S0-BUB-835

THE ENCYCLOPEDIA OF ANCIENT NATURAL SCIENTISTS

The Encyclopedia of Ancient Natural Scientists is the first comprehensive English-language work to provide a survey of ancient natural science, from its beginnings through to the end of late antiquity. A team of over 100 of the world's experts in the field have compiled this *Encyclopedia*, including entries which are not mentioned in any other reference work – resulting in a unique and hugely ambitious resource which will prove indispensable for anyone seeking the details of the history of ancient science.

Additional features include a *Glossary*, *Gazetteer*, and *Time-Line*. The *Glossary* explains many Greek (or Latin) terms difficult to translate, whilst the *Gazetteer* describes the many locales from which scientists came. The *Time-Line* shows the rapid rise in the practice of science in the 5th century BCE and rapid decline after Hadrian, due to the centralization of Roman power, with consequent loss of a context within which science could flourish.

Paul T. Keyser's publications include work on gravitational physics, computer science, stylometry, Greek tragedy, and ancient science. Formerly a teacher of Classics, he is currently crafting Java for IBM's Watson Research Center.

Georgia L. Irby-Massie is Assistant Professor at the College of William and Mary. Her research investigates reflections of science in literature and society, and includes publications on astrology, geography, natural philosophy in tragedy, and women scientists.

THE ENCYCLOPEDIA OF ANCIENT NATURAL SCIENTISTS

The Greek tradition and its many heirs

Edited by Paul T. Keyser and Georgia L. Irby-Massie

LONDON AND NEW YORK

First published 2008
by Routledge
2 Park Square, Milton Park, Abingdon, Oxon OX14 4RN

Simultaneously published in the USA and Canada
by Routledge
270 Madison Ave, New York, NY 10016

Routledge is an imprint of the Taylor & Francis Group, an informa business

Typeset in Baskerville by
RefineCatch Limited, Bungay, Suffolk
Printed and bound in Great Britain by
CPI Antony Rowe, Chippenham, Wiltshire

British Library Cataloguing in Publication Data
A catalogue record for this book is available from the British Library

Library of Congress Cataloging in Publication Data
A catalog record for this book has been requested

ISBN13: 978–0–415–34020–5 (hbk)
ISBN13: 978–0–203–46273–7 (ebk)

"... the person who is used to inquiry tries every possible pathway as he conducts his search and turns in every direction, and, so far from giving up the inquiry in the space of a day, does not cease his search throughout his life: directing his attention to one thing after another that is relevant to what is being investigated, he presses on until he attains his goal."

Erasistratos of Ioulis, *Paralysis* book 2

(in Galēn, *Habits* §1, *CMG* S.3 [1941] 12;

trans. by G.E.R. Lloyd, *Greek Science After Aristotle* [1973] 86, altered)

CONTENTS

ILLUSTRATIONS

INTRODUCTION

This work provides a synoptic survey of all "ancient," i.e., Greek and Greek-based, natural science, broadly defined, from its beginnings through the end of late antiquity, for the benefit of anyone interested in the history of science. Greek science is a central field for the understanding of antiquity – more of Greek science survives than does any other category of ancient Greek literature, and yet much of that is obscure even to classicists.

It is proper to describe the work of the people included herein as "science," with no more risk of anachronism than in using any modern term to refer to a corresponding ancient practice, because the ancient models of nature, whether correct or not, were indeed attempts at *models*. That is, they were created and debated as abstracted descriptions of phenomena, intended to give a naturalistic and self-consistent causal account, of a world viewed as regular or constant in its behavior. Their methods and aims were scientific, even when their theoretical entities or intellectual achievements are ones we now perceive as inadequate. Histories of science must be comprehensive, including all abandoned paths, since roads *not* taken seem evitable only in hindsight.

I. Scope. Natural science is a conceptual territory which cannot be precisely distinguished from other intellectual activities, and which resembles but is distinct from magic, philosophy, technology, and theology. Science borders on philosophy (we exclude metaphysics, ethics, and epistemology), lies near technology (we exclude writers who only record technological achievements, such as lists of manmade wonders or non-medical cookbooks), has affinity with magic (we exclude theurgy and all incantations, but we include astrology and alchemy), and touches on theology (we exclude divine cosmogonies and the theology of the soul). We prefer to err on the side of inclusion, so that readers working in or near the area of ancient science will be able to consult the work with profit, and thus we include writers and works whose topics lie on our margins, so long as it seems likely that they wrote on relevant topics. Most of the results and theories of ancient science would no longer pass muster as scientific (humoral medicine no more than astrology), but our principle of inclusion is to ask whether the endeavor was to understand or model some aspect of the natural world on the basis of investigation and reason, without recourse to hypotheses about purposive agents, and without reliance on tradition *per se*.

Texts of ancient writers often survive by accident or despite the ever-changing tastes of copyists and their patrons. Textual remains rarely represent what contemporaries would have agreed were the most important works: Strabōn for example seems almost unknown to his contemporaries, who cite and use works now lost. Many lost works, known to have been widely read and very influential during antiquity, perished only later. Thus, to present only or even primarily what is now extant or canonical is to distort. As a result, we include the obvious major authors, but also all the lost, fragmentary, obscure, and anonymous scientific writers whom we can. However, we include only authors known to have made some written or conceptual contribution, but not practitioners, such as architects and physicians, however famous or accomplished. Likewise, teachers, however important they may have been to their students, or for the transmission of ideas, we include only if they themselves made a contribution. Despite the plausible presumption that most of the adherents of the schools of Plato, Aristotle, Zēnōn, or Epicurus who made any contribution would also have made some contribution to natural science, we include only those who are explicitly attested to have done so. Anonymous and pseudonymous works receive a separate entry, in order to give them their due prominence – for example, the Hippokratic and Aristotelian corpora are each divided into about a dozen entries, whereas many other *anonymi* each receive individual entries.

We begin with Hēsiod and Homer and Thalēs, who represent key parts of the origin of Greek scientific thought, even if we can no longer or not yet assert with confidence any theory of that origin. (It may be that we should go further back: Arnott 1996.) The encyclopedia extends to *ca* 650 CE, rather than (say) to 529 (closing of the Academy), 410 (sack of Rome), or even 313 (edict of Milan), on the grounds that the whole "late antique" period represents a gradual transition from Mediterranean antiquity to the medieval or Byzantine periods. (Moreover, the century 650–750 CE is remarkably weak in science, whereas adequately to study science after 650 CE requires great familiarity with many languages, such as Arabic, Aramaic, Armenian, Coptic, etc., which the editors alas do not possess.) We include the authors of the "early Byzantine" period (330–650 CE) in order to ensure that the encyclopedia offer a synoptic view of ancient Greek science. Preferring errors of inclusion to those of exclusion, we also include authors of unknown or uncertain date, so long as there is a reasonable chance that they are prior to our terminal date.

Although nearly all of the entries concern works written in Greek, we include around 200 entries on authors or works in other languages, that in each case are based upon the Greek scientific tradition; we thus exclude the copious works of Chinese science, as well as most Babylonian, or Egyptian, or Sanskrit works. Classicists and historians are familiar with the Latin reception and transformation of Greek science; less familiar but of equal interest are a number of other scientific traditions also influenced by the Greek. Readers should consult the relevant indices for works whose language or author was Armenian, Celtic, Gothic, Egyptian, Persian, Sanskrit, or one of the Semitic languages (Arabic, Aramaic including Mandaic and Syriac, Babylonian, Hebrew, or Punic), as well as the index of well over a hundred Latin authors and works.

II. Names. Because the book primarily contains Greek scientists, Greek names are transliterated without prior Latinization. Moreover, names of other traditions (including Latin) are also directly transliterated, all according to the conventions of scholarship in those fields. Direct transliteration is no more arbitrary than any other system, more accurately preserves pronunciation and etymology, and more clearly signals culture. Of course, no system is wholly consistent: the standard "Anglo-Latinate" rendering of Greek names has Plato and Hero for Platōn and Hērōn, but Theon (not Theo) and Cleon (not Cleo), and oscillates over Dio or Dion; finally, even in that system, Nikē is never Nice. Even in Latin, if one writes "Pompey" and "Pliny" and "Livy" and "Antony", why then does one not also write "Tully" (as indeed in 18th c. English) or even "Porcy"? Moreover, even in English, we do not write

"Cafca" for the Czech novelist or "Cant" for the Prussian philosopher. The direct transliteration of certain very familiar Greek and Latin names might confuse the casual or novice reader, though surely not a classicist. Therefore, we use traditional Anglicized Latinized spellings for the two names "Alexander" and "Gregory" and for these 14 people: Aristotle, Chrysippus, Epicurus, Euclid, Galēn, Hēsiod, Homer, Plato, Pliny, Plutarch, Ptolemy, Pythagoras, Thucydidēs, and Zēnō of Elea (note macrons); a similar set of 19 exceptions for locations (many of which, like "Alexander," are names in common use in English) includes Alexandria, Antioch, Athens, Babylōn, Carthage, Corinth, Crete, Cyprus, Egypt, Ethiopia, Italy, Jerusalem, Libya, Macedon, Oxyrhynchos, Rhodes, Rome, Sicily, Spain, and Tyre. There is no reason to make this exception for unfamiliar names, whether of people or places, since they lack any obtrusive "familiar" transliteration. For names of ancient mythological people and places, we adopt a similar plan, thus, e.g., Achilles, Apollo, Athena, Atropos, Bacchant, Dionysus, Hadēs, Horus, Kronos, Lachesis, Muses, Odysseus, Ōriōn, Osiris, Ozymandias, Pandōra, Poseidōn, Promētheus, Sykeus, Thyestes, and Troy. Historical Greek and Byzantine figures outside the scope of the encyclopedia, but frequently cited or mentioned within (such as the historian Dio Cassius or the emperor Justinian or the theologian Clement of Alexandria), are listed with brief identifications in an index of 33 names (pp. 1037–8).

Transliteration and polyonomy dictate the need for cross-references. All binomial Latin names (e.g. "Tullius Cicero") are cross-referenced from *cognomen* to *nomen*. Late imperial Latin names also require cross-references, from the alternate name(s) to the diacritical name. Most of the Greek transliterations that are more direct (ē for *eta* and ō for *omega*, final -os not -us, and final -ōn rather than just -o) do not require any cross-reference, whereas the transliteration of omicron-upsilon as "ou" rather than as -u (so Euboulos etc.), and the transliteration of upsilon always as "u" rather than as -y, plus the transliteration of epsilon-iota as "ei" rather than as -i, does slightly affect the order of the entries in a few cases, all of which are handled with cross-references. The use of K rather than the Latin C, just as (in Latin names) the use of I rather than the medieval J (increasingly favored even by conservative Anglophone scholarship), does noticeably affect the order, which again is handled by cross-references.

As Roman rule absorbed Greek culture, some Romans took Greek *cognomina*, and some Greeks upon receiving Roman citizenship assumed Latin names; the earliest certain cases appear to be late Republican (1st c. BCE); *cf.* Salway 1994. Nearly all such bicultural names in this work appear to belong to Greek writers and are hence filed under their Greek name (without assertion as to a primary culture of their bearer), with their Latin name(s) given as an epithet (Dionusios Cassius, Kritōn Statilius, etc.), and cross-referenced. The few Greek writers who possess no known Greek name are simply filed under their Latin name (Aelianus, Arrianus, Rufus, Vettius, etc.). There are very few examples of Latin writers possessing both a Latin and a Greek name – in this work, perhaps only Fauonius Eulogios (filed under his Latin name). In late antiquity, many names lost their ethnic specificity: for example, Iōannēs, once a Hebrew/Jewish name (Yohannan, as at *I Macc.* 2.1–2, Iosephus *Ant. Iud.* 18 [116–119]), then became a Christian name, and by the 5th c. was in general use (e.g., Iōannēs of Stoboi). Likewise, many Greek names, Grēgorios, Hieroklēs, Isidōros, Palladios, Paulos, and Theodōros, e.g., became fully Latinized by the 4th or 5th c., whereas Latin names such as Marcellus and Marcianus/Martianus became fully Hellenized apparently by the 3rd c.

III. Gazetteer, Glossary, Time-Line, and Indices. Many terms used in describing the work of ancient scientists are Greek (or Latin) terms whose translation raises subtle and important questions. Any adequate treatment of such issues is outside the scope of this work, but we have provided a Glossary of commonly used terms, giving a brief discussion of their meaning, with some references. Entries refer to these terms in **bold**, or for terms transliterated rather than translated, in ***bold italic***.

Three indices or appendices, to aid consultation and cross-referencing, are provided: (1) topical or categorical index, listing all authors astronomical, geographical, medical, etc., in order to facilitate understanding the degree to which ancient science was understood by its practitioners as straddling multiple traditions; (2) the "Time-Line," a chronological index, to facilitate understanding the chronological development of ancient natural science; (3) the "Gazetteer," a geographical index, listing by place of author's origin all entries for which that is possible, to clarify the diverse origins of the scientists and the degree to which ancient science was conducted away from the two traditional intellectual centers of Athens and Alexandria.

The Gazetteer shows that scientists originated from a wide variety of locales: over 325 are listed for the almost 1,000 scientists (i.e., about half those in this encyclopedia) whose place of origin is attested or inferred (thus, an average of three scientists per locale). Alexandria (with *ca* 80) and Athens (with *ca* 50) indeed each produced more scientists than any other two places together; but 16 other sites also produced significant numbers (at least thrice the average): Samos (22), Kurēnē and Milētos (at least 17 each), Rhodes (16), Ephesos, Kōs, and Surakousai (at least 13 each), Pergamon, Smurna and Tarsos (at least 11 each), plus Buzantion, Khios, Knidos, Kuzikos, Taras, and Tralleis (at least ten each). The total for Alexandria or Athens, although many times the average, amounts to only 1/8 of the number of scientists whose place of origin is known (8% for Alexandria; 5% for Athens). Those two centers did produce (and attract) many scientists, as indeed one would expect for places that provided resources and an environment congenial to the practice of science. But prosperity, trade, and democracy seem also to promote the practice of science, or at least be correlated with it, as can be seen in the list of 16 cities above, which altogether provided at least 204 (*ca* 1/5) of the scientists whose place of origin is known. (That conclusion is based upon only half of our entries, but to alter it significantly would require establishing a place of origin for a large number of the unassigned scientists, which itself would be a welcome result.) The same can also be seen in the Gazetteer as a whole: scientists come primarily from prosperous places open to outside influences and which foster free discussion of ideas (*cf.* Keyser in Irby-Massie and Keyser 2002, c. 1).

The Time-Line shows that the practice of science was not uniform over time, as may well be expected; but the "classical" era of Hippokratēs, Plato, and Aristotle was **not** the most productive – rather the 4th to 3rd cc. BCE and the 1st c. BCE through 1st c. CE were (the "dip" in the 2nd c. BCE may be an artifact of fragmentary data). Periodization is always a scholarly imposition of discontinuity upon complex and continuous data, since every era is transitional between its own past and its own future. Nonetheless, using only those scientists who are relatively narrowly dated (about half those in this encyclopedia), we create **Fig. 1** of the Time-Line, showing the rapid rise in the 5th c. BCE and rapid decline after Hadrian. As previously argued (*ibid.*), the decline seems due to the centralization of political power under Hadrian and abolition of semi-autonomous polities throughout the Mediterranean world, with the consequent loss of a context within which science could flourish. No doubt these conclusions should be held somewhat tentatively, given the likelihood that many names have been entirely lost, especially in the latest periods (and perhaps also in the turbulent and less-well-known periods of the 2nd c. BCE and 3rd c. CE).

However, to alter it significantly would require establishing a narrower date-range for a large number of the poorly-dated scientists, or else would require finding many hitherto unknown scientists dating to post-Hadrianic times, either of which would be another welcome result.

IV. Creation of the Encyclopedia. The germ of this encyclopedia was sown two decades ago when Keyser began annotating the margins of his copy of the *OCD*, 2nd ed., with missing scientists. That seed fell on fertile ground during a memorable dinner with Richard Stoneman celebrating the publication of Irby-Massie and Keyser (2002), at the APA meeting of January 2002. The first contributor was recruited that very evening, and numerous scholars were contacted over the next 18 months. Almost all were supportive, and most were willing to participate; many others provided useful advice. Scholars were recruited in the first instance to compose groups of entries (e.g., on medical Empiricists, or Neo-Platonic mathematics, or Hellenistic and Greco-Roman agronomy); nevertheless many scientists were covered individually. In the end, slightly over half of our entries, and all the more important, were contributed by 119 scholars familiar with the relevant material (listed below); the balance were composed by the editors. One team of contributors contributed many entries on paradoxographers, and the entries on the Sanskrit authors were also planned as teamwork; other collaborations developed in the course of the work.

The editors specified the total lengths of sets of entries to be supplied; individual contributors were then free to adjust the relative lengths within their sets as they saw fit. We selected a few entries (Aristotle, Galēn, Ptolemy) to have the maximum length, of *ca* 2,000 words. For most of the entries, much lengthier pieces could have been written, and it is hoped that the texts here, together with their bibliographies, will serve as a useful *introduction.* For the better-known scientists (the three mentioned and many others), the bibliographies must perforce be very selective and serve only as prolegomena.

Anaximandros and Anaximenēs, similarly-named and both of Milētos, are nonetheless well-distinguished; there are cases far more problematic than that herein. Despite care and diligence, we cannot be sure to have made all distinctions correctly, and in some cases the entries discuss the problems quite explicitly: see esp. Aelianus, Apollōnios, Apuleius/Placitus Papyriensis, Dēmētrios, pseudo-Dēmokritos, Dionusios, Magnus, Olumpiodōros, Orpheus, Plutarch, and Stephanos.

Less than 25% of our entries are found in English-language reference works (such as *OCD3* or *DSB*), although the *BNP* when complete will contain about 40% of our entries; even in such works the coverage of Latin authors is almost twice that of the Greeks. Moreover, about 1/8 of our entries are not listed in **any** encyclopedia whatsoever, neither the famously capacious 85-volume *RE*, nor even the most complete list of medical authors heretofore, Fabricius (1726). We consulted not only modern encyclopedias, but also many ancient authors more generous than usual with explicit citation (esp. Aëtios of Amida, Galēn, Iōannēs of Stoboi, Oreibasios, Paulos of Aigina, Pappos, Pliny, Plutarch, Proklos, Simplicius, and Vitruuius). Finally, over 1/5 of our 2,043 entries were discovered during the writing of the originally-proposed 1,558 entries.

The proportion of entries herein not found in any encyclopedia, averaging 1/8, seems to rise from about 1/12 early in the English (or German) alphabet to over one-seventh at the end (See the last *Index*). Compared to the distribution of initial letters in the names of the *LGPN*, some of our initial letters seem underrepresented. Those observations, as well as the fact that many entries were discovered during composition, suggests that there are still items to be discovered, perhaps principally late in the alphabet, or from the early Byzantine period (330–650 CE), or among the papyri. It is therefore almost certain, although no less regrettable, that we have omitted some names. If the fates are kind and scholars diligent, we hope to include them in a revised edition.

V. Contributors and Supporters. Over half, and all the more important, of the entries of the *EANS* were composed by the following 119 contributors (we indicate in each case the category or categories within which they contributed, or list the entries):

Gianfranco Agosti (Biological poets)
Università di Udine, Udine
Eugenio Amato (Lithika authors)
Université de Nantes, France ***and*** Department of Classics, Fordham University, New York USA (from 2008)
Cosmin Andron (Neo-Platonists)
Department of Classics, Royal Holloway College, University of London, England, UK
Jacques Bailly (Theagenēs)
Classics Department, University of Vermont, USA
Han Baltussen (Neo-Platonists)
Classics, School of Humanities, University of Adelaide, Australia
Alain Bernard (Neo-Platonists)
Centre Université Paris 12 (IUFM Créteil), EHESS et PAI "Mathematiques et Histoire"
Sylvia Berryman (Stratōn)
Department of Philosophy, University of British Columbia, Vancouver, Canada
Gábor Betegh (Derveni Papyrus, Stēsimbrotos)
Department of Philosophy, Central European University, Budapest, Hungary
Richard Bett (Skeptics)
Department of Philosophy, Johns Hopkins University, Baltimore Maryland, USA
Siam Bhayro (Syriac authors)
Department of Theology, University of Exeter, Exeter, United Kingdom
Shane Bjornlie (Ambrose, Macrobius, Cassiodorus)
Assistant Professor of Ancient and Medieval History, Department of History, Claremont McKenna College, Claremont California, USA
Larry Bliquez (Hippokratic Corpus)
Classics, University of Washington, Seattle Washington, USA
István Bodnár (Aristotelian Corpus, Peripatetics)
Institute of Philosophy, Eötvös University/Department of Philosophy, Central European University, Budapest, Hungary
Jan Bollansée (Paradoxographers)
Faculteit Letteren, Katholieke Universiteit Leuven, Leuven, Belgium
Alan C. Bowen (Hellenistic astronomers)
Institute for Research in Classical Philosophy and Science, Princeton New Jersey, USA
István M. Bugár (Gorgias, Melissos, *On Melissos, Xenophanes and Gorgias*)
University of Debrecen, Department of Philosophy, Debrecen, Hungary
Stanley M. Burstein (Agatharkhidēs, Bērossos)
Department of History, California State University, Los Angeles California, USA

Brian Campbell (Agrimensores)
School of History and Anthropology, The Queen's University of Belfast, Northern Ireland, UK
Bruno Centrone (Neo-Pythagoreans)
Professor of Ancient Philosophy, Dipartimento di Filosofia, Università di Pisa, Italy
Elizabeth Craik (Hippokratic Corpus)
School of Classics, University of St. Andrews, Scotland, UK
David Creese (Harmonics authors)
Classical, Near Eastern and Religious Studies, University of British Columbia, Vancouver, Canada
Patricia Curd (Parmenidēs, Xenophanēs, Zēnō)
Department of Philosophy, Purdue University, West Lafayette Indiana, USA
Christophe Cusset (Astronomical poets)
Littérature Grecque, École Normale Superieure Lettres et Sciences Humaines, Université de Lyon, France
Mauro de Nardis (Metrologists)
Dipartimento di Discipline storiche, Università di Napoli "Federico II," Italy
Claudio De Stefani (Aemilius Macer, Aglaias, Ovid)
Venice International University, Venezia, Italy
Lesley Dean-Jones (Hippokratic Corpus)
Department of Classics, University of Texas, Austin Texas, USA
Leo Depuydt (Demotic and Coptic texts)
Norton, Massachusetts, USA
Keith M. Dickson (Byzantine medical authors)
Classical Studies, Purdue University, West Lafayette Indiana, USA
Cristiano Dognini (Androsthenēs, Arrian, Megasthenēs, Xenophōn)
Liceo Scientifico Elio Vittorini, Milano, Italy
Daniela Dueck (Hellenistic geographers)
Department of History, Department of Classical Studies, Bar Ilan University, Israel
Walter G. Englert (Atomists)
Department of Classics, Reed College, Portland Oregon, USA
Silvia Fazzo (Peripatetics)
Università di Trento, Italy – Université de Lille 3, France
Klaus-Dietrich Fischer (*Mulomedicina Chironis*)
Institut für Geschichte, Theorie und Ethik der Medizin der Johannes Gutenberg-Universität Mainz, Germany
Rebecca Flemming (Hippokratic Corpus)
University Lecturer in Ancient History and Fellow of Jesus College, Cambridge University, UK
Emma Gee (Firmicus, Iōannēs "Lydus")
School of Classics, University of St. Andrews, Scotland, UK
Daniel W. Graham (Pre-Socratics)
Department of Philosophy, Brigham Young University, Provo Utah, USA
Andrew Gregory (Plato)
Department of Science and Technology Studies, University College, London England, UK

Aurélie Gribomont (Hermēs, *Korē Kosmou*)
Faculteit Letteren, Katholieke Universiteit Leuven, Leuven, Belgium
Antonino Grillone (pseudo-Hyginus)
"Aglaia" – Dipartimento di studi greci, latini e musicali. Tradizione e modernità, Università di Palermo, Italy
Jean-Yves Guillaumin (Agrimensores; Boëthius, Capella)
Professeur de Langue et Littérature Latines, Université de Franche-Comté, Besançon, France
Karen Haegemans (Paradoxographers)
Faculteit Letteren, Katholieke Universiteit Leuven, Leuven, Belgium
David E. Hahm (Aristōn of Keōs)
Department of Greek and Latin, Ohio State University, Columbus Ohio, USA
Bink Hallum (Hellenistic alchemists)
Warburg Institute, University of London, UK
R. J. Hankinson (Galēn)
Department of Classics, University of Texas, Austin Texas, USA
Maury Hanson (Hippokratic Corpus)
Madison Virginia, USA
Stephan Heilen ("Lion Horoscope")
Department of Classics, University of Illinois, Urbana Illinois, USA
Oliver Hellmann (Aristotelian Corpus)
Klassische Philologie, Universität Trier, Germany
Thomas Noble Howe (Roman architectural authors)
Coordinatore Generale, Fondazione Restoring Ancient Stabiae, Brown Distinguished Research Professor, Southwestern University, Georgetown Texas, USA
Katerina Ierodiakonou (Aristotelian Corpus)
Associate Professor in Ancient Philosophy, Department of Philosophy and History of Science, University of Athens, Greece
Jean-Marie Jacques (Iologists)
Professeur émérité de l'Université Michel de Montaigne, Bordeaux, France
Alexander Jones (Astronomers, astrologers, and mathematicians)
Institute for the Study of the Ancient World (NYU), New York City, USA
Jacques Jouanna (Hippokratēs)
UFR de Grec – Université de Paris 4, Paris, France
Philip G. Kaplan (Classical geographers)
Department of History, University of North Florida, Jacksonville Florida, USA
George Karamanolis (Neo-Platonists)
Department of Philosophy and Social Studies, University of Crete, Rethimno, Greece
Toke Lindegaard Knudsen (Sanskrit authors)
Department of Mathematics, Computer Science, and Statistics, State University of New York, College at Oneonta, Oneonta, New York, USA
Kostis Kourelis (Byzantine architectural authors)
Art Department, Clemson University, South Carolina, USA
Andreas Kuelzer (Byzantine geographers)
Oesterreichische Akademie der Wissenschaften, Institut für Byzanzforschung, Wien, Austria

Volker Langholf (Hippokratic Corpus)
Classics Department, Universität Hamburg, Germany
Julie Laskaris (Hippokratic Corpus)
Department of Classical Studies, University of Richmond, USA
Peter Lautner (Neo-Platonists)
Inst. of Philosophy, Faculty of Arts, Pázmány Péter Catholic University, Hungary
Daryn Lehoux (Stoics; astrologers)
Classics and Ancient History, University of Manchester, UK
Robert Littman (Classical medical authors)
LLEA, University of Hawaii at Manoa, Honolulu Hawai'i, USA
Natalia Lozovsky (Late Roman geographers)
San Mateo California, USA
Daniela Manetti (Classical Medical authors)
Universita' degli Studi di Firenze, Facolta' di Lettere e Filosofia, Dipartimento di Scienze dell'Antichità, Italy
Maria Marsilio (Classical agronomists)
Department of Foreign Languages and Literatures, Saint Joseph's University, Philadelphia Pennsylvania, USA
Edward G. Mathews, Jr. (Armenian authors)
Tunkhannock Pennsylvania, USA
Thomas J. Mathiesen (Late Harmonics)
Center for the History of Music Theory and Literature, Jacobs School of Music, Indiana University, Bloomington Indiana, USA
Anne McCabe (Greek veterinarians)
Centre for the Study of Ancient Documents, Oxford England, UK
Jørgen Mejer (Doxographers)
Washington DC, USA ***and*** Copenhagen, Denmark
Claudio Meliadò (Biological poets)
Università degli Studi di Messina, Italy
Henry Mendell (Classical astronomers)
Philosophy Department, Cal. State University, Los Angeles California, USA
Margaret Miles (Greek architectural authors)
Department of Art History, University of California, Irvine California, USA
Ian Mueller (Euclid and Classical mathematicians)
University of Chicago, Illinois, USA
Bret Mulligan (Claudian)
Haverford College, Pennsylvania, USA
Trevor Murphy (Pliny)
Department of Classics, University of California, Berkeley California, USA
Reviel Netz (Hieroklēs, Pelagonius)
Department of Classics, Stanford University, Stanford California, USA
Jennifer Nilson (Hierocles, Pelagonius)
UW-Madison Classics Department, Madison Wisconsin, USA
Jan Opsomer (Middle Platonists)
Philosophisches Seminar, Universität zu Köln, Germany
Vincenzo Ortoleva (Vegetius)
Università di Catania, Dipartimento di Studi archeologici, filologici e storici, Italy

Antonio Panaino (Iranian authors)
Facoltà di Conservazione dei Beni Culturali, Ravenna, Italy
Holt N. Parker (Damastēs, Mētrodōra)
Department of Classics, University of Cincinnati, Ohio, USA
Carl Pearson (Iōannēs Philoponos)
Department of History, Rice University, Houston Texas, USA
Gerard J. Pendrick (Antiphōn)
Spelman College, Atlanta Georgia, USA
Christopher A. Pfaff (Greek architectural authors)
Florida State University, Department of Classics, Tallahassee Florida, USA
Kim L. Plofker (Sanskrit authors)
Department of Mathematics, Union College, Schenectady New York, USA
Peter E. Pormann (Paulos of Aigina)
Wellcome Trust Assistant Professor, University of Warwick, Coventry, UK
Annette Yoshiko Reed (Jewish authors)
Department of Religious Studies, University of Pennsylvania, Philadelphia Pennsylvania, USA
Eleonora Rocconi (Harmonics authors)
Università degli studi di Pavia, Facoltà di Musicologia, Cremona, Italy
Francesca Rochberg (Early astrologers)
Department of Near Eastern Studies, University of California, Berkeley California, USA
Robert H. Rodgers (Late agronomists)
Classics Department, University of Vermont, USA
Lucía Rodríguez-Noriega Guillén (Epikharmos)
Universidad de Oviedo, Facultad de Filología, Dpto. De Filología Clásica y Romànica, Oviedo, Spain
James Romm (Hērodotos)
Bard College, Annandale-on-Hudson New York, USA
Anne Roth Congès (Innocentius)
Centre Camille Jullian, CNRS, Aix-en-Provence, France
John Scarborough (Asklēpiadeans, Erasistrateans, Hērophileans, Methodists; Byzantine Medical authors; Hellenistic Pharmacists; Roman Medical authors)
School of Pharmacy, Departments of History and Classics, University of Wisconsin, Madison, Wisconsin, USA
Guido Schepens (Paradoxographers)
Faculteit Letteren, Katholieke Universiteit Leuven, Leuven, Belgium
Hermann S. Schibli (Pherekudēs)
Universität Passau, Rotthalmünster, Germany
Daniel Schwartz (Greek Fathers)
Bryn Mawr College, History Department, Bryn Mawr Pennsylvania, USA
Jacques Sesiano (Diophantos)
Département de Mathématiques, École Polytechnique Fédérale de Lausanne, Switzerland
José Solana Dueso (*Dissoi Logoi*, Iōn, Kritias)
Universidad de Zaragoza, Zaragoza, Spain
Michael G. Sollenberger (Theophrastos)

Foreign Languages and Literatures, Mount St. Mary's University, Emmitsburg Maryland, USA

Anna Somfai (Calcidius, Isidore of Hispalis)
Visiting Professor, Department of Medieval Studies, Central European University, Budapest, Hungary

Fabio Stok (Empiricist medical authors)
Dipartimento di Antichità e tradizione classica, Facoltà di Lettere e Filosofia, Roma, Italy

Peter Struck (Artemidōros of Daldis)
Classics Department, University of Pennsylvania, Philadelphia Pennsylvania, USA

Ioannis Taifacos (Klearkhos)
Faculty of Letters, University of Cyprus, Nicosia, Cyprus

Richard Talbert (*Itineraries*, *Peutinger Map*)
Department of History, University of North Carolina, Chapel Hill North Carolina, USA

Harold Tarrant (Early Platonists, Thrasullos)
School of Humanities and Social Science, University of Newcastle, New South Wales, Australia

Philip Thibodeau (Hellenistic and Roman agronomists)
Brooklyn College, Classics Department, Brooklyn New York, USA

Robert B. Todd (Damianos, Geminus, Hēliodōros of Larissa, Kleomēdēs, Themistios)
Emeritus Professor of Classics, University of British Columbia, Canada

Laurence Totelin (Hippokratic Corpus)
The Wellcome Trust Centre for the History of Medicine, University College, London England, UK

Alain Touwaide (Hellenistic pharmacists; pneumaticists; Byzantine medical authors)
Historian of Sciences, Department of Botany, National Museum of Natural History, Smithsonian Institution, Washington DC, USA

Simon Trépanier (Empedoklēs)
Lecturer in Classics, School of History, Classics and Archaeology, University of Edinburgh, Scotland, UK

Karin Tybjerg (Hellenistic mechanics authors)
Head of the Department of Astronomy, Kroppedal Museum, Copenhagen Denmark

Kevin van Bladel (Arabic)
Assistant Professor of Classics, Classics Department, University of Southern California, Los Angeles, California

Cristina Viano (Byzantine alchemy)
Université de Paris Sorbonne, Centre de recherches sur la pensée antique, Paris, France

Sabine Vogt (Physiognomers)
Institut für Klassische Philologie, Ludwig-Maximilians-Universität, München, Germany

J. M. Wilkins (Hippokratic Corpus)
Department of Classics and Ancient History, Queens Building, University of Exeter, UK

Malcolm Cameron Wilson (Aristotle)
Department of Classics, University of Oregon, Eugene Oregon, USA

Leonid Zhmud (Pythagoreans)
Institute for the History of Science, St. Petersburg, Russia

Arnaud Zucker (Biological authors)
Université de Nice, Faculté LASH, Nice, France

Many other scholars and friends assisted in the creation of this work, often by helping us find contributors (marked * in the list below). Much of the work was carried out in the libraries at Columbia University, at New York University, at the New York Public Library, and at the College of William and Mary whose Inter-Library Loan office is the very model of cheerful efficiency. We are grateful to those organizations, to Richard Stoneman for first seeing the value of the book, and to the 78 individuals listed below, notably Michèle Lowrie (whose questions always clarify) and Keith Massie (whose efforts included tracking down museum addresses, creating digital images, and building a secure web site which greatly facilitated the final editing process). Georgia L. Irby-Massie dedicates her entries and editorial efforts to her father, Richard E. Irby, an autodidact and polymath, who first set Georgia's feet on the path of knowledge, truth, and scientific inquiry.

Silvia Barbantani*
Andrew Barker*
Adam H. Becker*
Peter Bing*
Calvin Bower*
Ewen Bowie*
Laurel M. Bowman
Charles Burnett*
William M. Calder III*
Alan D.E. Cameron
Dee Clayman*
Shaye Cohen*
J.J. Coulton*
Serafina Cuomo*
Frans de Haas*
Denise Demetriou
Bruce Eastwood*
Michel Federspiel*
Simonetta Feraboli
William W. Fortenbaugh*
Dorothea Frede
Michael Frede
Karen Green
Jerry Heverly
Brooke Holmes
C.A. Huffman*
Richard Janko*
C.H. Kahn*
R.A. Kaster
Rachel L. Keyser
Ewald Kislinger*
Johannes Koder*
Nita Krevans*
Bob Lamberton*
M.J.T. Lewis*
Ian Lockey
Michèle Lowrie
Enrico Magnelli*
Jaap Mansfeld*
Keith Massie
Steve McCluskey*
Michael McCormick
Michèle Mertens*
Chris Minkowski*
David Mirhady
Phil Mitsis*
Tony Natoli*
Thomas Noble*
Jim O'Donnell*
Tim O'Keefe*
Michael Peachin*
Jim Porter*
John M. Riddle*
Tracey E. Rihll*
Duane Roller
David Runia*
Jacques Schamp*
David N. Sedley*
Robert W. Sharples*
David Sider*
Ian Simmonds
Lucas Siorvanes
P. Oktor Skjaervo*
David A. Smith
Richard Sorabji*
Liba Taub*
Bill Thayer
Teun Tieleman*
Philip van der Eijk*
Marlein van Raalte*

Evangelos Venetis*
Heinrich von Staden*
Bonna Wescoat*
Stephen Wheeler*
Craig Williams
Georg Woehrle*
Ehsan Yarshater*
Jim Zetzel

Needless to say, they should not be held accountable for any defects of this work.

Abbreviations and Bibliography

b. = born
c. = century (pl.: "cc.")
ca = circa
d. = died
ed. = edition/editor (pl.: "edd.")
f. = folio (pl.: "ff.")
fr. = fragment (pl.: "frr.")
mod. = modern
n. = note (pl.: "nn.")
n.d. = no date
# = number
ns = new series
pr. = preface/proem/prologue
repr. = reprinted
§ = section
s. = series
S. = Supplement
v. = volume (pl.: "vv.")

The bibliographic closing date of the *EANS* was 31 December 2007: items appearing after that date could not be taken into account; in a few cases contributors were aware of items forthcoming, and those are cited as such. Abbreviations of journal titles are according to *L'Année Philologique* (Paris 1924–).

Texts of authors in the encyclopedia:

Are cited according to the edition(s) given in the relevant entry (q.v.); see esp. GALĒN. Note also:

- Editions cited by an abbreviation of their title are also listed below under "Works (of reference and editions) cited by abbreviation".
- Editions cited by author-date (e.g., those in the Loeb and *CUF* series) are also listed below under "Frequently-cited" works.
- Editions cited by name of editor(s), without date or title, are listed here:

Athan. — P. Athanassiadi, *Damascius, The Philosophical History* (Athens 1999)
BDM — U.C. Bussemaker, Ch. Daremberg, and A. Molinier, *Oeuvres d'Oribase* 6 vv. (Paris 1851–1876)
Cornarius — J. Cornarius, *Aetii medici graeci contractae ex veteribus medicinæ Tetrabiblos*, etc. (Basel 1542)
DK — H. Diels and W. Kranz, *Die Fragmente der Vorsokratiker*, 2 vv. (Zürich and Berlin 1964: 11th ed.) **– cited by section and fragment number**
DR — Ch. Daremberg and Ch.É. Ruelle, *Oeuvres de Rufus d'Éphèse* (Paris 1879; repr. 1963)

Deichgr.	K. Deichgräber, *Die griechische Empirikerschule* (Berlin 1930; augmented repr. 1965)
EK	L. Edelstein and I.G. Kidd, *Posidonius*, 3 vv. in 4: Edelstein and Kidd, v. 1 Text, 2nd ed. (Cambridge 1989); Kidd, vv. 2.1, 2.2 Commentary (Cambridge 1988), Kidd, v. 3 Translation (Cambridge 1999) – **cited by fragment number**
FHSG	W.W. Fortenbaugh, Pamela Huby, Robert W. Sharples, and Dimitri Gutas, edd., *Theophrastus of Eresus. Sources for his Life, Writings, Thought and Influence*, 2 vv. (Leiden 1992) = *Philosophia Antiqua* 54; *Commentary*, vv. 2, 3.1, 4, 5, 8 (of 9 projected) (1995–2006) = *Philosophia Antiqua* 103, 79, 81, 64, 97 (respectively) – **cited by fragment number**
Fr.	G. Friedlein, *Procli Diadochi In primum Euclidis Elementorum librum commentarii* (Leipzig 1873)
Hultsch	Fr. Hultsch, *Pappi: Math. Collectiones quae supersunt*, 3 vv. (Berlin 1876–1878)
K.	C.G. Kühn, *Claudii Galeni Opera omnia*, 20 vv. in 22 parts (Leipzig 1821–1833; repr. Hildesheim 1964–1965; 1986); note: *CMGen* = *De Compositione Medicamentorum per Genera; CMLoc* = *De Compositione Medicamentorum secundum Locos*
KRS	G.S. Kirk, J.E. Raven, and Malcolm Schofield, *The Presocratic Philosophers*, 2nd ed. (Cambridge 1983)
Littré	É. Littré, *Oeuvres complètes d'Hippocrate*, 10 vv. (Paris 1839–1861; repr. Amsterdam 1973–1980)
MMH	J. Marquardt, I. Müller, and G. Helmreich, *Claudii Galeni Pergameni Scripta Minora* 3 vv. (Leipzig 1884, 1891, 1893; repr. Amsterdam 1967)
Nachm.	E. Nachmanson, *Erotiani Vocum Hippocraticarum collectio cum fragmentis* (Uppsala 1918)
Puschm.	Th. Puschmann, *Alexander von Tralles. Original-Text und Übersetzung* 2 vv. (Vienna 1878–1879; repr. Amsterdam 1963)
Speranza	F. Speranza, *Scriptorum Romanorum De Re Rustica Reliquiae* (Messina 1971)
Tecusan	M. Tecusan, *The Fragments of the Methodists*, v.1: *Methodism outside Soranus* (Leiden 2004) = *SAM* v.24/1
Thesleff	H. Thesleff, *The Pythagorean Texts of the Hellenistic Period* (Äbo 1965)
W.	Carle Wescher, *Poliorkētika kai poliorkiai diaphorōn poleōn. Poliorcétique des Grecs. Traités théoriques. – Récits historiques* (Paris 1867)
Wa.	C. Wachsmuth, *Ioannis Laurentii Lydi, Liber de ostentis* (Leipzig 1863; 2nd ed. 1897)
W.-C.	L.G. Westerink and J. Combès, *Damascius, Traité des premiers principes*, 3 vv. (*CUF*: Paris 1986–1991)
W.-H.	C. Wachsmuth and O. Hense, *Iohannis Stobaei Anthologium* 5 vv. (1884–1912) – often instead cited by "book.section"
Wu.	R. Wuensch, *Ioannis Laurentii Lydi, Liber de mensibus* (Stuttgart 1903; repr. 1967).
Wehrli	F. Wehrli, *Die Schule des Aristoteles*, 2nd ed., 10 vv. (Basel 1967–1969), Suppl. 2 vv. (Basel 1974, 1978)
Wellmann	M. Wellmann, *Fragmentsammlung der griechischen Ärzte, I: Die Fragmente der sikelischen Ärtze Akron, Philistion und des Diokles von Karystos* (Berlin 1901)

***Works (of reference and editions) cited by abbreviation of title*:**

ACA	*Ancient Commentators on Aristotle Series* (London and Ithaca 1989–), ed. R. Sorabji
ACT	*Astronomical cuneiform texts: Babylonian ephemerides of the Seleucid period for the motion of the sun, the moon, and the planets* (London 1955; repr. New York and Berlin 1983), ed. O. Neugebauer
AL	*Anthologia Latina*, 2 vv. in 5 (Leipzig 1893–1926; repr. 1964, 1973), ed. Fr. Bücheler (v. 2) and A. Riese (v. 1)
AML	*Antike Medizin: Ein Lexikon* (München 2005), ed. K.-H. Leven
ANRW	*Aufstieg und Niedergang der römischen Welt* (Berlin and New York 1972–), ed. H. Temporini
BAGRW	*Barrington atlas of the Greek and Roman world* (Princeton 2000), ed. R.J. Talbert
BBKL	*Biographisch-Bibliographisches Kirchenlexikon*, 26 vv. (Hamm 1975–2005), ed. F.W. Bautz
BEA	*Biographical Encyclopedia of Astronomers* (New York and Berlin 2007), ed. Thos. A. Hockey
BNJ	*Brill's new Jacoby*, ed. I. Worthington *et al.* (forthcoming); see http://www.brillsnewjacoby.com/description.html
BNP	*Brill's New Pauly* 11 vv. to date (Leiden 2002–), ed. H. Cancik and H. Schneider (the *NP* is cited instead of not-yet published volumes of the *BNP*, or when an entry is missing from the *BNP*)
BTML	*Bibliographie des textes médicaux latins. Antiquité et haut moyen âge*, edd. G. Sabbah, P.-P. Corsetti, and K.-D. Fischer (Saint-Étienne 1987 [1988]) = *Mémoires du Centre Jean Palerne* 6; and *Premier Supplement, 1986–1999* (2000), ed. K.-D. Fischer
CA	*Collectanea Alexandrina* (Oxford 1925; repr. 1970, 1981), ed. J.U. Powell
CAAG	*Collection des anciens alchimistes Grecs*, 3 vv. (Paris 1883–1888; repr. London 1963), ed. M. Berthelot and Ch.-Ém. Ruelle
CAG	*Commentaria in Aristotelem Graeca*, 23 vv. and three supplements, some in multiple parts (Berlin 1882–1909)
CCAG	*Catalogus Codicum Astrologorum Graecorum*, 12 vv. in 20 parts (Brussels 1898–1953), ed. D. Bassi, Fr. Boll, P. Boudreaux, Fr. Cumont, A. Delatte, J. Heeg, W. Kroll, E. Martini, A. Olivieri, Ch.-Ém. Ruelle, M.A.F. Šangin, St. Weinstock, and C.O. Zuretti
CAR	*Corpus Agrimensorum Romanorum* (Naples 1993–): 1. M. Clavel-Lévêque *et al.*, *Les conditions des terres, Siculus Flaccus* (1993); 2. *Présentation systematique de toutes les figures* (1996); 3. J.-Y. Guillaumin, *Balbus. Podismus et textes connexes* (1996); 4. M. Clavel-Lévêque *et al.*, *Hygin l'arpenteur, l'établissement des limites* (1996); 5. O. Behrends *et al.*, *Hygin. L'oeuvre gromatique* (2000)
CESS	*Census of the Exact Sciences in Sanskrit, Series A*, 5 vv. to date (Philadelphia 1970–1994), ed. D.E. Pingree
CHG	*Corpus Hippiatricorum Graecorum* 1: *Hippiatrica Berolinensia* (Leipzig, 1924) and 2: *Hippiatrica Parisina, Hippiatrica Cantabrigiensia, Additamenta Londinensia, Excerpta Lugdunensia* (Leipzig 1927); (both vv. repr. Stuttgart 1971), ed. E. Oder and C. Hoppe
CMAG	*Catalogue des manuscripts alchimiques grecs*, 8 vv. (Brussels 1924–1932), ed. J. Bidez *et al.*
CMG	*Corpus Medicorum Graecorum* (Berlin 1908–)

CML	*Corpus Medicorum Latinorum* (Berlin 1915–)
CTC	*Catalogus translationum et commentariorum*, 8 vv. to date (Washington 1960, 1971, 1976, 1980, 1984, 1986, 1992, 2003), ed. P.O. Kristeller *et al.*
CUF	*Collection des Universités de France* (i.e. the "Budé" series)
DPA	*Dictionnaire des Philosophes Antiques*, 4 vv. (Paris 1989, 1994, 2000, 2005) and Supplement (2003) to date, ed. R. Goulet
DSB	*Dictionary of Scientific Biography*, 14 vv. (New York 1970–1976, 2 vv. per year), v. 15 = S.1 (1978), v.16, index (1980), ed. C.C. Gillispie (vv. 17–18 = S.2–3 contain no relevant entries); see also *NDSB*
EAAE	*Encyclopedia of the Archaeology of Ancient Egypt* (Routledge 1999), ed. Kathryn A. Bard
ECP	*Encyclopedia of Classical Philosophy* (Westport 1997), ed. D.J. Zeyl
EI	*Encyclopaedia Iranica*, 12 vv. to date (London and Boston 1982–), ed. E. Yarshater
EJ2	*Encyclopedia Judaica*, 2nd ed. 22 vv. (Detroit 2007), ed. F. Skolnik and M. Berenbaum
FHG	*Fragmenta Historicorum Graecorum*, 5 vv. (Paris 1849–1884), ed. K.O. Muller
FLP	*The Fragmentary Latin poets* (Oxford 2003), rev. ed., E. Courtney
FGrHist	*Fragmente der griechischen Historiker* (Leiden 1923–), ed. F. Jacoby – **cited by number (not volume and page)**
GAS	*Geschichte des arabischen Schrifttums*, 12 vv. to date (Leiden 1967–), ed. F. Sezgin: v. 3 (Medizin, Pharmazie, Zoologie, Tierheilkunde, 1970), v. 4 (Alchimie, Chemie, Botanik, Agrikultur, 1971), v. 5 (Mathematik, 1974), v. 6 (Astronomie, 1978), v. 7 (Astrologie, Meteorologie und Verwandtes, 1979), vv. 10–12 (Mathematische Geographie und Kartographie, 2000)
GGL	*Geschichte der griechischen Literatur*, 5 vv. (München 1929–1948), ed. Wilhelm Schmid und O. Stählin – **cited by section**
GGLA	*Geschichte der griechischen Litteratur in der Alexandrinerzeit*, 2 vv. (Leipzig 1891–1892), ed. Fr. Susemihl – **cited by section**
GGM	*Geographi Graeci Minores*, 2 vv. (Paris 1855–1861; repr. Hildesheim 1990), ed. K. Müller
GGP	*Grundriß der Geschichte der Philosophie. Die Philosophie der Antike*, 4 vv. (Basel 1983–), ed. H. Flashar
GL	*Grammatici Latini* 8 vv. (Leipzig 1855–1880; repr. 1961; repr. 1981), ed. H. Keil
GLLM	*Geschichte der lateinischen Literatur des Mittelalters*, v.1 (München 1911), ed. M. Manitius
GLM	*Geographi Latini Minores* (Heilbrun 1878; repr. Hildesheim 1964), ed. A.L. Riese
GRL	*Geschichte der römischen Literatur*, 4 vv. (München 1896–1935), ed. M. Schanz and C. Hosius – **cited by section**
HGM	*Handbuch der Geschichte der Medizin*, v. 1, begründet von Th. Puschmann, ed. M. Neuburger and J. Pagel (Jena 1902; repr. Hildesheim 1971)
HGP	*History of Greek Philosophy*, 6 vv. (Cambridge 1965–1981), W.K.C. Guthrie
HLB	*Die hochsprachliche profane Literatur der Byzantiner*, 2 vv. (München 1978), Herbert Hunger
HLL	*Handbuch der lateinischen Literatur der Antike* 5 vv. (München 1989–), ed. R. Herzog and P.L. Schmidt – **cited by volume and section number**

HWPhil — *Historisches Wörterbuch der Philosophie* 12 vv. (Basel and Stuttgart 1971–2004)

KLA — *Künstlerlexikon der Antike* 2 vv. (Munich and Leipzig, 2004), ed. R. Vollkommer and D. Vollkommer-Gloekler

KP — *Der Kleine Pauly*, 5 vv. (Stuttgart 1969–1975), ed. K. Ziegler and W. Sontheimer

LGPN — *Lexicon of Greek Personal Names* 4 vv. (in 5) to date (Oxford 1987–), ed. P.M. Fraser and E. Matthews

MGG2 — *Die Musik in Geschichte und Gegenwart: Personenteil*, 2nd rev. ed., 17 vv. (Kassel, New York and Stuttgart 1999–2007), ed. L. Finscher

MRR — *Magistrates of the Roman Republic*, 2nd ed., 3 vv. (Atlanta 1984–1986), ed. T.R.S. Broughton

MSG — *Musici scriptores graeci* (Leipzig 1895; repr. Hildesheim 1962), ed. Karl von Jan

MSR — *Metrologicorum Scriptorum Reliquiae* (Stuttgart 1864–1866; reprint 1971), ed. Fr. Hultsch

NDSB — *New Dictionary of Scientific Biography* 8 vv. (Detroit 2007), ed. Noretta Koertge

NGD2 — *New Grove Dictionary of Music and Musicians*, 2nd rev. ed., 29 vv. (London 2001), ed. S. Sadie and J. Tyrrell

NP — *Der Neue Pauly*, 12 vv. (Stuttgart 1996–2003), ed. H. Cancik und Helmuth Schneider (the *BNP* is cited by preference when available, i.e., normally up through "Pr-"): vv. 10 (2001), 11 (2001), 12/1 (2002), 12/2 (2003)

OCD3 — *Oxford Classical Dictionary*, 3rd ed. rev. (Oxford 2003), ed. S. Hornblower and A.J.S. Spawforth – pages differ from the 3rd ed. only for Aristoklēs of Messēnē and Hēliodōros of Alexandria (med.)

ODB — *Oxford Dictionary of Byzantium*, 3 vv. (Oxford 1991), ed. A.P. Kazhdan (paginated continuously)

PCG — *Poetae Comici Graeci*, 8 vv. (New York and Berlin 1983–2001), ed. R. Kassel and C. Austin

PGR — *Paradoxographorum Graecorum Reliquiae* (*Classici Greci e Latini* 3, Milan 1966), ed. A. Giannini

PG — *Patrologiae cursus completus . . . series graeca* 161 vv. (Paris 1857–1866; repr. Athens 1988), ed. J.P. Migne

PL — *Patrologiae cursus completus . . . series latina* 221 vv. (Paris 1844–1891), ed. J.P. Migne

PIR2 — *Prosopographia Imperii Romani*, 2nd ed., (Berlin 1933–), ed. E. Groag and Arthur Stein – **cited by Letter + number (not volume and page)**

PLRE — *Prosopography of the Later Roman Empire*, 3 vv. (Cambridge 1971, 1980, 1992), ed. A.H.M. Jones, J.R. Martindale, and J. Morris

RAC — *Reallexikon für Antike und Christentum*, 20 vv. to date (Stuttgart 1950–), ed. Th. Klauser *et al.*

RBK — *Reallexikon zur byzantinischen Kunst* 6 vv. to date, issued in fascicles (Stuttgart v.1: 1963–1966; v.2: 1967–1971; v.3: 1972–1978; v.4: 1982–1990; v.5: 1991–1995; v.6: 1997–2005), ed. K. Wessel and M. Restle

RE — *Paulys Realencyclopädie der classischen Altertumswissenschaft*, 85 vv, incl. 15 supplements (Stuttgart 1893–1978), ed. G. Wissowa *et al.*; see also the *Paulys Realencyclopädie der Classischen Altertunswissenschaft: Gesamtregister* 1 (Stuttgart 1997), ed. Tobias Erler *et al.*

REP — *Routledge Encyclopedia of Philosophy*, 10 vv. (Routledge 1998), ed. Edward Craig

RUSCH — *Rutgers University Studies in Classical Humanities* 13 vv. to date (New Brunswick and London 1983–) ed. W.W. Fortenbaugh *et al.*

SAM	*Studies in Ancient Medicine*, 33 vv. to date (Leiden 1990–)
SDS	*Storia della scienza* (Rome 2001–), ed. S. Petruccioli
SEP	*Stanford Encyclopedia of Philosophy*, ed. E.N. Zalta, online: http://plato.stanford.edu/ (search under the given entry-title)
SH	*Supplementum Hellenisticum* (Berlin and New York 1983), ed. H. Lloyd-Jones and P. Parsons – **cited by fragment (not page)**
Souda	A. Adler, *Suidae Lexicon* 5 vv. (Leipzig 1928-1938; repr. 1967–1971; repr. 1989–2001) – **cited by Letter + number of entry**
SRMH	*Source Readings in Music History* (New York 1998), ed. W. Oliver Strunk, rev. ed. L. Treitler: v. 1: *Greek Views of Music*, ed. T.J. Mathiesen
SSR	*Socratis et Socraticorum Reliquiae* 4 vv. (Naples 1990), ed. G. Giannantoni
SVF	*Stoicorum veterum fragmenta*, 4 vv. (Leipzig 1905–1924; repr. Stuttgart 1968; Dubuque 1973), ed. H.F.A. von Arnim
TAM	*Tituli Asiae Minoris*, 5 vv. to date (Vienna 1901–), ed. E. Kalinka, R. Heberdey, and P. Herrmann
TLL	*Thesaurus Linguae Latinae*, 10 vv. to date (Leipzig 1900–)
TTE	*Trade Travel and Exploration in the Middle Ages* (New York 2000), ed. J.B. Friedman, K.M. Figg, *et al.*

Inscriptions, Manuscripts, and Papyri:

Papyri edited in a series are cited usually with both the volume number and papyrus number, e.g., *P.Hibeh* 1.27 or *P.Mich.* 3.148.

BNF	*Bibliothèque nationale de France*
CIL	*Corpus inscriptionum Latinarum*, 17 vv. to date, some in 2nd ed. (Berlin 1862–)
IBM	*The Collection of Ancient Greek Inscriptions in the British Museum* 4vv. (Oxford 1874–1916; repr. Milan 1977–1979): vv. 1–3, ed. T.C. Newton, E.L. Hicks, and Gustav Hirschfeld; v. 4 ed. Gustav Hirschfeld and F.H. Marshall
IG3	*Inscriptiones Graecae*, 3rd ed. (Berlin 1981–)
IGLSyr	*Inscriptions grecques et latines de la Syrie* 7 vv. (and parts of others) to date (Paris 1929–), ed. L. Jalabert, R. Mouterde, *et al.*
IGRR	*Inscriptiones Graecae ad res Romanas pertinentes*, v. 1 (Paris 1911) ed. R. Cagnat *et al.*; v. 3 (1906), ed. R. Cagnat and G. Lafaye; v. 4 (1927), ed. G. Lafaye [v. 2 never published]; all vv. (repr. Rome 1964)
ILS	*Inscriptiones latinae selectae* 3 vv. (Berlin 1892–1916; repr. 1954–1955; repr. Chicago 1979), ed. Hermann Dessau
OGIS	*Orientis Graeci Inscriptiones Selectae* 2 vv. (Leipzig 1903–1905; repr. Hildesheim 1960), ed. Wilh. Dittenberger
P.Ant.	*The Antinopolis papyri* 3 vv. (London 1950–1967)
P.Berol.	*Papyri graecae berolinenses* (Berlin 1911), ed. Wilhelm Schubart
PGM	*Papyri graecae magicae* 2 vv. (Leipzig 1928–1931), ed. K. Preisendanz; rev. ed. by Albert Henrichs (Stuttgart 1973–1974)
P.Hibeh	*The Hibeh Papyri*, 2 vv. (London 1906, 1955), ed. B.P. Grenfell and E.G. Turner

P.Lit.Lond.	*Catalogue of the Literary Papyri in the British Museum* (London 1927), ed. H.J.M. Milne
P.Lund.	*Aus der Papyrussammlung der Universitätsbibliothek in Lund* (Lund 1934/1935), ed. A.F. Wifstrand
P.Mich.	*Michigan Papyri* (individual volumes variously subtitled and edited) 19 vv. to date (Ann Arbor 1931–); see esp: v.2 (1933), ed. A.E.R Boak, *Papyri from Tebtunis, Part I* v.3 (1936), ed. J.G. Winter, *Papyri in the University of Michigan collection; miscellaneous papyri* v.5 (1944), ed. E.M. Husselman, A.E.R. Boak, and W.F. Edgerton, *Papyri from Tebtunis, Part II*
P.Mil.Vogl.	*Papiri della R. Università di Milano* (Milan 1937; repr. 1966), ed. A. Vogliano
P.Oslo.	*Papyri Osloenses* 3 vv. to date (Oslo 1925–), ed. S. Eitrem
POxy	*The Oxyrhynchus papyri*, 67 vv. to date (London 1898–), ed. B.P. Grenfell *et al.*
P.Ryl.	*Catalogue of the Greek {and Latin} Papyri in the John Rylands Library, Manchester* 4 vv. (Manchester 1911–1952): vv.3–4 have the augmented title
PSI	*Papiri Greci e Latini, Pubblicazioni della Società italiana per la ricerca dei papiri greci e latini in Egitto* 15 vv. (Florence 1912–1979; repr. 2004)
P.Tebtunis	*Tebtunis Papyri*, vv.1-3, ed. C.C. Edgar, E.J. Goodspeed, B.P. Grenfell, A.S. Hunt, and J.G. Smyly (London 1902–1938); v.4, ed. J.G. Keenan and J.C. Shelton (London 1976); see also several vv. of *P.Mich.*
P.Turner	*Papyri Greek and Egyptian edited by various hands in Honour of Eric Gardner Turner* (London 1981)
Pack	**Roger A. Pack,** ed., *The Greek and Latin literary texts from Greco-Roman Egypt*, 2nd ed. (Ann Arbor 1965) – **cited by number (not page)**
Samama	**Évelyne Samama**, *Les médecins dans le monde grec: Sources épigraphiques sur la naissance d'un corps medical* (Geneva 2003) – **cited by number (not page)**
SB	*Sammelbuch Griechischer Urkunden aus Ägypten* 26 vv. to date (Strassburg 1915)

***Frequently-cited works, cited by "Author (Date) pp."* (works entered here if cited thrice or more; forenames written out when needed for clarification):**

J.N. Adams, *The Latin Sexual Vocabulary* (London and Baltimore 1982).

Idem, *Pelagonius and Latin Veterinary Terminology in the Roman Empire* (Leiden 1995) = *SAM* 11.

B. Alexanderson, *Peri Kriseōn. Galenos. Überlieferung und Text* (Stockholm, Göteborg and Uppsala 1967).

I. Andorlini Marcone, "L'apporto dei papiri alla conoscenza della scienza medica antica," *ANRW* 2.37.1 (1993) 458–562.

Jacques André, *Les Noms de plantes dans la Rome antique* (Paris 1985).

R.G. Arnott, "Healing and medicine in the Aegean Bronze Age," *J Roy Soc Med* 89 (1996) 265–270.

Athanassiadi (1999): see "editions," above.

H. Bailey, *Zoroastrian Problems in the Ninth-Century Books* (Oxford 1943; new ed. 1971).

A.D. Barker, *Greek Musical Writings* 2 vv. (Cambridge 1984, 1989).

Idem, *The Science of Harmonics in Classical Greece* (Cambridge 2007).

I.C. Beavis, *Insects and Other Invertebrates in Classical Antiquity* (Exeter 1988).

A. Beccaria, *I codici di medicina del periodo presalernitano (secoli IX, X e XI)* (Rome 1956).

M. Berthelot, *Les origines de l'alchimie* (Paris 1885).
I. Bodnár and W.W. Fortenbaugh, edd., *Eudemus of Rhodes* (2002) = *RUSCH* 11.
A.C. Bowen, ed., *Science and philosophy in classical Greece* (New York 1991).
J. Bidez, "Plantes et pierres magiques d'après le ps. Plutarque de fluviis," in *Mélanges offerts à O. Navarre* (Toulouse 1935) 25–38.
Idem and F. Cumont, *Les Mages hellénisés* (Paris 1938; repr. New York 1975).
G. Björck, *Zum Corpus hippiatricorum graecorum. Beiträge zur antiken Tierheilkunde = Uppsala Universitets Årsskrift* (1932) # 5.
Idem, Apsyrtus, Julius Africanus, et l'hippiatrique grecque = Uppsala Universitets Årsskrift (1944) # 4.
J. Blänsdorf, *Fragmenta Poetarum Latinorum epicorum et lyricorum* (Leipzig 1995).
R.C. Blockley, *The Fragmentary Classicising Historians of the Later Roman Empire:. Eunapius, Olympiodorus, Priscus and Malchus* 2 vv. (Liverpool 1981–1982).
Peter Brain, trans., *Galen on Bloodletting* (Cambridge 1986).
Jean Rhys Bram, *Ancient Astrology Theory and Practice* (Park Ridge, NJ 1975; repr. 2005).
A.J. Brock, *Greek Medicine. Being Extracts Illustrative of Medical Writers from Hippocrates to Galen* (London, Toronto and New York 1929).
S.P. Brock, *A Brief Outline of Syriac Literature* (Kottayam 1997).
W. Burkert, *Lore and Science in Ancient Pythagoreanism*, trans. by E.L. Minar, Jr. (Cambridge Mass. 1972).
Bussemaker, Daremberg, and Molinier (1851–1876): see "editions", above.
Brian Campbell, *The Writings of the Roman Land Surveyors* (London 2000).
L. Casson, *The Periplus Maris Erythraei* (Princeton 1989).
R. Chartier, *Magni Hippocratis Coi et Claudii Galeni Pergameni Archiatron Universa Quae Extant Opera* 13 vv. (Paris 1639).
G.-A. Costomiris "Études sur les écrits inédits des anciens médecins grecs. Deuxième série," *REG* 3 (1890) 145–179.
S. Cuomo, *Pappus of Alexandria and the Mathematics of Late Antiquity* (Cambridge 2000).
M. Decorps-Foulquier, *Recherches sur les Coniques d'Apollonios de Pergé et leurs commentateurs grecs* (Paris 2000).
Deichgräber (1930): see "editions," above.
A. De Lazzer, *Plutarco. Paralleli minori* (Naples 2000).
Idem, Plutarco. Fiumi e monti (Naples 2003).
D.R. Dicks, *Early Greek Astronomy to Aristotle* (London 1970).
Keith Dickson, *Stephanus the Philosopher and Physician. Commentary on Galen's Therapeutics to Glaucon* (Leiden 1998) = *SAM* 19.
H. Diels, *Doxographi Graeci* (Berlin 1879; repr. 1929, 1958, 1965).
Idem, Die Handschriften der antiken Ärzte 2 vv. (Berlin 1905–1907); plus *Bericht über den Stand des interakademischen Corpus Medicorum Antiquorum und erster Nachtrag zu den in den Abhandlungen 1905 and 1906 veröffentlichten Katalogen: Die Handschriften der antiken Ärzte, I. und II. Teil. Zusammengestelt im Namen der Kommission der königl. preuss. Akademie der Wissenschaften = Abhandlungen der königl. preuss. Akademie der Wissenschaften, philosophisch-historische Klasse, Abhandlung* 2 (1907; repr. Berlin 1908); all three vv. repr. (Leipzig 1970).
Idem, Antike Technik, 2nd ed. (Leipzig and Berlin 1920).
F. Dietz, *Apollonii Citiensis, Stephani, Palladii, Theophili, Meletii, Damascii, Ioannis, aliorum Scholia in Hippocratem et Galenum* 2 vv. (Königsburg 1840; repr. Amsterdam 1963).
O.A.W. Dilke, *Greek and Roman Maps* (Ithaca 1985).
Aubrey Diller, *The Tradition of the Minor Greek Geographers* (Lancaster 1952).
John M. Dillon, *The Middle Platonists, 80 B.C. to A.D. 220*, 2nd ed. (Ithaca 1996).
Idem, The Heirs of Plato: a study of the Old Academy, 347–274 B.C. (Oxford 2003).
K. Dimitriadis, *Byzantinische Uroskopie* (Inaugural-Dissertation, Universität Bonn 1971).
B. Dodge, *The Fihrist of Al-Nadim: A Tenth-Century Survey of Muslim Culture* 2 vv. (New York 1970).
G. Downey, "Byzantine Architects: Their Training and Methods," *Byzantion* 18 (1948) 99–118.
A.G. Drachmann, *Ktesibios, Philon and Heron* (Copenhagen 1948).
Idem, Mechanical Technology of Greek and Roman antiquity (Munksgaard 1963).

W.L.H. Duckworth (ed. M.C. Lyons and B. Towers), *Galen. On anatomical procedures, the later books* (Cambridge 1962).

D. Dueck, *Strabo of Amasia: a Greek Man of Letters in Augustan Rome* (Routledge 2000).

I. Düring, *Porphyrios Kommentar zur Harmonielehre des Ptolemaios* (Göteborg 1932).

R.J. Durling, *A Dictionary of Medical Terms in Galen* (Leiden 1993) = *SAM* 5.

L. Edelstein, "Methodiker," *RE* S.6 (1935) 358–373, English trans. by C.L. Temkin as "The Methodists" in O. Temkin and C.L. Temkin, edd., *Ancient Medicine: Selected Papers by Ludwig Edelstein* (Baltimore 1967; repr. 1987) 173–191.

Edelstein and Kidd (1972-1999): see "editions," above.

É. Espérandieu, *Signacula medicorum ocularium* = *CIL* 13.3.2 (Berlin 1906).

W.C. Evans, *Trease and Evans' Pharmacognosy*, 14th ed. (London 1996).

Johann Albert Fabricius, *Bibliotheca Graeca*, v.13 (Hamburg 1726).

Cajus Fabricius, *Galens Exzerpte aus älteren Pharmakologen* (Berlin 1972).

K.-D. Fischer, *Pelagonii Ars veterinaria* (Leipzig 1980).

A.J. Festugière, *La révélation d'Hermès Trismégiste* 4 vv, 2nd ed. (Paris 1949–1953; repr. in 3 vv, 1983).

P.M. Fraser, *Ptolemaic Alexandria* 3 vv. (Oxford 1972).

M. Frede, "The Method of the So-Called Methodical School of Medicine," in J. Barnes *et al.*, edd., *Science and Speculation: Studies in Hellenistic Theory and Practice* (Cambridge, New York and Paris 1982) 1–23.

J.W. Fück, "The Arabic Literature on Alchemy According to An-Nadim (A.D. 987): A translation of the tenth discourse of the *Book of the Catalogue (Al-Fihrist)* with introduction and commentary," *Ambix* 4 (1951) 81–144.

D.J. Furley and J.S. Wilkie (w/trans. and comm.), ed., *Galen On Respiration and the Arteries* (Princeton 1984).

H. Gerstinger, *Codex Vindobonensis med. Gr. 1 der Österreichischen Nationalbibliothek* (Graz 1970).

A. Giannini, "Studi sulla paradossografia greca II," *Acme* 17 (1964) 99–140.

A. Gioè, *Filosofi medioplatonici del II secolo d.C. Gaio, Albino, Lucio, Nicostrato, Tauro, Severo, Arpocrazione* = *Elenchos* 36 (Naples 2002).

B.R. Goldstein and A.C. Bowen, "Meton of Athens and Astronomy in the Late Fifth Century B.C.," in E. Leichty, M. de J. Ellis, and P. Gerardi, edd., *A Scientific Humanist: Studies in Memory of Abraham Sachs* (Philadelphia 1988) 40–81.

Eidem, "The introduction of dated observations and precise measurement in Greek astronomy," *AHES* 43 (1991) 93–132.

H.B. Gottschalk, "Aristotelian philosophy in the Roman world from the time of Cicero to the end of the second century," *ANRW* 2.36.2 (1987) 1079–1174; in part reprinted as "The earliest Aristotelian commentators," in R. Sorabji, ed., *Aristotle Transformed. The ancient commentators and their influence* (London 1990) 55–81.

D. Gourevitch, "L'*Anonyme* de Londres et la médecine d'Italie du Sud," *HPLS* 11 (1989) 237–251.

Henry Gray, *Anatomy of the Human Body*, 27th ed., ed. Charles Mayo Goss (Philadelphia 1959).

H. Grensemann, *Knidische Medizin* (Berlin 1975).

C.L. Grotefend, *Die Stempel der römische Augenärzte* (Hannover 1867).

M.D. Grmek, *Diseases in the Ancient Greek World* (Baltimore and London 1989), trans. of *Les maladies à l'aube de la civilisation* (Paris 1983) by Mireille Muellner and Leonard Muellner.

Idem and D. Gourevitch, "Aux sources de la doctrine médicale de Galien: l'enseignement de Marinus, Quintus et Numisianus," *ANRW* 2.37.2 (1994) 1491–1528.

W. Gundel and H.G. Gundel, *Astrologumena: Die astrologische Literatur in der Antike und ihre Geschichte* (Wiesbaden 1966).

R. Halleux, *Les alchimistes grecs I: Papyrus de Leyde. Papyrus de Stockholm, Fragments de recettes* (Paris: *CUF* 1981).

Idem and J. Schamp, *Les lapidaires grecs* (Paris: *CUF* 1985).

R.J. Hankinson, *Galen On the Therapeutic Method Books I and II* (Oxford and New York 1991).

C.R.S. Harris, *The Heart and the Vascular System in Ancient Greek Medicine from Alcmaeon to Galen* (Oxford 1973).
T.L. Heath, *A history of Greek mathematics* (Oxford 1921; repr. New York 1981).
Idem, The thirteen books of Euclid's Elements, 3 vv., 2nd ed. (Cambridge 1926; repr. New York 1956).
E. Heitsch, *Griechischen Dichterfragmente der römischen Kaiserzeit* 2 vv. (Göttingen 1963–1964).
Jeffrey Henderson, *The Maculate Muse: Obscene Language in Attic Comedy*, 2nd ed. (New York and Oxford 1991).
Gustav Heuser, *Die Personennamen der Kopten* (Leipzig 1929).
R.G. Hoyland, "Theomnestus of Nicopolis, Hunayn ibn Ishaq, and the beginnings of Islamic veterinary science," in R.G. Hoyland and P.F. Kennedy, edd., *Islamic Reflections, Arabic Musings. Studies in Honour of Professor Alan Jones* (Cambridge 2004) 150–169.
C.A. Huffman, *Philolaus of Croton: Pythagorean and Presocratic* (Cambridge 1993).
Fr. Hultsch, *Griechische und römische Metrologie*, 2nd ed. (Berlin 1882; repr. Graz 1971).
Idem (1876–1878): see "editions," above.
J.L. Ideler, *Physici et medici graeci minores* 2 vv. (Berlin 1841–1842; repr. Amsterdam 1963).
A.M. Ieraci Bio, "La transmissione della letteratura medica greca nell'Italia meridionale fra x e xv secolo," in A. Garzya, ed., *Contributi alla Cultura greca nell'Italia meridionale* (Naples 1989) 133–257.
S. Ihm, *Clavis commentariorum der antiken medizinischen Texte* (Leiden 2002).
G.L. Irby-Massie and P.T. Keyser, *Greek Science of the Hellenistic Era* (Routledge 2002).
J.-M. Jacques, *Nicandre*, v.2 (Paris: *CUF* 2002); v.3 (2007).
F. Jacoby, "Die Überlieferung von ps.-Plutarchs *parallela minora* und die Schwindelautoren," *Mnemosyne* 8 (1940) 73–144.
Ian Johnston, *Galen on Diseases and Symptoms* (Cambridge 2006).
Alexander Jones, *Book 7 of the Collection: Pappus of Alexandria*, 2 vv. (New York 1986).
Idem, "Uses and Users of Astronomical Commentaries," in G.W. Most, ed., *Commentaries – Kommentare* (Göttingen 1999) 163–172.
H.W.S. Jones, *Hippocrates* vv.1–2 (Cambridge, MA: Loeb 1923), v.4 (1931).
D. Kahn and S. Matton, edd., *Alchimie: art, histoire et mythes. Actes du 1er colloque international de la Société d'Étude de l'Histoire de l'Alchimie, (Textes et Travaux de Chrysopoeia, I)* (Paris and Milan 1995).
S. Kapetanaki and R.W. Sharples, *Pseudo-Aristoteles (Pseudo-Alexander), Supplementa Problematorum* (Berlin and New York 2006).
R.A. Kaster, *C. Suetonius Tranquillus: De Grammaticis et Rhetoribus* (Oxford 1995).
A. Keller, *Die Abortiva in der römischen Kaiserzeit* (Stuttgart 1988).
W.R. Knorr, *The Evolution of Euclidean Elements* (Dordrecht 1975).
Idem, The ancient tradition of geometric problems (Boston 1986).
Idem, Textual studies in ancient and medieval geometry (Boston and New York 1989).
J. Kollesch, "René Chartier, Herausgegeber und Fälscher der Werke Galens," *Klio* 48 (1967) 183–198.
Idem, Untersuchungen zu den pseudogalenischen Definitiones Medicae (Berlin 1973).
J. Korpela, *Das Medizinalpersonal im antiken Rom: eine sozialgeschichte Untersuchung* (Helsinki 1987).
Fr. Kudlien, "Poseidonios und die Ärztesschule der Pneumatiker," *Hermes* 90 (1962) 419–429.
Idem, "Pneumatische Ärzte," in *RE* S.11 (1968) 1097–1108.
Kühn (1821–1833): see "editions," above.
J.H. Langenheim, *Plant Resins* (Portland and Cambridge 2003).
D.R. Langslow, *Medical Latin in the Roman Empire* (Oxford and New York 2000).
Fr. Lasserre, *De Léodamas de Thasos à Philippe d'Oponte: témoignages et fragments: edition, traduction et commentaire* (Naples 1987).
E. Leichty, M. de J. Ellis and P. Gerardi, edd., *A Scientific Humanist: Studies in Memory of Abraham Sachs = Occasional Publications of the Samuel Noah Kramer Fund* 9 (Philadelphia 1988).
Jean Letrouit, "Chronologie des alchimistes grecs," in Kahn and Matton (1995) 11–93.
Littré (1839–1861): see "editions," above.
A.A. Long, ed., *Cambridge Companion to Early Greek Philosophy* (Cambridge 1999).

Idem and D.N. Sedley, *The Hellenistic Philosophers* 2 vv. (Cambridge 1987).
J. Longrigg, *Greek rational medicine: Philosophy and medicine from Alcmaeon to the Alexandrians* (London 1993).
E. Maass, *Commentariorum in Aratum reliquiae* (Berlin 1898).
Anne McCabe, *A Byzantine Encyclopaedia of Horse Medicine: The Sources, Compilation, and Transmission of the Hippiatrica* (Oxford 2007).
R.J. Mainstone, *Hagia Sophia* (New York 1988).
D. Manetti and A. Roselli, "Galeno commentatore di Ippocrate," *ANRW* 2.37.2 (1994) 1529–1635.
J. Mansfeld and D. Runia, *Aetiana. The Method and Intellectual Context of a Doxographer I. The Sources* (Leiden 1996).
M.-H. Marganne, *Inventaire Analytique des Papyrus Grecs de Médecine* (Geneva 1981).
Eadem, L'ophthalmologie dans l'Égypte gréco-romaine (Leiden 1994) = *SAM* 8.
Eadem, "Les medicaments estampillés dans le Corpus Galénqiue," *Galen on Pharmacology*, ed. A. Debru = *SAM* 16 (1997) 153–174.
Eadem, La chirurgie dans l'Égypte gréco-romaine d'aprés les papyrus littéraires grecs = *SAM* 17 (1998).
Marquardt, Müller and Helmreich: see "editions," above.
E.W. Marsden, *Greek and Roman Artillery* 2 vv. (Oxford 1969, 1971).
Jean Martin, *Histoire du texte des Phénomènes d'Aratos* (Paris 1956).
T.J. Mathiesen, *Apollo's Lyre* (Lincoln 1999).
J. Matthews, *Western Aristocracies and Imperial Court, A.D. 364–425* (Oxford 1975; repr. w/postscript 1990; repr. 1998).
I. Mazzini and F. Fusco, *I testi di medicina latini antichi* (Rome 1985).
Jørgen Mejer, *Diogenes Laertius and his Hellenistic Background* (Wiesbaden 1978).
A. Meredith, *The Cappadocians* (London 1995).
M. Mertens, *Les alchimistes grecs* 4.1 (Paris: *CUF* 1995).
M. Michler, *Die Alexandrinischen Chirurgen = Die Hellenistische Chirurgie* 1 (Wiesbaden 1968).
J.I. Miller *Spice Trade of the Roman Empire* (Oxford 1969).
P. Moraux, *Der Aristotelismus bei den Griechen von Andronikos bis Alexander von Aphrodisias* 3 vv. (Berlin and New York 1973, 1984, 2001).
Ph. Mudry and J. Pigeaud, edd., *Les Écoles médicales à Rome: Actes du 2ème colloque international sur les textes médicaux latins antiques, Lausanne septembre 1986* (Geneva 1991).
Ian Mueller, "Greek arithmetic, geometry, and harmonics: Thales to Plato," in C.C.W. Taylor, ed., *The Routledge History of Philosophy* (1997) 1.271–322.
Nachmanson (1918): see "editions," above.
C.A. Nallino, "Tracce di opere greche giunte agli Arabi per trafila pehlevica," *A Volume of Oriental Studies Presented to Professor E.G. Browne*, T.W. Arnold and R. Nicholson, edd. (Cambridge 1922) 345–363, reprinted in *Idem, Raccolta di scritti editi e inediti* 6 (Rome 1948) 285–303.
R. Netz, "Classical Mathematics in the Classical mediterranean," *Mediterranean Historical Review* 12.2 (1997) 1–24.
O. Neugebauer, *A History of Ancient Mathematical Astronomy* (Berlin, Heidelberg and NewYork 1975).
Idem and H.B. van Hoesen, *Greek Horoscopes = Memoirs of the American Philosophical Society* 48 (1959).
V. Nutton, "Drug Trade in Antiquity," *JRoySocMed* 78 (1985) 138–145; repr. in *From Democedes to Harvey* (1988), #IX.
E. Oder, "Beiträge zur Geschichte der Landwirtschaft bei den Griechen," *RhM* 45 (1890) 58–99, 212–222; 48 (1893) 1–40 (3 parts of one article).
D.J. O'Meara, *Pythagoras Revived: mathematics and philosophy in late antiquity* (Oxford 1989).
A. Önnerfors, "Das medizinische Latein von Celsus bis Cassius Felix," *ANRW* 2.37.1 (1993) 227–392 and 924–937.
A. Panaino, "L'influsso greco nella letteratura e nella cultura medio-persiana," *Autori classici in Lingue del Vicino e Medio Oriente* (Rome 2001) 29–45.
W. Pape and G.E. Benseler, *Wörterbuch der griechischen Eigennamen* 2 vv. (Braunschweig 1884).

H.M. Parker, "Women Doctors in Greece, Rome, and the Byzantine Empire," in L.R. Furst, ed., *Women healers and physicians: climbing a long hill* (Lexington KY 1997) 131–150.
L. Pearson, *Lost Histories of Alexander* (New York 1960).
P. Pédech, *Historiens compagnons d'Alexandre* (Paris 1984).
J. Pigeaud, "Les fondements théoriques du méthodisme," in Mudry and Pigeaud (1991) 7–50.
J.R. Pinault, *Hippocratic Lives and Legends* (Leiden 1992) = *SAM* 4.
D.E. Pingree, "Classical and Byzantine Astrology in Sassanian Persia," *DOP* 43 (1989) 227–239.
Idem, *Yavanajātaka of Sphujidhvaja* 2 vv. (Harvard 1978).
Paul Potter, *Hippocrates* v.8: *Places in Man. Glands. Fleshes. Prorrhetic I. Prorrhetic II. Physician. Use of Liquids. Ulcers. Haemorrhoids. Fistulas* (Cambridge, MA: Loeb 1995).
R. Rashed, *Les catoptriciens grecs I: Les mirroirs ardents* (*CUF* 2000).
E. Rawson, *Intellectual Life in the Late Roman Republic* (London and Baltimore 1985).
A. Rehm, *Parapegmastudien* = *Abhandlungen der Bayerischen Akademie der Wissenschaften. Philosophisch-historische Abt.* N.F., Heft 19 (1941).
M. Riley, "A Survey of Vettius Valens," *ANRW* 2.37.5 (forthcoming), *cf.*: http://www.csus.edu/indiv/r/rileymt/pdf_folder/VettiusValens.pdf
C.A. Robinson, *The History of Alexander the Great*, v. 1 (Providence 1953).
D.W. Roller, *Scholarly Kings: The Writings of Juba II of Mauretania, Archelaos of Kappadokia, Herod the Great and the Emperor Claudius* (Chicago 2004).
M. Roueché, "The Definitions of Philosophy and a New Fragment of Stephanus the Philosopher," *JÖByz* 40 (1990) 107–128.
G. Sabbah, ed., *Mémoires VIII. Études de Médecine romaine* (Saint-Étienne 1988) = *Mémoires (Centre Jean Palerne)* 8.
H.-D. Saffrey, "Historique et description du Marcianus Graecus 299" in Kahn and Matton (1995) 1–10.
S.M.R. Sala, *Lexicon nominum Semiticorum quae in papyris Graecis in Aegypto repertis ab anno 323 a. Ch. n. usque ad annum 70 p. Ch. n. laudata reperiuntur* (Milan 1974).
B. Salway, "What's in a Name? A Survey of Roman Onomastic Practice from c. 700 B.C. to A.D. 700," *JRS* 84 (1994) 124–145.
J. Scarborough, "Roman Pharmacy and the Eastern Drug Trade: Some Problems Illustrated by the Example of Aloe," *PhH* 24 (1982) 135–143.
Idem, ed., *Symposium on Byzantine Medicine* (1985) = *DOP* 38.
Idem, "Early Byzantine Pharmacology," in *Idem* (1985a) 213–232.
Idem, "Criton, Physician to Trajan: Historian and Pharmacist," in J.W. Eadie and J. Ober, edd., *The Craft of the Ancient Historian: Essays in Honor of Chester G. Starr* (1985) 387–405.
Idem, "The Opium Poppy in Hellenistic and Roman Medicine," in R. Porter and M. Teich, edd., *Drugs and Narcotics in History* (1995) 4–23.
Idem and V. Nutton, "The *Preface* of Dioscorides' *Materia Medica*: Introduction, Translation, Commentary," *Transactions and Studies of the College of Physicians of Philadelphia* 4.3 (1982) 187–227.
I. Schlereth, *De Plutarchi quae feruntur parallelis minoribus* (Freiburg i. Br. 1931).
Wilh. Schulze, *Zur Geschichte lateinischer Eigennamen* (Berlin 1904; repr. 1966).
S.M. Sherwin-White, *Ancient Cos: an historical study from the Dorian settlement to the imperial period* (Göttingen 1978).
P.N. Singer, trans., *Galen: Selected Works* (Oxford 1997).
Wesley D. Smith, *The Hippocratic Tradition* (Ithaca and London 1979).
Idem, ed. and trans., *Hippocrates: Pseudepigraphic Works* (1990) = *SAM* 2.
H. Solin, *Die griechischen Personennamen in Rom: ein Namenbuch* 2nd ed., 3 vv. (Berlin and New York 2003).
Fr. Solmsen, "Greek philosophy and the discovery of the nerves," *MusH* 18 (1961) 150–197, repr. in his *Kleine Schriften* 1 (1968) 536–582.
Speranza (1971): see "editions," above.
Fr. Staab, "Ostrogothic geographers at the Court of Theodoric the Great," *Viator* 7 (1976) 27–58.
H. von Staden, *Herophilus* (Cambridge 1989).

Idem, "Rupture and Continuity: Hellenistic Reflections on the History of Medicine" in van der Eijk (1999) 143–187.
F. Steckerl, *The Fragments of Praxagoras of Cos and his School* (Leiden 1958).
M. Stern, *Greek and Latin Authors on Jews and Judaism* 3 vv. (Jerusalem 1974, 1980, 1984).
A.F. Stewart, "Nuggets: Mining the Texts Again," *AJA* 102 (1998) 271–282.
Malcolm Stuart, ed., *The Encyclopedia of Herbs and Herbalism* (London and Novara 1979).
H. Svenson-Evers, *Die griechische Architekten archaischer und klassischer Zeit* (Frankfurt am Main 1996).
S. Swain, ed., *Seeing the Face, Seeing the Soul. Polemon's* Physiognomy *from Classical Antiquity to Medieval Islam* (Oxford and New York 2007).
O. Temkin, *Geschichte des Hippokratismus im ausgehenden Altertum* = *Kyklos: Jahrbuch des Instituts für Geschichte der Medizin an der Universität Leipzig* 4 (1932).
Idem, Hippocrates in a World of Pagans and Christians (Baltimore 1991).
Thesleff (1965): see "editions," above.
D'A.W. Thompson, *A Glossary of Greek Fishes* (London 1947).
L. Thorndike and P. Kibre, *A Catalogue of Incipits of Mediaeval Scientific Writings in Latin*, 2nd ed. (London 1963).
C.O. Thulin, *Corpus agrimensorum Romanorum* (Berlin 1913; repr. Stuttgart 1971).
G.J. Toomer, *Diocles on Burning Mirrors* (Berlin and New York 1976).
A. Touwaide, "Medicinal Plants," *BNP* 8 (2006) 558–568.
M. Ullmann, *Die Medizin in Islam* (Leiden and Köln 1970).
Idem, Die Natur- und Geheimwissenschaften im Islam (Leiden and Köln 1972).
George Usher, *A Dictionary of Plants Used by Man* (London 1974).
J.T. Vallance, *The Lost Theory of Asclepiades of Bithynia* (Oxford 1990).
M.E. Vázquez Buján, ed., *Tradición e Innovación de la Medicina Latina de la Antigüedad y de la Alta Edad Media* (Santiago de Compostela 1994).
Ph. van der Eijk, ed., *Ancient Histories of Medicine: Essays in Medical Doxography and Historiography in Classical Antiquity* (Leiden 1999) = *SAM* 20.
B.L. van der Waerden, *Die Pythagoreer* (Zürich 1979).
J. Voinot, *Inventaire des cachets d'oculists gallo-romains* = *Conférences Lyonnaise d'ophtalmologie*, # 150 (Lyon 1981–1982).
Idem, Les cachets à collyres dans le monde romain (Montagnac 1999).
F.W. Walbank, *A historical commentary on Polybius* 3 vv. (Oxford 1957–1979).
John Warren, *Greek Mathematics and the Architects to Justinian* (London 1976).
Gilbert Watson, *Theriac and Mithridatium. A Study in Therapeutics* (London 1966).
Wehrli (1967–1969; 1978): see "editions", above.
M. Wellmann, *Die pneumatische Schule bis auf Archigenes* (Berlin 1895).
Idem, "Die Pflanzennamen des Dioskurides," *Hermes* 33 (1898) 360–422.
Idem (1901): see "editions," above.
Idem, "Pamphilos," *Hermes* 51 (1916) 1–64.
M.L. West, "Magnus and Marcellinus: Unnoticed Acrostics in the Cyranides," *CQ* 32 (1982) 480–481.
Idem, Orphic Poems (Oxford and New York 1983).
Westerink and Combès (1986–1991): see "editions," above.
M. Wichtl, ed., *Herbal Drugs and Phytopharmaceuticals*, 3rd ed. trans. from the 4th German ed. (Boca Raton and Stuttgart 2004).
E. Wickersheimer, "A Note on the *Liber de medicinis expertis* Attributed to Galen," *Annals of Medical History* 4 (1922) 323–327.
M.C. Wilson, "Hippocrates of Chios's Theory of Comets," *JHA* 39 (2008) 141–160.
E.T. Withington, *Hippocrates*, v.3 *Wounds in the Head. In the Surgery. Fractures, Joints, Instruments of Reduction* (Cambridge, MA: Loeb 1928, repr. 1984).
W. Wolska-Conus, "Stéphanos d'Athènes et Stéphanos d'Alexandrie. Essai d'identification et de biographie," *REByz* 47 (1989) 5–89.

H.C. Wood, Jr. and C.H. LaWall, *et al.*, edd., *The Dispensatory of the United States of America*, 21st ed. (Philadelphia and London 1926): later editions are less relevant.
L. Zhmud, *Wissenschaft, Philosophie und Religion im frühen Pythagoreismus* (Berlin 1997).
Idem, The Origin of the History of Science in Classical Antiquity (Berlin and New York 2006).

NOTE TO USERS

Entries whose inclusion in this work is doubtful have their lemma *italic*; entries with uncertain name have a "(?)" suffixed.

Entries for *anonymi* give the title of the work underlined; almost all papyri are filed under "Papyrus" with the papyrological citation.

Date-ranges in the lemmata are *termini post* and *ante* of the period of activity. Such a system is the only one that can be fairly applied to all cases. (For some entries only the *terminus post* or *ante* is known; a very few entries give only a single "*akmē*" date; the few known or inferred birth-dates are given *within* the lemma.) Precision is almost never possible, so most dates should be regarded as best estimates.

Homonyms are listed in this order: (1) bare names, in order by topic (mathematics, cosmology, astronomy, astrology, geography, mechanics, alchemy, biology, pharmacy, and medicine); (2) names provided with some accepted, usually ancient, epithet (e.g., Apollodōros "Dēmokritean" and then "*thēriakos*"); (3) names with known patronymics; (4) names with known ethnics (cities of origin or residence), in order by city-name. These 11 frequent names best show the system: Apollodōros, Apollōnios, Dēmētrios, Diodōros, Diogenēs, Dionusios, Hēliodōros, Hērakleidēs, Iōannēs, Mētrodōros, and Philōn.

Cross-references to other entries are indicated by SMALL CAPITALS (on their first occurrence within an entry).

Terms in the Glossary are marked in **bold** wherever they appear in an entry.

Bibliography at the end of an entry is intended to be initiatory and not complete (especially for entries such as Aristotle, Euclid, Galēn, Homer, or Plato); items in English have generally been preferred (e.g., *BNP* rather than *NP*), but not exclusively. Authors or works whose editions are cited within the encyclopedia (e.g., Aëtios of Amida, Pappos, Proklos, etc.) or else which would not readily be found through the initiatory bibliography cited, are given under "**Ed.**:" before other items. (*) indicates a person (or work) for which we could find no modern bibliography.

A

Aarōn ⇒ Ahrun

Abas (or Aias) (500 – 330 BCE)

Greek physician, quoted only by the Londiniensis medicus (8.45–9.4), who attributes diseases to discharges from the head through nose, ears, eyes and mouth. Health or disease depends on the quantity of these flows. *Cf.* Hippokratēs' *Loc. hom.* 1.10 (6.276 Littré); *Gland.* 11 (8.564 Littré); *Morb. sacr.* 3.17 (6.366 Littré).

RE S.1 (1903) 1–2 (#12), M. Wellmann; S.3 (1918) 13, H. Gossen; H. Grensemann, *Die hippokratische Schrift "Über die heilige Krankheit"* (1968) 30–31; *BNP* 1 (2002) 6, V. Nutton.

Daniela Manetti

Abaskantos of Lugdunum (10 BCE – 80 CE)

Andromakhos approves his remedy for ***phthisis***, composed of birthwort, saffron, ***euphorbia*** (*cf.* Iouba), gentian, henbane, mandrake, myrrh, opium, etc.: Galēn, *CMLoc* 7.2 (13.71 K.); he again cites Abaskantos for a colic remedy, involving Indian nard, myrrh, opium, pepper, etc. in boiled wine, *ibid.* 9.4 (13.278). Asklēpiadēs Pharm., in Galēn *Antid.* 2.12 (14.177 K.), cites him (with ethnic) for an antidote: castoreum, saffron, Illyrian iris, myrrh, opium, white pepper, germander, wild ***staphis***, etc., in wine. Andromakhos (13.71) gives him an apparently Roman *nomen*, *ΚΛΗΤΙΟΣ*, which may represent *GLITIVS: cf.* Pliny 7.39, Schulze (1904/1966) 232, n. 2, and *RE* S.3 (1918) 790–791 (#4).

RE 1.1 (1893) 20 (#8), M. Wellmann.

PTK

Abdaraxos (of Cyprus?) (330 – 25 BCE)

Writer on mechanics resident in Alexandria, listed by *P. Berol.* P-13044, col.8. The name is otherwise unattested, but compare Abdimilkos (4th c. BCE) and Abdubalos (5th c. BCE), both of Cyprus, *LGPN* 1.1: if Semitic, the prefix Abd- corresponds to *-doulos* (Sala 1974: 1–3). Perhaps the name derives from Abdēra, *cf.* Abdēriōn of Thrakē (*bis*, 4th and 3rd c. BCE) as well as Abdarakos of Tanais (3rd c. CE), *LGPN* 4.1.

Diels (1920) 30, n.1.

PTK

Abiyūn al-Biṭrīq (*ca* 630 CE)

"Apiōn the Patricius," at the time of the advent of Islam, mentioned by Ibn-al-Nadīm and Ibn-al-Qifṭī. He wrote a work "On Operating the Planispherical Astrolabe," at present unknown.

GAS 6 (1978) 103.

Kevin van Bladel

Abram (150 BCE – 150 CE)

Presumably pseudepigraphic astrological authority cited by VETTIUS VALENS (2.29–30) as a "most wondrous" authority on astrological prediction of a propensity to travel, and several times on various topics by FIRMICUS MATERNUS (4.*pr*, 4.17–18, and 8.3), who calls him "divine." The patriarch Abraham was regarded in early Jewish and Christian lore as a discoverer of astronomy (e.g. Iosephus, *Ant. Iud.* 1.156–157), but it is remarkable to find his reputation thus reflected in the "pagan" astrological tradition already in the 2nd c. CE.

RE S.1 (1903) 5 (#2), F. Boll; Riley (n.d.).

Alexander Jones

AC- ⇒ AK-

C. Acilius (155 – 140 BCE)

Wrote Roman history in Greek, and served as translator when KARNEADĒS, DIOGENĒS OF BABYLŌN, and KRITOLAOS OF PHASĒLIS addressed the Senate in 155 BCE. His annals explained Sicily as an island rent from the mainland in a prehistoric flood (F13).

FGrHist 813; *OCD3* 7–8, A.H. McDonald.

PTK

Acilius Hyginus of Kappadokia (20 – 55 CE)

Modified the colic remedy taught by ATIMĒTOS, substituting white pepper for black: SCRIBONIUS LARGUS 120 (MARCELLUS OF BORDEAUX 29.5 [*CML* 5, p. 502]). Presumably distinct from his contemporary, Acilius the rake: TACITUS, *Ann.* 13.19, 13.21–22.

RE S.3 (1918) 17 (#47a), W. Kroll; Korpela (1987) 164.

PTK

Adamantios (300 – 350 CE?)

Author of a *paraphrasis* of the physiognomy by POLEMŌN (whose Greek original is lost), taking into account also the ARISTOTELIAN CORPUS PHYSIOGNOMY, as he states in the foreword. The metaphor in the first sentence for his use of past physiognomic lore, that of setting up a holy statue of a god in a Pagan sacred precinct, hardly allows identifying him with ADAMANTIOS IOUDAIOS, which some have suggested (Rose; Wellmann; Nutton). Foerster (1.CII) deduces from style and language a date of 300–350 CE.

The treatise has two parts, the first of which contains a brief theoretical introduction on the methods of physiognomy (1.1–4) and long chapters on the significance of the eyes (1.5–23). The second briefly resumes the main areas of signs (2.1) and the significance of gender

(2.2–3) and then lists the signs in the unusual order from toes to head (2.4–30) as well as color, hair, stride, voice and the like (2.31–42). It ends with a list of character types and their signs (2.43–61), much in accordance with the equivalent list in the Aristotelian Corpus *Physiognomy*.

Ed.: I. Repath, "The *Physiognomy* of Adamantius the Sophist," in Swain (2007) 487–547.
V. Rose, *Aristoteles Pseudepigraphus* (1863) 697; *RE* 1.1 (1893) 343 (#1), M. Wellmann; *KP* 1.61, F. Kudlien; *PLRE* 2 (1980) 6; *BNP* 1 (2002) 133 (#1), V. Nutton.

Sabine Vogt

Adamantios of Alexandria, Ioudaios (*ca* 412/415 CE)

Jewish iatrosophist, who was expelled with other Jews from Alexandria by the patriarch Cyril (*cf.* KURILLOS) in *ca* 412 or 415 CE, went to Constantinople to be christened by the patriarch Atticus, and returned to live in Alexandria (Sōcr. *Hist. eccl.* 7.13.54–57). Sōcratēs calls him a "sophist of medical works" (*iatrikōn logōn sophistēs*). Given his interest in medicine, he might be the author of a metrology and recipes quoted by OREIBASIOS (*Syn.* 2.58; 3.24–25; 3.28–29; 3.35; 7.6; 9.57 = *CMG* 6.3, pp. 50–51, 73–77, etc.); if so, these must have been written before Oreibasios' death around 400 CE. Two recipes in AËTIOS OF AMIDA (8.29.1–47, *CMG* 8.2, pp. 438–440, for toothache and 15.6 [Zervos 1909: 23] for tumor of the throat) are also likely to be his.

Aëtios cites him as "Adamantios the sophist" in quoting the first of those recipes (8.29.2) and in an excerpt of a treatise "on the winds" (*peri anemōn*) (3.163). It has been doubted, however, whether the latter treatise was written by the same author, as it resembles **Peripatetic** meteorology and might stem rather from the 3rd c. CE (Rose 1.22, Nutton).

Ed.: V. Rose, *Anecdota Graeca et Graecolatina* (1864) 1.1–26 (introduction) and 1.27–52 (text).
RE 1.1 (1893) 343 (#1) M. Wellmann; *KP* 1.61, F. Kudlien; *PLRE* 2 (1980) 6 (#1); *BNP* 1 (2002) 133 (#1), V. Nutton.

Sabine Vogt

Adeimantos (325 BCE – 75 CE)

Listed by PLINY as an authority on "foreign" trees, such as cinnamon, and distinguished from medical authorities, 1.*ind.*12.

(*)

PTK

ADMĒTOS ⇒ ATIMĒTOS

Adrastos of Aphrodisias (60 – 170 CE)

Peripatetic philosopher. Two inscriptions of Aphrodisias (*ca* 110 and *ca* 185 CE) mention an Adrastos, but neither is identified as a philosopher (scholars have suggested identification, not proven). His commentary on the *Categories* – along with that of ASPASIOS – is mentioned by GALĒN. Semantic and metaphysical considerations play an important role in the passage from his commentary on the *Physics* quoted – through PORPHURIOS – by SIMPLICIUS. Besides some philological works on the history and the internal structure of the **Peripatetic** corpus, a commentary – or at least an extended discussion of the technically difficult passages and facets – of PLATO's *Timaeus* is attested. This commentary is quoted by Porphurios and transcribed by THEŌN OF SMURNA and CALCIDIUS. In his interpretation, Adrastos sets out the details, and the astronomical and musical issues operative in the Platonic text, but, characteristically, he brushes over the differences between ARISTOTLE's planetary theory of homocentric spheres and the later theory of **epicycles**; indeed, he claims that Plato already knew the theory of **epicycles**.

Moraux 2 (1984) 294–332; Gottschalk (1987) 1155–1156.

István Bodnár

Adrastos of Kuzikos (120 – 80 BCE)

Augustine, *City of God* 21.8.2, quotes VARRO following Kastōr of Rhodes saying that Adrastos and DIŌN OF NEAPOLIS computed the date of a portent of Venus.

RE S.1 (1903) 11–12 (#9), Fr. Hultsch.

PTK

AE- ⇒ AI-

Aeficianus (130 – 160 CE)

Stoicizing doctor who wrote commentaries on two books of the HIPPOKRATIC CORPUS, EPIDEMICS, was a student of QUINTUS, and taught GALĒN at Corinth *ca* 151–152 CE: Galēn, *Comm. in Hipp. Off.* 1.3 (18B.654 K.), *Comm. in Hipp. Epid. III* 1.40 (*CMG* 5.10.2.1, p. 59), *On the Order of my own Books* 3 (2.87 MMH).

Grmek and Gourevitch (1994) 1520–1521; Manetti and Roselli (1994) 1590–1591; *DPA* 1 (1989) 88, R. Goulet; Ihm (2002) #5–7.

PTK

AELIANUS ⇒ MAECIUS

Aelianus "the Platonist" (165/170 – 230/235 CE)

Author of a commentary on PLATO's *Timaeus*, a fragment of which survives in PORPHURIOS' commentary on PTOLEMY's *Harmonics*. Porphurios calls him Aelianus "the Platonist;" he is probably to be identified with the rhetorician and natural scientist CLAUDIUS AELIANUS, whose other lost works included an *Indictment of the Effeminate*, an *On Providence* and an *On Divine Manifestations*. His three extant works, *On the Nature of Animals, Varia Historia* (or "Miscellany") and *Rustic Letters*, show a concern to elucidate the workings of the divine in human and animal life.

Porphurios' quotations from Aelianus' commentary are limited to discussions of acoustics, harmonics, and musicological terminology; the tradition of comment on the *Timaeus* was a common forum for such discussions, not all of which were restricted to explanations of the harmonizing of the world soul (34b–36d). Porphurios' passing references to Aelianus show that in several other respects he adhered to the mathematical (rather than the Aristoxenian) musicological tradition: he followed Ptolemy in admitting six concords, rather than ARISTOXENOS' eight, and he discussed and explained musicological terminology peculiar to **Pythagorean** authors.

The four-page fragment from the second book of Aelianus' *Timaeus* commentary quoted by Porphurios (33.16–37.5 Düring) is concerned with the physical determinants of pitch-difference in musical notes. Aelianus subscribes to the traditional **Pythagorean** thesis (established by ARKHUTAS and codified in the EUCLIDEAN SECTIO CANONIS) that movement is the cause of all sound, and that sound is "air that has been struck" (*aēr peplēgmenos*, 33.21, a notion familiar from *Timaeus* 67b and the ARISTOTELIAN CORPUS ON SOUNDS). Differences in the speeds of the movements of the air cause differences in pitch: faster movement causes higher pitch, and slower movement causes lower pitch. Aelianus illustrates the theory with demonstrations on both wind and stringed instruments – demonstrations which, while they employ two different hypotheses about the causation of pitch, are unified in an attempt to explain how movement within an instrument is transferred to a movement of the surrounding "air that has been struck;" it is the relative speeds of the latter, in Aelianus' argument, which constitute the pitch differences we apprehend with our ears.

Aelianus discusses concord and discord, and defines concord as a blend (*krasis*) of two notes of different pitch, combined according to a principle of proportionality (*summetria*). He is explicit in his view (a logical consequence of his acoustic theory) that the two notes in a musical interval travel at different rates, and thus cover different distances in the same amount of time. (In the case of the 2:1 octave interval, Aelianus' theory demands that the higher note travel twice the distance of the lower note in the same amount of time). If empirical observation played a part in his investigations (as suggested by his instrumental demonstrations), it must therefore have been limited. Aelianus appears, from Porphurios' quotation, not to have been worried by the implications of this acoustic theory, about which ARISTOTLE had already expressed concern (*De Sensu* 448a).

Düring (1932); Barker (1989); Mathiesen (1999); *BNP* 1 (2002) 201 (#3), M. Baltes.

David Creese

Claudius Aelianus of Praeneste (*ca* 195 – *ca* 235 CE)

Born *ca* 170, Roman freedman and well-connected orator and priest, "honey-tongue" (*meliglōssos: Souda* AI-178), a canonical sophist who wrote in Greek (Philostratos *VS*). His lost treatises *On Providence* and *On Divine Manifestations* (perhaps the same work), based on a couple of fragments, may show **Stoic** ideas, which appear superficial or irrelevant in his two extant writings, which extol through exquisite anecdotes human morality and animal virtue. Besides a probably posthumous pamphlet *Indictment of the Effeminate* against Elagabalus, and 20 *Rustic Letters* (maybe apocryphal), he wrote a collection of edifying tales known as *Miscellany* (*Poikilē historia*, in 14 books) and a monumental compilation *On the Characteristics of Animals* (*Peri idiotētos zōiōn*, in 17 books), which is, after ARISTOTLE, the most important extant zoological opus in Greek. His subtly affected style delighted Byzantine scholars (see the numerous mentions in the *Souda*), and his work, surviving in many MSS, was abundantly imitated in the east and used in medieval bestiaries. The Constantinian animal anthology known as *Epitome of Aristophanēs of Byzantium* (10th c.) was primarily composed of an abstract of Aristotle and a wide choice of whole chapters from Aelianus.

Aelianus' work is a personal selection, made from numerous Greek authors (ALEXANDER OF MUNDOS, KTĒSIAS, DŌRIŌN, IOUBA, PAMPHILOS OF ALEXANDRIA, etc.), addressing all aspects of animals (mythology, ethology, biology, zootechnics, . . .), untidily dispatched in 808 chapters of uneven length. He records only three original observations (2.56, 5.47,

5.56), but his "personal contribution to science" (*prologue*) lies in literary achievement and scientific popularization. Mixing quasi-quotations and abrupt summaries, the book slips often into paradoxographical accounts, treating mythical animals, e.g., phoenix (6.58), basilisk (2.7), mantichore (4.21), gryphon (4.27), unicorn (16.20), amphisbaena (9.23). As usual in such collections, Aelianus included other natural ***paradoxa*** in chapters on springs and rivers (8.21, 9.29, 10.38, 12.36, 12.42, 14.19, 15.25) or on plants (9.31–33, 9.37, 14.27). Nevertheless, he preserves scientific information, e.g. DĒMOKRITOS on how deers' horns grow (12.18); the toxic action of venoms (4.36, 4.41); anatomy of cobra teeth (9.4). His testimony is especially worthy on Asiatic and African fauna, ichthyology, and angling (even more than OPPIANUS' *Halieutika*): Aelianus discusses butterfly-fish in minute detail (11.23), the otolith of some fishes (9.7), symbiosis of sponge and hermit crab (8.16), techniques of musical fishing (6.31–32, 17.18) and submarine hunting (4.58, 8.16), the subtle tactics of fishing-frogs (9.24), and gives the first reference to fly fishing (15.1, 15.10).

Ed.: A.F. Scholfield, *Aelian, On the characteristics of animals* 3 vv. (Loeb 1958–1959).
RE 1.1 (1893) 486–488 (#11), M. Wellmann; *DPA* 1 (1989) 79–81, S. Follet; *OCD3* 18, M.B. Trapp; *ANRW* 2.34.4 (1998) 2954–2996, J.F. Kindstrand.

Arnaud Zucker

AELIUS ⇒ PHLEGŌN

L. Aelius Gallus, praefect. Aegypti (45 – 5 BCE)

Although the chronology of Gallus' military expedition into Sabaean country (after 27 BCE) is debated (Jameson 1968), there is little doubt it was a disaster. Gallus was seduced by prospects of controlling the spice trade, which had enriched the Roman client kingdom of Nabatea, which received caravans and camel-loads of frankincense and myrrh from south-western Arabia, as well as many spices by then imported from India and south-east Asia. AUGUSTUS appointed Gallus Prefect of Egypt 27–25 BCE, years that witnessed his ill-fated attempts to control ports on the Red Sea and emporia further south. The Aelii were a late Roman Republic family of intellectuals, and Gallus' father, C. Aelius, was a legal lexicographer (Syme 1986: 308); the *gens* produced scholars known for varied interests, including science and medicine (*Ibid.*, 300). Aelius Gallus was patron and friend of STRABŌN, who was with Gallus (25 BCE) in Egypt (Syme 1995: 243, 322, 360).

Gallus' medical interests focused on pharmacology and toxicology. GALĒN (*Antid.* 2.17 [14.203 K.]) records a theriac against the stings of scorpions, a multi-ingredient drug that "...Gallus brought out of Arabia and gave to Caesar [Augustus], [and] many soldiers received cures from it." Probably Gallus was an "Asklēpiadean," since Galēn cites a "Marcus" Gallus, "follower of ASKLĒPIADĒS," as the inventor of a useful prophylactic aid (for use before luxurious meals), a compound of henbane seeds, roses, anise, celery seeds, old myrrh, and saffron crocus, boiled in wine and honey (*CMLoc* 8.5 [13.179–180 K.]). Galēn writes that an "...antidote of Aelius Gallus was employed by CAESAR and KHARMĒS against the lethal effects of poisons," which also aids women who have difficulty being purged, and it "...expels a fetus painlessly" (*Antid.* 2.1 [14.114–115 K.]). Gallus, a gourmand, also offered digestive "antidotes" for gluttony (Galēn, *Antid.* 2.10 [14.158–159, 161–162 K.]), containing myrrh and other costly, imported spices. ANDROMAKHOS SENIOR quotes from Gallus' books on cough syrups (Galēn quotes the quotes at *CMLoc* 7.2 [13.28–30 K.]), suggesting he had learned how useful were frankincense, myrrh, the two cinnamons, and other exotic ingredients in compounding effective cough drops and other medicines that soothed the windpipe.

RE 1.1 (1893) 492–493, P. von Rohden and M. Wellmann; S. Jameson, "Chronology of the Campaigns of Aelius Gallus and C. Petronius," *JRS* 58 (1968) 71–84; G.W. Bowersock, *Roman Arabia* (1983) 46–49; R. Syme, *The Augustan Aristocracy* (1986); *Idem, Anatolica: Studies in Strabo* (1995).

John Scarborough

Aelius Promotus of Alexandria (*ca* 140 – 190 CE?)

Physician, wrote *On Curative Remedies*, and *Natures and Antipathies* (*cf.* NEPUALIOS), the latter unedited. The first work (ed. Crismani) contains 130 chapters of remedies for such conditions as falling hair (1), ***duspnoia*** (30), fevers (39–40), ***anthrax*** (57), ***erusipelas*** (59), insomnia (85–90), eye disorders (96–99), and colic or dysentery (119–130). Nestled between the two attributed treatises, two MSS (*Vat. Gr.* 299, *Ambros. Gr.* S3) transmit an anonymous text *On Venomous Animals and Poisons*, attributed by scholars to Aelius, ARKHIGENĒS, or AISKHRIŌN OF PERGAMON (but PHILOUMENOS 14 shows that Arkhigenēs offered a different analysis of scorpion poisons than *Venomous* 15). It cites NOUMĒNIOS OF HĒRAKLEIA (14), AFRICANUS PHARM. (50), SŌRANOS (56), and EPAINETĒS (64–67, etc.); and is cited first by AËTIOS OF AMIDA (Book 13, *passim*). The treatise falls into two parts: poisoning caused by bites (snakes – *cf.* NIKANDROS, lizards, humans, ***hudrophobic*** dogs, cats, etc.), and from ingesting plants (aconite, hemlock, henbane, mandrake, etc.), minerals (**litharge**, mercury, ***psimuthion***, etc.), or small animals (leeches, etc.). Ihm divides the text into 79 chapters, each describing the poison, symptoms, and remedies. Though references to Arabia and Egypt may suggest a geographical link, the author's treatment of crocodiles and lions makes no particular Egyptian correlation. Ihm dates the core of the treatise to the era of Aelius, and subsequent additions before Aëtios; Touwaide considers this text a 14th century compilation.

Ed.: S. Ihm, *Der Traktat Peri ton iobolon therion kai deleterion pharmakon des sog. Aelius Promotus* (1995); D. Crismani, *Manuale della Salute* (2002).

OCD3 19, anon.; A. Touwaide, rev. of Ihm, *Medicina nei Secoli* 8 (1996) 306–307; *BNP* 1 (2002) 207, V. Nutton.

PTK and GLIM

Palladius Rutilius Taurus Aemilianus (*ca* 375 – *ca* 450 CE)

Latest surviving Latin agricultural writer, of uncertain date: he used the work of VINDONIUS ANATOLIOS, and his title *uir inlustris* postdates *ca* 375. Author of *Opus agriculturae* in 13 books: the first treats general matters (e.g. siting, water, building, poultry, beekeeping), while each of the following is devoted to the range of tasks appropriate for each calendar month. A 14th book discusses veterinary medicine. An elegiac poem on grafting, addressed to an unknown Pasiphilus, is appended as literary flourish in the manner of COLUMELLA's Book 10.

Palladius owned property near Rome (3.25.20) and in the area of Neapolis in Sardinia (4.10.16); he describes in detail a reaping machine used in the plains of Gaul (7.2.2–4). Primarily he follows literary sources: Columella on field crops, vines, livestock, and GARGILIUS MARTIALIS on gardens and fruit trees; these are supplemented by *Graeci* (i.e., Anatolios) and, for building topics, CETIUS FAUENTINUS' epitome of VITRUUIUS (not named). He expresses personal views and practices desultorily (e.g. 1.28.5, 2.9.1, 4.10.24). His is not the world of large (and absentee) landowners; agricultural slaves are mentioned but once (1.6.18). That "some fragment of column" will serve for rolling the threshing floor (7.1) implies agricultural recession. Lists of necessities to be kept ready (e.g. 1.42 implements, 14.3 medicinal plants) combine with straightforward organization to appeal to an audience of free tenants. He provides novel uses of wood: vine props made of winter oak (*aesculus*) and exposed structures made of Spanish chestnut (*castanea*): 12.15.2.

Language and style are characteristic of late Antiquity. There is an explicit aversion to rhetorical embellishment (1.1.1), belied to some extent by conscious application of both quantitative and accentual prose rhythms. Palladius' work was recommended by CASSIODORUS (*Inst.* 1.28.6) and used by ISIDORE (e.g. *Etym.* 17.10.8). Books 1–13 were transmitted as a unit and circulated widely from the 9th c. onward, eclipsing all similar works in the Latin Middle Ages. The *Carmen de insitione* was known to 15th c. readers, but Book 14 re-emerged only in the 20th c.

Ed.: Robert H. Rodgers (1975); concordance: J.F. Núñez (2003).

J. Svennung, "De auctoribus Palladii," *Eranos* 25 (1927) 123–178, 230–248; *Idem, Untersuchungen zu Palladius* (1935); *PLRE* 1 (1971) 23–24; F. Morgenstern, "Die Auswertung des opus agriculturae des Palladius . . .," *Klio* 71 (1989) 179–192; D. Vera, "Dalla 'villa perfecta' alla villa di Palladio," *Athenaeum* 83 (1995) 189–211, 331–356; *OCD3* 1101, M.S. Spurr; *BNP* 10 (2007) 393–394, K. Ruffing.

Robert H. Rodgers

Aemilius Hispanus (*ca* 100 BCE – *ca* 350 CE)

Cited by PELAGONIUS OF SALONA for a remedy for arthritic **glanders** (Pel. 23 = *Hippiatrica Parisina* 57 = *Hippiatrica Berolinensia* 4.14) and described as a *mango* or horse-dealer.

Fischer (1980) 23; Adams (1995).

Anne McCabe

Aemilius Macer of Verona (d. 16 BCE)

Wrote didactic poems. The *scholia Bernensia ad VERGILIUS Ecl.* 5.1 claim he was *Vergilii amantissimus*, and that Vergil disguised him as Mopsus and himself as Menalcas. Macer read his poems to OVID, his younger contemporary (*Trist.* 4.10.43–44). He died in Asia. Two titles and only fragments thereof survive: The *Generation of Birds* (*Ornithogonia*), in two or more books, *frr.* 1–6, and *Thēriaka*, in two books, *frr.* 7–11. Scholars assume the existence of another poem, to which Ovid (above) *legit . . . quae iuuat herba* seems to allude, as does MANILIUS (2.44). Fragments 12–14 probably belonged to the latter work, whose title was perhaps *Alexipharmaka* (alternatively, this poem on herbal remedies might simply have been *Thēriaka* 2). The *Generation of Birds* was based on Boios' poem with the same title, whereas *Thēriaka* followed NIKANDROS. LUCANUS seems to have drawn on Macer in his excursus on Lybian snakes (9.700–947); a *scholion* to Lucanus 9.701, in fact, cites Macer as a possible source.

W. Morel, "Iologica," *Philologus* 83 (1928) 345–389; R. Rau, "Ein Jugendwerk Ovids," *PhW* 52 (1932) 895–896; A.S. Hollis, "Aemilius Macer, Alexipharmaca?" *CR* 87 (1973) 11; H. Dahlmann, *Über Aemilius Macer* (1981); F. Brena, "Nota a Macro, fr. 17 Büchn.," *Maia* 44 (1992) 171–172; *FLP* 292–299 and 520; Blänsdorf (1995) 271–278; Jacques (2002) CXVII, n. 253.

Claudio De Stefani

Aethicus Ister (650 – 800 CE?)

The purported author, otherwise unknown, of a Latin *Cosmographia* written in the late 7th or 8th c. and containing later interpolations. It claims that a philosopher and traveler Aethicus wrote it in Greek, and that Jerome translated it into Latin (i.e., *ca* 400 CE). The work begins with the creation of the world and describes the author's travels through the ***oikoumenē***, including Taprobanē (Ceylon), Britain, Thule, Asia Minor, Greece, and many other real and imaginary places. Some of the material is taken from Isidorus of Hispalis, Iulius Solinus, and other earlier writers. The author emphasizes that much of his material is not mentioned in any other authority, and he obviously invents some place names. The names of the author and the translator are considered to be a mystification, and the work may have been a parody, missed by its medieval audiences. The *Cosmographia* was often used in the Middle Ages by geographical writers and mapmakers.

Ed.: O. Prinz, *Die Kosmographie des Aethicus* (1993).

RE 1.1 (1893) 697–699, H. Berger; *TTE* 4–5, M. Hamel; M.H. Herren, "The 'Cosmography' of Aethicus Ister: Speculations about Its Date, Provenance, and Audience," in A. Bihrer and E. Stein, edd., *Nova de veteribus: mittel- und neulateinische Studien für Paul Gerhard Schmidt* (2004) 79–102; D. Shanzer, "The Cosmographia Attributed to Aethicus Ister as Philosophen- or Reiseroman," in G.R. Wieland *et al.*, edd., *Insignis Sophiae Arcator: Medieval Latin Studies in Honour of Michael Herren on His 65th Birthday* (2006) 57–86.

Natalia Lozovsky

Aethicus, Pseudo (450 – 600 CE?)

The unknown author of a *Cosmographia* (different from that by Aethicus Ister but falsely attributed to him in some MSS). The first part is based on Iulius Honorius, and reports that Iulius Caesar as consul ordered a survey and measurement of the world. Then it lists geographical features, such as seas, rivers, and mountains, as well as provinces, towns, and peoples. The second part, drawn from Orosius, describes the three known parts of the world, Asia, Europe, and Africa. The work's focus on Italy suggests a Roman compiler.

Ed.: *GLM* 71–103.

RE 1.1 (1893) 697–699, H. Berger; *GRL* §1061; *PLRE* 2 (1980) 19; C. Nicolet and P.G. Dalché, "Les 'quatre sages' de Jules César et la 'mesure du monde' selon Julius Honorius: réalité antique et tradition médiévale," *Journal des savants* (1986) 157–218.

Natalia Lozovsky

Aethlios of Samos (350 – 200 BCE?)

Wrote a chronicle of Samos, giving geographical or botanical data: fruits that grow twice a year, pears of Keōs. For the very rare name, *cf.* Diogenēs Laërtios 8.89.

FGrHist 536 = Ath., *Deipn.* 14 (650d, 653f).

PTK

Aëtios (1st c. CE)

The name of an otherwise unknown writer of a survey of philosophical opinions (often called *Placita*). Since Diels (1879) he is assumed to be the source of the two extant specimens of doxography found in pseudo-Plutarch's *Epitome of the Opinions of the Philosophers* and

Iōannēs Stobaios' *Eclogae Physicae*. Aëtios presents various philosophical views in short thematic entries on cosmology and (meta)physics, meteorology, psychology and perception, and human physiology and embryology. There are no arguments and no context; many of the questions posed and answered seem to reflect the concerns of Hellenistic philosophy and not those of the original philosophers. As for the Pre-Socratic philosophers, many pieces of information derive from Aristotle and Theophrastos.

Ed.: Diels (1879) 267–444; L. Torraca, trans., *I Dossografi Greci* (1961); H. Daiber, *Aëtius Arabus. Die Vorsokratiker in arabischer Überlieferung* (1980).

J. Mansfeld "Chrysippus and the Placita," *Phronesis* 34 (1989) 311–342; *Idem*, "Physikai doxai and Problemata physica from Aristotle to Aëtius (and Beyond)," in W.W. Fortenbaugh and D. Gutas, edd., *Theophrastus; his Psychological, Doxographical and Scientific Writings* (1992) 63–111; Mansfeld and Runia (1996); *BNP* 1 (2002) 274–276 (#2), D.T. Runia.

Jørgen Mejer

Aëtios of Amida (500 – 550 CE)

A scholion to a MS of the *Tetrabiblos* (*CMG* 8.1, p. 8) terms the author a "*komētos tou opsikiou*," indicating that Aëtios was probably a court physician. Traditions and MSS uniformly suggest a *floruit* in the reign of Justinian (527–565 CE) and, given the unique preponderance of obstetrics and gynecology (and large number of contraceptive and abortifacient recipes) in Book 16 of the *Tetrabiblos*, it is also likely that Aëtios was a personal physician to the empress Theodōra (d. 548 CE), whose checkered career receives scurrilous if overdrawn detail in Prokopios' *Arcana*. Aëtios studied medicine in Alexandria (*Tetr.* 1.131; 1.132; 2.121; *cf.* 4.22 [*CMG* 8.1, pp. 65, 67, 197, 368]), and may have practiced for a time in Egypt before moving to Constantinople.

Olivieri collated 29 MSS to produce the *CMG* edition of Books 1–8 of the *Sixteen Books* (Grk. *Tetrabiblos*, so named for the usual subdivision into four blocks of four books each), and the large number of texts (showing widespread popularity) descending into the Renaissance generally militated against meticulous editing of the Greek, although good translations into Latin appeared in the 16th c. (Cornarius [1542] remains the only complete and fairly reliable edition of Aëtios' gigantic handbook). Dry in style but remarkably comprehensive, the *Tetrabiblos* reflects the teaching of medicine in 6th c. Alexandria: an authoritative text is quoted, *then* the practicing physician adds his own experiences, especially those recipes for drugs and surgical techniques found to be beneficial; probably Aëtios had at hand a well stocked medical library in Alexandria, as well as in Constantinople. Book 1 begins with a famous "drugs-by-degrees" summary, the theoretical constructs of pharmacy that predominated until medicinal chemistry in the 19th c. Significant are Aëtios' accounts of mastectomies, embryotomies, abortions (never after the third month, never before), and repair of inguinal hernias in Book 16, the toxicology in Book 13, general surgery in 14, and the rightly famous ophthalmology in Book 7. Phōtios includes a lengthy summary of Aëtios' work in the *Bibliotheca*, and is duly impressed, concluding, "Indeed, those who have chosen to demonstrate through their [medical] practice that [medical] attention can drive away diseases [or afflictions], should devote continual study and close attention to this work" (*Biblioth.*, 221.181a R. Henry).

Ed.: [Latin] J. Cornarius, *Aetii medici graeci contractae ex veteribus medicinæ Tetrabiblos*, etc. (1542): still essential for Books 9–16; A. Olivieri, *Aetii Amideni Libri medicinales I–IV, V–VIII* (1935, 1950) = *CMG* 8.1–2; J. Hirschberg, *Die Augenheilkunde des Aëtius aus Amida* (1899); Ch. Daremberg and Ch.É. Ruelle,

Aetiou tou Amidēnou Biblion IA, in *Oeuvres de Rufus d'Éphèse* (1879; repr. 1963) 85–131; S. Zervos, "Aetiou Amidēnou Logos Enatos," *Athēna* 23 (1911) 265–392; G.A. Kostomiris, *Aetiou logos dodekatos* (1892); S. Zervos, *Aetiou Amidinou* [sic] *Logos dekatos pemptos* in *Athēna* 21 (1909) 3–144; *Idem, Aetii Sermo sextidecimus et ultimus. Ersten aus Handschriften veröffentlicht* (1901); Brock (1929) 247–249 ("Aetius of Amida: Aneurysms" and "Fatty Tumours").

I. Bloch, "Aëtios von Amida," in *HGM* 529–535; Scarborough (1985b) 224–226; R. Masullo, "Problemi relativi alle fonti di Aezio Amideno nei libri IX–XVI: Filumeno, Areteo e altri medici minori;" A. Pignani, "Aezio Amideno L.XI: La considerazione delle fonti nella costituzione del testo;" and R. Maisano, "L'edizione di Aezio Amideno, IX–XVI," in A. Garzya, ed., *Tradizione e ecdotica dei testi medici tardoantichi e bizantini* (1992) 237–256, 271–274, and 350–353; John Scarborough, "Teaching Surgery in Late Byzantine Alexandria," in H.D.F. Horstmanshoff, ed., *Medical Education in Antiquity* (forthcoming).

John Scarborough

Aetna (55 BCE – 78 CE)

Latin didactic poem, of unknown authorship and date, treating the causes of volcanic activity in general by focusing on Sicily's Mount Etna in particular. It is generally seen as postdating LUCRETIUS for stylistic reasons, and predating the eruption of Vesuvius (79 CE) because of the poem's reference to the Naples area as having been long volcanically inactive (line 431). The poem's authorship has always been a question of some debate. Many of the poem's MSS are ascribed to VERGIL, though most (but certainly not all) recent commentators reject this ascription. Other candidates, including SENECA, MANILIUS, and PLINY, have been variously (and sometimes rashly) offered. The Augustan poet Cornelius Seuerus was long preferred, but more recently, C. Lucilius, to whom Seneca dedicated his *Naturales Quaestiones*, has been proposed. Nevertheless, the question is far from settled and the evidence does not strongly favor any candidate.

The poem's explanation of volcanism is similar to Seneca's explanation of earthquakes (*Q.Nat.* 6). Several other passages in the poem may also indicate a **Stoic** author, and POSEIDŌNIOS' influence is often hypothesized. A longish passage extolling the importance of studying both physics and astronomy also includes references to the divinity of the stars, and possibly to the **Stoic** end-of-the-world conflagration. The basic argument of the poem is that volcanic activity is caused by the powerful motion of wind through natural subterranean passages. Certain types of sulfurous stones, with which Aetna is prodigiously furnished, serve in combination with the subterranean winds to "feed" and "nourish" the flames of the volcano during an eruption. Much of the argument's detail is, however, obscure and the difficulty is compounded by the highly corrupted state of the text.

Ed.: J. Vessereau, *L'Etna* (1905; repr. 1961) with commentary; W. Richter, *Aetna* (1963), with German translation; F.R.D. Goodyear, *Incerti auctoris Aetna* (1965) with commentary.

Daryn Lehoux

AFRICANUS ⇒ IULIUS AFRICANUS

Africanus (Metrol.) (200 – 300 CE)

A short treatise *On measures and weights* is transmitted by the MSS under, among others, Africanus' name, which modern scholars ascribe either to IULIUS AFRICANUS or, alternatively, to this later Africanus.

P. De Lagarde, *Symmicta* 1 (1874) 166–176; *RE* 1.1 (1893) 715–716 (#7), M. Wellmann; *BNP* 1 (2002) 300–301 (#1), V. Nutton.

Mauro de Nardis

Africanus (Pharm.) (***ca*** **40 – 30 BCE**)

Both Africanus' name and work are attested only by manuscript tradition. In a Greek codex containing an excerpt on medical matters and antidotes from Aelius Promotus, Africanus, hypothetically a pharmacologist or a physician, is referred to as an eyewitness, "under king Antigonos" of how citron can act as antidote to any poison. Since Africanus is a name not commonly used before the 2nd c. BCE (Kajanto, *Latin Cognomina* [1965] 205–206), probably the reference is to Antigonos of Judea, who ruled 40–37 BCE.

E. Rohde, "Aelius Promotus," *RhM* 28 (1873) 287; *RE* 1.1 (1893) 715–716 (#7), M. Wellmann; *BNP* 1 (2002) 300–301 (#1), V. Nutton.

Mauro de Nardis

Aganis (520 – 550 CE)

Cited by Simplicius, *In Eucl. Elem. I* (preserved solely in Arabic), as a companion, and as following Ploutarkhos of Athens on defining the angle. The name seems to be Egyptian (Coptic), not Greek.

DPA 1 (1989) 60–62, R. Goulet and M. Aouad.

PTK

Agapētós (200 – 560 CE)

Alexander of Tralleis (2.529–531 Puschm.) and Paulos of Aigina 7.11.59 (*CMG* 9.2, pp. 312–313) record two versions of Agapētós' 15-ingredient gout remedy, both containing aloes, saffron, ***malabathron***, myrrh, pimpernel, peony, spikenard, etc., but differing in five ingredients. The rare name is likely Christian, though *cf.* *Iliad* 6.401.

RE 1.1 (1893) 734 (#2), M. Wellmann.

PTK

Agapios of Alexandria **(470 – 510 CE)**

Damaskios' *Life of Isidore* describes the very erudite and admired medical scholar, who migrated to Constantinople and there became wealthy from his work (Phōtios, *Bibl.* 242.298 [352a34–b2]; *Souda* A-158), perhaps to be distinguished from the homonymous coeval neo-**Platonist**, *Souda* A-157, who taught Iōannēs of Philadelpheia.

DPA 1 (1989) 63, R. Goulet.

PTK

Agatharkhidēs of Knidos (***ca*** **200 – 140 BCE**)

Agatharkhidēs was a historian and grammarian. Born in Knidos (Cnidus), he was raised in the household of a councilor of Ptolemy VI named Kineas and was the protégé of the historian and diplomat Hērakleidēs "Lembos," whom he served as personal secretary and reader. Nothing is known about his life except that he was a **Peripatetic** like his patron

Hērakleidēs and that he lived for some time in Athens, probably after being exiled by Ptolemy VIII in 145 BCE.

Only minor fragments survive of most of his works. According to Phōtios, he wrote seven works. These include an epitome of the 4th c. BCE poet Antimakhos of Kolophōn's *Ludē* ("*Lyde*"), a book on friendship, and a collection of excerpts from writers on remarkable natural and human phenomena. Agatharkhidēs was best known in antiquity, however, as a historian. His principal works were two large histories – *On Affairs in Asia* in 10 books and *On Affairs in Europe* in 49 books – which together surveyed world history up to his own time. His third historical work, the *On the Erythraean Sea* in five books, is better known thanks to the survival of an epitome of its first and fifth books by Phōtios and extensive excerpts in the third book of DIODŌROS OF SICILY.

The fifth book of the *On the Erythraean Sea* treated comprehensively the history and cultural geography of the Red Sea and its hinterlands based on the reports of 3rd c. BCE Ptolemaic explorers. Its ethnographic accounts were organized according to the **Peripatetic** theory that a people's interaction with its environment determined the nature of its culture. Although not a formal geographical work, the fifth book of the *On the Erythraean Sea* was the main source for later accounts of the geography and ethnology of the region, strongly influencing STRABŌN's *Geography*, PLINY's *Natural History*, and AELIANUS' *On the Nature of Animals*.

Ed.: *FGrHist* 86; Stanley M. Burstein, *Agatharchides of Cnidus, On the Erythraean Sea* (1989).
BNP 1 (2002) 311, K. Meister.

Stanley M. Burstein

Agatharkhidēs of Samos (250 BCE – 50 CE?)

Author of an *On stones* in at least four books (PSEUDO-PLUTARCH *De fluu.* 9.5 [1155D]). Many consider him fictive. (Schlereth, however, identifies him with the **Peripatetic** historian and geographer AGATHARKHIDĒS OF KNIDOS.) PLUTARCH quotes our author in *Parall. min.* 305E, and attributes to him a *Persika* in at least two books, possibly supporting the historicity of our Agatharkhidēs.

J. Geffken, *Geographie des Westens* (1892) 85, n.2; *RE* S.1 (1903) 22, G. Knaack; Schlereth (1931) 97–99; Jacoby (1940) 76; *FGrHist* 284; Giannini (1964) 124; *PGR* 144–145; De Lazzer (2003) 66–67.

Eugenio Amato

Agatharkhos of Samos (460 – 410 BCE?)

Son of Eudēmos, worked as a painter in Athens, and wrote a book on *skēnē*-painting for Aeschylus, or a revival of Aeschylus, offering a novel theory of perspective, that inspired DĒMOKRITOS and ANAXAGORAS (VITRUUIUS 7.*pr.*11; *cf.* PLATO *Rep.* 10 [602c–d]). Agatharkhos worked rapidly (PLUTARCH, *Per.* 13.2), and was compelled to paint Alkibiadēs' house (Andokidēs 4.17).

BNP 1 (2002) 311–312, N. Hoesch.

PTK

Agathēmeros son of Orthōn (400 – 600 CE)

Otherwise unknown author of a treatise *Geōgraphias hupotupōsis*, preserved only in later copies of the 9th c. codex Palatinus gr. 398. The text treats the history of geography from

ANAXIMANDROS to POSEIDŌNIOS and describes the continents, winds, seas, length of the ***oikoumenē*** (in stades), and perimeters of some Mediterranean islands between the Strait of Gibraltar in the west and Lesbos in the east. Written without literary pretensions, the work seems influenced by ERATOSTHENĒS. Agathēmeros quotes from DIKAIARKHOS and TIMOSTHENĒS, and draws from ARTEMIDŌROS OF EPHESOS and MENIPPOS OF PERGAMON for the passage on perimeters. The MS-tradition appended two anonymous works to the *Geōgraphias hupotupōsis*: the *Hupotupōsis geōgraphias en epitomē* was present already in the codex Palatinus gr. 398 (see EXPOSITIO GEOGRAPHIAE). Then in the 15th c. the *Diagnōsis en epitomē tēs en tē sphaira geographias* was added: see SUMMARIA RATIONIS GEOGRAPHIAE. Two further fragments concern distances (in stades) and islands: *GGM* 2.509–511.

A. Diller, "Agathemerus, *Sketch of Geography*," *GRBS* 16 (1975) 59–76.
KP 1.116, H. Gams; *HLB* 1.528; *BNP* 1 (2002) 312, K. Brodersen.

Andreas Kuelzer

Agathinos of Sparta (30 – 70 CE)

Greek physician from Sparta, active in Rome, probably distinct from Claudius Agathinos of Bithunia (1st c. BCE), but possibly in contact with the **Stoic** L. Annaeus Cornutus. Agathinos may have been a pupil of ATHĒNAIOS OF ATTALEIA (Wellmann 1895: 14), or just a member of his *circle* (GALĒN, *Dign. Puls.* 1, 8.787 K.). His students included ARKHIGENĒS and HĒRODOTOS (PNEUM.). Agathinos is generally considered a **Pneumaticist** (Galēn, *Diff. Puls.* 3, 8.674 K.), or even that school's founder (Gourevitch 1993: 135–136). PSEUDO-GALĒN DEFINITIONES (19.353 K.) further credited him with creating the *episunthetic* or eclectic school (*cf.* 7.359, 8.771 K.). He seems also to have borrowed **Empiricist** and **Methodist** elements, and (probably SŌRANOS in) CAELIUS AURELIANUS includes him among the **Methodists** (*ex nostris: Acut.* 2.57, *CML* 6.1.1, p. 166). The unfixed and preliminary state of the developing school, or else his own attempt to transcend contemporary medical sectarianism, may be the origin of the ambiguity of his affiliation. **Pneumaticism** indeed seems to have evolved rapidly to an open and flexible system, permeable to heterogeneous contributions.

In accord with **Pneumaticist** theory on the importance of the vascular system as the vehicle of ***pneuma***, Agathinos wrote on pulse (Galēn, *Diff. Puls.* 2 [8.593–594 K.], 4 [8.748–750, 753–754 K.], *Dign. Puls.* 1 [8.787 K.], 4 [8.931, 935–936, 953 K.]), fever (Galēn, *Febr. Diff.* 2 [7.367, 369, 373 K.], *De Typis* [7.469 K.]; 17A.120, 228, 942 K.), pharmaceuticals (Galēn, *CMGen* 5 [13.830 K.]), and hydrotherapy (OREIBASIOS, *Coll.* 10.7, *CMG* 6.1.2, pp. 49–53). He wrote a book on hellebore (Cael. Aur., *Acut.* 3.135, *CML* 6.1.1, p. 371; *cf.* Oreib., *Coll.* 8.2, *CMG* 6.1.1, p. 252), speculated on the definitions to be used in sphygmology (Galēn, *Diff. Puls.* 4 [8.750 K.]), and analyzed tertian fever, distinguishing an intermediary semi-tertian fever on which he might have written a treatise (Marganne 1981: 311–314).

Only fragments of Agathinos' works survive, with the exception of the papyrus fragment on semi-tertian fever attributed to him. Galēn respected him highly (*Adv. Typ.* 7.488 K.; *Dign. Puls.* 4 [8.938 K.]), and compared him with HĒROPHILOS for his efforts to improve medicine and for his capacity to reason (*Dign. Puls.* 1 [8.786–788 K.]), but also criticized Agathinos' unnecessary prolixity (*Diff. Puls.* 4 [8.750 K.]), enigmatic explanations (*Dign. Puls.* 4 [8.935 K.]), and excessive emphasis on terms rather than on facts (*Febr. Diff.* 2 [7.367 K.]).

RE 1.1 (1893) 745, M. Wellmann; *KP* 1.117 (#2), Fr. Kudlien; Kudlien (1968) 1098; *DSB* 1.74–75, *Idem; OCD3* 36, J.T. Vallance; *BNP* 1 (2002) 313–314, V. Nutton.

Alain Touwaide

Agathodaimōn, pseudo (250 BCE – 300 CE)

Pseudonymous authority first mentioned in the Greek alchemical corpus by ZŌSIMOS OF PANŌPOLIS, but also found in the CORPUS HERMETICUM. OLUMPIODŌROS OF ALEXANDRIA gives three mythical reports on the identity of this otherwise unknown author or authors (*CAAG* 2.79–80). His only extant Greek works are a *Demonstration and Commentary on the Oracle of Orpheus* (*CAAG* 2.268–271) and an aphorism (*CAAG* 2.115). Zōsimos (*CAAG* 2.193) cites his *Teaching on the Pretincture* and Olumpiodōros (*loc. cit.*) mentions his *Alchemical Book* (*Biblos Khēmeutikē*), both lost. He is also associated with the verse *Riddle of the Philosophical Stone of Hermēs and Agathodaimōn* (*CAAG* 2.267–268), an excerpt from the *Sibylline Oracles* (lines 141–146) which had been given an alchemical interpretation by, at least, the time of Olumpiodōros (*CAAG* 2.71). Texts attributed to Agathodaimōn are extant in Arabic.

Ullmann (1972) 175–177.

Bink Hallum

Agathodaimōn of Alexandria (after 178 CE)

Several MSS of PTOLEMY's *Geōgraphikē huphēgēsis* (including 13th c. Vat. Urb. gr. 82, 14th c. Florent. Laurent. XXVIII 49, and 15th c. Venet. Marc. gr. 516) contain at the end of Book 8 (where the world-map is divided into 26 regional maps) a small notice that one *Agathodaimōn Alexandreus mēkhanikos* has drawn the whole ***oikoumenē*** (*hupetupōsato, hupetupōsa* or *hupetupōse*). In spite of intensive research, it is still today impossible to decide whether this sentence means the drawing of all Ptolemaic maps or simply the drawing of the Ptolemaic world map.

H. von Mžik, *Denkschriften Akademie Wien* 59.4 (1916), appendix 2; *RE* S.3 (1918) 59, Jos. Fischer; *RE* S.10 (1965) 737–741, E. Polaschek.

Andreas Kuelzer

Agathoklēs (50 BCE – 75 CE)

Wrote a work *On Nutrition*, after DĒMĒTRIOS "KHLŌROS" (*Schol. Nik. Thēr.* 622), and is cited by ANDROMAKHOS, in Galēn *CMGen* 5.12 (13.832–833 K.), for an ***erusipelas*** remedy, and by PLINY 22.90 for an antidote to "bull's blood." Lucian's joke merely uses the name, *Kataplous* 7.

BNP 1 (2002) 317 (#12), V. Nutton.

PTK

Agathoklēs of Atrax (300 BCE – 175 CE)

Wrote *On fishes* (*Halieutika*) in prose (Ath. *Deipn.* 1 [13c] = *Souda* K-1596); *cf.* POSEIDŌNIOS OF CORINTH. Perhaps identical to the Agathoklēs whom PLINY includes as a foreign authority on geography (1.*ind.*4–6), if that is not a reference to the memoirs of the tyrant Agathoklēs of Surakousai.

RE 1.1 (1893) 759 (#33), E. Oder.

PTK and GLIM

Agathoklēs of Khios (325 – 90 BCE)

Authored a treatise on agriculture possibly covering cereals, livestock, poultry, viticulture, and arboriculture (*cf.* PLINY, 1.*ind.*8, 10, 14–15, 17–18) that CASSIUS DIONUSIOS excerpted (VARRO, *RR* 1.1.8–10, *cf.* COLUMELLA, 1.1.9).

RE 1.1 (1893) 759 (#32), E. Oder.

Philip Thibodeau

***Agathoklēs of Milētos* (250 BCE – 50 CE?)**

Cited for a ***paradoxon*** by PSEUDO-PLUTARCH, *On Rivers* 18.3, and likely fictive.

RE 1.1 (1893) 759 (#27), M. Wellmann.

PTK

AGATHŌN ⇒ BOOK OF ASSUMPTIONS BY AQĀṬUN

Agathōn of Samos (300 – 50 BCE?)

Anethnically credited with an accurate ***Periplous*** *of the Pontos* by a scholiast, and credited by PSEUDO-PLUTARCH (most of whose citations are fictive) with a *Skuthika*, whose sole fragment describes a marvelous plant, and a book *On Rivers*, whose two fragments describe marvelous plants.

FGrHist 801 (*Periplous*), 843 (*Skuthika, On Rivers*); *BNP* 1 (2002) 318 (#2), K. Brodersen.

PTK

***Agathosthenēs* (unknown date)**

Mentioned among paradoxographers discussing aquatic phenomena (Tzetzēs *Chil.* 8.144, 645). Whether he is identical to Aglaosthenēs, author of *Naxiaka* (Müller *FHG* 4.294), is uncertain.

RE 18.3 (1949) 1137–1166 (§24, 1159–1160), K. Ziegler.

Jan Bollansée, Karen Haegemans, and Guido Schepens

Agathotukhos (*ca* 325 BCE – *ca* 300 CE)

Wrote on veterinary medicine. Three fragments, quoted by THEOMNĒSTOS, are preserved in the *Hippiatrika*: a drench for fever (*Hippiatrica Parisina* 5 = *Hippiatrica Berolinensia* 1.25), a remedy for arthritic **glanders** (*Hippiatrica Parisina* 35 = *Hippiatrica Berolinensia* 2.24), and a description of symptoms of liver trouble with treatments (*Hippiatrica Parisina* 546 = *Hippiatrica Berolinensia* 32.4). These passages also figure in the Arabic translation of Theomnēstos.

CHG v.1; Hoyland (2004); McCabe (2007) 201.

Anne McCabe

Agennius Urbicus (390 – 410 CE)

The earliest MSS transmitting the most important remains of the ancient Latin prose treatises on surveying and/or similar topics, namely the two halves of the Arcerianus, contain also Urbicus' work *On land disputes* (*de controuersiis*). Owing to damage suffered by the Arcerianus, the preface and the conclusion of Urbicus' work are both lost: a text largely rearranged by Lachmann and Thulin survives.

After explaining technicalities of land surveying, Urbicus treats the globe and its four parts made by the Oceans, following a **Stoic** source. Since part of this land is under Roman rule, Urbicus aims at illustrating what were the "types" (*genera*), "conditions" (*statūs*) and "effects" (*effectūs*) of 15 different types of land disputes and how to settle them (the key issue), along with instructions on both Roman law (as for possession/ownership) and surveying technique (the nature and identification of any kind of boundary marker and land division) that trainees had to know. Lachmann argued Urbicus largely drew on FRONTINUS (as is clearly the case with *controuersia de proprietate*, "dispute about ownership": Frontinus, p. 15.1–7 La. and Agennius, p. 79.13–22 La.), whereas Thulin suggested a 1st c. CE unknown main source.

The same manuscript tradition of land-surveying texts also falsely attributed to Urbicus a commentary on the first two sections of Frontinus' handbook, which for textual reasons must be dated to *ca* 450–550 CE.

Ed.: F. Bluhme, K. Lachmann, A Rudorff, *Die Schriften der römischen Feldmsser*, 2 vols. (1848–1852); Thulin (1913).
RE 1.1 (1893) 773, W. Kubitschek; Campbell (2000) XXXI–XXXV.

Mauro de Nardis

Agēsias of Megara (250 BCE – 200 CE)

If not a mistake for HĒGĒSIAS OF MAGNESIA, this otherwise unknown author is cited for crane ethology by the PARADOXOGRAPHUS VATICANUS (§1).

RE 1.1 (1893) 795 (#3), E. Schwartz.

PTK

Agēsistratos (100 – 50 BCE)

Studied under APOLLŌNIOS OF ATHENS and tutored ATHĒNAIOS MECH., modified designs for spring-frames, using oval instead of round washers, creating more powerful catapults with greater range (Athēn. Mech. p. 8 W.; VITRUUIUS 7.*pr*.14). The name is especially common on Rhodes (*LGPN* 1.10), where his teacher worked.

Cichorius, *Römische Studien* (1922) 271–279.

GLIM

AGGELEUAS ⇒ ANQĪLĀWAS

Aglaias of Buzantion (40 – 70 CE)

Two 15th c. MSS, *Ambr.* A 162 sup./Gr. 58 and *Marc. gr.* Z. 480, transmit under the name of Aglaias from Buzantion 28 verses (14 elegiac couplets), containing a prescription for cataracts, invented by Aglaias himself (v. 4). Aëtios of Amida (7.101, *CMG* 8.2, p. 351) quotes in prose the same prescription (naming the inventor as Aglaidas). The verses must have been well known, as a scholion to Nikandros' *Alexipharmaca* 314 mentions one phrase. Aglaias calls himself a physician (verse 2) and addresses the poem to an unknown poet Dēmētrios (verse 3). The two MSS further reveal that Aglaias belonged to an illustrious family of Buzantion (the poet likewise attests this origin: verse 1), was a pupil of Alexander (probably Alexander Philalēthēs, the physician), and schoolmate and friend of Dēmosthenēs (apparently, the famous ophthalmologist), dating Aglaias to the reigns of Claudius/Nero. The style is obscure: in some cases the ingredients are expressed through mythological circumlocutions recalling Lykophrōn and Dosiadas (e.g. verses 10, 14, 15–6).

Ed.: Claudio De Stefani, "Aglaia di Bisanzio, *SH* 18: edizione critica e note," in G. Cresci Marrone and A. Pistellato, edd., *Studi in ricordo di Fulviomario Broilo. Atti del Convegno Venezia, 14–15 ottobre 2005* (2007) 266–275.

Claudio De Stefani

Aglaosthenēs ⇒ Agathosthenēs

Agnellus of Ravenna (*ca* 590 – 615 CE)

Gradually coming to light are lectures or perhaps lecture notes by a master physician and teacher (*iatrosophista*) in Byzantine Ravenna, probably based on similar commentaries then taught as part of a "medical curriculum" at Alexandria. Attempts to link the medical professor with St. Agnellus (*ca* 525–555) are not fruitful, nor is there firm evidence connecting medical lectures with Patricius Agnellus, sent to Africa by Theodoric between 507 and 511. Agnellus *iatrosophista* knew Greek and was aware of the sequence of topics taught over about two years in Byzantine Alexandria, a curriculum preserved in outline in Arabic and confirmed by surviving Greek texts of several elaborate commentaries; medical and exegetical tracts function on two levels: first, students receive commentaries on selected works from a "canon" of Hippokratēs and Galēn; then the professor interlards his own experiences as a practitioner within the commentary, a characteristic displayed by Agnellus in Latin and Stephanos of Athens, Iōannēs of Alexandria, and others in Greek. The Latin MSS are a tangle of attributions and misattributions, but diligence has begun to bring some order and restoration of both authors' names and the actual works, indicating a lively northern Italian medical and intellectual life in the late 6th and early 7th c.

Ed.: L.G. Westerink *et al.*, *Agnellus of Ravenna: Lectures on Galen's De sectis* (1981); N. Palmieri, *Agnellus de Ravenne. Lectures galéniques: le «De pulsibus ad tirones»* (2005); D. Irmer, *Palladius. Kommentar zu Hippokrate "De fracturis" und seine Parallelversion under dem Namen des Stephanus von Alexandria* (1977); C.D. Pritchett, *Iohannis Alexandandrini Commentaria In sextum librum Hippocratis Epidemiarum* (1975); *Idem, Iohannis Alexandrini Commentaria In librum De sectis Galeni* (1982); J.M. Duffy, *Stephanus the Philosopher. A Commentary on the Prognosticon of Hippocrates* (1983) = *CMG* 11.1.2; L.G. Westerink, *Stephanus of Athens. Commentary on Hippocrates' Aphorisms* (1985–1995 = *CMG* 11.1.3.1–3; Dickson (1998); J.M. Duffy, *Commentary on Hippocrates' Epidemics VI Fragments. Commentary of an Anonymous Author on Hippocrates' Epidemics VI Fragments* [and] T.A. Bell *et al.*, *John of Alexandria. Commentary on Hippocrates' On the Nature of the Child* (1997) = *CMG* 11.1.4.

O. Temkin, "Studies on Late Alexandrian Medicine. I. Alexandrian Commentaries on Galen's *De sectis ad introdocendos*," *BHM* 3 (1935) 405–435 = *The Double Face of Janus and Other Essays in the History of Medicine* (1977) 178–197; A.Z. Iskandar, "An Attempted Reconstruction of the Late Alexandrian

Medical Curriculum," *Medical History* 20 (1976) 235–258; M.E. Vázquez Buján, "El Hipócrates de los comentarios atribuidos al Circulo de Rávena," in J.A. López Férez, ed., *Tratados hipócraticos (estudio acerca de su contenido, forma e influencia)* (1992) 657–685; N. Palmieri, "Survivance d'une lecture alexandrine de l' 'Ars medica' en latin et en arabe," *Archives d'histoire doctrinale et littéraire du Moyen Age* 60 (1993) 57–102; *Eadem,* "Il commento latino-ravennate all'Ars medica di Galeno e la tradizione alessandrina" in Vázquez Buján (1994) 57–76; *Eadem,* " 'Practicon diuiditur in duo': mesures prophylactiques et mesures thérapeutiques chez Agnellus de Ravenne," in Fr. Gaide and Fr. Biville, edd., *Manus Medica. Actions et gestes de l'officiant dans les texts médicaux latins. Questions de thérapeutique et de lexique* (2003) 183–206.

John Scarborough

AGRIPPA ⇒ (1) IULIUS; (2) VIPSANIUS

Agrippa of Bithunia (92 CE)

PTOLEMY, *Synt.* 7.3, records Agrippa's observation of the occultation by the moon of part of the Pleiades.

BNP 1 (2002) 393 (#4), W. Hübner.

PTK

Ahrun ibn-Aʿyan al-Qass (*ca* 600 – 640 CE)

Known from Arabic sources as an Alexandrian physician living in the early 7th c. He wrote a medical compendium said to have been translated into Syriac by an unknown *GWSYWS* (Gessios? *cf. GAS* 3 [1970] 160–161) and later translated into Arabic in the late 7th or early 8th c. as al-Kunnāš. It discussed causes, symptoms, and treatments for diseases. Mined by later Arabic authors for material, some of its contents can be gathered from the numerous citations.

GAS 3 (1970) 166–168; Ullmann (1972) 23, 87–89; A. Dietrich, "Ahrun (Ahrūn) b. Aʿyan al-Ḳass," *Encyclopaedia of Islam*, 2nd ed., Supplement Vol. (1980) 52; *NP* 12/2.884–885, Chr. Schulze.

Kevin van Bladel

AIAS ⇒ ABAS

Aigeias of Hierapolis (*ca* 200 BCE – 460 CE)

Wrote an epitome of EUCLID's *Elements*, combining theorems, according to PROKLOS, *In Eucl.* p. 361 Fr. For the name, compare only *LGPN* 3A.17, Aigeia of Surakousai (3rd–5th cc. CE).

Netz (1997) #74.

PTK

Aigimios of Ēlis (325 – 300 BCE)

Greek physician, perhaps the first to write a work on pulse: GALĒN *Diff. Puls.* 1.2, 4.11 (8.498, 751–752 K.) knew a work *On throbbing* under his name; Aigimios thought diseases arise from residues (*perittomata*) and nourishment (LONDINIENSIS MEDICUS 13.21–14.3). The residues are normally eliminated through the bodily secretions, but under certain conditions become pathological, e.g. if the residues are not yet expulsed and an excess of nourishment intervenes. He was considered wrongly as a forerunner of the corpuscular theory of digestion.

RE 1.1 (1893) 964 and S.1 (1903) 36, M. Wellmann; W.A. Heidel, "Antecedents of Greek Corpuscular Theories," *HSCPh* 22 (1911) 111–172 at 165; *BNP* 1 (2002) 191, V. Nutton.

Daniela Manetti

Aineias Tacticus (370 – 350 BCE)

The earliest author on military topics. His identity is much debated, but he was probably identical to Aineias Stumphalos, general of the Arcadian League in 367 BCE, who helped overcome Euphrōn, the tyrant of Sikuōn (Xenophōn *Hell.* 7.1.44–46). Aineias wrote several treatises on military topics, but only the long extract, *Siege-Craft*, survives (Aelian *Tactics* 1.2). From the historical examples of sieges used in the treatise it can be dated to the mid-4th c. *Siege-Craft* deals with the preparations for and methods for countering sieges, although he sometimes switches briefly to the viewpoint of the attacker. Aineias refers to other treatises of his: *Siege Preparations* (*Paraskeuastikē*) (7.4, 8.5, 21.1, 40.8) and *Procurement* (*Poristikē*) (14.2).

Siege-Craft has no discernible structure. It may, however, be seen as falling broadly into three parts: preparing for a siege by an unknown threat (1–14); preparing for a siege by an enemy known to be on its way (15–31); and resisting an actual attack (32–40). Only a minor part of the treatise thus deals with the attack itself, concerning, for instance, techniques for protecting walls against attack, dealing with incendiary devices, countering attempts to undermine walls and making the defending forces seem as large as possible. The majority of the treatise is concerned with strategies for dealing with potential threats and dangers from within. Aineias discusses how to select troops, prepare defenses, keep up morale and discipline, estimate the approach of an enemy, avoid treachery at the gate and prevent enemies from communicating with sympathizers inside. The treatise thus gives a vivid picture of life in a Greek ***polis*** and the role of military technology on the scale of a small city state. Although siege technology and generalship were developing into forms of technical knowledge taught outside practical contexts, *Siege-Craft* appears to be written by someone with a measure of direct practical experience.

Ed.: D. Whitehead, *Aineias the Tactician: How to Survive under Siege* (1990).
KP 1.175 (#2), W. Sontheimer; *OCD3* 23, D. Whitehead; *BNP* 1 (2002) 221–222, L. Burckhardt.

Karin Tybjerg

Aineias of Gaza (*ca* 460 – 530 CE)

Christian orator of the Gaza school, Prokopios of Gaza's contemporary (465–528) and pupil of the **Platonist** Hieroklēs (*Theophrastos* 2.9, 2.20). Aineias wrote 25 letters and a dialogue entitled *Theophrastos*. The dialogue is between a **Platonist** called Theophrastos and two Christians, Aiguptos and Euxitheos, trying to defend Christian doctrine, especially the immortality of the soul and the resurrection of the body, rejecting **Platonist** doctrines incompatible with Christianity, like the pre-existence of the soul and the eternity of the world.

Ed.: M.E. Colonna, *Enea di Gaza Teofrasto* (1958); L.M. Positano, *Enea di Gaza Epistole* (1962).
DPA 1 (1989) 82–87, A. Segonds; *BNP* 1 (2002) 222 (#4), P. Hadot.

George Karamanolis

Aineios (of Kōs?) (10 BCE – 110 CE)

KRITŌN, in Galēn *CMLoc* 2.2 (12.589–590 K.) cites from Aineios a decongestant, containing beeswax, goat-fat, lye, natron, pitch, and soap, instilled nasally. The following three recipes, possibly also his, involve ***euphorbia*** (*cf.* IOUBA), or pepper, or both (a sternutatory). This form of the name is rare, cited only by STEPHANOS OF BUZANTION, s.v. Kōs, for a doctor (likely earlier), and *LGPN* 3B.18.

RE 1.1 (1893) 1022 (#3), M. Wellmann.

PTK

Ainesidēmos of Knōssos (100 – 50 BCE)

Initiated the skeptical movement known as Pyrrhonism, claiming inspiration from Pyrrho of Ēlis (*ca* 360–270 BCE). We have an informative summary by Phōtios, *Bibl.* §212, of his *Pyrrhonist Discourses* (*Purrōneioi Logoi*: in eight books), and several other works are attested. It is contestable whether the variety of Pyrrhonism espoused by Ainesidēmos was identical with that of SEXTUS EMPIRICUS. Whereas Sextus stresses the *undecidability* of the conflicts between incompatible arguments or impressions, Ainesidēmos seems rather to have stressed the *relativity* to circumstances, or to persons, of each of these arguments or impressions – the consequence being that none of them can be taken to capture the way things are intrinsically.

Like Sextus, Ainesidēmos appears to have applied his skeptical method to a great variety of topics, including scientific topics. The subjects addressed in *Pyrrhonist Discourses* included causes, effects, generation, destruction, motion and sense-perception. Ainesidēmos also discussed "signs" – observable phenomena that, according to non-skeptical philosophers, constituted evidence of non-observable states of affairs. Signs were an important aspect of the scientific methodology of particularly the Hellenistic period; not surprisingly, Ainesidēmos is reported to have argued that there are no such things.

Long and Sedley (1987) §§71–72; *ECP* 6–8, J. Allen.

Richard Bett

Aisara of Lucania (100 BCE – 100 CE?)

PYTHAGORAS' daughter according to Phōtios, *Bibl.* 249. IŌANNĒS OF STOBOI (1.49.27) transmits under her name *On the Nature of Man* (*Peri Anthrōpou* ***Phuseōs***), a spurious Dorian fragment conjecturally attributed to Aresas, a **Pythagorean scholarch**. The text probably belongs to a group of treatises ascribed to **Pythagorean** women philosophers and mainly treating the ethics of the household. Human nature is the criterion of law and justice; justice consists in harmonizing the parts of the soul, which occurs, in **Platonic** fashion, when the superior part (intelligence) rules the inferior (appetite), and the intermediate (spirit) rules the appetitive and follows the superior part. The best life results from a commingling of virtue and pleasure.

Thesleff (1965) 48.20–50.23; *DPA* 1 (1989) 348–349, Bruno Centrone.

Bruno Centrone

Aiskhinēs of Athens (350 BCE – 77 CE)

Wrote on medicine and recommended burnt excrement (in a remedy called *botruon*) for tonsil complaints, sore uvula, and carcinomata (PLINY 1.*ind.*28, 28.44).

RE 1.1 (1893) 1063 (#21), M. Wellmann.

GLIM

Aiskhriōn (325 – 90 BCE)

Wrote a work on agriculture possibly treating cereals, livestock, poultry, viticulture, and arboriculture (*cf.* PLINY, 1.*ind.*8, 10, 14–15, 17–18); excerpted by DIONUSIOS OF UTICA (VARRO, *RR* 1.1.9; *cf.* COLUMELLA, 1.1.10).

RE 1.1 (1893) 1064 (#9), E. Oder.

Philip Thibodeau

***Aiskhriōn of Pergamon* (100 – 150 CE)**

Empiricist physician, in old age teacher of GALĒN in Pergamon (*Simples* 12.1.24 [12.356 K.]); perhaps to be identified with the **Empiricist** teacher mentioned by Galēn in *Plen.* 9 (7.558 K.). Galēn praises Aiskhriōn for his pharmacological knowledge and especially for a (magical) remedy for **rabid** dog-bites based on ash of crayfish; another teacher of Galēn, PELOPS, did not refute but explained the recipe (12.357–359 K.). This remedy is quoted also by other authors, but without reference to Aiskhriōn; amongst them by the anonymous author of *On Poisonous Animals* commonly attributed to AELIUS PROMOTUS (which the 16th c. physician Antonius Possevinus, *Bibliotheca selecta* [Venetiis 1603] 2.163, attributed to "Aischrion Empiricus"). Galēn, however, does not refer to any writings by Aiskhriōn, who could have been a practitioner physician.

Ed.: Deichgräber (1930) 3, 215 (fragment).
RE 1.1 (1893) 1064 (#8), M. Wellmann.

Fabio Stok

Aiskhulidēs of Keōs (325 BCE – 200 CE)

Composed a work on agriculture in at least three books; the two fragments describe pears (Ath., *Deipn.* 14 [650d]) and sheep (AELIANUS, *HA* 16.32) of Keōs.

RE 1.1 (1893) 1064–1065 (#2), E. Oder

Philip Thibodeau

Aiskhulos (430 – 400 BCE)

Student of HIPPOKRATĒS OF KHIOS who with his mentor held that comets are planets, much slower than the sun, which move to the north and south of the tropics and that the tail arises when the planet is north of the tropic, away from the dry region of the sun, and results from moisture around the planet that causes reflection from the eye to the sun to the planet: ARISTOTLE, *Meteorology* 1.6 (342b36–3a20). For Aristotle's criticisms, *cf. ibid.* 1.6 (343a20–b7), 1.7 (344b8–17).

DSB 6.416B (s.v. Oinopides), I. Bulmer-Thomas; Wilson (2008).

Henry Mendell

AITHANARID ⇒ ATHANARID

Akesias of Athens (350 – 230 BCE)

Wrote on culinary art (Ath., *Deipn.* 12 [516c]). Proverbially, patients under his care declined, and so he was declared to have "healed (patients) for the worse" (*Souda* A-842; Zenob. 1.52; Diogenianus, 2.3). ARISTOPHANĒS OF BUZANTION (p. 238 Nauck) provides the *terminus ante quem*.

RE 1.1 (1893) 1163, M. Wellmann; *KP* 1.217, Fr. Hiller von Gärtringen; *BNP* 1 (2002) 67, V. Nutton.

GLIM

Akhaios (200 BCE – 80 CE)

ANDROMAKHOS, in GALĒN *CMLoc* 7.4 (13.79 K.), records a pill for blood-spitting (*cf.* ***phthisis***) from *ΑΚΑΚΙΟΣ*, containing several red ingredients (Samian and **Sinōpian earths**, red coral, and pomegranate-flower), for **sympathetic** effect, as well as henbane and opium. *ΑΧΑΙΟΣ*, common from the 5th to 2nd cc. BCE, could easily become *ΑΚΑΚΙΟΣ*, common from the 3rd c. CE (*LGPN*).

RE 1.1 (1893) 1140 (#1), M. Wellmann.

GLIM

Akhillās (120 BCE – 80 CE)

ANDROMAKHOS, in GALĒN *CMLoc* 7.5 (13.90 K.), describes him as a *parakentētos* (cataract-coucher or **dropsy**-tapper), and records his opium-based anodyne (containing also ***ammi***, cassia, Indian nard, and pepper), and in *CMGen* 5.12 (13.834 K.) lists his pill, containing aloes, alum, frankincense, ***khalkanthon, misu***, myrrh, etc., with the renowned ones of POLUEIDOS and ANDRŌN. This form of the name is not recorded before the 1st c. BCE (*LGPN*). (The *Akhilleios* of AËTIOS 15.15 [Zervos 1909: 67] is likely a brand-name.)

RE 1.1 (1893) 220 (#2), M. Wellmann.

PTK

Akhilleus (Tatius?) (200 – 300 CE)

Three MSS include an introduction to ARATOS' *Phainomena* deriving "from Akhilleus' <treatise> *On the Universe*" (*Univ*). Its 40 chapters, constituting an elementary introduction to astronomy, emphasizes underlying physical theories more than mathematics (spherical astronomy) and quotes an impressive range of sources: pre-Socratics, ARISTOTLE, PLATO, EPICURUS, Aratos, many **Stoic** and **Pythagorean** philosophers, mathematicians (astronomers) and ***grammatikoi***, whose divergences Akhilleus is quick to indicate. Since *Univ* quotes 2nd c. authorities (including PTOLEMY and ADRASTOS OF APHRODISIAS) and FIRMICUS MATERNUS cites *prudentissimus Achilles* as an authority in astrology (*Mathesis* 4.17.2, but see Neugebauer 1975: 950–952 for *Univ*'s meager astrological content), Akhilleus must have lived around the 3rd c. and may be identifiable with an homonymous ***grammatikos*** (Di Maria 2) – according with *Univ*'s style and content.

The *Souda* (A-4695) lists only one Akhilleus "Statios" (i.e. Tatios), author of *Leukippē and Kleitophōn*, attributing to him an additional three works: *On the Sphere* (of which *Univ* may be a chapter) and *Etumologiai* and *Historia summiktos* (both lost). Identification with the novelist is doubtful (the *Souda* seems to rely only on the novel's stylistic similarities to the last three works) and is now usually rejected (the earliest known fragment of the novel is dated *ca*

250 CE, but see Di Maria XI). Two MSS also attribute shorter tracts, *Life of Aratos* and *Peri exēgēseōs*, commenting on the first verses of Aratos and possibly part of a larger commentary (this plausible attribution is disputed, see Di Maria VII–XII).

Ed.: G. Di Maria, *Achillis quae feruntur astronomica et in Aratum opuscula* (1996).
Martin (1956); *DPA* 1 (1989) 48–49, P. Robiano; *BNP* 1 (2002) 96 (#2), K. Brodersen.

Alain Bernard

Akhinapolos (?) (*ca* 150 – 25 BCE?)

Devised a method of casting zodiacs from the time of conception rather than birth (VITRUUIUS 9.6.2). All MSS agree on *ACHINAPOLVS* (save two late Vaticani, 2767 and 1328, which read *ARCHINAPOLVS*); the name seems otherwise unattested. Rose emended to Athēnodōros.

RE 1.1 (1893) 248, E. Riess, s.v. Achinapolus.

GLIM

Akholios (400 – 500 CE?)

AËTIOS OF AMIDA 8.58 (*CMG* 8.2, p.506) records his cough medicine composed of pennyroyal, pepper, hyssop, etc. in **terebinth**, fresh butter, and honey. For the rare name, *cf.* *PLRE* 1 (1971) 9–10 (*ca* 400 CE) or Phōtios, *Bibl.* 257 (477a).

Fabricius (1726) 31.

PTK

Akrōn of Akragas (*ca* 450 – 400 BCE)

Son of a doctor with the same name (DIOGENĒS LAËRTIOS 8.63). A late tradition relates that Akrōn had some success in curing the plague of Athens by lighting fires (PLUTARCH, *Isis and Osiris* 80 [383 C–D], AËTIOS OF AMIDA, 5.95 [*CMG* 8.2, pp. 80–82]), a story also told of HIPPOKRATĒS, his slightly younger contemporary. According to PLINY 29.5 he founded the **Empiricists** (and was recommended by EMPEDOKLĒS), a foundation-legend rejected by PSEUDO-GALĒN *Introd.* 4 (14.683 K.).

Ed.: Wellmann (1901) 70, 73, fragments pp. 108–109; Deichgräber (1965) 40–41, 270.
Pinault (1992) 45–46, 55; *BNP* 1 (2002) 114, V. Nutton.

Robert Littman

ALBINUS ⇒ D. CLODIUS ALBINUS

Albinus (Encyclo.) (*ca* 320 – 345 CE?)

Latin encyclopedist, wrote on music (CASSIODORUS, *Inst.* 2.10), geometry, and dialectic (BOËTHIUS, *Inst. Mus.* 1.12, 26), all lost. Perhaps identifiable with one of the men named Ceionius Rufinus Albinus, and/or Albinus the poet of *De Metris* and *Res Romanae* (*FLP* 425–426); *cf.* MACROBIUS, *Sat.* 1.24.19.

PLRE 1 (1971) 33–34 (#4,5), 37–38 (#14,15); *OCD3* 50, R.A. Kaster; *BNP* 1 (2002) 431 (#2), L. Zanoncelli.

GLIM

Albinus of Smurna (130 – 170 CE)

Wrote a brief *Introduction to Plato's dialogues* (*Prologos* or *Eisagōgē*), preserved in the PLATO MS *Vindob. suppl. gr.* 7 and containing a theory of the dialogue genre, a classification of Plato's works and two distinct sequences for reading the dialogues. No longer extant are: (a) transcripts of GAIUS' lectures, a survey of **Platonic** doctrine, (b) a treatise on the incorporeal, and (c) possibly commentaries on Plato's *Timaeus, Republic* and *Phaedo*. Transcripts of Gaius' lectures and the survey of **Platonic** doctrine, still available in the 6th c., figured in the lost half of MS *Paris. gr.* 1962 (9th c.), the *pinax* of which is still extant. The *Prologos* could well be the introduction to the lecture transcripts. Albinus was Gaius' disciple and was considered important by later **Platonists**, such as PROKLOS (*cf. in Remp.* 2.96.10–13). For a long while credited with the *Didaskalikos* (*cf.* J. Freudenthal, *Der Platoniker Albinos und der falsche Alkinoos*, 1879), now re-attributed to ALKINOOS. GALĒN met Albinus in Smurna some time between 149 and 157.

Ed.: Burkhard Reis, *Der Platoniker Albinos und sein sogenannter 'Prologos'* (1999); Gioè (2002) 79–115.
DPA 1 (1989) 96–97, J. Whittaker; T. Göransson, *Albinus, Alcinous, Arius Didymus* (1995); *BNP* 1 (2002) 431–432, M. Baltes.

Jan Opsomer

ALC- ⇒ ALK-

Alexander (Geog.) (300 BCE? – 110 CE)

MARINOS OF TYRE in PTOLEMY *Geography* 1.14 cites an Alexander for the description of a voyage far to the east (so perhaps after *ca* 120 BCE).

RE S.6 (1935) 3–5 (#90a), W. Kubitschek.

PTK

Alexander (Med.) (400 – 600 CE)

Early Byzantine physician, wrote on sphygmology and urology (*Alexandrou iatrou peri diagnōseōs sphugmōn epi tōn puressōntōn kai peri ourōn aphorismoi*). The work survives in a 15th c. MS (Paris, BNF, *graecus* 2316, ff.207^{V}–214^{V}), there attributed to ALEXANDER OF TRALLEIS, and is cited by Iōannēs Aktouarios (*On urine*, Ideler 2 [1842/1963] 5). The work, perhaps a fragment of a larger unknown work (1.88 Puschm.), proffers a typical 5th/6th c. Alexandrian school synthesis of earlier knowledge. The work explores the causes of diseases as the basis for prognosis. The part on the pulse deals with fevers and several diseases classified *a capite ad calcem*. The part on urine is more aphoristic in nature. Both take for granted a good knowledge of the topics as they do not explain any of the notions they use. Two Medieval Latin translations are ascribed to GALĒN: Diels 1905–1907: 1.128, 132, 2.13; Beccaria (1956) 126, 137, 299, 327; L. McKinney, *Early Medieval Medicine with Special Reference to France and Chartres* (1937) 188–191. Through these, the text helped disseminate early-Byzantine uroscopical knowledge to the West.

Ed.: E.F. Farge, *Alexandre de Tralles, ms. latin du du X siècle: un livre inédit* (1891); E. Landgraf, *Program der königlichen Progymnasiums in Ludwigshafen am Rhein* (1895); B. Nosske, *Alexandri (Tralliani?) liber de agnoscendis febribus et [sic] pulsibus et urinis aus dem Breslauer Codex Salernitanus* (1919); H. Pohl, *"De pulsis et urinis omnium causarum" aus der Handschrift Nr.44 der Stiftsbibliothek zu St.Gallen: Ein Pseudogalentext aus dem frühen Mittelalter* (Inaug.-Diss. Leipzig 1922); H. Leisinger, *Die lateinischen Harnschriften Pseudo-Galens* =

Beiträge zur Geschichte der Medizin 2 (1925); M. Stoffregen, *Eine frühmittelalterliche lateinische Übersetzung des byzantinischen Puls- und Urintraktats des Alexandros. Text, Übersetzung, Kommentar* (Diss. Med. Berlin 1977).
Thorndike and Kibre (1963) 1004; Dimitriadis (1971) 28–29; F. Wallis, "Signs and Senses: Diagnosis and Prognosis in Early Medieval Pulse and Urine Texts," *Social History of Medicine* 13 (2000) 265–278; *BNP* 1 (2002) 485 (#30), V. Nutton.

Alain Touwaide

Alexander Sophistēs (400 – 600 CE?)

Two unpublished treatises are attributed to "Alexander Sophistēs": one on embryology (MS Paris, BNF, *suppl. gr.* 165, ff.116–117V: Costomiris 97–98); the other on sacred plants (MS Oxford, Bodleian Library, *Baroccianus* 150, ff. 67V–68) (Costomiris 98; see also Ch. Daremberg, *Notices et extraits des manuscrits médicaux grecs, latins et français. Ire partie: manuscrits grecs d'Angleterre*, 1853: 39), resembling the treatise published by M. Thomson, *Textes grecs inédits relatifs aux plantes*, 1955: 80–87. The distinctive qualification *Sophistēs* scarcely proves the same man composed both works. Moreover, there is reason to doubt the attribution of the sacred botany treatise, since a similar text is explicitly attributed to an "Alexander the King" (Oxford, Bodleian Library, Roe 15, ff.103, 105), and the author discusses plants (e.g., bryony and mandrake) characteristic of the largely anonymous and freely circulating corpus of iatromagics.

G.-A. Costomiris, "Études sur les écrits inédits des anciens médecins grecs. Troisième série," *REG* 4 (1891) 97–99; Diels 2 (1907) 11.

Alain Touwaide

Alexander of Aphrodisias (T. Aurelius Alexander) (*ca* 200 CE)

The most influential commentator on ARISTOTLE in antiquity, and the first whose works are well known to us. Son of an homonymous philosopher, Alexander acquired Roman citizenship through connection with the imperial family, and was appointed in Athens by Septimius Seuerus between 198 and 211 CE (most probably before 209) as professor and **scholarch** in Aristotelian philosophy. His commentaries are "continuous" in their devotion to careful sentence-by-sentence explanations of the whole of Aristotle's texts. He commented on most of Aristotle's logical and theoretical works, the latter on the basis of the former. He tends to reshape the treatises' contents into syllogisms and other forms of arguments described by Aristotle's *Organon*, reducing them to a standard terminology. By contrast with Aristotle's flexible way of thinking and lexical usages, Alexander produced a coherent and consistent system of thought, suitable for teaching. Alexander also wrote original treatises, more pedagogical in character, and a number of shorter discussions of various kinds, always focused on Aristotelian exegesis. When this procedure is difficult and where no convenient consensus has been reached, problems (*aporiai*) are openly discussed and more than one solution may be kept, either within the commentaries or in separate *opuscula* (*aporiai kai luseis*, the so-called *Quaestiones*). Altogether, Alexander is the main representative of a distinctively Aristotelian commentary tradition, which was to be the basis for subsequent exegesis by Neo-**Platonic**, Arabic, and Renaissance commentators.

Some theoretical assumptions seem original to Alexander or scarcely expressed before. He explained circular heavenly motions as due to desire of imitating the eternal perfection of the Unmoved Mover. As for the origin of soul, which is form and perfection of the living beings, it derives from a divine power (*theia dunamis*) exerted by the movement of the celestial bodies, which supervenes upon the physical mixture (*krasis*) of their organic components (see *De prouidentia* 148 Zonta, 75–77 Ruland; *Quaestio* 2.3). By this activity, the celestial power is the cause both for providence, which guarantees preservation and well-being of living species, and for individual fate. From fate, nonetheless, one is able to escape through education and exertion, so that moral life remains up to us (*eph' hēmin*). As for the active intellect (*nous poiētikos*, of Aristotle's *De an.* 3.5), this is identical with its peculiar object, namely the eternal first intelligible, and is therefore itself eternal. But since the first intelligible is alike for all rational beings, no place seems to remain in Alexander for the immortality of individual souls, and Alexander's Renaissance followers were charged with impiety for their position in this regard.

Extant commentaries and minor works of Alexander are mainly edited within the *CAG* volumes (1883–1901): 1. *Comm. in Metaph.* (Books 1–5 only are authentic); 2.1: *In An. pr.* I (M. Wallies, 1883); 2.2: *In Top.* (M. Wallies, 1891); 3.1: *In De sensu et sensato* (P. Wendland, 1901); 3.2: *In Meteor.* (M. Hayduck, 1899). Treatises and *opuscula: Suppl. Arist.* 2.1: *De anima* and *Mantissa* (I. Bruns, 1889); v. 2.2: *De fato, Quaestiones, De mixtione* (I. Bruns, 1892; also a further edition of *De fato* by P. Thillet, 1984; a revision of Bruns' *Quaest.* 1.10, 1.15, 2.3 by Silvia Fazzo, *Aporia e sistema,* 2002; and of *De mixtione* by R.B. Todd, 1976). We also have fragments of Alexander on *Cat.*, *De int.*, *An. pr. I*, *An. post.* (ed. P. Moraux, 1979), *Phys.*, *De caelo* (fr. on Book I ed. by A. Rescigno, 2004), *De gen. et corr.*, *De an.*, and *Met.* XII, and Arabic translations of Alexander's lost *De prouidentia* (ed. J.-H. Ruland, Diss. Saarbrucken, 1976; P. Thillet's Thèse d'état, Paris 1979; M. Zonta in Silvia Fazzo, ed., *Alessandro di Afrodisia. La provvidenza. Questioni sulla provvidenza,* 1999), *De principiis* (ed. C. Genequand, *Alexander of Aphrodisias On the Cosmos,* 2001), and of other minor writings. Many of Alexander's works have been translated into English, among others in the *ACA*.

The name "Alexander" has been abused, especially in the Middle Ages, as a generous label for different writings with some Aristotelian connection; texts preserved only in Arabic with no Greek parallel or counterpart must be handled with caution as sources for Alexander; and so also Greek texts, *opuscula*, or fragments, when no safe indication of authorship is given. Therefore, authenticity is sometimes controversial (e.g. of some *opuscula*, including *Quaestio* 2.21 and the famous *De intellectu* = *Mantissa* II), and some works are certainly spurious: the commentaries to *Soph. El.* and *Met.* 6–14 (both by Michael of Ephesos), various medical writings (edited by Ideler 1 [1841/1963], and by Kapetanaki and Sharples [2006]), some Arabic treatises or titles of treatises, most of which are polemical, against the thinkers denying *creatio ex nihilo* and against GALĒN.

R.W. Sharples, "Alexander of Aphrodisias: Scholasticism and Innovation," *ANRW* 2.36.2 (1987) 1176–1243 (good starting point on themes, problems, and bibliography); *DPA* 1 (1989) 125–139, R. Goulet and M. Aouad; Moraux v. 3 (2001), with a chapter on ethics and a general bibliography by R.W. Sharples, integrated and supplemented in *DPA* S. (2003) 61–67, Silvia Fazzo; *Eadem*, "Alexandre d'Aphrodise contre Galien: la naissance d'une légende," *Philosophie Antique* 2 (2002) 109–144; J. Sellars, "The Aristotelian Commentators: a Bibliographical Guide," in *Philosophy, Science and Exegesis in Greek, Arabic and Latin Commentaries*, edited by P. Adamson, H. Baltussen and M.W.F. Stone, *BICS* S.82.1 (2004) 244–245.

Silvia Fazzo

Alexander of Aphrodisias, pseudo, On Fevers (150 – 200 CE)

Otherwise unknown medical writer, whose treatise *On Fevers* defines and classifies them according to the **Pneumaticist** school, based on a lengthy discussion of the nature of causation. The author (§§26–30) divides causes into *prokatartika* (predisposing), *proēgoumena* (antecedent), and *sunezeugmena* (conjoint), as had ATHĒNAIOS OF ATTALEIA. The heat of fever arises from the innate heat of the heart (§2), but is none the less unnatural (§8). ***Pneuma*** and **humors** are not the primary agents of disease or fever (§13). In §§16.1, 24.5, 30.1, he appears to cite ARETAIOS OF KAPPADOKIA.

Ed.: Ideler 1 (1841/1963) 81–106; P. Tassinari, *Trattato sulla febbre* (1994).
P. Tassinari, "Il trattato sulle febbri dello ps. Alessandro d'Afrodisia," *ANRW* 2.37.2 (1994) 2019–2034.

PTK

Alexander of Ephesos, Lukhnos (75 – 45 BCE)

This rhetor was contemporary with CICERO who described him in mixed words such as *poeta ineptus, non inutilis* (*Att.*, 2.20.6; 2.22.7). He was known as Lukhnos ("The Light") and wrote an historical work on the Marsian War. STRABŌN (14.1.25) especially mentions him as an author of didactic poems on geography (*SH* 23–38: it seems that DIONUSIOS OF ALEXANDRIA took him as a model) and astronomy: he wrote *Phainomena* of which the remaining 26 hexameters show the influence of **Pythagorean** philosophy on his description of the harmony of spheres (*SH* 21).

BNP 1 (2003) 479 (#22), C. Selzer; *SH* 19–38.

Christophe Cusset

Alexander of Laodikeia on the Lukos, Philalēthēs (20 BCE – 25 CE)

Originally a follower of ASKLĒPIADĒS OF BITHUNIA, succeeded ZEUXIS as *arkhiatros* of the **Hērophilean** school in Asia, tutored ARISTOXENOS and DĒMOSTHENĒS PHILALĒTHĒS (GALĒN, *Puls. Diff.* 4.10 [8.744, 746 K.]). His views on lethargy, a sudden loss of reason accompanied by fever and impaired senses (CAELIUS AURELIANUS, *Acute* 2.5 [*CML* 6.1.1, p. 132]), digestion, a predisposition but not assimilation of nutriments (LONDINIENSIS MED. 24.27–35), and pores, all derive from Asklēpiadēs (*cf.* von Staden 1989: 532–534). Alexander posits invisible apertures ("apprehensible only by reason") through which corporeal matter enters and leaves the body (Londiniensis med. 35.21–9, 38.58–39.13). Alexander offered two definitions of the pulse (Galēn, *Puls. Diff.* 4.4–5 [8.725–727, 731 K.]): (1) involuntary contractions and distentions of heart and arteries ("objective," according to nature); (2) the throb resulting from the continuous involuntary motion of arteries against one's touch and its following interval ("subjective"). His "objective" definition is orthodox **Hērophilean**: *cf.* BAKKHEIOS, ZĒNŌN (HĒROPH.), and KHRUSERMOS. Alexander concurs with the **Hērophilean** theory that male seminal fluid arises in the blood (*fr.*9 von Staden). He wrote a *Gynecology* (at least two books), denying illnesses specific to women (SŌRANOS *Gyn.* 3.2: *CMG* 4, pp. 94–95; *CUF* v. 3, p. 47) and defining vaginal flux as a sanguineous flow over the uterus (*cf.* DĒMĒTRIOS OF APAMEIA). The ANONYMOUS OF BRUSSELS (1, p. 208 Wellmann) cites the first book of Alexander's *On Seed*. In *Opinions*, at least five books and filtered to Galēn through Aristoxenos, Alexander is connected with the doxographic tradition reaching back, perhaps, to HĒROPHILOS himself (von Staden 1989: 538; *cf.* Galēn, above).

von Staden (1989) 532–539; *OCD3* 61, *Idem; Idem* (1999) 164–165; *BNP* 1 (2002) 485, V. Nutton.

GLIM

Alexander of Lukaia (250 – 30 BCE)

Wrote a *Phainomena*, according to BOËTHOS OF SIDŌN (PERIPATETIC), entirely lost; HYGINUS, *Astr.* 2.21.3, mentions an Alexander who wrote on the Hyades, possibly ALEXANDER OF AITOLIA, ALEXANDER OF EPHESOS, ALEXANDER OF MILĒTOS, or this man.

DPA 1 (1989) 144, P. Robiano.

PTK

Alexander of Milētos, Cornelius, Poluhistōr (*ca* 80 – *ca* 40 BCE)

"Poluhistōr" for his wide range of interests. A pupil of KRATĒS OF MALLOS, Alexander was taken prisoner in Asia Minor during Rome's war with MITHRADATĒS VI EUPATŌR, brought to Rome as slave, and then tutored a certain Cornelius Lentulus. He was freed and granted Roman citizenship under Sulla *ca* 80 BCE, whereupon he adopted the clan name of his former Roman master, Cornelius. He also taught C. IULIUS HYGINUS, himself a prolific author. Alexander died in a fire in Laurentum when his house was destroyed; his wife Helēnē hanged herself (*Souda* A-1129). Alexander produced works in various fields, including literature (e.g. on the poetry of Alcman and Corinna), and philosophy, but most of the surviving fragments indicate his specific interest in geography and ethnography. To these belong, according to their order in *FGrHist*, excerpts of descriptions of Egypt, the Black Sea, Illyria, India, Italy, various parts of Asia Minor, Crete, Cyprus, Libya, Rome and Syria as well as an ethnographic treatise on the Jews and one on marvels (*thaumasia*).

Ed.: *FGrHist* 273.

L. Troiani, "Sull'opera di Cornelio Alessandro soprannominato Polistore," in: *Due studi di storiografia e religione antiche* (1988) 9–39.

Daniela Dueck

Alexander of Mundos (10 BCE – 40 CE)

Wrote a ***Periplous*** *of the Red Sea* of which AELIANUS, *NA* 17.1 preserves one fragment, a book on dream interpretation of which ARTEMIDŌROS OF DALDIS preserves three fragments (1.67, 2.9, 2.66), perhaps a book on theriac, of which a scholiast preserves one fragment, and a book *On Animals* of which over two dozen fragments are preserved, mainly in PLUTARCH *Marius* 17.3, Aelianus 3.23 on storks, 4.33 on chameleons, 5.27 on unusual goats, and Athēnaios, *Deipn.*, 5 (221b–d) and esp. 9 (387–398). The *Collection of Marvels* on animals, plants, rivers, and springs, excepted by Phōtios *Bibl.* 188, is probably his.

FGrHist 25; *DSB* 1.120–121, J. Stannard.

PTK

Alexander of Pleuron, Aitoleus (290 – 250 BCE)

Born at Pleuron *ca* 315, son of Saturos and Stratokleia, the poet and ***grammatikos*** Alexander of Pleuron was contemporary with ARATOS, KALLIMAKHOS and Theokritos. He lived both in Alexandria, Egypt (perhaps at different times of his life) and in Pella, Macedon. He seems to be the only known Aitolian poet. In Alexandria, he worked at the Library for Ptolemy II Philadelphos and undertook the *diōrthosis* (correction of copies, critical and exegetical commentary and classification) of tragedies and satyr plays. Around 276 he was called to the court of Antigonos Gonatas in Macedon along with Aratos and Antagoras of Rhodes: it is uncertain whether he had a special cultural task there or not.

He was best known as a tragedian, as well as one of the tragic Pleiad; however very little of this production has survived. He also wrote *epyllia* (like *The Fisherman*), epigrams and elegies (like *Apollo* and *The Muses*): several fragments survive. But, according to SEXTUS EMPIRICUS and *The Second Life of Aratos*, he is also said to have composed *Phainomena* on constellations like Aratos: the authenticity of this poem is disputed and no fragment has survived.

Ed.: E. Magnelli, *Alexandri Ætoli Testimonia et Fragmenta* (1999).
OCD3 60, K. Dowden; *BNP* 1 (2003) 478–479 (#21), F. Pressler.

Christophe Cusset

Alexander of Tralleis (*ca* 550 – 605 CE)

In his *Histories* (5.5–6 [171 Keydell]), Agathias of Murina (*ca* 535–*ca* 575 CE) sketches Alexander's family: born *ca* 525 CE, he was the youngest of five brothers, all distinguishing themselves in their chosen professions; ANTHĒMIOS, the eldest, became Justinian's architectural and engineering confidant in the rebuilding of the famous Hagia Sophia, the massive structure still admired in modern Istanbul; the second brother, MĒTRODŌROS, a prominent grammarian, was summoned by the emperor to Constantinople, where he taught the "...young sons of the ruling class ... the love of eloquence;" Olumpios, the third, was famed as a legal advocate and lawyer, and Dioskouros (the fourth) became a physician, practicing in Tralleis with honor and success. Alexander was a renowned traveling doctor, ending his career in Rome, where he "...had been called to hold a position of the highest distinction." Agathias says their mother was "...especially blessed to have borne such gifted children." Alexander's father, the physician STEPHANOS OF TRALLEIS, was probably the principal teacher of his two sons following the profession, and Alexander thanks him for an effective, multi-ingredient gargle. Perhaps the "Kosmās", to whom Alexander dedicates his *Books on Fevers* (1.289 Puschm.) and who was a friend of Stephanos, is the famous KOSMĀS INDIKOPLEUSTĒS; if so, he may have introduced some Far Eastern pharmaceuticals to the physicians, father and sons (the cloves present among the ingredients in fashioning emetic lozenges to expel black bile and in treatments for gout suggest a special connection with the spice trade: *Quartan Fevers*, 1.429, 431 Puschm.; *Podagra*, 2.231 Puschm.). One passage hints that Alexander studied for a time in Alexandria (*Colic*, 2.343 Puschm.)

Alexander traveled widely, settling in varying locales where he gathered folk traditions on drugs and therapies, and fused them into the venerated theoretical package of elements, qualities, and **humors**; most simples in Alexander's pharmaceutical recipes are known from earlier works, but some are new and some have different nomenclatures; from time to time, he allows amulets as useful for patients who find them powerful healing agents, and occasionally Alexander's comments reflect an ordinary use of magical remedies among the *pagani* wherever he might journey or settle – documented are stints in Italy, southern Gaul, Africa (presumably the recently re-conquered rims of coastal Tunisia and Algeria), and Spain – finally going to Rome in old age, when he says he lacks vitality and energy to continue the rigors of a full practice. In Rome, late in the 6th c., Alexander likely set down writings incorporating his lengthy experiences as a practitioner, and frequently there is blunt criticism of the classical authorities (especially GALĒN) when practice demonstrated clinical errors in the written texts. One receives the impression of a highly intelligent, innovative and warmly curious, kindly and sensitive physician, a clinician quite willing to listen to local "experts" as much as he might cite earlier medical authorities. Wanderlust touches Alexander as he combines current and local details drawn from districts of Asia Minor and Thrace as easily as those derived from Africa, Gaul, Spain, and Italy; his prose is clear and direct, and the numerous pharmaceutical recipes scattered throughout his books have precise weights, measures, and dosage forms, so that one could "test" ingredients in a modern laboratory. Alexander recommends the fern *thēlupterion* for flatworms (*Letter*, 2.595, 597 Puschm.; Brunet, 2.110–111) and opium poppy latex and meadow saffron in the treatment of gout (*Podagra*, 2.275, 563, 565 Puschm.; Brunet, 4.246–251). Ferns, esp. *Dryopteris filix-mas* (L.) Schott. (*cf.* Diosk. 4.184), but also *Pteris aquilina* L. (*cf.* Diosk. 4.185), remain in modern pharmacopeias as potent vermifuges, especially for tapeworms, and *Colchicum autumnale* L. (the autumn crocus or meadow saffron) contains the alkaloid colchicine, a prescription drug still one of the most effective against gout, rheumatism, eczema, and bronchitis.

Uniquely among Byzantine physicians, Alexander gains admiration from modern doctors, who sometimes call him the "third Hippokratēs," mirrored in enthusiasm for Alexander the Clinician by Puschmann (editor and German translator) and Brunet (French translator). Extant are Alexander's remarkable *Letter on Intestinal Worms* (the first parasitological tract worthy of the name), the detailed *Twelve Books on Medicine* (ailments and pathologies in the traditional "from head to heel"), and the minutiae-packed *Books on Fevers*. Arabic and Latin sources indicate Alexander composed lost works on gynecology and obstetrics, ophthalmology, pulse lore, and perhaps on toxicology.

Ed.: Th. Puschmann, *Alexander von Tralles. Original-Text und Übersetzung* 2 vv. (1878–1879; repr. 1963); *Idem*, *Nachträge zu Alexander Trallianus. Fragmente aus Philumenus und Philagrius nebst einer bisher noch ungreduckten Abhandlung über Augenkrankheiten* (1887; repr. 1963); F. Brunet, *Médecine et thérapeutique byzantines. Oeuvres Médicales d'Alexandre de Tralles* (1933–1937); M. Stoffregen, *Eine frühmittelalterliche lateinische Übersetzung des byzantinischen Puls- und Urintraktats des Alexandros* (1977); D.R. Langslow, *The Latin Alexander Trallianus. The Text and Transmission of a Late Latin Medical Book* (2006).

I. Bloch, "Alexandros von Tralles" in *HGM* 535–544; A. Cameron, *Agathias* (1970) 1–11; Scarborough (1985b) 226–228; Temkin (1991) 231–236; John Scarborough, "The Life and Times of Alexander of Tralles," *Expedition* 39.2 (1997) 51–60.

John Scarborough

Alexias (350 – 280 BCE)

Pharmacist, Thrasuas' student. Theophrastos acclaimed his skill and general knowledge of medical science (*HP* 9.16.8). The name, rare at Athens, is known widely throughout the Mediterranean from the 5th c. BCE on (*LGPN*).

RE 1.2 (1894) 1464 (#6), M. Wellmann.

GLIM

Alfius Flauus (*ca* 50 – 75 CE)

Child prodigy born *ca* 35, primarily a rhetorician (Sen., *Con.* 1.1.22, 1.7.7), cited in Pliny's index as a source for the story about a boy and a dolphin in the Lucrine lake (9.25).

RE 1.2 (1894) 1475 (#6), P. von Rohden.

Arnaud Zucker

Alkamenēs of Abudos (500 – 300 BCE)

Greek physician, who, according to the **Peripatetic** doxography in LONDINIENSIS MEDICUS (7.40–8.10), states that diseases are due to the residues of nourishments going to the head, like EURUPHŌN OF KNIDOS, from whom Alkamenēs differs in assigning a more specific role to the head as the origin of diseases.

RE S.1 (1903) 60–61 (#4b), M. Wellmann; S.3 (1918) 82 (#6), H. Gossen; *BNP* 1 (2002) 439, V. Nutton.

Daniela Manetti

Alkimakhos (250 – 50 BCE)

Cited by ALEXANDER OF APHRODISIAS, *Physical Problems* 4.181 (p. 36 Usener), for what the Celts say about the conception of the mule. The archaic name is hardly attested after the 1st c. BCE (*LGPN*: only 5 of 103).

RE S.1 (1903) 62 (#5b), M. Wellmann.

PTK

Alkimiōn (120 BCE – 25 CE)

Possibly identifiable with *Claudius Alcimus*, physician to Tiberius or Claudius (*IGRR* 1.283); twice cited with APOLLŌNIOS CLAUDIUS (GALĒN, *CMLoc* 7.2 [13.31–32 K.] and *CMGen* 5.12 [13.835 K.]). The name, a variant of the commonly cited Alkimos (6th c. BCE to 2nd c. CE: *LGPN*) is attested only in the 1st c. CE (*LGPN*). ANDROMAKHOS records his treatments for various ailments: roughness of the trachea and hoarseness (Galēn *CMLoc* 7.2 [13.31–32 K.]); ***duspnoia*** (*CMLoc* 7.6 [13.112 K.]: "Alkimios"); a multi-use green plaster involving Indian aloes (*CMGen* 2.2 [13.493–494 K.]); a cicatrizant (*CMGen* 2.15 [13.529 K.]); a green plaster attributed to Alkimiōn or NIKOMAKHOS, also involving Indian aloes, and effective especially for ulcers, rejoining bones, and dissipating scrofulous tumors (*CMGen* 5.5 [13.807–808 K.]), a multi-use lozenge (*CMGen* 5.12 [13.835 K.]); and a pill for skin lesions (*CMGen* 5.13 [13.841–842K.]). ASKLĒPIADĒS PHARM., in Galēn *CMGen* 7.6 (13.973–974), records his emollient compounded from beeswax, "Kolophōn" resin, ***ammōniakon*** incense, **galbanum**, myrrh, frankincense, ***opopanax***, bee-glue, vinegar, goat-dung, and olive oil.

RE 1.2 (1894) 1541, M. Wellmann; C.A. Forbes, "The Education and Training of Slaves in Antiquity," *TAPA* 86 (1955) 321–360 at 346.

GLIM

Alkinoos (100 – 200 CE?)

The **Platonic** handbook *Didaskalikos* is transmitted under the name of the otherwise unknown Alkinoos, long falsely identified with ALBINUS. The author begins with "dialectic," comprising epistemology, the theory of division, and an essentially **Peripatetic** syllogistic. Throughout the book arguments are often syllogistic. In the physical section, Alkinoos praises the usefulness of mathematics and discusses theology and cosmology, the formation of elements from geometrical shapes, astronomy, anthropology – including a description of bodily organs, an etiology of diseases, accounts of sense perception and the partition of the soul. Most of this is based on PLATO's *Timaeus*, supplemented by other Platonic works and more recent ideas (e.g. on the Great Year as measured by the conjunction of all the planets). The book's last part treats ethics.

Ed.: J. Whittaker and P. Louis, *Alcinoos. Enseignement des doctrines de Platon* (*CUF* 1990); J. Dillon, trans., *Alcinous. The Handbook of Platonism* (1993).
DPA 1 (1989) 96–97, J. Whittaker; *OCD3* 54, J. Dillon; *BNP* 1 (2002) 452 (#2), M. Baltes.

Jan Opsomer

Alkmaiōn of Krotōn (*ca* 500 – 480 BCE)

As a natural philosopher Alkmaiōn belonged to the **Pythagorean** school and as a doctor to the Krotonian medical school. His book *On Nature*, preserved in several fragments and testimonia, is addressed to three **Pythagoreans** (B1 DK), and several later **Pythagoreans** (Menestōr, Ikkos, Hippōn, Philolaos) shared his interest in natural science and medicine.

Following Xenophanēs, Alkmaiōn insisted that human cognition is limited and has to rely on the empirical evidence (B1). He revealed no interest in cosmogony and very little in cosmology. Apart from the old Ionian views, Alkmaiōn shared some ideas of **Pythagorean** astronomy, e.g. the motion of the planets from west to east, which he most likely conceived as circular (A4, 12). In Pre-Socratic thought he was a founder of the trend that focused on the human body, its nature and functioning. He transferred Pythagoras' idea of the qualitative opposite principles from cosmic to human realm ("most human affairs go in pairs," A1) and based on it his theory that health is kept due to "equality" of the forces (*isonomia tōn dunameōn*) – wet, dry, cold, hot, bitter, sweet, etc. – whereas domination (*monarkhia*) of one of them causes diseases (B4). (*Isonomia* is understood here in the spirit of the **Pythagorean** aristocracy, to which Alkmaiōn belonged politically, and not in the later democratic sense.) Alkmaiōn's theory also took into account external factors (character of place, water, etc.) and, contrary to the later schemes, did not fix the number of the contrarieties. To keep and restore their balance is the task of the rational dietetics that became a cornerstone of **Pythagorean** medicine and through the authors of the Hippokratic corpus exerted powerful influence throughout Greek medicine.

As a pioneer in anatomic research, Alkmaiōn discovered the optic nerves, visibly connected with the brain (A10). Hence his ingenious conclusion that all the senses (sight, hearing, smell, taste) are transferred through special channels (*poroi*) from organs of sense (eyes, ears, nose, tongue) to the brain, which is the center of consciousness. Alkmaiōn was the first to distinguish between intelligence and perception; both of them inhere in humans, animals possess only perception (A5–9). His embryology is understandably more primitive: semen comes from the brain, embryos from the mixture of male and female semen; sex of the child depends on whose semen was stronger (A13–14).

Alkmaiōn regarded the soul as immortal; like the immortal heavenly bodies it is in eternal circular motion (A12). Apparently, he distinguished between intelligence located in the brain and the soul located, probably, in the heart (A18; *cf.* similar theory of Philolaos, 44 A13 DK). The soul is responsible for locomotion and perception, which are inherent in animals as well. Humans die "because they cannot connect the beginning with the end" (B2), i.e. when circular motion of the soul ceases.

DK 24; G.E.R. Lloyd, "Alcmaeon and the early history of dissection," *Sudhoffs Archiv* 59 (1975) 113–147; Longrigg (1993); Zhmud (1997); L. Perilli, "Alcmeone di Crotone tra filosofia e scienza," *QUCC* 69 (2001) 55–79.

Leonid Zhmud

Alkōn (40 – 55 CE)

Surgeon working in Rome, famously wealthy and famously fined: PLINY 29.22, Iosephus, *Ant. Iud.* 19.157 (Alkuōn). Cited by Martial 6.70.6, 11.84.5, as an example of the surgeon (Kay, *Martial Book XI* [1985] *ad loc.*); *cf.* DASIUS. Pliny 1.*ind.*28 cites *BIALCON* as a source, perhaps a *Bios Alkontos*.

PIR2 A-493; Korpela (1987) 164.

PTK

Alupios (*ca* 300 – 400 CE)

His fragmentary *Introduction to music* preserves a short section, reflecting the later Aristoxenian tradition (KLEONEIDĒS, ARISTEIDĒS QUINTILIANUS, and GAUDENTIUS), followed by the most complete surviving tabulation of ancient Greek musical notation, arranged in two sets (one for text and one for instruments), according to a system of fifteen *tonoi* in the diatonic, chromatic, and enharmonic genera. The tables for the enharmonic genus, however, are incomplete and most probably defective. Each *tonos* (a basic two-octave scale with three alternative notes in the middle, for a total of 18 notes) following the lowest (the Hypodorian) is one semitone higher overall, resulting in a total range of three octaves and a tone between the lowest note of the lowest *tonos* and the highest note of the highest *tonos*. The Alupian system of symbols largely accords with the notation found in the surviving fragments of music, enabling them to be transcribed with reasonable confidence.

Alupios is mentioned by CASSIODORUS (*Institutiones* 2.5) in the list of important Greek musical authors. BOËTHIUS (*De institutione musica* 4.3–4) reproduces the Alupian notational symbols for the Lydian *tonos* in all three genera, but they are not attributed to Alupios and could have been derived from other sources. Nothing is known of his life. The date-range is assigned on the basis of the name (not otherwise attested prior to 300), the content of the opening prose section, and the general disregard for musical notation by writers securely dated prior to 200.

Ed.: *MSG* 367–406.
Mathiesen (1999) 593–607; *NGD2* 1.435–37.

Thomas J. Mathiesen

Alupios of Antioch (358 – 371 CE)

Possibly a Kilikian, educated in Antioch. He had a brother Caesarius and a son Hieroklēs, named after his uncle (LIBANIOS *Ep.* 324). In 358 when Alupius was *uicarius Britanniarum*, he formed a friendship with the later emperor Julian (Julian *Ep.* 9 Bidez and Cumont), to whom Alupios soon presented a map "which was better drawn than the older ones," together with an epigram (Julian *Ep.* 10 Bidez and Cumont). In 363 Alupios (maybe serving as a *comes*) was in charge of rebuilding the temple at Jerusalem. The project failed immediately (Ammianus Marcellinus 23.1.2; PHILOSTORGIOS *HE* 7.9; Rufin. *Hist.* 10.38). In 371, Alupios was tried for poisoning at Antioch, and condemned to exile, but soon pardoned (Ammianus Marcellinus 29.1.44).

RE 1.2 (1894) 1709 (#1) O. Seeck; *PLRE* 1 (1971) 46–47; *BNP* 1 (2002) 553, W. Portmann.

Andreas Kuelzer

Amarantos of Alexandria (20 BCE – 95 CE)

Grammarian, wrote a commentary on Theokritos (*Schol. Theokr.* 4.57, 7.154) and *On the Theater*, wherein he cited an epigram of IOUBA (Ath., *Deipn.* 8 [343f], 10 [414e]). ASKLĒPIADĒS PHARM. in GALĒN *CMLoc* 7.4 (13.84–85 K.) cites his recipe for a lozenge to expel blood: wine and Attic honey, heated and cooled, to which are added gum, pomegranate flowers, frankincense, and Samian earth, again heated and cooled. Explicitly naming Amarantos a grammarian, Galēn details his multi-ingredient remedy for sore feet, which Amarantos used himself, including (not exhaustively) Pontic rhubarb, white pepper, parsley, **shelf-fungus**, St. John's wort, yellow iris, eryngo, cardamom, shepherd's-purse, acacia, licorice juice, roses, butcher's broom, ginger, gentian, seeds of parsley and wild turnip, frankincense, Pontic nard, and cassia: *Antid.* 2.17 (14.208–209 K.).

RE 1.2 (1894) 1728–1729 (#3), G. Wentzel.

GLIM

M. Ambiuius (30 BCE – 20 CE)

Wrote a treatise known to COLUMELLA (12.4.2) on storing wine and other agricultural produce.

RE 1.2 (1894) 1804 (#2), E. Klebs; *GRL* §203.

Philip Thibodeau

AMBRŌN ⇒ HABRŌN

Ambrose (Ambrosius) of Milan (374 – 397 CE)

Ambrose Reproduced with kind permission of the Parrocchia di S.Ambrogio, Milan

Theologian and bishop of Milan, Ambrose (b. *ca* 340 CE) held senatorial rank and received a liberal education at Rome in Greek and Latin. Prior to his election as bishop in 374 CE, his career followed the traditional *cursus* for public life culminating in the governorship of Aemilia (approximately modern Emilia) and Liguria.

Only theological works written as bishop survive, and these are deeply indebted to the Classical tradition and neo-**Platonism**. His *De officiis* reflects a thorough reading of CICERO, while his *Hexaemeron* (*ca* 389 CE: on the days of creation) follows BASIL OF CAESAREA, although not without original contribution. Knowledge of the natural world found in the *Hexaemeron* is literary and homiletic, treating nature as a landscape for moral contemplation: *cf.* AELIANUS OF

PRAENESTE and the PHYSIOLOGUS. The work engages with interpretations of creation from PLATO, ARISTOTLE and DĒMOKRITOS (*Hexaemeron* 1.1–2). Ambrose's discussion includes properties of the heavens, oceans and land (1.6–3.5), the utility and typology of vegetation (3.6–17), phenomena pertaining to celestial bodies and the calendar (4.1–9), typology and physiology of sea life (5.2–11) and aerial (5.12–24) and terrestrial (6.2–5) creatures, and an excursus on human physiology (6.9) drawing from GALĒN.

Ed.: *PL* 14–17.

J.J. Savage, trans., *Hexameron, Paradise, and Cain and Abel* (1961); N.B. McLynn, *Ambrose of Milan* (1994); *OCD3* 71, P. Rousseau.

M. Shane Bjornlie

Ambrosios Sophistēs **(unknown date)**

Author of a remedy for horses preserved in the chapter on head infections in the C and L recensions of the *Hippiatrika* (*Hippiatrica Cantabrigiensia* 11.8).

CHG v.2; McCabe (2007).

Anne McCabe

Ambrosios of Puteoli, Rusticus (40 – 80 CE)

SCRIBONIUS LARGUS records the diuretic against kidney stones by Ambrosios of Puteoli, composed of seeds of anise, carrot, celery, cucumber, and parsley, plus myrrh, etc., to be taken for 40 days; on the seventh day the patient passes sandy residue: 152 = MARCELLUS OF BORDEAUX 26.10 (*CML* 5, p. 430); the preparer must use a wooden pestle and wear no iron ring. The same recipe and magical conditions are attributed to Rusticus by ANDROMAKHOS, in GALĒN *CMGen* 10.1 (13.325–326 K.), whom Andromakhos, in GALĒN *CMGen* 2.7 (13.507–508 K.), describes as an associate (*gnōrimos*) of ISIDŌROS OF ANTIOCH (who himself is an associate of Andromakhos, 13.834 K., *cf.* Fabricius 1972: 228). Andromakhos also records Rusticus' wound-plaster based on **litharge** and ***psimuthion*** (*ibid.*), and his ***hedrikē*** composed of human milk, poppy juice, two raw eggs, butter, honey, etc.: in Galēn, *CMLoc* 9.6 (13.309 K.). ASKLĒPIADĒS PHARM., in Galēn *Antid.* 2.14 (14.184 K.), quotes Rusticus' snake-bite antidote.

RE 1.2 (1894) 1812 (#3), M. Wellmann; *PIR2* R-230.

PTK

AMBROSIUS ⇒ MACROBIUS AMBROSIUS THEODOSIUS

Amelius Gentilianus of Etruria (*ca* 245 – 275 CE)

One of PLŌTINOS' most loyal students, joined Plōtinos' school in Rome in 246 – three years after its opening – staying until 269 (*Vit. Plot.* 3.38–42). Amelius had already been educated in philosophy by Lusimakhos and had learnt by heart almost all of NOUMĒNIOS OF APAMEIA's works (3.42–48).

A key figure in Plōtinos' school, Amelius prepared critical editions of Plōtinos' works (*Vit. Plot.* 19.22–23), kept notes from Plōtinos' seminars (3.46–48), and also refuted Plōtinos' rivals and critics, writing 40 books against ZOROASTER (16.12–14), trying to discredit the charge that Plōtinos had stolen his doctrines from Noumēnios (17.1–6). Amelius also wrote two treatises rebutting PORPHURIOS' view that the intelligibles are outside the intellect (18.11–17). Porphurios criticized Amelius for the "unphilosophical complexity" of his works (21.17; *cf.* 20.76–80) and his naïve religiosity (10.33–34), suggesting rivalry between Plōtinos' two most prominent students.

Amelius' own philosophical views, close to Plōtinos' by his own admission (*Vit. Plot.* 20.76–78), sometimes differed from them (17.41–42). Amelius upheld the existence of three divine ***hupostaseis***, but unlike Plōtinos maintained (based on PLATO, *Timaeus* 30c7–d1) that all are intellects. Amelius commented on John 1.1 arguing that the *logos* of which John speaks is the cause of all beings (EUSEBIOS, *PE* 11.18.26–19.1). SYRIANUS reports (*In Met.* 119.12–15) that Amelius also maintained that human souls accommodate the Forms which are present in nature as *logoi* and are the causes of everything.

L. Brisson, "Amélius: Sa vie, son oeuvre, sa doctrine, son style," *ANRW* 2.36.2 (1987) 793–860; *BNP* 1 (2002) 575–576, *Idem.*

George Karamanolis

AMIGERŌN ⇒ DAMIGERŌN

Ammōn (Metrol.) (395 – 405 CE)

One of the extant fragments (fr. 41 Blockley), preserved by Phōtios, *Bibl.* 80 (63a), from the history of OLUMPIODŌROS OF THĒBAI refers to Ammōn the *geōmetrēs*. His name is typically Egyptian, and he was not simply a "land surveyor," but probably an "architect," according to the three official titles listed in *Theodosian Code*, 13.4.3. Ammōn is said to have carried out the measurement of the perimeter of Rome's city walls when "the Goths were about to launch the first attack on her." Therefore, he may have surveyed the wall after the original circuit was completed and restored under Honorius 402–3 CE and before 410 CE, when Alaric plundered Rome. According to Olumpiodōros, Ammōn's survey was 21 Roman miles, but this record is obviously corrupt, since the actual circuit is only about 12½.

RE 1.2 (1894) 1857–1858 (#2), Fr. Hultsch; I.A. Richmond, *The City Wall of Imperial Rome* (1930; repr. 1971) 25, 35.

Mauro de Nardis

Ammōn (Astrol.) (*ca* 100 BCE – *ca* 400 CE?)

Poetical astrologer, from whose *Concerning Beginnings* (*peri* ***katarkhōn***) two fragments (19 hexameters) were preserved in the scholia to Tzetzēs, *Allegoriai Iliadis* 7.117; he is cited by name at 7.126. Ammōn interprets the moon in tropic signs as indicating the falsity of oracles and dreams, and swift homeward return for travelers (*cf.* MANETHŌN 6.359–360); in solid signs the moon presages illness, but slow return for travelers. The dedicatee of HERMĒS TRISMEGISTOS' ***Iatromathēmatika*** is an Ammōn, who if not the Egyptian god might be an astrological Ammōn, such as ours.

Ed.: A. Ludwich, *Maximi et Ammonis carminum* (1877) 52–54.
RE 1.2 (1894) 1858 (#3), E. Riess.

GLIM

Ammōnios, M. Annius (*ca* 40 – 85 CE)

PLUTARCH depicts his teacher Ammōnios (born *ca* 5 CE) as an impressive and stimulating intellectual figure, with **Pythagorean** leanings and special interests in theology, physical phenomena, mathematics, and astronomy: *De E* 391E–394C; *Quaest. conv.* 648B–F; 743C–748D. An Egyptian living in Athens as a citizen, Ammōnios held important offices including *stratēgos* (thrice), and received Roman citizenship.

C.P. Jones, "The Teacher of Plutarch," *HSPh* 71 (1967) 205–213; J. Whittaker, "Ammonius on the Delphic E," *CQ* 63 (1969) 185–192; *DPA* 1 (1989) 164–165, B. Puech; *BNP* 1 (2002) 589 (#5), M. Baltes and M.L. Lakmann.

Jan Opsomer

Ammōnios of Alexandria (50 – 10 BCE)

CELSUS 7.*pr*.3 credits him as an innovative surgeon, and specifies that he invented a reliable process for splitting and extracting bladder stones (7.26.3B). AËTIOS OF AMIDA 14.51 (p. 795 Cornarius) reports a blood-stanch: ***khalkitis***, quicklime, orpiment, and realgar. OREIBASIOS, *Ecl. Med.* 8 (*CMG* 6.2.2, p. 188), preserves his wound-cream, especially useful on eyes, containing aloes, antimony, **calamine**, saffron, **Indian buckthorn** and Indian nard, ***psimuthion***, etc. in rainwater; repeated by Aëtios 7.117 (*CMG* 8.2, pp. 393–394). A very similar, but unattributed, collyrium is given by ASKLĒPIADĒS PHARM., in GALĒN *CMLoc* 4.8 (12.761 K.), and repeated by PAULOS OF AIGINA, 7.16.23 (*CMG* 9.2, p. 339).

Michler (1968) 72, 115–116.

PTK

Ammōnios of Alexandria, son of Hermeias (*ca* 470 – after 517 CE)

Born *ca* 440 CE; Neo-**Platonist** philosopher from Alexandria, PROKLOS' student. DAMASKIOS, SIMPLICIUS, ASKLĒPIOS OF TRALLEIS, and IŌANNĒS PHILOPONOS attended his lectures. Preserved are these commentaries: *In Porphyrii Isagogen, In Aristotelis Categorias, In Aristotelis De Interpretatione, In Aristotelis Analytica Priora* (*CAG* 4.3–6). Asklēpios' *Comm. In Metaph.* and Philoponos' *In Anal. (I and II), De Gen.* and *De An.* derive heavily from his lectures. Damaskios describes Ammōnios (*Vita Is.* 79) as *philoponōtatos*, an expert in ARISTOTLE, geometry, and astronomy, and critical of Proklean metaphysics. He argued that god(s) know all of time, but that such knowledge does not constrain future events: they have knowledge of future contingents but not as future (Tempelis; *cf.* IAMBLIKHOS' suggestion that divine knowledge is definite but is about indefinites). Ammōnios observed planetary occultations or near-conjunctions (with his brother HĒLIODŌROS and his uncle), and Arcturus' longitude (with Simplicius), the latter to check PTOLEMY's value of the precession of the equinoxes (which he erroneously confirmed); his work on the use of the astrolabe has been rediscovered and published.

KP 1.306, H. Dörrie; *DSB* 1.137, Ph. Merlan; Neugebauer (1975) 1031–1041; Ch. Soliotis, "Unpublished Greek texts on the use and construction of the Astrolabe," *Praktika tēs Akadēmias Athēnōn* 61 (1986) 423–454; E. Tempelis, "Iamblichus and the School of Ammonius, Son of Hermias, on Divine Omniscience," *SyllClass* 8 (1997) 207–217; *ECP* 25–26, H.J. Blumenthal; *REP* 1.208–210, Chr. Wildberg; Athanassiadi (1999); *BNP* 1 (2002) 590–591 (#12), P. Hadot.

Cosmin Andron

Amōmētos (of Kurēnē?) (280 – 245 BCE)

Wrote geographical ***paradoxa***, including *Voyage up from Memphis* and *On the Attakori* (a mythical blessed people of the Far East), quoted by KALLIMAKHOS, PLINY 6.55, and AELIANUS *NA* 17.6. The name is rare, except in Kurēnē: *LGPN* 1.35.

FGrHist 645.

PTK

L. Ampelius (175 – 180 CE?)

Wrote a small encyclopedia or epitome on cosmology, geography, theology, and history, entitled *Liber Memorialis* ("aide-mémoire"), and addressed to his student Macrinus. The work is preserved in one 17th c. MS copied by Claude Saumaise. Scholars dispute the work's date, some preferring the 4th c., because of genre and language. The most recent editor argues that §47, on the greatest Roman victories, predates those of Septimius Seuerus in 194–198 CE. Consequently, the dedicatee, a student in Caesarea Mauretania (one of the few provincial cities mentioned: §38), could be the emperor Macrinus – hence perhaps explaining the work's survival.

The text begins with an elementary cosmology, describing the four elements and five zones of the earth (§1), then proceeds through fire, i.e., the stars (§2–3), air, i.e., the winds (§4–5), earth (§6), and water (§7). Ampelius describes many human ***paradoxa*** (§8), and offers a Euhemeristic theology (§9). About three-fifths of the work is a Romano-centric outline of history (§10–50).

Ampelius' description of the zodiac (§2) is based on NIGIDIUS, to which he adds the constellations Bears, Orion, Pleiades, Hyades, and Canis, plus the seven planets (§3). He correlates the 12 winds with the zodiacal signs (§4), then explains wind as the movement of air, saying there are four *generalis* winds, east, west, north, south, and various *specialis* winds of particular times or places (§5). He follows KRATĒS OF MALLOS' fourfold division of the globe, partitioning our quarter into the continents Asia, Europe, and Libya; he lists peoples, mountains, rivers, and isles of each continent (§6). His treatment of the sea (§7) resembles that in ON THE KOSMOS.

Ed.: M.-P. Arnaud-Linder, *Aide-mémoire* (*CUF* 1993).
OCD3 75, L.A. Holford-Strevens.

PTK

Amphilokhos of Athens (325 – 90 BCE)

Wrote a work on agriculture, possibly treating cereals, livestock, poultry, viticulture, and arboriculture (*cf.* PLINY, 1.*ind.*8, 10, 12–15, 17–18), that was excerpted by CASSIUS DIONUSIOS (VARRO, *RR* 1.1.8, *cf.* COLUMELLA, 1.1.9). One volume, entitled *On Moon-trefoil and Alfalfa*, discussed the cultivation of these two important fodder plants, and described their appearance, country of origin, and medicinal uses for both humans and animals (Pliny, 13.130–131, 18.144–145, *cf.* Schol. Nik. *Thēr.* 617).

RE 1.2 (1894) 1940–1941 (#7), M. Wellmann.

Philip Thibodeau

Amphinomos (365 – 325 BCE)

Mentioned four times by PROKLOS, twice with known followers of PLATO, but not in his sketch of the history of geometry (*In Eucl.* pp. 65–68 Fr.). These references suggest that Amphinomos was interested in metamathematical rather than mathematical issues: the circle of SPEUSIPPOS and Amphinomos held that, since geometry treats eternal entities, one should not call anything established in it a problem (77.15–78.8); Amphinomos denied that geometry explains why its results are true (202.9–11), and classified problems in terms of the number of solutions they admit (220.7–12); finally mathematicians "around" MENAIKHMOS and Amphinomos discussed the convertibility of propositions of the form "All A are B" (254.4–5).

DPA 1 (1989) 173, R. Goulet.

Ian Mueller

Amphiōn (250 BCE – 95 CE)

ASKLĒPIADĒS PHARM., in GALĒN *CMGen* 4.13 (13.736 K.), preserves his wound-plaster of copper flakes (DIOSKOURIDĒS 5.78–79), **litharge**, and ***khalkitis***, ground in olive oil and vinegar.

Fabricius (1726) 56.

PTK

Amuntas (Geog.) (*ca* 320 – *ca* 230 BCE)

Wrote an account of Alexander's journey, *Stages in Asia* or *Stages in Persia*. ARISTOPHANĒS OF BUZANTION *Epit.* 2.358 (*CAG* S.1.1 [1885] 106) cites him on the iron-ore-eating mice of Terēdōn in Babylonia, repeated by AELIANUS *NA* 5.14, who also transcribes his somewhat paradoxographical account of Caspian-region rats and foxes (*NA* 17.17). Athēnaios, *Deipn.* 2 (67a), quotes him on Persian mountain products (**terebinth**, squill, and walnuts), and, 11 (500d), on the "manna" exuded from oak leaves and used to make a sweet drink (*cf.* THEOPHRASTOS, *HP* 3.7.6). Although the name is common from an early date, especially in northern Greece (*LGPN*), he may be the same as the Amuntas involved in a conspiracy against Ptolemy Philadelphos, *ca* 250 BCE.

FGrHist 122; *KP* 1.322 (#1), G. Wirth.

PTK

Amuntas (Med.) (350 BCE – 200 CE)

Wrote on bandages, as recorded by OREIBASIOS, *Coll.* 48.31 (*CMG* 6.2.1, p. 278); similar bandages are described by pseudo-GALĒN, *de Fasciis* 58, 61, 89 (18A.805–806, 807–808, 818 K.).

Michler (1968) 88–89, 131.

PTK

Amuntas of Hērakleia Pontikē (365 – 325 BCE)

Among PLATO's pupils (PHILODĒMOS, *Pap. Herc.* 1021, col. 6: Gaiser 1988: 183). Many scholars think that Ibn al-Qifṭī (*Ta'riḫ al-Ḥukāmā'* (Lippert 1903: 24), AELIANUS (*VH* 3.19),

Diogenēs Laërtios (3.46), and Proklos (*In Eucl.* p. 67.8 Fr.) refer to this same pupil as Amuklas or Amuklos. That he is from Hērakleia Pontikē is inferred from Philodēmos' description of Amuntas and a Hērakleidēs as Hērakleians. Proklos, linking Amuklas with Menaikhmos and Deinostratos, says that they "made the whole of geometry more perfect (or complete)."

J. Lippert, *Ibn al-Qifti's Ta'rih al-hukama'* (1903); Lasserre (1987) 7; K. Gaiser, *Philodems Academica* (1988).

Ian Mueller

Amuntianos (160 – 180 CE)

Wrote a history of Alexander dedicated to Marcus Aurelius (*Schol. Bern. Verg. Georg.* 2.137), parallel lives of Dionysius (tyrant of Surakousai) and Domitian, of Philip of Macedon and Augustus, and a biography of Olympias (Phōtios *Bibl.* 131). In *On Elephants*, he claimed "all Ethiopian elephants, males and females alike, have 'tooth-horns'" in contrast to the female Indian elephant (*Schol. Pind. Ol.* 3.52).

FGrHist 1072.

GLIM

Amuthaōn (120 BCE – 80 CE?)

Prepared ointments containing **bdellium**, frankincense, and **galbanum**, in a beeswax and **terebinth** base. Asklēpiadēs Pharm., in Galēn *CMGen* 7.6 (13.967 K.), cites his ointment for joints etc., as above, plus henna oil, ***ammōniakon*** incense, and myrrh. Paulos of Aigina cites two of his remedies for stiff or distorted joints, found unattributed or otherwise attributed in Andromakhos in Galēn: (1) 4.55 (*CMG* 9.1, p. 381) = *CMGen* 7.6 (13.969 K.: Leukios') = 7.7 (13.977 K.), contents as above, plus ***opopanax***, pepper, etc.; and (2) a simpler version 7.17.33 (*CMG* 9.2, p. 355) = *CMGen* 7.7 (13.983 K.); also in Cassius Felix 42.8, 43.11 (*CUF* pp. 112, 121). Oreibasios, *Syn.* 3.57 (*CMG* 6.3, p. 83), and Aëtios of Amida 10.11 (p. 578 Cornarius) record his very similar recipe for splenetic disorders, cited as well known by the *Hipp. Cant.* 31.2 (2.166–167 ed. Oder-Hoppe). Caelius Aurelianus, *Chron.* 2.211 (*CML* 6.1.1, p. 672), prescribes his *malagma* for ***phthisis***, and Aëtios 8.63 (*CMG* 8.2, p. 512) mentions his remedy for ***orthopnoia***. The rare name is attested as Amutheōn and Amuthaoun (*LGPN* 2.26, 3B.28).

RE S.11 (1968) 29–30, M. Michler.

PTK

Anakreōn (Astron.) (300 – 100 BCE?)

Nothing is known about this poet of Alexandrian time. According to *The Second Life of Aratos*, he is said to have written an elegiac poem entitled *Phainomena* from which Iulius Hyginus (*Astr.* 2.6.2) quotes a single pentameter about the constellation Lyra (*CA* 130).

E. Maass, *Aratea* (1892) 150; *BNP* 1 (2002) 631, M. Di Marco.

Christophe Cusset

Anakreōn (Pharm.) (after 100 BCE?)

Wrote *On Root-Cutting* (*peri rhizotomikēs*), recording that some called "horse-celery" by the name *Smurneion* (*Schol. Nik. Thēr.* 597c). Besides the poet, a rare name, cited at Kition and Dēlos (3rd c. BCE: *LGPN* 1.35) and Athens (1st/2nd cc. CE: 2.28).

RE 1.2 (1894) 2050 (#2), M. Wellmann.

GLIM

Anania of Shirak (Arm.: Anania Širakacʿi) (*ca* 610 – *ca* 685 CE)

Distinguished mathematician and astronomer, known as the "father of the exact sciences" in Armenia. Born in the village of Anania in Shirak around the time HĒRAKLEIOS reopened the school of Constantinople, he traveled widely to satisfy his thirst for scientific learning, and studied under several masters before finding TUKHIKOS in Trebizond with whom he studied for eight years. Upon returning home he set up his own school, presumably near Shirak; he claims that until this time no Armenian knew philosophy nor were any books on science to be found. Samuēl of Ani (*ca* 1100–1180) names the following as his students: Hermon, Trdat, Azaria, Ezekiel, and Kirakos. The tradition that he was buried in the village of Anavank' is more likely to be an etiology for the name of the village.

Although more than half are now lost, more than 40 scientific works have been attributed to Anania in the fields of cosmology, astronomy, mathematics, geography, and chronology. Matʿewosyan has demonstrated that a number of these works once formed a textbook, completed in 666, known as the *Kʿnnonikon* (Gk. *Kanonikon*), and comprised all the major sciences known in the medieval curriculum. His work, *A History of the Universe*, based largely on the *Hexaemeron* of BASIL OF CAESAREA, served for centuries in Armenia as the basic textbook for science.

On Heaven and Earth is a compendium of what was known in the Classical Greek tradition: e.g., the earth was a sphere and that the ***kosmos*** turned eternally on its concentric spheres around the earth at its center. Curiously, despite his mathematical interests, Anania did not seem interested in calculating meridians, longitudes or latitudes. Unlike many of his fellow Christians, he adhered to a **Peripatetic** system of the world. Anania maintained belief in a transcendental universe, and was guilty of occasional flights into mystical speculation.

Of his astronomical works, *Concerning the Skies* is based on Basil, and a work called *On Clouds and Signs* seems rather a conglomeration of Basil and ARATOS. Recent scholarship has also attributed to Anania the Armenian translation of Aratos, *Phainomena*; both are certainly works of the HELLENIZING SCHOOL at whose height Anania was most active.

Seven mathematical treatises treat the fundamentals of addition, subtraction, multiplication, division, odd and even numbers. These works, complete with long lists of tables, are considered the oldest extant writings of their kind in the world, although they are of much less sophistication than Greek mathematical thought such as found in EUCLID's algebraic and geometrical imagination. Anania considered mathematics to be the "mother of all sciences" and only through it can the human mind apprehend that all natural phenomenon presents itself in the language of mathematics, under definitely structured forms, and develops or moves according to definite patterns or laws. Although a confirmed scientist, Anania was not free of the magical side of contemporary mathematics.

On Questions and Answers contains 24 arithmetical problems and their solutions, as does *Fun with Arithmetic* (Arm. *Xraxčanakankʿ*, lit., "Things for festal occasions"). These too are possibly the oldest extant texts of their kind in the world. Two short treatises, one on the numerology of the Old and New Testaments, and another on the allegorical significance or power of numbers also survive. A number of fragmentary works are considered to have been part of his lost mathematical textbook, *Arithmetic: Texts for Four Applications*.

One of the most important of Anania's works is his *Geography* (Arm., *Ašxarhacʿoycʿ*), which was once thought to be the work of MOSES OF XOREN. This work, which survives in a long and short recension, is explicitly based on the lost *Geography* of PAPPOS OF ALEXANDRIA, and also makes use of PTOLEMY. It is a rambling, epitomized descriptive geography, covering all the then-known world, from Spain to China, offering nothing new except for some few details about Asia Minor. Nearly one quarter is devoted specifically to the Caucasus, Armenia, and the Sasanian empire, providing much information from local otherwise unknown sources, and comprising an invaluable source for the historical geography of these areas, but especially Armenia. Another work, *On the Languages of the World*, lists and locates the speakers of the 72 known world languages.

Anania's chronological works survive mostly in fragmentary form. *The Book of Caesars*, up to the year 685/6, is essentially a poor translation of Greek sources. In addition to a short treatise entitled *On Calendars*, fragments of a treatise survive in which he seems to have worked out a universal schema of calendar calculation. Short pieces, on the date of Christmas and on the date of Easter, and a comparison of the Armenian and Hebrew calendars, also survive.

Other surviving works include a book on *Weights and Measures*, which is essentially a very loose, inaccurate translation of EPIPHANIOS OF CYPRUS. Two related opuscules, *On Measures* and *On Weights and Weighers*, also raise certain ethical issues. A short essay, *On the Names of Gems and their Colors*, also survives. A treatise *On Music* is of disputed authorship. Other untitled works on meteorology and on foretelling the weather remain unedited.

Anania's works served as the curriculum standard in the medieval Armenian academy, as well as the basis for many later Armenian writers who devoted themselves to science in the Middle Ages, of whom the more famous were Anania of Narek, Yovhannēs Kosern, Grigor Magistros, Yovhannēs Sarkavag, Samuēl Anecʿi, Yovhannēs Erznkacʿi, Grigor Datevacʿi, etc.

Ed.: Ašot Abrahamyan, *Anania Širakacʿu Matenagrutʿyunē* [The Works of Anania Širakacʿi] (1944); Robert H. Hewsen, *The Geography of Ananias of Širak (Ašxarhacʿoucʿ). The Long and the Short Recensions* (1992).

Hakob Manandyan, "Les mesures attribuées à Anania Širakacʿi convertées en poids et mesures actuels," *Revue des études arméniennes* 5 (1968) 369–419; Robert H. Hewsen, "Science in Seventh Century Armenia: Ananias of Širak," *Isis* 49 (1968) 32–45; Jean-Pierre Mahé, "Quadrivium et cursus d'études au VIIe siècle en Arménie et dans le monde byzantin d'après le *Kʿnnikon* d'Anania Širakacʿi," *Travaux et Mémoires* 10 (1987) 159–206; A.S. Matʿewosyan, "Anania Širakacʿu 'Ašxarhagrutʿyan' het kapvacʿ mi kʿani harcʿer [Some Questions relative to the *Geography* of Anania of Shirak]," *Lraber* 9 (1979) 73–86.

Edward G. Mathews, Jr.

Anastasios (200? – 540 CE)

AËTIOS OF AMIDA 12.47 (p. 681 Cornarius) preserves his gout remedy, containing **shelf-fungus**, aloes, cinnamon, gentian (*cf.* GENTHIOS), ***malabathron***, parsley, spikenard, etc. This Christian name is attested from *ca* 200 CE (*LGPN*, esp. 3A.36, 3B.33), but rare before the middle-Byzantine period: *ODB* 86–87, *Prosopographie der mittelbyzantinischen Zeit* 1.1 (1999) #228–341.

RE 1.2 (1894) 2067 (#4), M. Wellmann.

PTK

Vindonius Anatolios of Bērutos (*ca* 330 – *ca* 370 CE)

Agricultural writer (Vindanius in Phōtios, *Bibl.* cod. 163), perhaps but not certainly identifiable with the imperial official and friend of LIBANIOS who died in 360 (*PLRE* 1 [1971] 59–60, #3). His *Sunagōgē geōrgikōn epitēdeumatōn*, a collection from earlier Greek writers in 12 books, extant in the 9th c., was distinguished by Phōtios as useful and superior to similar treatises. From its preface Phōtios names DĒMOKRITOS, IULIUS AFRICANUS, HĒRAKLEIDĒS OF TARAS, APULEIUS, "Florentius" (=FLORENTINUS), VETTIUS VALENS, Leōn (perhaps LEONTINOS), PAMPHILOS, and the ***Paradoxa*** of DIOPHANĒS. Its contents embraced siting the farmstead; quality of soil and water; weather prognostication and remedies for natural phenomena such as storms and insects; culture and preservation of cereal crops, vines, olives, as well as the products of garden and orchard and their potential medicinal applications; livestock and their care. From what survives, Phōtios' judgment seems entirely valid: alongside empirical commentary of known and accepted validity is a rich miscellany of "irrational and incredible elements, reeking of pagan folly." The latter embrace folklore, apotropaic customs, and practices based on the "Democritean" dogma of **sympathy** and antipathy.

Anatolios' work achieved widespread circulation. In the west, PALLADIUS, whose references take the form *Graeci (auctores)*, used it. Probably in the 6th c., it was translated into Syriac (an abbreviation partially survives) and thence into Arabic (*Kitāb al-Filāḥa* of Yūniūs, preserving the author's name and apparently much of his format). Roughly contemporaneously, it was revised and expanded by CASSIANUS BASSUS, whose *Eklogai* not only were incorporated (largely wholesale) into the 10th c. GEŌPONIKA, but were in turn translated into Old Persian and thence a second time into Arabic. Complex relationships amongst the Middle Eastern and Islamic versions, themselves marked by a certain fluidity, are the subject of current studies, which will illuminate the combination of literary tradition and practical innovation characteristic of Arabo-Andalusian agronomical works.

Only fragments of the *Sunagōgē* survive in Greek. Independently transmitted are a chapter on hail, frost and pests in the vineyard (Paris, BNF *graecus* 2313, f. 49^{V}), and a short section on medical treatment of cattle (*P. Vindob.* G40302, 6th/7th c.). Sixteen excerpts in the late antique compilation of veterinary texts known as the *Hippiatrika* are securely attributed to Anatolios; these closely correspond to material on horses preserved in *Geōponika* Book 16, indicating that this part of Cassianus' compilation is derived, without much alteration, from Anatolios. Anonymous passages on the points of the horse and on breeding which preface the C recension of the *Hippiatrika* are attributed by Oder (1896) to Anatolios on account of their similarity to parts of *Geōponika* 16.

Oder (1890); *RE* 1.2 (1894) 2073 (#14), M. Wellmann; E. Oder, *Anecdota Cantabrigiensia* (1896); E. Fehrle, *Studien zu den griechischen Geoponikern* (1920); Robert H. Rodgers, "Hail, Frost, and Pests in the Vineyard," *JAOS* 100 (1980) 1–11; A. Papathomas, "Das erste antike Zeugnis für die veterinärmedizinische Exzerptensammlung des Anatolios von Berytos," *WSt* 113 (2000) 135–151; J. Carabaza Bravo, *Arabic Sciences and Philosophy* 12 (2002) 155–178; J. Hämeen-Anttila, "The Oriental

Tradition of Vindanius Anatolius of Berytus' *Sunagōgē geōrgikōn epitēdeumatōn*," *WZKM* 94 (2004) 73–108; McCabe (2007).

Anne McCabe and Robert H. Rodgers

Anatolios of Laodikeia (250 – 282 CE)

Christian polymath born in Alexandria, succeeded Eusebios as bishop of Laodikeia (268 CE). Because of his high reputation for learning (in arithmetic, geometry, astronomy, dialectics, physics, and rhetoric), Alexandrians asked him to establish a school of **Peripatetic** philosophy (Eus., *EH* 7.32.6–21). As testimony to Anatolios' intellectual scope, EUSEBIOS quotes (rather, misquotes: McCarthy 126–139) the second part of his *De ratione Paschali* – his only attested work of substantial length – a defense of an original method of computing the date of Easter based on a 19-year lunar cycle and drawing from various sources, Greek (PTOLEMY), Christian (mainly Origen), and Jewish (McCarthy 114–125). An hagiographic anecdote, underscoring Anatolios' attitude and virtues during the Roman siege of Alexandrian Pyrucheum in 264, suggests his political influence.

At least three other Anatolioi might be identifiable with our bishop: (1) most plausibly, the Neo-**Platonist** philosopher of Alexandria mentioned in EUNAPIOS as IAMBLIKHOS' master and PORPHURIOS' renowned contemporary, probably the author of the short neo-**Pythagorean** tract *On the Decade and the numbers within it* included in the THEOLOGUMENA ARITHMETICAE, perhaps part of the *Arithmetical introductions* mentioned by Eusebios (*EH* 7.32.20); (2) the source of an appendix to HĒRŌN's *Definitiones* (4.160.8–166.22 H.), proposing answers to general questions about the nature and parts of mathematics; (3) one "very learned Anatolios" mentioned in a letter of Michael Psellos discussing a nomenclature of "powers of the unknown" similar to, but more general than DIOPHANTOS', adding names for the 5th and 7th powers and proceeding to the 10th power instead of the 7th (*Diophantos* 2.37–39 Tannery). Tannery's conjectural identification of the latter with the bishop is weak.

Ed.: D.P. McCarthy, *The ante-Nicene Christian Pasch 'De ratione paschali'* (2003).
DPA 1 (1989) 179–183, R. Goulet.

Alain Bernard

Anaxagoras of Klazomenai (480 – 428 BCE)

Born *ca* 500, he came to reside in Athens during the golden age of that city, where he established himself as the leading philosopher-scientist of his day. He associated with the statesman Periklēs and other prominent figures. After spending 30 years in Athens (probably *ca* 480–450, though some argue for 460–430), he left under the threat of prosecution and set up a school in Lampsakos, where he died.

In the wake of PARMENIDĒS' criticisms of cosmological theories, Anaxagoras developed a radically pluralistic theory. According to this theory, nothing comes to be or perishes (a Parmenidean principle); everything is mixed in everything; matter is infinitely divisible; and whatever predominates in a mixture determines the character of that mixture. The theory is designed to explain how the appearance of radical change is possible without there being any radical change. For instance, he can explain how people add flesh to their bodies by eating bread: flesh, mixed in the bread, is extracted in the process of digestion. Anaxagoras seemingly posits an elemental reality corresponding to every kind of stuff in the world: water, earth, flesh, bone, wood, iron, and so on. A piece of wood, for instance, has portions of all others in it, but more elemental wood than anything else.

Anaxagoras' principle of infinite divisibility allowed him to account for change on any scale, even a microscopic one. His contemporary ZĒNŌ OF ELEA argued that this principle has absurd consequences.

In his cosmogony, Anaxagoras described how in the beginning there was a more or less uniform mixture of elements. "Mind" (*nous*) created a rotary motion expanding into a great whirl or vortex, which tended to separate like elements to like, as in a centrifuge. The present world is the result of this motion, with the vortex carrying the heavenly bodies around the earth. Mind plays a cosmic role in exercising some control over the world, at least in its origin. Anaxagoras described mind as an extended stuff with special properties, that does not mix with all things like the elements. It is, however, found in all living things.

Anaxagoras' most important scientific innovations came in astronomy. He held that the heavens consisted of heavy bodies held aloft by the vortex motion. When a large meteor fell near Aigospotamoi in northern Greece, he was said to have predicted it. Anaxagoras also correctly explained solar and lunar eclipses, perhaps for the first time in recorded human history. Probably drawing on Parmenidēs' insight that the moon gets its light from the sun, he explained eclipses as screenings of the sun's light. He also posited the existence of unseen bodies (asteroids in effect). He argued (probably against Parmenidēs) that the earth is flat; in this he was wrong, but he argued his case empirically, and his advances seem to qualify him as the first empirical astronomer.

Like other philosophers of his time, Anaxagoras also proposed a number of meteorological theories. He gave an essentially correct explanation of hail and conventional accounts of lightning and thunder. He explained the Nile floods as resulting from melting snows in southern mountains.

As the first philosopher-scientist to reside in Athens, he laid the foundations of an important intellectual tradition. His mechanistic views of the heavenly bodies seemed blasphemous to conservative Athenians and led to proceedings against him, yet he remained the most influential scientific thinker of the early 5th c.

Ed.: DK 59; D. Sider, *The Fragments of Anaxagoras*[2] (2005); P. Curd, *Anaxagoras of Clazomenae = Phoenix* S.44 (2007).

KRS 352–384; D. Panchenko, "Anaxagoras' Argument against the Sphericity of the Earth," *Hyperboreus* 3 (1997) 175–178.

Daniel W. Graham

Anaxikratēs (of Rhodes?) (325 – 315 BCE)

Wrote a ***Periplous*** *of the Red Sea*, based on his voyage south from Aila (mod. Aqaba), and cited by STRABŌN 16.4.4 from ERATOSTHENĒS; his expedition may be the one in THEOPHRASTOS, *HP* 9.4.2–6, *cf.* ARRIANUS, *Indika* 43.7. Perhaps the same as the writer of *Argolika* (*FGrHist* 307). Tzetzēs *Chil.* 7.177–180, refers to an Anaxikratēs *epistatēs* under Seleukos I, along with Asklēpiodōros the satrap of Persis (*cf.* DIODŌROS OF SICILY 19.48.5), who is likely different. The name is rare except at Athens and Rhodes (*LGPN*).

RE S.14 (1974) 44–47 (#8), W. Schmithenner.

PTK

Anaxilaïdēs (*ca* 350 BCE – *ca* 200 CE)

DIOGENĒS LAËRTIOS 3.2 cites Anaxilaïdēs, *On Philosophers* 2, for a tale about PLATO's mother Periktiōnē also told by SPEUSIPPOS and KLEARKHOS; at D.L. 1.107 the same man, or someone named Anaxilaos (hardly the alchemist ANAXILAOS OF LARISSA), along with SŌSIKRATĒS OF RHODES, is cited for variant data on an early Greek wise man, Musōn. (In the latter case, Dionusios of Halikarnassos, *A.R.* 1.1, may be related.)

FGrHist 1094–1095.

PTK

Anaxilaos of Larissa (40 – 20 BCE?)

Paradoxographer, interested in occultism and with **Pythagorean** tendencies, banned by AUGUSTUS from Italy, in 28 BCE. He is regarded as one of the chief sources of PLINY, who mentioned Anaxilaos several times: 19.20; 25.154; 28.181; 32.141; 35.175. He wrote *Physics* (*phusikà*), *On gilding and silvering* (*baphikà*), and *Trifles* (*paignia*).

DSB 1 (1970) 150, L. Taràn; *DPA* 1 (1989) 192, R. Goulet.

Bruno Centrone

Anaximandros of Milētos (*ca* 580 – 545 BCE)

Anaximandros © Rheinisches Landesmuseum Trier

Born *ca* 610, and a student of THALĒS, Anaximandros continued the kind of scientific theorizing his master had begun. He wrote a book, perhaps one of the first prose compositions and a paradigm for later theorists, describing how the world arose and its shape. Anaximandros said a seed-like substance separated from "the everlasting" or "boundless" (*apeiron*) and from it grew the earth surrounded by fiery stuff, which divided into rings surrounded by air. The earth was a disk surrounded by these fiery rings (the heavenly bodies), enclosed by air to make them invisible except at an opening through which the light shone. Thus, what are called the sun, moon, and stars are just the apertures of invisible rings. The sun is farthest from the earth, the moon below it, and the stars closest to the earth. Lunar phases and eclipses of the sun and moon result from the blocked openings.

Anaximandros said that periodic phenomena, apparently including day and night, summer and winter, are caused by some contrary matter (such as the hot) prevailing, and then paying a penalty to its opposite (such as the cold) "according to the ordering of time," for its trespass. Thus a kind of law-like balance was maintained which put a limit to the excesses of the contraries.

Anaximandros, further, gave natural explanations to meteorological phenomena such as lightning and thunder (caused by wind in clouds). Life emerged in the primeval seas, and some creatures were enclosed in shells, which burst open on land to give birth to terrestrial animals.

Anaximandros was a pioneer of naturalistic explanations. Although he seems to have ascribed divine properties to the boundless stuff from which the elements originate, he gave purely natural explanations to the events of the world, including the creation of the world itself, and what had seemed to be miraculous events, such as lightning (traditionally explained by Zeus throwing thunderbolts). He also drew the first Greek map of the world. His conception of the world as a product of natural forces acting in a law-like way inspired all later Greek science.

DK 12; C. H. Kahn, *Anaximander and the Origins of Greek Cosmology* (1960); KRS 100–142.

Daniel W. Graham

Anaximenēs of Milētos (*ca* 555 – 535 BCE)

The student of ANAXIMANDROS, Anaximenēs developed a cosmological theory like his teacher. He said that the world arose out of air, which by condensation or rarefaction changes into other things. When air is somewhat rarefied it becomes fire; when it is somewhat condensed it becomes wind; when it is more condensed it becomes cloud; when still more, water, then earth, then stones. Thus Anaximenēs posited a mechanism which by acting on air or its successor substances produces other substances, providing the first systematic theory of change. According to this theory, variations in pressure produced different states of air (on the standard account) or perhaps even completely different substances.

In the beginning, according to Anaximenēs, there was only air. As air was condensed, the earth – a flat, thin disk – arose, and then the heavenly bodies from terrestrial evaporations. The earth floats on a cushion of air, and the heavenly bodies – or probably just the sun, moon and planets – also float on the air like leaves; the fixed stars are attached to a dome like a felt cap, which circles around the earth. The heavenly bodies do not go under the earth, but are hidden behind an elevated region in the north. Anaximenēs explained meteorological phenomena such as rain, hail, and lightning on the basis of his physical theory. He also explained earthquakes as resulting from the drying out and cracking of the earth.

Anaximenēs' physical theory and meteorology influenced a number of later cosmologists including XENOPHANĒS, HĒRAKLEITOS and ANAXAGORAS; moreover, PARMENIDĒS seems to have refuted his theory of change. By positing a mechanism of change, Anaximenēs made the change of elemental bodies subject to determinate physical laws for the first time.

DK 13; KRS 143–162; G. Wöhrle, *Anaximenes aus Milet* (1993); Daniel W. Graham, "A New Look at Anaximenes," *History of Philosophy Quarterly* 20 (2003) 1–20.

Daniel W. Graham

Anaxipolis of Thasos (325 – 90 BCE)

Authored a work on agriculture which may have treated cereals, livestock, poultry, viticulture, and arboriculture (*cf.* PLINY, 1.*ind*.8, 10, 14–15, 17–18), and was excerpted by CASSIUS DIONUSIOS (VARRO, *RR* 1.1.8–10, *cf.* COLUMELLA, 1.1.9).

RE 1.2 (1894) 2098, M. Wellmann.

Philip Thibodeau

Andreas (of Athens?) (345 – 355 CE)

Brother of a bishop Magnus, composed a Paschal canon valid for 353–553 CE, and a work on Paschal canons. An Andreas "philosopher," likely the same man, is credited with a work on solar eclipse; and an Andreas of Athens is credited with a work on astrological predictions from conjunctions in Taurus. Little survives, and only in Armenian.

DPA 1 (1989) 196–197, J.-P. Mahé.

PTK

Andreas of Karustos (*ca* 250 – 217 BCE)

Andreas (*Vind. Med. Gr.* 1, f.3V) © Österreichische Nationalbibliothek

Personal physician to Ptolemy Philopatōr, mistaken for Ptolemy and murdered before the battle of Raphia (POLUBIOS 5.81.6). His writings include a pharmacopoeia and a history of medicine, *On Medical Genealogy*, wherein Andreas traced, e.g., HIPPOKRATĒS' "descent" from Asklēpios and Hēraklēs, and preserved anecdotes and traditions occasionally in conflict with other accounts of Hippokratēs' life (von Staden 1999: 150–157). In *On Poisonous Animals* and *Against False Beliefs* (Ath., *Deipn.* 7[312d–e]), he described the poisonous moray eel. The latter text seems to have treated popular belief and non-medical wonders rather than the history of scientific medicine. In *Casket*, he likened plant leaves to *skolopendra* (centipedes) in that their astringent properties aid victims of poisonous bites (An.25 von Staden). DIOSKOURIDĒS, although chiding Andreas and KRATEUAS for omitting useful roots and herbs (*pr.*1), conceded they were exact describers of plants. PLINY, citing Andreas as a medical authority for 14 books (1.*ind.*20–28, 31–35), reported his opinions that because poppy is adulterated in Alexandria, instant blindness does not ensue (22.200; *cf.* Dioskouridēs 4.64.6), and chickpeas, a digestive aid to women and the elderly, do not cause flatulence (22.102). In *To Sōsibios*, an epistolary work treating obstetrics, *inter alia*, and dedicated to Ptolemy Philopatōr's minister, Andreas concurred with the **Hērophilean** theory of difficult parturition, adding that the light weight of paralyzed or emaciated fetuses further complicates labor (SŌRANOS *Gyn.* 4.1 [*CMG* 4, p. 131]).

Andreas, employing both mechanical and pharmaceutical treatments, was widely cited for the latter. He observed that bandages could alleviate headaches (Ath., *Deipn.* 15 [675c]), and he developed a machine to set dislocated limbs described in detail by OREIBASIOS (*Coll.* 49.4–6 = *CMG* 6.2.2, pp. 6–12; *cf.* CELSUS 8.20.4). Celsus quotes three complex prescriptions: a multi-use emollient of costly exotics and aromatics (5.18.7); Andreas' invention for scrofulous tumors including nettle seed, **galbanum**, sea spurge seed, and unheated sulfur (5.18.14; *cf.* ASKLĒPIADĒS PHARM. in GALĒN *CMLoc* 10.1 [13.343 K.]); and a salve of gum, ***psimuthion***, antimony, and **litharge**, smeared on the forehead to treat eye complaints (6.6.16). Pliny preserves Andreas' prescription of ashed crabs applied with oil to leprous eye sores (32.87), and Dioskouridēs preserves his prescription of thistle applied over varicose veins to prevent pain (4.118). In Galēn, Andreas is cited as the source for NEILEUS' rose compound for painful boils and prolapses (An.32 von Staden); also preserved are a treatment for calloused ulcers (An.31 von Staden); a treatment for sciatica and arthritis, of fenugreek dissolved and cooked in vinegar and honey (13.345 K.); and several emollients similar to those in Celsus (An.29 von Staden).

J. Scarborough, "Nicander *Theriaca* 811. A note," *CPh* 75 (1980) 138–140; Beavis (1988) 10–19; von Staden (1989) 472–477; *OCD3* 88, A.J.S. Spawforth; von Staden (1999) 149–158; *BNP* 1 (2002) 680–681 (#1), V. Nutton.

GLIM

Andrias (120 – 80 BCE)

According to VITRUUIUS 9.8.1, the inventor, with THEODOSIOS OF BITHUNIA, of the "All-Latitude" (*pros pan klima*) sundial, which simultaneously provided directional orientation and time (contrast PARMENIŌN).

D.J.deS. Price, "Portable Sundials in Antiquity, including an Account of a New Example from Aphrodisias," *Centaurus* 14 (1969) 242–266.

PTK

Androitas of Tenedos (400 – 200 BCE?)

Wrote a ***Periplous*** *of the Propontis* of which one fragment survives in a scholiast.

FGrHist 599.

PTK

Androkudēs (Pythag.) **(1st c. BCE?)**

The Androkudēs cited by THEOLOGUMENA ARITHMETICAE (p. 52 de Falco), with ARISTOXENOS, HIPPOBOTOS, *et al.*, as a writer on the numerology of PYTHAGORAS' reincarnations may well be the same neo-**Pythagorean** (cited by Clement of Alexandria, *Strom.* 5.8.45.2, IAMBLIKHOS *VP* 145, etc.), who wrote *On Pythagorean Symbols.*

RE 1.2 (1894) 2149 (#2), J. Freudenthal; Burkert (1972) 167.

PTK

Androkudēs (Med.) (360 – 320 BCE)

Physician; contemporary of Alexander the Great, to be distinguished from ANDROKUDĒS THE PYTHAGOREAN. The apocryphal epistle which an *Androcydes sapientia clarus* addressed to the king to warn of the effects of immoderate drinking (PLINY 14.58) is insufficient proof to declare him Alexander's private physician. Some Androkudēs (Clement of Alexandria, *Strom.* 7.6.33.7) — the same as the **Pythagorean** (*Androkudēs ho Puthagorikos, ibid.* 5.8.45.2)? — used to say that "both wine and meat strengthen bodies but weigh down one's spirits" (*cf.* PLUTARCH, *Meat-Eating* 995E, *al.*). The epistolary Androkudēs may be identifiable with THEOPHRASTOS' Androkudēs who considered cabbage a remedy against drunkenness (*HP* 4.16.6, *unde* Pliny 17.240). Only Athēnaios (6 [258b]) titled Androkudēs a physician; but the etymology of *kolakeia* (flattery), from *kollasthai*, that he attributed to Androkudēs is better suited to a grammarian (*cf. Etymologicum Genuinum* = *Etymol. Magnum*, s.v. *kolax* [525.2]).

RE 1.2 (1894) 2149 (#1), M. Wellmann.

Jean-Marie Jacques

Andromakhos of Crete (Elder) (50 – 65 CE)

Greek physician from Crete, father of ANDROMAKHOS OF CRETE (YOUNGER). Perhaps a *peregrinus* (Korpela 1987: 164, #54), he practiced medicine in Rome as Nero's personal doctor (GALĒN, *Antid.* 1.1, 14.2 K.), probably specializing in pharmaceutical therapy. He altered the formula of the *Mithridateion* supposedly created by (or for) MITHRADATĒS VI (*ibid.*), recording his version in 87 elegiac couplets, twice given by Galēn: *Antid.* 1.6 (14.32–42 K.) and *Thēr. Pis.* 6–7 (14.233 K., but omitted by Kühn). Modifying both the list and proportions of ingredients, Andromakhos called the new 64–ingredient compound *galēnē* as effecting tranquility regarding health (Galēn *ibid.* 15, 14.270–271 K.). Andromakhos' formula resulted in a wide-spectrum medicine, instead of a supposedly universal antidote against all kinds of venoms and poisons. The medicine, later known as theriac, circulated independently in Byzantine MSS, was translated into Arabic (ed. L. Richter-Bernburg, *Eine arabische Version der pseudogalenischen Schrift De theriaca ad Pisonem*, Diss. Göttingen 1969), and continued to be prepared into the early 20th c.

Ed.: Ideler 1 (1841/1963) 138–143; Heitsch 2 (1964) 8–15.
RE 1.2 (1894) 2153–2154 (#17), M. Wellmann; Watson (1966) esp. 45–53; *KP* 5.1573, J. Kollesch; *BNP* 1 (2002) 685 (#4), V. Nutton.

Alain Touwaide

Andromakhos of Crete (Younger) (70 – 90 CE)

Greek physician, son of ANDROMAKOS OF CRETE (ELDER), physician to the imperial court at Rome: GALĒN *CMGen* 1.15 (13.427–429 K.). Andromakhos, perhaps specializing in pharmaceutical therapy like his father whose theriac recipe he is believed to have transcribed into prose (Wellmann), wrote on medicine in three works, each cited as a monograph: 1. medicines for external use; 2. medicines for internal use; 3. ophthalmologic medicines (Galēn, *CMGen* 2.1, 13.463 K.). Galēn claims that Andromakhos simply compiled them from his many sources without properly explaining the doses and uses of the medicines: *CMGen ibid.* and 1.16 (13.441–442 K.). Indeed, Andromakhos often claims to have employed or preferred a particular recipe – e.g. *CMLoc* 9.5 (13.299 K.) and *CMGen* 6.14 (13.930–931 K.), *cf.* Fabricius (1972) 174–179. Contrast *CMLoc* 9.5 (13.300 K.) and *CMGen* 7.13 (13.1037 K.), where Andromakhos gives detailed preparation directions (and claims to have used the second recipe). Maybe Galēn's criticism is due to his willingness to appear as the major authority in pharmacology, something that led him to sharply criticize his predecessors. In spite of that, however, Galēn reproduced from Andromakhos' works over 50 lengthy extracts.

RE 1.2 (1894) 2154 (#18), M. Wellmann; Watson (1966) 45, 55, 138; *KP* 5.1573, J. Kollesch; Fabricius (1972) 185–189, 201; *BNP* 1 (2002) 685–686 (#5), V. Nutton.

Alain Touwaide

Andrōn (Math.) (430 – 370 BCE)

Student of Oinopidēs of Khios and teacher of Zēnodotos, who distinguished *theorēma* (seeking to know the character of a matter under investigation) from *problēma*, which seeks conditions of existence (Proklos, *In Eucl.* p. 80.17 Fr.). Zhmud (2006: 178–179) identifies Proklos' source as Geminus not Eudēmos, and dates this Andrōn (with his teacher and student) to the Hellenistic era.

RE S.7 (1940) 39 (#18), K. von Fritz.

GLIM

Andrōn (Pharm.) (225? – 75 BCE)

Pharmacologist to be distinguished from Andreas of Karustos; prior to Hērakleidēs of Taras who mentioned Andrōn's remedy in his *Against Antiokhis* (Galēn *CMLoc* 6.8 [12.983.17–984.6 K.] = Hkld. *fr.* 205 Deichgr.). He had invented a ***trokhiskos***, the formula of which was modified by Hērakleidēs (*cf.* Asklēpiadēs Pharm. in Galēn *CMGen* 5.11 [13.825.16–826.1 K.]). It is described by Andromakhos the Younger (*ibid.* 5.12 [834.9–13 K.], *cf.* Asklēpiadēs, 5.11 [825.13–15 K.]) as follows: pomegranate flowers, oak-gall, myrrh, birthwort, vitriol, fissile alum and Cyprian ***misu*** macerated in sweet wine. This **pastille** became famous and is mentioned in various prescriptions not only by the Greeks (Galēn, Oreibasios, Aëtios, and Paulos of Aigina) but also by the Romans (Scribonius Largus [5 times], Celsus [thrice] and Caelius Aurelianus [*Acute* 3.18, *CML* 6.1.1, p. 302.31]). Athēnaios (15 [680d]) quotes Andrōn when writing about *akinos*, a coronary plant rather akin to wild basil.

RE 1.2 (1894) 2161 (#16), M. Wellmann.

Jean-Marie Jacques

***Andrōn of Rome* (*ca* 120 – 170 CE)**

Marcus Aurelius' childhood geometry and music teacher, afterwards honored by the emperor (*SHA* 2.2).

RE 1.2 (1894) 2159 (#10), P. von Rhoden; Netz (1997) #98.

GLIM

Andrōn of Teōs (350 – 300 BCE?)

Wrote a *Pontika*, of which traces on early myths are preserved in the scholia to Apollōnios of Rhodes; he is also cited by Arrianus, *Indika* 18.8.

FGrHist 802.

PTK

Andronikos (Paradox.) (250 BCE – 300 CE)

Otherwise unidentifiable author whom the PARADOXOGRAPHUS PALATINUS (§12) cites for a self-generating mineral from Spain.

(*)

PTK

Andronikos (Pharm.) (250 BCE – 80 CE)

ANDROMAKHOS, in GALĒN *CMLoc* 7.5 (13.114 K.), gives his recipe for ***orthopnoia***: anise, "Ethiopian" cumin, ***ammōniakon*** incense, castoreum, myrrh, and sulfur, formed into pills, to be taken at bedtime with water.

BNP 1 (2002) 688 (#6), W. Portmann.

PTK

Andronikos of Kurrhos (*ca* 150 – 125 BCE)

Astronomer; built at Athens the *Horologion* ("Tower of the Winds"), an octagonal marble tower with an internal water-clock and a sundial on each face (VARRO, *RR* 3.5.17). Topping the tower was a weather-vane in the shape of a Triton or sea-monster, whose rod held in the right hand hovered above the side carved with a representation of the prevailing wind (VITRUUIUS 1.6.4), thereby signifying wind direction.

RE 1.2 (1894) 2167–2168 (#28), E. Fabricius; C. Mee and A.J.S. Spawforth, *Greece: An Archaeologial Guide* (2001) 74–76.

GLIM

Andronikos of Rhodes (100 – 20 BCE)

Peripatetic philosopher, "the eleventh (**scholarch**) after ARISTOTLE" according to AMMŌNIOS, *De interpr.* 5.28–29; ancient sources connect him especially to an edition of ARISTOTLE's and THEOPHRASTOS' treatises, previously rather neglected if not partially lost: during the 1st c. BCE (in the first half: Moraux and Gottschalk; *ca* 40–20: Düring) he "published Aristotle's and Theophrastos' writings and prepared those catalogues (*pinakes*) of them" which were still used in PLUTARCH's time (*cf. Sulla* 26). Moreover, "he divided them into *pragmateiai*, having put together treatises on the same subject into one work" (PORPHURIOS, *Plot.* 24). Since all of Andronikos' works are lost (many lost already in late antiquity: e.g. there is no reason to believe that his *Peri diaireseōs* was available any longer to BOËTHIUS, *De diuisione, PL* 64.875D), it is unclear how responsible he is for the shape of the extant works of Aristotle (e.g. our *Metaphysics*). The catalogue of Aristotle's writings preserved in Arabic attributed to a Ptolemaios is regarded as deriving from Andronikos' list. But as far as we can see, his main contribution seems to concern the collection, constitution or completion of the Aristotelian corpus, which he made more available for further exegesis, so that he can be regarded as a main founder of the Aristotelian commentary tradition. He wanted to order the corpus into a curriculum, "starting with logic, for this is concerned with proof" (*cf.* IŌANNĒS PHILOPONOS *in Cat.* 5.18–23); logic in turn presupposes the *Categories*, a text scarcely influential hitherto, on which Andronikos wrote an explanatory paraphrase, starting the series of a long and dense exegetical tradition on this work. In his *pinakes*, five books at least, he recorded the titles (which he eventually discussed, rejecting e.g. a previous title for *Categories: pro tōn topōn*, i.e. "Before the Topics"), the beginnings, and the length of each treatise (e.g. he disregarded the final chapters, 10–15, of the *Categories*, as not belonging to the work). He possibly included a discussion of content, aim, and authenticity (e.g. the *De interpretatione* rejected as spurious); moreover Aristotle's legacy, possibly his life, and a collection of (spurious) correspondence. His attitude toward Aristotle was sometimes critical, and not exempt from **Stoic** influence.

Ed.: Fr. Littig, *Andronikos von Rhodes* (1890).

Spuria: Two works preserved in Greek have been wrongly attributed to Andronikos: a *Paraphrasis of Nichomachean Ethics* (*editio princeps*: D. Heinsius, Lugduni Batavorum 1617) and a treatise *On passions* (*peri pathōn*) (ed. Mullach, *Fr. Philos. Gr.* 3 [1881] 570–578).

M. Plezia, *De Andronici Rhodii studiis Aristotelicis* (1946); I. Düring, *Aristotle in the ancient Biographical tradition* (1957) 412–425; Moraux (1973) 1.45–141, with L. Taran's review, *Gnomon* 53 (1981) 725–742; Gottschalk (1987) 1084–1107, 1112–1116, 1129–1131; J. Barnes, "Roman Aristotle," in *Philosophia togata* II, ed. J. Barnes and M. Griffin (1997) 1–69.

Silvia Fazzo

Androsthenēs of Thasos (324 – 286 BCE)

Son of Kallistratos, oversaw the navigation of NEARKHOS' fleet to the mouths of the Tigris and Euphratēs, and supervised one of the three naval voyages to Arabia promoted by Alexander the Great (324 BCE). Covering a good section of the Arabian coast, presumably during the navigation, he made a stop near Tulos (modern Bahrein), where previous exploration had ceased.

He wrote a treatise on India, ***Periplous*** *tēs Indikēs* (*FGrHist* 711), discussing also naval voyages to Arabia, observations on rains, some fields, flora, and shellfishes. Androsthenēs' autoptical observations allow a precise dating of his arrival at the island Tulos in November, when abundant rains begin, and yet it is still possible to see the ripe fruits of tree-cotton (*Gossypium arboreum*).

F. Susemihl, *GGLA* 1 (1891) 653–654; *RE* 1.2 (1894) 2172–2173, H. Berger.

Cristiano Dognini

Androtiōn of Athens (385 – 355 BCE)

Son of Andrōn, Athenian politician and atthidographer, born *ca* 410 BCE from a wealthy and distinguished family and was Isokratēs' pupil before he entered political life *ca* 385. He was a member of the Council twice (before 378/377 and again in 356); a tax commissioner (perhaps in 374/373); a governor of the Athenian garrison in Arkesinē on Amorgos during the Social War (358–356); and a member of the embassy to Maussōllos of Karia (355/354), intended to prepare the way for war against Persia. His personal enemies accused him of an unconstitutional proposal in 354/353; Dēmosthenēs undertook the prosecution but was defeated. For reasons uncertain, Androtiōn was exiled to Megara (after 344/343), where he composed his *Atthis* (published *ca* 340), the only history of Athens written by an active politician, comprising eight books; 68 fragments survive. Like other *Atthides*, Androtiōn's work began with primeval time and the history of early kings and continued to at least 344/343 BCE. Its contents reveal that Androtiōn treated 4th c. history and his own time in far greater detail than past history. The last five books cover the events of 404–*ca* 340, a period Androtiōn considered critical for Athens and wherein his own political activities peaked.

Androtiōn's *Atthis*, chief source for ARISTOTLE's *Athenian Constitution* and Philokhoros' *Atthis*, was a standard work on Attic history until Philokhoros' *Atthis* surpassed it.

The authenticity of the *Geōrgikon* (composed in one book) ascribed to Androtiōn (also to the otherwise unknown figures Philippos or Hēgēmōn: F75) has been doubted, but the work's content suits our knowledge of Androtiōn's character. Furthermore, an agricultural work was also assigned to Androtiōn's predecessor KLEIDĒMOS, although Kleidēmos' work was more theoretical than Androtiōn's technical and practical *Geōrgikon*. Tree culture received thorough treatment. Athēnaios cites Androtiōn's *Geōrgikon* on kinds of fig trees (F75), apple trees (F77) and pears (F78). THEOPHRASTOS (F81) refers to Androtiōn's recommendation that olive and myrtle trees need the most pruning to promote growth and fruitfulness; olive, myrtle and pomegranate trees also need the most pungent manure, the heaviest watering, and the most complete pruning to avoid any underground disease (*HP* 2.7.2–3). Theophrastos (F82) also cites Androtiōn's view of the **sympathy** between olive and myrtle trees, which entwine their roots so the myrtle's fruit becomes tender and sweet when the olive shelters it from sun and wind (*CP* 3.10.4). The story (F76) of the Titan Sykeus, which (according to Athēnaios) the grammarian Truphōn in his plant history says was recorded in Androtiōn's *Geōrgikon*, looks to be Hellenistic, but from this we should not regard the *Geōrgikon* as a forgery (as did Wellmann).

FGrHist 324 F75–82; F. Jacoby, *Atthis* (1949); P.E. Harding, *Androtion and the Atthis* (1994); *OCD3* 89, P.E. Harding; *BNP* 1 (2002) 690, K. Meister.

Maria Marsilio

Ankhialos (*ca* 135 – 105 BCE)

Named by PANAITIOS OF RHODES with KASSANDROS as the best astrologers of his time: they did not practice predictive astrology (CICERO, *Div.* 2.88), perhaps instead focusing on signs indicating divine will. A rare name (usually with gamma), cited in the feminine at Samos (5th c. BCE) and variously spelled, especially at Boiōtia (Ankhialos with gamma or nu, plus *-aros* or *-alis*: 5th c. BCE – 2nd c. CE: *LGPN*).

RE S.3 (1918) 99, W. Kroll.

GLIM

M. Annaeus Lucanus (60 – 65 CE)

Born at Cordoba on 3 November 39 CE, studied in Rome and learned **Stoic** philosophy under his uncle L. ANNAEUS SENECA and L. Annaeus Cornutus. In 60 CE he was nominated *quaestor* by Nero, but in 65 CE joined the conspiracy of C. Calpurnius Piso and was arrested with the other conspirators. Having received the order to kill himself, he cut his veins and died reciting verses in which he described the similar death of a soldier (TACITUS, *Ann.* 15.70). He wrote several lost poetic works: *Iliacon* (on the end of the Trojan war, to the death of Hector and the ransom of his corpse); *Catachthonion* (a description of a journey in the Underworld); *Saturnalia; Epistulae ex Campania; De incendio urbis; Orpheus*; 14 *fabulae salticae*; epigrams; *Medea* (an unfinished tragedy). He was also the author of two *Controuersiae*, one for Octauius Sagitta and the other against him.

His best known works are the *Siluae* and the *Bellum Ciuile* (*Pharsalia*) in 10 books. In the latter he wanted to show his own negative vision of the civil war and the horrors caused by it. In Books I–III he recounts Roman history from the crossing of the Rubicon to

Caesar's arrival in Rome and the siege of Marseilles; in Books IV–VIII the civil war in Spain, Illyria and Africa, Caesar's arrival in Italy through Epirus, the siege against Pompeius at Dyrrachium, the battle of Pharsalus, Pompeius' flight into Egypt and his death. In Book IX Lucanus describes the withdrawal of Cato Uticensis' troops through the Libyan desert towards Leptis Magna, among dangers and sufferings. Lucanus includes a catalogue of 16 snakes, born in Libya from the blood streaming form the Medusa's cut head and describes the atrocious deaths inflicted on soldiers by their bites (9.607–733), probably based on NIKANDROS. Finally in Book X (incomplete), Lucanus speaks of Caesar's visit to the tomb of Alexander the Great and of a banquet given for him by KLEOPATRA VII.

Ed.: C.R. Raschle, Pestes Harenae. *Die Schlangenepisode in Lucans* Pharsalia *(IX 587–949)* (2001); C. Wick, *Marcus Annaeus Lucanus,* Bellum Civile, *liber IX* (2004).

OCD3 94–95; J. Radicke, *Lucans poetische Technik. Studien zum historischen Epos* (2004).

Claudio Meliadò

L. Annaeus Seneca (*ca* 40 – 65 CE)

Annaeus Seneca Bildarchiv Preussischer Kulturbesitz/Art Resource, NY

Born *ca* 4 BCE/1 CE, **Stoic** philosopher and Roman politician. Tutor and advisor to the emperor Nero. Seneca was born in Spain to a wealthy Italian family. His interest in philosophy began early, and he studied under both **Stoics** and Sextians (Seneca later described SEXTIUS' independent Roman school, a little misleadingly, as a kind of **Stoicism**). He was politically active from the age of 25, becoming *quaestor* in 31, and thereafter moving in influential circles, including that of the imperial family. In 41 he became ensnared in a move by Claudius' first wife Messallina to banish Caligula's sister Iulia Liuilla, ostensibly for adultery with Seneca. With the help of Claudius' second wife, Agrippina, Seneca was able to return to Rome in 49, when he also became tutor to Agrippina's 12-year-old son, Nero. After Nero's accession in 54, Seneca became one of the two chief advisors to Nero, exerting a powerful influence on the young emperor, and ensuring a remarkably smooth and scandal-free early reign. His co-advisor in this period was Sex. Afrianus Burrus, Nero's praetorian prefect. Over time, however, Seneca's beneficial influence on Nero began to wane as the emperor became more and more self-indulgent, and on Burrus' death Seneca attempted to retire. His political reputation had become compromised as a result of Nero's increasingly unacceptable behavior, including Nero's murder of his own mother. In 64, Seneca was implicated (perhaps unjustly) in the conspiracy of C. Calpurnius Piso and forced to commit suicide.

Seneca's main philosophical interests were in ethics and natural philosophy, and he also authored some celebrated and influential tragedies, nine of which are extant. As with his tragedies, Seneca's natural philosophy (to judge by the extant *Naturales quaestiones*) is deeply engaged with moral themes. It covers the field known as meteorology in antiquity, which included the study of earthquakes, volcanoes, and comets as well as winds and weather phenomena. Seneca also shows an interest in rivers, with a long discussion of the origins of rivers generally, and an entire book dedicated to the Nile. The *Q.Nat.* shows POSEIDŌNIOS' influence, mentioned repeatedly by name, but it is overall a much more original work than scholars usually deem it (Seneca's originality in ethics and literature has similarly been positively re-assessed in recent years). The physics underlying the *Q.Nat.* often deploys a theory that the earth is permeated by subterranean caverns through which winds, water, and fire can move and from which they can emerge in various ways, causing storms, earthquakes, and volcanic eruptions. In the *Q.Nat.*, Seneca generally introduces a topic, reviews various competing theories about it, and then adjudicates between different theories or offers novel solutions. But moral questions are never very far from Seneca's mind, and his introduction of ethical themes in the middle of an otherwise apparently straightforward text on physics can strike the modern reader as disconcerting or off topic. Recent work (Inwood 2005) has tried to see the *Q.Nat.* as part of Seneca's larger intellectual project, and this has the effect of bringing ethics and physics into a clearer relation to each other. This approach also avoids anachronistically exporting modern disciplinary boundaries and expectations onto Seneca's work.

Ed.: conveniently available in Loeb editions.
M. Griffin, *Seneca: A Philosopher in Politics* (1976); B. Inwood, *Reading Seneca* (2005).

Daryn Lehoux

ANNIUS ⇒ AMMŌNIOS ANNIUS

"Anonymous"

Most anonymi are filed within under some more specific name, in order to emphasize their specific nature (most pseudonymous works and those preserved on PAPYRI could well be labeled "anonymous"). See also the parts of the ARISTOTELIAN CORPUS and the HIPPOKRATIC CORPUS. Some anonymous works have no standard label as "Anonymous": those 47 we cross-reference here first; then seven cited elsewhere as "Anonymous X."

⇒ AELIUS PROMOTUS

⇒ "ANTIKYTHERA DEVICE"

⇒ ASTROLOGOS OF 379

⇒ "BĒRUTIOS"

⇒ BOOK OF ASSUMPTIONS

⇒ BOOK OF THE ZODIAC

⇒ CARMEN ASTROLOGICUM

⇒ CARMEN DE PONDERIBUS ET MENSURIS

⇒ Demotic Scientific Texts

⇒ Derveni Papyrus

⇒ Dimensuratio Prouinciarum and Diuisio Orbis Terrarum

⇒ Dissoi Logoi

⇒ Expositio geographiae

⇒ Expositio totius mundi

⇒ Geōponika

⇒ Hellenizing School

⇒ De Herbis

⇒ Isidōros of Milētos' Student

⇒ Itineraria

⇒ Keskintos

⇒ Korē Kosmou

⇒ On The Kosmos

⇒ "Lion Horoscope"

⇒ Massiliot Periplous

⇒ Medicina Plinii

⇒ On Melissos, Xenophanēs, and Gorgias

⇒ Mulomedicina Chironis

⇒ Paradoxographus Florentinus

⇒ Paradoxographus Vaticanus

⇒ Paradoxographus Palatinus

⇒ Paraphrasis eis ta Oppianou Halieutika

⇒ Periplus Maris Erythraei

⇒ Periplus Ponti Euxini

⇒ Peutinger Map

⇒ Physica Plinii

⇒ Physiognomista Latinus

⇒ Physiologus

⇒ De planetis

⇒ Pontica

⇒ Pracepta Salubria

⇒ Prolegomena to Ptolemy's Suntaxis

⇒ De Quaternionibus

⇒ Ravenna Cosmography

⇒ Stadiasmus Maris Magni

⇒ Summaria Rationis Geographiae

⇒ Theologumena Arithmeticae

⇒ Yavaneśvara

... Arabs ⇒ Arrabaios

... De rebus Bellicis ⇒ Bellicis, de Rebus

... of Brussels ⇒ Vindicianus

... Fuchsii ⇒ Parisinus medicus

... Londinensis ⇒ Londiniensis medicus

... Parisinus ⇒ Parisinus medicus

... in Theaetetum ⇒ P. Berol. 9782

Anonymous Alchemist "Christianus" (500 – 800 CE?)

Hard to date alchemical commentator, whose *On the Divine Water* (*CAAG* 2.399) was dedicated to "Sergius," probably the patriarch of Constantinople under Emperor Hērakleios

(Saffrey 1995: 6). The work (*CAAG* 2.395–421) consists of 30 chapters, with commentary, following the early authors (PSEUDO-HERMĒS, ZŌSIMOS, PSEUDO-DĒMOKRITOS) on precise themes and questions. Berthelot (*CAAG* 3.381) notes that this compilation follows the general system adopted by the Byzantines of the 8th and 10th centuries (e.g., Phōtios and Constantine VII Pophurogennētos), which consisted of drawing extracts and summaries from early authors. This method, although conserving fragments, also dismembered texts. Berthelot (*CAAG* 3.380) gives a comparative table of the chapters in different MSS and their distribution in his edition.

The author's fragments essentially concern the idea of the "divine water" and the science's methodology and its operations. As with SUNESIOS and other commentators, the obscurity of the language of the ancients is explained as having the double goal of fooling the jealous and training the minds of the adepts. The author insists upon the apparent discord of the ancients regarding the names for the "divine water" and especially about the meaning of its unity. As Zōsimos had already done, this author wanted to demonstrate the fundamental accord among the authors as to the unity of kind of the "divine water." In particular, he shows that pseudo-Dēmokritos speaks in general of a unique kind, and Zōsimos speaks of multiple material kinds. In reality, all multiplicity leads to the one unity.

Some reflections concern the method. Distinctions of materials and treatments show the influence of the descriptions of states of physical bodies (liquids, solids, compounds) and processes of transformations (cooking, melting, decomposition by fire or by liquids) from ARISTOTLE, *Meteor.* 4. The treatments (*oikonomiai*) are compared to plane geometrical figures, recalling PLATO's *Timaios* and the work of STEPHANOS. Finally, the author applies the "dialectic" method of dividing and reuniting by species and genera (originating with Plato) to explain the operations, for clarity.

Ed.: *CAAG* 2.395–421.
CAAG 3.378–382; Letrouit (1995) 2; Saffrey (1995) 6.

Cristina Viano

***Anonymous Alchemist Philosopher* (600 – 800 CE?)**

Berthelot collected the writings attributed to the "Anonymous Philosopher" into three works: *On Divine Water, On the Gold-making Procedure*, and *On Music and Alchemy* (*CAAG* 2.421–441); sometimes split into two persons, the first and second Anonymous (Letrouit 1995). Our author offers one of the oldest lists of alchemists: he summarizes the ANONYMOUS ALCHEMIST "CHRISTIANUS", mentions HERMĒS TRISMEGISTOS, PSEUDO-DĒMOKRITOS, SUNESIOS, ZŌSIMOS, IŌANNĒS ARCHPRIEST, and then "the famous ecumenical philosophers, the commentators on PLATO and ARISTOTLE, who employed dialectic, OLUMPIODŌROS and STEPHANOS" (*CAAG* 2.425).

In particular, the Anonymous examines the mixture of substances using liquids, and without fire, of which Olumpiodōros also speaks (*CAAG* 2.426). He is influenced by the Aristotelian theory of mixture, the basic composition of all natural bodies (*CAAG* 2.439). The Anonymous establishes a curious methodological analogy between musical instruments and the parts of alchemical science.

Ed.: *CAAG* 2.421–441.
CAAG 3.378–382; Letrouit (1995) 63–64; Saffrey (1995) 6.

Cristina Viano

Anoubiōn of Diospolis (1st c. CE?)

Astrologer featuring in the fictitious account of pseudo-Clement, *Homiliae* (4.6), as an associate of the famous wizard Simōn Magos who later repudiated this master. Anoubiōn was also the ostensible author of a widely diffused Greek astrological poem, in at least four books, that, exceptionally for its genre, was composed in elegiac couplets rather than hexameters. Firmicus Maternus (3.1), as well as Greek astrological compilations from late antiquity, cite Anoubiōn as a source of doctrines, and a few elegiac lines from the poem survive through the medieval manuscript tradition. Substantial fragments of elegiac verse on astrology, almost certainly belonging to Anoubiōn's opus, have been identified in several papyri dating from the 2nd c. CE and after. The surviving portions are devoted to specifics of interpreting horoscopes, and the poem, although betraying some literary pretensions, appears to have been intended as a practical handbook directed primarily to professional astrologers.

Ed.: *POxy* 66 (1999) 57–109; D. Obbink, *Anubio: Carmen astrologicum elegiadum* (2006).

Alexander Jones

Anqīlāwas (or Anqīlāʾus) (620 – 640 CE)

Probably the Aggeleuas mentioned by Stephanos in his commentary on Galēn's *Therapeutics to Glaukōn*. Arabic sources know him, together with the earlier Gessios of Petra and Marinos, as one of the early 7th c. Alexandrian physicians who abridged Galēn's works into a canonical medical curriculum surviving in Arabic as *Jawāmiʿ al-Iskandarāniyyīn* or *Summaria Alexandrinorum*. Ibn-Juljul says Anqīlāwas was the chief (*raʾīs*) of the Alexandrians, evidently the head of the school, and wrote books of medicine.

Ullmann (1970) 21, 65; *GAS* 3 (1970) 160; Dickson (1998) 77; D. Gutas, "The 'Alexandria to Baghdad' Complex of Narratives. A Contribution to the Study of Philosophical and Medical Historiography among the Arabs," *Documenti e Studi sulla Tradizione Filosofica Medievale* 10 (1999) 155–193.

Kevin van Bladel

Anthaios, Sextilius (25 – 75 CE)

Pliny lists him after Dēmokritos, Apollōnios "Mus," Melētos, and Artemōn (1.*ind.*28 and 28.7–8) as giving medicines *from* the human body, and records his quasi-magical remedy for ***hudrophobia*** (pills made from the skull of a hanged man). Scribonius Largus in Asklēpiadēs Pharm., in Galēn *CMLoc* 4.7 (12.764–765 K.), records his green plaster of acacia, **calamine**, saffron, myrrh, Indian nard, opium, etc., in gum and rain-water; *cf.* Aëtios of Amida 12.44 (p. 102 Kostomiris). Pliny also lists *Antaeus* as a Greek authority on trees, 1.*ind.*12–13.

RE 1.2 (1894) 2343 (#7), M. Wellmann.

PTK

Anthedius of Vesunnici (450 – 470 CE)

President of a poetical society, lectured on musicians, geometers, arithmeticians, and astrologers, and was, according to Sidonius Apollinaris' fawning account of his friend, an expert on astrology and astronomy (*Carmen* 22.*pr*.2–3, *cf. Ep.* 8.11.2), whose sources included FULLONIUS SATURNINUS, IULIANUS VERTACUS, IULIUS FIRMICUS, and THRASUBULUS.

PLRE 2 (1980) 93.

GLIM

Anthēmios of Tralleis (*ca* 500 – 558 CE?)

Byzantine mathematician, engineer, and architect, who was born at Tralleis in Lydia *ca* 475 CE, and died in Constantinople before 558 CE. His father, STEPHANOS, was a physician and his two brothers Dioskoros and ALEXANDER were physicians, his brother MĒTRODŌROS was a famous ***grammatikos*** and another brother Olumpios was a lawyer (Agathias, *Hist.* 5.6–9). Little is known about Anthēmios' education and training, but *mēkhanikos* (or *mēkhanopoios*), the term used for Anthēmios' profession, suggests that he was an engineer with some theoretical training in mathematics.

Agathias states that his fame for mathematical competence reached the emperor. On the one hand, this competence seems confirmed by EUTOKIOS' warm dedication to him of his commentary to APOLLŌNIOS' *Kōnika* (2.168.5, 290.3, 314.2, 354.2 Heiberg), in which Anthēmios appears either as Eutokios' young companion or disciple (Decorps-Foulquier 2000: 63–64), as well as by the excerpts of the work entitled in Greek *Mechanical Paradoxes* and in Arabic *On the Construction of Burning Mirrors* (Greek text in Rashed 343–359, to be completed with ʿUtārid's revision of its Arabic translations, *ibid.* 312–315). They contain the clever design of ellipsoidal and parabolic mirrors (the axial section of each being conceived and constructed through its tangents), and of a burning mirror such as the one attributed to ARCHIMĒDĒS through the combination of plane mirrors. On the other hand, another collection of catoptrical problems, *On Burning Mirrors and other Mirrors*, also appears in ʿUtārid under Anthēmios' name and apparently contains much material naively derived from HĒRŌN's *Katoptrika*, with little care for experimental likelihood (Jones 1987). Whether Anthēmios is also the author of the so-called Bobbio fragment edited by Heiberg (*Mathematici Graeci Minores* 87–92 = Huxley 53–58 with trans. 20–26) is disputed (see Rashed 264–271 *contra* Knorr 63–70); the Arabic evidence neither confirms nor disproves Anthēmios' authorship.

Anthēmios gained fame as the senior designer of Justinian I's Hagia Sophia in Constantinople, celebrated by the historian Prokopios (*Aed.* 1.1.24) and Paul Silentiarios (*Ekphrasis* 267, 552); see also ISIDŌROS OF MILĒTOS. Anthēmios consulted in the construction of a dam in Dara (Prokopios, *Aed.* 2.3.7) and, according to Constantine of Rhodes (870–931 CE), he designed Justinian's church of the Holy Apostles in Constantinople (no additional commissions are known). There are (diverging) speculations that the architects may have based their work on a geometrical project related to late neo-**Platonist** conceptions of mathematics and/or the ***kosmos*** (Dennert, *Sehepunkte* 6, Nr 7/8, 2006, summarizes the issue). Anthēmios also produced an artificial earthquake, loud noises, and blinding reflections to annoy his neighbor, the orator Zēnōn (Agath. 5.6–8).

Anthēmios continued to be respected by medieval mathematicians. Al-Kindī, Ibn Īsā or Ibn al-Haytham used his works, and Alhazen (11th c.) pairs Anthēmios with Archimēdēs as pioneers in paraboloid mirrors. The Byzantine poet John Tzetzēs (12th c.) quotes Anthēmios when describing Archimēdēs' use of mirrors to burn Marcellus' fleet. Finally, the Thuringian Vitello (13th c.) refers to Anthēmios in his *Perspectiva*, an optical treatise commissioned by the humanist bishop William of Moerbeke.

RE 1.2 (1894) 2368–2369, Fr. Hultsch; Heath (1921) 2.203; E. Darmstaedter, "Anthemios und sein 'künstliches Erdbeben' in Byzanz," *Philologus* 88 (1933) 477–482; Downey (1948) 112–114; G.L. Huxley, *Anthemius of Tralles* (1959); Soulis (review of Huxley), *Speculum* 35 (1960) 123–124; *RBK* 1 (1963) 177–178, M. Restle; Warren (1976); W.R. Knorr, "The Geometry of Burning Mirrors in Antiquity," *Isis* 74 (1983) 63–70; A. Jones, "On Some Borrowed and Misunderstood Problems in Greek Catoptrics," *Centaurus* 30 (1987) 1–17; Mainstone (1988) 157; *ODB* 109, M.J. Johnson and A. Kazhdan; *PLRE* 3 (1992) 88–89 (#2); Rashed (2000).

Kostis Kourelis and Alain Bernard

Anthemustiōn (unknown date)

Author of a recipe for a ***trokhiskos*** for snake-bite or bites of other venomous creatures, preserved in the C and L recensions of the *Hippiatrika* (*Hippiatrica Cantabrigiensia* 71.22).

CHG v.2; McCabe (2007).

Anne McCabe

Anthimus (of Constantinople?) (*ca* 475 – 525 CE)

Greek-speaking physician, born *ca* 455, wrote an extant food manual/cookery book in Latin, the *Epistula Anthimi De obseruatione ciborum*. The short booklet carries the full title *Viri inlustris comitis et legatarii ad gloriosissimum Theudoricum regem Francorum de obseruatione ciborum*, thus dedicated to the Frankish king, Theuderic. Malkhos of Philadelpheia (*fr.*15 Blockley) writes that Anthimus had been deeply involved in the complicated and treacherous bargaining by two Gothic chieftains (Theodoric Strabō, and Theodoric the Amal, son of Theodemir, who would become Theodoric the Great) as they attempted to manipulate advantage in making alliance with the emperor Zēnōn in Constantinople. In 478, letters addressed to Theodoric the Amal, signed by Anthimus and two other officials, were intercepted while negotiations were ongoing with Theodoric Strabō. Zēnōn arrested all three, had them whipped in public, and sentenced them to perpetual exile. Anthimus spent some years in Ravenna at the court of Theodoric the Amal (Strabō was killed in 481), and was sent as an ambassador by the Ostrogothic king to Theuderic, king of the Franks, sometime after 511. Dates of Anthimus' birth and death are unknown, but Grant (23–24) speculates that *De obseruatione* was written either in 516 or 523.

De obseruatione is essentially a "letter" to the king of the Franks about foodstuffs: some are good for maintaining health, others are not. Occasionally Anthimus suggests how to cook and serve foods according either to a "Greek" style or perhaps to what was customary in Ravenna. He occasionally reveals his native language in offhanded comments: e.g. §64 (*fit etiam de hordeo opus bonum, quod nos graece alfita, latine uero polentam, Gothi uero fenea*: "Also from barley is made a good (recipe) from what we call *alfita* in Greek, in Latin termed *polenta*, but in the foreign tongue of the Goths, *fenea*"), 78 (*oxygala graece quod latine uocant melca id est lac*: "curdled milk is called *oxygala* in Greek and *melca* in Latin"), etc. Anthimus' Byzantine medical background often shows through (specific dietetics on consumption of fowl, e.g. §23), and he is sometimes clearly impressed by Frankish foods that are both medicines and fit nourishment (§14: bacon eaten raw is an excellent vermifuge, and is likewise a superb wound-healer; §15: Frankish beer is of the finest quality as is mead "if the honey is good"). The Goths and Franks were very fond of venison, ox, pork, hare, lamb, beef, and boar: Anthimus (§3) suggests such might be even better for one's health if cooked with spices, including peppercorns, ***kostos***, spikenard, cloves, pennyroyal, celery, and fennel.

Ed.: V. Rose, *Die Diätetik des Anthimus an Theuderich König der Franken* in *Anecdota Graeca et Graecolatina*, v.2 (1870; repr. 1963) 41–102; E. Liechtenhan, *Anthimi De observatione ciborum ad Theodoricum regem Francorum epistula* (1963) = *CML* 8.1; Blockley 2 (1983) 401–462; M. Grant, *Anthimus De observatione ciborum. On the Observance of Foods* (1996): with trans. and comm.

E. Brandt, *Untersuchungen zum römischen Kochbuche. Versuch einer Lösung der Apiciusfrage* = *Philologus Supplementband* 19.3 (1927); G.M. Messing, "Remarks on Anthimus de observatione ciborum," *CPh* 37 (1942) 150–158; C. Deroux, "Une acception nouvelle pour le mot *lardum* (Anthimus, *De obs. Cib.* 20)," in Sabbah (1988) 33–38.

John Scarborough

Antigenēs (240 – 200 BCE)

Student of Kleophantos of Keōs (thus presumably an **Erasistratean**), who (in a work on care of infants?) advocated a Thessalian-style swaddling, in which the infant was tied into a padded board, analogous to the Native American "cradleboard" (Sōranos, *Gyn.* 2.83: *CMG* 4, pp. 60–61; *CUF* v. 2, p. 21; Temkin 1956: 84), and who in his book *On Fevers and Inflammations* described catalepsy as "deafness" (Caelius Aurelianus, *Acute* 2.56, *CML* 6.1.1, p. 164). Galēn, *In Hipp. Nat. Hom.* (*CMG* 5.9.1, pp. 69–70), lists him among other "early anatomists," suggesting a work on anatomy.

RE 1.2 (1894) 2399 (#12), M. Wellmann.

PTK

Antigonos (Med.) (270? – 80 BCE)

Asklēpiadēs Pharm. in Galēn, *CMLoc* 2.1 (12.557–558 K.) = 2.2 (12.580.12–17 K.), records the headache remedy of Antigonos, the military physician, and in 4.7 (12.773–774 K.) preserves his "*krokōdes leontarion epigraphomenon* (*sc. pharmakon*)," including white pepper and Falernian wine, especially useful for children. Marcellus of Bordeaux cites three collyria probably invented by the same physician: 8.11, 8.15, 8.124 (*CML* 5, pp. 116.20, 118.10, 144.15). In 8.15, he describes the collyrium as "*acharistum theudotium ab Antigono inuentum*," seemingly indicating Antigonos precedes Theodotos. Is he to be identified with Antigonos Nikaieus (see Antigonos of Alexandria)?

RE 1.2 (1894) 2422 (#22), M. Wellmann.

Jean-Marie Jacques

Antigonos of Alexandria (80 – 40 BCE)

Alexandrian scholar, one of the earliest commentators on Nikandros' *Thēriaka*; before Didumos, and a younger contemporary of Dēmētrios Khlōros whom he criticized. The Antigonos whose glosses are quoted eight times in Schol. Nik. *Thēr.*, often in conjunction with Dēmētrios', is identical to the author of a Hippokratēs *Lexicon* mentioned by Erōtianos (p. 5.19 Nachm.). Rohde, followed by Wellmann and Susemihl, wanted to identify him with an obscure physician from Nikaia quoted only once ("Aelius Promotus," p. 67.5 Ihm: *Antigonou tou Nikaieōs*). Some points of similarity in Erōtianos and the Nikandros *Scholia* may be explained by the use of Antigonos as their common source.

E. Rohde, *RhM* 28 (1873) 270, n. 5 = *Kl. Schr.* 1.387, n. 4; *GGLA* 2 (1891) 194–195, F. Susemihl; Jacques (2002) 2.cxxix.

Jean-Marie Jacques

Antigonos of Karustos (*ca* 290 – *ca* 240 BCE)

The *Antigonou Historiōn paradoxōn sunagōgē* contains 173 paradoxographical excerpts. Four sections can be discerned according to theme and author: (1) *Mirabilia de animalibus* drawn from various authors including Antigonos of Karustos; (2) *Mirabilia de animalibus* selected from PSEUDO-ARISTOTLE; (3) *Mirabilia de uariis rebus* (water, human physiology, botany etc.) from various authors; (4) *Mirabilia de aquis et de aliis rebus* derived from KALLIMAKHOS.

Modern judgment of this collection is critical: the structure is weak, the style is dry and its main contribution is to have preserved a large collection of otherwise unknown citations of ancient authors. This criticism brings us to the question of the identity of its author.

While for a long time the discussion concentrated on which Antigonos of Karustos wrote this compilation, Musso (1976: 1–10) convincingly suggested that the *Historiōn paradoxōn sunagōgē* was in fact a paradoxographical collection composed under Constantine VII Porphurogennētos (905–959 CE), in the first part of which Antigonos of Karustos' *Peri Zōiōn* features prominently and seems to have been the compiler's main source.

The question of the identity of "Antigonos" still remains. Several men of the same name and origin are known: a biographer, a sculptor, an art historian and a poet. In all likelihood, an Antigonos, born in Karustos (Euboia) in the 3rd c. BCE, was a famous biographer, sculptor, and art historian. This man – an acquaintance of Menedēmos of Eretria – lived in Athens for a long time and worked at the Pergamene court under Attalos I. (The 1st c. BCE saw a poet of the same name and origin, cited by several authors, among whom Athēnaios 3.82.)

The zoological work *Peri Zōiōn*, mentioned in *Antigonou Historiōn paradoxōn sunagōgē*, was probably created by the 3rd c. prose writer rather than the poet.

Ed.: *PGR* 31–109; O. Musso, *[Antigonus Carystius], Rerum mirabilium collectio* (1985); T. Dorandi, *Antigone de Caryste, Fragments* (*CUF* 1999).

RE 18.3 (1949) 1137–1166 (§10, 1145–49), K. Ziegler; Giannini (1964) 112–116; O. Musso, "Sulla struttura del cod. Pal. Gr. 398 e deduzioni storico-letterarie," *Prometheus* 2 (1976) 1–10; *OCD3* 106, A.F. Stewart; *BNP* 1 (2002) 751 (#7), H.A. Gärtner.

Jan Bollansée, Karen Haegemans, and Guido Schepens

Antigonos of Kumē (325 – 90 BCE)

Agricultural writer, whose work, which may have treated cereals, livestock, poultry, viticulture, and arboriculture (*cf.* PLINY, 1.*ind.*8, 10, 14–15, 17–18), was excerpted by CASSIUS DIONUSIOS (VARRO, *RR* 1.1.8–10, *cf.* COLUMELLA, 1.1.10).

RE 1.2 (1894) 2422 (#21), M. Wellmann.

Philip Thibodeau

Antigonos of Nikaia (125 – 175 CE)

Wrote a lost astrological treatise in Greek, comprising at least four books. The treatise, used by HEPHAISTIŌN OF THĒBAI and – almost certainly – by ANTIOKHOS OF ATHENS in his lost *Thēsauroi*, contained a biographical interpretation of the horoscope of an unnamed emperor easily identifiable as Hadrian. He might perhaps be identifiable with a physician Antigonos of Nikaia whose antidote against poison was cited by pseudo-AELIUS PROMOTUS (*cf.* ANTIGONOS OF ALEXANDRIA). Antigonos' astrological handbook illustrated its methods by means of specific horoscopes. Hadrian's horoscope, preserved by Hephaistiōn (2.18), is exceptionally elaborate.

Ed.: S. Heilen, *Hadriani genitura. Die astrologischen Fragmente des Antigonos von Nikaia. Edition, Übersetzung und Kommentar* (2006).

Alexander Jones

"Antikythera Device" (120 – 100 BCE)

Discovered in 1900–1901 by sponge divers from a shipwreck of *ca* 60 BCE off Antikythera island, this bronze and wood device (a box *ca* 10 by 20 by 30 cm) containing triangular-toothed gears was reconstructed by Price as a calendar computer for predicting lunar and (possible) solar eclipses; a crank was turned to drive the gears from day to day. The device displays the Babylonian "Saros" cycle, the 235 months of the 19-year cycle of METŌN, and the 76 years of the cycle of KALLIPPOS. Enough of the text survives to show that a ***parapēgma*** was inscribed on the bronze front of the box. Recent re-examination by Freeth *at al.* has recovered further text, containing the word *sphairion*, which Keyser argued was the name for such a device, although Freeth *et al.* see a reference to a (lost) indicator for the sun or moon. They also establish that the gearing was based on HIPPARKHOS' model of lunar motion.

D.deS. Price, *Gears from the Greeks* (1975); P.T. Keyser, "Orreries, the Date of [Plato] *Letter* ii, and Eudorus of Alexandria," *AGP* 80 (1998) 241–267; T. Freeth *et al.*, "Decoding the ancient Greek astronomical calculator known as the Antikythera Mechanism," *Nature* 444 (2006) 587–591.

PTK

Antimakhos (30 BCE – 80 CE)

ANDROMAKHOS, in GALĒN *CMGen* 7.13 (13.1034 K.), gives his ***akopon***, composed of fresh marjoram (*sampsukhon*), *libanōtis, aspalathos* (DIOSKOURIDĒS 1.20), savin juniper (named *sabina*, showing a Latin source), oak-galls, and beeswax in aged wine and must.

RE S.1 (1903) 91 (#26a), M. Wellmann.

PTK

Antimakhos of Hēliopolis (unknown date)

The *Souda* A-2682 records that Antimakhos wrote a *Kosmopoiia* in 3,700 hexameters.

RE 1.2 (1894) 2436 (#25), G. Knaack.

PTK

Antiokhis of Tlōs (95 – 55 BCE)

Honored with a statue by Tlōs for her medical expertise (*TAM* 2.2, #595), and by HĒRAKLEIDĒS OF TARAS with a book-dedication, she was perhaps the daughter of DIODOTOS. ASKLĒPIADĒS PHARM., in GALĒN *CMLoc* 9.2 (13.350 K.) = 10.1 (13.341), cites her **terebinth**-based ointment for arthritis, **dropsy**, and sciatica (as prepared by FAUILLA).

BNP 1 (2002) 761–762 (#2), V. Nutton.

PTK

Antiokhos, Paccius (20 BCE – 14 CE)

SCRIBONIUS LARGUS, *Compositiones*, 97 (p. 51 Sconocchia), says that *Paccius Antiochus* was a student (*auditor*) of PHILŌNIDĒS OF CATANA, and that after Paccius' death, his book on gout and pains in the flanks (deposited among the books of Tiberius' *bibliothecae publicae*) finally became known to fellow physicians. Scribonius seems a bit peeved that Paccius refused to share his recipe, since he profited handsomely from a compound formulated from well-known simples, not only hiding the formula from his own students, but preparing the drug in a barricaded room (*clusus*). GALĒN, *CMLoc* 4.8 (12.772 K.), calls him an "**Asklēpiadean**" in a section of recipes apparently extracted from Scribonius Largus' Greek books (12.764–790 K.), attesting again to the bilingual abilities of Sicilian physicians prominent in early imperial Rome (e.g. the collyrium of FLORUS, prepared for Drusus' mother Antonia [12.768–769 K.]).

Several pharmacal recipes devised by Paccius were extracted by Scribonius Largus, Galēn, AËTIOS OF AMIDA, and PAULOS OF AIGINA. Most of Paccius' drugs were analgesic, mild pain-relievers occasionally including the latex of the opium poppy, but not mandrake, henbane, and similar anesthetics: e.g. the Eye-Instillation (*enstakton*) of Paccius, quoted from THEMISŌN (12.782–783 K.), the general analgesic in a collection of "**Asklēpiadean**" recipes *CMLoc* 9.4 (13.284–285 K.), and the *Antidotos hiera Paccii Antiochi ad uniuersa corporis uitia, maxime ad lateris et podagram* which consumes three chapters of the *Compositiones* (pp. 51–53). Paccius' *Emplastrum album*, however, designed to treat breast cancer in women ("lumps" that have hardened: *cum in mammis mulierum alioue quouis durita fuerit. . . quam Gracei carcinoma aut cacoethes uocant*), does contain quantities of narcotics, if the *terrae mali* were, indeed, mandrake "apples" (*Compositiones*, 220 [p. 100]). This multi-ingredient plaster probably was a good transdermal anesthetic, certainly in no way curative. The "**Hiera** of Antiokhos" (Paulos 7.8.1 [*CMG* 9.2, pp. 286–287]) is a compaction of the recipes in *Compositiones*, 106–107 (pp. 57–58), and as a purgative includes saffron crocus (leaves), birthwort, "white" pepper, cinnamon, spikenard, honey, and myrrh.

RE 18.2 (1942) 2063, H. Diller, corrected by Fabricius (1972) 226, with n. 41.

John Scarborough

Antiokhos VIII Philomētōr (141 – 96 BCE)

Antiokhos VIII (inv. 1977.158.706) Courtesy of the American Numismatic Society

Born 141 BCE; king of Syria (125/121–96). His name is linked to a famous anti-venom compound (***opopanax***, bitter-vetch flour, clover and many other vegetable-garden plants). PLINY (20.264) says the theriac was "engraved in verse on stone in Aesculapius' temple in Cos": see the interpolated *prographē* quoting "Antiokhos' theriac which, according to Pliny, was inscribed near the doors of Asklēpios" (GALĒN *Antid.* 2.14 [14.183.6–8 K.]). Both ASKLĒPIADĒS PHARM. (*ibid.* 185.3–186.2 K.) and HĒRĀS OF KAPPADOKIA (2.17 [201.16–202.14 K.]) transmit a formula in verse, suitable for a metrical inscription, followed by a prose resolution. Pliny gives a prose version (without the *sphragis* mentioning Antiokhos), as the culmination of his book about garden plants. Commended against vipers (v.15–16), spiders and scorpions, this theriac is a sort of panacea (see NIKANDROS). The initial *sphragis* shows Antiokhos to be an inventor of remedies akin to ATTALOS III and MITHRADATĒS. The author of the metrical inscription is no longer thought to be Antiokhos; nor is it THEMISŌN's student EUDĒMOS, *pace* Wellmann (see *RE* 6.904.68–905.3), assuming it is the **Methodist** physician and not EUDĒMOS THE ELDER (Galēn 13.291.10 K.) that Asklēpiadēs had in mind in his *prographē* (*Antid.* 185.1 K.). The theriac and other recipes in verse were to be found in his work on pharmacology – that is all the *prographē* means. Pliny mentions that Antiokhos III the Great used it. Except for a possible confusion between Antiokhos III and VIII, one may suppose that it could well have been invented by Antiokhos III's private physician, APOLLOPHANĒS OF SELEUKEIA.

RE 1.2 (1894) 2483.7–12, M. Wellmann; *SH* 412A; *BNP* 1 (2002) 765 (#10), A. Mehl; Jacques (2002) 2.308–309 (see XLVI–XLVIII).

Jean-Marie Jacques

Antiokhos of Athens (30 BCE – 260 CE)

Greek astrological author of uncertain date (perhaps 2nd c. CE), but earlier than PORPHURIOS, who quotes him in his astrological *Eisagōgē* (38). Antiokhos wrote two treatises on astrology, both lost. A summary of his *Eisagōgika* is extant in a 15th c. codex (*Par. gr.* 2425), from which it is apparent that Porphurios appropriated the content of many of its chapters without acknowledgement. The *Thēsauroi*, a treatise of similar scope, was one of the principal sources of a complex family of astrological epitomes, some associated in the MSS with RHĒTORIOS. Antiokhos is often cited as an authority on **genethlialogy** and interrogatory astrology in the Arabic tradition, where his influence was probably by way of translations from the Greek compilations of late antiquity.

D.E. Pingree, "Antiochus and Rhetorius," *CPh* 72 (1977) 203–223.

Alexander Jones

Antiokhos of Surakousai (430 – 410 BCE)

Wrote chronicles of Italy and Sicily, describing the migrations and settlements of its peoples, cited often by STRABŌN (5.4.3, 6.1.1–15, and 6.3.2) and others. Gomme-Andrewes-Dover 4.198–205 present the argument for Antiokhos as a source of THUCYDIDĒS 6.2–5 (and probably 3.88.2–3, 3.90.1, 6.96.2, and 6.104.2).

FGrHist 555; A.W. Gomme, A. Andrewes, and K.J. Dover, *A historical commentary on Thucydides* v. 4 (1970).

PTK

Antipatros (Pharm.) (30 BCE – 80 CE)

Greek physician, perhaps contemporary with AUGUSTUS (Wellmann); the compound medicines linked with his name imply a *floruit* no earlier than the 1st c. CE. GALĒN mentions the **Methodist** ANTIPATROS postdating THESSALOS OF TRALLEIS (*Introd.* 4, 14.684 K.), perhaps the source of the following citations. CAELIUS AURELIANUS cites letters in at least three books, attributed to Antipatros, treating medical topics (*Chron.* 2.157 [*CML* 6.1.1,

p. 640.11–12], 2.187 [*CML* 6.1.1, p. 658.14]). Galēn mentions Antipatros' collection of recipes for compound medicines (*CMLoc* 9.5, 13.292 K.) and further quotes recipes via ANDROMAKHOS (from this collection or not) as used by Antipatros (*CMLoc* 3.3, 12.684 K.: nasal polyps; *CMLoc* 6.6, 12.936 K.: oral wounds; and 13.292 K.: dysentery and *tenesmus*). He might be the same Antipatros who created the recipes bearing his name in Galēn (*CMLoc* 9.2, 13.239 K.: draught for spleen disease; *CMLoc* 3.1, 12.630 K.: against pain in the ears). Antipatros composed a *Mithridateion* (*Antid.* 2.1, 14.108–109 K.) and a theriac whence the antidote to asp-bites Galēn himself used (*Antid.* 2.10, 14.160 K.). It is unknown (though typical of the 1st c. CE) whether the same physician, perhaps a specialist in pharmaceutical therapeutics (particularly compound medicines), authored all these works.

RE 1.2 (1894) 2517 (#32), M. Wellmann; Watson (1966) 37–43.

Alain Touwaide

Antipatros (Methodist) (*ca* 50 – 193 CE)

Physician listed among the **Methodists** post-dating THEMISŌN and THESSALOS (GALĒN, *MM* 1.7.5 [10.52–53 K. = p. 27 Hankinson], PSEUDO-GALĒN, INTRODUCTIO 14.684 K. = *frr.* 162, 283 Tecusan), but omitted from other **Methodist** catalogues. Tecusan (2004: 46) identifies as a **Methodist** the homonymous author of medical letters, in at least three books (CAELIUS AURELIANUS *Chron.* 2.157, 187 [*CML* 6.1.1, pp. 640, 658] = *frr.* 68, 72), which addressed the application of the number three to medical treatments, and the cautious use of bandaging and its concomitant dangers, in defense of a **Methodist** stance. But if this Antipatros' addressee "Gallus" was AELIUS GALLUS (as has been suggested), he would predate Thessalos. See also ANTIPATROS (PHARM.), whose interests accord with the **Methodist** profile. Galēn records an Antipatros in Rome, whose alarm, when fevered, at an erratic pulse led to his self-treatment of that fever with baths, exercise, and plain simple foods (*Affected Parts* 4 [8.293–294 K.] = *fr.* 153 Tecusan). That Galēn treats his contemporary respectfully and without disputation may render his **Methodism** unlikely (Tecusan 2004: 48).

Tecusan (2004) 45–51.

GLIM

Antipatros (of Tarsos?) (*ca* 200 – 100 BCE?)

The second Greek, after BĒROSSOS and before AKHINAPOLOS, to study astrology (VITRUUIUS 9.6.2); perhaps the **Stoic** ANTIPATROS OF TARSOS.

RE 1.2 (1894) 2517 (#34), E. Riess.

GLIM

Antipatros of Tarsos (*ca* 160 – 130 BCE)

Student and successor of DIOGENĒS OF BABYLŌN as head of the **Stoa** in *ca* 150 BCE, and teacher of PANAITIOS. Died at an old age in *ca* 130 BCE by drinking poison. He worked innovatively in many areas of philosophy and was noted as a particularly gifted logician. SEXTUS EMPIRICUS praised him highly as a philosopher. In opposition to the standard **Stoic** position, Antipatros claimed that there were such things as single-premise arguments (such as "if you are breathing, you are alive"). He is credited with introducing the adjective as a distinct class of word in Greek grammar (omitted by CHRYSIPPUS OF SOLOI and

Diogenēs of Babylōn). He also engaged Diodōros Kronos' infamous so-called *Master Argument*, denying (as Kleanthēs of Assos before him had done) one of Diodōros' premises in the *Argument*: that past truths are necessary. He modified (or at least tightened up) the earlier **Stoic** definition of the goal of life in response to the sharp criticisms of the **Academic** Karneadēs, with whom Antipatros debated at length on paper (*SVF* 3.244 = Plutarch, *Garrul.* 23 [514D]), but never, to Karneadēs' apparent consternation, face to face.

Cicero (*Div.* 1.6) cites Antipatros as having defended divination in two books on the subject, and arguing (as had Chrysippus and Diogenēs of Babylōn) from the premise that "there are gods" to the conclusion that "divination exists" (*ibid.* 1.9, 1.82–84). Our knowledge of his physics is clouded by the fact that he is in some sources confused with (or at least not sufficiently distinguished from) Antipatros of Tyre. Modern sources tend to accrete many unspecified "Antipatros" fragments onto Antipatros of Tarsos rather than his Tyrian namesake. One of the two argued that the voice of humans and animals is corporeal, that the soul is "warm ***pneuma***," both of which were common **Stoic** positions, and that "body is finite substance." One of them wrote books *On the Soul*, and *On Substance.*

Ed.: *SVF* 3.244–258.

Daryn Lehoux

Antipatros of Tyre (100 – 40 BCE)

Stoic who introduced Cato the Younger to philosophy. He wrote at least ten books *On the **Kosmos*** (Diogenēs Laërtios, 7.139–140), claiming that the ***kosmos*** is living, ensouled, and rational, with ***aithēr*** as its ***hēgemonikon***. He says that the substance of god is air-like (D.L., 7.148). In some references to "Antipatros'" arguments, it is unclear whether he or Antipatros of Tarsos is meant, but given that a work *On the **Kosmos*** is connected with them, a typically standard **Stoic** claim that the ***kosmos*** is spherical, harmonious, and unified is probably attributable to Antipatros of Tyre. He also seems to have discussed the generation and destruction of the ***kosmos*** (D.L., 7.142).

RE 1.2 (1894) 2516, H. von Arnim; Long and Sedley (1987) #47.O.

Daryn Lehoux

Antiphanēs of Dēlos (400? – 300 BCE)

Medical writer before the time of Theophrastos (see *De sudore* 17). He took a marked interest in dietetics: "The physician of Dēlos Antiphanēs said one of the causes of sicknesses stemmed from the variety of exotic dishes which were sought out of sheer snobbishness" (Clement of Alexandria, *Paedag.* 2.1.2). Andromakhos the Younger (Galēn *CMLoc* 5.5 [12.877.8–11 K.]) knew of one of his plasters for painful molars. Caelius Aurelianus (*Chron.* 4.114, *CML* 6.1.2, p. 838.25) tells us that he is the author of a book called *Panoptēs*, but there may be here some confusion with Antiphanēs, the comic poet: *Panoptēs* is the title of plays by two other comic dramatists, Kratinos (*frr.*158–170 *PCG*) and Euboulos (*fr.*71 *PCG*).

GGLA 1 (1891) 828, M. Wellmann.

Jean-Marie Jacques

Antiphōn of Athens (450 – 400 BCE)

The sophist, distinguished by Hermogenēs of Tarsos (*De ideis* 399.18–400.6 = DK 87 A 2) from the homonymous Athenian politician and logographer Antiphōn of Rhamnous. (Many, however, reject the distinction and identify the two.) Hermogenēs ascribes to the sophist Antiphōn the treatises *On Truth*, *On Concord*, and *Politicus*, and a dream-book of no fixed title. All the evidence bearing on Antiphōn's scientific and mathematical interests is plausibly ascribed to the two-volume treatise *On Truth*.

ARISTOTLE and related sources describe Antiphōn's contribution to the problem of the quadrature of the circle (DK 87B13). Antiphōn (it is reported) inscribed a regular polygon in a circle, then constructed isosceles triangles upon the sides of the polygon, and subsequently repeated this process until an inscribed polygon was produced whose sides coincided with the circumference of the circle. Here we find the first attested adumbration of the notion of exhausting a curvilinear figure, an idea taken up and perfected later by EUDOXOS OF KNIDOS and which represents an important early contribution to the development of calculus. Other fragments evidence Antiphōn's interests in subjects ranging from cosmology (DK 87B22–25) to the source of the moon's illumination and the cause of its eclipses (DK 87B27–28) and the nature and movements of the sun (DK 87B26). An important text transmitted in Arabic attests Antiphōn's interest in medicine and the **humoral** theory of disease (DK 87B29a=F29a Pendrick), while a papyrus fragment suggests he espoused the theory of sight by visual rays (*POxy* 52 [1984] 3647, col. III.6–8).

Taken together, the evidence shows Antiphōn's thorough engagement in conventional topics of Pre-Socratic natural inquiry. Whether he espoused a general theory of nature or ***phusis*** is unknown. Aristotle (DK 87B14) reports a sort of thought experiment about a buried bed, which he claims shows that Antiphōn defined nature generally as matter. But the inference is suspect; and the larger import of Antiphōn's views on nature, including its relation to morality, remains uncertain.

DPA 1 (1989) 225–244 (#209), M. Narcy; *Corpus dei papiri filosofici greci e latini* I.1* (1989) 176–236, F. Decleva Caizzi and G. Bastianini; M. Gagarin, *Antiphon the Athenian: Oratory, Law and Justice in the Age of the Sophists* (2002); G.J. Pendrick, *Antiphon the Sophist: The Fragments* (2002).

G.J. Pendrick

Antisthenēs of Athens (*ca* 425 – *ca* 365 BCE)

Antisthenēs (born *ca* 445 BCE), one of the associates of Sōcratēs, who praised his bravery at the battle at Tanagra 424 BCE where both fought (DIOGENĒS LAËRTIOS 6.1) and whose death he attended (PLATO, *Phaidōn* 59b), became one of the most prominent Socratic teachers in Athens and was later regarded as the founder of the Cynic tradition; he died *ca* 365 BCE. The work titled *Peri tōn sophistōn phusiognōmonikos* (D.L. 6.15.15) or *Phusiognōmonikos* (Ath., *Deipn.* 14 [656f]) quite certainly is not a technical text on the art of physiognomy, but rather an anti-Sophistic text in the same tradition as that of other Socratic writers, making use of the art of physiognomy in much the same way as Plato in the *Sumposion* 215a4–217a2 has Alkibiadēs compare Sōcratēs' features to that of satyrs, and Phaidōn of Ēlis in his treatise *Zōpuros* (*cf.* ZŌPUROS): to contrast the external appearance of Sōcratēs and his internal soul and character, and thus to refute the possibility of physiognomical inferences from body to soul.

Ed.: *SSR* 2.163–164 (fragments) and 4.281–283 (commentary).

Sabine Vogt

Antisthenēs (of Rhodes) (1st c. BCE?)

Wrote a *Successions of Philosophers*, known only from references in DIOGENĒS LAËRTIOS. The preserved fragments contain the usual bits and pieces of biographical material. He seems to have lived in the 1st c. BCE, but we know nothing about him.

J. Janda, "D'Antishène, auteur des Successions des philosophes," *Listy Filologické* 89 (1966) 341–364; Mejer (1978) 62–64.

Jørgen Mejer

Antoninus of Kōs (30 BCE – 80 CE)

GALĒN, *CMLoc* 7.3 (12.843–844 K.), preserves ASKLĒPIADĒS PHARM.'s record that TIMOKRATĒS used Antoninus' plaster; Asklēpiadēs also gives his anti-venom made from the herb ***alussos***, *Antid.* 2.11 (14.168–169 K.). This man and Arrius Antoninus, *PIR2* A-1086, *cos. suff.* 69 CE, are the earliest bearers of this imperial-era cognomen.

RE 1.2 (1894) 2572 (#16), M. Wellmann.

PTK

Antonius (170 – 190 CE)

GALĒN's *Affections and Errors of the Soul* (trans. Singer, 1997, pp. 100–149) was directed against Antonius' *Control of Individual Affections*, which presented diagnoses and cures, that Galēn asserted inadequately distinguished "affection" (*pathos*), caused by irrational internal powers, from "error" (*hamartēma*), caused by false opinions. The affections discussed by Galēn include anger (*orgē*), fear (*phobos*), grief (*lupē*), and envy (*phthonos*), which Antonius probably also covered. PSEUDO-GALĒN DE PULSIBUS is directed against an apparently distinct Antonius.

DPA 1 (1989) 258 (#222), R. Goulet.

PTK

Antonius Castor (10 – 75 CE)

PLINY considered him the most authoritative pharmacist of his era, and used to visit his garden, in which he worked up the age of 100 (apparently deceased when Pliny published): 25.9; he is cited as an authority 1.*ind.*20–27. Antonius prescribed ferula-root for eyesight (20.261) and root of *potamogiton* (perhaps *Hippuris vulgaris* L.) for *strumae* (*cf.* CELSUS 1.9.6); he carefully describes *potamogiton, piperitis* (20.174, for the mouth), and butcher's-broom (23.166). Like other botanists, he distinguished similar plants, such as the two kinds of horehound, one of which he prescribed for abscesses and dog bites (20.244). Possibly the same as ANTONIUS "THE ROOT-CUTTER."

GRL §495.3.

PTK

Antonius Musa (40 – 20 BCE)

One of four "**Asklēpiadean**" physicians known to AUGUSTUS (PLINY 29.6), thus linked with the teachings of ASKLĒPIADĒS OF BITHUNIA; Musa is frequently quoted by Roman and Byzantine medical authors (e.g. Pliny 29.141; GALĒN *CMLoc*, 7.2 [13.57 K.]; AËTIOS OF AMIDA, *Tetrabiblon*, 8.56 and 75; and PAULOS OF AIGINA, *Epitome of Medicine*, 7.12.18), for his drug-prescriptions, especially multi-ingredient pills, noted for their mildly narcotic properties, well diluted in wine or water. Famed for his cure of the emperor (23 BCE) with cold-water baths (Suetonius *Aug.*, 59; Dio Cassius 53.30), Musa was one of a number of physicians in the late Roman Republic and early Empire who advocated various forms of "hydrotherapy" accompanied by mild-acting pharmaceuticals. Pliny (29.6) associates Musa with THEMISŌN OF LAODIKEIA, thereby putting him in the direct heritage of Asklēpiadēs. Musa's brother, EUPHORBOS, was a physician to IOUBA (Pliny 25.77); the genus *Euphorbia* commemorates Iouba's doctor and their research on medical plants, discovering one species (*E. dendroides* L.) still used to stun fish. Galēn (*CMGen*, 6.15 [13.935 K.]) cites ANTONIUS "THE ROOT-CUTTER," perhaps the same as the author of the extant *De herba botanica* (*CML* 4 [1927] 3–11).

RE 1.2 (1896) 2633–2634, M. Wellmann; M. Michler, "Principis medicus: Antonius Musa," *ANRW* 2.37.1 (1993) 757–785; D.W. Roller, *The World of Juba II and Kleopatra Selene* (2003) 159–160, 178–179; *Idem* (2004) 103–107.

John Scarborough

ANTONIUS MUSA (PSEUDO) ⇒ APULEIUS (PSEUDO)

ANTONIUS POLEMŌN ⇒ POLEMŌN OF LAODIKEIA

Antonius "root-cutter" (100 BCE – 95 CE)

ASKLĒPIADĒS PHARM. (as preserved in GALĒN) four times cites an Antonius, thrice calling him "root-cutter" (*rhizotomos*), and once "druggist" (*pharmakopōlos*), nearly synonymous. Evidently a practical man, and considered "very experienced" (*CMLoc* 2.2 [12.580 K.]), he may well be identical with ANTONIUS CASTOR. The root-cutter compounded a henbane- and poppy-juice-based vegetal remedy for headache (*ibid.* 2.1 [12.557]), and a compound of mined salt and **litharge** in gum, laced with **galbanum**, for joint conditions; as druggist he prescribed a vegetal remedy for gassy colic: *CMLoc* 9.4 (13.281–282 K.).

RE S.1 (1903) 96 (#14a), M. Wellmann.

PTK

Antullos (100 – 260 CE)

Greek physician, surgeon, therapist, and pharmacologist who lived after ARKHIGENĒS OF APAMEIA and before OREIBASIOS. His treatment of disease with purgation is a typically **Pneumaticist** therapeutic principle, but he also assimilated Hippokratic elements into his practice: he studied the influence of the environment on human health, recommended physical exercise, and considered patients' housing and diet. He recommended therapeutic thermal baths, particularly mineral waters (*cf.* ANTONIUS MUSA), and wrote on pharmacology in the manner of 1st c. CE therapists.

Antullos' major contribution was in surgery. He performed difficult interventions (including laryngotomy, fistulas, several eye interventions, and head abscesses), and wrote a major work *Kheirourgoumena* (at least two books), quoting HĒLIODŌROS, and another on hydrocephalic newborns.

In the field of pharmaceutical therapy, his *Peri Boēthēmatōn* discussed external medicines (Book 1); cathartic medicines (Book 2); diet (Book 3); and gymnastics (Book 4). A work *On the Preparation of Medicines*, a large compilation of formulas from **Pneumaticist** physicians including ATHĒNAIOS OF ATTALEIA, HĒRODOTOS (PNEUM.), APOLLŌNIOS OF PERGAMON, and Arkhigenēs, plus others such as DIOKLĒS OF KARUSTOS and RUFUS OF EPHESOS, is probably a segment of his work on pharmaceutical therapy.

Antullos' works are known only as fragments in Byzantine encyclopedists (Oreibasios and PAULOS OF AIGINA), in Arabic works, and in a commentary on the HIPPOKRATIC CORPUS *Humors* dubiously ascribed to GALĒN, but probably dating to the Renaissance. His work on gymnastics was the main source of Girolamo Mercuriale (1530–1606).

P. Nicolaides, *Antylli, veteris chirurgi, ta leipsana* (1799); *RE* 1.2 (1894) 2644–2645, M. Wellmann; Wellmann (1895) 104–114; I. Bloch, "Griechische Aerzte des dritten und vierten (nach-christlichen) Jahrhunderts," in *HGM* 483–488; R.L. Grant, "Antyllus and his medical works," *BHM* 34 (1960) 154–174; *KP* 1.415–416, F. Kudlien; Marganne (1981) 99; *OCD3* 117, J.T. Vallance; Marganne (1998) XVII–XIX, 5, 11, 78; *BNP* 1 (2002) 810–811, V. Nutton.

Alain Touwaide

Apeimantos (280 – 250 BCE)

GALĒN, *On Venesection, Against Erasistratos* 2 (11.151 K. = p. 18 Brain), lists him, and STRATŌN (ERASI.), as students of KHRUSIPPOS OF KNIDOS (II). Both, like their fellow-student ERASISTRATOS, eschewed venesection due to the danger of excessive bleeding. (Kühn prints "*Apoi-*", unattested, instead of *Apei-/Apē-*, common from the 5th to 2nd cc. BCE.)

Fabricius (1726) 73.

PTK

Apellās of Kurēnē (350 BCE – 465 CE)

Apellās *ho Kurēnaios* is cited by MARCIANUS OF HĒRAKLEIA as one of his sources for his epitome of MENIPPOS OF PERGAMON: *GGM* 1.565. Perhaps to be identified with OPHELLĀS OF KURĒNĒ.

RE 1.2 (1894) 2686 (#7), H. Berger; *RE* 19.1 (1937) 849, F. Gisinger; *RE* 18.1 (1939) 630, E. Honigmann.

Andreas Kuelzer

Apellās of Laodikeia (*ca* 150 – 350 CE?)

Addressee – real or fictitious – of one of the letters that make up APSURTOS' veterinary treatise. The letter, about dislocated joints, is preserved in the *Hippiatrika* (*Hippiatrica Parisina* 182 = *Hippiatrica Berolinensia* 26.3). Apsurtos calls Apellās *hippiatros*, "horse-doctor," but does not specify which of the cities named Laodikeia was his correspondent's home.

CHG v.1; McCabe (2007).

Anne McCabe

Apellēs (of Thasos?) (250 BCE – 20 CE)

An Apellēs is qualified as the "founder of botany" by the 15th-century Byzantine teacher and copyist Michael Apostolios (III.60c). An Apollās wrote *On Herbs* (*Peri Botanōn: Schol. Nik. Thēr.* 559a, p. 214 Crugnola). PLINY cites "Apelles" as a foreign authority on drugs obtained from animals, especially aquatic (1.*ind*.28, qualified as "physician;" 31–32). Pliny cites his theory that a decoction of land crocodile (*skinkos*) flesh is an effective antidote to arrow poison (28.120) and to poisonous honey (32.43). ASKLĒPIADĒS PHARM., in GALĒN *CMGen* 5.14 (13.853 K.), reports Apellēs' treatment and preventive for ulcers prescribed also for dysentery, compounded from ashed papyrus, roasted lead, roasted copper, orpiment, iron scales, and raw sulfur, applied with honey or rose oil. APOLLŌNIOS "MUS," in Galēn *Antid.* 2.8 (14.148 K.), records Apellēs' antidote comprising dittany, *polion*, long pepper, wild rue, and ***skordion***, taken with honey, effective against pleurisy and as a menstrual emmenagogue. Wellmann identifies the man in Pliny and Galēn with Apollās. Fabricius (1726) 72 assigns the ethnic (not supplied by Pliny or Galēn).

RE 1.2 (1894) 2687 (#1), J. Kirchner, 2688 (#11), M. Wellmann; Jacques (2002) LV–LVI.

GLIM and Alain Touwaide

Apellis (260 – 120 BCE?)

Invented a *trispaston* (triple pulley) for hauling ships by winch (*ergatēs*), producing its mechanical advantage through a cascading series of pulleys that drew two paired ropes inward to the machine when a third single rope was hauled outward. The device is described by OREIBASIOS *Coll.* 49.22 (*CMG* 6.2.2, p. 33) probably from HĒLIODŌROS OF ALEXANDRIA; compare PASIKRATĒS. The rare name is attested through *ca* 100 BCE (*LGPN* 3A.48, 4.33).

Drachmann (1963) 178–180.

PTK

Aphrodās (90 BCE – 80 CE)

Student or follower of MOSKHIŌN (see ANDROMAKHOS, in GALĒN, *CMLoc* 7.2 [13.30–31 K.]), and thus dated. He is often cited by Andromakhos: for a ***khalkitis***-based bloodstanch (*CMLoc* 3.3 [12.695 K.]), henbane- and opium-laced toothache remedies (*ibid.* 5.5 [12.878]), a complex and costly aromatic ***artēriakē*** (*ibid.* 7.2 [13.30–31]: ***amōmon***, cassia, cinnamon, frankincense, ***kostos***, myrrh, nard, roses, saffron, etc.), an opium- and henbane-based anodyne (*ibid.* 7.5 [13.94–95]), a colic remedy involving opium, aloes, etc. (*ibid.* 8.2 [13.135–136]), and an ***akopon*** (*CMGen* 7.3 [13.1035 K.]). Andromakhos cites several of his wound plasters, two "green" (copper-salts-based) – one with **verdigris** (*ibid.* 2.2 [13.494–495 K.]), the other with ***misu***, copper flakes, roasted copper, and much else (*ibid.* 2.20 [13.551]) – and a bitumen plaster admired by Andromakhos (*ibid.* 2.22 [13.555–556]); ASKLĒPIADĒS PHARM. gives another "green" plaster from Aphrodās (*ibid.* 4.13 [13.738]: not only ***khalkitis, khalkanthon***, and ***misu***, but also **litharge** and **Sinōpian earth**). Aphrodās also compounded antidotes, one used by Andromakhos containing over two dozen ingredients (including ***amōmon***, cinnamon, Indian nard, ***kostos***, myrrh, and saffron: Galēn, *Antid.* 2.2 [14.111–112 K.]), the other cited by Galēn himself (*ibid.* 2.17 [14.207–208]) for ***hudrophobia***, a rose-water potion of only *lathuris* (*Euphorbia lathyris* L., *cf.* Galēn, *Simples* 7.11.2 [12.56 K.]; Durling 1999: 217).

RE 1.2 (1894) 2725, M. Wellmann.

PTK

Aphrodisis (50 BCE – 95 CE)

ASKLĒPIADĒS PHARM., in GALĒN, *CMGen* 7.12 (13.1013 K.), records his ***akopon*** containing rose oil, henna oil, iris, myrrh oil, **terebinth**, balsam, and beeswax, in must and honey. The rare name is widely attested, beginning in the 1st c. CE (*LGPN*), and an emendation to APHRODĀS is otiose.

Fabricius (1726) 72.

PTK

Aphros (250 BCE – 540 CE)

AËTIOS OF AMIDA 7.114 (*CMG* 8.2, p. 390) records his "Black Phoenix" collyrium, including cuttlefish bone, ground pumice, ***ammōniakon*** incense, **verdigris**, burnt deer-horn, myrrh and opium, plus notably burnt "phoenix" bones, in honey and water. The name ("foam") is otherwise unattested, and an emendation to APHRODĀS or even APHRODISIS seems likely.

(*)

PTK

Aphthonios of Rome (250 – 350 CE)

OREIBASIOS, *Ecl. Med.* 75.18 (*CMG* 6.2.2, p. 246), calling him a *dikologos* (advocate), records his arthritis ointment, compounded from beeswax, butter, **galbanum**, lanolin, lard, pine-resin, **terebinth**, etc. (For this rare late form of Aphthonētos, *cf. PLRE* 1 [1971] 81–82, 2 [1980] 110.)

(*)

PTK

APIŌN ⇒ ABIYŪN

Apiōn of Oasis, Egypt (20 – 50 CE)

Son of Poseidōnios, migrated to Alexandria, studied under Didumos "Khalkenteros," and became head of the grammatical school there; member of the Alexandrian embassy to Caligula in 40 CE, denouncing the Alexandrian Jews, thereafter remaining in Rome. Wrote on HOMER, Egypt, and other topics, and was the object of Iosephus' *Against Apiōn.* PLINY 1.*ind.*30–32 (medicine from animals: *cf.* AELIANUS *NA* 11.40), and 35–37 (minerals: presumably his *De Metallica Disciplina*) lists him as a source (*cf.* 37.19). ASKLĒPIADĒS PHARM., in GALĒN *CMGen* 5.15 (13.856 K.), gives his mineral remedy for ***anthrax***: copper ore, twice-roasted ***khalkanthon***, red natron, orpiment, and realgar, ground in vinegar, dried and ground again.

FGrHist 616=1057; *OCD3* 121, N.G. Wilson; *EJ2* 2.256, A. Schalit.

PTK

Apios Phaskos (100 BCE – 110 CE)

KRITŌN, in GALĒN *CMLoc* 5.3 (12.841–842 K.), records his skin-treatment, composed of copper flakes, orpiment, realgar, squirting cucumber, black hellebore, and the abdomens of cantharides beetles, ground into cedar oil. The name is puzzling, apparently "Pear-tree Sage-apple" (*cf.* Theophrastos, *HP* 3.8.6), although Apios is attested as a name (*LGPN* 1.49, 3A.50). Perhaps emend the first name to the Roman "Appius," or more likely emend *ΑΠΙΟΣΦΑΣΚΟΣ* to *ΑΠΙ<ΩΝ>ΟΣΟΑΣΕΩΣ* (i.e., Apiōn of Oasis).

RE 1.2 (1894) 2810, M. Wellmann.

PTK

Apollinarios (Pharm.) (*ca* 160 – 260 CE?)

Marcellus of Bordeaux lists "Apollinaris" after Celsus, Pliny, Apuleius, and before Largius Designatianus, among the ancient medical authorities writing in Latin whose work he examined (*pr.*2 [*CML* 5, p. 3]). Oreibasios preserves the recipe for Apollinarios' eye-salve containing ***psimuthion***, **calamine**, roasted copper, myrrh, aloe, saffron, acacia, tragacanth, meal, opium, gum, and rain-water (*Syn.* 3.118 [*CMG* 6.3, p. 98]). Not an uncommon name, both variants of which are known from the 1st–3rd cc. CE (*LGPN*).

RE 2.1 (1895) 2844 (#2), P. von Rohden.

GLIM

Apollinarios of Aizanoi (30 – 180 CE)

Astronomer mentioned in a handful of ancient sources (Galēn; Vettius Valens 6.4.8 [p. 239], 9.12.10 [p. 339]; Porphurios of Tyre; Paulos of Alexandria), and quoted at some length in two MSS of Hephaistiōn of Thēbai. He was seemingly reputed an important astronomer in antiquity. Surviving evidence allows the attribution of a 248-day scheme for lunar motion based on Babylonian models, with the solstitial and equinoctial points at 8° of their respective signs. Surviving fragments contain an account of lunar motion of some sophistication.

Ed.: A. Jones, "Ptolemy's First Commentator," *Transactions of the American Philosophical Society* 80.7 (1990).

A. Jones, "The Development and Transmission of 248-Day Schemes for Lunar Motion in Ancient Astronomy," *AHES* 29 (1983) 1–36; *NDSB* 1.82–83, *Idem.*

Daryn Lehoux

Apollodōros (Med.) (325 – 150 BCE)

In his register of sources Pliny lists three medical writers named Apollodōros, in addition to Apollodōros the *thēriakos*: (1) doctor Apollodōros of Kition, 1.*ind.*20–27, (2) doctor Apollodōros of Taras, 1.*ind.*20–27, and (3) the author of *On Perfumes and Chaplets*, 1.*ind.*12–13, of which Athēnaios (*Deipn.* 15 [675e]) quotes a substantial fragment. One of these men may be identical to the Apollodōros who wrote to an unspecified "King Ptolemy" instructing him which wines to drink, in an era when Italian wines were still unknown to the Greeks (Pliny 14.76).

RE 1.2 (1894) 2895 (#70), M. Wellmann.

Philip Thibodeau

Apollodōros Dēmokritean (150 – 80 BCE)

PLINY 24.167, listing him before KRATEUAS, records that to PSEUDO-DĒMOKRITOS' list of Persian plants, he added *aiskhunomenē* (*Mimosa asperata* L.) and *krokis*, allegedly a spider-bane. Wellmann, noting that of the sources Pliny lists for Book 27, only "Apollodorus" is not cited explicitly within, argues that this Apollodōros is Pliny's primary source there. He also identifies this Apollodōros with the Apollās cited in the *Schol. Nik. Thēr.* (but *cf.* APELLĒS OF THASOS), as well as with Apollodōros *On Perfumes* (*cf.* APOLLODŌROS (MED.)); finally he dates this Apollodōros to the 1st c. CE. Jacques (2002) *fr.*15 identifies the man of 24.167 with APOLLODŌROS THE *THĒRIAKOS*.

M. Wellmann, "Beiträge zur Quellenanalyse des älteren Plinius. II," *Hermes* 68 (1933) 93–105.

PTK

Apollodōros the _thēriakos_ (280 – 240 BCE)

Physician and naturalist, possibly active in Alexandria, though there is no persuasive evidence of links with Egypt, except for a possible but unverifiable hypothesis that Apollodōros was the author of a wine handbook for a Ptolemy (PLINY 14.76). Pliny's *Apollodorus adsectator Democriti* (24.167) refers to the *thēriakos* (*fr.*15; *cf.* Jacques [2002] 2.XXXVI with n. 59). Perhaps the *thēriakos* wrote a general work on pharmacology (*frr.*16, 18), but his only work to have been clearly confirmed is his iological work. His book on venom treatments is quoted as *thēriakos logos* (*fr.*1), *Peri thēriōn* (*frr.*4, 8, 10, *cf.* Pliny 1.*ind.*11: *Apollodoro qui de bestiis uenenatis*). The venomous species mentioned are *khershudros* (*fr.*1), *pareias* (2), *tuphlōpes* (3), spiders (4) and scorpions (5). Specific chapters are probably tripartite: see *frr.*2, 5 (descriptive elements), 1 (symptomatology), 6–10 (therapy). Despite the lack of explicit testimonies referring to an *alexipharmakos logos*, the mention of poison renders its actual existence probable: *toxikon* (11), mushrooms (12), toad (13), **litharge** (14), henbane and salamanders (16). Apollodōros is concerned with zoological (2, 3) and botanical (10, 17) nomenclature. His work is influenced by the **Peripatos**: see *frr.*4 (ARISTOTLE) and 1 (THEOPHRASTOS). It has been said that NIKANDROS merely put Apollodōros' work into verse – an opinion which has become a dogma – and that it could be reconstructed on the basis of parallel passages in Nikandros' and Pliny's works and those of iologists before or after Nikandros' time. However, there are several divergences between Apollodōros and the rest of iological literature, and disagreements between Apollodōros and Nikandros.

Ed.: O. Schneider, *Nicandrea* (1856) 181–201; Jacques (2002) 2.285–292 (see XXXIII–XXXVI, XLIX–LII).

RE 1.2 (1894) 2895 (#69), M. Wellmann; *Idem*, *Hermes* 43 (1908) 379, n.1; Jacques (2007) 3.301 (see Index V, s.v. Apollodoros).

Jean-Marie Jacques

Apollodōros of Artemita (*ca* 130 – 50 BCE?)

Parthian Greek, wrote *Parthika*, in at least four books, the primary resource for STRABŌN's evidence on Central Asia, Asiatic Skuthia, Iran, Armenia, India, and POMPEIUS TROGUS' *Historiae Philippicae* (Books 41 and 42: Nikonorov 108). PLINY lists an unspecified Apollodōros among his foreign experts on Central Asian geography (1.*ind.*6), perhaps his source on Margiana (6.46–47) and possibly our author (Nikonorov 112). Apollodōros determined the borders of Hurkania and Baktria more accurately than others (Str. 2.5.12), estimated distances between major cites (11.9.1), states, and natural landmarks (11.11.7), claimed (erroneously) that the Araxes separates the Armenians from Pontos and Kolkhis (1.3.21), and referred often to the Okhos – unmentioned by other ancient authors – implying that it flows continually through Parthia (11.7.3). He argued that the Parthians, conquering Ariana and India, subdued more tribes than Alexander (11.11.1). He described the unusual *philadelphum*, able to unite and grow onto other *philadelpha*, the roots of which are planted as impenetrable garden fences (Ath. *Deipn.* 15 [682c]).

Ed.: *FGrHist* 779.

V.P. Nikonorov, "Apollodorus of Artemita and the Date of His Parthica Revisited," in E. Dabrowa, ed., *Ancient Iran and the Mediterranean World* (1998) 107–122; *NP* 12/2.897 (#8a), H.A. Gärtner.

GLIM

Apollodōros of Athens (150 – 110 BCE)

Epicurean scholarch in Athens and author of many works (Diogenēs Laërtios 10.25), all lost, including a *Summary of Doctrine* (D.L. 7.181) and a defense of Epicurus against Chrysippus (D.L. 1.60).

GGP 4.1 (1994) 280–281, M. Erler.

PTK

Apollodōros of Athens, pseudo (80 BCE – 10 CE)

Stephanos of Buzantion makes over two dozen citations from Apollodōros' *On the Earth*, Book 2, including iambs on the Oritans of India (F313), the Psēssoi (F318), the Hulleis (F321–322), and the Iberians (F324). Strabōn 14.5.22 attributes an iambic *Circuit of the Earth* (***periodos*** *gēs*) to Apollodōros of Athens, the grammarian and chronographer, which scholars reject, since the cited fragments seem to post-date pseudo-Skumnos.

Ed.: *FGrHist* 244 F313–330.

RE S.6 (1935) 8–10, F. Atenstädt.

PTK

Apollodōros of Damaskos (100 – 120 CE)

Architect and military engineer (*arkhitektōn*) under Trajan and Hadrian. He is celebrated as the designer of Trajan's forum in Rome (Dio Cassius 69.4) and the emperor's bridge over the Danube, pictured on Trajan's column (Prokopios, *De aedificiis*). According to an improbable account, he insulted Hadrian, mocking the emperor's architectural interests in vaulted structures by calling them "pumpkins," and was banished and later executed because he criticized Hadrian's temple of Venus and Rome.

Excerpts from an illustrated work on *Siege Craft* (*Poliorkētika*) survive. A letter to the emperor, most likely Trajan, offering help for upcoming campaigns, introduces the treatise. The author notes his involvement with earlier campaigns and with this treatise he sets out to assist current campaigns; he also offers to send a man who has observed the construction of the devices. Apollodōros describes his designs as effective, safe, light and speedy to manufacture from available materials with the manpower at hand (137–138 S.). The *Poliorkētika* describes a series of devices roughly in the order of the progress of a siege of a hill fort. It starts with the approach and devices for protecting troops against objects rolled down the hills; then screens for protecting miners and advice on undermining the walls, as well as a drill and a covered ram for attacking the walls; he covers a siege tower with attachments such as an assault bridge, a sweep for throwing enemies off the wall and fire extinguishing equipment, and assault ladders with a number of attachments; and lastly he describes an armored floating bridge. Some of the devices are simple and light as advertised in the introduction, while others are complex and hardly realistic.

It is debated whether the complex material is the work of another, later author, or whether it simply demonstrates the lack of a clear boundary between realistic and imaginative devices found in many ancient military treatises. Blyth (145–154) doubts the identification of Apollodōros as the author of *Siege Craft*.

Ed.: R. Schneider, *Griechischer Poliorketiker, I: Apollodorus* (*AbhGöttingen* N.F. 10.1, 1908).

RE 1.2 (1894) 2896 (#73), E. Fabricius; P.H. Blyth, "Apollodorus of Damascus and the Poliorcetica," *GRBS* 33 (1992) 127–158; *OCD3* 124, N. Purcell; *BNP* 1 (2002) 862–863 (#14), C. Höcker.

Karin Tybjerg

Apollodōros of Kerkura (170 – 130 BCE?)

Theorized that tidal ebb and flux was due to the reflux (*palirrhoia*) of the Ocean, according to AËTIOS 3.17.8, who lists Apollodōros with KRATĒS and SELEUKOS OF SELEUKEIA (although Aëtios' list is not chronological).

DPA 1 (1989) 274–275, R. Goulet.

PTK

Apollodōros of Kition (325 BCE – 75 CE)

Doctor whose writings were known to PLINY, 1.*ind*.20–27; he recommended crushed radish in water as an antidote to mistletoe poisoning, 20.25.

RE 1.2 (1894) 2895 (#70), M. Wellmann.

Philip Thibodeau

Apollodōros of Kuzikos (350 BCE – 200 CE)

Dēmokritean mathematician (*arithmētikos*) or calculator (*logistikos*), cited for traditions that PYTHAGORAS sacrificed a hecatomb to celebrate his discovery of the "Pythagorean" theorem (Ath. *Deipn.* 10 [418f–419a]; DIOGENĒS LAËRTIOS 1.25, 8.12), and that DĒMOKRITOS was acquainted with PHILOLAOS (D.L. 9.38).

FGrHist 1097.

GLIM

Apollodōros of Lēmnos (450 – 335 BCE)

Named with KHARETIDĒS OF PAROS by ARISTOTLE as a writer of (lost) manuals on agriculture, treating both crops and fruits (*Politics* 1.11 [1258b39–1259a2]). VARRO names him in a catalogue of farmers who have written treatises on agriculture (*RR* 1.1.8); and PLINY 1.*ind*.8, 10, 14–15, 17–18 mentioned his work.

RE 1.2 (1894) 2895–2896 (#71), M. Wellmann.

Maria Marsilio

Apollodōros of Seleukeia (Tigris) (175 – 125 BCE)

A student of DIOGENĒS OF BABYLŌN, called Ephēlos from his cataracts, Apollodōros wrote an *Introduction to the Doctrines* (which included *Ethics* and *Physics*), a comprehensive and systematic defense of orthodox **Stoic** logic and epistemology, preserved by AREIOS DIDUMOS and DIOGENĒS LAËRTIOS. Apollodōros argues for a completely material and continuous, living, sensible, and rational ***kosmos***, surrounded by an infinite void (D.L. 7.142–143). He used geometry to define fundamental physical concepts: so bodies have three-fold extension, while surfaces, which limit bodies, have only two-fold extension (D.L. 7.135). He believed change and continuity in place or shape explained motion and rest (Ar. Did. *fr.*24). Time he thought to be the "extension of cosmic motion" (Ar. Did. *fr.*26), and vision he explained as light between viewer and object stretching into a cone, extending from its base at the object seen to the apex at the eye. Air stretching between viewer and object relays visual data to the viewer (D.L. 7.157). See THEŌN OF ALEXANDRIA (STOIC).

DPA 1 (1989) 276–278, M.-O. Goulet-Cazé; *GGP* 4.2 (1994) 635, P. Steinmetz; *ECP* 44–45, S.A. White.

GLIM

Apollodōros of Taras (325 BCE – 75 CE)

Doctor whose writings were known to PLINY, 1.*ind.*20–27; he preferred straight radish juice as an antidote to mistletoe poisoning, 20.25.

RE 1.2 (1894) 2895 (#70), M. Wellmann.

Philip Thibodeau

Apollōnidēs (100 – 50 BCE)

Wrote a ***Periplous*** *of Europe* which STRABŌN cites thrice (7.4.3 MITHRADATĒS' war against the Skuths, 11.13.2 northern Media, and 11.14.4 mythical Median snow-worms) and PLINY once (7.17 Skuthian women's evil-eye). The *scholia* to Apollōnios of Rhodes offer further fragments.

BNP 1 (2002) 867 (#1), K. Brodersen.

PTK

Apollōnidēs of Cyprus (*fl. ca* 150 CE)

Called a surgeon by ARTEMIDŌROS OF DALDIS, *Onirocriticon*, 4.2 (p. 245 Pack = p. 188 White), and listed among the excoriated **Methodists** by GALĒN, *MM* 1.7 (10.54 K. = 1991: 27, and PSEUDO-GALĒN, INTRODUCTIO 14.684 K.). He may well be the Appius Apollōnidēs addressed in a letter from Fronto (*Ad amicos*, 1.2: ed. Naber, p. 174 = ed. and trans. Haines v. 1 [Loeb 1919] 286–289; *cf.* Pack 1955: 285–286). Apollōnidēs taught IULIANUS the **Methodist**, and studied under OLUMPIAKOS OF MILĒTOS (Galēn 10.54 K.). Galēn, further, accuses Apollōnidēs of mangling descriptions of the pulses, presuming he could understand them without actual clinical observations, dressing his false depictions in pedantic and elaborate terminologies (*Causes of Pulses* 3.9 [9.138–139 K.]).

Ed.: Tecusan (2004) *frr.* 19, 108, and 162.

RE 2.1 (1895) 121 (#33), M. Wellmann; R.A. Pack, "Artemidorus and his Waking World," *TAPA* 86 (1955) 280–290.

John Scarborough

Apollōnios (Paradoxographer) (150 – 100 BCE?)

Wrote *Historiai thaumasiai* in 51 chapters. The first six sections focus on six thaumatographers (EPIMENIDĒS, Aristeas, Hermotimos, Abaris, PHEREKUDĒS, PYTHAGORAS). Chapters 7 to 51 make up a paradoxographical collection of brief anecdotes, mainly treating botanical, anthropological, physical and ethnographic curiosities.

Very often Apollōnios mentions his sources, often highly authoritative authors, such as ARISTOTLE (*Phusika* 7; 9; 21–23; 37; 51; *Zōika* 27; 28; *Peri Zōiōn* 44; etc.), THEOPHRASTOS (*Peri phutōn* 16; 29; 31–34; 41–43; 47–48; 50; etc.), ARISTOXENOS (30; 40), Theopompos (1; 10), KTĒSIAS (17, 20), Phularkhos (4; 18), perhaps BŌLOS and many others.

When Apollōnios lived cannot be determined with certainty. That most of his sources flourished at the turn of the 3rd c. BCE, while none is later than the first half of the 2nd c. BCE, suggests a slightly later date. Attempts to identify the paradoxographer with any other known homonym remain futile.

Ed.: *PGR* 119–143.

RE 18.3 (1949) 1137–1166 (§14, 1152–55), K. Ziegler; Giannini (1964) 122–123; *OCD3* 127, R.L. Hunter; *BNP* 10 (2007) 506–509 (I.B.1, 508–509), O. Wenskus.

Jan Bollansée, Karen Haegemans, and Guido Schepens

Apollōnios, Claudius (40? – 80 CE)

ANDROMAKHOS twice cites an Apollōnios via ALKIMIŌN, in GALĒN, *CMGen* 5.12 (13.835 K.) for a wound remedy of **litharge**, ***psimuthion***, etc., and for a cough drop of saffron, licorice, myrrh, white pepper, tragacanth, etc., *CMLoc* 7.2 (13.31–32 K.). Wellmann identifies this man with the Apollōnios Claudius cited by ASKLĒPIADĒS, in Galēn, *Antid.* 2.11 (14.171–172 K.) for a ***hudrophobia*** remedy involving ashed crabs, clover, licorice, etc., in Falernian wine. Wellmann (followed by Korpela) supports this by reading *arkhiatēr tou autokratoros* for the garble *ΑΡΧΙΣΤΡΑΤΩΡ* in the first passage.

RE 2.1 (1895) 150 (#105), M. Wellmann; Korpela (1987) 166.

PTK

Apollōnios Glaukos (250 BCE – 100 CE)

Discussed the expulsion of round worms from the anus in *Internal Diseases*. Empty dead worms indicate recovery; live, full, bloody worms denote trouble (SŌRANOS in CAELIUS AURELIANUS, *Chron.* 4.113, *CML* 6.1.2, p. 838). Listed after, and probably later than, HĒROPHILOS.

RE 2.1 (1895) 151 (#107), M. Wellmann.

GLIM

Apollōnios "Ophis" (Snake), "Organikos," "Thēr" (Beast) (225 – 25 BCE)

ERŌTIANOS *pr.* (p. 5 Nachm.) lists his predecessors apparently in chronological order, placing Apollōnios Ophis after BAKKHEIOS and before DIOSKOURIDĒS PHAKĀS; and A-103 (p. 23 Nachm.) says that Apollōnios Thēr explained *ambē* (HIPPOKRATIC JOINTS 7 [4.88 Littré]) as "projection." (If *ophis* is not a precision of *thēr*, it may stand for "snake-bald," *ophiasis*, thus, like *phakas*, "warty," indicating a physical distinction.) ASKLĒPIADĒS PHARM., in GALĒN *CMGen* 5.15 (13.856 K.), cites a mineral-based recipe of Apollōnios

Organikos for ***anthrax***; a machine-maker (*organikos*) might well have remarked upon *ambē*. OREIBASIOS, *Coll.* 48.41 (*CMG* 6.2.1, p. 282), cites Apollōnios Thēr for the "Monōps" bandage; Michler 121 suggests "Organikos" might be APOLLŌNIOS OF ANTIOCH; von Staden (1989) 549 suggests that *thēr* might mean "mouse," i.e., that he may be APOLLŌNIOS "MUS."

Michler (1968) 82–83, 119–122.

PTK

Apollōnios (of Alexandria?) (200 – 150 BCE?)

APOLLŌNIOS OF PERGĒ's son, entrusted with delivering *Conics* Book 2 (*Con.* 2.*pr.*)

Netz (1997) #10.

GLIM

Apollōnios of Alexandria, "Mus" (*ca* 50 BCE – 30 CE)

Apollōnios "Mus" (*Vind. Med. Gr* 1, f.3[V]) © Österreichische Nationalbibliothek

Hērophilean physician, KHRUSERMOS' student, HĒRAKLEIDĒS OF ERUTHRAI's fellow pupil (STRABŌN 14.1.34), wrote *On the School of Hērophilos* (29+ books, apparently a comprehensive treatment of physiological and pathological theory), cited extensively by medical writers including CELSUS, ANDROMAKHOS, ASKLĒPIADĒS PHARM., SŌRANOS, GALĒN, PHILOUMENOS, OREIBASIOS, AËTIOS OF AMIDA, and the Hippokratic commentators IŌANNĒS OF ALEXANDRIA and PALLADIOS (see further von Staden [1989]; Oreibasios and Aëtios also often cite an unspecified Apollōnios); also by PLINY (1.*ind.*28, 28.7), Athēnaios, and PLUTARCH (*Quaest. Nat.* 3 [912D–E]). The well-attested but ambiguous nickname may mean "mouse," "muscle," or "mussel."

Like other **Hērophileans**, Apollōnios' interests included pulse theory, where he concurred with Khrusermos and Hērakleidēs: the pulse occurs through the agency of a (dominant) vital and psychic faculty (AM.4 von Staden). Other fragments, from *On Common Remedies* (several volumes), address garden variety ailments, toothaches, earaches, skin disorders, dandruff, treatable with ingredients from the mundane to the bizarre: e.g. bull or camel urine or turtle blood against dandruff (GALĒN considered turtle blood impractical: 12.475–482 K.), donkey urine for a sore throat, which remedy Galēn finds astonishingly repugnant (12.979–983 K.). Although Apollōnios distinguishes, for example, categories of headaches – those caused by heatstroke, chills, intoxication, blows to the head, falls (AM.12–16 von Staden) – Galēn criticizes Apollōnios for prescribing remedies generically without properly noting causes and symptoms, an approach with potential for harm (*ibid.*). Preserved are Apollōnios' recommendations for using leeches (AM.30 von Staden), his antidote compounded from rue, walnuts, salt, and *iskhas* (AM.31 von Staden), and, from the *Euporista*, Book 1, his tooth-whiteners, one consisting of mineral salt roasted with honey on a shell, then ground with myrrh (AM.24b von Staden). Sōranos (*Gyn.* 3.2 [*CMG* 4, pp. 94–95; *CUF* v. 3, p. 3]) cites him among physicians denying illnesses specific to women. CAELIUS AURELIANUS (*Acute* 2.88–89 [*CML* 6.1.1, p. 186]) criticizes his definitions of pleurisy as pleonastic and omitting essential details (e.g., the presence of fever).

Athēnaios refers at length to Apollōnios' *On Perfumes and Unguents* (15.688e–689b): recounting the best quality varieties of many; and commenting on the changing fortunes of quality control: the excellent tradition of perfume in Ephesos is in decline, but royal interest in Adramuttion has yielded an improved dropwort perfume.

Pack #2386 = *POxy* 234; Fabricius (1972) 180–183; von Staden (1989) 540–558; *OCD3* 127, *Idem; Idem* (1999) 166–169; *BNP* 1 (2002) 882–883 (#17), V. Nutton.

GLIM

Apollōnios of Antioch (200 – 150 BCE)

Apollōnios "Biblas" (175 – 125 BCE)

Empiricist physicians, father and son, both from Antioch. The Elder is quoted by CELSUS (*pr.*10) as the second exponent of the **Empirical** School after SERAPIŌN; PSEUDO-GALĒN INTRODUCTIO 14.683 K. quotes both the Apollonii after PHILINOS and Serapiōn. The Elder carried on the polemical tradition that had been typical of the first empirical physicians: we know about his polemics against the **Epicureans** about the foundations of sensible experience (*cf.* DĒMĒTRIOS OF LAKŌNIKA, *P. Herc.* 1012, *frr.*23, 58, 71 Puglia: criticizes particularly Apollōnios' exegesis of the HIPPOKRATIC CORPUS EPIDEMICS 6.9), and against the **Hērophilean** ZĒNŌN, who in a work entitled *On the Marks* had attributed to HIPPOKRATĒS himself the marks or symbols (*kharaktēres*) in the Alexandrian copies of Book III of the Hippokratic *Epidemics*; Apollōnios challenged Zēnōn's thesis, denying that those marks went back to Hippokratēs (GALĒN *Hipp. Epid.*: *CMG* 5.10.2.1, p. 86). As we can infer from ERŌTIANOS (p. 23.17 Nachm.), he wrote also a Hippokratic lexicon. His polemic against Zēnōn was taken up after Zēnōn's death by his son, Apollōnios "Biblas" ("the Bookworm"). Biblas' references to Hippokratic MSS suggest that he spent some time in Alexandria. From SŌRANOS (*Gyn.* 2.87 [*CMG* 4, p. 65; *CUF* v. 2, pp. 26–27]) we know that he was also interested in gynecology.

Ed.: Deichgräber (1930) 171–172 (fragments), 256–257.

RE 2.1 (1895) 149 (#101), M. Wellmann; M. Gigante, *Scetticismo e epicureismo* (1980) 170–175; J. Nollé, "Die 'Charaktere' im 3. Epidemienbuch," *Epigr. Anat.* (1983) 85–98; E. Puglia, ed., *Demetrio Lacone, Aporie testuali ed esegetiche in Epicuro (PHerc. 1012)* (1988) 217–219, 286, 311; von Staden (1989) 501–502; Ihm (2002) #12.

Fabio Stok

Apollōnios of Aphrodisias (265 – 195 BCE)

Wrote a *Karika* describing the land and its history, of which STEPHANOS OF BUZANTION preserves over a dozen fragments. Stephanos, s.v. *Agkura*, provides the *terminus post*; the *Souda* A-3424 gives the ethnic and his works. Stephanos, s.v. *Lētous polis*, establishes an Egyptian connection (not origin, despite *RE* 2.1 [1895] 134–135 [#73], E. Schwartz), which suggests the *terminus ante*. (He also wrote on ORPHEUS and his rites, of which cult he was high priest.)

FGrHist 740; *PLRE* 2 (1980) 120 [impossibly dating him to *ca* 400 CE].

PTK

Apollōnios of Athens (130 – 70 BCE)

Teacher of AGĒSISTRATOS, designed siege machines, making practical rather than theoretical contributions, according to ATHĒNAIOS MECH. (p. 12 W.), and helped defend Rhodes in the siege of 88–87 BCE.

C. Cichorius, *Römische Studien* (1922) 271–279.

PTK and GLIM

Apollōnios of Kition (90 – 60 BCE)

Empiricist physician, pupil of ZŌPUROS OF ALEXANDRIA, perhaps one of the two surgeons Apollōnii mentioned by CELSUS at 7.*pr*. We have a commentary in three books on (or rather an up-to-date revision of) the HIPPOKRATIC ON JOINTS, dedicated to Ptolemy XII Aulētēs (or to Ptolemy of Cyprus). The illustrations contained in the 10th c. Laurentian codex are probably modifications of the original ones, to which Apollōnios refers in the proems to the three books. This treatise is important not only for the history of orthopedic surgery, but also as a testimony to the tradition of the Hippokratic text, and as a linguistic document. In his commentary Apollōnios polemizes against the **Hērophileans** BAKKHEIOS and HĒGĒTŌR. In lost works about Hippokratic exegesis, he dealt with lexicography and criticized the *Lexeis* by Bakkheios; in another polemic treatise, in 18 books, he shielded Bakkheios from the criticisms leveled against him by the **Empiricist** HĒRAKLEIDĒS OF TARAS (this fact has called into question his actual belonging to the **Empiricist** "school"). We also know (from CAELIUS AURELIANUS *Chron.* 1.140 [*CML* 6.1.1, p. 512]) about a treatise *On epileptics* (whence a fragment in ALEXANDER OF TRALLEIS 1.15 [1.559, 561 Puschm.]).

Ed.: J. Kollesch and F. Kudlien, *Apollonios of Kition. Kommentar zu Hippokrates über die Einrenken der Gelenke* (1965) = *CMG* 11.1.1; Deichgräber (1930) 206–209 (fragments), 262–263.

J. Kollesch and F. Kudlien, "Bemerkungen zum Peri arthrōn Kommentar des Apollonios von Kition," *Hermes* 89 (1961) 322–332; J. Blomqvist, *Der Hippokratestext des Apollonios von Kitios* (1974); Smith (1979) 212–222; P. Potter, "Apollonius and Galen on 'Joints'," *AGM* 32 (1993) 117–123; A. Roselli, "Tra pratica medica e filologia ippocratica," in *Sciences exactes et sciences appliquées à Alexandria*, ed. Argoud-Guillaumin (1998) 217–231; *OCD3* 127, H. von Staden; *BNP* 1 (2002) 881–882 (#16), V. Nutton; Ihm (2002) #13; *AML* 69–70, K.-H. Leven.

Fabio Stok

Apollōnios of Laodikeia (*ca* 180 – 380 CE)

Wrote a work arguing against the system of computing zodiacal rising times used by "Egyptian," i.e., Alexandrian, astrologers, according to PAULOS OF ALEXANDRIA, *pr.*, who mentions him "in addition to" PTOLEMY and APOLLINARIOS.

RE 2.1 (1895) 161 (#115), E. Riess.

PTK

Apollōnios of Memphis (250 – 200 BCE)

Student of the **Erasistratean** STRATŌN; he explained the pulse as ***pneuma*** from the heart filling the arteries (GALĒN, *Diff. Puls.* 4.17, 8.759–761 K.), and also wrote on anatomy (PSEUDO-GALĒN, INTRODUCTIO 10, 14.699–700 K.; PSEUDO-GALĒN, DEFINITIONES *pr.*

19.347 K.), including a work *On Joints* (ERŌTIANOS, A-103 [p. 23 Nachm.]). CAELIUS AURELIANUS attests to his work on pathology, classifying kinds of **dropsy** (*Chron.* 3.101–102, *CML* 6.1.2, p. 740), and offering prognoses based on intestinal worms (4.114, p. 838). The *Schol. Nik. Thēr.* 52c preserve his explanation of a rare plant name in NIKANDROS. ASKLĒPIADĒS PHARM., in Galēn *Antid.* 2.14 (14.188–189 K.), preserves HUBRISTĒS' record of his antidote, and AËTIOS OF AMIDA records two prescriptions: 6.84 (*CMG* 8.2, p. 230) a mineral and incense ***trokhiskos*** for fleshy overgrowth in the ears, and 7.22 (p. 270) a collyrium including sun-dried blood from a donkey's heart and a boy's urine. Aëtios records many other recipes attributed to an anethnic Apollōnios, which may belong to this man, or to a later Apollōnios, such as AP. CLAUDIUS, AP. OF PERGAMON, AP. OF PITANĒ, AP. OF PROUSIAS, or AP. OF TARSOS, or especially AP. "MUS": vegetal remedy for ears 6.79 (pp. 223–224), vegetal remedy for nasal polyps 6.91 (p. 238), vegetal remedy for cataracts 7.101 (p. 353), and mineral and incense ocular wound plaster with saffron 7.109 (p. 375).

Michler (1968) 43, 96; Jacques (2002) 298–299.

PTK

Apollōnios of Mundos (120 – 80 BCE)

Studied with the "Chaldeans," i.e., Babylonian astrologers, and cast horoscopes; he also wrote a work explaining comets as long-period planets of elongated shape and on non-circular orbits (SENECA, *QN* 7.4, 7.17). For periodic comets, compare also the Talmud, *Horayoth* 10a (in 95 CE, R. Joshua said "a certain star rises once in 70 years and leads the sailors astray").

P.T. Keyser, "On Cometary Theory and Typology from Nechepso-Petosiris through Apuleius to Servius," *Mnemosyne* 47 (1994) 625–651 at 648.

PTK

Apollōnios of Pergē (*ca* 220 – *ca* 170 BCE)

Chronology. In the introduction to *Conics* II, Apollōnios mentions (a) having introduced his dedicatee Eudēmos to PHILŌNIDĒS OF LAODIKEIA "the Geometer," (b) sending the new work via his son, also APOLLŌNIOS. Philōnidēs is known to have been active in the mid-2nd c. BCE, so that Apollōnios would have been mature in the early 2nd c. BCE. (PAPPOS' claim that Apollōnios studied with "the students of EUCLID" seems to be pure fiction).

Works. The *Conics*, Apollōnios' major work, originally in eight books, is mostly extant: Books I–IV in Greek (EUTOKIOS' commentary also survives), Books V–VII in what appears to be the fairly close Arabic translation by the Banū Musā (Toomer). *Cutting Off of a Ratio* survives in Arabic only. Pappos' discussion of books on analysis (*Collection*, Book VII) offers a detailed survey of the aforementioned work together with five others by Apollōnios, no longer extant (*Cutting off of an Area, Determinate Section, Inclinations, Tangencies, Plane Loci*). All appear to be detailed surveys of various combinations arising from a geometrical problem with several parameters. Pappos' *Collection* II (unfortunately fragmentary) is dedicated to a single work by Apollōnios, unattested otherwise, where the letters of an hexameter line are considered as Greek numerals and multiplied (!). This calculatory tour de force may resemble ARCHIMĒDĒS' *Sand-Reckoner*; another lost work, the "*Quick Delivery*," offered (to use modern notation) an approximation of π closer than Archimēdēs' estimate in the *Measurement of the Circle*. A desire to compete against Archimēdēs is easy to imagine. Also attested are a study comparing dodecahedra and icosahedra (an obvious attempt to further Euclid's

Elements XIII), as well as a study in the cylindrical helix (a response to Archimēdēs' *Spiral Lines*?), a study of "unordered irrationals" (a response to Euclid's *Elements* X?), and a "general treatise" of unknown significance.

The Conics. Apollōnios himself considered the opus, or at least its first four books, as "Elements of Conics." It was certainly not the first of its kind, and Apollōnios' originality is never certain. While, in part, "Elementary," *Conics* – not an axiomatic sequence in the manner of Euclid's *Elements* – instead forms a diverse collection of nearly independent treatises, with certain unifying themes and goals. Book I is the most obviously "Elementary" in character, defining the main conic sections, deriving their most useful proportions and leading up to their construction from given points and lines. This appears also to be the least original to Apollōnios. Furthermore, this book inspires the main modern scholarly debates concerning the *Conics*: did Apollōnios define the conic sections primarily as (a) the results of a geometrical cut, or as (b) the locus satisfying a certain proportion? (a) is consistent with a thoroughly geometrical interpretation of Greek mathematics, (b) – with a more modernizing and algebraic one. Books II–III form a certain continuity (Book III is the only extant book not to carry any introduction), building up various surprising equalities arising from conic sections. The great historical significance of that sequence is that it provides the tools (as asserted by Apollōnios himself) for the problem of a three and four line locus, a major inspiration for the algebraization of geometry down to Descartes and beyond. Book IV studies the intersections of conic sections. Its more qualitative character allows for less spectacular results than those of Books II–III, and it has therefore been relatively neglected by scholarship. Fried makes the case for its importance. Book V, the most ambitious among the extant books, studies the shortest and longest lines drawn to conic sections from given points. It can be taken as an example of the precise, systematic and advanced character of Apollōnios' geometry. Once again, the question whether Apollōnios' shortest and longest lines should be considered as "normals" impinges on the question of the overall interpretation of the *Conics* as geometric or algebraic in character. Book VI studies the similarities between conic sections, and suffers the same critical fate as Book IV, for similar reasons. Book VII returns to the spirit of Books II–III, producing a remarkable succession of theorems concerning, specifically, conjugate diameters. It may have served as preparation for the problems of Book VIII, now lost.

For survey and study of the *Conics*, see Fried and Unguru (2001). Zeuthen (1886), usually rejected today for its extreme modernizing interpretations, remains a classic.

Ed.: G.J. Toomer, Conics: *Books V to VII/Apollonius: the Arabic Tradition* (1990); M. Fried, *Apollonius of Perga:* Conics *Book IV* (2002).

H.G. Zeuthen, *Die Lehre von den Kegelschnitten im Altertum* (1886); M. Fried and S. Unguru, *Apollonius of Perga's* Conica: *Text, Context, Subtext* (2001); *NDSB* 1.83–85, F. Acerbi.

Reviel Netz

Apollōnios of Pergamon (Agric.) (325 – 90 BCE)

Wrote a work on agriculture which may have treated cereals, livestock, poultry, viticulture, and arboriculture (*cf.* PLINY, 1.*ind.*8, 10, 14–15, 17–18), and was excerpted by CASSIUS DIONUSIOS (VARRO, *RR* 1.1.8–10, *cf.* COLUMELLA, 1.1.9). In Wellman's opinion he is to be distinguished from the homonymous medical writer.

RE 2.1 (1895) 150 (#104), M. Wellmann.

Philip Thibodeau

Apollōnios of Pergamon (Med.) (15 BCE – 182 CE)

Greek physician quoted for scarification of the legs as a therapeutic method which he had successfully employed on himself during a plague in Asia (OREIBASIOS *Coll.* 7.19 = *CMG* 6.1.1, p. 218; *Eustath.* 1.14 = 6.3, pp. 12–13, and *Eunap.* 1.9.7 = 6.3, p. 325, with ethnic); for the method *cf.* GALĒN, *De hirudinibus* (11.322 K.). Scarification was a typically **Pneumaticist** therapy, attempting to restore the body's natural balance (*eukrasia*) by eliminating blood to reduce excessive bodily heat which caused the plague. ALEXANDER OF TRALLEIS (1.15 = 1.559, 1.561 Puschm.) twice cites an anethnic Apollōnios on epilepsy, who must be APOLLŌNIOS OF KITION (*cf.* CAELIUS AURELIANUS *Chron.* 1.140 = *CML* 6.1.1, p. 512.22), refuting Wellmann's attribution to this Apollōnios. Likewise, the anethnic Apollōnios, author of *Euporista*, at Oreibasios, *Eunap., pr.* (*CMG* 6.3, p. 318), must be APOLLŌNIOS "MUS." This Apollōnios is distinct from the homonymous agronomist.

RE 2.1 (1895) 150 (#104) M. Wellmann; von Staden (1989) 548–550.

Alain Touwaide

Apollōnios of Pitanē (350 BCE – 77 CE)

Physician who treated cataracts and *albugines* (opaque white spots on the eye) with honey and dog's (rather than hyena's) gall (PLINY 29.117).

RE 2.1 (1895) 151 (#108), M. Wellmann.

GLIM

Apollōnios of Prousias (30 BCE – 120 CE)

Wrote on childbirth, advising attendants to grasp the projecting part of the *khorion* (afterbirth) to draw it out (SŌRANOS, *Gyn.* 4.14 [*CMG* 4, pp. 144–145; *CUF* v. 2, p. 11]), as had EUĒNŌR and SŌSTRATOS. Listed after and probably later than Sōstratos.

RE 2.1 (1895) 151 (#110), M. Wellmann.

GLIM

Apollōnios of Tarsos (250 BCE – 80 CE)

ANDROMAKHOS, in GALĒN *CMGen* 5.13 (13.843 K.), records his mineral remedy for hemorrhoids: alum, copper and its flakes, ***diphruges***, roasted orpiment, ***khalkanthon, khalkitis***, **litharge**, and roasted ***sōru***.

RE 2.1 (1895) 151 (#109), M. Wellmann.

PTK

Apollōnios of Tuana, pseudo

Cited for amulets in Byzantine times, and famous in Arabic tradition as "Master of Talismans" under the name Balīnās (and the like, whence Medieval Latin Belenus). Several works attributed to him survive in Arabic, most famous of which is *Kitāb Sirr al-khalīqa*. This synthetic work, composed perhaps around 900 CE, utilizes NEMESIOS' *De natura hominis* and includes the famous alchemical *tabula smaragdina* attributed to HERMĒS TRISMEGISTOS. Other extant works attributed to Apollōnios in Arabic, on talismans, await careful investigation.

Ed.: U. Weisser, *"Buch über der Schöpfung und die Darstellung der Natur" von Pseudo-Apollonios von Tyana* (1979). *GAS* 4 (1971) 77–91; Ullmann (1972) 378–381; *GAS* 6 (1978) 102–103, 7 (1979) 64–66, 227–229; U. Weisser, *Das "Buch über das Geheimnis der Schöpfung" von Pseudo-Apollonios von Tyana* (1980); F.W. Zimmermann, review of Weisser 1980, *Medical History* 25 (1981) 439–440; M. Dzielska, *Apollonius of Tyana in legend and history* (1986) 99–108; P. Travaglia, *Una Cosmologia Ermetica* (2001).

Kevin van Bladel

Apollophanēs of Nisibis (280 – 220 BCE)

From Antioch "Mugdonia," i.e., Nisibis (Stephanos of Buzantion, s.v. Antioch, #3). Entitled a treatise, *Aristōn*, in honor of his friend and teacher the **Stoic** Aristōn of Khios (Ath. *Deipn.* 7 [281d]), emphasizing Aristōn's love of pleasure. Apollophanēs also wrote *On Physics* arguing that there is no void inside the ***kosmos*** (Diogenēs Laërtios 7.140), and hypothesized that the soul has nine parts (rather than the usual **Stoic** eight: five senses, generation, speech, reason), *SVF fr.* 405.

DPA 1 (1989) 296–297, C. Guérard; *GGP* 4.2 (1994) 561, P. Steinmetz.

PTK and GLIM

Apollophanēs of Seleukeia "Pieria" (223 – 187 BCE)

Court physician in service to Antiokhos III ("the Great") of the Seleukid monarchy, also probably wrote a history of his own day (Brown 1961), an important source for some events in Polubios' *Histories*. Apollophanēs is a major figure in foiling a plot against Antiokhos in 220 BCE: "Apollophanēs, a physician for whom the king had great regard, discerned that Hermeias [Antiokhos' chief minister] had no scruples in his attempt to seize power, and began to be concerned about the king" (Polubios 5.56.1). Warned of the looming treachery, Antiokhos counter-plotted to entrap Hermeias, publishing a diagnosis of dizziness which prevented the usual business schedule, and for which court doctors prescribed walks in the cool morning hours announced to everyone. At the appointed time, Hermeias appeared, but since the king had taken his stroll some hours earlier, Hermeias was slain with the daggers of his presumed co-conspirators (5.56.13). Apollophanēs' crucial speech regarding the coming campaign against Ptolemy IV is quoted verbatim (5.58.3–8). Apollophanēs' prominence is attested by a contemporary honorific inscription from Kōs (Samama #133), in which the king praises his doctor for supererogatory deeds, who is numbered among the king's "friends" (*philoi*), an elite Hellenistic title (Mastrocinque 147–149; *cf.* Welles #44; Sherwin-White 1978: 131–132).

Apollophanēs was not content with merely political power: as an **Erasistratean** (Caelius Aurelianus, *Acute*, 2.173, 175 [pp. 250, 252 Drabkin; *CML* 6.1.1, pp. 248, 250]), Apollophanēs would certainly have been a royal rival to Andreas, and equally interested in fashioning drugs and antidotes for use at court, including a cooling emollient plaster suitable for use against burns (2.136 [p. 222 Drabkin; *CML* 6.1.1, p. 224]). An anodyne salve for flank-pains (*ad laterum dolores*) consisted of mastic, powdered frankincense, four ounces of myrrh, four ounces of ***ammōniakon***, four ounces each of mistletoe and the kidney-fat from a calf or a goat. In recording this recipe, Celsus (5.18.6) says it "softens calluses, eases all sorts of pain, and is only slightly warming." Galēn, quoting Arkhigenēs in *CMLoc* 8.9 (13.220 K.), registers a similar Apollophanean compound, but augmented with ground-down iris bulb. For hemorrhoids, he prescribed suppositories fashioned from antimony-ore, acacia-gum, **verdigris**, the latex of the opium poppy, frankincense, myrrh, and the gum from Babul acacia (*Acacia arabica* Lam.), softened with salted wine (Galēn, *CMGen* 5.11 [13.831 K.]). Apollophanēs' pharmacology retained its value well into the Byzantine era, illustrated by re-quotations (from Galēn and Oreibasios) of the anodyne salve, now employed for pains in the liver (Alexander of tralleis, *Liver Inflammations*, 1 [2.386 Puschm.]; Aëtios of Amida 13.18 [p. 617 Cornarius], Paulos of Aigina, 7.18.20 [*CMG* 9.2, p. 373]; *cf.* 3.46.6 [9.1, p. 253]). Appropriately enough, Apollophanēs had his own version of a theriac, useful against snakes, scorpions, and poisons, of unknown composition (Pliny 22.59; *Scholia on* Nikandros, *Thēriaka*, 491 [p. 197 Crugnola]). The strangely corrupted text in Galēn, *Antid.*, 2.14 (14.183 K.) says that Antiokhos had a theriac made from snake-meat, which might be a faint echo of Apollophanēs' theriac and its main ingredient. (See also Pliny 20.264 and Antiokhos VIII.)

RE 2.1 (1895) 165–166, M. Wellmann; C.B. Welles, *Royal Correspondence in the Hellenistic Period* (1934; repr. 1974); Walbank 1 (1957) 584–585; T.S. Brown, "Apollophanes and Polybius, Book 5," *Phoenix* 15 (1961) 187–195; A. Mastrocinque, "Les médecins des Séleucides," in Ph. van der Eijk *et al.*, edd., *Ancient Medicine in its Socio-Cultural Context*, 2 vols. (1995 [*Clio Medica* 27]) 1.143–151.

John Scarborough

Apsurtos of Klazomenai (*ca* 150 – 350 CE?)

Wrote on the care and medical treatment of horses, donkeys, mules, and cattle; probably the most influential veterinary author of Late Antiquity. According to the *Souda* (A-4739), Apsurtos was a native of Prousa or Nikomēdeia; however, evidence in his text suggests he was from Klazomenai. Apsurtos explains in his preface (which dedicates the work to an unknown Asklēpiadēs) that he was a soldier, a statement corroborated by his frequent references to elements of military life, and to Thrakian and Sarmatian horses and horsemen encountered on the Danube frontier. The treatise takes the form of a collection of letters purporting to answer questions from Apsurtos' friends and acquaintances. The letters are addressed to over 60 individuals, including soldiers of various ranks and more than 20 horse-doctors. Apsurtos' letters are preserved in the *Hippiatrika*: they served as the compilation's armature, onto which excerpts from other texts were added. Apsurtos' treatise predates that of Theomnēstos (early-mid 4th c.), in which it is quoted: there is no solid evidence for a more precise dating, although Björck proposed 150–250 CE. Apsurtos cites written sources including "Magōn of Carthage" (presumably Cassius Dionusios' reworking of Magōn's agricultural manual), Eumēlos, and others; he also presents recipes found in texts on human medicine. Twenty-one spells appear under Apsurtos' name in the M recension of the *Hippiatrika*; it is unclear whether these were excerpted from the veterinary manual or from a separate book of magical cures. Also preserved in the *Hippiatrika* is the preface to a treatise or chapter on cows. Passages from Apsurtos appear in Latin translation in the Mulomedicina Chironis, which contains moreover two chapters attributed to Apsurtos not extant in Greek. The Latin author Vegetius seems to have known this translation, but Pelagonius, who used Apsurtos both as a source of content and as a model of literary style, worked from the Greek text. Hieroklēs also used Apsurtos, reworking the text into higher style. Apsurtos' name appears among the false attributions in Geōponika 16, and also (along with Simōn and Xenophōn) in the title of the collections of hippiatric texts in three late MSS.

CHG vv.1–2 *passim*; Björck (1932) 64–87; *Idem* (1944); *BNP* 1 (2002) 916 (#2), A.-M. Doyen-Higuet; McCabe (2007) 122–155.

Anne McCabe

Apuleius Celsus of Centuripae (*ca* 20 – 40 CE)

SCRIBONIUS LARGUS (§94) names Apuleius Celsus as his mentor alongside TERENTIUS VALENS, his fellow-apprentice in medicine and pharmacology. Celsus was born and practiced in the Sicilian town of Centuripae (§171), a city on a hill south-west across a valley facing Mt. Aetna, where Celsus was known for his ***rabies*** cure, "since there are numerous rabid dogs in Sicily." The drug became famous in Crete, after a shipwreck and a successful cure, but ZŌPUROS OF GORTUNA countermanded a better and pricier drug fashioned from hyena-hides (§172). Scribonius (or his teacher) replied he would await an opportunity to verify for himself the effectiveness of such an exotic ingredient.

The "cure" principally entailed waiting a month to ensure that the victim lost his fear of water, since most bite victims would not necessarily have rabies. Meanwhile, the "Antidote of Celsus," (§173) prepared in bulk, was administered to treat common ailments such as diarrhea, runny eyes, and colic, and even as an antidote for snake bites. The 19-ingredient compound and its preparation display the complex pharmacological technology characteristic of recipes in Scribonius: it consisted of three parts each of Syrian nard, saffron, myrrh, ***kostos***, cassia, cinnamon, camel-grass oil (*Cymbopogon schoenanthus* [L.] Spreng.), "white" pepper, "long" pepper, beaver castor, **galbanum, mastic**, and the latex of the opium poppy; two parts of "white" henbane seeds and flowers; one part of anise seeds; six parts each of celery seeds and tragacanth gum – all to be mixed with Attic honey and Falernian wine. The tragacanth gum and opium latex were to be steeped in wine for a day before assembly; and on the following day, the remaining substances were crushed and mixed with honey; meanwhile, the **galbanum** and **mastic** were heated over charcoal, reduced to a hot powder, and some honey added to ensure a waxy consistency; the mixture was boiled until it attained a saffron-yellow color; then the ingredients steeped in wine were added. "The antidote is put away for storage in a glass container;" and a small quantity administered with water.

The "Antidote of Celsus" required a series of complicated and quite technical pharmacological stages for its manufacture (impressive in themselves), and the compound would have been extraordinarily expensive (due to the saffron and beaver castor); the inclusion of henbane suggests an anesthetic property, as does opium, and beaver castor (naturally rich in salicylates) combined with mildly analgesic celery seeds suggest that the "Antidote of Celsus" was given as a month-long calming pill that could enable a much-worried patient to wait out the 30 days.

RE 1.2 (1894) 259 (#20), P. von Rohden.

John Scarborough

L. Apuleius of Madaurus (150 – 170 CE)

Platonic philosopher and writer belonging to the Second Sophistic, born *ca* 125 CE to a wealthy family, and educated in Carthage (*Florida* 18) and Athens (*Apol.* 72). Apuleius traveled widely, visiting Samos, Phrugia, and Rome (*Flor.* 15.49; 17.77). He was initiated into the mysteries of Isis. In the winter of 156, on his way to Alexandria, at Oea (mod. Tripoli) he married his friend Pontianus' mother Pudentilla, whose relatives accused him of seducing her by magic. His *Apologia* (*De magia*), our only source for this episode, recounts the trial at Sabratha in 158 or 159. In the 160s he delivered epideictic speeches and philosophical lectures in Carthage, received a statue thanks to his friend Aemilianus Strabo, and possibly was made a priest of the imperial cult in Africa. Two of his works are dedicated to his son Faustinus.

Metamorphoses (*The Golden Ass*) is Apuleius' best known work. The rhetorical *De deo Socratis* analyzes Sōcratēs' ***daimonion***. The authenticity of two other works, *De Platone et eius dogmate* and *De mundo*, is questioned, though not conclusively. *Peri Hermeneias* (in Latin, despite its title), a work on **Peripatetic** logic citing also THEOPHRASTOS and Aristōn of Alexandria (*ca* 70 BCE), is generally considered spurious. Certainly spurious are *Asclepius* (a HERMETIC treatise), *Herbarius* (see next entry), *De remediis salutaribus* and *Physiognomonia*. Among the significant lost works are *Quaestiones naturales* in Greek; a work on ichthyology, presumably adapted from works by ARISTOTLE, Theophrastos, EUDĒMOS OF RHODES and LUKŌN OF ALEXANDRIA TROAS (*Apol.* 36); *De arboribus* and *De medicinalibus* (*frr.*14–17); a work on astronomy and meteorological *miracula* (*frr.*22–25); a translation of NIKOMAKHOS OF GERASA (CASSIODORUS *Inst. div.* 2.4.7); *De republica* (*fr.*13); a translation of PLATO's *Phaedo* (*frr.*9–10); a handbook on music (Cassiod., *Inst. div.* 2.5.10).

Apuleius inserted into his *De mundo*, a free translation of ON THE KOSMOS, FAUORINUS' account of the winds (13–14: based on Aulus Gellius 2.22). *De Platone et eius dogmate* is an arid and superficial exposition of **Platonic** physics and ethics, supplementing Plato's medical views with Aristotelian elements. The *Apologia* contains sections on the mechanism of vision (15) and ichthyology (29–36; 40–42).

Apuleii fragmenta, in W.A. Oldfather, H.V. Canter and B.E. Perry, *Index Apuleianus* (1934) IX–XIII; *DPA* 1 (1989) 294–297, J.-M. Flamand; Dillon (1996) 306–340; *OCD3* 131–132, S.J. Harrison; *Idem*, *Apuleius. A Latin Sophist* (2000); *BNP* 1 (2002) 905–909, M. Zimmermann.

Jan Opsomer

Apuleius, pseudo, Herbarius (500 – 530 CE)

About 20 Latin MSS in two distinct families have herbals under names Apuleius Platonicus, ANTONIUS MUSA, and SEXTUS PLACITUS, with an Anonymus occasionally but inconsistently attached. An ancestral archetype (what Howald and Sigerist call *Herbarius pseudo-Apulei genuinus*) seems to have emerged sometime in the first decades of the 3rd c., soon augmented by a synonym-list in the mid 4th c., a hypothetical *Herbarius synonymis aliis addit. auctus*. About 500 CE, this text becomes conjoined with three more herbals, one presumably authored by a Latinated PSEUDO-DIOSKOURIDĒS (the *De herbis femininis*), another pseudo-Apuleius, and an otherwise unknown Sextus Placitus. In turn, *ca* 600 CE, this collection splits into two separate traditions, with various MSS omitting some passages, others supplemented with apparent borrowings from MARCELLUS' *De medicamentis* of about 400 CE, or yet other texts attributed to Antonius Musa, or still others without a definite author. Intertwining are echoes of PLINY's *Natural History*, and the MSS edited by Howald and Sigerist bear some affinities to similar traditions known as the MEDICINA PLINII and the PHYSICA PLINII, but were transmitted independently from them.

"Antonius Musa" prefaces his *De herba uettonica liber* with an obviously confused *Antonius Musa M. Agrippae salutem*, followed by *Caesari Augusto, praestantissimo omnium mortalium, sed et iudicibus proximum esse auxilium quodque artis meae carissimum Caesaris iudicio* (*CML* 4, p. 4); in the succeeding lines (p. 5, esp. line 30), the editors signal a parallel to Marcellus, 30.106, and the first of the very clipped 47 chapters (*Ad capitis fracturam*) is a Latin translation of the Greek in DIOSKOURIDĒS 2.170. Many of the MSS carry fictive illuminations, with only a few remotely resembling the plants under discussion. The *Pseudo-Apulei Platonici herbarius* begins with the fanciful *quem accepit a Cirone centauro, magistro Achillis, et ab Aesculapio,* then comes the obligatory *Apuleius Platonicus ad ciues suos* (*CML* 4, p. 15); this *Herbarius* has numerous illuminations throughout its 131 chapters (*CML* 4, pp. 22–225), with (e.g.) 1.1–24 single lines (or sometimes two) suggesting for what *herba plantago* might be used (the artist got the tassels right, but has ballooned the leaves beyond recognition: *CML* 4, p. 22). Surprising is the folklore-free chapter 131, *Effectus herbae mandragorae* (*CML* 4, p. 222), which is as sober as the account in ISIDORUS OF HISPALIS, *Etym.* 27.9.30, or the pseudo-Dioskouridean *De herbis femininis* 15 (599–600 Kästner). The next page (folio?), however, provides a classic illustration of the man-like *mandragora,* bound at the ankles with a rope extending to a dog's collar (the dog is stretching the rope as it sniffs a ball-like object, presumably a culinary reward for pulling up the mandrake roots). Howald and Sigerist believe the mandrake is an interpolation (*CML* 4, p. XVIII).

The very brief and anonymous *De taxone* (*CML* 4, pp. 229–232) purports to be a letter to AUGUSTUS (version No. 2 [the text is in two columns]: *Partus rex Aegyptiorum Octauio Augusto salutem*), and contains some generalities about remedies derived from badgers (or perhaps "bacon-fat" [Isid., *Orig.* 20.2.24]). There are no illuminations, and one can doubt if this tiny text has an Egyptian pedigree. And the *Liber medicinae Sexti Placiti Papyriensis ex animalibus pecoribus et bestiis uel auibus* (*CML* 4, pp. 235–298) carries no pictures, and the 130 chapters are closer to the Pliny-excerptors than the remainder of these texts; the editors adduce frequent parallels to Pliny, often gained through Marcellus. Much *Dreckapotheke* is interwoven, e.g. 5: *De capra*, 24: *Ad luxum* (*CML* 4, p. 255): *Idem stercus facit ad luxum et tumores discutit et non patitur postmodum consurgere*, or 26 (p. 256): *Ad carbunculos: idem stercus caprae cum melle conmixtum et superpositum carbuncolos, qui in uentre nascuntur, discutit.* The author suggests a mix of goat manure, beaver-castor, myrrh, and honey, and made into little **pastilles** and inserted into the mouth of the womb as an excellent abortifacient (5.41: *Ad aborsum* [*CML* 4, pp. 255–256]).

Ed.: H.F. Kästner, "Pseudo-Dioscoridis De herbis femininis," *Hermes* 31 (1896) 578–636; E. Howald and H.E. Sigerist, *Antonii Musae De herba vettonica liber. Pseudoapulei herbarius. Anononymi De taxone liber. Sexti Placiti Liber medicinae ex animalibus etc.* (1927) = *CML* 4.

H.E. Sigerist, "Zum Herbarius Pseudo-Apuleius," *Sudhoffs Archiv* 23 (1930) 197–204; *Idem*, "The Medical Literature of the Middle Ages," *BHM* 2 (1934) 26–52; *Idem*, "Materia Medica in the Middle Ages," *BHM* 7 (1939) 417–423; L.E. Voigts, "The Significance of the Name Apuleius to the *Herbarium Apulei*," *BHM* 52 (1978) 214–227; J.M. Riddle, "Pseudo-Dioscorides' *Ex herbis femininis* and Early Medieval Medical Botany," *JHB* 14 (1981) 43–81; Önnerfors (1993) 318.

John Scarborough

Aquila Secundilla (10 BCE – 95 CE)

ASKLĒPIADĒS twice cites her: for a **terebinth**-based ointment containing myrrh, gentian (*cf.* GENTHIOS), sulfur, and white pepper, GALĒN, *CMGen* 7.6 (13.976 K.), and for an ***akopon*** potion containing myrrh, ***euphorbia*** (*cf.* IOUBA), ***malabathron***, etc., *ibid.* 7.12 (13.1031 K.). The name Aquila/*Akulas* is first attested in the mid-1st c. BCE: Suetonius, *Iulius* 78.2; *LGPN* 3A.23, 3B.21; *PIR2* A-979.

Parker (1997) 145 (#48).

PTK

Arabic Translations (of Greek scientific works not extant in the original)

Numerous Greek medical, scientific, and philosophical works available in Late Antiquity were translated into Arabic in the 8th–11th cc. An unknown number of these came by way of Syriac and, to a lesser degree, Pahlavi Translations. In Arabic tradition, these originally Greek works were much used, commented upon, cited, criticized, corrected, and built upon in continuity of scientific tradition. Many Arabic authors explicitly tell us that they saw themselves as continuing the sciences of the ancients in their research.

The Greek authors whose works had the greatest impact and were the most numerous in Arabic translation are those studied most extensively in Late Antique schools and hence most easily available and recently copied. Foremost were Aristotle as interpreted by the Neo-**Platonists** and Galēn as systematized in a 6th c. Alexandrian medical curriculum. On the other hand, because the reception of Greek science and philosophy in Arabic occurred at the end of Late Antiquity, the earliest Greek philosophers' views (such as those of Thalēs and Pythagoras) are mostly transmitted in Arabic in gnomologia and doxographies, just as in Greek and Latin tradition.

Among extant translations are many not surviving in their original Greek. Therefore Arabic MSS contain genuinely ancient scientific works of great interest to historians of ancient Greek science, even to those who have no concern whatever with Arabic tradition as such. Examples of texts surviving only in Arabic include Proklos' first proposition on the eternity of the world and Alexander of Aphrodisias' refutation of Galēn's criticisms of Aristotle's physics, as well as many other ancient traditions. In the area of doxography, too, Arabic works preserve doctrines of the ancients not preserved elsewhere: for example, 18% of the sayings of Diogenēs the Cynic survive only in Arabic.

Listed here are Greek scientific authors *some* of whose known works survive *only* in Arabic in part or in their entirety, either complete or fragmentary (citations). This incomplete list of 38 provides a fair picture of authors or works surviving only in Arabic, in fragments or more: Agathodaimōn (Alch.), Alexander of Aphrodisias, Apollōnios of Pergē, pseudo-Apollōnios of Tuana, pseudo-Archimēdēs, pseudo-Aristotle, Bōlos, Dioklēs (Math.), Diophantos, Dōrotheos of Sidōn, "Dtrums," Euclid (five works), Eutokios, Galēn (at least 27 works), Hermēs Trismegistos, Hērōn, pseudo-Hippokratēs, Iōannēs Philoponos, Kritōn of Hērakleia, Menelaos, Nikolaos of Damaskos, Olumpiodōros of Alexandria, Palladios, Pappos, Paulos of Alexandria, Paulos (Music.), Porphurios, Proklos, Ptolemy, Rufus of Ephesos, pseudo-Sōkratēs, Thalēs, Themistios, Theodosios (on burning mirrors), Theomnēstos of Magnesia, Theophrastos, Vindonios Anatolios, and Zōsimos of Panōpolis.

Most of these Greek works surviving only in Arabic have not yet been studied in detail. In some cases (as with pseudo-Sōkratēs), Arabic works are clearly forgeries attributed to ancient authorities. But even then, it is often unclear whether these counterfeits are faithful translations from lost Greek forgeries or pseudepigraphs composed in Arabic.

Excluded from the list above are numerous other Arabic translations of texts surviving in the original Greek that are not "scientific" by the criteria used in this volume, numbering in the hundreds. Research in Arabic literature is still in the early stages of providing a complete account of this entire Greco-Arabic tradition. There are millions [*sic*] of Arabic MSS extant today, found from western Africa to south-east Asia, not to mention in European and American collections. This ocean of MSS is partly uncatalogued, and many of the catalogued MSS are imperfectly registered, so that more Arabic translations of hitherto unknown Greek works will surely be discovered. Yet it is already clear that in the fields of science, medicine, and philosophy, Arabic far surpasses ancient and Medieval Latin in importance as a successor to Greek tradition.

Arabic translations of extant Greek originals must be considered witnesses to the Greek text and used in editions of the Greek. Often the Arabic translations were made from MSS centuries older than the oldest extant Greek MSS of a given work. Some of the works translated into Arabic from ancient Greek were later translated from Arabic into other languages including Persian, Hebrew, Latin, and even Byzantine Greek, adding further witnesses to the ancient texts.

For more complete coverage of the Greco-Arabic translations of scientific works, one may consult the introductory bibliography given below.

GAS (especially v. 3 *Medizin, Pharmazie, Zoologie, Tierheilkunde*, v. 4 *Alchimie, Chemie, Botanik, Agrikultur*, v. 5 *Mathematik*, v. 6 *Astronomie*, v. 7 *Astrologie, Meteorologie und Verwandtes*, vv. 10–12 *Mathematische Geographie und Kartographie*); Ullmann (1970) and (1972); G. Endress, "Die wissenschaftliche Literatur," in *Grundriss der arabischen Philologie* (1987–1992) 2.400–506, 3.3–152; D. Gutas, "Pre-Plotinian Philosophy in Arabic (Other than Platonism and Aristotelianism): A Review of the Sources" in *ANRW* 2.36.7 (1994) 4939–4973; *Idem, Greek Thought, Arabic Culture* (1998); *Encyclopedia of the History of Arabic Sciences* 3vv. (1996) ed. R. Rashed; H. Daiber, *Bibliography of Islamic Philosophy*, 2 vv. (1999).

Kevin van Bladel

Arabas of Thēbai ⇒ Arrabaios

Arabianus ⇒ Arrabaios

Aratos of Soloi (Kilikia) (290? – 240 BCE)

Aratos (inv. 1968.244.38) Courtesy of the American Numismatic Society

Born around 300, Aratos was the son of Athēnodōros and Letophila. He had three brothers among whom was Athēnodōros II, disciple of Zēnōn of Kition. Aratos attended the lessons of Zēnōn with him in Athens and came under the influence of **Stoicism**. Aratos also had good relationships with the philosopher Menedēmos of Eretria who warmly received other poets for the purpose of discussion of literary questions. Aratos lived under the reigns of both Ptolemy II Philadelphos and Antigonos Gonatas and was called to the court of the latter in 276. Antigonos Gonatas is even said to have invited Aratos to compose his *Phainomena* by giving him a copy of Eudoxos' prose treatise on the same subject: this anecdote, highly improbable and legendary, at least suggests that Aratos wrote his *Phainomena* after 276 in Pella, even if he already thought about it in Athens. In Pella, he also may have had close relationships with the **Stoic** Persaios, the tragedian Antagoras of Rhodes and the poet Alexander of Pleuron. He also could have been a disciple of the astronomer Aristotheros. Finally Aratos is said to have been in contact with Dionusios of

Hērakleia, another disciple of Zēnōn, either as a disciple or as a master, but surely as a friend.

Indeed Aratos' work was varied. He not only produced an edition of the *Odyssey* but he also composed a study on the *Iliad* at the request of Antiokhos I; he wrote epigrams, funeral laments (*Epikedia*), hymns (e.g. the *Hymn to Pan*, perhaps composed for Antigonos' victory at Lusimakheia/Lysimachia in 277) and even several scientific treatises like the *Canon* on the planets or others in the areas of pharmacology or anatomy. All of this work is more or less lost today.

Thus, it is as the author of *Phainomena* that Aratos is known to us and achieved fame (as shown by several translations into Latin at different times). This long poem begins with a Hymn to Zeus as Proem in the manner of the **Stoic** Kleanthēs. The first part (19–461) presents the constellations as well as the method to recognize them. The second part (462–757) deals with the passage of time and how to estimate it by observing the constellations as well as the moon and the sun. After these astronomical subjects, the poet then turns to meteorology (758–1141) and explains local weather signs which are observable in natural phenomena and the behavior of animals. The poem ends with a short conclusion (1142–1154).

By putting into verse such an unpoetic subject, Aratos tries to emulate Hēsiod and to bring up to date the early calendar of *Works and Days*. Aratos is indeed borrowing from Hēsiod the shape of his poem and is rewriting in his own style the myth of ages. Moreover, Kallimakhos (*AP* 9.507) considered him to be a new Hēsiod due to this influence.

Ed.: D. Kidd, *Aratus. Phenomena* (1997); J. Martin, *Aratos. Les Phénomènes* 2 vv. (*CUF* 1998).
OCD3 136–137, G.J. Toomer; *BNP* 1 (2002) 955–960, M. Fantuzzi.

Christophe Cusset

Arbinas of Indos (Lukia) (120 BCE – 80 CE)

Andromakhos, in Galēn *Antid.* 2.1 (14.109–111 K.), recites his 56-ingredient antidote and abortive, including birthwort, gentian, ginger, poppy-juice, mandrake, pepper, and wild rue seeds. Aëtios twice cites an unnamed doctor of Indos, presumably the same man, for much simpler remedies, 9.49 (p. 556 Cornarius) and 11.11 (p. 608 Cornarius). Kühn prints *OPBAN-*, a rare name (*LGPN*), but probably we have to do with the Lukian name Arbinas, *cf.* *Cambridge Ancient History*, 2nd ed., v. 6 (1994) 214–215 and *SEG* 28.1245.

Fabricius (1726) 254, 451.

PTK

Arbitio (350 – 400 CE?)

Wrote in Latin a geographical work on Roman possessions, cited by the Ravenna Cosmography: 4.3, 4.6–7, 4.9. The name is attested *ca* 350–400 CE: *PLRE* 1 (1971) 94–95. *Cf.* Castorius, Lollianus, and Marpēssos.

(*)

PTK

Arc- ⇒ Ark-

Archimēdēs of Surakousai (*ca* 250 – 212 BCE)

The most important scientist of antiquity.

1. Biographic Evidence. Archimēdēs' death is dated, most securely on POLUBIOS' authority (Book 8, *frr.*4–6, 37), to the fall of Surakousai in the Second Punic War; he likewise describes the scientist then as *presbutēs*, old. (The often repeated statement that he was 75 when he died is based on the worthless authority of the Byzantine poet Tzetzēs.) Archimēdēs refers, in the introductions to several of his major treatises, to the death of KONŌN, known to have been alive in 245 (when Konōn named the "Lock of Berenice"). It follows that at least some of Archimēdēs' major works were likely written during the 230s–220s. Knorr (1978) offers the most ambitious attempt to offer a chronology of Archimēdēs' life and works.

EUTOKIOS refers (p. 228 H.) to an ancient *Life of Archimēdēs* written by a certain Hērakleios or HĒRAKLEIDĒS. Since Archimēdēs' introduction to *Spiral Lines* mentions an associate named Hērakleidēs, it is likely that this *Life* was written by a knowledgeable, if partisan, author. While none of this work survives save for the two comments quoted by Eutokios (both of a strictly mathematical significance), the possibility remains that at least some of Archimēdēs' ancient biographical tradition stems ultimately from such a reliable source. It remains impossible to say which parts of the tradition are fictional, and which are historical. The familiar stories (such as that of the forged crown problem solved in the bath, or launching the giant ship while uttering "give me where to stand and I shall move the earth"), having the ring of legend, derive from late authorities.

Archimēdēs refers in the *Sand-Reckoner* to an astronomer, "PHEIDIAS, my father." Two onomastic comments suggest themselves. (a) From the late-5th c. onwards, "Pheidias" was a name traditionally given in artistic and artisanal families, (b) the name "Archimēdēs" is effectively a *hapax*, apparently modeled on "Diomēdēs," or "the mind of Zeus," meaning roughly "the mind of the *Archē*." This clearly suggests a religion motivated by **Platonic** or **Stoic** metaphysics. The sum total of our evidence is that Archimēdēs' grandfather was likely an artist, his father an astronomer and follower of contemporary, **Platonic** or **Stoic**, currents of metaphysical thought.

Reliable historical evidence on Archimēdēs' death portrays him as heroically, to some extent single-handedly, and very effectively, contributing to the defense of Surakousai, through the construction of original war engines. Nothing in the extant corpus bears on the problem of war engines, probably the simple product of necessity. Mention of the sands of Sicily (in the introduction to the *Sand-Reckoner*), as well as the choice of setting for the *Cattle Problem*, both suggest a patriotic devotion consistent with his wartime conduct.

2. Bibliographic Evidence. As mentioned above, some works ascribed to him start with a letter of introduction, whose first words typically are "Archimēdēs to X, greetings." The sober, mathematical character of those letters suggests authenticity. If so, several extant works are definitely authentic (titles, however, always unreliable, are provided here for reference, followed by the abbreviation to be used below): *Sphere and Cylinder I* (SC I) and the independent but closely-related *Sphere and Cylinder II* (SC II); *Spiral Lines* (SL); *Conoids and Spheroids* (CS); *Quadrature of Parabola* (QP); *Sand-Reckoner* (Arenarius); *Method* (Meth.). The first five – SC I, SC II, SL, CS, QP – all addressed to DŌSITHEOS, can be seen as the core of Archimēdēs' achievement.

The Greek manuscript tradition further includes the following works: *Measurement of the Circle* (DC); *Planes in Equilibrium* I, II (PE I, II) (two rolls of one work); *Floating Bodies* I,

II (CF I, II) (also a single work divided into two rolls); *Stomakhion* (Stom.); *Cattle Problem* (Bov.).

DC contains obvious mistakes, alongside some inspired mathematics, and it is often assumed that the extant version is a corruption of an original work by Archimēdēs (Knorr 1989, part 3). Similar doubts were raised concerning PE I (Berggren). Still, the presence of Doric dialect in PE, CF, as well as some ancient testimonies connecting the contents of DC, Stom., and Bov. to Archimēdēs, make us believe that indeed all the works extant in Greek are by Archimēdēs himself – even if in corrupt form. To this should be added the treatise *On Polyhedra* (Poly.) which Pappos (*Coll.* 5.19, pp. 2.352–358 H.) describes in detail.

Ancient testimony mentions several more works (in a few cases, inside the works of Archimēdēs himself), but evidence is meager and allows no firm conclusions. The mention of a work on Optics, though, is especially intriguing, as this field – otherwise not very well attested in Archimēdēs' time – would provide ample opportunities for Archimēdēs' genius in mathematical physics (Knorr 1985 connects Archimēdēs and the *Catoptrics* ascribed to Euclid).

Many Arabic treatises are ascribed to Archimēdēs, but most go beyond the Greek corpus. Four of these are usually taken to be authentic (even if in a more or less mediated form: Sesiano): *Construction of the Regular Heptagon* (Hept.); *Tangent Circles* (Tan.); *Lemmas* (Lem.); *Assumptions* (Assum.).

With the exception of Bov. (surviving through collections of epigrams), all Greek works are transmitted through one or more of three early Byzantine MSS. One of these, a collection of various mechanical works, only some by Archimēdēs, was lost and is known only through Moerbeke's Latin translations of PE and CF, partly based on this MS. The two remaining MSS seem to form a kind of "collected works of Archimēdēs" (Medieval scientific MSS are arranged typically by subject matter rather than author, so that those "collected works" are an exception testifying to Archimēdēs' stature). One of these two, containing SC I, SC II, DC, CS, SL, PE I-II, Aren., QP, became lost in the Renaissance, after serving as another source for Moerbeke's translation as well as for numerous copies. The second, "The Archimēdēs Palimpsest," contains PE II (end only), CF I-II, Meth., SL, SC I, SC II, DC, Stom. (beginning only). This 10th c. Byzantine MS, turned in the 13th c. into a prayer book, was rediscovered by Heiberg in 1906. Again lost, it resurfaced in an auction in 1998 and has been recently the subject of intensive study, giving rise to notable changes in the received text.

3. Major Areas of Discovery. The most significant body of Archimēdēs' work concerns measuring curvilinear objects, based on the technique of "exhaustion," which prefigures the calculus (SC I, SC II: sphere; CS: conoids of revolution; SL: spirals, QP: parabolic segment, Meth.: various figures). In the most general terms, the method of exhaustion works as follows (Fig.). The curvilinear object is bound (from above, below, or both) by a complex rectilinear object whose difference, or ratio, of volume or area, from the given curvilinear object, can be made indefinitely small. Typically, this involves dividing the curvilinear object into an indefinitely large number of sections, each of which is circumscribed or inscribed by a respective section of the complex rectilinear object. Certain measurements are then made for the rectilinear objects and, based on these measurements, one shows through contradiction that the curvilinear object must possess the specified measure (or else for instance it can be made *smaller* than a rectilinear object it *circumscribes*, etc.). Euclid had already applied the same technique for the measurement of the cone (*Elements* 12.10) – a result ascribed to Eudoxos of Knidos, partly on the authority of Archimēdēs himself (in the introduction to Meth.). Archimēdēs has put his signature on this technique, through his wide ranging and elegant application of it. It was following on his lead that modern mathematicians strove to transform this technique into the project of measuring, systematically, all curvilinear objects – a project giving rise to the calculus.

In both PE and CF, Archimēdēs applies mathematics rigorously to the physical world. Starting from simple assumptions – e.g. that equal weights balance at equal distances (PE) or that the columns of a liquid at rest all press down with equal force (CF) – he derives by pure logic the main principles of Statics and Hydrostatics, respectively – Archimēdēs' Laws of the Lever and Buoyancy. From these, Archimēdēs derives special results such as the finding of centers of weight, ultimately deriving complex results of a more geometrical character, once again having to do with curvilinear objects: e.g. the center of weight of a parabolic segment (PE II) and the hydrostatic properties of certain conoids of revolution (CF II). The rigorous application of mathematics to physics appears to have been original to Archimēdēs and served as major inspiration to the scientific revolution.

Meth. combines Archimēdēs' interest in measuring curvilinear objects with mathematical physics, in subtle and surprising ways. In most propositions of this treatise, plane figures (or solid figures) are sliced by parallel lines (or parallel planes), and results are obtained for the center of gravity of each of the resulting linear segments (or planar segments). Those results are then summed up as a result for the center of gravity of the plane or solid figure as a whole, giving rise to its measurement. This technique prefigures Cavalieri's indivisibles (1635). Unfortunately this treatise was discovered only with the appearance of the Palimpsest in 1906 and so did not contribute to the scientific revolution. In 2001, a new reading revealed Archimēdēs' use of actual infinity, in the course of applying proportion theory to a geometrical arrangement involving indivisibles (Netz, Saito and Tchernetska). This appears to be unique in the extant Greek corpus.

While Archimēdēs' most remarkable achievements are qualitative in character (in either pure geometry or in mathematical physics) many of his works involve detailed calculation. His bounds for the value of π (to use the modern notation), $3^1/_7 \geq \pi \geq 3^{10}/_{71}$, are obtained in DC based on a whole set of numerical results, including an approximation of $\sqrt{3}$. Aren. states the number of grains of sand it takes to fill up the universe; Bov. is a staggeringly difficult numerical puzzle; Poly., at least in the form reported by Pappos, appears to have been primarily a numerical study in the faces, edges and vertices of solids; finally, it has been suggested recently that Stom. formed a study in geometrical combinatorics, counting the number of ways in which a certain jigsaw puzzle can be put together (Netz, Acerbi and Wilson).

4. Scientific Personality. The subject matters chosen by Archimēdēs all revolve around the surprising: curved objects are equal to straight ones; physical objects obey geometrical laws; apparently impossible calculations are executed. Such results are always shown through elegant and surprising routes. In a typical work, Archimēdēs builds up an arsenal of apparently unrelated results which then unexpectedly combine to yield the main result of the treatise. No allusion is ever made, within the works themselves, to any extra-mathematical interests, and it appears that Archimēdēs saw himself (at least in his persona of a scientific author) as a pure mathematician, dedicated to the pure pursuit of proofs.

Archimēdēs' *manner* of proof is more difficult to study. While a substantial part of Archimēdēs' work does survive, the works as transmitted contain what appears to be obvious later glosses, in places quite substantial. It is thus to some extent a matter of conjecture to say which of the text is by Archimēdēs and which by later commentators, considerably limiting our ability to judge Archimēdēs' scientific personality. Assuming most of those apparent glosses are indeed late, the emerging personality is that of a very precise, and yet somewhat impatient author. While no mistakes are ever made, and careful attention is given to many subtle points of logic, Archimēdēs can be quite cavalier about details he considers obvious. (This, indeed, may be the reason why later readers felt the urge to add in their glosses.)

In many of the introductions to his works, Archimēdēs mentions previously sent "enunciations," apparently challenges distributed so as to test Archimēdēs' contemporaries. In one case (the introduction to SL) he specifically mentions a false enunciation, i.e. one meant to tease his contemporaries into proving a falsehood. The overall tone of the introductions is of supreme self confidence. In the very choice of scientific subject matter, in the spirit of "intellectual tournament" sustained through his correspondence, and finally in the subtle and yet cavalier manner of his writing, Archimēdēs radiates a consistent persona – subtle, self-confident, playful.

J.L. Berggren, "Spurious Theorems in Archimedes' Equilibrium of Planes, Book I," *AHES* 16 (1976–1977) 87–103; W.R. Knorr, "Archimedes and the Elements: Proposal for a revised Chronological ordering of the corpus," *AHES* 19 (1978) 211–290; *Idem*, "Archimedes and the pseudo-Euclidean Catoptrics," *AIHS* 35 (1985) 28–105; *Idem* (1989); J. Sesiano, "Un Fragment attribué à Archimède," *MH* 48 (1991) 21–32; Reviel Netz, K. Saito, and N. Tchernetska, "A New Reading of Method Proposition 14: Preliminary Evidence from the Archimedes Palimpsest," *SCIAMVS* 2 (2001) 9–29, 3 (2002) 109–125; Reviel Netz, F. Acerbi, and N.W. Wilson, "Towards a Reconstruction of Archimedes' Stomachion," *SCIAMVS* 5 (2004) 67–99; *NDSB* 1.85–91, F. Acerbi.

Reviel Netz

Areios Didumos (100 BCE – 200 CE)

A number of passages on **Peripatetic** and **Stoic** ethics and on **Stoic** physics in the church father Eusebios and in Iōannēs Stobaios are attributed to Areios, to Didumos or once to Areios Didumos. It is a modern assumption that all these passages refer to one and the same person called Areios Didumos, and that this person was identical with Emperor Augustus' philosophical friend Areios. This identification is far from assured: all we can say is that these texts seem to have been written before the end of the 2nd c. CE. In any case, these excerpts seem to belong to that kind of doxographical literature where the philosophy of each of the schools was dealt with in separate sections on logic, physics and ethics.

Ed.: Fragments of physical doxography in Diels (1879) 447–472; fragments of ethical doxography in A.J. Pomeroy, trans., *Epitome of Stoic Ethics* (1999).

Moraux (1973) 1.259–443; W.W. Fortenbaugh, ed., *On Stoic and Peripatetic Ethics, The Work of Arius Didymus* (1983); D. Hahm "The Ethical Doxography in Arius Didymus," *ANRW* 2.36.4 (1990) 2935–3055; T. Göransson, *Albinus, Alcinous, Arius Didymus* (1995); Mansfeld and Runia (1996) 238–266; *NP* 1 (1996) 1041–1042, D.T. Runia (not in *BNP*).

Jørgen Mejer

Areios of Tarsos, Laecanius (54 – 77 CE)

Dioskouridēs dedicates his *Materia Medica* to an Areios (*Pr.*1), an instructor of medical botany and mineralogy then resident in Tarsos, and likely one of Dioskouridēs' primary mentors in pharmacology. In *CMLoc* 4.8, and 5.3 (12.776 and 829 K.) Galēn ascribes

Asklēpiadean connections to Areios, but given Dioskouridēs' criticism of **Asklēpiadean** theories (*Pr.*2), Areios was not a strict adherent of the sect. He was an accomplished physician and pharmacologist in his own right, and a client of the consular C. Laecanius Bassus (*consul* 64 CE: *CIL* 5.698; TACITUS *Annals*, 15.33; PLINY 26.5 and 36.203). Areios wrote a handbook on pharmacology, known to Galēn through quotations in ANDROMAKHOS' compilation (*CMGen* 5.13 [13.840 K.]), and Dioskouridēs contributed a styptic compound included by Areios in that work (Galēn, *CMLoc* 5.15 [13.857 K.]). He also wrote a *Life of HIPPOKRATĒS*, mentioned by SŌRANOS (*Vita Hipp.* 1 [CMG 4, p. 175]).

If the formulas and recipes quoted by Galēn are representative, Areios was an expert compounder of drugs fashioned from ores and minerals, and Book 5 of Dioskouridēs' *Materia Medica* plausibly reflects this emphasis on medical mineralogy and concomitant technologies of smelting and refining in 1st c. CE Roman metallurgy and mining. Not only would that variety of drugs have indefinite "shelf-lives," many were superb styptics quite suitable for staunching the wounds commonly suffered by soldiers and gladiators, and kindred pharmaceuticals were quite effective in the treatment of the widespread ophthalmic ailments of the day. Ingredients are exemplified in one of Areios' styptic *collyria* and include copper "flakes," iron pyrite, fissile alum, **calamine**, and **verdigris**, compounded with the latex of the opium poppy and thickened with acacia-gum (Galēn, *CMLoc* 4.8 [12.776 K]).

RE 2.1 (1895) 626, M. Wellmann; R. Syme, "People in Pliny," *JRS* 58 (1968) 135–151; J.F. Healy, *Mining and Metallurgy in the Greek and Roman* World (1978) 246 with nn.133–154; Scarborough and Nutton (1982) 198–199, 206–208; R. Syme, "Eight Consuls from Patavium," *PBSR* 51 (1983) 102–124; John Scarborough, "Introduction" to L.Y. Beck, trans., *Dioscorides of Anazarbus De materia medica* (2005) XIII–XXI at XV.

John Scarborough

Aretaios of Kappadokia (150 – 190 CE?)

Greek physician from Kappadokia of uncertain date; PSEUDO-ALEXANDER OF APHRODISIAS, ON FEVERS, quoting Aretaios (16.1, 24.5, 30.1 Tassinari), provides a clear *terminus ante quem*. GALĒN, who does not quote Aretaios explicitly, fails to mention his source in reporting a case of ***elephantiasis*** (*Subfiguratio Empirica* 10 [pp. 75–79 Deichgr.]; *Simples* 11.*pr.* [12.312 K.]), which corresponds to Aretaios, *Morb. Chron.*, 4.13. Galēn mentions that this episode happened when he was young, living in Asia minor (in the 140s): thus Aretaios might have lived during Galēn's lifetime rather than a century earlier as suggested by Kudlien and Oberhelmann.

Aretaios' major work analyzes causes, signs, and therapy of acute and chronic diseases in two groups of four books each (*On causes and signs of acute and chronic diseases*, and *On therapy of acute and chronic diseases*). He composed four other treatises, all lost, three known only through Aretaios' own testimonia: (1) *On Fevers* (*Acute*, 3.*pr.*; possibly cited by ps.-Alexander Aphrodisias, *De Febr.*, 16 and 30); (2) *On Gynecology* (*Acut.*, 3.3); (3) *On Surgery* (*Chron.*, 3.2); (4) *On Preventive Medicines* (?) (*Peri Phulaktikōn*: ps.-Alexander Aphrodisias, *De Febr.*, 24).

Aretaios' theoretical leanings are debated. Although often presented as a **Pneumaticist** strongly influenced by Hippokratic medicine (Kudlien), Oberhelmann's recent re-evaluation shows him more **Pneumaticist** than Hippokratic. Aretaios defined four major categories of diseases according to elemental imbalance (*duskrasia*), those of (1) dry and cold, (2) cold and wet, (3) dry and warm, and (4) warm and wet. Therapy consists, first, in eliminating the excess of ***pneuma*** by bleeding, cupping, and possibly also rubefaction. Simultaneously, the patient is assisted with psychotherapy, physiotherapy, and diet to recover strength and to prepare the body for pharmacotherapy, based on the principle of allotherapy and aiming to restore *eukrasia*. Aretaios' conservative *materia medica* includes animal, vegetable, and mineral products, prepared as in antecedent literature, and administered according to use (internal or external) and the organ to be treated. According to Oberhelmann, **Pneumaticism** in Aretaios is the "orthodox" theory at its zenith, before it evolved toward eclecticism. Aretaios' Hippokraticism seems more literary than medical: like the Hippokratic physicians, he wrote in Ionic Greek.

An abundant Byzantine MS tradition transmitted Aretaios' extant work (Diels 2 [1907] 17–19), first printed in Greek in 1554 (Paris, edition by Jacques Goupyl), and contributing to the development of Morgagni's (1682–1771) anatomico-pathological theory.

Ed.: C. Hude, *Aretaeus*, 2nd ed. = *CMG* 2 (1958); F. Adams, trans., *The extant works of Aretaeus, the Cappadocian* (1856).

RE 2.1 (1895) 669–670, M. Wellmann; *Idem* (1895); Kudlien (1962); J. Stannard, "Materia Medica and Philosophic Theory in Aretaeus," *Sudhoffs Archiv* 18 (1964) 27–53 (reprinted in *Pristina Medicamenta* [1999], #V); Fr. Kudlien, *Untersuchungen zu Aretaios von Kappadokien* (1963); *Idem* (1968) 1098; *KP* 1.529, *Idem; DSB* 1.235–236, *Idem*; A.D. Mouroudes, "Aretaios o Kappadokes; Analutikē bibliografia," *Ellênika* 36 (1986) 26–68; S. Oberhelmann, "On the Chronology and Pneumatism of Aretaeus of Cappadocia," *ANRW* 2.37.2 (1994) 941–966; G. Weber, *Areteo di Cappadocia. Interpretazioni e aspetti della formazione anatomo-patologica del Morgagni* (1996); *OCD3* 152–153, W.D. Ross; *BNP* 1 (2002) 1051–1052, V. Nutton.

Alain Touwaide

ARETHAS ⇒ ḤĀRITH

Ariobarzanēs (1st c. BCE)

Medical writer, whom Philostratos (*VS* 1.19), identifying as a Sophist, calls a "Kilikian," probably the same *Ariobarzanios* credited with the invention of a special plaster against cancers and sclerodermas, according to HĒRĀS in GALĒN, *CMGen* 4 (13.439 K.) and 14 (13.750–751 K.); the composition of this plaster is given by ALEXANDER OF TRALLEIS (2.108–111, 388–389 Puschm.) and quoted by AËTIOS OF AMIDA (6.89 [*CMG* 8.2, p. 234]) and PAULOS OF AIGINA (4.23.13 [*CMG* 9.1, p. 193], 7.17.22 [9.2, p. 353]).

Fabricius (1726) 82.

Antonio Panaino

Aristagoras (of Milētos?) (380 – 340 BCE)

Wrote a geographical work *On Egypt* cited by PLINY 36.79, PLUTARCH *Isis* 5 (352F), AELIANUS *NA* 11.10, DIOGENĒS LAËRTIOS 1.11, and STEPHANOS OF BUZANTION.

FGrHist 608.

PTK

Aristaios (350 – 250 BCE)

PAPPOS OF ALEXANDRIA (*Collection* 7.3), in discussing analysis and synthesis, refers to a set of works by EUCLID, APOLLŌNIOS, Aristaios "the elder," and ERATOSTHENĒS, supplementing the "common elements" of geometry as tools for solving geometric problems.

Pappos lists Aristaios' *Solid Loci* in five books (also called *Conic Elements*: 7.29) after Apollōnios' *Conics*, the last work discussed in detail, and before the final two treatises in the catalogue, Euclid's *Surface Loci* and Eratosthenēs' *On Means*. The order suggests Pappos thought Aristaios best studied after Apollōnios, and, indeed, he says that Aristaios' work was written rather succinctly as if for readers already competent. However, he clearly thinks Apollōnios was chronologically later since he tells us (7.30) that Apollōnios introduced the words "ellipse," "parabola," and "hyperbola" for what his predecessors and Aristaios called sections of acute-angled, right-angled, and obtuse-angled cones (7.30), and Euclid published a (lost) work on *loci* after Aristaios' *Solid Loci* (7.34). Attempts to reconstruct Aristaios' contribution to the study of conics are inferences from this information and what else we know about Greek studies in the field.

HUPSIKLĒS (in the so-called Book 14 of Euclid's *Elements*) says that in his *Comparison of the Five Figures*, Aristaios proved that the same circle circumscribes the pentagonal face of the regular dodecahedron and the triangular face of the regular icosahedron (Heiberg-Stamatis 1977: 4.4–7). Hupsiklēs gives his own proof of this result as proposition 3. Doubts have been raised about whether *Solid Loci* and *Comparison of the Five Figures* have the same author, but these seem to depend on more specific chronological assumptions than the evidence warrants.

Heath (1926) 3.513–515; J.L. Heiberg and E.S. Stamatis, edd., *Euclidis Elementa* (1977) v. 5.1; Jones (1986) 573–591.

Ian Mueller

Aristanax (330 BCE – 120 CE)

Greek physician criticized by SŌRANOS, *Gyn.* 2.48 (*CMG* 4, p. 87; *CUF* v. 2, p. 57), for recommending that female infants be weaned six months later than males, based on his generalizing assumption that females are weaker. Listed after, and probably later than, MNĒSITHEOS (OF ATHENS? OF KUZIKOS?). This Doric form of the name is esp. common on Rhodes (*LGPN*).

RE 2.1 (1895) 859 (#2), M. Wellmann.

GLIM

Aristandros of Athens (240 – 90 BCE)

Wrote a work on agriculture, possibly treating cereals, livestock, poultry, viticulture, and arboriculture (*cf.* PLINY, 1.*ind*.8, 10, 14–15, 17–18), that was excerpted by CASSIUS DIONUSIOS (VARRO, *RR* 1.1.8, *cf.* COLUMELLA, 1.1.8). He had a special interest in botanical "portents" or anomalies, such as trees bearing fruit on their trunks, or fruit but no leaves, and trees which altered in color or genus (Pliny, 17.241–243). A *terminus post quem* is provided by his reference to the city of Laodikeia, which was founded by Antiokhos II sometime between 261 and 247.

RE S.1 (1903) 131 (#6a), M. Wellmann.

Philip Thibodeau

Aristarkhos of Samos (*ca* 280 – 270 BCE)

Particularly renowned for proposing a heliocentric theory that inspired Copernicus, but his one extant work is a study *On the Distances of the Sun and the Moon*. Prior to Aristarkhos, there were theories proposing the motion of the Earth (Philolaos, Hērakleidēs). According to Plutarch (*Platonic Questions*, 1006C, and *The face on the Moon*, 923A, *cf.* pseudo-Plutarch, *De placita phil.* 891A), Aristarkhos merely hypothesized with demonstrations that the Earth revolved around the Sun on an inclined circle and also rotated so that the sphere of the fixed stars would not move. Seleukos of Seleukeia later maintained this view. We may surmise that the account included other basic elements of a heliocentric theory. Unfortunately, no extant testimony provides any motivation for his heliocentric theory or even details. Indeed, our other principal source, Archimēdēs' *Sand-Reckoner*, 1.4–7, only mentions that such a universe must be larger than a geocentric one. Even here, we must infer a motivation for this claim, that otherwise we would expect an observable stellar parallax, i.e. a change in the angle between stars in different seasons. Indeed, it is easier to speculate on why the theory was not accepted than why it was proposed. Does Kleanthēs' accusation of impiety represent a common judgment among ancient scientists?

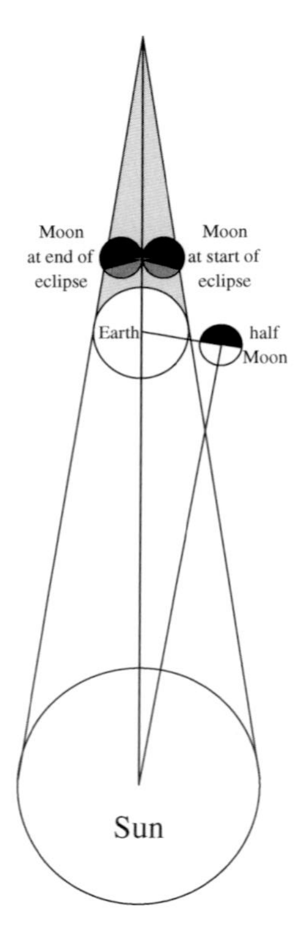

Aristarkhos of Samos: distances and sizes of Sun and Moon © Mendell

The treatise on the measurement of the distances and sizes of the Sun and Moon was profoundly influential in the ancient world. Interestingly it assumes a geocentric universe (*cf. prop.* 6). Using four basic phenomena in addition to the equal apparent or angular size of the Moon and Sun, Aristarkhos constructed a valid geometrical analysis of the relative distances of the Sun and Moon, based on the insight that the triangle formed by the Sun, Moon, and observer must form a right triangle at half moon, with the right angle at the Moon. One phenomenon is that the shadow of the Earth on the Moon during a lunar eclipse is two lunar diameters: the actual ratio varies and is larger (about 2.6). Aristarkhos also treated the distances of the Moon and Sun as constant, as would most contemporary astronomers, and ignored parallax. Only these assumptions affect the geometry of his argument. Although Archimedes ascribes to him the correct angular size of the sun as ½°, here it is 2° (by inference from eclipses and his value for the moon as $^{1}/_{45}$ right angle). These three observational errors facilitate the geometry and calculation but do not affect the calculation seriously. Quite different is his assumption that the angle of the Sun-Earth-Moon at half moon is $^{1}/_{30}$ less than a right angle (87° in a later system). The actual angle, *ca* 89°51', is not observable without sophisticated instruments. At 87° the Moon is *ca* 47.5% illuminated, about 4 to 5 hours before true half Moon. So his approximation makes the Sun much nearer. The principal results are that the distance of the Sun's distance from the Earth and diameter are respectively between 18 and 20 times the Moon's distance and diameter, and that the Sun's diameter to the Earth's is between 19:3 and 43:6, while the Earth's to the Moon's is between 108:43 and 60:19. The upper and lower ranges are typical of Hellenistic science. However, they are purely trigonometrical and not based on the phenomena. In modern discussions, it is sometimes suggested that this exercise should be regarded as a thought experiment. It may be significant that no known ancient reader ever said so.

In other matters, Aristarkhos reports views of THALĒS and HĒRAKLEITOS on eclipses and the month (*POxy* 3710). He observed the summer solstice of 280 BCE. VITRUUIUS 9.8.1 credits him with a hemispherical sun dial and a level disk sun dial. The work on the phases of the Moon as described by Vitruuius 9.2.3–4 is fairly trivial, but may reflect more sophisticated work. Aristarkhos appears to have constructed a solar/lunar/eclipse cycle of 2,434 years (CENSORINUS 19.2, as corrected), so that it seems that he added $^{1}/_{1623}$ day to the 365 $^{1}/_{4}$ day solar year of KALLIPPOS.

Ed.: T.L. Heath, *Aristarchus of Samos* (1913); B. Noack, *Aristarch von Samos: Untersuchungen zur Überlieferungsgeschichte* (1992).

Neugebauer (1975) esp. 621, 634–643, 697–698; J.L. Berggren and N. Sidoli, "Aristarchus's *On the Sizes and Distances of the Sun and the Moon*: Greek and Arabic Texts," *AHES* 61 (2007) 213–254.

Henry Mendell

Aristarkhos of Sikuōn (60 BCE – 60 CE)

Wrote a description of Greece and lands to the north, cited by PLINY 1.*ind*.5, and used by the RAVENNA COSMOGRAPHY, 4.8–11 on Greece, and 4.14 on Dacia – i.e. after 60 BCE. *Cf.* HULAS and SARDONIUS.

J. Schnetz, *SBAW* (1942), # 6, pp. 81–84.

PTK

Aristarkhos of Tarsos (30 – 70 CE?)

Physician to be distinguished from Polyainos' Aristarkhos (*Strateg.* 8.50.1.14), a physician living at Berenikē's court, wife of Antiokhos II Theos (261–246 BCE). Prior to ASKLĒPIADĒS PHARM. who quotes (in GALĒN *CMGen* 5.1 [13.824.13 K.]) his "multi-purpose pastilles" (***trokhiskoi***), and (*CMLoc* 7.5 [13.103.7–104.6 K.]) the formula of "an admirably potent" antidote called *Paulina*, with various applications in such conditions as ***duspnoia*** and poisonous and venomous infections; KRITŌN (*CMLoc* 5.1 [12.818.3–8 K.]) quotes his ointment for the treatment of black eyes.

RE 2.1 (1895) 873 (#24), M. Wellmann.

Jean-Marie Jacques

Aristeidēs (Mech.) (440 – 400 BCE?)

Pausanias 6.20.14 reports that he improved the *hippaphesis* of KLEOITAS; Fabricius suggests he may be the student of the sculptor POLUKLEITOS.

RE 2.1 (1895) 896 (#28), E. Fabricius

PTK

Aristeidēs (Paradoxographer) (250 BCE – 25 BCE)

Among other authors discussing geographical properties and qualities of water, VITRUUIUS (8.3.27) enumerates Aristeidēs. Whether this man should be identified with the homonymous geographer (PLINY 4.70) is doubtful.

RE S.5 (1932) 46 (#23a), W. Kroll.

Jan Bollansée, Karen Haegemans, and Guido Schepens

Aristeidēs Quintilianus (*ca* 270 – 330 CE)

Author of a treatise *On music*, arranged in three books (in Greek). The overall vocabulary and style of the treatise are Neo-**Platonic**, reflecting or suggesting specific notions found in the writings of PLŌTINOS, PORPHURIOS, and IAMBLIKHOS. In addition, Aristeidēs almost certainly drew on such 2nd c. authors as THEŌN OF SMURNA, PTOLEMY, PLUTARCH, and Hephaistiōn, and on other authors of less certain date such as KLEONEIDĒS and GAUDENTIUS. Whole sections of MARTIANUS CAPELLA's *On the Marriage of Philology and Mercury* have been identified as derived from his treatise.

On music weaves together in rigorous, systematic, and highly complex language a wide range of materials – musical, philosophical, medical, grammatical, metrical, and literary – to create a unified philosophical discourse in which music serves as a paradigm for the order of the soul and the universe. Book I, largely following the Aristoxenian model, defines the science of music (*mousikē*), conjoining the treatments of harmonics (1.6–12), rhythmics (1.13–19), and metrics (1.20–29) by means of vocabulary and the development of definitions. Various notational diagrams are included, one of which (1.9) purports to preserve scales of "the exceedingly ancient peoples." Another diagram (1.11) illustrates the 15 *tonoi* (*cf.* ALUPIOS) laid out "akin to a wing." The treatments of rhythmics and metrics exhibit apparent *loci paralleli* with Hephaestion's *Handbook* and Dionysius of Halicarnassus' *On literary composition* (1st c. BCE). Book II, conceived in three sections, applies the definitions of the first book to larger considerations: the soul, the influence of music on character, and ethnic stereotypes and the use of music in the Roman empire (2.1–6); the development of ethical notions through music, the relationship of souls and bodies (human and otherwise), and the association of masculine, feminine, and medial natures with the technical details of music (2.7–16); and the affective power of instruments, exercised through their association with the soul, the Muses, and the gods (2.17–19). Book III reveals music as a paradigm for cosmic order in two sections: the first reviewing mathematical–musical affinities (3.1–8); the second, as Aristeidēs states, "making quite plain the similarity of each particular to the universe altogether" (3.9–27), in which nearly every particular of the preceding material is related in a grand Neo-**Platonic** cosmology based not only on PLATO (especially *Republic* and *Timaeus*) and ARISTOTLE (*On the heavens, Physics, Metaphysics*, and *History of Animals*) but also on Plōtinos, Ptolemy (*Tetrabiblos*), Porphurios, and Theōn of Smurna.

Ed.: R.P. Winnington-Ingram, *De musica* (1963); Thomas J. Mathiesen, trans., *Aristides Quintilianus On Music* (1983); *SRMH* 1.47–66 (with diagrams).
MGG2 1 (1999) 917–922; Mathiesen (1999) 521–582; *NGD2* 1.905–907.

Thomas J. Mathiesen

Aristeidēs (of Knidos?) (360 – 325 BCE?)

Cited twice by PLINY for names of islands: 4.64 Euboia is "Long Island" (Makra), and 4.70 Mēlos is *MIMBLIS*; Pliny 1.*ind*.4 lists him between EPHOROS and ARISTOTLE, possibly indicating the date-range. Jacoby tentatively identifies with the writer of Knidiaka.

FGrHist 444.

PTK

Aristeidēs of Samos (360 – 50 BCE)

VARRO, *Hebd.* 1 (in Gellius 3.10.6), reports that he cited the 28-day lunar month as evidence of the power of seven in nature (*cf.* ANATOLIOS, *On the Decade*). Scholars suggest emending to ARISTARKHOS OF SAMOS, but the more obscure person is the *lectio difficilior.*

RE 2.1 (1895) 896 (#26), G. Kauffmann.

PTK

Aristiōn, father and grandson (Mech.) (200 – 160 and 140 – 80 BCE?)

Aristiōn (or Kharistiōn), the father of the physician and pharmacologist PASIKRATĒS (OREIBASIOS *Coll.* 49.24, 26 = *CMG* 6.2.2., pp. 36–38, 41–43), was a medical engineer who may have developed the balance (*kharistion*) employing a ratio between the motor power, the weight moved, and the space traveled, which SIMPLICIUS credits to ARCHIMĒDĒS (*in Phys.* 7.5: *CAG* 10 [1895] 1110). He designed a triple-pulley (*trispaston*) described by Oreibasios (*Coll.* 49.15–27, pp. 26–43), which the grandson then altered. One of the two developed also a plaster for fractures, if "Aristos" is to be read as Aristiōn (SCRIBONIUS LARGUS 209).

P. Duhem, *Origins of Statics* (1905–1906; translated, 1991) 65–66, 70–73; Drachmann (1963) 181–183; Michler (1968) 87–88, 130–131.

PTK and GLIM

Aristippos of Kurēnē (225 – 175 BCE?)

Author of a doxographical account *On the Natural Philosophers*, known only for explaining PYTHAGORAS' name (DIOGENĒS LAËRTIOS 8.21). The author is probably the philosopher of the new **Academy**, student of LAKUDĒS (Classen 180; *cf.* EUSEBIOS, *PE* 14.7.14). D.L. 8.60 also cites him on the love of EMPEDOKLĒS for PAUSANIAS. The student of Sōcratēs who taught and wrote that pleasure was the goal of life (*ca* 410 – *ca* 360 BCE) does not seem probable; nor his grandson Aristippos (D.L. 2.83; *cf.* STRABŌN 17.3.22 and AELIANUS, *NA* 3.40).

C.J. Classen, "Bemerkungen zu zwei griechischen, Philosophiehistorikern'," *Philologus* 109 (1965) 175–181; *SSR* 4.155–168, esp. 164; *BNP* 1 (2002) 1103–1104, K.-H. Stanzel.

PTK

Aristoboulos (250 BCE – 50 CE?)

Wrote an *On stones* quoted by PSEUDO-PLUTARCH (*De fluu.* 14.3 [1158C]) reporting a fragment from the first book, treating a stone similar to crystal, common in the river Tanais. He is to be distinguished from ARISTOBOULOS OF KASSANDREIA and from the homonymous Aristoboulos, often quoted as a source by pseudo-Plutarch in his *Parallela minora*. He could, however, possibly be identified with a homonymous author whom Giannini lists among the non-specialist paradoxographical writers.

RE S.1 (1903) 133 (#14), F. Knaack; Schlereth (1931) 105; *FGrHist* 830 F1; A. Giannini, "Studi sulla paradossografia greca I," *Rend. Ist. Lomb. Sc. Lett.* 97 (1963) 247–266 at 265; De Lazzer (2003) 71.

Eugenio Amato

Aristoboulos of Kassandreia (334 – 301 BCE)

Serving in Alexander's army, he had no known military role, but his technical skills were such that Alexander commissioned him to restore Cyrus the Great's desecrated tomb at Pasargadae. Alexander may have commissioned him to restore the water systems around Babylōn. Late in his long life he wrote an account of Alexander's expedition, which does not survive; but ARRIANUS used him as a source, along with Ptolemy. He gave a largely eye-witness description of the geography, flora and fauna, and ethnography of the regions through which the expedition passed. His observations on Mesopotamian and Central Asian rivers were particularly detailed. He discussed the cause of the Indus' flooding, which he ascribed, along with that of the Nile, to summer rains. He described the Hindu Kush, which he called the Caucasus. His remarks on the climate of Afghanistan and Pakistan were accurate for the regions through which he actually traveled; but he maintained erroneously that the Indian plains are desert. He also discussed the Indus crocodiles, its abundant fish, and numerous venomous serpents of India. Furthermore, he described in detail the banyan tree, and gave the first known Greek account of the banana and the cultivation of rice. He mentioned various customs he heard about at Taxila such as the exposure of the dead, the throwing of the elderly to dogs, the sale of daughters, and the sati. His observations were often more sober than those of ONĒSIKRITOS and NEARKHOS: Aristoboulos gave more modest figures for the size of serpents in India, and for the extent of the shade of the banyan tree, and contradicted Onēsikritos' assertion that there are hippopotami in the Indus river. He gave a graphic account of Alexander's trek through the desert of Gedrosia. Aristoboulos' account of the lands of Mesopotamia, Iran, and India was a reasonable first-hand report; but it was unable to displace the tradition of eastern wonders begun by SKULAX, HĒRODOTOS, and KTĒSIAS.

Ed.: *FGrHist* 139.
Robinson (1953) 1.205–243; Pearson (1960) 150–187; P. Brunt, "Notes on Aristobulus of Cassandria," *CQ* 26 (1974) 65–69; Pédech (1984) 331–406.

Philip Kaplan

Aristodēmos (250 BCE – 175 CE)

Perhaps the grammarian from Nusa (active *ca* 90–40 BCE), STRABŌN's teacher, or his younger relative (*ca* 80–30 BCE), teacher of Pompey's sons. Cited by AËTIOS OF AMIDA 13.86 (p. 713 Cornarius), probably from CALPURNIUS PISO's work *On Animals*, on the domesticated weasel's ability to sniff out medicinal roots. The name is very frequent through the 1st c. CE, quite rare in the 2nd/3rd c. CE, and unattested thereafter: *LGPN*.

BNP 1 (2002) 1114–1115 (#7), F. Montanari (the grammarian).

PTK

Aristogeitōn (of Boiōtian Thēbai?) (60 – 75 CE)

PLINY 27.31 (*cf.* 1.*ind*.27), listing him after HIKESIOS, records that he prescribed the Skuthian herb *anonymus* for wounds. After the 1st c. BCE, the name is attested only from Boiōtian Thēbai.

Fabricius (1726) 83.

PTK

Aristogenēs of Knidos (260 – 240 BCE)

The *Souda* A-3911 credits him with curing Antigonos "Gonatas," King of Macedon (reigned 284–239 BCE). Scholars generally equate him with the widely cited anethnic doctor, but *cf.* ARISTOGENĒS OF THASOS. The ARISTOTELIAN CORPUS, BREATH confutes Aristogenēs' theory of breath and respiration, that air is somehow "digested" via the lungs which also excrete some residue, 2 (481a28–30), that respiration extends only to the lungs (i.e., not to the whole body as EMPEDOKLĒS had taught), 2 (481b17–18), and that the respiratory vessels grow like other body parts and when larger contain more air. PLINY mentions Aristogenēs as an authority on drugs from animals and minerals (1.*ind*.29–30, 33–35), and CELSUS, who refers to a "student of Chrysippus at the court of Antigonus" (presumably KHRUSIPPOS OF KNIDOS (II)), cites Aristogenēs' emollient of natron, squill, sulfur, etc. in **terebinth**, bovine suet, and beeswax, 5.18.27. GALĒN records that he and MĒDEIOS, students of Khrusippos of Knidos (II), abjured phlebotomy as did ERASISTRATOS: *On Venesection, Against the Erasistrateans in Rome* 2 (11.197 K. = p. 43 Brain), *Treatment by Venesection* 2 (11.252 K. = p. 68 Brain), *cf.* probably *CMG* 5.9.1, pp. 69–70.

RE 2.1 (1895) 932–933 (#5), M. Wellmann.

PTK

***Aristogenēs of Thasos* (unknown date)**

The *Souda* A-3910 credits him with 24 works, including *Biting Beasts, Diet, Health, Semen,* and an *Epitome of Natural Remedies*. Usually identified with ARISTOGENĒS OF KNIDOS.

(*)

PTK

ARISTOKLĒS ⇒ PLATO

Aristoklēs (120 BCE – 80 CE)

Pharmacologist, three of whose recipes ANDROMAKHOS preserves in GALĒN. His remedy for oral infections, later used by ANTIPATROS (the one cited by ANDROMAKHOS or else Galēn's contemporary the **Methodist**), included oak-gall cooked in vinegar until tender, myrrh, Indian nard, rhubarb, 20 peppercorns, Attic honey, and mandrake seeds (*CMLoc* 6.6 [12.936 K.]). His remedy for the liver and internal ailments consisted in pepper, myrrh, saffron, ***kostos***, *meōn*, yellow iris, nard, carrot seeds, parsley, ***skordion***, cinnamon, cassia, mixed with sufficient honey (*CMLoc* 8.6 [13.205 K.]). Aristoklēs' emollient was compounded of pitch, beeswax, resin, ***terebinth, ammōniakon*** incense, and **galbanum** (*CMGen* 7.6 [13.977 K.]).

RE 2.1 (1895) 937 (#20), M. Wellmann.

GLIM

Aristoklēs of Messēnē (Sicily) (1st c. CE?)

Peripatetic philosopher who wrote a treatise *On Philosophy* in ten books, transmitted by EUSEBIOS in *Praeparatio Euangelica* Books 11, 14 and 15. All the fragments come from Books 7 and 8 of Aristoklēs' treatise and deal mainly with the epistemology of the Skeptics, Cyrenaics, PRŌTAGORAS, MĒTRODŌROS OF KHIOS, **Epicureans**, and Eleatics. The purpose of Aristoklēs' treatise as such is not known but he seems to have had extensive knowledge of the history of philosophy.

M.L. Chiesara, *Aristocles of Messene: testimonia and fragments* (2001).

Jørgen Mejer

Aristokratēs (30 – 80 CE)

Grammarian, possibly identifiable with Cornutus' friend Petronius Aristocrates (*RE* 19.1 [1937] 1214 [#30], O. Stein: *ca* 35–65 CE). ANDROMAKHOS in GALĒN *CMLoc* 5.5 (12.878–879 K.) preserves his toothache remedy compounded from poppy, ***sagapēnon, silphion***, pepper, *sphondulion* (DIOSKOURIDĒS 3.76), myrrh, **galbanum**, ***purethron***, and saffron, made glutinous with honey. Andromakhos then gives Aristokratēs' gingivitis remedy.

RE 2.1 (1895) 941 (#27), M. Wellmann.

GLIM

Aristokreōn (*ca* 300 – 250 BCE?)

Cited with BIŌN and DALIŌN as a foreign authority on geography and ethnography (PLINY 1.*ind*.5–6). In his *Aithiopika*, he estimated the country's dimensions (6.183), placed Elephantis 750 Roman miles from the Mediterranean, presumably following the Nile (5.59), and placed Tolles five days from Meroe (towards Libya), a further 12 days being needed to reach Aesar (which town Biōn calls Sapēs: 6.191). HERMIPPOS cites Aristokreōn's witness to an "Ethiopian" tribe whose king was a dog (AELIANUS, *HA* 7.40; *FGrHist* 667). The name is rare, attested on Cyprus and at Kōs (2nd to 1st cc. BCE: *LGPN* 1.70), so he may be the same as, or an ancestor of, the homonymous **Stoic**, active 230–185 BCE, the nephew and student of CHRYSIPPUS (DIOGENĒS LAËRTIOS 7.185).

RE 2.1 (1895) 941–942 (#1), H. Berger.

GLIM

Aristolaos (250 BCE – 80 CE)

ANDROMAKHOS, in GALĒN *CMLoc* 9.5 (13.296 K.), notes that he employed an enema similar to that of ATHĒNAIOS, which itself contained minerals (lime, realgar, etc.), plus sour grapes and ashed papyrus, in myrtle wine.

Fabricius (1726) 83.

PTK

Aristomakhos of Soloi (325 – 25 BCE)

Wrote a treatise on beekeeping (*Melittourgika*) which according to PLINY, 11.19, was based on 58 years of first-hand experience. He recommended feeding bees on moon-trefoil, and adding new bees to a colony that had grown old (Pliny, 13.131–132, COLUMELLA, 9.13.8–9). His work also dealt with trees, the cultivation of radishes, and wine-making (Pliny 1.*ind.*11–15, 19, 14.120, 19.84). His *floruit* falls somewhere between ARISTOTLE, whom he appears to surpass in apiary expertise, and IULIUS HYGINUS, who cited him.

RE 2.1 (1895) 946 (#20), M. Wellmann.

Philip Thibodeau

Aristombrotos (350 BCE – 50 CE)

Attributed with a brief Doric fragment from a pseudo-**Pythagorean** treatise *On sight* (IŌANNĒS OF STOBOI 1.52.21), treating the properties and the interrelations of air, sight and light.

Thesleff (1965) 53.23–54.7.

Bruno Centrone

Aristomenēs (325 – 90 BCE)

Agricultural writer whose work was used by CASSIUS DIONUSIOS (VARRO, *RR* 1.1.9–10).

RE 2.1 (1895) 949 (#14), M. Wellmann.

Philip Thibodeau

Aristōn (I) (450 – 400 BCE)

Physician, pupil of PETRŌN (or PETRONAS) OF AIGINA. GALĒN (15.455.15 K., *cf.* 18A.9.1 K.) includes him among the *palaioi*, to whom the *Regimen in Health* was attributed. He was an upholder of the thesis which the author of the HIPPOKRATIC CORPUS ON SACRED DISEASE objected to, according to which intelligence was located in the diaphragm, so that, in his opinion, the diaphragm played an important role in mental diseases.

RE 2.1 (1895) 959 (#58), S.1 (1903) 135, M. Wellmann.

Jean-Marie Jacques

Aristōn (II) (50 – 10 BCE?)

Not the same as ARISTŌN (I), but a later homonym whom CELSUS 5.18.33 quotes concerning an ointment for gout. Aristōn (II) may be the same Aristōn whose medication ANDROMAKHOS THE YOUNGER (in GALĒN *CMLoc* 9.4 [13.281.4 K.]) mentions as being an excellent soothing remedy for intestinal disorder.

RE 2.1 (1895) 959 (#58), S.1 (1903) 135, M. Wellmann.

Jean-Marie Jacques

Aristōn of Ioulis on Keōs (*fl. ca* 225 BCE)

Student of LUKŌN OF TROAS and his successor as **scholarch** of the **Peripatos** from *ca* 225; author of *Erotic Examples* and other works, probably including a dialogue *Lukōn*, and possibly also ethical works *On Old Age* and *On Relieving Arrogance.* Lack of specific identification in ancient sources often makes it impossible to distinguish his works and ideas from those of other writers named Aristōn. He played a role in preserving and transmitting the wills of earlier **Peripatetic scholarchs** and may have written about HĒRAKLEITOS, Sōcratēs and/or EPICURUS. **Stoics** in the next century, wishing to distance their school from the Cynic tendencies of earlier **Stoics**, attempted unsuccessfully to reassign to him the books of the **Stoic** Aristōn of Khios (*fr.*8 = DIOGENĒS LAËRTIOS 7.163). Though no scientific writings have been attributed to him, he is credited with incidental observations on mitigating hangovers (*frr.*10–11) and on the deleterious physical and mental effects of drinking water from certain springs (*fr.*17).

BNP 1 (2002) 1119–20 (#3), R.W. Sharples; P. Stork *et al.*, "Aristo of Ceos: The Sources, Text and Translation," David E. Hahm, "In Search of Aristo," and R.W. Sharples, "Natural Philosophy in the Peripatos after Strato," in Fortenbaugh and White, *RUSCH* 13 (2006) 1–177, 179–215 and 312–20.

David E. Hahm

Aristōn of Khios (100 – 60 BCE)

Peripatetic who argued with EUDŌROS OF ALEXANDRIA over the priority of their published theories of the rise of the Nile (STRABŌN 17.1.5).

KP 1.571–572 (#3), H. Dörrie.

PTK

ARISTŌNUMOS ⇒ HEKATŌNUMOS

Aristophanēs (250 BCE – 100 CE)

PAULOS OF AIGINA 7.17.34 (*CMG* 9.2, p. 356) records his emollient of beeswax, liquid pitch, ***panax***-juice, and vinegar (*cf.* 4.55 [9.1, p. 380]). The name is almost unattested after the 1st c. CE (*LGPN*).

(*)

PTK

Aristophanēs of Buzantion (*ca* 230 – 180 BCE)

Born *ca* 257 BCE; prominent scholar and head of the Alexandrian library (195–180), wrote numerous critical editions and treatises on classical poets, and lexicographical compilations (*Lexeis* or *Glōssai*). He also abridged the Aristotelian zoological corpus (*Epitome by Aristophanes of the writings of Aristotle on living creatures*), partly preserved in a Byzantine zoological *Sylloge* dedicated to Constantine VII Porphurogennētos (10th c. CE), with additional extracts mainly borrowed from AELIANUS, TIMOTHEOS, KTĒSIAS and the PSEUDO-ARISTOTELIAN DE MIRABILIBUS AUSCULTATIONIBUS. Originally in four books, the *Epitome* excerpted, summarized and reorganized **Peripatetic** zoological material, mainly but not exclusively from *History of Animals*, and including later material (e.g., THEOPHRASTOS: 1.98, etc.), to provide a practical handbook. This digest demonstrates an original structure. Aristophanēs treats first general questions on mammals (Book 1), then systematically describes principal mammalian species and ovoviviparous fishes (Book 2); he seems to have posited general questions on ovoviviparous animals in Book 3 and then described them specifically in Book 4 (both lost), rejecting the remaining animals as unworthy (*Epit.* 2.2–3).

Aristophanēs' didactic and synthetic introduction focuses on zoological classification

(1.1–27), generation (1.27–97), and sensational peculiarities of men (1.98–113) and animals (1.114–154). Book 2, altered in the sole surviving MS, presents 26 monographic files on mammals (beginning with fissipeds: man, elephant, lion. . .) systematically arranged and including the following headings: complete anatomy, mating, gestation, reproduction, number of offspring, lifestyle, ethology, longevity (1.155, 2.1). This new arrangement of zoological knowledge, separating clearly the (general) theory and the (singular) concrete description leads to a significant distortion of the meaning and aim of Aristotelian inquiry. This kind of naturalist guide, with a general introduction and systematic monographic files, was apparently the standard format of "Aristotelian" zoology transmitted through ancient and Byzantine times. Scholars have erroneously considered this *Epitome* the enigmatic Aristotelian *Zoika*, but the complex composition and intertwined citations suggest rather the result of a well-developed early **Peripatetic** tradition of editing, reorganizing and rewriting the broad Aristotelian zoological corpus and information disseminated (see *Epit.* 2.1) by his inquiry. Of a lexicographical treatise entitled *On the names [of animals] according to their age*, 182 brief fragments survive.

Ed.: S.P. Lambros, *Excerptorum Constantini De Natura Animalium Libri Duo: Aristophanis Historiae Animalium Epitome*, Supplementum Aristotelicum, 1.1 (1885); Arnaud Zucker, *Recueil zoologique de Constantin* (*CUF*, forthcoming).

RE 2.1 (1895) 994–1005 (#14), L. Cohn; O. Hellmann, "Peripatetic biology and the *Epitome* of Aristophanes of Byzantium," *Aristo of Ceos, RUSCH* XIII (2006) 329–359.

Arnaud Zucker

Aristophanēs of Mallos in Kilikia (325 – 90 BCE)

Agronomist whose writings may have treated cereals, livestock, poultry, viticulture, and arboriculture (*cf.* PLINY, 1.*ind.*8, 10, 14–15, 17–18); CASSIUS DIONUSIOS excerpted from his work (VARRO, *RR* 1.1.8–10, *cf.* COLUMELLA, 1.1.7). Pliny gives Milētos as Aristophanēs' homeland, but Varro, somewhat closer to the source, has it as Mallos.

RE 2.1 (1895) 1005 (#15), M. Wellmann.

Philip Thibodeau

Aristophilos of Plataia (350 – 280 BCE)

Pharmacologist, used his knowledge of antaphrodisiac medicines to punish and reform his slaves (THEOPHRASTOS *HP* 9.18.4).

RE 2.1 (1895) 1005, M. Wellmann.

GLIM

Aristotelēs of Mutilēnē (180 – 205 CE)

Peripatetic teacher of ALEXANDER OF APHRODISIAS, who promulgated a **Stoic** theory of mind (*In de Anima: CAG* S.2.2 [1887] 110) and contributed to the debate on circular motion (SIMPLICIUS *In de Caelo: CAG* 7 [1894] 153–154). GALĒN, *Peri Ethōn* (2.11–12 MMH), gives the ethnic and records how he died from lack of proper care.

P. Moraux, "Aristoteles, der Lehrer Alexanders von Aphrodisias," *AGP* 49 (1967) 169–182.

PTK

ARISTOTELĒS OF STAGEIRA ⇒ ARISTOTLE

Aristotle (355 – 322 BCE)

Aristotle © Kunsthistorisches Museum, Vienna

Aristotelēs; born 384 BCE in Stageira, son of NIKOMAKHOS, a Macedonian court physician; joined PLATO's **Academy** in Athens at age 17 and stayed until Plato's death in 348/7 BCE; left Athens for the Troas and Lesbos (perhaps in association with fellow Academician THEOPHRASTOS); was tutor to Alexander the Great (343–340); returned to Athens and founded the **Lyceum** (*ca* 335); fled anti-Macedonian Athens after Alexander's death (323); died in Khalkis (322). Aristotle is the most influential observer and recorder, philosopher and systematizer of antiquity. Though his most voluminous contribution was in biology, he is best known for his physical theory, dominant in the Middle Ages and overturned only in the early modern period. This article treats the contents and contribution of his treatises in systematic order.

Aristotle's physical treatises (originally lecture notes and catalogues edited and rearranged) form a hierarchical unity of discrete but related disciplines subject to a variety of methods. Aristotle divides rational thought, according to its object, into theoretical, practical and productive areas (*Metaphysics* 6.1 [1025b25]). *Theoria* is further divided into physics, mathematics and theology depending on the materiality and changeability of its subjects. The three domains remain closely related: physics studies embodied form subject to change, theology provides the final cause of physics, and mathematics is the study of unchanging quantity inhering in physical substance.

Aristotle's natural science arises from Greek speculative philosophy. PARMENIDĒS denied the possibility of change (since what is not cannot be), and EMPEDOKLĒS, ANAXAGORAS, and DĒMOKRITOS, among others, responded by reducing genesis and alteration to the less problematic locomotion (all change is the movement of elements, homoeomeries or atoms). Aristotle starts from these problems, and his discussion assumes a philosophical and deductive rather than empirical tone.

Nature is "the principle of change and rest in a thing": what makes a thing act and react the way it does (*Physics* 2.1 [192b13–15]). Not only is the nature of earth to fall to the center of the universe; plants and animals also have their own complex internal sources of change governing growth and behavior. Though each complex and simple nature is a principle of change, complex natures subsume simple natures hierarchically: the nature of complex bodies cannot be reduced to the nature of their constituent elements like a clockwork mechanism.

The *Physics* begins with the most general conditions of change: an unchanging substrate (matter) upon which a change (locomotion, genesis and destruction, growth and decay, alteration) occurs from privation to form. Parmenidēs' problem is solved: we can deny *ex nihilo* genesis while accounting for change and motion, and without reducing all change to locomotion. Aristotle also adapts Plato's **demiurge** (*Timaios*) and his Forms into a dynamic conception of the good. Natural, like artificial, change is purposive, and tends regularly and through its own agency toward some end, its perfected form. Aristotle's four causes (formal, final, efficient and material) are thus accounted for: the form, the good, and the agent are identified and made immanent in each material substance. The substrate-form response to Parmenidēs is melded with the final cause to generate the potentiality-actuality distinction: change is the fulfillment of potentiality. The changer is actually, what the changed thing is potentially.

The first half of the *Physics* discusses the existence and definition of the fundamental principles concerning change: infinity (material stuff is a continuum infinitely divisible, but its extent is finite: Aristotle uses this principle throughout the physical treatises), place (the innermost boundary of the containing thing, having direction up and down: useful in cosmology and biology), void (does not exist: *antiperistasis*, mutual replacement of material, explains locomotion), time (the measure of motion, a continuum on which the now is a point: used throughout the physical works).

The second half of the *Physics* applies these principles (along with continuity and contact) to prove facts about motion. Aristotle uses the infinite to demonstrate that Zēnōn's paradoxes of motion and time are misdirected: motion, time and magnitude are all continuous and infinitely divisible. The theorems culminate in a set of arguments for the first unmoved mover. All alteration, growth and decay, and terrestrial locomotion occur between some terminal contraries. All such motion and change come to an end. Yet change and motion are eternal: for how, if it ceased, would it ever get going again? Moreover, there cannot be an infinite series of moved movers, and so there must be eventually an unmoved mover, which is the source of the other finite movements. This unmoved mover will have no magnitude and will cause movement by desire (*Met.* 12).

On the Heavens (*de Caelo*) uses the conclusions of the *Physics* as principles and studies locomotion in its specific kinds: the natural motions of the elements, the heavy (earth and water), the light (air and fire) and the fifth element (***aithēr***) whose natural motion is circular. Again Aristotle's method is philosophical and deductive, constantly engaging the theories of Empedoklēs, Anaxagoras, Dēmokritos and Plato. He discusses the roles of the elements as heavy and light. He avoids mathematical methods: his frequent use of inverse proportion (among speed, distance, weight, resistance) is intended merely to prove the impossibility of situations involving zero and infinity, and not to establish finite mathematical relations. He is keenly aware how tenuous his conclusions are where direct knowledge is impossible.

The world is a unique, single, finite, uncreated and everlasting thing, comprising all matter. The earth is spherical, about 50,000 miles in circumference (2.14 [298a15–17]). It is unmoving and heavy things, like earth, tend to its center, which is identical with the center of the universe. Light things, like fire, have an opposite natural tendency to seek the periphery. Such is the world below the level of the moon. The heavens are occupied by a finite but much greater amount of an element – ***aithēr*** – whose natural motion is circular, and which cannot transmute into other elements. The animate planets and stars are fixed into, and made of the same stuff as, the spheres that revolve regularly from the right (south pole is up). In *Metaphysics* 12, Aristotle adapts the homocentric model of Eudoxos and Kallippos to explain the complex motions of the planets.

Generation and Corruption considers differences among changes (alteration, growth, and generation and corruption) and the lower four elements. Change among contrary powers (hot-cold, wet-dry) accounts for change among the elements. Terrestrial changes are driven ultimately by the northward and southward oscillation of the sun, and generation and corruption will be continuous, the closest possible approximation to eternal being.

Meteorologica concerns the phenomena below the lowest sphere of the moon, and deduces them from the material causes (the dry and the moist exhalations from the earth) and the efficient cause (the sun). More specific causes include burning (comets, milky way), ejection (lightning), condensation (dew, rain), reflection (rainbows). In absence of the final cause, relative place above the earth (or below in the case of earthquakes and mineral formation) is a fundamental ordering principle. *Meteorologica* 4 is a separate treatment of chemical transformations (concoction, parboiling, etc.)

Aristotle's biological works, the most extensive part of his natural writings, focus on zoology. As he moves to more specific principles, his interpretation of matter and form is adapted to living things. The principle of life is the soul, the form, or first actuality, of the potentially living body, likened to the ability of the axe to cut (*de Anima*). Soul consists of a series of ultimate functions: nutrition and reproduction, sensation and locomotion, intellect. These are both the form and the final cause of a living thing. With the exception of the intellect these functions cannot exist apart from the body (*cf.* Plato's **Pythagorean** soul). Each of the ultimate soul functions is characterized in relation to some external object: the reception of food by digestion, the reception of the form of sensitive and intelligible objects. The sense organs are affected and in turn affect the heart, the common sensorium, which discriminates and unifies the sense perceptions. The heart is also the seat of *phantasia*, the primary intentional faculty. The passive intellect contemplates its objects through the illumination of the active intellect. The *Short Natural Treatises* (*Parva Naturalia*) discusses some general life functions from other perspectives. *Sense and Sensibilia* discusses what makes the objects of sense sensible. Other treatises concern memory and recollection, dreams, and aging (all of which concern the physical principle, time). Of this same series are the *Movement of Animals* and the *Progression of Animals*. The first is a discussion of the general conditions of movement: the moved and moving parts of the body, desire, etc.; the second is a specific treatment of the forms movement takes in various animal groups (flying, crawling, swimming) and how many and of what form are the appendages (points of motion).

The *History of Animals* is a sprawling collection of data in the Ionian tradition, probably a collaboration with Theophrastos and other members of the **Lyceum**. It introduces a new, empirical approach, appropriate here where data are available from extensive dissection and reports from fishermen, hunters and farmers. Since the physicist's task is to study both the matter and the form, the *History* is a description of the various parts of animals, and their activities and functions. Aristotle isolates the universal nature of each animal kind without reference to incidentals like the animals' utility to man (in contrast to Theophrastos' parallel work on plants). The differences among animals are studied in accordance with the basic principles of the *Physics*: privation and contrariety of properties. Humans are studied first because of their familiarity, then the basic kinds of blooded (viviparous quadrupeds, oviparous quadrupeds, birds, fishes, selachia and cetacea) and bloodless animals (crustacea, testacea, insects) in their internal and external parts and their behavior.

The *History* provides the facts for which the *Parts of Animals* gives the causes. The form, the ultimate functions of the soul, provides the final cause, and the parts and tissues are the material cause. These two causes, inherited from Plato's *Timaios*, act together through hypothetical necessity: if an animal is to discharge a certain function, it must have such and such a material constitution. Aristotle arranges his subjects as in the *History*: animal groups that belong together, e.g. birds, fish, are treated together; parts common to many groups (especially internal organs) are treated together and given single explanations. Broadly Aristotle moves from highest life form to lowest, from most general to most specific, and from the head down. He rejects the dichotomous form of division popular in the **Academy** in favor of a system of multiple differentiae, e.g. birds have several general essential features in which they vary from one another by having more or less (longer/shorter beaks).

Finally (apart from spurious works) the *Generation of Animals* investigates the material and efficient cause, though final and formal play subsidiary roles. Differences among sexual organs, modes of propagation and causes of the differences among animal groups are treated in the manner of the *Parts*. In procreation the male semen, a concoction of blood, provides the form and efficient cause; female menstrual fluid provides the matter. Imperfect mastering of the menstrual fluid by the semen results in a female offspring. Resemblances to other relatives on the mother's side and father's occur by "relapse." Aristotle refutes the theory of pangenesis, according to which the seed is gathered from and composed of all parts of the body. Lower animal forms, including testacea, are generated spontaneously by the vital heat present in the air forming a kind of froth in mud. Finally, some parts and qualities (e.g. eye color) are not for a purpose, but have a material cause.

Ed.: I. Bekker, *Aristotelis Opera* (1831); Oxford Classical Texts for *Physics, de Caelo, de Anima*; H. Joachim, *de Generatione et Corruptione* (1926); F.H. Fobes, *Meteorologica* (1919); Loeb for biological works.
Trans.: *The Complete Works of Aristotle*, ed. J. Barnes (1984).
Studies: *DSB* 1.250–281, G.E.L. Owen *et al.*; G. Freudenthal, *Aristotle's Theory of Material Substance* (1995); L. Judson, *Aristotle's Physics* (1991); J.G. Lennox, *Aristotle's Philosophy of Biology* (2001); Fr. Solmsen, *Aristotle's System of the Physical World* (1960); *NDSB* 1.99–107, J.G. Lennox.

Malcolm C. Wilson

Aristotelian Corpus On Breath (*ca* 270 – 230 BCE)

The short Pseudo-Aristotelian treatise *On Breath* (481a1–486b4) is commonly believed to be a product of the mid-3rd c. **Peripatos**. Some have connected it with THEOPHRASTOS, STRATŌN OF LAMPSAKOS and ERASISTRATOS.

The treatise starts by investigating the preservation of the innate breath (*emphuton* ***pneuma***) and its increase (§1–2), polemizes by name against ARISTOGENĒS (§2), and discusses various physiological questions all in some way connected with the function of ***pneuma***. The author treats respiration (§3), alluding (482a28–31) to ARISTOTLE *On Respiration* 1 (470b6–9), movement of ***pneuma*** through the vessels (§4–6), nature and functions of the bones (§ 7), locomotion (§8), and the function of heat in biological processes (§9). Jaeger considers §9 a later **Stoic** polemic against Stratōn; Roselli defends it as part of the original work.

The author very rarely gives solutions to the problems presented to the reader. The aporetic nature of the discussion, combined with its brevity and the corrupt status of the text, makes *On Breath* a rather difficult text. It may be interpreted as a **Peripatetic** reaction to 3rd c. medical discoveries. The author refers clearly to Erasistratos' theory of skin composition (483b5–19), but elsewhere the unnamed referent is difficult to determine.

Ed.: W.S. Hett, *Aristotle, On the Soul, Parva Naturalia, On Breath*, rev. ed. (Loeb 1957); A. Roselli, [Aristotle], *De spiritu* (1992).
W.W. Jaeger, "Das Pneuma im Lykeion," *Hermes* 48 (1913) 29–74.

Oliver Hellmann

Aristotelian Corpus, de Coloribus (*ca* 320 – 250 BCE?)

Brief treatise, textually corrupt in many places, generally thought to have been authored not by ARISTOTLE, but by another **Peripatetic**. There is no unanimity, however, regarding the author's identity; scholars recently have attributed the treatise to THEOPHRASTOS, though STRATŌN also had been suggested, while its unpolished style suggests a student's lecture notes.

The author claims that the basic colors are those of the elements: fire by nature is yellow, while air, water and earth intrinsically are white. When heated, though, air and water become black; in addition, things appear black when reflecting little or no light. All other colors are produced by mixing elemental colors; an object's specific color depends not only on the colors mixed but also on proportion and intensity. For instance, violet is produced at sunrise and sunset by the mixture of the sun's rays, then weak, with the then shadowy colored air.

The treatise includes numerous illustrations of color phenomena which the author tries to explain, especially on artificial dyeing and the colors of plants and animals. The last part of this work treats particular cases exemplifying color changes of plants and animals, due either to exsiccation or to the earth's absorption of liquids. For instance, human hair changes color because it acquires through the skin different degrees of moisture at different ages.

There are clear deviations from Aristotle's color theory; e.g. elements are said to be colored, the "transparent" is not mentioned, light is treated as a material substance. But these deviations, together with the treatise's method, are not foreign to the **Peripatetic** school; for similar doctrines and the preference for observed phenomena over abstract generalizations can be found in post-Aristotelian **Peripatetic** accounts.

Having been included in the Aristotelian corpus, this work was widely read and paraphrased in medieval times. Michael of Ephesos commented on it in the 12th c., and his comments as well as the text itself were later translated into Latin.

Ed.: K. Prantl, *Aristoteles über die Farben* (1849); M.F. Ferrini, *Pseudo Aristotele, I colori* (1999).
H.B. Gottschalk, "The *De coloribus* and its author," *Hermes* 92 (1964) 59–85; G. Wöhrle, *Aristoteles, De coloribus. Aristoteles Werke in deutscher Übersetzung* (1999).

Katerina Ierodiakonou

Aristotelian Corpus On the Flood of the Nile (*ca* 340 – 328 BCE)

A Medieval Latin translation preserves, under ARISTOTLE's name, this short treatise, whose scientific aim is to investigate why the Nile is the only river that floods in summer. The text discusses and refutes some 12 different explanations of this, some made by Greek intellectuals or famous philosophers from the 6th to 4th cc. BCE. The author's conjecture, namely that heavy rains in "Ethiopia" cause the Nile's rise, is strikingly accurate, but not original.

Although modern scholars have raised doubts about both Aristotle's authorship of this short treatise and its original structure, it is likely that the Latin version comes from a Greek original written by Aristotle, since both in this treatise and in his *Meteōrologikà* the Red Sea and the Persian Gulf are not yet considered branches of the Indian Ocean, as became standard after Alexander's expedition to India (327–325 BCE).

FGrHist 646 F 1; M. de Nardis, "Aristotelismo e doxographia," *Geographia antiqua* 1 (1992) 89–108.

Mauro de Nardis

Aristotelian Corpus, Historia animalium 10 (*ca* 350 – 270 BCE)

As stated in its first sentence, the main topic of *HA* 10 (633b12–638b37) is sterility and its causes in women and men. In fact, female sterility is treated in much greater detail. §1 discusses the condition and position of the uterus and menstruation, §2 the condition and position of the mouth of the uterus, §3 the uterus after menstruation, emissions during sleep, flatulences in the uterus, and wind pregnancy, §4 spasms of the uterus. Male sterility is briefly discussed in §5 – testable, according to the author, by intercourse with a different woman – followed by a theory of simultaneous emission in men and women as well as further details about the female role in reproduction (female seed); §6 addresses reproduction in animals, §7 *mola uteri.*

HA 10, transmitted in some MSS as the last book of the *HA*, missing in others, may be identical with the work *On sterility* listed in the Catalogues of ARISTOTLE's work by DIOGENĒS LAËRTIOS and in the *Vita Hesychii*. Perhaps ANDRONIKOS OF RHODES added it to the *HA*. Today some scholars do not believe *HA* 10 to be a genuine work of Aristotle but an addition presumably by a later **Peripatetic** author with medical knowledge or by a Hippokratic physician. Besides language, style, and its anthropocentric perspective, *HA* 10 differs from *HA* and the other biological treatises mainly in the doctrine of female seed (rejected by Aristotle in *GA* 1.17–23 [721a30–731b14]), of ***pneuma*** drawing the seed into the uterus, and because it lacks the concept of form (*eidos*) and matter (*hulē*). In the last decades, Balme and van der Eijk, in different ways, have defended it as a genuine work.

Ed.: P. Louis, *Aristote, Histoire des animaux* III (1969); D. Balme and A. Gotthelf, *Aristotle, History of Animals, Books VII–X* (1991); *Eidem, Aristotle, Historia animalium*, v. 1, *Books I–X: Text* (2002).

G. Rudberg, *Zum sogenannten zehnten Buche der Aristotelischen Tiergeschichte* (1911); D. Balme, "Aristotle Historia animalium Book Ten," in J. Wiesner, ed., *Aristoteles, Werk und Wirkung* I (1985) 191–206; S. Föllinger, *Differenz und Gleichheit* (1996) 143–156; P. van der Eijk, "On Sterility ('*HA* X'), A Medical Work By Aristotle?" *CQ* 49 (1999) 490–502.

Oliver Hellmann

Aristotelian Corpus On Indivisible Lines (330 – 300 BCE)

Preserved in the Aristotelian corpus in a rough and often unintelligible form, inspiring much philological work. Sometimes ascribed to THEOPHRASTOS in antiquity, and today generally agreed not to be by ARISTOTLE, but assignable to the **Peripatos** of the later 4th c. BCE. Evidence suggests that PLATO and his follower XENOKRATĒS maintained a doctrine of indivisible lines, as much metaphysical as mathematical, although the treatise stresses its mathematical aspect, i.e., its inconsistency with accepted mathematics. Mathematically the doctrine holds that the primary entity of geometry is minimal, indivisible (atomic) lines, with every line being a (apparently finite) sum of such lines. The argumentation of the treatise is fundamentally Aristotelian. It begins (968a2–b21) with five arguments in support of indivisibles, refutes the arguments (968b21–969b28), and then (969b9–971a5) offers further considerations against indivisible lines. Some related claims are then made: a line is not a concatenation of points (971a6–972a13); a point cannot be added to or subtracted from a line (972a13–30); a point is not a minimum component of a line (972a30–b24). The treatise concludes (972b25–33) by arguing that a point is not an indivisible connector (*arthron*). Unfortunately our knowledge of the ideas refuted here comes only from such refutations, making it difficult to understand why such ideas were propounded.

Ed.: H.H. Joachim, *De Lineis Insecabilibus*, in W.D. Ross, ed., *The Works of Aristotle* v. 6 (Loeb 1913); M. Timpanaro Cardini (with Italian trans.), *Pseudo-Aristotele, De Lineis Insecabilibus* (1970).

O. Apelt, *Beiträge zur Geschichte der Griechischen Philosophie* (1891) 255–286; *RE* S.7 (1940) 1542–1543, O. Regenbogen.

Ian Mueller

ARISTOTELIAN CORPUS ON MELISSOS, XENOPHANĒS, AND GORGIAS ⇒ MELISSOS, . . .

Aristotelian Corpus Mēkhanika (Problēmata mēkhanika) (320 – 200 BCE)

This collection of problems (or a similar one) apparently has been part of the Aristotelian corpus from early on: a work with the title *Mēkhanikon* features in DIOGENĒS LAËRTIOS' list of ARISTOTLE's works. Internal characteristics also place the work quite early: its mathematical terminology is close to EUCLID's, and it is not acquainted with ARCHIMĒDĒS' contributions to the study of mechanics. This, however, does not necessarily exclude that the work was written after Archimēdēs.

The key concepts in the *Mēkhanika* are the force (*iskhus* or *dunamis*) and the load (*baros*): a force has to be equal to hold a load, or it has to exceed the load to be able to lift it or move it. These fundamental relations are apparently not observed when a mechanical device is operative: little forces are able to hold or move a much bigger load. The *Mēkhanika* uses a balance-lever model to explain how such mechanical devices work. The aspect of the balance addresses cases of equilibria, whereas the aspect of the lever accounts for cases when the device described is in motion. The *Mēkhanika* opens with a general explanation why lesser forces are able to move greater loads with the help of a lever. This is possible because of the amazing features of the circle, which combines in itself the opposites of motion and rest, and of two component motions – one centripetal, another tangential – which for no extended period of time remain in the same relation to each other. On account of these, the author argues (§8), circles have an intrinsic tendency to move. This is also why the circular motion of the balance and the lever is able to make the small force produce a greater effect. It is important to note that the author does not formulate this enhancing capacity of the lever, or of circles in general, in terms of explicit proportionalities. Nevertheless the idea of the proportionality among the distances of the force and the load from the fulcrum, and of the magnitude of the force and the load themselves is expressed repeatedly (for the most unequivocal formulation see §3).

The authorship of the *Mēkhanika* is still being debated. Among other indications, the most compelling one against an Aristotelian authorship is the way §§32–33 give a rather unskillful recapitulation of Aristotle's account of projectile motion (*Physics* 8.10). As the title *Mēkhanikon* also occurs in Diogenēs Laërtios' list of STRATŌN's works (5.59), it has been repeatedly suggested that the work is by Stratōn. The fact, however, that Stratōn is credited with a work on mechanics does not require that he should be identified with the author (or indeed, any of the authors) of these mechanical problems.

Ed.: O. Apelt, *Aristotelis quae feruntur De plantis, De mirabilibus auscultationibus,. . .* (1888); M.E. Bottecchia,

Aristotele, Mechanica (1982: provides the most detailed apparatus; however, Bottecchia also integrates in her lemmata a MS of Book 12 of Geōrgios Pachymeres' *Epitome of Aristotelian philosophy*).

Th. Heath, *Mathematics in Aristotle* (1949; reprint: 1998) 227–254; István Bodnár, "The mechanical principles of animal motion," in: A. Laks and M. Rashed, edd., *Aristote et le mouvement des animaux* (2004) 137–147.

István Bodnár

Aristotelian Corpus Physiognomy (320 – 280 BCE)

This text transmitted in the Corpus Aristotelicum was believed until the 17th c. to have been written by Aristotle himself. Since then, his authorship has been rightly doubted, and it is now assumed to have been composed shortly after Aristotle's death and based on his own interest in the subject, *cf. Anal. Pr.* 2.27 (70b7–38), a passage absorbed into the argument of *Phgn.* 805b10–806a18, and the occasional passage especially in the biological works (Vogt 1999: 120–145).

The treatise consists of four parts: (1) a discourse on the theory and methods of the subject (805a1–807a30); (2) a catalogue of 21 character types, listing their bodily signs (807a31–808b10); (3) again, and slightly different from before, an introduction to the subject, focusing on the prevalent distinction between the genders (808b11–810a13); (4) another catalogue, this time in order of body traits and qualities, listing their correlations to character types (810a12–814b9). The different focus in the two methodical parts has led to the assumption of two separate treatises A (1 and 2) and B (3 and 4) by two different authors (*cf.* Boys-Stone 64–75), but such strong separation is not necessary; the differences can be explained by a more practical attitude in the second half.

The methodological considerations became the model for physiognomists for centuries, adopted not only by Loxos, Polemōn, the Physiognomista Latinus and Adamantios, but also by modern physiognomies that cared about the theory and method (e.g. Gianbattista della Porta 1535–1615).

Ed.: I. Bekker, *Aristotelis Opera Omnia* I (1831) 805–814; R. Foerster, *Scriptores Physiognomonici* (1893) 1.2–90; G. Raina, *Pseudo-Aristotele, Fisiognomica. Anonimo Latino, Il trattato di fisiognomica* (1993).

M.M. Sassi, *La scienza dell'uomo nella Grecia antica* (1988); Sabine Vogt, *Aristoteles, Physiognomonika. Aristoteles Werke in deutscher Übersetzung* 18.6 (1999); G. Boys-Stones, "Physiognomy and Ancient Psychological Theory," in Swain (2007) 19–124.

Sabine Vogt

Aristotelian Corpus Problems (*ca* 270 – 230 BCE)

The *Problēmata phusika* is the third-largest work in the Aristotelian Corpus (859a1–967b27). In 38 topical sections of different length, the text treats nearly 900 scientific problems (with about 200 repetitions) in question-and-answer form. The questions are always introduced by a characteristic "Why is it that . . ." (*Dia ti*); the causal explanation is usually given as a rhetorical question "Is it because . . ." (*ē oti*). In many cases, alternative answers are added.

Section 1 treats medical questions, 2–9 different human phenomena (sweat, wine and drunkenness, sexual intercourse, fatigue, position, **sympathy**, frost and shivering, skin), 10 zoology, 11 voice, 12–13 odors, 14 mixtures, 15 mathematics, 16–17 animate and inanimate things, 18 philology, 19 music, 20–22 botany, 23–26 waters, air and wind, 27–30 ethics and mental faculties (including the influential discussion of melancholy: 30.1), 31–35 sense organs (eyes, ears, nose, mouth, touch), 36–38 human body (face, body and color of the skin).

This treatise clearly postdates ARISTOTLE although he authored a work by the same title (e.g. *PA* 3.15 [676a18]; *GA* 2.8 [747b5]). How much of Aristotle's work has been incorporated into the extant *Problems* remains a matter of dispute: Louis accepts a greater proportion as genuinely Aristotelian than Flashar. Other sources include THEOPHRASTOS (sections 2, 5, 12–13, 23, and 26 bear the titles corresponding to his *On Sweating, On Tiredness, On Odors, On Waters*, and *On Winds*), and medical writings (HIPPOKRATĒS, DIOKLĒS OF KARUSTOS). There is great variety in the treatment of sources, from wholesale quotations to loose parallels.

Flashar considers the work a **Peripatetic** handbook collecting and summarizing knowledge in the fields of medicine and science. While the author utilizes the four Aristotelian qualities warm and cold, wet and dry as fundamental explanatory principles, there are significant conceptual differences: for example, the concept of *vacua* shows clear parallels to STRATŌN OF LAMPSAKOS.

It is generally agreed that the extant work is the result of several redactions, as evidenced by contradictions and repetitions. Flashar convincingly argues that most of the material dates to the mid 3rd c. BCE, though there are later additions, common with this type of literature. A collection of problems known today as *Supplementa Problematorum* was attributed to Aristotle or Alexander of Aphrodisias in antiquity: Kapetanaki and Sharples (2006).

Also surviving is Ḥunain ibn Isḥāq's Arabic translation of a version of the text composed after 200 CE. Moses ibn Tibbon translated the Arabic into Hebrew in 1264.

Ed.: W.S. Hett, *Aristotle, Problems*, 2 vols., rev. ed. (Loeb: 1970 and 1965); P. Louis, *Aristote, Problèmes*, 3 vols. (*CUF* 1991–1994); L.S. Filius, *The Problemata Physica Attributed to Aristotle. The Arabic Version of Ḥunain ibn Isḥāq and the Hebrew Version of Moses ibn Tibbon* (1999); M.F. Ferrini (with Italian trans.), *Aristotele, Problemi* (2002).

H. Flashar, *Aristoteles, Problemata Physica. Aristoteles Werke in deutscher Übersetzung* (1991); A. Blair, "The *Problemata* as a Natural Philosophical Genre," in: A. Grafton and N. Siraisi, edd., *Natural Particulars* (1999) 171–204.

Oliver Hellmann

Aristotelian Corpus On Sounds (322? – 269? BCE)

A short document *On Sounds* (*De Audibilibus* = *Peri akoustōn*) quoted by PORPHURIOS in his commentary on PTOLEMY's *Harmonics* (67.24–77.18 Düring), and attributed there to ARISTOTLE (51.1, 67.17). The treatise survives in no other ancient source, and in his introduction Porphurios confesses that he quotes only "some of it, abridging it on account of its length." Its Aristotelian authorship has been doubted since the 19th c., but on this issue there is still no consensus; inconclusive arguments have been put forward in favor of HĒRAKLEIDĒS PONTIKOS, THEOPHRASTOS and STRATŌN.

The text, as it stands, is concerned with the generation and transmission of sounds (vocal sounds in particular), and offers explanations of the causes of their pitches and qualities which differ in important details from other 4th c. accounts. Porphurios' ten-page abridgement opens with the general thesis, articulated already by ARKHUTAS (*fr.* 1), that sounds are the result of impacts (*plēgai*) between bodies or between a body and the surrounding air. But where other theorists (both earlier [e.g. Arkhutas] and later [e.g. AELIANUS "THE PLATONIST"] had imagined that the impacts cause the air itself to move, the author of the *De*

Audibilibus proposes instead that the impacts which constitute a sound are transmitted from the point of origin to the point of perception through the stationary, yet flexible, medium of the air. (Examples of bronze statues resonating under the file, and ships' masts being tapped to detect cracks, though not adduced in defense of this pulsation-theory, show a concern to account for the transmission of sound through solid objects as well as through the air.) Impacts, thus transmitted, are diffused widely, each portion of the air conveying to the next by its movement the timbre, as well as the pitch, of the sound. This is effected not by the air being "shaped" (*skhēmatizesthai*), but by each portion of the air being identically moved (*kineisthai*) by a neighboring portion of air. Each of these portions is momentarily contracted and expanded by the pulse of sound which travels through it, but is not pushed or shunted to a new position, as in the acoustic theory of the ARISTOTELIAN PROBLEMS (11.6).

While this theory of the propagation and transmission of sound attempts to improve on earlier ones, it does not, in the *De Audibilibus,* underpin a thesis that higher pitch is caused by a greater frequency of impacts. The author makes an important improvement on Arkhutan acoustics by separating force and speed as distinct variables which cause different qualities in a sound (73.23–24). But by maintaining the view that velocity of transmission is the determinant of pitch, the author falls short of a theory in which differences between frequencies of impact are directly responsible for differences between pitches. Greater frequency of impact at the point of perception will be a consequence of greater velocity of travel, and so, accidentally, higher pitches will be constituted by a greater number of impacts in an equal amount of time; but this higher frequency is a consequence rather than a cause. Aristotle's own objections to the pitch-velocity theory (*De Sensu* 448a) are therefore not avoided in the present text, as they are in the EUCLIDEAN SECTIO CANONIS.

Düring (1932); *Idem, Ptolemaios und Porphyrios über die Musik* (1934); H.B. Gottschalk, "The *De Audibilibus* and Peripatetic acoustics," *Hermes* 96 (1968) 435–460; Barker (1989); Mathiesen (1999).

David Creese

Aristotelian Corpus Situations and Names of Winds (*ca* 300 – 200 BCE?)

This brief and fragmentary text (973a1–b25) lists 11 winds with various local names in the Mediterranean region and some brief etymological explanations: the name of one wind (north) is missing. The list agrees almost exactly with TIMOSTHENĒS' system of winds. A (now lost) drawing of a "wind rose" (illustrating the situations of the winds) is promised at the end of the text. According to the MSS, the text is an excerpt of ARISTOTLE *On Signs*, an attribution accepted by Sider and Brunschön but doubted by other modern scholars, who suggest THEOPHRASTOS' *On Signs* or another unknown **Peripatetic** author as source of the excerpt. Regenbogen assumed that the text was possibly the mutilated end of Theophrastos' *On Winds*, whereas Rehm denied a **Peripatetic** origin.

Ed.: W.S. Hett, *Aristotle, Minor Works* (Loeb, 1936); V. D'Avella, "[Aristotle] On the Locations and Names of the Winds," in D. Sider and C.W. Brunschön, *Theophrastus On Weather Signs = Philosophia Antiqua* 104 (2007), 221–225.

A. Rehm, *Griechische Windrosen* (1916) 94–103; *RE* S.7 (1940) 1412, O. Regenbogen; *RE* 8A.2 (1958) 2350–2351, R. Böker; J.F. Masselink, *De Grieks-romeinse Windroos* (1956).

Oliver Hellmann

pseudo-Aristotle, De Mirabilibus Auscultationibus (250 BCE – 200 CE)

From Athēnaios (12 [541a–b]), and throughout late Antiquity, there is evidence for a *De mirabilibus auscultationibus* (*On Marvelous Things Heard*) circulating under the name of ARISTOTLE. In its present state, it contains 178 chapters probably resulting from merging three pre-existing collections of excerpts, the nucleus of which might be traced back to the 3rd c. BCE. Notwithstanding, internal evidence suggests a date well after Aristotle, so the work's attribution to the Stagirite (possibly due to the similarity between its opening paragraphs, 1–2, 5–6, and 8, and parts of Book 9 of the *Historia Animalium*) is definitely spurious – although it probably contributed greatly to its survival.

In keeping with its composite nature, *De mirabilibus auscultationibus* is structurally muddled. More than one principle of organization is used to order heterogeneous data. Chapters 1–77 and 139–151 display a random thematic arrangement, whereas §§78–136 are clearly organized along geographical lines, albeit with a few stray chapters; §§152–178 – the latest addition to the corpus – are arranged topically. The collection's content varies widely, featuring scattered clusters on zoological marvels (the predominant theme), fire-related ***paradoxa***, curious stones and ore, and wondrous rivers; botanical phenomena, however, are conspicuous by their near-complete absence.

True to its paradoxographical nature, *De mirabilibus auscultationibus* acknowledges numerous sources. In §§1–151, the limited number of sources is of the highest quality: including Aristotle and THEOPHRASTOS on natural phenomena, TIMAIOS on ***paradoxa*** from the west and Theopompos on those from Greece and the eastern part of the world. In the late addendum, §§152–178, the quality declines, with references to PSEUDO-PLUTARCH, DE FLUUIIS, PHILOSTRATOS, and Herodian.

Ed.: *PGR* 221–313; G. Vanotti, *Aristotele, De mirabilibus auscultationibus* (1997).
RE 18.3 (1949) 1137–1166 (§13, 1149–1152), K. Ziegler; Giannini (1964) 133–135; H. Flashar, *Aristoteles, Opuscula 2. Mirabilia* (1990, 3rd ed.); *BNP* 10 (2007) 506–509 (I.B.1, 508), O. Wenskus.

Jan Bollansée, Karen Haegemans, and Guido Schepens

Aristotelian Corpus, Translations into Pahlavi (200 – 900 CE)

The impact of Greek philosophy on Sasanian Iran was particularly significant directly through Pahlavi translations of Greek originals or through Syriac versions of Christian translators; terms like "philosopher," "physicist," and "sophist" are attested with loanwords in Pahlavi sources. ARISTOTLE's influence was seminal not only in physics, but also in later Zoroastrian ethics, especially with regard to *mesótēs*. According to the *Anecdota Syriaca*, Paulus Persa (6th c.) offered a Syriac synthesis of Aristotle's dialectics and logic to King Xusraw I. Aristotelian concepts and categories such as "movement," "time," "space," "nature," "becoming," "change," and "increasing" are well attested in the 9th c. Zoroastrian encyclopedia *Dēnkard*. They seem to be directly inspired by Aristotle's *Generation and Corruption*, translated also in Syriac by Ḥunain (809–876). The doctrine of primal matter, that of the principal elements, and other related concepts were known; all these Greek elements were more or less adapted to the Zoroastrian framework, and had a certain impact on Sasanian astronomy, astrology, and culture.

J.P.N. Land, ed., *Anecdota Syriaca* 4 (1875) 1–30; L.C. Casartelli, *La philosophie religieuse du Mazdéisme sous les Sassanides* (1884); Bailey (1943; 1971) 82–119; R.Ch. Zaehner, *Zurvan. A Zoroastrian Dilemma* (1955; 1971^2); J. de Menasce, *Le troisième livre du Dēnkart* (1973); Panaino (2001).

Antonio Panaino

PSEUDO-ARISTOTLE, On the Kosmos ⇒ Kosmos

Aristotheros (*ca* 250 BCE?)

According to Simplicius, *In de cael.* 2.1.2 (*CAG* 7 [1894] 504.16–505.19), Aristotheros had a dispute with Autolukos, which Simplicius cites as proof that Autolukos failed in his attempts to explain the apparent variation in the distance of the planets from the Earth by means of hypotheses (here, models: *sc.* homocentric spheres). Simplicius does not say what the dispute was about, and we cannot confirm his account in any of its details; indeed, given that Simplicius here reconstructs the past by retrojecting later astronomical theory, it is probably a thorough misrepresentation. Still, from this text one might infer that Aristotheros was at least a contemporary of Autolukos. But, even were Simplicius right in this limited respect, Autolukos' dates are uncertain, so this would hardly help to identify Aristotheros. There is also a report in the anonymous *Vita Arati IV* that Aristotheros was an astronomer and the teacher of Aratos, but this is contradicted in the same text which also says that Aratos was the student of Persaios of Athens as well as by reports elsewhere that he was the pupil of Dionusios of Hērakleia (*Vita Arati I*) or of Menekratēs of Ephesos (*Souda* A-3745). Given the current state of our sources, there is no sensible way to adjudicate this disagreement.

SDS 1.806–839, Alan C. Bowen; *Idem*, "Simplicius and the Early History of Greek Planetary Theory," *Perspectives on Science* 10 (2002) 155–167; *Idem*, "Simplicius' Commentary on Aristotle, *De caelo* 2.10–12: An Annotated Translation (Part 2)" *SCIAMVS* 9 (2008: forthcoming).

Alan C. Bowen

Aristoxenos (*ca* 25 – 50 CE)

Hērophilean physician, Alexander Philalēthēs' student (Galēn, *Puls. Diff.* 4.10 [8.746 K.]), wrote a polemical *On the School of Hērophilos*, attacking even prominent members of his own school for their illogical or defective understanding of medicine and their imprecise, redundant and superfluous definitions, especially of sphygmology, e.g., Bakkheios, Zēnōn, Khrusermos, Apollōnios "Mus," and Hērakleidēs of Eruthrai (Galēn, *Puls. Diff.* 4.7, 4.8, 4.10 [8.734–735, 738–740, 744–747 K.], *Puls. Dign.* 4.3 [955 K.]). Galēn, who may have relied on the text for substantial portions of his *Puls. Diff.* (von Staden 1999: 170–171), praises Aristoxenos' theory of the pulse, which distinguished between distention and contraction, and identified the pulse as a function of the arteries and heart (8.734–5, 955 K.). He prescribed purgatives to patients to maintain **humoral** balance (Caelius Aurelianus, *Acute* 3.134 [*CML* 6.1.1, p. 372]).

von Staden (1989) 559–563; *Idem* (1999) 170–176; *BNP* 1 (2002) 1155 (#2), V. Nutton.

GLIM

Aristoxenos of Taras (350 – 310 BCE)

Traditionally regarded as the major musical authority of the ancient world (hence simply called "the musician"), he was born in Taras to Spintharos, or Mnēsias, a nickname that some scholars think derived from the verb *mimnēskō* ("to remember"), according to the habit widespread among **Pythagoreans** of using epithets related to the sphere of memory. From his father he received his first training in music which, according to the sources, continued with Lamprōs of Eruthrai (perhaps during his stay in Mantinea, whence he moved to Corinth and then to Athens), the **Pythagorean** Xenophilos and, from about 330 BCE,

ARISTOTLE. His failure to gain the **Lyceum**'s headship after his master's death is prominent in the biographical sources because of the harsh resentment it seems to have aroused in him. The *Souda* ascribes 453 writings to Aristoxenos on many different topics: music, history, philosophy and education. However this exceptional number is more likely the total number of book-rolls comprising different works than as independent titles. Among his writings were biographies, including lives of PYTHAGORAS, ARKHUTAS, Sōcratēs, and PLATO; a large number of musicological works (*On Music, On Musical Listening, On Melodic Composition, On Tonoi, On Instruments, On Auloi, On the Boring of Auloi, On Aulos Players, On Tragic Dancing*); and ethical and political writings (*Educational Nomoi, Political Nomoi, Pythagorean Sentences, Customs of Mantineans*). Of all this material only titles or fragments survive, the plurality of which belong to the *Elementa Harmonica*, a treatise which was very influential and became paradigmatic for musical theory in antiquity (not only for the "Aristoxenian" tradition), and to the *Elementa Rhythmica*.

The conventional division of the *Elementa Harmonica* into three books (the third incomplete) has nowadays become almost unanimously rejected thanks to the correct reading, in the earliest MSS, of the title as "Before the (*pro tōn*) *Harmonic Elements*," corrupted throughout the manuscript tradition to "The First Book (*prōton*) of the *Harmonic Elements*." In fact, PORPHURIOS ascribes to Aristoxenos a preliminary treatment of the subject entitled *On principles* (*Peri arkhōn*), which proposed his criteria for a theoretical enquiry on music, perception and reason: that suggests that Book 1 of the *Harmonica* – as its generic content also seems to show – belongs to this separate and more introductory work, while the traditional second book is probably the original beginning of the *Elements*.

According to Aristoxenos, harmonics is a theoretical science (*theōrētikē epistēmē*) concerning audible *melos*, an element which exists in nature as a continuous becoming. Thus, in accordance with the Aristotelian grounds of his methodology, harmonics is a "physics" concerned with melody, but it is only a part of a wider and multifaceted science, as are the sciences of rhythm, meter and instruments. Harmonics, in particular, has the purpose of picking out musical facts – like notes, intervals and scales – grasped by perception (*aisthēsis*), and then of discovering – by means of reason (*dianoia*) – the principles governing the ways in which these elements are combined to form melodic or unmelodic sequences. For the comprehension of music, Aristoxenos states that the harmonic scientist should also use the memory (*mnēmē*) to perceive the *melos* as a process of coming to be, remarking the distance between the **Pythagorean** theory (whose mathematical representation of intervals conceived them only as relations between immovable pitches) and his dynamic approach. The conception of melodic movement of the voice with respect to "place" (*kinēsis kata topon*) is actually one of his most original and lasting concepts, and the description of musical structures as combinations of conjoined and disjoined tetrachords to form bigger arrangements (as the "Great Perfect System," shown in the **diagram**) is the first full account of an extensive scalar system in antiquity. His scientific approach to the subject, overemphasized by himself as absolutely new and innovative, was also directed against earlier empiricists, faulted for having merely sought to catalogue different forms of scales without investigating the principles on which they were constructed. Aristoxenos lists seven subjects of study in harmonics: genera (*genē*, i.e. different arrangements of tetrachords, depending on the tuning of movable notes); notes (*phthongoi*, conceived as dimensionless points lying on a spatial continuum); intervals (*diastēmata*, lit. "distances" between two points in the continuum); scales (*sustēmata*, lit. "combinations of intervals"); *tonoi* (somewhat like "keys" in which scales are placed when they occur in melody); modulations (*metabolai*, variations between systems, genera, keys and so on); and melodic composition (*melopoiïa*, basically the use to which the notes are put in melodies). The discussion of the last three topics has not been preserved in the treatise, but can be reconstructed from later Aristoxenian writers such as KLEONEIDĒS, BAKKHEIOS and GAUDENTIUS, who in their handbooks gave a scholastic exposition of the master's doctrines.

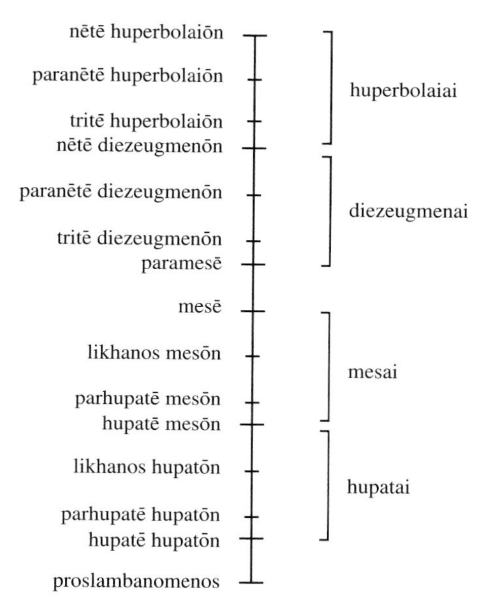

Aristoxenos: the Great Perfect System © Rocconi

Of the *Elementa Rhythmica* we have only a section of Book 2, but we can infer the main topics of its missing portion from later sources such as Bakkheios, ARISTEIDĒS QUINTILIANUS and the Byzantine scholar Michael Psellos, as well as from *POxy* 34 (1968) 2687 plus *POxy* 1 (1898) 9 (identified by scholars as an Aristoxenian source). In this treatise, Aristoxenos asserted that rhythm is a temporal structure imposed on a medium susceptible of rhythmic formation (*to rhuthmizomenon*, i.e. speech, melody, or bodily movement) to which rhythm gives a particular arrangement of *khronoi*. To be "rhythmic," these arrangements need a clear ratio between arsis (*anō*, the up-beat) and thesis (*katō*, the down-beat), and they are defined by reference to a minimal time-length, different from the syllable, to which the performer will match each of his notes, the "primary duration" or *prōtos khronos*. Thus, for the first time in antiquity, rhythm is described as something independent, not inherent in the syllabic structure, therefore no longer subject to verbal prosody. Confirming its importance, Aristoxenos devoted an entire work (*On the Primary Duration*) to this topic, as attested by Porphurios.

Ed.: R. Da Rios, *Aristoxeni Elementa Harmonica* (1954); F. Wehrli, *Aristoxenus* (1967); Barker (1989) 119–189; L. Pearson, *Aristoxenus, Elementa Rhythmica. The fragment of book II and the additional evidence for Aristoxenean rhythmic theory* (1990).

L. Laloy, *Aristoxène de Tarente, disciple d'Aristote, et la musique de l'antiquité* (1904; repr. 1973); A. Bélis, *Aristoxène de Tarente et Aristote: le Traité d'Harmonique* (1986); M. Litchfield, "Aristoxenus and empiricism: a reevaluation based on his theories," *JMT* 32 (1988) 51–73; A.D. Barker, "Aristoxenus' harmonics and Aristotle's theory of science," in Bowen (1991) 188–226; Mathiesen (1999) 294–344; A. Visconti, *Aristosseno di Taranto. Biografia e formazione spirituale* (1999); A.D. Barker, *The science of harmonics in classical Greece* (2007) 113–259.

E. Rocconi

Aristullos (300 – 265 BCE)

Astronomer cited in PTOLEMY, *Alm.* 7.3 for the undated measurement of the declinations of six fixed stars. Ptolemy (*Alm.* 7.1) reports that HIPPARKHOS had access to a few such measurements by Aristullos and TIMOKHARIS and that, on such a basis, he conjectured the precession of the equinoxes. It is doubtful that Aristullos' measurements were made with this question in mind. A better hypothesis would be that they were part of an effort to improve on contemporary accounts of the heavens (*cf.* ARATOS, *Phain.*) by mapping the celestial sphere scientifically (in prose: *cf.* PLUTARCH, *De Pyth.* 18) and, perhaps, to construct a precisely marked celestial globe.

E. Maass, *Aratea* 2 (1892) 123, 151; Goldstein and Bowen (1991).

Alan C. Bowen

Arkadios (200? – 500 CE)

A commentator on PTOLEMY's *Almagest* criticized by EUTOKIOS in his commentary on *De Sphaer. et Cyl.* 2.4 (3.120.8 H.) for his account of compound ratio. Knorr conjectured that this account is found at the end of the PROLEGOMENA TO PTOLEMY'S SUNTAXIS, but the argument, though not impossible, is weak.

Knorr (1989) 168.

Alain Bernard

Arkhagathos of Lakōnika (*fl.* 219 BCE)

PLINY 29.12–13 preserves the tale, by the historian Cassius Hemina (*ca* 150 BCE), of Arkhagathos as the first Greek doctor to come to Rome, his receipt of citizenship, and practice at public expense of surgery and cautery in the centrally-sited "Crossroads of Acilius." His plaster of ***misu***, burnt copper, litharge and ***psimuthion***, in **terebinth** resin, is recorded by CELSUS, 5.19.27, used by THEMISŌN, and still in use in the time of SŌRANOS (CAELIUS AURELIANUS *Chron.* 4.7 [*CML* 6.1.2, p. 778]).

BNP 1 (2002) 975–976 (#3), V. Nutton.

PTK

Arkhebios/Arkesios (190 – 25 BCE)

Wrote on Corinthian proportions, the Ionic temple of Asklēpios at Tralleis whose construction he may have overseen (VITRUUIUS 7.*pr.*12: where the MSS have *ARGEL(L)IVS* – *cf.* 7.5.5 regarding Apaturios of Alabanda at Tralleis – and argued against using the Doric order in temple construction because of faulty and inconsistent proportions (*ibid.* 4.3.1: where the MSS have *(T)ARCHESIVS*). A better restoration than the rare *Arkesi-* (for which *cf.* only *LGPN* 1.80, 2.64) might be Arkhebios. Tralleis was the seat of the Pergamene governor 189–133 BCE (POLUBIOS Book 21, *fr.*46.1–10), and its Asklēpieion may date to this period.

RE 2.1 (1895) 1169, E. Fabricius.

PTK and GLIM

Arkhedēmos (Veterin.) (before *ca* 300 CE?)

Author of a recipe for an ointment for foals quoted by APSURTOS, preserved in the *Hippiatrika* (*Hippiatrica Parisina* 837 = *Hippiatrica Berolinensia* 130.14). Apsurtos describes Arkhedēmos as *hippotrophos*, "horse-keeper" or "breeder."

CHG v.1; McCabe (2007).

Anne McCabe

Arkhedēmos of Tarsos (175 – 125 BCE)

Arkhedēmos studied **Stoic** philosophy under DIOGENĒS OF BABYLŌN and ZĒNŌN OF TARSOS at Athens and subsequently opened a school in Seleukeia. He is credited with several titles including *On Voice* which defined "voice" as body (*soma*: DIOGENĒS LAËRTIOS 7.55), *On Elements*, a discussion of the active and passive principles and the four elements (D.L. 7.134, 136), and *On the Master Argument*, probably an attack on DIODŌROS OF IASOS. In some work he taught that the ***hēgemōn*** of the ***kosmos*** was fiery and resided in the center of the Earth, thus explaining volcanoes (*SVF frr.*15–16).

DPA 1 (1989) 331–333, R. Goulet; *GGP* 4.2 (1994) 634–635, P. Steinmetz; *ECP* 48–49, Tiz. Dorandi.

GLIM

Arkhelaos (Geog.) (320 – 50 BCE)

Described the lands traversed by Alexander, according to Dēmētrios of Magnesia in DIOGENĒS LAËRTIOS 2.17. SOLINUS 52.18 cites an *Archelaus* on India, as does PSEUDO-PLUTARCH, *Rivers* 1.3 (1150B).

(*)

PTK

Arkhelaos (Lithika) **(60 BCE – 10 CE?)**

Wrote *On rivers*, in at least 13 books, and *On stones* (PSEUDO-PLUTARCH *De Fluu.* 1.3 [1150B], 9.3 [1155D]; IŌANNĒS STOBAIOS 4.36.17). His identity, as well as his very existence are uncertain. Bidez considered him authentic, while Jacoby, collecting the fragments under ARKHELAOS OF KAPPADOKIA, thought them spurious or not attributable to our Arkhelaos. However, since PLINY quotes the Kappadokian in relation to amber, his identification with our author could be accepted. Finally nothing excludes identifying our man with ARKHELAOS OF KHERSONĒSOS.

Bidez (1935) 31; *FGrHist* 123; Giannini (1964) 111–112; *KP* 1.503 (#6), H. Volkmann; De Lazzer (2003) 70; Roller (2004); I. Ramelli, in G. Reale *et al.*, edd., *Diogene Laerzio. Vite e dottrine dei più celebri filosofi* (2005) 1330–1331, n.67.

Eugenio Amato

ARKHELAOS (ALCH.) ⇒ HĒLIODŌROS

Arkhelaos (Med.) (200 – 700 CE)

Credited with medical fragments. One, in the 15th c. MS of Bologna, Biblioteca Universitaria, 3632, ff. 43–45 (Baffioni), discusses pediatrics and seems to come from a larger treatise based on GALĒN. Another appears in *On intestines* by Iōannēs bishop of Prisduana (supposedly of the 12th century), contained in manuscript Paris, BNF, *graecus* 2286, f.127^{V}, written by the mid-14th c. monk, philosopher and physician Neophutos Prodromēnos in Constantinople. In the title of the *On intestines*, Arkhelaos is listed between PALLADIOS and STEPHANOS OF ALEXANDRIA, and these three names are followed by *kai diaphorōn palaiōn iatrōn*. These two facts suggest that Arkhelaos might have been a *palaios iatros* and that he was a member of the same group as Palladios and Stephanos of Alexandria, that is, the School of Alexandria. Significantly, the Paris MS comes from the most important hospital in

14th c. Constantinople, the Kralē, with which the Bolognese MS is linked, suggesting that Arkhelaos' works were still in use at that time.

Ed.: G. Baffioni, "Inediti di Archelao da una codice Bolognese," *Bollettino del Comitato per la preparazione dell' edizione nazionale dei classici greci e latini* ns 3 (1954) 57–76.
Diels 2 (1907) 16.

Alain Touwaide

Arkhelaos (Veterin.) (*ca* 100 – 300 CE?)

Author of a recipe for a ***trokhiskos***; the recipe is preserved in the *Hippiatrika* as a quotation in the treatise of APSURTOS (*Hippiatrica Parisina* 772 = *Hippiatrica Berolinensia* 129.27).

CHG v.1; McCabe (2007).

Anne McCabe

Arkhelaos of Athens (*ca* 460 – 440 BCE)

A student of ANAXAGORAS and teacher of Sōcratēs, he propounded a cosmology broadly similar to that of Anaxagoras, positing material stuffs as basic principles. Hot and cold caused these to differentiate into the four elements recognized by EMPEDOKLĒS: earth, water, air and fire. Moisture gathered in the center of the world, where heat caused some of it to turn into air and rise, while another part solidified into earth. The earth is a concave disk, which originally was swampy in the middle so as to produce living things. Mind is found in all animals, and it mixes with the stuffs of the world. Arkhelaos seems also to have reflected on ethics and political theory but apparently made no significant advances in scientific theory or observation.

DK 60; KRS 385–389; V. Tilman, "Archélaos d'Athènes," *Revue de Philosophie Ancienne* (2000) 65–107.

Daniel W. Graham

Arkhelaos (of Hērakleia Salbakē?) (40 – 95 CE)

ASKLĒPIADĒS PHARM., in GALĒN *CMLoc* 9.6 (13.312 K.), preserves his recipe to treat anal prolapse, compounded from lead-slag, rose blossoms, rhubarb, and myrrh. Uncertain is the identification of our Arkhelaos with the son of Euneikos, physician, priest, gymnasiarch, and stephanophoros whose virtue and skill is acclaimed in an inscription from Hērakleia Salbakē, *ca* 50 CE (Robert and Robert 2.177, #70).

J. and L. Robert, *La Carie II: Le Plateau de Tabai* (1954); *RE* S.14 (1974) 56 (#37a), J. Benedum.

GLIM

Arkhelaos of Kappadokia (36 BCE – 17 CE)

Was installed by M. Antonius as king of Kappadokia, which he ruled until his death. PLINY makes him the author of writings on animals, agriculture, and stones (1.*ind*.8–9, 17–18, 37). However, the anecdotes on animals (8.202, 218) are copied from Varro, *RR* 2.3.5, 3.12.4, and Varro's source is clearly ARKHELAOS OF KHERSONĒSOS (3.16.4; note that Varro mentions Arkhelaos several times but does not cite him in his index of sources). It would appear that Pliny's claims that the king wrote on animals, and perhaps agriculture as well, are based on a mistaken inference from Varro's text. Since Jacoby, scholars have identified the king with ARKHELAOS (GEOGRAPHER); the fragments on *lithika* from Pliny are assigned to this work (37.46, 95, 104, 107). But as DIOGENĒS' source was Dēmētrios of Magnesia, *fl.* 50 BCE, this identification too must be rejected. Perhaps the most economical hypothesis is to make one person of the geographer, Pliny's writer on stones, the Arkhelaos of SOLINUS 52.18 (on India), and ARKHELAOS (LITHIKA).

RE 2.1 (1895) 451–452 (#15), U. Wilcken and H. Berger; *FGrHist* 123; *OCD3* 144, S. Harrison.

Philip Thibodeau

Arkhelaos of Khersonēsos (270? – 180? BCE)

Mysterious Egyptian author (ANTIGONOS, *Mir.* 19) of *Special Natural Phenomena* (*Idiophuē*, *cf.* Ath. *Deipn.*, 9 [409c]; DIOGENĒS LAËRTIOS 2.17). He seemingly transposed into elegiacs a prose collection of marvels (Antigonos, *Mir.* 89) for Ptolemy III. Arkhelaos was more poet and paradoxographer (AELIANUS, *NA* 2.7) than *phusikos*. His only preserved epigram tells how some animals arise from corpses of other animals, following the principle of **sympathy** or similitude: scorpion from crocodile, wasp from horse, snake from human spinal column (Antigonos, *Mir.* 19).

GGLA 1 (1891) 465–467; *RE* 2.1 (1895) 453–454 (#34), R. Reitzenstein.

Arnaud Zucker

Arkhestratos (250 – 150 BCE)

Musical theorist, quoted by PORPHURIOS on the authority of DIDUMOS' writings and recalled by PHILODĒMOS as a scholar whose approach to harmonics was based more on reason than perception, hence not Aristoxenian. According to Athēnaios (*Deipn.* 14 [634d]), he wrote *On Aulos Players*.

RE 2.1 (1895) 459 (#13, 14), C. von Jan; *BNP* 1 (2002) 984–985 (#3), F. Zaminer; A.D. Barker, "Diogenes of Babylon and Hellenistic Musical Theory," in C. Auvray-Assayas and D. Delattre, *Cicéron et Philodème. La polemique en philosophie* (2001) 353–370.

E. Rocconi

ARKHIAS ⇒ MOSKHIŌN

Arkhibios (50 – 75 CE)

Empiricist physician, lived after ASKLĒPIADĒS OF BITHUNIA (he knows a pharmaceutic remedy by him: GALĒN *CMGen* 13.849 K.) and before ASKLĒPIADĒS PHARM. (from whom Galēn *Antid.* 14.159–160 K. derives an antidote by Arkhibios). Besides being a pharmacologist, he was a surgeon: his activity is documented by the surgical procedure in OREIBASIOS *Coll.* 46.11.31 (*CMG* 6.2.1 p. 222: from HĒLIODŌROS) and *P. Berol.* 9764 (Pack^2 #2354) on the teaching of surgery, against **Dogmatic** medicine (perhaps also a commentary on HIPPOKRATIC CORPUS, APHORISMS 1.1). The name of Arkhibios is mentioned amongst the sources of Book 18 of PLINY, but what Pliny says about him at 18.294 is surprising: Arkhibios suggested to one king Antiokhos of Syria that to avert bad weather, one buries in a field a pot with a stolen frog inside. It is not certain that the physician Arkhibios in Lucian *Gall.* 10 alludes to this Arkhibios.

Ed.: Deichgräber (1930) 21, 209–210, 407 (fragments).

RE 2.1 (1895) 466 (#5), M. Wellmann; Manetti and Roselli (1994) 1536; Marganne (1998) 13–34; Ihm (2002) #17.

Fabio Stok

Arkhidēmos (350 – 290 BCE)

Galēn, *Simples* 2.5 (11.471–474 K.), cites Dioklēs of Karustos' work *Arkhidēmos*, in which Dioklēs rejected Arkhidēmos' view that oil-massages hardened the skin, which absorbed the olive oil and blocked the pores, thus impeding the normal flow of secretions and ***pneuma*** through the flesh; he therefore preferred dry massage, *ibid.* 2.6 (11.477). Pliny, 1.*ind.*12–13 (trees), 29–30 (animal-based drugs), and 33–35 (metals and pigments) cites the doctor *Archedemus*, who may be the same person.

van der Eijk (2000–2001) *fr.*185.

PTK

Arkhigenēs of Apameia (95 – 115 CE)

Greek physician from Apameia, possibly son of Philippos of Rome. Arkhigenēs studied medicine with Agathinos, practiced in Rome under Trajan, and died aged 63, his name nearly synonymous with "physician" (Juvenal 6.236, 13.97, 14.251). He either taught medicine, or had a group of followers (Galēn, *CMLoc* 7.1 [13.14 K.]). Though fundamentally a **Pneumaticist**, he also incorporated elements from contemporary medical thought, especially the four **humors** and the Hippokratic *kairos*. As a result, he was considered eclectic as early as pseudo-Galēn Introductio 4 (14.684 K.), and was even credited with the foundation of an "eclectic" school, just like Agathinos.

A productive writer, Arkhigenēs worked on physiology, pathology, and therapy. In physiology, he followed mainly the **Pneumaticist** system, and particularly explored sphygmology. He refined Agathinos' definition of the pulse (*sustolē* and *diastolē*), adding that each phase is a movement that is "natural" (***phusikē***), i.e., involuntary. He also classified different types of pulse according to qualities (with eight major types, Harris 1973: 251–257). His work on sphygmology (Galēn, *Diff. Puls.* 2.4 [8.576 K.]), and Galēn's seven-book commentary (*Febr. Diff.* 2.8 [7.365 K.]), are lost (Ihm 2002: #89). Galēn preserves abundant passages, yet criticizes Arkhigenēs' many distinctions of pulse quality (*Diff. Puls.* 2.10 [8.625–635 K.]) as well as his opacity (*Febr. Diff.* 2.8 [7.365 K.]).

Arkhigenēs wrote two works on pathology: *Peri Topōn* and *Peri Peponthotōn Topōn* in three books: Galēn considered the latter "the best of all works" previously written on the topic (*Cris.* 2.8 [9.670, 672 K.]). Arkhigenēs wrote on fevers (*Peri Puretōn Sēmeiōseōs: ibid.* [9.668–669, 672 K.]), on the development of diseases, i.e. the Hippokratic *kairos* (*Peri tōn en tais Nosois Kairōn*: Galēn, *De totius morb. temp.* [7.461 K.]), and on chronic diseases (*Tōn Khroniōn Pathognōmonika*: Galēn, *De Locis Affect.* 3 [8.203 K.]). He collected his letters of medical advice to friends and colleagues (*ibid.* 3.5 [8.150 K.]).

Arkhigenēs compiled an overview of surgery (*Sunopsis tōn Kheirourgoumenōn*: Oreibasios *Coll.* 45.29 [*CMG* 6.2.1, p. 190], with scholia *ad locum*), influenced by Leōnidas of Alexandria. He supposedly wrote on acute and chronic diseases, known through Arabic sources (Ullmann 1972: 69–70); though Oreibasios (*Coll.* 8.1, *CMG* 6.1.1, p. 247) cites it as a work only on chronic diseases. Arkhigenēs also wrote on *materia medica*, devoting at least an entire book to castoreum (Galēn, *Simples* 20.15 [13.337 K.]) and another to hellebore (Galēn, *in Hipp. de Humor.* 1 [16.124 K.]). Arkhigenēs also composed a treatise on medicines by types (*kata genos pharmaka*: Galēn, *CMLoc* 1.8 [12.468 K.]), heavily extracted by Galēn, and perhaps on medicines (*Peri Boēthēmatōn*, Wellmann 1895: 485). Contrary to Wellmann (1895: 486), Arkhigenēs did not write a text on toxicology, but seems rather to have included this topic in his pharmaceutical treatise. According to the extant fragment (pseudo-Aelius Promotus, §58, pp. 30–31 Ihm), the work listed medicines and detailed the symptoms for which they were prescribed.

Arkhigenēs' works survive only in fragments preserved by Galēn (Fabricius 1972: 198–199), Oreibasios, Aëtios of Amida, and Paulos of Aigina (for the latter two, Brescia), and in some Byzantine MSS (Olivieri; Brescia; and Calabrò). Some works, intact or fragmentary, were still known in 14th c. Constantinople, in the Prodromou monastery. His treatises on pathology and medicines by types were used by Neophytos Prodromenos in his work on dental pathologies (contained in MSS Paris, BNF, *graecus* 2286 and Athens, BN, 1481), and in the anonymous treatise on toxicology contained in MS *Vat. graec.* 299 (i.e., pseudo-Aelius Promotus).

Ed.: A. Olivieri, "Frammenti di Archigene," *Memorie della Real Accademia di Archeologia, Lettere e Belle Arte della Società Reale di Napoli* 6 (1938) 44–46; C. Brescia, *Frammenti medicinali di Archigene* (1955); G. Larizza Calabrò, "Frammenti inediti di Archigene," *Accademia Nazionale dei Lincei, Bolletino del Comitato per la preparazione della Edizione nazionale dei classici greci e latini* ns 9 (1961) 69–72.

Wellmann (1895) 19–22; *RE* 2.1 (1895) 484–486, *Idem*; Diels 2 (1907) 16–17; Kudlien (1968) 1099; *KP* 1.507, *Idem; DSB* 1.212–213, J. Stannard; *GAS* 3 (1970) 61; Ullmann (1972) 159; A.D. Mauroudes, "O iatros Archigenes," *Ellēnika* 36 (1985) 278–285; *OCD3* 145, V. Nutton; *BNP* 1 (2002) 989–990, *Idem*.

Alain Touwaide

Arkhutas (350 – 90 BCE)

Cassius Dionusios excerpted from a work on agriculture attributed to Arkhutas of Taras (Varro, *RR* 1.1.8–10, *cf.* Columella, 1.1.7), while Diogenēs Laërtios (8.82), following Dēmētrios of Magnesia's discussion of homonymous persons, considered the agronomist and the philosopher to be different people. It is reasonable to assume that all the sources knew a pseudo-Arkhutan work on farming, which may have discussed cereals, livestock, poultry, viticulture, and arboriculture (*cf.* Pliny, 1.*ind*.8, 10, 14–15, 17–18).

RE 2.1 (1895) 602 (#6), M. Wellmann.

Philip Thibodeau

Arkhutas of Taras (400 – 360 BCE?)

Arkhutas/Archytas (b. *ca* 435 BCE), the last prominent representative of ancient **Pythagoreanism**, is a rare example of a brilliant mathematician and an original thinker who achieved success in ruling a state. He was elected seven times in succession as a *stratēgos* of Taras, at that time one of the most powerful cities of Greece; as a *stratēgos-autokratōr* he headed the union of the Greek cities in Italy (A1–2 DK). Some of Archytas' original works, e.g. *On mathematical sciences*, *Diatribae*, and *Harmonics*, are preserved in several fragments and indirect testimonies. Most writings bearing his name belong to the late Hellenistic pseudo-**Pythagorean** literature, in which Archytas, judging by the number of the forged treatises (45), was even more popular than Pythagoras.

Archytas actively and fruitfully took up all the sciences of **Pythagorean** *quadrivium* (arithmetic, geometry, harmonics, astronomy), which he regarded as akin (B1). He was the first to solve the famous problem of doubling the cube (A14) by having found two mean proportionals between two given lines ($a{:}x = x{:}y = y{:}2a$; $x^3 = 2a^3$). His remarkable stereometrical construction that for the first time introduces movement in geometry employed the intersection of the cone, the torus and the half-cylinder, which produced the necessary curve. Archytas' discoveries might have prompted his pupil EUDOXOS OF KNIDOS to develop a similar kinematic theory for the motion of the heavenly bodies. Archytas' arithmetic was closely related to harmonics. He demonstrated that between numbers in the ratio $(n+1){:}\ n$ there is no mean proportional (A19), hence the basic harmonic intervals, e.g. the octave (2:1), the fourth (4:3) and the fifth (3:2) cannot be divided in half. His researches in acoustics combined mathematics with empirical observations and experiments, though not always with correct results: following HIPPASOS, he considered the pitch of a sound to depend on the velocity of its propagation (B1); Ciancaglini questioned this standard interpretation. These and other studies of Archytas (A16–17, B2) completed Pythagorean harmonics, which was further advanced by the EUCLIDEAN SECTIO CANONIS. In astronomy, contrary to the subsequently dominant scheme, he argued for an unlimited universe (A24).

In physics Archytas developed the mathematical approach characteristic for **Pythagoreanism**: any motion occurs according to proportion (*analogia*). In "natural," circular motion it is "the proportion of equality," for "it is the only motion that returns to itself" (A 23a), as in the circular motion of the heavenly bodies. The causes of mechanical motion are the unequal and uneven (A23), e.g., unequal arms of the lever. The ARISTOTELIAN CORPUS MĒKHANIKA drew upon Archytas' discoveries. There are grounds to regard him as a founder of mechanics and, possibly, of optics (A1, 25).

Following PHILOLAOS, Archytas was engaged in philosophical analysis of mathematics, in particular of its epistemological potential (B1, 3–4). He taught that arithmetic promotes consent and justice in the society and even improves morality (B3). PLATO's first trip to Italy (388 BCE) started his long acquaintance with Archytas. Though their relationship was not devoid of rivalry, it was Archytas' intervention that made possible Plato's return from his trip to Surakousai (361 BCE), where he was kept by the tyrant Dionysius II. Archytas was an important source of Plato's knowledge of Pythagoreanism and stimulated many of his general ideas: on the ruler-philosopher, on beneficial influence of mathematics on the soul, on mathematical sciences as a threshold of dialectic, etc. Mathematics, in which Archytas was the main expert in his generation, served as a model for Plato's theory of ideas and for Aristotle's logic. Aristotle devoted two special works to Archytas' philosophy (A13); the **Peripatetic** ARISTOXENOS, whose father was close to Archytas, wrote his biography.

DK 47; Thesleff (1965); F. Krafft, *Dynamische und statische Betrachtungsweise in der antiken Mechanik* (1970); van der Waerden (1979); Barker (1989); G.E.R. Lloyd, "Plato and Archytas in the Seventh Letter," *Phronesis* 35 (1990) 159–173; C.A. Ciancaglini, "L'acustica in Archita," *Maia* 50 (1998) 213–251; M. Burnyeat, "Archytas and optics," *Science in context* 18 (2005) 35–53; C.A. Huffman, *Archytas of Tarentum: Pythagorean, Philosopher and Mathematician King* (2005); Zhmud (2006).

Leonid Zhmud

Arrabaios (of Macedon?) (250 BCE – 25 CE)

ASKLĒPIADĒS PHARM. cites his "Pontic" recipe for blood-spitting, GALĒN, *CMLoc* 7.4 (13.83 K.): reduce bear-berry (*arkou staphulos*) by one third, boiling in rainwater. Known only from Macedon, *LGPN* 4.48 (5th–2nd cc. BCE; *cf.* Errabaios, 4.127; see Krahe, *Lexicon altil-*

lyrischer Personnamen [1929]), the name is given by Kühn as "Arrabianus," presumably due to "Arabianus," attested from the 2nd c. CE (*LGPN* 2.48, 4.40); *cf.* Arabaiās (Solin 2003: 1.667, 2nd/3rd c. CE). Galēn, *Antid.* 2.12 (14.179–180 K.) records the scorpion antidote, containing **galbanum, terebinth**, myrrh, raw sulfur, etc., of *APABAΘHBAIOY*, probably *Arabbaiou*, with a marginal "*AΘH*" taken into the text. CELSUS 5.18.16 records the emollient of an *anonymus Arabs*, probably a mistake by Celsus for "Arabaios." The complex kidney-pill cited by ANDROMAKHOS, in Galēn *CMLoc* 10.1 (13.324 K.), from "a Macedonian" more likely belongs to THEODŌROS or ZŌILOS.

Fabricius (1726) 78, 85.

PTK

ARRIANUS ⇒ FLAUIUS ARRIANUS

Arrianus (210 – 220 CE)

Poet who translated VERGIL's *Georgics* into Greek. His other works included an epic about Alexander the Great, and a panegyric for "Attalos of Pergamon," presumably a late Attalos like Claudius Attalos Paterculianus, governor of *Thracia* and *Cyprus* under Elagabalus.

FGrHist 143F15; *SH* 207; S. Swain, "Arrian the epic poet," *JHS* 111 (1991) 211–214.

Philip Thibodeau

Arruntius Celsus (200 – 350 CE)

PRISCIANUS OF CAESAREA quotes a short fragment in Latin (*de fig. num.*, 9) from an unnamed work (metrological or philological) of an Arruntius, on the etymology of *sestertius*. He is probably Arruntius Celsus, a grammarian who wrote on Terence and VERGIL. Priscian derives the word *sestertius* from *semis tertius*, based on Arruntius' claim that the *sestertius* "long ago" was worth two asses and a half (*dupondius et semis*), "when the *denarius* was ten asses."

RE 2.1 (1895) 1265 (#16), G. Goetz; *PLRE* 1 (1971) 194; *BNP* 2 (2003) 30 (#II.9), P. Gatti.

Mauro de Nardis

Arsenios (300 – 400 CE)

Greek physician, prescribed **pessary** laxatives used by Arsinoë and Saluina (pseudo-THEODORUS p. 338.4 Rose). In his *Letter to Nepotianus*, in scope similar to the HIPPOKRATIC OATH, but Christianizing in tone, Arsenios described the qualities and duties of the ideal physician, vigilant in study, modest in character and appearance.

RE S.3 (1918) 162 (#2a), R. Ganschinietz; E. Hirschfeld, "Deontologische Texte des frühen Mittelalters," *AGM* 20 (1928) 353–371; *BNP* 2 (2003) 33–34, V. Nutton.

GLIM

ARSINOË ⇒ MARSINUS

Artemidōros (Astron.) (210 – 215 CE)

Wrote a commentary on PTOLEMY's *Almagest*, a fragment of which is extant: *CCAG* 8.2 (1911) 129–130.

Neugebauer (1975) 948–949.

PTK

Artemidōros Capito (115 – 135 CE)

Greek physician, relative of DIOSKOURIDĒS OF ALEXANDRIA (*CMG* 5.9.1, p. 113), together frequently cited by GALĒN. Artemidōros, following Dioskouridēs' re-attribution of many Hippokratic works, published an edition of the entire HIPPOKRATIC CORPUS which Hadrian valued highly (15.21–22 K., 18B.631 K.). Galēn criticizes Artemidōros and Dioskouridēs for greatly emending text and modernizing language (17B.104 K., 19.83 K.), but preserves Artemidōros' recipe for treating scars left by tumors (12.828–829 K.). Stratōn, in a series of epigrams satirizing "types" of (mostly fictive) physicians, lampoons Artemidōros' eye-salve which destroyed the vision of keen-sighted Khrusēs (*Anth.* 11.117).

J. Ilberg, "Die Hippokratesausgaben des Artemidoros Kapiton und Dioskourides," *RhM* 45 (1890) 111–137; *RE* 2.1 (1895) 1332 (#34), M. Wellmann; *KP* 1.618 (#5), F. Kudlien; Smith (1979) 234–240; Manetti and Roselli (1994) 1617–1625; *BNP* 2 (2003) 62 (#8), V. Nutton.

GLIM

Artemidōros of Daldis (*ca* 150 CE)

Of Ephesos, but called himself "of Daldis" in deference to his mother's birthplace, where Apollo, god of divination, was the principle deity. His five-book *Oneirokritikon*, the sole extant example of the popular ancient genre of dream interpretation, consists mostly of a copious catalogue of dreams and the results they portend. The *Souda*, GALĒN, and pseudo-Lucian in the *Philopatris* mention him as a famous dream interpreter.

His work shares characteristics with travel literature, encyclopedias, empirical medical tracts, and the other cataloguing genres thriving during his period. The first three books are addressed to "Cassius Maximus," likely Maximus of Tyre (*ca* 125–185 CE). Books 1 and 2 organize dreams topically and chronologically according to the life of a Roman male – from dreams of birth to dreams of death. The third book is presented as a supplement, adding anything omitted from the first two. Books 4 and 5, addressed to his son, also named Artemidōros, appear more rudimentary and pedagogical. These two books are full of fatherly advice to the practicing dream interpreter, much of it consonant with the tips for aspiring itinerant doctors in the HIPPOKRATIC CORPUS AIRS, WATERS, PLACES.

The *Oneirokritikon* opens with a discussion of theory, though the multiple schematic formulas introduced are an overlapping and sometimes redundant succession of taxonomies more than a coherent synthesis. Though his interest in systematization is somewhat half-hearted, the real engine of the work is Artemidōros' acute interest in his subjects' personal details, local customs, and peculiarities, resulting in a guide of unmatched usefulness to the personal lives of ancient dreamers. His method is decidedly empirical, showing some influences from skepticism and **Epicureanism**. He endorses broad travel and wide reading of details. Aspects of **Stoicism** are also recognizable, since most of his past authorities are influenced by that school, and his notions of the soul and of how dreams are produced resonates with **Stoicism** as well.

Ed.: R.A. Pack, *Artemidori Daldiani Onirocriticon* (1963); trans. R.J. White, *The Interpretation of Dreams. The Oneirocritica of Artemidorus* (1975).

C. Blum, *Studies in the Dream-Book of Artemidorus* (Diss. Uppsala, 1936); S.R.F. Price, "The Future of Dreams: From Freud to Artemidorus," *P&P* 113 (1986) 3–37; J. Winkler, *The Constraints of Desire* (1990) 17–44.

Peter Struck

Artemidōros of Ephesos (104 – 101 BCE)

Greek geographer, author of an 11-book geographical description of the world preserved in an epitome by Marcianus of Hērakleia. The work, divided into three sections (Europe, Libya and Asia), included distances between sites and measurements of geographical features along the lines of a traditional ***periplous*** perhaps partially based on his own travels. Part of it was devoted to *Ionic Notes* (*upomnēmata*), which may have been a separate work. The work also included calculations of the measurements of the inhabited world. Artemidōros started his description of the ***oikoumenē*** with the Iberian peninsula as did also Skulax of Karuanda and Skumnos of Khios before him and Strabōn, Pliny and Dionusios of Alexandria after him. Book 1 was introductory, Book 2 described Spain and Lusitania, Book 3 Gallia, and so forth around the Mediterranean. Artemidōros' sources were mainly Agatharkhidēs, Megasthenēs and the geographers of Alexander the Great, and he himself became an important source for later geographers including Strabōn. Artemidōros' work had been known only through literary references, but recently a papyrus excerpt of the text was discovered including a *prooimion* saying that geography is a branch of philosophy, and a description of Spain – its name, political situation, coasts and distances between sites. Micunco casts doubt on the authenticity of the animal drawings on the *verso*, Canfora on the text of Artemidōros, in that papyrus.

GGM 1.574–576; *FGrHist* 438; C. Gallazi and B. Kramer, "Artemidor in Zeichensaal. Eine Papyrusrolle mit Text, Landkarte und Skizzenbüchern aus späthellenistischer Zeit," *APF* 44 (1998) 189–208; S. Settis and C. Gallazzi, *Le tre vite del Papiro di Artemidoro: Voci e sguardi dall'Egitto greco-romano* (2006); St. Micunco, "Figure di animali: il *verso* del papiro di Artemidoro," *Quaderni di Storia* 64 (2006) 5–43; L. Canfora, "Postilla Testuale Sul Nuovo Artemidoro," *ibid.* 45–60.

Daniela Dueck

Artemidōros of Parion (70 – 50 BCE)

Wrote an account of the ***kosmos*** collecting opinions of Anaxagoras, Apollōnios of Mundos, and others; entitled *Phainomena*, if Boëthos of Sidōn (Peripatetic) refers to the same man, or if he is the same as the writer on Aratos (Robiano). Seneca, *QN* 1.4.3–4, describes Artemidōros' explanation of rainbows as specular reflections from clouds, and 7.13 his theory of multiple normally unseen orbiting bodies, under a heaven congealed from atoms (Goulet sees an **Epicurean**), and with apertures admitting occasional extra-cosmic fire.

DPA 1 (1989) 604, P. Robiano and 614, R. Goulet; P.T. Keyser, "On Cometary Theory and Typology from Nechepso-Petosiris through Apuleius to Servius," *Mnemosyne* 47 (1994) 625–651 at 649–650.

PTK

Artemidōros of Pergē, Cornelius (75 – 70 BCE)

Kritōn, in Galēn *CMLoc* 5.3 (12.828–829 K.), preserves his pill for facial growths (copper flakes, ***khalkanthon***, and alum, in clear carpenter's glue and vinegar). Cicero, *Verr.* II 3, provides the ethnic and *nomen* (54), and scurrilous anecdotes (*ibid.* 69–60, 117, 138). Diels (1905–1907) 2.19 records a British Museum MS, 16C XVI (16th c.), f.8, of his work *On Urines* (in epitome?). *Cf.* perhaps Cornelius (Pharm.).

RE 2.1 (1895) 1332 (#33), M. Wellmann; Korpela (1987) 157–158.

PTK and GLIM

Artemidōros of Sidē (90 – 30 BCE)

Follower of ERASISTRATOS, who wrote on pathology, explaining ***hudrophobia*** as an affection of the upper gastro-intestinal tract, and denied that there could be any new diseases, including ***hudrophobia***: CAELIUS AURELIANUS *Acute* 3.113, 118 [*CML* 6.1.1, pp. 358, 362]); he is there listed before ARTORIUS, supplying the likely *terminus ante*. He also defined cardiac disease as an inflammation in the region of the heart: *ibid.*, 2.163 (p. 242).

RE 2.1 (1895) 1332 (#32), M. Wellmann.

PTK

Artemisius Dianio (200 – 400 CE)

MARCELLUS OF BORDEAUX ascribes two recipes to an Artemius (13.17, *CML* 5, p. 228) or Artemisius (36.54, p. 614) Dianio, whose name seems somehow related to the small island Dianium, near the Etruscan coastline, called Artemisium by the Greeks (*cf.* PLINY 3.81 and others), today Giannutri, where remain important ruins of a 1st/2nd c. BCE Roman villa. The first recipe is a toothpaste against the gnashing of teeth; the second one is a cure for gout (*podagra*).

RE 2.2 (1896) 1445 (#5), M. Wellmann.

Fabio Stok

Artemōn (Epicurean) (*ca* 240 – 180 BCE)

Teacher of PHILŌNIDĒS, who in turn wrote on Artemōn's commentary on EPICURUS' *On Nature*.

DPA 1 (1989) 615, T. Dorandi.

PTK

Artemōn (Med.) (20 BCE – 25 CE)

PLINY lists him, apparently in chronological order, after DĒMOKRITOS, APOLLŌNIOS "MUS," and MELĒTOS, and before ANTHAIOS SEXTILIUS (1.*ind.*28 and 28.7–8) as giving medicines *from* the human body, and records his quasi-magical remedy for epilepsy (night-drawn spring-water drunk from a dead man's exhumed skull: *cf.* HĒRODOTOS 4.65). SCRIBONIUS LARGUS in ASKLĒPIADĒS PHARM., in GALĒN *CMLoc* 4.7 (12.780 K.), records the "Artemonion" collyrium used by IULIUS BASSUS, containing antimony, saffron, myrrh, ***psimuthion***, white pepper, etc. in gum and wine.

RE 2.2 (1896) 1447 (#21), M. Wellmann.

PTK

Artemōn of Kassandreia (250 – 150 BCE?)

Certainly lived after the mid-3rd c. BCE, because of the reference he makes, in one of his fragments, to the grammarian Dionusius Skutobrakhiōn (Ath. *Deipn.* 12 [515d–e]); he wrote treatises on several topics, *On collecting of books*, *On the use of books* (probably a part of the same work previously quoted) and *On the Guild of Dionusos*, whose title seems to refer to guilds of theatrical artists – musicians as well as poets and actors – called "Artists of Dionusos" (*Dionusiakoi tekhnitai*), active in several parts of Greece from the late 3rd c. BCE. Of this musical work we have two fragments preserved by Athēnaios, our main source on Artemōn: the former is concerned with the musician Timotheos of Milētos, accused by Lacedaemonians of fitting too many strings on his instrument, the latter with Puthagoras of Zakunthos.

FHG 4.340–343; *BNP* 2 (2002) 69–70 (#1), F. Montanari.

E. Rocconi

Artemōn of Klazomenai (450 – 430 BCE)

Designed rams and tortoises for Periklēs in the siege against Samos (440–439 BCE: Ephoros *fr.*194 = Diodōros of Sicily 12.28 and Plutarch *Periklēs* 27.3–4). Hērakleidēs of Hērakleia Pontikē linked this Artemōn with a man (described by the poet Anakrēon, *frr.*372, 388) who rode in litters (*Periphorētos*), was notorious for his cowardice, and who had slaves hold shields over his head during sieges. Aelianus cites the *Annals of Klazomenai* by another Artemōn describing a destructive winged pig (*HA* 12.38).

RE 2.2 (1896) 1445 (#1), J. Toepffer.

PTK and GLIM

M. Artorius (55 – 27 BCE)

Brought a warning to the future emperor Augustus before the battle of Philippi (42 BCE, Vell. Pat. 2.70.1). Caelius Aurelianus, *Acut.* 3.113 (*CML* 6.1.1, p. 358), assigns him to the sect of Asklēpiadēs of Bithunia, and describes his theory of ***hudrophobia***: the stomach is the affected part, causing the hiccups, vomiting, and thirst. He also wrote *On Long Life* (Clement of Alexandria, *Paid.* 2.2.23). Inscriptions reveal that he benefited Dēlos, and died in a shipwreck, 27 BCE.

BNP 2 (2003) 81, V. Nutton.

PTK and GLIM

Āryabhaṭa (*ca* 500 CE)

Āryabhaṭa (born 476 CE) lived in Pāṭaliputra (modern Patna in Bihar, India), authored two works, the *Āryabhaṭīya* (*ca* 500 CE), the origin of the Āryapakṣa school of astronomy, and a now lost work, the origin of the Ārdharātrikapakṣa school of astronomy. The Āryapakṣa, which became influential in south India, has a dawn epoch, whereas the Ārdharātrikapakṣa, which influenced north-west India and Iran, has a midnight epoch, but otherwise the two schools differ only in certain parameters.

While both schools differ from the Brāhmapakṣa and the *Āryabhaṭīya* has a different structure from other Indian astronomical treatises, the Paitāmahasiddhānta, the founding text of the Brāhmapakṣa, was among Āryabhaṭa's sources, and he states in the *Āryabhaṭīya* that his astronomical system was revealed by Svayambhū, i.e., Brahmā.

The planetary model used by Āryabhaṭa is derived from a pre-Ptolemaic Greek model, which sought to preserve the Aristotelian principle of concentricity. The mean planet moves in a circle around the Earth, and centered around the mean planet are one or two **epicycles**, depending on whether the planet is one of the two luminaries or a star-planet. Pingree (*DSB* 15.590) believes that the mean motions of the planets in the *Āryabhaṭīya*, apparently unrelated to those of the Brāhmapakṣa, were derived from a Greek table of mean longitudes corresponding to noon on 21 March 499 CE.

In the *Āryabhaṭīya*, Āryabhaṭa speaks of the diurnal rotation both as a rotation of the earth and as a rotation of the fixed stars. Mathematically, these are equivalent, but other Indian astronomers, including VARĀHAMIHIRA, rejected that the earth is rotating, for physical reasons. Āryabhaṭa also divided the world ages untraditionally, for which later Indian astronomers criticized him.

DSB 1.308–309, 15.590–602 (Āryapakṣa), 15.602–608 (Ārdharātrikapakṣa), D.E. Pingree; *CESS* A.1.50–53, A.2.15, A.3.16, A.4.27–28, A.5.16–17; D.E. Pingree, *Jyotiḥśāstra: astral and mathematical literature* = *A History of Indian Literature* 6.4 (1981); *Idem*, "Āryabhaṭa, the Paitāmahasiddhānta, and Greek astronomy," *Studies in History of Medicine and Science* ns 12.1–2 (1993) 69–79.

Kim Plofker and Toke Lindegaard Knudsen

Asaf ha-Rofe, Asaf the Jew, Asaf ben Berekhiah **(300 – 900 CE)**

Jewish physician associated with the oldest extant medical text in Hebrew, *Sefer Asaf ha-Rofe* or *Sefer Refu'ot* ("Book of Remedies"), not completely published or translated and sorely understudied. Its date and provenance remain debated. Among current theories, an 8th/9th c. compilation in Byzantine Italy seems most plausible. *Sefer Asaf* records the teachings of Asaf and his colleagues Yohanan ben Zabda and Yehudah ha-Yarhoni. Some MSS identify Asaf with Asaf ben Berekhiah, mentioned briefly in *I Chron.* 15:17 and associated with Solomon in Jewish and Islamic folklore. The book is prefaced by an account of the origins of medicine in revelations by the angel Raphael (lit. "God heals") to Noah, transmitted to Shem, progenitor of the Jews. From the Jews, medicine was taught to Indians, Greeks, Egyptians, and Mesopotamians. In listing famous physicians, *Sefer Asaf* groups "Asaf the Jew" with HIPPOKRATĒS, GALĒN, and DIOSKOURIDĒS. The body of the text is an eclectic compendium of medical traditions, covering all areas but surgery. Its anatomy and embryology reflect Jewish tradition. Hippokratic influence is marked; its medical aphorisms are essentially Hebrew paraphrases of the HIPPOKRATIC CORPUS, APHORISMS, and the oath that Asaf and Yohanan require of their students stands in a close relationship to the HIPPOKRATIC OATH. Galēn's influence is minor, but lists of pharmacological plants derive largely from Dioskouridēs. In presenting a Jewish interpretation of Hippokratic medicine, in particular, Asaf and his colleagues may stand in the tradition of earlier Jewish physicians such as RUFUS OF SAMARIA. No evidence suggests influence from Arabic medicine.

S. Pines, "The Oath of Asaph the Physician and Yohanan Ben Zabda," *Proceedings of the Israel Academy of Sciences and Humanities* 9 (1975) 223–264; E. Lieber, "Asaf's Book of Medicines," *DOP* 38 (1984) 233–249; *EJ2* 2.543–544, S. Muntner.

Annette Yoshiko Reed

Asamōn **(unknown date)**

Wrote about the rise of the Nile. The name appears to be Egyptian for "eagle" (Heuser 1929: 13), but *cf.* also *LGPN* 3A.78 (2nd c. BCE, Ēlis).

RE 2.2 (1896) 1515 (#3), H. Berger.

PTK

Asarubas or Asdrubas (55 – 75 CE)

Wrote on electrum, relaying that the mud of lake Cephesis – known to the Mauri as "Electrum" – produced electrum when dried by the sun (PLINY 37.37, see also 1.*ind*.3). Some scholars read Asdrubas, i.e., the Punic name Hasdrubal.

Fr. Buecheler, "Zwei Gewährsmänner des Plinius," *RhM* 40 (1885) 304–307; *RE* 2.2 (1896) 1518, P. von Rohden; *RE* S.1 (1903) 151, G. Knaack.

Eugenio Amato

Asinius Pollio of Tralleis (40 – 10 BCE)

Freedman of the historian Asinius Pollio, described by the *Souda* Pi-2165 as both a sophist and a philosopher. Besides writing various historical works, he made an epitome of DIOPHANĒS OF NIKAIA's *Geōrgika*, reducing it from six books to two, and wrote ten books "against ARISTOTLE on animals."

RE 2.2 (1896) 1589 (#23), E. Schwartz.

Philip Thibodeau

Asklatiōn (Astrol.) (50 – 535 CE)

Astrologer (IŌANNĒS "LYDUS," *Ost.* p. 6.24 Wa.). Dubiously identifiable with Domitian's astrologer Ascletario whom, upon predicting his own rending by dogs, the emperor executed as an object-lesson in the mendacity of astrology. Dogs mangled the corpse (Suet. *Dom.* 15.3).

RE 2.2 (1896) 1622 (#2), E. Riess; Gundel and Gundel (1966) 158–159.

GLIM

Asklatiōn (Med.) (250 BCE – 65 CE)

Commentator on HIPPOKRATĒS, mentioned by ERŌTIANOS (A-103.9, p. 23.10 Nachm.), probably distinct from the homonymous astrologer.

RE 2.2 (1896) 1662 (#1), M. Wellmann; Ihm (2002) #25.

Alain Touwaide

Asklēpiadēs Pharmakiōn (*ca* 90 – 100 CE)

Greek pharmacologist, distinct from ASKLĒPIADĒS OF BITHUNIA, wrote a pharmacological work quoting ANDROMAKHOS OF CRETE (YOUNGER) and cited by ARKHIGENĒS OF APAMEIA; scholars view the citation at Pliny 14.183 as interpolated. Perhaps, like many contemporary pharmacologists, he lived in Rome. He studied medicine under LEUKIOS OF TARSOS (as did KRITŌN); he also cites MOSKHIŌN, and seems to have read SCRIBONIUS LARGUS in a Greek edition. Asklēpiadēs authored ten books of recipes, perhaps a single compilation, more likely two works of five books each; he also composed works on theriac and gynecology (whose precise nature is uncertain: GALĒN, *CMGen* 1.16–17, 13.441–442 K.). His ten books treated medicines for external use (called, and possibly dedicated to a, Marcella) and for internal use (called, and dedicated to a Mnasōn), organized by place; he often provides detailed preparations. Galēn, highly praising Asklēpiadēs for his careful cataloguing of recipes, quotes him firsthand in over 50 lengthy extracts (Fabricius), more than from HĒRĀS, ANDROMAKHOS, and KRITŌN together.

RE 2.2 (1896) 1633–1634, M. Wellmann; Watson (1966) 8–10, 15–16, 60–61; Fabricius (1972) 192–198, 246–253; *BNP* 2 (2003) 99 (#9), V. Nutton.

Alain Touwaide

Asklēpiadēs Titiensis (*ca* 100 BCE?)

Cited after HIPPOKRATĒS, DIOKLĒS, PRAXAGORAS, and before "Dēmētrios," for identifying apoplexy with paralysis (SŌRANOS in CAELIUS AURELIANUS, *Acute* 3.55 [*CML* 6.1.1, p. 324]). The text may originally have read *Apollōnios Kitiensis* (*Citiensis* in Latin), emended to *Titiensis* by a copyist (referring to an obscure Bithunian town), who further confused Apollōnios with Asklēpiadēs (both common medical names). Earlier editions of Caelius Aurelianus (Sichardus, Rovillius, Amman) proffer *ASCLEPIADES TITIENSIS*; Wellmann followed by Bendz corrected Titiensis to Citiensis, and Drabkin restored Apollōnios Citiensis. Furthermore, Caelius Aurelianus may be citing authors chronologically (the first three are *ca* 400, *ca* 300, and *ca* 200 BCE respectively). DĒMETRIOS may be OF APAMEIA or some other medical Dēmētrios (e.g., "KHLŌROS"). Most likely, our author is APOLLŌNIOS OF KITION. (*Cf.* AUIDIANUS, SEXTUS OF APOLLONIA.)

RE 2.2 (1896) 1632 (#37), M. Wellmann.

GLIM

Asklēpiadēs of Bithunia (in Rome, *ca* 120 – 90 BCE)

PLINY 26.12 relates the story of Asklēpiadēs turning to medicine from rhetoric since he was not making a good living in a very crowded profession, but our polymath garbles the chronology, setting the rhetorician-turned-physician in the time of Pompeius Magnus. Rawson demonstrates that Asklēpiadēs was dead by 91 BCE (CICERO, *De or.* 1.62), but she advances flaccid arguments against Pliny's switch of professions, fairly common in an age long before legally sanctioned medicine of any particular outlook. Pliny is ambivalent about Asklēpiadēs: he is brilliant (*sagacis ingenii*), but reduced medicine to a discovery of causes and diagnostics into guesswork (*medicinam ad causas reuocando coniecturae fecit*); as a gifted speaker, he persuaded patients that diseases were cured by simple means, but used the lies ordinary among magicians (26.18: *adiuuere eum magicae uanitates*). CELSUS, 4.26.4, says Asklēpiadēs advocated giving cold rainwater mixed with wine (LONDINIENSIS MEDICUS 34.30 calls him the "wine giver"), and his practice was marked by huge success and many prominent patients, who valued his advice on dietetics, moderation in personal habits, mild exercise, and careful employment of drugs – and then only rarely. Many of his students became prominent physicians in their own right, and his pseudo-mechanistic medical philosophy fit well into the general popularity of EPICURUS in the late Republic, suggested by the famous Library at Herculaneum, from which have emerged unknown works of PHILODĒMOS. "Asklēpiadean" physicians attended the emperors from AUGUSTUS through Nero, and ANTONIUS MUSA used the "cold water treatment" to save Augustus in 23 BCE.

Debated are origins of his medical theories: Asklēpiadēs taught the body is formed from "fragile corpuscles" (*anarmoi ongkoi*), but, as Vallance (1990: 7–43) warns, certainly not the "corpuscles" as understood in modern hematology. *Ongkos* in medico-philosophical context could be a "lump," and the *ongkoi* presumably passed through channels (*poroi*) throughout the living body. If the *ongkoi* were blocked or the motion became too easy, disease occurred. The mechanical nature of Asklēpiadēs' theories suggests Epicurus, the **Platonist** HĒRAKLEIDĒS OF HĒRAKLEIA PONTIKĒ, the **Peripatetic** STRATŌN OF LAMPSAKOS, or possibly all three, but Vallance thinks Asklēpiadēs was rejecting the then-popular notions (also mechanical) of ERASISTRATOS. None of Asklēpiadēs' writings survives, although he is cited as late as AËTIOS OF AMIDA's *Tetrabiblos*, and GALĒN's vicious attacks against Asklēpiadēs and his "followers," especially THEMISŌN and THESSALOS, rather well doomed the works to obscurity. Galēn's major objection was Asklēpiadēs' denial of teleology.

Ed.: C.G. Gumpert, *Asclepiadis Bithyniae Fragmenta* (1794) = trans. by R.M. Green, "Fragments from Asclepiades of Bithynia" in *Asclepiades: his Life and Writings* (1955); J.T. Vallance (in preparation).

N.W. DeWitt, "Epicureanism in Italy" and "Epicureanism in Rome" in *Epicurus and his Philosophy* (1954) 340–344; H.B. Gottschalk, "The Theory of *anarmoi ongkoi*" in *Heraclides of Pontus* (1980) 37–57; E. Rawson, "The Life and Death of Asclepiades of Bithynia," *CQ* 32 (1982) 358–370 = repr. in *Roman Culture and Society: Collected Papers* (1991) 427–443; Vallance (1990); *Idem*, "The Medical System of Asclepiades of Bithynia," *ANRW* 2.37.1 (1993) 693–727; R. Polito, "On the Life of Asclepiades of Bithynia," *JHS* 119 (1999) 48–66; W.R. Johnson, "A Secret Garden: *Georgics* 4.116–148;" M. Gigante, "Vergil in the shadow of Vesuvius;" D. Delattre, "Vergil and Music, in Diogenes of Babylon and Philodemus;" and F. Cairns, "Varius and Vergil: Two Pupils of Philodemus in Propertius 2.34?," in D. Armstrong *et al.*, edd., *Vergil, Philodemus and the Augustans* (2004) 75–99, 245–263, and 299–321; D. Sider, "Philodemus and his Texts" in *The Library of the Villa dei Papiri at Herculaneum* (2005) 78–95.

John Scarborough

Asklēpiadēs of Murleia (*ca* 90 – 60 BCE)

Son of Diotimos; taught in Turdetania (inland north of Gadēs). Before leaving his homeland, wrote its history, of which a few fragments survive, notably Ath., *Deipn.* 2.35 (50d-e), on the soporific and headache-inducing berries of the *khamaikerasos* ("ground cherry") of Bithunia. While in Turdetania wrote a ***periēgēsis*** of that land, Galicia, and Catalonia (perhaps of all Iberia?), which STRABŌN cites (3.4.3, 19). Also composed a commentary on ARATOS, a monograph on the Pleiades, and a work on Nestor's Cup in HOMER, of which Athēnaios, *Deipn.* 11 (489c–494b), preserves much: Asklēpiadēs argues that it metaphorically reflects the ***kosmos***. The confused notice in the *Souda* A-4173, and the frequency of his name, have spawned modern scholarly debates about his identity with the grammarian Asklēpiadēs used by SEXTUS EMPIRICUS. He is called *doctus ac diligens* by MACROBIUS 5.21.5.

FGrHist 697; *BNP* 2 (2003) 98–99 (#8), F. Montanari.

PTK

ASKLĒPIADĒS OF PROUSIAS ⇒ ASKLĒPIADĒS OF BITHUNIA

Asklēpiodotos of Alexandria (460 – 510 CE?)

Born in Alexandria, where, as a boy, devoting himself to science and crafts (paints and dyes, rocks, and especially biology), he invented mechanical devices for religious ritual use. He studied medicine with IAKŌBOS PSUKHRESTOS, and later revived the use of white hellebore. PROKLOS was his mentor in Athens, where *ca* 470 CE Asklēpiodotos met and befriended DOMNINOS (later quarreling over a mathematical dispute), and became acquainted with DAMASKIOS, whose fragmentary *Life of Isidore* is our primary source for Asklēpiodotos. SIMPLICIUS called him Proklos' best student (*Coroll. Time* = *CAG* 9 [1882] 795). He left Athens for study in Seleukeia of Syria, married Damianē of Aphrodisias, and taught in Aphrodisias. Initially childless, the couple moved to Alexandria to entreat Isis for a child, and later they reared daughters. He also taught in Alexandria and apparently returned to Aphrodisias *ca* 485 CE. He wrote a lost commentary on PLATO's *Timaios* and attempted to reconstruct the "enharmonic" scale. He was a careful and expert farmer, and a devoted pagan, reviving pagan worship in Aphrodisias, at great personal expense, so that despite his wealth, he bequeathed his estates encumbered by debt.

PLRE 2 (1980) 161–162; Athanassiadi (1999) 212–216.

PTK

Asklēpiodotos (of Nikaia?) (40 BCE – 30 CE)

A student of POSEIDŌNIOS, whose meteorological theories SENECA cites on the volcanic isle Hiera (*QN* 2.26.6), on thunder and lightning produced by the collision of solid bodies (2.30.1), on the discovery of subterranean lakes (5.15.1), and on earthquakes generating winds and transmitting shocks (6.17.2–3, 6.22.1). He is perhaps identical with the tactician whose treatise *Tactical Summary* details technical aspects of the organization and disposition of an ideal phalanx.

GGP 4.2 (1994) 709, P. Steinmetz; *OCD3* 187, J.B. Campbell; *BNP* 2 (2003) 100 (#2), L. Burckhardt.

PTK and GLIM

***Asklēpion/Asklēpios (Med.)* (250 – 75 BCE)**

GALĒN, *Diff. Morb.* 9 (6.869 K.), mentions that a certain Nikomakhos of Smurna became immobile because of obesity, but "Asklēpios cured him" – god or man? GEŌPONIKA 20.6, citing HĒRAKLEIDĒS OF TARAS, ascribes works on nutrition from fish to Asklēpios, MANETHŌN, PAXAMOS, and PSEUDO-DĒMOKRITOS. *Hipp. Cant.* 31.2 (2.166–167 ed. Oder-Hoppe) mentions Asklēpiōn's spleen-remedy.

RE 2.2 (1896) 1698 (#8), M. Wellmann.

PTK

Asklēpios (Pharm.) (*ca* 515 – *ca* 565 CE)

ASKLĒPIOS OF TRALLEIS (*in Metaphys.* 995b20 = *CAG* 6.2 [1888] 143) cites a homonymous pharmacist, fellow-student of AMMŌNIOS and later a teacher of pharmacy. He composed a commentary on the HIPPOKRATIC APHORISMS (Westerink, *CMG* 11.1.3.1 [1985] 17–23).

RE 2.2 (1896) 1698 (#6), J. Freudenthal, (#9), M. Wellmann.

PTK

Asklēpios of Tralleis (Math.) (515 – 565 CE)

Neo-**Platonic** mathematician and philosopher, student of AMMŌNIOS, wrote a commentary on NIKOMAKHOS' *Arithmetica*, derived from Ammōnios' lectures and surviving as lengthy *scholia*. Asklēpios also wrote a commentary (*CAG* 6.2) on ARISTOTLE's *Metaphysics A–Z*, valuable primarily for *testimonia* of Ammōnios as well as extracts from ALEXANDER OF APHRODISIAS.

RE 2.2 (1896) 1697–8 (#5), A. Gerke; L. Tarán, Asclepius of Tralles: *Commentary to Nicomachus' Introduction to Arithmetic* (1969).

GLIM

Aspasia (120? – 540? CE)

Cited by AËTIOS OF AMIDA, Book 16, for gynecological remedies and practices, more often than SŌRANOS. She cites HIPPOKRATIC CORPUS, APHORISMS 5.31 (*apud* Aëtios 16.18 [Zervos 1901: 21–22]; *cf.* 16.99 [pp. 147–148]), employs gentian (*cf.* GENTHIOS) in an abortifacient (16.18), and may cite ASKLĒPIADĒS (16.94 [p. 141]: of Bithunia? Pharmakiōn?), providing a *terminus post quem*. That Sōranos (usually careful to cite predecessors) nowhere mentions her suggests a later terminus. Her care of the gravida (16.12 [p. 12]) and advice on abortives (16.18) resemble the corresponding sections of Sōranos *Gyn.* (1.46 and 1.64–65 [*CMG* 4, pp. 32–34, 47–49; *CUF* v. 1, pp. 43–46, 62–65]), whereas she intervenes less in difficult births (16.15 [p. 16] contrasted with Sōranos *Gyn.* 4.7–8 [*CMG* 4, pp. 136–139; *CUF* v. 4, pp. 11–16]), but more in cases of tilted womb (16.73 [pp. 112–115] contrasted with Sōranos *Gyn.* 3.50 [*CMG* 4, pp. 127–128; *CUF* v. 3, pp. 54–55]). Since she also advises on the care of the woman after embryotomy (16.25 [p. 36]), whereas Sōranos focuses on the embryotomy itself (*Gyn.* 4.9–13 [*CMG* 4, pp. 140–144; *CUF* v. 4, pp. 16–22]), she seems on the whole to evince greater care for the mother than did Sōranos. Her prescriptions for various uterine disorders (corresponding to lost sections of Sōranos, *Gyn.*), 16.94 and 99 (above), plus 102 (p. 150), 104 (pp. 151–152), and 108 (p. 155), offer a variety of relatively simple recipes, plus venesection and some radical surgeries. Her unapologetic stance on abortives and embryotomy might suggest a date before Constantine (or Theodosius I), but Aëtios likely cites her as an authority for his actual practice, and not merely out of antiquarian interest, showing that although officially condemned, such procedures continued to be employed by women and their gynecologists and midwives.

H. Fasbender, *Geschichte der Geburtshülfe* (1906) 58–61; Parker (1997) 138 (#54).

PTK

Aspasios (Perip.) (*ca* 100 – 130 CE)

Aristotelian commentator. GALĒN studied in Pergamon with one of his students, when the student had returned home after a long sojourn elsewhere – either from Aspasios' school, or some other city where the student had been active as a teacher. Wherever Aspasios was active, he was not simply an instructor at a local **Peripatetic** school. It is certain that Aspasios' commentaries were standard works, but the evidence we have about them only gives us a glimpse of the state of the textual tradition of the Aristotelian corpus in the 2nd c. CE, not about Aspasios' positions. Galēn took Aspasios' (and ADRASTOS') commentaries as his starting point when writing his more extensive (lost) commentary on the *Categories*, and later interpreters – HERMINOS, ALEXANDER OF APHRODISIAS, even PLŌTINOS and BOËTHIUS – had access to, and used one or another of Aspasios' commentaries.

References in later authors attest that Aspasios wrote commentaries on the *Categories*, the *De interpretatione*, the *Physics*, the *De sensu*, and the *Metaphysics*. In one instance, Alexander of Aphrodisias refers to Aspasios' interpretation of a passage of the *De caelo*, about which he learnt from his teacher, Herminos – apparently this was not part of a commentary to which Alexander had access. Only the commentary on Books I–IV and VII–VIII of the *Nicomachean Ethics* is extant (*CAG* 19.1).

Moraux 2 (1984) 226–293; Gottschalk (1987) 1156–1158.

István Bodnár

Aspasios (Pharm.) (250 BCE – 90 CE)

Asklēpiadēs Pharm., in Galēn *CMLoc* 9.5 (13.302 K.), records his remedy for dysentery: parsley, pomegranate, heath-fruit, and opium, reduced in myrtle. Possibly Asklēpiadēs' reference was originally to Aspasia.

Fabricius (1726) 92.

PTK

Asterios (120 BCE – 540 CE)

Aëtios of Amida 7.117 (*CMG* 8.2, p. 398) records his collyrium, an opium-laced mixture of minerals (antimony, **calamine**, and ***psimuthion***) and aromatics (cassia, myrrh, and spikenard).

Fabricius (1726) 92.

PTK

Astrampsukhos (*ca* 1st – 9th c. CE?)

Legendary Persian magus; the name appears in a list given by Diogenēs Laërtios (*pr.*2) of Zoroaster's successors in the period before the Persians were defeated by Alexander. Several works of occult nature circulated under Astrampsukhos' name in antiquity: the popular *Sortes Astrampsychi*, a set of oracular questions and answers (in some MSS prefaced with a dedication by "Astrampsukhos the Egyptian" to an unknown Ptolemaios, also falsely attributed in one MS to Leōn the Wise; 3rd c. CE papyri provide a *terminus ante quem* for the work); a dream-book in verse (dated by Oberhelman between the 6th and 9th centuries CE); and a spell for the prosperity of a workshop (*PGM* 8.1–63, labeled a love-charm). According to the *Souda* A-4251, a book on the healing of donkeys was also attributed to Astrampsukhos; this, however, is not preserved.

Ed.: N. Rigault *et al.*, *Artemidori Daldiani et Achmetis Sereimi f. Oneirocritica, Astrampsychi et Nicephori versus etiam oneirocritici* (1603); G.M. Browne, *Sortes Astrampsychi*, v.1 (1983), R. Stewart, v.2 (2001).

RE 2.2 (1896) 1796–1797, E. Riess; P. Tannery, "Astrampsychus," *REG* 11 (1898) 96–106; S. Oberhelman, "Prolegomena to the Byzantine *Oneirokritika*," *Byzantion* 50 (1980) 489–491; R. Stewart, "The Textual Transmission of the *Sortes Astrampsychi*," *ICS* 20 (1995) 135–47; *BNP* 2 (2003) 121–122, C. Harrauer; McCabe (2007) 5.

Anne McCabe

Astrologos of 379 (379 CE)

Anonymous author of a brief text "prognostications from the positions of the fixed stars," forming part of a long compilation of astrological texts in Greek ascribed to "Palkhos" but actually the work of the 14th c. Byzantine astrologer Eleutherios Eleios. The author claims Egyptian ancestry, and to be writing in a location having the latitude of Rome, giving his date of writing as the consulate of Olybrius and Ausonius (379 CE). His text catalogues astrological influences determined by a list of bright stars, the selection of which was clearly influenced by Ptolemy's *Phaseis*, though the positions of the stars derive from the *Almagest* with adjustment for precession. This short catalogue of stars was repeatedly reworked with updated precessional corrections in late antiquity and the Byzantine period.

Ed.: *CCAG* 5.1 (1904) 196–206.

D.E. Pingree, "The Astrological School of John Abramius," *DOP* 25 (1971) 189–215; S. Feraboli, "L'evoluzione di un catalogo stellare," *Maia* 45 (1993) 269–273.

Alexander Jones

Astunomos (350 – 100 BCE?)

Wrote a book on islands (i.e., likely after EUDOXOS) or a geographical gazetteer, cited by PLINY 1.*ind.*4 and 5.129, and by STEPHANOS OF BUZANTION. The archaic name was rare before 100 BCE, and almost unattested thereafter (*LGPN*).

RE 2.2 (1896) 1872 (#2), E. Schwartz.

PTK

Athanarid (496 – 507 CE)

Wrote in Gothic a geography of Europe, covering Finland to Spain, listing towns according to the rivers on which they stood, and cited extensively by the RAVENNA COSMOGRAPHY, Book 4. See also HELDEBALD and MARCOMIR.

Staab (1976); *DPA* 1 (1989) 639, R. Goulet; *BNP* 1 (2002) 408, A. Schwarcz.

PTK

Athēnagoras (Agric.) (325 – 90 BCE)

Agricultural writer whose work was used by CASSIUS DIONUSIOS (VARRO, *RR* 1.1.9; *cf.* COLUMELLA 1.1.10).

RE 2.2 (1896) 2021 (#11), M. Wellmann.

Philip Thibodeau

Athēnagoras (Med.) (400 – 600 CE?)

Credited with a Latin treatise on pulse and urine (*Incipit liber Athenagore de pulsis et urinis. Quoniam medicus peritissimus debet esse . . .*), although the author's name seems to indicate a tract written in Greek and translated into Latin. Citations suggest either a work discussing only urine (Diels 1907: 2.21) or two separate treatises (Thorndike and Kibre 1963: 1285, 1610). Medical content in the classical, especially Galēnic tradition, and the earliest extant manuscript (Paris, BNF, *latinus* 7028, 10th/11th c.) predate the post-Constantinian activity of Salerno school (12th c.), suggesting that the Latin version might date to the period of the early medieval translations of Greek medical works into Latin in the medical schools of northern Africa and Italy (Ravenna).

RE 2.2 (1896) 2021 (#10), M. Wellmann; Cam. Vitelli, "Studiorum Celsianorum particula prima," *SIFC* 8 (1900) 450–476 at 467; Beccaria (1956) 155; E. Wickersheimer, *Manuscrits latins de médecine du haut moyen âge dans les bibliothèques de France* (1966) 85.

Alain Touwaide

Athēnagoras son of Arimnēstos (365 – 350 BCE)

Hypothesized that the Red Sea and the Ocean outside the Pillars of Hēraklēs were connected, according to the ARISTOTELIAN CORPUS ON THE FLOOD OF THE NILE, which places Athēnagoras at the Persian court before the expedition against Egypt by Artaxerxēs III "Ōkhos," 357 BCE.

RE S.5 (1931) 46 (#12), W. Kroll.

PTK

Athēnaios Mechanicus (30 – 20 BCE?)

Wrote a work on siege machinery, *On Machines*. Its dating is uncertain, but current consensus places it in the second half of the 1st c. BCE and identifies the dedicatee of the treatise, "Marcellus," as M. Claudius Marcellus (42–23 BCE), AUGUSTUS' nephew and son-in-law. The treatise was perhaps prepared for his departure to Spain with Augustus. Athēnaios himself is identified as a **Peripatetic** philosopher who lived and held public positions in Rome in this period (STRABŌN 14.5.4).

On Machines opens with an introduction on the need to speak briefly and avoid theoretical digressions (3.1–7.7). The main part of the treatise contains descriptions of a number of devices for attacking cities: DIADĒS' portable siege towers, rams and tortoises, i.e. sheds for covering and moving siege machinery; the probably impossible monumental ram-tortoise by HĒGĒTŌR; and EPIMAKHOS' ***helepolis*** (7.8–27.6). These descriptions are followed by a shorter section on more unfortunate devices: some that are scaled wrongly, and KTĒSIBIOS' sea-saw tube for scaling walls, which Athēnaios regards as pure imagination (27.7–29.2). Last he includes devices of his own invention, such as devices for making a tortoise change direction, so it is less easy to hit (31.6–38.13).

The first section on machines is closely related to VITRUUIUS' descriptions of siege machinery in *On Architecture* (10.13–16). Both are thought to have drawn on the artillery specialist AGĒSISTRATOS; Athēnaios states that he is "relating in full" everything he learned from him. The style of the treatise is seemingly nuts-and-bolts and includes details and measurements, but the material does not appear organized. Athēnaios' descriptions are often hard to interpret, because of his vagueness and frequent *hapax legomena* in describing his own inventions. It is likely that Athēnaios, despite his emphasis on practical application, drew his material mainly from technical literature and teaching rather than direct experience.

Ed.: D. Whitehead and P.H. Blyth, *Athenaeus Mechanicus, On Machines* (2004).
Marsden (1969); *BNP* 2 (2003) 243–244 (#5), D. Baatz.

Karin Tybjerg

Athēnaios of Attaleia (or Tarsos?) (30 – 70 CE)

Greek physician from Attaleia (GALĒN, *Ars medica* [1.306 K.], *Elem. Hipp.* 1 [1.457 K.], *Temp.* 1 [1.522 K.]; *Dign. Puls.* 1 [8.787 K.]; etc.; PSEUDO-GALĒN, DEFINITIONS 19.347, 356, 392 K.) or Tarsos (CAELIUS AURELIANUS, *Acut.* 2.6 [*CML* 6.1.1, p. 134], reading Tharsus). Traditionally dated to the 1st c. CE, Kudlien (1962) dates him to the 1st c. BCE on the basis of his doctrine, likely derived from POSEIDŌNIOS OF APAMEIA (see Galēn, *Morb. Diff.* 6.842 K.), whose student he may have been (Galēn, *On Cohesive Causes* 2 = *CMG S. Orient.* 2, pp. 54–57; *cf.* p. 134), or whose philosophy he simply may have followed (Galēn, *Elem. Hipp.* 1 [1.469 K.]; PSEUDO-GALĒN, INTRODUCTIO 14.698 K.). The 1st c. CE date would better explain CELSUS' omission of the **Pneumaticists** in his overview of contemporary philosophical approaches to medicine (1.*pr.*), and accord with the date of Athēnaios' first known follower (i.e., AGATHINOS).

Athēnaios, practicing in Rome, opened a new avenue in medical thinking, loosely and variously defined by Galēn, most explicitly as *pneumatikē* (*Diff. Puls.* 3, 4 [8.646, 756 K.]) and its members as *pneumatikoi* (*ibid.* [8.674, 749 K.]), on the basis of the role of ***pneuma*** in their physiological system. According to Galēn, such physicians were especially well-qualified in general medicine, particularly with regard to fever (*Febr. Diff.* 1 [7.295 K.]). Among those deserving his deepest esteem was Athēnaios himself (*Sympt. Caus.* 2 [7.174 K.]).

According to Athēnaios, reasoning from theories of nature alone is the basis of medicine (Galēn, *Elem. Hipp.* 1 [1.457–486 K.]; ps.-Galēn, *Intro.* 14.676–677, 698 K.; ps.-Galēn, *Def.*, 19.356 K.), contrary both to those relying on tradition, and to Asklēpiadēs of Bithunia, who valued reason and experience (*logos* and *peira*). Athēnaios' theory was based on **Stoicism**, particularly regarding the role of fire both in the ***kosmos*** and in physiology. Again contrary to Asklēpiadēs, Athēnaios opted for an incorporeal biological theory. He believed that the body is made of four elemental properties (*stoikheia*: heat, cold, dry and moist), which Galēn found unclearly defined, but Athēnaios claimed were evident (*enargē*) and not requiring demonstration. Athēnaios variously called the *stoikheia* "qualities" (*poiotētes*), "powers" (*dunameis*), and "bodies" (*sōmata*). Galēn found it unclear whether the bodies composed of these four elements were **homoiomerous**. Two of these elementary qualities are active (*poiētika*: heat and cold), two are material (*hulika*: dry and wet). A fifth element, ***pneuma***, holding the qualities together and contained in the blood, generates cardiac movement. It circulates through the heart and the arteries, stimulating their expansion, a natural and involuntary movement. Heat thus moves from the heart and returns to it (Galēn, *Diff. Puls.* 4 [8.755–756 K.]). As a result, the source and directing principle of human life (***hēgemonikon***) is located in the heart (Galēn, *MM* 13 [10.929 K.]).

Athēnaios defined health as an equilibrium (*eukrasia*) of ***pneuma*** and the four elements (*MM* 7 [1.523 K.]). The equilibrium between the ***pneuma*** and four elements is created by the tension (*tonos*) between them (Galēn, *Diff. Puls.* 3 [8.646 K.]). The ***pneuma*** is thus responsible for both health and disease (ps.-Galēn, *Intro.* 14.699 K.), the latter being a disequilibrium (*duskrasia*). Diseases are caused by substances altering the quantity or the quality of either the ***pneuma*** or of the elements.

Athēnaios dedicated no specific work to articulate his theory (ps.-Galēn, *Def.*, 19.347 K.), and none of his writings, or even their titles, survives. Galēn refers to the 24th book of an unnamed treatise (*Sympt. Caus.* 2 [7.165 K.]), and quotes an enema (through Andromakhos, in *CMLoc* 9.5 [13.296 K.]), and a medical formula (through Asklēpiadēs Pharm., in *CMGen* 5.3 [13.847 K.]). Galēn connects Athēnaios' work on embryology with Aristotle (Galēn, *De Semine*, *passim*).

Galēn admired Athēnaios (*Tremor* [7.609 K.]), and agreed with him except on quotidian fever (*Febr. Diff.* 1 [7.295 K.]). Nevertheless he criticized him, principally for his vagaries regarding the four *stoikheia* (*Elem. Hipp.* 1 [1.457, 460 K.]). Athēnaios' theory reached its zenith with Aretaios, but evolved very early toward a more synthetic system absorbing elements from other contemporary schools, leading to the so-called *episunthetic* or eclectic school (ps.-Galēn, *Def.* 19.353 K.) supposedly created by Agathinos, and followed by Arkhigenēs and Leōnidas of Alexandria.

Wellmann (1895); *RE* 2.2 (1896) 2034–2036 (#24), *Idem*; Kudlien (1962); *Idem* (1968) 1097–1098; *KP* 1.703, *Idem; DSB* 1.324–325, J.S. Kieffer; Harris (1973) 237–242; Smith (1979) 231–234; *OCD3* 203, V. Nutton; *BNP* 2 (2003) 244–245, *Idem*.

Alain Touwaide

Athēnaios of Kuzikos (390 – 345 BCE)

Mathematician and geometer, and member of PLATO's **Academy** (PROKLOS *In Eucl.* p. 67 Fr.).

RE 2.2 (1896) 2025 (#18), P. Natorp.

GLIM

Athēniōn (of Athens?) (50 – 10 BCE)

CELSUS 5.25.9 records his ***trokhiskos*** against cough: castoreum, myrrh, poppy "tears" and pepper, ground separately, then mixed (take two in the morning and two at bedtime). SŌRANOS, *Gyn.* 3.2 (*CMG* 4, p. 94; *CUF* v. 3, pp. 2–3), classing him as **Erasistratean**, indicates that he, like MILTIADĒS, argued in favor of the existence of diseases special to women. In this period, the name is especially Athenian: *LGPN.*

RE 2.2 (1896) 2041 (#9), M. Wellmann.

PTK

Athēnippos (120 BCE – 40 CE)

Compounded a collyrium that became known as "Athēnippion"; recorded by SCRIBONIUS LARGUS 26–27 (opium, white pepper, ***pompholux***, roasted copper, etc.), by ASKLĒPIADĒS PHARM. in GALĒN, *CMLoc* 4.7 (12.789 K., *cf.* 774) as "universal" (*pankhrēston*), and by MARCELLUS OF BORDEAUX 8.6 (*CML* 5, p. 114.26).

RE 2.2 (1896) 2042 (#2), M. Wellmann.

PTK

Athēnodōros (Med.) (50 – 100 CE)

Physician and philosopher, contemporary with PLUTARCH. He wrote *On Epidemics* (*peri epidēmiōn*) claiming that ***hudrophobia*** and ***elephantiasis*** first appeared in the time of ASKLĒPIADĒS OF BITHUNIA, cited as a witness against the conjecture of PHILŌN OF HUAMPOLIS that the diseases were newly discovered (*Quaest. Symp.* 8.9.1 [731A]).

RE 2.2 (1896) 2046 (#23), M. Wellmann.

GLIM

Athēnodōros (of Rhodes?) (250 BCE – 50 CE)

DIOGENĒS LAËRTIOS four times cites the 8th book of Athēnodōros' *Peripatoi*, apparently a biographical account of philosophers: 3.3 (PLATO and Diōn), 5.36 (THEOPHRASTOS' father's occupation), 6.81 (Diogenēs the Cynic), and 9.42 (DĒMOKRITOS' visual acuity). Grammarians record the remarks of some Athēnodōros on prosody and melody, and Quintilian 2.17.15 assigns that man to Rhodes.

GGP 3 (1983) 585, Fr. Wehrli.

PTK

Athenodōros of Tarsos (*ca* 60 – 20 BCE)

Athēnodōros "Caluus" from the village of Kana near Tarsos (CICERO *Att.* 16.11.4, 14.4), son of Sandōn, was, like AREIOS DIDUMOS, a court philosopher and tutor to AUGUSTUS. STRABŌN mentions Athēnodōros as one of his companions and a source for government at Petra (16.4.21). Athēnodōros' writings include a work on tides (Strabōn 1.1.9, 1.3.12, 3.5.7), in which he argued that ebb and flux are analogous to breathing, and that sub-oceanic springs may exist whose flux raises the tide. Late in life, he returned to Tarsos to expel the despot Boëthos who was placed in power there by M. Antonius (Strabōn 14.5.14). He died at the age of 82. Sometimes confused with a coeval philosopher of the same name, also from Tarsos (Athēnodōros Kordulion); one of the two argued that divination is reliable and a skill (DIOGENĒS LAËRTIOS 7.149).

FGrHist 746; *GGP* 4.2 (1994) 711–712, P. Steinmetz; *OCD3* 203, J. Annas; *ECP* 100, I. Vasiliou; *BNP* 2 (2003) 252–253 (#3), K.-H. Hülser.

GLIM

Atimētos (10 – 40 CE)

Taught SCRIBONIUS LARGUS (120) the colic remedy of CASSIUS, which included Indian nard, opium, black pepper, etc., in honey. ASKLĒPIADĒS PHARM., in GALĒN *CMLoc* 4.7 (12.771 K.: emending from *ATIMHTP-*), cites him for a collyrium containing ***psimuthion, khalkanthon, pompholux***, opium, and saffron. He was associated with the emperor Tiberius, and may be the same as the *ocularis* Attius Atimetus, known from his **collyrium stamps**. See ACILIUS HYGINUS.

RE 2.2 (1896) 2253 (#10), M. Wellmann; S.3 (1918) 17, W. Kroll; Korpela (1987) 180.

PTK

Attalos (Med.) (130 – 170 CE)

GALĒN, *MM* 13.15 (10.909–916 K.) describes his elder contemporary Attalos, a student of SŌRANOS; Galēn disparages his treatment of the Cynic Theagenēs via plasters of honey-bread, affusions of warmed olive oil, and a diet of porridge. He is perhaps the same as Statilius Attalus, personal physician to Antoninus Pius and M. Aurelius. AELIUS PROMOTUS, *Dyn.* 72.5, refers to the "plaster we received from Attalos" (more likely a contemporary than ATTALOS III OF PERGAMON), also to be found in THEODŌROS OF MACEDON, Book 6. The preface to a late-Latin commentary on OREIBASIOS lists "Attalion" (perhaps our man) among earlier commentators on the HIPPOKRATIC CORPUS, APHORISMS.

RE 3A.2 (1929) 2186 (Statilius #11), F.E. Kind; Korpela (1987) 180; Ihm (2002) #32.

PTK

Attalos III of Pergamon, Philomētōr (138 – 133 BCE)

Last monarch of the Pergamene kingdom, which he famously bequeathed to Rome in his will. He probably served as patron to NIKANDROS OF KOLOPHŌN, and authored a treatise on agriculture known to CASSIUS DIONUSIOS (VARRO, *RR* 1.1.8–10; *cf.* COLUMELLA 1.1.8); to judge from references in PLINY, 1.*ind.*10–11, 14–15, 17–18, it discussed beekeeping, cereals, viticulture, and arboriculture. He reportedly devoted his final years to gardening, pharmacology, and bronze-smithing (Justin 36.4.1–5; *cf.* Pliny 1.*ind.*33). HĒRĀS, in

GALĒN *CMGen* 1.13 (13.414–416 K.), *cf.* 1.17 (13.446–447 K.), records his wound plaster containing white pepper, **litharge**, and ***psimuthion*** in **terebinth** and beeswax; *cf.* ANDROMAKHOS in Galēn, *CMGen* 1.14 (13.419–427 K.). CELSUS 5.19.11 offers a wound plaster from Attalos composed of copper flake, frankincense soot, and ***ammōniakon*** incense in ***terebinth***, bull-fat, vinegar and olive oil, and 6.6.5B a collyrium of aloes, antimony, **calamine**, myrrh, saffron, etc. Pliny, 1.*ind.*28, 31, cites him for medicine from animals. Galēn praises the pharmacological work of "our Attalos," *Simpl. Med.* 10.1 (12.251 K.), *Antid.* 1.1 (14.2 K.). (*Cf.* perhaps ASKLĒPIADĒS PHARM., in Galēn *CMLoc* 8.3 [13.162–163 K.], apparently used by MANTIAS.)

J. Hopp, *Untersuchungen zur Geschichte der letzen Attaliden* (1977); *OCD3* 211, R.M. Errington.

Philip Thibodeau

Attalos of Rhodes (*ca* 150 – 125 BCE)

HIPPARKHOS (*In Eudoxi et Arati Phaenomena*) frequently castigates a commentary on ARATOS' *Phainomena* by Attalos, his local contemporary whom he calls *mathēmatikos* (i.e. astronomer); the designation "of Rhodes" is found only in an anonymous list of commentators on Aratos. Hipparkhos' criticisms and 14 brief quotations show that Attalos for the most part sought to exonerate Aratos from charges of astronomical inaccuracy.

Maass (1898) 1–24; J. Martin, *Histoire du texte des Phénomènes d'Aratos* (1956) 22–27.

Alexander Jones

ATTICUS ⇒ (1) IULIUS; (2) LICINIUS

Atticus (*ca* 150 – 200 CE)

Platonic philosopher, wrote a polemical tract *Against those who interpret Plato's teachings through Aristotle's*, cited extensively by EUSEBIOS (*PE* 11 and 15), and commentaries on Platonic dialogues, cited by PROKLOS (often grouping Atticus with PLUTARCH). Evidence exists for commentaries on *Timaeus*, *Phaedo* (?) and *Phaedrus* (?). Eusebios (*Chron.* p. 207 Helm²) places his *floruit* in 176–180 CE. He taught HARPOKRATIŌN OF ARGOS.

Regarding ontology and cosmology Atticus often sides with Plutarch against the **Platonic** mainstream, teaching the temporal creation of the ***kosmos***. Atticus' anti-Aristotelianism is noticeable in his rejection of the fifth element (***aithēr***), criticism of the astronomical views expressed in *De caelo*, and of the *Categories*, for which he follows his near-contemporary Nikostratos the **Platonist**. Atticus' works were read in PLŌTINOS' school. He moreover seems to have influenced GALĒN and LONGINUS.

Ed.: E. des Places, *Atticus. Fragments* (*CUF* 1977).
Moraux (1984) 2.564–582; *DPA* 1 (1989) 664–665, J. Whittaker; Dillon (1996) 247–258; *BNP* 2 (2003) 325–326, M. Baltes.

Jan Opsomer

Attius (30 BCE? – 75 CE)

Among the first astrologers to write in Latin, listed last (after MAMILIUS SURA) with agricultural authorities (PLINY 1.*ind.*18). In *Praxidikē*, Attius suggests that the best time to sow is when the moon is in Aries, Gemini, Leo, Libra, or Aquarius (Pliny 18.200): masculine and diurnal signs (Sagittarius omitted: MANILIUS 1.150–154, 2.358–384, *cf.* ARISTOTLE *Met.* 1.5 [986a]) in accord with the **Pythagorean** theory that odd numbers are masculine. Praxidikē, exactor of justice, is identified with Persephone in the Orphic tradition (Paus. 9.33.3).

(*)

GLIM

T. Aufidius of Sicily (*ca* 100 – 50 BCE)

Listed by PHILŌNIDĒS OF DURRAKHION among the medical notables resident in Durrakhion (STEPHANOS OF BUZANTION, *Ethnika* s.v. "Durrakhion" [Meineke, p. 245]), a *sectator* of ASKLĒPIADĒS, likely the *Titus Asclepiadis sectator* in CAELIUS AURELIANUS, *Acute* 2.158, and *Chronic* 3.78 (pp. 239 and 761 Drabkin; *CML* 6.1, pp. 238, 726). Nothing else survives of his books on chronic diseases or his two-volume *De anima* (*sc. Peri psukhēs*, "mental illness"). Aufidius thought it beneficial to flog a mental patient (here afflicted with *mania*), or to put him in chains, starve and deny him water, then after a time entice him with wine and the prospect of sex (Cael. Aur., *Chron.*, 1.179 [*CML* 6.1, p. 536]: *tunc uino corrumpi, uel in amorem induci*); sex is again recommended in treating jaundice (*Chron.*, 3.78 [above]), since it "relaxed the flesh" (*laxationem carnis faciendum*). It is little wonder that Asklēpiadēs and his students were popular among their Roman patients.

RE 2.2 (1896) 2290 (#13), M. Wellmann.

John Scarborough

AUGUSTINUS ⇒ AURELIUS AUGUSTINUS

AUGUSTUS ⇒ IULIUS CAESAR OCTAUIANUS

Auidianus (200 – 650 CE)

IŌANNĒS OF ALEXANDRIA, *In Galeni Sect.* (pp. 15–16 P., *cf.* DIOKLĒS OF KARUSTOS *fr.*13c van der Eijk), lists **Methodists**: "THEMISŌN, THESSALOS, DIONUSIOS, *MANASEVS* (sc. MNASEAS), PHILŌN, *OLIMPICVS* (i.e., OLUMPIAKOS), MENEMAKHOS, *AVIDIANVS*." Although the name is attested in the mid-3rd c. CE, *Cod. Iust.* 9.2.6 (*RE* 2.2 [1896] 2378, P. von Rohden), if for *-ID-* we restore *-REL-* we have *AVRELIANVS*, i.e. a reference to CAELIUS AURELIANUS. (*Cf.* ASKLĒPIADĒS TITIENSIS, SEXTUS OF APOLLŌNIA.)

(*)

PTK

Postumius Rufius Festus Auienus of Volsinii (340 – 380 CE)

Roman aristocrat and poet, whose work includes translations of the astronomical poem *Phainomena* by ARATOS and of the geographical poem by DIONUSIOS PERIĒGĒTĒS, *Descriptio orbis terrae.* He also wrote a poem *De ora maritima,* in which he follows the model of a ***periplous*** and describes regions in the order of travel along the coastline. Only the part describing regions of the Atlantic Ocean and the Mediterranean Sea from Brittany to Marseille survives. Scholars debate whether the poem is based on older ***periploi*** or later compilations.

Ed.: *GGM* 2.177–189; P. van de Woestijne, *La descriptio orbis terrae d'Avienus* (1961).
KP 1.788–789, M. Fuhrmann; *PLRE* 1 (1971) 336–337; *OCD3* 226, J.H.D. Scourfield; *BNP* 1 (2003) 426–427, J. Küppers.

Natalia Lozovsky

AURELIANUS ⇒ CAELIUS AURELIANUS

AURELIUS ⇒ (1) ALEXANDER OF APHRODISIAS; (2) HĒRAKLEIDĒS; (3) NEMESIANUS; (4) CASSIODORUS SENATOR

Aurelius (*ca* 155 – 200 CE?)

GALĒN himself records Aurelius' dentifrice in *CMLoc* 5.5 (12.892 K.), compounded from alum, roasted and then quenched with dry wine, to which were added **mastic**, frankincense, ***malabathron***, and *kuperos*, mixed and applied. He is conceivably identifiable with the military physician Aurelius Artemōn, attested at Moesia Inferior (155 CE: *CIL* 3.7449).

F. Cramer, *Anecdota Paris.* 1 (1839) 394; *RE* S.1 (1903) 229–230 (#60a), A. Stein.

GLIM

Aurelius Augustinus (*ca* 385 – 430 CE)

Born 354 CE in Thagaste, he was concerned with transmitting the classical intellectual disciplines arrayed by VARRO until his conversion in 386 to the blend of neo-**Platonism** and Christianity taught by AMBROSIUS; he was bishop of Hippo from 395. In addition to scores of theological works, wrote a treatise *On music* in six books. Its famous definitions of "music" (1.2–3) as "*scientia bene modulandi*" and "*scientia bene mouendi*" ground it in Roman rhetorical tradition rather than Greek music theory. Books 1–5 (completed *ca* 388 CE), on rhythm, develop the primacy of music over grammar in understanding the movement of sound in language, illustrate proportions of time (following **Pythagorean** traditions) expressed in rhythm and meter, establish that number provides the basis for true knowledge of music, and provide examples of various meters and verse types. In the sixth book (*ca* 391 CE), on musical metaphysics, number and proportion are expanded from the corporeal to the incorporeal. Numbers in rhythm – found as well in light, color, dance, and celestial harmony – are heard and exist in the memory but are also eternal. Genera of number exist in: sound (*sonus*), the sense of hearing (*sensus audientis*), the act of presentation (*actus pronuntiantis*), memory (*memoria*), and discernment (*iudicium*). Augustine concludes that when organized according to the numerical principles of proportion, music can stimulate the soul to imitate celestial harmony and lead it to a love of God.

Ed.: M. Jacobsson (with English trans.), *Aurelius Augustinus De musica liber VI* (2002).
R. Catesby-Taliaferro, trans., *Saint Augustine on Music* (1947); Mathiesen (1999) 619–622; *NGD2* 1.173–174.

Thomas J. Mathiesen

Iulius Ausonius of Vasates (*ca* 315 – 378 CE)

Born *ca* 290 CE; father of the poet Decimus Magnus Ausonius, was born at Vasates and practiced medicine at Burdigala, where was member of the *curia*; in old age became *praefectus* of Illyria (Aus. *Epic.* 2; *Lect.* 5–14; *Parent.* 3). He spoke Greek better than Latin (*Epic.* 2.9), possibly because he had studied in a medical school of Massalia. He is listed by MARCELLUS OF BORDEAUX (*pr.*2) as a source, who also refers to a remedy of Ausonius for sciatica and arthritis (25.21).

RE 2.2 (1896) 2562 (#2), F. Marx; M.K. Hopkins, "Social Mobility in the Later Roman Empire," *CQ* 11 (1961) 239–249; *PLRE* 1 (1971) 139 (#5); Matthews (1975) 81–82.

Fabio Stok

Autolukos of Pitanē (*ca* 300 BCE)

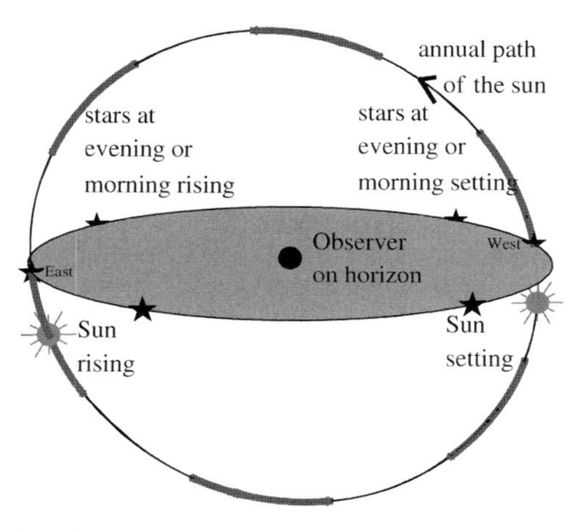

Autolukos: On Risings and Settings © Mendell

Taught Arkesilaos with whom he then traveled to Sardēs (DIOGENĒS LAËRTIOS, 4.29). Two works of his survive, *On Rotating Spheres* and *On Risings and Settings*. Although earlier Greek mathematical texts may be found in ARISTOTLE and the fragments of EUDĒMOS, these are probably the earliest extant complete Greek mathematical texts; yet they display a formal presentational style similar to other extant treatises of the 3rd c. BCE. The 12 propositions of *On Rotating Spheres* concern a sphere attached to a fixed horizon, where it rotates obliquely to it, and consider the properties of visible and invisible parts of the sphere. In *On Risings and Settings* (two books), Autolukos defines the basic phases of fixed stars, evening rising and setting, morning rising and setting, distinguishes the apparent phases from the true, and examines the conditions for the phases to occur for stars north of, on, and south of the **ecliptic**. He assumes the arc determining visibility is ½ a zodiacal sign or a 24th part of the zodiacal circle, i.e. if the sun is at least ½ a sign from the **ecliptic** point on the horizon, then stars above the horizon can be visible. The second book elaborates in detail some of the theorems of the first book. He attempted, in a dispute with ARISTOTHEROS, to explain variations in the brightness of planets, which seem to indicate changes in their distances from the earth (SIMPLICIUS, *In de caelo, CAG* 7 [1894] 504: *cf.* POLEMARKHOS).

Ed.: J. Mogenet, *Autolycus de Pitane: histoire du texte* (1950); F. Bruin and A. Vondjidis, trans. and ed., *The Books of Autolykos: On a Moving Sphere and On Risings and Settings* (1971).
Neugebauer (1975) 747–767.

Henry Mendell

Auxanōn (before *ca* 350 CE)

Described as *hippiatros* by APSURTOS, who quotes Auxanōn on cures for diarrhea (*Hippiatrica Parisina* 103 = *Hippiatrica Berolinensia* 35.1).

McCabe (2007) 141–142.

Anne McCabe

Axios (45 BCE – 100 CE)

Ocularis of the Roman British fleet. KRITŌN OF HĒRAKLEIA preserves two recipes: his collyrium based on cinnabar, in GALĒN *CMLoc* 4.8 (12.786 K.), and his treatment for ***leikhēn***, *ibid.* 5.3 (12.841 K.: emending from *AΞIOPI-*), compounded of copper flakes, ***khalkanthon***, realgar, cantharides, and white hellebore. The name could be Latin, the plebeian *nomen* Axius.

RE 2.2 (1896) 2633 (#2), M. Wellmann.

PTK

Azanitēs (1st c. BCE)

Created a special plaster against various types of ulcers, wounds, and other diseases, mentioned by HĒRĀS and approved by GALĒN (*CMGen* 5 [13.784–785 K.]). Azanitēs' pharmacological formulae were popular, quoted by OREIBASIOS (*Synopsis*, 3, p. 43), AËTIOS OF AMIDA (14.34, p. 781 Cornarius; 15.21, p. 854 Cornarius, where Zervos 1909: 123 reads "Ananias"), and PAULOS OF AIGINA (*CMG* 9.2, p. 376; *cf.* also *Hipp. Berol.* 130.126–127, p. 424 ed. Oder-Hoppe).

RE 2.2 (1896) 2640, M. Wellmann.

Antonio Panaino

B

Babylonian Astronomy (*ca* 1800 BCE – *ca* 100 CE)

A tradition of celestial divination, the oldest cuneiform record of which goes back to *ca* 1800 BCE, produced systematic observation of the Moon, Sun, planets, and fixed stars. In addition to celestial divination, early astronomical texts, such as the Astrolabes, a tradition mainly concerning fixed stars originating in the middle of the 2nd millennium BCE (Hunger and Pingree 50–57), reflect knowledge of the seasonal heliacal rising of fixed stars as well as the change in the length of the day with the north-south progress of the Sun's rising over the eastern horizon in the course of an ideal year (12 30-day months). The compendium "MUL.APIN," composed *ca* 1000 BCE, systematizes astronomical phenomena such as stellar risings, settings, and culminations, intervals of visibility between first and last appearances of the superior planets, intervals of invisibility between last and first appearances of the superior planets, as well as periods of visibility and invisibility between synodic appearances of the inferior planets, and intercalation schemes based on the ideal calendar. With recognition of the periodic nature of some of the ominous phenomena, methods to predict such phenomena were developed, first within the 7th c. Neo-Assyrian court, then after 500 BCE in the scribal centers of Babylōn and Uruk.

Observational reports extant from 709–649 record phenomena considered ominous and interpreted according to the canonical omen series *Enūma Anu Enlil*. The more comprehensive Babylonian astronomical diaries contain systematic and continuous nightly observations from the 7th to 1st cc., though evidence points to their origin already in the 8th c. Celestial data focused on the Moon's progress through the fixed stars, eclipses both lunar and solar, planetary phenomena, meteors, comets, and various weather reports. The diaries were utilized by PTOLEMY: for example, *Alm.* 9.7 dates observations of Mercury "according to the Khaldaeans," and gives positions by means of the cubit and the **ecliptical** stars used in the diaries.

Late Babylonian astronomy was an exact science characterized by mathematical models of the longitudinal progress of synodic lunar and planetary phenomena. These models underlie the computation of lunar and planetary **ephemerides**, which tabulate the dates and longitudes of the synodic phenomena. Hellenistic authors (e.g., STRABŌN, PLINY, and VETTIUS VALENS) associated this science with the names KIDĒNAS, SUDINĒS, and NABURIANOS. Greek papyri from Roman Egypt, containing sequences of sexagesimal numbers forming "zig-zag" and "step" functions with Babylonian astronomical parameters, attest to the transmission of Babylonian astronomy to the Greeks. Babylonian predictive methods were therefore fully integrated in Greco-Roman astronomy until the 5th c. CE, and Greek awareness of their Babylonian inheritance is indicated by mention of *Orchenoi* (*P.Oxy.* 4139, line 8: Jones [1999]), "people of Uruk," whom Strabōn identified as "astronomical Khaldaeans" (16.1.6).

The Babylonian paradigm for calculating the rising times of the zodiacal signs (Greek *anaphorai*) significantly influenced Hellenistic astronomy. The evidence for zodiacal rising times is embedded in the column for generating length of daylight in Babylonian **ephemerides**. The calculation for the length of daylight is derived from the sum of the rising times for the appropriate half of the zodiac rising on the day in question beginning with the position of the Sun. The direct connection between the position of the Sun in the **ecliptic** and the length of daylight is therefore expressed. Two such schemes are attested, and their values are constrained by a 3:2 ratio of longest to shortest daylight, assumed in Hellenistic astronomy as the canonical value for the *klima* of Babylōn (latitude 32.5°).

Furthermore, the ancient Mesopotamian celestial science of **genethlialogy** significantly shaped Hellenistic astrology. Natal omens are attested in cuneiform texts of the mid- to late first millennium. By *ca* 500 BCE the celestial signs visible at birth were noted, and such divination was soon followed by the earliest horoscope, i.e., a collection of the positions of the planets, Moon, and Sun at the moment of birth. Most Babylonian horoscopes come from the city of Babylōn. Others are known from Uruk and one from Nippur. They date from the 3rd to the 1st c. BCE, excepting two 5th c. BCE documents. Because it was necessary to obtain the positions of all planets at the arbitrary moment of birth, methods to compute these positions were critical. As horoscopes begin to appear, so do a variety of methods to compute astronomical data. It is unknown whether the schemes attested in the **ephemerides** were used for this purpose. Certainly the astronomical diaries and almanacs were a source for the scribes who prepared horoscopes.

ACT; H. Hunger, *Astrological Reports to Assyrian Kings* (1992); *Idem* and D.E. Pingree, *Astral Sciences in Mesopotamia* (1999); Francesca Rochberg, *Babylonian Horoscopes* (1998); *Eadem*, *The Heavenly Writing: Divination, Horoscopy, and Astronomy in Mesopotamian Culture* (2004); *Eadem*, "A Babylonian Rising-Times Scheme in Non-Tabular Astronomical Texts," in Ch. Burnett, J. Hogendijk, K. Plofker, and M. Yano, edd., *Studies in the History of the Exact Sciences in Honour of David Pingree* (2004) 56–94.

Francesca Rochberg

Baitōn (335 – 305 BCE)

Recorded Alexander's itinerary, giving data on peoples, plants, and the heavens along the route, preserved in PLINY 6.61–62, 6.69, and 7.11; *cf.* DIOGNĒTOS OF ERUTHRAI and PHILŌNIDĒS OF KHERSONĒSOS. The apparently unattested name Baitōn may be Egyptian, meaning "hawk" (Heuser 1929: 14, 20). Compare perhaps Batōn in DIOGENĒS LAËRTIOS 6.99 (3rd c. BCE), *LGPN* 3A.89 and 3B.85 (4th c. BCE), and Baitis of Larissa (*LGPN* 3B.84), 3rd c. BCE.

FGrHist 119.

PTK

Bakkheios Gerōn (*ca* 300 – 400 CE?)

Author of a small musical catechism preserved under the title of *Introduction to the art of music*, usually (though not always) followed in the MSS by a second distinct treatise. The MSS regularly apply the same title and author to the second treatise, due to the inadvertent connection of the end-title of the first treatise with the beginning of the second, but the second is entirely different from the first in approach, style, and content. It is almost certainly not by Bakkheios. The second treatise is followed in most (but not all) MSS by an epigram referring to a certain Dionusios and the emperor Constantine. The epigram has commonly been taken to refer to the second treatise, the attribution of which is accordingly modified (even in some of the MSS) to Dionusios, but this, too, is hardly certain, nor is it certain which of the several possible emperors Constantine is intended or whether Bakkheios is contemporary with Dionusios. In the end, the epigram is of no use in dating either treatise, and only the first can be reasonably assigned to Bakkheios.

The treatise, presented as a series of questions and answers, mixes definitions and theories from various early traditions. The first 88 questions define common terms and concepts in harmonics; questions 89–101 are devoted to rhythmics. Some of the answers (11, 13–18, 29–34 and 38–42) employ musical notation, recognizable from the tables of ALUPIOS. The treatise represents nothing completely new, but several of the answers, especially in the section on rhythmics, clarify or confirm other sources. The unassuming character, routine content, and style of the treatise suggest a date no earlier than 300.

Ed.: *MSG* 292–316.

O. Steinmayer, trans., "Bacchius Geron's 'Introduction to the Art of Music'," *Journal of Music Theory* 29 (1985) 271–298; *NGD2* 2.293–294; Mathiesen (1999) 583–593.

Thomas J. Mathiesen

Bakkheios of Milētos (325 – 90 BCE)

Agronomist whose work, which may have treated cereals, livestock, poultry, viticulture, and arboriculture (*cf.* PLINY, 1.*ind.*8, 10, 14–15, 17–18), was excerpted by CASSIUS DIONUSIOS (VARRO, *RR* 1.1.8–10, *cf.* COLUMELLA, 1.1.9).

RE 2.2 (1896) 2790 (#9), M. Wellmann.

Philip Thibodeau

Bakkheios of Tanagra (250 – 200 BCE)

Physician, resident in Alexandria, among the rare central Greek immigrants. Bakkheios was a "**Hērophilean**" in medical practice, valuing anatomy, pharmacology, and knowledge of the pulses in diagnosis and prognosis, but he is generally cited for his lexicographical studies of Hippokratic terminologies. ERŌTIANOS records his glosses on at least 18 works of the HIPPOKRATIC CORPUS (including EPIDEMICS III, SACRED DISEASE, APHORISMS, WOUNDS IN THE HEAD, and *Joints, Instruments of Reduction, Fractures*: see SURGERY), demonstrating the circulation of "Hippokratic" writings by the 3rd c. BCE. Probably the predominance of so-called "Koan" treatises (as contrasted to those presumably from Knidos) led later commentators to make what modern scholarship has determined to be a false dichotomy among the Hippokratic tracts. In addition to the extensive fragments of Bakkheios' *Hippokratic Lexicon* (*Lexeis*), later authors cite him for his work in pulse theory (GALĒN, *Diff. Puls.* 4.6, 10 [8.732–733, 748–749 K.]; MARCELLINUS, *On Pulses* 3 [p. 457 Schöne]; and Galēn, *Dign. Puls.* 4.3 [8.955 K.]), and pharmacology (Galēn, *CMGen* 7.7 [13.987 K.]). Also mentioned is his *Memoirs of HĒROPHILOS and Those from his House* (Galēn, *CMG* 5.10.2.2, p. 203).

W.D. Smith, "Galen on Coans versus Cnidians," *BHM* 47 (1973) 569–585; *Idem* (1979) 202–204; von Staden (1989) 484–500; *OCD3* 230, *Idem; Idem* (1999) 158–160.

John Scarborough

Bakkhulidios (250 BCE – 500 CE)

Cited by pseudo-Aelius Promotus 51 (p. 66 Ihm) for an antidote; the name as such is unattested, although the archaic "Bakkhulidēs" was in use until Aelius' era (*LGPN* 1.98, 2.86). *Cf.* perhaps Aiskhulidēs of Keōs.

(*)

PTK

Bakōris of Rhodes (*ca* 405 – 350 BCE?)

Wrote a ***Periplous*** of unknown scope, which Auienius, *OM* 42–50, cites with Phileas of Athens and others. The otherwise unattested name might be Egyptian: either from Pakōris ("he of the snake," *cf.* Heuser 1929: 34) or a version of Bokkhōris. Compare also Bakō(n) of Athens (*LGPN* 2.86), 5th/4th c. BCE, and Bakos of Tauris (*LGPN* 4.64), 4th c. BCE.

(*)

PTK

Balbillos (Barbillos), Ti. Claudius (40 – 80 CE)

Roman court astrologer of the 1st c. CE and *praefectus Aegypti*, 55–59 CE. Cichorius identified Balbillos as the son of Thrasullos, astrologer for Tiberius and Claudius. Balbillos served in the courts of Claudius, Nero, and Vespasian. Fragments of his work survive (*CCAG* 8.4, 232–238 and 240–244), including *sunkephalaiōsis* (*CCAG* 8.3, 103), and he is mentioned by Seneca (*QN* 4.2.13), Tacitus (*Annals* 15.47), Suetonius (*Nero* 36) and Dio Cassius (65.9.2). His *Astrologumena* is dedicated to Hermogenēs (*CCAG* 8.3, pp. 103–104).

A proponent of deterministic astrology in an era when the nature of astral influences was in debate, Balbillos utilized a "method concerning the length of life from starter and destroyer" (*CCAG* 8.4, p. 232), in which a linear arithmetical scheme is given to predict the month of a person's death (*CCAG* 8.4, p. 243). The numbers of the scheme are simple substitutions for the days of a lifetime, modeled loosely on schemes for the change in the length of daylight through the year. The same treatise preserves the two earliest Greek literary horoscopes, dated 72 BCE Jan. 21 or 16, and 42 BCE December 27 (Neugebauer-van Hoesen 1959: 76–78).

After serving as *praefectus*, Balbillos' career is obscure until the accession of Vespasian, when he again rose to prominence, perhaps through his relation by marriage to Vespasian's ally Antiokhos IV Epiphanēs of Kommagēnē. Dio Cassius mentions games in Balbillos' honor established at Ephesos. Other Latin inscriptions attest to these games from 90 CE until the 3rd c., referring to them either as the *Balbillea* or the *Barbillea*.

C. Cichorius, *RhMus* 76 (1927) 102–105, contra, see *PIR2* C-813; Pingree (1978) 423; *BNP* 2 (2003) 471, W. Hübner; *BNP* 3 (2003) 397–398 (#II.15), W. Eck.

Francesca Rochberg

Balbus (102 – 106 CE?)

Nothing is known of this man's life. If the beginning of his treatise *does* refer to Trajan's expedition to Dacia, it can be dated to between 102 and 106. The *Expositio et ratio omnium formarum* has come to us in mutilated form; this is why, contrary to what its title promises, it does not deal with all figures. This handbook of geometry was written by an *agrimensor*: the author therefore always keeps in mind the relationship between geometry and a surveyor's work. Beside definitions concerning the categories of Roman land management (*rigor, extremitas, decumanus, cardo, ager arcifinius*), the geometrical definitions (point, line, parallel, area, and so on) to be found in the extant part of the *Expositio* testify that EUCLID's *Elements* had already been translated into Latin, at least Books 1 to 3, when Balbus wrote, therefore a long time before such translations as attested by MARTIANUS CAPELLA or attributed to BOËTHIUS. The fortunes of this handbook, to be found in many MSS and still used in the medieval *Demonstratio artis geometricae* edited by Lachmann, are precisely due to its offering only definitions together with very elaborate figure classifications. In this respect it is comparable with the Greek *Definitiones* attributed to HĒRŌN OF ALEXANDRIA, actually apocryphal but whose substance may date back to him.

K. Lachmann, *Die Schriften der Römischen Feldmesser*, v.1 (1848); French translation and commentary: Guillaumin, *Balbus* (1996) = *CAR* 3.

Jean-Yves Guillaumin

Bardaisan of Edessa (174 – 222 CE)

Bardaisan was born in Edessa in 154 CE, and grew up in the court of Abgar VIII, the Great. A pagan convert to Christianity, his distinct ideas diverged enough from orthodoxy that later writers counted him a heretic. He travelled through Armenia, about which he wrote a history that is known in Greek translation. He was also an accomplished poet whose skill was admired even by his opponents. He died at the fortress of Anium in 222 CE. Almost unique among Christian writers of his cultural milieu, Bardaisan did not embrace sexual abstinence but stated that sexual activity was beneficial, particularly for women. He thus had a son, Harmonius, who was also a poet.

His one major surviving work is the *Book of the Laws of the Countries* (BLC) that, although ascribed to him, was probably written by his pupil Philip. Written in Syriac, BLC contains a philosophical argument about freewill and determination. This entails discussions of the customs of various regions, hence its title, but also much cosmological speculation and a discernible anthropological discourse. The structure of the ***kosmos***, with its various (probably seven) spheres of influence, is thus reflected in the individual, whose various constituent parts are duly influenced by the corresponding cosmological element. BLC is extant in its Syriac version, and a Greek translation is quoted by later writers including EUSEBIOS.

H.J.W. Drijvers, *Bardaisan of Edessa* (1966); *KP* 1.824, K. Wegenast; Brock (1997) 15; *BNP* 2 (2003) 507–508, *Idem*.

Siam Bhayro

Basil of Caesarea (Kappadokia) (*ca* 365 – 379 CE)

Born in Caesarea around 330 in a Christian upper-class family, older brother to GREGORY OF NUSSA. Basil was educated first in his native city and then at Antioch and Athens by LIBANIOS, Himerios, Prohairesios, and was a fellow student with his friend GREGORY OF

Nazianzos (Sōcratēs *HE* 4.26). Basil left a career as a rhetorician in Caesarea to become a monk; in 364 he was elected bishop of Caesarea where he died in 379. His writings greatly influenced the development of Christian theology, and his argument in favor of reading pre-Christian Greek literature encouraged the preservation of texts. He is received as a saint by the Orthodox and Roman churches.

Basil demonstrated his familiarity with contemporary science in his *Homilies on the Six Days of Creation* (*Homilies in Hexaemeron*: "*HH*") where he criticized prevailing allegorical interpretations of Genesis in favor of a more scientific approach (*HH* 3.9, pp. 234–238 Giet). The Homilies seek to articulate a systematic account of creation inspired by *Genesis* against rival accounts proposed by pagans, Manicheans, and Gnostics, deploying scientific knowledge to articulate Christian salvation-history. Basil opposed the view of Plato's *Timaeus*, that God was a craftsman of disordered and eternal matter, since that limited God's freedom. He argued instead that the ***kosmos*** was created by God *ex nihilo* through goodness (*HH* 1.2, 7.7 = pp. 96, 464 Giet), which continues to be exercised after the world's creation through God's providence (*HH* 7.5, p. 416 Giet). Basil accepted the two **Platonic** worlds, one invisible and the other, later-created, visible world composed of the standard four elements (*HH* 1.7, pp. 116–118 Giet). The universe was created for the education and edification of human souls (*HH* 1.5, p. 106 Giet). Humans are the sole being created not by divine command, but by God's own hands (*HH* 6.1, p. 328 Giet), and belong to both the invisible (spiritual) and visible (material) world. Basil accepted the Platonic tripartite soul, incorporeal, but localized after joining the body (*Hom. Att. tibi ipsi* 7, p. 35 Rudberg), the rational part, which is characterized by free will, being the image of God (*Hom. in Psalm.* 48.7, 1.184E–185A Garnier). Basil also had considerable medical knowledge and was one of the first bishops to use the resources of the church to found hospitals for the care of the sick and poor.

Ed.: *PG* 29–31; S. Giet, *Basile de Césarée. Homélies sur l' Hexaéméron* (1950); B. Sesboüe, *Basile de Césarée Contre Eunome*, 2 vv. (1982–1983).

RAC 1 (1950) 1261–1265, G. Bardy; Y. Courtonne, *Un témoin du IVe siècle oriental. Saint Basile et son temps d'après sa correspondence* (1973); N.G. Wilson, *Saint Basil on the value of Greek Literature* (1975); M.A. Orphanos, *Creation and Salvation according to St. Basil of Caesarea* (1975); P.J. Fedwick, ed., *Basil of Caesarea: Christian, Humanist, Ascetic* 2 vv. (1981); *ODB* 269–270, B. Baldwin, A.P. Kazhdan, and N.P. Ševčenko; Ph. Rousseau, *Basil of Caesarea* (1994); Meredith (1995).

George Karamanolis and Daniel L. Schwartz

Basileidēs (225 – 175 BCE)

Epicurean philosopher, fourth **scholarch** of the **Garden** after Epicurus (Diogenēs Laërtios 10.25). He collaborated with the Alexandrian mathematician and astronomer Hupsiklēs on the work of the mathematician Apollōnios of Perga, and contributed to a debate about the nature of anger, arguing against the views of the **Epicureans** Nikasikratēs and Timasagoras (*cf.* Timagoras).

BNP 2 (2003) 516–517 (#1), T. Dorandi.

Walter G. Englert

Basilis (*ca* 300 – 115 BCE?)

Cited with Hekataios as treating the eastern quadrant of the world, wrote an *Indika*, in at least two books, wherein he described partridge-riding dwarfs warring against cranes

(Ath., *Deipn.* 9 [390b]) and perhaps an *Aithiopika*. He is cited with DALIŌN, ARISTOKREŌN, BIŌN, the younger SIMŌNIDĒS, and XENOPHŌN as a foreign authority on geography and ethnography (PLINY 1.*ind*.6), estimating the size of "Ethiopia" (6.183). The name is rare (2nd c. BCE – 2nd c. CE: *LGPN*).

FGrHist 718.

GLIM

BASSUS ⇒ (1) CASSIANUS BASSUS; (2) IULIUS BASSUS; (3) POMPONIUS BASSUS

Bathullos (?) (100 BCE? – 10 CE)

ASKLĒPIADĒS PHARM., drawing on HĒRĀS, in GALĒN, *Antid.* 2.11 (14.173–174 K.), gives his anti-***hudrophobia*** plaster, compounded from ***psimuthion*** and **litharge**, plus myrrh, etc., reduced in olive oil to a paste. Kühn prints *ΒΑΦΟΥΛΑΟΣ*, unattested; if not "Bathullos," perhaps "Babulos" (*LGPN* 3B.84, Delphi), BOUPHANTOS, or FAUILLA.

Fabricius (1726) 100.

PTK

BELKHIONIOS ⇒ VELCHIONIUS

De Rebus Bellicis (365 – 375 CE)

The anonymous author of this treatise details a proposed reformation of provincial, military, legal, and financial policy for late Roman imperial administration. The author describes military machines and equipment which he considers essential to the apparatus of the Roman army and for which meticulous colored drawings are provided, but his accounts lack the technical rigor of Hellenistic writers on siege-machines. Precise measurements and material specifications are lacking, and the author refers the reader to illuminations for clarification (7.1).

The author emphasizes current frontier dangers and the need for state-of-the-art equipment in attacking walled cities (6.3). He includes design specifications and instructions for deploying and utilizing a ballista whose arrows are propelled by a windlass rather than torsion (7), a wall-scaling mechanism (*tichodifrus*: 8, "wall-chariot"), short-range fletched javelins to enable speed (*plumbata tribulata*: 10), fletched shield-piercing weapons (*plumbata maillata*: 11), and several scythed-combat vehicles (12–14). The author also recommends a calf-skin device to facilitate crossing rivers (*ascogefyrus*: 16, "skin-bridge"), and, recognizing human limitations, ingeniously suggests adapting oxen-driven watermills into a paddle-wheel system to ease the operation of massive warships (*liburna*: 17), seemingly the first, perhaps theoretical, attempt to propel a ship without oars or sails.

Ed.: R. Ireland, *Anonymi auctoris De rebus bellicis* (1984); A. Giardina, *De rebus bellicis* (1989).

GLIM

Bērossos of Babylōn (*ca* 330 – 280 BCE)

Bērossos was a Chaldaean and a priest of Bel (=Marduk) in Babylōn. His life spanned the period from the reign of Alexander the Great to that of Antiokhos I Sōtēr. He is reported to have spent his final years in exile, teaching astrology on the island of Kōs in Ptolemaic territory. Only one work by Bērossos is known, the *Babyloniaca*, a history of Babylōn in three books that was dedicated to King Antiokhos I. It was written in Greek and intended to correct Greek misconceptions about the history of Babylōn. The *Babyloniaca* does not survive in its original form, but extensive fragments are preserved in the *Against* APIŌN of Iosephus and the *Chronicle* of EUSEBIOS. These reveal that Bērossos organized his work according to traditional Babylonian views of man and his place in the world.

Book 1 treated the origins of Babylōn and the gift of culture to the first men by the demigod Oannes. Book 2 recounted the history of Babylōn from the appearance of Oannes 432,000 years before the flood to the reign of Nabunasir in the 8th c. BCE. Book 3 dealt with the Neo-Babylonian and Persian periods. Bērossos' claim that he used cuneiform sources preserved in the temple of Bel at Babylon is confirmed by the fragments. These reveal that he used a version of the Babylonian creation epic in Book 1, a flood story and a king-list similar to the *Sumerian King List* in Book 2, and chronicle texts related to extant neo-Babylonian chronicles in Book 3.

Bērossos' influence on Hellenistic Greek culture in general and science in particular was negligible. Thus, despite claims that Bērossos was responsible for the introduction of Babylonian astrology to the Greeks, the fragments reveal no knowledge of mathematical astronomy or horoscopic astrology. Instead, they contain only an account of the motion of the Moon based on that found in the fifth tablet of the Babylonian creation epic and a description of the "Great Year."

FGrHist 680; Stanley M. Burstein, *The Babyloniaca of Berossos* (1978); A. Kuhrt in *Eadem* and S. Sherwin-White, *Hellenism in the East* (1987) 32–56; *BNP* 2 (2003) 608–609, B. Pongratz-Leisten.

Stanley M. Burstein

"Bērutios" (350 – 500 CE?)

Wrote a work on Asia Minor, cited by the RAVENNA COSMOGRAPHY 2.16, as "Purit(i)os": *cf.* 5.7 where the city "Bērutos" is similarly misspelled. Cited with IAMBLIKHOS and PORPHURIOS; perhaps *cf.* VINDONIOS ANATOLIOS or the orator Celsinus, both of Bērutos.

J. Schnetz, *SBAW* (1942) #6, p. 62.

PTK

BIALCON ⇒ ALKŌN

Billaros (of Thessalia?) (105 – 75 BCE)

Made a globe displayed in Sinōpē that was carried off by Lucullus in 74 BCE: STRABŌN 12.3.11. (Hultsch believed the globe demonstrated the diurnal heavenly rotation.) For the name, *cf. LGPN* 3B.86, Bilaros of Thessalia (20 BCE), 3B.422, Philērō of Thessalia (2nd c. BCE), or 4.71, Bilarra of Macedon (4th c. BCE).

RE 3.1 (1897) 472, Fr. Hultsch.

PTK

Biōn Caecilius (100 BCE – 77 CE)

Wrote *On Potencies* (*Peri Dunameōn*), listed by PLINY after BIALCON and before ANAXILAOS OF LARISSA as a Greek medical authority (1.*ind*.28). He learned a spleen remedy (dried horse tongue administered in wine) from *barbari* (Pliny 28.200). His identification with Caecilius Medicus is uncertain.

Fabricius (1726) 103.

GLIM

Biōn of Abdēra (330 – 270 BCE?)

Hypothesized six-month days and six-month nights in polar regions (thus likely not long after Putheas), according to Dēmētrios of Magnesia in Diogenēs Laërtios 4.58, who identifies him as a *mathēmatikos* (sc. "astrologer"), and seems to place him before Biōn of Soloi. According to Poseidōnios (*fr.*137 E–K in Strabōn 1.2.21), Biōn the *astrologos*, along with Aristotle and Timosthenēs, hypothesized a richer wind-rose than Homer.

DK 77.

PTK

Biōn of Soloi (325 – 250 BCE)

Biōn the agronomist is most likely identical to the historian of the same name whose multi-book account of the geography and ethnography of "Ethiopia" was based on first-hand travels. His agricultural interests – whether recorded in a separate treatise, it is unclear – included cereals, livestock, poultry, viticulture, and arboriculture (Varro, *RR* 1.1.8, Pliny 1.*ind.*8, 10, 14–15, 17–18). A title in Pliny, 1.*ind.*28, implies that he also wrote about the properties of drugs made from animal products. His travels probably postdate Alexander's conquest of Egypt, but as his figures for the dimensions of "Ethiopia" are cruder than Eratosthenēs', he presumably precedes him.

Ed.: *FGrHist* 668.
RE 3.1 (1897) 483 (#9), E. Schwartz.

Philip Thibodeau

Bithus of Durrakhion (80 BCE? – 75 CE)

Although Pliny 28.82 only records Bithus' magical claims about mirrors and menstruation, Bithus appears in his index (1.*ind.*28) among Latin medical writers. Andromakhos, in Galēn *CMGen* 5.12 (13.836–837 K.), records the ***trokhiskos*** which *Bithunos* made in Sicily, containing pomegranate-peel, **calamine**, birthwort, frankincense, Illyrian iris, oak-galls, ***khalkanthon, khalkitis, misu***, etc. (prepared when Sirius shines); Galēn himself approves it, *Simples* 10.2.13 (12.276 K.). The ***trokhiskos*** is prescribed, with wine, by Aëtios of Amida 4.12 (*CMG* 8.1, p. 365), and for auricular phlegm by Paulos of Aigina, 7.12.28 (*CMG* 9.2, p. 319), slightly altered. The name is Thrakian, *cf.* Plutarch, *Aratos* 34.1; *CIL* 3.703, 12391, 12395 and *LGPN* (esp. v.4).

(*)

PTK

Bitōn (170 – 160 BCE?)

Wrote a treatise on catapults and siege devices, *The Construction of War Machines and Artillery*, dedicated to King Attalos, so it must be dated in the reigns of Attalos I–III, i.e. 241–133

BCE. Athēnaios (*Deipn.* 14 [634]) refers to the work. Bitōn also refers to a treatise entitled *Optics*, which covers the use of the ***dioptra*** (52.7–53.2 W.).

The treatise describes four non-torsion catapults, i.e. catapults whose force derives from a bow: a large and a small stone thrower by Kharōn of Magnesia and Isidōros of Abudos (44.8–51.5 W.) respectively; and two belly-bows, i.e. glorified crossbows, by Zōpuros of Taras (61.2–67.3). It also includes a ***helepolis***, by Poseidōnios of Macedon (51.6–56.8); and a ***sambukē***, by Damis of Kolophōn (57.1–61.1). The treatise is technical and detailed in style with exact measurements of the different parts and advice on materials. It is often hard to interpret, perhaps because the diagrams are not preserved.

Bitōn is the best source on the early non-torsion catapults, little used after the invention of torsion catapults around 350 BCE, and otherwise only described briefly by Hērōn of Alexandria. Because Bitōn described technology that was almost certainly outdated in his time, his date is much debated. Marsden (61) placed him as early as possible in the 3rd c., arguing that non-torsion catapults were still in use in some places; Lewis places him in the 2nd c. ascribing the use of non-torsion catapults to an emergency lack of materials for the springs, and arguing that Damis' ***sambukē***, which uses a screw, cannot be too soon after Archimēdēs, the screw's inventor. Perhaps the descriptions of old catapults are simply the result of technical interest, rather than necessity.

Marsden (1971); M.J.T. Lewis, "When was Biton?", *Mnemosyne* 52 (1999) 159–168; *BNP* 2 (2003) 682–683, H. Schwerteck.

Karin Tybjerg

Blastos (30 BCE – 80 CE)

Galēn, *CMLoc* 7.2 (13.17–21 K.), quotes Andromakhos' record of three throat lozenges from Blastos: Andromakhos approves the second, containing aloes, saffron, and **mastic**; the third has frankincense, myrrh, saffron, etc. in honey; and the first has 18 ingredients, including frankincense, **mastic**, myrrh, saffron, plus Indian nard, cassia, ***kostos, malabathron***, etc., boiled in honey wine. Korpela tentatively identifies with Ti. Claudius Blastus of *CIL* 6.9571 (*ca* 40–60 CE).

Korpela (1987) 180.

PTK

Blatausis (before 700 CE)

Known from a single mention in the anonymous Ravenna Cosmography 3.1, among many otherwise unnamed philosophers (*phylosophi*) who geographically surveyed Egypt (*Egypti patrie descriptores*). Blatausis and Cynchris were portrayed as of Egyptian nationality (*genere Egyptios*) and as surveyors of southern Egypt (*meridiane partis descriptores*). But the cosmography prefers citing Lollianus "cosmographer of the Romans" to Blatausis and Cynchris as a source for the names of Egyptian cities designated variously by different surveyors.

J. Schnetz, *Itineraria Romana*, v.2: *Ravennatis Anonymi Cosmographia et Guidonis Geographica* (1990; reprint of the edition of 1940 with index by M. Zumschlinge), 33.

Leo Depuydt

Bocchus ⇒ Cornelius

Anicius Manlius Seuerinus Boëthius (500? – 524 CE)

This Roman aristocrat, born in Rome around 480, senator, consul (510), prime minister (*magister officiorum*) to the Ostrogoth king Theoderic (522), but sentenced to death without trial (523) under pretence of conspiracy with the court of Constantinople and executed in Pavia in 524 (after several months spent in jail, where he wrote his masterpiece, the *Consolation of Philosophy*), was both an intellectual and a politician. Together with his father-in-law, Symmachus, he meant to foster a revival of Greek literary pursuits in Italy during Theoderic's 30-year peaceful reign and planned to translate into Latin not only Aristotle's logical corpus but Greek scientific works he deemed fundamental as well.

Still in his youth, he worked out an adaptation of NIKOMAKHOS OF GERASA's *Introduction to arithmetic*, under the title *Institutio arithmetica*. Starting from basic definitions (even and odd numbers, compound, prime, perfect ones and so on) and going through a presentation of multiples and proportions, this summary of the "**Pythagorean**" doctrine rises to the study of "means" (notable proportions), enabling one to elucidate the composition of the World-Soul as described in PLATO's *Timaeus*. Boëthius coined the word *quadriuium* to mean the Neo-Platonic conjunction of the four mathematical sciences (*Inst. ar.* 1.1.1 and 7) and wrote, according to CASSIODORUS (*Variae* 1.45.4), a treatise on each (arithmetic, music, geometry, astronomy). However, no trace of his *Astronomy* exists, and his *Geometry* has survived only in the pseudonymous medieval *Geometries* (compilations from the 9th and 11th c.). A Latin translation of EUCLID's works is also ascribed to him, extracts of which have survived. Only his *Institutio musica* (mutilated at its end) and his *Institutio arithmetica* have survived. The latter ought to be read first since it deals with numbers "in themselves," while music deals with them "in relation" (musical relationships being numerical ratios).

There is nothing original in Boëthius' scientific writings, since they simply transmit the Greek tradition, but he is important because he provided a bridge between ancient and medieval times. The western world of the Middle Ages drew its body of knowledge from him, until it could benefit from the contribution of the Arabs. The teachings of his *Institutio arithmetica* provided the basis for the medieval game "rhythmomachy" (actually "arithmomachy," a contest of numbers), a kind of arithmetical chess game still widely played at the time of the Renaissance.

Ed.: *Geom.* I and translation from Euclid: K.Lachmann, *Die Schriften der Römischen Feldmesser*, v. 1 (1848); *Inst. Mus.* and *Geom.* II: G.Friedlein (1867); *Inst. Ar.*: Jean-Yves Guillaumin, *Institution arithmétique: Boèce* (*CUF* 1995).

H. Chadwick, *Boethius* (1981).

Jean-Yves Guillaumin

Boëthos of Sidōn (Peripatetic) (1st c. BCE)

Peripatetic scholar, is recorded in late sources as a pupil of ANDRONIKOS OF RHODES (IŌANNĒS PHILOPONOS, *in Cat.* = *CAG* 13.1 [1898] 5.18–19), and as **Peripatetic scholarch** (AMMŌNIOS, *in An. Pr.* I = *CAG* 4.6 [1899] 31.12). With Andronikos, he seems to be at the origin of that peculiar form of Aristotelianism, the **Peripatetic** commentary tradition which will culminate with ALEXANDER OF APHRODISIAS. Born in the first half of the 1st c. BCE, he flourished either in the middle or second half of the century, depending on Andronikos' dates, which are uncertain and controversial. Later scholars preserve some of his views on *Cat.*, *An. Pr.*, *Phys.*, psychology and ethics. Pre-eminent was his commentary on ARISTOTLE's *Categories*, where Boëthos discussed among others such categories as Time,

Action, Passion, not treated in detail in Aristotle, and retained in the *Categories* the final chapters (10–15), the so-called *Post-Praedicamenta,* although they had been rejected by Andronikos. He devoted a book on the category of relation (*pros ti*), partly in polemic against the **Stoics**. Boëthos' theoretical tendencies have often been regarded as materialistic or naturalistic: he wanted to start the curriculum with physics (and not with logic, as Andronikos did) because this is the most familiar field of knowledge for us (Philop., *in Cat.: ibid.*, 5.16–18); he maintained that only matter and compounds are substances, since they are not said of substrate (*kath' hupokeimenou*) nor they are in a substrate (*en hupokeimenōi, cf.* Simplicius, *in Cat.* = *CAG* 8 [1907] 78.4–20), whereas forms belong to other categories, such as quality or quantity; he regarded universal genera (including Plato's ideas) as posterior to individuals (see Dexippos, *in Cat.* = *CAG* 4.2 [1888] 45.12–28, Syrianus, *in Met.* = *CAG* 6.1 [1902] 106.5–7); he distinguished matter and substrate, thus preparing Alexander's theory of a matter (see Simpl., *in Phys.* = *CAG* 9 [1882] 211.13–23, and *cf. e.g.* Alex., *DA* 3 = *CAG* S.2.1 [1887] 21–4.4). According to Gottschalk (vs. Moraux), Boëthos the Peripatetic (not Boëthos of Sidōn the Stoic) is the philosopher criticized in Porphurios' lost books *Against Boethus on the Soul* (fr. in Eusebios *Praep. Ev.* 11.28, pp. 62.25ff. Mras) for having rejected the proof for the immortality of the soul as held in Plato's *Phaedo* 79–81.

E. Zeller, *Die Philosophie der Griechen in ihrer geschichtlichen Entwicklung* 3.1 (1865) 624–627; K. Prantl, *Geschichte der Logik im Abendlande* (1955) 1.540–544; Moraux (1973) 1.143–179, with L. Taran's review, *Gnomon* 46 (1981) 732–734; H.B. Gottschalk, "Boethus' psychology and the Neoplatonists," *Phronesis* 31 (1986) 243–257; *Idem* (1987) 1107–1110, 1116–1119; *DPA* 2 (1994) 126–130, J.-P. Schneider.

Silvia Fazzo

Boëthos of Sidōn (Stoic) (175 – 125 BCE)

Stoic, student of Diogenēs of Babylōn, held several rather unorthodox opinions for a Stoic, including denial of the living ***kosmos*** and the end-of-the-world conflagration. He also argued that the ***kosmos*** was eternal and incorruptible, that the substance of god is the sphere of the fixed stars (or ***aithēr***), and that everything happens according to Fate. He investigated the causes of meteorological phenomena, and he seems also to have said that soul is a mixture of air and fire. Books *On Fate, On Nature*, and a commentary on Aratos' *Phainomena* are attested.

Ed.: *SVF* 3.265–7.

Daryn Lehoux

Bolās (250 BCE – 540 CE)

Aëtios of Amida, 7.106 (*CMG* 8.2, p. 371), records his collyrium, a compound of ***psimuthion***, saffron, opium, fresh starch, and gum acacia, in water. For the rare name, *cf.* only *LGPN* 4.73.

Fabricius (1726) 103.

PTK

Bōlos of Mendēs (*ca* 250 – 115 BCE)

Paradoxographer and author on magic, thought coeval with or after Kallimakhos and before the paradoxographer Apollōnios.

The *Souda*'s two entries for Bōlos, one labeled "DĒMOKRITOS [perhaps *Dēmokritean*] philosopher" (B-481), and the other "Mendēsian **Pythagorean**" (B-482), generally believed to refer to the same person, ascribe to him the following (lost) works: *Scientific Inquiry and Medical Art; Concerning Things from the Reading of the Histories that Lead us to Pause* [*in Thought*]; *Concerning Wonders; Naturally Potent* [*Drugs?*]; *On Sympathies and Antipathies* (the *Souda* adds *of Stones*, probably the vestige of another title); and *On Signs from the Sun, Moon, Ursa Maior, Lamps and the Rainbow*.

He was also said to have authored the *Kheirokmēta* (*Things Wrought by Hand*) falsely ascribed to DĒMOKRITOS (COLUMELLA 7.5.17 and contested by PLINY 24.160). A fragment preserved by Columella, and illustrative of the text's genre, concerns a method for curing diseased sheep using magical **sympathies** and antipathies. It is unlikely, however, that Bōlos is the source of some or indeed any of the alchemical recipes and treatises attributed to Dēmokritos (*CAAG* 2.41–56; see P. HOLMIENSIS) as has sometimes been claimed.

DK 78; *DSB* 2.256–257, J. Stannard; *OCD3* 249, D.J. Furley and J.T. Vallance.

Bink Hallum

Book of Assumptions by Aqāṭun (Hekatōn? Agathōn?) (200 – 600 CE?)

The Arabic translation of a Greek treatise containing 43 demonstrated geometrical theorems pertaining to the geometry of triangles and circles (chords and tangents). This collection of lemmas drawing from various sources (but with no explicit reference to other works) is probably of late antique origin. The first half contains 19 propositions also found (with slight differences) in the MS "the Book by ARCHIMĒDĒS on the Elements of Geometry." The third proposition recalls *prop.* 10 of Archimēdēs' *liber assumptorum*, and some others are (sometimes strongly) similar to propositions found in PAPPOS, probably suggesting common sources rather than derivation. The names "Hekatōn" or "Agathōn" are only possible guesses, and no Greek geometers with such names are known.

Jones (1986) 2.603–605; Y. Dold-Samplonius, "Some Remarks on the 'Book of Assumptions by Aqāṭun,'" *JHAS* 2.2 (1978) 255–263.

Alain Bernard

Book of the Signs of the Zodiac (after *ca* 200 CE?)

The *Asfar Malwašia*, or "Book of the Signs of the Zodiac" (AM), is a compilation of texts dealing with astrology and omens. It is written in Mandaic, an eastern Middle Aramaic dialect associated with the Mandaeans, a Gnostic sect historically native to Mesopotamia (from at least the 2nd c. CE) but which exists today mainly through its diaspora in Europe, North America and Australia.

In addition to its astronomical basis, there is a clearly discernible anthropological discourse in AM, particularly in chapters 1–20, presenting an astrological analysis of humankind. There is much evidence for the influence of Babylonian science in AM, particularly astronomy and astrology. The most likely Babylonian sources are *Enuma Anu Enlil*, on celestial omens, *Iqqur Ipuš*, on hemerology, and *Šumma alu* on terrestrial omens. Thus the oldest parts of AM probably date to *ca* 200 CE, when the central Babylonian temples, the depository of the sciences (medicine, astronomy, astrology, incantations), were still functioning. Although it is possible that the Babylonian elements apparent in AM derive from a Hellenistic intermediary, Rochberg has presented compelling evidence for a direct reception of Babylonian cuneiform traditions in Mandaic sources and Müller-Kessler has argued convincingly for a Sasanian context for this reception.

F. Rochberg, "The Babylonian Origins of the Mandaean Book of the Zodiac," *Aram* 11–12 (1999–2000) 237–247; C. Müller-Kessler, "The Mandaeans and the Question of their Origin," *Aram* 16 (2004) 47–60.

Siam Bhayro

Bothros (150 BCE – 500 CE?)

Magico-medical "Philosopher" whose letter to a Persian king details medical applications of vultures. A fragment, surviving in two versions (*CCAG* 8.3), explicates **sympathetic** and magical medicine: the skull heals headaches, blood mixed with Syrian cedar oil cures skin diseases, a feather on the belly of a pregnant woman eases parturition (f.100), vulture eyes placed in piglet hides treat ophthalmologic ailments (f.153). Dissection is conducted ritually by invoking three messengers: Adamaēl (*adam*: man), Elkhōē, and Abrak (*ab*: Father) (f.100), or else Adamanēl, Elōēl (*ēl*: god), and Babriēl (*bab*: gate) (f.153). These Semitic names may suggest an eastern-Mediterranean origin. Otherwise unattested as a proper name, Bothros, "a hole or pit dug into the ground" (and therefore contextually relevant to magic, e.g. *defixiones*), may be a Hellenized version of a Semitic name or, more likely, a pseudonym.

RE 3.1 (1897) 792, E. Riess; *CCAG* 8.1 (1929) 47: Paris 2180, f.100; 8.3 (1912) 126–127: Paris 2419, f.153.

GLIM

Botrus (350 – 270 BCE)

Medical writer, reviled by TIMAIOS as shameless (*FGrHist* 566 F 35 = POLUBIOS Book 12, *fr*.13.1), cited as a foreign authority on trees (PLINY 1.*ind*.12–13), on drugs obtained from animals (1.*ind*.29–30), and on the properties of copper (1.*ind*.34–35). ASKLĒPIADĒS PHARM. records his treatment for auricular hemorrhaging in GALĒN *CMLoc* 3.1 (12.640 K.), compounded from blackberry juice and oak-galls, cooked in vinegar, poured into the ear. This not altogether common name, attested from the 5th c. BCE into the imperial era, is more frequently known in the 3rd/2nd cc. BCE (*LGPN*).

RE 3.1 (1897) 794 (#4), M. Wellmann.

GLIM

Bōtthaios (?) (400 – 300 BCE?)

Wrote a ***periplous*** giving distances in days, cited with SKULAX OF KARUANDA by MARCIANUS OF HĒRAKLEIA 1.2. The name seems otherwise unattested, and Pape-Benseler emend to *BOYΘHP-* (an epithet of LUKOS OF RHĒGION). Better might be *BOHΘAI-* (which became *BOΘΘAI-*, and was "corrected" to *BΩTΘAI-*); Boēthos is attested in the 5th/4th c. BCE (and later): *LGPN* 1.102–103, 2.89, 3B.87. Also possible is the Macedonian ethnic *BOTTIAI-* (THUCYDIDĒS 2.99.3).

RE 3.1 (1897) 794, H. Berger.

PTK

Boupha(n)tos (120 BCE – 565 CE)

Alexander of Tralleis (2.577 Puschm.) quotes his remedy for gout, composed of anise, ***silphion***, ginger, pepper, ***kostos***, and salt. The form Bouphantos is attested: *LGPN* 3B.88 (Boiōtia, 3rd c. BCE); *cf.* perh. Bathullos.

(*)

PTK

Boutoridas (*ca* 100 BCE – 20 CE)

Wrote a geographical or paradoxographical work on Egypt, cited once by Pliny 36.79. The name is otherwise unattested (though *cf.* Boutadas, *LGPN* 3A.94), but probably derives from the Egyptian city Boutos, Strabōn 17.1.18.

FGrHist 654.

PTK

Brenitus (120 BCE – *ca* 90 CE)

Asklēpiadēs in Galēn, *CMLoc* 6.4 (13.288 K.), cites his laxative containing opium, Indian nard, white pepper, gentian (*cf.* Genthios), etc.; and 10.1 (13.330–331), his pain-killer for nephritis; Arkhigenēs in Galēn *CMLoc* 9.3 (13.266) cites the laxative of 6.4 as a remedy for **dropsy**. The presence of Indian nard and white pepper suggests the *terminus post* of *ca* 120 BCE. Both 9.3 and 10.1 read *BIENNIOY* (*cf.* the Latin *nomina* Biennius and Brennius: Schulze [1904/1966] 133), only 6.4 preserving *BPENITOY*, the spelling of this probably Celtic name (*cf.* Breniton, a place in Burgundy, Ravenna Cosmography 4.26, and the Celtic chieftains named Brennos, Strabōn 4.1.13 and Livy 5.38.3, 48.8). Perhaps *cf.* Brentēs of Thasos, 7th c. BCE (*LGPN* 1.104)?

Fabricius (1726) 102, 104.

PTK

Bromios of Athens (315 – 305 BCE)

Constructor of the earliest double-purpose artillery piece, capable of shooting both stones and bolts: IG^2 2.1487.B, lines 84–90.

Marsden (1969) 70.

PTK

Brusōn of Hērakleia Pontikē (380 – 350 BCE)

Aristotle twice refers to Brusōn's attempt to square the circle: it is a proof from immediate truths, but provides only accidental knowledge of its result because its principles are too general and apply to other subject matters (*APo.* 1.9 [75b37–76a3]); moreover, it is "sophistic, even if the circle is squared, because it is not in accordance with its subject" (*Soph. Ref.* 11 [171b16–18]) and it is contentious or eristic because it does not proceed from principles specific <to geometry> and would only work with people not knowledgeable in geometry (172a2–7). Ancient commentators on Aristotle offer several reconstructions of Brusōn's argument, all involving the inscription and circumscription of one or more polygons in and about a circle, and the taking of some intermediate polygon. Apparently Brusōn relied on some general principle of continuity to infer the existence of this intermediate.

Aristotle refers to a sophist Brusōn as holding obscenity to be impossible (*Rh.* 3.2 [1450b8–10]), and twice identifies him as the son of Hērodōros (of Hērakleia Pontikē: *Hist. An.* 6.5 [563a7] and 9.22 [615a9–10]). Other sources identify a Brusōn as a pupil of Kleinomakhos of Thurii, of Euclid of Megara (disputed), and of Sōcratēs (generally rejected), and as a teacher of Pyrrho of Ēlis, of Polyxenos, and of Theodōros of Kurēnē. Theopompos (Athēnaios, *Deipn.* 11 [508c–d]) accused Plato of plagiarizing from Brusōn of Hērakleia. It is an open question which if any of these Brusōns are identical with the man who attempted to square the circle.

K. Döring, *Die Megariker* (1972) 62–67, 157–166; Ian Mueller, "Aristotle and the quadrature of the circle," in N. Kretzmann, ed., *Infinity and Continuity in Ancient and Medieval Thought* (1982) 146–164; R. Muller, *Les Mégariques* (1985) 67–71, 174–179.

Ian Mueller

Burzōy (550 – 600 CE?)

Sasanian physician, son of Azdhar, generally but highly improbably identified with Wuzurgmihr, vizir of Xusraw I (531–579). According to Islamic sources (in Arabic and Persian), Burzōy traveled to India, translating from Sanskrit into Pahlavi some Buddhist stories and the *Pañcatantra*. His work *Kalīlag ud Damnag* – in Arabic translation (*Kalīlah wa Dimnah*) – was the origin of subsequent versions in Greek and other European languages. Burzōy probably knew and practiced Indian medicine and pharmacology: the introduction to the *Kalīlah wa Dimnah* refers to Indian medical concepts such as embryology.

Diels (1905–1907) 2.81, s.v. Perzoë; Th. Nöldeke, *Burzōes Einleitung zu dem Buche Kalīla waDimna, übersetzt und erlautert* (1912); G. Sarton, *Introduction to the History of Sciences* 1 (1927) 449; F. De Blois. *Burzōy's Voyage to India and the Origin of the Book of Kalīlah wa Dimnah* (1984); *EI* 4 (1990) 381–382, D.K. Motlagh (s.v. Borzūya); *PLRE* 3 (1992) 991, s.v. Perzoë; Antonio Panaino, *La novella degli Scacchi* (1999) 105–119.

Antonio Panaino

C

C- ⇒ K-

CAECILIUS ⇒ (1) BIŌN; (2) LACTANTIUS

Caecilius "Medicus" (100 BCE – 77 CE)

Listed by PLINY after SEXTIUS NIGER and before Metellus Scipio as a (Latin) medical authority (1.*ind*.29). In his *Commentarii* (presumably written in Latin), he described a contraceptive amulet containing two worms from a large-headed hairy spider, a type of *phalangium* (29.85). Possibly bilingual, his identification with BIŌN CAECILIUS is uncertain.

RE 3.1 (1897) 1188 (#3), G. Wissowa.

GLIM

Caelius Aurelianus of Sicca (425 – 460 CE)

A superscription in the Leiden fragment and in the Lorsch catalogue of MSS (9th c.) reads *Caelius Aurelianus methodicus Siccensis*, informing us that the author of a Latin *Acute and Chronic Diseases* was from Sicca in Roman Numidia, and a **Methodist**. He claims to have translated SŌRANOS' (lost) books *On Acute and Chronic Diseases* from the Greek (*Acute* 2.8, 2.65). Rose (1882) collated the Leiden Greek fragment of Sōranos' *Acute and Chronic Diseases*, and showed Caelius' Latin translation was literal, but much abridged and paraphrased. Caelius occasionally mentions his own writings (e.g. a *Responsiones* portion of a work on surgery: *Chron.*, 2.27–28 and 4.3; a *Medicamina: Chron.*, 2.78; others including a *Fevers*: Amman 710, Urso 125–149; and a work in Greek titled *Letters to Praetextatus: Chron.* 2.60), all lost, save a battered six folios from a 10th c. *Questions and Answers*, a medical catechism (Rose 1870: 183–192). The *Letters* may indicate Caelius' contacts among the Senatorial class, if the addressee is Rufus Praetextatus Postumianus, *consul* in 448 CE (*CIL* 6.1761), a date consistent with Caelius' Latin, similar to the medical Latin of CASSIUS FELIX. Caelius had students (e.g. "Bellicus": *Acute* 1.*pr*.1) and a circle of friends who knew Greek (*Acute* 1.*pr*.2; "Lucretius"), unless the names are spurious (Urso 138–143).

The *editio princeps* of the *Chronic Diseases* appeared in 1529 (Basel), by Johannes Sichart, based on a Lorsch MS which shortly vanished, and *Acute* was edited and published by Winter (Joannes Guinterius) von Andernach at Paris in 1533, founded on a Paris MS that also quickly disappeared. Each printed text was probably produced from a single MS and, with the re-discovery of three leaves of the Lorsch, Sichart's edition was shown to be an accurate transcription, excepting the use of 16th c. orthography (Bendz, v. 1, pp. 12–14; Drabkin, p. XII). All subsequent edited texts are thus based on the first printed editions, since actual MSS are no longer extant.

Methodists took medical doxography as essential to the understanding of the principles in diagnosis, prognosis, and therapeutics. *Acute and Chronic Diseases* is a lengthy catalogue of individuals and their suggested treatments for various ailments (usually firmly criticized), often beginning with ASKLĒPIADĒS (probably of Bithunia). This is a medical history in the service of current medical practice, and the work emphasizes the many mistakes of prior physicians: e.g. PRAXAGORAS OF KŌS, whose surgeries for hernias are likened to "murder with the hands": *Acute* 3.165).

Besides doxographical accounts of *phrenitis*, lethargy, ***sunankhē***, stupor, catalepsy, pleurisy, pneumonia, heart ailments, etc., Caelius/Sōranos follow the custom of ancient medical and philosophical writing by disputing with the "living presence" of their predecessors. *Acute* 1 closes with "replies" to DIOKLĒS, ERASISTRATOS, Asklēpiadēs, THEMISŌN OF LAODIKEIA, and HĒRAKLEIDĒS; and 2 has "replies" to those plus PRAXAGORAS and HIPPOKRATĒS. Of particular interest is *Acute* 1.10 and 11, "Venesection" and "Cupping," the latter the most complete account of the procedure to survive from antiquity. The complex ancient debate about ***hudrophobia*** (*Acute* 3.9–16) attests to the quandary of patients presenting symptoms of this supposedly always-fatal illness, but who survived without ill effect.

Ed.: J.C. Amman, with appendix by T.J. Almeloveen, *Caelii Aurelianii Siccensis medici vestusti, secta Methodici, De morbis acutis & chronicis libri VIII* (1755); V. Rose, *Caeli Aureliani de salutaribus preceptis* in *Anecdota Graeca et Graecolatina* 2 vv. (1864–1870; repr. in 1 v. 1963) 2.181–192; *Idem, Sorani Gynaeciorum vetus translatio Latina . . . cum additis textus reliquiis a Dietzio repertis atque ad ipsum codicem Parisiensem* (1882); I.E. Drabkin, with English trans., *Caelius Aurelianus On Acute Diseases and On Chronic Diseases* (1950); G. Bendz, with German trans. by I. Pape, *Caelii Aureliani Celerum passionum libri III. Tardarum passionum libri V* 2 vv. (1990–1993) = *CML* 6.1.

G. Bendz, *Studien zu Caelius Aurelianus und Cassius Felix* (1964), rev. by R. Browning in *CR* 15 (1965) 230–231; J. Pigeaud, "Pro Caelio Aureliano," in G. Sabbah, ed., *Mémoires III. Médecins et Médecine dans l'Antiquité* (1982) 105–117; J. Pigeaud, "Les origines du méthodisme d'après *Maladies aiguës* et *Maladies chroniques* de Caelius Aurélien," in Mazzini and Fusco (1985) 321–338; John Scarborough, "The Pharmacy of Methodist Medicine: The Evidence of Soranus' *Gynecology*," in Mudry and Pigeaud (1991) 203–216; Önnerfors (1993) 301–317; A.M. Urso, *Dall'autore al traduttore. Studi sulle Passiones celeres e tardae di Celio Aureliano* (1997); P.J. van der Eijk, "Antiquarianism and Criticism: Forms and Functions of Medical Doxography in Methodism (Soranus and Caelius Aurelianus)," in van der Eijk (1999) 397–452; B. Maire, "Le triangle méthodique: Soranos, Caelius Aurelianus et Mustio," in N. Palmieri, ed., *Rationelle et irrationnel dans la médecine ancienne et médiévale* (2003); *BNP* 2 (2003) 894–895 (#II.11), V. Nutton; Tecusan (2004) 10–14.

John Scarborough

Caepio (15 – 35 CE)

Roman author of a treatise discussing roses – including one variety said to have a 100 petals (PLINY, 21.18). Pliny specifies that he wrote during the reign of Tiberius, and he may be identical to the *quaestor* Caepio Crispinus, a notorious informant (TACITUS *Ann.* 1.74).

W. Smith, *A Dictionary of Greek and Roman Biography and Mythology* (1867) 535; *RE* 3.1 (1897) 1280 (#2), A. Stein, (#3), E. Groag; *GRL* §495.4.

Philip Thibodeau

CAESAR ⇒ C. IULIUS CAESAR

Caesarius of Nazianzos (Kappadokia) (*ca* 355 – 368 CE)

Born *ca* 330, younger brother of GREGORY OF NAZIANZOS, whose funeral oration (*Or.* 7) and epigrams (*Epi.* 7, 12–18, 21 = *Anth. Gr.* 77, 78, 85, 86, 88–98, 100) are our principal biographical sources. His higher education in Alexandria focused primarily on geometry, mathematics, astronomy, and medicine. He enjoyed considerable fame as a physician from *ca* 355 in Constantinople, and was offered senatorial status, though he reputedly declined a position at Constantius II's court (337–361) to return to Kappadokia *ca* 358. Soon returning to the capital, he served perhaps as *arkhiatros* under Constantius and Julian (361–363), who unsuccessfully attempted to convert Caesarius to paganism. Under Valens (364–378), Caesarius was awarded a financial office (*comes sacrarum largitionum*) in Bithunia. He escaped the 11 October 368 earthquake at Nikaia, but died not long afterwards, having received clinical baptism. Accounted a saint by the Orthodox and Catholic churches.

KP 1.1006, A. Lippold; *PLRE* 1 (1971) 169–170; *NP* 2 (1997) 925–926, H. Leppin; *BNP* 2 (2003) 918–919 (#2), W.A. Portmann.

Keith Dickson

Caesennius (*ca* 50 BCE – 75 CE)

Roman author of a treatise on gardening (*Kepourika*) employed by PLINY (1.*ind.*19). Schulze (1904/1966) 135–137 notes that the name is first found in the era of CAESAR and CICERO. COLUMELLA (10.1.1) suggests that Roman horticultural writers did not predate the age of AUGUSTUS.

RE 3.1 (1897) 1306 (#1), A. Stein.

Philip Thibodeau

Calcidius (*ca* 400 CE)

Wrote his Middle-**Platonic** *Commentary* to elucidate his Latin translation of the first two-thirds of PLATO's *Timaeus* (17a–53c). Throughout the Middle Ages, Plato's name was associated almost exclusively with this portion of the *Timaeus* and Calcidius' *Commentary* established itself as the primary source for its interpretation.

Calcidius provided evidence for his life only in his preface to Osius, friend and possibly patron, either bishop of Cordova (d. 357), or a Roman living in Milan *ca* 400 CE. Calcidius' medieval readership considered him Christian, though he probably was not. His Latin makes no facile reading, and it is possible that his native tongue was Greek. His primary audience is likely to have been an educated, Latin-speaking group of scholars whom he wanted to provide with a summary of Plato's philosophy and its Greek commentary tradition. Beyond Plato and Aristotle, Calcidius drew on various Greek **Academic, Peripatetic**, and **Pythagorean** texts. Of extant works, the *Commentary* shows knowledge of THEŌN OF SMURNA's commentary on Plato (itself based on ADRASTOS OF APHRODISIAS). Calcidius followed the Greek mathematical commentary tradition in considering mathematical disciplines the basis for grasping the *Timaeus* and in using lettered mathematical diagrams.

Calcidius, dividing both translation and *Commentary* into two parts (*Timaeus* 17a–39e and

39e–53c respectively), focused on cosmology. The first part, describing the creation and elemental structure of the universe's body and soul, contains 25 diagrams. The first diagrams explain the workings of mathematical proportions present in the binding of the four cosmic elements and in the musical and planetary intervals constituting the soul's harmonious structure. The remaining diagrams and text explicate astronomy focusing on planetary motion and providing an astronomical manual. The *Commentary*'s second part, building on the mathematical introduction, treats the nature of created beings and matter, *silua*, combining Platonic and Aristotelian concepts and terminology.

Calcidius' original contributions include the concept of *analogia* as the hermeneutic axis of his *Commentary* and his theory of elements that brought together in a coherent system the Platonic and Aristotelian element concepts.

Calcidius (with *ca* 130 extant medieval MSS) was one of the chief sources during the Middle Ages for the study of **Platonic** philosophy and cosmology as well as for mathematics and astronomy, and his text was excerpted for encyclopedias.

Ed.: J.H. Waszink, *Timaeus a Calcidio translatus commentarioque instructus* (1962, repr. 1975) = *Plato Latinus* 4.

Dillon (1996) 242–245, 401–408; Anna Somfai, "Calcidius's Commentary to Plato's Timaeus and its place in the commentary tradition: the concept of analogia in text and diagrams," in P. Adamson, H. Baltussen, and M.W.F. Stone, edd., *Philosophy, Science and Exegesis in Greek, Arabic and Latin Commentaries* = *BICS* S.83 (2004) 1.203–220.

Anna Somfai

CALLI- ⇒ KALLI-

Calpurnius Piso (I) (90 – 130 CE)

Wrote an elegiac *Constellations*, contemporary with the younger Pliny (*Ep.* 5.17) who praised the *Katasterismoi* after a recitation. He may be identifiable with C. Calpurnius Piso, consul of 111 (*PIR2* C285) or his brother.

OCD3 280, A.J.S. Spawforth; *BNP* 2 (2003) 1000 (#II.14), W. Eck.

GLIM

L. Calpurnius Piso (II) (175 – 200 CE)

Consul in 175 CE, the dedicatee of GALĒN's *Theriac* and, after his retirement, much given to philosophy and learning: 14.210–214 K. AËTIOS OF AMIDA 13.86 (p. 713 Cornarius) preserves one fragment from his work *On Animals*: "a partridge we raised was accustomed to call out, run in circles, and claw at the partridge-hutch, when any drug or poison was being prepared in the house" – probably part of a case for the intelligence of animals. The Byzantine scholar Psellos, *Epitaph. Xiphil.*, credits the work to Galēn, presumably its dedicatee (K.N. Sathas, *Mesaiōnikē Vivliothēkē* [1886; repr. 1972] v. 4, p. 462). See also ARISTODĒMOS and SIMŌNIDĒS.

Fabricius (1726) 370; E. Orth, "Eine unbekannte Schrift Galens," *Philologische Wochenschrift* 54 (1934) 846–848; *PIR2* C-295.

PTK

CALUENA ⇒ MATIUS

CALUUS ⇒ (1) ATHĒNODŌROS OF TARSOS; (2) LICINIUS

Campestris (or Campester) (100 BCE – 400 CE)

Wrote a lost work on astral omens, known, at least by reputation, to several late antique Greek and Latin authors. The most substantial reports of its contents are in IŌANNĒS "LYDUS," *De Ostentis* (pp. 24 and 35 Wu.). These passages concern the appearance of the Sun and, especially, comets as omens relating to forthcoming events on a national scale. Campestris held the number of comets small and each associated with a planet. His doctrines were supposed to follow pseudo-PETOSIRIS.

BNP 2 (2003) 1027, Klaus Sallmann.

Alexander Jones

Candidus (30 BCE – 80 CE)

GALĒN quotes ANDROMAKHOS' record and approval of Candidus' recipe for a ***diaphorētikē*** based on **terebinth**: *CMGen* 6.14 (13.926 K.), *cf.* Fabricius (1972) 174–179. AËTIOS OF AMIDA, 7.117 (*CMG* 8.2, p. 393), records his collyrium, containing **calamine**, ***psimuthion***, antimony, saffron, myrrh, opium, etc. The non-Republican cognomen is attested from the mid-1st c. CE: *PIR2* C-1257 (*ca* 45 CE), *cf. BNP* 2 (2003) 1047.

Fabricius (1726) 107.

PTK

Martianus Minneius Felix Capella of Carthage (*ca* 430 CE?)

Everything concerning the life of Capella is a matter of conjecture; some date him to the last decades of the 5th c. He may have been a rhetor, and wrote a curious work entitled *De nuptiis Philologiae et Mercurii*. When the god Mercury wished to marry, Jupiter approved the choice of Philology as bride. The marriage ceremony is the subject of the nine books of *De nuptiis*. After the first two books have dealt with the mythos, depicting Philology's apotheosis, a necessary condition for her union with a god to be made possible, seven books are devoted to the three literary disciplines (i.e. Book III: Grammar; Book IV: Dialectic; Book V: Rhetoric) and to the four mathematical ones (i.e. Book VI: Geometry; Book VII: Arithmetic; Book VIII: Astronomy; Book IX: Harmony).

Every one of these is personified as a maiden who will attend the new bride. Capella's originality lies in his uniting Roman encyclopedism, going back to VARRO, with the Neo-**Platonic** doctrine that knowledge, mainly mathematical, can save the soul. The technical contents of the books dealing with the four sciences can be traced back in large part to ancient sources known to the author through school compendia written during the preceding centuries. His Geometry is made up of a geography derived mostly from PLINY with borrowings from IULIUS SOLINUS, and *in fine* of Euclidian fundamentals, mostly definitions. The book on Arithmetic starts with an arithmology, goes through elements present in NIKOMAKHOS OF GERASA's *Introduction to Arithmetic*, and ends with a statement of several EUCLID's arithmetical propositions: these last are valuable evidence of the translation of part of the *Elements* into Latin, even before BOËTHIUS. The originality of Capella's book on astronomy is twofold: first in his two-part plan dealing with cosmography to start with and then with the planets, secondly in the semi-heliocentric system he describes, where Mercury and Venus revolve round the Sun, whereas the other planets and the Sun itself turn round the Earth. Book IX largely copies ARISTEIDĒS QUINTILIANUS by means of intermediaries unknown to us.

Beside the works of BOËTHIUS, CASSIODORUS, and ISIDORUS OF HISPALIS, the *De nuptiis* became a basic textbook with Carolingian schools, because it provided a genuine encyclopedia. Many Carolingian MSS from the 9th c. bear witness to its systematic use in schools, and the pictures of the seven maidens standing for the seven disciplines are plentiful in medieval and Renaissance iconography.

Ed.: J. Willis, *De nuptiis Mercurii et Philologiae* (1983); Book VI: G. Gasparotto, *Geometria: De nuptiis Philologiae et Mercurii, liber sextus* (1983); VII: Jean-Yves Guillaumin (2003); VIII: A. Le Boeuffle, *Martianus Capella, Astronomie* (1998); IX: L. Cristante, *Martiani Capellae De nuptiis Philologiae et Mercurii. Liber IX* (1987).

W.H. Stahl, R. Johnson, E.L. Burge, *Martianus Capella and the Seven Liberal Arts* (1971); S. Grebe, *Martianus Capella* (1999); M. Bovey, *Disciplinae cyclicae* (2003).

Jean-Yves Guillaumin

CAPITO ⇒ FONTEIUS CAPITO

Capito (Kapitōn) of Lukia (500 – 550 CE)

Historian, born in Lukia, wrote an *Isaurika* in eight books (not 18, a misconception resulting from a copyist's mistake), a translation of EUTROPIUS' *Breuiarium*, and a treatise *Peri Lukias kai Pamphulias*, *Souda* K-342. The very few fragments are quoted in STEPHANOS OF BUZANTION'S *Ethnika*, one of which mentions a certain Konōn of Psimatha in Isauria, bishop of Apameia in 484, thus dating Capito to the early 6th c.

FHG 4.133–134; *RE* 3.2 (1899) 1527, E. Schwartz; *PLRE* 2 (1980) 259–260.

Andreas Kuelzer

Carmen Astrologicum (100 – 500 CE)

Dubiously attributed to THEŌN OF ALEXANDRIA. Thirteen dactyls correlate the planets, Moon and Sun to various emotions and physical needs: e.g., hated Kronos is weeping; Hēlios is laughter; Moon is sleep. The poet identifies Zeus as the primal force whence ***phusis*** emerges.

Heitsch 2 (1964) 43–44.

GLIM

Carmen de ponderibus et mensuris (280 – 510 CE)

Some 30 MSS preserve this metrological poem of 208 hexameters under the name of REMMIUS FAUINUS or PRISCIAN OF CAESAREA. Its arrangement and subject suggests didactic purposes. The first section treats the Greco-Roman weight system and the *as*-pound subdivision; then follows a discussion of capacity measures both for grain and liquids and their equal (sub-)units, including the *artaba* (chief Egyptian grain-measure, *ca* 38 liters). Section three covers the specific gravity of fluids and how an areometer can determine it, plus, how to measure the silver and gold content of alloyed objects (possibly through MANTIAS' work quoted by Menelaos). Still disputed is the poem's authorship and the date: in addition to Priscian, Remmius Palaemon or Dardanos (as a partial Latin version of his work), it has been recently assigned either to Fauinus or to an unknown poet of the 6th c. CE.

Ed.: *MSR* 2.24–32, 88–100: text #120.

RE 3.2 (1899) 1593–1594, F. Hultsch; K.D. Raios, *Recherches sur le Carmen de ponderibus et mensuris* (1983); S. Grimaudo, "Metrologia e poesia nel tardoantico: struttura e cronologia del *Carmen de ponderibus et mensuris*," *Pan* 10 (1990) 87–110; *BNP* 2 (2003) 1112, J. Gruber.

Mauro de Nardis

Carminius (*ca* 300 – 400 CE?)

Wrote a grammatical work *On Expressions* (cited by Seruius *ad Aen.* 5.233, 7.445) and a geographical work *On Italy*, cited by Macrobius *Sat.* 5.19.3 on the use of bronze in early Italy.

BNP 2 (2003) 1114–1115 (#6), P.L. Schmidt

PTK

Carn- ⇒ Karn-

Cass- ⇒ Kass-

Cassiodorus ⇒ Cassiodorus Senator

Cassianus Bassus (500 – 600 CE?)

Of uncertain origin (his title *skholastikos* predates *ca* 620), author of a Greek agricultural compendium, *Peri geōrgias eklogai*. The work, surviving only in the Geōponika, was compiled, apparently with little modification, from the *Sunagōgē geōrgikōn epitēdeumatōn* of Vindonios Anatolios and the *Geōrgika* of Didumos of Alexandria. Cassianus' name and title are preserved in one MS of the *Geōponika* (Marcianus gr. 524), along with vestiges that the work was written for "my dear son Bassus" (7.*pr.*, 8.*pr.*, 9.*pr.*). First-person references in the *Geōponika* seem to belong to Cassianus (e.g. 1.*pr.*, 10.73.1, 13.1.1); among them, we find personal acquaintance with the didactic poetry of Nestōr of Laranda (12.16.1, 15.1.11, 32). The author speaks of his property in a district called Marōton (5.6.6, perhaps in Syria). The *Eklogai* were translated into Old Persian (*Warz-nāmag*) and thence into Arabic, the *Filaha farisiyya* (with its author's name variously corrupted to Quṣṭūs or Kasinūs).

Oder (1890, 1893); *RE* 3.2 (1899) 1667–1668 (#10), M. Wellmann; C.A. Nallino, in T.W. Arnold and R. Nicholson, edd., *Festschrift Edward G. Browne* (1922) 346–351; J. Carabaza Bravo, *Arabic Sciences and Philosophy* 12 (2002) 155–178.

Robert H. Rodgers

Cassius ⇒ Dionusios of Utica

Cassius (10 BCE – 30 CE)

Physician of the time of Augustus and Tiberius (Celsus *pr.*69: "the most ingenious practitioner of our generation"; Pliny 29.7), famous for a specific for the relief of colic (*kolikon*) that was used also by Tiberius (Scribonius Largus 120; Celsus 4.21.2; 5.25.12;

P.Harris 46). According to Celsus, he also cured a case of intoxication (*pr.*69); an antidote by him is described by GALĒN *CMLoc* 4.8 (12.738 K.) and Scrib. L. 176; a ginger-based drug by CAELIUS AURELIANUS *Chron.* 4.99 (*CML* 6.1.2, p. 830). As for his methodology, he tended possibly towards earlier **Empiricism**, purged from **Rationalist** accretions: Galēn *Subf. emp.* 4 ascribes to a "Pyrrhonean" Cassius a whole book on the "transition to the similar" (the third part of the **Empiricist** "tripod"), in which he upheld the thesis "that the **Empiricist** does not make use of this kind of transition." The same Cassius is mentioned by DIOGENĒS LAËRTIOS (7.32–34) in his list of skeptic philosophers.

Ed.: Deichgräber (1930) 210–212 (fragments), 264.

RE 3.2 (1899) 1678–79 (#3), M. Wellmann; *Idem, A. Cornelius Celsus. Eine Quellenuntersuchung* (1913) 123–131; I. Andorlini, "Una ricetta del medico Cassio. P.Harris 46," *BASP* (1981) 97–100; Fabio Stok, "La scuola medica empirica a Roma," *ANRW* 2.37.1 (1993) 632–633; *ECP* 122–123, M. Schofield; H. von Staden, "Was Cassius an Empiricist? Reflections on Method," in *Synopia. Studia humanitatis Antonio Garzya dicata*, edd. U. Criscuolo and R. Maisano (1997) 939–966; R.J. Hankinson, *Cause and Explanation in Ancient Greek Thought* (1998) 313–314.

Fabio Stok

Cassius Felix (*ca* 400 – 450 CE)

Renowned physician in Roman north Africa. In the *De Miraculis Sancti Stephani protomartyris* (*PL* 41.833–854), he is a widely admired and honored *archiater* of Carthage, rendering a pessimistic diagnosis in a case of facial paralysis. That corresponds with a funerary inscription from Cirta (*CIL* 8.7566) attesting an ancestral home, where he probably became a respected practitioner before going to Carthage. His *De medicina* is dedicated (according to the 13th c. Codex Parisinus Latinus 6114) to the consuls Artaburus and Calepius, i.e. 447 CE. Moreover, AUGUSTINE's account of Innocentius (*Civ. Dei.* 22.8.3) reflects physicians at Carthage not only among the rich and powerful, but also frequently in contact their counterparts in Alexandria (. . .*nisi ut adhiberet Alexandrinum quondam, qui tunc chirurgus mirabilis habebatur*. . .).

Cassius was one of several prominent physicians in Roman Africa eminent in 5th c. politics (others include VINDICIANUS and THEODORUS PRISCIANUS), and Cassius' *De medicina* reflects local Punic dialects, as well as common knowledge of Greek in learned circles (HIPPOKRATĒS, GALĒN, etc. are sources of sections in the *De medicina*). Perhaps trilingual, Cassius uses Semitic terms only incidentally in transliteration, revealing Latin as the language of medicine, law, and politics. He probably maintained communication with physicians and medical teachers in Alexandria, citing Greek texts of the *Commentaries on the Aphorisms of Hippocrates*, set down in the mid 4th c. by MAGNUS OF NISIBIS (*Med.* 29.1: . . .*omnia haec quinque secundum expositionem Magni iatrosophistae super memoratus senior Hippocrates cum discretione*; also 76.3). Cassius composed his *De medicina* as a practical guide based on Greek sources but, as is also true in the late Alexandrian commentaries, he infuses his translations and citations with his own personal experiences as a medical practitioner.

Ed.: A. Fraisse, *Cassius Felix De la médecine* (*CUF* 2002).

A. Köhler, "Handschriften römischer Mediziner II: Cassius Felix," *Hermes* 18 (1883) 392–395; O. Probst, "Biographisches zu Cassius Felix," *Philologus* 67, n.f. 21 (1908) 319–320; E. Wölfflin, "Über die Latinität des Afrikaners Cassius Felix" in *Ausgewählte Schriften* (1933) 225–230; G. Bardy, "Saint Augustin et les médecins," *Année Théologique Augustinienne*, 13 (1953) 327–346 esp. 331–332; B.H. Warmington, *The North African Provinces from Diocletian to the Vandal Conquest* (1954) 103–111; G. Bendz, "Textkritisches zu Cassius Felix" in *Studien zu Caelius Aurelianus und Cassius Felix* (1964)

91–122; G. Sabbah, "Observations préliminaires à une nouvelle edition de Cassius Felix," in Mazzini and Fusco (1985) 279–312; E. Giuliani, "Note su alcuni calchi nel *De medicina* di Cassio Felice," *Ibid.*, 313–320; Önnerfors (1993) 336–342; G. Sabbah, "Le *De medicina* de Cassius Felix à la charnière de l'Antiquité et du Haut Moyen Age," in Vázquez Buján (1994) 11–28; M. Conde Salazar and A. Moreno Hernández, "Estudio del léxico tardio de los tratados latonos africanos de los siglos IV y V," *Ibid.*, 241–252; D.R. Langslow, *Medical Latin in the Roman Empire* (2000) 513 (index entries, s.v. Cassius Felix).

John Scarborough

Cassius Iatrosophist (200 – 240 CE)

Otherwise unattested author of a "*problēmata*"-work (*cf.* ARISTOTELIAN CORPUS PROBLEMS), an oft-used format whose exemplars elude precise dating. The rough chronology of "early" adaptations is established by starting from the **Peripatetic** works, and assuming revisions, accretions, and recombinations, reaching works ". . .at the beginning of the second century by . . . SŌRANOS, [and] about a century later Cassius the Iatrosophist, and still later by ps.-Alexander Aphrodisias" (Lawn p. 3). Cassius' work is probably an early 3rd c. CE collection. A similarly complicated history characterizes a parallel series of compilations attributed to Alexander of Aphrodisias: see Kapetanaki and Sharples (2006).

The latest editors of Cassius' work conclude that its vocabulary and style bear affinities to the works of THEOPHULAKTOS SIMOKATTĒS, but Cassius incorporates questions reflecting Hellenistic and early Roman imperial medical practice, e.g. opinions of **Hērophileans** and ASKLĒPIADĒS OF BITHUNIA (*Prob.* 1 = Garzya and Masullo pp. 35–37 = Ideler pp. 144–146, *cf.* von Staden 1989: 411–412), and a quotation from Asklēpiadēs' lost *On Wounds* (*Prob.* 40 = Garzya and Masullo p. 55 = Ideler pp. 157–158 = Vallance 1990: 87). Appearance of **Methodist** opinions (e.g. *Prob.* 8 = Garzya and Masullo pp. 39–40 = Ideler, pp. 147–148 = Tecusan 2004: 269, 271) suggests a flourishing medical sect in full competition with other medical philosophies, pointing to composition in the time of GALĒN or a little later, as do the several mentions of Alexander of Aphrodisias (Wellmann 1899: 1679).

Questions in Cassius' *Problēmata* indicate he was a practicing physician, e.g. the old quandary of why circular wounds take so long to heal (*Prob.* 1; *cf.* the diagrams in Majno 1975: 156), or why stubbing one's toe does not cause swelling unlike the swollen results from blows to the feet or legs (*Prob.* 40: Cassius approves Asklēpiadēs' analogies to water seeking its own level unless a depression intervenes; *cf.* Vallance 1990: 87–88). Cassius resolves disputes about links between sleep and pleasure, by observing that nature sometimes produces divergent results from a single cause (*Prob.* 8; *cf.* THEOPHRASTOS *HP* 9.18.4; Scarborough 2006: 19).

Ed.: Ideler 1 (1841/1963) 144–167; A. Garzya and R. Masullo (with trans. and comm.), *I Problemi di Cassio Iatrosofista* (2004).

RE 3.2 (1899) 1679–1680, M. Wellmann; B. Lawn, *The Salernitan Questions* (1963); G. Majno, *The Healing Hand: Man and Wound in the Ancient World* (1975); John Scarborough, "Drugs and Drug Lore in the Time of Theophrastus: Folklore, Magic, Botany, Philosophy and the Rootcutters," *AClass* 49 (2006) 1–29.

John Scarborough

Cassius Longinus (*ca* 240 – 272/3 CE)

Born *ca* 210; renowned more for his polymathy than philosophical skills, so fond of the classical authors (PORPHURIOS, *Vit. Plot.* 14.18–20) that EUNAPIOS called him "a living library and a walking museum" (*Vit. Soph.* 4.1.3), widely known as "the critic" (PROKLOS *In Tim.* 1.14.7), or "the greatest critic of his time" (*Vit. Plot.* 20.1–3). PLŌTINOS maintained that Longinus was "a lover of learning and speech (*philologos*) rather than a philosopher in any way" (*Vit. Plot.* 14.19–20). Longinus, studying with Origen under Ammōnios Saccas (20.17–25), later opened his own school in Athens, where Porphurios studied before joining Plōtinos. *Ca* 267, Longinus left Athens to serve as advisor to queen Zenobia of Palmyra, possibly bringing with him his rich library perhaps used later by Christians like EUSEBIOS. He was executed after Zenobia's revolution failed.

A prolific author, his titles include *On principles*, read in Plōtinos' seminar (*Vit. Plot.* 14.18–20), *On impulse* (17.10–12), *On final end* (20–21), *On life in accordance with nature* (*Souda* Lambda-645), and a polemical work against the **Stoics** on the soul (Eusebios, *PE* 15.20.8–21.3). Additionally, he wrote several works on language, style, and rhetoric, most famously the (extant) *Art of Rhetoric*, and presumably also *On the Sublime.*

Longinus, concerned with interpreting PLATO's dialogues, focused more on the letter of the Platonic text (Proklos, *In Tim.* 1.59.10–19, 1.94.4–14) as a way of finding the sense of Plato's words (1.83.19–25). As a philosopher, Longinus disagreed with Plōtinos on two main issues: 1) he denied the existence of any intelligible entity higher than the divine demiurge, which he identified with the Form of the Good; 2) regarding how the Forms relate to the Intellect (i.e., the divine demiurge), he maintained that the Forms exist outside the Intellect and are subordinate to it (Proklos, *In Tim.* 1.322.18–26), thus upholding the absolute metaphysical primacy of the divine intellect.

Ed.: M. Patillon and L. Brisson, *Longin: Fragments, Art rhétorique, Rufus* (*CUF* 2001).

M. Frede, "La teoria de las ideas de Longino," *Methexis* 3 (1990) 183–190; L. Brisson and M. Patillon, "Longinus Platonicus Philosophus et Philologus, I. Longinus Philosophus," *ANRW* 2.36.7 (1994) 5214–5299; *BNP* 7 (2005) 808–810 (#1), M. Baltes and F. Montanari; *OCD3* 300, D.A. Russell; L. Brisson and M. Patillon, "Longinus Platonicus Philosophus et Philologus, I. Longinus Philologus," *ANRW* 2.34.4 (1998) 3023–3108; P. Kalligas, "Traces of Longinus' library in Eusebius' *Preparatio Evangelica*," *CQ* 51 (2001) 584–598; *DPA* 4 (2005) 116–125, L. Brisson.

George Karamanolis

Castorius (480 – 520 CE?)

Wrote in Latin a geographical work giving a detailed treatment of the regions he covered. It was the primary source cited by the RAVENNA COSMOGRAPHY on all of Asia (Book 2) and much of Africa (3.1, 3.5–8, 3.11), as well as Burgundy (*post* 480 CE), Italy, and Spain (4.26–30, 4.42) He covered all of Europe, and Asia east to India and Baktria, referring not to Sasanians but to long-gone Parthians. The *Ravenna Cosmography* cites him with ARBITIO and LOLLIANUS, but prefers LIBANIOS on regions around Constantinople (4.3, 4.6–7), ARISTARKHOS OF SIKUŌN on Macedon (4.9), and MARCOMIR on Pannonia (4.19). See also ATHANARID, "BĒRUTIOS," HELDEBALD, HULAS, IAMBLIKHOS GEOG., MARCELLUS GEOG., MARPĒSSOS, MAXIMINUS, MELITIANUS, PENTHESILEUS, PORPHURIOS GEOG., PROBINUS, and SARDONIUS. The name is primarily Christian: AUGUSTINE, *Epist.* 69; CASSIODORUS, *Variae* 3.20.2; *PLRE* 1 (1971) 185, 2 (1980) 271.

BNP 2 (2003) 1183, K. Brodersen.

PTK

Castricius (*ca* 30 BCE)

Roman author of a treatise on gardening (*Kepourika*) utilized by Pliny (1.*ind*.19); quite possibly identical to the "C. Castricius T. f. Caluus" of *CIL* 11.600 (from Forum Livi), a decorated military officer from the triumviral period, and a devotee of agriculture.

RE 3.2 (1899) 1776 (#1), A. Stein.

Philip Thibodeau

Castus (50 – 80 CE)

Andromakhos, in Galēn *CMGen* 7.13 (13.1037–1038 K.), cites his **terebinth**-based heating ***akopon***, containing pine-pitch, olive oil, Illyrian iris, and *Holarrhena*-bark (*xulomaker phloiou*, *cf.* Casson 1989: 125–126; Kühn prints *xusmatos ploiou*, "ship scrapings"). Andromakhos, *ibid.* 6.14 (13.931), preserves his version of Xenokratēs of Aphrodisias' and Polustomos' anodyne, ***aphronitron*** and ***psimuthion***, in beeswax, aged olive oil, and **terebinth**. Asklēpiadēs, in Galēn *CMGen* 4.13 (13.739 K.), cites his remedy for gangrenous wounds, composed of **litharge, verdigris**, myrrh, pine-bark, and root of black chameleon (Dioskouridēs 3.9), in olive oil and vinegar. The name is first attested in the 1st c. CE, *TLL Onomasticon* 2 (1909) 251–252, *cf. PIR2* C-547.

Fabricius (1726) 109.

PTK

Cato ⇒ M. Porcius Cato

Caystrius of Sicily (before *ca* 350 CE)

Quoted by Pelagonius of Salona for a remedy for **glanders** (Pel. 9 = *Hippiatrica Parisina* 44 = *Hippiatrica Berolinensia* 4.5) and described as a *mango* or horse-dealer.

Fischer (1980); Adams (1995); McCabe (2007) 167.

Anne McCabe

Ce- ⇒ Ke-

Celer the Centurion (10 BCE – 95 CE)

Asklēpiadēs Pharm., in Galēn *CMGen* 7.12 (13.1030–1031 K.), records a complex ***akopon***, prepared by (or for) this officer, good for sciatica, arthritis, and other ills, containing over two dozen ingredients, among which ***amōmon***, cardamom, cassia, ***euphorbia***, **malabathron** perfume, myrrh from the Trogodutae, saffron, and finest nard. Asklēpiadēs insists this "royal" ointment is "very good."

Nutton (1985) 145.

PTK

Celsinus of Kastabala (300 – 380 CE?)

Son of Eudōros, and wrote, according to the *Souda* K-1305, a *Collection of the opinions of each philosophical Sect* (lost). However, if Celsinus is to be identified with the Celsinus whom Augustine mentions, and with the Celsus whom he at the end of his life mentions as author of a doxographical work (*De haeresibus, prol.: PL* 42, col.23), he cannot be earlier than the 4th c. CE. It is worth quoting Augustine's evaluation since he uses this work as a source elsewhere: "A man named Celsus summarized the opinions of all the philosophers who founded philosophical schools, down to his own time (that's all he could do) in six rather large volumes. He did not criticize anyone, but only described their views; his style was so concise that it left no room for praise or criticism, nor confirming or defending any views, only for presenting and describing them. He mentioned almost 100 philosophers by name, not all of whom had founded their own philosophical sects since he did not think that he should omit those who followed their teacher without any dissent."

P. Courcelle, *Late Latin Writers and their Greek Sources*, trans. H.E. Wedeck (1969) 192–194; *DPA* 2 (1994) 252–253, R. Goulet and 253, G. Madec.

Jørgen Mejer

CELSUS ⇒ (1) APULEIUS; (2) ARRUNTIUS; (3) CORNELIUS

Censorinus (I) (180 – 220 CE)

ALEXANDER OF APHRODISIAS, *Quaest.* 1.13, rejects this earlier Censorinus' claim that EPICURUS' theory of color was similar to that of other schools.

R.W. Sharples, *Alexander of Aphrodisias, Quaestiones 1.1–2.15* (1992) *ad loc.; DPA* 2 (1994) 261–262, S. Follet.

PTK

Censorinus (II) (*ca* 230 – 250 CE)

Grammarian, author of *On Accents* (*De accentibus*; PRISCIANUS OF CAESAREA 1.4.17; CASSIODORUS, *Inst.* 2.1.1), now lost, and (most probably) *On birthday* (*De die natali*), partially extant, wherein Censorinus examines astrological aspects of the birthday in the first part, and chronological issues (division of time into days, months, years, etc.) in the second. Referring frequently to Greco-Roman philosophical and scientific views, Censorinus draws especially from VARRO and Suetonius. Several folios from the 7th c. MS which alone preserves Censorinus' work are lost, so we lack the end of *On Birthday* and the beginning of the so-called *Fragmentum Censorini*, a scientific miscellany treating the position of heavens, the stars, the rhythm of names, music, and numbers – i.e., more or less the *quadriuium* of Varro.

Ed.: N. Sallman, *De die natali liber* (1983).
RE 3.2 (1899) 1908–1910 (#7), G. Wissowa; *BNP* 3 (2003) 105 (#4) and 5 (2004) 536, K. Sallmann.

George Karamanolis

CETIUS ⇒ FAUENTINUS

CH- ⇒ KH-

Chrysippus of Soloi (Kilikia) (*ca* 250 – *ca* 205 BCE)

Khrusippos; born *ca* 280 BCE, student of KLEANTHĒS and third head of the **Stoa** (from 232 BCE). Briefly also studied under Arkesilaos at the **Academy**. Teacher of Zēnōn of Tarsos and DIOGENĒS OF BABYLŌN. Chrysippus was an exceptionally powerful and original thinker who framed his project as a preservation and elaboration of ZĒNŌN's ideas. It is difficult now to determine to what extent he faithfully followed Zēnōn and to what extent he was critical. Although some ancient sources seem to indicate considerable divergence, solid evidence of a sharp break is not always obvious. In any case, Chrysippus' **Stoicism** was to become the standard orthodoxy for the whole early period of the school. Chrysippus was a prolific author, with over 700 books attributed to him. DIOGENĒS LAËRTIOS (7.181) relates that Chrysippus is supposed to have written 500 lines a day, though his books are said to be padded with unnecessarily long quotations from other authors.

He was a considerable logician, and Diogenēs Laërtios (7.180) reports a "common" saying that "if there were dialectic among the gods, it would be Chrysippean." It is certainly true that much of the logic among the **Stoics**, at least, was Chrysippean. He divided logic into dialectic and rhetoric. Under dialectic, he contributed to the theory of meaning, the logic of signs, and modal logic, among other subjects. Rhetoric he defined as "the science of speaking correctly" (Quintilian, *Inst.* 2.15.34).

For Chrysippus, the ***kosmos*** is a finite and unified sphere, containing no void within, but surrounded by void. It is subject to periodic conflagration and eternal recurrence. It consists of the four Aristotelian elements, of which the primary element is fire. ***Pneuma***, composed of fire and air, sustains and unifies the ***kosmos*** and is the source of the tension (*tonos*) that keeps the ***kosmos*** together both as a whole and in all of its parts, through the dynamic interaction of necessarily inherent heat and cold. ***Pneuma*** also provides the "tenor" or character (*hexis*) of what we might now be tempted to call natural kinds. Only bodies exist. All causes are bodies, and the causal agent is that "because of which" a state of affairs comes about. The ***kosmos*** is reducible, at one level of description, to two coextensive bodily principles: matter (passive) and god (active, shaping and structuring matter). Chrysippus put forth a complex theory of mixtures (best preserved in ALEXANDER OF APHRODISIAS' *De mixtione*) to understand the nature of the co-extensiveness of these two bodies. That an eternal self-mover (god) permeates the ***kosmos*** means that the ***kosmos*** itself must be both rational and divine.

Chrysippus argued for strict determinism still allowing for individual moral responsibility. The important inter-relations between physics and ethics are shown by Chrysippus' position that ethics is best grounded in "universal nature and the governance of the world" (PLUTARCH, *St. rep.* 1035C–D), an important landmark in the development of natural-law theories of ethics.

GALĒN and others report that Chrysippus situated the controlling part of the human soul in the heart. Like all **Stoics**, Chrysippus argued for the corporeality of the soul, and he claimed the soul of individuals to be mortal, though the souls of the wise survive the deaths of individuals and persist until the conflagration (this in contrast to Kleanthēs: D.L., 7.157).

Several stories are told of his death: one that he died after falling dizzy while drinking. Another story has him dying in a fit of laughter after seeing a donkey eat some figs. He called out to an old woman to give the donkey some wine to wash them down, and then laughed himself to death.

Ed.: *SVF* v.2.

E. Brehier, *Chrysippe et l'ancien stoicisme* (1951); J.B. Gould, *The Philosophy of Chrysippus* (1970); S. Bobzien, *Determinism and Freedom in Stoic Philosophy* (1998); B. Inwood, ed., *Cambridge Companion to the Stoics* (2003).

Daryn Lehoux

CI- ⇒ KI-

CICERO ⇒ M. TULLIUS CICERO

Claudianus (Alch.) (1st – 7th c. CE?)

Cited in the list of *poiētai* (makers of gold, *CAAG* 2.26), but no treatise survives. The name is derived from Claudius (meaning "from Claudius") and, according to Festugière (1.225), he must be after the emperor Claudius. The name Claudianus is also found in a magical papyrus, associated with a sacrifice of incense to the Moon (*PMG* 7.862). Otherwise, the name refers to *klaudianos/-on*, the alloy of lead and silver mentioned by PSEUDO-DĒMOKRITOS (*CAAG* 2.44).

CAAG 2.26; Festugière (1944/1950) 1.225, n.3, 240, 323 n.4.

Cristina Viano

Claudian, Claudius Claudianus (395 – 404 CE)

Acclaimed Latin poet, Claudian was born (*ca* 370 CE) and educated in Egypt, perhaps Alexandria, before migrating to Italy (*ca* 394 CE). There he secured the patronage of the imperial court under the regent Stilicho. His major works – panegyrics, invectives, epithalamia, and an unfinished epic on the rape of Prosperina – exhibit medical, natural, philosophical, and scientific flourishes, although many derive from rhetorical or literary commonplaces. The panegyric for MALLIUS THEODORUS (*carm. mai.* 17) is of particular interest for its discussion of natural phenomena and philosophers. More than a third of his *carmina minora* treat animals and natural phenomena: e.g. hedgehogs (9), trained mules (18), lobsters (24), the geothermal lake at Aponus (26), the phoenix (27), the source and flooding of the Nile (28), magnetism (29), water encased in ice crystals (33–39), the torpedo-fish (49: derived from OPPIANUS *Hal.* 2.56–85 and 3.149–155), and orreries (51). Titles of other animal poems, now lost, are preserved in his *Appendix*. A few Greek works survive, including two similar to *c.m.* 33–39. His silence on Stilicho's second consulship strongly suggests his death or retirement before 405 CE.

Ed.: M.-L. Ricci, *Carmina Minora* (2001).
A. Cameron, *Claudian* (1970); *OCD3* 337, J.H.D. Scourfield; A. Prenner, *Quattro studi su Claudiano* (2003).

Bret Mulligan

CLAUDIUS ⇒ (1) AELIANUS; (2) APOLLŌNIOS; (3) BALBILLOS; (4) DAMONIKOS; (5) GALĒN; (6) MENEKRATĒS; (7) PTOLEMY; (8) THRASULLOS

CLE- ⇒ KLE-

CLEMENS ⇒ (1) FLAUIUS CLEMENS; (2) SERTORIUS CLEMENS

Cloatius Verus (*ca* 1 – 10 CE)

Roman grammarian whose *Ordinata Graeca* might be taken to mean "Greek Categories." The fragments quoted by MACROBIUS (*Sat.* 3.19.2, 6, 20.1) are mainly alphabetical lists of names for nuts, apples, pears, and figs, both Greek and Italian; his commentary on the names drew on the botanical works of Theophrastos (3.18.8). Verus knew varieties of apple named after Augustus and C. Matius, and his grammatical works were quoted by the Augustan grammarian Verrius Flaccus.

*GRF frr.*6–8; Rawson (1985) 141.

Philip Thibodeau

Clodius (Asklēpiadean) (80 BCE – 120 CE)

Asklēpiadean physician or perhaps a medical historian. He described drinking **silphium** juice mixed with beeswax, and other remedies involving acrid substances, as means used by some to treat **tetanus** (Caelius Aurelianus *Acute* 3.96 [*CML* 6.1.1, p. 348]), and gave an account of a cure of an effeminate man (interpreted by Sōranos/Caelius as a cure for ascarides: *Chron.* 4.134 [*CML* 6.1.2, p. 850]).

Fabricius (1726) 122–123.

PTK and GLIM

Clodius Tuscus (30 BCE – 15 CE)

Composed a Latin ***parapēgma***, which Iōannēs "Lydus," *Ost.* 58–70 (pp. 117–158 Wu.) preserves in Greek. Clodius indicates meteorological and astronomical events for each Julian date, and often agrees with the calendars of Ovid, *Fasti*, and of Columella, *RR* 11.2. Tuscus' **ephemeris** may be Ovid's source (*Fasti*, ed. R. Merkel, pp. LXVI–LXXIV).

Rehm (1941).

PTK

D. Clodius Albinus of Hadrumetum (d. 197 CE)

Roman senator, born at Hadrumetum in Africa, governor of Britain, colleague then rival of Septimius Seuerus, who defeated him at the battle of Lugdunum. Reportedly an expert in agronomy who composed a *Geōrgika* or treatise on farming (*SHA* 11.7).

OCD3 351, A.R. Birley.

Philip Thibodeau

Clodius of Naples (50 BCE ? – 200 CE)

Porphurios, *Abstinence* 1.17.2, cites him for an anecdote about the servant of the doctor Krateros being saved by eating viper-flesh (in a tract against vegetarianism, extracted in 1.13–17). Clearly not the Sicilian orator of Suetonius, *Gramm.* 29 (Kaster 1995: 308–309).

(*)

PTK

Cly- ⇒ Klu-

Co- ⇒ Ko-

Columella ⇒ Iunius Columella

Constantinus (250 – 360 CE)

Greek physician, whom OREIBASIOS cites as the author of a soap formula made of several plant ingredients, alum and exotic (perfumed) substances amalgamated with "soap from Gaul" (Latin translation of the *Synopsis*, 3.164 [5.881 BDM]), repeated in later medical encyclopedias (AËTIOS OF AMIDA 6.54, *CMG* 8.2, p. 197.20–27]; PAULOS OF AIGINA *CMG* 9.2, p. 237). It is also known in some Latin manuscripts, probably from the Latin translation of Oreibasios but differing from it: MS Laon, Bibliothèque communale, 426bis, f.117 (9th c.), specifying the soap for drying excessive humor in the head; Paris, BNF, *lat.* 11219, f.104^{V} (9th c.), where it is followed by the recipe for the preparation of an oil also attributed to a Constantinus not definitively identifiable with our man; Vendôme, Bibliothèque municipale, 175, f.126^{V} (11th c.). A MS also credits Constantinus with an unpublished (?) Latin *De coitu* (Diels 2.24) that might derive from the treatise on the same topic by Constantinus the African (11th c.), or one of its sources.

Diels 2 (1907) 24; E. Wickersheimer, *Manuscrits latins de médecine du haut moyen âge dans les bibliothèques de France* (1966) 39, 119, 188.

Alain Touwaide

CORNELIUS ⇒ ARTEMIDŌROS

Cornelius (120 BCE – 80 CE)

ANDROMAKHOS in GALĒN, *CMLoc* 9.5 (13.292 K.), records his ***trokhiskos*** against "dysentery" and blood-spitting, composed of opium, myrrh, aloes, **Indian buckthorn**, etc.; *cf.* GARGILIUS MARTIALIS, *Med.* 30. Cited also by the *Antidotarium Brux.* 38 (THEODORUS PRISCIANUS pp. 373–374 R.), for a **dropsy** remedy, composed of squill steeped in aged Aminaian wine and stored in glass, and 40 (pp. 374–375 R.), for a chest-pain remedy, containing calamint, ginger, hartwort, juniper, lovage, parsley, pennyroyal, pepper, and Spanish thyme, in honey, also stored in glass. *Cf.* perhaps ARTEMIDŌROS CORNELIUS.

Fabricius (1726) 128.

PTK

Cornelius Bocchus (120 BCE – 75 CE)

Expert in Spanish geography, cited by PLINY (16.216; 37.24, 97 and 127) regarding minerals and precious stones from Spain. His identification with the homonymous historian, author of a *Chronicle* ending with the 207th Olympiad (49 CE), and mentioned by IULIUS SOLINUS 1.97; 2.11 and 18, is much debated. L. Cornelius Bocchus, *flamen* and military tribune, is attested by a 1st c. CE inscription from Lusitania (*PIR2* C-1333; *ILS* 2920–2921); the name "Bocchus" enters Latin from SALLUST, *Iug.* 70–83, 100–104.

RE 3.1 (1897) (#3), W. Henze; *GRL* §503.Prosa.2, 8 (pp. 835, 863); H. Bardon, *La littérature latine inconnue* 2 (1956) 148–149; *BNP* 3 (2003) 836–837, M. Meier and M. Strothmann.

Eugenio Amato

A. Cornelius Celsus (15 – 35 CE)

Life. Encyclopedist, worked during the Tiberian age (at *pr.*69 he states that he knew Cassius, the **Empiricist** physician who had cured Tiberius). The *nomen* Cornelius is frequent in Gallia Narbonensis and north Spain, but Celsus certainly worked in Rome (it may be he who is attested in *CIL* 6.4/2.36285, close to his wife's name, Sabina). From Quintilian (*Inst.* 10.1.124) we know that Celsus had been a follower of the philosophical school of the SEXTII, from which he could have inherited his interest in science and medicine. But at the time he wrote his work he was not following the vegetarian doctrine of this school (in the *De medicina* he proposes meat-based recipes); furthermore, he never mentions SEXTIUS NIGER, son of the founder of that school and medical writer.

It is a debated question whether Celsus actually practiced medicine. He was probably of noble family, hence it is unlikely that he practiced a profession that at Rome was typical of freedmen or immigrants; moreover, his encyclopedic interests, extending to other fields than medicine, also render it unlikely. Especially in the surgical section, however, the *De medicina* suggests that the author had a practical experience, and that that section is addressed to practicing physicians, able to perform complex operations. Medical practice is suggested also by his references (2.17.1; 3.21.6) to hot baths near Baiae (Campania), where he perhaps owned a villa used for thermal therapy.

Works. His encyclopedia was divided into disciplinary sections: agriculture (five books), medicine (eight books), rhetoric (perhaps seven books), philosophy (perhaps six books), military science, and perhaps jurisprudence. Only the section on medicine (*De medicina*) is preserved; there are fragments of other sections (agriculture; rhetoric).

Structure of De medicina. The *De medicina* comprises eight books divided into three sections, dietetics (Books 1–4), pharmacology (Books 5–6) and surgery (Books 7–8). The section on dietetics comprises subsections on hygiene (Book 1) and semiotics (2.1–8). The present *Prooemium* probably combines the original *proem* to the entire work, containing a brief history of the medicine since its beginnings up to THEMISŌN (1–11), and the *proem* to the dietetics (12–75). The other two sections, pharmacology and surgery, are also introduced by brief *proems*.

Celsus and the Medical Schools. The *proem* to dietetics comprises an outline of the principles of the **Rationalist** (13–26) and the **Empiricist** medicine (27–44), followed by an outline of Celsus' own position (45–53): on the one hand, he appropriates the **Empiricist** principle that therapy must be based on experience, on the other hand he states that medicine is "an art based on conjecture" (48). His approach is avowedly a probabilistic one, based on the pursuit of the probable (*ueri similis*) and of statements "as may seem nearest to the truth" (45). Moreover, his decision to confront the positions of the different schools is inspired by a probabilistic approach (*disputatio in utramque partem*).

Celsus' own position has received conflicting interpretations: in the past he has been viewed as a **Rationalist**, an **Empiricist**, a follower of ASKLĒPIADĒS OF BITHUNIA. The *proem* is certainly influenced by **Empiricism** (probably especially by the moderate **Empiricism** of HĒRAKLEIDĒS OF TARAS, who introduced into the **Empiricist** tradition rational elements). **Empiricist** arguments can be seen in Celsus' criticism of ERASISTRATOS and of the **Methodist** school (54–73) and in his disapproval of Alexandrian vivisection (74–75); as for vivisection, however, Celsus disapproved of it not only on theoretical grounds (as the **Empiricists** did) but on humanitarian ones as well.

The influence of **Empiricism** can be seen also in the structure of *De Medicina*, for this work deals only with therapeutics, and not with anatomy or etiology. In the individual books, however, Celsus mainly uses **Dogmatic** authors (the most frequently mentioned authors are HIPPOKRATĒS, Asklēpiadēs of Bithunia and Erasistratos; among **Empiricists**, the only one he mentions quite often is Hērakleidēs of Taras), and sometimes he even opens up to **Dogmatic** theories (e.g. to the Hippokratic theory of **humors** in Book 2). As for these issues, Celsus's "probabilism" cannot be associated with known positions from the **Empiricist** tradition.

Sources. Celsus uses different sources (the hypothesis that there was a single Greek source for *De Medicina* is now totally discarded): probably an **Empiricist** doxography in the *proem*; Hippokratēs in Books 2 (*Progn.*, *Prorrh.*, *Aphor.* and other works) and 8 (surgical treatises); Alexandrian anatomy in Book 7; Asklēpiadēs in Books 2–4. It seems certain, even if not precisely verifiable, that he was influenced by several Hellenistic sources. He possibly uses also some Latin sources (at least the medical section of VARRO's encyclopedia).

Dietetics. Following hygiene (Book 1) and semiotics (2.1–8), the remaining part of Book 2 deals with dietetic cures in general, namely therapies that can be used for different diseases: bleeding (that he deems useful for almost every disease, even if he urges always taking into consideration the general state of the patient), cups, purgatives, vomit, massages, rocking (*gestatio*), abstinence and fasting, and sweating. This section comprises also a treatment of the characteristics of foods and drinks, in connection with the characteristic of patients and to the effects they have on human body.

Book 3 deals with the therapies of the diseases affecting the whole body: fevers (3–17), madness, "cardiac" affection, lethargy, **dropsy**, wasting disease, epilepsy, jaundice, "***elephantiasis***," and apoplexy (18–27). Celsus focuses on symptoms and therapies; there are only scattered references to causes of diseases (e.g. 3.18.17, where he says that melancholy "is apparently caused by black bile"). Book 4 deals with the diseases affecting body parts in the *a capite ad calcem* order ("from head to foot"). The treatment is preceded by a concise description of human anatomy connected to the following exposition, that comprises diseases affecting both external parts of the body (eyes, oral cavity, genitals, etc.) and internal organs (stomach, lungs, liver, etc.).

Pharmacology. In his brief *proem* to pharmacology Celsus mediates between the pharmacological tradition (mentioning Erasistratos, HĒROPHILOS and his school, the Empiricists) and the rejection of the use of drugs proposed by Asklēpiadēs of Bithunia, by urging integration of drugs with dietetic therapy. Book 5 deals with drugs and remedies in general. The single, mostly vegetable, remedies are classified according to their therapeutic effects (anesthetics, laxatives, etc.); recipes are subdivided into emollients, plasters, **pastilles**, and **pessaries**. Specific sections are dedicated to pharmaceutical (but also surgical) treatment of wounds, poisonings and skin diseases. Book 6 deals with remedies for the diseases affecting body parts, according to the *a capite ad calcem* order (special attention is given to ophthalmic diseases: Celsus knows about 30 of them, and quotes many collyria recipes). In this section Celsus mentions several inventors of specific remedies and recipes, mostly otherwise unknown.

Surgery. The third section opens with a brief history of surgery, from pre-Hippocratic age to Alexandrian medicine and to the surgeons working at Rome. The *proem* also comprises a portrait of the ideal surgeon, with observations about surgical deontology. The first part of Book 7 (1–5) is devoted to surgical operations affecting the entire body: dislocations (*luxata*), wound-caused suppurations and ulcerations, extraction of missiles which have entered the body, with specific directions according to the different kinds of missiles (here Celsus attests Roman military surgical techniques). The second part of Book 7 (6–33) deals with specific surgical treatments in the *a capite ad calcem* order. Ample space is also given to ophthalmology, e.g. to the surgical treatment of cataract (7.7.14). Other surgical operations described by Celsus include catheterization (26.1), embryotomy (29), tonsillectomy (30.3), treatment of varicose veins (31). Book 8 is devoted to orthopedics and to the treatment of fractures, dislocations, and other orthopedic pathologies. The book opens with a general description of skeleton and bones (1).

Transmission. The *De Medicina* is scarcely known in the ancient medical tradition. Pliny lists it amongst his sources; the two epistles MARCELLUS OF BORDEAUX ascribes to Celsus are apocryphal (the two lost sections on agriculture and rhetoric had more success, and were used by COLUMELLA and Quintilian). Little known in the Middle Ages, Celsus' work was rediscovered in the humanistic age. The *editio princeps* was published in Florence in 1478. The text in the lacuna at 4.27 (on vesical calculus) was discovered in the early 1970s in a MS of the Capitular Library of Toledo (Spain).

Ed.: F. Marx *CML* 1 (1915; repr. 2002); W.G. Spencer, *Celsus* (1935–1938); **4.27**: U. Capitani, "Il recupero di un passo di Celso in un codice del *De medicina* conservato a Toledo," *Maia* 26 (1974) 161–212, D. Ollero Granados, "Dos nuevos capítulos de A. Cornelio Celso (*De medicina* IV, 27,1 D)," *Emerita* 41 (1973) 99–108; Ph. Mudry, *La Préface du De medicina de Celse* (1982); G. Serbat, *De la médecine: Celse* (1995), Books 1–2; R.M. Cuilla, *Celso, De medicina libro IV* (1990); I. Mazzini, *La chirurgia: libri VII e VIII del De medicina* (1999); S. Contino, *De Medicina Liber VIII* (1988); Wm.F. Richardson, *A Word index to Celsus: De medicina* (1982).

Studies: *RE* 4.1 (1900) 1273–1276, M. Wellmann; *DSB* 3.174–175, F. Kudlien; *KP* 1.1102, F. Kudlien; *OCD3* 392–393, J.T. Vallance; *ECP* 123–125, H. von Staden; *BNP* 3 (2003) 74–75 (#7), K. Sallmann; *AML* 189–191, C. Oser-Grote; *NDSB* 2.81–84, I. Mazzini.

K. Barwick, "Die Enzyklopädie des Cornelius Celsus," *Philologus* 104 (1960) 236–249; U. Capitani, "A. C. Celso e la terminologia tecnica greca," *ASNP* 5 (1975) 449–518; H.D. Jocelyn, "The new chapters of the ninth book of Celsus' *Artes*, V," *Papers of Liverpool Latin Seminar* 5 (1985) 299–336; Ph. Mudry, "Le 1er livre de la Médecine de Celse," in Mazzini and Fusco (1985) 141–150; Ph. Mudry, "Le 'De medicina' de Celse. Rapport bibliographique," and W. Deuse, "Celsus im Prooemium von 'De Medicina'," *ANRW* 2.37.1 (1993) 787–818 and 819–841; Önnerfors (1993) 233–250; G. Sabbah and Ph. Mudry, edd., *La médecine de Celse* (1994); H. von Staden, "Author and Authority. Celsus and the construction of a scientific self," in Vásquez Buján (1994) 103–117; Fabio Stok, "Natura corporis. Costituzioni e temperamenti in Celso," in S. Sconocchia and L. Toneatto, edd., *Lingue tecniche del greco e del latino II* (1997) 151–170; Ch. Schulze, *Aulus Cornelius Celsus: Artz oder Laie?* (1999); H. von Staden, "Celsus as historian?" in van der Eijk (1999) 251–294; Ch. Schulze, *Celsus* (2001): with bibliography; Ph. Mudry, *Medicina soror philosophiae* (2006) 307–316, 317–332 (repr. of Mudry 1985 and 1993, above).

Fabio Stok

Cornelius Nepos of Transpadana (*ca* 80 – 24 BCE)

Earliest extant Latin biographer, born *ca* 110 BCE, in Rome by 65, friend of CICERO and Atticus, and Catullus' dedicatee. Nepos' prolific output, mostly lost, includes biographies of famous men (24 of 400+ survive), a three-book universal history (Catullus 1), *Anecdotes* (Gellius 6 [7].18.11), and light verse (probably never published: Pliny Jr. *Ep.* 5.3.6). Nepos' geography, perhaps "universal," is cited by PLINY on the Danube and its tributaries (3.127),

Mela on the properties of Ocean surrounding the world (3.5), and by both on the navigable route between the Arabian Gulf and Gadēs (Mela 3.9; Pliny 2.169). Neither Mela nor Pliny give a book title.

KP 4.62–63, G. Wirth; *BNP* 9 (2006) 659–660 (#2), U. Eigler; *OCD3* 396, A.J.S. Spawforth.

GLIM

Cornelius Tacitus (98 – ca 120 CE)

Born *ca* 55–58 CE in southern Gaul, Tacitus married the daughter of Cn. Iulius Agricola *ca* 76 CE, was *praetor* in 88, *consul* 97, and *proconsul* 112/113; he died early in Hadrian's reign. His first work, *Agricola*, and his second, *Germania*, were published in 98 CE, the mostly lost *Histories* (composed by *ca* 110 CE) covered Roman history from 69 CE, and the *Annales* (composed after the *Histories*) covered Tiberius to Nero (although Caligula's reign and the first half of Claudius' are lost); Tacitus also wrote a dialogue on oratory dedicated to Fabius Iustus, *consul* 102 CE.

Tacitus' geography in the *Agricola*, the *Germania*, and the *Histories* is always ancillary to ethnography. The *Agricola*, a laudatory biography of his father-in-law, narrates Agricola's governorship of Britain and conquests there. Tacitus' excursus (*Agr.* 10–12) on the people of Britain includes geographical notes on the surrounding sea (following Putheas of Massalia and recent observations), on the land and climate, and sites the island between Ocean, Germany, Gaul, and Spain (following Caesar, *BG* 5.13). ***Paradoxa*** about peculiar pearls and the odd behavior of sun and sea serve to emphasize Britain's extreme position and nature. The *Germania* is an ethnographic monograph (drawing on Poseidōnios, Caesar, Pliny, and others) that includes geographical notes, on the site of "Germany" bordered by Ocean, two great rivers, and mountains (§1), and on the cold and wet character of the land which lacks metals (§5). The second half forms a kind of ***periēgēsis*** (28–46), concluding again with a pair of ***paradoxa***, the same odd behavior of sea and sun (45.1), and the genre-bending nature of amber (45.5). The *Histories* reflect Itineraries and ***periploi***, and offer one extant ethnographic excursus, on the Jews (5.2–7). Their dry but fertile land is sited and sketched, and the odd behavior of its chief ***paradoxon***, the Dead Sea, is portrayed, with its genre-bending product, bitumen (5.6), and the dead plain whose sole product is glass-makers' sand (5.7).

Tacitus' works were rediscovered in the 15th c., and in the 16th–20th cc. the *Germania* deeply influenced German nationalism and ethnic exceptionalism, whereas the *Annales* and *Histories* informed modern thought on tyranny.

Ed.: R.M. Ogilvie and I. Richmond, *Cornelii Taciti de Vita Agricolae* (1967); A.A. Lund, *P. Cornelius Tacitus Germania* (1988); R. Oniga, *Opera omnia: Tacito* 2 vv. (2003).

DLB 211 (1999) 306–313, R. Mellor; J.B. Rives, *Tacitus Germania* (1999); *NP* 11.1209–1214 (#1), E. Flaig; R.S. Bloch, "Geography without territory," in J.U. Kalms, ed., *Internationales Josephus-Kolloquium* (2000) 38–54; K. Clarke, "An Island Nation," *JRS* 91 (2001) 94–112; F. Mittenhuber, "Die Naturphänomene des hohen Nordens," *MH* 60 (2003) 44–59; A.G. Pomeroy, "Center and Periphery in Tacitus' *Histories*," *Arethusa* 36 (2003) 361–374.

PTK

Cornelius Valerianus (*ca* 36 CE)

Mentioned by Pliny in the index for Books 8, 10, 14–15 (on plants and animals), as the source for a phoenix sighting in 36 CE (10.5), and on climbing vines (14.11).

RE 4.1 (1901) 1591 (#400), A. Stein.

Arnaud Zucker

Cosm- ⇒ Kosm-

Cr- ⇒ Kr-

Crispus ⇒ (1) Iunius; (2) Sallustius

Ct- ⇒ Kt-

Cy- ⇒ Ku-

Cynchris ⇒ Blatausis

D

Dadis (325 – 90 BCE)

Agronomist whose work was known to CASSIUS DIONUSIOS (VARRO, *RR* 1.1.9–10; *cf.* COLUMELLA, 1.1.11).

RE 4.2 (1901) 1978 (#2), M. Wellmann.

Philip Thibodeau

Daimakhos of Plataia (280 – 260 BCE)

Daimakhos served Antiokhos I as ambassador to the Mauryan king Bindusāra (ruled 297–272 BCE), and wrote an *Indika*, criticizing MEGASTHENĒS, and dismissed by ERATOSTHENĒS (in STRABŌN 2.1.19). He also authored a work on siege warfare (*Poliorkētika*), according to STEPHANOS OF BUZANTION (s.v. Lakedaimōn) and ATHĒNAIOS MECH. (p. 5 W.), as well as one on piety (*Peri Eusebias*), of which the sole fragment (PLUTARCH, *Lysander* 12.6–7) attributes the fall of a meteorite to a comet (*cf.* ANAXAGORAS).

FGrHist 716; *BNP* 4 (2004) 40 (#2), K. Meister.

PTK and GLIM

Daliōn (Geog.) (*ca* 325 – 275 BCE?)

Cited with BIŌN and ARISTOKREŌN as a foreign authority on geography and ethnography (PLINY 1.*ind*.6), wrote an *Aithiopika*. Daliōn, sailing beyond Meroë, was the first to estimate the country's dimensions (6.183) and described several peoples living along the southern Nile, including the Vacathi, who used only rainwater (6.194). For the rare name see *LGPN* 1.112 (2nd/1st cc. BCE); compare Dalion and Dalōn (*LGPN*), and perhaps Daïleōn and DĒÏLEŌN.

FGrHist 666.

GLIM

Daliōn (Med.) (350 BCE – 77 CE)

Greek physician, cited with "Damiōn" by PLINY as a medical authority (1.*ind*.20–23) – Damiōn alone is listed 1.*ind*.24–27. Wellman is confident that both names indicate the same man and that Pliny, finding both versions in his sources, faithfully copied them into his indices. Damiōn advises hyacinth bulbs mixed with honey-wine, and other remedies, to treat wounds (20.103–104). Daliōn "the herbalist" prescribes a poultice of anise and parsley as well as an anise-dill drink for women in labor (20.191) and, as an aphrodisiac, to drink a potion of ass-genitals ashed, or bull's urine produced after copulation (28.262). Distinct from Daliōn the geographer. The name Daliōn is five times rarer (*LGPN* 1.112, 3B.96) than Damiōn (*LGPN* 1.115, 2.99, 3A.110, 3B.99).

RE 4.2 (1901) 2022 (#2), M. Wellmann.

GLIM

Damas (*ca* 280 – 250 BCE)

Simplicius, *in Arist. Phys.* 6 (*CAG* 10 [1895] 924), cites Damas' *Life of Eudēmos*, discussing the arrangement of the works of Aristotle; probably Eudēmos' student, hence the date range.

FGrHist 1101.

PTK

***Damaskēnos* (*ca* 800 – 857 CE)**

Attested in the unpublished Greek translation of the *Ephodia* (*Zād al Musāfir*) by the Arabic physician ibn al-Jazzār of Kairouan (d. 979/980), and in some medical recipes contained in 15th c. manuscripts (Paris, BNF, *graecus* 2194, ff. 455 and 462–462^{V}, and Bologna, Biblioteca Universitaria, 3632, f. 189^{V}). In the *Ephodia*, Damaskēnos is credited with 17 medical recipes to treat white spots on the face, affections of the throat, cough, wounds in kidneys and bladder, cardiac arrhythmia, intestinal wounds due to yellow bile, amenorrhea, wounds due to heat and fractures, as well as an aphrodisiac. The same text further credits Iōannēs Damaskēnos with eight treatments (against hemorrhage in the lung, shivering, excess of heat in the throat, hemorrhage in the stomach, swellings, spots on the skin, plus a cathartic, and one using squirting cucumber). The Vatican manuscript of the *Ephodia* (= *graecus* 300) and others preserve a short treatise on cathartic medicines ascribed to a Iōannēs Damaskēnos (ff. 273^{V}; 284), to whom other manuscripts (e.g. Escorial, T.II.12, ff. 182^{V}; 183^{V}, and Madrid, Biblioteca Nacional, Vitr. 26.1, ff. 212, 213^{V}) likewise ascribe medical recipes. Both Damaskēnos and Iōannēs Damaskēnos are very probably identifiable with the well-known Arabic physician Abū Zakarīyā' Yūhannā ibn Māsawayh (b. *ca* 777 CE), also known in the medieval West as "Mesue." He wrote 42 medical treatises treating wide-ranging topics (e.g., *materia medica*, ophthalmology, diet, cathartic medicines), several of which were translated into Latin. In the Latin West, the treatise on cathartic medicines, medical recipes and collections circulated under the name of Mesue or John Damascenus (the 7th/8th c. theologian).

Diels 2 (1907) 25; Thorndike and Kibre (1963) 85, 128, 415, 833, 1493; *GAS* 3 (1970) 231–236; S. Lieberknecht, *Die Canones des Pseudo-Mesue. Eine mittelalterliche Purgantien-Lehre. Übersetzung und Kommentar* (1995) 212–215.

Alain Touwaide

Damaskios (*ca* 485 – after 538 CE)

Born *ca* 460 CE; Neo-**Platonic** philosopher, the last *diadokhos* of the **Academy** in Athens. Originally from Damascus, Syria, Damaskios studied and taught in Alexandria, Athens and – after the Academy's closure in 529 – in Mesopotamia and Syria. He studied rhetoric with

Theōn of Alexandria (not the astronomer), geometry and other sciences with MARINOS, and philosophy with Zēnodotos and AMMŌNIOS; his greatest intellectual influence, however, was Isidōros of Gaza. Some works and fragments survive: a treatise on metaphysics *Peri Protōn Arkhōn*, a treatise on *Number, Place and Time*, a *Commentary to Parmenides*, lecture notes from his commentaries on PLATO's *Phaedo* and *Philebus*, fragments from a commentary on ARISTOTLE's *de Cael.* (possibly *Meteor.*) (preserved in IŌANNĒS PHILOPONOS, *In Meteor.*), fragments from either a biography of Isidōros or a philosophical history commonly cited as *Vita Isidori*. Above all, Damaskios was a metaphysician who, unlike his predecessors, was less preoccupied to harmonize Platonic and Aristotelian doctrines, but more concerned to explain soundly the tenets of Neo-**Platonism**. His philosophy is characterized by painstakingly detailed analysis, and he is not shy to come into conflict with orthodox doctrines, mainly Proklean. His metaphysics strives to take on Neo-**Platonist** ontology in its relationship to how it is comprehended, thus converting it into a very interesting and challenging philosophy of the mind.

Ed.: M.C. Galperine, *Damascius, Des premiers principes. Apories et résolutions* (1987); Westerink and Combès (1986–1991); Athanassiadi (1999).

DPA 2 (1994) 541–593, Ph. Hoffmann; Cosmin Andron, "Damascius on knowledge and its object," *Rhizai* 1 (2004) 107–124.

Cosmin Andron

Damastēs (200 – 150 BCE)

Mentioned by SŌRANOS (*Gyn.* 2.18 [*CMG* 4, p. 65; *CUF* v. 2, pp. 26–27]) as a medical writer on pediatrics. Damastēs thought mothers should breastfeed immediately and appealed to nature apparently in a type of Aristotelian teleological argument which Sōranos found unacceptable. Sōranos mentions APOLLŌNIOS OF ANTIOCH (BIBLAS) as among "those who agree with him," giving a *terminus ante quem.*

The Biblioteca Laurenziana in Florence preserves an 11th c. MS (*Laur.* 74.2, f.381^{V}, lines 3–26) with a short excerpt under "Dam{n}astes." The chapter "Concerning those who are able to conceive and carry to term" is a calendar outlining stages in fetal development, from his book *On the Care of Pregnant Women and Infants*, combining obstetrics and child care. The calendar outlines the steps of foam-blood-flesh-shape-motion-birth for seven-, eight-, nine-, and ten-month children, based on **Pythagorean** models but considering viable the eighth-month child.

Holt N. Parker, "Greek Embryological Calendars and a Fragment from the Lost Work of Damastes, *On the Care of Pregnant Women and of Infants*," *CQ* 49 (1999) 515–534.

Holt N. Parker

Damastēs of Sigeion (440 – 410 BCE)

A prose writer immediately preceding THUCYDIDĒS, considered a student of Hellanikos. He wrote an *Events in Greece*, possibly covering the Persian Wars; a genealogical work, *On the ancestors of those who fought at Troy; On Poets and Sophists*, a work of literary criticism; and a geographical work, variously titled *Catalogue of Peoples and Cities, On Peoples*, or ***Periplous***. This was considered derivative of HEKATAIOS OF MILĒTOS, and criticized as faulty: Damastēs thought the Arabian Gulf was a lake; he accepted a story of the Athenian ambassador Diotimos that it was possible to sail up the Kudnos river in Kilikia to the

Khoaspes and thence to Susa; and he placed the Hyperboreans north of the Arimaspians, beyond the Rhipaean mountains.

Ed.: *FGrHist* 5.

RE 4.2 (1901) 2050–2051, E. Schwartz; *OCD3* 427, K. Meister.

Philip Kaplan

Damianos of Larissa (400 – 600 CE?)

Damianos (probably the son of HĒLIODŌROS OF LARISSA) is known only as the author of the *Optica* (the conventional name for the title in the Greek MSS, *Summaries of Optical Principles*), an elementary treatise in which 14 propositions on optics and the theory of vision are stated and then discussed. The date of the treatise is uncertain, but passages identical with sections of THEŌN OF ALEXANDRIA's recension of EUCLID's *Optics* made in the second half of the 4th c. CE render it likely that Damianos wrote in the 5th or even 6th c. CE. He broadened his discussion into philosophical topics, and in particular the **Platonic** extromission theory of vision, in connection with which he referred to "the great PLATO" as having shown that sight was the most "sunlike" of the senses (*cf.* Plato *Republic* 6 [508b3–4]). This suggests that the treatise belongs to milieux such as we find at Alexandria and Athens of the 5th and 6th cc. CE in which Platonic exegesis was central to philosophical education. Otherwise, the *Optica* merits Knorr's assessment of it as "unsystematic and non-technical"; but while it might not be consulted by serious students of optics, it deserves note as part of the corpus of elementary pedagogical texts that are integral to the transmission of the ancient scientific tradition.

Ed.: R. Schöne, *Damianos Schrift über Optik* (1897).
W.R. Knorr, "Archimedes and the Pseudo-Euclidean *Catoptrics*," *AIHS* 35 (1985) 28–105 at 32–33 and 89–96; *DPA* 2 (1994) 594–597, Robert B. Todd, *CTC* 8 (2003) 1–6, *Idem; NDSB* 2.233–234, F. Acerbi.

Alan C. Bowen and Robert B. Todd

Damigerōn (325 BCE – 160 CE)

Listed with other eastern magi (APULEIUS *Apol.* 90.6, Tertullian *De anim.* 57.1, and Arnobius *Adu. nat.* 1.52). It is not clear if he is the same figure attributed – in some medieval MSS, although in a corrupt form (Amigerōn) – with the Latin lapidary also ascribed to EUAX. The current opinion is that Damigerōn is the Greek name of the author of an Alexandrian Greek lapidary used as a model for the surviving Latin version. Damigerōn is also cited in the GEŌPONIKA as a source of information regarding the preservation of cereals (2.30–31), viticulture and wine (5.21–22, 37; 7.13, 24), olive oil production (9.18, 26), etc., without any reference to magic and mineralogy.

Ed.: Halleux and Schamp (1985) 193–290.
RE 4.2 (1901) 2055–2056, M. Wellmann.

Eugenio Amato

DAMIŌN ⇒ DALIŌN

Damis of Kolophōn (195 – 185 BCE?)

Designed a ***sambukē***, perhaps for Kolophōn's successful emergency defense against Antiokhos III's circumvallation (Livy 37.26.5–8, 28.4, 31.3). Damis' ***sambukē***, described by BITŌN, *Belop.* 5 (pp. 57–61 W.), rolled into position on four wheels, with the ladder horizontal, which was then elevated to the top of the wall, by *kokhlias* (screw) and *ergatēs* (winch) rotating the ladder and counterweight (the exact mechanism is disputed). The ladder was screened and held about ten fully armed men at its extremity, balanced by several metric tons of lead. The name is rare outside Kurēnē, after POLUBIOS Book 21, *fr.*31.6 (*LGPN*).

Irby-Massie and Keyser (2002) 162–163.

PTK

Dāmokratēs, Seruilius (*ca* 70 – 80 CE)

PLINY (25.87) mentions a physician Seruilius Dāmokratēs, who "recently" in Spain had discovered a herb called *iberis*. This with Dāmokratēs' quotation of the younger ANDROMAKHOS (GALĒN, *CMGen* 6.12, 13.920 K.) fixes his period of activity. According to Pliny (24.43), Dāmokratēs cured the daughter of the former consul M. Seruilius, suggesting that he gained the rights of Roman citizenship from this patrician. The cure was purely pharmaceutical, as are all his surviving 1,650 iambs. These 48 recipes (from nine to 173 verses in length) are cited exclusively by Galēn who keeps praising Dāmokratēs for his exactitude, succinctness, and usefulness, and quotes three titles: *Puthikos, Philiatros*, and *Klinikos*.

Ed.: U.C. Bussemaker, *Poetarum de re physica et medica reliquiae* (1851) 99–132; Sabine Vogt, *Servilius Damokrates. Die iambischen Pharmaka im Corpus Galenicum* (in preparation).

H. von Staden, "Gattung und Gedächtnis: Galen über Wahrheit und Lehrdichtung," in W. Kullmann, J. Althoff, and M. Asper, edd., *Gattungen wissenschaftlicher Literatur in der Antike* (1998) 65–94; *BNP* 4 (2004) 64, E. Bowie; Sabine Vogt, "'. . . er schrieb in Versen, und er tat recht daran.' Lehrdichtung im Urteil Galens," in Th. Fögen, ed., *Antike Fachtexte/Ancient Technical Texts* (2005) 51–78.

Sabine Vogt

Damōn (Geog.) (250 BCE – 77 CE)

Wrote a geographical or paradoxographical work from which PLINY 7.17 cites a ***paradoxon*** about the healing sweat of "Ethiopian" folk.

(*)

PTK

Damōn of Athens (465 – 425 BCE)

Son of Damonidēs, he was born in the Athenian district of Oa and was greatly admired by contemporaries as "the most accomplished of men not only in music" (PLATO, *Lach.* 180d). According to Plato, he was pupil of Prodikos of Keōs – thus, some scholars think, close to the sophists' environment, if not even a sophist himself – and later on teacher and adviser of Periklēs, probably suggesting to him the construction of the public building for musical performances called Ōdeion. He was ostracized around 440 BCE (four *ostraka* were found with his name) and then, ten years later, came back to Athens for teaching. The hypothesis that he delivered a speech in front of the Council titled *Areopagiticus* has recently been set aside as erroneous: echoes of his writings survive through secondary sources, the most important of which are Plato, PHILODĒMOS and ARISTEIDĒS QUINTILIANUS.

Damōn's main theoretical interests were the ethical effects of music which – as reported by Plato *Republic* Book 3 – he seems to have categorized according to affinities between musical structures (rhythms and *harmoniai*) and types of characters: hence the political and social relevance of music, basic for the formation of citizens' character. He also described music and dance as the product of a special kind of movement in the soul, and seemed to base the intimate relation between characters and musical forms on "qualitative" features – in terms of male-female dichotomy – inherent in the elements out of which scales were built (according to Aristeidēs Quintilianus, who cites Damōn but may have added other material).

Ed.: DK 37; F. Lasserre, *Plutarque. De la Musique* (1954) 53–79.

Barker (1984) 168–169; R. Wallace, "Damone di Oa ed i suoi successsori: un'analisi delle fonti," in R. Wallace and B. MacLachlan, edd., *Harmonia mundi: musica e filosofia nell'antichità* (1991) 30–53; *BNP* 4 (2004) 65–66 (#3), R. Harmon; Z. Ritoók, "Damon. Sein Platz in der Geschichte des ästhetischen Denkens," *WSt* 114 (2001) 59–68.

E. Rocconi

Damōn of Kurēnē (225 – 185 BCE)

DIOGENĒS LAËRTIOS 1.40 cites his *On the Philosophers*, which criticized all the wise men (*cf.* HERMEIAS), especially the Seven (THALĒS *et al.*). According to a Herculaneum papyrus (Classen 179–180), he was a student of LAKUDĒS.

C.J. Classen, "Bemerkungen zu zwei griechischen, Philosophiehistorikern'," *Philologus* 109 (1965) 175–181.

PTK

Damonikos, Claudius (40 – 60 CE?)

Nestled among numerous sources and authorities in the "drug books" of ASKLĒPIADĒS PHARM., as quoted by GALĒN, are two recipes by a "Damonikos," with the second as Claudius Damonikos, probably the same. Perhaps a specialist in compounding drugs to heal inveterate and macrobiotic wounds and ailments, Claudius Damonikos formulates pharmaceuticals to soften long-hardened fistulas (*suringai*) and leathery sores known as *phagedainika* (Galēn, *CMGen.*, 4.13 [13.739–740 K.]), and to alleviate long-standing ear-purulency (Galēn, *CMLoc.*, 3.1 [12.637 K.]). Ingredients are sensible and appropriate, including beeswax, *propolis* ("bee-glue"), bitumen (*asphaltos*), boiled olive oil combined with hot pine-pitch, **terebinth**-resin, heated fissile alum and **litharge** combined with just-liquified aloe-sap and **galbanum** – cooled before application – to treat hardened fistulas and *phagedainika*. For long-standing purulence oozing from the ears, Claudius fashions a compound from saffron oil, myrrh, fissile alum, frankincense, Syrian spikenard, Egyptian natron, and the pounded-smooth and purified "meat" of 30 walnuts, all mixed in sharp vinegar; one "drips" this into the ear as further diluted in more vinegar; the saffron, frankincense, myrrh, and natron constitute a mild antibiotic, but the saffron oil as an ingredient makes this remedy phenomenally expensive.

Fabricius (1726) 121.

John Scarborough

Damostratos or Dēmostratos (of Apameia?) (80 BCE – 20 CE)

Roman senator writing in Greek on history and fishes (*Souda* Delta-51). Renowned and important was his *On fishing* (*Halieutika*, 20 books), indirectly used by AELIANUS (see, e.g., *NA* epilogue) who considered him a prominent scientific authority and a brilliant stylist (*NA* 15.5, 29). Aelianus preserves verbatim a fragment describing Dēmostratos' personal experience of dissection and embalming of a fish (maybe the oarfish, *Regalecus glesne* Ascanius, *NA* 15.9). Dēmostratos' book treated many related themes, including aquatic divination (*Peri enudrou mantikēs: Souda, ibid.*). Identification with a Demostratus (who wrote on amber: PLINY 37.34) and a Dēmostratos of Apameia (*sic*), author of *On Rivers* (at least two books, PSEUDO-PLUTARCH, *Fluv.* 9.2), is then highly probable.

RE 4.2 (1901) 2080–2081 (#5), M. Wellmann.

Arnaud Zucker

Daphnis of Milētos (*ca* 300 BCE)

Architect who, with PAIŌNIOS OF EPHESOS, built the Temple of Apollo at Milētos (at Didyma, begun *ca* 300 BCE: Vitruuius 7.*pr*.16). This Apolloneion replaced an earlier temple burnt by the Persians in 494 BCE; work continued well into the Roman period but was never finished.

BNP 4 (2004) 83–84, C. Höcker; *KLA* 1.160, R. Vollkommer.

Margaret M. Miles

Dardanos or Dardanios (*ca* 360 – 410 CE)

IŌANNĒS "LYDUS" (*Mens.*, 4.9 Wu.) quotes an etymology of the name of *miliarísion*: "from a thousand *oboloi*" from the work of Dardanos *On weights*. This silver coin was introduced under Constantine and so called because it was in fact equal to one-thousandth of a pound of gold (see *MSR* 1.307). PRISCIANUS OF CAESAREA cites two short passages (*de fig. num.*, 10 and 14) where Dardanos lists the equivalence rates (in two cases, wrong) of obol, drachma, ounce, pound and mina, "light" and "heavy" talent. Nevertheless, Priscianus neither refers to the *miliarísion* etymology nor specifies Dardanos' work. Moreover Priscianus strikingly quotes Dardanos in Latin. The evidence being too scanty, the most likely conclusion is either that Priscianus quoted Dardanos from a Latin metrological collection or his Greek work was translated into Latin *ca* 450–550 CE, and transmitted as CARMEN DE PONDERIBUS.

MSR 2 (1866) 83, 85; Hultsch (1882) 7–8, 201; *RE* 4.2 (1901) 2163 (Dardanios); 2180 (Dardanos #14), F. Hultsch; G. Mercati, "Il περι σταθμων di Dardano tradotto anticamente in latino?," *RIL* 41 (1909) 149–156; J.-P. Callu, "Les origines du «miliarensis»: le témoignage de Dardanius," *RN* 22 (1980) 120–130.

Mauro de Nardis

DAREIOS ⇒ DASIUS

Dasius (120 BCE – 80 CE)

Cited by Martial 6.70.6, along with ALKŌN and Symmachus (*cf.* 5.9, 7.18.10), presumably as an instance of some medical genus other than surgeon. ANDROMAKHOS twice cites a Dareios, who may be the same person: GALĒN *CMLoc* 7.3 (13.69 K.), a febrifuge of saffron, henbane, myrrh, poppy juice, pepper, etc.; and *CMGen* 5.12 (13.832), a suppository of acacia, myrrh, poppy juice, pepper, etc.

RE 4.2 (1901) 2219 (#3), A. Stein; Korpela (1987) 186.

PTK

DAVID (PSEUDO) ⇒ ELIAS (PSEUDO)

Dēïleōn (250 BCE – 95 CE)

ASKLĒPIADĒS PHARM., in GALĒN *CMGen* 4.13 (13.744–745 K.), records his remedy for infected head-wounds: copper flakes (DIOSKOURIDĒS 5.78–79), frankincense, and dry resin, ground fine into vinegar, and dried away from light, applied with beeswax. For the rare name, *cf.* Apollōnios of Rhodes 2.955–956 (of Trikka), Daïleōn (*LGPN* 1.112, 2.98, 3B.95), and perhaps Dēïlleōs (*LGPN* 1.123).

Fabricius (1726) 136.

PTK

Deinōn of Kolophōn (360 – 330 BCE)

The "Diōn of Kolophōn" whom both VARRO, *RR* 1.1.8, and COLUMELLA, 1.1.9, regard as an agricultural authority is probably identical to the Deinōn/Dinōn of Kolophōn, who wrote a history of Persia in three books; the evidence for this comes from PLINY's equivocation on the spelling of the name in his index, 8, 10, 12, 14–15, 17–18. He is often cited as an expert on wild trees, and he seems to have given special attention to cultivation of myrtle. His work may also have included discussion of livestock, poultry, and cereals among the Persians. He was the father of the historian Kleitarkhos.

Ed.: *FGrHist* 690.
RE 5.1 (1903) 708 (#20), M. Wellmann, with 654 (#2) E. Schwartz.

Philip Thibodeau

Deinostratos of Prokonessos (365 – 325 BCE)

Pupil of EUDOXOS and associate of PLATO, who, with his brother MENAIKHMOS OF PROKONESSOS and AMUKLAS OF HĒRAKLEIA, "made the whole of geometry more perfect (or complete)" (PROKLOS *In Eucl.* p. 67.8–12 Fr.). PAPPOS, saying that Deinostratos, NIKOMĒDĒS, and some later people used a line called the quadratix (*tetragōnizousa grammē*) for squaring the circle, proceeds to describe the line and its use (*Collection* 4.30 [pp. 250–252 Hultsch]). No other ancient source associates Deinostratos with the quadratix. SIMPLICIUS (*in Cat.* = *CAG* 8 [1907] 190.20–21, quoting IAMBLIKHOS; *cf. in Phys.* = *CAG* 9 [1882] 60.12–13) links it with Nikomēdēs, and Proklos (*In Eucl.* p. 272.7–8 Fr.; *cf.* p. 356.6–11 Fr.)

says that some people used the quadratices of Nikomēdēs and Hippias (usually thought to be HIPPIAS OF ĒLIS) to trisect an angle. The bibliography lists three different accounts of Deinostratos' treatment of the quadratix.

Heath (1921) 1.225–230; *DSB* 4.103–105, I. Bulmer-Thomas; Knorr (1986) 80–86.

Ian Mueller

Dēmarkhos (300 BCE – 100 CE)

CAELIUS AURELIANUS, *Chron.* 1.140 (*CML* 6.1.1, p. 512), lists him, among other "ancients" (PHULOTIMOS to TRUPHŌN), as having propounded ineffective cures for epilepsy.

Fabricius (1726) 136.

PTK

Dēmētrios (Math.) (250 – 300 CE)

Platonic geometer who discussed the properties of odd and even numbers; teacher of PORPHURIOS (PROKLOS, *in Plat. Remp.* 2.23.14 Kroll), who met at Athens with CASSIUS LONGINUS, Porphurios, and others, to celebrate PLATO's birthday (EUSEBIOS, *Pr. Ev.* 10.3.1). Chronology and similarity of interests suggest, though not definitively, that our Dēmētrios may be identifiable with DĒMĒTRIOS (MUSIC).

PLRE 1 (1971) 247.

GLIM

Dēmētrios (Music) (before *ca* 300 CE)

Author of a *Peri logou sunaphēs* (perhaps *On the Composition of Ratio*, if it was a mathematical treatise), of which only the title survives, cited by PORPHURIOS in his commentary on PTOLEMY's *Harmonics* (92.25–26). Porphurios mentions Dēmētrios several times in a discussion on the correct use of the terms "ratio" (*logos*), "excess" (*huperokhē*, i.e. the amount by which the greater term exceeds the lesser) and "interval" (*diastēma*). He calls Dēmētrios a mathematical scientist (*mathēmatikos*), and names him in a list of writers who used the term "interval" to mean "ratio," just as PLATO did when he referred to hemiolic, epitritic and epogdoic *intervals* (rather than *ratios*) at *Timaeus* 36a–b. The list also includes PANAITIOS THE YOUNGER, ARKHUTAS, DIONUSIOS (OF HALIKARNASSOS?), EUCLID, "and many other canonic theorists (*kanonikoi*)."

RE 4.2 (1901) 2487 (#110), C. von Jan; Düring (1932).

David Creese

Dēmētrios (Pythag.) (200 – 100 BCE)

Authored a treatise wherein he explained why the number four, one of Hēraklēs' prerogatives (*herculaneus numerus quaternaries*), is particularly important for the healing art (PLINY 28.64). Evidence is too meager to consider him a physician.

RE 4.2 (1901) 2850 (#119), Fr. Hultsch.

Bruno Centrone

Dēmētrios (Astrol.) (unknown date)

Astrologer to whom "Palkhos" (pseudonym for Eleutherios, 1388 CE [Pingree 1978: 437]) ascribes an *Astrologoumena*, one fragment of which (*Peri drapeteuontōn*) details signs for fugitives. The zenith of various constellations and planets indicates escape, and the nadir presages capture; a fugitive's sign also indicates the timeframe for capture (e.g., the Bull means capture will occur within a year). Another fragment gives a nautical ***melothesia***, on analogy with human anatomy (e.g., the Bull represents the keel, the Crab the rudder, Gemini, connected to the human ribs, governs the *pleura* or "ribs" of ships, while Pisces, associated with feet, protects the submerged parts of the ship). The same fragment lists lunar portents for sailors: the Moon in Aries or Cancer impedes travel, and in Aquarius endangers the ship (*contra* HEPHAISTIŌN 3.17 where those "swimming" signs *favor* sea travel). Emphasized also is the dire nature of retrograde motion, especially of the malefic Mars, as in VALENS, PTOLEMY, and Hephaistiōn.

Ed.: *CCAG* 1 (1898) 104–106; *CCAG* 8.3 (1912) 98–99.
Pingree (1978) 426; J. Komorowska, "Seamanship, sea-travel, and nautical astrology: Demetrius, 'Rhetorius' and Naval Prognostication," *Eos* 88 (2001) 245–256 at 246–252.

GLIM

Dēmētrios (Geog.) (*ca* 100 BCE – 60 CE?)

Wrote a geographical or paradoxographical work *On Egypt*, cited by PLINY 36.79, for the pyramids, and Athēnaios, *Deipn.* 15 (680a–b) for acacia-trees.

RE 4.2 (1901) 2850 (#120), Fr. Hultsch.

PTK

Dēmētrios Khlōros (90 – 50 BCE)

Grammarian; probably a Pergamene scholar, NIKANDROS' earliest commentator. He is mentioned in our *Scholia in Thēriaka* nine times (once as Meneklēs' son, thrice nicknamed *khlōros*, "pale"), and once more by STEPHANOS OF BUZANTION (s.v. *Koropē*) who quotes *Thēr.* 613–614 with remarks drawn from *Scholia* richer than ours. He is our sole authority for Alkibios' story (545–549). His comments, centered on *realia*, are sometimes based on *falsae lectiones* or unfounded speculations with erroneous or even absurd glosses (*Schol. Thēr.* 377–378a, 622c). ANTIGONOS OF ALEXANDRIA sometimes contradicts him (585a, *cf.* 748); in 781b, the Scholiast (after PLUTARCH?) dismisses both of them.

GGLA 2 (1891) 220; Jacques (2002) 2.CXXIX–CXXX (and n. 296).

Jean-Marie Jacques

Dēmētrios "physicus" (1st c. BCE)

Among the authorities consulted by PLINY (1.*ind*.8) who in particular records a remarkable story about a panther showing gratitude to a man who saved her cubs (8.59–60).

RE 4.2 (1901) 2849 (#114), M. Wellmann.

Jan Bollansée, Karen Haegemans, and Guido Schepens

Dēmētrios of Alexandria (200 BCE – 100 CE)

Wrote *Linear Considerations* (*Grammikai Epistaseis*: the third of the three types of geometrical problems: plane, solid, linear) on the geometrical curves, which he discovered in attempting to trisect rectilinear angles by studying the interaction of various surfaces (e.g., *plektoids*/spiral curves with other surfaces), and which exhibit "astonishing properties" (n.b., the quadratrix of DEINOSTRATOS). Dēmētrios discovered many complex solutions including, perhaps, spirals, quadratrices, conchoids, and cissoids. Dēmētrios predates MENELAOS, who called one of his, or PHILŌN OF TUANA's, curves "paradoxical": PAPPOS, *Coll.* 4.36 (p. 270 H.)

RE 4.2 (1901) 2849 (#116), Fr. Hultsch.

GLIM

***Dēmētrios of Amisos* (*ca* 250 BCE?)**

Son of Rathēnos, mathematician noteworthy for his knowledge: STRABŌN 12.3.16.

RE 4.2 (1901) 2849 (#117), Fr. Hultsch; Netz (1997) #30.

GLIM

Dēmētrios of Apameia (*ca* 200 – 100 BCE?)

Hērophilean physician extensively interested in pathology and gynecology; cited mostly in SŌRANOS (and CAELIUS AURELIANUS), but also HĒRAKLEIDĒS OF TARAS, whereas he cites ANDREAS, yielding the date-range. Whether he hails from Apameia in Bithunia or on the Orontēs river in Syria is uncertain. In *On Diseases*, he described indications of **dropsy** distinguishable from diabetes (12 books: Cael. Aur. *Chron.* 3.99, 102 [*CML* 6.1, pp. 738, 740]) and differentiated causes of hemorrhaging including cutting, open pores, diffusion, bodily weakness, and openings at the ends of vessels (*Chron.* 2.122–123 [p. 618]). He considered apoplexy and paralysis identical (*Acute* 3.55 [pp. 324–326]), distinguished bodily convulsions from tremors (*Acute* 3.71–72 [p. 334], *Chron.* 2.64 [p. 582]), and defined mania as a strain on the mind (*Chron.* 1.150 [p. 518]). Caelius Aurelianus discounts his theory that ***hudrophobia*** is a slow and long-lasting ailment (*Acute* 3.106 [p. 354]) and criticizes his omission of fever from his discussion of pneumonia (*Acute* 2.141 [p. 226]). Nonetheless, Dēmētrios included fever as a symptom of *phrenitis* (*Acute* 1.4 [p. 24]), lethargy (*Acute* 2.4 [p. 132]), and cardiac disease (*Acute* 2.173 [p. 248]). Regarding gynecology, Dēmētrios argued for diseases particular to women, in stark contrast to other **Hērophileans** (HĒROPHILOS, ALEXANDER OF LAODIKEIA). Dēmētrios recognized six varieties of vaginal flux which he defined as the flow of fluid matter (bloody and other) through the uterus (Sōr. *Gyn.* 3.43.2 [*CMG* 4, p. 122]). He described the causes and cures of difficult childbirth, including the parturient's emotional state and ignorance of the process, uterine shape and size, and the embryo's size and position (Sōr. *Gyn.* 4.2–5 [*CMG* 4, pp. 131–135]).

von Staden (1989) 506–511; *OCD3* 450–451, *Idem.*

GLIM

Dēmētrios (of Athens?) (300 – 270 BCE)

Geometer, wrote *On the Difficulties of* POLUAINOS, preserved in badly damaged papyrus fragments (*P. Herc.* 1429, 1061), arguing the incompatibility of the Euclidean doctrine of indivisibles (*atoma*) with the fundamental principles of mathematics. Dēmētrios, not citing EUCLID explicitly, discusses bisecting straight lines and angles (1061: col.9–11; *cf.* Euclid 1.*prop*.9–10), and the properties of circles (1061: col.8.9–17: *cf.* Eucl. 1.*def*.15).

J.-L. Heiberg, "Quelques papyrus traitant de mathématiques," *Oversigt over det Kgl. Danske Videnskabernes Selskabs Forhandlinger* 2 (1900) 147–171; *RE* 4.2 (1901) 2849 (#115), Fr. Hultsch; Netz (1997) #45.

GLIM

Dēmētrios of Kallatis (215 – 145 BCE)

Historian, wrote a geography *About Asia and Europe* in 20 books (DIOGENĒS LAËRTIOS 5.83), treating Sarmatian history and topography, and referring to the death of Hierōn II of Surakousai (F3: 216 BCE). Dēmētrios is cited with DIOPHANTOS as treating the world's northern quadrant (AGATHARKHIDĒS OF KNIDOS, Book 5, *fr*.65). His account of all the earthquakes occurring in Greece, of which STRABŌN provides a lengthy synopsis (F6: 1.3.20) was, presumably, part of the larger history.

Ed.: *FGrHist* 805.
KP 1.1467 (#20), H. Gärtner; *BNP* 4 (2004) 252 (#30), K. Meister.

GLIM

Dēmētrios of Lakōnika (150 – 80 BCE)

Epicurean student of PRŌTARKHOS OF BARGULIA and younger associate of ZĒNŌN OF SIDŌN. The Herculaneum papyri contain numerous fragments of Dēmētrios' works on poetry, physics, mathematics, theology, and philology. Some of his views can be reconstructed from these fragments and later reports of his doctrines. He defended EPICURUS' views on the size of the Sun, the nature of the gods, the infinity of space, the infinite number of atoms, inference from similarities (PHILODĒMOS, *On Signs* 45–46), the nature of time ("an accident of accidents"), the **Epicurean** theory of minima, and, against KARNEADĒS, he defended the existence of proof. He also considered the possibility that some apparent shortcomings in **Epicurean** philosophy were caused by scribal errors in earlier MSS.

Ed.: V. de Falco, *Demetrio Lacone* (1923).
RE 4.2 (1901) 2842 (#89), H. von Arnim; C. Romeo, "Demetrio Lacone sulla grandezza del sole (PHerc. 1013)," *CrErc* 9 (1979) 11–35; Long and Sedley (1987) §7C; *OCD3* 450, D. Obbink; *ECP* 178, J.S. Purinton; *BNP* 4 (2004) 250 (#21), T. Dorandi.

Walter G. Englert

Dēmodamas of Milētos (290 – 260 BCE)

General of Seleukos and Antiokhos, crossed the Iaxartes river, and wrote an account of those lands, which PLINY followed: 1.*ind*.6, 6.49; repeated in IULIUS SOLINUS 49.1–6 and MARTIANUS CAPELLA 6.692. STEPHANOS OF BUZANTION, s.v. *Antissa*, notes an Indian island of that name in "Dēmodamas the Milesian."

R. Hennig, *Terrae Incognitae* 1 (1936) 222–223.

PTK

Dēmokedēs of Krotōn (*ca* 560 – 500 BCE)

According to HĒRODOTOS (3.131–137), Dēmokedēs was the most famous doctor of his time. At high salary he was invited to Aigina, then to Polukratēs of Samos, where he was taken as a captive by the Persians and served as a doctor to King Darius. He managed to flee and to return to his native Krotōn (*ca* 518), where he joined the political community of the **Pythagoreans**. His father, the doctor Kalliphōn, probably was **Pythagorean** too (Hērodotos 3.125). PLINY mentions Dēmokedēs' medical writing, but its authenticity is dubious.

DK 19; M. Michler, "Demokedes von Kroton," *Gesnerus* 23 (1966) 213–229; Zhmud (1997).

Leonid Zhmud

Dēmokleitos (200 – 160 BCE)

With KLEOXENOS invented a binocular ***dioptra*** and a cipher, for use as a nocturnal telegraph, which POLUBIOS improved (Book 10, *fr.*45–46). This form of the name is otherwise unattested, but compare the common names Dēmoklēs and Dēmokleidēs (*LGPN*).

Diels (1920) 85–87.

PTK

Dēmoklēs (200 – 25 BCE)

Wrote on machinery (VITRUUIUS 7.*pr.*14) and was cited an authority on metallurgy and *lithika* (PLINY 1.*ind.*34–35). He also described severe earthquakes in Lydia and Ionia (STRABŌN 1.3.17). Possibly he is Damoklēs of Messēnē, the student of PANAITIOS.

RE 5.1 (1903) 133 (#13), E. Fabricius.

PTK

Dēmokritos (Neo-Platonist) (*ca* 200 – 270 CE)

Listed by his contemporary LONGINUS in the preface of *On final end* among **Platonists** who expounded their views in writing (PORPHURIOS, *Vit. Plot.* 20.31), but basically upheld their predecessors' doctrines (*Vit. Plot.* 20.60). Dēmokritos commented on PLATO's *Alcibiadēs* (OLUMPIODŌROS, *In Alc.* p. 70.16 Westerink), the *Phaedo* (DAMASKIOS, *In Phaed.* 1.503.3 Westerink) and perhaps also the *Timaeus* (PROKLOS, *In Tim.* 2.33.13).

That Dēmokritos' views, insofar as they survive, are hardly original confirms Longinus' testimony of Dēmokritos' attachment to tradition. Dēmokritos reportedly identified the soul's faculties – a favorite topic among **Platonists** like DIONUSODŌROS SEUERUS, Longinus, PLŌTINOS, and Porphurios – with the soul's substance (IŌANNĒS STOBAIOS, 1.370.1–2 W.-H.), a view similar to Longinus' (Porphurios, *On the faculties of the Soul* in Stobaios 1.351.14–19 W.-H.). SYRIANUS (*In Met.* 105.36–39) reports that Dēmokritos along with PLUTARCH and ATTICUS believed that the Forms always exist in the soul of the divine **demiurge**, answering a question of concern to contemporary **Platonists**, namely how the Forms relate to the **demiurge**. Dēmokritos sided with those maintaining that the divine **demiurge** is the primary cause of everything that exists, while the Forms are subordinate to the **demiurge**, and thus secondary causes. **Platonists** like Dēmokritos preferred to distinguish between soul and intellect within the divine

demiurge, a less radical distinction than that between divine ***hupostaseis*** upheld by Noumēnios of Apameia and Plōtinos. Hence the **demiurge** remains the primary cause of the universe, and yet his intellect is more distanced from the material world which because of matter can taint and divide.

L. Brisson in M.O. Goulet-Gazé, R. Goulet, and D.J. O'Brien, *Porphyry: La vie de Plotin* 1 (1982) 78–79; *BNP* 4 (2004) 269–270 (#2), M. Baltes and M.-L. Lakmann.

George Karamanolis

Dēmokritos of Abdēra (440 – 380 BCE)

Greek **atomist** and student of Leukippos, he was responsible for developing **atomism** into a coherent philosophical system. Evidence about his life, travels, and teachings is sparse and inconsistent, but he was most likely born about 460 BCE and died sometime after 380. He is reported to have traveled widely, including to Egypt and Athens. We have the titles of over 60 works, all of them lost. Apart from about 300 extant ethical sayings that have been preserved, we must rely on testimonia in Aristotle and others to reconstruct his philosophical system. Dēmokritos greatly influenced Epicurus in the development of his own atomic system.

Leukippos and Dēmokritos developed **atomism** in response to the challenge posed by the Eleatic philosopher Parmenidēs, who had argued that the apparent multiplicity of the world around us is illusory, and that all that exists is Being, one and motionless. Motion, plurality, and "not being" cannot and do not exist. Addressing this paradox, Leukippos and Dēmokritos divided the world into being and non-being, defining being as "the full" (i.e., without any empty space) and non-being as "the empty." Being, or "the full," was composed of indestructible "atoms" (literally, "uncuttables"), and non-being, or "the empty," was described as empty space. "Being," that is, each individual atom, was "one," but could move, thanks to the "existence" of non-being, or void. Leukippos and Dēmokritos thus explained how and in what sense being and non-being, and motion and plurality, could exist while still meeting many of Parmenidēs' objections.

Leukippos and Dēmokritos taught that atoms were entities that were indivisible, indestructible, and homogenous in substance. Atoms are infinite in number, and they come in an infinite number of different shapes. Atoms thus possess the properties of size and shape, but probably do not possess weight (which is apparently a quality that Epicurus later ascribed to the atom). Atoms move through the void eternally in all directions, traveling in one direction until they collide with another atom and rebound. As these atoms move and collide eternally in the void, they at times fall into groupings called "vortexes." Worlds or ***kosmoi*** are formed as atoms in these vortexes are sorted out into various arrangements, gradually forming elements and then compound bodies of various degrees of complexity. An infinite number of worlds, including our own, are constantly coming into and out of existence in the infinite void.

Within a world, all compound bodies, including inanimate objects, plants, animals, and humans are made up of atoms jostling back and forth in the void. Both body and soul, including the human soul, are made up of atoms. Dēmokritos held that the human soul, the source of human life, sensation, and thought, is made up, like fire, of tiny, smooth, spherical atoms that move quickly at the slightest impulse. The soul is mortal, disintegrating along with the body at death. Sensation occurs when atoms from external objects come in contact with our sense organs. Vision, for example, results from the impact on our eyes of eidola, or thin atomic images, that all objects are constantly shedding. Perceptible qualities (e.g., color, taste, temperature) of compound bodies did not belong to individual atoms, but were the result of the particular shapes and arrangements of atoms that struck and produced impressions in our sense organs. These secondary qualities existed "by convention."

Little is known about his views on the gods, though Dēmokritos seems to have thought humans developed their belief in gods as a result of receiving images of huge and powerful anthropomorphic beings. His ethical system is also difficult to assess, though he is said to have made *euthumia* ("contentment") the goal of life.

Ed.: DK 68.
DSB 4.30–35, G.B. Kerferd; KRS 402–433; *OCD3* 454–455, D.J. Furley; *ECP* 169–172, J.S. Purinton; *REP* 2.872–878, C.C.W. Taylor; *BNP* 4 (2004) 267–269 (#1), I. Bodnár.

Walter G. Englert

Dēmokritos, pseudo (Lith.) (250 BCE – 50 CE)

PLINY refers to Dēmokritos on the virtues and properties of gems and precious stones: Macedonian and Persian emeralds (37.69); Arabian *aspisatis* (perhaps a variety of coal) and the Leukopetrian silvery stone, effective respectively against spleen diseases and hysteria (37.146); Erbilian *belum* (37.149: a kind of agate?); *erotylos* (or *amphicomos* or *hieromnemon*), valued for divining (37.160); and finally Median *zathene* (37.185: a kind of amber?). Dēmokritos' identity is entirely uncertain. THRASULLOS OF MENDĒS assigns a book on stones to DĒMOKRITOS OF ABDĒRA (see DIOGENĒS LAËRTIOS 9.47), but this study very probably treated magnets exclusively (*cf.* DK 68 A 165 = ALEXANDER OF APHRODISIAS *Quaest.* 2.23). Pliny's Dēmokritos is more likely a late homonym or rather pseudonym, perhaps concealing BŌLOS OF MENDĒS.

RE S.4 (1924) 219–223, I. Hammer-Jensen.

Eugenio Amato

Dēmokritos, pseudo (Alch.) (200 BCE – 250 CE)

Pre-eminent amongst authorities cited in the Greek alchemical corpus where he is often called simply "the Philosopher," esteemed not for his **atomist** doctrine, but rather for the magical and alchemical pseudepigrapha circulating under his name and largely believed genuine (PLINY 30.8–11; Gellius 10.12). At least one of pseudo-Dēmokritos' magical works, the *Kheirokmēta* (*Things Wrought by Hand*), is thought to have been written by BŌLOS OF MENDĒS (COLUMELLA 7.5.17) and some of the alchemical or proto-alchemical recipes ascribed to Dēmokritos (P. HOLM. recipe 2; DK B300, 25; SENECA *Ep. Mor.* 90.33) were written by *anonymi* perhaps as early as the last few centuries BCE. However, the texts for which alchemists revered him, or at least those that are extant, appear to have been composed after the mid-1st c. CE (Letrouit 1995: 74).

These texts, cited throughout the Greek alchemical corpus, which (on the sole testimony of OLUMPIODŌROS, *CAAG* 2.102) together formed the lost *Principles*, were four in number: *On Gold* (aka *The Yellow* or *Gold-making*), *On Silver* (aka *The White* or *Silver-making*), *On Stones* and *On Purple* (Letrouit 1995: 75–80). The ANONYMOUS ALCHEMIST PHILOSOPHER tentatively suggests a fifth book, *On Pearls* (*CAAG* 2.433).

Berthelot published two alchemical treatises ascribed to Dēmokritos: the so-called

Physica et mystica (*CAAG* 2.41–53) and *Book 5 Addressed to* LEUKIPPOS (*CAAG* 2.53–56). The first, a compilation of unrelated fragments, can be divided as follows: a) recipes concerning purple (41–42); b) the story of a student in Egypt (presumably Dēmokritos) who recalls from Hades his unnamed teacher (42–43); c) polemic against a group of "new" alchemists; d) ten recipes for making gold (43–46); e) three further recipes for gold (48–49); f) nine recipes for ***asēmos*** (49–53) and g) a conclusion on gold- and silver-making (53).

The tale in section b) should probably be interpreted in relation to the tradition first preserved in SUNESIOS' dialogue on *The Book of Dēmokritos* (*CAAG* 2.57), but perhaps derived from HERMIPPOS OF SMURNA's *On the Mages* (Bidez and Cumont 170–171), that the Persian mage OSTANĒS initiated Dēmokritos in the temple at Memphis. At the end of this section is found the oft-repeated alchemical maxim "Nature is delighted by nature; nature conquers nature; nature rules nature," which is said to bring together the whole of alchemical teaching. The texts edited under the title *Physica et mystica,* although not the four books of the *Principle*, appear to contain material from those books. *Book 5 Addressed to Leukippos*, containing instructions for a single process resulting in *khrusokorallos* (gold-coral), is, however, not the *On Pearls*. Not referred to elsewhere in the alchemical corpus, it appears to be a late composition.

Alchemical texts ascribed to Dēmokritos are extant in both Syriac (Berthelot and Duval 1893: edition 10–60 and partial translation of this and other texts 19–106; 267–293) and Arabic (Ullmann 1972: 159–160), but their relations with the Greek texts remain unclear.

Berthelot (1885) 145–163; *Idem.* and Duval: *La chimie au moyen âge*, v. 2, *L'alchimie syriaque* (1893) 19–104 and 267–293; Bidez and Cumont (1938) 1.170–171; J.P. Hershbell, "Democritus and the Beginnings of Greek Alchemy," *Ambix* 34.1 (1987) 5–20.

Bink Hallum

Dēmokritos, pseudo (Agric.) (*ca* 250 – 50 BCE)

Numerous opinions regarding agricultural doctrine and procedure are attributed to DĒMOKRITOS. Some may go back to the philosopher's work *On Farming* (DIOGENĒS LAËRTIOS 9.48); COLUMELLA (11.3.2) specifically ascribes an opinion on the expense of building a wall around a garden to that treatise. But the majority ought to be classified as apocryphal; Columella identifies BŌLOS OF MENDĒS as the author of the pseudo-Dēmokritean *Kheirokmēta*, a collection of recipes which included such things as advice on how to prevent the spread of ***erusipelas*** in a flock of sheep (bury an infected animal before the threshold of the stall: 7.5.17). A treatise *On* ***Sympathies*** *and Antipathies*, if distinct from the *Kheirokmēta*, seems to have contained precepts of a magical character, e.g. to eliminate caterpillars from a field, have a menstruating woman walk around it (11.3.64). The Dēmokritean material in the GEŌPONIKA and the Arabic agricultural tradition often veers into paradoxography, e.g. giving instructions for "wild" grafts (Wellman *fr.*41). Various lists of weather-signs are assigned to Dēmokritos, some of which may be genuine (*cf.* PTOLEMY *Phaseis* 27), but one that links weather and climate phenomena to the position of the planets in different zodiacal houses will postdate 100 BCE (Wellmann *fr.*4).

Ed.: *DK* 68 B300.1–20; M. Wellmann, "Die Georgika des Demokritos," *Abhandlungen der Preussischen Akademie der Wissenschaften, Phil.-Hist. Kl.* (1921) #4.
RE S.4 (1924) 219–223, I. Hammer-Jensen.

Philip Thibodeau

Dēmokritos, pseudo (Pharmacy) (150? – 80 BCE)

PLINY quotes *Democritus* on herbs, once giving the title, *De Effectu Herbarum* (25.23, *cf.* 25.13), describing magical properties of over a dozen Armenian, Persian, or Indian plants (24.160–166, 25.14, 26.18–19, 27.141); APOLLODŌROS THE DĒMOKRITEAN augmented the list (so this man may predate APOLLODŌROS *THĒRIAKOS*). The eastern plants are listed approximately in Greek alphabetical order (24.160–166), but none can be reliably identified: 24.163 *aethiopis* is not the sage variety of 27.11 and the *therionarca* of 24.163 is not the oleander of 25.113 (the narcotic *onothuris* of 26.18 may recur at 26.111, 146). Pliny quotes Dēmokritos' potion for ensuring good and pretty children (24.166); presumably the book included more such; *cf.* GEŌPONIKA 7.32, recording Dēmokritos' potion to cure excessive desire for alcohol. Dietary properties of turnips (20.19) and radishes (20.28) may derive from another work, perhaps from BŌLOS. Pliny's citation of Dēmokritos – followed apparently in chronological order by APOLLŌNIOS "MUS," MELĒTOS, ARTEMŌN and ANTHAIOS SEXTILIUS – as the source of medicines derived from humans, 1.*ind*.28, 28.7–8, apparently refers to a similar pseudepigraphon. AELIUS PROMOTUS, *Dyn.* 39.1, 46.1 cites other remedies from a pharmaceutical work attributed to Dēmokritos. The 5th–6th c. CE Latin work *Liber Medicinalis* purports to be a translation.

RE S.4 (1924) 219–233, I. Hammer-Jensen; K.-D. Fischer, "Der *Liber Medicinalis* des Pseudo-Democritus," in Vázquez Buján (1994) 45–56.

PTK

Dēmokritos, pseudo (Medicine) (*ca* 150 – *ca* 50 BCE)

CELSUS *pr.*7–8 describes PYTHAGORAS, EMPEDOKLĒS, and DĒMOKRITOS as founders of medicine, the last of whom taught HIPPOKRATĒS (*cf.* Gellius 4.13.2). Celsus seems to be relying on a tradition found also in the 1st c. BCE pseudo-Hippokratic letters (Smith 20–34), which represent Dēmokritos as a medical expert excelling Hippokratēs. *Letter* 17.3 (9.356 Littré) makes gall the cause of madness, proven by animal dissection, 18 (9.380–384 Littré) concerns the use of hellebore, 19 (9.304–306 Littré) pretends that Dēmokritos composed the HIPPOKRATIC CORPUS SACRED DISEASE, and 23 (9.392–398) is a précis of human anatomy. RUFUS, in OREIBASIOS *Coll.* 45.28.1 (*CMG* 6.2.1, p. 184) rejects, and SŌRANOS/ CAELIUS AURELIANUS, *Chron.* 4.4 (*CML* 6.1.2, p. 776), hesitantly cites, Dēmokritos' book on ***elephantiasis***; Sōranos/Caelius says the author prescribed a plant found in Syria and Kilikia. Rufus also cites him on bubonic plague: Oreib. *Coll.* 44.14.1 (pp. 131–132). ANON. PARIS. 51.1 records Dēmokritos' theory that ***elephantiasis*** arose from phlegm blocking surface veins. Caelius also cites Dēmokritos' work on ***hudrophobia***, describing it as an affection of the ***neura***, and prescribing marjoram-potion, drunk from a hemi-spherical cup (*Acute* 3.132–133 [*CML* 6.1.1, p. 372], *cf.* 3.112, 120 [pp. 358, 364]). DIOGENĒS LAËRTIOS 9.48 attributes works on prognosis, diet, and regimen to Dēmokritos.

The medical pseudo-Dēmokritos "listened to bird's voices" (*Letter* 10.1 [9.322 Littré]), and Pliny says *Democritus* claimed that mixing blood from various birds generated snakes, the consumption of which conferred comprehension of birds' speech (10.137, 29.72). He appears also to have written on the anatomy (11.80), ethology (8.61), and medicinal value (28.153, 32.49) of animals (*cf.* XĒNOKRATĒS OF APHRODISIAS). Two lengthy paraphrases are preserved, on the chameleon (Pliny 28.112–118, rejected by Gellius 10.12.1, 6), and on the basilisk of Libya (DK 68 B300.7a), described per NIKANDROS, *Thēr.* 396–398, said to be antipathic to the household weasel, and whose bite is cured by the Psulloi (herpetologists first in Nikandros, *apud* AELIANUS, *NA* 16.28). GEŌPONIKA 20.6 mentions Dēmokritos' book on the dietary properties of fish.

Diels (1905–1907) 2.26–27, 3.26 records MSS containing medical Dēmokritean works: Vatican 299 (15th c.), ff.309–314^V, 329, 366^V, 391, Vatican 1174 (14th/15th c.), ff.1, 32^V, 33^V, Vatican 2304 (15th/16th c.), f.6, and Florence Laurent. App.2 (15th c.), ff.340^V, 356, 359.

RE S.4 (1924) 219–223, I. Hammer-Jensen (some citations misprinted); DK 68 B300.1–20; Smith (1990).

PTK

Dēmophilos (500 – 400 BCE?)

Greek architect, painter or sculptor, mentioned by VITRUUIUS (7.*pr*.14) as a second-tier author (i.e. less famous than those previously listed) of a treatise with precepts on symmetry. He may be identifiable with the homonymous painter and sculptor in clay working in Rome in the 5th c. BCE (PLINY 35.154), or the painter from Himera (*ca* 450–400 BCE, Pliny 35.61).

KLA 1.167, R. Vollkommer.

Margaret M. Miles

Dēmosthenēs Philalēthēs (*ca* 50 BCE – 25 CE)

"The Truth Lover," one of the last members of the "**Hērophilean** School" founded by ZEUXIS. Traditions attached to that school suggest much internecine quarreling, a late Hellenistic version of "medical politics" that doomed its influence, but did not diminish the striking accomplishments of some of its members. In fact, Dēmosthenēs left his mark on ophthalmology from his day through excerpts by AËTIOS OF AMIDA, as well as scattered references in sometimes fragmentary Latin texts dating to as late as the 13th c. (von Staden 572–573). Aëtios likely preserved the organizational principles of Dēmosthenēs' *Ophthalmikos*, even while adding details of his own experiences with eye diseases: Bk. 7 of the *Tetrabiblos* has numerous references to Dēmosthenēs' lost *Ophthalmikos*, and it is obvious that great advances came in the last century BCE in treating common ailments that occasionally threatened one's sight, including cataracts, attested by CELSUS 6.6.12 (*CML* 1, p. 266).

Illustrative is Dēmosthenēs' "On Cancerous Ulcers of the Eye" (Aëtios 7.33, *CMG* 8.2, pp. 283–284), which indicates keen clinical observation, a good employment of case histories, and sensible advice on regimen, diet, and appropriate drugs. "When ulcers that do not heal are at the back of the eye, and are small and painful, they have small blood vessels, and are termed "cancerous" when they turn hard. Sometimes they will appear as if they are healing, but they fall apart again without any obvious cause . . . the victims lose their appetites . . . their great pain becomes worse if any physician were to apply caustic salves . . . the disease is common in old men after a long history of inflammation of the eye, and in women whose menstruals have ceased. . . ." After careful enumeration of treatments including rubbing the body with sweetened olive oil, or good quality rose oil, or mild juice made from the skins of grapes, the patient is to attach a small green patch over the eye, go for a walk where there are shady trees, few people chattering, and later to take some moderately boiled thinned milk ("this deadens the pain of the caustic flowing from the eye"); hunger is to be satisfied by drinking two eggs, then taking a long snooze, preferably for 24 hours. At the end of the section, "pain in the temples . . . is alleviated with compresses made from small poppy-heads . . . and some added saffron and women's milk." Dēmosthenēs' ophthalmology remained the most meticulous and successful type of treatment available until the 14th c., and even in the late 19th and early 20th cc., Dēmosthenēs still gained praise from medical practitioners (Hirschberg 1919).

Fragments (46): von Staden (1989) 576–578, enumerated but not edited; extensive extracts in Aëtios 7 (*CMG* 8.2, pp. 250–399); German translation (1–90 only, of 117): J. Hirschberg, *Die Augenheilkunde des Aëtius aus Amida* (1899).

M. Wellmann, "Demosthenes' *ΠΕΡΙ ΟΦΘΑΛΜΩΝ*," *Hermes* 38 (1903) 546–566; J. Hirschberg, "Die Bruchstücke der Augenheilkunde des Demosthenes," *AGM* 11 (1919) 183–188; von Staden (1989) 570–578.

John Scarborough

DĒMOSTRATOS ⇒ DAMOSTRATOS

Dēmotelēs (*ca* 100 BCE – 20 CE?)

Wrote a geographical or paradoxographical work on Egyptian antiquities, cited by PLINY 36.79, for the pyramids, and 36.84, for the labyrinth.

FGrHist 656.

PTK

Demotic Scientific Texts (650 BCE – 450 CE)

The Demotic stage of Egyptian is the fourth of five stages distinguished in the history of ancient Egyptian (a language attested from *ca* 3000 BCE to after 1000 CE) and is written in a cursive hieroglyphic variant analogous to shorthand. Texts are inscribed typically either on papyrus or *ostraka*. Writings date from *ca* 650 BCE to *ca* 450 CE, a period in which Greek became an Egyptian language and a rival linguistic medium used especially in the upper classes. The Demotic period is divided into Early (Saïte and Persian) Demotic (*ca* 650 BCE to *ca* 300 BCE), Ptolemaic Demotic (*ca* 300 BCE to *ca* 30 BCE), and Roman Demotic (*ca* 30 BCE to *ca* 450 CE). The vast majority of Demotic texts remain unpublished.

Scientific Demotic texts are often short, hardly ever completely preserved, and always anonymous, and neither distinct authors nor the place and time of composition are recognizable. Moreover, due to the Persian conquest (*ca* 525 BCE) and then Greek conquest and immigration (332 BCE and after), Babylonian and Greek influence on Egyptian science is both plausible and rarely verifiable. When hieroglyphic writing was dominant, Egyptian science never significantly lagged behind any other nation's science. A certain level of sophistication is absent from Demotic scientific texts: but a proper historical perspective, namely the fact that Greek had become Egypt's preferred linguistic vehicle for scientific discourse, shows that it would be wrong to expect such a presence.

The most basic expression of knowledge is collecting and listing objects by name (onomastics), and such lists are preserved in Demotic. E.g., *P. Cairo* CG 31168+31169 (found at Saqqara and dating to Ptolemaic times) lists place-names and gods, the early Roman *ostrakon Ashmolean Museum* D.O. 956 lists southern Egyptian place-names, and *P. Carlsberg* 230 lists plants. Lists of words suggest that the script was not learned sign by sign but word by word. Thus, *P. Saqqara* 27 lists birds in an alphabetical order that begins with *H*.

Concerning architecture, the so-called *Book of the Temple* describes how a temple ought to be built and its cult organized. As regards pharmacology, the efficacy of surviving remedies such as the application of mouse-dung and wine to an ailing ear recommended in *P. Vienna* D-6257 seems questionable. Among the medical texts is a dentist's manual, *P. Vienna* D-12287.

Several mathematical texts survive. They evidence a degree of sophistication that progressed somewhat beyond the great mathematical texts of the first half of the second millennium BCE, namely *P. Rhind* and the mathematical *P. Moscow*. Examples of problems treated are as follows: (a) divide 100 by $15^{2}/_{3}$; (b) if a circular piece of land is 100 square cubits large, give the diameter; (c) if a pyramid has a height of 300 cubits and a square base whose side is 500 cubits, give the distance from the center of any side to the apex.

Among astronomical texts, the Stobart Tables record the motions of the planets over a number of years. *P. Berlin* 13146+13147 contains a canon of lunar eclipses. The demotic and hieratic *P. Carlsberg* 1 and 1a of the 2nd c. CE comment on a hieroglyphic astronomical text surviving in two copies dating to about 1300–1150 BCE in the Osireion at Abydos and in the rock tomb of Ramses IV in the Valley of the Kings at Thebes. *P. Carlsberg* 9 provides a simple rule for optimally distributing 309 calendrical lunar months of 29 or 30 days over a cycle of 25 Egyptian years, there being about 9,124.95 days in 309 astronomical lunar months and precisely 9,125 days in both 25 Egyptian years (25 × 365) and 309 calendrical lunar months of which 164 have 30 days and 145 have 29 (164 × 30 + 145 × 30 = 9,125). Horoscopes, of which about eight in Demotic have been published so far (far fewer than Greek specimens from Egypt), are not attested before the Roman period (that is, before 30 BCE). The zodiac was presumably imported from Babylonia not long before the earliest known horoscopes. Much more unpublished astronomical material exists in the Carlsberg Collection in Copenhagen and in *ostraka* excavated at Medinet Madi or ancient Narmuthis, now at the Cairo museum, as well as at the British Museum in London and in Lille and Florence.

For decades, Otto Neugebauer (1899–1990) was the principal and sometimes the sole student of mathematics and astronomy as transmitted in Demotic (on Neugebauer, see *Proceedings of the American Philosophical Society* 137 [1993] 139–65). All his writings on the subject are essential (for a complete bibliography up to the late 1970s, see *Centaurus* 22 [1979] 257–80).

R.A. Parker, *Demotic Mathematical Papyri* (1972); E.A.E. Reymond, "From an Ancient Egyptian Dentist's Handbook. P. Vindob. D. 12287," in: *Mélanges Adolphe Gutbub* (1984) 183–199; A. Jones, "The Place of Astronomy in Roman Egypt," *Apeiron* 27 (1994) 25–51 (with bibliography); Leo Depuydt, "The Demotic Mathematical Astronomical Papyrus Carlsberg 9 Reinterpreted," *Egyptian Religion; The Last Thousand Years* (1998) 1277–1297; Fr. Hoffmann, *Ägypten: Kultur und Lebenswelt in griechisch-römischer Zeit; Eine Darstellung nach den demotischen Quellen* (2000) 103–137, 271–277.

Leo Depuydt

Derkullidēs (*ca* 50 BCE – 120 CE)

Commentator on Plato, apparently specialized in mathematical and astronomical passages. Simplicius (*CAG* 9 [1882] 247–248, 256), quoting Porphurios, mentions the 11th book of Derkullidēs' *The Philosophy of Plato*, where Derkullidēs cites Hermodōros on Plato's "categories." Theōn of Smurna (*Expos.* 198.11–204.21 Hiller; 202.7–204.21 may be Theōn not Derkullidēs) quotes from *The Spindle and the Whorl in Plato's Republic*, an exegesis of *Rep.* 616c–617d which may have been a part of the work mentioned by Simplicius.

Derkullidēs emphasizes the regularity of planetary motion, and rejects eccentrics and **epicycles** introduced, he claims, by ARISTOTLE and the mathematicians MENAIKHMOS and KALLIPPOS. PROKLOS (*in Remp.* 2.24.6–15; 25.14–26) refers twice to Derkullidēs regarding the nuptial number in *Rep.* 546a–d.

ALBINUS *Prol.* 4 links Derkullidēs with THRASULLOS as developing the tetralogic division of Plato's dialogues. Derkullidēs, listed first in Albinus, was usually assumed to be the older of the two. This supposition, together with the interpretation of VARRO *LL* 7.37 as a reference to the tetralogical division, has led scholars to date Derkullidēs to the 1st c. BCE. Tarrant has shown both assumptions to be questionable, meaning Theōn's work is the *terminus ante quem*, as given.

H. Tarrant, *Thrasyllan Platonism* (1983) 11–13, 72–84; *DPA* 2 (1994) 747–8, J. Dillon; *BNP* 4 (2004) 311–312, M. Baltes and M.-L. Lakmann.

Jan Opsomer

Derkullos (250 BCE – 50 CE?)

Lapidary author, considered authentic by Bidez and Schlereth, but fictive by Jacoby, who distinguishes him from the homonymous historian from Argos. PSEUDO-PLUTARCH, however, preserves some fragments from Derkullos' *On stones* 1 (*De fluu.* 19.4 [1162D]), from *On mountains* 3 (*ibid.* 1.4 [1150C] and 8.4 [1155B]), from *Saturika* 1 (*ibid.* 10.3 [1156C]) and from *Aitolika* 3 (*ibid.* 22.5 [1164C]). His name is also mentioned in pseudo-Plutarch *Parall. min.* 17A (*Ktiseis*) and 38B (*Italika*) and in IŌANNĒS "LYDUS," *Mens.* 3.11 (p. 21 Wu.) in relation to the *lukhnis* (ragged robin).

Ph.J. Maussac, "Annotationes in Plutarchum *De fluviis*," in J. Hudson, *Geographiae veteris scriptores Greci minores* 2 (1703) 15; Schlereth (1931) 113; Bidez (1935) 28–29, 31; *FGrHist* 288; De Lazzer (2000) 63; De Lazzer (2003) 80–81.

Eugenio Amato

Derveni papyrus (400 – 300 BCE?)

Found in 1962 at Derveni near Thessalonikē in the remains of a funeral pyre, the date and authorship of the text is debated. (The pyre itself is dated late 4th to early 3rd c. BCE.) Suggested authors are the Epigenēs of DK 36B2 (*cf.* perhaps PLATO, *Phaidōn* 59b), Euthuphrōn of Athens, STĒSIMBROTOS OF THASOS, and Diagoras of Mēlos, but the text was most probably written by an unknown Orphic priest.

In the first extant columns, the author gives a rationalizing interpretation of certain ritual practices. In the main part of the text, he expounds his cosmological theory as a running commentary on a poem – attributed to the mythical poet Orpheus – treating successive divine generations. The theory shows the influence of HĒRAKLEITOS OF EPHESOS, ANAXAGORAS, DIOGENĒS OF APOLLŌNIA, and ARKHELAOS OF ATHENS. **Stoic** influence, also suggested, would however require a significantly later dating.

The author identifies the poem's different divine beings with various cosmic functions of a divine Mind (*cf.* ***kosmos***), the physical manifestation of which is air. The cosmogonic processes are described as the dynamic interplay between intelligent air and fire's brute force. In the pre-cosmic stage, fire completely mingled with the other elements, and its excessive heat did not let independent entities form. When Mind wanted to create the present cosmic order, it removed much fire, forming the Sun from it. As the Sun would have been still too large, Mind dispersed surplus fire into the sky; consequently the stars were born. A sentence seems to say that Mind put the Sun "in the center," but it remains unclear whether we should attribute to the author some version of heliocentrism. Alternatively, the sentence may mean that air keeps the Sun under control by encircling it. The stars are kept in their places by "necessity" so that they do not join the Sun driven by the force of like to like. The Moon derives from a different material, which is not hot. Its function is also teleological, for without the Moon people could not calculate the seasons and winds.

Ed.: R. Janko, "The Derveni Papyrus: an Interim Text," *ZPE* 141 (2002) 11–62; Gábor Betegh, *The Derveni Papyrus* (2004); Th. Kouremenos, G.M. Parássoglou, and K. Tsantsanoglou, *The Derveni Papyrus. Edited with Introduction and Commentary = Studi e testi per il "Corpus dei papiri filosofici greci e latini"* 13 (2006).

Gábor Betegh

Dexios (120 BCE – 25 CE)

Physician, recommended a salve of lime, ***psimuthion***, pine-resin, peppercorns, beeswax and wine, for hardening over the joints (CELSUS 5.18.36). The uncommon name is known from the 5th c. BCE into the 2nd c. CE without geographical or chronological concentration (*LGPN*).

RE 5.1 (1903) 287 (#3), M. Wellmann.

GLIM

Dexippos of Kōs (400 – 360 BCE)

Greek physician, pupil of HIPPOKRATĒS (*Souda* Delta-238; GALĒN *In Hipp. Acute Morb.* 15.478 K.), invited by Hekatomnos, Karian king (*ca* 390 BCE), to treat his sons. He wrote a treatise *On medicine* in one book and *On prognosis* in two books. He defended the theory that drinks and liquefied food go to the lungs and was criticized by ERASISTRATOS (PLUTARCH, *Quaest. Conv.* 7.1.3; Gellius, 17.11) for not understanding the function of the epiglottis, which, according to Dexippos, simply directed the liquids in part to the stomach and in part to the lungs. Erasistratos also criticized him for prescribing fasting and minimal liquids to feverish patients (Galēn *Opt. Sect. Thras.* 1.144 K.; *On Venesection, Against Erasistratos* 9 [11.182 K. = p. 35 Brain]). Dexippos attributes the cause of diseases to digestive residues (LONDINIENSIS MEDICUS 12.8–36), that is to bile and phlegm: for him they are a physiological product that, in continuous change, is transformed into other substances, like sweat serum, mucus, fat etc., but can be impaired by various circumstances, such as unsuitable or excessive food or excessive heat or cold. From these alterations, particularly if blended with blood, there derive a series of four pathological humors: *cf.* Hippokratic treatises, such as *Morb.* I and *Aff.*

RE 5.1 (1903) 294–295, M. Wellmann; Grensemann (1975) 209–214; A. Thivel, *Cnide et Cos?* (1981) 111–114; *BNP* 4 (2004) 330, V. Nutton.

Daniela Manetti

Diadēs (330 – 310 BCE)

Wrote on mechanics and with KHARIAS studied under POLUIDOS, traveled with Alexander for whom he developed a massive "wall-borer" (*truganon*), moved by eight cylinders, featuring an iron-tipped yard-arm with battering ram operated by a system of pulleys and windlasses; he also describes movable towers and a ram-tortoise (Athēnaios Mech. pp. 10–15 W.; Vitruuius 10.13). For the rare name, *cf.* the founder of the Lukian city Dias: Stephanos of Buzantion, s.v.

Irby-Massie and Keyser (2002) 164.

GLIM

Diagoras of Cyprus (220 – 180 BCE?)

Dioskouridēs, *MM* 4.64.6, recounts that he restricted the use of opium because it weakened the senses of sight and hearing (he is listed between Erasistratos and Andreas, yielding the date-range estimate); Pliny 20.200 repeats this, but also records his directions for extracting opium, 20.198, and lists him in his index, 1.*ind.*12–13, 20–21, 34–35. Oreibasios, *Syn* 3.158 (*CMG* 6.3, p.106), and Aëtios of Amida 7.110 (*CMG* 8.2, pp. 375–376), record his rose-collyrium, starting with rose-petals stripped of their whitish bases, antimony, **calamine**, copper flakes, saffron, Indian nard, myrrh, and opium, in gum and rainwater; evidently popular, as Asklēpiadēs, in Galēn *CMLoc* 4.8 (12.767–768 K.), Alexander of Tralleis (2.63 Puschm.), and Paulos of Aigina 7.16.37 (*CMG* 9.2, p. 342), record very similar recipes. Erōtianos Pi-37 (p. 71 Nachm.) records that he called the voluntary nerves *peronas* ("pins").

RE 5.1 (1903) 311 (#3), M. Wellmann.

PTK

Diagoras of Mēlos ⇒ Derveni Papyrus

Dicaearchus ⇒ Dikaiarkhos

Didumos "the music theorist" (*ca* 60 CE)

Musical writer whose book *On the Difference Between the Aristoxenians and the* ***Pythagoreans*** (*Peri tēs diaphoras tōn Aristoxeneiōn te kai Puthagoreiōn*) is quoted by Porphurios in his commentary on Ptolemy's *Harmonics* (26.2–29, 27.17–28.6 Düring). Ptolemy himself preserves and critiques a set of tetrachordal divisions worked out by Didumos (*Harm.* ii.13–14); these are important for their strict adherence both to certain core **Pythagorean** mathematical principles and to key structural features of Aristoxenos' tetrachordal divisions. Ptolemy also reports that Didumos made certain improvements to the monochord to facilitate the playing of melodies on the instrument (one not well suited to musical performance); his intention may have been to demonstrate in a melodic context the scales of Aristoxenos' age, newly "translated" into ratios (see Barker 1994).

Didumos' work may have been both Ptolemy's and Porphurios' immediate source for the harmonic writings of Arkhutas, Eratosthenēs (for both of whom Ptolemy transmits tetrachordal divisions) and Ptolemaïs of Kurēnē (whom Porphurios quotes immediately before his excerpts from Didumos) – indeed, Porphurios goes so far as to claim that Ptolemy plagiarized the greater part of his material from Didumos' book. Although Porphurios' own quotations from Didumos do not adequately support the accusation, the fact that Porphurios could make it at all suggests that Didumos' work was both substantial and ambitious, and it is clear that in many respects Ptolemy's treatise owes much to Didumos.

Like Ptolemaïs, Didumos divided his predecessors according to the methodological criterion of relative emphasis on either sense-perception (*aisthēsis*) or reason (*logos*), with the **Pythagoreans** inclined to favor the latter when the two appeared to disagree, and the Aristoxenians requiring that all conclusions be brought to the former for final approval. In light of this and of his improvements to the monochord, it is noteworthy that one of Ptolemy's central criticisms of Didumos' tetrachordal divisions was that they would not stand the test of musical perception.

He may be identical with the Neronian Didumos of *Souda* Delta-875, son of Hērakleidēs (perhaps of Hērakleia Pontikē, Junior), a grammarian, practicing musician and composer. He may also be same as both the Didumos whose book *On* ***Pythagorean*** *Philosophy* (*Peri Puthagorikēs philosophias*) is mentioned by Clement of Alexandria (*Strom.* 1.16), and the Didumos whose definition of rhythm is quoted in the musical treatise of Bakkheios (313.9 Jan).

RE 5.1 (1903) 473–474 (#11), L. Cohn; I. Düring, *Die Harmonielehre des Klaudios Ptolemaios* (1930); *Idem* (1932); Barker (1989); M.L. West, *Ancient Greek Music* (1992); A.D. Barker, "Greek Musicologists in the Roman Empire," in T.D. Barnes, ed., *The Sciences in Greco-Roman Society* = *Apeiron* 27 (1994) 53–74; *OCD3* 468, A.D. Barker; Mathiesen (1999); A.D. Barker, *Scientific Method in Ptolemy's "Harmonics"* (2000); *BNP* 4 (2004) 398 (#1), F. Zaminer and 399 (#4), F. Montanari; *NDSB* 2.284–286, E. Rocconi.

David Creese

Didumos of Alexandria (I: Metrol.) (30 BCE – 30 CE)

The few fragments of Didumos' metrological work are mostly transmitted as Hērōn's. Didumos' *On the Measures of Marble and Wood of All Sorts* is a concise technical guide to estimating the volume of blocks of stone or timber, along with a table of correspondence of both Ptolemaic and Roman basic units of measurement, to be used for surface areas.

MSR v.1 (1864) 21–23, 180, v.2 (1866) 22–23; Hultsch (1882) 9, 609–610; *RE* 5.1 (1903) 474 (#12), *Idem.*

Mauro de Nardis

Didumos of Alexandria (II: Agric.) (*ca* 350 – 450 CE)

Author of *Geōrgika* (in 15 books, known from the *Souda* Delta-876), used extensively by Cassianus Bassus: parts of his *Eklogai* taken from Didumos concern astronomy and medicine as well as fables of metamorphoses, making likely an identification with the homonymous man whose *Phusika* is cited by Alexander of Tralleis (*Therap.* 7.13); there are similarities as well with the *Iatrika* of Aëtios of Amida.

Oder (1890) 212–222.

Robert H. Rodgers

Didumos of Knidos (250 BCE – 100 CE)

Wrote a commentary on Aratos (*FGrHist* 1026 T19), entirely lost.

(*)

PTK

Dieukhēs (300 – 200 BCE)

Greek physician, Noumēnios of Hērakleia's teacher (Ath., *Deipn.* 1 [5b]). Galēn often associated Dieukhēs with Mnēsitheos and Dioklēs of Karustos as one of the great historical **Dogmatic** physicians (*MM* 1.3.13 [10.28 K. = p. 15 Hankinson], *On Venesection, Against Erasistratos* 5 [11.163 K. = p. 25 Brain], etc.). A votive inscription seemingly confirming the association between Dieukhēs and Mnēsitheos quotes both their names (or of members of their families: *IG* II2 1449, 350 BCE). Dieukhēs applied to the human body the "Hippokratic" four qualities theory (Galēn *MM* 7.3 [10.462 K.]), practiced bloodletting (Galēn 11.163 K.) and used cataplasms of hellebore (OREIBASIOS *Coll.* 7.26.196 [*CMG* 6.1.1, p. 245]). PLINY quotes him five times as an authority for many treatments (20.31, 78, 191, 23.60, 24.145). Oreibasios also quotes Dieukhēs (from a work perhaps entitled *On the preparation of bread*) prescribing special food for patients and maritime travelers, recommending vomiting and light food.

Ed.: J. Bertier, *Mnésithée et Dieuchès* (1972).
BNP 4 (2004) 404–405, V. Nutton.

Daniela Manetti

Dikaiarkhos of Messēnē (Sicily) (340 – 290 BCE)

Son of Pheidias, student of ARISTOTLE and student or colleague of THEOPHRASTOS, dwelt in the Peloponnesus, wrote on the Spartan constitution, *Life of Greece* (a "biography" of the Greek people explaining civilization on naturalistic grounds), doxographical biographies including THALĒS, PYTHAGORAS, and PLATO, plus literary criticism and political philosophy. In *On the Soul* he denied there was a soul separable from the mortal and material body.

Ca 305 BCE, supported by kings Kassandros of Macedon and Ptolemy of Egypt, and employing the method of EUCLID, *Optics* 19, he measured the heights of mountains, to show that even tall peaks did not significantly affect the sphericity of the Earth. (Probably he was refuting EPICURUS' revival of ANAXIMENĒS' hypothesis that the Sun sets by moving behind high mountains at the edge of the flat Earth; such high peaks were still part of Platonic and Aristotelian spherical-Earth models: *Meteor.* 1.13 [350a28–35].) An un-attributed measurement of the Earth's circumference as 30 myriad stades (ARCHIMĒDĒS, *Aren.* 1.8, KLEOMĒDĒS 1.5.57–75) was probably part of the same work. In *Circuit of the Earth*, he drew maps of the oblong known inhabited world and established a non-equatorial central latitude, running through Sardinia, Sicily, the Peloponnesus, Karia, Lukia, the Tauros Mts., and some western part of the Himalayas; his data west of Sicily were not very reliable. In the same book, he explained the rise of the Nile as an influx from the Atlantic (via a west African river) – much as had HOMER, *Iliad* 21.194–197, and EUTHUMENĒS – presumably assuming the Atlantic was higher than the Mediterranean (*cf.* Aristotle, *Meteor.* 1.14 [352b22–31], *HA* 8.13 [598b15–18], STRATŌN in STRABŌN 1.3.4–5, and PSEUDO-SKULAX 20). He advocated a solar-attraction theory of tides.

VARRO's *Life of the Roman People* probably owed much to Dikaiarkhos' *Life of Greece*; Varro praised him as "most learned," as did CICERO and PLINY, both of whom utilized his works. PHILODĒMOS used him for doxography, PLUTARCH and PORPHURIOS for his record of Greek ways.

W.W. Fortenbaugh and E. Schütrumpf, edd., *Dicaearchus of Messana: Text, Translation, and Discussion* = *RUSCH* 10 (2001), esp. P.T. Keyser, "The Geographical Work of Dikaiarchos," 353–372.

PTK

Dimensuratio Prouinciarum and Diuisio orbis terrarum (400 – 500 CE?)

Two brief geographical works in Latin giving the names, boundaries, and dimensions of the provinces of the Roman Empire and of some regions beyond it, such as India. The *Dimensuratio* also lists islands. The works are independent of one another but based on the same sources. Most scholars believe that they ultimately go back to AGRIPPA's survey.

Ed.: *GLM* 9–20.
RE 5.1 (1903) 647 and 1236–1237, G. Wissowa; *KP* 2.33–34, A. Lippold; *BNP* 4 (2004) 418, K. Brodersen.

Natalia Lozovsky

DINŌN OF KOLOPHŌN ⇒ DEINŌN OF KOLOPHŌN

DIOC- ⇒ DIOK-

Diodōros (Astron.) (150 BCE – 250 CE)

Geometer and astronomer, possibly identifiable with a commentator on ARATOS named Diodōros, and with a Diodōros of Alexandria who wrote on astronomical topics. The geometer Diodōros authored a treatise *On Analemma* concerning the theory underlying sundials. PAPPOS wrote a lost commentary on this work, and PROKLOS speaks of Diodōros as one of the earlier writers on sundials. Diodōros' method of determining the cardinal directions from three measured shadows (a problem equivalent to finding the axis of a hyperbola given three points on the curve and one on the axis) is reported by HYGINUS GROMATICUS, Abu Sa'id ad-Darir, and al-Biruni. We also know from Pappos' reference that *On Analemma* contained a construction requiring the trisection of a given angle. According to al-Nairizi, the commentator on EUCLID's *Elements*, Diodōros attempted a geometrical demonstration of Euclid's fifth postulate, perhaps in a different book.

Diodōros, the commentator, is cited intermittently in the *scholia* to Aratos' poem; he criticized **Stoic** interpreters such as KRATĒS as well as HIPPARKHOS, and was in turn attacked by an otherwise unknown Dōsitheos. Diodōros of Alexandria figures, together with METŌN, EUDOXOS, Hipparkhos, and LASOS, in an anonymous list of astronomers, and he is cited by AKHILLEUS and MACROBIUS on the distinction between mathematical and physical astronomy, the meaning of the words ***kosmos*** and "star" (*astēr*), and the nature of the Milky Way.

Ed.: D.R. Edwards, *Ptolemy's Περὶ ἀναλήμματος – An Annotated Transcription of Moerbeke's Latin Translation and of the Surviving Greek Fragments with an English Version and Commentary* (1984) 152–182.
Neugebauer (1975) 840–843; *NDSB* 2.304–305, J.L. Berggren.

Alexander Jones

Diodōros (Metrol.) (350 – 410 CE)

All that survives of Diodōros' possibly metrological work *On weights* is a very short table of the equivalence rates between a talent and its parts, namely mina, drachma, obol and *khalkós*, the bronze coin. Diodōros' table possibly reflects contemporary variations in exchange rates between imperial coins.

Ed.: *MSR* 1 (1864) 156–157, 299–300.
Hultsch (1882) 8, 339–340; *RE* 5.1 (1905) 712, *Idem; BNP* 4 (2004) 445 (#19), M. Folkerts.

Mauro de Nardis

Diodōros (Empir.) (1st c. BCE)

The Diodōros mentioned by Galēn in his list of **Empirical** physicians in *MM* (10.142 K.) is probably to be identified with the Diodōros author of *Empirica* mentioned by Pliny for a property of basil (20.119), and as a source for Books 29–30 (and for a remedy at 29.142). Other remedies are mentioned by Galēn at *CMLoc* 5.3 (12.834 K.: via Kritōn), 9.2 (13.248 K.) and 10.3 (13.361 K.: via Asklēpiadēs Pharm.) and *CMGen* 5.15 (13.857 K.). It is possible that he belonged to the school of Hērakleidēs of Taras.

Ed.: Deichgräber (1930) 203–204 (fragments), 261.
RE 5.1 (1901) 708 (#50), M. Wellmann.

Fabio Stok

Diodōros of Antioch ⇒ Diodōros of Tarsos

Diodōros of Ephesos (*ca* 400 BCE – *ca* 200 CE)

Wrote on Empedoklēs' imitation of Anaximandros' pomposity: Diogenēs Laërtios, 8.70.

FGrHist 1102.

PTK

Diodōros of Eretria (*ca* 400 – 350 BCE)

Hippolutos, *Ref.* 1.2.12–13, cites Diodōros, along with Aristoxenos, for the tale that Pythagoras learned dualism from Zoroaster. Probably, Aristoxenos cited Diodōros.

FGrHist 1103.

PTK

Diodōros of Iasos, "Kronos" (*ca* 320 – 284 BCE)

Lived in both Athens and Alexandria (the nickname Kronos, "Old Fool," inherited from his teacher Apollōnios Kronos), was the teacher of Zēnōn of Kition and the logician Philo, and was known as a brilliant dialectician. Recently some have seen him as a leading member of a school called Dialectical, in contrast to his traditional placing among the Megarian school; but the evidence is inconclusive. He was important for a view of the truth-conditions for conditionals, for his arguments against motion, and for the so-called Master Argument, which defines the possible as what is or will be the case, blurring the distinction between a fixed or necessary past and a future open with possibilities; it was highly influential on Hellenistic debates on necessity and possibility. Regarding motion, he argued that it can never truly be said that something *is moving*, but only that something *has moved* (Sextus *Adv Math.* 10.85–87); the arguments employ a conception of motion and time as consisting of minimal partless units, as opposed to continuous and in principle infinitely divisible (as Aristotle held: *Physics* 6.1–2 [231a21–233b32]).

Long and Sedley (1987) 1.44, 51–52, 209–210, 230–236; *OCD3* 472, D.N. Sedley; *ECP* 185–187, M. White.

Richard Bett

Diodōros of Priēnē (325 – 90 BCE)

Wrote an agricultural work, possibly treating cereals, livestock, poultry, viticulture, and arboriculture (*cf.* PLINY, 1.*ind*.8, 10, 14–15, 17–18), excerpted by CASSIUS DIONUSIOS (VARRO, *RR* 1.1.9–10, *cf.* COLUMELLA, 1.1.9).

RE 5.1 (1903) 708 (#49), M. Wellmann.

Philip Thibodeau

Diodōros of Samos (*ca* 100 BCE – 50 CE?)

Wrote a geographical work giving astronomical data about the sea-voyage to India, according to MARINOS OF TYRE in PTOLEMY, *Geog.* 1.7.

RE 5.1 (1903) 704–705 (#39), H. Berger.

PTK

Diodōros of Sicily (*ca* 80 – *ca* 20 BCE)

Greek historian from Agurion, Sicily, composed a universal history entitled *Bibliothēkē* some time in the late 40s. He said that he worked on the project for 30 years, for this purpose visiting extensive regions of Asia and Europe. He met certain Romans in Sicily, learned Latin, and thus could study Roman historical records (1.4.1–4). The first six books of Diodōros' history, treating the time before the Trojan War, comprise three books on the barbarians and three on the Greeks; the next 11 books explicate events from the Trojan War to the death of Alexander; the last 23 books survey events up to the Gallic War and CAESAR's arrival in the British Isles (1.4.6–7). Books 1–5 and 11–20 survived intact, but the rest are fragmentary. According to Greek historiographical tradition, Diodōros sets his universal history against a geographical and ethnographical background, describing regions and their inhabitants, particularly Greece, Sicily, Rome and the surrounding areas. Some modern scholars seem to have misjudged him as a mere compiler, for he relies heavily on whole parts of earlier works, including HEKATAIOS OF ABDĒRA on Egypt, MEGASTHENĒS on India, and AGATHARKHIDĒS on the Arabian Gulf. However, the vastness of his project with its clear editorial plan resulted in an important even if partially preserved work recording significant and unusual historical and geographical information. Stylistically, Diodōros tended towards the grandiose and the fantastic when referring to fauna, flora, and other natural phenomena.

K.S. Sacks, *Diodorus Siculus and the First Century* (1990); D. Ambaglio, *La Biblioteca Storica di Diodoro Siculo: Problemi e Metodo* (1995), esp. 59–82; P.J. Stylianou, *A Historical Commentary on Diodorus Siculus Book 15* (1998).

Daniela Dueck

Diodōros of Tarsos (365 – 393 CE)

Monk, born in Antioch to a noble family, lived under Julian and Valens, studied philosophy at Athens, banished to Armenia (372), was appointed bishop of Tarsos and Kilikia (378), and taught IŌANNĒS KHRUSOSTOMOS among others. An ally of BASIL OF CAESAREA,

Diodōros wrote on theology and scriptural exegesis (fragments of his commentaries in *PG* 33.1546–1628); he attacked PLATO and PORPHURIOS.

Phōtios (*Bibl.* 223), calling Diodōros knowledge-loving and wise and his arguments clear and rigorous, epitomized his *Against Destiny*, directed against BARDAISAN, including his theories of cosmogony, astrology, astronomy, ethnology, and zoology. Diodōros, Book 2, contended that the entire ***kosmos*** and its matter were created, perishing and coming to be again and again, and that all is governed by the creator (209a–210a). In Book 3, he argued against a spherical heaven (210ab), and against geographical astrology, since each sign would affect all regions alike (210b–211a); he disparaged astrological explanations of planetary retrograde motion, the variable apparent size of the Moon, and the relative sizes of the Sun, Moon, and Saturn (211b). Diodōros, in Book 4, used the Sun to explain the Earth's climatic zones, discounting the effects of astral movement [212b], and considered Greek astrology inadequate to explain the diversity of colors, figures, and qualities of living things [213b], much less (Book 5) explain the course of a human life [214a] or even the color-changing chameleon that harmonizes itself to its environment [215b]. In Book 6 (217b–218b) Diodōros advanced arguments regarding causality, Book 7 (218b–220b) concerned the problem of evil, and Book 8 (220b) attempted a non-spherical cosmology based on Hebrew scripture (*cf.* LACTANTIUS and KOSMĀS).

The *Souda* (Delta-1149) preserves numerous other titles surviving in scant fragments, including the following nine "scientific" treatises (many of which may be parts of a larger work on natural philosophy): *On the Sphere and the Seven Zones and of the Contrary Motion of the Stars, On Hipparkhos' Sphere, On Nature and Matter, That the Unseen Natures are not from the Elements but Were Made from Nothing along with the Elements, Against Aristotle concerning Celestial Body, How Hot is the Sun, Against Those Who Say the Heaven is a Living Being, On the Question of How the Creator is Forever but the Created is Not, Against Porphurios about Animals and Sacrifices.*

ODB 626–627, B. Baldwin and A. Kazhdan.

PTK and GLIM

Diodotos (Astr. I) (95 – 60 BCE?)

Wrote a lost commentary on ARATOS (*FGrHist* 1026 T19); perhaps the brother of BOËTHOS OF SIDŌN (PERIP.) (STRABŌN 16.2.24), or else the **Stoic** teacher of CICERO.

Maass, *Aratea* (1892) 159; *RE* 5.1 (1903) 715 (#11), von Arnim, (#12) E. Martini.

PTK

Diodotos (Astr. II) (190 – 230 CE)

Called the foremost *astrologos* of his time by ALEXANDER OF APHRODISIAS, who reports his denial that reflection from the **Peripatetic** exhalation could produce an image of the Sun that would appear as a comet in the southern sky (*In Meteor.* 1.6, *CAG* 3.2 [1899] 28).

RE 5.1 (1903) 715 (#16), F. Boll; Wilson (2008).

PTK

Diodotos (Pharm.) (10 – 30 CE)

Among the **Asklēpiadeans** mentioned by DIOSKOURIDĒS (*pr.*2) distinguished by their "vain prating about causation [who] have explained the actions of an individual drug by differences among particles [and] confusing one drug for another." Diodotos wrote in iambic trimeters on medical botany and ointments rich in exotic oils, but the title of the poem is uncertain. PLINY (20.77) calls it *Anthologoumena*, while ERŌTIANOS cites Book 2 of a *Murologiōn* [N-4, s.v. *niōpon*, p. 62 Nachm.]). Erōtianos carefully separates the writings of PETRONIUS from those by Diodotos, but Pliny (20.77, 25.110) confuses the two, causing much perplexity among scholars who have presumed a "Petronius Diodotos," when they are two separate but roughly contemporary writers.

Scarborough and Nutton (1982) 205–206.

John Scarborough

Diogās "the anointer" (30 BCE – 95 CE)

ASKLĒPIADĒS PHARM., in GALĒN, *CMLoc* 7.5 (13.104 K.), records that Diogās the *iatraleiptēs* employed the "panacea" of ANTONIUS MUSA. He may merely have been a practitioner; the name is rare, attested from the Cimmerian Bosporos (*LGPN* 4.98).

Nutton (1985) 145.

PTK

Diogenēs (Geog.) (*ca* 50 BCE – 50 CE?)

Described his return from India and subsequent voyage down the east coast of Africa, as recorded by MARINOS OF TYRE in PTOLEMY, *Geog.* 1.9; *cf.* DIOSKOROS and THEOPHILOS.

RE 5.1 (1903) 763–764 (#41), H. Berger.

PTK

Diogenēs (Pharm.) (10 BCE – 30 CE)

ASKLĒPIADĒS PHARM., in GALĒN *CMLoc* 3.1 (12.686 K.), records his remedy based on lanolin, **terebinth**, and rose oil for nasal ulcers, and, *ibid.* 9.7 (13.313–314 K.), a recipe for hemorrhoids (reduce, in a bronze pot, cyclamen juice and honey to the consistency of beeswax). CELSUS 5.19.20 (*cf.* 5.27.1A) credits presumably the same man with a black bite-plaster (bitumen, beeswax, pitch, **litharge**, and olive oil). AËTIOS OF AMIDA 3.111 (*CMG* 8.1, pp. 301–302) records a purifying phlegmagogue of ***euphorbia*** (*cf.* IOUBA), pepper, and sal ammoniac in raw egg. Aëtios 2.30 (*CMG* 8.1, p. 166) cites "Diogenēs" in (scribal?) error for DAMIGERŌN.

Fabricius (1726) 142.

PTK

Diogenēs Laërtios (150 – 250 CE)

The most important example of historiography of philosophy from antiquity is Diogenēs Laërtios' *Compendium of the Lives and Opinions of Philosophers*. The author is not otherwise known, and his work can only be dated from a combination of the latest personalities he mentions, and the fact that he was not yet influenced by Neo-**Platonism**. His work is divided into ten books: (1) introduction and various wise men (including THALĒS), The Ionian Tradition (Books 2–7): Ionian physicists, Sōcratēs and the minor Socratics (2), PLATO (3), the **Academy** down to KLEITOMAKHOS (4), ARISTOTLE and the **Peripatetics** down to LUKŌN (5), ANTISTHENĒS and the Cynics (6), ZĒNŌN OF KITION and the **Stoics** (7: possibly down to the 1st c. CE: the end of the book is lost); the Italic Tradition (Books 8–10): PYTHAGORAS and his early successors, and EMPEDOKLĒS (8), HĒRAKLEITOS, the Eleatics, the **Atomists**, PRŌTAGORAS, DIOGENĒS OF APOLLŌNIA, and Pyrrho (9), and finally EPICURUS (10).

Diogenēs' book is basically a compilation of excerpts from a large number of sources and often provides us with the main evidence for the Hellenistic tradition. The qualities and the structure are very uneven: some lives are nothing but anecdotes and aphorism while others are primarily doxographies. Some lives have important sections on philosophy (e.g. the Lives of Zēnōn in Book 7 and Pyrrho in Book 9) while others are of no help in reconstructing the philosophy of a thinker (e.g. Plato in Book 3 and Aristotle in Book 5; in both cases, however, they exemplify how later generations interpreted their predecessors). For the Pre-Socratics, Diogenēs used a doxographical source of a type also found in other late sources (e.g. HIPPOLUTOS).

Most of Diogenēs' biographies included a number of items like birth, parents, name, appearance, relations to other philosophers, travels, life style, and manner of death; many lives also contain bibliographies and some pieces of documentary evidence. The comprehensiveness of a life depends on the number of anecdotes available, but the factual information must always be viewed with skepticism and there are many obvious mistakes.

Ed.: M. Marcovich, *Diogenes Laertius Vitae Philosophorum* (1999); M.-O. Goulet-Cazé, *Diogène Laërce: Vies et doctrines des philosophes illustres* (1999), with intr. and commentary.

Mejer (1978); *Elenchos* 7 (1986) and *ANRW* 2.36.5–6 (1991–1992) contain many articles on Diogenēs; R. Goulet, *Études sur les vies de philosophes dans l'antiquité tardive. Diogène Laërce, Porphyre de Tyr, Eunape de Sardes* (2001).

Jørgen Mejer

Diogenēs of Apollōnia (*ca* 445 – 425 BCE)

Followed the long-deceased ANAXIMENĒS in making air the basic reality. Often thought eclectic, he was influenced both by ANAXAGORAS and LEUKIPPOS. He may, however, have given the first argument for material monism, the view that all things are made up of one kind of matter (*fr.*2): if there were not some common principle in different things, they would not be able to interact. Diogenēs held that air is the ultimate reality and the source of intelligence in living things, as can be seen by the fact that animals must breathe to live. The intelligence in air, responsible for the orderly nature of the world, controls all things and is to be identified with God. Diogenēs developed a cosmology similar to Anaxagoras', with stony bodies carried around the Earth. He gave detailed account of the circulatory system as a means of distributing air throughout the body, and an account of perception as resulting from air. His views seem to have been popular among intellectuals and to have been parodied by Aristophanēs in *The Clouds* of 423 BCE. Some scholars have claimed he stressed the purposiveness of the world and invented the first argument for the existence of God from design. But he may have used the orderliness of the world only to argue for the ascendancy of air.

DK 64; KRS 434–452; A. Laks, *Diogène d'Apollonie* (1983).

Daniel W. Graham

Diogenēs of Babylōn (*ca* 200 – 150 BCE)

Born *ca* 240 BCE, fifth head of the **Stoa**; student of CHRYSIPPUS and Zēnōn of Tarsos, teacher of PANAITIOS and ANTIPATROS OF TARSOS, among others. Diogenēs was instrumental in introducing **Stoicism** to Rome during his visit in 156–155 BCE. He contributed to political theory, ethics, logic, and the theory of language, among other subjects. His work on music was largely concerned with questions of moral education. GALĒN (*PHP* 2.5.9, 2.8.40 = *CMG* 5.4.1.2, pp. 130, 138) cites Diogenēs' defense of Chrysippus' theory that the ***hēgemonikon*** is located in the heart. Diogenēs wrote a work *On Divination* in which he argued for the Chrysippean position that if there are gods, then there is divination, a position also held later by Antipatros of Tarsos. In his *On Voice*, Diogenēs claimed that sound was corporeal and that it was a percussion of the air, one effected by impulse in animals but by reason in people, at least after age 14. Diogenēs followed Chrysippus in dividing language into only five parts: names, common nouns, verbs, conjunctions, and articles, though Diogenēs may have given a fuller explanation of these terms than Chrysippus did insofar as DIOGENĒS LAËRTIOS (7.58) cites Diogenēs of Babylōn rather than Chrysippus for definitions of the parts of speech. New work on the Herculaneum papyri (as-yet unpublished) is raising the possibility that Diogenēs may have been responsible for other important developments in **Stoicism**.

Ed.: *SVF* 3.210–243.

D. Obbink and P.A. Vander Waerdt, "Diogenes of Babylon: the Stoic sage in the city of fools," *GRBS* 32 (1991) 355–396; J. Brunschwig, "Did Diogenes Invent the Ontological Argument?" in *Idem*, ed., *Papers in Hellenistic Philosophy* (1994) 170–189.

Daryn Lehoux

Diogenēs of Oinoanda (*ca* 120 – 200 CE)

Epicurean philosopher: little is known of his life, other than what he tells us in the massive **Epicurean** inscription he erected for his fellow citizens sometime in the 2nd c. CE. He was near the end of his life when he had the inscription erected in a prominent *stoa* in Oinoanda; the cost of such an undertaking suggests he was wealthy. Fragments of the inscription first came to light in 1884. Only about one third of the whole has been recovered (now about 200 fragments, but more pieces continue to come to light), and the reconstruction of many fragments and their original arrangement on the wall is still disputed. The inscription, arranged in a number of rows and columns on a wall some 80–100 meters in length, was erected by Diogenēs, he tells us, as a benefit for the citizens of Oinoanda and for visitors to the city. Topics Diogenēs treated were **Epicurean** ethics (on the lowest level), **Epicurean** physics (on the next level), and old age (on the top level). He also included letters to friends (*Letter to Antipatros*, *Letter to Dionusios*), *Maxims*, and *Directions to Family and Friends*. In the section on ethics, he treats the goal of life, the nature of happiness, pleasure, pain, fear, the nature of death, mortality of the soul, dreams, necessity and freedom of action, and an **Epicurean** golden age. In the section on physics he attacks the physical theories of earlier philosophers, including DĒMOKRITOS, and discusses images, dreams, the development of human society, astronomy, meteorology, and the nature of the gods. The inscription provides additional evidence for many of EPICURUS' doctrines known from other sources, and a unique glimpse into the state of **Epicurean** philosophy in the 2nd c. CE.

Ed.: M.F. Smith, *Diogenes of Oinoanda: The Epicurean Inscription* (1993); *Idem, Supplement to Diogenes of Oinoanda: The Epicurean Inscription* (2003).

D. Clay, "The Philosophical Inscription of Diogenes of Oenoanda," *ANRW* 2.36.4 (1990) 2446–2559, 3231–3232; P. Gordon, *Epicurus in Lycia* (1996); *OCD3* 474, D. Konstan; *ECP* 191–192, D. Clay; *REP* 3.89–90, M. Erler; *BNP* 4 (2004) 455–456 (#18), T. Dorandi.

Walter G. Englert

Diogenēs of Tarsos (*ca* 150 – 100 BCE)

Epicurean philosopher, probably identical with the Diogenēs of Tarsos mentioned in Strabōn (14.5.15), and thus to be dated to the second half of the 2nd c. BCE. Little is known about his life, but he is said to have traveled widely giving lectures on **Epicurean** topics. These were gathered together in a collection of at least 20 books entitled *Select Lectures* (*epilektoi skholai*), and were a source for Diogenēs Laërtios' account of **Epicurean** philosophy. Topics in the *Select Lectures* included possible causes of lunar eclipses (D.L. 10.97), the non-divine nature of love (D.L. 10.118), and the existence of static and kinetic pleasures in the body and soul (D.L. 10.136). He also argued, following Epicurus, that one should choose virtue for the sake of pleasure (D.L. 10.119), and that the **Epicurean** sage will feel grief (D.L. 10.119). In another of his works, an *Epitome of Epicurus' Ethical Doctrines*, he asserted that the **Epicurean** sage would not have illicit sexual relations (D.L. 10.118). He also composed poetry, for the most part tragedies (Strabōn 14.5.15).

OCD3 474, D. Konstan; *ECP* 196, S.A. White; *BNP* 4 (2004) 452 (#16), T. Dorandi, and 456 (#20) F. Pressler.

Walter G. Englert

Diognētos (of Eruthrai?) (335 – 305 BCE)

Recorded the itinerary of Alexander, giving data on peoples and plants along the route, preserved only in Pliny 6.61; *cf.* Baitōn and Philōnidēs of Khersonēsos.

FGrHist 120.

PTK

Diognētos of Rhodes (310 – 300 BCE)

Engineer at Rhodes displaced by Kallias of Arados, then rehired, after which he succeeded in stopping the ***helepolis*** of Epimakhos of Athens, by soaking the ground in its path (Vitruuius 10.16.3–7).

RE 5.1 (1903) 786 (#19), E. Fabricius

PTK

Diokleidēs of Abdēra (285 – 220 BCE)

Described the ***helepolis*** built by Epimakhos of Athens for Dēmētrios besieging Rhodes, according to Moskhiōn in Ath., *Deipn.* 5 (206d). Since Moskhiōn also cites the historians Timaios of Tauromenion, Hierōnumos of Kardia, and Polukleitos of Larissa, perhaps Diokleidēs was a historian, not a technical writer (omitted by Jacoby, *FGrHist*).

RE 5.1 (1903) 791 (#3), E. Schwartz.

PTK

Dioklēs (*ca* 200 – 175 BCE)

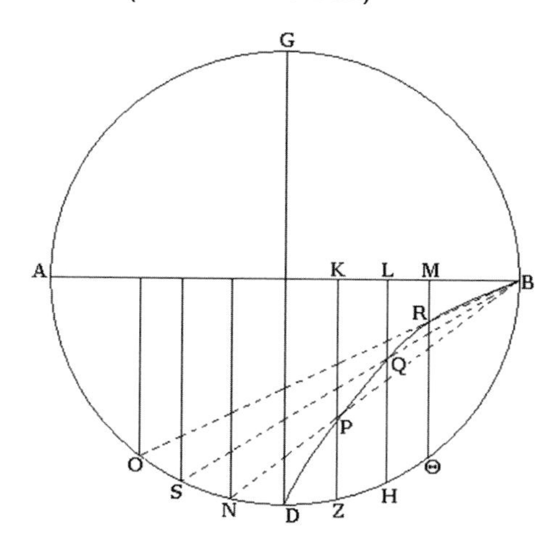

Dioklēs' Cissoid. Given: circle ABDG with perpendicular diameters AB and DG, where arc DZ=ZH=HΘ=DN=NS=SO, and KZ, LH and MΘ are perpendicular to AB. Line DPQRB is a cissoid, and ZK and KB are two mean proportionals between AK and KP, so that AK : ZK = ZK : KB = KB: KP. We can then (if we like) work out that for the special case where AK is twice the length of KP, then a cube whose sides equal line KP will be doubled by building a cube whose sides equal line KB. © Lehoux and Massie

Mathematician, probably lived in Arkadia for a while, as he mentions (in the introduction to his one surviving work) another mathematician visiting him there. Fragments of Dioklēs had long been known from EUTOKIOS' commentary on ARCHIMĒDĒS' *On the Sphere and Cylinder*, but until Toomer's 1976 publication of the then-newly-found Arabic translation of Dioklēs' *On Burning Mirrors*, no full text survived. *On Burning Mirrors* treats generally, but not exclusively, the geometry of conic sections. Dioklēs presents some original work on focal properties of parabolas, including the construction of a parabola with any given focal length (leading Toomer to attribute to Dioklēs the first construction of a parabola from focus and directrix). Dioklēs next turns to a problem posed by Archimēdēs: dividing a sphere such that the two segments bear a given ratio to each other, which Dioklēs solves by means of intersecting an ellipse and an hyperbola. Finally, he addresses the classic question of doubling the cube, solved in two ways. His first solution employs the intersection of two parabolas and his second a cissoid, to find two mean proportionals between two given magnitudes. *Cf.* "DTRUMS" (probably later).

Ed.: Toomer (1976); Rashed (2000) 3–151.
W.R. Knorr, "The Geometry of Burning Mirrors in Antiquity," *Isis* 74 (1983) 53–73.

Daryn Lehoux

Dioklēs of Karustos (400 – 300 BCE)

Life. Son of Arkhidamos, one of the most celebrated physicians and medical writers in antiquity: regularly described by ancient sources as a **Dogmatist** (*fr.*2: CELSUS, and later). He is sometimes said to be HIPPOKRATĒS' pupil or follower (*frr.*3, 40), to have come second in age and fame only to Hippokratēs (*fr.*4), and was called a younger Hippokratēs by the Athenians (*fr.*3). Since Dioklēs is thought to have written the first systematic handbook on anatomy (*fr.*17), Wellmann dated him to the first half of the 4th c. BCE. W. Jaeger's later date, *ca* 340–260, making Dioklēs ARISTOTLE's younger contemporary (and his pupil), rests on controvertible evidence (e.g. a probably spurious dietetic letter transmitted by PAULOS OF AIGINA, *fr.*183a). Thus, it is only certain that Dioklēs lived after Hippokratēs and presumably somewhat earlier than HĒROPHILOS and ERASISTRATOS. His relationship to his predecessors and Aristotle or the **Peripatos** remains obscure: the Dioklēs quoted in THEOPHRASTOS' *On Stones* (*fr.*239a) is not necessarily the Karustian, and is not mentioned elsewhere in Aristotle's or Theophrastos' treatises. On the other hand similarities exist between some of Dioklēs' ideas and some of those developed by Aristotle and his circle as well as by some Hippokratic writings of MNĒSITHEOS and PRAXAGORAS.

Writings. Dioklēs' writings are encyclopedic in range: at least 20 titles (some in four or more books) on subjects as varied as prognostics (*On prognosis*), physiology (*On digestion*), diagnostics and therapy (*Affection, Cause, Treatment* and *On treatments*), gynecology (*Matters related to women*), bandages (*On bandages*), surgery (*On things in the surgery*), regimen in health (*Matters of Health to Pleistarkhos*, his most influential work), food and wines, herbs (*On Rootcutting*), olive oil for massage (*Arkhidamos*), drugs, poisons (*On lethal drugs*), etc. Among many strictly technical treatises devoted to particular fields of medicine (e.g., surgical instruments or bandages), some of his works (*Matters of Health to Pleistarkhos* and *Arkhidamos*) apparently addressed a larger non-specialist audience. His books, in Attic dialect, long remained available: GALĒN read first-hand *Arkhidamos* and *Affections, Cause, Treatment*, plus *Matters of Health*, of which he knew variant editions in circulation (*fr.*188); *On things in the surgery* circulated in Galēn's time under various titles (*fr.*160a). Even OREIBASIOS probably compiled his medical encyclopedia copying directly from Dioklēs' works.

Doctrine. Dioklēs generally agrees with Hippokratēs on most important issues (he uses notions such as ***krisis***, "resolution," "concoction," **humors**), and his doctrines are frequently conflated with those of other ancient authorities, making it difficult to reconstruct Dioklēs' own ideas. Precisely aware of methodological questions, Dioklēs tends to relate his medical views to more general views on nature. Like Hippokratēs, he believes that treatment of a body-part cannot be effective without considering the body as a whole (*fr.*61) or the essence of the disease, and he privileges external climatic factors as causes of disease (*frr.*54–55). He uses inference from signs and refers to hidden causes (*frr.*56, 177), and was accordingly classified among the **Dogmatists** (*frr.*13–16). However, believing in an alliance between reason and experience, Dioklēs is always concerned with empirical confirmation both for pathological inner processes (*frr.*60, 137, 176) and anatomy (repeated dissections in *fr.*24). Dioklēs is interested in comparative animal anatomy and physiology (*frr.*24, 39) and in teleological explanations of bodily structures and processes (*fr.*23). SŌRANOS preferentially quotes Dioklēs' detailed description (*frr.*22–23) of the anatomy of the female reproductive organs, including breast-like offshoots inside the uterus (the *kotuledons* of, e.g., the HIPPOKRATIC CORPUS, APHORISMS 5.45). He argues against the Aristotelian view that semen is concocted blood (*fr.*40), claiming instead that semen is ***pneuma***, which, unlike blood, has the power of self-movement and is generated directly out of nutrition: semen has its origin ultimately in the brain (*frr.*40, 41a). The psychic ***pneuma***, a kind of breath different from the air inhaled by respiration for the purpose of cooling the innate heat (*fr.*31), seems to originate, however, in the heart and is responsible for distributing consciousness and voluntary movement over the body (*frr.*78, 80, 98). The ***hēgemonikon*** is primarily located in the heart, whence the ***pneuma*** moves upwards to the head and the brain: the blood probably carries the ***pneuma*** through the vascular system. The ANONYMOUS OF BRUSSELS (*fr.*40) attributes to him the four-**humor** doctrine, each produced by nutriment, although their physiological and/or pathological role does not emerge clearly from the other fragments. Obstructions to the ***pneuma*** in passages or vessels through which it is distributed are caused by phlegm (*frr.*95, 98, occasionally by bile, *fr.*108) and cause affections such as lethargy, paralysis, epilepsy (sharing Praxagoras' view). In fact what causes fever is "blockage at the end of the veins" (*fr.*63) and blockages feature prominently in his pathology. Similarly, coagulation and corruption of the blood cause disease (*frr.*34, 59). In his *Affection, Cause, Treatment,* he addressed diseases in a systematic way, ordering descriptions by symptoms, causal explanations and therapeutics. Convinced that therapy is by opposites, Dioklēs tends to differentiate it according to a patient's age and constitution (*frr.*73, 92, 125). His remedies include bathing, venesection, emetics, walks, fomentations, drugs and pills, dieting. His fame in antiquity derived mostly from his regimen in health, wherein he focuses on the balance between nutriment and physical movement, as in Hippokratic and even more ancient traditions. In dietetics he is keenly interested in the "powers" of foodstuffs and takes into account all variables to determine them correctly, specifying the influence of the mode of preparation on their effects (*frr.*187, 196).

Ed.: Wellmann (1901); P.J. van der Eijk, *Diocles of Carystus. A Collection of the Fragments with Translation and Commentary*, 2 vv. (2000–2001) = *SAM* 22–23.

RE 5.1 (1903) 802–812 (#53), M. Wellmann; W. Jaeger, *Diokles von Karystos. Die griechische Medizin und die Schule von Aristoteles* (1938); *Idem*, "Vergessene Fragmente des Peripatetikers Diokles von Karystos," *ABAW* (1938) #3, pp. 1–46; F. Kudlien, "Probleme um Diokles von Karystos," *AGM* (1963) 456–464; *DSB* 4.105–106, K.H. Dannenfeldt; *KP* 2.52–53 (#7), F. Kudlien; H. von Staden, "Jaeger's 'Skandalon der historischen Vernunft': Diocles, Aristotle and Theophrastus," in W.M. Calder III, *Werner Jaeger reconsidered* (1992) 227–265; *OCD3* 470, J.T. Vallance; *DPA* 2 (1994) 772–774, R. Goulet; P.J. van der Eijk, "Diocles and the Hippocratic Writings on the method of dietetics and the limits of causal explanation," in R. Wittern and P. Pellegrin, *Hippokratische Medizin und antike Philosophie* (1996) 227–259; *Idem*, "The systematic status of therapy in the Hippocratic Corpus and in the work of Diocles of Carystus," in I. Garofalo, *et al.*, edd., *Aspetti della terapia nel Corpus Hippocraticum* (1999) 389–404; *BNP* 4 (2004) 424–426 (#6), V. Nutton.

Daniela Manetti

Dioklēs of Khalkēdōn (250 BCE – 95 CE)

Asklēpiadēs Pharm., in Galēn *CMLoc* 7.4 (13.87 K.), records his opium-based pill for blood-spitting (***phthisis***?), containing acacia juice, *hupokistos* juice, pomegranate flower, red coral, Samian earth, etc.

Fabricius (1726) 141.

PTK

Dioklēs of Magnesia (1st c. BCE?)

If he was the friend of the poet Meleager, Dioklēs belongs to the 1st c. BCE. His *Compendium of Philosophers* is quoted in Diogenēs Laërtios 7.49–53 on **Stoic** logic, and some sayings are attributed by him to Antisthenēs (D.L. 6.12–12). Otherwise, information from his works is biographical and concerns mainly Cynic and **Stoic** philosophers, preserved in Diogenēs Laërtios.

V. Cellupríca "Diocle di Magnesia come fonte della dossografia stoica in Diogene Laerzio," *Orpheus* 10 (1989) 58–79; Jørgen Mejer "Diogenes Laertius and the Transmission of Greek Philosophy," *ANRW* 2.36.5 (1992) 3580.

Jørgen Mejer

Diomēdēs (250 BCE – 95 CE)

Asklēpiadēs Pharm. in Galēn *CMLoc* 4.7 (12.759 K.) recommends Diomēdēs' two collyria taken with rainwater and an egg: one composed of ***pompholux, psimuthion,***

verdigris, myrrh, frankincense, and opium in gum; and the other substituting ***khalkanthon*** for the **verdigris**, and saffron for the opium, and providing immediate pain relief (12.771 K.).

RE 5.1 (1903) 829 (#15), M. Wellmann.

GLIM

Diōn (Med.) (120 BCE – 120 CE)

Sōranos, *Gyn* 4.14 (*CMG* 4, p. 144; *CUF* v. 2, p. 11), records that he recommended potions of *elelisphakos*, myrrh, or celery seed, to expel afterbirth; he is there listed after Hippokratēs and Euruphōn, among other writers later than them. Oreibasios, *Syn.* 3.138 (*CMG* 6.3, p. 103), and Aëtios of Amida, 7.30 (*CMG* 8.2, pp. 276–277), record a collyrium (containing acacia, frankincense, Indian nard, myrrh, poppy-juice, etc., and to be applied in egg-white) by Diōn, who despite the frequency of the name may be the same man.

RE 5.1 (1903) 877 (#21), M. Wellmann.

PTK

Diōn of Kolophōn ⇒ Deinōn of Kolophōn

Diōn of Neapolis (120 – 80 BCE)

Censorinus 18.14 records that Diōn computed the "Great Year" as 10,884 years; Augustine, *City of God* 21.8.2, quotes Varro following Kastōr of Rhodes saying that Diōn and Adrastos of Kuzikos computed the date of a portent of Venus.

RE 5.1 (1903) 877 (#23), Fr. Hultsch.

PTK

Diōnidēs (350 BCE – 400 CE)

Cyril of Alexandria, *Dict.* (Dioklēs, fr.12 van der Eijk) – *cf.* Kurillos – lists many outstanding doctors (Galēn and earlier, save for Philagrios and the presumably intrusive Alexander of Tralleis), among whom the otherwise unknown and possibly corrupt Diōnidēs.

(*)

PTK

Dionusios (Astron.) (*fl.* 285 BCE)

Astronomer who lived at Alexandria, according to a *scholion* to the *Almagest.* In his *Almagest,* Ptolemy cites seven observations of the apparent positions of Mercury (9.7, 9.10) and Mars (9.9) relative to stars, dated by a calendar "according to Dionusios," ranging from 272 to 241 BCE; it is not known whether any of them were made by Dionusios himself. His calendar used solar years beginning at the summer solstice, divided into 12 months named after signs of the zodiac. Years were counted sequentially from a year 1 beginning with the solstice of 285 BCE, chosen perhaps because it immediately preceded Ptolemy II Philadelphos' first regnal year. The precise structure of the calendar, probably intended for astronomical, not civil, use, is disputed.

G.J. Toomer, *Ptolemy's Almagest* (1984); B. L. van der Waerden, "Greek Astronomical Calendars. III. The Calendar of Dionysios," *AHES* 29 (1984) 125–130; Alexander Jones, "A Posy of Almagest Scholia," *Centaurus* 45 (2003) 69–78.

Alexander Jones

Dionusios (Geog.) (285 – 245 BCE)

Sent by Ptolemy Philadelphos to India, and may have written an *Indika*, according to PLINY 6.58, who also cites a Dionusios for data on Euboia (4.64) and on the pyramids (36.79).

FGrHist 653, 717.

PTK

Dionusios (Lithika) (70 – 200 CE)

Together with SŌKRATĒS (LITH.), attributed with a prose lapidary, in the 14th c. *Vaticanus graecus* 578. This lapidary was probably written in Egypt during the early Roman imperial age. What is ascribable to Dionusios and what to Sōkratēs is difficult to determine, but Dionusios is thought to have been the illustrator. It is also debated whether our author should be identified with DIONUSIOS OF ALEXANDRIA, to whom ancient sources also attribute an *On stones*, surviving only in fragments.

Ed.: Halleux and Schamp (1985) 139–144, 166–167 (text); G. Giannakis, *Orpheōs Lithika, Kērugmata – Sōkratous kai Dionusiou peri lithōn* (1987).

RE 5.1 (1903) 977 (#133), M. Wellmann; K.W. Wirbelauer, *Antike Lapidarien* (1937) 31–42; Eugenio Amato, *Dionisio di Alessandria. Descrizione della Terra abitata* (2005) 69–73.

Eugenio Amato

Dionusios (Med.) (*ca* 340 – 300 BCE)

Known for his treatment of wounds. He advocated using pressure, particularly through tight bandages to staunch the flow of blood (CAELIUS AURELIANUS, *Chron.* 2.186 [*CML* 6.1.1, p. 658]). Several references in other writers might refer to this Dionusios, though because the name was so common, it is far from certain: CELSUS 6.6.4, 6.18.9C, SCRIBONIUS LARGUS 212 (called a surgeon), and PLINY 20.19, 219, and 22.67. Pliny (20.113–115) cites perhaps this Dionusios, in conjunction with KHRUSIPPOS OF KNIDOS (I), on various remedies from male parsley. Wellmann and others have suggested identification with DIONUSIOS SON OF OXUMAKHOS.

RE 5.1 (1903) 976 (#132), M. Wellmann; von Staden (1989) 424; *BNP* 4 (2004) 114 (#24), V. Nutton.

Robert Littman

Dionusios (Methodist) (*ca* 50 – 75 CE)

Physician, listed among the **Methodists** post-dating THEMISŌN and THESSALOS (GALĒN 10.53 K. = *fr.*162 Tecusan; PSEUDO-GALĒN, *Introd.* 14.684 K. = *fr.*283; IŌANNĒS OF ALEXANDRIA, *pr.*2 = *fr.*219), probably identifiable with the author whom MNASEAS cites on the "naturalness" of certain unhealthy sanguineous constrictions and laxities (*fr.*305). Of the numerous medical Dionusioi, Tecusan suggests that the **Methodist** may be identifiable with either (a) the pharmacologist whose recipes treated eye-sores and hemorrhoids (CELSUS 6.6.2, 4, 6.18.9 = *frr.*104, 107; etc.); (b) the

Hippokratic exegete (Galēn, *In Hipp. Aph.* 4.49 [17B.750–51 K.] = *fr.*209); (c) the botanist who described clover (Seruius, *Ad Geor.* 1.215 = *fr.*302); or, most likely according to Tecusan 2004: 55–56, (d) the Dionusios listed by Pliny (1.*ind.*4, 8, 10–17, 27, 31, 33–36 = *fr.*256) as a foreign medical authority. Pliny's Dionusios recommended turnips for joint pains (20.18–19 = *fr.*258), believed that eating parsley caused sterility and epilepsy in suckling infants (20.112–114 = *fr.*259), and wrote on the correct preparation and dangers of orache (20.219 = *fr.*260), and the properties and benefits of asphodel (22.67 = *fr.*61). Pliny's **Methodist** is probably identical to Seruius' botanist, but it is unclear how many medical Dionusioi Pliny may have cited. *Cf.* Dionusios (Med.) and Dionusios (of Milētos?).

Tecusan (2004) 53–59.

GLIM

Dionusios, Sallustius (100 bce – 75 ce)

Pliny describes his cure for toothache or loose teeth: eat a frog boiled in vinegar; to the weak of stomach, he instead offered frog-saliva brewed in vinegar (1.*ind.*31, 32.80–81).

RE 5.1 (1903) 976 (#131), M. Wellmann.

PTK and GLIM

Dionusios, son of Diogenēs (*ca* 210 – 90 bce?)

Credited by Marcianus 1.4 with determining the circumference of the Earth, and finding a value agreeing with that of Eratosthenēs.

RE 5.1 (1903) 992 (#145), Fr. Hultsch.

PTK

Dionusios son of Kalliphōn (*ca* 100 – 87 bce)

Greek, possibly Athenian, geographer, author of a description (*anagraphē*) of Greece in trimetric iambics addressed to an unknown Theophrastos and written *ca* 100–87 bce. Earlier attributed to Dikaiarkhos, the poem revealed the author's name and epithet in the acrostics of verses 1–23. The work adheres to the ***periplous*** form, describing coast-lines, giving distances measured in stades and sailing days, and using specific terminology of relative positions of sites and of toponymy. Dionusios' sources were Philētas, an unknown Athenian historian, Apollodōros of Athens, and Artemidōros of Ephesos. Manifesting some **Stoic** tendencies, the extant 150 verses describe western and central Greece, Crete and the Aegean islands.

Ed.: *GGM* 1.238–243; D. Marcotte, *Le poème géographique de Dionysios fils de Calliphon* (1990).

Daniela Dueck

Dionusios son of Oxumakhos (300 – 250 bce)

Cited by Rufus of Ephesos, *Onom. Anthr. Mor.* 205–208 (pp. 162–163 DR), as having coined the term *epanthismos* (used by Eudēmos of Alexandria for "vein"), which Dionusios used for a "vein-like vessel" (distinguished from both vein and artery: i.e., after Praxagoras). This Dionusios is perhaps contemporary, and thus possibly identical, with

Dionusios of Ephesos; Wellmann and others have suggested identification with Dionusios (Med.). The father's name is very rare: *LGPN* 1.353 (Crete), 3A.344 (Argos), 3B.327 (Amphissa and Eruthrai).

C.G. Kühn, *Additamenta ad elenchum medicorum ueterum* 14 (1827) 7; *RE* 5.1 (1903) 976 (#132), M. Wellmann; *BNP* 4 (2004) 486 (#24), V. Nutton.

PTK

Dionusios of Aigai (200 – 300 CE?)

Empiricist physician and Skeptic who probably lived after Galēn; von Staden (1999) has suggested that he was perhaps not an **Empiricist**. He wrote a work called *Diktuaka* (whose date and provenance is not certain), twice excerpted by Phōtios, *Bibl.* 185 and 211, containing 100 chapters, consisting of 50 pairs of medical thesis and antithesis (listed), with arguments (lost). For example, §50: "the capacity to think is located in the area of the central vesicle of the brain," or §42: "the liver is the source of the veins." The work can be divided into five sections: A = 1–4: spermatogenesis; B = 5–26: digestion and nutrition; C = 27–52: pathology; D = 52–69: therapy; E = 70–100: anatomy and cognition. He often presented several theoretical alternatives, and how each could be refuted. His work focused on Hellenistic debates, but also contained traces of theories of the 2nd c. CE, including those of Galēn, such as antithesis 42 (first attested in Galēn *PHP* 6.3–6 = *CMG* 5.4.1.2, pp. 372–404). Antithesis 50 apparently reflects the works of Nemesios and Poseidōnios (Med. II), which may indicate that Dionusios is later than usually assigned. A number of the counter-hypotheses include doctrines attributed to Erasistratos, with whom Dionusius at times agrees and at other times, disagrees.

RE 5.1 (1903) 975 (#124), H. von Arnim; Deichgräber (1965) 340; von Staden (1989) 389; *Idem* (1999) 177–185.

Robert Littman

Dionusios (of Alexandria?) (ca 240 – 260 CE?)

Addressed by Diophantos, *Arithmetica Pr.*1, as an enthusiastic beginning student of mathematics. Hultsch rejects identification with the contemporary bishop of Alexandria.

RE 5.1 (1903) 993 (#147), Fr. Hultsch.

GLIM

Dionusios of Alexandria, Periēgētēs (130 – 138 CE)

Greek grammarian and geographer, son of Glaukos, living under Hadrian. Dionusios was director of the imperial libraries and secretary in charge of correspondence and embassies. There is some confusion regarding Dionusios' period. However, he himself hints at his origin and time in the acrostics of some verses (109–134; 513–532) of his poetic description (***periēgēsis***) of the ***oikoumenē***, consisting in 1,186 hexameters. The work describes the ocean and the three continents: Africa, Europe and Asia, based mainly on Eratosthenēs.

Ed.: *GGM* 2.xv–xl, 103–176; K. Brodersen, *Dionysios von Alexandria: Das Lied von der Welt* (1994); A.A. Raschieri, *Dionigi di Alessandria: Guida delle terre abitate* (2004); E. Amato, *Dionisio di Alessandria. Descrizione della Terra abitata* (2005).

C. Jacob, *La Description de la terre habitée de Denys d'Alexandrie ou la leçon de géographie* (1990); I.O. Tsavari, *Histoire du texte de la description de la terre de Denys le Périégète* (1990); H. White, "On the date of Dionysius Periegetes," *Orpheus* 22 (2001) 288–290; E. Amato, "Per la Cronologia di Dionisio el Periegeta," *RPh* 77 (2003) 7–16; E. Bowie, "Dénys d'Alexandrie. Un poète grec dans l'empire romain," *REA* 106 (2004) 177–186; R. Hunter, "The Periegesis of Dionysius and the traditions of Hellenistic poetry," *REA* 106 (2004) 217–232.

Daniela Dueck

Dionusios of Alexandria (300 – 220 BCE)

Designed a repeating catapult for discharging arrows, possibly constructed during the siege of Rhodes by Dēmētrios Poliorkētēs. PHILŌN (*Belop.* p. 73 W.), who saw the weapon in action, describes the catapult and its operation: a winch was pulled back and forth to volley arrows as quickly as the men could turn the handles. Philōn criticizes the catapult which could not be re-sighted between shots.

Irby-Massie and Keyser (2002) 160–161.

GLIM

Dionusios of Buzantion (120 – 180 CE?)

Composed the extant *Anaplous of the Bosporos*, describing the sail up (*anaplous*) and back down the Thrakian Bosporos. The work is preserved in one MS, missing a middle folio, representing one-third of the work (§57–96), for which we rely on the 16th c. Latin of Pierre Gilles, translated from a lost Greek MS. Dionusios' detailed description of Buzantion seems to predate the city's razing by Septimius Seuerus in 196 CE, and the language suggests the 2nd c. CE. Dionusios refers to or imitates HĒRODOTOS, THUCYDIDĒS, XENOPHŌN, POLUBIOS, STRABŌN, PLINY, and ARRIANUS, possibly not using any directly. The work gives an overview of the Bosporos (§1–6), describes sites in and around Buzantion (§7–34), proceeds up the European shore of the Bosporos to the Black Sea (§35–87), and returns down the Asian shore to Khalkēdōn (§87–112). Dionusios' unusual ***periplous*** gives extents only for the Maiotis, Bosporos, *Keras*, and walls of Buzantion (§2–6). He describes monuments, anchorages, and fishing, and for all names provides *aitia*, which include myths (§7, 24, 45) and ***paradoxa*** (§24, 42, 70, 95). The monuments include many temples, Philip's siege-bridge (§27), Dareios' throne (§57), and the ruined lighthouse *Timaion* (§77). His greatest practical interest lies in anchorages, described for 16 sites, and fishing, described for 15 sites, plus oyster-beds (§37).

Talbert (2000) #53; *BNP* 2 (2003) 733–735 (Bosporus #1), E. Olshausen; *BNP* 4 (2004) 487 (#28), K. Brodersen.

PTK

Dionusios of Corinth (265 BCE – 75 CE)

This epic poet is of uncertain date but he surely lived after KALLIMAKHOS. According to the *Souda* Delta-1177, he wrote *Hupothekai, Meteōrologoumena*, and *Aitia* from which only a small fragment remains (PLUTARCH, *Amat.* 761B = *SH* 388). He is likely to have also written a prose commentary on HĒSIOD. However all this remains uncertain because the *Souda* attributed to him a ***Periēgēsis*** owing to confusion with DIONUSIOS OF ALEXANDRIA.

BNP 4 (2004) 490 (#43), M. Di Marco.

Christophe Cusset

Dionusios of Ephesos (290 – 250 BCE?)

Physician, wrote a *Record of Physicians* wherein he reports that NIKIAS OF MILĒTOS was ERASISTRATOS' fellow student.

FGrHist 1104.

GLIM

Dionusios (of Halikarnassos?) (200 BCE – *ca* 300 CE)

One or perhaps two authors of this name are cited by PORPHURIOS in his commentary on PTOLEMY's *Harmonics*.

Dionusios "the musicologist" (*ho mousikos*) is introduced early in Porphurios' work as the author of a *Peri homoiotētōn* (*On likenesses*, 37.16). Porphurios preserves a brief quotation from the first book of this treatise, in which Dionusios relates four basic doctrines of the "canonic theorists" (*kanonikoi*): (1) the nature of rhythm and melody is nearly one and the same; (2) what is high-pitched is fast and what is low-pitched is slow; (3) attunement is the commensurability (*summetria*) of certain speeds; and (4) well-tuned intervals (*diastēmata*) are in ratios (*logoi*) of numbers.

Much later in Porphurios' commentary a Dionusios "of Halikarnassos" is mentioned in a different context (92.28); he is named in a list of authorities including PLATO, DĒMĒTRIOS, ARKHUTAS and EUCLID, all of whom used the term "interval" (*diastēma*) in place of "ratio" (*logos*) – i.e., they referred to (e.g.) the "epitritic interval" rather than the "epitritic ratio." This Dionusios agreed with ARISTOXENOS and ERATOSTHENĒS "and many others" in admitting eight concords, as opposed to AELIANUS and Ptolemy, who only admitted six (96.11). His authority is invoked to corroborate the statement that the octave interval does not differ in function (*kata dunamin*) from a single note, and therefore that any interval combined with an octave will have the same melodic function as the uncompounded interval, like the numbers under ten when added to ten (104.14).

The second Dionusios ("of Halikarnassos") may be the same man as the first Dionusios (*ho mousikos*), notwithstanding his use of both terms *logos* and *diastēma* in the same sentence (37.19–20). (The context here requires the use of both terms, and the discussion at 92.28 is of a somewhat different point.)

Porphurios' *Dionusios ho mousikos* has therefore been identified by some as Dionusios of Halikarnassos – not the author of the *De compositione verborum*, who lived in the time of AUGUSTUS, but a Hadrianic musicologist for whom the *Souda* (Delta-1171, which gives him both epithets) lists three further titles: *Rhythmics* (in 24 books), *Musical History* (in 36 books) and *Science of Music* (in 22 books). He may be identical with Aelius Dionusios, the Atticist lexicographer of Hadrian's time (so Düring, following Scherer).

Karl Scherer, *De Aelio Dionysio musico qui vocatur* (Diss. Bonn 1886); *RE* 5.1 (1903) 986–991 (#142), L. Cohn; Düring (1932); *BNP* 4 (2004) 484 (#20), F. Montanari.

David Creese

Dionusios of Kurēnē (160 – 110 CE)

A student of ANTIPATROS OF TARSOS and of DIOGENĒS OF SELEUKEIA, Dionusios was an acclaimed **Stoic** geometer (according to a Herculaneum papyrus, *Index Stoicorum*, col. 52) who wrote against POLUAINOS, and was attacked by DĒMĒTRIOS OF LAKŌNIKA (*P. Herc.*

1642, *fr.*4). He insisted that induction must be based upon what is always and everywhere observed.

GGP 4.2 (1994) 641–642, P. Steinmetz; *BNP* 4 (2004) 476 (#10), B. Inwood.

PTK and GLIM

Dionusios Kurtos or Dionusios of Kurtos (100 BCE? – 50 CE)

Physician from Egypt also named after his homeland Kurtos (though one would then expect *Kurtitēs*), not because he was actually *kurtos* ("hunchbacked"), as we are told by STEPHANOS OF BUZANTION (s.v. *Kurtos; cf. Schol. Oribas.* III.687 BDM), citing "HERENNIUS PHILŌN's book *On the Physicians.*" Dionusios was used by RUFUS OF EPHESOS and before him by ANDROMAKHOS THE YOUNGER, which indicates Nero's period as *terminus ante quem.* Andromakhos (in GALĒN *CMGen* 6.14 [13.928.7–11 K.]) describes one of his vesicatory plasters (*Kurtou epispastikē*), whereas Rufus of Ephesos in OREIBASIOS, *Coll.* 44.14 (*CMG* 6.2.1, pp. 131–132: *Dionusion ton kurton*) cites him regarding a pestilential bubo specific to Libya, Egypt and Syria.

RE 5.1 (1903) 976 (#132), M. Wellmann; *RE* 12.1 (1924) 206 (#2), F.E. Kind.

Jean-Marie Jacques

Dionusios of Milētos (460 – 430 BCE)

Wrote histories of Persia and mythographical works (*Souda* Delta-1180), plus a *Guide to the World,* of which a few fragments are preserved by scholiasts.

FGrHist 687; *OCD3* 478, K. Meister.

PTK

Dionusios (of Milētos?) (75 – 35 BCE)

GALĒN at *CMLoc* 5.3 (12.835 K.: following ASKLĒPIADĒS PHARM.) ascribes a dermatological recipe to a Dionusios schoolfellow (*summathētēs*) of HĒRAKLEIDĒS OF TARAS: he is possibly to be identified either with the Dionusios of Milētos mentioned at *CMLoc* 4.7 (12.741–742 K.) and *Antid.* 2.11 (14.171 K.) or with the DIONUSIOS OF SAMOS mentioned at *CMGen* 6.16 (13.938 K.), or with the Dionusios mentioned at *In Hipp. Aph.* 17B.751 K. A certain *Dionysius* is mentioned by PLINY for different remedies at 1.*ind.*20, 19.113, 219; 22.67 and 25.8.

RE 5.1 (1903) 976 (#132), M. Wellmann.

Fabio Stok

Dionusios of Philadelpheia (140 BCE? – 20 CE?)

Enigmatic figure sometimes identified with DIONUSIOS OF ALEXANDRIA, PERIĒGĒTĒS. Authored a poem *On Bird-catching* (*Ixeutika*) or *On Birds* (*Ornithiaka*), originally in two or three books, traditionally but wrongly attributed to OPPIANUS, whose substance is well preserved in a Byzantine paraphrase previously attributed to the sophist EUTEKNIOS. Dionusios' text shares many parallels with those of Athēnaios and AELIANUS. The paraphrase, preserving typical rhythmic endings and special vocabulary, is probably a prosaic transcription, very close to the original. More folkloric than technical treatise, the text mentions prey as well as domestic and mythic fowls, and presents "the names, residences and customs, talents, forces and desires of birds, and the ways of catching them" (1.1). The first book treats land-birds (beginning with eagle and finishing with phoenix), the second with water-birds (beginning with water-eagle and finishing with swan), the third with bird-catching, with birdlime (*ixos*: 3.1–6) and various other means and sometimes subtle traps (3.7–28).

Ed.: A. Garzya, *Dionysii Ixeuticon* (1963).
RE 5.1 (1903) 925 (#96), G. Knaack; *OCD3* 478 (#9–10), J.S. Rusten.

Arnaud Zucker

Dionusios of Rhodes (265 BCE – 200 CE)

Souda Delta-1181 mentions that some attribute to this historian the *Guide to the Earth* (***periēgēsis*** *gēs*) by Dionusios of Alexandria; the same work is also attributed to Dionusios of Corinth. Contrast the epigrammatist of *BNP* 4.488 (#33).

(*)

PTK

Dionusios of Samos (250 BCE – 95 CE)

Asklēpiadēs Pharm., in Galēn *CMGen* 4.13 (13.745–746 K.), records his recipe for a lotion compounded of Eretrian earth (a grey clay), copper flake, ***ikhthuokolla***, frankincense, **verdigris**, myrrh, and vinegar. He is probably distinct from other homonymous medical authors; *cf.* Dionusios (of Milētos?).

RE 5.1 (1903) 976 (#132), M. Wellmann.

GLIM

Dionusios of Utica, Cassius (90 BCE)

Agronomist who translated Mago the Carthaginian's encyclopedic work on agriculture into Greek, reducing it to eight books while adding a further 12 books of excerpts from the Greek agricultural writers listed by Varro (*RR* 1.1.8–10). The complete volume was dedicated to the praetor P. Sextilius, governor of Africa in 89 or 88 BCE. Dionusios discussed cattle-breeding in his first book, leeks in the seventh (Mago *frr.*42, 63, Speranza). If the alteration of Stephanos of Buzantion, s.v. *Itukē*, is correct, Cassius also composed a work on botanical medicine (*Rhizotomika*) with illustrations of the flora discussed (Pliny 25.8), including rape, parsley, orache, and asphodel (20.19, 113, 219, 22.67; *cf. Schol. Nik.* 519). Like the agricultural work, this collection may have been partly an anthology, drawing on writers such as Dioklēs of Karustos and Khrusippos of Knidos (I). "Dionusios" suggests a Greek-speaking freedman, perhaps from the household of Cassius Longinus, the praetor (111 BCE) who escorted Iugurtha from Africa to Rome.

Ed.: Speranza (1971) 75–119.
Rawson (1985) 135; *BNP* 2 (2003) 1172, C. Hünemörder.

Philip Thibodeau

Dionusodōros (Pharm.) (300 BCE – 115 CE)

Arkhigenēs, in Galēn *CMLoc* 1.2 (12.409–410 K.), cites his *alōpekia* remedy, composed of ashed raw-hide in sharp vinegar (optionally add *thapsia*-juice), as a scalp-rinse to exfoliate the skin and produce hair-growth. Wellmann tentatively equates him with the oculist "C. Iulius Dionysodorus" known from a **collyrium stamp**: Grotefend (1867) #43; but the name is exceedingly common (*LGPN*).

RE 5.1 (1903) 1004–1005 (#17), M. Wellmann.

PTK

Dionusodōros, Maecius Seuerus (100 – 200 CE)

Platonist, wrote a commentary on the *Timaeus*, to which PROKLOS repeatedly refers (e.g. *in Tim.* 1.204.16–18), and *On the Soul* (possibly part of the same commentary), from which EUSEBIOS (*PE* 13.17) preserves an extract. "Seuerus," as he is known in the later tradition, is probably identical with "Flauius Maecius Se[. . .] Dionusodōrus, Platonic philosopher and counselor" honored in an inscription from Antinoē, *IBM* IV 1076 = *SB* III 6012. A Dionusodōrus mentioned by DIOGENĒS LAËRTIOS (2.42) may well be the same person. Seuerus was read in PLŌTINOS' school. SYRIANUS (*in Metaph.* 84.23–5) censures Seuerus for misusing mathematics to study nature.

Ed.: Gioè (2002) 379–433.

RE 2A.2 (1923) 2007–2010, K. Praechter; P. Cauderlier and K.A. Worp, "SB III 6012 = IBM IV 1076: Unrecognised Evidence for a Mysterious Philosopher," *Aegyptus* 62 (1982) 72–79; Dillon (1996) 262–264; *NP* 11.484–485, M. Baltes and M.L. Lakmann.

Jan Opsomer

Dionusodōros (of Kaunos?) (*ca* 200 BCE)

We know of three geometers named Dionusodōros: (1) of Amisēnē, mentioned by STRABŌN (12.3.16), (2) of Mēlos, mentioned by Strabōn (*ibid.*) and PLINY (2.248), who relates a foolish anecdote about his funeral inscription, and (3) of Kaunos, one of the teachers of PHILŌNIDĒS, and thus a member of a circle of intellectuals including the mathematicians APOLLŌNIOS OF PERGĒ, Eudēmos of Pergamon (otherwise unknown), ZĒNODŌROS, and probably DIOKLĒS. This milieu makes Dionusodōros of Kaunos the most credible candidate for the authorship of two sophisticated geometrical results attributed to an unspecified Dionusodōros.

EUTOKIOS (*In Arch. Sph. Cyl.* pp.152–160 H.) quotes an alternative solution by Dionusodōros to the problem of dividing a sphere by a plane such that the two segments are in a given ratio, which ARCHIMĒDĒS reduced to a complex division of a line segment in *Sphere and Cylinder* 2.4. Dionusodōros' construction solves the problem by finding the intersections of a parabola and a hyperbola. HĒRŌN (*Metrika* 2.13) reports that Dionusodōros' *On the Torus* contained a formula effectively relating the volume of a torus to the diameters of the generating circle and the circle of revolution. The proof, which probably resembled Archimēdēs' procedures in *Sphere and Cylinder* and other works, is lost. VITRUUIUS (9.8) attributes the invention of the conical sundial to a Dionusodōros.

DSB 4.108–110, I. Bulmer-Thomas; Knorr (1986) 263–277; R. Netz, *The Transformation of Mathematics in the Early Mediterranean World* (2004) 29–38.

Alexander Jones

Diophanēs of Nikaia (85 – 60 BCE)

Compiled a six-book epitome of CASSIUS DIONUSIOS' translation of Mago's agricultural work, dedicated to Deiotaros, tetrarch of Galatia; *cf.* VARRO, *RR* 1.1.8–10. The popular epitome eventually superseded the original translation; Varro treats it as well-known *ca* 55 BCE (1.9.7). A collection of "paradoxes" was also ascribed to Diophanēs (Phōtios *Bibl.* 163), though it may be the case that these were simply items culled from his agricultural work.

Ed.: Speranza (1971) 75–119.
RE 5.1 (1903) 1049 (#9), M. Wellmann, S.6 (1935) 27, W. Kroll; 18.3 (1949) 1137–1166 (§22, 1159), K. Ziegler; *KP* 2.85, H.G. Gundel.

Philip Thibodeau

Diophantos (Geog.) (325 – 150 BCE)

AGATHARKHIDĒS OF KNIDOS notes that Diophantos wrote about the north-lands; scraps are preserved in the *scholia* to Apollōnios of Rhodes and in STEPHANOS OF BUZANTION, s.v. *Abioi* and *Libustinoi.*

FGrHist 805.

PTK

Diophantos of Alexandria (*ca* 250 CE)

Greek mathematician, author of a sizeable and influential algebraic treatise, the *Arithmētika.* A small fragment of a treatise on polygonal numbers is also attributed to him. Diophantos lived in Alexandria, the main scientific center of antiquity. Not mentioned before the 4th c., he is thought to have lived in the 3rd, and the Dionusios addressed in the introduction to the *Arithmētika* may have been Saint Dionusios of Alexandria (d. *ca* 264 CE). An arithmetical epigram from the *Anthologia Graeca* (14.126), retracing some events of his life (marriage at 33, birth of his son at 38, death of his son four years before his own at 84), seems contrived.

The *Arithmētika* has not been preserved in its entirety; only ten of its 13 books (*biblia*) have been transmitted, at different times and in a different form. Six in Greek (numbered 1–6) reached Europe in the 15th c. Four more, in a 9th c. Arabic translation, (numbered 4–7) were discovered in 1968; since this latter numbering turned out to be correct, the last three Greek books must follow them, probably as Books 8–10, while Books 11–13, still missing, must be considered lost. The Arabic version (originally covering Books 1–7) is notably more prolix than the extant Greek text, for it completes computations and verifies that solutions indeed fulfill the equations. Certainly Greek in origin, it must be the commentary (*hupomnēma*) which the *Souda* Y-166 (as emended by Tannery 1895: 36) attributed to HUPATIA, the daughter of THEŌN OF ALEXANDRIA.

In the introduction, Diophantos provides general instructions regarding treating equations, and he defines relevant symbols, with signs for the unknown and its powers, as well as for a few operations, the first known algebraic symbolism. Like its late medieval successors, Diophantos' symbolism originated in scribal abbreviations of commonly repeated words. Then Diophantos proceeds with the problems (some 250 survive). In Book 1, the problems are of the familiar kind seen by the student in school and solved, with an early form of algebra, by applying identities already in use in Mesopotamia; but Diophantos treats these problems in his new, algebraic way. Books 2–3 teach and apply some fundamental methods which are then extended to further problems in Books 4–7; for their scope is, as Diophantos says, "experience and skill" (introductions to Books 4 and 7, pp. 87, 156). (Thus no really new methods are taught here, perhaps explaining why the lacuna in the middle of the six Greek books escaped notice.) The last three Greek books contain problems of a notably higher level. All the problems require that the solution be a rational and positive number.

The *Arithmētika*'s historical importance is twofold. First, it is the only surviving testimony of higher algebra in antiquity. Secondly, the extant Greek books initiated the first modern studies on number theory, beginning principally with the observations of the French mathematician Pierre de Fermat (1601–1665) in his copy of Diophantos. (Fermat's note on problem 2.8 is well-known: the equality $x^n+y^n=z^n$ with n any integer larger than 2 is impossible in rational numbers; this assertion was to occupy mathematicians for more than three centuries until proved in 1992–1995.)

Most of Diophantos' problems are indeterminate ones of the second degree, that is, with more unknowns than equations. Now such problems may be soluble or not. Since Diophantos had few general methods at his disposal, only through skillful assumptions were his problems made determinate and reduced to an already known problem or method; departing form Diophantos' assumptions may lead to quite another situation. Furthermore, Diophantos states, when necessary, that certain numbers cannot be represented as the sum of two squares, or as the sum of three squares, but without offering any proof. So it is hardly surprising that later mathematicians found in Diophantos an incentive for further research to the extent that Diophantos' name is now associated with various fields of modern mathematics quite foreign to the content and spirit of his *Arithmētika.*

Ed.: P. Tannery, *Diophanti Alexandrini Opera omnia* 2 vv. (1893–1895, repr. 1974); Jacques Sesiano, *Books IV to VII of Diophantus'* Arithmetica *in the Arabic translation attributed to Qusṭā ibn Lūqā* (1982).

Th. Heath, *Diophantus of Alexandria* (1910, repr. 1964); Jacques Sesiano "An early form of Greek algebra," *Centaurus* 40 (1998) 276–302.

Jacques Sesiano

Diophantos of Lukia (40 – 10 BCE)

Physician and surgeon, probably identifiable with C. Iulius Diophantos, son of C. Iulius Hēliodōros of Ludē in Lukia, responsible for inscribing a remedy on the base of an Asklēpios statue in the Ludean agora (*JHS* 10 [1889] 59, #11); the father may have been among IULIUS CAESAR's freedmen (*ibid.*, #8). Diophantos' colic remedy, admired by ARISTŌN (II), included centaury sap (either *Centaurium nemoralis* Jord., native to Spain and Portugal, or common or lesser centaury, *C. erythraea* Rafin.: Durling 1993: 199), plus castoreum, squill, both white and long pepper, myrrh, rue, hyssop, wormwood, Illyrian iris, saffron, ***ammōniakon*** incense, yellow iris root, Pontic nard, ginger, and black hellebore, administered with ***oxumel*** (GALĒN *CMLoc* 9.4, 13.281 K.). KRITŌN, in Galēn *CMLoc* 5.3 (12.845 K.), records his salve for burns and intertrigo, effective also against ***erusipelas***, compounded from **litharge**, stag marrow, ***psimuthion***, beeswax, **terebinth**, frankincense and olive oil, prepared as needed. ANDROMAKHOS records three remedies: Diophantos' *Aphra* (*sc.* "foaming"?) quince-yellow emollient for drawing and squeezing out to heal joints in Galēn *CMGen* 2.7 (13.507 K.); and two antidotes for scorpion and spider bites in *Antid.* 2.12, 13 (14.175–176, 181 K.), the second of which treats also all serpent bites.

Asklēpiadēs Pharm., in Galēn *CMGen* 5.4 (13.805 K.), discussing the uses of dittany, includes Diophantos' antidote to any poison, used by Phulakos.

RE 5.1 (1903) 1051 (#17), M. Wellmann; S.14 (1974) 113–114, J. Benedum.

GLIM

Diophil- (150 BCE – 50 CE?)

An astronomical poem of disputed title (*en tō epigraphomenō Prok[. . .]ō*) survives in eight corrupt verses (*POxy* 20 [1952] 2258C *fr.*1, *ad Callimimachi Comam*). The scholiast attributes the verses to a Diophil- (i.e., of uncertain gender). The fragment, following Aratos and Kallimakhos (*frr.*110, 387), describes a constellation of seven stars, near Virgo, Leo, Boötes, and the Bear – likely the triangular "Coma Berenices." The titular expansion *Prok[omi]ō* ("lock") suggests an encomium to Berenice. The extremely rare name (*LGPN* 3B) could be corrupted from the less rare Dinophil- (*LGPN* 3B) or common Diphil- (*LGPN*).

L. Lehnus, "Notizie Callimachee V," *Acme* 54.3 (2001) 283–291 at 285.

GLIM

Dioskoros (Geog.) (*ca* 50 BCE – 80 CE?)

Described a voyage down the east coast of Africa near the equator, according to Marinos of Tyre in Ptolemy, *Geog.* 1.9, 1.14. *Cf.* Diogenēs and Theophilos.

RE 5.1 (1903) 1086 (#4), H. Berger.

PTK

Dioskoros (Alch.) (300 – 390 CE)

Found in the list of *poiētai* (makers of gold, *CAAG* 2.25), but no specific work attributed to him is found in the Greek alchemical corpus. Dioskoros priest of Serapis at Alexandria is the dedicatee of Sunesios' *Commentary on the Book of Dēmokritos* (*CAAG* 2.56–69); Berthelot (1885: 186) considers our Dioskoros historical. The 10th c. catalogue of books, *Kitāb al-Fihrist*, mentions a *Book of Dioskoros about the Art.*

Berthelot (1885) 131, 156, 186, 190–191; Fück (1951) 94 (#1, 5), 122.

Cristina Viano

Dioskoros (Pharm.) (120 BCE – 80 CE)

Andromakhos, in Galēn *CMLoc* 8.7 (13.204–205 K.), records his hepatic antidote, of cassia, ***kostos***, licorice, nard, saffron, etc., in honey.

Fabricius (1726) 144.

PTK

Dioskouridēs (Metrology) (*ca* 60 – 200 CE?)

A treatise *On Weights and Measures* (*peri metrōn kai stathmōn*), surviving as a table in two parts, appears to be attributed to the pharmacist Dioskouridēs. One defines various measurements of weight, including three standards of *mina* (Attic, Italian, Alexandrian). The other, in three sections, treats Roman liquid quantities (of wine, olive oil, and honey), including quantities for the amphora (*keramion*), urn (*ourna*), and congium (*chous*). Dioskouridēs gives weights of an amphora of wine, oil, and honey as 80, 72, and 70 pounds, agreeing with OREIBASIOS.

Ed.: *MSR* 1.132–133, 239–244.
RE 5.1 (1903) 974 (#122), Fr. Hultsch.

PTK and GLIM

Dioskouridēs Phakās (80 – 45 BCE)

Hērophilean physician, whose epithet means "warty-faced," resident in Alexandria during the first years of the joint reign of Ptolemy XIII and KLEOPATRA VII, likely court doctor and roving ambassador under Ptolemy XII Aulētēs (80–51 BCE). CAESAR (*BC* 3.109.3–6) is ambiguous about his fate as an emissary of Ptolemy XIII to Akhillās threatening civil war (48 BCE): "[Akhillās] ordered them [Dioskouridēs and Serapiōn] arrested and killed, but one of them was simply wounded and was quickly rescued by his friends and borne away as if he were dead . . ." If Dioskouridēs survived, he would have been an elderly and wily court physician "associated with Kleopatra in the time of Antony" (*Souda* Delta-1206). He wrote 24 books on medicine (*ibid.*) as well as tracts on strange Hippokratic terminology (ERŌTIANOS, *Pr.*, and *fr.*5 [pp. 5, 91 Nachm.]). In his *Strange Diseases*, RUFUS OF EPHESOS reports that (probably this) Dioskouridēs had composed a work on a nodular-swelling ("bubonic") plague of uncertain era ravaging Libya (excerpted in OREIBASIOS *Coll.* 44.14.2 [*CMG* 6.2.1, p. 132]). PAULOS OF AIGINA (4.24 [*CMG* 9.1, p. 345]) quotes directly from "Dioskouridēs of Alexandria" on skin diseases, providing a careful description: "Dioskouridēs of Alexandria says that *terminthoi* are protuberances formed in the skin, that are round and colored dark green, like the fruit of the **terebinth**-tree" (*cf.* pseudo-GALĒN *Commentary on the Hippocratic Humors* 3.26 [16.461 K.]). This small bit of evidence, if typical, suggests an expertise in pharmacology, rather necessary at the Ptolemaic court.

RE 5.1 (1903) 1129–1130 (#10), M. Wellmann; von Staden (1989) 519–522 (incl. 7 fragments enumerated but not edited).

John Scarborough

Dioskouridēs of Alexandria (100 – 120 CE)

Wrote commentaries on the HIPPOKRATIC CORPUS, APHORISMS, EPIDEMICS 2, 3, 6, PROGNOSIS, etc., and a glossary, all much used by GALĒN; only PAULOS OF AIGINA 4.24 (*CMG* 9.1, p. 345) preserves the ethnic. Galēn often cites him with ARTEMIDŌROS CAPITO, and says he imitated the text-critic Aristarkhos, altering or athetizing passages (*In Hipp. Epid VI* [*CMG* 5.10.2.2, pp. 415, 464, 483]), and was accustomed to rewrite passages for clarity (pp. 4, 232, 400); he re-attributed many Hippokratic works, to HIPPOKRATĒS' grandson (p. 55: NATURE OF MAN), THESSALOS OF KŌS (p. 76: *Epidemics* 2 and 6), or DRAKŌN (Galēn, *Difficulty Breathing* 2.8 [7.854–855 K.]: *Epidemics* 5). Galēn cites him extensively in his own *Hippokratic Glossary* (19.63–64, 83, 88–89, 97, 105–106, 109, 140–142, 148, 152, etc. K.).

Manetti and Roselli (1994) 1617–1633; Ihm (2002) #45–46.

PTK

Dioskouridēs of Anazarbos (*ca* 40 – 80 CE)

Dioskouridēs of Anazarbos (*Vind. Med. Gr.* 1, f.3^{V}) © Österreichische Nationalbibliothek

The *De materia medica* (Greek: *Peri hulēs iatrikēs*) is one of the most influential works of its kind, but its gifted and energetic author is almost unknown, biographically. Dioskouridēs' birthplace was the small city of Anazarbos, about 100 km east north-east from Tarsos on a major highway in the Roman province of Cilicia. Comparing passages in PLINY's *Natural History* with similar extracts in the *De materia medica* reveals both quoting independently from SEXTIUS NIGER, so probably Dioskouridēs was born sometime in the reign of Tiberius or Caligula, and set down his observations in the same decade as Pliny composed his encyclopedia. In the *Preface* to the *De materia medica*, Dioskouridēs intimates he studied herbal pharmacology at Tarsos, and that here were teachers of medical botany, and medicaments fabricated from animal products and minerals; an early and respected instructor was AREIOS OF TARSOS, to whom Dioskouridēs dedicates the *De materia medica*. GALĒN notes that Areios was a famous teacher in Tarsos in the right decade, and the *De materia medica* and other texts reflect teaching centers in the eastern half of the Roman Empire, cities with reputations for given subjects available for instruction. Alexandria in Egypt long remained a center for medical learning, and other urban clusters of medical education existed in Laodikeia, Ephesos, and probably Smurna.

Dioskouridēs traveled widely in the Greek-speaking parts of the empire: prominent are his citations of herbal lore and pharmacology in Egypt, Syria, Palestine, various provinces in Asia Minor, Greece and the Islands, and he visited Greek communities in Sicily, southern Italy, and southern Gaul. He was not part of his contemporary elite, although Areios' connections with the consular Bassus suggest intermittent if occasionally important contacts. *Preface* 4 (*oistha gar hēmin stratiōtikon ton bion*) does not mean that Dioskouridēs was a military physician, but perhaps he had served in an eastern legion for short periods as a civilian doctor, a common custom in the western legions. "My soldier's life" likely says that Dioskouridēs lived as a soldier as he journeyed from region to region, listening to the inhabitants and surviving on the minimum of food, drink, and clothing. Perhaps he made his living as an itinerant physician in the manner of the medical travelers recorded in the works under the name of HIPPOKRATĒS.

Dioskouridēs arranges his material into five books, writing in the *Preface* that his way of organization is superior to previous compilations of drugstuffs, but he never explicitly explains his new scheme. Clues are the linking of "similars" in each book, or as he writes in *Preface* 3, ". . . [not] using the alphabetical arrangement which splits *materia medica* and their properties from those which they are closely related." Drugs will be classed according to the *dunameis* (almost always "properties" to Dioskouridēs) they evince as pharmaceuticals, as they "act" in or on the body of a patient. Sense-perceptions (quite probably adapted from THEOPHRASTOS) are central: smells linked with tastes identify drugs in Book 1 (aromatic oils, salves, trees, and shrubs, and the strongly fragrant liquids, gums, and fruits produced by them); Book 2 takes up animals and parts of animals, pharmaceuticals fashioned from various insects, crustaceans, arthropods, reptiles, and larger animals, both wild and domestic, and then follow cereals, pot herbs, and others which are "sharper"; Book 3 continues with more roots, juices, and seeds, and Book 4 provides roots and herbs not included previously; finally in Book 5 are details of wines and mineralogy, disclosing that Dioskouridēs knew well the vintner's technologies since ancient wine production struggled to produce a beverage that did not become "sour" (viz. turn into vinegar), and the "additives" in Book 5 are priceless listings of ingredients used to flavor wines, or were substances used in hopes of controlling what we call fermentation. In Book 5 Dioskouridēs considers subjects far beyond what moderns expect in a work on pharmacology: here are the technologies of quicklime, the important properties of minerals as manufactured into drugstuffs and other products, and why one has to know the best sites of mining and smelting of fine ores, so that the physician can procure good mineral pharmaceuticals; knowledge of such metallurgical technologies enables the doctor to recognize the best remedies, as contrasted to some common poisons, also derived from minerals.

Several of Dioskouridēs' descriptions became standard, appearing repeatedly in accounts of medicines in later Greek, Latin, Arabic, and Armenian, coming down to the first printed editions of the Renaissance. Illustrative are the opium poppy (4.64), mandrake (4.75), willow (1.104), the chaste tree (1.103), the two blister beetles (2.61), sea creatures and similar land animals (2.1–33: n.b. beaver castor, 2.24, standard to the 18th c.), the numerous milks (2.70), rennets (2.75), fats (2.76), honeys, beeswax, and bee glue (2.82–84), the flavorings of wines (5. 34–73), minerals and ores (5.76–162), and many others. The complete *De materia medica* is a compaction of over 700 items fused into more than 2,000 recipes and formulas, and its bulk guaranteed it would be modified and augmented according to local requirements. The original work did not have illustrations, but when codices replaced papyrus rolls, scribes and artists produced handsome versions of the *De materia medica*, with the Codex Juliana Anicia of 512 CE the earliest, extant exemplar. First to employ Dioskouridēs in a rearranged, alphabetical format was OREIBASIOS, physician and friend of Julian the Apostate (reigned 361–363 CE), but papyri as early as 150 CE show recensions, as do allusions to PAMPHILOS OF ALEXANDRIA's alphabetical *Herbs* (*ca* 100 CE) nestled in Galēn (*Simples* 6. *pr* [11.792–796 K.]). The *De materia medica* achieved immediate popularity reflected in the citation by ERŌTIANOS, who mentions Dioskouridēs' synonyms for leopard's bane (31 [p. 51 Nachm.] = Diosk. 4.76).

Ed.: H. Stadler, "Dioscorides Longobardus (Cod. Lat. Monacensis 337)," *Romanische Forschungen* 10 (1897) 181–247, 369–445; 11 (1899) 1–121; 13 (1902) 161–243; 14 (1903) 601–635; M. Wellmann, *Pedanii Dioscuridis Anazarbei De materia medica libri quinque* (1906–1914; repr. 1958): **the edition cited**; C. Bonner, "A Papyrus of Dioscurides in the University of Michigan Collection," *TAPA* 53 (1922) 142–168; H. Mihăescu, *Dioscoride Latino Materia medica libro primo* (1938); C.E. Dubler and E. Terés, *La "Materia Médica" de Dioscórides. Transmisión Medieval y Renacentista* (1953–1959); Fr. Rosenthal, "Pharmacology" in *The Classical Heritage in Islam* (1975) 194–197; Scarborough and Nutton (1982) 187–227; A. Dietrich, *Dioscurides Triumphans. Ein anonymer arabischer Kommentar (Ende 12. Jahr. N. Chr.) zur Materia medica* (1988); *Idem*, *Die Dioskurides-Erklärung des Ibn al-Baitār* (1991); R. Flemming and A.E. Hanson, "2. *Dioscorides*, De materia medica *II 76.2 and 76.7–18*," in I. Andorlini, ed., *Greek Medical Papyri* I (2001) 9–35; M. Aufmesser, *Pedanius Dioscurides aus Anazarba Fünf Bücher über die Heilkunde* (2002); L.Y. Beck, *Pedanius Dioscorides of Anazarbus De materia medica* (2005).

Wellmann (1898); *Idem, Die Schrift des Dioskurides Peri Haplōn Pharmakōn* (1914); *Idem* (1916); H. Gerstinger, *Dioscurides Codex Vindobonensis Med. Gr. 1 der Österreichischen Nationalbibliothek. Kommentarband zu der Faksimileausgabe* (1970); *DSB* 4 (1971) 119–123, J.M. Riddle; M. Ullmann, "Pharmaceutics," in *Islamic Medicine* (1978) 103–106; J.M. Riddle, "Dioscorides," *CTC* 4 (1980) 1–143; O. Mazal, *Pflanzen, Wurzeln, Säfte, Samen. Antike Heilkunst in Miniaturen des Wiener Dioskurides* (1981); M.M. Sadek, *The Arabic Materia Medica of Dioscorides* (1983); A. Touwaide, "L'authenticité et l'origine de deux traits de toxicologie attributés à Dioscoride," *Janus* 70 (1983) 1–53; J.M. Riddle, "Byzantine Commentaries on Dioscorides," in Scarborough (1985b) 95–102; J.M. Riddle, *Dioscorides on Pharmacy and Medicine* (1985); A. Touwaide, "Un Recueil grec de pharmacologie du Xe siècle illustré au XIVe siècle: Le *Vaticanus Gr.* 284," *Scriptorium* 39 (1985) 13–56 with plates 7–8; Scarborough (1995); A. Touwaide, "Tradition and Innovation in Mediaeval Arabic Medicine. The Translations and the Heuristic Role of the Word," *Forum* 5.2 (1995) 203–213; *OCD3* 483–484, J.M. Riddle; A. Touwaide, "La thérapeutique médicamenteuse de Dioscoride à Galien: du *pharmaco-centrisme* au *medico-centrisme*," in A. Debru, ed., *Galen on Pharmacology* (1997) 255–282; M. Aufmesser, *Etymologische und wortgeschichtliche Erläuterungen zu De materia medica des Pedanius Dioscurides Anazarbeus* (2000); J.E. Raven, "Lecture 4: Primitive Medicine. The Rhizotomists and Druggists. Crateuas and the Illustration of Plants. The Codex Vindobonensis. Dioscorides' Herbal, its Nature and Influence," in F. Raven *et al.*, edd., *Plants and Plant Lore in Ancient Greece* (2000) 33–40.

John Scarborough

Diphilos (200 – 25 BCE)

Writer on machines listed by VITRUUIUS 7. *pr.*14, and to be distinguished from Q. TULLIUS CICERO's architect (CICERO, *ad Q. fr.*3.1.1).

RE 5.1 (1903) 1156 (#19–20), E. Fabricius.

PTK

Diphilos of Laodikeia (40 BCE – 180 CE)

Uncertainly dated grammarian; wrote on NIKANDROS' *Thēriaka*. Wellmann dated him to the High Roman Empire, solely because most commentaries on Nikandros were composed then. The only two testimonies about Diphilos do not concern Nikandros directly. (1) *Schol. Theokritos* 10.1–3b, Diphilos quotes *boukaios* as a proper name (*cf.* Theokritos *Id.* 10.1); but in Nik. *Thēr.* 5 and *fr.* 90 *boukaios* = *boukolos*, herdsman. (2) Athēnaios, *Deipn.* 7 (314d), *Diphilos ho Laodikeus* speaks of the torpedo-fish: referring to its efficacy even through a solid body, as for the Basilisk? If so, he augmented Nikandros' teachings which neither attribute this power to the Basilisk (*Thēr.* 396–410) nor mention the Torpedo. The *Schol. Nik. Thēr.* may have used Diphilos; this is however impossible to prove.

RE 5.1 (1903) 1155 (#18), M. Wellmann; Jacques (2002) 2.CXXXI (and n. 300).

Jean-Marie Jacques

Diphilos of Siphnos (300 – 250 BCE)

Greek physician, active at LUSIMAKHOS' court (306–281 BCE: Ath., *Deipn.* 2 [51a]). In *On diet for ill and healthy people*, discussing a wide variety of vegetables, fruits and other foods (including fishes), Diphilos describes the effects of single foods on human health and gives instructions for preparing them: Ath. 8 (355a).

RE 5.1 (1903) 1155, W. Schmid; J. Scarborough, "Diphilus of Siphnos and Hellenistic Medical Dietetics," *JHM* 25 (1970) 194–201; *KP* 2.97, F. Kudlien; *BNP* 4 (2004) 527, V. Nutton; *AML* 230, M. Stamatu.

Daniela Manetti

Dissoi Logoi (*ca* 400 BCE)

Short anonymous treatise, written in Doric and transmitted in the MSS on the *folii* following the text of SEXTUS EMPIRICUS, currently known by the initial words, *Dissoi Logoi* (*Double Arguments*). H. Estienne published it in 1570 under the title *Dialexeis*; Diels included it in the Early Sophistic.

According to most scholars, the treatise was composed about 400 BCE (*ca* 403–395 according to Robinson). The first five of the treatise's nine brief chapters treat moral (good/bad, beautiful/ugly, just/unjust), epistemological (true/false) and ontological questions (being/not being). The next four chapters refer to topics discussed by the sophists, such as whether virtue and wisdom can be taught, the assignment of offices by lot, the ideal of the wise man, and a short praise of memory.

The fact that the author could have been a student summarizing the controversy between two sophists expounding opposite viewpoints on the same topic would account for its imperfect literary form.

Although philosophically controversial, the *Dissoi Logoi* shows some interesting scientific aspects. The first five chapters expound two theses, the first of which, like PRŌTAGORAS' work, could be described as relativistic for two reasons: firstly, because it makes use of ethnographic accounts depicting the variation and opposition between ways of life and cultural and moral values in different societies or different social groups. Secondly, because it employs a language of dyadic predicates (good, beautiful *for* . . .) in one case and of complex propositions (just, true *if* . . .) in the other case. The second thesis, Socratic in character, develops arguments against the relativistic thesis. Therefore, the controversy concerns both anthropology and logic.

In the dispute between the defenders of both theses, some refined discursive devices show the high level attained in the art of criticism, such as the use of thought experiments (2.18, 6.12) or the distinction between the premises and the conclusion of an argument (6.13).

Ed.: *DK* 90; T.M. Robinson, *Contrasting Arguments. An Edition of the Dissoi Logoi* (1979).
DPA 2 (1994) 888–889, M. Narcy; M. Untersteiner, *I Sofisti* 3 (1967) 148–191.

José Solana Dueso

DIUISIO ORBIS TERRARUM ⇒ DIMENSURATIO PROUINCIARUM

Doarios (325 – 540 CE)

AËTIOS OF AMIDA 12.47 (p. 681 Cornarius) cites Doarios the bishop for a gout-remedy containing **shelf-fungus**, parsley, gentian, etc. The name seems otherwise unattested, and is likely corrupt: besides "Dareios," or else an ethnic based on Euhēmeros' mythical land Dōa (DIODŌROS OF SICILY 5.44.6–7), perhaps most likely is Daorsios, from the Hellenized Illyrian town of Daorsi, *cf.* POLUBIOS Book 32, *fr.*9.2, STRABŌN 7.5.5, and *BNP* 4 (2004) 78–79.

Fabricius (1726) 145.

PTK

Domitius Nigrinus (*ca* 10 BCE – *ca* 90 CE)

ASKLĒPIADĒS in GALĒN, *CMGen* 7.12 (13.1021 K.), cites his powerful ***akopon***, containing, among much else, mandrake and ***euphorbia*** (i.e., *post* IOUBA).

PIR2 D-155.

PTK

Domninos of Larissa (*ca* 430 – *ca* 475 CE)

Neo-**Platonist** philosopher and mathematician, studied under SYRIANUS with PROKLOS, who reports two of his theories (*In Tim.* I.109.30 and 122.18). DAMASKIOS' *Vita Isidori* describes him as "a philosopher," "of Syrian origin" (which, added to another anecdote concerning Domninos' disrespect for "the Syrian Law," may indicate he was a Jew), "from Laodikeia and Larissa, a Syrian city" (Laodikeia may refer to his residence in Thessalia). MARINOS calls him "philosopher and successor" (*Vita Procli* 26), implying that he may have been Syrianus' successor, which is not improbable but reliant upon meager evidence. Both Proklos and Damaskios respected him as an able mathematician, despite their strong opposition to his philosophical opinions; Syrianus, in contrast, held Domninos and Proklos in the same respect.

Domninos wrote a *Manual of introduction to arithmetic* (extant), wherein his *Elements of arithmetic* (lost) is attested. To Domninos is also attributed the tract *How one is to subtract a ratio from a ratio*. The first text points to a lucid and competent treatment of ancient arithmetic, based mainly on NIKOMAKHOS and EUCLID, with a preference for the latter. The first two treatises were meant to introduce the reading of PLATO, following THEŌN OF SMURNA's tradition. The third one, together with Proklos *In Tim.* 122.18, intimate Domninos' interest in mathematical astronomy.

Ed.: F. Romano, *Domnino di Larissa* (2000).
DPA 2 (1994) 892–896, A. Segonds.

Alain Bernard

Domnus (*ca* 450 – 500 CE)

Jewish physician, taught and was superseded by GESSIOS OF PETRA (STEPHANOS OF BUZANTION, s.v. *Gea; Souda* Gamma-207), listed as a commentator on the HIPPOKRATIC CORPUS, APHORISMS in pseudo-OREIBASIOS commentary.

RE 5.1 (1903) 1526, M. Wellmann; Stern 2 (1980) 678–679; P. Kibre, *Hippocrates Latinus* (1985) 31; Ihm (2002) #48.

Annette Yoshiko Reed

Dōriōn (Mech.) (200 – 25 BCE)

Writer on machines and inventor of the *lusipolemos*, listed by *P. Berol.* P-13044, col.8.

Diels (1920) 30, n.1.

PTK

Dōriōn (Biol.) (1st c. BCE)

Compiled gastronomical and dietetic treatises and authored a book *On Fishes*, where he gave names, descriptions and main characteristics of different species, apparently with great detail (Ath., *Deipn.* 7 [306e]). Dōriōn was concerned with lexicology and synonyms (*Deipn.* 7 [282c, 285a, 304c, 315f, etc.]) and fond of cookery books (e.g., Euthudēmos, *On Pickles*; EPAINETOS, *On Cookery*). He advised on culinary preparations (*Deipn.* 7 [287c, 300f], and 7 [309f]: seasoning garfishes) and offered one technical recommendation (using the juice of a fish called *gnapheus*, unfortunately not identified, against stains: 7 [297c]). Dōriōn, quoted 34 times, is considered Athēnaios' main source, although indirect, for Book 7 on fishes.

GGLA 1 (1891) 850; *RE* 5.2 (1905) 1563 (#3), M. Wellmann.

Arnaud Zucker

Dōrotheos of Athens (325 BCE? – 79 CE)

Author of a medical poem quoted by PLINY (22.91) for a herb called *condrion* that could be helpful for stomach and digestive ailments. Dōrotheos is also cited among Pliny's sources: 1.*ind.*12 (on the nature of trees) and 1.*ind.*13 (on foreign trees). He is probably different from DŌROTHEOS OF HĒLIOPOLIS.

FGrHist 145; *RE* 5.2 (1905) 1571 (#19), M. Wellmann.

Claudio Meliadò

Dōrotheos of Hēliopolis (250 BCE – 95 CE)

Consulted by ASKLĒPIADĒS PHARM. in GALĒN *Antid.* 2.14 (14.183, 187 K.) on cures for snake bites. Perhaps identical to Dōrotheos *medicus*, possibly from Egypt, whom PHLEGŌN OF TRALLEIS mentioned (*On Marvels* 26). Identification of this doctor with the Dōrotheos quoted by PLINY (22.91) is doubtful.

RE 5.2 (1905) 1571 (#19), M. Wellmann.

Jan Bollansée, Karen Haegemans, and Guido Schepens

Dōrotheos of Khaldaea **(250 BCE – 50 CE?)**

Wrote *On stones*. PSEUDO-PLUTARCH *De fluu.* 23.3 (1165A) preserves a single fragment of the second book, regarding the stone *sikuonos*. Among various identifications proposed are DŌROTHEOS OF SIDŌN, and the homonymous author of the *Pandektē* quoted by Clement of Alexandria, *Str.* 1.21.133. According to Jacoby, he is entirely fictive.

E. Hiller, "Zur Quellenkritik des Clemens Alexandrinus," *Hermes* 21 (1886) 126–133 at 129; *GGM* 1.LIII; *RE* 5.2 (1905) 1571 (#15), E. Schwartz; Schlereth (1931) 114–115; Jacoby (1940) 95–96; *FGrHist* 289; Halleux and Schamp (1985) XXXVI, n.8; De Lazzer (2000) 64–66; De Lazzer (2003) 81–82.

Eugenio Amato

Dōrotheos of Sidōn (50 – 100 CE)

Authored a widely influential astrological poem in Greek hexameters, comprising five books, addressed to "his son, Hermēs." Only brief excerpts of the original text survive in quotations by later authors, but an Arabic translation of a lost Pahlavi version of the whole is extant (see PAHLAVI, TRANSLATIONS INTO). At the beginning of the poem, Dōrotheos, calling himself an Egyptian, claims to have traveled through Babylōn as well as Egypt, but these are presumably fictions. That Dōrotheos was active in the 1st c. CE is established by eight horoscopes dating from 7 BCE to 43 CE and included for illustrative purposes. Dōrotheos' work, notwithstanding its use of verse as a medium, is a practical handbook. The first four books address interpreting personal horoscopes to determine the individual's character and the course of his life; the fifth concerns ***katarkhic*** astronomy.

Ed.: D.E. Pingree, *Dorothei Sidonii Carmen Astrologicum* (1976).
Irby-Massie and Keyser (2002) 93–96 (partial trans.).

Alexander Jones

DŌSITHEOS (ASTRON. II) ⇒ DIODŌROS (ASTRON.)

Dōsitheos (pharm.) (30 BCE – 540 CE)

AËTIOS OF AMIDA 8.70 (*CMG* 8.2, p. 530), records his opium-based pill for "blood-spitting" (*cf.* ***phthisis***), containing also frankincense, ***lukion***, myrtle, saffron, roses, etc.; and PAULOS OF AIGINA 7.11.45 (*CMG* 9.2, p. 308), his liver-pill, containing aloes, ***kostos, malabathron***, **mastic, shelf-fungus**, etc.

Fabricius (1726) 146.

PTK

Dōsitheos of Pēlousion (250 – 210 BCE)

Student of KONŌN and a correspondent of ARCHIMĒDĒS. He wrote and observed in Alexandria, and perhaps on Kōs. The name, meaning "god-given," is typically Jewish, so it may translate Nathaniel. After Konōn died, Archimēdēs addressed four works to Dōsitheos, providing requested proofs, while acknowledging Dōsitheos' familiarity, not expertise, with geometry, although according to DIOKLĒS, *On Burning Mirrors* 1, he was the first to discover the focal property of the parabola. His astronomical contributions chiefly concerned the calendar, on which he wrote three works: *Appearances of Fixed Stars* (rising and setting dates), *Weather-signs* (seasonal weather-predictions based on astronomical phenomena), and *On the* ***Oktaetēris*** *of* EUDOXOS (all lost). Notes from the first and second are preserved in the calendar appended to GEMINUS' *Introduction*, in PLINY, and in PTOLEMY's *Phaseis*. A work entitled *To Diodōros* (an exceedingly common name) apparently gave information on the life of ARATOS.

R. Netz, "The First Jewish Scientist?" *SCI* 17 (1998) 27–33; *BNP* 4 (2004) 695 (#3), M. Folkerts.

PTK

Douris of Samos (*ca* 340 – 260 BCE)

Greek historian and tyrant of Samos, claimed descent from Alkibiadēs, probably born in Sicily after his family's banishment from Samos when Athens captured the island from Persia in 366 BCE. His father, Kaios, an Olympic boxing victor, tyrant of Samos, had three sons: Douris who inherited the Samian tyranny, Lunkeos, a comic poet and friend of Menander, and Lusagoras involved in Samian politics. In about 304–302 BCE Douris and his brother Lunkeos studied under THEOPHRASTOS in Athens, returning to Samos in 300. Douris composed several historiographical works including a biography of Agathoklēs the Sicilian tyrant (at least four books), a history of Macedon (at least 23 books), and a local history of Samos (at least two books). He also wrote various (lost) works on tragedy, art, laws and competitions. His interest in Macedon and Samos had some geographical undertones; AGATHARKHIDĒS expressed appreciation of Douris' work. His style followed the Hērodotean tradition, emphasizing fascination and amusement.

Ed.: *FGrHist* 76; P. Pédech, *Trois historiens méconnus: Théopompe, Duris, Phylarque* (*CUF* 1989) 255–389; F. Landucci Gattinoni, *Duride di Samo* (1997).

R.B. Kebric, *In the Shadow of Macedon: Duris of Samos* (1977); D. Knoepfler, "Trois historiens hellénistiques: Douris de Samos, Hiéronymos de Cardia, Philochore d'Athènes," in *Histoire et Historiographie dans l'Antiquité* (2001) 25–44.

Daniela Dueck

Drakōn of Kerkura (80? – 120 CE)

Wrote an *On stones* (PLUTARCH, *QR* 22.41 and Ath., *Deipn.* 15.46 [692d]). Drakōn may postdate PLINY to whom he was apparently unknown. However, his interest in the Janus legend (to whom the invention of the crown, rafts, boats and bronze coinage are attributed) might suggest the Augustan age.

RE 5.2 (1905) 1663 (#16), M. Wellmann.

Eugenio Amato

Drakōn of Kōs (400 – 350 BCE)

GALĒN, commenting on two Hippokratic treatises, mentions Drakōn the son of HIPPOKRATĒS, and brother of THESSALOS OF KŌS, and suggests that some claimed that Drakōn authored them: *In Hipp. Nat. Hom* 2.1 (*CMG* 5.9.1, p. 58) and *In Hipp. Prorrhet. I* 2.17 (*CMG* 5.9.2, p. 68). SŌRANOS, *Vita Hipp.* 15 (*CMG* 4, p. 178), describes the family; *Souda* Delta-1497 distorts that account.

Von Staden (1989) 64; van der Eijk (2000–2001) *fr.*13.

PTK

"Dtrums" (230 – 30 BCE?)

Wrote a Greek work on burning mirrors surviving only in Arabic, unknown beyond his text itself, which has only one internal reference, to an anonymous *Katoptrika*. Rashed renders the author's original name, distorted beyond recognition in Arabic transliteration and subsequent tradition, according to an ad-hoc transliteration of the Arabic characters used for the name: *DTRWMS*. The treatise's level and contents are comparable to DIOKLĒS', and may suggest a Hellenistic date, but the methods used indicate no dependence of one treatise on the other. There is, furthermore, no clear dependence or influence on ANTHĒMIOS, Didumos (also edited by Rashed 2000, and post-Anthēmios), or the Bobbio fragment (see Rashed 1997). The Arabic translator has explicitly replaced the first two parts, treating proprieties of conic sections, with excerpts of APOLLŌNIOS' *Kōnika*. Only the third part is translated from Dtrums' Greek; it first addresses the properties of the parabolic mirror, including a skillful and original point by point construction of the parabola, given its axis and diameter (*prop.* 12 and 13). The end discusses the burning properties of the spherical mirror (*prop.* 14 and 15) and includes an original discussion of the path of reflected sunrays, coming to meet the axis after more than one reflection on the mirror.

Ed.: R. Rashed, *Œuvres philosophiques et scientifiques d'al-Kindi*, v. 1, *L'optique et la catoptrique* (1997) 117–120; Rashed (2000) 153–213.

Alain Bernard and Kevin van Bladel

Dulcitius (180 – 360 CE)

OREIBASIOS, *Ecl. Med.* 114.8 (*CMG* 6.2.2, p. 289), cites his remedy against warts of all kinds (*thumoi, murmēkiai, akrokhordonai*): fava beans, bruised, pounded, and applied. The name is attested from the late 2nd c. CE to the era of Oreibasios: *CIL* 3.7088, 3.12030, *LGPN* 4.111 (Doulkitios), *PLRE* 1 (1971) 273–274, esp. Libanios, *Orat.* 42.24.

(*)

PTK

E

Egnatius (of Spain?) (*ca* 100 – 50 BCE?)

Wrote a poem *De Rerum Natura* in at least three books, of which MACROBIUS (*Sat.* 6.5.2, 12) preserves two very short passages: in *fr.* 1 Blänsdorf, Egnatius speaks about metal working, and in *fr.* 2, he describes the Moon (Phoebē) setting or disappearing at dawn. Egnatius lived between Accius and VERGIL (150–50 BCE), and probably was a contemporary of LUCRETIUS. Bergk and Baehrens identified him with the Egnatius Celtiber mentioned by Catullus (*Carm.* 37 and 39), a rather unlikely conjecture. It is also impossible to ascertain if Egnatius were an imitator of Lucretius or wrote independently.

Ed.: N. Marinone, "I frammenti di Egnazio," in *Poesia Latina in frammenti* (1974) 179–199; *FLP* 147–148.

BNP 4 (2004) 842 (#I.4), P.L. Schmidt.

Claudio Meliadò

Eirēnaios (250 BCE – 25 CE)

Pharmacist whose remedy for *uitiligo* (psoriasis) comprised ***alkuoneion***, natron, cumin, and dried fig leaves, pounded with vinegar, to be applied under sunlight and washed off to prevent corrosion (CELSUS 5.28.19C).

RE 9.2 (1916) 2032 (#3), H. Gossen.

GLIM

Ekhekratēs of Phleious (400 – 360 BCE)

Student of PHILOLAOS and of EURUTOS (DIOGENĒS LAËRTIOS 8.46; IAMBLIKHOS *VP* 251, 267), he described Sōcratēs' last day to Phaedo, and sympathized with the view that the soul "is a kind of harmony" (PLATO, *Phaedo* 57a, 88d–e). A later legend suggested that Plato visited Ekhekratēs at Lokri (pseudo-Plato *Epist.* 9 [358b]; CICERO *Fin.* 5.87; Val. Max. 8.7. *ext.*3).

DK 53; *BNP* 4 (2004) 781 (#2), C. Riedweg; *OCD3* 501 C. Roueché.

GLIM

Ekphantos of Surakousai (400 – 350 BCE?)

Ekphantos belongs to a group of later **Pythagoreans** active in Surakousai in the first part of the 4th c. (DK 50–51, 55). As distinct from other later **Pythagoreans**, Ekphantos' theories are described in some detail in the doxographical tradition (DK 51 A1–5), implying that he wrote a treatise on natural philosophy that was available to THEOPHRASTOS. In the catalogue of **Pythagoreans**, compiled by ARISTOXENOS, his birthplace is given as Krotōn (DK 59 A1), but otherwise he is from Surakousai (A1–2, 5). He must be earlier than HĒRAKLEIDĒS OF HĒRAKLEIA PONTIKĒ, who accepted his theory that the Earth rotates around its own axis. Ekphantos could have been a follower of PHILOLAOS, though not necessarily his pupil.

As a philosopher Ekphantos is an example of an eclectic, typical of late Pre-Socratics. Following DĒMOKRITOS he taught that the world consists of indivisible bodies (*adiaireta sōmata, atoma*) and void (A2), but is governed by "divine power, which he calls 'mind' and 'soul'," as ANAXAGORAS believed, and not by necessity (A1, 4). The idea that he identified these *adiaireta sōmata* with "**Pythagorean** *monadas*", i.e. arithmetical units (A2), which gave rise to the **Pythagorean** "number atomism," is unattested in the other testimonia on Ekphantos and comes most probably from doxographers. Ekphantos' skepticism ("it is not possible to attain a true knowledge of things," A1) is close both to the epistemological stance of some **Pythagoreans** (ALKMAIŌN, Philolaos) and of Dēmokritos. His astronomical hypothesis (attested also in his contemporary HIKETAS) develops Philolaos' theory that the Earth rotates around the Central Fire in 24 hours. Ekphantos abandoned Philolaos' ideas on the Central Fire and the Counter-Earth, returned the Earth to the central place in the universe, and asserted that it moves about its own center from west to east (A1, 5), in order to explain the apparent diurnal rotation of the heavens. Copernicus mentioned both Ekphantos and Hērakleidēs in the preface to his *De revolutionibus*.

DK 51; T.L. Heath, *Aristarchus of Samos* (1913); Dicks (1970).

Leonid Zhmud

Elephantinē/Elephantis (100 BCE – 75 CE)

PLINY 28.81 cites her for quasi-magical abortifacients, and SŌRANOS, in GALĒN *CMLoc* 1.1 (12.416–420 K.) lists her, with ASKLĒPIADĒS, HĒRAKLEIDĒS, and MOSKHIŌN, as providing recipes for *alōpekia*. The *Souda* A-4261 blames her or a homonym for a work on sexual positions.

RE 5.2 (1905) 2324–2325 (#3), O. Crusius; Parker (1997) 145 (#43).

PTK

Eleutheros (900 – 1450 CE)

Physician, credited with a treatment for sciatica compounded from the juice of wild figs and radish mixed with olive oil and applied externally or injected internally as an enema, and preserved in MS Antinori 102 of Florence, Medicea Laurenziana, *ca* 1460 CE, f. 358. The MS probably comes from the collection of a medical library in Constantinople. It excerpts canonical writers such as HIPPOKRATĒS, DIOSKOURIDĒS, GALĒN, and PAUL OF AIGINA, plus physicians of the mid- to late-Byzantine period: Theophanēs Khrusobalantēs "Nonnos" (mid-10th c.), Sumeōn Sēth (mid-11th c.), Nikēphoros Blemmidēs (mid-13th c.), etc., thus suggesting the date-range.

Diels 2 (1907) 35.

Alain Touwaide

Pseudo-Elias (Pseudo-David) **(600 – 726 CE?)**

Anonymous collection of 51 lectures, replete with medical learning, on PORPHURIOS' *Isagoge* (lectures 1–7 are lost), which the MS tradition connects to commentaries by Elias (on the *Isagoge* and ARISTOTLE's *Categories, CAG* 18.1) and David (on the *Isagoge, CAG* 18.2). The author seems Christian and probably taught at Constantinople. He cites GALĒN by name (pp. 17.22, 24.12, 28.27–8, 35.3): e.g., ginger, pepper, and ***purethron*** exhibit similarity in difference in degree, "as Galēn writes" (p. 14.4–5; *cf.* Galēn *Simples* 6.6.2 [11.880–882 K.], 8.16.11 [12.97 K.], 8.16.41 [12.110 K.]). The author distinguishes corporeal and incorporeal bodies, simple and composite bodies, and composite bodies in equilibrium or dominated by one property (e.g., wet, cold: pp. 35.2–4; *cf.* Galēn, *Bones for Beginners, pr.* [2.733 K.]; contrast David, *CAG* 18.2 [1904] 151.18–28, who makes only the first distinction). The author employs medical technical terminology (pp. 18.5: *epidiaresis*; 29.29: *antembainein*; 45.13: *analōsis*), examples (p. 19.4: finger as a continuous quantity), and metaphors (p. 13.23: suffering is to the soul as painful surgical cuts are to the ill). Westerink (p. XV) surmises the author may be "a professor of medicine giving an elementary course in logic." Our author considered himself a philosopher, but misunderstood PLATO and basic Aristotelian logic. Differences in presentation, style, emphasis, and approach to Porphurios' text militate strongly against identifying the author with either Elias or David.

Ed.: L.G. Westerink, *Pseudo-Elias (Pseudo-David): Lectures on Porphyry's Isagoge* (1967).

GLIM

Emboularkhos (?) (30 BCE – 540 CE)

AËTIOS OF AMIDA 16.142 (Zervos 1901: 171) cites his fumigation recipe, containing **bdellium**, cassia, cinnamon, saffron, ***malabathron***, myrrh, spikenard, fresh and dried roses, ***sturax***, etc. The name is otherwise unattested and seems incorrectly formed; Boularkhos is attested through the 1st c. BCE (*LGPN*), and perhaps Euboularkhos, though unattested, is correct; alternatively, perhaps emend *ΕΜΒΟΥΛ-* to *ΠΟΛΥ-* (*cf.* POLUARKHOS, cited for gynecological remedies).

Fabricius (1726) 148.

PTK

Emeritus (Hemeritos) (100 BCE – *ca* 400 CE?)

Author of remedies quoted in PELAGONIUS, who calls Emeritus *mulomedicus*, "horse-doctor." The remedies are for cough (85, 99, 110); dysury (153); *opisthotonos* (272, 274); and colic (290). Three, translated into Greek, figure in the *Hippiatrika*: on pneumonia (Pel. 72 = *Hippiatrica Berolinensia* 7.5), cough (Pel. 85 = *Hippiatrica Parisina* 564), and a caustic ointment for shoulders and hips known in Latin only from the Einsiedeln MS (XXXII.519, Corsetti, 53–54 = *Hippiatrica Parisina* 963 = *Hippiatrica Berolinensia* 96.23).

Fischer (1980); P.-P. Corsetti, "Un nouveau témoin de l'*Ars veterinaria* de Pelagonius," *Revue d'histoire des textes* 19 (1989) 31–56; *CHG* vv.1–2; McCabe (2007).

Anne McCabe

Empedoklēs of Akragas (*ca* 460 – 430 BCE)

Philosopher-poet and natural scientist, born *ca* 483 BCE, author of one or two lost didactic epics, the *On Nature* and *The Purifications*. His prominent family secured victories in the chariot-race at the Olympics, and retained its position after the fall of the tyranny in Akragas. Diogenēs Laërtios records Empedoklēs' involvement in early struggles for the democratic regime (8.64–66), which may have some independent basis, for his source, Timaios (*FGrHist* 566 F 2), remarks that Empedoklēs' democratic leanings seem at odds with his lordly and conceited posture in his poetry. This presumptuous tone, however, probably inspired his colorful figure in the biographical tradition, including the tale of his leap into the flaming caldera of Aetna.

Empedoklēs' poetry survives mainly from citations in later authors, especially Aristotle and Simplicius, but a recently-reconstructed papyrus containing 74 lines of four continuous sections (a, b, c, d) brings the extant total to *ca* 490 lines. DK divide our fragments between two works, following the thematic affiliations of the two titles. Thus, *On Nature* discussed natural science, while *The Purifications* told of the exile of the soul and its struggle to regain its place over several reincarnations. Some recent scholarship, however, prefers a single poem, combining both themes. The debate continues. Only Diogenēs Laërtios (8.77) gives both titles, but even he, perhaps considering them a unit, provides a single verse-sum for both. Other authors mention either no titles or only one. The opening of section d of the Strasbourg papyrus, omitting a title, overlaps with a number of lines which Simplicius records from *On Nature*, and contains a discussion of reincarnation, including the previously known fragment B 139, cited from *The Purifications*. The second half of section d shifts to the origin of life, material suited to the *On Nature*. This does not eliminate the possibility of two original works, but now it seems that *On Nature* also discussed the after-life.

Empedoklēs' most lasting influence on Western science remains his theory of the four elements – earth, air (sometimes ***aithēr***), fire and water – the permanent building blocks of the universe, adopted, with modifications, by most subsequent ancient philosophical schools except the **Atomists**. Less historically influential, but equally central to Empedoklēs' physical system, was his doctrine of the cosmic cycle driven by two equal and opposite moving/volitional powers, Love and Strife, sharing dominion over the elements, Love combining and Strife separating them. Each power always eventually achieves, in alternation, full sway over the elements.

Thus, the universe oscillates between two extreme states, during which no world can come to be, because of the exclusive predominance of Love or Strife over the elements. Under the rule of Love, all four elements become harmoniously fused into one all-embracing super-organism, which Empedoklēs calls the *Sphairos* god, while under the rule of Strife the four elements either separate into different places, or perhaps slide into chaos (the evidence is unclear). Only in the middle periods do worlds like ours occur.

The apparent motivation for the theory seems to be a commitment to non-emergence (i.e., no state has ontological priority to any other), and through it, an attempt to respond to Parmenidēs' critique of change. Aristotle provides an important hint (*GC* 1.1 [315a19–20]), wondering if one ought not to consider the *Sphairos*-god as having an equal claim to be a first-principle, alongside the elements. That is, perhaps neither elements nor *Sphairos* are prior to each other, but merely extreme limits of the two-way never-ending process of becoming. Thus, becoming as a whole might acquire eternal and invariant limits, like Parmenidēs' Being.

Within this framework, Empedoklēs aimed to be as encyclopedic as possible. Both the fragments and doxography include passages on physics/cosmology, botany, zoology, physiology, reproduction and sense-perception. Also attested is a critique of Greek religion and ritual, especially animal sacrifice, based on Empedoklēs' **Pythagorean** belief in reincarnation.

The over-all unity of Empedoklēs' thought remains perplexing, although, since both reincarnation and physics appear in a single passage, it can no longer be denied. The problem is whether or not Empedoklean physics and reincarnation can be accommodated in the same system. At a minimum, many difficulties can be avoided if Empedoklēs' transmigrating soul is not anachronistically identified with PLATO's immortal soul.

Ed.: DK 31; M.R. Wright, *Empedocles* (1981); A. Martin and O. Primavesi, *L'Empédocle de Strasbourg, P. Strasb. Gr. 1665–1666* (1999); B. Inwood, *The Poem of Empedocles*, 2nd ed. (2001).

D. O'Brien, *Empedocles' Cosmic Cycle* (1969); P. Kingsley, *Ancient Philosophy, Mystery and Magic* (1995); *DPA* 3 (2000) 66–88, R. Goulet; Simon Trépanier, *Empedocles, an Interpretation* (2004); *NDSB* 2.395–398, *Idem*.

Simon Trépanier

Q. Ennius of Rudiae (*ca* 205 – 169 BCE)

Ennius © Rheinisches Landesmuseum Trier

Latin poet, born 239 BCE in Rudiae (near Lecce), had a Greek cultural formation. In 204 he came from Sardinia to Rome with CATO, where he taught Greek. A member of Scipio's retinue, he obtained Roman citizenship in 184. Only fragments of his diverse works survive. Ennius wrote tragedies (*Andromache*, *Medea*, *Telamon*, inspired by Greek models), poems, comedies, and *saturae*. In the *Euhemerus*, Ennius expounded the successive reigns of Sky, Saturn, and Jupiter, stressing particularly, in agreement with Euhēmēros' theories, their human characteristics. In the *Epicharmus*, Ennius identified gods with primordial elements, whence the ***kosmos*** arose (*cf.* EPIKHARMOS). In the epic *Annales*, he celebrated Roman history from her origins to his own time. The **Pythagorean** theory of metempsychosis is suggested in Ennius' claim to be HOMER's reincarnation.

J. Vahlen, *Ennianae poesis reliquiae*, 3rd ed. (1928); E.H. Warmington, *Remains of Old Latin*, v. 1 (Loeb 1935); O. Skutsch, *Ennius* (1972); *Idem, The Annals of Ennius* (1985).

Bruno Centrone

Epagathos (100 BCE – 80 CE)

ANDROMAKHOS, in GALĒN *CMLoc* 9.5 (13.300–301 K.), cites his enema for "dysentery," composed of orpiment, realgar, and wild pomegranate flower (*balaustion*), in old dry wine. The Greek name is frequent from the 1st c. BCE, and unattested prior; his seeming *nomen* "Deletius" is unexplained (perhaps "Dēmētrios"?). If the remedy for blood-spitting, also containing *balaustion*, cited *ep' agathou kathēgētou* by Andromakhos, *ibid.* 7.4 (13.79), is by the same man, perhaps emend *ΔΗΛΗΤΙΟΥ* to *ΚΑΘΗΓΗΤΟΥ* ("teacher"; and thus set the *terminus post* as *ca* 40 CE).

Fabricius (1726) 136.

PTK

Epainetēs (100 BCE – 100 CE?)

Greek toxicologist, who wrote on iology (*Thēriaka*), often mentioned by pseudo-AELIUS PROMOTUS (*On Venomous Animals and Poisons*), who calls him an herbalist (*rhizotomos*) and presents under his name various remedies for intoxication: leopard's bane (53), hemlock (63), mandrake (65), opium poppy (64), henbane (66), deadly mushrooms (67), a plant called black chameleon (70), bull's blood (71), gypsum (72) and sea-hare (79). See EPAINETOS.

BNP 4 (2004) 1011 (#1), V. Nutton.

Arnaud Zucker

Epainetos (*ca* 90 BCE)

Writer of an *On Cookery* (*Opsartutika*) often mentioned by Athēnaios (esp. *Deipn.* 12 [506c]), who preserves a fragment giving a recipe for *muma* (*Deipn.* 14 [662d]) and repeatedly quotes him for lexical remarks on food, strongly suggesting that Athēnaios knew Epainetos' book through a grammarian (see *Deipn.* 9 [387e]). The titles *On Vegetables* and *On Fishes*, if not erroneous, must have been chapters of *On Cookery*. The identification of Epainetos with EPAINETĒS, formerly accepted (and still plausible since gastronomy, dietetics and toxicology are closely related), appears now at least doubtful.

RE 5.2 (1905) 2672–2673 (#9), L. Cohn; *BNP* 4 (2004) 1011 (#2), G. Binder.

Arnaud Zucker

Epaphroditos (Meteor.) (unknown date)

Wrote a "Commentary on ARISTOTLE's Discussion of the Halo (of the Moon) and the Rainbow," as noted by Ibn-al-Nadīm from writing of the Aristotelian Yaḥyā ibn-ʿAdī (d. 974). Thābit ibn-Qurra's (*ca* 826–901) Arabic translation has not yet been found in Arabic MSS.

GAS 7 (1979) 230.

Kevin van Bladel

Epaphroditos and Vitruuius Rufus (200 – 300 CE?)

A collection of geometrical problems to be found in Latin gromatic MSS (i.e. collections of texts about land surveying) has survived with these two otherwise unknown names attached to it; but Lachmann did not include them in his edition of the corpus. Following the same order as that in the works attributed to HĒRŌN OF ALEXANDRIA (*Metrika* I, authentic, and *Geōmetrika*, considered apocryphal), whose influence is obvious, the calculations of perimeters and areas of triangles, of quadrangular figures, regular polygons, and of the circle and its segments are all dealt with practically, with detailed figures but no attempt at demonstration, which is a great difference from the *Metrika*. Surprisingly, the polygonal areas (pentagon and so on up to dodecagon) are here dealt with arithmetically, not geometrically; they are looked at in the **Pythagorean** manner as sums, not products. The origin of these developments ought to be looked for in DIOPHANTOS' treatise *Polygonal Numbers*, which provides evidence for dating. As they show similarities with the *Podismus* (Lachmann, pp. 295–301), Epaphroditos' and Vitruuius' excerpts may bear some link with the calculation of triangular, trapezoidal, and pentagonal *subseciua* (minor areas of a centuriation not allotted to any owner), such as presented by IUNIUS NIPSIUS (Lachmann p. 290).

Ed.: N. Bubnov, *Gerberti opera mathematica* (1899); *CAR* 3 (1996).

Jean-Yves Guillaumin

Epaphroditos of Carthage (25 – 80 CE)

ANDROMAKHOS records that he used – as patient or doctor? – an antidote from ZŌILOS as a once-a-year prophylactic: birthwort, clover, ***herpullos***, myrrh, ***opopanax***, pimpernel, ***skordion***, germander, etc., plus bitumen and sulfur, in wine: GALĒN, *Antid.* 2.12 (14.178–179 K.).

RE S.9 (1962) 36 (#7), J. Kollesch.

PTK

Ephoros of Kumē (360 – 330 BCE)

Wrote a *History of Kumē*, a work *On Words*, and a work *On Inventions*. His *Histories* in 30 books – lost but frequently cited by later writers – traced the history of the inhabited world from the return of the Hērakleidai to the siege of Perinthos in 340 BCE (the final book was written by his son Dēmophilos). Arranging his history by nation (*kata genos*), he took a particular interest in geography. In Books 4–5, he gave a geographical overview of the ***oikoumenē***, covering Europe and Asia respectively. A particularly important fragment, preserved by STRABŌN (1.2.28) and KOSMĀS INDIKOPLEUSTĒS, represents the Earth as a flat rectangle, bordered to the north by Skuthians, to the east by Indians, to the south by Aithiopians (Ethiopians), and to the west by Kelts, an advance over the older Ionian view of the world as a circle surrounded by the River Ocean. Ephoros likely arranged his conspectus of the lands along the standard lines, following the Mediterranean coast from western Europe, covering Greece and the Pontos, moving down the eastern Mediterranean to north Africa, and ending with the African coast outside of the Straits of Gibraltar. Ephoros showed particular interest in the historical geography of places, their early inhabitants, and foundation accounts of cities. He speculated on the origins of the Nile flood, proposing that the ground soaks up water like a sponge in the cool months and sweats it out in the hot months. He believed that the Tanais (Don) river originated in a sea of unknown extent. He attributed the cause of the great earthquake and flood that destroyed Helikē and Boura in Achaia in 373 to a comet seen before the earthquake, which split into two planets.

Ed.: *FrGrHist* 70.

W.A. Heidel, *The Frame of the Early Greek Maps* (1937) 16–17; Chr. van Paassen, *The Classical Tradition of Geography* (1957) 246–253; G.L. Barber, *The Historian Ephorus* (1985); G. Schepens, "The Phoenicians in Ephorus' Universal History," in *Studia Phoenicia* 5 (1987) 315–330.

Philip Kaplan

Epic- ⇒ Epik-

Epicurus of Samos (310 – 270 BCE)

Epicurus © Roma, Musei Capitolini, Archivo Fotografico dei Musei Capitolini

1. Life and Writings. Epicurus (Epikouros) was an Athenian citizen born on the island of Samos. He founded the **Epicurean** school, called the **Garden** (*kēpos*), in Athens around 307 BCE, having taught previously at Mutilēnē on Lesbos and at Lampsakos. Epicurus developed the atomic theory of Leukippos and Dēmokritos (which he had studied with the atomist Nausiphanēs) and wrote prolifically: 300 books are recorded, most lost. Extant works include three "epitomes" discussing physics (*Letter to Herodotus*), ethics (*Letter to Menoeceus*), and meteorology and astronomy (*Letter to Pythocles*). In addition, there are two collections of short sayings (the *Principal Doctrines* and *Vatican Sayings*), and fragments of other works, most notably his major work in 37 books, *On Nature*.

2. Physics. In his physics, Epicurus adapted earlier **atomism** to meet the criticisms of Aristotle and others. He taught that there exist indestructible atoms and the void (empty space), and that all other objects in the world are compound bodies made up of atoms moving in the void. The universe is infinite in all directions, and there are an infinite number of variously shaped atoms moving constantly through empty space. He claimed that while the number of atoms in the universe was infinite, the sizes and shapes they could take were not. He denied that individual atoms could ever be so large as to be visible to the naked eye. He posited that there are three types of atomic movement: (1) a natural motion downward caused by the weight of the atom (How Epicurus defined the direction "down" in an infinite void is not fully understood); (2) forced motion in all directions caused by collisions with other atoms, and (3) a minimal, completely random motion of the atom he called the "swerve" (Greek *parenklisis*; Latin *clinamen*). He posited the random swerve in his physics in order to explain how atoms, falling naturally downward at the same high speed, can cross each other's path and collide. At times atoms move about separately, but at other times they come together to create different worlds comprised of various compounds. Even in compound bodies, however, atoms are in ceaseless motion, traveling at a constant and incredibly high speed. Compound bodies and the worlds of which they are a part are transient, coming into being and passing away. Only atoms and the void are eternal and indestructible, having no beginning and no end. Epicurus also posited several physical theories criticized by later ancient philosophers, including that the Earth is flat and rests on a gradually less and less dense foundation, and that the Sun and stars are very small, in fact about the size that they appear to us.

Like all compound bodies, humans consist of atoms. Epicurus taught that both body and soul were corporeal: the body was made of relatively large, dense atoms, and the soul, responsible for sensation and thinking, of several types of small, light, and mobile atoms. Perceptions arise when images (*eidōla*) flow off of physical objects and strike the sense organs. Sight, for instance, occurs when thin, swift moving images fly off of objects and strike the eyes. Thought is caused by even thinner images directly striking the mind, which Epicurus located in the chest near the heart. He held that there are an almost infinite number of different images flying around us at any time on which our thoughts can focus. The process of thinking is thus a focusing of the mind on one external image after another. At death, the soul atoms escape from the body and disperse. Epicurus taught that there is no afterlife, since the soul does not survive after death, and held that therefore we should fear neither death nor punishment in the afterlife. Although a strict materialist, Epicurus was not an atheist. He held that the gods existed, but were completely blessed creatures who lived lives of perfect pleasure and had nothing to do with our world.

3. Scientific Method. Epicurus shunned traditional logic, substituting what he labeled "canonic" (from the Greek word *kanōn*, "rule, standard"), his term for his theory of knowledge that he connected closely to physics. Epicurus was an empiricist, teaching that knowledge was possible and derived from sensation. He held that there were three criteria of truth: sensation, general concepts, and feelings. Sensation was the primary criterion of truth. He said "all sensations are true," a claim which at first sight appears implausible. Epicurus, though, carefully distinguished sensations themselves from the judgments that people make about them. In the case of an optical illusion like an oar appearing bent when partially submerged in water, Epicurus would say that the image of the oar that reaches our eyes is true: we see an image made up of certain sizes, shapes, and colors. Error occurs when we add false judgments to our perceptions, such as "this oar is bent." Sensation has not fooled us, but our interpretation of the sensation that has reached our eyes. Our knowledge of the world is ultimately based on sensations, and the judgments we make on the basis of sensation must be scrutinized for possible error. An important way to avoid making errors of judgment and attain knowledge is by attending to "general concepts" (*prolēpseis*). Epicurus maintained that general concepts could function as a criterion of truth. He believed that humans form general concepts by generalizing from their sensations. From such general concepts, people make statements that are true and false about objects in the world. Epicurus' third criterion of truth was "feelings" (*pathē*). He taught that our actions must be judged by the primary feelings of pleasure and pain, and took them to be the criterion of ethical truth. All our actions must be directed to maximizing our pleasure and minimizing our pain in the long run.

Relying on these criteria of truth, Epicurus argued that we could gain knowledge not only of the visible world, but also of the microscopic world of atoms and the distant movements of the heavens. When we are investigating the visible world directly accessible to us, Epicurus taught that we should accept as true things verifiable by direct and clear observation, and false what we cannot so verify. But when we are investigating the underlying principles of matter (e.g., atoms and the void) or the heavens, realms that we cannot examine directly, he argued that we must make use of analogies with the physical world, and take as true "uncontested" views and as false those that are "contested." For example, Epicurus argues that the only view that can explain the workings of the physical world around us is **atomism**, because it alone accounts for and does not conflict with the facts of the world as we see them. Similarly, when discussing the movements of the heavens, Epicurus posits explanations that are not contradicted by the evidence. At the microscopic level, though, only one theory, **atomism**, fits all the facts, whereas in astronomy and meteorology there are often several hypotheses that are not contradicted by the phenomena. For example, Epicurus posited a number of possibilities for why the Moon waxes and wanes, all of which he says may be true. Only one of the possibilities will in fact be true for our Moon, but that does not stop the other explanations from being true of other similar phenomena somewhere else in the universe.

4. Ethics. In ethics, Epicurus taught that the highest good is pleasure, defined as freedom from pain in the body (*aponia*), and freedom from anxiety and disturbance in the mind (*ataraxia*). Epicurus identified two types of pleasure, static and kinetic. Static pleasure is the state an organism feels when it suffers no pain and is functioning well. Kinetic pleasure is what an organism feels when it is physically or mentally stimulated. Kinetic pleasure apparently occurred in two ways: either in the process of satisfying a want and returning an organism to its static state of pleasure, or when an organism's experience of static pleasure is "varied" by the addition of kinetic pleasure. Epicurus taught that static pleasure is the highest possible for a human being. Kinetic pleasure does not increase pleasure, but only varies it.

Epicurus taught that human beings often fail to achieve happiness because they do not distinguish among three types of desires: (1) natural and necessary desires, i.e., desire for things that are necessary for life; (2) natural and non-necessary desires, i.e., desires for things that are not necessary for life but help to "vary" our pleasure; and (3) desires that are neither natural nor necessary, i.e., desires for things like honor and political office. Epicurus advocated leading a simple life, taking pleasure in easily satisfying our natural and necessary desires. He also taught that not all pleasures should be chosen, nor all pain avoided. Humans often must give up pleasure now to avoid greater pain later, and chose some pain now to attain greater pleasure later. If an action promotes long-term freedom from pain and anxiety, it should be chosen, otherwise not. Epicurus also taught techniques for maintaining mental *ataraxia* even when the body was feeling great pain. He maintained that physical pain could be endured, his reasoning captured later in a memorable Latin phrase: *si grauis, breuis; si longus, leuis* ("Pain is short if it is strong, light if it is long"). Pleasure was also the basis for evaluating virtue and ethical behavior. According to Epicurus, it is important to be virtuous not because the virtues are valuable in themselves, but because the virtues are the means to the most pleasant life. He maintained that human beings, although they had minds and souls made up of atoms and void, had freedom of action because of the swerve of atoms. How he thought the swerve preserved the freedom of living creatures, and what kind of freedom he thought it preserved, have been the subject of intense scholarly debate.

DSB 4.381–382, D.J. Furley; G. Arrighetti, *Epicuro, Opere* (1973); E. Asmis, *Epicurus' Scientific Method* (1984); Walter G. Englert, *Epicurus on the Swerve and Voluntary Action* (1987); Long and Sedley (1987) §4–25; *OCD3* 532–534, D.J. Furley; *ECP* 214–219, E. Asmis; *REP* 3.350–351, D.N. Sedley; *BNP* 4 (2004) 1075–1084, M. Erler.

Walter G. Englert

Epidauros (?) (120 BCE – 80 CE)

ANDROMAKHOS, in GALĒN *CMGen* 7.7 (13.985 K.), gives his ointment for circumcision: *thapsia* root, pepper, veal fat, frankincense, balsam-wood, resin, and beeswax. The rare name is attested in the 3rd–1st cc. BCE: *LGPN* 1.156, 2.148. (For toponyms as personal names, *cf.* EUPHRATĒS and KTĒSIPHŌN.)

Fabricius (1726) 150.

PTK

Epidikos (300 BCE – 500 CE)

Taught that the ***kosmos*** was caused by nature (***phusis***): IŌANNĒS STOBAIOS 1.21.6, Phōtios, *Bibl.* 167 (p. 114a). The name is rare, attested in Akhaia: *LGPN* 3A.146, and in Plautus' eponymous comedy.

DPA 3 (2000) 182, R. Goulet.

PTK

EPIGENĒS ⇒ DERVENI PAPYRUS

Epigenēs (Med.) (*ca* 390? – 310 BCE)

Physician, claimed that rancid water purified seven times would not putrefy again (PLINY 31.34), from THEOPHRASTOS *On Water* (Wellmann 1900). In *On Fatigue* (p. 398 W.), Theophrastos may cite an Epigenēs arguing that fatigue occurs primarily in veins and tendons. The MS, however, is corrupt, and the name is Furlanus' (reasonable) restoration for *ΕΠΙΓΟΝ*, so perhaps *cf.* EPIGONOS.

M. Wellmann, "Zur Geschichte der Medicin im Altertum," *Hermes* 35 (1900) 349–384 at 354–358; *RE* 6.1 (1907) 66 (#18), *Idem*; W.W. Fortenbaugh, R.W. Sharples, and M.G. Sollenberger, *Theophrastus of Eresus: On Sweat, On Dizziness, On Fatigue* (2003) 279–280.

GLIM

Epigenēs of Buzantion (120 – 30 BCE?)

PLINY (7.160) cites Epigenēs (who "studied the stars"), together with BĒROSSOS and PETOSIRIS, as astrological authorities on the destined length of life; he also invokes Epigenēs as a source on the antiquity of Babylonian astronomical observations (7.193), saying they went back 720,000 years.

Epigenēs claimed that the maximum possible human lifespan was less than 112 years, a value that may be evidence for the astrological application of a Babylonian style scheme for the rising times of the zodiacal signs, adapted for Alexandria where the longest to shortest day ratio is 7:5, the longest day (M) is 3,30;0° (14 hours), and the constant difference is 3;20°. The longest life, derived from the quadrant with the greatest rising time, is found from ½M + 2d, hence 111;40° for Alexandria. Epigenēs' value might represent a rounding of this result.

RE 6 (1907) 65–66 (#17), A. Rehm; Honigmann in *P. Mich.* 3 (1936) 310–311; Neugebauer (1975) 721.

Francesca Rochberg

Epigenēs of Rhodes (285 – 90 BCE)

Agronomist whose work may have treated cereals, livestock, poultry, viticulture, and arboriculture (*cf.* PLINY, 1.*ind.*8, 10, 14–15, 17–18). CASSIUS DIONUSIOS excerpted from his writings (VARRO, *RR* 1.1.8–10, *cf.* COLUMELLA, 1.1.9). Pseudo-PLUTARCH, *Nobil.* 20 (7.269 Bern.), reports that Epigenēs advanced numerous arguments to prove that humans lived in the countryside long before they lived in cities (*cf.* Varro, *RR* 3.1); this sort of speculative anthropology seems to have been popularized by DIKAIARKHOS.

RE 6.1 (1907) 65 (#19), E. Fabricius.

Philip Thibodeau

Epigonos (250 BCE – 10 CE)

HĒRĀS records that some attributed the "ISIS" plaster to Epigonos; it "drew out poison," and contained aloes, alum, birthwort, **galbanum**, copper flakes, myrrh, **verdigris**, etc. in aged olive oil and vinegar: GALĒN, *CMGen* 5.2 (13.774–778 K.). ANDROMAKHOS credits him with a "green" plaster, of almost identical ingredients, in an olive oil and "Kolophōn" resin base: *ibid.* 2.2 (pp.492–493); Galēn himself cites Epigonos' plaster as exemplary, *Rat. Cur. ad Glauk.* (11.126 K.). *Cf.* also GLUKŌN and HERMŌN.

RE 6.1 (1907) 66 (#21), M. Wellmann.

PTK

Epikharmos of Surakousai (*fl.* 488 – 485 BCE)

Sicilian comic poet, known through several hundred testimonia and fragments. Most evidence about his life is obscure, but he undoubtedly lived and wrote in Surakousai in the times of Gelōn and Hierōn (491–467), and died after 458 (perhaps as late as 438). Fragments containing either satires against contemporary thinkers or sententious maxims, taken out of context, shaped the idea of Epikharmos as philosopher and "wise man," later augmented by his alleged relationship with PYTHAGORAS. Consequently, other writers ascribed to him their own philosophical or quasi-scientific works, most of them linked to the **Pythagorean** school, written in trochaic tetrameters, and in a dialect which tried to imitate Epikharmos' Sicilian Doric. According to Athēnaios, *Deipn.* 14 (648d), the spuriousness of these writings (the *Pseudepikharmeia*) was known to some authors from the late 4th c. BCE, thus ARISTOXENOS, Philokhoros (early 3rd c. BCE), and Apollodōros of Athens (2nd c. BCE), but many continued treating them as genuine, and at least one of them, the *Antenor*, seems to have been forged after Aristoxenos, and may be the latest. They addressed philosophy or physics (so the *Republic*, written by a flute-player called Khrusogonos; the *Kanōn*, by a certain Axiopistos; and the *Antenor*), general truths and rules of conduct (the *Maxims*, also by Axiopistos), and medicine and veterinary medicine.

The first writer to connect Epikharmos with medical subjects is DIODŌROS OF SICILY, and many others did so afterwards (PLINY 20.89, 20.94, etc., COLUMELLA 7.3.6, PAMPHILOS OF ALEXANDRIA, DIOGENĒS LAERTIOS, CENSORINUS 7.5–6, IAMBLIKHOS). By asserting that Epikharmos was a native of Kōs, Diogenēs Laertios (8.78) might mean to connect him with that island's medical school. In a rather obscure passage, Iamblikhos (*VP* 241) also links a certain MĒTRODŌROS (allegedly his son) with Epikharmos' theories on medicine. In all likelihood, the source for pharmaceutical prescriptions allegedly coming from Epikharmos was the poem *Kheirōn* (*Chiron*), which probably included the culinary treatise also attributed to him. The real author and date of the *Kheirōn* are unknown, but it might have been written as early as the 4th–3rd cc. BCE, if the papyrus fragment *295 *PCG* = *335 R–N is confirmed to be a part of the poem.

Ed.: Lucía Rodríguez-Noriega, *Epicarmo de Siracusa. Testimonios y fragmentos. Edición crítica bilingüe* (1996). R. Kerkhof, *Dorische Posse, Epicharm und Attische Komödie* (2001).

Lucía Rodríguez-Noriega

Epiklēs of Crete (130 – 30 BCE)

Abridged BAKKHEIOS' Hippokratic glossary (ERŌTIANOS p. 5.5 Nachm.), alphabetizing (Erōtianos p. 7.23 Nachm.), revising (A-8, A-58, A-66 [pp. 13.3, 19.3, 20.2 Nachm.]), and mediating Bakkheios (A-4, A-8, A-58, A-66, A-69, A-73, B-8, B-30, and fragment 42 [pp. 10.16–17, 13.2–4, 19.2–5, 20.1–2, 20.12–13, 21.10–11, 28.10–14, 37.9–10, and 112.2–7 Nachm.]). Erōtianos, drawing comparisons and contrasts between Bakkheios and his successors, cites Epiklēs by name 23 times, quoting his glosses all but twice. See APOLLŌNIOS "OPHIS."

RE 6.1 (1907) 117 (#5), M. Wellmann; H. von Staden, "Lexicography in the III B.C.: Bacchius of Tanagra, Erotian, and Hippocrates" in J.A. López Férez, ed., *Tratados Hipocraticos* (1992) 549–569; Ihm (2002) #50.

GLIM

Epikouros (250 BCE – 80 CE)

Pharmacist: quoted by GALĒN (*CMGen* 5.5, 13.807 K.), from ANDROMAKHOS, for his recipe of a plaster for the cure of scars, containing aloes, **galbanum**, myrrh, **verdigris**, etc.

Deichgäber (1930) 408 (attributed to EPIKOUROS OF PERGAMON); *RE* S.9 (1962) 64 (#5), F. Kudlien; Fabricius (1972) 226–227.

Fabio Stok

Epikouros of Pergamon (120 – 180 CE)

Empiricist physician, teacher of GALĒN, author of a commentary on the HIPPOKRATIC CORPUS, EPIDEMICS, Book 6 (Gal. *Hipp. Epid. CMG* 5.10.2.2, p. 412), and probably also of other exegeses of Hippokratic works.

Deichgräber (1930) 408 (fragment); *RE* S.9 (1962) 64 (#5), F. Kudlien; Ihm (2002) #51; A. Anastassiou and D. Irmer, *Testimonien zum Corpus Hippocraticum* 2.1 (1997) 486.

Fabio Stok

EPIKOUROS OF SAMOS ⇒ EPICURUS

Epikratēs of Hērakleia (325 – 25 BCE)

Writer on machines listed by *P. Berol.* P-13044, col.8, as having constructed war-machines in Rhodes, perhaps for the siege in 88 BCE; contrast EPIMAKHOS OF ATHENS.

Diels (1920) 30, n. 1.

PTK

Epimakhos of Athens (310 – 300 BCE)

Epimakhos designed a giant ***helepolis*** (siege-tower) for Dēmētrios Poliorkētēs at the siege of Rhodes (305–304 BCE). This costly and elaborate siege machine was *ca* 40m high, *ca* 20m wide, weighed *ca* 160 metric tons, and could withstand the impact of a 160-kg stone thrown by a ballista (Athēnaios Mech. p. 27 W.; Vitruuius 10.16.4).

RE 6.1 (1907) 160 (#3), E. Fabricius.

GLIM

Epimenidēs of Crete (650 – 520 BCE)

Wise man credited with wonders including a 57-year nap (Diogenēs Laërtios 1.109). His "hunger-banishing" recipe, allegedly based on Hēsiod, *WD* 41, is cited by Theophrastos, *HP* 7.12.1 (contains squill), Hermippos of Smurna (*FGrHist* 1026 T8e = Proklos, *in Hes. Op.* 41), and Plutarch, *Conv. Sap.* 157D–E (*cf. Fac. Orb.* 940), among others. Aelianus 12.7 = *fr.* B2 preserves three lines of verse which he interprets as a claim that the Nemean lion fell from the inhabited Moon.

DK 3.

PTK

Epinomis ⇒ Philippos of Opous

Epiphanēs (?) (400 BCE – 300 CE)

The "Laurentian" list of medical writers (MS *Laur. Lat.* 73.1, f. 143^{V} = *fr.*13 Tecusan) includes Epiphanēs, more likely as the *epiklēsis* of a king or god than a proper name (although attested: *LGPN*). Perhaps Epigenēs or Epainetēs is meant, or else Antiphanēs or Aristophanēs of Buzantion. *Cf.* also Hipposiadēs, Lupus, and Philippos of Kōs.

(*)

PTK

Epiphanios (Meteor.) (unknown date)

Author of an unedited work *On Thunder and Lightning*, conceivably a work similar to Vicellius'.

RE 6.1 (1907) 196 (#11), A. Rehm.

PTK

Epiphanios of Eleutheropolis/Salamis (*ca* 365 – 403 CE)

Born in a Jewish family of Eleutheropolis, after conversion became bishop of Salamis. Besides his Christian dogmatic works, Epiphanios wrote also a metrological treatise whose title, *On measures and weights*, seems a later addition. The work survives abridged in Greek and Georgian, but complete only in Syriac. It seems clear the work was basically didactic, and contained much Biblical and historical material, including the metrological terms. What survives is an unsystematic exposition of Biblical units, giving the meaning of their Hebrew name, comparison with the measures used in the Greco-Roman world, along with entries about the currency units.

Ed.: *MSR* 1 (1864) 140–142, 259–276, 2 (1866) 100–106; J.E. Dean, *Epiphanius' Treatise on weights and measures. The Syriac Version* (1935); M.-J. van Esbroeck, *Les versions géorgiennes d'Épiphane de Chypre. Traité des poids et des mesures* (1984).
BNP 4 (2004) 1119–1120 (#1), C. Markschies.

Mauro de Nardis

Erasistratos (Astrol.) (200 – 300 CE?)

CCAG 1 (1898) 81–82 prints a Greek translation from Mash'allah al-Misri (*ca* 760 CE), who lists his 11 sources as: PTOLEMY, HERMĒS, PLATO (six books), DŌROTHEOS (11 books), DĒMOKRITOS (14 books), ARISTOTLE (ten books), ANTIOKHOS (OF ATHENS) (seven books), (VETTIUS) VALENS (ten books), "Erasistratos" (11 books), "Stokhos" (sc. "Eustokhios"? "Stoikos"?) (six books), and "the Persians" (44 books). Antiokhos, Dōrotheos, Ptolemy, and Valens are genuine, and an otherwise unknown astrologer Erasistratos probably is too, *cf.* Al-Bīrūnī, *Book of Instruction in the Elements of the Art of Astrology* (1029 CE), §453 (p. 265, ed. R.M. Wright, 1934).

(*)

PTK

Erasistratos of Ioulis on Keōs (*ca* 260 – 240 BCE)

Erasistratos (b. *ca* 315 BCE) may or may not have been a "colleague" of HĒROPHILOS at the Museion in Ptolemaic Alexandria, but ancient *testimonia* attest to his presence as a "younger contemporary" and that he also performed systematic dissections (and less likely vivisections) on human cadavers *ca* 260–250 BCE. Ancient sources also tell us that Erasistratos had links with the **Peripatetic** School in Athens (but not a student of THEOPHRASTOS: Scarborough 1985), and that he served for a time as a court physician to one of the Seleukid or Antigonid monarchs (Wellmann 1907: 333–334), before moving on to Alexandria. Fraser asserts (1969, 1972: 1.347) that Erasistratos spent his entire career in Antioch, refuted by Lloyd (1975). Biographical details are at best confused and confusing in our ancient testimonies, and no work survives intact.

Erasistratos' connections to the **Peripatetics** are well-documented in DIOGENĒS LAËRTIOS 5.57 and GALĒN, *Blood in the Arteries* 7 (4.729 K. = Furley and Wilkie 174), and it is likely that the mechanical and corpuscular theory espoused by Erasistratos owed much to STRATŌN OF LAMPSAKOS. Thereby Erasistratos differed greatly from Hērophilos regarding what we would term "physiological functions": Erasistratos employed mechanistic principles fused with an Aristotelian notion of teleology, occasionally verifying hypotheses by means of experiment.

From contemporary mechanics and physics, he derived a major mechanistic principle: substances move in nature by "going toward that which is being emptied" (*pros to kenoumenon akolouthia*: Galēn, *Natural Faculties* 1.16 [2.62–63 K.] = Garofalo, *fr.* 74; *cf. frr.* 93–96, 109, 110, 198). In *General Principles* (*kath'holou logoi* [Garofalo, *frr.* 74–152]), Erasistratos combined veins and arteries, nerves, muscles, the function of appetite, and digestion into a unified template of physiology. The system was: air enters the lungs via the trachea and bronchi while the thorax expands after exhalation: some of the "breath" (***pneuma*** [*frr.* 101–108 Garofalo]) in the lungs then moves via the "vein-like artery" (our pulmonary vein) into the left ventricle of the heart, when this cavity expands after contraction; meanwhile the ***pneuma*** in the left ventricle of the heart is "refined" or "thinned" into a "vital" (*zōtikon*) ***pneuma***, and thence pushed into the arteries when the heart contracts; meanwhile, excess air in the lungs has absorbed or "sucked up" some of the superfluous heat of the body and the heart, and is exhaled when the thorax contracts. Following the basic principle ("an empty space fills up"), new breath (***pneuma***) rushes into the expanding thorax. Thus Erasistratos explained how the breathing-cycle both cools the body and provides the arteries with essential ***pneuma*** (Scarborough 1998: 175) To explain why the arteries are empty but the veins are full of blood, Erasistratos argued that when an artery is severed, it spurts blood because there are (theoretical) extremely minute blood vessels, the *sunastomōseis* (invisible to the eye), that connect blood-filled veins to arteries (Galēn, *Blood in the Arteries* 2 [4.709 K.] = Furley and Wilkie 150 = Garofalo, *fr.*109 [*paremptōsis*]), and therefore blood rushed into the severed artery. Erasistratos' theory led to the capillaries, not demonstrated until the 17th c. by Malpighi in *De pulmonibus observationes anatomicae* (1661): using a microscope was the key (Major 1954: 1,511). Erasistratos' postulation of *sunastomōseis* allowed his system to "work," and his dissections gave a physiology founded on anatomy. Telling is the absence of blood in a "normal" artery (which Galēn condemned in *Venesection against Erasistratos*, and proved wrong in *Blood in the Arteries*): no observed arterial blood (esp. in the aorta) in the living human, thus no vivisection of humans (Scarborough 1976).

Works known by title and shorter and longer quotations in AËTIOS OF AMIDA, the LONDINIENSIS MEDICUS, CELSUS, DIOSKOURIDĒS, Galēn, MARCELLINUS, OREIBASIOS, RUFUS OF EPHESOS, SŌRANOS, and others (see Garofalo, "Index Fontium" and "Index auctorum et locorum") are *Fevers* (*frr.*194–226 Garofalo), *Expectoration of Blood, Paralysis,* ***Dropsy***, *Podagra* (*viz. Gout* [*frr.*267–269]), *The Abdominal Cavity*, and *Divisions*. In the last, Erasistratos enunciated the famous dictum that "every organ is supplied by an artery, a vein, and a nerve," and he confirmed Hērophilos' observations of the "two parts" of the brain, re-emphasizing that human cranial convolutions were far more complex than those in animals, which proved higher human intelligence (geometry easily demonstrated a "greater surface area"). Erasistratos rejected the Hippokratic notions of a **humoral** pathology, teaching that blood in the veins and two kinds of ***pneumata*** were essential for life. Detailed dissection of the heart yielded description of the semi-lunar valves and the tricuspid valve, which he named and understood prevented reflux of blood. The list of accurate descriptions (many from *Abdominal Cavity* [*frr.*258–269 Garofalo]) is impressive: the aorta, the pulmonary artery, the intercostal arteries, hepatic artery, the arteries of the stomach, pulmonary vein, vena cava, the azygos vein, the milk-white vessels of the mesentery (lymphatic vessels [Gray, *Anatomy*, 800–803]), and the complicated courses of the hepatic veins.

The function of the nerves also followed the same basic principles. The nerves also carried "vital" (*zōtikon*) ***pneuma***, "pushed" through the arteries from the left ventricle of the heart to the brain, where the ***pneuma*** gains further refinement into a "psychic" (*psukhikon*) ***pneuma***, lacking any notion of "soul"; this then is "pulled" throughout the body by means of the two kinds of nerves (we call them "sensory" and "motor"). Sight requires the most "psychic" ***pneuma***, and thus the optic nerve has the greatest "psychic" ***pneuma***, and was tubular or hollow for the ***pneuma*** directed at the eye. Meanwhile, appetite and digestion gave the liver liquid nourishment so that it could "process" food into blood, then "pushed" into the veins by his principle "all empty spaces are filled up." Arteries + nerves, therefore, contain only ***pneuma***, and the veins have blood, "pushed" as nutriment to all parts of the body. Also, therefore, all parts, muscles, and organs, to live and grow must have "triple-woven" (*triplokiai*) ingrowths of veins, arteries, and nerves; ***pneuma*** "pushed" to muscles by arteries and nerves enables them to contract and relax, displaying "voluntary" motion. This theory of "eating + digestion + manufacture of blood by the liver + ready-made food (blood) from the liver to the rest of the body" was adapted by Galēn in his *Natural Faculties*, and held until 1833 (Scarborough 1998: 222).

For Erasistratos, the laws of probability governed symptoms of disease and any applied therapeutics, so that he opposed phlebotomy (Galēn excoriates him for this), and any harsh treatments. The etiology of disease emerges from the classification of matter (blood, ***pneuma***, other life-supporting liquids), usually absolutely (somehow) separate, but mixed in disease; thus one has a *plethora* = too-much-blood-as-food in the veins, causing inflammation, in turn causing fevers, and swollen limbs (***Dropsy*** [*frr.*248–257 Garofalo]), unhealthy states in the liver and the stomach, the Falling Sickness (epilepsy), and many more. The mechanics of pathology: excess venous blood undergoes a "spill-over" (*paremptōsis*) into the arteries through the invisible *sunastomōseis*, which lessens the arterial "push" of "vital" ***pneuma***. Women do not have pathologies peculiar to females, except for matters obstetrical (Hērophilos had said the same), so that in Erasistratos' *Hygiene* (*frr.*115–167 Garofalo) he urges a healthy life-style (regimen) to prevent *plethora*, and mild intervention to restore displaced matter.

The *Londiniensis medicus* 33 (ed. Diels, 62–63 = Jones, 126–127) records an experiment by Erasistratos to determine weight-loss in a fasting animal: "If one were to take a creature, such as bird or something of the sort, and were to place it in a pot for some time without giving it any food, and then were to weigh it with the excrement that visibly had been passed, he will find that there has been a great loss of weight plainly because, perceptible only to the reason, a copious emanation has taken place" (trans. Jones; *cf.* von Staden 1975).

Erasistratos' writings were long available for discussion and citation, indicated by the rather precise account of the epiglottis, esophagus, and trachea cited by Gellius 17.11. The learned elite in 2nd c. Rome (in PLUTARCH's *Table-Talk*) continued to debate PLATO's assertion that food and drink went into the lungs, and from the comments in Gellius, Erasistratos' correct description based on dissection remained controversial. Plato (and Plutarch) were "authorities," and Erasistratos' medical mechanics fades before Gellius, ". . . who allows the last word to the defense of Plato" (Holford-Strevens, 303).

Ed.: I. Garofalo, *Erasistrati Fragmenta* (1988) [incomplete]; see also the following: A.J. Brock, *Galen on the Natural Faculties* (Loeb 1916); J.F. Dobson [trans., selected passages], "Erasistratus," *Proceedings of the Royal Society of Medicine* 20 (1927) 21–27 [= 825–832]; Wehrli 5 (1969); Furley and Wilkie (1984); Brain (1986).

RE 6.1 (1907) 333–350, M. Wellmann; R. Major, *A History of Medicine*, 2 vols. (1954); L. Wilson, "Erasistratus, Galen, and the Pneuma," *BHM* 33 (1959) 293–314; Solmsen (1961); P.M. Fraser, "The Career of Erasistratus of Ceos," *Rendiconti del Istituto Lombardo* 103 (1969) 518–537; Fraser (1972) 1.347–348, 2.503–504; *DSB* 4 (1972) 382–386, J. Longrigg; G.E.R. Lloyd, "A Note on Erasistratus of Ceos," *JHS* 95 (1975) 172–175; John Scarborough, "Celsus on Human Vivisection at Ptolemaic Alexandria," *CM* 11 (1976) 25–38; H. von Staden, "Experiment and Experience in Hellenistic Medicine," *BICS* 22 (1975) 178–199; W.D. Smith, "Erasistratus' Dietetic Medicine," *BHM* 56 (1982) 399–409; John Scarborough, "Erasistratus, Student of Theophrastus?," *BHM* 59 (1985) 515–517; *Idem*, Medical and Biological Terminologies, 2nd ed. (1998); L. Holford-Strevens, *Aulus Gellius: An Antonine Scholar and his Achievement*, rev. ed. (2003).

John Scarborough

Erasistratos of Sikuōn (250 BCE – 95 CE)

ASKLĒPIADĒS PHARM., in GALĒN *CMLoc* 10.3 (13.356–358 K.), records two **humor-** extracting remedies. For gout, draw out the phlegm from blisters raised by pouring on chilled feet a solution of caper-root, henbane-seed, hemlock, mandrake, etc., heated with dried lees, and then bandaging the feet with thin vinegar-soaked cloth for two hours. Blood is extracted by an analogous procedure, rather than by cupping vessels or leeches. *Cf.* THEOXENOS.

(*)

PTK

Eratoklēs (450 – 390 BCE)

Musical theorist quoted by ARISTOXENOS as representative of the school of *harmonikoi*, earlier empiricists who rejected the **Pythagorean** description of notes as quantities and conceived them as dimensionless points lying on a linear continuum they called the "diagram." He and his followers seem to have attempted a distinction of conjunct from disjunct tetrachords and to have enumerated arrangements of octaves interpreting *harmoniai* as approximations to "octave species" (*eidē tou dia pasōn*). However, according to Aristoxenos, they made no serious attempt to explain the principles governing the melodic phenomena: thus their results are described by him as incomplete.

A.D. Barker, "*Hoi kaloumenoi harmonikoi*: the predecessors of Aristoxenos," *PCPhS* 24 (1978) 1–21; *Idem* (1989) 124–125; *OCD3* 553, *Idem*.

E. Rocconi

Eratosthenēs of Kurēnē (*ca* 240 – 194 BCE?)

Polymath, wrote works on a wide range of subjects including mathematics, harmonic theory, geography, chronology, grammar, and literary criticism, as well as composing poetry. Scarcely any of this *oeuvre* remains extant except for a mathematical epigram and quotations from his most important work, *Geographica*. Various ancient sources report conflicting biographical details. The *Souda* (E-2898) credibly states that he was born in the 126th Olympiad (276–272 BCE) and died at age 80 during the reign of Ptolemy V. Son of Aglaos, during his youth in Kurēnē he supposedly received a literary education from the philologist Lusanias and poet KALLIMAKHOS. During a sojourn at Athens, he associated with prominent philosophers, the **Peripatetic** Aristōn of Khios and **Academic** Arkesilaos. From Athens he was invited to Egypt by Ptolemy III. A Roman-period papyrus (*POxy* 10.1241) asserts that Eratosthenēs succeeded Apollōnios of Rhodes as head of the Alexandrian library, and that in turn Aristophanēs of Buzantion succeeded him. Probably during the earlier part of his Alexandrian period, ARCHIMĒDĒS communicated to him a lost collection of geometrical propositions asserted without proofs, as well as the extant *Method Concerning Mechanical Theorems*, though in his other surviving prefatory letters (to DŌSITHEOS OF PĒLOUSION), which must have been written while Eratosthenēs was still alive, Archimēdēs singles out KONŌN as the Alexandrian mathematician for whom Eratosthenēs had greatest respect.

1. Mathematics and Harmonics. Our evidence for Eratosthenēs' mathematical work is severely limited. PAPPOS (*Collection* 7.3) includes a formal geometrical treatise by Eratosthenēs called *On Means*, apparently in two books, as one of the writings making up the "Treasury of Analysis" (*topos analuomenos*), a corpus of resources for the solution of geometrical problems by analysis, but he does not describe the work's contents. It was likely in *On Means* that Eratosthenēs discussed "loci on means," which according to an obscure statement of Pappos' (7.22) seem to have comprised straight lines and circles, conic sections, and other curved lines. We do not know the context in which Eratosthenēs presented the so-called "sieve," an algorithm for finding prime numbers (Nikomakhos, *Introductio Arithmetica* 1.13).

Eutokios (*In Arch. Sph. Cyl.* pp. 88–96 Heiberg) quotes what purports to be a letter addressed by Eratosthenēs to "King Ptolemy," describing a geometrical and instrumental solution of the problem of finding two mean proportionals between two given rectilinear magnitudes; that is, given linear magnitudes A and D, to find magnitudes B and C such that $A : B = B : C = C : D$. The letter originates the problem in the story of how the Delians consulted the "geometers around Plato" on how to obey an oracle commanding the doubling of a cubical altar. Eratosthenēs' geometrical solution is to erect A and D as perpendiculars to a base line, and to construct three similar right triangles adjacent to one another on this base such that the first has the end point of A as its vertex and the other two have their vertices collinear with the end points of A and D (**Fig.**); the heights of these latter triangles are the mean proportionals. The solution is to be implemented mechanically by an arrangement of rigid triangles sliding along grooves. According to the letter, Eratosthenēs made a votive dedication of a bronze specimen of this *mesolabon* ("mean-obtainer") accompanied by a proud epigram in elegiacs, reproduced at the end of the letter, asserting the superiority of Eratosthenēs' solution to those of Arkhutas, Eudoxos, and Menaikhmos. (Nikomēdēs would in turn castigate Eratosthenēs' approach as both unmechanical and ungeometrical; *cf.* Eutokios p. 98 Heiberg.) Modern scholarship has, for the most part, followed Wilamowitz in considering the letter spurious but the epigram authentic, though Knorr has argued that the whole is genuine. It seems plausible in any case that Eratosthenēs did commemorate his discovery through a votive object and inscription.

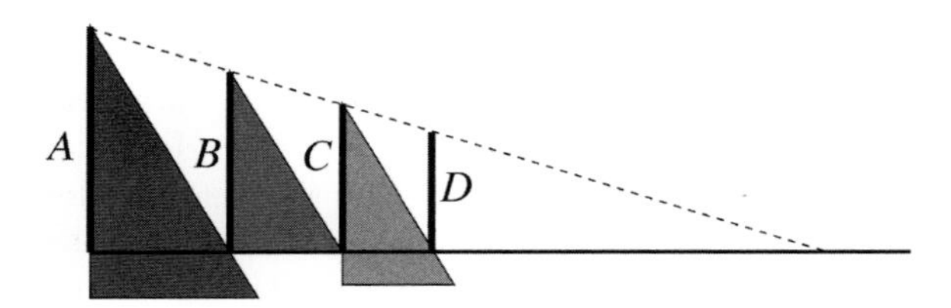

Eratosthenēs' mechanical method of finding two mean proportionals between magnitudes *A* and *D*. The left and right triangles are slid along grooves until by trial and error the four vertices are collinear © Jones

Theōn of Smurna (p. 2, Hiller) reports that Eratosthenēs gave a similar account of the Delians' efforts to double their altar in *Platōnikos*, "the Platonist." This seems to have been a discursive book on the philosophy of mathematical objects and relations, and Theōn's several citations of Eratosthenēs on the topic of ratios probably come from it. Theōn (p. 142) also ascribes to Eratosthenēs a discussion of the harmonies of the celestial spheres that was partly in verse and contained an etiological myth for the origins of the celestial tuning.

A more technical work on harmonic theory seems to lie behind schemes that PTOLEMY attributes to Eratosthenēs (*Harmonics* 2.14) specifying numbers associated with the pitches of three tuning systems: on Ptolemy's understanding, these numbers represent the lengths of a uniformly tensed string that would sound at the corresponding pitches. Two of the schemes coincide perfectly with schemes that Ptolemy associates with ARISTOXENOS; hence it appears that Eratosthenēs was attempting somehow to reconcile Aristoxenos' theory of tonal "distances" with the Pythagorean model of musical intervals as whole number ratios.

2. Geography. HĒRŌN (*Dioptra* 35) refers to Eratosthenēs' work "on the measurement of the Earth," seemingly independent of his *Geographica* and in which he presented a geometrical deduction of the length of the spherical Earth's circumference from ostensibly empirical data. KLEOMĒDĒS (1.7) reports a summary of Eratosthenēs' approach, which he characterizes as following a *geōmetrikē ephodos*, a phrase that could mean a method involving surveying or, more likely, deductive argument expressed in the manner of the geometers. The assumptions are (1) that the Sun is effectively at infinite distance from the Earth, so that shadows cast in all localities are parallel, (2) that Alexandria is situated 5,000 stades north of Suēnē as measured along a meridian, (3) that for an observer at Suēnē the Sun passes through the zenith at noon on the summer solstice, and (4) that for an observer at Alexandria the Sun is 1/50 of the meridian circle south of the zenith at the solstitial noon. Of these data, (3) was probably derived from common report, and is accurate, and the interval (2) between the two cities – not in fact on the same meridian – has the appearance of a round estimate. Kleomēdēs, supported by MARTIANUS CAPELLA's (6.596–598) dubious testimony, states that (4) was measured using a spherical sundial, though this may be a didactic simplification. The resulting value of the Earth's circumference, 250,000 stades, is often cited in ancient sources, but not always attributed to Eratosthenēs. Eratosthenēs himself is likely to be responsible for the well attested "rounding" of this number to 252,000, allowing a convenient equation of 700 stades with one degree of the meridian. (Eratosthenēs apparently employed a division of the meridian into 60 units, however, rather than into degrees.) In the same work, Eratosthenēs may have treated related questions of mathematical geography, including estimates of the latitudes of Alexandria and other cities derived from the ratio of a gnomon to its noon shadow on an equinox, and an estimate of the obliquity of the **ecliptic** (or equivalently, the latitude of Suēnē), which Ptolemy (*Almagest* 1.12) says was very near his own value, 11/83 of a semicircle.

The *Geographica*, in three books, was a treatise on the construction of a map of the ***oikoumenē***. Eratosthenēs may have coined the word *geōgraphia* (in the sense of "world-cartography") and terminology derived from it, reflecting a new emphasis on setting map-making on a rational and quantitative scientific basis. Eratosthenēs thus initiated a genre that was to lead, by way of MARINOS OF TYRE, to Ptolemy's *Geography*. Though no longer extant, the *Geographica* is often mentioned in ancient authors, in particular STRABŌN, who reports many specific details. Strabōn had direct access to Eratosthenēs' work and also drew extensively from HIPPARKHOS' lost polemic against it, and thus we can recover from Strabōn the general structure and character of the *Geographica*. Book 1 contained a critical review of earlier geographical authors and cartographers, a list from which Eratosthenēs significantly excluded HOMER. Book 2 appears to have addressed methodology and the situation and dimensions of the ***oikoumenē***. Book 3 provided the detailed discussion of the dimensional and positional data necessary for drawing a map of the ***oikoumenē***, employing a division of the continents into large geometrically defined regions with rectilinear borders (*sphragides*, "seals"). It is not clear whether the text was meant to be accompanied by an actual map. Notwithstanding Hipparkhos' criticisms, Eratosthenēs' conception of the general shape and layout of the known world remained the basis for verbal and pictorial portrayals of the world well into the Roman period.

3. Astronomy and Chronology. Although Eratosthenēs' geodesy had an ostensibly astronomical empirical foundation, his direct contributions to astronomy were slight. GEMINUS (8.24) refers to a work on the ***oktaetēris***, in which Eratosthenēs explained how the 365-day year of the Egyptian calendar meant that Egyptian festivals gradually shifted backwards in relation to the natural seasons. A work for which the *Souda* offers the alternative titles *Astronomia* and *Katastērigmoi* ("constellations") retailed myths relating to the constellations; an extant anonymous book containing such material may represent an adaptation or digest of Eratosthenēs' work.

Eratosthenēs is often credited as the principal founder of Greek chronography, chiefly on the basis of his *On Chronographers* and *Olympic Victors*. It is unclear, however, whether this reputation is wholly deserved. *On Chronographers* (an alternative version of the title, *Chronography*, is less likely to be correct) appears to have been more a critical review of earlier writings pertaining to chronology rather than an original study, though it did propose a framework of specific intervals of years between landmark dates from the Trojan War up to the death of Alexander. The *Olympic Victors*, in at least two books, was not primarily a chronological catalogue – such works in any case had allegedly been compiled already by HIPPIAS OF ĒLIS, ARISTOTLE, and Philokhōros – but a gathering of general information relating to the Olympic Games from literary sources.

Ed.: H. Berger, *Die geographischen Fragmente des Eratosthenes* (1880).
P.M. Fraser, "Eratosthenes of Cyrene," *PBA* 56 (1970) 175–207; K. Geus, *Eratosthenes von Kyrene. Studien zur hellenistischen Kultur- und Wissenschaftsgeschichte* (2002).

Alexander Jones

ERGOTELĒS ⇒ PURGOTELĒS

Erōtianos (60 – 80 CE)

Dedicates his extant *Hippokratic Lexicon* to ANDROMAKHOS, "*arkhiatros*." GALĒN indicates that both Erōtianos and Andromakhos were at Nero's court (54–68 CE). Two physicians named Andromakhos (likely father and son) are attested, and Galēn often cites the Elder for pharmaceutical recipes. Erōtianos frequently cites BAKKHEIOS OF TANAGRA's lost *Hippokratic Glossary*, one of several such compilations attempting to "explain" the often-obscure Hippokratic medical terms. As Smith (1979) illustrates in translation (203, n. 31), Erōtianos' purpose in setting forth a new collection of readings was straightforward enough, and similar to previous glossographers: "HIPPOKRATĒS is important . . . because he is useful for literary instruction. He is useful for physicians especially because in reading him they can learn new things and test the ones they already know."

Entries are arranged alphabetically, most likely following the template devised by ARISTOPHANĒS OF BUZANTION, with each term explicated by citations from other writers, including near-contemporaries (Erōtianos is the first known witness to DIOSKOURIDĒS' *Materia Medica* [*Lexicon*, K, 31.85, p. 51 Nachm.]), and there are comparisons of medical terminologies employed by earlier Hellenistic physicians including APOLLŌNIOS OF

Kition, Hērakleidēs of Taras, and several others; most of Erōtianos' appositions, however, are drawn from drama and poetry, with numerous selections from Homer, Menander, Aeschylus, Sophoklēs, Aristophanēs, and similar "classics." Erōtianos' mentions of authors providing him with glosses are a valuable listing of medical writers circulating in Rome in the 1st c. CE: noteworthy are Praxagoras of Kōs, Nikandros of Kolophōn, Lukos of Naples, Sextius Niger, Dioskouridēs, as well as many of the works in the Hippokratic *corpus*. Erōtianos himself probably was not a physician.

Ed.: Nachmanson (1918).

K. Strecher, "Zu Erotian," *Hermes* 26 (1891) 262–307; *RE* 6.1 (1907) 544–548, L. Cohn; E. Nachmanson, *Erotianstudien* (1917); M. Wellmann, *Hippokratesglossare* (1931); Smith (1979) 202–204; E.M. Craik, "Medical References in Euripides," *BICS* 45 (2001) 81–95.

John Scarborough

Erukinos (before 250 CE)

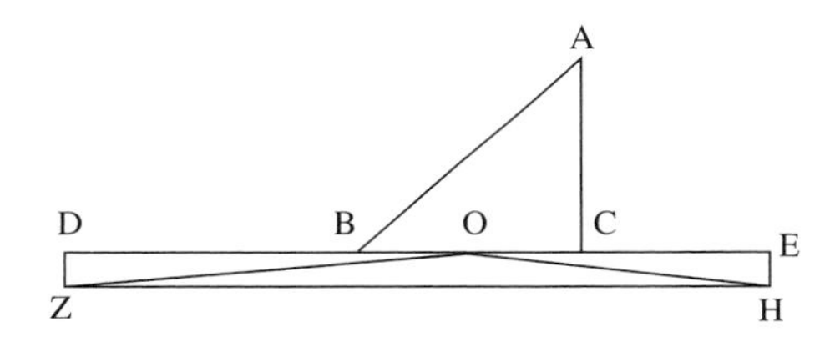

One of Erukinos' (?) paradoxes: AC=CE, AB=DB, DH is constructed so as to have the same area as ABC: thus OHZ has a lesser area than ABC but greater sides. © Bernard

Pappos (*MC* 3, pp. 104–130 H.) explains and demonstrates 15 paradoxical constructions of triangles and quadrilaterals, starting from the "well-known paradoxes" of Erukinos (106.8). Seven of them concern triangles drawn inside given triangles, the paradox being that two sides of the inner triangle can be made greater than the corresponding sides of the outer triangle. From those are derived five more that propose the same kind of paradox for quadrilaterals. The last three concern the areas of triangles or parallelograms, the areas being in inverse relation to the lengths of the sides of the corresponding figures. In many places (e.g. 130.5), Pappos seems to have added his own constructions to Erukinos' so as to reinforce the "paradoxical effect" of the latter. The whole order of exposition follows Pappos' style, so that it is plausible that only some of these theorems are taken from Erukinos. Moreover, the construction submitted by one of Pandrosion's students to Pappos triggering his discussion (104.15–23), is not among the latter and therefore in Erukinos. On the other hand, this same construction is retrieved by Eutokios in his discussion of Archimēdēs' postulates about the relative size of lines having the same extremities (in *Sph. and Cyl.* 3, pp.12–14 Heiberg) and in Proklos' comments on *Elem.* 1.21, in which he clearly refers to it as a "mathematical paradox" (*In Eucl.* p. 326.24–25 Fr.). Proklos (397) also mentions *Elem.* 1.25–27 as belonging to the "treasury of paradoxes" worked out by "mathematicians" and Pappos repeatedly mentions "paradoxes" as a recognized genre, to which Erukinos' text therefore probably belonged.

Heath (1921) 365–368.

Alain Bernard

Eruthrios (*ca* 350? – 640 CE)

Paulos of Aigina 7.18.10 (*CMG* 9.2, p. 371) records his ointment of two dozen ingredients, including three compounds, plus clove-flowers, saffron, cyclamen, nard, propolis, rose oil, spikenard, ***sturax***-gum, **terebinth**, and absinthe wormwood, in honey and Falernian wine. The archaic name (*cf. LGPN* 3B.145) was seeing a late-antique revival: *cf. LGPN* 1.283–284, 2.401–402.

(*)

PTK

Esdras (100 – 500 CE?)

Greek physician of dubious historicity as he is identified in the MSS as a "great prophet" (below) and a "teacher." Whatever the case, the formulas for medicines with which he is credited are of the same type as many antidotes from the 1st to 4th cc. CE. They might thus date back to this period, and have been attributed to a possibly mythical Esdras at a later period. So far, two formulas for compound medicines are known under his name: a 40 ingredient antidote made mainly of vegetals, with also brimstone, and used for the treatment of a wide range of pathologies, from venoms and poisons to difficult childbirth, headache, delirium, cough, fever, swellings, edema, gout and sciatica, for example (three Greek MSS: Bologna, Biblioteca Universitaria, 1808, early 14th c.; Oxford, Bodleian Library, *Baroccianus* 150, 15th c., and Roe 14, 15th c.; and one Latin: Monte Cassino, Archivio della Badia, V.225, 11th c. [Beccaria 1956: 304], ascribed to *Esdre*). The other medicine, which is a shorter version of the prior, was made of 27 ingredients, mostly vegetals but plus also castoreum and dog's flesh, and was prescribed against **dropsy** and cold diseases (MSS: München, *graecus* 72, 16th c.; Roma, Biblioteca Angelica, 4, 15th c.; Wien, Österreichische Nationalbibliothek, *med. gr.* 31, 16th c., and 41, 14th c.).

Esdras *prophētēs* is cited in astrological texts (*CCAG* 8.3 [1912] 13, 26–27, 34, 64–65, 76, 88; *CCAG* 6 [1903] 51, 56), and in Venezia, Biblioteca Nazionale Marciana, *appendix graeca* IV.46 (E. Mioni, *Codices graeci manuscripti Bibliothecae Divi Marci Venetiarum* I.2 [1972] 236). Latin astrological texts circulated in the medieval West under Esdras' name (Thorndike and Kibre 1963: 427, 603, 739, 805, 837, 1444, 1451, 1453), but whether the medical and astrological writers are the same man is uncertain.

Diels 2 (1907) 27, 37–38, Suppl. (1908) 50; M. Formentin, *I codici greci di medicina nelle tre Venezie* (1978) 50, 81.

Alain Touwaide

Euagōn of Thasos (325 – 90 BCE)

Authored a work on agriculture which may have treated cereals, livestock, poultry, viticulture, and arboriculture (*cf.* PLINY, 1.*ind*.8, 10, 14–15, 17–18). It was used by CASSIUS DIONUSIOS (VARRO, *RR* 1.1.8–10, *cf.* COLUMELLA, 1.1.9).

RE 6.1 (1907) 820 (#2), M. Wellmann.

Philip Thibodeau

Euainetos (250 BCE – 100 CE?)

Wrote a commentary on ARATOS, entirely lost; there may have been two such men (*FGrHist* 1026 T19). The name is common before *ca* 100 CE and unattested thereafter (*LGPN*).

(*)

PTK

Euangeus (?) (250 BCE – 80 CE)

ANDROMAKHOS, in GALĒN *CMGen* 5.5 (13.806 K.), gives his "green" plaster, containing aloes, birthwort, frankincense, **galbanum**, ***ikhthuokolla***, myrrh, ***opopanax***, **verdigris**, etc. in an olive oil, vinegar, and **terebinth** base. The name is otherwise unattested (Pape-Benseler; *LGPN*), and if we do not emend to "Euangelos," *ΕΥΑΓΓΕΩΣ* may be a garbled brand name rather than a possessive.

Fabricius (1726) 154.

PTK

Euax (400 – 500 CE)

A name of probably pseudepigraphic origin. Together with DAMIGERŌN, he is credited with a Latin lapidary tract, of uncertain date and composition, consisting, in its current state, in two introductory letters, two very short astrological lapidaries (a planet corresponds to each stone), and the description and find-places of 80 stones with magical properties, most likely translated from Greek originals and then synthesized into one text. Such fusion can be dated to *ca* the 5th/6th c. CE, that is to the same period when, in Italy (particularly in Ravenna and in Cava de' Tirreni) but also in Vandal Africa, medical-scientific works, such as those by DIOSKOURIDĒS, OREIBASIOS, and GALĒN, were translated into Latin.

It is impossible to determine how much of the text is due to Damigerōn, and how much to Euax. Likely the dual Euax-Damigerōn authorship reflects two stages of tradition. Euax probably refers to more recent revisions, while Damigerōn may refer to an older edition, probably the original Alexandrine Greek text used as a model. The name Euax, completely unknown in the ancient world (in Latin *euax* is an interjection of joy, while the Greek suffix *-ax* forms several proper names, e.g. Hierax, Phaiax, Skulax), appears only at the beginning of the second introductory letter, addressed to the emperor Tiberius (in some codices, however, both letters, as well as the lapidary itself, are attributed only to Euax). In it, Euax is characterized as "king of the Arabs," possible evidence of the lapidary's presumed original date (commercial relationships between Romans and Arabs are attested from the imperial age on).

RE 6.1 (1907) 849–850, M. Wellmann; Halleux and Schamp (1985) 193–290.

Eugenio Amato

Euboulidēs (*ca* 200 BCE – *ca* 250 CE)

ANATOLIOS, *On the Decade*, as preserved in the THEOLOGUMENA ARITHMETICAE (p. 52 de Falco), cites Euboulidēs, ANDROKUDĒS, ARISTOXENOS, HIPPOBOTOS, and NEANTHĒS, as writers on PYTHAGORAS and his rebirths. BOËTHIUS, *Inst. Mus.* 2.19, cites (the same?) Euboulidēs with HIPPASOS on the order and generation of the harmonies from the **Pythagorean** *tetraktus*, i.e., the number ten and its representation as 1 + 2 + 3 + 4. (The name is more frequent in Athens and areas under Athenian influence: *LGPN*.)

FGrHist 1106.

PTK

Euboulos (Agric. and Veterin.) (325 – 90 BCE)

Author of a remedy for *opisthotonos* in horses quoted by PELAGONIUS (Pelagonius 271). HIEROKLĒS attributes the same remedy to unknown authorities (*henioi*): *Hippiatrica Parisina*

325 = *Hippiatrica Berolinensia* 34.10). A Euboulos appears in VARRO's list of Greek writers on agriculture (*RR* 1.1.9) added by CASSIUS DIONUSIOS to Magōn's agricultural treatise (*cf.* COLUMELLA, 1.1.11). Pelagonius and Hieroklēs may have used sources derived from Cassius Dionusios.

RE 6.1 (1907) 879 (#18), M. Wellmann; Fischer (1980); *CHG* v. 1; McCabe (2007) 159, 168, 236–237.

Anne McCabe and Philip Thibodeau

Euboulos (Pharm.) (250 BCE? – 80 CE)

ANDROMAKHOS quotes two of Euboulus' recipes: an enema for dysentery compounded from realgar, copper, acacia, etc., in myrtle wine and infused with warm, diluted wine (GALĒN *CMLoc* 9.5, 13.297 K.); and a *phaia*, possibly against venoms, compounded from **litharge**, roast copper, **verdigris**, beeswax, **terebinth**, ***ammōniakon*** incense, ***opopanax***, etc. (*CMGen* 6.1, 13.911–912 K.; *cf.* 3.9, 13.650 K.). *Phaia* (dark) plasters are so-called probably because of their colorful mineral ingredients.

RE 6.1 (1907) 879 (#19), M. Wellmann.

Alain Touwaide

EUC- ⇒ EUK-

Euclid of Alexandria (300 – 260 BCE)

We have remarkably little personal information about Euclid (Eukleidēs), arguably the most influential mathematician who ever lived. PAPPOS (*Collection* 7.35, p. 678.10–12 H.) says that APOLLŌNIOS OF PERGĒ studied with Euclid's students in Alexandria, suggesting a *floruit* in the middle of the 3rd c. BCE. PROKLOS (*In Eucl.* p. 68.10–11 Fr.) makes Euclid a contemporary of the first Ptolemy (d. 282), but his evidence does not inspire confidence. The standard edition of Euclid's works (Heiberg and Menge) includes the following complete texts in Greek: *Elements,* 13 books (vv. 1–4) plus a 14th book written by HUPSIKLĒS and a 15th book at least in part due to a pupil of the elder ISIDŌROS OF MILĒTOS (v. 5); *Data* (v. 6); *Optics* (in two recensions) and *Catoptrics* (v. 7); *Phenomena, Sectio Canonis*, and *Introductio Harmonica* (v. 8). Volume 8 also contains textual evidence relating to non-extant works ascribed in ancient sources to Euclid: *On Divisions, Fallacies, Porisms, Conics,* and *Surface Loci.* Arabic evidence indicates that Euclid also wrote on mechanics.

Mathematical texts are especially vulnerable to "improvements," inserted "explanations," and recasting, as is shown, for example, by the 14th and 15th books of the *Elements* and the two recensions of the *Optics*. Most Greek MSS of the *Elements* and all early printed versions derive from an edition by THEŌN OF ALEXANDRIA, whereas the standard printed edition purports to be pre-Theonine. There are considerable variations between our Greek text and Arabic translations and also among the Greek MSS themselves. Because the texts are so subject to tampering, it is really not possible to speak about exactly what Euclid wrote, but only about whether a work is based on something Euclid could have written. Of the complete works published in the standard edition, only the *Introductio Harmonica* is universally rejected as non-Euclidean in this sense. Older scholars tended to consider the *Catoptrics* and *Sectio Canonis* spurious, but both works have been defended as Euclidean in more recent years (the *Sectio* is treated in a separate entry). Only the other surviving works which can be considered Euclidean, all characterized by apparently rigorous, stylized deduction from first principles, are discussed here.

Data ("Things Given") is mentioned first by Pappos (*Collection* 7.3, p. 636.18–19 H.) in his list of works useful for analysis, that is, the finding of solutions to problems and of proofs of propositions, by supposing that what has to be done is accomplished or that what is to be proved is true, and asking what else must be accomplished or true as a result: the idea is that when one reaches things one knows how to accomplish or prove, one will be able to reverse the steps and produce a solution or proof for what is sought.

Optics is essentially a treatise on monocular perspective. It is assumed that vision is a matter of the emission of rectilinear rays from the eye which strike an object and form a cone with vertex in the eye and base a plane figure determined by the shape of the object seen, and that the relative apparent size of an object is determined by the size of the angle "under which" it is seen and its relative apparent position by the relative position of the rays under which it is seen; the rays are treated as discrete straight lines, so that an object will not be seen if it falls between rays.

Catoptrics takes the same approach to mirror vision, treating plane, convex, and concave mirrors.

Phenomena is an essay in very elementary geometric astronomy, the main point of which seems to be showing that certain astronomical appearances can be represented and understood geometrically. In the prologue simple astronomical data are invoked to justify the claim that the sphere of the fixed stars rotates uniformly about a fixed axis and that the eye of an observer is at the center of the sphere, and geometrical definitions are given of such astronomical terms as "horizon," and "meridian." Among the theorems proved are the assertion that if two stars lie on a great circle which has no point in common with the arctic circle (the circle including all stars that are never seen to set), the one which rises earlier sets earlier (*prop.* 4).

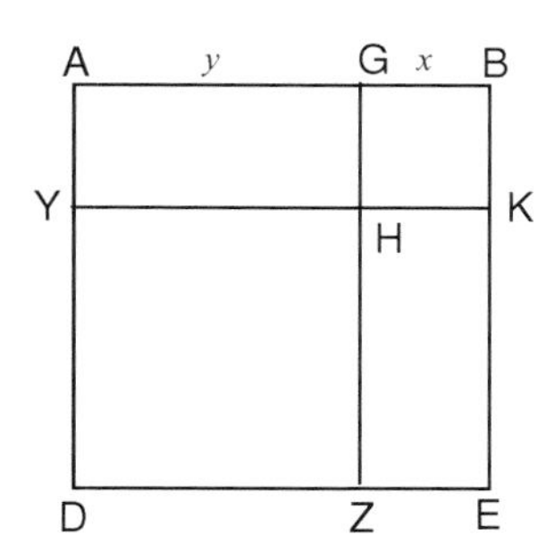

Euclid's geometric algebra (1. *prop.* 2) © Mueller

The name of "Euclid" is associated first and foremost with the *Elements*, apparently a single treatise in which propositions are derived from principles labeled as "definitions," "postulates," and "common notions" (the last frequently called axioms). Careful scholarship of the last century has made clear that the work is a compilation based on several sources. The subject of book 1 is the geometry of plane rectilinear figures. The book is noteworthy for avoiding the use of proportions and for postponing the use of the parallel postulate until it is required. Book 2 introduces what is now frequently called geometric algebra in a series of geometric propositions corresponding to what we know as algebraic equations; for example, proposition 2, which corresponds to "$(x+y)^2 = x^2 + y^2 + 2xy$," says that if AGB is a straight line, the square with side equal to AB [**SQ**(AB)] is equal to **SQ**(AG) plus **SQ**(BG) plus two times the rectangle with sides equal to AG and BG.

Book 3 treats circles and their relations to straight lines and angles, Book 4 the inscription in circles and circumscription about circles of rectilinear figures. Book 5 brings in proportionality, developing a theory based on a definition which says of four magnitudes A, B, C, D that $A{:}B :: C{:}D$ if and only if for any multiples $m{\cdot}A$, $n{\cdot}B$, $m{\cdot}C$, $n{\cdot}D$ of those magnitudes, if $m{\cdot}A$ is greater than, equal to, or less than $n{\cdot}B$, $m{\cdot}C$ is accordingly greater than, equal to, or less than $n{\cdot}D$, and that $A{:}B > C{:}D$ if and only if for some m and n, $m{\cdot}A > n{\cdot}B$ and $m{\cdot}C \leq n{\cdot}D$. Euclid's theory deals only with proportionalities among magnitudes, not with ratios between pairs of magnitudes, but it is a simple matter to reformulate the theory of Book 5, by treating ratios $A{:}B$ as "cuts" in the system of positive fractions m/n. In Book 6 Euclid applies the theory of proportion to geometric entities and develops the notion of similarity. Books 7–9 introduce numbers as objects of study using a separately developed theory of proportion. The major topic of the very difficult Book 10 is a classification of straight lines A which are called irrational relative to a given straight line R if both A and R and **SQ**(A) and **SQ**(R) are incommensurable.

Book 11 develops basic ideas of solid geometry. Book 12 uses a method, which is called the "method of exhaustion," to prove a series of sophisticated results, the simplest of which is *prop.* 2: if C and C' are circles with diameters d and d', then $C{:}C'$ :: **SQ**(d):**SQ**(d').

In Book 13 Euclid constructs the five regular solids, triangular pyramid, octahedron, cube, icosahedron, and dodecahedron, circumscribes spheres around them, and characterizes their edges relative to the diameters of the circumscribing spheres using in the last three cases the classification of Book 10.

It is clear from Proklos (*In Eucl.* pp. 65–68 Fr.) that Euclid's *Elements* had more than one predecessor, starting with a work of Hippokratēs of Khios. It is also clear that much of the contents of the *Elements* is based on the work of others, most clearly Eudoxos (Books 5 and 12) and Theaitētos (Books 10 and 13). Nevertheless, Euclid's *Elements* is an outstanding achievement which replaced all of its predecessors and sources, and became both an inspiration and a foil for much of the subsequent history of Western mathematics.

Ed.: J.L. Heiberg, and H. Menge, *Euclidis Opera Omnia*, 9 vv. (1883–1916);
P. Ver Eecke, trans., *Euclide, L'optique et al catoptrique* (1938); *DSB* 4.414–437, I. Bulmer-Thomas; B. Vitrac, trans., *Euclide, Les Elements* 4 vv. (1990–2001); J.L. Berggren and R.S.D. Thomas, trans., *Euclid's Phaenomena* (1996); *DPA* 3 (2000) 252–272, B. Vitrac; C.M. Taisbak, trans., *Dedomena* (2003).

Ian Mueller

pseudo-Euclid, Elements 15 ⇒ Isidōros of Milētos' student

Euclidean Sectio Canonis (300 – 260 BCE?)

"Division of the Monochord" (= *kanonos katatomē* = *Sectio Canonis*), a short text on mathematical harmonics ascribed in most MSS to Euclid. Fragments are quoted by Porphurios (title and authorship: *In Ptolemaei Harmonica Commentarium* 98.19 Düring; preface: 90.7–22; props.1–16: 99.1–103.25) and Boëthius (*De Institutione Musica* iv). Its authorship, date and unity of composition have been long debated: a logical error in *prop.* 11 has been used as evidence against Euclid's authorship, but arguments for dating it substantially later than Euclid, and for excising the preface and two (or four) final propositions as late accretions, have not met with consensus.

The text as we now have it is comprised of five types of material: (1) a discursive preface attempting to derive mathematical harmonics from a physical acoustics which can account for the behavior of strings (an essential connection in order for the monochord to be used demonstratively in props.19–20); (2) nine purely mathematical propositions demonstrating the properties of the simplest multiple ($mn:n$) and epimoric ($(n+1):n$) ratios; (3) seven subsequent propositions (10–16) wherein the properties of simple musical intervals are shown to be analogous to those of the ratios of props.1–9; (4) two propositions (17–18) locating the "movable" notes in the scale by the method of concordance; (5) a final two propositions (19–20) introducing the monochord and marking on it the bridge-positions corresponding to the notes of a two-octave scale-system.

The *Sectio* owes as much to 4th c. developments in acoustics and harmonics as it does to Euclidean mathematics. In the preface, the basic assumptions of Arkhutan acoustics (e.g., that sound is caused by impact, *plēgē*) are adopted and modified, apparently with the aim of allowing mathematical propositions to be demonstrated on strings, in ways that suggest the influence of theories akin to those expressed in the ARISTOTELIAN ON SOUNDS and the ARISTOTELIAN PROBLEMS (11.6, 19.12, 19.23, 19.39). The *Sectio* is also a polemical text; certain propositions (e.g. 16, 18) are clearly intended to refute not only the conclusions but also the basic assumptions of Aristoxenian harmonics.

Ed.: *MSG*; H. Menge, *Euclides Phaenomena et scripta musica* (1916).

Düring (1932); A.D. Barker, "Methods and aims in the Euclidean *Sectio Canonis*," *JHS* 101 (1981) 1–16; A. Barbera, "Placing *Sectio Canonis* in historical and philosophical contexts," *JHS* 104 (1984) 157–161; Barker (1989); A. Barbera, *The Euclidean Division of the Canon* (1991); A.D. Barker, "Three approaches to canonic division," *Apeiron* 24 (1991) 49–83; A.C. Bowen, "Euclid's *Sectio canonis* and the history of Pythagoreanism," in Bowen (1991); O. Busch, *Logos Syntheseos* (1998); Mathiesen (1999); S. Hagel, "Zur physikalischen Begründung der pythagoreischen Musikbetrachtung," *WS* 114 (2001) 85–93; Barker (2007) ch. 14, 364–410.

David Creese

Eudēmos (Methodist) (*ca* 21 – 31 CE)

The irony-infused episode in TACITUS, *Hist.* 4.3 and 11, and the gossipy notice in PLINY 29.20 name Eudēmos the **Methodist** as a personal physician to Tiberius' son Drusus "the Younger" and his wife, Liuilla (Liuia Iulia). Implicated long after Drusus' death in 23, Eudēmos was put on the rack in 31, "confessing" to murder by poisoning, a charge stemming from a letter written by Apicata to Tiberius after the disgrace and execution of her ex-husband Seianus; she conveniently committed suicide once the letter was sent. Liuilla's adultery with Eudēmos, as recorded by Pliny, ". . . was an easy frill" (Levick, 279 n. 151). Only Tacitus' innuendo suggests Eudēmos was named in Apicata's letter; more likely it simply stated that Seianus and Liuia had poisoned Drusus (whose death was quite likely to have been "natural"). "[A doctor's] professional duties at the time of the alleged murder would ensure that the unfortunate Eudēmos . . . stood at the head of the list of candidates for the rack" (Seager, 156).

Eudēmos was THEMISŌN's student (CAELIUS AURELIANUS, *Acute* 2.219 [Drabkin, p. 286; *CML* 6.1.1, p. 278]), and under Eudēmos' name is preserved a poetic version of a theriac invented by ANTIOKHOS VIII PHILOMĒTŌR (GALĒN, *Antid.* 2.14 [14.185–186 K.]; Tecusan, p. 339, rough translation), perhaps indicating that Themisōn, who had emigrated from Syria to Italy, passed along to Eudēmos some of Antiokhos' detailed knowledge of pharmacology and toxicology, especially antidotes against snake bites, scorpion stings, and spiders (*cf.* Galēn, *Antid.* 2.17 [14.201–202 K.]). Eudēmos was one of the first physicians to write about ***hudrophobia***, as contracted from dog bites (PHILOUMENOS, *Poisonous Animals* 1.4 [*CMG* 10.1.1, p. 5]), but assumed that ***hudrophobia*** was the same as *melankholia*, a diagnosis refuted by Caelius Aurelianus (*Acute* 3.107–108 [Drabkin, p. 368; *CML* 6.1.1,

p. 354]); therapy for ***hudrophobia***, according to Eudēmos, consisted of venesection, administration of hellebore (type not specified), and application of cupping vessels (bodily position also not designated [*ibid.*, 3.134–135; Drabkin, p. 386; *CML* 6.1.1, p. 372]). Eudēmos recorded the anecdote of a physician who had contracted ***hudrophobia***; recognizing his unhappy and painful fate, weeping, he dropped to his knees but seeing the tears dripping onto his body, "he leaped up and tore his clothes to pieces" (*ibid.*, 3.105 [Drabkin, pp. 366–367; *CML* 6.1.1, p. 354]). To alleviate "cardiac disease" (a severe and constricting pain in the upper chest, our *angina pectoris*), Eudēmos recommended an enema of cold water (*ibid.*, 2.219 [Drabkin, p. 286; *CML* 6.1.1, p. 278]).

Ed.: Tecusan (2004) 83, 91, 98, 103, and 107 ("Eudemus: Thematic Synopsis").
RE 6.1 (1907) 904–905, M. Wellmann; B. Levick, *Tiberius the Politician* (1976); R. Seager, *Tiberius*, 2nd ed. (2005).

John Scarborough

Eudēmos "the Elder" (250 – 30 BCE)

Cited once by ANDROMAKHOS, in GALĒN *CMLoc* 9.5 (13.291 K.), for a ***trokhiskos*** against "dysentery" (compounded from saffron, "tubes" of cassia, nard, myrrh, alum, and poppy juice). Designated "the Elder" presumably to distinguish him from the then-recent EUDĒMOS (METHODIST). The name is very frequent, and there is no need to identify with any other medical Eudēmos.

(*)

PTK

Eudēmos of Alexandria (285 – 235 BCE)

Greek anatomist, often quoted by GALĒN, together with his younger contemporary HĒROPHILOS (Galēn, *In Hipp. Aph.* 18A.7 K.), as among the great historical anatomists (*De Semine* 2.6.13 [*CMG* 5.3.1, p. 200], *In Hipp. Nat. Hom.* 15.134 K., etc.). He seems to have worked on bones (RUFUS, *Onom. Anthr. Mor.* 73 [p. 142 DR]), arteries and veins (Galēn, *UP* 3.8 [Helmreich 1907: 148–149]), joints of hands and feet (*ibid.*), and the embryonic vascular system (SŌRANOS *Gyn.* 1.57.4 [*CMG* 4, p. 42; *CUF* v. 1, pp. 56–57]). He apparently wrote on the nervous system (Galēn, *On My Own Books* 3 [2.108 MMH]).

RE 6.1 (1907) 904 (#17), M. Wellmann; *KP* 2.405, F. Kudlien; *BNP* 5 (2004) 147 (#4), V. Nutton; *AML* 280, K.-H. Leven.

Daniela Manetti

EUDĒMOS OF ATHENS ⇒ EUTHUDĒMOS OF ATHENS

Eudēmos of Athens (380 – 300 BCE)

Drug merchant to be distinguished from later homonymous physicians (see ANTIOKHOS VIII). He may have been active as early as the beginning of 4th c. BCE, if one can identify him with Eudamos (Aristophanēs, *Plut.* 884), which is hardly certain. In order to demonstrate that drugs have different effects according to the person, THEOPHRASTOS (*HP* 9.17.2–3) contrasts Eudēmos, who, "after making a wager that he would experience no after-effect before sunset, drank a quite modest dose" of hellebore and could not withstand this purgative, with another *pharmakopōlēs*, EUNOMOS OF KHIOS, who took a draught of hellebore with impunity.

RE 6.1 (1907) 903–904 (#16), M. Wellmann.

Jean-Marie Jacques

EUDĒMOS OF PERGAMON ⇒ PHILŌNIDĒS OF LAODIKEIA

Eudēmos of Rhodes (330 – 285 BCE)

Student of ARISTOTLE, founded a philosophical school in Rhodes. DAMAS' vita of Eudēmos is lost (*fr.*1 Wehrli). From biographical sources on other members of the **Lyceum** we can glean that Eudēmos, already a mature scholar by 322 BCE, was a candidate to succeed Aristotle (*fr.*5 Wehrli); that, upon THEOPHRASTOS' designation as **scholarch**, Eudēmos left for his native Rhodes and set up a school there; furthermore, that he remained in correspondence with Theophrastos about matters of Aristotelian philosophy (*fr.*6 Wehrli).

Eudēmos is credited with some works on logic (*frr.*7–24 Wehrli, these testimonies almost always mention Eudēmos together with Theophrastos, suggesting there was no specific contribution by him which would have set him apart from Theophrastos), *On angle* (*fr.*30 Wehrli), a *Physics* (*frr.*31–123 Wehrli), a collection of data on animal behavior (*frr.*125–132 Wehrli), and histories of the mathematical sciences (geometry: *frr.*133–141 Wehrli, arithmetic: *fr.*142 Wehrli, astronomy: *frr.*143–149 Wehrli), and perhaps one of theology (*fr.*150 Wehrli). The testimony from *On angle* situates angles in terms of Aristotelian ontology, as belonging to the category of quality. The *Physics* – which must have been composed for purposes of Eudēmos' own school – follows the discussion of Aristotle's *Physics* in a linear fashion, omitting Book VII. The collection of data on animal behavior continues Aristotle's investigations in the *History of Animals*.

Practically everything we know about early Greek mathematics and astronomy comes from Eudēmos' histories. These histories must have belonged to the category of collections of data, known as hypomnematic works in the Aristotelian corpus. Nevertheless, such hypomnematic works were not just loose collections: rather, Eudēmos' histories rested on what could be termed a framework of the rational reconstruction of the development of these disciplines, in terms of a sequence of crucial discoveries, each attributed to a first discoverer (*prōtos heuretēs*), and contributing to the perfection of the discipline – a perfection which either has already been achieved by Eudēmos' contemporaries, or which can be expected to be achieved soon.

When assessing Eudēmos' sources and methods in writing his histories, we can with some confidence assume that he had access to his predecessors' works, at least beginning with OINOPIDĒS and HIPPOKRATĒS OF KHIOS, and for earlier authors he relied on collections, like e.g. the collection (*Sunagōgē*) of HIPPIAS. These works, however, must have been less rich in detail about earlier mathematicians than Eudēmos' histories, hence Eudēmos almost certainly had some further material at his disposal to supplement these earlier collections.

Ed.: Wehrli, v. 8; H. Baltussen, "Wehrli's edition of Eudemus of Rhodes," in Bodnár and Fortenbaugh (2002) 127–156.

Bodnár and Fortenbaugh (2002); Zhmud (2006).

István Bodnár

Eudikos (250 BCE – 75 CE)

PLINY 31.13 cites him for two springs near Hestiaia: one which blackens and one which whitens the skin of drinkers. Pliny, 1.*ind.*31, lists him among early sources, such as KTĒSIAS and THEOPHRASTOS, explicitly distinguishing him from EUDOXOS, and the name Eudikos is more frequent than Eudoxos through the 1st c. CE (*LGPN*): contrast Gisinger, *Eudoxos* (1921) 123–124. Perhaps the same as, or confused by Pliny with, the (neo)-**Pythagorean** of Lokroi (IAMBLIKHOS, *VP* 267).

(*)

PTK

Eudōros of Alexandria (*ca* 60 – 35 BCE)

An "**Academic**" (IŌANNĒS STOBAIOS 2.24.7–8 W.-H.) considered the founder of Middle **Platonism**. His doctrine of principles has a Neo-**Pythagorean** outlook. The "elements" Monad and Dyad are transcended by a higher principle "the One." Eudōros is reported to have ventured an emendation of *Metaph.* 1 (988a11), where ARISTOTLE discusses PLATO's first principles (ASPASIOS in ALEXANDER OF APHRODISIAS *CAG* 1 [1891] 58–59). A subdivision of ethics survives from Eudōros' classification of philosophy. PLUTARCH (*Anim. Procr.* 1013B; 1019E; 1020C) refers to a work on the *Timaeus*, wherein Eudōros upheld a non-literal interpretation of Plato's cosmogony and calculated the numbers of the soul. In a work on the *Categories*, Eudōros raised detailed objections against Aristotle (SIMPLICIUS, *in Categ. CAG* 8 [1907] 159). A work on the heavens or on the world seems to have been a principal source for AKHILLEUS. Drawing on DIODŌROS OF ALEXANDRIA, Eudōros discussed among other things the division of the Earth into five zones and argued that the torrid (equatorial) zone is inhabited. Eudōros also wrote on the Nile (STRABŌN 17.1.5).

Ed.: C. Mazzarelli, "Raccolta e interpretazione delle testimonianze e dei frammenti del medioplatonico Eudoro di Alessandria," *Rivista di filosofia neo-scolastica* 77 (1985) 197–209, 535–555.

Moraux (1984) 509–527; Dillon (1996) 115–135; *DPA* 3 (2000) 290–293, *Idem; BNP* 5 (2004) 149–150 (#2), M. Baltes and M.-L. Lakmann.

Jan Opsomer

Eudoxos of Knidos (*ca* 365 – *ca* 340 BCE)

Son of Aiskhinēs, born *ca* 395–390 BCE; mathematician, astronomer, and geographer. DIOGENĒS LAËRTIOS (8.86–91) provides our principal biographical evidence. At age 23, though impoverished, Eudoxos visited Athens for two months with the otherwise unknown Theomedōn, a physician who funded him. After returning to Knidos, he visited Egypt with KHRUSIPPOS OF KNIDOS (I), and stayed for 16 months, where he studied astronomy. The doxographical tradition also claims that he studied mathematics with ARKHUTAS OF TARAS and medicine with PHILISTIŌN OF LOKROI. He then proceeded to the Hellespont where he lectured and gained many followers, especially from Kuzikos (KALLIPPOS, HELIKŌN, POLEMARKHOS). He then returned to Athens, where he associated with the **Academy**. Having returned home, he did legislative work for Knidos, probably after 347, and died in his 53rd year, well honored by his city.

Eudoxos of Knidos © Budapest Museum

Eudoxos completed the generalization of proportion theory, one of the principal intellectual efforts of the previous 50 years, and developed fundamental techniques for comparing figures by approximating figures leading to a *reductio*. They are among the most enduring achievements of ancient Greek mathematics. Our knowledge of his work on mathematics comes principally from four sources: PROKLOS' claim that Eudoxos expanded the number of general theorems; scholia to EUCLID's *Elements* claiming Eudoxos as the author of Book 5 on proportion theory and to *Elements* 12.2 (circles are as the square on their diagonals) and 12.10 (a cone is $^1/_3$ a cylinder with the same height and base); ARCHIMĒDĒS' comment in the introductions to *On the Sphere and Cylinder* and *Method* that Eudoxos proved that the pyramid is $^1/_3$ a prism with the same height and base (= *Elements* 13.3–7) and the cone/cylinder theorem; and finally ERATOSTHENĒS' claim that Eudoxos produced a solution to the double mean proportion problem: given a, b, to find x, y so that a : x = x : y = y : b, using curved lines, which Eratosthenēs found impractical and EUTOKIOS (our source for Eratosthenēs) found too garbled to reproduce.

From these sources, a general understanding of 4th c. BCE mathematics, and traces especially in ARISTOTLE, Archimēdēs, and THEODOSIOS, we can reconstruct some of Eudoxos' mathematical ideas. The method by which theorems from Euclid, *Elements* 12 are proved, inappropriately called "the method of exhaustion," approximates the compared figures by inscribed figures whose relations are known. Then it proves by contradiction that the approximated figures must be in the same relation. The *reductio* builds on two implicit principles: (1) given two comparable magnitudes A, B, $A > B$ or $A = B$ or $A < B$ (connectivity), (2) given comparable magnitudes, A and B, and a magnitude C, there is an X such that $A : B = C : X$ (existence of 4th proportional), and one fundamental theorem: (3) given A, B, if $A > B$ and more than half is taken away from A, and so continuously from the remainder, there will eventually be left a magnitude X, such that $X < B$ (a bisection principle proved in *Elements* 10.1). He also uses a theorem based on (2): (4) if $X > B$ and $A : B$ is a ratio, then there is a Y, such that $Y < A$, $X : A = B : Y$ (*cf. Elements* 5.14). Principles (2) and (4) are not used in the cone/cylinder theorem.

As an example of the structure of the method, for which there are several other forms, suppose that one needs to prove that $A : B = C : D$. The proof involves two theorems. In the first, one proves for an approximating class of figures, a, b of A, B, that $a : b = C : D$. In the second, one assumes that $A : B \neq C : D$, in which case, by (2), there is an X, $A : X = C : D$, where $X < B$ or $X > B$, by (1). For the first part of the proof, suppose $X < B$. One now finds, by construction, a_i, b_i, such that $a_i < A$ and $b_i < B$, where $a_i : b_i = C : D$, by the first part of the proof, and $B\text{-}b_i < B - X$ by (3), so that $X < b_i < B$. But since $a_i : b_i = C : D = A : X$ and $a_i < A$, it follows that $b_i < X$. This is a contradiction, so that $A : B \nless C : D$. We can take this as a general theorem about magnitudes A, B. In the second part of the proof, one now assumes that $X > B$. Here this case is reduced to the first, since, by (4), there will be a Y such that $Y < A$ and $D : C = X : A = B : Y$, which contradicts the first case. Hence, $X \ngtr Y$. So, by (1), $A : B = C : D$.

There is some evidence that Eudoxos also used this method for proving general theorems in proportion theory, where the first case would be for commensurable magnitudes and the second for incommensurable magnitudes. If so, then *Elements* 5, on proportion theory, would represent a later reworking of his theory and proofs. Here, there is a general definition of "same ratio" (*Elements* 5. def. 5), eliminating the need for separate cases:

$$A : B = C : D \text{ iff } \forall\, n, m: (n \times A) \gtreqless (m \times B) \text{ iff } (n \times C) \gtreqless (m \times D).$$

Eudoxos was the first astronomer to attempt a general geometrical model to explain apparent motion of planetary stars, the sun, Moon, and five visible planets. The model assumed that all celestial bodies are spheres with the Earth as their center and whose motion is regular: circular about an axis through the center of the Earth. Each planetary star has a system of concentric spheres where the axis of one inner sphere is fixed to the next outer sphere. In this way, Eudoxos could create apparent irregular motions. For example, each planetary system consisted of an outer sphere whose poles would be the poles of the celestial equator and which rotated daily east/west. Fixed to it were the poles of a sphere contained in it with the same center. The poles of the fixed sphere would be perhaps $^1/_{15}$ circle (the obliquity of the **ecliptic**) from the poles of the first sphere with the second sphere rotating slowly west/east, i.e., with the zodiacal period of the planetary star, where the net motion produced is a spherical spiral. For the five planets, further variations in their motion would then be explained by two

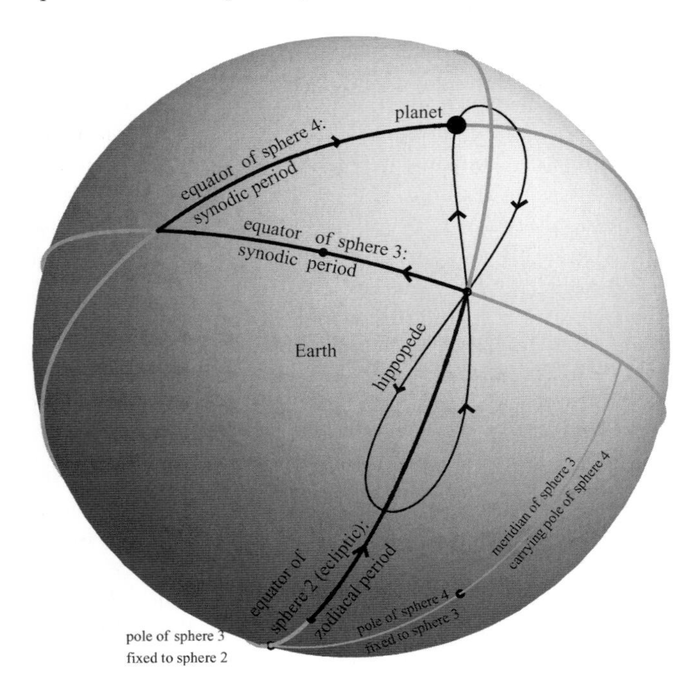

Eudoxos of Knidos: Hippopede © Mendell

more spheres rotating oppositely with the synodic period of the planet, one with poles on the equator of the second sphere, and the last, carrying the planet, with poles at an angle to this. The motivation for the extra two spheres is uncertain, although the spheres for Mercury and Venus certainly provided elongations from the Sun. Other suggested phenomena include retrograde motion, invisibility periods, and latitudes.

On the other hand, the Sun and Moon only required one additional sphere each. Here the second and third sphere of the Sun moved east/west, although the purpose of the third sphere is not known, except that it provided a latitudinal motion for the Sun, rightly criticized by HIPPARKHOS two centuries later. The best understood model is that of the Moon. The outer sphere moved with the daily motion, while the second moved west/east with the longitudinal motion plus the latitudinal motion, and the third moved oppositely with the latitudinal motion, but at an angle perhaps of 1/12 or 1/15 of a right angle to the second sphere. Here the latitudinal motion is, e.g., 1 cycle per interval from downward node to downward node, while the longitudinal motion is 1 cycle per interval between conjunctions with the same star (the zodiacal motion).

Eudoxos' other work in astronomy includes notably the systematic organization of fixed stars, which he described in two works, the *Mirror* and later the *Phainomena*, which survive in ARATOS' versification and Hipparkhos' commentary on Aratos. In the first work the ratio of longest day to shortest night for Greece is given as 5 : 3, and in the other as 12 : 7, as also in PHILIPPOS OF OPOUS. His division of the year, consisting of three seasons of 91 days and autumn of 92 days, makes the seasons as equal as possible and would seem to be an explicit rejection of METŌN and EUKTĒMŌN. Eudoxos may have endorsed an ***oktaetēris*** calendar (*cf.* KLEOSTRATOS), also a rejection of Metōn and Euktēmōn, but the attribution of the book ***Oktaetēris*** to him was doubted in antiquity (esp. by Eratosthenēs, *cf.* AKHILLEUS, *Introduction* 19). Traces of his ***parapēgma***, possibly erected on Dēlos, survive in GEMINUS and PTOLEMY.

Eudoxos also wrote an extensive geography in seven books, *The Circuit of the Earth*, and may have been the first to divide the Earth into regions according to projections of celestial circles: equatorial, tropical, arctic and antarctic. He held that the inhabited world, from India to the Iberian Peninsula, was twice as long as it was wide. The work included ethnologies.

In other matters, Eudoxos defined the good as "what all things aim at" and identified it with pleasure (Aristotle, *Nicomachean Ethics* 1.1.12 and 10.2–3). He also thought that Forms were immanent in things (*cf.* Aristotle, *Metaphyics* A.9).

Ed.: Fr. Lasserre, *Die Fragmente des Eudoxos von Knidos* (1966).

Knorr (1975); Neugebauer (1975) 596, 620–621, 662, 675–689; W.R. Knorr, "Archimedes and the Pre-Euclidean Proportion Theory," *AIHS* 28 (1978) 183–244; I. Mueller, *Philosophy of Mathematics and Deductive Structure in Euclid's Elements* (1981); A.C. Bowen and B.R. Goldstein, "A New View of Early Greek Astronomy," *Isis* 74 (1983) 330–340; R.M. Dancy, *Two Studies in the Early Academy* (1991); Henry Mendell, "The Trouble with Eudoxus," in P. Suppes, J. Moravcsik, and Henry Mendell, edd., *Ancient and Medieval Traditions in the Exact Sciences: Essays in Memory of Wilbur Knorr* (2001) 59–138; Henry Mendell, "Reflections on Eudoxus, Callippus and their Curves: Hippopedes and Callippopedes," *Centaurus* 40 (1998) 177–275; I. Yavetz, "On the Homocentric Spheres of Eudoxus," *AHES* 51 (1998) 221–278.

Henry Mendell

Eudoxos of Kuzikos (120 – 110 BCE)

Greek navigator whose adventures and discoveries were documented by POSEIDŌNIOS OF APAMEIA (*On the Ocean*, F 49 E–K), later transmitted and denounced by STRABŌN (2.3.4–5). According to these reports, Eudoxos was a sacred ambassador and peace herald at the festival of Persephone. Coming to Egypt under Ptolemy VIII Euergētēs II (145–116 BCE), he formed an association with the king and his ministers regarding voyages up the Nile. An Indian sailor, found alone and half dead on his stranded ship at the Arabian Gulf, and brought before the king, declared that he had come from India; whereupon Ptolemy sent an expedition including Eudoxos to explore the route.

Kleopatra III, Ptolemy's wife and successor, sent Eudoxos to sail the same route. Returning to Egypt, he drifted to southern "Ethiopia" where he contacted the inhabitants and found a horse-shaped wooden prow of a ship from Gadēs that had sailed beyond the Lixos river (in Morocco) but had not returned. Eudoxos concluded that it was possible to sail around Africa, and set sail from Kuzikos with all his property and a large entourage. But the ship sank.

Eudoxos built another ship and set sail again. Arriving in Maurousia (Mauretania) he traveled on foot to the court of Bogos whose advisors opposed the exposure of their country to foreigners. Eudoxos, fleeing to Roman territory, crossed over to Iberia where he built two ships, one for sailing along the coast and the other for the open sea. He equipped the ships with supplies and carpenters and set sail once again, but never returned.

Eudoxos' achievements reflect two maritime routes: from Egypt to India, and around Africa from west to east. Both journeys occurred around 120–110 BCE. Eudoxos' journeys were probably not along the coast (***periploi***), but followed the monsoon sailing with the monsoon winds across the open sea between Egypt and India. In the second route, around Africa, Eudoxos' failures were probably due to ERATOSTHENĒS' misconception of the size and shape of Africa as a right angled triangle, the right angle being in Egypt.

J.H. Thiel, *Eudoxus of Cyzicus, A chapter in the History of the Sea-route round the Cape in ancient times* (1966): Strabōn's text with commentary.

Daniela Dueck

Eudoxos of Rhodes (*ca* 275 – 200 BCE)

Historian (DIOGENĒS LAËRTIOS 8.90), who incorporated ***periploi*** into his *Histories*; the fragments suggest paradoxography. He described birds larger than oxen beyond the Pillars of Hēraklēs (F3 = AELIANUS, *HA* 17.14) and how the Galatians charmed birds to vanquish locust swarms (F4 = *HA* 17.19). He noted a lack of sunlight in Celtic climes (F2).

FGrHist 79; *OCD3* 566, anon.

GLIM

Euelpidēs (15 – 35 CE)

Greek ophthalmologist contemporary with CELSUS (6.6.8A) who preserves several of Euelpidēs' recipes for eye pathologies, all containing poppy juice and minerals, in gum; all but one contain saffron. The *trygodes*, resembling wine lees (*trux*), he compounded from **calamine**, antimony, ***lukion***, myrrh, etc. (6.6.8A [2.196.12–19 Spencer]); the *memigmenon* salve he "mixed" from white peppercorn, roasted copper, etc., without saffron (6.6.17 [2.210.3–6 Spencer]); the *pyrrona*, red from roasted copper, contained also myrrh and white pepper (6.6.20 [2.210.24–28 Spencer]); the ball-shaped *sphaerion* he compounded from hematite, peppercorns, **calamine**, and myrrh pounded in Aminian wine (6.6.21 [2.212.1–6 Spencer]); the *pyxinum* he kept in a box-wood case, having compounded it from ***ammōniakon*** incense, **calamine**, and ***psimuthion*** (6.6.25C [2.214.12–17 Spencer]); the royal *basilicon* he formed also from ***psimuthion***, white pepper, **calamine**, etc. (6.6.31A [2.218.23–27 Spencer]). Scribonius Largus 215 (p. 99 Sconocchia) quotes the plaster of the surgeon Euelpistos, whom Wellmann equated with Euelpidēs. Since the plaster treats skin abrasions, contrasting with Euelpidēs' ophthalmologic interests, the identification seems unlikely.

RE 6.1 (1907) 951, M. Wellmann.

Alain Touwaide

Euelpistos, Terentius (30 BCE – 10 CE)

Celsus 7.*pr*.3 names him as a prominent surgeon operating in Rome between Truphōn and Megēs. Scribonius Largus 215 preserves his ointment of **litharge** and pine resin in wax and olive oil; whereas Marcellus of Bordeaux, *Epist. Hipp. ad Maec.* 11 (*CML* 5, p. 32), states that he wrote a book on the powers of herbs, describing the influence thereon of the lunar cycle.

Michler (1968) 74, 117; Korpela (1987) 169.

PTK

Euēnōr of Argos (Akarnania) (350 – 290 BCE)

Greek physician, lived mostly in Athens, where he probably achieved considerable success: he was honored as benefactor in 322 for giving the city a great sum of money and received citizenship in 307–303 BCE (*IG* II–III, 374). He wrote *On Therapy* in at least five books (Caelius Aurelianus, *Chron.* 3.122 [*CML* 6.1.2, p. 752]) and was interested also in gynecology (Sōranos *Gyn.* 1.35.3 [*CMG* 4, p. 24; *CUF* v. 1, p. 32]; 4.36.8 [*CMG* 4, p. 149; *CUF* v. 4, p. 25]). He thought that *pleuritis* concerned lungs (Cael. Aur., *Acute* 2.96 [*CML* 6.1.1, p. 194]) and that an abnormal increase of innate heat caused fevers. He cured **dropsical** patients and thought highly of water therapy, particularly that connected with Amphiaraus' sanctuary in Eretria (Ath., *Deipn.* 2 [46d]).

RE 6.1 (1907) 972–973, M. Wellmann; *KP* 2.411, F. Kudlien; *BNP* 5 (2004) 247 (#2), V. Nutton.

Daniela Manetti

Euēnos (250 BCE – 95 CE)

Asklēpiadēs Pharm., in Galēn *CMLoc* 8.5 (13.178 K.), cites his stomach compress, of amber-filings, melilot (*cf.* Theophrastos *HP* 7.15.3, Nikandros *Thēr.* 897, Dioskouridēs 3.40, Durling 1993: 233), *oinanthē* (dropwort, *cf.* Theophrastos *HP* 6.8.1–2, Nikandros *Thēr.* 898, Dioskouridēs 3.120, Durling 1993: 250), dried roses, and saffron, pounded and sieved, mixed with myrtle wine to a waxy consistency, and topped off with date-nut meat, to be applied in a linen bandage. The name, though rare, is widely attested (*LGPN*), and he is surely distinct from Euēnōr.

RE 6.1 (1907) 977 (#9), M. Wellmann.

PTK

Eugamios (250 BCE – 300 CE?)

The *Antidotarium Brux.* 39 (THEODORUS PRISCIANUS p. 374 Rose) cites his (or her?) remedy against **dropsy**: ashed dove, feathers and all, savin juniper, pounded and sifted, and myrrh, in pure African wine, warmed. The name is only attested in the feminine (*LGPN* 2.165), but Eugamos is found (*LGPN*).

RE 6.1 (1907) 984 (#2), M. Wellmann.

PTK

Eugeneia (120 BCE – 80 CE)

ANDROMAKHOS, in GALĒN *CMLoc* 7.6 (13.114–115 K.), records her remedy for lung and other disorders, containing saffron, **galbanum**, ***kostos***, laurel, licorice, ***misu***, white and long pepper, opium, and ***terebinth***, in gum and honey. The use of pepper might suggest a *terminus post* of *ca* 120 BCE. Kühn prints *-ΕΙΟΣ*, but the name is primarily feminine until the 2nd c. CE (*LGPN*). *Cf.* EUGĒRASIA and ŌRIGENEIA.

RE 6.1 (1907) 988 (#11), M. Wellmann.

PTK

Eugenios (Alch.) **(300 – 800 CE?)**

Extant is a short treatise entitled *Eugenios' On the Doubling* (i.e., doubling the quantity of a metal, *CAAG* 2.39). In the early table of MS *Marcianus gr.* 299, a short treatise *On the Sacred Art* is attributed both to HIEROTHEOS and to Eugenios. But in the alchemical corpus, the treatise appears only under Hierotheos' name. The 10th c. catalogue of books, *Kitāb al-Fihrist*, mentions the Eugenios' name among the authors of alchemy.

Ed.: *CAAG* 2.39
Berthelot (1885) 131, 176; Dodge 2 (1970) 852, 983; Letrouit (1995) 83.

Cristina Viano

Eugērasia (?) (120 BCE – 90 CE)

ASKLĒPIADĒS PHARM., in GALĒN *CMLoc* 9.2 (13.244 K.), preserves her spleen remedy: squill, boiled and strained, bryony, Cretan-carrot-seed, iris, cedar-berry, myrrh, ***panax***, parsley, pepper, and ground bitter vetch, in vinegar and Falernian wine (famed since the mid-2nd c. BCE: CICERO, *Brut.* 287; PLINY 14.55, 76); stored away from light. Her name seems otherwise unattested, but *cf.* the later Eugēros (*LGPN* 1.172) and EUGENEIA.

Fabricius (1726) 156; Parker (1997) 145 (#50).

PTK

Euhēmeros (200 BCE – 25 CE)

Cited four times by SCRIBONIUS LARGUS in ASKLĒPIADĒS PHARM. in GALĒN *CMLoc* 4.7, for eye-medicines (12.774, 777–778, 788 K.). Three of the four use saffron and opium, with various minerals; two of the four use "Italian" or Falernian wine, rendering a date after 200 BCE more likely.

RE 6.1 (1907) 972 (#4), M. Wellmann.

PTK

Euhēmeros/Himerios (*ca* 150 – 350 CE?)

Euhēmeros (in MS Parisinus gr. 2322) or Himerios (in MS Phillipps 1538); addressee of APSURTOS' letter on ailments of the spleen, preserved in the *Hippiatrika* (*Hippiatrica Parisina* 549 = *Hippiatrica Berolinensia* 40.1). Euhēmeros is described as *hippiatros*, horse-doctor.

McCabe (2007).

Anne McCabe

Eukleidēs "Palatianus" (100 BCE – 80 CE)

ANDROMAKHOS in GALĒN, *Antid.* 2.10 (14.162–163 K.), records his viper anti-venom, useful also for quartan fevers. Apparently a Roman citizen of the tribe "Palatinus": VARRO, *LL* 5.56.

Fabricius (1726) 155.

PTK

EUKLEIDĒS ⇒ EUCLID

Euktēmōn of Athens (440 – 410 BCE)

Astronomer and geographer. With METŌN, he observed the summer solstice of 13 Skirophorion, 432 BCE, and established the 19-year soli-lunar cycle. He also set up a ***parapēgma***. Although tracing seasons and weather patterns through fixed star phases is a part of Greek culture already in HĒSIOD, and of Babylonian and other cultures, a public tracking device may well be a late 5th c. invention. Our principal sources, (pseudo?) GEMINUS, *Calendarium* and PTOLEMY, *Phaseis*, mention three 5th c. BCE ***parapēgma***-authors, DĒMOKRITOS, Euktēmōn, and Metōn. Of these, Euktēmōn's is the most elaborately preserved in these works as well as in the Anonymous in MS *Vindob. Gr. philos.* 108 and one of two ***parapēgmata*** from Milētos. Euktēmōn divided the seasons: summer (90 days), fall (90 days), winter (92 days), spring (93 days). If he divided the year into 12 parts of 30 or 31 days each, and started the year at about the summer solstice by assigning all months 30 days, then to get the total to 365 days, he gives the last five months 31 days each, accounting for this division of seasons. Perhaps, this also comes with some knowledge that spring is, in fact, the longest season. Euktēmōn is probably the author to whom AUIENUS, *De ora marit.*, 47, 337–380, attributes a geographical work that at least discussed the western Mediterranean. If so, Euktēmōn also resided in the Athenian colony of Amphipolis, sometime in 437–424 BCE.

A. Rehm, "Das Parapegma des Euktemon," *Sitzungsberichte der Heidelberger Akad. der Wissenschaft. phil.-hist. Kl.* 1913; *RE* 18.4 (1949) 1295–1366 (s.v. "Parapegma"), *Idem; Idem* (1941); R. Hannah, "Euctemon's Parapegma," in C.J. Tuplin and T.E. Rihll, *Science and Mathematics in Ancient Greek Culture* (2002) 76–132.

Henry Mendell

Eumakhos of Kerkura (25 BCE – 75 CE)

Wrote *Root Gathering* wherein he states that the narcissus is called *akakallis* and *krotalon* (Ath., *Deipn.* 15 [681e]); Wellmann guesses the date-range; indeed the name is rare after the 1st c. CE, but is attested from the 6th c. BCE (*LGPN*).

RE 6.1 (1907) 1073 (#5), M. Wellmann.

GLIM

Eumēlos of Thēbai (before *ca* 350 CE?)

Author of a treatise on the medical treatment of horses and cows, preserved in excerpts as one of the seven principal sources of the *Hippiatrika*. Striking instances of near-verbatim correspondence with Columella, Pelagonius, and Vindonius anatolios illustrate Eumēlos' dependence upon the agricultural tradition, and imply that he copied his source or sources uncritically. Through Apsurtos, Eumēlos' advice reappears in the treatises of Theomnēstos and Hieroklēs, as well as in the Latin Mulomedicina Chironis. Apsurtos' use of Eumēlos provides a mid-4th c. CE *terminus ante quem*. Unlike the agricultural manuals, which cover a wide array of subjects, Eumēlos' treatise apparently focused on veterinary treatments. Apsurtos calls him *hippiatros megalos*, implying that Eumēlos was not simply a compiler but also a practitioner. Apsurtos also refers to him as *Thēbaios*, but without specifying which of the numerous cities named Thēbai was Eumēlos' home. Hieroklēs, following Apsurtos, mentions Eumēlos by name; but no other authors do so.

CHG vv.1–2 *passim*; J.N. Adams, "Pelagonius, Eumelus, and a lost Latin veterinary writer," *Mémoires du Centre Jean Palerne* 5 (1984) 7–32; McCabe (2007) 98–121.

Anne McCabe

Eunapios of Sardēs (375 – 420 CE)

Greek historian (b. *ca* 345 CE?) who wrote an account of the period 270–404 CE (now lost) and a *Lives of the Sophists*, dealing mainly with contemporary Neo-**Platonists** and intellectual life in Constantinople and Asia Minor.

Ed.: Blockley v. 2 (1983); W.C. Wright, *Lives of the Sophists* (1922).

R.J. Penella, *Greek Philosophers and Sophists in the Fourth Century A.D.: Studies in Eunapius of Sardis* (1990); R. Goulet, *Études sur les vies de philosophes dans l'antiquité tardive. Diogène Laerce, Porphyre de Tyr, Eunape de Sardes* (2001).

Jørgen Mejer

Eunomos Asklēpiadean (1 – 50 CE)

So called by Asklēpiadēs Pharm. in Galēn *CMGen* 5.14 [13.850.17 K.]). Asklēpiadēs describes three of his dry healing ointments (see *ibid.* 851.1–2, 11–15; 852.8–11 K.).

RE 6.1 (1907) 1133 (#9), M. Wellmann.

Jean-Marie Jacques

Eunomos of Khios (380 – 300 BCE)

Drug merchant active in Athens before Theophrastos' time. Apollōnios the paradoxographer (*Mirab.* 50) tells the same anecdote as Theophrastos (*HP* 9.17.2–3) using the name Eunomos, whereas Theophrastos calls him Eudēmos (§3), though here he seems to be mistaken, having mentioned in §2 Eudēmos, another *pharmakopōlēs* who could be one of Eunomos' contemporaries. Apollōnios explains Eunomos's ability to resist the effects of hellebore by his progressive addiction to the drug, in conformity with one of Theophrastos' teachings in the chapter quoted. Theophrastos however states that he used an antidote.

RE 6.1 (1907) 904 (#16), M. Wellmann.

Jean-Marie Jacques

Eupalinos of Megara (550 – 500 BCE)

Son of Naustrophos, architect and engineer, built a water supply system for Samos noted by HĒRODOTOS (3.60) as one of the three greatest achievements in Greek building and engineering. The system, dating to *ca* 550–530 BCE, *ca* 2.5 km long, included a reservoir at the spring's source, a covered pipeline leading to a water conduit tunneled 1,036 m through a large hill rising up behind the city, and a second pipeline that brought the water into the city. The work may be associated with the patronage of the tyrant Polukratēs in its later phase (see also PHAIAX for another impressive hydraulic project). The attribution of a springhouse in Megara of *ca* 500 BCE to Eupalinos is uncertain.

H.J. Kienast, *Die Wasserleitung des Eupalinos auf Samos* (1995); *BNP* 5 (2004) 176, 177–8 (illus.), C. Höcker; *KLA* 1.227–228, M. Weber.

Margaret M. Miles

Euphēmios of Sicily **(1000 – 1200 CE)**

Greek physician cited with PHILIPPOS XĒROS OF RHĒGION in the *Book containing compound medicines, brought together and tried by Euphēmios of Sicily the most commendable [physician], and Philippos Xēros of Rēgion, commendable physicians* (Paris, BNF, *graecus* 2194, ff. 454–464[V]). Euphēmios' association with Philippos Xēros (known also from the 12th c. *Vat. graec.* 300) suggests that he also came from Reggio. Euphēmios was probably a member of a family of physicians, as his son composed pharmaceutical recipes (one quoted in Euphēmios' work: the Parisian MS, f.454[V]). Since some of the formulae in the Parisian MS are introduced *apo phōnēs* ("from oral presentation"), Euphēmios and Philippos Xēros may have taught medicine at a school in the area of Reggio (Ieraci Bio 228).

Costomiris (1890) 170–171; Diels 2 (1907) 38; *S.* (1908) 51; G. Mercati, *Notizie varie di antica letteratura medica e di bibliografia* (1917) 12; Ieraci Bio (1989) 226–227.

Alain Touwaide

Euphorbos (40 – 20 BCE)

Brother of ANTONIUS MUSA, and like him a follower of ASKLĒPIADĒS OF BITHUNIA, became the personal physician and traveling companion of IOUBA, with or for whom he discovered the plant named for him (***euphorbia***), and widely used in ancient pharmacy: PLINY 5.16, 25.77–79.

M. Michler, "*Principis medicus*: Antonius Musa," *ANRW* 2.37.1 (1993) 757–785 at 760–764.

PTK

Euphoriōn of Khalkis (275 – 220 BCE)

A scholar-poet in the manner of KALLIMAKHOS, who studied at Athens under LAKUDĒS OF KURĒNĒ, enjoyed the patronage of the wife of Alexander of Euboia, and towards the end of his life served as librarian for Antiokhos III. His poetry, notorious for its erudition and obscurity, became an important source for later lexicographers and technical writers like STEPHANOS OF BUZANTION, who used its material on toponyms extensively. His *Mopsopia* apparently contained a discussion of perfect numbers (*SH* 417; *cf.* Lightfoot). Among his prose works was a glossary to the Hippokratic corpus in six books (ERŌTIANOS, *Pr.* [p. 5 Nachm.], B-8 [p. 28 Nachm.], *fr.*29 [p. 107 Nachm.]). That Euphoriōn also wrote a prose treatise on agriculture excerpted by Cassius Dionusios seems doubtful (Varro, *RR* 1.1.9–10; *cf.* Columella, 1.1.11); perhaps Cassius found material relevant to agronomy in a poem like the *Hēsiodos*.

Ed.: B.A. van Groningen, *Euphorion* (1977); *SH* 413–453.
RE 6.1 (1907) 1189–1190, O. Skutsch; J.L. Lightfoot, "An early reference to perfect numbers? Some notes on Euphorion, *SH* 417," *CQ* 48 (1998) 187–194; *OCD3* 570, F. Williams; Ihm (2002) #52.

Philip Thibodeau

Euphranōr (Music) (400 – 350 BCE?)

Pythagorean musical theorist quoted by Athēnaios, together with Arkhutas and Philolaos (*Deipn.* 6 [184e]), as devoted to the art of the *aulos*; he also wrote a treatise titled *On Auloi* or *On Aulos Players*.

RE 6.1 (1907) 1190–1191 (#5), E. Wellmann.

E. Rocconi

Euphranōr (Pythag.) (*ca* 150 – 50 BCE)

Second-generation **Pythagorean** musician who, together with Muōnidēs, some time after Eratosthenēs, discovered four new means (*mesòtēs*), added to the six already known (Iamblikhos *in Nikom.* 2.28.6–11 [p. 116]).

M. Timpanaro Cardini, *I Pitagorici. Testimonianze e frammenti* (1962) 2.436–439.

Bruno Centrone

Euphranōr (Arch.) (*ca* 364 – 325 BCE)

Sculptor, painter, wrote on theories of art. Quintilian singles out for praise Euphranōr's talents in sculpture, painting, and the other arts (*Inst.* 12.10.3). Vitruuius notes his treatises on proportions and color (7. *pr.*14). Ancient authors describe many of his statues and paintings (e.g. Pliny 34.77, 35.128–129, Pausanias 1.3.3–4), of which the torso of the colossal marble cult statue of Apollo Patroös (Athenian Agora) has been excavated and is generally accepted as a genuine work of Euphranōr.

O. Palagia, *Euphranor* (1980); C. Hedrick, "The Temple and Cult of Apollo Patroos in Athens," *AJA* 92 (1988) 185–210; *KLA* 1.229–230, W. Müller.

Margaret M. Miles

Euphranōr (Pharm.) (200 BCE – 95 CE)

Asklēpiadēs Pharm., in Galēn *CMGen* 2.14 (13.525 K.), records his cicatrizing ointment composed of **calamine**, ***khalkitis, diphruges***, lead, roasted ***misu***, in beeswax, "Kolophōn" resin, myrtle oil, and Italian wine. The last-named ingredient renders a date after *ca* 200 BCE more likely; the name is very common on Rhodes, and rare after the 1st c. BCE: *LGPN*.

RE 6.1 (1907) 1191 (#7), M. Wellmann.

PTK

Euphratēs (160 – 180 CE)

Procurator a rationibus, who provided the simples used in the preparation of the antidote favored by the emperors Antoninus and M. Aurelius, and recommended GALĒN as compounder: *Antid* 1.1 (14.4–5 K.); *cf.* M. Aurelius 10.31.

RE 2.2 (1896) 2491 (#126), P. von Rohden, 6.1 (1907) 1216 (#3), A. Stein.

PTK

Euphrōnios of Amphipolis (325 – 90 BCE)

Wrote a work on agriculture, possibly treating cereals, livestock, poultry, viticulture, arboriculture (PLINY, 1.*ind*.8, 10, 14–15, 17–18), excerpted by CASSIUS DIONUSIOS (VARRO, *RR* 1.1.8–10, *cf.* COLUMELLA, 1.1.9).

RE 6.1 (1907) 879 (#18), M. Wellmann.

Philip Thibodeau

Euphrōnios of Athens (325 – 90 BCE)

Wrote a work on agriculture, possibly treating cereals, livestock, poultry, viticulture, arboriculture (PLINY, 1.*ind*.8, 10, 14–15, 17–18), in which he discussed winemaking (14.120), and, if the emended text of COLUMELLA 9.2.4 is correct, the origin of bees on Mt. Hymettos in Athens. It was excerpted by CASSIUS DIONUSIOS (VARRO, *RR* 1.1.8–10, *cf.* Columella, 1.1.9).

RE 6.1 (1907) 1221 (#8), M. Wellmann.

Philip Thibodeau

Euphutōn (325 – 90 BCE)

Agricultural writer whose work was used by CASSIUS DIONUSIOS (VARRO, *RR* 1.1.9–10); *Euphutōn*, "Goodplanter," was presumably his *nom de plume*.

RE 6.1 (1907) 1170, M. Wellmann.

Philip Thibodeau

Euruōdēs (?) of Sicily (400 BCE – 100 CE)

Surgeon who like HIPPOKRATĒS understood how to cut for kidney stones so as to allow rapid post-operative healing (RUFUS, *CMG* 3.1, p.112). Wellmann suggests Empedoklean influence: Euruōdēs' method of rapid post-operative healing may have been an attempt to reconcile the mandates of ritual purity with the normally bloody practice of surgery (*cf.* PYTHAGORAS and HIPPOKRATIC CORPUS, OATH). The name, elsewhere unattested, may be a corruption of the archaic and classical Eurumēdēs/das (*LGPN*) or Hērōidēs, cited from the 5th c. BCE, esp. in the West (*cf.* XENOPHŌN, *HG* 3.4.1; *LGPN*).

RE 6.1 (1907) 1341, M. Wellmann.

GLIM

Euruphōn of Knidos (460 – 400 BCE)

SŌRANOS (*Vita Hipp.* 5 [*CMG* 4, p. 176]) says that Euruphōn accompanied his younger contemporary HIPPOKRATĒS to the court of the Macedonian king Perdikkas II, to treat his son. His fame must soon have spread widely, if the comic poet Plato (425–385 BCE) mentioned him; he is credited with the saying that his teacher was Time (IŌANNĒS OF STOBOI 1.8.40a). Ancient scholars ascribed to him the *Knidiai Gnomai* (GALĒN, *In Hipp. Epid. VI* [*CMG* 5.10.2.2, p. 55]) and also some works attributed to Hippokratēs such as *On diet* (Galēn, *Alim. Fac.* 1 [6.473 K.]; *In Hipp. Reg. Acute* 15.455 K. = *CMG* 5.9.1, p. 135). Listed among ancient anatomists (Galēn, *Uteri Diss.* 2.900 K.), he named the *labia minora* "cliffs (*kremnoi*)" (ERŌTIANOS; RUFUS, *Onom. Anthr. Mor.* 112, p. 147 DR) and thought that *pleuritis* concerned lungs (CAELIUS AURELIANUS, *Acut.* 2.96 [*CML* 6.1.1, p. 194]). He cured ***phthisis*** by making patients suck woman's milk (Galēn, *Prob. Prav. Alim. Succ.* 6.775 K.; *MM* 7.6 [10.474 K.]) and used cautery frequently (Plato, *fr.*200 *PCG*). In gynecology, he utilized both drugs (Sōr. *Gyn.* 1.35.3 [*CMG* 4, p. 24; *CUF* v. 1, p. 32]; 4.14.2 [*CMG* 4, p. 144; *CUF* v. 2, p. 11]) and mechanical tools such as a ladder, by which, in order to expel the placenta, he shook women, or, to cure uterine prolapse, he suspended them by the feet (Sōr. *Gyn.* 4.14.3 [*CMG* 4, pp. 144–145; *CUF* v. 4, p. 25]; *cf.* 4.36.7 [*CMG* 4, p. 149; *CUF* v. 4, p. 12]). In general, Euruphōn thinks that diseases arise if the abdomen fails to evacuate, and then digestive residues rise to the head (LONDINIENSIS MEDICUS 4.31–40).

RE 6.1 (1907) 1342–1344 (#5), M. Wellmann; *KP* 2.455, F. Kudlien; Grensemann (1975) 4–15; J. Jouanna, *Hippocrate. Maladies II* (*CUF* 1983) 40–48; *BNP* 5 (2004) 218–219, V. Nutton.

Daniela Manetti

Eurutos of Krotōn or Taras (400 – 375 BCE)

Student of PHILOLAOS and teacher of EKHEKRATĒS, **Pythagorean** either from Krotōn (IAMBLIKHOS *VP* 148), Taras (DIOGENĒS LAËRTIOS 8.45; Iamblikhos *VP* 267) or maybe Metapontion (Iamblikhos *VP* 266, 267). By using pebbles to outline shapes, Eurutos determined the number which governed various objects, e.g., "man," "horse" (ARISTOTLE *Metaphys.* 14.5 [1092b8–13]; THEOPHRASTOS, *Metaphys.* 11 [6a19]).

DK 45; *BNP* 5 (2004) 223 (#2), C. Riedweg.

GLIM

Eusebios of Caesarea, pseudo (500 – 560 CE?)

A MS attributes an *Abridged choice selection on weights and measures* to Eusebios bishop of Caesarea and church historian (*ca* 290–340 CE). This fragment lists in Greek some 60 short entries: a list of capacity measures both for grain and liquids and their equivalents, followed by equivalence-exchange rates of weights and coins. This table is a conflation of quotations from several metrological sources.

MSR 1 (1864) 149–151, 276–278.

Mauro de Nardis

Eusebius son of Theodorus (380 – 400 CE)

His father, THEODORUS PRISCIANUS, addressed *Physica* to him (4.2), and the *Antidotarium Bruxellensis* 49 (p. 377 Rose) attributes a remedy for dysentery to him (sour wild grapes, pips and all, crushed and dried, and stored in glass, to be given with aged wine). Eusebius the *archiater* was active in Rome *ca* 382 CE (Symm., *Epist.* 2.18 and 5.36–37: *RE* 6.1 [1907] 1369

[#11], O. Seeck), who may be the same man. The Christian name is first attested in the 3rd c. CE, although Eusebēs/Eusebis is found from the 2nd c. BCE: *LGPN*.

Fabricius (1726) 158; Korpela (1987) 207.

PTK

Euskhēmos the Eunuch (100 BCE? – 90 CE)

Asklēpiadēs Pharm., in Galēn *CMLoc* 9.4 (13.287–288 K.), preserves his colon-remedy, including cassidony, Illyrian iris, hazelwort, hemlock seed, myrrh, pepper, Indian nard, ***kostos***, mandrake, opium, etc. in boiled honey. The name is attested only from *ca* 100 BCE, in Illyria and S. Italy: *LGPN* 3A.176.

Fabricius (1726) 157.

PTK

Euteknios (250 – 450 CE)

Sophist of unknown date and uncertain origin, attributed with various prosaic paraphrases of didactic poems: Nikandros' *Alexipharmaka* and *Thēriaka*; the *Kunēgetika* of Oppianus of Apameia, but probably not of the *Halieutika* of Oppianus of Kilikia (see Paraphrasis eis ta Oppianou Halieutika), and certainly not the *Ixeutika* of Dionusios of Philadelpheia, as formerly thought. Based on a wide corpus of scholia, these prosaic transcriptions, equal in length to the originals (*e.g.* Opp., *Kun.* 13,583 words *vs* 13,800 for the *Paraphrase*), sometimes clarifying poetical expressions, but not exempt from small errors, are rhetorical exercises occasionally explicating mythology rather than interpreting zoology. Nevertheless, these paraphrases, especially for Nikandros (since Marianus' metrical paraphrasis is lost), are of great value for the manuscript tradition (*e.g.* *Alex.* 616–628 on envenimation through mushrooms is missing).

Ed.: M. Papathomopoulos, *Eutekniou Paraphraseis eis ta Nikandrou Theriaka kai Alexipharmaka* (1976).
RE 6.1 (1907) 1491, L. Cohn; *HLB* 2.265, 272; *BNP* 5 (2004) 231–232, S. Fornaro.

Arnaud Zucker

Euthudēmos of Athens (350 – 50 BCE)

Known only from Athēnaios, *Deipn.*, whose quotations scholars trace to Pamphilos of Alexandria, and ultimately to Dōriōn and Hērakleidēs of Taras. Two titles are known, *On Vegetables* (*lakhana*) – (58f), 3 (74b), and 9 (369e, 371a) – and *On Pickles* (*tarikha*), salted fish: 3 (116a–d, 118b), 7 (307b, 308e, 315f, and 328d); both appear to be works on nutrition. The name is much more frequent before *ca* 200 BCE: *LGPN*.

BNP 5 (2004) 234–235 (#5), V. Nutton.

PTK

Euthukleos (250 BCE – 25 CE)

Formulated an emollient for general joint and bladder pain and a poultice specific to fingers. Celsus (5.18.28) preserves the recipes, both involving ***ammōniakon*** and **galbanum**.

RE 6.1 (1907) 1502, M. Wellmann.

GLIM

Euthumenēs of Massalia (*ca* 550 – 510 BCE?)

Wrote a ***periplous*** of the Atlantic coast of Africa, describing the mouth of the Senegal river, and its fauna, and suggested that the reflux of the Ocean up its estuary drove the rise of the Nile; *cf.* HĒRODOTOS 2.20–21 and MASSILIOT PERIPLOUS.

BNP 5 (2004) 235, K. Brodersen.

PTK

EUTHUPHRŌN OF ATHENS ⇒ DERVENI PAPYRUS

Eutokios of Askalōn (*ca* 510 – 530 CE)

Wrote commentaries (extant) on ARCHIMĒDĒS' *Sphere and Cylinder* (*inSC*), *Plane Equilibria* and *Measure of the Circle*, and APOLLŌNIOS' *Kōnica* (*inCo*), the latter conceived together with a new edition of the first four books of Apollōnios (*inCo* 176.17–22). Eutokios' *scholia* on PTOLEMY's *Almagest* (*inCo* 218.11–12 Heiberg) are lost. He most probably worked in Alexandria under AMMŌNIOS whom he may have succeeded among the late Neo-**Platonist** commentators on ARISTOTLE (Decorps-Foulquier 65). ANTHĒMIOS OF TRALLEIS, named as a companion, was Eutokios' perhaps younger contemporary (*inCo* 168.5 Heiberg). Some have speculated that Eutokios may have studied under ISIDŌROS OF MILĒTOS (the elder), but evidence shows only that the anonymous pupil of Isidōros (see ISIDŌROS OF MILĒTOS' STUDENT), who may also be responsible for part of pseudo-EUCLID *Elements* Book XV, edited Eutokios' commentary (Jones 170–172; Decorps-Foulquier 62 n.8, *contra* Cameron 1990).

Eutokios claimed to have been the first in his time to write a "valuable treatise" on Archimēdēs (*inSC* 2.2–3 Heiberg), seemingly regarding the following:

- Clarity (*saphēneia*) by which Eutokios either refers to clarifying difficult points or elliptic explanations, or to rewriting, correcting or selecting manuscript readings considered better according to his mathematical judgment.
- Authority of classical authors: Eutokios attributes his "clearer" discovered or reconstructed versions to recognized authors, so that his attempts to *clarify* often *renovate* their past lessons. By contrast, his opinion of the intermediate tradition of textual transmission is often poor.
- Invention (*heuresis*): Eutokios emphasizes in his foreword to *inSC* that reading Archimēdēs requires both precision and imagination. He twice provides his reader a detailed list of constructions and demonstrations both filling a gap in Archimēdēs' explanations in *SC* II, and displaying the method of invention (*tropos heureseōs*) of their inventors (*inSC* 66.2–126.3, 152.3–208.6), whereby Eutokios displays his own talent as their imitator.

Eutokios' fidelity to late Neo-**Platonist** principles of philosophical and mathematical exegesis, recalled in his foreword to *inSC*, may plausibly explain these characteristics. Eutokios both appeals to divine inspiration and to Ammōnios' *epistēmonikē theōria* as the ultimate guarantee of his commentary's value. Again, in his brief allusion to the kinship of arithmetic and geometry as part of "mathematics" in general and the use of proportions in particular (*inCo* 220.18–25), he probably refers to the late Neo-**Platonist** emphasis on "general mathematics" and its specific contents, as found in Ammōnios' mentor PROKLOS or in MARINOS OF NEAPOLIS.

DSB 4.488–491, I. Bulmer-Thomas; A. Cameron "Isidore of Miletus and Hypatia: On the editing of mathematical texts," *GRBS* 31 (1990) 103–127; Jones (1999); Decorps-Foulquier (2000).

Alain Bernard

Eutonios (250 BCE – 365 CE)

Arkhiatros whose sciatica remedy was compounded from aged olive oil, wild cucumber cooked therein, beeswax, ***purethron***, raw sulfur, ***terebinth, euphorbia, staphis***, barley, *thapsia*, dittany, and marsh-salt (Dioskouridēs 5.119) (Oreibasios, *Syn.* 3.91, *CMG* 6.3, p. 93). An extremely rare name, known only at Athens (4th c. BCE: *LGPN* 2.184).

RE 6.1 (1907) 1519, M. Wellmann.

GLIM

Eutropius of Bordeaux (350 – 390 CE)

Physician mentioned among his sources by Marcellus of Bordeaux, who also says that he worked recently, and was an eminent citizen of Bordeaux (*pr.*2). He is possibly the Eutropius addressee of eight letters of Symmachus (*epist.* 3.46–53: 377–387 CE), a political personage who is generally identified with the historian author of the *Breuiarium* (369 CE).

RE 6.1 (1907) 1520 (#3), J Seeck; *PLRE* 1 (1971) 317 (#2); Matthews (1975) 8–9, 72–73.

Fabio Stok

Eutychianus (200 – 400 CE)

Marcellus of Bordeaux attributes to the *archiater* ("imperial physician") Eutychianus the recipe of a pill against various diseases (14.70: *CML* 5, p. 246); Oreibasios that of a salve for cicatrization (*Ecl. Med.* 95: *CMG* 6.2.2, p. 272). He is identifiable with the Terentius Eutychianus quoted by the *Antidotarium Bruxellense* for a laxative recipe (Theodorus Priscianus p. 368 Rose).

RE 6.1 (1907) 1532 (#8), M. Wellmann.

Fabio Stok

Expositio geographiae (9th c. CE?)

The author is more likely to be from the circle of the 9th c. patriarch Phōtios of Constantinople than to be the much earlier Prōtagoras. His geographical compendium *Hupotupōsis geōgraphias en epitomē* is connected in 9th c. and later MSS with Agathēmeros' *Geōgraphias hupotupōsis*. The text (14 chapters and 53 paragraphs), starting with reflections on the Earth's circumference, according to Strabōn, presents different passages from Strabōn and Ptolemy, sometimes *verbatim* (§§46–53 = Strabōn 2.5.18–25), and emphasizes geographical more than astronomical matters. There are numerous arithmetical errors and inaccuracies, possibly suggesting two redactions.

Ed.: *GGM* 2.494–509.

RE 1.1 (1893) 743, H. Berger; A. Diller, "The Scholia on Strabo," *Traditio* 10 (1954) 29–50 at 49–50; *RE* S.10 (1965) 800–805, E. Polaschek; *HLB* 1.508–509.

Andreas Kuelzer

Expositio totius mundi (*ca* 360 CE)

Greek geographical treatise, *ca* 360, whose anonymous author was possibly a pagan from Syria; preserved only in two distinct Latin versions under the title *Expositio totius mundi*. The work begins with a description of the paradise Eden in the Far East, followed by remarks on India and Persia. This introductory part resembles Greek *hodoiporeia*, the fabulous guidebooks to Eden. The author then describes "our land" (*terra nostra*), the Byzantine Empire, starting with Syria and Egypt. From Asia Minor, he comes to Thrakē and Macedon, Greece and countries in the western Mediterranean, ending with the description of some famous islands including Cyprus and Britannia. He gives distances in *mansiones*. The second part of the treatise contains data on climate, commerce and treaties; the author may have been a widely traveled merchant.

Ed.: *GGM* 2.513–528; J. Rougé, *Sources Chrétiennes* 124 (1966).

RE 6.2 (1909) 1693–1694, H. Berger; A.A. Vasiliev, *Seminarium Kondakovianum, recueil d'études* 8 (1936) 1–39; N.V. Pigulevskaja, *Byzanz auf den Wegen nach Indien* (1969) 46–51; *HLB* 1.515; *ODB* 771, A. Kazhdan.

Andreas Kuelzer

F

Faber ⇒ Tektōn

M. Cetius Fauentinus (*ca* 300 ce?)

Authored a late antique manual on private architecture, usually known in the MSS as *De Diuersis Fabricae Architectonicae*, but more properly titled *M. Ceti Fauentinus artis architectonicae priuatis usibus adbreuiatus liber*. The material is almost completely a reduced recension of Vitruuius, with some additions from the author's own experience, and explicitly states that its goal is to make Vitruuius more accessible.

The text proceeds in logical manner, eliminating Vitruuius' "scientific" explanations. It opens with Vitruuius' "principles," then it discusses winds, finding and conducting water including building wells, cisterns and water pipe sizes, mortar, bricks, timber, siting components of a villa and a town house, major rooms and baths, vaults and pavements, stucco finishes, heating, pigments and the use of the square, and two sun-dials *pelecinum* (double axe with gnomon) and *hemicyclium*. The work, representing practices of a later period than Vitruuius, features critiques and modifications of Vitruuius' formulae; *opus testaceum* is taken for granted, and the author recommends higher, more open hypocausts and a different formula for mortar for cisterns: one part of lime to two of sand as opposed to two of lime to five of sand.

H. Nohl, "Palladius und Faventinus in ihrem Verhältnis zu einander und zu Vitruvius," in *Commentationes Philologae in honorem Theodori Mommseni* (1877) 64–74; H. Plommer, *Vitruvius and the Later Roman Building Manuals* (1973).

Thomas Noble Howe

Fauilla (?) of Libya (*ca* 30 bce – *ca* 90 ce)

Asklēpiadēs Pharm. in Galēn, *CMLoc* 9.2 = 10.1 (13.250 = 341 K.), cites her for two aromatic anti-sciatic ointments, the first her preparation of Antiokhis' **terebinth**-based recipe, the second containing white pepper. The apparently Latin name ("Ashe") may conceal a writer on cosmetics (*cf.* Ovid, *Ars* 3.203), or could perhaps be Berber; *cf.* also Pha(o)ullos (*LGPN*: rare in this period), or even Baphullos/Bathullos?

Fabricius (1726) 158.

PTK

Fauinus ⇒ Remmius

Fauonius Eulogius (*ca* 380 – 420 CE)

Municipal Rhetor in Carthage and student of AUGUSTINE, who mentions Eulogius in his *De cura pro mortuis gerenda* 11.13, which tells of Augustine unraveling for Eulogius in a dream (probably in 386 or 387) an obscure passage in a rhetorical text of CICERO. This may possibly refer to the composition of Eulogius's *Disputation on Cicero's Dream of Scipio* (generally dated between 390 and 410), written in Latin and dedicated to Superius, consul of *Byzacium* (modern-day Tunisia and Libya), which is arranged in two sections: the first (§§2–20) reviews the characteristic **Pythagorean** arithmology of the first nine numbers in order to explain why Cicero regarded the age of 56 as perfect; the second (§§21–27) deals with the musical intervals, harmonic ratios, and the harmony of the spheres. It is likely, though not certain, that Eulogius' treatment predates the far more extensive commentary of MACROBIUS.

Ed.: R.-E. van Weddingen, *Disputatio de Somnio Scipionis* (1957), with French trans.
RE 6.2 (1909) 2077, G. Wissowa; *PLRE* 1 (1971) 294; *BNP* 5 (2004) 375, J. Flamant.

Thomas J. Mathiesen

Fauorinus of Arelate (110 – 150 CE)

Roman intellectual who traveled all over the Roman world and knew many of the leading men. He wrote in Greek, and of his numerous works, mostly lost, fragments of his *Memoirs* and his *Miscellaneous History* are of particular interest for the history of philosophy and science. He seems to have been mainly interested in biographical details; most fragments are preserved in DIOGENĒS LAËRTIOS. He argued against astrology, Gellius 14.1.

Ed.: E. Mensching, *Favorin von Arelate* I (1963): all published; A. Barigazzi, *Favorino di Arelate, Opere* (1966); E. Amato and Y. Julien, *Favorinos d'Arles, Oeuvres* 1 (*CUF* 2005).
Mejer (1978) 30–32; *BNP* 5 (2004) 375–376, E.-G. Schmidt.

Jørgen Mejer

***Faustinus* (*ca* 100 BCE – *ca* 80 CE)**

ANDROMAKHOS in GALĒN, *CMLoc* 9.5 (13.296 K.), records the "Faustinian" enema for "dysentery," attributed to a Faustinus by AËTIOS OF AMIDA 14.50 (p. 790 Cornarius), and still in use by ALEXANDER OF TRALLEIS (2.427 Puschm.) and PAULOS OF AIGINA, 7.12.24 (*CMG* 9.2, p. 318); Aëtios also refers to the ***trokhiskos*** of Faustinus, 9.42 (Zervos 1911: 386). KRITŌN OF HĒRAKLEIA in Galēn, *CMLoc* 7.2 (13.36 K.), cites the "Faustinian" cough-drop (*ekleigma*) of wine, honey and rue, which Paulos attributes to "Faustinus." These may just be "lucky" medicines, as Galēn, *Antid.* 1.3 (14.20 K.), but the name Faustinus/a is attested from the 1st c. BCE (*LGPN* 1.456, Crete; Martial 1.25, etc.; CIL^2 2.5.268, 615), and is a common cognomen among such *gentes* as Aelius, Caecilius, Iulius, Pompeius, and others.

Fabricius (1726) 159.

PTK

Iulius Firmicus Maternus (334 – *ca* 357 CE)

Wrote the *Mathesis* (*Matheseos Libri*), probably between 334 and 337 (*PLRE* 1 [1971] 568), which treats practical astrology, and may represent a "popular" rather than "scientific" viewpoint (*DSB* 4.622). Its importance is twofold: firstly, it is the only substantially extant ancient handbook on astrological practice, as opposed to theory (*cf.* PTOLEMY's *Tetrabiblos*);

secondly, it uses, and in some cases invents, Latin astrological terminology. Firmicus strives to make technical, sometimes eastern, material available to Romans (*Math.* 4.*pr.*5; 5.*pr.*6). His work purports to mirror the celestial system it describes, seven books symbolizing the seven planets (*Math.* 8.33.1).

The *Mathesis* illustrates the eclectic, but above all **Platonic** and **Stoic**, philosophical orientation of the 4th c. intelligentsia. Firmicus invokes the **Stoic** concept of ***sumpatheia*** at *Math.* 1.5.10–12, 3.*pr.*2–4 (Barton, *Power and Knowledge* 90). Firmicus' work also contains Neo-**Platonic** elements (Hoheisel 523): he describes astrology in terms of a mystery religion (*Math.* 1.6, 2.30.13–15, 7.1.1), and cites ORPHEUS, PLATO, PYTHAGORAS, and PORPHURIOS (*Math.* 7.1.1). Scholars have differed as to the strength of his Neo-**Platonism** (e.g. Bram 312, n. 79; Barton, *Power and Knowledge* 172). Firmicus cites a plethora of sources, some quasi-mythical, some accessible (e.g. *Math.* 2.29.2, DŌROTHEOS OF SIDON). MANILIUS, although not cited, is certainly among Firmicus' sources (Housman XLIII–XLVI; G.P. Goold, ed., *Manilius, Astronomica* [1977] XIV, XCVI–VII).

The *Mathesis* was printed early (*editio princeps*: Bivilaqua, Venice 1497), and "sparked the astrological enthusiasm of the Renaissance" (Bram 4). As well as being associated with astrological works such as Ptolemy's *Tetrabiblos* or Manilius' *Astronomica* (in the edition of Pruckner, Basel 1551, for example), the *Mathesis* was sometimes printed alongside ARATOS and his Roman translators (e.g. in the Aldus edition, Venice 1499), together with whom he furnishes an accessible picture of celestial geography, and the predictive value of the stars as set out by divine providence; these are the concepts that underlie Firmicus' astrology.

Ed./Trans.: P. Monat, *Firmicus Maternus: Mathesis,* 3 vv. (*CUF* 1992, 1994, 1997); W. Kroll and F. Skutsch, *Iulii Firmici Materni Matheseos Libri VIII,* 2 vv. (1897 and 1913, repr. 1968); Bram.

A.E. Housman, ed., *M. Manilii Astronomicon V* (1937); *PLRE* 1 (1971) 567–568; *DSB* 4.621–622, W.H. Stahl; *KP* 2.554, E. Berneker; F. Fontanella, "A proposito di Manilio e Firmico," *Prometheus* 17 (1991) 75–92; T.S. Barton, *Ancient Astrology* (1994); T.S. Barton, *Power and Knowledge: Astrology, Physiognomics and Medicine under the Roman Empire* (1994); *OCD3* 598, D.S. Potter; *BNP* 5 (2004) 434–435, K. Hoheisel; J.-H. Abry, "Manilius et Julius Firmicus Maternus," in N. Blanc and A. Buisson, edd., Imago Antiquitatis: *Religions et iconographie du monde romaine; Mélanges offerts à Robert Turcan* (1999) 35–45.

Emma Gee

Firmius (50 BCE – 75 CE)

Roman author of a treatise on gardening (*Kepourika*) used by PLINY (1.*ind.*19). COLUMELLA (10.1.1) suggests that Roman horticultural writers did not predate the age of AUGUSTUS.

RE 6.2 (1909) 2382 (#3), A. Stein.

Philip Thibodeau

Flauianus of Crete (60 BCE – 80 CE)

ANDROMAKHOS, in GALĒN, *CMLoc* 7.3 (13.72–73 K.), records his cough-drop for the tubercular: poppy juice, henbane seed, mandrake pith, white and black pepper, anise, etc., in honey. The use of a Roman *cognomen* on Crete provides the *terminus post*; he was probably a Flavian-era freedman.

Fabricius (1726) 158, s.v. Fabianus.

PTK

Flauius (?) (270 – 305 CE)

Fellow-teacher with LACTANTIUS in Nikomēdeia, and author of a poetic *De Medicinalibus* (*cf.* Q. SERENUS and MEDICINA PLINII), entirely lost. The "name" Flauius was by this time becoming a title, and *FLAVIVS* may conceal some other name (Hosius suggests Fabius; *cf.* perhaps FAUINUS).

RE S.5 (1931) 224 (#83a), C. Hosius.

PTK

L. Flauius Arrianus of Nikomēdeia (120 – 170 CE)

Arrian(us), born *ca* 85–95 CE, Roman citizen of high status, interested since youth in hunting, military exercises and scholarship. Arrian published his mentor Epiktētos/Epictetus' lectures (*Diatribai*) epitomized in *Enkheiridion*. He became proconsul of Hispania, suffect consul (129 or 130), *legatus Augusti* in Cappadocia, and eponymous archon in Athens in 145/6. Arrian's writings include a lost *Meteorology* (wherein he tried to show that comets are atmospheric phenomena), *On the Hunt* (*Kunēgetikos*), *Ektaxis* and *Taktikē* (on military art), *Anabasis of Alexander*, and *Indikē* (an historical and geographical treatise on India). Arrian, greatly admiring XENOPHŌN, seems to have based his methodological criteria on two principles: mathematical scientific nature and autoptical observation. Arrian assembled the best sources (in his opinion), reading most of them directly, not second-hand, citing authors often, but not always explicitly, never confusing them. Occasionally synthesizing sources, he often quoted them verbatim. The *Ektaxis* and *Taktikē* reveal Arrian's deep tactical comprehension, regarding especially the position of two infantry bulwarks anchoring the wings, anticipating the T battle-formation; the modern concept of draw concentration from the circumference to the center; and the principle of the mass applied to the draw.

Cristiano Dognini, *L' "Indikê" di Arriano. Commento storico* (2000); A.G. Roos and G. Wirth, *Flavii Arriani, Quae extant omnia* 2 vv. (1967–1968); H. Tonnet, *Recherches sur Arrien. Sa personnalité et ses écrits atticistes* 2 vv. (1988); G. Wirth and O. von Hinüber, *Der Alexanderzug. Indische Geschichte* (1985).

Cristiano Dognini

Flauius "the boxer" (30 BCE – 80 CE)

GALĒN, *CMLoc* 9.5 (13.294 K.), preserves ANDROMAKHOS' record of Flauius' powder for "dysentery": myrtle, roses, ***malabathron***, juniper-berries, etc., taken with diluted wine. For a boxer as medical writer, *cf.* MOUSAIOS or THEŌN OF ALEXANDRIA (MED. I), or perhaps the scholar M. Pomponius Porcellus (Suet., *Gram.* 22.3).

Fabricius (1726) 161.

PTK

Flauius Clemens (30 BCE – 90 CE)

ASKLĒPIADĒS PHARM. in GALĒN, *CMGen* 7.12 (13.1026–1027 K.), records Flauius Clemens' ointment for relief of gout in hand or foot, re-compounded by VALERIUS PAULINUS. The non-Republican cognomen (see also SERTORIUS CLEMENS) is first attested in the Augustan period: *PIR2* C-1134, I-270; *CIL* II^2.5.106 (Voltinia); and *RE* S.9 (1962) 1856 (#5): L. Volusenus.

PIR2 F-241.

PTK

T. Flauius Vespasianus (70 – 78 CE)

The Roman emperor "Titus" (reigned 79–81 CE) is credited by PLINY 2.89 with a work on the comet of 76 CE, and by GALĒN *CMLoc* 10.3 (13.360 K.), probably from the pharmacist ASKLĒPIADĒS, with a recipe for a plaster.

OCD3 1532–1533, J.B. Campbell.

PTK

Florentinus (200 – 250 CE)

Author of *Geōrgika*, a comprehensive work on agriculture, in at least 11 books, the most influential of its age written in Greek, although only fragments survive, in the GEŌPONIKA. Among his sources were the QUINTILII. The work mostly conveyed traditional practices rather than superstitious customs, although innovation, especially through grafting, was of interest. His homeland is unknown, but he traveled widely: he reports seeing a giraffe at Rome (*Geōpon.* 16.22.8) and, in the garden of Marius Maximus, an olive grafted to a vine and bearing both fruits (9.14.1). His inclusion of therapeutic characteristics of plants and fruits, not hitherto much addressed by Greek writers, was representative of the age (*cf.* GARGILIUS MARTIALIS).

Oder (1890) 83–87; *PIR2* P-454a; *BNP* 5 (2004) 469 (#2), P.L. Schmidt.

Robert H. Rodgers

Florus (20 BCE – 20 CE?)

ASKLĒPIADĒS PHARM., in GALĒN *CMLoc* 4.7 (12.768–769 K.), records his cure of Antonia, "almost mutilated by other doctors," using a collyrium of saffron, henbane, mandrake, myrrh, opium, roses, etc., in Falernian wine and rainwater; repeated by AËTIOS OF AMIDA 7.110 (*CMG* 8.2, p. 376), and mentioned by *Hipp. Berol.* 62.6 (pp. 254–255 ed. Oder-Hoppe). Galēn and Aëtios both read "Antonia, of Drusus the mother," which is incorrect; restore "Antonia, of Drusus <the wife and of Germanicus> the mother," one line of 20 letters having fallen out of Asklēpiadēs' text (*gunēs te kai Germanikou*).

RE 6.2 (1909) 2760 (#1), A. Stein.

PTK

Fonteius Capito (50 – 30 BCE)

Born *ca* 80 BCE, antiquarian, member of the coterie of NIGIDIUS FIGULUS and VARRO. Citations of "Fonteius," "Capito," and "Fonteius Capito" by IŌANNĒS "LYDUS" probably all refer to the same man, M. Antonius' supporter, *pontifex* after 44, suffect consul in 33 (Horace *Sat.* 1.5.32–33; PLUTARCH *Ant.* 36.1; Weinstock 44). Iōannēs cites Fonteius on astrology, the calendar (the beginning of the day in Babylōn, Umbria, Athens, and Rome: *De Mens.* 2.2 [pp. 18–20 Wu.]; the Earth's warming in May: *ibid.* 4.80 [p. 132 Wu.]), and religion. Iōannēs also attributes to Fonteius dire predictions from thunder when the Moon is in Capricorn, including threats to the *pax Romana* (*De Ost.* 39–41 [pp. 88–91 Wa.]). Although geographical references, indicating Egypt as the text's country of origin, and language (reference to *pax Romana*), do not support Fonteian authorship, Weinstock (47–48) proposes, nonetheless, that the attributions may be genuine: Fonteius may have attempted to follow Nigidius who couched contemporary politics within a Graeco-Etruscan astrological model, but Fonteius' political references are oblique or absent. Bram 1975: 305, n. 16 suggests that Firmicus' citation of FRONTO (ASTROL.) belongs to this man.

St. Weinstock, "C. Fonteius Capito and the Libri Tagetici," *PBSR* ns 5 (1950) 44–49; Bram (1975); *BNP* 5 (2004) 491 (#I.6), K.-L. Elvers, (#I.9), Fr. Graf.

GLIM

FRONTINUS ⇒ IULIUS FRONTINUS

Fronto (Astrol.) (120 BCE – 350 CE)

Latin astrologer who published rules for forecasting by stars and copied HIPPARKHOS' *antiscia* theory, though not usefully, according to FIRMICUS MATERNUS 2.*pr.*2, 4. (Firmicus favored the rare aspect *antiscia*, i.e., "opposite shadows," which relates planets to signs equidistant from Mid-Heaven or *Imum Caelum*.) Bram 1975: 305, n.16 suggests that "Fronto" may be corrupted from FONTEIUS CAPITO.

RE 7.1 (1910) 112, Fr. Boll; St. Weinstock, "C. Fonteius Capito and the Libri Tagetici," *PBSR* ns 5 (1950) 44–49.

GLIM

Fronto (Agric.) (100 – 450 CE)

The GEŌPONIKA preserves five extracts (three substantial) from an otherwise unknown agronomist Fronto; Wellmann suggested emending *ΦΡΟΝΤΟΝΟΣ* to *ΦΡΟΝΤΙΝΟΥ*, i.e., IULIUS FRONTINUS, not elsewhere attested to have written agronomy. The Latin name Fronto, known from the Augustan era (*PIR2* F-485; *LGPN* 1.476, 3B.436), is concentrated in the 2nd c. CE. Fronto advised against intercropping white- and red-grape vines (5.15, pp. 139–140 Beckh), offered 29 ways to preserve wine (7.12, pp. 196–198: mix in salt, or gypsum, or oak chips, immerse a hot sword, etc.), and advised on the care and judging of dogs (19.2, pp. 502–504); see also 7.22 (clarifying wine, p. 208) and 12.10 (vegetables benefit from intercropping with arugula, p. 355).

RE 7.1 (1910) 112 (#13), M. Wellmann.

PTK

Fufi(ci)us (*ca* 100 – *ca* 50 BCE?)

Listed by VITRUUIUS (7. *pr.*14) as the first Roman to have written a short book (presumably one scroll, *liber*) on architecture. Called Fuficius in MSS BM Harleanius 2767 and Wolfenbüttel, Gudianus 69. Possibly the same as Q. Fufius Calenus, tribune of the plebs in 61 BCE and legate of CAESAR in Gaul and Spain.

BNP 5 (2004) 570–571 (#3), U. Egelhaaf-Gaiser

Thomas Noble Howe

L. Fullonius Saturninus (300 – 470 CE)

Eminent and profound astrologer, used by ANTHEDIUS (Sidonius Apollinaris *Ep.* 8.11.10; *Carmen* 22.*pr.*3).

PLRE 1 (1971) 808.

GLIM

M. Fuluius Nobilior (*ca* 190 – 179 BCE)

Son of Marcus, grandson of Seruius, successful as *praetor* 193 BCE in Spain, then as consul in 189 against Aitolian Ambrakia, a lucrative conquest celebrated by ENNIUS, but reproached by CATO as exaggerated, censor 179. Constructed the temple of "Hercules of the Muses" at Rome, wherein he placed his commentary on the *Fasti* (Roman civil and legal calendar), which included folk-etymologies of Latin month-names: March (Mars), April (Aphroditē), May (*maior*, older), June (*iunior*, younger): VARRO, *LL* 6.33, CENSORINUS 20.2–4, 22.9, MACROBIUS THEODOSIUS, *Sat.* 1.12.16, 1.13.21. He is also said to have advocated astral studies as a means to comprehend the divine (IŌANNĒS "LYDUS," *de Ost.* 16 [p. 47 Wa.]), perhaps influenced by ARATOS, or the sole Aitolian poet, ALEXANDER OF PLEURON.

GRL 1, §77; P. Boyancé, "Fulvius Nobilior et le Dieu Ineffable," *RevPhil* s.3, v.29 (1955) 172–192; *DPA* 3 (2000) 434, M. Ducos; *BNP* 5 (2004) 572–583 (#I.15), K.-H. Elvers.

PTK

G

Gaius (Platonist) (100 – 140 CE)

Active in Asia Minor, renowned among later **Platonists**. Scholars long believed in a school of Gaius as the main center of **Platonic** activity in the 2nd c., arguing from similarities between the *Didaskalikos*, falsely attributed to ALBINUS, and APULEIUS' *De Platone et eius dogmate*. Gaius, however, was probably not the common source of these works. GALĒN studied under two of Gaius' disciples, of whom only Albinus is known by name (*De aff. dign.* 8.3: *CMG* 5.4.1.1, p. 28.9–15; *De prop. libr.* 2.97 MMH). Albinus published transcripts of Gaius' lectures, probably consisting partly of textual exegesis of PLATO's works. Gaius' works were read in PLŌTINOS' school (PORPHURIOS, *Vit. Plot.* 14.10–14). PROKLOS (*in Remp.* 2.96.10–15) speaks positively of Gaius' interpretation of the myth of Er. It is unclear whether Gaius himself wrote commentaries. The distinction between an exposition aiming at likelihood and one resulting in knowledge (Plato, *Tim.* 29b–c), concurring with a dogmatic interpretation of Plato, seems to have played an important role in Gaius' (and Albinus') exegesis of the dialogues (Proklos, *in Tim.* 1.340–342).

Ed.: Gioè (2002) 47–76.
DPA 3 (2000) 437–440, J. Whittaker; *BNP* 5 (2004) 642, M. Baltes and M.L. Lakmann.

Jan Opsomer

Gaius (Hēroph.) (70 – 90 CE?)

Hērophilean physician, wrote *On* ***Hudrophobia*** wherein he argued that the disease affected the brain and its meninges, where nerves controlling voluntary actions and those connected to the esophagus originated (CAELIUS AURELIANUS, *Acute* 3.113–114 [*CML* 6.1.1, p. 360]). He is probably distinct from GAIUS OF NAPLES, but identification with a homonymous oculist (ASKLĒPIADĒS in GALĒN *CMLoc* 4.8 [12.771 K.]) and "godlike" Gaius (ANDROMAKHOS in Galēn *CMLoc* 3.1 [12.628 K.]) is possible.

von Staden (1989) 566–569; *BNP* 5 (2004) 642 (#II.1), V. Nutton.

GLIM

Gaius of Neapolis (10 – 70 CE)

Pharmacist and oculist. Of his 19 preserved recipes, PAULOS OF AIGINA (3.22.16, *CMG* 9.1, p. 177) attests one, GALĒN preserves the others, quoted from ANDROMAKHOS (*CMLoc* 3.1, 12.628 K.) and ASKLĒPIADĒS PHARM. He is called Gaius the "Neapolitan" (*CMGen*

5.11, 13.830 K.), the "oculist" (*CMLoc* 4.8, 12.771 K.), "godlike" Gaius (*CMLoc* 3.1, 12.628 K.) and, more frequently, simply "Neapolitan." He is perhaps identifiable with the oculist "Gallio" quoted by Galēn, from Asklēpiadēs, in *CMLoc* 4.8 (12.766 K.), where the name could be textually corrupt. Six of Gaius' recipes are ophthalmic. The others concern various pathologies: pills, e.g. for rheumatic suppurations (*CMLoc* 7.4, 13.86–87 K.), emollients for internal affections (*CMLoc* 8.5, 13.183 K) and against *podagra* (*CMGen* 7.12, 13.1020–21 K.), etc. Galēn used and approved Gaius' recipe against throat inflammations (*CMLoc* 6.8, 12.986 K). AQUILA SECUNDILLA used the cataplasm of *CMGen* 7.7 (13.976 K); IULIUS AGRIPPA used the tonic of *CMGen* 7.12 (13.1030 K).

J. Diehl, *Sphragis* (Diss. Giessen, 1938) 140; von Staden (1989) 566–569 (the three recipes for which Galēn quotes "Gaius" are doubtfully included under the fragments of GAIUS THE HĒROPHILEAN); Marganne (1997) 164–165.

Fabio Stok

Galēn of Pergamon (155 – 215 CE)

Galēn (*Vind. Med. Gr.* 1, f.3^{v}) © Österreichische Nationalbibliothek

Galēnos, a Greek from Pergamon in Asia Minor: sometimes mistakenly called "Claudius Galēn." Born in September 129 CE, he probably died *ca* 215 CE (the tradition putting his death in 199, although still frequently repeated, is based on worthless late testimony). He was a doctor by profession, a teacher by avocation, and a philosopher and grammarian by inclination and self-assessment. He wrote prolifically, on philosophy as well as medicine, and much of his medical (although little of the more strictly philosophical) *oeuvre* survives; indeed, his are the largest individual literary remains of antiquity, stretching to some 10,000 pages in the far from complete edition of C.G. Kühn (1821–1833), which, for all its faults, remains the only modern version of much of Galēn's writing (equally, only a minority, albeit a steadily growing one, of his works have been translated into modern languages). Moreover, such was his reputation in subsequent centuries, that much of his work was translated into Arabic (usually by way of Syriac); and much that has perished in Greek survives in Arabic (or Hebrew, or Armenian, or Latin). Indeed, "new" works of Galēn are still being edited and published, and superior texts of "old" ones generated by making use of the indirect tradition.

We are unusually well-informed about Galēn's life and practice, since he peppers his writing with autobiographical anecdotes; and while self-serving and self-advertising, they are also frequently amusing and savagely polemical. His father was a wealthy architect, NIKŌN; and Galēn writes affectionately about his educational, moral and salutary influence (Nikōn prescribed him a regimen which kept him disease-free as a youth, unlike his more acratic contemporaries: *On Good and Bad* ***Humor*** 6.755–756 K.). In a late work *On My Own Books* (wherein Galēn catalogues his genuine works to distinguish them from widely-circulating forgeries), he tells us that he was sent as a boy to study **Stoic** logic (§11, 2.119 MMH). In *On the Order of My Own Books* 4 (2.88 MMH) he lauds Nikōn for having provided him with an excellent education in mathematics and grammar, and he appeared destined for a career in philosophy (he also studied with **Platonists, Peripatetics** and **Epicureans**: *On the Affections of the Soul* 5.41–42 K.). When Galēn was 16, however, Nikōn had a dream which "persuaded him to make me study medicine as well," and after his father's death in 148–9 (*Good and Bad* ***Humor*** 6.756 K.), Galēn continued his studies in various places (Smurna, Corinth, Alexandria) to sit at the feet of various masters of various different persuasions. Consequently, Galēn "did not declare allegiance to any school" (*Affections of the Soul* 5.43) but rather resolved to take only what was best from each tradition, medical and philosophical, to create a sound system which would meet all reasonable empirical tests. This eclectic approach was to serve him throughout his long and active life.

He returned to Pergamon in 157, where he became physician to the gladiatorial school; in an age when it was hard to gain practical experience of human anatomy, this gave him a great opportunity for detailed observation, of which he took full advantage. In 162, he visited Rome for the first time, rapidly gaining a reputation as an effective and combative physician (he records several of his more impressive, and socially advantageous, cures in *On Prognosis*), eventually moving into the imperial circle itself. It was at this time too that he gave several spectacular public demonstrations of anatomical knowledge and surgical ability, including a demonstration of the function of the recurrent laryngeal nerve, which he had recently discovered (*On Anatomical Procedures* 2.663–666 K.). Four years later, however, he left Rome for Pergamon under somewhat obscure circumstances (he instances the jealousies of his enemies – his *On Slander* is lost– but he also seems to have been avoiding an epidemic), returning again at the behest of Marcus Aurelius to join the imperial army at Aquileia. Invited to accompany the German expedition as physician to the army, he politely declined, citing an admonitory dream from his patron Asclepius (*On My Own Books* 2: 2.97–99 MMH); he was allowed to return to Rome to supervise the emperor's son Commodus.

As far as we know, he remained in Rome until his death; hereafter solid facts about his life are harder to establish. He completed his systematization of medicine, and stopped giving public lectures and demonstrations (*ibid.* pp. 96, 99), an embargo he relaxed only once, after the publication of *On the Function of the Parts of the Body* (his vast compendium of teleological, functional anatomy) and *Anatomical Procedures* (the summation of his anatomical knowledge enlivened by accounts of his triumphant discoveries and displays), to refute the slanders of his enemies. Many of his writings were destroyed by fire with the Temple of Peace (which functioned as a public depository) in 192. Some he rewrote, some survived by other means. At the very end of his life he wrote a "philosophical testament," *On My Own Opinions*, which largely vindicates his claim that his views underwent little substantial change in the latter 50 years of his life.

In medicine, although he resisted association with any of the "schools" (in his view they promoted uncritical acceptance of ill-founded dogma), he is himself to a large extent responsible for the tendency to think of later Greek medicine in terms of sectarian affiliation; for he exploits and deploys CELSUS' tripartite distinction between **Empiricists, Rationalists** and **Methodists**. Although undoubtedly crude (the **Rationalist** category is particularly generic in form), it is still serviceable and not wholly misleading. The distinction is made in terms of theoretical commitment (or the lack of it). **Rationalists** believe that medicine must have a sound theoretical basis; doctors must understand the etiology of diseases, in terms of the disruptions they represent to proper physical functioning, in order to be able to treat them. **Empiricists**, by contrast, think all such theorizing about the "hidden conditions" of the body to be both ill-founded and pointless (they point to the "undecidable disputes" of the theoreticians as evidence of this): doctors can derive all the knowledge they need by carefully observing and recording similar sets of symptoms and antecedent circumstances, and determining by trial and error what works and what does not in the case of each syndrome. **Methodists** suppose that all diseases fall into one of three phenomenally-determinable general types, constricted, relaxed, and mixed, which indicate in themselves appropriate therapies. Galēn has no time for **Methodism** (it has no recognizable method at all, he says); but while he is committed, as the **Rationalists** are, to there being a true physical account of the circumstances underlying disease, he is quite prepared to allow that **Empiricism**, within limits, can be perfectly successful (*On Sects* 1.72–73 K.); and he holds that all medical claims must pass examination at the tribunal of experience, *peira*.

But although Galēn professes independence from sectarian affiliation (which he likens to slavery), he holds that the proper medical method was discovered by his great hero HIPPOKRATĒS. Galēn's "Hippokratēs" is a complex, semi-mythical figure; and Galēn's "Hippokratism' is thus a good deal more original than Galēn himself often allows. Still, he contends that Hippokratēs (pre-eminently in *Nature of Man*) showed both what sort of physical theory was required for successful medicine and also how to establish it. He rebuts the **Empiricists**' contention that the **Rationalists**' disputes exemplify the poverty of their method by asserting roundly that it simply indicates the incompetence of most **Rationalists**. Moreover, this is *logical* incompetence: they can neither construct sound arguments themselves nor recognize and refute them when propounded by others: so they fall hopelessly into error and sophistry. Galēn believes that one can aspire to proper theoretical practice only given talent for and constant practice in the "logical methods," the formal logics and demonstrative theories of ARISTOTLE and the **Stoics**, the **Platonic** method of analysis by division, and the "method of the geometers." Galēn wrote a massive work *On Demonstration*, numerous texts on **Stoic** and **Peripatetic** logic, now lost, as a well as a surviving *Introduction to Logic* showing him to be aware, uniquely, of the non-syllogistic nature of most mathematical reasoning. Armed with all of this, the doctor must first discover, by analysis of ordinary conceptions, the basic meaning of such terms as "health" and "disease." He must then seek to discover the true physical theory underlying physiology. He claims, in *Elements according to Hippokratēs*, that Hippokratēs demonstrated the fundamental nature of Hot, Cold, Wet and Dry, physiologically associated with the four **humors**: yellow bile (Hot-Dry), black bile (Cold-Dry), blood (Hot-Wet) and phlegm (Cold-Wet); see also *On Mixtures*. Galēn utilizes Hippokratean physics (allied with certain conceptual truths such as "opposites cure opposites" and "nothing occurs without a cause") to determine what, as a matter of physical fact, is discordant with any particular distemper; this involves empirical testing (the ultimate criterion for all qualitative analysis is perception), on the basis of which he will derive a "therapeutic indication" of what ought to be done.

Galēn outlines this method at great length in *On the Therapeutic Method* – but it is also evident throughout his clinical and diagnostic *oeuvre*. The empirical component accounts for the importance he accords anatomical knowledge, always confirmed on the basis of personal dissective experience. His anatomy (preserved in *Anatomical Procedures* and some smaller works) is very impressive; but his human anatomy is vitiated by the fact that he was rarely, if ever, able to dissect human cadavers (a fact he laments: he recommends the novice anatomist to take advantage of any lucky chance, such as the fortuitous exposure of a riparian graveyard by flooding, to gain knowledge of the human skeleton). Consequently his human anatomy relies on inference from comparison with primates and other mammals, which he both dissected and vivisected to uncover inner workings and determine particular functions of parts (such as the recurrent laryngeal nerve). All of this knowledge contributed to his boundless admiration for nature's technical skill, which in turn informs his powerful natural teleology: no-one versed in anatomy can seriously doubt that animals' parts are providentially constructed, ordered and arranged, a faith most clearly expounded in *On the Function of the Parts*, but evident elsewhere as well.

This teleological bent makes him unremittingly hostile to **atomism** (although he also considers **atomism**, as a sort of monism, to have been refuted by Hippokratēs' *Nature of Man* (s.v. HIPPOKRATIC CORPUS): nothing which did not involve qualitative intermixture could feel pain; we do feel pain, so we involve qualitative intermixture). Equally, he is hostile to what he sees (rightly) as the excessively reductionistic mechanism of the **atomists** in philosophy, and their medical counterparts, in particular ERASISTRATOS and ASKLĒPIADĒS. Mere mechanical principles such as that of *horror vacui* cannot on their own account for the fluid dynamics of the body; and this, as well as the falsity of other **Erasistratean** views, such as that the arteries in their normal condition contain no blood, can be shown by experience and reason. Rather we must posit (although not as a scientific last word: this is merely the beginning of the account) the existence of certain natural powers (attraction, retention, alteration, expulsion) possessed by the various parts of the body (the kidneys naturally attract urine, for instance: *On the Natural Powers*). His great avowed philosophical debt is to PLATO, a debt he details in *On the Doctrines of Hippokratēs and Plato* (often cited as *PHP*), a long work which seeks to show that in all important respects his two great authorities were in agreement, not just about physics, but also about the nature of the soul, which Galēn views, *contra* the **Stoics**, as being demonstrably tripartite, although he is notably and consistently agnostic when it comes to its actual nature, refusing to commit himself to **Platonic** immaterialism. His overall teleology is also **Platonic** (the marvelous complexity and adaptiveness of biological structures is clear evidence, he thinks, of intelligent design), although the fine details owe more, as he acknowledges, to Aristotle (with whom he disagrees on important points, however, rejecting cardiocentric psychology on the basis of neurological investigation, and ridiculing the notion that the female contributes no developed form in generation: *On Semen*).

In the end, Galēn's system is powerful, synthetic, but not crudely eclectic. It allows for both reason and experience; and it is optimistic – excessively so – about its ability to deliver theoretical and practical knowledge. Its obvious methodological strengths (insistence on logic, empirical testing, the ultimate criterial role of the senses) are undercut by evident shortcomings, notably Galēn's unduly sanguine belief that he has, in fact, provided solid, irrefutable demonstrations of physical hypotheses (such as four-quality and four-**humor** theory). Yet he is also capable of caution; he repeatedly notes that, while he has shown that the origin of voluntary motion and the receptor of sensation is the brain and that these psychological faculties are mediated by the nervous system, the actual nature of the soul is unknown. Evidence clearly shows it to be susceptible to material effects (*The Powers of the Soul follow the Mixtures of the Body*), suggesting that materialism is true; but he cannot rule out the possibility of Plato being right after all about the soul's immateriality.

This survey is necessarily brief and partial, and the bibliography is merely indicative. I have been unable even to sketch the breadth of Galēn's other work, including hundreds of pages of learned commentary on Hippokratēs, more hundreds on pharmacology, as well as four massive (and two shorter) works on the pulse, the theory of which as a diagnostic and prognostic tool Galēn did much to advance. His influence too was remarkable and enduring. Largely as a result of the success of his synthetic achievement, the disputes between the schools fade in the succeeding centuries, while the general outlines of his **humoral** physiology, pathology and therapeutics were preserved and institutionalized, at first in the Arab world and then in the Christian West, where they reigned supreme until the 16th c., and were highly influential still as late as the 19th. They linger still in our vocabulary of moods: sanguine, bilious, choleric, melancholic.

Ed.: Chartier (1639) for some works; Kühn (1821–1833; repr. 1986); Marquardt, Müller and Helmreich (1884–1893); *CMG* 5 (1914–); Alexanderson (1967); Furley and Wilkie (1984); for other works edited individually, see esp. Hankinson (1991) 238–247; I. Magnaldi, *Claudii Galeni pergameni Peri psychēs kai hamartēmatōn* (1999); V. Boudon *et al.*, *Galien* 3 vv. to date (*CUF* 2000–); Chr. Otte, *Galen de Plenitudine* (2001).

Wickersheimer (1922); Brock (1929); Duckworth (1962); Fabricius (1972); K. Schubring, "Bibliographische Hinweise zu Galen," in Kühn (1986 reprint) v. 20, pp. XVII–LXII; Brain (1986); Hankinson (1991); Durling (1993); *ANRW* 2.37.2 (1994) 1351–2080 (over a dozen articles on Galen); Singer (1997); *DPA* 3 (2000) 440–466, V. Boudon; O. Powell, *Galen on the Properties of Foodstuffs* (2003); *BNP* 5 (2004) 654–661, V. Nutton; Johnston (2006); *NDSB* 3.91–96, R.J. Hankinson.

R.J. Hankinson

Galēn, pseudo, An Animal (260 – 320 CE?)

Preserved among the works of GALĒN, but not possibly his, is a work entitled *Whether what is in the womb is alive*. The author, apparently responding to PORPHURIOS' *Pros Gauron*, argues for ensoulment at conception, and asks the emperor to legislate against abortion. The author's crabbed style matches IAMBLIKHOS' (*cf.* EUNAPIOS 5.1.2–4), and many turns of phrase are paralleled only in Iamblikhos; the author is, however, more likely his student or reader. He assumes a unitary soul, which can only be wholly present or wholly absent, and his four-part argument deploys mainly analogies, teleology, and proof-texts without context. Since God put soul into the ***kosmos*** from its beginning (PLATO, *Timaios* 35–37), soul must be in rational creatures from their beginning (assumed to be conception, §1); soul is an efflux of the ***kosmic*** soul (§4.7), *cf.* Iamblikhos, *In Tim.*, *fr.*82. Fetuses receive food and breath through their mouths (§3.1), according to the HIPPOKRATIC CORPUS, *Nature of the Child* 17 (which invokes "breath" as the efficient cause of bone-formation and of the articulation of the large tubes and passages through the body, later in pregnancy). Fetal sense-organs prove fetuses have sensation (§4.1); and fetal breathing proves the presence of soul (§4.5). The author claims mere mortals cannot understand how Nature causes a fetus to become a living being (§5.1–2), *cf.* Iamblikhos *In Tim.*, *fr.*88. His request for legislation defines the purpose of laws as the prohibition of evil, and the preservation and promotion of good (§5.5), *cf.* Iamblikhos, *Letter to Agrippa* (IŌANNĒS STOBAIOS, *Ecl.* 4.5 [223–224]). The superscription *ΓΑΛΗΝΟΥ* should perhaps be emended to *ΓΑΛΛΙΗΝΩ* ("to the [emperor] Gallienus," reigned 253–268, admirer of PLŌTINOS, *cf.* Porphurios, *V.Plot.* 12), or the later and thus perhaps more likely *ΓΑΛΗΡΙΩ* ("to Galerius," tetrarch from 293; emperor 305–311).

Ed.: H. Wagner, *Galeni qui fertur Libellus ΕΙ ΖΩΙΟΝ ΤΟ ΚΑΤΑ ΓΑΣΤΡΟΣ* (1914); C.M. Colucci, *Galeno, se ciò che è nell' utero è un essere vivente* (1971).

PTK

Galēn, pseudo, Definitiones Medicinales (100 – 150 CE)

The text attributed to GALĒN in several Byzantine MSS (Diels 1 [1905] 111; S. [1908] 36), under the title *Oroi Iatrikoi* (*Medical Definitions*), contains 487 medical definitions. Galēn does not include it in *On My Own Works*. Not strictly limited to medical science, it includes all disciplines integrated into medicine and its doxographies, as well as topics related to its history.

The treatise proceeds methodically from the notion of definition itself (1), the concept of medicine (9), the parts of medicine (10–11), and different philosophical approaches with their methods (14); the definition of man (17), and the elements of physiology (18), and anatomy (24), including the **humors** (65–69). Among the physiological processes analyzed (95), pulse and the cardio-vascular system receive much attention (110). Senses are studied (116), as are memory (124) and sleep (127). The concept of health (129) and disease (133) open a new section, devoted to pathology, including the causes (154–163) and signs of diseases (164). Treated at length are fevers (185), cardiac movement (205) and pulse (208–233). The *Definitions* then addresses all types of diseases; general (*phrenitis* 234), those of the nose (252), mouth and throat (204), and respiratory system (258), heart (265) and bile (266), digestive system (267), liver and gall bladder (274), **dropsy** (279), the bladder (283) and urine (284), and joint diseases (290). After skin diseases (295), the author analyzes gynecology (299). The *Definitions* seem to start again, with pathologies of hair (306), skin and skin pigmentation (315), and wounds, eyes and sight (340), pathologies of the nose (370), skin (373), swellings of all kinds (384), gangrenous and similar wounds (392), abnormal anatomical excrescences (396), including genital (413) and anal (419). After some definitions of skin diseases (436), the author discusses generative organs and their possible troubles (439), generation, obstetrics, and newborn care (443), including abnormalities. He ends with hemorrhages (460) and evacuations (462), followed by some 24 definitions, perhaps additions to the collection.

Wellmann dates the work to the late 1st c. CE, and attributes it to the **Pneumaticist** school (1865: 66). Its explanation of the cardio-vascular system, pulse, fever and its types, and its use of purgations, are **Pneumaticist**, as is the only personage quoted by name in the definitions dealing with Roman-imperial medical schools (12–17), AGATHINOS. The work, however, also contains elements of other contemporary medico-philosophical schools (*cf.* ATHĒNAIOS OF ATTALEIA), and need not be ascribed to known **Pneumaticists**. The treatise reflects rather the medical milieu of the 1st c. CE, and likely dates to the early 2nd c. (*cf.* PSEUDO-GALĒN, INTRODUCTIO). Fragments of a similar text were introduced into pseudo-SŌRANOS, *Isagoge* or *Quaestiones medicinales* (*BTML* S.1: 30).

The *Definitions*, translated into Arabic at some point, are mentioned by Ḥunayn ibn Isḥāq (808–873 CE, Ullmann 1970: 38; *GAS* 3 [1970] 138–139). A new Latin translation was published in 1528 in Paris by a *Johannes Philologus* (tentatively identified as Johann Gunther von Andernach, i.e., Johannes Guinterius, 1505–1574), while the Greek text was not printed before 1537 (Basel).

Ed.: 19.346–462 Kühn; V. Rose, *Anecdota graeca et graecolatina* 2 (1870) 241–280.

M. Wellmann (1895) 65–104; *RE* 7.1 (1910) 590, J. Mewaldt; Kollesch (1967); Kudlien (1968) 1101; Kollesch (1973); *Idem*, "Eine hippokratische Krankheitseinteilung in den pseudo-Galenischen

Definitiones medicae," in P. Potter, G. Maloney and J. Desautels, edd., *La maladie et les maladies dans la Collection hippocratique. Actes du VIe Colloque international hippocratique (Québec, du 28 septembre au 3 octobre 1987)* (1990) 255–264; A.D. Mauroudês, "Pseudo-galênikoi Oroi ston kôdika Vaticanus Palat. gr. 199 ekdedomenoi ôs apospasmata tou Archigenê," *Epistêmonikê epetêrida tês filosofikês scholês tou Aristoteleiou Panepistêmiou Thessalonikês, teuchos tmêmatos filologias* 4 (1994) 203–223; K.-D. Fischer, "Beiträge zu den pseudosoranischen Quaestiones medicinales," in K.-D. Fischer, D. Nickel and P. Potter, edd., *Text and Tradition. Studies in Ancient Medicine and its Transmission presented to Jutta Kollesch* = *SAM* 18 (1998) 1–54.

Alain Touwaide

Galēn, pseudo, Historia Philosopha (100 – 400 CE)

Doxographical survey from late antiquity, preserved with the works of the doctor Galēn and falsely attributed to him. It consists of 133 short chapters; §§25–133 are simply excerpts from pseudo-Plutarch's *Epitome, cf.* Aëtios; chapters 1–25 are also derived from the doxographical tradition but are more independently shaped. The author claims no independent effort and simply tried to put together a compendium of previous literature for those who are "fond of learning." The result is of no value as a source for ancient philosophy and contains much misinformation.

Mansfeld and Runia (1996) 141–156.

Jørgen Mejer

Galēn, pseudo, Introductio (130 – 170 CE)

The *Eisagōgē ē Iatros* (*Introduction or the Physician*), attributed to Galēn, is contained in some 40 Byzantine MSS (Diels 1 [1905] 100–101, and S. [1908] 34). It has been identified with the *Galēnou Iatros*, sold in a bookshop in Galēn's time: *On My Own Works Pr.*1 (ed. Boudon [2007]; Mewaldt read *Galēnos Iatros*). Wellmann (1903: 546 and 547, n.1) suggested Hērodotos (Pneum.) as author; against which see Kudlien (1963: 253–254, and 1968: 1102).

It is a doxographical and partly historical introduction to medicine. After a brief chapter on the discovery of medicine and its heuristic principles, the work presents the **Rational, Empiric** and **Methodist** medical schools, and their main historical figures. The following epistemological section discusses the scientific or practical nature of medicine, its very definition, and its constitutive parts. The elements constituting the body are then analyzed, and the names and definitions of bodily parts are catalogued. The author then discusses the internal parts of the body (anatomy), bones (osteology), and physiological fluids (**humors**, etc.) with their function (physiology) and dysfunction (pathology). The author also surveys diseases and therapeutics, starting with purgation and continuing with a general presentation of several medical formulas. The author returns to the examination of pathology, with the specific diseases of the head, the skin, and conditions to be treated by surgery, including bandages.

The work is encyclopedic, a compilation resulting from three different methods of collection and organization: definitions, catalogue, and history, with some **Pneumaticist** material. In presenting the historical figures of the three schools, the text lists a fourth, the **Pneumaticist** school, here divided into the *episunthetic* and eclectic schools, ascribed to Leōnidas and Arkhigenēs of Apameia respectively (14.684 K.; contrast pseudo-Galēn, Def. Med.). The author mentions Athēnaios' theory on physiology, recognizably that of Athēnaios of Attaleia (14.689, 698 K.). Similarly, the author's concepts of pulse and fever (14.729–730 K.), and phlebotomy and catharsis (14.733 and 759 K. respectively) strongly resemble those of Athēnaios. The work is best understood as a mirror of the intellectual and philosophical activity of the medical milieu in Rome before Galēn (not mentioned). The most recent physicians named are Leōnidas and Arkhigenēs, yielding the date-range (*cf.* pseudo-Galēn, *Def. Med.*).

Fragments of a 6th/7th c. CE Latin translation are contained in one MS (*BTML* p. 87). The work was translated into Arabic, possibly by Ḥunayn ibn Isḥāq (808–873 CE, Ullmann 1970: 139), and into Latin by Niccolò da Reggio (*ca* 1308–1345, Thorndike 1946: 226), and again by Johann Gunther von Andernach (Johannes Guinterius, 1505–1574). Its Greek text was published in 1537 in Basel.

Ed.: 14.674–797 Kühn.

Wellmann (1895); *Idem*, "Demosthenes *ΠΕΡΙ ΟΦΘΑΛΜΩΝ*," *Hermes* 38 (1903) 546–566; *RE* 7.1 (1910) 590.11–17, J. Mewaldt; L. Thorndike, "Translations of Works of Galen from the Greek by Niccolo da Reggio (c. 1308–1345)," *Byzantina-Metabyzantina* 1 (1946) 213–235; F. Kudlien, "Die Datierung des Sextus Empiricus und des Diogenes Laertius," *RhM* 106 (1963) 251–254; *Idem* (1968) 1102; Kollesch (1973) 30–35; V. Boudon, *Galien. Tome I* (*CUF* 2007) 176–177, n. 1.

Alain Touwaide

Galēn, pseudo, De Pulsibus ad Antonium (220 – 650 CE)

A few Byzantine MSS contain a text attributed to Galēn (Diels 1 [1905] 113, S. [1908] 37), entitled *Peri Sphugmōn pros Antōnion Philomathē kai Philosophon* (*De Pulsibus ad Antonium Disciplinae Studiosum et Philosophum*), considered spurious since omitted from Galēn's *On My Own Works*. The work provides a brief synthesis of sphygmology containing material mainly from Galēn and also from **Pneumaticist** physicians, for whom pulse was of primary importance as a diagnostic tool (*cf.* Athēnaios of Attaleia). Temkin showed that it closely resembles Philaretos' treatise (1932: 56–66); it is extracted in several MSS under various authors: *Vat. graec.* 280 in a commentary by "Iōannēs of Alexandria" whose name is followed by a partial erasure rewritten as *tou epiklēthentos philoponos* (sc. Iōannēs of Alexandria, Philoponos); Paris, BNF, *graec.* 1884 and 2316, Gregory of Nussa; Paris, BNF, *graec.* 2224, Meletios the monk (9th c. CE?); and Paris, BNF, *graec.* 2324. These attributions, although highly problematic, suggest the date-range. Widely circulated in the early-Byzantine world and possibly receiving commentary in Alexandria, the work was transmitted also in the West through Philaretos' Latin translation, and via its assimilation into the *Articella*, the major Western medical manual in the post-Salernitan and early-Renaissance periods.

Ed.: 19.629–642 Kühn.

H.A. Lutz, *Leitfaden der Pulse dem Galēn zugeschriebenen. Galēns Schrift über die Pulse an Antonius, den Freund der Wissenschaften und den Philosophen (Übersetzung und Erläuterung)* (1940).

Alain Touwaide

Gallus ⇒ (1) Aelius Gallus; (2) Sulpicius Gallus

Gamaliel VI (d. 425 CE)

Jewish patriarch (*nasi*) and physician. "Gamaliel the patriarch," most likely Rabban Gamaliel VI, is cited by Marcellus of Bordeaux (23.77, *CML* 5, p. 408) as a source for a remedy for spleen disorders. In his youth, he appears to have corresponded with LIBANIOS OF ANTIOCH; the letters attest his knowledge of Greek and openness to Greco-Roman learning. In the classical Rabbinic literature, no mention is made of his medical skills, but remedies for the spleen feature prominently in Talmudic lists of remedies (esp. *b. Gittin* 69b).

Stern 2 (1980) 678–679; F. Kudlien, "Jüdische Ärzte im Römischen Reich," *MHJ* 20 (1985) 49–50; P.W. van der Horst, *Japheth in the Tents of Shem* (2002) 27–36.

Annette Yoshiko Reed

Q. Gargilius Martialis (220 – 270 CE)

Author of a Latin work on horticulture, likely to be identified with a homonymous man from Auzia in Mauretania Caesariensis (*CIL* 7.9047, 260 CE). Surviving in a 6th c. palimpsest (Naples, A.IV.8, from Bobbio) is a fragment on gardens and orchards (*De hortis*), treating of apples, peaches, quinces, almonds, and chestnuts. The same work apparently embraced also therapeutic properties and applications of plants and fruits, anonymously transmitted under the title *Medicinae ex holeribus et pomis* (a section on quinces overlaps with *De hortis*), as Book 4 of MEDICINA PLINII. Of unlikely attribution are a life of Alexander Seuerus and extracts on tending cattle (*Curae boum*). Gargilius' primary sources were PLINY and DIOSKOURIDĒS; he refers also to CORNELIUS CELSUS, IUNIUS COLUMELLA and the QUINTILII. He was abreast of developments in arboriculture, e.g. the peach, scarcely noted by earlier writers. His work, like his Greek contemporary FLORENTINUS, represents a renewed circulation of old-fashioned medical lore and traditions addressed to the landed middle class in the century after GALĒN. Extensively used as a source by PALLADIUS AEMILIANUS (for gardens and fruit-trees), the treatise won praise for its practicality and elegance of style (CASSIODORUS *Inst.* 1.28.5). It is assumed, perhaps without adequate study, that Gargilius is the "Martial" or "Marsial" frequently named by the Hispano-Arabic writer Ibn al-Awwam.

Ed.: S. Condorelli, *Gargilii Martialis quae exstant* 1 (1977: incomplete, *Hort.* only); I. Mazzini, *Q. Gargilii Martialis De hortis*, 2nd ed. (1988); B. Maire, *Les remèdes tirés des legumes et des fruits* (*CUF* 2002); *Eadem, Concordantiae Gargilianae* (2002).

PIR2 G-82; J.H. Riddle, *Quid pro quo* (1992); *OCD3* 224, M.S. Spurr; *BNP* 5 (2004) 700 (#4), E. Christmann.

Robert H. Rodgers

Gaudentius (*ca* 200 – 400 CE)

Author of an *Harmonic introduction* in Greek, a mixture of Aristoxenian and **Pythagorean** theory, together with a treatment of notation. CASSIODORUS knew the treatise in a Latin translation credited to MUCIANUS and clearly made use of it in his own treatment of consonances (*Institutiones* 2.5); he also specifically cites Gaudentius as one whose treatise "will open to you the courts of this science."

The treatise begins as if Gaudentius were an Aristoxenian, moves in the middle section to the story of PYTHAGORAS' discovery of harmonic phenomena, returns to a discussion of the consonant intervals, and concludes with a description of ancient Greek musical notation, which breaks off in the middle of the Hypoaeolian *tonos*. As the treatises survive today, only the tables of ALUPIOS provide a more complete representation of the notation found in surviving pieces of Greek music, although notational symbols also appear in the treatises of Aristeidēs Quintilianus and Bakkheios Gerōn.

Gaudentius' treatments parallel other treatises for the most part, but a few unique or unusual features include: the definition of paraphonic notes (§8), distinct from those of Bakkheios and Theōn of Smurna; the possibility of combining individual species of the fourth and the fifth into 12 different species of the octave (§19), though only the traditional seven species of the octave are accepted as "melodic and consonant"; acceptance of the eleventh as a consonance (§§9–10), unusual in a treatise showing some adherence to the Pythagorean tradition; and explanations (§20) of the purpose of musical notation and the necessity of multiple signs for each note-name.

MSG 327–356; *SRMH* 1.66–85; Mathiesen (1999) 498–509; *NGD2* 9.576; *MGG2* 7 (2002) 619–621.

Thomas J. Mathiesen

Gemellus (50 BCE – 80 CE)

Andromakhos in Galēn *CMLoc* 9.5 (13.299 K.) records Gemellus' mineral-based enema in diluted wine, with approval (Fabricius 1972: 174–179). The word is first attested in Caesar, *BC* 3.4.1, and the name first in Iosephus, *Ant. Iud.* 16.241–243 = *PIR2* G-138; *cf.* also *LGPN* 1.106 and 4.78.

Fabricius (1726) 167.

PTK

Geminus (1st c. BCE)

Greek writer with astronomical, mathematical, and philosophical interests, who probably came from Rhodes. His only surviving work, *Eisagōgē eis ta phainomena* (*Introductio astronomiae*) is an introduction to the visible heavens as observed with the naked eye. Its 18 chapters deal with the celestial sphere (its articulation and observed motion), mathematical geography, calendrical cycles, solar and lunar motions, and the names of the five planets. This treatise was the first explicit attempt to introduce and, thus, to define astronomy; and it was composed when recently received Babylonian ideas and techniques, which Geminus tries to explain and adapt to traditional Greek concerns, demanded assimilation.

Four chapters of the *Introductio* (reordered as 4, 5, 15, and 13) were excerpted in the late 14th c., given the title *Sphaera*, and mistakenly attributed to Proklos. This treatise proved extremely popular and was not recognized as Geminus' until the late 16th c. Geminus probably also wrote on mathematics, to judge from citations by Pappos and Eutokios, as well as from references by Proklos in his commentary on Book 1 of Euclid's *Elements* (to which there are also parallels in excerpts on optics preserved in several MSS of Damianos' treatise).

None of this material nor the content of the *Introductio* readily fixes Geminus' philosophical allegiance. Even his use of **Stoic** concepts and language in the *Introductio* may just indicate a reliance on commonly-used terminology and vocabulary. However, Geminus did prepare an epitome of the **Stoic** Poseidōnios' *Meteōrologika*, from which an important passage is summarized in Simplicius' commentary on Aristotle's *Physics*. Its central idea, that astronomical theorizing is subordinated to, and integrated with, philosophical speculation about the nature and constitution of the heavens and its underlying causal processes, fits with Geminus' own procedure in the *Introductio* and is reflected in later **Stoicizing** literature, notably KLEOMĒDĒS' astronomical handbook.

In particular, this Poseidōnian passage may illuminate the relation between the *Introductio* and the *Calendar* (or ***parapēgma***), appended in the MSS. The *Calendar* lists for a complete year those days on which, according to specified authorities, certain constellations rise or set or cross the meridian. It begins with the winter solstice, and proceeds by zodiacal month (i.e., the time the Sun takes to go through a given zodiacal sign or 30°-segment of the **ecliptic** that is named after a zodiacal constellation). Added to these stellar phenomena are statements about the winds and rains that begin or stop. Some scholars today deny Geminus' authorship of the *Calendar*, but alleged disparities between the *Introductio* and the *Calendar* can be effectively resolved by considering the purpose and scope of the *Introductio*.

So while in §17 Geminus may attack the very idea of a calendar, his key target is non-experts (e.g., his students and readers) who assume that the stellar risings and settings are the causes of the meteorological phenomena associated with them. In this chapter, he not only takes such risings and settings simply as signs correlated with the meteorological phenomena through sustained observation, but shows that he regards such correlations as true only "for the most part," while urging that for surer predictive knowledge of the weather astronomers should develop calendars based on causes rooted in natural processes. The *Introductio* thus confirms that the traditional calendar is a legitimate part of astronomy, although lacking the certainty that Geminus identifies as the goal of astronomical theorizing, and regards as attained, for example, in the prediction of eclipses. Moreover, Geminus respects Poseidōnios' position on the relation between astronomy and natural philosophy by assigning greater cognitive value to meteorological predictions based on a theory of natural causes than to predictions based on observed correlations. Geminus may, therefore, be legitimately labeled a post-Poseidonian **Stoic**, since for him those predictions based on physical causation, where the Sun is the leading cause, supersede predictive correlations based solely on observation.

Ed.: G. Aujac, *Géminos: Introduction aux phénomènes* (*CUF* 1975).

Heath (1921) 2.222–231; Neugebauer (1975) 578–589; Alan C. Bowen and B.R. Goldstein, "Geminus and the concept of mean motion in Greco-Latin astronomy," *AHES* 50 (1996) 157–185; *CTC* 8 (2003) 7–48, Robert B. Todd; J. Evans and J.L. Berggren, *Geminos's Introduction to the Phenomena: A Translation and Study of a Hellenistic Survey of Astronomy* (2007).

Alan C. Bowen and Robert B. Todd

Gennadios (250 BCE – 95 CE)

ASKLĒPIADĒS PHARM., in GALĒN *CMLoc* 4.7 (12.760 K.), records his collyrium of antimony, copper flakes (DIOSKOURIDĒS 5.78–79), ***psimuthion***, gum acacia, myrrh, and opium in rainwater.

Fabricius (1726) 167.

PTK

Genthios, King of Illyria (180 – 168 BCE)

Son of King Pleuratos and Eurudikē, bribed into alliance with Perseus of Macedon against Rome in 169/168 BCE, and taken captive: Livy 43.19–20, 44.23, 27, 29–30.

Dioskouridēs 3.3.1, followed by Pliny 25.71, credits him with introducing gentian to the pharmacopoeia.

BNP 5 (2004) 763–764, L.-M. Günther.

PTK

Geōponika (*ca* 950 CE)

Byzantine encyclopedia on agriculture, compiled under Emperor Constantine VII (913–959). It represents, with some modifications, the *Eklogai* of Cassianus Bassus. Ancient authorities cited within the text appear to be reliably transmitted, while those in the chapter-headings are severely problematical. It is organized into 20 books: astrological weather lore (1), siting, cereals and legumes (2), monthly calendar (3), viticulture and wine (4–8), olives and olive oil (9), garden and fruit-trees (10), ornamental and medicinal plants (11), vegetables (12), pests and vermin (13), poultry (14), bees (15), horses (16), cattle (17), sheep and goats (18), dogs, swine and game (19), fishes (20).

Ed.: H. Beckh (1895); M. Meana, J. Cubero, P. Sáez, *Geopónica*, trans. and comm. (1998).

Oder (1890, 1893); J.L. Teall, "The Byzantine Agricultural Tradition," *DOP* 25 (1971) 35–59; *ODB* 834, A. Kazhdan; *BNP* 5 (2004) 780–783, J. Niehoff and E. Christmann.

Robert H. Rodgers

Geōponika, Translation into Pahlavi (*ca* 700 – 900 CE)

One of the two Arabic translations of Cassianus Bassus' Geōponika derived from an anonymous "Persian" version, i.e., probably Pahlavi (*cf.* Nallino). The Arabic title *Warz-nāmah* ("The Book of Agriculture") followed the Pahlavi (*Warz-nāmag*). The later Arabic text was translated from Greek by Sergios (Sarjis ibn Hiliyyā ar-Rūmī), and entitled *al-Filāḥah ar-rūmiyyah* "The (Roman, i.e.) Greek Agriculture."

Nallino (1922; 1948); Pingree (1989) 236–237, correcting Bidez and Cumont (1938) 2.173–197; Panaino (2001) 38–39.

Antonio Panaino

Geōrgios of Cyprus (600 – 620 CE)

Obscure geographer, born in Lapithos on Cyprus, who wrote a description of the Byzantine Empire similar to Hieroklēs' *Sunekdēmos*, presenting the secular administrative divisions of its single districts. Starting with the eparchy of Italy, he transmits unique information on different spheres of influence of Buzantion and of the Lombards there. Next, he treats Africa, Egypt, some parts of Asia Minor, Syria and Mesopotamia, Arabia and Cyprus. Around the middle of the 9th c., an editor, probably the Armenian Basil of Ialimbana, connected the text with a *notitia*, a list of episcopal sees, concentrated on the diocese of Constantinople.

Ed.: H. Gelzer, *Descriptio orbis romani* (1890); E. Honigmann, *Le Synekdèmos d'Hiéroklès et l'opuscule géographique de Georges de Chypre* (1939) 51–70.

E. Honigmann, "Die Notitia des Basileios von Ialimbana," *Byzantion* 9 (1934) 205–222; V. Laurent, "La «Notitia» de Basile l'Arménien," *EO* 34 (1935) 439–472; *HLB* 1.531–532; *Tusculum-Lexikon* (1982) 276; *ODB* 837–838, A. Kazhdan.

Andreas Kuelzer

Geōrgios of Pisidia (*ca* 610 – *ca* 634 CE)

Born *ca* 580, deacon and archivist of Saint-Sophia, a remarkable and admired poet, wrote various historical works and a long elaborate poem (1,910 trochaic trimeters) on the creation: *Hexaemeron* (or *Kosmourgia*; after 630). Following the hexaemeronic tradition (see BASIL, GREGORY OF NUSSA, EPIPHANIOS, and PHILŌN OF ALEXANDRIA, *de opificio mundi*), in this apologetic and lyrical poem, Geōrgios develops relevant observations on natural history in a clear and classical style, juxtaposing biblical tradition with "pagan" sciences, emphasizing **Stoic** themes such as the animated ***kosmos*** and universal **sympathy** (lines 1397, 1679). Especially influenced by the two Kappadokians Gregory and Basil, he treats with erudition, but loosely ordered, questions of astronomy, human anatomy and psychology, and pharmacology, sometimes degenerating to popular marvels (such as vulture parthenogenesis: 1136–1153). In the pharmacological sections (lines 636–680, 1353–1440, 1512–1624) his medical terminology is rich (probably from a handbook, and partly directly from GALĒN). He mentions the silkworm (*skōlex Serikōn*: i.e., *bombyx mori*), introduced to Constantinople (*ca* 554), producing unworthy clothes, but proof of the resurrection (lines 1293–1302).

Ed.: *PG* 92.1425–1580.

G. Bianchi, "Note sulla Cultura a Bisanzio all'Inizio del VII secolo in Rapporto all *Esamerone* di Giorgio di Pisidia," *Rivista di studi bizantini e neoellenici* 2–3 (1965–1966) 137–143; *HLB* 2.269–270; *BNP* 5 (2004) 788 (#6), I. Vassis.

Arnaud Zucker

GERMANICUS ⇒ GERM. IULIUS CAESAR

Gessios of Petra (475 – 520 CE?)

Greek physician active in Alexandria where he studied and then taught medicine, with IŌANNĒS OF ALEXANDRIA seemingly among his students. A highly praised medical teacher (*iatrosophistēs*), Gessios commented on the HIPPOKRATIC CORPUS (e.g. *De natura pueri*) and perhaps also Galēnic treatises, e.g. *De sectis* (Latin translations of commentaries attributed in most MSS to Iōannēs of Alexandria or AGNELLUS OF RAVENNA are credited to Gessios in MS Città del Vaticano, Biblioteca Apostolica Vaticana, *Palatinus latinus* 1090). In Arabic sources, Gessios is considered creator of the *Summaria Alexandrinorum*, a collection of 16 Galēnic treatises with commentary, constituting the basis of late-Alexandrian and Arab medical teaching, and whose interpretation is still disputed. Gessios was forcedly converted to Christianity by the Byzantine emperor Zēnōn (474–491) and baptized. Nevertheless, he also received "extraordinary honors" from the emperor, as well as large amounts of money. Sophronios (possibly the patriarch of Jerusalem, 630–638) in his account of the therapeutic miracles performed by Holy Healers (miracle 30 [302–306 Fernandez Marcos]) reports that Gessios, a known pagan, denied that the Holy Healers Kuros (Cyrus) and Iōannēs (venerated in Menuthis, near Alexandria) obtained their knowledge of medicine directly from God as according to legend, for their therapeutic methods and treatments were strictly Hippokratic. In punishment, his back and neck were paralyzed, and all "pagan" medicine (Hippokratic and Galēnic) failed to cure Gessios, who was then forced to seek a cure from the Saints.

Ed.: C.D. Pritchet, *Iohannis Alexandrini commentaria in libru De sectis Galeni. Recognovit et adnotatione critica instruxit* (1982); N. Palmieri, *L'antica versione latina del «De sectis» di Galeno (Pal. lat. 1090)* (1989).

Temkin (1932); *Idem*, "Studies on Late Alexandrian Medicine. I. Alexandrian Commentaries on Galen's De Sectis Ad Introducendos," *BHM* 3 (1935) 405–430 = *The Double Face of Janus and Other Essays in the History of Medicine* (1977) 178–197; N. Fernandez Marcos, *Los Thaumata de Sofronio. Contribución al estudio de la incubatio cristiana* (1975); V. Nutton, "From Galen to Alexander, Aspects of Medicine and Medical Practice in Late Antiquity," B. Baldwin, "Beyond the House Call: Doctors in Early Byzantine History and Politics," and J. Duffy, "Byzantine Medicine in the Sixth and Seventh Centuries: Aspects of Teaching and Practice," in Scarborough (1985a) 1–14, 15–19, and 21–27; Wolska-Conus (1989); I. Mazzini and N. Palmieri, "L'école médicale de Ravenne. Programmes et méthodes d'enseignement, langue, hommes," in Mudry and Pigeaud (1991) 285–317; *BNP* 5 (2004) 824–825, V. Nutton.

Alain Touwaide

Gildas of Britain (540 – 550 CE)

Gildas "Sapiens" (504–569 CE) composed a narrative history of Britain from the Roman occupation to his own day. *De excidio et conquestu Britanniae*, written *ca* 547 CE (Higham prefers 479–485 CE), is a vitriolic denunciation of contemporary rulers and clergy whom Gildas blamed for the island's troubles after the Roman withdrawal (§1). In the tradition of Classical historians, Gildas included a brief geographical description of the island (§3) quoted from PAULUS OROSIUS (*Historiae* 1.2.76–7) who followed PTOLEMY (*Geography* 2.1–2). Several errors are apparent, including the width (200 Roman miles) and the number of cities (Gildas' 28 British cities, reproduced by Nennius, reflect a scribal error for Ptolemy's 38 cities south of Hadrian's wall). Gildas' description of topography (wide plains: *campis late pansis*) and geology (white stones: *niueas ueluti glareas pellentibus*) is consistent with the southern lowlands.

Ed.: M. Winterbottom, with English trans., *The ruin of Britain, and other works: Gildas* (1978).
N.J. Higham, *English Conquest* (1994).

GLIM

Glaukias of Taras (195 – 155 BCE)

Empiricist physician, contemporary of APOLLŌNIOS OF ANTIOCH (CELSUS *pr.*10), author of a lost work *Tripod* (*Trípous*: GALĒN *Subf. Emp.* 11), in which he improved the elaboration of the three main principles of the school (already developed by SERAPIŌN OF ALEXANDRIA): experience (*empeiria*), reports of others (*historia*) and analogical reasoning (*tou homoíou metábasis*: "transition to the similar"). In the field of Hippokratic exegesis, he wrote a Hippokratic lexicon in alphabetical order (ERŌTIANOS p. 8.5 Nachm.) containing the relevant passages of HIPPOKRATĒS (a few lemmata remain, attested by Erōtianos), and commentaries on single works: we know about those on *Epidemics* 2 and 6 (Galēn, *Hipp. epid.*: *CMG* 5.10.1 p. 230 and 5.10.2.2 p. 3); doubtful on *De humoribus* (pseudo-Galēn, *Hipp. hum.* 16.1 K.) and on *De alimento* (pseudo-Galēn, *Hipp. alim.* 15.409 K.). He had a tendency to modify Hippokratēs' text in order to support his own interpretation. Other fragments (attested by PLINY, Galēn, Athēnaios, OREIBASIOS) are concerned with pharmaceutical, therapeutic (Galēn *Fasc.* 18A.790 K.: a technique of bandaging), and dietetic matters.

Ed.: Deichgräber (1930) 168–170 (fragments), 257–258.
RE 7.2 (1912) 1399 (#8), H. Gossen; *KP* 2.809, F. Kudlien; P. Manuli, "Lo Stile del Commento: Galeno e la Tradizione Ippocratica," in *La scienza ellenistica*, edd. G. Giannantoni and M. Vegetti (1985) 375–394 at 391; *BNP* 5 (2004) 867 (#3), V. Nutton; Ihm (2002) #105–109.

Fabio Stok

Glaukidēs (350 BCE – 100 CE)

Physician, thinking quinces, *phaulia*, and *strouthia* the three best fruits, distinguished between the varieties, in contrast to the grammarian Nikandros of Thuateira who claimed all quinces are called "*strouthia*" (Ath., *Deipn.* 3 [81a, d] = *FGrHist* 343). Meineke thinks he is likely identifiable with GLAUKIAS OF TARAS, but "Glaukiadēs" is attested thrice in the 4th/3rd cc. BCE (*LGPN* 1.107, 2.93, 3B.91), as is "Glaukidēs/as" five times in the 5th/4th cc. BCE and once in the 1st c. BCE/1st c. CE (*LGPN* 1.108, 2.93, 3A.99), and even "Glaukudēs," once in the 3rd c. BCE (*LGPN* 1.108).

RE 7.1 (1910) 1401 (#2), H. Gossen.

GLIM

Glaukōn/Glaukos (Med.) (120 BCE – 77 CE)

PLINY 22.77 records that Glaukōn praised the medicinal benefits of the unidentified plant *boupleuron*, as did NIKANDROS (i.e., *Thēr.* 586, in an anti-venom: Jacques 2002: 170), who prescribes the seeds; Glaukōn employed the root in wine to the same end, but the leaves in wine for afterbirth-expulsion, and for swollen lymph-nodes. ASKLĒPIADĒS PHARM., in GALĒN *CMLoc* 4.7 (12.743 K.), credits Glaukos with a pain-relieving collyrium containing aloes, saffron, **Indian buckthorn**, myrrh, fresh roses, and opium, reduced in wine, formed into pills, and stored away from light. (The disputatious Glaukos of PLUTARCH, *Precepts of Health* 1 [122BC, 124D] is likely distinct and later.) *Cf.* GLUKŌN or perhaps APOLLŌNIOS GLAUKOS.

RE 7.1 (1910) 1403 (#9), H. Gossen; 1421 (#40), *Idem.*

PTK

Glaukos (Geog. I) (*ca* 200 – 100 BCE?)

Wrote an *Arabian Antiquaries* in four books, a ***periēgēsis*** with historical and ethnographical data, describing the coastline, as well as cities and peoples in Arabia, and Parthia along the Euphrates (F3).

FGrHist 674.

GLIM

Glaukos (Geog. II) (300 BCE – 220 CE)

The sole surviving fragment of his *Pontika*, treating the sea's left bank, describes *melugion*, a drink more inebriating than wine, made from honey boiled with water and "a certain herb" (perhaps the Skuthian hemp, *cf.* HĒRODOTOS 4.75). The land produces much honey and beer made from millet.

FGrHist 806.

GLIM

Glaukos of Khios (*ca* 620 – 560 BCE)

Metal-worker who invented welding (or perhaps soldering: HĒRODOTOS 1.25). He was famous for the iron stand he created for a silver *kratēr* dedicated by Aluattēs II (617–560) at the temple of Apollo at Delphi (Paus. 10.16).

RE 7.1 (1910) 1421–1422 (#46), C. Robert; *KP* 2.812 (#8), H. Marwitz.

Bink Hallum

Glukōn (250 – 25 BCE)

SCRIBONIUS LARGUS 206–207 records his two plasters, the "ISIS," best of its kind, suitable for trepanation and belly-incisions, the other useful for gladiators. Despite the rarity of the name at this period (*LGPN*), the Glukōn suspected of hastening Vibius Pansa's death with poison (CICERO, *ad Brut.* 1.6.2) is probably distinct (*contra* Korpela). *Cf.* also EPIGONOS and HERMŌN.

Michler (1968) 86–87, 129; Korpela (1987) 158.

PTK

Gorgias of Alexandria (100 – 50 BCE)

Surgeon whom CELSUS cites on navel lesions (7.*pr.*3, 7.14.2). Gorgias gave three causes (not preserved by Celsus) and stated that breath (*spiritus*, i.e., ***pneuma***) occasionally ruptures into such lesions. Gossen suggested a 2nd c. BCE date.

RE 7.2 (1912) 1619 (#11), H. Gossen; Michler (1968) 61, 105.

GLIM

Gorgias of Leontinoi (*ca* 460 – 380 BCE)

Born *ca* 480, a celebrated orator of the Sophistic movement, who composed a (lost) pamphlet on Eleatic philosophy offering arguments (reproduced by SEXTUS EMPIRICUS and by the ON MELISSOS, XENOPHANĒS, GORGIAS) that (1) nothing exists, or, (2) if anything exists, it cannot be known, or else, (3) it cannot be communicated. The significance of the three arguments can be summarized as follows. (1) By reusing and ridiculing Eleatic arguments, it promoted research into logic and questions of ontology. (2) More inspiring are his considerations concerning problems of knowing, probably drawing on contemporary discussions (*cf.* PRŌTAGORAS) on the relationship between sensation and reflection. He asserts that all sense data and arguments have an equal claim for truth. Mansfeld adduces evidence from his other speeches that Gorgias, nevertheless, allowed for personal experience as a criterion for a restricted validity. (3) The difficulties Gorgias raises concerning the communication of information are of considerable interest. First, he emphasizes the radical difference of media and information, which jeopardizes decipherable correspondence. Second, he adds the problem of interpretation, *vs.* understanding, which, he claims, is subjective, and different with each individual. Further, he is said to have touched upon the theory of perception (B4) and of fire (B5) using the Empedoklean theory of pores, but in what context remains unclear.

J. Mansfeld, "Historical and Philosophical Aspects of Gorgias' *On What is not*," in L. Montoneri and F. Romano, edd., *Gorgia e la Sofistica: Atti del convegno internazionale (Leontini – Catania 12–15 dic. 1983)* = *Siculorum Gymnasium* 38 (1985) 243–271; P. Woodruff in Long (1999) 290–310.

István M. Bugár

Granius (120 BCE – 75 CE)

Listed among medical sources whom PLINY consulted (1.*ind.*28), and cited for a marvel about bladder-stones excised by iron and child-birth, 28.42.

RE 7.2 (1912) 1819 (#11), H. Gossen.

Jan Bollansée, Karen Haegemans, and Guido Schepens

Grattius Faliscus (30 BCE – 8 CE)

Author of a *Cynegeticon* in 541 hexameters. Mentioned in OVID, who might have considered the poet among his friends (*Ex Ponto* 4.16.34: *aptaque uenanti Grattius arma daret* – where *Grattius* is Bücheler's correction of *gratius* or *gracius*). If Bücheler's conjecture is correct, we can fix the poem's *terminus post quem* at about 30 BCE and the *terminus ante quem* at 8 CE (the date of Ovid's death). This range is further supported by Grattius' deep knowledge of VERGIL. After a proem (1–23), the poet writes about hunters' equipment: nets, spears, dogs and horses (24–541). Subjects sometimes alternate with brief narratives: the myths of the huntsmen Derkulos (95–125) and Hagnōn (213–252); and digressions: e.g., the harmful effects of luxury (310–325), the description of a cave at the foot of Aetna (430–460) and of a sacrifice to Diana (480–496). The final part of the poem is incomplete, probably due to an accident of transmission. Grattius seems to have influenced MANILIUS, Calpurnius, AURELIUS NEMESIANUS, and many others.

C. Formicola, "Rassegna di studi grattiani," *BSL* 24 (1994) 155–186; *OCD3* 647–648, A. Schachter; *NP* 12/2.981–982, C. Schindler.

Claudio Meliadò

Grēgorios (Pharm.) (150 – 500 CE)

Pseudo-AELIUS PROMOTUS *On Venomous Creatures* §40 (p. 62 Ihm) cites him for a remedy against lion, leopard, and bear bites. The name was first used by Christians, from *ca* 150 CE: *LGPN* 3A.103 (170 CE; see also 4.83, *ca* 200 CE); Solin (2003) 2.826–828.

(*)

PTK

Grēgorios (before *ca* 400 CE)

Author of two remedies for horses preserved in the *Hippiatrika* as quotations in HIEROKLĒS: a remedy for cough (*Hippiatrica Parisina* 483 = *Hippiatrica Berolinensia* 22.26) and a ***trokhiskos*** *dusenterikos* (*Hippiatrica Berolinensia* 130.183; *cf. Hippiatrica Parisina* pinax 1219).

McCabe (2007) 227.

Anne McCabe

Gregory of Nazianzos (*ca* 370 – 389 CE)

Born in Kappadokia around 329/330 in an upper-class Christian family, he studied in Alexandria and Athens, with Himerios and Prohairesios, along with his friend BASIL OF CAESAREA (Sōcratēs *HE* 4.26). He become a monk in 361 and bishop of Sasima a decade later, but remained at Nazianzos until 379 when he was summoned to Constantinople. Gregory played a major role in the Council of Constantinople in 381, and he became bishop of the city from 379 to 381, when he composed most of his orations. He was received as a saint by the Orthodox and Roman churches, and entitled "Theologian" from 451.

Gregory maintained that the human intellect (*nous*) was created in accordance with God's image but, though master of man, is far from being perfect (*Letter* 101.43–49). He stressed the mystery of God and the purity required in order to approach God, being the first to talk about *theōsis* (deification). Gregory's theological and autobiographical verse exhibits considerable poetic talent. In his panegyric for Basil he articulated a doctrine of nature which gave a high position to the natural world and the study of nature as an earthly indicator which pointed the careful observer to the divine creator (*Or.* 43.11). The brother of a famous physician, CAESARIUS, Gregory also received some systematic medical education and shows considerable interest in medical theory and practice (*Or.* 2; 7).

Ed.: *PG* 35–38.

R.R. Ruether, *Gregory of Nazianzus: Rhetor and Philosopher* (1969); D. Winslow, *The Dynamics of Salvation: A Study in Gregory of Nazianzus* (1979); *RAC* 12 (1983) 793–863, B. Wyß; *ODB* 880–881, B. Baldwin, A. Kazhdan, R.S. Nelson, N.P. Ševčenko; Meredith (1995).

George Karamanolis and Daniel L. Schwartz

Gregory of Nussa (Nyssa) (*ca* 370 – *ca* 395 CE)

Born at Kappadokian Caesarea around 330, Gregory, unlike his older brother BASIL, was married, and did not study systematically but was self-taught in rhetoric, philosophy and science; he was a friend of GREGORY OF NAZIANZOS. Initially he became a rhetorician and was later consecrated bishop of Nyssa in Kappadokia in 371 (hence his traditional ethnic). He participated in several church councils, especially at Constantinople in 381, where his arguments so impressed the emperor Theodosius that he considered communion with him a mark of orthodoxy (*Cod. Theod.* 16.1.3, Sōcratēs *HE* 5.10). He died *ca* 395, and is received as a saint by the Orthodox and Roman churches.

Gregory argued that the term "God" signifies not an individual person, but a substance which corresponds to a genus (*Ad Graecos* 176–177 M., 184–185 M.), while he also stressed the infinite nature of God which allows infinite participation (*Against Eunomius* 1.291). Like Basil, Gregory wrote on the creation of the ***kosmos*** in his *Hexaemeron*, demonstrating considerable familiarity with contemporary science. Gregory was the first Christian to argue that it is man, not God, who is the author of all arts and sciences, being endowed by God with an inventive intellect (*Against Eunomius* 2.184–190). In *On the Making of Man*, Gregory speaks in detail of human physiology and mental activities, defending the immaterial nature of the intellect which permeates the entire body (§12.3) fed by the senses (§13.5). Gregory maintained that the soul is created by God (not before the body), as an immaterial substance with the capacity to enliven bodies and perceive (*On the soul and resurrection* 12, 21). Gregory adopted PLATO's tripartite soul, but maintained that strictly the soul is the rational part, the one godlike part. For Gregory the purpose of the incarnation is the deification of human nature as a whole including the body. While lacking special medical training, Gregory of Nyssa acquired substantial medical knowledge and was conversant in anatomy, physiology, surgery, and pharmacology. (*Cf.* PSEUDO-GALĒN, DE PULSIBUS.)

Ed.: *PG* 44–46; W. Jaeger *et al.*, *Gregorii Nysseni Opera*, 8 vv. to date (1952 ff).

H. Cherniss, "The Platonism of Gregory of Nyssa," *University of California Publications in Classical Philology* 11 (1934) 1–92; J. Danielou, *Platonisme et théologie mystique: doctrine spirituelle de saint Grégoire de Nysse* (1944); W. Jaeger, *Gregor von Nyssa's Lehre vom heiligen Geist* (1966); *RAC* 12 (1983) 863–895, H. Dörrie; *ODB* 882, A. Kazhdan, B. Baldwin, and N.P. Ševčenko; Meredith (1995).

George Karamanolis and Daniel L. Schwartz

Gregory of Tours (570 – 594 CE)

The powerful bishop and historian of Tours also wrote a brief cosmological work, *de Cursu Stellarum Ratio*. He describes seven human-made wonders §2–8 (Noah's ark, Babylōn,

Solomon's temple, the Mausoleum, the Colossus of Rhodes, the Alexandrian Pharos, and the theatre of some Hērakleia, perhaps Latmos, an Episcopal see in his era, though none of the many Hērakleiai are known for their theatre: *BNP* 6 [2005] 150–155). Those are excelled by the seven perpetually-renewed divine "miracles" (§9–16): tides, terrestrial fertility, the Phoenix (quoting LACTANTIUS), Aetna (quoting VERGIL and IULIUS TITIANUS), the hot spring at Grenoble (quoting HILARIUS OF ARLES), the Sun, and the Moon. He gives 15 hours as the longest day (§18), as did HIPPARKHOS for the latitude of Massalia (*cf.* STRABŌN 2.5.40), correct within ¼ hour. Then, using epichoric (*rusticitas nostra*), not mythological, names (§16), he describes (§19–33) the monthly rising or setting times of bright constellations serving as nocturnal chronometers, including Arcturus (*robeola*), Corona Borealis (*sigma*), Cygnus (*crux maior*), Delphinus (*crux minor*), Auriga (*signum Christi*), Gemini (*anguis*), Pleiades (*butrio*), and Orion (*falx*). He describes comets and explains them as omens (§34), mentioning the comets of 565 and 574 CE. He concludes with a month-by-month account of the chronometers, from September through August (§35–47).

Ed.: B. Krusch, *Scriptores Rerum Merovingicarum* 1.2 (1885; repr. 1969) 404–422.
TTE 238–239, Robt. Penkett; *BNP* 5 (2004) 1030 (#4), U. Eigler.

PTK

H

Habrōn (100 – 200 CE)

One of the sources named by THEOPHULAKTOS, at the end of his *Quaestiones Physicae.* The name is especially common at Athens, and seems unattested after the 2nd c. CE (*LGPN*). *Cf.* HIEROKLĒS OF ALEXANDRIA, IMBRASIOS (PARADOX.), and SŌTIŌN, also named as sources.

RE 1.2 (1894) 1808, E. Oder.

PTK

***Hagnodikē of Athens* (290 – 260 BCE)**

Disguised herself as a man in order to learn better midwifery from HĒROPHILOS (in Alexandria?), and, after practicing in Athens, was tried on the Areopagos for impropriety. She revealed herself a woman, whereupon the Athenians modified their laws to allow free-born women to study medicine: HYGINUS, *Fabulae* 274.10–13, who includes her among mythical "first discoverers." The name seems otherwise unattested (Pape-Benseler; *LGPN*). Most other midwives credited with remedies are later (ASPASIA, ELEPHANTIS, etc.), but *cf.* perhaps SŌTEIRA. A recipe for skin disorders (composed of oak-gall, myrrh, lead, and ***psimuthion***) attributed to "the midwife" by ANDROMAKHOS, in GALĒN *CMGen* 5.13 (13.840 K.), could belong to Hagnodikē.

von Staden (1989) 38–41 and T8; Parker (1997) 146.

PTK

Halieus (250 – 10 BCE)

HĒRĀS, in GALĒN *CMGen* 2.2 (13.785–786 K.), records his ointment for wounds and scorpion-stings, containing frankincense, **galbanum**, litharge, and **Sinōpian earth**, in a beeswax, olive oil, and **terebinth** base; repeated by ASKLĒPIADĒS PHARM., *ibid.* 3.9 (pp. 645–646) = 5.4 (p. 802). ANDROMAKHOS, *ibid.* 7.13 (p. 1032), records his ***akopon*** potion, containing ***aphronitron***, frankincense, **galbanum, verdigris**, etc. in a vinegar and **terebinth** base; two other ***akopa*** were revised by VALERIUS PAULINUS. AËTIOS OF AMIDA 12.41 (p. 672 Cornarius) and 14.53 (p. 797 Cornarius) cites plasters. The name is almost unattested (*cf. LGPN* 3A.27), but may represent the occupational epithet "Fisherman" transformed into a proper name, *cf.* PEPHRASMENOS, TEKTŌN, or TURANNOS.

RE 7.2 (1912) 2252 (#2), H. Gossen.

PTK

Hanno of Carthage (*ca* 480 BCE)

King (*suffete*) of Carthage in the early 5th c., probably a relative of the Himilkōn who commanded Carthaginian forces at the battle of Himera. His expedition through the Pillars of Hēraklēs and down the coast of Africa is mentioned in the PSEUDO-ARISTOTELIAN DE MIRABILIBUS AUSCULTATIONIBUS, POMPONIUS MELA, PLINY and ARRIANUS. An MS in the 9th c. *Codex Palatinus graecus* 398 purports to be a Greek translation of his account, posted in the "temple of Kronos" (Baʿal Ḥaman) in Carthage. The text describes an expedition with 60 ships and 30,000 men and women to found cities and explore the coast. The expedition founded a series of settlements up to the Lixos river (mod. Wadi Loukkos, Morocco?). Beyond this point the expedition encountered "Ethiopians" and Troglodytes, as well as savage men dressed in animal skins. The coastal topography is described in some detail, noting islands, bays, rivers, mountains, fragrant forests, and elephants, crocodiles and hippopotami. The account culminates in a description of a volcanic region and a high mountain visible from the sea called the Chariot of the Gods. Sailing past this mountain, the expedition reached a bay called the Horn of the South; on a large island they encountered, captured and skinned "wild men" whom their interpreters called "gorillas" – possibly humans, western lowland gorillas, chimpanzees or baboons. Shortly afterwards, the expedition ran out of provisions and turned back. The account lacks mythological or fantastic references, and so suggests a real voyage; but there is no consensus on what part of the African coast was reached. Most commentators accept the beginning as an authentic account of the Carthaginian settlement of the Moroccan coast. The route beyond the Lixos is harder to plot. Some see the rest of the account as a late fabrication. Others locate the entire journey along the coast of Morocco as far as the Canary Islands. A common view is that the expedition made it as far as the coast of Sierra Leone and Sherbro Island, with the Chariot of the Gods being Mount Kakoulima, visible from the sea but not a volcano. Some identify the volcano as Mount Cameroon, the only active volcano on the west African coast; if so, the island of the gorillas could be Bioko (Fernando Po); but one must posit major elisions in the account to explain such an extended journey.

Ed.: J. Ramin, *Le Périple d'Hannon/The Periplus of Hanno* (1976).
J. Blomqvist, *The Date and Origin of the Greek Version of Hanno's Periplus* (1979); E. Lipiński, *Itineraria Phoenicia* (*Studia Phoenicia* 18) (2004) 435–476.

Philip Kaplan

al-Ḥārith ibn-Kalada al-Thaqafī (*ca* 620 – *ca* 680 CE)

Born in Ṭāʾif, al-Ḥārith ibn-Kalada, a physician who studied medicine in Iran around the time of the prophet Muḥammad (b. *ca* 570, d. 632). It is difficult to sift reality from legend: he is said to have lived to the time of Muʿāwiya (reigned 661–680) but also to have participated in learned exchanges with the Persian emperor Xusraw I (reigned 531–579), preserved as a dialogue by the great historian of medicine Ibn-Abī-Uṣaybiʿa (d. 1270). Albeit, the name al-Ḥārith ibn-Kalada is usually registered as the earliest of Arab physicians known to later Arabic historians, the beginning of a very long Hellenistic tradition of medicine. Several other early Arab physicians' names are known but little else remains.

GAS 3 (1970) 203–204; Ullmann (1972) 19–20.

Kevin van Bladel

Harpalos (Astron.) (500 – 400 BCE)

Proposed an ***oktaetēris***, with intercalated months differing from KLEOSTRATOS', as well as a year of 365 days 13 equinoctial hours (CENSORINUS *De die natali* 18–19). If correct, Harpalos would have expressed parameters in a different form, e.g., 24 years of 8,773 days. AUIENUS (*Arati phen.* 1366–1370) seems to place him before METŌN. The shared name and technical professions suggested to Diels (1904) that our Harpalos may be identifiable with the Harpalos attested in a Hellenistic papyrus as among the architects who built the pontoon/cable bridge across the Hellespont for Xerxēs (480 BCE).

H. Diels, *Laterculi Alexandrini* (1904) 8; DK 6A4, n.12.

Henry Mendell

Harpalos (Pharm.) (120 BCE – 80 CE)

ANDROMAKHOS records three of Harpalos' treatments, a compound for auricular inflammation and two plasters. The ear compound, according with PRUTANIS', contained myrrh, nard, saffron, burnt copper, opium, castoreum, and alum, taken with must when the ears are runny, when painful with rose oil (GALĒN *CMLoc* 3.1 [12.627–628 K.]). The first plaster for extraction was compounded of ***ammōniakon*** incense, beeswax, iris, frankincense granules, pepper, raw sulfur, pumice, ***terebinth***, and olive oil. The second, containing ***terebinth***, pumice, natron, ***ammōniakon*** incense, beeswax, and a little olive oil, was mixed with vinegar and red ochre for color (Galēn *CMGen* 6.4 [13.928–929 K.]). Galēn also preserves his long-lasting remedy for quartan fevers, comprised of myrrh, white or long pepper, opium, castoreum, cardamom, and ***sagapēnon***, taken with wine by the mortally ill, decocted with spring water and administered with hydromel to feverish patients (*Antid.* 2.10 [14.167 K.]). A rare name, attested most often in northern Greece, 4th c. BCE to 1st c. CE (*LGPN*).

RE 7.2 (1912) 2401 (#6), H. Gossen.

GLIM

Harpokrās of Alexandria (250 BCE? – 80 CE)

Traditionally identified with the *Harpocras iatroliptes* (physiotherapist) from Egyptian Memphis who treated Pliny the Younger, who in turn petitioned Trajan to grant the physician Roman citizenship (*Epist.* 10.5, 10.6, 10.7, 10.10). However, the fact that our evidence comes from ANDROMAKHOS (in GALĒN) sets an earlier terminus, and the recipes quoted therein are not physiotherapeutic. Andromakhos attributes six recipes to Harpokrās, one of which is clearly his own: against pain in the ears, compounded from spikenard, myrrh, saffron, opium, etc. (*CMLoc* 3.1, 12.631 K.: "Harpokratēs"). Other formulae are "according to" Harpokrās (i.e., possibly from a collection by him): against sciatica, compounded from burnt swallow nestlings, honey, green myrtle sap, and myrrh (*CMLoc* 6.6, 12.943 K.), an unguent comprising fenugreek, parsley seeds, cardamom, natron, ***panax***, iris, **terebinth**, ***ammōniakon*** incense, etc., in bull-fat, beeswax, honey, and vinegar (*CMGen* 7.7, 13.978–979 K.), and three powders of realgar, malachite, and orpiment: against overgrowth of flesh (*CMGen* 4.8, 13.729 K.), against bleeding (*CMGen* 5.13, 13.838 K.), and to close wounds (*ibid.*, 13.840–841 K.). References to compound medicines under HARPOKRATIŌN may be so-called in honor of Harpokrās, as the Theodotion was for THEODOTOS.

RE 7.2 (1912) 2410 (#4), H. Gossen and A. Stein; Fabricius (1972) 226.

Alain Touwaide

Hekataios of Milētos (*ca* 520 – 490 BCE)

Mythographer and geographer, of whom HĒRODOTOS made use. Hekataios traveled to Egypt, the Black Sea, and probably elsewhere in Asia and Greece. He took part in the councils at the start of the Ionian Revolt (499–494), at which time he had substantial knowledge of the Persian Empire. He improved on the first map of the inhabited world created by ANAXIMANDROS. He wrote a mythographical work in four books, later called *Genealogies, Histories* or *Heröology*, which to some degree rationalized the Greek myths, by setting them in plausible geographical contexts. His major geographical work, ***Periodos Gēs*** or ***Periēgēsis***, was a catalogue of places, divided into two scrolls, Europe and Asia. Many brief fragments survive, although doubts were raised about their authenticity in antiquity. The opus, arranged as an itinerary with basic directional and topographical indicators, probably followed the order of the earliest ***Periploi***, tracing the Mediterranean shore from the Pillars of Hēraklēs east along the European coast, and returning west along the African Coast. His treatment of the interiors of Egypt and Asia was limited. Hekataios claimed to have visited Thebes in Egypt, but his information about the east may have derived from Persian sources or predecessors such as SKULAX OF KARUANDA. Hekataios recorded toponymy and tribal names, and the location of rivers, mountains, plains, capes and gulfs, along with some data concerning mythology, ethnography and natural history.

Ed.: *FGrHist* 1.

C. van Paassen, *The Classical Tradition of Geography* (1957) 65–71; S. West, "Herodotus' Portrait of Hecataeus," *JHS* 111 (1991) 144–160.

Philip Kaplan

HEKATŌN ⇒ BOOK OF ASSUMPTIONS BY AQĀṬUN

Hekatōnumos (?) of Khios (50 – 250 CE)

Cited by PSI INV.3011 on the medicinal properties of bitumen. Only "]tōnumos of Khios" is preserved, but of the three possible names, Aristōnumos although by far the most frequent is not attested on Khios (*LGPN*), whereas Hekatōnumos is (*LGPN* 1.148); the archaic name Kleitōnumos is very rare (*LGPN* 1.260, 2.265). The papyrus also cites an "*–os*" of Thessalia and an "*–ēs*" of Milētos, who could perhaps be the agronomist ARISTOPHANĒS.

(*)

PTK

Heldebald (500 – 600 CE)

Wrote in Gothic a geography of Europe, covering Denmark to Spain, sketching the physical geography, and cited extensively by the RAVENNA COSMOGRAPHY, Book 4. See also ATHANARID and MARCOMIR.

Staab (1976); *DPA* 3 (2000) 707–708, R. Goulet.

PTK

Helenos **(before *ca* 950 CE)**

Author of a recipe for ointment for cysts in horses preserved in the B recension of the veterinary compilation *Hippiatrika* (*Hippiatrica Berolinensia* 77.14). Helenos is described in the lemma as *hippiatros*, a horse doctor.

McCabe (2007).

Anne McCabe

Hēliadēs (250 BCE – 540 CE)

AËTIOS OF AMIDA 7.114 (*CMG* 8.2, pp. 385–386), records his collyrium for ***leukōmata***, composed of various minerals, ground cuttlebone, flax-seed, and the Egyptian incense *kommi*. Listed after OREIBASIOS, and before PETROS OF CONSTANTIA, so perhaps *ca* 380–440 CE.

Fabricius (1726) 175.

PTK

Helikōn of Kuzikos (375 – 350 BCE)

Follower of EUDOXOS and of students of Isokratēs and BRUSŌN, as well as an associate of PLATO (pseudo-Plato, *Epistle* 13). He predicted a partial solar eclipse in Surakousai while there with Plato, for which the tyrant Dionusios gave him a talent (PLUTARCH *Dion* 19.6), perhaps 12 May 361 BCE (or 29 Feb. 357 BCE), and may have contributed a solution to the problem of duplicating a cube (Plutarch *De genio Socratis* 579C). The former is probably either fable or fortune; as to the latter, even the speaker in the dialogue is uncertain.

Lasserre (1987) 139–133, 347–352, 573–576.

Henry Mendell

Hēliodōros (Stoic) (10 – 50 CE)

Wrote a commentary on ARATOS (*FGrHist* 1026 T19b), entirely lost, and informed against his pupil L. Iunius Silanus (Juv. 1.34).

DPA 3 (2000) 532, M. Ducos.

PTK

Hēliodōros (Astrol.) (350 – 370 CE)

"Horoscope Reader" (*fatorum per genituras interpretes*), instrumental in trials for treason and magic at Antioch under Valens. The courtier Fortunatianus accused Hēliodōros, with Palladius, of attempting to poison him. Palladius lodged the more serious counter-charge that ex-governor Fistudius secretly employed divination to ascertain Valens' successor. Hēliodōros, coddled and employed at court to reveal what he knew or had fabricated regarding plots against Valens, accused many nobles of treason (Ammianus Marecllinus 29.1.5, 2.6, 2.13), and died mysteriously in 372 (*incertum morbo an quadam excogitata ui*: Ammianus Marcellinus 29.2.13).

RE 8.1 (1912) 42 (#19), Fr. Boll; *CCAG* 1 (1898) 57.

GLIM

Hēliodōros of Alexandria (Astron.) (475 – 510 CE)

Born *ca* 445 CE to Hermeias and Aidēsia, a close relative of SYRIANUS. After Hermeias died, Aidēsia took Hēliodōros and his older brother AMMŌNIOS to Athens for study with PROKLOS. Hēliodōros proved to be the less talented and studious of the brothers, according to DAMASKIOS. While in Athens, he observed the Moon occulting Venus (475 CE). The brothers returned to Alexandria in 485 CE. Hēliodōros cast horoscopes in 492–493 CE, preserved in the commentary on PAULOS OF ALEXANDRIA attributed to him (§16, 22; *cf.* Boer and Pingree pp. 149–150), which may include other Hēliodōran material. In Alexandria, he observed the Moon occulting Saturn (503 CE, with Ammōnios) or a star (509 CE), and conjunctions of Jupiter with Mars (498, 509 CE), Venus (510 CE), or a star (508 CE), probably observational attempts to confirm their relative geocentric distances.

Ed.: A. Boer and D.E. Pingree, *Heliodori ut dicitur in Paulum Alexandrinum commentarium* (1962); A. Jones, "Ptolemy's Canobic Inscription and Heliodorus' Observation Reports," *SCIAMVS* 6 (2005) 53–97.

Neugebauer (1975) 1038–1041; *DPA* 3 (2000) 534–535, H.D. Saffrey.

PTK

Hēliodōros of Alexandria (Pneum.) (70 – 110 CE)

Greek surgeon supposedly from Egypt (Alexandria), perhaps practiced in Rome and contemporary with Juvenal who accused him of castrations (*Sat.* 6.370–373). His **Pneumaticist** leanings suggest the given date-range (contrary to scholarly opinion making him Hellenistic). LEŌNIDAS OF ALEXANDRIA may have influenced him, and HĒRAKLĀS may have been his pupil; moreover, he may have influenced ANTULLOS.

Hēliodōros wrote four works. His treatise on surgery (*Kheirourgoumenōn Hupomnēma*), probably in five books (not 11, per the *scholia ad* OREIBASIOS, *Coll. med.*, 44.11.4), with interventions arranged *a capite ad calcem*, demonstrates a search for safe procedures. The work survives in fragments in one papyrus and in extracts in Oreibasios. He also wrote *On Luxations* (*Peri Olisthēmatōn*), *On Joints* (*Peri Arthrōn*), and *On Bandages* (*Peri Edesmatōn*), all lost.

Several small tracts in Greek (Diels 2 [1907] 41–42) or Latin (*BTML* pp. 93–94) have been attributed to Hēliodōros, including the *Cirurgia Eliodori*, the translation of an Hellenistic questionnaire (Marganne 1986). Recent scholars have credited Hēliodōros with some papyrus texts (Marganne 1981 and Andorlini Marcone 1993), sometimes on the basis of nothing more than his reputation as a surgeon from antiquity to the Renaissance (Fausti; Marganne 1994: 139, 164–165).

Wellmann (1895) 14–19; *RE* 8.1 (1912) 41–42 (#18), H. Gossen; Drachmann (1963) 171–172, 183–184; Kudlien (1968) 1099–1100; Michler (1968) 7, 104, 106, 130, 148, 151; *KP* 2.998 (#8), F. Kudlien; Marganne (1981) #75, 77, 87, 103, 153, 168; D. Manetti, "P.Coln. inv. 339," in A. Carile, *Die Papyri der Bayerischen Staatsbibliothek München* (1986) 19–25; M.-H. Marganne, "La Cirurgia Eliodori et le P. Genève inv. 111," in *Études de Lettres* (1986) 65–73; *Eadem*, "Le chirurgien Héliodore. Tradition directe et indirecte," in Sabbah (1988) 107–111; D. Fausti, "P. Strasb. inv. gr. 1187," in *Annali della Facoltà di Lettere e Filosofia di Siena* 10 (1989) 157–169; M.-H. Marganne, "Un témoignage unique sur l'incontinence intestinale: P. Monac. 2.23," in D. Gourevitch, ed., *Maladie et maladies, histoire et conceptualisation (Mélanges Grmek)* (1992) 109–121; Andorlini Marcone (1993) #9, 54, 57, 70, 75, 98; *OCD3* 675–676, V. Nutton; Marganne (1998) *passim; BNP* 6 (2005) 71–72 (#5), Alain Touwaide.

Alain Touwaide

Hēliodōros of Athens (250 BCE – 95 CE)

A tragic poet and paradoxographer, who wrote a didactic poem *Apolutika pros Nikomakhon*, discussing remedies against disease. A citation from Hēliodōros in Iōannēs of Stoboi's *Anthology* (4.36.8 W.-H.), about wells near mount Gaurus in Italy with curative effects on eye disorders, can be linked to a passage in Pliny (31.3) on medicinal springs said to have arisen on Cicero's estate near Puteoli in the time after his death, but closer examination shows that this connection does not provide a conclusive *terminus post Ciceronem* for the author. Thus, the quotation by Asklēpiadēs Pharm. in Galēn (*Antid.* 2.7 [14.145 K.]), offers the only chronological clue regarding his life, pointing to 95 CE as a *terminus ante quem*.

RE 8.1 (1912) 15 (#10), E. Diehl; A.A.M. Esser, "Zur Frage der Lebenszeit Heliodors von Athen," *Gymnasium* 54/55 (1943/1944) 114–117; Fabricius (1972) 203; *BNP* 6 (2005) 71 (#4), B. Zimmermann.

Jan Bollansée, Karen Haegemans, and Guido Schepens

Hēliodōros of Larissa (400 – 600 CE?)

Hēliodōros is known only by his presence in the title of the *Optica* of Damianos, where Damianos is said to be "of Hēliodōros," which probably means that Hēliodōros was Damianos' father. Were he the author of the *Optica* that Damianos had later edited, this status would probably have been indicated more explicitly. Nonetheless, Hēliodōros is identified as the author of the *Optica* in all editions prior to the most recent.

DPA 3 (2000) 544–546, Robert B. Todd.

Robert B. Todd

Hēliodōros (pseudo?) et alii (700 – 800 CE)

Four iambic poems *On the Divine Art* are preserved in MS *Marcianus gr.* 299, attributed respectively to Hēliodōros, Theophrastos, Hierotheos, and Arkhelaos. These very mystically-inspired poems contain litanies on gold and parallel Stephanos in style and content. Attributed to Hierotheos is the extant *On the Sacred Art* (*CAAG* 2.450–451), credited also to Eugenios in the early table of MS *Marcianus gr.* 299. Goldschmidt (1923: 11–15) considers all these names as referring to one person, probably Hēliodōros, said to have addressed his poems to Theodosios (probably the emperor Theodosios III, reigned 716–717 CE). Goldschmidt explains the pseudonyms thus: "Theophrastos" for his interest in natural philosophy, "Hierotheos" as being the teacher of Dionusios the Areopagite, and "Arkhelaos" as having been considered the teacher of Sōcratēs. Hēliodōros is either the real name of the author (Goldschmidt), or a forger trying to pass as Hēliodōros of Emesa, the 2nd to 4th c. CE novelist (Berthelot 1885: 202).

Ed.: Ideler 2 (1842/1963) 382–352; Goldschmidt (1923).

Berthelot (1885) 121–122, 201–202; C.A. Browne, "Rhetorical and religious aspects of Greek alchemy. Including a commentary and translation of the poem of the philosopher Archelaos upon the sacred art," *Ambix* 2 (1938) 129–137; 3 (1948) 15–25; *DPA* 1 (1989) 334, R. Goulet; Saffrey (1995) 5; Letrouit (1995) 82–83.

Cristina Viano

Hellenizing School (Arm., *Yunaban Dproc̒*; *ca* 570 – *ca* 730)

This term has been given to a group of translators, many unknown, who were responsible for the translation of numerous Greek, predominantly philosophical, texts into Armenian, betraying an overwhelming interest in things Greek on the part of the Armenians during this period. These Armenian translators seem all to have been associated with the school in Constantinople, and the translations are characterized by an ever-increasing tendency to provide literal translations, even to the point of rendering Greek verbal prefixes by a single corresponding Armenian prefix. Beginning, most likely, with the translation of the grammar of Dionusios Thrax, and other such works, the "corpus" includes works of certain contemporary ecclesiastics, works of and commentaries on PLATO, ARISTOTLE, PORPHURIOS, and especially on PHILŌN OF ALEXANDRIA. In addition, there were also a number of scientific works translated into Armenian during this period, including: the HERMETICA; the *De Animalibus* of PHILŌN OF ALEXANDRIA; pseudo-Aristotle, *De Mundo* (ON THE KOSMOS); ARATOS, *Phainomena*; NEMESIOS OF EMESA, *De natura hominis*; GREGORY OF NUSSA, *De hominis opificio*; and two anonymous treatises, *On Nature*. Some original works in Armenia, largely based on classical sources, were also composed during this period, most notably the works of ANANIA OF SHIRAK and the commentaries on the works of Philōn. The translations from this period are very important for the later development of Armenian thought, but in not a few cases are also of importance as the Greek original has been lost.

H. Manandyan, *Yunaban Dproc̒ě ew nra zargac̒man Šrǰannerě* [The Hellenizing School and the (chronological) Limits of its Activity] (1928); A. Terian, "The Hellenizing School: Its Time, Place, and Scope of Activities Reconsidered," in N.G. Garsoïan *et al.*, edd., *East of Byzantium* (1982) 175–186.

Edward G. Mathews, Jr.

HELUIUS ⇒ VINDICIANUS

HEMERITOS ⇒ EMERITUS

Hephaistiōn of Egyptian Thēbai (420 – 450 CE)

Egyptian author of an astrological treatise in Greek, *Apotelesmatika*, in three books addressed to one Athanasios. He used his own birth-date, November 26, 380 CE, to demonstrate a technique of retrocalculating a date of conception. Large parts of Hephaistiōn's work are paraphrased from PTOLEMY's *Tetrabiblos*, but he acknowledges other sources, including frequent citations of DŌROTHEOS OF SIDŌN and ANTIGONOS OF NIKAIA. Of particular value for the history of earlier Greek astrology and its relations to first-millennium BCE Egyptian and Mesopotamian astral divination are two chapters (1.21 and 1.23) wherein he summarizes methods of the "old Egyptians" of making prognostications for entire geographical regions on the basis of eclipses, comets, and the rising of Sirius.

Ed.: D.E. Pingree, *Hephaestionis Thebani Apotelesmaticorum Libri Tres* (1973).

Alexander Jones

Hēraiskos of Egypt (*ca* 480 – 495 CE)

Studied Neo-**Platonic** philosophy under PROKLOS to whom he dedicated a work on the general doctrine of the Egyptians (DAMASKIOS, *De princ.* 3.167.20–21 W.-C.). He may have taught Isidōros (Damaskios, *Vita Isid. fr.*160 Zintzen). Hēraiskos' interests centered on philosophical explanation of religious phenomena and practices. He formulated a theory of the elements with reference to Egyptian mysteries (Damaskios, *De princ.* 3.167.1–24 W.-C.). The origin of all is the unknowable Darkness, whence arise water and sand, giving birth to the first Kméphis, an Egyptian god who mated with his mother, symbolic of cyclical regeneration. He engenders the second Kméphis which in turn produces the third. They populate the intelligible ***kosmos***. Named after his father and grandfather, the third is in fact the Sun (which Damaskios interprets as the intelligible intellect). Hēraiskos may also divide the intelligible world according to divine features.

RE 8.1 (1912) 421–422, K. Praechter; *PLRE* 2 (1980) 543–544, 1326; *DPA* 3 (2000) 628–630, R. Goulet; *BNP* 6 (2005) 183, M. Tardieu.

Peter Lautner

Hēraklās (110 – 140 CE)

Greek physician (surgeon?), considered a **Pneumaticist** and pupil of HĒLIODŌROS OF ALEXANDRIA (whence the date-range). OREIBASIOS, *Coll.* 48.1–8 (*CMG* 6.2.1, pp. 262–268), quotes him on bandages, from a work hypothetically entitled *Peri Edesmatōn.*

RE 8.1 (1912) 423 (#1), H. Gossen.

Alain Touwaide

***Hērakleianos of Alexandria* (*ca* 125 – 160 CE)**

Physician, son of NUMISIANUS, whose *Anatomy* he epitomized (if *ΑΙΛΙΑΝΟΣ* in Kühn's text is thrice emendable to *ΗΡΑΚΛΕΙΑΝΟΣ*: GALĒN does not cite the father by name: *Musc. Diss.* 18B.926–927, 935 K.). Hērakleianos, whom Galēn met in Alexandria in 151 (*CMG* 5.9.1, p. 70 [15.136 K.]), refused Galēn's later request to see Numisianus' works, burning them shortly before his death (Galēn, *Admin. Anat.* 14.1 [pp.183–184 D.]). See also MAECIUS AELIANUS.

BNP 6 (2005) 155, V. Nutton.

GLIM

Hērakleidēs "Kritikos" (*ca* 270 – 230 BCE)

Greek geographer, possibly of a Cretan school (he is known as Krētikos, traditionally emended to Kritikos), author of a prose work *On the* ***poleis*** *in Greece*, probably based on personal travels. Fragments of the work were found together with the text of DIONUSIOS SON OF KALLIPHŌN's poetic ***periplous*** and were at first attributed to DIKAIARKHOS. The description includes lists of sites, distances, details on scenery and inhabitants; it incorporates poetic citations.

Ed.: *GGM* 1.97–110; *FHG* 2.154–164.

F. Pfister, *Die Reisebilder des Herakleides* (1951); E. Perrin, "Héracleidès le Crétois a Athènes: Les Plaisirs du tourisme culturel," *REG* 107 (1994) 192–202; T. Ballati, "Nota al *Peri tōn en tē Helladi poleōn* di Eraclide Critico: Ellade e Peloponneso," in: S. Bianchetti *et al.*, edd., *Ποικίλμα*: *Studi in onore di Michele R. Cataudella* (2001) 1.49–62.

Daniela Dueck

Hērakleidēs "Lembos" ⇒ Hērakleidēs of Kallatis

Hērakleidēs "Pontikos" ⇒ Hērakleidēs of Hērakleia Pontikē

Hērakleidēs of Athens, Aurelius (150 – 190 CE)

Prominent **Stoic**, possibly an imperially-appointed public teacher, awarded with Roman citizenship *ca* 170; his full Roman name "Aurelius Heraclides Eupyrides" is attested: *IG* II[2] 3801. He wrote on the fifth element (***aithēr***), to which Alexander of Aphrodisias replied (*fr.*2 Vitelli).

R.W. Sharples, *Alexander of Aphrodisias, Quaestiones 1.1–2.15* (1992) *ad loc.; DPA* 3 (2000) 559, B. Puech.

PTK

Hērakleidēs of Ephesos (75 – 50 BCE)

Member of the **Erasistratean** school of Hikesios of Smurna, according to Dēmētrios of Magnesia in Diogenēs Laërtios 5.94. Oreibasios, *Coll.* 49.4.45–50 (*CMG* 6.2.2, p. 9), records that he constructed a variant of Tektōn's reduction machine, and transmits the recipe for a wound-cleansing salve, alum, copper-flake, ***misu***, and frankincense, ground with vinegar and formed into ***trokhiskoi***, applied with beeswax and **terebinth** or resin: *Ecl. Med.* 98.22 (pp. 278–279), possibly belonging to Hērakleidēs of Taras.

Michler (1968) 89, 132–134.

PTK

Hērakleidēs of Eruthrai (*ca* 30 BCE – 30 CE)

Khrusermos' most famous student, respected **Hērophilean** physician (Galēn, *Ars Med.* 1.305 K.), pupil with Apollōnios "Mus," and Strabōn's contemporary (14.1.34). Wrote commentaries on the Hippokratic Corpus, Epidemics III and VI, and probably also II (Galēn, *In Hipp. Epid. II, III,* and *VI* = *CMG* 5.10.1, p. 130; 5.10.2.1, p. 80; and 5.10.2.2, pp. 3–4, 212, 243). Galēn, while conceding Hērakleidēs' usual sensibility, criticizes these as inaccurate and inappropriately explicated (*In Hipp. Epid. VI* 5.15 [pp. 304, 306], 6.14 [p. 378]). He elucidated the *sigla* introduced by Mnēmōn of Sidē, apparently nearly ending the feud between **Empiricists** and **Hērophileans** regarding their authenticity by suggesting the symbols were post-Hērophilos interpolations (Galēn, *In Hipp. Epid. III* = *CMG* 5.10.2.1, pp. 75–77, 86–94). In *On the Hērophilean Sect* (seven books), Hērakleidēs, censuring his teacher's definition of pulse theory, largely on semantic grounds, explained the contraction and dilation of veins and arteries through vital and psychic power (Galēn, *Puls. Diff.* 4.10 [8.743–746 K.]).

von Staden (1989) 555–558; *Idem* (1999) 169–170; *BNP* 6 (2005) 173 (#26), V. Nutton.

GLIM

Hērakleidēs of Kallatis, "Lembos" (150 – 100 BCE)

Politician and intellectual active in and around Alexandria. He seems to have belonged to the **Peripatetic** tradition and wrote, in addition to a large historical work, a number of epitomes of earlier biographies by Sōtiōn, Saturos, and Hermippos of Smurna; most of the fragments are preserved in DIOGENĒS LAËRTIOS. Epitomes of ARISTOTLE's *Politeia* and *Nomina Barbarica* have been transmitted separately.

Ed.: M. Dilts, *Heraclidis Lembi excerpta Politiarum* (1971); S. Schorn, *Satyros aus Kallatis. Sammlung der Fragmente mit Kommentar* (2005).
Mejer (1978) 40–42, 62–72.

Jørgen Mejer

Hērakleidēs of Hērakleia Pontikē, "Pontikos" (*ca* 365 – *ca* 320 BCE)

Son of Euthuphrōn, born between 390 and 380 into a prominent family of Hērakleia on the Black Sea. In Athens, according to SŌTIŌN (*apud* DIOGENĒS LAËRTIOS 5.86 [*fr.*3 Wehrli]), Hērakleidēs first encountered SPEUSIPPOS, listened to the **Pythagoreans**, and joined PLATO's **Academy**, where he was among those who recorded Plato's lecture *On the Good*, and served as **scholarch** during one of Plato's visits to Sicily, probably in 361/360 (*Souda* H-461 [*fr.*2 W.]). After Plato's death (347), Hērakleidēs remained in the **Academy** under the **scholarchate** of Speusippos (*fr.*9). At that time (347–*ca* 334), ARISTOTLE left Athens, so if it is true, as Sōtiōn adds, that Hērakleidēs also studied under Aristotle, this must have happened during Plato's lifetime, when both were members of the **Academy**. This therefore does not imply that later in his life Hērakleidēs became a **Peripatetic**, although he shared literary and historical interests with the **Peripatos**. In Athens, his obesity and affected dress and manner encouraged the transformation of his toponymic "Pontikos" into the nickname "Pompikos" (*fr.*3), i.e., "inclined to a life of luxury." After Speusippos' death (339), Hērakleidēs competed with XENOKRATĒS for the **scholarchate** and was defeated by only a few votes; he then returned to Hērakleia (*fr.*9), where he is known to have had pupils (D.L. 7.166 = *fr.*12). But we do not whether he opened a regular school.

None of Hērakleidēs' numerous works survives. His wrote many dialogues, usually set in the past, sometimes in a comic, sometimes in a tragic style, and employing myths rather than dialectical arguments. Diogenēs Laërtios (5.86 = *fr.*22), listing Hērakleidēs' works, roughly classifies them under the headings "Ethics," "Physics," "Grammar," "Music" (including literary criticism), "Rhetoric," and "History." Only a small portion of the topics reflected scientific interests, and his theories were largely embedded in mythical contexts. He believed in the immortality and transmigration of souls, and in divine intervention in human affairs. He explained e.g. the inundation and subsidence, which destroyed the town of Helikē in 373, as an act of divine vengeance. He was famous in antiquity not as a scientist nor as a philosopher but as a literary writer, and as such he continued to have a wide audience well into the Roman era.

According to Hērakleidēs, matter is composed of "jointless particles" (*anarmoi onkoi, fr.*199a-b) endowed with quality and subject to change (unlike DĒMOKRITOS' and EPICURUS' atoms), but different in quality from the bodies composed of them, a theory followed by later medical thinkers, especially ASKLĒPIADĒS OF BITHUNIA (influential in his own right), and well-known to GALĒN. Hērakleidēs' main contributions to the history of science concern the theory of planetary movements. He explained the daily movement of the fixed stars by a rotation of the Earth on its own axis, as opposed to the traditional explanation by a movement of the celestial spheres around the Earth – his main predecessor being possibly PHILOLAOS, who explained it by the daily motion of the Earth around a central fire. Hērakleidēs was probably the first to develop in detail an astronomical theory hypothesizing infinite space (a belief he shared with others, like Dēmokritos and some **Pythagoreans**). Moreover, he carefully described the movements of Venus and Mercury as both morning and evening stars endowed with a maximum elongation from the Sun (50° for Venus, less for Mercury). But there is no reason to believe that he had these two planets rotating around the Sun rather than around the Earth. Furthermore, there is no proof that he thought the Earth circulated around the Sun (as Aristarkhos of Samos was later to suggest), and no hint at all that he attributed such a revolution to the superior planets (Mars, Jupiter, Saturn).

Ed.: Wehrli v. 7 (1953); this will be soon superseded by W.W. Fortenbaugh and E. Schütrumpf, edd., *Heraclides of Pontus* = *RUSCH* 14 (forthcoming 2008), with a collection of essays (on Hērakleidean astronomy, see especially those by A.C. Bowen and R.B. Todd, and by P.T. Keyser; on physics, R.W. Sharples).

RE S.11 (1968) 675–686, F. Wehrli; H.B. Gottschalk, *Heraclides of Pontus* (1980); *DPA* 3 (2000) 563–568, J.-P. Schneider.

Silvia Fazzo

Hērakleidēs of Hērakleia Pontikē, Junior (1st c. CE)

Author of a *Musical Introduction* (*Mousikē eisagōgē*), from which a fragment on acoustics is preserved by Porphurios in his commentary on Ptolemy's *Harmonics* (30.1–31.21 Düring). It is possible that he is to be identified with the famous 4th c. BCE Hērakleidēs of Hērakleia Pontikē, but more probably Porphurios is quoting from the 1st c. CE author of the same name, who studied under Didumos of Alexandria ("Khalkenteros") and later lived in Rome during the reigns of Claudius and Nero. This Hērakleidēs may have been the father of Didumos "the music theorist," whose work is also quoted by Porphurios.

The fragment of Hērakleidēs is concerned with the physical causes of pitched sound. It begins with a quotation from Xĕnokratēs about Pythagoras' discovery of the numerical basis of musical intervals, and briefly discusses his investigations of concord and discord, in which sound was linked to movement, movement to quantity, and thus quantity to sound.

Hērakleidēs' main argument, similar in many respects to the acoustic theories of the Euclidean Sectio Canonis and the Aristotelian Corpus On Sounds, stops short of making the anticipated link between speed of movement and pitch of note. Musical notes are made up of discrete impacts, each of which has no duration in time, but which are perceived in succession as a single pitch because of the weakness of our hearing, just as a single dot of white on a spinning cone appears to the eye as a solid line.

Hērakleidēs demonstrates his theory with the example of a stretched string. (He makes no appeal to wind instruments as Aelianus does, and thus avoids the complications which such instruments introduce.) The backward and forward movement of the string produces discrete impacts on the air; between the impacts are silences, which are so brief as to be imperceptible to the ear. The impacts thus give the appearance (*phantasia*) of a single continuous sound. One difficulty is that Hērakleidēs considers the individual impacts as "notes" (*phthongoi*); his theory does not apparently deal with the inevitable question of how the impacts themselves acquire their pitch. The theory is therefore not in the strictest sense a kind of acoustic **atomism**.

RE 8.1 (1912) 487–488 (#49), H. Daebritz and G. Funaioli; Düring (1932); *KP* 2.1043 (#19), H. Gärtner; Barker (1989); Mathiesen (1999); *BNP* 6 (2005) 171–172 (#21), S. Fornaro.

David Creese

Hērakleidēs of Taras (Mech.) (220 – 200 BCE)

The bright son of a craftsman who became an *arkhitektōn* (engineer) but was exiled on suspicion of treason and fled to the Romans, then defected to Philip V of Macedon (Polubios Book 13, *fr*.4.4–8; Diodōros of Sicily 28.2, 28.9), and served on his staff (*Syll.* 552; Livy 31.16.3, etc.). Moskhiōn says Hērakleidēs invented the ***sambukē***.

M.J.T. Lewis, "When was Biton?" *Mnemosyne* 52 (1999) 159–168.

PTK and GLIM

Hērakleidēs of Taras (Med.) (95 – 55 BCE)

Hērakleides of Taras (*Vind. Med. Gr.* 1, f.2V) © Österreichische Nationalbibliothek

The **Empiricist** physician who in ancient times had the most renown (praised by Galēn *Hipp.Artic.* 18A.735 K. and Caelius Aurelianus *Acut.* 1.166 [*CML* 6.1.1, p. 114]) and the most fortune (almost 100 fragments survive). He was a pupil of Mantias in Alexandria, but afterwards he joined the **Empiricist** school (under the influence of Ptolemaios of Kurēnē, if he is to be identified with Hērakleidēs pupil of Ptolemaios and teacher of the philosopher Ainesidēmos mentioned by Diogenēs Laërtios 9.115–116 in his catalogue of the Skeptics). As an **Empiricist**, he probably still worked in Alexandria (as his dissection of human bodies suggests). His date, once controversial, is guaranteed by his use of the work of the **Erasistratean** Hikesios, and that he was studied by Apollōnios of Kitiōn; for Celsus he lived "somewhat later" than Apollōnios and Glaukias (*pr.*10). The images of Mantias and Hērakleidēs contained in the Vienna codex of Dioskouridēs (6th c.) probably come from the *Hebdomades* of Varro, who mentioned Hērakleidēs in his *Menippeae* (*fr.*445 Ast).

Compared to earlier **Empiricists**, he theorized a more sustained use of rational and causal argumentation (*logos*), thus toning down the contrast with the rival schools. He outlined **Empirical** doctrine in a work *On the Empiricist Sect*, from which Galēn extracted a lost *Synopsis* in seven books (*On My Own Books* 9 [2.115 MMH]). Hērakleidēs was probably used by Celsus for his exposition of **Empirical** doctrine (or perhaps for his own exposition) in the *proem* of *De medicina*, and by Galēn in *On medical experience.*

Hērakleidēs' interest in pharmacology is influenced by the work of Mantias, who had also distinguished himself in that field. One of his pharmaceutical works was dedicated to Antiokhis of Tlōs; another was dedicated "to Astudamas" (both were used by Galēn in *CMGen* and *CMLoc*). Specific works by him concerned theriatrics and military pharmacopoeia. His *Symposium* (used by Athēnaios in *Deipn.*) was devoted to dietetics, another discipline already studied by Mantias. He also wrote two treatises on therapeutics, one dealing with external diseases, of surgical interest too, and the other, largely used by Caelius Aurelianus, dealing with internal diseases. Interest in ophthalmic surgery is testified by P. CAIRO CRAWFORD 1.

Following the **Empirical** tradition, he dealt also with Hippokratic exegesis, writing the first commentary ever on all the works of HIPPOKRATĒS (certainly on *Aph.*, *De art.*, *Epid.* 2–4 and 6, *De off. med.*, doubtful *De hum.*). He also wrote a work in three books against the Hippokratic interpretations of the **Hērophilean** BAKKHEIOS. Hērakleidēs also dealt with other traditional topics of the **Empiricist** polemic, such as the polemic against HĒROPHILOS' *On Pulses* and against the **Hērophilean** ZĒNŌN about the marks contained in the Alexandrian copies of Book III of the HIPPOKRATIC CORPUS EPIDEMICS.

Two of his recipes (for the treatment of fractures and against chronic warts) are attested by the *Hippiatrica Cantabrigensia* (62.5 from MOSKHIŌN p.194.13–19, and 67.3 p. 199.4–11 Oder-Hoppe: the first one also in PAULOS OF AIGINA 7.17.87 [*CMG* 9.2, pp. 367–368], and given by ASKLĒPIADĒS PHARM. in Galēn *CMGen* 2.17 [13.537–539 K.] but without reference to Hērakleidēs; the second one also by AËTIOS OF AMIDA 16.6 = Guardasole *fr.*22a). Perhaps his are also two fragments attested by *Hipp. Berolinensia*, both from HIEROKLĒS and ascribed to "Tarentinus": the anecdote of the old Athenian mule (1.13, p. 5.23), also known to ARISTOTLE *HA* 6.24 (577b) and others, and a recipe against shrew bites (87.2, p. 314.21). Both fragments were ascribed by Oder and by Georgoudi to an agricultural writer "Tarentinus" quoted by Phōtios *Bibl.* 163 and frequently in GEŌPONIKA 3–4; but Hērakleidēs too is sometimes referred to as "Tarentinus": Galēn *Antid.* 2.13 (14.181 K.) and by *Etym. magn.* s.v. *elinuein*, on the Hippokratic *Epidemics* 6.1 (omitted by editors, connected with *fr.*352 D. against Bakkheios). All the fragments attested by veterinarians came likely from Hērakleidēs' pharmaceutical works, and not from "the first attested veterinary work," as stated by Gossen (1913) 1714 (*contra* Deichgräber, 260 and Björck).

Ed.: Deichgräber (1930) 172–204 (fragments), 258–261; A. Guardasole, *Frammenti* (1997).

Oder (1890) 89–90; *RE* 8.1 (1912) 493–496 (#54) and 8.2 (1913) 1713–1715, H. Gossen; Björck (1932) 38–39; *KP* 2.1044 (#23), F. Kudlien; Fabricius (1972) 200; Smith (1979) 211–212; M. Frede "The Empiricist attitude towards reason and theory," in R.J. Hankinson, ed., *Method, Medicine and Metaphysics* = *Apeiron* 21.2 (1988) 79–97 at 91–94; S. Georgoudi, *Des chevaux et des boeufs dans le monde grec* (1990) 55–56; Marganne (1994) 147–167; Manetti and Roselli (1994) 1595–1597; *OCD3* 687, H. von Staden; *ECP* 258–259, *Idem; BNP* 6 (2005) 173–174 (#27), V. Nutton; Ihm (2002) #114–123; *AML* 401–402, A. Guardasole.

Fabio Stok

Hērakleios Imp. (610 – 640 CE)

Byzantine emperor (reigned 610–640 CE), found in the list of *poiētai* (makers of gold, *CAAG* 2.25). The early table in MS *Marcianus gr.* 299 attributes to him three treatises not preserved in the corpus: *On Alchemy*, *Eleven Chapters on the Making of Gold*, and *Collection Concerning the Study of the Sacred Art by Philosophers*. The first was addressed to Modestus (patriarch of Jerusalem, 614–630 CE). The 10th c. catalogue of books, *Kitāb al-Fihrist*, mentions: "of Hērakleios the larger book, fourteen chapters."

Berthelot (1885) 132; Fück (1951) 95 (#42), 124; *ODB* 916–917, W.E. Kaegi *et al.*

Cristina Viano

Hērakleitos (Math.) (450 – 150 BCE?)

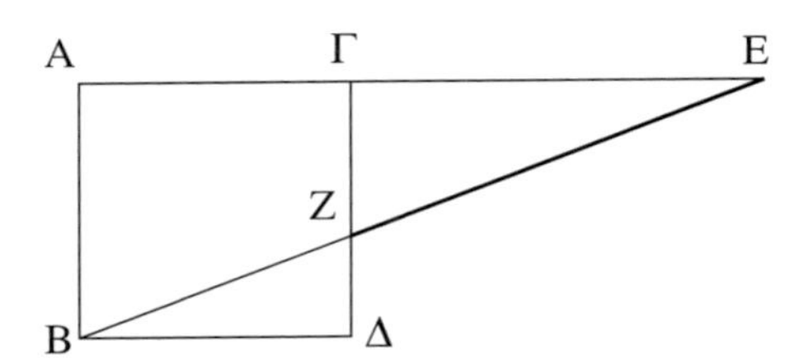

Hērakleitos' *neusis* in a given square **AΔ**: the given length EZ should verge toward point B. © Bernard

In commenting on APOLLŌNIOS OF PERGĒ's lost *Neuseis*, PAPPOS (*Coll.* 7.128 Jones) quotes Hērakleitos' ***neusis***-construction for the square (*see figure*), which depends on the previous lemma (7.127 Jones), probably Pappos' own contribution but perhaps also his. Since, as mentioned by Pappos (p. 203.18–19 Jones), the problem solves a particular case of Apollōnios' ***neusis*** in a rhombus (problems 8 and 9 in his *Neuseis* I, evoked by Pappos at 7.27 and again 7.126 Jones) and since Pappos criticizes Apollōnios' (alleged) disdain for his predecessors' efforts (7.34–35 Jones), Hērakleitos is most probably earlier than Apollōnios. EUTOKIOS (p. 228 H.) cites a Hērakleidēs who wrote a life of ARCHIMĒDĒS, (probably the same as the Hērakleios mentioned in the first lines of Eutokios' commentary on the *Conics*) whom Heiberg and Knorr unconvincingly identified with Archimēdēs' companion mentioned in his *On Spirals* and with Hērakleitos, although Eutokios' Hērakleidēs is most probably a later biographer (Decorps 2000: n.5 p.10), perhaps the same as the biographer and doxographer HĒRAKLEIDĒS OF KALLATIS.

Jones (1986) 436; Knorr (1986) 294–302.

Alain Bernard

Hērakleitos of Ephesos (*ca* 510 – 490 BCE)

Known for his dark sayings and caustic invectives against his predecessors, Hērakleitos has an anomalous position in the history of science. On the one hand, he seems to continue the cosmological theories of the philosophers of Milētos; on the other, he calls into question many of their assumptions and turns attention to human concerns. Hērakleitos criticized famous wise men for pursuing information without a unifying theory. He seemingly admired the natural philosophers for trying to explain the world in a systematic way. Stressing the unity underlying the natural phenomena of the world, he criticized people for not recognizing that unity.

Hērakleitos posited fire as the source of all things. The main forms of matter were fire, water and earth. There is a constant interchange of matter such that portions of fire turn into water, while portions of water return into fire and other portions transform into earth. A given amount of one element becomes a proportionate amount of the other. Overall, the relative amounts of the three kinds of element remain fixed in the world, so that a balance is maintained. Hērakleitos first envisaged a "law of conservation of matter." Although a given portion of matter can change its form, it remains some kind of matter, and changes in a law-like way.

Earlier cosmologists saw cosmic unity arising from a single stable substance such as water or air, but Hērakleitos seems to have chosen fire for its instability. Fire is matter and also a symbol of the changeability of matter. What makes the world possible is balanced cycles in the interchange of elements: a hidden harmony. Hērakleitos seems to have identified this harmony with the *Logos*, which is at once a message about the world and its underlying structure. His own sayings are often cast in the form of riddles whose solution manifests the unity of different meanings, as though in imitation of the world's structure.

Hērakleitos of Ephesos © Reproduced by kind permission of the Archaeological Museum of Hērakleion

Unlike earlier cosmologists, who saw the world as arising out of a uniform state of matter, Hērakleitos seems to have rejected cosmogony: the world always is and was and will be because of the balance of its contrary changes (some scholars however think he believed in periodic conflagrations in which everything turned into fire). The heavenly bodies are composed of bowl-like structures which trap fiery vapors from the Earth. The phases of the Moon and eclipses of the Sun and Moon are caused by tilting of the bowls.

Hērakleitos is famous for allegedly having said that everything flows, so that one could not step twice into the same river. Hērakleitos' original statement, however, implied something rather different: "On those stepping into rivers staying the same, other and other waters flow" (*fr.*12). Although (and perhaps because) the waters of a river are constantly changing, the river remains the same. There is flux in matter, but there is also regularity and constancy in the world.

Ed.: *DK* 22; M. Marcovich, *Heraclitus: Greek text with a short commentary* (1967; 2nd ed., 2001); S.N. Mouraviev, *Heraclitea: édition critique complète des témoignages sur la vie et l'oeuvre d'Héraclite d'Éphése et des vestiges de son livre* (1999–).

C. H. Kahn, *The Art and Thought of Heraclitus* (1979); KRS 181–212.

Daniel W. Graham

Hērakleitos of Rhodiapolis (*ca* 60 – 140 CE)

Physician and author of medical and philosophical works, attested in a single inscription (*TAM* II.2.910) accompanying a statue dedicated to him by his hometown, Rhodiapolis (Lukia), because of his donations. This document tells us that Hērakleitos, the Homer of medical poetry, for his works received the same honors from Alexandrian, Rhodian and Athenian citizens and the Areopagos as accrued to **Epicurean** philosophers.

BNP 6 (2005) 179 (#6), V. Nutton.

Claudio Meliadò

Hērakleitos of Sikuōn (250 BCE – 50 CE?)

Wrote an *On stones* in at least two books. Pseudo-Plutarch, *De fluu.* 13.4 (1158A) transmits a single fragment regarding the stone *kruphios*. Jacoby considers him fictive.

FHG 4.426; *RE* 8.1 (1912) 510 (#14), F. Jacoby; De Lazzer (2003) 83.

Eugenio Amato

Hērakle(i)odōros (370 – 320 BCE?)

LONDINIENSIS MEDICUS 10, heavily damaged, includes him, after ALKAMENĒS, TIMOTHEOS OF METAPONTION and ABAS, and before HĒRODIKOS OF SELUMBRIA, in a series of accounts of the causes of diseases: the prior three blame the head, the latter diet. The rare name is first attested in the 4th c. BCE (*LGPN*), and possibly our man is the **Platonist** of (pseudo?) Dēmosthenēs *Letter* 5, on whom see *DPA* 3 (2000) 552, T. Dorandi.

(*)

PTK

Hērās of Kappadokia (20 BCE – 20 CE)

Greek physician or pharmacologist practicing medicine in Rome as a *peregrinus*, perhaps contemporary with AUGUSTUS (although traditionally dated 100 BCE – 40 CE). Considered an **Empiricist** by modern scholars (solely because he used recipes from HĒRAKLEIDĒS OF TARAS), he followed probably the current pragmatico-synthetist trend. His pharmacological treatise, known through a papyrus (*ca* 300 CE, and overlapping with GALĒN, *CMLoc* 1.2, 12.430 K., remedy against hair-loss), and about two dozen lengthy extracts in Galēn (Fabricius), was variously designated by Galēn's sources (Fabricius 183, n.8), *cf.* *CMGen* 1.14 (13.416 K.), esp. as *Narthēx* (*CMLoc* 1.1, 1.2 [12.398, 430 K.]) or *Pharmakitis* (*CMLoc* 2.3 [12.593 K.], *CMGen* 7.6, 7.14 [13.969, 1042 K.]). The work was a compilation of recipes for compound medicines listed according to two different principles: topographical (affected parts of the body) arranged *a capite ad calcem*; and pharmaceutical (*per genera*). *Materia medica* of animal, vegetable, and mineral kinds were used; and Hērās seems to use the word *antidotos* in its earlier toxicological sense. The hair-loss remedy contained *ladanon* (resin of some *Cistus* sp. Galēn, *Simples* 7.10.28 (12.28–29 K.); *cf.* Durling 1993: 220–221) and maidenhair, macerated in dry wine and myrtle oil to a honey-like consistency, and applied after the bath. Galēn in *CMGen* preserves numerous topical plasters some of which seem collected from local usage (barbarian: 2.22 [13.557 K.]; Mēlian white: 2.10 [13.511 K.]; Kuzikēnian: 5.7 [13.814–818 K.]; Hellespontian: 6.11 [13.914–915 K.]). Other compounds treat ***hudrophobia*** (*CMGen* 1.16, 13.431–432 K.), dysentery (an enema: *CMLoc* 9.5, 13.297 K.), sciatica, endorsed but perhaps not created by Hērās (*CMGen* 7.7, 13.986–987 K.), and bruises (*CMLoc* 5.1, 12.819 K.). OREIBASIOS *Syn.* 3.96 (*CMG* 6.3, p. 94) cites "Hēra" (with feminine article, implying a woman, but perhaps an error for our pharmacologist) and preserves a treatment for herpes compounded from saffron, myrrh, and **oxymel**.

RE 8.1 (1912) 529 (#4), H. Gossen; *KP* 2.1053, F. Kudlien; Fabricius (1972) 183–185, 242–246; I. Andorlini, in *Atti e Memorie dell'Accademia Toscana di Scienze e Lettere "La Colombaria"* 46 (1981) 41–45 and nn.36–37; Marganne (1981) 134–135; Korpela (1987) 169, #81; Andorlini-Marcone (1993) #10 (p. 478); *BNP* 6 (2005) 183–184, Alain Touwaide.

Alain Touwaide

DE HERBA VETTONICA ⇒ APULEIUS, PSEUDO

De herbis/De viribus herbarum (*ca* 200 – 300 CE)

Anonymous 216–line Ionic Greek hexameter poem in highly mannered style treating curative properties of herbs. The text is preserved in a number of MSS of DIOSKOURIDĒS, on whom (together with NIKANDROS) it seems largely based. Characteristic of its age, the poem blends medical with purely magical pharmacology, offering herbal remedies for fever alongside love potions.

Ed.: Heitsch 2 (1964) #64.
G. Kaibel, "Sententiarum Liber Quintus," [§VII] *Hermes* 25 (1890) 103–109; *RE* 1.2 (1894) 2327, M. Wellmann; *KP* 5.1573, J. Kollesch; *BNP* 1 (2002) 710–711, V. Nutton.

Keith Dickson

HERENNIUS PHILŌN ⇒ PHILŌN OF BUBLOS

Hermarkhos of Mutilēnē (300 – 240 BCE)

Hermarkhos © Budapest Museum

Epicurean philosopher who studied under EPICURUS at Mutilēnē on the island of Lesbos *ca* 311, and moved to Athens in 307 when Epicurus founded his philosophical school, the **Garden**. Along with Epicurus, MĒTRODŌROS, and POLUAINOS, he was considered one of the co-founders of the Epicurean school. He became **scholarch** of the school when Epicurus died in 271. His writings, considered authoritative by later **Epicureans**, included *Essays in Letter Form, On the Sciences, Against* EMPEDOKLĒS, *Against* PLATO, and *Against* ARISTOTLE. The few remaining fragments deal with the development of justice and homicide laws (PORPHURIOS, *De abstinentia* 1), and the nature of the gods.

Ed.: Long and Sedley (1987) §22M-N; F. Longo Auricchio, *Ermarco, Frammenti* (1988).
OCD3 689–690, D. Obbink; *ECP* 262–263, D.N. Sedley; *BNP* 6 (2005) 208–209, T. Dorandi.

Walter G. Englert

Hermās *thēriakos* (300 BCE – 180 CE)

PHILOUMENOS 8 (*CMG* 10.1.1, p. 13) records his blood-stanch (*iskhaimon*): ***khalkitis***, *melanteria* (DIOSKOURIDĒS 5.101), and spider-webs; AËTIOS OF AMIDA 8.49 (*CMG* 8.2, p. 476) records his remedy for oral disorders: heath-fruit and henbane in honey. The simplicity of the remedies argues for an early date.

RE 8.1 (1912) 722 (#3), H. Gossen.

PTK

Hermeias (Math.) (40 – 100 CE)

One of PLUTARCH's interlocutors in *Table Talk* 9.3 (738D–739A), a geometer who addressed why the Greek alphabet contains 24 letters. His solution rested upon perfect numbers (for which he provided two definitions), squares, and cubes. The number of letters of the alphabet are 3x8 (the first perfect number with a beginning, middle, and end *times* the first cube) or 6x4 (the first perfect number equal to the aliquot sum of its factors *times* the first square).

RE 8.1 (1912) 732 (#12), C.R. Tittel.

GLIM

Hermeias (Doxogr.) (150 – 250 CE)

Otherwise unknown Christian author of a satire of philosophical doctrines, commonly known as *Irrisio gentilium philosophorum*. This short treatise is based mainly on doxographical information and contains some close parallels to known texts. It contains nothing of importance not found elsewhere, but it demonstrates how varied doxographical texts were, and how important it was for some early Christian writers to minimize the pagan philosophers.

Ed.: R.P.C. Hanson and D. Joussot, *Hermias: Satire des philosophes paiens = Sources chrétiennes* 388 (1993).
Mansfeld and Runia (1996) 314–317.

Jørgen Mejer

Hermeias (Astrol.) (150 BCE – 150 CE)

Astrologer, whose name (though very common) was perhaps adopted in allusion to Hermēs. The heading VETTIUS VALENS' *Anthologiai* 4.27 states that the following discussion of a scheme of determining planetary lords for temporal intervals in an individual's life are "from Seuthēs *On Years*." Confusingly, immediately following are the words, "a lecture from Hermeias." Hermeias refers to himself by name in the first person in 4.29, making the three successive chapters appear a direct quotation of his lecture. It is not clear whether Seuthēs was an anthologizer or an auditor recording the lecture at first hand.

Riley (n.d.).

Alexander Jones

Hermeias (Geog.) (325 BCE – 540 CE)

Wrote a ***periēgēsis***, cited once by STEPHANOS OF BUZANTION, s.v. "Khalkis."

RE 8.1 (1912) 731 (#7), F. Jacoby.

PTK

Hermeias (Ophthalm.) (250 BCE – 95 CE)

Ophthalmologist whose eyewash ASKLĒPIADĒS PHARM., in GALĒN, *CMLoc* 4.8 (12.754 K.), records – aloes, **calamine**, frankincense, myrrh, roasted copper, saffron, and opium in gum, egg-white, and Mendesian wine, applied every 3–4 hours; he is also cited as an authority on inverted eyelashes, *ibid.* (12.801).

RE 8.1 (1912) 832 (#12), H. Gossen.

PTK

Hermēs Trismegistos, pseudo (*ca* 100 BCE – *ca* 400 CE)

Several texts have been ascribed to the Greco-Egyptian god of *gnōsis*, Hermēs Trismegistos, most of which seem to have been written by unknown authors between the 1st c. BCE and the 4th c. CE, and were revised at later dates (late antiquity to the Middle Ages). Modern scholars traditionally divide the literary works attributed to Hermēs Trismegistos into two groups: the "philosophical Hermetica" are the religious and philosophical texts (KORĒ KOSMOU) and the "technical Hermetica," addressing "more practical" matters related to magic, alchemy, astrology and natural sciences. This division, however, is not absolute and its limits are not easily drawn. Both philosophical and technical work share two major characteristics, among others: universal ***sumpatheia***, and the *topos* of revelation – true knowledge can only be transmitted by a revelation coming either from Hermēs Trismegistos or one of his messengers. Without it, knowledge remains ineffective (for example, *CCAG* 8.3 [1912] 134–138).

The major works of the "technical Hermetica" are, among others, a *Brontologion*, which explains the significance of thunder for every month of the year, and the *Peri Seismōn*, where the significance of earthquakes is explained in relation to the zodiac (*cf.* VICELLIUS). There are also several texts of ***iatromathēmatika*** based on the ***melothesia***, e.g., the ***Iatromathēmatika*** *from Hermēs Trismegistos to Ammōn the Egyptian* which explains not only the relationship between the planets and the different parts of the body but also the importance of the hour and day of the beginning of the illness and their astrological significance for its treatment (*cf.* IMBRASIOS). There are also astrological herbals, a group of texts associating a plant with the seven planets, the 12 signs of the zodiac and the 36 **decans** (*cf.* THESSALOS OF TRALLEIS). Finally, the *Sacred Book of the* ***Decans***, which explains how to make amulets with the help of a plant, a stone and the figure of each **decan**, and the KURANIDES, devoted to describing the capacities allotted to plants, birds, animals, fishes and stones. Most alchemical texts attributed to Hermēs Trismegistos have been lost. However, fragments have been preserved as quotations in the work of authors such as ZŌSIMOS OF PANŌPOLIS, OLUMPIODŌROS or SUNESIOS. One of the longest quotations comes from the ANONYMOUS ALCHEMIST PHILOSOPHER, which attributes to Hermēs Trismegistos a recipe for the preparation of silver.

Festugière (1950); *Dictionary of Gnosis and Western Esotericism* (2005) 487–499, R. Van den Broek; *ODB* 920, J. Duffy; Aurélie Gribomont, "La pivoine dans les herbiers astrologiques grecs," *Bulletin de l'Institut Historique Belge de Rome* 74 (2004) 6–59 (the author regrets the errors of Greek caused by electronically-generated misprints).

Aurélie Gribomont

Herminos (of Pergamon?) (*ca* 160 – 180 CE)

Aristotelian commentator, ALEXANDER OF APHRODISIAS' teacher (SIMPLICIUS, *in De Caelo* = *CAG* 7 [1894] 432.32) and perhaps ASPASIOS' student. Alexander addressed a short tract to GALĒN, extant only in Arabic, responding to Galēn's criticism of a **Peripatetic** theory originally directed to a certain *'RMNWS*, probably Herminos who may also have been the student of Aspasios whom Galēn heard *ca* 145 (Moraux 362). Some fragments of his commentaries on ARISTOTLE's *Categories, Prior Analytics, On interpretation, Topics* and *On the Heavens* survive. According to Simplicius, Herminos participated in an extended debate about the cause of the eternal movement of the Heavens.

Moraux (1984) 2.361–398; *BNP* 6 (2005) 225, H.B. Gottschalk.

Alain Bernard

Hermippos of Bērutos (105 – 165 CE)

Born a slave in an inland Bērutos (*Souda* E-3045) but eventually freed, studied under PHILŌN OF BUBLOS. Hermippos was a grammarian and mathematician whose titles include *Interpreting Dreams* (five books), *On the Number Seven*, and *About Slaves Eminent in Learning*. Only fragments survive (*FHG* 3.35–36, 51–52), not always securely assignable to our Hermippos (*cf.* HERMIPPOS OF SMURNA).

OCD3 692 (#3), (anonymous).

GLIM

Hermippos (of Smurna?) (250 – 200 BCE)

It seems likely that references to Hermippos, whether he is called "Kallimakhean," "**Peripatetic**" or "of Smurna," all refer to one and the same person. He was mainly known for his biographies of philosophers and statesmen. His biographies were epitomized late in the 2nd c. BCE and remained popular down to the 3rd c. CE. They contained many picturesque anecdotes, but also lists of books and pupils; he is one of DIOGENĒS LAËRTIOS' main sources and is often referred to when Diogenēs is describing a philosopher's death. Hermippos is not more nor less reliable than other ancient biographers. He is also said to have written a *Phainomena*, of which little survives.

Ed.: *FGrHist* 1026; *SH* 485–490.
J. Bollansée, *Hermippos of Smyrna and his Biographical Writings, A Reappraisal* = *Studia Hellenistica* 35 (1999).

Jørgen Mejer

Hermodōros of Alexandria (*ca* 300 – 340 CE)

PAPPOS' pupil and dedicatee of the seventh book of the *Mathematical Collection* (7.*pr.*).

Jones (1986) 379–380; Netz (1997) #25.

GLIM

Hermodōros of Surakousai (365 – 325 BCE)

Wrote and promulgated in Sicily the biography of his mentor PLATO (*frr.*1–5 I-P: CICERO, *ad Att.* 13.21.4; DIOGENĒS LAËRTIOS 2.106, 3.6; *Souda* Lambda-661), important studies of Plato's ideas about matter (*frr.*7–8 I-P: SIMPLICIUS, *in Phys.* 1.9 = *CAG* 9 [1882] 247–248, 256–257), and a book *On Mathematics* (*peri mathēmatōn*) of which one fragment survives: *fr.*6 I-P: D.L. 1.2.

Ed.: M. Isnardi Parente, *Senocrate – Ermodoro Frammenti* (1982) 157–160, 261–263, 437–444.
BNP 6 (2005) 231 (#2), K-H. Stanzel.

PTK and GLIM

Hermogenēs of Alabanda (200 – 150 BCE)

Architect and architectural theorist, cited often by VITRUUIUS, who seems to have relied heavily on Hermogenēs' treatises for the Ionic order. Hermogenēs considered the Doric order unsuitable for temples because of the difficulty of arranging its frieze over a peristyle with harmonious proportions. He changed the Temple of Dionysus at Teos – on which he wrote a treatise – from Doric to Ionic, apparently in mid-construction (Vitr. 3.3.1, 3.3.8, 4.3.1). He also wrote on his Temple of Artemis at Magnesia (Vitr., 3.2.6, 3.3.6–9, 7.*pr.*12). Other buildings stylistically attributed to him include a Temple of Zeus at Magnesia, and monumental altars at Priēnē and Magnesia. He used modular grids as a method of design. His interest in proportions, favoring eustyle, can be found in some of his attributed works; he also revived the pseudodipteral plan, little used since the archaic period, and 5th c. BCE atticisms. His temples show some influence from PUTHEOS' Temple of Athena at Priēnē. Hermogenēs combined an appreciation of historical precedent with new ideas about space, proportion and efficient building; his writings seem to have had a didactic and theoretical aspect. Debate continues over dating buildings attributed to him and the man himself: some place him earlier in the late 3rd c. BCE, or later in the mid 2nd c. BCE. Hermogenēs' version of the Ionic order became canonical through Vitruuius, and still endures after revival in the Renaissance.

J. Pollitt, *Art in the Hellenistic Age* (1986) 242–247; W. Hoepfner and E.L. Schwander, edd., *Hermogenes und die hochhellenistische Architektur* (1990); *KLA* 1.305–310, W. Hoepfner; *BNP* 6 (2005) 232–234, H. Knell.

Margaret M. Miles

Hermogenēs of Smurna (30 – 70 CE?)

Pharmacist cited by GALĒN, *Simples* 1.29 (11.432 K.), as exemplary of the **Erasistratean** school, as HĒRODOTOS was of the **Pneumaticists**, MĒTRODŌROS (PHARM.) was of the Asklēpiadeans, and ZĒNŌN was of the **Hērophileans**. OREIBASIOS, *Ecl. Med.* 109.2 (*CMG* 6.2.2, p. 287), cites his plaster against infections of the extremities: grind frankincense, copper-flakes, iron-rust, and honey together in the sun until pale yellow, wash the affected spot with wine, apply with olive oil. Another is preserved by AELIUS PROMOTUS, *Dyn.* 63.4: lime and orpiment, but omit the ***khalkitis***, he says. Despite the frequency of the name (*LGPN*), the **Hērophilean** pharmacist has been identified with the historian doctor of *CIG* 3311 = *Inschr. Smyrn.* 1 (1982) #536, who died aged 77 having written a like number of medical scrolls, a *History of Smurna*, *Stadiasmoi* of Asia and Europe, and other works. The father of the epigraphic Hermogenēs is Kharidēmos, identified with the **Hērophilean** KHARIDĒMOS. The Hermogenēs of Smurna, however, whose wife was Melitinē (*CIG* 3350) is 2nd c. BCE: *Inschr. Smyrn.* 1 (1982) #118, whereas Hadrian's physician (Dio Cassius 69.22.3) is 138 CE. The epigraphic doctor is perhaps the target of epigrams by Nikarkhos of Alexandria (*Anth. Pal.* 11.114) and the Neronian-era Lucilius (*Anth. Pal.* 11.131, 257), as surgeon (257), and slayer of the astrologer Diophantos (114, 131). Galēn's citation does not greatly restrict the date of Hermogenēs, who may be either of the Smurnian doctors, or neither.

RE 8.1 (1912) 877–878 (#23), H. Gossen; *FGrHist* 579.

PTK

Hermolaos (Geog.) (*ca* 530 – 560 or *ca* 690 – 710 CE)

Grammatikos at Constantinople, wrote the extant epitome of STEPHANOS OF BUZANTION's *Ethnika*. *Souda* E-3048 states he dedicated his work to the emperor Justinian, meaning either Justinian I (525–565) or II (685–695, 705–711). Likewise uncertain is the inclusion of contemporary notes while making the epitome.

RE 8.1 (1912) 891 (#2), A. Gudeman; *RE* 3A.2 (1929) 2374–2375, E. Honigmann; *HLB* 2.37; *PLRE* 2 (1980) 1032, 3 (1992) 593; *ODB* 1954, A. Kazhdan.

Andreas Kuelzer

Hermolaos (Pharm.) (120 BCE – 450 CE)

CASSIUS FELIX 29.9 (*CUF*, p. 62) and AËTIOS OF AMIDA 7.104 (*CMG* 8.2, p. 364) record two versions of his water-based collyrium, one with ***pompholux***, copper, saffron, myrrh, aloes, acacia, and the Egyptian incense *kommi*; the other adds spikenard, opium, and **Indian buckthorn**. The same recipes are repeated by ALEXANDER OF TRALLEIS (2.21 Puschm.); and MAXIMIANUS uses both.

Fabricius (1726) 182.

PTK

Hermōn of Egypt (120 BCE – 10 CE)

An Egyptian temple-scribe (*hierogrammateus*) who published collyria, one recorded by HĒRĀS, in GALĒN *CMGen* 5.2 (13.776–777 K.), involving **galbanum** and **terebinth**, which others attributed to EPIGONOS; and another by CELSUS 6.6.24, containing aloes, antimony, cassia, cinnamon, saffron, ***kostos***, myrrh, nard, poppy "tears," ***psimuthion***, etc. AËTIOS OF AMIDA 15.13 (Zervos 1909: 39–40) records Hermōn's ISIS-plaster, also attributed by some to Epigonos. *Cf.* also GLUKŌN, and perhaps NAUKRATITĒS MEDICUS.

RE 8.1 (1912) 894 (#10), H. Gossen.

PTK

Hermophilos (120 BCE – 95 CE)

ASKLĒPIADĒS PHARM., in GALĒN *CMLoc* 4.7 (12.781 K.), records his "*thalasseros*" collyrium, containing **calamine**, white pepper, **verdigris**, etc. in gum and water; repeated by AËTIOS OF AMIDA 7.114 (*CMG* 8.2, p. 388) and PAULOS OF AIGINA 7.16.46 (*CMG* 9.2, p. 344), unattributed. The name is more often spelled Herm<u>a</u>philos: *LGPN*.

von Staden (1989) 583–584.

PTK

Hermotimos of Kolophōn (360 – 310 BCE)

Extended the work of EUDOXOS and THEAITĒTOS, discovered many elementary propositions, and wrote on *loci* (PROKLOS *In Eucl.* p. 67.20–23 Fr.).

Lasserre (1987) 17.

Ian Mueller

Hērodikos of Knidos (440 – 400 BCE)

Greek physician of uncertain date, earlier than HIPPOKRATĒS but slightly later than EURUPHŌN OF KNIDOS with whom he is associated regarding special therapies: the use of woman's milk in curing ***phthisis*** (GALĒN *Prob. Prav. Alim. Suc.* 6.775 K.), of stomach purging, vomiting, steam baths etc. (CAELIUS AURELIANUS, *Chron.* 3.139 [*CML* 6.1.2, p. 762]). Hērodikos (LONDINIENSIS MEDICUS 4.40–5.34) agrees with Euruphōn that diseases arise from digestive residues, but his view is more elaborated: poor digestion occurs when movement and diet are imbalanced, fitting with the suggestion that Hērodikos invented the concept of diet (*Sch. Hom. Iliad* 11.515c). When food is not processed, residues are generated; these in turn generate harmful acidic or bitter liquids. Different diseases arise if one or the other liquid predominates. Because of the mention of physical exercise, our Hērodikos has been confused in many sources with Hērodikos of Selumbria.

RE 8.1 (1912) 979, H. Gossen; Grensemann (1975) 12–14; A. Thivel, *Cnide et Cos?* (1981) 363–64; *BNP* 6 (2005) 264, A. Touwaide; Daniela Manetti, "Medici contemporanei a Ippocrate: problemi di identificazione dei medici di nome Erodico," in Ph. van der Eijk, ed., *Hippocrates in context* = *SAM* 31 (2005) 295–313.

Daniela Manetti

Hērodikos of Selumbria (*ca* 500 – *ca* 425 BCE)

PRŌTAGORAS' contemporary, originally from Megara, settled in Selumbria, trained athletes for many years, best known through references in PLATO and HIPPOKRATĒS, and possibly the latter's pupil, at least according to SŌRANOS, *V. Hipp.* 2 (*CMG* 4, pp. 175–178). Modern scholars, such as Smith, tend to discount such a relationship between Hippokratēs and Hērodikos. The HIPPOKRATIC CORPUS, EPIDEMICS (6.3.18) complains "Hērodikos used to kill fever-patients with running, much wrestling and with vapor baths: bad policy" (but Hērodikos' doctrines may have influenced the work). Plato criticizes him for mixing gymnastic training and medicine (*Resp.* 408d), attributes to him a treatment of observation and waiting (*Resp.* 406a-d), and conveys his prescription of brisk walks (*Phdr.* 227d3–4). The LONDINIENSIS MEDICUS 9.20–36 ascribes to him an elaborate theory concerning diet, especially in regards to athletes. Hērodikos, interested in dietetics, was considered one of its inventors: PORPHURIOS, *Homeric Enquiries* (*Iliad* 11.515) says that Hērodikos began dietetic medicine, later perfected by Hippokratēs, PRAXAGORAS and KHRUSIPPOS (sc. of KNIDOS (I)). Contrarily, Eustathios remarks that dietetics was begun by Hippokratēs and completed by Hērodikos, Praxagoras and Khrusippos (*Commentaries on Homer's Iliad* 11.514). Modern scholars have wrongly attributed to him the authorship of the HIPPOKRATIC CORPUS, REGIMEN.

RE 8.1 (1912) 978–979 (#2), H. Gossen; K. Deichgräber, *Die epidemien und das Corpus Hippocraticum*, in *Abhandlungen der Preussischen Akademie der Wissenschaften* (1933) 162–163; R. Joly, *Hippocrate, Du régime* (*CUF* 1967); E.D. Phillips, *Aspects of Greek Medicine* (1973) 190; Grensemann (1975); W.D. Smith, "Notes on Ancient Medical Historiography," *BHM* 63 (1989) 73–109 at 87; Pinault (1992) 7, 18–23, 31; *BNP* 6 (2005) 265, A. Touwaide.

Robert Littman

HĒRODOTOS ⇒ HEIRODOTOS

Hērodotos (Mech.) (230 – 180 BCE?)

Made minor improvements to the traction machine of NEILEUS, according to HĒLIODŌROS in OREIBASIOS *Coll.* 49.8 (*CMG* 6.2.2, p. 14); see PASIKRATĒS.

Drachmann (1963) 174–175; Michler (1968) 87, 130.

PTK

Hērodotos of Halikarnassos (445? – 420? BCE)

Hērodotos of Halikarnassos © Biblioteca Nazionale "Vittorio Emanuele III", Naples, Italy. Reproduced with permission of the Ministero per i Beni e le Attività Culturali

Born *ca* 485 BCE, author of the long prose work that has come to be called *Histories*, based on the opening sentence which identifies the work as "the display of the inquiry (*historiē*)" of the author. The modern title should therefore not be interpreted as denoting an exclusively historical focus; *Inquiries* might be more accurate. While taking the events of *ca* 550–479 BCE as his general framework, i.e. the rise of the Achaemenid Persian Empire and its escalating conflict with the Greek cities of Europe, Hērodotos makes frequent and sometimes lengthy excursuses into questions of geography, ethnography, and natural science, exploring in discursive fashion the topics that most intrigued him and his contemporaries. The extremely broad scope of his work and the range of attitudes and methodologies found there make the *Histories* a rich but often frustratingly complex source of insight into the evolution of scientific thinking in Greece of the mid-5th c. BCE.

About Hērodotos' life little is known, but if his own account of his travels is believed (as most scholars do), he ranks as one of the Classical world's great explorers. He visited upper Egypt, the Black Sea coast, southern Italy, the Levant, and perhaps even the city of Babylōn. At least some of his travels seem to have been undertaken for the purpose of historical, geographical and anthropological research. His visit to Egypt provided the material for an extremely long excursus filling all of the second book of the *Histories*, almost 15% of the work's total length, which investigates matters as diverse as the source of the Nile and the causes of its annual flooding, the geology of the Nile valley, local flora and fauna, and the religious practices of the inhabitants. When discussing the many mysteries of the Nile, Hērodotos shows great independence of mind and acute powers of observation, as he rejects the myth-based or speculative accounts of his Ionian predecessors in favor of deductions grounded in empirical evidence. At one point (2.12) he cites five first-hand observations supporting his thesis that the land of Egypt had been formed from layers of silt deposited by the Nile in a gulf of the Mediterranean. Elsewhere in his discussion of the Nile, as well as in his discussion of global geography in his fourth book, Hērodotos rejects the idea of a river "Ocean" encircling the landmass of Asia, Africa and Europe, simply on the basis of lack of empirical support (2.23, 4.8, 4.45).

The scientific method adopted by Hērodotos in the *Histories* is by no means uniform or consistent, however, and he himself occasionally reverts to the very mythic or non-empirical kinds of explanation he condemns in the Ionians. His religious conservatism led him to assert, for example, that the gods, rather than natural forces, had caused a severe storm which damaged the Persian navy as it prepared to attack mainland Greece (8.13). (In discussing another storm, however, which also caused great harm to the Persians, Hērodotos seems unwilling to choose between natural and divine causes: "At long last the Magi priests stopped the wind after three days, by offering sacrifices and shouting incantations, as well as by sacrificing to Thetis and the Nereids; or else it was otherwise and the wind stopped by itself, as it is wont to do" [7.191]). He often supports, or at least is unwilling to contradict, a traditional piety which saw in the natural world a pattern familiar from contemporary tragic drama, of excess or transgression leading to divine retribution. Thus, while Hērodotos sometimes approaches scientific problems in a way that anticipates the empirical method of ARISTOTLE, at other times he relies on mythic notions inherited from Homer and the archaic world, and it is hard to know at any one point in his complex narrative why he leans one way or the other.

Though the excursus on Egypt is by far Hērodotos' longest, other important discussions deal with the lands and peoples of "Skuthia" or north-eastern Eurasia (4.1–82), the geography of the ***oikoumenē*** (contained within the Skuthian account, 4.36–45), and the contrast between the edges of the Earth and its central regions (3.106–116). Framed by this last passage is a fascinating discussion of the reproductive patterns of lions and snakes (3.108–109), which shows Hērodotos making bold yet naive forays into the then-nascent field of biology. A brief dialogue between King Xerxēs and his uncle concerning the origin of dreams (7.14–16) contains a theory (ultimately rejected) which anticipates modern psychological explanations.

OCD3 696–698, J.P.A. Gould; J.S. Romm, *Herodotus* (1998); *BNP* 6 (2005) 265–271, K. Meister; C. Dewald & J. Marincola, *The Cambridge companion to Herodotus* (2006).

J.S. Romm

Hērodotos (Pneum., of Tarsos?) (70 – 100 CE)

Pneumaticist, (preferred?) student of AGATHINOS, who dedicated to him a work (perhaps *On Pulse*); became famous in Rome (GALĒN, *Diff. Puls.* 4 [8.751 K.]), where he might have taught medicine (OREIBASIOS, *Ecl. Med.* 73[74].22); criticized all other sects' approaches to medicine (Galēn, *Simples* 1.19 [11.432 K.]). Despite the chronological difficulty, identified by some scholars with the Skeptic Hērodotos of Tarsos, son of Areios, student of MĒNODOTOS, in DIOGENĒS LAËRTIOS 9.116 (the father being groundlessly identified with DIOSKOURIDĒS' homonymous dedicatee), Hērodotos may have been SEXTUS EMPIRICUS' teacher (Kudlien 1963: 252–253). Nevertheless, he has also been identified as a **Methodist** and an Eclectic (Scarborough 1969: 45, 155), and was perhaps influenced by **Empiricists**: he seems to have professed an etiological nihilism, at least regarding fevers (PSEUDO-GALĒN, HIST. PHIL. 19.343 K.).

Hērodotos probably worked mainly on therapeutics. His activity in this field does recall the school of Tarsos: Galēn, when he criticizes Hērodotos for his **Pneumaticist** method which proceeded by invalid logic (*Simples* 1.34, 36, 11.442–443 K.), associates him with Dioskouridēs. Oreibasios extracts three pharmacological books: (1) *Active Remedies* (*Peri tōn*

Poioumenōn Boēthēmatōn); (2) *Cathartic Remedies* (*Peri tōn Kenoumenōn Boēthēmatōn*); (3) *External Remedies* (*Peri tōn Exōthen Piptontōn Boēthēmatōn*). Galēn, referring to a second book on *Materia Medica* (*Alim. Fac.* 1 [6.516 K.]), draws material from one or more treatises, perhaps on simple medicines and compound medicines (*CMGen* 5.3 [13.788–789, 801 K.]). A papyrus contains a further fragment of his work (Marganne 1981). Hērodotos has been identified as the author of PSEUDO-GALĒN, INTRODUCTIO, probably incorrectly (Kudlien 1963: 253–254), and was also thought to have written the treatise now known as the PARISINUS MEDICUS, and a glossary containing Hippokratic data (Gossen).

M. Wellmann (1895); *RE* 15.1 (1931) 990–991 (#12), H. Gossen; J. Steudel, "Die physikalische Therapie des Pneumatikers Herodot," *Gesnerus* 19 (1962) 75–82; F. Kudlien, "Die Datierung des Sextus Empiricus und des Diogenes Laertius," *RhM* 106 (1963) 251–254; *Idem* (1968) 1098–1099; J. Scarborough, *Roman medicine* (1969); *KP* 2.1103, F. Kudlien; M.-H. Marganne, "Un fragment du médecin Hérodote: P. Tebt. II 272," *Proceedings of the Sixteenth International Congress of Papyrology* (1981) 73–78; *OCD3* 698, V. Nutton; *BNP* 6 (2005) 271, Alain Touwaide.

Alain Touwaide

Hērōn (Math.) (*ca* 410 – 460 CE)

PROKLOS' mathematics teacher, who initiated him into some mystery cult (MARINOS OF NEAPOLIS, *Vita Procli* 9; *Souda* H-552). Our Hērōn may be identifiable with HĒRŌNAS.

RE 8.1 (1912) 1080 (#6), C.R. Tittel; Netz (1997) #25.

GLIM

Hērōn (Med.) (100 – 50 BCE)

CELSUS 7.*pr*.3 lists him among surgeons, after PHILOXENOS and before AMMŌNIOS; and 7.14.2 records his four-fold categorization of tumors around the navel. ASKLĒPIADĒS PHARM., in GALĒN *CMLoc* 4.7 (12.745–746 K.), records the "Parrot" collyrium of Hērōn the oculist, containing saffron, ***glaukion***, mandrake, opium, ***sarkokolla***, and tragacanth, in rainwater; and SŌRANOS, *Gyn.* 2.5.3–4 (*CMG* 4, pp. 53–54; *CUF* v. 2, p. 7), disputes his advice that the midwife stand in a pit to assist delivery. P. CAIRO CRAWFORD 1 prefers the method of treating ocular flux (by localized incisions) used by Hērōn (and others).

Michler (1968) 63, 108–109.

PTK

Hērōn of Alexandria (*ca* 62 CE)

Date: Hērōn's dates have been the topic of extended discussion; the only explicit markers cover a 500-year span: he quotes ARCHIMĒDĒS, and is quoted by PAPPOS. Neugebauer settled the question, observing that Hērōn, in his ***Dioptra***, described a lunar eclipse visible in both Rome and Alexandria. The only eclipse fitting Hērōn's data occurred in 62 CE. Neugebauer argued that Hērōn was earlier than PTOLEMY, as he did not make use of his results, and one of his devices is described as a "new invention" by PLINY (15.5).

Work: Very little is known of Hērōn's life, but a large number of his treatises on mechanical and mathematical topics have been preserved. The *Pneumatics* (*Pneumatika*), one of Hērōn's best-known works, concerns the construction of devices driven by the properties of water, air and steam. Hērōn explains how to construct novelty drinking cups, mechanical singing birds, water-organs, a pump, medical instruments and many other devices. A long introduction gives a theoretical account of the constitution and properties of matter, and Hērōn argues that air can be compressed and expanded because it consists of small particles separated by pockets of void. This theory has been associated with STRATŌN and **Epicureanism** as well as with ERASISTRATOS, but the arguments are inconclusive.

Other treatises describe devices for entertainment and show. The *Automaton Construction* (*Automatika*) describes two automatic theaters – one stationary and one moving. At the pull of a string, the theaters deliver shows featuring moving figures and effects such as lightning or flames on an altar. One of the shows may reproduce imagery from religious processions, for instance Dionysos and dancing maenads and the pouring of libations; the other is a modified version of a show that Hērōn ascribes to PHILŌN OF BUZANTION.

Catoptrics (*Katoptrika*), which discusses reflection in mirrors, is only preserved in a Latin translation and was first thought to be PTOLEMY's *Optics*. Its authenticity has been questioned, but it almost certainly belongs to an author of Hērōn's school. Catoptrics concerns the construction of mirror devices such as a street mirror and a trick mirror showing visitors in a temple an image of a goddess where they expect to see their own reflections. A theoretical introduction explains that visual beams are emitted from the eyes and demonstrates that they are reflected at equal angles in mirrors. Hērōn describes how a mirror is manufactured and proves geometrically how beams are reflected by different mirrors: plane, convex, concave, and various cut mirrors. While offering many of the same cases as pseudo-EUCLID's *Optics*, Hērōn uses a language that indicates he is dealing with actual mirrors as well as geometrical cases.

A close relationship between geometry and mechanics is also a feature of the *Mechanics* (*Mēkhanika*), only preserved in an Arabic translation. It opens with a description of the *baroulkos* ("weight-lifter"), a box with geared wheels, which can lift a large weight with a small power; this opening section may derive from an independent treatise on the *baroulkos*. A first book concerns various mechanical principles such as geared wheels, the parallelogram of forces and other problems also treated in the ARISTOTELIAN CORPUS MĒKHANIKA. It also deals with enlarging or reducing areas proportionally and includes the famous geometrical problem of doubling the cube, i.e. finding the length of the sides of a cube that has

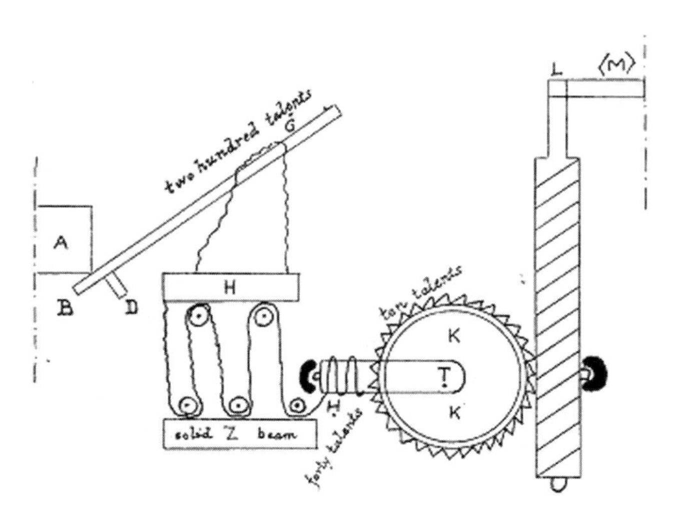

Hērōn of Alexandria: the *baroulkos* © Reproduced from Drachmann (1963) 90

double the volume. This problem cannot be solved just with ruler and compass, and Hērōn uses a special sliding ruler, thus combining geometrical and mechanical methods. The first book also contains an analysis of the center of gravity related to Archimēdēs. The second book covers the five "simple machines", i.e. mechanical principles for lifting a heavy weight with a small power: windlass, lever, pulley, wedge, screw, and combinations thereof. Lastly, Book three considers practical problems of lifting weight and applying power in, e.g., presses and cranes. Throughout the treatise geometrical, mechanical and practical solutions and advice are combined.

Artillery Construction (*Belopoiika*) describes the development from the simple belly-bow to the torsion catapult. The account is historical, and the newest catapults described were 300 years old when the treatise was written. The function of the account is therefore to describe how mechanics responds to demands and needs, rather than to provide "recipes" for artillery. In the introduction, Hērōn argues that mechanics rather than philosophy can offer a tranquil life and thus presents mechanics as a competitor to philosophy. The treatise ends with a solution to the problem of the doubling of the cube also found in the *Mechanics*, but this time introduced as a measure for scaling catapults up or down. It is used in a similar way by Philōn. A fragment of an additional treatise on hand-operated catapults, *Kheiroballista*, is also preserved.

Several of Hērōn's treatises concern measurement. ***Dioptra*** describes a surveying instrument that can measure angular distances and heights. The introduction lists many uses of the instrument ranging from astronomy and geography, over harbor and aqueduct construction, to measuring the height of a wall before a siege. The main part of the treatise primarily concerns problems of water transport, for instance how to construct a tunnel so that the teams digging from either side will meet in the middle. Hērōn also considers instruments and techniques for long distance measurement such as a road measurer and an astronomical method for measuring the distance from Alexandria to Rome.

Metrika is a more geometrical treatise on the measurement of two-dimensional figures (Book I), three-dimensional figures (Book II) and division of areas (Book III). The third book links division of areas directly to land division and states that geometry secures both equality and justice. The techniques employed in the treatises combine geometrical proofs with numerical calculations, and the treatise thus demonstrates a connection between practical mathematics and geometry. Additionally, the treatise *On Measurements* compares Egyptian, Greek and Roman standards.

A group of more purely geometrical treatises, *Geometry* (*Geōmetrika*) and *Stereometry* (*Stereometrika*), discusses two and three dimensional geometry, and *Definitions* (*Horismoi*), a geometrical handbook, gives definitions for an array of geometrical objects. These have not survived in their original form and it is hard to separate Heronic material from later additions.

Lastly a few fragments are preserved from a treatise on *Water-Clocks* (*Hudria Horoskopeia*) and a commentary on Euclid's *Elements*. There are reports of treatises on balances, vaults and astrolabes, but nothing further is known about these works.

Sources and Character of Work: Hērōn's work has, in the past, been used in a fragmentary fashion mainly as a source for a few selected technologies or for authors considered more significant. Thus Diels assigned large parts of the introduction of Hērōn's *Pneumatics* to Stratōn; whereas Drachmann and Heiberg associated most of the discussion on centers of gravity in *Mechanics* with Archimēdēs. More generally Hērōn is often viewed as a compiler, who simply preserved material from his predecessors (such as Philōn and KTĒSIBIOS), rather than proving himself a creative thinker. But fragments from other authors can only be verified in a few brief sections. Hērōn always combines material from a range of contexts in an original synthesis that supports his presentation of mechanics. He employs philosophical vocabulary in his treatises, but it is impossible to pin him down as associated with a particular author or philosophical school.

Hērōn wrote within an established tradition of mechanical writings, drawing on the work of Philōn of Buzantion and Ktēsibios who cover a similar range of topics. But while treating a recognized area of knowledge, he pushed the boundaries with other disciplines such as philosophy and geometry. The contents of Hērōn's treatises demonstrate his interest in mixing theoretical and practical approaches. He addresses both questions akin to those discussed by contemporary philosophers and geometers, and problems of constructing military engines, buildings and entertainment devices. Hērōn should thus not be seen as a purely technical writer. His treatises have many features in common with introductory works on philosophy and geometry (*eisagōgai*) and should be seen in this context as well as in the context of the craftsmanship also evident in the treatises. Hērōn blurs the boundary between theory and practice and shows how mechanical devices play a central role in resolving philosophical questions and producing a complete geometry. This is seen for instance in his claim in *Artillery Construction* that catapults can secure the philosophical aim of tranquility and in his combination of geometrical, arithmetical and mechanical methods in the *Metrika* and *Mechanics*.

Ed.: Wilhelm Schmidt, L. Nix, H. Schoene, and J.L. Heiberg, *Heronis Alexandrini Opera* 5 vv. (Leipzig 1899–1914).

RE 8.1 (1912) 992–1080, C.R. Tittel; Drachmann (1948); *Idem, The Mechanical Technology of Greek and Roman Antiquity* (1963); Marsden (1971); *DSB* 6.310–315, A.G. Drachmann and M.S. Mahoney; *OCD3* 689–699, G.J. Toomer; Karin Tybjerg, "Hero of Alexandria's Geometry of Mechanics," *Apeiron* (2004) 29–56.

Karin Tybjerg

Hērōnas (100 – 530 CE)

EUTOKIOS (*in Archimedis De Sphair.* 3.120 H.), discussing ratios (EUCLID 6.*def.*5) cites Hērōnas' commentary on NIKOMAKHOS' *Arithmētika*.

RE 8.1 (1912) 1080, C.R. Tittel.

GLIM

Hērophilos of Khalkēdōn (*ca* 280 – 260 BCE)

Physician and anatomist, born *ca* 330 BCE, resident in Alexandria *ca* 280–260 BCE, student of PRAXAGORAS, famous for his pulse-theories. Probably Hērophilos was a "medical apprentice" as documented in the HIPPOKRATIC CORPUS *Decorum* 17, in which "apprentices" are "initiated" into the profession, in contrast with mere "laymen." Hērophilos may have practiced in Athens (von Staden 1989: 38–41) before his tenure in Alexandria, and he was probably connected in some way with the *Mouseion* (Museum), whose complex of buildings included the famed library (Fraser 1972: 1.312–335 with texts, 2.467–461). Most medical tracts from the *Mouseion* were commentaries on earlier writings, earning the jibe that Alexandrian scholars were "well-fed bookish birds in a gilded cage" (TIMŌN OF PHLEIOUS, in Ath., *Deipn.*, 1d [epitome]). A strong contrast was the career of Hērophilos, whose dissections of the human cadaver at the *Mouseion* (Longrigg 1993: 179), along with those by ERASISTRATOS, established details about the human body that were standard until the European Renaissance. A single ancient source (all others are derivative) claims that the Alexandrian anatomists vivisected living men (CELSUS, *pr.*23–26), which leaves this aspect of Alexandrian anatomy and physiology in the realm of controversial uncertainty (Scarborough 1976: 11; von Staden 1989: 29–30, 144–153).

None of Hērophilos' writings has survived intact, so one depends on excerpts (sometimes lengthy, sometimes short) in GALĒN, RUFUS OF EPHESOS, SŌRANOS, and others, diligently gathered, edited, translated, and analyzed by von Staden. Quotations from six or seven works survive: (1) *On Anatomy* in at least four books (T63–129 and 136–142 von Staden), which considered dissection techniques, the liver, anatomy of the brain and features of the skull, the nerves and their function after generally exiting from the brain and spinal column, the eye and its chiasmatic pair of optic nerves, bones and cartilages of the larynx including the hyoid bone and styloid process, the viscera (stomach, parts of the small intestine, the large intestine and its omentum), male and female reproductive organs, anatomy of the veins and arteries, and perhaps the bones in general (represented by a short fragment from Rufus' *Anatomical Nomenclature*); (2) *On Pulses* ("vascular physiology": T144–188); (3) *Midwifery* or *Gynecology* (T193–204 with 105–114 [female reproductive anatomy] and 247 [abortion]); (4) *Against Common Opinions* (203–204), an innovative collection of doxographical and historiographical information about earlier Greek medico-philosophical tenets (von Staden 1999: 144–149); (5) *Regimen* (T230, probably linked with the *Gymnastics* of T227–229); *Therapeutics* (T231–259, with 248–259 compound drugs as devised by Hērophilos); and (6) *On Eyes* (T260 from AËTIOS OF AMIDA 7.48). Hērophilos may have written an uncertainly-titled *Hippokratic Exegesis* (T261–275).

Hērophilos' discovery of the anatomy and function of the nerves ranks as a signal achievement in medical history: not only did he demonstrate structural affinities among parts of the brain and the sprouting from the brain of ten of the cranial nerves, he was also the first to distinguish between what we call "sensory" and "voluntary" or "motor" nerves (Solmsen 1961: 185); he was the first to observe the 4th ventricle of the brain, and the surface convolutions ("inward foldings") of the cerebrum, noting that humans had many more than animals; he was the first to describe accurately the human liver and its accompanying vessels; he discovered the ovaries in the female, likening them by analogy to the male testicles, and he seems to have observed the Fallopian tubes but did not deduce their reproductive function; he traced a good portion of the vascular system, described the valves in the heart as well as its four chambers, described the venous cavities of the skull (the largest of the sinuses at the base of the skull still bears his name), and his observations, dissections (discovering that the eye had four membranes) and theories about the eye, its twin optic nerves that crossed each other at a point we still call the *optic chiasma*, led Hērophilos to think that sight and vision were conducted by the ***pneuma*** in these large, possibly tubular, ***neura*** – a "transmission" theory of sensation that had millennially long influence (the ***pneuma*** ultimately derived from breathing). Having observed that the lungs contract and dilate in a quadripartite cycle, he could posit that respiration occurred from the normal habit of the lungs to do so. In *On Pulses*, Hērophilos formulated the notions of *diastolē* ("dilation") and *sustolē* ("contraction") among the arteries, resulting from the *dunamis* ("power" or "faculty") of the heart which thereby pulled into the arteries a mix of blood and ***pneuma*** (Harris 1973: 180), while the veins had only blood.

Hērophilos had a genius for coining anatomical names, and some are still standard in anatomy today. He named part of the small intestine the *dōdekadaktulon*, Latinized as *duodenum*, i.e., named "...from being about equal in length to the breadth of twelve fingers (25 cm.)" (Gray 1959: 1,278). He called another part of the small intestine the *nēstis* or "fasting" intestine, *ieiunium* in Latin, the "empty" bowel, learned by modern medical students as the *jejunum* (*cf.* "jejune"). Finely woven nets resembled cobwebs, so that when Hērophilos described the back of the eye as being *arakhnoeidēs* ("like a cobweb") or *khitōn amphiblēstroeides* ("a net-like tunic"), this becomes Latin's *retina*, which retains Hērophilos' analogy to fish-nets. This passion for terminologies and labeling is characteristic of Hellenistic taxonomy as applied to medicine, zoology, botany, philosophy, rhetoric and educational theory, so that nomenclature in medicine provides its relation to structure ("what does it look like?"), function ("what does it do?"), and treatment ("how can one restore a healthy state?"). Categories, analogies, naming things, ordering things, relating things, and dividing things pervaded the works of ARISTOTLE (and *cf.* PLATO *Soph.* and SPEUSIPPOS), emphasizing relationships of parts to one another, genus and species, classifications by differences and similarities (the zoological books), and the context of models in correlation joined with analogy (Elements, Qualities, and **Humors**). Hērophilos' knowledge of "healthy things" mirrors the notion presented in the Hippokratic Corpus, *Nourishment*: food has *dunamis*, good or bad only in relation to something; and **Stoic**-type terminologies are definitive for Hērophilos, who divided the Art of Medicine into three parts: knowledge of the disease, knowledge of health, and neutral things, and thus semiotics became *trikhronos sēmeiōsis* (a "three-phased inference from symptoms [or signs]").

Vascular physiology likewise employed such analogies, using poetic meter to suggest the relation between diastole and systole as reflecting the age of a patient: the pyrrhic, trochaic, spondaic, and iambic rhythms correspond respectively to infant, child, adult, and elderly, each with an up-beat and down-beat (*arsis* and *thesis*), analogous to rhythms in music and poetry, beginning with a *prōtos khronos aisthētos* ("a first perceivable unit of time"), the dilation of an artery in a newborn (von Staden 1989: T174, 183). Diagnostics received colorful names for pulse-rates ("meters" as "frequencies"), each suggestive by analogy to the disease and age of the patient: a eunuch's pulse is *dorkadizōn*, Latinized as *caprizans*, "leaping like a gazelle," another's pulse is *murmēkizōn*, Latin *formicans*, "crawling like an ant" (T163a, 163b, with 169, 170, and 180). Striking is Hērophilos' clinical application of his pulse-theories by his construction of a small portable water clock (a *klepsudra*) as a combination thermometer and adaptable timer, assuming the correlation of the pulse rates to a patient's presentation of fever (viz. the greater the frequency, the higher the temperature). Hērophilos used his water clock when he felt the pulse of feverish patients, then adjusted the clock for the patient's age, "...by as much as the movements of the pulse exceeded the number that is natural for filling up the recalibrated clock, by that much he also stated the pulse too frequent, viz. the patient had either more or less of a fever" (T182; trans. von Staden, with minor changes; *cf.* Longrigg 1993: 204). Modern authorities in the history of technology generally agree that Hērophilos' *klepsudra* is mechanically feasible (Thompson 1954: 37–38; Brumbaugh 1966: 68–73; Fraser 1972: 2.518, n. 113; West 1973).

Ed.: von Staden (1989).

A. Souques, "Que doivent à Hérophile et à Erasistrate l'anatomie et la physiologie du système nerveux?" *Bulletin de la Société française d'Histoire de la Médecine* 28 (1934) 357–365; *Idem*, "Connaissances neurologiques d'Hérophile et d'Erasistrate," *Revue Neurologique* 63 (1935) 145–176; H.A. Thompson, "Excavations in the Athenian Agora: 1953," *Hesperia* 23 (1954) 31–67; R.S.

Brumbaugh, *Ancient Greek Gadgets and Machines* (1966); Harris (1973) 177–233; S. West, "Cultural Exchange over a Water-Clock," *CQ* 23 (1973) 61–64; *DSB* 6 (1975) 316–319, J. Longrigg; H. von Staden, "Experiment and Experience in Hellenistic Medicine," *BICS* 22 (1975) 178–199; John Scarborough, "Celsus on Human Vivisection at Ptolemaic Alexandria," *CM* 11 (1976) 25–38; P. Potter, "Herophilus of Chalcedon: An Assessment of his Place in the History of Medicine," *BHM* 50 (1976) 45–60; Longrigg (1993) 177–219; von Staden (1999); V. Nutton, "Alexandria, Anatomy and Experimentation" in *Ancient Medicine* (2004) 128–139.

John Scarborough

Hēsiod of Askra (*ca* 750 – 650 BCE)

Linked with HOMER as a composer of early epic. In *Theogony* he names himself as "Hēsiod" (*Hēsiodos*) and describes his visitation by the Muses while pasturing his sheep on Mount Helikon (22–34). The Muses help the poet sing the genealogy of the gods, especially Zeus' birth and his consolidation of power over gods and humans. Hēsiod addresses *Works and Days* to his brother, Persēs, who allegedly cheated Hēsiod of his share of their father's inheritance (37–41). Now destitute, Persēs pursues a legal dispute with Hēsiod and receives guidance about the need for justice and hard work. *Works and Days* also relates that Hēsiod's father hoped for prosperity as a professional sea trader. However, poverty compelled him to leave Aiolian Kumē and to settle in Askra in Boiōtia where he resigned himself to the difficult but more reliable occupation of farming (633–640). Unlike his father, Hēsiod took only one sea voyage from Aulis to Khalkis, where he won a poetic competition at the funeral games of Amphidamas (654–662). Ancient testimony and modern archaeological evidence have placed Amphidamas' funeral and Hēsiod's victory in 730–700 BCE. Hēsiod's claim of poetic victory in *Works and Days* may also be the source of the *Certamen Homeri et Hesiodi*.

Homer and Hēsiod drew from a common tradition of Ionian hexameter poetry. However, Hēsiod's poems differ in content and perhaps incorporated separate traditions, such as Near-Eastern succession myths and didactic "wisdom" texts. *Theogony* and *Works and Days* are regarded as genuine compositions of Hēsiod. *Theogony* begins with a hymn to the Muses (1–104) and explains the origin of the physical world. It details the relationship of Zeus and the Olympian gods to the earlier Titan gods and other primordial divinities and describes Zeus' birth and ascent to power, establishing the nature of his rule over gods and mortals as one based upon justice and mutual alliances. *Works and Days* recounts the world's decline and the reason why mortals now must endure painful labor for survival. It stresses the importance of just and independent work in accordance with Zeus' designs. Hēsiod's poems influenced the Pre-Socratics.

Works and Days reprimands Persēs for his idleness and dishonesty in bribing "gift-devouring" kings to ratify his theft of Hēsiod's property. Persēs must abandon bad strife and adhere to good strife, which promotes just and honest work, productive envy of others' successes, and healthy competition (11–26 "corrects" *Theogony* 225 and marks *Works and Days* as the later poem). Hēsiod's moral instruction incorporates myths (Promētheus' crimes and Pandōra's creation establish the need for work among mortals, 42–105; the Myth of Ages charts the world's decline to the present Iron Age, 106–201); fable (the story of the hawk and nightingale elevates humans above animals by the presence of justice, 202–212); injunctions (warning Persēs and the "kings" about *dikē* and *hubris*, 213–285); and allegory (on the importance of hard work, 286–334). There are additional injunctions about respect for the gods, honesty, friendship, and reciprocity (335–382). Hēsiod's advice on farming (the poem's "Works") is addressed to Persēs and the peasant farmers of Boiōtia. The poet instructs them on fall woodcutting and plowing (383–492); winter protection from cold (493–563); and spring vine-pruning, harvest, threshing, and summer drinking and relaxation (564–617). Hēsiod emphasizes timeliness and perception in his agricultural calendar, incorporating many elements of the natural world: stars (Sirius, Pleiades, Orion, Arcturus), winds (Boreas, Notus, Zephyr), heat, cold, rain, drought, birds (crane, cuckoo, swallow, crow), insects (cicada), oxen, crops, and fields. Sailing is a dangerous and risky occupation which the farmer should pursue only in summer after amassing a surplus of livelihood (618–694). The poem ends with a list of auspicious and inauspicious "Days" (765–828) for given tasks. Hēsiod's *Works and Days* is unique in its focus upon the ordinary Boiōtian farmer, whose agricultural toil, while necessary for survival, enables him to gain a special understanding of Zeus' justice and to find his path toward wealth and economic self-sufficiency. Recent scholarship focusing upon the unity and artistry of *Works and Days* has been fruitful.

Ed.: M.L. West, *Theogony* (1966); *Idem, Works and Days* (1978); Gr. Arrighetti, *Esiodo: Opere* (Turin 1998). G.P. Edwards, *The Language of Hesiod* (1971); W.J. Verdenius, *A Commentary on Hesiod Works and Days vv. 1–382* (1985); R. Lamberton, *Hesiod* (1988); R. Hamilton, *The Architecture of Hesiodic Poetry* (1989); R.M. Rosen, "Poetry and sailing in Hesiod's Works and Days," *ClAnt* 9 (1990) 99–113; J.C.B. Petropoulos, *Heat and Lust* (1994); *OCD3* 700, M.L. West; *ECP* 267–269, R. Lamberton; *REP* 4.412–413, G.W. Most; S. Nelson, *God and the Land* (1998); Maria Marsilio, *Farming and Poetry in Hesiod's Works and Days* (2000); E.F. Beall, "The Plow that Broke the Plain Epic Tradition: Hesiod *Works and Days*, vv. 414–503," *ClAnt* 23.1 (2004) 1–32; J.S. Clay, *Hesiod's Cosmos* (2003); A.T. Edwards, *Hesiod's Ascra* (2004); *BNP* 6 (2005) 279–284, Gr. Arrighetti; E.F. Beall, "An Artistic and Optimistic Passage in Hesiod: *Works and Days* 564–614," *TAPA* 135 (2005) 231–247.

Maria Marsilio

Hestiaios of Perinthos (365 – 325 BCE)

Student of PLATO, according to DIOGENĒS LAËRTIOS (3.46) and PHILODĒMOS. IŌANNĒS STOBAIOS calls him a physicist, and gives his view on the basis (*ousia*) of time (AËTIOS 1.22.3), *viz.* the movement of heavenly bodies relative to one another. Stobaios 4.13.5 also reports Hestiaios' view on the mechanism of sight. Here he opted for an explanation apparently indebted to EMPEDOKLĒS, combining the Pre-Socratic idea of representations (*eidōla*) emanating from objects (4.13.1) with the **Academic** idea of light-rays (*aktines*) stemming from the eye, and reverting to it (4.13.3). He spoke of ray-representation (*aktineidōla*). The significance of his innovations is unclear. THEOPHRASTOS (*Metaphysics* 13) speaks as if Hestiaios went some way towards generating the wider universe from mathematical principles, a project of XENOKRATĒS.

BNP 6 (2005) 287, K.-H. Stanzel.

Harold Tarrant

Hēsukhios of Damaskos (*ca* 410 – 470 CE)

Physician, father of IAKŌBOS PSUKHRESTOS (*Souda* I-12), traveled widely to Rhodes, and Argive Drepanon where he married Iakōbos' mother, then leaving his family, he spent 19 years in Alexandria and Italy, his family thinking him dead. His son, joining him in Constantinople, became his student (DAMASKIOS, *Vit. Isid.* in Phōtios, *Bibl.* 242, §120, and

Phil. Hist. 84A-D, pp. 206–208 Athan.). Hēsukhios scorned the physicians at Constantinople (121), prescribed purges, cold baths, and strict diets, avoiding surgery, phlebotomy, and cautery (122) – apparently a **Pythagoreanizing** doctor.

RE 8.2 (1913) 1317 (#6), O. Seeck; *PLRE* 2 (1980) 554 (#8).

GLIM

Hic- ⇒ Hik-

Hierax of Thēbai (250 BCE – 25 CE)

Asklēpiadēs Pharm., in Galēn *CMLoc* 4.8 (12.775–776 K.), records Hierax' treatments for trachoma, comprised of ***misu***, saffron, opium, "hematite," roasted copper, myrrh, and gum applied in very sour vinegar. Celsus records a simpler "efficacious" treatment, compounded of myrrh, ***ammōniakon*** incense, and **verdigris** filings (6.6.28). Designating Hierax as "Theban," Asklēpiadēs also recounts his remedy efficacious against various skin ailments including pimples, night pustules, nasal sores, chapping, and scars: containing ***psimuthion***, **litharge**, alum, *halikababon* ("winter cherry"), ***khalkanthon***, **Sinōpian earth**, mixed with vinegar (*CMGen* 5.11 [13.829 K.], *cf. CMLoc* 1.8 [12.489 K.] on skin diseases). The "Theban" to whom is attributed a medicament compounded of **litharge**, old olive oil, ***khalkanthon***, white chameleon, birthwort, **galbanum**, and frankincense (by Asklēpiadēs Pharm. in Galēn *CMGen* 4.13 [13.739 K.]) may be our pharmacologist. The name *hierax*, a homonym of both a hawk and a type of bandage (*cf. hierakion*: 12.783 K.), is attested from the 4th c. BCE to the 3rd c. CE (*LGPN*).

RE 8.2 (1913) 1411 (#12), H. Gossen.

GLIM

Hierios (of Alexandria?) (*ca* 285 – 362 CE)

Philosopher and friend of Pappos, who sought Pappos' opinion regarding a "plane" method devised by an unnamed geometer to solve the problem of the two mean proportionals between two straight lines (Pappos, *Coll.* 3.43.3). Probably identifiable with the homonymous student of Iamblikhos and teacher of Maximus (*cf.* Ammōnios of Alexandria, *In Prior Analytics, CAG* 4.6 [1899] 31), and the philosopher mentioned by Libanios (*Ode* 14.7, 32, 34): Jones 1986: 4, n. 9.

RE 8.2 (1913) 1458–1459 (#8–9), K. Praechter; Netz (1997) #21.

GLIM

Hieroklēs (Geog.) (*ca* 450 – 535 CE)

Author of the *Sunekdēmos*, a list of 64 provinces (*eparkhiai*) and 923 cities belonging to the Eastern Empire in the early Byzantine era. The names are arranged geographically, and the text reflects errors and lacunae. The provincial organization strongly suggests a date before the reforms of Justinian I in 535/536: only one of the 27 cities renamed *Ioustinianoupolis* or *Ioustiniana* in the emperor's honor is here mentioned. However, the work, seemingly a revision of a secular administrative document from the mid 5th c., includes much information from the time of Theodosius II (408–450). The work, comparable with Geōrgios of Cyprus' *Descriptio orbis romani*, was one of the most important sources for Constantine VII Pophurogennētos' 10th c. *De thematibus*.

A.H.M. Jones, *The Cities of the Eastern Roman Provinces* (1937) 514–521; *KP* 2.1133–1134, H. Gärtner; *HLB* 1.531; *Tusculum-Lexikon* (1982) 338; *ODB* 930, T.E. Gregory; *PLRE* 3 (1992) 597.

Andreas Kuelzer

Hieroklēs (Veterin.) (300 – 400 CE?)

Wrote an unnamed veterinary text, included in the *Hippiatrika*, later reconstructed and published. Hieroklēs was, according to his preface, a solicitor wishing to compile a book on equine veterinary medicine. The resulting discourse consists of two books, of which both prefaces have survived intact. Hieroklēs relies heavily on Apsurtos, although he also cites Hierōn, Kleomenēs the Libyan, and Stratonikos, as well as making use of Aristophanēs of Buzantion and the Quintilii. Hieroklēs does not, however, cover as much equine ground as Apsurtos, stating in the first preface that he will only discuss equine diseases and cures, as his reader already knows about breeding and anatomy.

While the body of Hieroklēs' text straightforwardly restates Apsurtos (littered with such sayings as: "Apsurtos recommends"), the prefaces reveal literary pretensions. His style throughout is Atticizing, and each book of the treatise is adorned with a preface whose rhetorical flourishes contrast with the plainer style used in the body of the text. Hieroklēs discards Apsurtos' epistolary form which, by its nature, contains information irrelevant to the topic at hand – the care and curing of horses. He includes no information based on first-hand experience, but faithfully relays the opinions of his sources. The resulting style is smoother and more unified than Apsurtos and other parts of the *Hippiatrika.*

This smoothness and the appeal of the style probably inspired the reconstitution of Hieroklēs' work by later students of equine medicine, which was the primary conduit of Greek equine medicine to the west. This work, smaller than the *Hippiatrika* and more accessible than Apsurtos, was the basis for several medieval veterinary expositions, the most notable of which is that of Giordano Ruffo, a 13th c. resident of the Sicilian court, later translated into six languages. A Latin translation of Hieroklēs was made by Bartholomew of Messana for King Manfred of Sicily (reigned 1258–1266).

Björck (1944); K.-D. Fischer, "A horse! A horse! My kingdom for a horse! Versions of Greek Horse Medicine in Medieval Italy," *MHJ* 34 (1999) 123–138; McCabe (2007) 208–244.

Jennifer Nilson

Hieroklēs of Alexandria (*ca* 100 – 120 CE)

Stoic philosopher with geometrical approach to ethics, frequently cited by the **Platonist** Tauros for opposition to **Epicurean** hedonism (Gellius 9.5.8). A papyrus fragment (*PBerol.* 9780) seems to preserve the introduction to his *Elements of Ethics*, further extracted by Iōannēs Stobaios. The papyrus illustrates how the **Stoics** sought the foundation of ethics in animal instinct for self-preservation (*oikeiōsis*). Observing how animals behave in relation to themselves and their environment proves self-perception: animals must perceive body parts (wings or horns) to use them effectively, and must assess strength or vulnerability of body parts (a tortoise withdraws into its shell for protection). Hieroklēs further treats how one species behaves towards another (lions are circumspect regarding bulls' horns but unconcerned with the rest of the body). Animals' behavior towards other species seems to imply a comparative valuation (perception) of assets and weakness relative to other species. Hieroklēs subscribed to the common **Stoic** view that the soul enters the body at birth, extending this to animals as well. Stobaios' excerpts (4.672 W.-H.) treat actions proper to human nature, including behavior towards parents, reverence of the gods and homeland. In his most striking passage, Hieroklēs offers an image of concentric circles to illustrate the expansion of concern for others.

KP 2.1133 (#5), Günther Schmidt; G. Bastianini and A.A. Long, *CPF* I.1.3 (1992) 268–451; A.A. Long, "Hierocles on oikeiōsis and self-perception," *Stoic Studies* (1996) 250–263; *OCD3* 704–705, J. Annas; *ECP* 269–270, A.A. Long; *DPA* 3 (2000) 686–688, R. Goulet.

GLIM

Hierōn (Veterin.) (250 BCE – *ca* 400 CE)

Source of a remedy for ***elephantiasis*** preserved in the *Hippiatrika* (*Hippiatrica Berolinensia* 3.3). The remedy appears, without attribution, in the Einsiedeln MS of PELAGONIUS, E 529 *bis* (Corsetti 55).

P.-P. Corsetti, "Un nouveau témoin de l'Ars veterinaria de Pelagonius," *Revue d'histoire des textes* 19 (1989) 31–56; J.N. Adams, "Notes on the Text, Language, and Content of Some New Fragments of Pelagonius," *CQ ns* 42 (1992) 490–493; McCabe (2007) 102, 109, 168.

Anne McCabe

Hierōn of Soloi (Cyprus) (*ca* 325 – 320 BCE)

Pilot (*kubernētēs*) of Alexander, who sailed down the Persian gulf and along the Indian Sea-coast of Arabia, as recorded by ARISTOBOULOS OF KASSANDREIA *fr.*55.20.7–8 = ARRIANUS *Indika* 7.20.7–8.

RE S.4 (1924) 743 (#17a), H. Berve.

PTK

Hierōn II of Surakousai (*ca* 270 – 216 BCE)

Enlightened despot of Surakousai who was the patron of ARCHIMĒDĒS and sponsor of his engineering projects. He paid close attention to Sicily's agricultural production and established laws regulating the tithing of grain which were still in force over a century later. VARRO (*RR* 1.1.8) knew, through CASSIUS DIONUSIOS, an agricultural treatise under his name, which may have discussed cereals, livestock, poultry, viticulture, and arboriculture (*cf.* PLINY 1.*ind*.8, 10, 14–15, 17–18).

H. Berve, *König Hieron II* (1956); *OCD3* 705–706, B.M. Caven.

Philip Thibodeau

Hierōnumos of Kardia (325 – 250 BCE)

STRABŌN thrice records geographic observations or descriptions from an anethnic Hierōnumos, generally identified as the Kardian historian. That man served Antigonos "One-Eyed" (316–301 BCE), then Dēmētrios "Besieger" (301 to after 291), and wrote a lost history from the death of Alexander to the death of PURRHOS; he died aged 104 years. Strabōn's Hierōnumos described Corinth as had EUDOXOS OF KNIDOS (Strabōn 8.6.21 = *FGrHist* F16) described Thessalia, as preserved in ARTEMIDŌROS OF EPHESOS (Strabōn 9.5.22 = *FGrHist* F17), and gave the size of Crete, likewise (10.4.3 = *FGrHist* F18).

FGrHist 154; *BNP* 6 (2005) 316–317 (#6), K. Meister.

PTK

Hierōnumos of Rhodes (260 – 230 BCE)

Scholarch of the **Peripatos**, and writer of biographies. PLUTARCH, *Quaest. Conv.* 1.8 (626 A–B) credits him with a theory of vision based upon corpuscular (non-**atomic**) emission.

T.S. Ganson, "Third-century Peripatetics on Vision," and P. Lautner, "The Historical Setting of Hieronymus *fr.*10 White," in Fortenbaugh and White, *RUSCH* 12 (2004), 355–362 and 363–374; *BNP* 6 (2005) 317 (#7), H.B. Gottschalk.

PTK

Hierophilos Sophistēs (550 – 1050 CE)

Wrote a dietetic calendar, often dated to the 12th c; however, he might be earlier (7th/9th c.). A poem on the same topic by Theodōros Prodromos (*fl. ca* 1130) shows close similarities to Hierophilos' work and establishes a terminus, whereas a Syriac calendar also shows close parallels, allowing a date as early as the 6th c., when Syriac medical literature was first translated from Greek. Although the epithet *sophistēs* (physician) recalls the late antique medical milieu, both name and epithet may be apocryphal, added to a previously anonymous work to recall HĒROPHILOS, whose work included dietetics. Whatever its period and origin, this typically Byzantine text is mainly based on GALĒN's system of health preservation, with the dietary properties of food and all the human activities that might affect health: bathing, unguents, and sexual intercourse.

Ed.: J.Fr. Boissonade, in *Notices et extraits des manuscrits de la blibliothèque du roi et autres bibliothèques* 11 (1827) 178–273; *Idem*, in *Anecdota Graeca e codicibus regiis* 3 (1831) 409–421; Ideler 1 (1841/1963) 409–417; A. Delatte, *Anecdota atheniensia et alia* 2 (1939) 456–466.

Ch. Daremberg, in *Archives des missions scientifiques et littéraires* 3 (1854) 19–20; Diels 2 (1907) 49; L. Oeconomos, *Actes du VIème Congrès International d'Études Byzantines* (1950) 1.169–179; M. Formentin, *I codici greci di medicina nelle tre Venezie* (1978) 83.

Alain Touwaide

HIEROTHEOS (ALCH.) ⇒ HĒLIODŌROS

Hikatidas (200 BCE – 75 CE)

PLINY 28.83–84 records that Hikatidas claimed sex relieved quartan fevers (*cf.* AUFIDIUS) – so long as the woman is commencing menstruation. The name is recorded once elsewhere, *LGPN* 1.234. *Cf.* perhaps HIPPOSIADĒS or HURIADAS.

Fabricius (1726) 253, s.v. Icetidas.

PTK

Hikesios (325 BCE – 75 CE)

Known to PLINY (1.*ind.*14–15, 14.120) as the author of a technical treatise on winemaking. Pliny seems to distinguish him from the doctor of Smurna, but an identification cannot be ruled out.

RE 8.2 (1913) 1593–1594 (#5), H. Gossen.

Philip Thibodeau

Hikesios of Smurna (120 – 80 BCE?)

In his terse summary of the medical instructors assembled at Mēn Karou (founded as a school by ZEUXIS), STRABŌN (12.8.20) adds as an aside, ". . . the **Erasistratean** teaching center was established by Hikesios in Smurna." Athēnaios, *Deipn.* 3 (87b), notes that Hikesios was a "follower of ERASISTRATOS," and Strabōn's clipped notice (the text seems badly corrupted in part [Syme 1995: 344–347]) indicates that the Smurnean school did not last very long. DIOGENĒS LAËRTIOS 5.94 lists the eighth Hērakleidēs, *iatros tōn apo Hikesiou*, probably "a student of Hikesios" (number nine being the famous **Empiricist** physician, HĒRAKLEIDĒS OF TARAS). Scattered in Athēnaios' *Deipnosophists* are numerous fragments of Hikesios' *Peri hulēs*, rendered by Gourevitch as *On Materials for Health* (2000: 490–491), i.e., foods as remedies, especially fish; but many of these remnants as quoted are fused with other bits, often difficult to separate from one another. PLINY's references are condensed (Gourevitch 2000: 484), and the single notice of Hikesios as a surgeon occurs in Tertullian's well-known condemnation of pagan anatomists for their destructive practices with embryos and fetuses (*De anima* 25.6 Waszink).

Though known for his medical dietetics, later physicians more highly respected Hikesios' pharmaceutical compounds, evinced by Hikesios' *Melaina* ("The Black One": HĒRĀS in GALĒN, *CMGen* 5.2 [13.780–781 K.]) and an all-inclusive, multiple-use plaster (KRITŌN in Galēn, *CMGen* 5.3 [13.787–788 K.; *cf.* 13.809–810: "ANDROMAKHOS' Hikesian *Melaina*"]), on which Galēn offers his extended critical commentary, *viz.* on changes by Hērās and Kritōn, and the properties of the ingredients (13.788–794 K.). Hikesios as a good **Erasistratean** was a keen student of herbal, mineral, and entomological pharmacology, and Galēn's sometimes acidic remarks about alterations in Hikesios' recipes demonstrate the long term respect for **Erasistratean** drug lore. PAULOS OF AIGINA 3.64 (*CMG* 9.1, p. 281) recommends the "Hikesian Plaster" (without formula) in the treatment of external hardening of the uterus; then among extracts from ANTULLOS (7.17.1, *CMG* 9.2, p. 347) he combines the extant recipes, above, into the "Plaster of Hikesios: For Scrofulas, Abscesses, the Spleen, Joints, and Ailments of the Hips/Sciaticas" (7.17.45, *CMG* 9.2, p. 359]). Paulos' 6th c. streamlining of the recipes still carrying Hikesios' name reduced the 19 substances debated in the 2nd c. to twelve: dropped from the Hērās-Andromakhos-Kritōn revisions are bitumen, Ampelitidian earth, alum, powdered frankincense, and honey, whereas retained are **litharge**, old olive oil, *propolis* ("bee glue"), beeswax, vinegar, and **verdigris**; Paulos augments with pine bark, pseudo-mastic gum (*Atractylis gummifera* L.), horseheal (*Inula helenium* L.), ***purethron, euphorbia***, and the juice of the parasitic *hupokistis* (*Cytinus hypocistis* L.). The bee glue, beeswax, and pseudo-mastic made this plaster sticky, while the horseheal, ***verdigris, euphorbia*** and *Cytinus* provided properties that were bactericidal, and the ***purethron*** was lightly anesthetic and insecticidal. Especially effective against microorganisms of many kinds was/is the *propolis*, one of the most hypertonic natural substances known.

RE 8.2 (1913) 1593–1594 (#5), H. Gossen; R. Syme, *Anatolica: Studies in Strabo* (1995); Dueck (2000) 142; D. Gourevitch, "Hicesius' Fish and Chips," in D. Braund and J. Wilkins, edd., *Athenaeus and his World: Reading Greek Culture in the Roman Empire* (2000) 483–491.

John Scarborough

Hiketas of Surakousai (*ca* 400 – 350 BCE)

Belongs to a group of later **Pythagoreans** active in Surakousai in the first part of the 4th c. (DK 50–51, 55). An astronomical hypothesis, ascribed to him, is identical to that of his countryman Ekphantos: the Earth rotates about its own axis in 24 hours, whereas the diurnal rotation of the heavens is only apparent (A1). This was a modification of Philolaos' system; Hiketas could have been his follower, though not necessarily a pupil. Theophrastos refers to Hiketas' theory which therefore must have been put into writing.

DK 50; T.L. Heath, *Aristarchus of Samos* (1913); Dicks (1970).

Leonid Zhmud

Hilarius of Arles (425 – 450 CE)

The nobly-born and liberally-educated bishop of Arles, among other writings, composed verses on the hot spring at Grenoble; Gregory of Tours preserves one quatrain.

FLP 454; *OCD3* 706–707, P. Rousseau.

PTK

Hildebald ⇒ Heldebald

Himerios (Veterin.) ⇒ Euhēmeros (Veterin.)

Himilkōn (of Carthage?) (520 – 480 BCE?)

Wrote a ***periplous*** of his voyage up the Atlantic coast of Iberia to the Kassiterides, fragments of which describe the shallow Ocean: Pliny 2.169 and Auienus, *OM* 117–129, 380–389, 402–415.

BNP 6 (2005) 332 (#6), L.-M. Günther.

PTK

Hipparkhos (Veterin.) (before *ca* 350 CE)

Quoted by Pelagonius of Salona on evaluating stallions for stud. The passage is preserved in Greek in the *Hippiatrika* (Pel. 3 = *Hippiatrica Parisina* 85 = *Hippiatrica Berolinensia* 14.10).

Fischer (1980) 3.

Anne McCabe

Hipparkhos of Nikaia (*ca* 140 – 120 BCE)

Astronomer, astrologer, geographer, without a doubt the most important of Ptolemy's predecessors, and central in incorporating Babylonian astronomy into Greek mathematical astronomy. Unfortunately eclipsed by Ptolemy's *Almagest*, most of Hipparkhos' works have not survived. His short commentary on the *Phainomena* of Aratos and Eudoxos survives, but all other evidence is secondary. He wrote at least one book (possibly two) on the fixed stars, including material excerpted in later ***parapēgmata***, and he is also known to have compiled a star catalogue, long thought the basis of Ptolemy's but now generally seen as independent. Indeed, parts of the *Almagest* are so indebted to Hipparkhos that it is sometimes difficult to distinguish Hipparkhos' material from Ptolemy's. Consequently, some scholars have underemphasized the scale and nature of Ptolemy's achievement, but generally a considerable degree of difference is now acknowledged between Ptolemy and Hipparkhos, in spite of important lines of dependence. Clearly Hipparkhos seems not to have written a systematic astronomical treatise to prefigure the *Almagest*, but he instead did much foundational work that Ptolemy would later use in his systematization of mathematical astronomy. Considerable evidence suggests that Hipparkhos used both (Greek) geometrical and (Babylonian) arithmetical methods in his astronomical calculations. Hipparkhos is believed to have published a collection of eclipse observations spanning 600 years (thus going back to the 8th c. BCE) and including much Babylonian material that Hipparkhos himself may have collected in Babylōn. Indeed, when Kugler discovered in 1900 that Hipparkhos' very precise value for the mean synodic month (in sexagesimal notation: 29; 31,50,08,20 days) was actually borrowed from the Babylonian "System B" for the Moon, only then did we recognize the deep indebtedness of Greek mathematical astronomy to Babylon. Hipparkhos' published eclipse records were an invaluable resource for Ptolemy's lunar theory. Ptolemy's solar theory is also deeply indebted to Hipparkhos, including the wholesale use of his values for the lengths of the seasons and solar year.

Hipparkhos is perhaps best known for his discovery of the precession of the equinoxes, the very slow (and very difficult to observe) movement of the equinoctial points relative to the fixed stars. His fundamental work on parallax allowed accurate prediction of solar eclipses for the first time. He is also known to have been innovative in developing astronomical instruments, including a *dioptra*, an accurate star globe, and possibly even the plane astrolabe.

Later writers praise Hipparkhos for his skill in astrology, and PLINY (2.95) reports that no-one had done so much as Hipparkhos to establish clearly the connection between human souls and the stars. Unfortunately, however, virtually nothing of the details of his astrology survive. He also seems to have written something on combinatoric logic, criticizing CHRYSIPPUS.

His attested lost works include: *On the Movements of the Solsticial and Equinoctial Points, On the Length of the Year, On Intercalary Months and Days, On the Risings of the Twelve Zodiacal Signs, Treatise on Simultaneous Risings, On Sizes and Distances, On the Moon's Monthly Motion in Latitude, On Things Carried down by their Weight*, and *Against the Geography of ERATOSTHENĒS* (most of our knowledge of which comes from STRABŌN). Hipparkhos also wrote on chords in 12 (implausibly long) books, according to THEŌN OF ALEXANDRIA (*In Ptol. Synt.* 1.10). Two other titles commonly found in modern sources, *On the Length of the Month* and *On Matters Pertaining to Straight Lines in the Circle*, are essentially fabrications of Albert Rehm.

Ed.: K. Manitius, *Hipparchi in Arati et Eudoxi phaenomena commentariorum libri tres* (1894).

Neugebauer (1975) 274–343; G.J. Toomer, "Hipparchus and Babylonian Astronomy," in E. Leichty *et al.* (1988) 353–362; A. Jones, "The Adaptation of Babylonian Methods in Greek Numerical Astronomy," *Isis* 82 (1991) 441–453.

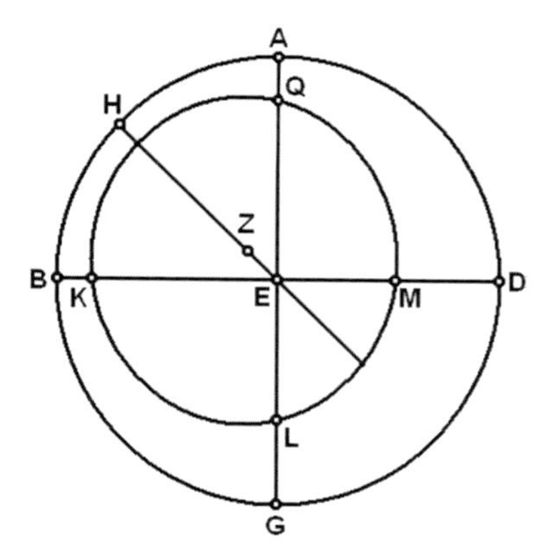

Ptolemy's version of Hipparkhos' model for solar motion © Lehoux and Massie. With the Earth at the center of the ***kosmos***, E, and looking out at the four cardinal divisions of the heavens (vernal equinox at A, summer solstice at B, autumnal equinox at G, and winter solstice at D), we know that the Sun takes different times to travel through each of the seasons (viz., 94 ½ days from A to B, 92 ½ days from B to G, 88 1/8 days from G to D, and 90 1/8 days from D back to A again). The problem then becomes how to model this mathematically. The elegant solution adopted by Hipparkhos is to assume that the Sun does not actually move on circle ABDG, but in fact moves uniformly on a different circle (QKLM), one that is not centered on the Earth but is instead centered on Z. Looking out from the Earth at E, what was an apparent motion of the Sun from A to B is actually a real motion from Q to K on the Sun's own smaller circle, and this should obviously take longer than the motion from L to M (since arc QK is longer than arc LM), as it in fact does. (Adapted from Toomer, *Ptolemy's Almagest* (1984) 154.)

Daryn Lehoux

Hippasios of Ēlis (before *ca* 400 CE?)

Author of a remedy for horses or cows preserved in the *Hippiatrika* (*Hippiatrica Parisina* 1148 = *Hippiatrica Berolinensia* 130.160). The remedy is attributed to Hippasios in the treatise of HIEROKLĒS. Hippasios is called "Ēleios"; according to STEPHANOS OF BUZANTION there were three cities called Ēlis.

CHG v.1; McCabe (2007) 227.

Anne McCabe

Hippasos of Metapontum (520 – 480 BCE)

Student of PYTHAGORAS. Evidence of his natural philosophical doctrines is very scanty; if he put them in writing this book was lost very early. Hippasos made fire the first principle (18 A7 DK), regarded soul as fiery (A9) and the ***kosmos*** as finite and ever moving (A1). The doxographical tradition frequently combines philosophical doctrines of Hippasos and HĒRAKLEITOS, therefore it is difficult to say what exactly belongs to whom. Hippasos' principle was reflected in PHILOLAOS' theory, who made all heavenly bodies rotate around the Central Fire. Hippasos discovered irrational magnitudes, which left a profound trace in Greek mathematics (the legend that he "disclosed" this **Pythagorean** secret arose from a double meaning of the word *arrētos*: "inexpressible in numbers" and "secret"). In solid geometry he is connected with a construction of the dodecahedron. Along with Pythagoras, Hippasos was one of the founders of mathematical harmonics; he developed the theory of proportions in its application to harmonics (A14–15) and carried out acoustic experiments (A12–13). To already known harmonious intervals (octave, fifth and fourth) he added a double octave (4:1) and a twelfth (3:1). Hippasos, probably, was the first to connect the pitch of a sound with frequency of vibration (A13).

DK 18; K. von. Fritz, "The discovery of incommensurability by Hippasos of Metapontum," *Annals of Mathematics* 46 (1945) 242–264; van der Waerden (1979); Zhmud (1997); Zhmud (2006).

Leonid Zhmud

Hippias of Ēlis (440 – 400 BCE)

Best known to us from PLATO's dialogues, where he is represented as a pretentious polymath, public performer, and "sophist." It appears from Plato's *Protagoras* (318d7–e5) that Hippias taught what came to be called the *quadrivium* (arithmetic, geometry, astronomy, and music). Most scholars have accepted that PROKLOS refers to this Hippias when he says that some people have carried out angle trisection on the basis of the *quadratices* (*tetragōnizousai grammai*) of Hippias and NIKOMĒDĒS (*In Eucl.* p. 272.1–10 Fr.) and that Hippias gave the characterizing feature (*sumptōma*) of the same curve (356.6–11). The name "*quadratix*" derives from the use of the curve to square a circle, but it has a simpler application to dividing an angle in a given ratio. Scholars disagree about whether Hippias of Ēlis used the *quadratix* for squaring the circle as well as for dividing an angle in a given ratio and about whether he knew anything about the *quadratix* under any name. There is no other known Hippias to whom such knowledge can be ascribed.

DK 86; R.K. Sprague, ed., *The Older Sophists* (1972) 94–105; *DSB* 6.405–410, I. Bulmer-Thomas; Knorr (1986) 80–86.

Ian Mueller

Hippobotos (200 – 180 BCE)

Wrote two books on ancient philosophers. Most of the fragments are preserved in DIOGENĒS LAËRTIOS and offer biographical information. Non-biographical information seems to come from his *On the Sects*, dealing only with Hellenistic ethical thought. (For the rare name compare only *LGPN* 2.237.)

M. Gigante, "Frammenti di Ippoboto. Contributo alla storia della storiografia filosofica," *Ommagio a Piero Treves* (1983) 151–193; Mejer (1978) *passim*.

Jørgen Mejer

Hippokratēs (Veterin.) (before *ca* 500 CE?)

Wrote on the medical treatment of horses, mules, and other beasts of burden; one of the seven principal sources of the *Hippiatrika*. Concealed behind the famous name, his identity is elusive, but apparently distinct from the Hippokratēs to whom APSURTOS addresses two letters, the Hippokratēs cited by Ibn al-Awwām, and the so-called "Ipocras Indicus," and certainly from the Hippokratēs to whom, along with GALĒN, the *Epitomē* of the *Hippiatrika* is attributed in MSS. Although he does not name any sources, the author of the Hippokratēs-extracts in the *Hippiatrika* apparently drew upon the CASSIUS DIONUSIOS-Magōn tradition, without depending on any other known author as an intermediary. His text is notable for its introductory material on bloodletting, as well as for its lack of literary pretension.

CHG vv.1–2 *passim*; G. Björck, "Griechische Pferdeheilkunde in arabischer Überlieferung," *Le monde oriental* 30 (1936) 1–12; McCabe (2007) 245–258.

Anne McCabe

Hippokratēs of Khios (440 – 400 BCE)

In his history of geometry, PROKLOS mentions Hippokratēs of Khios together with THEODŌROS OF KURĒNĒ as distinguished people in geometry, and tells us that Hippokratēs was the first person said to have written elements of geometry (*In Eucl.* p. 66.4–8 Fr.). EUTOKIOS tells us that Hippokratēs reduced the problem of constructing a cube twice the size of a given one to the finding of two mean proportionals between two given straight lines x and y (*In Arch.* 88.17–23), and Proklos says that he was the first person to reduce outstanding geometric questions to other propositions (*In Eucl.* p. 212.24–213.11 Fr.). It appears that he also concerned himself with questions of natural philosophy, since ARISTOTLE tells us (*Mete.* 1.6 [342b35–343a20]) that those around Hippokratēs and also his pupil AISKHULOS gave an account of the tail of a comet as an optical illusion and explained the rareness with which one appears; Aristotle also implies (1.8 [345b10–12]) that they also considered the Milky Way to be an illusion.

However, the most mathematically interesting material relating to Hippokratēs concerns quadrature. In the *Physics* (1.2 [185a14–17]), Aristotle mentions an attempt to square the circle "by means of segments" as a false inference from true geometrical principles. The ancient commentators on the *Physics* passage all attribute this attempted quadrature to Hippokratēs of Khios. The commentators express uncertainty about what the quadrature was, but it is now generally accepted that SIMPLICIUS provides our best information about it in his comment on the Aristotle passage (*In Phys.* = *CAG* 9 [1882] 60.22–68.32) in which Simplicius adds his own explanatory material to EUDĒMOS' reworking of Hippokratēs' argument. The argument has several problematic features, but I shall discuss only the three major ones after giving my own formulation of Hippokratēs' quadratures. The first problematic aspect of the argument is Hippokratēs' assumption that:

> If a and b are similar segments of circles on the bases α and β, then a is to b as the square with side α [**SQ**(α)] is to **SQ**(β).

Simplicius implies that Hippokratēs proved this principle, as he could have, from an equivalent of EUCLID's *Elements* 12, *prop.* 2, according to which circles are to one another as the squares on their diameters. However, Simplicius also says that Hippokratēs proved this proposition, which Euclid proves by the method of exhaustion, a method almost certainly unavailable to Hippokratēs.

Hippokratēs applies his principle in squaring three "lunes," the shaded plane figures in figures 1, 2, and 3a, which are contained by arcs of two circles. In particular, in figure 1, the lune $ABCD$ is contained by the semicircle ACB and the arc ADB, which is similar to the arcs cut off by the equal straight lines AC and CB; in figure 2 arc $ADCB$ is greater than a semicircle, DC is parallel to AB, and **SQ**(AB) = **SQ**(AD) + **SQ**(DC) + **SQ**(CB); and in figure 3a $ADCB$ is less than a semicircle, DC is parallel to AB and equal to AD and CB, and such that AC and DB intersect at E with **SQ**(AE) = $3/2 \times$ **SQ**(AD). In order to carry out the construction of this third quadrilateral *ADCB* Hippokratēs starts from a semicircle *FGD'* with diameter *FC'D'* and center *C'*. He bisects *C'D'* at *H* and draws *HJ* perpendicular to *C'D'*. He then determines *B'* on the semicircle *FGD'* and *E'* on *HJ* by using a so-called *neusis* (verging) construction in which *B'E'* is placed in such a way that it "verges" toward *D'* and so that **SQ**(*B'E'*) is 3/2 × **SQ**(*B'C'*). He draws *B'K* parallel to *D'F* and connects *C'* to *B'* and to *E'*. He extends *C'E'* to meet *B'K* at *A'*, and connects *D'* to *E'* and *A'*. Then *A'D'* is equal to *B'C'*, which is equal to *C'D'*. So *A'B'C'D'* is the desired quadrilateral. Simplicius does not discuss this *neusis* construction so we can only conjecture how Hippokratēs carried it out. One might think of the verging argument as a matter of marking a line or ruler LN at a point M so that **SQ**(*LM*) = 3/2 × **SQ**(*B'C'*), then moving the line around until a position is found in which *L* lies on the circumference of the semicircle *FGD'*, *M* lies on *HJ* and the line passes through *D'*. The construction can also be carried out using a fairly complicated application of areas.

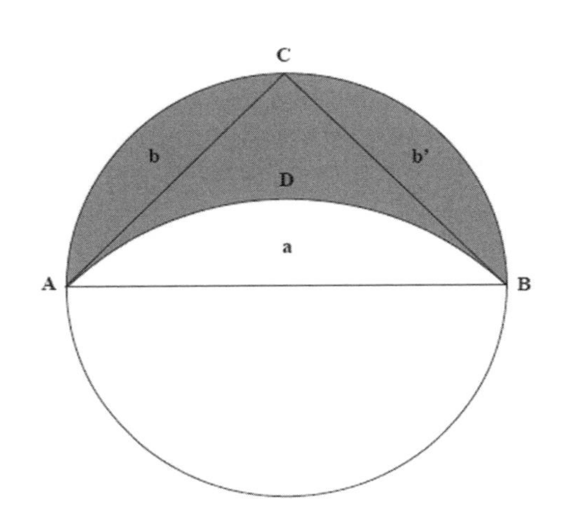

Hippokrates of Khios, 1 © Mueller

Having squared these three lunes, Hippokratēs shows how for any circle *A'B'C'-D'E'F'* one can construct a square equal to it plus a lune *AGCB* which is constructed as follows. *ABCDEF* and *A'B'C'D'E'F'* are taken as circles with center *O* arranged as in figure 4 with *A'B'C'D'E'F'* a regular hexagon, with **SQ**(*AD*) = 6 × **SQ**(A'D'), and with the circular segment *AGC* similar to the segment *a* on *A'B'*. Scholars in general have been dubious about whether Hippokratēs believed or even claimed that he had squared the circle when he had shown how to square any circle plus a particular lune and how to square particular members of three classes of lunes into which every lune must fall.

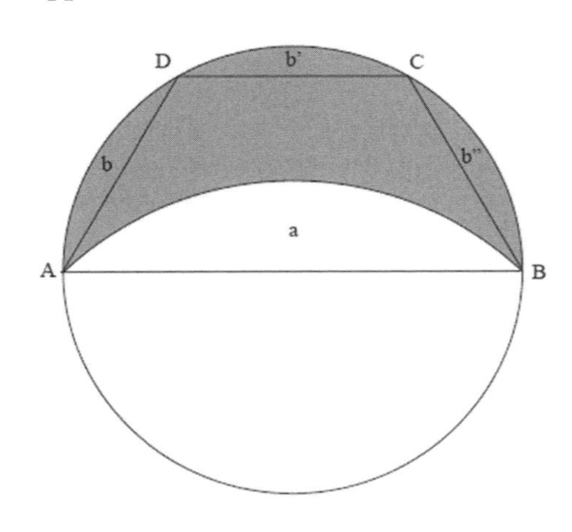

Hippokrates of Khios, 2 © Mueller

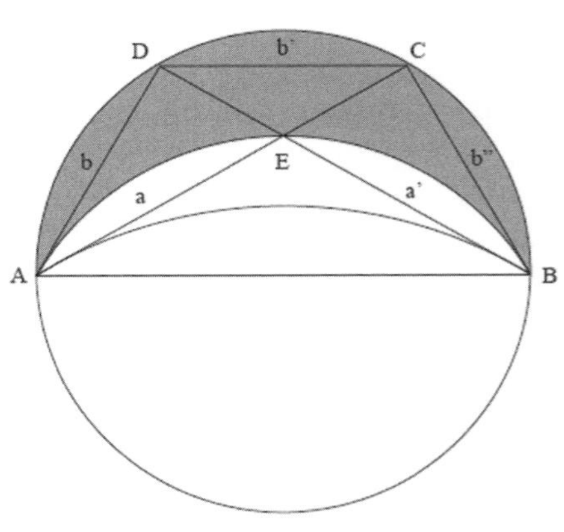

Hippokrates of Khios, 3a © Mueller

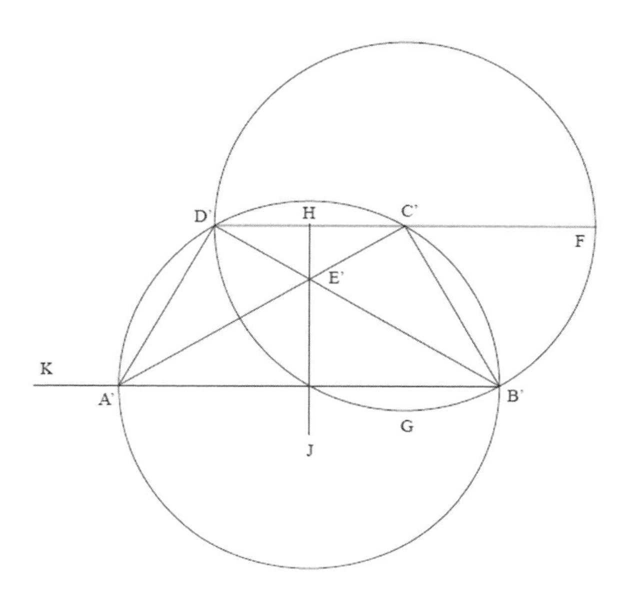

Hippokrates of Khios, 3b © Mueller

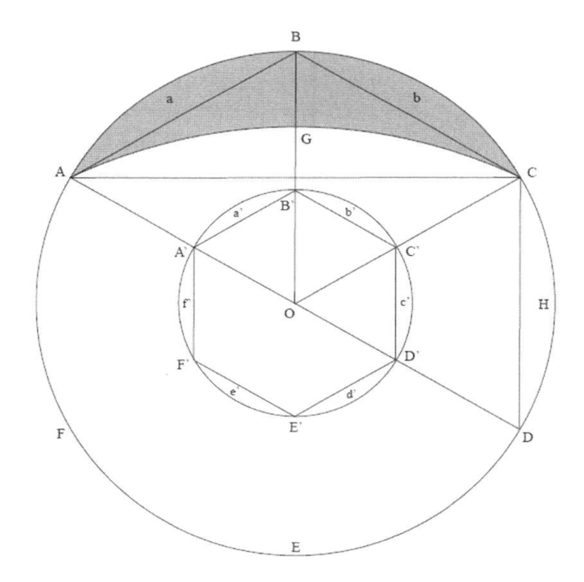

Hippokrates of Khios, 4 © Mueller

Ed.: F. Rudio (with trans.), *Der Bericht des Simplicius über die Quadraturen des Antiphon und des Hippokrates* (1907).

Heath (1921) 182–202; *DSB* 6.410–418, I. Bulmer-Thomas; Mueller (1997) 304–309; Wilson (2008).

Ian Mueller

Hippokratēs of Kōs (*ca* 440 – *ca* 370 BCE)

Hippokratēs is considered the father of medicine, just as HĒRODOTOS is considered the father of history.

Life: It is not always easy to untangle the historical and the legendary in the life of this doctor of the 5th/4th c. BCE, who was famous during his lifetime. According to PLATO (in the *Protagoras* and the *Phaidros*), he was the best doctor, in his teaching as in his method of medical discovery. According to ARISTOTLE, Hippokratēs was great, not in size but in talent.

Born on Kōs in 460 BCE, Hippokratēs belonged to the aristocratic family of Asklēpiads. Famed for its medical knowledge, the family traced its lineage to Asklēpios, prince of Trikka in Thessaly in the Homeric era (later the god of medicine). Asklēpios' two sons, Makhaōn and Podaleirios, were the best-known doctors among the Greeks at Troy. Podaleirios, upon his return after the war, fell in Karia. Three extensive branches of the family each in a specific area (Rhodes, Kōs, Knidos) traced their descent from him; the Rhodian branch rather soon died out. The best-known, due to Hippokratēs, was that of Kōs. The Knidian branch included both EURUPHŌN, to whom is attributed the *Knidian Sentences*, in fact a compilation of the Knidian doctors, as well as KTĒSIAS, doctor to Artaxerxēs and author of a Persian history.

Hippokratēs was considered the 17th (or 18th or 19th) descendant of Asklēpios in the male line. In the Koan branch of the family, the best-known doctor before Hippokratēs was Nebros (summoned by the Pythia at the time of the "first sacred war," *ca* 590 BCE). Son of Hērakleidas, Hippokratēs was named after his grandfather; his ancestry was also traced to Hēraklēs, probably through his mother, Phainaretē. Medicine was taught in each of the branches of the family by father to son. Hippokratēs had two sons, THESSALOS and DRAKŌN, who were also doctors, and a daughter who married one of his students, POLUBOS. Hippokratēs' notoriety came about because that familial instruction was opened to unrelated students, instructed for a fee, under the terms of the HIPPOKRATIC CORPUS OATH.

The reputation of the doctor from Kōs is shown also by two widely-told tales popularized by the *Hippokratic Letters*:

1 Hippokratēs and DĒMOKRITOS: summoned by the residents of Abdēra to treat Dēmokritos, whom they believed stricken with madness, Hippokratēs established the wisdom of the philosopher, who laughed at others' madness.
2 Hippokratēs and the King of Persia: Artaxerxēs I attempted to hire Hippokratēs, who refused to serve barbarians.

Leaving his student Polubos on Kōs, Hippokratēs went to mainland Greece, arrived in Thessalia and traveled, probably with his students, in northern Greece (Thessalia, Macedon, the Thrakian coast, esp. Abdēra and Thasos, and the Propontis, esp. Perinthos), as the provenance of the patients in the HIPPOKRATIC CORPUS EPIDEMICS attests.

Two other anecdotes belong to this period:

1 Summoned before Perdikkas, king of Macedon, believed stricken with ***phthisis*** after the death of his father, Hippokratēs is said to have diagnosed love-sickness.

2 Hippokratēs is said again to have refused to treat barbarians during a plague (not the Athenian plague); on the contrary, he sent his disciples to treat Greeks.

He died (aged 85 to 109) at Larissa in Thessaly. After his death, he received honors and a public cult as a hero at Kōs, and his image appeared on their bronze coins.

Works: Under the name of Hippokratēs, Renaissance scholars collected some 60 treatises. This mass of treatises, when examined closely, cannot possibly derive from a single person (Hippokratēs), nor even from a single school (that of Hippokratēs, called the school of Kōs), nor even from a single era (5th–4th cc. BCE). Some treatises are later than Aristotle, such as the HIPPOKRATIC CORPUS HEART, notable for its anatomical understanding. The sole treatise of the corpus to which one can reasonably attach an author's name is the *Nature of Man*, a work of his student Polubos, known for his theory of the four **humors** (blood, phlegm, yellow bile, and black bile).

Besides treatises of the school of Hippokratēs, whose medicine was environmental and for which the patient was at the center of observation (the Hippokratic Corpus *Epidemics*, the HIPPOKRATIC CORPUS AIRS, WATERS, PLACES, the HIPPOKRATIC CORPUS PROGNOSTIC, etc.), some treatises are ascribed to the school of Knidos, where the concept of the disease was primary (HIPPOKRATIC CORPUS, NOSOLOGICAL WORKS: *Diseases*; HIPPOKRATIC CORPUS, GYNECOLOGY: *Diseases of Women*). Others are philosophical, basing their understanding of diseases upon a prior understanding of human nature (*Fleshes*, HIPPOKRATIC CORPUS REGIMEN, HIPPOKRATIC CORPUS SEVENS).

All aspects of medicine are represented in the totality of the corpus: semiology, prognostic, etiology, therapy by surgery and by pharmacy, regimen, and deontology. Nevertheless, these treatises, diverse in origin, subject, and audience, demonstrate sufficient coherence, especially in their rational spirit of a medicine detached from magic, that they could be read as the work of one man.

Reception: Hippokratēs enjoyed in the history of medicine a reputation comparable to that which Plato or Aristotle had in the history of philosophy.

In antiquity, it was GALĒN who contributed more than anyone else to Hippokratēs' reputation, in re-interpreting him the better to make himself his continuator. Thus, Galēn attributed to Hippokratēs himself the theory of the four **humors** (blood, phlegm, yellow bile, and black bile), which was the theory of his student Polubos. This four-fold theory, relating the **humors** to the seasons and life-stages, became received as the teaching of Hippokratēs in western thought. Even in the Byzantine era, Hippokratēs was credited with the theory of four **humors**, as augmented with the theory of the four temperaments. This Byzantine Greek medicine translated into Latin (in VINDICIANUS' *Letter to Pentadius*), and also into Arabic, was to have a decisive influence on medieval thought, especially through the medical school of Salerno.

After the Renaissance rediscovery of the Greek text of the Hippokratic Corpus, Hippokratēs continued to excite admiration in the West for his observations (the "hippokratic face") up to the 19th c., including Laënnec (d. 1826), the inventor of indirect auscultation (the stethoscope), who had found in Hippokratēs the practice of direct auscultation. The two best-known treatises that remained attached to the name of Hippokratēs were the *Oath* and the HIPPOKRATIC CORPUS APHORISMS.

Ed.: É. Littré, *Oeuvres complètes d'Hippocrate* 10 vv. (1839–1861); *CUF* (partial: see individual entries).
Smith (1979); Pinault (1992); Jacques Jouanna, *Hippocrate* (1992; Engl. trans. 1999); V. Nutton, *Ancient Medicine* (2004).

Jacques Jouanna

Hippokratic Corpus, Airs, Waters, Places (430 – 400 BCE)

Intended to aid itinerant doctors traveling to unfamiliar places, *Airs, Waters, Places* discusses the effects of the natural environment, including astronomical phenomena, on constitutions, illnesses, and ethnic traits. The treatise's date is debated, but it is doubtless among the earliest of the extant texts on these topics.

The environmental elements influencing constitutions and illnesses are the city's position with respect to the Sun and prevailing winds; the source(s) and qualities of drinking water; the seasons; and such astronomical phenomena as solstices and equinoxes. Two constitutions are identified (bilious and phlegmatic) and more than 30 illnesses. Variations may occur according to individual constitution, regimen, sex, or age.

Environmental effects on ethnic groups are discussed more broadly, especially with regard to Asia Minor (sections on Egypt and Libya are missing). The entire region has an unchanging, moderate climate producing people of greater beauty and size than elsewhere; individual variation plays little role and there is no mention of constitutions. Asians' environment also makes them lack courage and spirit, in contrast to Europeans, whose changeable and more extreme climates produce an array of physiques and a bellicose temperament. Political institutions often reinforce such tendencies, but they may also over-ride them.

The customs of several European ethnic groups are described in detail, perhaps reflecting contemporary *nomos*-***phusis*** debates. A relationship with HĒRODOTOS has been suggested, particularly because both texts discuss the Skuthians and their sacred disease; relative dates are debated.

Because *Airs, Waters, Places* and ON THE SACRED DISEASE discuss some topics, most notably "sacred disease," with similar ideas and language, a single author, or closely-related authors, has been posited. There are also points of dispute, including whether bile can cause the sacred disease.

The recommendation that doctors study astronomy (*contra* ON ANCIENT MEDICINE) may echo contemporary inquiries concerning the impact of the environment (or of the ***kosmos***) on individual nature and health. Ideas similar to the text's are found variously in *Epidemics I–III*, ***Humors***, APHORISMS, REGIMEN, PLATO, and ARISTOTLE and his school.

BAKKHEIOS may have known the treatise. ERŌTIANOS and GALĒN provide some glosses, and consider it a genuine work of HIPPOKRATĒS. Galēn's commentary survives in Arabic (now in German).

Littré v. 2; J. Jouanna, *Airs, Eaux, Lieux* (*CUF* 1996).

Julie Laskaris

Hippokratic Corpus, Anatomy and Physiology (*ca* 430 – 370 BCE)

Nature of Man (Littré 6). This work, written in an agonistic debating style, is most celebrated for its exposition of the nature of the four **humors** – blood, phlegm, yellow bile and black bile – which must be in proper balance to ensure bodily health. It is the only work of the Corpus which explicitly takes this stance. The treatise is sometimes attributed to HIPPOKRATĒS' son-in-law POLUBOS, as ARISTOTLE quotes (*HA* 3.3 [512b13–513a8]), with that ascription, a passage from it in which the vascular system is described. In all ancient MSS, *Nature of Man* and *Regimen in Health* are transcribed as a single work; the two contain similar provisions for attaining and maintaining a state of health. (J. Jouanna, *CMG* 1.1.3, 2nd ed. [2002])

Breaths (Littré 6). This is a treatise of unusual content, affirming that all diseases are caused by "winds" or "breaths" (*phusai*), affecting the body or the atmosphere. This view resembles that apparently attributed to the historical Hippokratēs by a pupil of Aristotle (MENŌN, on the evidence of the LONDINIENSIS MEDICUS): accordingly, the work has some claim to be considered truly Hippokratic. (J.L. Heiberg, *CMG* 1.1 [1927] pp. 91–101; J. Jouanna, *Hippocrate* v. 5.1 [*CUF* 1988].)

Places in Man (Littré 6). This long work deals with a large number of subjects in anatomy, physiology and pathology, as well as enunciating various medical precepts and doctrines. Throughout, there is much stress on bodily balance as a factor in health, and on flux of excess or noxious matter from the head to other parts of the body (eyes, chest etc.) as a causative agent in disease; in therapy cauterization is favored. There is an excursus on gynecology. The work seems to be early and may have a west Greek origin (Craik 1998).

Glands (Littré 8). This work deals with the function of glands, believed to be situated in parts of the body where moisture gathers, usually associated with places where hair grows. Among these the brain has an important place, as it is the starting point for flux of disease-inducing fluids; the theory of seven fluxes resembles that of *Places in Man*.

Flesh(es) (Littré 8). This treatise gives an elegant scientific account of the formation of the ***kosmos***, and of the body and its parts. It is envisaged that two types of matter (cold and gluey or viscous on the one hand; hot and fatty or slippery on the other) underlie the process. The formation of the body, including eyes and other sense organs, is described. The work ends with an excursus on the importance of the number seven in both embryology and nosology.

Bones (Littré 9). Despite the title, probably drawn from the first words, this treatise deals with the (blood) vessels, bones being the subject only of the first section. Various views of the vascular system, with similarities and differences alike apparent, are presented. Although the accounts are confused and fanciful in details, a salient common supposition is that the vessels originate in the head. Different parts of *Bones* can be traced to different sources: to a work described by GALĒN as "Appendix to *Mokhlikon*" (Galēn 19.128 K.), to the obscure SUENNESIS OF CYPRUS, and to similar passages in *Nature of Man* and in *Epidemics 2* (C.R.S. Harris, *The Heart and the Vascular System in Ancient Greek Medicine* [1973]; Duminil 1998).

Anatomy (Littré 8). This very short piece, comprising a single page in the modern printed text, is an account, with some reference to comparative anatomy, of the internal configuration of the human trunk. It seems to be a late pastiche, incorporating material both from Hippokratic sources and from the work of DĒMOKRITOS. (Duminil 1998; ed.: Elizabeth Craik, "The Hippocratic Treatise *On anatomy*," *CQ* 48 [1998] 135–167.)

Vision (Littré 9). This is a short surgical manual prescribing treatment, most commonly purging and cautery, for various diseases affecting the eyesight and the eyelids; these can be plausibly identified as cataract, trachoma and other common conditions (Elizabeth Craik, *Two Hippocratic treatises on sight and on anatomy* = *SAM* 33 [2006]).

Jones v. 2 (1923); Jones (1931); R. Joly, *Hippocrate* v. 13 (*CUF* 1978); Potter (1995); Elizabeth Craik, *Hippocrates: Places in Man* (1998); M.-P. Duminil, *Hippocrate* v. 8 (*CUF* 1998).

Elizabeth Craik

Hippokratic Corpus, On Ancient Medicine (430 – 380 BCE)

On Ancient Medicine denies the value of assumptions (*hupotheseis*) as the foundation of medicine, or of any discipline, claiming that their appeal to invisible or non-existent substances makes them unverifiable and impedes communication between doctor and patient. Medicine's traditional method – reasoning based upon observation – is sufficient. Philosophical theories are attacked, especially those narrowing etiologies to one or two of such unmixed substances as the hot, cold, wet, and dry. The author, however, employs *hupotheseis* himself (Lloyd), asserting that innumerable substances (e.g., bitterness, sweetness), properly mixed, constitute the healthy body (*cf.* Alkmaiōn). Illness results from the separation of one of the **humors**, or from the body's own structures.

Singling out Empedoklēs, the author also refutes those asserting that medicine requires first a knowledge of Nature, claiming the converse: Nature is understood only via medicine. Doctors require such knowledge, but it is subordinate to medicine.

The defense of medicine's status as a *tekhnē* echoes sophistic debates (*cf. On the Art, On Breaths*, On the Sacred Disease). Ideally, a *tekhnē* has a theoretical foundation and is unfailingly successful, in contrast with chance (*tukhē*). Here, however, medicine's inevitable (though infrequent) fallibility does not diminish its technical status (it surpasses chance), whose attainment is attributed to empirically-based reasoning requiring no *hupotheseis*.

The author traces medicine's origins to primitive man's discovery that many illnesses were prevented by eating mild, cooked foods rather than the strong, raw ones suited to animals. Expertise in treating illness came later. Early doctors' regimens of restricted food intake, weak gruels, and liquid nourishment followed the same reasoning as the primitive discoveries. Good doctors know that individual constitutions have individual dietary requirements.

Erōtianos attributed the treatise to Hippokratēs. Modern scholars have suggested various schools and figures as the author's specific targets. The particular sense of "*hupothesis*" – otherwise known only in Plato (*Meno*) and later authors (and rare in any sense before Plato [Lloyd]) – clouds this issue and the dating of the text.

Ed.: Littré v. 1; J. Jouanna, *Hippocrate: De l'Ancienne médecine* (*CUF* 1990).
G.E.R. Lloyd, "Who is Attacked in On Ancient Medicine?" *Phronesis* 8 (1963) 108–126; repr. in *Methods and Problems in Greek Science* (1991) with new introduction; M.J. Schiefsky, trans., *Hippocrates on ancient medicine* = *SAM* 28 (2005).

Julie Laskaris

Hippokratic Corpus, Aphoristic Works (*ca* 430 – 370 BCE)

Aphorisms (Littré 4) and *Coan Prognoses* (Littré 5). These two compilations are alike: both comprise lengthy collections of disjointed sayings, conveying useful information for the doctor. The content of the latter is somewhat more restricted, dealing primarily with prognostic guidance, though this relates to a wide range of diseases and conditions; it is also more clearly organized by subject matter. The guidance ranges from common-sense rules of thumb to superstitious observations. There is some overlap in content, but differences in vocabulary and modes of expression suggest that the collections had different origins; *Coan Prognoses* is related to *Prorrhetic* 1. Both collections, especially *Aphorisms*, were subject to much reprinting, being cited and followed by practicing physicians until the 19th c.

Humors (Littré 5). The title, based on a passing reference to **humors** in the first sentence, is misleading. The work is a collection of aphorisms on miscellaneous subjects, notably on signs and symptoms to be observed by the physician. Particular attention is paid to the nature of body fluids and evacuations, and to signs which indicate medical crisis; some attention is given to seasonal factors in causing disease. The expression is frequently obscure and elliptical, with recourse to simile and metaphor. Where case notes are recorded, the content overlaps with that of EPIDEMICS 6.

Dentition (Littré 8). This work consists of a set of 32 aphorisms, relating to feverish illnesses which typically beset infants at the time of teething. The expression is concise and the content condensed.

Nutriment (Littré 9). This short collection of aphorisms deals with the importance of nourishment to all parts of the body, though it is clearly perceived that individual needs differ. The style is contorted, and there is much riddling antithesis in the manner of HĒRAKLEITOS OF EPHESOS.

Crises and *Critical Days* (Littré 9). These two works are "late" and derivative selections of aphoristic material from a variety of Hippokratic sources.

Jones (1923, 1931); R. Joly, *Hippocrate* vv. 6.2, 13 (*CUF* 1972, 1978).

Elizabeth Craik

Hippokratic Corpus, Epidēmiai (*ca* 430 – 350 BCE)

Seven anonymous books transmitted in Ionic Greek under the name of HIPPOKRATĒS as part of the *Corpus Hippocraticum*. Their title is perhaps old (4th c. BCE?) but probably not original; its meaning is uncertain: "arrival" or "sojourn" of persons (of itinerant physicians and/or patients) or diseases (but not only of epidemics and endemics, since the books describe others as well, e.g. injuries) or both.

The seven books form three groups: (1) *Epidēmiai* 1 and 3 (*ca* 410 BCE; extant commentaries by GALĒN) contain observations on weather conditions of particular years and on concomitant diseases; individual case descriptions; and aphorisms. They are, in parts, carefully composed. (2) *Epidēmiai* 2, 4, 6 (between 427/426 and 373/372 BCE) are, in content, similar to 1 and 3 (plus a description of an anatomical dissection, 2.4.1–2) but read more like notebooks and are composed largely chaotically (commentaries are extant by Galēn on 2, by Galēn, PALLADIOS, and IŌANNĒS OF ALEXANDRIA on 6). (3) *Epidēmiai* 5 and 7 (*ca* 375–350 BCE) consist mostly of individual case descriptions.

The "literary" or in parts subliterary form of the *Epidēmiai*, characterized by inexplicitness and unpolished (elliptic, "telegraphic") diction, is remarkable. With hundreds of records about individual patients or groups of patients and with sometimes obscure generalizing texts ("aphorisms"), the *Epidēmiai* were obviously written for informal use, perhaps as notes which medical teachers would expand upon during lessons, or as internal materials for a professional group. They seem to have served didactic as well as practical purposes and represent an advanced stage of Greek medical development as compared to certain other Hippokratic treatises. The patients, whose names, professions, and addresses are often revealed, lived throughout the Aegean and belonged to both sexes, all age-groups (perhaps except very young infants) and social strata. Modern attempts to relate the *Epidēmiai* to specific "medical schools," in particular a hypothetical "school of Kōs," have failed. The technical vocabulary comprises many terms absent from earlier, more traditional texts. At first sight, one might conclude that the *Epidēmiai* were designed as a database of observational raw data; but a more thorough analysis makes it clear that observations have been "filtered": that they presuppose (i) elaborate methods of prognosis, (ii) nosological doctrine (i.e. lore about particular diseases conceived of as entities), (iii) sophisticated theoretical assumptions about the healthy body (e.g. its physiological "type"; its "**humors**") and the diseased organ(ism) (e.g. the processes of "ripening" and "***krisis***" during a malady), and

(iv) about the environment (including weather, food etc.). Observation was, therefore, guided by dogma, "facts" became "facts" through dogma. All medical fields (i–iv) were considered interdependent; but since data collected within this framework are (in modern terms) incommensurate, "there is," as one of the texts admits, "a difficulty of evaluation, even if one knows the method" (*Epidēmiai* 6.8.26). The theoretical activity of the physicians who wrote the *Epidēmiai* may aptly be called "research," and a spirit of criticism is not absent. This skeptical attitude, however, concerns only details. Fundamental dogmas remain unchallenged. In the *Epidēmiai* pre-existing theories or methods are never submitted to criticism by questioning their value or by restricting their range of applicability; instead, there is a marked tendency to ask only constructive theoretical questions which extend the validity of existing doctrines and prompt an affirmative answer.

Apart from their high value as documents of Greek medical history, the *Epidēmiai* are, more generally, monuments of the evolution of Greek thought in the "classical" period and as such hitherto little explored.

Ed.: Jones (1923) [*Epidēmiai* 1 and 3]; W.D. Smith, *Hippocrates* v. 7 [*Epidēmiai* 2, 4–7] (Loeb 1994).

G. Baader and R. Winau, edd., *Die hippokratischen Epidemien* (1989); Volker Langholf, *Medical theories in Hippocrates. Early texts and the "Epidemics"* (1990); L.A. Graumann, *Die Krankengeschichten der Epidemienbücher des Corpus Hippocraticum. Medizinhistorische Bedeutung und Möglichkeiten der retrospektiven* Diagnose (2000).

Volker Langholf

Hippokratic Corpus, Gynecological Works (*ca* 470 – 370 BCE)

The gynecological works of the Hippokratic Corpus comprise eight works: (1) *Diseases of Women* I and II (*Mul.* I and II) contain the bulk of material treating female anatomy, physiology and pathology as well as many issues of reproduction. (2) *On Sterile Women* (*Steril.*) continues *Mul.* II, but focuses primarily on causes and treatment for infertility. Some of the material in these books may date from the first half of the 5th c., but their ancient compiler also inserted later material, including an independent treatise on women's diseases written by the author of *On Seed* or *On Generation* (*Genit.*) and *On the Nature of the Child* (*Nat. Puer.*). (3) *Nature of Women* (*Nat. Mul.*) contains descriptions of female diseases and remedies corresponding to what are considered the earlier sections of *Mul.* I and II and *Steril.* (4) *Superfetation* (*Superf.*) also includes some material found in *Mul.* I and II and *Steril.* The first part of the treatise focuses on the problems of pregnancy and childbirth and the second part on sterility. The treatise is named for the topic of the first chapter: the rare occurrence of a second conception in an already pregnant woman. (5) *Excision of the Fetus* (*Exc.*), a very short treatise on childbirth and its attendant problems, also takes its title from the topic of the first chapter. (6) *On the Seven Month Fetus* (*Sept.*) and *On the Eight Month Fetus* (*Oct.*), which deal with embryology and the problems of premature births, are a single treatise most often cited as *Oct.*, though the title *Sept.* is sometimes retained when referring to the chapters of what was traditionally thought to be a self-contained work. Similarly, (7) *Genit./Nat. Puer.* form one continuous treatise on conception, gestation and parturition, with *Genit.* indicating the first 11 chapters. There is also a brief treatise, maybe a fragment of a treatise on epilepsy, (8) *On the Diseases of Young Girls* (*Virg.*).

Genit./Nat. Puer., and therefore the latest sections of *Mul.* I and II and *Steril.*, is the most reliably datable of the treatises. *Diseases* IV (*Morb.* IV), by the same author, aims to reduce the **humors** into a tetrad schema similar to that of *Nature of Man* – the watery **humor** *hudrops* replaces black bile – implying a date of *ca* 420–380 BCE. The other gynecological treatises too are thought to have taken their present shape around this time, although they probably enshrine much older material, such as perhaps the extended series of pharmacological recipes of *Mul.* I and II, *Steril.*, *Nat. Mul.* and *Superf.*, rare elsewhere in the Corpus.

The model of the female body emerging from these diverse treatises is generally consistent, and coheres with female physiology and pathology in the general works of the Corpus, when differentiated from male physiology. However, because the word the ancient medical writers used for "patient" (*anthropos*) is an unmarked term, it is possible that some Hippokratics assimilated women's bodies to men's to a greater extent than the gynecological authors. The author of *Mul.* I castigates some doctors for "treating women as if they had men's diseases" (63). No evidence exists that any Hippokratic physician would have specialized as a gynecologist.

Women were differentiated from men by the nature of their flesh. Female flesh was spongier than the male and absorbed excess blood produced in the woman's stomach from the nourishment which her smaller, weaker body and less active life style could not consume. The blood was stored in her flesh for a month to act as nourishment for a fetus should the woman conceive. If she was not pregnant at the end of a month, the womb would draw the blood from all over the body and evacuate it through the cervix. If the woman's passages were all open and the mechanism functioned properly, this acted as a very efficient purge and prophylactic. However, the benefits of menstruation were offset by the fact that the mechanism was liable to malfunction causing a retention of menses, potentially leading to a variety of pathological conditions. Other specifically female conditions were caused by the tendency of the womb to move from its position. Both the wandering womb and menstrual retention were most easily averted by regular intercourse. This moistened and warmed the womb, keeping the cervix and the passages throughout a woman's body open. Thus, unmarried girls approaching puberty were thought to be at increased risk of disease. The healthiest result of intercourse for a woman was pregnancy. The fetus anchored the womb in place and consumed the woman's excess blood. It also drew the blood to itself in the womb steadily throughout the month, thereby avoiding pain and discomfort caused by the menses being drawn through the narrow passages to the womb all at once. Once a woman had given birth, the abundant lochial flow broke down her body, opening up the passages and making her menstrual mechanism more reliable in the future, another reason why it was considered healthy for a pubescent girl to marry.

If a woman did fall ill, the attendant physician would often try to stimulate the menses with an emmenagogue, administered as a drink or a **pessary**. Pessaries and fumigations were also used to try to attract the uterus back to its normal position. Regimen, bleeding and cauterization seem to have been employed less often to cure women than men. The death rate for female patients of the Hippokratics is comparable to that of male.

The Hippokratics believed that a woman contributed seed to conception much as a man did, and argued that a child resembled one parent more than the other in those characteristics for which they had received more or stronger seed. The theories of the embryological treatises appear to be largely guesswork, though there is some evidence that some authors had seen aborted fetuses. Normal childbirth is not mentioned in the gynecology of the Corpus. Apparently, a Hippokratic doctor would only be called to attend a problematic parturition.

Ed.: ***Mul.* I and II, *Steril.***: Littré 8; some sections, with German trans., in H. Grensemann, *Knidische Medizin* 1 (1975), *Hippokratische Gynäkologie* (1982), *Knidische Medizin* 2 (1987); ***Nat. Mul.***: Littré 7; H. Trapp, (Diss. Hamburg, 1967); ***Superf.***: Littré 8; C. Lienau, *CMG* 1.2.2 (1973), with German trans.; ***Foet. Exsect.***: Littré 8; ***Oct.*** (and ***Sept.***): Littré 7; H. Grensemann, *CMG* 1.2.1 (1968), with German trans.; R. Joly, *Hippocrate* 11 (*CUF* 1970); ***Genit./Nat. Puer.***: Littré 7; Joly (1970); ***Virg.***: Littré 8.

Trans.: ***Mul.* I**: A.E. Hanson, "Hippocrates: *Diseases of Women* I," *Signs* 1 (1975) 567–584 (selected chapters); ***Genit./Nat. Puer.***: I.M. Lonie, *The Hippocratic Treatises "On Generation"; "On the Nature of the Child"; "Diseases 4"* (1981); ***Mul.* I and II, *Nat. Mul., Oct., Virg.*** (excerpts): in M. Lefkowitz and M. Fant, *Women's Life in Greece and Rome* (1992) 230–243; ***Virg.***: A.E. Hanson and R. Flemming, "Hippocrates' *Peri parthenion* (*Diseases of young girls*): Text and Translation," *Early Science and Medicine* 3 (1998) 241–252.

Lesley Dean-Jones, *Women's Bodies in Classical Greek Science* (1994); N. Demand, *Birth, Death and Motherhood in Ancient Greece* (1994); H. King, *Hippocrates' Woman* (1998).

Lesley Dean-Jones

Hippokratic Corpus, On Head Wounds (*ca* 400 BCE)

Included by GALĒN among the genuine and most useful of the Hippokratic works (17[1].577 K.), this treatise contains vocabulary, dialectal forms and grammar consistent with a date of composition *ca* 400 BCE. Other Hippokratic treatises have similarities: in particular, EPIDEMICS V presents case histories illustrating its advice, and *On Ulcers* (see HIPPOKRATIC CORPUS SURGERY) contains almost identical language.

On Head Wounds survives in nine MSS, the most authoritative being the beautiful 10th c. codex, Laurentianus Gr. 74.7. The treatise begins with a description of cranial anatomy, then lists the types of skull injury, discusses clinical evaluation of the patient, and concludes with advice on treatment. It falls short of modern knowledge in some aspects of anatomy, in the use of the neurological examination, and especially in the indications for surgery. However, some of the anatomical description is accurate; there is clear evidence of the emergence of a technical medical vocabulary (*bregma, diploē*, suture [*raphē*], linear fracture [*rōgmē*], depressed fracture [*esphlasis*]); the relevance of brain function (state of consciousness, contralateral paresis/paralysis) is acknowledged; the importance of a good trauma history is stressed; and excellent advice is given on examination of the wound and on surgical technique. The examining surgeon is warned to distinguish between sutures and fractures, and in cases where fracture is suspected but not seen to apply a black solution which will enter a fracture but not normal skull, and to enlarge by incision any wound too small to allow adequate visualization of the injury. Trephination is described in detail with emphasis on recognizing when the inner table of the skull is perforated and in allowing for the relative thinness of children's skulls.

The advice in this treatise, itself only a small remnant of a surgical tradition ancient at the time of its composition, clearly represents the experience not just of one talented surgeon but of many generations of surgeons. Following its translation into Latin in the 16th c. (Calvus, 1525; Vidius, 1544), it remained a surgical reference work well into the 19th c. (Littré 1.IX and 3.150–261). It is a good example of a practical surgical handbook.

Maury Hanson, *CMG* 1.4.1 (1999): Greek text, English translation, commentary; Withington (1928) 6–51.

Maury Hanson

Hippokratic Corpus, Heart (*ca* 350 – 250 BCE)

Although GALĒN does not question its authenticity, it is unlikely that this work was written by HIPPOKRATĒS. PLUTARCH *Quaest. conv.* 7.1 (699 C–D), our earliest reference, cites Hippokratēs as having mentioned that some things that are drunk pass down the windpipe into the lungs. The author argues that the heart is the center of the vascular system. Since ARISTOTLE *HA* 3.3 (513a16–22) claimed to have known this, the mid 4th c. most probably is a *terminus post quem* for the work. Fredrich and others, however, dated the work before Aristotle. Abel, on philological grounds, dates the work in the 3rd c. None of the arguments for dating are entirely persuasive. A 3rd c. date would correlate with the knowledge of anatomy of the heart. The author does not appear to be HĒROPHILOS' or ERASISTRATOS' pupil.

De Corde describes the heart and vascular system, as well as the atrio-ventricular valves. The heart is described as being like a pyramid and dark red, it is a strong muscle because of its thickness and texture. The treatise discusses the inlet valves and semi-lunar valves. The author knows that there is blood-flow from the right to the left side of the heart. Given the level of sophistication, the author most probably gained his knowledge through dissection, either by him or others. Possibly he could have been one of the earlier **Pneumaticists**.

Ed.: Littré 9.80–93.

C. Fredrich, *Hippokratische Untersuchungen* (1899) 15, 77; K. Abel, "Die Lehre vom Blutkreislauf im Corpus Hippocraticum," *Hermes* (1958) 192–219, esp. 196; C.R.S. Harris, *The Heart and the Vascular System in Ancient Greek Medicine* (1973) 83–96; I.M. Lonie, "The Paradoxical Text 'On the Heart'," *Medical History* 27 (1973) 1–15, 136–153.

Robert Littman

Hippokratic Corpus, Nosological Works (*ca* 450 – 380 BCE)

Five nosological treatises, overlapping much in content:

(1) *Internal Affections* (*Int.*) describes 54 diseases, classified in "head to toe" order, starting with afflictions of the chest. The compiler divides several diseases into a number of varieties (e.g. 4 jaundices, 4 typhuses, 3 **tetanuses**).

(2) *Affections* (*Aff.*) in two parts: a nosological section (§§1–38), wherein illnesses are classified from head to toe, and which refers several times to a lost recipe book entitled *Pharmakitis*; and a dietetic section (§§39–61), lacking a clear organizing principle. The compiler claims (§1) that he is writing for laymen (*idiōtai*); but the technicality of some chapters seems to indicate that this work was addressed to physicians.

(3) *Diseases* I (*Morb.* I) in two sections: the first, comprising general remarks on the medical art, is intended to enable the physician to defend his views in debates with colleagues; while the second describes internal diseases ("suppurating" and acute diseases).

(4) *Diseases* II (*Morb.* II), not the continuation of *Morb.* I, contains two sub-treatises: the first (§§1–11; *Morb.* II-1) describes 14 diseases of the head and throat; the second (§§12–75; *Morb.* II-2) addresses the same afflictions, and also lists diseases of the nose, chest and back.

(5) All the descriptions of diseases in *Diseases* III (*Morb.* III) have parallels in *Morb.* II and *Int.* However, *Morb.* III also includes a collection of recipes for cooling remedies (§17), which has no parallel in the nosological treatises.

Each description of disease in the nosological works includes the identification of the disease and its symptomatology. Other elements figuring in the nosological descriptions are: details of treatment, both dietetic and pharmacological (found in *Int.*, *Aff.*, *Morb.* II-1, and *Morb.* III); prognosis (found in *Int.*, *Morb.* II-1, and *Morb.* III); and etiology (found in *Aff.*, *Morb.* I, and *Morb.* II-2). The etiology in *Aff.* and *Morb.* I is centered on the **humors**: bile and phlegm.

Numerous parallel redactions of material in these treatises indicate that their compilers exploited the same source(s), one of which may be the *Cnidian Sentences*, a lost fifth-century treatise. The relative chronology of the nosological treatises is disputed, but a date in the second half of the 5th c. can be advanced for *Int.*, *Morb.* II and *Morb.* III, whilst *Aff.* I and *Morb.* I, with their systematic bi-**humoral** etiology, were more likely composed at the beginning of the 4th c. BCE.

Ed.: Littré 6 (*Diseases* I; *Affections*) and 7 (*Internal Affections; Diseases* II and III); R. Wittern, *Die hippokratische Schrift De Morbis I* (1974); P. Potter, *CMG* 1.2.3 (1980); J. Jouanna, *Hippocrates. Maladies* II (*CUF* 1983); P. Potter, *Hippocrates* v. 5 (1983) (*Diseases* I and II; *Affections*) and v. 6 (1988) (*Diseases* III; *Internal Affections*).

J. Jouanna, *Hippocrate: Pour une archéologie de l'école de Cnide* (1974).

Laurence M.V. Totelin

Hippokratic Corpus, Oath (350 – 100 BCE)

Short Hippokratic treatise of uncertain origin. The oath falls into two parts. In the first, the oath-taker vows to Apollo, Asklēpios, and the gods of healing to hold his teacher equal to his own parents, to make him partner in his livelihood, to share with him his own goods, to impart oral instructions only to his own sons, to the teacher's sons and to those pupils who have taken the oath. The second part of the oath contains specific deontological prescriptions; the physician will use medical treatments only to the advantage of the sick, abstaining from any injury and wrongdoing; it is forbidden to administer poison, to perform abortion, to operate using the knife, to divulge professional secrets.

Edelstein argued that the oath displays features that are **Pythagorean** in tone or seem to echo the precepts of PYTHAGORAS (*cf.* DIOGENĒS LAËRTIOS 8.34–35); other scholars are less certain. **Pythagoreans** were unique in prohibiting suicide (PLATO, *Phaedo* 61d-62b), in considering embryos animate from the moment of conception (D.L. 8.28–29), and in their 4th c. BCE disputes over blood sacrifice (D.L. 8.13). Furthermore, **Pythagorean** stipulations of purity may have contributed to prohibitions against using the knife (Edelstein 1943: 32–33).

It is uncertain whether the oath was required by some physician's guild or merely a normative moral and ethical guide. ERŌTIANOS *Pr.* (p. 9 Nachm.) considered the oath genuinely Hippokratic.

W.H.S. Jones, *The Doctor's Oath* (1924); L. Edelstein, *The Hippocratic Oath* (1943); G. Harig and J. Kollesch, "Der hippokratische Eid," *Philologus* 122 (1978) 157–176.

Bruno Centrone

Hippokratic Corpus, Prognostic Works (*ca* 450 – *ca* 370 BCE)

"I hold that it is an excellent thing for a physician to practice *pronoia* (forethought/foresight)." So opens the Hippokratic treatise *Prognostic* (*Prog.*), in which the author offers a threefold argument about the benefits of medical prognosis for both doctor and patient, and then provides detailed practical instructions about how most accurately and effectively to read signs offered by the bodies of the acutely sick. Nor is this the only Hippokratic text dedicated to medical forecasting. It is joined by *Prorrhetic* (*Prorrh.*) 1 and 2, and *Coan Prenotions* (*Coac.*); though none of these achieved the same canonical status as *Prog.*, which was always counted amongst the writings of HIPPOKRATĒS himself in antiquity.

There is some further variation within this prognostic grouping. Both *Prorrh.* 1 and *Coac.* are, in contrast to the more polished and synthetic prose of *Prog.* and *Prorrh.* 2, aphoristic and disconnected (though not disorganized); and their content overlaps considerably. For example, *Prorrh.* 1.55 states that, "The loss of speech arising from exertion brings a bad death," which follows a similarly terse sentence about a different (but also fatal) form of speechlessness, in a longer sequence of bad signs. The same aphorism appears at *Coac.* 244, with the preceding entry also identical, though the longer sequence in this treatise is dedicated more specifically to loss of speech or speech-related symptoms. *Coac.* contains about 90% of the 170 aphorisms in *Prorrh.* 1 in some form – more often contracted or otherwise amended than replicated verbatim – collected and arranged together with almost 500 additional, and more heterogeneous, segments making a total of about 640 aphorisms in all. Compare HIPPOKRATIC CORPUS, APHORISTIC WORKS.

On the other hand, though both of the other two works open with programmatic statements about the importance, and basis, of medical prognosis, and share a certain style and literary vocabulary, they advocate distinct programs and their technical terminology differs. While the author of *Prog.* promotes the practice of medical prediction as beneficial to the physician's authority and reputation and to his success in treating the sick, with the two combining to produce the "good doctor," *Prorrh.* 2 concentrates heavily on the former. Sound forecasting here brings success in competition with other physicians, and there is no mention of healing. *Prorrh.* 2 is, however, more explicit about predictive methods and their limitations than *Prog.* The former text defines the careful and methodical observation and interpretation of medical signs it advocates and describes, against a more prophetic or divinatory form of prognosis; while the latter blurs the distinction between the two.

Thus the opening sentence of *Prog.* paraphrases HOMER's description of the famous seer Kalkhas (*Iliad* 1.69–70) in explicating what *pronoia* means in a medical context: i.e., "foreknowing and foretelling, in the presence of the sick, the present, the past, and the future." In his introductory sequence, however, the author of *Prorrh.* 2. rejects "prophecy about the past and present," stating "I will not divine in this way; rather I will record the signs from which one must judge which persons will become well and which will die." The same author also uses only words of "foretelling" while eschewing entirely the "foreknowing" that accompanies prediction in *Prog.*

These points make it tempting to think that *Prorrh.* 2 was written in response to *Prog.*, and consciously attempts to promote an alternative view of what constitutes "the best medical prognosis," an interpretation strengthened by recent suggestions that *Prorrh.* 2 is early 4th c., rather than sharing a late 5th c. date with *Prog.* More caution is perhaps needed about the traditional ordering of *Prorrh.* 1 (usually placed in the mid-5th c.) and *Coac.* (4th c.). Although the latter could directly depend on the former (and *Prog.* and *Prorrh.* 2 might also have borrowed vocabulary from the same source), it is increasingly accepted that large parts of Hippokratic material were held, roughly speaking, in common; and so might be multiply drawn on, rearranged and modified, without establishing a clear, vertical, line of textual succession.

Ed.: Littré 5.588–733; B. Alexanderson, *Die hippokratische Schrift Prognostikon: Überlieferung und Text* (1963); H. Polack, *Textkritische Untersuchungen zu der hippokratischen Schrift Prorrhetikos I* (1976); Potter (1995) 167–293.

L. Edelstein, *Ancient Medicine* (1967) 65–85; V. Langholf, *Medical Theories in Hippocrates: Early Texts and the "Epidemics"* (1990) 232–254; T. Stover, "Form and function in *Prorrhetic 2*," in P. van der Eijk, ed., *Hippocrates in Context* (2005) 345–361.

Rebecca Flemming

Hippokratic Corpus, Protreptic Works (*ca* 420 – 100 BCE)

The Art (Littré 6). This carefully worked treatise sets out to demonstrate that there really is a *tekhnē* ("art," "craft," "science") of medicine. Various arguments to the contrary are set out and rebutted, for instance the contention that medical cures arise from *tukhē* ("luck") rather than from *tekhnē*. The author, perhaps a sophist rather than a practicing doctor, is at home with techniques of literary prose and rhetorical expression (Jones v.2, 1923; Heiberg 1927: 9–19; J. Jouanna, *Hippocrate* v. 5.1 [*CUF* 1988]).

Precepts (Littré 9). This little work is made up of a disjointed amalgam of notes and remarks, where much is individually and collectively obscure. The vocabulary is recondite and the style self-consciously arresting. It describes the precepts to be followed by the ideal, high-principled physician. In date, it is regarded as "late": at least Hellenistic and possibly Roman (Jones v.1, 1923; Heiberg 1927: 30–35).

Law (Littré 4). The *Law* is frequently linked with the Hippokratic Corpus, Oath, but is more reflective in character. Debate on the *tekhnē* of medicine centers on the qualities required for medical expertise and understanding: innate ability, proper instruction and diligence. There is in conclusion a reference to the peripatetic nature of the profession and to the "sacred" character of its knowledge (Jones v.2, 1923; Heiberg 1927: 7–8).

Decorum (Littré 9). The author argues that personal ***phusis*** is necessary for progress in medical wisdom, as the prerequisites of this cannot be taught; indeed teaching in general is suspect. The work is idiosyncratic in vocabulary and contorted in expression; on the basis of this it is commonly regarded as "late." The content, however, accords with matters debated in the 5th c. (Jones v.2, 1923; Heiberg 1927: 25–29).

Physician (Littré 9). In this tract, the qualities of appearance and character desiderated in the ideal doctor are first outlined: health, dignity and trustworthiness are important. The essential elements of basic medical education are then set out: particular attention is paid to the orientation of the surgery, to proper ways of bandaging and to appropriate types of instrument. The work with its practical tenor has an appealing immediacy. (Potter 1995; Heiberg 1927: 20–24).

J. L. Heiberg, *CMG* 1.1.1 (1927).

Elizabeth Craik

Hippokratic Corpus, Regimen (*ca* 430 – 370 BCE)

Regimen I–IV: a series of four treatises placing the human being, both physically and psychologically, in the ***kosmos***. The first, much influenced by the Pre-Socratics, draws analogies between the ***kosmos*** and human microcosm, both constituted from fire and water, which govern diet, health, sickness, and even reproduction. The second establishes the impact of location (with significant differences from Hippokratic Corpus, Airs,

WATERS, PLACES), food and drink, and lifestyle on the body. Food and drink are presented authoritatively but summarily, hence the need for GALĒN's supplements in *On the Powers of Foods*. The third develops the role of bathing, exercise and daily regime in maintaining health, while recognizing the impact of work and limited resources on the majority of the population. The fourth reviews the production and significance of dreams. The integration of diet, health and cosmology is comparable to Chinese and Indian medicine. The action of bodily heat and fluids (or "humors") on fluids of ingested plants and animals underpins the Hippokratic system of **humors** (which varies between different groups of treatises). These treatises are powerfully located in the thought of the late 5th c. BCE and reflect the importance of lifestyle in maintaining good health, in preference to treatment by drugs (themselves often essences of foods) or surgery. The point is reinforced in *Regimen in Acute Diseases* (next) where the patient's life is under serious threat. Joly (1960) and Joly and Byl (1984) present major reviews of the scholarly debate.

Regimen in Acute Diseases: this work has been much discussed, not least by Galēn in an important commentary, because of its explicit attack from the very first sentence on the *Knidiai Gnomai* of the "Cnidian School," its links with HĒRAKLEITOS OF EPHESOS, and its possible relationship to *Regimen I*, ANCIENT MEDICINE, some of the EPIDEMICS, and *Fractures* and *Joints* (see HIPPOKRATIC CORPUS, SURGERY), among other treatises. The treatise, apparently written in the later 5th c. BCE, is designed for use by professional physicians in critical cases, particularly fevers. The key treatment is varied preparations of barley water (with careful regulation of food), from which the patient progresses to stronger liquids (honey and water, vinegar and water, and wine). An Appendix (of disputed authenticity) discusses certain conditions, prognostics and therapy. An important Arabic translation preserves *Regimen* but not the Appendix. Kühlewein edited the text (1894), and summaries of the scholarly debate appear in Jouanna 1992: 559–560, Joly 1972 (with French translation), and Jones 1931 (with English translation).

Regimen in Health: a short treatise transmitted with *Nature of Man* and normally considered with it, *Regimen in Health* sets out dietary requirements according to season, bodily state, and exercise taken by the ordinary person (men who take moderate exercise) – women, children, and athletes are considered as special cases. Particular attention is given to emetics and clysters. The treatise may date to the late 5th c., like *Nature of Man*, though some doubt surrounds its integrity, not least the quotations from other treatises at the end.

Use Of Liquids: this work complements *Regimen I–IV*, *Regimen in Acute Diseases* and other treatises in concentrating on external applications of fresh water, salt water, vinegar and wine. Heating and cooling are major issues, along with moistening and drying, and cleansing and softening – all according to the medical condition of the patient and consequent state of the skin. The treatise, difficult to read and to date, contains many obscurities as if the text were merely reference notes – but is closely related to APHORISMS 5 and is cited frequently in ERŌTIANOS and Galēn. Heiberg edited the text (1927) and Joly (1972) and Potter include useful editorial comments, translations and summaries of the scholarly debate.

Ed.: J.L. Heiberg, *CMG* 1.1.1 (1927); Jones (1931); R. Joly, *Hippocrate* v. 6.2: *Du regime des maladies aigues, Appendice, De l'Aliment, De l'Usage des liquides* (*CUF* 1972); R. Joly and S. Byl, *Hippocrate: du Regime* = *CMG* 1.2.4 (1984); H. Kühlewein, *Hippocrates: Opera Omnia* 1 (1894); Potter (1995).

R. Joly, *Recherches sur le traité pseudo-hippocratique du regime* (1960); J. Jouanna, *Hippocrate* (1992).

J.M. Wilkins

Hippokratic Corpus, On the Sacred Disease (430 – 400 BCE)

On the Sacred Disease attacks magicians and priests as impious charlatans for blaming the illness on the gods and prescribing ritual cures. The illness, it maintains, is no more sacred than others, since all are divine – yet subject to human expertise. The illness develops *in utero* from an excess of phlegm; phlegmatics are particularly susceptible. Symptoms include convulsions, nightmares, hallucinations, hunchback and, apparently, the epileptic "aura." The illness might disappear in childhood or become chronic; victims may be "distorted" or have no obvious vestiges. Scholars have equated the illness with epilepsy, though stroke, schizophrenia, and tuberculosis may also be indicated. "Sacred disease" is a topic of AIRS, WATERS, PLACES; *On Breaths*; and *Diseases of Young Girls*.

The author also argues for and against the views of philosophers and doctors. Bile cannot cause the illness (*contra Airs, Waters, Places*); constitution is familial (*contra Airs, Waters, Places* which posits that winds and environmental factors determine regional constitutions). Reproduction occurs by pangenesis, and unhealthy seed can be inherited. The importance of air to cognition may show DIOGENĒS OF APOLLŌNIA's influence. The author may be following ALKMAIŌN (and anticipating PLATO and DĒMOKRITOS) in considering the brain the locus of intelligence, emotion, and perception, and not the diaphragm (HOMER, etc.) or the heart (ARISTOTLE, etc.). *On the Sacred Disease* shares enough with *Airs, Waters, Places* that many assert single authorship; there is no consensus on relative dating. There are also significant differences between the texts (as above).

BAKKHEIOS knew the treatise; ERŌTIANOS deems it a genuine work of HIPPOKRATĒS. The pseudo-Hippokratic *Letter* 19 on madness incorporates a section. GALĒN glossed some words and considered its author "noteworthy," though inferior to Hippokratēs, and wrote no commentary. Others providing testimonia assume Hippokratic authorship: HĒRODOTOS (PNEUM.), SŌRANOS in CAELIUS AURELIANUS, and Theodōrētos (the 5th c. bishop of Cyprus).

On the Sacred Disease did not greatly influence the etiology or treatment of the sacred disease. Plato and Galēn attributed the illness to black bile. DIOKLĒS, PRAXAGORAS, Aristotle, THEOPHRASTOS, and SERAPIŌN variously recommend for it such typical substances of the *materia magica* as genitals, blood, and excrement; Galēn advocates concocting pulverized human bone and wearing an amulet. Caelius Aurelianus reports that some doctors thought magicians' aid helpful in treatment.

Littré v. 6; J. Jouanna, *Hippocrate* v. 2.3: *La Maladie Sacrée* (*CUF* 2003).

Julie Laskaris

Hippokratic Corpus, Sevens (440? – 50 BCE)

First known in a Latin translation, then in a fragmentary Greek version (the complete text survives in Arabic), it treats cosmology and pathology by applying the pattern of the number seven (hebdomadic principle). The ***kosmos*** divides into seven parts as do all of the things in it; the outermost part is Olympos, then the stars, Sun and Moon, the sublunary region, with air and waters over the Earth, then at the seventh place the Earth itself. The Earth and the outer region are stationary, but the other five parts revolve eternally around the Earth, moved by themselves and the immortal gods. There are seven stars, and seven seasons, seven winds, seven "seasons" or "ages" in the human life; the seven parts of the world are associated with the seven parts of the body and each part can itself be divided into seven; the soul is also a mixture of seven substances, the Earth's surface divides into seven parts, which correspond to the parts of the body. The second part of the work, discussing the causes and treatment of fevers, makes little reference to arithmology and the number seven. Chronologies proposed swing between the 5th and 1st cc. BCE, but the presence of postclassical features in the tract can be taken for granted.

J. Mansfeld, *The Pseudo-Hippocratic Tract* Περὶ ῾εβδομάδων *and Greek Philosophy* (1971); M.L. West, "The Cosmology of 'Hippocrates', *de hebdomadibus*," *CQ* 21 (1971) 365–388 (text and commentary).

Bruno Centrone

Hippokratic Corpus, Surgery (*ca* 430 – 370 BCE)

Fractures (Fr), Surgery (S): Littré 3; Joints (J), Mochlikon (M): Littré 4; Fistulas (F), Hemorrhoids (H), Ulcers (U): Littré 6

Along with ON HEAD WOUNDS, Fr and J constitute the major Hippokratic surgical treatises. Both address dislocations and fractures, their descriptions, treatment (including diet and purges) and the consequences of non-treatment. Their common approach, similarity of language and cross references (*cf.* Fr 31 and 13 and J 67 and 72) suggest a once unitary work. M ("Instruments of Reduction," garbled toward the end) epitomizes Fr and J, and some passages (7–19, 27–31) were introduced verbatim into J (17–29, 82–87). M, largely following the traditional tendency to proceed from head to foot (except in the introductory chapter on bones where, curiously, it reverses the sequence), possibly reflects the original order of the now hodge-podge arrangement of topics in Fr and J. In antiquity Fr and J were almost universally attributed to Hippokratēs. If by one author, he was a surgeon experienced with bones, muscles, tendons and major blood vessels, presenting himself as a practitioner, not a theorizer, and describing cases he witnessed or attended (e.g. Fr 1–3). He manifests the adversarial attitude found elsewhere in the Corpus (J 1 attests to a public dispute), but, to his credit, considers it fitting for a good surgeon to admit and describe personal failures (J 48). A few of his views, if we correctly understand the text, have puzzled modern readers: e.g. that the fibula is longer than the tibia (Fr 12 and 37).

Reduction devices described in Fr, J and M range from simple (leather balls for shoulder dislocations) to complex ("Bench of Hippokratēs": J 72–73). APOLLŌNIOS OF KITION's commentary preserves illustrations of some mechanisms and maneuvers detailed in J.

S falls into two parts: the first (1–6) treats necessary equipment and conditions as well as personal appearance, positioning and movements of surgeons and assistants. Remaining chapters describe bandaging (types and modes of application with attendant problems): bandages are both knotted and sutured into position, both with and without splints and supports (7–25); only linen is mentioned (11, 12, 22). It is debated whether these condensed and sometimes obscure notes represent an instructional outline (e.g. for opening an "office") to be filled in later, or an abbreviated summary like M.

The author of U, positing that moisture promotes lesions, provides a mine of pharmaceutical information, as he favors non-surgical cures promoted by purges, plasters, and styptics, emphasizing desiccating ingredients. Numerous concoctions include vegetable, mineral and animal products such as clover, lentils, oak gall, myrrh, blister beetles, copper and lead by-products and, of course, hellebore. The final chapters, treating bleeding and cupping (25–27), may be later additions.

The brief treatises H and F focus on maladies of the anal tract and associated conditions like strangury. Their language is similar and, like Fr and J, since antiquity have often been thought to have originally constituted a unitary work. After identifying heated and/or accumulated blood as the cause of piles, fistula in ano, and condyloma, both recommend treatment by medication and fomentation but favor surgery, though apparently not for piles in women (H 7).

These surgical tracts refer to many metal instruments: cupping vessels (J 48, M 38, U 27); knives with fine sharp blades (U 24); probes (U 10, 24, J 11, 37, F 5), of tin (F 4) and lead (F 6); iron cauteries that are slender (J 11), obeliskoid (H 2), and passed through a tube (H 6); iron reduction levers (Fr 31, J 68, M 25, 33, 42); and, for anal dilation, a *katoptēr* (H 4–5; F 3), possibly the familiar rectal speculum of the Roman Empire. These Greek tools were often not regularly professionally prepared, but created *ad hoc.*

Translations (with informative introductions): Fr, J, M, and S: Withington (1928); Fr reprinted in G.E.R. Lloyd, ed., *Hippocratic Writings* (1983); U, H, and F: Potter (1995).

Lawrence J. Bliquez

Hippokratic Corpus, in Pahlavi texts

The deep influence of Greek medicine in Iran is attested from the Achaemenid period when Greek physicians such as Dēmokedēs of Krotōn, Ktēsias of Knidos or Apollōnidēs of Kōs (Ktēsias, *FGrHist* 688 F14.34, 44) were active at the Persian court. Moreover, a Greek doctor, Stephanos of Edessa, cured the Sasanian king Kāwād I (Prokopios, *Bell. Pers.* 2.26). Pahlavi sources clearly refer to the Hippokratic tradition, e.g., **humoral** theory (*Wizidagīhā ī Zadsprām* 29–30). The 9th c. Zoroastrian encyclopedia *Dēnkard* (3.157) attests the Hippokratic distinction between medicine of the body and of the soul. The Christian (Nestorian) school of medicine, where Greek and Indian doctrines intermingled, surely represents a center of diffusion of Western medicine in Iran. A ***melothesia*** in *Iranian Bundahišn* 28.3–5 (e.g., the eyes relate to the Sun and Moon) seems to derive from the Hippokratic Corpus Sevens.

L.C. Casartelli, *La philosophie religieuse du Mazdéisme sous les Sassanides* (1884); *Idem*, "Un traité pehlevi sur la médecine," *Le Muséon* 5 (1886) 296–316, 531–558; H. Fichtner, *Die Medezin im Avesta* (1924); Bailey (1943; 1971) 104–108; E. Benveniste, "La doctrine médicale des Indoeuropéens," *RHR* 130 (1945) 5–12; R.Ch. Zaehner, *Zurvan. A Zoroastrian Dilemma* (1955; 1971[2]); A. Götze, "Persische Weisheit in griechischem Gewande," *Zeitschrift für Indologie und Iranistik* 2 (1963) 60–98, 167–174; J. de Menasce, *Le troisième livre du Dēnkart* (1973); Panaino (2001).

Antonio Panaino

Hippolutos of Rome (200 – 236 CE)

Controversial Christian father (b. *ca* 170 CE) who disputed the status of the official bishop of Rome and had to go into exile on Sardinia. His main work, *Refutation of all Heresies*, is an attempt to derive Christian ideas from earlier Greek philosophy and contains numerous fragments of Pre-Socratics and other philosophers. Book 1 has a number of doxographical reports similar to Diogenēs Laërtios while later books contain many fragments of, in particular, Hērakleitos and Empedoklēs.

C. Osborne, *Rethinking Early Greek Philosophy. Hippolytus and the Presocratics*, (1987); J. Mansfeld, *Heresiology in Context. Hippolytus'* Elenchos *as a Source of Greek Philosophy* (1992); I. Mueller, "Heterodoxy and Doxography in Hippolytus' 'Refutation of All Heresies'," *ANRW* 2.36.6 (1992) 4309–4374; K. Alt "Hippolytos als Referent Platonischer Lehren," *Jahrbuch für Antike und Christentum* 40 (1997) 78–105.

Jørgen Mejer

Hippōn of Krotōn (450 – 430 BCE?)

Hippōn (b. *ca* 475 BCE) continued a line of early **Pythagorean** natural philosophy and Italian medicine (ALKMAIŌN, MENESTŌR, EMPEDOKLĒS); his theories concern mainly physiology, embryology and botany. Of his two books only one literal fragment and about 20 testimonia are preserved. Kratinos in his comedy *Panoptai* ("All-seers" *fr.*167 *PCG* = DK 38A2; *ca* 435–431 BCE) derides Hippōn as a reprobate, which seems to be the origin of his (hardly deserved) reputation as an atheist. Hippōn's activity is connected with traditional centers of **Pythagoreanism** in Italy (Krotōn, Metapontion, Rhēgion), whereas Samos as his birthplace (ARISTOXENOS *fr.*21) is probably a mistake. In any case, he cannot be regarded as an epigone of the Ionian school: his principle, moisture (*to hugron*) seems to be microcosmic rather than macrocosmic and is not identical to THALĒS' water. He believed that there is a moisture in the body due to which it feels and lives; the lack or surplus of moisture, e.g. because of an excessive cold or heat, leads to illness and death. The soul has a moist nature, as does male seed; the latter comes not from the brain (as Alkmaiōn thought), but from marrow (this thesis Hippōn tried to prove "experimentally"). Many embryological views of Hippōn are naive (sex of the child depends on what seed appeared stronger, male or female; twins are born, if seed was more than it is necessary for one child), though some of them survived up to the 19th c. Hippōn's materialistic monism seemed primitive and vulgar to ARISTOTLE, but the idea that health depends on a balance of liquids in an organism became standard for ancient medicine.

DK 38; E. Lesky, *Die Zeugungs- und Vererbungslehren der Antike und ihr Nachwirken* (1950); *HGP* 2.354–358; Zhmud (1997).

Leonid Zhmud

HIPPONAX ⇒ HIPPŌN

Hipponikos (of Athens?) (*ca* 285 – 250 BCE)

Proficient but lackluster geometry teacher whose lectures in Athens the **Platonist** Arkesilaos attended, and whom he restored to health in his own house (DIOGENĒS LAËRTIOS 4.32).

Netz (1997) #47.

GLIM

Hipposiadēs (?) (400 BCE – 300 CE)

The "Laurentian" list of medical writers (MS *Laur. Lat.* 73.1, f.143^{V} = *fr.*13 Tecusan) includes this otherwise unattested name. The list repeats no names, so this entry cannot be an error for *Hippo<kratēs . . . Asklēp>iadēs*. Perhaps the same as HIKATIDAS or HURIADAS; or else perhaps we should restore some name like Iasiadēs (*LGPN* 2.231: one), Dosiadēs (*LGPN* 1.145, 2.135: six; *cf.* the historian, *FGrHist* 458), Sōsiadēs (*LGPN* 1.420, 2.415, 3A.412, 3B.392: eight; *cf.* the non-medical author cited by IŌANNĒS OF STOBOI 1.90), or most likely Pasiadēs (*LGPN* 2.361, 3A.354, 3B.337, 4.274: 18). *Cf.* also EPIPHANĒS, LUPUS, and PHILIPPOS OF KŌS.

(*)

PTK

Homer (750 – 700 BCE)

Associated with the two earliest surviving Greek epics, the *Iliad* and the *Odyssey*, along with a number of shorter hymns and several lost epics of later date. The ancients believed Homer (*Homēros*) to be a blind poet from Khios or another eastern Greek city. Modern scholarship largely accepts the view of Parry and his school that the poems were recorded in the second half of the 8th c., following a long oral tradition. Still disputed is whether the poems preserve traditions of their Late Bronze age setting or reflect the cultural background of the 9th or 8th c.

The later **Stoic** view, followed by STRABŌN (throughout), saw Homer as an entirely accurate guide to the world, particularly in geography; indeed, Strabōn put Homer first among reliable geographers. Geographic and toponymic references in the *Iliad*, concentrated in the Catalogue of Ships and Trojan Allies (2.493–877), reflect knowledge of mainland Greece and western Anatolia. Beyond these regions, geographic data in the poem are scarcer; it is uncertain, for example, whether the poetic tradition knew of the Black Sea. Simpson and Lazenby attribute the geography of the Catalogue to the Aegean Bronze Age, arguing from the prominence of Bronze Age sites later abandoned; this view has been challenged. The old debate about Troy's location has been resolved in favor of Hisarlık in the Troad, first identified as Troy by Calvert and excavated by Schliemann. Excavations have revealed a substantial Bronze Age settlement reoccupied in the Archaic, Classical, Hellenistic and Roman periods after a hiatus. Topographical references in the *Iliad* suggest a familiarity with the Troad and the environs of the settlement.

The *Odyssey*'s geography proves even more contentious. The poem's references to Egypt, Cyprus and Phoenicia suggest an 8th c. worldview. Descriptions of Ithaca and surrounding Ionian islands have proven difficult to reconcile with geographical facts. But the most hotly disputed issue has been the geography of Odysseus' journey to fantastic lands. ERATOSTHENĒS rejected all attempts to locate the journey, but the view that placed the fantastic lands in the central and western Mediterranean prevailed.

R.H. Simpson and J.F. Lazenby, *The Catalogue of Ships in Homer's Iliad* (1970); J.K. Anderson, "The Geometric Catalogue of Ships," in *The Ages of Homer* (1995) 181–191; M. Dickie in *Homer's World: Fiction, Tradition, Reality* (1995) 29–56; J.V. Luce, *Celebrating Homer's Landscapes: Troy and Ithaca Revisited* (1998); C. Dougherty, *The Raft of Odysseus: The Ethnographic Imagination of Homer's Odyssey* (1999); J. Latacz, *Troy and Homer. Towards a Solution of an Old Mystery* (2004).

Philip Kaplan

HONORIUS ⇒ IULIUS

Hostilius Saserna and son (125 – 60 BCE)

The father and son pair wrote about agriculture based in part on their experience with a farm in Cisalpine Gaul (VARRO, *RR* 1.18.6). Theirs was the second oldest Latin treatise on agriculture, after CATO's. TREMELLIUS SCROFA and Varro frequently criticized its recommendations, while COLUMELLA praised it for its detail and expertise. It offered formulae for staffing a farm (Columella 1.7.4, 2.12.7; Varro 1.16.5, 18.2, 6, 19.1), folk remedies (1.2.25–28, 2.9.6), advice on growing vines (Columella 3.3.2, 3.12.5, 3.17.4, 4.11.1; PLINY 17.199), fertilizing crops (Columella 2.13.1), and operating clay, stone, and sand pits (Varro 1.2.22–23). Their land-holdings appear to have been extensive (*ca* 100 ha); thus, like other known Sasernae, they may have been senatorial.

The Sasernae apparently referred to HIPPARKHOS' discovery of the precession of the equinoxes (Columella 1.1.4–5). They are thus the earliest of the very small number of ancient authors to acknowledge the discovery (the others are PTOLEMY, THEŌN OF ALEXANDRIA, and PROKLOS). Their suggestion that the spread of viticulture northward was a sign of climate change brought on by precession constituted an ingenious, if ultimately mistaken, speculation. Finally, the recorded dates for Hipparkhos' observations provide a *terminus post quem* for the Sasernas' work of 127 BCE (but *ca* 70 BCE if they knew Hipparkhos only through DIOPHANĒS OF NIKAIA).

GRL §81; Speranza (1971) 33–45; *OCD3* 1358, A.J.S. Spawforth; *NP* 11.98, K. Ruffing; *HLL* §196.3.

Philip Thibodeau

Hubristēs of Oxyrhynchos (120 – 100? BCE)

ASKLĒPIADĒS PHARM., in GALĒN *Antid.* 2.14 (14.188–189 K.), preserves Hubristēs' record of the prescription of APOLLŌNIOS OF MEMPHIS for "all bites," containing 16 ingredients including Massilian hartwort (DIOSKOURIDĒS 3.53–54), Indian nard, pepper, *kangkhru*, rue, and St. John's wort. The name is unattested after 100 BCE (*LGPN*).

Fabricius (1726) 249.

PTK

Hugiēnos, the "Hippokratic" (100 BCE – 10 CE)

Physician, known for various topical treatments: KRITŌN, in GALĒN *CMLoc* 1.8 (12.488–489 K.), describes his quince-yellow plaster effective against all fluxes; ASKLĒPIADĒS PHARM., in Galēn *CMLoc* 4.8 (12.788), details his collyrium for inflammation at the corner of the eyes and scabs; ARKHIGENĒS, in Galēn *CMLoc* 10.2 (13.353–354 K.), records his topical for sciatica and chills, to remain on the pained area for three hours, after which a bath is prescribed; and HĒRĀS, in Galēn *CMGen* 2.10, 4.14 (13.512, 747 K.), declares his plaster for cicatrization and whitlow the best. The physician's name, curiously recalling the Greek for "health," is attested, along with variants Hugianos, Huginianos, and Huginos (*cf.* the Latinized name HYGINUS), from the 1st c. BCE to 3rd c. CE (*LGPN*).

RE 9.1 (1914) 97, H. Gossen.

GLIM

Hulas (*ca* 60 BCE – 430 CE)

Wrote a geographical work, cited by the RAVENNA COSMOGRAPHY, which treated Macedon (4.9), Sarmatia (4.11: i.e. before 430 CE) and Dacia (4.14: i.e. after 60 BCE). *Cf.* perhaps the Hulas cited by PLINY 10.38 on five ominous Greek birds (for the name *cf. PIR2* H-240, 242); see ARISTARKHOS OF SIKUŌN and SARDONIUS.

(*)

PTK

Hupatia of Alexandria (*ca* 380 – 415 CE)

Lived, was educated and taught in Alexandria. The daughter of THEŌN OF ALEXANDRIA (*PLRE* 2 [1980] 439), she is known, through the correspondence of her famous student

Sunesios of Kurēnē, to have mentored at least 15 students, most of them Christians, all socially and politically elite (Dzielska 27–46). Since Sunesios studied under her until 398 and she probably collaborated with her father, she was probably born *ca* 355. Already elderly when she was murdered, she was the victim of a political conflict between the bishop Cyril (*cf.* Kurillos) and the augustal prefect Orestēs (see Dzielska 83–100 and 1–26 for literary and historical fictions derived from this episode).

Several sources describe her proficiency in mathematics (astronomy and astrology included), as surpassing her father's. The (notoriously unreliable) *Souda*, in particular, credits her with *The astronomical canon*, commentaries on Diophantos (perhaps his *Arithmētika*) and Apollōnios' *Kōnika*, all of which, if they ever existed, are lost or at best survive as anonymous fragments (Cameron 44–48). The heading of Theōn's commentary on the 3rd book of Ptolemy's *Almagest* (2.807 Rome) only shows that Hupatia proofread it for her father (see Jones 1999: 170–172, *contra* Cameron 1993). Cameron 1993 conjectured, on weak evidence, that her "astronomical canon" refers to an edition of Ptolemy's *Handy Tables* and that she is responsible for some interpolations found in the MSS of the Almagest. Qusta's Arabic translation of Books 4–7 of Diophantos' *Arithmētika* may have been based on a Greek text that already included interpolated commentaries perhaps due to Hupatia, but this point is again disputed. Moreover, the study of the direct transmission does not allow positive conclusions on *scholia* to Diophantos written before the 13th c. Knorr (1989: 765–770) plausibly argued that part of Hupatia's hypothetical commentary on Apollōnios may be found in the material used for Eutokios' commented edition of the *Kōnika*. But Knorr's own attempt to circumscribe part of it is weak. His attribution to her of a reworking of Archimēdēs' *De dim. circ.*, likewise, is highly conjectural (*ibid.* 771–780).

The *Souda* notice is partly based on Damaskios' *Life of Isidore*, in which Hupatia's reputation in mathematics is used to belittle her proficiency in philosophy (Cameron 41–43), somewhat contradicting the well-informed and enthusiastic testimony of her student Sunesios, whose letters, although deliberately allusive regarding the content of Hupatia's teaching, clearly show that she considered astronomy "a divine science" leading to philosophy (*Ad Paeonium de dono*, 310c-311a), and one of Euclid's common notions liable to an ethical interpretation (*Epist.* 93). She therefore probably considered mathematics one stage in a philosophical and "psychagogic" curriculum (Dzielska 54–56). This does not imply that she was not competent in mathematics (*cf.* Sunesios' letter to Paeonius), nor that her philosophical obedience was to Iamblikhean Neo-**Platonism** (Dzielska 62–64 *contra* Cameron 49–58). But it may bring her teaching close to what is found some decades later in Proklos' *Hupotuposis*. Probably following her father's interests, she also taught astrology (see Sunesios' allusion to an astrological hydroscope in *Epist.* 15, Dzielska 74–79). She also showed interest in music and musical instruments (Cameron 60). These aspects of her teaching may have contributed to Alexandrian hostility leading up to her death (Dzielska 91).

A. Cameron, *Barbarians and Politics at the Court of Arcadius* (1993) 42–60; M. Dzielska, *Hypatia of Alexandria* (1995); *BNP* 6 (2005) 627–628, P. Hadot; *NDSB* 3.435–437, F. Acerbi.

Alain Bernard

Hupatos (1000 – 1250 CE?)

Some MSS contain a lexicon of terms designating the parts of the body, the title of which includes the word *hupatos*, traditionally interpreted as the author's name, a supposed physician, Hupatos. It was, however, a professorial title in use at the university in Constantinople (from the 11th c. onward); it may also be an adjective expressing the distinct quality of the author whose name has been lost in the MS tradition. The *hupatos* (*tōn philosophōn*) Iōannēs Pediasimos (*fl. ca* 1250), writer of a medical work on obstetrics, might very well be our author, particularly because all the MSS of the work are recent. This lexicon is attributed to HIPPOKRATĒS in some MSS (Diels 1905: 1.43).

Diels 2 (1907) 50; *RE* 9.1 (1914) 251 (#6), H. Gossen; F. Fuchs, *Die höheren Schulen von Konstantinople im Mittelalter* (1926) 50–54; G. Björck, "Remarques sur trois documents médicaux de la Bibliothèque universitaire de Leyde," *Mnemosyne* 3 (1938) 139–150 at 141–145; C.N. Constantinides, *Higher education in Byzantium in the thirteenth and early fourteenth centuries (1204–ca. 1310)* (1982) 113–132.

Alain Touwaide

Hupsiklēs of Alexandria (150 – 100 BCE)

Mathematician and astronomer, later than APOLLŌNIOS OF PERGĒ and roughly contemporary with HIPPARKHOS. Hupsiklēs is perhaps best known as the author of a treatise that survives as Book 14 of EUCLID's *Elements*. This treatise, which concerns the ratio of a regular dodecahedron and icosahedron inscribed in the same sphere, is addressed to PRŌTARKHOS OF BARGULIA. His other surviving work, the *Anaphorikos*, is remarkable for its introduction of the division of the circle into 360 degrees of arc and of the day into 360 degrees of time, as well as for its quantitative and arithmetical approach to a problem that is treated qualitatively and geometrically in Euclid's *Phainomena*. This treatise addresses the question of the time-intervals required for the individual zodiacal signs (that is, the 30°-segments of the **ecliptic** named after the zodiacal constellations) to rise at a given latitude (Alexandria), and uses an arithmetical scheme for computing such rising-times that is known to be Babylonian in origin to answer it. (Some mistakenly infer that Hupsiklēs' use of such a scheme dates him before Hipparkhos.) DIOPHANTOS (*De polyg. num.*) attributes a definition of polygonal number to Hupsiklēs, which some speculate belonged to a treatise on polygonal numbers that has been lost. AKHILLEUS TATIUS indicates that Hupsiklēs also wrote a treatise (not extant) on the harmony of the spheres.

Ed.: V. De Falco, M. Krause, and O. Neugebauer, *Hypsikles: Die Aufgangszeiten der Gestirne* (1966).
Maass (1898) 43; Heath (1921) 1.419–421; Fraser (1972) 2.612, n.381; Neugebauer (1975) 712–733.

Alan C. Bowen

Hupsikratēs of Amisos (30 – 10 BCE)

Wrote a geographical work cited by STRABŌN, describing the Crimean region (7.4.6), the Amazons of the Caucasus (11.5.1), and the western "Ethiopians" (17.3.5). He attained an age of 92, and also wrote history and grammar.

FGrHist 190.

PTK

***Huriadas* (400 – 300 BCE)**

Listed by THEOPHRASTOS (*Sweat* 17) with ANTIPHANĒS OF DĒLOS on disorders related to sweat: uncertain whether a dietician like Antiphanēs or perhaps an athlete or trainer. (The name is otherwise unattested.) *Cf.* perhaps EURUŌDĒS, HIKATIDAS, HIPPOSIADĒS, or Hurradios (father of Pittakos: DIOGENĒS LAËRTIOS 1.74).

(*)

PTK

HYGINUS ⇒ (1) ACILIUS; (2) IULIUS

Hyginus (Agrimensor) (*ca* 100 – 120 CE)

One of two writers named Hyginus in the *Corpus Agrimensorum Romanorum*, the compilation (*ca* 4th c. CE) of texts concerned with land survey and various aspects of measurement. Hyginus refers to a recent distribution of land in Pannonia to veterans of Trajan, i.e., *post* 102 CE. A professional surveyor with substantial field experience, including work at Kurēnē in north Africa, and Samnium, where he investigated changes of ownership in lands allocated to veteran soldiers by Vespasian, he also produced a collection of imperial edicts and decisions on land. Hyginus expounds the role of *limites*, which, because of their specified width and status, were the crucial elements in dividing land into units (*centuriae*) for distribution. He carefully describes the erection of stones appropriately inscribed to designate each *centuria*. In general, Hyginus offers practical guidance to surveyors, advising on methods for recognizing and interpreting boundaries, and emphasizing the importance of using wide ranging evidence. He notes important regional variations in expressing an area of land, such as the *uersus* (8,640 square feet) in Dalmatia. In Kurēnē the Ptolemaic foot (25/24 Roman feet) was in use, and in Germany the Drusian foot (9/8 Roman feet). Notably, in a wide-ranging discussion of land-holding conditions, Hyginus insists on the relevance and importance of local practices, and that each land-holding community should be judged on its own terms.

Thulin (1913); *CAR* 5 (2000); Campbell (2000) 76–101.

Brian Campbell

Hyginus Gromaticus (100 – 300 CE?)

The second of the two writers named Hyginus in the *Corpus Agrimensorum Romanorum* (see HYGINUS [AGRIMENSOR]), and often referred to as "Gromaticus" on the basis of the rather confused MS headings. He refers to the poet LUCANUS, but otherwise makes no datable references. His approach is partly historical in that he discusses the foundation of colonies, but he also describes the procedures of land survey in a way that offers guidance to other surveyors. He is particularly informative on the establishment of *limites*, the dimensions of land division units (*centuriae*), and their proper designation with inscribed stones so that plots of land could be found easily and without ambiguity. Hyginus describes methods of orientation and the alignment of *limites*, using a sundial and the measurement of shadows, and a more complex method based on solid geometry. He also outlines a method for measuring parallel lines using similar right-angled triangles. Hyginus sets out the best methods of land division starting from the principle that the two main *limites*, aligned north-south and east-west, intersected in the middle of the settlement and extended through four gates. Although this could rarely be achieved, surveyors with their professional, scientific approach worked with the administrative bureaucracy to overcome and exploit physical terrain. In a way, they represented the power of the Roman state to control natural resources.

Thulin (1913); *CAR* 4 (1996); Campbell (2000) 134–163.

Brian Campbell

Hyginus, pseudo, de Metatione Castrorum (*ca* 200 – 212 CE)

A military geometer of good theoretical training and practical experience (§45, 47), who wrote probably in the beginning of the 3rd c. CE, but not later than 212 (edict of Caracalla: Grillone 1987: 407–411). *De metatione castrorum* is a more suitable title than the commonly-accepted *de munitionibus castrorum*, proposed by a copyist: the author treats fortifications only briefly at the end (§48–58), where however he expends no small attention on geometrical matters, *coxae* and *clauiculae* (§54–55). *Coxae* round and thus strengthen the angles of the *castra; clauiculae* form a vertical quarter-cylinder, extending from the door's right jamb until the point corresponding to the central point of the opening part of the wall reserved to the door (width = 60 feet: §14,49; Grillone 2000: 378–379). *Clauiculae* and small *fossae* (§50: *titula*) aim to impede frontal attacks, to defend retreating soldiers, and to allow defenders to hit assailants everywhere.

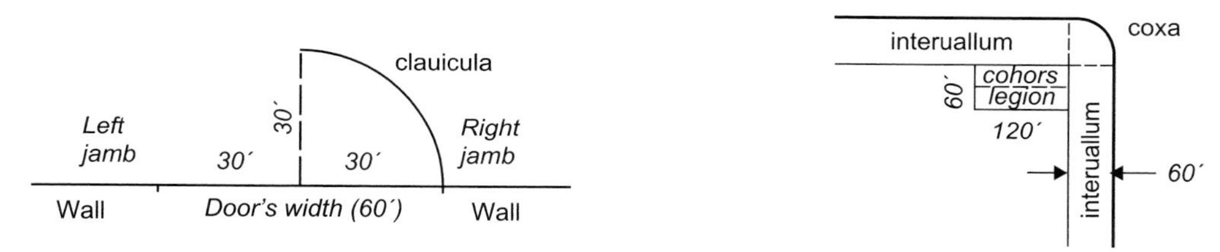

Plan view of Clauicula and plan view of Coxa © Grillone

The rest of the booklet (§1–47), mutilated at the beginning, addresses only *metatio*, i.e., how a camp's surface is distributed between the units of an army of three legions in three parts, to the front (*praetentura*), in the middle (*latera praetorii*), to the back (*retentura*); *cohortes* partly are symmetrically disposed along the four sides of the *castra* (8+8 on left and right hand [§36], 4+4 to the front and to the back [§44]: Grillone 1984: n. 25), partly in *praetentura* and in *latera praetorii* (4+2; §3,9; Grillone 1984: n. 24). In calculating the area necessary for any unit, the geometer allots $1^1/_5$ foot for each infantryman, and three feet for each horseman (width fixed at 30 feet for arms, animals . . .; §1). *Cohortes legionariae* and other units – *auxiliarii* and *gentes* (*nationes* and *symmachares*) – differ in that *cohortes legionariae* take up quarters according to a fixed plan, also if they have less than 600 soldiers (720×30 feet: §1–2), while other troops have an area corresponding to the number of soldiers (i.e., *cohors peditata*: 600 men = 720×30 feet: §27–28), and sometimes the usable area accords to the circumstances (for *gentes*, if they are less or more people: §40).

Ed.: Antonino Grillone, *Hygini qui dicitur de metatione castrorum liber* (1977); M. Lenoir, *Pseudo-Hygin, Des fortifications du camp* (*CUF* 1979).

Antonino Grillone rev. of Lenoir, in: *Gnomon* 56 (1984) 15–26; *Idem*, "Problemi tecnici e datazione del de metatione castrorum dello ps.-Igino," *Latomus* 46 (1987) 399–412; *Idem*, "Soluzioni tecniche e linguaggio di un geometra militare del III secolo: lo pseudo-Igino," in *Atti del IV Seminario Internazionale di studi sulla letteratura scientifica e tecnica greca e latina (Messina 29–31 ottobre 1997)* (2000) 365–395; *Idem*, "Lessico ed espressioni della gromatica militare dello ps. Igino," in *Atti del*

Congresso Internazionale "Les vocabulaires techniques des arpenteurs romains," Besançon (19–21 Septembre 2002) (2005) 125–136.

Antonino Grillone

HY- ⇒ HU-

I

Iakōbos Psukhrestos (457 – 474 CE)

DAMASKIOS, *Philosophical History* 5.84 (pp. 206–213 Athan.), attests most fully to the political, intellectual, and medical prominence of Iakōbos "The Cooler" (from his habit of prescribing cold baths for a number of diseases) in the reign of Leo the Thrakian (457–474 CE); the *Souda* (I-12, 13), John Malalas' *Chronicle*, Marcellinus' *Chronicle*, the *Chronichon Pascale 284–628 AD*, and ALEXANDER OF TRALLEIS add details regarding the remarkable career of Iakōbos and his father HĒSUKHIOS. An avowed pagan, closely associated with Neo-**Platonists** in Athens (including PROKLOS OF LUKIA, whom he treated for a stomach ailment: Dam., 84J), Iakōbos was so renowned for his medical skills, the equal of Asklēpios, that the sculptor Zeuxis produced idealized statues of him. "Iakōbos persuaded his wealthy patients to alleviate the poverty-stricken: he took no payment for his services, being quite satisfied with his salary as *arkhomenos*" (Dam. 84G). In 462 CE, Iakōbos, summoned to the emperor's bedside to cure Leo's high fever, seated himself without the proper signal from Leo, and laid his "healing hands" on his royal patient, scandalizing observers. Returning later, he explained "that he had not acted arrogantly but had done this in accordance with the practices of the ancient founders of his discipline" (Marcellinus, *Chronicle: Leonis Aug. II Solius* [*sic; viz.* 462 CE]: Croke 1995, pp. 23–24). When the wealthy and learned Isokasios was accused of paganism (467 CE), Iakōbos' sensational defense achieved Isokasios' acquittal through his close association with the emperor (that Isokasios underwent baptism may have helped: Malalas, *Chronicle*, and *Chronicon Paschale*).

Fame did not assure good preservation of biographical detail; our sources suggest Iakōbos was born either in Damascus, or Alexandria, or at Argive Drepanon. For two decades he studied the Art of Medicine under his father before going to Constantinople, where father and son prescribed baths, diet, and purgatives, generally avoiding cautery, surgery, and phlebotomy (Dam. 84D).

Alexander of Tralleis has enormous respect for Iakōbos Psukhrestos, writing that he was a "great man possessed of the most divine virtues in the practice of the Art" (5.4 [*On Coughs*] = 2.163 Puschm.), even while praising how Iakōbos had improved the traditional composition of the "Secret Cough-Medicine" (*cf.* NIKĒRATOS' *Secret Remedy*), judiciously combining licorice (*Glycyrrhiza glabra* L.), tragacanth-gum (*Astragalus gummifer* Labill.), high-grade flour, and lettuce-juice. Highly significant are two dual-ingredient recipes to treat gout (Alex. Trall., *Twelve Books* 12: *Podagra* = 2.565, 571 Puschm.), both among the simplest of the recipes in *Podagra*, and both including *hermodaktulon*, the autumn crocus, *Colchicum autumnale* L., the source of *colchicine*, the fundamental drug of modern gout-therapy. Iakōbos was also a skilled pharmacological technologist, as evinced by the five multi-staged, multi-ingredient compounds recorded in the Latin translation of Oreibasios' *Syn.* 7.22 (6.160–161 BDM) to treat nerve-pain. The ingredients of one of them, to be ground in a mortar and mixed with beeswax, olive oil, and butter, include beaver castor, **terebinth** oil, ***opopanax***, chamomile, and other substances, to produce a narcotic salve (*unguentum*) called a *Bromios* (sc. the god Bacchus, thus a drug acting like strong wine), "a pain-killer (*anodinus*) good for luxations and wounded nerves."

O. Holder-Egger, "Die Chronik des Marcellinus Comes und die oströmischen Fasten," *Neues Archiv* 2 (1877) 59–109 at 107; *RE* 9.1 (1914) 622–623, H. Gossen; E. Jeffreys *et al.*, trans. *The Chronicle of John Malalas* (1986); M. and M. Whitby, trans., *Chronicon Paschale 284–628 AD* (1989); Temkin (1991) 214–215 and 222; B. Croke, ed., trans., comm., *The Chronicle of Marcellinus* (1995); *Idem, Count Marcellinus and his Chronicle* (2001) 260.

John Scarborough

Iamblikhos (of Syria?) (*ca* 50 BCE – 450 CE?)

Wrote a geographical work used by the Ravenna Cosmography 2.16–19 on Asia Minor and 4.1–3 on Europe, the Black Sea, and the Bosporos (see also 1.5). The Syrian name Iamblikhos transliterates *ya-mliku*, "God rules," and is attested from *ca* 50 BCE: Cicero, *Fam.* 15.1.2, Strabōn 16.2.10. *Cf.* Libanios Geog. and Porphurios Geog.

(*)

PTK

Iamblikhos (Alch.) (200 – 800 CE)

Two alchemical recipes are ascribed to Iamblikhos (*CAAG* 2.285–287): a procedure for tincturing a metal and another for making gold. Whether falsely attributed to the Neo-**Platonic** philosopher Iamblikhos or actually written by an homonymous author is unknown.

(*)

Bink Hallum

Iamblikhos of Constantinople (*ca* 300? – *ca* 540 CE)

Leontios Skholastikos praises him as a virginal old man, who taught and practiced medicine without fee (*AP* 16.272). Paulos of Aigina, 3.48.4 (*CMG* 9.1, p. 258), records that he prescribed a diet for **dropsy**; such a diet is described in some detail by Alexander of Tralleis (2.455–461 Puschm.). Iamblikhos' probable contemporary Aëtios of Amida records his *digestif* salt, like that of Marcellus (Pharm.), but substituting for the anise and seeds of elecampane and *nasturcium* instead arugula and thistle seeds: 9.24 (p. 507 Cornarius; omitted by Zervos 1911: 324–325).

RE 9.1 (1914) 651 (#5), H. Gossen.

PTK

Iamblikhos of Khalkis (Syria) (300 – 327 CE)

Studied Neo-**Platonic** philosophy with Anatolios of Laodikeia, then with Porphurios of Tyre and subsequently established his own school in Apameia. He wrote numerous works, among them a treatise in ten books on **Pythagoreanism**, commentaries on Plato

(*Timaeus, Parmenides, Phaedrus* and, possibly, *Alcibiades I, Phaedo, Philebus* and *Sophist*) and ARISTOTLE (*Categories, Prior Analytics*). He is credited with the titles, *Pythagorean Way of Life, Protrepticus, On the General Principles of Mathematics, Introduction to the Arithmetic of* NIKOMAKHOS OF GERASA, *On Mysteries* (the title comes from Ficino), a response to Porphurios on the use of mysteries and religion in general, but also containing the core of his metaphysics, *On the Soul*, which draws not only on Plato but Aristotle as well.

Influenced by **Pythagorean** doctrines, Iamblikhos' philosophy is a complex version of the teaching of PLŌTINOS and Porphurios. He introduced triadic schemata into each level of being below the first One, completely unspeakable, and the second One, not related to the triadic structure of the intelligible realm. Below the second One are the Limit and the Unlimited making up the One Existent (DAMASKIOS, *De Principiis* 2, pp. 25.15–26.8 W.-C.). Then come seven triads constituting the intelligible and intellective realms, of which the first member is the One Existent, and the last, called Zeus, plays the role of the **Demiurge** (PROKLOS, *In Tim.* 1, p. 308.17–23). Other members of the triads were also identified as gods, which shows the attempt to integrate traditional religion into Neo-**Platonic** metaphysics. The realm of the soul also has a threefold structure, with a transcendent soul differing both from the world-soul and from individual souls. In contrast to Plōtinos, Iamblikhos denies that any part of the individual human soul does not descend into body: when connected, the whole human soul pervades the body. One consequence is the need for theurgy to set the soul free of the pollution coming from bodies, the other is that the rational element shows itself in each psychic activity of men, and even in the arrangement of the human body (SIMPLICIUS, *In De Anima: CAG* 11 [1882] 187.35–188.3). He is also credited with establishing the curriculum followed later in the Neo-**Platonic** schools at Athens and Alexandria. For generations of later Neo-**Platonists** he was the authoritative philosopher after Plato and Aristotle.

Ed.: (cited in works below)

B.D. Larsen, *Jamblique de Chalcis* (1972); J. Dillon, "Iamblichus of Chalcis (c. 240 – 325 A.D.)," *ANRW* 2.36.2 (1987) 862–909; *RAC* 16 (1994) 1244–1259, G. O'Daly; *NP* 5 (1998) 848–852, L. Brisson; *BNP* 6 (2005) 666–670, M. Fusillo and L. Gallo.

Peter Lautner

Ianuarinus (*ca* 250 – 400 CE)

MARCELLUS OF BORDEAUX 23.24 (*CML* 5, p. 398) records his spleen-poultice, composed of ben-nut oil (CELSUS 6.2.2), cardamom, mustard-seed, nettle-seed, pepper and ***purethron***, ground into vinegar, and placed over the spleen after washing the skin with natron-water. For the name, *cf. PLRE* 1 (1971) 452–453.

Fabricius (1726) 252.

PTK

Iasōn of Nusa (*ca* 80 – 10 BCE)

He succeeded POSEIDŌNIOS as head of the **Stoic** school in Rhodes; son of Menekratēs and of Poseidōnios' daughter, Iasōn was his grandfather's student. He wrote two biographical works, *Lives of Famous Men* and *Successions of Philosophers*, but no fragments of these or any other works have been preserved.

RE 9.1 (1914) 780–781 (#1), F. Jacoby; *GGP* 4.2 (1994) 709, P. Steinmetz.

Jørgen Mejer

Ic- ⇒ Ik-

Idios (250 BCE – 80 CE)

Galēn (*CMLoc* 9.5 [13.297 K.]) quotes Andromakhos' record of his enema, compounded of quicklime, roast copper, realgar, and burnt papyrus, reduced in myrtle wine (or rose-water) and dry wine. Asklēpiadēs Pharm., in Galēn *CMLoc* 9.2 (13.245 K.), gives his spleen remedy (emending *ΙΔΙΩΤΟΥ* to *ΙΔΙΟΥ*): oak mistletoe, reduced in a pottery vessel, then add quicklime, and apply, leaving in place until it falls off of its own accord. For the rare name, *cf. LGPN* 1.231, 2.232, 4.172, or perhaps *cf.* Ioudaios.

Fabricius (1726) 253.

PTK

Idomeneus of Lampsakos (300 – 270 BCE)

Epicurean philosopher who met Epicurus in Lampsakos when Epicurus founded a school there *ca* 310–307. When Epicurus departed to found his school in Athens, Idomeneus remained in Lampsakos as **scholarch**, and kept in touch with Epicurus in a series of letters, fragments of which remain. It is debated whether he is to be identified with the Idomeneus who was active as a politician in Lampsakos during the same period. If so, his works also include *On the Socratics* (Diogenēs Laërtios 2.19, 2.60, 3.36), *On Demagogues*, and a *History of Samothrakē*.

FGrHist 338 (Samothrakē); A. Angeli, "I frammenti di Idomeneo di Lampsaco," *CrErc* 11 (1981) 41–101; *BNP* 6 (2005) 717 (#2), T. Dorandi, and (#3), K. Meister.

Walter G. Englert

Ikkos of Taras (*ca* 470 – 440 BCE)

The **Pythagorean** Ikkos (b. *ca* 500) was in his youth a famous athlete and an Olympic victor. As a doctor and a trainer he practiced gymnastics and dietetics and possibly wrote a book on dietetics as a basis for athletes' training. Approvingly mentioned by Plato, Ikkos was known for his moderate way of life, which included strict diet and abstention during athletic competitions.

DK 25; W. Fiedler, "Sexuelle Enthaltsamkeit Griechischer Athleten und ihre medizinische Begründung," *Stadion* 11 (1985) 137–175; Zhmud (1997).

Leonid Zhmud

Iktinos (465 – 410 BCE)

Architect and author, famous for the Parthenon in Athens, the Temple of Apollo at Bassae, and the Telesterion at Eleusis (Strabōn 9.1.12, 16; Vitruuius 7.*pr*.12, 16; Pausanias 8.41–7–9), co-wrote (with Karpiōn) a treatise on the Parthenon (Vitr. 7.*pr*.12). Plutarch (*Pericles* 13) states that Kallikratēs collaborated with Iktinos on the Parthenon, and names other architects participating in the Telesterion's construction, perhaps in a later phase. Iktinos faced substantial challenges in all three projects.

The current Parthenon was constructed (449–432 BCE) on a platform intended for

its narrower and longer predecessor of *ca* 485 BCE, burnt by the Persians in 480/479 BCE. Iktinos designed a wider temple, with 8 × 17 columns allowing a larger interior, and re-used many blocks and column drums from the older building, some of them re-cut. By emphasizing the proportion 4:9, using many "refinements" (deviations from the horizontal and vertical), including Ionic features, and an elaborate sculptural program, Iktinos created a superlative temple. M. Korres has shown that the Parthenon had windows in the door wall between the cella and pronaos, an interior service staircase to the attic in the width of the same wall on its north side, and included an earlier shrine in the north peristyle.

At Eleusis, Iktinos sought a spacious interior to accommodate initiates into the Eleusinian Mysteries in privacy. He chose a nearly square plan with a Doric exterior and an Ionic, many-columned interior, more than doubling the space of its predecessor. Others completed this work when Iktinos went to Bassae *ca* 429–427 BCE, where he repeated an archaic predecessor's plan for continuity, but introduced the first engaged Ionic columns into the interior, the first Corinthian capital, and an interior sculptured frieze. Iktinos was outstanding for his innovations, adaptability, and skillful engineering.

M. Korres in P. Tournikiotis, ed., *The Parthenon* (1994) 56–97, 138–161; F. Cooper, *The Temple of Apollo Bassitas* I (1996) 369–379; Svenson-Ebers (1996) 157–211; *BNP* 6 (2005) 708–709, H. Knell; *KLA* 1.338–345, M. Korres; J. Neils, ed., *The Parthenon* (2005).

Margaret M. Miles

Imbrasios (Paradox.) (100 – 550 CE)

One of the sources named by THEOPHULAKTOS, at the end of his *Quastiones Physicae*. The rare name is attested (Markovich). *Cf.* HABRŌN, HIEROKLĒS OF ALEXANDRIA, and SŌTIŌN, also named as sources.

M. Markovich, "Supplement to *RE*: A New Paradoxographer," *CP* 54 (1959) 260; *RE* S.10 (1965) 328, *Idem.*

PTK

Imbrasios of Ephesos (300 BCE – 650 CE?)

Putative author of a short work on **iatromathematics**, *Prognostica de decubitu ex mathematica scientia*, attributed to GALĒN in most MSS but to Imbrasios in a single codex seemingly independent of the rest of the manuscript tradition. Weinstock argued for the authenticity of the ascription to Imbrasios and further speculated that this Imbrasios was a pseudepigraphical writer identifiable with a legendary Egyptian priest-magician Iambrēs or Ambrēs. The work is of exceptional interest as one of the few extant on **iatromathematics**. After a preface addressed to an Aphrodisios and invoking the **Stoics**, HIPPOKRATĒS, and DIOKLĒS OF KARUSTOS, the main body of the text works systematically through forecasts for a sick person based on astronomical conditions in effect at the time the patient is bedridden, with particular emphasis on zodiacal position, apparent speed, and latitude of the Moon.

S. Weinstock "The Author of Ps.-Galen's *Prognostica de Decubitu*," *CQ* 42 (1948) 41–43.

Alexander Jones

IN THEAETETUM AUCTOR ⇒ P. BEROL. 9782

Innocentius (350 – 360 CE?)

Vir perfectissimus and *auctor*, known as the writer of a tract entitled *On explaining legal records and signs* (*De litteris et notis iuris exponendis*), extracted from Book 12 of a work otherwise lost, probably devoted to surveying. The name, Innocentius, of this *agrimensor*, a high-level imperial administrator of equestrian rank, is associated with five lists (*Casae litterarum*), recording the boundaries of 107 properties (*fundi*), each identified by a letter of the Latin (lists 1, 4, 5) or Greek alphabet (lists 2, 3). Critical analysis shows that only the 2nd, 5th and half the material in list 1 are original, referring to genuine *fundi*, some along the *uia Flaminia*, perhaps near Rome; the 4th list is interpolated and list 3 is just a catalogue of symbols. It is uncertain if Innocentius authored the whole collection, the first delimitation, or just the revision as interpolations to list 1 (in part) and list 4 (entire) intimate. The dates of the lists appear to vary, but the emperor Arcadius (395–408 CE), speaking about demarcation of properties and letters, quotes a "12th Book," surely Innocentius'. He is probably (1) the surveyor who advised Constantius II near the Danube in 359 CE (Amm. Marc., 19.11.8); but he could be (2) the Innocentius associated with a Paulus (probably the jurisconsult, Praetorian prefect *ca* 218–219 CE) in Gisemundus' *Ars gromatica*; or even (3) the presumed author of the *Ius respondendi*, written late in the 3rd c., before Diocletian's reign (*RE* 9.2 [1916] 1558–1559). However, the late-antique language and Christian allusions indicate a more recent date: Constantius II's surveyor would have written the 1st and perhaps 2nd and 5th lists.

Ed.: K. Lachmann, *Die Schriften der römischen Feldmesser* 1 (1848) 310–338.

Å. Josephson, *Casae litterarum. Studien zum Corpus Agrimensorum Romanorum* (1950); L. Toneatto, "Note sulla tradizione del *Corpus agrimensorum Romanorum*. I. Contenuti e struttura dell'Ars gromatica di Gisemundus (IX sec.)," *MEFRM* 94 (1982) 191–313 at 223; *Idem, Codices artis mensoriae. I manoscritti degli antichi opuscoli latini d'agrimensura (V–XIX sec.)* (1994–1995) 1002; J. Peyras, "Ecrits d'arpentage et hauts fonctionnaires géomètres de l'Antiquité tardive," *DHA* 21 (1995) 166–186, 29 (2003) 160–176, 30 (2004) 166–182; St. Del Lungo, *La pratica agrimensoria nella tarda Antichità e nell'alto Medioevo* (2004) 569–637; A. Roth Congès, "Nature et authenticité des Casae litterarum d'après l'analyse de leur vocabulaire," in *Les vocabulaires techniques des arpenteurs latins, Actes du colloque international de Besançon (19–21/09/2002)* (2006) 71–124.

A. Roth Congès

***Iōannēs Archpriest* (700 – 800 CE?)**

Cited as "from the divine Euagia" in the list of *poiētai* (makers of gold, *CAAG* 2.25). In the treatise bearing his name, *On the Sacred Art* (*CAAG* 2.263–267), he invokes in Gnostic fashion celestial and ***dēmiourgic*** natures, Unity, and the Triad; he cites PSEUDO-DĒMOKRITOS and ZŌSIMOS. The ANONYMOUS ALCHEMIST PHILOSOPHER (*CAAG* 2.424) cites him as "Iōannēs made archpriest in Euagia of the Tuthia and of the sanctuaries included." Berthelot notes that Euagia can either be a place-name or mean "sanctity"; whereas Tuthia could indicate a location, or **calamine**, or even the place to prepare that substance (Berthelot 1885: 118; *CAAG* 3.406, note). Berthelot considers Iōannēs historical and notes that the name is Christian, but his function seems to recall some Egyptian institution (1885: 186).

Ed.: *CAAG* 2.263–267.

Berthelot (1885) 186–187; *ODB* 55, s.v. Alchemy.

Cristina Viano

Iōannēs Esdras (1100 – 1200 CE)

Credited with an unpublished treatise on urine preserved in a Greek MS (Città del Vaticano, *Reginensis graecus* 182, f.4). The text strongly resembles that of the 12th c. Byzantine physician Iōannēs bishop of Prisduana contained in several MSS (Diels 2.55; Zervos in *EEBS* 10 [1933]: 362–382; see Dimitriadis 1971: 41–42).

Diels 2 (1907) 53; *RE* 9.2 (1916) 1800 (#32), H. Gossen; Dimitriadis (1971) 43.

Alain Touwaide

Iōannēs Iakōbos (1200 – 1400 CE?)

Wrote several medical treatises known in Latin MSS; the orthography of the name suggests a Greek origin, yet no Greek text seems extant. He probably lived after 1000 CE, as none of his work is known in pre-Salernitan Latin MSS. Furthermore, one of the texts attributed to him seems to be a Latin translation of compound medicines from Avicenna's *Qanūn* (*ca* 1000). Similarly, *de pestilentia*, known in numerous MSS under different titles, might be related to the plague of 1348. His name is also attached to texts on head wounds, stones, fever, and female sterility.

Diels 2 (1907) 53; *RE* 9.2 (1916) 1800 (#33), H. Gossen; Thorndike and Kibre (1963) 45, 521, 1421, 1502, 1693, 1709 (plague); 227 (formulas from Avicenna); 1028, 1081 (head wounds); 1214 (stones); 1341 (fever); 1506 (female sterility).

Alain Touwaide

Iōannēs Iatrosophist (400 – 650 CE?)

Known solely from a treatise on therapeutics apparently contained in only one 15th c. Byzantine MS (Paris, BNF, *graecus* 2316). The text closely follows the ancient version of the therapeutic collection by Iōannēs *arkhiatros* (date unknown) attested through several MSS and different versions. The therapeutic work of the MS might be attributed to this Iōannēs. In any case, Iōannēs' title (*iatrosophistēs*), the iatrosophic nature of the work along the lines of the *Alphabetum empiricum* ascribed to DIOSKOURIDĒS and STEPHANOS OF ALEXANDRIA, and the presence of a commentary on the HIPPOKRATIC CORPUS APHORISMS preceding Iōannēs' text in the same MS, all suggest a late antique date and a location in such a medical school as Alexandria or Ravenna. Iōannēs might be identifiable with other better-known Iōannēs but is probably distinct from IŌANNĒS OF ALEXANDRIA.

Diels 2 (1907) 54; *RE* 9.2 (1916) 1800 (#34), H. Gossen; Temkin (1932) 66; Ihm (2002) #283–284; *BNP* 6 (2005) 897, V. Nutton.

Alain Touwaide

Iōannēs Matthaios (*ca* 1450 CE?)

In one Latin manuscript credited with the work *Consilia medicinalia*. The ascription might be a partially truncated form of the name of the late medieval Italian physician Giovanni Matteo Ferrari de Gradi (d. 1472), who authored similar *Consilia* and commentaries on Avicenna's *Qanūn*, GALĒN's *Tegni* (the Latin translation of the Arabic version) and Rāzī's *Nonus Almansoris*.

Diels 2 (1907) 54; *RE* 9.2 (1916) 1800 (#35), H. Gossen.

Alain Touwaide

Iōannēs of Alexandria (500 – 700 CE?)

Physician (*iatrosophistēs*) and teacher in the Alexandrian school; his name is typically Christian. He refers vaguely to his teacher, assumed without secure evidence to have been GESSIOS OF PETRA. He authored commentaries on at least two treatises of the HIPPOKRATIC CORPUS: EPIDEMICS VI and *Nature of the Child.* The works are typical of the Alexandrian school in their Galēnic interpretation of Hippokratic medicine. Iōannēs might have written other works, lost in Greek, but preserved in Arabic, such as a commentary on GALĒN's *Theriac.* This work and other commentaries on Galēnic treatises such as *De pulsibus* attributed in the Arabic versions to an unspecified Iōannēs may be best ascribed to this man (or a homonym?), all the more because this man is often confused with several Arabic authors including Iōannēs ***grammatikos***. It is unlikely that our Iōannēs of Alexandria was responsible for the manual on nosology and therapeutics contained in a unique manuscript of Paris as by an otherwise unknown homonym.

Ed.: C.D. Pritchet, *Iohannis Alexandrini Commentaria in librum de sectis Galeni* (1982); J.M. Duffy, *John of Alexandria, Commentary on Hippocrates' Epidemics VI. Fragments. Commentary of an Anonymous Author on Hippocrates' Epidemics VI. Fragments* (1997) = *CMG* 11.1.4.

Diels 2 (1907) 51; *RE* 9.2 (1916) 1800 (#25), H. Gossen; Temkin (1932) 66–71; *KP* 2.1430 (#13), F. Kudlien; Ullmann (1970) 89–91; I. Garofalo, "La tradizione araba del commento di Ioannes grammatikos al De pulsibus di Galeno," in A. Garzya and J. Jouanna, edd., *I testi medici greci. Tradizione e ecdotica. Atti del III Convegno Internazionale Napoli 15–18 ottobre 1997* (1999) 185–218; P.E. Pormann, "Jean le grammarien et le De sectis dans la littérature médicale d'Alexandrie," in I. Garofalo, A. Roselli, *Galenismo e medicina tardoantica: fonti greche, latine e arabe* (2003) 233–263; *BNP* 6 (2005) 897, V. Nutton.

Alain Touwaide

Ioannēs of Alexandria, Philoponos, Grammatikos (*ca* 510 – 570 CE)

Born *ca* 490; studied at the **Academy** in Alexandria under AMMŌNIOS OF ALEXANDRIA (SON OF HERMEIAS) and probably taught there, although he never held the chair of philosophy. A Christian Neo-**Platonist**, his name probably indicates his association with a group of lay Christians, the *philoponoi.* His earliest surviving works are Neo-**Platonic** commentaries on ARISTOTLE, as well as more elementary works, including a treatise on the astrolabe and an introduction to NIKOMAKHOS' *Arithmetic.* Four of these commentaries (on the *Prior* and *Posterior Analytics, De anima* and *De generatione et corruptione*) derive from Ammōnios' lectures, edited and augmented by Philoponos. Commentaries on the *Physics*, the *Categories* and the *Meteorologica* also survive. Starting around 529, when Justinian closed the Athenian **Academy**, Philoponos wrote a series of anti-eternity polemics, including *Against Proklos on the Eternity of the World* and the fragmentary *Against Aristotle on the Eternity of the World.* His natural philosophical corpus culminated in *De opificio mundi*, written in the 540s (although some dispute this date), an attempt to harmonize pagan natural philosophy with the account of creation in *Genesis.* For the remainder of his career he focused his formidable intellectual talents on Christian theological matters, including developing a doctrine of the trinity based on a rigorous application of Aristotle's definition of substance. This doctrine, called "tritheism" by its opponents, ultimately led the Church to anathematize Philoponos in 681.

The commentaries written with Ammōnios are traditional, Neo-**Platonic** exercises aiming to construct a harmonized and systematic philosophy from the writings of PLATO and

Aristotle. Later commentaries on the *Physics* (517) and the *Meteorologica* (after 529), however, display a marked shift toward a more critical approach to Aristotelian physics. Philoponos considers void space a theoretical possibility. He proposes an alternative to Aristotelian forced motion, suggesting an impetus force is somehow imparted to the moved thing by the mover. In the *De opificio mundi* (1.12, pp. 28–29 Reichardt), Philoponos even suggests God placed an impetus force in the heavens at creation. Most Christian natural philosophers denied the heavens a divine status and Philoponos also denied them a soul, so a primitive impetus theory could provide a natural – as opposed to a supernatural or psychic – explanation for their motion. He also refines the concepts of prime matter and place.

His best known and most widely influential contributions to natural philosophy were his arguments against eternity. The extent of these polemics is quite broad. All arguments in defense of eternity are attacked in detail, and, while some of these have little philosophical force on their own, as a whole the polemics make for a compelling dossier against the philosophical case for eternity. Moreover, some of his arguments are quite novel and powerful, particularly those using puzzles about infinity. Their effectiveness is evident in SIMPLICIUS' response in his commentary on the *De caelo*. He attacks not only the arguments, but also Philoponos' character and his Christianity. Nonetheless, Philoponos' arguments against eternity spread widely, particularly in the Islamic Middle Ages, and subsequently in the Latin West.

While the anti-eternity polemics are primarily a negative critique, Philoponos' final natural philosophical work aimed to construct a Christian natural philosophy through a literal reading of *Genesis*. The *De opificio mundi*, primarily striving to reconcile Moses' account with Greek science, was written in response to anti-pagan Christian natural philosophers, such as KOSMĀS INDIKOPLEUSTĒS, whose *Topographia Christiana* ridiculed Christians who failed to abandon Greek philosophy when they forsook pagan religion. Philoponos' reply shows that Greek rationalism is not only useful for Christians but necessary. Moreover, he argues that the *Genesis* narrative prefigures and even influences later Greek cosmology. The main points of contention between Kosmās and Philoponos focus on the shape of the world, the materiality of angels and the anthropology implied by being made in the image of God.

Ed.: H. Hase, *De usu astrolabii eiusque constructione* (1839); R. Hoche, *Eis to prōton [kai deuteron] tēs Nikomakhou Arithmētikēs eisagōgēs* (1864); *CAG* 13–17 (1887–1909); H. Rabe, *De aeternitate mundi* (1890); G. Reichardt, *De opificio mundi libri VII* (1897); R. Sorabji, ed., *ACA* (1987–); C. Scholten, trans., *De opificio mundi* (1997).

RE 9.2 (1916) 1764–1795 (#21), W. Kroll; *DSB* 7.134–139, S. Sambursky; R. Sorabji, ed., *Philoponus and the Rejection of Aristotelian Science* (1987); *REP* 7.371–378, C. Wildberg; *BNP* 11 (2007) 89–91, K. Savvidis and C. Wildberg; *NDSB* 4.51–52, Carl Pearson.

Carl Pearson

Iōannēs of Antioch, arkhiatros (1200 – 1500 CE?)

A collection of compound medicines is preserved under the name of a Iōannēs of Antioch in a Byzantine manuscript now in Paris, BNF, *graecus* 2315 partially copied by Zakharias Kalliergēs (d. after 1524). The texts in the manuscript seem to reproduce a collection created in a late Byzantine hospital, perhaps in Constantinople, as they include the Byzantine translation of Avicenna's *De pulsibus* (*ca* 1000) and other treatises circulating among contemporary practicing physicians. If the collection does derive from such a milieu, it might represent a hospital's accumulated recipes: other similar collections make explicit reference to hospitals and that mode of gathering recipes, and the epithet *arkhiatros* provides some confirmation.

RE 9.2 (1916) 1800 (#27, 28), H. Gossen; Diels 2 (1907) 2.51.

Alain Touwaide

Iōannēs of Antioch, Khrusostomos ("Chrysostom") (*ca* 380 – 407 CE)

Born *ca* 350, student of DIODŌROS OF TARSOS and of LIBANIOS, priested 386, bishop of Constantinople 397/398, deposed 403, exiled 404, died 407. Wrote numerous ethical treatises, commentaries on Christian scriptures, sermons, letters, and speeches (e.g., those *Against the Jews*, 386–387: trans. Harkins, 1999). A work *Demonstration of the Construction of the Human Body*, attributed in the margin to "Khrusostomos," is preserved in MS *Ambros.* Q94 Sup. (undated), f.364V (Diels 1907: 2.23). Three Paris MSS – *Coislin.* 78 (11th c.), f.199, 79 (11th/12th c.), f.79, and *Parisin.* 912 (14th c.), f.266 (Diels 1907: 2.52) – preserve a work *On Diseases and Doctors* attributed to the bishop, possibly in error for IŌANNĒS OF ANTIOCH, *ARKHIATROS* (*cf.* Diels 1907: 2.51–52, *Parisin.* 2315, 15th c., f.117, extracts from DIOSKOURIDĒS, and *Therapy of Various Diseases*, in many MSS). From the 6th c., Byzantine texts describe him as "golden-tongued" (*khruso-stomos*) for his sermons, and numerous works were ascribed to him; received as a saint by the Orthodox and Roman churches.

RE 9.2 (1916) 1800 (#29), H. Gossen; *OCD3* 329, W. Liebeschütz (no mention of medical writings); *BNP* 6 (2005) 890–892 (#4), J. Rist (*ditto*).

PTK

Iōannēs of Philadelpheia, "Lydus" (*ca* 540 – *ca* 561 CE)

Imperial bureaucrat and scholar, who served most of his career under Justinian. Perhaps in 543, he was appointed to a chair at the imperial school in Constantinople. Of Lydus' three extant works, *De mēnsibus* (*Peri mēnōn*), *De ostentis (Peri diosēmeiōn)*, and *De magistratibus*, only the first two are scientific. *De mens.* and *De ost.* share calendrical interests, and a consistent, if understated, engagement with philosophical issues. The concept of the motion of heavenly bodies as a chronological mechanism underlies both treatises. *De mens.* contains passages of **Pythagorean** numerology, probably drawn from a contemporary compendium, while *De ost.* is concerned with exegesis of Ptolemaic, and ultimately Platonic and Aristotelian, world-systems. These concerns are apt for a student (in 511) of the Neo-**Platonic** philosopher AGAPIOS (*cf. De Mag.* 3.26).

Lydus' natural scientific works, particularly *De ost.*, stand in a tradition of calendrical and meteorological astrology which continued from the ***parapēgmata*** of METŌN, EUKTEMŌN and EUDOXOS, through ARATOS, VERGIL and OVID, to the astronomical didactic writings of the Renaissance.

The *De mens.* gives information about the week and months. Book 1 treats the old Roman calendar; the second book, days of the week; the third, the months; the fourth gives a ritual calendar similar to OVID's *Fasti.* Of primary natural scientific interest is Book 1, where Lydus describes Numa's institution of the solar year.

The *De ostentis* has been characterised as an astrological compilation, as confirmed by Lydus' own list of sources, *De ost.* 2 (pp. 4–5 Wa.), citing authors as diverse as

(PSEUDO)-ZOROASTER, PETOSIRIS, ANTIGONOS, ARISTOTLE, HĒLIODŌROS, ASKLATIŌN, ŌDAPSOS OF THĒBAI, POLLĒS OF AIGAI, and PTOLEMY; among his Roman sources are PLINY and VARRO (*cf.* Wachsmuth, pp. XVII–XXVIII). The *De ost.* is also valuable for the study of important but poorly-attested authors, such as NIGIDIUS FIGULUS and CLODIUS TUSCUS. What looks like a miscellany, from the standpoint of *Quellenforschung*, actually presents a coherent system, based on synchronic signs, months and dates. The desire for such unity seems evident in Lydus' own program at *De ost.* 4 (pp. 6–7 Wa.).

Of special interest are Lydus' various systems of dating, indicating divergences between sources, or competing contemporary systems. For example, days of the month listed in the **ephemeris** of chapters *De ost.* §§27–38 are numbered consecutively, whereas, in the ephemeris of Clodius Tuscus (§§59–70), this system is combined with a Greek version of the old Roman system of counting back from the fixed points of the month. In *De mens.* Book 2, Lydus follows the seven-day "planetary" week, rather than the nundinal cycle appropriate to his antiquarian material.

Ed.: R. Wuensch, *Ioannis Laurentii Lydi, Liber de mensibus* (1903; repr. 1967); C. Wachsmuth, *Ioannis Laurentii Lydi, Liber de ostentis* (1863; 2nd edition 1897).

A.K. Michels, *The Calendar of the Roman Republic* (1967); *KP* 3.801–2 (#2), T.F. Carney; *PLRE* 2 (1980) 612–615; M. Beard, "A complex of times: no more sheep on Romulus' birthday," *PCPS* (1987) 1–15; M. Maas, *John Lydus and the Roman Past: Antiquarianism and Politics in the Age of Justinian* (1992); *OCD3* 899, L.M. Whitby; *BNP* 8 (2006) 14–15, F. Tinnefeld.

Emma Gee

Iōannēs of Stoboi (400 – 440 CE)

Iōannēs from the Macedonian city Stoboi (often cited as "Stobaios") collected a large number of "excerpts, sayings, and precepts" from more than 500 Greek authors, from HOMER to THEMISTIOS (who is the *terminus post quem*). He dedicated this collection to his son. It was divided into four books on physics (1), logic and ethics (2–3), and political theory and practice and various practical matters (4). In the medieval tradition it was split into two different volumes, Books 1–2 called *Eclogae physicae et ethicae*, Books 3–4 *Florilegium*. Stobaios has preserved many quotations from Greek authors otherwise lost; the text was arranged in thematic chapters, e.g. "Is the universe one?," "No one is willingly evil," or "On virtue." In Book 1 he used the doxographical collection of AËTIOS, in Books 2, 3 and 4 we find many of DĒMOKRITOS' ethico-political statements. Stobaios is also an important source for many (Neo)**Pythagoreans** and Neo-**Platonists**, not to mention poets like Euripidēs and Menander.

Ed.: C. Wachsmuth and O. Hense, *Iohannis Stobaei Anthologium* 5 vv. (1884–1912).

Mansfeld and Runia (1996) 196–271; *NP* 11.1006–1010, R.M. Piccione and D.T. Runia; *DPA* 3 (2000) 1012–1016, R. Goulet.

Jørgen Mejer

Iollas of Bithunia (150 – 110 BCE?)

Physician prior to HĒRAKLEIDĒS OF TARAS, both of whom are quoted by DIOSKOURIDĒS (*MM* 1.*pr.*1); wrote a lacunose pharmacological work of unknown title. NIKANDROS' *Scholia in Thēriaka* twice refer to somebody called Iolaos, who must be the same person as Iollas: (1) *ad* verse 683, concerning the herb named *puritis*/***purethron***; (2) *ad* verse 523, on the Peloponnesian phytonym *rhutē* (*Ruta graveolens* L.), from another work *On the Peloponnesian cities*. Celsus 5.22.5 describes one of his compound medicines, a cauterizing powder.

GGLA 1 (1891) 826, M. Wellmann; *RE* 9.2 (1916) 1855 (#2), H. Gossen.

Jean-Marie Jacques

Iōn of Khios (*ca* 460 – before 421 BCE)

Writer of lyric poetry and tragedy, born in Khios *ca* 490, also known as Xouthos. His tragedies were performed during the 82nd Olympiad (452–449). He wrote prose works, including a history of the foundation of Khios (*Khiou Ktisis*) and a book of memoirs (*Hupomnēmata* or *Epidēmiai*), where Iōn recounts his meetings with and opinions about great men such as Kimōn, Aeschylus, Sophoklēs, Periklēs or Sōcratēs.

The earliest known testimonies about Iōn are found in Aristophanēs and Isokratēs. The former (*Peace* 832–837, presented in 421 BCE), dramatically assuming Iōn's recent death, calls him "morning star," alluding to the first words of one of his dithyrambs. Isokratēs (*Antidosis* 268), including Iōn among the "old sophists," together with Empedoklēs, Alkmaiōn, Parmenidēs, Melissos and Gorgias, recalls their theories about the number of the first elements.

Iōn authored a philosophical book entitled *Triagmos* or *Triagmoi*, a word of dubious meaning interpreted as "tripartition" or "triad." The treatise opens as follows: "This is the beginning of my discourse: all things are three and not more or less than these three. The virtue of each singular thing consists of a triad, intelligence, power and fortune." (DK 36B1). Iōannēs Philoponos says that Iōn postulated fire, earth and air as the material elements (DK 36A6), which some scholars interpret as a cosmological version of the triad-theory. The scholion to Aristophanēs' *Peace* (832) quotes a book entitled *Kosmologikos*, probably a different title for the same work. According to Aëtios, "about the nature of the Moon, Iōn believes that it is partly a translucent and transparent body, partly an opaque one" (DK 36A7). Ancient scholars considered Iōn to be not only a poet but also a natural scientist.

Ed.: DK 36; *FGrHist* 392; A. Leurini, *Ionis Chii. Testimonia et Fragmenta* (1992).
DPA 3 (2000) 864–866, L. Brisson.

José Solana Dueso

Iōnikos of Sardēs (*ca* 380 – 400 CE?)

Philosopher, physician, rhetorician, and poet, son of a physician and student of Zēnōn of Cyprus, knowledgeable in all aspects of medicine, especially theory and anatomy, a highly admired teacher, esteemed for his practical therapy, pharmacology, bandaging, and surgery. He was also skilled in medical prognostication and divination (Eunapios, *Vit. Phil.* 499), but no publications are attested.

BNP 6 (2005) 1078, V. Nutton.

GLIM

Iordanes (*ca* 550 CE)

Historian of Gothic descent who most probably worked in Constantinople and wrote in Latin. He compiled a world chronicle, *De Summa Temporum Vel Origine Gentis Romanorum* (known as the *Romana*) and a history of the Goths, *De Origine Actibusque Getarum* (known as the

Getica). The *Getica* is based on the lost history of the Goths by CASSIODORUS, and scholars debate the extent of Iordanes' borrowings. The *Getica* contains much geographical information, beginning with a geographical introduction and including geographical digressions. Iordanes' geographical descriptions highlight the places important in Gothic history. His introduction focuses on Scandza, according to Iordanes an island in the Northern Ocean and the place of the origin of the Goths. Iordanes cites PTOLEMY and POMPONIUS MELA, but the sources for most of his description of Scandza have not been identified. Some scholars have suggested that he may have relied on Gothic historical and geographical writers, but a consensus has not been reached.

Ed.: Th. Mommsen, *Monumenta Germaniae historica. Auctores antiquissimi* v. 5.1 (1882) 53–138.
C.C. Mierow, *The Gothic History of Jordanes* 2nd ed. (1915; repr. 1960, 2006); *RE* 9.2 (1916) 1908–1929, A. Kappelmacher; *KP* 2.1439 (#1), M. Fuhrmann; *PLRE* 3 (1992) 713–714; *OCD3* 798, P.J. Heather; Natalia Lozovsky, *"The Earth Is Our Book": Geographical Knowledge in the Latin West ca. 400–1000* (2000); *BNP* 6 (2005) 917–918 (#1), P.L. Schmidt.

Natalia Lozovsky

Iouba II of Mauretania, C. Iulius (*ca* 20 BCE – 24 CE)

Iouba II (inv. 1944.100.81120) © Courtesy of the American Numismatic Society

King of Mauretania and Libya, an eminent scholar who wrote in Greek (PLINY 5.16). Son of Iouba I and prisoner of CAESAR in 46, along with his pro-Pompeian father, Iouba II became a friend and client of AUGUSTUS and a Roman citizen, renamed C. Iulius. He married Kleopatra Selēnē, daughter of M. Antonius and KLEOPATRA, and was restored to his father's throne in 25 BCE. Thoroughly Greco-Roman by his education among the Roman oligarchy, and moreover Punic-speaking, he developed this double culture in his kingdom. With his extensive resources, he sent exploratory missions to the Canary islands (*fr.*44), to seek the source of the Nile which he supposed was in the Atlas mountains (*fr.*38a). Iouba discovered the plant ***euphorbia*** (DIOSKOURIDĒS 3.82.1) and developed and established the "Getulian purple" industry from orchil (indigenous to Mogador).

A prolific author (*Souda* I-399), aided by numerous collaborators, Iouba wrote many compilations: *On theatre* (17 books), *On painting* (eight books), *On the history of Rome* (two or four books), and on cultural comparative history *On Similitudes* (15 books), treating parallel customs, manners, words, etc. between different peoples, especially Greek vs. Roman. This broad natural-anthropological perspective seems typical of Iouba's cultural conception as shown in the remains of his three ethnological works *Libuka* (at least three books), *Assuriaka* (two books derived from BERŌSSOS) and *Arabika* (on southern countries from Egypt and "Ethiopia" up to India). He collected many data on natural history and treated zoology (*frr.*3, 40, 58, 70, 71), botany (*frr.*2, 62–69), and mineralogy (*frr.*72–79) in later works. Unfortunately only 100 fragments survive, despite his immeasurable influence on Greek (as PLUTARCH, Athēnaios, and AELIANUS who copies him in maybe 50 chapters of *NA*) and Latin writers (Pliny quotes him 37 times – e.g., 8.4 on the nature of elephant tusks: horn [Iouba] or tooth [HĒRODOTOS]). Mentioned among the *auctores externi* for 16 books, he is probably the main source of Pliny and later encyclopedists for natural history in exotic countries. His work was lost and unknown to Latin early medieval writers, but "Iorach," often quoted as a scientific authority in Arabic literature and influential on medieval encyclopedias (e.g. Arnoldus Saxo), is very probably Iouba II himself.

GGLA 2 (1892) 402–414; *OCD3* 799, K.S. Sacks; *DPA* 3 (2000) 940–954, J.M. Camacho Rojo and P.P. Fuentes Gonzales; I. Draelants, "Le dossier des livres sur les animaux et les plantes de Iorach: tradition occidentale et orientale," in I. Draelants *et al.*, edd., *Occident et Proche-Orient: contacts scientifiques au temps des croisades* (2000) 191–276; D.W. Roller, *Scholarly Kings: The Writings of Juba II of Mauretania, Archelaos of Kappadokia, Herod the Great and the Emperor Claudius* (2004).

Arnaud Zucker

Ioudaios (250 BCE – 25 CE)

Physician whose plaster for skull fractures consisted in salt, red copper scales, roasted copper, ***ammōniakon*** incense, frankincense soot, dried resin, "Kolophōn" resin, calf suet, vinegar and olive oil (CELSUS 5.19.11B). Celsus also preserves Ioudaios' skin powder of lime, red natron, and a young boy's urine, recommending that the area to be treated be moistened occasionally (5.22.4). Ioudaios, attested only once, 2nd c. BCE (*LGPN* 3B.207), is perhaps a corruption of the ethnic Ioudas, Ioudion, or Ioudiōn known from the 1st c. CE (*LGPN*).

RE 9.2 (1916) 2461, H. Gossen.

GLIM

IOUNIAS ⇒ IUNIA

Iriōn (?) (250 BCE – 80 CE)

ANDROMAKHOS, in GALĒN *CMGen* 6.10 (13.913 K.), records this man's *phaia* (dark) plaster, containing **litharge**, roast copper, and **verdigris**, plus birthwort, **galbanum**, and ***opopanax***, in a beeswax-and-resin base. The name *ΙΡΙΩΝ* seems otherwise unattested, and perhaps we ought to read *ΕΙΡΗΝΙΩΝ* (*LGPN* 2.139) or else *῞ΗΡΙΩΝ* (*LGPN* 3B.183–184). Alternatively, it may be a brand-name, as are the immediately preceding ("Phtheirograph") and following ("Hellespontian") plasters.

Nutton (1985) 145.

PTK

Isidōros (300 – 500 CE?)

Found in the list of philosophers "of the science and of the sacred art," at the beginning of MS *Marcianus gr.* 299 (f.7V), and probably identifiable with PETASIOS, an Egyptian synonym of "Isidōros" (gift of Isis).

(*)

Cristina Viano

Isidōros the Younger (*ca* 510 – 563 CE)

Nephew of ISIDŌROS OF MILĒTOS, who repaired Hagia Sophia, designed by his uncle, after the dome collapsed during the earthquake of 558 CE (Prokop. *Aed.* 2.8.25). Consecrated in

563, the new dome was 6 m deeper, more stable, but less "awe inspiring" (Agathias, *History* 5.9). Isidōros and John of Constantinople, both young men, built fortifications, churches, barracks and baths in Zenobia, Mesopotamia (Prokop. *Aed.* 2.8.25). Constantine of Rhodes (9th c.) names Isidōros as co-designer of Justinian's church of the Holy Apostles in Constantinople. Although not an academic, Isidōros seems to have matched his uncle's reputation. The illustrious *mēkhanikōs* celebrated in two house inscriptions at Qinnesrīn, Syria, (*ca* 550 CE) has been associated with Isidōros the Younger.

RE 9.2 (1916) 2081, E. Fabricius; *IGLSyr* 2, #348, #349; Downey (1948) 105; W. Emerson and R.L. van Nice, "Haghia Sophia, Istanbul," *AJA* 47 (1947) 403–436; *eidem*, "Haghia Sophia: The collapse of the first dome," *Archaeology* 4 (1951) 94–103; *RBK* 3 (1975) 508–510, M. Restle; Warren (1976) 10–12; Mainstone (1988) 215–217; *ODB* 1017, W. Loerke and M.J. Johnson; *PLRE* 3 (1992) 724–725 (#5); R. Taylor, "A Literary and Structural Analysis of the First Dome on Justinian's Hagia Sophia, Constantinople," *JSAH* 55 (1996) 66–78.

Kostis Kourelis

Isidōros of Abudos (170 – 160 BCE?)

Designed a large stone-throwing catapult at Thessalonikē, probably for its successful emergency defense against the Romans (Livy 44.10.5–7). Isidōros' stone-thrower, described by BITŌN, *Belop.* 3 (pp.48–51 W.), was a mechanically-assisted bow (i.e., *gastraphetēs*), cocked by a winch (*kokhlias*), that shot stones of *ca* 20 kg (40 *minae*).

Marsden (1971) 68–69, 82–84.

PTK

Isidōros of Antioch (50? – 80 CE)

Traditionally considered GALĒN's student and friend (Gossen), but the reference is by ANDROMAKHOS OF CRETE (YOUNGER) in Galēn (*CMGen* 5.12, 13.834–835 K.); he might have practiced in Rome (Fabricius). Andromakhos quotes five compound recipes authored or used by Isidōros: a lozenge for dysentery compounded from yellow orpiment, realgar, copper scales, saffron, etc, mixed with sweet wine (*CMLoc* 9.5, 13.295–296 K.); ***trokhiskoi*** against *aphthae* (*CMGen* 5.12, 13.833–835 K.); plaster for wounds (gangrenous and malign: *CMGen* 6.6, 13.885 K., giving the ethnic); and a plaster for dermatological affections (*ibid.*, 908 K.).

RE 9.2 (1916) 2080 (#29), H. Gossen; Fabricius (1972) 228.

Alain Touwaide

Isidōros of Kharax (*ca* 40 – 1 BCE)

Greek geographer of Kharax Spasinou (later Antioch), an important mercantile center in southern Mesopotamia on the Persian Gulf, author of *Stathmoi Parthikoi*, an itinerary of the caravan trail from Zeugma to the borders of India, naming the supply stations maintained by the Parthian authorities for the convenience of merchants and containing some description of local traits. The work includes names of stations and intervening distances indicated in *skhoinoi*, a Persian unit of measure. Other fragments attributed to Isidōros deal with long-lived people, pearl fisheries in the Persian Gulf, measurements of the ***oikoumenē*** based on ERATOSTHENĒS and records of distances given by PLINY. Isidōros, identified as "Dionusios," is probably also the author of a survey of the east commissioned by AUGUSTUS before Gaius' expedition to Armenia, against the Parthians and Arabs, 1 BCE (Pliny 6.141).

Ed.: *GGM* 1.244–256; *FGrHist* 781; W.H. Schoff, *Parthian Stations by Isidore of Charax* (1914).

M.L. Chaumont, "Études d'histoire parthe. V. La route royale des Parthes de Zeugma à Séleucie du Tigre d'après l'itinéraire d'Isidore de Charax," *Syria* 61 (1984) 63–107; A. Luther, "Zwei Bemerkungen zu Isidor von Charax," *ZPE* 119 (1997) 237–242.

Daniela Dueck

Isidōros of Memphis (250 BCE – 540 CE)

AËTIOS OF AMIDA 7.110 (*CMG* 8.2, p. 387) cites his collyrium: grind ***ammōniakon*** incense, cuttlefish ink, ***opopanax***, **silphium**, **verdigris**, ***sagapēnon***, and gum in water, and pour into a mixture of fennel-juice and honey.

Fabricius (1726) 303; *RE* 9.2 (1916) 2080 (#30), H. Gossen.

PTK

Isidōros of Milētos (*ca* 500 – 558 CE)

Architect, mathematician, and academic. In 532 CE he collaborated with ANTHĒMIOS OF TRALLEIS in the design of Hagia Sophia, Constantinople, and he advised Justinian I in the dams of Dara (Prokop. *Aed.* 1.1.24, 2.3.7). Isidoros edited mathematical texts, particularly ARCHIMĒDĒS and EUCLID, and was probably professor of geometry in Constantinople. He also wrote a commentary on HĒRŌN OF ALEXANDRIA's lost treatise *On Vaulting*. Among Isidoros' students was EUTŌKIOS OF ASKALON, who notes his teacher's invention of a device for drawing parabolas. Scholars have celebrated Anthēmios and Isidoros as mathematical theorists akin to the architects of antiquity and the Renaissance. Although their scientific interest is irrefutable, their editorial activities served the practical needs of their profession rather than the search for higher mathematical principles.

RE 9.2 (1916) 2081, E. Fabricius; Downey (1948) 99–118; *RBK* 3 (1975) 505–508, M. Restle; Warren (1976); Mainstone (1988) 157; Alan Cameron, "Isidore of Miletus and Hypatia: On the Editing of Mathematical Texts," *GRBS* 31 (1990) 103–127; *ODB* 1016, M.J. Johnson and W. Loerke; *PLRE* 3 (1992) 724 (#4).

Kostis Kourelis

Isidōros of Milētos' student (author of *Elements* Book XV) (520 – 580 CE)

The end of the pseudo-Euclidean "Book 15" of the *Elements* (*Elementa* 5.1, pp. 29–38 Heiberg), extant in Greek but missing from known Arabic translations, treats the following question: how to find by geometrical construction the inclination between adjacent faces of the five regular solids. The constructions detailed therein are explicitly attributed to "Isidōros our great teacher" (29.21 Heiberg), later called "the most glorious man previously mentioned" (30.26 Heiberg); the five "instrumental constructions" are given first and each is then carefully justified through demonstrations, including analyses through *data*.

Four similar mentions of "Isidōros the Milesian *mēkhanikos*, our teacher" are found in addenda to EUTOKIOS' commentaries on ARCHIMĒDĒS: three (48.30, 224.9, 260.12 Heiberg) allude to Isidōros' proofreading of Eutokios' commentaries (Decorps 2000: 62, n.8), and the last (84.8–11) mentions a compass drawing parabolas invented and described by Isidōros in his commentary to HĒRŌN's (lost) *Kamarika*. The precise references to

APOLLŌNIOS' *Kōnika* in Eutokios' commentaries probably derive from Isidōros' revision (Decorps 2000: 82). The Isidōros in question may be either the uncle or the nephew, since both were famous *mēkhanikoi.*

Heath (1926) 3.519–520.

Alain Bernard

Isidorus (Isidore) of Hispalis (Seville) (*ca* 610 – 636 CE)

Encyclopedist, historian, theologian. Isidore was born (possibly in Cartagena, Spain *ca* 560) to a noble family in Visigothic Spain and was educated by his brother, Leander, whom he succeeded as Bishop of Seville in 600; he died April 4th, 636. His works (extant in several hundred medieval MSS), covering Biblical exegesis, canon law, theology, history, philosophy and science, served throughout the Middle Ages as handbooks for various disciplines. They preserved philosophical and scientific ideas current in late ancient Rome that had ultimately derived from Greek sources. Isidore was among the late ancient encyclopedists (CALCIDIUS, MACROBIUS, MARTIANUS CAPELLA, BOËTHIUS, and CASSIODORUS) whose works contained both texts and diagrams, setting the model for the genre of the medieval encyclopedia and serving as its sources.

Isidore's chief scientific works include *Etymologiae* or *Origines, De natura rerum, De ordine creaturarum* (brief explanations of various natural phenomena), and *De differentiis uerborum* and *De differentiis rerum* (concepts and distinct nature of difference present in words and in things respectively).

The encyclopedic *De natura rerum* includes the division of time, and the description of the planetary system and Earth with its parts and connected astronomical and natural phenomena. Isidore's natural philosophy centers on his theory of elements, visualized in a cubic diagram and a series of circular (*rota*) diagrams that became the standard visual means of depicting elemental concepts during the Middle Ages. His theory of elements relies on Calcidius' *Commentary* on PLATO's *Timaeus* (combining Aristotelian and Platonic concepts) and on medical sources (connecting elemental qualities with **humors** of the human body and temperaments based on them). He linked the *mikrokosmos* (man) with the *makrokosmos* (universe) through their parallel elemental structure and described atoms conceptually as the smallest invisible particles present in bodies and time or even in numbers or letters.

In *Etymologiae* (20 books), Isidore organized a large body of diverse encyclopedic knowledge around the etymology of words on the principle that the name of a thing is key to its nature. Though his etymologies are often farfetched and misleading, they represent a new approach towards organizing knowledge. The first two books discuss the *trivium* (*grammatica, rhetorica, dialectica*), the third book the *quadrivium* (*arithmetica, musica, geometria, astronomia*). Books 4, 11, and 12 discuss medicine, man, and the animal world respectively. Books 13 and 14, describing the parts of the universe and natural phenomena, provide a theory of elements and atoms; Book 16 treats stones and metals.

By using Christian as well as pagan sources, Isidore secured not only survival but also broad acceptance for ancient concepts.

Ed.: *Opera omnia* in *PL* 81–84; W.M. Lindsay, *Etymologiarum sive originum libri XX* (1911); J. Fontaine, *Traité de la nature* (1960).
J. Fontaine, *Isidore de Séville et la culture classique dans l'Espagne Wisigothique* vv. 1–2 (1959), v. 3 (1983).

Anna Somfai

Isigonos of Nikaia (50 BCE – 70 CE)

Author of high standing (Aulus Gellius 9.4), composed a work of *Apista*, of which fragments survive in *Cod. Vatic.* 12. As the codex mentions a second book, this work consisted of at least two books. Different themes were discussed: ethnography, zoology, hydrography. Isigonos seems to have relied on excellent sources, such as ARISTOTLE, ANTIGONOS OF KARUSTOS, THEOPHRASTOS and NUMPHODŌROS. Some disagreement remains regarding chronology. Albeit, PLINY 7.12, 7.16 functions as the ultimate *terminus ante quem*. The *terminus post quem* depends on whether we can count VARRO among Isigonos' sources, combined with Isigonos' possible role as source for POSEIDŌNIOS and NIKOLAOS OF DAMASKOS. In all probability, the author was active towards the end of the 1st c. BCE.

Ed.: *PGR* 146–148.

RE 18.3 (1949) 1137–1166 (§17, 1155–56), K. Ziegler; Giannini (1964) 124–125; *KP* 2 (1967) 1463, W. Spoerri; *OCD3* 768, J.S. Rusten; *BNP* 10 (2007) 506–509 (I.B.1, 508–509), O. Wenskus.

Jan Bollansée, Karen Haegemans, and Guido Schepens

Isis, pseudo (Alch.) (175 – 225 CE)

An alchemical text entitled *Of Isis the Queen of Egypt and Wife of Osiris Concerning the Sacred Art, Addressed to her son Hōros* survives in two redactions (*CAAG* 2.28–35; for date see Mertens [1988] 4). Both begin with variations of a myth in which Isis receives knowledge of alchemy from the angel Amnaēl and end with almost identical procedures for the "whitening of all bodies."

Festugière (1950) 253–256; M. Mertens, "Une scène d'initiation alchimique: la *Lettre d'Isis à Horus*," *RHR* 205 (1988) 3–23.

Bink Hallum

***Isis, pseudo (Pharm.)* (250 – 10 BCE)**

SCRIBONIUS LARGUS 206 explains that GLUKŌN's excellent plaster was known as "Isis"; according to HĒRĀS, in GALĒN *CMGen* 5.2 (13.774–775 K.), EPIGONOS' plaster was so-named (*cf.* 5.3, p. 794); Galēn distinguishes MAKHAIRIŌN's, Epigonos', and "the one called Isis," *Ad Glauk. Meth. Med.* 2.10 (11.126 K.), 2.11 (p. 138). Galēn cites several remedies thus inscribed – *CMGen* 4.13 (13.736–737, 747), as does PAULOS OF AIGINA 4.19.2 (*CMG* 9.1, p. 339, also Makhairiōn), 4.40.3 (p. 360, also MANETHŌN and the "Athēnē" drug), 4.43.3 (p. 362), 4.45.5 (p. 366), and 7.17.39–40 (*CMG* 9.2, pp. 356–357, also "Athēnē"). But at 4.48.2 (p. 369), Paulos appears to refer to "drugs of Isis and of Makhairiōn," as if "Isis" were a person. *Cf.* ISIS, PSEUDO (ALCH.).

Fabricius (1726) 303–304.

PTK

Iskhomakhos of Bithunia (70 BCE – 60 CE)

Physician, wrote *On the School of Hippokratēs* (*CMG* 4, p. 175), suggesting that HĒRAKLEIDĒS OF EPHESOS (perhaps rather HĒRAKLEIDĒS OF TARAS) attributed *Regimen* to HIPPOKRATĒS. ERŌTIANOS censures Iskhomakhos and KUDIAS OF MULASA for their alternate orthography of *iktar* (I-20 [p. 47.2 Nachm.]) and quotes our author with GLAUKIAS OF TARAS and an unidentifiable Hippōnax for their explanation of *kokhonē* (*fr.* 17 [p. 103.15 Nachm.]).

FGrHist 1058; Ihm (2002) #153.

GLIM

Itineraries (from *ca* 100 CE)

Although Greeks listed notable points on coastal voyages (***periploi***), the compilation of equivalent documents for land travel (*itineraria*, from Latin *iter* = journey) was primarily a Roman development stemming from construction of roads along which milestones were placed. Even so, claims that reliance upon itineraries caused Romans' worldview to be linear rather than spatial are extreme. The content of a typical *itinerarium* is minimal, comprising a start- and end-point, the names of intermediate stopping-points and the distance between each, and a total figure for the entire distance. Point-to-point distances rarely exceed 20–25 miles (thus furnishing successive overnight stops after a day's journey), although the fullest itineraries may include additional intermediate points. Distances are usually recorded in Roman miles (sometimes even half-miles), except in Gaul where the local *leuga* (one Roman mile and a half) is often preferred. Itineraries seldom include any reference to the nature and condition of road surfaces, the character of terrain to be traversed, the relative importance of stopping-points, or other circumstances of concern to travelers.

Itineraries were produced by both public and private initiative, and recorded by various means; there is as yet no indication that they were linked to maps. Inscribed stone tablets erected at city gates offer onward itineraries to neighboring communities and even to Rome. Small silver beakers survive listing over 100 intermediate points on the 1,840–mile journey from Gadēs (modern Cadiz) through Spain and across the Alps to Rome. The trip made by a privileged lawyer from Hermopolis Magna (Egypt) to Antioch (Syria) and back *ca* 320 is detailed on papyrus. An unnamed Christian pilgrim writes a notably full record of travels between Gaul and Jerusalem in 333. No doubt some sets of itineraries were assembled for reference by provincial administrators and imperial couriers, but how comprehensive such collections were, and how widely available, is far from clear. The one surviving collection, the misnamed *Antonine Itinerary* of *ca* 300, is a raw, confusing assemblage of routes (not all of them direct), seemingly the work of an anonymous individual enthusiast. The maker of the PEUTINGER MAP too, who is so dependent upon itineraries, evidently needed to gather and organize them as he seems to have lacked access to a full, collated collection.

Ed.: B. Löhberg, *Das* Itinerarium Provinciarum Antonini Augusti: *Ein kaiserzeitliches Strassenverzeichnis des Römischen Reiches – Überlieferung, Strecken, Kommentare, Karten* 2 vv. (2006).

Richard Talbert, "Author, audience and the Roman empire in the *Antonine Itinerary*," in R. Haensch and J. Heinrichs, edd., *Der Alltag der römischen Administration in der Hohen Kaiserzeit* (2007) 256–270.

Richard Talbert

Iuliana (485 – 527/8 CE)

Cited only in the list of philosophers "of the science and of the sacred art," at the beginning of MS *Marcianus gr.* 299 (f.7V). Berthelot identifies her with Iuliana Anicia (b. 462, d. 527/8), daughter of Olybrius (Western emperor 472), for whom the illustrated MS of DIOSKOURIDĒS was produced.

CAAG 1.122; Letrouit (1995) 57.

Cristina Viano

Iulianus (Pharm.) (520 – 540 CE)

AËTIOS OF AMIDA 11.12 (p. 609 Cornarius), giving remedies employing goat's blood, notes that he is a contemporary deacon, and records his antidote, containing saffron,

"Ethiopian" cumin, myrrh, parsley, two kinds of pepper, spikenard, etc., in dried goat's blood and honey.

PLRE 2 (1980) 638.

PTK

Iulianus Imp. (330 – 363 CE)

Cited in f. 242 of MS *Parisinus gr.* 2327: "Thus is accomplished the precept of the Emperor Iulianus." Berthelot finds this significant, since Iulianus consorted with magician students of IAMBLIKHOS and himself practiced theurgy.

Berthelot (1885) 145.

Cristina Viano

L. Iulianus Vertacus (300 – 470 CE)

Writer on arithmetic and astrology, used by ANTHEDIUS (Sidonius Apollinaris, *Ep.* 8.11.10; *Carmen* 22.*pr.*3).

PLRE 1 (1971) 952.

GLIM

Iulianus (of Alexandria?) (*ca* 140 – 160 CE)

GALĒN met Iulianus ("Julian"), the **Methodist** physician, some time during his youthful sojourn in Alexandria: ". . .more than twenty years ago, since when he has written handbook upon handbook, always changing them and altering them, never content with what he has written. . .' (*MM* 1.7.6 [10.53 K.] = Hankinson 1991: 27). Iulianus had studied under APOLLŌNIDĒS OF CYPRUS. Thanks to Galēn's acidic logic and nuanced condemnation, little remains of Iulianus' writings, even though one can, through painstaking reading, discern the main outlines of his works on the definitions of health and disease. Galēn's *Against Iulianus* so completely demolishes **Methodism**'s medical logic that Tecusan simply edits and translates the entire tract to suggest the involuted and precise philosophical sarcasm applied to **Methodist** doctrine, also explicated by Hankinson (1991: 145–160).

Despite his scorn for the **Methodists**, Galēn (*CMGen* 2.21 [13.557 K.]) preserves the complicated recipe, suggesting an expertise in pharmacology, for Iulianus' *enaimos* – a thick, adhesive, styptic plaster that "sealed wounds shut," to avoid stitches (*cf.* HIPPOKRATIC CORPUS, *Fractures* 24; THEOPHRASTOS *HP* 4.7.2). The *enaimos*, prepared in bulk, probably was an ordinarily available plaster to treat wounds suffered by gladiators; it had a long "shelf-life," since it included 50 parts each of **litharge** and Dead Sea bitumen (*asphaltos*), copper flakes (12 parts), and ***khalkitis*** (four parts). The beeswax (50 parts), carefully roasted pine-resin (15 parts), and the finest Bruttian pine-pitch (50 parts), ensured the *enaimos*' adhesive properties. Finally, smaller quantities of frankincense, myrrh, two kinds of birthwort (*Aristolochia* spp.), and aloe-latex (prob. the "best," viz. *Aloe perryi* Baker from Socotra) gave the plaster a mild analgesic and antibiotic quality, the latter augmented with oak-gall (*kēkis*). Those 13 ingredients, plus **galbanum**, were compounded in "old olive oil."

Ed.: E. Wenkebach, *Galeni Adversus Lycum et Adversus Iulianum libelli* (1951) = *CMG* 5.10.3, pp. 33–70; Tecusan (2004) 290–331 (*fr.*111), with trans.

RE 10.1 (1918) 11–12 (#4), H. Gossen; Frede (1982); Scarborough (1982); R.J. Hankinson, "Methodism" in *Cause and Explanation in Ancient Greek Thought* (1998) 318–321.

John Scarborough

Iulianus of Askalon (*ca* 530 – 535 CE)

Architect from Askalon, known exclusively as author of a treatise composed around 531–533 CE. *On the laws or customs [nomoi . . . ethē] of Palestine* was transmitted as an appendix in the *Book of the Eparch* (9th/10th c.) and incorporated in Harmenōpoulos' *Hexabiblos* (14th c.). The treatise prescribes codes for building in Askalon and encapsulates local customs as well as the influence of Beirut's law school. Iulianos, revealing his interests in natural philosophy, organizes the work around the four elements of fire, air, water, and earth.

Scholars have inconclusively attempted to link him with other architects by the same name. A Iulianos *arkhitektōn* who built a *noria* is addressed in a letter by AINEIAS OF GAZA (*PLRE* 2 [1980] 639 #16), and another *Iulianos* supervised repairs in an aqueduct at Sardica (*PLRE* 3 [1992] 738 #21). Inscriptions referring to a *Iulianos* at Qasr al-Brad, Syria, have also been associated with Iulianos of Askalon or his hypothetically homonymous father.

J. Geiger, "Julian of Ascalon," *JHS* 112 (1992) 31–43; B.S. Hakim, "Julian of Ascalon's Treatise of Construction and Design Rules from Sixth-Century Palestine," *JSAH* 60 (2001) 4–25; *ODB* 1079–1080, M.Th. Fögen; M.Ja. Sjuzjumov, "O tractate Juliana Askalonita," *ADSV* 1 (1960) 3–34.

Kostis Kourelis

Iulianus of Laodikeia (*ca* 500 CE?)

Astrological author mentioned often in late antique and Byzantine Greek astrological literature. A chapter transmitted as part of the compilations of astrological texts associated with RHĒTORIOS, in fact a reworking of the text on influences of fixed stars by the ASTROLOGOS OF 379, identifies "Iulianus the polyhistor" as its source, and since the positions of the stars cited in this version have been corrected to fit a date within a year or two of 360 years after PTOLEMY's star catalogue in the *Almagest* (whose epoch is 137 CE), it is plausibly supposed that Iulianus was active about 500 CE. Another section of the Rhētorian corpus comprises a series of ten chapters headed "Useful Selections from the Discoveries of Iulianus of Laodikeia on ***katarkhai***." Other texts, both astrological and astronomical, came to be falsely ascribed to him (in particular in the form of a pseudo-treatise called *Astronomical Episkepsis*) through the chaotic processes by which the astrological literature was selected and reordered in the Byzantine manuscript tradition.

CCAG 8.4 (1921) 244–253.

Alexander Jones

Iulianus of Tralleis (100 – 180 CE?)

SIMPLICIUS, in *De Caelo* 2.1 (*CAG* 7 [1894] 379–380), quotes ALEXANDER OF APHRODISIAS refuting Iulianus' theory that the cause of the regularity and right-handedness of the motion of the heavens is "soul." He is presumably distinct from the *theourgos* of *ca* 160–180 CE, who placed the Sun midmost of the planets (PROKLOS, *In rem Publ.* 2, p. 220 = *In Tim.* 1, pp. 63, 132) – always called "*theourgos*" and never assigned to Tralleis: *BNP* 6 (2005) 1045 (#4–5), S.I. Johnston.

RE 10.1 (1918) 9 (#1), H. von Arnim.

PTK

IULIUS ⇒ (1) AUSONIUS; (2) FIRMICUS; (3) IOUBA

Iulius Africanus (*ca* 190 – 235 CE)

Born *ca* 160 CE; in addition to his five books on the world chronology, both pagan and Christian, from early ages to his own time, his main work is a technical encyclopedia entitled *Kestoì* ("*Embroideries*"), written between 227/8 and 232/3 and presented to the emperor Alexander Seuerus. No firm hypothesis can be made about the actual framework of this book, since only excerpts and fragments have come down to us. Since a papyrus preserves the end of Book 18, the original number of books was likely 24 (*Souda* A-4647).

Vieillefond divides this material into these main sections, exhibiting a variety of interests and approaches: extracts from Book 7: on warfare, on horse diseases, on weights and measures, lyric fragments; extracts from Book 13: on cinnamon, on dyeing. The metrological chapters, in five recensions, appear as a somewhat muddled conflation of lemmas. In descending order, Africanus explained the main weights, liquid- and grain-measures used in the Mediterranean, along with each sub-multiple. Some recensions record the corresponding weight of the Roman currency system in use later than Africanus' time, or assign this extract to HĒRŌN or DIDUMOS, suggesting this section should perhaps be credited to a different writer.

RE 10.1 (1917) 116–125 (#47), W. Kroll; J.-R. Vieillefond, *Les "Cestes" de Julius Africanus* (1970); H. Chantraine, "Der metrologische Traktat des Sextus Iulius Africanus, seine Zugehörigkeit zu den κεστοί und seine Authentizität," *Hermes* 105 (1977) 422–441; *OCD3* 778, J.F. Matthews; T. Rampoldi, "I 'κεστοί' di Giulio Africano e l'imperatore Severo Alessandro," in *ANRW* 2.34.3 (1997) 2451–2470; *RAC* 19 (2001) 508–518, F. Winkelmann; *NP* 11 (2001) 494–495 ("Sextus" #2), J. Rist.

Mauro de Nardis

Iulius Agrippa (10 BCE – 90 CE)

Recompounded recipes by earlier pharmacists, as recorded by ASKLĒPIADĒS in GALĒN: *CMGen* 7.12 (13.1030–1031 K.), an ***akopon*** potion including ***euphorbia*** (*cf.* IOUBA), ***malabathron***, etc., revising GAIUS OF NAPLES; and *CMLoc* 8.5 (13.185–186 K.), stomach ointment including **bdellium**, cardamom, cassia, cinnamon, ***malabathron***, myrrh, Indian nard, pepper, etc., in **terebinth**, from POLUARKHOS. The use of ***euphorbia*** and ***malabathron***, plus the citation by Asklēpiadēs, yield a date-range consistent with either of two homonymous descendants of King Herod, though an identification is not substantiated.

PIR2 I-128 to I-132.

PTK

Iulius Atticus (10 – 30 CE)

From a prominent family in Gallia Narbonensis, an older contemporary of COLUMELLA (1.1.4), author of a monograph on viticulture. Columella, calling him an expert in the field, cites some of his recommendations with approval (4.2.2, 3.11.9), while criticizing him for e.g. preparing trenches for vine plants too deep (3.16.3, 4.1.1–6, 4.2.2). Atticus considered the shade of elm trees noxious (PLINY 17.90). *Cf.* IULIUS GRAECINUS.

GRL §497.1; *OCD3* 779, M.S. Spurr; *BNP* 6 (2005) 1080 (#IV.3), E. Christmann.

Philip Thibodeau

IULIUS AUSONIUS ⇒ AUSONIUS

Iulius Bassus (*ca* 10 – 40 CE)

Friend of SEXTIUS NIGER (CAELIUS AURELIANUS, *Acut.* 3.135 [Drabkin, p. 386; *CML* 6.1.1, p. 372]; the MS has *TVLLIVS*), who appears in the listing of "less-than-accurate" **Asklēpiadeans** in DIOSKOURIDĒS, *pr.*2 (Beck p. 2), and as one of PLINY's Greek *auctores* (1.*ind.*20–27), but among "medical writers" (1.*ind.*33–34). Caelius Aurelianus (*ibid.*) cites Bassus as prescribing sternutatories and enemas in treating ***rabies***, instead of the **Methodist** therapy of alternating remedies (*metasyncritica*). SCRIBONIUS LARGUS, *Comp.* 121 (ed. Sconocchia, pp. 63–64 = ANDROMAKHOS in GALĒN, *CMLoc* 9.4 [13.280–281 K.]) records his "wonderful remedy for intestinal colic," which "gives relief quickly and then counters the bloated state of the lower bowel along with all of the other parts of the body." Among the ingredients are spikenard oil (*Nardostachys jatamansi* DC.), white pepper, black pepper (viz. the unshelled peppercorns), henbane-root (*Hyoscyamus niger* L.), myrrh, frankincense, cabbage seeds, the latex of the opium poppy (*Papaver somniferum* L.), and beaver-castor; such a compound would engender a mild narcotic effect. Andromakhos adds mandrake root-bark, and hemlock seeds (*Conium maculatum* L.; the dried, unripe "fruits" are a potent sedative and narcotic); *cf.* Bassus' clipped formulas quoted by Galēn from Andromakhos in *CMLoc* 7.2 and *CMGen* 7.13 (13.60 and 1033 K.). If the quotations are representative, Bassus was adept at devising effective anodynes and narcotics for chronic illnesses affecting the digestive tract.

RE 10.1 (1918) 180–181, M. Wellmann; Scarborough and Nutton (1982) 205.

John Scarborough

C. Iulius Caesar (77 – 44 BCE)

Iulius Caesar Courtesy of the Vatican Museums

Roman statesman, historian, orator, accomplished military general, politician and dictator, born 100 BCE to an ancient but recently undistinguished patrician family. He saw military service in Asia in the 70s, defeating an advance force of MITHRADATĒS VI and receiving the *corona ciuica* for service at the sack of Milētos. He published legal orations and eulogies, was elected *tribunus militum* in 73, served as *quaestor* and *praetor* in Further Spain, and consul in 59. As governor of Illyricum, Cisalpine and Transalpine Gaul for an unprecedented ten years, he launched campaigns against Helvetian uprisings, resulting in economic depletion of his provinces, deaths of one million Gauls, enslavement of another million (by his own account), and conquest of Gaul. Caesar also engaged in civil war against Pompeius Magnus and senatorial forces from 48–47, and the Alexandrine war to avenge

Pompey's execution on Ptolemy XIII's orders. Caesar reigned in Rome until his assassination by a senatorial mob in 44.

In Caesar's two treatises *Gallic Wars* (on the campaigns during his governorship) and *Civil Wars* (48–47 BCE), geography is ancillary but essential to military success. His *Gallic Wars* famously opens with a description of the three provinces, their demarcating bodies of water, and cultural and linguistic distinctions (*BG* 1.1). As a field general, Caesar emphasizes rivers, especially as landmarks (*BG* 1.2, 1.12, 1.38, 2.5), mountain ranges (*BG* 1.2), ease and length of marching routes (*BG* 1.8, 1.10), distances (*BG* 1.48, 2.6), supply lines, and battlefield topography (*BG* 1.26, 2.9, 2.23: especially vivid are his descriptions of the Hecyrnian forest: *BG* 6.24–26, Alesia: *BG* 7.69, and Dyrrhachium: *BG* 3.44–46). In his excursus on Britain, informing TACITUS', Caesar notes the Channel's frequent but small tidal activity necessitating adaptations in ship design (*BG* 5.1). He discusses ethnography, natural resources, climate, the island's shape, distances, and the surprising behavior of the midwinter sun: regarding which the locals were unable to provide information, but Caesar's own exact water measurements (presumably with a *klepsudra*) showed that British summertime nights were shorter than on the continent (*BG* 5.12–13).

Caesar's authorship of the accounts of the Alexandrine, African, and Spanish Wars is currently regarded as dubious. Caesar's calendar, executed by SŌSIGENĒS (and on which Caesar published the *De Astris*: PLINY 18.212; MACROBIUS *Sat.* 1.16.39), was not substantially revised until 1582. For the world map commissioned by Caesar, see IULIUS HONORIUS.

Dilke (1985) 39–41; Rawson (1985) 109–114, 259–263; *OCD3* 780–782, E. Badian; *DLB* 211 (1999) 109–117, C.B. Champion; *BNP* 2 (2003) 900–912, J. Rüpke.

GLIM

Germanicus Iulius Caesar (10 – 19 CE)

Born on May 24, 15 BCE. He was the son of Nero Claudius Drusus and Antonia, and then nephew of Tiberius and great-nephew of AUGUSTUS. He was nicknamed Germanicus after his father's death. But, when he became one of the closest male relatives of Augustus, he was adopted by Tiberius. He also married Agrippina under the influence of Augustus in 5 CE. He took part between 11 and 16 in the German campaigns so that he celebrated a triumph in 17 (TACITUS *Ann.* 2.41). He then left for the eastern lands where he died at Antioch in October, 19 CE; his ashes were brought back to Rom (Tac. *Ann.* 3.1–4). He received a very good literary, rhetorical and philosophical education (Suet. *Calig.* 3.1): he was clever, cultured and excelled in rhetoric (OVID *Pont.* 2.5.53; Tac. *Ann.* 2.83.5). He delivered many defense speeches before the courts or the emperor. He also wrote several comedies in Greek and different kinds of poems: only two epigrams survive (*AL* 708–709 Riese).

Finally, he rendered in Latin ARATOS' *Phainomena* during his stay in Rome in 16–17 CE. In all likelihood he already knew Ovid's *Fasti* and MANILIUS' *Astronomica*. Germanicus was not a specialist in astronomy, but he was very fond of it and wanted to popularize the science. He dedicated his work not to Zeus like Aratos, but to his father (*genitor*) who may be Augustus himself. Germanicus' poem does not correspond entirely to Aratos', but only to its astronomical part: so, after the prologue (1–16), we find first one long description, of the constellations (17–445), and then another shorter one, of the different circles of the heavens (446–572); thirdly Germanicus explains how to estimate the passage of time according to the rise of zodiacal constellations (573–725). Besides this poem, we also have six fragments of varying length on the zodiac, planets and meteorology. It is not the adaptation of the second part of Aratos' *Phainomena* (ordinarily called *Diosēmeiai*), but these fragments may have been part of a large poem dealing with astronomy, astrology and meteorology.

Ed.: A. Le Boeuffle, *Germanicus. Les Phénomènes d'Aratos* (*CUF* 1975).
OCD3 783, B.M. Levick; *BNP* 5 (2004) 812–814, Werner Eck.

Christophe Cusset

C. Iulius Caesar Octauianus, Augustus (31 BCE – 14 CE)

The emperor Augustus, born in Rome (Suet., *Aug.* 5); according to Suetonius, *Aug.* 85, he wrote a possibly geographical poem on Sicily, and is also attested to have "completed" the map of AGRIPPA.

OCD3 217–218, N. Purcell.

PTK

IULIUS FIRMICUS ⇒ FIRMICUS

Sex. Iulius Frontinus (*ca* 90 – 103/104 CE)

Born *ca* 40, Roman senator, possibly from southern Gaul, with a distinguished active career (70 CE: *praetor urbanus* and assisted in repressing the Iulius Ciuilis revolt; consul, 72 or 73; governor of Britain, 73/74–77; proconsul of Asia, 87; *curator aquarum* under Nerva, 97; suffect consul, 98; and consul, 100), and authored *De aquis urbis Romae* and *Strategemata*.

Frontinus writes on the aqueducts of Rome as the Roman senatorial administrator (*curator*) deeply cognizant of his department's technology. He cites technical reports from engineers, senatorial decrees and known abuses of the public water system (e.g. illegal tapping). He credits M. VIPSANIUS AGRIPPA and his architect VITRUUIUS for having introduced the use of standard pipe sizes (the only other reference in antiquity) based on the measurement of the *quinaria* (meaning either a five-digit lead sheet rolled into a pipe, or a pipe five *quadrantes* – quarter digits – in diameter), but he gives slightly different and more complicated measurements than Vitruuius, probably indicating further evolution of the system (*De aquis*, 25, 26–34, Vitr. 8.6.4). *Strategemata* is divided into four books: before battle; during and after battle; sieges; and generalship. The authenticity of the fourth book, though questioned, is probably genuine, and was likely meant to be a manual on military practice to assist the education of the Roman senatorial elite in their potential roles as field commanders. He is also possibly the author of certain sections of the *Corpus Agrimensorum*.

Ed.: C. Thulin, *Corpus Agrimensorum Romanorum* 1/1 (1913); C.E. Bennet and M.B. McElwain, *Stratagems, The Aqueducts of Rome* (Loeb 1925); P. Grimal, *Les Aqueducs de la ville de Rome* 2nd ed. (*CUF* 1961; repr. 2003); R.H. Rodgers, *De aquaeductu urbis Romae* (2004).
O.A.W. Dilke, *The Roman Land Surveyors* (1971; repr. 1992).

Thomas Noble Howe

L. Iulius Graecinus (30 – 50 CE)

Roman senator from Forum Iulii, and the father of the Iulius Agricola immortalized by TACITUS (*Agricola*); Caligula executed him for refusing to participate in a show-trial (Tac., *Agr.* 4). He wrote in Latin a treatise *On Vineyards* (*de Vineis*) in two books whose style

COLUMELLA (1.1.14) praised. He ascribed the decline of viticulture in his day to an ignorance of good practice on the part of growers, since his calculations showed that the income from viticulture ought to exceed outlays even on poor land (*cf.* Columella 3.3.4–7, 4.3.1, 6). He described the best soil for vines as slightly warmer and looser than average, recommended dates for various activities, and maintained that vines can have a life-span of up to 600 years (3.12.1, 4.28.2–29.1; PLINY 16.241). Columella (1.1.14) calls him a "student, as it were," of IULIUS ATTICUS, and Pliny 14.33 reports that he closely followed CELSUS.

GRL §497.2; *DPA* 3 (2000) 493, M. Ducos; *BNP* 6 (2005) 1082 (#IV.9), E. Christmann.

Philip Thibodeau

Iulius Honorius (300 – 450 CE?)

A teacher who wrote a geographical treatise in Latin for the purpose of instructing students. He says his text included a map (not extant). The text lists geographical objects (seas, islands, mountains, rivers), as well as administrative divisions, cities, and peoples. In some MSS the text begins with the report of a survey and measurement of the world made by four Greeks, which continued from "the consulate of IULIUS CAESAR and Marc Antony" until the time of AUGUSTUS. The four surveyors explored the east, the west, the north, and the south, and produced a description, which supposedly served as the basis for this treatise, and probably its lost map. Modern scholars often connect this information to the survey made by AGRIPPA by the order of Augustus. The story of the survey of the world was repeated in later geographers (such as PSEUDO-AETHICUS) and sometimes represented on maps.

Ed.: *GLM* 21–55.

GRL §1060; *RE* 10.1 (1918) 614–628 (#277), W. Kubitschek; *PLRE* 2 (1980) 569; C. Nicolet and P. Gautier Dalché, "Les 'quatre sages' de Jules César et la 'mesure du monde' selon Julius Honorius: réalité antique et tradition médiévale," *Journal des savants* (1986) 157–218.

Natalia Lozovsky

C. Iulius Hyginus (*ca* 30 BCE – *ca* 10 CE)

A learned, Greek-speaking slave from Spain or perhaps Alexandria who was brought by IULIUS CAESAR to Rome (*ca* 45?), where he became a student of the scholar ALEXANDER OF MILĒTOS (Suet. *Gram.* 20). After Caesar's death he passed into the possession of the emperor AUGUSTUS, who eventually freed him, and appointed him overseer of the Palatine library (28 BCE or later). He became a friend of the poet OVID, and of the consular historian Clodius Licinus, who supported him after he lost his post and fell into poverty. (Ovid probably did not address him in the *Tristia*: Kaster 1995: 212.)

Most of his numerous writings, many cited by Gellius, Seruius, and MACROBIUS, were devoted to topics of interest to the Augustan nobility, such as the genealogy of Italian families (*de familiis Troianis*), the history of religious practice at Rome (*de diis penatibus; de proprietatibus deorum*), and the customs of the Italic peoples (*de origine situque urbium Italicarum*). He also wrote commentaries on VERGIL and Heluius Cinna, and fragments from a biographical collection have survived. A work dealing with the geography of Greece, Italy and perhaps other parts of the world was used as a source by PLINY, 1.*ind.*3–6. (Contrast the works on surveying attributed to one or another, later, HYGINUS.)

Despite much controversy, a mythological compendium entitled *Genealogiae* (or *Fabulae*), based on Greek sources and apparently interpolated, and an astronomical treatise in four books are probably by C. Iulius Hyginus. *De Astronomia* (probably not its original title), dedicated to M. Fabius, certainly belongs to the Augustan period as proved by the lack of astrological speculations, the lack of reference to GERMANICUS' translation of ARATOS' *Phainomena* (whereas the author knows CICERO's translation), the agreement with ERATOSTHENĒS, *Catasterisms*, and the fact that the author never uses the words *astronomia* nor *astronomus* (well-attested at Nero's time: Le Boeuffle [1983] XXXVIII). The *De Astronomia* covers: Book I, summary of cosmography and basic definitions for astronomy; Book II, catasterisms and legends about 42 constellations, planets and the Milky Way; Book III, the position and composition of these constellations; Book IV, in a more varied way, studies the circles of heaven, spheres, nights and days, risings and settings of stars, and planets. This treatise does not really innovate in the field of astronomy: it is not the work of an astronomer but of a more or less enlightened compiler. It pretends to be a companion to initiate one to astronomy and could take the place of Aratos' *Phainomena*.

Hyginus certainly wrote a treatise on agriculture which is known to us through numerous citations in COLUMELLA and Pliny; the citation *de apibus* may refer to part of this work, or a separate work. The work seems to have been arranged as a doxography, with opinions drawn from ancient authors grouped according to topic, such as the best soil for vines, the feeding of oxen, the origin of bees, and treatments for apiary illnesses. Hyginus clearly relied on books from the library which he oversaw for his information; nevertheless, his treatise was regarded seriously, and Columella, 3.11.8, 11.3.62, reports testing a few of his agricultural precepts, finding some that worked, and some that did not. His most original contribution was a calendar of tasks for beekeepers, praised by Columella, 9.14, in which he stressed the need to keep the hive clean, and listed various methods for eliminating pests. Hyginus claims to have taken the dating system for his calendar from EUDOXOS and METŌN, but in this he was deceived, since the Babylonian convention he used of placing the solstices at the eighth-degree of their respective zodiacal signs did not reach Greece before the Hellenistic era. Moreover, *Astronomia* 4.2 refers to the same eighth-degree convention, providing another connection between the two books.

Ed.: A. Le Boeuffle, *Hygin: L'astronomie* (*CUF* 1983); G. Viré, *Hygini De astronomia* (1992).

J. Christes, *Sklaven und Freigelassene als Grammatiker und Philologen im antiken Rom* (1979); Kaster (1995); *BNP* 6 (2005) 606–607, P.L. Schmidt and Helmuth Schneider.

Christophe Cusset and Philip Thibodeau

Iulius Secundus (*ca* 10 BCE – *ca* 90 CE)

ASKLĒPIADĒS in GALĒN, *CMGen* 7.12 (13.1029 K.), cites his pore-relieving potion (***akopon***) useful for sciatica, arthritis, and headaches, containing ***euphorbia*** (i.e., *post* IOUBA) as well as myrrh, balsam-tree sap, etc., and labeled as "metasyncritic" (pore-altering), a **Methodist** term.

RE 10.1 (1918) 803 (#471), H. Gossen.

PTK

C. Iulius Solinus (230 – 240 CE)

Author of the *Collectanea rerum memorabilium*, a compendium of geographical information borrowed mainly from PLINY and POMPONIUS MELA. The work begins with a lengthy account of the foundation and history of Rome and goes on to describe Italy and other regions of the three known parts of the world. Solinus emphasizes wonders and marvels and records much of the lore on monstrous races, and unusual animals and plants that passed on to the Middle Ages. The *Collectanea* was widely read in the following centuries (at least 350 MSS survive) and used as a source for compilations on geography, ethnography, and natural philosophy.

Ed.: Th. Mommsen, *C. Iulii Solini Collectanea rerum memorabilium* (1895; repr. 1958).
RE 10.1 (1918) 823–838 (#492), E. Diehl; *KP* 5.260–261, Kl. Sallmann; *TTE* 566–567, Z.R.W.M. von Martels; *OCD3* 786, E.H. Warmington.

Natalia Lozovsky

Iulius Titianus (145 – 175 CE)

Among other works, wrote a description (*chorographia*) of the provinces of the Roman Empire, cited by Seruius, *Ad Aen.* 4.22, 11.651, and by GREGORIUS OF TOURS, on Aetna.

NP 12/1.628 (Titianus #1), M. Zelzer.

PTK

Iunia/Iounias (30 BCE? – 80 CE)

OREIBASIOS, *Ecl. Med.* 136.3 (*CMG* 6.2.2, p. 299), from ARKHIGENĒS, preserves Iounias' two recipes for mammary abscesses, using **litharge** and ***psimuthion***, frankincense and "Kolophōn" resin (*cf.* GALĒN, *CMGen* 2.2 [13.475 K.]), in olive oil and chicken fat. PHILOUMENOS, in AËTIOS OF AMIDA 16.37 (Zervos 1901: 55), preserves fuller versions of the pair, adding flour, linseed oil, and beeswax. The name *ΙΟΥΝΙΑΣ* is otherwise unattested (*LGPN*, Solin 2003), and the genitive *ΙΟΥΝΙΑΔΟΣ* found in Oreibasios and Aëtios is unexpected (contrast AKHILLĀS, APELLĀS, HĒRĀS, and HERMĀS, e.g.). Perhaps emendable to *ΙΟΥΝΙΑ[ΔΟ]Σ*, i.e., *Iuniae*. (*Cf.* Epp in Dunaux, ed., *New Testament Textual Criticism and Exegesis*, 2002, pp. 227–292.)

(*)

PTK

L. Iunius Moderatus Columella of Gadēs (*ca* 40 – *ca* 70 CE)

Born 4 CE; acquaintance of SENECA and his brother Gallio. To an unknown P. Siluinus he addressed *De re rustica*, a systematic treatment of agriculture, in 12 books: siting of property, labor force (1), cereal crops (2), viticulture and fruit-trees (3–5), animals and their care (6–7), poultry and bees (8–9), horticulture (10, hexameters, after VERGIL), manager's duties, astronomical and meteorological calendar (11), household duties, preservation of produce (12). Rustic enterprise requires technical knowledge, resources for investment, willingness to work (1.1.1). Financial profit is compared to interest on loans; Columella's attitude anticipates capitalism. He describes an intensive system with slave labor, integrating field crops with animal husbandry, yet deplores absentee landlords and comments on land leasing by *coloni*. Among numerous literary sources are the Carthaginian Mago, CATO, CORNELIUS CELSUS, especially Vergil's *Georgics*. He criticizes earlier and contemporary writers on philosophical views and specific techniques (in turn he is criticized by his contemporary PLINY). His uncle was an experimental farmer in Baetica; he reports practices from Syria (where he was military tribune in 35); he owned properties near Rome and repeatedly relates personal experience. His Latin style is conscientious and elegant, especially in lexical and syntactical variety. The work was highly influential (Gargilius Martialis, Palladius Aemilianus), but he was not the Yūniyūs of Arabo-Andalusian medieval writers.

Neither his *Aduersus astrologos* (11.1.31) nor a projected tract on farmers' religion (2.21.5) has survived. *De arboribus*, transmitted in some MSS, may be a later abridgement of the *De re rustica* (Richter) rather than, as conventionally assumed (Goujard), part of an earlier work by Columella himself.

Ed.: V. Lundström, Å. Josephson, S. Hedberg (1897–1968); concordance: G.G. Betts and W.D. Ashworth (1971).

PIR2 4.340–341 (I-779); W. Richter, *The "Liber de arboribus" und Columella* = *SBAW* (1972) # 1; R. Martin, "État présent des études sur Columelle," *ANRW* 2.32.3 (1985) 1959–1979; R. Goujard, "Encore à propos de l'authenticité du *De arboribus*," *Latomus* 45 (1986) 612–618; J.I. García Armendáriz, *Agronomía y tradición clásica* (1995); *OCD*3 367, M.S. Spurr; E. Noè, *Il progetto di Columella* (2002); *BNP* 3 (2004) 584–585, E. Christmann.

Robert H. Rodgers

Iunius Crispus (*ca* 70 – 80 CE)

Andromakhos in Galēn cites "Crispus": *CMLoc* 7.3 (13.67 K.: "freedman") for cough-drops, *CMGen* 5.13 (13.841 K.) for a powder (*xēron*), 7.7 (13.984 K.) for an ointment (*malagma*), and *CMGen* 9.4 (13.276 K.), for his preparation of the *kōlikē* of Cassius. Kritōn of Hērakleia in Galēn, *CMLoc* 5.3 (12.831 K.), cites him as a *philos* (hence a contemporary?) and gives his remedy for facial ***leikhēn***. Marcellus of Bordeaux 23.9 (*CML* 5, p. 394) gives his *nomen* and recipe for the intestinal drug "Ambrosia."

PIR2 I-747.

PTK

Iunius Nipsus (200 – 400 CE?)

Three works in the *Corpus Agrimensorum Romanorum* (see Hyginus), *Fluminis uaratio, Limitis repositio*, and *Podismus*, can possibly though not certainly be ascribed to Iunius Nipsus, whose name appears at the end of the Podismus fragment. Nipsus on the basis of his Latinity was probably writing in the later Roman Empire. The *Fluminis uaratio* describes a method for indirectly calculating the width of a river without actually measuring, by using the surveying instrument (*groma*) to establish identical right-angled triangles. The *Limitis repositio* discusses the resiting of *limites* in surveyed land, a method for correcting a measured distance, how to plot a *limes* when there was an obstruction, and also analyses inscriptions on *centuria* stones, which helped the surveyor find the location of *centuriae* within the system, and check details of ownership. The *Podismus* examines definitions of types of measurement and angles, and ways of measuring figures.

Ed.: J. Bouma, *Marcus Junius Nypsus. Fluminis Varatio, Limitis Repositio: Introduction, Text, Translation, and Commentary* (1993); *CAR* 3 (1996) 120–138.

A. Roth Congès, "Modalités pratiques d'implantation des cadastres romains: quelques aspects (Quintarios claudere. Perpendere. Cultellare. Varare: la construction des cadastres sur une diagonale et ses traces dans le Corpus Agrimensorum)," *MEFRA* 108 (1996) 299–422.

Brian Campbell

D. Iunius Silanus (*ca* 146 BCE)

Led the group of translators whom the Roman Senate commissioned to render into Latin the 28 books on agriculture by Mago of Carthage (PLINY 18.22–23). Pliny cites him as a source for his books on cereals, viticulture, and arboriculture (1.*ind.*14–15, 17–18).

RE 10.1 (1918) 1088–9 (#160), F. Münzer.

Philip Thibodeau

IUSTINUS ⇒ POMPEIUS TROGUS

Iustinianus Imp. (600 – 800 CE?)

Two lost works are attributed to Iustinianus in the early table in MS *Marcianus gr.* 299: *Letter* and five *Chapters on the Divine Art and Discussion Addressed to the Philosophers*. A fragment entitled *Procedure of the Emperor Iustinianus* survives (*CAAG* 2.384–387), and added to MS *Marcianus gr.* 299 in a 15th c. hand is a text written in an almost barbarous dialect ending with the words "thus is accomplished, with the aid of God, the procedure of Iustinianus" (*CAAG* 2.104–105). Since nowhere else is any alchemy attributed to Iustinianus, Letrouit suggested a pseudepigraphon; whereas Berthelot (*CAAG* 1.176) suggested the 7th/8th c. emperor Iustinianus II.

Ed.: *CAAG* 2.104–105, 384–387.
CAAG 1.176; Letrouit (1995) 57.

Cristina Viano

Iustinus (Pharm.) (30 BCE – 115 CE)

AËTIOS OF AMIDA 11.12 (pp. 609–610 Cornarius) records that ARKHIGENĒS prescribed Iustinus' antidote for stone, compounded of cassia, castoreum, cinnamon, ***kostos***, myrrh, saffron, spikenard, etc., that was later prescribed by OREIBASIOS. The non-Republican cognomen is first attested in the 1st c. CE: Martial 1.71.1, 11.65.1, and the potter, *RE* 10.2 (1919) 1337 (#2); *cf.* also the 2nd c. writers *PIR2* I-713, I-871.

Fabricius (1726) 306.

PTK

Iustus the Pharmacologist (30 BCE – *ca* 150 CE)

Three multi-ingredient pharmaceutical recipes are recorded under the name of a Iustus, who apparently lived around the time of RUFUS and ARKHIGENĒS: *cf.* MARCELLUS OF SIDĒ in AËTIOS OF AMIDA 6.11 (*CMG* 8.2, pp. 151–152). One is by MARCELLUS OF BORDEAUX, 25.32 (*CML* 5.2, pp. 422–424), using peppercorns added to four ingredients (wind rose, vervain, sorrel, the roots of a mullein [*Verbascum thapsus* L.; *cf.* André 1985: 40]) in a sitzbath for the relief of sciatic pains in the hips.

The second is a ***hiera*** (Aëtios of Amida 3.117 [*CMG* 8.1, p. 306] = slightly rearranged in PAULOS OF AIGINA 7.8.3 [*CMG* 9.2, p. 287]) containing 23 ingredients (two kinds of *aristolokhia* [birthwort], "white" and "black" pepper, each counted as a single component) + salted honey-water administered in increasing dosages. The compound also includes thyme, germander, black hellebore, **shelf-fungus**, roasted squill, pulp from a gourd, aloe, saffron, ***opopanax***, cassia and cinnamon, pennyroyal, and myrrh, among the 23, quite obviously a drastic purge with aromatic properties (saffron, myrrh, thyme, cinnamon) rendering this solution palatable to the patient quaffing it. Iustus' ***hiera*** in Oreibasios' *Collection*, 8.47.21 (*CMG* 6.1.1, p. 300) is a "short-version" of the more elaborate recipe recorded by Aëtios and Paulos.

The third recipe (a milder cathartic) occurs in Oreibasios' *Collection* 8.47.8 (*CMG* 6.1.1, p. 298), with the heading, "A Purge of Iustus, which easily moves down the waste in the stomach, and, at the same time takes away the heaviness in the chest and in the head," suggesting that Oreibasios had successfully prescribed it. About a dozen harsh ingredients include **shelf-fungus**, the two peppers, and the squills, softened and sweetened with saffron, the pulp of the gourd, germander, and gentian, triturated with **bdellium**, and administered in a drink fortified with ***opopanax*** and honey. Saffron (*Crocus sativus* L.) is the costliest ingredient by far, but Iustus appears to favor myrrh, cinnamon, and pepper among the exotic imports.

BNP 6 (2005) 1143 (#4), V. Nutton.

John Scarborough

Iustus the Ophthalmologist (160 – 180 CE)

Galēn, *MM* 14.19 (10.1019 K.) mentions his contemporary, Iustus, an eye-doctor (*ophthalmikos*) whose achievement was to heal ulcerative and pussy suppurations resulting from blows to the head. Iustus' simple technique involved draining the pus over a few hours, while the patient sat upright on a stool, with his head cocked slightly to the right or the left. Galēn does not mention a written text on Iustus' technique, but he claims to have witnessed the procedure.

BNP 6 (2005) 1143 (#4), V. Nutton.

John Scarborough

J- ⇒ I-

John ⇒ Iōannēs

Juba ⇒ Iouba

K

Kaikalos (?) of Argos (400 – 300 BCE?)

Author of a poem about fishing (*Halieutika*), lived before OPPIANUS OF KILIKIA (Ath., *Deipn.* 1 [13b] and *Souda* K-1596). His name is uncertain: Athēnaios' MSS have *Kaiklon*, while *Souda* calls him *Kikilios*. On this basis, Birt conjectured *Kikinos* and Meineke, more plausibly, *Kaikalos*.

SH 237; A. Zumbo, "Ateneo 1, 13b–c e il 'canone' degli autori alieutici," in P. Radici Colace and A. Zumbo, *Atti del Seminario Internazionale di Studi "Letteratura scientifica e tecnica greca e latina"* (2000) 163–170; *BNP* 3 (2003) 871, S. Fornaro.

Claudio Meliadò

KALLANEUS ⇒ KALYĀṆA

Kallianax (*ca* 280 – 230 BCE?)

Hērophilean physician whose treatise on early Hērophileans is quoted by ZEUXIS in GALĒN *In Hipp. Epid. VI* (*CMG* 5.10.2.2, p. 203). Kallianax recited HOMER and the tragedians to patients experiencing anxiety before death to underscore its inevitability to everyone, including heroes (e.g., Patroklos); only the Immortals escape death.

von Staden (1989) 478–479; *BNP* 2 (2003) 960, V. Nutton.

GLIM

Kallias of Arados (310 – 300 BCE)

Replaced DIOGNĒTOS OF RHODES, to construct an anti-catapult crane, but failed to counter the ***helepolis*** of EPIMAKHOS OF ATHENS, and was dismissed in favor of Diognētos (VITRUUIUS 10.16.3–7).

(*)

PTK

Kalliklēs (200 BCE – 150 CE)

Empiricist mentioned by Galēn, *MM* 2.7.23 (10.142–143 K. = p. 71 Hankinson), in a non-chronological list including SĒRAPIŌN, MĒNODOTOS, etc. (the name "Kalliklēs" is very common).

Fabricius (1726) 106.

PTK

Kallikratēs (Astrol.) (50 – 150 CE?)

In a collection of epitomes of such astrologers as THRASULLOS, KRITODĒMOS, BALBILLOS, and ANTIOKHOS OF ATHENS, *CCAG* 8.3 (1912) 102–103 includes a paragraph on the *Treasury* of Kallikratēs, addressed to an otherwise unknown Timogenēs (both names are very rare after the 2nd c. CE). He began with the Moon, turned to the Sun, and proceeded thence through Saturn, Mars, Jupiter, Venus, and finally Mercury: i.e., first the luminaries, next the two malefic planets, then the others. Kallikratēs described each planet's **sympathies** and configurations with the others, and with each of the signs, and explained what its oppositions and other aspects prognosticated.

(*)

PTK

Kallikratēs (Arch.) (*ca* 450 – 425 BCE)

Architect connected with various projects in Athens in the 5th c. BCE. According to PLUTARCH (*Perikles* 13.4), he and IKTINOS built the Parthenon. According to the same source (*Perikles* 13.5), he took up the contract to build the (middle) Long Wall that connected Athens with its port at Piraeus. An inscription (*IG* I³ 45) specifies that Kallikratēs should provide the *sungraphē* (specifications) for some kind of construction to secure the Acropolis from runaway slaves and thieves. Another inscription (*IG* I³ 35) records that Kallikratēs should provide *sungraphai* for a door to the sanctuary of Athena Nikē on the Acropolis and for a temple in the same sanctuary. This literary and epigraphical evidence has generated considerable debate about the role and significance of Kallikratēs. McCredie has argued that Kallikratēs's involvement in mundane projects, such as the Long Wall, shows that Kallikratēs was more of a builder or contractor than a designing architect. Although *IG* I³ 35 clearly connects Kallikratēs with a temple of Athena Nikē, McCredie argues that there is no compelling reason to link the temple mentioned in the inscription with the well-known marble building constructed in the 430s and 420s. He notes that both Bundgaard and Mark connect the inscription with the simple limestone predecessor building that would have required only a minimal design. A more generous view of the career of Kallikratēs is taken by most scholars, who believe that he designed the marble temple of Athena Nikē and had a role in the design of some phase of the Parthenon. The temple of Athena Nikē is a small but finely crafted amphiprostyle temple that embodies the newly developed Athenian version of the Ionic order. Because the inscription that links Kallikratēs with the building (*IG* I³ 35) was dated on the basis of letter forms to the 440s BCE it was long thought that the execution of the project was delayed some 15 to 20 years and that the original plan of the building needed to be shortened in response to the reduction in the size of the sanctuary that resulted from the encroachment of the new Propylaia, built earlier in the 430s BCE. More recent studies by Wesenberg and Mattingly have lowered the date of the inscription to the date of the commencement of the building, thus eliminating the awkward gap between them. Although many scholars have accepted Kallikratēs' role in the design of the Parthenon, the nature of that role has been subject to a variety of interpretations. Whereas some scholars, such as Dinsmoor and Martin, see Kallikratēs and Iktinos as collaborators in the design of the famous Periklean building, others have attempted, without very convincing evidence, to assign Kallikratēs and Iktinos to separate phases of the building. Carpenter assigned Kallikratēs to a hypothethical pre-Iktinian phase of the building dated to the time of Kimōn (460s); Svenson-Evers, by contrast, has assigned Kallikratēs to a post-Iktinian phase dated to the time of Periklēs (440s-430s). Those who see Kallikratēs as a major designer of the 5th c. BCE have increased his oeuvre on the basis of characteristics that are directly or indirectly linked to the Temple of Athena Nikē. As early as 1908, Lethaby connected the small Ionic "temple on the Ilissos" in Athens with Kallikratēs and soon thereafter Studniczka suggested that this temple preserved the plan originally intended for the temple of Athena Nikē. Subsequently, Dinsmoor added to Kallikratēs's oeuvre the Doric temple of the Athenians on Delos, and I. M. Shear added the Erechtheion on the Athenian Acropolis. Carpenter, while leaving aside the Erechtheion, added the Hephaisteion and the temple of Ares in Athens, the temple of Poseidon at Sounion, and the temple of Nemesis at Rhamnous. Although Carpenter attempted to demonstrate that such a large number of buildings could have been created by a single man, it would seem that few other scholars have followed him. Martin rejects Carpenter's additions, but expands Kallikratēs's oeuvre in yet another direction; he argues that Kallikratēs, as a master of decorative design in the Ionic style, was responsible for the innovative Ionic-Corinthian interior of the Temple of Apollo at Bassai.

R. Carpenter, *The Architects of the Parthenon* (1970) 83–109; W.B. Dinsmoor, *Architecture of Ancient Greece*, 3rd ed. (1950) 148, 159, 183–187; G. Gruben, *Die Tempel der Griechen*, 4th ed. (1986) 149–151, 163–178, 188–193; R. Martin, "L'atelier Ictinos-Callicratès au temple de Bassae," *BCH* 100 (1976) 427–442; H.B. Mattingly, "The Athena Nike Temple Reconsidered," *AJA* 86 (1982) 381–385; J.R. McCredie "Architects of the Parthenon," in *Studies in Classical Archaeology* (1979) 69–73; I.M. Shear, "Kallikrates," *Hesperia* 32 (1963) 375–424; Svenson-Evers (1996) 214–236; B. Wesenberg, "Zur Baugeschichte des Niketempels," *JdI* 96 (1981) 28–54.

Christopher A. Pfaff

Kallimakhos (of Bithunia) (*ca* 275 – 205 BCE)

Hērophilos' student, whom Erōtianos, citing with Bakkheios of Tanagra and Philinos, chides for calling the plague "divine" because of superstitions regarding it (*fr.*33, p. 108 Nachm.). Polubios referred to Kallimakhos as one of the two eponymous founders (with Hērophilos) of a **Rationalist** school of medicine in Alexandria, but asserted that the Kallimakheans' focus on disease theory (to the exclusion of dietetics, surgery, and pharmaceutics) was detrimental to patients (Book 12, *fr.*25d). Rufus of Ephesos affirmed Kallimakhos' emphasis on symptoms in diagnosis (*Quaest. Med.* 3.21). Kallimakhos wrote an Hippokratic lexicography (Erōtianos *Pr.*, p. 4.26 Nachm.) and an exegesis of the Hippokratic Corpus, Epidemics 6 (and possibly also *Prognosis*), often chiding earlier **Hērophilean** commentators, including Hērophilos (Galēn *In Hipp. Epid. VI* [*CMG* 5.10.2.1, p. 21]). Pliny, calling him a medical writer (1.*ind.*21–27), reports that Kallimakhos wrote on toxic and pharmaceutical effects of botanicals, including fragrant wreaths (21.12), and called *ērigerōn* by the name *akanthis* (25.167–168).

von Staden (1989) 480–483; *OCD3* 277–278, *Idem; BNP* 2 (2003) 978–979 (#5), V. Nutton.

GLIM

Kallimakhos of Kurēnē (*ca* 285 – *ca* 245 BCE)

Kallimakhos of Kurene (identification disputed) Photo: Ole Haupt, Ny Carlsberg Glyptotek

Born *ca* 310 BCE in Kurēnē and perhaps a descendant of the formerly ruling Battiad family, who spent the greater part of his life at the Ptolemaic court in Alexandria, during the reigns of Philadelphos (285–246 BCE) and Euergetēs (246–221 BCE). A prominent member of the *Mouseion* (Museum), he was highly regarded as a scholar and a learned poet. His prolific pen produced more than 800 works in prose and verse (*Souda* K-227), but besides several hundred fragments only six hymns and some 60 epigrams survive intact.

The most famous part of Kallimakhos' scholarly activity concerned the organization of the great library of Alexandria, resulting in the publication of 120 books of *Pinakes* or *Lists of People who have Distinguished themselves in all Fields of Learning, and their Writings*, but he also wrote a number of scientific treatises. In addition to authoring lesser-known books *On Winds* (F 404), *On Birds* (F 414–428) and *On The Rivers of the World* (F 457–459), he was probably the first to compile a collection of *thaumasia* or marvelous phenomena occurring in the natural world. The topics chosen for inclusion by the founding-father of ancient paradoxography would remain strong favorites throughout the genre's history: wondrous waters (rivers, springs and lakes), animals, plants, stones and fire (F 407–411). The detailed title transmitted in the *Souda* (K-227), *Collection of the Wonders Happening all around the World, Arranged by Locality*, perhaps not entirely authentic, at least suggests a geographical organization; one sub-section is known to have gathered the wonders found in the Peloponnesus and Italy. This kind of arrangement would remain one of four basic types commonly used in paradoxographical treatises, the other three being topical or thematic, bibliographic (according to the excerpted sources) and alphabetical.

Kallimakhos' working method was essentially bookish, resting exclusively on the wealth of information accumulated in the Alexandrian library. Judging from the long series of excerpts contained in ANTIGONOS' *Historia mirabilium* (§129–173), he made a point, in his collection of wonders, of substantiating all reported curiosities by carefully acknowledging his written sources, many of whom were reliable historical authorities (EUDOXOS, Theopompos, ARISTOTLE, THEOPHRASTOS). This served to heighten credibility and, thus, to increase the audience's sense of amazement. This emphasis on documentation and trustworthiness became a common feature of ancient paradoxography.

Ed.: R. Pfeiffer, *Callimachus* I (1949); *PGR* 15–20.

RE 18.3 (1949) 1137–1166 (§1, 1140–1141), K. Ziegler; Giannini (1964) 105–109; Guido Schepens and K. Delcroix, "Ancient Paradoxography," in: O. Pecere and A. Stramaglia, edd., *La Letteratura di Consumo nel Mondo Greco-Latino* (1996) 373–460 (380–409); *BNP* 10 (2007) 506–509 (I.B.1, 508), O. Wenskus.

Jan Bollansée, Karen Haegemans, and Guido Schepens

Kallimakhos Jr. of Kurēnē (260 – 230 BCE)

The son of Stasēnōr and Megatimā, the latter the sister of KALLIMAKHOS, senior. He wrote a geographical – perhaps paradoxographical – work *On Islands: Souda* K-228.

SH 309; *BNP* 2 (2003) 978 (#4), S. Fornaro.

PTK

Kallimorphos (*ca* 150 – 170 CE)

Surgeon of the 6th Pikemen, participated in Verus' Parthian campaign, wrote a *History of Parthia* (Lucian, *How to Write History* 16 [*FGrHist* 210]), claiming that "it is suitable for a physician to write history, if Asklēpios is the son of Apollo, Apollo being leader of the Muses and of all education." Lucian criticizes his inconsistent style and use of dialects. Gossen questions Kallimorphos' existence, but the name is attested in Lucian's era: *LGPN* 3A.232, 4.183.

RE 10.2 (1919) 1648–1649 (#1), F. Jacoby; and 1649 (#2), H. Gossen.

GLIM

Kallinikos (Pharm.) (20? – 90 CE)

ASKLĒPIADĒS PHARM., in GALĒN *CMGen* 7.7 (13.984 K.), records his ointment, based on (or recorded among?) those of ANTIOKHOS PACCIUS: clear **bdellium**, fresh beeswax, gentian (see GENTHIOS), henbane seed, pepper, saffron, **terebinth**, etc.

Fabricius (1726) 107.

PTK

Kalliphanēs (250 BCE – 75 CE)

Possibly a writer of *Thaumasia*, consulted by PLINY as an authority on Libyan hermaphrodites (1.*ind*.3, 5.7.15).

RE 10.2 (1919) 1655, F. Jacoby.

Jan Bollansée, Karen Haegemans, and Guido Schepens

Kallippos of Kuzikos (350 – 320 BCE)

Follower of EUDOXOS and perhaps also POLEMARKHOS. He composed a ***parapēgma***, preserved in GEMINUS, *Calendarium*, and PTOLEMY, *Phaseis*, which divides the year symmetrically into 12 months named after the zodiac with Sagittarius having 29 days and its six adjacent signs having 30 days each, and with Gemini having 32 days and its four adjacent signs having 31 days each. He divided the year (P. PARISINUS GRAECUS 1 for the first three numbers): summer (92 days), fall (89 days), winter (90 days), spring (94 days). With summer beginning with the start of Cancer and every season being exactly three months, the two systems are coordinated. He also established a 76 year cycle, which reduces the solar year of METŌN of 365 $^{5}/_{19}$ days (or 6,940 days in 19 years) to 365 ¼ or 6,939 $^{3}/_{4}$ days (hence four times the period will yield a whole number, 27,769 days). The Metonic system yielded a 19-year cycle of 6,940 days resulting in an error of one day every four cycles or 1 ¼ days in 95 years. Kallippos began the calendar on 1 Hekatombaion (June 28, 330 BCE), when the summer solstice fell on the same day as the lunar conjunction and a day before the visible new moon. Kallippos' cycle was used by HIPPARKHOS and others for astronomical dating. ARISTOTLE (*Met.* 12.8 [1073b32–38]) credits Kallippos with revising Eudoxos' celestial models. Where Eudoxos used four spheres to produce the motions of Mars, Venus, and Mercury, Kallippos used five. SIMPLICIUS says that people objected to the latitudes in Eudoxos' models, so Kallippos' additions possibly repaired these aspects. However, other defects are apparent in Eudoxos' planetary models, especially for Venus and Mercury. Additionally, Kallippos added two spheres to Eudoxos' three each for the Sun and Moon. Again, we do not know what was being improved. However, if Simplicius (*In de caelo*, = *CAG* 7 [1894] 497.18–22) accurately represents his source, EUDĒMOS, the extra spheres for the Sun seemingly improved Eudoxos' models to make them consistent with the periods for EUKTEMŌN and Metōn, and not his own. Hence, the added pair of spheres would create an anomaly with a hippopede where the Sun accelerates 2½ days in summer/fall, and slowed down 2½ days in winter/spring. Aristotle seems to have doubted the need for the extra solar/lunar spheres.

RE S.4 (1924) 1431–1438, A. Rehm; *Idem*, "Parapegma," *RE* 18.4 (1949) 1295–1366; *KP* 3.83–84, J. Mau; *DSB* 3.21–22, J.S. Kieffer; Goldstein and Bowen (1988); *OCD3* 278, G.J. Toomer; Henry Mendell, "Reflections on Eudoxus, Callippus and their Curves: Hippopedes and Callippopedes," *Centaurus* 40 (1998) 177–275; I. Yavetz, "On the Homocentric Spheres of Eudoxus," *AHES* 51 (1998) 221–278; A. Jones, "Calendrica I: New Callippic Dates," *ZPE* 129 (2000) 141–158; Henry Mendell, "The Trouble with Eudoxus," in P. Suppes, J. Moravcsik, and Henry Mendell, edd., *Ancient and Medieval Traditions in the Exact Sciences: Essays in Memory of Wilbur Knorr* (2001) 59–138; *BNP* 1 (2002) 985–986 (#5), W. Hübner.

Henry Mendell

Kallisthenēs of Olunthos (356 – 327 BCE)

Relative and student of ARISTOTLE, born in Olunthos. Kallisthenēs made his reputation as a writer with a history of Greece from the King's Peace (386) to the beginning of the Third Sacred War (356), along with several lesser works including a ***periplous***, fragments of which concern the Pontic and southern coasts of Anatolia. He accompanied Alexander as official court historian of the expedition. His account tended towards obsequiousness, with frequent allusions to Alexander's divine ancestry and favor. Nevertheless, Kallisthenēs refused to participate in the rituals derived from Persian court ceremonial, including the *proskunesis*. Falling out of favor, Kallisthenēs was implicated in a plot to assassinate Alexander, and was either executed immediately, or died in 327 after being carted around in chains for several months. His unfinished account, *Deeds of Alexander*, survived to be consulted by later writers often on geographical matters. Kallisthenēs was particularly interested in historical and mythological associations of places, as well as natural phenomena. He speculated that the Nile flood was caused by the summer rains, and that earthquakes were caused by air trapped in caverns beneath the Earth. He reportedly sent astronomical data home from Babylōn to Aristotle. He paid some attention to the hydrography of Central Asia, and so must have discussed the Baktrian campaign to some extent. He also noted interesting flora and fauna. He reported credulously on portents and accepted the oracles at Delphi, Dodona, Brankhidai and Siwah. The extant fragments preserve far less novel data than do the other Alexander historians, due to his curtailed participation in the expedition.

Ed.: *FGrHist* 124.
Robinson (1953) 1.45–77; Pearson (1960) 22–49; A.B. Bosworth, "Aristotle and Callisthenes," *Historia* 19 (1970) 407–413; Pédech (1984) 15–70.

Philip Kaplan

Kallistratos (350 – 25 BCE)

Wrote a book on mechanics, and made a scaling error (involving a "famous triangle") in designing a machine for transporting stones to the temple at Ephesos (Athēnaios Mech. p. 28 S.), thus perhaps datable to 350–320 BCE, or else perhaps a contemporary of Paconius.

(*)

PTK

Kallixeinos of Rhodes (210 – 150 BCE)

Athēnaios, *Deipn.*, cites Kallixeinos five times, preserving two lengthy passages from his paradoxographical *On Alexandria* (*cf.* 9 [387c], 11 [474c], 15 [677d]), a work of at least four books, perhaps arranged topically. One long fragment relates a triumphal procession of Ptolemy II Philadelphus, in which a cart carried an automaton (*Deipn.* 5 [198f]). Another long fragment describes two ships of Ptolemy IV Philopatōr. The account of the warship (203a–204d), the "Forty," includes measurements, equipment (four steering oars), armaments, adornments, manpower (4,000 oarsmen, 2,850 marines), and mechanics: e.g., the oars, although long and heavy, were properly balanced for ease of use; it was launched from a "cradle" and pulled into the water by a team of men (*cf.* Plutarch, *Demetr.* 43.4–5, who describes the warship as monstrous and unmaneuverable). Rice 1983: 142 speculates that "Forty" may indicate a catamaran, two "Twenties" lashed together, with 20 oarsmen distributed along three banks of oars per hull. In the style of a ***periēgēsis***, Kallixeinos describes Ptolemy IV's shallow riverboat or barge (204d–206c), a "cabin-cruiser" (*thalamēgos*), inspired by Egyptian architecture and powered partly by sail, with a double bow and stern. Kallixeinos includes the barge's measurements, materials, and function, plus digressions on Egyptian flora. The barge was also probably a catamaran (Rice 1983: 146); such "double-boats" were used in Ptolemaic Egypt to transport especially heavy loads, and the tall, top-heavy superstructure of the barge would require the additional stability of a broad hull and shallow keel.

E.E. Rice, *The Grand Procession of Ptolemy Philadelphus* (1983) 134–179; *OCD3* 279, K. Meister.

GLIM

Kalyāṇa (before 100 BCE)

References to an Indian astronomer named Kallaneus, the Greek transcription of the name Kallāṇa, i.e., Kalyāṇa, are found on a fragment of a ***parapēgma*** from Milētos (Diels and Rehm 1904). Kalyāṇa apparently composed a treatise which gave the times of the heliacal risings and settings of certain fixed stars. Pingree (*CESS* A.2.24) points out that Kalyāṇa's existence, or identity with Kalanos, a gymnosophist at the time of Alexander the Great, is uncertain.

CESS A.2.24 (Kalyāṇa's date wrongly recorded), A.4.47; H. Diels and A. Rehm, "Parapegmenfragmente aus milet," *SBAW* (1904) 92–111; A. Rehm, "Weiteres zu den milesischen Parapegmen,"

Sitzungsberichte der Prussischen Akademie der Wissenschaften zu Berlin 23 (1904) 752–759; D.E. Pingree, "The Indian and Pseudo-Indian Passages in Greek and Latin Astronomical Texts," *Viator* 7 (1976) 141–195.

Kim Plofker and Toke Lindegaard Knudsen

Karneadēs of Kurēnē (190 – 130 BCE)

Karneades © Antikenmuseum Basel und Sammlung Ludwig

A member and later head of the **Academy**, perhaps representing the climax of its skeptical phase. He wrote nothing; our knowledge of his arguments depends ultimately on reports of his younger contemporaries, particularly KLEITOMAKHOS OF CARTHAGE and Philo of Larissa. His arguments were directed especially against the **Stoics**, but also against the **Epicureans** and others. It is much disputed whether his procedure was purely dialectical – intended to demonstrate what these *other* philosophers should conclude given their own intellectual commitments – or whether a skeptical stance of suspension of judgment, resulting from the irreconcilability of opposing arguments, was something he personally adopted. He developed a theory of decision based on the notion of "persuasive appearances," a form of fallibilism designed to operate in the absence of certainty; again, it is debatable whether he intended this for his own use or as a dialectical device.

The topics of Karneadēs' arguments range widely over epistemology and ethics, but also include some issues of more direct relevance to science. He attacked both **Stoics** and **Epicureans** on determinism, fate and causation; the **Epicureans**, he held, did not need their notorious atomic swerve in order to preserve human freedom, and the **Stoics** need not infer that everything is fated from the fact that everything is caused. He also addressed issues in theology (considered part of physics in the Hellenistic period), in particular arguing that there is no clear boundary between the divine and the non-divine and that divination has no clear purpose.

Long and Sedley (1987) §§68–70; *OCD3* 293–294, G. Striker; *SEP* "Carneades," J. Allen.

Richard Bett

Karpiōn (450 – 430 BCE)

Presumably an architect, co-author with IKTINOS of a book on the Parthenon (VITRUUIUS 7.*pr.*12). There has been speculation about his role in the Parthenon's construction, but, in keeping with ancient tradition, modern discussions usually credit Iktinos with the design.

KLA 1.404–405, M. Korres.

Margaret M. Miles

Kárpos of Antioch (10 BCE – 40 CE?)

"*Ho mēkhanikos*," worked in mathematics, astronomy, and mechanics. IAMBLIKHOS, *In Categ.*, mentions his attempt to square the circle using a curve of "double motion" (SIMPLICIUS *In Categ.* 7, *CAG* 8 [1907] 192, = *In Phys.* 1.2, *CAG* 9 [1882] 60). PROKLOS cites him for defining angles as quantities, i.e., the distance between the enclosing lines or planes (*In Eucl.* pp. 125.25–126.7 Fr.), and for speculating that Venus is larger than the Earth (*In Rem Pub.* 2, p. 218 Kr.). Proklos (*In. Eucl.* pp. 241.18–244.1 Fr.) quotes from Kárpos' *Astrologikē Pragmateia* an argument that *problēmata* have priority over theorems, since *problēmata* possess evident constructions, and the method of analysis always finds their solution, whereas no-one up to his time had found a uniform way of dealing with theorems; moreover, he criticized GEMINUS, for saying a theorem is more perfect than a problem. (Given that PAPPOS, *Coll.* 7.6, records HĒRŌN's claim to deal uniformly with theorems, Segonds argues for Kárpos' priority.) Kárpos' *Mēkhanikos* applied geometry to practical aims, citing ARCHIMĒDĒS' *Sphere-making* (Pappos, *Coll.* 8.3 [p. 1026 Fr.]), and he is credited with a type of level similar to the *alpharion* or *diabētēs* (THEŌN OF ALEXANDRIA, *In Ptol. Alm.* 1, p. 524 R.).

G. Sarton, *History of Science* (1959) 2.360, 362; *DPA* 2 (1994) 228–230, A. Segonds.

PTK and GLIM

Kassandros (*ca* 135 – 105 BCE)

Named by PANAITIOS OF LINDOS with ANKHIALOS as the best astrologers of his time: they did not practice predictive astrology (CICERO, *Div.* 2.88). Probably distinct from the historian Kassandros of Salamis (*FHG* 4.359), our Kassandros believed that the Great Year occurred every 180,000 years (CENSORINUS 18.11).

RE 10.2 (1919) 2314 (#9), F. Jacoby.

GLIM

Kēphisophōn (of Athens?) (325 BCE – 120 CE)

SŌRANOS, *Gyn.* 3.38 (*CMG* 4, p. 118; *CUF* v. 3, p. 41), refers to Kēphisophōn's remedy for uterine mole (along with that of POLUARKHOS); OREIBASIOS, *Ecl. Med.* 53.5 (*CMG* 6.2.2, p. 215) records his pitch-plaster (*drōpax*) of bitumen, sulfur, and marsh-salt (DIOSKOURIDĒS 5.119), and refers to his ointment, *Ecl. Med.* 59.3 (p. 224). CAELIUS AURELIANUS repeats Sōranos' prescriptions of his remedies: *Acute* 2.153 (*CML* 6.1, p. 236), *Chron.* 2.34, 3.55 (pp. 564, 710). The name, like all names in Kēphiso-, is almost solely Athenian, and common there before the Roman period.

(*)

PTK

***Keras of Carthage* (450 – 350 BCE?)**

Improved the ram design of PEPHRASMENOS OF TYRE, by adding a rolling platform with protective covering, and was the first to use the term "ram tortoise" (ATHĒNAIOS MECH. pp. 9–10 W.; VITRUUIUS 10.13.2). Although the name is attested for an Olympic victor (*LGPN* 3A.240: Argos, 300 BCE), the mechanist may be a legendary "*prōtos heurētēs*."

(*)

PTK

Keskintos, Inscription of (150 – 50 BCE)

The surviving lower half of an anonymous Greek votive inscription recording a list of planetary periodicities, discovered in 1894 at Keskintos, near Lardos, Rhodes, and now in the Antikensammlung, Staatliche Museen, Berlin. Each planet is assigned four periods, and the table records how many of each kind of period occur in 29,140 and 291,400 solar years. The underlying planetary theory was apparently geometrical though radically different from PTOLEMY's, and the long cycles relate the inscription to the "Great Years" of Greek philosophy and astrology as well as to the Yugas (enormous astronomical cycles) of Indian astronomy. The inscription also defines units of arc, the "degree" (*moira*) as 1/360 of a circle, and the "point" (*stigmē*) as 1/9,720 of a circle.

Ed.: Alexander Jones, "The Keskintos Astronomical Inscription: Text and Interpretations," *SCIAMVS* 7 (2006) 3–41.

P. Tannery, "L'Inscription astronomique de Keskinto," *REG* 8 (1895) 49–58; Neugebauer (1975) 698–705.

Alexander Jones

Khaireas of Athens (325 – 175 BCE)

Was the author of an agricultural work which distinguished different varieties of wine (Ath., *Deipn.* 1 [32c]), and devoted special attention to the properties of thistle, according to PLINY, 20.263 (ed. *CUF*), quoting the opinion of GLAUKIAS OF TARAS. VARRO, *RR* 1.1.8 and COLUMELLA, 1.1.8, knew his work, which, to judge from Pliny, 1.*ind*.8, 10, 14–15, 17–18, may also have treated cereals, livestock, poultry, and arboriculture.

KP 1.1121 (#4), Kl. Stiewe; *RE* 3.2 (1899) 2023 (#8), M. Wellmann.

Philip Thibodeau

Khairēmōn (25 – 75 CE)

Egyptian priest (*hierogrammateus*) adhering to **Stoicism**, among the emperor Nero's tutors. Khairēmōn wrote an account of Egyptian hieroglyphics, perhaps as part of his lost *Aiguptiakē Historia*. EUSEBIOS reports in his *Praeparatio Evangelica* (3.4) that Khairēmōn maintained that Egyptian gods were the planets, zodiacal signs, and constellations, referring to the SALMESKHOINIAKA for authority. Origen, *Contra Celsum* (1.59) asserts that Khairēmōn wrote *On Comets* discussing their significance as astral omens, but the testimonia do not suggest that he was an astrologer in his own right.

Ed.: P.W. van der Horst, *Chaeremon: Egyptian Priest and Stoic Philosopher. The Fragments Collected and Translated with Explanatory Notes* (1984).

M. Frede, "Chaeremon," *ANRW* 2.36.3 (1989) 2067–2103.

Alexander Jones

Khairesteos of Athens (325 – 90 BCE)

Wrote a work on agriculture used by CASSIUS DIONUSIOS (VARRO, *RR* 1.1.8–10). COLUMELLA (1.1.8) gives his name as "Chrestus," and PLINY (1.*ind*.14, 15, 17, 18) as "Chaerestus," both perhaps confusing him with KHAIREAS OF ATHENS. To judge from Pliny's index, he discussed cereals, viticulture, and arboriculture.

RE 3.2 (1899) 2029, M. Wellmann.

Philip Thibodeau

Khalkideus (250 BCE – 95 CE)

ASKLĒPIADĒS PHARM., in GALĒN *CMGen* 5.4 (13.803–804 K.), records his wide-spectrum plaster ("on the whole its power is amazing"): boil **litharge** and copper-flake in aged olive oil, then add ***diphruges***, beeswax, and pine resin; remove from heat and add ***ammōniakon*** incense (and the **galbanum**?), when cool, add birthwort, and apply. The rare name is unattested after the 2nd c. BCE (*LGPN*).

(*)

PTK

Kharetidēs of Paros (450 – 335 BCE)

Linked by ARISTOTLE with APOLLODŌROS OF LĒMNOS as a writer of (lost) manuals on agriculture, treating both crops and fruits (*Politics* 1.11 [1258b39–1259a2]). Like Apollodōros, Kharetidēs may be counted among the earliest Greek technical writers on agriculture.

RE 3.2 (1899) 2131–2132, M. Wellmann.

Maria Marsilio

Kharias (330 – 310 BCE)

The mechanicians Kharias and DIADĒS were students of POLUIDOS OF THESSALIA, whose improvements on ram-tortoise design they continued (ATHĒNAIOS MECH. pp. 5, 10 W.; VITRUUIUS 10.13.3); they accompanied Alexander on campaign. The name Kharias is widespread, but especially frequent at Athens (*LGPN*).

RE 3.2 (1899) 2133 (#11), Fr. Hultsch.

GLIM

Kharidēmos (50 BCE – 120 CE)

Mentioned by CAELIUS AURELIANUS, *Acute* 3.118 (*CML* 6.1.1, p. 362), as denying that ***hudrophobia*** specifically was a new disease, while rejecting ARTEMIDŌROS OF SIDĒ's stronger claim that no disease was new. On the basis of *CIG* 3311 = *Inschr. Smyrn.* 1 (1982) #536, identified as the father of the **Erasistratean** HERMOGENĒS OF SMURNA. Although Kharidēmos is much rarer after the 1st c. BCE (*LGPN*), the identification is uncertain.

RE 3.2 (1899) 2138 (#7), M. Wellmann.

PTK

Khariklēs (120 BCE – 80 CE)

ANDROMAKHOS, in GALĒN *CMLoc* 7.5 (13.94 K.), records his anodyne, based on henbane and opium, and including Cretan-carrot seed; whereas ASKLĒPIADĒS PHARM., in Galēn, records three recipes, for ***duspnoia*** (*CMLoc* 7.6 [13.109 K.]), for colic (*ibid.* 9.4 [13.282]) and nephritis (*ibid.* 10.1 [13.329]), the second and third involving Cretan-carrot seed, celery seed, and white pepper. He is also credited with two headache compresses, both involving castoreum, laurel, roses, and sulfurwort root, dissolved in vinegar and rose oil; soak a linen bandage in the result, shave the head, and wrap; leave on for a day: *ibid.* 2.1 (12.556–557, 558) = 2.2 (12.579, 581). The name is rare after the 1st c. BCE (*LGPN*), but he may be the doctor who diagnosed Tiberius' impending demise, Suet. *Tib.* 72, Tacitus, *Ann.* 6.50.

RE 3.2 (1899) 2140 (#7), M. Wellmann; Korpela (1987) 166.

PTK

Kharistiōn ⇒ Aristiōn

Kharitōn (325 BCE – 95 CE)

Asklēpiadēs Pharm., in Galēn, *Antid.* 2.13 (14.180 K.) records his simple remedy for spider bites: fruit of *sphondulion* (Dioskouridēs 3.76) and calamint, given in wine; repeated by Aëtios of Amida, 13.18 (p. 693 Cornarius). His book of remedies *Pharmaka* appears to be preserved in whole or part, in MS BNF *Parisin.* 2240 (16th c.), f. 1.

Fabricius (1726) 111; Diels 2 (1907) 23.

PTK

Kharixenēs (30 BCE? – 95 CE)

Greek pharmacologist datable by inclusion in Asklēpiadēs Pharm., in Galēn, *CMLoc*; he did not quote recipes of Antonius Musa (contra Wellmann). He may have lived in Rome as did other contemporary pharmacologists. Asklēpiadēs cites Kharixenēs' collection of compound medicines only for remedies for the head and breathing organs. It probably included recipes Kharixenēs used (*CMLoc* 7.4, 13.85 K., against blood-spitting, including opium) and others, bearing his name, that he might have created: 3.1 (12.635 K.), also with opium, against open abscesses; 3.1 (12.638 K.) against *aphthae* and halitosis; 3.3 (12.685 K.) for surgical treatment of nasal polyps; 7.6 (13.48–50 K.), against *aphonia* and pathologies of the throat; 7.4 (13.82–83 K.), against blood-spitting (?); 7.5 (13.102 K.), with mandrake and henbane, against respiratory affections, blood-spitting and coughs; 7.6 (13.108–109 K.), against ***duspnoia***; and Aëtios of Amida 8.56 (*CMG* 8.2, pp. 492.22–493.2).

RE 3.2 (1899) 2172, M. Wellmann.

Alain Touwaide

Kharmandros (*ca* 350 – 200 BCE?)

In his doxography on comets, along with Kallisthenēs and Aristotle, Seneca includes Kharmandros' book (*QN* 7.5.3), which discussed a comet seen by Anaxagoras. In discussing Apollōnios' now-lost *Plane Loci*, Pappos, *Coll.* 7.23–24 (pp. 104–109 Jones), credits a Kharmandros with three simple plane loci, as first among the loci "additional" to the "ancient" loci. The name is very rare (hardly known aside from Plato's accuser: Diogenēs Laërtios 3.19), making an identification plausible, though not necessary.

Jones (1986) 104–109, 395–397; *DPA* 2 (1994) 299, R. Goulet.

PTK

Kharmēs of Massalia (30 – 60 CE)

Achieved fame and fortune by prescribing cold-water baths; practiced in Rome from 55 CE and bequeathed his fortune to his hometown: Pliny 29.10, 29.22. Damokratēs, in

Galēn, *Antid.* 2.4 (14.126–129 K.), versified his antidote (priced at 1,000 drachmas the dose), which had been employed by Aelius Gallus and Iulius Caesar: 2.1 (14.114–115). Asklēpiadēs, in Galēn *CMGen* 5.15 (13.855–856 K.; *cf.* Aëtios of Amida 14.58, p. 804 Cornarius), records the remedy donated to Massalia for ***anthrax***, of "the Massaliote" – Kharmēs, or of Dēmosthenēs Philalethēs, or possibly a brand-name? He also studied bird-behavior, according to Aelianus, *NA* 5.38 (nightingale).

BNP 3 (2003) 202, V. Nutton.

PTK

Kharōn of Carthage (*ca* 300 – 150 BCE?)

The *Souda* Khi-136 attributes geographical works to the historian Kharōn of Lampsakos (*ca* 430 BCE), which scholars assign to Kharōn of Carthage, to whom the *Souda* Khi-137 attributes only biographies. The four works, all entirely lost, are: *Aithiopika, Krētika, Libuka,* and a ***Periplous*** *beyond the Pillars of Hēraklēs.*

FGrHist 262, 1077.

PTK

Kharōn of Magnesia (220 – 200 BCE?)

Designed a stone-throwing catapult at Rhodes, perhaps for its successful defense of Khios in 201 BCE (Polubios Book 16, *frr.* 2–9). Kharōn's catapult, described by Bitōn, *Belop.* 2 (pp. 45–48 W.), was a mechanically-assisted bow (i.e., *gastraphetēs*), cocked by a winch, that shot stones of *ca* 2.5 kg (5 *minae*).

Marsden (1971) 66–69, 78–82.

PTK

Khēmēs or Khumēs (250 BCE – 300 CE)

Pseudepigraphic alchemical authority first mentioned by Zōsimos of Panōpolis (*CAAG* 2.169, 172, 182–183). He may be identifiable with the angel Khēmeu, who, according to Zōsimos (*apud* Synkellos *Ec.Chron.* 14, ed. Mosshammer), revealed the alchemical arts to humans and from whom alchemy (Greek *khumeia/khēmeia*) took its name. He was most famous for the aphorism "all is one and through it all has come into being; all is one and if all does not have all, all has not come into being," found repeatedly throughout the Greek alchemical corpus with minor variations, and with or without attribution.

(*)

Bink Hallum

Khersiphrōn of Knōssos (570 – 520 BCE)

Architect, inventor, built the great Temple of Artemis at Ephesos, with his son Metagenēs of Knossos (Strabōn 14.1.22; Vitruuius 3.2.7; 7.*pr.*12, 16; Pliny 7.125; 36.95), and wrote a treatise on the temple, among the earliest architectural tracts. Begun *ca* 560 BCE, the dipteral Ionic temple was one of the largest of its time. Special preparations (charcoal, sheepskins for a damp course) for the foundations were needed because of the flat and marshy site. Khersiphrōn, a pioneer in practical mechanics, invented a rolling framework allowing the large column-drums to be moved to the construction site easily and quickly (Vitr. 10.2.11). He also used sand to help settle the upper courses into place.

Svenson-Evers (1996) 67–99; A. Bammer and U. Muss, *Das Artemision von Ephesos* (1996); J. Healy, *Pliny the Elder on Science and Technology* (1999) 161–164; *KLA* 1.139, R. Vollkommer.

Margaret M. Miles

Kh(o)ios **(250 BCE – 100 CE)**

SŌRANOS *Gyn.* 1.12 (*CMG* 4, p. 9; *CUF* v. 1, p. 12) records that *ΧΙΟΣ* propounded the existence of a suspender (*kremastēr*) muscle on the ovaries, where scholars suppose the name corrupt. "Khios" is, however, attested (1st c. BCE – 1st c. CE: *LGPN* 3A.477, Illyria and S. Italy), as is "Khoios" (Hellenistic: 3B.444, Thessalia), and OREIBASIOS, *Coll.* 24.31.20 (*CMG* 6.2.1, pp. 43–44), confirms the text. If "Khios" is to be emended, perhaps BAKKHEIOS OF TANAGRA (von Staden 1989: 214–215).

(*)

PTK

Khrusanthos Gratianus (*ca* 100 BCE – *ca* 80 CE)

ANDROMAKHOS in GALĒN, *CMLoc* 3.1 (12.631–632 K.), cites his remedy for aural inflammations and wounds. The Greek name is first attested as a *cognomen* in south Italy: Q. Iunius Chrysanthus (freedman) of Herculaneum, five men from Pompeii, and others (*LGPN* 3A.479; Solin 2003: 1.174–176), although *cf.* Khrusantas of Kōs, *ca* 200 BCE (*LGPN* 1.487).

Fabricius (1726) 115.

PTK

Khrusermos of Alexandria (*ca* 70 – 30 BCE?)

Hērophilean physician, mentor of HĒRAKLEIDĒS OF ERUTHRAI and APOLLŌNIOS "MUS." Controversial was his amplified definition of the pulse as distention and contraction of arteries through vital and psychic power, as the arterial layer expands and shrinks (GALĒN *Puls. Diff.* 4.9 [8.741–743 K.]). Accepted only by his students, Khrusermos' theory was rejected by ARISTOXENOS and Galēn, among others. ASKLĒPIADĒS in Galēn *CMLoc* 9.2 (13.243–244 K.) preserves Khrusermos' diuretic belly-soothing ***trokhiskos*** for the spleen and **dropsy** compounded from squill, carrot seeds, anise, hartwort, ***panax*** root, iris, nettle seeds, cedar berries, myrrh, bitter vetch, tragacanth, and fragrant wine. PLINY, citing Khrusermos as a medical authority (1.*ind.*22), recounts two treatments with decocted asphodel root: in wine for parotid abscesses, and added to parched barley in wine for scrofulous swellings (22.71).

von Staden (1989) 523–528; *OCD3* 329, *Idem.*

GLIM

Khrusēs of Alexandria (500 – 520 CE?)

Architect employed by Justinian I in Dara, capital of Mesopotamia, according to Prokopios (*Aed.* 2.3.1–23). Khrusēs' most challenging design involved a dam protecting the city.

According to archaeological data, Dara's urban projects were completed by Emperor Anastasios I following the establishment of a military stronghold in 505–507 CE. Prokopios' fabrication of Justinian's activities makes the true identity of Khrusēs difficult to ascertain.

Warren (1976) 8; B. Croke and J. Crow, "Procopius and Dara," *JRS* 73 (1983) 143–159; *PLRE* 3 (1992) 314.

Kostis Kourelis

Khrusippos (Agric.) (250 – 50 BCE)

Wrote *On Agriculture*; listed after Erasistratos' student Khrusippos of Knidos (II) (Dēmētrios of Magnesia in Diogenēs Laërtios 7.186). If the silence of Dionusios of Utica in Varro is significant, perhaps after 90 BCE.

RE 3.2 (1899) 2511 (#20), M. Wellmann.

PTK and GLIM

Khrusippos (Med.) (*ca* 100 BCE – 100 CE)

Follower of Asklēpiadēs of Bithunia, wrote a medical work, in Book 3 of which he considers the statements of predecessors including Hippokratēs and Dioklēs on worms (Caelius Aurelianus *Chron.* 4.114–115 [*CML* 6.1.2, p. 838]). His own theory on worms was that the expulsion of dead worms is a dire sign only in dangerous cases, indicating extreme weakness. In the last part of his book on catalepsy, he discussed the signs of an imminent attack, which he distinguished from lethargy (Cael. Aurel. *Acute* 2.57, 64, 82 [*CML* 6.1.1, pp. 164–166, 170, 182]; *Chron.* 2.86 [p. 596], with Nikēratos); he prescribed sharp ointments, such as henna with pepper and natron or brine with sulfur and bitumen, to treat numb or trembling limbs. Caelius Aurelianus lists Khrusippos (or a homonym) among those "ancients" whose treatments for epilepsy failed (*Chron.* 1.140 [p. 512])

RE 3.2 (1899) 2511 (#19), M. Wellmann.

GLIM

Khrusippos of Knidos (I) (*ca* 385 – 335 BCE)

Physician, son of Erineos, student of Philistiōn and Eudoxos (Diogenēs Laërtios 8.89–90), listed among Pliny's foreign authorities on drugs (1.*ind.*20–27, 29, 30). He is probably the author of *On Vegetables* (*Schol. Nik. Thēr.* 845; *cf.* Pliny 26.10), to which his volume on the benefits of cabbage by body part (presumably "head-to-toe") may have belonged (20.78). Khrusippos suggested that cabbage heals flatulence, biliousness, and wounds if applied with honey and left on for seven days; the brassicid also cures scrofula and fistulas (20.93). Pliny reports Khrusippos' opinions that *kaukalis* aids in conception (22.83), and that gourds are bad to eat despite consensus on their benefit to the stomach and to intestinal and bladder ulcers (20.17). He distinguished female from male parsley as hard, with curlier leaves, thick stem, and a hot, sharp taste (20.113). He prescribed wild asparagus, parsley, and cumin seeds for hematuria (Pliny 20.111). Bad for **dropsy**, sexual arousal, and the bladder unless boiled in water (the juice kills dogs), but asparagus root, boiled in wine, cures toothache. Khrusippos condemned basil as harmful to the stomach, urine, and eyesight, causing madness, "comas," and liver complaints, hence explaining why goats avoid it (Pliny 20.119). CELSUS preserves his emollient for joint pain including liquid pitch, realgar, and pepper, fused with a little beeswax (5.18.30).

RE 3.2 (1899) 2509–2510 (#15), S.1 (1903) 299, M. Wellmann.

GLIM

Khrusippos of Knidos (II) (280 – 250 BCE)

Son of Aristagoras, student of Aethlios (DIOGENĒS LAËRTIOS 8.89, *cf.* 7.186), and teacher of ERASISTRATOS OF KEŌS, STRATŌN (ERASI.), APEIMANTOS, ARISTOGENĒS OF KNIDOS, and MĒDEIOS; he eschewed the practice of venesection (as dangerous and ineffective), and advocated redistributing blood using tourniquets: GALĒN, *On Venesection, Against Erasistratos* 1, 2, 7 (11.148, 151–152, 175–176 K. = pp. 16, 18, 31–32 Brain), *Venes. Rome* 2, 7 (11.197, 230 K. = pp. 43, 58 Brain). Theorized, as did Erasistratos, that all fevers were caused by undigested residues of food (Galēn, *In Hipp. Epid. VI = CMG* 5.10.2.2, p. 44), and was renowned as an anatomist (Galēn, *In Hipp. Nat. Hom.* 15.136 K., listed with DIEUKHĒS, PHULOTIMOS, PLEISTONIKOS, PRAXAGORAS, etc.). In therapy, he avoided strong purgatives (*Venes. Rome* 9 [11.245 K. = p. 65 Brain]), prescribed wine in cold water for cholerics (*Venes. Eras.* 7 [11.171 K. = pp. 29–30 Brain]), and steam baths for **dropsy** (Galēn, *Use of Respiration* 4 [4.495–496 K.]). Treated a patient who believed she had swallowed a snake by prescribing an emetic, then slipping a snake into the vomit-basin (*In Hipp. Epid. II = CMG* 5.10.1, pp. 207–208). Galēn had trouble obtaining his works (*Venes. Rome* 5 [11.221 K. = p. 54 Brain]), and they probably did not survive the 3rd c. CE. Often confused by scholars with KHRUSIPPOS OF KNIDOS (I).

RE 3.2 (1899) 2510–2511 (#16) + S.1 (1903) 299, M. Wellmann; *BNP* 3 (2003) 293–294 (#3), V. Nutton.

PTK

KHRUSIPPOS OF SOLOI ⇒ CHRYSIPPUS OF SOLOI

Kidēnas (Kidinnu) of Babylōn (*ca* 150 – 50 BCE?)

According to STRABŌN (16.1.6), a *mathēmatikos* (astronomer) from Babylōn. The name Kidin(nu) appears in the colophons of two cuneiform **ephemeris** tables (*ACT* 122 and 123a). The tablet (*ACT* 122) is designated "*tersētu* of Kidin . . . concerning (the years) 208 to 210" (Seleukid Era, i.e., 104–102 BCE). The table, part of which provided the key for Kugler's pioneering analysis of the Babylonian astronomical "System B," published in his *Babylonische Mondrechnung* (1900), concerns new moons for the years mentioned. In the colophon of the second tablet, *ACT* 123a, the table is called a "*tersētu* of Kidinnu," concerning new and full moons for two years.

VETTIUS VALENS 9.12 says he used "HIPPARKHOS for the Sun, SUDINĒS and Kidēnas and APOLLŌNIOS for the Moon, and again Apollōnios for both types (solar and lunar)." The methods he refers to here are not given, so any comparison with actual Babylonian eclipse prediction schemes is impossible. PLINY (2.38–39) gives values for the maximum elongations of the inner planets from the Sun, Venus (46°) and Mercury (22°), based on TIMAIOS for Venus and Kidēnas and SŌSIGENĒS (I) for Mercury.

KP 3.207, B.L. van der Waerden; Neugebauer (1975) 611, 804.

Francesca Rochberg

Kimōn (250 BCE – 30 CE)

ASKLĒPIADĒS PHARM., in GALĒN *CMLoc* 3.1 (12.637 K.), records his remedy for purulent ears, compounded of castoreum, alum, saffron, myrrh, frankincense, to be taken with must, or myrtle-wine, or with honey-wine if the infection is recent, with vinegar if old. The name, known from the 6th c. BCE to the 1st c. CE, is slightly concentrated in the 4th/3rd cc. BCE (*LGPN* 1.255, 2.261, 3A.241, 3B.230).

RE 11.1 (1921) 454 (#9), F.E. Kind.

GLIM

Kineas of Thessalia (*ca* 325 – 277 BCE)

The very eloquent ambassador (*ca* 282–277) of King PURRHOS OF ĒPEIROS, Kineas was a student of the orator Dēmosthenēs, and a follower of EPICURUS (PLUTARCH, *Pyrrh.* 14–22). He wrote a history of Thessalia (*FGrHist* 603), and abridged AINEIAS TACTICUS and the king's own work (Aelianus, *Taktika* 1.2).

BNP 3 (2003) 342 (#2), J. Engels.

PTK

Kleandros of Surakousai (350 BCE – 200 CE?)

Wrote *On the Horizon*, a geographical work explaining that concept, cited only by the scholiast at *Iliad* 5.6. *Cf.* perhaps KLEŌN OF SURAKOUSAI.

(*)

PTK

Kleanthēs of Assos (*ca* 290 – 230 BCE)

Born *ca* 330; student and successor of ZĒNŌN OF KITION as head of the **Stoa** from 262 to 230. Ancient sources depict Kleanthēs as hard working, if a bit of a dullard. He is said to have been a boxer before going to Athens and studying under Zēnōn, and he reputedly financed his study with manual labor. He appears to have had some trouble defending **Stoic** doctrines against attacks by Arkesilaos the **Academic** Skeptic, and by the dissident **Stoic** Aristōn of Khios. Kleanthēs' student CHRYSIPPUS is said to have eventually entered the fray to defend more vigorously and successfully the **Stoic** position. DIOGENĒS LAËRTIOS (7.174–175) attributes 50 works to him, with focuses on ethics, theology, and physics. Kleanthēs wrote on the **Stoic** doctrine of the end-of-the-world conflagration, on time, on sense perception, on HĒRAKLEITOS, and on Zēnōn's natural philosophy, as well as the *Hymn to Zeus*. He also composed a book sharply critical of ARISTARKHOS OF SAMOS' heliocentrism. He used the human body as an analogy for the ***kosmos***, with the Sun as the ***hēgemonikon*** (soul) of the universe. He also elaborated Zēnōn's concept of ***pneuma***. He is the first **Stoic** to answer the Master Argument of DIODŌROS KRONOS by denying the necessity of past truths, and he may have introduced the category of *lekta* ("sayables") into **Stoic** logic.

Ed.: *SVF* 1.103–139.

Daryn Lehoux

Klearkhos of Soloi (*ca* 330 – 290/280 BCE)

Born *ca* 370/360 BCE; ARISTOTLE's pupil, **Peripatetic** philosopher, possibly from Soloi of Cyprus based on the first-person plural testimonies (*par' hēmin*) he gives for Cypriot customs in his *Gergithios* as quoted by Athēnaios (6 [256c]). His dedication of a collection of Delphic dicta in the Aï Khanoum sanctuary of Kineas provides further evidence for his life. That he was Aristotle's pupil (Ath. 6 [235a]), plus Aristotelian influence on his writings, suggests the date-range. Aristotle was the main figure in the lost dialogue *On Sleep* (*Peri Hupnou*), where Klearkhos described the adventures of the soul in the manner of HERAKLEIDĒS OF HĒRAKLEIA PONTIKĒ.

None of Klearkhos' works survived the Second Sophistic. Only fragments remain, mainly preserved by Athēnaios, from about 20 named works, the most important being those *On Lives* (*Peri Biōn*), *On Proverbs* (*Peri Paroimiōn*), and *On Riddles* (*Peri Griphōn*), unique in classical literature. Even in these works, the selection of Athēnaios, insisting on anecdotes, does not permit a clear idea of his philosophy. Other titles show **Peripatetic** interests: *On Love* (*Erōtika*), *On Friendship* (*Peri Philias*), *On Education* (*Peri Paideias*), etc.

No more than 12 "scientific" fragments survive; five titles of works on physics and anatomy are given: *On Water-animals* (*Peri tōn Enudrōn*), *On Creatures that dwell in Water, On Anatomy* (*Peri skeletōn*) in at least two books, *On Sand-Banks* (*Peri Thinōn*) and *On the Electric Ray* (*Peri Narkēs*). The fragmentary evidence shows his emphasis on descriptions, but not theory. That Klearkhos was interested in peculiarities and detailed descriptions is evident from his work in general. Klearkhos described in detail some phenonema – "the sleeper-out" fish or the "sacred octopus" – but only short notices have survived, especially in fragments not ascribed to named works, such as a certain river-fish making a sound, the colors of various fluids, stones that give birth. (This last and some other items seem to come from a *Mirabilia*.) Athēnaios (7 [314c]) omits Klearkhos' explanation of the electric ray (*torpedo*). In other fragments we read his opinion that the Moon is bright and made of ***aithēr***, and the markings on its disk reflect the Ocean; he insisted that there are 26 bones in the hand, and that men, bats, and elephants have breasts.

Ed.: Wehrli 3 (1969), with commentary.

J.B. Verraert, *Diatribe academica inauguralis de Clearcho Solensi, Philosopho Peripatetico* (1828); M. Weber, *De Clearchi Solensis vita et operibus* (Diss. Frankurt Oder, 1880); O. Stein, "Klearchos von Soloi," *Philologus* 40 (1931) 258–259; P. Moraux, "Cléarque de Soles, disciple d'Aristote," *LEC* 8 (1950) 22–26; P. Pédech, "Cléarque le Philosophe," *Au miroir de la culture antique: Mélanges R. Marache* (1992) 385–391; *DPA* 2 (1994) 415–420 (#141), R. Schneider.

Ioannis Taifacos

Kleëmporos (350 BCE – 77 CE)

Physician, wrote on medicinal uses of plants. He asserted that the white *sonkhos* was suitable for pharmaceutical preparations (e.g., for earaches) but the black, disease-causing, *sonkhos* should be avoided (PLINY 22.90). Pliny dismissed the claim that a book on botanicals circulating under PYTHAGORAS' name was ascribable to Kleëmporos, since our author published known volumes under his own name (24.159); *cf.* also PAMPHILOS OF ALEXANDRIA. An uncommon name, three citations attest Kleëmporoi at Athens in the 4th c. BCE (*LGPN* 2.263), others elsewhere from the 3rd to 1st cc. BCE: *LGPN* 3A.244.

RE 11.1 (1921) 591, H. Gossen.

GLIM

Kleidēmos of Athens (380 – 340 BCE)

Considered the oldest atthidographer (Pausanias 10.15.5, who with a few later sources names him "Kleitodēmos"), and supplanted by ANDROTIŌN's *Atthis* for the historical period. Athēnaios 14 (660ab) = *fr.*5 and Harpokratiōn, s.v. *ΠΥΚΝΙ* = *fr.*7, called his *Atthis*, composed *ca* 350 BCE, *Protogonia* – probably the authentic title and an indicator of Kleidēmos' antiquarian interests. Likely intended as an historical work also claiming literary merits, only 25 fragments, from four books, remain. The fragments from the first two books refer to the mythological period and regal history down to 683/2 BCE, from the third book to the reforms of Kleisthenēs, and the rest continue to the Peloponnesian War with his last recorded event in 415 BCE (F 10). Most fragments contain references to cults, constitutional issues, descriptions of the country, and details about localities. Kleidēmos' writings showed support of the democratic constitution. His scientific method is evident in his rationalization of myth and his application of etymology. Kleidēmos, an interpreter of and expert in ceremonial ritual (*exēgētēs*), also wrote an *Exēgētikon*.

Scientific fragments quoted by ARISTOTLE and THEOPHRASTOS (F 31–36) are probably also the work of the atthidographer (but Kroll dates Kleidēmos the agricultural writer to *ca* 440 BCE and distinguishes him from the atthidographer). These fragments may not belong to one work. Aristotle explains Kleidēmos' view that lightning has no objective existence but is merely an appearance as due to his ignorance of the theory of reflection (*Meteorologica* 2.9.18 = F31). Theophrastos mentions Kleidēmos on sensory perceptions with polemic against ANAXAGORAS (*De Sensu* 7.38 = F32). Fragments 31–32 would suit a book *peri* ***phuseōs***. Fragments 33–36 may belong to a *Geōrgikon*; a work having this title is ascribed to Androtiōn also. In these fragments Theophrastos cites Kleidēmos concerning fruit and vines and the reasons why vines would fail to bear fruit. He also gives advice for pruning vines, the proper time for sowing (at the setting of the Pleiades), and on the elemental composition and differences between plants and animals.

Ed.: DK 62; *FGrHist* 323 F31–36 with commentary.

RE S.7 (1940) 321, W. Kroll; G. Huxley, "Kleidemos and the Themistokles decree," *GRBS* 9 (1968) 313–318; J. McInerny, "Politicizing the past," *ClAnt* 13.1 (1994) 17–37; *OCD3* 343, P.E. Harding; *BNP* 3 (2003) 417, K. Meister.

Maria Marsilio

Kleinias of Taras (390 – 350 BCE)

Contemporary of PLATO, who with AMUNTAS allegedly prevented Plato from burning the collected works of DĒMOKRITOS OF ABDĒRA (DIOGENĒS LAËRTIOS 9.40). Kleinias was reputed for philanthropy (see PRŌROS) and exemplary character (Ath., *Deipn.* 14 [624a]; IAMBLIKHOS, *VP* 197–198). His treatise on piety survives in two fragments (IŌANNĒS STOBAIOS 3.1.75, 76). One fragment of his work on numbers describes how the first four numerals, setting arithmetic and geometry in motion, provide the foundation of harmony and astronomy (PSEUDO-IAMBLIKHOS, THEOL. ARITHM. [p. 21 de Falco]). The other discusses the pre-eminence of the numeral one (SYRIANUS, *In Metaphys.: CAG* 6.1 [1902] 168).

Thesleff (1965) 107–108; *BNP* 3 (2003) 417–418 (#6), Chr. Riedeweg.

GLIM

KLEITODĒMOS OF ATHENS ⇒ KLEIDĒMOS OF ATHENS

Kleitomakhos (Hasdrubal) of Carthage (155 – 110 BCE)

Son of Diognētos, born 187/186, taught philosophy at Carthage in Punic, studied at Athens from 147/146 under KARNEADĒS (DIOGENĒS LAËRTIOS 4.67). Kleitomakhos founded a school in Palladion (140/139), and then returned to the **Academy** (129/128), whose head he became (127/126). Although Diogenēs Laërtios attributes 400 works to him, only five titles are known, including *On the Sects*, in which he denied the utility of physics and logic, since *ataraxia* can be attained by study of ethics (D.L. 2.92); and *On Withholding Assent* (*Peri epokhēs*) in which he discussed perception and probability (CICERO, *Acad.* 2.98).

BNP 3 (2003) 421–422 (#1), K.-H. Stanzel.

PTK and GLIM

KLEITŌNUMOS ⇒ HEKATŌNUMOS

Kleoboulos (Geog.) (60 BCE – 60 CE?)

Wrote a ***periplous*** or the like in which he gave the name "Khia" for Khios (PLINY 5.136). For Khios as feminine, *cf.* Eupolis of Athens, *Poleis fr.*246 *PCG.*

RE 11.1 (1921) 672 (#5), F. Jacoby.

PTK

Kleoboulos (Pharm.) (250 BCE – 95 CE)

ASKLĒPIADĒS, in GALĒN *CMGen* 5.14 (13.854 K.), records his wound-powder, containing copper flakes, oak-gall, frankincense, ***khalkanthon***, myrrh, orpiment, ashed papyrus, and realgar.

RE 11.1 (1921) 672 (#6), F.E. Kind.

PTK

Kleoitas (Mech.) (480 – 440 BCE)

Son of Aristoklēs, invented the mechanical *hippaphesis* (horse-race starting gate) used at Olympia, according to Paus. 6.20.14. The device was later improved by ARISTEIDĒS.

RE 11.1 (1921) 675–676, G. Lippold.

PTK

Kleomēdēs (*ca* 50 BCE – *ca* 200 CE)

Stoic philosopher and teacher. His date has been inferred from the fact that his sole surviving treatise, the *Caelestia*, appears to include an account of the equation of time (1.4.72–89), which is usually viewed as an original discovery of PTOLEMY. This account, however, follows GEMINUS, *Intro. Astr.* 6.1–4, in failing to separate the contribution of latitude to the variation in the length of a full day during the course of the year, and so shows no dependence on Ptolemy. Fortunately, Kleomēdēs can be dated effectively by his polemics against the followers of ARISTOTLE and of EPICURUS, which characterize debates between **Stoics** and other philosophers during the 1st and 2nd cc. CE but which largely cease by the early 3rd c. CE. Attempts to date Kleomēdēs to the 4th c. CE on the basis of an astronomical observation reported at *Cael.* 1.8.46–56 are not warranted by the text.

The *Caelestia* is an astronomical digression, and the only surviving part of a series of lectures on all aspects of **Stoic** philosophy by its author. It conveys more about contemporary **Stoicism** by its program of following Poseidōnios in defining astronomy as a science that operates within first principles derived from physical theory and cosmology than it does about current astronomical theory. Indeed, the astronomy it presents is elementary and mostly limited to the Sun, Moon, and celestial sphere. After an introductory section on cosmology, which provides important evidence on the **Stoic** theory of the void, Kleomēdēs deals with the following topics: the division of the world into zones, seasonal and climatic differences (1.1–4); the sphericity and centrality of the Earth (1.5–6); the absence of parallax in observations of the Sun and beyond (1.8); the sizes of the heavenly bodies (2.1–3) – specifically, the claim by Epicurus that they are the size they appear to be – the illumination and phases of the Moon (2.4–5); and lunar eclipses (2.6). There is a brief appendix (2.7) listing reliable values for planetary latitudes and elongations. Underlying this presentation is, of course, **Stoic** cosmology but also a methodology of arguments (or "procedures," *ephodoi*) that represent, probably through Poseidōnios' influence, the extension of earlier **Stoic** epistemology into the realm of the philosophy of science. In this regard, the treatise often addresses optics (possibly based on Euclid), especially when discussing the illusions involved in observations of the heavenly bodies. Indeed, the polemic against Epicurus is largely an excuse to demonstrate the nature of such observations, and the possibilities for integrating them in calculations of the size of the Sun and the Moon.

Historians of astronomy have valued the *Caelestia* mainly for offering two geometrical arguments estimating the size of the Earth (1.7), one attributed to Eratosthenēs, the other to Poseidōnios. The presentation of these arguments, however, is plainly governed by Kleomēdēs' goal of illustrating that aspect of Poseidōnios' philosophy of science that allowed for the structuring of data so as to permit inferences regarding unobservables. It is, therefore, difficult to assess the historicity of these accounts, and in particular the calculation attributed to Eratosthenēs, especially when the value for the circumference of the Earth ascribed to Eratosthenēs differs from the one reported in numerous earlier sources.

Ed.: Robert B. Todd, *Cleomedis Caelestia* (1990); R. Goulet, *Cléomède: Théorie élémentaire* (1980); Alan C. Bowen and Robert B. Todd, trans., *Cleomedes' Lectures on Astronomy* (2004).

CTC 7 (1992) 1–11, Robert B. Todd; *DPA* 2 (1994) 436–439, R. Goulet; *ECP* 147, Robert B. Todd; *BNP* 3 (2003) 431–432, W. Hübner; Alan C. Bowen, "Cleomedes and the Measurement of the Earth: A Question of Procedures," *Centaurus* 45 (2003) 59–68.

Alan C. Bowen and Robert B. Todd

Kleomenēs the Libyan (before *ca* 400 CE?)

Author of remedies for horses and other beasts of burden. The remedies are preserved in the *Hippiatrika*, cited in Hieroklēs' text (for abrasions of the neck, *Hippiatrica Berolinensia* 23.1; for ***orthopnoia***, *Hippiatrica Parisina* 457 = *Hippiatrica Berolinensia* 27.2; for nephritis, *Hippiatrica Berolinensia* 30.2, attributed to "Kleomenēs the Lindian"). Hieroklēs may have used Kleomenēs' work via an agricultural compilation related to Cassius Dionusios' reworking of Magōn.

CHG v.1; McCabe (2007) 234–236.

Anne McCabe

Kleōn (of Kuzikos?) (100? – 20 BCE)

CELSUS 6.6.5 reports two of his collyria, one with saffron, poppy-juice and rose oil in gum, the other of roasted copper, **litharge**, and *squamae aeris quod stomoma appellant* (but *stomōma* are iron flakes), in gum. OREIBASIOS, *Syn.* 3.137 (*CMG* 6.3, p. 102), is still using a variant of the latter (***spodion***, lead and sulfur roasted together, and iron flakes in gum), and AËTIOS OF AMIDA 7.109 (*CMG* 8.2, p. 375), says Kleōn's recipe was known to DĒMOSTHENĒS PHILALĒTHĒS and APOLLŌNIOS "MUS" (Kind therefore suggested Kleōn was a Hērophilean); *cf.* also PAULOS OF AIGINA 7.16.36 (*CMG* 9.2, p. 342), using ***pompholux*** for the ***spodion***, and adding saffron. Other ophthalmologic recipes are preserved by Oreibasios, *Syn.* 3.146 (p. 104), and Paulos 3.22.21 (*CMG* 9.1, p. 179), 7.16.36, 58 (9.1, pp. 342, 346). ASKLĒPIADĒS PHARM., in GALĒN *CMLoc* 3.1 (12.636 K.), cites an ear-remedy, composed of aloes, frankincense, ***misu***, myrrh, and poppy-juice, in vinegar. Perhaps from Kuzikos, if the same as the Kleōn of Kuzikos credited with speculations about the salamander as fire-extinguisher by (pseudo?)-AELIUS PROMOTUS §74 (p. 75 Ihm).

RE 11.1 (1921) 719–720 (#11), F.E. Kind.

PTK

Kleōn of Surakousai (*ca* 350 – 270 BCE?)

Wrote *On Harbors* cited by PAUSANIAS OF DAMASKOS 118, AUIENUS *OM* 42–50, MARCIANUS 1.2, and STEPHANOS OF BUZANTION. *Cf.* perhaps KLEANDROS OF SURAKOUSAI?

RE 11.1 (1921) 718–719 (#8), F. Jacoby.

PTK

Kleoneidēs (100 – 200 CE?)

Author of the most important of a number of handbooks of the Imperial period which summarized the musical theory of ARISTOXENOS (others include those of BAKKHEIOS, the Dionusios who composed the second half of Bakkheios' treatise, and GAUDENTIUS). His *Introduction to Harmonics* (*Eisagōgē harmonikē*), ascribed also to EUCLID, Zōsimos, or PAPPOS OF ALEXANDRIA in some MSS but now accepted as the work of Kleoneidēs, is a précis in 14 short chapters of the principal harmonic doctrines of Aristoxenos. Its structure is simple, transparent, and uncomplicated: the seven parts of harmonics are treated in turn (notes [§4], intervals [§5], genera [§§3, 6–7], scale-systems [§§8–11], *tonos* [§12], modulation [§13] and melodic composition [§14]). Consisting mainly of terms and their definitions, the epitome is thus devoid of the richness and sophistication of its original. Much of Kleoneidēs' material was in turn appropriated by Manuel Bryennius in his *Harmonics* (*ca* 1300).

Ed.: *MSG*.

KP 5.1622, D. Najock; J. Solomon, *Kleōneidēs, Eisagōgē harmonikē* (Diss. Chapel Hill, 1980); L. Zanoncelli, *La manualistica musicale greca* (1990); *BNP* 3 (2003) 437, D. Najock; *SRMH* 1; Mathiesen (1999).

David Creese

Kleopatra of Alexandria (Queen of Egypt, 51 – 30 BCE)

Kleopatra VII (inv. 1967. 152.567) © Courtesy of the American Numismatic Society

Born 69 BCE. In *On the Manufacture of Medicaments*, GALĒN (12.403–405, 432–434, 492–493 K.) records three of Kleopatra's recipes from the work she allegedly wrote *On Cosmetics* (*tò kosmētikòn*): for *alōpekia*, for scalp rash, and a mixture for hair loss. The last is also quoted, under Kleopatra's name, in AËTIOS OF AMIDA's medical books (6.56: *CMG* 8.2, p. 205), where a body unguent is added (8.6: p. 408, with ***amōmon***, cassia, ***kostos, malabathron***, etc.). Galēn expressly states (12.445–446 K.) that such recipes were part of a collection of medicaments assembled by KRITŌN, from which he probably quoted them.

In addition, an extract *On Weights and Measures* under Kleopatra's name has survived which claims to be drawn from her work *On Cosmetics*. Basically a table, the extract lists the average weight of *mina* and its submultiples, and of what and how many fractions each of them is made. Eight measures for fluids are then displayed according to the same pattern. Kleopatra herself may have added such a table as a useful appendix to which her readers could turn. Nevertheless, since, in some cases, the text records the corresponding standard value in the Roman currency system, valid for the 1st c. CE and afterward, the extract may have been manipulated in the course of time, or an anonymous compiler may have added the famous Egyptian Queen's name to make his table on weights and measures more authoritative.

Furthermore, according to Arabic tradition, Kleopatra is credited with both a late Greek *Dialogue* between herself, as the mistress or as an important member of an Egyptian school of alchemy, and some philosophers, on transformative dynamics in nature, and a book *On Poisons*.

Ed.: *MSR* 1 (1864) 108–129, 233–236; J. Lindsay, *The Origins of Alchemy in Greco-Roman Egypt* (1970) 253–277; G. Marasco, "Cléopatre et les sciences de son temps," in G. Argoud and J.-Y. Guillaumin, edd., *Sciences exactes et sciences appliquées à Alexandrie* (1998) 39–53.

Mauro de Nardis

Kleophanēs (450 – 325 BCE)

PSEUDO-GALĒN, HIST. PHIL. 32 (19.324 K.), attributes the theory that males are generated by the right testicle, and females by the left, to "Kleophanēs," citing ARISTOTLE. Presumably a scribal mistake for LEOPHANĒS, for whom the doctrine is attested: Aristotle, *GA* 4.1 (765a25). (In the same paragraph, the text has "Hippōnax" for HIPPŌN.)

(*)

PTK

Kleophantos (80 BCE – 80 CE)

Greek physician, identified (though not definitively) with a physician mentioned in a poison case of 74 BCE (CICERO *Clu.* 47). ANDROMAKHOS, in GALĒN *Antid.* 2.1 (14.108–109 K.), cites Kleophantos with ANTIPATROS as the authors of a multi-ingredient *Mithridateios* antidote compounded from myrrh, varieties of nard, saffron, opium, ***storax***, castoreum, cinnamon, ginger, ***kostos***, wormwood, rock salt, cassia, frankincense, balsam, ***opopanax***, **galbanum, terebinth**, etc., administered with Attic honey, not wine. Two more of Kleophantos' recipes are likewise recorded: a treatment for **dropsy** compounded from beeswax, resin, ***aphronitron***, **galbanum**, propolis, ***ammōniakon*** incense, henna oil, and vinegar (*CMLoc* 9.3, 13.262 K.) and a ***hedrikē***, used by Andromakhos himself, compounded from myrtle oil, **mastich**, rose, ***psimuthion***, **litharge**, wine, and beeswax (*CMLoc* 9.6, 13.310 K.).

Watson (1966) 37–43; *KP* 3.251, F. Kudlien; *BNP* 3 (2003) 447 (#3), V. Nutton.

Alain Touwaide

Kleophantos of Keōs (270 – 240 BCE)

The son of Kleombrotos, a physician (RUFUS OF EPHESOS, *Kidney and Bladder Diseases*, 4.32 = *CMG* 3.1, p. 128), and the brother of ERASISTRATOS, both of whom were students of KHRUSIPPOS OF KNIDOS (II) (Gossen and Kind; *cf.* pseudo-SŌRANOS in V. Rose, ed., *Anecdota*, 2.226–227). Kleophantos founded his own medical group, whose students included MNĒMŌN OF SIDĒ and ANTIGENĒS (GALĒN, *In Hipp. Epid. III* 2.4, *CMG* 5.10.2.1, p. 77; CAELIUS AURELIANUS, *Acute* 2.56 [Drabkin, p. 158; *CML* 6.1.1, p. 164]; and Wellmann 1891: 814–815). It is likely that Kleophantos was in Alexandria while HĒROPHILOS and Erasistratos were conducting dissections of human cadavers. Kleophantos composed a work on the medical benefits of wine, prescribing it, chilled, after soaking the head in hot water, for tertian fever (CELSUS 3.14.1; Caelius Aurelianus, *Acute* 2.230–231 [p. 284]; *cf.* PLINY 26.14). Celsus (*ibid.*), identifying him as one of the *antiqui medici*, clearly distinguishes this Kleophantos from the later homonym.

Pliny 20.30–31, in his typical manner of "...parallel sources piled one on another" (Scarborough 1986: 75), has a multi-layered account of *staphylinus*, usually Latin's *pastinaca* (Grk. *staphulinos* = DIOSKOURIDĒS 3.52), that includes Kleophantos (along with DIEUKHĒS and PHILISTIŌN) as sources for the uses of this "wild carrot" (*Daucus carota* L.), which Kleophantos recommends for chronic dysentery. He also wrote an 11-book *Gynecology*, respectfully quoted by Sōranos, who nonetheless faults the earlier author for not detailing "all the causes of difficult labor" (*Gynecology*, 4[1].53.3: *CMG* 4, pp. 129–130; v. 4, p. 3 *CUF*), and appends similar observations from Hērophilos' *Midwifery*, and from DĒMĒTRIOS OF APAMEIA. Kleophantos was a respected authority on drugstuffs and gynecology, but is not firmly associated with either **Hērophileans** or **Erasistrateans**.

M. Wellmann, "Die Medicin bis in die zweite Hälfte des zweiten Jahrhunderts" in *GGLA* 1.777–828; *RE* 11.1 (1921) 790, H. Gossen and F.E. Kind; J. André, *Pline l'Ancien Histoire naturelle Livre XX* (*CUF* 1965) 138; John Scarborough, "Pharmacy in Pliny's *Natural History*," in R. French and F. Greenaway, edd., *Science in the Early Roman Empire: Pliny the Elder, his Sources and Influence* (1986) 59–85.

John Scarborough

Kleostratos of Tenedos (530 – 470 BCE)

After ANAXIMANDROS (PLINY 2.31). He may have been the first to introduce the ***oktaetēris***, an eight year cycle consisting of 99 lunar months, five years of 12 months and three years of 13 months (Censorinus 18.5, p. 37). Assuming he gave a length of the year in days, the actual number is unknown (the 365 ¼ days mentioned by Censorinus is probably a later correction). He wrote an astronomical work entitled *Phainomena* (Anon., *In Arati Phaenomena* [ed. Maass, 324.10–14]) or *Astrologia* (Ath., *Deipn.* 7.7 [278b]). The one surviving fragment suggests that that work was in verse (*Schol. in Eurip. Rhes.* 528) and that it discussed phases (Scorpio's rising). Kleostratos introduced one star pair, the Kids, in Auriga (Hyginus, *Astron.* 2.13), mentioned the constellations on the zodiac, including Aries, Sagittarius, and Scorpio (Pliny, *ibid.*), and observed solstices from Mt. Ida (pseudo-Theophrastos, *De signis* 4). He may well have been the first Greek to translate the constellations of the zodiac from Babylonian astronomy into Greek. Objections to this suggestion rest on inferring invalidly from Pliny's description of Kleostratos as "next after Anaximander" that he lived in the 6th c. and on skepticism about early Greek adoption of the Babylonian zodiacal constellations. But he could have lived into the mid 5th c., when calendrical systems and the concept of the zodiacal belt as a region of solar and lunar motion were being developed, and he could have been a significant figure in their development. Censorinus implies that Kleostratos precedes Harpalos.

DK 6; Dicks (1970) 87; Neugebauer (1975) 620–621.

Henry Mendell

Kleoxenos (200 – 160 BCE)

The primary inventor of the binocular ***dioptra*** and cipher conceived with Dēmokleitos.

(*)

PTK

Kloniakos (?) (250 BCE – 80 CE)

Andromakhos, in Galēn *CMGen* 7.7 (13.987–988 K.), records an ointment of orpiment, alum, and quick-lime, dissolved in vinegar, then mixed with resin and beeswax dissolved in olive oil. For *ΚΛΟΝΙΑΚΟΥ* (otherwise unattested as a name), Kind suggests *Kleōnik-* and compares Kleōn, but Pape-Benseler accept the reading, since *ΚΛΟΝ-* means agitate- (Plutarch, *Table-talk* 5.7 [681A]; Galēn, *Caus. Puls.* 2.6 [9.76 K.]; *Souda* K-1825); *cf.* also Klonios of Boiōtia, *Iliad* 2.494–495. Perhaps a brand-name, to be read *ΚΛΟΝΙΑΚΟΝ*.

RE 11.1 (1921) 876, F.E. Kind.

PTK

Klutos (200 BCE – 80 BCE)

Andromakhos, in Galēn *CMGen* 7.13 (13.1036–1037 K.), cites Philoxenos "the grammarian" for a recipe from "Glutos" for a complex **terebinth**-based ***akopon***, including **bdellium**, frankincense, **galbanum**, myrrh, ***storax***, three perfumes, etc. One expects the slightly earlier pharmacist Philoxenos of Alexandria rather than the grammarian Philoxenos of Alexandria (1st c. BCE). "Glutos" is unattested, but Klutē/Klutos is widespread, though not Athenian (*LGPN*).

(*)

PTK

Kōdios Toukos **(250 BCE – 95 CE)**

Asklēpiadēs Pharm., in Galēn *Antid.* 2.7 (14.147 K.), cites him for an antidote (also used by Krateros) composed of *rhamnos* root-bark, horehound, wild rue, ***skordion***, and "sacred plant," in honey, or for deadly poisons in honeyed wine and olive oil. The name is unattested and presumably corrupt, and we might read Kōdinos, Kōdalos (*cf.* Ath., *Deipn.* 14 [624b]), Kodros, Kotus, Clodius Tuscus, Clodius Tucca, or Clodius of Kōs. Perhaps interpretable as a syncopated pseudonym, like "Kodamos" for Nikomēdēs IV in Andromakhos in Galēn, *CMGen* 6.14 (13.929 K.). If Krateros is not the *terminus ante*, perhaps emendable to Modius Asiaticus.

Fabricius (1726) 123.

PTK

Koiranos (*ca* 45 – 65 CE)

Teacher and *delator* of Rubellius Plautus, executed 62 CE (Tacitus, *Ann.* 13.19, 14.22, 14.59); Pliny 1.*ind.*2 cites Koiranos, evidently as a writer of cosmology.

RE 11.1 (1921) 1061 (#6), H. von Arnim.

PTK

Kōlōtēs of Lampsakos (280 – 240 BCE)

Epicurean philosopher who studied with Epicurus when Epicurus established a school at Lampsakos *ca* 310–307 BCE. In his most famous work, *Concerning the Fact that One is not Able to Live in accordance with the Doctrines of Other Philosophers*, he attacks the views of earlier philosophers including Dēmokritos, Parmenidēs, Melissos, Empedoklēs, Sōcratēs, Plato, Stilpo, and Arkesilaos (who became head of the skeptical **Academy** *ca* 268). The work was especially concerned to refute the skeptical doctrine of "withholding assent" (*epokhē*) championed by Arkesilaos. Kōlōtēs' attack was later answered by Plutarch's *Against Kōlōtēs*. Kōlōtēs wrote other polemical works, including *Against Plato's Lysis*, *Against Plato's Euthydemus*, *Against the Gorgias*, and *Against the Republic*. Some fragments from letters that Epicurus wrote to him survive.

RE 11.1 (1921) 1120–1122 (#1), H. von Arnim; Long and Sedley (1987) §22R, 68H, 69A; P.A. van der Waerdt, "Colotes and the Epicurean Refutation of Skepticism," *GRBS* 30 (1989) 225–267; *OCD3* 366, D. Obbink; *BNP* 3 (2003) 583 (#2), M. Erler.

Walter G. Englert

Komerios/Komarios/Kōmarios of Egypt **(1st c. CE/7th c. CE?)**

Probably pseudonymous author of a lost alchemical *Discourse to Kleopatra* (*CMAG* 2.20). Komerios is discussed in the *[Book] of the Philosopher and High-Priest Komarios Teaching Kleopatra the Divine and Sacred Art of the Stone of Philosophy* (*CMAG* 4.400–403). Komerios has long been placed near the beginning of the Hellenistic alchemical tradition, as early as the 1st c. CE (Taylor 1930: 116), but Letrouit (1995: 83–84), believing he post-dates Stephanos of Alexandria, argues for a 7th c. or later date.

F.S. Taylor, "A Survey of Greek Alchemy," *JHS* 50 (1930) 109–139.

Bink Hallum

Kommiadēs (325 BCE – 75 CE)

Greek author of a treatise on winemaking read by PLINY (14.120). The name is otherwise unattested, but *cf.* Komniadēs of Corinth (*LGPN* 3A.254), or more likely Kosmiadēs of Dēlos (*LGPN* 1.270, 3rd/2nd c. BCE).

RE 11.1 (1921) 1194, W. Kroll.

Philip Thibodeau

Konōn of Samos (*ca* 250 – 200 BCE)

Mathematician and astronomer, courtier to Ptolemy III Euergetēs, ARCHIMĒDĒS' friend and correspondent. Ptolemy (*Phaseis*) reports that Konōn made astrometeorological observations in Italy and Sicily, but he seems to have spent much of his time at Alexandria. KALLIMAKHOS (*Aet., fr.* 110) and Catullus (66) celebrated Konōn for identifying the constellation *Coma Berenices*, an opportune bit of courtiership. According to the charming story, Ptolemy's new queen, Berenikē, dedicated a lock of her hair (Latin: *coma*) in a temple for her husband's safe return from war. To her distress, the hair disappeared, but Konōn claimed to have found it transposed into the sky, between Virgo, Leo, and Boötes. Konōn worked on spirals, and on the intersection of conic sections, criticized by APOLLŌNIOS OF PERGĒ, and is also reported to have written seven books on *Astrologia*. We know from reports and attributions that he contributed to the ***parapēgmata*** tradition.

RE 11.2 (1922) 1338–1340, A. Rehm.

Daryn Lehoux

Korē Kosmou (*ca* 100 BCE – *ca* 400 CE)

HERMETIC text known under the Greek title *Hermou trismegistou ek tēs hieras biblou epikaloumenēs korēs kosmou* (*From Hermēs Trismegistos' Sacred Book called the Pupil of the World*), one of the 40 Hermetic fragments and extracts collected, among others, by IŌANNĒS STOBAIOS in his *Anthologium*. In this text, the goddess Isis, instructed by Hermēs, describes to her son Horus the cosmogony, the creation of souls and their order, zoogony, and finally how the human body was created to punish and imprison souls after their revolution. It also explains how Isis and Osiris, educated by Hermēs Trismegistos, came to rule the world and to create harmony by teaching crafts and science. Small parts of the text are devoted to astrology, zoology and alchemy.

The *Korē Kosmou* is a representative work of the philosophical Hermetica. Other important hermetic philosophical and religious works include the *Corpus Hermeticum*, a Byzantine collection of 17 Hermetic texts and the *Asclepius*, an extensive Hermetic compendium made by the collection and abbreviation of several hermetic treatises. The original Greek text of this compendium has been lost except for a few fragments and its content is mainly known from its rather free Latin translation.

Ed.: A.D. Nock and A.-J. Festugière, edd., *Corpus Hermeticum* (1946–1954).

Festugière (1949–1953); B.P. Copenhaver, trans., *Hermetica* (1992); *Dictionary of Gnosis and Western Esotericism* (2005) 487–499, R. Van den Broek.

Aurélie Gribomont

Kosmās of Alexandria, Indikopleustēs (530 – 570 CE)

Merchant who wrote a *Christian Topography* between 535 and 547. Kosmās aimed to produce a thoroughly Christian view of the world. He attacked the theory of a spherical Earth as pagan and argued that, according to the Bible, the Earth was flat and rectangular and the shape of the world resembled the Tabernacle of Moses. In Kosmās' view, the Earth is divided into two parts, covered by heaven as if by a roof. The part on which we now live is surrounded by ocean. The other part of the Earth, which people inhabited at one time before the Flood and which contains Paradise, encircles the ocean. These ideas were not universally accepted in the Byzantine Empire. Judging from the number of surviving MSS, Kosmās' work had a limited circulation in the Greek world and remained inaccessible to the Latin West. Kosmās' work was, however, translated into Slavic languages (Russian, Bulgarian, and Serbian).

Ed.: W. Wolska-Conus, *La topographie chrétienne de Cosmas Indicopleustès* 3 vv. (1968–1973).
KP 3.315–316, F. Lasserre; *HLB* 1.528–530; *ODB* 1151–1152, B. Baldwin and A. Cutler; *PLRE* 3 (1992) 355–356; *OCD3* 404, S.J.B. Barnish; *TTE* 129–131, W. Wolska-Conus; *BNP* 3 (2003) 861–862 (#2), K. Brodersen.

Natalia Lozovsky

KOSMIADĒS ⇒ KOMMIADĒS

Kosmos (80 – 100 CE)

MARCELLUS OF BORDEAUX cites remedies from Kosmos: (a) ***trokhiskos*** for the eyes, with myrrh, saffron, and other aromatics or imports, such as **bdellium**, cardamom, or ***kostos*** (8.8, 14: *CML* 5, pp. 115–116), (b) for pain, also with myrrh and saffron, plus cassia (14.45, p. 240), and (c) a complex ***trokhiskos*** for intestinal disorders and serpent bites, again with myrrh and saffron, plus aromatics and imports including ***malabathron*** (20.19, p. 332). Kosmos also revised ANTIGONOS' collyrium (8.11, p. 115) and offered a purgative (30.28, p. 528). ASKLĒPIADĒS PHARM., in GALĒN *CMLoc* 7.5 (13.100 K.), cites Kosmos' remedy for excess wetness: myrrh and ground pepper in Attic honey boiled to gumminess (Kühn prints *ΚΟΣΟΥ*). Probably the same Kosmos is ridiculed by Martial for his use of "leaf" (11.18.9, 14.146.1: i.e., ***malabathron***) and other aromatics (3.55.1, 9.26.2, 14.59.2), in remedies made into *pastillae*.

RE 11.2 (1922) 1499 (#2), A. Stein, (#4), F.E. Kind.

PTK

On the Kosmos (80 – 20 BCE)

This anonymous work addressed to an Alexander, "best of rulers," depends on CHRYSIPPUS and POSEIDŌNIOS, but a few scholars attribute it to ARISTOTLE. The author described the ***kosmos*** as a "system of heaven and earth and the elements contained in them." He added ***aithēr*** to the Aristotelian elements fire, air, water, and earth, and arranged them in four concentric spherical shells around a central spherical Earth. Phenomena of the planets, Moon and Sun, whose endless uniform circular motions showed them to be composed of material fundamentally distinct from mundane matter, occurred in ***aithēr***, while irregular events such as comets, haloes, or meteors were fiery. The planets orbited on the "different" geocentric circles of APOLLŌNIOS' **epicyclical** model, in the order, from Earth outwards: Moon, Sun, Venus, Mercury, Mars, Jupiter, Saturn. The author named Saturn

Phainōn ("bright"), Jupiter *Phaëthōn* ("shining"), Mars *Puroeis* ("fiery"), and Mercury *Stilbōn* ("scintillating"), names attested in Alexander of Ephesos and Geminus, although Aristotle denied that planets scintillated, *On Heaven* 2.8 (290a17–24). The planetary periods were one year for Venus and Mercury, two years for Mars, 12 years for Jupiter, and two and a half times that for Saturn, as in Eudoxos of Kuzikos. The planets produce by their motion the resounding harmony hypothesized by Philolaos of Kroton, but rejected by Aristotle, *On Heaven* 2.9 (290b12–291a27). The known world is divided into Europe, Asia, and Libya, with unknown inhabited continents likely. Two exhalations arose from the earth: the wet produced precipitation, the dry caused other aerial phenomena. He classified precipitation, wind, lightning, comets, and earthquakes.

The ***kosmos*** maintained its eternal order, despite the opposed powers of the four mutable elements, through *harmonia* enforced by one pervading Power that ruled them all and in balance bound them. All pairs of mundane powers constituted the whole, and planetary movements were beautifully constant and ordered. The author claimed to "theologize" about the ***kosmos***, and the work amounts to a cosmological argument for god, concluding with a sermon on the 37 names of God. In contrast to Aristotle, who had viewed the relation between the supreme god and the ***kosmos*** in metaphysical terms, this author perceived a religious relation and a single god. In antiquity, it seems to have been read primarily by non-philosophers and Christians: Philodēmos probably, and Proklos certainly, denied that it was by Aristotle, and Aristotelian commentators such as Alexander of Aphrodisias and Simplicius say next to nothing about this work. It was translated into Latin (by Apuleius around 160 CE), and into Syriac (around 500 CE), later into Armenian, and thrice into Arabic. In the Middle Ages and the Renaissance it exerted tremendous influence, especially on Roman Catholic philosophers such as Ficino and Pico della Mirandola; it was Daniel Heinsius (in 1609) who first argued at length that it could not be by Aristotle.

Ed.: G. Reale and A.P. Bos, *Il trattato sul cosmo per Alessandro attribuito ad Aristotele* 2nd ed. (1995).
Gottschalk (1987) 1132–1139; J. Mansfeld, "ΠΕΡΙ ΚΟΣΜΟΥ: A note on the history of the title," *Mnemosyne* 46 (1992) 391–411.

PTK

Krantōr of Soloi (Kilikia) (*ca* 335 – 275 BCE)

Academic philosopher, old enough to have been Xenokratēs' pupil, he worked mainly under Polemōn, and formed a close relationship with the young Arkesilaos. Apart from Diogenēs Laërtios' *vita* 4.24–27, Krantōr is known to us chiefly from his widespread influence on consolation-literature through his *On Grief*, and from references in Plutarch and Proklos to his commentary on Plato's *Timaios*. Proklos (*in Tim.* 1, p. 76.1) describes him as the first commentator, making him a key figure in the interpretation of Platonic physics. Proklos says Krantōr treated the Atlantis-story as simple historical narrative (*historia psilē*) rather than as a mythical construction revealing some deeper meaning. Krantōr took the story as a rebuff to those accusing Plato of stealing his constitution (in the *Republic*) from Egypt, making the Egyptians in turn indebted to an early Athenian state! It is usually assumed that he thought the story true, but Proklos' evidence comes from Porphurios' discussions of the genre rather than the historical status of the tale.

More importantly, Krantōr was already suggesting how Plato (*ibid.* 1, p. 277.8) could describe a universe (***kosmos***) without a beginning as generated. Krantōr held that it owed its being to an outside entity upon which it depended. Plutarch likewise attests that he adopted a non-literal reading of the creation-process like Xenokratēs and others (*Anim. Procr.* 1013A–B), and saw the psychogony as a means of expounding various powers eternally inherent in the World-Soul. The construction of the World-Soul was from Same, Different, Divided, and Undivided since these were the fundamentals of the universe that it would need to apprehend, and cognition involved internal elements that matched its external objects (1012D–13F). He arranged the numbers that were the basis of the Soul's numerical ratios (*Tim.* 36a–b) in a lambda shape, with 2, its square, and its cube on one side, and 3, its square, and its cube on the other; he also used the number 384, in lieu of one, to illustrate the filling in of the intervals without using fractions (1020C).

H.J. Mette, "Zwei Akamediker heute. Krantor von Soloi und Arkesilaos von Pitane," *Lustrum* 26 (1984) 7–94; Dillon (2003) 216–231; Harold Tarrant, ed., *Proklos: Commentary on Plato's Timaios* v.1 (2006): intro.

Harold Tarrant

Krateros of Antioch (50 – 25 BCE)

ANDROMAKHOS, in GALĒN *CMLoc* 7.5 (13.96 K.), quotes Krateros' analgesic for patients suffering from ***phthisis*** and blood-spitting compounded from myrrh, mandrake, opium, henbane, saffron, frankincense, etc., and taken with honey-wine. ASKLĒPIADĒS PHARM., in GALĒN *Antid.* 2.8 (14.147 K.), preserves the prophylactic antidote of KŌDIOS TOUKOS used by some Krateros. The pharmacist is traditionally identified with the Greek physician who cured Attica in 45 BCE (CICERO *Att.* 12.13.1, 12.14.4), and was known to Horace (*Sat.* 2.3.161); although the *arkhiatros* (130–115 BCE) of Antiokhos VII and *tropheus* of Antiokhos VIIII might be meant.

RE 11.2 (1922) 1622 (#4), H. Gossen and F.E. Kind; Watson (1966) 24, 83; *KP* 3.327, F. Kudlien; *BNP* 3 (2003) 915 (#4), V. Nutton.

Alain Touwaide

Kratēs (Geom.) (*ca* 150 – 130 BCE)

According to DIOGENĒS LAËRTIOS 4.23 (following Dēmētrios of Magnesia), an otherwise unknown Kratēs wrote a *Geōmetrika* (*cf.* STRABŌN 2.5.2, 2.5.4, 2.5.6, and HĒRŌN, *Metr.* 1.*pr*), between KRATĒS OF MALLOS and Kratēs of Tarsos.

Netz (1997) #128.

PTK

***Kratēs (Agric.)* (325 – 90 BCE)**

Agricultural writer whose work was excerpted by CASSIUS DIONUSIOS (VARRO, *RR* 1.1.9–10); the name in the MSS should perhaps be corrected to KRATEUAS, the doctor whose illustrated herbal PLINY (25.8) lists before Cassius'.

RE 11.2 (1922) 1634 (#15), W. Kroll.

Philip Thibodeau

Kratēs (Med.) (50 BCE – 30 CE)

THEODŌROS OF MACEDON, Book 63, in PHILOUMENOS 4.15 (*CMG* 10.1.1, p. 9), cites his ***hudrophobia***-remedy: apply rust, salt, and calf-fat to the spleen; retain the patient's urine in a (transparent?) glass vessel for testing. Since not in DIOGENĒS LAËRTIOS 4.23, presumably post 50 BCE; the anonymous epigram against Krateas is probably not relevant: *Anth. Pal.* 11.125.

(*)

PTK

Kratēs of Khalkis (335 – 325 BCE)

Miner (*metalleutēs*) under Alexander, who cleaned out the drains of Lake Kopais, until stopped by Boiōtian strife, recorded in his letter to Alexander, cited by STRABŌN 9.2.18; called a *taphrōrukhos* by DIOGENĒS LAËRTIOS 4.23; STEPHANOS OF BUZANTION, s.v. *Athēnai*, cites him for a canal to the sea.

RE 11.2 (1922) 1642 (#21), E. Fabricius.

PTK

Kratēs of Mallos (*ca* 170 – *ca* 120 BCE)

Kratēs of Mallos: the four inhabited *oikoumenai* Reproduced with kind permission from Thames & Hudson, O.A.W. Dilke, *Greek and Roman Maps* (1985) 36

Greek polymath, from Mallos, Kilikia, son of Timokratēs, head of the library in Pergamon, nicknamed *Homērikos* and *Kritikos* because of his intense engagement in grammatical and poetic discourses, and teacher of PANAITIOS. In 168, or perhaps 159, the king of Pergamon sent Kratēs to the Roman senate. He fell into a sewer opening in the Palatine quarter, broke his leg and had to stay in Rome. Kratēs, lecturing frequently, was the first to introduce grammar and literary criticism into Rome (Suet., *Gram.* 2). Kratēs was mainly interested in Homeric and linguistic investigations which became the basis of his scientific endeavors. Believing HOMER was the founder of geography (STRABŌN 2.5.10), he composed a textual analysis of the *Iliad* and *Odyssey* in nine books and constructed a large terrestrial globe at least three meters in diameter to illustrate Odysseus' wanderings and to solve the mystery of the two nations of "Ethiopians" in the Homeric epics (*Od.* 1.22–23). Combining the geometric and scientific thinking of ERATOSTHENĒS with his own interpretation of Homer, Kratēs represented four inhabited ***oikoumenai*** on the surface of his globe (***Figure***), displayed in Pergamon in about 150 BCE. Kratēs' presentation of the Earth was referred to in antiquity as *Sphairopoiia* or *Sphairikos logos*, although recognition of the Earth as a spheroid occurred as early as the 5th c. BCE.

Ed.: M. Broggiato, *Cratete di Mallo: I frammenti* (2001).

H.J. Mette, *Sphairopoiia: Untersuchungen zur Kosmologie des Krates von Pergamon* (1936); Dilke (1985) 36–37; J.B. Harley and D. Woodward, *A History of Cartography* 1 (1987) 162–164.

Daniela Dueck

Krateuas (100 – 60 BCE)

Krateuas (*Vind. Med. Gr.* 1, f.3[V]) © Österreichische Nationalbibliothek

Herbalist (DIOSKOURIDĒS *MM* 1.*pr.*1: *rhizotomos*) belonging to the entourage of MITHRADATĒS VI EUPATŌR; among PLINY's physicians (1.*ind.*20–27). He wrote a *Rhizotomikon* ("*Herbal*") in alphabetical order and most probably owed to DIOKLĒS OF KARUSTOS more than just the title. The entries on plants included their synonyms, description and a list of their medicinal properties. The plant he named *Mithridatia* (25.62) in honor of the king is unidentified, an "antidote against all poisons and magical practices" (25.127), unknown to any other but Pliny and (following Pliny) PSEUDO-APULEIUS (66.12 = *CML* 4, p. 123). He dealt with remedies extracted from metals (*metallika pharmaka*: Diosk. *l.c.*, *cf.* GALĒN 15.134.17 K.) which, as in the case of his writings about medicinal botany, may have been part of an extended pharmacological work. He also wrote a popular alphabetical Herbal in which descriptions were replaced by colored plates: see Pliny 25.8 *pinxere* (sc. Krateuas *et alii*) . . . *effigies herbarum atque ita subscripsere effectus*. Dioskouridēs' illustrated alphabetical revision, which has passed down to us through *Vindob. med. gr.* 1 (late 5th c. CE), includes several extracts from this Herbal and his portrait (f. 3[V]). These are the only direct fragments that we have, besides the testimonies of Dioskouridēs, Pliny, and the Scholia to Theokritos and NIKANDROS.

Ed.: M. Wellmann, *Diosk. Mat. med.*, v. 3 (1914) 139–146.

M. Wellmann, *AGGW philol.-hist. Kl.* 2 (1897) 3–32; *Idem, Festgabe für Fr. Susemihl* (1898) 1–31; *RE* 11.2 (1922) 1644–1646 (#2), F.E. Kind; *KP* 3 (1969) 329, F. Kudlien.

Jean-Marie Jacques

Kratippos (100 BCE – 80 CE)

Wrote a *Narthēx* ("Casket"), and kept hounds. ANDROMAKHOS, in GALĒN *CMLoc* 6.6 (12.946 K.) = 6.7 (12.959), cites his gargle, of alum, saffron, pine-nuts, roses, and starch, in honey; ASKLĒPIADĒS, in Galēn *Antid.* 2.11 (14.170 K.), cites his antidote for ***hudrophobia***: ashed crabs, plus saffron, gentian (*cf.* GENTHIOS), myrrh, and white pepper, in wine.

RE 11.2 (1922) 1659 (#4), F.E. Kind.

PTK

Kratistos (of Athens?) (*ca* 430 – 485 CE)

PROKLOS (*In Eucl.* p. 211.16 Fr.) cites Kratistos' natural talent in arriving at desired results of mathematical problems from the fewest possible first principles.

Netz (1997) #48.

GLIM

Kratōn (Pharm.) (120 BCE – 25 CE)

CELSUS records his remedy for ear-infection: aloes, cassia, ***lukion***, myrrh, and nard, in honey and wine (6.7.2CD, *cf.* 6.18.2E). The Vatican MS, 1470 (13th c.), f.158, contains (extracts of?) the *Prognostica de infirmorum uita ac morte* by this man, or by KRATŌN OF ATHENS.

Diels 2 (1907) 25.

PTK

Kratōn (of Athens?) (80 – 120 CE)

PLUTARCH records a medical relative by marriage who, along with ZĒNŌN (OF ATHENS?), advised the sick to consume fish, since it was easier to digest: *Table-Talk* 1.4.1 (620A), 4.4.3 (669C). *Cf.* perhaps the Athenian homonym in *IG* II(2).5925.

RE 11.2 (1922) 1660 (#2), F.E. Kind.

PTK

Kratulos of Athens (*ca* 420 – 400 BCE)

A follower of HĒRAKLEITOS' theory, which, radically modified, he taught to PLATO. Whereas Hērakleitos had said one cannot step twice into the same river, Kratulos maintained that one cannot step into it even once. (His interpretation of Hērakleitos seems to be mistaken, but it was influential.) Kratulos also held that each thing has a proper name that is natural to it, which may differ from the conventional name – a view difficult to reconcile with his belief in radical flux. According to ARISTOTLE, Plato got his views about the instability of the sensible world from Kratulos.

DK 65; *ECP* 158–159, T.M. Robinson.

Daniel W. Graham

Krinas of Massalia (25 – 50 CE)

Physician, included in PLINY's entertaining catalogue of fashionable doctors of the early Empire (29.9). Krinas earned an enormous fortune, partly spent on public works in Massalia, by practicing **iatromathematics** at Rome. His technique was based on consulting **ephemerides**, many fragments of which are extant among the Greek astronomical papyri from Roman Egypt.

BNP 3 (2003) 943, V. Nutton.

Alexander Jones

Kritias of Athens (*ca* 430 – 403 BCE)

Born in Athens of a noble and wealthy family *ca* 455, first cousin of Periktionē, PLATO's mother. Close friend of Alkibiadēs and pupil of Sōcratēs, in 404 he led the government of the Thirty Tyrants, of which he was the most radical and violent member (XENOPHŌN, *Hellenica* 2.3–4). He died in the battle of Mounikhia in 403. Kritias was a poet, author of elegies (*To Alkibiadēs*), tragedies (*Tennes, Rhadamantis*, and *Pirithous*) and the satirical drama *Sisyphus*, although many scholars maintain a Euripidean authorship for the dramas. He also wrote in prose (*Constitutions* and *Conversations*). Diels (DK), like Philostratos (2nd–3rd c. CE; *Lives of the Sophists* 501–503), included him among the Older Sophists.

Although Kritias did not engage in natural science, his works include references to contemporary scientific issues, among which the following are worth noting: (a) The thesis that time (*Chronos/Khronos*) is a metaphysical reality, implied in fragments B18 and B19 of *Pirithous*, where time is shown as first principle and qualified as "self-generated" (*autophuē*). Untersteiner (295) notes that the god Chronos is always connected with Orphic and other mysteries. Other scholars interpret those fragments as a mixture of Orphic speculation and Pre-Socratic physics. (b) According to ARISTOTLE (*De anima* 1 [405b5]), Kritias stated that the soul is blood, which scholars consider typical of EMPEDOKLĒS. (c) Fragment B2 presents a catalogue of inventions useful for humanity, both in the technical and the social spheres (lawgivers). Kritias ascribes these inventions to peoples and not to individuals. (d) Among the useful inventions, Kritias makes religion a special case, claiming that a wise man invented the gods to prevent humans from breaking laws in the absence of witnesses (B25, *Sisyphus*).

DK 88; M. Untersteiner, *I Sofisti* 4 (1967); R.K. Sprague, *The Older Sophists, a complete translation* (1972); *DPA* 2 (1994) 512–520, R. Goulet.

José Solana Dueso

Kritodēmos (50 BCE – 50 CE)

A Greco-Egyptian authority on astrology and one of the earliest Greek astrological writers. His work *Horasis* (*Vision*, i.e., a visionary divine revelation) is cited by VETTIUS VALENS (*Anth.* 3.12, p. 150 K.), for whom he was an important source. PLINY 1.*ind*.2, 7 also cites him as a source. His alleged role as transmitter of Babylonian astrology to the Greeks is no longer given much credence, and the "Babylonian" chronological schemes with which he was associated are similarly in doubt. HEPHAISTIŌN OF THĒBAI (2.10) quotes him on stillborn children, and his name is still found cited by FIRMICUS MATERNUS (4.*pr.*, 1.196 K.-S.) and by RHĒTORIOS (*CCAG* 8.4.199–202). The latest reference to Kritodēmos is found in the 8th c. work of Theophilos (*CCAG* 1.129–131), where he is said to have normed the zodiac at Aries 0°, which Neugebauer and van Hoesen (1958: 185) dismiss.

Pliny claimed (7.193) that Kritodēmos was a student of BĒROSSOS and had direct access to Babylonian sources. Bērossos and Kritodēmos are cited as saying Babylonian astronomical observations go back 490,000 years. Babylonian influences may be traced in *Horasis*, specifically the *sunkephalaiōsis* (*CCAG* 8.3, 102). Valens (*Anth.* 3.7, p. 142.28 K.) attributes to Kritodēmos the theory of *antiscia*, in which points equidistant from the equinoctial or solstitial axis are opposed and paired, perhaps having to do with sun dials or at least the length of daylight, as day/night length are equal at the *antiscia*. Valens (Book 8) presents Kritodēmos' method for calculating the length of life, attributing to him a correlation of times with division of zodiacal signs into six parts.

W. Kroll, "Aus der Geschichte der Astrologie," *Neue Jahrbücher* 7 (1901) 559–577 at 572–573; Neugebauer-van Hoesen (1958) 185–186; Gundel and Gundel (1966) 106–107; *KP* 3.350, E. Boer; Pingree (1978) 424–426; *BNP* 3 (2003) 947, W. Hübner.

Francesca Rochberg

Kritolaos of Phasēlis (175 – 135 BCE)

Peripatetic scholarch (in Athens). Kritolaos lived 82 years (*fr.*6 Wehrli) and, together with KARNEADĒS and DIOGENĒS OF BABYLŌN, was sent in 156/155 BCE on an embassy to Rome to plead against the fee exacted on Athens for the ransacking of Ōrōpos. Of Kritolaos' students, we know Aristōn the younger and Diodōros of Tyre. Kritolaos argued for the eternity of the world from the eternity and immutability of the human race – both orthodox Aristotelian doctrines – remarking that an eternal human race can only be housed in an eternal world (*fr.*13 Wehrli). Another argument apparently relied on the principle of the synonymy of cause and effect, which postulates that the cause has to possess the feature for the emergence of which it is responsible. Accordingly, the cause of health cannot be sickly, the cause of being awake is being awake itself, and the cause of the eternal subsistence of the animal-kinds, the world, also has to be eternal itself (*fr.*12 Wehrli). Taking up some loose Aristotelian suggestions, Kritolaos submitted that the divine intellect itself (*fr.*16) and the souls (*frr.*17 and 18) are constituted from the special element of the celestial bodies (*cf.* also Cicero's testimony about Aristotle: *Acad.* 1.7.26).

Ed.: Wehrli v.10 (1969), some further testimonia from papyri, mostly on Kritolaos' teachings about rhetoric.

István Bodnár

Kritōn of Hērakleia Salbakē, T. Statilius (80 – 120 CE)

Martial, in an epigram dated to December 96 CE (11.60.2–7), features a physician named Kritōn, a medical professional skilled in curing *ulcus tendere* (*saturiasis*: pseudo-Galēn *Def. Med.* 19.426 K.; Adams). Martial's Kritōn, with medical abilities superior to those of the goddess Hygeia, could cure this disease, which often became full-blown priapism in men. Two years later, Kritōn appears as court physician to Trajan, perhaps recommended by his remarkable practice among the fashionable and sensual senatorial classes of Rome. Beneath the flashy practice discerned in Martial's acidic lines is a physician of high abilities, who studied with the accomplished Leukios of Tarsos. Kritōn was a member of a medical dynasty (inscriptions record numerous honors for the family: Benedum), served as procurator (*epitropos*), was a benefactor of Ephesos and his hometown, and apparently among the fashionable physicians plying their trade *ca* 80–100 CE (Wellmann).

Kritōn was also a known expert on *kosmētika* (broadly "The Art of Dress and Bodily Ornament") on which he had written four books, extracted by Galēn according to topic, with detailed table of contents; it described normal (Books 1–2) and diseased (Books 3–4) bodily conditions, discussing first the head (Books 1 and 3) and then the rest of the body (Books 2 and 4): Galēn, *CMLoc* 1.3 (12.446–450 K.). The *Kosmētika* incorporated topical "make-ups" but also treated *kommōtikē*, the "Art of Embellishment" that sought by means of artistic arrangement of facial and body ornaments to enhance the attractiveness of one's natural appearance. The work was a careful collection of numerous recipes for plasters, ointments, hair dyes, depilatories, and salves with known and beneficial properties, and drew on the best written sources, including Dioskouridēs. If Galēn has accurately excerpted, it began not simply with instructions for the "preservation of hair" (*CMLoc* 1.2 [12.435 K.]), but with eight recipes carefully quoted from Hērakleidēs of Taras (12.435–438 K.) and one from Andromakhos (12.438–439 K.). Galēn adds that the first book included drugs to de-louse hair and scalp (12.450 K.), which he does not extract, but instead records its depilatories (*psilōthra*: 12.453–455 K.). One included quicklime, orpiment, and the medicinal earth from Selinus in Sicily: Galēn's remark that "Kritōn advised close attention to the treatment" suggests his awareness of the ointment's efficacy as well as its latent danger. Another was "The Depilatory of Paris the Dancer," a famous contemporary actor (Suet. *Dom.* 3.1; Mart. 11.13.7; Juv. 6.87). The *Kosmētika* enjoyed a long life, circulating widely in Byzantine times, with Arabic translations appearing sometime after 850 CE (*GAS* 3 [1970] 60–61; Ullmann 1972: 69–70).

Galēn also extracts at length Kritōn's pharmacological writings: the *Pharmakitis* (*CMLoc* 6.4 [12.883–884 K.]) cited SERAPIŌN's plaster for all sorts of ailments resulting from sexual overindulgence, and was arranged topographically, by affected body parts (head in Book 1: *CMLoc* 2.2 [12.587 K.]), and pharmaceutically, by types of medicines: theriac in Book 3 (*Antid.* 1.17 [14.103 K.]), and plasters in Book 4 (*CMGen* 4.6 [13.708–716 K.]). Another tract, in five books (*CMGen* 4.6, 5.3, 6.1–2 [13.708–716, 786–801, 859–882 K.]) was entitled either *Peri tēs tōn pharmakōn suntheseōs* (*Compounding Drugs* [13.786 K.]), or *Peri tōn haplōn pharmakōn* (*Simples* [13.862 K.]).

Kritōn accompanied his royal patient on the Dacian campaigns (101–102 and 105–106 CE), and composed an account of those wars, the *Getika*, apparently a striking eyewitness narrative (peculiar words, such as those describing Dacian defenses, Dacian felt-caps, and arable land, can derive only from an eyewitness: Scarborough 390–393). His medical vocabulary probably obscured his history for later readers, as suggested by the *Souda*'s extracts of weird words; and Lucian may have aimed KALLIMORPHOS at Kritōn's unfortunate style (*Hist. concsr.* 16; *cf.* Baldwin), which doomed the *Getika* to extinction, once exploited by Dio Cassius. Kritōn may also have participated in Trajan's Mesopotamian campaign (115–117 CE). Some well-informed medical details preserved about Trajan's death suggest a witness acquainted with 2nd c. CE prognostics, perhaps again Kritōn.

Ed. [*Getika*]: *FGrHist* 200.

Wellmann (1895) chs. 3–4 ("Theodorus, Magnus, Herodot, Leonidas"); F.A. Lepper, *Trajan's Parthian War* (1948; repr. 1993) 198–201 ["Trajan's Health"]; R. Syme, *Tacitus* (1958) 1.221 [on the Dacian campaigns]; Fabricius (1972) 190–192; B. Baldwin, *Studies in Lucian* (1973) 36–40; *RE* S.14 (1974) 216–220, J. Benedum; J.N. Adams, *The Latin Sexual Vocabulary* (1982) 40–41; Scarborough (1985c); *NP* 11.921–922, Alain Touwaide.

John Scarborough and Alain Touwaide

Kritōn of Naxos (335 – 250 BCE?)

Wrote an ***oktaetēris*** sometimes attributed to EUDOXOS (*Souda* K-2454); PLINY 18.312 (*cf.* 1.*ind.*18) appears to cite him.

RE 11.2 (1922) 1935 (#6), A. Rehm.

PTK

Kronios (130 – 170 CE)

Friend of NOUMĒNIOS (PORPHURIOS, *de antro Nymph.* 21), listed among the neo-**Platonic** authors read in PLŌTINOS' school (Porph., *Vita Plot.* 14), and perhaps identifiable with the Kronios to whom Lucian dedicates *Death of Peregrinus* (165 CE). He wrote on the nuptial number in PLATO, *Rep.* 546c (asserting that fire is incapable of destroying all matter, evidence for which is the "Carystian" stone, asbestos, and explaining that the male is a myriad and the female 7500: PROKLOS, *In Remp.* 2.22–23 Kroll), and the myth of Er in Plato, *Rep.* 614–621 (declaring that Er was a historical teacher of ZOROASTER: Proklos, *In Remp.* 2.110 Kroll). In *On Reincarnation* (*peri paliggenesias*), his only titled work, Kronios argued that "souls always remain rational" (NEMESIOS, *Nat. Hom.* 2.116–7 M.).

BNP 3 (2003) 958–959 (#1), M. Frede.

PTK and GLIM

Ktēsias of Knidos (405 – 390 BCE)

Son of Ktēsiarkhos, served as Artaxerxēs II's doctor for nearly two decades before returning to Knidos where he wrote a multi-volume *History of Persia*; a one-volume *Account of India; On the Tributes of Asia; Voyages* (or ***Periplous*** or *Descriptions*); and a *Medical Treatise*, preserved only in fragments, including lengthy summaries in NIKOLAOS OF DAMASKOS and in Phōtios' *Bibliotheka*, and extensive citations by DIODŌROS OF SICILY. His Persian history was full of historical and geographical error: e.g., he located Nineveh on the Euphrates. His description of Babylōn, although based on eye-witness with authentic details, also contained inaccuracies. He described Darius' inscription at Bisitun, but credited it to Semiramis. He gave a list (now lost) of places and distances in the Persian Empire. He was also interested in bizarre and fantastic natural phenomena. The few extant fragments of his ***periplous*** suggest that it covered Asia, Libya, and Italy; it was a mixture of geography and ethnography, with fantasies such as the Skiapods (Shade-Foots). He claimed that his geographical and ethnographical account of India was based on eye-witness; at best, it was based on Persians' tales and riddled with fancies. He claimed that the Indus river is 200 stades wide and contains only worms. His descriptions of animals range from exotic, to exaggerated, to utterly fantastic: elephants, monkeys, the parrot, poisonous snakes with two different kinds of venom, as well as the *martikhora*, the unicorn and the griffon. He described the Indians as just and perfectly healthy; he reported on "Pygmies" and their diminutive livestock; and he gave ethnographical detail about the "dog-headed people." He also reported marvelous physical phenomena, such as the "unquenchable fire" at Mount Khimaira. The marvels he recorded became standard items in later accounts of eastern wonders.

Ed.: *FGrHist* 688; F.W. König, *Die Persika des Ktesias von Knidos* (1972); J. Auberger, trans., *Ctésias, Histoires de l'Orient* (*CUF* 1991).
J.S. Romm, *The Edges of the Earth in Ancient Thought* (1992) 86–88.

Philip Kaplan

Ktēsibios of Alexandria (290 – 250 BCE)

Dated by an epigram of Hedulos, mentioning a pneumatic drinking horn with figurines set up in a temple to honor Ptolemy II's wife Arsinoē (Ath., *Deipn.* 11 [497d]). Ktēsibios wrote on mechanical topics, but no treatise survives. PHILŌN OF BUZANTION refers to two catapults driven by bronze springs and air pressure respectively (*Artillery Construction* 56, 67–73, 77–78) and ATHĒNAIOS MECH. describes a seesaw tube for scaling walls (29). VITRUUIUS probably had access to Ktēsibios' work and presents him as the discoverer of the principles of pneumatics (9.8.2, 10.7.4); he describes a number of Ktēsibios' inventions: water clocks with moving figurines, a water-organ, a water-pump and catapults, but leaves out song-bird automata as frivolous. Ktēsibios begins the mechanical tradition continued by Philōn and HĒRŌN OF ALEXANDRIA, but it is impossible to judge the extent to which they drew on his work.

Drachmann (1948); *DSB* 3.491–492, AG. Drachmann; *BNP* 3 (2003) 971–973, F. Krafft.

Karin Tybjerg

Ktēsiphōn (250 BCE – 25 CE)

PSEUDO-GALĒN, INTRODUCTIO, discussing the medicinal efficacy as ***diaphorētikai*** of certain ingredients, cites Ktēsiphōn's use of natron (14.764 K.). ANDROMAKHOS records his plaster for breaking up, drawing in, and cleansing, in Galēn *CMGen* 6.14 (13.927 K.). Asklēpiadēs Pharm. records, in Galēn *CMGen* 6.16 (13.936–937 K.), Ktēsiphōn's beeswax plaster – the most effective medicament against various types of tumors – compounded of **terebinth**, beeswax, aged olive oil, and natron, dissolved in water, heated in a ceramic pot over a fire until it stops rumbling, to which then liquids are added. The mixture, shaken violently until it stops smearing, poured into a mortar, is then pounded before use. Celsus, reading *CLESIPHON*, recommends his treatment for joint pain (compounded from Cretan beeswax, **terebinth**, olive oil, and the "reddest natron," pounded for three days) also for parotid swelling, diseased growths, scrofulous tumors, and for mollifying accrued **humors** (5.18.31). Two botanical treatises, *On Plants* and *On Trees*, are attributed to the "historian" Ktēsiphōn, perhaps conflated with our pharmacologist (*FGrHist* 294.3–4). Attestations of Ktēsiphōn, known into the 3rd c. CE, are concentrated in the 4th/3rd cc. BCE (*LGPN*).

RE 11.2 (1922) 2079–2080 (#4), F.E. Kind.

GLIM

Kudias (of Kuthnos?) (370 – 340 BCE)

Painter, who developed a method of making ruddle ("red ochre") from yellow ochre (hydrated iron oxide) by roasting it, after observing the accidentally-caused effect: Theophrastos, *Stones* 53. Theophrastos gives no ethnic, but the name is rare and especially Athenian at this period (*LGPN* 2.276 and 1.277–278; contrast 3A.260, 3B.250, 4.204), so probably identical to the mid-4th c. BCE painter from Kuthnos: Pliny 35.130.

BNP 3 (2003) 1045 (#3), N. Hoesch.

PTK

Kudias of Mulasa (250 – 100 BCE)

Hērophilean physician, probably engaged in Hippokratic exegesis, censured by Lusimakhos of Kōs in three books (Erōtianos *Pr.*, p. 5 Nachm.); Erōtianos himself censures Kudias and Iskhomakhos (I-20, p. 47.2 Nachm.).

von Staden (1989) 564–565; *BNP* 3 (2003) 1045 (#4), V. Nutton.

GLIM

Kuranides (50 – 200 CE?)

A Byzantine compilation of six books of Egyptian-Syrian origin. The work discusses magical powers of stones, plants, and animals. Authorship is uncertain: Book 1 is attributed to the Persian king Kuranos, augmented from a similar book by Harpokratiōn of Alexandria (*pr.* 1.1.75, 1.1.128), while Books 2–6 are simply named *Kuranides bibloi* (the compiler lamenting the loss of the corresponding books of Harpokratiōn). The work, explicitly referring to Hermēs (*pr.*7, 1.4), relies on the broad Hermetic literary tradition and is also related to the pseudo-**Pythagorean** and pseudo-Dēmokritean occult and popular tradition (see Bōlos, Nepualios, . . .).

The treatise's extant title is "book on the natural faculties, **sympathies** and antipathies." Book 1 (also called *Kuranis*, possibly derived from the Egyptian word for "stele") has 24 chapters, giving for each Greek letter the name of a plant, a bird, a stone, and a fish. Often homonyms (for *K* the four items are called *kinaidios*; for *A*: vineyard, eagle, eagle-stone, sea-eagle), the four objects are described and occasionally combined into a single magical preparation (a powerful amulet), expressing a kind of "alphabetic **sympathy**." But the implicit program of the description (localization, denomination, anatomy, medical and magical recipes) is unfulfilled, most of the animals being only mentioned. The following books, also in alphabetic order but more therapeutic than magical, explicate separately and specifically 46 land animals (2), 54 birds (3), 77 fishes (4), five plants (5) and nine stones (6); the last book ends after the third letter (only 280 words). The treatment is equally uneven in Books 2–6, focusing on medical uses and detailing complex recipes. Due to the relative independence of the books and flexibility of the classifications (in Book 2 are treated sea-flea, ant, spider, viper, . . ., and in Book 3 bee and glow-worm), there are repetitions: thrice introduced is the seal, and the benefic power of his mustache and skin thrice mentioned (1.21, 2.41, 4.67), the peony twice (1.3, 5.3). The book (even when addressing such rare animals as giraffe or halcyon) offers hardly an original observation, but incredibly rich pharmacological data, corresponding to PLINY *NH* 28–32, chapters of MARCELLUS' *Iatrika*, and DIOSKOURIDĒS (for Book 5).

Ed.: D. Kaimakis, *Die Kyraniden* (1976).

RE 12.1 (1924) 127–134, R. Ganszyniec; M. Wellmann, *Marcellus von Side als Arzt und die Koiraniden des Hermes Trismegistos* (1934); West (1982); *OCD3* 421, J. Scarborough; D. Bain, "Some textual and lexical notes on Cyranides 'Books Five and Six'," *C&M* 47 (1996) 151–168.

Arnaud Zucker

Kurillos (of Jerusalem? Alexandria?) (350 – 540 CE)

The earlier archbishop ("Cyril"), of Jerusalem, consecrated *ca* 350, and died 386/387 aged *ca* 70; he defended the primacy of Jerusalem, and advocated the sanctity of places. The later man, born *ca* 378, succeeded his uncle Theophilos as archbishop of Alexandria in 412, instigated the lynching of HUPATIA, actively suppressed pagan and Jewish practice, wrote against DIODŌROS OF TARSOS, and sparked the monophysite/dyophysite controversy; he died 444. AËTIOS OF AMIDA 9.24 (p. 508 Cornarius; omitted by Zervos 1911: 324–325) credits "Kurillos the archbishop" with a *digestif*, composed of laurel-leaves, ***kostos, malabathron***, pennyroyal, St. John's wort, salt, plus ground seeds of celery, coriander, and fennel, macerated for a few hours in vinegar. Aëtios 9.50 (p. 557 Cornarius; Zervos' edition ends earlier) credits "Kurillos" with a plaster for dysentery (containing acacia, aloes, frankincense, myrrh, oak-galls, roses, etc. in myrtle oil, pitch, beeswax, and date-wine). Diels 2 (1907) 25 records two MSS attributing medical works to Kurillos: *Parisin.* 1389 (16th c.) f. 387^{V}, *De mensuris & ponderibus* (*cf.* ADAMANTIOS), and *Coislin.* 335 (15th c.) f. 1, medical extracts. All the attributions may be scribal errors for (e.g.) the *arkhiatros* KUROS OF EDESSA, but *cf.* BASIL OF CAESAREA and the metrology of EPIPHANIOS. Both archbishops are received as saints by the Orthodox and Roman churches.

Fabricius (1726) 134; *RE* 12.1 (1924) 175 (#6), F.E. Kind; *OCD3* 422–423, J.F. Mathews, and 423, E.D. Hunt.

PTK

Kuros of Edessa (50 – 540 CE)

AËTIOS OF AMIDA 6.91 (*CMG* 8.2, p. 237) entitles him *arkhiatros*, and records his pill for nasal polyps (*cf.* GALĒN, *CMGen* 3.3 [12.678–679 K.]), composed of almost 20 ingredients, including **calamine, litharge**, ***psimuthion***, ruddle, acacia, aloes, frankincense, and spikenard, in a vinegar base.

RE 12.1 (1924) 191 (#16), F.E. Kind.

PTK

KURTOS ⇒ DIONUSIOS OF KURTOS

L

L. Caecilius Firmianus Lactantius of Sicca Veneria (*ca* 270 – 315 CE?)

Before his conversion to Christianity (*ca* 300 CE), Lactantius composed an extant Latin poem *Phoenix*, describing the home of the bird at the eastern edge of the flat Earth (1–30), and her life and singing for Apollo (31–58). The phoenix lives a millennium, then flies to a Syrian palm (*phoenix*), where she builds a nest of incense, on which she bursts into flame, from the ashes of which a worm forms and becomes an egg, out of which hatches a new phoenix (59–122). Lactantius concludes with an *ekphrasis* of the phoenix (123–170). After his conversion, he composed among others a work on the construction of the human body, *De Opificio Dei*, arguing from function to design. He argued in his systematic theology, *Divine Institutes* 3.24, that the spherical-Earth theory was absurd and ill-founded.

Ed.: E. Rapisarda, *Il carme "De ave phoenice" di Lattanzio* (1959); M.F. McDonald, trans., *Lactantius: Minor Works* (1965) 213–220.

P.A. Roots, "The De opificio dei. The workmanship of God and Lactantius," *CQ* 37 (1987) 466–486; *DPA* 4 (2005) 65–71, Chr. Ingremeau.

PTK

LAECANIUS ⇒ AREIOS

LAETORIUS ⇒ LITORIUS

LAETUS ⇒ OFELLIUS LAETUS

Laïs (100 BCE – 77 CE)

Female physician listed (after SŌTEIRA and before ELEPHANTIS) as a foreign authority on drugs obtained from animals (PLINY 1.*ind.*28). She disagreed with Elephantis regarding the efficacy of various abortifacients (28.81), but agreed with SALPĒ regarding treating ***hudrophobia*** and fevers magically with wool from a black ram (28.82). A relatively common name from the 5th c. BCE (*LGPN*).

RE 12.1 (1924) 516 (#4), F.E. Kind; Parker (1997) 145 (#44)

GLIM

LAITOS ⇒ OFELLIUS LAETUS

Lakudēs of Kurēnē (245 – 205 BCE)

Son of Alexander, one of CHRYSIPPUS' teachers (DIOGENĒS LAËRTIOS 7.183), succeeded Arkesilaos as **scholarch** of the **Academy** (CICERO, *Acad.* 2.16). His students included ARISTIPPOS OF KURĒNĒ and DAMŌN OF KURĒNĒ. Although credited with founding the New **Academy**, he simply emphasized skepticism, already present in Arkesilaos. Lakudēs' writings include the lost *On Nature* (*Souda* Lambda-72). Resigned the **scholarchate** five years before his death. Best known for sealing his storeroom and tossing the signet inside through a hole, to prevent theft, but his slaves did the same, hence perplexing Lakudēs, who interpreted the result via Arkesilaos' doctrine of incomprehensibility (D.L. 4.59–61, EUSEBIOS, *PE* 14.7).

KP 3.462, E.G. Schmidt; *OCD3* 811, W.D. Rouse; *BNP* 7 (2005) 161, K.-H. Stanzel; *DPA* 4 (2005) 74–5, T. Dorandi.

GLIM

Lampōn of Pēlousion (120 BCE – 80 CE)

Physician, treated nasal polyps and ***aigilōps***, using red copper, alum, ***ammōniakon*** incense, and sharp vinegar (GALĒN *CMLoc* 3.3, 12.682–683 K.). ANDROMAKHOS, in Galēn *CMLoc* 8.1 (13.133–134 K.), records his universal remedy compounded from cinnamon, black cassia, and aromatic reed, balsam-wood, and rush blossoms, plus fir, mixed with rainwater and Indian aloe, set in the summer sun until dried; the results were mixed with saffron, myrrh, and **mastic**; ingested with water, it is useful for sprains, fractures, and internal wounds, rib, lung, and stomach pains, digestion, and blood-spitting. The name, most frequently cited in the 3rd c. BCE, is attested perhaps into the 1st c. CE (*LGPN*).

RE 12.1 (1924) 581–582 (#5), F.E. Kind.

GLIM

Laodikos (250 BCE – 80 CE)

ANDROMAKHOS, in GALĒN *CMLoc* 3.1 (12.626 K.), records and assiduously used the earache remedy of "king" Laodikos containing castoreum, poppy juice, ***opopanax***, and froth of ***lukion***, to be taken tepid in must. A rare name, attested at Eretria (4th/3rd cc. BCE: *LGPN* 1.282); the more common feminine form is known from the same period (*LGPN*), and Andromakhos may have intended the wife of Antiokhos II (Laodikē I) or of Seleukos II (Laodikē II). The name may instead have been corrupted from Laodokos, a son of Apollo (Apoll. 1.7.6), or perhaps refers to a king of Laodikeia (*cf.* DIPHILOS, PAPIAS, THEODĀS).

RE 12.1 (1924) 726 (#5), F.E. Kind.

GLIM

Largius Designatianus (200 – 350 CE?)

Author of a lost medical work (probably a collection of remedies) in Latin, used by MARCELLUS OF BORDEAUX (*pr.*2). Marcellus also gives, among the prefaces to his *De medicamentis*, a letter of Largius to his sons and an epistle of HIPPOKRATĒS to a King

Antiokhos which Largius translated from Greek and dedicated to his sons. The epistle offers dietetic prescriptions for diseases of the four main parts of the body, and to be used during the 12 months of the year. A Greek version of the epistle is reported by PAULOS OF AIGINA (1.100, *CMG* 9.1, pp. 68–72), for whom however the epistle was written by DIOKLĒS OF KARUSTOS and addressed to Antigonos Gonatas.

RE 12.1 (1924) 836 (#2), F.E. Kind; C. Opsomer and R. Halleux, "La lettre d'Hippocrate à Mécène et la lettre d'Hippocrate à Antiochus," in Mazzini and Fusco (1985) 339–364; *BNP* 7 (2005) 250, V. Nutton.

Fabio Stok

Lasos of Magnesia (300 – 250 BCE?)

Astronomer, wrote a work on fixed stars which an anonymous life of ARATOS calls *Phainomena*, but a scholion to BASIL OF CAESAREA, *Hexaemeron* cites as "on the distance of the fixed stars." Lasos' work may have contained quantitative measurements of the angular separation of bright stars. An early Hellenistic date is suggested by the order of the astronomical authors listed in the two sources.

RE 12.1 (1924) 888, A. Rehm; *BNP* 7 (2005) 260, W. Hübner.

Alexander Jones

LENAEUS POMPEIUS ⇒ POMPEIUS LENAEUS

Leōdamas of Thasos (390 – 350 BCE)

A geometrician and contemporary of PLATO, ARKHUTAS and THEAITĒTOS (PROKLOS, *On the First Book of Euclid's Elements*, p. 66 Fr.), who is reported by Proklos (p. 211) to have made many discoveries by the method of analysis, there defined as tracing back the desired result to an acknowledged principle. This method is said to have been taught to him by Plato (*cf.* DIOGENĒS LAËRTIOS 3.24: a suspect tradition, since it tries to credit Plato with several discoveries).

BNP 7 (2005) 395 (#3), M. Folkerts.

Harold Tarrant

Leōn (385 – 345 BCE)

Chronologically intermediate between EUDOXOS OF KNIDOS and LEŌDAMAS OF THASOS, he, and his teacher NEOKLEIDĒS, added to geometric knowledge, enabling Leōn to compile a book of *Elements* more attentive to the number of theorems and their usefulness than its predecessors. Leōn also "discovered" *diorismoi*, that is, conditions under which a problem is solvable or not solvable (PROKLOS, *In Eucl.* pp. 66.18–67.2 Fr.).

DSB 8.189–190, I. Bulmer-Thomas; Lassere (1987) 6.

Ian Mueller

Leōnidas (Geog.) (300 – 150 BCE?)

Wrote a geographical work *On Italy* of which one fragment, on net-floats made from pine-tree bark, is preserved by a scholiast. This Spartan name is rare outside Lakōnika before 300 BCE (*LGPN*).

FGrHist 827.

PTK

Leōnidas of Alexandria (Astr.) (50 – 80 CE)

This poet lived at the time of Nero and Vespasian. He is particularly known to have written three books of epigrams: only 42 brilliant pieces remain, nearly all isopsephic; they are addressed mainly to the imperial family who are praised on many anniversary feasts. But according to his own statement (*AP* 9.344), he was primarily an astronomer: unfortunately nothing else is known about his scientific work.

BNP 7 (2005) 403 (#4), M.G. Albiani.

Christophe Cusset

Leōnidas of Alexandria (Med.) (80 – 120 CE)

Called *episunthetic* by PSEUDO-GALĒN, INTRODUCTIO (14.684 K.), and CAELIUS AURELIANUS explicitly following SŌRANOS (*Acut.* 2.7–8 [*CML* 6.1.1, p. 134]), a school whose creation PSEUDO-GALĒN, DEFINITIONS, attributes to AGATHINOS. Hence scholars have surmised that Leōnidas studied under Agathinos (at a time when **Pneumaticists** began to incorporate elements from other sects, e.g., **Methodism** and **Empiricism**). No work by Leōnidas has survived; extant fragments in AËTIOS OF AMIDA and PAULOS OF AIGINA show clearly that he was a surgeon, practicing in Alexandria. His explanation of lethargy as the obstruction of brain canals (Cael. Aurel., *ibid.*) might result from anatomical explorations; his surgical work could have influenced ARKHIGENĒS, HĒLIODŌROS, and ANTULLOS.

RE 12.2 (1925) 2034 (#18), F.E. Kind; *KP* 3.569 (#8), F. Kudlien; *Idem* (1968) 1099; *BNP* 7 (2005) 402 (#3), V. Nutton.

Alain Touwaide

Leōnidas of Buzantion (150 – 50 BCE)

Son of the ichthyologist MĒTRODŌROS OF BUZANTION, wrote in Greek prose a famous *On fishing* (*Halieutika*) used by Athēnaios (*Deipn.* 1 [13c]) and probably also by PLUTARCH and OPPIAN (see Wellmann, *Hermes* 30 [1895] 161–176). AELIANUS (*NA*, epilogue) mentions Leōnidas, along with his father and DĒMOSTRATOS, among the main authorities on sea-fishing. Aelianus' explicit references concern (personal?) observations made while traveling (2.6; 2.50), technical explanations of special baits (12.42), and a mysterious poisonous fish (a kind of globe-fish) of the Red Sea (3.18).

GGLA 1 (1891) 851; *RE* 12.2 (1925) 2033–2034 (#16), W. Kroll.

Arnaud Zucker

Leōnidas of Naxos (350 – 325 BCE)

Son of Leotēs, architect and perhaps sculptor, built a large eponymous guest-house at Olympia, adjacent on the south-west to the sacred precinct of Zeus, *ca* 330 BCE (Pausanias 5.15.1). Pausanias also saw an honorary statue of Leōnidas set up by the people of Arcadian Psōphis (6.16.5). Fragments from the epistyle of the Leōnidaion preserve a dedicatory inscription, repeated at least twice and possibly on three sides of the building, naming Leōnidas as builder and donor. The Leōnidaion is nearly square, 74.80 x 81.08 m, with an

Ionic colonnade around the exterior, and an inner courtyard with a Doric peristyle, and includes an extensive series of guest-rooms. An homonymous sculptor may be our architect (VITRUUIUS 7.*pr.*14).

A. Mallwitz, *Olympia und seine Bauten* (1972) 246–254; Svenson-Ebers (1996) 380–387; *KLA* 2.12–13, W. Müller.

Margaret M. Miles

Leontinos (Agric.) (200? – 330 CE)

Listed by Phōtios, *Bibl.* 163, among the sources of VINDONIUS ANATOLIOS. The MSS give Leōn, but Oder plausibly emends to Leontinos, more typical of the period.

Oder (1890) 92–93.

PTK

Leontios (600 – 650 CE)

Wrote two short astronomical tracts. In his commentary on ARATOS' *On the Construction of the Sphere*, addressed against THEODŌROS (MECH.), Leontios described Aratos' division of the stars into three parts, the division of the celestial sphere into six parts, the placement of the constellations, including the zodiac, to those six circles inscribed on the heavenly sphere, and the relationship between the celestial sphere and the five zones of the Earth. The treatise also details practical applications (navigation). His *On the Circle of the Zodiac* describes the zodiac, its connections to the tropics, the path of the Sun and how it affects the seasons, and the necessity of the seasons. The tract cites PTOLEMY and recalls Neo-**Pythagoreanism**, comparing the zodiac circle to the **Demiurge** – lacking a beginning and end – and explaining the 12 parts of the zodiac in terms of musical theory.

E. Maass, *Commentariorum in Aratum reliquiae* (1958) LXXI, 561–570.

GLIM

Leophanēs (470 – 430 BCE)

Named by AËTIOS 5.7.5 (Diels 1879: 420) between ANAXAGORAS and LEUKIPPOS, providing an approximate date. THEOPHRASTOS quotes him as commending black soil: it absorbs both heat and water and therefore is able to withstand both rain and drought (*CP* 2.4.12). ARISTOTLE refers to Leophanēs' view that males who copulate with the right testicle bound up will produce male progeny, while those with the left testicle bound will produce female offspring (*GA* 4.1 [765a23–25]). Wellmann (*Abh. Akad. Wiss. Berlin* 1921: 22) thinks that Leophanēs' regulation, which would have generated some popular superstition, might have recommended his work to BŌLOS.

RE 12.2 (1925) 2057, W. Kroll.

Maria Marsilio

Lepidianus (30 BCE – 360 CE)

OREIBASIOS, *Ecl. Med.* 75.21 (*CMG* 6.2.2, p. 246), records a potion for gout and arthritis composed of *khamaidrus* (germander), *aristolokhia* (birthwort), gentian (*cf.* GENTHIOS), and rue-seed, among other ingredients. For the cognomen *cf. CIL* 9.2693 and *RE* S.15 (1978) 124 (#316a), W. Eck: *procos. Syria-Palestina*, 186 CE.

(*)

PTK

Leptinēs (I) (260 – 240 BCE)

Astrologer (*mathematicus*) at the court of Seleukos, who diagnosed his son Antiokhos' love for his step-mother Stratonikē (a diagnosis also attributed to ERASISTRATOS): Val. Max. 5.7.*ext*.1.

RE 12.2 (1925) 2074 (#6), C.O. Thulin.

PTK

Leptinēs (II) (200 – 100 BCE?)

On the recto, in column 24 of P. PARISINUS GRAECUS 1, the last surviving column and perhaps the end of this astronomical text written sometime in the 2nd c. before 165 BCE, there is drawn an annulus divided into twelfths that are inscribed with the Greek names of the zodiacal signs. In the same hand is written within the annulus the two-line caption, "Celestial Circle/Oracles of Sarapis." Beneath the annulus is the sentence, "Work men so that you may work no longer," which is followed by the lines, "To the kings/celestial/instruction/oracles of Sarapis/xxxx of Leptinēs/oracles of Hermes." Of these six lines "To the kings/celestial/instruction/xxxx of Leptines" are written in the same hand as the rest of the papyrus; whereas the other two are in the hand that inscribed the annulus. Such is the meager and unpromising basis on which some have supposed that Leptinēs is the author or redactor of the papyrus, or that the papyrus is a copy of an *Instructio caelestis* by an otherwise unknown Leptinēs.

Neugebauer (1975) 686–687; *NDSB* 4.271–272, A. Jones.

Alan C. Bowen

Leukios (of Tarsos?), Kathēgētēs (50 – 85 CE)

Greek physician, the teacher of ASKLĒPIADĒS PHARM., (see GALĒN, *CMGen* 2.17 [13.539 K], 3.9 [648], and *CMLoc* 7.6 [972]) and of KRITŌN OF HĒRAKLEIA (*CMLoc* 5.3 [12.827–828 K.]), both of whom usually qualify him as teacher (*kathēgētēs*), and so probably best-known as a professor of pharmacology. ARKHIGENĒS, in Galēn *CMLoc* 3.1 (12.623 K.), attributing therapeutic innovations, does not call him "teacher." ANDROMAKHOS OF CRETE (YOUNGER), in Galēn *CMLoc* 9.5 (13.292–293 K.), refers to a Loukios (Lucius), probably the same man (Leukios is also once *Loukios* in Asklēpiadēs, *CMLoc* 4.7 [12.787 K.], possibly to be emended; *cf.* Solin 2003: 2.749), and records a book of remedies by him, including one for dysentery compounded from opium, saffron, acacia, oak-gall, etc. (13.292); Andromakhos also gives the ethnic (295), with a remedy for flux and *empneumatōsis*, composed of opium, henbane, and various seeds. Diverse remedies are transmitted under Leukios' name by Asklēpiadēs Pharm.: e.g., a rose collyrium (*CMLoc* 4.7, 12.767–768 K.); a **calamine**-plaster for cicatrization (*CMGen* 2.14, 13.524 K.); an acne treatment with alum, **Sinōpian earth**, and Corinthian **verdigris**, in vinegar (*CMGen* 5.3, 13.829 K.); an herbal plaster of pimpernel, opium, henbane, aloe, ***ammōniakon*** incense, beeswax, aged olive oil, diligently beaten and dissolved in vinegar (*CMGen* 4.13, 13.746–747 K.); a "dry," i.e., fat-free, head plaster (*CMGen* 5.3, 13.846 K.); several styptics (*CMGen* 5.14 [13.850–854 K.], 5.15 [13.857 K.]) composed of **calamine**, ***khalkitis***, orpiment, sulfur, or other minerals; a compound prevailing against all sinew ailments of **bdellium**, ***ammōniakon*** incense, Illyrian iris, ***opopanax***, **galbanum**, ***storax***, frankincense, pepper, beeswax, and **terebinth** (*CMGen* 7.6, 13.969 K.). Kritōn (*loc. cit.*) transmits one of Leukios' remedies, a face-powder of **litharge** and ***misu***.

RE 13.2 (1927) 1652–53 (#7), F.E. Kind; *KP* 3.756, F. Kudlien; *BNP* 7 (2005) 854–855 (Lucius #1), V. Nutton.

Alain Touwaide

Leukippos of Abdēra (460 – 420 BCE)

First **atomist** philosopher, he may have been born in Milētos or Elea and then later moved to Abdēra. He taught DĒMOKRITOS, and together they developed the atomic theory of matter. Leukippos seems to have been its inventor, positing **atomism** as a response to PARMENIDĒS' philosophical doctrines. Although it is difficult to distinguish Leukippos' and Dēmokritos' respective contributions to the atomic system, the most likely view is that Leukippos worked out the main outlines of **atomism**, and that Dēmokritos elaborated the system in greater detail. If true, Leukippos would be responsible for the brilliant response to Parmenidēs' arguments that ancient **atomism** represents. Parmenidēs had argued that according to the rules of logic it can be deduced that all that exists is Being, one and motionless. Motion, plurality, change, and "not being" are mere illusions and cannot and do not exist. Moreover, coming into being and passing away are impossible. Addressing this paradox, Leukippos responded that Parmenidēs' objections could be met by positing an infinite number of indivisible and indestructible atoms traveling forever in infinite void or space. The atoms come together to form compound bodies, and depart as the compound bodies disintegrate. Thus, the ultimate principles of reality are "being" (the atoms) and "non-being" (the void), and all compound bodies are made up of them. The atoms and void are eternal, but the compound bodies composed of them are not. Two works attributed to Leukippos are *The Great World System*, and *On Mind*, neither of which survives. In *The Great World System*, Leukippos explained how worlds (***kosmoi***) are formed. As the atoms move and collide eternally in the void, they at times fall into patterns called "vortexes," developing into worlds as the atoms sort themselves out.

DK 67; C. Bailey, *The Greek Atomists and Epicurus* (1928); KRS 402–433; *DSB* 8.269, G.B. Kerferd; *OCD3* 848, D.J. Furley; *ECP* 298, J.S. Purinton; *BNP* 7 (2005) 447 (#5), I. Bodnár.

Walter G. Englert

Libanios (*ca* 350 – 450 CE?)

Wrote a geographical work on Macedon and the lands around Constantinople, which the RAVENNA COSMOGRAPHY used as a principle source on the Bosporos (4.3), and on Dardania, Thrakē, and Musia (4.5–7); cited also on Macedon (4.9). *Cf.* IAMBLIKHOS GEOG. and PORPHURIOS GEOG. All attested bearers of the name seem to be related to the orator (*PLRE*); the name is derived from the Semitic *liban*, "white" (Sala 1974: 25).

(*)

PTK

Libanios of Antioch, pseudo (300 – 500 CE?)

A 13th c. Greek MS, Paris, BNF, *graecus* 2894, contains a work on *De hominis generatione* attributed to "Libanios of Antioch," presumably distinct from the homonymous rhetor. It seems probable that the work was spuriously ascribed to him either by an accident of transmission or with the intent to include it in the great late antique literary tradition. Albeit, the work is typical of the period, represented also by GREGORY OF NAZIANZOS, *De humana natura*; NEMESIOS OF EMESA, *De natura hominis* and BASIL OF CAESAREA, *On the origin of man*, which recast classical Greek medical anthropology into Christian theological terms (*cf.* also LOUKĀS and MAKARIOS). As a result, bodily processes (both in health and disease) are no longer considered as physiological phenomena resulting from material causes, but as gifts, punishments, or trials of divine origin, thus reintroducing supernatural explanations of diseases, as in earlier times (see the HIPPOKRATIC CORPUS SACRED DISEASE). Consequently, pathology is merely descriptive and does not investigate the possible material causes affecting human health.

Diels 2 (1907) 57.

Alain Touwaide

Licinius Atticus (*ca* 100 BCE – 90 CE)

ASKLĒPIADĒS PHARM. in GALĒN, *CMLoc* 8.5 (13.182 K.), records that Licinius Atticus prescribed NEILOS' aromatic ointment for stomach disorders. The cognomen is attested from the 1st c. BCE (*PIR2* A-1333; *TLL* 2.1135–1138: most famously, CICERO's friend; *cf.* also IULIUS ATTICUS). A. Manlius Torquatus (*cos.* 241 BCE) is also called "Atticus" in Augustan-era sources, probably in error (*cf.* Syme, *Roman Papers* 3 [1984] 1430–1431; Rübekeil, *Suebica* [1992] 156–157).

RE 13.1 (1926) 232 (#37), F.E. Kind.

PTK

C. Licinius Caluus (60 – 47 BCE?)

Possibly the politician, orator, and neoteric poet, CICERO's rival, Catullus' friend, born 82 BCE (PLINY 7.165). Martial (14.196) celebrates his lost prose *Use of Cold Water*, perhaps detailing sources and types of waters, on which Charisius (*GL* 1.81) comments that the stomach is unable to endure very sweet food.

GRL §100; *KP* 3.850–851 (Macer #2), P.L. Schmidt; *OCD3* 857, E. Courtney; *FLP* 201–211; *BNP* 7 (2005) 532–533, P.L. Schmidt.

GLIM

LICINIUS MAECENAS ⇒ MAECENAS LICINIUS

Licinius Mucianus (70 – 75 CE)

Roman statesman, legate of *Lycia ca* 57 CE, of *Syria* 68/69 CE; in the civil war, he sided first with Otho, then with Vespasian whom he assisted to the throne, was consul for the second time in 70 CE, and for a third time 72 CE; he died *ca* 76 CE. PLINY 32.62 and TACITUS (*Hist.* 1.10, 1.76, 2.5–7, 2.74–84, 3.46–53) give biographical data. In the last years of his life, he wrote history, published his letters, and compiled observations made on his travels into a paradoxography or *commentarius*. Pliny preserves *ca* three dozen extracts, mostly on marvelous springs or the intelligence of animals (translated by Williamson 247–252). There was a fresh-water spring at Arados below the sea, 5.128; a spring on Andros produced wine-flavored water, 31.16; and one at Kuzikos curbed lust, 31.19. Many animals display almost human intelligence: elephants on gangplanks 8.6, goats crossing streams 8.201, apes playing *latrunculi* and worshipping the waxing moon 8.215.

BNP 7 (2005) 539–540 (#II.14), W. Eck; G. Williamson, "Mucianus and a Touch of the Miraculous," in J. Elsner and I. Rutherford, edd., *Pilgrimage in Graeco-Roman and Early Christian Antiquity* (2005) 219–252.

PTK

Lingōn (10 BCE – 95 CE)

Asklēpiadēs Pharm., in Galēn, *CMLoc* 9.4 (13.286 K.), records his anodyne, based on henbane and opium, and including Cretan-carrot seed (*cf.* Khariklēs), ***euphorbia*** (*cf.* Iouba), ***purethron***, and saffron. The name seems derived from the Gallic tribe, *cf.* Polubios 2.17.7, Caesar *BG* 1.26.5–6, 1.40.11, Strabōn 4.3.4, 4.6.11, and J.-H. Billy, *Thes. Linguae Gallicae* (1993).

RE 13.1 (1926) 714, F.E. Kind.

PTK

Linos, pseudo (*ca* 250 – 150 BCE?)

"Linos" was the lamented, or a song for the lamented, but is listed by Hippobotos as a sage (Diogenēs Laërtios 1.42); Diogenēs Laërtios credits him with a verse cosmogony, entitled *On the Nature of the World*, of which Iōannēs Stobaios quotes two excerpts: 1.10.5 on the essential unity and perpetual flux of all things, and 3.1.70 on avoiding gluttony; *cf. Souda* Lambda-572 (giving Thebes as the ethnic). He taught a Great Year of 10,800 years, Hērakleitos' value (Censorinus 18.11), that the four elements were held together by three "bonds" (Theologumena Arithmeticae, p. 67 de Falco), and the special significance of the seven planets and seven days (Aristoboulos in Clement, *Strom.* 5.107.4).

Ed.: West (1983) 56–67.
DPA 4 (2005) 107, B. Centrone.

PTK

"Lion Horoscope" of Kommagēnē (109 – 62 BCE)

Stone relief on top of Mt. Nemrud in east Anatolia (37°59' N, 38°45' E, at 2206m elev.), discovered in 1890 as part of the western terrace of the *Hierothesion* of Antiokhos I of Kommagēnē (fragments of a twin copy have been found on the east terrace). It represents a conjunction of Mars, Mercury, Jupiter, and the Moon in the constellation Leo with the main emphasis on the application of the Moon to the "royal" bright fixed star Regulus. This seems to give the ruler cult of Kommagēnē a cosmic dimension. Neugebauer's long accepted dating of the conjunction to July 6 or 7, 62 BCE, has recently been questioned by Crijns who argues for July 14, 109 BCE. Depending on the difficult interpretation of the unique iconography and a number of indispensable hypotheses for its astronomical dating,

"Lion Horoscope" Reproduced from K. Humann and O. Puchstein, *Reisen in Kleinasien und Nordsyrren* (1890)

both dates are plausible but neither is cogent. The relief has been variously yet inconclusively interpreted as the horoscope of Antiokhos' conception, birth, coronation, or apotheosis, of the foundation or inauguration of this site, or as the coronation horoscope of Antiokhos' father Mithradatēs I Kallinikos. The state of preservation has much deteriorated since 1890.

Neugebauer and van Hoesen (1959) 14–16; M. Crijns, "The Lion horoscope: proposal for a new dating," in: E.M. Moormann and M.J. Versluys, "The Nemrud Dağ Project: first interim report," *BABesch* 77 (2002) 73–111 at 97–99; Stephan Heilen, "Zur Deutung und Datierung des 'Löwenhoroskops' auf dem Nemrud Dağı," *EA* 38 (2005) 145–158; B. Jacobs and R. Rollinger, "Die 'Himmlischen Hände' der Götter. Zu zwei neuen Datierungsvorschlägen für die Kommagenischen Reliefstelen," *Parthica* 7 (2005) 137–154.

Stephan Heilen

Litorius of Beneventum (100 BCE – *ca* 350 CE?)

Quoted by PELAGONIUS OF SALONA for a remedy for **glanders**; the passage is preserved in Greek translation in the *Hippiatrika* (Pel. 6.1 = *Hippiatrica Berolinensia* 4.15). Part of the remedy is copied from APSURTOS, who may be Pelagonius' true source. Pelagonius describes Litorius, perhaps anachronistically, as *uir clarissimus* (a late-antique title meaning "of senatorial rank").

Fischer (1980); Adams (1995); McCabe (2007) 167.

Anne McCabe

Lobōn of Argos (200 BCE? – 200 CE)

Mentioned at most three times in ancient literature (DIOGĒNES LAËRTIOS 1.34: on THALĒS, 1.112: on EPIMENIDĒS; and possibly *Vita Sophoclis* §16), wrote a book *On Poets*, at some unknown date; all other fragments attributed to him by modern scholars are hypothetical and doubtful. The name is very rare, attested also at Athens 410/409 BCE: *LGPN* 2.285.

SH 504–526.
C. Farinelli "Lobone di Argo, ovvero la psicosa moderna del falso antico," *Annali dell' Istituto Universitario Orienteale di Napoli (filol.)* 22 (2000) 367–379; *DPA* 4 (2005) 111–112, R. Goulet.

Jørgen Mejer

Logadios (20 BCE – 450 CE)

CASSIUS FELIX 73.2 (*CUF*, p. 196), repeated by AËTIOS OF AMIDA 3.113 (*CMG* 8.1, p. 302) = PAULOS OF AIGINA 7.8.2 (*CMG* 9.2, p. 287), preserves his ***hiera*** of aloes, black hellebore, cinnamon, ***euphorbia*** (see IOUBA), gentian, myrrh, pepper, squill, etc. Paulos lists him with GALĒN and earlier pharmacists; the name is otherwise unattested (Pape-Benseler; *LGPN; PIR; PLRE*), though *logades* are the whites of the eyes (NIKANDROS, *Thēr.* 292); it may be a mistake for Lagodius (attested 409 CE: *RE* 12.1 [1924] 457, O. Seeck), or else derived from "Lugh" (Irish sun-god), as *Lugaid* (attested from the 3rd c. CE). Diels 2 (1907) 58, and 3 (1908) 35, lists MSS containing extracts from Logadios, including Brit. Mus. *Harl.* 5626 (15th c.) f. 2, Oxford *Barocc.* 150 (15th c.) f. 3b, Vienna *Med.* 31 (15th c.) ff. 133^{V}–136^{V} and *Med.* 41 (14th–15th c.), ff. 93^{V}–96.

RE 13.1 (1926) 990, F.E. Kind.

PTK

Lollianus (480 – 520 CE?)

Wrote in Latin a geographical work on current and former Roman possessions, which the RAVENNA COSMOGRAPHY follows on Egypt (3.2, 3.8), and cites often in Book 4 on Europe. Although the name is attested primarily earlier (*Souda* Lambda-670; *BNP* 7 [2005] 802–803; *PIR2* Q-52; *PLRE* 1 [1971] 511–512), his reference to Burgundia, 4.26–27, postdates 480 CE. *Cf.* ARBITIO and CASTORIUS.

(*)

PTK

Londiniensis medicus (80 – 100 CE)

The unknown author of the text in *Papyrus London inv.* 137, first published by Diels who viewed it as a text consisting of notes on an introductory medical course, badly copied by a scribe or an uneducated pupil, written under Domitian or Trajan. The text is, in fact, autographous, originating probably in an instructional context: the scribe was at the same time "composing" the text. Clearly incomplete, the papyrus breaks off abruptly halfway down col. 39 and is perhaps only a rough draft. The contents can be divided into three sections. The first part defines fundamental medical concepts, such as "affection," "condition," "disease," etc. Then follows an extensive section (4.18–21.9), treating causes of disease, derived from "ARISTOTLE": it is rich in unique testimony on Pre-Socratic doctors and philosophers of 5th/4th cc. BCE. Diels traced it back to Aristotle's pupil MENŌN, through ALEXANDER PHILALĒTHĒS' doxographical work, cited by the author, but his thesis remains dubious. Finally, there is a physiological section (cols. 21–39), a lengthy discussion of the theory of digestion and assimilation of food. The author seems to know and manipulate a wide range of doxographical material: in his section on definitions he uses **Stoic**-oriented manuals, but Aristotle's doxography on the causes of disease appears to trace back to the early **Peripatos**, while in the physiological discussion the author probably draws on Alexander

Philalēthēs' doxographical work, at least for the doctrines of ASKLĒPIADĒS OF PROUSIAS, but he knows also discussions for and against single doctrines of HĒROPHILOS and ERASISTRATOS. The author freely selects and criticizes his sources; in the section of definitions he sides with the **Peripatetics** against the **Stoics**, he corrects "Aristotle" on HIPPOKRATĒS' theory of diseases, and emphasizes his account of PLATO's pathology much more than the doxographical source. By his choice of the Aristotelian doxography addressing the "ancients" (nothing is said about any Hellenistic theory of diseases), he seems to locate himself purposely in a **Platonic-Peripatetic** tradition; as for the physiological discussion he favors the **Hērophilean** tradition, against the "mechanistic" doctrines of Erasistratos and Asklēpiadēs.

Ed.: H. Diels, *Anonymi Londinensis ex Aristotelicis Iatricis Menoniis et aliis medicis eclogae* = *CAG* S.3.1 (1893); W.H.S. Jones, *The medical writing of Anonymus Londinensis* (1947).

Daniela Manetti, "Autografi e incompiuti. Il caso dell'Anonimus Londinensis, *Pap. Lit. Lond.* 165," *ZPE* 100 (1994) 47–58; Daniela Manetti, "Aristotle and the role of doxography in the Anonymus Londiniensis (*P.Br.Libr. inv.* 137)," in van der Eijk (1999) 95–141; *BNP* 1 (2002) 712–713, V. Nutton; *AML* 52–53, H. Flashar.

Daniela Manetti

LONGINUS ⇒ CASSIUS LONGINUS

Loukās, pseudo (Alch.) (300 – 700 CE)

Listed among philosophers "of the science and sacred art," at the beginning of MS *Marcianus gr.* 299 (f.7^{V}).

(*)

Cristina Viano

Loukās, pseudo (Med.) (350 – 550 CE)

Byzantine MSS contain some brief texts supposedly by Saint Luke: an *alation* (salt; ed. Ideler 1 [1841/1963] 297); an *Epistle on the Taxis of the Human Body* (ed. Rose in *Theod. Prisc.*, p. 463); and unpublished *remedia*. These texts, none of which is authentic, are typical of the early-Byzantine period. The *alation*, resembling a theriac, is compounded from 15 substances (mainly vegetal) and supposedly treated a wide range of ailments. However, it is more a miraculous treatment in curing the ailments of the elderly. A similar *alation* (ed. Ideler 2 [1842/1963] 297–298) is attributed to a Gregory identified only as a *Saint* and a *Theologian* (GREGORY OF NUSSA?). The *Epistle*, extant only in Latin (perhaps its original language), resembles the writings on Christian anthropology by such authors as BASIL OF CAESAREA, GREGORY OF NAZIANZOS, and NEMESIOS; *cf.* also PSEUDO-LIBANIOS and MAKARIOS. The Latin text might result from an adaptation in the West of Byzantine texts (e.g., the late 8th c. medical volume of the German Abbey of Lorsch known as the *Lorsch Arzneibuch*) including a Christian anthropology, a history of Greek medicine and other short texts summarizing Greek written data. The works supposedly by Loukās and other Byzantine writings witness the reformulation of medicine occurring in the east in the early Christian centuries until Justinian's reign, aimed at absorbing ancient medical art and changing its inspiration in a way that is best illustrated by the saints Kosmās and Damianos (Latin texts, translations and adaptations or original works, reflect the same process, although dating to the 8th c.).

Diels 2 (1907) 58; G.A. Lindeboom, "Luke the Evangelist and the ancient Greek writers on medicine," *Janus* 52 (1965) 143–148; G. Del Guerra and A. Scapini, "S. Luca era medico?" *Scientia veterum* 106 (1967) 3–78.

Alain Touwaide

Loxos (400 – 350 BCE?)

Mentioned by Origen *contra Celsum* 1.53 as physiognomist (together with ZŌPUROS and POLEMŌN) and eight times by the anonymous PHYSIOGNOMISTA LATINUS (§1) when he refers to "Loxos the physician" as one of his three sources, together with "ARISTOTLE the philosopher and Polemōn the rhetor." The Origen quotation provides a *terminus ante quem* of *ca* 220 CE; and the fact that the Physiognomista Latinus, when he mentions his sources in order (§1, 48, 80), puts Loxos before "Aristotle" (meaning, the ARISTOTELIAN CORPUS PHYSIOGNOMY), has led to the assumption (e.g. by Misener) that he was older than Aristotle. Boys-Stone (58–64) argues that Loxos "is a writer of clear **Peripatetic** affiliation" (59).

The Physiognomista Latinus employs Loxos for the chapters on the eyes and pupils (§81) and hair of the ears and nose (§ 82) and especially for the animal analogies (§118–131). He quotes Loxos for a physiological explanation of the relation between body signs and character traits: the blood is the seat of the soul, and its greater or lesser fluidity and free or obstructed passages cause the differences in signs of the body and its parts, signifying different character types (§2, 12). And he attributes to Loxos and Polemōn the belief that physiognomy can be a method of divining the future (§133).

G. Misener, "Loxus, Physician and Physiognomist," *CP* 18 (1923) 1–22; Jacques André, *Traité de physiognomonie: anonyme latin* (*CUF* 1981) 24–26; G. Boys-Stones, "Physiognomy and Ancient Psychological Theory," in Swain (2007) 19–124.

Sabine Vogt

LUCANUS ⇒ (1) ANNAEUS LUCANUS; (2) OCELLUS LUCANUS

T. Lucretius Carus (65 – 55 BCE)

Roman poet and **Epicurean**, author of the philosophical poem *De Rerum Natura* (*On the Nature of Things*), one of our main sources for details of EPICURUS' **atomic** theory. Little is known about Lucretius' life. He was born sometime in the 90s BCE and probably died *ca* 50 BCE. His poem is addressed to Memmius, probably to be identified with C. Memmius, the patron of the poet Catullus.

De Rerum Natura, approximately 7,400 lines in length, and divided into six books, is a didactic epic, designed to instruct the reader in a memorable poetic style about the elements of Epicurus' **atomic** system. Lucretius' poetic model for versifying natural philosophy was the Pre-Socratic philosopher and poet EMPEDOKLĒS, whom he praised explicitly (1.705–741). His chief, and perhaps sole, philosophical source was Epicurus, from whom he received his philosophical inspiration and doctrines. Lucretius drew heavily on Epicurus' writings, particularly his major work in 37 books, *Peri* ***Phuseōs*** (*On Nature*), which is now preserved only in fragments. Lucretius tried to present Epicurus' doctrines as accurately and memorably as he could.

The six books of Lucretius' poem describe the **atomic** nature of the world, moving gradually from the **atomic** level to the workings of the universe as a whole. Books 1 and 2 examine the nature of the atom, Books 3 and 4 explore the nature of the soul, and Books 5 and 6 explicate the nature of the world. Book 1, after an invocation to the goddess Venus, praise of Epicurus, and an attack on traditional religion, sets out the arguments for the existence of atoms and the void, and criticizes the views of HĒRAKLEITOS, Empedoklēs, and ANAXAGORAS; the book ends by describing the infinite nature of matter, space, and the universe. Book 2 describes the motions, shapes, and characteristics of atoms, and explains how worlds are created and destroyed. Book 3 sets out the nature of the soul in **atomic** terms, arguing that the soul is mortal and thus that death is nothing to fear. Book 4 treats the nature of **atomic** images and their role in perception and thinking, as well as the topics of digestion, locomotion, sleeping, dreaming, and the evils of passionate love. Book 5 treats the birth and growth of our world, the nature and motion of the heavenly bodies, and the origins of life and development of human society. Finally, Book 6 treats meteorology and geology, explaining thunder and lightning, clouds and rain, earthquakes, volcanoes, magnets, and plagues; the book ends with a detailed description of the great plague at Athens based on THUCYDIDĒS' account. The poem is unfinished, although scholars disagree to what extent. As the summary of its contents indicates, the poem primarily treats **Epicurean** physical theory, and omits an explicit account of **Epicurean** ethics, although Lucretius clearly intended the poem to have consequences for how people lived their lives. It has been argued that fragments of the poem can be identified among the charred papyrus rolls from the **Epicurean** library buried at Herculaneum during the eruption of Vesuvius in 79 CE.

C. Bailey, *Lucretius* 3 vv. (1947); *KP* 3.759–764, G. Schmidt; *DSB* 8.536–539, D.J. Furley; D. Clay, *Lucretius and Epicurus* (1983); K. Kleve, "Lucretius in Herculaneum," *CrErc* 19 (1989) 5–27; M. Gale, *Myth and Poetry in Lucretius* (1994); *OCD3* 888–890, P.G. Fowler and D.P. Fowler; *ECP* 309–311, Walter G. Englert; *REP* 5.854–856, M. Erler; D.N. Sedley, *Lucretius and the Transformation of Greek Wisdom* (1998); *BNP* 7 (2005) 860–864, K. Sallmann.

Walter G. Englert

Lukomēdēs (120 BCE – 80 CE)

ANDROMAKHOS, in GALĒN *CMLoc* 7.5 (13.92 K.), records two anodynes, based on henbane and opium, the second also containing, e.g., Indian nard, parsley, pomegranate-flower, rose-petals, saffron, and ***storax***. The name is rarer after *ca* 100 BCE (*LGPN*), whereas the Indian nard makes a date after *ca* 120 BCE more likely.

RE 13.2 (1927) 2300 (#14), F.E. Kind.

PTK

Lukōn of Iasos (335 – 270 BCE)

Pythagorean critic of Aristotle (DIOGENĒS LAËRTIOS 5.16; Ath. *Deipn.* 10 [418e]; EUSEBIOS *PE* 15.2.8–10), whom Dēmētrios of Magnesia, in D.L. 5.69, seems to place before LUKŌN OF TROAS. (The Lukōn cited by ANTIGONOS OF ALEXANDRIA in *Schol. Nik. Thēr.* 585, and the **Pythagorean** Lukos of Ath., *Deipn.* 2.80 [69e], are probably LUKOS OF NEAPOLIS).

FGrHist 1110; *DPA* 4 (2005) 200–203, B. Centrone and C. Macris.

PTK

Lukōn of Troas (*ca* 280 – 225 BCE)

Student of STRATŌN, and his successor as **scholarch** of the **Peripatos**; celebrated for eloquence; died at 74 of gout. APULEIUS, *Apol.* 36, credits him with investigating fish (along with ARISTOTLE, THEOPHRASTOS, and EUDĒMOS OF RHODES), probably an error for LUKOS OF RHĒGION. The fragment quoted by the grammarian Herodian (*ca* 220 CE) on salt, "dug up sweet or foul smelling," might refer to natural philosophy.

W.W. Fortenbaugh and S.A. White, *RUSCH* 12 (2004) *frr.* 13, 15; *BNP* 7 (2005) 924 (#4), R.W. Sharples; *DPA* 4 (2005) 197–200, J.-P. Schneider.

PTK

Lukos of Macedon (130 – 160? CE)

Student of QUINTUS and bête-noire of GALĒN, who never met him, but sought to refute his work on anatomy (*On My Own Books*, 2.101 MMH). He taught the doctrine of ERASISTRATOS that urine is a residue of food (Galēn, *Nat.Fac.* 1.17, 3.152 MMH), and argued that all forms of heat are the same, a long fragment being preserved in Galēn's sophistic refutation (*CMG* 5.10.3, pp. 19–23). His commentaries on the HIPPOKRATIC CORPUS, EPIDEMICS, Books 3 and 6 (and others?), were criticized by Galēn as excessively interventionist, and as displaying **Empiricist** tendencies (*CMG* 5.10.2.2, pp. 225, 239 and 5.10.2.1, pp. 16–17, respectively). Recipes of a Lukos, this man or the Neapolitan, are found in OREIBASIOS, *Coll.* 8.25 (*CMG* 6.1.1, pp. 278–282) and 8.43 (p. 293) against dysentery and for "downward purges" (*cf.* also *CMG* 6.1.2, p. 28, 6.2.1, p. 46, 6.3, pp. 22, 88, 119); and PAULOS OF AIGINA 5.3.1 (*CMG* 9.2, pp. 7–8) and 5.13.4 (p. 17), antidotes against ***hudrophobia*** and viper bites.

Manetti and Roselli (1994) 1582–1589; Ihm (2002) #159–165; *BNP* 7 (2005) 939–940 (#13), A. Touwaide.

PTK

Lukos of Neapolis (130 – 70 BCE)

The Lukos included by GALĒN *MM* 2.7.23 (10.142–143 K. = p. 71 Hankinson) in a list of **Empiricist** physicians is certainly to be identified with the *Licus Neapolitanus* whose recipe PLINY mentions at 20.220 (and whom he mentions amongst the sources of Books 20–27). Some remedies (also of gynecological interest) mentioned by OREIBASIOS and PAULOS OF AIGINA, and also the dietetic fragment preserved by the Homeric scholia *Townl. Il.* 6.260, might derive from the same work on pharmacology (although in these cases the author might also be LUKOS OF MACEDON). He dealt with exegesis of the HIPPOKRATIC CORPUS: we know about a commentary in at least two books on *De locis*, from which ERŌTIANOS and Phōtios derived some lemmata.

Ed.: Deichgräber (1930) 20, 204–205 (fragments), 261.
RE 13.2 (1927) 2407–2408, F.E. Kind; *KP* 3.819 (#13), F. Kudlien; Ihm (2002) #166; *BNP* 7 (2005) 938–939 (#10), A. Touwaide.

Fabio Stok

Lukos of Rhēgion (*ca* 360 – 280 BCE)

Greek historian, adoptive father of the tragedian Lykophrōn of Euboian Khalkis, wrote a history of Libya and a work on Sicily, both containing geographical and ethnographical information. Together with TIMAIOS OF TAUROMENION, Lukos was considered an authority on the western part of the ***oikoumenē***. Fragments of his works, many preserved in the paradoxographical compilation of ANTIGONOS OF KARUSTOS, include, for instance, information on the mixture of cold and hot water in certain seas, on solar effects in Libya, and on the long-lived honey-eating inhabitants of Corsica.

FGrHist 570.

Daniela Dueck

Lunkeus (250 BCE – 95 CE)

ASKLĒPIADĒS PHARM., in GALĒN *CMLoc* 4.7 (12.778 K.), records his scar- and callous-softener composed of aloes, **calamine**, roasted copper, **galbanum**, hematite, myrrh, opium, **verdigris**, etc., in rainwater.

(*)

PTK

Lupus (of Thēbai?) (100 BCE – 400 CE)

The "Laurentian" list of medical writers (MS *Laur. Lat.* 73.1, f.143[V] = *fr.*13 Tecusan) includes "Lupus Pelopis," usually emended to "Lupus of Thebes," but perhaps "Lupus, student of PELOPS OF SMURNA" is meant; if so, the date-range would be *ca* 150–190 CE (*cf.* also LUKOS OF MACEDON, student of QUINTUS). *Cf.* also EPIPHANĒS, HIPPOSIADĒS, and PHILIPPOS OF KŌS.

(*)

PTK

Lusias (*ca* 100 BCE – 35 CE)

Greek physician, probably a follower of ASKLĒPIADĒS OF BITHUNIA. CELSUS (5.18.5) preserves Lusias' multi-use emollient good for abscesses, parotid swellings, joints, painful heels, and digestion compounded from ***opopanax, storax***, **galbanum**, ***ammōniakon*** incense, **bdellium**, beeswax, beef suet, dried iris, barley, and peppercorns, pounded with iris ointment. ASKLĒPIADĒS PHARM., in GALĒN *CMLoc* 7.5 (13.49–50 K.), quotes his ***arteriakē*** compounded from saffron, myrrh, licorice, frankincense, cassia, and peppercorns, in Cretan must and Attic honey. SŌRANOS, *Gyn.* 3.1 (*CMG* 4, p. 94; *CUF* v. 3, p. 3) refers to a *Chronic Diseases* written by a follower of Asklēpiadēs of Bithunia: editors print *ΛΟΥΚΙΟΣ* but the MS has *(Ε)ΛΑΙΟΥΣΙΟΣ*, very likely identifiable with our physician. CAELIUS AURELIANUS cites a *LVCIVS* (sometimes emended to *LYSIAS*), probably a **Methodist** (*Chron.* 2.59 [*CML* 6.1.2, p. 578.27]), as author of a *Chronic Diseases* in at least four books (4.79 [*CML* 6.1.2, p. 818.12]). He recommended patients suffering from throat inflammation take, before eating, dry-parched figs soaked in wine or sip hot wine and water mixed with realgar and an egg-white or colts-foot root juice (*Chron.* 2.111 [*CML* 6.1.1, p. 610.17]). He also prescribed induced vomiting after meals for patients with stomach or bowel ailments (*Chron.* 4.79 [above]).

KP 3.836 (#7), F. Kudlien; *BNP* 8 (2006) 36 (#9), V. Nutton.

Alain Touwaide

Lusimakhos (325 – 90 BCE)

Author of a treatise on agriculture excerpted by CASSIUS DIONUSIOS (VARRO, *RR* 1.1.8–10, *cf.* COLUMELLA, 1.1.11); to judge from PLINY's index, it discussed cereals, livestock, poultry, viticulture, and arboriculture (1.*ind.*8, 10, 14–15, 17–18). There is no reason to identify him with the Lusimakhos of Pliny 25.72, *pace* Gudeman.

RE 14.1 (1928) 32 (#19), A. Gudeman.

Philip Thibodeau

Lusimakhos of Kōs (*ca* 280 BCE – 60 CE)

A doctor who wrote commentaries on the HIPPOKRATIC CORPUS. One work in 20 books dealt with obscure Hippocratic terminology, another in three books attacked KUDIAS THE HĒROPHILEAN, and one in four books attacked a certain Dēmētrios (ERŌTIANOS *Pr.*, B-8, T-13 [pp. 5, 28, 85 Nachm.]; *cf.* *Schol. Nic. Alex.* 376) – see SALIMACHUS. If this Dēmētrios is identical to the **Epicurean** DĒMĒTRIOS OF LAKŌNIKA (so Kind and Nutton), Lusimakhos will date *ca* 100 BCE; if to another, e.g. DĒMĒTRIOS OF APAMEIA, he may be earlier.

RE 14.1 (1928) 32 (#19), A. Gudeman, and 39 (#21), F.E. Kind and W. Kroll; *BNP* 8 (2006) 42 (#7), V. Nutton.

Philip Thibodeau

Lusimakhos of Macedon (*ca* 335 – 281 BCE)

Born *ca* 360, officer of Alexander the Great and later *diadoch* king of Thrakē and Asia Minor. PLINY (25.71–72) seems to regard him as the discoverer of a willow-like shrub called *lusimakhia*, whose therapeutic powers were praised by ERASISTRATOS (Pliny *ibid.; cf.* 25.100, 26.131, 141, 147, GALĒN *Simpl. Med.* 8.21 [12.64 K.], DIOSKOURIDĒS 4.85).

RE 14.1 (1928) 39 (#21) W. Kroll; *OCD3* 902, A.B. Bosworth.

Philip Thibodeau

M

Mac- ⇒ Mak-

Macer ⇒ Aemilius

Macharius (of Rome?) (395 – 400 ce)

Rufinus, *Apol.* 1.11.19, records that Macharius, a Christian, was composing an attack upon astrology in 397. The unusual name presumably derives from Makharēs, as in Plutarch, *Luc.* 24.1, App. *Mithr.* 67–83.

DPA 4 (2005) 226, R. Goulet.

PTK

Macrobius ⇒ Macrobius Theodosius

Maecenas Licinius (30 bce – 15 ce)

Authored a treatise on storing wine and agricultural produce known to Columella, (12.4.2), who dates him roughly to the age of Augustus. The MSS name the writer variously as Bascenas or Mecenas Licinius. Since Maecenas' household drew other writers interested in agricultural topics (Melissus, *cf.* Vergil, Sabinus Tiro), the first name is plausibly corrected to Maecenas. The author would then be his freedman (Licinius is not uncommon as a slave's name).

GRL §203.

Philip Thibodeau

C. Maecenas Melissus of Spoletium (30 – 10 bce)

Free-born foundling raised a slave, instructed in literature, and given as a gift to Maecenas, who quickly recognized his talents. When his true identity was discovered, he was manumitted and assigned by Augustus to organize the library in the *Porticus Octauiae*; he lived to at least 60. Though better known as a writer of plays and joke-books, he is presumably identical to the Roman Melissus who wrote about zoology (Pliny, 1.*ind.*9–11), physiology (1.*ind.*7; *cf.* 28.62) and bees (Serv. *ad Aen.* 7.66).

RE 15.1 (1931) 532–534, P. Wessner; Kaster (1995) *ad* §21.

Philip Thibodeau

MAECIANUS ⇒ VOLUSIUS MAECIANUS

Maecius Aelianus (100? – 155 CE)

The oldest man among GALĒN's teachers, outstanding in experience and kindness, who once, when a plague afflicted Italy, saved people using his antidote (*Theriac for Pamphilianus* 14.298–299 K.); the same man is credited with an epitome of muscular anatomy (*Diss. Musc.* 18B.926–927, 935, 986 K.), based on his unnamed father's work. Kühn reads *ΜΕΚΚΙΟΣ*, an otherwise unattested name (*LGPN, MRR, PIR2*), but the old republican *gentilicium* Maecius is sometimes rendered *ΜΑΚΚΙΟΣ* (Geffcken). (Grmek and Gourevitch [1994] 1498, n. 13, to explain the father as NUMISIANUS, suggest reading "HĒRAKLEIANOS," whom Galēn's description does not fit.)

RE 1.1 (1893) 488 (#2), M. Wellmann; 14.1 (1928) 233 (#2), J. Geffcken.

PTK

MAECIUS DIONUSODŌROS ⇒ DIONUSODŌROS MAECIUS

Maēs Titianus of Macedon (50 – 110 CE)

Composed a report on the overland trade with China, preserved by MARINOS OF TYRE in PTOLEMY 1.11. The name Maēs is unusual (*LGPN* 1.295, 2.296, 3B.268: mostly 2nd/1st c. BCE); and Pape-Benseler suggest *maeitai* ("babble") as the etymology. Instead, it may be an error for the Latin *nomen* Maesius (*cf.* Maesius Titianus: *ILS* 1.1083, *ca* 150 CE), or an attempt to transcribe a Hebrew name usually rendered "Maasia" (as in Ezra 10.18, Nehemiah 3.23, etc.); if the latter, Maēs may have been from the Jewish community of Thessalonikē. The non-Republican cognomen is attested from the early 1st c. CE: *CIL* 6.5194 ("Augustan"), *LGPN* 2.434.

RE S.11 (1968) 1365, K. Ziegler.

PTK

Magistrianus (*ca* 60 – 100 CE?)

AËTIOS OF AMIDA 16.39 (p. 883 Cornarius = Zervos 1901: 57) quotes Magistrianus' brief remedy for abscesses of the breasts (chopped earthworms mixed with barley meal and quaffed), and thrice more cites him: 13.126 (p. 742 Cornarius), 13.134 (p. 753 Cornarius), and 14.55 (p. 800 Cornarius), once within a quotation from ARKHIGENĒS. Probably Magistrianus predates Arkhigenēs by a generation or so, and presumably writes in Greek. The name seems otherwise unattested (*LGPN, CIL, PIR*).

Fabricius (1726) 313.

John Scarborough

Magnēs or Magnus (200 BCE? – 460 CE)

Published *Logistika*, a treatise on numerical computations cited by EUTOKIOS in his commentary on ARCHIMĒDĒS' *Measurement of a Circle* (302.3 Heiberg), criticizing Magnēs' calculation of circumferences of circles via multiplication and division of myriads as difficult to follow. Orinksy suggests that Magnēs is possibly identifiable with THEUDIOS OF MAGNESIA (PROKLOS, *In Eucl.* p. 67 Fr.). *Cf.* PHILŌN OF GADARA; NIKOLAOS (MATH.).

RE S.6 (1935) 237 (s.v. Magnes), K. Orinsky; *DPA* 4 (2005) 245, R. Goulet.

GLIM

Magnus *arkhiatros* (90 – 130 CE)

GALĒN, *Thēr. Pis.* 12–13 (14.261–263, 267 K.), records his antidote (employing ***amōmon***, cinnamon, ***kostos***, ***malabathron***, and Indian nard), revising an antidote from ANDROMAKHOS and DAMOKRATĒS (Magnus is not Galēn's contemporary, the practitioner Dēmētrios, 14.261 K., *contra* Kroll). He may be the same as MAGNUS OF TARSOS.

RE 14.1 (1928) 494 (#29), W. Kroll.

PTK

Magnus of Emesa (*ca* 300 – 400 CE)

Wrote treatises on prognostics, fevers, and urines; his identification with the contemporary Alexandrian *iatrosophistēs* MAGNUS OF NISIBIS is plausible but still unproved. Magnus' *De Urinis* survives in Arabic and partly in revised and excerpted Greek (GALĒN 19.574–601 K.; Ideler; Moraux 68–74). The treatise, although largely restating content from HIPPOKRATĒS and Galēn, presents one of the first arrangements of that material into a systematic compendium of types of urine and their differences, based on color, consistency, and sedimentation, marking a genuine advance, since examination of urine plays a relatively small part in Galēnic medicine. To the extent to which Magnus' work thereby helped elevate urine to the far more central place it subsequently enjoyed in medical practice, *De urinis* had a major influence on diagnostic uroscopy in later Byzantine antiquity and the medieval Latin West. It was a principal source and model for Theophilos Protospatharios (9th c.) in his own *De urinis*, who nonetheless criticizes Magnus' incomplete account (Ideler 1 [1841/1963] 261–262); it is also cited in the work of the same title by Iōannēs Aktuarios (13th c.; Ideler 2 [1842/1963] 5) with much the same complaint. Magnus' *De urinis* was translated into Arabic, extensively excerpted, and included in later Byzantine compilations (Baader).

Ed.: Ideler 2 (1842/1963) 307–316.
Diels 2 (1907) 59–60; G. Baader, "Early medieval Latin adaptations of Byzantine medicine in Western Europe," *DOP* 38 (1984) 251–259; P. Moraux, "Anecdota Graeca Minora VI: Pseudo Galen, de Signis ex urinis," *ZPE* 60 (1985) 63–74; *BNP* 8 (2006) 175 (#1), V. Nutton.

Keith Dickson

Magnus of Ephesos (50 – 100 CE)

Greek physician from Ephesos (CAELIUS AURELIANUS, *Acut.* 3.114 [*CML* 6.1.1, p. 360]), after ATHĒNAIOS OF ATTALEIA (GALĒN, *Diff. Puls.* 3 [8.674 K.]). He preceded AGATHINOS and ARKHIGENĒS (*ibid.* and Cael. Aurel., *Acut.* 2.57–58 [*CML* 6.1.1, p. 166]), who refuted his theories (Galēn, *Diffic. Resp.* 1 [7.763 K]; *Diff. Puls.* 3 [8.640, 642, 646, 648, 650, 674 K.]; *Caus. Puls.* 1 [9.8, 18, 21, 22 K.]). Moreover, Galēn (*Diff. Puls.* 3 [8.641 K.]) credits Magnus with writing a work "on the [medical] discoveries after Themisōn's time," probably THEMISŌN OF LAODIKEIA, whereas ASKLĒPIADĒS PHARMAKION (in Galēn) preserves one of his medical formulae. Nevertheless, the name is very common, and compare his near-contemporaries MAGNUS *ARKHIATROS*, MAGNUS OF PHILADELPHEIA, and MAGNUS OF TARSOS, some of whose recipes might belong to this Magnus. Although Galēn cites Magnus as a **Pneumaticist** (above), Caelius Aurelianus considered him a **Methodist** (above). However, his theories and their refutation by Arkhigenēs seem clearly to indicate his **Pneumaticism**. If the chronology in Galēn and Caelius Aurelianus is correct, Magnus was a first generation **Pneumaticist** who might have refined concepts left ambiguous by Athēnaios.

In accord with **Pneumaticist** theory, Magnus studied in great detail the function of the heart and vascular system and took great care to define pulse (Galēn, *Diff. Puls.* 3 [8.674 K.]). Arkhigenēs' refutations imply that Magnus attached great importance to the intensity (*sphodrotēs*) of the pulse – which does not result from ***pneuma*** and *stoikheia* (*ibid.* 8.638 K.), its amplitude (*megethos*) and fullness (*plērotēs: ibid.* 8.640 K.), its quality (*poiotēs: ibid.* 8.650 K.), its strength (*rhōmē*: Galēn, *Diffic. Resp.* 1 [7.762–763 K.]), and its speed (*takhos*: Galēn, *Caus. Puls.* 1 [9.8 K.]). To this end, he paid great attention to technical terms and created neologisms (*Diff. Puls.* 3 [8.640–642 K.]). Magnus is also credited with medical letters, in at least two books (Cael. Aurel., *Acut.* 3.114: *CML* 6.1.1, p. 360).

RE 14.1 (1928) 494 (#28), W. Kroll; *KP* 3.887 (#8), F. Kudlien; *Idem* (1968) 1098.

Alain Touwaide

Magnus of Nisibis (350 – 400 CE)

A pagan student of Zēnōn of Cyprus and Oreibasios' schoolmate, Magnus of Nisibis taught and practiced rhetoric and medicine in Alexandria. Libanios' letter (*Ep.* 843) confirms Magnus' presence in Egypt in 364 and 388; other letters (*Ep.* 1208, 1358) contain less than flattering character details. Eunapios (*Vit. Soph.* 20.5) claims, on evidence of Magnus' popularity as a teacher, that the Alexandrians assigned him his own public lecture-hall (*didaskaleion koinon*), to which students from throughout the eastern empire were drawn. Philostorgios also mentions Magnus (*Hist. eccl.* 8.10).

Eunapios implies that Magnus excelled more in eloquence than healing, a common complaint against Alexandrian medical professors, though presumably not without some basis. Theophilos Protospatharios (9th c.) levels virtually the same criticism (Ideler 1 [1841/1963] 261) against a contemporary Alexandrian, Magnus of Emesa, perhaps but not definitively identifiable with our Magnus. Lending some plausibility to their identification is that Theophilos refers to his Magnus as *iatrosophistēn*, just as Eunapios styles his Nisibian Magnus; the latter may well be the subject of a satirical epigram (addressed *eis Magnon iatrosophistēn*) attributed to Palladas (*Anth. Gr.* 11.281), celebrating Magnus' powers to raise the dead. The ubiquity of the name and generic character of the contrast between rhetoric and experience still leave open the possibility that they are distinct.

There is additionally a chance that Magnus of Nisibis wrote verses, wherein the name Magnus appears acrostically, preserved in the Kuranides attributed to Harpokratiōn of Alexandria (West). Finally, he may also have written an epigram on Galēn (*Pal. Anth.* 16.270).

RE 14.1 (1928) 494 (#34), W. Kroll; *KP* 3.887 (#9), F. Kudlien; *PLRE* 1 (1971) 534; West (1982); *BNP* 8 (2006) 176 (#5), V. Nutton.

Keith Dickson

Magnus of Philadelpheia (100 BCE – 80 CE)

Greek physician from Philadelpheia in Egypt. Andromakhos in Galēn quotes his recipe against blood-spitting, compounded from coral, Samian earth and *polugonon: CMLoc* 7.4

(13.80 K.) and an enema of minerals including alum, quicklime, orpiment, and realgar: *ibid.* 9.5 (13.296 K.). ASKLĒPIADĒS PHARM., in Galēn, preserves a ***trokhiskos*** against several skin affections made of alum, iron, burnt copper, and **Sinōpian earth**: *CMGen* 5.11 (13.829 K.). Galēn quotes various compound medicines attributed to "Magnus": otherwise unidentified (13.831, 14.262, 14.263 K.), qualified as clinician (12.829 K.), and *periodeutēs* (12.844 K.); he is also credited with a recipe preserved by AËTIOS OF AMIDA (7.107 [*CMG* 8.2, p. 372.11–15] = PAULOS OF AIGINA 7.16.33 [*CMG* 9.2, p. 341]). It is unknown if the same Magnus is the author of all these recipes.

RE 14.1 (1928) 494 (#30), W. Kroll (see also #29 and 31–33).

Alain Touwaide

Magnus of Tarsos (60 BCE – 95 CE)

ASKLĒPIADĒS PHARM., in GALĒN, *CMLoc* 9.7 (13.313 K.), records his recipe for relief of hemorrhoids (grind pepper with natron, apply and remove before applying other medicaments). He may be MAGNUS *ARKHIATROS*.

RE 14.1 (1928) 494 (#31), W. Kroll.

PTK

MAGO ⇒ DIONUSIOS OF UTICA

Maiorianus (*ca* 350 – 540 CE)

AËTIOS OF AMIDA 12.48 (p. 682 Cornarius) cites his wound-powder, of alum and pumice (store in glass); to apply, first wash the wound in wine, then sprinkle on, for eight days (Aëtios approves). The name is first attested in the mid-4th c. CE: *PLRE* 1 (1971) 537–538, 2 (1980) 702–703.

Fabricius (1726) 314.

PTK

Makarios of Magnesia (300 – 400 CE?)

Byzantine medical MSS contain two texts, *On the soul* and *On urine*, under the names "Makarios Maximos" and "Makarios Mangens" respectively, neither precisely identified. We can exclude the Christian apologist Makarios Magnēs (*ca* 350–400 CE), not known to have written on medicine. "Makarios Mangens" might be corrupted from MAGNUS OF EMESA, usually identified in the MSS as *Magnos Emesinos* (or *Aimesinos*), but sometimes also called *Makaritēs Magnēs*. The adjective *Makarios* may have been transformed into a noun and the noun *Magnos* into an adjective in confusion with Magnus' Magnesian origins. Furthermore, the text *On urine* in the MS Paris, BNF, *graecus* 2316, somewhat resembles Magnus of Emesa's and might be a later re-arrangement of it. *On the soul* seems typical of 4th c. Christian anthropology, in which bodily processes are attributed to the action of God rather than to physiological transformations: *cf.* PSEUDO-LIBANIOS and LOUKĀS.

Diels 2 (1907) 59, 60; Dimitriadis (1971) 44 (Makarios Magnēs), 29–33, 44, 47–50, 55, 59 (Magnus of Emesa).

Alain Touwaide

Makhairiōn (100 BCE – 110 CE)

KRITŌN OF HĒRAKLEIA, in GALĒN, *CMGen* 5.3 (13.796–797 K.), cites his **terebinth**-based wound plaster, good enough even for ***hudrophobia***, containing birthwort, **galbanum**, myrrh, ***opopanax***, etc. (*cf.* AËTIOS OF AMIDA 15.13 [Zervos 1909: 41]), which Aëtios 10.11 (p. 579 Cornarius) prescribes for the spleen, and PAULOS OF AIGINA 7.17.67 (*CMG* 9.2, p. 364) for sciatica; Paulos also prescribes other remedies from Makhairiōn, sufficiently familiar to omit ingredients: 3.49.2, 4.19.2, 4.48.2 (*CMG* 9.1, pp. 259, 339, 369), *cf.* Aëtios 12.41 (p. 672 Cornarius). Galēn cites Makhairiōn's plaster, plus EPIGONOS' and ISIS', as exemplary, *Meth. Med. ad Glauk.* 2.10 (11.126 K.), 2.11 (p. 138), and *CMGen* 2.5 (13.499 K.).

Fabricius (1726) 312–313.

PTK

MALKH- ⇒ PORPHURIOS OF TYRE

Mallius Theodorus of Milan (385 – 395 CE)

Christian of humble origin whose family included a brother Lampadius (*PLRE* 1 [1971] 493, #3) and a son Theodorus. He rose from advocate to a series of offices *ca* 376–382 CE, after which he retired for about a decade. Under Stilicho he returned to political service, as *praefectus praetorio* of Illyria, Italy, and Africa, and was consul *posterior factus* in 399, with Eutropius, after whose overthrow he remained sole consul. During his midlife retirement, he wrote several works, including a philosophical treatise (entirely lost), which according to CLAUDIAN, *Panegyric on Theodorus* 67–112, illuminated the obscure tenets of Greek cosmology as taught by KLEANTHĒS, CHRYSIPPUS, DĒMOKRITOS, PYTHAGORAS, and others, explaining their theories on elements, the motions of the stars and planets, the lunar cause of tides, and meteorology including comets.

PLRE 1 (1971) 900–902.

PTK

Mamerkos (of Italy) (600 – 560 BCE)

Brother of the poet Stēsikhoros, cited variously as Mamerkos (HIPPIAS DK B12 = PROKLOS *In Eucl.* p. 65.12 Fr.), Mamertinos (*Souda* Sigma-1095), or Mamertios (pseudo-HĒRŌN, *Def.* 136.1). The name strongly suggests Italian lineage, and Stēsikhoros' family plausibly hailed from Lokroi or its colonies. Mamerkos studied geometry and influenced Hippias.

RE 14.1 (1928) 950–951 (#1), W.A. Oldfather; Morrow (1970) 52.

PTK and GLIM

Mamilius Sura (55 BCE?)

Agricultural author whom PLINY (18.143) cites for the view that *ocinum* is the name for a mixture of fava-bean and vetch-seeds sown to make cattle-fodder. Pliny lists him between CATO and VARRO, and elsewhere describes as "ancient" those who use the term *ocinum* (17.198). Mamilius' work apparently discussed cereals, livestock, fowl, viticulture, arboriculture, beekeeping, and garden plants (1.*ind.*8, 10–11, 17–19). Since the name is rare for this period, he may be identical to the plebian tribune Mamilius who in 55 BCE helped to pass a bill designed to remedy defects in CAESAR's agrarian legislation of 59 (*cf.* Varro, TREMELLIUS SCROFA). His family, the *gens Mamilia*, had long-standing ties to Tusculum.

Ed.: Speranza (1971) 66–68.
M. Cary, "Note on the legislation of Julius Caesar," *JRS* 19 (1929) 113–119; *HLL* §196.4.

Philip Thibodeau

Mandroklēs of Samos (525 – 500 BCE)

Engineer, built a bridge of ships across the Bosporus, for Darius I's invasion of Skuthia. HĒRODOTOS (4.87–89) states that with Darius' rewards, Mandroklēs commissioned a painting of his construction work, and dedicated it in the Hēraion at Samos, one of the earlier known documentary paintings.

Svenson-Ebers (1996) 59–66; *KLA* 2.50, K. Hornig.

Margaret M. Miles

Mandrolutos of Priēnē (585 – 525 BCE)

APULEIUS, *Florida* 18.30–35 (DK 11 A 19), records that THALĒS' solar theories were published by his student, *MANDRAVTVS*. The name Mandrolutos is attested (Goulet), and names in Mandro- are frequent (*LGPN*).

RE 14.1 (1928) 1041 (#2), O. Kern; *DPA* 4 (2005) 248, R. Goulet.

PTK

Manethōn (Astrol.) (120 – 140 CE)

Authored an astrological poem *Apotelesmatika* in Greek hexameters in six books. The name Manethōn was associated in Egypt with magical revelations, making its occurrence in connection with the *Apotelesmatika* likely pseudepigraphic; ultimately it appears to recall the historical MANETHŌN OF SEBENNUTOS. The complete *Apotelesmatika* survived in a single medieval copy, but two 3rd c. papyrus fragments preserve passages from Book 4; quotations by HEPHAISTIŌN OF THĒBAI (2.4) and IŌANNĒS PHILOPONOS (*De Opificio Mundi* 4.20 and 6.2) also attest to the work's popularity. The separate books appear to have been composed by several authors, Books 2 and 3 (depending heavily on DŌROTHEOS), and perhaps 6, forming the original kernel (Koechly reordered the books, making these three Books 1–3, but we refer here to the MS sequence). From a horoscope in Book 6 dating to 80 CE, it would appear that these oldest books were written during the 2nd c. CE. The *Apotelesmatika* is a practical reference rather than a literary effort.

Ed.: A. Koechly, *Manethonis Apotelesmaticorum qui Feruntur Libri VI* (1858); R. Lopilato, *The Apotelesmatica of Manetho* (1998).

Alexander Jones

Manethōn (Pharm.) (300 BCE – 400 CE)

The "Laurentian" list of medical writers (MS *Laur. Lat.* 73.1, f.142^{V} = *fr.* 13 Tecusan) includes Manethōn, according to Wellmann's emendation, followed by *Nechepso* (*sc.* PETOSIRIS) and

KLEOPATRA, among Egyptian doctors. PAULOS OF AIGINA 7.13.4 (*CMG* 9.2, p. 324) records his wound-ointment composed of **calamine**, burnt lees, frankincense, etc. (*cf.* 4.40.3, *CMG* 9.1, p. 360). GEŌPONIKA 20.6 credits him with a book on the dietary properties of fish.

RE 14.1 (1928) 1101–1102 (#1), F.E. Kind.

PTK

Manethōn of Sebennutos (*ca* 280 – *ca* 260 BCE)

Egyptian priest who, under Ptolemy II Philadelphos, wrote a history of the Egyptian kingdoms (*Aiguptiaka*), from their "beginning" to the end of Persian rule, employing a 30–dynasty scheme still in use (the work is lost). He also wrote *Phusiologika* (*cf. Souda* M-143), a collection of doctrines cited by DIOGENĒS LAËRTIOS, *pr.*10, and EUSEBIOS, *PE* 3.2.6–3.3.10, for Egyptian cosmological beliefs (as in DIODŌROS OF SICILY, 1.11.1, 1.11.5–6, 1.12.1–9, 1.13.1–2). He claimed sow's milk caused skin diseases (AELIANUS, *NA* 10.16), and solar eclipses afflicted the head and stomach (IŌANNĒS "LYDUS," *Mens.* 4.87 [p. 136 Wu.]); he or a homonym described the compounding of *kuphi* (PLUTARCH, *Isis and Osiris* 80 [383E-384C], *cf. Souda* M-142; *RE* 12.1 [1924] 52–57, R. Ganschinietz).

OCD3 917, A.B. Lloyd and N. Hopkinson.

PTK

M. Manilius (10 – 30 CE)

Wrote *Astronomica*, a didactic poem in five books of Latin hexameters on astrology. Book 1 alludes (898) to Arminius' defeat of Varus in 9 CE, and Book 2, with its praise of Capricorn as AUGUSTUS' birth sign, must have been composed before Augustus' death in 14 CE, whereas Book 4 obliquely refers to Tiberius as the current emperor. Whether the poem (aside from a long gap in Book 5) is complete is controverted.

The *Astronomica* is the earliest extant comprehensive exposition of the fundamentals of Greco-Roman astrology, but unlike the slightly later astrological poems in Greek by DŌROTHEOS and MANETHŌN, it seems to have sought a literary rather than a professional audience; hence both the purple passages prefacing each book and the tour-de-force versifications of technical subjects, such as the table of rising times of zodiacal signs in 3.275–300. Manilius writes as a determinist and **Stoic**, intermittently combating **Epicurean** teachings in language that shows the imprint of LUCRETIUS.

The first book treats the cosmological and astronomical underpinnings of astrology, introducing the celestial sphere, the zodiac, and the other constellations. Books 2 through 4 chiefly concern the signs and further astrologically significant subdivisions of the zodiac, and its interaction with the local horizon. The concluding book is primarily astronomical, listing the constellations that rise simultaneously with each part of the zodiac. Surprisingly, Manilius says little about the Sun, Moon, and planets, and their influences, addressed possibly in the lost section of Book 5.

Ed.: G.P. Goold, *Manilius: Astronomica* (Loeb 1977).
DSB 9.79–80, D.E. Pingree.

Alexander Jones

Mantias (*ca* 15 CE)

Wrote *Concerning the Phenomena Observed during the Rarefaction and Condensation of Bodies*, of which one fragment survives in an Arabic translation of MENELAOS OF ALEXANDRIA's *Concerning the Technique by which the Amount of Each of a Number of Mixed Bodies May be Known* (Madrid, Escurial MS árabe 960, ff. 43a.18–43b.29; Würschmidt 1925: 381–382). The names of both author, Mantias, and dedicatee, Germanicus the King (= Germanicus Iulius Caesar?, d. 19 CE), suggesting Mantias' *floruit*, are conjectural reconstructions of Greek names garbled in the Arabic (Würschmidt 1925: 380–381).

Menelaos cites Mantias as the source of a solution to the problem of ARCHIMĒDĒS' water-test of the purity of the gold wreath of HIERŌN II OF SURAKOUSAI. However, Mantias did not himself claim that Archimēdēs devised his test. Unlike VITRUUIUS' (9.9–12) improbable explanation of Archimēdēs' water-test as based solely upon comparing the volume of water displaced by Hierōn's wreath with that displaced by an equal weight of pure gold, Mantias provides a description of a test based on Archimēdēs' principles of buoyancy and of the lever. Mantias' test goes further than evaluating the purity of a given metal and can be used to calculate the relative proportions of the constituents of any two-metal alloy provided that the identities of the two metals are known.

Mantias' test for a gold-silver alloy first calls for the construction of a calibrated balance: (1) place gold and silver in a proportion of 1:1 in one scale of a balance and place an equal weight of pure silver in the other scale; (2) Submerge the scales in water and watch the balance tip towards the side of the gold and silver; (3) Move the scale with the gold and silver along the beam of the balance and mark the position of the scale at which the beam becomes horizontal; (4) Repeat this procedure increasing the proportion of gold to silver (2:1, 3:1, 4:1 etc.) and marking the final position of the scale each time. To use the balance, an object (e.g. Hierōn's wreath) made of gold and silver in unknown proportions is placed in one scale and an equal weight of pure silver is placed in the other. The scales are then submerged in water and when the balance tips towards the alloy, its scale is moved along the beam until it becomes horizontal. A reading is taken at that point and the proportions of gold and silver in the alloy are ascertained. Mantias places limitations on his method, pointing out that differences in the waters in which the scales are submerged during each use of the balance can affect its readings. This method was later described in the CARMEN DE PONDERIBUS (lines 124–162).

Text: Madrid, Escurial MS árabe 960 (item 3), ff. 43a-50b; *GAS* 5 (1974) 164.
Trans.: J. Würschmidt, "Die Schrift des Menelaus über die Bestimmung der Zusammensetzung von Legierungen," *Philologus* 80 (1925) 377–409.

Bink Hallum

Mantias (Hēroph.) (150 – 100 BCE)

"**Hērophilean**" physician, teacher of HĒRAKLEIDĒS OF TARAS (GALĒN, *CMGen* 2.1 [13.462 K.], *CMLoc*, 6.11 [12.989 K.: "a true **Hērophilean** from the beginning"]). By naming one of his multi-ingredient anti-diarrhea compounds an "Attalikē" (Galēn *CMLoc* 8.3 [13.162 K.]), Mantias was probably honoring ATTALOS III OF PERGAMON, confirming the king's reputation as an investigator of drugs and poisons, and perhaps signaling personal association. Famed for works on pharmacology, Mantias was, if Galēn is right, one of the first physicians to devise effective multi-ingredient compounds to treat specific ailments, and

Mantias (*Vind. Med. Gr.* 1, f.2[V]) © Österreichische Nationalbibliothek

was noted, moreover, for his excellent dietetics and regimen (Galēn, *CMGen* 2.1 [13.462 K.]).

Mantias, however, did not compose books on pharmacology in the manner of Dioskourides: Mantias' "drug books" focused on single diseases and the compounds employed to treat them, e.g. the "Attalikē" with its 11 ingredients including saffron, spikenard, henbane seeds, aloe-latex, pomegranate flowers, tragacanth-gum (*Astralagus* spp.), "white" pepper, acacia-gum, Pontic rhubarb, and one cooked Syrian pomegranate – all mixed with rose oil and dry wine, then boiled and, when cooled, fashioned into ***trokhiskoi***. As Galēn/Asklēpiadēs Pharmakion says, there is no better astringent purge. The "Attalikē" may have been one of the recipes in Mantias' *Dunameis*. Another tract by Mantias perhaps carried the title *Druggist* or *In the Physician's Office* (*frr.*15–16 von Staden), and it too contained prescriptions for compounds. There are traces of writings on gynecology, including the infamous gynecological ailment known as "uterine suffocation," preserved by Sōranos, *Gyn.* 3.4.29 (*CMG* 4, pp. 109–113; *CUF* v. 3, pp. 30–31; *fr.*11 von Staden), and on afterbirth expulsion, *ibid.* 4.14.5 (=1.71.5) (*CMG* 4, p. 145; *CUF* v. 2, p. 11; *fr.*12 von Staden).

RE 14.1 (1928) 1257, F.E. Kind; von Staden (1989) 515–518.

John Scarborough

Marcellinus (Pharm.) (30 BCE – 80 CE)

Andromakhos, in Galēn *CMLoc* 7.5 (13.90 K.), records his anodyne, based on henbane and poppy-juice, also containing ***amōmon***, anise, myrrh, celery, dried roses, and saffron, ground in water and taken at bedtime. Alexander of Tralleis (2.357 Puschm.) appears to mention this Marcellinus. Perhaps identical to Marcellus (Pharm.).

RE 14.2 (1930) 1489 (#53), F.E. Kind.

PTK

Marcellinus (Med.) (140 – 160 CE?)

Four MSS attest a Greek *On Pulses* by an otherwise unknown Marcellinus, whose *floruit* seems to be some time in the middle of the 2nd c. CE (Schöne 450). In 1895, Olivieri reported on a 15th c. MS (Codex gr. Bononiensis bibl. Univ. 3632) presenting two series of portraits of physicians, accompanied by their names; among those in the second set (f. 213) appears a "Markelēnos," presumed by Schöne to be "Markellinos," the author of this *On Pulses.* Seven "pulse lore" writings are known in the Galēnic corpus, three more (probably spurious) are in Kühn, v.19 (*cf.* pseudo-Galēn, De Pulsibus); the Daremberg-Ruelle edition of Rufus of Ephesos (1879) contains a Greek account believed to be a pseudo-Rufus, and the *Anecdota Græca et Græcolatina* (ed. V. Rose, 2 [1870] 263–266 and 275–280) includes the *De pulsibus* and *Peri sfigmon* by a pseudo-SŌRANOS. Although all these works generally overlap one another, Marcellinus' version is a valuable "history" of pulse-lore, running from HIPPOKRATĒS (13 [463 Schöne]) to the middle of the 2nd c. Featured are, of course, the famous names for pulses coined by HĒROPHILOS and ERASISTRATOS, e.g. *tis ho dorkadizōn sphugmos* (31 [Schöne 468–469]), *tis ho murmēkizōn sphugmos* (32 [Schöne 469]), and *tis ho skōlēkizōn sphugmos* (33 [*ibid.*]).

Ed.: H. Schöne, "Markellinos' Pulslehre. Ein griechisches Anekdoton," in [no ed.] *Festschrift zur 49. Versammlung Deutscher Philologen und Schulmänner in Basel im Jahre 1907* (1907) 448–472.

John Scarborough

Marcellus (Geog.) (*ca* 300 – 400 CE?)

Wrote an *Aithiopika* cited by PROKLOS *In Tim.* 1.177, 1.181, for Atlantis, and a work on Illyria and Dalmatia cited by the RAVENNA COSMOGRAPHY 4.15–16.

BNP 8 (2006) 298 (#3), P.L. Schmidt

PTK

Marcellus (Mech.) (350 – 450 CE?)

Assisted QUIRINUS in writing a *Mēkhanikē*, according to Leōn, *Anth. Gr.* 9.200.

Netz (1997) #132.

GLIM

Marcellus (Pharm.) (50 – 70 CE)

Compounded Nero's *digestif*, according to MARCELLUS OF BORDEAUX 20.84 (*CML* 5, pp. 349–350), and perhaps the wound-cream attributed to Nero, containing **litharge**, myrrh, ***opopanax***, and ***psimuthion***: PAULOS OF AIGINA 7.17.46 (*CMG* 9.2, p. 359). Nero's Marcellus also prescribed, as a *digestif* and febrifuge, a cathartic salt (***ammi***, anise, celery-seed, ginger, ***malabathron***, marjoram, parsley, pepper, **silphium**, thyme, and seeds of elecampane and *nasturcium*, plus sal ammoniac and salt, all finely ground: Marcellus 30.51 [*CML* 5, pp. 532–534]), and a skin treatment of pumice, ***psimuthion***, and rose oil in butter, goat-fat, and beeswax: Paulos 4.11.2 (*CMG* 9.1, p. 331). The Galēnic *Euporista* 2.21 (14.459 K.) cites Marcellus for a spleen remedy, of cardamom in squill-vinegar taken while fasting.

Fabricius (1726) 315.

PTK

Marcellus of Bordeaux, "Empiricus" (375 – 425 CE)

Life: Wrote the treatise *De medicamentis* ("*On medicaments*") after the birth (401) or maybe after the accession (408) of Theodosius II: in the *inscriptio* of his dedicatory epistle he mentions Theodosius I (under whom he was chief of the chancellery, *magister officiorum*) as *Theodosius senior*. This is confirmed by *Codex Theodosianus* (16.5.29; 6.29.8 = *Cod. Iustin.* 12.22.4), whence it appears that he was *magister officiorum* in 394–395 (i.e., under Arcadius, as confirmed by the *Souda* M-203). The botanical nomenclature and the language of his work prove that he lived in Gaul; Bordeaux as his possible place of origin is suggested by the fact that he mentions as his elder fellow-citizens SIBURIUS, EUTROPIUS and IULIUS AUSONIUS (*pr.*2). He was certainly a Christian (see esp. *pr.*3; 21.2; 23.29; 25.13); but he is unlikely to be the "illustrious person in the employ of Theodosius," from Narbonne, met by OROSIUS (7.43.4) at Bethlehem in 415 together with Jerome. It is also uncertain whether he is the Marcellus addressed in two epistles of 399 by Symmachus (9.11, 9.23), and mentioned in a letter of 395 (2.15); this Marcellus was a politician as well and a landowner in Spain.

The title of "Empiricus" given to Marcellus is a modern one, and was suggested by the title that was given to his work by the *editio princeps* by Janus Cornarius (Johann Haynpul), published in Basel in 1536. In fact, Marcellus' work has no connection at all with the **Empiricist** "school," which ended around 200 CE. The title was suggested by the *incipit* of the dedicatory epistle, where Marcellus defines his work as *libellus de empiricis* (as well as by his frequent use of the words *empiricus, expertus, experimentum*, etc., especially in the phrase, recurring in titles of chapters, *remedia diversa physica et rationabilia de experimentis*, with reference to testing and checking the efficacy of the remedies).

Genre of *De medicamentis*: He does not seem to have been a professional physician: in the dedicatory letter to his sons he states (*pr.*3) that he wrote his work to enable them to heal themselves without turning to physicians (but at *pr.*5 he warns that drugs must be prepared carefully and under a physician's supervision). Thus his work enters the genre of the *eupórista* or *parabilia* ("remedies easy to prepare") that went back at least to the lost *Euporista* by APOLLŌNIOS "MUS," and became a great success in late antiquity. Likewise, Marcellus's work takes its place in the Roman tradition of the medicine of the *paterfamilias* that went back to CATO THE CENSOR and was carried on in the Imperial age by PLINY and by GARGILIUS MARTIALIS. The revival of this tradition is probably to be connected with the traditionalist culture typical of the milieu of Symmachus (if Marcellus was actually associated with him). But Marcellus also justifies his choice by appealing to Christian charity: in the dedicatory letter, he says that, thanks to his advice, his sons will be able to heal wayfarers (*pr.*3–4).

Prefatory epistles: Besides the dedicatory letter, the treatise is introduced by seven more letters: (1) by LARGIUS DESIGNATIANUS to his sons, introducing (2) by HIPPOKRATĒS to *Antiochus*, (3) to Maecenas (another translation of epistle 2, perhaps attributable to ANTONIUS MUSA), (4) by *Plinius Secundus* to his friends (it is the preface to the MEDICINA PLINII), (5) by CORNELIUS CELSUS to Iulius Callistus and (6) to Pullius Natalis (the first one is the preface to Scribonius Largus' *Compositiones*; the second one is certainly apocryphal: it is the introduction to a translation of a collection of *Compositiones* in two books), and (7) by VINDICIANUS to the emperor Valentinianus (probably the preface to a collection of pharmaceutical recipes). The choice of these epistles on the one hand follows the sources used by Marcellus, on the other hand the choice has been clearly suggested by their consonance with the author's program, mainly devoted to collecting ready-to-use remedies; but the epistle attributed to Hippokratēs deals with subjects (**humoral** theory, etiology of the diseases, influence of the seasons) that are absent from Marcellus's work.

Sources: The main sources Marcellus demonstrably uses are Scribonius Largus' *Compositiones* (two thirds of which are reproduced by Marcellus) and the *Medicina Plinii*. This latter work is sometimes supplemented with Pliny's *Naturalis historia*; the "two Plinies" (*uterque Plinius*) mentioned among the sources listed in the prefatory epistles probably refer to these two works. The other sources are APULEIUS (that is PSEUDO-APULEIUS, *Herbarius*, which seems in fact to have been used by Marcellus), CELSUS (i.e., evidently, Scribonius Largus, as for the preface to the *Compositiones*; Marcellus does not know Celsus' *De medicina*), and other authors otherwise unknown: APOLLINARIOS, Designatianus (the author of the epistle) and his three fellow-citizens above mentioned (only Ausonius is mentioned later in the work, in reference to a remedy exposed in 25.21), and also "rural and popular remedies checked by experience" (*pr.*2).

Content: The work comprises 36 chapters, in which the remedies (about 2,500) are offered for the body part to be cured, according to the *a capite ad calcem* order ("from head to foot"). As a rule, each chapter deals with all the diseases affecting single body parts or organs (8 eyes; 9 ears; 12 teeth; 14 throat and trachea; 20 stomach; 22 liver; 26 kidneys and bladder etc.); sometimes the material is arranged according to the typology of disease (1 headaches; 2 hemicranias; 3 vertigo; 4 dermatological diseases and head parasites; 5 *alōpekia* and hair problems; 16 coughs and blood expectorations; 28 diseases caused by intestinal parasites; 36 *podagra* and *chiragra*, etc.); some chapters deal with not specifically medical subjects: 7 on hair dye and treatment; 12 on toothpastes. Compared with Scribonius Largus' *Compositiones* and with the *Medicina Plinii*, Marcellus uses in a more rigid way the *a capite ad calcem* pattern: the fact that he does not reproduce certain parts from these sources seems to be due to his difficulty in inserting them in that scheme (this is true for most of Book III of *Medicina Plinii*, and also for diseases treated by Scribonius such as epilepsy or **dropsy**). Sometimes, however, Marcellus omits topics that had been treated by Scribonius but were not significant in his own socio-cultural milieu: e.g. the section on antidotes and the treatment of poisonous snake bites, a topic more important in an African or Mediterranean area. Marcellus mostly limits himself to indicating the trouble or the disease; single symptoms are mentioned in reference to proposed remedies that are specifically intended for those same symptoms; he does not make any suggestions about the causes of diseases or to theoretical or doctrinal problems. Among therapies he proposes both simple, mainly vegetable, remedies, and compound recipes.

Marcellus' treatise is accompanied by a bilingual handbook (Greek and Latin) about the weights and measures used in the recipes. Knowledge of Greek on Marcellus' part (not surprising, if we consider the office he held under Theodosius) is confirmed by the presence of about 40 Greek words (and of many other transliterated terms).

Besides the remedies taken from Scribonius and other medical sources, we find 266 magical remedies (this is one of the features connecting Marcellus to the tradition of Cato the Censor, Pliny, and QUINTUS SERENUS): precepts about gestures to be made (1.54), formulas to be recited while administering medicines, also of the *Ephesia grammata* kind (e.g. 18.30 against *paronychia*: touch a wall and say "*pu pu pu*"), and actual incantations (e.g. 21.2–3).

The treatise is closed by a poem (*carmen de speciebus*) in dactylic hexameters (78 lines): after a brief outline of the history of medicine, extending from mythical physicians Chiron and Machaon to Hippokratēs, Marcellus dwells on the ingredients used in the preparation of their remedies. There are similarities, but no direct relationship, with Quintus Serenus' *Liber medicinalis*.

Transmission. The *De medicamentis* is used in the *Liber medicinae ex animalibus* of SEXTUS PLACITUS PAPYRIENSIS. The three known MSS attest a limited knowledge of it in the Middle Ages. It entered modern culture with the *editio princeps* in 1536.

Ed.: M. Niedermann and E. Liechtenhan, *CML* 5 (1968^2; with German translation); *Concordantiae* ed. S. Sconocchia (1996).

R. Heim, *Incantamenta magica Graeca Latina* = *JCPh* S.19 (1893), 463–576; *RE* 14.2 (1930) 498–503 (#58), F.E. Kind; *KP* 3.993–994 (#14), F. Kudlien; J.F. Matthews, "Gallic supporters of Theodosius," *Latomus* 30 (1971) 1073–1099 at 1083–1087; *PLRE* 1 (1971) 551–552; C. Opsomer and R. Halleux, "La lettre d'Hippocrate à Mécène et la lettre d'Hippocrate à Antiochus," in Mazzini and Fusco (1985) 339–364; *Eidem*, "Marcellus ou le mythe empirique," in Mudry and Pigeaud (1991) 159–178;

A. Önnerfors, "Marcellus, De medicamentis. Latin de science, de superstition, d'humanité," in *Le latin médical*, ed. Sabbah (1991) 397–405; Önnerfors (1993) 319–330; *Lexikon des Mittelalters* 6 (1993) 221–222, K.-D. Fischer; W. Meid, *Heilpflanzen und Heilsprüche: Zeugnisse gallischer Sprache bei Marcellus von Bordeaux* (1996); *AML* 591–592, K.-D. Fischer; *BNP* 8 (2006) 300–301 (#8), A. Touwaide.

Fabio Stok

Marcellus of Sidē (*ca* 140 – 160 CE)

Widely reputed and imperially recognized physician and poet who lived under Antoninus Pius. He wrote an immense compilation of 40 books in hexameters, *On Medical Matters* (*Iatrika*) or, more poetically (*AP* 7.158) *Daughters of Chiron* (*Chironides*). A preserved fragment of *On Werewolves* (*Peri lukanthropias*; see *Souda* M-205), perhaps a part of the *Iatrika*, presents this disease as a form of melancholia, gives clinical symptoms, and prescribes bloodletting, baths, and an antidote used against viper bites (Aëtios of Amida 6.11 = *CMG* 8.2, pp. 151–152, citing Rufus and Arkhigenēs). Contemporary with Galēn, with whom he was probably personally in contact, Marcellus is mentioned by major later physicians, including Aëtios and Paulos of Aigina who borrowed medical treatments from him. He supplied numerous medico-magical recipes similar to those given by Kuranides. He also wrote a poem *On Fishes*, preserved in a lengthy fragment (101 verses) which offers in Homeric diction a long catalogue of 91 so-called fishes (*ikhthus*) including shellfish and dolphins.

RE 14.2 (1930) 1496–1498 (#56), W. Kroll; M. Wellmann, *Marcellus von Side als Arzt* (1934); *OCD3* 922, A.J.S. Spawforth.

Arnaud Zucker

Marcianus (of Africa?) (10 BCE – 15 CE)

Prepared an antidote for Augustus, which Scribonius Largus (177) also used, containing over 40 ingredients, including such imports as "Ethiopian" cumin, African **silphium** and ***ammōniakon*** incense, and Indian cinnamon, ***kostos***, Celtic and Indian nards, and pepper, plus fresh duck blood. Oreibasios, *Ecl. Med.* 74.9 (*CMG* 6.2.2, p. 243) gives Marcianus' ***akopon***, employing African ***euphorbia*** (*cf.* Iouba) and Indian **galbanum**, cardamom, and pepper. Aëtios of Amida cites probably the same man six times for relatively simple recipes, once calling him African (11.11, p. 608 Cornarius: recipe for kidney and bladder stones): a collyrium including rue, fennel, and coriander (7.110, *CMG* 8.2, p. 387), an emetic for ***sunankhē*** with ***aphronitron*** and bull gall, etc. (8.50, p. 485), two compresses for intestinal disturbances, both involving rue and fenugreek (9.27, Zervos 1911: 331), and a potion for those who cannot keep food or water down, based on the bark of the Libyan *lōtos* tree (9.42, p. 389 Zervos = 9.48, pp. 550–551 Cornarius, with further recipes; on the tree, *cf.* Theophrastos, *HP* 4.3.1–2); *cf.* 12 (p. 28 Kostomiris). The prevalence of African ingredients (***ammōniakon***, cumin, ***euphorbia***, **silphium**, and *lōtos*) accords with an African origin. Marcianus seems an imperial-era name (often corrupted to *martianus* in Latin minuscules).

Fabricius (1726) 320, 322.

PTK

Marcianus of Hērakleia Pontikē (*ca* 300 – 430 CE)

Geographer who tells us he wrote an *epitomē* of Artemidōros of Ephesos' *Geographia* (*GGM* 1.574–576), and of Menippos of Pergamon's ***Periplous*** *tēs entos thalassēs* (the

Mediterranean) in three books (*GGM* 1.563–573), of which only fragments are extant. His ***Periplous*** *tēs exō thalassēs*, however, is preserved almost completely (*GGM* 1.515–562). After a *prooimion* with some general deliberations about the structure of the world and references to Ptolemy and Prōtagoras (Geog.), his primary sources, a first book describes the world from the Gulf of Aqaba (*Arabios kolpos*) to the Indian Ocean, from the Persian Gulf to the "Gulf of the Chinese" (*kolpos tōn Sinōn*). A second book describes the coasts of the Atlantic Ocean from Spain to Britain.

RE S.6 (1935) 271–281, F. Gisinger; *RE* S.10 (1965) 772–789, E. Polaschek; *PLRE* 1 (1971) 555; *HLB* 1.528; *ODB* 1302, A. Kazhdan.

Andreas Kuelzer

Marciōn ⇒ Markiōn

Marcomir (500 – 600 CE)

Wrote in Gothic a geography of Europe, covering Denmark to Spain, giving data about tribes unknown to earlier geographers, and cited extensively by the Ravenna Cosmography, Book 4. See also Athanarid and Heldebald.

Staab (1976); *DPA* 4 (2005) 268–269, R. Goulet.

PTK

Maria (100 BCE – 250 CE?)

Jewish, among the earliest alchemists in Hellenistic Egypt, highly regarded by later alchemists for descriptions of furnaces and other apparatus, many of which are thought to be her own inventions, given in her *Descriptions of Furnaces*, first mentioned by Zōsimos of Panōpolis (*CAAG* 2.240; see Festugière 1950: 365) and perhaps identical with *On Furnaces and Apparatus* (Mertens 1995, §1.2). Presumably in this work Maria gave her instructions, often quoted by later alchemical authors, for making and using various chemical equipment including stills, the *kērotakis* reflux device, furnaces and baths for slow, constant heating (*CAAG* 2.224–227). A hot water bath in culinary use today, the bain-marie, bears testament to her. Zōsimos also attributes to her the *Procedures for the Making of a Little Image* (*CAAG* 2.157). None of her works survives in the original Greek, but a few short and possibly apocryphal treatises and fragments exist in Arabic (Ullmann 1972: 181–183) one of which, *The Crown and the Nature of Creation*, was thought by its modern translator (Holmyard 1927: 162) to be a genuine translation from the Greek, if not an authentic work of Maria.

E.J. Holmyard, "An Alchemical Tract Ascribed to Mary the Copt," *Archeion* 8 (1927) 161–167; R. Patai, "Maria," *Ambix* 29 (1982) 177–197.

Bink Hallum

Marianus (490 – 520 CE)

Erudite poet, perhaps an epigrammatist (if identifiable with Marianus Skholastikos, *Anth. Graec.*, 9.668–669, etc.), of Roman patrician origin, lived under Anastasios (491–518). He metrically paraphrased numerous Alexandrine poems, including epics (Apollōnios of Rhodes, Kallimakhos, and Theokritos: *Souda* M-194) and transposed various didactic, dactylic poems such as ARATOS' *Phainomena* (in 1,140 verses) and NIKANDROS' *Thēriaka* (in 1,370 verses) into iambic meters, all completely lost.

RE 14.1 (1928) 1750, J. Geffcken; *GGL* §973; *BNP* 8 (2006) 353 (#1), G. Damschen.

Arnaud Zucker

Marinos (Med.) (70 – 120 CE)

GALĒN generously preserves the memory of Marinos (the teacher of QUINTUS), who after the "ancients" (HIPPOKRATĒS and HĒROPHILOS), "in the time of my grandparents," revived anatomical study, and gave his whole life to its study, based on dissections of apes and other animals. His *Anatomy* comprised 20 books, of which Books 1–2 covered the **homoiomerous** parts, 3–4 the tubes and vessels, 5–6 the bones, 7–10 the muscles, 11–15 internal organs, 16–19 the head, nerves, and ***hēgemōn*** (Galēn's *On My Own Books* 3 is damaged, leaving the contents of Book 20 unknown). He taught that the glands have two uses, to stabilize vessels at junctions, and to secrete liquids to moisten and soften parts (*On Seed* 2.6.14–21, *CMG* 5.3.1, pp. 200–202). He wrote commentaries on the HIPPOKRATIC CORPUS, APHORISMS 7, and EPIDEMICS 2 and 7, from which Galēn cites. ANDROMAKHOS, in Galēn *CMLoc* 7.2 (13.25 K.), cites his ***artēriakē***, composed of saffron, gum, and tragacanth, boiled in honey.

Marquardt, Müller and Helmreich 2 (1891) 104–108; Grmek and Gourevitch (1994) 1493–1503; Manetti and Roselli (1994) 1580–1581; Ihm (2002) #170–172; *BNP* 8 (2006) 357 (#I.2), V. Nutton.

PTK

Marinos of Neapolis (Palestine) (460 – 495 CE)

Studied Neo-**Platonic** philosophy under PROKLOS who dedicated to him an essay on a theme in PLATO's *Republic*, the myth of Er (Proklos, *In Remp.* 2, p. 96.2 K.), and subsequently became head of the philosophical school at Athens. He is credited with the title *Life of Proklos*, a biography containing a discussion of the virtues. He wrote an introduction and commentary on EUCLID's *Data*, starting with definitions, following ARISTOTLE's model in discussing scientific material, on Plato's *Philebus* and *Parmenides*, and on Aristotle (*Prior Analytics, Posterior Analytics, De Anima*).

Marinos was considerably influenced by Aristotelian ideas. He emphasized the need for definitions and analyses of terms. When asking what the data are, he surveyed the relevant mathematical material in APOLLŌNIOS, DIODŌROS OF ALEXANDRIA, and PTOLEMY, examined complex definitions and disputed the explanation of PAPPOS. He tried to attach philosophy to mathematics, with an emphasis on exactness through definitions and consistent use of terms (PSEUDO-DAVID = Elias, *Prolegomena: CAG* 18.1 [1900] 28.9–29.5). In psychology he connected Neo-**Platonic** theories with Aristotelian ones. The distinction between six grades of virtue has precedents in PLŌTINOS and PORPHURIOS. Those possessing lower grade virtues do not necessarily possess higher virtues, whereas those having higher virtues retain the lower ones as well. The lowest grade includes good birth and education, followed by the virtues of character, civic virtues, purificatory virtues – cleansing the soul from bodily influences – theoretical and, in the end, theurgical virtues, the importance of which is due to IAMBLIKHOS' influence.

Ed.: J. Fr. Boissonade, *Marini Vita Procli* (1814); H. Menge, *Euclidis Data cum commentario Marini et scholiis antiquis* (1896); H.D. Saffrey and A.Ph. Segonds, *Marinus: Proclus ou Sur la bonheur* (*CUF* 2001).

RE 14.2 (1930) 1759–1767, O. Schissel von Fleschenberg; *DPA* 4 (2005) 282–284, H.D. Saffrey.

Peter Lautner

Marinos of Tyre (100 CE)

Greek geographer known only through hostile criticism in PTOLEMY's *Geography*, author of a *Correction (diorthōsis) of the World Map*. On the basis of astronomical observation and the duration of land and sea journeys, Marinos calculated coordinates of regions and sites on the globe and attempted to modify existing maps. He employed records of travelers and merchants, Greek and Roman alike, converting voyage duration from days into stades. Living in Tyre, a busy Phoenician port, Marinos could meet people able to supply such information. He adopted a rectilinear projection of the world and incorporated his unscientific measurements into his text. Ptolemy thought this was difficult to work from without a map at hand but nevertheless borrowed some features. Ptolemy says that Marinos never drew a map to illustrate his claims, although Arabic geographers mention maps attributed to him. Marinos dealt with two major cartographic problems confronting mapmakers: (1) The size and position of the inhabited world: according to him the ***oikoumenē*** occupied more than a quarter of the terrestrial globe, lying mostly in the northern hemisphere but drawn in both hemispheres. Ptolemy contested the width and length of Marinos' map. (2) Map projections: regarding the problem of representing a portion of the globe on a plane, Marinos, like ERATOSTHENĒS and STRABŌN, adopted a rectangular projection in which parallels and meridians were drawn as straight parallel lines, in Marinos' version at regular distances from each other. Further problems arise from uncritical copying of geographical detail from written commentaries. Ptolemy thus rejected Marinos' work as a cartographer and considered its information incoherent and impractical. However, Marinos' importance in the history of cartography still lies in his critical approach to existing maps.

Dilke (1985) 72–86; J.B. Harley and D. Woodward, *A History of Cartography* 1 (1987) 178–180; R. Wieber, "Marinos von Tyros in der arabischen Überlieferung," in M. Weinmann-Walser, ed., *Historische Interpretationen* (1995) 161–190; *NDSB* 5.27, A. Jones.

Daniela Dueck

Marius Victorinus (*ca* 340 – 370 CE)

Grammarian, philosopher, rhetorician, and theologian, born *ca* 300 CE in Africa. Moving to Rome, he taught rhetoric under Constantius (337–361). His works divide into grammar, rhetoric, philosophy, and theology, which was his focus after conversion to Christianity later in life (AUGUSTINUS, *Conf.* 8.2.1.–8.3.5). His philosophical works include a translation with commentary on ARISTOTLE's *Categories* and *De interpretatione*, a commentary on CICERO's *Topica* and *De inuentione*, and a translation of PORPHURIOS' *Isagoge*, all lost. Extant is Victorinus' *Ars Grammatica*, his anti-Arian *Ad candidum Arianum, De generatione uerbi diuini ad candidum*, three hymns *De trinitate*, and commentaries on Paul's letters to Ephesians, Galatians, and Philippians. Influenced by Porphurios' version of **Platonism**, Victorinus inherited his scheme of three divine ***hupostaseis***: the One, the ultimate source of Being; Intellect which is life; and soul, the source of thinking. Victorinus seems to have identified the Christian triad of Father, Son, and Holy Spirit with the Porphurian triad.

Ed.: *PL* 8.993–1310; I. Mariotti, *Ars Grammatica* (1967); P. Henry-P. Hadot, *Opera Theologica, CSEL* 83.1 (1971); A. Locher, *Marii Victorini Commentarii in epistolas Pauli ad Galatas ad Philippenses ad Ephesios* (1972); F. Gori, *Opera Exegetica, CSEL* 83.2 (1986).

RE 14.2 (1930) 1840–1848, P. Wessner; P. Hadot, *Porphyre et Victorinus* 2 vv. (1968); *idem, Marius Victorinus. Recherches sur sa vie et ses oeuvres* (1971); *OCD3* 1597–1598, S. Hornblower; *BNP* 8 (2006) 371–372 (#II.21), Chr. Markschies.

George Karamanolis

Markianos (before 11th c.)

The 11th c. MS of Florence, Biblioteca Medicea Laurenziana 75.3, probably of Italian origin (Calabria-Campania), contains several medical texts of a practical nature, among which is Markianos' compound *Medicine to Relax Nerves*. Markianos is identified as *Rhakendutes* (wearer of rags), an adjective seemingly referring to monastic status (an homonymous commissioner, who *owned* the late 13th c. codex of Venice, Biblioteca Nazionale Marciana, *graecus* 294 [coll. 288], is there qualified as a physician by the copyist Theophilos *Rhakendutes*: but this later Markianos must be distinct). Markianos' recipe is added to the last folio of the manuscript, probably by one of the owners and also users of the manuscript: this might suggest a south-Italian provenance for Markianos.

Diels 2 (1907) 61; E. Trapp, ed., *Prosopographische Lexicon der Palaiologenzeit*, fasc. 7 (1985) 16985; Ieraci Bio (1989) 169–170, 190, 235, 237, 239.

Alain Touwaide

Markiōn of Smurna (30 BCE – 77 CE)

Wrote on the virtues of simples, cited after ANDREAS and before AISKHINĒS as a foreign authority on drugs obtained from animals (PLINY 1.*ind*.28). Pliny reports his observation that sea scolopendrae burst if spat upon (28.38: see also OFELLIUS and SALPĒ). Unattested before the 1st c. CE (*LGPN* 1.298: *cf.* Markios, 2nd–3rd cc. CE: *LGPN* 2.298, 3A.288–289, 3B.270, 4.222), Markiōn might be a Romano-Greek name postdating the battle of Actium.

Fabricius (1726) 302.

GLIM

Marpēssos (300 BCE – 500 CE)

Wrote a work on Kolkhis, cited by the RAVENNA COSMOGRAPHY, 4.4, as *MARPESIVS* (*cf.* PENTHESILEUS). The name when personal is otherwise attested solely as feminine: Pausanias 4.2.7, 5.18.2, 8.47.2; OROSIUS 1.15.4–5; Pape-Benseler s.v. *Marpēss-*.

J. Schnetz, *SBAW* (1942), # 6, pp. 58–59, 61–62.

PTK

Marsinus of Thrakē (500 – 565 CE)

According to ALEXANDER OF TRALLEIS (1.565 Puschm.), giving a series of recipes he himself collected, Marsinus prescribed for epilepsy seven doses of the ashes of a rag bloodied by an executed man, taken in wine. For the rare Latin name *cf.* Schulze (1904/1966) 189; or perhaps emend to Arsinoë (see ARSENIOS) or MARINOS.

Fabricius (1726) 322.

PTK

MARTIALIS ⇒ GARGILIUS

Martialius/Martianus (150 – 190 CE)

Erasistratean who wrote on anatomy, and whom GALĒN had in mind as the object of his attack on the **Erasistrateans** at Rome; he attended Eudēmos, whom Galēn claims to have cured: *Progn.* 3.6–7, 4.1–2 (*CMG* 5.8.1, pp. 84, 88), *On My Own Books* 1 (2.94–95 MMH). The name Martialius is rare before *ca* 200 CE, and *MAPTIAΛIO-* differs very slightly from *MAPTIANO-*.

RE 14.2 (1930) 2003 (Martianus #1), W. Kroll; Korpela (1987) 198 #240; *DPA* 4 (2005) 286–288, V. Boudon-Millot.

PTK

MARTIANUS CAPELLA ⇒ CAPELLA

Massiliot Periplous (520 – 350 BCE?)

Scholars have identified a ***periplous*** written in Massalia that described the Iberian coast, Atlantic and Mediterranean, and which is preserved in the verse paraphrase of AUIENUS, *OM*. Auienus claims to use an ancient document (*OM* 9, 17), and over 450 lines of his poem, from line 85, are credited to the Massiliot ***Periplous***. Ascribed to other sources are lines 115–129, 266–283 on the Carthaginians, 336–389, 406–413 explicitly from DAMASTĒS, EUKTEMŌN, and HIMILKŌN, and 323–334, 390–405, 645–679, 689–698 explicitly from DIONUSIOS OF ALEXANDRIA (331) and PHILEAS (695). In view of Auienus' citation of BAKŌRIS and THUCYDIDĒS in lines 42–50, some scholars date the Massaliot ***Periplous*** to *ca* 400–350 BCE. See also EUTHUMENĒS.

L. Antonelli, *Il periplo Nascosto* (1998).

PTK

C. Matius Caluena (45 BCE – 5 CE)

Equestrian friend of CAESAR, CICERO, and AUGUSTUS who introduced the cultivation of dwarf shrubs at Rome (PLINY, 12.13), and developed a new variety of apple, the *malum Matianum*, grown near Aquileia, presumably on his estates there (PLINY 15.49; Ath., *Deipn.* 3 [82c]). COLUMELLA (12.4.2, 46.1) credits him with three books ("The Cook," "The Fish Dealer," and "The Pickler"), wherein he discussed the storage of wine and agricultural produce.

RE 14.2 (1930) 2210 (#2), A. Stein; *KP* 3.1080 (#2), H. Gundel; *BNP* 8 (2006) 479 (#2), Tho. Frigo; *OCD3* 937, R.J. Seager.

Philip Thibodeau

Matriketas of Mēthumna (before 320 BCE)

Astronomer who observed solstices from Mt. Lepetumnos on Lesbos (PSEUDO-THEOPHRASTOS, *De signis* 4).

(*)

Henry Mendell

Maximianus (*ca* 300 – 565 CE)

ALEXANDER OF TRALLEIS (2.57 Puschm.) records his collyrium, composed of one part each of the two collyria of HERMOLAOS, and two parts of the "swan" collyrium. Surely distinct from the poet, *PLRE* 2 (1980) 739–740.

Fabricius (1726) 327.

PTK

Maximinus (*ca* 350 – 550 CE?)

Wrote in Latin a geographical work that treated at least Illyria and Dalmatia, and was followed by the RAVENNA COSMOGRAPHY 4.15–16. *Cf.* MARCELLUS and PROBINUS.

J. Schnetz, *SBAW* (1942), # 6, pp. 80–81.

PTK

Maximus (300 – 400 CE?)

Wrote an astrological poem in Greek hexameters entitled *Peri Katarkhōn*, treating ***katarkhic*** astrology but surviving incomplete (a later prose version of the entire poem is extant). The *Souda* (M-174), with what authority one cannot say, identifies the author of *Peri Katarkhōn* as the philosopher Maximus who taught the emperor Julian, and says that he was from either Ēpeiros or Buzantion.

Ed.: A. Ludwich, *Maximi et Ammonis carminum de actionum auspiciis reliquiae* (1877).
BNP 8 (2006) 517 (#2), W. Hübner.

Alexander Jones

Mēdeios (320 – 270 BCE)

Maternal uncle of ERASISTRATOS and student of KHRUSIPPOS OF KNIDOS (II), who like his teacher rejected phlebotomy (GALĒN, *On Venesection, Against Erasistratos* 2 [11.196–197 K. = p. 43 Brain]; *Treatment by Venesection* 2 [11.252–253 K. = p. 68 Brain]). He appears to have been a grandson of ARISTOTLE and to have attended THEOPHRASTOS in his last illness: SEXTUS EMPIRICUS *Math.* 1.258 (with Kroll 1932), DIOGENĒS LAËRTIOS 5.53, 72. Galēn refers to him in a list of early anatomists: *In Hipp. Nat. Hom.* (*CMG* 5.9.1, pp. 69–70). CELSUS 5.18.11 preserves the recipe of his ointment of alum, copper-flakes, roasted lead, and ***panax*** in beeswax, and PLINY (who cites him as an authority, 1.*ind.*20–27) his prescription of radishes for blood-spitting and to promote lactation, 20.27.

RE 15.1 (1931) 106 (#5), 15.2 (1932) 1482–1483 (#26), W. Kroll; Brain (1986).

PTK

Medicina Plinii (200 – 240 CE)

By the first decades of the 3rd c. CE, an unknown student of PLINY had extracted a compilation of the "medical sections" of the *Natural History*, especially Books 20–32, on pharmaceuticals derived from plants and animals. Although philologists debate when *auctor ignotus* assembled the collection of extracts, the borrowings in SERENUS' *Liber medicinalis* seem decisive (Önnerfors [1963] esp. 62–83), although firm proof remains elusive. In his edition, Önnerfors' earliest MS is the Codex Sangallensis 752 (9th c.), and his proposed stemma (p. XXX) suggests a double origin in the 4th c. and what are labeled *alii fontes uarii generis*. The text itself is a tangle of quotations, snippets, and abridgements mostly from Pliny, but discerned are subsumed bits from earlier authors as varied as CELSUS, SCRIBONIUS LARGUS, and a Latin version of DIOSKOURIDĒS (pp. XXXI–XXXII), sometimes shadowed in the Latin texts under the name of Sextus Placitus and others. MARCELLUS OF BORDEAUX in the preface of his *De medicamentis liber* (*CML* 5.1, p. 2) says he has lifted things from "both" Plinys (*cui rei operam uterque Plinius*), indicating that the *Medicina Plinii* was in common circulation by 400 CE.

Ed.: A. Önnerfors, *Plinii Secundi Iunioris qui feruntur De medicina libri tres* (1964) = *CML* 3.

A. Köhler, "Handscriften römischer Mediciner. 1. Pseudoplinii medicina," *Hermes* 18 (1883) 382–392; *GRL* §523.3; R. Laux, "Ars medicinae. Ein frühmittelalterliches Kompendium der Medizin," *Kyklos* 3 (1930) 417–434; *RE* 15.1 (1931) 81–85, E. Steier; A. Önnerfors, *In medicinam Plinii Studia Philologica* (1963); *Idem*, "Die mittelalterlichen Fassungen der Medicina Plinii," *Berliner Medizin* 16 (1965) 652–655; Ch.G. Nauert, "Caius Plinius Secundus, Spurious Work: *Medicina Plinii*," *CTC* 4 (1980) 422; Önnerfors (1993) 277–280; Langslow (2000) 64.

John Scarborough

Mēdios (Stoic) (*ca* 240 – 270 CE?)

Older contemporary of CASSIUS LONGINUS, compiled the writings of earlier philosophers, making no original contributions (PORPHURIOS, *Vit. Plot.* 20). Longinus defended the soul's unity against Mēdios' traditional **Stoic** division of the soul into eight parts (PROKLOS *in Plat. Rep.* 1.233.29–234.30).

H. Dorrie, *Porphyrios' "Symmikta Zetemata": Ihre Stellung in System und Geschichte des Neuplatonismus nebst einem Kommentar zu den Fragmenten* (1959) 104–107; *BNP* 8 (2006) 588 (#3), B. Inwood.

GLIM

Megasthenēs (*ca* 320 – 290 BCE)

Born *ca* 350 in Asia Minor, ambassador of Seleukos I Nikatōr (or Siburtios, satrap of Arakhosia) near the court of Chandragupta Maurya in eastern India. In his *Indiká* (four books: *FGrHist* 715), he described geography, fauna, flora (Book 1), customs, towns and administration (Book 2), society and philosophy (Book 3), archaeology, myth and history of India (Book 4). Megasthenēs supplemented personal observations with data from earlier Greek authors and information from Indian scholars whom he met. Modern scholars often debate Megasthenēs' credibility, although some consider his "the most reliable account of India produced account in antiquity." Of especial value are the long descriptions of techniques for capturing, training, and utilizing elephants in hunting and warfare. Not very reliable, however, is his sociological treatment of the caste system and Indian society, since Megasthenēs seems more interested in presenting India as a social model than realistic description.

Ed.: *FGrHist* 715.

J. Timmer, *Megasthenes en de Indische maatschappij* (1930); *RE* 15.1 (1931) 232–233, O. Stein; T.S. Brown, "The Reliability of Megasthenes," *AJPh* 76 (1955) 18–33; A. Zambrini, "Gli Indiká di Megastene," *ASNP* 12 (1982) 71–149; A.B. Bosworth, "The Historical Setting of Megasthenes' Indiké," *CPh* 91 (1996) 113–127; *DPA* 4 (2005) 367–380, J.M. Camacho Rojo and P.P. Fuentes González.

Cristiano Dognini

Megēs of Sidōn (10 BCE – 30 CE)

Surgeon from Sidōn (*kheirourgos*: GALĒN, *CMLoc* 5.3 [13.845 K.]; *ho Sidōnios*: Galēn, *MM* 6.6 [10.454 K.]), THEMISŌN's student (scholion to OREIBASIOS, *Coll.* 44.21, title = *CMG* 6.2.1, p. 142). He emigrated to Rome where he attained fame and presumably fortune from his skilful surgical procedures for bladder stones (CELSUS 7.26.2N) and fistulas, as well as carefully compounded *collyria* (here, "surgical tents": glutinous pastes rolled into rods to dilate fistulas), for treating fistulas – very common and troublesome (abnormally-open tubular passages between epithelial surfaces; modern diagnostics names *ca* 100 varieties). Megēs' *collyrium* for hardened fistulas was simple and rapidly effective, consisting of **verdigris**, ***ammōniakon*** incense, and vinegar (Celsus 5.28.12K). KRITŌN in Galēn (*CMLoc* 5.2 [12.845 K.]) records another of Megēs' famous plasters (also for dissolving calluses), a rather harsh one, compounded from ***psimuthion***, beeswax, **terebinth**-resin, **litharge**, olive oil, and water. Celsus 7.*pr*.3 rates Megēs as "most learned" of surgeons who have practiced in Rome, and Oreibasios (*ibid.*, pp. 142–144) cites with enormous respect Megēs' *On Fistulas*. (Even at the end of the 19th c., Gurlt mirrored a professional esteem afforded to Megēs' scrupulous surgical techniques.) Unusually Megēs appears to have studied human anatomy: *On Fistulas* contains meticulous descriptions, and Celsus, recording detailed bladder anatomy, credits Megēs with inventing a "straight blade, a knife bordered widely on its upper part but semicircular below" (7.26.2N), for use in operations to remove rough bladder stones.

E. Gurlt, *Geschichte der Chirurgie und ihrer Ausübung* (1898) 1.332–333 ("Meges"); *RE* 15.1 (1931) 328, H. Raeder; J.D. Grainger, *Hellenistic Phoenicia* (1991) 185.

John Scarborough

Megethiōn (of Alexandria?) (*ca* 285 – 320 CE)

Dedicatee of the fifth book of PAPPOS' *Mathematical Collection* (5.*pr*.).

Netz (1997) #22.

GLIM

Megethios of Alexandria (*ca* 530 – 540 CE)

SIMPLICIUS, *In de Caelo* 3.3 (*CAG* 7 [1894] 602), on the potential presence of elements in substances cites both THEOPHRASTOS, *fr.*281 FHSG (fire excreted from the eyes, *cf.* *De Sensu* 26), and his own contemporary the doctor Megethios, who showed that fire was excreted by the flesh of a man with sciatica. For the rare name, *cf.* only *LGPN* 4.226 (2nd/3rd c. CE).

Fabricius (1726) 328.

PTK

MEKKIOS ⇒ MAECIUS

MELA ⇒ POMPONIUS MELA

Melampous (300 – 200 BCE?)

Physiognomist and astrologer; three brief seemingly complete treatises survive. (I) *Divination by Birthmarks* (*peri elaiōn tou sōmatos*) details signs indicated by birthmarks – probably moles (*elaiōn*) – on various parts of the human body. Reading like a horoscope in predicting the course of a human life, the essay proceeds from head to toe. Melampous' signs reflect Hellenistic ***melothesia***. Moles on the nose, for example, suggest sexual insatiability (3): the nose is governed by Aphrodite (Vettius Valens 1.1). Birthmarks seem to enhance the function of the body part to which they bring attention. Melampous' predictions for moles on the belly (gluttonous behavior: 15), the spleen (sickliness: 16) and genitalia (parents of same-sex children) follow anatomical function. (II) *On Bodily Tremors* (*Peri palmōn mantikēs*), addressed to Ptolemaïs, likewise descends from head to toe, right to left, detailing the signs indicated by trembling of quite specific body parts often connected to deities, e.g., each ear, the tip of the nose (right side and left) and each finger and toe (and all parts between). For example, the third finger of the right hand, governed by Kronos, indicates glory for some, subjugation for slaves, and illness for virgins; a palpitation of the left knee presages great unhappiness for all. Melampous offers predictions especially for slaves, virgins, and widows, and refers to named and generalized sources (Phēmōn, Antiphōn, and the "Egyptians": 461 Franz). (III) In *Prognostication by the Moon* (*peri tōn tēs selēnēs prognōseōn*), the Moon in various signs together with weather (thunder, clouds, wind) offers signs for political and agricultural ventures: e.g., a blood-red Moon in Ram portends a fruitful grain crop; with Moon in Taurus, winds portend flock destruction and noises from the sky indicate civil war; if an earthquake occurs with Moon in Gemini, war is evident; with Moon in Cancer, thunder presages crop-destruction.

Ed.: J.G.F. Franz, *Scriptores Physiognomoniae ueteres* (1780) 451–508; *CCAG* 4 (1903) 110–113.
RE 15.1 (1931) 404–405 (#9), W Kroll; H. Diels, *Beiträge zur Zuckungsliteratur des Okzidents und Orients* 2 vv. (1908–1909; repr. 1970); *OCD3* 952, anonymous.

GLIM

Melampous of Sarnaka (500 – 25 BCE)

Listed among minor artisans and artists; compiled rules of architectural symmetry (Vitruuius 7.*pr*.14).

RE 15.1 (1931) 405 (#10), G. Lippold.

GLIM

Meleagros (*ca* 350 BCE – *ca* 200 CE)

Diogenēs Laërtios 2.92 cites Meleagros' *On Philosophical Opinions*, Book 2, on Aristippos of Kurēnē (Sōcratēs' student); Meleagros doubtless included other philosophers. Perhaps the same as the Cynic philosopher and poet Meleagros of Gadara (*ca* 100 BCE).

OCD3 953 (the Cynic), A.D.E. Cameron.

PTK

Melētos (20 BCE – 25 CE)

Pliny lists him after Dēmokritos and Apollōnios "Mus," and before Artemōn and Anthaios Sextilius (1.*ind*.28 and 28.7–8) as giving medicines *from* the human body, and cites him as claiming human gall cures cataracts. Andromakhos, in Galēn *CMLoc* 6.6 (12.946–947 K.), indicating that he wrote a multi-volume work on pharmacy, cites two gargles, one with saffron, ***kostos***, roses, and sumac, the other with rush-flower, alum, cassia, saffron, Illyrian iris, Indian nard, and myrrh, both in honey.

RE 15.1 (1931) 504 (#4), R. Hanslik.

PTK

Melior (d. 144 CE)

Calculator who wrote notebooks (*commentarii*) of everything he knew. Melior, possibly a home-bred slave (Russell 214), died at age 13 and was honored by his grieving teacher (Sex. Aufustius Agreus) with an epitaph (*ILS* 7755: Ostia) proclaiming his recall and knowledge (*scientia*; he apparently had mastered the names of all things from antiquity to the day of his death), which would "fill a volume rather than an inscription." Russell (214) hesitates over authenticity, probably in view of the unusual and precise death-date.

D.A. Russell, "Arts and Sciences in Ancient Education," *G&R* 36 (1989) 210–225.

GLIM

Melissos of Samos (*ca* 480 – 430 BCE)

Born *ca* 500, admiral who defeated the Athenian fleet at Samos in 441 BCE. The argument of his lost work can be reconstructed from extensive fragments, the brief account of ARISTOTLE, and a more detailed paraphrase by the anonymous ON MELISSOS, XENOPHANĒS AND GORGIAS. Melissos revives the arguments of PARMENIDĒS in prose, with some change. By insisting that the only Being is infinite, argues Aristotle (*Physics* 1.2), he gave Parmenidēs a materialistic interpretation. He is especially interested in the theory of motion, void (a necessary condition of motion according to Melissos) and mixture, excluding the possibility of all three. Sense perception and "common mind" are delusory. A late source adduces a fragment claiming incorporeality for his only being, but the authenticity and the context of the statement is highly problematic.

D.N. Sedley in Long (1999) 390–441; *DPA* 4 (2005) 391–393, R. Goulet.

István M. Bugár

On Melissos, Xenophanēs, and Gorgias ("MXG") (300 – 50 BCE)

Transmitted by MSS under the name of ARISTOTLE or THEOPHRASTOS, consisting of three treatises dedicated to a pre-Socratic philosopher each. Each treatise first provides a reconstruction of the main tenets and arguments of the philosopher in question, followed by a detailed analysis of the validity of their reasoning. The three philosophers treated are related to the Eleatic school, the central figure of which was PARMENIDĒS.

The textual tradition is very poor, and the text badly needs a critical edition, since the century-old editions of Apelt and Diels are over-emended, while that of Cassin follows the most puzzling readings. There is already confusion about the titles in the MSS. Some supply the following title: *On Xenophanēs, Zēnōn and Gorgias*, others reverse the order of the first two names. However, the first treatise begins with an accurate account of MELISSOS' philosophy, and the second can only be related to XENOPHANĒS. The transmitted title and some cross-references in the work – which support a single author-editor for the three treatises – make it likely that a treatise on ZĒNŌ OF ELEA has been lost.

Like the account of Melissos, the GORGIAS treatise appears more reliable on doctrine and terminology than later sources. Since, from the Hellenistic catalogue of the works of Aristotle preserved by DIOGENĒS LAËRTIOS, we know that he had composed a monograph on Melissos and Gorgias, it seems reasonable that those two treatises of the *MXG* are based on the lost Aristotelian prototypes, with some shifts of interest. The treatise on Xenophanēs largely departs from the evidence on the author known from other witnesses, such as Aristotle. The *MXG* attributes to Xenophanēs the argument *e gradu entium* for the existence of God (known from *fr.*16 of Aristotle's *On philosophy*). Further, it misinterprets Theophrastos' brief account on Xenophanēs presented in his *Doctrines of the physicists*, claiming that Xenophanēs did not say whether God was limited or unlimited, moving or unmoved, and elaborates it into a "negative theology": God is neither limited, nor unlimited (*cf.* Aristotle, *Physics* 8.10), neither moved, nor unmoved. In this, as in other parts of the work, the influence of PLATO's *Parmenides* is considerable.

Like that dialogue, the *MXG* also appears to be a dialectical exercise, but in the Aristotelian fashion. In this respect, the author may also be influenced by the Megarians, or the skeptical **Academy** (Diels, 10–12). Other suggestions are less likely (Neo-Pyrrhonist influence: Mansfeld; sophistic movement: Cassin). However, typical Skeptical vocabulary is totally absent from the treatises, especially the *On Gorgias*, where the author has *agnōston* ("unknowable") for SEXTUS' *akatalēpton* ("incomprehensible"), which becomes standard in epistemology from Arkesilaos onwards. At one point (977a4–10, *cf.* 975a6–7), the author calls a physical theory mentioned but finally rejected by Aristotle (*GC* 1.10 [327b30–328a18]) "probable" – compare the probabilism of the Skeptical **Academy**.

Nevertheless, the author is strikingly ignorant of some crucial Aristotelian passages. Metaphysical issues especially seem alien to him: e.g. he cannot conceive that the divine is without magnitude, although has also exegetical and eristic reasons for excluding this solution (978a16–20, contra *Physics* 8.10 [267b19–24]). The dominance of dialectic, as well as some trace of probabilism, is attested in Aristotle's school after STRATŌN and before the revival under ANDRONIKOS OF RHODES (*cf.* STRABŌN 13.1.54 and CICERO *Tusc.* 2.3, 2.9, *De finibus* 5.10, *De oratore* 3.80, *Orator* 14). For dialectic, the exemplary work for Cicero was Aristotle's *Topics*, from which *MXG* offers a wide range of reminiscences. For other elements of Aristotelian learning, the author seems interested only in such physical theories as that of empty space (976b14–19) and mixture (977a4–11), which were widely discussed among **Peripatetics** and **Stoics**, from Stratōn (*frr.*54–67 W.) to ALEXANDER OF APHRODISIAS.

Thus, the author appears to be a dialectician in the tradition of the school of Aristotle, working some time between the composition of the early catalogue of Aristotle's writings and the rediscovery of the Aristotle of his school-works, reflecting some developments of contemporary physics and theology.

Ed.: H. Diels, *Aristotelis qui fertur de M.X.G. libellus = Abhandlungen der Königlichen Preussischen Akademie der Wissenschaften* (1900); B. Cassin, *Si Parménide: le traité anonyme De Melisso, Xenophane, Gorgia* (1980).

J. Mansfeld, "De Melisso Xenophane Gorgia. Pyrrhonizing Aristotelianism," *RhM* 131 (1988) 239–276; István M. Bugár, "How to Prove the Existence of a Supreme Being?" *Acta Antiqua Academiae Scientiarum Hungaricae* 42 (2002) 203–215 at 205–206.

István M. Bugár

MELISSUS ⇒ MAECENAS MELISSUS

Melitianus (350 – 500 CE?)

Wrote in Latin a geographical work on Africa, cited by the RAVENNA COSMOGRAPHY 3.5, and there said to be an African. The name is otherwise unattested, but *cf.* Melinianus (*PLRE* 1 [1971] 594), or Melitius/Meletius, common 350–500 CE.

(*)

PTK

Melitōn (250 BCE – 80 CE)

ANDROMAKHOS, in GALĒN *CMGen* 5.13 (13.843 K.), cites his wound-powder, containing lime, orpiment, pumice, and realgar, ground fine in water for 30 days, then dried for an entire day. *PIR2* suggests identification with Ti. Claudius Meliton (on whom see Korpela), physician to some Germanicus, perhaps Tiberius' adopted heir (s.v.).

PIR2 M-451; Korpela (1987) 167 #69.

PTK

Menaikhmos of Prokonessos (365 – 325 BCE)

A pupil of EUDOXOS and associate of PLATO, who together with his brother DEINOSTRATOS and AMUKLAS OF HĒRAKLEIA, also an associate of Plato, "made the whole of geometry more perfect (or complete)" (PROKLOS *In Eucl.* p. 67.8–12 Fr.). Proklos associates Menaikhmos with metamathematical questions, telling us that he gave an account of the word "element" (pp. 72.23–73.14 Fr.), that the mathematicians "around" him considered everything proved in geometry to be a "problem" rather than a theorem – although he allowed that some problems seek to determine a feature of some defined thing – (p. 78.8–13 Fr.), and that the mathematicians around him and Amphinomos dealt with questions concerning the convertibility of propositions of the form "All A are B."

But Proklos also reports (p. 111.20–23 Fr.) that Menaikhmos "conceived" (*epinoeisthai*, apparently meaning "discovered") the conic sections and cites a line of poetry by ERATOSTHENĒS: "Don't section the cone with the triads (standardly taken to be the curves we call parabola, hyperbola, and ellipse) of Menaikhmos." The line is from an epigram which Eratosthenēs attached to his mechanical solution to the problem of producing a cube double a given one and urges the reader not to follow the solution of Menaikhmos. The epigram is quoted by EUTOKIOS in his commentary on ARCHIMĒDĒS' *On the Sphere and Cylinder* (p. 96.10–27 H.). In the commentary Eutokios describes a number of solutions, including one which is ascribed to Menaikhmos (78.13–80.24 H.). In it, there is an analysis and a synthesis. The procedure depends upon the reduction of cube duplication to the problem of finding two mean proportionals between straight lines a and b, a reduction due to HIPPOKRATĒS OF KHIOS (p. 88.17–21 H.); for if $b = 2a$, and $a{:}x :: x{:}y :: y{:}b$, then, in algebraic terms, (i) $x^2 = ay$, (ii) $y^2 = 2ax$, and (iii) $xy = 2a^2$, and any two of these equations yield that $x^3 = 2a^3$, i.e., geometrically, the cube with side x is double the cube with side a. Eutokios' presentation of Menaikhmos' solutions uses the terms "parabola" and "hyperbola," which were introduced by APOLLŌNIOS OF PERGĒ to replace his predecessors' "section of a right-angled cone" and "section of an obtuse-angled cone" (Eutokios, *in Apol.* pp. 168.12–170.24 H.; *cf.* Pappos, *Collection* 7.30). We cannot determine exactly how Menaikhmos proceeded and how much he knew about conics. Most scholars assume that he knew quite a bit, but it has been argued that Menaikhmos only used point-wise constructions of curves which were later determined to be definable as conic sections (Knorr 1982).

Menaikhmos' solution uses a parabola and a hyperbola. Immediately after describing it, Eutokios (pp. 82.1–84.7 H.) gives under the heading "in another way" (*allōs*) a very similar one, using two parabolas. Menaikhmos has usually been credited with this solution, but a very similar argument is found in the Arabic translation of DIOKLĒS (186–207), raising doubts about the attribution to Menaikhmos. If that attribution is moot, then so must be the attribution to Menaikhmos of a mechanical solution, ascribed by Eutokios (pp. 56.13–58.14 H.) to Plato, the configuration of which is very like the alternative solution. However, PLUTARCH (*Quaest. Conv.* 718E) says that Plato reproached Eudoxos, ARKHUTAS, and Menaikhmos for trying to reduce the duplication of the cube to a matter of mechanical constructions.

THEŌN OF SMURNA (pp. 201.22–202.2) connects Menaikhmos with the astronomical theory of homocentric spheres, commonly attributed to Eudoxos.

DSB 9.268–277, I. Bulmer-Thomas; Toomer (1976) 90–96, 169–170; W.R. Knorr, "Observations on the early history of the conics," *Centaurus* 26 (1982) 1–24; Jones (1986) 573–577; Lasserre (1987) 12.

Ian Mueller

Menandros Iatrosophist (600 – 1200? CE)

Under the name of *Menandros iatrosophistēs*, the 14th c. MS, Paris BNF, *graecus* 1630, contains a fragment of a work on gynecology in the vein of the undated MĒTRODŌRA, and the many works from southern Italy that constituted the so-called 12th-century *Trotula* (in fact a collection of treatises rather than a single author). The author of the Paris fragment is probably not the same as the MENANDROS cited by PLINY; the epithet *iatrosophistēs* might confirm a late-antique date.

Diels 2 (1907) 64 (*De Mulieribus*).

Alain Touwaide

Menandros of Hērakleia (325 – 90 BCE)

Wrote a treatise on agriculture excerpted by CASSIUS DIONUSIOS (VARRO, *RR* 1.1.8–10). To judge from references in PLINY (1.*ind*.8, 11), he discussed livestock and bees. Pseudo-PLUTARCH, *Nobil.* 20 (7.269 Bern.) reports his claim that farmers were the last remnant of the Saturnian race.

RE 15.1 (1931) 764–765 (#19), Ernst Diehl, and S.6 (1935) 297, W. Kroll.

Philip Thibodeau

Menandros (of Pergamon?) (*ca* 175 – 155 BCE)

Physician, medical authority on drugs (PLINY 1.*ind*.30), possibly the same as the Menandros from Pergamon attested at Athens, and a companion of King Eumenēs (*Syll.* 655). He prescribed eating beetroot roasted on hot coals to neutralize "garlic breath" (19.113). NIKOSTRATOS reused his enema. Identification with Pliny's non-medical authority who wrote *Necessities for Life* (*BIOXPHΣTA*: 1.*ind*.19–27) is tenuous, for whom better see M. OF HĒRAKLEIA or OF PRIĒNĒ.

RE S.6 (1935) 297, W. Kroll.

GLIM

Menandros of Priēnē (325 – 90 BCE)

Wrote a work on agriculture excerpted by CASSIUS DIONUSIOS (VARRO, *RR* 1.1.8–10, *cf.* COLUMELLA, 1.1.9). To judge from references in PLINY (1.*ind*.8, 11), he wrote about livestock and bees.

RE 15.1 (1931) 764–765 (#19), Ernst Diehl, and S.6 (1935) 297, W. Kroll.

Philip Thibodeau

Mēnās (350? – 540 CE)

Cited by AËTIOS OF AMIDA 10.5 (p. 567 Cornarius) for a remedy involving **bdellium**, cassia, saffron, myrrh, spikenard, etc. Aëtios 7.42 (p. 351 Cornarius) and 7.110 (p. 391 Cornarius) appear to refer to the collyrium of "Monus," where Olivieri (*CMG* 8.2) reads respectively *ΠΙ* (7.44, p. 297) and "Nonnos" (7.114, p. 382). The name Mēnās is frequent in the early Byzantine period (*PLRE* 3 [1992], *cf. LGPN*), e.g., the patriarch of Constantinople 536–552 CE (*ODB* 1339–1340, A. Kazhdan), under the influence of the Egyptian cult of the mythical martyr Mēnās (*ODB* 1339, A. Kazhdan and N.P. Ševčenko).

Fabricius (1726) 329, 341.

PTK

Menekratēs, Ti. Claudius (10 – 40 CE)

Syll. 803 (Rome) records that he wrote 156 books establishing his own medical system, and was an imperial physician; the Latin *nomen* accords with GALĒN's confused notice placing him after HĒRĀS, before ANDROMAKHOS, and contemporary with ANTONIUS MUSA, *CMLoc* 6.9 (12.989 K.), if the emperor could be Tiberius. His pharmaceutical work, *Written-in-Full Emperor*, gave recipes whose quantities were in words not numerals, but even those became corrupted, which led DAMOKRATĒS to put recipes into verse: Galēn, *CMGen* 2.6 (13.502–503 K.), 7.9 (13.994–996), *Antid.* 1.5 (14.31–32 K.). A few of his recipes are preserved – ASKLĒPIADĒS PHARM., in Galēn *CMGen* 6.14 (13.937–938), and KRITŌN, in Galēn *CMLoc* 5.3 (12.846) – but he did not write on theriac: Galēn, *Theriac* (14.306). SŌRANOS, as preserved in CAELIUS AURELIANUS, approves the internal-abscess drug of some Menekratēs, *Chron.* 5.126 (*CML* 6.1, p. 930), and disparages the epilepsy-treatment of Menekratēs *ZEOPHLETENSIS*, *Chron.* 1.140 (p. 512), plausibly emended by Drabkin to refer to MENEKRATĒS ZEUS OF SURAKOUSAI, famed for epilepsy treatments.

BNP 3 (2003) 410 (#IV.2), V. Nutton.

PTK

Menekratēs of Elaious (330 – 300 BCE)

A student of XENOKRATĒS OF KHALKĒDŌN, and approved by MĒTRODŌROS OF SKEPSIS. Credited with two works, *Foundations* (of cities), which claimed that the Ionian coast and neighboring isles were originally Pelasgian (STRABŌN 13.3.3), and ***Periodos** of the Hellespont*, which explained the Halizones of *Iliad* 2 as a mountainous tribe near Murleia

(Strabōn 13.3.22–23). He also explained the name "Musian" as the Ludian word for the beech tree, common in Musia (*idem* 12.8.3).

RE 15.1 (1931) 801 (#25), P.E. Göbel; *DPA* 4 (2005) 442–443, R. Goulet.

PTK

Menekratēs of Ephesos (330 – 270 BCE)

A philologist perhaps best known as the teacher of the astronomical poet ARATOS. He was also the author of a poem in the Hēsiodic style called *Works*, in which the discussion of bee varieties was apparently based on ARISTOTLE's *Historia Animalium* (*cf.* VARRO, *RR* 3.16.18, with *HA* 5.21 [553a25], 9.40 [624b21]). He seems thus to have been a pioneer in the early-Hellenistic revival of scientifically-informed didactic poetry. He may also be identical to the statesman who led a rebellion at Ephesos after the death of the *diadoch* LUSIMAKHOS in 281 BCE (Polyain. 8.57).

Ed.: *SH* 542–550.
RE 15.1 (1931) 798 (#6), 800 (#16), P.E. Göbel; *OCD3* 958, J.S. Rusten; *DPA* 4 (2005) 443, R. Goulet.

Philip Thibodeau

Menekratēs of Surakousai (*ca* 350 BCE)

Greek physician who carried out a detailed study of the qualities (*poiŏtētes*) of bodies in his work *On medicine* (LONDINIENSIS MEDICUS 19.18–20.1): bodies are formed of four elements, two of which are hot and two cold (blood and bile, phlegm and air). Good or bad mixtures (*krasis*) result in health or disease; alterations produce flows or other secondary substances such as red or black bile. Menekratēs had enormous success in curing epilepsy and demanded his patients obey him like slaves: contemporary comic poets mocked him for calling himself Zeus and giving his patients names of other gods (Ath., *Deipn.* 7 [289a–b]). Philip II (359–336 BCE) and Agesilaos (445–359 BCE), to whom he wrote letters, also ridiculed him.

RE 15.1 (1931) 802 (#29), H. Raeder; O. Weinreich, *Menekrates Zeus und Salmoneus = Tübinger Beiträge zur Altertumswissenschaft* 18 (1933); *RE* S.9 (1962) 401, K. Deichgräber; D. Gourevitch, "Médecins fous," *Evolution psychiatrique* 47 (1982) 1113–1118; Gourevitch (1989) 246–248; G. Squillace, "Le lettere di Menecrate/Zeus ad Agesilao di Sparta e Filippo II di Macedonia," *Kokalos* 46 (2000) 175–191; *AML* 604, G. Marasco; *BNP* 8 (2006) 672–673 (#3), V. Nutton.

Daniela Manetti

Menekritos (350 BCE – 80 CE)

HĒLIODŌROS OF ALEXANDRIA, in OREIBASIOS, *Coll.* 48.53 (*CMG* 6.2.1, p. 286), describes the winding of his *exakros* hand-bandage. H.G. Liddell, R. Scott, and H.S. Jones, *A Greek-Emglish Lexicon* (1968), *s.v.*, identify as MENEKRATĒS, but the name Menekritos is attested, *LGPN* 3B.279, 4.229.

Fabricius (1726) 334.

PTK

Menelaos (Pharm.) (100 BCE? – 95 CE)

ASKLĒPIADĒS PHARM., in GALĒN *Antid.* 2.11 (14.173 K.), records Menelaos' salve to treat ***hudrophobia***, also effective against ***aigilōps*** and attacks of all serpents, consisting in red natron, goat-suet, olive oil, beeswax, charred lees, and ***ammōniakon*** incense ground in water until glutinous. The ingredients are mixed, dissolved, heated, and then softened in a mortar.

RE 15.1 (1931) 835 (#17), K. Deichgräber.

GLIM

Menelaos of Alexandria (*ca* 90 – 100 CE)

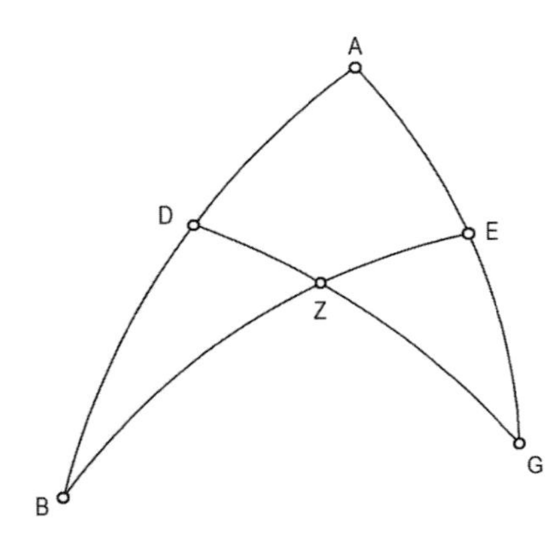

Ptolemy's version of part of the theorem of Menelaus. Given: great circles AB and AG on the face of a sphere, cut by great circles GD and BE, which meet at Z, and where each of the arcs is less than a semicircle. Then the chord (CRD) of arc 2GE : CRD arc 2EA is combined from (CRD arc 2GZ : CRD arc 2ZD) and (CRD arc 2 DB : CRD arc 2BA), where "A is combined from B and C" can be treated in modern terminology as "A = B × C". © Lehoux and Massie

Mathematician and astronomer, to whom PTOLEMY attributes two observations of lunar positions relative to fixed stars, both made in Rome in January 98 CE (*Almagest* 7.3, pp. 30 and 33 H.). Menelaus also appears as a bystander in PLUTARCH's *On the Face that Appears in the Moon*, set in the late 1st c. Two mathematical works are extant in Arabic translations: the *Sphairika* and *On Specific Gravities* (MS *Escurial* 960/3, ff.43–50, 742 H., dedicated to Domitian; *cf.* MANTIAS). The *Sphairika* was translated into Latin several times, first by Gerard of Cremona, and later by Edmund Halley. Menelaos' most important contributions are to spherical trigonometry, which field he pioneered, with immediate applications in spherical astronomy. He wrote six (lost) books on chords, usually taken to include a chord table. His *Sphairika* includes a general proof of what has come to be known as the "Theorem of Menelaus" (actually two closely related theorems), allowing one to solve for triangles on a spherical surface. The theorem has wide-ranging astronomical applications including conversions between spherical coordinate systems, the calculations of rising times of oblique arcs, and (hence) the determination of the length of daylight at any given latitude, for example.

Ed.: M. Krause, *Die Sphärik von Menelaos aus Alexandrien* (1936).
A.A. Björnbo, *Studien über Menelaos' Sphärik* (1902); Neugebauer (1975) 26–27.

Daryn Lehoux

Menemakhos of Aphrodisias (30 – 90 CE)

Physician, listed with THEMISŌN and SŌRANOS among the **Methodists** (PSEUDO-GALĒN INTRODUCTIO 14.684 K.; MS *Laur. Lat.* 73.1, f.143^{V} = *fr.*13 Tecusan), probably not before THESSALOS (Tecusan 2004: 15–16, 65), disagreed with predecessors, sometimes vituperatively (GALĒN *MM* 1.7.5 [10.54 K. = p. 27 Hankinson]). ANDROMAKHOS, in Galēn *CMLoc* 3.1 (12.625 K.), records (and uses) his ear remedy compounded from myrrh, frankincense,

aphronitron, saffron, poppy-juice, bitter almonds, **galbanum**, and vinegar. OREIBASIOS *Coll.* 7.22 reports Menemakhos' instructions for applying and removing leeches (*CMG* 6.1.1, pp. 220–221 = *fr.*226 Tecusan), and a depilatory method involving scratching the scalp and carefully applying pitch plaster (*drōpax: Coll.* 10.14: *CMG* 6.1.2, p. 58 = *fr.*227 Tecusan). CAELIUS AURELIANUS, *Acut.* 1.8 (*CML* 6.1.1, p. 134), conveys his definition of lethargy, augmented by SŌRANOS, as a swift or acute pressure accompanied by acute (but not always continuous) fever. Tecusan doubts our **Methodist** is as early as the Menemakhos attributed by CELSUS with a multi-ingredient toothache remedy whose active component was ***purethron*** (6.9.5 = *fr.*106 Tecusan; *cf.* p. 64).

RE 15.1 (1931) 838 (#6), H. Raeder; Tecusan (2004) 63–65.

GLIM

Menenius Rufus (30 BCE – 90 CE)

ASKLĒPIADĒS PHARM. in GALĒN, *CMGen* 7.12 (13.1010–1011 K.), records his complex recipe for a potion against gout, calling for over two dozen ingredients concocted over three days in three stages. Kühn reads *MENI-*, which *PIR2* accepts, but Fischer points out that *Menenius* is far more likely as a *nomen* (*cf. RE* 15.1 [1931] 838–844, Fr. Münzer, and Catullus 59 on Menenius husband of Rufa).

PIR2 M-256.

PTK

Menestheus of Stratonikeia (150 BCE? – 50 CE)

GALĒN *Hipp. Gloss.* cites his *Names of Drugs* twice (where Fabricius and Kühn print *MENHΘEYΣ*, otherwise unattested): s.v. *boukeras* (19.89 K.) and *Indikon* (19.105–106 K.), there agreeing with ANDREAS and XENOKRATĒS OF APHRODISIAS that it is ginger. ERŌTIANOS A-103 (p. 23.12–13 Nachm.) supplies the ethnic, recording his opinion that in the HIPPOKRATIC CORPUS, JOINTS, 7 (4.88 Littré), *ambē* means "leverage." ASKLĒPIADĒS PHARM. in Galēn, *CMGen* 5 (13.830 K.), records his ***trokhiskos*** for skin disorders (chapping, callosities, etc.), of aloes, alum, and saffron, in wine. Perhaps *cf.* Galēn, *In Hipp. Epid. VI* 4.11 (*CMG* 5.10.2.2, p. 212), *ad* 4.8 Littré, where the Arabic records a *MNSNUS* among other commentators.

Fabricius (1726) 335; *RE* 15.1 (1931) 852 (#6), K. Deichgräber.

PTK

Menestōr of Subaris (460 – 440 BCE?)

A **Pythagorean** natural philosopher and the earliest Greek botanist. His botanical treatise is lost, and our knowledge of his theories rests on several references to them, preserved in THEOPHRASTOS' works on plants. Theophrastos reckons him among the ancient *phusiologoi* (32 A7 DK) and says that he sided with (A5) an opinion of EMPEDOKLĒS. Menestōr is usually regarded as a contemporary of Empedoklēs, most probably slightly older than him. His name occurs in the list of the **Pythagoreans** compiled by ARISTOXENOS (A1).

Following ALKMAIŌN, whose theory of health was based on the idea of qualitative opposite principles (cold/hot, wet/dry, etc.), Menestōr transferred this explanatory pattern to the realm of plants. He believed that the moisture, or the juice of plants (*hugron, khumos*, A2, 7), bears life and heat. There are infinite number of such juices of plants, which are distributed in pairs: bitter/sweet, harsh/oily etc.; a plant consists of their mixture (A7). Menestōr methodically divides all plants into cold and warm and derives from their balance the most important qualities of plans, taking into account external factors as well (climate, soil, etc., A4, 6). An excessive cold or heat reduces the moisture of plants, so that they either freeze or dry out. The warm plants bear fruits, the cold do not. The warmer a plant is, the faster it grows and the earlier it bears fruits. The evergreen plants have more inner heat than others, which shed their leaves in winter due to the cold (A5). Plants can live only in places with opposite climate: the warm in cold, the cold in warm.

DK 32; W. Capelle, "Menestor redivivus," *RhM* 104 (1961) 47–69; C. Viano, "Théophraste, Ménestor de Sybaris et la summetria de la chaleur," *REG* 105 (1992) 584–592; Zhmud (1997).

Leonid Zhmud

Menestratos (I) (*ca* 400 – 250 BCE?)

CENSORINUS 18.5 lists HARPALOS, NAUTELĒS, the otherwise unknown Menestratos, and DŌSITHEOS as writers of works on the ***oktaetēris***. *Cf.* perhaps MENESTRATOS (II).

(*)

PTK

Menestratos (II) (325 – 90 BCE)

Authored a treatise on agriculture excerpted by CASSIUS DIONUSIOS (VARRO, *RR* 1.1.9–10). He is perhaps identical to MENESTRATOS (I), whose ***oktaetēris*** may have been a farmer's calendar.

RE 15.1 (1931) 856 (#9), W. Kroll.

Philip Thibodeau

Menippos (100 BCE? – 95 CE)

ASKLĒPIADĒS PHARM., in GALĒN *Antid.* 2.11 (14.172–173 K.), records his simple salve against ***hudrophobia***, used by Galēn's teacher PELOPS, compounded of Bruttian pine pitch, ***opopanax*** and vinegar, heated, but not to boiling, administered to the wound with a compress. The wound should be watched for 40 days.

RE 15.1 (1931) 894 (#12), K. Deichgräber.

GLIM

Menippos of Pergamon (*ca* 80 – *ca* 20 BCE)

Greek geographer, author of a ***Periplous*** *of the Inner Sea* (Mediterranean) in three books and possibly also one of the Black Sea, relying on the work of ARTEMIDŌROS OF EPHESOS. Menippos is known mainly through an epitome by MARCIANUS OF HĒRAKLEIA. The extant fragments contain the *prooimion* and the descriptions of the Black Sea, Bithunia, Paphlagonia, Pontos, Bosporos, Propontis and Europe. The work has a clear navigational inclination concentrating on coasts, ports and distances between coastal points. An allusion by the Greek Augustan epigrammatist Krinagoras of Mutilēnē to Menippos, as his friend and author of a circular tour, determines Menippos' date.

Ed.: *GGM* 1.563–573.
Diller (1952) 151–164; F.J. González Ponce, "El *Periplo* Griego antiquo: Verdadera Guí a de viajes o mero género literario? El ejemplo de Menipo de Pérgamo," *Habis* 24 (1993) 69–76.

Daniela Dueck

MENIUS RUFUS ⇒ MENENIUS RUFUS

Mēnodōros of Smurna (85 – 35 BCE)

Friend of the **Erasistratean** HIKESIOS OF SMURNA, and offered dietary advice including remarks on squashes and their preparation: Ath., *Deipn.* 2 (59a). ANDROMAKHOS, in GALĒN *CMLoc* 7.3 (13.64 K.), preserves RIPALUS' preparation of a cough-syrup, for sufferers from ***phthisis***, named after Mēnodōros. OREIBASIOS, *Coll.* 46.11 (*CMG* 6.2.1, p. 222), records his work as a surgeon, and his procedure in cases of skull fracture (to excise all damaged bone). P. CAIRO CRAWFORD 1 records his practice in eye-surgery. Michler (1968a) identifies MĒNOPHILOS with this man, but all the names Mēnodōros, Mēnodotos, and Mēnophilos are sufficiently common (*LGPN*) to render such equation otiose.

RE S.9 (1962) 402, J. Kollesch; S.11 (1968) 934–935, M. Michler; *Idem* (1968) 71, 113–114.

PTK

Mēnodotos (Astr.) (250 BCE – 100 CE?)

Wrote a commentary on ARATOS (*FGrHist* 1026 T 19), entirely lost.

(*)

PTK

Mēnodotos of Nikomēdeia (105 – 145 CE)

Empiricist physician, pupil (with THEODĀS OF LAODIKEIA) of the Skeptic philosopher Antiokhos of Laodikeia, and teacher of the Skeptical Hērodotos of Tarsos (in the catalogue of DIOGENĒS LAËRTIOS 9.115; it is doubtful whether the physician is the Mēnodotos to whom Diogenēs Laërtios ascribes the catalogue itself; it is also controversial whether to read the name of Mēnodotos in SEXTUS EMPIRICUS *Pyrrh. hyp.* 1.222 regarding PLATO's skepticism). Of his works we only know that one of them, in several books, was dedicated to a certain Seuerus (perhaps the Cn. Claudius Seuerus, **Peripatetic**, interlocutor of the emperor M. Aurelius: SHA, *Marc.* 3.3 and M. Aur. *ad se ipsum* 1.14), and that about it GALĒN wrote a lost work in 11 books (*On My Own Books* 2.115 MMH; *Subf. emp.* 11); it is still uncertain whether we should read the name of Mēnodotos in the title of Galēn's lost *Protreptikos* (*On My Own Books, ibid.*).

The fact that Mēnodotos is frequently mentioned by Galēn in *The Outline of Empiricism* and in other works had suggested the idea that he was the main source of Galēn for **Empirical** doctrine, and more generally that he played the role of an innovator in the development of **Empirical** doctrine (Favier went so far as to view him as a forerunner of modern experimental science). It is unclear, however, the real extent of Mēnodotos' innovations in the elaboration of the **Empiricist** doctrine created by HĒRAKLEIDĒS OF TARAS (who was used by Galēn as well): certainly Mēnodotos, as well as Hērakleidēs, has a tendency to give more importance to the "rational" element in the doctrine of the "school." In his layout of the three principles of the **Empiricist** "tripod," Mēnodotos separated the different types of "experience" (Galēn *Subf. emp.* 2–3), and he maintained that the third element of the tripod, the "transition to the similar," is not a true criterion but only a criterion of what is possible (*Subf. emp.* 9). Probably Mēnodotos introduced the concept of *epilogismos*, namely the possibility of rational inferences on the basis of the data coming from the experience – distance from **Dogmatic** medicine remains assured by the fact that the *epilogismos* is an inference directed toward visible things, whereas the *analogismos* is the conclusion pointing to invisible things (Galēn, *Med. exp.* 24).

For other fields of Mēnodotos' production, we know that, according to Galēn (who basically agreed with him: *Nat. fac.* 3.71 MMH; perhaps *Caus. resp.* 4.475 K.), he polemized against Asklēpiadēs of Bithunia in an excessively violent manner (*Subf. emp.* 11; Galēn gives a similar opinion also about the criticisms Mēnodotos made against other **Empiricists**); that he deemed it legitimate for the physician to seek fame and money (Galēn *PHP: CMG* 5.3.1.2, p. 764); that he used phlebotomy only in case of *plethora*, that is excessive increase in the blood mass (Galēn, *Cur. rat. ven. rom.* 11.277, 285 K.; *Hipp. ac. mor. vic.* 15.766 K.; *Hipp. art.* 18A.575 K.).

Ed.: Deichgräber (1930) 212–214 (fragments), 264–265.

A. Favier, *Un médecin grec du deuxième siècle ap. J.C., précurseur du la méthode expérimentale moderne: Ménodote de Nicomédie* (1906); *RE* 15.1 (1931) 901–916 (#2), W. Capelle; 916 (#3), H. Raeder; *KP* 3.993–994, F. Kudlien; L. Perilli, *Menodoto di Nicomedia* (2004); *BNP* 8 (2006) 695 (#2), V. Nutton; *DPA* 4 (2005) 476–482, V. Boudon-Millot.

Fabio Stok

Menoitas/Menoitios (250 BCE – 10 CE)

Hērās, in Galēn *CMGen* 2.10 (13.511–512 K.: *Menoitios*), and Andromakhos, *ibid.* 2.8 (13.509 K.: *Menoitas*), cite two versions of his *mēlinē*, containing beeswax, **litharge**, clear **terebinth**, olive oil; Andromakhos adds frankincense, **galbanum**, and **verdigris**, providing a multi-step preparation. The epic name appears in both forms (e.g., the herdsman of Hadēs and the father of Patroklos), as for historical figures, among whom Menoitas is usually Doric and more widely used, being especially frequent in Aitolia: *RE* 15.1 (1931) 918–922, K. Keyßner; *LGPN*.

Fabricius (1726) 335.

PTK

Menōn (350 – 300 BCE?)

A pupil of Aristotle, known to Galēn (15.24 K.) as the author of a medical doxography, *Medical Collection*, circulating under Aristotle's name, probably the same work quoted by Plutarch (*Quaest. conv.* 8.9.3) as *Menoneia* (i.e. "work by Menōn"). Diels considered Menōn the source of the doxography about the causes of diseases preserved in the first part of the Londiniensis medicus. It is impossible to determine his actual role, whether he wrote the *Medical Collection*, or was a later editor of **Peripatetic** material, a reviser, or merely one who "possessed" a copy of the work, used by early imperial Aristotelian scholars.

H. Diels, "Ueber die Excerpte von Menons Iatrika in dem Londoner Papyrus 137," *Hermes* 28 (1893) 407–434; *KP* 3.1223, F. Kudlien; D. Manetti, *CPF* I.1 (1989) 345–351; *OCD3* 960, J.T. Vallance.

Daniela Manetti

Mēnophilos (120 BCE – 25 CE)

CELSUS 6.7.2C describes his ear medicine: pepper, myrrh, saffron, poppy "tears," pomegranate peel, almonds, etc., in honey and very sour vinegar. The name is unattested before 300 BCE (*LGPN*), and the use of pepper suggests the *terminus post*, when Indian trade made it more available.

Fabricius (1726) 336.

PTK

MESSALLA ⇒ VALERIUS

MESTRIUS ⇒ PLUTARCH

Metagenēs of Knōssos (550 – 500 BCE)

With his father KHERSIPHRŌN, began the great Temple of Artemis at Ephesos, and wrote about it in one of the earliest known architectural treatises (VITRUUIUS 7.*pr.*12, 16). Metagenēs also invented a rolling framework for moving large rectangular epistyle blocks of the temple, an extension of his father's invention for the column drums (Vitr. 10.2.12, 13).

Svenson-Ebers (1996) 67–99; *KLA* 2.78–79, A. Bammer.

Margaret M. Miles

Metōn of Athens (440 – 410 BCE)

Astronomer who, with EUKTĒMŌN, observed the summer solstice on the morning of 13 Skirophorion (probably 27 June 432 BCE) on the Pnyx using a *hēliotropaion* (an instrument of disputed nature) and devised a 19-year calendar, whose first period presumably would have begun on the next new moon. The period was 235 months, which required 12 years with 12 months and 7 years with 13 months. This part of the system was certainly based on the Babylonian 19-year system and probably distributed months in the same way. Moreover, the period was 6,940 days, implying a year of 365 5/19 days and an average synodic month of 29 25/47 days. The distribution of hollow (29 days) and full months (30 days) might have used a scheme like that reported by GEMINUS, *Elem. Astron.* 8. To yield 235 months in 6,940 days, the system treats all months as having 30 days (making 7,050 days), but then drops a day every 64th day (i.e. after the 63rd day), with the month being hollow, to bring the total back down to 6,940 days. The period was called "Metōn's cycle" (DIODŌROS OF SICILY 12.36.2–3). Whatever Metōn's purpose was in devising the calendar, it was used as the basis of astronomical observation, especially in its revised form by KALLIPPOS. Metōn may have begun the practice of erecting public ***parapēgmata***, traces of which survive in Geminus and PTOLEMY (also, *Schol.* *ARATOS* 752). Metōn and Euktēmōn parceled out the seasons (P. PARISINUS GRAECUS 1): summer (90 days), fall (90 days), winter (92 days), and spring (by inference, 91 days). His appearance in two comedies in 414 BCE, as a cloudy architect and geometer in Aristophanēs, *The Birds*, and as a well-maintainer in Phrunikhos, *The Recluse* (*Schol. Aristoph. Birds* 997), probably has more to do with his attempts to avoid military service in Sicily the previous year than to his work as a mathematician (PLUTARCH, *Alkibiadēs* 17.5, *Nikias* 13.6; AELIANUS, *VH* 13.12).

DSB 9.337–340, G.J. Toomer; B. Goldstein and A.C. Bowen, "Meton of Athens and Astronomy in the Late Fifth Century B.C.," in Leichty *et al.* (1988) 40–81; R. Hannah, "Euctemon's Parapēgma," in C.J. Tuplin and T.E. Rihll, *Science and Mathematics in Ancient Greek Culture* (2002) 76–132.

Henry Mendell

Mētrodōra (50 – 400 CE?)

Preserved in a single MS (Florence, Biblioteca Laurenziana, *Pluteus* 75.3, 4[V]–19[R]). Dating is difficult, since Mētrodōra mentions no names, apart from a cosmetic used by "Berenikē called Kleopatra" (a confused reference, possibly an interpolation). Use of the vaginal speculum argues for a date beginning *ca* 1st c. CE, and the text cites neither SŌRANOS nor encyclopedias, and displays no Galenism, which places it probably before the fifth.

The title, *From the Works of Mētrodōra*, indicates a selection from a corpus of at least two books. The preserved text, entitled "Concerning the Feminine Diseases of the Womb," contains 63 chapters in seven well-organized sections. 1: Introduction; 2–19: General conditions of the womb (inflammation, suppuration, hardness, cancer, discharges, hemorrhages, prolapses, coldness, and inflation); 20–25: Diseases caused by excessive moisture (**dropsy**, cleansing of ulcers, recipes to restore the appearance of virginity); 26–28: Conception and contraception (fertility, female and male children, cures for sterility, three recipes for contraception); 29–32: Childbirth; 33–39: Sexual recipes (tests for virginity, aphrodisiacs), 40–55: Diseases of the breasts; 56–63: Cosmetics and general preparations. (Four sets of mainly pharmaceutical extracts following in the MS are probably not Mētrodōra's). The earlier chapters are fuller and the text may have been abbreviated at some point. There is no mention of obstetrics; the work was not confined to midwifery, but focuses on pathology.

Mētrodōra is an interesting figure in the history of medicine for reasons independent of gender. More than an anthologist like OREIBASIOS or an encyclopedist like AËTIOS OF AMIDA, she does not depend on the growing secondary literature of the handbooks but reaches directly back to HIPPOKRATĒS, quoting, paraphrasing, synthesizing, and gathering symptoms missed by others.

Mētrodōra takes sides in several medical controversies over symptomatology and etiology (e.g., inflammation of the womb). She formulates an individual classification of various vaginal discharges, a hotly debated topic. She makes several seemingly original contributions to theory and etiology (e.g., linking certain vaginal discharges to irritation of the adjoining rectum produced by intestinal worms). Some of her compounds became part of the ancient medical common stock, but the vast majority appear only in her work. In clinical practice Mētrodōra employs both digital examination and the vaginal speculum, providing a unique and detailed description of pathology based on its use. These are indications of individual scholarship of a high level, backed by experience.

A Latin translation was made in late antiquity (probably 5th/6th c.), and portions of the material circulated under the names of KLEOPATRA, THEODORUS PRISCIANUS, and in other early medieval sources, notably the *Liber de causis feminarum* (ed. Egert 1936). Through these the material passed to Caspar Wolf's *Harmonia Gynaeciorum* (1566), the first Renaissance encyclopedia of gynecology.

Ed.: A.P. Kousis [Kuzes], "Metrodora's work 'On the feminine diseases of the womb' according to the Greek codex 75, 3 of the Laurentian Library," *Praktika tēs Akademias Athenōn* (1945 [1949]) 20, 46–68: *editio princeps* and unreliable; G. Del Guerra, *Il Libro di Metrodora* (1953), repr. with Italian trans.:

Metrodora: Medicina e cosmei ad uso delle donne (1994): unreliable; Holt N. Parker, *Metrodora: The Gynecology* = *SAM* (forthcoming).

Holt N. Parker

Mētrodōros (Astr. I) (*ca* 150 – 50 BCE?)

PTOLEMY's *Phaseis* records that Mētrodōros observed in Italy and Sicily (p. 67 H.), and cites him, with other ***parapēgmatists*** from DĒMOKRITOS to HIPPARKHOS and "Caesar" (SŌSIGENĒS I) for over a dozen weather-signs: Phaōphi 5: "rain" (p. 18), Athur 13: "tempest and thunderstorm" (p. 22), Mekhir 15: start of spring (p. 38), Mechir 30: "the swallow appears" (p. 39), Pakhōn 17: start of summer (p. 50), Epiphi 27: start of fall (p. 60), etc. IŌANNĒS "LYDUS," *Mens.* cites him *ad* March 15, September 17, and October 27.

Rehm (1941) 82, n.2; *BNP* 8 (2006) 838 (#7), W. Hübner.

PTK

Mētrodōros (Astr. II) (*ca* 10 – 300 CE?)

Seruius, *ad Georg.* 1.229 (3.1.185 Th.-H.), cites Mētrodōros' work on the zones, which also defended VERGIL's astronomy, and "Probus" *ad Georg.* 2.224 (3.2.371 Th.-H.) cites him for the geographical tidbit that the River Clanius near Mount Vesuvius is named for a giant. Goulet identifies with MĒTRODŌROS (ASTR. I).

RE S.7 (1940) 449 (#24a), W. Kroll; *DPA* 4 (2005) 504, R. Goulet.

PTK

Mētrodōros (Arch.) (20 BCE – 77 CE)

Listed among the non-Roman authorities on painting, pigments, and drugs derived therefrom consulted by PLINY (1.*ind.*35). He wrote *On the Science of Architecture* (*de Architectonice*), but is omitted from VITRUUIUS 7.*pr*, perhaps providing a *terminus post.*

RE 15.2 (1932) 1483 (#29), W. Kroll.

GLIM

Mētrodōros (Pharm.) (100 BCE – 60/75 CE)

Wrote *Epitome of Rootcutting*, recommending *peplis* (a *Euphorbia* sp.) after delivery, to ease expulsion of the *khorion* (PLINY 20.214). Illustrations accompanied exegeses of botanical properties, as for DIONUSIOS (OF MILĒTOS?) and KRATEUAS (25.8). He was presumably the doctor listed after TLĒPOLEMOS and before SOLŌN (1.*ind.*20–27). Our pharmacist, perhaps identifiable with the Hippokratic commentator cited by ERŌTIANOS under *Epidemics* 5.26, "caul" (*fr.*19, p. 105 Nachm.), may also be the homonymous **Asklēpiadean** pharmacist cited by GALĒN, *Simples* 1.29, 35 (11.432, 442 K.).

RE 15.2 (1932) 1483 (#27), W. Kroll.

GLIM

Mētrodōros son of Epikharmos, pseudo (200 BCE – 100 CE?)

EPIKHARMOS' son, credited by IAMBLIKHOS (*VP* 241) with a medical treatise (probably pseudepigraphical, *cf.* Thesleff 1965: 121–122) where supposedly PYTHAGORAS' teachings were applied.

DPA 4 (2005) 502–503 (#143), Bruno Centrone and C. Macris.

Bruno Centrone

Mētrodōros of Alexandria (*ca* 130 – 170 CE)

SABINUS' student, PHILISTIŌN OF PERGAMON's teacher (GALĒN *CMG* 5.10.1, p. 401), wrote commentaries on the HIPPOKRATIC CORPUS, EPIDEMICS, and was acclaimed with Sabinus as more accurate than previous Hippokratic scholars (*CMG* 5.10.2.1, pp. 17–18). Galēn sharply criticizes numerous interpretative errors, chiding Sabinus and his followers for their unique view of the dangers of pustules (5.10.2.2, pp. 46–47).

Smith (1979) 151, n.71, 152, n.73; Ihm (2002) #176–178; *BNP* 8 (2006) 838–839 (#8), V. Nutton.

GLIM

Mētrodōros of Buzantion (180 – 80 BCE)

Father of LEŌNIDAS OF BUZANTION, mentioned among famous ichthyologists by AELIANUS (*NA*, epilogue).

RE 15.2 (1932) 1482 (#25a), W. Kroll.

Arnaud Zucker

Mētrodōros of Khios (400 – 350 BCE)

Atomist philosopher and student of DĒMOKRITOS. His major work *On Nature* (*peri* ***phuseōs***) combined skeptical views about the possibility of knowledge with an atomic analysis of the nature of reality. Following Dēmokritos, he taught that everything was made up of atoms and the void, and that there are an infinite number of worlds (***kosmoi***). He also discussed meteorology and astronomy.

Ed.: DK 70.

RE 15.2 (1932) 1475–76 (#14), W. Nestle; *KP* 3.1280 (#4), H. Dörrie; Long and Sedley (1987) §1D; *OCD3* 977, W.D. Ross; *ECP* 342, J.S. Purinton; *DPA* 4 (2005) 506–508, R. Goulet; *BNP* 8 (2006) 836–837 (#1), I. Bodnár.

Walter G. Englert

Mētrodōros of Lampsakos (305 – 278 BCE)

Metrodōros of Lampsakos Reproduced with kind permission of the National Archaeological Museum, Athens

Epicurean philosopher who studied under EPICURUS at Lampsakos *ca* 310–307, and moved to Athens with him in 307. Along with Epicurus, HERMARKHOS, and POLUAINOS, he was considered one of the four founders of the Epicurean school. Epicurus dedicated some of his works to him, and he wrote extensively. His works included: *Against the Physicians, On the Senses, Against the Dialecticians, Against the Sophists, Against Dēmokritos*, and *On Change* (DIOGENĒS LAËRTIOS 10.24).

RE 15.2 (1932) 1477–80 (#16), W. Kroll; *KP* 3.1280 (#6), H. Dörrie; Long and Sedley (1987) §21G; *OCD3* 977, D. Obbink; *ECP* 342–343, D.N. Sedley; *DPA* 4 (2005) 514–517, B. Puech and R. Goulet; *BNP* 8 (2006) 837–838 (#2), T. Dorandi.

Walter G. Englert

Mētrodōros of Skēpsis (*ca* 100 – *ca* 70 BCE)

Greek rhetorician and historian, son of KARNEADĒS' disciple Mētrodōros. An impoverished Skēpsian, also interested in philosophy, Mētrodōros, marrying well in Khalkēdōn, became an intimate friend of MITHRADATĒS VI. Appointed a senior judge, Mētrodōros was called the king's father. Some time between 73 and 71 BCE, Eupatōr sent Mētrodōros as an ambassador to Tigranēs of Armenia to ask for military aid against the Romans. Mētrodōros betrayed Eupatōr and died shortly afterwards, probably by the king's order: STRABŌN 13.1.55; PLUTARCH *Luc.* 22. Renowned for his excellent rhetorical style, Mētrodōros wrote on diverse subjects, including a biography of Tigranēs, and treatises on history, habits and gymnastic training. A scholium on LACTANTIUS (F 16) calls Mētrodōros "Periegeticus" suggesting a lost ***periēgēsis***. Fragments of his lost works contain various ethnographic and geographical data on Italy, Greece, Pontos and Kappadokia, and PLINY (7.89, 8.36) used him for geographical and mineralogical information. Possible anti-Roman undertones earned him the nickname "Misoromaios" (Roman-hater).

Ed.: *FGrHist* 184.

J.-M. Alonso-Núñez, "Un historien antiromain: Métrodore de Scepsis," *DHA* 10 (1984) 253–258; P. Pédech, "Deux grecs face à Rome au Ier siècle av. J. C.: Métrodore de Scepsis et Théophane de Mitylène," *REA* 93 (1991) 65–78; *DPA* 4 (2005) 515, T. Dorandi.

Daniela Dueck

Mētrodōros of Tralleis (*ca* 550 – 600 CE)

Grammatikos who compiled or penned 30 epigrams presenting arithmetical puzzles of aliquot parts (*Greek Anthology* 14.116–146). He lived some time after DIOPHANTOS (*cf.* 14.126), and is probably the ***grammatikos*** brother of ANTHĒMIOS and ALEXANDER OF TRALLEIS.

BNP 8 (2006) 839 (#9), M.G. Albiani.

PTK and GLIM

Mikiōn (100 – 40 BCE?)

His *Rhizotomoumena* is cited after PETRIKHOS by PLINY 20.258, as prescribing *hippomarathron* for snake bite, apparently in reference to NIKANDROS, *Thēr.* 596. (For *hippomarathron* Durling 1993: 185 suggests either *Prangos ferulacea* [L.] Lindl. or *Cachrys ferulacea* [L.] Calest.; *cf.* DIOSKOURIDĒS *MM* 3.71; GALĒN, *Simpl.* 7.12.5 [12.67–68 K.].) Mikiōn is also cited by the *Schol. Nik. Thēr.* 617, on *tithumallos* ("petty spurge" and other names, i.e., *Euphorbia peplus* L.: Dioskouridēs *MM* 4.164; Galēn, *Simples* 8.19.7 [12.141–143 K.]; Durling 1993: 311). The name, rarely spelled with -kk-, is almost unknown after the 1st c. BCE (*LGPN*).

RE 15.2 (1932) 1555 (#5), W. Kroll.

PTK

Milēsios (280 BCE – 120 CE)

Wrote on seminal ducts, denying physiological distinction between nocturnal emissions resulting from dreams of coitus wherein semen is completely discharged (*oneirōgmos*) and is not (*oneiropolēsis*: probably SŌRANOS, in CAELIUS AURELIANUS, *Chron.* 5.82 [*CML* 6.1.2, p. 904]). He also believed that weakness in seminal ducts results in discharging blood rather than semen during coitus (*Chron.* 5.87 [p. 906]). The rare name is attested from the 3rd c. BCE: *LGPN* 1.314, 3A.301, 3B.286, 4.337.

Fabricius (1726) 338.

GLIM

Milōn (450 – 300 BCE?)

IŌANNĒS OF STOBOI 1.29.3 records Milōn's theory that lightning is produced when water is "broken" (*rhag-*), diurnally by the sun, and nocturnally by the stars. The latter claim suggests an early date, when the stars were imagined as nearby. The name is most frequent in the 4th–3rd cc. BCE, but is attested as late as the 3rd c. CE: *LGPN* 1.314, 2.315, 3A.301.

RE 15.2 (1932) 1677–1678 (#7), W. Kroll and A. Modrze; *DPA* 4 (2005) 522, R. Goulet.

PTK

Miltiadēs (250 BCE – 120 CE)

Physician, perhaps **Erasistratean**, argued that some diseases are exclusive to women (SŌRANOS, *Gyn.* 3.2 [*CMG* 4, p. 94; *CUF* v. 3, pp. 2–3]). The name is very common at Athens (*LGPN* 2.314–315), but rare elsewhere (*LGPN* 1.314, 3A.301).

RE 15.2 (1932) 1705 (#7), K. Deichgräber.

GLIM

Mīnarāja (*ca* **300 – 325 CE**)

Mīnarāja was a *yavanādhirāja*, i.e., person of authority in the settlements of Greeks under the western Kṣatrapas in what is now Gujarat and Rajasthan in western India. He wrote a long astrological compendium, the *Vṛddhayavanajātaka*, covering every subject of astrology, in 71 chapters. The work is based on SPHUJIDHVAJA's *Yavanajātaka* and a lost work of Satya. Pingree suggests that the first part of his name, "mīna," is a designation of the Śakas, i.e., Indo-Skuthians.

CESS A.4.427–429, A.5.310; Pingree (1978) 1.24, n. 75.

Kim Plofker and Toke Lindegaard Knudsen

Minius Percennius of Nola (200 – 150 BCE)

Agronomist who "demonstrated" a superior method for sowing seed of the Tarentine cypress (PORCIUS CATO, 151). It is unclear how Cato learned of his method: whether from personal contacts, or from a treatise, written perhaps in Latin, Greek, or even Minius' native tongue, Oscan.

RE 19.1 (1937) 588 (#1), F. Münzer; Speranza (1971) 11–13.

Philip Thibodeau

Minucianus (10 – 80 CE)

ANDROMAKHOS approves his recipe for scrofula (beeswax, **galbanum**, propolis, **terebinth**, and mistletoe from oak, add lees and natron, set on coals and add olive oil, wild cucumber root, gladiolus bulb, and "Asian flower"): GALĒN, *CMGen* 6.14 (13.930–931 K.). Minucianus also preserved an antidote from ZĒNŌN OF LAODIKEIA, see Galēn, *Antid.* 2.20 (14.163 K.).

RE 15.2 (1932) 1988 (#4), K. Deichgräber.

PTK

Minuēs (*ca* **500 BCE –** *ca* **200 CE**)

DIOGENĒS LAĔRTIOS 1.27 cites Minuēs for the tale that THALĒS associated with Thrasuboulos, tyrant of Milētos; probably, like SŌTIŌN, he is *ca* 200 BCE. The name is otherwise unattested (*LGPN*) but accepted by Pape-Benseler, and may mean "a Minyan" (i.e., from Orkhomenos, destroyed in 368 and 346 BCE, and mostly abandoned after 85 BCE), like the eponymous hero of the Minyans, Paus. 9.36.4, or else "an informer" (*mēnuēs*).

FGrHist 1111.

PTK

Mithradatēs VI, King of Pontos (*ca* **115 – 63 BCE**)

Born 132 BCE in Sinōpē, son of King Mithradatēs V (d. 120 BCE); he deposed his regent mother Gespaepuris *ca* 115 BCE, and expanded his realm, allying with Armenia. By *ca* 95 he had come into conflict with Roman interests, after which he conquered Bithunia and allied with many Greek cities, including Ephesos, Milētos, Pergamon, and Athens, defecting from Rome in 90. More or less continuous warfare ensued for 25 years, Mithradatēs representing himself as the savior of Hellenism, until his defeat by Pompey and suicide in 63; his treasury immoderately enriched the Roman Republic. Like contemporary kings (ANTIOKHOS, ATTALOS, and NIKOMĒDĒS), he practiced pharmacy, with the legendary intent of immunizing himself against all poisons, by often sampling each: PLINY 25.5–7; GALĒN *Antid.* 1.1 (14.2–5 K.). POMPEIUS LENAEUS translated his Greek into Latin, and numerous pharmacists of the 1st c. CE record antidotes alleged to be his, from a simple one, "in his own hand," of walnuts, figs, and rue (Pliny 23.49; *cf.* Pliny Jr. 3.33.4, Gellius 17.16), through the earliest known, of 37 ingredients, CELSUS 5.23.3, to one of 54 ingredients, Pliny 29.24 (no recipe). The more complex recipes all include cinnamon (usually with cassia), ***kostos***, myrrh, pepper, and saffron, and most add frankincense, parsley, and ***skordion***: ANTIPATROS and KLEOPHANTOS in Galēn, *Antid.* 2.1 (14.108), Celsus, DAMOKRATĒS *ibid.* 2.2 (14.115–117), XENOKRATĒS *ibid.* 2.10 (14.164–165), ANDROMAKHOS, *ibid.* 2.1 (14.107), 2.7 (14.148), and 2.9 (14.152–155), and ASKLĒPIADĒS PHARM. in Galēn *CMLoc* 10.1 (13.329–330 K.). The last two pharmacists also credit him with throat-remedies, in Galēn *CMLoc* 7.2 (13.23–25, 52–56), all containing cinnamon, cassia, frankincense, myrrh, and saffron. KRATEUAS named a plant *mithradatia* (Pliny 25.62), and agrimony was known as *eupatoria* in his honor (25.65; *cf.* DIOSKOURIDĒS 4.41). He was the subject of a play by Racine (1673), an Italian opera by Mozart (1770), a poem by Housman (1896), and several 20th/21st c. English novels.

Fr. de Callataÿ, *L'histoire des guerres mithridatiques vue par les monnaies* (1997) 235–388.

PTK

Mnaseas (Method.) (54 – 68 CE)

Physician listed among the **Methodists**; SŌRANOS accepts him as one of "his" sect: "Mnaseas says that some [women] are by nature healthy, but others are by nature less than healthy, and among those who are less than healthy some are more constricted (*stegnoteron*) than not, some are more 'flowing' (*rhoōdesteron*, i.e. 'lax' or 'unconstricted') than not" (*Gyn.* 1.6.29 [*CMG* 4, p. 19; *CUF* v. 1, p. 24]). Similarly **Methodist** is Mnaseas' bipartite diagnosis of lethargy: one kind is from a state of stricture, another kind from a state of laxity (*solutio*: CAELIUS AURELIANUS, *Acut.* 2.24 [Drabkin, p. 134; *CML* 6.1.1, p. 144]); and he thinks that paralysis is caused by contraction (*paraleipsis*) saying that sometimes paralysis is a constriction (here *extentio*) and sometimes a loosening (*solutio; Chron.* 2.16 [Drabkin, p. 574; *CML* 6.1.2, p. 554]). Later, PSEUDO-GALĒN, INTRODUCTIO 4, lists him among the **Methodists**: "after THESSALOS OF TRALLEIS, then Mnaseas, DIONUSIOS, PROKLOS, ANTIPATROS" (14.684 K.). Once, Sōranos compares (or contrasts?) him with HĒROPHILOS: "Hērophilos and Mnaseas – although basing their opinions on differing doctrines – both state that in some women, menstruation is health-producing, in others it is not" (*Gyn.* 1.6.27 [*CMG* 4, p. 17; *CUF* v. 1, p. 22]). Nevertheless, perhaps because SEXTUS EMPIRICUS, *Pyrrh.* 1.34 (esp. 1.34.236–237), describes a physician who combined **Methodism** with Skepticism, some modern scholars have ranked Mnaseas with the Skeptics (Deichgräber 1930/1965: 267, n.2; *cf.* Tecusan, pp. 60–61). Mnaseas is typical of **Methodists**, who were rarely rigidly sectarian.

Mnaseas' effective and simple plaster is recommended by several authorities, **Methodist** and not. PAULOS OF AIGINA, 7.27.21 (*CMG* 9.2, p. 353), gives its basic recipe of five common and easily compounded ingredients: one *litra* each of beeswax and pig's fat ("lard"), six ounces of scammony-resin (*Convolvulus scammonia* L.), two *litrai* of **litharge**, mixed with four *litrai* of good wine. As Paul says, this is an excellent "diaphoretic," i.e., a "discutient" (a common property in salves and plasters before 1920, when they were sometimes termed "resolvents," drugs that could dissipate pus in a wound). Galēn harshly criticizes the plaster as overly-simplistic; his typically legalistic attack seemingly demolished the theoretical usefulness of such a homely and ordinary five-ingredient drug (*CMGen* 7.5 [13.962–966 K.], *cf.* 1.4 [13.392 K.]: "Mnasaios"). Significantly, Galēn's criticism does not address the scammony. Asklēpiadēs Pharm., in Galēn *CMGen* 1.17 (13.445 K.), records another plaster of "Mnasaios": 100 *drachmai* of **litharge** and ***psimuthion***, 50 of beeswax, 25 each of **terebinth** and frankincense, 12 of alum, in two cups of olive oil.

Ed.: Tecusan (2004) 85–86, 99, 104–105 ("Thematic Synopsis: Mnaseas").
RE 15.2 (1932) 2247 (s.v. Mnasaios), H. Raeder; 2252–2253 (#7), K. Deichgräber.

John Scarborough

Mnaseas of Milētos (90 – 40 BCE)

Wrote a treatise on agriculture known to Varro, *RR* 1.1.9. According to Columella, 12.4.2, he discussed the preservation of foodstuffs, "following Mago" – presumably he read his source in Cassius Dionusios' translation.

RE 15.2 (1932) 2253 (#8), R. Laqueur and W. Kroll.

Philip Thibodeau

Mnaseas of Patara (215 – 175 BCE)

A student of Eratosthenēs, wrote a compilation of myths and *thaumasia*, probably entitled ***Periplous*** or ***Periēgēsis***, organized geographically, three chapters being entitled "On Europe," "On Asia," and "On Libya." In addition, Mnaseas authored *Peri khrēsmōn*. He tries to explain mythical stories rationally and genealogically. Judging from extant fragments, compared with parallel traditions, Mnaseas seems to have followed his sources quite faithfully and added few inventions of his own.

Ed.: P. Cappelletto, *I frammenti di Mnasea: Introduzione testo e commento* (2003).
POxy 13 (1919) #1611; H.J. Mette, *Lustrum* 21 (1978) 39–40; *OCD3* 992, K.S. Sacks; *BNP* 9 (2006) 93 (#2), G. Damschen.

Jan Bollansée, Karen Haegemans and Guido Schepens

Mnēmōn of Sidē (245 – 220 BCE)

Student of Kleophantos who brought to Alexandria from Sidē a copy of the Hippokratic Corpus, Epidemics 3, annotated with marks whose interpretation exercised generations of Alexandrian commentators; Galēn doubts their authenticity: *In Hipp. Epid. III* (*CMG* 5.10.2.1, pp. 77–80, 87, 157). They may be notes written in the epichoric Sidetan script.

Ihm (2002) #179.

PTK

Mnēsarkhos of Athens (*ca* 110 – 90 BCE)

Taught by Diogenēs of Babylōn, Antipatros of Tarsos and Panaitios (Cicero *De Or.* 1.45–46), Mnēsarkhos, son of Onēsimos of Athens, was the **Stoic scholarch** at Athens (Cic. *Acad. Pr.* 2.69). He taught that the primary substance of the universe (*prōtē ousia*) was located in ***pneuma*** (Iōannēs Stobaios 2.29.24 = Diels 1879: 303), and that language and procreation were not rational faculties but sensory faculties shared by all animals (PSEUDO-GALĒN HIST. PHIL. 24 = Diels 1879: 615).

GGP 4.2 (1994) 661–662, P. Steinmetz; *ECP* 349, T. Dorandi; *DPA* 4 (2005) 538–542, R. Goulet.

PTK and GLIM

Mnēsidēmos (200 – 120 BCE)

DIOSKOURIDĒS, *MM* 4.64.6, cites ERASISTRATOS, DIAGORAS OF CYPRUS, and ANDREAS, (all *ca* 250–200 BCE), on the use of the opium-poppy, after whom Mnēsidēmos restricted it to sleeping-draughts. The name is attested up through the mid-2nd c. BCE (*LGPN*).

RE 15.2 (1932) 2275 (#2), K. Deichgräber.

PTK

Mnēsidēs (300 BCE – 77 CE)

Cited as a foreign authority on scents from trees (PLINY 1.*ind*.12–13), an expert on medicines from botanics (20–27), and an expert on metals (33–35). Pliny cites only his opinion that henbane seed is the best preservative of opium (20.203). Mnēsidēs may be corrupted from Mnēisidēs (Athens, 3rd c. BCE), Mnēsiadēs (cited eight times at Athens, 6th–3rd cc. BCE: *LGPN* 2.316, and four times at Dēlos, late 3rd c. BCE: *LGPN* 1.317), or Mnasiadēs: (3rd–2nd cc. BCE: 3A.303, 3B.288). Our author is possibly identifiable with MNĒSIDĒMOS, who wrote on opium preparation.

RE 15.2 (1932) 2275, K. Deichgräber.

GLIM

Mnēsimakhos of Phasēlis (400 – 200 BCE)

Composed a work on Skuthia, which recorded myths about the north, and assigned the region to Europe; the few fragments are preserved in the scholia to Apollōnios Rhodios.

FGrHist 841; *BNP* 9 (2006) 101 (#2), M. Baumbach.

PTK

Mnēsitheos of Athens (370 – 330 BCE)

Greek physician, mentioned, together with DIEUKHĒS, in an Athenian votive inscription dedicated to Asklēpios (350 BCE: *IG* II^2 1449: citing either our physicians or their families), and quoted in Alexis' comedy *Foster Brothers* (370–280 BCE: *fr.*219 *PCG* = Ath., *Deipn.* 10 [419b]). Mnēsitheos is usually included among the **dogmatic** physicians with DIOKLĒS (whom he postdates: GALĒN, 17B.608 K). Pausanias mentions his grave in Athens, not far from the Kēphisos, near the altar of Zeus Meilikhios (1.37.4).

Mnēsitheos followed, but also innovated, "Hippokratic" humoral etiology and developed the difference between **humors** (*khumoi*) and savors (*khuloi: frr.*12–15). He approved the theory of the innate heat and ***pneuma*** and tried to systematize his medical theories using the diairetical **Platonic** method. Much interested in dietetics, he said that health is maintained through similes and disease is cured by opposites (*fr.*11). He treated specialized subjects: his *Letter to Lukiskos* (an Athenian archon of 344/343 BCE), devoted to infants' care, seems a polemical answer to PLATO's regimen of children (*Laws*, VII). His *Letter on Tippling* reveals the importance given to wine both as nutriment and as drug (*frr.*41, 45–47). More general texts are *On Edibles* or *On the Properties of Foods*, whence Athēnaios, Galēn and OREIBASIOS preserve many literal quotations. He addressed morpho-pathology in *On the Construction of the Body*, examining the proportion of individual body parts and their predisposition to disease. He wrote also on therapy discussing the use of hellebore and of clysters. A work *Pathology* is mentioned (PSEUDO-GALĒN *Def. Med.* 19.457 K.).

J. Bertier, *Mnésithée et Dieuchès* (1972); G. Wöhrle, *Studien zur antiken Gesundheitslehren* (1990) 160–169; *AML* 623–624, R. De Lucia; *BNP* 9 (2006) 102, V. Nutton.

Daniela Manetti

Mnēsitheos of Kuzikos (200? – 160 BCE)

Composed a work on the virtues of cabbage, OREIBASIOS *Coll.* 4.4 (*CMG* 6.1.1, p. 100), a version of which appears in CATO, *Agric.* 156–157 and PLINY 20.80–81. He eschewed hellebore as dangerous, Oreib., *Coll.* 8.9 (p. 261), and described the testing of human milk, for color, smell, taste, and even viscosity, by storing it overnight in a glass, horn, or shell (i.e., non-reactive) vessel, *ibid.* inc.32 (6.2.2, pp. 124–126). He, or his Athenian homonym, also wrote on anatomy, *cf. ibid.* inc.7 (p. 84) and 8.38 (6.1.1, pp. 288–290).

R.M. Grant, *Dieting for an Emperor* (1997) 300–302.

PTK

MODERATUS ⇒ IUNIUS MODERATUS COLUMELLA

Moderatus of Gadēs (*ca* 25 – 75 CE)

Neo-**Pythagorean** philosopher, wrote *Lectures on Pythagoreanism* in ten or 11 books. PLUTARCH's portrayal of Moderatus' student Lucius as an interlocutor in *Quaestiones Conviviales* (8.7–8) establishes Moderatus' approximate date as well, perhaps, as his observance of strict **Pythagorean** asceticism. The *Lectures*, quoted in PORPHURIOS' *Vita Pythagorae* (48–53), were a source from which Porphurios and other authors of late antiquity apparently drew much information on **Pythagorean** teachings. IŌANNĒS STOBAIOS, *Anthologium* 1 (p. 21 W.-H.) reports Moderatus' metaphysical definitions of number and monad.

Dillon (1996) 344–351.

Alexander Jones

M. Modius Asiaticus (30 – 90 CE)

An inscribed honorific bust from Smurna (*CIG* 3283 = Kaibel #306 = *fr.*12 Tecusan) records this **Methodist**; he may be the pharmacist whose name is transmitted as KŌDIOS TOUKOS.

G. Kaibel, *Epigrammata Graeca* (1878/1879; repr. 1965); Schefold (1997) #204.

PTK

Molpis (250 – 50 BCE?)

Listed (with PHULOTIMOS, EUĒNŌR, NEILEUS, and NUMPHODŌROS) by HĒRAKLEIDĒS OF TARAS, in GALĒN, *Comm. in Hipp. Artic.* 4.40 (18A.735–736 K.), as having reduced dislocations of the thigh. He, or PERIGENĒS, invented the *thaïs* bandage: pseudo-Galēn, *de Fasciis* 16 (18A.789 K.).

Michler (1968) 47, 98; van der Eijk (2000–2001) *fr.* 164.

PTK

Monās (350 – 270 BCE)

Mentioned by THEOPHRASTOS, *Sweat* 12, as theorizing about sweat. (The rare name is attested from Epidauros, *LGPN* 3A.305, and Egyptian Thēbai, *CIG* 4951.)

(*)

PTK

MOPSEATĒS ⇒ SALLUSTIUS

***Mō(u)sēs* (600 – 800 CE)**

Alchemist, perhaps pseudonymous and alleging to be the prophet Moses. A brief late alchemical recipe for doubling the weight of gold (*The Doubling of Mōsēs: CAAG* 2.38–39) may in fact be the work of the alchemist PAPPOS (Letrouit 1995: 87). A *Domestic Chemical Treatise* is ascribed to the prophet Mōusēs in an anonymous Byzantine discussion on dying stones (*CAAG* 2.353). The text entitled *Chemistry of Moses* by Berthelot (*CAAG* 2.300–315) is an acephalic collection of alchemical recipes, following a brief introduction (paraphrasing *Exodus* 31.1–5) in which God tells Mōusēs that he gave Beseleēl mastery over metal, stone and woodworking (*CAAG* 2.300). However, the rest of the text refers neither to Mōusēs nor this story.

(*)

Bink Hallum

Moses of Xoren (Arm., Movsēs Xorenac'i) (traditionally 400 – 500 CE: disputed)

Known as the "father of Armenian history" (Arm. *patmahayr*), there is next to nothing known of his life. Explicit references to him or to his work are not found before the mid-9th c. His *History of Armenia* traces the beginnings of the Armenian people from a descendant of Noah to the middle of the 5th c. He provides the most detailed account of pre-Christian Armenia and makes use of much – some otherwise unattested – archival material, as well as extensive use of Greek historical and scientific works. In addition to his *History*, Moses, who claims to have worked as a translator (III.65), was long considered to have been the author of a *Geography of Armenia*, which is clearly dependent on Greek sources, such as PAPPOS and PTOLEMY. This attribution, however, is found only in late MSS (post 17th c.); the work is now generally ascribed to ANANIA OF SHIRAK.

J. Marquart, *Ērānšahr nach der Geographie des Ps. Movsēs Chorenac'i* (1901); see also sources cited on Anania of Shirak.

Edward G. Mathews, Jr.

Moskhiōn or Moskhos (220 – 180 BCE)

Moskhiōn's treatise *Mēkhanika* described every aspect of the construction of HIERŌN II's massive ship *Surakousia* overseen by Arkhias, ARCHIMĒDĒS, and Phileas of Tauromenion: sources and preparation of materials, workmen and workmanship, adornments, including artwork and a private library, launching the ship, battlements, siege engines, defensive leather "shields," masts, and a screw for pumping out bilge water (Ath., *Deipn.* 5 [206d-209e]). Moskhiōn also described the invention of the ***sambukē*** by HĒRAKLEIDĒS OF TARAS (*idem*, 14 [634b]).

RE 16.1 (1933) 356 (#7), K. Orinsky.

GLIM

Moskhiōn (Pharm.) (90 BCE – 80 CE)

Within a seriatim listing of pharmaceutical recipes culled from ASKLĒPIADĒS PHARM., GALĒN inserts a formula from the books of the "very familiar" or "celebrated" (*gnōrimos*) Moskhiōn, an always-reliable compound that removed calluses and heavy scar-tissue (*CMGen* 2.14 [13.528–529 K.]), made from notably caustic simples, including **litharge**, ***psimuthion***, and quicklime (*asbestos*), fashioned into a plaster using deer marrow, beeswax, and myrtle oil (*cf.* KRITŌN, *ibid.* 5.3 [13.787–794]); such ingredients were typical in the pharmaceutical cosmetics of the day, and one notes similar substances especially in the treatment of *alōpekia* (*cf.* Moskhiōn [emended from Moskhos] in Galēn, *CMLoc* 1.2. [12.401 K.]: sea urchins + ashed shells, and *ibid.*, [12.416 K.] cat or crocodile dung, bear fat, ashed frog, sharp vinegar, white hellebore, among several). Galēn there indicates that Moskhiōn was one of a group of pharmacologists whose collection of recipes he has consulted, including also Asklēpiadēs "the Pharmacist" and HĒRAKLEIDĒS OF TARAS. Galēn's excerpts from this handbook show Moskhiōn and others specializing in wound treatments, the manufacture of collyria, and ***artēriakai***. Moskhiōn understood the narcotic properties of opium latex and mandrake, illustrated by a collyrium-formula, also noted as invented by Moskhiōn *gnōrimos* (Galēn, *CMLoc* 4.8 [12.745 K.]). Moskhiōn's styptic wound-clotter (Asklēpiadēs Pharm. in *CMGen* 2.17 [13.537–539 K.; *cf.* 13.528 and 646–647]), good for fractures, hemorrhoids, and other bleeding skin-lesions, is a complex, 14–ingredient, multi-staged preparation, altered somewhat as "fashioned by our mentor LEUKIOS (*ho hēmeteros kathēgētēs Leukios*)," and includes **litharge**, decocted pine-pitch, frankincense, beeswax, and fig-juice, to be applied with wine and sharp vinegar. Moskhiōn followed ASKLĒPIADĒS OF BITHUNIA on pulsation as arising from the heart, veins, arteries, and the brain, to emerge as a single pulsation via the meninges (Galēn, *Puls. Diff.* 16 [8.758–759 K.]), and is thereby grouped with those called **Asklēpiadeans**, even though Galēn is unusually mild with his criticism in these passages. APHRODĀS (known to ANDROMAKHOS) followed Moskhiōn's work, providing the *terminus ante* of 80 CE; if we emend the *MOSCHI* of CELSUS 5.18B.10 to *MOSCHIONIS*, his *terminus ante* could even be 40 CE. In either case, he is probably the man cited by PLINY 19.87, for a book on the radish.

RE 16.1 (1933) 349–350 (#9), K. Deichgräber; *BNP* 9 (2006) 227 (#4), V. Nutton.

John Scarborough

Mousaios "the boxer" (350 BCE – 75 CE)

Prescribed the rubbing of decapitated *muloikon* beetles on the skin, for *lepra*, according to PLINY 29.141 (*cf.* MOSKHIŌN). For a boxer as medical writer, *cf.* FLAUIUS or THEŌN OF ALEXANDRIA (MED. I). The *Musaeus* cited by Pliny 1.*ind*.21–27, with HOMER, HĒSIOD, and Sophoklēs the tragedian, presumably intends the early Greek prophet.

(*)

PTK

Mucianus (560 – 590 CE)

Translated into Latin GAUDENTIUS' treatise on musical theory (CASSIODORUS *Inst.* 2.5.2), as well as 34 homilies of IŌANNĒS KHRUSOSTOMOS on the Christian *Letter to the Hebrews* (1.8.3).

RE 16.1 (1933) 411 (#3), W. Enßlin.

PTK and GLIM

Muia, pseudo (250 BCE – 150 CE)

Neo-**Pythagorean**; daughter of PYTHAGORAS and wife of Milōn of Krotōn according to PORPHURIOS, *V.Pyth.* 4 and IAMBLIKHOS, *V.Pyth.* 267. An apocryphal letter to Phyllis has been transmitted under her name. The letter indicates how to hire a wet-nurse and particularly emphasizes measure and balance in the child's upbringing.

Ed.: Thesleff (1965) 123.5–124.8.

H.J. Snyder, *Woman and the Lyre* (1989) 110–111 (trans.); A. Städele, *Die Briefe des Pythagoras und der Pythagoreer* (1980) 267–281 (comm.); *DPA* 4 (2005) 573–574, Bruno Centrone.

Bruno Centrone

Mulomedicina Chironis (***ca*** **300 CE?)**

Two closely related Latin MSS, both from the second half of the 15th c., transmit the single most comprehensive work on equine medicine that has survived from antiquity. Probably the main source of VEGETIUS' *Digesta artis mulomedicinalis*, its redaction goes back to the 4th if not the 3rd c. CE. A number of passages preserved in Greek within the Greek collection of veterinary writers (*Hippiatrika*) provides close parallels, making it almost certain that the *Mulomedicina* is mainly based on Greek writings now partly lost, but preserved here in Latin translation. Among its ten books, the structure of Books 3 and 4 (§§114–421) closely resembles the structure of the *Hippiatrika* in presenting extracts from a number of writers excerpted in sequence. Similarly, a collection of recipes (starting in §796 and extending to the very end = §999) concludes the treatise. Book 1 is devoted to phlebotomy and cauterization, Book 8 (§§741–774) to reproduction, while the books in between roughly follow the order *a capite ad calcem* (from head to hoof). This order seems to have been disturbed at an early time during the transmission, because even Vegetius complains about it in his preface. It is unclear how often the text was redacted or augmented; the presentation echoes that of therapeutic manuals on human medicine from Imperial times, and it is evident how much the authors (apart from APSURTOS, parts of whose work survive within the *Hippiatrika*, other names – e.g. SŌTIŌN, PHARNAX, and – a clear pseudonym – Chiron the Centaur, cannot be linked to known fragments) wished to achieve a standard of diagnosis and therapy on a par with human medicine. Accordingly, medical historians must pay more attention to this source than has hitherto been the case. Book 1 and some other passages obviously derive from the doctrine of the **Methodist** school of medicine and have not been exploited adequately. Nevertheless, after the first (and to date only complete) edition in 1901, the language of the *Mulomedicina* has been studied intensively by Latinists for whom it constituted a very important source of vulgar Latin (inspired by remarks in Vegetius' preface). While this view deserves to be challenged or bolstered with fresh arguments, the *Mulomedicina* remains one of the most important (and often puzzling) sources of technical Latin and veterinary expertise in late antiquity.

Ed.: E. Oder, *Claudii Hermeri Mulomedicina Chironis* (1901); other editions and translations in *BTML* 409–422.

K. Hoppe, *Die Chironfrage* (1933); *RE* 16.1 (1933) 503–513, K. Hoppe; Klaus-Dietrich Fischer, *HLL* §513; W. Sackmann, "Eine bisher unbekannte Handschrift der Mulomedicina Chironis aus der Basler Universitätsbibliothek," *ZWG* 77 (1993) 117–119; Önnerfors (1993) 370–380; Adams (1995).

Klaus-Dietrich Fischer

Muōnidēs (*ca* 150 – 50 BCE)

Neo-**Pythagorean** musician who some time after Eratosthenēs, together with Euphranōr, discovered four new means (*mesòtēs*), added to the six already known (Iamblikhos *in Nikom.* 2.28.6–11 [p. 116]). The name is apparently attested only on Rhodes, in the 1st c. BCE: *LGPN* 1.323.

M. Timpanaro Cardini, *I Pitagorici. Testimonianze e frammenti* (1962) 2.436–439; *DPA* 4 (2005) 575, Bruno Centrone

Bruno Centrone

Murōn (250 BCE – 25 CE)

Celsus records two dermatological recipes from Murōn, against ***leikhēn*** (5.28.18B), containing raw sulfur, red natron, **terebinth**, pine pitch, frankincense, etc., and against *alphos* (5.28.19D), containing sulfur, natron, alum, and myrtle.

RE 16.1 (1933) 1115 (blind cross-reference).

PTK

Mursilos of Mēthumna (300 – 250 BCE)

Wrote a local history of his native island Lesbos (*Lesbiaka*) and a paradoxographical treatise (*Historika* ***Paradoxa***). The former was cited by Antigonos of Karustos (as indicated in the first part of *Antigonou Historiōn* ***paradoxōn*** *sunagōgē*, at 5; 15.3; 117–118), which establishes an early 3rd c. date for Mursilos. Later writers to use his works include Dionusios of Halikarnassos (*A.R.* 1.23.1–5; 1.28.4), Strabōn (1.3.19; 13.1.58), Pliny (3.85; 4.65), Plutarch (*Arat.* 3.5; *De soll. anim.* 36 [984E]), Athēnaios (*Deipn.* 13 [609f-610a]), and Clement of Alexandria (*Protr.* 2.31). Hardly any of the scanty surviving fragments can be assigned with certainty to either of the two known treatises, as they seem to have featured the same mix of historical, etymological, and paradoxographical data. Mursilos was perhaps the earliest author to collect *mirabilia* of contemporary life alongside natural (botanical and ornithological) wonders.

Ed.: *FGrHist* 477; *PGR* 29–30.

RE 18.3 (1949) 1137–1166 (§7, 1143), K. Ziegler; Giannini (1964) 116–117; S. Jackson, *Myrsilus of Methymna: Hellenistic Paradoxographer* (1995); *BNP* 9 (2006) 422 (#2), K. Meister.

Jan Bollansée, Karen Haegemans, and Guido Schepens

MUSA ⇒ ANTONIUS

Muscio/Mustio (440 – 460 CE)

Otherwise unknown physician, resident in Roman north Africa; extant is an extended Latin catechism (*viz.* "question-and-answer" format) on women's diseases and midwifery titled *Gynaecia* or *De muliebribus passionibus*, generally based on the *Gynecology* of SŌRANOS OF EPHESOS. The MSS couple this work with a Latin *Genesia*, attributed to a "Kleopatra," probably of the 4th or 5th c. The author had access to texts varying from those cited by CAELIUS AURELIANUS and other, near-contemporary, medical writers in north Africa, illustrated by the mention of SŌSTRATOS as physician to KLEOPATRA VII (26.78: *Apollonius et Sostratus et Filoxenus adseuerant. . .* [ed. Rose, p. 106]). Perhaps circulating was the lost "medical journal" of Sōstratos, a physician attending Kleopatra in 30 BCE, and witness to the famous suicide. Later copyists fused some of Mustio with *Gynaeciae* produced by Caelius Aurelianus and Kleopatra, as well as several other writers on surgery and gynecology, in a 13th c. MS luckily recovered in 1948 through a Zurich antiquities sale catalogue.

Ed.: V. Rose, *Sorani Gynaeciorum vetus translatio Latina nunc primum edita cum additis Graeci textus reliquiis a Dietzio repertis atque ad ipsum Codicem Parisiensem* (1882) 3–167; M.F. Drabkin and I.E. Drabkin, *Caelius Aurelianus Gynaecia. Fragments of a Latin Version of Soranus' Gynaecia from a Thirteenth Century Manuscript* = *BHM* S.13 (1951).

J. Ilberg, *Die Überlieferung der Gynäkologie des Soranos von Ephesos* (1910); J. Medert, *Quaestiones criticae et grammaticae ad Gynaecia Mustionis pertinentes* (1911); Önnerfors (1993) 331–336.

John Scarborough

N

Naburianos (Naburimannu) of Babylōn (*ca* 50 BCE)

Known to Greeks as a Babylonian *mathēmatikos* (astronomer), together with KIDĒNAS and SUDINĒS (*cf.* STRABŌN 16.1.6). Naburianos is assumed to be the Greek version of the Babylonian name Nabū-rimannu or Nabū-rimanni appearing in the colophon of a Babylonian astronomical cuneiform tablet (*ACT* #18, lower edge of reverse 1). The tablet is broken, however, so the reading is uncertain. The colophon designates the tablet as a *tersētu* or "computed table" of Nabū-rimannu, giving dates and positions in the **ecliptic** of new and full moons for the year 49–48 BCE and is among the youngest extant cuneiform lunar **ephemerides** of System A. Consequently, the report of Naburianos being an inventor of Babylonian astronomy is unfounded.

ACT p. 23.

Francesca Rochberg

Naukratēs (200 – 180 BCE)

Geometer who encouraged APOLLŌNIOS OF PERGĒ to study conic sections when he visited Alexandria (*Kōnika* 1.*pr.*), and received an uncorrected, unrevised copy of the *Kōnika* before setting sail.

RE 16.2 (1935) 1954 (#4), K. Orinksy.

GLIM

Naukratitēs medicus (250 BCE – 25 CE)

SCRIBONIUS LARGUS in ASKLĒPIADĒS PHARM., in GALĒN *CMLoc* 4.7 (12.764 K.), records a collyrium containing **calamine**, copper flakes, iron flakes, roasted lead, rose juice, acacia, gum, myrrh, nard, saffron, and opium, credited to a "Naukratitēs medicus." No medical writer is known from Naukratis (and only one scientist, STAPHULOS), although two anethnic Egyptian pharmacists, HERMŌN and NEILAMMŌN, compounded collyria, Hermōn's also including myrrh, nard, saffron, and opium (plus MANETHŌN, known for a wound-ointment). The text may conceal *NEAΠOΛITAN-* (i.e., GAIUS OF NAPLES) or an otherwise unknown physician "Naukratēs." SŌRANOS, *Gyn.* 3.32.7 (*CMG* 4, p. 115; *CUF* v. 3, p. 35), and Galēn, *Sanit.* 4.5.12, 4.7.18, 6.7.18, 6.10.23–35 (*CMG* 5.4.2, pp. 117, 125, 182, 188–189), record a "Diospolitikos" ointment, as if from Diospolis; perhaps likewise the collyrium of the "Naukratite" physician is simply from Naukratis.

Fabricius (1726) 344.

PTK

Nausiphanēs of Teōs (340 – 320 BCE)

Dēmokritean philosopher and teacher of the **atomist** EPICURUS, he was influenced by the Skepticism of Pyrrho and wrote an epistemological work called the *Tripod*. Epicurus' major work on epistemology, the *Canon*, was partially a response to it. Nausiphanēs' interests included physics, mathematics, ethics, music, and rhetoric.

DK 75; Long and Sedley (1987) §1B; *OCD3* 1029, D.N. Sedley; *ECP* 352, D. Konstan; *DPA* 4 (2005) 585–586, R. Goulet; *BNP* 9 (2006) 552–553, I. Bodnár.

Walter G. Englert

Nautelēs (*ca* 400 – 250 BCE?)

CENSORINUS 18.5 lists HARPALOS, the otherwise unknown Nautelēs, MENESTRATOS (I), and DŌSITHEOS as writers of works on the ***oktaetēris***.

(*)

PTK

Neanthēs of Kuzikos (330 – 30 BCE?)

Greek historian who wrote a collection of biographies, *On famous Men*, that mainly dealt with the lives of philosophers up to the generation of PLATO.

FGrHist 84; *RE* 16.2 (1935) 2108–2110, R. Laqueur; *DPA* 4 (2005) 587–594, P.P. Fuentes González.

Jørgen Mejer

Nearkhos (60 BCE – 80 CE)

ANDROMAKHOS, in GALĒN *CMLoc* 8.7 (13.204 K.), gives his liver-pill recipe, containing agrimony (after MITHRADATĒS: *cf.* DIOSKOURIDĒS 4.41), arugula seed, elecampane, eryngo, gentian (*cf.* GENTHIOS), hart's tongue (THEOPHRASTOS, *HP* 9.18.7), *polion* (Diosk. 3.110), juniper, ***kostos***, madder, pepper, and nine other ingredients.

Fabricius (1726) 344.

PTK

Nearkhos of Crete (315 – 295 BCE)

Originally from Crete, lived in Amphipolis, one of the boyhood companions of Alexander of Macedon. He accompanied Alexander on his expedition and was made satrap of Lukia and Pamphulia in 334/3. In 329/8 he rejoined Alexander in Baktria and was made a Khiliarkh of the Hypaspists. When the fleet was built on the Hydaspes river, Alexander appointed Nearkhos admiral of the fleet, sharing responsibility with ONĒSIKRITOS, the chief pilot of Alexander's ship. Alexander charged Nearkhos with guiding the fleet back to the Persian Gulf and exploring the coast along the way. The half-year journey ended successfully with Nearkhos arriving at the mouth of the Euphrates; Nearkhos then sailed the fleet up the Pasitigris (Karun) river to Susa and was awarded a gold crown by Alexander. Shortly before Alexander's death, he and Nearkhos were planning an expedition to Arabia. Afterwards, Nearkhos served under Antigonos Monophthalmos. Nearkhos' lost account of India and the coasting expedition was used extensively by ARRIANUS in the latter part of the *Anabasis Alexandrou* and in the *Indika*. Nearkhos' account was skeptical of superstitions and full of observed detail. He described in detail the topography and climate of the lands through which he passed, including distances, harborages, islands, and water sources. He witnessed ocean tides, and speculated on the alluviation of major rivers and the cause of the flood of the Indus river. His observations of the flora and fauna of India and the sea voyage contained some misinformation and exaggeration. His astronomical comments probably derived from speculation or hearsay, rather than observation: he noted the absence of shadows at midday when he sailed out to sea and described sailing to a region where shadows pointed south, but it is unlikely that he made it south of the Tropic of Cancer. He also evidently reported that both Dippers could be seen to set in India, which could only be observed near the equator.

Ed.: *FGrHist* 133.
Robinson (1953) 1.100–149; Pearson (1960) 112–149; E. Badian, "Nearchus the Cretan," *YClS* 24 (1975) 147–170; A.S. Sofman and D.I. Tsibukidis, "Nearchus and Alexander," *AncW* 16 (1987) 71–77.

Philip Kaplan

Nechepso ⇒ Petosiris

Neilammōn (250 bce – 540 ce)

Paulos of Aigina 3.21 (*CMG* 9.2, p. 179) claims that the best of all anodynes is Neilammōn's, contraindicated for chronic use because too narcotic. Aëtios of Amida 7.106 (*CMG* 8.2, p. 370), repeated by Paulos 7.16.16 (*CMG* 9.2, p. 338), records his collyrium of **calamine**, ***pompholux, psimuthion***, tragacanth, gum acacia, and opium, in rainwater. The Egyptian name is not so rare as to require identification with the medical deacon, *PLRE* 2 (1980) 784; *cf.* perhaps Naukratitēs medicus.

(*)

PTK

Neileus (255 – 215 bce)

Neileus (or Neilos), son of Neileus, was a surgeon and pharmacist, who developed recipes for muscle relaxation (Celsus 5.18.9), inflammation of the eyes (Celsus 6.6.8–9) – both often repeated later, an antidote recorded by Andromakhos (Galēn, *Antid.* 2.10 [14.165 K.]), and a spleen remedy in Asklēpiadēs Pharm. (Galēn, *CMLoc* 9.2 [13.239 K.]), both connected to Antipatros. Sōranos in Caelius Aurelianus repeatedly prescribes his remedies: *Acute* 2.153 (*CML* 6.1, p. 236); *Chron.* 2.34 (p. 564), 5.13 (p. 862). A renowned authority on dislocated joints, especially the thigh (Celsus 8.20.4), Neileus developed a spanner for setting bone fractures, an improvement on the Hippokratic bench (*cf. Joints* 72–73). The device was an oblong quadrangle, with holes bored through the centers of the longer boards to accommodate an axle with a peg and handles on the projecting ends to maintain tension (Hēliodōros in Oreibasios, *Coll.* 49.8, 49.23 [*CMG* 6.2.2, pp. 13–15, 32–33]). The apparatus was lashed to a bench or a ladder to keep the fractured bone immobile. See Hērodotos (Mech.).

Drachmann (1963) 174; Michler (1968) 45, 97; *BNP* 9 (2006) 619 (#2), V. Nutton.

GLIM

Neilos (*ca* 250 – 300 CE)

Alchemist and member of THEOSEBEIA's alchemical milieu. ZŌSIMOS OF PANŌPOLIS, addressing Theosebeia, calls Neilos "your priest" (*CAAG* 2.191) and urges her to disassociate from him. Elsewhere Zōsimos refers to "the pseudo-prophet of yours" (Festugière 1950: 367), almost certainly an allusion to Neilos, in connection with an astrological/alchemical doctrine employing astral ***daimones*** in alchemical procedures which Zōsimos considers dangerous. Zōsimos' diatribe against reliance on the use of astrologically opportune moments in alchemy (Mertens 1995, §1) should also be read as tacitly directed against Neilos. Although none of Neilos' writings survives, *Chapters of Neilos*, now lost, are announced in the index of the alchemical miscellany, codex *Marcianus gr.* 299 (*CMAG* 2.21).

D. Stolzenberg, "Unpropitious Tinctures. Alchemy, Astrology & Gnosis According to Zosimos of Panopolis," *AIHS* 49 (1999) 3–31; K.A. Frazer, "Zosimos of Panopolis and the Book of Enoch: Alchemy as Forbidden Knowledge," *Aries* 4.2 (2004) 125–147.

Bink Hallum

NEKHEPSO ⇒ PETOSIRIS

Nemesianus, M. Aurelius Olympius, of Carthage (*fl.* 284 CE)

Wrote three didactic poems in Latin on hunting and fishing: *Halieutica, Cynegetica, Nautica* (perhaps rather: *Ixeutica*?), and five pastoral eclogues. Only 325 lines of the *Cynegetica* remain, discussing rearing dogs (103–238), training horses (238–298) and nets and traps (299–320). Nemesianus also alludes, mimetically and conventionally, to different hound breeds and their main diseases (scabies and ***rabies***). The truncated hexameter poem, inspired by VERGIL and probably by GRATTIUS and OPPIANUS, ends before the description of the hunt. Two fragments on bird-catching (*de aucupio* vel *Ixeutica*) in 28 hexameters (the woodcock and the little bustard) are spurious.

KP 4.47–48, R. Herzog; *OCD3* 1033–1034, J.H.D. Scourfield.

Arnaud Zucker

Nemesios of Emesa (*ca* 360 – 430 CE)

Bishop of Emesa in Syria, brilliant author of the philosophical and scientific *On the Nature of Man* (*Peri* ***phuseōs*** *anthrōpou*), whose title is borrowed from the HIPPOKRATIC CORPUS, wherein Nemesios contributes to establishing Christian anthropology (following Origen, and GREGORY OF NUSSA's *On the Creation of man*). Based on pagan scientific tradition rather than Christian literature, this text, probably unfinished, tries to reconcile Christianity and neo-**Platonism**. Asserting the eternity of the world, the pre-existence of the soul, and a subtle union (without blending) between soul and body, in the manner of PORPHURIOS, Nemesios assumes the body's natural limitations condition man's spiritual life. Although he never mentions a personal practice, Nemesios' exceptional medical education and current physiological knowledge permitted him to discuss and even refute GALĒN (on the anatomy of the tongue: §30; on female semen: §42). He was apparently aware of the circulatory system and the functions of bile (§§24, 28). He is, in fact, especially renowned for a "ventricular theory" of the mind (§§6–13). Galēn asserted that reasoning is localized in ventricles (*Loc. Aff.* 4.3 [8.232 K.]) and Gregory claimed that "the cerebral membrane . . . forms a foundation for the senses" (*Opif.* 12.3). Nemesios states that *all* mental faculties lie specifically located in the three brain ventricles (*koiliai tou enkephalou*): the intellect (*aestimativa* or *cogitativa*) in the middle ventricle, imagination (*phantasia*) in the front (= union of the two lateral ventricles, sometimes plural, e.g., §27), and memory in the posterior (or *cerebellum*, §§30–32; AUGUSTINE, *de Genesi ad litteram* 7.18, where a similar ventricular doctrine appears, also based on observing brain-lesions in humans). This "Nemesian" theory, perhaps originated by HĒROPHILOS, and first detailed by POSEIDŌNIOS (MED. II) (in AËTIOS OF AMIDA 6.2 [*CMG* 8.2, pp. 125–128]), was widely accepted, translated and reformulated (Johannes Damascenus, Meletius, Al-Razi, Avicenna. . .) until the 16th c. (Vesalius).

RE S.7 (1940) 562–566, E. Skard; *DSB* 10.20–21, C.D. O'Malley; *REP* 6.763–764, J. Bussanich; *DPA* 4 (2005) 625–654, M. Chase; *BNP* 9 (2006) 630–631, L. Brisson.

Arnaud Zucker

Neokleidēs (of Athens?) (390 – 350 BCE)

Younger mathematical contemporary of LEŌDAMAS, ARKHUTAS, and THEAITĒTOS; and the teacher of LEŌN (Proklos, *In Eucl.* p. 66 Fr.). The name is rare except in Athens (6th–4th cc. BCE: *LGPN* 2.328), and possibly indicates an Athenian origin.

RE S.7 (1940) 566–567 (#4), K. von Fritz.

GLIM

Neoklēs of Krotōn (300 – 50 BCE)

Aelianus, *NA* 17.15,records that the doctor Neoklēs claimed toads had two livers, one poisonous, one healthful (Wellmann assigned the fragment to DIOKLĒS OF KARUSTOS). Athēnaios, *Deipn.* 2 (57f), records that Neoklēs of Krotōn claimed the Moon was inhabited (and that Helen's egg came thence); Bicknell thus assigns the remark in *Schol. Ap. Rhod.* 1.498 about the Nemean lion being from the Moon to Neoklēs; *cf.* also EPIMENIDĒS DK B2.

RE 16.2 (1935) 2422 (#7), K. Deichgräber; P.J. Bicknell, "Lunar Eclipses and Selenites," *Apeiron* 1.2 (1967) 16–21.

PTK

Neoptolemos (325 – 25 BCE)

Greek author of a treatise on beekeeping (*Melittourgika*) of which PLINY knew (1.*ind.*11), as in all likelihood did IULIUS HYGINUS (*cf.* ARISTOMAKHOS OF SOLOI).

RE 16.2 (1935) 2470 (#12), W. Kroll.

Philip Thibodeau

Nephōn (unknown date)

Source of a remedy for arthritic **glanders** cited by THEOMNĒSTOS, preserved in the *Hippiatrika* (*Hippiatrica Parisina* 34–35 = *Hippiatrica Berolinensia* 2.23–34). The passage is preserved in the Arabic translation of Theomnēstos.

Hoyland (2004) 162; McCabe (2007) 201.

Anne McCabe

NEPOS ⇒ CORNELIUS NEPOS

Nepualios or Neptunianus (100 – 200 CE?)

Authored a treatise *On Antipathy and* ***Sympathy*** or *Phusika* (IULIUS AFRICANUS, *Kest.* 2.4), preserved in a Byzantine epitome (86 sentences). Pretending to reject vulgar marvels and following the pseudo-Dēmokritean tradition (see BŌLOS), the book describes treatments used by animals (1–26), prophylactics against enemies, and wide-spread **sympathies** (lion fears cock, magnet attracts iron, salamander does not burn, etc.). Flagrant parallels can be found with TIMOTHEOS OF GAZA and PSEUDO-ZOROASTER (in GEŌPONIKA 15.1).

Ed.: W. Gemoll, *Nepualii fragmentum Peri tōn kata antipatheian kai sumpatheian & Democriti Peri sumpatheiōn kai antipatheiōn, Städtisches Realprogymnasium zu Striegau* (1884) 1–3.
RE 16.2 (1935) 2535–2537, W. Kroll; *BNP* 9 (2006) 663, C. Hünemörder.

Arnaud Zucker

Nestōr of Laranda, Septimius (195 – 210 CE)

Father of Peisandros the epic poet, and dwelt for a time in Nikaia of Bithunia; was honored in his lifetime by statues in Paphos, Ephesos, Kuzikos, Ostia, and Rome (*Souda* N-261). Wrote didactic and epic verse in the tradition of NIKANDROS OF KOLOPHŌN, especially a *Metamorphoses* of which a few fragments survive in the *Greek Anthology*: 9.129 the dragon Python drinking up the River Kēphisos, 9.536 the Alphaios flowing sweetly through the salt sea, and 9.364, 537. His *Alexikēpos* ("antidote garden") is cited by CASSIANUS BASSUS in the GEŌPONIKA, 12.16.1 and 12.17.16–17 (on the antipathy of cabbage and grape-vine), as is Nestōr's *Panakeia* ("Heal-all"): 15.1.11, the hyaena's attack, and 15.1.32, the paradoxical properties of lignite (*gagatēs*), as in PLINY 36.141.

BNP 9 (2006) 683 (#3), J. Latacz.

PTK

NEXARĒS ⇒ XEN(OKH)ARĒS

NIGER ⇒ (1) SEXTIUS; (2) TREBIUS; (3) TURRANIUS

P. Nigidius Figulus (70 – 45 BCE)

Influential Roman politician and scholar, born *ca* 100 BCE. Senator and supporter of Pompey in the civil war, he died in exile 45 BCE. His friend CICERO (*Timaeus* 1) describes him as a hard-working researcher and the renewer of the ancient *disciplina pythagorica* (though the extent of this revival remains uncertain). Nigidius Figulus was a very learned and versatile scholar with a wide range of interests, compared by some, e.g., Aulus Gellius 19.14.3, to VARRO. The extant fragments and the titles of his works suggest that he devoted himself to the study of natural sciences (*De uentis, De hominum natura*), zoology (*De animalibus*), astronomy (*De sphaera*), grammar, occultism and divination. In his *Commentarii grammatici*, he treated questions of phonetics and morphology and displayed a deep interest in speech. He maintained the natural origin of language, according thereby a pivotal role to etymology. His work *On Gods* (*De diis*) was the first comprehensive study on Roman divinities. He built up a society (*sodalicium*), of uncertain nature, perhaps a philosophical school or secret society.

Some anecdotes speak of his interest in divination: at the birth of Octauius (who became AUGUSTUS), Nigidius is said to have predicted that the newborn would become the ruler of the universe; he also practiced dish-divining (*lekanomanteia*) and was deeply interested in astrology (*de extis, de augurio priuato*). To him was attributed a brontoscopic calendar. Although connections to **Pythagoreanism** remain possible, Nigidius' extant doctrines do not display typical **Pythagorean** features.

Ed.: A. Swoboda, *P. Nigidii Figuli operum reliquiae* (1889; repr. 1964); D. Liuzzi, *Nigidio Figulo, "astrologo e mago": testimonianze e frammenti* (1983).
RE 17.1 (1936) 200–212, W. Kroll; A. Della Casa, *Nigidio Figulo* (1962).

Bruno Centrone

Nikagoras of Cyprus (375 – 335 BCE)

Wrote a geographical or paradoxographical work cited by KALLIMAKHOS for mineral salt from Kition and by the ARISTOTELIAN CORPUS ON THE FLOOD OF THE NILE for the theory that the rise of the Nile is caused by trans-equatorial rainfall.

BNP 9 (2006) 705 (#3), Fr. Lasserre.

PTK

Nikandros (Nicander) of Kolophōn (150 – 110 BCE)

Nikandros (*Vind. Med. Gr.* 1, f.3[V]) © Österreichische Nationalbibliothek

Physician and poet who wrote *Thēriaka* (958 lines), *Alexipharmaka* (630) and other *epē* (*Souda*); son of Damaios (*fr.* 110), hereditary priest of Clarian Apollo (*cf. Alex.* 11 ~ *Thēr.* 958), he was Aitolian by origin, and spent much time in Aitolia (*Nikandrou Genos*); according to the *Vitae* of Theokritos, ARATOS, and Lykophrōn (= test. C.I–V Gow-Scholfield), a contemporary of Aratos (C.I–III), or of Ptolemy V (204–181) (C.IV–V); according to *Souda* and *Genos*, of ATTALOS III (138–133). A *proxenia* decree honoring "Nikandros, son of Anaxagoras, Kolophōnios, *epeōn poētēs*" (*SIG*^{3} 452: Delphi, *ca* 210 BCE) compels us to distinguish Nikandros (I), the epic poet honored in Delphi, from Nikandros (II), his grandson or great-nephew, author of the iological poems and a eulogy to Attalos III (*fr.* 104) – not Attalos I (241–197), *pace* Cazzaniga (*PP* 27 [1972] 369–396) and Cameron. The reverse combination (Nikandros [I], the iologist: Cameron, after Bethe, *Hermes* 53 [1918] 110–112) as opposed to the *Genos* attestation, the most reliable authority (Theōn?) on Nikandros' biography (also to be rejected is C.V's dating: *cf.* Fantuzzi, *BNP* 9 [2006] 706: 200 BCE), can no longer be supported by the Delphian decree (formerly dated *ca* 250 BCE) but only by the early dating of *Vitae* which is as questionable as the exchange of poems between Aratos (writer of *Thēriaka*) and Nikandros (of *Phainomena*), a legend condemned by his own sources (C.IV–V). Nikandros (I) may be the author of works ascribed to Nikandros (II), the only one recognized in literary tradition.

This could even be true for *Ophiaka, Iaseōn sunagōgē*, and the epic transposition of HIPPOKRATIC CORPUS PROGNOSIS.

We are only concerned here with *Thēriaka* and *Alexipharmaka*, the oldest monuments of a science that flourished during the Hellenistic period. The study of scientific and literary parallels (Jacques [2002] 2.XXV–LVI, CXI–CXIV; *cf.* [2007] 3.LIX–LXVII) shows that they may be the work of a poet-physician belonging to Attalos III's entourage. These poems, between a *proem* to a relative/friend (possibly a physician) and a *sphragis* (*cf.* the acrostic, *Thēr.* 345–353/*Alex.* 266–274), deal with venoms/poisons and their antidotes, beginning with the most dangerous (cobra/aconite). However similar they are in presentation, language and style, there are differences. In *Alexipharmaka*, 22 vegetable, animal and mineral poisons are the subject of tripartite articles (description, symptomatology and therapy) following each other without any general preamble or any other rule but a sense of variety. *Thēriaka* divides poisonous creatures into two groups (1/snakes 2/arachnids and miscellaneous [description and symptomatology only]), each followed by a collective therapy including *hapla* and *suntheta pharmaka*. The whole is both preceded and followed by general precepts, first on prophylaxis and, at the end, on other methods of treatment (among which leeches are quoted for the first time for medical use), and then crowned by an *antidotos polumigmatos*, a panacea anticipating the great antidotes to come (*Mithridateion, Galēnē*). Some of the descriptions or symptomatologies are remarkable, i.e. proteroglyph fangs (*Thēr.* 182–185), side-winding progression (*Thēr.* 264–270, *cf.* Jacques 2004: 120–121), viper and hemlock poisoning (*Thēr.* 235–257, *Alex.* 195–206).

Nikandros' medical competence was not questioned in antiquity. His name appears in a list of physicians in an MS of CELSUS (Wellmann, *Hermes* 35 [1900] 370). He is among the medical *auctores* mentioned in PLINY's index of 17 books (e.g. those on medicinal plants, 1.*ind.*20–27); there are more parallels between Nikandros and Pliny/DIOSKOURIDĒS than the latter's explicit references. In Dioskouridēs' *Vindob. med. gr.* 1 and some other MSS, EUTEKNIOS' *paraphraseis* of Nikandros replace pseudo-Dioskouridēs' iological books (his portrait, f.3[V]). Far from being APOLLODŌROS' versifier, Nikandros treats him freely, as a professional pharmacologist, as he does his other predecessors (see NOUMĒNIOS); and the iologists that followed Nikandros (see PHILOUMENOS) sometimes used him tacitly (Jacques [2002] 2.LX–LXIV). The concept of Nikandros, versifier of a subject alien to him (Schneider), does not take medical poetry (of which he is a representative) into account. All known iologists were physicians including those who expressed themselves in verse such as Noumēnios and PETRIKHOS. Polypharmacy manifests Nikandros' **Empiricist** tendency.

Ed.: O. Schneider (1856); Gow and Scholfield (1953); Jean-Marie Jacques v. 2 (2002): *Les Thériaques, Fragments iologiques antérieurs*; 3 (2007): *Les Alexipharmaques, Lieux parallèles du livre XIII des Iatrika d'Aétius*. Scholia: *Theriaka*: Crugnola (1971); *Alexipharmaka*: Geymonat (1974).

G. Pasquali, "I due Nicandri," *SIFC* 20 (1913) 55–111; A. Cameron, *Callimachus and his Critics* (1995) 194–207; G. Massimilla, "Nuovi elementi per la cronologia di Nicandro," in: R. Pretagostini, ed., *La Letteratura ellenistica* (2000) 127–137; Jean-Marie Jacques, "Médecine et poésie: Nicandre de Colophon et ses poèmes iologiques," in J. Jouanna and J. Leclant, edd., *La médecine grecque antique = Cahiers Kérylos* 15 (2004) 109–124; Jean-Marie Jacques, "Situation de Nicandre de Colophon," *REA* 109 (2007) 99–121.

Jean-Marie Jacques

Nikanōr of Samos (118 – 131 CE?)

Wrote *On rivers*, in at least two books, quoted by PSEUDO-PLUTARCH, DE FLUUIIS 17.2 (1160C) as a source of information about the stone *thrasudeilos* ("audacious-cowardly") found along the Eurotas river. Among the Nikanōrs mentioned by Müller (*FHG* 3.632–634) and Jacoby, our Nikanōr is perhaps identifiable with the one (perhaps from Kurēnē) who lived under Hadrian, and to whom the *Metonomasiai* are attributed; consequently, possibly the Nikanōr quoted by STEPHANOS OF BUZANTION, who says that he loved to Hellenize Barbarian names (according to Jacoby, however, he should be an unidentifiable author before VARRO).

RE 17.1 (1936) 277 (#28), C. Wendel; *FGrHist* 146; De Lazzer (2003) 85.

Eugenio Amato

Nikēratos (of Athens?) (10 – 40 CE)

A 3rd c. papyrus cites a "Nikēratos of Athens" and his suggestions for the use of liquid bitumen to treat mange in dogs (Gazza 1955: 96–97), probably the Nikēratos mentioned by DIOSKOURIDĒS, *pr.*2, as an **Asklēpiadean** pharmacologist who lacked precision in his description of medicinals. Like IULIUS BASSUS, PETRŌNIOS, SEXTIUS NIGER, and DIODOTOS, these **Asklēpiadeans** did not ". . .measure the activities of drugs experimentally, and in their vain prating about causation, they have explained the action of an individual drug by differences among particles, as well as confusing one drug for another" (Scarborough and Nutton 1982: 196, with comm., 205).

Dioskouridēs' criticism notwithstanding, Nikēratos' is quoted with respect by SCRIBONIUS LARGUS 39 (unnamed, attribution established from ASKLĒPIADĒS PHARM. in GALĒN, *CMLoc* 3.1 [12.633–634 K.]; *cf.* Wellmann 1914: 44, n. 1), listing simples to treat unulcerated but painful ears, including small millipedes or pillbugs (*oniokon tōn katoikidiōn; cf.* Scarborough 1980) boiled in oil, then inserted into the external auditory meatus. Nikēratos' coral-based ***trokhiskoi*** incorporated two kinds of "earths" (Samian and Lemnian), henbane seeds, the latex of the opium poppy, pomegranate flowers, high-grade flour, broom (*Cytinus* spp.), and plantain juice (Asklēpiadēs in Galēn, *CMLoc* 7.1 [13.87 K.]), an effective narcotic compound for raw windpipes that produced bloody sputum. His *Secret Pain-Killer* (*ibid.*, 96) – good for "consumption" (***phthisis***), coughs, bowel pains, diarrhea, and catarrhs of all sorts – consisted of saffron, a double-measure of henbane, opium poppy latex, beaver castor, the rhizomes of European wild ginger (*Asarum* spp. [a good emetic]), and ***storax***, to be administered with honey. Among the *ekleikta* (lozenges manufactured to "melt in the mouth" or medicinals "made into a linctus," *viz.* an "electuary," a medicine to be licked from a spoon), Nikēratos' intended his *Pharyngeal Linctus/Lozenge* to be a galactagogue for the new mother suffering from suppurations, difficult breathing, persistent coughs bringing up glutinous or sticky phlegm, and whose infant was not receiving sufficient milk (*ibid.*, 98): she was given fresh horehound-leaf juice (*prasion*: likely *Marrubium vulgare* L.), liberally mixed with "Falernian" wine and Attic honey, augmented with white pepper, frankincense, and myrrh (horehound syrup remains a common remedy for sore throats). Nikēratos' little pills (*katapotia*) for difficult breathing and panting (*asthmatikos*) combined beaver castor, the gum of the Libyan giant fennel (*Ferula marmarica* L.), lavender cotton (probably the oil from the leaves of *Santolina chamaecyparissus* L.), wormwood oil (*Artemisia* spp.), and "Ethiopian" ***ammi***, mixed with vinegar and administered as pills the size of chickpeas (*ibid.*, 110), and his two recipes for medicines to treat jaundice (Asklēpiadēs in Galēn, *CMLoc* 9.1 [13.232–233 K.]) employ quantities ("handfuls") of chickpeas, asparagus, rosemary, and common fennel, mixed with wine. Embedded in the medical poetry of Damokratēs (Galēn, *Antid.* 2.15 [14.196–201 K.]) is Nikēratos' multi-ingredient antidote against poisons and the bites of **rabid** animals. As probably recorded by Sōranos, Caelius Aurelianus, *Chron.* 2.86 (Drabkin, p. 620; *CML* 6.1.1, p. 596), cites and approves Nikēratos' tract *Katalēpsis* ("seizure"). The remnants of Nikēratos' medical, and especially pharmaceutical works, indicate a prominent practitioner whose pharmacological expertise included the full range of drugs fashioned from animals (e.g. coral, beaver castor, and pill-bugs), common foodstuffs, and several gum-exudates that ensured successful administration as pills and **pastilles** to patients for a number of diseases.

M. Wellmann, *Die Schrift des Dioskurides Peri haplōn pharmakōn: Ein Beitrag zur Geschichte der Medizin* (1914); *RE* 17.1 (1936) 314, K. Deichgräber; V. Gazza, "Prescrizione mediche nei papiri dell'Egitto greco-romano," *Aegyptus* 35 (1955) 86–110, and 36 (1956) 73–114; John Scarborough, "Nicander *Theriaca* 811," *CPh* 75 (1980) 138–140; Beavis (1988) 13–19 ("Arthropoda: Diplopoda and Isopoda").

John Scarborough

Nikesios of Maronea ⇒ Hēgēsias of Magnesia

Nikētēs (of Athens?) (250 BCE – 90 CE)

Asklēpiadēs Pharm., in Galēn *CMLoc* 4.8 (12.765 K.), records his collyrium prescribed for great pain and eye-infections, composed of **calamine**, roasted copper, ***spodos***, **galbanum**, myrrh, and opium, in water. The name is rare outside Athens (*LGPN*); emendation to *NIKH<PA>TOY* (i.e., Nikēratos) is possible but unnecessary.

Fabricius (1726) 346.

PTK

Nikias of Mallos (125 BCE – 75 CE?)

Lapidary writer whose *On stones* is cited by pseudo-Plutarch, De Fluuiis 20.4 (1163A), regarding a stone similar to sardonyx. It is debated, however, if the Nikias Maleōtēs, quoted by pseudo-Plutarch; (*Parall. min.* 13A: on Hēraklēs' attempt to seize Iolē), is to be identified with our author: the ethnic hinders such an interpretation. Moreover, Pliny 37.36 also mentions, in discussing electrum, a certain Nikias and an Homeric scholium about Helen's rape by Alexander (*FHG* 4.463–464), which is similar in content to the *Parallela minora*: presumably the homonymous Homeric grammarian.

Ed.: *FGrHist* 60.

J. Tolkiehn, *Philologische Streifzüge* (1916) 11–19; F. Atenstadt, "Zwei Quellen des sogenannten Plutarch de fluviis," *Hermes* 57 (1922) 219–246, esp. 237–238; Schlereth (1931) 118–120; Bidez (1935) 31; Jacoby (1940) 129, n. 1; De Lazzer (2003) 85–86.

Eugenio Amato

Nikias of Milētos (*ca* 300 – 250 BCE)

Poet and physician, friend of Theokritos, who addressed him in *Idd.* 11 (asserting poetry was the only remedy for love) and 13, described a cedar statue which Nikias dedicated to Asklēpios (*Ep.* 8), and also composed *Id.* 28 to accompany a distaff for Nikias' wife Theugenis.

The *hupothesis* to *Id.* 11 recounts that Nikias replied to Theokritos with a short poem in hexameters; in the opening, preserved in the scholia, he said that *Erōtes* inspired many poets previously *amousoi* (*SH* 566). The same source informs us that Nikias was a fellow student of Erasistratos, according to Dionusios of Ephesos, and also the author of epigrams. Therefore, it is probable that he is to be identified with the homonymous poet defined by Meleagros in his *Garland* (*AP* 4.1.19) as *khloeron sisumbron*, a plant sacred to Aphroditē and with great healing properties, in which we may see an allusion to Nikias' experience in amatory matters, consistent with Theokritos' account. Eight epigrams are transmitted under his name in the *Anthologia Palatina* and in the *Planudean*: three were written for the dedication of objects to Athena (*AP* 6.122), Artemis (*AP* 6.127) and Eileithuia (*AP* 6.270); two are inscriptions for a statue of Hermēs (*APl.* 188) and one of Pan (*APl.* 189); two have insects as subjects (*AP* 7.200 and 9.564); and one is an inscription for a fountain built by Simos for the grave of his son (*AP* 9.315).

W. Schott, *Arzt und Dichter: Nikias von Miletos* (1976); A. Lai, "Il *chloeron sisumbron* di Nicia, medico-poeta milesio," *QUCC* 51 (1995) 125–131; *BNP* 9 (2006) 720 (#4), M.G. Albiani.

Claudio Meliadò

Nikias of Nikaia (1st c. BCE?)

Author of a *Successions of Philosophers* quoted by Athēnaios, *Deipn.* 4 (162e), 6 (273d), 10 (437e), 11 (505b, 506c), 13 (591f), but not by Diogenēs Laërtios.

Mejer (1978) 63–64; *DPA* 4 (2005) 666–667, R. Goulet.

Jørgen Mejer

Nikolaos (Math.) (350 BCE – 460 CE)

Wrote a political interpretation of numbers, especially the marital number, in Plato's *Republic*, that was paraphrased by Magnus: Proklos *In Plat. Rep.* (2.25–26 K.).

(*)

PTK

Nikolaos (Pharm.) (150 BCE – 95 CE)

Asklēpiadēs Pharm., in Galēn *CMGen* 5.11 (13.831 K.), records his plaster, composed of acacia, antimony, myrrh, opium, and **verdigris**, in gum and wine. Paulos of Aigina 7.17.44 (*CMG* 9.2, pp. 358–359) records his 40-ingredient blood-stanch (*enaimos*), including various oxides and minerals, aloes, **bdellium**, various resins, mandrake, and opium; *cf. Idem*, 4.37 (*CMG* 9.1, p. 358).

Fabricius (1726) 346–347.

PTK

Nikolaos of Damaskos (*ca* 40 BCE – *ca* 10 CE?)

Born 64 BCE, major political and intellectual personality of Judea, Nikolaos was minister and personal counselor to Herod the Great, king of Judea, before going with Herod Arkhelaos to Rome, where he settled, in the court of Augustus, and tutored the children of M. Antonius and Kleopatra VII. Distinguished encyclopedic scholar and polygraph, he wrote comedies, tragedies, a comprehensive compilatory *Universal history* (144 books) from the beginning of the time until Herod, a fundamental source for Flauius Josephus and STRABŌN, two biographies (CAESAR and Augustus), and an autobiography *On my own Life and Education* (*Souda* N-393). He is now chiefly known as a philosopher and commentator of ARISTOTLE. Besides a collection, in **Peripatetic** style, of data *On strange Manners and Customs* of 50 nations (***Paradoxōn** ethōn sunagōgē*), he wrote many commentaries and paraphrases of Aristotle's philosophical and natural historical treatises. Nikolaos' *On the Philosophy of Aristotle* (in many books with numerous full extracts) is often mentioned and celebrated by later philosophers such as SIMPLICIUS or PORPHURIOS.

Concerning natural sciences, he wrote *On Meteorology* (*Peri meteōrōn*), treating, among other things, the origin of springs and rivers, an *Epitome of the Historia Animalium* of Aristotle, and *On Plants* (two books), which played – as did Nikolaos' work in general – a decisive role in Syriac and Arabic culture, since Aristotle's *On plants* disappeared early and THEOPHRASTOS' *Historia Plantarum* was never translated in the East. Nikolaos' treatise, probably a patchwork of extracts and commentaries based on Aristotle's lost *On plants* and Theophrastos' broader botanic corpus, was translated into Syriac (*ca* 870) – only fragments of the first book survive – then into Arabic (*ca* 1000), and again both into Hebrew (*ca* 1280), and, independently into Latin by Alfred of Sareshel (*ca* 1200). This Latin version was translated back into Greek by an unknown Byzantine scholar (*ca* 1300), perhaps Maximus Planudēs or Manuel Holobolos, whose text was still included by Bekker (1831) and Hett (1936) in the Aristotelian corpus, after the ARISTOTELIAN CORPUS, PHYSIOGNOMY. This retroversion, less reliable than the Latin text, despite being complete, presents an unsatisfactory text which cannot be emended by other versions. Chaotic and full of internal contradictions (e.g. the sex of plants; the definition of plant life), this patchwork of epitomized extracts formally describes (in seven chapters of Book 1) theoretical and biological questions, and (in ten of Book 2) more heteroclite matters (including the paradoxographical, digressions on floating stones, strange perfumes, etc.). This opuscule, almost always attributed to Aristotle himself (in place of the lost *De Plantis*), was a primary reference for Aristotelian and generalized botany in the late Middle Ages.

DSB 10.111–112, J. Longrigg; H.J. Drossaart Lulofs and E.L.J. Poortman, *Aristoteles semitico-latinus. Nicolaus Damascenus "De plantis." Five translations* (1989); *OCD3* 1041–1042, Kl. Meister; *DPA* 4 (2005) 669–679. J.-P. Schneider; *BNP* 9 (2006) 725–728 (#3), Kl. Meister.

Arnaud Zucker

Nikomakhos (Pharm.) (250 BCE – 80 CE)

ALKIMIŌN's green plaster, ANDROMAKHOS in GALĒN *CMGen* 5.5 (13.807 K.), is alternatively attributed to Nikomakhos, possibly NIKOMAKHOS OF STAGEIRA. Galēn, *Diff. Morb.* 9 (6.869 K.), mentions that a certain obese Nikomakhos of Smurna was cured by ASKLĒPIOS.

Fabricius (1726) 348.

PTK

Nikomakhos of Athens (*ca* 320 – 310 BCE)

ARISTOTLE's son, by his concubine Herpullis, a minor at his father's death, and who died young in battle, probably at Mounukhia harbor in 307 BCE (DIODŌROS OF SICILY, 20.45.5–7); the extant *Ethica Nicomachea* and a lost work on his father's *Physics* are attributed to him: CICERO, *Fin.* 5.12, DIOGENĒS LAËRTIOS 8.88, and *Souda* N-398.

RE 17.1 (1936) 462–463 (#19), K. von Fritz; *DPA* 4 (2005) 694–696, J.-P. Schneider.

PTK

Nikomakhos of Gerasa (100 – 150 CE)

Neo-**Pythagorean** philosopher; the Gerasa whence he came is likely to have been the one in Palestine. Nikomakhos composed two surviving treatises, *Introductio Arithmetica* (two books) on the philosophy of number and number theory, and *Harmonicum Enchiridion* on the **Pythagorean** theory of pitches and tuning systems. In the latter (11) he cites THRASULLOS, while CASSIODORUS, *Institutiones* (p. 140 Mynors) writes that APULEIUS translated the *Introductio Arithmetica* into Latin, thus bracketing his date. The contents of two further lost works are known in great part. *Arithmētika Theologoumena* ("arithmetic subjected to theology"), an exposition of **Pythagorean** number symbolism, was one source of the THEOLOGUMENA ARITHMETICAE, and Phōtios also summarized it (*Bibl.* cod.187). A work on the life of PYTHAGORAS is cited by PORPHURIOS, *Vita Pythagorae* (20, 59) and was also exploited without acknowledgement by IAMBLIKHOS in his *De Vita Pythagorica*. Nikomakhos himself alludes to a lost *Introduction to Geometry* in *Introductio Arithmetica* 2.6.

The *Introductio Arithmetica* moves fairly rapidly from discussing the ontology of numbers to exposing elementary number-theoretic classifications of numbers, e.g. into even and odd, prime and composite. Other prominent topics are ratio equalities and inequalities and figurate numbers. The presentation is discursive and eschews proofs. In the *Harmonicum Enchiridion*, Nikomakhos presents in a comparably discursive manner a **Pythagorean** theory of the celestial and numerical foundations of musical pitch, while tacitly incorporating elements from ARISTOXENOS.

DSB 10.112–114, L. Tarán; Barker (1989) 245–269; Dillon (1996) 352–361; *DPA* 4 (2005) 686–694, G. Freudenthal.

Alexander Jones

Nikomakhos of Stageira (*ca* 410 – *ca* 370 BCE)

The father of ARISTOTLE, who died in his son's childhood; according to the *Souda* N-399, he wrote a *Iatrika* in six books and a *Phusika* in one.

RE 17.1 (1936) 462 (#18), Kurt von Fritz.

PTK

Nikomēdēs (Hērakleitean) (230 – 50 BCE)

Interpreter of HĒRAKLEITOS, later than SPHAIROS OF BORUSTHENĒS, according to Dēmētrios of Magnesia in DIOGENĒS LAËRTIOS 9.15; see also PAUSANIAS.

RE 17.1 (1936) 500 (#13), R. Laqueur.

PTK

Nikomēdēs (*ca* 225 – 200 BCE)

Mathematician, credited by later authors with inventing an easily constructed group of mechanical curves, conchoids (*On Conchoid Lines*, lost), which could be used to solve two important classical geometrical problems: doubling a cube (by finding two mean proportionals), and trisecting an angle. PAPPOS, *Coll.* 4 (pp. 242–246, 274 Hultsch), PROKLOS (p. 272 Fr.), and IAMBLIKHOS (in SIMPLICIUS *In Categ.*, *CAG* 8 [1907] 192.19–24) also credit him with using a *quadratrix* in solving the squaring of a circle. EUTOKIOS, *In Archim. circ. dim.* (3.114 H.), reports that Nikomēdēs sharply criticized ERATOSTHENĒS' solution to the problem of two mean proportionals.

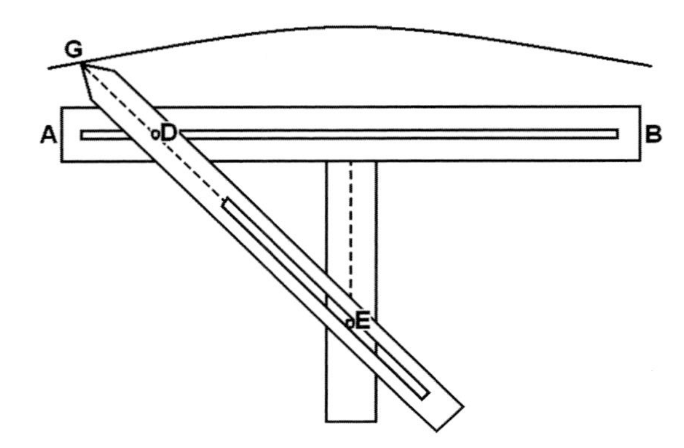

Nikomēdēs' "first" conchoid. Point D is allowed to slide along fixed line AB. The length of line DG is constant, but it is always oriented toward point E, fixed below line AB (the Greek term for a line that verges toward a distant point like this was a *neusis*). Point G then traces out a conchoid. © Lehoux and Massie

Knorr (1986) 219–220.

Daryn Lehoux

Nikomēdēs Iatrosophist (900 – 1200 CE?)

Wrote a lexicon of plant names (typical of post-classical medical literature), attested by a 15th c. manuscript in the Iviron Monastery in Mount Athos (4271.151) and an early 16th c. one (Paris, BNF, *graecus* 2224: the basis of Delatte's edition). Nikomēdēs' lexicon, containing neither magical terms nor plant names borrowed from Arabic, pertains to a pre-13th/14th c. period and perhaps even to the 10th c. The current text is clearly augmented with references to other manuscripts and literary explanations, probably first added in some manuscript as *scholia*, later integrated into the main text. Such a lexicon, diffuse in genre, was probably aimed at connecting the medical practices of non-learned healers with learned technical texts.

Ed.: A. Delatte, *Anecdota Atheniensia* 2 (1939) 302–318.

Diels 2 (1907) 69; *RE* 17.1 (1936) 500 (#15), H. Diller; M. Thomson, *Textes grecs inedits relatifs aux plantes* (1955) 176–177; Alain Touwaide, "Lexica medico-botanica byzantina. Prolégomènes à une étude" in *Tês filiês tade ta dôra. Miscellánea léxica en memoria de Conchita Serrano* (1999) 211–228 at 214.

Alain Touwaide

Nikomēdēs IV of Bithunia (100 – 74 BCE)

ANDROMAKHOS, in GALĒN *CMGen* 6.14 (13.929 K.), records that some king Nikomēdēs, under the syncopated nickname "Kodamos," published a plaster (***ammōniakon*** incense, red natron, propolis, and oak mistletoe, in beeswax, lye, and resin). Presumably the same Nikomēdēs is the author of the head-compress recorded by ASKLĒPIADĒS PHARM., in Galēn *CMLoc* 2.1 (12.556 K.), containing sulfurwort (hog-fennel), rue, mint, and other herbs in rose oil. Probably we should read Nikomēdēs for the "Nikodēmos" in SCRIBONIUS LARGUS in Asklēpiadēs Pharm. in Galēn *CMLoc* 9.7 (13.314 K.), author of a mineral-based ***hedrikē*** in myrtle oil, good wine, and butter. The last-cited practiced in Rome, suggesting either Nikomēdēs II (who was in Rome 167 BCE: POLUBIOS Book 32, *fr.*16.4, Livy 45.44.4–18), or better Nikomēdēs IV, who dwelt in Rome in the 80's BCE (STRABŌN 12.3.40). That would explain the presence of such northerly ingredients as oak mistletoe, butter, and sulfurwort.

BNP 9 (2006) 736–737 (#6), M. Schottky.

PTK

Nikōn of Akragas (80 – 40 BCE)

Physician, Sextus Fadius' mentor, wrote *On Overeating*: CICERO, *ad Fam.* 7.20.3, calling him "pleasant" (*O medicum suauem*). CELSUS describes his emollients for scrofulous tumors (5.18.14) and for relaxing, cleaning, and opening pores (5.18.26). He is probably the same Nikōn whom PHILŌN HERENNIUS includes among ASKLĒPIADĒS' students (STEPHANOS OF BUZANTION, s.v. Durrakhion) and possibly the one claiming the best rennet comes from young deer, then hares, then goats (*Schol. Nik. Thēr.* 577a).

RE 17.1 (1936) 506–507 (#17), H. Diller.

GLIM

Nikōn of Pergamon, Aelius (120 – 150 CE)

Architect and geometer, father of GALĒN (*Souda* Gamma-32), who does not name his father but thanks him for his grounding in mathematics and logic (2.116.22–26, 119.2–9 MMH). Galēn's father is probably the Aelius Nikōn who erected isopsephic inscriptions at Pergamon (*IGRR* 4, #502–506; Schlange-Schöningen). Using $\pi = 22/7$, Nikōn compares the volumes of a cone, cylinder, and sphere, all with a common given radius (that radius equal also to the height of the cylinder and cone), and compares the surface areas of a cube (superposed over a cone), of a cylinder, and of a sphere, likewise with a common radius, yielding a proportion of 42 : 33 : 22 (#503).

H. Schlange-Schöningen, *Die römische Gesellschaft bei Galen: Biographie und Sozialgeschichte* (2003) 45–54; *DPA* 4 (2005) 696–698, V. Boudon-Millot; *BNP* 9 (2006) 740 (#4), M. Folkerts.

GLIM

Nikōnidēs of Thessalia (75 – 70 BCE)

An engineer of MITHRADATĒS VI, Nikōnidēs designed siege engines used at the siege of Kuzikos (PLUTARCH, *Luc.* 10; Appian, *Mithr.* 73–75): rams, towers, and a novel and amazing boarding bridge extended from a ship-mounted tower.

BNP 9 (2006) 740, W.H. Groß.

GLIM

NIKOPHŌN ⇒ NIKŌN OF AKRAGAS

Nikostratos (Pharm.) (50 – 80 CE)

ANDROMAKHOS in GALĒN cites eight of his recipes: gout-ointment, *CMGen* 7.7 (13.985 K.); stomach-remedy *CMLoc* 8.2 (13.139 K.); his "god-like" (*isotheos*) colic-remedy, including **galbanum**, ***kostos***, mandrake, myrrh, pepper, saffron, *ibid.* 9.4 (13.279–280 K.), *cf.* 7.3 (13.65–66 K.); his enema derived from MENANDROS OF PERGAMON, *ibid.* 9.5 (13.299–300 K.); lanolin and beeswax-based ***hedrikē*** including **Indian buckthorn**, *ibid.* 9.6 (13.308 K.); his *Mithridateion* derived from XENOKRATĒS OF APHRODISIAS, involving **galbanum**, ***kostos***, white pepper, saffron, etc., *Antid.* 2.10 (14.164–165 K.); his ***hudrophobia***-treatment, of **Indian buckthorn**, gentian (*cf.* GENTHIOS), and burnt crabs in honey, *ibid.* 2.17 (14.208 K.); and his best antidote, including cinnamon, gentian, ginger, ***kostos***, licorice, ***malabathron***, long pepper, white pepper, etc., *ibid.* 2.1 (14.112–114 K.).

Fabricius (1726) 350–351.

PTK

Nikotelēs of Kurēnē (240 – 220 BCE)

Writer on conic sections and circles who responded to KONŌN's work (APOLLŌNIOS *Conica* 4.*pr.*).

DPA 4 (2005) 702–703, P.P. Fuentes González.

GLIM

NIPSUS ⇒ IUNIUS NIPSUS

Ninuas of Egypt (400 – 300 BCE)

Physician quoted only by the LONDINIENSIS MEDICUS (9.37), Ninuas distinguished congenital and non-congenital diseases: the latter are caused by heat, which, if the nourishment remains blocked, generates dangerous residues. The theory is actually common in ancient Egyptian medicine.

BNP 9 (2006) 770 (#2), V. Nutton.

Daniela Manetti

Nonnos (200 – 540 CE)

AËTIOS OF AMIDA, 7.114 (*CMG* 8.2, p. 382), records his collyrium for trachoma (roasted copper, hematite, **calamine**, opium, etc.), said to be suitable for children.

(*)

PTK

Nonnosos (525 – 540 CE)

Member of a Jewish family of envoys, dispatched on a diplomatic mission to central and southern Arabia and "Ethiopia" by the emperor Justinian I in 530/531. His grandfather Euphrasios (in 502) and his father Abram (in 524 and later) performed similar duties. He wrote an account of the embassy in Greek, now lost but still known to Phōtios in the 9th c.

(*Bibl.* 3 = *FHG* 4.178–180). In a lively style, Nonnosos describes elephants and pygmies, as well as the Arabian language and religion. John Malalas and Theophanēs the Confessor used this narrative.

RE 17.1 (1936) 920–921, R. Laqueur; I. Kawar, "Byzantium and Kinda," *ByzZ* 53 (1960) 57–73; *HLB* 1.303; *ODB* 1492–1493, B. Baldwin; *BNP* 9 (2006) 812, A. Berger.

Andreas Kuelzer

Noumēnios of Apameia (*ca* 150 – 180 CE)

Pythagoreanizing Platonist, often associated with KRONIOS. His works were read in PLŌTINOS' school (PORPHURIOS, *Vit. Plot.* 14.10–14) and he had a considerable influence on the development of **Platonism**. Seven works are known by title: *On Place, On Number, The Hoopoe* ("*Epops*," a pun on *epopteia*, the mystical vision), *On the Indestructibility of the Soul, On the Secret Doctrines of Plato, On the Unfaithfulness of the Academics toward Plato, On the Good* (in at least six books, in dialogue form). EUSEBIOS preserves longer fragments of the last two: *PE* 14.5, 14.7–9; and *PE* 9.7, 9.17, 9.21–22, 11.8–10, 11.17–18, and 11.21–22. His hierarchy of three gods foreshadows the neo-**Platonic *hupostaseis***. Noumēnios believed in an original wisdom preserved in eastern religions, especially Judaism, and in the teachings of PYTHAGORAS and PLATO.

Ed.: E. des Places, *Numénius. Fragments* (*CUF* 1973).
M. Frede, "Numenius," *ANRW* 2.36.2 (1987) 1034–1075; Dillon (1996) 361–379; *OCD3* 1054–1055, D.J. O'Meara; *DPA* 4 (2005) 724–740, P.P. Fuentes González; *BNP* 9 (2006) 895–898, M. Frede.

Jan Opsomer

Noumēnios of Hērakleia (270 – 230 BCE)

Physician and poet, student of DIEUKHĒS, who wrote a *Deipnon* under his influence (Ath., *Deipn.* 1 [5a]), and a didactic poem on fishing, *Halieutika*, often quoted by Athēnaios (Noumēnios the *Hērakleōtēs*: Ath. 1 [5a, 13a], 7 [282a, 306d]). The scholia to NIKANDROS' *Thēriaka*, and later iological treaties (PHILOUMENOS, "AELIUS PROMOTUS," AËTIOS OF AMIDA), refer to Noumēnios' *Thēriaka* (*cf.* Nikandros' homonymous poem and, prior to Nikandros, PETRIKHOS' *Ophiaka*), without giving his ethnic *Hērakleōtēs* or the poem's title (*thēriakos* in *frr.*3 [= *SH* 593] and 5 [= *SH* 594], however, alludes to the poem). The iologists quote Noumēnios in prose, but a few verses have come down to us in Nikandros' scholia (*fr.*1 = *SH* 590 ~ Nik. *Thēr.* 237; *fr.*2 = *SH* 591 ~ *Thēr.* 257–258). Poetically, Nikandros is influenced by Noumēnios (*Schol.* [Theōn?] 237a *metapepoiēke*, 257a *memnētai*), but his treatment of scientific facts is independent (*cf. Thēr.* 643–4 ~ Noumēnios *fr.*6 = *SH* 589). The fragments of Noumēnios' *Thēriaka* are concerned with symptomatology (*frr.*1–2) and therapy (3–6). Noumēnios' therapy, unlike Nikandros', deals with anti-venoms separately (*fr.*3: cobra; *fr.*5: gecko). Did he also write a book entitled *Therapeiai*? CELSUS quotes two of his compound medicines that may be derived from it, one treating gout (5.18.35) and the other inflammation of the womb (5.21.4).

Ed.: *SH* 568–596; *Thēriaka* in Jacques (2002) 2.304–306 (see XLIV–XLV).
GGLA 1 (1891) 812–813, M. Wellmann; *RE* S.7 (1940) 663–664 (#7a), H. Diller; *KP* 4 (1972) 192, R. Keydell; *BNP* 9 (2006) 895 (#1), S. Fornaro (relation Noumēnios/Nikandros inverted); Jean-Marie Jacques, "Situation de Nicandre de Colophon," *REA* 109 (2007) 99–121 at 115–117.

Jean-Marie Jacques

Numisianus (130 – 150 CE)

Student of QUINTUS, father of HĒRAKLEIANOS, and teacher of PELOPS. Wrote commentaries on the HIPPOKRATIC CORPUS, APHORISMS, EPIDEMICS 2, and perhaps other works. GALĒN traveled to Corinth in 152 CE to hear him, the "most famous" student of Quintus, and then on to Alexandria with the same goal: *Anat. Admin.* 1.1 (2.217–218 K.). Galēn reports the doctrines of Pelops and Numisianus usually without distinction.

Grmek and Gourevitch (1994) 1513–1518; Manetti and Roselli (1994) 1581–1582; Ihm (2002) #180–184; *BNP* 9 (2006) 906, V. Nutton.

PTK

Numius **(300 BCE – 540 CE)**

AËTIOS OF AMIDA 13.20 (p. 694 Cornarius) cites him for the use of wild marjoram as an antidote for asp bites. The name Numius is otherwise unattested, though Nummius is found from the 3rd c. CE (*PIR2* N-225 to 241), and Numisius is an old Republican *nomen* (*MRR* 1.398, 435, *PIR2* N-207 to 220); probably read NOUMĒNIOS OF HĒRAKLEIA or perhaps NUMISIANUS?

Fabricius (1726) 351.

PTK

Numphis of Hērakleia Pontikē (*ca* 280 – 245 BCE)

Greek historian, son of Xenagoras, born *ca* 310 BCE, author of a local history of Hērakleia (13 books) and a universal history of Alexander and his successors (24 books). His 1st c. CE compatriot, Memnōn (7.3, 16.3) presented Numphis in his own Hērakleian history as a leader of the Hērakleians expelled by Seleukos I. Numphis convinced his fellow exiles to renounce restoration of the property taken from their parents and return to the city. They did so and were received with pleasure. Later Numphis led the Hērakleian embassy to the Celts who had devastated the territory of the Hērakleians because of their alliance with Mithradatēs I. By paying 5,000 gold pieces to the army of the Celts as a whole and 200 pieces to each leader, Numphis persuaded them to withdraw from the country. Numphis' work was a source for both Memnōn and POMPEIUS TROGUS. Surviving fragments reveal no particular geographical orientation other than the traditional historiographic allusions to toponyms and to ethnographic details. Numphis is sometimes confused with NUMPHODŌROS OF SURAKOUSAI who in his ***periplous*** of Asia referred to Sappho's love affair with Phaōn (Ath., *Deipn.* 13 [596e]).

FGrHist 432 (Numphis); *FGrHist* 434 (Memnōn).

Daniela Dueck

Numphodōros (240 – 200 BCE)

OREIBASIOS (*Coll.* 49.21–22: *CMG* 6.2.2, pp. 30–33) preserves the "chest" of Numphodōros, a square spanner to which a patient was tied for treating dislocated legs (*cf.* also VITRUUIUS 7.*pr.*14; CELSUS 8.20.4; GALĒN *UP* 7.14 [3.572–575 K.] and *MM* 6.5 [10.442–443 K.]). A screw-thread engaged a wheel, on each side of which were eyelets for ropes. The axle and wheel turned together drawing in the ropes to stretch the patient. Probably the same man composed recipes for a wound poultice preserved by PLINY 34.104 and a skin plaster quoted by ANDROMAKHOS (in Galēn *CMGen* 4.14 [13.926 K.]).

RE S.11 (1968) 1020–1022, M. Michler; Michler (1968) 48, 98–99.

GLIM

Numphodōros of Surakousai (230 – 190 BCE)

Composed ***Periploi*** and *Peri tōn en Sikeliai thaumazomenōn*. The former work was divided into sections, one of which was entitled ***Periplous*** *tēs Asias*. A selection *Peri tōn en Sardiniai thaumazomenōn* possibly completed his *oeuvre*. Numphodōros perhaps composed his paradoxographical writings from the most interesting paradoxographical material in his travel compilation. The few extant fragments treat anthropological and zoological themes.

Ed.: *FGrHist* 572; *PGR* 112–115.
RE 18.3 (1949) 1137–1166 (§12, 1149), K. Ziegler; Giannini (1964) 119–120; *KP* 4 (1972) 217 (#1), W. Spoerri; *OCD3* 1055–1056, K. Meister; *BNP* 9 (2006) 927–928 (#1), H.A. Gärtner.

Jan Bollansée, Karen Haegemans, and Guido Schepens

O

Ocellus Lucanus (200 – 50 BCE)

Early **Pythagorean** (IAMBLIKHOS *V.Pyth.* 267) under whose name is preserved an apocryphal treatise deeply influenced by **Peripatetic** doctrines: the universe is eternal, indestructible and ungenerated; it undergoes no change, by remaining always identical with itself. The whole divides into a celestial or superlunary region, which, inhabited by divinities, is in perpetual motion and governs the constantly changing sublunary realm, inhabited by mortals. This part of the world, wherein generation occurs, is the realm of the four elements, neither corrupted nor generated. The human species, which as a co-subsistent part of an eternal world must be perpetual, attains immortality through the continuity of generation, where intercourse should occur not for pleasure, but only for the procreation of children. Under Ocellus' name, we also have a brief fragment from an apocryphal *On laws*, where law is described as the cause of concord in the family and the city, since God is the cause of harmony in the world.

Ed.: R. Harder, *Ocellus Lucanus. Text und kommentar* (1926, repr. 1966); Thesleff (1965); K.S. Guthrie, *The Pythagoran Sourcebook and Library* (1987) 203–213; *DPA* 4 (2005) 746–750, Bruno Centrone and C. Macris.

Bruno Centrone

Ōdapsos of Thebes (150 – 400 CE?)

Authored an astrological work cited several times by HEPHAISTIŌN OF THĒBAI for associations between parts of the zodiacal signs and geographical regions (1.1.65, 123, 163, 221). IŌANNĒS "LYDUS" mentions Ōdapsos in his *De Ostentis* 2 (p. 6 Wa.) as an authority on astral omens, indicating that he postdated PTOLEMY.

RE 17.2 (1937) 1881–1883, W. Kroll.

Alexander Jones

Ofellius Laetus (*ca* 50 – 95 CE)

In his *Quaestiones naturales*, PLUTARCH cites a certain Laitos on the effect of rainfall on plants (911F) and the harmful effect of dew on human skin (913E). Plutarch's wording suggests personal encounters with Laitos. He is probably identical with the **Platonic** philosopher Ofellius Laetus, known from two 1st c. inscriptions, from Ephesos (J. Nollé, *ZPE* 41 [1981] 197–206) and Athens (*IG* II^2 3816). Ofellius Laetus is the author of a hymn extolling the heavens and is called a theologian. He probably belonged to the Ephesian Ofellius family. He is possibly also identical with Laetus the author of lives of philosophers, and the translator of historiographies of Phoenicia by Theodotos, Hupsikratēs and Mōkhos (Tatianos *ad Graec.* 37; *FGrHist* 784). The "opening of heaven" mentioned in *IG* II^2 3816 could then be an allusion to a cosmogony attributed to Mōkhos by Damaskios: *Prim. Princ.* 3 (p. 166.11–20 W.-C.). Ofellius Laetus may also be the same person as the physician Ofillius/Ofilius consulted by Pliny regarding the benefits of saliva against snakes (1.*ind.*28; 28.38).

G.W. Bowersock, "Plutarch and the Sublime Hymn of Ofellius Laetus," *GRBS* 23 (1982) 275–279; *DPA* 4 (2005) 79, B. Puech.

Jan Opsomer

Oinopidēs of Khios (450 – 420 BCE)

According to Eudēmos (DK 41A7), the first to identify the region of the zodiacal circle and the cycle of the Great Year, attributed to many others and plausibly adopted from Babylonian astronomy. However, his particular innovation may have been to distinguish geometrically the west/east revolution of the Sun on the **ecliptic** circle from the daily east/west motion of the entire ***kosmos***, the basic model of most subsequent Greek astronomy. As to the Great Year, this was probably a terminological innovation to name a minimum integer of years and synodic months. His Great Year was 59 years (Aelianus *VH* 10.7), and more dubiously a solar year of 365 22/59 days (Censorinus 19). Proklos credits him with two geometrical problems, the construction of a perpendicular, which he called "gnomonwise," according to Proklos from the gnomon of a sundial (*In Eucl.* pp. 283–283 Fr.; *cf.* Euclid, *Elements* 1.12), and, according to Eudēmos (Proklos, *ibid.* p. 333), the construction of an angle equal to a given rectilinear angle on a given line at a given point (*cf.* Euclid, *Elements* 1.23), both of which exhibit the growing contemporary interest in geometrical constructions.

Oinopidēs regarded fire and air as the basic principles of the physical ***kosmos*** and considered god the soul of the ***kosmos***. This model suggests his account of the seasons. When the Sun is below the horizon, it generates moisture below the earth and heats it; when above, it heats the Earth. Hence, the water below is cold in summer and hot in winter. Oinopidēs explained the rise of the Nile according to this model. The Milky Way, a previous path of the Sun, moved to its present route and changed direction in disgust at Thyestes' feast, a view akin to one Aristotle attributes to **Pythagoreans**: *Meteor.* 1.8 (345a16–18). It may be significant here that one doxographical tradition (DK 41A7) accuses Oinopidēs of appropriating the discovery of the zodiac from Pythagoras.

DK 41; *DSB* 10.179–182, Bulmer-Thomas; *DPA* 4 (2005) 761–767, I. Bodnár: revised with new collection of fragments at http://www.mpiwg-berlin.mpg.de/Preprints/P327.PDF

Henry Mendell

Okianos (1000 – 1400 CE?)

A manuscript of the Iviron monastery on Mount Athos (4271.151) contains a collection of medical recipes presented as the fifth book of an otherwise unknown Okianos (not *Okeanos*). However, the recipes show an Arabic influence, thus suggesting the date-range, when this type of formulary was typical.

Diels 2 (1907) 70.

Alain Touwaide

Olumnios of Alexandria (400 – 650 CE)

Olumnios' Alexandrine origin indicates a *terminus ante quem* and suggests his participation in the Alexandrian school (provided that he did not move to another center such as Ravenna, which reproduced the Alexandrian model). Two MSS (the more recent perhaps a copy of the other) contain two fragments of medical works: Paris, BNF *graecus* 2289 (mid 14th c.), and New Haven, Yale University, History of Medicine Library, 34 (ex Phillipps 6763) (1540s). One discusses the critical days, that is, the evolution of diseases in the Hippokratic and Galēnic model; the other fragment describes the evolution and treatment of a clinical case. It is unclear if Olumnios is likewise responsible for the **iatromathematical** texts that follow in the MSS. Significantly, both works seem still to have been used in 14th c. Constantinople: the Parisian codex was written and owned by the monk, philosopher, and physician Neophutos Prodromēnos who possibly used Olumnios' work in his own medical practice or in teaching at the hospital of the Kralē, which was adjacent to the Monastery of the Prodromos, also hosting a school.

Diels 2 (1907) 70; Boudreaux in *CCAG* 8.3 (1912) 23–27.

Alain Touwaide

Olumpiakos of Milētos (*ca* 80 – 150 CE)

Physician cited as Olumpikos (GALĒN *MM* 1.7, 1.9 = 10.54–57, 67–68 K. = pp. 28–29, 34–35 Hankinson = *frr.*162, 165 Tecusan) and Olumpiakos (PSEUDO-GALĒN INTRODUCTIO 14.684 K. = *fr.*283), taught APOLLŌNIDĒS OF CYPRUS, and listed among the **Methodists** (*frr.*11, 162, 219, 283). Disagreeing in some points with predecessors, he was considered arrogant and foolish by Galēn who sharply criticizes Olumpiakos' definition of illness as a bodily change from a state in accord with nature into one beyond nature as simplistic and naïve (*frr.*162, 165). Galēn further reprimands Olumpiakos' failure to discriminate between "affection" and "symptom" (*frr.*165–166). PAULOS OF AIGINA preserves a recipe for a treatment called *Olumpiakon* (thus perhaps by Olumpiakos), or *Olumpos* (*cf.* OLUMPOS), compounded from 20 (pricey and exotic) ingredients including frankincense, myrrh, spikenard, saffron, and **Indian buckthorn** (7.16.24 [*CMG* 9.2, p. 339] = *fr.*248), recommended for prolapses and various protuberances (warts, staphylomata: 3.22.22 [*CMG* 9.1, pp. 179–180] = *fr.*239). Olumpiakos is probably distinct from OLUMPIKOS.

RE 18.1 (1939) 199 (#2), K. Deichgräber; Tecusan (2004) 63.

GLIM

Olumpias of Thebes (325 BCE – 77 CE)

PLINY, citing her among medical authorities (1.*ind.*20–28), records her emmenagogue **pessary** of bull's gall, lanolin, and natron on wool, 28.246, as well as her abortifacient **pessary** of mallows in goose-fat, 20.226 (Pollux 10.12 [p. 192 Bethe] cites her for the danger of mallows). She claimed to cure parturition-induced barrenness with a vaginal ointment of bull's gall, snake-fat, and **verdigris** in honey, 28.253: copper-salts are however contraceptive. Alleged parallels to DIOSKOURIDĒS seem weak.

BNP 10 (2007) 109–110 (#2), V. Nutton.

PTK

Olumpikos (300 BCE – 77 CE)

Wrote *On stones* (PLINY 1.*ind*.37). Wellmann has unconvincingly identified him with OLUMPIAKOS OF MILĒTOS. Our Olumpikos is more likely a technical writer, who lived at the beginning of Vespasian's reign.

M. Wellmann, "Der Verfasser der *Anonymus Londinensis*," *Hermes* 57 (1922) 396–429, esp. 406, n.4; *RE* 18.1 (1939) 199 (#2), K. Deichgräber.

Eugenio Amato

Olumpiodōros (*ca* 525 – 565 CE)

Platonist philosopher, active in Alexandria. Of his commentaries on PLATO there survive those on *Alcibiades I* and *Gorgias*; that on the *Phaidros* is lost. Several fragments of commentaries on the *Phaedo*, as well as a commentary on the *Philebos*, attributed to Olumpiodōros, are of dubious authorship, arguably authored by DAMASKIOS. Olumpiodōros' work on ARISTOTLE includes the *Prolegomena to the Organon*, and commentaries on the *Categories* and *Meteorologica*. Olumpiodōros also authored a biography of Plato and a (lost) commentary on PORPHURIOS' *Isagoge* upon which depended the *Isagoge* exegeses of his students Elias and David. Olumpiodōros' commentaries derive from school lectures, as suggested by the recurrence of the term *praxis*, probably indicating school hours. The fairly detailed, sometimes repetitive commentaries are structured according to *lemmata*, divided into *theoria* (interpretation of an issue) and *lexis* (interpretation of a specific *lemma*). Olumpiodōros relied considerably on ALEXANDER OF APHRODISIAS regarding the *Meteorologica*, and on Porphurios for Aristotle's logic.

Ed.: L.G. Westerink, *Olympiodori In Platonis Gorgiam Commentaria* (1970); W. Norvin, *Olympiodori In Platonis Phaedonem Commentaria* (1913); L.G. Westerink, *The Greek Commentaries on Plato's Phaedo* (1976); A. Busse, *Olympiodori Prolegomena et in Categorias Commentarium* = *CAG* 12.1 (1902); G. Stüve, *Olympiodori in Aristotelis Meteora Commentaria* = *CAG* 12.2 (1900).
RE 18.1 (1939) 207–227, R. Beutler; *SEP* "Olympiodorus," Chr. Wildberg.

George Karamanolis

Olumpiodōros of Alexandria (Alch.) (530? – after 565 CE)

Wrote a commentary (*CAAG* 2.69–106) on a lost treatise of ZŌSIMOS entitled *The Alexandrian Philosopher Olumpiodōros on the Book "Kat' Energeian" by Zōsimos and on the Sayings of Hermēs and the Philosophers* (*kat' energeian* is "On the Action" or "According to the Action"). Olumpiodōros distinguishes himself among the Alexandrian alchemists by his exegesis treating both the principles of transmutation and the philosophical models of those principles. In the alchemical corpus, Olumpiodōros is mentioned with STEPHANOS among "the ecumenical masters celebrated everywhere, the new interpreters of PLATO and ARISTOTLE" (*CAAG* 2.425). The title of his treatise indicates his Alexandrian origin, and, according to some MSS, he dedicated his work to PETASIOS.

Olumpiodōros explicitly presents his commentary as both exegetical and doxographical. His originality consists in explicitly vindicating Greek philosophy, notably pre-Socratic, as the epistemological foundation of transmutation. In fact, near the middle of the commentary, Olumpiodōros sets forth the opinions of nine pre-Socratics (MELISSOS, PARMENIDĒS, THALĒS, DIOGENĒS, HĒRAKLEITOS, HIPPASOS, XENOPHANĒS, ANAXIMENĒS, and ANAXIMANDROS) on the monistic principle of the Universe and then sketches a comparison between their theses and those of the principle masters of the art of transmutation (Zōsimos,

KHĒMĒS, AGATHODAIMŌN, and PSEUDO-HERMĒS: *CAAG* 2.79–85). Moreover, the structure of this treatise remains on the whole disconnected and lacunose: lacking a preface and conclusion, it begins and ends abruptly. The most coherent part is at the beginning where the author comments upon Zōsimos on the extraction of gold, in the typical structure of Neo-**Platonic** exegesis originating precisely from the school of Olumpiodōros. Otherwise, the treatise consists in a collection of excerpts from early authors concerning the principle notions and operations of alchemy, accompanied with commentary, among which are found further extracts of Zōsimos.

The question of the alchemist Olumpiodōros' identity has attracted much scholarly attention. Formerly, he was identified with OLUMPIODŌROS OF THĒBAI, the historian (Berthelot 1885); later opinion was split between attributing the treatise to the Neo-**Platonic** commentator, or to a homonym, or even to a pseudonym. In fact, reasons in favor of attributing the work to the Neo-**Platonist** are numerous. Firstly, the tradition attributes the alchemical work to an Alexandrian Olumpiodōros interpreter of Plato and Aristotle. Moreover, the author demonstrates familiarity with Platonic and Aristotelian philosophy. His commentary presents characteristic traits of Alexandrian Neo-**Platonic** exegesis, such as the apparent obscurity of philosophic language, the identification of the goal of research with a unitary principle, and even the structure of the doxography about that principle. Finally, one can discover evident doctrinal and terminological convergences, whether with the commentary on Aristotle's *Meteōrologika*, or whether with other works of the Neo-**Platonist** Olumpiodōros. Recently, Brisson (1992) decided in favor of the identification with the Neo-**Platonic** commentator, relying on a passage in the *Commentary on the Phaedo* (1.3, pp. 42–43 Westerink), where Olumpiodōros seems to describe the generation of human beings as an alchemical operation based on the sublimates of the vapors rising from the bodies of the thunder-struck Titans.

To conclude: given its discontinuous and composite state, one can imagine that our text consists in extracts of a lost alchemical work of the Neo-**Platonist** Olumpiodōros (perhaps a full commentary on the *Kat' Energeian* of Zōsimos), arranged by a copyist, or perhaps that the copyist copied the work of Olumpiodōros up to a point, and then added a series of unstructured notes on the principal alchemical operations, accompanied by excerpts from other alchemists and, probably, other works of Olumpiodōros himself.

Ed.: *CAAG* 2.69–106 (text), 3.75–115 (trans.); *CAAG* 2.79.11–85.5 ed. and trans. in Cristina Viano, "Olympiodore l'alchimiste et les Présocratiques. Une doxographie de l'unité (*De arte Sacra*, §18–27)," in Kahn and Matton (1995) 95–115.

Berthelot (1885) 191–195; J. Letrouit, "Datation d'Olympiodore l'Alchimiste," *Emerita* 58 (1990) 289–292; L.G. Westerink, *The Greek Commentaries on Plato's Phaedo*, v.1, *Olympiodorus* (1976) 20–32; L. Brisson, "Le corps 'dionysiaque.' L'anthropogonie décrite dans le *Commentaire sur le Phédon de Platon* (1, par. 3–6) attribuée à Olympiodore est-elle Orphique?" in *Sophiēs Maiētores. Hommage à Jean Pépin*, ed. M.O. Goulet-Cazé, G. Madec, and D. O'Brien (1992) 481–499; *DPA* 4 (2005) 768–771, H.D. Saffrey; Cristina Viano, *La matière des choses: Le livre IV des Météorologiques d'Aristote, et son interprétation par Olympiodore* (2006) 197–206; *NDSB* 5.338–340, *Eadem*.

Cristina Viano

Olumpiodōros of Thēbai (Egypt) (405 – 425 CE)

Served as an envoy to the Huns, 412 CE, and has been tentatively identified with the neo-**Platonist** and envoy "to barbarians," anethnic dedicatee of Hieroklēs (in Phōtios, *Bibl.* 214 [p. 171b]; but the name is very common, esp. in Athens: *LGPN*). He wrote a 22-book lost "material for history" covering 407–425 CE, excerpted by Phōtios, *Bibl.* 80, and used by

PHILOSTORGIOS and others. A few geographical remarks are preserved: *fr.*32 Blockley (Phōtios, *Bibl.* pp. 179–180) on the Oasis in Egypt; *fr.*35 (p. 182) on the emerald mines of the Blemmues, and *fr.*42 (p. 186) on the geography of HOMER's *Odyssey*.

Ed.: Blockley 2 (1982) 152–220.
PLRE 2 (1980) 798–799; *BNP* 10 (2007) 112 (#3), W. Portmann; *NDSB* 5.338–340, C. Viano.

PTK

Olumpionikos (?) (200 BCE – 90 CE)

ASKLĒPIADĒS PHARM., in GALĒN *CMLoc* 4.8 (12.753 K = AËTIOS OF AMIDA 7.104 [*CMG* 8.2, p. 364]), preserves a multi-ingredient *phaia*-plaster (*cf.* EUBOULOS) attributed to *Olumpionikos*, immediately effective against numerous severe afflictions including eye-ailments (*khēmōseis*), and compounded from minerals (**calamine**, antimony, both burnt and washed) and botanicals (acacia, Indian aloe, saffron, myrrh, opium poppy, gum), taken with water, or used with an egg, to which Asklēpiadēs adds ***pompholux*** and frankincense. The name is otherwise unattested, and is perhaps to be emended to Olumpikos, Olumpikhos, or Olumpiōn (*LGPN*).

(*)

GLIM

Olumpos of Alexandria (35 – 25 BCE)

Physician to KLEOPATRA VII, assisted in her suicide, and published an account of her end (PLUTARCH, *Ant.* 82.3–5 [*FGrHist* 198]). AELIANUS (*NA* 9.61) explains that asp poison spreads quickly, disappearing from the skin and thereby is difficult to detect, as in her death. GALĒN (*CMLoc* 9.3 [13.261 K.]) preserves Olumpos' recipe for an ointment of anise, cardamom, licorice, Celtic nard, and ***panax*** in a beeswax, bovine fat, honey, ***terebinth***, and perfumed wine base. PAULOS OF AIGINA proclaimed his ability to heal prolapses and all wounds with olive juices (3.22.22 [*CMG* 9.1, p. 180]) and preserves a remedy (perhaps by OLUMPIAKOS) compounded from 20 ingredients (including acacia, myrrh, saffron, **verdigris**, rose, grapes, dates, spikenard, copper, antimony, ***psimuthion***) mixed with rainwater, standing for three days and nights (7.16.24 [*CMG* 9.2, p. 339]).

RE 18.1 (1939) 324 (#32), H. Diller; *BNP* 10 (2007) 118–119 (#15), V. Nutton.

GLIM

Onēsidēmos (*ca* 40 BCE? – *ca* 90 CE)

ASKLĒPIADĒS, in GALĒN *CMLoc* 10.1 (13.327–328 K.), cites his recipes for kidney ailments, between ANTONIUS MUSA and HĒRAKLEIDĒS OF TARAS: so perhaps *post ca* 40 BCE. For the rare name, *cf.* *LGPN* 1.351 (of Ioulis, *ca* 300 BCE).

Fabricius (1726) 353.

PTK

Onēsikritos of Astupalaia (315 – 295 BCE)

A student of Diogenēs of Sinōpē, accompanied Alexander's expedition to Asia, and was sent as an envoy to the gymnosophists of Taxila. He was later appointed the pilot of

Alexander's ship on its trip down the Indus river, and was made chief pilot of the fleet in its journey from the Indus to the Euphrates. He quarreled with Nearkhos, under whom he served, but was crowned by Alexander upon reaching Susa. He wrote an idealized account of Alexander's life, entitled *How Alexander Was Raised*, which included the expedition in India and the fleet's return through the Arabian Sea; only citations of the work survive. His account of India was heavily influenced by his own philosophical training: he presented the gymnosophists, Brahmin priests, as Cynics. His descriptions of the lands through which the expedition passed were full of detail, and were used by Pliny and others; but he favored the marvelous, and relied on reports of places he did not see, so that was accused of flattery and exaggeration by later writers. Despite his role as pilot, his estimates of size and distance were exaggerated: he made India a third part of the Earth, put the width of the Indus delta at 2,000 stades, and claimed that the Indus is 200 stades across. He reported on the customs of the people of Baktria, Sogdia, Cathaia, and the land of Mousikanos, transmitting utopian exaggerations about the peoples of the East. Similarly, in describing the exotic flora and fauna of India, he recorded distortions and exaggerations. He claimed that the Sun reaches zenith and the northern half of the sky in India in places north of the Tropic of Cancer, perhaps based on reports of southern India. He was the first Greek to know of Taprobanē (Sri Lanka), and the islands between it and the mainland, and described an animal that might be the dugong.

Ed.: *FGrHist* 134.

T.S. Brown, *Onesicritus* (1949); Robinson (1953) 1.149–167; Pearson (1960) 83–111; Pédech (1984) 71–158.

Philip Kaplan

Onētidēs/Onētōr (300 BCE – 80 CE)

Andromakhos in Galēn, *CMLoc* 7.6 (13.115 K.) cites his remedy for ***orthopnoia*** (castoreum and ***ammōniakon*** incense). Kühn reads *ΟΝΗΤΡΟΥ*, so perhaps *ΟΝΗΤΙΔΟΥ*, the rare name Onētidēs (as in *LGPN* 2.353, Athens, 390–375 BCE). *Cf.* also *CMGen* 5.13 (13.840 K.) *ΟΝΗΤΩΡΕΙΣ* (perhaps read *ΟΝΗΤΟΡ<ΟΣ>ΕΙΣ*: "one of Onētōr's"), composed of oak-gall, iris, myrrh, **litharge** and ***psimuthion***, and "good for everything."

Fabricius (1726) 353.

PTK

Onētōr (*ca* 50 – 150 CE?)

Platonist author of *On Mathematical Analogy*, cited once by Proklos, and possibly another work, cited by Damaskios, *in Phaed.* 100.3, where Onētōr is coupled with Atticus. Diogenēs Laërtios cites probably the same man, 2.114 (on Nikarētē and Stilpōn of Megara) and 3.9 (for a work *Whether a Wise Man will Make Money*). The name is rare except in 5th/4th cc. BCE Athens, but could have been revived (*cf.* e.g., Adrastos, Akhillēs, Hulas, Menelaos, Pelops, and Penthesileus).

Ed.: *FGrHist* 1113.

DPA 4 (2005) 781–782, R. Goulet.

PTK

Opil(l)ius ⇒ Ofellius

Ophellās of Kurēnē (320 – 310 BCE)

Ruler of Kurēnē 322–308 BCE (DIODŌROS OF SICILY 18.21, 20.40–42), composed a ***periplous*** of the exterior, Oceanic, coast of Libya, considered unreliable by STRABŌN 17.3.3, but perhaps used by MARCIANUS 1.2 (see APELLĀS).

RE 18.1 (1939) 630 (#1), E. Honigmann; *OCD3* 1068, S. Hornblower.

PTK

Ophiōn (280 – 250 BCE?)

Listed after DIOKLĒS OF KARUSTOS and before HĒRAKLEIDĒS – SC. OF TARAS? – as a foreign authority on pharmacology (PLINY 1.*ind*.20–21) and linked with ERASISTRATOS (22.80). He believed that parsnip was both a diuretic and an aphrodisiac (20.34). Diller suggests that Ophiōn used Dioklēs and lived perhaps before Erasistratos.

RE 18.1 (1939) 646 (#3), H. Diller.

GLIM

Oppianus of Anazarbos in Kilikia (176 – 180 CE)

Wrote the didactic poem *On Fishing* (*Halieutikà*), dedicated to two Antonini, probably M. Aurelius and Commodus (176–180 CE: *Hal.* 1.3, 66, 78; 2.41, 683; 4.4–5), distinct from OPPIANUS OF APAMEIA author of *On Hunting*. Athēnaios, *Deipn.* 1 (13b), alludes to Oppianus as his older contemporary. We have *Lives* preserved in Byzantine MSS, but they are not reliable: for example, the *Vita A* tells that Oppianus – already 30 – accompanied his father, whom Septimius Seuerus (193–211 CE) banished, to an island in the Adriatic sea, where he composed *Halieutikà*; he later went to Rome, obtaining from Caracalla (211–217 CE) his father's restoration and some money.

The poem (3,506 hexameters) divides into five books: 1. Habitat and habit of various fishes (the author shows remarkable ichthyologic knowledge); 2. The struggle for survival among sea creatures; 3. The fisherman and fishing seasons; various methods whereby fishes deceive fishermen; 4. Fish reproduction and ways to capture them; 5. Sea-monsters (dolphins, whales, Cetaceans). Oppianus appears to believe in the **Stoic** doctrine of Divine Principle (1.409–411), but also in the constant fight between Hate and Love (2.11–39); his style reflects the influence of popular **Stoic** philosophical discourse, e.g. in the use of developed similes. He finds parallels between the struggle for life in the sea and among human beings; he exalts the Caesars as divine peacemakers. Oppianus' *Halieutikà*, perhaps the best poem from the first two centuries of the Imperial Age, is highly refined metrically and stylistically, especially in his use of similes and rhetorical devices. The sources can be traced back to ARISTOTLE and THEOPHRASTOS and to Hellenistic scholars, including LEŌNIDAS OF BUZANTION and ALEXANDER OF MUNDOS.

Ed.: A.W. Mair, *Oppian, Colluthus, and Tryphiodorus* (Loeb 1928), F. Fajen, *Oppianus, Halieutica. Einführung, Text, Übersetzung in deutscher Sprache, ausführliche Kataloge der Meeresfauna* (1999).
Conc.: F. Martín García and Á. Ruiz Pérez, *Oppiani Cilicis Halieuticorum concordantiae* (1999).
RE 18.1 (1939) 698–703 (#1), R. Keydell; N. Hopkinson, *Greek Poetry of the Imperial Period. An Anthology* (1994) 185–186; E. Rebuffat, *"Poiêtês epeôn." Tecniche di composizione poetica negli Halieutica di Oppiano* (2001); S. Martínez, "Opiano en la poesía bizantina. Lección y leyenda," *Prometheus* 29 (2003) 259–268; T. Silva Sánchez, "Opiano, ¿un poeta o dos?" *AC* 72 (2003) 219–230; A.N. Bartley, *Stories*

from the Mountains, Stories from the Sea. The Digressions and Similes of Oppian's Halieutica and the Cynegetica (2003); *BNP* 10 (2007) 164–165, S. Fornaro.

Gianfranco Agosti

Oppianus of Apameia (198 – 217 CE)

Wrote the didactic poem *On Hunting* (*Kunēgetikà*), distinct from OPPIANUS OF KILIKIA author of *On Fishing*. Our Oppianus dedicated his poem to Caracalla (212–217 CE). The allusion to the capture of Ktēsiphōn by Seuerus (1.31) dates the poem securely after 198 CE; it is likely to have been completed after 212 CE or after Caracalla's tour of Syria in 215 CE. Oppianus tells he was from Apameia-on-the-Orontes at 2.125–127. His sources include ARISTOTLE's writings on zoology and cynegetic authors (XENOPHŌN, ARRIANUS and others). The poem (2,144 hexameters) falls into four books: 1. Types of hunting, fishing and fowling; the hunter and his equipment, horses and dogs; 2. Inventors of hunting; horned animals; 3. Wild animals; 4. Seasons of hunting and the hunter's weapons; lion and bear hunting. The lack of an epilogue and hints about animal instincts in the fourth book suggest an unfinished work. A major feature of the poem is the importance attached to myths and paradoxography (e.g., leopards can be captured with water and wine, since they originate in the Bacchants who killed Pentheus: 4.230–353), and the great number of epic similes (91), which often tend to the grotesque. In general the style is more forced and obscure than Oppianus of Kilikia's (whose work was known to our Oppianus), with considerable neologisms and occasional metrical inaccuracies.

Ed.: A.W. Mair, *Oppian, Colluthus, and Tryphiodorus* (Loeb 1928); M.A. Papathomopoulos, *Oppianus Apameensis, Cynegetica. Eutecnius Sophistes, Paraphrasis metro soluta* (2003).

Conc.: F. Fajen and M. Wacht, *Concordantia Oppianea. Konkordanz zu den Halieutika des Oppian aus Kilikien* (2002).

Comm.: W. Schmitt, *Kommentar zum ersten Buch von Pseudo-Oppians Kynegetika* (1970).

RE 18.1 (1939) 698–703 (#2), R. Keydell; A.S. Hollis, "[Oppian], *Cyn.* 2, 100–158 and the mythical past of Apamea-on-the-Orontes," *ZPE* 102 (1994) 153–166; N. Hopkinson, *Greek Poetry of the Imperial Period. An Anthology* (1994) 197–198; T. Silva Sánchez, *Sobre el texto de los Cynegetica de Opiano de Apamea* (2002); G. Agosta, "Due note testuali al proemio dei Cynegetica (I 26, 32–34)," *Eikasmos* 14 (2003) 133–160; E. Giomi, "Ps. Oppiano, 'Cynegetica' III 53–55 e la zampa 'narcotizzante' del leone," *Maia* 55 (2003) 537–543; A.N. Bartley, *Stories from the Mountains, Stories from the Sea. The Digressions and Similes of Oppian's Halieutica and the Cynegetica* (2003); *BNP* 10 (2007) 163–164, S. Fornaro.

Gianfranco Agosti

Oppius (100 BCE – 15 CE)

Wrote "On Woodland Trees" (*De Siluestribus Arboribus*); fragments discuss the chestnut (MACROBIUS, *Sat.* 3.18.7) and citrons (19.4). There is no way to be sure with which, if any, of the several better-known Oppii our author may be identified; but he may be cited in PLINY's book on bees, insects, and anatomy (1.*ind.*11). Other authors quoted by Macrobius in Book 3 date no later than the age of AUGUSTUS.

Speranza (1971) 69–72.

Philip Thibodeau

ORBANUS ⇒ ARBINAS

ORBIKIOS ⇒ URBICIUS

Oreibasios of Pergamon (*ca* 350 – *ca* 400 CE)

Born *ca* 325 CE, well-educated member of the non-Christian aristocracy in Asia Minor, who from his earliest years associated with orators, writers, and various intellectuals, ranging from classically-trained physicians to neo-**Platonic** philosophers. Having studied medicine in Alexandria with the famous Zēnōn of Cyprus (fellow-pupils included IŌNIKOS OF SARDĒS and perhaps MAGNUS OF ANTIOCH), Oreibasios took up the practice of medicine as a *iatrosophistēs*, probably in his home city of Pergamon, famed in its own right for the ornate and massive temple precincts dedicated to Asklēpios. Some time before 350 CE, he became friend and confidant of Julian (then in forced internal exile by Constantius II), and when the emperor appointed Julian general in Gaul (355), Oreibasios accompanied him as personal physician. Another friend of Oreibasios was the sophist and historian EUNAPIOS OF SARDĒS, who has given a short account of Oreibasios' life and career in *Lives of the Philosophers* (488–489 Wright [Loeb]). Eunapios asserts that Oreibasios was fundamental in Julian's assuming the purple in 361. Oreibasios accompanied Julian on his last campaign into Persia (363) and was present when Julian died from wounds received in battle (the fragments of PHILOSTORGIOS' *Church History* reflect details that can emerge only from a physician's journal). Exiled to residence among barbarian tribes north of the Danube after 364 CE, he was restored to his property and titles *ca* 380, and lived in honor until his death.

While serving Julian in Gaul, Oreibasios composed summations of GALĒN (now lost), and after Julian became emperor, Oreibasios prepared an encyclopedia of medicine, judiciously selecting the best pagan medical works from ALKMAIŌN to his contemporaries PHILAGRIOS and ADAMANTIOS (Greek only) in conjunction with the short-lived pagan revival; about a third of this *Collectiones medicae* survives (25 books of 70 or 72, with many of the rest deduced from the extant *Synopsis ad Eustathium* and *Ad Eunapium*), handbooks intended for use by laypersons (Eunapios being his friend the doxographer). Often denigrated for slavish quotation of earlier authorities, Oreibasios may be responsible for the survival of Galēn's reputation (quite uncertain is the circulation of Galēn's writings throughout most of the 3rd c.), and in *Collectiones medicae*, Oreibasios devises a careful system of editing, summarizing, and fusing much of the self-contradictory, scattered, and obtuse work of Galēn, editorial mechanics generally followed thereafter by medical encyclopedists. More importantly, Oreibasios inserts his own professional experiences among the quotations, so that his own practice becomes a commentary on the great classics of medicine.

Ed.: Bussemaker, Daremberg, and Molnier (1851–1876): the most useful edition, scholia, Latin trans. texts, full indices; J. Raeder, *Oribasii Collectionum medicarum reliquiae* (1928–1933) = *CMG* 6.1–2, *Synopsis ad Eustathium. Libri ad Eunapium* (1926) = *CMG* 6.3; Mark Grant, *Dieting for an Emperor. A Translation of Books 1 and 4 of Oribasius' Medical Compilations with Introduction and Commentary* (1997); C.L. Day, *Quipus and Witches' Knots. The Role of the Knot in Primitive and Ancient Cultures with a Translation and Analysis of Oribasius' De laqueis* (1967); J.W. Humphrey, J.P. Oleson, and A.N. Sherwood, "Machine-Screws and Nuts in Bone-Setting Devices: Oribasius *Compendium of Medicine* 49.4.52–58, 5.1–5, 7–9," *Greek and Roman Technology* (1998) 54–55.

RE S.7 (1940) 797–812, H.O. Schröder; A. Sideras, "Aetius und Oribasius," *ByzZ* 67 (1974) 110–130; B. Baldwin, "The Career of Oribasius," *AC* 18 (1975) 85–97; Scarborough (1985b) 221–224; M. Michler, "Zu einer Konjectur in Heliodors Verbandslehre bei Oreibasios," *Hermes* 114 (1986)

252–255; J. Bertier, "Reflets et inflexions des tendances théoriques de la médecine des enfants dans l'Oribase latin," in Mudry and Pigeaud (1991) 269–283.

John Scarborough

Orestinos (100 – 50 BCE)

Hērakleidēs of Taras, in Galēn *CMLoc* 1.1 (12.402–403 K.), records his three remedies for baldness, compounded, e.g., from **sympathetically** magical ingredients including bear hair, maidenhair, reed roots, fig leaves, in equal amounts, all singed, mixed with cedar-resin and bear-fat, heated in a vapor-bath, applied daily. Attested as a cognomen from the 1st c. BCE (Solin 2003: 1.552), this rare diminutive derives from the infrequently attested Orestēs.

RE 18.1 (1939) 1017, H. Diller.

GLIM

Orfitus (10 BCE – 90 CE)

Asklēpiadēs Pharm., in Galēn *CMGen* 7.12 (13.1029–1030 K.), records the ***akopon*** prepared by (of for) Orfitus, very similar in composition and use to Iulius Secundus'. The non-Republican cognomen is first attested in the Augustan era (Pliny 7.39 = *PIR2* O-139); *cf.* also *PIR2* P-18 (Paccius), and perhaps *LGPN* 1.354, Orphetēs of Kurēnē.

PIR2 O-143 (suggesting identification with C-1448, *cos.* 178 CE).

PTK

Ōrigeneia (150 BCE – 90 CE)

Asklēpiadēs Pharm., in Galēn *CMLoc*, quotes three of her remedies: throat lozenge (boiled licorice, myrrh, saffron, roasted nettle- and flax-seed, in honey), 7.2 (13.58 K.); for blood-spitting (***phthisis***): gentian (*cf.* Genthios), licorice, myrrh, saffron, etc., 7.3 (13.85 K.); and two versions of a stomach-remedy involving henbane, 8.3 (13.143–144 K.). This feminine form of Ōrigenēs seems otherwise unattested (*LGPN*), but *cf.* Eugeneia.

Fabricius (1726) 354; Parker (1997) 145 (#51).

PTK

Ōriōn of Bithunia (250 BCE – 80 CE)

Cited as a hairdresser; four recipes are known: Andromakhos gives his multi-ingredient aromatic and greasy ***akopon*** including almond oil, beeswax, lanolin, olive oil, **terebinth**, etc. (Galēn, *CMGen* 7.13 [13.1038 K.]); Asklēpiadēs Pharm. records two sciatica remedies (Galēn, *CMLoc* 9.3 [13.260 K.)]), ***ammōniakon*** incense, laurel, red natron, etc.; and Aëtios of Amida (giving the ethnic) reports a remedy for **dropsy** and splenetic disorders, which looks very like a compaction of the first half of the first sciatica remedy with the second half of the second: 10.22 (p. 590 Cornarius). For the rare name, *cf.* *LGPN* 1.488, 3A.481–482, 4.360.

Fabricius (1726) 354.

PTK

Ōros of Mendēs (300 BCE – 75 CE)

PLINY, listing him as a doctor providing drugs from animals (1.*ind.*29), cites his remedy for weasel bites (37.138): crush the *iritis* stone (unidentified), and roast it. ASKLĒPIADĒS PHARM. in GALĒN *Antid.* 2.7 (14.144–145 K.) names him, as younger than ORPHEUS, and before (?) HĒLIODŌROS OF ATHENS and ARATOS, as a writer of verses giving antidotes. AËTIOS OF AMIDA 7.35 (*CMG* 8.2, p. 286) refers to his collyrium against *muokephalon* (eye-disease characterized by growths in the shape of a fly's head), and 15.27 (Zervos 1909: 124) gives his "nine-ingredient" wound-cream, good also as a **pessary** and a ***hedrikē***: goose-fat, calf-fat, beeswax, **terebinth**, butter, deer-marrow, rose oil, castor oil, and honey. Notwithstanding *Iliad* 11.303, the name is probably Egyptian: MANETHŌN in Iosephus *Ap.* 1.96, Paus. 2.30.6.

RE 18.1 (1939) 1183 (#5, 7), W. Kroll.

PTK

Paulus Orosius (*ca* 385 – 420 CE)

Christian historian and theologian. His most famous work, *Historiae aduersum Paganos*, written at the suggestion of AUGUSTINE, covers the history of the world from the Creation to approximately 416 CE. Orosius began his universal history with a long and detailed geographical description of the entire world, the earliest attested Roman historian to thus open his history. Orosius' geographical description conforms to the classical tradition, but his precise sources are hard to identify. He may have used information which ultimately traces back to a survey of the world, produced by AGRIPPA. Orosius began with the tripartite division of the world into Asia, Europe, and Africa and described the provinces of the three continents with their cities, rivers, and mountains. Orosius' geographical survey became very influential in the Middle Ages (more than 200 MSS survive), and was sometimes transmitted separately from the rest of Orosius' book. It was used by later writers, such as ISIDORUS OF HISPALIS, and by mapmakers. Orosius' history was translated into Old English in the 9th c. on the initiative of King Alfred the Great, and geographical information therein was updated and complemented by contemporary accounts about northern Europe, provided by travelers.

Ed.: M.-P. Arnaud-Lindet, *Orose. Histoires: contre les paiens* 3 vv. (*CUF* 1990–1991).

R.J. Deferrari, trans., *The Seven Books of History Against the Pagans* (1964); *KP* 4.350–351, B.R. Voss; *PLRE* 2 (1980) 813; Y. Janvier, *La géographie d'Orose* (*CUF* 1982); *OCD3* 1078, E.D. Hunt; *TTE* 462–463, Z.R.W.M. von Martels; Natalia Lozovsky, *"The Earth Is Our Book": Geographical Knowledge in the Latin West ca. 400–1000* (2000); *BNP* 10 (2007) 240–242, U. Eigler.

Natalia Lozovsky

Orpheus, pseudo (Astrol.) (*ca* 200 BCE – 200 CE?)

Legendary poet whose syncretistic mystery cult propounded metempsychosis, was influenced by **Pythagoreanism**, and informed Neo-**Platonism** and Dionysian mystery rites. Concerned primarily with theogony, Eleusinian and Dionysian mythology, eschatology, and ritual prescriptions, the Orphic Corpus includes also astrological, medical, and lapidary fragments (see PSEUDO-ORPHEUS, LITHIKA).

In an *Argonautika*, an autobiographical account of Orpheus' participation in the Argo's voyage heavily influenced by Apollōnios of Rhodes' version, the poet inventories his interests and writings, including divination by birds, beasts, entrails, dreams, and stars (1.33–39:

#224 Kern). The Orphic *Dōdekaetēris* treats Jupiter's 12-Year Cycle wherein, allotting 120 years to the Great Year, the author describes years governed by zodiac signs and their effects, often agricultural: the year of the Archer presages war from the beginning and the falling of wagons on harvest-days (#249–270 Kern). *On Earthquakes*, also attributed to HERMĒS TRISMEGISTOS, explains the connection between earthquakes and the zodiac (#285 Kern); *cf.* VICELLIUS. *On the Rise of Heavenly Bodies* details signs presaged by zodiac constellations in relation to the Sun, Moon, and planets (#286–287 Kern). Other titles include *On Lucky and Unlucky Days* (#271–279 Kern); a *Georgics* with signs for planting and harvesting; and verses common with MAXIMUS' *Georgics* (#280–284 Kern).

Ed.: O. Kern, *Orphica Fragmenta* (1922) Astrol.: #249–288, 318; *CCAG* 8.3 (1912).
RE 18.2 (1942) 1338–1341, 1341–1417 (esp., §23: 1400–1406), R. Keydell; 18.3 (1949) 1137–1166 (§4, 1142), K. Ziegler; *KP* 4.358, K. Ziegler; West (1983) 32–33, 37; *OCD3* 1078–9, F. Graf (s.v. "Orphic literature"), *DPA* 4.843–858, L. Brisson.

GLIM

Orpheus, pseudo (Lithika) (100 – 150 CE)

Mythical Thrakian bard and founder of the religious Orphic sect, to whom have been attributed about 50 titles of extant little poems, together with some works written in hexameters, whose language and style suggest a date not before the 6th c. BCE. Among them is the *Lithika*, a poem on the magical properties of precious stones in 774 hexameters, dated on the basis of language, style and meter. Internal evidence is lacking, and interpretations of verses 71–74 as a reference to a historical prosecution against philosophers are bound to fail.

The lapidary can be divided into four parts: an introduction (verses 1–90), wherein the poet remembers his encounter with HERMĒS, who gave him the mission of showing men the wonders concealed in the god's cave; a short bucolic tale (91–164), wherein the narrator recalls to the wise Theiodamas (whom he met along the way to the sanctuary of Hēlios) and some episodes of his childhood, and explains why he was making a pilgrimage to that divine place; the strictly mineralogical section (172–761); the envoy (762–774).

Ed.: O. Kern, *Orphica Fragmenta* (1922) 249–287, 318.
RE 18.2 (1942) 1338–1341, R. Keydell; 18.3 (1949) 1137–1166, §4 (1142), K. Ziegler (s.v. Paradoxographoi); *HLB* 2.277; Pingree (1978) 437; J. Schamp, "Apollon prophète de la pierre," *RBPh* 69 (1981) 29–49; G.N. Giannakis, *Orpheōs Lithika* (1982); West (1983); Halleux and Schamp (1985) 125–177; F. Vian, "La nouvelle édition des Lithica orphiques," *REG* 99 (1986) 161–170; *DPA* 4 (2005) 843–858, L. Brisson.

Eugenio Amato

Orpheus, pseudo (Med.) (*ca* 200 BCE – 50 CE?)

The corpus of the legendary poet includes astrological and lapidary verses (see PSEUDO-ORPHEUS, ASTROLOGY and LITHIKA), but all medical fragments are in prose, suggesting distinct origin. PLINY lists Orpheus' *Idiophue* among his sources on drugs obtained from animals (1.*ind*.28). Orpheus wrote on plants (25.12), advised blood-letting as a treatment for *angina* (***sunankhē***: 28.43), believed carrots aphrodisiacal (20.32: *cf.* 28.232, COLUMELLA 10.168), and suggested how to use arrows as a love-charm (28.34; *cf.* APULEIUS *Apol.* 30). Likewise quasi-magical was Orpheus' treatment for epilepsy: a drink of *strukhnos* (nightshade?) root plucked under a waning moon, one portion on the first day, then two up to 15 consecutively (ALEXANDER OF TRALLEIS 1.15 [1.565 Puschm.]). GALĒN *Antid.* 2.7

(14.144 K.), torn between disapproval and necessity of comprehensiveness, mentions Orpheus "Theologus," with ŌROS OF MENDĒS (emended to BŌLOS by Kern) and HĒLIODŌROS OF ATHENS, as compounders of poisons. AËTIOS OF AMIDA 1.175 (*CMG* 8.1, p. 80), in discussing calamint, records Orpheus' burn ointment (grind calamint juice and rose oil with ***psimuthion*** to gummy consistency), and 1.139 (p. 70) preserves Orpheus' treatment for consumption compounded from spikenard, ginger, *elelisphakos* seeds, long pepper, given before breakfast and at bedtime, with pure water.

Ed.: O. Kern, *Orphica Fragmenta* (1922) #319–331.
RE 18.2 (1942) 1338–1341, 1341–1417 (esp. §23: 1400–1406), R. Keydell; 18.3 (1949) 1137–1166 (§4, 1142), K. Ziegler (s.v. Paradoxographoi); *KP* 4.358, *Idem*; West (1983) 32–33, 37; *OCD3* 1078–1079, F. Graf (s.v. "Orphic literature"); *DPA* 4 (2005) 843–858, L. Brisson.

GLIM

Orthagoras (330 – 310 BCE)

A companion of NEARKHOS OF CRETE, who apparently composed his own description of the coastal voyage from the Indus to Mesopotamia, cited by STRABŌN 16.3.5 and AELIANUS, *NA* 16.35 and 17.6 (very large whale).

FGrHist 713; *BNP* 10 (2007) 259 (#2), K. Karttunen.

PTK

Orthōn of Sicily (120 BCE – 100 CE)

KRITŌN, in GALĒN *CMLoc* 1.2 (12.403 K.), records Orthōn's recipe against *alōpekia* (mange), including white hellebore and white pepper, not to mention ashed frogs and mouse-dung. The use of white pepper suggests a date after Indian trade made it more available. The unusual name is common in Sicily and south Italy, 3rd–1st cc. BCE: *LGPN* 3A.345; but attested only thrice elsewhere, 1.354. (Kühn, followed by *LGPN* 3A.338, prints "Otho" the *cognomen* attested since the late Republic: *cf.* L. Roscius Otho, in CICERO, *Murena* 40.)

Fabricius (1726) 354, s.v. Otho.

PTK

Ostanēs, pseudo (*ca* 50 BCE – 50 CE?)

Disputed legendary figure, first cited by HERMODŌROS (DIOGENĒS LAËRTIOS, *pr.*2), placing Ostanēs among the early Magi, priests of Zoroaster (dated to 5,000 years before the fall of Troy). PLINY (30.3–11) raised doubts about this tradition, but assumed that the first to write about Zoroastrian "magic" was a certain Ostanēs who followed Xerxēs into Greece. Pliny 30.9 assumed that these magical books would have been so intriguing that PYTHAGORAS, EMPEDOKLĒS, DĒMOKRITOS and PLATO would have desired to learn this science. Such a legend (and the writings attributed to Ostanēs) probably developed within the Alexandrian intellectual framework some years before Pliny (Smith), who attributed to the Persian sages necromancy and various forms of divination. In addition, Pliny (28.5–7; 30.14, 69, 256, 261) wrote that Ostanēs was an expert in the use of human and animal body parts for medical purposes. Among the works falsely attributed to Ostanēs, PHILŌN OF BUBLOS (*FGrHist* 790 F4.52 = EUSEBIOS *PE* 1.10.52) cites the *Oktateukhos*, wherein the Persian author seemingly teaches non-Iranian doctrines (e.g., a hawk-headed god). Sources from the end of the 1st c. CE and later refer to Ostanēs (sometimes with PSEUDO-ZOROASTER) as an expert in astrology, magical amulets, herbology, and in the medical use of stones: *cf.* e.g. DIOSKOURIDĒS. According to later alchemists, Ostanēs initiated PSEUDO-DĒMOKRITOS into alchemy (*CAAG* 1.57).

Berthelot (1885) 163–167; *CAAG* (1887) 1.10–12, 3.261–262; R. Reitzenstein, *Die hellenistische Mysterienreligion* (19273) 172; Bidez and Cumont (1938) 1.165–212, 2.265–356; *RE* 18.2 (1942) 1610–1642, K. Preisendanz; R. Beck, "Thus Spake not Zarathustra," in M. Boyce *et al.*, edd., *A History of Zoroastrianism* 3 (1991) 491–565; *BNP* 10 (2007) 279–280, L. Käppel; M. Smith, "Ostanes," *EI* (http://www.iranica.com/articles/sup/Ostanes.html).

Antonio Panaino

Ouranios (500 – 540 CE?)

Wrote a geographical treatise *Arabika*, in at least five books, of which only a few fragments are extant in STEPHANOS OF BUZANTION and John Tzetzēs (*FHG* 4.523–526), of disputed and controversial date, possibly early 6th c. (Bowersock). Among other things, Ouranios gives some useful information on the Nabataeans.

RE 9A.1 (1961) 947 (#4), H. Papenhoff; *RE* S.11 (1968) 1278–1292 (#4), H. von Wissmann; G.W. Bowersock, *Aporemata* 1 (1997) 173–185; *NP* 12/1 (2002) 1025 (#3), H.A. Gärtner.

Andreas Kuelzer

"Ovid": P. Ouidius Naso of Sulmo (*ca* 20 BCE – 17 CE)

Born 43 BCE, famous and gifted poet, belonged to the circle of Messalla Coruinus and was a friend of Horace and Propertius. Wrote several works on love and mythology and, at the height of fame, was banished to Tomi (Black Sea), thus leaving unfinished an antiquarian poem (*Fasti*). His *relegatio* has never been adequately explained. In exile, he wrote various works, including important books of elegies. He died at Tomi.

Two extant technical works are attributed to Ouidius: *Cosmetics for Women* (*Medicamina Faciei Feminae*) and *Fishing* (*Halieutica*). *Med.*, mentioned by Ouidius (*Ars* 3.205–206), and probably written before the composition of *Ars* 3 (Pohlenz), is a short didactic poem (100 verses). The disproportionately long proem (50 verses) indicates that part of *Med.* is missing, and abrupt transitions suggest lacunae. The work offers four prescriptions for facial cures, in a typical didactic style recalling NIKANDROS' prescriptions against poisons.

Hal. (134 verses) is attributed to Ouidius by PLINY (32.11–12) and its manuscripts. Ouidian authorship, doubted since the 16th c. (first Muretus, then Vlitius), may be undermined by prosodic oddities (Housman). The text, full of lacunae and highly corrupted, describes how the world (*mundus*) – or *deus* or *natura* – gave to each animal the way to defend itself and escape death (1–9); fishes defend themselves in various ways (9–48); *natura* impels sylvan beasts to attack hunters or to flee from them (49–65); horses proudly compete in the circus and dogs pursue game (66–82); the open sea is unsuitable for fishing (83–91); each fish has its own habitat (93–134).

From Ouidius' lost works there survive two fragments (five hexameters total) of *Phaenomena*, a partial translation of ARATOS' *Phainomena* (through v. 451). One fragment describes the Pleiades, the other ends the description of fixed stars. Pliny (30.33) expounds a medical remedy against angina, attributing it to Ouidius (*fr.* 13): if Pliny's attribution is accurate and the text uncorrupted, the prescription might have belonged to the lost part of *Med.*

Med.: E.J. Kenney, *Amores; Medicamina faciei femineae; Ars amatoria; Remedia amoris*, 2nd ed. (1994).
Hal.: T. Birth, *De Halieuticis Ovidio falso adscriptis* (1878); J.A. Richmond, *The Halieutica, ascribed to Ovid* (1962); F. Capponi, *P. Ovidii Nasonis Halieuticon* (1972); E. de Saint-Denis, *Halieutiques* (*CUF* 1975).
Phaenomena and other fragments: *FLP* 308–309, 312, 521; Blänsdorf (1995) 285, 288; M. Ciappi, "Nota al frg. 1 Blänsdorf (= 1 Courtney, 3 Lenz) dei «Phaenomena» di Ovidio," *RhM* 146 (2003) 365–371.

Claudio De Stefani

P

PACCIUS ⇒ ANTIOKHOS

Paconius (80 – 25 BCE)

To transport a marble pediment intended to replace the cracked base of the colossal statue of Apollo at Ephesos, Paconius designed a machine described by VITRUUIUS 10.2.13–14, whose design caused it to swerve off the path (*cf.* KALLISTRATOS). The family was Oscan and mercantile.

RE 18.2 (1942) 2123–2124, Fr. Münzer.

PTK and GLIM

Paetus (50 – 100 CE?)

Physician, a greybeard and a supporter of the false prophet and paradoxographer Alexander of Abonouteikhos (Lucian *Alex.* 60.8). A physician with this name is the addressee of the first "Hippokratic" letter, wherein King Artaxerxēs requests some remedy against a plague; the second letter, containing the reply of "Paetus," refers the king to HIPPOKRATĒS. Supposed neo-**Pythagorean** elements in the correspondence make possible a connection with our Paetus, who may have forged the letters to increase his prestige.

RE 10 (1965) 473–474 (#3), Fr. Kudlien.

Bruno Centrone

Pahlavi, Translations into (200 – 900 CE)

A number of Greek astronomical and astrological works translated into Pahlavi, later lost, and partly surviving only in Arabic translation. See also PSEUDO-ZOROASTER.

PTOLEMY's (*Ptalamayus*) *Mathematical Syntaxis* (*Almagest*) was translated into Pahlavi: this work is mentioned in the *Dēnkard* (4.428.15–429.8) with the title *Megistīg ī hrōmay* "the *Megístē* of the Romans" (i.e., "Greeks"). The redactors of the ZĪG consulted the Pahlavi *Almagest*; Manuščihr's *Epistle* (2.2.9–11) discusses mathematical parameters introduced by Ptolemy, as compared with those derived from Indian astronomical works. It remains unclear if Rabbān al-Ṭabarī's mid-9th c. Arabic translation derived from the Pahlavi version.

VETTIUS VALENS' *Anthologies*, already translated in the 3rd c., underwent a new recension by WUZURGMIHR. Arabic sources claim that the Sasanian astrologer Zādānfarrūx al-Andarzaghar was a great admirer of VALENS' *Bizīdaj* (*Anthologies*). Zādānfarrūx wrote a *Kitāb al-mawālīd*, a large section of which survives in al-Dāmaghānī's astrological "Collection" (written 1113 CE), in Hugo of Santalla's Latin *Liber Aristotelis*, in the Jewish astrologer Sahl ibn Bišr's *Kitāb al-mawālīd* (9th c.), and in Ibn Hibintā's *Kitāb al-mughnī* (*ca* 950), which is quoted in a Byzantine MS (Vat.Gr. 1056) as *Moúgnē*.

The *Paranatéllonta toîs dekanoîs*, only partially ascribable to Teukros of Babylōn, were translated under Xusraw I, perhaps *ca* 542 (von Gutschmid 88, and Boll 416), but the precision of this date is still questioned (some scholars suggest a separate earlier translation). Nonetheless, other works, e.g. Abū Ma'šar's *Greater Introduction* (9th c.) – describing the Persian system of the *Paranatéllonta* and mentioning numerous Pahlavi terms (especially names of constellations) – confirm the existence of a Pahlavi translation. Ibn Hibintā's *Kitāb al-mughnī* also refers to the Pahlavi Teuker, whom the Arabic sources name variously as Ṭīnkalūs, Ṭīnkarūs, Tinkalūša, or (with reference to an archaizing forgery) Tankalūša. Boll (415–439) argued that Abū Ma'šar utilized the Sasanian text, in combination with the Indian iconographic tradition of the **decans** from Varāhamihira via the astrological text *Yavanajātaka* of Sphujidhvaja, while Warburg and Saxl argued for the presence of the Egyptian iconography of the **Decans**, through Indian and then Persian (Sasanian) intermediation. Abū Ma'šar's work was embedded in the *Astrolabium planum* of Pietro d'Abano (*ca* 1300), who there describes the *sphaera barbarica*.

'Umar ibn al-Farruxān al-Ṭabarī's (*ca* 800) Arabic version of Dōrotheos of Sidōn's *Astrological Poem* was based on a Pahlavi translation. This treatise survived in Greek until the 7th c., while some prose paraphrases were attested at Buzantion in the 9th c.; few fragments survived, but some excerpts were inserted into Hephaistiōn's *Apotelesmatiká*. Other chapters of Dōrotheos survived through a recension based on the Pahlavi version, preserved in Māšā'allāh's *Kitāb al-mawālīd* and *Kitāb al-mawālīd al-kabīr* "The Great Book of the nativities," extant in Latin translation.

Māšā'allāh refers to parts of the Pahlavi Dōrotheos missing from al-Farruxān's translation. Thus, Māšā'allāh was probably still able to consult another Pahlavi annotated version, apparently earlier and lengthier than al-Farruxān's later Arabic translation. The University of Leiden Arabic MS (Oriental 891, ff.1–28) contains further material of the Pahlavi Dōrotheos, including some horoscopes, also attested in the monumental Byzantine astrological compilation, *Introduction and Foundation to Astrology*, attributed to Aḥmad the Persian. A few chapters deriving from Māšā'allāh's redaction are preserved in Hugo of Santalla's *Liber Aristotilis*.

A planetary ***melothesia*** is attested in the framework of the relations between microcosm and macrocosm (*Wizīdagīhā ī Zādspram* 30.1–13). The pattern partly follows that attested in Sphujidhvaja's *Yavanajātaka*, (1.123–126), but is partly Greek. Another ***melothesia***, *Iranian Bundahišn* 28.3–5, perhaps reflects the influence of the Hippokratic Corpus, Sevens.

The description of the World horoscope (*thema mundi*) attested in the *Iranian Bundahišn* 5 and 6B reveals further Greek and Indian influences.

A. von Gutschmidt, "Die nabatäische Landwirtschaft und ihre Geschwister," *ZDMG* 15 (1861) 1–110; K. Dyroff in F. Boll, *Sphaera: neue griechische Texte und Untersuchungen zur Geschichte der Sternbilder* (1903) 482–539, see also Boll pp. 412–439; Nallino (1922) 352, 361 = (1948) 292, 301; Fr. Saxl, *Verzeichnis astrologischer und mythologischer illustrierter Handschriften des lateinischen Mittelalters, 2., Die Handschriften der National-Bibliothek in Wien = Sitzungsberichte der Heidelberger Akademie der Wissenschaften, Philosophisch-Historische Klasse* 2 (1925–1926); A. Götze, "Persische Weisheit in griechischem Gewande," *Zeitschrift für Indologie und Iranistik* 2 (1963) 60–98, 167–174; D.E. Pingree, "The Indian Iconography of the

Decans and Horās," *Journal of the Warburg and Courtauld Institutes* 26 (1963) 223–254; *Idem, The Thousands of Abū Ma'shar* (1968); *Idem*, "Indian Influence on Sassanian and Early Islamic Astronomy and Astrology," *The Journal of Oriental Research*, 34–35 (1973) 118–126; *Idem*, "The Greek Influence on Early Islamic Mathematical Astronomy," *JAOS* 93, 1 (1973) 32–43; *Idem* (1978) 2.251–252; *GAS* 6 (1978) 109–110, 7 (1979) 71–72, 81–86, 139–151; A. Warburg, *La rinascita del paganesimo antico* (1980) 253–257; Pingree (1989) 227–230, 237; Ch. Burnett and A. al-Hamdi, "Zādānfarrūkh al-Andarzaghar on Anniversary Horoscopes. Edition and Translation," *Zeitschrift für Geschichte der arabisch-islamischen Wissenschaften* 7 (1991–1992) 294–399; Ph. Gignoux and A. Tafazzoli, *Anthologie de Zādspram* (1993) 96–99; P. Kunitzsch, "The Chapter on the Fixed Stars in Zarādusht's *Kitāb al-mawālīd*," *Zeitschrift für Geschichte der arabisch-islamischen Wissenschaften* 8 (1993) 241–249, esp. 241; Ch. Burnett and D.E. Pingree, *The* Liber Aristotilis *of Hugo of Santalla* (1997) 151, 196; D.E. Pingree, *From Astral Omens to Astrology. From Babylon to Bīkāner* (1997); Antonio Panaino, *Tessere il cielo* (1998) 38–40, 211–212; E. Raffaelli, *L'Oroscopo del mondo* (2001).

Antonio Panaino

Paiōnios of Ephesos (350 – 300 BCE)

Architect, credited by VITRUUIUS (7.*pr*.16) with completing (together with Dēmētrios, a temple-slave) the archaic Temple of Artemis at Ephesos (begun by KHERSIPHRŌN and METAGENĒS), and with building (together with DAPHNIS OF MILĒTOS) the Temple of Apollo at Milētos (at Didyma, begun *ca* 300 BCE). These accomplishments can be resolved chronologically only if Vitruuius intended by "completion" of the Artemision its reconstruction after destruction by fire in 356 BCE. STRABŌN (14.1.23) assigns the reconstructed Artemision to Kheirokratēs (sc. Deinokratēs?). Both temples were Ionic, very large in scale, and required lengthy construction, probably with a series of architects.

Svenson-Ebers (1996) 100–102; *BNP* 10 (2007) 335 (#2), C. Höcker; *KLA* 2.174–175, A. Bammer.

Margaret M. Miles

Paitāmahasiddhānta (*ca* 425 CE?)

The *Paitāmahasiddhānta* or "Treatise of Brahmā (Pitāmaha)," known as the *Paitāmahasiddhānta* of the *Viṣṇudharmottarapurāṇa* to distinguish it from similarly-named works, appears to be the inspiration for much of the classical *siddhānta* tradition in Indian mathematical astronomy. The *siddhānta* is a standard treatise format that explains universal computations for all significant astronomical phenomena. The core *siddhāntas* of the two earliest major schools or pakṣas of Indian astronomy (upon which the later schools are based) – namely, the *Āryabhaṭīya* of ĀRYABHAṬA (*ca* 500 CE) in the Āryapakṣa and the *Brāhmasphuṭasiddhānta* of Brahmagupta (628 CE) in the Brāhmapakṣa – both claim to follow a treatise of Brahmā. Similarities in content strongly indicate this *Paitāmahasiddhānta* as the treatise referred to in both cases. It is considered to be the founding text of the Brāhmapakṣa, although its original version has long been lost.

Based on the dates of the *siddhāntas* it inspired and some of the parameters it uses, the *Paitāmahasiddhānta* is thought to have been composed in the early 5th c. CE. It was incompletely absorbed into a large non-astronomical collection called the *Viṣṇudharmottarapurāṇa*, probably in the 7th c. Fragmented and corrupt, especially in its technical details, this surviving version preserves its original format of a dialogue between the sage Bhṛgu and the god Brahmā, who instructs the sage in astronomy.

The extant form of the *Paitāmahasiddhānta* still contains many of the basic features of classical Indian astronomy that were apparently derived from Hellenistic spherical astronomy models. These include large-integer period relations used to calculate mean celestial positions, planetary **epicycles** and equations for correcting mean positions on the assumption of circular orbits, and orbital sizes and geocentric distances. Their details reflect a rather chaotic mix of (among other things) Babylonian and Aristotelian notions invoked by various early Hellenistic theories that fell into oblivion after PTOLEMY. Indian astronomers combined these concepts with other parameters and techniques in their astronomical tradition to produce the cosmological and computational models that became standard in *siddhāntas*.

D.E. Pingree, "The *Paitāmahasiddhānta* of the *Viṣṇudharmottarapurāṇa*," *Brahmavidyā* 31–32 (1967–68) 472–510; *Idem*, "The recovery of early Greek astronomy from India," *JHA* 7 (1976) 109–123; *CESS* 4.259; *DSB* 15 (1978) 555–564, D.E. Pingree; *Idem*, "Āryabhaṭa, the *Paitāmahasiddhānta*, and Greek astronomy," *Studies in History of Medicine and Science* ns 12.1–2 (1993) 69–79.

Kim Plofker and Toke Lindegaard Knudsen

Palladios of Alexandria (*ca* 500 – 600 CE)

Physician and lecturer on medicine (*sophistēs/iatrosophistēs*). Transcripts of his lectures on GALĒN's *De sectis* and on the HIPPOKRATIC CORPUS EPIDEMICS VI survive, the former (Dietz 1840: 2.VII) in fragments, the latter (Dietz 2.1–204) nearly intact; both are designated *skholia apo phōnēs Palladiou* (*cf.* Richard) in the MSS. The mutilated redaction of a commentary on HIPPOKRATIC CORPUS *De fracturis* appears in a single MS under the name of STEPHANOS OF ALEXANDRIA, and a lecture on his *Aphorisms* survives only in fragments in Arabic. All four texts, and presumably also others known only through citation, reflect the standard format of late 6th c. Alexandrian exegesis, with its ordered approach to the text in terms of (1) lexical issues, (2) general explication of the lemma, and (3) discussion of earlier interpretations. Palladios is also credited with a fragmentary text on diet (*Peri brōseōs kai poseōs*: Dietz 2.VII). The authorship of a synoptic work on fevers (*Peri puretōn suntomos sunopsis*: Ideler 1 [1841/1963] 107–120) is contested among Palladios, STEPHANOS OF ATHENS, and Theophilos Protospatharios (9th c.); Stephanos and Theophilos each probably later reworked an earlier Palladian text.

Ed.: D. Irmer, *Palladius Alexandrinus. Komm. zu Hippokrates De fracturis und seine Parallelversion unter dem Namen des Stephanus von Alexandria* (1977).

Diels 2 (1907) 75–76; R. Walzer, "Fragmenta graeca in litteris arabicis: 1. Palladios and Aristotle," *JRAS* (1939) 407–422; *RE* 18.3 (1949) 211–214 (#8), H. Diller; M. Richard, "*Apo phōnēs*," *Byzantion* 20 (1950) 191–222 at 204–205; G. Baffioni, "Scoli inediti di Palladio al de Sectis di Galeno," *Boll. dei Classici Graeci e Latini* ns 6 (1958) 61–78; *KP* 4.433 (#5), F. Kudlien; *HLB* 2.292, 301; G. Endress in W. Fischer, *Grundriss der arabischen Philologie* 3 (1992) 120, n. 24; *PLRE* 3 (1992) 962; *BNP* 10 (2007) 393 (#I.5), A. Touwaide.

Keith Dickson

PALLADIUS ⇒ AEMILIANUS

Pammenēs (Alch.) (50 – 250 CE)

Alchemist mentioned in the *Physical and Mystical Things* of PSEUDO-DĒMOKRITOS where he is said to have demonstrated his knowledge to the priests of Egypt (*CAAG* 2.49). ZŌSIMOS OF PANŌPOLIS (*CAAG* 2.148) says Dēmokritos introduced (invented?; *eisagei*) Pammenēs, while Synkellos (*Ec.Chron.* 279–278, ed. Mosshammer) claims he was criticized for speaking plainly about alchemy. Probably not the same as the alchemist PHIMENAS OF SAÏS mentioned in P. LEIDENSIS X (recipe 82; ed. Halleux [1981]), but maybe identifiable with either or both PAMMENĒS (BIOLOGY) and/or the homonymous astrologer (*fl.* 66 CE, TACITUS *Ann.* 16.14).

Berthelot (1885) 170; *RE* 18.4 (1949) 303 (#4), W. Nestle.

Bink Hallum

Pammenēs (Biology) (350 BCE – 235 CE)

Biologist, wrote *Concerning Wild Animals* cited by AELIANUS (*HA* 16.42) for Egyptian winged scorpions with double stings (which Pammenēs claims to have seen first-hand) and two-headed "bipedal" serpents. Probably distinct from the homonymous astrologer; identification with the alchemist is uncertain. The name may be Egyptian ("he of Min," or "he of Amōn": Heuser 1929: 15), but is attested in Greek from the mid 4th c. BCE: DIODŌROS OF SICILY 15.94.2–3.

RE 18.4 (1949) 303 (#4), W. Nestle.

GLIM

Pamphilos of Alexandria (60 – 80 CE)

Pamphilos of Alexandria (*Vind. Med. Gr.* 1, f.2V) © Österreichische Nationalbibliothek

The *Souda* Pi-141 jumbles Amphipolis, Sikuōn, and Nikopolis as ethnics for a writer on painting, grammar, and *Geōrgika*. The painter Pamphilos (*ca* 350–300 BCE), who worked at Sikuōn, is labeled Macedonian by PLINY 35.75–77, 123, explaining Amphipolis. *Souda* Pi-142 says the Alexandrian Pamphilos composed *Leimōn* (*Meadow*: typical name for a miscellany), a dictionary (E to Ω) – epitomized before *ca* 135 CE by Diogenianus (*Souda* Delta-1140, *cf.* O-835) – and other grammatical works. Wellmann (1916) has argued that the *Leimōn* lies behind many of the animal stories of AELIANUS OF PRAENESTE, such as 1.35–38 (cited from Pamphilos in GEŌPONIKA 15.1), 3.5–6 (*ibid.*), 5.40–51, etc. The Alexandrian's *Geōrgika* is the likely source of the botanical data cited from Pamphilos in *Geopōnika* 2.20 (sowing), 5.23 (vine-pruning), 7.20 (perfuming and sweetening wine), 10.39–40 (damson-plums), 10.86 (sowing), 13.15 (fleas), and 14.14 (birds). Athēnaios, *Deipn.* preserves dozens of lexicographical notes, mostly in Books 2–3 and 11 (on cups and vessels), from his *Glōssai* and other works, seven involving plants: 2 (62d, 69d), 3 (77a, 82d, 85c), and 14 (650d, 653b).

In addition, GALĒN, *Simples* 6.*pr* (11.792–798 K.), 7.10.31 (12.31 K.), and *Hipp. Gloss.* (19.63–64, 69 K.), mentions a "younger" Pamphilos, who wrote an alphabetical *Herbs* (probably using data from Dioskouridēs: Wellmann 1898 and ed. Dioskouridēs, v.3, pp. 327–329), not listed by the *Souda*. Galēn complains that Pamphilos was inclined to old wives' tales and foolish Egyptian sorcery, but confesses he offers useful data, including Egyptian names of herbs, and describes him as a grammarian writing without autopsy (i.e., not an herbalist), and inferior to Krateuas, Hērakleidēs of Taras, Sextius Niger, and Dioskouridēs. Nevertheless, Andromakhos, in Galēn *CMLoc* 7.3 (13.68) cites his febrifuge: cardamom, saffron, ***euphorbia***, henbane, Illyrian iris, ***kostos***, myrrh, poppy-juice, white pepper, and sulfur, in honey. Asklēpiadēs Pharm., in Galēn *CMGen* 1.17 (13.446–447 K.) + 2.12 (13.527), records the cicatrizing plaster named for him. Kritōn, in Galēn *CMLoc* 5.3 (12.839–842) = Aëtios of Amida 8.16 (*CMG* 8.2, pp. 426–428), describes him as profitably curing ***leikhēn*** at Rome.

Wellmann (1898); *Idem* (1916); *RE* 18.3 (1949) 334–336 (#24), W. Stegemann, 336–349 (#25), C. Wendel, and 350–351 (#28), H. Diller; *BNP* 10 (2007) 413–414 (#6), R. Tosi; Ullmann (1972) 394; van der Eijk (2000–2001) *fr.*145.

PTK

Pamphilos of Bērutos (270 – 309 CE)

Christian bishop of Caesarea in Palestine, teacher of Eusebios of Caesarea, and executed under Galerius and Licinius. While imprisoned he wrote a five-volume defense of the Christian **Platonist** Origen (d. *ca* 250 CE), of which the first volume survives in a Latin rendition. During his tenure as bishop (from *ca* 300), he created a vast Christian library. Aëtios of Amida 16.122 (Zervos 1901: 171) appears to credit him with a complex gynecological fumigation, containing roses, plus other aromatics such as ***amōmon***, **bdellium**, cassia, ***kostos, malabathron***, spikenard, and ***storax***; apparently confirmed by MS Bonon. 1808 (15th c.), f. 53: Diels 2 (1907) 76. Perhaps a Byzantine Christian scribe added *episkopos* to a citation of Pamphilos of Alexandria?

BNP 10 (2007) 413 (#4), Chr. Markschies.

PTK

Panaitios the Younger (135 BCE – 300 CE)

Mathematician and music theorist known only from Porphurios' commentary on Ptolemy's *Harmonics*. Porphurios provides no biographical information about Panaitios except to refer to him as "the Younger" (*ho neōteros*, 65.21 Düring), presumably to distinguish him from the more famous **Stoic** philosopher Panaitios of Rhodes. Of his work we know no more than what is preserved by Porphurios, who quotes briefly from Panaitios' book *On the Ratios and Intervals in Geometry and Music* (*Peri tōn kata geōmetrian kai mousikēn logōn kai diastēmatōn*, 65.21); no other titles survive. A single-sentence paraphrase giving the rationale for the analysis of musical notes by means of mathematical proportion (88.5–7) appears to have been drawn from the same work, and three subsequent references (92.20, 92.24, 94.24), almost certainly to the same Panaitios, link him with Dēmētrios on a point of scientific vocabulary.

The only substantial quotation is the first (65.26–66.15 according to Düring, but probably extending at least as far as 67.8; Barker translates the passage to 67.10), an argument intended to prove that the term "semitone" (*hēmitonion*) is an invalid term of reference, because sense-perception, on the one hand, is not sufficiently accurate to divide musical intervals exactly in half, and because "canonic theory," on the other hand, denies that such a division is mathematically possible in the first place. For this conclusion Panaitios relies on several premises: (1) that the intervals in music can be shown to correspond to certain mathematical ratios, which he demonstrates by means of a brief canonic division; (2) that from this division it is evident that the ratio corresponding to the tone (*tonos*) is 9:8; (3) that the 9:8 ratio cannot receive a geometric mean expressible in a ratio of whole numbers (relying on a proof attributed by Boëthius to Arkhutas and spelled out at Euclidean Sectio Canonis *prop.* 3). From this he concludes that the tone cannot be divided into two equal intervals (*cf. Sect. can. prop.* 16), and that the term *hēmitonion* is consequently as much a misuse of language as the term *hēmionos* (mule, lit. "half-ass").

The fragment is also noteworthy for its mention of sympathetic vibration of strings. The phenomenon was noted by other ancient authors (Adrastos, the Aristotelian Corpus Problems, Aristeidēs Quintilianus), but Panaitios is the only extant author to connect it with the discovery of the concord-ratios.

Düring (1932); *RE* 18.3 (1949) 440–441 (#6), K. Ziegler; Barker (1989); Mathiesen (1999).

David Creese

Panaitios of Rhodes (Lindos) (*ca* 150 – 109 BCE)

Born *ca* 185 BCE, student of Diogenēs of Babylōn and Antipatros of Tarsos and successor of Antipatros as head of the **Stoa** from 129–109 BCE. Active in Rome from the 140s onwards, he later divided his time between Rome and Athens, and his ethics, informed by upper-class Roman interests, shows a marked practicality. He was a member of Scipio Africanus the Younger's circle (*cf.* Cicero *Somn.*). Panaitios is generally taken to mark the division between the "early" and "middle" **Stoa**. He emphasizes the role of an individual's own nature and dispositions in contrast to the earlier **Stoics**' grounding of natural-law ethics in a "universal nature." His ethics underpins much of Cicero's discussion in the *De officiis*. Panaitios was also more eclectically influenced than most earlier **Stoics**, using (among other targets of earlier **Stoic** critique) both Plato and Aristotle as authorities and sources. In particular he preferred an eternal ***kosmos*** to the traditional **Stoic** conflagration. He was also, according to Cicero, the only **Stoic** to reject astrology, and his arguments against astrology and divination are a source for some of the discussion in Cicero, *Div.* 2.88. Seneca (*Q.Nat.* 7.30.2) reports that Panaitios thought comets were "false," rather than "real," stars.

Ed.: M. van Straaten, *Panaetii Rhodii fragmenta* (1952), with a rather over-enthusiastic idea of what should count as a "fragment."

Daryn Lehoux

Pandrosion, and anonymous students (*ca* 285 – 320 CE)

The female teacher of mathematics to whom Pappos addresses the tract forming the first part of what later became the third book of the *Mathematical Collection* (1.30–131 Hultsch), a long and skillful response to a challenge set to him by (at least) three of Pandrosion's students, seeking his opinion about some geometrical constructions (30.17–22). The first one (1.32), a clever (though erroneous) construction, perhaps derived from Eratosthenēs' *mesolabē* (Knorr 1989: 63–69) and was meant to find two geometrical means between two given lines. The second one (68.17–25), seemingly not fully understood by Pappos himself, is also an elegant solution to the problem of finding in the same figure the geometrical, harmonic and arithmetical means between two given lines. The last one (114.14–20) is a paradoxical theorem akin to ERUKINOS' paradoxes, as Pappos remarked. Despite Pappos' (calculated) claim that his mathematical knowledge is superior to Pandrosion's and her students', the tone and content of his response, as well as the level of their achievements, show their mathematical competence. A feminine name, perhaps of Athenian origin, the diminutive of Pandrosos, Kekrops' dewy daughter, Pandrosion was never common (*LGPN*).

Cuomo (2000) 127–128, 170; Jones (1986) 4, n.8; Alain Bernard, "Sophistic aspects of Pappus's Collection," *AHES* 57 (2003) 93–150.

Alain Bernard

Pankharios (140 – 380 CE)

Iatromathematical astrologer. HEPHAISTIŌN preserves several brief fragments from his commentary on PTOLEMY's *Tetrabiblos*: Pankharios discussed the importance of the Moon over the Sun to those born at night (2.11.26–30), criticized Ptolemy for not determining the "starter" (for determining length of life) by proceeding from the degree nearest the descending degree (2.11.46–50), argued that allotting a 30° interval to the three places around Midheaven was not always necessary (2.11.63–64), and explained how to determine a horoscope when the signs did not fall neatly on the centers, i.e., Midheaven, the Horoscopus, etc. (2.11.83–86). MARCELLUS OF BORDEAUX reports Pankharios' upset stomach remedy containing pepper, hartwort, ginger, ***ammi***, anise, *libysticum* (i.e., *ligusticum*), and Spanish juniper (20.88 = *CML* 5.1, p. 350). Pankharios also wrote *Epitome Concerning Bed-Illnesses* (*peri katakliseōs nosouontōn epitomē*), similar to HERMĒS TRISMEGISTOS' *Iatromathēmatika*; Pankharios' work survives in whole or part in MS BNF *Parisin.* 1991 (15th c.) f.29^V, *Parisin.* 2139 (17th c.) f.70^V, *Barberin.* I.127 (16th c.) f.197^V, and *Vatic.* 1444 (15th c.) f.235^V.

RE 18.3 (1949) 495 (#1), K. Preisendanz; Diels 2 (1907) 76–77.

GLIM

Pankratēs of Alexandria (125 – 140 CE)

Appointed a member of the Museion by Hadrian, wrote hexameters on flowers, preserved in *POxy* 1085, *P. Brit. Mus.* 1109, and Ath., *Deipn.* 15 (677d-f = *FGrHist* 625). Ulpianus (*apud* Athēnaios) speaks about two types of lotus flower, blue and red; the latter was used to plait the so-called "Antinoos garland." According to Pankratēs the red lotus was to be called "Antinoeios" in honor of Hadrian's favorite, since it had grown from the blood of the lion killed by Antinoos hunting near Alexandria, celebrated in the *tondi* on Constantine's Arch (originally a part of a Hadrianic hunting monument). The four lines preserved by Athēnaios describe several flowers, echoing *Iliad* 14.347–349, before the Antinoos lotus arose. The papyri allow reconstruction of 50 hexameters, 30 of which are complete, with traces of a prologue and a description of the hunting scene where the lion attacks and Hadrian saves Antinoos. Echoes of HOMER and the Hēsiodic *Shield* are evident. The epyllion probably dates to soon after 130 CE. It is difficult to state with certainty if this Pankratēs is distinguishable from, or identifiable with, the wizard Pankratēs from Hēliopolis, mentioned by the Paris magical papyrus (*PGM* 4.2441–55: he showed a magical sacrifice to Hadrian); perhaps our man is the same Pankratēs cited in Lucian's *Philopseudes* 34.

Ed.: Heitsch (1963–1964) 1.51–54.

RE 18.3 (1949) 615–619 (#5), F. Stoessl; A. Garzya, "Pankrates," *Atti XVII Congresso Internazionale di Papirologia* (1984) 2.319–325; E.L. Bowie, "Greek Poetry on the Antonine Age," in D.A. Russell, *Antonine Literature* (1990) 53–90 at 81–82; *BNP* 10 (2007) 430–431 (#3), S. Fornaro.

Gianfranco Agosti

Pankratēs of Argos (*ca* 300 – *ca* 100 BCE)

Wrote a didactic poem *Sea Works*, of which only three fragments remain about different types of fishes (Ath., *Deipn.* 7 [283a, 305c, 321e]): *pompilos* called "sacred" fish, the *khiklē* and its names, and the *salpē* (Thompson 1947: 208–209, 116–117, 224–225). This Pankratēs might be the inventor of pancretian meter (Seruius, *GL* 4.459 Keil). An identification with the author of *Bokkhoreis*, a poem about the Egyptian king Bokkhoris, attributed by some to PANKRATĒS OF ALEXANDRIA, is possible.

Ed.: Heitsch (1963–1964) 1.54; *SH* 598–603.

RE 18.3 (1949) 612–614 (#3), F. Stoessl; *BNP* 10 (2007) 430 (#2), S. Fornaro.

Gianfranco Agosti

PANSĒRIS ⇒ PAUSĒRIS

Pantainos (250 BCE – 25 CE)

CELSUS 5.18.12 records his recipe for a "dispersing" ointment, similar to MĒDEIOS': quicklime, ground mustard, fenugreek, and alum in ox-fat; and ASKLĒPIADĒS PHARM., in GALĒN *CMLoc* 7.2 (13.57–58 K.), records his reduction of honey-wine and tallow, seasoned with rue, for ulcers and infections. (Kühn prints "Peteinos," *cf.* Claudius' wife Petinē: Iosephus, *AntJ* 20.150, *BJ* 2.249, Suet. *Claud.* 26.2–3.)

Fabricius (1726) 357 (s.v. Panthemus), 360 (s.v. Petinus).

PTK

Papias of Laodikeia (300 BCE – 90 CE)

The doctor of "Autolukos" – not the Athenian politicians, since the Asiatic-Greek name Papias is hardly attested before 300 BCE (*LGPN*), and no Laodikeia was founded before 300 BCE. Perhaps the astronomer AUTOLUKOS, or more likely the Rhodian pilot who went down with his ship at the battle of Khios (POLUBIOS Book 16, *fr.*5.1–2), 201 BCE. ASKLĒPIADĒS, in GALĒN, *CMGen* 4.7 (12.799–800 K.), cites Papias' remedy for inverted eyelids (*trikhiasis*).

Fabricius (1726) 357.

PTK

Papirius Fabianus (*ca* 35 BCE – *ca* 30 CE)

A rhetorician and a philosopher (Seneca Senior, *Controu.* 2.4), he paid great attention to physical science, and is called *rerum naturae peritissimus* by PLINY (36.15), who refers to him in the indices of 13 books (esp. cosmology, botany, and zoology). A prolific author (SENECA, *Epist.* 100.1) who deeply inspired Seneca (*QN* 3.27), he wrote on Physics (*Libri Causarum Naturalium*, at least three books), and his *De Animalibus* (at least two books: Charisius *GL* 105.14 and 146.28, 4th c. CE) seems to have been akin to Greek and especially Aristotelian literature (Pliny 12.20 and 28.54). Pliny's numerous references suggest real scientific inquiry (on tide: 2.224, on winds: 2.121, on corn rust: 18.276) as well as plain curiosity (9.24; 23.62).

RE 18.3 (#54) 1056–1059, W. Kroll; *DPA* 3 (2000) 413, M. Ducos; *BNP* 10 (2007) 489 (#II.3), B. Inwood.

Arnaud Zucker

Pappos of Alexandria (*ca* 285 – 320 CE)

Influential polymath, astronomically dated to 320 CE; although a marginal note places him under Diocletian. He wrote on theoretical and computational astronomy, classical geometry, practical arithmetic, geography, and perhaps astrology. Many of his works are known only through later quotations (in PROKLOS, MARINOS, EUTOKIOS or *scholia* to PTOLEMY's *Almagest*), heavily interpolated commentaries (e.g., on EUCLID's *Elements*, Book 10 = *IE*), and the collection of originally separate treatises later known as the *Mathematical Collection* (*MC*), probably compiled after the 6th c. CE (Decorps 47–51) and much interpolated. His geography (*khōrographia* ***oikoumenikē***) is known through an Armenian translation (see Jones 3–15), and Books 5 and 6 of his commentary on the *Almagest* are extant (*IA*). Pappos' scientific contribution consists not in any substantial innovation but in the way he used, organized and compared an impressive mass of scientific texts. He claims originality usually only for variations on traditional inventions, according to his own values (see his revealing criticism of APOLLŌNIOS' alleged attitude toward Euclid, *MC* 7, pp. 119.16–120.12 Jones). It is therefore necessary to outline Pappos' social and intellectual context as well as the scope of his sources to assess his key interests and contributions.

Biography and Intellectual Context: Pappos' tracts addressed various audiences, interested either in philosophy (*MC* 5; Cuomo 57–90), mechanics or architecture (*MC* 8; Cuomo 91–103), astronomy (*MC* 6, *IA*), geometry (*MC* 3, 4, 5, 7). In *MC* 3, Pappos addresses his competitor PANDROSION, her students and some of his own friends (including HIERIOS "the philosopher," perhaps among IAMBLIKHOS' followers, Jones 5) and thereby tries to attract new students by displaying his mathematical knowledge and skill, plausibly implying he worked as a private teacher. He also cleverly shows Pandrosion's students how to improve their own propositions and consequently their geometrical knowledge and skill, especially in analysis. In general Pappos seems to situate himself as a professional mathematician or at best as a teacher of liberal arts, seemingly confirmed by the scope and variety of his interests, his care for learning and his Atticist language.

Pappos' Sources: In computational and theoretical astronomy, Pappos utilized Ptolemy's *Almagest, Geography, Planispherium*, the lost *Meteōroskopeion*, the *Handy Tables*, repeatedly alluded to in *IA*, and introductory works belonging to the corpus of "little astronomy," some of which are criticized or amplified in *MC* 6. Repeated allusion to the *Handy Tables*, lost works on the interpretation of dreams as well as the building of a hydroscope suggest (plausibly but conjecturally) astrological interests.

In arithmetic and logistic, Pappos paid interest to practical calculations (*IA, MC* 3 and 8) and Apollōnios' system of notation for large numbers (*MC* 2). Neo-**Pythagorean** hints contained in *MC* 3 and *IE* are probably interpolated: Pappos' approach to *mesotai* seems predominantly geometrical, in the tradition of THEAITĒTOS and ERATOSTHENĒS.

In practical and theoretical mechanics, Pappos heavily uses HĒRŌN (esp. in *MC* 8; see also *IA* and *MC* 3); ARCHIMĒDĒS, KÁRPOS, PHILŌN OF BUZANTION and Ptolemy are also mentioned.

In geometry, Pappos relies on a considerable number of works, many known only through his allusions in *MC* 7 (also *MC* 3, 4 and 5), several from Euclid and Apollōnios. Pappos includes them in the "field of analysis" described as "useful material" for the invention of geometrical problems (Jones 66–70).

Key Interests and Contribution to Ancient Science: In modern commentaries, Pappos is sometimes characterized as only interested in geometry (e.g. Heath 2.358). But in late antiquity, Pappos was mainly known for *IA* (to which *MC* also refers) and anthologies of geometrical and mechanical problems, well reflecting the content and structure of many of his works (mainly *MC* 3, 4, 7 and 8). His exposition is often structured by *problems* or *series of problems*, for which he provides various approaches: fully articulated demonstrations, analyses that open to new problems, mechanical devices, missing lemmas, or calculations. In some cases unsolved problems or questions are presented with a variety of solutions (e.g. the treatment of the "squaring" curve in *MC* 4). In one famous case, the generalization of Euclid's problem of three and four lines, Pappos proposes a generalization of the problem itself (*MC* 7, pp. 120–123 Jones). This taste for the variety of problems or solutions is related to Pappos' interest in problem-solving and mathematical heuristic and to his endeavor to provide his students *treasuries of solutions* as a resource for their own efforts as well as *guidance* through the use of these works.

This core feature of Pappos' work is related to his other characteristics: his emphasis on the dichotomy between invention and demonstration (*MC* 3.30 Hultsch and 7.1, p. 83 Jones); his tendency to *classify* and *generalize* problems according to kinship, either formally or according to the solutions used; his mixture of mechanics and geometry (*MC* 8); his mixture of calculation and geometrical reasoning, akin to the techniques used in the *Almagest*; his complex use of the tradition, simultaneously reverential and critical (Cuomo 186–199).

Heath (1921); *RE* 18 (1949) 1084–1106, K. Ziegler; Jones (1986); Cuomo (2000); Decorps-Foulquier (2000); Alain Bernard, "Sophistic aspects of Pappus's Collection," *AHES* 57 (2003) 93–150.

Alain Bernard

Pappos (II) (600 – 800 CE)

The early table in MS *Marcianus gr.* 299 mentions a treatise entitled *By Pappos the Philosopher, On the Divine Art*; in the same MS is found a short treatise *By Pappos the Philosopher* (*CAAG* 2.27–28), which begins with a sermon steeped in Christian metaphysics and consisting in an enigmatic description of the alchemical task. The author's abridgment of a recipe of STEPHANOS provides the *terminus post.*

Ed.: *CAAG* 2.27–28.
Letrouit (1995) 61.

Cristina Viano

Papyri (Overview)

Numerous papyri survive, mostly in small pieces, containing scientific texts. They are primarily texts of practical science, with alchemy, astrology, pharmacy, and medicine being the most heavily represented; mathematical, astronomical, and other scientific texts are also found. Papyri constitute a kind of snapshot of what was in circulation and used – for example, there are more papyri of HOMER than of any other author or text. We have included only a small selection of the scientific papyri, those of the greatest length (e.g., the commentary in P. BEROL. 9782), or which are not represented by authorial entries. Many papyri are not copies of "published" texts, but are notes made by practitioners (e.g., the LONDINIENSIS MEDICUS), or are carefully-produced works for a limited audience (e.g., the horoscope by PITENIUS), or had other contexts, not now clearly discernible (e.g., the alchemical P. HOLM. or the astronomical P. PARISINUS GRAECUS 1). Such items, described as "sub-literary" by papyrologists, bring us to the border of what constitutes a "work" or a "contribution," but are valuable as holographs, unique evidence of practice and belief (e.g., P. MICHIGANENSIS 17.758). We have excluded tables of data (numerous though they are), whether astrological or pharmaceutical. We have also had to omit items that are inadequately published, although some appear to be of great interest: e.g., *P.Tebt.* II.676 (Pack 2366: on surgical abortion?). Many texts are so fragmentary that we can hardly explain them: see for example *P.Ant.* 3 (1967) #123 (*Materia medica*, similar to DIOSKOURIDĒS), 124 (on diet and fever), 126 (encyclopedia entry on tonsils, similar to AËTIOS OF AMIDA 8.51 [*CMG* 8.2, pp. 485–486]), 140 (magico-medical recipes employing the blind mole rat, *Spalax typhlus* Nordmann), and 141 (lunar astrology similar to VETTIUS VALENS 1.4–5).

H. Harraurer and P.J. Sijpesteijn, edd., *Medizinische Rezepte und Verwandtes* (1981); Marganne (1981); W. Cavini *et al.*, *Studi su papiri greci di logica e medicina* (1988); Andorlini Marcone (1993); *Eadem, Trattato di medicina su papiro* (1995); A. Jones, *Astronomical papyri from Oxyrhynchus* (1999); I. Andorlini Marcone, *Greek Medical Papyri* 1 (2001).

PTK and GLIM

Papyrus Aberdeen 11 (100 – 200 CE)

Problēmata work, whose preserved fragment concerns ***pterugeion***, as in the HIPPOKRATIC CORPUS, *Prorrh.* 2.20 (9.48 Littré) and CELSUS 7.7.4.

Pack #2342; Marganne (1994) 104–111.

PTK

Papyrus Akhmīm (500 – 800 CE)

Found in six codex folios in the late 19th c. in the newly discovered necropolis of Akhmīm (ancient Panōpolis), which had also hosted a Christian cemetery. Its author (or copyist) was Christian; Baillet palaeographically estimated its date. The first two folios present 20 tables giving multiples of nth-parts of unity (n between 2 and 20) and of two-thirds; the last four folios exhibit 50 problems either using, or asking for, calculations with numbers expressed as parts or sums of integers and parts, for which the tables are indeed useful. The large variety of techniques displayed for this may indicate this was the main focus of this booklet, although some of the problems are also presented within a (pseudo-) practical context.

J. Baillet, *Le papyrus mathématique d'Akhmîm* (1892); B. Vitrac, *Histoire de Fractions, fractions d'histoire* (1992) 149–172.

Alain Bernard

Papyrus Ashmolean Library (200 – 100 BCE)

Fragment on ***glaukōma***, stating that it is sometimes fatal, and sometimes induced by injury.

Pack #2344; Marganne (1994) 97–103.

PTK

Papyrus Ayer (70 – 140 CE)

This papyrus contains fragments of a work on mensuration, giving computations of areas of irregular quadrilaterals, performed by slicing them into triangles (often right) and regular quadrilaterals, whose more-easily computed areas are summed. The text uses *aroura* as an abstract areal unit (not in its Hellenistic sense of a concrete unit of land area); contrast HĒRŌN OF ALEXANDRIA, who uses *monas* for abstract units (*Metr.* 1.1, linear and areal; even volumetric: e.g., *Metr.* 2.11). Likewise, the papyrus considers a parallelogram to be any quadrilateral with at least *one* pair of sides parallel; whereas Hērōn follows EUCLID in requiring both pairs parallel. Other terminology (*koruphē* for the "upper side" of a quadrilateral) and the procedure are similar to Hērōn, *Metr.* A similar document is *P. Cornell* inv. 69 (Buelow-Jacobson and Taisbak).

E.J. Goodspeed, "The Ayer Papyrus: A Mathematical Fragment," *AJPhilol* 19 (1898) 25–39; A. Buelow-Jacobson and Ch. M. Taisbak, "P. Cornell inv. 69: Fragment of a Handbook in Geometry," in A. Piltz *et al.*, edd., *For particular reasons: studies in honour of Jerker Blomqvist* (2003) 54–70.

PTK

Papyrus Berol. 9782 (*Anonymous in Theaetetum*) (45 BCE – 150 CE)

PBerol. 9782 (papyrus dated to *ca* 150 CE), discovered in Hermupolis Magna in 1901, contains 75 columns and fragments of a running commentary on PLATO's *Theaetetus*. The extant part starts with preliminary questions about the dialogue and ends with the commentary on *Theaet.* 153d; fragment C covers 157e4–158a2. The author's familiarity with AINESIDĒMOS' skepticism proves that the commentary was written after 45 BCE. Attempts to identify the author have failed to convince. The author mentions commentaries on the *Timaeus* and the *Symposium* as other works of his, and promises one on the *Phaedo*. He shows familiarity with Middle **Platonic** doctrines, distances himself from **Academic** skepticism, yet avoids radically dogmatic interpretations; he is critical of **Stoics** and **Epicureans**. He uses Aristotelian logical tools (syllogistic, the theory of the definition, the categories), but not accurately. His mathematical knowledge is rather rudimentary, as appears from his discussion of *Theaet.* 147d3–148b2.

Ed.: G. Bastianini and D.N. Sedley, *Commentarium in Platonis Theaetetum*, in *Corpus dei papiri filosofici* 3 (1995) 227–562.

Moraux (1984) 2.481–493; *ECP* 543–544, D.N. Sedley; *BNP* 1 (2002) 712, K.-H. Stanzel.

Jan Opsomer

Papyrus bibl. univ. Giss. IV.44 (100 – 80 BCE)

Discusses the autoplastic surgical repair of mutilation (*kolobōma*) of the lips, as in CELSUS 7.9.2–5.

M.-H. Marganne, "Un témoignage antérieur à Celse sur l'opération du coloboma," *CE* 66 (1991) 226–236.

PTK

Papyrus Cairo Crawford 1 (200 – 300 CE)

Fragment on eye-surgery to cure *rheumatismos*, advocating procedures recorded by PAULOS OF AIGINA 6.6 (*CMG* 9.2, p. 49) or SEUERUS (*ibid.* 6.7, p. 50). The author claims to follow PHILOXENOS and "those around" HĒRAKLEIDĒS, HĒRŌN, MĒNODŌROS, and SŌSTRATOS, so s/he must have been writing after *ca* 10 CE. Marganne (1981) 140–143 suggests the author was HĒLIODŌROS.

Marganne (1994) 146–172.

PTK

Papyrus Fayumensis (Apoplexy) (200 – 100 BCE)

Problēmata work, whose preserved fragment concerns apoplexy.

Pack #2370.

PTK

Papyrus Florentinus (before 400 CE)

Fragment of a recipe for dyeing the skins of living animals; the papyrus offers analogies to P. HOLM. and P. LEID. V.

Ed.: C. Gallavotti, "Tre Papiri Fiorentini," *RFIC* 67, ns 17 (1939) 252–260; Pack #2000; Halleux (1981) 160–163.

Cristina Viano

Papyrus Geneva inv. 259 (100 – 200 CE?)

This papyrus contains three problems of increasing complexity, each seeking to solve an integer right triangle (of sides 3, 4, 5) with two givens: (a) the hypotenuse plus one side, (b) the sum of the hypotenuse and one side plus the other side, and (c) the hypotenuse plus the sum of the two sides. The method is algebraic, making use of the "**Pythagorean**" formula.

J. Rudhardt, "Trois problèmes de géometrie, conservés par un papyrus genevois," *MusHelv* 35 (1978) 233–240, pl. 7.

PTK

Papyrus Hibeh 1.27 (*ca* 320 – *ca* 280 BCE)

This papyrus from the ruins of El-Hibeh was wrapped about the same mummies as the drafts of two letters (*P.Hibeh* 1.34, 73) written in the early reign of Ptolemy III, probably before 240 BCE. Its purpose was to provide a calendar of the major festivals or days of obligation in the Egyptian wandering year of 365 days.

After a brief introduction (first hand) in which the author claims to have learned what follows from a wise man in the Saïte nome, he presents the calendar (second hand, corrected by a third or perhaps the first). The typical entry gives a date, the risings and setting of certain stars and constellations, the changes in the weather and condition of the Nile, the length of daytime and nighttime, and the religious festival to be observed. The date itself is a month and day number when the Sun is in a new zodiacal constellation (*not* a zodiacal sign) or simply a day number. Computation of the lengths of daytime (the interval from sunrise to sunset) assumes that the ratio of maximum (M) to minimum (m), where $M + m = 24$ and vary with latitude, is 14:10 equinoctial hours (which is consistent with the reference to Saïs). The scheme starts with Thoth 1; it supposes that the length of daytime increases daily for 180 days by $^1/_{45}$ hour, stays the same for four days, decreases daily for 180 days by $^1/_{45}$ hour, and then stays the same for three days before increasing again. The day *after* the days of maximum or minimum daytime is called a solstice (*tropē*). In essence, *P.Hibeh* 1.27 adapts a Babylonian linear zigzag scheme for lengths of daytime and night-time throughout a year of 360 days such as is found in MUL.APIN (see BABYLONIAN ASTRONOMY) to the Egyptian year of 365 days, by flattening the extremes in order to accommodate the five extra days (see P. PARISINUS GRAECUS 1).

The astronomical evidence, though schematic, suggests that the calendar was composed for the interval of a few decades around 300 BCE; and the handwriting is consistent with the papyrus' being written then.

P.Hibeh 1 (1906) 138–157; Neugebauer (1975) 599–600, 706; Alan C. Bowen and B.R. Goldstein, "Hipparchus' Treatment of Early Greek Astronomy," *PAPS* 135 (1991) 238–245.

Alan C. Bowen

Papyrus Hibeh 2.187 (325 – 240 BCE)

The exiguous remains of *P. Hibeh* 2.187 describe methods for increasing the yield and sweetness of almonds; in both cases drilling the trunk is the primary recommendation. The close resemblance this text bears to THEOPHRASTOS, *HP* 2.7.6–7, suggests that it is an abridgement or a source of Theophrastos. (This work appears to be the sole technical treatise on agriculture among the Egyptian papyri: *PSI* 6.624 is not a manual on viticulture, but a report written by Zēnōn, secretary of Apollōnios, about the work performed by vine-dressers on one of his estates.)

Pack #1985.

Philip Thibodeau

Papyrus Hibeh (Ophthalmology) (300 – 250 BCE)

Fragment of a work deploying ***pneuma*** to explain the mechanism of vision and the pathology, etiology, and therapy of eye-diseases.

Pack #2343; Marganne (1994) 37–96.

PTK

Papyrus Holmiensis (230 – 350 CE)

A compendium of alchemical recipes on loose papyrus folios of unknown provenance, but probably written by the same scribe or group of scribes as P. LEIDENSIS X. The papyrus contains 155 separate alchemical recipes: 1–9 concern gold and mostly ***asēmos***; 10–88 involve stones and 89–159 treat materials; some recipes are identical with or similar to those in *P. Leidensis X* (see concordances in Halleux 1981: 14–15).

The papyrus cites a certain *Aphrikianos* (recipes 116 and 141), possibly IULIUS AFRICANUS.

Based on this and what he sees as deference to legislation against forgery enforced at the beginning of the 4th c., Halleux dates *P. Holmiensis* and *P. Leidensis X* to the reign of Constantine (306–337), which would make them roughly contemporary with Zōsimos of Panōpolis and the supposed burning of Egyptian alchemical books by the emperor Diocletian (John of Antioch *fr.*165 [*FHG* 4.601], *Souda* Delta-1156 and Khi-280). Recipe 2, at least, is attributed to pseudo-Dēmokritos and a certain Anaxilaos (perhaps Anaxilaos of Larissa) is cited as the source of this attribution.

The recipes preserved in *P. Holmiensis* are purely practical in nature containing none of the theoretical passages, elaborate *decknamen* or mystical references found in the Greek alchemical corpus. However, a reference to the *kērotakis* device (recipe 31), associated in the Greek alchemical corpus with Maria, provides a link with the wider alchemical tradition.

Ed.: Halleux (1981) 110–151.

Bink Hallum

Papyrus Iandanae 85 (75 – 125 CE)

Two fragmentary alchemical recipes preserved on a single folio of a papyrus roll or perhaps codex of unknown provenance. Similarities with P. Holmiensis indicate that both recipes address imparting a red tincture; the second perhaps for coloring stones.

Ed.: Halleux (1981) 158–160.

Bink Hallum

Papyrus Johnson (Antinoensis) (350 – 400 CE)

Fragment of an illustrated herbal, describing *sumphuton* and *phlommos*, which are synonyms of *helenion* (elecampane), according to Dioskouridēs 1.28.

Pack #2095.

PTK

Papyrus Laur. Inv. 68 (400 – 500 CE)

Fragment of a toxicology on the *lepus marinus* (sea hare, an *Aplysia* sp.) and antidotes thereto, similar to Asklēpiadēs Pharm., in Galēn *Antid.* 2.7 (14.139 K.): human, cow, or goat milk, or pennyroyal in must, or cyclamen root in wine, or mallows; Pliny 20.223 mentions mallows (9.155 describes the animal), and Scribonius Largus 186 milks and mallows.

I. Andorlini, "Una trattazione 'sui veneni e sugli antidoti' (PL 68)," *Analecta Papyrologica* 3 (1991) 85–101.

PTK

Papyrus Leidensis V (300 – 350 CE)

A bilingual magical miscellany (*P. Leid.* V = *PGM* XII) of unknown provenance containing *inter alia* a single alchemical recipe. The section wherein the recipe is found dates to *ca* 300–350 CE and its appearance in this context demonstrates the early association of alchemy and magic. The recipe, for the rusting (*iōsis*) of gold, calls for the application of chlorides and sulfurous compounds to a leaf of gold. Halleux believes that these ingredients could have formed sulfuric or hydrochloric acid and that the resulting dross left on the gold could have been considered its "rust."

Ed.: Halleux (1981) 163–166; *PGM* 2.71.

Bink Hallum

Papyrus Leidensis X (230 – 350 CE)

A compendium of alchemical recipes contained in a papyrus codex of unknown provenance but probably written by the same scribe or group of scribes as P. HOLMIENSIS. The codex contains 99 alchemical recipes plus ten sections from DIOSKOURIDĒS' *De materia medica*. Recipes 1–88 concern gold, silver and methods for writing with gold and silver, while recipes 89–99 address coloring metals, stones and wool. Three recipes (31, 42 and 43) are for the assaying of metals, but the vast majority is for creating alloys of between two and five metals. The codex contains some recipes identical or similar to those in *P. Holmiensis* (concordances in Halleux 1981: 14–15).

The recipes are similar to those in the Greek alchemical corpus, but with none of the theoretical passages, elaborate *decknamen* or mystical speculation also found there. The recipes seemingly are of a purely practical nature aimed at creating cheap imitations of precious goods. Yet the fact that the papyrus is in the form of a codex written in an elegant book-hand and that it shows no stains or signs of frequent thumbing suggests that it was read in a library and not a laboratory.

Ed.: Halleux (1981) 84–109.

E.R. Caley, "The Leyden Papyrus X. An English Translation with Brief Notes," *Journal of Chemical Education* 3.10 (1926) 1149–1166.

Bink Hallum

PAPYRUS LIT. LOND. 165 ⇒ LONDINIENSIS MEDICUS

Papyrus Lit. Lond. 167 (100 BCE – 100 CE)

Careful and precise description of the bones of the foot, similar to RUFUS OF EPHESOS, *Bones* 38 (p. 193 DR); GALĒN, *Bones for Beginners* 24 (2.776–777 K.); and PSEUDO-GALĒN, INTRODUCTIO 12 (14.724–725 K.).

M.-H. Marganne, "Une description des os du tarse," *BASP* 24 (1987) 23–34.

PTK

Papyrus London 98 (Coptic Part) (*ca* 110 – 190 CE)

Contains a connected text written in both Greek and Coptic. The Greek portion at the beginning presents the calculation of a horoscope (13 April 95 CE). The Coptic part consists of non-mathematical statements of magical purport.

J. Černý, P.E. Kahle, and R.A. Parker, "The Old Coptic Horoscope," *JEA* 43 (1957) 86–100; Neugebauer and van Hoesen (1959) 28–38.

Leo Depuydt

PAPYRUS LONDON INV. 137 ⇒ LONDINIENSIS MEDICUS

PAPYRUS LOUVRE 2329/2388 ⇒ P. PARISINUS GRAECUS 1

Papyrus Louvre inv. 7733 (300 – 200 BCE)

This fragmentary papyrus, in a 3rd c. BCE hand, discovered in Egypt in 1869, discusses optical distortions and illusions and their explanations. The work has been attributed to DĒMOKRITOS, EPICURUS and the milieu of EUCLID; but indications are too slender for confidence. A Skeptical author from the circle of Pyrrho has also been suggested. But the text is not endeavoring to cast doubt, in Skeptical fashion, on our knowledge of things; rather, it attempts scientific explanations of why we do not always see things accurately. The phenomena discussed include the apparent motion and apparent lack of motion of objects, as well as variations in their apparent size and their degree of visibility.

Richard Bett, "Sceptic Optics?" *Apeiron* 40 (2007) 95–121.

Richard Bett

Papyrus Lund I.7 (200 – 400 CE)

Fragment of an anatomical catechism, on the caecum (*tuphlon enteron; cf.* PSEUDO-GALĒN, *Aff. Ren.* 1 [19.646 K.]) and rectum (*apeuthusmenon; cf.* RUFUS OF EPHESOS, *Anat.* 48 [p. 180 DR]).

M.-H. Marganne, "Un questionnaire d'anatomie," *CE* 62 (1987) 189–200.

PTK

Papyrus Michiganensis 3.148 (1st c. CE)

Two columns of an astrological treatise perhaps entitled *On Lunar Conjunctions*, and perhaps written *ca* 160–70 BCE (judging from its copious references to pirates). Prognostications are made on the basis of the conjunctions.

F.E. Robbins in *P. Mich.* 3 (1936).

PTK

Papyrus Michiganensis 3.149 (100 – 200 CE)

Unknown author of an astrological prose work in Greek of which there survive in the 2nd c. papyrus substantial fragments, amounting to parts of 22 columns of text. The work is exceptional within the corpus of Greco-Roman astrological texts for its integration of elements of contemporary astronomical modeling into its astrological doctrines, which are themselves largely non-standard. The most interesting and idiosyncratic passage comes at the beginning: here the varying apparent speeds of the Sun, Moon, and planets are explained in terms of their revolving on **epicycles**, spoken of as spheres. Unlike in PTOLEMY's planetary models, but paralleling aspects of PLINY's obscure discussion (2.68–76), the planets are said to produce their fastest apparent motion when they are nearest the Earth on their **epicycles**. Specific values for the radii of the epicycles are given, expressed in degrees and minutes such that the radius of the circle bearing the **epicycle**'s center is 60°. These radii are made the numerical basis of a scheme of ***melothesia***, and from this scheme in turn the author derives astrologically significant characterizations of divisions of the zodiac.

A. Aaboe, "On a Greek Qualitative Planetary Model of the Epicyclic Variety," *Centaurus* 9 (1963) 1–10; Neugebauer (1975) 805–808.

Alexander Jones

Papyrus Michiganensis 17.758 (350 – 370 CE)

Sometime in the middle of the 4th c., a practicing physician in Roman Egypt hired a local scribe (perhaps in Oxyrhynchos, perhaps Antinoopolis: the papyrus' provenance is uncertain) to extract pharmaceutical recipes from circulating Greek-language medical books. The nameless physician and his anonymous scribe were both quite literate, evinced by corrections in two hands: Youtie believes that addenda written more rapidly with abbreviations ("ligatures") are those of the doctor, but the scribe also inserted some of his own corrections; the texts copied and augmented are usually familiar from formal pharmacological works, whether extant (DIOSKOURIDĒS), or as known through later compactions (e.g., HĒRĀS in GALĒN, OREIBASIOS, etc.), with simples and compounds suggesting consistency across the long history of ancient drug lore. Occasionally, there are quotations of otherwise unknown formulas (e.g. the rue-plaster from Book 2 of some Dionusios: Youtie 22–23). The papyrus includes full dosages and timing of application as usual in papyri and in Galēn, etc., but it also shows a crudely applied "common knowledge" of "how much" of X or Y should be used. Youtie meticulously matches the substances with the better known and much more copious accounts of the philosopher-physicians, indicating a kind of "filtering down," a variety of cookbookery; and yet this and other papyri also reveal a "filtering up" of information derived from folk medicine (paralleled with sparse data in the *PGM*). "Drugs" include substances of multiple employment, e.g. *P.Mich. Inv. 21G* (Youtie 56–58) with its *kollēs*: the "plaster" is sticky (as it should be) but the term suggests the "flour-paste, made from the best wheaten-flour and the finest meal to glue books . . . is helpful for those who spit blood, when the flour-paste is diluted with water, warmed, and administered to the patient a spoonful at a time" (Dioskouridēs *MM* 2.85.3, as Youtie adduces).

Ed.: L.C. Youtie, *The Michigan Medical Codex (P. Mich. 758 = P. Mich. Inv. 21)*, with introd. by A.E. Hanson (1996) = *P.Mich.* 17.
Andorlini Marcone (1993).

John Scarborough

Papyrus Mil. Vogl. I.14 (100 – 200 CE)

Fragment discussing the nerves and how they carry ***pneuma*** to organs as needed, and how that explains diseases.

M.-H. Marganne and P. Mertens, "Medici et Medica," in B. Mandilaras, ed., *Proc. of the XVIII Inter. Congr. of Papyrology* (1988) 1.105–146 at 123, #2361.

PTK

Papyrus Mil. Vogl. I.15 (100 – 200 CE)

Fragment of a medical catechism, with two questions on apoplexy and one on ***elephantiasis***, adjacent as in CELSUS 3.25–26, and PSEUDO-GALĒN, DEF. MED. (19.346–347 K.).

Tecusan (2004) *fr.* 14.

PTK

PAPYRUS MIL. VOGL. VIII.309 => POSEIDIPPOS

Papyrus Oslo. 72 (100 – 130 CE)

Fragment on the treatment of epilepsy and *paraplexy* by diet; the etiology given is the brain "and what comes from it" (*cf.* Aretaios 5.4–5).

Pack #2384.

PTK

Papyrus Osloensis 73 (*ca* 150 BCE – 50 CE?)

This papyrus of *ca* 100 CE contains one column describing the use of a ***dioptra*** to measure the apparent solar diameter (*cf.* Archimēdēs, *Sand-reckoner* 1.10), and a water-clock to find the rising time of the sun (from first to last contact with the horizon); the two measurements both yield an angular measure of the apparent solar diameter as ½° (*cf.* Kleomēdēs 2.1).

P. Osloenses 3 (1936) #73.

PTK

Papyrus Oxyrhynchos 3.467 (75 – 125 CE)

Two fragmentary alchemical recipes preserved in *POxy* 3, #467 (= Pack #1999), dated to the end of the 1st or beginning of the 2nd c. CE. The first recipe is for coloring silver to appear like gold, while the second is for purifying ***asēmos*** by cupellation.

Ed.: Halleux (1981) 155–158.

Bink Hallum

Papyrus Oxyrhynchos 3.470 (300 BCE – 280 CE?)

This papyrus of the 3rd c. CE describes the construction of a *pesseutērion* (marked with the "House of Horus," *Phoror*, and the "House of Beauty," *Phernouphis*), i.e., a calendar-abacus similar to the Egyptian "Senet" game, and secondly of a water-clock.

POxy 3 (1903) #470; W. Decker, *Sports and Games of Ancient Egypt* (1992).

PTK

Papyrus Oxyrhynchos 13.1609 (*ca* 300 BCE – 100 CE)

This papyrus of *ca* 125 CE contains part of a column rejecting the "efflux" theory of vision held by Dēmokritos, Empedoklēs, and Epicurus, and referring to the author's commentary on Plato's *Timaios*.

POxy 13 (1919) #1609.

PTK

Papyrus Oxyrhynchos 15.1796 (*De plantis Aegyptiis*) (100 BCE – 100 CE?)

A fragment in 22 wholly readable hexameters from a poem on Nilotic countryside preserved in the second column of *POxy* 1796. The fragments, treating cyclamen or, maybe, sycamore (vv. 1–11), and *persea* (now called *Mimusops* by botanists: *cf.* Pliny 13.60, 15.45) and its flourishing (vv. 12–22), derive from the tradition of Alexandrian didactic poetry: for example, we know that Nikandros had written a *Georgics* mentioning cyclamen (*fr.*74 G.), and one of Kallimakhos' fragments treats the Egyptian origin of *persea* (*fr.*655 Pf.). Our fragment, probably part of a longer poem, is refined in style and technique, with descriptive images and variations (e.g., *persea* fruit derives moisture from the Nile to survive the dry season). The author possesses a good knowledge of epic models; metrical features suggest the date.

Ed.: D.L. Page, *Select Papyri. III. Poetry* (Loeb 1941; repr. 1992) #124; Heitsch (1963–1964) 1.60.

D. Bonneau, *La cru du Nil* (1964) 49–50; A. Zumbo, "Considerazioni sul P. Oxy. 1796: *De Plantis Aegyptiis*," *Analecta Papyrologica* 4 (1992) 41–47; D. Fausti, "Il POxy XV 1796 *verso*: nuovi contribute interpretativi," in I. Andorlini *et al.*, edd., *Atti XXII Congr. Intern. di Papirologia* (2001) 1.443–455.

Gianfranco Agosti

Papyrus Parisinus graecus 1 (now P. Louvre 2388 Ro + Paris, Louvre 2329 Ro) (*ca* 200 – *ca* 165 BCE)

This opisthographic papyrus, found near Gizeh, has on the verso 12 lines of iambic trimeter, the first letters of which form an acrostic *ΤΕΧΝΗ ΕΥΔΟΞΟΥ* (meaning *Art of Eudoxos*), which is often treated as the name of the papyrus. Around this poem are administrative documents (in later hands) indicating composition before 165 BCE. The assertion (col. 22) that, according to EUDOXOS and KALLIPPOS, the winter solstice falls on Athyr 19 or 20 holds for *ca* 190 BCE. The Dionysius addressed in the latest letter, perhaps a *strategos* of Memphis, may have owned the papyrus.

On the recto, in the same hand as the poem and also containing sections in iambic trimeter, are preserved 24 columns (including the end but not the beginning) of what amounts to a rudimentary handbook in astronomy. The subjects broached include the day-intervals between stellar phenomena; the course of the Sun; the annual variation in the length of daytime – the author describes the scheme found in P. HIBEH 1.27 but has two (should be three) days of longest daytime or summer solstice and three (should be four) days of shortest daytime or winter solstice – the course of the Moon through the zodiacal signs; the planets, their names and periods; the celestial sphere, its layout and motion; the risings and settings of the fixed stars; the relation between the celestial sphere and the observer's latitude; the Moon, its shape and illumination; the ***oktaetēris***; the *arcus visionis*; lunar and solar eclipses; the relative sizes of the Sun, Moon, and Earth; and the lengths of the seasons.

The papyrus does not derive from Eudoxos; moreover, the poem's location and lack of explicit connection to the text on the recto make the title "*Art of Eudoxos*" unlikely. Some suppose that its author or final redactor was LEPTINĒS (II). The technical errors and the repetition of passages columns apart suggest a careless compilation of two prose versions of an original perhaps written in verse during the 3rd c. BCE.

A.-J. Letronne, with W. Brunet de Presle, *Notices et extraits des manuscrits de la Bibliothèque Impériale* 18.2 (1865) 25–76; F. Blass, *Eudoxi ars astronomica* (1887) 138–157; Neugebauer (1975) 686–689, 706.

Alan C. Bowen

Papyrus Ross. Georg. 1.20 (140 – 160 CE)

Problēmata work on ophthalmology, mentioning ***glaukōma*** (distinguished from cataract as RUFUS in OREIBASIOS, *Syn.* 8.49, *CMG* 6.3, pp. 266–267), ***staphulōma*** (*cf.* CELSUS 7.7.11), and ***pterugeion*** (*cf.* P. ABERDEEN 11).

Marganne (1994) 112–132.

PTK

Papyrus Rylandensis 27 (*ca* **250 – 300 CE**)

Gives rules for computing lunar latitudes and longitudes at lunar apogee, probably for computing those quantities on an arbitrary day (omitted or lost). The zero-point of the zodiac used is that of the Babylonian sidereal zodiac, and the epoch is 32 BCE, June 30/July 1, when the Moon was at apogee. The computation employs Babylonian numerical methods adapted to the Egyptian 25-year calendar cycle.

Neugebauer (1975) 808–817.

PTK

Papyrus Ryl. III.529 (200 – 300 CE)

Fragment on the mechanical setting of compound fractures; the author cites his own *Tekhnikos Logos* (*Practical Treatise*) and refers to the "Alexandrian position" as inferior to the recumbent.

Pack #2376.

PTK

Papyrus Ryl. III.531 (300 – 200 BCE)

Fragment of a pharmaceutical treatise, possibly gynecological. Dried otter kidney is prescribed for hysterical suffocation, testicular pain, and as a womb-enema; a compound of realgar, unfired sulfur, and almonds in wine (following the HIPPOKRATIC CORPUS, DISEASES OF WOMEN 200 [8.382 Littré]) is prescribed for coughing and choking; oak-gall, pomegranate, and alum form part of a fragmentary recipe for a contraceptive (*atokeion*).

Pack #2418.

PTK

Papyrus Strassbourg Inv. Gr. 90 (130 – 170 CE)

Fragment giving both etiologies of various eye-diseases (*psōr-*, *xēr-*, *sklēr-*, and *lag-ophthalmia*, plus *pheimōsis*, *onukhion*, etc.), apparently following DĒMOSTHENĒS, and recipes for collyria.

Marganne (1994) 133–146, 173–176.

PTK

PSI 6.624 ⇒ P. HIBEH 2.187

PSI inv. 3011 (250 – 300 CE)

Fragment on the medical properties of bitumen, citing NIKĒRATOS OF ATHENS (as in DIOSKOURIDĒS 1.73) and HEKATŌNUMOS OF KHIOS.

G.A. Gerhard, "Frammento medico: sulle proprietà terapeutiche dell' asphalto," *SIFC* ns 12 (1935) 93–94; Pack #2388.

PTK

Papyrus Tebtunis 679 (100 – 200 CE)

Fragment of an illustrated treatise on medicinal properties of plants.

Pack #2094.

PTK

Papyrus Turner. 14 (150 – 200 CE)

Fragment of a medical catechism, with questions on olive oil, and the best moment for a *katabrokhē* (soaking) in cases involving paroxysms ("at the beginning"). The terminology, of constriction, relaxation, and dispersion, seems **Methodist**.

L.C. and H.C. Youtie, in *P.Turner.*

PTK

Papyrus Vindob. 19996 (50 – 100 CE)

This papyrus from the Fayum, Egypt, contains several columns of stereometrical calculations, computing the volumes of parallelepipeds, pyramids (triangular and quadrangular), cylinders, and truncated cones. The procedures are similar to those of HĒRŌN OF ALEXANDRIA, *Metr.*

H. Gerstinger and K. Vogel, "Eine stereometrische Aufgabensammlung in Papyrus Graecus Vindobonensis 19996," *Griechische Literarische Papyri* v. 1, ed. H. Gerstinger *et al.* (1932) 11–76.

PTK

Paradoxographus Florentinus (100 – 200 CE)

Assembled 43 extracts concentrating on the theme of water. Following STEPHANOS OF BUZANTION's attribution of the collection to SŌTIŌN, the anonymous author is also known as pseudo-Sōtiōn. The work divides into two sections, one on springs, the other on lakes and rivers. The stories are not arranged in a specific geographical order. Many different sources were used, most of them indirectly.

Ed.: *PGR* 315–329.
RE 18.3 (1949) 1137–1166 (§31, 1161–62), K. Ziegler; Giannini (1964) 135–136.

Jan Bollansée, Karen Haegemans, and Guido Schepens

Paradoxographus Palatinus (200 – 300 CE?)

Anonymous author, who passed on a collection of 21 *mirabilia*, dubbed *Palatinus Paradoxographus* by Öhler. The topics in this compilation range from animals, over water and stones, to medicinal plants. Numerous authors, many not cited first-hand, provided the stories, among whom ARISTOTLE (10), KALLIMAKHOS (15), Theopompos (19), TIMAIOS (13), ANTIGONOS PARADOX. (20), AGLAOSTHENĒS (7), ARTEMIDŌROS OF EPHESOS (11), ANDRONIKOS PARADOX. (12), CATO (21), and finally Athēnaios (18), providing the *terminus post quem*.

Ed.: *PGR* 354–361.
RE 18.3 (1949) 1137–1166 (§33, 1163–64), K. Ziegler; Giannini (1964) 138.

Jan Bollansée, Karen Haegemans, and Guido Schepens

Paradoxographus Vaticanus (14 – 200 CE)

Also known as *Paradoxographus Rohdii*, after its first editor E. Rohde. An anonymous compilation of 67 paradoxographical excerpts surviving in a 15th c. mixed Vatican MS (*Vat. gr.* 12, ff. 211–215). The material is arranged in three groups: ten unusual zoological phenomena open the collection, after which 16 water-wonders (11–14, 17–23, 34–36, 38 and 39) and 32 ethnographical curiosities (25–30, 41–43, 45–67) alternate, interspersed with additional isolated topics (metamorphoses, geological marvels, etc.). The compilation relies heavily on NIKOLAOS OF DAMASKOS' *Collection of Customs* (which provides the vague *terminus post quem* of 14 CE) and further contains numerous second- and third-hand citations of sources (via ANTIGONOS PARADOX.), ranging from the well-known (Theopompos *FGrHist* 115 F11, 16; ARISTOTLE *HA* 1.1 [487a28–32]; ARISTŌN OF KEŌS) to the obscure (the otherwise unknown AGĒSIAS and POLITĒS, plus Polukleitos of Larissa and DALIŌN). Its attribution to ISIGONOS, suggested in the *editio princeps*, is rejected by most scholars as lacking proof.

Ed.: *PGR* 331–351.
RE 18.3 (1949) 1137–1166 (§32, 1162–1163), K. Ziegler; Giannini (1964) 137–138.

Jan Bollansée, Karen Haegemans, and Guido Schepens

Paraphrasis eis ta Oppianou Halieutika (200 – 500 CE)

Paraphrase of the poem by OPPIANUS OF APAMEIA, sometimes attributed to EUTEKNIOS, and partially preserved (from Book 3.605 to the end). Appearing in almost all manuscripts along with the paraphrases of Euteknios, and of similar linguistic features, it shows more interest in stylistic elaboration than in clear transcription.

Ed.: M. Papathomopoulos, *Anônumou Parafrasis eis ta Oppianou Halieutika* (1976).

Arnaud Zucker

Parisinus medicus (of Crete?) (70 – 180 CE)

Anonymous text (named for its MS), rediscovered in 1844, though not completely edited until 1997; a citation of MNASEAS the **Methodist** (50.3.10) provides the *terminus post*, and an extract in PHILOUMENOS the *terminus ante*; the author may be Cretan (12.3.7). The work covers acute (§1–16) and chronic (§17–51) diseases, each division proceeding from head to foot. It defines each disease, offers a doxographic and often aporetic etiology citing ERASISTRATOS, PRAXAGORAS, DIOKLĒS, and HIPPOKRATĒS, usually in that (approximately reverse chronological) order, and thirdly prescribes therapy, a structure resembling P. MIL.VOGL. I.15 and PSEUDO-GALĒN INTRODUCTIO. Garofalo (1997: XI–XIII) rejects the scholarly tendency to identify the author as HĒRODOTOS (PNEUM.), while accepting parallels of doctrine and diction. Although the therapeutics are often **Methodist** in character, the author's own etiologies in §34 (cirrhosis) and 40 (bladder-paralysis) refer to **humors** and unobservable entities (not possibly **Methodist**), and §16 on satyriasis cites no authorities, a suppression of THEMISŌN (contrast CAELIUS AURELIANUS *Acute* 3.185–186 [*CML* 6.1.1, p. 400]) precluding **Methodist** authorship. The author is careful to distinguish between what "the four" (*cf.* §22.1) say and what s/he infers, and their etiologies are recast to focus on the affected part(s). Pharmaceutical prescriptions include few animal products other than ordinary food: only castoreum and bull-gall; and the sole Indian import is cardamom (note also Arabian products acacia, aloes, and frankincense, and African products ***ammi***, ***ammōniakon***, **bdellium** and ***euphorbia***).

Ed.: I. Garofalo, *Anonymi Medici de Morbis Acutis et Chroniis* (1997) = *SAM* 12.
van der Eijk (1999) 295–331; *BNP* 1 (2002) 713–714, V. Nutton.

PTK

Parmenidēs of Elea (*ca* 490 – *ca* 450 BCE)

Born *ca* 520 BCE, and perhaps the most influential Pre-Socratic philosopher. In a hexameter poem, Parmenidēs puts into the mouth of an unnamed goddess a criticism of "mortal" thought and a set of claims about what truly is: coming-to-be and passing-away are both impossible, what-is is complete, whole, perfect, unchanging, and of a single kind. Mortals go astray in assuming that what-is can not be (i.e. can come to be, pass away, or change). A fundamental problem in interpreting Parmenidēs is determining the subject of Parmenidēs' discourse. The metaphysical arguments have been seen by some as rejecting altogether the possibility of inquiry such as that practiced by the Milesians (THALĒS, ANAXIMANDROS, and ANAXIMENĒS) and recommended by XENOPHANĒS. These interpretations take Parmenidēs to be claiming that there exists only Being: a single, unmovable, unchangeable entity. Yet, Parmenidēs can also be understood as offering a corrective to earlier scientific inquiry: on this view, Parmenidēs inquires into the nature of that which genuinely is and so is genuinely knowable. In the Alētheia section (B2–B8.50) Parmenidēs gives an analysis of the nature or essence of a thing, explaining what it is to be such an entity. Only explanations of experience grounded in fundamental entities of the right sort (meeting the criteria for what-is) can hope to succeed; this offers a solution to the problem posed by Xenophanēs' rejection of divine revelation as a source of knowledge. Thus, Parmenidēs' arguments allow for a rational science grounded in metaphysically acceptable entities. Parmenidēs himself claims that one who understands his account will "know all things" including cosmological claims, and will be able to evaluate and reject unsuccessful accounts of what-is. He himself explains the sensible world (in the Doxa section) and was arguably the first to assert that the Moon lacks its own light but reflects light from the Sun. Despite apparently allowing for properly grounded cosmological explanation, he insists on the fundamental role in knowledge and explanation of self-justifying thought uncontaminated by sense experience.

Ed.: DK 28; L. Tarán *Parmenides* (1965); D. Gallop, *Parmenides of Elea: Fragments* (1984).
D.J. Furley, "Parmenides of Elea," in P. Edwards, ed. *The Encyclopedia of Philosophy* (1967) 6.47–51; A.P.D. Mourelatos, *The Route of Parmenides* (1971); *ECP* 363–369, *Idem*; Patricia Curd, *The Legacy of Parmenides: Eleatic Monism and Later Presocratic Thought* (1998; corrected paper edition with new introductory chapter, 2004); A. Hermann, *To Think Like God* (2004); *SEP* "Parmenides," John Palmer; R. McKirahan, "Signs and Arguments in Parmenides B8," in Patricia Curd and D.W. Graham, edd., *The Oxford Handbook of Presocratic Philosophy* (2008: forthcoming).

Patricia Curd

Parmeniōn (*ca* 310 – 280 BCE)

Architect of the Serapeion in Alexandria and the Iasōneion in Abdēra, and cited by VITRUUIUS 9.8.1 for having invented a type of sundial, the *pros ta historoumena*, usable at preset latitudes (contrast ANDRIAS).

RE 18.4 (1949) 1567–1569 (#5), H. Riemann; D.J.deS. Price, "Portable Sundials in Antiquity, including an Account of a New Example from Aphrodisias," *Centaurus* 14 (1969) 242–266.

PTK

Parmeniskos of Alexandria (120 – 80 BCE)

Grammarian who commented on Homer and Euripidēs, and also on Aratos, from which astronomical explanations are cited by Pliny 18.312 and Iulius Hyginus 2.2, 2.13.

RE 18.4 (1949) 1570–1572 (#3), C. Wendel; *FGrHist* 1026 T19.

PTK

Pasikratēs (of Sidōn?) (170 – 100 BCE?)

Reported on the machines of Apellis, Archimēdēs, Neileus, and Numphodōros, according to Oreibasios *Coll.* 49.7, 49.13, and 49.22 (*CMG* 6.2.2, pp. 14, 23–26, 34–35). Pasikratēs, working in Sidōn, improved the *trispaston* (triple-pulley) of Apellis, by adding a winch to give it greater traction, and, by adding a locking mechanism to maintain constant traction, improved the traction machine of Neileus, used to reduce dislocations and set fractures. Erōtianos *fr.*40 (p. 111 Nachm.) records that he wrote a commentary on Hippokratēs' *Mokhlikon*; Asklēpiadēs Pharm. in Galēn *CMLoc* 8.8 (13.213–214 K.) preserves his diuretic, including anise, carrot-seed, cassia and cinnamon, hazelwort, Indian nard, root of Pontic rhubarb, and saffron. Compare Pasikratēs' father and son, both Aristiōn, as well as Hērodotos (Mech.) and Tektōn.

RE S.9 (1962) 799–800, J. Kollesch; Drachmann (1963) 174–175, 180–181, Michler (1968) 87–88, 130–131.

PTK

Pasiōn (250 BCE – 10 CE)

Aëtios of Amida 15 (Zervos 1909: 89–90) cites from Galēn (not in Kühn) and Hērās the widely-useful ***trokhiskos*** of Pasiōn: **litharge**, pine-resin, and beeswax in olive oil and aged dry wine; Galēn often approves it: *MM* 5.6 (10.330 K.), *MM Glauk.* 2.3, 2.11 (11.87, 136–137 K.), and *Simples* 10.2.13 (12.276 K.), along with those of Andrōn and Polueidēs. Andromakhos, in Galēn *CMGen* 2.2 (13.493 K.), describes his "green" plaster of alum, sal ammoniac, frankincense, **verdigris**, etc. Oreibasios, *Syn.* 3.102 (*CMG* 6.3, p. 95), cites his very similar ***trokhiskos***, made by grinding copper flakes, roast copper, sal ammoniac, alum, and **verdigris**, in vinegar, under the sun, then adding frankincense; repeated by Aëtios 14.50 (p. 792 Cornarius) and Paulos of Aigina 7.12.22 (*CMG* 9.2, p. 318), the last adding aloes. Asklēpiadēs, in Galēn *CMGen* 5.14 (13.854 K.), cites a wound ointment (with orpiment, ***khalkanthon***, myrrh, realgar, etc.) of "Prasiōn" (an attested name: *LGPN* 3B.362). Aëtios 14.58 (p. 803 Cornarius) = Paulos of Aigina 4.25.2 (*CMG* 9.1, pp. 346–347) lists Pasiōn, with Andrōn and Polueidēs, as among those offering remedies for **anthrax**; Aëtios 14.53 (p. 797 Cornarius) mentions his *melinon*, and Halieus', again citing Galēn. The name is rare after the 1st c. CE: *LGPN*.

Fabricius (1726) 358; *RE* 22.2 (1954) 1699, H. Diller.

PTK

Paterios (250 – 400 CE)

In his explanations of the "geometrical number" governing human generations in Plato's *Republic* (546b3–c7), Proklos exposes two geometrical methods to find the numbers 27, 36, 48 and 64, in continuous proportion with epitrite ratio (4:3). The second method

(*In Resp.* 2.40.25–42.10 Kroll) uses elementary constructions within the right triangle with sides 3,4,5 as well as basic calculations on fractions and is attributed to a certain Paterios, probably the exegete of Plato's *Phaedo* whom Proklos approvingly mentions in explaining the myth of Er (2.134.10). DAMASKIOS uses Paterios' exegesis of the *Phaedo* to solve a difficulty raised by HARPOKRATIŌN (*In Phaed.* p. 137 Westerink, on *Phaedo* 68c1–3). Paterios is thus either a middle- or Neo-**Platonist** who commented on Plato between Harpokratiōn and Proklos.

RE 18.4 (1949) 2562–2563, R. Beutler.

Alain Bernard

Patroklēs of Macedon (*ca* 312 – 261 BCE)

Macedonian explorer and navigator, general in the armies of Seleukos I Nikatōr and Antiokhos I Sōtēr, author of geographical work(s) now lost, used by ERATOSTHENĒS, censured by HIPPARKHOS OF NIKAIA, and regarded as trustworthy by STRABŌN. As the admiral of the fleets of Seleukos and Antiokhos, Patroklēs sailed around the Hyrcanian and Caspian seas, was appointed governor of these regions and based his work on personal experience. Memnōn of Hērakleia (*FGrHist* 434 F 9.1) reports that Antiokhos sent Patroklēs with his army to Asia Minor where Patroklēs appointed Hermogenēs of Aspendos to attack Hērakleia and the other cities. VITRUUIUS (9.8.1) says that Patroklēs invented the Dovetail or the Axe (*Pelikinon*) sundial. His work included records of distances, measurements of countries and regions, outlines of sailing routes and descriptions of various sites.

FGrHist 712.

Daniela Dueck

Patroklos (50 BCE – 10 CE)

ASKLĒPIADĒS PHARM., in GALĒN *CMGen* 7.13 (13.1019–1020 K.), describes him as a freedman of CAESAR (AUGUSTUS?), and quotes his gout ointment, containing frankincense, myrrh, white pepper, etc. For the rare name, *cf. LGPN* 2.363, 3A.356, 3B.339, 4.276.

PIR2 P-163.

PTK

PAULINUS ⇒ VALERIUS

Paulos (Music) (*ca* 610 – *ca* 640 CE)

Ordered by the emperor HĒRAKLEIOS (reigned 610–641) to compile ancient philosophers' sayings on music. This work is preserved only in an Arabic translation attributed to Isḥāq ibn-Ḥunayn (d. 910/911). The unedited collection, more gnomology than scientific treatise, nevertheless presents ancient traditions on music and harmony attributed to PYTHAGORAS, THEOPHRASTOS, and other such figures.

Fr. Rosenthal, "Two Graeco-Arabic Works on Music," *PAPhS* 110.4 (1966) 261–268.

Kevin van Bladel

Paulos of Aigina (*ca* 630 – 670 CE?)

Lived in Alexandria "at the beginning of Islam" and was known as "the obstetrician (*al-qawābilī*)," according to Arabic bio-bibliographical literature. This chimes well with the latest author whom he quotes being AËTIOS OF AMIDA, and his dealing extensively with gynecology and pediatrics. His own works suggest that he was not only an active practitioner, but also a teacher of surgery. Heiberg (1919: 270) surmised that he was a Christian, but apart from his name and a variant reading, there is little evidence to support this claim. His epithet *Aiginētēs* ("of Aigina") is the sole evidence for his origin. He is mostly known for his medical handbook in seven books which had a great influence on the Byzantine and Arabic medical tradition. Paulos is also the author of a pediatric monograph which only survives through quotations in later medical writers such as Damastēs and al-Baladī. Other works attributed to Paulos (*On Uroscopy, On the Diseases of Women, On Lethal Drugs*) are spurious.

Paulos' handbook, which he himself calls *pragmateía* and *hupómnēma*, was inspired by contemporary jurists' manuals: he wanted to provide a comprehensive, yet portable, work for practical needs. It is divided into seven books: I: hygiene, prophylactics and diet; II: fevers; III: diseases from tip to toe; IV: external ailments and worms; V: poisonous animals; VI: surgery; VII: *materia medica*. It is often based on OREIBASIOS, as he himself states (*CMG* 9.1, p. 1.27), but also on other medical writers such as SŌRANOS (especially in the sections dealing with gynecology and pediatrics), DIOSKOURIDĒS (particularly in Book VII), the inescapable GALĒN, as well as Aëtios and ALEXANDER OF TRALLEIS. At times, however, Paulos displayed some independence, especially in Book VI, on surgery (translated by the French surgeon Briau, explicitly aiming to improve surgical practice; *cf.* Salazar). Book III on diseases from tip to toe was translated into Latin in 11th c. south Italy. Paulos' other work, his pediatric treatise, is the only Classical Greek monograph on the subject apart from RUFUS OF EPHESOS' *Therapy of Children*; Paulos uses Rufus as well as the relevant pediatric chapters in Aëtios.

Paulos had an extraordinary influence on subsequent medical tradition. The Greek text of his *pragmateía* survives in numerous manuscripts, a result of intense interest in the Byzantine world. Moreover, the Alexandrian tradition in general, and Paulos in particular, had a profound impact on medieval Syriac and Arabic medicine. His views on gynecology, surgery and pediatrics were incorporated into the writings of such luminaries as Ibn Sarābiyūn (*fl.* late 9th c.), ar-Rāzī (d. *ca* 925), al-Baladī (*fl. ca* 970s), Az-Zahrāwī (or "Albucasis," *fl. ca* 1000). Albucasis quietly borrowed many surgical procedures, and became in his turn a rich source of inspiration for medieval surgeons such as Guy de Chauliac (Guido de Cauliaco, d. 1368). By perfecting and promoting the genre of the encyclopedia, Paulos had an enduring impact on medical writings, shaping the work of authors such as al-Maōūsī (Haly Abbas, d. before 995) and Ibn Sīnā (Avicenna, 980–1037). These encyclopedias became core curriculum in the nascent European universities, where Paul's impact was felt both directly and indirectly.

Ed.: J.L. Heiberg, *CMG* 9.1–2 (1921–1924).

F. Adams, trans., *The Seven Books of Paulus Ægineta* (1844–1847); R. Briau, *La Chirurgie de Paul d'Égine* (1855); J.L. Heiberg, *Pauli Aeginetae libri tertii interpretatio antiqua* (1912); *Idem*, "De codicibus Pauli Aeginetae observations," *REG* 26 (1919) 268–277; C.F. Salazar, "Getting the point: Paul of Aegina on arrow wounds," *Sudhoffs Archiv* 82 (1998) 170–187; P.E. Pormann, *The Greek and Arabic Fragments of Paul of Aegina's Therapy of Children* (Diss. Oxford, 1999); *Idem, The Oriental Tradition of Paul of Aegina's Pragmateia* (2004).

P. E. Pormann

Paulos of Alexandria (350 – 400 CE)

Wrote an introduction to astrology in Greek prose, *Eisagōgika*, approximately datable by its inclusion of a worked example of a computation for 378 CE. The extant version is a revised edition dedicated to one Kronamōn (an Egyptian name), whom Paulos addresses as "dear son," probably a pupil; Kronamōn had detected errors in the earlier edition. The *Eisagōgika* is an elementary handbook introducing fundamental concepts of Greek astrology in clear language, but with little engagement with astrological practice or astronomy. It served as the basis for a series of astrological lectures delivered by OLUMPIODŌROS in Alexandria in 564.

Ed.: E. Boer, *Pauli Alexandrini Elementa Apotelesmatica* (1958).
DSB 10 (1974) 419, D.E. Pingree.

Alexander Jones

Paulos (of Italy) (*ca* 50 – 350 CE)

OREIBASIOS, *Ecl. Med.* 108.6 (*CMG* 6.2.2, p. 287), records that "our Paul" prescribed henbane juice rubbed on chilblains. The name is attested from the mid-1st c. CE (*LGPN*), and is usually Christian, so Oreibasios' "our" may be meant to specify "pagan." Diels (1905–1907) 2.81 records an Oxford MS, Barocc. 88 (15th/16th c.), f. 47, with extracts from a Paulos of Italy, possibly the same man. If, however, "our" means "of Pergamon," as in GALĒN, *Simpl. Med.* 10.1 (12.251 K.: ATTALOS III), perhaps we should read *ΑΤΤΑΛΟΣ* for *ΠΑΥΛΟΣ*.

RE 18.4 (1949) 2397 (#24), H. Diller.

PTK

PAULUS ⇒ OROSIUS

Pausanias "Hērakleiteios" (200? – 50 BCE)

Interpreter of HĒRAKLEITOS, later than NIKOMĒDĒS, according to Dēmētrios of Magnesia in DIOGENĒS LAËRTIOS 9.15.

RE 18.4 (1949) 2405 (#19), W. Nestle.

PTK

Pausanias of Damaskos (125 – 95 BCE)

Syrian-Greek geographer of Syria, author of a poetic composition in iambic trimeters dedicated to Nikomēdēs III Euergetēs king of Bithunia (127–94 BCE). The work, including descriptions of the Mediterranean and Greece, does not specify an author and was variously attributed to MARCIANUS OF HĒRAKLEIA, SKUMNOS OF KHIOS, PSEUDO-SKUMNOS and APOLLODŌROS OF ATHENS. Pausanias emerged as an option on the basis of a reference in the work of the 10th c. Byzantine emperor Constantine VII Pophurogennētos. The extant text, comprising 747 verses and some fragments, describes the Mediterranean world starting from the Pillars of Hēraklēs (Gibraltar). The last parts, originally including the Asiatic and African coastal regions, are missing. The author says that he chose iambic trimeters for their brevity and clarity (*cf.* DIONUSIOS SON OF KALLIPHŌN), and emphasizes that he relies specifically on ERATOSTHENĒS, EPHOROS and TIMAIOS OF TAUROMENION. The text has three parts: *prooimion* 1–138; description of the European coasts from Gadēs to the mouth of the Danube on the Black Sea; description of the coast of the Asiatic Black Sea coast. The author declares he has personally seen Greece, the Asian cities, Tyrrhenia and Sicily, almost all of Libya and Carthage (109–138). He mentions the traditional four large nations of the ends of the ***oikoumenē***: the Celts in the west, Indians in the east, Skuths in the north, and "Ethiopians" in the south.

Ed.: *GGM* 1.196–237; Diller (1952); D. Marcotte, *Les Géographes Grecs:* v. 1, *Pseudo-Scymnus, Circuit de la Terre* (2002); M. Korenjak, *Die Welt-Rundreise eines anonymen griechischen Autors* (2003).

A. Diller, "The Authors named Pausanias," *TAPA* 86 (1955) 268–279 at 276–279; D. Marcotte, *Le poème géographique de Dionysios fils de Calliphon* (1990) 40–44.

Daniela Dueck

Pausanias of Gela (*ca* 460 – 430 BCE?)

Son of Ankhitos (Iamblikhos, *VP* 113), from Gela (Diogenēs Laërtios 8.61), follower of and admired by Empedoklēs of Akragas, to whom Empedoklēs dedicated his work (D.L. 8.71). Aristippos of Kurēnē and Saturos of Kallatis claim he was Empedoklēs' boy lover (D.L. 8.60); Hērakleidēs of Hērakleia Pontikē declares Empedoklēs related to Pausanias the tale of reviving a dead woman (*ibid.*), and gives him a prominent role in the narrative of Empedoklēs' disappearance (D.L. 8.67–69). Wright 1981: 75–76 surmises that Empedoklēs encouraged Pausanias to study the art of healing and elevate his thoughts to improve the constitution and mixture of his soul. He is very likely the same Pausanias listed by Galēn *Meth. Med.* 1.1 (10.6 K.; Hankinson 1991: 5; Inwood 2001: 162–163) with Empedoklēs and Philistiōn of Lokroi as Italian physicians. Empedoklēs' epigram on Pausanias (D.L. 8.61) is almost certainly spurious (Wright 1981: 160; *cf. Anth. Gr.* 7.508).

RE S.14 (1974) 368–372 (#28), M. Michler; M.R. Wright, *Empedocles: The Extant Fragments* (1981) 11–19, 159–161; B. Inwood, *The Poem of Empedocles: A text and translation with a commentary* (2001).

GLIM

Pausēris (50 – 300 CE)

An interlocutor with Hermēs in a lost alchemical dialogue cited by the Anonymous Alchemist "Christianus" (*CAAG* 2.281). In a fragment of the same dialogue preserved by Olumpiodōros of Alexandria, Hermēs is once mistakenly replaced by Pelagios, as shown by the citation of this and one other passage from the dialogue by Zōsimos of Panōpolis in his writings preserved in Arabic (Hallum 2008: 209–211). For the Egyptian name, "he of Osiris," *cf.* Hērodotos 3.15 and Polubios, Book 22, *fr.* 17.4.

Berthelot (1885) 170; Bink Hallum, *Zosimus Arabus: the Arabic/Islamic Reception of Zosimos of Panopolis* (Diss. London, 2008).

Bink Hallum

Pausimakhos of Samos (440 – 400 BCE?)

Wrote a ***periplous*** of unknown scope, cited by Auienus, *OM* 42–50, along with Phileas of Athens and others dated to the 5th c. BCE. The name is almost unknown in the Greco-Roman period: *LGPN* 1.366–367, 2.364. Gisinger compares Promathos and dates Pausimakhos to "before 500 BCE."

RE 18.4 (1949) 2423 (#9), Fr. Gisinger.

PTK

Pausistratos of Rhodes (200 – 190 BCE)

Commander of the Rhodian fleet, betrayed by Antiokhos' admiral Poluxenidas (Livy 36.45, 37.9–12; Poluainos 5.27). Pausistratos used a funnel-shaped iron basket manipulated by iron chains and suspended from a ship's prow, on long poles (to clear his own ships), to hurl fire at enemy ships in frontal and front lateral attacks (App. *Syr.* 24; *Inscr. Lind.* 264; POLUBIOS Book 21, *fr.*7.1–4: Walbank 3 [1979] 97). Appian renders the name as Pausimakhos.

RE 18.4 (1949) 2423–2425 (#29), Thos. Lenschau; *BNP* 10 (2007) 654, L.-M. Günther; S.T. Teodorsson, "Pausistratos' Fire Basket" *SO* 65 (1990) 31–35.

GLIM

Paxamos (90 – 30 BCE)

Greek author (though his name is Egyptian) who wrote a treatise on agriculture in two books (*Souda* Pi-253). COLUMELLA (12.4.2) states that in his work he "followed Mago" – presumably using CASSIUS DIONUSIOS' translation – and wrote before the age of AUGUSTUS. He discussed the cultivation of pistachios (GEŌPONIKA 10.12.3) and probably much else (attributions in the *Geōponika* are unreliable, however). Other works ascribed to him by the *Souda* include an alphabetically-organized cookbook, two books on dyes (*Baphika*), a sex manual, and a history of Boiōtia (*Boiōtika*) in two books – though perhaps the title should be emended to read *Botanika*, "Herbal Remedies." He may also be mentioned by the alchemist ZŌSIMOS OF PANŌPOLIS in his *On the Evaporation of the Divine Water which Fixes Mercury* (*Mém. Auth.* 8.5 in Mertens [1995]; contrast *CAAG* 3.140).

H. Beckh, *Geoponica* (1895) *passim; RE* 18.4 (1949) 2436–2437, W. Morel; *FGrHist* 377.

Philip Thibodeau

Pēbikhios or Pibēkhios (50 – 300 CE)

Alchemist of Egyptian descent, judging by his name, a transliteration of Egyptian "he of the hawk" (i.e., "Hierax"); first mentioned by ZŌSIMOS OF PANŌPOLIS (*CAAG* 2.155, 158, 169, 182 and 196), who cites a work in which Pēbikhios addresses a discussion of yellow washes to "The Philosopher" (viz. PSEUDO-DĒMOKRITOS; *CAAG* 2.184–185). He is later mentioned by SUNESIOS (*CAAG* 2.63, and *apud* OLUMPIODŌROS, *CAAG* 2.91), STEPHANOS OF ALEXANDRIA (Ideler 2 [1842/1963] 236) and the ANONYMOUS ALCHEMIST PHILOSOPHER (*CAAG* 2.220). Preisendanz suggests that he may be identifiable with the Egyptian magician Pibēchis – to whom is attributed an invocation against epilepsy exhibiting a strong Jewish influence (*PGM* 4.3007–3086) – and, less certainly, with the magician Apollobēx (APULEIUS *Apol.* 90; *PGM* 12.121 see P. LEIDENSIS V) or Apollobeches (PLINY 30.9). If this last identification is indeed true, his *floruit* must be before the mid-1st c. CE. Syriac *Letters of Pēbikhios* survive addressed to the otherwise unknown Osron the Mage (Berthelot and Duval [1893] summary XXXVIII–XXXIX and partial translation 309–312).

M. Berthelot and R. Duval, *La chimie au moyen âge*, v. 2: *L'alchimie syriaque* (1893) 309–312; *RE* 20.1 (1941) 1310–1312 (s.v. Pibechis), K. Preisendanz.

Bink Hallum

Peithōn (of Antinoeia?) (330 – 390 CE?)

Geometer, contemporary with SERENUS OF ANTINOEIA, who preserves Peithōn's definition of parallel lines (p. 96, ed. Heiberg).

RE S.7 (1940) 836 (#6), M. Kraus.

GLIM

Pelagios (300 – 520 CE)

Mentioned by OLUMPIODŌROS, who cites Pelagios' work addressed to PAUSĒRIS (*CAAG* 2.89). In the Arabic ZŌSIMOS material the line is more sensibly attributed to HERMĒS. Pelagios' treatise *On This Divine and Sacred Art* (*CAAG* 2.253–261) seems to address the gilding and silvering of metals like copper and iron; it cites Zōsimos, providing the *terminus post*.

Ed.: *CAAG* 2.253–261.
Festugière (1944) 1.240, 247; Letrouit (1995) 46–47.

Cristina Viano

Pelagonius of Salona (350 – 400 CE)

Wrote the *Ars Veterinaria*, a Latin treatise on veterinary medicine, partly surviving in 6th c. fragments from Bobbio, and two MSS, *R* and *E*, the former probably reconstituted from *testimonia* and a mutilated original. The recently identified *E* contains only a partial text, but does explain many discrepancies between *R* and VEGETIUS, whose *Mulomedicina* draws on Pelagonius extensively. The two manuscripts vary widely, at times appearing more like separate works than different versions of the same opus.

Pelagonius employed several earlier authors including – primarily – APSURTOS whose epistolary form he followed; COLUMELLA, whose writing style he tried to imitate; CELSUS; and EUMĒLOS (perhaps only via Apsurtos). Typical of ancient veterinary writers, Pelagonius concerned himself almost exclusively with the horse and its care, for a presumably upper-class audience using horses for racing and riding. The *Ars Veterinaria* originally consisted of 35 chapters each in the form of a letter to a friend or patron. A brief introductory letter praises horses and states the scope of the work. The second letter contains general information about the horse, describing points of good equine conformation and how to determine age. Subsequent letters discuss common equine diseases and afflictions, such as lethargy, colic, and **glanders**. Nearly all remedies are pharmacological, but a few are overtly magical, possibly interpolated. Arguing from his errors of translation and terminology, some scholars have doubted whether Pelagonius was a practicing veterinarian. Moreover, he may have contributed little first-hand experience, and the surviving work appears heavily redacted.

Ed.: Fischer (1980).
K.D. Fischer, "The first Latin treatise on horse medicine and its author Pelagonius Saloninus," *MHJ* 16 (1981) 215–226; Önnerfors (1993) 380–381; Adams (1995); *BNP* 10 (2007) 691–692, K.D. Fischer.

Jennifer Nilson

Pelops (Med.) (30 BCE – 75 CE)

PLINY, 1.*ind*.31–32, lists him as a foreign source, and 32.43 cites his prescription for honey overdose: consume a tortoise boiled without its head and extremities. The Latin translation of the HIPPOKRATIC CORPUS, APHORISMS by "Pelops" mentioned by (the Latin) pseudo-OREIBASIOS *In Hipp. Aph., pr.*, and attributed to PELOPS OF SMURNA (*RE* S.10 [1965] 531 [#5], Fr. Kudlien), more likely belongs to a medical writer known to the Latin tradition (Pliny). Fabricius (1726) 360 anachronistically identifies this man with Galen's teacher. The name is very rare (*LGPN*), rendering possible a family relation with Pelops of Smurna, perhaps a grandson.

(*)

PTK

Pelops of Smurna (140 – 160 CE)

Student of NUMISIANUS, teacher of GALĒN; wrote commentaries on the HIPPOKRATIC CORPUS, EPIDEMICS 2 and 6, a *Hippokratic Introduction*, and an *Epitome on the Muscles*, but many more writings, unpublished, perished in a house-fire. Pelops (a **Rationalist**) came to Pergamon to debate the **Empiricist** physicians; Galēn in 149 CE then traveled to Smurna for study with Pelops. He distinguished bodily constitutions by signs, e.g., red indicated the warm mixture, a thin nose and small eyes the dry mixture: *In Hipp. Epid. II* (*CMG* 5.10.1, pp. 347–348). In his commentaries, he sought to explain obscurities by rearrangements of text: *In Hipp. Epid. VI* (*CMG* 5.10.2.2, p. 291). Although he taught that the brain is the source of all vessels, his anatomy commenced from the liver: *Opinions of Plato and Hippokrates* 6.5.23 (*CMG* 5.4.1.2, p. 392). He tried to explain how burnt river-crabs could cure ***hudrophobia*** because of their watery nature, said when combusted to absorb the poison that caused the diagnostic symptom: *Simples* 11.24 (12.356–359 K.; *cf.* AISKHRIŌN); but he also used MENIPPOS' potion, without river-crabs: *Antid.* 2.11 (14.172–173 K.). A fragment of his anatomy of the bovine tongue is preserved in *Dissection of the Muscles* (18B.959 K.), translated by Goss; and PAULOS OF AIGINA, 3.20.1 (*CMG* 9.1, pp. 167–168), preserves his explanation of ***tetanos***: the muscles around the spine fill with ***pneuma*** that is thick and cloudy.

C.M. Goss, "On the Anatomy of Muscles for Beginners by Galen of Pergamon," *Anatomical Record* 143 (1963) 477–501; Grmek and Gourevitch (1994) 1521–1522; Manetti and Roselli (1994) 1591–1635; Ihm (2002) #196–197; *BNP* 10 (2007) 713 (#5), V. Nutton.

PTK

Penthesileus (325 BCE – 300 CE)

Wrote a work on Kolkhis, cited by the RAVENNA COSMOGRAPHY 4.4. *Cf.* MARPĒSSOS. The masculine form of the name seems otherwise unattested.

J. Schnetz, *SBAW* (1942), # 6, pp. 58–59, 61–62.

PTK

Pephrasmenos of Tyre (500 – 350 BCE?)

Is said to have invented a type of ram wherein a cross-beam, suspended from a transverse beam, can be thrust back and forth violently, for an otherwise unattested siege at Gadēs (ATHĒNAIOS MECH. p. 9 W.; VITRUUIUS 10.13.2), although such designs are Assyrian. KERAS OF CARTHAGE later improved the device. Pephrasmenos ("Designer") seems otherwise unattested as a name (Pape-Benseler; *LGPN*), although HALIEUS, TEKTŌN, and TURANNOS seem similarly formed.

RE 19.1 (1937) 560, K. Orinsky.

PTK and GLIM

PERCENNIUS ⇒ MINIUS

Periandros **(*ca* 360 – *ca* 335 BCE)**

A good doctor who wrote bad verse – on unspecified topics (PLUTARCH, *Spartan Sayings* 218F).

RE 19.1 (1937) 717 (#2), W. Kroll.

PTK

Perigenēs (200 BCE – 50 CE)

Perigenēs "*Organikos*" (ERŌTIANOS A-103, p. 23 Nachm.; *cf.* APOLLŌNIOS) was a surgeon and engineer. His medical apparatus contributed to refinements in bone surgery. Perigenēs wrote a *Mēkhanika* in which he described three bandages: one called "*thais*," also ascribed to MOLPIS (PSEUDO-GALĒN *De Fasciis* 16 [18A.789 K.]), a second called the "helmet bandage," seemingly his own invention (*ibid.* 35 [18A.797 K.]), and a "cranesbill" bandage for a luxated humerus (*ibid.* 80 [18A.814 K.]). ANDROMAKHOS preserves three remedies for breathing disorders (GALĒN, *CMLoc* 7.2–3 [13.33–34, 69–70, and 73 K.]).

RE S.11 (1968) 1054–1055 (#7), M. Michler; *Idem* (1968) 89, 132.

GLIM

Periklēs (150 BCE – 300 CE)

Wrote a lost commentary on the *Cutting-off of a Ratio* by APOLLŌNIOS OF PERGĒ, according to PAPPOS, *Coll.* 7.6.

Jones (1986) 386, 511; Netz (1997) #133.

PTK

Periklēs of Ludia (*ca* 430 – 480 CE)

Philosopher associated with PROKLOS (MARINOS OF NEAPOLIS, *Vit. Pr.* 29) whose *Theologia Platonica* was dedicated to him (*Theol. Plat.* 1.1, p. 5.7 S-W). In his interpretation of PLATO's *Parmenidēs* 131d–e, he refers to the idea of the Small, concluding that Smallness is not divisible (Proklos, *in Parm.* 872.18–32). He claims that the very first matter is body without qualities, a view he attributes to Plato, ARISTOTLE, and the **Stoics** (SIMPLICIUS, *in Phys.* = *CAG* 9 [1882] 227.23–26). Periklēs may also have participated in theurgic rituals (Marinos, *Vit. Pr.* 29).

RE S.7 (1940) 899 (#8), R. Beutler; *PLRE* 2 (1980) 860.

Peter Lautner

Periplus Maris Erythraei (40 – 70 CE)

An Egyptian Greek, author of a ***periplous*** of the Erythraean Sea, as the Greeks called the Indian Ocean and its branches including the Red Sea and the Persian Gulf. Drawing on personal experience, the author described the African route down to the ancient town of Raphta (situated at the mouth of the Pangani river in present-day Tanzania), and the

Arabian-Indian route at least down to Cape Comorin at the southern tip of India. He probably spent some time in India: Indian names and words are transcribed accurately into Greek. He may have been a merchant who decided to write a handbook for traders between Roman Egypt and eastern Africa, southern Arabia and India, unlike the traditional ***periploi***, primarily guides for seamen. The author concentrates on trading routes, ports and products, and indicates local friendliness or hostility. The *PME* provides information on the rank and sometimes the name of the local ruler of each port and specifies goods which can be sold to the ruler and his court. The text also alludes to historical events and includes anthropological and natural historical information: description of the tides along India's north-western coast, unusual animals, distinctive appearance of locals, their dwellings, language, eating habits and dress. As he was a businessman and not a scholar, the author's language is mainly functional and technical. Unlike ERATOSTHENĒS, MARINOS OF TYRE, and PTOLEMY, who relied on second hand information, the writer of the *PME* is the only author with personal acquaintance of the Indian Ocean whose work has survived.

Ed.: *GGM* 1.257–305; G.W.B. Huntingford, *The Periplus of the Erythraean Sea* (1980); Casson (1989).

Daniela Dueck

Periplus Ponti Euxini (575 – 600 CE)

Collection of sailing instructions for the Black Sea by an unknown author. The text, containing much information on coast-lines, harbors, rivers and cities, begins at the Thrakian Bosporos, working counter-clockwise from Bithunia to Paphlagonia, Pontos, the Caucasus and the Tauric Khersonēsos to Thrakē. Sources include ARRIANUS' ***Periplus*** *Ponti Euxini*, MENIPPOS OF PERGAMON's ***Periplous*** *tēs entos thalassēs* and minor treatises (including PSEUDO-SKULAX). More than 40 references to "current" (*nun*) cities and people, to Alans and Goths, and to the Turkish invasion of the Crimea in 576, suggest a date of not earlier than the last quarter of the 6th c.

Ed.: *GGM* 1.402–423; Diller (1952) 118–138.
Diller (1952) 102–146; *HLB* 1.528; *ODB* 1629, A. Kazhdan.

Andreas Kuelzer

Perseus (250 – 50 BCE?)

According to PROKLOS (*In primum Euclidis Elementorum librum* pp. 111–112, 356 Fr.), Perseus investigated the properties of curves generated by the intersection of a plane with the surface of a torus (the solid of revolution of a circle about a straight line not passing through the circle's center). Proklos quotes an epigram by Perseus, indicating that he distinguished five cases of intersection and three kinds of curve. By analogy with the Greek study of conic sections, Perseus is likely to have demonstrated a *sumptoma* or characteristic property of each curve. His work clearly belongs to the tradition of Hellenistic geometry; if, as generally supposed, Proklos derived his information from a lost work of GEMINUS, Perseus lived before the middle of the 1st c. BCE.

DSB 10.529–530, I. Bulmer-Thomas; Knorr (1986) 267–272.

Alexander Jones

Persis (325 – 90 BCE)

Agricultural writer whose work CASSIUS DIONUSIOS excerpted (VARRO, *RR* 1.1.9–10). *Persis* (a woman's name) is the only known female agronomist from classical antiquity.

RE 19.1 (1937) 1030 (#2), W. Kroll.

Philip Thibodeau

Petasios, pseudo? (300 – 400 CE)

Alchemist sometimes falsely called "King of Armenia" and first cited by OLUMPIODŌROS OF ALEXANDRIA concerning "our lead" (*CAAG* 2.95, 97). "Petasios" is a Hellenized version of the Egyptian for "given by Isis" (Heuser 1929: 49, 61); *cf.* ISIDŌROS. The title of an alchemical treatise ascribed to OSTANĒS claims Petasios as its addressee (*CAAG* 2.261), but he is not mentioned in the text. An anonymous alchemical treatise preserves a fragment of Petasios said to be from the (his?) *Dēmokritean Commentaries* (*CAAG* 2.356). The ANONYMOUS ALCHEMIST "CHRISTIANUS" attributes to Petasios two aphorisms concerning unspecified instruments (*CAAG* 2.278, 282), and he is mentioned in an alchemical lexicon compiled perhaps *ca* 8th–9th c. (*CAAG* 2.15). Texts attributed to Petasios may be extant in Arabic (Ullmann 1972: 188). Letrouit (1995: 48) takes Petasios to be contemporary with SUNESIOS, but an earlier date may be established if we equate him with the Peteēsios cited in some versions of DIOSKOURIDĒS as an authority on *khalkanthes* (also ***khalkanthon*** and *khalkanthos; MM* 5.98). In turn, this Peteēsios may perhaps be identified with the priest and magician Petēsios (*ca* 99 BCE) found in *P. Leid. G, H, I,* and *K.* However, the fact that ZŌSIMOS OF PANŌPOLIS does not mention Petasios suggests three separate people, or that alchemical doctrines were ascribed to this name only in the 4th c.

Berthelot (1885) 168–169; *RE* 19.1 (1937) 1125, W. Kroll; *BNP* 10 (2007) 864, J. Quack.

Bink Hallum

Petosiris, or Nekhepso-Petosiris (*ca* 150 – 100 BCE)

At Hermopolis the tomb of Petosiris ("whom Osiris has given"), a high priest of Thoth, has been dated to *ca* 300 BCE (G. Lefebvre, *Le tombeau de Petosiris,* three vols. [1923–24]), and the Egyptian royal name Nechepso/Nekhepso (possibly from "Nekho the King," referring to Nekho II of the Saïte dynasty) was included as a forerunner to the 26th or Saïte dynasty (664–525 BCE) in one of the recensions of MANETHŌN's compilation of Egyptian kings. Neither of these figures can be connected to the Greek works on celestial omens and astrology that circulated under the pseudepigraphic authorship of Petosiris and Nekhepso, generally assumed to be compositions produced in Ptolemaic Egypt of the 2nd c. BCE. Fragments of this material are collected in Riess (1892), but more are now known. THRASULLOS is the first dateable source to cite either Nekhepso or Petosiris, but by late antiquity, e.g., in FIRMICUS MATERNUS, Petosiris and Nekhepso were widely associated with divination, astrology, and the HERMETIC tradition. They are the creation of the interaction between the Greek and Egyptian cultural realms of the Hellenistic period.

Textual fragments of the Petosiris tradition include celestial omens in the Babylonian style of *Enūma Anu Enlil,* transmitted to Egypt during the Achaemenid period and no doubt a development from the tradition represented in a DEMOTIC papyrus concerning lunar and eclipse omens (R.A. Parker, *A Vienna Demotic Papyrus on Eclipse and lunar-omina* [1959]).

The Petosiris-Nekhepso fragments (*frr.*6–12 Riess) include omens from phenomena such as eclipses with their colors and the direction of the winds blowing, heliacal risings of Sirius, and comets, and they are preserved in the works of HEPHAISTIŌN OF THĒBAI, PROKLOS, and IŌANNĒS "LYDUS". Fragment 12 bears relation to another demotic papyrus (G.R. Hughes, "A Demotic Astrological Text," *JNES* 10 [1951] 256–264) concerning the heliacal rising of Sirius, the positions of planets and the directions of winds. Another set of fragments deals with the date of conception for the purpose of computing the length of a native's life by means of the rising times of the zodiacal signs between the ascendant and mid-heaven at birth.

Another group of texts ascribed to Petorisis-Nekhepso are the "mysteries" revealed to Nekhepso in a vision and recorded in 13 books, many fragments and passages of which are given by VETTIUS VALENS, *Anthologies* (e.g., 3.16) and by Firmicus Maternus (4.22.2), who calls him "the most just ruler of Egypt and an exceedingly good astrologer," as well as simply "the king." Here again, the length of life of the native is computed, as is the Lot of Fortune and other times in a person's life found to be good or bad, or related to travel, injury, children and death, on the basis of horoscopic methods, such as the lord of the year.

Nekhepso's knowledge of the healing power of plants and stones in **sympathy** with the zodiac comes down to us in the autobiographical epistolary prologue of THESSALOS OF TRALLEIS. Thessalos, engaged in the study of the miraculous, comes across a book by Nekhepso, but the remedies are not good and he seeks more direct knowledge from the god Asclepius (Imhotep) himself.

Finally, numerological treatises addressed to Nekhepso by Petosiris are attested (e.g., *CCAG* 7 [1908] 161–162).

E. Riess, "Nechepsonis et Petosiridis fragmenta magica," *Philologus*, S.6 (1892) 327–394; C. Darmstadt, *De Nechepsonis-Petosiridis Isagoge quaestiones selectae* (1916); *RE* 16 (1935) 2160–2167, 19 (1938) 1165, W. Kroll; Gundel and Gundel (1966) 27–36; *DSB* 10.547–549, D.E. Pingree; *DPA* 4 (2005) 601–615, P.P. Fuentes Gonzalez.

Francesca Rochberg

Petrikhos (200 – 100 BCE)

Wrote a didactic poem on poisons and remedies (*Ophiaka*), a physician cited by PLINY (and scholiasts) as source for Books 20–27 on plants. Petrikhos indicates as remedies for snake bites hen's brain, oregano, and fennel (20.258), and recommends small bur-parsley for wounds caused by poisonous marine animals (22.83).

RE 19.1 (1938) 1189–1190, W. Kroll; Jacques (2002) XLV–XLVI.

Arnaud Zucker

Petrōn(as) of Aigina (500 – 400 BCE)

Teacher of an Aristōn (PARISINUS MEDICUS 10, p. 72.13), supposed author of the HIPPOKRATIC ON DIET, recognized by GALĒN as "ancient" (*CMG* 5.9.1, p. 135): he was known for giving feverish patients roasted pork, wine and cold water (CELSUS 3.9 = ERASISTRATOS *fr.*213 Garofalo, *cf. frr.*214, 217). His theory about health and disease is quoted by the LONDINIENSIS MEDICUS (20.2–24): bodies are composed of two elements (heat and cold), each of which has a corresponding element (*antistoikhon*), hot corresponding to dry, and cold to moist. Diseases can arise from the elements themselves and from digestive residues.

Finally he is said to be closer to PHILOLAOS because he too regarded bile as an effect of disease rather than a cause.

RE 19.1 (1937) 1191, K. Deichgräber; Daniela Manetti, "Doxographical Deformation of Medical Tradition in the Report of the Anonymous Londinensis on Philolaus," *ZPE* 83 (1990) 219–233 at 223; *BNP* 10 (2007) 874–875, A. Touwaide.

Daniela Manetti

Petrōn of Himera (450 – 410 BCE?)

The sources of PLUTARCH, *de defectu oraculorum* 22 (422B), claim that Petrōn hypothesized a triangular universe consisting of 183 discrete ***kosmoi*** arranged with 60 worlds along each side, the remaining three at the corners, each ***kosmos*** in contact with its two neighbors and revolving "as in a dance" (*cf.* ARISTOTELIAN CORPUS, MĒKHANIKA 848a20–37): Huxley argues that the total number must be even, i.e., 180. The interior "Plain of Truth" provided the common hearth to all, wherein lay "Eternity," and whence "Time" flowed to ***kosmoi***.

DK 16; G.L. Huxley, "Petronian Numbers," *GRBS* 9 (1968) 55–57; *BNP* 10 (2007) 874, C. Riedweg.

PTK and GLIM

Petrōnios Musa (*ca* 10 – 40 CE)

Designated by GALĒN (*CMGen* 2.5 [13.502 K.]) as one of the best pharmacologists in the handbooks on the subject, and by DIOSKOURIDĒS (*MM pr.*2 [Wellmann, 1.1]) as an "**Asklēpiadean**," along with SEXTIUS NIGER and DIODOTOS. He wrote a *Hulika*, i.e., *(Medical) Materials* (ERŌTIANOS N-4, p. 62.12 Nachm.), lost except for short quotations. PLINY cites Petrōnios as a source for Books 20–27, and specifically mentions Petrōnios' accounts of endives (20.77) and carrots (25.110): both compacted with a tract by Diodotos. The passage in Galēn closely resembles Dioskouridēs' in grammatical structure, indicating transmission of Petrōnios through the collection of ASKLĒPIADĒS PHARM.

Galēn (*CMGen* 5.11 [13.831–832 K.]) provides the formula of Petrōnios' "excellent" lozenge-suppository, to be compounded for a rapidly pain-killing treatment of hemorrhoids and probably fistulas in the anus. The drug would have a good "shelf-life," being roughly half mineral (antimony, roasted and washed lead, fissile alum, and copper sulfate), added to acacia-gum, ashed henna-flowers, seeds of the tree-heath, frankincense, myrrh, and the latex of the opium poppy: "compound [these ingredients] in wine and fashion into suppositories. Administer/insert the suppositories using grape-syrup." Not only would this compound relieve pain (the opium latex), it would also be a reasonably good bactericide (the copper sulfate, frankincense, myrrh, henna, probably the acacia-gum), certain to promote healing.

RE 19.1 (1937) 1193–1194, K. Deichgräber; Fabricius (1972) 226, 243; Scarborough and Nutton (1982) 205–206.

John Scarborough

Petros (of Constantia?) (*fl.* 449 CE)

Christian *arkhiatros*, wrote a book about astrology, and is cited by AËTIOS OF AMIDA 7.114 (*CMG* 8.2, p. 386) for an eyewash against ***leukōmata***.

PLRE 2 (1980) 865–866

PTK

Peutinger Map (300 – 330 CE?)

Large maps of the Roman world were made from AUGUSTUS' time onwards, but only one survives – as a copy produced in 11 segments *ca* 1200 CE, and missing its western end (*Codex Vindob.* 324). Copying slips are detectible, together with minimal modifications reflecting Christian belief, but overall this survival seems a faithful reproduction of the original. There is no clue to the identity or work-site of the mapmaker(s). His date is also uncertain, although the absence of Christian influence and the ample reflection of Late Roman tastes suggest the early 4th c. There could be no finer articulation of the Tetrarchy's ideals in map form: a seamless, stable, united world – with Rome conspicuously at its center – under Roman sway and readily accessible everywhere overland. In his presentation of physical landscape and his placement of principal settlements, the mapmaker demonstrates a geographical awareness which must derive from the Hellenistic tradition inspired by ERATOSTHENĒS. Within this landscape he deftly integrates a comprehensive assemblage (how and where obtained?) of route data from the Roman Empire and even far beyond to the East; about 2,700 places and associated distance figures are marked.

Whether this combination of elements is innovatory remains unclear. The same applies to the bold choice of map frame, no more than 34 cm tall by perhaps 850 long. It was only possible to span Britannia to Taprobane thus by virtual abandonment of a north-south dimension, compression of the Mediterranean and Black Seas into narrow channels, and manipulation of principal landmasses to appear at different scales (the heartland of Italy especially large, Persian territory eastwards small). The ingenious mapmaker accomplishes these feats, and handles line-work, symbols, lettering, and palette, with a mastery that recalls Rome's Marble Plan (*ca* 200 CE), and presupposes a mature cartographic tradition. With its indifference to up-to-date information or direct routes, the map appears not so much a practical guide for travelers as a colorful display piece for, say, a palace wall, possibly just one component of a greater artwork. It is inviting on various levels. From afar, the main regions and their names stand out; close up, the unfamiliar worldview and the richness of detail both intrigue and delight the learned.

Richard Talbert, "Cartography and taste in Peutinger's Roman map," in *Idem* and K. Brodersen, edd., *Space in the Roman World: its Perception and Presentation* (2004) 113–141.

Richard Talbert

Phaeinos (460 – 430 BCE)

Astronomer, a metic in Athens, who observed solstices from Mt. Lukabettos and taught METŌN astronomy (PSEUDO-THEOPHRASTOS, *De signis* 4), as well as, perhaps, a Babylonian version of the 19-year Metonic cycle, but the text does not require this suggestion.

DSB 9.339, G.J. Toomer.

Henry Mendell

Phaiax (490 – 470 BCE)

Civil engineer in Akragas, designed and supervised the construction of an elaborate water distribution and canal system at Akragas, built (among other projects) by prisoners of war from the Greek victory over the Carthaginians at Himera (480 BCE: DIODŌROS OF SICILY

11.25.3–4). The conduits, called *phaiakoi* after Phaiax, with 23 separate branches totaling 14.6 km, emptied into and helped drain a large artificial lake or basin once stocked with fish and waterfowl. Completed under the tyrant Thērōn (*ca* 530–472 BCE), it may be compared to other large hydraulic projects encouraged by Greek tyrants (see EUPALINOS).

KLA 2.208, W. Müller; D. Mertens, *Städt und Bauten der Westgriechen* (2006) 319–321.

Margaret M. Miles

Phaidros (250 BCE – 90 CE)

ASKLĒPIADĒS PHARM., in GALĒN *CMLoc* 4.7 (12.736-737 K.), preserves his nasal remedy (*rhinion*) composed of ***ammōniakon***, **verdigris**, roasted ***alkuoneion***, and "flower of Asian stone" (*cf.* Galēn, *Simpl. Med.* 9.2.9 [12.202 K.]). He taught that the umbilical vessels grew into the heart: SŌRANOS *Gyn.* 1.57 (*CMG* 4, p. 42; *CUF* v. 1, p. 56). The name (with its variants) is especially Athenian (*LGPN*).

Fabricius (1726) 363.

PTK

Phainias of Eresos (340 – 310 BCE)

Student of ARISTOTLE, correspondent of THEOPHRASTOS, writer of biographical accounts about tyrants (*FGrHist* 1012), and of a work against DIODŌROS OF IASOS. Composed a popular treatise on plants, in at least five books, often cited with, and apparently similar to, Theophrastos' extant work *History of Plants*. PLINY lists Phainias among *medici*, but with the epithet *physicus*, 1.*ind*.21–26, and cites him on the benefits of nettle-root, 22.35–36. Athēnaios, *Deipn.*, preserves a dozen fragments, including remarks on the absence of flowers and seeds in fungi and ferns, 2 (61f), the seeds of umbellifers, 9 (371c–d), differing preparations of *ōkhros*, broad-beans, and chickpeas, 2 (54f), and the art of wine-making, 1 (29f, 31f–32a).

Ed.: Wehrli (1967–1969) v. 9; *FGrHist* 1012.
BNP 10 (2007) 901–902, H.B. Gottschalk; *GGP* 3.588–590.

PTK

Phanias (250 BCE – 80 CE)

ANDROMAKHOS, in GALĒN *CMGen* 5.13 (13.840 K.), records his *enkathisma* ("sitz-bath") for hemorrhoids: grind alum, ***khalkanthon, khalkitis***, "raw" ***misu***, and realgar, then mix with aged pickled (*tarikhēr-*) urine of a man; apply for seven days. (*Cf.* the alchemical use of pickled urine on stones, P. HOLM. §29, p. 119 Halleux.) The name is very rare after the 2nd c. BCE (*LGPN*).

(*)

PTK

Phanokritos (of Thasos?) (*ca* 330 – *ca* 200 BCE)

Athēnaios, *Deipn.* 7.4 (276f), cites Phanokritos' work on EUDOXOS OF KNIDOS, which apparently discussed his theory of pleasure. The name is rare, except on Thasos, and is hardly attested anywhere after *ca* 200 BCE (*LGPN*).

FGrHist 1114.

PTK

Phaōn (420 – 350 BCE)

Listed by GALĒN, *In Hipp. Reg. Acute* (15.455 K. = *CMG* 5.9.1, p. 135), with EURUPHŌN, PHILISTIŌN, and ARISTŌN, as a putative author of the HIPPOKRATIC CORPUS, REGIMEN IN HEALTH. Besides Sappho's ferryman, the name is attested through the 2nd c. BCE (*LGPN*).

(*)

PTK

Pharnax (75 – 125 CE?)

Parthian physician. A medical formula derived from his *herbarium* was considered marvelous against epathical diseases (GALĒN, *CMLoc* 8.7, 13.204 K.).

Fabricius (1726) 363.

Antonio Panaino

Phasitas (Phaeitas?) of Tenedos (400 – 300 BCE)

Greek physician quoted by the LONDINIENSIS MEDICUS (12.36–13.9): he thinks that diseases arise from exhalations of **humors**, if they concentrate inopportunely in some body parts or from excrements themselves. The name, uncertain, was probably wrongly copied: Phasitas has been connected to the doctor Phaïdas of <Te>nedos (on an epitaph found at Paphos: W. Peek, *Griechische Vers-Inschriften* 1 [1955] #902, *ca* 300 BCE).

U. v. Wilamowitz, "Lesefrüchte," *Hermes* 33 (1898) 513–533 at 519.

Daniela Manetti

Pheidias (of Surakousai) (300 – 250 BCE)

Astronomer who estimated the size of the Sun as 12 times the Moon, according to his son ARCHIMĒDĒS, *Sand-reckoner* 1.9.

RE 19.2 (1938) 1918–1919 (#1), W. Kroll.

PTK

Pherekudēs of Suros (*fl.* 544/1 BCE)

Semi-legendary figure, reputed to have taught PYTHAGORAS, which probably stems from his influence on **Pythagorean** theories of the soul. According to Theopompos in DIOGENĒS LAËRTIOS, Pherekudēs was the first to write about nature and the gods. HĒSIOD had already written about the gods and THALĒS about nature, but Pherekudēs' account is a fascinating "theo-cosmogony," a mixture of theogonical and physical speculation. His book, surviving in fragments, counts as the first prose work of Greek literature, appearing to predate even ANAXIMANDROS' book. (This Pherekudēs is distinct from the later homonymous genealogist from Athens.)

The narrative begins with the existence from eternity of three primary deities: Zās (Zeus), Khronos (the personification of Time), and Khthoniē (the Earth in its Urform). Khronos, without a consort, from his own seed produces three elements: fire, breath (or air), and water. From these elements, deposited, presumably in various mixtures, in five nooks or hollows, arises a numerous second generation of gods, the five-nook generation (problematic, since Theopompos entitles Pherekudēs' book *The Seven-Nook Mingling of the Gods or Birth of the Gods*). This auto-erotic act can be seen as the first stage of creation, the genesis of the larger ***kosmos***. The second stage occurs when Zās weds Khthoniē and presents her with a robe embroidered with Gē (Earth) and Ogēnos (Ocean). By the investiture of the robe, Khthoniē becomes Gē, Earth actualized. This earth-robe is also depicted as cloaking a winged oak tree, possibly to explain the suspension of the Earth in the ***kosmos***. Ophioneus, a snake-god, threatens the created order but is defeated and cast into the Ocean. The account concludes with a division of cosmic portions among the gods, similar to Homer.

Aristotle (*Metaphysics* 14 [1091b6–10]) names Pherekudēs as one of the early theologians who did not explain everything "in myth." Although Pherekudēs clearly draws from Hēsiod and Greek mythology, he also reveals rudimentary notions of a philosophical and natural-scientific nature about eternity and time, first principles and causes, and demiurgic creation.

Hermann S. Schibli, *Pherekydes of Syros* (1990); *OCD3* 1157 (#1), J.S. Rusten; *BNP* 10 (2007) 951 (#1), L. Käppel.

Hermann S. Schibli

Philagrios of Ēpeiros (300 – 340 CE)

The *Souda* (Phi-295) says Philagrios was a physician from Ēpeiros (or the Lukian island of Makra), a student of Naumakhios, and that Philagrios practiced in Thessalonikē. Long rejected is the tradition of Philagrios' "brother-doctor" Poseidōnios (Temkin 1931); but reference works attempt to match our Philagrios with the family described by Philostorgios, *Ecclesiatical History* 8.10 (p. 111): ". . .in the days of Valens and Valentinian [I], there lived a Philostorgios, the most famous physician of his time [who] had two sons, Philagrios and Poseidōnios [of whom] Poseidōnios was also famed as a doctor. . .." Philostorgios the historian thus refers to Philostorgios the doctor, and his likely non-physician son Philagrios, of *ca* 365–375 CE. Our physician Philagrios, however, is first extracted by Oreibasios, establishing a *terminus ante* of roughly the mid-4th c. CE (Masullo, *Frammenti*, 19). Perhaps initially identifying himself as a **Pneumaticist** physician, Philagrios seems to have become more eclectic through the years (Wellmann 1895). The *Souda* also tells us Philagrios wrote 70 books, including commentaries on Hippokratēs. The *Souda* entry is supplemented by six from Arabic *testimonia* (Rhazes, Avicenna, ibn ad-Nadīm's *Fihrist*, others) whence we have recovered most of the titles and subjects of Philagrios' lost writings (Masullo, Test. 2–7).

Philagrios gained fame for his skills in treatment (surgery and application of plasters) of kidney and bladder ailments, including urinary obstruction and commonly occurring stones (Masullo, *fr.*95 = Aëtios of Amida 11.5); he improved the techniques of Antullos and Hēliodōros of Alexandria in curing *ganglia* (swellings or tumors of nerves or tendons occurring from blows, especially to the joints and articulations of the hands and feet), particularly those of fingers, toes, wrists, and ankles (Masullo, *fr.*210 = Aëtios 15.9); he refined procedures for surgical correction of damaged ligaments and tendons in the leg, especially those providing articulation of the tibia and bones of the ankle (Masullo, *fr.*211 = Aëtios 15.13), accompanied by careful cicatrization and application of wine-and-honey soaked plasters to assure healing and prevention of inflammation. Significant fragments of Philagrios' works emerge from multi-ingredient preparations of invigorating or soothing potions (*pōmata*) administered in conjunction with phlebotomy, cupping, and precise cautery (esp. e.g. Masullo, *frr.*2–7), and substantial passages are quoted with approval and in detail by Oreibasios (prominent are *melikraton* [honey + water], quince-juice, the employment of the full heads of ripe poppies, and a honey + rose oil mix]). Details of Philagrios' account of gout (*podagra* [Masullo, *frr.*8–34]) display an accurate clinical picture, as well as detailed treatment, with a mineral-heavy, multi-ingredient drink made quaffable with large quantities of olive oil and vinegar (Masullo, *fr.*8c = Aëtios 12.66; *cf. fr.*29 = Aëtios 12.68: *Antidotes* or *Remedies for Gouty Conditions*). The majority, however, of the texts of Philagrios' *On Gout* are embedded in Rhazes' Arabic quotations, followed by an important fragment from Philagrios' *On Sciatica* (Masullo, *fr.*35) also given by Rhazes. Other Greek, Arabic, and Latin fragments collected by Puschmann and Masullo are from *On* ***Phthisis***, *Pain in the Chest, Colic, Sweet-Urine Disease* (*viz. Diabetes*), ***Dropsy***, *Cancer, The Spleen, On the Suffocation of the Womb, Bites of* ***Rabid*** *Dogs, Arthritis, Bladder and Kidney Stones, Jaundice*, and others. One also finds traces of Philagrios' therapies for ailments of the eyes, ears, migraine headaches, nocturnal emissions, and the restoration of hair on the head, as well as indications of works on gynecology and obstetrics. The 10th c. list of Philagrios' treatises (*Fihrist* 292 = Dodge 1970: 2.687–688) adds *To Those without a Physician, Making an Antidote for Salt, Impetigo* (or *Ringworm*), and *What Befalls the Gums and the Teeth*, all lost.

Ed.: Th. Puschmann (with German trans.), *Nachträge zu Alexander Trallianus. Fragmente aus Philumenus und Philagrius nebst einer bischer noch ungedructen Abhandlung über Augenkrankheiten* (1887; repr. 1963) 74–129 [Latin translations of *Diseases of the Spleen, The Swollen Spleen, The Inflamed Spleen, The Hardened Spleen*]; R. Masullo (with Italian trans.), *Filagrio Frammenti* (1999).

M. Steinschneider, "Die toxicologischen Schriften der Araber bis Ende des XII. Jahrhunderts," *Virchows Archiv* 52 (1871) 340–503; Wellmann (1895) 63; O. Temkin, "Das 'Brüderpaar' Philagrios und Poseidonios," *AGM* 24 (1931) 268–270; *Idem* (1932) 30–32 and 41; *RE* 19.2 (1938) 2103–2105, E. Bernert; R. Masullo, "*Prolegomena* all'edizione critica di Filagrio," in A. Garzya and J. Jouanna, edd., *Histoire et ecdotique des textes médicaux grecs/Storia e ecdotica dei testi medici greci. Actes du IIe Colloque International (Paris-Sorbonne, 24–26 mai 1994)* (1996) 319–335; D.R. Langslow, *The Latin Alexander Trallianus* (2006) = *JRS Monograph* 10, pp. 19 and 25–26.

John Scarborough

Philaretos (50 – 300 CE?)

Alchemist listed in *Names of Philosophers of the Divine Science and Art* (*CAAG* 1.111). He is the addressee of an alchemical work ascribed to DĒMOKRITOS (*CAAG* 2.159). Although the name is attested elsewhere (*LGPN* 2.446, 3B.421 and 4.343), Letrouit (1995: 36) has recently suggested that a discrepancy in the text preserving this fragment indicates that *philaretos* is here not a proper name but simply means "lover of virtue" and that no such alchemist ever existed.

(*)

Bink Hallum

Philaretos (Med.) (700 – 1000 CE)

Byzantine physician, confused in modern scientific literature until the end of the 19th c. with Theophilos Protospatharios, and also with PHILAGRIOS, whose name could be corrupted to Philaretos, a well-attested name (*LGPN*). Our Philaretos wrote *On pulses*, a treatise on the diagnostic method typical of the late antique and Byzantine periods, probably post-dating the 7th c. The book is largely based on and abbreviated from the **Pneumaticist** PSEUDO-GALĒN, DE PULSIBUS, revised and expanded several times. Some passages possibly come from the Arabic physician Rāzī (9th c.) whose text was translated very early into Greek. Philaretos' work was translated into Latin in the early years of the School of Salerno (*ca* 1000?). Together with the Latin version of Theophilos' *On urine* and other texts translated from Arabic to Latin, it formed the so-called *Articella*, the major manual for learning medicine in Salerno and, eventually, in all late-medieval Europe, particularly Paris.

Ed.: J.A. Pithis, *Die Schriften ΠΕΡΙ ΣΦΥΓΜΩΝ des Philaretos* (1983).
BNP 11 (2007) 13 (#1), V. Nutton.

Alain Touwaide

PHILEAS ⇒ MOSKHIŌN

Phileas of Athens (440 – 400 BCE?)

Wrote a *Trip Around the Earth* (***periodos*** *gēs*) describing the Mediterranean from west to east, from which a dozen fragments survive on Greece and Asia Minor, explaining foundations and myths. AUIENUS, *OM* 691–696, claims he divided Europe and Libya at the Rhône. *Cf.* BAKŌRIS, KLEŌN, and PAUSIMAKHOS OF SAMOS, plus DAMASTĒS OF SIGEION, HEKATAIOS OF MILĒTOS, and SKULAX OF KARUANDA.

BNP 11 (2007) 14 (#1), H.A. Gärtner.

PTK

Philemōn (5 – 25 CE)

Wrote a geographical work on the northern ocean, including Ireland and the amber isles, cited by PLINY and MARINOS OF TYRE. He discussed local names for the northern seas (Pliny 4.95), the nature and source of amber (Pliny 37.33, 37.36), and the size of Ireland (Marinos in PTOLEMY *Geog.* 1.11).

BNP 11 (2007) 17 (#6), H.A. Gärtner.

PTK

PHILĒTAS ⇒ DIONUSIOS SON OF KALLIPHŌN

Philinos of Kōs (280 – 220 BCE)

Physician, worked in Alexandria and was the founder of the **Empiricist** "school" (so PSEUDO-GALĒN, INTRODUCTIO 14.683 K.; CELSUS *pr.*10 mentions as the founder of the school the later SERAPIŌN OF ALEXANDRIA). A pupil of HĒROPHILOS, he broke with his teacher: we know he found fault with the theory of the pulse (MARCELLINUS p. 455.15 Sch.), but probably he disagreed with the whole Hērophilian etiology (perhaps he also polemized against ERASISTRATOS, who in pseudo-DIOSKOURIDĒS *De ven.an.* 6.49 K. = fr.35 Gar. criticizes some **Empiricist** views). It is unclear how much of the later theorization about the principles of medical **Empiricism** can be attributed to Philinos. Already

Philinos, however, initiated the two major genres of the school, i.e., HIPPOKRATIC exegesis and pharmacology. As for Hippokratic exegesis, he wrote a work in six books against the Hippokratic lexicon of the **Hērophilian** BAKKHEIOS (but the three remaining glosses, attested by ERŌTIANOS A-4, A-103, *fr.*33 (pp. 10.17, 23.9, and 108.17 Nachm.), do not differ from those given by Bakkheios). His pharmaceutical work is quoted, for single remedies, by PLINY and GALĒN (via ANDROMAKHOS (YOUNGER)). It is uncertain whether he is the same Philinos who wrote about theriac (PHILOUMENOS, *CMG* 11.1.1, p. 10.19).

Ed.: Deichgräber (1930) 163–164 (frgg.), 254–255.
RE 19.2 (1938) 2193–2194, H. Diller; *DSB* 10.581, F. Kudlien; *OCD3* 1160, H. von Staden; Ihm (2002) #199; *AML* 694–695, K.-H. Leven; *BNP* 11 (2007) 22 (#4), V. Nutton.

Fabio Stok

Philippos (Astron.) (150 – 90 BCE?)

Named in unclear contexts in two inscriptions from Dēlos. He is designated simply as *astrologos*, at this period signifying an astronomer, not an astrologer. A proposed identification with PHILIPPOS OF MEDMA is highly dubious; more probably our Philippos was a local expert of no wide fame.

RE 19.2 (1938) 2558–2560 (#71), P. Treves.

Alexander Jones

PHILIPPOS ⇒ ANDROTIŌN

Philippos of Egypt (100 – 170 CE)

GALĒN records the teaching of a philosopher from Egypt, who at the age of 40 published a book prescribing a regimen for eternal youth, by maintaining the wetness of one's bodily **humors**; he lived to be a frail 80, publishing a second work, *Amazing Agelessness*, explaining that for success, one had to start the diet in infancy: *Sanitate* 1.12.14–15, 6.3.45 (*CMG* 5.4.2, pp. 29, 176), *Marasmos* 2, 4–7, 9 (7.670–671, 678, 685–686, 689, 694, 701 K.), *Caus. Puls.* 4 (9.176–177 K.), *Praesag. Puls.* 1 (9.246–247 K.), and *MM* 7 (10.495 K.). He attributed the wasting of age to disorders of the pulse and excessive fevers. AËTIOS OF AMIDA 4.97 (*CMG* 8.1, p. 407) repeats parts, and gives the ethnic. (The contemporary Ignatius of Antioch, 105–110 CE, *Epist. Ephes.* 20, refers to the "drug of immortality, the antidote of undying.")

RE 19.2 (1938) 2369–2370 (#51), H. Diller.

PTK

Philippos of Kōs (400 BCE – 300 CE)

The "Laurentian" list of medical writers (MS *Laur. Lat.* 73.1, f.143^{V} = *fr.*13 Tecusan) includes *PHILIPPVS COVS*. Most likely a mistake for PHILINOS OF KŌS, not elsewhere on the list (which includes, *inter alia*, HĒROPHILOS, ERASISTRATOS, SERAPIŌN, GLAUKIAS, and PLEISTONIKOS: i.e., other school-founders and **Empiricists**); less likely a reference to PHILIPPOS (OF PERGAMON?), or an otherwise unknown medical writer. *Cf.* EPIPHANĒS, HIPPOSIADĒS, and LUPUS.

(*)

PTK

Philippos of Macedon (120 – 10 BCE?)

ASKLĒPIADĒS PHARM. in GALĒN *Antid.* 2.8 (14.149–150 K.) credits "Philippos of Macedon" with a complex antidote, including cassia, cinnamon, saffron, "Ethiopian" cumin, ***kostos***, pepper, the Egyptian incense *kuphi*, etc. (Diller suggests he was royal). AËTIOS OF AMIDA 9.49 (pp. 553–554 Cornarius) records that HĒRĀS altered Philippos' simple dysentery remedy of acacia, lime, orpiment, and realgar, in plantain juice or dry wine; *cf.* PAULOS OF AIGINA 3.42.3 (*CMG* 9.1, p. 234), 7.12.47 (9.2, p. 317). Those two must be distinct from PHILIPPOS OF ROME; PLINY 1.*ind.*29–30 cites a Philippus for medicine from animals, and OREIBASIOS, *Syn.* 9.5.2 (*CMG* 6.3, p. 277), records a simple asthma remedy by Philippos, who may also be our man. The Philippos of Ēpeiros whose prognosis CELSUS 3.21 records, and shown ignorant by an unnamed student of KHRUSIPPOS OF KNIDOS (II) (perhaps ARISTOGENĒS), must, however, be a distinct earlier person.

RE 19.2 (1938) 2367–2370 (#49–51), H. Diller.

PTK

Philippos of Medma (*ca* 350 – 200 BCE?)

A "remarkable man," born to wealthy and well-known family. He wrote *On Winds*, describing winds and their activities by region (STEPHANOS OF BUZANTION, s.v. *Medma*), presumably similar to THEOPHRASTOS' treatment. Our meteorologist is plausibly, though not certainly, identifiable with the homonymous astronomer from Opous (von Fritz).

RE 19.2 (1938) 2558 (#70–71), P. Treves; 2351–2367 (#42), K. von Fritz.

GLIM

Philippos of Opous (365 – 335 BCE)

Reputed to have edited PLATO's *Laws* and to have written the supplement, *Epinomis*. He wrote on many topics, arguing, from its appearance as the observer moves from right to left, that the rainbow is due to reflection, and that lunar eclipses arose from the interposition of either the Earth or counter-Earth, as in some late **Pythagoreans**. He endorsed the 19-year solar/lunar cycle of METŌN and EUKTĒMŌN and composed his own ***parapēgma*** (*cf.* Euktēmōn). He may have produced an annuary table of midday shadows for Greece. His writings on geometry apparently concerned issues important for Plato's philosophy, including critiques of contemporary formulations of theorems. The *Epinomis* may be the earliest extant Greek work to name all the planets, and identifies Egypt/Syria as a source for awareness of Hermēs' Star (Mercury) and the name of Aphroditē's Star (Venus).

L. Tarán, *Plato, Philip of Opus, and the Pseudo-Platonic Epinomis* (1975); Neugebauer (1975) 574, 739–740; Lasserre (1987) 157–188, 365–393, 591–659.

Henry Mendell

Philippos (of Pergamon?) (130 – 190 CE)

Empiricist physician, exegete of the HIPPOKRATIC CORPUS (GALĒN, *In Hipp. Epid. VI*: *CMG* 5.10.2.2, p. 412). Galēn, in his juvenile work *On medical experience* (150/151 CE), expounds the quarrel about the principles of medicine that arose between Philippos and the **Dogmatic** PELOPS, later teacher of Galēn at Smurna (§8–30: identification of the characters is supplied by *On My Own Books* 2, 9 [2.97, 115 MMH]).

Ed.: Deichgräber (1930) 400–406, 408.
RE 19.2 (1938) 2369–2370 (#51), H. Diller; *KP* 4.752 (#23), F. Kudlien; *BNP* 11 (2007) 41 (#I.33), V. Nutton.

Fabio Stok

Philippos (of Rhēgion?) Xēros (1000 – 1100 CE?)

Greek physician cited in the text and scholia of the *Ephodia tōn apodēmountōn* (Greek translation of ibn al-Gazzār's *Zād al musāfir, ca* 896–979 CE) in MS *Vat. graec.* 300 (*ca* mid-12th c., S. Italy), ff. 90^V^–91^R^, 126^RV^, 230^V^, 292^R^, 300^RV^, and probably also 17^R^, and other MSS. The MS Paris, BNF, *graecus* 2194, ff. 454^R^–464^V^ contains a tract under his name (Costomiris 1890: 170–171). Mercati (1917) and Ieraci Bio (1989: 223) considered Philippos the owner and copyist-annotator of the Vatican MS. Unlikely to have been either (*cf. CMG* 11.1.4, p. 14), Philippos Xēros predated the codex and may have lived during the 11th c. CE; he practiced medicine in southern Italy. In the Paris MS, he is qualified as *ΡΙΓΙΝΟΣ*, i.e., of Rhēgion, and is associated with EUPHĒMIOS OF SICILY.

G. Mercati, *Notizie varie di antica letteratura medica e di bibliografia* (1917); A.M. Ieraci Bio, "La medicina greca nello Stretto (Filippo Xeros ed Eufemio Siculo)," in F. Burgarella and A.M. Ieraci Bio, edd., *La cultura scientifica e tecnica nel'Italia meridionale bizantina* (2006) 109–123.

Alain Touwaide

Philippos of Rome (45 – 95 CE)

Physician often mentioned by GALĒN with ARKHIGENĒS, as together having followers (*In Hipp. Progn.* [16.684 K.], etc.). Since the *Souda* (A-4107) names Arkhigenēs' father as Philippos, scholars assume that this Philippos is the father of the better-known **Pneumaticist**; Kudlien (1968: 1099) prefers a close friendship between the two men. Galēn's inclusion of Philippos among the *neōteroi* (*Febr. Diff.* 2 [7.347 K.]) does not preclude Philippos being Arkhigenēs' father. Galēn mentions students of Philippos (*CMGen* 3 [13.642 K.]; *cf.* Juvenal 13.125, of some doctor Philippos in Rome), who might thus have been a teacher. Since Arkhigenēs practiced in Rome, Philippos is usually located there, despite lack of evidence beyond Juvenal. Galēn mentions an ophthalmic compound medicine used by Philippos in Caesarea of Kappadokia (*CMLoc* 4.8 [12.735 K.]); if that is the same man, and he did not simply travel there, he may have been a native (note Arkhigenēs' origin from Apameia).

Scholars assume Philippos shared Arkhigenēs' medico-philosophical orientation; and indeed, Galēn's several references imply Philippos' **Pneumaticism**. Galēn states that "the group around Philippos and Arkhigenēs studied the question of repletion" (*Plenit.* 7.530 K.), which might be linked to the **Pneumaticist** theory of ***pneuma*** and *stoikheia: cf.* ATHĒNAIOS OF ATTALEIA. Moreover, their group worked on *katokhē* (catalepsy: Galēn, *In Hipp. Epid. III* [17A.640 K.]), resulting from an obstruction of physiological organs according to **Pneumaticist** theory (the reading attributing such a theory to Philippos in CAELIUS AURELIANUS, *Acute* 2.57 [*CML* 6.1.1, p. 166] is a Renaissance paradiorthosis). Although Kudlien (1968: 1099) distinguishes the PHILIPPOS OF EGYPT who wrote on *marasmos*, his theories, emphasizing the diagnostic role of pulse, are not inconsistent with **Pneumaticism**. He defined senescence as a bodily state that does not resemble "burning coal," but is already like "ashes of a fading fire" (Galēn, *Caus. Puls.* 4 [9.176–177 K.], *Praesag. Puls.* 1 [9.246–247 K.], and *MM* 7 [10.495 K.]); Galēn criticizes the author's recommendation to avoid bathing (*Marasmos* 6 [7.690 K.]; also *MM* 10 [10.706–707, 722 K.]).

Philippos wrote profusely on medicines (Galēn, *CMLoc* 7 [13.14 K.], *CMGen* 2 [13.502 K.]: "those around Arkhigenēs and Philip"), though all works are lost, and Galēn quotes several compounds (through Asklēpiadēs Pharmakion): for dysentery (intestinal pain) and blood-spitting (*CMLoc* 7 [13.88 K.]); for ***phthisic*** patients and those spitting blood (*CMLoc* 7 [13.105 K.]); and an analgesic against chronic dysentery (*CMLoc* 7 [13.304 K.]). Some preparations are also quoted in Aëtios of Amida (9.48, p. 552 Cornarius [omitted by Zervos 1911] = Paulos of Aigina 7.12.17), and in Paulos of Aigina (3.42.3 and 7.12.7 [*CMG* 9.1, p. 234; 9.2, p. 315]). If we take Juvenal (above) literally, Philippos taught his students to bleed patients, corresponding to **Pneumaticist** therapeutic methods.

RE 19.2 (1938) 2367–2368 (#50), H. Diller; *KP* 4.752 (#24), F. Kudlien; *Idem* (1968) 1099; *BNP* 11 (2007) 41 (#I.33), V. Nutton.

Alain Touwaide

Philiskos of Thasos (325 – 25 BCE)

An authority on beekeeping, who earned the nickname "Wild-man" (*Agrios*) for setting up his apiary in a remote part of the countryside. His work *Melittourgika* was known to Pliny (1.*ind.*11, 11.19) and probably Hyginus as well.

RE 19.2 (1938) 2389 (#14), W. Kroll.

Philip Thibodeau

Philistidēs of Mallos (110 BCE – 70 CE?)

Cited by Pliny 4.58 (Crete) and 4.120 (Gaddir/Gadēs), with Kratēs of Mallos, for a work on islands.

BNP 11 (2007) 45, W. Ax.

PTK

Philistiōn of Lokroi (370 – 340 BCE)

Physician of Dionysios II of Surakousai (Plato *Ep.* 2.314d: 364 BCE), quoted in Kallimakhos' *Pinakes* as Eudoxos of Knidos' teacher (Diogenēs Laërtios 8.86). Perhaps he came to Athens, if a fragment of the comic playwright Epikratēs (*fr.*10 *PCG*) can be interpreted as referring to Philistiōn.

The titles of Philistiōn's works are lost, but he surely wrote on surgery (*fr.*15 W.), dietetics (*fr.*9 W.) and pharmacology (*fr.*10–12 W.); his "dialectal" name for the temporal vein is recorded ("eagle": *fr.*8 W.). Perhaps because of his fame in dietetics, he was thought by some to have authored the Hippokratic Corpus, Regimen (*fr.*14 W.). His doctrine about the causes of disease is described by Londiniensis medicus (20.24–50): man is constituted by four ***kosmic*** elements, each with a single property, fire/hot, air/cold, water/moist, earth/dry. The origin of diseases is complex, but three series of causes are distinguishable: the imbalance of elements in the body, external causes (physical trauma etc.), and the disposition of bodies, wherein good circulation and transpiration of ***pneuma*** is the essential condition for good health. It has been inferred from this account, compared with Galēn 4.471 K. (*fr.*6 W.), that in Philistiōn's opinion respiration (and transpiration) had the function of moderating innate heat. His doctrine is clearly connected with Empedoklēs and probably influenced the pathological theory of Plato in *Timaios* 70a–d, 82a–b, 84d. Traces of his doctrines are discernible in the HIPPOKRATIC CORPUS, HEART and in DIOKLĒS OF KARUSTOS.

Ed.: Wellmann (1901) 65–93, *frr.*109–116.
RE 19.2 (1938) 2405–2408, H. Diller; T.J. Tracy, *Physiological Theory and the Doctrine of the Mean in Plato and Aristotle* (1969) 28–32; *KP* 4.756, F. Kudlien; Gourevitch (1989) 248–251; *OCD3* 1163, J.T. Vallance; *BNP* 11 (2007) 46–47 (#1), V. Nutton.

Daniela Manetti

Philistiōn of Pergamon (180 – 190 CE)

Student in Alexandria of MĒTRODŌROS (himself SABINUS' student), who read the HIPPOKRATIC CORPUS, EPIDEMICS, 2 (5.138 Littré), as prescribing the consumption of cephalopods to cure barrenness, on the grounds that the clinginess of their feet would induce the womb to cling to the semen. GALĒN rejects this, and records how Philistiōn's wealthy and fastidious patient fired and shamed Philistiōn, *In Hipp. Epid. II* (*CMG* 5.10.1, pp. 401–403).

BNP 11 (2007) 47 (#2), V. Nutton.

PTK

Philodēmos of Gadara (85 – 40 BCE)

Epicurean poet, teacher, and philosopher, and one of the most important figures for the transmission of **Epicurean** philosophy to the Romans in the 1st c. BCE. He was born in Gadara in Syria *ca* 110 BCE, and studied with the **Epicurean** teacher ZĒNŌN OF SIDŌN in Athens. He moved to Italy *ca* 75 BCE, where the Roman noble L. Calpurnius Piso Caesoninus became his patron. Philodēmos taught and wrote poetry and philosophy at Herculaneum and Neapolis, and his students included the Roman poets VERGIL and Horace. Philodēmos was famous in contemporary Roman society for his poetry, but he also wrote a number of philosophical treatises that have survived in fragments among the Herculaneum papyri. These papyri formed part of the library of Piso's villa in Herculaneum and were buried when Mt. Vesuvius erupted in 79 CE (about 120 years after Philodēmos' death). Excavations at Piso's villa (also known as the "Villa of the Papyri") began in the 18th c., and with painstaking effort many of the papyri, numbering about 1800, have been gradually deciphered and edited. Among Philodēmos' works that have come to light are a *History of Philosophers, On Epicurus, On Rhetoric, On Plain Speaking, On the Good King According to Homer, On Signs, On Poems, On Music, On Anger, On Death, On the Gods*, and *On Piety*.

KP 4.759–763, G. Schmidt; Long and Sedley (1987) §18F-G, 23H, 25J, 42G-H, J; *OCD3* 1165–1166, D. Obbink; *Idem, On Piety*, Part I (1996); *BNP* 11 (2007) 68–73, T. Dorandi.

Walter G. Englert

Philogenēs (unknown date)

Cited by Tzetzēs, *In Lykophr.* 603, which Philogenēs refers to Lokroi, and 1085, on the Italian River Lamētos, west of Krotōn (*RE* 12.1 [1924] 544, H. Philipp; *BAGRW* 46-D4); perhaps a geographer.

RE 19.2 (1938) 2483 (#2), W. Kroll.

PTK

Philokalos (*ca* 100 BCE – *ca* 90 CE)

ASKLĒPIADĒS, in GALĒN, *CMLoc* 10.2 (13.349 K.), records his ointment (*malagma*) for sciatica. The name is very rare before the 1st c. BCE (of 37 in the *LGPN* only one, 2.454, is earlier, 349/348 BCE).

Fabricius (1726) 367.

PTK

Philoklēs (250 BCE – 80 CE)

GALĒN quotes ANDROMAKHOS' record of his ***akopon***: wild cucumber root, frankincense, **galbanum**, and *sampsukhon* (marjoram), in beeswax, deer marrow, goose-fat, olive oil, and wine: *CMGen* 7.13 (13.1034–1035 K.).

Fabricius (1726) 367.

PTK

Philokratēs (250 BCE – 25 CE)

CELSUS 5.19.14 (*cf.* 5.26.35C) transmits his wound-plaster: sal ammoniac, birthwort, **galbanum**, iris-root, and **litharge**, said to be especially good for deep wounds.

Fabricius (1726) 367.

PTK

Philolaos of Krotōn (*ca* 430 – after 400 BCE)

The **Pythagorean** philosopher and scientist, born *ca* 470 BCE. Because of the anti-**Pythagorean** revolt in Italy *ca* 450 he fled to Thebes, where he lived and taught for a long time; at the end of his life he probably moved to Taras. His book is preserved in several dozens of fragments and testimonia. Relying on Burkert's fundamental study, Huffman considers as authentic B1–7, 13, 17; A7a, 9, the beginning of 16, 17–24, 27–29; the rest goes back to the pseudo-**Pythagorean** literature. Contrary to the late legendary tradition, Philolaos was not the first to publish the "**Pythagorean** teaching" that before him allegedly was oral and/or secret. Earlier **Pythagoreans** also wrote books, in which they like Philolaos set forth their own views, not never-existent "general **Pythagorean** teaching."

Under the influence of the Eleatics, who asserted that Being cannot be generated, Philolaos modified PYTHAGORAS' cosmogonic principles, limit and unlimited. His ***kosmos*** arose from and consists of the unlimited and limiting things, which are eternal (as is Being) and fitted together by cosmic "harmony" (B1–2, 6). Educated in **Pythagorean** mathematics, Philolaos was first among them to place number and mathematics in a philosophical, above all an epistemological context (A29). Like ALKMAIŌN, he believed that human knowledge is limited (B6), yet tried to rely not on empirical evidence, but on mathematics and related sciences. He asserted that "if all things are unlimited, there will not be anything that is going to be known" (B3); "all the things that are known have number, without which it is impossible to understand or to know anything" (B4), i.e. things are cognizable to the extent that they can be expressed in numbers. Philolaos' number is not an ontological principle, but a function of the limiting things which bring certainty in the world and make it cognizable.

As a scientist Philolaos was most original in astronomy. His system incorporated early **Pythagorean** ideas of sphericity of the Earth and uniform circular motion of the planets, but placed in the center a Central Fire, Hestia ("hearth"), which is the first thing that was fitted together out of the unlimited and limiting things (B7). Around Hestia rotate counter-Earth, Earth, Moon, Sun, five planets and the heavenly sphere. The bodies closer to the center rotate faster: the Earth makes a revolution in a day, the Moon in a month, etc. The Sun is a glass-like body and reflects the light of Hestia that, like the counter-Earth, is invisible to us, since we live on the opposite hemisphere. The counter-Earth was, probably, introduced in order to explain why lunar eclipses are more frequent than those of the Sun, and not in order to bring the number of moving bodies to the "perfect" number 10. (In total, there are 11 celestial bodies in Philolaos' system, not 10.)

Philolaos also discussed arithmetic (B5), geometry (A7a), and harmonics; he gave a mathematical expression of musical intervals from an octave to a semitone (A6a). In physiology Philolaos followed Alkmaiōn (consciousness is located in the brain, B13) and the other **Pythagoreans** (Hippōn, Menestōr); he explained vital functions of an organism by interaction of warm and cold. The soul is "harmony" of the opposite elements of the body (A23) and, hence, dies with it (*cf.* Plato *Phaid.* 86b–c, 88d). Diseases are caused by the influence of external factors on blood, bile and phlegm (A27).

DK 44; Burkert (1972); Huffman (1993); Zhmud (1997); *Idem,* "Some Notes on Philolaus and the Pythagoreans," *Hyperboreus* 4 (1998) 243–270.

Leonid Zhmud

Philomēlos (325 bce – 105 ce)

Rufus, *Ren. Ves. Morb.* 6.7 (*CMG* 3.1, pp. 136–138), records that he pioneered assisted urination in cases of partially-blocked urethra, by applying direct pressure. Galēn, *Loc. Aff.* 1.1 (8.9 K.), refers to the procedure. The archaic name is almost unattested after the 1st c. bce: *LGPN.*

RE S.15 (1978) 308 (#8), H. Gärtner.

PTK

Philōn (Geog.) (*ca* 300 – 250 bce?)

Wrote an *Aithiopika* inspired by his expedition (under Ptolemy I or II) reaching Meroë and the Red Sea. His astral data, e.g., that the Sun was in its zenith 45 days before the summer solstice and the relation of gnomons to shadows in the solstices and equinoxes (Strabōn 2.1.20), were significant in estimating the Earth's circumference. Cited by Antigonos of Karustos and Eratosthenēs; *cf.* perhaps Philōn of Hērakleia.

Ed.: *FGrHist* 670.
RE 20.1 (1941) 51 (#44), R. Laqueur; *BNP* 11 (2007) 51 (#I.5), W. Ameling.

GLIM

Philōn (Meteor.) (*ca* 200 bce – *ca* 200 ce)

Author of a lost *On Metals,* known only through a brief reference in Athēnaios, *Deipn.* (7 [322a]), to a certain fish called a *strōmateus* ("patchwork") found in the Red Sea and mentioned in Philōn's book (Thompson 1947: 253).

(*)

Bink Hallum

Philōn (Methodist) (*ca* **65 – 120** CE)

Physician, listed among the **Methodists** post-dating THEMISŌN and THESSALOS (GALĒN *MM* 10.53 K. = Tecusan, *fr.*162; etc.), perhaps identifiable with PHILŌN (METEOR.) or PHILŌN OF HUAMPOLIS (Tecusan 2004: 52–53).

RE 20.1 (1941) 60 (#60), H. Diller; Tecusan (2004) 51–53.

GLIM

Philōn of Alexandria (*ca* **10 – 55** CE)

Platonic philosopher, political leader and influential theological writer, born *ca* 20 BCE to a prominent Jewish family, Philōn (often called Philōn the Jew, *Judaeus*) wrote in Greek and perhaps did not even know Hebrew. His literary production, listed by EUSEBIOS (*HE* 2.18), is traditionally divided between biblical exegesis and commentaries, (apologetic) history, and philosophy. The bulk of it (about three quarters) concerns the interpretation and depth of the Pentateuch and his writings, valued by the first Eastern Church fathers, exerted significant influence on Christian theology and Alexandrian allegory (Clement and Origen) and on Neo-**Platonism**. His 36 surviving works, among which the strictly philosophical ones (*On Providence, On the Eternity of the World*) are of minor importance, attempt to conciliate and syncretize Jewish spirituality and Greek philosophy.

Of special interest to the zoological tradition is *Alexander* or *On Whether Brute Animals Possess Reason* (as given in Eusebios, *HE* 2.18.6), written *ca* 38 CE on the model of PLATO's *Phaedrus*. In this text, only preserved in an Armenian version, Philōn, who sharply condemns Egyptian zoolatry (*Decal.* 76–80) states an anthropocentric vision of the universe (§73–100) and sustains, against the position of his nephew Alexander, that animals are deprived of both forms of *logos* (reason and language). It would be unjust, he argues, to introduce morality in our relation with animals as equals – as Alexander demands (§10) – since they are not equal (§100). In the first part of the dialogue, Alexander, stressing arguments of the New **Academy**, based on traditional examples (also in AELIANUS *NA*, PLUTARCH *On the Intelligence of Animals*, and PLINY) of some 70 animals, awards to them language (12–15), intelligence (16–29) and morality (30–71). Elsewhere, Philōn develops his cosmologic conception, of a mitigated **Stoicism** (e.g. in *Leg.II* 9), especially, in *On the Creation of the World*, of a world created on an intelligible and prior model (§19) and resumes the **Platonic** idea (see *Tim.* 91), discordant with Jewish theology, of a continuity of beings from fishes to men (*Opif.* 65–66).

S. Sandmel, "Philo Judaeus: An Introduction to the Man, his Writings, and his Significance," *ANRW* 2.21.1 (1984) 3–46; *OCD3* 1167–1168, T. Rajak; *BNP* 11 (2007) 55–61 (#I.12), D.T. Runia.

Arnaud Zucker

Philōn of Bublos, Herennius (115 – 135 CE)

Born 50 CE, and a client of Herennius Seuerus, *cos. suff.* 128 CE (*Souda* Phi-447). He wrote learned works on *Phoenician History* (claiming to use thousand-year-old Phoenician texts: *cf.* BĒROSSOS), *Paradoxes, Cities*, and others. The section "On Doctors" of his *Getting and Choosing Books* circulated separately (*FGrHist* 790 F52–53), as did the section "On Stages of Life" of his *Different Meanings of Words* (§42, pp. 150–152, 235–236 Palmieri, *cf.* Diels 1907: 85).

Ed.: V. Palmieri, *De diversis verborum significationibus* (1988) 17–48.
BNP 6 (2005) 199–201, S. Fornaro.

PTK

Philōn of Buzantion (240 – 200 BCE)

Wrote on mechanics and lived shortly after KTĒSIBIOS on whose work he draws. Philōn is quoted by HĒRŌN OF ALEXANDRIA and EUTOKIOS, and included by VITRUUIUS in a list of writers on mechanics also including ARKHUTAS, ARCHIMĒDĒS, Ktēsibios, NUMPHODŌROS, and later authors (7.*pr.*14). Otherwise little is known of his life. His own work, referring to travels to Alexandria and Rhodes, indicates that he may have worked for a patron, perhaps the unknown Aristōn to whom his treatises are dedicated.

Philōn appears to have written a collection of treatises on mechanics, the *Mechanical Collection* (*Mēkhanikē Suntaxis*). Internal references suggest that there were nine books: 1. *Introduction*, 2. *The Lever* (*Mokhlika*), 3. *Harbor Construction* (*Limenopoiika*), 4. *Artillery Construction* (*Belopoiika*), 5. *Pneumatics* (*Pneumatika*), 6. *Automaton Construction* (*Automatopoiika*), 7. *Siege Preparations* (*Paraskeuastika*), 8. *Siege Craft* (*Poliorkētika*) and 9. *Strategems* (*Stratēgēmata*). Of these only *Artillery Construction*, and parts of *Siege Preparations* and *Siege Craft*, are preserved in Greek, while *Pneumatics* is preserved in an abbreviated Latin and a modified Arabic translation. By assembling this particular range of topics under the heading of mechanics, the *Collection* presents mechanics as a well-defined discipline, and covers similar ground as Ktēsibios had, and as Hērōn would, three centuries later.

Artillery Construction is introduced by methodological considerations (49.1–56.8). Philōn emphasizes that artillery construction depends on both theoretical principles and practical trial and error. The construction and scaling of standard torsion catapults is systematized by introducing a fundamental measure – the size of the hole through which the springs are drawn. To find the diameter of the hole for a catapult that is double the size, it is necessary to solve the famous geometrical problem of doubling the cube, i.e. finding the side of a cube with double the volume of a known cube. This problem cannot be solved by the standard geometrical methods of ruler and compass, and Philōn uses a sliding ruler, thus mixing geometrical and mechanical methods (Hērōn offers a similar solution). Philōn then describes a number of more advanced catapults: an arrow-firing engine (56.8–67.27), a bronze-spring engine inspired by Ktēsibios (67.28–73.20), DIONUSIOS OF ALEXANDRIA's repeating catapult (73.21–77.8) and Ktēsibios' air-spring engine (77.9–78.26).

The *Pneumatics* begins with theoretical chapters on the interaction of water and air (1–5). On the basis of simple experiments air is shown to be a body. Water cannot enter a vessel filled with air unless the air can escape. Moreover there can be no void. Even water, which is heavy, moves upwards in a vessel if the air is sucked out, as if the water and air were glued together. The main part of the treatise consists of a series of chapters describing pneumatic devices such as novelty drink dispensers, constant level bowls, washstands, pumps, simple automata and water lifting devices (6–16). Both the introduction and the devices have been related to Ktēsibios' lost work(s).

The lost book on *Automaton Construction* that may have followed the *Pneumatics* is known from Hērōn's work of the same name, which includes a show ascribed to Philōn about Nauplius and the return of the Greeks from Troy (20–30).

Philōn's treatises on *Siege Craft* and *Siege Preparations* are printed as one work in current editions; it consists of short chapters that can be divided into four sections. The first two

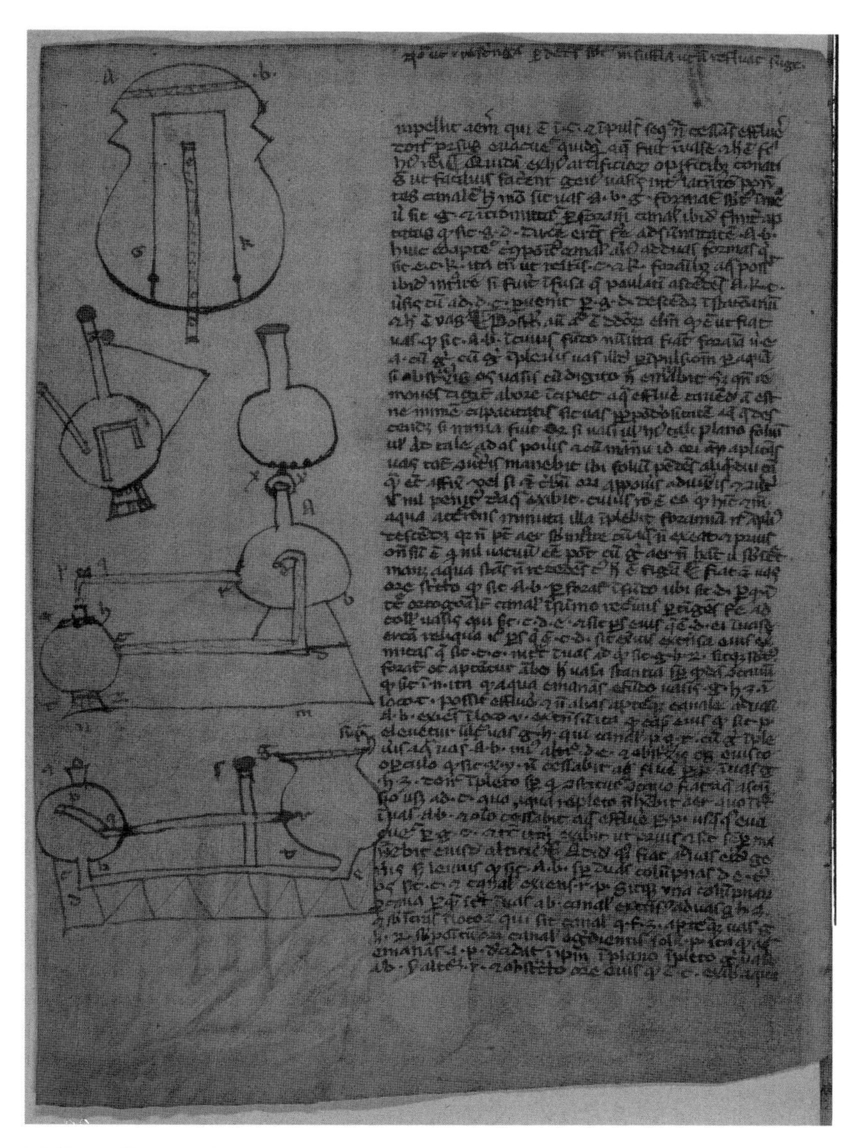

Philon of Buzantion © Bayerische StaatsBibliothek

parts may be identified as *Siege Craft* and the second two with *Siege Preparations*. The first section deals with fortification techniques, such as the construction of walls, towers and moats, as well as the positioning of catapults; the second treats provisions for a town under siege, such as positioning of storerooms and foodstuffs. The third section concerns defense against attacks from land and sea, the necessity of doctors and of pensions and burials for the wounded and dead. The last section discusses how to conduct a siege using catapults, starvation, bribery, and secret messages.

With his focus on principles, practical experience and non-standard devices, Philōn presents himself as a practically and theoretically grounded specialist in mechanics – especially its military applications.

Drachmann (1948); Marsden (1971); *DSB* 10.586–589, A.G. Drachmann; Y. Garlan, *Recherches de poliorcétique greque* (1974); F. Prager, *Philo of Byzantium: Pneumatica* (1974); *OCD3* 1166–1167, G.J. Toomer; *BNP* 11 (2007) 53 (#I.7), M. Folkerts.

Karin Tybjerg

Philōn of Eleusis (350 – 305 BCE)

Named by Vitruuius (7.*pr*.12) as the author of a treatise on the symmetries (proportions?) of temples and on the arsenal at Piraeus (the harbor of Athens); the same source (7.*pr*.17) adds that Philōn was responsible for adding columns to the front of the "cella of Ceres and Proserpina" (the Telesterion) at Eleusis.

That Philōn was the architect of the naval arsenal (*skeuothekē* or *hoplothekē* in Greek and *armentarium* in Latin) at Piraeus (*Peiraieus*) is confirmed by numerous literary sources: Strabōn 9.1.15; Pliny 7.12; Plutarch *Sulla* 14.7; Philodēmos *Rhet.* I, p. 192 ed. Sudhaus; Cicero *de oratore* 1.14.62; Valerius Maximus 8.12.2. These accounts, which were all written after the destruction of the building by Sulla's forces in 86 BCE, show that the fame of the building long outlasted the actual structure, probably as a result of Philōn's treatise. Pliny goes so far as to count the building among the wonders of the ancient world, though he does not specify the criteria by which the building qualified for such a high distinction. A number of the literary accounts take less interest in the excellence of the building than in the rhetorical skill with which Philōn is said to have discussed his building before the Athenians. An inscription contemporary with the building (*IG* II[2] 1668) names both Philōn and a certain Euthudomos of Melitē as the men responsible for the *sungraphē* (building specifications) inscribed on the stone. That the later tradition recognized Philōn alone as the architect of the building may reflect his more significant role in the design, but it may, on the other hand, reflect the fact that Philōn was responsible for the treatise by which the building was known to later generations. Until recently, virtually all information concerning the arsenal was derived from the remarkably detailed *sungraphē*, and for many years the chief interest of scholars was in producing restorations of the building based on it. In 1988 and 1989 portions of the building were excavated in the north-west corner of Zea harbor. The remains are consistent with the *sungraphē* and show that the building was a three-aisled structure, 131 m long and 18 m wide, provided with storage space in its side aisles for the sails and tackle of Athenian warships. The design was admirably functional, providing easy access to the stored material, as well as much needed ventilation. Architectural embellishment was restricted, but, as the presence of a Doric frieze around the exterior shows, it was not altogether ignored. Historical evidence indicates that the arsenal was constructed between 357/6 and 325/4 BCE, when Athens made its final attempt to restore its historic naval power.

The second project with which Philōn is associated, according to Vitruuius, is the columnar porch of the Telesterion at Eleusis. This porch, called the *prostoon* in 4th c. BCE building accounts, was an addition to the earlier 5th c. BCE hall of the mysteries. It comprises 12 Doric columns stretched across the 55 m. expanse of the façade with one additional column on the return at each end. One of the building accounts (*IG* II[2] 1673) shows that Philōn was not, in fact, responsible for either the original design or the initial construction of the porch; the foundations, *krepis*, and at least a portion of the columns were completed by a certain

Athēnodōros of Melitē before Philōn took over completion of the project (between 317 and 307 BCE). Rebuilding of the colonnade in Roman times has apparently obliterated most, if not all, of the portions of the project completed under Philōn.

F. Noack, *Eleusis* (1927) 112–139; K. Jeppesen, *Paradeigmata: Three Mid-Fourth Century Main Works of Hellenic Architecture Reconsidered* (1958) 69–101, 109–131; K. Clinton, "Inscriptions from Eleusis," *ArchEph* (1971) 107–113; J.J. Coulton, *Greek Architects at Work* (1977) 25, 54–55, 57–58; A. Linfert, *Die Skeuotheke des Philon im Piräus* (1981); G. Steinhauer, "La decouverte de l'arsenal de Philon," *Tropis* 4 (1996) 471–479; Svenson-Evers (1996) 301–315.

Christopher A. Pfaff

Philōn of Gadara (150 – 270 CE)

EUTOKIOS (*In Arch. Circ.dim.* 4.162.18–24), following Philōn's student SPOROS OF NIKAIA, explains that Philōn found a better numerical approximation of the circumference of the circle than ARCHIMĒDĒS'. Eutokios next apparently refers to lost works of both APOLLŌNIOS and Philōn, reproaching both for using excessively complex algorithms for calculating with myriads.

BNP 11 (2007) 62 (#I.15), Gr. Damschen.

Alain Bernard

Philōn of Hērakleia (295 – 235 BCE)

To be distinguished from the Byzantine Philōn (*Peri tōn hepta theamatōn*); composed a *mirabilia*-collection (IŌANNĒS OF STOBOI, *Ecl. Phys.* 1.52.48).

Ed.: *PGR* 110–111.
RE 18.3 (1949) 1137–1166 (§6, 1142–1143), K. Ziegler.

Jan Bollansée, Karen Haegemans, and Guido Schepens

Philōn of Huampolis (50 – 90 CE)

Physician, interlocutor in PLUTARCH's *Quaest. Conv.* He argues that animals partaking in simple and uniform diets are healthier than humans with mixed and richly flavored diets (4.1.2 [661B-E]). Despite favoring a simple diet, his pharmaceuticals included many mineral, vegetable, and animal products of land and sea (4.1.3 [663C]). Philōn suggests that oil is hostile to most plants, which perish when grafted onto cypress and pine, which are very oily (2.6 [640B-D]). He claims that thirst arises not from a deficiency of drink but a change in the shape of bodily channels (*poroi*: 6.2.1 [687B-C] – suggesting **Methodist** or **Asklēpiadean** leanings). He also conjectured that some diseases, e.g., ***elephantiasis***, were newly discovered since no ancient treatises existed, against which the doctor ATHĒNODŌROS was cited as a witness; Philōn further adds the long existence of these diseases would convict ancient medical writers of either negligence or ignorance (8.9.1 [731A–B], 8.9.3 [732B]).

RE 20.1 (1941) 60 (#61), H. Diller; *BNP* 11 (2007) 61–62 (#I.14), V. Nutton.

GLIM

Philōn of Tarsos (10 – 35 CE)

Greek physician, in whose honor, perhaps, the *Philoneia* medicines were named. CELSUS preserves Philōn's collyrium compounded from ***psimuthion, spodion***, gum, and poppy, each pounded separately and then mixed together with liquid added gradually (6.6.3). ANDROMAKHOS, in GALĒN *CMLoc* 9.4 (13.267–269 K.), quotes Philōn's 13 elegiac couplets detailing an analgesic of white pepper, henbane, opium, saffron, ***purethron***, ***euphorbia*** (*cf.* IOUBA), and spikenard in Attic honey, for gastro-intestinal, urinary tract, respiratory and neurological use. Galēn, considering this recipe the first of its kind, explicates the couplets, pharmaceutical preparation, and efficacy (269–276 K.). The verses were rendered in prose in early Byzantine medical encyclopedias (OREIBASIOS *Syn.* 3.182 = AËTIOS OF AMIDA 9.32 = PAULOS OF AIGINA 7.11.13), and are transmitted independently in some MSS.

Diels 2 (1907) 85, Suppl. (1908) 63; *RE* 20.1 (1941) 52–53 (#47), 60 (#59), H. Diller; *KP* 4.776 (#13), F. Kudlien; Fabricius (1972) 202; Scarborough and Nutton (1982) 193, n. 24; *BNP* 11 (2007) 61 (#13), Alain Touwaide.

Alain Touwaide

Philōn of Tuana (200 BCE – 100 CE)

Together with DĒMĒTRIOS OF ALEXANDRIA, discovered linear curves in efforts to trisect rectilineal angles. Philōn probably predates MENELAOS who called one of his or Dēmētrios' curves "paradoxical": PAPPOS *Coll.* 4.36 (270 H.; *cf.* Pappos 4.33–34: pp. 259–265 H.).

RE 20.1 (1941) 55 (#51), K. Orinsky.

GLIM

Philōnidēs of Catina (10 BCE – 25 CE)

Physician from Catina (Sicily), known from brief quotations in later medical authorities; GALĒN (*Puls. Diff.* 4.10 [8.748 K.]) has at hand Philōnidēs' *On Medicine* Book 18 (*Peri iatrikēs*) which discusses arterial pulsation, and is cognizant of Philōnidēs' multi-ingredient compound emollient plaster as quoted by ANDROMAKHOS (*CMGen* 8.7 [13.978 K.]). MARCELLUS OF BORDEAUX (*Med.* 29.38, *CML* 5, p. 514]) says he has used an analgesic compound against lower abdominal pains, a remedy for colic ". . .from the book of Philōnidēs," which includes celery seeds, myrrh, opium latex, saffron, and spikenard among ten ingredients, *graues et inueteratos dolores mitigans, mihi quoque experimentis notissima.* DIOSKOURIDĒS (*MM* 4.148.3) writes simply (on white hellebore) that ". . . I completely concur with Philōnidēs, the Sicilian from Enna . . . regarding dosage, administration, and consequent dietetic regulation"; and ERŌTIANOS (E-16 and T-8 [pp. 36, 84 Nachm.]) cites Philōnidēs for descriptions of reddish or flushed skin conditions and terms for unhealthy life-styles, probably from the *On Medicine* noted by Galēn. Philōnidēs' most famous student was ANTIOKHOS PACCIUS. Philōnidēs' books were probably not translated into Latin, so it is probable that he was bilingual, quite characteristic of physicians prominent in Roman Sicily (*cf.* Scribonius Largus). Philōnidēs may have been one of the first Sicilian doctors to produce tracts in both Latin and Greek.

Ihm (2002) #202.

John Scarborough

Philōnidēs of Durrakhion (80 – 60 BCE?)

STEPHANOS OF BUZANTION (s.v. Durrakhion [p. 245 Meineke]) quotes HERENNIUS PHILŌN's *Physicians from Durrakhion* noting that Philōnidēs was a student of ASKLĒPIADĒS, and author of a work about the practice of medicine in his home city in 45 books. Nothing more is known about these books on Durrakhian doctors, unless the Philōnidēs cited by Athēnaios, *Deipn.*15 (675a-e: in turn quoting an ANDREAS, 676c, and 691f–692b), is the same writer who produced the quoted *On Perfumes and Wreaths*. If so, the physicians of Durrakhion were known for their drinking-parties, remedies for hangovers, and how one oiled the head with redolent perfumes and unguents for "medical reasons." "[The imbibers] were persuaded to oil the head since the ill-effects of the wine would be lessened if they [did so] before they began their drinking-bout" (Ath., *Deipn.* 15 [692a-b]). Philōnidēs explains that the application of unguents prevented partly-charred elements taken into the stomach from contributing to fevers engendered in the "dry" head, empty when drinking began.

PIR2 P-378.

John Scarborough

Philōnidēs of Khersonēsos (335 – 305 BCE)

Bematist of Alexander, from the Khersonēsos region of Crete (F1); he measured the Sikuōn to Ēlis road (PLINY 2.181, 7.84). Pliny 1.*ind.*5 and 5.129 cites an anethnic Philōnidēs, probably the same man, for *ACAMANTIS* as an alternate name of the Kilikian Sea. *Cf.* BAITŌN and DIOGNĒTOS OF ERUTHRAI.

FGrHist 121.

PTK

Philōnidēs of Laodikeia "ad Mare" (*ca* 210 – 150 BCE)

Philosopher, mathematician, and statesman. The primary source for Philōnidēs' life is a biography fragmentarily preserved in *PHerc.* 1044 whose title and author are unknown; PHILODĒMOS is a possible candidate. Philōnidēs is also mentioned in the preface to Book 1 of APOLLŌNIOS' *Conics*, and, together with his father Philōnidēs and brother Dikaiarkhos, in two inscriptions from Athens and Delphi. Philōnidēs was born, probably in the last years of the 2nd c. BCE, at Laodikeia "on the Sea," in southern Syria. His family was politically important, and Philōnidēs and his brother followed their father in completing diplomatic missions between Greek cities and Seleukid kings, including Antiokhos IV Epiphanēs and Dēmētrios I Sōtēr; in recognition of their services all three were awarded Athenian citizenship and honored by the priests of Eleusis. In his youth, Philōnidēs met Apollōnios, who introduced him at Ephesos to another mathematician, Eudēmos of Pergamon (otherwise unknown), who became his teacher. Philōnidēs also had contacts with other mathematicians including DIONUSODŌROS OF KAUNOS and ZĒNODŌROS. His training as an **Epicurean** philosopher appears to have begun under one ARTEMŌN, whose commentary on Books 1–33 of EPICURUS' *On Nature* was later the target of one of Philōnidēs' critical writings; during two sojourns in Athens he also studied with two other Epicureans, BASILEIDĒS OF TYRE and Thespis (otherwise unknown). In addition to the attack on Artemōn, Philōnidēs composed his own commentary on Book 8 of *On Nature*. His chief historical significance, however, is for the light cast on chronology, prosopography, and social history of Hellenistic intellectual circles by his apparently amicable relations with contemporary mathematicians and philosophers, including adherents of rival sects such as the **Academic** KARNEADĒS and the **Stoic** DIOGENĒS OF BABYLŌN.

I. Gallo, *Frammenti Biografici da Papiri*, v.2 (1980) 23–166.

Alexander Jones

PHILOPONOS ⇒ IŌANNĒS PHILOPONOS

Philostephanos of Kurēnē (260 – 210 BCE)

In all probability a student of KALLIMAKHOS, who followed his mentor in entertaining antiquarian, mythological and paradoxographical interests. He mainly stayed in Alexandria, where he may have been court poet of Ptolemy IV Philopatōr. His titles – including *On Cities of Asia, On Cities of Europe, On Islands, On Wonderful Rivers, On Springs, On Inventions, Hypomnemata* – may not all denote separate works, but rather chapters within broader treatises. Philostephanos was used at least indirectly as a source by MNASEAS, ERATHOSTHENĒS (*Schol. Lykophr.* 447), PLUTARCH, Athēnaios *Deipn.* 8 (331d), PLINY, and Aulus Gellius 9.4.2, among others.

Ed.: *PGR* 21–23; *SH* 691–693.

RE 18.3 (1949) 1137–1166 (§2, 1141), K. Ziegler; Giannini (1964) 110–111; Fraser (1972) 1.522–4, 777–8, 2.752–778, 1085; *OCD3* 1171, J.S. Rusten; *BNP* 10 (2007) 506–509 (I.B.1, 508), O. Wenskus; *BNP* 11 (2007) 112–113 (#1), S. Fornaro.

Jan Bollansée, Karen Haegemans, and Guido Schepens

Philostorgios of Borissos (*ca* 390 – *ca* 439 CE)

Well-educated ecclesiastical historian, born *ca* 368 at Borissos in *Cappadocia Secunda*. At the age of 20, he moved to Constantinople, where he spent much of his life, and became an adherent of his compatriot Eunomios. He traveled to Palestine and Syria. His *Ecclesiastical History* to 425, in 12 books, each beginning with a letter of his name, continued EUSEBIOS OF CAESAREA's account. Only fragments survive, primarily in the *Artemii Passio* of John of Rhodes (before 668), and in an *epitomē* by Phōtios (*Bibl.* 40), undoubtedly because of its extreme Arianism. Geographical digressions (concerning Daphnē near Antioch, the source of the River Jordan etc.) are based on Philostorgios' own observations. He also wrote an *enkōmion Eunomiou*, a refutation of the philosopher PORPHURIOS OF TYRE, and a vita of the martyr and saint Lucian of Antioch (d. *ca* 362).

Ed.: J. Bidez and F. Winkelmann, *Philostorgius: Kirchengeschichte* (1972).

RE 20.1 (1941) 119–122 (#3), G. Geutz; F. Winkelmann, "Kirchengeschichtswerke," in *Berliner Byzantinistische Arbeiten* 55 (1990) 211; *ODB* 1661, B. Baldwin; *OCD3* 1171, A.M. Nobbs; *BNP* 11 (2007) 113–114, C. Markschies.

Andreas Kuelzer

Philostratos of Lēmnos, the eldest (180 – 200 CE?)

Souda Phi-422 attributes a (lost) *Lithognōmikón* (or *-kós*) to the eldest Philostratos, father of Flauius Philostratos. Part of its content (scientific, according to Münscher) probably survives in his son's *Life of Apollōnios of Tuana* (2.4; 3.7–8.46) wherein numerous stones with magical properties are discussed (*cf.* APOLLŌNIOS OF TUANA, PSEUDO).

K. Münscher, "Die Philostrate," *Philologus* S.10.4 (1907) 543–546; *RE* 20.1 (1941) 135 (#9), F. Solmsen; Halleux and Schamp (1985) XXI–XXII.

Eugenio Amato

Philōtas of Amphissa (35 BCE – 30 CE)

Born *ca* 55 BCE; one of the young medical attendants serving M. Antonius Antyllus (b. 47/46 BCE, Antony's elder son by Fuluia) in the early 30s BCE, he later returned to Amphissa after completing medical studies in Alexandria. At the age of about 75 (Oldfather 1924: 177), Philōtas was honored in a Delphic inscription (*SEG* 1.181; Pelling 1988: 195) for his many years of service. PLUTARCH's grandfather, Lamprias, passed on the stories of garrulous old Philōtas (Plut., *Ant.* 28.3). Scholars have often noted that oral sources are important to Plutarch (Jones 1971: 10; Millar 1984: 23–24 with n. 97; Pelling 1988: 29, 195), and Lamprias was "his most eloquent and resourcefully clever self while imbibing, saying that since frankincense becomes vaporous fumes from heat, thus he was made so by wine" (Plut., *Quaest. Conv.* 1.5 [622E]).

Philōtas learned some typical medical theory while a student at Alexandria, dimly mirrored in this third-hand quotation from Lamprias: during an evening meal with M. Antonius Antyllus and his attendants, the young Philōtas challenged an annoying older physician with "To someone who is slightly feverish one must administer something cold; and anyone who displays a fever is slightly feverish; therefore everyone who has a fever should be given cold [water]" (Plut., *Ant.* 28.5). Philōtas valued complicated and specialized compounds, suitable for a late Hellenistic royal court, and perhaps useful for a military physician serving one of the doomed generals contesting the Roman takeover of the Eastern Mediterranean. Recorded under Philōtas' name is a *kephalikon* (in the class of compounds called *rhaptousai*: that "sew up" or "seal" a wound; usually prepared as plasters), especially useful for skull fractures (CELSUS 5.19.7 = ASKLĒPIADĒS PHARM. in GALĒN, *CMGen* 4.13 [13.745 K.]). Not only does Philōtas' *kephalikon* contain the expected beeswax, myrrh and frankincense, agglutinative "Eretrian" earth combined with vinegar, four variants of copper flakes and **verdigris**, the gummy "juice" of birthwort, raw alum, oil of roses, and olive oil, but also ***ikhthuokolla***. One is tempted to place Philōtas' multi-ingredient eye-salve (Galēn, *CMLoc* 4.8 [12.752 K.]: an *aphroditarion, viz.* "darling") in the context of Antony and KLEOPATRA's court, and the inclusion of 12 drachmas of saffron crocus would make this special collyrium fittingly costly; the 12 drachmas of opium poppy latex combined with 24 drachmas of **calamine**, 12 drachmas of washed Cyprian copper dust, 36 drachmas of acacia-gum, all to be mixed in pure rainwater and spread on with an egg (then washed off) could engender the advertised copious flowing (of tears, presumably).

W.A. Oldfather, "A Friend of Plutarch's Grandfather," *CPh* 19 (1924) 177; C.P. Jones, *Plutarch and Rome* (1971); E.G. Huzar, *Mark Antony: A Biography* (1978) 70–71; F. Millar, "The Mediterranean and the Roman Revolution: Politics, War and the Economy," *P&P* 102 (1984) 3–24; C.B.R. Pelling, ed., with commentary, *Plutarch: Life of Antony* (1988).

John Scarborough

PHILOTIMOS ⇒ PHULOTIMOS

Philoumenos of Alexandria (150 – 190 CE)

The author of an extensive pharmacological compilation modeled on ARKHIGENĒS' *Peri tōn kata genos pharmakōn*, its main source. There are several excerpts in OREIBASIOS' *Collectiones medicae* (see *Index* s.v. Philumenus in *CMG* 6.2.2), who was his earliest user, followed by AËTIOS OF AMIDA. The iological part (*De uenenatis animalibus eorumque remediis*), inspired by Arkhigenēs (Book V), was directly transmitted (*Vaticanus gr.* 284, 11th c.), though in somewhat abbreviated excerpt form. One of its antidotes, also quoted by "Aelius Promotus" (p. 54.19–26 Ihm), is tacitly derived from Nikandros' panacea (Jacques 138–145; Jacques 2002: v. 2, n. 119). Other sources mentioned include physicians who were active in Rome under Trajan, Hadrian, and the Antonines such as Theodōros of Macedon (**Pneumaticist** like Arkhigenēs) and the **Methodist** Sōranos of Ephesos. These borrowings do not establish Philoumenos' membership in a medical school. The fact that he was a compiler explains the absence of references to his work in Galēn. Fragments of Philoumenos' work found in Alexander of Tralleis' Latin version show that Philoumenos had also used Marcellus of Sidē from whom he quotes a ***trokhiskos*** against dysentery (*De medicina* 2.138.15).

Ed.: M. Wellmann, *De uenenatis animalibus* = *CMG* 10.1.1 (1908).
M. Wellmann, "Philumenos," *Hermes* 43 (1908) 373–404; *RE* 20.1 (1941) 209–211 (#7), H. Diller; Jean-Marie Jacques, "Nicandre de Colophon poète et médecin," *Ktema* 4 (1979) 133–149; *BNP* 11 (2007) 126–127, A. Touwaide.

Jean-Marie Jacques

Philoxenos of Alexandria (120 – 90 BCE?)

Celsus 7.*pr*.3, assigning him a leading role in the systematization and development of surgery, describes his book as "most diligent"; P. Cairo Crawford 1 is similar. Asklēpiadēs Pharm., in Galēn, *CMLoc* records several recipes: mineral-based remedy for tumors, 3.3 (12.683–684 K.); three collyria, the last containing henbane, mandrake, and opium, 4.7 (pp. 731, 735–736, 743–744); and two wound-ointments, both containing frankincense and birthwort, in *CMGen* 4.13 (13.738–739 K.) and 4.13 (13.742–743) = 5.8 (13.819–820). Galēn, adding the probably extraneous *nomen* Claudius, praises and records ointments he himself used, *CMGen* 2.17, 3.9 (13.539–540, 645 K.).

Michler (1968) 58–60, 104–105; *BNP* 11 (2007) 125 (#7), V. Nutton.

PTK

Phimenas of Saïs (before 230 – 350 CE)

Alchemist mentioned in P. Leidensis X as the source of a recipe for the making of Egyptian ***asēmos*** (recipe 82; ed. Halleux 1981). The name is Egyptian, meaning roughly "Mēnodoulos." There is no reason to suppose that he is the alchemist Pammenēs (Halleux 1981: 103, n.2).

(*)

Bink Hallum

Phlegōn of Tralleis, P. Aelius (*ca* 117 – 138 CE)

A freedman of the emperor Hadrian who moved in the imperial literary circle. Apart from a chronographical compendium, *Olympiads*, and geographical and topographical treatises, of which at best a few fragments remain, he compiled two shorter books, *On Marvels* (*Peri thaumasiōn*) and *Long-Lived Persons* (*Peri makrobiōn*), surviving more or less complete. Although they appear side by side in the *mirabilia*-section of the Heidelberg MS Palat. gr. 398 (10th c. CE) containing additionally the works of Antigonos of Karustos and Apollōnios, and although the *Souda* (Phi-527) also mentions both books together, they probably were not different parts of a single composition, and only *On Marvels* was a truly paradoxographical compilation.

Moreover, the work heralds a new step in the genre's development. Omitting the staple zoological, botanical, and mineralogical subjects, it focuses exclusively on wonders pertaining to the human world. Stories featuring revenants and hermaphrodites, and reports on living hippocentaurs and births from males outnumber those concerning unusual natural phenomena (giant bones, multiple births). Tellingly, the scientific underpinning of reported curiosities (through detailed source-citations) has largely passed from fashion, the work rather exhibiting a penchant for pure and unadulterated sensationalism. Even so, like the wonder-books of Kallimakhos and Philostephanos catering to the Ptolemies' interest in ***paradoxa***, Phlegōn's *On Marvels* may well have appealed to his patron Hadrian, *omnium curiositatum explorator* (Tert. *Apol.* 5.7).

Ed.: *FGrHist* 257; *PGR* 169–219.

RE 18.3 (1949) 1137–1166 (§19, 1157–1159), K. Ziegler; Giannini (1964) 129–130; A. Stramaglia, "Sul *Peri thaumasion* di Flegonte di Tralle," *SCO* 45 (1995) 191–234; Guido Schepens and K. Delcroix, "Ancient Paradoxography," in: O. Pecere and A. Stramaglia, edd., *La Letteratura di Consumo nel Mondo Greco-Latino* (1996) 373–460 at 430–433, 449–451; W. Hansen, *Phlegon of Tralles' Book of Marvels* (1996).

Jan Bollansée, Karen Haegemans, and Guido Schepens

Phoibos, Ulpius (100 BCE – *ca* 350 CE)

Quoted by Apsurtos for a remedy for sore back in horses (*Hippiatrica Parisina* 223, reappearing without attribution in *Hippiatrica Berolinensia* 26.34) and described as *eparkhikos*, the equivalent of *consularis*.

McCabe (2007).

Anne McCabe

Phokos of Samos (before 330 BCE?)

Probably the author of a *Nautical Astronomy* (also attributed to Thalēs), which presumably described the phases of some fixed stars and constellations (Diogenēs Laërtios 1.23).

DK 5; DK 11A18; O. Wenskus, *Astronomisches Zeitangaben von Homer bis Theophrast* = *Hermes* S.55 (1990), p. 53.

Henry Mendell

Phulakos (?) (10 BCE – 95 CE)

Asklēpiadēs Pharm., in Galēn *CMGen* 5.4 (13.805 K.), indicates that Phulakos employed the dittany-containing anti-venom of Diophantos of Lukia. The archaic name (*cf.* Homer, *Iliad* 2.695, 6.35; Hērodotos 8.39, 8.85) is not attested after the 2nd c. BCE (*LGPN*), but Phulax was in use in south Italy in the imperial era (*LGPN* 3A.469), the probable date of this possible pharmacist. *Cf.* perhaps Pharnax?

Fabricius (1726) 370.

PTK

Phulotimos (330 – 270 BCE)

Greek physician, PRAXAGORAS' pupil together with HĒROPHILOS (GALĒN *Alim. Fac.* 3.30.8 [*CMG* 5.4.2, p. 374]), mentioned, as son of Timolukos, in two inscriptions of Kōs (300–260 BCE: R. Herzog, *Heilige Gesetze von Kos* = *Abh. Ak. Berl.* 1928.6, n° 14, 37–38), quoted usually with Praxagoras in lists of **Dogmatic** physicians (Galēn, *Sanit.* 4.6.22 [*CMG* 5.4.2, p. 122]; *On Venesection, Against Erasistratos* 5, 6 [11.163, 169 K. = pp. 25, 28 Brain]; etc.), often erroneously cited as "Philotimos."

It is difficult to discern Phulotimos' positions. He shared his teacher's main anatomical and physiological theories and terminology, practiced phlebotomy (Galēn *Anat. Womb* 2.890 K., *Ven. Sect.* 11.163 K. = p. 25 Brain) and gave treatments for epilepsy and pleuritis. He is quoted more rarely alone, for his treatment with hellebore, the anatomical explanation of a rare word in HOMER (*Schol. Hom. Iliad.* 11.424d: not in Steckerl), and an anecdote about his answer to someone dying of consumption who asked him to cure a sore finger. He wrote *Art of cookery* (Ath., *Deipn.* 7 [308f]) – perhaps a part of the influential *On (the properties of) food* expanding his teacher's text (Galēn *Alim. Fac.* 1.13.2 [*CMG* 5.4.2, pp. 234–235]) – describing and classifying the properties of foods and frequently quoted, for cereals, nuts and different fishes (*frr.*6–20 Steckerl). A surgeon (CELSUS 8.20.4), he wrote *On matters concerning surgery* (Galēn *In Hipp. Offic.* 18B.629 K.).

Ed.: Steckerl (1958).
RE 20.1 (1941) 1030–1032, H. Diller; Sherwin-White (1978); *BNP* 11 (2007) 215, V. Nutton.

Daniela Manetti

Physica Plinii (*ca* 450 – 500 CE)

Collection of "extracts" from PLINY, even more truncated and mangled than the earlier MEDICINA PLINII. The extractor, another *Ignotus*, has sandwiched passages from the *Medicina Plinii* (thus an extract of an extraction), plus bits from the *Natural History* itself, MARCELLUS' *De medicamentis*, the PSEUDO-APULEIUS tracts, and what Önnerfors calls *alii fontes*. From this abridgment of abridgements was produced an even more truncated set of excerpts (12 folios in an 8th/9th c. MS), and another Italian excerptor (6th/7th c.) had independently created a longer set of excerpts, about 120 folios re-copied into a 9th/10th c. Italian MS labeled by Önnerfors as Q, the so-called Bamberg Pliny. Not surprisingly, other enterprising scribes were busily doing separate extractions, and another set of folios, called the Plinii Florentino-Pragensis, has survived to be added to our knowledge of this obviously most popular genre that circulated in the Latin West in the Middle Ages. The very bulk of Pliny's original 37-book *Natural History* encouraged the production of summaries, synopses, outlines, and abridgements, much as we find many more digests and epitomes of Gibbon, Frazer, or Toynbee, than we do the complete *Decline and Fall of the Roman Empire* (seven vols.), *The Golden Bough* (14 vols.), or *History of the World* (ten vols.). The *Physica Plinii*, however, is best compared with the all-too-popular *Reader's Digest* series of abridged novels.

Ed.: A. Önnerfors, *Physica Plinii Bambergensis (Cod. Bamb. Med. 2, fol. 93v–232r)* (1975); J. Winkler, *Physicae quae fertur Plinii Florentino-Pragensis Liber Primus* (1984); W. Wachtmeister, *Physicae Plinii quae fertur Florentino-Pragensis Liber secundus* (1985); A. Önnerfors, *Physica Plinii Sangallensis*, vv. 1–3 (2006–2007): *non vidi.*
H.E. Sigerist, *Studien und Texte zur frühmittelalterlichen Rezeptliteratur* (1923; repr. 1977); J. Jörimann,

Frühmittelalterliche Rezeptarien (1925); H. Fischer, *Mittelalterliche Pflanzenkunde* (1929; repr. 1967); A. Önnerfors, *Pliniana* (1956); B. Löfstedt, *Studien über die Sprache der langobardischen Gesetze* (1961).

John Scarborough

Physiognomista Latinus (350 – 400 CE)

Composer of a book on physiognomy, based on and extensively translating and paraphrasing POLEMŌN, using also LOXOS and "Aristotle" (i.e. the ARISTOTELIAN CORPUS PHYSIOGNOMY), as he states in the first sentence. An attribution by Albertus Magnus to APULEIUS was once discussed (*cf.* Rose 77–86), but has been rejected by André (31–34) and Repath (549–550), who, because of its language, instead suggest the date we give. The treatise has four main parts: 1–15: an introduction on the theory and method of physiognomy which closely follows the methodical introduction in the Aristotelian Corpus *Physiognomy*; 16–89: the signs from head to feet and the characters they signify, according mostly to Polemōn, especially for the eyes (20–43); 90–117: several character types who bear combinations of the signs, including a brief chapter on the importance of the "overall impression" (*epiprépeia*), following the Aristotelian Corpus *Physiognomy* and Polemōn; finally, 118–133: the characteristics of animals, according mostly to Loxos.

Ed.: V. Rose, *Anecdota Graeca* (1864) 1.59–102 (introduction) and 103–169 (text); J. André, *Anonyme Latin: Traité de physiognomonie* (*CUF* 1981); I. Repath, "Anonymus Latinus, *Book of Physiognomy*," in Swain (2007) 549–635.

Sabine Vogt

Physiologos (100 – 400 CE)

Greek Christian anonymous collection of brief animal portraits, originated in Alexandria or Palestine, widely distributed throughout the Greco-Roman world until late Middle Ages. With a wide range of texts of variable length and material, the *Physiologos* (i.e. "Expert-in-nature") has no standard form and is rather a genre than a work. It usually presents a twofold description of one or several features of an animal, first naturalistically, and then allegorically and symbolically, with regular scriptural quotations, revealing how nature (***phusis***) itself expresses Christian realities and spiritual truths. On the basis of some 80 MSS, Sbordone distinguished three main Greek recensions: ancient (immediately post-dating the gospels), Byzantine (5th/6th c.), and a so-called Basilean (10th/11th c.), erroneously attributed to BASIL OF CAESAREA. Traditionally ascribed to various heterogeneous authors (e.g., the *Christian* bishop EPIPHANIOS, the *pagan* naturalist ARISTOTLE, or the *Hebrew* king Solomon), this very popular syncretic digest of Egyptian lore, Greek natural history and Judeo-Christian exegesis was early translated into the main ancient Eastern languages (Syriac, Coptic, Armenian, Georgian, etc.), and Latin. All medieval Bestiaries (books of beasts) or Aviaries (in Latin as in vernacular versions) originate from one of the numerous Latin recensions (B, Y, A, C, etc., produced before 500 CE). This extensible collection with an average of 45 chapters (cumulatively treating some 80 different creatures in the ancient versions) is not even strictly zoological, including also plants (sycamore and the *peridexion* tree) and stones (*adamas*, magnetite, fire-flints, etc.). This cultural medley abounds in popular beliefs and ethology (with many parallels in Aristotle, AELIANUS and KURANIDES), sometimes misconceived animal behavior (the beaver's autocastration, the fox's simulating death, the crow's monogamy, the snake's hibernation, etc.), and theological interpretations: the ichneumon covering himself with mud to kill the snake becomes thus a natural allegory of Christ assuming mortal nature to defeat the evil spirit. The selected animals are mostly savage (and often exotic), and some are mythical: onocentaur, siren, phoenix, ant-lion (fantastic creature generated by a free translation in the Septuagint of a rare Hebraic word for *lion*). However, *Physiologos*, containing no original naturalistic data, offers rather a series of peculiar behaviors or powers (called "natures") turning to popular themes (curative properties of the bird *kharadrios* – not the ordinary plover, the horror of the wolf in front of a naked man, the tears of the anthropophagous crocodile, etc.); and it gathers the main moral figures of medieval imaginary and Romanesque architecture (the fireproof salamander, abstinent elephant, resuscitating phoenix, heroic ichneumon, savage unicorn tamed by a virgin, eagle renewed by the sunlight of the truth and the water of baptism, etc.). The apologetical function and homiletical use of the text is obvious, but *Physiologos* often occurs in the manuscript tradition with zoological (and not Christian) writings and was read and treated as such. Medieval zoology (from Isidorus of Hispalis to Bartholomaeus Anglicus, 13th c.) relies in fact amply on *Physiologos*' moralizing, myths, and erroneous assumptions.

Ed.: A. Zucker, *Physiologos. Le bestiaire des bestiaires* (2004).

M. Wellmann, *Der Physiologos. Eine religionsgeschichtlich-naturwissenschaftliche Untersuchung* = *Philologus* S.22.1 (1930) 1–116; F. Sbordone, *Ricerche sulle fonti e sulla composizione del Physiologus greco* (1936); *RE* 20.1 (1941) 1074–1129, B.E. Perry; N. Henkel, *Studium zum Physiologus im Mittelalter* (1976); J.H. Declerck, "Remarques sur la tradition du Physiologus grec," *Byzantion* 51 (1981) 148–158; A. Scott, "The Date of the Physiologus," *Vigiliae Christianae* 52 (1998) 430–441; *BNP* 11 (2007) 227–228, K. Alpers.

Arnaud Zucker

Piso ⇒ Calpurnius

T. Pitenius (*ca* 100 CE)

Astrologer, wrote an elaborate horoscope on a papyrus roll for Hermōn, born on April 1, 81 CE, presumably in Lower Egypt (*P. Lond.* 1.130). The positions of the heavenly bodies were computed by the "Eternal Tables," mentioned also by Ptolemy (*Almagest* 9.2) and Vettius Valens (6.1).

Neugebauer and van Hoesen (1959) 21–28.

Alexander Jones

Sextus Placitus Papyriensis (400 – 450 CE)

In the corpus constituted of (1) pseudo-Antonius Musa, *De herba uettonica*, (2) pseudo-Apuleius, Herbarius, (3) the anonymous *De taxone*, and (4) pseudo-Dioskouridēs, *De herbis feminis*, there is a *Liber medicinae ex animalibus*, ascribed to this man. Each of its 34 chapters treats an animal, describing its products used as *materia medica* (e.g., from deer, fox, rabbit and wild goat to eagle, vulture and other birds). It borrows material from Marcellus of Bordeaux and the Plinian tradition, and its illustrations may be based on Hellenistic models (Grape-Albers 1977: 27, 35). Both the text and its illustrations, originating probably in the first half of the 5th c. CE, are known through two recensions (text: Howald and Sigerist 1927; illustrations: Talbot and Unterkircher 1971–1972; Grape-Albers 1977: 23–25), probably resulting from independent rearrangements of an original nucleus, rather than from two authors (Howald and Sigerist 1927: xxi). The compiler has sometimes been identified with Sextus Platonicus. The date of the English translation is still debated, D'Aronco (2007: 38) preferring the late 10th c. Constantine the African (d. after 1081 CE) rearranged the treatise (Ackermann 1788: 1–112), and it saw several Renaissance printings.

Ed.: E. Howald and H.E. Sigerist, *Antonii Musae De herba vettonica liber. Pseudo-Apuleius Herbarius. Anonymi De taxone liber. Sextii Placiti Liber medicinae ex animalibus* = *CML* 4 (1927).

J.G. Ackermann, *Parabilium medicamentorum antiqui* (1788); C.H. Talbot and F. Unterkircher, *Medicina antiqua. Codex Vindobonensis 93 der ÖNB. Facsimile & Kommentarband* (1971–1972); H. Grape-Albers, *Spätantike Bilder aus der Welt des Arztes* (1977); H.J. De Vriend, *The Old English Herbarium* (1984); M.P. Segolini, *Libri medicinae Sexti Placiti Papyriensis ex animalibus pecoribus et bestiis vel avibus concordantiae* (1998); M.A. D'Aronco, "The Transmission of Medical Knowledge in Anglo-Saxon England: the Voices of Manuscripts," in P. Lendinara *et al.*, edd., *Form and Content of Instruction in Anglo-Saxon England in the Light of Contemporary Manuscript Evidence* (2007) 35–58.

Alain Touwaide

De Planetis (200 – 300 CE?)

The author explains the powers of single planets, of planetary conjunctions, and of the planetary figures trine, quadrature, and opposition; the work is sufficiently similar to FIRMICUS 6.3–27 (on those figures) that Kroll thought a common source likely (*cf.* also F. Boll in *PSI* 3.158, a 3rd c. CE papyrus from Oxyrhynchos). The author apparently exploited a hexameter poem, of which a few lines remain embedded in the prose (*cf.* perhaps MANETHŌN or ANOUBIŌN). The text proceeds systematically from Kronos (Saturn: pp. 160–168) inward through Zeus (Jupiter: pp. 169–173) and Ares (Mars: pp. 173–176) to the Sun (p. 176), then Aphroditē (Venus: pp. 177–178) and Hermēs (Mercury: pp. 178–179), and ending with the Moon (pp. 179–180). Each planet's role in ***melothesia*** is given (citing PTOLEMY and VETTIUS VALENS); then its effects in conjunctions with more inward planets; then its effects with more inward planets in the three figures (omitting impossible figures). For the inner planets, Aphroditē, Hermēs, and Moon, the effects of "superiority" are recorded (*cf.* Manethōn 6.279); for Hermēs, the effects of conjunctions with all preceding planets are repeated.

W. Kroll, *CCAG* 2 (1900) 159–180.

PTK

Platōn (Med.) (50 BCE – 95 CE)

Wrote *On Phlebotomy*, of which a Latin translation apparently survives, in MS *Monac.* 8.2 (16th c.). ASKLĒPIADĒS PHARM. in GALĒN *CMLoc* 7.2 (13.60 K.) records two cough remedies, both containing ***sturax***, opium, and myrrh, and providing immediate relief. Since not mentioned in DIOGENĒS LAËRTIOS 3.109, he probably postdates 50 BCE.

Diels 2 (1907) 86; *RE* 20.2 (1950) 2542 (#10), Johanna Schmidt.

GLIM

Plato (*ca* 390 – 348/347 BCE)

Platōn; born at Athens (or Aigina?) 427 BCE, and died in Athens 348/347 BCE. Plato's attitude and contribution to ancient natural science are both difficult to judge and have been the subject of considerable controversy. As Plato wrote dialogues, whose often complex arguments are sometimes inconclusive, and never appears in person in these works, there

Plato © Fitzwilliam Museum, Cambridge

has been considerable debate concerning Plato's actual thoughts. This is exacerbated by the fact that Plato does not appear to write to record his own doctrines, but to engage in or illustrate the nature of philosophical debate, or perhaps even to provoke his readers to examine their own opinions. The order of his works and the possible development of his thought are also areas of contention. The consensus on groupings of early, middle and later works is broad, but the position and significance of individual works can still be hotly contested. The key work for Plato's views on natural science is the *Timaeus*, now generally agreed to be late, but its relation to other late works and to the development of Plato's thought is unclear. Sources for Plato's biography include DIOGENĒS LAËRTIOS, whose own sources vary in reliability, and ARISTOTLE, Plato's pupil. Some letters in Plato's name give interesting information, but their provenance is open to considerable doubt.

Plato gives us the first thoroughgoing teleological account of the ***kosmos***, its formation and the origins of humans and animals in the *Timaeus*. This work was hugely influential in astronomy and cosmology, and significantly affected attitudes to explanation down to the 17th c. Why does Plato adopt this teleology? Plato's critics argue that his motivation here is some sort of overspill from his programs in ethics and epistemology, both dominated by an absolute conception of the good. They argue this was a reaction against materialist science preceding Plato, and had a malign effect on subsequent thought.

Plato found contemporary materialist explanations crude, implausible and inadequate, a reasonable conclusion given the lack of sophistication of these accounts at this stage of their development. His alternative was to postulate a craftsman God, the demiurge, who organized all things out of chaos, always with the best arrangement in mind. Where LEUKIPPOS and DĒMOKRITOS had an unlimited number of worlds occurring by accident, an unlimited number of sizes and shapes of atoms, and EMPEDOKLĒS had a multiplicity of biological accidents before viable species are formed, Plato was adamant that there was one well-designed ***kosmos***, a small number of well-designed basic particles and unitary, well-designed species.

That Plato criticized many theories of Pre-Socratic *phusiologoi* is sometimes taken as evidence of a negative attitude towards natural science. Here it is important to distinguish between Plato's attitude to the *phusiologoi* and his own conception of how natural phenomena should be explained. When he is critical of materialist accounts of Sōcratēs remaining in jail, or why the Earth has its shape and position, or why one person is taller than another, it is not that he believes these issues are not worthy of investigation, but rather that materialist accounts of these issues are inadequate, either because they do not refer to the good, and so are not teleological in the sense required, or do not refer to Plato's forms.

Plato contrasted his unchanging, intelligible, knowable forms with the changing, perceptible physical world, the subject of opinion only. Modern interpretations of Plato downplay the extent to which these should be seen as two separate worlds and emphasize that both participate in investigation for Plato. There is no reason to suppose that Plato thought natural science solely concerned with forms, and so entirely non-empirical, or solely concerned with the physical and so unable to constitute knowledge.

Plato has been much criticized for appearing to denigrate the role of observation and experiment in science. Plato has Sōcratēs say (*Republic* 530b6–c1):

> "It is by means of problems, then, that we shall proceed with astronomy as we do geometry, and we shall leave the things in the heavens alone, if we propose by really taking part in astronomy to make useful instead of useless the understanding that is by nature in the soul."

The context and the conditional nature of this passage are critical here. Plato prescribes a curriculum for the intellectual development of the guardians of his ideal state, not offering a methodology for astronomy, nor does this passage have any implication for such a methodology. It says that if we are to use astronomy to educate the guardians, then we use it in this specific manner. The *Timaeus* (47b6–c5), more concerned with method, tells us in contrast that:

> "God devised and gave to us vision in order that we might observe the rational revolutions of the heavens and use them against the revolutions of thought that are in us, which are like them, though those are clear and ours confused, and by learning thoroughly and partaking in calculations correct according to nature, by imitation of the entirely unwandering revolutions of God we might stabilize the wandering revolutions in ourselves."

If Plato's *Timaeus* supports the idea that the motions of the Sun, Moon and planets can be resolved into combinations of regular circular motions, then this is probably Plato's most important contribution to contemporary Hellenic science. While (apparent) motions of the fixed stars were easy to model, motions of the other heavenly bodies were not. They were commonly referred to as "wanderers," as their motion appeared to defy simple laws.

The problem in ascribing regular, circular motion to Plato is that the astronomical model of the *Timaeus* is very crude, using only two circular motions each for Sun, Moon and five planets, and so can only reproduce very few phenomena. This appears to produce a dilemma. Either Plato is ignorant of the phenomena, or his model must be able to account for more of the phenomena, by using motions which are not regular and circular, if Plato believes his model can reproduce all the phenomena of which he is aware. However, Simplicius (*in De Caelo* = *CAG* 7 [1894] 504.17–20) tells us that authors proposed models that could not account for all the phenomena of which they were aware. It may well be that Plato considered his model of the *Timaeus* a prototype, not able to account for all the known phenomena but showing the way in terms of regular circular motion. If so, Simplicius' comment (*in De Caelo* = *ibid.*, 488.18–20) makes sense:

> "Plato posed the following problem for those engaged in these studies: 'Which hypotheses of regular and ordered motion are able to save the phenomena of the planets?'"

It matters little whether Plato or Eudoxos, his associate, originated the idea as it is the *Timaeus* which popularizes it. Eudoxos greatly improved on Plato's model using a more complex array of regular circular motions. Once the paradigm of regular circular motion is established, an enormously successful research program ensues, resulting in one of the finest products of ancient science, Ptolemaic astronomy, and it is not until *ca* 1600 that Kepler first questioned that astronomy should be done in this fashion.

A further part of Plato's legacy was the **Academy**, a school of intellectuals researching in Athens down to its closure by Christian authorities in 529 CE. We are by no means certain of the nature of the school's activities, though it is reported that "Let no one ignorant of geometry enter here" was written above the door.

G.E.R. Lloyd, "Plato as Natural Scientist," *JHS* 28 (1968) 78–92; J.P. Anton, ed., *Science and the Sciences in Plato* (1981); Andrew Gregory, *Plato's Philosophy of Science* (2000).

Andrew Gregory

Platusēmos (?) (100 BCE – 360 CE)

Oreibasios, *Ecl. Med.* 86.6 (*CMG* 6.2.2, p. 263), records his blood-stanch (*iskhaimon*) of lime, orpiment, realgar, and sulfur. The word seems otherwise unattested as a Greek name (*LGPN*, Pape-Benseler), but represents the Latin *laticlauia* (Strabōn 3.5.1), the senatorial stripe, or its rank; compare the late Roman name Senator, *PLRE* 2.989–991, esp. Cassiodorus Senator. (Or perhaps *cf.* Platulaimos in Alkiphrōn 1.23.)

(*)

PTK

Pleistonikos (300 – 240 BCE)

Greek physician, Praxagoras' pupil (Celsus 1.*pr.*20), cited by Galēn mostly with his teacher and other **Dogmatic** physicians, especially Dioklēs (4.732, 10.28, 10.110 K., etc). Although his place of origin is unknown, he probably practiced in Kōs. His opinions on physiology and anatomy must have been similar to Praxagoras', but it is difficult to distinguish Pleistonikos' theories in Galēn's general lists: certainly Pleistonikos described and analyzed the **humors** (Galēn, *Atra Bile* 1.2 [*CMG* 5.4.1.1, p. 71], *PHP* 8 [*CMG* 5.4.1.2, p. 510]). He believed that air entered the arteries not only from the heart but also from the entire body (Galēn *Blood Arter.* 8.1, pp. 176–177 Furley and Wilkie) and approved phlebotomy (Galēn *On Venesection, Against Erasistratos* 5 [11.163 K. = p. 25 Brain]). To him alone is attributed the opinion that digestion is a process of putrefaction (*sepsis*: Celsus 1.*pr.*20). He claimed that water is a better aid to digestion than wine (Ath., *Deipn.* 2 [45d]), treated some illness with radish (Pliny 20.26), and used hellebore in a peculiar way, employing it as a **pessary** and making patients smell it to induce vomit: Oreibasios *Coll.* 7.26.194 (*CMG* 6.1.1, p. 245).

Ed.: Steckerl (1958).

KP 4.925, F. Kudlien; *BNP* 11 (2007) 379–380, V. Nutton.

Daniela Manetti

***Plentiphanēs* (500 – 90 BCE)**

Agricultural writer whose work was known to Cassius Dionusios (Varro, *RR* 1.1.9–10). Since "Plentiphanēs" is not a plausible Greek name, one may infer textual corruption of e.g. Leophanēs.

RE 21.1 (1951) 226, K. Ziegler.

Philip Thibodeau

C. Plinius Secundus of Novum Comum (43 – 79 CE)

Wrote *Naturalis Historia* (*NH*), a 37-volume compendium of knowledge about the natural world, medicine, technology, and art, offering a universal index of the world as known and imagined by the educated classes of early imperial Rome.

Life: Born in Novum Comum in 23 or 24 CE, Pliny had a career typical of the wealthy equestrian class to which he was born. As a young man, he served in the army (47–52 CE) as military tribune and commander of a cavalry unit; a decorative roundel (*phalera*) bearing his name has been recovered at Xanten. He participated in campaigns in the frontier provinces of Upper and Lower Germania, under Domitius Corbulo against the Chauci (47 CE), and against the Chatti under Pomponius Secundus (50–51 CE). In military life, he befriended the future emperor T. FLAVIUS VESPASIANUS, the future dedicatee of *NH*. In civilian life, he acted as a forensic orator and wrote prolifically. Beside *NH*, his works included a biography of Pomponius Secundus; a manual on throwing javelins from horseback; a history of Rome's wars with the Germanic tribes (inspired by a dream-vision of Nero Claudius Drusus, hero of AUGUSTUS' German campaigns); a book on the education of orators; a book on linguistic problems (an apolitical choice dictated by the dangers of life under Nero); and a history of his times.

After Nero's death and the Flavians' ascent to power (69 CE), Pliny became a man of importance. Between 70–76 CE, Pliny took procuratorships in several provinces including *Hispania Tarraconensis*, *Africa*, and probably both *Gallia Narbonensis* and *Belgica*. Eventually he was recalled to Vespasian's court as an imperial adviser (*amicus principum*), and finally was appointed commander of the Roman fleet at Misenum on the Bay of Naples. From his house at Misenum, as described by his nephew (Pliny the Younger, *Letters* 6.16), Pliny saw the eruption of Vesuvius of 79 CE, August 24. Having taken a galley to Stabiae to observe the eruption and rescue others in the neighborhood, he died when asphyxiated by volcanic gas (some scholars prefer to adduce a heart-attack).

Pliny the Younger described his uncle's work-habits in detail (*Letters* 3.5). Reducing his sleep to an austere minimum, he spent his waking hours either at official duties or studying. While listening to a reader, he dictated whatever caught his interest to scribe, keeping reader and scribe employed even at meals; rather than suspend note-taking while walking, he traveled by litter. No book was so bad, he said, that some part of it might not be somehow useful. In this way he produced the raw materials of his enormous books: 20 volumes on Rome's wars against the *Germani*, 31 volumes of contemporary history, and the 37 volumes of *NH*.

Work: Pliny is of major significance as a pioneer in the encyclopedic tradition. For historians of science, Pliny's importance lies more in the concept and outline of Nature implicit in the structure of his book than in original theories or first-hand observations, of which there are few.

Pliny dedicates *NH* (*Pr.*1) to Titus, Emperor Vespasian's son and co-ruler – a measure of Pliny's ambitions. There follows an extended table of contents or index-list, setting out his topics by book and subsection; for each book Pliny gives a total sum of facts contained (consistently undercounted) as well as authors consulted, listing Roman authors separately from foreigners. *NH* begins with cosmology, including astronomy and geology (Book 2); there follow a geographical gazetteer of the known world (Books 3–6); man (7); creatures of land, sea, and air (8–10); insects and comparative anatomy (11); botany (12–19); medicine and pharmacology (20–32); and finally minerals, including long subsections on pigments, painting, sculpture, architecture, and gems (33–37).

As the Hellenistic period saw the methodical collection of knowledge in royal libraries, Rome's ascendancy was marked by the appearance of books aiming at comprehensive syntheses of Greek scholarship and traditional Roman culture. Designed to embrace *enkuklios paideia*, "general culture," these books represent the beginning of the encyclopedic tradition. What *enkuklios paideia* meant was not yet fixed: VARRO's *Disciplinae* covered dialectic, rhetoric, grammar, arithmetic, geometry, astronomy, music, architecture, and medicine; CELSUS' *Artes* comprised agriculture, military science, medicine, oratory, jurisprudence, and philosophy. Unlike these older encyclopedias, the goal of *NH*, and its prime structural criterion, was not tuition in skills valued by Roman society, but investigation into nature.

The index-list of *NH*, a novel device intended to let readers find particular facts without lengthy browsing, represents another innovation. A universal taxonomy in miniature, the index-list also demonstrates how to fit the world into a referential shape. This instrument of reference, as well as Pliny's sums of facts recorded, set a standard and issued a challenge to later encyclopedic authors.

Pliny is sometimes attacked for excessive credulity, since *NH* abounds in the surprising and the marvelous (*mirabilia*): fantastic animals, astonishing springs, and oddly-shaped peoples. But from an ancient perspective, *mirabilia* serve not only the recognized literary end of entertainment, they also illustrate the variety and power of Pliny's chosen subject, Nature. Since the normal is understood by contrast with the strange, Pliny's *mirabilia* work as limit-cases, demarcating the realm of accepted knowledge by tracing its periphery.

Pliny's book, which often reads like an inventory of things available to his contemporaries, is not simply collected data given referenceable form, it is knowledge collected for Roman use and made accessible, as Pliny himself says, by the spread of Roman authority. As with the treasures displayed in a triumphal procession, one witnesses in *NH* Rome's power at work subduing and taxonomizing Nature.

Ed.: R. König and G. Winkler, ed. and trans., *Naturkunde: Lateinisch-Deutsch/C. Plinius Secundus der Ältere* 27 vv. (1973–2004); H. Rackham and W.H.S. Jones, ed. and trans., *Pliny: Natural History* 10 vv. (Loeb: 1938–1963: complete, though not always reliable, English translation with Latin text).

M. Beagon, *Roman Nature: The Thought of Pliny the Elder* (1992); Trevor Murphy, *Pliny the Elder's Natural History* (2004); V. Naas, *Le Projet Encyclopédique de Pline l'Ancien* (2002), with other bibliography; *NDSB* 6.116–121, A. Doody.

Trevor Murphy

Plōtinos (254 – 270 CE)

The most important **Platonist** philosopher in late antiquity, whose life is well documented in his pupil PORPHURIOS' detailed biography. Born 204 CE in Egypt, Plōtinos studied philosophy, with both the Christian and the **Platonist** Origen, under Ammōnios Saccas at Alexandria for 11 years (*Vit. Plot.* 3). In 242–3 he joined Gordian's expedition to Persia to learn about Persian and Indian philosophy, but without success (*Vit. Plot.* 3). In 244, Plōtinos moved to Rome where he opened his own school, seemingly quite popular, attracting students from abroad – Porphurios from Athens, Roman senators (*Vit. Plot.* 7), and women (*ibid.* 9) – and enjoying the emperor Gallienus' favor (*ibid.* 12). Porphurios provides a good impression of Plōtinos' seminars, consisting in reading the exegetical works of DIONUSODŌROS SEUERUS, NOUMĒNIOS, GAIUS, ATTICUS, ALEXANDER, ASPASIOS, and ADRASTOS OF APHRODISIAS (*ibid.* 14), presumably to elucidate PLATO's and ARISTOTLE's philosophy, while his pupils raised questions and argued for their views (*ibid.* 13, 18). Suffering from a serious disease, in 270 Plōtinos retired to Campania to die.

Porphurios divided Plōtinos' works into three periods: early, those recorded before Porphurios' arrival (263); middle, during Porphurios' stay in Rome (until 268); and late, written after Porphurios' departure (*Vit. Plot.* 4–6). Porphurios reports that Plōtinos, who started to write at age 50, composed carelessly (*ibid.* 8) and his writings needed editorial care, provided by both AMELIUS GENTILIANUS (*ibid.* 19–20) and Porphurios, Plōtinos' most loyal students (*ibid.* 24). Plōtinos' work survives today in the arrangement of Porphurios' edition published *ca* 300–305. Porphurios arranged Plotinos' treatises into six groups of nine treatises (*Enneads*), because he regarded the numbers six and nine as perfect, symbolizing the perfection of Plōtinos' philosophy. Pedagogically the arrangement guides the reader to the heights of philosophy, the vision of the ultimate divine entity, the One. The first *Ennead* treats ethics, the second physics, the third cosmology, the fourth the soul, the fifth the intellect, and the sixth the One. Yet this division does not correspond to the treatises as Plōtinos wrote them, since sometimes Porphurios gathered his mentor's notes (*Enn.* 3.9) but more often divided longer treatises into smaller pieces (e.g. *Enn.* 3.8, 5.8, 5.5, 2.9)

Plōtinos intended to elucidate and expound Plato's philosophy, not to create a new one (*Enn.* 5.1.8.10–14). His understanding of Plato is much indebted to earlier **Platonists** especially PLUTARCH and Noumēnios, but also to Aristotle and **Peripatetics** like Alexander. Plōtinos tried to systematize various ideas in Plato's work, defending them against **Peripatetic, Stoic**, and other critics. For Plōtinos only what subsists of its own is a substance, a ***hupostasis***, and as such only intelligible entities qualify; but given that intelligible entities have different degrees of unity and simplicity, there are higher and lower entities representing different degrees of reality. Plōtinos maintained the existence of three divine ***hupostaseis***, the One, the Intellect, and the Soul, from which everything else results. Inspired by Plato's *Parmenides*, Plōtinos, like Noumēnios, postulated the existence of the One, which he identified with the Form of the Good of *Republic* 6 and which he considered the ultimate cause of everything in the intelligible and the sensible world. The One is claimed to be above the Intellect, or the divine **demiurge**, first because the Intellect acting under constraints, such as matter, is incompatible with the unlimited freedom that the highest God merits; and second because an intellect implies dualism, since it has thoughts, while the first principle must be utterly simple and united. The Intellect is characterized by non-discursive thinking (*noesis*), while the Soul displays discursive or *dianoetic* thinking. Below the Soul lies Nature maintained by the higher ***hupostaseis***. The ***hupostaseis*** play a role also in Plōtinos' cosmology. Plōtinos argues for the everlastingness of the universe, the heavens, and the heavenly bodies, which means that all of them persist and retain their individual identity over time because they are ultimately ontologically dependent on the World-Soul, which in turn is dependent on the Intellect. The crux of Plōtinos' philosophy is his psychology. Plōtinos seems to approach the question of how the intelligible realm relates to the sensible one by investigating the relation between soul and body. Plōtinos' preoccupation with the soul was both metaphysical and ethical. He distinguished between inner and outer man, and he identified man's self with the former which is the soul. By "soul" is not intended the embodied soul which enlivens the body, but rather the transcendent, intellective one, from which the embodied emanates. Man's aim, according to Plōtinos, is to achieve unity with the One (*Enn.* 1.4.3, 1.4.10).

Plōtinos' philosophy exerted enormous influence on later generations of **Platonists**, leading historians of philosophy to consider Plōtinos the founder of a distinct version of

Platonism, Neo-**Platonism**, a label to be used with caution: first because Plōtinos did not aim to create a new interpretation of Plato, "Neo-**Platonism**," and second because much of this is anticipated by earlier **Platonists** including Noumēnios.

Ed.: P. Henry and H.-R. Schwyzer, *Plotini Opera* (1964–1982).
A. Armstrong, *The Architecture of the Intelligible Universe in the Philosophy of Plotinus* (1940); *RE* 21.1 (1951) 471–592, H.-R. Schwyzer; H. Blumenthal, *Plotinus' Psychology. His Doctrine of the Embodied Soul* (1971); E. Emilsson, *Plotinus on Sense Perception* (1988); D.J. O'Meara, *Plotinus. An Introduction to the Enneads* (1993); L. Gerson, *Plotinus* (1994); *OCD3* 1198–1200, J. M. Dillon; P. Hadot, *Plotin ou la simplicité du regard* (1997); J. Wilberding, *Plotinus' Cosmology. A Study of Ennead II.1 (40)* (2006); *BNP* 11 (2007) 395–403, P. Hadot.

George Karamanolis

Ploutarkhos of Athens, son of Nestorios (d. 432 CE)

Neo-**Platonist** philosopher from Athens, taught (in his own house) Hieroklēs of Alexandria, SYRIANUS, PROKLOS, and his own daughter Asklēpigeneia; was acquainted with DOMNINOS. He wrote commentaries on PLATO (*Gorgias, Phaedo* and *Parmenides*) and ARISTOTLE (at least on *De Anim.* 3), of which only fragments are preserved in later commentators. His successors, especially Proklos, esteemed him highly, and his work's main focus seems to have been harmonizing Platonic and Aristotelian doctrines, partly based on IAMBLIKHOS. Ploutarkhos was the main source of the revival of **Platonism** in Athens.

RE 21.2 (1952) 962–975 (#3), R. Beutler; D.P. Taormina, *Plutarco di Atene, L'Uno, l'Anima, le Forme* (1989); *ECP* 429–430, H.J. Blumenthal; *BNP* 11 (2007) 426–427 (#3), H.D. Saffrey.

Cosmin Andron

PLOUTARKHOS OF KHAIRONEIA ⇒ PLUTARCH

Plutarch of Khairōneia, L. Mestrius (*ca* 80 – 120 CE)

Born *ca* 46, biographer and **Platonic** philosopher. At the time of Nero's visit to Greece (66/67 CE), Plutarch was not older than 20 (*De E* 385B). Born to a wealthy family, he held various public offices: a mission to the proconsul of Achaia (*Praec. ger. reip.* 816B), *agoranomos* and eponymous archon in Khairōneia (*Quaest. conv.* 642F; 693F), Boeotarch, and probably several times president of the Amphictyony. Hadrian entrusted the government of Greece to Plutarch (119 CE: EUSEBIOS, *Chron.* 2135 *ab Abr.*). Trajan elevated him to consular status (*Souda* Pi-1794). Plutarch counted influential Romans, such as Sosius Senecio, Fundanus, and Mestrius Florus, among his friends. Plutarch's *nomen gentilicium* Mestrius indicated Roman citizenship (*CIG* 1713). With his wife Timoxena he had four sons and a daughter, who died young, like two of her brothers. Plutarch himself died between 119 and 127.

Plutarch studied under AMMŌNIOS ANNIUS in Athens but resided mostly in Khairōneia, where he established a philosophical school, and in Delphi, where he held a priesthood (at the latest from the beginning of Hadrian's reign – *cf. CIG* 1713 – but probably already long before: *An seni* 792F). He traveled to Egypt (*Quaest. conv.* 678C), Asia Minor (*An. an corp.* 501E) and several times to Rome (*Demosth.* 2.2). His extant writings include 50 biographies (23 parallel lives, *Vitae*, and lives of Otho, Galba, Aratos and Artaxerxēs) and various other works belonging to different genres (in modern editions known as *Moralia*). The dialogues portray Plutarch's circle of friends and students.

Plutarch's philosophy of nature is largely based on PLATO's *Timaeus* but is also influenced by the skepticism of the Hellenistic **Academy** and especially by the fallibilism characterizing its final phase. Any enquiry into the physical world and its physical causes can merely attain probability (*cf. De def. or.* 435E–436A; *Ti.* 68e–69a; *Phaed.* 97b–99d). The philosopher, however, should also look for final, i.e. teleological, causes. The fallible character of any inquiry on the level of material causes enables Plutarch to give serious consideration not only to Plato's views, but also to physical doctrines of the **Peripatetic, Stoic** and **Epicurean** schools. Even if their specific doctrines find provisional acceptance, they remain subordinated to Plutarch's overall **Platonism**.

In *De facie in orbe lunae*, Plutarch, citing HIPPARKHOS and ARISTARKHOS, discusses astronomy, geography, and catoptrics. Lunar phenomena, Plutarch argues, show that the Moon's constitution is earthy. He mentions the theory that the Moon's velocity prevents it from falling, rejects the Aristotelian doctrine of natural motion, discusses distances between heavenly bodies, size, position and shape of the Earth, the existence of the "antipodes," lunar phases, solar and lunar eclipses, the habitability of the Moon, lunar vegetation, the apparent face in the Moon (the great ocean reflected in the Moon, according to an Aristotelian speaker). In the introduction to the concluding myth a trans-Atlantic continent and islands westward of Britain are mentioned. Plutarch rejects motion of the Earth (*Quaestiones Platonicae* 1006C–E; citing Aristarkhos, SELEUKOS, THEOPHRASTOS). He also treats the parts of speech (1009B–1011E), and *antiperistasis*, a Platonic theory to explain the properties of magnets and amber, the motion of projectiles and thunderbolts, the working of cupping-instruments, the perception of consonance (1004D–1006B; *cf. Ti.* 79e–80c). In *De animae procreatione*, Plutarch discusses arithmetical problems related to Plato's harmonic division of the soul. In *Quaestiones naturales*, he discusses various issues, including agriculture, zoology, medicine, meteorology, fishing, hunting, cooking, properties of sea water, many previously addressed by ARISTOTLE or Theophrastos or "Laitos" (probably OFELLIUS LAETUS). Plutarch often offers original solutions, probably of his own making. *Quaestiones Convivales* address medicine, botany, zoology, physics in general, and astronomy. Plutarch considers comparable subjects in some of the fragments of his commentaries on HOMER (*fr.*127), HĒSIOD (*fr.*75–76, 80–81, 102, 104) and NIKANDROS' *Thēriaka* (*fr.*113–114). The scholia (*fr.*13–20) preserve excerpts from notes on ARATOS' *Diosemiae*. In *De primo frigido*, Plutarch searches for the primarily cold element: not air (the **Stoics**), not water (EMPEDOKLĒS, STRATŌN OF LAMPSAKOS), but earth. The essay closes with an appeal to suspend judgment. In *De sollertia animalium* and *Bruta animalia ratione uti*, Plutarch upholds the intelligence of animals against the **Stoics**. *De tuenda sanitate praecepta* provides dietary advice.

Plutarch occasionally addresses scientific theories in the anti-**Stoic** works *De Stoicorum repugnantiis* (rejecting the **Stoic** view on the role of air in the animation of the fetus, and of air as primarily cold: 41–43), and *De communibus notitiis* (on mixture, the divisibility of body, the continuum, the structure of matter: 37–43; 49–50). Plutarch inserts a treatment of the **Stoic** hypothetical syllogism and speculations on the number five in *De E Delphico. De Pythiae oraculis* opens by discussing the atmospheric conditions in Delphi giving bronze a peculiar patina and then moves on to exhalations as material causes for the oracle. This issue also features in *De defectu oraculorum*, which, moreover, describes the lamps at the shrine of Ammōn consuming less and less olive oil every year (410B).

The handbook *On music* and the geographical work *On rivers* (*De fluuiis*) are spurious: see the next two entries.

RE 21.1 (1951) 636–962, K. Ziegler; P.L. Donini, "Science and metaphysics. Platonism, Aristotelianism, and Stoicism in Plutarch's *On the face in the moon*," J.M. Dillon and A.A. Long, edd., *The Question of "Eclecticism"* (1988) 126–144; P.L. Donini, "I fondamenti della fisica e la teoria delle cause in Plutarco," in I. Gallo, ed., *Plutarco e le scienze* (1992) 99–120; *OCD3* 1200–1201, D. Russell.

Jan Opsomer

Plutarch (?), *On Music* (*ca* 150 CE)

The dialogue *On music*, included among the *Moralia* by tradition, is rejected by current scholarship as an authentic work of PLUTARCH. Nevertheless, a number of authentic treatises contain important information on **Pythagorean** mathematics and music (*On the generation of the soul in the Timaeus*), the ethical effect and value of music in society (*Table-Talk*), and the history of musical instruments (*Ancient customs of the Spartans; Life of Crassus; On progress in virtue; On the control of anger*).

Regardless of its author, *On music* is in a sense the earliest "history" of Greek music and a prime source of information on ancient Greek musical life, including historical material on **Pythagorean** music theory, the "invention" of musical forms, and the development of early musical scales. Some of this material is attributed to now-lost works by ALEXANDER OF MILĒTOS, ARISTOXENOS, Glaukos of Rhēgion, and HĒRAKLEIDĒS OF HĒRAKLEIA PONTIKĒ, JUNIOR. The two primary speakers in the dialogue, Lysias and Soterichus, represent respectively the practical and theoretical viewpoints of music and its development. After describing various musico-poetic forms and attributing them to early "inventors," Lysias explains the construction of the enharmonic genus, its relationship to the other genera, and a special "spondeion" scale, the precise structure of which remains obscure. Soterichus expands on Lysias' practical presentation, correcting and augmenting his descriptions of the musico-poetic forms and the spondeion scale. He subsequently turns his attention to the realm of **Pythagorean** mathematics and music, concluding that music should be elevating, instructive, and useful. Modern musical innovations have led music to its present low estate, aptly represented by the famous fragment from the *Cheiron* of Pherecratēs. Music must be restored to its proper place by copying the ancient style, following the guidance of philosophy. Reviewing the principles of harmonics and rhythmics, Soterichus recognizes that this knowledge is insufficient alone for the creation or judgment of musical art and yields to the precentor Onēsicratēs, who provides the philosophical capstone of the dialogue: PYTHAGORAS, PLATO, and ARKHUTAS have revealed that music is of value because the revolution of the universe is based on music and god has arranged everything to accord with harmonia (*kath' harmonian*).

K. Ziegler, *Plutarchi Moralia* 6.3 (1966); Barker (1984) 1.205–257; *NGD2* 19.931–932; Mathiesen (1999) 355–66; *MGG2* 13.698–699.

Thomas J. Mathiesen

Plutarch, pseudo, De Fluuiis (300 CE?)

The anonymous collection *De Fluuiis* is divided into 25 chapters containing etiological myths about the names of as many streams in Greece, Gaul, Asia and Egypt, with added information about unusual or wonderful stones, metals and plants found in those rivers or on nearby mountains. The work survives in a single MS preserving several other paradoxographical treatises (*Palat.* gr. 398). Numerous source-citations, underscoring the collection's credibility, further place it squarely in the *mirabilia*-tradition. However, while some of the acknowledged writers (46) and books (65) *peri potamōn, peri lithōn, peri orōn* etc. are possibly real sources (AGATHŌN OF SAMOS, ARKHELAOS OF KAPPADOKIA, PUTHOKLĒS OF SAMOS,

Sōstratos of Nusa, and Thrasullos of Mendēs), the great majority of them have long since been exposed as figments of the author's imagination, only to be found in *De Fluuiis* and another mediocre Pseudo-Plutarchan writing obviously coming from the same pen (*Parallela minora*). As such, *De Fluuiis* is merely pseudo-paradoxographical, a gratuitous concoction for a gullible, sensation-seeking audience.

Ed.: N. Bernardakis, *Plutarchos, Moralia* 7 (1896) 282–328.
RE 18.3 (1949) 1137–1166 (§34, 1164), K. Ziegler; *RE* 21.1 (1951) 636–962 (#2; III.10f-g, 867–871), *Idem; BNP* 11 (2007) 424 (#2, IV.A), E. Olshausen.

Jan Bollansée, Karen Haegemans, and Guido Schepens

Podanitēs (250 BCE – 80 CE)

Andromakhos, in Galēn *CMLoc* 7.4 (13.115 K.), quotes his remedy for ***duspnoia***: dissolve mustard, salt, and natron in water, drink often. The name seems otherwise unattested: *cf.* Podanikos (*LGPN* 1.374), Podanemos (*LGPN* 3A.365), or perhaps Podarēs (*ibid.*).

Fabricius (1726) 375.

PTK

Polemarkhos of Kuzikos (360 – 330 BCE)

Follower of Eudoxos, classmate or teacher of Kallippos (Simplicius, *In de caelo*, = *CAG* 7 [1894] 492). He noted the apparent variation in the brightness of planets but dismissed it as evidence for variation in the distances of planets from the Earth, on the grounds that the distance is not really observable, and so defended Eudoxos' theory of homocentric spheres (*ibid.*, p. 505).

Simplicius, *On Aristotle's "On the Heavens 2.10–14,"* trans. Ian Mueller (*ACA* 2005).

Henry Mendell

Polemōn of Athens (*ca* 345 – 270/269 BCE)

Followed his teacher Xenokratēs as **scholarch** of the **Academy** in 314/313 BCE until his death in (or near) 270/269 BCE. Sources other than Diogenēs Laërtios (4.16–20) and Cicero are limited. Until recently, Polemōn had been considered mainly a moral philosopher (following D.L. 4.18), but Cicero's mentor Antiokhos of Askalon stressed wide-ranging connections between Polemōn and Zēnōn the **Stoic**, once his pupil. Sedley has now persuasively argued that the account of Platonic physics offered by Cicero (*Acad.* 1.24–29), attributing to Plato a single organic universe with two principles, active and passive, operating in close conjunction within it, and hitherto seen as Antiokhos' own stoicizing contribution, is in fact an account of **Academic** physics under Polemōn. Like Zēnōn, Polemōn had idealized "life according to nature (***phusis***)," and written a book on the subject (Clement, *Strom.* 7.6.32.9). The passage identifies matter with the passive principle, while the other principle seems to be given a sentient nature and perfect reason, called the "soul of the world," god, or providence. It also makes much use of the idea of bodies as *qualified* matter, using earth, air, fire and water as "elements," two more active and two more passive, mentioning in addition the fifth astral body of Aristotle. These are distinguished from compound bodies. While the attribution to Polemōn is only hypothetical, it accords with trends already found in Xenokratēs and can be reconciled with early **Academic** non-literal readings of the *Timaios*' creation-process.

As an ethicist, Polemōn is associated with the idea that, whereas virtue is the most important object of pursuit, being sufficient for happiness, humans must also give thought to the provision of the "first things according to nature," various desirable things promoting life from its very first stages. His definition of love, associating it with service to the gods in looking after the youth, is preserved by PLUTARCH, *Uneducated Ruler* 3 (780D), and the **Academy** under him was characterized by a number of prominent male-to-male relationships, including his own with Xenokratēs and then his successor as **scholarch**, Kratēs, causing Tarrant (*JHP* 43 [2005] 131–155) to postulate a relationship with the pseudo-Platonic *Theagēs*, combining the divinely inspired Sōcratēs with the erotic one.

Ed.: M. Gigante, *Polemonis Academici Fragmenta* (1977).
Dillon (2003) 155–177; D.N. Sedley, "The Origins of the Stoic God," in M. Frede and A. Laks, edd., *Traditions of Theology: Studies in Hellenistic Theology, its Background, and Aftermath* (2002) 41–83.

Harold Tarrant

Polemōn of Ilion (190 – 160 BCE)

Wrote a set of ***periēgētic*** works on various Greek lands, explaining myths and customs, a number of scholarly works, and paradoxographical works. Over 100 fragments survive, in authors from STRABŌN to MACROBIUS, and he was described by PLUTARCH, *Q.Conv.* 5.2 (675B), as "of wide learning, tireless, and accurate." From his paradoxographical work *Rivers in Sicily*, Macrobius, *Sat.* 5.19.26–30, preserves a long fragment; see also Ath., *Deipn.* 7 (307b).

BNP 11 (2007) 458–459 (#2), A.A. Donohue.

PTK

Polemōn of Laodikeia on the Lukos ("Antonius Polemo") (*ca* 110 – 144 CE)

Born *ca* 88 CE from an influential family, Polemōn was a rhetor and prominent politician in Smurna and a representative of the Second Sophistic (Philostratos, *Vit. Soph.* 1.23). He enjoyed privileged access to power through his friendship with Hadrian, but also Trajan and Antoninus Pius; he died in 144 CE. From his rhetorical work, only two short declamations and several fragments have survived, showing him as preferring the "Asian" style of brief sentences and rhetorical *tropoi*.

He is reported to have been extraordinarily conscious of comportment and self-representation, so it is not surprising that he also was a physiognomist. His written work on physiognomy can be reconstructed from one brief fragment, an Arabic translation and a Greek paraphrase by ADAMANTIOS. It appears to have been silently based on the ARISTOTELIAN CORPUS PHYSIOGNOMY, and interspersed with anecdotes from Polemōn's own travels. Most prominent in his physiognomical method is the observation of signs in the eyes, comprising one-third of the text (in the Arabic version). Polemōn also reviews the character traits of 92 animals, signs of the various parts of the body, ethnographic differences, the color of skin and eyes, the significance of hair on head and body, and features of comportment such as body movement, gait, gesture and voice. As in the Aristotelian Corpus *Physiognomy*, Polemōn considers it essential to "seek an overall impression (*epiprépeia*) so that you may apply it to the body the way a signet ring is applied to material on which it is to print" (1 [1.168 F.] Arabic; *cf.* 2.1 [1.348–9. F.] Adamantios).

Ed.: R. Hoyland, "A New Edition and Translation of the Leiden Polemon," and A. Ghersetti, "The Istanbul Polemon (TK Recension): Edition and Translation of the Introduction," in Swain (2007) 328–463 and 465–485.

KP 4.972–973 (#5), H. Gärtner; *OCD3* 1204 (#4), D.A.F.M. Russell; *BNP* 11 (2007) 460–461 (#6), E. Bowie; M.W. Gleason, *Making Man. Sophists and Self-Representation in Ancient Rome* (1995); S. Swain, "Polemon's Physiognomy," in Swain (2007) 125–201.

Sabine Vogt

Politēs (250 BCE – 200 CE)

Cited by the PARADOXOGRAPHUS VATICANUS 3 for the claim that the tuna-fry in the Pontos are generated from mud (*pēlos*), hence their name "*pēlamus*." The name is rare after the 1st c. BCE: *LGPN* (-tas, -tēs, and -tis).

(*)

PTK

Pollēs (Med.) (120 – 365 CE)

OREIBASIOS, *Syn.* 3.13 (*CMG* 6.3, p. 65), cites his beeswax-based "crane" remedy composed of **verdigris**, realgar, *sandux* (minium, i.e., red lead [lead tetroxide], prepared by roasting **litharge** or ***psimuthion*** in air, below 500 °C: DIOSKOURIDĒS, *MM* 5.88), **terebinth**, and notably a crane wing, combusted in a sealed ceramic jar, all powdered. His recipes for a chest-wound plaster and for an ointment for headaches and migraines, *ibid.* 3.15–16 (pp. 67–68) contain primarily botanicals, but the *Schol. Oreib. Coll.* 45.21.1 (*CMG* 6.2.1, p. 177) says he used a mixture of pigeon dung, barley-chaf fand water on scrofula. AËTIOS OF AMIDA preserves his *digestif* salt, containing calamint, chamomile-flower, eryngo-root, *konuza* (*cf.* PLINY 21.58; GALĒN, *Simples* 7.10.42 [12.35–36 K.]; André 1985: 74; Durling 1993: 209: either *Inula viscosa* Aiton ["fleabane"] or *Inula graveolens* Desf.), marjoram, pepper, and ***silphion***, plus roasted salt: 9.24 (p. 507 Cornarius; omitted by Zervos 1911: 324–325), and seven complex softening ***diaphorētikē*** plasters (15.15, Zervos 1909: 73–75, 77–80, 82–83). PAULOS OF AIGINA, 4.16 (*CMG* 9.1, p. 334), places him after ARKHIGENĒS.

(*)

PTK

Pollēs of Aigai, pseudo (80 – 120 CE)

Wrote three extensive volumes *Concerning antipathies and* ***sympathies*** (*Souda*, O-163), including a fragment *Roadside Augury* (*Enodion Oïōnisma*). Along with MELAMPOUS, Pollēs was considered infallible (*Souda*, M-448; *cf.* MARINOS, *V. Prokl.* 10).

RE 21.2 (1952) 1410–1411 (#1), K. Scherling; Ullmann (1972) 394.

GLIM

POLLIO ⇒ (1) ASINIUS; (2) VITRUUIUS

Pollis (or Pollēs?) (300 – 25 BCE)

Compiled rules of architectural symmetry and proportion (VITRUUIUS 7.*pr.*14), perhaps identifiable with the homonymous sculptor (*post* 300 BCE) of athletes, hunters, warriors, and men offering sacrifices, whom PLINY (34.91) lists among minor sculptors.

RE 21.2 (1952) 1417–1418 (#4), G. Lippold.

GLIM

Poluainos of Lampsakos (310 – 275 BCE)

Epicurean philosopher, he became EPICURUS' student when Epicurus was teaching at Lampsakos *ca* 310–307, and moved to Athens with him in 307. Epicurus was said to have turned Poluainos' interests from mathematics to philosophy (CICERO, *Lucullus* 106). Along with Epicurus, HERMARKHOS, and MĒTRODŌROS, he was known as one of the four founders of the **Epicurean** school. He wrote a number of works, including *On Definitions*, *On Philosophy*, *Against Aristōn* (target uncertain), and *Aporiai* (*Puzzles*).

Ed.: A. Tepedino Guerra, *Polieno: Frammenti* (1991).
OCD3 1209, D. Obbink; *ECP* 445–446, D.N. Sedley; *BNP* 11 (2007) 494–495 (#1), T. Dorandi.

Walter G. Englert

Poluarkhos (30 BCE – 35 CE)

Widely-cited pharmacist. CELSUS 5.18.8 (*cf.* 8.9.1D) gives his softening *malagma* (resin, beeswax, cardamom, *kuperos*, etc.); ANDROMAKHOS, in GALĒN *CMGen* 7.7 (13.981 K.), offers his ointment; and ASKLĒPIADĒS PHARM., in Galēn *CMLoc* 8.5 (13.184–185 K.), reports two internal remedies, one with **bdellium**, saffron, cinnamon-wood, Indian nard, etc., the second revised by IULIUS AGRIPPA. MARCELLUS OF BORDEAUX 20.149 (*CML* 5, p. 372) reports another: cardamom, cassia, galingale, ***malabathron***, roses, etc. SŌRANOS, *Gyn.* 3.32, 3.38 (*CMG* 4, pp. 115, 118 = *CUF* v. 3, pp. 35, 41; along with KĒPHISOPHŌN's ointment), OREIBASIOS, *Syn.* 9.43.19 (*CMG* 6.3, p. 303), AËTIOS OF AMIDA 8.63 (*CMG* 8.2, p. 512), and PAULOS OF AIGINA 3.74.3, 7.18.4–5 (*CMG* 9.1, p. 292; 9.2, pp. 369–370) prescribe Poluarkhos' remedies.

RE 21.2 (1952) 1439–1440, H. Diller.

PTK

Polubios of Megalopolis (*ca* 180 – 118 BCE)

Son of Lukortas, one of the leaders of the Akhaian Confederation.

Biography: In 182 Polubios buried the ashes of Philopoimen, well-known general of the Confederation. Two years later, Polubios was appointed envoy to Alexandria and in 170 served as general of cavalry of the Confederation. After the Roman victory over Perseus of Macedon in 168, Polubios was deported to Rome (due to insufficiently good relations with the Roman occupier), together with a thousand elite Akhaians. Polubios became friend and mentor of P. Cornelius Scipio Aemilianus. In his years as a political prisoner of Rome (167–150 BCE), Polubios made several journeys: through Africa, Spain, Gaul and the Alps. On his 151 BCE visit to Spain, accompanying Scipio, he probably visited New Carthage. In 149, released from exile, he was asked to come to Lilubaion in Sicily. He arrived in Kerkura/Corcyra, was informed that Carthage had accepted Roman terms and returned home to Arkadia. When the war resumed, Polubios came to Carthage and after its destruction in 146 BCE he went on a voyage of discovery through the Pillars of Hēraklēs (Gibraltar), up the coast of Portugal and back along the African coast as far as the river Lixos (in Morocco), attempting to locate Mount Atlas more accurately than had yet been done. Later Polubios visited Asia Minor. He was also at Alexandria and probably Buzantion. He may have visited Scipio's camp at Numantia in 133 BCE. Polubios died at the age of 82 after a fall from a horse.

Works: (1) *Histories* in 40 books beginning chronologically where Timaios left off (264 BCE) and concentrating on the swift ascendance of Rome as a world power; lost works: (2) biography of Philopoimen; (3) *On Tactics*, a history of the Numantine War; (4) on living conditions in the equatorial region.

Contribution: Polubios expanded the role of geography within historiography by reducing the traditional use of geographical digressions and assigning a separate section of his *Histories* to geography. The idea probably came from Ephoros, whose Books 4–5 were exclusively geographical. Polubios concentrated his geographical descriptions and discussions in *Histories* Book 34, surviving only as paraphrases in Strabōn, Pliny and Athēnaios *Deipnosophists*; it contained a geographical survey of the entire ***oikoumenē*** and a detailed description (*khorographia*) of Europe and Africa. Polubios' pragmatic attitude towards geography made him less interested in scientific and theoretical discussions of geography and topography and more in information aimed at increasing his readers' knowledge of remote and little known regions. Much of the book comprised attacks on previous geographers, discussions of theory, practical details concerning distances and topography. Like his contemporary, Kratēs of Mallos, Polubios emphasized the role of Homer in geographical tradition, drew geographical information from the *Iliad* and *Odyssey* and dealt at length with locating Odysseus' wanderings. Polubios wrote of latitudinal climatic zones and their influence on the character of their human inhabitants and on animals and plants. He divided the globe into six zones unlike Strabōn who preferred Poseidōnios of Apameia's five-zone division. Polubios used astronomical methods to measure the length of the ***oikoumenē*** and a system of triangulation to describe the main outlines of Italy (2.14.4–6). He did not make a serious contribution to the scientific study of geography but recorded topographical details as well as assembling and comparing distances.

F.W. Walbank, "The geography of Polybius," *C&M* 9 (1948) 155–182; P. Pédech, "La géographie de Polybe: Structure et contenu du livre XXXIV des Histoires," *LEC* 24 (1956) 3–24; Walbank v.3 (1979).

Daniela Dueck

Polubos (420 – 350 BCE)

Greek physician, son of Apollōnios, credited by Aristotle (*HA* 3.3 [512b–513a]) with the description of blood vessels preserved also in the Hippokratic On the Nature of Man 11 (and *On the nature of bones* 9). He argues that all vessels originate in the head. A later biographical tradition (*Vita Hp. Bruss.* 1; Hipp., *Letter* 27) considers him Hippokratēs' pupil and son-in-law; Polubos remained at Kōs, heading the school after Hippokratēs. Accordingly, Polubos was credited with the authorship of Hippokratic works (*On the Nature of Man* partly, *Nature of the Child, On Birth in the Eighth Month*): see Galēn 4.653, 18A.8 K., *CMG* 5.9.1, p. 8; pseudo-Plutarch *Plac.* 5.18. But he was probably connected with Hippokratēs only later, because the "Aristotelian" doxography in the Londiniensis medicus 19.2–18

K. separates them neatly: Polubos admits four qualities (hot, cold, dry, moist) and four substances (yellow bile, black bile, phlegm and blood); diseases arise from alterations in the manner in which substances are blended and are differentiated according to where **humors** derive and where they end up: a humoral pathology similar to *Nature of Man* 3–4.

H. Grensemann, *Abh. Ak. Wiss. Mainz, geist. u. sozialw. Kl.* (1968) 2; J. Jouanna, "Le médecin Polybe est-il l'auteur de plusiers ouvrages de la Collection Hippocratique?" *REG* 82 (1969) 552–562; *Idem, Hippocrate. La nature de l'homme* (*CMG* 1.1.3) 56; *KP* 5.1639 (#4), J. Kollesch; *RE* S.14 (1974) 428–436, H. Grensemann; *OCD3* 1211, J.T. Vallance; *AML* 723–724, C. Oser-Grote; *BNP* 11 (2007) 504–505 (#6), V. Nutton.

Daniela Manetti

Poludeukēs (250 BCE – 565 CE)

ALEXANDER OF TRALLEIS (2.15 Puschm.), in a series of collyria containing saffron, ***glaukion***, and ***sarkokolla***, gives his recipe additionally including fresh roses, gum, and opium. *Cf.* the "Parrot" collyrium of HĒRŌN (MED.), and SERGIUS OF BABYLŌN.

(*)

PTK

Polueidēs (250 BCE – 25 CE)

The "*sphragis*" of Polueidēs – composed of aloes, alum, ***khalkanthon***, myrrh, pomegranate flowers, and bull gall, ground in dry wine – is repeatedly prescribed for wounds by pharmacists from CELSUS 5.20.2 and ANDROMAKHOS, in GALĒN *CMGen* 5.12 (13.834 K.), through PAULOS OF AIGINA 7.12.21 (*CMG* 9.2, p. 318). PHILOUMENOS 17.9 (*CMG* 10.1.1, p. 24) cites his remedy for snake bite: drink *alkibiadion* juice and apply the mash to the wound; Paulos mentions his cream for ***anthrax***, 4.25.2 (9.1, pp. 346–347). Diller argued he was merely a brand-name, for the mythical early doctor, as at PSEUDO-GALĒN, INTRODUCTIO (14.675 K.), but Galēn himself cites this Polueidēs as an individual, and places him after ANDRŌN: *CMGen* 3.3 (13.612 K.), *CMLoc* 3.1, 3.3 (12.611, 690–191 K.).

RE 21.2 (1952) 1661–1662 (#12), H. Diller.

PTK

Poluidos of Thessalia (360 – 320 BCE)

Poluidos, whose students included KHARIAS and DIADĒS, designed siege machines for Philip II for the siege of Buzantion in 340–339 BCE (VITRUUIUS 10.13.3), and wrote *de machinationibus* (ATHĒNAIOS MECH. p. 10 W.; Vitruuius 7.*pr.*14), of which nothing survives.

BNP 11 (2007) 527 (#4), M. Folkerts.

GLIM

Polukleitos of Argos or Sikuōn (*ca* 460 – 415 BCE)

Polukleitos was born in Argos or Sikuōn around 480 BCE and died probably in 415 BCE. He sculpted almost exclusively in bronze and preferably standing virile naked youths. The main motif in his sculpted work is the contrapost position which had already been invented and which Polukleitos brought to perfection. According to GALĒN (*PHP* 5.3.16 = *CMG* 5.4.1.2, p. 308), he wrote a treatise entitled *Kanōn*, arguing that "beauty lies in the proportion (*sum-*

metria) of the members: of finger to finger, of all the fingers to the palm and wrist, of these to forearm, of forearm to upper arm, and of all to all." Hippocratic medicine can be regarded as an inspiration for the way in which Polukleitos conceptualized and organized the body, in the tract as well as in sculpture. In a conjectural reconstruction of the outline of the *Kanōn* from all known and attributed sources (Stewart 1998: 273–275), it emerges that Polukleitos obsessively insists on exactitude at all levels in order to achieve perfection and beauty.

Ed.: DK 40.
J.J. Pollitt, "The Canon of Polykleitos and Other Canons," in: W.G. Moon, ed., *Polykleitos, the Doryphoros, and Tradition* (1995) 19–24; *OCD3* 1211–1212, A.F. Stewart; *BNP* 11 (2007) 511–513 (#1), R. Neudecker.

Sabine Vogt

Polukleitos of Larissa (*ca* 360 – 300 BCE?)

Father of Olumpias (mother of Antigonos), captured in Alexander's campaigns in Kurēnē (*FGrHist* 128T1), among STRABŌN's geographical sources, and cited as a foreign authority on trees (PLINY 1.*ind.*12–13). His *Historiae* described Persia, Mesopotamia, and India, and treated Alexander the Great's luxurious lifestyle (Ath., *Deipn.* 12 [539a]; PLUTARCH, *Alex.* 46). Polukleitos claimed that the Euphratēs did not overflow, due in part to the mountains' low altitudes and shallow snow cover, and its drainage into the Tigris and flooding of the plains; dismissed as absurd by Strabōn (16.1.13). Polukleitos described the mountainous terrain between Susa and Persis and the river Choaspēs, which flowed through Susis, meeting the Tigris and Eulaeos in a lake subsequently emptying into the sea (Str. 15.3.4). He proffered the opinion that the serpent-producing Caspian with its sweet water was in fact Lake Maeotis, into which the Tanais poured (Str. 11.7.4). He reported large, polychrome, dappled Indian lizards, soft to the touch (AELIANUS, *NA* 16.41) and tortoises from the Ganges whose shells could hold 5 *medimnoi* (PARADOXOGRAPHUS VATICANUS 10). It is likely that "Polukleitos of Liparis," Pliny's authority on the river Liparis in Soloi near Kilikia (31.17), is identifiable with our geographer.

FGrHist 128.
RE 21.2 (1952) 1700–1707 (#7–8), Fr. Gisinger; Pearson (1960) 70–77; *KP* 4.999 (#3), H. Gärtner; *OCD3* 1211, A.B. Bosworth; *BNP* 11 (2007) 514 (#4), E. Badian.

GLIM

Polukritos of Mendē (400 – 350 BCE)

Historiographer whose treatise on water-*mirabilia* survives in two fragments. He wrote a *Sikelika* in verse with an obvious leaning towards natural sciences and paradoxography.

Ed.: *FGrHist* 559.
RE 21.2 (1952) 1760–1761 (#7, 8), K. Ziegler.

Jan Bollansée, Karen Haegemans, and Guido Schepens

Polustomos (250 BCE – 80 CE)

ANDROMAKHOS, in GALĒN *CMGen* 6.14 (13.931 K.), records his recipe for gout (*podagra*): ***aphronitron*** and ***psimuthion*** in beeswax, aged olive oil, and **terebinth**. The name seems otherwise unattested, and may be distorted from, e.g, Polustratos (though no pharmacist by that name is otherwise known) or perhaps PLATUSĒMOS.

(*)

PTK

Polustratos (275 – 225 BCE)

Epicurean philosopher and student of EPICURUS, he became the third **scholarch** of the **Garden** upon the death of HERMARKHOS (DIOGENĒS LAËRTIOS 10.25). Little is known of his life, but two titles are attested: *On Philosophy* and *On the Unfounded Contempt of Commonly Held Beliefs*. The latter is preserved in fragments among the Herculaneum papyri, and attacks skeptical philosophers who denied that knowledge could be based on the senses.

Ed.: G. Indelli, *Sul Disprezzo Irrationale* (1978).
RE 21.2 (1952) 1833 (#7), H.J. Mette; Long and Sedley (1987) §7D; *OCD3* 1213, D. Obbink; *ECP* 446, D. Clay; *BNP* 11 (2007) 533–534 (#2), T. Dorandi.

Walter G. Englert

POLY- ⇒ POLU-

POMPEIUS ⇒ THEOPHANĒS

Pompeius Lenaeus (70 – 40 BCE)

A learned freedman of Pompey Magnus who at the latter's request translated into Latin MITHRADATĒS VI's pharmacological writings, which PLINY (1.*ind.*14–15, 20–27, 25.5–7) utilized. A fragment describes *mustax* – a variety of laurel with pale, drooping leaves – distinguished from the more familiar Delphic and Cyprian laurels mentioned by CATO; another sketches the appearance and medicinal properties of a plant from Pontos called *scordotis* or *scordion* (15.127, 25.63). The famous story of Mithradatēs' efforts to immunize himself against poisons by consuming them in minute quantities comes from this work (25.6; Gellius 17.16).

Speranza (1971) 63–65; *KP* 3.556, W. Richter; *OCD3* 1215, J.W. Duff; *BNP* 11 (2007) 386 (#2), Ed. Courtney.

Philip Thibodeau

Pompeius Sabinus (90 – 110 CE)

Prepared for Aburnius Valens a complex herbal antidote: ASKLĒPIADĒS PHARM., in GALĒN *CMGen* 7.12 (13.1021–1023, 1027 K.). *PIR2* identify the pharmacist as the procurator of Ēpeiros (*CIL* 3.12299), and the patient as L. Fuluius Aburnius Valens (*PIR2* F-526), the latter of whom is later than Asklēpiadēs; probably ancestors of those men are involved.

PIR2 P-649.

PTK

Pompeius Trogus (30 BCE – 10 CE)

Roman historian of Gallic descent living in the Augustan age, author of a 44-book historical work in Latin titled *Historiae Philippicae* surviving as a 2nd (?) c. CE epitome made by M. Iunianus Iustinus. According to his own testimony, Pompeius was of the Vocontii, a Gallic tribe in Gallia Narbonensis conquered and incorporated into a Roman province in the 120s BCE. His grandfather, receiving Roman citizenship from Pompey, served under him in Spain against Sertorius. His uncle was a cavalry commander under Pompey in the Mithridatic War in Asia. His father was in charge of correspondence and embassies under Julius Caesar. These origins explain Pompeius' double name – his Roman name from his patron, and Trogus, a Gallic name. Pompeius' work, relying on the historiographic model of Theopompos of Khios in his *Philippika*, treats universal history, focusing on the Near East and Greece. As the subtitle (*totius mundi origo et terrae situs*) indicates, Trogus includes ethnographic and geographic digressions, for instance a description of Skuthia and its inhabitants (Iust. 2.2.1–15); an allusion to Egypt and its customs (Iust. 2.1.5–9); an excursus on the local history and foundation of Kurēnē (Iust. 13.7); a digression on Hērakleia Pontikē (Iust. 16.3–5) based on NUMPHIS and Memnōn of Hērakleia; an excerpt on the Jews including a geographical description of Judaea (Iust. 36.3.1–7); and a description of Parthia and the Parthians (Iust. 41.1.11–41.3.10). Pompeius' scientific interests probably inspired his lesser known work *On Animals*, quoted by PLINY, and based on the earlier works of ARISTOTLE and THEOPHRASTOS.

R. Develin, "Pompeius Trogus and Philippic History," *Storia della Storiografia* 8 (1985) 110–115; H.D. Richter, *Untersuchungen zur hellenistischen Historiographie. Die Vorlagen des Pompeius Trogus für die Darstellung der nachalexandrischen hellenistischen Geschichte (Iust. 13–40)* (1987); J.M. Alonso-Núñez, "Trogue-Pompée et l'impérialisme romain," *BAGB* (1) (1990) 72–86; R. Develin and J.C. Yardley, *Justin: Epitome of the Philippic History of Pompeius Trogus* (1994); W. Heckel and J.C. Yardley, *Justin: Epitome of the Philippic History of Pompeius Trogus Books 11–12* (1997).

Daniela Dueck

Pomponius Bassus (65 – 95 CE)

ASKLĒPIADĒS PHARM., in GALĒN *CMLoc* 4.8 (12.781–782 K.), preserves his collyrium compounded from **calamine**, ***euphorbia***, long and white peppers, opium, cinnamon, ***opopanax***, etc, moistened with fennel sap. Asklēpiadēs describes him as "companion" (12.780 K.).

RE 21.2 (1952) 2420 (#108), H. Diller.

GLIM

Pomponius Mela of Tingentera (*ca* 30 – 60 CE)

From Hispania Baetica (2.96, probably Iulia Traducta; *cf.* Iulia Soza, STRABŌN 3.18; Romer, 1), the first extant systematic Roman geographer, composing under Claudius whose British triumph the work was perhaps intended to celebrate (3.49–52). PLINY lists him as an authority for nine books (1.*ind.*3–6, 8, 12–13, 21–22), but never later cites him by name.

Vat. Lat. 4929 (9th c.: our source for the text) gives the title *De Chorographia*, changed by copyists to *De Cosmographia*, the former suggesting regional geography, the latter a description of the entire Earth. Mela, broadly treating the known world, selectively includes places considered well-known, important, or interesting, and omits lesser-known sites and features.

Positing a spherical Earth without much considering the mathematical ramifications of sphericity, he divides the world into two longitudinal hemispheres (east and west) and, following ERATOSTHENĒS, five latitudinal zones, all habitable (*cf.* POLUBIOS). Mela suggests the existence of the Antikhthonēs, occupying the habitable but unknown and unexplored southern zone (1.4, 1.54). Naming his sources with greater frequency in Book 3, Mela drew from HOMER and HANNO, but especially CORNELIUS NEPOS (3.45).

Mela declares his purpose to describe the known world and trace the complex arrangement of peoples and places (1.1–2). Mela overviews the continents from east to west: Asia, Europe, Africa (1.3–23), but proceeds unusually counter-clockwise, detailing sites and peoples in Africa (1.25–48), Asia (1.49–117), Europe (Book 2), outer coasts and islands (Book 3). Mela's interests included anthropological curiosities (3.75: hairy, fish-skin wearing nomadic Carmanii of the Persian Gulf), ***paradoxa*** (1.39: the diurnal temperature cycle of the Katabathmian fountain, boiling at midnight and freezing at midday; 1.94: XENOPHANĒS' sun on Trojan Mt. Ida), and topography (1.35: gulf of Syrtis; 3.70: Taprobane), plus mythology (1.37: Lotus-Eaters; 2.120: Calypso's island Aeaee) and history (2.32: Xerxēs' invasion of Hellas). He exhibits some skepticism, especially regarding mythical and legendary creatures of the African interior (1.23). No evidence suggests that maps accompanied the text.

Ed.: A. Silberman (*CUF* 1988).

DSB 11.74–76, Ed. Grant; *OCD3* 1218, N. Purcell; P. Berry, *Pomponius Mela: De Chorographia* (1997); F.E. Romer, *Pomponius Mela's description of the world* (1998).

GLIM

Pontica (250 – 400 CE?)

Latin hexameter poem of which only the first 22 lines survive, in MSS of IULIUS SOLINUS, who is probably not the author. It describes the Black Sea, and assigns it as the province of Venus.

GRL §523.3; *RE* 22.1 (1953) 26, K. Ziegler.

PTK

M. Porcius Cato of Tusculum (185 – 149 BCE)

Born 234 BCE to a plebeian family in the Latin countryside outside Rome, and died 149 BCE one of the most important political and cultural figures of his day (for his life, we are primarily indebted to PLUTARCH, *Cato Maior*). Early on his talents won him the nickname Cato, "the Shrewd." For the first half of his career political advancement and military achievement alternated. Elected *quaestor* for 204, he served under Scipio Africanus in Sicily and Africa, where he criticized his superior for allowing the Roman troops to indulge in Greek ways. As governor of Sardinia he made his administration a model of justice and frugality. A consulship in 195 saw him campaigning in Spain, where he put down a rebellion and opened up gold and silver mines to Roman exploitation; for his efforts there he was awarded a triumph. He played a key role in the defeat of Antiokhos III at the battle of Thermopylae, leading his troops around the same short cut once taken by the Persians. His censorship in 184 became the stuff of legend thanks to his efforts to clamp-down on luxurious living and corruption among senators and commoners alike; during his term he also made much-needed improvements to the city's infrastructure, in particular the sewer system. He died before Rome, in the third Punic War, could carry out his famous demand that *Karthago delenda est.*

Cato's most significant contributions to Roman culture was Latin prose, which he essentially invented as a literary form. His *Origins* was the first history of Rome composed in Latin, while his works *To my Son* and *Recital on Conduct* were the first Latin ethical treatises. He also wrote about civil law and military affairs, and was the earliest Roman orator known to have published his speeches (he reportedly never lost a case in court). In his day Rome was already possessed by an enthusiasm for Greek culture, hence Cato's purpose was not so much to introduce the Romans to literature as to make them less reliant on foreigners for it.

The one product of Cato's pen to survive complete is his *De Agricultura.* The audience for this work consisted of wealthy landowners like himself who owned several large estates in different parts of Italy and were interested in acquiring more. This growing interest seems to have created a demand for better information on farming; shortly after Cato's death, the Roman Senate ordered that the 28 volumes on agriculture ascribed to Mago the Carthaginian be translated into Latin (*cf.* DIONUSIOS OF UTICA). Cato describes individual estates which are between 60 and 150 acres in size; he seems to have owned at least six different ones.

The first noteworthy feature of his work is its organization, or lack of such. After a brief piece of praise for the life of the citizen-farmer, Cato launches into his subject with no obvious plan. Topics are frequently repeated and discussions are broken up, e.g. advice on how to process olives occurs in chapters 3, 31, 52, 54, 55, 64–69, and 93. Given this disorganization, the conclusion has sometimes been drawn that the work represents, wholly or in part, a compilation made after Cato's death; yet lack of flow would hardly present any difficulty to the energetic reader Cato seems to imagine himself addressing. Also distinctive is the treatise's explicitness about numbers: Cato prescribes exactly how many slaves one should assign to particular tasks, details the size of the rations to be distributed to workers and animals, and even lists the number of tools each building on the farm should have (*cf.* 10 ff.). He also offers several specimen contracts for the letting out of work on the farm. Such specifics make the work a crucial document for our understanding of the ancient economy.

Cato devotes little space to the production of cereals; vines, olives, and orchards absorb most of his attention, either because they were the most profitable forms of agriculture, or because there was a greater demand for technical knowledge on these subjects. For these crops Cato details at length the efforts required, describing the specialized equipment which wine- and oil-making demand, incorporating a complete calendar of annual tasks, and explaining the techniques of transplanting, grafting, and layering. He gives a long list of practical uses for *amurca*, a viscous by-product of olive oil production, and offers directions for making six varieties of flavored wine. Towards the end of the treatise he includes numerous medical recipes, in many of which the magical or superstitious element is pronounced; cabbage is praised, highly and at great length, for its medicinal virtues (*cf.* MNĒSITHEOS OF KUZIKOS).

Cato's treatise is of some interest for the history of technology. His recommendations for olive oil production include a detailed set of instructions for building a press, the most innovative feature of which is its levered drum, which when turned pulls a rope that lowers the press-beam onto the fruit (18). Working models of the "Catonian press" have been constructed based on his account, and examples have even been uncovered at Pompeii. Also described in considerable detail is a rotary olive-crusher called a *trapetum*; Cato even gives the cost of shipping and handling (20). His treatise may also contain the earliest mention of the donkey mill – the first mill of any kind to dispense with human labor as its main power source (10).

There is little to Cato's work that by Greek standards might be deemed scientific: only the barest of descriptions of plants and animals, for example, and almost nothing about physical causes. Yet it treats certain aspects of the farm, particularly the use of slaves, animals, and equipment, with a degree of detail found in no other surviving author. His main importance was as a pioneer, who, in the words of Columella, 1.1.12, "taught agriculture to speak Latin."

E. Brehaut, *Cato the Censor on Farming* (1933); K.D. White, *Roman Farming* (1970); A. Astin, *Cato the Censor* (1978); *OCD3* 1224–1225, M.S. Smith; *BNP* 1 (2002) 368–372 (§B.1, 369–370), E. Christmann; *BNP* 3 (2003) 20–23 (#1), W. Kierdorf.

Philip Thibodeau

Porphurios (Geog.) (*ca* 350 – 450 CE?)

Wrote a geographical work on Asia Minor and the lands around Constantinople, cited by the Ravenna Cosmography, 2.16 (Asia Minor), 4.3–4.7 (Bosporos, Dardania, Thrakē, and Musia). The *Ravenna Cosmography* calls him *miserus* and *nefandissimus*, confusing him with the anti-Christian Neo-**Platonist**. *Cf.* Iamblikhos Geog. and Libanios Geog.

(*)

PTK

Porphurios (Med.) **(300 – 1000 CE?)**

In iatrosophic therapeutic collections contained in two late-Byzantine MSS (Oxford, Bodleian, *Barocc.* 150 and Paris, BNF, *suppl. gr.* 1202), and in the *scholia* of another late-Byzantine MS (Paris, BNF, *graecus* 2183; mid 14th c.), credited with information on some *materia medica* and medicines (for example, in Paris, BNF, *suppl. gr.* 1202, f. 16, a fragment on *oxuphoinikon*). Paris, BNF, *graecus* 2183 was probably copied and used in the Kralē hospital in Constantinople. The formularies preserving Porphurios' fragments were probably developed in the context of Byzantine hospitals, especially common after 1200, and amalgamate formulae extracted from authors such as Dioskouridēs, Galēn, and the encyclopedias of Oreibasios, Aëtios of Amida and Paulos of Aigina, or anonymous physicians. Thus, although Porphurios of Tyre supposedly wrote on vegetarianism and included medical considerations in his philosophical works, he is probably not our author.

Diels 2 (1907) 86.

Alain Touwaide

Porphurios of Tyre (*ca* 260 – 305 CE)

Platonist philosopher. Born *ca* 234 in Tyre in Phoenicia, he changed his original Semitic name "Malkhos" (king) to "Porphurios" to celebrate his native city famous for purple (*porphura*). He studied with Longinus in Athens before joining Plōtinos in Rome (263–269), becoming one of his most loyal students. On Plōtinos' advice, Porphurios left for Sicily to overcome depression; his later activity is poorly documented. His students, for whom he wrote some texts, included the Roman aristocrat Chrysaorius and perhaps also Iamblikhos, but whether he had established a school is unclear.

Porphurios was a man of formidable talent and learning, as his surviving work shows. He wrote commentaries on PLATO's dialogues including *Sophist* and *Timaeus*, ARISTOTLE's *Categories, On Interpretation, Physics.* The nature of his writings on Aristotle's *Ethics, Sophistic Refutations, Prior Analytics*, and *Metaphysics* 12 remains unclear. Other titles include *Isagoge* (an introduction to Aristotle's *Categories*), a history of philosophy, *Starting points leading to intelligibles* or *Sententiae* (a philosophical handbook), *On Abstinence from eating food from animals* (a treatise on vegetarianism), comments on HOMER, the *Cave of the Nymphs* (an allegorical interpretation of Odysseus' hiding the Phaeacean gifts in a cave in Ithaca: *Od.* 13.102–112), *On how the embryos are ensouled*, a polemical *Against the Christians*, a commentary on PTOLEMY's *Harmonics* (underscoring Porphurios' commitment to the **Pythagorean** view that music manifests the work of reason in the world), and works on rhetoric and grammar. Particularly important is Porphurios' edition of Plōtinos' writings divided into six books of nine treatises each (*Enneads*), prefaced with his *Life of Plotinus*.

Remaining philosophically close to Plōtinos, Porphurios departed from his mentor, though never expressly. Porphurios agreed with Plōtinos on the structure of the intelligible world, acknowledging three divine ***hupostaseis*** (the One, the Intellect, and the Soul), but seemingly disagreed on some aspects of the relation between the intelligible and the sensible world. Like Plōtinos, Porphurios identified the human soul with the intellect but apparently maintained contrarily that the human soul was a manifestation of the ***hupostasis*** soul, rather than a power stemming from it. Porphurios also held that Plato regarded immanent Forms as versions of the transcendent ones, the latter being thoughts of the divine intellect instantiated into matter. Porphurios seemed to ascribe the same view to Aristotle, affecting the evaluation of Aristotle's ontology. Further, Porphurios did not object to the priority of particulars over universals in Aristotle's *Categories*, as did **Platonists** until Plōtinos: first because, for him, Aristotle's treatise examines significant expressions signifying particular substances, and secondly because Porphurios interpreted "particulars" as the whole class of entities (e.g. men), which are prior to the universal term (e.g. man). Porphurios saw ethics as the end of philosophy and agreed with Plōtinos' division of levels of virtue, but apparently differed in believing that happiness is not obtained only at the ultimate level of virtue but, in a different degree, also at the first.

Porphurios' impact on later generations was huge, especially regarding his example of writing commentaries on Aristotle and his appreciation of Aristotle's logic, integrated then into the **Platonist** philosophical curriculum. Later **Platonist** commentators on Aristotle draw much from him, and in turn commented on his *Isagage.*

Ed.: J. Bouffartique and M. Patillon, *Porphyre, De l'abstinence* 4 vv. (*CUF* 1977–1996); A. Smith, *Porphyrii Philosophi Fragmenta* (1993).

RE 22.1 (1953) 275–313, R. Beutler; P. Hadot, *Porphyry et Victorinus* (1968), vols. 1–2; A. Smith, *Porphyry's Place in the Neoplatonic Tradition. A Study of post-Plotinian Neoplatonism* (1974); J. Barnes, *Porphyry Introduction* (2003); *BNP* 11 (2007) 646–652, R. Harmon; George Karamanolis, "Porphyry, the First Platonist Commentator of Aristotle," in P. Adamson, H. Baltussen, and M. Stone, edd., *Science and Exegesis in Greek, Arabic and Latin Commentaries* = *BICS* S. 83 (2004) 79–113; George Karamanolis and A. Sheppard, edd., *Studies on Porphyry* = *BICS* S. 98 (2007).

George Karamanolis

Poseidippos of Pella (290 – 240 BCE)

Friend and disciple of Asklēpiadēs of Samos, writer of epigrams and a famous member of the Ptolemaic court in Alexandria in Egypt. Born *ca* 310 BCE, he makes reference (*Epigr.*

112) to the Olympic Games in 240, suggesting a long life. Before moving to Alexandria, he lived in Athens (where his friends included the **Stoics** ZĒNŌN and KLEANTHĒS) and in Kōs (where he met important poets like Philetas, Asklēpiadēs and Theokritos, participated in heated contemporary cultural debates, and polemicized KALLIMAKHOS).

Only 23 authentic Poseidippean epigrams were preserved until the discovery (in 1990) of the Milan papyrus *Vogliano* 1295 dating to the second half of the 3rd c. BCE (nearly contemporary with our author). This MS transmits about 100 compositions, nearly 600 new verses. Among the most interesting elements emerging from this important literary document are the thematic distribution of epigrams (not according to an alphabetic order) and their nature in relation to a specific client's request. The MS groups epigrams into nine categories, each with its own subject heading (a tenth section may lurk in the tattered remains of the end of the roll). Some of these categories are familiar, such as "poems on tombs" (*epitumbia*); other sections are more exotic and almost bizarre, e.g., poems about stones (*lithika*), omens (*oionoskopika*), statue-making (*andriantopoiïka*), even a group of funerary epigrams with the enigmatic title "turnings" (*tropoi*). The first and longest section of the papyrus, the lithika, treats gems and other noteworthy stones in a tour de force of geographical, cultural and literary references. The ecphrastic content has a practical end, as in the case of an epigram ordered by a suitor to accompany the gift of a precious stone to his beloved. The entire section reads like a gazetteer of the Hellenistic world: beginning far in the east with the Indian river Hydaspes, it proceeds through Persia and Arabia to the island of Euboia, as it details the provenances of the stones and the distances they have traveled. Women often serve as the final destination, the recipients of the precious objects. The lithika displays in miniature the ambitious scope of the entire collection, its ability to weave together literary and material culture, the powerful and the humble.

Ed.: G. Bastianini *et al.*, *Posidippo di Pella Epigrammi (P. Mil. Vogl. VII 309)* (2001); C. Austin and G. Bastianini, *Posidippi Pellaei quae supersunt omnia* (2002).

RE 22.1 (1953) 428–444, W. Peek; *KP* 4 (1972) 1075–1076, R. Keydell; K. Niatas, "A poetic gem. Posidippus on Pegasus," *Pegasus* 40 (1997) 16–17; W. Luppe, "Weitere Überlegungen zu Poseidipps Lithika-Epigramm Kol. III 14ff.," *APF* 47 (2001) 250–251; G. Bastianini, ed., *Un poeta ritrovato: Posidippo di Pella. Giornata di studio Milano 23 novembre 2001* (2002); *Idem* and A. Casanova, edd., *Il papiro di Posidippo un anno dopo.* (2002) 1–5; W. Luppe, "Poseidipp, Lithika-Epigramm II 23–28," *Eikasmos* 13 (2002) 177–179; *Idem*, "Zum Lithika Epigramm Kol. III 28–41 Poseidipps (P.Mil.Vogl. VIII 309)," *AC* 71 (2002) 135–153; B. Acosta-Hughes *et al.*, edd., *Labored in Papyrus Leaves. Perspectives on an Epigram Collection Attributed to Posidippus (P.Mil.Vogl. VIII 309)* (2004); R. Casamassa, "Posidippo fra arte e mito. La gemma di Pegaso (Posidipp. ep. 14 A–B)," *Acme* 57 (2004) 241–252; E. Lelli, "I gioielli di Posidippo," *QUCC* 76 (2004) 127–138; M. Di Marco *et al.*, edd., *Posidippo e gli altri. Il poeta, il genere, il contesto culturale e letterario. Atti dell'incontro di studio, Roma, 14–15 maggio 2004* (2005); V. Garulli, "Rassegna di studi sul nuovo Posidippo (1993–2003)," *Lexis* 22 (2004) 291–340; K.J. Gutzwiller, ed., *The New Posidippus. A Hellenistic Poetry Book* (2005); *BNP* 11 (2007) 671–672 (#2), M.G. Albiani.

Eugenio Amato

Poseidōnios (Med. I) (70 – 30 BCE?)

Student of ZŌPUROS OF ALEXANDRIA, along with APOLLŌNIOS OF KITION (*CMG* 11.1.1, p. 12), and co-authored with DIOSKOURIDĒS PHAKĀS a work on the bubonic plague in Libya (high fever, terrible pain, widespread buboes): RUFUS in OREIBASIOS, *Coll.* 44.14.2 (*CMG* 6.2.1, p. 132); compare the similar work by DIONUSIOS OF KURTOS.

He is hardly likely to be identifiable with the homonymous **Stoic** (contrast Kudlien; *cf.* T113 EK).

F. Kudlien, "Poseidonios und die Ärzteschule der Pneumatiker," *Hermes* 90 (1962) 419–429.

PTK

Poseidōnios (Med. II) (400 – 440 CE)

Denied that demons caused any disease or mental illness (Philostorgios, *HE* 8.10), and explained them on the basis of displacement of **humors**, to the head in the case of mental illness. Aëtios of Amida gives extensive extracts, showing that he prescribed phlebotomy, regimen, and mainly vegetal drugs, containing few exotics (aloes: *CMG* 8.2, pp. 139, 147; ginger, pp. 150–151; **Indian buckthorn**: p. 167; and **silphium**: p. 129); he followed Antullos on hellebore: 3.122 (*CMG* 8.1, pp. 309–310). His explanations of mental illnesses, which show some affinity with pseudo-Galēn, Introductio 13 (14.732–733, 741 K.), consistently invoke **humoral** pathology: three forms of *phrenitis* come from phlegm, Aëtios 6.2 (*CMG* 8.2, pp. 125–128); cold and wet **humors** cause *karos*, 6.5 (p. 133), wet and warm cause *kōma*, 6.6 (pp. 133–134); chilling of the head causes *mōrōsis* and *lēros*, 6.22 (pp. 159–160); *cf.* also on melancholy, 6.9 (pp. 141–143) and epilepsy, 6.13 (pp. 153–155), including three recipes 6.19–21 (pp. 158–159). Not even nightmares, long attributed to spirits, were demonic, but rather caused by indigestion, 6.12 (pp. 152–153): *cf.* Sōranos in Caelius Aurelianus *Chron.* 1.55 (*CML* 6.1.1, p. 220). He follows Arkhigenēs on *lēthargia* (6.3, pp. 128–131), *katalēpsia* as the mean between *phrenitis* and *lēthargia* (6.4, pp. 131–133), ***skotōma*** (6.7, pp. 134–136), and *mania*, caused by blood rising to the head (6.8, pp. 136–141). He cites Galēn, *Simples* 11.24 (12.356–357 K.), on ashed crabs as an antidote for ***hudropobia***, though primarily following Rufus' method of treatment, 6.24 (pp. 163–169). See also Paulos of Aigina, 7.3, 7.20.26, 7.21.2, 7.22.4 (*CMG* 9.2, pp. 196, 387, 392, 394).

PLRE 1 (1971) 717; *BNP* 11 (2007) 682 (#1), V. Nutton.

PTK

Poseidōnios of Apameia (*ca* 110 – *ca* 51 BCE)

Born in Apameia on the Orontes, *ca* 135 BCE, student of Panaitios and founder of a **Stoic** school at Rhodes which superseded the school at Athens after Panaitios' death. Poseidōnios wrote broadly on physics, logic, and ethics. As Rhodian ambassador to Rome in 87–86 BCE, Poseidōnios became close to influential Romans, including Cicero and Pompey. Precise details of his philosophy have been much debated, compounded by methodological divisions among modern historians of philosophy regarding which sources should count as fragments (the most secure collection is that of Edelstein and Kidd, confining itself to sources mentioning Poseidōnios by name, a good antidote to the kind of pan-Poseidonianism found in some earlier accounts).

Much of Poseidōnios' philosophy built on and developed standard **Stoic** fare: the unity of the kosmos; its finite and spherical nature; its governance by divine reason; the division into active and passive principles; the soul as warm ***pneuma***; the doctrine of fate; and the efficacy of divination (though Poseidōnios seems to have emphasized astrology in particular). Poseidōnios is said to have written five books on divination (Diogenēs Laërtios, 7.149; Cicero, *Div.* 1.6), and Cicero (*ibid.* 2.35) reports that he followed Chrysippus and

Antipatros (of Tyre?) in claiming that a providential deity – leading the diviner to select the particular victim whose internal organs will correspond with the situation under enquiry – guides the choice of sacrificial victim in extispicy (entrail divination). It is debated whether Poseidōnios broke up the older **Stoic** unity of god-fate-nature into its individual components and placed them hierarchically with god at the top, fate below, and nature in the third tier, although this may only have been done for the purposes of grounding a particular argument in defense of divination (Reydam-Schils). Poseidōnios accused Epicurus of atheism, on the grounds that Epicurus' inclusion of "gods" in his ***kosmos*** (gods who did not and could not interact or interfere with the workings of the world) was no more than a token gesture designed to make his philosophy less unpalatable to the masses.

Poseidōnios also made novel and important contributions in many areas, including history, geography, meteorology, cosmology, psychology, and mathematics. Galēn (*PHP* 8.1.14: *CMG* 5.4.1.2, p. 482) calls him the "most scientific" of the **Stoics**. Strabōn (Books 1–3) discussed his work in geography at length. Poseidōnios is reported to have estimated the circumference of the Earth at 180,000 stades, a significantly smaller figure than Eratosthenēs' 252,000 stades. In meteorology, he wrote on the causes of hail, snow, winds, storms, thunder, lightning, rainbows, earthquakes, halos and parhelia. He argued influentially that tides were caused by winds affected by the Moon (*cf.* Seleukos of Seleukeia). Poseidōnios seems to have gone to some length to clarify the methodological differences between philosophy generally and particular sciences like astronomy. He is reported to have constructed a celestial globe. He estimated the sizes and distances of the Sun and Moon, and a few fixed-star observations are credited to him. For Poseidōnios, the Sun is larger than the Earth, a sphere, and composed of pure fire. The Moon was composed of a mixture of fire and air, and its opacity was caused only by extreme thickness. He also wrote on eclipses and comets. In mathematics, he contributed to the foundations of geometry, and composed a book (quoted at some length by Proklos in his *Commentary on the First Book of Euclid's Elements*) to counter Zēnōn of Sidōn's critique of geometry.

Galēn reports (*PHP* 5.7.1–10: *CMG* 5.4.1.2, pp. 336–338) that Poseidōnios broke with earlier **Stoics** and followed Plato and Aristotle in dividing the faculties of the soul into three: thinking, desiring, and being angry. But Poseidōnios, on Galēn's account, broke with Plato in situating all three faculties in the heart rather than in separate locations in the body. The extent and depth of Poseidōnios' use of Plato is still debated.

Ed.: Edelstein and Kidd (1972–1999).

G. Reydams-Schils, "Posidonius and the Timaeus: Off to Rhodes and Back to Plato?" *CQ* 47 (1997) 455–476; A.D. Nock, "Posidonius," *JRS* 49 (1959) 1–15.

Daryn Lehoux

Poseidōnios of Corinth (325 BCE? – 175 CE?)

Mentioned by Athēnaios, *Deipn.* 1 (13b), in a catalogue of authors of *Halieutika*. He lived probably before Oppianus of Kilikia.

RE 22.1 (1953) 826 (#5), R. Keydell; *SH* 709; A. Zumbo, "Ateneo 1, 13b–c e il 'canone' degli autori alieutici," in P. Radici Colace and A. Zumbo, *Atti del Seminario Internazionale di Studi "Letteratura scientifica e tecnica greca e latina"* (2000) 163–170.

Claudio Meliadò

Poseidōnios of Macedon (335 – 325 BCE)

Designed a *helepolis*, described in Bitōn, *Belop.* 4 (pp. 51–56 W.), for Alexander the Great, perhaps in 335 BCE. Poseidōnios' rolling siege-tower, over 20 m (50 cubits) tall and made of light flexible wood coated with flame-retardant, was padded against missiles and supported on a wheeled oaken platform about 18 m (60 feet) square. It was self-propelled, driven by a capstan operated by the soldiers inside, and had floors corresponding to the heights of the walls to be assaulted.

Marsden (1971) 70–73, 84–90.

PTK

Potamōn (300 BCE – 80 CE)

Andromakhos, in Galēn *CMGen* 2.2 (13.473, 488–489 K.), mentions Potamōn's "green" plaster. The name is rare before *ca* 300 BCE (*LGPN*).

RE 22.1 (1953) 1028 (#5), H. Diller.

PTK

Potamōn of Alexandria (*ca* 40 – *ca* 10 BCE)

Described as an "eclectic" by Diogenēs Laërtios *pr.*21; the *Souda* Pi-2126 records that he wrote on *Elements*; his *arkhai* were matter, quantity, quality, and space. Simplicius, *In de Caelo* 3.4 (*CAG* 7 [1894] 607), says that he defined mathematical *arkhai* through quantity, starting from the monad.

OCD3 1235, J. M. Dillon.

PTK

Praecepta Salubria (100 – 400 CE)

Iambic poem giving advice on regimen, purportedly based on Asklēpiadēs and Dioskouridēs; its pharmaceutical use of beer (*zumē* lines 88, 97) suggests an Egyptian (or Mesopotamian) origin. The invocation of an unnamed single deity, if not merely formulaic ("with God", line 27), and emphasis on repressing libido (62–63 and 72–75), may suggest a Christian origin. Moderation is urged (1–5) and a diet prescribed that balances the qualities hot/cold and wet/dry by solar and lunar cycles (7–46); therapeutic interventions include sleeping on one's right-hand side (6, 45), bathing (14–16), purges and phlebotomy (30–32), and fumigations (47–52). Those are followed by six simple prescriptions: raw honey for long life (57–61), chicory (*intubion*) and lentils to repress libido (62–63: on chicory, contrast Galēn *Properties of Foodstuffs* 2.40–41 [6.624–628 K.], and Arkhigenēs, in Galēn [?], *Eupor.* 1.1 [14.321 K.]), boiled oregano taken at the new moon for good memory (64–67), Thasian almonds to prevent inebriation (68–71), lettuce-seed in water for sexual restraint (72–75), garlic or cinnamon to clear the throat (76–79 – perhaps another encratic therapy, *cf.* pseudo-Dēmokritos in Pliny 20.28). The poem closes with four pest-control potions: wormwood, absinthe wormwood, or boiled fig-juice to banish fleas and bedbugs (80–85), pellets formed of iron filings in beer and fat to slay mice (86–90), an ointment of mercury simmered in fat for delousing (91–94), and aconite or realgar boiled in beer and fat to exterminate rodents (95–100).

Ed.: U.C. Bussemaker, *Poetae Bucolici et Didactici* (1862) 132–134.

PTK

PRASIŌN ⇒ PASIŌN

Praxagoras of Kōs (325 – 275 BCE)

Greek physician, from a highly reputed medical family of the Asclepiad tradition active in Kōs (including his father Nikarkhos and an earlier Praxagoras, pupil of HIPPOKRATĒS). His pupils included HĒROPHILOS, PLEISTONIKOS, PHULOTIMOS and XENOPHŌN. His name is mentioned for an ointment in a papyrus of *ca* 250 BCE (*SB* 9859d). GALĒN, citing him frequently with DIOKLĒS as an important member of the **Dogmatic** medical tradition, dates him a little after Hippokratēs (*Tremor Palp.* 7.584 K., *Diff. Puls.* 4.3 [8.723 K.]). A more ancient tradition credits Praxagoras, together with Hippokratēs and KHRUSIPPOS OF KNIDOS (I), with perfecting dietetics (*Schol. Hom. Iliad.* 11.515).

He is credited with many treatises, revealing the broad spectrum of his medical activity, some surviving to Galēn's era (Galēn wrote a polemical essay against Praxagoras' **humoral** doctrine). Praxagoras wrote on physiology and anatomy (*Phusika*, at least two books, and *Anatomy* apparently in many books), as well as on pathology (*On diseases*, at least three books, and *On differences in acute diseases*), prognosis (*On the concurrent signs*, two books, and *On the supervening affections* [or *signs*]), and therapy (*Ways of therapy*, four books, and *Causes, affections and therapies*).

Having developed and expanded "Hippokratic" **humoral** physiology and pathology, Praxagoras probably also wrote a treatise on **humors**, wherein he distinguished ten **humors** (the most-often cited among them being the "glassy" one, *hualodes*) plus blood. Distinguishing some as kinds of phlegm, others according to taste, consistency or quality, he claimed they are formed in the veins by nutriment transformed by heat, determining health and disease (Galēn *Sympt. Caus.* 1.6, 1.7 [7.124, 137 K.], PSEUDO-GALĒN, INTRODUCTIO 9 [14.698–699 K.]). Fever, for example, is caused by the putrefaction of **humors** (*MM* 2.4.13 [10.101 K. = p. 51 Hankinson).

Praxagoras was apparently the first to distinguish the anatomical structure and physiological function of veins and arteries. Veins contain blood, and arteries air (***pneuma***) introduced through respiration (and from gaseous digestive byproducts) and nourishing the soul. He places the seat of the soul in the heart, thus making it the ***hēgemonikon***, arguing that the brain is just an extension of the spinal marrow. Praxagoras did not believe in innate heat, but thought that bodily heat was drawn in from the outside. No details of his theory of digestion remain, but he thought that blood was the product of good digestion, becoming flesh through the veins. (It is wrong to ascribe to Praxagoras the theory of his pupil Pleistonikos that digestion is a process of putrefaction, sepsis: CELSUS 1.*pr.*20). He surely shared with his pupils the idea that sperm comes not only from the brain but from the entire body (PSEUDO-GALĒN, DEFINITIONS 19.449 K.), and asserted the filtering role of kidneys (Galēn *Nat. Fac.* 1.13 [2.30 K.]). His anatomical doctrines (that arteries become more and more subtle finishing as nerves, and pulse independently of the heart) are sometimes criticized by Galēn. Regarding therapy, he approved bloodletting, used emetics, and discussed the utility of fasting. Many of his opinions and treatments of single diseases are preserved by the PARISINUS MEDICUS and in CAELIUS AURELIANUS, while Athēnaios (*Deipn.* 2 [32d, 41a, 46d], 3 [81c]) records some of his opinions in the use of wine, water and other foods, a field that Praxagoras and his school explored in detail.

Ed.: Steckerl (1958).
RE 22.2 (1954) 1735–1739, K. Bardong; *OCD3* 1241–1242, J.T. Vallance; *BNP* 11 (2007) 782–783, V. Nutton.

Daniela Manetti

Primiōn (100 BCE – 80 CE)

ANDROMAKHOS records and approves Primiōn's remedy to cicatrize severe wounds, in GALĒN *CMGen* 4.5 (13.695–696 K.). The balm is compounded from ***sōru***, alum, pomegranate peel, unslaked lime, frankincense, oak-gall, beeswax, calf-suet, and old olive oil. Although Prīmos is a far more common variant, Primiōn is not unique, known from the 1st c. BCE to the 2nd c. CE (*LGPN* 2.380, 3A.376, 3B.362, 4.290).

RE 22.2 (1954) 1974, H. Diller.

GLIM

PRISCIANUS ⇒ THEODORUS PRISCIANUS

Priscianus (*ca* 300 – 365 CE)

OREIBASIOS, *Ecl. Med.* 54.10 (*CMG* 6.2.2., p. 218), records his enema for dysentery, composed of ashed papyrus, lime, orpiment, and realgar, to be used like NUMPHODŌROS' (pp. 217–218). Unless the recipe is a later insertion (*cf.* ADAMANTIOS), this Priscianus predates THEODORUS PRISCIANUS. The name is attested from *ca* 300 CE: *PLRE* 1 (1971) 727–728, but perhaps *cf.* PROCLIANUS.

(*)

PTK

Priscianus of Caesarea (Mauretania) (500 – 525 CE)

The widely-used grammarian of Latin, who taught in Constantinople. He also composed two or three small works, *De Figuris, Description of the World-Globe*, and perhaps a brief poem *On the Stars*. The first gives a false theory of the Roman numerals (§1–8), an accurate account of the Roman weights and coins (§ 9–18), and the conjugation of the Latin numeral-words (§19–32); ed. Keil, *GL* 3 (1859) 406–417. The geographical poem is a free interpretation of the poem of DIONUSIOS OF ALEXANDRIA, omitting most pagan references: Priscianus describes in hexameters the parts of the Earth (vv. 1–36), the ocean and its inlets (37–66), the Mediterranean from the Pillars to the Pontos (67–159), Libya, its peoples and lands (160–258), Europe, its peoples, lands, and isles (259–567), the isles of the Ocean (568–613), and Asia, its peoples and lands (614–1034): ed. Paul van de Woestijne, *La périégèse de Priscian* (1953). A mnemonic poem listing the constellations, northern, zodiacal, and southern, may be his: ed. A. Riese, *Anthologia Latina* (1906) #679.

BNP 11 (2007) 868–870, P.L. Schmidt.

PTK

Priscianus of Ludia (*ca* 530 CE)

Neo-**Platonic** philosopher and colleague of SIMPLICIUS active in Athens when Justinian's new laws forbade pagan philosophers to teach (529 CE). Little is known about his life or his works. His contribution to scientific writing lies solely in the incomplete *Metaphrasis [paraphrase] of Theophrastos' On Sense-Perception*, which discusses ARISTOTLE's psychology from a Neo-**Platonic** perspective, and specifically inquires into what THEOPHRASTOS contributes to the subject in his *Physics* (Books 4–5). Together with THEMISTIOS' summary version of Aristotle's *On the soul*, Priscian's *Metaphrasis* is a major source on Theophrastos' psychology. Steel attributes to Priscian a commentary on Aristotle's *On the soul*, but this is still disputed. Priscian's *Solutions to King Chosroes' scientific questions* (*Solutiones eorum de quibus dubitavit Chosroes Persarum rex* – only in Latin translation, *CAG* S.1.2), presumably written in Persia, belongs to the *problēmata*-genre, covering without originality soul, sleep, astronomy, lunar phases, the four elements, animal species, and motion.

RE 22.2 (1954) 2348 (#9), W. Enßlin; C.G. Steel, *The Changing Self. A Study on the Soul in Later Neoplatonism: Iamblichus, Damascius and Priscianus* (1978); C.G. Steel and P.M. Huby, *Priscian, On Theophrastus' on Sense-Perception with 'Simplicius' On Aristotle On the Soul 2.5–12* (*ACA* 1997); P.M. Huby, *Theophrastus of Eresus. Sources for His Life, Writings, Thought and Influence. Commentary Volume 4: Psychology* (1999); *BNP* 11 (2007) 870, L. Brisson.

Han Baltussen

Priskos of Panion (*ca* 445 – 480 CE)

Rhetorician and historian, born in Panion no later than 420. Primarily a teacher of rhetoric at Constantinople, he was among Theodosios II's envoys to Attila the Hun in 449. In 450, he stayed at Rome; in 452/453 he was in Syria and Egypt, visiting Damaskos, Alexandria and the Thebaid. Around 456 he served as *assessor* to the *magister officiorum* Euphēmios. In addition to lost declamations and letters, he wrote an eight-book history, probably entitled *Historia Buzantiakē*, covering from at least 433 to 472, classicizing in style and rich in ethnographic detail, and one of the most important sources for the Huns in the time of Attila. Unfortunately only long fragments, incorporated in the 10th c. *Excerpta de legationibus* of Constantine VII Pophurogennētos, are extant. Narratives of Euagrios Skholastikos and Theophanēs the Confessor preserve some fragments, other quotations survive in the *Chronicon paschale* and the *Souda* (Pi-2301 and Z-39, s.v. Zerkōn). Apparently, Priskos was an influential author in Buzantion.

Ed.: *FHG* 4.69–110; Blockley 1 (1981) 48–70, 113–123, 2 (1982) 222–400.
HLB 1.282–284; B. Baldwin, "Priscus of Panium," *Byzantion* 50 (1980) 18–61; *ODB* 1721, *Idem; OCD3* 1248, R.J. Hopper; *NP* 10 (2001) 343, K.-P. Johne.

Andreas Kuelzer

Probinus (*ca* 350 – 450 CE?)

Wrote in Latin a geographical work, cited by the RAVENNA COSMOGRAPHY, and which treated at least Africa (3.5: where he is called African), and Illyria and Dalmatia (4.15–16). The name is common 350–450 CE: *PLRE* 1 (1971) 734–735, 2 (1980) 909–910.

(*)

PTK

Proclianus (*ca* 180 – *ca* 400 CE)

MARCELLUS OF BORDEAUX cites two remedies from Proclianus for liver ailments: 22.34, 37 (*CML* 5, pp. 388, 390). The rare name is first attested 182 CE, Proklianē of Thessalonikē

(*LGPN* 4.29; see also 3A.377), and primarily from the 4th c.: *PLRE* 1 (1971) 741, 2 (1980) 914. *Cf.* perhaps Protlius.

Fabricius (1726) 380.

PTK

Prodikos (Med.) ⇒ Hērodikos

Proëkhios (?) (120 BCE – 365 CE)

Oreibasios, *Ecl. Med.* 90.1 (*CMG* 6.2.2, p. 270), records his remedy for scrofula, composed of barley, **galbanum**, ***ammōniakon*** incense, oak mistletoe, natron, pigeon-dung, propolis, and pyrites, in **terebinth**; Paulos of Aigina, 7.16.22 (*CMG* 9.2, p. 339), records his blood-stanch of antimony, **calamine**, saffron, ***khalkanthon, misu***, opium, balsam, white pepper, and **verdigris**, in gum and water. The name is otherwise attested only for an obscure bishop from Arsinoë, in the acts of the council of Khalkēdōn (451 CE): *RE* 23.1 (1957) 104. *Cf.* perhaps Prosdokhos.

Fabricius (1726) 380, s.v. Prosechius.

PTK

Proklos the Methodist (*ca* 27 BCE – 30 CE)

Caelius Aurelianus (*Chron.* 3.8.100) calls Proklos a "follower" of the early **Methodist** Themisōn (*cf.* Galēn *MM* 1.7.4 [10.52 K.]), placing Proklos in the reign of Augustus; pseudo-Galēn, Introductio 4 (14.684 K.) puts him slightly later, listing Proklos after Thessalos, Mnaseas, and Dionusios (Method.). Caelius Aurelianus (*Chron.* 3.100–101 [*CML* 6.1.2, p. 738]) records Proklos' theories regarding developmental stages of edemas: first are the beginnings of **dropsies** characterized by small changes in the flesh (*leucophlegmatia*); the most severe state *tympanites* occurring when abdominal swelling and tightness are at their worst, followed by a lessening phase, *ascites* (Caelius further claims that Proklos ". . . strays from [the sect's] true doctrine"). Oreibasios, *Synopsis*, 3.103 (*CMG* 6.3, pp. 95–96), strongly commends Proklos' recipe for the treatment of gout (*podagra*) which if consumed for one year ". . . cures gout, sciatica, and generally any sort of ailment and pain in the joints. It enables the [five] senses to be more acute since the compound cleanses mildly through urination, engendering a more healthy state in the entire body. It also cures epilepsies and hardened swellings in the liver and spleen." The nine ingredients in stepwise reduced quantities are 9 ounces of germander (*Teucrium chaemaedrys* L.), 8 ounces of the full twig – fruits and all – of the white centaury (*Centaurium umbellatum* Gilib.), 7 ounces of "long birthwort brought from the mountains" (*Aristolochia longa* L.), 6 ounces of "great" or "yellow" gentian (*Gentiana lutea* L.), 5 ounces of *huperikon* ("St. John's wort"; probably *Hypericum crispum* L.), 3 [sic]ounces of parsley, 3 ounces of valerian (*Valeriana phu* L.), and a single ounce of **shelf-fungus**, mixed with honey. Proklos grinds in a mortar each ingredient separately, fashioning the compound into ***trokhiskoi***, to be taken daily in water some three hours after a bowel movement. Proklos' "Medicine for Gout" was a mild analgesic, anti-depressant, and diuretic, but any benefits for the patient would have been counteracted in the long term by a gradual poisoning of the kidneys from the substantial total quantity of birthwort consumed.

RE 23.1 (1957) 247, H. Diller; Wichtl (2004) 630–634 [valerian], 305–308 [St. John's wort].

John Scarborough

Proklos of Laodikeia (Syria), "Proklēios" (150 – 480 CE)

Son of Themisōn and hierophant, according to the *Souda* Pi-2472. He wrote a commentary on NIKOMAKHOS, some *geōmetrika*, and other works, all lost, save one citation by DAMASKIOS, *In Philebum* 19.

PLRE 1 (1971) 742.

PTK

Proklos of Lukia, *diadokhos* (*ca* 430 – 485 CE)

Life: Proklos' life is amply recorded in the edifying encomium by his student MARINOS OF NEAPOLIS, composed the year after his mentor's death (*Vita Procli* = *VP*), and in DAMASKIOS' *Vita Isidori*. Born in Constantinople in 412 to a Lukian noble family, Proklos began his studies with a ***grammatikos*** in Lukian Xanthos and continued them in Alexandria. His father Patricius, a high-ranking advocate, had practiced in the capital and wanted his son to learn Roman Law. Proklos also began studying rhetoric under the Sophist Leōnas, whom Proklos accompanied on an embassy to Constantinople (*VP* 8–9). Returning to Alexandria, Proklos studied philosophy (especially ARISTOTLE's) and mathematics under HĒRŌN (MATH). *Ca* 430, Proklos went to Athens to study with PLOUTARKHOS OF ATHENS and SYRIANUS, the latter connected with Athenian Sophists, especially Lakharēs and his student Nikolaos who welcomed Proklos on his arrival (*VP* 10–11). Among his fellow students were DOMNINOS OF LARISSA and Hermeias. After Syrianus died, *ca* 437, Proklos became his "successor" (*diadokhos*), and, for the rest of his life, lived and taught in Athens, except during one year when Christian threats forced him into exile in Lydia (*VP* 15). Several of his students would become influential in government and/or philosophy, including AMMŌNIOS OF ALEXANDRIA and his brother HĒLIODŌROS, Marinos and Isidōros of Alexandria (who successively succeeded Proklos). His students also included high-ranking notables of the late empire, many of them Christians (see Saffrey and Westerink, edd., *Theologie Platonicienne: Proclus* I.XLIX–LIV). Proklos himself was an influential political figure, extending patronage to many contemporaries (*VP* 16–17). He was also a devoted pagan who scrupulously observed traditional rites (*VP* 18–19).

Nature of his scientific works and activities: Marinos describes Proklos as a hard worker, devoting time to courses, lectures and discussions which he subsequently recorded in commentaries (*VP* 22). Some of his extant works greatly influenced philosophy, theology, and science. Of particular significance are his commentaries on PLATO's *Timaeus* (*IT*) and *Republic* (*IR*), on EUCLID's first book of the *Elements* (*IE*), his *Outline of astronomical hypotheses* (or *Hupotuposis*), and his *Elements of Physics*.

Proklos' written works derive from a method of reading and discussion that could be qualified as *mystagogical, eclectic, conciliatory,* and *agonistic*. (a) *Mystagogical*: Proklos remained faithful to Syrianus' idea that preparatory readings (like the study of Aristotle) should lead one to Plato's mystagogy (*VP* 13), i.e., to the idea that Plato's dialogues (especially *Timaeus* and *Parmenides*) were designed to lead their readers to higher ***hypostases***. Proklos thus considered it his duty to imitate Plato by providing his own "guidance" into Neo-**Platonist** metaphysics, e.g., a teaching both inspired and inspiring. In particular, he considered reading Euclid, Plato's *Timaeus* or PTOLEMY as steps along the same path. In general, he considered commentary in itself as a kind of religious performance, akin to prayer and theurgy. (b) *Eclectic*, since Proklos chose from among his extensive literary knowledge everything relevant to attain "true reality" (*pragmata*) and Platonic theology. The literature upon which Proklos relied included important mathematical works like Euclid's *Elements*, HĒRŌN's or PAPPOS' commentaries thereon, Neo-**Pythagorean** arithmetic works, GEMINUS' encyclopedia of mathematical science, and astronomical works (especially Ptolemy's *Almagest* and *Hypotheseis*). (c) *Conciliatory*, since Proklos also tried to build a harmony (*sumphōnia*) between different kinds of reasoning or theories – e.g., Euclid's proofs and Aristotle's theory of demonstration (revised by Syrianus). (d) *Agonistic*: the interpretation of texts was discussed in a closed circle, some of them sometimes raising valid objections (e.g. *IE* 29–30). This, in turn, was consistent with Proklos' view that teaching should awaken the souls of his listeners, controlled by, and directed toward, higher levels of cognition.

Main scientific works and influence: *IT*, Proklos' favorite work (*VP* 38), is an ambitious attempt to reconcile Plato's dialogue with Aristotelian physics and cosmology, which Proklos substantially criticized and modified. Likewise, his *Hupotuposis* attempts to criticize Ptolemy's cosmology by emphasizing the artificiality of his hypotheses as compared with the simplicity and the independence from human needs characterizing natural processes, ideas also explored in *IR* (2.213–236 Kroll). The 13th dissertation of *IR* also includes a long discussion, in which Proklos discusses various issues pertaining to astrology or Neo-**Pythagorean** arithmetic. He addresses in particular side and diagonal numbers and confronts the Neo-**Pythagorean** procedure with a geometrical proof drawn from Euclid. Proklos' *IE* contains an original theory of mathematical activity and invention, derived from Syrianus' own projectionist theories about the activity of the soul (*IE* 49–57). In its first Prologue, Proklos also developed IAMBLIKHOS' earlier idea of "general mathematics" (*holē mathēmatikē*) by expressing it according the late Neo-**Platonist** metaphysics (*IE* 5–10). Proklos' *Elements of Physics*, as well as his *Elements of Theology*, show his eagerness to adapt the Euclidean paradigm of demonstration to other subjects, such as Aristotelian physics and Neo-**Platonist** theology. Proklos' immediate influence is seen in the interest that some of his pupils took in ancient science, particularly Ammōnios and Marinos.

DSB 11.160–162, G.R. Morrow; A.-Ph. Segonds, "Proclus: astronomie et philosophie," in J. Pepin and H.D. Saffrey, edd., *Proclus, lecteur et interprète des anciens* (1987) 319–334; O'Meara (1989) 142–208; L. Siorvanes, *Proclus, Neo-Platonic Philosophy and Science* (1996); *ECP* 452–454, D.J. O'Meara; H.D. Saffrey and A.-Ph. Segonds, *Marinus: Proclus ou sur le bonheur* (*CUF* 2001).

Alain Bernard

Prolegomena to Ptolemy's Suntaxis (*ca* 450 – 500 CE?)

Some 25 MSS of PTOLEMY's *Suntaxis* (early 9th c. and later) have a long introduction consisting of a preliminary chapter, to which alone the title *prolegomena* legitimately applies (ed.: Hultsch 1878: 3.XVII–XIX) and three independent studies, probably not from the same author: (1) on isoperimetric figures, deriving from an earlier treatment traditionally ascribed to ZĒNODŌROS (Knorr 1989: 725 and 738–741; ed.: Hultsch 1878: 3.1138–1165); (2) on the calculation of the volume of the Earth, perhaps deriving from PAPPOS with different, sometimes erroneous, calculations (ed.: Hultsch 1878, 3.XX–XXI); (3) on various calculation techniques: multiplication, division, extraction of square roots, division of ratios (partial ed.: Tannery, *Diophantos* 3–15 and *Mémoires Scientifiques* II.447–450; Knorr 1989: 185–210 and 787–793). This last study explicitly praises SYRIANUS, is close to DOMNINOS' style, and was plausibly written by a member of the Neo-**Platonist** circle in Athens around the middle of the 5th c. (Knorr 1989: 168).

Knorr convincingly refuted Mogenet's previous attribution to EUTOKIOS and proposed instead ARKADIOS, but the argument is weak. Mogenet conjectured that the text is a clandestine edition of notes on a public course on the *Almagest.* Later MSS (14th c. onwards) attribute the text to THEŌN OF ALEXANDRIA or DIOPHANTOS.

J. Mogenet, "L'introduction à l'Almageste" (1956); Knorr (1989) 155–211, 689–751, 787–793; Decorps-Foulquier (2000) 66, nn.27–30.

Alain Bernard

Promathos of Samos (460 – 430 BCE?)

Cited by the ARISTOTELIAN CORPUS ON THE FLOOD OF THE NILE, as the source of ANAXAGORAS' opinion that the Nile rises when snow melts on the mountain at the headwaters of the Khremes river, an opinion rejected by HĒRODOTOS 2.22, but echoed by ARISTOTLE, *Meteor.* 1.13 (350a14–350b23). The name is rare, *cf.* only *LGPN* 3A.378 (7th c. BCE).

RE 23 (1957) 1285–1286, Fr. Gisinger.

PTK

Prōros of Kurēnē (370 – 340 BCE?)

Said by IAMBLIKHOS to be from Kurēnē (*VP* 127, 267), he became impoverished in a political upheaval but was restored to fortune by KLEINIAS OF TARAS (DIODŌROS OF SICILY Book 10, *fr.*4.1; Iamblikhos, *VP* 239). Diodōros (16.2.1) and Pausanias (10.2.3) name Prōros of Kurēnē victor in the stadion sprint of the 105th Olympiad. His only known work, *On the Heptad*, of which two brief *testimonia* survive (pseudo-Iamblikhos, THEOL. ARITH. 7 [p. 57 de Falco]), discussed the holiness of the number seven.

Thesleff (1965) 154–155.

PTK and GLIM

Prosdokhos (200 BCE – *ca* 400 CE)

MARCELLUS OF BORDEAUX 29.55 (*CML* 5, p. 518) refers to his collection of recipes, and cites a quasi-magical recipe for *ikhneumōn*-urine and black-cow milk as a remedy for colon-troubles. The name is attested once elsewhere, Prosdokhē of Edessa (*LGPN* 4.292, *ca* 100 CE). *Cf.* perhaps PROËKHIOS.

Fabricius (1726) 380.

PTK

Prōtagoras (200 – 300 CE?)

Geographer, working in the tradition of PTOLEMY. He lived after him but long before MARCIANUS OF HĒRAKLEIA, who regarded him as one of the *arkhaioi andres* (ancient authors) and used his material extensively. Prōtagoras wrote a work in six books, entitled either *Geōmetria* (Phōtios, *Bibl.* 188) or *Geōgraphia tēs oikoumenēs* (*GGM* 1.543), preserved only in fragments. The first five books, written in a poor style (Phōtios), described Asia Minor, Libya, and Europe in general; the last book mentioned *doxologoumena* (marvelous things) of the ***oikoumenē***. Some material was taken from older sources, some based on autopsy

(John Tzetzēs, *Chil.* 7.647). Distances along coast-lines are given in stades (with the exception of north-eastern Europe). The compendium *Hupotupōsis geōgraphias en epitomē* (EXPOSITIO GEOGRAPHIAE), however, occasionally attributed to Prōtagoras, belonged most probably to the circle of the 9th c. patriarch Phōtios of Constantinople.

RE 18.3 (1949) 1160–1161, K. Ziegler; *RE* 23.1 (1957) 921–923 (#5), F. Gisinger; *NP* 10.458, H.A. Gärtner.

Andreas Kuelzer

Prōtagoras of Abdēra (*ca* 460 – 420 BCE)

The oldest and perhaps most important of the 5th c. sophists whose educational activities and intellectual interests belong more to the social sciences than the natural sciences. But one central doctrine of Prōtagoras, that "a human being is measure of all things," has important consequences for all kinds of inquiry. We have only a single sentence of Prōtagoras' own words on this topic. But as explained by both PLATO (*Tht.* 152a–c) and SEXTUS EMPIRICUS (*PH* 1.216–219), the doctrine entails that, for any individual in a given set of circumstances, the way things appear is the way they are. (Whether this is equivalent to relativism, as that term is normally understood, is another question.) Prōtagoras thus erases any distinction between appearance and reality, which seems seriously to undermine the impulse to scientific inquiry.

Plato connects Prōtagoras' "measure" doctrine with an ontology of radical flux. It is not clear, however, whether he means to attribute this to Prōtagoras himself, or suggest that Prōtagoras *should have* taken this as a corollary of the doctrine. What does seem to have been connected with the measure doctrine was a suspicion of claims about matters falling outside ordinary experience. Prōtagoras' famous expression of religious agnosticism is one of several indications of this attitude. ARISTOTLE *Metaph.* 3 (998a1–4) also reports that he took issue with geometers about whether a line touches a circle only at a point. It looks as if his point was that this is clearly not the case for *visible* straight and circular objects; the implication seems to be that any other kinds of objects, such as those of pure mathematics, are not even worth considering. Other indications suggest he was dismissive of mathematics, explained, presumably, in his *On Mathematics*.

ECP 455–458, P. Woodruff.

Richard Bett

Prōtagoras of Nikaia (100 BCE – 350 CE)

Authored a lost astrological work entitled *Sunagōgai*, part of which HEPHAISTIŌN OF THĒBAI paraphrases for doctrines relating journeys of individuals to planetary motions (3.30 and 3.47). The same work contained material on astrological medicine, for which it is cited, along with similar texts attributed to HERMĒS and PETOSIRIS, in an anonymous **iatromathematical** chapter in an 11th c. Byzantine astrological codex. The character of the doctrines indicates a date no earlier than the 1st c. BCE. DIOGENĒS LAËRTIOS (9.56) refers to an *astrologos* Prōtagoras who lived about 200 BCE, probably distinct from our Prōtagoras, and probably an astronomer rather than an astrologer.

Pingree (1978) 2.438–439.

Alexander Jones

Prōtarkhos (Mech. and Pharm.) (220 – 180 BCE?)

Named only by CELSUS, as one of those who, like NUMPHODŌROS and HĒRAKLEIDĒS OF TARAS (MECH.), built devices for the reduction of dislocations of the thigh (8.20.4); also cited for an ear-ointment (5.18.18) and a remedy against *scabies* (5.28.16–18).

Michler (1968) 49, 99.

PTK

Prōtarkhos of Bargulia (150 – 120 BCE)

Epicurean mathematician whose student was DĒMĒTRIOS OF LAKŌNIKA (STRABŌN 14.2.20), and to whom HUPSIKLĒS addressed his appendix of EUCLID.

RE 23.1 (1957) 924 (#5), W. Aly.

GLIM

Prōtarkhos of Tralleis (160 – 60 BCE)

Cited by IULIUS HYGINUS in MACROBIUS, *Sat.* 1.7.19, on early Italy, and by STEPHANOS OF BUZANTION on the Hyperboreans who live in and beyond the Alps (which appear first in POLUBIOS 2.14).

RE 23.1 (1957) 923–924 (#4), K. Ziegler.

PTK

Prōtās of Pēlousion (120 BCE – 80 CE)

ANDROMAKHOS, in GALĒN, *CMLoc* 10.2 (13.338 K.), cites his remedy against sciatica and headache. ASKLĒPIADĒS, in Galēn, *CMLoc* 4.7 (12.787–788 K.), cites the "Proteus" collyrium (a brand-name? emend to "*ΠΡΩΤᾹ*"?), containing **calamine**, ***khalkitis, psimuthion***, saffron, opium, white pepper, etc., in rainwater; AËTIOS OF AMIDA 7.114 (*CMG* 8.2, p. 388), ALEXANDER OF TRALLEIS (2.47 Puschm.), and PAULOS OF AIGINA 7.16.43 (*CMG* 9.2, p. 343) repeat the prescription. The use of white pepper suggests the *terminus post.*

RE 23.1 (1957) 924, H. Diller.

PTK

Prothlius/Protlius (?) (150 – 378 CE?)

"Only physician saved by war," came to Germany as a captive. He invented a plaster called *captiuum* to treat the stricken daughter of a "king" (M. Aurelius is the only emperor known to have been in Germany with a daughter, but her illness is otherwise unattested), and for his success was rewarded and freed together with his fellow-captives. The plaster – compounded from ocean water, natron, pure beeswax, roasted resin, sal ammoniac, ***opopanax***, **galbanum**, the beak of a dove, old olive oil, birthwort, ***psimuthion***, and the dung of a white dog – was efficacious against scrofulous tumors, abscesses, punctures, and calluses: "wherever you will have used it, you will praise it" extols Nicholas Myrepsus (1.202). Kühn reads the name as "Protlius." Neither variant is Greek or Roman, but perhaps emendable to Procilius or Procilianus.

C.G. Kühn, *Additamenta ad elenchum medicorum ueterum* 25 (1837) 5.

GLIM

Proxenos (120 – 30 BCE)

Used by ANTONIUS MUSA and commended by GALĒN, his "harmonious" remedy for long-standing coughs and fevers consisted in white pepper, opium, cardamom, saffron, raw sulfur, myrrh, white henbane seeds, and honey, administered with hydromel (Galēn, *CMLoc* 7.2 [13.61 K.]). This name is attested from the 5th c. BCE to 1st c. CE (*LGPN*).

RE 23.1 (1957) 1034 (#15), H. Diller.

GLIM

Prutanis (250 BCE – 80 CE)

ANDROMAKHOS, in GALĒN *CMLoc* 3.1 and 7.3, records two remedies of Prutanis. His compound to treat auricular inflammation, according with HARPALOS', contained myrrh, nard, saffron, burnt copper, opium, castoreum, and alum, taken with must when the sore is running, when painful with rose oil (12.627–628 K.). Prutanis' white pill for ***phthisis*** was compounded of myrrh, henbane seed, opium, ***sturax***, taken with ***staphis***, must or a date (13.73 K.). This rare name, attested from the heroic era into the 2nd c. CE (*LGPN*), primarily in the Black Sea area where eight Prutaneis are known, is a cognomen of the Augustan era and later (Solin 2003: 1.1090).

RE 23.1 (1957) 1158 (#6), H. Diller.

GLIM

PSEUDO-<NAME> ⇒ <NAME>

Ptolemaios (Pharm.) (70 – 90 CE)

ASKLĒPIADĒS PHARM., in GALĒN, describes a pharmaceutical Ptolemaios as an acquaintance, and records: **(a)** eye-ointment, with ***khalkitis***, ***misu***, realgar, cassia, ***malabathron***, myrrh, ***omphakion***, opium, pepper, and saffron, in gum and rainwater (*CMGen* 4.7, 12.789 K.), **(b)** pain-killer in cases of blood-spitting (***phthisis***?), of henbane, mandrake, opium, with saffron, cassia, etc. in Aminian wine (*CMLoc* 7.5, 13.101 K.), and **(c)** wound-powder of pine-bark, **calamine**, copper-flake, roasted deer-antler, etc. (*CMGen* 5.14, 13.849–850 K.). Shortly thereafter, Asklēpiadēs cites probably the same Ptolemaios, without epithet, for another wound-powder, of roasted lead, orpiment, copper-flake, ashed papyrus, and unfired sulfur (13.852–853 K.), and probably again, for a headache remedy (good for ***skotōmatics*** and epileptics) based on white hellebore: *CMLoc* 2.2 (12.584 K.). (Michler 1968: 122–125 wrongly equates this pharmacist with CELSUS' surgeon.) *Cf.* perhaps PTOLEMAIOS (ERASISTRATEAN).

Fabricius (1972) 224, 228.

PTK

Ptolemaios (Med.) (250 BCE – 25 CE)

Surgeon whose recipe to treat ear ulcers is preserved in CELSUS (6.7.2B–C), immediately after one by ERASISTRATOS. Diller distinguishes him from the homonymous **Erasistratean** physician, because Ptolemaios utilizes completely different ingredients than the Erasistratean; i.e., **mastic**, oak gall, ***omphakion***, and pomegranate juice; Michler however argues that Celsus is here giving recipes from **Erasistratean** sources.

RE 23.2 (1959) 1863 (#81), H. Diller; Michler (1968) 83–84, 122–125.

GLIM

Ptolemaios (Erasi.) (250 BCE – 100 CE?)

Erasistratean who taught that **dropsy** originated in the "hardness" of the liver, concluding that *parakentesis* (direct fluid extraction through an inserted tube) treated only the symptom not the cause of the disease: CAELIUS AURELIANUS *Chron.* 3.125, 130 (*CML* 6.1.2, pp. 754, 756). The name is much more frequent before the Roman conquests (*LGPN*), suggesting an early *terminus ante*, and our **Erasistratean** may be identical with CELSUS' surgeon or ASKLĒPIADĒS' pharmacist.

Michler (1968) 83–84, 122–125; Fabricius (1972) 224, 228.

PTK

Ptolemaios Platonikos (*ca* 50 – *ca* 250 CE)

Philosopher cited by IAMBLIKHOS OF KHALKIS (IŌANNĒS STOBAIOS 1.378.1–11 W.) and PROKLOS (*in Tim.* 1.20.7–9 D.). Ptolemaios may have commented on PLATO's *Timaeus*, since Proklos (*ibid.*) claims that Ptolemaios believed the fourth, missing person in the dialogue was Kleitophōn. Influenced by both Platonic and Aristotelian theories, Ptolemaios thinks that the soul is always in a body and moves from bodies of a rare nature into the oyster-like body, the vehicle of the soul. It can reside in the sensible world, inhabiting solid bodies (Stobaios 1.378.1–11 W.). Ptolemaios may also have compiled a list of ARISTOTLE's writings, along with his biography and testament (ELIAS, *in Cat.: CAG* 18.1 [1900] 107.11–14 [referring to a certain Ptolemaios Philadelphos], 128.6–7). In this context, his name is likewise mentioned by Arabic sources.

A. Dihle, "Der Platoniker Ptolemaios," *Hermes* 85 (1957) 314–325; *RE* 23.2 (1959) 1859–1860 (#69), A. Dihle; *PLRE* 1 (1971) 753 (#1); Moraux 1 (1973) 60–94; *NP* 10.571 (#68), M.-L. Lakmann.

Peter Lautner

PTOLEMAIOS OF ALEXANDRIA ⇒ PTOLEMY

***Ptolemaios of Kurēnē* (125 – 75 BCE)**

That the skeptic philosopher Ptolemaios of Kurēnē was an **Empiricist** physician as well is an idea suggested by the catalogue of skeptic philosophers in DIOGENĒS LAËRTIOS (9.115–116), where it is said that Ptolemaios, besides having relaunched skepticism after a period of decline, was the teacher of Hērakleidēs. The identification of this Hērakleidēs with HĒRAKLEIDĒS OF TARAS suggested the hypothesis that the **Empiricist** Ptolemaios was his teacher after his breaking with the **Hērophilean** MANTIAS. But this identification appears to be highly hypothetical, and it finds little support in the two fragments Deichgräber decided to ascribe to him (thinking that they had been transmitted by Hērakleidēs): the recipe against auricular ulceration that CELSUS (6.7.2B) ascribes to a *Ptolemaeus chirurgus* (PTOLEMAIOS (MED.)), and those against headache that GALĒN ascribes to a *Ptolemaios*; but Celsus' *Ptolemaeus* is possibly to be identified as PTOLEMAIOS (ERASI.), whereas Galēn's *Ptolemaios* is probably the same as the *Ptolemaios* of ASKLĒPIADĒS PHARM.: see PTOLEMAIOS (PHARM.).

Ed.: Deichgräber (1930) 20, 172 (fragments), 258.
RE 23.2 (1959) 1861 (#72), A. Dihle, 1863 (#80–82), H. Diller; C.A. Viano, "Lo scetticismo antico e la medicina," in G. Giannantoni, ed., *Lo Scetticismo antico* = *Elenchos* 6 (1981) 2.563–656 at 640; J. Barnes, "Ancient Skepticism and Causation," in M. Burnyeat, ed., *The Skeptical Tradition* (1983) 149–203 at 189–190 (n. 14).

Fabio Stok

Ptolemaios of Kuthēra (100 – 120 CE)

The *Souda* Pi-3032 says he wrote a didactic poem on the power and marvel of the plant *psalakanthē*: "power" (*dunamis*) could be pharmacological or magical. To be distinguished from his contemporary, the polymathic paradoxographer Ptolemaios "Khennos" ("Chennus"), son of Hephaistiōn, who mentioned the same plant (Phōtios, *Bibl.* 190, p. 150a20–37), citing a possibly-fictive line of Euboulos, *fr.*27 *PCG*.

RE 23.2 (1959) 1859 (#68), A. Dihle.

PTK

Ptolemaïs of Kurēnē (*ca* 50 BCE? – *ca* 50 CE?)

Musicologist, the only surviving fragments of whose catechetic manual *Pythagorean Elements of Music* (*Puthagorikē tēs mousikēs stoikheiōsis*) are preserved by PORPHURIOS in his commentary on PTOLEMY's *Harmonics* (22.22–24.6, 25.3–26.5 Düring). Porphurios' source for Ptolemaïs' writings may have been DIDUMOS (25.3–6).

Ptolemaïs presents the different types of musical theorists in a spectrum, arranged according to the importance they placed on either reason or perception. On one end of the spectrum are certain **Pythagoreans** who regarded reason as an autonomous criterion and excluded sensory data altogether; on the other are the "instrumentalists" (*organikoi*), who based their conclusions solely on the evidence of perception. The latter, she says, were followers of ARISTOXENOS, though she takes care to place Aristoxenos himself more centrally on account of his more balanced treatment of the necessary cooperation of the two faculties.

Ptolemaïs also discusses *kanonikoi*, "canonic theorists," who practiced a mathematical harmonic theory which she calls "canonic science" (*hē kanonikē pragmateia*), in which the monochord (*kanōn*) had a central role in demonstrating the numerical ratios of musical intervals to the ear. She locates canonic science at the meeting point between reason and perception; its fundamental postulates are drawn from the hypotheses of both the musicians (1–3 below) and the mathematicians (4–5): (1) that there are concordant and discordant intervals, (2) that the octave is made up of a fourth and a fifth, (3) that the tone is the excess of a fifth over a fourth, (4) that intervals are in ratios of numbers, and (5) that a note consists of numbers of collisions.

Ptolemaïs is an important source for our understanding of the range of approaches to harmonic science between ERATOSTHENĒS and THRASULLOS, and for the development of specific terminology within the discipline. She may in fact be the earliest extant author to use the term *kanonikē* to indicate mathematical harmonics, a label which gained common currency among contemporary or later authors (e.g. PHILŌN OF ALEXANDRIA, HĒRŌN OF ALEXANDRIA, GEMINUS, PANAITIOS THE YOUNGER, PLUTARCH, DIONUSIOS "THE MUSICOLOGIST").

Düring (1932); Barker (1989); Mathiesen (1999); *NP* 10.571–572, R. Harmon; *NDSB* 6.172–173, E. Rocconi.

David Creese

Ptolemy ("Claudius Ptolemaeus," 127 – after 146 CE)

Ptolemy (*Ptolemaios*) was the most important author working in the mathematical and physical sciences during the Roman Empire. His extant writings are devoted to astronomy, astrology, cartography, harmonic theory, and optics. A central concern of his work was the deduction of systems of models representing physical causes of various categories of phenomena, whether in the heavens or in our more immediate environment. From his own works we know that he made astronomical observations between 127 and 141 CE at Alexandria, and erected an inscription reporting the numerical details of his astronomical models at Kanobos ("Canopus," a suburb of Alexandria) in 146 or 147 CE. The order of several of his books is known from cross-references, and most were completed after the inscription. Authentic tradition may be behind OLUMPIODŌROS OF ALEXANDRIA's assertion that Ptolemy lived for 40 years in an isolated place called the "Wings" at Kanopos; the few medieval sources attesting Ptolemy's biography are untrustworthy or fictitious.

Astronomy: Among Ptolemy's several works on astronomy, occupying a central place is the *Almagest* – the medieval nickname derived from the Greek *megistos* ("greatest") by way of Arabic and Latin; Ptolemy entitled it *Mathematical Composition* (*Suntaxis Mathēmatikē*). This treatise in 13 books attempts to use mathematics – by which Ptolemy means the rational study of shape, number, size, position, and time in physical bodies – to establish models for the motions of the Sun, Moon, planets, and stars. The fundamental assumption is that these motions are combinations of uniform circular revolutions representing the spinning of spherical bodies of ***aithēr***. Starting from appropriately selected observations, subjected to mathematical analysis, Ptolemy demonstrates first the qualitative arrangement and then the quantitative details such as radii and rates of revolution in the various circles.

The opening chapters of Book 1 give empirical arguments for Ptolemy's basic cosmological framework, most of which would not have been controversial among contemporary astronomers. The Earth is spherical, stationary, and located at the center of the ***kosmos***. The Earth's size is negligible relative to the heavens, which taken as a whole revolve uniformly in an east-to-west direction around the Earth, causing the daily risings and settings of the visible heavenly bodies. The complex secondary motion of the Sun, Moon, and planets occurs from west to east along the **ecliptic circle**.

To account for the apparent irregularity in the motions of the heavenly bodies, Ptolemy employs two devices, introduced into Greek astronomy by the time of HIPPARKHOS OF NIKAIA: eccentric motion and **epicycles**. A uniform circular motion, when seen from a point off center, appears to vary in speed, and is called an eccenter; likewise a uniform circular motion, the center of which is carried uniformly in a circular path around the observer, will appear non-uniform, and is called an **epicycle**. Any periodic variation in apparent speed explainable by an eccentric model can equivalently be explicated by an epicyclic model, though the circles involved have different physical meanings. Ptolemy uses a simple eccenter for the Sun, but his models for the Moon and planets combine the two principles since the apparent motions of these bodies exhibited two intertwined periodicities. Moreover, in his models for the Moon and planets, Ptolemy considers that a motion is uniform if it sweeps out equal angles as seen from some fixed point which need not be the

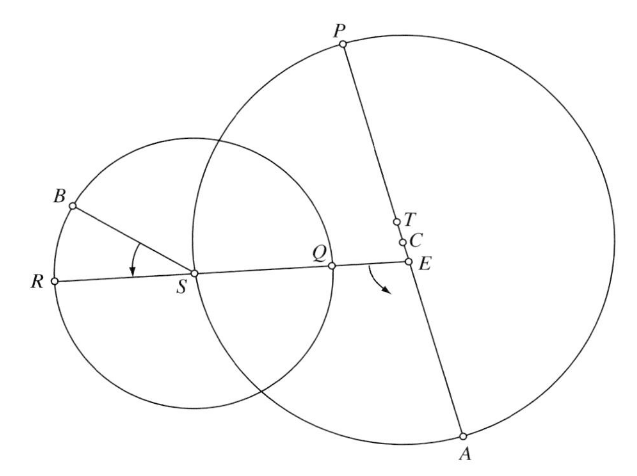

Ptolemy's model for the motion of Mars © Jones

circle's center. The model for Mars illustrates these concepts (***figure***). Planet *B* is assumed to revolve uniformly around an epicycle *RQ*. The center of this epicycle *S* revolves around an eccenter *AP*, the center of which *C* is displaced from the Earth *T*. The motion of *S* around the eccenter is uniform as seen, not from *C*, but from an "equant" point *E* which in this model is twice as far from the Earth as the eccenter's center.

The central argument of the *Almagest* begins, after various mathematical and geographical preliminaries, in Book 3 with the theory of the Sun, chosen because it can be established without recourse to assumptions about the motions of other heavenly bodies. In this and subsequent sections, Ptolemy follows a recurring pattern. First, the basic structure of the model is supported by very general observed facts. Then Ptolemy applies a geometrical and trigonometrical analysis to a small number of dated observations of the positions of the body to acquire numerical details. In his solar theory the deduction is straightforward and final, but for the other bodies Ptolemy must correct the initial data in the light of his first approximate results, so that the arguments are recursive and depend on convergence.

In discussing solar and lunar theory (Books 4–5), Ptolemy acknowledges Hipparkhos, who had developed some of the same deductive methods and arrived at some of the same results. However, Ptolemy explicitly takes credit for the discovery that the Moon has two periodicities. Book 6 applies the solar and lunar models to the study and prediction of eclipses. In Books 7–8, where Ptolemy again draws on Hipparkhos, Ptolemy shows that the stars, while maintaining configurations relative to each other, make a gradual revolution ("precession") around the poles of the **ecliptic**, and he presents a catalogue of 1022 stars. Books 9–13 are devoted to the five planets known to antiquity (Mercury, Venus, Mars, Jupiter, and Saturn). Ptolemy asserts that Hipparkhos contributed little to planetary theory; and it is striking that he acknowledges no debt in the *Almagest* to any astronomers during the intervening centuries except as observers.

The *Almagest* does not profess to be a historical account of discovery, and internal analysis of its details reveals that many of its results cannot have been obtained originally in the way that Ptolemy deduces them. Moreover, some of Ptolemy's reports of his own observations – perhaps all – have been adjusted or fabricated to agree closely with the theories. How much

Ptolemy appropriated from his predecessors is difficult to determine because of Ptolemy's silences and the dearth of other Greek technical literature on the subject, and there is no modern consensus.

A byproduct of Ptolemy's determining celestial models in the *Almagest* is a set of tables, interspersed among the chapters of the *Almagest*, enabling computation of a full range of phenomena including the positions of heavenly bodies at any given date. Subsequently, Ptolemy published a revised and expanded set of tables as the *Handy Tables*, used extensively in antiquity and the Middle Ages, especially by astrologers. Ptolemy gave a more physical account of the models in *Planetary Hypotheses*, in two books, surviving complete only in Arabic translation. He returned here to a question that he had regarded as inconclusive in the *Almagest*, the distances of the planets, proposing a system of nested and contiguous systems of etherial spheres in the order (outward from the Earth): Moon, Mercury, Venus, Sun, Mars, Jupiter, Saturn, stars. By way of Islamic astronomy this became the standard cosmological model until the 16th c.

Ptolemy's other astronomical writings are relatively minor. *Phaseis* is a ***parapēgma***, arranged according to the solar year and the principal latitudes of the Greco-Roman world. *Planispherium* (extant only through Arabic translation) is a study of stereographic projection, a mathematical technique for representing circles on the celestial sphere by circles in a plane, the basis of the primary astronomical instrument of late antiquity and the Middle Ages, the plane astrolabe. *Analemma*, extant in fragments in Greek and a more or less complete Medieval Latin translation, concerns the mathematical theory underlying sundials. Ptolemy also appears to have written monographs, now lost, on the theory of visibility of stars and of the planets Venus and Mercury.

Other sciences: The Harmonics, in three books, deduces models for systems of tuning employed by Greek musicians. Probably one of his earliest treatises, it contains discussions of scientific epistemology that have bearing on Ptolemy's work in other sciences. Ptolemy situates his own harmonic theory in relation to two faulty theories: that of the **Pythagoreans**, which modeled the intervals in Greek scales by means of a highly restricted set of ratios of whole numbers, and that of the Aristoxeneans, which was ostensibly empirical and eschewed ratios. These theories complementarily exemplify reason insufficiently controlled by the senses, and empiricism insufficiently controlled by reason, though Ptolemy's solution falls closer to the **Pythagoreans** by embracing a more flexible system of whole-number ratios. Ptolemy's central claim is that the general constraints he proposes for the ratios of musical intervals within scales lead to a finite set of possible scales that is almost coextensive with the scales employed by contemporary musicians. Experimental apparatus plays an important role in this complex work.

The *Optics*, in five books, is a study of the phenomena of visual perception, including long treatments of binocular vision and the appearances of objects seen reflected in mirrors or refracted through the interfaces between different transparent media. It unfortunately survives only in a defective Medieval Latin translation of a lost Arabic translation, lacking the whole of Book 1 and end of Book 5. Ptolemy's model for vision assumes that it is effected through a cone-shaped visual ray with its vertex at the eye; perception occurs along straight lines emanating from the cone's vertex. Ptolemy probably thought of the ray as an alteration of the exterior environment caused by the human soul. In contrast to the model of visual rays in EUCLID's *Optics*, there are no gaps between potential lines of sight in Ptolemy's cone, and it has some ability to perceive the distance of objects. Again Ptolemy makes appeal to experimental apparatus, most impressively on binocular vision and measuring angles of refraction at boundaries between different media (air, water, and glass).

The treatise in four books on astrology known as the *Tetrabiblos* (Ptolemy's title for it is not known) makes a case for astrology as a legitimate though inexact science, primarily physical rather than mathematical, and grounded in the interplay of reason and tradition rather than reason and observation. Ptolemy divides his subject into a more reliable "general" astrology treating the influence of celestial bodies on entire geographical regions (Book 2) and a personal astrology covering influences on individual lives and characters (Books 3 and 4).

In the *Geography* (properly *Guide to Drawing Maps of the World*) Ptolemy addressed the problems of how best to determine positions on the globe of localities throughout the known world from the source materials, mostly unscientific, available to an ancient geographer; how to present this information in an image conveying the impression of the Earth's spherical shape; and how to transmit this picture accurately from copy to copy. Of the eight books, almost six consist of a list of about 8,000 localities with their assigned latitudes (in degrees north or south of the equator) and longitudes (in degrees east of the meridian through the Isles of the Blest, i.e. the Canaries). Ptolemy explains how to construct a map from these data on a large globe or on a plane surface, employing sophisticated grids of circular arcs and straight lines representing parallels and meridians. Ptolemy's map was closely modeled on MARINOS OF TYRE, though he incorporated new information especially concerning east Africa and south Asia.

On the Kriterion and ***Hēgemonikon*** is a brief work presenting an eclectic Hellenistic approach to the classification of epistemological standards. The authenticity of this work has been disputed. SIMPLICIUS (*Comm. in Aristotelem de caelo* = *CAG* 7 [1894] pp. 20 and 710) cites what appears to be a single lost work by the differing titles *On Weights* and *On the Elements*. This book replaced Aristotle's theory of the natural motion of the five elements earth, water, air, fire, and ***aithēr*** with a theory resembling that of XENARKHOS.

F. Boll, "Studien über Claudius Ptolemaeus," *Jahrbücher für Classische Philologie*, S.21 (1894) 51–244; *RE* 23.2 (1959) 1793–1831, 1839–1853, 1858–1859, B.L. van der Waerden; *DSB* 11.186–206, G.J. Toomer; *Idem, Ptolemy's Almagest* (1984); A. Lejeune, *L'Optique de Claude Ptolémée dans la version latine d'après l'arabe de l'émir Eugène de Sicile* (1989); A. Murschel, "The Structure and Function of Ptolemy's Physical Hypotheses of Planetary Motion," *JHA* 26 (1995) 33–61; A. Barker, *Scientific Method in Ptolemy's Harmonics* (2000); J.L. Berggren and Alexander Jones, *Ptolemy's Geography: An Annotated Translation of the Theoretical Chapter* (2000); Alexander Jones, "Ptolemy's Canobic Inscription and Heliodorus' Observation Reports," *SCIAMVS* 6 (2005) 53–97; *NDSB* 6.173–178, Alexander Jones.

Alexander Jones

Publius of Puteoli (60 – 80 CE)

Cited thrice by ANDROMAKHOS, for various remedies (in GALĒN *CMLoc* 9.4 [13.281 K.], *CMGen* 2.15 [p. 533], 5.13 [p. 842]), and once by ASKLĒPIADĒS PHARM., *ibid.* 5.14 (13.852 K.) who calls him his teacher.

RE 23.2 (1959) 1936 (#3), H. Diller.

PTK

Puramos (250 BCE – 95 CE)

Asklēpiadēs Pharm. records in Galēn *CMLoc* 4.7 (12.777–778 K.) his eye-salve, treating scars and swelling/calluses, and all protuberances, compounded from Cyprian **calamine**, hematite, roasted copper, roasted ***misu***, saffron, ***ammōniakon*** incense, copper scales, heated and cooled, to which are then added opium, myrrh, gum and very sour vinegar. This rare name, known from the 5th c. BCE to 3rd c. CE, is concentrated in western Greece and Magna Graecia (*LGPN*).

RE 24 (1963) 11 (#4), H. Diller.

GLIM

Purgotelēs (or Ergotelēs?) (280 – 260 BCE)

Son of Zōēs, master ship builder for Ptolemy II Philadelphos, for whose services Philadelphos erected in the temple of Aphrodite at Paphos a statue whose inscription (*OGIS* 39) cites Purgotelēs' two ships, one a "thirty" the other a "twenty" (*cf.* Athēnaios, *Deipn.* 5 [203c–d], and see Casson pp. 99–116). The inscription is damaged, and only . . .*ΟΤΕΛΗΣ* is preserved. Purgotelēs is attested once at Dēlos and four times at Rhodes (2nd c. BCE: *LGPN* 1.396); the sole alternate supplement, Ergotelēs, is more common, attested 30 times from the 5th–2nd cc. BCE (*LGPN* 1.162, 2.155, 3A.151, 3B.142), though also not on Cyprus. For Zōēs, see *LGPN* 1.195 (Cyprus and Lesbos).

RE 24 (1963) 49 (#2), E. Fabricius; L. Casson, *Ships and seamanship in the ancient world* (1971; repr. 1986).

GLIM

Purrhos of Ēpeiros (and Macedon) (295 – 275 BCE)

Athēnaios Mech. p. 5 W. lists Purrhos "of Macedon" (after Daimakhos, Kharias, and Diadēs) as a writer on siege engines (*cf.* Vitruuius 7.*pr*.14). He also wrote so well on mining and tunneling that Athēnaios had nothing to add (p. 31 W.). Plutarch, *Pyrrh.* 8, Aelianus, *Taktika* 1.2, and Cicero, *ad Fam.* 9.25.1, record that King Purrhos of Ēpeiros wrote a *Taktika*, abridged by his minister Kineas. Purrhos was king of Macedon for less than a year (287 BCE: Plut., *Pyrrh.* 11).

OCD3 1283, P.S. Derow.

PTK

Purrhos of Magnesia (250 BCE – 100 CE)

Wrote a commentary on Aratos (*FGrHist* 1026 T19), entirely lost.

(*)

PTK

Puthagoras (Med.) (450 – 50 BCE)

Physician, wrote a work *On Squill* (*skillēs: cf.* Epikharmos), or less likely *On Hernia* (*kēlēs*), and perhaps a book about Homer: Dēmētrios of Magnesia, in Diogenēs Laërtios 8.47 (later than Puthagoras of Rhēgion, the sculptor); pseudo-Galēn, *Eupor.* 3 (14.567 K.) cites Galēn for "Pythagoras on squill." Two fragments in Arabic concerning urine are attributed to a Pythagoras of Alexandria (Ullmann 1970: 82). Aelianus' notice regarding Puthagoras' interest in medicine may refer to our author rather than the philosopher (*VH* 9.22), if it does not refer to the pseudo-Puthagoras of Dioskouridēs; *cf.* also Pamphilos of Alexandria. Ziegler distinguishes the physician from the author of the Homeric treatise.

RE 24 (1963) 305 (#12b), H. Diller and K. Ziegler.

PTK and GLIM

Puthagoras (of Alexandria?) (275 – 265 BCE)

Officer under Ptolemy Philadelphos, whose lost *On the Red Sea* described stones (Iouba in Pliny 37.24), musical instruments of the Trogodutes (Ath., *Deipn.* 4 [183f], 14 [634a]), and the long-tailed monkey (Aelianus, *NA* 17.8).

NP 10.654 (#6), H.A. Gärtner.

PTK

Puthagoras of Samos ⇒ Pythagoras

Puthagoras of Samos, pseudo (Astrol.) (*ca* 1 – 150 CE)

Three extant treatises on occult sciences are attributed to Pythagoras of Samos. Two are astrological, *The Pebble* (*CCAG* 11.2.124–125), describing planetary influences on character (from Saturn inward to the Moon), and *On the Forms and Indications of the 12 Signs of the Zodiac* (*CCAG* 11.2.135–138), which has been shown to depend on Dōrotheos of Sidōn (Pingree 1978: 292–299). The third is an untitled work on numerological onomancy, which takes the form of a letter addressed to his son Tēlaugēs (*CCAG* 11.2.139). Hippolutos of Rome cited this text *in extenso* (*Haer.* 4.13.1–14.20).

Diels (1905) 87–88; P. Tannery, "Notice sur des fragments d'Onomatomancie arithmétique" (1886), repr. in *Mémoirs scientifiques* 9 (1929) 17–50.

Bink Hallum

Puthagoras of Zakunthos (500 – 450 BCE?)

Professional *kitharodos* and theoretician probably active in the mid-5th c. BCE (according to Diogenēs Laërtios 8.46), he is also recalled by Aristoxenos for his theoretical interests in describing different forms of scales without achieving a complete enumeration of their possible structures and without investigating the general principles governing them (*El. Harm.* 36.33). As stated by a fragment of Artemōn of Kassandreia in Athēnaios, *Deipn.* 14 (637b–f), he became famous for the invention of an instrument resembling the Delphic tripod (hence called *tripous*), a triple kithara where three sets of strings were put on a revolving base with a big common sound-box, so that the Dorian, Phrygian, and Lydian *harmoniai* could all be played.

M.L. West, *Ancient Greek Music* (1992) 226.

E. Rocconi

Putheas of Massalia (*ca* 320 – 305 BCE)

Explorer and geographer of uncertain date; Dikaiarkhos' criticism (Strabōn 2.4.2) seems our earliest reference. Polubios described him as a private individual and a poor man (*ibid.*). His motivation may have been commercial, to find northern sources of tin and amber, but his interest in science was genuine, and he may have been a student of EUDOXOS OF KNIDOS. Although later writers such as Polubios and Strabōn made him out to be a liar, ERATOSTHENĒS, HIPPARKHOS, PLINY, and others used his observations, and Putheas' contributions to descriptive and mathematical geography and astronomy are fundamental.

In one or more works, entitled *On the Ocean* and *Circuit of the Earth*, Putheas described his journey from Marseilles to northern Europe via the Atlantic coast, either in a Massilian ship out the Pillars of Hēraklēs around the Iberian peninsula to the British Isles, or on foot and on native craft. He described Britain as three-sided and estimated its circumference with reasonable accuracy. He also mentioned the island of Thule, six days beyond Britain, whether referring to the islands north of Scotland, or to Norway, Jutland, or even Iceland, is a matter of longstanding debate. He described the enigmatic Congealed Sea, like a sea-lung (jellyfish?), wherein earth, air and water were mixed together; perhaps a first- or second-hand report of pack or slush ice. He mentioned finding the source of amber on an island Abalus, in an estuary on the northern shore of Europe; perhaps west of Jutland, or even east into the Baltic. A reference to reaching the Tanais (the Don) must be discounted.

The perspicacity of Putheas' observations is astonishing. He calculated the latitude of Marseilles fairly exactly using a gnomon. He established, through careful observation, that the celestial North Pole was occupied not by a star, but by empty space bordered by four stars. He also observed ocean tides, and proposed that their amplitude depends on lunar phases. He reported an improbably high tide in northern Britain – likely an exaggerated account of surge tides of the Pentland Firth or regions around the Scottish islands. As he journeyed north, he recorded the lengthening of the solsticial day, although his claim to have witnessed the midnight sun north of the Arctic Circle is doubtful.

Ed.: C.H. Roseman, *Pytheas of Massalia: On the Ocean* (1994); S. Bianchetti, *Pitea di Massalia. L'Oceano* (1998).

DSB 11. 225–226, A. Diller; B. Cunliffe, *The Extraordinary Voyage of Pytheas the Greek* (2001).

Philip Kaplan

Putheos of Priēnē (*ca* 370 – 330 BCE)

Architect, sculptor and author, collaborated with SATUROS OF PAROS on the Maussōlleion, and designed the Temple of Athena at Priēnē. VITRUUIUS (7.*pr*.12) credits him with a commentary on the Maussōlleion (written with Saturos), and one on the Temple of Athena that included comments on the architect's proper education and on the Doric order's disadvantages (Vitr. 1.1.12, 4.3.1). The Maussōlleion (funerary monument for Maussōllos, d. 352 BCE) was famed for its great height and extensive, high-quality sculptural decoration. Putheos may be the "Puthis" mentioned by PLINY (36.30) as the sculptor (or designer) of the marble four-horse chariot on the top of the Maussōlleion. The Temple of Athena at Priēnē, financed and dedicated by Alexander the Great *ca* 334 BCE, admired for its proportions, became a standard model for the Ionic order. Putheos is assumed to have taken great interest in theories of proportion and design and is believed to have influenced later theoreticians such as HERMOGENĒS, and eventually Vitruuius.

Svenson-Ebers (1996) 116–150; K. Jeppesen, *The Maussolleion at Hallikarnassos* 2 (1986) 52–113; 5 (2002) 29–42; *KLA* 2.334–338, W. Hoepfner.

Margaret M. Miles

Puthiōn (Pharm.) (50 – 30 BCE)

ASKLĒPIADĒS PHARM., in GALĒN *CMGen* 2.17 (13.536–537 K.), preserves his fracture-compound, used by the freedman Helenos: asphalt, Bruttian pine pitch, beeswax, frankincense, and copper flakes in olive oil and vinegar. AUGUSTUS' freedman C. Iulius Helenos held Sardinia for him in 40 BCE: *BNP* 6 (2005) 67 (#3).

RE 24 (1963) 568 (#6), H. Diller, 1429 (#2), R. Hanslik.

PTK

Puthiōn of Rhodes (325 – 90 BCE)

Author of a treatise on agriculture excerpted by CASSIUS DIONUSIOS (VARRO, *RR* 1.1.9–10).

(*)

Philip Thibodeau

Puthiōn of Thasos (*ca* 255 – 235 BCE)

Wrote a lost letter to KONŌN proposing the problem of finding a mirror surface that reflects solar rays to meet the circumference of a circle: DIOKLĒS, *On Burning Mirrors* 1.

Toomer (1976) 138.

PTK

Puthios (325 BCE – 80 CE)

ANDROMAKHOS, in GALĒN *CMLoc* 5.5 (12.879–880 K.), gives his treatment for loose teeth and gingivitis (alum roasted in papyrus, mixed with ruddle). For the rare name, probably not a mistake for PUTHIŌN, see *LGPN* 1.392, 2.386, 3A.380, 4.295.

Fabricius (1726) 382.

PTK

Puthoklēs (450 BCE – *ca* 350 BCE)

Cited in the HIPPOKRATIC CORPUS, EPIDEMICS, 5.56 = 7.75 (5.238, 434 Littré), for curing patients through use of water or watered milk. (For the name, *cf.* Phaidros' father in PLATO, *Phdr.* 244a.)

Fabricius (1726) 382.

PTK

Puthoklēs of Samos (325 BCE – 200 CE)

Was the author of a treatise on agriculture, according to pseudo-PLUTARCH, *Para. Min.* 41. The *Parallels* contains much fabricated information, but the *History of Italy* which it ascribes to Puthoklēs (14) is independently attested (Clement of Alexandria, *Strom.* 1.135).

Ed.: *FGrHist* 833.
RE 24 (1963) 601 (#10), K. Ziegler; *KP* 4.1279–1280, O. Dreyer.

Philip Thibodeau

Pythagoras of Samos (*ca* 570 – 495 BCE)

In the ancient tradition Pythagoras (Grk. *Puthagoras*) is presented as a philosopher, scientist, religious reformer and politician. The old debate on the reliability of this image is still unresolved. Some scholars accept evidence for Pythagoras' scientific and philosophical activities, but disagree on what can be safely ascribed to him; others regard him only as a religious leader and moral reformer. Like THALĒS and Sōcratēs, Pythagoras wrote nothing, whereas his pupils and followers, unlike Sōcratēs' pupils, did not take care to expose his ideas. Absence of direct sources is only partially compensated by a very extensive indirect tradition, which can be used only as far as it goes back to the 5th/4th centuries.

Pythagoras left Samos *ca* 530 because of Polukratēs' tyranny and moved to Krotōn. Owing to his talents and charisma he found here many supporters and founded a political community. A special way of life and cultivation of friendship contributed to the rallying of the Pythagoreans; many of Pythagoras' ethical rules had a religious basis and were supported by belief in his god-like nature. The Pythagoreans' influence increased after the defeat of Subaris by a Krotonian army under Pythagorean command (*ca* 510). Shortly after this, an opposing faction of the Krotonian elite organized an anti-Pythagorean revolt; Pythagoras fled to Metapontion, where he soon died.

Pythagoras' teaching has to be considered in a context of the Ionian natural philosophy and science. A cosmogony that might go back to him explains the origin of the world by the interaction of two principles, "limit" and "unlimited." The "unlimited" is identified with an empty space and with an infinite ***pneuma*** that surrounds ***kosmos***. It is inhaled into the ***kosmos*** and, limited by "limit," begins to separate individual things from each other. Opposite principles of a different kind play a further and important role both in the **Pythagoreans** (ALKMAIŌN, MENESTŌR, PHILOLAOS) and in other Italian philosophers (PARMENIDĒS, EMPEDOKLĒS).

Pythagoras' contributions to cosmology and astronomy are hard to discern. Such important discoveries as the sphericity of the Earth, a division of heavenly and terrestrial spheres into zones, an identification of the Evening and Morning star with Venus, are ascribed both to Pythagoras and Parmenidēs. Independent planetary movement from west to east and on circular orbits are first attested in the **Pythagorean** Alkmaiōn (24 A4, 12 DK). It is possible that relying on ANAXIMANDROS' concept of "geometrical ***kosmos***," Pythagoras transferred to the planets the circular motion inherent in the Sun, the Moon and stars in Anaximandros' system. According to the early **Pythagorean** theory of "heavenly harmony," the circular motions of all the heavenly bodies produce sounds; their pitch depends on the speed of motion, which, in turn, corresponds to the relative positions of the heavenly bodies: the farther from the Earth the greater speed of rotation. The speeds correspond to each other as the harmonious intervals, so that common circular movement of all the bodies generates harmonious sound.

The search for the heavenly harmony was undoubtedly prompted by Pythagoras' discovery of the numerical expression of harmonic intervals: octave (2:1), fourth (4:3) and fifth (3:2). Most probably, he obtained these results by dividing the string of a monochord; further scientific experiments in acoustics were carried out by Pythagoras' student HIPPASOS. Pythagoras' discovery laid the foundations of the mathematical harmonics and contributed to the formation of the mathematical quadrivium: geometry, arithmetic, astronomy and harmonics (ARKHUTAS 47 B1 DK). The theory of proportions valid for commensurable magnitudes became a link between all the four sciences. It is very probable that Pythagoras knew the geometric, arithmetic and harmonic means. In geometry, where he continued the line of Thalēs, the deductive proof of Pythagoras' theorem is ascribed to him (empirical formulas for some of the "Pythagorean triplets" – 3, 4, 5, etc. – were known already in Babylōn). Pythagoras further applied the technique of deductive proof to arithmetic; to him must go back one of the earliest samples of the theoretical arithmetic – the theory of even and odd numbers (Euclid, *Elem.* 9.21–34) using an indirect proof. The method of deductive proof was further transferred from mathematics to philosophy by Parmenidēs, who had a **Pythagorean** teacher.

DK 14; *RE* 24 (1963) 171–203, K. von Fritz; *HGP* v. 1; Burkert (1972); L. Navia, *Pythagoras: An Annotated Bibliography* (1990); Zhmud (1997); C. H. Kahn, *Pythagoras and the Pythagoreans: A Brief History* (2001).

Leonid Zhmud

Q

Quadratus (*ca* 100 BCE – 80 CE)

ANDROMAKHOS in GALĒN, *CMGen* 7.13 (13.1034 K.), records Quadratus' ***akopon***, of a baker's dozen ingredients. The cognomen is recorded from the late Republic – *MRR* 1.196 (L. Ninnius, *tr. pl.* 58 BCE), 2.239=3.224, and plausibly restored in *LGPN* 2.271 – and later, e.g., TACITUS, *Ann.* 6.7 and *PIR2* I-507.

PIR2 Q-4.

PTK

De Quaternionibus (180 – 300? CE)

Neo-**Pythagorean** author who cites PTOLEMY, *Almagest* 8.4, in arguing that the **Pythagorean** *tetraktus* rules and orders the ***kosmos***. He cites the seasons, the lunar phases, the elements, the **humors**, and the dimensions (including "point") as instances of cosmological tetrads, and adds the four astrological centers (Horoscopus, Midheaven, Setting, and Nadir), and the quarters of the year (two equinoxes and two tropics); *cf.* THEŌN OF SMURNA (pp. 93–99 H.) and ANTIOKHOS OF ATHENS (*CCAG* 1 [1898] 143, 146; 8.3 [1912] 105). His astrological contribution is to argue that the planets' powers shift as they pass the centers, and that they are more effective when in quadrature than in trine.

CCAG 9.1 (1951) 172–174.

PTK

Sex. Quintilii, Condianus and Valerius Maximus, of Alexandria Troas (*ca* 140 – *ca* 182 CE)

Brothers whose lifelong harmony was proverbial: they were *consules ordinarii* in 151, both held influential offices under Marcus Aurelius, and were executed under Commodus, their sumptuous villa outside of Rome confiscated. Authors of a comprehensive agricultural treatise which was cited by GARGILIUS MARTIALIS and VINDONIUS ANATOLIOS, and of which echoes survive in the *Hippiatrika* (*Hipp. Berol.* 1.18) and the GEŌPONIKA. The calendar attributed to them (*Geōpon.* 3.1) is independently known (Boll).

F. Boll, "Der Kalender der Quintilier," *SB Heidelberg* 1 (1911) 3–18; *PIR2* Q-21, 27, W. Eck; *OCD3* 1291, A.R. Birley; *NP* 10.702 (#II.1, 6), W. Eck.

Robert H. Rodgers

Quintus (of Pergamon?) (*ca* 115 – *ca* 145 CE)

Student of MARINOS, and praised by GALĒN (*Prognosis* 1 [*CMG* 5.8.1, pp. 70–72]) as the best doctor of his era, he made anatomical discoveries that he taught to his students (including LUKOS OF MACEDON and SATUROS OF SMURNA), but did not publish (Galēn, *Anat. Admin.* 14.1 = Duckworth 1962: 183). He also commented upon the HIPPOKRATIC CORPUS, EPIDEMICS, perhaps published (*In Hipp. Epid. III* [*CMG* 5.10.2.1, pp. 14–17, 59], *In Hipp. Epid. VI* [*CMG* 5.10.2.2, pp. 212, 314]). He practiced in Rome under Hadrian, but was banished thence on a charge of malpractice, and died (in Pergamon?) in Galēn's youth (*Anat. Admin.* 1.2 [2.224–225 K.]). Galēn wrote a (lost) work in support of Quintus' criticism of the four qualities (*GAS* 3 [1970] 167), and describes him as being like an **Empiricist**, but not of that school (*In Hipp. Epid. I, CMG* 5.10.1, pp. 6, 17, 52). Galēn explains that Quintus substituted Pontic nard (*karpēsion*) for cinnamon (*Antid.* 1.14 [14.69–72 K.]); OREIBASIOS, *Syn.* 3.192 (*CMG* 6.3, p. 115), preserves Quintus' henbane- and opium-based anodyne.

Grmek and Gourevitch (1994) 1503–1513.

PTK

Quirinus (350 – 450 CE?)

With the assistance of MARCELLUS, wrote a *Mēkhanikē*, according to Leōn, *Anth. Gr.* 9.200. Leōn praises technical works of the 4th–5th centuries CE (9.201–202), but also of the late 3rd c. BCE (9.578), so if Marcellus is the taker of Surakousai, Quirinus could instead be contemporary with PHILŌN. (Note that ATHĒNAIOS MECH. dedicated his *On Machines* to Marcellus, usually taken to be AUGUSTUS' nephew.) The name may be a pseudonym, since Quirinus is the Sabine god of war.

PLRE 2 (1980) 933 (#1); Netz (1997) #136.

PTK

R

Rabirius (*ca* 150 BCE – 75 CE)

PLINY records his advice that human milk benefits the bowels and serves as an emmenagogue: 28.74 (note 1.*ind*.28). To be distinguished from CICERO's two clients, and unlikely to be the epic poet contemporary with OVID (*Ex Ponto* 4.16.5; *FLP* 332); the name is not rare enough to identify him with the **Epicurean** Rabirius in Cicero, *Acad.* 1.5, much less the Rubrius of Pliny 29.7; *cf.* also *LGPN* 1.398, *MRR* 2.35, 2.273.

RE 1A.1 (1914) 23 (#2), H. Gossen.

PTK

Ravenna Cosmography **(600 – 720 CE?)**

A Latin work composed in Ravenna. It begins with an introduction that contains biblical and patristic references and places the geographical account in the framework of Christian knowledge about the world created by God. The work combines late Roman and biblical traditions. Thus, following a pattern already established by earlier Christian writers, the author uses the notion of a tripartite division of the Earth, which became traditional in classical geography, and supplies from the Bible the names of the sons of Noah who settled in each continent. The work describes the Earth's three continents and contains lists of geographical names partially arranged in the order of Roman provinces. Most of the sources used by the anonymous author are now lost (*cf.* CASTORIUS). Similarities to the PEUTINGER MAP suggest that both works ultimately go back to a common exemplar, probably a Roman road map.

Ed.: J. Schnetz, *Ravennatis Anonymi Cosmographia = Itineraria Romana* 2 (1990) 1–110.
RE 1A.1 (1914) 305–310, G. Funaioli; *KP* 4.1343, Fr. Lasserre; *NP* 4.934, K. Brodersen.

Natalia Lozovsky

Remmius Fauinus (300 – 400 CE)

Some MSS attribute the CARMEN DE PONDERIBUS ET MENSURIS to Remus or Rem(m)ius Fauinus (or Fauianus). He has been identified with Dunamius, alias Flauinus, a poet friend of Ausonius (*ca* 310–400? CE).

PLRE 1 (1971) 325; K.D. Raios, *Recherches sur le Carmen de ponderibus et mensuris* (1983) 27–45.

Mauro de Nardis

Rhēginos (*ca* 65 – 180 CE)

Physician, listed among the **Methodists** post-dating Themisōn and Thessalos (Galēn *MM* 2.7.5 [10.52–53 K. = p. 27 Hankinson]), omitted by Tecusan 2004.

Edelstein (1935) col. 358 = (1967) p. 173.

GLIM

Rhētorios (600 – 700 CE?)

Extremely shadowy figure, apparently participated in forming large compilations of Greek astrological texts now found in many Byzantine codices. His authentic contribution is obscured by misattributions by later Byzantine scholars and by frequently speculative identifications of authors on the part of the editors of the *CCAG* (1898–1940), still the only published repository for the majority of the relevant texts. Pingree hypothesized that Rhētorios assembled a lost enormous anthology of astrological chapters that was the common ancestor of selections preserved in two MSS now in Paris.

Ed.: D.E. Pingree, *Rhetorii Aegyptii compendium astrologicum [. . .], imprimendum curavit* S. Heilen (Teubner, forthcoming).

D.E. Pingree, "Antiochus and Rhetorius," *CPh* 72 (1977) 203–223.

Alexander Jones

Rhoikos of Samos (550 – 500 BCE)

Son of Phileus, sculptor and architect of an archaic Temple of Hēra at Samos (begun *ca* 530 BCE, Hēraion IV), then the largest temple in Greece (*ca* 55 × 110 m). Hērodotos (3.60) names Rhoikos the first architect of this temple; later authors link him with Theodōros of Samos for this and other projects. Vitruuius (7.*pr.*12) names them as co-authors of a book on the temple. Pausanias (10.18.5) cites Rhoikos as sculptor of a statue at the Artemesion at Ephesos. Under Rhoikos' supervision, the problem of stable foundations for the temple at Samos was solved.

Svenson-Ebers (1996) 7–49; *KLA* 2.351–352, H.J. Kienast.

Margaret M. Miles

Ripalus (50 BCE – 80 CE)

Andromakhos in Galēn, *CMLoc* 7.3 (13.64 K.), records his cough-syrup for ***phthisis*** and recurrent fevers, composed of nard, myrrh, cinnamon, black and white pepper, poppy juice, henbane seed, etc., in honey-wine. He called his recipe *Ambrosia* or *Mēnodōrios*, presumably from Mēnodōros of Smurna.

Fabricius (1726) 383.

PTK

Romula (180? – 400? CE)

Aëtios of Amida 16.141 (Zervios 1901: 171) records her uterine fumigation, compounding cinnamon, saffron, ***kostos***, myrtle, spikenard, ***sturax***, etc. (*cf.* Theopompos). He describes her as *kurias*, an honorific for women over 40, in use by the late 2nd c. CE, and probably out of use once Christianity became dominant: *cf.* Williger in *RE* 12.1 (1924) 176–183.

(*)

PTK

Rufīnos of Antioch (*ca* 402 CE)

Architect from Antioch summoned by Bishop Porphurios in 402 CE to build the cathedral of Gaza. Mark the Deacon (*V. Porph.* 78.1), calling him an *arkhitektōn*, describes the process of construction in great detail. The empress Eudoxia prescribed the plan which Rufīnos marked on the ground during a public ceremony.

Downey (1948) 104; *Idem, Gaza in the Early Sixth Century* (1963) 26–29; *ODB* 157, M.J. Johnson *et al.; PLRE* 3 (1992) 952 (#4).

Kostis Kourelis

RUFINUS ⇒ VIBIUS RUFINUS

RUFIUS FESTUS AUIENUS ⇒ AUIENUS

RUFUS ⇒ (1) MENENIUS; (2) VALGIUS; (3) VITRUUIUS

Rufus of Ephesos (*ca* 70 – 100 CE)

Rufus of Ephesos (*Vind. Med. Gr.* 1, f.3^{V}) © Österreichische Nationalbibliothek

Despite fame in his own era, Rufus' historical contexts and milieu are not certain. The *Souda* (Gamma-241) says that he and KRITŌN were physicians under Trajan (98–117 CE). GALĒN, however, in quoting DAMOKRATĒS' didactic poem on the Egyptian triplicate-use incense *kuphi* (*Antid.* 2.2 [14.117–119 K.]), reveals that Damokratēs cites Rufus for *his* work on *kuphi*, and since Damokratēs worked under Nero and Vespasian (54–77 CE), Rufus presumably lived a full generation before the decades given by the *Souda*, which often skips the Flavians.

His birthplace was the thriving commercial center of Ephesos, and the fuzzy traditions in later Byzantine and Arabic sources suggest he practiced medicine in his home city; no evidence suggests that Rufus ever was in Rome, although he seems to have traveled widely in the eastern half of the Roman Empire. Two sources probably suggest that Rufus studied medicine in Alexandria: at *Anatomical Nomenclature*, 133 (p. 151 DR), Rufus mentions that the Egyptians have their own names for the parts of the body, and two passages in *Interrogation of the Patient* (12.67–68, 70 [pp. 44, 46 Gärtner]) are striking observations on guinea-worm infestations in Egypt. By Rufus' day, human dissection was no longer an option for medical students; in fact, dissection of animals ". . ..most closely like a human being. . ." was the accepted norm, whereas "in ancient times, [the internal parts] were learned from a man" (*Nomenclature* 9–10, p. 134 DR).

Rufus was principally a clinician, and his talents in diagnostics are sharply revealed in the *Case Histories* (preserved in an Arabic translation: Ullmann 1978); even without direct anatomical observations, Rufus' *Kidney and Bladder Diseases* (Sideras 1977) discloses consummate skills in treatment of common urological ailments, and long experience with many patients brought forth the incisive *Jaundice* (extant in Latin and Arabic: Ullmann 1983). Galēn has enormous admiration for Rufus, frequently citing tracts (occasionally quoting them *in extenso*) with obvious approval, e.g. a four-book *Herbs* in hexameters (*Simples pr.* [11.796 K.]), *Black Bile* (*Atra bile* 1 [5.105 K. = *CMG* 5.4.1.1 (1937) 71]), *On the Gum-Resin Labdanum* (*CMLoc* 1 [12.425 K.], perhaps from the verses of *Herbs*), and the remarkable *Pain-Alleviating Potion* (*CMLoc* 7.5 [13.92–93 K.]). This last is a truly anesthetic compound, combining with precision the root-bark of mandrake, frankincense, "white" pepper, saffron crocus, the seeds of henbane, the latex of the opium poppy, myrrh, spikenard, and the outer rinds of cassia; Rufus (and Galēn) probably employed such drugs as they performed surgeries or cauteries.

Other writings, known either *in toto*, in extracts, or by title alone, display his considerable intellectual attentions: e.g. *Satyriasis and Gonorrhea, Purging Drugs, Bones* (probably spurious), *Pulses, Diseases of the Joints, Aphrodisiacs, Melancholia, On Rabies, Glaucoma and Diseases of the Eyes, Fevers, Urines, Commentary on the Hippocratic Airs Waters Places, Andrapodismus, Acute and Chronic Diseases, Therapeutics* (from which *Pain-Alleviating Potion* is extracted by Galēn), and many more.

Widely cited, quoted, extracted, recopied and summarized in later Roman, Byzantine, and Arabic medicine, Rufus of Ephesos' influence slowly became swamped in the long shadows cast by Galēn's prescient synthesis, but it is little surprise that Rufus appears as a "standard" authority in the famous "Seven Physicians" Folio of the 6th c. Codex Juliana Anicia (notably absent is HIPPOKRATĒS). As late as the 9th c., Rufus' status was unquestioned as one of the "Four Silencers of Disease," to approximate the baroquely piquant Byzantine Greek phrase (Gossen 1212).

Ed.: Daremberg and Ruelle (1879/1963); H. Gärtner, *Rufus von Ephesos. Die Fragen des Artzes an den Kranken* (1962) = *CMG* S.4; A. Sideras, *Rufus von Ephesos. Über die Nieren- und Blasenleiden* (1977) = *CMG* 3.1; M. Ullmann, *Rufus von Ephesos Krankenjournale* (1978 [Arabic]); *Idem, Die Schrift des Rufus von Ephesos über die Gelbsucht* (1983 [Arabic and Latin]); P.E. Pormann, *Rufus of Ephesus On Melancholy* = *SAPERE (Scripta Antiquitatis Posterioris ad Ethicam REligionemque pertinentia)* 13 (2008 [Arabic, Greek, and Latin]); Brock (1929) 112–129: partial translations of *Interrogation of the Patient* and *Anatomical Nomenclature*.

RE 1A.1 (1914) 1207–1212, H. Gossen; J. Ilberg, *Rufus von Ephesos. Ein griechischer Arzt in trajanischer Zeit* (1930); A. Sideras, *Textkritische Beiträge zur Schrift des Rufus von Ephesos De renum et vesicae morbis* (1971); *DSB* 11 (1974) 601–603, F. Kudlien; Scarborough (1985c); *NDSB* 6.290–292, V. Nutton.

John Scarborough

Rufus of Samaria (*ca* 100 CE)

Jewish physician who lived in Rome and wrote in Greek. GALĒN refers to Rufus' commentaries on the sixth book of HIPPOKRATĒS' *Epidemics* in his commentary on the same work, noting Rufus' Jewish ethnicity negatively, even as he uses his commentary: *CMG* 5.10.2.2, pp. 213, 289, 293, 413.

F. Pfaff, "Rufus aus Samaria: Hippokrateskommentator und Quelle Galens," *Hermes* 67 (1932) 356–359; *RE* S.6 (1935) 646 (#18a), L. Edelstein; S. Muntner, "Rufus of Samaria," *Israel Medical Journal* 17 (1958) 273–275; *NP* 10.1158, V. Nutton; *EJ2* 17.527–528, S. Muntner.

Annette Yoshiko Reed

S

Sabinius Tiro (35 – 10 BCE)

Author of a book on gardening (*Kepourika*) which he dedicated to Maecenas. In it he stated that rue, savory, mint, and basil were harmed by contact with iron (PLINY 19.177).

GRL §213, 225, 356.7; *RE* 1A.2 (1920) 1601–1602 (#33) s.v. Sabinus, G. Funaioli.

Philip Thibodeau

SABINUS ⇒ POMPEIUS SABINUS

Sabinus (15 BCE – 15 CE)

Poet friend of OVID who left unfinished at his death (*ex Pont.* 4.16.15–16) a didactic poem on the calendar like the *Fasti*.

RE 1A.2 (1920) 1598–1599 (#21), Fr. Vollmer; *OCD3* 1342, E. Courtney.

Philip Thibodeau

Sabinus (Med.) (*ca* 100 – 120 CE)

Commentator of Hippokratic treatises frequently referred to by GALĒN, traditionally dated to Hadrian's era. He was the teacher of STRATONIKOS OF PERGAMON and MĒTRODŌROS. He wrote commentaries on the HIPPOKRATIC CORPUS, AIRS, WATERS, PLACES, NUTRIMENT; APHORISMS, EPIDEMICS II, III and VI, and *On the Nature of Man*; perhaps also *Humors*. A fragment is preserved in OREIBASIOS *Coll.* 9.12 (*CMG* 6.1.2, pp. 15–16) on geomedicine, which summarizes such Hippokratic theories as those of *Airs, Waters, Places*. Though Galēn held him in high esteem, he also criticized his ignorance of anatomy and overly-teleological exegeses.

RE 1A.2 (1920) 1600 (#25), H. Gossen; Deichgraber (1930) 25–28, 29 n.1; *KP* 4.1483, F. Kudlien; Smith (1979) 64–72, 132–133, 149–154, 162–163, 171–172, 245–246; *NP* 10.1189, V. Nutton; Ihm (2002) #220–227.

Alain Touwaide

Salimachus

CAELIUS AURELIANUS (probably mostly from SŌRANOS), in *Acute* 3.138 (*CML* 6.1.1, p. 376), cites *SALIMACHVS* recording that **Pythagorean** physicians in Sicily called *ileus* by the name *phragmos*. The name is unattested and Kind emends to *LVSI-* (i.e., LUSIMAKHOS OF KŌS); *cf.* SILIMACHUS.

RE 3A.1 (1927) 61, F.E. Kind.

PTK

SALLUSTIUS (NEO-PLAT.) ⇒ SALOUSTIOS

Cn. Sallustius (65 – 45 BCE)

A friend of CICERO, whom he followed into banishment in 58 BCE. He is likely to be the author of the *Empedoclea* (Cicero, *ad Q. fratr.* 2.9.3), presumably a philosophical poem treating EMPEDOKLĒS' theories (or simply a Latin translation?).

GRL §110.

Bruno Centrone

C. Sallustius Crispus of Amiternum (45 – 35 BCE)

Born 87–86 BCE, *tr. pl.* 52 BCE, expelled from the Senate 50 BCE by the *censor* Ap. Claudius Pulcher on a possibly trumped-up charge of adultery, then entered military service under IULIUS CAESAR; *praetor* 47 BCE, and governor of *Africa Noua* 46 BCE. He retired from politics and composed two historical monographs, the *Catilina* (events of 64–62 BCE) and the *Iugurtha* (events of 112–105 BCE), then began an annalistic history of Rome from 78 BCE, unfinished at his death *ca* 35 BCE.

The two monographs offer evaluative narratives of moral decay in Roman society, presented through speeches and letters, and focused on a single enemy of Rome. They display an increasing occupation with geography, from the brief excursus on Rome (*Cat.* 6) to an extensive one on Africa (*Iug.* 16–19). Like the speeches and letters, the excursuses set the stage, and mark turns, in the narrative; they evince a pragmatic approach to geography (*cf.* POLUBIOS Book 34; or STRABŌN 1.1.1, 1.1.16, 1.1.18, 1.1.22, 2.5.8, 2.5.13). Sallust describes each site, then its people, with Hērodotean flourishes.

Sallust completed four books (to 72 BCE), and part of Book 5, of the *Historiae*, which survive in extracts (speeches and a letter), quotations by grammarians and others, and some scraps on papyrus and parchment; six excursuses can be detected among the fragments of Books 1–4. Sertorius is said to have met sailors at Gadēs returning from a voyage to a pair of Blessed Isles remote in the Atlantic (probably Madeira and Porto Santo), which Sallust briefly described (Book 1, *frr.*100–102 M.; Keyser 1993). In recording the pirate war and the conquest of Isauria by Seruilius, Sallust described (probably in Book 1) the region of his activity: Lukian coast, Pamphulia, and Lukaonia (Keyser 1997). In connection with the revolt of Lepidus, Sallust described the islands of Sardinia and Corsica, offering data on the fertility of Sardinia and its deadly "parsley" (Book 2, *frr.*1–8, 10–12); again for the pirate war, he provided an excursus on Crete (again treating fertility: Book 3?). As his scene shifted from the Aegean to the Black Sea, he included a counter-clockwise ***periplous*** of that sea, delineating coastal sites and rivers and addressing hydrography (20 fragments of Book 3). Probably to introduce Spartacus, Sallust described the Sicilian strait, giving a naturalistic etiology of its myths and currents (Book 4).

Sallust emphasized the value of local traditions (*Iug.* 17.7), resorted to geographical determinism (*Hist.* Book 3, *frr.*74, 78 M.), introduced physical etiologies (ANAXAGORAS' atmospheric earthquake-theory, Book 2, *fr*.28 M.), and took an interest in the natural products of his regions (Book 4, *frr*.61, 72 M.). Scholars suggest various sources, notably TIMAIOS, POSEIDŌNIOS, and VARRO. Probably used by VERGIL, MELA, and PLINY (on the Sicilian strait), Sallust was received as the Roman THUCYDIDĒS (V.P. 2.36.2; Quint. 10.1.101–102), and Martial records (14.191) that he was ranked the foremost Roman historian by scholars. Zēnobios translated him into Greek (under Hadrian), and he is cited by Latin grammarians on a par with CICERO, Horace, Terence, and VERGIL. The ***periplous*** of the Pontos was admired by AUIENIUS (*OM* 32–37) and used by Ammianus Marcellinus (22.8); whereas ISIDORUS OF HISPALIS exploited him widely in his geography (9.2.119–122 + 18.1–10: *Iugurtha*; 13.18: Sicily; 14.6–7: Sardinia). The *Historiae* were barely lost, surviving to have lengthy extracts made in the 9th c.

Ed.: B. Maurenbrecher, *C. Sallusti Crispi Historiarum Reliquiae* (1891–1893; repr. 1967); L.D. Reynolds, *Catilina, Iugurtha, Historiarum Fragmenta Selecta* (1991).

R. Syme, *Sallust* (1964; French trans. 1982; repr. 2002); P. McGushin, *Sallust The Histories* 2 vv. (1992–1994); P.T. Keyser, "From Myth to Map," *AncW* 24 (1993) 149–168; *OCD3* 1348–1349, C.B.R. Pelling; P.T. Keyser, "Sallust's Historiae, Dioskorides and the sites of the Korykos captured by P. Servilius Vatia," *Historia* 46 (1997) 64–79; *DLB* 211 (1999) 267–276, R.W. Ulery, Jr.; *NP* 10.1254–1258 (#II.3), P.L. Schmidt.

PTK

SALLUSTIUS DIONUSIOS ⇒ DIONUSIOS SALLUSTIUS

Sallustius Mopseatēs (15 – 40 CE)

Wrote a *Iatrika* at the time of Tiberius: *Souda* Sigma-61. The odd Greek name is attested for a Kallippos Mopsiatēs at Athens, *ca* 162–170 CE (*LGPN* 2.322).

Fabricius (1726) 390.

PTK

Salmeskhoiniaka (200 – 100 BCE?)

Lost early astrological text in Greek, probably composed in Egypt during the Ptolemaic period from Egyptian sources of the last millennium BCE. The strange title may have been a transcription of an Egyptian phrase meaning "traveling of the influences." Testimonia to the *Salmeskhoiniaka* in IAMBLIKHOS OF KHALKIS *De Mysteriis* (8.4) and EUSEBIOS *Praeparatio Evangelica* (3.4), both deriving from the lost letter of PORPHURIOS to Anebo, show that it was very close in doctrine to a 4th c. BCE Egyptian inscription on the so-called Saft el-Henna Naos; similar material is found also in a 2nd c. CE Greek papyrus, *POxy* 3.465. The *Salmeskhoiniaka* set out a scheme of astrological attributes and predictions associated with the **decans**.

S. Heilen, *Hadriani genitura. Die astrologischen Fragmente des Antigonos von Nikaia. Edition, Übersetzung und Kommentar* (2006) 470–474.

Alexander Jones

Saloustios (*ca* 350 – 370 CE)

Author of *On gods and the world* (as called by its first editor), presumably written during Julian's reign (361–363 CE) with the aim to restore the pagan religion. Probably identifiable with Julian's friend Saturninus Sallustius Secundus, he resembled Julian and IAMBLIKHOS intellectually. Saloustios treats divine attributes, myths and how to interpret them (1–4), classes of gods, namely those living outside the world, those immanent in it, and the world itself (5–7), the soul, providence and fate, ethics and politics (8–11), and the problem of evil, said to come not from the gods, humans, or even ***daimones***, but merely the absence of goodness, as darkness is the absence of light (12). Saloustios finally discusses the soul's fate after death and endeavors to refute atheism (13–21), examining in particular the essence and causes of atheism, under which heading he included also Christianity (18).

Ed.: A.D. Nock, *Sallustius Concerning the Gods and the Universe* (1926); G. Rochefort, *Des Dieux et du Monde* (1960).
RE 1A.2 (1920) 1960–1967 (#37), K. Praechter; *NP* 10.1270 (#2), L. Brisson.

George Karamanolis

Salpē (of Lesbos?) (100 BCE – 77 CE)

Midwife (*obstetrix*), wrote on women's diseases, listed as a foreign authority on drugs from animals (PLINY 1.*ind*.28 [with ELEPHANTIS and OLUMPIAS OF THĒBAI], 32). She agreed with LAÏS in treating ***hudrophobia*** and fevers magically with wool from a black ram (28.82), and offered other **sympathetic** remedies, e.g. restoring sensation to numbed limbs (28.38), strengthening eyelids and ameliorating sunburn (28.66), an aphrodisiac (28.262). With her tuna-based depilatory (32.135), she prepared enslaved boys for market by making them appear less sexually mature and therefore more costly (*cf.* RUFUS OF EPHESOS, *On the Sale of Slaves* pp. 469–470 DR; Bain p. 267, n. 44). Athēnaios, *Deipn.* 7 (321f-322a), quotes NUMPHODŌROS OF SURAKOUSAI who mentions an homonymous Lesbian authoress of *Paignia*, trifles or "playful tricks" of an alchemical-magical nature, perhaps the same writer (Pliny's *obstetrix* is anethnic); if so, she must predate 190 BCE, but the identification is doubtful. Salpē, otherwise unattested as a name (but *cf.* Salpis, of Rhēgion, 1st c. CE: *LGPN* 3A.387), is a beautiful and supposedly aphrodisiac fish (KURANIDES 1.18; Thompson 1947: 225; Bain p. 268).

D. Bain, "Salpe's *ΠΑΙΓΝΙΑ*: Athenaeus 322A and Plin. *H.N.* 28.38," *CQ* 48 (1998) 262–268.

GLIM

Samithra/Tanitros (?) (100 BCE – 40 CE)

ANDROMAKHOS, in GALĒN *CMLoc* 9.6 (13.310 K.), records a ***hedrikē*** (composed of **litharge**, ***psimuthion***, ***khalkitis***, and ***misu*** in **terebinth**, olive oil, and water) under the otherwise unattested name *ΣΑΜΙΘΡΑ* (*LGPN*, Pape-Benseler). Galēn himself records that the **Asklēpiadean** *ΤΑΝΙΤΡΟΣ*, before DIOSKOURIDĒS, very accurately described plants but not the causes of their effects. Ta-nitr- might be Egyptian for "she of the god(dess)" (Heuser 1919: 93–94, 109); another possible Egyptian parallel is the formation Psam(m)e/i-, as in HĒRODOTOS 1.105, 2.2, etc. Perhaps emend to the typically feminine name Psamatha/ē (*LGPN* 2.481, 3A.481) or Saminthos (THUCYDIDĒS 5.58). A Persian or Indian connection, however, seems very unlikely.

Fabricius (1726) 390; Parker (1997) 145 (#52).

PTK

SAMMONICUS ⇒ SERENUS SAMMONICUS

Samuel of Nehardea, Mar Samuel (d. *ca* 254 CE)

Jewish legal scholar, physician, astronomer, and head of a Rabbinic academy in Nehardea. The Babylonian *Talmud* preserves many of his teachings and many traditions about him. He is famous for curing eye diseases (*'Abodah Zarah* 28b; *Shabbat* 78a, 108b). He also advises on a variety of other ailments, including medical complications of circumcision (*Shabbat* 133b–134b, 137a–b; *Ketubbot* 110b; *Nedarim* 37b, 41a, 54b; *Gittin* 70a; *Baba Batra* 146a; *'Abodah Zarah* 28a). Samuel is also celebrated as an astronomer (*Berakhot* 58b; *cf. Shabbat* 156b). Some connection between his medical and astronomical interests is suggested by Samuel's comments on bloodletting, which include astrological elements (*Shabbat* 129a–b). Most stressed, however, is his knowledge about lunar, solar, and planetary cycles essential to calculate the luni-solar calendar and intercalation of months (*'Erubin* 56a; *Rosh Hashanah* 20b; *Sanhedrin* 12b; *Arakhin* 9b). Jewish tradition associates him with one of the two traditional methods for calculating solstices and equinoxes. The earliest Hebrew astronomical work, *Baraita di-Shmuel* (*ca* 8th c.), circulated in his name, was widely read by medieval Jewish astronomers (e.g. Shabbatai Donnolo, Abraham bar Hiyya, Abraham ibn Ezra). Analysis of traditions associated with Samuel has led some modern scholars to speculate about the development of a lunar visibility theory by ancient Jews. The *Talmud* also preserves a tradition about Samuel Yarhina'ah (astronomer, lunar expert), physician to the Palestinian Jewish sage R. Judah the Patriarch, who treated his eye disease (*Baba Metzia* 85b). Since few traditions connect Mar Samuel to Palestine, some scholars question whether this tradition refers to the same Samuel.

J. Preuss, *Biblical and Talmudic Medicine*, trans. F. Rosner (1978) *passim*; E. Beller, "Ancient Jewish Mathematical Astronomy," *AHES* 38 (1988) 51–66; *EJ2* 17.757–758, M. Beer.

Annette Yoshiko Reed

Sandarius / Sardacius (550 – 1300 CE)

Known only in Nicolaus Myrepsus (1.202) who attributes to *SARDACIVS* an "extraordinary" multi-ingredient treatment for irritated bowels compounded from roses, **mastic**, cinnamon, **bdellium**, cardamom, several varieties of pomegranate (including Syrian), ginger, gum Arabic, acacia sap, rosemary seeds, quince, and Armenian *symphutum bulbosum*. Kühn, reading the name as "Sandarius," is skeptical of his authenticity. If Sardacius, the name may derive from Sardēs, suggesting a Ludian connection. *Cf.* Sardanapalus, king of Assyria, 7th c. BCE (Aristophanēs, *Birds* 1021, ARISTOTLE, *Eth. Nic.* 1095b19–22, Ath., *Deipn.* 8 [335b-337a], 12 [528e-530c]). *Cf.* also Sandrocottus (Chandragupta), king of India, 3rd c. BCE (Ath. *Deip.* 1 [18d–e]). Greek names in *ΣΑΡΔ-* are rare but attested, names in *ΣΑΝΔ-* are even rarer (*LGPN*). Neither variant is Latin.

C.G. Kühn, *Additamenta ad elenchum medicorum ueterum* 24 (1836) 7.

GLIM

Sardonius (*ca* 60 BCE – 430 CE)

Wrote a geographical work, followed by the RAVENNA COSMOGRAPHY, that treated at least Sarmatia (4.11) – i.e. before 430 CE, and Dacia (4.14) – i.e. after 60 BCE; *cf.* ARISTARKHOS of Sikuōn and Hulas. The name may be from the Greek Sardo (Sardinia), or see Aurelius Victor, *Caes.* 13.3, on the Dacian king Sardonius, *ca* 105 CE.

(*)

PTK

Sarkeuthitēs/os (250 BCE – 80 CE)

Andromakhos, in Galēn *CMLoc* 6.4 (13.927 K.), records an "extractive" plaster (of **bdellium**, ***ammōniakon*** incense, and natron in beeswax, olive oil, salt water, and **terebinth**) under this otherwise unattested name (*LGPN*, Pape-Benseler). Various emendations of the name are possible, perhaps the most likely being Theudās "Sarkophagos" (cited just 27 lines above); other possibilities include an ethnic of the city Arkeuthēs of Lukaonia, an Illyrian name "Skarditheutēs" (*cf.* names in Skerd-/Skord-, Polubios 2.5.6, and those in Theut-/Teuth-, Polubios 2.6.3), or even a doubly-divine Egyptian name, Serkhou-Thōth (Heuser 1929: 57–58).

Fabricius (1726) 390.

PTK

Saserna ⇒ Hostilius Saserna

Saturos (Lithika) (325 BCE – 77 CE)

Lapidary writer of uncertain identity whose *On stones* is cited by Pliny (37.91 and 94) regarding onyx and *syrtite.* Whether the homonymous poet, quoted again by Pliny 37.31 on electrum, together with Aeschylus, Philoxenos, Euripidēs and Nikandros of Kolophōn, is our man is also disputed.

RE 2A.1 (1921) 235 (#20), A. Gudeman; *SH* 717–719.

Eugenio Amato

Saturos of Kallatis (200 – 150 BCE?)

Wrote a series of biographies of which only part of his *Life of Euripides* has been transmitted in a papyrus fragment. His biographies of philosophers are mainly preserved in Diogenēs Laërtios. Saturos seems not to have offered much information about philosophical ideas.

Ed.: S. Schorn, *Satyros aus Kallatis. Sammlung der Fragmente mit Kommentar* (2005).
NP 11.123–125, G. Arrighetti.

Jørgen Mejer

Saturos of Paros (360 – 340 BCE)

Son of Isotimos of Paros. He was a sculptor, architect, and author who collaborated with Putheos on the Maussōlleion at Halikarnassos, one of the seven ancient wonders of the world. Vitruuius cites their commentary on the building (7.*pr.*12–13). Saturos is also known from an inscribed statue base at Delphi that supported images of Ada and Idrieus, younger sister and brother of Maussōllos and Artemisia (ruled Halikarnassos 351–344 BCE), which he sculpted. His name (probably the same person) appears in building accounts for the sanctuary of Asklēpios at Epidauros.

Svenson-Ebers (1996) 116–142; G.B. Waywell, "The sculptors of the Mausoleum at Halicarnassus," in I. Jenkins and G.B. Waywell, *Sculptors and sculpture of Caria and the Dodecanese* (1997) 60–67; *KLA* 2.366–367, G.B. Waywell.

Margaret M. Miles

Saturos of Smurna (130 – 160 CE)

His student GALĒN praises him as the best of QUINTUS' students, and mentions his presence at Pergamon during the ***anthrax*** epidemic of 146–147 CE. Aristidēs, *Sacred Discourse* 3.8–11, records his treatment in Pergamon by Saturos. Galēn preserves Saturos' words on "phrenetics," *In Hipp. Prorrhet.* (*CMG* 5.9.2, p. 20), and indicates that he wrote (or taught) commentaries on the HIPPOKRATIC CORPUS, EPIDEMICS: *In Hipp. Epid. III* (*CMG* 5.10.2.1, p. 59), *In Hipp. Epid. VI* (*CMG* 5.10.2.2, pp. 287, 412–413). Probably Saturos was Galēn's primary source for the teachings of Quintus.

Grmek and Gourevitch (1994) 1519–1520; *OCD3* 1362, V. Nutton; Ihm (2002) #228–229.

PTK

Scribonius Largus (*ca* 25 BCE – 55 CE)

A Sicilian-born, bi-lingual physician, one of several medical professionals in service at the courts of Caligula and Claudius (PLINY 29.7); the "Dedicatory Epistle" to Scribonius' extant *Compositiones* (*Prescriptions* or *Recipes*) reveals his patron as C. Iulius Callistus, a conspicuously wealthy and powerful freedman (*Comp.*, *pr.*1, with Pliny, 36.60) who went to Britain (43 CE) to assist in supervising arrangements for Claudius' celebration of conquest. Scribonius says he too went north (*Comp.* 163: *cum Britanniam peteremus cum deo nostro Caesare*), attending to the needs of the traveling court. He also mentions locales and teachers that identify him as Sicilian, quite likely a member of the prominent Greco-Roman farmer intelligentsia characteristic of the new imperial order established by Augustus after 27 BCE, and the bicultural intellectual heritage is patent in the *Compositiones*: he cites Greek authorities in medicine, known through their writings (HIPPOKRATĒS, HĒROPHILOS, ASKLĒPIADĒS PHARM.) as well as contemporary or near-contemporary practitioners from both traditions (Greek: AMBROSIOS OF PUTEOLI, ZŌPUROS OF GORTUN, TRUPHŌN OF GORTUN, PHILŌNIDĒS OF CATINA, THRASEAS, ANDRŌN, MEGĒS OF SIDŌN, and others; Roman: ANTONIUS MUSA, VETTIUS VALENS, PACCIUS ANTIOKHOS, APULEIUS CELSUS, IULIUS BASSUS, and MARCIANUS).

Scribonius knows his plants (especially those native to Sicily and the western Mediterranean), minerals, and animal products, is experienced in some of the surgeries and treatments of grievous injuries suffered in the arena (a number of plasters and poultices are "designed" for gladiators), the complex technologies of drug-preparations, and is a remarkably talented and precise compounder of complicated, multi-ingredient, multiple-stage pharmaceuticals. Repeatedly, the *Compositiones* provides specifics for the spontaneous preparation of remedies or for storage in their fractions, which then could be mixed as the requirements arose: e.g. *Comp.* 70: the remedy for *angina* (i.e., "quinsy"; *cf.* ***sunankhē***); the "For Treatment of a Choking Quinsy" contains 15 ingredients (11 botanicals, one mineral ["½ ounce of fissile alum"], two animal products [Attic honey used to skim the compounded ingredients, and "one ounce of the ashes of a young wild swallow"], and "five medium-sized ground oak galls"), all producing a rather effective anodyne lictus applied directly to the swollen tonsils. "The Augusta always has this compound at hand": the "Augusta" is probably Antonia Minor, the mother of Claudius (so perhaps Scribonius had been a court physician well before the British campaign). He divided and sub-divided the *Compositiones* into a traditional "head-to-heel" arrangement, followed by antidotes (163–199), and finally (200–271) ***acopa***, plasters, poultices, and salves to soothe wounds, especially employed by surgeons, eventually retailing 271 recipes that use 242 botanicals, 36 minerals, and 27 medicines derived from animals.

The "Dedicatory Epistle" addressed to Callistus bewails the general lack of standards among physicians, and Scribonius firmly adheres to a "Hippokratic" ideal, showing that the *pagani* often contribute useful remedies in contrast to the professional *medici* who perform gratuitous surgeries and pander worthless drugs. Scribonius touts no theoretical constructs, making the *Compositiones* a clear, practical manual for the preparation of compounds generally useful and often effective for common ailments. Quotations by GALĒN (e.g. *CMLoc*, 7.3 [13.67 K.] and *CMGen*, 6.14 [13.930 K.]) indicate Scribonius had also written tracts on drugs in Greek, and among later authorities writing in Latin, only MARCELLUS OF BORDEAUX borrows goodly chunks of the *Compositiones* for his own *De medicamentis* (*CML* V [1968]). Recent partial translations of the *Compositiones* into German and English are unfortunately marred by a lack of technical expertise.

Ed.: S. Sconocchia, *Scribonii Largi Compositiones* (1983 [supersedes the edition of Helmreich (1887), esp. with readings from the Codex Toletanus]); German translation (from the Helmreich ed.): W. Schonack, *Die Rezepte des Scribonius Largus* (1913).

F. Rinne, "Das Receptbuch des Scribonius Largus," *Historische Studien aus dem Pharmakologischen Institute der Kaiserlichen Universität Dorpat* 5 (1896) 1–99; *RE* 2A.1 (1921) 876–880, F.E. Kind; S. Sconocchia, *Per una nuova edizione di Scribonio Largo* (1981); Önnerfors (1993) 250–258; S. Sconocchia, "L'opera di Scribonio Largo e la letteratura medica del 1 sec. D.C.," *ANRW* 2.37.1 (1993) 843–922; V. Nutton, "Scribonius Largus, the Unknown Pharmacologist," *Pharmaceutical Historian* 25.1 (1995) 5–8.

John Scarborough

SCROFA ⇒ TREMELLIUS

Sebosus Statius (*ca* 20 BCE – 60 CE?)

Wrote in Latin a ***periplous*** or paradoxography cited by PLINY: distance between Suēnē and Meroë (6.183), the blessed isles of IOUBA distinct from the earlier-known Madeiran archipelago (6.202), and giant blue worms of the Ganges (9.46: *cf.* KTĒSIAS on the Indus).

NP 11.929 (s.v. Statius #II.5), Kl. Sallmann.

PTK

SECUNDILLA ⇒ AQUILA

SECUNDUS ⇒ (1) IULIUS; (2) PLINIUS

Secundus **(ca 150 – 350 CE?)**

Addressee – real or fictitious – of two letters in APSURTOS' veterinary treatise. These are preserved in the *Hippiatrika*: one on cough (*Hippiatrica Parisina* 458 = *Hippiatrica Berolinensia*

22.1), the other on ***orthopnoia*** (*Hippiatrica Parisina* 456 = *Hippiatrica Berolinensia* 27.1). In both, Apsurtos addresses Secundus as *hippiatros*, "horse-doctor."

CHG v.1; McCabe (2007).

Anne McCabe

M. Seius ⇒ Sueius

Seleukos of Alexandria (*ca* 10 BCE – *ca* 30 CE)

Taught and wrote in Rome; compelled to commit suicide by the emperor Tiberius (Suet., *Tib.* 56). Diogenēs Laërtios twice cites Seleukos' *On Philosophy*: 3.109 for Platōn of Rhodes, a student of Panaitios, and 9.11–12 for the transmission of Hērakleitos' book to mainland Greece.

FGrHist 1056 F2–3.

PTK

Seleukos of Seleukeia on Tigris (165 – 135 BCE)

Studied with Babylōnian astronomers and astrologers (Strabōn 16.1.6), responded to Kratēs of Mallos, and preceded Hipparkhos. Seleukos argued the ***kosmos*** was infinite since no boundary could exist (*cf.* Arkhutas of Taras), and hypothesized a mechanism for the long-known lunar influence on tides, addressing two objections to heliocentrism, the apparent absence of stellar parallax and terrestrial rotation. His adherence to heliocentrism is inferred from Plutarch, *Plat.Q.* 8.1. According to Seleukos, the rotation of the Earth and the orbital motion of the Moon disturbed the intervening ***pneuma***, thus swelling the Ocean. Tidal irregularity was proportional to the Moon's distance from the Earth's equatorial plane, but tides differed from sea to sea, and the monthly tidal cycle had seven phases. He may have proposed the system recorded by Theōn of Smurna 3.33, in which the spheres of Venus, Mercury, and the Sun are nested, and together orbit the Earth on a common hollow sphere.

P.T. Keyser, "Seleukos," *BEA* (2007).

PTK

Seleukos of Tarsos (200 – 100 BCE?)

Wrote *On Fishing* (*Halieutikos logos*). Athēnaios describes Seleukos as a competent specialist (*Deipn.* 1 [13c]), who argues that the parrot-fish (*skaros*) never sleeps and cannot be caught by night (*Deipn.* 7 [320a]).

NP 11.364–365 (#10), C. Hünemörder.

Arnaud Zucker

Sēmos of Dēlos (300 – 150 BCE?)

Geographer and antiquarian, wrote *Nēsias*, *On Paros* (one book), *On Pergamum* (one book), ***Periodoi*** (two books, lost), and a treatise on islands (*Souda* Sigma-327, calling him ***grammatikos***). Athēnaios quoted extensively from Sēmos' *Dēlias* (eight books) treating geography, antiquities, culture, religion, and curiosities of Dēlos (*Deipn.* 3 [109e–f], 8 [335a–b],

11 [469c], 14 [618a], 14 [645b], 15 [676f-177a]). In *Nēsias*, he reported the summertime practice of chilling water underground at Kimolos (3 [123d]). He described Dionysian rites in *On Paean* (*Deipn.* 14 [622a–d]; *Souda* Sigma-327) and Thesmophorian rituals (*Souda* Chi-43). The rare name, found at Dēlos 3rd/2nd cc. BCE (*LGPN* 1.404), may be a variant of the commonly attested Sīmos.

FGrHist 396; *KP* 5.98, H. Gärtner; *OCD3* 1383, K.S. Sacks; *NP* 11.383–384, S. Fornaro (erroneously dating Sēmos to 200 CE).

GLIM

Magnus Aurelius Cassiodorus Senator (507 – 585 CE)

Born 485 CE in Scylletium; a statesman of the Ostrogothic government of Italy and, later, founder of the monastery Vivarium in his native Calabria, Cassiodorus (d. 585 CE) wrote a variety of political, philosophical, religious and pedagogical works. He held office as quaestor, consul, master of offices and praetorian prefect. When the Ostrogothic court surrendered Ravenna to Justinian's forces in 540 CE, Cassiodorus' official public career ended and he seems to have relocated to Constantinople, later returning to Italy and retiring at Vivarium *ca* 552.

Among Cassiodorus' political pieces, the *Variae* survive as a collection of administrative letters containing disquisitions on the liberal arts, natural history and certain technologies. The text is traditionally dated at 538 CE, although recent research suggests a later publication during the 540s CE. Digressions within individual letters include material pertaining to animal typology (*Var.* 1.35, 2.14, 3.48, 5.34, 8.31, 10.30, 11.40), astronomy (11.36), engineering and land surveying (1.45, 3.52), environment (2.21, 2.39, 3.47, 3.48, 4.50, 8.32, 8.33, 11.14, 11.39, 12.12, 12.14, 12.15, 12.22, 12.24, 12.25), epidemiology (10.29), mathematics (1.10, 3.52), music (2.40) and different modes of production (1.2, 5.1, 5.16, 11.38). These digressions evince continued interest in natural and technical sciences in a 6th c. audience and highlight the importance of libraries in the transmission of such topics. Cassiodorus deployed this material epideictically and derivatively, illustrating wide reading rather than genuine inquiry. AMBROSE OF MILAN (e.g., *Hexaemeron*) and BOËTHIUS (e.g., *De arithmetica*) number among his sources. Neo-**Platonic** conceptions of law and nature form a unifying theme for the encyclopedic interest of the *Variae*. The text's *varietas* was intended to demonstrate traditionalism at the Ostrogothic court by attaching legal decisions to a universal understanding of nature.

Sometime after 562 CE, Cassiodorus began the *Institutiones diuinarum et saecularium litterarum*, wherein he outlined the tenets to a liberal Christian education, attempting to integrate the Classical tradition of letters with the study of Christian scriptures and to provide a point of departure for those interested in pursuing secular and divine topics. The treatment of Classical learning in the *Institutiones* is neither exhaustive nor entirely systematic, but it reveals a willingness to expand the precepts of education to a wider epistemological conception of past and present learning. Book 2 addresses the liberal arts with references to diverse authors, perhaps illustrative of scientific texts available at Vivarium – 2.4 on arithmetic (PYTHAGORAS, NIKOMAKHOS, APULEIUS, Boëthius), 2.5 on music (GAUDENTIUS, CENSORINUS, ALUPIOS, EUCLID, ALBINUS, Apuleius, AUGUSTINE), 2.6 on geometry (VARRO, Censorinus, Euclid, APOLLŌNIOS OF PERGĒ, ARCHIMĒDĒS, Boëthius, SENECA), and 2.7 on astronomy (Varro, PTOLEMY, BASIL OF CAESAREA, Augustine). Each section pertaining to the Classical tradition provides an etymological background, a broad description of the topic as treated by previous commentators, usually an outline of the basic principles, and an orientation with scriptural interpretation.

OCD3 298–299, S.J.B. Barnish; J.W. Halporn and M. Vessey, *Cassiodorus: Institutions of Divine and Secular Learning* (2004); M. Shane Bjornlie, *The Variae of Cassiodorus Senator* (Diss. Princeton, 2006).

M. Shane Bjornlie

SENECA ⇒ L. ANNAEUS SENECA

SEPTIMIUS ⇒ NESTŌR

P. Septimius (*ca* 100 – 60 BCE?)

One of the few Romans before VITRUUIUS' time to have written on architecture (7.*pr*.14). He wrote two books and may have been Vitruuius' source for the teachings of Hermodōros of Salamis, the first Greek architect to build a marble temple in Rome (*ca* 146 BCE for Q. Caecilius Metellus Macedonicus), who was in turn the likely source for the system of columnar proportions of the Greek HERMOGENĒS OF ALABANDA.

P. Gros, "Hermodorus et Vitruve," *MEFRA* 85 (1973) 137–161.

Thomas Noble Howe

SEQUESTER ⇒ VIBIUS SEQUESTER

Serapiōn (Astron.) (150 – 360 CE)

Astronomer whose method for using PTOLEMY's tables to calculate the equation of time (the difference between mean and observed noon) THEŌN OF ALEXANDRIA describes in his "Great Commentary" on Ptolemy's *Handy Tables*. The passage's textual difficulties are due in part to Theōn's habit of updating his own works, but Theōn is clearly referring to our Serapiōn's manual of instructions for the use of the *Handy Tables*, thus an early example of the same genre of commentary as Theōn's "Little Commentary." Modern scholars have frequently conflated him with other homonyms, including the astrologer SERAPIŌN OF ALEXANDRIA and the geographer SERAPIŌN OF ANTIOCH, both of whom, however, preceded Ptolemy.

J. Mogenet and A. Tihon, *Le "grand commentaire" de Théon d'Alexandrie aux Tables faciles de Ptolémée. Livre I* (1985) 288–300.

Alexander Jones

Serapiōn of Alexandria (Astrol.) (100 BCE – 100 CE?)

Astrologer whose lost work or works on ***katarkhic*** astrology are partially summarized in several chapters of the astrological compilation falsely ascribed to "Palkhos" (but actually the work of Eleutherios Eleios, a 14th c. Byzantine astrologer). Other citations of Serapiōn are scattered through the chaos of Byzantine astrological codices, but their authenticity is doubtful. The ASTROLOGOS OF 379 asserts that PTOLEMY postdated Serapiōn. It seems unlikely, therefore, that our astrologer was the Serapiōn "the Egyptian" executed in 217 CE for forecasting Caracalla's imminent death (Dio Cassius 78.4).

Pingree (1978) 2.440–441.

Alexander Jones

Serapiōn (or Sarapiōn) of Alexandria (240 – 200 BCE)

Physician, the second exponent of the **Empiricist** "school" after PHILINOS OF KŌS in the catalogue in PSEUDO-GALĒN INTRODUCTION 14.683 K.; CELSUS *pr.*10 speaks of him as the founder of that "school," perhaps echoing a claim by Serapiōn himself, whom GALĒN ironically calls "the new Asclepius" (*Subf. Emp.* 11). As Philinos before him, he polemized against the **Hērophileans**, but also against other medical schools, as the title *Against the Sects* shows (CAELIUS AURELIANUS *Acute* 2.32 [*CML* 6.1.1, p. 148]; about this work Galēn wrote two lost books: *On my own Books* 9 [2.115 MMH]). Perhaps in a work entitled *Through three* (*Dià triōn*: Galēn, *Subf. Emp.* 11) he formalized the doctrine of the school, elaborating its three basic concepts, experience (*empeiria*), the reports of others (*historia*), and analogical reasoning (*tou homoíou metábasis*: "transition to the similar"). According to Galēn, it was controversial whether this third principle was, for Serapiōn, a constitutive part of the medicine, or whether it was only a heuristic principle (*Subf. Emp.* 4 and *Part. Art. Med.*: H. Schöne, *Galeni de partibus artis medicativae* [1911] 32). The greater part of the remaining fragments comes from the *Therapeutics* in three books (testified particularly by Caelius Aurelianus). They are concerned with single remedies, related to different pathologies, which probably included the *malagma* that Serapiōn took up from the **Hērophilean** ANDREAS (ASKLĒPIADĒS PHARM. in Galēn, *CMLoc* 10.2 [13.343–344 K.]) and the so-called "Serapiōn's emplastrum," used for dermatological diseases (KRITŌN in Galēn, *CMGen* 6.6 [13.883 K.]).

Ed.: Deichgräber (1930) 164–168 (fragments), 255–256.
RE 2A.2 (1923) 1667–1668 (#9), H. Gasser; *OCD3* 1392, von Staden; *Idem* (1999) 160–163; Ihm (2002) #230.

Fabio Stok

Serapiōn of Antioch (100 – 60 BCE)

Mathematical geographer, called "most recent" by CICERO, *ad Att.* 2.4.1, and a critic of ERATOSTHENĒS, *ad Att.* 2.6.1. PLINY, 1.*ind.*2, 4–5 cites him as a writer on gnomons and geography.

NP 11.444 (#1), W. Hübner.

PTK

Serenus (Pharm.) (50 – 540 CE)

AËTIOS OF AMIDA 6.16 (*CMG* 8.2, p. 157) quotes Serenus' drug for epileptics, compounded of castor (the plant), black hellebore, scammony, ***opopanax***, Theban cumin, natron, sulfur, wild rue, and wormwood-seeds, among other potent ingredients, to be drunk in vinegar-water. Emendation to "SŌRANOS" may be ruled out, as he cautioned against even penny-royal (1.62), and no Greek text is attributed to SERENUS (SAMMONICUS). The name is attested from the 1st c. CE: *LGPN* 1.404, 2.396; *PLRE* 1 (1971) 826, 2 (1980) 993.

RE 2A.2 (1923) 1677 (#8), F.E. Kind.

John Scarborough

Q. Serenus Sammonicus (*ca* 190 – 212 CE)

Preserved under the name of Quinctus (or Quintus) Serenus (or Serenius) is a *Liber medicinalis*, a poem (1,107 lines including subtitles: Vollmer) that treats various medical topics, especially remedies prescribed for particular ailments. The MSS link this writer with a Q. Serenus Sammonicus, known through citations as the author of a work on antiquities, *Res reconditae* in five books (MACROBIUS, *Saturnalia*, 3.9.6; 3.16.6–9; Sid. Apoll., *Carm.*, 14.3; Arnobius, *Adv. Gent.*, 6.7; etc.), a not unreasonable association, given the subject-matter in the *Res reconditae*, an esoteric series of odd characteristics displayed by animals. A Serenus Sammonicus appears among the thicket of bogus and quasi-fictive names in the *SHA*, and even though this Serenus Sammonicus (or two of them: father and son) is said to have donated 60,000 books to an emperor, and been murdered by Caracalla, the sum of literary evidence suggests a role at the courts of Septimus Seuerus and Caracalla (thus 193–212 CE).

The *Liber medicinalis* derives from PLINY, and quite significantly from the passages extracted from the *Natural History* known as the MEDICINA PLINII, probably indicating that the "abbreviated Pliny" was already in circulation. Serenus adds words and phrases occasionally similar to the works of GARGILIUS MARTIALIS, whereas MARCELLUS OF BORDEAUX and THEODORUS PRISCIANUS sometimes borrowed terms from Serenus' poem ("Remediorum fontes uel testes," Vollmer pp. 53–64). *Liber medicinalis* probably owes its survival to its redaction into poetry of numerous diseases and therapeutics, and "medical poetry" in classical antiquity has a long pedigree, extending back to at least NIKANDROS' *Thēriaka* and *Alexipharmaka*.

Ed.: Fr. Vollmer, *Quinti Sereni Liber medicinalis* (1916) = *CML* 2.3; re-ed. with French trans., R. Pépin, *Quintus Serenus (Serenus Sammonicus) Liber medicinalis (Le livre de medicine)* (1950).

R. Syme, *Ammianus and the Historia Augusta* (1968) 186; *Idem, Emperors and Biography* (1971) 279; *Idem, Historia Augusta Papers* (1983) 23; J.H. Phillips, "The Incunable Editions of the Liber medicinalis Quinti Sereni," in Mazzini and Fusco (1985) 215–236; J.H. Phillips, "Liber medicinalis Quinti Sereni, XLI.775–776 [podagrae depellendae]: seminecisue hirci reserato pectore calces insere," in Sabbah (1988) 157–160; Önnerfors (1993) 274–277.

John Scarborough

Serenus of Antinoeia (200 – 230 CE?)

Wrote various geometrical treatises, *The section of the cylinder* (*SCy*), *The section of the cone* (*SCo*), both edited by Heiberg (French translation by Ver Eecke), a lost commentary on APOLLŌNIOS' *Kōnika* to which *SCy* 17 alludes, and geometrical *lēmmata* from which a small addendum to THEŌN OF SMURNA's treatise is borrowed (Heiberg XVIII). MSS cite him as a [**Platonist**] philosopher, as a geometer, or according to his birthplace, long considered Antissa (Lesbos), now Antinoeia according to Heiberg's plausible correction to the corrupt subscription of MS *Vat. gr.* 206 (Heiberg XVII, 116, 120). MS *Par. gr.* 1918 indicates that "Sirinos the geometer" followed the views of HARPOKRATIŌN OF ARGOS, implying perhaps that Serenus was temporally close to Harpokratiōn and (less plausibly) that he was himself a **Platonist** (Decorps-Foulquier 2000: 19).

SCy and *SCo* are both dedicated to a certain Kuros, and the last four propositions of *SCy* are presented in defense of his friend PEITHŌN's views on parallel lines (Heiberg 96). Both treatises, of respectable length, imitate Apollōnios' *Kōnika*, less by the results on which they rely (elementary properties of Apollōnios' ellipse for *SCy* and of right and oblique cones for

SCo) than by their structure: a skillful combination of theorems and problems, the first progressively leading the reader to conceive solutions to the second.

Ed.: J.L. Heiberg, *Sereni Ant. Opuscula* (1896).
P. Ver Eecke, *Le Livre de la section du cylindre, et le livre de la section du cône* (1929); Decorps-Foulquier (2000) 33–39.

Alain Bernard

Sergius of Babylōn (250 BCE – 90 CE)

Asklēpiadēs Pharm. records two gum and rainwater collyria from Sergius, one for use over several days, containing saffron, ***glaukion***, hematite, and ***sarkokolla***, in Galēn *CMLoc* 4.7 (12.746 K.: *cf.* Hērōn (Med.) and Poludeukēs), the other a "one-day" cure, compounded from saffron, acacia, burnt copper, **calamine**, opium, and ***pompholux***, applied in egg-white, *ibid.* (12.751 K.). The name is probably Semitic rather than Latin, as Trajan was the first Roman to conquer Babylōn.

RE 2A.2 (1923) 1689 (#5), F.E. Kind.

PTK

Sergius of Rešʿaina (500 – 536 CE)

Educated in medicine and theology at Alexandria, and appointed Arkhiatros of Rešʿaina, Sergius was a priest, theologian, doctor and diplomat who towards the end of his life turned against, and thus incurred the wrath of, his fellow Monophysites. He was sent to Rome by Ephraim, bishop of Antioch, on a mission to Pope Agapetus, to secure support for Ephraim's persecution of the Monophysites, and died in Constantinople in 536 CE, whilst returning from Rome.

According to the 13th c. scholar Bar Hebraeus, Sergius was the first to translate the works of Galēn from Greek into Syriac. In addition to translating the entire Alexandrian medical syllabus into Syriac, Sergius also composed his own medical treatises, including one concerning **dropsy**. He also composed and translated other philosophical, logical, theological and astronomical works, including a treatise on the *Categories* of Aristotle, in which he refers to the work of Bardaisan. Several of Sergius' works are dedicated to his pupil and assistant Theodore.

Although later, mainly 13th c., commentators were critical of the quality of his translations, such criticisms should probably be understood as revisionist propaganda that, motivated by the cataclysmic consequences of the Mongol invasions, sought to promote the accomplishments of the early ʿAbbasids (*ca* 750–900 CE) and did so at the expense of the pre-Islamic translation movement.

PLRE 3 (1992) 1123–1124; *NP* 11.453–454, S.P. Brock; Siam Bhayro, "Syriac Medical Terminology: Sergius and Galen's Pharmacopia," *Aramaic Studies* 3.2 (2005).

Siam Bhayro

Sertorius Clemens (30 BCE – 80 CE)

Andromakhos records and approves his ***akopon***: chop fresh laurel, myrtle, *libanōtis*, rue, and sage; steep for a night and a day in a mixture of aged olive oil, laurel oil, myrtle oil, etc., then boil with resin, lamb-fat, Pontic beeswax, and **galbanum**; remove from heat and mix in **verdigris**: GALĒN *CMGen* 7.13 (13.1037 K.); another recipe, for a ***diaphorētikē***, *ibid.* 6.14 (13.926–927 K.). Since sage is given as *salbia* (i.e., Latin), Sertorius presumably wrote in Latin, or used Latin sources (such as SCRIBONIUS LARGUS or the Latin synonyms from DIOSKOURIDĒS). For the non-Republican *cognomen, cf.* FLAUIUS CLEMENS.

Fabricius (1726) 121.

PTK

SERUILIUS ⇒ DAMOKRATĒS

Seuerianus of Gabala **(d. before 430 CE)**

One MS (Wien, Österreichische Nationalbibliothek, *medicus graecus* 27, 16th c.) attributes a short text *peri tēs prosēgorias anthrōpou* – better known as *De hominis nomine* (ed. in *PG* 56, 473.16–474.10) – to a Seuerianus, most likely the bishop of Gabala in Syria. The work is philological and religious rather than medical: it explains biblical terms, as typical of 4th c. Christian literature, especially regarding the transformation of medical anthropology (*cf.* also PSEUDO-LIBANIOS and MAKARIOS). Seuerianus also authored several exegetic and homiletic treatises and a short work on the names of God.

Diels 2 (1907) 91 and *Suppl.* (1908) 65; H. Hunger, *Katalog der griechischen Handschriften der Österreichischen Nationalbibliothek* 2 (1969) 74–75; *ODB* 1883–1884, B. Baldwin.

Alain Touwaide

SEUERUS ⇒ DIONUSODŌROS

Seuerus Iatrosophista (500 – 520 CE)

This "Seuerus" is distinct from any homonym mentioned or quoted in GALĒN, AËTIOS OF AMIDA, OREIBASIOS, ALEXANDER OF TRALLEIS, or PHILOUMENOS, and Dietz himself (p. VI) placed him *ca* 600. The addressee Timotheos is probably the zoologist TIMOTHEOS OF GAZA. Seuerus adds two "appendices": the first claiming to summarize Galēn's views on the "Seven-Month Child" (pp. 45–46), and the second listing 97 names for surgical tools (pp. 46–48).

The tract is similar in structure to known pseudo-Galēnic texts, e.g. PSEUDO-GALĒN INTRODUCTION, which has its own synopsis of surgery and surgical operations (780–791 K.), as well as purges (759–761 K.). Perhaps Seuerus noted the absence of enemas in contemporary synopses, so set down a summary of his own, expanded from lines and paragraphs in Galēnic writings: e.g. *ibid.*, 675–676, or Galēn, *Venesection against* ERASISTRATOS 6 (11.168 K. = Brain p. 28), ". . .the Egyptian bird [ibis] imitating the enema," both based on HĒRODOTOS 2.76–77, compared with Seuerus, pp. 1–2, whose bird-lore on the Egyptian Ibis converges reasonably with Timotheos' *On Animals* (50.3: Bodenheimer and Rabinowitz, p. 48).

The nomenclature of surgical instruments bears some affinities with the text analyzed by Bliquez (pp. 195–197), but adds names of its own, and the synopsis on the Seven-Month Child has some overlap with PSEUDO-GALĒN, HISTORY OF PHILOSOPHY 34, and PSEUDO-GALĒN DEFINITIONS 451 (19.331, 454 K.), but is not derivative. "Simple" enemas (water) precede the paragraphs on particular substances, all familiar from earlier Greco-Roman writings on botanical pharmacology: beets, centaury, colocynths, mints, leeks, wormwoods (the "oil," good for worms, says Seuerus), rose-water, opium-poppy capsules, etc. Seuerus' *Enemas* is not so much a physician's handbook, as a student's guide to the major substances used in enemas, and since the contexts are decidedly Egyptian, this might have been a "Cliff's Notes" for students attending medical lectures in Alexandria.

Ed.: Fr.R. Dietz, *Severi Iatrosophistae De clysteribus liber* (1836).
RE 2A.2 (1923) 2011, F.E. Kind; Kollesch (1973) 86–87; L.J. Bliquez, "Two Lists of Greek Surgical Instruments and the State of Surgery in Byzantine Times," *DOP* 38 (1984) 187–204.

John Scarborough

Seuerus the Ophthalmologist (20 – 40 CE?)

In Book 7 (ophthalmology) of his *Tetrabiblos*, AËTIOS OF AMIDA features Seuerus, a well-known physician and eye-doctor in the first decades of the 1st c. CE. Seuerus became an important authority alongside his contemporary DĒMOSTHENĒS for diagnostics, prognoses, and treatments of the numerous ophthalmologic afflictions common (then and now) in the Mediterranean world. As a practicing physician, Aëtios respectfully summarizes Seuerus' "Standard Treatment of Ulcerous Eyes" (*CMG* 8.2, pp. 267–268), who sensibly notes that ". . .initially, one observes the whole body so that overabundances are to be reduced by phlebotomy or purges or enemas . . . [one must] restore the proper proportion (*eukrasia*) to the ill: one must promote regular bowel function, massage the legs vigorously, [have the patient] take pleasant and frequent walks each day, drink water, and bathe only infrequently; the mildest eye-salves are to be employed in treatment of an ophthalmic ulcer . . . especially that prepared from the sap of the fenugreek." The juice of *Trigonella foenum-graecum* L. seeds is a strong demulcent, quite suitable to treat crusted ulcers on the eye. Particularly famous was Seuerus' "powdered eye-medicine," simply known by its eponym "The Seuerian" (GALĒN, *CMLoc* 4.7 [12.734 K.]; Aëtios 7.45, 100 [*CMG* 8.2, pp. 296, 344]; PAULOS OF AIGINA 3.22, 7.16 [*CMG* 8.1, pp. 182–183, 8.2, p. 337]; etc.), and appears among **collyrium stamps** (*cf.* Voinot 1981–1982: #87, 100, and 105, and Marganne 1994: 154, fig.12). Seuerus was a skilled surgeon (*Eadem*, 155–159), and favored finely-ground and powdered drugs exemplified by Aëtios 7.87, "Treatments for Lachrymal Fistula" (*CMG* 8.2, pp. 331–334).

Ed.: J. Hirschberg, *Die Augenheilkunde des Aëtius aus Amida* (1899), esp. 107–115; Aëtios, *CMG* 8.2, pp. 207–300 (*passim*); T.H. Shastid, trans. (Aëtios 7.45), "History of Ophthalmology" in C.A. Wood, ed., *The American Encyclopedia and Dictionary of Ophthalmology* 11 (1918) 8662–8664.
RE 2A.2 (1923) 2010–2011, F.E. Kind; H. Nielsen, *Ancient Ophthalmological Agents* (1974); E. Savage-Smith, "Hellenistic and Byzantine Ophthalmology: Trachoma and Sequelae," *DOP* 38 (1984) 169–186 at 178–179, 185.

John Scarborough

SEUTHĒS ⇒ HERMEIAS (ASTROL.)

Severus Sebokht of Nisibis (630 – 667 CE)

Appointed bishop of Qennešre, the site of a monastery and famous school that promoted Greek science and literature in Syriac. In addition to writing on theology, logic, grammar and mathematics, he excelled in philosophy, geography, astronomy and astrology. Of his surviving scientific treatises, one concerns the astrolabe, and another the constellations.

There are also some extant fragments concerning geography. In his astronomical writings, he refers to the work of BARDAISAN, particularly regarding his astronomical estimate that the lifespan of the ***kosmos*** was to be 6,000 years. He is significant for promoting Indian sciences, particularly mathematics and astronomy, in addition to Greek learning, and may indeed be the first external writer to refer to the Indian numerical system. Severus continued to write and work until at least 665 CE, just two years before his death in 667 CE.

W. Wright, *A Short History of Syriac Literature* (1894) 137–141; *PLRE* 3 (1992) 1105; Brock (1997) 53, 222–223.

Siam Bhayro

SEXTILIUS ⇒ ANTHAIOS

L. Sextilius Paconianus (26 – 35 CE)

Praetor in 26, strangled in prison in 35, for verses against Tiberius. He is probably the author of an astronomic poem, of which there survive four hexameters describing the cardinal winds (Diomēdēs, *de art. Gramm.*: *GL* 1.500.1–4).

GRL §503, 813, 863; H. Wieland, "Pacon. Carm. Frg. 3," *MH* 31 (1974) 114–116.

Bruno Centrone

Q. Sextius (5 – 40 CE)

Roman philosopher and scientist, father of SEXTIUS NIGER. Renouncing his state offices to pursue philosophy, Sextius founded a philosophical sect which, according to SENECA *QN* 7.32.2, despite its "Roman vigor," was extinct at its very beginning. Sextius was a moral philosopher influenced by **Stoic** and **Pythagorean** ideas. Nonetheless, his practice of vegetarianism was not grounded in the doctrine of transmigration of souls, but was rather justified by hygiene and the contention that practicing butchery forms a habit of cruelty. Reflecting **Stoic** dogma is his simile of the wise man with an army marching in a hollow square, his fighting qualities deployed on every side, so as to be able to withstand any attack.

RE 2A.2 (1923) 2040–2041 (#10), H. von Arnim.

Bruno Centrone

Sextius Niger (30 – 50 CE)

Writer of handbooks on pharmacology (in Greek), excoriated by DIOSKOURIDĒS OF ANAZARBOS (*MM, Pr.*3) as a woefully inadequate armchair herbalist, a would-be botanist who thought aloe was "mined" in Judea (Sextius Niger was probably describing the prepared form of aloe-juice, which was set out in trays to harden in the sun, then cut into lozenges to be re-melted or dissolved as needed), and could not differentiate spurges from spurge-olives. Despite his condemnation, Dioskouridēs employs – as does PLINY – Sextius Niger's (lost) works fairly frequently. E.g., Pliny (32.26) names Sextius as his source on beaver castor, writing that Sextius is *diligentissimus medicinae*, but what follows is sheer folklore, slightly corrected by Dioskouridēs (2.24). Pliny/Sextius (29.76) prescribes a gutted, decapitated, and dismembered salamander preserved in honey as an aphrodisiac, while Dioskouridēs (2.62)

Sextius Niger (*Vind. Med. Gr.* 1, f.2[V]) © Österreichische Nationalbibliothek

offers the same preparation for use as a hair remover; Sextius had stated that the salamander did not quench fire, a "fact" replicated by both Pliny and Dioskouridēs. CAELIUS AURELIANUS, *Acute* 3.134 (Drabkin, p. 386; *CML* 6.1.1, p. 372), notes Sextius was a "friend" (*amicus*) of a "Tullius" Bassus, probably the IULIUS BASSUS also mentioned by Dioskouridēs (*MM Pr.*2), and GALĒN (*Simples* 11.797 K.) lists Sextius Niger alongside Dioskouridēs, HĒRAKLEIDĒS OF TARAS, and KRATEUAS, as essential reading for the pharmacologist-in-training. If Pliny's excerpts are representative, Sextius Niger specialized in harsh and dangerous drugs made from minerals and various animals (e.g. the *kantharides* [blister beetles]: *NH* 29.93–96, and *MM* 2.61) employed as aphrodisiacs, depilatories, and slowly-acting poisons. Wellmann's collected fragments are all that remain from Sextius Niger's writing (cross-quotations in Pliny, ERŌTIANOS, EPIPHANIOS, Dioskouridēs, and Galēn).

Ed.: M. Wellmann, "II. Sextius Niger, A. Testimonia vitae doctrinae" in *Pedanii Dioscuridis Anazarbei De materia medica*, v.3 (1914) 146–148.

M. Wellmann, "Sextius Niger. Eine Quellenuntersuchungen zu Dioskorides," *Hermes* 24 (1889) 530–569; John Scarborough, "Some Beetles in Pliny's Natural History," *Coleopterists Bulletin* 31 (1977) 293–296; *Idem*, "Remedies: The Blister Beetles" in "Nicander's Toxicology II: Spiders, Scorpions, Insects and Myriapods, pt. 2," *PhH* 21 (1979) 73–80; *Idem* (1982); *Idem* and Nutton (1982) 206 [Sextius Niger] and 210–212 [aloe]; M. Davies and J. Kithirithamby, *Greek Insects* (1986) 91–94 (*bouprestis* and *kantharis*); Beavis (1988) 168–175 (*kantharis* and *bouprestis*).

John Scarborough

Sextus Empiricus (*ca* 100 – 200 CE)

The only ancient Greek skeptic of whom complete works survive. Virtually the only thing known about him is that he was a doctor. DIOGENĒS LAËRTIOS 9.116, and others, refer to him as a member of the **Empiric** school, as his name suggests. In one puzzling passage Sextus expresses a preference for the **Methodic** school over the **Empiric**; but his criticism may be only of one particular form of **Empiricism**. Sextus belonged to the Pyrrhonist skeptical tradition, whose method, as he explains it, was as follows.

The skeptic assembles opposing arguments and impressions on any given topic. These arguments and impressions are found to exhibit *isostheneia*, "equal strength"; each of them appears no more or less persuasive than any of the others. Given this situation, the skeptic suspends judgment. And this suspension in turn is supposed to yield *ataraxia*, "freedom from worry." The Pyrrhonist skeptic does not claim that knowledge of things is impossible; that too is a topic about which he suspends judgment. Rather, the skeptic refrains from all pretensions to knowledge – or even to belief – about how things really are. There is considerable dispute about what falls under the heading of "how things really are." But it is at least clear that the findings of natural science are among the matters on which the skeptic suspends judgment.

Sextus applied this method, unrestricted as to subject-matter and clearly intended to be employed globally, to the central topics of ancient physics in *Against the Physicists* (part of a comprehensive but incomplete work), and in the complete but more synoptic *Outlines of Pyrrhonism* ("*PH*"). In addition, Sextus' third surviving work, *Pros Mathēmatikous* (*Against the Learned*), discusses six specialized fields of study, of which several are scientific: grammar, rhetoric, arithmetic, geometry, astrology, and music (i.e., musical theory).

OCD3 1398–1399, G. Striker; *ECP* 488–490, J. Allen; *NP* 12/2.1104–1106 (#2), M. Frede.

Richard Bett

SEXTUS PLACITUS ⇒ (1) APULEIUS, PSEUDO; (2) PLACITUS PAPYRIENSIS

Sextus of Apollōnia

Some later versions of the catalogue of **Empiricist** physicians (PALLADIOS, *Comm. on Galen's On the Sects* p. 77 Baffioni = DIOKLĒS *fr.*13e van der Eijk; *cf.* also *fr.*7a Deichgr.) include an apocryphal "Sextus of Apollōnia" whose name results from a textual conflation between SEXTUS EMPIRICUS and either APOLLŌNIOS OF ANTIOCH or APOLLŌNIOS BIBLAS (named last in other versions of the catalogue: *cf. fr.*7b Deichgr.).

Deichgräber (1930) 40–41.

Fabio Stok

Siburius of Bordeaux (350 – 390 CE)

Physician mentioned among his sources by MARCELLUS OF BORDEAUX, who also says that he worked recently, and was an eminent citizen of Bordeaux (*pr.*2). He is possibly the Siburius addressee of eight letters of Symmachus (*Epist.* 3.43–45: 375–380 CE) and of one of Libanius (*Epist.* 963: 390 CE; mentioned in *epist.* 973, 982, 989), who was a follower of archaistic style (Symm. *Epist.* 3.44), *magister officiorum* of Emperor Gratian about 379 and *Praefectus praetorio Galliarum* in 379 (*Cod. Theod.* 11.31.7).

RE 2.A2 (1923) 2072–2073 (#1), J. Seeck; *PLRE* 1 (1971) 839 (#1); Matthews (1975) 72–73.

Fabio Stok

Siculus Flaccus (100 – 200? CE)

One of the authors whose work appears in the *Corpus Agrimensorum Romanorum* (see HYGINUS). He refers to a decision of Domitian on unsurveyed land, but we know nothing of his life. Since he refers to "our profession," he was presumably a practicing surveyor. Writing in a didactic tone, as if offering advice to surveyors, but in a clear and coherent style, he analytically deploys a wealth of technical knowledge and shows considerable pride in the achievements and role of land surveyors. Flaccus establishes the history and context of land holding in Italy, Rome's gradual acquisition of more territory, and the foundation of settlements. Within his three categories of land – "occupied" (i.e., without formal division), "quaestorian" (sold off by the state), "divided and allocated" (land formally surveyed for setting up colonies) – he describes boundary marking techniques, boundary disputes, and the principles for conducting a survey. He discusses *limites* in allocated lands (those facing east and west were *decumani* and those facing north and south *kardines*), describing how units of land division (*centuriae*) were normally 20 *actus* square with an area of 200 *iugera* (50.4 hectares). However, he recognizes many variations in the layout of *centuriae*, borne out by modern archaeological investigation.

Thulin (1913); *CAR* 1 (1993); Campbell (2000) 102–133.

Brian Campbell

Silaniōn of Athens (360 – 320 BCE)

Wrote on symmetry (PLINY 34.51; VITRUUIUS 7.*pr.*14). He sculpted portraits in bronze, including a Iokastē (PLUTARCH, *Table Talk* 5.1 [674A]), whose deathly facial pallor was achieved by "mixing silver into the bronze": perhaps a silvered surface. Pliny remarks that Silaniōn was a severely self-critical perfectionist who earned the nick-name "madman" (*insanum*) for breaking his statues to pieces after completion. His *Plato* (DIOGENĒS LAËRTIOS 3.25), known only from Roman copies, provided the paradigm for the seated, plainly dressed, contemplative philosopher.

A.F. Stewart, *Greek Sculpture* (1990) 179–180; *NP* 11.545, R. Neudecker; *OCD3* 1406, A.F. Stewart; *Idem* (1998) 278–280.

GLIM

SILANUS ⇒ IUNIUS SILANUS

Silēnos (*ca* 400 – *ca* 30 BCE)

Listed as the first after ANAXAGORAS who wrote on Doric proportions (VITRUUIUS 7.*pr.*12), so perhaps not long after 400.

RE 3A.1 (1927) 56 (#3), M. Fluß.

GLIM

Silimachus

CAELIUS AURELIANUS (probably mostly from SŌRANOS), in *Chron.* 1.57 (*CML* 6.1.1, p. 462), cites *SILIMACHVS*, a follower of HIPPOKRATĒS, who recorded that *incubus* carried off many people at Rome, infected by contagion. The name is unattested and Kind emends to *LVSI-* (i.e., LUSIMAKHOS OF KŌS); *cf.* SALIMACHUS.

RE 3A.1 (1929) 61, F.E. Kind.

PTK

Silo (*ca* 120 BCE – 40 CE)

ANDROMAKHOS, in GALĒN *CMGen* 6.4 (13.928 K.), records Silo's ***diaphorētikē***; he also records, *CMLoc* 9.4 (13.285 K.), an analgesic intestinal remedy, containing white pepper, "white" henbane, opium, mandrake, etc., from *ΣΙΓΩΝ*, sc. *ΣΙΛΩΝ*, used by *Valens*, presumably TERENTIUS VALENS. The use of white pepper, an Indian import, suggests a terminus of *ca* 120 BCE. The name is rare but widespread (*LGPN*); Pape-Benseler derive from the Latin *Silo* (as at Catullus 103.1).

RE 2A.2 (1923) 2455, F.E. Kind; 3A.1 (1927) 103, anon.

PTK

Simmias the Stoic (125 – 145 CE)

Among QUINTUS' students, and wrote an exegesis of the senses followed by AEFICIANUS (GALĒN, *In Hipp. Off.* 18B.654 K.). Kühn read *Simiou tou Stōikou*, for differing MS readings (*Paris. Gr.* 1849: *sēmainomenou stōikou; Marc. Gr.* 279: *sēmeiou*; see DK 88B39). Distinct from SIMMIAS SON OF MĒDIOS.

Ihm (2002) #231.

GLIM

Simmias son of Mēdios or Mēdeios (260 – 220 BCE?)

The name Simmias is linked to two potions against the sting of phalanx-spiders: (1) an antidote presented as being of his own creation (ASKLĒPIADĒS PHARM. in GALĒN *Antid.* 2.13 [14.180.10 K. *Simmia tou Mēdiou*, but *Knidiou* Laur. gr. 74.5]); (2) a remedy to be drunk or used as a plaster, good for all venomous snake bites, which he is cited for having been known to use (*ibid.* 182.15 K. *Simmias ho okhlagōgos*). If *Mēdiou* is the *uera lectio*, one might identify his father as Mēdeios, a physician in Olunthos, who knew how to cure Libyan cobra bites (POSEIDIPPOS, *Epigr.* 95.5 A.–B.), whose son probably received an excellent medical education; so the hypothesis is less attractive if Gossen is right about him being the same person as "the quack."

RE 3A.1 (1927) 158 (#8), H. Gossen.

Jean-Marie Jacques

Simmias (of Macedon?) (245 – 220 BCE)

Friend and elephant-driver of Ptolemy III Euergetēs, sent to report on the land of the fish-eaters and the elephant-hunting there, according to AGATHARKHIDĒS Book 5, *fr.*41, in DIODŌROS OF SICILY 3.18.4. The name appears primarily as Simías, the form Simmías being mostly Boiōtian and Thessalian (*LGPN*).

GGL §468; *RE* 3A.1 (1927) 142–143 (Simias #2), A. Klotz, and 144 (Simmias #3), P. Schoch.

PTK

Simōn of Athens (*ca* 470 – 400 BCE)

Equestrian and prestigious author of a lost *Art of Horsemanship* (*Peri Hippikēs*). He was perhaps a hipparch (Aristophanēs, *Knights* 242) and is supposed to have dedicated a bronze horse in the Athenian Eleusinium, on which pedestal he inscribed his deeds (XENOPHŌN *Eq.* 1.1). Xenophōn is widely and explicitly inspired by Simōn with whom he always agrees, and pretends only to have filled his omissions (Xenophōn, *ibid.*; ARRIANUS, *On hunting* 1.5). The *Hippiatrika* (*Hipp. Cant.* 92), giving the complete title as *On Kinds and (good) Choice of Horses*, preserves a fragment of Simōn, still influential on late Greek hippiatric literature (e.g. HIEROKLĒS). Simōn gave much practical riding advice and a detailed portrait (e.g. on hoof angles) of the perfect horse type (well balanced: *summetros*), to help, as Xenophōn said, a buyer avoid being cheated. He probably also treated horse diseases, although a special book on the subject (*Hippoiatrikos logos* according to *Souda* T-987) is dubious.

NP 11.570 (#2), H. Schneider; McCabe (2007) 195–197.

Arnaud Zucker

Simōn of Magnesia (*ca* 350 – 250 BCE)

Physician perhaps in the time of Seleukos Nikanōr (358–281 BCE: DIOGENĒS LAËRTIOS 2.123), wrote *Midwifery*. HĒROPHILOS cites him (providing the *terminus ante quem*) as having observed difficulties in labor encountered by women who experienced troublesome pregnancies with three to five fetuses (SŌRANOS, *Gyn.* 4.1[53]: *CMG* 4, p. 130; *CUF* v. 4, p. 3; Temkin 1956: 176, 214). Other ancient sources who discuss multiple births include ARISTOTLE, *HA* 9(7).4 (584b) and PLINY, 7.33.

von Staden (1989) 367–368; V. Dasen, "Multiple births in Graeco-Roman Antiquity," *OJA* 16 (1997) 439–463.

Robert Littman

Simōnidēs (Geog.) (300 – 250 BCE)

Dwelt five years in Meroë where he wrote an *Aithiopika* cited only by PLINY 6.183, for the extent of "Ethiopia."

Ed.: *FGrHist* 669.
RE 3A.1 (1927) 197 (#5), A. Klotz.

PTK

Simōnidēs (Biol.) (400 BCE – 175 CE)

Cited by AËTIOS OF AMIDA 13.86 (p. 713 Cornarius), probably from CALPURNIUS PISO's work *On Animals*, for the idea that peacocks will detect drugs, run towards them, cry, display, and even scatter them or dig up drugs hidden in the ground. The archaic name, found through the 1st c. BCE, is revived in the 2nd c. CE (*LGPN* esp. 2.399) and the 4th (Ammianus Marcellinus 29.1.37–39, *Souda* Sigma-445).

(*)

PTK

Simos of Kōs (350 BCE – 20 CE)

Famous physician from Kōs after HIPPOKRATĒS (STRABŌN 14.2.19), medical writer whom PLINY (1.*ind*.21–27) cites after PHAINIAS and before TIMARISTOS (21–22) or after Timaristos and before Hippokratēs (23–27) as an authority on drugs from botanicals. Simos proclaimed the toxicity of clover contrary to popular belief in its efficacy against snake bites: decocted clover applied to wounds effects a burning sensation similar to snake bite (21.153). Simos recommended a decoction of asphodel in wine as a remedy for kidney stones (22.72).

RE 3A.1 (1927) 203 (#9), H. Gossen and F.E. Kind.

GLIM

Simos of Poseidōnia (360 – 340 BCE)

"The musician," Simos removed a bronze pillar (erected by Arimnēstos, son of PYTHAGORAS, but the story is a fiction by the historian DOURIS OF SAMOS, *FGrHist* 76 F23), upon which were engraved the seven skills, and published them as his own. Listed among the **Pythagoreans** from Poseidōnia (IAMBLIKHOS, *VP* 267).

RE 3A.1 (1927) 201–202 (#5), H. Hobein.

GLIM

Simplicius of Kilikia (530 – 538 CE)

Pupil of DAMASKIOS and AMMŌNIOS in Alexandria, who wrote several long commentaries on ARISTOTLE's works. Upon Justinian's closure of the school in 529 CE, Simplicius and some colleagues fled to King Chosroes of Persia, reputed for enlightened rule and an interest in philosophy (Agathias *Histories* 2.28.1 Keydell). Simplicius most probably wrote his commentaries after 532 (it is disputed where, but provided with a sizeable library, given the range of writers he uses). He preserves important material from early sources on astronomy and mathematics (Eudēmos, Eudoxos), and meteorology (Poseidōnios from Geminus' summary), and enhances our understanding of work in ancient physics by Aristotle and others.

With Plōtinos, the focus of **Platonists** became otherworldly, but without fully rejecting nature. While the physical world is of secondary importance, their analysis of physics is anything but irrelevant. Their perspective is religious as well as philosophical: a deeper understanding of, and concomitant respect for, the creation was a form of worshipping God, and an aid to achieving their ultimate goal, the "return" to God.

In explicating Aristotle's philosophy, Neo-**Platonists** use commentaries as a vehicle for philosophical and scientific thought, and studying Aristotle prepared students in the Neo-**Platonist** curriculum for studying the work of Plato. Simplicius paraphrases and clarifies Aristotle's dense prose, and further develops problems and themes from his own **Neo-Platonic** perspective, harmonizing Plato and Aristotle when possible. His claim that he adds little is partly a *topos*, partly a matter of respect and acknowledgement of belonging to a tradition: it does not exclude originality.

On scientific issues Simplicius does think that advances are being made (e.g. *Physics* commentary, *Corollary on Place: CAG* 9 [1882] 625.2, *cf.* 795.33–35). He himself significantly alters the cosmological account of Aristotle with full use of post-Aristotelian reactions inside and outside the **Peripatos**. The rotation of the sphere of fire is called "supernatural." Starting from criticisms by the **Peripatetic** Xēnarkhos and a suggestion by Origen (the 3rd c. **Platonizing** Christian) he makes the fifth element (***aithēr***) influence the motion of fire, while Aristotle considered fire to rotate according to the natural inclination of the fifth element. He also refers to an objection, found in Alexander of Aphrodisias, that their rotation on transparent spheres could not explain the occasional closeness of some planets. Like his teacher Ammōnios he made Aristotle's thinking-god into a creator-god (Plato *Timaeus*). He famously polemicizes against Philoponos about the eternity of the world.

His most original contribution is on time and place. On place, a two-dimensional surface for Aristotle, Simplicius follows the criticism of Theophrastos who wants a dynamic instead of static concept, and with Damaskios he gives place the power to arrange the parts of the world (which is viewed as an "organism" with "members"). Iamblikhos already had postulated that place holds things together, giving each thing a unique place which moves with it. Simplicius and Damaskios hold that the power to arrange members of an organism is assigned to a place (e.g. *Corollary on Place*, pp. 636.8–13, 637.25–30), but Simplicius disagrees with Damaskios' idea that measure – a kind of mould (*tupos*) "into which the organism should fit" (*ibid.*) – gives things size and arrangement. Each thing has a unique place (*idios topos*) which moves along with it (*Corollary on Place* p. 629.8–12).

A second excursus (to Book 4 of the *Physics* commentary: *CAG* 9, pp. 773–800), on time, responds to Aristotle's plain rejection of the paradoxes on whether time exists at all (according to Aristotle its parts do not, so time itself cannot), and whether an instance can cease to exist. The Neo-**Platonists** posit higher and lower time, the former being "above change" (Iamblikhos): the higher kind is immune to paradox, while the lower kind is a stretch of time between two instants. Simplicius reports Damaskios' solution, but merely agrees that time exists as something which continuously comes into being, divisible in thought only. In the discussion on the continuum (*Phys.* 6) he adds his own solution that time is infinite (without beginning or end), if viewed as a cycle.

Some evidence exists that Simplicius wrote a commentary on a Hippocratic work, found in the Fihrist (work not specified) and in Abu Bakr al-Razi, al-Hawi (v. 13, p. 159.9) who gives Simplicius as commentator on the *On Fractures* (*Peri agmōn*), in Arabic "Kitāb al-Kasr" or "Kitāb al-Jabr" ("On Setting [Bones]").

Ed.: *CAG* 7 (1894), 8 (1907), 9 (1882), 10 (1895), 11 (1882); translation: *ACA* (1992 etc.): 21 volumes so far: eight on *Physics*; four on *Categories*; five on *Heavens*; two on *Soul*.

RE 3A.1 (1927) 204–213 (#10), K. Praechter; *DSB* 12.440–443, G. Verbeke; R. Sorabji, *Time, Creation and the Continuum* (1983); I. Hadot, "The Life and Works of Simplicius in Greek and Arabic Sources," in Sorabji (1991) 275–303; R. Sorabji, ed., *Aristotle Transformed* (1991); K.A. Algra, *Concepts of space in Greek thought* (1995); *OCD3* 1409–1410, R. Sorabji; *NP* 11.578–580, I. Hadot; R. Sorabji, *The Philosophy of the Commentators 200–600 AD* 3 vv. (2004).

Han Baltussen

Skopinas of Surakousai (200 – 150 BCE)

Inventor of the sundial-type called *plinthion* or *lacunar*, of which an example was in the Circus Flaminius (Regio IX of Rome) in VITRUUIUS' time (9.8.1); *cf.* PLUTARCH, *Oracles* 3 (410E). Vitruuius 1.17 ranks him with PHILOLAOS, ARKHUTAS, ARISTARKHOS OF SAMOS, ARCHIMĒDĒS, ERATOSTHENĒS, and APOLLŌNIOS OF PERGĒ, as a writer of works on machines and sundials; all of his works are lost. The name seems otherwise unattested, but compare the very common Skopas (*LGPN*).

S.L. Gibbs, *Greek and Roman Sundials* (1976) 61; Netz (1997) #108.

PTK

Skulax of Halikarnassos (140 – 90 BCE)

In a passage treating the validity of divination in **Stoicism**, CICERO (*Div.* 2.88) mentions PANAITIOS' intimate friend Skulax who, *excellens in astrologia*, headed the government at Halikarnassos and, sharing in Panaitios' disdain for astrological prophecy, renounced the Chaldaean method of prognostication.

RE 3A.1 (1927) 646 (#3), H. von Arnim.

GLIM

Skulax of Karuanda (*ca* 510 – 500 BCE)

Karian sailor, commissioned by Darius I to explore the Indus river. According to HĒRODOTOS (4.44), he embarked from Kaspatyrus (more likely Kaspapyrus, thought to be on the Kabul near Peshāwar) east to the sea, and took 30 months to reach the north end of the Red Sea near Suez. The circumnavigation of Arabia under Darius is confirmed by inscriptions on stelai erected near the canal dug by the Persian king from the Mediterranean to the Red Sea; claims for control of India are made in the royal inscriptions of Darius and Xerxēs. Skulax wrote a report of his voyage for Darius, perhaps part of the ***Periplous*** *of the Sea outside the Pillars of Hēraklēs* or the *Circuit of the Earth* credited to him in the *Souda* (Sigma-710). He was used by HEKATAIOS OF MILĒTOS and Hērodotos, and cited by ARISTOTLE, Athēnaios (although he is uncertain about the attribution), Harpokratiōn, Philostratos and Tzetzēs, for information about the east, some of it fanciful – Skulax is credited with descriptions of Troglodytes, Shade-foots and Winnowing-fan-ears. Several citations of Skulax in STRABŌN, Schol. Apoll. Rhodes, and Constantine VII Porphurogennētos refer to peoples and places in the Mediterranean, and seem to come from a variant of the later ***Periplous*** of the Mediterranean wrongly attributed to him (PSEUDO-SKULAX); parts of which, some have argued, date to an original authored by Skulax.

Ed.: *FGrHist* 709; *BNJ* 709 (Kaplan).

RE 3.1 (1927) 619–635, F. Gisinger; D. Panchenko "Scylax' circumnavigation of India and its interpretation in early Greek geography, ethnography and cosmography," *Hyperboreus* 4 (1998) 211–242.

Philip Kaplan

Skulax of Karuanda, pseudo (362 – 335 BCE)

MS Paris, BNF, *graecus* 443 contains a ***periplous*** purportedly by SKULAX OF KARUANDA, which begins at the Pillars of Hēraklēs and describes the north coast of the Mediterranean, the Black Sea, the African shore, and Morocco beyond the Pillars; a series of measurements of distances along parallels in the Mediterranean follows; then a list of Mediterranean islands. Although usually taken to be a sailing manual, it is unlikely to have been of practical use; rather, it is a compendium of geographical data from different eras, providing a verbal map of the world known to the Greeks. Details seem conflated from various sources, and the text has suffered extensive corruption. The text gives distances throughout, either in days' sail or in stades; these are often inaccurate. It describes topographical features significant for sailing, such as islands, gulfs, promontories, harbors, rivers, lakes, and mountains. It also gives incidental details about man-made structures such as cities, emporia, temples, fortifications, and shipyards. The overall view of terrestrial geography is conventional, deriving from HĒRODOTOS and earlier Ionian speculation. Particular attention is paid to the flora and inhabitants of the coastal regions, and some ethnographical detail is provided as well, particularly regarding Libya (Africa), along with occasional references to historical events and Greek myth. The text's date is uncertain and debated. In its current state it is certainly not by Skulax, although portions may preserve his work. Internal evidence of geographical references suggests to some the years 362–357; others place it later, 338–335; it has also been thought a Byzantine pastiche. It is most likely an accretion of different strata, ending in the late 4th c. BCE; the measurements along parallels are Hellenistic additions.

A. Peretti, *Il periplo di Scilace* (1979); E. Lipiński, *Itineraria Phoenicia* (*Studia Phoenicia* 18) (2004) 337–434.

Philip Kaplan

Skumnos of Khios (175 – 145 BCE)

Greek geographer, author of a description (*periēgēsis*) of Asia and Europe in Ionic prose in at least 16 books, mistakenly identified as the author of a later poetic *periēgēsis* (PAUSANIAS OF DAMASKOS). Skumnos was the son of Apellēs, a *proxenos* to Delphi in 185 BCE. The 19 extant fragments of his work show an interest in foundation stories, mythology and botany.

RE 3A.1 (1927) 661–687, Fr. Gisinger.

Daniela Dueck

Skuthinos of Teōs (420 – 350 BCE?)

Versified HĒRAKLEITOS, according to HIERŌNUMOS OF KARDIA in DIOGENĒS LAËRTIOS 9.16, of which one fragment is preserved in IŌANNĒS STOBAIOS 1.8.43, citing from *On nature*. Skuthinos also wrote a *Historia* which described the deeds of Hēraklēs (*FGrHist* 13).

NP 11.656, E. Bowie.

PTK

Sminthēs (500 – 250 BCE?)

Nothing is known about this author of an astronomical poem entitled *Phainomena*, which AUIENIUS (*Ph.* 582–584) alluded to on the Pleiades. Sminthēs is also known thanks to *The Second Life of Aratos* and to a list of astronomical authors contained in MSS *Vatic. gr.* 191 and 381. He perhaps lived before ARATOS and seems to have been known by ERATOSTHENĒS. The name is rare, and usually spelled "Sminthis": *cf. LGPN* 1.409 (Rhodes and Thasos, *ca* 400 BCE) and 3A.398 (Megalopolis, *ca* 365 BCE).

RE 3A.1 (1927) 725–726, L. Wickert; *SH* 729–730.

Christophe Cusset

Sōkratēs (junior) (400 – 360 BCE)

Student of Sōkratēs, THEAITĒTOS' coeval and close companion (PLATO, *Soph.* 218b; *Thaeat.* 147d). Plato includes him as an interlocutor in the *Politikos* (257c). ARISTOTLE, who met Sōkratēs at the **Academy** (pseudo-ALEXANDER OF APHRODISIAS, *CAG* 1 [1891] 514; ASKLĒPIOS OF TRALLEIS *CAG* 6 [1888] 2, p. 420), opposes the younger Sōkratēs' analogy reducing sensate animals to mathematical properties (*Met.* 1036b24–32). Aristotle, merely hinting at Sōkratēs' correlation, claims that it "makes one assume that a man can exist without his parts, like circle without bronze."

RE 3A.1 (1927) 890–891 (#6), E. Kapp; *NP* 11.686–687 (#4), K.-H. Stanzel.

GLIM

Sōkratēs (Lithika) (70 – 200 CE)

Lapidary author, whose lost paradoxographical *On boundaries [?], places, fire and stones* Athēnaios, *Deipn.* 9 (388a) = *FGrHist* 310 F17, cites regarding *attagas* ("francolin"). The MS reading *horōn* "boundaries" is probably wrong; Casaubon suggests the emendation *horōn* "seasons" or *aërōn* "climates," while Müller posits *orōn* "mountains." Fragments of a prose *On stones* attributed to Sōkratēs and a certain DIONUSIOS appear together in manuscripts, unattributed except in the 14th c. Vaticanus Graecus 578, transmitting the so-called *Orphei lithika kērugmata* (occupying §§26–53). Since it is difficult to distinguish how much of this compilation derives from Sōkratēs and how much from Dionusios (the former is considered the author of the descriptions, the latter the illustrator), it is impossible to ascertain which surviving fragments are to be attributed to which author. According to Wirbelauer, Sōkratēs might be a corruption of Xenokratēs and this treatise, therefore, should be attributed to XENOKRATĒS OF EPHESOS.

The treatise describes about 30 stones (the hyacinth, the "rare" stone, the Babylonian stone, the chrysolite, etc.), all endowed with magical properties and some accompanied by engravings representing figures which were originally astrological (planetary, zodiac or **decans**) as well as magical figures (alphabets, secret formulas). Such characteristics (typical of Egyptian-Greek lapidaries), together with some historical references, scattered in the text (particularly in §26 the definition of "Neronian," given to a kind of emerald), suggest an eastern composition, probably in Egypt, during the imperial age.

Ed.: Halleux and Schamp (1985) 139–144, 166–167 (text).
RE 3A.1 (1927) 810–811 (#4), R. Laqueur; K.W. Wirbelauer, *Antike Lapidarien* (1937) 31–42.

Eugenio Amato

Sōkratēs (Med.) (10 BCE – 100 CE)

GALĒN, *Eupor.* 3 (14.501 K.), cites his "famous pill" for headaches and migraines, containing *thapsia* juice, ***euphorbia*** (*cf.* IOUBA), ginger, opium, and ***opopanax***, in vinegar, the patient to be rubbed on the forehead or fumigated therewith. SŌRANOS (?) in CAELIUS AURELIANUS, *Chron.* 3.151 (*CML* 6.1.2, p. 770), listing him with ASKLĒPIADĒS OF BITHUNIA and THEMISŌN, describes his **dropsy**-cure: multiple incisions are cauterized to induce spasm. Diels 2 (1907) 92 lists a Paris MS, BNF 1202 (13th c.), f.16, containing excerpts.

RE 3A.1 (1927) 893 (#11), F.E. Kind.

PTK

Sōkratēs of Argos **(300 – 50 BCE)**

Wrote a ***periēgēsis*** of Argos (DIOGENĒS LAËRTIOS 2.47), possibly geographical, although the sole secure fragment (*FGrHist* 310 F1) concerns religion; other fragments trace genealogies, treat religious and funerary practices, and mythology (F3–6, from PLUTARCH, concern Argos). This man, or a homonymous grammarian from Kōs, wrote on religious topics. The Sōkratēs credited with *On Boundaries, Places, Fires and Stones*, is probably SŌKRATĒS (LITH.).

RE 3A.1 (1927) 804–810, esp. 806 (#3), A. Gudeman; *NP* 11.687 (#7), A.A. Donohue.

PTK and GLIM

Sōkratiōn (250 BCE – 110 CE)

KRITŌN records, in GALĒN *CMLoc* 5.3 (12.835–6 K.), his refined lotion for ***leikhēn***, compounded from asphodel and ***alkuoneion***, reduced in vinegar, in which are dissolved ***ammōniakon*** incense, myrrh, frankincense, olive oil, salted meal, raw sulfur, ***misu***, ***khalkitis***, Kimolian earth, alum, and ***aphronitron***. Sōkratēs and its variations (Sōkratidās, Sōkratidēs) are not uncommon; for Sōkratiōn *cf.* Catullus 47.1.

RE 3A.1 (1927) 901 (#2), F.E. Kind.

GLIM

SOLINUS ⇒ IULIUS SOLINUS

Solōn of Smurna (250 BCE – 75 CE)

Listed by PLINY 1.*ind.*20–27 as an authority on botanical medicines, cited on orache (hard to grow in Italy: 20.220) and on the unidentified plant *bulapathum* (prescribed in wine for dysentery: 20.235). ANDROMAKHOS, in GALĒN *CMLoc* 3.1 (12.630 K.), calling him a dietician, cites his ear-medicine: reduce alum, castoreum, saffron, frankincense, myrrh, and opium in honeyed wine to honey-like viscosity.

RE 3A.1 (1927) 979 (#7), F.E. Kind.

PTK

Sophar/Sōphar the Persian **(before 8th/9th c. CE)**

Alchemist, pseudonymous if not entirely fictitious. The ANONYMOUS ALCHEMIST PHILOSOPHER claims that Sophar was discussed in a lost work attributed to OSTANĒS (*CAAG* 2.120–121; Bidez and Cumont 1938: 329).

(*)

Bink Hallum

Sōranos of Ephesos (98 – 138 CE)

Prominent **Methodist**, ranked with HIPPOKRATĒS and GALĒN for astute contributions to the practice of medicine, not simply to gynecology and obstetrics. His talents are displayed in his extant *Gynecology*: with access at least to books on human anatomy in Alexandria, Sōranos referred to HĒROPHILOS' dissections of the aorta and vessels of the liver (*Gyn.* 1.17[57].4 [*CMG* 4, p. 42) and uterus (*Gyn.* 4.36[85].2–3 [*CMG* 4, p. 148]). Sōranos practiced in Alexandria (*Gyn.* 2.6[70b].4 [*CMG* 4, p. 55]) before migrating to Rome to practice "pediatrics" (*Gyn.* 2.44[113].1–2 [*CMG* 4, p. 85]). The *Souda* (Sigma-851, 852) dates him to the reigns of Trajan and Hadrian. CAELIUS AURELIANUS, *Acute* 2.130 (*CML* 6.1.1, p. 220; Drabkin, pp. 218–219), says that in Rome he used phlebotomy to treat pleurisy, and MARCELLUS EMPIRICUS, 19.1 (*CML* 5.1, pp. 310–312), creates an amusing garble from PLINY 26.4 (Manilius Cornutus, legate in Aquitania), reflecting Sōranos' later fame.

In addition to the *Gynecology*, extant are (a) *Signs of Fractures* (*CMG* 4, pp. 153–158), (b) *Bandages* (pp. 159–171), and (c) a *Life of Hippokratēs* (pp. 173–178; Pinault 1992: 6–18; *FGrHist* 1062 F2). The first two are probably surviving fragments of Sōranos' *Surgical Operations* (*Kheirourgoumena: cf. Gyn.* 2.7[76].4 = *CMG* 4, p. 56). Most scholars now favor a mistaken attribution in the Hippokratic MSS as one of the biographies of *Lives of Physicians* (Kind 1927: 1115–1116, titles of biographies; *cf.* Smith 1990: 49 with n. 2, 51 n. 1, and 53 n. 3; Temkin 1991: 52–57; Mansfeld 1994: 182–183 with n. 329; van der Eijk 1999: 401–402; but *cf.* Radicke in *FGrHist* 1062).

Known by title and fragmentary quotations are a dozen other lost writings demonstrating Sōranos' wide-ranging interests. (1) Four books *On the Soul* (Tert., *De anima* 6, ed. Waszink [1947] 22*–40*; Polito 1994; van der Eijk 1999: 402–403). (2) *Commentaries on Hippokratēs* in an unknown number of books (Kind 1927: 1116–1117). (3) An *Etymology of Human Anatomy* (Scholia on RUFUS' *Anatomical Nomenclature*, pp. 237–246 DR, and lexicographers). (4) *Acute and Chronic Diseases*, essentially translated into Latin by Caelius Aurelianus, with a few Greek fragments. (5) *On Causes* (*Aitiologoumena*), quoted in Caelius Aurelianus, *Chron.* 1.55 (*CML* 6.1.1, p. 220; Drabkin, pp. 474–475), ". . . a nightmare [*incubus*] is not a god or godlet or Cupid." (6) *Communities of Diseases* (*Gyn.* 1.6.29 = *CMG* 4, p. 19), Book 2 of which criticizes DIONUSIOS (METHOD.) for thinking certain pathologies were "natural," whereas they were truly "diseased states" (*Gyn.* 3.4 = *CMG* 4, p. 96). Caelius Aurelianus, *Chron.* 4.5 (*CML* 6.1.2, p. 776; Drabkin, pp. 816–817), in citing *Communities* Book 2, indicates Sōranos had reduced the number of pathologies to three: *status strictus, status laxus, status mixtus*, contrasted to earlier **Methodists**. (7) *On Fevers*: Caelius Aurelianus, *Acute* 2.177 (*CML* 6.1.1, p. 250; Drabkin, pp. 254–255); *cf.* MUSTIO/MUSCIO, 2.2.23 (Rose, p. 57). (8) *Principles of Health* (*Hugieinon*): *Gyn.* 1.7.32, 1.10.40 (*CMG* 4, pp. 21, 28]). (9) *On Remedies* or *On Therapeutics* (*Peri boēthēmatōn*) cited in the famous account of diagnosis, prognosis, and therapeutic indications for "suffocation of the uterus," or "hysterical suffocation" (*husterikē pnix*), *Gyn.* 3.4.26–29 (*CMG* 4, pp. 109–113). There, many remedies are suggested, e.g., dry cupping in the groin, moistening the genitals with sweet olive oil, swinging in a hammock, and mild vaginal suppositories, optionally followed by an olive oil enema. Following the *metasyncritic* therapies common among **Methodists** (i.e. progressing gradually from mild drugs and therapeutic measures to harsher methods until the disease is alleviated: Scarborough 1991), Sōranos indicates that the patient is made to choke on white hellebore (*elleboros leukos: Veratrum album* L.) and induced to vomit with radishes. Notably, Sōranos rejected the theory of the animate uterus, "...although he admits that in some ways it behaves as if it were" (King 1998: 223; *cf.* Gourevitch 1984: 121–126; Dean-Jones 1991: 122, 135–136, n. 55). (10) A book either titled *A Drug from Poppy-Heads* (Galēn, *CMLoc* 7.2 [13.42–43 K.]) or *Philiatros* (Mustio/Muscio, Rose, p. 3: *sicut in opthalmico et chirurgumeno filiatro etiam boethematico legisti*), perhaps intended as a tersely worded counterpoint to DAMOKRATĒS: Sōranos' prosaic instructions begin "take 150 poppy-heads, 20 *sextarii* of water..." (Scarborough 1995: 6). (11) A probable five-book tract on drugs-as-cosmetics attached to a work that arranged pharmaceuticals into "communities," but which Galēn confusedly says is "four books on dandruff-treatments . . . and a single book on drug-actions" (*CMLoc* 1.8 [12.493–496 K.]); the listing that follows is a kind of "head-to-heel" catalogue, partly reminiscent of KRITŌN's *Kosmētika*, but Sōranos has grouped his medicinals as botanicals and minerals. The probable title is *Peri pharmakeias* (Scarborough 1985c: 394–397; 1991: 207–216). (12) A tract entitled *Ophthalmos*, cited once by CASSIUS IATROSOPHIST, *Problēmata* 27 (Garzya and Masullo 2004: 49).

Lloyd (1983: 168–200) and Gourevitch (1988: XXX–XLVI) provide succinct surveys of Sōranos' theoretical constructs and his critiques of earlier practices in gynecology and obstetrics. He has little patience for "superstition" (a debatable term, Scarborough 2006: 12–15), and he is well aware that gynecology is an aspect of medicine wherein popular beliefs fuse with the more "rational" outlooks of physicians equipped with a knowledge of anatomy and the elements of physiology, and whose increased knowledge of female vs. male physiologies eased the separation of the two sexes (Hanson 1991). Yet Sōranos' authoritative stance occasionally obscures the role of the midwife (literate or not) in prenatal care, and in birthing, with the necessary follow-up in care of the newborn.

Striking is Sōranos' discussion of contemporary ethical controversies as they necessarily impinged on the practice of medicine, and the tabulation of contraceptives and abortifacients in *Gyn.* 1.19.60–65 (*CMG* 4, pp. 45–49); Burguière *et al.* (1988: 1.59–65) very questionably insert the more elaborate account of such drugs as contained in AËTIOS OF AMIDA 16.17, 21 (Zervos 1901: 18–20, 25–26). Sōranos provides a priceless list of recommendations: of the two dozen substances, about three-quarters are chemically effective (Keller 1988; Riddle 1992: 25–30). Sōranos' only "mechanical" means, plugging the cervix with wool, is still used as a cheap – and fairly effective – contraceptive, although condoms were common (Scarborough 1969: 101 with n. 50, 209). He also records the "heated dispute" between doctors who forbid abortifacients, citing the HIPPOKRATIC CORPUS, OATH, and those who prescribe them "with great care" (only for the health of the woman); Sōranos prefers prevention, and prescribes contraceptives. Later term abortions too will be occasionally recommended, albeit rather more risky than the simples prescribed for contraception (Riddle 1992: 46–56).

Like many **Methodists**, Sōranos did not adhere rigidly to any particular "doctrine," other than how a physician theorized the origins of disease and maintenance of health. To Sōranos the simplest "explanation" bereft of most philosophical terminology was to assume that a "healthy body" was a freely-flowing one, but not to the extremes observed in profuse sweating or severe diarrhea; likewise, a "sick" body blocked the "flow," as in the often-diagnosed cases of constipation; thus a "healthy" body showed what Caelius Aurelianus called a *status mixtus*. Some **Methodists** valued anatomy, and Sōranos' skills in surgery suggest he had performed dissections during his training (or, like Galēn, had recorded observations of wounded bodies), but the anatomy displayed by Sōranos indicates that he followed what he had found in Hērophilos (e.g. the uterus and bladder are connected).

Sōranos' dependence on earlier medical traditions and dogmas is apparent as he composes his medico-historical doxography. Both Sōranos and Caelius Aurelianus retail previous written authorities in describing and treating specific diseases, and occasionally add harsh criticism – even of Hippokratēs. Sōranos seems one of the more "practical" in his approach, but even with his clear adaptation of much midwifery, he imposes the necessity of Greek literacy for any woman to be a success in what was recognized as the "common knowledge of women."

Ed.: V. Rose, *Sorani Gynaeciorum vetus translatio Latina nunc prima edita cum additis Graeci textus reliquiis a Dietzio repertis atque ad ispum codicum Parisiensem* (1882); J. Ilberg, *Sorani Gynaeciorum libri IV. De signis fracturarum. De fasciis. Vita Hippocratis secundum Soranum* (1927) = *CMG* 4; O. Temkin, trans.; *Soranus' Gynecology* (1956; repr. 1991); Pinault (1992) 8–18; *FGrHist* 1062 (J. Radicke); P. Burguière, D. Gourevitch, and Y. Malinas, *Soranos d'Éphèse: Maladies des femmes* 4 vv. (*CUF* 1988–2003).

RE 3A.1 (1927) 1113–1130, F.E. Kind; John Scarborough, *Roman Medicine* (1969); *DSB* 11 (1975) 538–542, M. Michler; G.E.R. Lloyd, "The Critique of Traditional Ideas in Soranus' Gynecology" in *Science, Folklore and Ideology* (1983) 168–200; D. Gourevitch, *Le mal d'être femme. La femme et la médecine à Rome* (1984); L. Dean-Jones, "The Cultural Construct of the Female Body in Classical Greek Science," in S.B. Pomeroy, ed., *Women's History and Ancient History* (1991) 111–137; John Scarborough, "The Pharmacy of Methodist Medicine," and A.E. Hanson, "The Restructuring of Female Physiology at Rome," in Mudry and Pigeaud (1991) 204–216 and 255–268; J.M. Riddle, *Contraception and Abortion from the Ancient World to the Renaissance* (1992) 25–30, 46–56; J. Mansfeld, *Prolegomena. Questions to be Studied before the Study of an Author, or a Text* (1994); R. Polito, "I quattri libri sull'anima di Sorano e lo scritto *De anima* di Tertulliano," *Rivista di Storia della Filosofia* 3 (1994) 423–468; H. King, *Hippocrates' Woman* (1998); Ph.J. van der Eijk, "Antiquarianism and Criticism: Forms and Functions of Medical Doxography in Methodism (Soranus, Caelius Aurelianus)," in van der Eijk (1999) 397–452; John Scarborough, "Drugs and Drug Lore in the Time of Theophrastus: Folklore, Magic, Botany, Philosophy and the Rootcutters," *AClass* 49 (2006) 1–29.

John Scarborough

Sōranos of Kōs (350 BCE – *ca* 120 CE)

One of Sōranos' sources for his biography of Hippokratēs (*Vita Hipp.* 3–5: *CMG* 4, pp. 175–176). The *Souda* Sigma-852 mentions Sōranos of Mallos, acclaimed by Asklēpiodotos of Athens as an excellent doctor (perhaps *ca* 460 CE, if not a confusion for Sōranos of Kōs or of Ephesos).

RE 3A.1 (1927) 1130, F.E. Kind; Pinault (1992) 7, 11, 83.

PTK

C. Sornatius (before 75 CE)

Roman author, cited by Pliny as an authority on drugs made from aquatic animals (1.*ind.*31–32, 32.68). He is hardly to be identified with the legate who accompanied Licinius Lucullus during the third Mithridatic war (73–68 BCE: *MRR* 2.621).

RE 3A.1 (1927) 1137–1138, Fr. Münzer.

Jan Bollansée, Karen Haegemans, and Guido Schepens

Sōsagoras (250 BCE – 25 CE)

Physician whose remedy for joint pain contained equal parts of roasted lead, poppy "tears," henbane-bark, ***sturax***, sulfurwort, suet, resin, and beeswax (Celsus 5.18.29). A rare name,

Sōsagoras, cited from the 3rd–1st cc. BCE, is known at Amorgas, Nisuros, and at Maroneia in Thrakē (*LGPN* 1.420, 4.323).

RE 3A.1 (1927) 1144, F.E. Kind.

GLIM

Sōsandros (Geog.) (120 BCE – 50 BCE)

A pilot (*kubernētēs*) who wrote a ***periplous*** of or to India, cited only by MARCIANUS OF HĒRAKLEIA in his epitome of MENIPPOS OF PERGAMON.

FGrHist 714.

PTK

Sōsandros (Pharm.) (250 BCE – 95 CE)

ASKLĒPIADĒS PHARM. records in GALĒN, *CMLoc* 4.7 (12.733–734 K. = AËTIOS OF AMIDA 7.78 [*CMG* 8.2, p. 328]), his collyrium for *milphosis* (eyelashes falling off), enduring disorders, and lesions at the corner of the eyes, compounded from **calamine**, antimony, ***khalkitis***, ***misu***, honey, roasted together and then soaked with wine, dried and used. The name is attested from the 4th–1st cc. BCE (*LGPN*).

RE 3A.1 (1927) 1145 (#2), F.E. Kind.

GLIM

Sōsandros (Veterin.)

Mythical inventor of the discipline of horse-medicine. According to the 12th c. CE chronicle of Geōrgios Kedrēnos (ed. I. Bekker, 1.213), Sōsandros was the brother of HIPPOKRATĒS: the relation of human and veterinary medicine is thus symbolically expressed as a fraternal one. Kedrēnos' notice echoes an epigram in the Planudean anthology (*AP* 16.271, ed. H. Beckby [1958] 446), which elaborates a pun on the names Hippokratēs, "lord of horses," and Sōsandros, "savior of men." In the 14th c. allegorical poem of Melitēniotēs (ed. E. Miller, "Poème allégorique de Meliténiote," *Notices et extraits des manuscrits de la Bibliothèque impériale et autres bibliothèques* 19.2 [1862] 71), a statue of Sōsandros appears among those of pagan poets, philosophers, sages, and sorcerers. The names Osandros and Sōstratos in the title of the *Epitome* of the *Hippiatrika* in 15th c. manuscripts (*Par. gr.* 2091 and 1995, respectively) may refer to this myth; *cf.* E. Miller, "Notice sur le Ms. grec 2322," *Notices et extraits* 21 (1865) 5–6.

McCabe (2007) 11–12.

Anne McCabe

Sōsigenēs (I) (*ca* 75 – 25 BCE)

An Alexandrian astronomer credited by PLINY (18.210–212) with helping CAESAR in his reform of the Roman calendar in 46 BCE, a reform that involved abandoning the quasi-lunar calendar then in use and adopting a solar calendar of 365 days with an intercalary day (the bissextile) every fourth year. Pliny also reports that Sōsigenēs aided Caesar in preparing a ***parapēgma*** and wrote three commentaries on it.

R. Hannah, *Greek and Roman Calendars* (2005) 113–114.

Alan C. Bowen

Sōsigenēs (II) (*ca* **125 – 190** CE)

Peripatetic philosopher, teacher of ALEXANDER OF APHRODISIAS, author of *On Vision* (now lost). Sōsigenēs' critical commentary on ARISTOTLE's use of winding and unwinding homocentric spheres in counting the number of celestial motions (*Metaphysics* 12.8 [1073a14–1074b14]) apparently drew on EUDĒMOS' *History of Astronomy*; it survives only in citations by pseudo-Alexander, PROKLOS, and SIMPLICIUS. Proklos, *Hyp. astr.* 4.97–99, asserts that Sōsigenēs explained that annular solar eclipses are observed when the Sun is at perigee. Modern scholars have mistakenly concluded that Sōsigenēs observed an annular eclipse, which they date to 164 CE. (No dated observational reports describing annular eclipses survive from Greek or Latin antiquity.) In context, however, the point is only that, for Sōsigenēs, annular total solar eclipses are possible – a compromise between PTOLEMY's denial that such eclipses occur and the claim found, e.g., in P. PARISINUS GRAECUS 1 col. 19.16–17 (*cf.* KLEOMĒDĒS, *Cael.* 2.4.108–115), that all total solar eclipses are annular. According to Proklos, Sōsigenēs constructed a Complete (or Perfect) Year of 648,483,416,738,640,000 years in which all the heavenly bodies return to their original positions, using Babylonian and Egyptian parameters.

Neugebauer (1975) 606; Alan C. Bowen, "Eudemus' History of Early Greek Astronomy: Two Hypotheses," in Bodnár and Fortenbaugh (2002) 307–322 at 315–318; *Idem*, "Simplicius' Commentary on Aristotle, *De caelo* 2.10–12: An Annotated Translation (Part 2)" *SCIAMVS* 9 (2008: forthcoming).

Alan C. Bowen

Sōsikratēs (250 BCE **– 80** CE)

ANDROMAKHOS in GALĒN *CMLoc* 7.6 (13.114 K.) records his pill for ***orthopnoia***, compounded from ***opopanax***, myrrh, pepper-corns, and rue, administered with water for fevers, otherwise, with wine. A common name, known from the 4th c. BCE into the imperial era, but attested more frequently during the Hellenistic era (*LGPN*).

RE 3A.1 (1927) 1166 (#6), F.E. Kind.

GLIM

Sōsikratēs of Rhodes (300 – 150 BCE**?)**

Geographer and doxographer, wrote *Krētika* and a chronological history of philosophers, whose citations in DIOGENĒS LAËRTIOS preserve strictly biographical, chronological, and political details (1.38, 1.49, 1.68, 1.75, 1.95, 1.101, 1.106–107, 6.13, 8.8), maintaining, e.g., that an Aristippos of Kurēnē and Diogenēs the Cynic left no writings (2.84, 6.80). In his *Krētika*, considered reliable by DIODŌROS SICULUS (5.80.4) and extracted by APOLLODŌROS OF ATHENS (the one upon whose testimony STRABŌN relied), Sōsikratēs gave 2,300 stades as the island's length, with a circuit exceeding 5,000 stades, larger than dimensions recorded by ARTEMIDŌROS OF EPHESOS and HIERŌNUMOS OF RHODES (Str. 10.4.3). In other fragments, Sōsikratēs provides ethnographic data regarding Crete.

Ed.: *FGrHist* 461.

GLIM

Sōsimenēs (350 BCE – 77 CE)

Cited as a foreign authority on medicines from plants, after DALIŌN and before TLĒPOLEMOS (PLINY 1.*ind.*20); recommended anise in vinegar for all ailments and as a remedy for fatigue, especially for travelers (20.192). The name, cited 4th c. BCE to 1st c. CE, is concentrated in the 4th/3rd cc. BCE (*LGPN*).

RE 3A.1 (1927) 1166, F.E. Kind.

GLIM

Sōstratos of Alexandria (*ca* 70 BCE – 10 CE)

Important Greek physician and surgeon, he was concerned with bandages (GALĒN, *De Fasc.* 18A.826 K.) and gave relevant advice on delivery (SŌRANOS, *Gyn.* 4.12, 14 = *CMG* 4, pp. 143, 144–145; *CUF* v. 4, p. 22 and v. 2, p. 11). He was chiefly a brilliant zoologist, perhaps the greatest after ARISTOTLE, and wrote a treatise *On Animals* or *On the Nature of Animals* in two or four books (Ath., *Deipn.* 7 [303b, 312e]), whose second book was on fishes, often mentioned in late literature. He also wrote *On Creatures which Strike and Ones which Bite* (*cf.* THEOPHRASTOS), well informed in scientific iology, and another one treating bears, maybe with a Theophrastean theme or title *On Animals which Live in Holes* (*fr.* 11). The few preserved fragments discuss beavers, blackbirds, gadflies, eels, bears, and tuna.

M. Wellmann, "Sostratos," *Hermes* 26 (1891) 321–350; *RE* 3A.1 (1927) 1203–1204 (#13), H. Gossen; *OCD3* 1427, W.D. Ross.

Arnaud Zucker

Sōstratos of Knidos (*ca* 300 – 250 BCE)

Son of Dexiphanēs. Architect prominent at the Ptolemaic court, he built the famous lighthouse on Pharos at Alexandria, one of the seven wonders of the ancient world, *ca* 290–280 BCE, although some sources present Sōstratos as a diplomat and courtier who paid for and dedicated the lighthouse (STRABŌN 17.1.6). The lighthouse, *ca* 100 m tall, was in three sections, decorated with sculpture. The mechanisms for providing fuel for its flame and any optical devices to magnify it (noted in late sources) are open to speculation. Cited by numerous Arab chroniclers, it stood until the 15th c. and some remains have been recovered in underwater excavations. Sōstratos is credited as the first architect to build a boardwalk supported by piers, at Knidos (PLINY 36.83).

H. Thiersch, *Pharos* (1909); T.L. Shear, Jr., *Kallias of Sphettos* = *Hesperia*, S.17 (1978) 22–25; F. Goddio and A. Bernand, *Sunken Egypt, Alexandria* (2004); *KLA* 2.414–415, W. Mueller.

Margaret M. Miles

Sōstratos of Nusa (100 – 50 BCE)

A grammarian, the brother of STRABŌN's teacher Aristodēmos, whose father was Menekratēs (14.1.48), and cited by PSEUDO-PLUTARCH and IŌANNĒS OF STOBOI for paradoxographical works. The works may be forgeries, or else one of the few genuine works cited by pseudo-Plutarch.

RE 3A.1 (1927) 1200–1201 (#7), E. Bux.

PTK

Sōtakos (320 – 270 BCE)

Lapidary author (*On stones*) according to APOLLŌNIOS PARADOX. 36, and whom PLINY 1.*ind*.36–37 includes among foreign authors. Pliny, citing Sōtakos on magnetite (36.128–129), hematite (36.146–148), amber (37.35), sardonyx (37.86), onyx (37.90), *keraunia* (37.135) and *drakonite* (37.158), quotes him among the most ancient writers; this might suggest that our author dates to the late 4th c. BCE. Sōtakos probably used PUTHEAS OF MASSILIA for information about Britain, reported by Pliny 37.35. Pliny 1.*ind*.37 lists Sōtakos with Putheas and other authors who probably spoke about northern European amber.

GGLA 1 (1891) 860–861; *RE* 3A.1 (1927) 1211, F.E. Kind; K.G. Sallmann, *Die Geographie des älteren Plinius in ihrem Verhältnis zu Varro. Versuch einer Quellenanalyse* (1971) 82–83; Ullmann (1972) 96–98; J. Kolendo, *À la recherche de l'ambre baltique. L'expédition d'un chevalier romain sous Néron* (1981) 68; A. Grilli, "La documentazione sulla provenienza dell'ambra in Plinio," *Acme* 36 (1983) 5–17; S. Bianchetti, *Plōta kai poreuta. Sulle tracce di una Periegesi anonima* (1990) 78–82; *Eadem, Pitea di Massalia. L'Oceano* (1998) 204.

Eugenio Amato

Sōteira (300 BCE – 77 CE)

Midwife (*obstetrix*) listed with LAÏS, ELEPHANTIS, and SALPĒ as a foreign authority on drugs obtained from animals (PLINY 1.*ind*.28). She recommended smearing the soles of the feet with a patient's own menstrual blood to treat fevers and epilepsy (28.83).

RE 3A.1 (1927) 1239 (#3), F.E. Kind; Parker (1997) 145 (#47).

GLIM

Sōtiōn (200 BCE – 65 CE)

Said by Phōtios (*Bibl.* 189) to have written a booklet containing "miscellaneous marvelous stories about rivers, springs and lakes" (*ta sporadèn paradoxologoumena peri potamōn kai krēnōn kai limnōn*). That the short work treated exclusively various kinds of miraculous waters (unlike other paradoxographical collections covering different parts of the physical world) elicited the suggestion that this Sōtiōn was the anonymous PARADOXOGRAPHUS FLORENTINUS, but this hypothesis cannot be supported. Likewise, it is impossible to tell whether the paradoxographer cited by Phōtios is identical with the homonymous author of a treatise on agricultural wonders (***Paradoxa*** *peri geōrgias*) mentioned in the GEŌPONIKA (*arg.* 1, v.1, p. 7 Nicl.), or with any other of the dozen of homonymous writers known from the early 2nd c. BCE until the first half of the 1st c. CE.

Ed.: *PGR* 167–168.
RE 18.3 (1949) 1137–1166 (§29, 1161), K. Ziegler; Giannini (1964) 128; *NP* 11.754–755 (#1), R.W. Sharples.

Jan Bollansée, Karen Haegemans, and Guido Schepens

Sōtiōn of Alexandria (200 – 150 BCE)

Wrote the earliest comprehensive survey of philosophers. On the basis of **Peripatetic** notions, he organized all philosophers in "schools," one philosopher succeeding another, and one school of philosophy linked up with another school. Sōtiōn obviously wanted to impose the organization of the four Hellenistic schools in Athens on the previous philosophical tradition, hence the title of his book *Successions of Philosophers*. The two main lines of schools were the Ionic and the Italic. There is no evidence that philosophical schools (except perhaps for the **Pythagoreans**) existed before PLATO. Both Sōtiōn and other writers of similar works seem to have concentrated on the biographical material and only dealt with philosophical ideas to the extent that they first appeared with a particular philosopher. Most of the fragments of Sōtiōn's book come from DIOGENĒS LAËRTIOS, the structure of whose work seems to reflect Sōtiōn's.

Ed.: Wehrli, S. 2 (1978).

W. von Kienle, *Die Berichte über die Sukzessionen der Philosophie in den hellenistischen und spaetantiken Literatur* (1961); Mejer (1978) 40–42 and 62–71; F. Aronadio, "Due fonti laerziane: Sozione e Demetrio di Magnesia," *Elenchos* 11 (1990) 203–255.

Jørgen Mejer

Spendousa (*ca* 100 BCE? – *ca* 80 CE)

ANDROMAKHOS in GALĒN, *CMLoc* 3.1 (12.631 K.), cites her *ōtikē* for infected ears: heat honey and barrow-fat in a glass vessel. The glass vessel may suggest a *terminus post* of *ca* 50 BCE. The rare name is attested from the 1st c. BCE (*LGPN*).

RE 3A.2 (1929) 1610, F.E. Kind.

PTK

Speusippos of Alexandria (250 – 50 BCE)

Hērophilean physician, distinct from PLATO's nephew and successor, the only homonym cited by DIOGENĒS LAËRTIOS (4.5). An uncommon name, attested six times in Athens from 5th–2nd cc. BCE (*LGPN*).

von Staden (1989) 585.

GLIM

Speusippos of Athens (*ca* 380 – 339 BCE)

PLATO's nephew, born *ca* 410 BCE; involved in Academic politics in Sicily during Plato's lifetime (PLUTARCH, *Dio* 17.22 = T29–30), and head of the **Academy** from Plato's death in 347 BCE until his own death. He wrote on a wide variety of philosophical topics, and an *Epistle to Philip II* survives, dating from *ca* 342 (if genuine). Largely independent intellectually, demonstrating **Pythagorean** tendencies, Speusippos showed a fascination for number and physical theory, wherein mathematical objects have a place of honor, and denied the existence of Platonic Ideas. He was involved in an elaborate classificatory project, resulting in ten books of *Homoia* (*Things Similar*), and regarding ethics he was anti-hedonistic. There is considerable useful (but hostile) evidence for his thought in ARISTOTLE, an unreliable biography in DIOGENĒS LAËRTIOS, a large fragment on the decad from a work *On Pythagorean Numbers*, a tantalizing chapter (4) in IAMBLIKHOS' *De Communi Mathematica Scientia* seemingly reflecting only Speusippean metaphysics, and a variety of other material (of uneven value and reliability). The comic fragment of Epikratēs (*fr.*10 *PCG* = T33), describing the close inspection of a pumpkin in a bid to determine its natural genus, further substantiates the classificatory project.

Speusippos postulates mathematical entities distinct from sensibles, but no Ideas. Aristotle (*Metaph.* 7.2 = F29a) reports that his theory of first principles begins with a One (not identical either with the Good or with Mind, as in XENOKRATĒS), and proceeds to Numbers, Geometrical Magnitudes, Souls, and Bodies, each with its own manifestation of the One (monad, point etc.) and its own *distinct* substrate (multiplicity, extension etc.). Aristotle (*Metaph.*

12.10 = T30) describes this system as "episodic" (i.e., like dramas consisting of a series of disconnected episodes). It seems that, while beauty appears at the mathematical levels, anything describable as "good" only appears at the level of soul. Speusippos explained the creation in the *Timaios* as simply a didactic device, so we must suppose his picture of successive stages of generation offered something other than a temporal account of how the world came to be as it is. Speusippos' explanation of soul in terms of the "Idea of the all-extended" may also relate to the *Timaios*, but its significance is unclear. He seems to have explained god as a soul-power governing all things, or an intellect (Aëtios 1.7.20 = T58), again avoiding any transcendent divinities (Cicero, *Nat. D.* 1.32 = T56).

In epistemology, Speusippos spoke of the "scientific" (*epistēmonikos*) sensation expected of experts in the use of the senses (Sextus Empiricus, *Adv. Math.* 7.145–6). We must infer that Speusippos was keen to show that scientific knowledge could be based on sensation, *pace* Plato. A further innovation seems to be the claim, connected with the classificatory project, that in order to know A, one must know also B, C, D, etc. from which A differs, since one needs to know all the differentiae of A (Arist. *An. Po.* 2.13 = T63).

In ethics, he clearly tried to avoid the hedonistic argument that if pain is bad, its opposite should be good, by appealing to something with a superficial resemblance to the doctrine of the mean: extremes are opposed to the middle, as well as to one another (Arist. *EN* 7.12, *cf.* 10.3 = T80–81). This suggests that the good is in fact a stable neutral state, rather than any deviation from or return to it.

Other idiosyncrasies in the areas of mathematic and logic suggest an original thinker, more scientific than Aristotle would allow, the loss of whose works is to be lamented.

Ed.: L. Tarán, *Speusippus of Athens* (1981).
Dillon (2003) 30–88.

Harold Tarrant

Sphairos of Borusthenēs (260 – 210 BCE)

Stoic philosopher and student of Zēnōn of Kition and Kleanthēs of Assos. Advisor or teacher to Kleomenēs III of Sparta, he later participated in the courts of Ptolemy IV Philopatōr (Diogenēs Laërtios, 7.177) and of at least one of Ptolemy's immediate predecessors, Ptolemy III Euergētēs, or Ptolemy II Philadelphos (D.L., 7.185, is unspecific and, given the dates of the participants, the Ptolemy who wrote to Kleanthēs could have been either Ptolemy II or III). Diogenēs Laërtios (7.178) ascribes 31 titles to him, including a set of lectures on Hērakleitos and books on a range of topics in ethics, physics, and logic. Very little else is known about his philosophy, although a few reports offer some hints. Cicero (*Tusc.* 4.53) says that the **Stoics** considered Sphairos particularly good at definitions, and Diogenēs Laërtios (7.177) recounts a story about Ptolemy IV tricking Sphairos with a wax pomegranate to show that a philosopher might assent to a phantasm (where Sphairos had argued that the wise man would not assent to an opinion). Sphairos countered that he had only assented to the fact that the pomegranate appeared real, not that it actually was real.

Ed.: *SVF* 1.620–630.

Daryn Lehoux

Sphujidhvaja (269/270 CE)

Composed in 269/270 CE a Sanskrit verse adaptation, entitled *Yavanajātaka* (*YJ*) or *Greek Horoscopy*, of a prose translation from Greek by an anonymous YAVANEŚVARA ("lord of the Greeks"). SPHUJIDHVAJA, who claimed (*YJ* 79.62) to bear the title "rājā" or "king," presumably referring to a similar elevation among Indian Greeks under the Śakas, wrote this work during the reign of the western Kṣatrapa monarch Rudrasena II. The *YJ*s 79 chapters primarily concern **genethlialogy** or horoscopic astrology (chapters 1–51), with some discussion of other astrological branches (interrogations, **katarkhic** astrology, and military astrology) and mathematical astronomy (chapter 79). Sphujidhvaja's presentation of this material, presumably similar to that of the prose translation, clearly reveals its Greek origin, including many transliterated Greek technical terms. However, it also bears witness to some "naturalization" within Indian traditions. The significance of the various astrological concepts is described in terms of Indian deities and culture, and some of the topics, techniques and parameters are apparently Indian rather than Hellenistic. Most of the standard subjects in the subsequent development of pre-Islamic Indian horoscopy are based on those of the *YJ*.

Pingree (1978); *Idem, Jyotiḥśāstra: astral and mathematical literature = History of Indian Literature* 6.4 (1981).

Kim Plofker and Toke Lindegaard Knudsen

Sporos of Nikaia (200 – 300 CE)

Six fragments and three *testimonia*, hard to synthesize, have reached us under this name. (F1) EUTOKIOS paraphrases his solution to the duplication of the cube (*In Arch. Circ. dim.* 4.57–58 Mugler). (F2) PAPPOS approvingly reports his criticism to the *quadratrix* curve (*Math. Coll.* 1.252–256 Hultsch): its generation requires determining the ratio of the circumference of a circle to its radius, although it is meant to find it. Attributed to Sporos are (F3–5) three nominal *scholia* to ARATOS' *Phainomena* (*Scholia in Aratum Vetera* 541.40–46, 881.21–27, 1093.1–8 Martin) giving physical explanations for natural phenomena (end of the visual ray pointing to the sky, parhelia, comets), and (F6) a short excerpt in one Aratos MS explaining why Aratos began with boreal constellations, introduced by the mention "*Hipparkhou Sporos.*"

Additionally, (T1) Eutokios (*In Arch. Circ. dim.* 4.162.18–24) mentions that *Poros ho Nikaieus* blamed ARCHIMĒDĒS for his vague approximation of the circumference of a circle, contrary to his teacher PHILŌN OF GADARA's more precise estimation, as reported in his *Kēria* (*Honeycombs*). (T2) This work may be the same as the *Aristotelika Kēria* mentioned by Eutokios in the same commentary (4.142.21), with no mention of author but as well-known to his readers (Aulus Gellius, *Pr.*1.6, signals *keria* as an example of a curious book-title). (T3) LEONTIOS reports that "Sporos the commentator" (3.6) excused Aratos' lack of precision, since his work was aimed only at navigators. Modern commentators strongly diverge on the positive conclusions to be drawn from such weak and disparate bases. (T3) and (F3–6) show that Sporos probably commented on Aratos; (F2) and (T1–2) might indicate that he wrote a compilation entitled *Kēria*, containing critical discussions of solutions to classical problems of geometry; (F4) and (T2) might indicate Sporos' relative obedience to ARISTOTLE.

Martin (1956) 205–209; *DSB* 12.579–580, M. Szabo; Knorr (1989) 87–93.

Alain Bernard

Stadiasmus Maris Magni (200 – 300 CE)

Description of the coast-lines (***periplous***) of the Mediterranean, based on older sources. Only the passages referring to the African shores between Alexandria and Utica, and between Arados and Milētos, as well as *data* concerning Cyprus and Crete, survive. *Stadiasmos*, "calculation of distances in stades," was first used as a book title by TIMOSTHENĒS OF RHODES.

Ed.: *GGM* 1.427–514.
KP 5.336, F. Lasserre; *OCD3* 1141, N. Robertson; *NP* 11.886, E. Olshausen.

Andreas Kuelzer

Staphulos of Naukratis (200 – 50 BCE)

Wrote histories of the Ailioans, Arkadians, Athenians, and Thessalians, telling myths, recounting migrations, explaining place-names, and describing customs. Scholiasts, STRABŌN 10.4.6, PLINY 5.134, and others preserve a few fragments. The masculine form of this rare name is attested only from 200 BCE; in archaic times, it was feminine (*LGPN*).

FGrHist 269.

PTK

STATILIUS ⇒ KRITŌN STATILIUS

STATIUS SEBOSUS ⇒ SEBOSUS STATIUS

Stephanos of Alexandria (*ca* 580? – 640? CE)

Greek philosopher and teacher, to whom are attributed works on alchemy (see STEPHANOS OF ALEXANDRIA (ALCH.)), astrology, astronomy, and philosophy. Stephanos is important as one of the last representatives of the Alexandrian tradition on the verge of the Islamic conquest, thus a significant figure in the transmission of Greek philosophy and science to the medieval world. Born around the mid-6th c. CE, he was trained in Alexandria, where he may have been a student of the Neo-**Platonic** philosopher Elias; he is often grouped (along with Elias and David) among the Christian members of the school of OLUMPIODŌROS. According to John Moskhos (*PG* 87.3:2929), he was active as "sophist and philosopher" in Alexandria in the 580s, where he taught courses and authored commentaries, and was involved (on both sides, apparently) in the monophysite controversy. Whether at the invitation of the Emperor HĒRAKLEIOS or for other reasons, he relocated to Constantinople soon after Hērakleios' accession in 610, thereby bridging late Alexandria and the medieval Byzantine world. There he assumed the title of professor (*oikoumenikos didaskalos*) at the recently reopened Imperial **Academy**; his teaching reportedly included courses on PLATO and ARISTOTLE, the *quadrivium*, alchemy, and astrology. Among his pupils at Constantinople were most probably the philosopher now designated PSEUDO-ELIAS (unless the two are in fact identical), as well as TUKHIKOS OF TREBIZOND, himself in turn the teacher of the Armenian mathematician and astronomer ANANIA OF SHIRAK. Works (falsely or otherwise) attributed to Stephanos during his career at Constantinople include both astronomical and astrological writings, as well as a series of alchemical lectures (*cf.* STEPHANOS OF ATHENS). Stephanos died some time before the emperor Hērakleios' own death in 641 CE.

Among philosophical works are commentaries in the Alexandrian tradition on ARISTOTLE's *De Interpretatione* and the third book of his *De Anima*; the latter is preserved in MSS as Book 3 of the *De Anima* commentary of IŌANNĒS PHILOPONOS, with whose group he was associated. Despite Stephanos' professed Christianity, his text offers neither refutation of such traditional doctrines as the eternity of the world, the existence of a fifth substance (***aithēr***), and the soul's pre-existence, nor any overt attempt at revision or reconciliation of pagan with Christian beliefs. This is not unusual; the same tendency is apparent in both Elias and David. Neo-**Platonic** influences perhaps deriving from AMMŌNIOS OF ALEXANDRIA via ASKLĒPIOS OF TRALLEIS are evident in the work on *De Anima*. Only a fragment survives of a *Prolegomenon Philosophiae* that served to introduce the *Eisagōgē* of PORPHURIOS.

Stephanos wrote a commentary on PTOLEMY's *Prokheiroi kanones* (*Handy Tables*), adapting them to Christian reckoning and thus contributing to the transmission of Alexandrian astronomy to the Byzantines; also his is an introduction to the commentary of THEŌN OF ALEXANDRIA on the same book. Spurious astrological works include a horoscopic prophecy concerning the Arabic peoples for the year 775 CE, and a treatise on the conjunction of Saturn and Jupiter: *CCAG* 2 (1900) 181–186.

RE 3A.2 (1929) 2404–2405 (#20), F.E. Kind; *KP* 5.360 (#9), F. Kudlien; *DSB* 13.37–38, K. Dannenfeldt; H. Blumenthal, "John Philoponus and Stephanus of Alexandria: Two Neoplatonic Christian Commentators on Aristotle?" in D.J. O'Meara, ed., *Neoplatonism and Christian Thought* (1982) 54–63; Wolska-Conus (1989); Roueché (1990); *ODB* 1953, A. Kazhdan; *BBKL* 10.1406–1409, A. Lumpe; *NP* 11.960 (#9), V. Nutton; *NDSB* 6.516–518, M. Papathanassiou.

Keith Dickson

Stephanos of Alexandria (Alch.) (*ca* 580 – *ca* 640 CE)

Belonging to the generation of alchemical commentators, Stephanos is the author of nine *Praxeis* (Lessons/readings) *On the Divine and Sacred Art*, and of a *Letter to Theodōros* (Ideler). *Praxis* 9 is addressed to the emperor HĒRAKLEIOS, thus datable to his rule; astronomical data further pinpoint his work to 617 (Papathanassiou).

In the alchemical corpus, Stephanos is mentioned with OLUMPIODŌROS among the "ecumenical masters everywhere celebrated, the new exegetes of PLATO and ARISTOTLE" (*CAAG* 2.425). In fact, he was named by Hērakleios as "ecumenical professor," i.e., professor of the imperial school in Constantinople. Modern scholarship tends to consider this Stephanos of Alexandria identifiable with the Neo-**Platonic** commentator on Plato and Aristotle (see STEPHANOS OF ALEXANDRIA). He may have also interpreted the *Handy Tables* of THEŌN OF ALEXANDRIA, and may have written an *Apotelesmatic Treatise*, explaining the horoscope of Islam and addressed to his student Theodōros (Papathanassiou 113). However, his identification with the Hippokratic commentator STEPHANOS OF ATHENS remains problematic.

In his alchemical work, Stephanos comments on the early alchemists in a highly rhetorical style and links alchemy to medicine, astrology, mathematics, and music. He declares alchemy compatible with Christianity and defines it as a "mystic" knowledge, inserted into a cosmology founded on the principles of unity and universal ***sumpatheia***. Alchemical transformations are considered natural and enter into the close analogy and correspondence between the micro-***kosmos*** and the macro-***kosmos***, the human body and the four elements, heavenly bodies and earthly bodies.

Methodologically, Stephanos aims to create a new system through critical comparison of theories and admission of the differences. This form of the *status quaestionis* of existing theories constitutes one of the most scientific aspects (in the modern sense) of Stephanos' work. Aristotelian, **Platonic**, and Neo-**Platonic** doctrines play a fundamental role in his concept of alchemy. In particular, his concept of the nature and transformation of metals is one of the most interesting in the alchemical corpus, because it seems to rely both upon the geometrical theory of Plato's *Timaios* and the theory of "exhalations" in the Aristotelian *Meteorologika*.

Stephanos' corpus was well-known to the Arabs. According to the Arabo-Latin work transmitted in the *Morienus* (Stavenhagen), one of Stephanos' students, the monk "Morienus" (i.e., Marianus), spread alchemy in the Arab world by initiating the Ummayad prince Khalid ibn Yazid, *ca* 675–700.

Ed.: Ideler 2 (1842/1963) 199–253; repr. with Engl. trans. (*praxeis* 1–3), F.S. Taylor, "The alchemical works of Stephanus of Alexandria," *Ambix* 1 (1937) 116–139, 2 (1938) 39–49.

L. Stavenhagen, ed., *A testament of alchemy. Being the revelation of Morienus to Khalid ibn Yazid* (1972); Wolska-Conus (1989); M. Papathanassiou, "Stephanus of Alexandria: pharmaceutical notions and cosmology in his alchemical work," *Ambix* 37 (1990) 121–133; *Eadem*, "Stephanus of Alexandria: on the structure and date of his alchemical work," *Medicina nei secoli* 8.2 (1996) 247–266; *Eadem*, "L'œuvre alchimique de Stéphanos d'Alexandrie: structures et transformations de la matière, unité et pluralité, l'énigme des philosophes," in Cristina Viano, ed., *L'alchimie et ses racines philosophiques. La tradition grecque et la tradition arabe* (2005) 113–133; *NDSB* 6.516–518, M. Papathanassiou.

Cristina Viano

Stephanos of Athens (*ca* 540 – 680 CE?)

Greek Christian physician and professor of medicine, born in Athens, studied in Alexandria under "Asklēpios" (i.e., OF TRALLEIS?), and later taught there. Three works survive under his name, given in some MSS as "Stephanos the Philosopher." These are commentaries on (1) the *Aphorisms* of HIPPOKRATĒS (*CMG* 11.1.3.1–2), (2) Hippokratēs' *Prognostics* (*CMG* 11.1.2) and (3) Book 1 of GALĒN's *Therapeutics to Glaukon* (Dickson 1998). All reflect traditional Alexandrian pedagogy in their division into "lectures" (*praxeis*) and "discussions" (*theoriai*), a format originally developed in the philosophical school of AMMŌNIOS OF ALEXANDRIA. There also survives a tract on uroscopy (*Peri ourōn*) and the redaction (attributed to "STEPHANOS OF ALEXANDRIA") of a commentary by PALLADIOS on the Hippokratic text *De fracturis*. Lost works include *On pulses*. His position is strongly Galēnic. His familiarity with philosophy, not deep but sounder than most, supports the hypothesis that he pursued both vocations: not unusual in this era. Although not implausible, the perennially argued identity between Stephanos of Athens and the philosopher STEPHANOS OF ALEXANDRIA, whose dates and sphere of activity roughly match those of the Athenian, still awaits satisfactory demonstration (Wolska-Conus; Roueché).

Ed.: J. Duffy, *Stephanus the Philosopher. A Commentary on the Prognosticon of Hippocrates* (1983) = *CMG* 11.1–2; L.G. Westerink, *Stephanus of Athens. Commentary on Hippocrates' Aphorisms* 2 vv. (1985/1992) = *CMG* 11.1.3.1–2; Dickson (1998).

RE 3A.2 (1929) 2404–2405 (#20), F.E. Kind; *KP* 5.360 (#9), F. Kudlien; *DSB* 13.37–38, K. Dannenfeldt; Wolska-Conus (1989); Roueché (1990); *ODB* 1953, A. Kazhdan; L. Angeletti and B. Cavarra, "The *Peri ouron* Treatise of Stephanus of Athens: Byzantine Uroscopy of the 6th–7th Centuries AD," *American Journal of Nephrology* 17 (1997) 228–232; *BBKL* 10.1406–1409, A. Lumpe; *NP* 11.960 (#9), V. Nutton; *NDSB* 6.516–518, M. Papathanassiou.

Keith Dickson

Stephanos of Buzantion (525 – 565 CE)

Greek grammarian, probably from Constantinople and a contemporary of Justinian I. Author of the *Ethnika*, an alphabetical list of geographical names together with information on etymologies, foundation-legends, changes of names, historical anecdotes, etc. Initially containing 55 to 60 books, it might have been a compendium for the officials and soldiery of the recently enlarged Byzantine Empire. Constantine VII Pophurogennētos (10th c.) was perhaps the last scholar to see the original *Ethnika* intact. The *Souda*, Eustathios of Thessalonikē, and others used HERMOLAOS' abridgement, extant in several MSS. Stephanos, not entirely uncritically, drew from grammarians and philologists, historians and geographers. He knew the works of PTOLEMY, STRABŌN, and Pausanias, but, in spite of his Christianity, seldom quoted Christian writings.

Ed.: A. Meineke (1849, repr. 1958, 1992); M. Billerbeck, *Stephani Byzantini Ethnica* v.1 (A–Γ) (2006).
RE 3A.2 (1929) 2369–2399 (#12), E. Honigmann; A. Diller, "The Tradition of Stephanus Byzantius," *TAPA* 69 (1938) 333–348; *HLB* 1.530–531; *ODB* 1953–1954, A. Kazhdan; *OCD3* 1442, R. Browning.

Andreas Kuelzer

Stephanos of Tralleis (475 – 525 CE?)

The father of ALEXANDER, ANTHĒMIOS, Dioskoros (doctor), MĒTRODŌROS, and Olympios (lawyer), of Tralleis, himself a practicing physician (Agathias 5.6.3–6), cited by his son, Alexander (2.139 Puschm.), for a sore-throat remedy, containing Egyptian acanthus, bran, "Nikolaos" dates (*cf.* PLINY 13.45, Ath., *Deipn.* 14.22 [652a–b]), iris, licorice, and dried roses, all boiled, and used as a gargle, hourly.

PLRE 2 (1980) 1030 (#9).

PTK

Stēsimbrotos of Thasos (*ca* 470 – *ca* 420 BCE)

References in XENOPHŌN (*Symposium* 3.6) and PLATO (*Ion* 530d) indicate that Stēsimbrotos was one of the foremost interpreters of HOMER in 5th c. Athens, able to reveal the poems' hidden meanings (*huponoia*) and to offer solutions to assumed textual puzzles. The nature of Stēsimbrotos' interpretations is debated, but probably he sometimes used physical allegory, as for example in his solution to the following assumed inconsistency: Homer says first that *everything* was divided into three to be distributed among Zeus, Poseidōn, and Hadēs (*Iliad* 15.189), but then he states that Earth and Olympus remained common. Stēsimbrotos' solution seems to be based on an identification of the gods with physical elements. (*Cf.* THEAGENĒS OF RHĒGION.) In *On Initiations* Stēsimbrotos discussed the Samothrakian mysteries, and interpreted mythical stories allegorically and divine names etymologically. Because of his interest in mystery cults and allegorical interpretation of traditional poetry, Stēsimbrotos has been suggested as the author of the DERVENI PAPYRUS. PLUTARCH preserves about a dozen fragments from a political pamphlet by Stēsimbrotos containing mainly gossip about leading Athenian politicians and generals.

Ed.: *FGrHist* 107, 1002 (biography).
NP 11.975–976, M. Baumbach.

Gábor Betegh

Stobaios ⇒ Iōannēs Stobaios

Strabōn of Amaseia (*ca* 30 BCE – 24 CE)

Greek geographer and historian, born *ca* 64 BCE in Amaseia, Pontos.

Biography: Strabōn's ancestors on his mother's side were intimates of Mithradatēs V Euergetēs and Mithradatēs VI Eupatōr, kings of Pontos, but during the Mithridatic War they supported the Romans. Strabōn was born and raised in Amaseia. As a boy he had several renowned teachers: Aristodēmos of Nusa, a historian and Homeric scholar; Xēnarkhos of Seleukeia; and Turannion of Amisos, a grammarian. As an adult Strabōn visited and lived in the intellectual centers of Rome, Alexandria, Nusa and possibly Smurna and Athens. Strabōn traveled to Rome (44 and 29 BCE, and possibly other times) where he established social relationships with Roman notables and Greek intellectuals. In 25 BCE he accompanied Aelius Gallus on his mission as Roman governor of Egypt. Although reluctant to mention his visits to particular sites, he seems to have traveled widely, eastward as far as the border between Pontos and Armenia; westward, Turrhenia; northward, Sinōpē and Kuzikos; and southward to Suēnē on the border of "Ethiopia." Strabōn probably composed his geographical work between 18 and 23 CE and died in Rome or perhaps Asia Minor.

Works: (1) Earlier historiographical work(s): *Alexander's Deeds, Historical Notes* and a sequel to Polubios' *Histories* in 43 books, surviving in very few fragments (*FGrHist* 91) (2) A 17-book geographical work describing the entire ***oikoumenē***: introductory remarks on geographical theory and Strabōn's predecessors from Homer to Poseidōnios of Apameia (Books 1–2); Iberia and the surrounding islands (Book 3); Gaul, the British Isle, Ireland, Thule (= the Shetlands) and the Alps (Book 4); Italy, Sicily and the islands between Sicily and Libya (Books 5–6); German tribes, Pannonia, Illyria, Macedon and Thrakē (Book 7); Greece (Books 8–10); Asia Minor (Books 11–14); India and Persia (Book 15); Mesopotamia, Syria, Phoenicia, Judaea, the Persian Gulf and Arabia (Book 16); Egypt, "Ethiopia" and Libya (Book 17).

Sources: Strabōn based his geographical notions on the works and ideas of several prominent predecessors. Regarding Homer as the founder of scientific geography, his description of the world follows the Homeric notion of ***oikoumenē*** as an island surrounded by Ocean, and specific geographical and toponymic details in the epics (see Kratēs of Mallos and Polubios). Strabōn both challenged and followed particularly Eratosthenēs, Hipparkhos of Nikaia, Artemidōros of Ephesos, and Poseidōnios and adopted Polubios' pragmatic considerations. He aimed at a task-oriented readership, particularly politicians and generals who in his time were mostly Roman. Numerous other sources of information – written works, reports of navigators, hearsay and visual media (perhaps the lost map of Agrippa) – also feature in the *Geography*.

Geographical ideas: Strabōn demanded philosophical qualities from the ideal geographer and himself adhered particularly to the **Stoic** school. He accepted the traditional division of the ***oikoumenē*** into three continents, Europe, Asia and Libya and the division of the globe into climatic latitudinal zones: the torrid on both sides of the equator, two temperate on both sides of the torrid and arctic zones at either pole. On this structural basis, Strabōn presented his awareness of the political and social changes apparent in the ethnic structure of the inhabited world. The work should be considered Augustan in reflecting the political ideas of the age and the official Roman image of the identity between the borders of the Roman Empire and the boundaries of the ***oikoumenē***. Strabōn's *Geography* in focus, terms and methodology, belongs with the descriptive branch of traditional Greek geography, which did not employ exact calculations and empirical research but presented descriptions of sites, their appearance and nature, their topographical, botanical and zoological traits.

Contribution: The *Geography* was the first attempt to collect all contemporary geographical knowledge and to compose a general treatise on geography which until then had featured only as an appendix to historical surveys. The work was not well known in antiquity but forms an important milestone in attitudes towards geography and offers an encyclopedic collection of otherwise lost information.

Ed.: H. Jones, *The Geography of Strabo* (Loeb 1969–1982, repr.); G. Aujac, F. Lasserre, and R. Baladié, Strabon. *Géographie* (*CUF* 1966–1996); S.L. Radt, *Strabons Geographika* vv. 1–10 (2002–). The historical work(s): *FGrHist* 91.

G. Maddoli and F. Prontera, edd., *Strabone: Contributi allo Studio della Personalita e dell' Opera* 2 vv. (1984–1986); K. Clarke, "In Search of the Author of Strabo's Geography," *JRS* 87 (1997) 92–110; *Eadem, Between Geography and History* (1999); J. Engels, *Augusteiche Oikumenegeographie und Universalhistorie im Werk Strabons von Amaseia* (1999); Daniela Dueck, "The Date and Method of Composition of Strabo's Geography," *Hermes* 127 (1999) 467–478; S. Pothecary, "Strabo the Geographer: his Name and its Meaning," *Mnemosyne* 52 (1999) 691–704; Dueck (2000); *Eadem*, S. Pothecary, and H. Lindsay, edd., *Strabo's World History: A Kolossourgia of a Work* (2005).

Daniela Dueck

Stratōn (Med.) (350 – 325 BCE)

Physician, cited by ARISTOTLE (DIOGENĒS LAËRTIOS 5.61), perhaps the Stratonikos who remarked that odors enjoyed for their own nature, such as flowers, smell beautiful (*kalon*) while others, e.g., foods, smell pleasant (*hēdu: Eth. Eud.* 3 [1231a11]), in turn possibly identifiable with Athēnaios' clever kitharist (*Deipn.*, 8 [347f, 348d–352d]).

RE 4A.1 (1931) 315 (#18), F.E. Kind.

GLIM

Stratōn (Erasistratean) (265 – 245 BCE)

Student and close connection of ERASISTRATOS, teacher of APOLLŌNIOS OF MEMPHIS, he performed all cures without phlebotomy, arguing that the procedure risked death (by fear or excess bleeding) and that practitioners might err and open an artery: GALĒN, *On Venesection, Against Erasistratos* 2 (11.151 K. = p. 18 Brain), *On Venesection, Against the Erasistrateans in Rome* 2 (11.196–197 K. = p. 43 Brain), and *Diff. Puls.* 4.17 (8.759 K.). He is described as "writing from the house" (11.151 K.) or "being a fosterling" (DIOGENĒS LAËRTIOS 5.61) of Erasistratos. In some unknown context he explained the *ambē* in HIPPOKRATIC CORPUS, JOINTS as a "round lever," ERŌTIANOS A-108 (p. 23.8 Nachm.). SŌRANOS, *Gyn.* 4.14 [1.71] (*CMG* 4, p. 145 = v. 2, p. 11 *CUF*), preserves his fumigation to expel the afterbirth, the herbs (including cassia, dittany, horehound, spikenard, tree wormwood, and oils of lily and rose) being heated in a silver or tinned-bronze vessel, with a spout directing the fumes into the vagina; he treated uterine prolapse with ***spodion*** and castoreum, *ibid.* 4.36[85] (*CMG* 4, pp. 149–150 = v.4, p. 26 *CUF*). PHILOUMENOS preserves his remedies for bites: human or dog (5.2, *CMG* 10.1.1, p. 9), snakes (21.6 [p. 28]; 23.4 [p. 30]; 33.1–3 [p. 36]), and stingrays or moray eels (37.3, p. 40); pseudo-AELIUS PROMOTUS, *Iobol.* 30

(p. 58 Ihm) records his remarks on the field-mouse (*mugalē*), and on the poisonous plant *ephēmeron* (§56, p. 69 Ihm). Rufus of Ephesos, in Oreibasios *Coll.* 45.28 (*CMG* 6.2.1, p. 184), says that he was the first to describe ***elephantiasis***, and called it *kakokhumia* (bad-humor).

NP 11.1043–1044 (#4), V. Nutton; Jacques (2002) 295–297.

PTK

Stratōn of Bērutos (90 BCE – 50 CE)

Greek physician. Asklēpiadēs Pharm., in Galēn *CMLoc* 4.7 (12.749 K.), preserves his *Bērution* collyrium, providing immediate relief, and compounded from roasted copper, ***pompholux***, acacia, opium, gum, saffron, and myrrh, in rainwater; administered with an egg(-white). Stratōn is perhaps also the *Bērutan* whose compounds for stomach ailments are preserved in Galēn *CMLoc* 9.5 by Andromakhos (13.290–291 K.) and Asklēpiadēs Pharm. (13.303 K.). Alexander of Tralleis cites perhaps this Stratōn, who followed Orpheus, for three prescriptions to treat epilepsy (1.15 [1.563.6, 1.565.11, 1.571.3 Puschm.]), one of which (1.571.3) Alexander also attributes to Moskhiōn, providing the probable *terminus post*.

RE 4A.1 (1931) 317 (#20), F.E. Kind.

Alain Touwaide

Stratōn of Lampsakos (*ca* 295 – 268 BCE)

The work of Stratōn of Lampsakos, third head of Aristotle's school, is known only through secondhand reports, as no work of his survives. He was particularly known in antiquity for his focus on natural philosophy and for denying any appeal to the divine in accounting for the natural world. It is likely that he knew of the work of doctors and scientists associated with Ptolemaic Alexandria, since he served as tutor to the young Ptolemy Philadelphus for some time before he took over the leadership of the **Peripatos**. Aristarkhos of Samos was a pupil of his, and he is said to have had ties to Erasistratos. From Theophrastos' death in 286 BCE, he was head of Aristotle's school in Athens, until his own death in 268 BCE.

Stratōn was known in antiquity as *ho phusikos*, "the natural philosopher" or "the physicalist," possibly because of his insistence on separating the study of the natural world from theological intervention. He reportedly ascribed all natural events to forces of weight and motion. He rejected Aristotle's doctrine of the fifth element, and also the idea that lighter bodies have a natural tendency to move upward, claiming instead that they are squeezed out by the fall of heavy bodies. He considered the natural world to contain the causes of all change.

He seems to have held a basically Aristotelian view of the nature of matter, inasmuch as he stressed the role of hot and cold in effecting change. Nonetheless, he altered the doctrine of void, holding that it is at least possible within the ***kosmos***. Some reports suggest that Stratōn thought void is only a conceptual possibility, and that it is coextensive with space. He supported Aristotle in denying that void is needed to account for magnetism or for the motion of bodies. One report, however, claims that he held that matter has passageways to allow the passage of light and heat. Some scholars take this report as evidence that Stratōn took void to be part of the microstructure of matter. Much controversy surrounds the relationship between Stratōn's view of the matter and void and the theory of Hērōn of

ALEXANDRIA. Scholars dispute the degree to which the introduction to Hērōn's *Pneumatics*, which evidently borrows a proof from Stratōn, is evidence for Stratōn's view.

He wrote a number of works on medical topics, and seems to have offered a naturalistic account of the soul. In this, he may have been following new medical theories that ascribed the functions of the soul to a substance, ***pneuma***, carried in passageways throughout the body. Although Aristotle ascribes some functions to ***pneuma***, it was assigned a greater role in the theories of the Hellenistic doctors who took the newly discovered nervous system to be the pathway for perception and motor functions. Stratōn located the centre of the soul's activity between the eyebrows, rejecting Aristotle's view that the heart is the center. He regarded rationality as a kind of causal change, and hence apparently part of the natural world; he offered lists of objections to PLATO's arguments for the immortality of the soul. A report by GALĒN may refer to Stratōn of Lampsakos: it concerns a figure who held that both male and female parents produce seed, as indeed new medical discoveries would suggest.

Scholars have long speculated that works in the Aristotelian corpus thought not to be by Aristotle may be written by Stratōn. In some cases, like *Meteorologica* Book 4, the arguments against attributing these to Aristotle are now generally rejected. His name is still often mentioned in connection with the ARISTOTELIAN MĒKHANIKA, but this is only speculation. He does seem to have had an interest in empirical investigations. His best known contributions to natural philosophy include attempts to prove the downward acceleration of falling bodies by the greater impact of bodies dropped from higher points, and by the breaking up of a continuous stream of water as it falls further. Other scientific contributions include a theory of the formation of seas by analogy to rivers, and an account of sound as impact transmitted through air.

Ed.: H.B. Gottschalk, "Strato of Lampsacus: Some Texts," *Proceedings of the Leeds Philosophical and Literary Society* 9 (1965) 95–182; F. Wehrli, *Straton von Lampsakos* 2nd ed. (1969); R.W. Sharples, ed. of fragments, *RUSCH* (forthcoming).

M. Gatzemeier, *Die Naturphilosophie des Straton von Lampsakos: Zur Geschichte des Problems der Bewegung im Bereich des frühen Peripatos* (1970); L. Repici, *La natura de l'anima: Saggi su Stratone de Lampsaco* (1988); D.J. Furley, "Strato's Theory of Void," in *Cosmic Problems: Essays on Greek and Roman Philosophy of Nature* (1989) 149–160; *NDSB* 6.540, Sylvia Berryman.

Sylvia Berryman

Stratonikos (Veterin.) (before *ca* 400 CE)

Quoted by HIEROKLĒS: on fever, *Hippiatrica Berolinensia* 1.18; sore throat, *Hippiatrica Berolinensia* 19.4; cholera, *Hippiatrica Parisina* 642 = *Hippiatrica Berolinensia* 75.5; and shrew-mouse bites, *Hippiatrica Parisina* 705 = *Hippiatrica Berolinensia* 87.2. Hieroklēs may have used Stratonikos in a compilation related to that of CASSIUS DIONUSIOS OF UTICA.

McCabe (2007) 227, 234.

Anne McCabe

Stratonikos of Pergamon (110 – 150 CE)

Student of both QUINTUS and SABINUS, possessed of a good bedside manner, taught GALĒN in Pergamon; leaving nothing in writing, he advanced novel interpretations of the HIPPOKRATIC CORPUS, EPIDEMICS, according to Galēn, *in Hipp. Epid. VI* (*CMG* 5.10.2.2,

p. 303). Galēn (*ibid.* p. 287) preserves his interpretation of *Epidemics* 6.5.15 (5.320 Littré), that the **humor** black bile indeed tracks the outbreak of hemorrhoids, and describes how he diagnosed an excess of black bile in the blood extracted during phlebotomy, and cured the patient with black-bile-expelling drugs and a balanced diet: *Atra Bile* 4.12 (*CMG* 5.4.1.1, p. 78, repeated by Oreibasios, *Coll.* 45.20.3, *CMG* 6.2.1, pp. 176–177).

RE 4A.1 (1931) 327 (#3), F.E. Kind.

PTK

Stuppax of Cyprus (460 – 430 BCE)

Inventor of a starting-gate at Olympia, according to *P. Berol.* P-13044, col.8 (anethnic). Lippold identified him with the sculptor Stuppax of Pliny 22.44, 34.81, whose "Entrail-Roaster" bronze depicted a favorite slave of Periklēs. The name seems otherwise unattested.

RE 4A.1 (1931) 454–455, A. Lippold.

PTK

Sudinēs (*fl. ca* 240 BCE)

Although Sudinēs is mentioned by Strabōn (16.1.6) as a Babylonian *mathēmatikos*, together with Kidēnas and Naburianus, no cuneiform evidence for his existence is extant. The Babylonian equivalent of the name is also a puzzle, although an Akkadian name with the common ending *-iddin* "he has given" is possible. A Sudinēs was named as a diviner (*bārû*) by Polyaenus 4.20: the *bārû* interpreted omens from extispicy, as Sudinēs supposedly did for King Attalos I of Pergamon before fighting and defeating the Gauls *ca* 235 BCE. While Babylonian astronomers were frequently also celestial diviners and experts on ritual and magic, the combination of astronomy and extispicy is not so common. Evidence that Sudinēs wrote on the properties of stones comes exclusively from Pliny, whose information is limited to Sudinēs' alleged knowledge of the provenance of onyx (36.59), rock-crystal (37.25), amber (37.34), *nilios* (37.114), and comments on the color of pearls (9.115), onyx (37.90) and *astrion* (37.133).

Pliny also mentions Sudinēs (9.115; 36.59; 37.25, 34, 90, 114, 153) as a "Khaldaean astrologer." Consistent with this designation is the papyrus fragment written in the 3rd c. CE, purportedly summarizing a commentary on Plato's *Timaeus* by the **Stoic** Poseidōnios (*P. Gen.* inv. 203). Here the influences of the five planets, Sun and Moon are enumerated in terms of Aristotelian qualities (warm, moist, dry) and further indications are given for the planets Saturn, Jupiter, Mars, and Venus as the "destroyers" of men and women, young and old. Venus is the destroyer of women "according to Sudinēs."

Vettius Valens lists parameters for the length of the year according to Greek and Babylonian astronomy (9.11). There Sudinēs is associated with year length of 365 ¼ + 1/3 + 1/5 days, which makes neither numerical nor astronomical sense. Valens adds that he used Sudinēs (and Kidēnas and Apollōnios of Pergē) to compute lunar eclipses and that he normed the equinoxes and solstices at 8° of their signs (9.12.10). This originally Babylonian norm for the cardinal points of the year was established perhaps *ca* 300 BCE for a zodiac in which degrees were not counted from the vernal point, however, but from the sidereally fixed zodiacal signs beginning with Aries ("The Hired Man" in the Babylonian zodiac). The 8° of Aries as the vernal point underlies much of Hellenistic astrological texts and continued in use throughout late antiquity.

RE 4A.1 (1931) 563, W. Kroll; F. Lasserre, "Abrégé inédit du commentaire de Posidonios au Timée de Platon (Pap. Gen. inv. 203)," *Protagora, Antifonte, Posidonio, Aristotele* (1986) 71–127; W. Hübner, "Zum Planetenfragment des Sudines (Pap. Gen. inv. 203)," *ZPE* 73 (1988) 33–42; *Idem*, "Nachtrag zum Planetenfragment des Sudines. P. Gen. inv. 203," *ibid.* 109–110.

Francesca Rochberg

Sueius (70 – 40 BCE)

Minor Roman poet contemporary with CICERO whose corpus included two agricultural poems. A quotation from the first, the *Peasant Salad* (*Moretum*), offers a learned history of Alexander the Great's introduction of the walnut from Persia to Greece. The title of the second work, *Chicks* (*Pulli*), suggests an identification of the author with the equestrian M. *Seius* famous for his innovations in raising chickens, geese, ducks, pigeons, cranes, and peacocks (VARRO, *RR* 3.2.7–14, PLINY 10.52) – though the variant spellings of the name may preclude this connection.

Ed.: Speranza (1971) 56–57; *FLP frr.*1–4.
RE 4A.1 (1931) 580–581, W. Kroll, w/2A.1 (1921) 1121–1122, Fr. Münzer; *NP* 11.1081, P.L. Schmidt.

Philip Thibodeau

Suennesis of Cyprus (*ca* 420 – 350 BCE)

HIPPOKRATĒS' pupil, among the first to describe the vascular system. According to his theory, there are two vessels, one originating from the eye, running along the eyebrow and through the lung under the breast. The other originates in the left eye and runs through the liver down to the right kidney and testicle. Another vessel runs from the right breast to the left buttock, and still another from the left buttock to the right breast (ARISTOTLE, *HA* 3.2.3 [511b 24–30]; Hippokratēs *Concerning the Nature of Bones* 8 [9.174–176 Littré]). Most likely, this theory originated from observations of the neurological system, that is an injury on one side of the head might affect the other side of the body.

RE 4A.1 (1931) 1024 (#2), F.E. Kind; E.D. Phillips, *Aspects of Greek Medicine* (1973) 190.

Robert Littman

C. Sulpicius Gallus (170 – 150 BCE)

Learned Roman consul. He wrote a book on Greek astronomy known to VARRO (in PLINY 2.53). Sulpicius described the celestial spheres (CICERO *Rep.* 1.23) and, on the occasion of a solar eclipse, mollified Paulus' army by explaining its causes (Pliny 2.9).

KP 5.424–425 (#I.17), M. Deißmann-Merten; J. Evans, *The History and Practice of Ancient Astronomy* (1998) 80, 82, 455 n.9; *NP* 11.1100 (#I.4), T. Schmitt.

Bruno Centrone

Summaria rationis geographiae in sphaera intelligenda (after 150 CE)

In the tradition of PTOLEMY, treated the proportions of the ***oikoumenē*** and the globe in stades and degrees, the connection between ***oikoumenē*** and the Sun's path at the equinox, the northern parallels and the *klimata*.

Ed.: *GGM* 2.488–493.
A. Diller, "The Anonymous *Diagnosis* of Ptolemaic Geography," *Classical Studies in Honor of W.A. Oldfather* (1943) 39–49; *RE* S.10 (1965) 794–800, E. Polaschek; *HLB* 1.512.

Andreas Kuelzer

Sunerōs (of Campania?) (200 BCE – 95 CE)

ASKLĒPIADĒS PHARM., in GALĒN *CMLoc* 4.8 (12.774–776 K.), records two of his collyria, one for conjunctivitis, containing **calamine**, roast hematite, myrrh, opium, pepper, ***pompholux***, and saffron, ground into aged olive oil and Italian wine, applied with egg-(white) – repeated by AËTIOS OF AMIDA 7.112 (*CMG* 8.2, pp. 377–378) – and the other for trachoma, containing roast copper, hematite, saffron, and opium, in gum and very sharp vinegar. The name in this era is attested almost solely from Campania: *LGPN* 3A.406–407, contrast 1.416, 2.410, 3B.388, 4.320.

RE 4A.2 (1932) 1362 (#2), F.E. Kind.

PTK

Sunesios (300 – 390 CE)

Unknown to ZŌSIMOS but cited by OLUMPIODŌROS, Sunesios opens the era of alchemical commentators; he wrote a commentary on the *Phusika and Mustika* of PSEUDO-DĒMOKRITOS as a dialogue entitled *Sunesios to Dioskoros, Commentary on the Book of Dēmokritos* (CAAG 2.56–69). His identification of DIOSKOROS as a priest of Serapis at Alexandria provides a *terminus ante* for the work, before the destruction of the Serapeion (391 CE). Berthelot (1885: 188–191) identifies this Sunesios with Bishop SUNESIOS OF KURĒNĒ, but a dedication to a pagan priest renders that rather difficult. Lacombrade (1951: 71) suggests an identification with Sunesios of Philadelpheia (*ca* 250 CE) mentioned by the *Souda* A-2180, but he has no connection with alchemy.

In his work, Sunesios declares his exegetical intention: one must explore the writings of Dēmokritos to learn his thoughts and the order of his teachings. The obscurity professed by Dēmokritos about procedures and substances is interpreted, in a *locus classicus* of alchemy, as a means of protection against outsiders and an exercise for the intelligence of adepts. The explanations of general principles show a strong Aristotelian influence: notions of mixture, putrefaction, qualitative change, matter and form, potential and actuality, all applied to procedures. The goal of gold-making is identified with agents of material transformation. The cause of transmutation is what one calls divine water, mercury, ***khrusokolla***, unfired sulfur, and it acts in effecting a dissolution of the bodies. Mercury appears simultaneously as agent, common substrate, and principle of liquidity.

Ed.: *CAAG* 2.56–69.

Berthelot (1885) 188–191; C. Lacombrade, *Synésios de Cyrène, hellène et chrétien* (1951) 64–71; Letrouit (1995) 47.

Cristina Viano

Sunesios of Kurēnē (*ca* 395 – 413 CE)

Christian **Platonist** philosopher and poet, born (*ca* 370) to an apparently aristocratic family. It is unclear whether he was born pagan or Christian. Sunesios studied philosophy and science with HUPATIA in Alexandria until 398. Sunesios participated in a Libyan mission to the emperor in Constantinople in 399 and he was elected bishop of Ptolemais in the province of Pentapolis in 411. Sunesios regarded himself as a philosopher (*Ep.* 79, 105) and rejected Christian doctrines of the world's corruption and soul's resurrection (*Ep.* 105) that conflict with his **Platonism**. Sunesios considered Christian priesthood complementary to (*Ep.* 41, 62), and a step towards, philosophy (*Ep.* 11), since he considered philosophy the guide to the divine (*Ep.* 105, 142) and strove towards virtue, contemplation and a life according to intellect (*Ep.* 137, 140, *Hymn* 1). Sunesios' work includes nine Doric hymns (a tenth is spurious) and treatises including *On reign, Egyptian discourses or on providence, Dion, On dreams*, and several letters rich in biographical detail and important evidence for the history, economy, and culture of contemporary Pentapolis. Most interesting is his *Dion*, wherein Sunesios defends his way of life dedicated to literature and philosophy and his endeavors to ascend to the intellect.

Ed.: A Garzya, *Sinesio di Cirene Opere* (1989); A. Garzya and D. Roques, *Synésius de Cyrène, Correspondance* (2000) 2 vols.

RE 4A.2 (1932) 1362–1365 (#1), H. von Campenhausen; Chr. Lacombrade, *Synésius Hellène et chrétien* (1951); *OCD3* 1463, P.J. Heather; *NP* 11.1147–1148 (#1), J. Rist.

George Karamanolis

Sura ⇒ Mamilius Sura

Suros (*ca* 125 – *ca* 150 CE)

Dedicatee of Ptolemy's *Almagest, Handy Tables, Planetary Hypotheses, Analemma, Planispherium*, and *Tetrabiblos*. Other writings of Ptolemy with extant beginnings, namely the *Harmonics, Geography*, and *Phaseis*, bear no dedications. A scholiast to the *Tetrabiblos* reports a tradition that Suros was a physician. Ptolemy addresses him as an intellectual peer. Two brief texts on astrological weather forecasting (*CCAG* 1.131–134 and 171–172) attributed to "*Suros tinos*," – probably "a Syrian" rather than someone named Suros – in any case have no known connection with Ptolemy's associate.

F. Boll, "Studien über Claudius Ptolemaeus," *Jahrbücher für Classische Philologie* S.21 (1894) 51–244, esp. 67.

Alexander Jones

Symmachus ⇒ Alkōn

Syrianus of Alexandria (before 432 – 437 CE)

Born in Alexandria (*Souda* Sigma-1662), the son of the otherwise unknown Philoxenos (*Vita Procli* 11), Syrianus was akin to his younger compatriot Aidesia, who married his student Hermeias of Alexandria (*Souda* A-79), and to the ***grammatikos*** Ammōnianus (*Souda* A-1639). Together with Hieroklēs and the sophists Lakharēs and Nikolaos, Syrianus studied with Ploutarkhos of Athens until the latter's death in 432, and succeeded him as the head of the Athenian Neo-**Platonic** school. At Ploutarkhos' bequest, Syrianus mentored Ploutarkhos' grandson Arkhiadas, and Proklos, who became his most prolific and devoted disciple. Syrianus' other famous students are Domninos of Larissa, Hermeias and Marinos of Neapolis.

Syrianus was probably the most innovative and influential late Athenian Neo-**Platonist**. His original elaboration of Neo-**Platonist** metaphysics and his theories and exegetical methods are reflected in Proklos (*Theol. Plat.* 1.6 Saffrey). Despite Syrianus' important written production, only two of his works survive: his commentary on Aristotle's *Metaphysics* BΓMN (*IM*; for its completeness see O'Meara 1989: 120–122); and a commentary on Hermogenēs' rhetorical treatises *On Style* and *On Issues*. Hermeias' commentary on Plato's

Phaidros derived from Syrianus' lessons, and Proklos' two major commentaries *in Tim.* and *in Parm.* are strongly indebted to Syrianus. Additionally, later commentators, especially SIMPLICIUS reproducing Syrianus' views on place, body and movement, frequently allude to Syrianus.

Syrianus explicitly followed IAMBLIKHOS' views on the role of mathematics as a path to **Pythagorean** theology and on the structure of reality in various levels. It led him to important theories later influencing exegetical literature on mathematics, natural philosophy, and psychology. Especially important is his theory of geometrical activity as the projection of the innate *logoi* of the soul onto the imagination, similar to the projection of the cosmic soul's ideas onto the screen of incorporeal space permeating all bodies like a beam of intangible light (*IM* 84–86). This implies an analogy between natural phenomena and mathematical and psychic reasoning as well as an original conception of space and the soul. Equally important is the relationship between his metaphysical system and his exegetical method, probably not to be separated from Syrianus' interests in rhetoric (see Praechter col.1744).

Meager evidence indicates that Syrianus' interest in mathematics exceeded epistemological theories. Thus, the invention of an elementary method of division of numbers of different sexagesimal orders is attributed to Syrianus in the anonymous PROLEGOMENA TO PTOLEMY'S SUNTAXIS (Knorr 1989: 167 and n.78). This endeavor in astronomical logistics, perhaps in relation to astrology (Proklos *in Remp.* II.64 Kroll), may be confirmed by allusions in Theodōros Melitēniotēs' *Tribiblos* preface (1.163 Leurquin) – that he (possibly) used Syrianus' writings, along with THEŌN's and PAPPOS'.

RE 4A.2 (1932) 1728–1775 (#1), K. Praechter; O'Meara (1989) 128–141.

Alain Bernard

T

Tacitus ⇒ P. Cornelius Tacitus

Tanitros ⇒ Samithra

L. Tarutius of Firmum Picenum (75 – 30 bce)

Roman astrologer, wrote *On Stars* in Greek, consulted by Pliny for astral weather signs (1.*ind.*18). More remarkable were his investigations of historical astrology made at his friend Varro's behest in the interests of establishing the chronology of Rome's beginnings (Cicero, *Div.* 2.98; Plutarch, *Rom.* 12.3–6). Tarutius claimed to have determined on astrological grounds the dates of Romulus' conception (synchronized with a supposed solar eclipse in 772 bce), his birth, and the founding of Rome (fixed by finding a horoscope fitting the character of the event in 754 bce).

A.T. Grafton and N.M. Swerdlow, "Technical Chronology and Astrological History in Varro, Censorinus and Others," *CQ* 35 (1985) 454–465.

Alexander Jones

Tauros of Bērutos, L. Caluenus (130 – 160 ce)

Student of Plutarch, then independent scholar or tutor, whose student Aulus Gellius provides vivid glimpses into the contemporary academic world: details of the formal curriculum, including the study of Plato's dialogues (*NA* 17.20) and Aristotle's scientific treatises (especially the *Problems: NA* 19.6), open class discussion (*NA* 1.26) as well as the social milieu including dinner parties where students brought topics for after-dinner discussion (*NA* 7.13). Tauros' literary output included a book like Plutarch's on **Stoic** contradictions (*NA* 12.5.5), an exegesis of the differences between Aristotle and Plato defending the **Academy** against contemporary charges of syncretism (*Souda*, T-166), and a commentary on Plato's *Timaeus*, cited by Iōannēs Philoponos, *De Aet. Mundi.* Tauros argued that the *Timaeus* describes an apparent "temporal creation" merely for clarity of instruction and to forestall *asebeia* and pride (Philoponos *De Aet. Mundi* 6.21 [pp. 186–187 Rabe]). Tauros distinguished four meanings of *genetos*: of the same genus as things that are created; composite things; in process of generation; dependant for existence on an outside source (Philoponos *De Aet. Mundi* 6.8 [pp. 145–148 Rabe]). Tauros also rejected Aristotle's fifth element, and to resolve the association of senses to elements, proposed a quasi-element between air and water, related to Aristotle's ***aithēr***, to account for smell (Philoponos *De Aet. Mundi* 13.15 [pp. 520–521 Rabe]).

Dillon (1996) 237–247; *NP* 12/1.59 (#1), M.-L. Lakmann.

PTK and GLIM

DE TAXONE LIBER ⇒ APULEIUS, PSEUDO

Tektōn (150 – 75 BCE?)

Designed a machine like a chair with armatures for attaching limbs which reduced a variety of dislocations and fractures; the moving part was a beam running in grooves for regular and smooth traction. Described by OREIBASIOS *Coll.* 49.4, 49.24–25 (*CMG* 6.2.2, pp. 6–9, 36–41), presumably following HĒLIODŌROS; mentioned also by GALĒN, *In Hipp. Artic.* 4.47 (18A.747 K.), and by CELSUS 8.20.4, as "Faber." The device was improved by HĒRAKLEIDĒS OF EPHESOS. For the occupational name *cf. Iliad* 5.59, and Tektōr of Dēlos (*LGPN* 1.431, 1st c. BCE); *cf.* also HALIEUS and PEPHRASMENOS.

(*)

PTK

Telamōn (250 BCE – 90 CE)

ASKLĒPIADĒS PHARM., in GALĒN *CMGen* 2.14 (13.528 K.), records his detailed instructions for preparing a cicatrizing ointment, good also for ***anthrax***: pine-resin, beeswax, quicklime, **litharge**, and ***psimuthion***, in olive oil. The simplicity of the recipe favors an early date; *cf. LGPN* 3B.402, 4.329.

Fabricius (1726) 430.

PTK

Telephanēs (250 BCE – 80 CE)

ANDROMAKHOS, in GALĒN *CMGen* 2.15 (13.532 K.), gives his "white" plaster of **litharge**, ***psimuthion***, frankincense, myrrh, and ***sagapēnon***, in olive oil. For the rare name *cf. LGPN* 3A.424.

Fabricius (1726) 430.

PTK

TELKHINAIOS ⇒ VELCHIONIUS

TERENTIUS ⇒ EUELPISTOS

M. Terentius Valens (25 – 40 CE)

SCRIBONIUS LARGUS includes among his teachers "Valens," renowned for pharmaceutical compounds designed as anodynes and anesthetics fundamental in surgeries (Scribonius alludes to treatments of gladiators, and possibly soldiers: 91, 94). Among anesthetic recipes for colicky intestinal pains, GALĒN cites via ANDROMAKHOS a Terentius Valens (*CMLoc* 9.4 [13.279 K.]; again at 9.5 [13.292]), who is likely the Terentius twice cited through ASKLĒPIADĒS PHARM., at *CMLoc* 4.7 (12.766 K.), for a rose-based collyrium, and at *CMGen* 5.11 (13.827 K.). Textual corruption most likely explains Asklēpiadēs' citation of

"M. Telentius" (an unattested *nomen*) at *CMGen* 7.6 (13.973 K.). All five of these are probably the same as Scribonius' teacher "Valens" (for whom scholars have also suggested VETTIUS VALENS).

"Marcus my teacher (*kathēgētēs*)" in Asklēpiadēs, in *CMLoc* 4.7 (12.750–751 K.), repeated by OREIBASIOS *Ecl. Med.* 117.8 (*CMG* 6.2.2, p. 292), is the same man, if Asklēpiadēs as often is quoting Scribonius' Greek. Citations of a "Valens" occur in Andromakhos, in Galēn *CMLoc* 7.4 (13.115 K.), *CMLoc* 9.4 (13.285 K.), and in CAELIUS AURELIANUS, *Acute* 3.2 "*physicus*" (Drabkin, p. 298; *CML* 6.1.1, p. 392), as the author of a work *curationes* (Book 3 considered ***sunankhē***).

Terentius Valens' recipes are consistently anesthetic, emollient, or styptic. *CMLoc* 9.4 (13.279 K.) prescribes for colicky pains of the bowels good measures of the leaves of the stone parsley, white pepper, henbane seeds, the latex of the opium poppy, saffron, and the peeled bark of mandrake-root, to be combined with honey and north Arabian balsam; not only are the stone parsley and Arabian balsam effective treatments for stomach and intestinal complaints, the henbane, opium, and mandrake all are powerful narcotics, and the saffron oil likewise is analgesic. Gladiators, soldiers, and civilians alike would have appreciated the styptic actions of the compound suggested in *CMGen* 5.11 (13.827 K.), that stopped bleeding from hemorrhoids and wounds in general: using a finger, the physician applied the drug made from realgar, ***misu, khalkanthon***, fissile alum, myrrh, copper flakes, aloes, rock alum, pomegranate calices, rose petals, oak galls, acacia gum, and pomegranate peelings. Other formulas include spikenard, beeswax, common rue, bitter-vetch seeds, birthwort, and clover.

Fabricius (1726) 431, 440.

John Scarborough

P. Terentius Varro of Narbo, "Atax" (*ca* 60 – 30 BCE)

Geographer and meterological poet born in Narbo or in the Atax valley of Gallia Narbonensis, 82 BCE. Learning Greek at age 35 (Jerome, *Chronicle* 151), he translated Apollōnios Rhodios' *Argonautika* into Latin. He wrote an historical panegyric epic on CAESAR's campaign against Ariovistus in 58 (*Bellum Sequanicum*), satires (Horace, *Sat.* 1.10.46), and erotic poetry addressed to a Leucadia (Propertius 2.34.85, OVID *Tr.* 439). His didactic *Chorographia*, in three books, possibly deriving from ALEXANDER OF EPHESOS, discussed the geography of Europe, Africa, and Asia, detailing the limits of bodies of water, describing vegetation, and using astral data to depict locations and explain climate. His ***Ephemeris***, a verse treatment of weather forecasting relying on ARATOS, influenced VERGIL, *Georgics* 1.375–397.

Ed.: *FLP* 235–253.
KP 5.1140 (#2), P.L. Schmidt; *OCD3* 1485, E. Courtney; *NP* 12/1.1144–5 (#3), P.L. Schmidt.

GLIM

M. Terentius Varro of Reate (81 – 27 BCE)

Ancient Rome's greatest scholar, born 116 BCE, came from the Sabine territory to the north-east of the city. Varro studied first at Rome under the philologist and antiquarian L. Aelius Stilo, then at Athens with the **Academic** philosopher Antiokhos of Askalon. A partisan of Pompey the Great, he was active in politics, being elected tribune, aedile, and quaestor, and serving several times as a naval and army commander. He was also a member of a 20-man commission set up by IULIUS CAESAR in 59 to redistribute public land in southern Italy. During the civil war he led forces for Pompey in Spain, and saw his property confiscated when the latter was defeated. He was given a pardon by Caesar, who selected him to oversee the establishment of Rome's first public library. After Caesar's death the project was abandoned, and Varro's property was once again targeted for confiscation, its owner marked out for death; only the protection of powerful friends ensured his survival. He devoted the rest of his life to scholarship. In his will he requested that he be buried in accordance with **Pythagorean** custom.

During his lifetime Varro composed some 620 books under 74 different titles. His efforts included humorous essays on human nature (*Saturae Menippeae*, 150 books), dialogues on philosophical topics (*Logistorici*, 76 books), a massive encyclopedia of Roman antiquities (*Antiquitates rerum humanarum et diuinarum*, 41 books), and his 700 collected portraits of famous Greeks and Romans (*Hebdomades uel de imaginibus*, 15 books). From this vast corpus only nine books under two titles survive intact, the works *De Rebus Rusticis*, and *De Lingua Latina*.

Surviving Works: The characteristic features of Varro's writings are prodigious erudition, keen interests in terminology, etymology, and numerology, close attention to the organization of his material, and occasional brilliant insights. These traits are all on display in the one work of his which has come down to us complete, the *De Rebus Rusticis* (*RR*). Divided into three books, the treatise deals with all the activities of cultivation that might be found on a typical large Roman estate in the 1st c. BCE. The first book covers the essentials of farming and the raising of plant crops, the second focuses on animal husbandry, while the third deals with *uillatica pastio*, or the raising of specialty products such as birds, bees, rabbits, and fish. It was apparently published in stages, with an early version of Book 1 appearing before 55 BCE, Book 2 composed somewhat later, and Book 3 added to complete the trilogy at the time of final publication in 37 BCE. Much of the material is clearly based on first-hand observation, but Varro introduces his treatise with a list of 52 authors whose writings he claims to have read.

The first book opens with an introduction of the *dramatis personae* (each book is presented as a dialogue among prominent Roman land-owners), and a lengthy debate as to the subjects that fall under the purview of agriculture proper. A guide to different types of land and soil is followed by directions for preparing vineyards and farm-buildings, procuring the right staff and equipment for the farm, and a discussion of the most profitable crops. Varro then inserts a farmer's calendar with tasks arranged according to an eight-fold division of the solar year, and concludes by describing how to sow, care for, and harvest various field crops. The book's advice (18.8) that the farmer combine imitation of his predecessors with systematic experimentation was to inspire many a later agronomist, and there are several interesting reports on agricultural technology, e.g. that in Spain farmers used a riding thresher on which the driver sat while it cleaned the grain (52.1). Nevertheless, the author's grasp of agricultural technique is uneven in this book. Varro is particularly weak on grafting, for which he paraphrases and in many cases misunderstands his source (*cf.* 40–41, and THEOPHRASTOS, *Caus. Pl.* 1.6).

The second book, by contrast, stands out for the depth and accuracy of its treatment of stock-breeding – a fact no doubt connected to Varro's possession of large-scale sheep-ranches in Apulia and mule-farms near his hometown of Reate. The book as a whole is structured according to a notional matrix: it has 81 subdivisions, to cover nine different varieties of animal and nine different kinds of animal care in every possible combination – that is, everything from "sheep, feeding of" to "dogs, health problems of." The longest section is devoted to swine (4), and detailed accounts are given of wool-shearing (11.5–12), the breaking of horses (7.12–14), and the "points" that breeders were to look for in cattle, horses, and donkeys (5.7–8, 7.5, 9.3–5). His discussions of transhumance (1.17, 2.9), and his notices regarding the primacy of mules and donkeys as sources of power for ground transport in ancient Italy (6.5, 8.5), are important texts for our understanding of the ancient economy. In a surprise attestation of rural literacy, he claims to have copied out by hand texts on veterinary health from Mago the Carthaginian and assigned them to his head-herdsmen to read (2.20, etc.).

The third book, on *uillatica pastio*, focuses mainly on the economics of animal breeding for urban niche-markets, although it also offers observations on the social organization of bees (16.4–9), patterns of bird migration (5.7) and learned behavior in birds, stags, boars, and fish (7.7, etc.). Varro exposes the ingenuity that went into the construction of Roman coops and bird houses, which were built to accommodate the special requirements of each breed and featured running water as well as elaborate networks of perches and roosts (5.2–6, etc.). A true *tour de force* is Varro's description of the aviary at his villa in Casinum, which included a stage for ducks to walk across, a lazy-susan bird feeder, and a planetarium (5).

The other Varronian work of which a substantial part survives intact is his *De Lingua Latina* (*DLL*). Published in the late forties BCE, it originally consisted of 25 books: one book of introduction, followed by six books on etymologies and meanings, six on inflectional morphology, and 12 dealing with syntax and grammaticality. Only Books 5–10 have survived complete. The first of these considers etymologies for the names of locations, the second for names of periods of time, and the third for the words poets use. The lacunose texts of Books 8–10 document Varro's major contribution to theoretical linguistics, which was a coherent account of variation among Latin roots and inflections that harmonized the contrary principles of anomaly and analogy.

Lost Works: Among Varro's lost works are many known to have provided impetus to the study of science at Rome. Perhaps the most important was his *Disciplines*, an encyclopedia of the *artes liberales* – the technical disciplines which it was felt proper for a free man to pursue – in nine books, one for each art: grammar, rhetoric, dialectic, arithmetic, geometry, astronomy, music, medicine, and architecture. This classification of sciences exerted a lasting influence on later scholarship, which dropped the last two fields due to their banausic character and consolidated the remainder into what would eventually become the *trivium* and *quadrivium* of medieval tradition. Varro's known contributions to the scientific subjects of the *quadrivium*, as well as medicine and geography, can be summarized as follows.

Geometry: Varro traced the origins of *geometria* to the needs of surveying and to a primitive interest in the size of the Earth (CASSIODORUS *Inst.* 2.6.1). He translated into Latin EUCLID's definitions of "plane," "solid," "line" and "cube" (*ibid.*, 1.20), revealed that in a circle the shortest distance from center to circumference is a line (*RR* 1.51.1), classified optics as a geometrical subject, and explained the causes of various optical illusions (Gell. 16.18). The elementary character of this material is patent, yet in *RR* 3.16.5 Varro betrays a familiarity with isoperimetry problems.

Arithmetic: Varro devoted a good deal of attention to Latin number terminology, seeking for instance to establish the precise difference in meaning between *secundum* and *secundo* (Gell. 10.1); a subtle examination of the "rule of nines" can be found at *DLL* 9.86–88. Beyond that, a deep and abiding interest in arithmology pervades the Varronian corpus. This comes across in his habit of working elaborate binary and ternary classificatory schemes into his treatises, as in *RR* bk. 2, with its 81 subtopics of animal husbandry, or in the lost *De*

Philosophia, which established as a theoretical possibility that 288 distinct schools of philosophy could exist (AUGUSTINE *CD* 19.1–3). The *Tubero de Origine Humana* sought to explain the viability of fetuses after seven months by dividing the 210–day period of gestation into 35–day chunks, then analyzing those sub-periods as compounds of harmonic ratios; a similar rationale was given for nine-month pregnancies (CENSORINUS *De Die Nat.* 9, 11). Finally, Varro's *Hebdomades* included a long catalogue of entities that come in groups of seven (Gellius 3.10), and there apparently existed a catalogue for threes as well (Ausonius 15.*pr* Green).

Astronomy: The eighth book of MARTIANUS CAPELLA's *De Nuptiis Mercurii et Philologiae* is thought to constitute a rough facsimile of the book on astronomy from the *Disciplines*. Martianus explains the nature of the poles, polar circles, colures, the **ecliptic**, the constellations, *sunanatolai* (simultaneous constellation risings), and the planets; the dimensions of the planetary orbits betray their Varronian provenance by their numerological invention (8.861). Elsewhere Varro observes the distinction between the sideral month (FAUONIUS EULOGIUS *In Somn. Scip.* 17.1) and the synodic month (Gell. 3.10), though in the interest of numerology he rounds the figures for their respective lengths to 27 days and 28 days.

Music: Varro divided harmonics into three sub-fields dealing respectively with rhythm, melody, and meter (Gell. 16.18); it was probably in the book on music from the *Disciplines* that he described the major modes, from hyperlydian to hypodorian (Cassiod. *Inst.* 2.5.8). He records numerous observations about animals attracted to music, such as the pigs he trained to respond to a horn (*RR* 2.4.20, etc.), and shows in other texts (*cf.* Mart. Cap. 9.926–929) a **Pythagorean**'s fascination with the inner connection between music and psychology.

Medicine: In his book on medicine Varro traced the science back to practices at the temple of Asklēpios at Kōs, and noted HIPPOKRATĒS' association with that temple; he seems also to have assembled a large collection of herbal remedies for diseases (PLINY 20.152, 22.114, 141). A number of the *logistorici* were devoted to medical topics such as diaetetics and mental health, and in the *Tubero de Origine Humana* he gave a detailed if theoretical account of fetal gestation (Censor. *De Die Nat.* 9, 11). Varro observes in passing at *RR* 1.12.2–4 that diseases could be caused by microorganisms, and accordingly recommends that country-houses built near swamps not have windows facing in the direction of the prevailing winds.

Geography: Varro's survey of geography appears to have concentrated on such subjects as ethnography (Pliny 3.8), the primary exports of different cities (4.62), the lengths of coasts and rivers (4.78; *cf.* Gell. 10.7), and the etymologies of place names (3.8); he made particular use of information Pompey Magnus gathered while on campaign near the Caspian Sea in the second Mithridatic War (6.38, 51–52). He also wrote a ***periplous*** of the Mediterranean (*De Ora Maritima*), a treatise on weather signs for use by sailors (VEGETIUS *De Re Militari* 4.41), and one on tides (*De Aestuariis*) which was probably based on POSEIDŌNIOS' pioneering work.

Such fragments represent only the tip of an iceberg. Much of the technical material in Pliny, Augustine, Martianus Capella and other late writers must also derive from Varro, although we are no longer in a position to determine precisely its extent. Placed in the broadest perspective, Varro's contributions to ancient science were twofold. Like Cicero, he played a crucial role in standardizing Latin technical terminology and digesting bodies of Hellenistic science and philosophy with a view towards transmitting them to later generations. In addition, he performed original work in several fields: first and foremost in the theory of grammar and linguistics, in the classification of the sciences, and in the development of the **Pythagorean** insight that numbers and patterns are crucial to our understanding of the world.

Ed.: D. Flach, *Marcus Terentius Varro. Gespräche über die Landwirtschaft* 3 vv. (1996–2002).

RE S.6 (1935) 1172–1277, H. Dahlmann; J.E. Skydsgaard, *Varro the Scholar* (1968); K.D. White, *Roman Farming* (1970); W.H. Stahl, *Martianus Capella and the Liberal Arts* (1971); K.D. White, "Roman Agricultural Writers I: Varro and his Predecessors," *ANRW* 1.4.1 (1973) 439–497; B. Riposati, ed., *Atti del Congresso Internazionale di Studi Varroniani* (1976); Rawson (1985); E.B. Cardauns, *Marcus Terentius Varro* (2001); *OCD3* 1582, R.A. Kaster.

Philip Thibodeau

Teukros of Egyptian Babylōn (*ca* 30 – 100 CE)

His astrological text, *Paranatellonta tois dekanois*, was instrumental in the transmission of the Hellenistic system of the **decans**. The earliest reference to Teukros is by PORPHURIOS OF TYRE in connection with the **decans**. Teukros is also important for transmitting the *paranatellonta*, or constellations rising on the horizon simultaneously with each **decan**. Given the Egyptian origin of the **decans**, it is not surprising to find references to the *sphaera barbarica*, i.e., the names of the stars and constellations known from the non-Greek world of Egypt or Mesopotamia, preserved in fragments of Teukros. Considering his date and his preservation of certain Greco-Egyptian astrological practices, the Babylōn of his epithet probably refers to the Egyptian Babylōn rather than the by then greatly diminished city in Mesopotamia. Egyptian Babylōn was already cited by DIODŌROS OF SICILY and by STRABŌN as having resident former prisoners from Babylōn in Mesopotamia. During the 3rd c., or perhaps later, Teukros' text was translated into PAHLAVI (q.v. for further developments).

F. Boll, *Sphaera* (1903) 16–21, 41–52, 416 n.2, 380–391, 545; *CCAG* 7 (1908) 194–213; *RE* 5A.1 (1934) 1132–1134 (#5), W. Gundel; *KP* 5.635–636 (#4), E. Boer; Pingree (1978) 442–443.

Francesca Rochberg

Teukros of Carthage (400 – 50 BCE)

DIOGENĒS LAËRTIOS 8.82 (in a list of homonyms derived from Dēmētrios of Magnesia) records that a work *Peri Mēkhanēs*, which began "From Teukros of Carthage I learned these things," was attributed either to ARKHUTAS OF TARAS, or a later homonym.

(*)

PTK

Teukros of Kuzikos (80 – 20 BCE)

Contemporary with MITHRADATĒS VI and Pompey, an author with a wide range of interests whose titles include *On the Goldbringing Earth, On Byzantium, On Mithradatēs' Deeds, On Tyre, Arabika, Jewish History*, etc.

Ed.: *FGrHist* 274.

OCD3 1488, K.S. Sacks; *NP* 12/1 (2002) 205 (#3), K. Meister.

Jan Bollansée, Karen Haegemans, and Guido Schepens

Thaïs (250 BCE – 300 CE)

PAULOS OF AIGINA, 3.25.7 (*CMG* 9.1, p. 198), among facial remedies, records her reddener, of saffron, frankincense, madder-root, myrrh, and orchil, in calf-fat and **mastic** oil. The name is primarily Hellenistic (*LGPN*), and need not refer solely to the *hetaira* of Alexander and Ptolemy I (Diodōros of Sicily 17.72).

(*)

PTK

Thalēs of Milētos (*ca* 600 – 545 BCE)

Renowned as the founder of philosophy, mathematics, and science. Born *ca* 624 BCE, he is said to have visited Egypt (a reasonable possibility since Milētos had a colony, Naukratis, there) and to have brought back a knowledge of geometry and astronomy. Thalēs said all was water and explained natural phenomena on the basis of the material composition of things. He is supposed to have predicted the solar eclipse of 585 BCE, May 28, and to have used geometry to measure the height of pyramids in Egypt by their shadows and the distance of ships at sea by triangulation. He was also reputedly a brilliant engineer who made the Halus river fordable by the army of Kroisos/Croesus, king of Ludia, by digging a second channel to divert some of the water.

Since Thalēs left no writings, we cannot verify his alleged accomplishments, and many seem anachronistic. He may have brought back a knowledge of practical astronomy from Egypt (determining the time of year from the stars), as well as techniques of land surveying (the root meaning of "geometry"). Eclipses were not predictable in this period, though Thalēs could possibly have used some scheme of periodic recurrences to anticipate an eclipse. Thalēs' isolationist political policy makes it unlikely that he helped Kroisos start a war by his engineering ability. One accomplishment more plausibly attributed to him is identifying the constellation of Ursa Minor as reliable for navigation (a technique probably borrowed from the Phoenicians). He seems also to have offered an interesting theory of why the Nile river floods in Egypt in the summer: the etesian (summer northerly) winds cause the river to back up.

Thalēs also seems to have started the tradition of explaining phenomena on the basis of the properties of natural substances, thus replacing mythological tradition with quasi-scientific speculation. He seems to have envisaged the Earth as a flat disk floating on a cosmic sea; ripples in the sea cause earthquakes. He thus pioneered natural explanations of phenomena and naturalistic cosmology. His student Anaximandros gave impetus to this tradition by publishing a written cosmology. Thalēs' legendary successes provided a kind of ideal paradigm for later theorists.

DK 11; KRS 76–99; P.F. O'Grady, *Thales of Miletus* (2002).

Daniel W. Graham

Thamuros (200 BCE – 80 CE)

Andromakhos, in Galēn *CMLoc* 9.5 (13.300 K.), records his detailed instructions for an enema containing alum, **calamine**, lime, yellow orpiment, burnt papyrus, and burnt cork from a Falernian wine-jar (famed since the mid-2nd c. BCE: Cicero, *Brutus* 287; Pliny 14.55, 76). The name is rare (*LGPN* 2.210, 3A.198), and probably Thrakian, *cf.* Homer *Iliad* 2.595–596, Strabōn 10.3.17.

Fabricius (1726) 431.

PTK

Tharseas/Thraseas/Tharrias (170? – 100 BCE)

Physician whose name is spelled in three different ways. (1) **Tharseas** is used by ASKLĒPIADĒS PHARM. in GALĒN *CMGen.* 4.13 [13.741.13 K.] to name the "surgeon" who invented "Indian plaster" and who appears in a list of physicians compiled in a manuscript of CELSUS (see NIKANDROS); this spelling is the most likely. (2) The plaster which Asklēpiadēs describes in the cited passage (741.14–18 K.) is identical with SCRIBONIUS LARGUS's *emplastrum nigrum* ***Thraseae*** *chirurgi* (208); he also gives the formula of "the surgeon **Thraseas**' green plaster" (204). (3) Lastly, there are good reasons to recognize as his the medical prescriptions concerning *lethargus* and **dropsy** which CELSUS 3.20.2, 21.14 has attributed to **Tharrias**, because they imply surgical knowledge which characterizes the **Erasistratean** sphere of influence.

Michler (1968) 84, 125–128.

Jean-Marie Jacques

Theagenēs of Rhēgion (530 – 500 BCE)

Reported to have written in Rhēgion about the life, text, and interpretation of HOMER: no fragments and few testimonia survive (DK has just four entries describing Theagenēs' thought only generically). Antiquity considered Theagenēs the founder of philology and allegorical interpretation. PORPHURIOS, in discussing *Iliad* 20.67 (descent of the gods to fight on the plain at Troy), explains that, in response to criticism (perhaps XENOPHANĒS') that Homer said inappropriately unseemly things about gods, Theagenēs originated allegorical interpretations of the following type (DK *fr.*2): Apollo, Hēlios, and Hephaistos = fire, Poseidōn and Skamander = water, Artemis = the Moon, Hēra = air, Athena = wisdom, Arēs = folly, Aphroditē = lust, Hermēs = reason. Hence, conflict of gods = conflict of elements, divine immortality = continued existence of elements in spite of occasional destruction. Details of Theagenēs' particular interpretations are not recoverable, and the nature of his innovation is unclear. Allegorical interpretation lies very near the surface of HĒSIOD's *Theogony* and is very nearly present in PHEREKUDĒS as well (*cf.* also STĒSIMBROTOS); Homer and Hēsiod both personify natural forces and elements, and offer etymologizing narrative (e.g. air/Hēra). Perhaps Theagenēs' allegorical thoughts were merely better, more extensive, or the first to be explicitly recorded.

A.L. Ford, *The Origins of Criticism* (2002) 68–72.

Jacques Bailly

Theaitētos of Athens (d. 369 BCE)

PLATO's dialogue *Theaitētos* is set shortly after Theaitētos was wounded, apparently fatally, in a war between Athens and Corinth, assumed to be that of 369. The body of the dialogue is the reading of the record of a conversation involving Sōcratēs, Theaitētos, and THEODŌROS OF KURĒNĒ occurring shortly before Sōcratēs' death (399), when Theaitētos was a *meirakion*, that is, what we might call a male of college-age. Theaitētos describes how he and a friend were being shown by Theodōros something about *dunameis* (powers), apparently what we would call the irrationality of the square roots of integers starting with 3 and ending at 17, where Theodōros stopped. Theaitētos and his friend thought that, since these powers appeared to be infinite, they would introduce a general characterization; they did this by distinguishing between square numbers (perfect squares) and the rest, which they called oblong; given a geometric representation of these numbers as quadrilaterals they called the sides of squares equal to them "lengths" and "powers" respectively. Plato's representation of this accomplishment has been taken to foreshadow Theaitētos' subsequent mathematical achievements.

In his history of geometry, PROKLOS mentions Theaitētos together with LEŌDAMAS OF THASOS and ARKHUTAS OF TARAS immediately after praising Plato's accomplishments, and says that the three of them increased the number of theorems and brought them into a more scientific arrangement (*In Eucl.* p. 66.14–18 Fr.). Later he tells us that EUCLID completed or perfected many of Theaitētos' results. According to the *Souda* Theta-93, Theaitētos was the first to describe (*graphein*) the five regular solids, information which the first scholion on Book 13 of the *Elements* presents more precisely by saying that the cube, pyramid, and dodecahedron are **Pythagorean**, but the octahedron and icosahedron belong to Theaitētos. The scholion credits Euclid with extending the elemental ordering to this subject, making it plausible to suppose that Book 13 is based on Theaitētos' work. Book 13 depends on Book 10, which develops an elaborate classification of irrational straight lines (lines which are incommensurable with a given straight line and such that a square with one of them as side is incommensurable with the square with the given straight line as side), focusing on lines called medials, binomials, and apotomes. PAPPOS (*In Eucl. X* 1.1; *cf.* 2.17) tells us that Theaitētos not only made the distinction ascribed to him in Plato's *Theaitētos*, but also made the distinction between the three fundamental lines of Book 10. Although Pappos' account of Theaitētos' characterization of these lines differs from the characterization found in the *Elements*, there is again good reason to suppose that Book 10 is based on Theaitētos' work. Histories of Greek mathematics now commonly ascribe to Theaitētos another achievement, the first theory of proportion applicable to both commensurable and incommensurable magnitudes. The theory which Euclid develops in Book 5 is thought to be dependent on the work of EUDOXOS, but a passage in ARISTOTLE (*Top.* 8.3 [158b24–35]; *cf.* ALEXANDER OF APHRODISIAS *ad loc.*) suggests the existence of an earlier theory in which (to state matters anachronistically) the ratio of two magnitudes x and y was expressed in terms of the application of the Euclidean algorithm of alternating subtraction to them (*Elem.* 7.2; *cf.* 10.1–3) and one said that $x{:}y :: z{:}w$ if and only if application of the algorithm to each pair produces the same result. No ancient text associates such a theory with Theaitētos; but even if he did not develop it, his accomplishments more than justify Theodōros' high praise of him in Plato's dialogue.

O. Becker, "Eine voreudoxische Proportionenlehre und ihre Spuren bei Aristoteles und Euklid," *Quellen und Studien zur Geschichte der Mathematik, Astronomie, und Physik* B.2 (1933) 311–333; B.L. van der Waerden, *Science Awakening* (trans. Arnold Dresden), (1963) 165–179; W.C. Waterhouse, "The discovery of the regular solids," *AHES* 9 (1972–3) 212–221; M.F. Burnyeat, "The philosophical sense of Theaetetus' mathematics," *Isis* 69 (1978) 489–513; *DSB* 13.301–307, I. Bulmer-Thomas; Lasserre (1987) 3; Mueller (1997) 277–285.

Ian Mueller

Theanō, pseudo (200 BCE – 100 CE?)

PYTHAGORAS' wife (or daughter, according to some traditions) and daughter of Brontinus of Krotōn, the popular idealized wife and mother. To her were ascribed various writings (*On virtue, Exhortations to women, On Pythagoras*) and apophthegms. IŌANNĒS OF STOBOI (1.10.13) transmits under her name a spurious Dorian fragment *On Piety* (*peri eusebeias*). Equally spurious are Theanō's letters, of uncertain date (proposed chronologies range from the 3rd c. BCE to the 2nd c. CE) which mainly treat the correct behavior of women toward husbands, infants and nurse-maids.

Ed.: Städele (1980) 166–185 (text); 288–335 (commentary).
Thesleff (1965) 193.17–201.9.

Bruno Centrone

Themisōn of Laodikeia (Syria) (*ca* 90 – 40 BCE)

From the Syrian city (PSEUDO-GALĒN, INTRODUCTIO 4 [14.684 K.]), and in Rome by *ca* 100 BCE, Themisōn studied under ASKLĒPIADĒS OF BITHUNIA (Tecusan, p. 14). Shortly before, Themisōn had emigrated to somewhere near Milan, seemingly becoming an expert on troublesome cases of *saturiasis* in both sexes (CAELIUS AURELIANUS, *Acute* 3.186 [Drabkin, p. 416; *CML* 6.1.1, p. 400]). Tecusan (*ibid.*) calculates that Themisōn was born *ca* 120 BCE, and that EUDĒMOS became his student *ca* 45 BCE (*cf. frr.*98 and 264). Tecusan's dates for the early **Methodist** figures rely on Rawson's chronology for Asklēpiadēs (prior scholars, e.g., Sepp, Wellmann, Deichgräber, Edelstein 1935/1967, offered a later chronology). Themisōn's physiological theories may be faintly reflected in VERGIL, *Georgics* 1.84–93 and 415–423 (so Pigeaud); but the supposed *archiater* C. Proculeius Themison (identified with our Themisōn by Roemer) is too late, viz. 7 CE. Tecusan's *only* firm date, however, is for Eudēmos' possible role in Drusus Caesar's assassination, 23 CE (TACITUS, *Ann.* 4.3, 4.8–11; PLINY 29.20–21). Moreover, Deichgräber (1934: 1633; *cf.* Sepp, pp. 119–120) has linked Themisōn with the obscure skeptic-**empiricist** philosopher-doctor, THEODĀS OF LAODIKEIA ON LUKOS, who is early 2nd c. CE (*cf.* GALĒN, *On My Own Books* 9 [2.115 MMH = Singer 1997: 17 = Boudon, *CUF* v.1, p. 163]).

Themisōn (or less likely THESSALOS) emerges from our often-contradictory texts as the "founder" of the **Methodist** "sect" (*hairesis*) of medicine (but *vide* Edelstein 1935/1967: 174–176). Themisōn was the first to distinguish "chronic" ailments in his (lost) three-book work on the subject (Caelius Aurelianus, *Chron.* 1.*pr.*1–3 [*CML* 6.1.1, pp. 426–428; *fr.*50 Tecusan]), and the separation of "chronic" from "acute" diseases became somewhat canonical among the **Methodists** (theoretical constructs later carefully elaborated by SŌRANOS). Doctrinal heritages from Asklēpiadēs' "**atomism**" are very weak (Tecusan, p. 14; *cf.* Vallance 1990: 141–142), and mechanical notions in Asklēpiadēs' theories are distinct from Themisōn's (Vallance 1990: 123–130). Occasionally, our sources attribute the notion of *koinōtetes* ("communities" viz. of diseases) to Themisōn (*frr.*63, 111, 161 Tecusan), but most sources "sandwich" citations of Themisōn among Thessalos and other later **Methodists** (Tecusan, pp. 82–83). Themisōn's remnants demonstrate his interest in pharmacology: e.g., *fr.*88 Tecusan (Caelius Aurelianus, *Chron.* 4.39 [*CML* 6.1.2, p. 796]), Themisōn's recipe for a beeswax-based remedy for bowel pains, whose ingredients also included acacia-gum, dried rose petals, and rose oil, triturated in wine once the beeswax has been melted; *cf. fr.*105 (CELSUS 6.7.1F), for ear problems, a recipe containing beeswax, opium poppy latex, and raisin wine. Themisōn authored four works known by title: *Acute Diseases, Chronic Diseases, Hygiene* (or *Rules of Health*), and a collection of *Letters*; two others, *The Method* and *Periodic Fevers*, Tecusan (p. 107) classes among her "dubia."

Ed.: Tecusan (2004) "Thematic Synopsis: Themison," pp. 82–83, 90–91, 97, and 101–102); F.P. Moog, *Die Fragmente des Themison von Laodikea* (Diss. Giessen, 1994 [cited by Tecusan, p. 13, n. 16]: *non vidi*).

S. Sepp, *Pyrrhonëische Studien* (1893); M. Wellmann, "Asklepiades aus Bithynien von einem herrschenden Vorurteil befreit," *NJb* 21 (1908) 684–703; *RE* 5A.2 (1934) 1632–1638, K. Deichgräber; E. Rawson, "The Life and Death of Asclepiades of Bithynia," *CQ* 32 (1982) 358–370, repr. in *Roman Culture and Society: Collected Papers* (1991) 427–443; J. Pigeaud, "Virgil et la médecine," *Helmantica* 33 (1982) 539–560; C. Roemer, "Ehrung für den Arzt Themison," *ZPE* 84 (1990) 81–88.

John Scarborough

Themistios of Buzantion (*ca* 340 – *ca* 385 CE)

A major Greek commentator on ARISTOTLE, born *ca* 317 at Buzantion, the son of a philosopher, Eugēnios. After a traditional education in Greek culture, he established a philosophical school at Constantinople (as Buzantion had by then become) and prepared paraphrases on several Aristotelian works. After about 350 he became involved in the political life of the eastern empire, and served several emperors as an ambassador, administrator and adviser, a phase of his career richly documented in his orations, some of which reflect his philosophical interests.

Themistios' extant paraphrases (of Aristotle's *De anima, De caelo, Metaphysics* Book 12, *Physics*, and *Posterior Analytics*) generally follow the Aristotelian text closely and are designed to facilitate study. He was clearly influenced by the work of the great **Peripatetic** commentator ALEXANDER OF APHRODISIAS, while also being thoroughly versed in the **Platonic** tradition. His interest in science is largely limited to responses to the Aristotelian text into which he never incorporates material from scientific writers, as did later commentators, notably IŌANNĒS PHILOPONOS, with the exception of his polemical reaction to GALĒN's views on void, place, and time in his paraphrase of Book 4 of the *Physics*. In general, however, he clarifies the essentials of Aristotelian scientific method and physical theory without developing original interpretations. He was well respected by later Aristotelian commentators in the Arabic, Hebrew and western Latin traditions, and during the Renaissance.

DSB 13.307–309, G. Verbeke; J. Vanderspoel, *Themistius and the Imperial Court: Oratory, Civic Duty and Paideia from Constantine to Theodosius* (1995); Robert B. Todd (trans.), *Themistius on Aristotle on the Soul* in *ACA* (1996); *OCD3* 1497, R. Browning; *ECP* 549–550, H.J. Blumenthal; *REP* 9.324–326, J. Bussanich; *CTC* 8 (2003) 57–102, Robert B. Todd; *Idem* (trans.), *Themistius on Aristotle Physics 4* in *ACA* (2003); *Idem* (trans.), *Themistius on Aristotle Physics 5–8* in *ACA* (2008: forthcoming).

Robert B. Todd

THEMISTOGENĒS OF SURAKOUSAI ⇒ XENOPHŌN OF ATHENS

THEOC- ⇒ THEOK-

Theodās of Laodikeia on Lukos (100 – 150 CE)

Empiricist physician, pupil (with MĒNODOTOS OF NIKOMĒDEIA) of the skeptic philosopher Antiokhos of Laodikeia on Lukos (DIOGENĒS LAËRTIOS 9.116: "Theiōdas"). He wrote: an *Introduction* to **Empiricist** medicine, on which GALĒN wrote a commentary in five books (*On My Own Books* 9 [2.115 MMH]); *Outlines* (*P. Petersburg* 13), on which there were commentaries by Galēn (in three books) and by THEODOSIOS; *On the parts of medicine* (Galēn *Subf. emp.* 5). As Mēnodotos already had, Theodās emphasized the rationalistic element of **Empiricism**, as is shown by the definition of "reasonable experience" he gives in order to explain the third principle of the **Empiricist** "tripod," the "transition to the similar" (Galēn *Subf. emp.* 4). He distinguished the three constitutive parts of the medicine, that coincide with the **Empiricist** "tripod" (experience, reports of others and "transition to the similar"), from its three final parts, that is semiotics, therapeutics and hygiene (Galēn *Subf. emp.* 5–6).

Ed.: Deichgräber (1930) 214–215 (fragments), 265–266.
RE 5A.2 (1934) 1713–1714, W. Capelle; *KP* 5.682, F. Kudlien; *NP* 12/1.308, V. Nutton.

Fabio Stok

Theodore, pupil of Sergius (525 – 545 CE?)

Theodore was a friend and pupil of SERGIUS OF REŠʿAINA, with whom he worked on translating the medical works of GALĒN from Greek into Syriac. Some of Sergius' works are dedicated to him. The identification of Theodore is not certain. Following a note in Assemani's 1725 CE edition of the *Catalogue* of 'Abdišu of Nisibis (1316 CE), he was traditionally identified with the Theodore who in 540 CE was appointed Nestorian bishop of Merv (capital of Sassanid Khorasan). Theodore of Merv produced a number of theological works, including a *Solution to Ten Questions of Sergius*. Based upon the 9th c. Ḥunayn's discussion of Galen's works, Brock identifies him with the Theodore who was bishop of Karkh Juddan, on the banks of the Tigris close to where Samarra was to be located.

Brock (1997) 43, 201–204; H. Hugonnard-Roche, "Note sur Sergius de Resʿaina, Traducteur du Grec en Syriaque et Commentateur d'Aristote," in G. Endress and R. Kruk, edd., *The Ancient Tradition in Christian and Islamic Hellenism* (1997) 121–143 at 124 (n. 13).

Siam Bhayro

Theodōrētos (30 BCE – 300 CE?)

Credited with two antidote formulae: one in MS *Berolinensis Phill. gr.* 1571 (16th c.) and the other in MS *Bodleianus Barroc.* 150 (15th c.). The latter is made of 11 vegetal ingredients (dodder [*epithumon*], spikenard, clove-tree [*kariophullon*], rush [*skhoinos*], Pontic rhubarb, **shelf-fungus**, balsam, balsam-perfume, aloes, saffron, and cassia, mixed with honey) for the treatment of ten medical conditions (headache, chest pain, liver and spleen ailments, long diseases, melancholia, **dropsy**, kidney and lung complaints, and gout). The ingredients, as well as their broad spectrum of uses, are typical of the "Classical/Golden" period of antidotes (1st–3rd cc. CE). *Theodōrētos* may name the physician who created these formulae, or simply describe them as "divine gifts" (for other such likely "brand names," compare FAUSTINUS and ISIS).

Diels 2 (1907) 100; *RE* 5A.2 (1934) 1803 (#9), K. Deichgräber.

Alain Touwaide

Theodōros (Mech.) (*ca* 435 – 485 CE)

Philosopher and engineer, addressee of PROKLOS' *de Prouidentia et fato et eo quod in nobis* (surviving only in William of Moerbeke's Latin translation). Proklos calls his friend "best among mechanicians" and summarizes Theodōros' belief in mechanical determinism: the Universe moves in a necessary motion like a machine on wheels and pulleys built by an engineer whose skill Theodōros imitates. Theodōros' work does not survive.

RE 5A.2 (1934) 1860–1863 (#41), K. Ziegler (prints relevant passages); *KP* 5.692–693 (#II.4), *Idem; PLRE* 2 (1980) 1091 (#29); *NP* 12/1.332 (#28), L. Brisson (cites recent editions of Proklos).

GLIM

Theodōros of Asinē (300 – 360 CE)

Studied Neo-**Platonic** philosophy under PORPHURIOS OF TYRE, then may have joined IAMBLIKHOS OF KHALKIS. He is credited with two titles: *On Names*, a discussion based on the myth of the *Phaedrus* about the first heaven, to be identified with the first principle (Test. 8 = PROKLOS, *Theol. Plat.* 4, pp. 68.24–69.25 S–W.), *That the Soul is All the Forms* addressing the transmigration of souls (Test. 37 = NEMESIOS, *De nat. hom.* 115.5–116.2, 117.1–4 Matthaei), culminating in the thesis that human souls can reside in animal bodies. He may also have composed commentaries on some of PLATO's dialogues (*Timaeus, Phaedrus, Phaedo*).

He taught a system of different realms of being, revealing a triadic structure at each level except for the first one, beyond human grasp and called "unspeakable," thus anticipating DAMASKIOS. The One, itself displaying a triadic structure, comes only second to the "unspeakable." The members of each triadic structure were named after the gods of the traditional Greek mythology, a procedure recalling Proklos. Theodōros' sympathy for arithmological symbolism led him to identify numbers and letters (Test. 6 = Proklos, *In Tim.* 2, pp. 274.10–277.26 D.). These considerations, linked to his theory of the soul, have their origin in the interpretation of Plato's view on the generation of the world-soul. Numbers are also symbols of the soul, since composed of ratios.

Ed.: W. Deuse, *Theodoros von Asine. Sammlung der Testimonien und Kommentar* (1973).
RE 5A.2 (1934) 1860–1863, K. Ziegler.

Peter Lautner

Theodōros of Gadara (30 BCE – 10 CE)

Rhetor and sophist who taught the emperor Tiberius and wrote learned works, including the probably geographical *On "Hollow" Syria*, all entirely lost.

Ed.: *FGrHist* 850.

PTK

Theodōros of Kurēnē (*ca* 470 – 400 BCE)

Known as a mathematician of the **Pythagorean** school, a friend of PROTAGORAS and a teacher of THEAITĒTOS (44 A4 DK). According to EUDĒMOS' *History of Geometry* (*fr.*133 Wehrli), he was a contemporary of HIPPOKRATĒS OF KHIOS. His name occurs in the list of the **Pythagoreans** compiled by ARISTOXENOS (A1). Theodōros figures in several Platonic dialogues (*Theaetetus, Sophist, Politicus*); he must have been one of PLATO's teachers in mathematics (DIOGENĒS LAËRTIOS 2.103 = A3) and spent considerable time in Athens. Theodōros was the first mathematician of whom we know who taught professionally all four sciences of the **Pythagorean** quadrivium – geometry, arithmetic, astronomy and harmonics (A4). Since the contemporary **Pythagorean** PHILOLAOS was also familiar with these sciences, and the **Pythagorean** HIPPASOS with all but astronomy, the formation of the quadrivium must go back to the **Pythagorean** school of the early 5th c.

Plato (*Tht.* 147d = A 4) ascribes to Theodōros a proof of irrationality of the magnitudes between $\sqrt{3}$ to $\sqrt{17}$, which means that Theodōros relied on the proof of the irrationality of

$\sqrt{2}$ that was found by Hippasos. Theodōros' discoveries were further incorporated in the general theory of irrational magnitudes, developed by his student Theaitētos and set forth in Book 10 of Euclid's *Elements*.

DK 43; Knorr (1975); van der Waerden (1979); H. Thesleff, "Theodoros and Theaetetus," *Arctos* 24 (1991) 147–159; Zhmud (2006).

Leonid Zhmud

Theodōros (of Macedon?) (70 – 150 CE)

Pharmacologist, wrote a compilation of compound medicines in at least 76 books (PHILOUMENOS 36 [*CMG* 10.1.1, p. 39]), now lost, and title unknown. Compounds are quoted by Philoumenos, AËTIOS OF AMIDA, ALEXANDER OF TRALLEIS, PAULOS OF AIGINA, and the 14th c. toxicological compilation, pseudo-AELIUS PROMOTUS, which identifies Theodōros as from Macedon. He is said to have acquired medical formulae from ANTIGONOS OF NIKAIA.

Scholars consider Theodōros a **Pneumaticist**, because DIOGENĒS LAËRTIOS (2.103) lists *Theodōros* (a very common name) as a physician student of an Athēnaios, perhaps ATHĒNAIOS OF ATTALEIA. Furthermore, one of Theodōros' recipes in Aëtios (14.48, p. 789 Cornarius) is attributed by Paulos (4.42 [*CMG* 9.1, p. 361]) to ARKHIGENĒS. Recipes attributed to Theodōros seem consistent with 1st c. CE methods and practice, although pseudo-DIOSKOURIDĒS *On Poisons* and *On Venoms* (both incorporated into Dioskourides' corpus before the 9th c.) never cite him. PLINY attributes two recipes (20.103, 24.186) to a Theodōros. These data, if all applicable to the same man, suggest the date-range, which remains troubling, as GALĒN never cites him.

Wellmann (1895) 13; *RE* 5A.2 (1934) 1865–1866 (#45), K. Deichgräber.

Alain Touwaide

Theodōros of Phōkaia (*ca* 380 BCE)

Author of a treatise on the Tholos at Delphi, according to VITRUUIUS (7.*pr*.12). On the basis of this single testimony it is generally accepted that Theodōros was the designing architect of the Tholos constructed around 380 BCE in the sanctuary of Athena Pronaia (the Marmaria) at Delphi. This sumptuous marble building, consisting of a cylindrical cella surrounded by a Doric peristyle, was probably a kind of temple, though the precise function is debated. The plan is an exceedingly rational radial scheme in which the position of nearly every element is related to the rays of the circle that determine the axes of the exterior columns. Innovations, such as the use of a new set of proportions in the elevation of the exterior order and the first securely attested use of a Corinthian colonnade in the interior, would suggest that Theodōros was a pioneering designer. Some features of the Tholos, including the Pentelic marble from which it was constructed, show a definite Athenian influence, though the ethnic, *Phocaeus*, that Vitruuius gives Theodōros should indicate that he was from Asia Minor. The idea that the name of the architect of the Tholos is a corruption of Theodotos, the architect named in the building accounts of the 4th c. temple of Asklēpios at Epidauros, has received some consideration but is beyond proof.

J. Charbonneaux and K. Gottlob, *Fouilles de Delphes*, II, *La Tholos* (1925); G. Gruben, *Die Tempel der Griechen*, 4th ed. (1986) 97–99; F. Seiler, *Die griechische Tholos* (1986) 57–71; G. Roux, "La tholos d'Athéna Pronaia dans son sanctuaire de Delphes," *CRAI* (1988) 290–309; Svenson-Evers (1996) 320–329.

Christopher A. Pfaff

Theodōros of Samos (590 – 530 BCE)

Son of Teleklēs, architect of an early archaic Temple of Hēra at Samos, also a sculptor, gem-cutter, metal smith, and inventor. Built, *ca* 575–550 BCE, the huge (*ca* 52.5 x 105 m) Ionic dipteral temple which had two rows of columns in the interior and a deep pronaos (Hēraion III). Because of weak foundations, it had to be replaced by a second dipteral temple 40 m to the west (Hēraion IV, with deep and carefully laid foundations, 530–500 BCE), associated with the tyrant Polukratēs of Samos (d. 525 BCE). Hērodotos (3.60) mentions the replacement as the largest temple in Greece, with Rhoikos as architect. Later sources link the two architects on these and other projects and attribute to Theodōros numerous works of art and the invention of useful tools and devices (Pliny 7.198). Vitruuius cites them as co-authors of a book on the Hēraion (7.*pr.*12).Work on the great temple continued for two centuries but was never completed.

Svenson-Ebers (1996) 7–49; H.J. Kienast, "Der Niedergang des Tempels des Theodoros," *MDAI(A)* 113 (1998) 111–131; *Idem* in M. Stamatopoulou and M. Yeroulanou, edd., "Topography and architecture of the Archaic Heraion at Samos," *Excavating Classical Culture* (2002) 317–325; *KLA* 2.445–447, S. Ebbinghaus.

Margaret M. Miles

Theodōros of Soloi (Kilikia) (*ca* 300 BCE – 100 CE)

Wrote a commentary on Plato's *Timaios*. He described the five regular Platonic solids, arguing against a common material origin because of the varying complexity (and number of triangles) comprising each element – less complex elements seemingly would be completed first. He also contended for a division and separation of matter into the five "worlds" of regular "elements." Theodōros explained elemental transformation as a change in environment, e.g., the compression and condensation of two units of fire being extinguished yield one unit of air (Plutarch, *de defect orac.* 427A–428B). In response to Plato, *Timaios* 35c2–36a5, the construction of the World Soul according to geometrical principles, Theodōros arranged numbers in a single line (not double or triple) arguing from the cleavage of matter lengthwise and, furthermore, that such an arrangement protects against disorder and confusion as the first power of 3 is transposed from the first power of 2 (Plutarch, *Anim. Proc.* 1022D). Proklos (*In Eucl.* p. 118 Fr.) disputes Theodōros' assumption that mixed (helical) lines are a blending of straight and curved.

RE 5A.2 (1934) 1811 (#30), K. von Fritz.

GLIM

Theodorus ⇒ Mallius

Theodorus Priscianus (364 – 375 CE)

Vindicianus' student, who had enormous respect for his medical mentor (Theod. Prisc., *Physica, Pr.* [Rose, p. 251]). Theodorus was probably also a member of the social and economic levels associated with the royal house in the western empire, perhaps, like his mentor, an *arkhiatros* in his own right. The medical masters and their apprentices in 4th c. Gaul, Italy, and north Africa knew their Greek texts, and when Alexander of Tralleis quotes from Theodorus Priscianus' (lost) book on epilepsy, there is no indication of a Latin original, suggesting a possible Greek version (Alex. Tr. 1.559 Puschm.). Two passages in

CAELIUS AURELIANUS' *Acute and Chronic Diseases* appear to be direct adaptations from Theodorus' *Logicus* 8 (on ***hudrophobia*** [Rose, p. 125]) and *ibid.* 20 (on "catarrh" [Rose, p.158]; *cf.* Cael. Aur., *Acute* 3.11.102, and *Chron.* 2.7.94 [Drabkin, 364 and 626]). Not only were Theodorus' works known and cited in the Greek-speaking east, but the medical writings of Latin medical writers were widely circulated and studied in the Roman west. Partially underpinning Theodorus' *Logicus* and *Gynaecia* are GALĒN and SŌRANOS, but the *Euporiston Faenomenon* is pointedly based on "natural remedies" of decidedly local origins (*Eup.* 1.4 [Rose, p. 4]), and there are traces of PLINY, SCRIBONIUS LARGUS, and GARGILIUS MARTIALIS who expressed similar grumpiness about exotic and complicated pharmacology.

Theodorus organizes his medical botany according to the "place on the body" to be treated, the traditional "head-to-heel," beginning with one's hair and skull. Farmer's lore abounds, e.g. *Eup.*, 1.5.12 (Rose, pp. 12–13), a tried-and true therapy for head and body lice, including bathing in a goodly mixture of powdered oak galls and pellitory roots (*hoc pyrethri et gallarum puluis ex aequo commixtus in balneis adhibitus facit*); *cf.* ***purethron***. Significantly, Theodorus does not suggest surgery for the very common childhood hernias, recommending a number of simples to be made into plasters (*Eup.*, 28, 79 [Rose, pp. 83–85]). Much of the *Gynaecia* is taken up with pharmacology, and Theodorus acknowledges the professional expertise of a woman named Victoria *medica*, whose status is honored without question (*Gyn.*, 1.1, 5.13 [Rose, pp. 222, 233]). Her specialty was the prescription of drugs ensuring pregnancies, a midwifely activity recognized in Roman law (Marcian in the *Digest*, 48.3.2–3). And even though Hippokratēs had advised against a *medicus* or *medica* prescribing drugs for abortions, *Gyn.*, 6.23–27 (Rose, pp. 240–244), provides five recipes for abortifacients, saying that ". . .occasionally these are necessary," much as farmers understand with their cattle.

Ed.: V. Rose, *Theodori Prisciani Euporiston Libri III cum physicorum fragmento et additamentis Pseudo-Theodoreis . . . accedunt Vindiciani Afri quae feruntur reliquiae* (1894 [includes the *Logicus* as Bk. II and *Gynaecia* as Bk. III]); Th. Meyer, *Theodorus Priscianus und die römische Medizin* (1909; repr. 1967).

E.H.F. Meyer, "Theodorus Priscianus" in *Geschichte der Botanik* v.2 (1855; repr. 1965) 286–299; *RE* 5A.2 (1934) 1866–1869, K. Deichgräber; Önnerfors (1993) 288–301; M.C. Salazar and A.M. Hernández, "Estudio del lexicon tardio de los tratados latinos africanos de los siglos IV y V," in Vázquez Buján (1994) 241–251; M.C. Salazar, "Grupos binarios de sinónimos en Theodoro Prisciano" in B. García Hernández, ed., *Latin vulgar y tardio: homenaje a Veikko Väänänen 1905–1997* (2000) 257–262; A. Fraisse, "Médecine rationelle et irrationelle dans le livre I des *Euporista* de Théodore Priscien" in N. Palmieri, ed., *Rationnel et irrationnel dans la médecine ancienne et médiévale* (2003) 183–192.

John Scarborough

Theodos of Alexandria, ha-Rofe (before 200 CE)

Jewish physician. Of the six people granted the title "physician" (*rofe* or *asya*) in classical Rabbinic literature (i.e., Theodos/Theodoros, Tobiya, Bar Ginte, Minyomi/Benjamin, R. Ammi, and Bar Nathan), we know most about Theodos. He is mentioned in the Mishnah (*Bekhorot* 4.4) as an expert on the ritual slaughter of animals, associated with Alexandria. The Tosefta (*Ohalot* 4.2) and Talmud (*Nazir* 52a; *Sanhedrin* 33a; *Bekhorot* 28b) depict Rabbis using his osteological expertise as a basis for Jewish legal decisions. Some scholars have sought to identify him with a THEUDĀS mentioned by (ANDROMAKHOS in) GALĒN, but the name was very common, and such connections highly speculative.

J. Preuss, *Biblical and Talmudic Medicine*, trans. F. Rosner (1978) 19–20; S. Kottek, "Alexandrian medicine in the Talmudic corpus," *Koroth* 12 (1996–1997) 85–87.

Annette Yoshiko Reed

Theodosios (Empir.) (150 – 210 CE)

Physician, mentioned by GALĒN together with SERAPIŌN and MĒNODOTOS as an exponent of the **Empiricist** "school" (*Med. Exp.* 29), and included in the list of **Empiricists** contained in the MS *Hauniensis Lat.* 1653 f.73 (following Mēnodotos and THEODĀS OF LAODIKEIA). He is probably to be identified with the skeptic philosopher to whom DIOGENĒS LAËRTIOS (9.70) ascribes the opinion that the Skepticism cannot be called "Pyrrhonism," and a work entitled *Skeptical Chapters* (also in *Souda* Theta-132, mentioning also the title *Comment on Outlines of Theodas*, and mistakenly identified with THEODOSIOS OF BITHUNIA). If his name goes back to the original version of Galēn's juvenile treatise, Theodosios was already active by the mid 2nd c. CE.

Ed.: Deichgräber (1930) 219 (fragments).
RE 5A.2 (1934) 1929–30 (#3), K. von Fritz; *KP* 5.699 (#1), H. Dörrie; *NP* 12/1.339–40 (#2), M. Frede.

Fabio Stok

Theodosios (of Bithunia) (200 – 50 BCE)

Mathematician, wrote three extant treatises on mathematical astronomy and a lost commentary on ARCHIMĒDĒS' *Method*, which establishes the only sure *terminus post* for his career. STRABŌN lists him (together with his unnamed sons) as a noteworthy Bithunian mathematician, and VITRUUIUS 9.8.1 as the inventor of a kind of sundial. An entry on Theodosios in the *Souda*, Theta-142, which ascribes philosophical and poetic works to him and states that he came from Tripolis, apparently confuses him with two other homonymous men.

The *Spherics*, in three books, was much studied in later antiquity (at least from the time of PAPPOS, who commented on it in Book 6 of his *Collection*). It is an elementary work on spherical geometry, with applications to astronomical problems that though obvious are never mentioned in the text; Theodosios' contribution was primarily to edit and organize material already known in the 3rd c. BCE if not earlier. Underlying the work are the conventional assumptions of contemporary astronomy, that the Earth and heavens are both spherical and concentric and that the Earth has a point-like magnitude in relation to the celestial sphere. The most advanced theorems demonstrate inequalities subsisting among the arcs of the horizon or the celestial equator corresponding to equal rising arcs of the **ecliptic** circle for observers situated either on or away from the terrestrial equator; for example these theorems allow comparison of the length of time required for successive signs of the zodiac to cross one's horizon. The treatise lacks theorems on configurations of great circle arcs ("MENELAOS' Theorem") by which PTOLEMY derives numerical values for quantities in spherical astronomy in *Almagest* Books 1–2.

On Habitations is a collection of 12 theorems concerning risings and settings of stars and length of daylight for different locations on the Earth; since most of the situations discussed are either close to the equator or near the poles, the book is clearly an intellectual exercise, not related to real observing conditions. The two books of *On Days and Nights* are similarly impractical, dealing with such questions as criteria for having day and night exactly equal at an equinox, taking into account the Sun's small movement along the **ecliptic** during the day in question.

O. Schmidt, *On the Relation Between Ancient Mathematics and Spherical Astronomy* (1943); *DSB* 13.319–321, I. Bulmer-Thomas.

Alexander Jones

Macrobius Ambrosius Theodosius (410 – *ca* 435 CE)

Imperial official and author who seems to have originated in north Africa, known to contemporaries as Theodosius, now generally identified with the praetorian prefect of Italy in 430 CE and possibly the proconsul of Africa in 410 CE. Macrobius wrote the *Saturnalia* and *In Somnium Scipionis*, both some time after 430 CE, and a treatise on Greek and Latin verbs surviving as a later-medieval epitome.

Macrobius addressed the *Saturnalia* to his son, Eustachius. As the title suggests, the work takes the form of a Platonic dialogue held over the course of the three-day festival by a small host of public and literary luminaries from the 4th c. pagan circle of Praetextatus and Q. Aurelius Symmachus (dramatic date 384 CE). The *Saturnalia* explicitly recalls the dialogues of CICERO's *Rep.* and Athēnaios' *Deipn.* Macrobius exploited this genre to portray the elite cultural attachments of the preceding generation within the ambit of *otium* and *conuiuium* and, by extension, to claim continuity with that culture in his own day. The text, now incomplete, presents a miscellany of antiquarianism, beginning with the Roman calendar and pagan religion (1.12–23) and leading to the main topic, a commentary on VERGIL and exposition of the poet's suitability as a cipher of antique lore (1.24, 3.1–12, 4.1–6.9). The last book, returning to the role of philosophy and science in a broader cultural context, contains a speculative discussion on diet, gender, physiology and medicine (7.4–9), and, although not explicit, forms a natural transition to the *In Somnium Scipionis*, also addressed to Eustachius. The *Saturnalia* excerpts also from HOMER, LUCRETIUS, AULUS GELLIUS, PORPHURIOS OF TYRE and SERENUS SAMMONICUS.

The commentary *In Somnium Scipionis* draws upon Cicero's famous *Somnium Scipionis* to depict (anachronistically) Neo-**Platonic** precepts. As such, the treatise does not form a traditional literary commentary of Cicero; instead, it serves as an exercise in secular exegesis, extrapolating selectively (rather than systematically) from passages of Cicero. The text depends heavily upon Porphurios for its philosophical doctrine, and Macrobius may have written in consultation with other commentaries on Cicero's *Somnium*, such as the briefer version by AUGUSTINE's pupil, FAUONIUS EULOGIUS. Similarly, Macrobius' use of Greek scientific authors probably derived from intermediary sources, providing insights into the transmission of philosophical and scientific learning. Macrobius employed long excurses in the late-Latin encyclopedic tradition. Topics include a classification of dreams (1.3), numerology and the **Pythagorean** decad (1.5–6), the nature and descent of the soul (1.8–14), a treatment of astronomy containing some original aspects (1.14–22), music and its relation to the harmony of creation (2.1–4), geography and a possibly original theory explaining the tides (2.5–9).

W.H. Stahl, trans., *Macrobius: Commentary on the Dream of Scipio* (1952); A. Cameron, "The date and identity of Macrobius," *JRS* 56 (1966) 25–38; P.V. Davies, trans., *Macrobius: The Saturnalia* (1969); R.A. Kaster, "Macrobius and Servius: Verecundia and the grammarian's function," *HSPh* 84 (1980) 219–262; *OCD3* 906–907, L.A. Holford-Strevens.

M. Shane Bjornlie

Theodotos (120 – 80 BCE)

Eye specialist from whom a type of remedy was called *Theodotion*: it cured abcesses of the eyelid and various eye conditions, more particularly the *muio-kephalon pathos* (a type of inflammation where the uvea juts out into the shape of a fly-head). One may determine his date because, according to CELSUS 6.6.5B, he added some ingredients to *Attalium*, an eye-wash named after ATTALOS III, and because of the fact that Celsus 6.6.6 provides the earliest testimony for a similar remedy, the famous eye-wash that Theodotos called *akharis-ton*, as its rapid efficiency favored *ingratitude*. The *Theodotia* were used until the end of antiquity, with some alterations in their formulae (see MARCELLUS OF BORDEAUX, PAULOS OF AIGINA, and ALEXANDER OF TRALLEIS, *passim*). In the seventh book of his *Iatrika* about eye conditions, AËTIOS OF AMIDA refers eight times to the *Theodotion* formulated by the eye specialist SEUERUS IATROSOPHIST (e.g. 7.36 = *CMG* 8.2, p. 287.28).

RE 5A.2 (1934) 1959–1960 (#24), H. Diller; *NP* 12 (2002) 348 (#7), V. Nutton.

Jean-Marie Jacques

Theokhrēstos (250 BCE – 77 CE)

Mentioned by a *scholion* on Apollōnios of Rhodes 4.1750 as the author of a *Libyka*, probably identical to the paradoxographer quoted by PLINY (1.*ind.*37, 37.37). The *Souda* Theta-166 attributes a Libyan history to a Theokritos. Although the latter is said to have been from Khios, the fact that he is only mentioned in relation with local Libyan tradition suggests Libyan origins. Identity between the two authors is uncertain, but probable.

RE 5A.2 (1934) 1704 (#3), R. Laqueur.

Jan Bollansée, Karen Haegemans, and Guido Schepens

Theoklēs (400 BCE – 200 CE)

Wrote a work on animals, which included (in the fourth book) an account of sea monsters larger than triremes near Syrtis (AELIANUS, *HA* 17.6). Aelianus cites him with AMŌMĒTOS, ONĒSIKRITOS, and ORTHAGORAS, as if he too were 4th/3rd cc. BCE.

(*)

GLIM

Theokritos (250 BCE – 80 CE)

ANDROMAKHOS, in GALĒN *CMGen* 6.5 (13.885 K.), records that he added Eretrian earth, ***khrusokolla***, orpiment, and sal ammoniac, to an otherwise herbal plaster. Besides the bucolic poet, the name is well-known, though more frequent before the Greco-Roman period (*LGPN*).

(*)

PTK

Theokudēs (500 – 25 BCE)

Listed early in an approximately chronological catalogue of minor artisans and artists who compiled rules of architectural symmetry (VITRUUIUS 7.*pr.*14). *Cf.* the homonymous 6th–5th c. BCE sculptor of Akraiphia (Boiōtia) (*LGPN* 3B.192).

RE 5A.2 (1934) 2030, E. Fabricius.

PTK and GLIM

Theologumena arithmeticae (*ca* 330 – 350 CE?)

Ascribed to IAMBLIKHOS OF KHALKIS, but with uncertainty. The treatise is a collection of passages from a lost work of the same title by NIKOMAKHOS OF GERASA, and from the extant *On the Decade and the Numbers within It* by ANATOLIOS (OF LAODIKEIA?). It might be a portion or summary of the work Iamblikhos planned for the seventh book in his series of **Pythagorean** treatises (Iamblikhos, *In Nic.* p. 125.15–25 Pistelli), though the assumption is disputed since newly recovered excerpts from Iamblikhos' *On Pythagoreanism* VII witness to a more elaborated stage of Neo-**Platonic** metaphysics missing from the anonymous *Theol. arithm.*

The author/compiler connects mathematics to physics, theology and ethics. The basic principle is the monad which can generate other numbers without underlying change (1.6–8). It also contains potentially all properties that show up in numbers, explaining its power to unite. Therefore the monad can be considered god and, as an organizing principle, the **Demiurge** (4.1–12). Numbers and gods are linked because the former also have generative force. Reflecting on theories in PLATO's *Republic* and *Timaeus*, the author presents the Ideas as numbers or characteristics of numbers with the intention to reduce the Ideas, the archetypes of the physical world, to numbers or relations between numbers. Thus he makes mathematics the highest science serving as a basis for other kinds of sciences and knowledge and from which all sorts of knowledge can be derived. Mathematical and ethical principles parallel each other: virtue is connected to knowledge – the highest of which is mathematics – and thus it can be examined mathematically; for example, the analysis of justice (36.20–40.19) recalls **Pythagorean** presumptions.

Ed.: V. de Falco, *[Iamblichi] Theologumena arithmeticae* (1922; rev. ed. by U. Klein 1975).
O'Meara (1989).

Peter Lautner

Theomenēs (300 BCE – 75 CE)

PLINY 37.38 records his explanation of the origin of amber: poplar sap drips into the pool called Electrum in the Hesperidēs along the Libyan coast.

(*)

PTK

Theomnēstos (Med.) (400 BCE – 400 CE)

The "Laurentian" list of medical writers (MS *Laur. Lat.* 73.1, f.143[V] = *fr.*13 Tecusan) includes this name, common before 300 CE (*LGPN*) and rare thereafter (*LGPN* 3A.204). The list includes no veterinarians, so this man must be distinct from the homonymous veterinarian.

(*)

PTK

Theomnēstos (of Nikopolis?) (313 – 650 CE)

Author of a text on horse care and veterinary medicine, one of the principal sources of the *Hippiatrika*. An allusion to an imperial marriage in Milan, apparently that of Licinius in 313

CE, provides a *terminus post quem* for Theomnēstos' work. Theomnēstos appears to have been acquainted with medical theory and also with practice in the field. His is the only treatise in the *Hippiatrika* that includes case-studies and detailed instructions for grooming and early training, and that does not contain magic. Also noteworthy is Theomnēstos' expression of affection for his patients. Theomnēstos' treatise includes quotations from Apsurtos, Agathotukhos, Nēphōn, and "Cassius" (possibly Cassius Dionusios). Echoes of Xenophōn and Simōn of Athens are present, but Theomnēstos does not mention these authorities by name, and may have used their work through a compilation such as that of Cassius Dionusios. An Arabic translation preserved in two MSS indicates that Theomnēstos was a native of Nikopolis, but there were numerous cities with that name.

CHG vv. 1–2 *passim*; Björck (1932) 54–55; *Idem*,"Griechische Pferdeheilkunde in arabischer Überlieferung," *Le monde oriental* 30 (1936) 1–12; *RE* S.7 (1940) 1353–54, K. Hoppe; Hoyland (2004); *NP* 12/1.373, V. Nutton; McCabe (2007) 181–207.

Anne McCabe

Theōn (Astr.) (127 – 132 CE)

Ptolemy, *Syntaxis*, records Theōn's observations of Mercury (9.9) and Venus (10.1–2), giving an elongation for Mercury of 26° 15', distinctly larger than the maximum elongation of 20° accepted by Theōn of Smurna (3.13, 3.30).

RE 5A.2 (1934) 2067–2068, K. Ziegler.

PTK

Theōn of Alexandria (Stoic) (15 BCE – 15 CE)

The **Stoic** Theōn, living at the time of Augustus, wrote a commentary on the physics section of the *Introduction to the Doctrines* of Apollodōros of Seleukeia (*Souda* Theta-203), as well as *On the Arts of Rhetoric*.

GGP 4.2 (1994) 714, P. Steinmetz.

GLIM

Theōn of Alexandria (Astr.) (*ca* 360 – 385 CE)

Active in Alexandria, according to his record of three astronomical events dated 360, 364 and 377 CE; father of Hupatia. Three of his commentaries on Ptolemy's works are almost extant, the most famous being the "Little Commentary" (*LC*), a practical guide to the use of the *Handy Tables* (*HT*) without theoretical justification. Less famous, but still influential, was his commentary on the *Almagest* (*IA*: only the section on Book 11 is lost). Apparently much less known were the five books of his "Great Commentary" on *HT* (*GC*: only Books 1–3 and part of 4 are extant), in which the correspondence between the *Almagest* and *HT* is examined. These texts contain the most reliable information on Theōn.

Despite their different purposes, the three commentaries address a composite audience, as can be seen from their respective prefaces. Theōn's foreword to *IA* states that his auditors urged him to explain certain difficulties in Ptolemy. Some (if not most) of these students were mainly interested in making use of Ptolemy's tables, most probably for astrological purposes like the calculation of horoscopes. Theōn's *LC* directly fulfils this demand, providing detailed guidance through the *HT* with no theoretical explanation but illustrated by fully calculated examples. At the same time, Theōn repeatedly complains about his students' insufficient skills in geometry and calculation and therefore urges them to turn themselves to theoretical studies, especially geometrical proofs underlying Ptolemy's calculations and tables. One of Theōn's purposes was thus to turn the *Almagest* into either an initiation to, or a consolidation of, his students' geometrical knowledge (best seen from his commentary on *Alm.* 1, *cf. IA* 319.2–3 Rome). He therefore emphasizes its mathematical interest and its contribution to liberal education (*IA* 321.10–13 Rome). Theōn saw this interpretative stance as the continuation of Ptolemy's own work as a commentator of the ancients, and urged the most able of his companions to go the same way (*IA* 319.6–10 Rome). It seems that Theōn shared with them a real veneration for Ptolemy, as an epigram preserved under his name reveals (Dzielska, 75). In spite of this, Theōn's project was essentially different from Ptolemy's, since Theōn showed no interest in checking or improving Ptolemy's models or calculations through new astronomical observations. Even the extant part of *GC*, in which such reflections could have been found, shows no effort in this direction.

Theōn's *IA* might betray some influence from PAPPOS' own commentary on the *Almagest.* But, although the *Souda* notice erroneously makes them contemporaries under Theodosius I, nothing precise is known about their exact dependence – in particular Theōn never mentions Pappos. Theōn may have derived his interest in classical geometry and liberal education from him, but he does not seem to have shared Pappos' special interest in mathematical heuristics and Hērōnian mechanics (HĒRŌN OF ALEXANDRIA); he was apparently more inclined toward accurate descriptions of computational procedures. Theōn's ambiguous statement about the study of philosophy in *IA* (319.20–22 Rome) shows either his lack of interest or his contempt for such studies, perhaps out of fidelity to Ptolemy's preference for mathematical studies over philosophical debates (*Alm.* 6.15–19 Heiberg). The scarce testimonies given by the *Souda* (A-205) and John Malalas' *Chronographia*, may point toward his (plausible) interest in astrological Hermetism (Dzielska, 74–77). Besides Hupatia, who seems to have proofread *IA* 3, Theōn mentions other collaborators: one Epiphanios to whom he dedicates *LC, IA* and *GC* 4 (most probably one of his able *akroatai*) and two "companions" Ōrigenēs and Eulalios, to whom *GC* 1–3 is dedicated.

In one passage of *IA* (492.6–8) Theōn mentions an additional case to *Elements* 6.33 published in his own edition of EUCLID's *Elements.* Heiberg, by comparing the Greek manuscripts of the *Elements*, thought he could determine the precise style and extent of Theōn's revisions on Euclid. But more recent research shows the comparison unreliable, partly due to Heiberg's neglect of the Arabic and Latin translations of Euclid as well as the complexity of the direct transmission itself (Vitrac, 27–30). Theōn's intervention was apparently a standardization, perhaps motivated by the need to provide better support for geometrical studies.

Heiberg's attribution of a revision of Euclid's *Optics* to Theōn has been shown to rely on negligible evidence, although it is consistent with Theōn's quotation of Euclid's *Optics* in *IA* as well as with his putative knowledge of Pappos' commentary. But it is not explicitly attested and is thus nothing more than a plausible guess. Similarly, much doubt has been cast on Heiberg's attribution to Theōn of revised editions of Euclid's *Catoptrics* and *HT.* Neugebauer (*Isis* 40, 1949) conjectured that the contents of Theōn's treatise on the "small astrolabe" correspond to Yaqubi's summary of Ptolemy's treatise on the plane astrolabe

(875 CE) and is the source of SEVERUS SEBOKHT's treatise (6th c., ed. Nau 1899), but this has been criticized (Sezgin, *GAS* 5 [1974] 180–186; Tihon, *Physis* 32, 1995: 239) and is at least questionable: Theōn never mentions it and neither PHILOPONOS nor SUNESIOS seem to know it. On the whole, the generous attribution of many works to Theōn seems to derive from the increasing celebrity he earned for his commentaries on Ptolemy in the Byzantine Middle Ages and Renaissance. He thus became a major figure of the commentary tradition, but this only partly reflects his actual work.

Ed.: A. Rome, *Commentaires de Pappus et de Théon d'Alexandrie sur l'Almageste*, v.3 *Théon d'Alexandrie* (1943); A. Tihon, *Le "Grand commentaire" de Théon d'Alexandrie aux "Tables faciles" de Ptolémée: Livre I* (1985), . . . *Livre II, III* (1991).

DSB 13.321, G.J. Toomer; M. Dzielska, *Hypatia of Alexandria* (1995); Jones (1999); *OCD3* 375, M. Folkerts; B. Vitrac, "À Propos des Démonstrations Alternatives et Autres Substitutions de Preuve Dans les Éléments d'Euclide," *AHES* 59 (2004) 1–44.

Alain Bernard

Theōn of Alexandria (Med. I) (130 – 160 CE)

Autodidact ex-athlete who wrote a work on exercise, and a longer work *Gymnastrion*, both known only from GALĒN, *Hygiene* (*CMG* 5.4.2), who praises him as wiser than other such writers, but chides him for thinking he knew better than HIPPOKRATĒS about massage, *Hygiene* 2.3–4 (pp. 44–53); *cf.* 3.3 (p. 80) on exercise, 3.8 (pp. 91–94) on bathing; *cf.* *Thras.* 47 (5.898 K. = 3.99 MMH). For athletes as medical writers, *cf.* FLAUIUS or MOUSAIOS.

NP 12/1.376 (#7), V. Nutton.

PTK

Theōn of Alexandria (Med. II) (*ca* 300 – 500 CE)

Alexandrian physician (*arkhiatros*), wrote the practical handbook *Anthropos* (Man), dedicated to a certain Theoktistos (a rare name attested from the 3rd c. CE: *LGPN*), surviving only in Phōtios' description (*Bibl.* 220). A list of therapeutic procedures for afflicted body parts presented in "head to foot" order is followed by a brief section on simple and compound medicines, along with a compilation of prescriptions from earlier medical writers. He is probably the same Theōn whose purgative recipe AËTIOS OF AMIDA quotes (3.58, *CMG* 8.1, p. 287). The epitomizing character of his handbook, which Phōtios compares to OREIBASIOS' *Iatrikai sunagogai*, suggests a date no earlier than the 4th c. His relation to the Theōn mentioned by EUNAPIOS (*Vit. Soph.* 499) as a physician successful in Gaul and a contemporary of IŌNIKOS OF SARDĒS is indeterminable.

RE 5A.2 (1934) 2082 (#17), K. Deichgräber; *KP* 5.716 (#7), K. Ziegler; *NP* 12/1.378 (#9), V. Nutton.

Keith Dickson

Theōn of Smurna (*ca* 100 – 130 CE)

Theon of Smurna Roma, Musei Capitolini, Archivo Fotografico dei Musei Capitolini

Philosopher. A surviving bust of Theōn the "**Platonist** philosopher," dedicated by his son Theōn "the priest," is dated stylistically to *ca* 135 CE. Theōn wrote several works to facilitate reading PLATO, of which the only one partially extant in Greek, the *Mathematical Things Useful for the Reading of Plato*, is of scientific interest. Theōn, whose most important source was ADRASTOS, gives a mostly elementary and entirely derivative exposition of topics in number theory, harmonic theory, and astronomy. Sections on harmonic theory and especially astronomical modeling are of much historical value. Theōn of Smurna has sometimes, implausibly, been identified with a THEŌN who supplied PTOLEMY with several astronomical observations, of which five dating from 127–132 CE are cited in the *Almagest*.

RE 5A.2 (1934) 2067–2075 (#14), K Ziegler; Neugebauer (1975) 949–950; *DSB* 13.325–326, G.L. Huxley; Barker (1989) 209–229.

Alexander Jones

Theophanēs of Hērakleopolis (Egypt) (before 530 CE)

Mentioned as ***phusikos*** by STEPHANOS OF BUZANTION (*Ethnika*, s.v. *Hērakleoupolis*, p. 304.7–9 Meineke).

RE 5A.1 (1934) 2127 (#2), W. Capelle.

Arnaud Zucker

Theophanēs of Mutilēnē, Cn. Pompeius (before 88 – after 36 BCE)

Greek historian, son of Hieroitas, of the Mutilēnean upper class, assumed the office of *prutanis*, and came to Rome probably around 88 BCE after MITHRADATĒS' expulsion of Asians protesting his slaughter of 80,000 Romans. Theophanēs met CICERO and became an intimate friend of Pompey whom he accompanied on his campaigns in the east. Pompey, stopping in Lesbos on his return to Rome, liberated Mutilēnē for Theophanēs' sake, earning honorable titles and divine respect. Receiving Roman citizenship from Pompey no later than 61, Theophanēs assumed the name Cn. Pompeius, according to custom. In 59, Theophanēs made a diplomatic trip to Egypt to convince Ptolemy XIII Auletēs to ally himself with Rome. In 48, Theophanēs again followed Pompey to the east after his defeat by IULIUS CAESAR at Pharsalos. Pompey was assassinated in Alexandria, perhaps due to Theophanēs' bad advice. After Caesar's murder in Rome (44 BCE), Theophanēs requested an interview with Cicero to discuss his situation. His end is unknown. Theophanēs composed the history of the third Mithridatic War on the basis of his experience with Pompey.

The work, with its propagandistic undertones, was finished in 62 BCE; of seven extant fragments five, preserved in STRABŌN, describe countries traversed by Pompey's army, including the sources of the Tanais (Don) (F 3); the position of the country of the Amazons (F 4); and the size of Armenia (F 6). Theophanēs was the first and the only Greek visitor known to have recorded impressions of Albania, Asian Iberia, and the Caucasus. His work thus served as a source for POSEIDŌNIOS and Strabōn.

Ed.: *FGrHist* 188.
B.K. Gold, "Pompey and Theophanes of Mytilene," *AJPh* 106 (1985) 312–327; P. Pédech, "Deux grecs face à Rome au Ier siècle av. J. C.: Métrodore de Scepsis et Théophane de Mitylène," *REA* 93 (1991) 65–78; V.I. Anastasiadis and G.A. Souris, "Theophanes of Mytilene: A new inscription relating to his early career," *Chiron* 22 (1992) 377–383.

Daniela Dueck

Theophilos (Geog.) (120 BCE – 110 CE)

Cited by MARINOS OF TYRE in PTOLEMY, *Geog.* 1.14, on the sea-voyage from the east coast of Africa 20 days' sail to the spice-lands (i.e., post Hippalos): if STRABŌN's silence (16.4.14) is reliable, he is post-20 BCE. *Cf.* the geographers DIOGENĒS and DIOSKOROS.

RE S.9 (1962) 1393–1394 (#7a), Fr. Gisinger.

PTK

Theophilos (Lithika) (250 BCE – 50 CE?)

Lapidary writer whose *On stones* PSEUDO-PLUTARCH *De fluu.* 24.1 (1165D) mentions with regard to the Tigris. Pseudo-Plutarch attributes two further tracts to Theophilos: *Italika* (*Parall. min.* 13B) and *Peloponnesiaka* (32A). Laqueur and Jacoby consider Theophilos fictitious. On the contrary, Schlereth argues that our man corresponds to Theophilos Zenodoteus, mentioned in a *scholion* to NIKANDROS OF KOLOPHŌN (*Thēr.* 12a [pp. 39–40 Crugnola]) regarding a story similar to pseudo-Plutarch's testimony on the Tigris.

Ed.: *FGrHist* 296.
Schlereth (1931) 123–124; *RE* 5A.2 (1934), 2139 (#11), R. Laqueur; De Lazzer (2003) 88–89.

Eugenio Amato

Theophilos (Agric.) (325 – 90 BCE)

Agricultural author whose work was excerpted by CASSIUS DIONUSIOS (VARRO, *RR* 1.1.9–10). Perhaps identical to the THEOPHILOS who wrote on geography or else the THEOPHILOS who wrote on *lithika*.

RE 5A.2 (1934) 2138 (#12), W. Kroll.

Philip Thibodeau

Theophilos (Pharm.) (120 BCE – 540 CE)

AËTIOS OF AMIDA 7.114 (*CMG* 8.2, p. 382) cites his collyrium, good for children and trachoma: **calamine**, copper, opium, and **verdigris** in gum and water, apply in egg-white; another in ALEXANDER OF TRALLEIS (2.19 Puschm.) substitutes acacia, saffron, Indian nard, and myrrh for the **verdigris**. Aëtios also cites his wound-cream: roasted copper and roasted ***misu***, with myrrh, ***omphakion***, and saffron, in Khian or other dry old wine and Attic honey, 7.45 (p. 299). The name is too common to risk identification with any homonym.

RE 5A.2 (1934) 2148 (#15), K. Deichgräber.

PTK

Theophilos son of Theogenēs (250 BCE – 300 CE)

Jewish Egyptian mineralogist and alchemist said by Zōsimos of Panōpolis to have written about "all the gold-mines of the *Chōrographia* (*Description of the Country*)" (Festugière 1950: 365). Stephanos of Alexandria preserves a small fragment of his work (Ideler 2 [1842/1963] 246) and he may be the same Theophilos of whose work Zōsimos preserves a fragment (*CAAG* 2.198).

(*)

Bink Hallum

Theophrastos (Alch.) ⇒ Hēliodōros

Theophrastos of Eresos (*ca* 340 – 287/6 BCE)

Born in Eresos, in 372/1 or 371/0 BCE, where he studied under Alkippos, went to Athens as a young man, and is said to have studied under Plato. Theophrastos probably met Aristotle at the **Academy**, whom, after Plato's death in 348/7, he accompanied to Assos. Theophrastos may have persuaded Aristotle to move to Mutilēnē on Lesbos in 345/4. When Philip summoned Aristotle to Macedon in 342 to tutor his son Alexander, Theophrastos apparently accompanied him. After seven years, both returned to Athens where Aristotle began to teach in the **Peripatos**. When Aristotle, fearing anti-Macedonian sentiments, fled Athens to Khalkis where he soon died, Theophrastos assumed leadership of the school in Athens, remaining **scholarch** for 36 years until his death at age 85 in 288/7 or 287/6 BCE.

Theophrastos' intellectual pursuits were as wide-ranging as Aristotle's. In his *vita Theophrasti*, Diogenēs Laërtios includes a catalogue of 225 titles of Theophrastos' works, some monographs, others in several volumes. Despite some duplication of titles and splitting of larger works, the list reveals Theophrastos' astonishing output; it also reveals how much has perished. From this enormous *corpus Theophrasteum*, all that survives are the two large works on botany (*Research on Plants* and *Plant Explanations*), nine *opuscula* treating various aspects of natural science (*On Sense-Perception, On Stones, On Fire, On Odors, On Winds, On Fatigue, On Dizziness, On Sweat, On Fish*), *Metaphysics, Meteorology*, six summaries from scientific works made by Phōtios, Patriarch of Constantinople in the 9th c. CE, and the well-known *Characters*. We can recover some idea of many of Theophrastos' scientific works from numerous fragments, recently published (Fortenbaugh *et al.*).

The two large botanical works, *Research on Plants* (nine books) and *Plant Explanations* (six books), complement one another. In general, the latter explains physiology and underlying causes and developments of plants and discusses common or unique characteristics of plants described in the former. Theophrastos follows Aristotle's division in his zoological works, ascertaining the different features (*diaphorai*) which characterize species and genera, moving from fundamental universal principles (*kath' holon*) to individual matters (*kath' hekasta*). Theophrastos' botanical researches so far eclipsed his predecessors' that he may rightly be entitled "the father of botany." Moreover, his attention to the relation between plant organisms and their environments shows him to have been an ecologist. Preserved by Phōtios is an excerpt (*fr.*435 FHSG) from another botanical work, *On Honey*, wherein Theophrastos noted that honey is produced in three ways: 1) in flowers, 2) from the air, when moisture is concocted and falls from the air, and 3) in reeds.

Theophrastos' zoological *On Fish* is an odd essay describing exotic and unusual fishes, especially two types: fish which venture onto land and live in air, and fish which bury themselves in the ground and survive there, both living without water normally required for cooling. Three excerpts preserved by Phōtios also depict strange or extraordinary phenomena. *On Creatures which Change Color* discusses the octopus, the chameleon, and the horned animal *tarand(r)os*, the elk or reindeer. The first two change color in their skins due to a change in "breath" (***pneuma***). The *tarand(r)os* changes color in its fur, but not due to *pneuma*. *On Creatures Appearing in Swarms* concerns creatures arriving in large numbers (e.g., frogs, locusts, snakes, mice) due to favorable climatic or environmental conditions and how they are exterminated by natural causes or human intervention. In *On Creatures Said to Be Grudging*, Theophrastos attributes jealousy to animals; to spite humans, they destroy or hide parts of their bodies useful to humans: e.g., geckoes swallow their skins used for epilepsy, the mare bites off the fleshy growth (*hippomanes*) on her newborn's forehead used as a love charm, and lynxes bury their urine which turns into a precious stone.

Theophrastos' *opuscula*, largely etiological, are composed in question-and-answer format like the ARISTOTELIAN CORPUS PROBLEMS, with which they have much in common. *On Sense-Perception* is primarily doxographical, recounting the opinions of natural philosophers down to Plato according to categories of problems, within which chronologically according to schools. *On Stones* and *On Fire* both consider phenomena associated with the combination and alteration of elemental substances and not with earth and fire as elements themselves. *On Fire* is structured much as the *opuscula* generally are – first there is a section on general questions followed by individual problems. Theophrastos observes that fire is unlike the other three elements, air, earth, and water, in that it alone requires an underlying substrate, a fuel, in order to exist. Of the elements only fire can be created, e.g., by striking stones together or from rubbing or friction. Several sections of this work address the operation of the Sun, the status of its fire and the heat it generates. *On Stones* primarily treats non-fusible stones, earths, and gems rather than metals or common rocks. He considers them composed of earth, produced by some sort of filtering (*diēthēsis*) or conflux (*surroē*), which purifies them, and solidified by fire or some sort of heat. While Aristotle (*Mete.* 3.6 [378a17–27]) claimed that stones are formed by dry "exhalation" (*anathumiasis*), Theophrastos makes little use of this process. In his closely related *On Metals* (lost), Theophrastos apparently claimed that metals (e.g., gold, silver, copper, etc.) come from water without mentioning the vaporous "exhalation" as Aristotle had done (*Mete.* 3.6 [378a27–b4]). In his lost *On Waters*, Theophrastos discussed the various qualities and powers of different types of waters (e.g., density, color, taste) which he ascribed to differences in temperature and admixture of earth. He also explained the Nile's annual flood as partly due to compression (*pilēsis*) of rain clouds on mountains. On *Odors*, no isolated treatise but most probably Book 8 of *Plant Explanations*, primarily deals with odors produced by art and design (*kata tekhnēn kai epinoian*), i.e., human intervention. Theophrastos' extant meteorological works, *Meteorology* (lost in Greek; preserved only in Syriac and Arabic translation) and *On Winds*, complement one another in many respects. Theophrastos' *Meteorology* gives causes of thunder, lighting, thunder without lightning, lightning without thunder, thunderbolts, clouds, different types of rains, snow, hail, dew and the like, as well as the Moon's halo and earthquakes. In general, Theophrastos attempts to correlate varieties of phenomena with their different causes. In *On Winds,* Theophrastos discovers multiple causes of winds and adopts no one as the primary cause. While he generally defines wind simply as movement of air (considered inadequate by Aristotle, *Mete.* 1.13 [349a17–32]), he asks whether winds move in order to restore equilibrium in the air. This imbalance is, in part, deductively attributed to the Sun and its heat. The notion of the restoration of air's equilibrium brings up the question of whether Theophrastos admitted the idea of *horror vacui. On* (*Weather*) *Signs*, attributed to Aristotle in manuscripts but to Theophrastos in modern editions, has been conclusively shown not to have been authored by either: *cf.* PSEUDO-THEOPHRASTOS.

Of the three extant physiological works, *On Fatigue* begins with a discussion of seats of fatigue in the blood vessels, tendons, joints, and even bones. He moves to their causes, symptoms, and therapies. Individual cases of fatiguing exertions follow, after which Theophrastos returns to therapies, concluding with remarks on constitutional dispositions toward fatigue. He presents no classification of different types of fatigue; the treatise appears disorderly. He does, however, present his favored explanation of the general cause of fatigue: colliquescence (*suntēxis*), the product of liquefaction of bodily wastes arising due to the motion of bodily parts in exercise or exertion. These fluids are not excreted like other natural bodily wastes (*perittōmata*) but permeate the body and settle in various places, e.g., joints, especially sinewy ones. One symptom upon which he dwells is the feeling of being weighed down by the *suntēxis*, essentially a hydraulic explanation leading to a hydraulic therapy – remove excess fluid and fatigue disappears. In *On Sweat,* rather than answer fundamental questions about the occurrence of sweating, Theophrastos turns his attention to certain qualities of sweat, e.g., saltiness and bad odor. He concludes that saltiness is due to the secretion of unconcocted matter not natural in the body; foul odor is due to imperfect concoction due to bodily condition, age, and eating certain foods. The sweat of young people, he says, smells worse than older people's due to their sexual drive, open pores, and continued bodily change, i.e., their bodies are less stable. A brief excursus on eruptions or ulcers of the skin due to sweating explains that skin sores may result if exercise fails to remove impurities along with sweat,. Various briefly-considered problems associated with sweating follow but are not conclusively answered. *On Dizziness* deals with a sensation involving disequilibrium and lack of coordination between vision and bodily position occurring from rotational movement, looking at moving objects, looking down from elevations. This dizziness can be accompanied by blurred vision and, in extreme instances, unconsciousness. Although Theophrastos recognizes multiple causes for dizziness, he still attempts to settle on one explanation – separation or imbalance of fluid in the head caused by some interference with that fluid's natural condition. Phōtios preserves excerpts from two other physiological works. In *On Paralysis,* Theophrastos explains that interruption of the flow of breath (***pneuma***) by pressure causes paralysis. The breath becomes trapped, triggering cooling and loss of heat in the afflicted area. In *On Fainting,* he considers fainting due more to the effects of hot and cold than breath. It happens from loss of heat occurring for various reasons: sudden cooling of the body can occur with excessive blood loss or when external heat overpowers the body's inner heat.

There is no clear evidence that Theophrastos seriously criticized Aristotle's scientific methods of inquiry, but rather basically accepted his mentor's hierarchical division of nature. He often develops, refines, and improves ideas already present in Aristotle's writings: resolving loose ends, continuing discussions initiated by Aristotle, or clarifying ideas which

Aristotle left implicit. Theophrastos willingly entertained multiple explanations for certain phenomena. For the most part, he raises problems and indicates various difficulties rather than offering a systematic theory. He adheres to a uniform physical system in which he emphasizes the hot, active element versus the three passive ones, putting the Sun in the center of activity as the pre-eminent heat. Theophrastos does not wholly oppose teleological explanations of all natural phenomena, but rejects them in several instances, e.g., ocean tides, droughts, male breasts, beards (*cf. Metaph.* 10a28–b16), and domestication of plants, which presents a conflict with their natural goals (e.g., *Plant Explanations* 1.16). Theophrastos maintains Aristotle's doctrine of the eternity of the universe, which depends on reciprocity of the four elements as they change into one another. Theophrastos nowhere mentions the Aristotelian notion that the heavens are composed of a fifth element, ***aithēr***, but he does consider the heavens ensouled yet self-moving, rather than moving through their longing for a transcendent Unmoved Mover. While he may have allowed supra-sensible principles, Theophrastos emphasized the limitations of human understanding and the need to start from what is accessible to us, which of course are the phenomena of the natural world to which he devoted so much of his intellectual energy.

Ed.: P. Steinmetz, *Die Physik des Theophrast* (1964); D. Eichholz, *On Stones* (1965); V. Coutant, *On Fire* (1971); V. Coutant and V. Eichenlaub, *On Winds* (1975); B. Einarson and G. Link, *Plant Explanations* (1976–1990); S. Amigues, *Researches on Plants* (1988–2003); H. Daiber, *Meteorology* (*RUSCH* 1992); Testimonia and fragments in W.W. Fortenbaugh *et al.*, *Theophrastus of Eresus* (1992); U. Eigler and G. Wöhrle, *On Odors* (1993); R.W. Sharples, *On Fish* (*RUSCH* 1992); A. Laks and G. Most, *Metaphysics* (1993); R.W. Sharples, *Theophrastus of Eresus*, Commentary v. 3.1: Biology (1998) and Commentary v. 5: *Physics* (1995); W.W. Fortenbaugh, R.W. Sharples, and Michael G. Sollenberger, *On Sweat, On Dizziness, On Fatigue* (2003).

G. Stratton, *Theophrastus and the Greek Physiological Psychology before Aristotle* [*On Sense-Perceptions*] (1917); A. Hort, *On Weather Signs* (1926); *RE* S.7 (1940) 1354–1562 (#3), O. Regenbogen; *OCD3* 1504–1505, R.W. Sharples.

Michael G. Sollenberger

Theophrastos, pseudo (330 – 300 BCE?)

Wrote *On Signs of rains, winds, storms, and clement weathers*, addressing each of these topics, mostly for predicting imminent weather within the seasonal patterns. Few of the signs are astronomical, although a miscellany of irregular astronomical signs (e.g., comets) at the end predict seasonal variations. Apparently a **Peripatetic** treatise and seemingly used by ARATOS, it could be a later compilation, but must be no earlier than 430 BCE, as it mentions METŌN's calendar, and probably much later, as it mentions Hermēs' Star (Mercury).

Ed.: D. Sider and C.W. Brunschön, *On Weather Signs* = *Philosophia Antiqua* 104 (2007).

P. Cronin, "The Authorship and Soures of *Peri sēmeiōn* Ascribed to Theophrastus," in W.W. Fortenbaugh and D. Gutas, *Theophrastus: His Psychological, Doxographical, and Scientific Writings* (1992) 307–345; D. Sider, "On *On Signs*," in W.W. Fortenbaugh and G. Wöhrle, *On the Opuscula of Theophrastus* (2002) 99–111.

Henry Mendell

Theophulaktos Simokattēs (610 – 645? CE)

Theophulaktos the "snub-nosed cat" (perhaps a physically descriptive epithet) was a "sophist" (*Souda* Theta-201; Sigma-435) and Egyptian civil servant (*Hist.* 7.16.10). Probably educated in rhetoric at Alexandria (Whitby 1986: XIII), he moved to Constantinople to study law before 610 (*Hist.* 8.12.3–7; perhaps the judge attested in an inscription from Aphrodisias: Grégoire #247). He wrote four works: *History, Problems of Natural History, Ethical Epistles,* and *On Predestined Terms of Life* (favoring a synergism between predestination and random fate). His eight-book moralizing "world history" continues Menander "Protector" and treats the reign of Maurice (582–602), whom Theophulaktos eulogized *ca* 610; he described Maurice's Persian and Balkan wars, and included modest ethnographical and geographical discussion of peoples from the Balkans to China: especially peoples along the Ister, their cities, military histories, rivalries, and strength (*Hist.* 1.3.1–7.6). His *Problems of Natural History* is cast in the form of a Platonic dialogue, wherein the fictional characters Antisthenēs and Polukratēs assess various explanations of 19 ***paradoxa***: e.g., why iron does not burn, why elephants stir water before drinking, why olive oil calms the ocean, why vultures gestate for three years, why goat's blood softens steel, why ravens do not drink in summer, why the frogs of Seriphos are mute. Many of these wonders are noted in AELIANUS from whom Theophulaktos drew deeply throughout his *oeuvre.* His description of Tempe (*Hist.* 2.11.4–8) relies on Ael. *Var. Hist.* 3.1 and examples in his *Ethical Epistles* derive from Ael. *Nat. An.* (Pignani). He lists some 18 "predecessors" (including Aelianus), from canonical authors (ARISTOTLE, DĒMOKRITOS, GALĒN, PLATO, PLUTARCH, and THEOPHRASTOS), to **neo-Platonists** (DAMASKIOS, IAMBLIKHOS, PLŌTINOS, PROKLOS), and notably the paradoxographers BŌLOS, HABRŌN, IMBRASIOS, and SŌTIŌN, plus the geographer TIMAGENĒS. His "Hieroklēs" is almost certainly HIEROKLĒS OF ALEXANDRIA. His "Alexander" is likely to be ALEXANDER OF MUNDOS or ALEXANDER OF MILĒTOS (read as a paradoxographer), although ALEXANDER OF APHRODISIAS cannot be ruled out.

Ed.: Ideler 1 (1841) 168–183; L. Massa Positano, *Simocatta, Theophylactus: Questioni naturali* (1965).

H. Grégoire, *Recueil des inscriptions grecques-chrétiennes d'Asie mineure* (1922); *KP* 5.725–726, H. Gärtner; A. Pignani, "Strutture compositive delle epistole 'morale' di Teofilatto Simocata," *Univ. di Napoli, Annali Fac. lett. e filos.* 22 (ns10) (1979–1980) 51–59; Michael and Mary Whitby, trans., *The History of Theophylact Simocatta* (1986); Michael Whitby, *The Emperor Maurice and his historian: Theophylact Simocatta on Persian and Balkan warfare* (1988); *ODB* 1900–1901, B. Baldwin; *PLRE* 3 (1992) 1311.

GLIM

Theopompos (120 BCE – 300 CE)

AËTIOS OF AMIDA 16.122 (Zervos 1901: 171) records his uterine fumigation, employing ***storax, kostos***, **mastic**, roses, etc. Diels 2 (1907) 106 records a Bologna MS, 1808 (15th c.), f.32^{V}, containing extracts from Theopompos. The name is unattested after 300 CE (*LGPN*). *Cf.* ROMULA.

Fabricius (1726) 435.

PTK

Theosebeia (*ca* 250 – 300 CE)

Alchemist and correspondent or even "sister" (*Souda* Z-168) of ZŌSIMOS OF PANŌPOLIS. Although none of her letters to Zōsimos (Mertens 1995, §1.19) survives, she practiced alchemy as part of a coterie with whom Zōsimos sometimes worked (Mertens 1995, §8.1). At some point she joined a group of alchemists including NEILOS and "the virgin Taphnoutia" (*CAAG* 2.190). The fact that Zōsimos addresses Theosebeia as "purple-robed lady," (*CAAG* 2.246), implies that she was of patrician if not imperial status, while a Latin term rarely used in Greek literature but applied to a member of her entourage suggests that she may have been of Roman lineage (Mertens 1995: §8.3 and note *ad loc.*).

(*)

Bink Hallum

Theosebios (100 – 300 CE?)

A 15th c. MS (Bologna, Biblioteca Universitaria, 3632) contains a five-ingredient formula (opium, myrrh, castoreum, *hupokistos* juice, and ***storax*** mixed with wine) to treat intestinal ailments. The formula's simple design reflects the early stages of compound medicines (1st–3rd cc. CE), although Theosebios does not seem to have been mentioned by (the sources of) GALĒN. The name appears Christian which perhaps renders the 3rd c. most likely.

Diels 2 (1907) 106.

Alain Touwaide

Theotropos (65 – 90 CE)

ASKLĒPIADĒS PHARM., in GALĒN *CMGen* 5.14 (13.852 K.), repeats from AREIOS OF TARSOS a medication given by Theotropos for ulcers of several kinds, containing yellow orpiment, **litharge**, etc. The name seems otherwise unattested (Pape-Benseler; *LGPN*).

Fabricius (1726) 435.

PTK

Theoxenos (300 BCE – 25 CE)

CELSUS 5.18.34 records his remedy for gout: smear the foot with kidney suet and salt, sheath it, and pour on a vinegar solution. *Cf.* ERASISTRATOS OF SIKUŌN.

Fabricius (1726) 435, s.v. Theosenus.

PTK

THESPIS ⇒ PHILŌNIDĒS OF LAODIKEIA

Thessalos of Kōs (*ca* 420 – 350 BCE)

Physician like his father HIPPOKRATĒS and brother DRAKŌN. The sons of both brothers were each called Hippokratēs (GALĒN, *In Hipp. De natura hominis* 2.1, *CMG* 5.9.1, p. 58) and, like their fathers and grandfather, were physicians, as was Hippokratēs' son-in-law POLUBOS. Inscriptional evidence suggests the continuing family tradition in medicine at Kōs (Benedum; Sherwin-White).

Thessalos, who may have worked at the Macedonian court (*Embassy*, 9.418, 428 Littré), contributed to the HIPPOKRATIC CORPUS, editing and publishing EPIDEMICS II, IV and VI from notes made by his father (Galēn, *In Hipp. Epid. VI* [*CMG* 5.10.2.2, pp. 13, 76, 156, 272]). Several Hippocratic works in antiquity were assigned alternately to Hippokratēs or Thessalos (Galēn, first reference above). For example, *On Nutriment* was assigned to Hippokratēs, Thessalos or even to HĒROPHILOS (Galēn, *De septimestri partu* 2 [Walzer]; *Schol. M* in Hippokratēs *On Nutriment*, MS *Marcianus graecus* 269 (11th c.) [*CMG* 1.1, p. 79]).

Thessalos and Drakōn in antiquity were held to be the founders of the **Dogmatists**, who believed that they should investigate not only the obvious, but also underlying and hidden causes of disease. In a letter concerning mathematics, supposedly from Hippokratēs to Thessalos (but generally believed to date much later), Hippokratēs explains the importance of geometry so that the physician can know better the location of the bones in the body, to correct them when they are twisted; he also tells him how to calculate changes in fevers (9.392 Littré). Thessalos is the supposed author of the *Embassy* (9.404–428 Littré), although it is clearly not by him, but a later fiction. See also SŌRANOS, *Vita Hippocratis* 15 (*CMG* 4, pp. 175–178); *Souda* I-564 ("Hippokratēs"); Tzetzēs, *Historiarum variarum chiliades* 7.968–973.

R. Walzer, "Über die Siebenmonatkinder," *Rivista di studi orientali* 15 (1935) 323–357 at 345; *RE* 6A.1 (1936) 165–168 (#5), H. Diller; A. Nikitas, *Untersuchungen zu den Epidemienbuchern II IV VI des Corpus Hippocraticum* (Diss. Hamburg, 1968); J. Benedum, "Griechische Artzinschriften aus Kos," *ZPE* 25 (1977) 265–276 at 272–274; Sherwin-White (1978) 262, 278; von Staden (1989) T16a–b, 36a–b; W.D. Smith, *Hippocrates. Pseudepigraphic Writings* (1990) 2, 4–5, 10, 39–40, 101, 111; Pinault (1992) 8–9, 11, 18, 19, 22, 25, 37, 39, 48, 75, 83, 85; *NP* 12/1.454–455 (#5), V. Nutton.

Robert Littman

Thessalos of Tralleis (*ca* 20 – 70 CE)

One of the reputed founders of the **Methodist** medico-philosophical sect, a claim buttressed by his letter to Nero: "I have founded a new sect, which is the only true one, as none of the earlier doctors propounded anything advantageous either for the preservation of health or the curing of disease" (GALĒN, *MM* 1.2.1 [10.7–8 K.]; Hankinson 1991: 6). Galēn avers that Thessalos, raised in "the women's quarters, was the son of a lowly wool-carder" (*On Crises* 2.3 [9.657 K. = Alexanderson 1967: 136–137]), but the scattered and often contradictory details regarding Thessalos' "life, times, and doctrines" allow no firm conclusions (Edelstein 1935: 358–363/1967: 173–179; Frede 1982: 15, 23; Vallance 1990: 132; and Tecusan pp. 9–16). That some Thessalos composed a tract addressed to some emperor (probably either Tiberius or Claudius, although one MS has "Germanicus Caesar") on medical astrology (linking pharmaceutically useful plants with planets and zodiac constellations) is attested by the MSS (Boudreaux pp. 134–165; Friedrich pp. 45–273), but scholarly opinion is neatly divided on its authorship: "by" Thessalos (Friedrich; Smith pp. 172–189; Fowden pp. 162–165), "probably" by Thessalos (Diller col. 180–182; Scarborough pp. 155–156), or total forgery (Pingree pp. 83–86). Tecusan "eliminated from the start every possibility that the famous zodiac produced under his name. . .could have been written by Thessalos or any other **Methodist**" (pp. 61–62). One could, however, easily associate medical astrology with the simple and simplistic theories espoused by Thessalos.

Galēn repeatedly excoriates for Thessalos "frivolity" (e.g. *Crises* 2.3, above). Thessalos' low status, recent date, and medical daring merits Galēn's deepest scorn: ". . .Thessalos not only especially cultivated the wealthy in Rome, but also promised to teach the art in six months, and thus readily attracted many pupils. For if those who wish to become doctors have no need of geometry, astronomy, logic, or music, or any of the other noble disciplines, as our fine friend Thessalos promised, and they do not even require long experience and familiarity with subject-matter of the art in question, then the way is clear to anyone who wants to become a doctor without any expenditure or effort. . ." (*MM* 1.1.5 [10.4–5 K.], Hankinson 1991: 4–5; *cf.* Diller col. 169).

Galēn's vitriol obscures much of Thessalos' doctrine, which derives ultimately from a rather modified **Epicureanism** (*cf.* entries ASKLĒPIADĒS, THEMISŌN, and later SŌRANOS). Grouping diseases and therapies into "communities" (*koinōtetes*) allowed distinguishing "acute" from "chronic" illnesses, thereby also giving superficial precision to recommended treatments (e.g., Tecusan *frr.*67 [surgery/bleeding] and 180 [chronic wounds]), as well as circumscribing the ailments (*frr.*46 [ileus], 54 [epilepsy], 62 [paralysis], 65 [excessive "flowings"], 70–73 [hemorrhagia], 85 [**dropsy**], 95 [gout/podagra], 146 [fevers], and 310 [uterine prolapse]), which in turn suggested applicable drugs and compounds. Galēn and CAELIUS AURELIANUS ascribe about nine titles to Thessalos: *The Canon, Commentary on Hippocrates' Aphorisms, The Communities, Letter to Nero, On Medicines, The Method, Regimen, Surgery*, and *Syncritics* (Tecusan pp. 107–108).

Ed.: P. Boudreaux in *CCAG* 8.3 (1912) 132–165; H.-V. Friedrich, *Thessalos von Tralles. griechisch und lateinisch* (1968); Tecusan (2004) "Thematic Synopsis: Thessalus," pp. 84–85, 91–92, 98, and 103.

Edelstein (1935/1967); *RE* 6A.1 (1936) 168–182, H. Diller; D.E. Pingree, "Thessalus Astrologus," *CTC* 3 (1976) 83–86; J.Z. Smith, "The Temple and the Magician," in *Map is Not Territory: Studies in the History of Religion* (1978) 172–189; Frede (1982); G. Fowden, *The Egyptian Hermes* (1986); John Scarborough, "The Pharmacology of Sacred Plants, Herbs, and Roots," in C.A. Faraone and D. Obbink, edd., *Magika Hiera: Ancient Greek Magic and Religion* (1991) 138–174.

John Scarborough

Theudās "Sarkophagos" (30 BCE – 80 CE)

Called "flesh-eater" or "coffin" (*sarkophagos*), and quoted approvingly, by ANDROMAKHOS, in GALĒN *CMGen* 6.14 (13.925–926 K.), for a plaster used on cancers, fistulas, etc., and containing **terebinth**, **litharge** (for which "some" substitute orpiment), copper flakes, **verdigris**, and frankincense, in beeswax. The more common form of the name (*LGPN* 1.222, 2.224, 3A.208, 4.168); contrast Theodās (*LGPN* 1.213, 4.162). Prior to THEODĀS OF LAODIKEIA; and hardly the same as THEODOS OF ALEXANDRIA; but *cf.* SARKEUTHITĒS.

(*)

PTK

Theudios of Magnesia (365 – 325 BCE)

Mentioned by PROKLOS after AMUKLAS, MENAIKHMOS, and DEINOSTRATOS and before ATHĒNAIOS OF KUZIKOS, all of whom worked together in PLATO's **Academy**, he excelled in mathematics and other parts of philosophy, and put the elements of geometry in good order, generalizing many results (*In Eucl.* p. 67.8–21 Fr.).

RE 6A.1 (1936) 244–246, K. von Fritz.

Ian Mueller

THOUKUDIDĒS OF ATHENS ⇒ THUCYDIDĒS

Thrasualkēs of Thasos (well before 350 BCE)

Held that there are only two winds, the North and the South (STRABŌN 1.2.21), and was ARISTOTLE's source for the view that flooding of the Nile was caused by summer rains in the far south (*ibid.* 17.1.5: discussed by Aristotle in the fragments of *On the Rising of the Nile*; *cf.* ARISTOTELIAN CORPUS ON THE FLOOD OF THE NILE).

Ed.: V. Rose, *Aristotelis Fragmenta* (1886), *frr.*246–248; DK 35.

Henry Mendell

Thrasuandros (300 – 30 BCE)

Aëtios of Amida mentions his pill for dysentery: 9.35, 42 (Zervos 1911: 363, 385). The rare name is unattested after the 1st c. BCE: *LGPN* 1.226.

Fabricius (1726) 437.

PTK

Thrasuas (350 – 280 BCE)

Theophrastos, *HP* 9.17.2, records his theory that poisons can become tolerated and mastered; he cited evidence that the same stuff was poisonous to some but not others, and made "clever" distinctions among constitutions. Theophrastos proceeds to relate stories about Eudēmos of Athens and Eunomos of Khios.

Fabricius (1726) 437.

PTK

Thrasubulus (220 – 470 CE)

Writer on astrology used by Anthedius (Sidonius Apollinaris *Ep.* 8.11.10). He is, possibly, the same astrologer Thrasubulus who advised Seuerus Alexander (*SHA Alex. Sev.* 62.2).

RE 6A.1 (1936) 577 (#11) A. Stein.

GLIM

Thrasudaios (*ca* 250 – 200 BCE)

Cited by Apollōnios of Pergē (*Kon.* 4.*pr.*) as the addressee of Konōn of Samos' work on conic sections. The rare name is mostly Doric: *LGPN*.

RE 6A.1 (1936) 577 (#3), K. Ziegler; Netz (1997) #119.

GLIM

Thrasullos, Ti. Claudius (of Mendēs?) (4 – 36 CE)

A polymath and scholar, who became the emperor Tiberius' astrologer, and best known for his tetralogical arrangements of the works of Plato and Dēmokritos, the former surviving in the manuscript tradition. Material on the Platonic corpus at Diogenēs Laërtios 3.47–66 (= T22) follows as introduction to the reading of Plato in the Thrasullan tradition. His work on the corpus involved interpretation, but surviving fragments clearly mark him as much more than a scholar. Though never explicitly referred to as a **Pythagorean**, it is agreed that he leans in this direction, and Porphurios' *Life of Plōtinos* 20–21 (= T19a-b), following Longinus, includes him in a list of Pythagorizers who treated the first principles of Pythagoras and Plato. A passage in Porphurios' *Commentary on Ptolemy's Harmonics* p. 12 (= T23) speaks of a *logos*, involving analogical relations and embracing all physical reality, which is imitated by human reasoning, informs matter, and is employed cognitively by the Universal Leader-God. Thrasullos is said to have called this "the logos of the forms" and sees its influence as penetrating to all levels. The information may have come from *On the Heptachord*, a harmonic writing of Thrasullos, cited by Porphurios (p. 91 = T15a, p. 96 = T15b), which included the octave along with the fourth and fifth as harmonic intervals and defined such terms as interval and harmony. THEŌN OF SMURNA (*Exposition of Plato's Mathematics*) preserves material offering definitions of such terms as "enharmonic" or "harmony," and of "symphonic" and "diaphonic" intervals, and explaining the differences between arithmetic, geometric, and harmonic rations (T13, 14a). Thrasullos also discussed astronomy in a mathematical context, tackling the size of the Sun, as well as influencing the astrological tradition in various ways (T24–28). PSEUDO-PLUTARCH, ON RIVERS, mentions a Thrasullos of Mendēs writing works *On Stones, Thrakian Matters*, and *Egyptian Matters*.

Barker (1989) 2.209–213, 226–229; Harold Tarrant, *Thrasyllan Platonism* (1993), with fragments.

Harold Tarrant

Thrasumakhos of Sardēs (430 – 330 BCE)

Greek physician known only from the doxography of the LONDINIENSIS MEDICUS (11.42–12.8): he attributes the origin of diseases to blood, considered a residue of food. Blood, modified through excessive heat or cold, produces bile, phlegm or sepsis, pathological **humors**. KRITIAS, according to ARISTOTLE, *de Anima* 1.2 (405b6), thought that soul is blood and Thrasumakhos' theory can well represent one of those against which the HIPPOKRATIC NATURE OF MAN polemizes, wherein bile and phlegm are **humors** parallel to blood and not produced by it: *cf.* PLATO *Timaios* 82e–83a.

K. Fredrich, *Hippokratische Untersuchungen* (1899) 27, n.1; M.P. Duminil, *Le sang, les vaisseaux, le coeur dans la collection hippocratique* (1983) 251–252.

Daniela Manetti

Threptos (100 BCE – 90 CE)

ASKLĒPIADĒS PHARM., in GALĒN *CMGen* 5.11 (13.828 K.), records his remedy for a wide variety of ulcers: alum, ***khalkanthon***, aloes, birthwort, frankincense, myrrh, oak-gall, and pomegranate peel. The name is unattested before the 1st c. BCE: *LGPN* 2.230, 3A.213.

Fabricius (1726) 439.

PTK

Thucydidēs of Athens (430 – 400 BCE)

Thoukudidēs, historian, son of Oloros, Athenian citizen, served as general in the war against Sparta in 424 BCE. He was exiled for failing to stop Brasidas from taking Amphipolis and spent the rest of the war gathering information and writing an account of it. His *History of the Peloponnesian War* ends abruptly in 411 but shows signs of work after 404.

Although not explicitly concerned with geographical questions, he understood better than any other ancient historian the importance of geography in interstate relations. In his introduction to the Sicilian Expedition, he himself notes, and endeavors to correct, the general ignorance of Athenians about the geography of Sicily. His descriptions of sites of various conflicts in the war are detailed and, where evaluation is possible, highly reliable. In several instances, such as the campaigns at Pulos, Amphipolis and Surakousai (Syracuse), even if he did not witness the events, he may very well have visited, studied the sites and interviewed eyewitnesses extensively. Elsewhere his brief geographical descriptions of the

Thucydidēs of Athens © Holkham Hall

various theaters of war around the Aegean suggest the language of ***periploi*** and ***periodoi*** *gēs*. He shows interest in the origins of place names, historical geography, and topographical detail, even when not relevant to his narrative.

Thucydidēs shows the influence of Hippokratēs in his description of the plague at Athens in 430 (2.47.3–54.5), which he himself contracted, and for which he provides an epidemiology and detailed prognosis. Thucydidēs' description of symptoms suggests adherence to contemporary medical doctrine rather than exact observation, while his language is not technical in the manner of medical writers, leaving key details ambiguous. The possibility that the disease has altered in its course, symptoms and virulence over the intervening millennia must also be considered. This has led to extensive debate about the nature of the disease: many candidates have been proposed, including typhus, influenza with toxic shock syndrome, and smallpox. Recent DNA analysis of remains found in mass burial pits in the Kerameikos cemetery points to typhoid fever as the cause of the plague. Although doubts about the accuracy of Thucydidēs' description remain, his recognition of the corrosive effects of epidemics on social order, and the long-term implications for political and military affairs, is unparalleled among ancient historians.

L. Pearson "Thucydides and the Geographical Tradition," *CQ* 33 (1939) 49–54; F. Sieveking, "Die Funktion geographischer Mitteilungen im Geschichtswerk des Thukydides," *Klio* 42 (1964) 73–179; J. Scarborough, "Thucydides, Greek medicine and the plague at Athens. A summary of possibilities," *Episteme* 4 (1970) 77–90; T.E. Morgan, "Plague or poetry?" *TAPA* 124 (1994) 197–207; *OCD3* 1516–1521, H.T. Wade-Gery *et al.*; M.J. Papagrigorakisa *et al.*, "DNA examination of ancient dental pulp incriminates typhoid fever as a probable cause of the Plague of Athens," *International Journal of Infectious Diseases* 10 (2006) 206–214.

Philip Kaplan

Thumaridas (of Paros?) (400 BCE – 200 CE)

In his *Life of Pythagoras* (33.239.7–240.2; *cf.* 36.267.34–35, where "Eumaridas" is listed as a well-known **Pythagorean** from Paros), Iamblikhos gives, as an illustration of friendship, a story about a man collecting money and sailing to Paros to give it to the Parian **Pythagorean** Thumaridas, who had fallen into poverty. Elsewhere (28.145.4–5) Iamblikhos mentions a **Pythagorean** Thumaridas from Tarentum, and he lists (28.104.7) some Thumaridas as a pupil of Pythagoras himself. In his commentary on the *Introduction to Arithmetic* of Nikomakhos, Iamblikhos ascribes to a Thumaridas the definition of the arithmetical unit as limiting quantity (*perainousa posotēs*: 11.2–3) and a characterization of prime numbers as rectilinear (*euthugrammikos*, 27.4), i.e., perhaps, not representable as rectangular arrays of points, but only as straight lines. Most strikingly Iamblikhos (62.18–63.2) attributes to Thumaridas his "Bloom" (*epanthēma*), which formulates in complex prose the idea that if $x + y_1 + \ldots + y_{n-1} = a$ and $x + y_i = b_i$, then:

$$x = \frac{b_1 + \ldots + b_{n-1} - a}{n - 2}.$$

Iamblikhos then demonstrates how other equations can be reduced to it. In most recent accounts, Thumaridas is assigned to Paros, and said to be PLATO's contemporary or perhaps even earlier. But what is said about Thumaridas in the *Life of Pythagoras* is legendary tradition rather than history, and some scholars assign the mathematical ideas associated with him in the commentary on Nikomakhos to the common era.

Heath (1921) 1.94–96; I. Bulmer-Thomas, *Selections Illustrating the History of Greek Mathematics* (Loeb 1939; rev., 1981) 1.139–141; DK, 1.447, n. on-line 3; Burkert (1972) 442, n. 9.

Ian Mueller

Tiberianus (*ca* 300 – 330 CE)

Possibly a prefect of Rome in 303–304, wrote poems, including a prayer to the **Platonic Demiurge** (*carmen* 4), following **Platonic** doctrines and influencing BOËTHIUS (*Consolatio* 3, *carmen* 9.22). Tiberianus prays for knowledge to a divinity who is unique and many in itself, a cause of the world. But he equates it with the whole nature and considers this world a home of both men and gods, reflecting **Stoic** views.

Ed.: S. Mattiacci, *I carmine e di frammenti di Tiberiano* (1990); *FLP* 429–446.
RE 6.A.1 (1936) 766–777, F. Lenz; *PLRE* 1 (1971) 911–912 (#1 and maybe #4); *NP* 12/1, 529, K. Smolak.

Peter Lautner

Tiberius (*ca* 150 CE – *ca* 500 CE?)

Wrote on the medical treatment of horses and cows. Tiberius' name belongs to late antiquity; however, there is no evidence in his text for a precise date. Excerpts are preserved in the 10th c. B recension, the L recension, and the RV recension of the *Hippiatrika*. Tiberius is related to the agricultural writers: his text contains parallels with VINDONIUS ANATOLIOS and IULIUS AFRICANUS. No treatments for cows appear in B, but a list of them is appended to L, and some appear anonymously in RV.

CHG vv.1–2; G. Björck, "Le *Parisinus grec* 2244 et l'art vétérinaire grec," *REG* 48 (1935) 505–524 at 513–515; *Idem* (1944) 16–17; McCabe (2007).

Anne McCabe

Timagenēs of Alexandria (*ca* 75 – *ca* 25 BCE)

Greek historian and rhetorician, apparently impulsive, witty and sharp, son of a royal moneychanger. Timagenēs arrived in Rome in 55 BCE as A. Gabinius' prisoner. Sulla's son Faustus liberated him. Seneca reports that "from captive he became a cook, from cook a chair-carrier, from chair-carrier a friend of AUGUSTUS" (Sen. Sr., *Contr.* 10.5.22). AUGUSTUS, angry over Timagenēs' remarks about the emperor and his family, banished Timagenēs from his house; in response Timagenēs burned parts of his histories relating to the emperor's deeds (Sen. *ibid.*). Timagenēs then moved to live with the Roman historian C. Asinius Pollio. Later he traveled a bit and died in Albania, having written many books. Solid information exists only regarding a work on kings and a universal history; the *Souda* (T-589) ascribes to him also a ***periplous*** of the sea in five books. Euagoras of Lindos composed a now lost biography.

Ed.: *FGrHist* 88.
M. Sordi, "Timagene di Alessandria: uno storico ellenocentrico e filobarbaro," *ANRW* 2.30.1 (1982) 775–797.

Daniela Dueck

Timagētos (*ca* 400 – 350 BCE?)

Wrote a ***periplous*** entitled *On Harbors*, preserved in the scholia on Apollōnios of Rhodes and in STEPHANOS OF BUZANTION. He described a river rising among the Celts, flowing into a lake (probably Lake Geneva), and thence bifurcating into the Rhône and the Istros (Danube). For the name *cf. LGPN* 3A.427, of Argos (3rd c. BCE).

NP 12/1.573–574, H.A. Gärtner.

PTK

Timagoras (*ca* 200 – 100 BCE)

Epicurean philosopher who disagreed with some of the teachings of the school, especially on the topic of sense perception (CICERO, *Lucullus* 80). If, as seems likely, he is identical to the **Epicurean** whose name is given as Timasagoras in PHILODĒMOS' *On Anger*, he and another Epicurean Nikasikratēs maintained, against the more orthodox **Epicurean** view, that anger was to be completely avoided in all its forms.

RE 6A.1 (1936) 1073–1074 (#5) – *cf.* Timasagoras: 17.1 (1936) 281–283 (s.v. Nikasikratēs), R. Philippson; *NP* 12/1.582 (Timasagoras), T. Dorandi.

Walter G. Englert

Timaios (Astrol.) (75 BCE – 79 CE)

Astrological doctrines are ascribed to Timaios by VETTIUS VALENS, commenting on the obscure vocabulary (*Anthologiai* 9.1), and in isolated chapters of the great Byzantine astrological anthologies. That Timaios discussed less conventional topics in his lost works is suggested by PLINY's references to "Timaeus mathematicus" as an authority on the influence of Scorpio causing leaves to fall off trees in autumn, on the causes of the Nile flood, and on the limits of Venus' elongation from the Sun (2.38, 5.55, and 16.82).

RE 6A.1 (1936) 1228 (#9), W. Kroll.

Alexander Jones

Timaios (Pharm.) (250 BCE – 25 CE)

Wrote *Mineral Drugs*, cited as a foreign authority on metals after IOUBA and before HĒRAKLEIDĒS (PLINY 1.*ind.*33). CELSUS preserves his remedy for a burning sensation of the skin (*ignis sacer*) and *cancer*, compounded of myrrh, frankincense, ***khalkanthon***, realgar, orpiment, copper scales, oak galls and roasted ***psimuthion***, applied dry or with honey (5.22.6).

Fabricius (1726) 438.

GLIM

Timaios of Lokris, pseudo (100 BCE – 100 CE)

The Timaios of PLATO's eponymous dialogue has been credited with an apocryphal tract in Doric prose (*On the Nature of the Soul and of the World*), an epitome of the Platonic dialogue which, for the most part, merely reiterates the Plato's content but also was deeply influenced by middle-**Platonic** doctrines. A two-principle theory, which sees mind (*nous*) and necessity (*anankē*) as causes of the universe, is combined with a three-principles doctrine which reproduces Aristotelian hylomorphism: the imposition of form on passive ("female") matter, thereby producing perceptible things. The universe, which is one, perfect, spherical and endowed with soul, was molded by the **Demiurge**, who reduced it to order by imprinting a definite form onto an undefined matter. By attuning the world soul according to harmonic ratios, the author, unlike in Plato's *Timaios* (35b–36d), starts from the number 384 to avoid fractions. The author explains the origin of the elements by a reduction to geometric figures and, in addressing physiological questions, also follows the Platonic model.

Ed.: W. Marg, *Timaeus Locrus, De natura mundi et animae* (1972); T.H. Tobin, *Timaios of Locri: On the nature of the World and the Soul* (1985).
M. Baltes, *Timaios Lokros: Über die Natur des Kosmos und der Seele* (1972); K.S. Guthrie, *The Pythagorean Sourcebook and Library* (1987) 287–296.

Bruno Centrone

Timaios of Tauromenion (*ca* 335 – 260 BCE)

Author of three works: a treatise on Olympic victors, perhaps based on his study of inscriptions in Olympia (ERATOSTHENĒS had also composed an *Olumpionika*); a *History* of events in Italy, Sicily and Libya; a work about PURRHOS OF ĒPEIROS. Timaios' father, Andromakhos, founded Tauromenion as a city of refuge for the people of Naxos when Dionusios I, the tyrant of Surakousai, destroyed their city (403 BCE). Andromakhos continued as their dynast for many years and welcomed the expedition of Timoleōn of Corinth in 345 BCE. Timaios moved to Athens as a very young man (339–329 BCE) and remained there for 50 years because *ca* 315 the Sicilian tyrant Agathoklēs officially banished him. Sometime between 289–279 BCE, during the reign of HIERŌN II, Timaios returned to Sicily (probably to Surakousai), where he died at the age of 96. While at Athens, Timaios studied rhetoric under Philiskos of Milētos, a pupil of Isokratēs, and wrote his historical works. His *History*, probably in 38 books, introduced the system of chronology by Olympiads and devoted special attention to colonies, foundations and peoples. Timaios considered geography an integral part of history, accepted the conventional division of the ***oikoumenē*** into three parts (Asia, Libya and Europe), and was particularly interested in islands. He approved of the work of PUTHEAS OF MASSALIA but did not have the mathematical training to appreciate some themes in geographic theory. Many of Timaios' preserved geographical notices lack a context and are too brief to enable a proper evaluation. According to the *Souda* (T-600) Timaios traveled very little and made only one expedition from Corinth to Surakousai, but POLUBIOS, starting his own work chronologically where Timaios ended (264 BCE), says that Timaios made a special journey to the Lokrians of Greece to get information. Polubios' *Histories* include many attacks and criticisms of Timaios' supposedly childish and illogical approach.

Ed.: *FGrHist* 566.
T.S. Brown, *Timaeus of Tauromenium* (1958); L. Pearson, *The Greek Historians of the West: Timaeus and his*

Predecessors (1987); R. Vattuone, "Timeo di Tauromenio," in R. Vattuone, ed., *Storici greci d'Occidente* (2002) 177–232.

Daniela Dueck

Timaris (325 – 90 BCE)

Allegedly a queen, to whom PLINY 37.178, following MĒTRODŌROS OF SKĒPSIS, attributes a poem devoted to Venus and referring to *panerōs*, presumably a type of amethyst, thought to foster fertility. Susemihl considers her historicity questionable, and the name fictitious, since Mētrodōros makes no reference to the stone or Timaris.

GGLA 1 (1891) 864–865; *RE* 6A.1 (1936) 1239, E. Diehl; *SH* 774.

Eugenio Amato

Timaristos (325 BCE? – 79 CE)

Cited among the sources of PLINY's books: 1.*ind*.21 (on the nature of flowers and garlands); 1.*ind*.22 (on the importance of herbs); 1.*ind*.23 (on medicines deriving from cultivated plants); 1.*ind*.24 (on medicines deriving from wild plants); 1.*ind*.25 (on the nature of spontaneous plants); 1.*ind*.26 (on other medicines divided into genera); 1.*ind*.27 (on other kinds of herbs and on medicines deriving from them). Pliny (21.180), treating a plant named *halicacabus*, says that it was celebrated by Timaristos in a poem.

Fabricius (1726) 438.

Claudio Meliadò

TIRO ⇒ SABINIUS

TITIANUS ⇒ (1) IULIUS; (2) MAĒS

TITUS, IMP. ⇒ FLAUIUS VESPASIANUS

Timokharis (300 – 265 BCE)

Astronomer, active in Alexandria, cited by PTOLEMY for the undated measurement of the declinations of 12 fixed stars (*Alm.* 7.3), for observing some undated lunar eclipses (one datable to 284 Mar 17), as well as the Moon's occultation of four fixed stars during the period from 295 to 283 (*Alm.* 7.3), and its overtaking η Virginis in 272 (*Alm.* 10.4). The first set of measurements may have been part of the same project as those measurements ascribed by Ptolemy to ARISTULLOS, a project with the goal of describing the heavens scientifically (in prose: *cf.* PLUTARCH, *De Pyth.* 18) and, perhaps, constructing a precisely marked celestial globe. Whatever their purpose, they were apparently used by HIPPARKHOS to discover the fact of precession (*cf. Alm.* 7.1). Likewise unknown is why Timokharis observed lunar eclipses, although they were used, according to Ptolemy, by Hipparkhos to quantify the rate of precession (*Alm.* 7.3). The observations of the lunar occultation of four fixed stars are the earliest known dated (as opposed to datable) observations by a Greek. It is difficult to say what the purpose of these observations was. Though there are some parallels between these observations and those recorded in the Babylonian *Astronomical Diaries*, they do not help identify the purpose. Perhaps Timokharis was investigating the length of the sidereal month (the period of the Moon's return to a fixed star). The observation of the Moon's overtaking η Vir concerns a conjunction, not an occultation. But, again, the purpose of this observation is unknown.

B.R. Goldstein and Alan C. Bowen, "On Early Hellenistic Astronomy: Timocharis and the First Callippic Calendar," *Centaurus* 32 (1989) 272–293; Goldstein and Bowen (1991).

Alan C. Bowen

Timokleanos (?) (10 BCE – 365 CE)

OREIBASIOS, *Ecl. Med.* 73.32 (*CMG* 6.2.2, p. 240), records his remedy for paralysis, containing ***euphorbia*** (*cf.* IOUBA), Chian **mastic**, pepper, spikenard, ***sturax***, etc. in beeswax. The name seems otherwise unattested, although names related to Timoklēs are common.

(*)

PTK

Timokratēs (30 BCE – 95 CE)

After ANTONINUS (q.v.) and before ASKLĒPIADĒS (PHARM.). GALĒN records his dentifrice, for gingivitis, loose teeth, etc., compounded by roasting salt, honey, and *perdikias* (probably *Convolvulus arvensis* L., *cf. helxinē* in Galēn *Simples* 6.5.10 [11.874–875 K.]; Durling 1993: 150, 263), in an almost-sealed vessel, until it just fumes, then mixing that with alum, celery seed, Illyrian iris, lanolin, mint, myrrh, ***purethron***, pennyroyal, white and black pepper, and pumice, all dried in the sun and pounded: *CMLoc* 5.5 (12.887 K.), possibly from KRITŌN (Fabricius 1972: 147). ANTULLOS, in PAULOS OF AIGINA 7.24.12 (*CMG* 9.2, p. 400), records his way of preparing pharmaceutical bitumen, "I boil it in olive oil" (*cf.* HEKATŌNUMOS OF KHIOS).

RE 6A.1 (1936) 1271 (#16), K. Deichgräber.

PTK

Timōn (250 BCE – 77 CE)

PLINY, discussing the efficacy of fenugreek against uterine and intestinal complaints, preserves Timōn's recipe for a drink of fenugreek seed with must and water as an emmenagogue (24.187; *cf.* DIOKLĒS in 24.185).

(*)

GLIM

Timōn of Phleious (290 – 240 BCE?)

The foremost exponent of the ideas of Pyrrho of Ēlis (*ca* 360–270 BCE), who inspired the later Pyrrhonist skeptical movement. Pyrrho was renowned for extraordinary tranquility, associated with some form of skeptical stance (the details are controversial). One aspect of this, according to Timōn, was his refusal to trouble himself with scientific inquiry, apparently because such inquiry is pointless and doomed to frustration.

There is reason to believe, however, that Timōn himself did not entirely adhere to this attitude concerning science. Titles attributed to him include *On the Senses* and *Against the Physicists*, the latter suggesting a critical rather than a constructive work, but at least indicating detailed engagement with scientific ideas. Very little is known of the content of these works. Just one sentence survives from *On the Senses*; in the mold of the later Pyrrhonists,

Timōn declines to posit anything about the real nature of things, while accepting their appearances for practical purposes. More specifically scientific interests are suggested for *Against the Physicists*. Timōn is reported as stressing the importance of the question whether anything should be assumed by hypothesis – a notion originally employed in geometry, but later without restriction as to subject-matter; it is a fair guess that Timōn's answer to the question was negative. Also probably from the same work is the claim that no process divisible into temporal parts can take place in an indivisible time; the point of this remark, in the absence of context, is unclear.

Long and Sedley (1987) §§1–3; *SEP* "Timon of Phlius," Richard Bett.

Richard Bett

Timosthenēs of Rhodes (270 – 240 BCE)

Admiral of Ptolemy Philadelphos, sailed west to the Tyrrhenian Sea and east to the lower Red Sea. He wrote *On Harbors*, a ***periplous*** covering Asia, Europe, and Libya, used extensively by ERATOSTHENĒS (see STRABŌN 2.1.40), as well as a *Summary of Distances*, and – according to Strabōn 9.3.10 – the song for the Pythian games. His wind-rose had 12 parts (Strabōn 1.2.21 and AGATHĒMEROS); 40 or so other fragments include citations by Strabōn (3.1.7, 13.2.5, 17.3.6), PLINY (5.47, 6.15, 6.163, 6.183, 6.198), and PTOLEMY, *Geog.* 1.15.

NP 12/1.595 (#2), H.A. Gärtner.

PTK

Timotheos (250 BCE – 100 CE)

Wrote a commentary on ARATOS (*FGrHist* 1026 T19), entirely lost.

(*)

PTK

Timotheos of Gaza (*ca* 490 – 510 CE)

Born *ca* 460, enigmatic grammarian who supposedly had "written in epic meter a book on quadrupeds, and Indian, Arabian, Egyptian and Libyan animals . . . and four books on exotic birds and on reptiles" (*Souda* T-621). Considering the *reliquiae* (a Byzantine epitome of 56 + 10 monographic chapters, and an anthology of 32 fragments preserved in a zoological *Sylloge* attributed to Constantine VII Pophurogennētos; see ARISTOPHANĒS OF BUZANTION), he was a Christian who, in addition to a political memorandum (on the chrysargyron-tax), composed a zoological compilation in mannered and rhythmical prose. Wellmann's argument failed to prove that Timotheos followed a lost book *On Animals* written by the apologist Tatian, but the Syro-Egyptian origin of the book is clear from the text itself. The zoological material was probably geographically dispatched in at least four books, and the work was highly esteemed in Byzantine times, mainly due to originality and stylistic refinement. Timotheos was considered a major zoological writer, listed with AELIANUS, OPPIANUS and LEŌNIDAS (Tzetzēs, *Chiliades*, 4.166). A significant proportion of the animals mentioned are exotic (tiger, bison, giraffe, hyena), but not restricted to land (griffin, seal, . . .). Timotheos possibly treated all macrofauna (including fishes). There are many parallels with Aelianus (whom he surely used), but Timotheos is often richer and more complex. The work offered seemingly complete monographs on animals, mixing anatomical and ethological remarks with various mythological, lexicographical, medical, magical and paradoxographical data. Timotheos, showing vivid interest in special powers of animals, insisted also on the themes of **sympathy** and hybridism.

Ed.: F.S. Bodenheimer and A. Rabinowitz, *Timotheus of Gaza, On animals* (1949).
M. Wellmann, "Timotheos," *Hermes* 62 (1927) 179–204; *RE* 6A.2 (1937) 1339–1341 (#18), A. Steier; *KP* 5.851 (#8), Th. Wolbergs.

Arnaud Zucker

Timotheos of Metapontion (500 – 400 BCE)

Greek physician, who, according to the LONDINIENSIS MEDICUS (8.10–34), supposes that nutrients are distributed throughout the body starting from the head. When pathways are obstructed, the digestive residues that have risen to the head remain blocked and are transformed into an acid and saline liquid, which travels to other parts of the body. Outcomes vary according to situations. When, for example, liquid concentrates in the larynx, sudden death follows. The head may fall prey to diseases, because of excessive heat or cold, or blows to the head. See EURUPHŌN OF KNIDOS, ABAS, and ALKAMENĒS.

Gourevitch (1989) 238–241; *NP* 12/1.596, V. Nutton.

Daniela Manetti

Tlēpolemos (of Dēlos?) (300 – 30 BCE)

PLINY 20.194 records his remedy for quartan fevers: anise-seed and fennel in honey-vinegar. The name is especially Dēlian, and scarce or unattested after *ca* 50 BCE (*LGPN*).

Fabricius (1726) 435 (s.v. Theopolemos).

PTK

Trebius Niger (150 – 130 BCE)

A companion of L. Lucullus (proconsul of *Hispania Baetica* in 150 BCE) who wrote a work on natural history in Latin which included observations made by the author and his general in Spain. One fragment from PLINY (9.89–93) describes at considerable length a giant octopus which harassed the garum-works near the Straits of Gibraltar; others discuss the remora (9.80), the swordfish and the cuttlefish (32.15) and the woodpecker (10.40). Scholarly attempts to date him to a later period founder on the clear evidence of Pliny's text: Trebius is one of the earliest Roman scientific writers.

GRL §495.5; *RE* 6A.2 (1937) 2272 (#5), Fr. Münzer.

Philip Thibodeau

Cn. Tremellius Scrofa (*fl.* 59 BCE)

Roman senator of praetorian rank, one of the 20 land-commissioners mandated by CAESAR's agrarian legislation (59 BCE). Regarded as the foremost authority on agriculture in his day, he offered personal instruction (VARRO, *RR* 2.1.2) and composed an agricultural treatise, noted for its stylistic polish (COLUMELLA 1.1.12). This contained advice on the tending of grape-vines (3.11.8, 3.12.5, PLINY 17.199) and trees (Columella 5.6.2), gave times and directions for sowing seed (2.8.5, 2.10.8), and observed that the soil of freshly-cleared forest-land rapidly declines in fertility (2.1.5). He read and critiqued his predecessors' writings (1.1.6, 3.3.2), and supplemented that reading with experience gained running his wife's farm in the Sabine country, as well as his own estate, located on the slopes of Mt. Vesuvius (Varro 1.15.1).

GRL §202; Speranza (1971) 46–55; P.A. Brunt, "Cn. Tremellius Scrofa the Agronomist," *CR* ns 22 (1972) 304–308; *KP* 5.937 (#4), M. Deißmann-Merten; *NP* 12/1.780 (#3), J. Fündling; *OCD3* 1549, E. Badian.

Philip Thibodeau

Tribonianus of Sidē (540 – 580 CE?)

A jurist distinct from the homonymous contributor to Justinian's code (*Souda*, T-957). This Tribonianus wrote a verse commentary on Ptolemy's *Kanōn*, and several works on astrology, as well as works on poetic diction and on Homer, all lost.

PLRE 3 (1992) 1339–1340.

PTK

Trogus ⇒ Pompeius Trogus

Trophilos **(220 BCE – 420 CE)**

Iōannēs of Stoboi's *Anthology* (4.36.24–28 W.-H.) purports to contain four excerpts about animals from the *Collection of Wonderful Reports* (*Sunagōgē akousmatōn thaumasiōn*) of a certain Trophilos. Elsewhere in the *Anthology* a *bon mot* about the accomplished doctor is attributed to Trophilos (4.36.9 W.-H.), but scholars prefer to change the rare name (not listed in any of *LGPN* vv. 1–5A) to that of the famous physician, Hērophilos. Because the four paradoxographical fragments correspond almost *verbatim* (albeit with omissions) with four paragraphs in the pseudo-Aristotelian De Mirabilibus Auscultationibus (from which there is one more fragment at 4.36.15 W.-H.), a textual corruption has also been suspected in this case, to the effect that the actual quotation from Trophilos – perhaps to be emended to Hērophilos again (so Roeper 569–570) or Pamphilos (so Giannini 131–132) – and the name of Aristotle (as the source-citation of the ensuing excerpt from the *Auscultationes*) would have dropped out of the Stobaios MSS. Whatever the case may be, nothing suggests that Trophilos was the true compiler of the pseudo-Aristotelian treatise, given its longstanding attribution to Aristotle.

Ed.: *PGR* 392–393.

Th. Roeper, "Joannis Stobaei Florilegium," *Philologus* 10 (1855) 569–571; *RE* 18.3 (1949) 1137–1166 (§30, 1161), K. Ziegler; Giannini (1964) 131–132.

Jan Bollansée, Karen Haegemans, and Guido Schepens

Truphōn of Alexandria (220 – 210 BCE)

Helped Illyrian Apollōnia successfully resist the siege of Philip V of Macedon (214 BCE: Livy 24.40), by detecting the besieger's excavations, using resonant vessels (*cf.* Hērodotos 4.100), and flooded the miners with heated water, pitch, sand, and dung (Vitruuius 10.16.9–10).

RE 7A.1 (1939) 745–746 (#31), H. Riemann.

PTK

Truphōn of Gortun (*ca* 15 BCE – 20 CE)

One of SCRIBONIUS LARGUS' colleagues, acknowledged as mentor or teacher (*praeceptor: Comp.* 175), from Gortun (GALĒN, *CMLoc* 9.2 [13.253 K]). Frequently termed "surgeon" (*Comp.* 201; 203; etc.), Truphōn was famed for his plasters, designed to aid knitting of broken bones – *Comp.* 201 ("Pale Green Plaster . . . effective for skull-fracture," *cf.* Galēn, *CMGen*, 4.13 [13.745 K.]), wounds sustained by gladiators – *Comp.* 203 ("Green Plaster . . . for fresh wounds . . . and for the wounds of gladiators"), as well as for his multiple-ingredient *collyria* – e.g. Galēn, *CMLoc*, 4.8 (12.784 K., "Spherical") – and antidote-plasters deemed effective against animal bites, particularly dogs (*Comp.* 175). Truphōn's plaster for dog bites saw use in the court of an "Augusta" (likely Antonia Minor, Claudius' mother), and Scribonius Largus says that she always had some at hand. Generally Truphōn favored minerals in his compounds (ensuring long shelf-lives), but he employed in a sophisticated way botanicals and animal products, suggested by the formulas for "The Antidote Plaster" and "The Pale Green Plaster." The former combined powdered iris rhizome, beaver castor, wild fig juice, the fat and blood of a black dog, Khian **terebinth** resin, hare's rennet, rock salt, ***silphion*** (either from Libya or Syria), Pontic beeswax, olive oil, and squill-flavored vinegar, all carefully ground in vinegar, mixed and melted into a consistency of honey, then stored in glass containers. "The Pale Green Plaster," good for broken bones (even old ones, all scabbed and corroded), mixed flakes of copper with frankincense, Libyan fennel-gum, Bruttian pine-resin, Khian **terebinth** resin, calf's fat, beeswax, olive oil, and vinegar, blended and heated, and then made into plasters called *magdalia*, viz. (Grk.) "lumps of bread used for wiping the hands at table." The "Green Plaster" for gladiators (*Comp.* 203), made in quantity, was intended to stop bleeding and engender quick healing: combining roasted copper, alum, rock salt, frankincense, **verdigris**, beeswax, resin, and olive oil, one notes not only the long shelf-life (most useful in the arena) but the styptic (alum), bactericidal and antibiotic (frankincense and **verdigris**), all "packaged" in the green of the **verdigris**. Truphōn's recipes were available in both Greek and Latin, and in some instances, Scribonius Largus translates directly into Latin from Greek originals. Like Scribonius, Truphōn was probably bilingual.

RE 7A.1 (1939) 745 (#28), H. Diller.

John Scarborough

Tukhikos of Trapezous (Arm., Tiwkʿikos; 500 – 600 CE)

Mathematician and philosopher known only from his short biography which comprises part of the *Autobiography* of ANANIA OF SHIRAK. He served the Byzantine army in Armenia during the reigns of Tiberius (578–582) and Maurice (582–602). After being wounded, he devoted the rest of his life to study, traveling successively to Antioch, Jerusalem, Alexandria, Rome, and eventually to Constantinople, where he studied with an unnamed "doctor of the city of philosophers." Upon the death of this last teacher, Tukhikos was nominated to succeed him, but declined and returned to his native Trapezous. Anania studied with Tukhikos for eight years and while there he thoroughly mastered mathematics and became learned in many other fields, for Tukhikos possessed many books "secret and esoteric, ecclesiastical and profane, scientific and historical, medical and chronological." Apparently a very learned man in all branches of learning, who taught a large number of students, Tukhikos has left no writings of his own.

F.C. Conybeare, "Ananias of Shirak (A.D. 600–c.650)," *Byzantinische Zeitschrift* 6 (1897) 572–574; A. Abrahamyan, *Anania Širakac'u Matenagrut'yunĕ* [The Works of Anania Širakac'i] (1944) 206–209; H. Berbérian, "Autobiographie d'Anania Širakac'i," *Revue des études arméniennes* 1 (1964) 189–194.

Edward G. Mathews, Jr.

M. Tullius Cicero (80 – 43 BCE)

M. Tullius Cicero Reproduced with permission of the Soprintendenza speciale per il Polo Museale fiorentino

Born 106 BCE; **Academic** philosopher, but with eclectic allegiances including a strong sympathy for **Stoic** ethics. His philosophical dialogues are a seminal source for **Stoicism** generally and for **Stoic** physics and theology in particular. He was instrumental in bringing Greek philosophy into Latin, inventing some of what would become the basic Latin vocabulary for discussing philosophy (most famously coining the words *essentia, qualitas*, and *moralis*, the roots of our essence, quality, and moral).

Cicero's reputation as a philosopher in his own right has fluctuated considerably: he played a prominent role in Early Modern and Enlightenment philosophy and political theory (it is now evident, for example, that Cicero was used as a resource for those who wanted to argue against moral skepticism – like that apparent in Hobbes and Mandeville – and his *De natura deorum* was a model and inspiration for Hume's *Dialogues Concerning Natural Religion*). Nevertheless, in the 20th century, his importance was downplayed, and he has often been mined only as a source for Hellenistic philosophy – unfortunate, as this seriously underestimates the force and originality of Cicero's thinking.

Cicero's trilogy of *De natura deorum, De diuinatione*, and *De fato* represent sophisticated treatments of contemporary theology (particularly **Stoic**), as well as the physics and logic of divination, causation, and free will. What is usually seen as a straightforward rationalist skepticism of superstition in *De diuinatione* can better be read as the insistence on causal (as opposed to indicative) accounts of the relationships between signs and predictions, further confirmed by the emphasis on particular logical questions in the *De fato*.

Cicero's *Dream of Scipio*, together with MACROBIUS' substantial *Commentary*, was an important source for (particularly early) medieval cosmology. Originally written as part of Cicero's *Republic* (corresponding to the myth of Er in PLATO's *Republic*), the *Dream*, cleaved and circulated as a text in its own right during the Middle Ages, describes a dream reported by P. Cornelius Scipio Africanus the Younger in which his dead grandfather, Scipio Africanus the Elder, takes him up through the spheres of the stars to see the structure of the ***Kosmos*** and to hear the music of the heavens. They look down upon the Earth, and the grandfather reflects on the futility of worldly glory. Although the point of the Dream is ultimately ethical, the story made a profound impression on the medieval cosmological imagination, and formed the model for the heavenly journey in Dante's *Paradiso*.

Ed.: Cicero's works are conveniently available in the Loeb Classical Library series, although better Latin texts are available in Teubner and Budé (*CUF*) editions. A.S. Pease, *M. Tvlli Ciceronis De divinatione* (1920–1923), ed. and comm.: highly recommended.

R. Philippson, "Cicero: Philosophische Schriften," *RE* 7A.1 (1939) 1104–1192; T.A. Dorey, ed., *Cicero* (1965); P. MacKendrick, *The Philosophical Books of Cicero* (1989); G. Striker, "Cicero and Greek Philosophy," *HSPh* 97 (1995) 53–61; J. Leonhardt, *Ciceros Kritik der Philosophenschulen* (1999).

Daryn Lehoux

Q. Tullius Cicero (55 – 43 BCE)

Younger brother of CICERO, Quintus Tullius was born 102 BCE, and similarly educated in Athens, but did not equal his brother's genius. He was a good soldier and had a respectable *cursus honorum*. He governed Asia from 61 to 58, served under CAESAR's command in Gaul from 54 to 51 and the next year under Marcus' in Cilicia. He joined Pompey during the Civil War. After Pharsalus, he returned to Rome in 47. In 43 he died with his son, betrayed by his own slaves. Only four short letters of Quintus Tullius survived: one to M. Tullius Cicero, three to M. Tullius Tiro. But he is said to have written four tragedies. We are also able to read a poem which is made up of 20 hexameters and is doubtfully attributed to him; this poem was transmitted by Ausonius (*Ecl.* 25) in order to compare his own poetry; it deals with zodiacal signs, but the astronomical description is poor. The zodiac is only used to illustrate in a poetical way the calendar of seasons.

Ed.: Blänsdorf (1995) 181–183; *FLP* 179–181.
RE 7A.2 (1948) 1286–1306 (#31), F. Münzer.

Christophe Cusset

Turannos (*ca* 100 BCE – *ca* 80 CE)

ANDROMAKHOS in GALĒN, *CMLoc* 9.6 (13.310 K.), preserves his mineral- and beeswax-based ***hedrikē***. The word is first attested as a name in the 1st c. BCE (*LGPN* 3A.437, Pompeii; 4.336, Buzantion). *Cf.* perhaps *CIL* 6.3985, Liuia's slave doctor (Korpela 1987: 176).

Fabricius (1726) 440.

PTK

Turpillianus (*ca* 30 BCE – 90 CE)

ASKLĒPIADĒS PHARM., in GALĒN *CMGen* 4.13 (13.736 K.), records his plaster, "The Philosophers'," based on **litharge** and ***khalkitis***, for the most infected wounds. The non-Republican cognomen (also spelled Turpilienus) is attested in the 1st c. CE: *PIR2* P-315, Petronius Turpillianus (*cos.* 61 CE), *cf.* Schulze (1904/1966) 246.

Fabricius (1726) 440.

PTK

Turranius (50 – 10 BCE)

Wrote a handbook on agriculture in at least two books (Diomēdēs, *GL* 1.368.24). The Latin of the sole fragment has an archaic flavor, and Diomēdēs ranges him with Plautus, CICERO, and CAESAR. If he is to be identified with a known individual, two plausible candidates are TURRANIUS GRACILIS or Turranius Niger, a rancher from Campi Macri in Cisalpine Gaul to whom VARRO dedicated the second book of his *Res Rusticae* (*pr.*6). The author may have introduced the variety of pear known as Turraniana (COLUMELLA, 5.10.18; MACROBIUS, *Sat.* 3.19.6).

Ed.: Speranza (1971) 60–62.
RE 7A.2 (1948) 1442–1443 (#7), W. Kroll, with 1443 (#10), Fr. Münzer.

Philip Thibodeau

Turranius Gracilis (10 BCE – 10 CE)

Wrote one or more geographical or agricultural works on Spain and Africa, cited by PLINY: pillars of Hēraklēs (3.3), monstrous fish at Gadēs (9.11), barley-drink of Andalusia and Africa (18.75).

RE 7A.2 (1948) 1442–1443 (#7), W. Kroll.

PTK

TURTAMOS ⇒ THEOPHRASTOS

TUSCUS ⇒ CLODIUS TUSCUS

U

Ulpianus (*ca* 475 – 500 CE)

Brother of the late Neo-**Platonist** Isidōros of Alexandria (*Souda* O-914 [*Oulpianos*], deriving partly from DAMASKIOS' *Vita Isidori*). His natural talent for solving mathematical problems, noted in particular by Syrianus the younger (= Syrianus 4 in *PLRE* 2 [1980] 1051–1052; *RE* 4A.2 [1932] 1775 [#2], K. Praechter), made him famous at Athens; by contrast, he produced no philosophical arguments of any worth (a commonplace in Damaskios), nor are any titles attributed to him. He died young and never married.

PLRE 2 (1980) 1181.

Alain Bernard

***Ulpianus of Emesa* (*ca* 300 – 330 CE)**

Sophist, born in Askalon, taught rhetoric at Emesa and at Antioch, to LIBANIOS, Proairēsios and Makedonios; his successor in the post was Zēnobios. His own rhetorical works and declamations (*Souda* O-911) are lost. The suggested authorship of the *scholia* to 18 speeches of Dēmosthenēs, with some geographical material, is doubtful (*FGrHist* 676).

RE 9A.1 (1961) 569 (#3), A. Lippold; *KP* 5 (1975) 1044 (#2), H.A. Gärtner; *OCD3* 1570, N.G. Wilson.

Andreas Kuelzer

ULPIUS ⇒ PHOIBOS

URANIUS ⇒ OURANIOS

Urbicius (*ca* 490 – *ca* 520 CE)

Author of *Epitēdeuma* (*Invention*), addressed to Anastasios, as well as *Tactica*, and perhaps an extant *Kunēgetika*. The *Invention* describes two innovations (*cf.* DE REBUS BELLICIS): bundles of spiked poles for rapid construction of an anti-cavalry fence (§4–7), and ballistae mounted upon carts as mobile artillery (§8, 14–16; already attested on Trajan's column).

Ed.: G. Greatrex, H. Elton, and R. Burgess, "Urbicius' *Epitedeuma*: An Edition, Translation and Commentary," *ByzZ* 98 (2005) 35–74.

PTK

URBICUS ⇒ AGENNIUS

V

VALENS ⇒ (1) M. TERENTIUS VALENS; (2) VETTIUS VALENS

VALERIANUS ⇒ CORNELIUS VALERIANUS

M. Valerius Messalla Potitus (45 – 15 BCE)

Suffect consul in 29 BCE, and author of a treatise on gardening (*Kepourika*) known to PLINY (1.*ind.*19). The emended text of Pliny (14.66) makes him the creator of a variety of wine known as Potitana. The *Messalla* of Pliny 14.69, who advertised the health-giving effects of a wine known as Lagarina, will either be this man or M. Valerius Messalla Coruinus, the famous orator and general.

RE 8A.1 (1955) 165–166 (#267), R. Hanslik.

Philip Thibodeau

Valerius Paulinus (*ca* 30 BCE – 90 CE)

ASKLĒPIADĒS PHARM. in GALĒN – *CMLoc* 8.8 (13.211–213 K.) for a liver remedy, and *CMGen* 7.12 (13.1025–1027 K.) for several ***akopa*** based on HALIEUS or FLAUIUS CLEMENS – twice cites this man, whose name and *terminus ante* match the Vespasianic *praefectus Annonae* and *praefectus Aegypti* (*PIR2* P-173). (*Cf.* also the *Paulina* prepared by ARISTARKHOS OF TARSOS.) But the *cognomen* is known from the 1st c. BCE: Paulinos (*ca* 100 BCE: *LGPN* 3A.356), Paulinos (1st c. BCE–1st c. CE: *LGPN* 4.276), Paulina (20 CE: Iosephus, *Ant. Iud.* 18.65–80), Paulina (38 CE: Dio Cassius 59.12.1, 59.23.7), etc.

Fabricius (1726) 440.

PTK

C. Valgius Rufus (45 – 5 BCE)

Roman senator, suffect consul (12 BCE), student of and translator for AUGUSTUS' teacher Apollodōros (Quint. 3.1.18, 3.5.17), friend of Horace who respected his literary judgments (*Sat.* 1.10.82) and consoled him over the loss of a beloved slave *Mystes* (*Carm.* 2.9). Valgius wrote a grammatical treatise (*de rebus per epistulam quaesitam*: Gell. 12.3.1), epigrams, elegies, and a panegyric for Messalla (*FLP* 287–290). PLINY, describing Valgius' erudition, cites him as only the second Latin author, after POMPEIUS LENAEUS, to write on herbal medicine. His unfinished treatise was dedicated to Augustus and prefaced with a prayer that the emperor heal all human evils (25.4–5).

GRL §273–274; *OCD3* 1581, E. Courtney; *NP* 12/1.1118–9 (#2), P.L. Schmidt.

GLIM

Varāhamihira (*ca* 550 CE)

A descendant of Zoroastrian immigrants from Iran to India and a resident of the area near Ujjain, and a prolific writer, whose works cover all aspects of traditional Indian astrology and astronomy. His *Pañcasiddhāntikā* is a summary of five astronomical works current at his time, but now lost: the *Paitāmahasiddhānta*, which expounds astronomy influenced by Mesopotamia (contrast the PAITĀMAHASIDDHĀNTA, the founding text of the Brāhmapakṣa); the *Vasiṣṭhasiddhānta*, the *Pauliśasiddhānta*, the *Romakasiddhānta* and the *Sūryasiddhānta*, which all expound Indian versions of Greco-Babylonian astronomy. The *Pañcasiddhāntikā* is an important work both in shedding light on the Indian astronomical tradition prior to 500 CE, and in recording pre-Ptolemaic Greek astronomy from which the Indian tradition borrowed. Varāhamihira authored three works on divination. The *Bṛhatsaṃhitā* is a large collection of omens in 106 chapters, based on adaptations of Mesopotamian omen series by earlier Indian writers. Two other works on divination, the *Samāsasaṃhitā* and the *Vaṭakaṇikā*, are now lost. On **genethlialogy**, Varāhamihira authored two works, the *Bṛhajjātaka* and the *Laghujātaka*, both based on the Indian adaptation of Greek material in the works of SPHUJIDHVAJA and others. On military astrology ("yātrā"), Varāhamihira composed three works, the *Bṛhadyātrā*, the *Yogayātrā*, and the *Ṭikaṇikāyātrā*, the earliest separate treatises on the topic. Varāhamihira's remaining work, the *Vivāhapaṭala*, deals with astrology applied to marriage.

DSB 13.581–583, D.E. Pingree; *CESS* A.5.563–595.

Kim Plofker and Toke Lindegaard Knudsen

VARRO ⇒ M. TERENTIUS VARRO

Vegetius Renatus (*ca* 445 – 450 CE)

Vir illustris, credited with three technical treatises: a compendium of military warfare (*Epitoma rei militaris*) in four books; a work on horse medicine (*Digesta artis mulomedicinalis*) in three books; a tract on bovine diseases (*De curis boum epitoma*) in one book.

Scholars have long debated the author's dates and name. The year of Gratianus' death (383 CE) provides a secure *terminus post quem*, since Vegetius calls this emperor *diuus* (*Mil.* 1.20.3). A secure *terminus ante quem* is 450 CE, given by the subscription of a corrector named Fl. Eutropius, who worked at Constantinople. The name of the emperor in the inscription varies in the MSS. One passage (4.*pr.*7) seems to allude to a datable historic event, the hurried reconstruction of Constantinople's walls (which had been destroyed by an earthquake) early in 447, due to the Huns having crossed the Danube border. Thus, the *Epitoma rei militaris* was probably written *ca* 447–448 under Theodosius II. Vegetius wrote the *De curis boum* while preparing to write the *Digesta*, for use in combating an epidemic in bovines (*Cur. boum pr.*1–2). Vegetius' statements (*Dig.* 3.6.1) about his travels through the empire suggest that he wrote his veterinary treatises after retiring from public life.

The author's name poses another problem. Authoritative MSS of the veterinary works report the author as *Publii*; those of the *Epitoma* have *Flauii*. Reeve 2004: vii suggests the imperial service of *Publius* entitled him to use the late-antique status-indicator *Flauius*. Although possible, this implies that the *Epitoma* was composed after the veterinary works, contrary to expectation. Additionally, the form *Vegeti* suggests the common nominative *Vegetus* rather than the rare form *Vegetius*.

All three of Vegetius' works are compilations. In the *Epitoma*, the author conflates lost military and strategic sources: at 1.8.10–11 he claims to have used Cato's *De disciplina militari*, Celsus, Frontinus and Tarruntenus Paternus (author *ca* 180 of a treatise on military law of which only two fragments remain, transmitted at *Digesta Imp. Iustiniani* 49.16.7 and 50.6.7: see *RE* 4A.2 [1932] 2405–2407). Vegetius' *Digesta* are almost entirely based on Pelagonius and the Mulomedicina Chironis, the latter elegantly modified by Vegetius. Some passages, whose source is unknown, provide interesting vocabulary and information about the anatomy or breeding of horses. The *De curis boum* almost entirely derives from Columella's *Res rustica* Book 6.

The *Epitoma*, popular in the Middle Ages and later (more than 200 MSS still survive), has been translated into several languages: most famously those of Jean de Meun (1284, old French) and Bono Giamboni (1286, old Italian). The two veterinary works, less widely diffused, survive in about 20 MSS, usually preserving the two treatises together. Nevertheless, Italian versions also exist. Theodericus Borgognoni used the *Digesta* extensively for his *Medela equorum* in the 13th c., and Dino Dini for his 14th c. vernacular work on horse medicine. Also noteworthy is Giovanni Brancati's Italian translation of both works (*ca* 1470).

Ed.: E. Lommatzsch, *P. Vegeti Renati Digestorum artis mulomedicinae libri* (1903) (with *De curis boum* erroneously as the fourth book); M.D. Reeve, *Vegetius, Epitoma rei militaris* (2004).

Vincenzo Ortoleva, *La tradizione manoscritta della «Mulomedicina» di Publio Vegezio Renato* (1996); M.B. Charles, *Vegetius in Context. Establishing the Date of the Epitoma rei Militaris* (2007).

Vincenzo Ortoleva

Velchionius (50 – 30 BCE)

Asklēpiadēs Pharm., in Galēn *Antid.* 2.11 (14.170–171 K.), records that *Belkhionios* said that a recipe of Aelius Gallus was used by Iulius Caesar. Although *ΒΕΛΧΙΟΝΙΟΣ* might conceal a Greek name such as *ΤΕΛΧΙΝΑΙΟΣ* (attested at Kurēnē, 1st c. CE: *LGPN* 1.433), given the Latin context, a name derived from the Etruscan *Velkhi-* seems more likely: Schulze (1904/1966) 99, 377–378.

(*)

PTK

P. Vergilius Maro of Mantua (42 – 19 BCE)

The man who would become arguably the greatest poet of the Latin language was born 70 BCE on a country estate in the village of Andes outside Mantua. Vergil trained as an orator in Milan and Rome and studied there with various prominent scholars and poets, including Parthenios of Nikaia. After moving to Naples he joined an **Epicurean** school led by Sirōn and became acquainted with Philodēmos. During the veteran-resettlement program of 42–40 BCE his family's estate was confiscated, then apparently restored through the intervention of his patron, Asinius Pollio. It was at this time that he began publishing versions

Vergilius Reproduced with permission, Musée national du Bardo

of the *Idylls* of Theokritos and other pastoral poems, which he released as a collection, entitled the *Bucolics*, in 37. Entering the circle of poets patronized by Maecenas, AUGUSTUS' most trusted political advisor, Vergil spent the next eight years of his life working on a didactic poem about agriculture, the *Georgics*; that he read to Augustus in person during the summer of 29. He devoted the rest of his life to the task of creating a Roman equivalent for Homer's *Iliad* and *Odyssey*, the result being a new national epic, the *Aeneid*. The poem had yet to receive its final touches when Vergil died while returning from vacation in Greece in 19.

Of his three poems, the *Georgics* best illustrates Vergil's place in the scientific tradition at Rome. The poem is in four books, the first treating cereal crops and weather signs, the second vines and orchards, the third animal husbandry, and the fourth bees. Its language is that of Latin lyric and epic, and for euphony and vividness it ranks among the most polished works of Latin literature. But aesthetic demands also forced the poet to be selective: typically Vergil will only relate a small set of precepts on a given topic, leaving it to the reader to fill in the rest (*cf.* his short list of wine varieties [2.89–108], or Book 3, which gives instructions for raising cattle, horses, and sheep but not donkeys, mules, or pigs). This compression can sometimes result in statements which are confused or simply incorrect (as often happens in the section on weather signs [1.351–460], and with the apiary lore of Book 4). On the other hand, Vergil's elisions encouraged later scholars to fill in the gaps; in doing so they were contributing to the promulgation of technical knowledge. Entire books were written to expand on or explain brief portions of the poem; *cf.* COLUMELLA bk. 10, and Seruius *ad Georg.* 1.231.

For the most part, Vergil derived his lore from other sources; the *Georgics* draws upon a wide range of authorities, most notably ARISTOTLE (for data on bees and animal sexual behavior), THEOPHRASTOS (for botany), ARATOS (for weather-signs), ERATOSTHENĒS (for an account of climate zones), Bōlos of Mendēs (for the *bugonia*), Lucretius (for methods of inference), and Varro (for much of the practical agricultural knowledge of Books 2–4). There are only a few particular items which, if not original to Vergil, at least receive their earliest mention in his poem. In Book 1 these include an explanation for the beneficial effects of burning stubble that shows the influence of **Epicurean** physics (84–93); advice to sow broad beans in the spring, which, though criticized by Seneca (*Epist.* 86.15), reflects the custom of the area in northern Italy where Vergil grew up (215; *cf.* Pliny 18.120); and an account of the physical causes of weather signs, of particular interest for its suggestion that abnormal animal behavior can be traced to internal perceptions of atmospheric density (415–423). Other Vergilian "firsts" include the mention of the Epirote breed of horses (1.59, 3.21), Crustumnian pears (2.85), a potent wine called *lagois* (2.93), a flower, the *amellus*, described as a panacea for apiary illnesses (4.271–280), and a recommendation to place spiked halters on kids to encourage early weaning (3.398–399).

More distinctive is the poet's holistic vision of agriculture, whereby the pedestrian characteristics of soils, plants, animals, and bees are traced back to broader cosmic trends and laws. Thus Vergil draws an analogy between the diversity of soils on a large estate and the diversity of products exported by countries around the ***oikoumenē*** (1.50–63). He uses the tendency of seeds to decline in fertility over time to illustrate a generalized principle of entropy (1.197–203). The favorable features of the climate in spring are said to reproduce the conditions which obtained when life first appeared on Earth (2.315–345). Farm animals' susceptibility to disease is worked up into an illustration of the death drive which affects all living beings (3.440–566). The rational, collective behavior of bees is offered as evidence for the existence of a world-soul (4.219–227). Even the peasant is portrayed as an ersatz philosopher-scientist, who combines practical knowledge of the natural world with a temperate lifestyle (2.475–494). The later ideal of the gentleman farmer who dabbles in scientific and philosophical speculation owes much to this poem, as do many organic and Romantic conceptions of Nature.

In the *Georgics*, Vergil often plays with the identification of known flora and fauna in ways that assume a fairly detailed knowledge of both. Although the *Bucolics* are not didactic poetry, they resemble the *Georgics* in this regard. In the *Aeneid*, this sort of erudite game is further expanded, so that it takes in such fields as geography, astronomy, and medicine. Vergil also engaged the traditions of Homeric allegoresis, which saw the epics as repositories of insight into the nature of the ***kosmos***, and generally identified the gods with the elements and other forces of nature, by incorporating allusions to **Stoic** physical theory into his own descriptions of the universe and the gods. Yet only once in the epic does he deal explicitly with cosmology; this is in Anchises' speech in Book 6 (724–751), where the vision of the ***kosmos*** bears no small resemblance to that presented by Plato in the Myth of Er, with touches of Cicero's *Somnium Scipionis* and other **Stoic** and **Pythagorean** doctrines added in.

Vergil's poetry wears its erudition lightly. Yet the reputation for learning and wisdom which he acquired during his lifetime continued to grow after his death. Subsequent generations of scholars felt challenged to interpret, expand upon, and even correct his work with a view towards establishing or refuting some scientific or philosophical dogma (*cf.* Pliny, Aulus Gellius, Macrobius, etc.). After a few centuries Vergil the scientific dilettante disappeared completely from view, having been replaced by the figure best known from Dante's *Divina Commedia*, who stood as the living embodiment of the entire pagan tradition of rational knowledge and wisdom.

T.F. Royds, *The Beasts, Birds and Bees of Virgil* (1918); J.J. Sargeaunt, *The Trees, Shrubs and Plants of Virgil* (1920); D. Comparetti, *Vergil in the Middle Ages* (1929); L.P. Wilkinson, *The Georgics of Virgil. A Critical Survey* (1969); *KP* 5.1190–1200 (#5), K. Büchner; P.R. Hardie, *Virgil's Aeneid. Cosmos and Imperium* (1986); R.A.B. Mynors, *Virgil. Georgics* (1990); *NP* 12/2.42–60 (#4) W. Suerbaum; *OCD3* 1602–1607, D.P. and P.G. Fowler.

Philip Thibodeau

Vettius Valens (Med.) (*ca* **35 – 48** CE)

Lover of Messalina, and founder of a new medical sect, which apparently died with him upon his execution in 48 by Claudius (TACITUS, *Ann.* 11.30, 35; PLINY 29.8, 20). Identified with the "Valens," teacher of SCRIBONIUS LARGUS (Moog), but the imperial-era *cognomen* Valens is very frequent, and there is no other reason to equate them; see instead M. TERENTIUS VALENS.

F.P. Moog, "Kaiserlicher Leibarzt und einziger römische Schulgründer," *Würzburger medizinhistorische Mitteilungen* 20 (2001) 18–35; *NP* 12/2.151–152.

PTK

Vettius Valens of Antioch (150 – 180 CE)

Wrote a Greek astrological treatise in nine books, the *Anthologiai*. The numerous horoscopes of unnamed individuals and details of their lives reveal the author as a working astrologer with an extensive practice as well as a teacher of his science. The birth years deducible from the horoscopes range from 50–150 CE; if we add to each birth year the greatest age of the individual that Valens reports, we find a great concentration through the 150s and until the 160s, which presumably reflects the interval during which he was hardest at work on his opus, but he continued to add new material into the 170s. He repeatedly adduces a horoscope cast for February 8, 120 CE, plausibly identified as his own birth-date.

The earlier books of the *Anthologiae* show some effort to cover basic topics of horoscopic astrology systematically. As the work progresses, however, it becomes increasingly devoted to specialized topics such as the precise forecasting of length of life. While drawing (sometimes without acknowledgement) on earlier authorities, Valens frequently claims elements of interpretative technique as his own inventions. He often criticizes unidentified contemporary astrologers, and occasionally pronounces on broader philosophical issues, for example arguing for a hard-line deterministic view of horoscopic predictions. Professing to aim at clear presentation, Valens was not successful; his Greek style is characterized by a penchant for rare vocabulary, not invariably used with precision. The authorial obscurities were exacerbated, moreover, by extensive textual corruption and tampering in later transmission. Nevertheless the *Anthologiae* is enormously valuable for its focus on astrological practice, without rival in the surviving Greco-Roman astrological literature. It is also an important source on contemporary astronomical resources, ranging from crude rules of thumb to arithmetically structured theories of ultimately Babylonian origin, comparable to methods known from Roman-period papyri and from early Indian astronomy.

Ed.: D.E. Pingree, *Vettii Valentis Antiocheni Anthologiarum Libri Novem* (1986).

O. Neugebauer, "The Chronology of Vettius Valens' Anthologiae," *HThR* 47 (1954) 65–67; Neugebauer and van Hoesen (1959); Neugebauer (1975) 793–801, 823–829; J. Komorowska, *Vettius Valens of Antioch: An Intellectual Monography* (2004); Riley (n.d.).

Alexander Jones

C. Vibius Rufinus of Tusculum (45 – 65 CE)

Latin authority on trees, plants, and flowers (PLINY 1.*ind.*14–15, 19, 21). Vibius Rufinus, a suffect consul of 21 or 22 with M. Cocceius Nerva (the emperor's grandfather), was possibly our botanist's father; Vibius Rufinus, the proconsul of Asia, *ca* 36/37, and legate of Germania Superior 42–45, was perhaps our botanist. Our Rufinus seems later than the addressee of two Ovidian epistles replete with medical imagery (*ex Pont.* 1.3 and 3.4) whom Syme argues was not a Rufinus (1434, n.94). Nonetheless an interest in botany accords equally with a friend of OVID or a provincial governor.

RE 8A.2 (1958) 1981 (#49), R. Hanslik; R. Syme, "Vibius Rufus and Vibius Rufinus," *Roman Papers* 3, ed. A.R. Birley (1984) 1423–1435 at 1430–1435; *NP* 12/2.177 (#II.14), W. Eck.

GLIM

Vibius Sequester (300 – 500 CE?)

Wrote a geographical work in Latin, *De fluminibus fontibus lacibus nemoribus paludibus montibus gentibus per litteras*. This is a list of geographical names which occur in Latin poets, such as VERGIL, LUCANUS, and OVID, arranged in alphabetical order. The list contains both real and mythological toponyms.

Ed.: *GLM* 145–159.
RE 8A.2 (1958) 2457–2462 (#80), W. Strzelecki; *KP* 5.1251–1252, F. Lasserre; *PLRE* 1 (1971) 823; *NP* 12/2.177–178 (#II.19), K. Sallmann.

Natalia Lozovsky

Vicellius (100 BCE? – 150 CE?)

Roman writer known solely from IŌANNĒS "LYDUS," *de Ost.*, who describes him as prior to APULEIUS (*cf.* perhaps *M. VIGELLIVS*, known solely as PANAITIOS' **Stoic** house-mate: CICERO, *De Or.* 3.78; *RE* 8A.2 [1958] 2130–2131[#1], H. Gundel). Iōannēs quotes or paraphrases Vicellius' *Seismologium*, which predicts, based on the sun-sign in which a quake occurs, catastrophes in the regions from India to Hispania which are ruled by that sign: §55–58 (pp. 110–117 Wa.). Compare the omen-literature of writers such as PETOSIRIS, and contrast *Seismologia* that predict type not place of trouble: *CCAG* 5.4 (1940) 155–163, 7 (1908) 167–171. Iōannēs §23–26 (pp. 57–62) also quotes or paraphrases a work predicting misfortunes in regions from India to Hispania, based on the sun-sign in which thunder occurs, which some scholars have attributed to Vicellius. Such *Brontologia* (or *Tonitrualia*) usually predict the type not place of trouble, based on the sun-sign: *CCAG* 4 (1903) 128–131, 8.3 (1912) 123–125, 9.2 (1953) 120–123, and PSEUDO-HERMĒS; but *CCAG* 7 (1908) 163–167 combines a lunar *Brontologion*, predicting type of trouble, with the Vicellian *Tonitruale*.

HLL §409.3.

PTK

Victorius of Aquitania (445 – 465 CE)

Mathematician, calculated paschal dates. In *Calculus*, his elementary arithmetical text, Victorius discussed the properties of numbers, conventions of arithmetical expression, and process of multiplication and division. MSS contain numerous multiplication and division charts, and tables of standard weights and measures (oils, honeys, lengths, liquid and dry measures).

Ed.: *MSR* 2.87–88; G. Friedlein, "Der Calculus des Victorius," *ZMP* 16 (1871) 42–79.
GRL §1229.8; F.K. Ginzel, *Handbuch der mathematischen Chronologie* (1914) 3.245–247; *RE* 8A.2 (1958) 2086–2087 (#8), W. Enßlin.

GLIM

Heluius Vindicianus (*ca* 350 – 410 CE)

Prominent politician and physician, perhaps receiving literary and medical education in Roman Gaul, Vindicianus appears as a gifted and crusty rhetorician of advanced years in AUGUSTINE's *Confessions* (4.3.5; 7.6.8; *cf. Ep.* 138.3), then resident in Carthage, holding the rank of *comes* and likely *archiater.* Augustine's youthful studies on astrology may have attracted the attention of Vindiciater, who detested such irrational approaches to diagnosis and prognosis: *uir sagax, acutus senex, magnus ille nostrorum temporum medicus,* so says Augustine of Vindicianus. Formerly court physician to Valentinian, Vindicianus' medical skills were outstanding enough to win the emperor's extension of privileges to loyal court doctors and their families, especially to those like Vindicianus who had attained the rank of *comes* (*Cod.Theod.*, 12.3.12 [14 Sept., 379]; *cf. Cod.Theod.*, 11.31.7 [3 Dec., 379]). Likely, too, *archiatri* functioned as teachers, and THEODORUS PRISCIANUS and CASSIUS FELIX were both proud to be Vindicianus' students. That medical education was fairly widespread in the 4th c. is indicated by Symmachus' *Relationes* (# 27: a *perfectissimus* demands the salary of *archiater* in Rome), as well as the medically informed poetry of Ausonius, whose father IULIUS AUSONIUS was court physician to Valentinian I.

Remnants of Vindicianus' medical writings demonstrate a respect for the Greek medical classics, esp. GALĒN and SŌRANOS, but with a practical twist emphasizing folk remedies, careful analysis of pharmaceuticals, and avoidance of gratuitous surgeries. An extant *Epistula Vindiciani* to Valentinian evinces expertise in medicinals to ease constipation, and the *Epistula Vindiciani ad Pentadium* (ed. Rose, pp. 484–492) briefly advises a nephew regarding the then-canonical doctrine of the four **humors** derived from Greek tracts (HIPPOKRATĒS and Galēn), which he says he has translated into Latin. The epitomes of Vindicianus' *Gynaecia* suggest close attention to anatomical structures, but details are almost certainly derived from Sōranos' *Gynecology*, not from actual dissection. The 13th c. *Codex Monacensis* (Clm 4622, ff.40^R-45^R) preserves Vindicianus' "Medical Etymology" compiled, as he says, since one cannot ". . .dissect corpses . . . because it is prohibited, and . . . an account is given of the joints, bones, limbs, and blood vessels of which we consist" (Cilliers 2005: 167). Debru (1996, 1999) and Cilliers (2005) have concluded that the attribution by Wellmann (1901) of the *Codex Bruxellensis* f.48^R to Vindicianus is no longer acceptable.

Ed.: V. Rose, *Vindiciani Afri Expositionis membrorum quae reliqua sunt. . . I. Gynaecia quae vocantur; II. Epitoma uberior altera adhaeret Epistula Vindiciani ad Pentadium* [in] *Theodori Prisciani Euporiston* (1894) 425–492; M. Niedermann, re-ed. E. Liechtenhan, *Epistula Vindiciani Comitis Archiatrorum ad Valentinianum Imperatorem* 1.46–53 in *Marcelli De medicamentis* (1968; 2 vols.) = *CML* 5; R.H. Barrow, *Prefect and Emperor: The Relationes of Symmachus A.D. 384* (1973) 148–151 (# 27); L. Cilliers, "Vindicianus's *Gynaecia*: Text and Translation of the Codex Monacensis (Clm 4622)," *Journal of Medieval Latin* 15 (2005) 153–236.
K. Sudhoff, "Zur Anatomie des Vindicianus," *Zeitschrift für Wissenschaftgeschichte* 8 (1915) 414–423; Th. Haarhof, ["Medicine"] in *Schools of Gaul* (1920) 87–89; *RE* 9A.1 (1961) 29–36, W. Enßlin and

K. Deichgräber; Matthews (1975; repr. 1990, 1998) 68, 72–73, and 399–400 [postscript to 1990 repr.]; V. Nutton, "*Archiatri* and the Medical Profession in Antiquity," *PBSR* 45 (1977) 191–226; *Idem*, "Continuity or Rediscovery: The City Physician in Classical Antiquity and Mediaeval Italy," in A.W. Russell, ed., *The Town and the State Physician in Europe from the Middle Ages to the Enlightenment* (1981) = *Wolfenbütteler Forschungen* 17; M.E. Vázquez Buján, "Vindiciano y el tratado «De natura generis humani»," *Dynamis* 2 (1982) 25–56; P. Migliorini, "Problemi Testuali in Vindiciano (Paris. Lat. 7027, cc. 3r–13v)," in Mazzini and Fusco (1985) 237–252; Önnerfors (1993) 281–288; A. Debru, "L'Anonyme de Bruxelles: un témoin latin de l'hippocratisme tardif," in R. Wittern and P. Pellegrin, edd., *Hippokratische Medizin und antike Philosophie*, v.1 (1996) 311–327; A. Debru, "Doctrine et tactique doxographie dans l'Anonyme de Bruxelles: Une comparaison avec l'Anonyme de Londres," in van der Eijk (1999) 453–471; L. Cilliers, "Vindicianus' Gynaecia and Theories on Generation" in H.F.J. Horstmanshoff and M. Stol, edd., *Magic and Rationality in Ancient Near Eastern and Graeco-Roman Medicine* (2004) 343–367.

John Scarborough

VINDONIUS ⇒ ANATOLIOS

M. Vipsanius Agrippa (40 – 12 BCE)

Born 64/63 BCE to an obscure but wealthy family, lifelong friend and doubly son-in-law to AUGUSTUS. He fitted and trained the fleet, orchestrated naval victories against Sex. Pompeius – wherein his improved grapnel proved effective – and M. Antonius, served in political positions of authority and distinction, was consul 37 BCE, then held extraordinary grants of *imperium* and *tribunicia potestas*. Traveling widely, Agrippa governed Gaul, represented Augustus' interests broadly in the east, and quelled a rebellion in Spain (20 BCE). His munificent building program at Rome (Pantheon, aqueducts, an expanded sewer, granary, etc) and in the provinces (roads from Lugdunum) earned him enduring popularity.

Lost are his autobiography, *Commentarius de Aquis*, geography, and map of the empire, intended to perfect IULIUS CAESAR's efforts. It is unclear if the geographical treatise represented a continuous commentary or supplementary notes. Whether Agrippa's map resembled more closely ERATOSTHENĒS' or the PEUTINGER MAP continues to be debated, but its display on a colonnaded wall in the Porticus Vipsania suits better a rectangular deployment. The map, completed by Augustus himself after Agrippa's death, probably represented the entire inhabited world. PLINY cites Agrippa almost exclusively for quantitative data (distances, lengths, circumferences) occasionally criticizing their veracity (3.16–17, 4.91, 4.102) or comparing with numbers from other geographical authors (4.45, 4.60, 4.77). Agrippa also described topography (5.9–10, 6.39). STRABŌN may have relied on Agrippa for figures for Italy, Corsica, Sardinia, and Sicily.

GRL §332–333; Dilke (1985) 41–53; *OCD3* 1601–1602, B.M. Levick; *TTE* 8, R.T. Macfarlane; *BNP* 1 (2002) 391–392 (#1), D. Kienast.

GLIM

M. Vitruuius Pollio (*ca* 30 – 20 BCE)

Professional architect and engineer, born *ca* 85 BCE, author of *De Architectura Libri Decem*, the only comprehensive summary of architecture to survive from antiquity, and possibly the only one ever written. The work presents the best panorama of what a broadly educated Roman professional with a "liberal arts" education would have known of the works of the leading scientists and mechanical authors of antiquity.

The *nomen* Vitruuius ("Vitruvius") is the only name known with certainty from most MSS, the *cognomen* Pollio comes from a single MS of FAUENTINUS, the *praenomen* is variously reported as Aulus, Lucius and most commonly Marcus. The *gens Vitruuia* is well attested on gravestones centering around Formia between Naples and Gaeta, and it seems likely from *De Architectura* that Vitruuius was raised and trained either in Campania or Rome, or both. He is almost certainly distinct from L. Vitruuius Cerdo, a freedman recorded in the arch of the Gauii in Verona, since the author was not a freedman, and from CAESAR's *praefectus fabrum* Mamurra, though the author served many years with Caesar and then AUGUSTUS as an artillery engineer and staff architect. He received a pension from Augustus and then his sister Octauia as a reward for service, and he is very likely the same Vitruuius credited with standardizing water pipe sizes in Rome while working as Agrippa's staff architect on the *cura aquarum* (FRONTINUS, *De Aquis Urbis Romae* 25.1), a position he also may have received as a reward for service or for his writing.

The treatise was written in the decade immediately after Actium (31 BCE), a period of tremendous renewed building activity after the civil wars. A literary hybrid, common in the last century of the Republic, *De Architectura* is a technical handbook with literary pretensions, aimed at the highly literate Roman elite, i.e., senators and equestrians who directed building projects, either private or public. Vitruuius seems unknown in contemporary accounts, but is mentioned later in ways suggesting that his writing remained the most comprehensive Roman building compendium: PLINY 33.87, 33.91, 36.171–172; FRONTINUS 25.1; FAUENTINUS; Seruius, *Ad Aen.* 6.43 (4th c. CE); Sidonius Apollinarius, *Epistulae* 4.3.5 (mid-5th c. CE), and indirectly PALLADIUS AEMILIANUS.

Vitruuius received a liberal arts education before training as an architect. His handbook in part presents architecture as a liberal art, whose practice had to be based on a mastery of those fundamentals of liberal knowledge common to many disciplines. In addition to the standard seven subjects (mathematics, music, geometry, astronomy; grammar, rhetoric, logic), he also lists draftsmanship, knowledge of painting and sculpture, law, and philosophy. Most architects were not trained that way. Vitruuius includes numerous supportive discourses based on generally understood principles of science: the four-element theory explained the properties of mortar (2.5.2), building stones (2.7.2), types of timber (2.9.1) and strength and weakness of *opus reticulatum* (2.8.2); retrograde motion of the planets as explained by the attraction of heat (9.1.12); latitude determining human physiology and justifying window placement (6.1.1–11); the variety of springs being due to the variety introduced into nature by the "inclination of the heavens" (*inclinatio mundi*), i.e., the inclination of the Earth on its axis.

Book 1 treats purported theoretical principles, orientation and siting of cities, and surveying, 2 addresses building materials, 3 and 4 the so-called orders and their proportional principles (Vitruuius never uses the term "orders," but rather calls them *genera*, or "types," of column), 5 public buildings, including bath construction, with a discourse about music theory, 6 private building, 7 finishes, 8 water distribution and surveying, 9 astronomy and the geometry of sundials and clocks (***Figure***), and 10 a variety of mechanical devices, including cranes, levers, water lifters, water wheels, pumps, pneumatic organs, catapults (scorpions and ballistae), and siege engines. The meticulous description of catapults and organs relies on the same type of modular geometry and arithmetic as his earlier description of the proportions of the orders, sundials, theaters, house proportions, boat design and proportions in nature (human anatomy). The section on siege-craft (10.13–15) is virtually identical to chapters in ATHĒNAIOS MECHANICUS' *Peri Mēkhanēmatōn* (9.4–10.4); they are

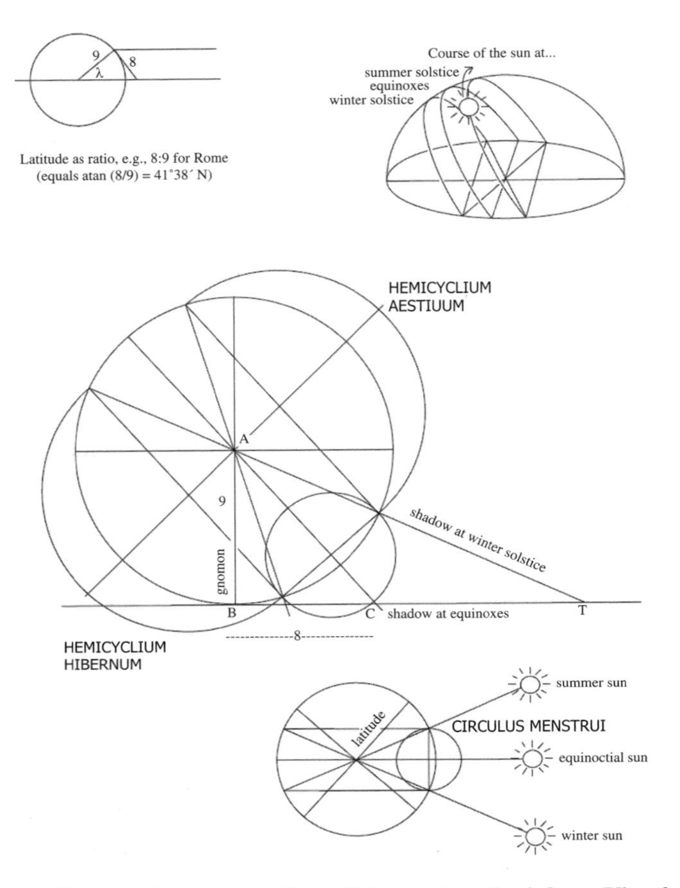

Analemma (geometry of sundial construction) from Vitr. 9.7.1–7 © Howe

probably contemporaries, and their common source was probably Agēsistratos, whom he cites (7.*pr.*14).

Vitruuius is often more "prescriptive" than descriptive, that is, arguing for innovations based on experience and critical evaluation rather than summing up current standard practice. The *chorobates* (surveyors' level, 8.6.1–3) is otherwise unattested in antiquity, and his recommendations for items such as polygonal fortification towers (1.5.1–8), sounding vessels in theaters (5.5.1–8), a peculiar form of *castellum aquae* (8.6.1–2), and fire-resistant larch (2.9.15) were not then standard Roman practice.

A. Boethius, "Vitruvius and the Roman Architecture of his Age," in *Dragma Martin Nilsson (Acta Ist. Sue Rom. 1*) (1939) 114–143; H. Knell, *Vitruvs Architekturtheorie* (1991); P. Fleury, *La méchanique di Vitruve* (1993); P. Gros, A Corso, and E. Romano, *Vitruvio, De Architectura* (1997); I. Rowland and Thomas Noble Howe, *Vitruvius, Ten Books on Architecture* (1999).

Thomas Noble Howe

Vitruuius Rufus ⇒ Epaphroditus

L. Volusius Maecianus (*ca* 140 – 170 CE)

Important member of the "court society" of the 2nd c. CE, assigned to high imperial offices, including the government of Egypt. As a legal expert, Maecianus wrote treatises on public trials and on trusts. Because of his competence in law, he also became advisor of Antoninus Pius and legal tutor to Marcus Aurelius (*SHA, MA* 3.6). A Caesar (deputy emperor), probably Marcus Aurelius, is the addressee of his *Distributio* (*Division*). Maecianus offers a didactic and exhaustive work on units of minted coins and their reckoning, dealing with the basic units of the Roman denominational system, both Republican and Imperial, their subdivisions and exchange rates, according to a peculiar system of symbols, useful to indicate sums of money. It ends with a short appendix on capacity measures both for grain and liquids, and their equal (sub-)units.

Ed.: *MSR* 2.17–22, 61–71.

RE 9A.1 (1961) 904–906 (#7), T. Mayer-Maly; *HLL* 4.130–133; *NP* 12/2 (2002) 323 (#II.1), T. Giaro; *OCD3* 1612, T. Honoré.

Mauro de Nardis

W

Wuzurgmihr (*ca* 531 – 579 CE)

Persian astrologer, active during the kingdom of Xusraw I (531–579), generally identified with the famous vizir, son of Bōxtag, of the same Sasanian *šāh*, an association debated by scholars (*cf.* BURZŌY). According to Arabic sources, Wuzurgmihr (also Buzurǰmihr and other spellings) translated VETTIUS VALENS' *Anthologies* into Pahlavi, augmenting the Greek with Indian and Iranian sources concerning ***katarkhic*** and interrogative astrology, "continuous" horoscopy and **genethlialogy**. The Pahlavi title of this important astrological treatise must have been **Wīzīdag* "Selections" (*al-Bizīdaj* in Arabic). The Pahlavi original was used by Māšā'allāh in his *Kitāb al-mawālīd* "The book of the nativities" (partly incorporated into Hugo of Santalla's *Liber Aristotilis*), and referred to in other works by Māšā'allāh. The Cod. Vat.Gr. 1056, ff.81^{V}–82 directly attests "Porzozómchar" (i.e., Wuzurgmihr). See also PAHLAVI, TRANSLATIONS INTO.

Nallino (1922) 351–357 = (1948) 291–296; D.E. Pingree, "The Indian and Pseudo-Indian Passages in Greek and Latin Astronomical Texts," *Viator* 7 (1976) 170, 187; *Idem* (1989); *EI* 4 (1990) 427–429, D.K. Motlagh (s.v. Bozorgmehr); Ch. Burnett and D.E. Pingree, *The* Liber Aristotilis *of Hugo of Santalla* (1997); Antonio Panaino, *La novella degli scacchi* (1999) 107–123.

Antonio Panaino

X

Xanitēs (?) (250 BCE – 80 CE)

ANDROMAKHOS in GALĒN, *CMLoc* 9.6 (13.311 K.), records the "very useful" *ΞΑΝΙΤΗΣ* ***hedrikē***, composed of beeswax, butter, goose-fat, deer-marrow, a little rose-water, etc. The name, otherwise unattested, may be a brand-name, a distortion of AZANITĒS, PODANITĒS, or else emendable to Naxitēs (*cf.* NEXARĒS), a remedy from Naxos (*cf.* NAUKRATITĒS).

Fabricius (1726) 452; Parker (1997) 145 (#53).

PTK

Xanthos (of Sardēs?) (480 – 440 BCE)

Son of Kandaulēs, a Ludian who wrote in Greek; EPHOROS names him as an older contemporary of HĒRODOTOS, who used him as a source. He wrote a *History of Ludia*, retailing myths, as well as the history of the Ludian royal family, enlivened with lurid details. He paid special attention to the toponymy and topography of Ludia, and included as well some herbal lore and ethnography of the Ludians and of the Persian Magi, to whom he may have devoted a separate work, along with a biography of EMPEDOKLĒS. He also anticipated Hērodotos in his observations of fossils found far inland as evidence for geological upheavals in the past. His work, or an epitome by MENIPPOS, was used by DIONUSIOS OF HALIKARNASSOS, STRABŌN, Athēnaios, and STEPHANOS OF BUZANTION, although Athēnaios raises doubts about its authenticity.

Ed.: *FGrHist* 765.

P. Kingsley, "Meetings with Magi: Iranian themes among the Greeks, from Xanthus of Lydia to Plato's Academy," *JAS* 5 (1995) 173–192; *OCD3* 1627, K. Meister.

Philip Kaplan

Xenagoras son of Eumēlos (*ca* 200 – 180 BCE?)

Measured the height of Mt. Olympus and recorded his precise results, ten stades and 96 feet, in an epigram addressed to the king (Philip V or Perseus) of Macedon, copied by P. Cornelius Scipio Nasica in 168 BCE, and quoted by PLUTARCH, *Aemil.* 15.9–11.

NP 12/2.606 (#2), H.A. Gärtner.

PTK

Xenagoras (of Hērakleia Pontikē?) (330 – 210 BCE?)

Wrote a chronology and a work *On Islands* (*cf.* the work of Eudoxos of Knidos) covering Cyprus (Pliny 5.129) and the Pithekousai (Dionusios of Halikarnassos, *A.R.* 1.72.5), about which he told etymological tales.

FGrHist 240; *NP* 12/2.606 (#1), H.A. Gärtner.

PTK

Xenarkhos of Seleukeia (Kilikia) (45 BCE – 17 CE)

Strabōn 14.5.4 records that this **Peripatetic** taught him and others in Alexandria and Athens; at Rome he joined the court of Augustus as a protégé of Areios Didumos; he died in 17 CE, aged over 90. His work *Against the Fifth Substance* argued the non-existence of the Aristotelian ***aithēr***, partly on the basis that "simple" motion need not be rectilinear and that circular motion must always be forced; Simplicius preserves a handful of fragments, e.g. *In De Caelo: CAG* 7 (1894) 13–14, 21–24, 286.

NP 12/2.608–609 (#4), A. Falcon; Irby-Massie and Keyser (2002) 67–69.

PTK

Xen(okh)arēs (500 – 25 BCE)

Compiled rules of architectural symmetry (Vitruuius 7.*pr.*14), listed first in an approximately chronological order, so perhaps 5th/4th c. BCE. Fabricius defends the unusual name *ΝΕΞΑΡΗΣ* on analogy with Drouarēs, but Xenarēs is found at Lokris (2nd c. BCE) and Thessalia (3rd c. BCE, *LGPN* 3B.314), at Sparta (5th c. BCE: *RE* 9A.2 [1967] 1435–1436), and Kerkura (6th c. BCE: *LGPN* 3A.333); whereas from the 5th c. BCE, Xenokharēs is frequent and widespread (*LGPN*).

RE 17.1 (1936) 163, E. Fabricius.

PTK and GLIM

Xenokratēs of Aphrodisias (Kilikia) (*ca* 50 – 70 CE)

Physician who wrote pharmaceutical works, famous until at least the 7th c. CE, but blamed by Galēn (*Simples* 9.1 [12.248–251 K.]) – although he utilizes the work abundantly – for making use of disgusting remedies, such as human brains, flesh, liver, or even urine and excrement (see also Pliny 32.144), in his *On Useful Things from Living Beings* (animals and humans equally serviceable as Wellmann rightly indicated). This text, a compilation partly based on Diphilos of Siphnos' treatise and used by Pliny in Books 20–30, which suggested treatments probably not more superstitious than many other pharmacologists (he proposes, e.g., remedies to steal an opponent's voice), is to be differentiated from a book with lexicographical relevance named *On Vegetal Remedies* (*Peri botanikōn pharmakōn*). Oreibasios (*Coll.* 2.58: *CMG* 6.1.1, pp. 47–57) preserved a large fragment of a possibly independent book, *On the Food Given by Aquatic Animals*, addressing shellfish, cetaceans (distinguished from fishes), and fishes, divided, as in Dioklēs, between fishes with hard flesh and those with soft flesh. It gives a great number of marine zoönyms and detailed remarks on dietetics and gastronomy (kinds and uses of scallops: 56–71, preparations of pinna: 98–104, Egyptian pickles: 148–151).

Xenokratēs of Aphrodisias (*Vind. Med. Gr.* 1, f.2[V]) © Österreichische Nationalbibliothek

M. Wellmann, "Xenokrates aus Aphrodisias," *Hermes* 42 (1907) 614–630; *RE* 9A.2 (1967) 1529–1531 (#8), F. Kudlien; *OCD3* 1628, W.D. Ross.

Arnaud Zucker

Xenokratēs of Ephesos (50 – 70 CE)

Son of Zēnōn, lived under Nero, and wrote at least a treatise on stones (*Lithika*), read and much appreciated in his own time and later. Pliny 37.38 speaks about Xenokratēs as still living and considers his work worthy of the greatest admiration. Origen, *In Ps.* 118.127, transmitting a description of topaz, calls our author *lithognōmōn*, "expert on stones." Xenokratēs' fragments mostly survived via the Arab tradition. According to Wellmann, the alphabetic catalogue of gems, included by Theodōros Melitēniotēs in his poem *On moderation* (second half of the 15th c.), probably derives from Xenokratēs.

Xenokratēs' lapidary was a compilation of various traditions, including detailed descriptions, classifications and geographical documentation, in the tradition of Theophrastos and Sōtakos, medical material in the tradition of Dioskouridēs, and magical properties in the traditions of Zoroaster or Bōlos, as well as medical questions related to minerals. Consequently Pliny sometimes quotes Xenokratēs among medical experts, but does not imply equating him with the homonymous physician Xenokratēs of Aphrodisias.

M. Wellmann, "Xenokrates aus Aphrodisias," *Hermes* 42 (1907) 614–629; *RE* 7.1 (1910) 1052–1115 at 1052, O. Rossbach; R. Cadiou, "L'île de Topaze. Le fragment du «lithognomon» de Xénocrate d'Éphèse," in *Mélanges Desrousseaux* (1937) 27–33; *RE* 9A.2 (1967) 1529 (#7), K. Ziegler; Ullmann (1972) 98–100; *Idem*, "Das Steinbuch des Xenokrates von Ephesos," *Medizinhistorisches Journ.* 7 (1972) 49–64; *Idem*, "Neues zum Steinbuch des Xenokrates," *Medizinhistorisches Journ.* 8 (1973) 59–76; *KP* 5.1416 (#4), C.J. Classen; *RE* S.14 (1974) 974–977 (#7), M. Ullmann; Halleux and Schamp (1985) xviii–xxi; *NP* 12/1.623–624 (#5), Chr. Hünemörder.

Eugenio Amato

Xenokratēs of Khalkēdōn (*ca* 375 – 314/3 BCE)

Student of PLATO and Head of the **Academy** for 25 years after the death of SPEUSIPPOS, probably also a candidate for the position on Plato's death; said to have died at the age of 82. Known as a systematizer, he organized philosophy into its three branches of physics, ethics and logic, and tried to integrate all kinds of reality into his ***kosmos***. While closer to Plato than Speusippos had been, Xenokratēs helped remove **Platonism** from any transcendent features. He is usually associated with dogmatic teaching, but it should be noted that mythical features abound in extant material, suggesting that he often communicated somewhat indirectly. At times, this makes the reconstruction of a supposedly systematic philosophy somewhat difficult.

This trend is observable in the epistemological fragment (SEXTUS EMPIRICUS *Adv. Math.* 7.147–9 = *fr.*83), where the intelligible objects (above the heavens) are linked with truth and scientific knowledge, sensory objects (below the heavens) with some qualified truth and sensation, and mixed or opinable objects (in the heavens) with both either truth or falsehood and with belief. The three fates, Atropos, Lachesis, and Clōthō, are associated with the three realms respectively.

The triads, observable here, recur elsewhere (e.g. PLUTARCH *Def. Orac.* 416C–E = *fr.*222; *Fac. Orb.* 943E–4A = *fr.*161), and Xenokratēs produced a variation on Plato's triad of Ideas, Mathematicals, and Sensibles, according to which the Ideas (treated as the patterns behind naturally occurring species) were themselves a kind of number (*fr.*103), superior to mathematical numbers. The definition of Ideas illustrates a further difficulty, that of knowing whether Xenokratēs speaks for himself or as an interpreter of Plato, in which capacity he appears in Plutarch (*Anim. Proc.* 1012D–13D = *fr.*188) and later commentators. His "theology" (*fr.*213) begins with the quasi-Platonic Monad (odd, male, father, supra-heavenly ruler, Zeus, intellect) and Dyad (even, female, mother, sub-heavenly ruler, universal soul). Doubts arise with regard to the last three attributes (Dillon 2003: 103), particularly for those who emphasize the systematic nature of Xenokratēs' philosophy. The theology continues with Heaven, heavenly bodies (= Olympians), and **daimonic** powers pervading air, water, and earth (= Hadēs, Poseidōn, Dēmētēr).

A famous physical-mathematical doctrine is that of the existence of indivisible lines, meeting with typical hostility from the **Peripatetic** tradition, as observed in the ARISTOTELIAN CORPUS ON INDIVISIBLE LINES. Various fragments of Xenokratean ethics and logic survive, but do not, on the whole, set him apart from the early **Academic** tradition. This may reflect the influence of Antiokhos of Askalon who had minimized differences. It may, however, indicate Xenokratēs' profound influence over it.

Ed.: M. Isnardi Parente, *Senocrate–Ermodoro: Frammenti* (1982).
Dillon (2003) 89–155.

Harold Tarrant

Xenokritos of Kōs (325 – 275 BCE)

Hippokratic commentator prior to KALLIMAKHOS OF BITHUNIA, cited twice by ERŌTIANOS, *pr.* (p. 4.24 Nachm.) and A-5 (p. 12.7 Nachm.).

RE 9A.2 (1967) 1533 (#4), M. Fuhrmann.

PTK

Xenophanēs of Kolophōn (540? – 478? BCE)

Born *ca* 570 BCE, an itinerant Greek bard and philosopher, criticized traditional claims about the gods as inconsistent with the concept of the divine. Rather than accepting that gods are like humans and behave in all-too-human ways, he claimed god was a single divine force: "One god is greatest among gods and men, not at all like mortals in body or in thought . . . whole he sees, whole he thinks, and whole he hears . . . but completely without toil he shakes all things by the thoughts of his mind" (B23, 24, 25). Xenophanēs rejected the notion that there is divine communication to human beings, claiming instead that through inquiry humans can "discover better" (B18). Recognizing possible skeptical consequences of this claim, he nevertheless offered naturalistic explanations of meteorological phenomena based on a theory about clouds: the rainbow and St. Elmo's fire are not divine messages: both are explainable as kinds of cloud (fragment B32: "She whom they call iris (rainbow), this too is by nature cloud, purple, red, and greenish yellow to see"). Indeed the Sun, Moon and all luminous celestial phenomena are clouds in various states. The Earth is flat (like a column drum) and extends unlimitedly outwards and downwards. Thus the Sun does not travel under the Earth as earlier theories had claimed: it is new each day and is cloud fed by exhalations from the Earth, traveling across the sky until it expires. Both the content of his scientific theories (he made claims instrumental for later Greek discoveries in astro-physics) and the epistemological problems generated by his rejection of divine intervention and warrant for knowledge influenced later thinkers, especially HĒRAKLEITOS and PARMENIDĒS.

Ed.: DK 21, J.H. Lesher, *Xenophanes of Colophon* (1992).

ECP 570–573, J.H. Lesher; *SEP* "Xenophanes," *Idem*; A.P.D. Mourelatos, "La Terre et les étoiles dans la cosmologie de Xénophane," in A. Laks and C. Louguet, edd. *Qu'est-ce que la Philosophie présocratique?* (2002) 331–350; A.P.D. Mourelatos "Xenophanes' Contribution to the Explanation of the Moon's Light," *Philosophia* 32 (2002) 47–59; *Idem*, "The Cloud-astrophysics of Xenophanes and Ionian Material Monism," in Patricia Curd and D.W. Graham, *The Oxford Handbook of Presocratic Philosophy* (forthcoming).

Patricia Curd

Xenophilos of Khalkidikē (375 – 325 BCE)

Musical theorist of the last generation of **Pythagoreans**, who is cited by different sources as teacher of ARISTOXENOS. According to DIOGENĒS LAËRTIOS 8.46, he was born in Khalkis of Thrakē (whose real existence, however, has been put in doubt by some archaeologists), and was active, probably in Athens, in the mid-4th c. BCE. According to PLINY 7.168, he reached an age of 105 years.

KP 5.1421 (#3), R. Engel; *NP* 12/2.632 (#2), R. Harmon.

E. Rocconi

Xenophōn of Athens (400 – 355 BCE)

Born *ca* 430–425 BCE, soldier, mercenary, known especially for his historical writings. He studied under Sōcratēs to whom he dedicated his *Apology, Memorabilia, Oeconomicus,* and *Symposium.* Among the Ten Thousand Greeks supporting Cyrus II's rebellion against his brother the Persian king Artaxerxēs II, Xenophōn commanded the rearguard. Although the Greeks were victorious at Kunaxa (401 BCE), Cyrus was killed, and the mercenaries, without leadership deep in hostile territory, elected new leaders, including Xenophōn, and traveled back to Greece. Xenophōn was later exiled from Athens for his association with Cyrus, or because he fought under the Spartan king Agesilaus against Athens at Koronea (394 BCE). The Spartans gave him property at Skillous, near Olympia in Ēlis, where he composed his works. Xenophōn's *Anabasis* ("The Expedition" or "The March Up Country"), to which he prefixed the pseudonym Themistogenēs of Surakousai for greater credibility, records the expedition of the Ten Thousand and the journey home, Notably, Alexander, during the early phases of his expedition into Persia, used the *Anabasis* as a field guide. Xenophōn's main historical work is the seven-book *Hellenika* treating events from 411 to 362, continuing THUCYDIDĒS' history, and underscoring Xenophōn's optimal knowledge of military art, his interest in prominent personalities, and skill at psychological analysis. Xenophōn was particularly interested in analyzing the character traits of leaders, as in the biography of his friend the Spartan king Agesilaus, the *Cyropaedia* and *Hiero*, and in his reflections on the *Constitution of Sparta*. Moreover his treatises *On Horsemanship* and *The Cavalry General* both address military cavalry art. He also composed *Hunting with Dogs*, and *Ways and Means* wherein he addresses economic problems of reorganizing Athenian finances.

M. Sordi, "I caratteri dell'opera storiografica di Senofonte nelle Elleniche," *Athenaeum* 28 (1950) 3–53 and 29 (1951) 273–348; É. Delebecque, *Essai sur la vie de Xénophon* (1957); *RE* 9A.2 (1967) 1569–2051, H.R. Breitenbach; L. Canfora, *Tucidide continuato* (1970); C.J. Tuplin, *The Failings of Empire. A Reading of Xenophon Hellenica 2.3.11–7.5.27* (1993); J. Dillery, *Xenophon and the History of his Times* (1995).

Cristiano Dognini

Xenophōn (of Kōs?) (330 – 270 BCE)

Greek physician, PRAXAGORAS' pupil. (The Xenophōn of Kōs in DIOGENĒS LAËRTIOS 2.59 is surely the physician of the emperor Claudius.) We do not know the origin of Praxagoras' pupil, although a medieval MS (*Laur. Lat.* 73.1, f.143[V]) quotes a "Xenophon Alexandrinus" in a list of physicians, after Praxagoras and HĒROPHILOS, referring perhaps merely to his place of activity. He studied malign tumors and tumors called *terminthoi* (OREIBASIOS *Coll.* 44.15 [*CMG* 6.2.1, p. 132]), and wrote *On cancers* (*ibid.* 45.11 [p. 166]). Most peculiarly, in referring to the word *theion* (divine) in the HIPPOKRATIC CORPUS, ON THE SACRED DISEASE 1, Xenophōn considered somewhat divine the phenomenon of the "critical day" of the disease's evolution and compared it to the Dioskouroi appearing to shipwreck victims (ERŌTIANOS *fr.*33 [p. 108 Nachm.]): the Hippokratic Corpus utterly denies the divine origin of disease. CAELIUS AURELIANUS 2.186 (*CML* 6.1.1, p. 658) mentions the uterine therapy and ligations for therapy of hemorrhage of a Xenophōn (as well as of DIONUSIOS (MED.)), whose identity is more doubtful.

Steckerl (1958); *RE* 9A.2 (1967) 2089–2092; F. Kudlien; *KP* 5.1430–1431, J. Kollesch; *NP* 12/2.643, V. Nutton.

Daniela Manetti

Xenophōn of Lampsakos (*ca* 100 – 60 BCE)

Greek geographer, author of two lost works on Syria and a ***periplous*** which described coasts beyond the Mediterranean, of northern and western Europe, and of Africa. He mentioned an enormous island, Balcia, three days' sail from the Skuthian coast (PLINY 4.95): probably Scandinavia. Xenophōn's fragments, as they are preserved for instance in

Pliny, give measurements in days of sail, indicating their navigational origin and purpose. ALEXANDER OF MILĒTOS used Xenophōn as a source.

NP 12/2.643–644, H.A. Gärtner.

Daniela Dueck

XOUTHOS ⇒ IŌN OF KHIOS

Xouthos (450 – 350 BCE)

ARISTOTLE, *Physics* 4.9 (216b22–27), cites his clever turn of phrase in his discussion of "micro-voids" interspersed within matter allowing for compression and rarefaction of matter yielding to other matter, without which "the universe would billow (*kumanei*)." Perhaps of Krotōn, if IAMBLIKHOS, *VP* 267 "Bouthos" should read "Xouthos." SIMPLICIUS, *ad loc.* (*CAG* 7 [1882] p. 683.24), calls him a **Pythagorean**.

DK 33.

GLIM

Y

Yavaneśvara (149/150 CE)

Anonymous author in 149/150 CE of a Sanskrit prose translation of an unidentified Greek (probably Alexandrian) text on horoscopy, a translation known only from the surviving Sanskrit verse version *Yavanajātaka* composed by SPHUJIDHVAJA in 269/270. The title Yavaneśvara then meant literally "lord of the Greeks," evidently a high position among the Greek residents of western India under the western Kṣatrapa rulers in the Śaka or Skuthian dominion of the area. This Yavaneśvara worked in the reign of Rudradāman I (and probably at his court in Ujjayinī, modern Ujjain). His text as known through the *Yavanajātaka* became the chief inspiration for Indian horoscopic astrology.

CESS A.5.330; Pingree (1978).

Kim Plofker and Toke Lindegaard Knudsen

Z

Zakhalias of Babylōn (*ca* 120 – 63 BCE?)

Babylonian physician, possibly Jewish. PLINY (37.169) cites his views about the medicinal and magical qualities of precious stones, noting that his books were dedicated to "King Mithradates" (probably MITHRADATĒS VI). Our Zakhalias may be identical to the otherwise unattested "Zalakhthes" in the summary of ARKHIGENĒS' comments on amulets for epilepsy preserved by ALEXANDER OF TRALLEIS (1.567 Puschm.); "Zalakhthes" is there credited with knowledge of the properties of jasper.

Real-Encyclopädie der classischen Alterthumswissenschaft, ed. A.F. von Pauly, v. 6.2 (1839) 2813, C. Cless; *RE* 9A.2 (1967) 2210, K. Ziegler; *EJ2* 13.720–729 at 723, S. Muntner; Stern 1 (1974) 467; M.W. Dickie, "The learned magician and the collection and transmission of magical lore," in D.R. Jordan *et al.*, edd., *World of Ancient Magic* (1999) 163–193 at 176.

Annette Yoshiko Reed

Zarathuštra (before *ca* 600 BCE?)

It is problematic if not impossible to construct an historical biography of the founder of Zoroastrianism. Zarathuštra's homeland and dates are deeply debated: traditionally Zarathuštra is dated *ca* late 7th c. BCE. Some scholars, mainly on linguistic grounds, date him to the end of the 2nd millennium BCE. Classical authors called him Zoroástrēs and considered him a magician and expert in astral sciences. According to the scholia to PLATO's *Alcibiades* (I.222a), the name should be interpreted as *astrothútēs* "sacrificer to the stars"; Dinōn, HERMODŌROS OF SURAKOUSAI (DIOGENĒS LAËRTIOS, *pr.*), and pseudo-Clement (*Recognitiones*, 4.27–29) considered Zoroaster an astrologer. Contrarily, Avestan sources do not confirm that early Zoroastrians were directly involved in astrology, *per se*, but refer only to astral conceptions and mythological speculations. During the Sasanian period (3rd–6th cc. CE) and after the Arab invasion of Iran, some astrological treatises were attributed to PSEUDO-ZOROASTER.

C. Clemen, *Fontes Historiae Religionis Persicae* (1920) 96; Bidez and Cumont (1938) 2.23–24, 66–67; J. Duchesne-Guillemin, *The Western Response to Zoroaster* (1958); *KP* 5.1561–2, W. Röllig; Antonio Panaino, *Tištrya* (1995); *OCD3* 1639–1640, H. Sancisi-Weerdenburg; Gh. Gnoli, *Zoroaster in History* (2001).

Antonio Panaino

Zēmarkhos of Kilikia (565 – 575 CE)

Native of Kilikia, *magister militum per Orientem*. In early August 569, Justin II (565–578) sent him on an embassy to Dizabulos, the khan of the Turks and ruler of Sogdia, accompanying a returning Turkish embassy under Maniach (Menander Prot., *fr.*19, *FHG* 4.227). A fully detailed report on the journey, lasting two to three years, is preserved in Menander Prot. *fr.*20–22, *FHG* 4.227–230. Zēmarkhos had to accompany the Turks on a military campaign against Persia. His return journey to Constantinople was long and dangerous; on this occasion he also spent some time with the Alans.

R. Hennig, "Die Einführung der Seidenraupenzucht ins Byzantinerreich," *ByzZ* 33 (1933) 302–305; *RE* 9A.2 (1967) 2500 (#4), A. Lippold; *KP* 5 (1975) 1490, *Idem; PLRE* 3 (1992) 1416–1417; *NP* 12/2 (2003) 728–729, K.-P. Johne.

Andreas Kuelzer

Zēnariōn (50 – 120 CE or 910 – 980 CE)

Cast a horoscope whose date computes either to December 22 of 57 CE, or else to December 3rd–16th of 911 CE. With the Sun in Sagittarius, Mercury's given position in Aquarius is impossible, but the actual position is closer to Aquarius in 911 (Capricorn) than in 57 (Sagittarius). The name, though typically Byzantine, is attested in the feminine (*-ion*) from 1st c. BCE to 1st c. CE (*LGPN* 3A.187: Lilybaeum).

CCAG 1 (1898) 128–129.

PTK

ZĒNŌ ⇒ ZĒNŌN

Zēnō of Elea (*ca* 470? – 430? BCE)

Perhaps born *ca* 490, Zēnō (Zēnōn) was a younger companion and follower of PARMENIDĒS, and constructed arguments exploiting conflicts between sensory evidence and claims supported by reason, with particular emphasis on problems of plurality, space, and time. He argued against plurality, and that all things are one, despite sensory evidence that things are many. Various arguments against plurality (mentioned in PLATO's *Parmenides* 127d6–128e4) all show that assuming either a plurality of entities or a plurality of predicates within a single entity entails logical contradiction. Similarly, Zēnō argued against the possibility of motion. There are no surviving texts of the arguments, but four are reconstructed from ARISTOTLE's discussions in the *Physics* and further discussions in the ancient Aristotelian commentators. Despite our belief (based on sensory experience) that things move, Zēnō argues that this is impossible: motions cannot be begun (the Achilles argument), or, if begun, cannot be completed (the Stadium, also called the Dichotomy). Further, according to "the Moving Blocks," relative motion entails contradictions, and "the Arrow" shows that any motion is indistinguishable from rest. Two other arguments, "the Paradox of Place" and "the Millet Seed," explore common-sense notions of the place wherein a thing rests, and the idea that things are composed of parts. Some ancient thinkers tended to treat Zēnō as a master of eristic rather than a philosopher raising serious problems about the natures of space, time, and the infinite; Aristotle, who called Zēnō "the Father of Dialectic" (and devoted part of the *Physics* to a study of Zēnō), was among those recognizing his importance. Twentieth-century mathematicians, physicists, and philosophers of science have found the study of Zēnō's arguments fruitful for analyses of space, time, and the infinite.

Ed.: DK 29, H.P.D. Lee, *Zeno of Elea* (1967).

G. Vlastos, "Zeno," in P. Edwards, ed. *The Encyclopedia of Philosophy* (1967) 8.369–379; A. Grünbaum, *Modern Science and Zeno's Paradoxes* (1968); W.C. Salmon, *Zeno's Paradoxes* (1970); *ECP* 579–573, Patricia Curd; *REP* 843–853, S. Makin; R. McKirahan, "Zeno," in Long (1999) 134–158; R. McKirahan, "Zeno of Elea," in D.M. Borchert, ed., *Encyclopedia of Philosophy* 2nd ed. (2005) 9.871–879; *SEP* "Zeno's Paradoxes," N. Huggett; *SEP* "Zeno of Elea," John Palmer.

Patricia Curd

Zēnodōros (*fl.* 200 BCE?)

Geometer whose only known writing, *On Isoperimetric Figures*, was a study of polygons and polyhedra, demonstrating various inequalities between polygons of equal perimeter, and polyhedra of equal surface area, aiming – unsuccessfully – to prove that the circle has a greater area than any polygon of equal perimeter and that the sphere has the corresponding property among solids. THEŌN OF ALEXANDRIA in his commentary to PTOLEMY's *Almagest* reports part of Zēnodōros' mathematical argument as does PAPPOS in *Collection* Book 5, without attribution, and the anonymous author of the PROLEGOMENA TO PTOLEMY'S SUNTAXIS. (The connection to the *Almagest* is Ptolemy's statement without proof [1.3] that the circle and sphere are the greatest of isoperimetric figures.)

The character of Zēnodōros' mathematics appears to fit the time of ARCHIMĒDĒS or soon after, but a definite dating to about 200 BCE depends on his identification with a Zēnodōros mentioned twice in the fragmentary anonymous *Life of Philōnidēs* as an associate of that **Epicurean** philosopher. Since PHILŌNIDĒS' encounters with this man were at Athens and the name Zēnodōros was rare except in Attica and the Near East, Zēnodōros probably was Athenian. Zēnodōros may also be the "astronomer" mentioned by DIOKLĒS in his *On Burning Mirrors* as having challenged Dioklēs to solve a problem in mirror optics, but the name is corrupt in the extant Arabic text.

G.J. Toomer, "The Mathematician Zenodoros," *GRBS* 13 (1972) 177–192.

Alexander Jones

Zēnodotos (Math.) (*ca* 390 – *ca* 350 BCE)

According to EUDĒMOS OF RHODES (in PROKLOS, *In Eucl.* p. 80.17 Fr.), student of ANDRŌN, with whom he distinguished *theorēma* from *problēma*. *Cf.* his contemporary AMPHINOMOS. Zhmud (2006: 178–179) identifies Proklos' source as GEMINUS not EUDĒMOS, and dates Zēnodotos (with his teacher) to the Hellenistic era.

RE 17.2 (1937) 2267–2271 (s.v. Oinopides), K. von Fritz.

PTK

***Zēnodotos (of Mallos?)* (*ca* 170 – *ca* 120 BCE?)**

Scholar who commented on HOMER and ARATOS (probably not astronomically), and perhaps identical to Zēnodotos the **Stoic** student of "Diogenēs" (perhaps DIOGENĒS OF BABYLŌN): DIOGENĒS LAËRTIOS 7.30.

FGrHist 1026 T19; *NP* 12/2.740 (#3), M.G. Albiani.

PTK

ZĒNŌN ⇒ ZĒNŌ

Zēnōn (Med.) (*ca* **200 – 150 BCE?**)

Physician, perhaps identical to ZĒNŌN OF LAODIKEIA, wrote on pharmacology and Hippokratic lexicography; cited by CELSUS 5.*pr.*1 and GALĒN as exemplary among **Hērophileans** for pharmacy (Zn.4 von Staden). Participating in the **Hērophilean** debate on pulse theory, Zēnōn argued for contraction and dilation of the arteries producing an harmonious sequence, with variation in type of pulses and timing, effecting both equal and unequal beats, analogous with respiration; ARISTOXENOS criticized the definition for its redundancies, and GALĒN noted Zēnōn's omission of the heart in his discussion of the pulse (Zn.1 von Staden), probably reflecting acceptance of BAKKHEIOS' definition of "arterial parts" as including both the arteries and the left ventricle (von Staden 1989: 504; Ba.2). Zēnōn attributed the *sigla* found, by MNĒMŌN OF SIDĒ, in MSS of the HIPPOKRATIC CORPUS, EPIDEMICS 3, to HIPPOKRATĒS himself. Zēnōn read the symbols as indicating the diagnosis, the length of the illness, and the outcome (health or death: Zn.5–6 von Staden). APOLLŌNIOS "BIBLAS" claimed he could not find Zēnōn's readings in any of the three versions he had examined personally, and further accused Zēnōn of emending the *sigla* where conceivable resolutions were lacking (Zn.6 von Staden; *cf.* He.5). Zēnōn explained the *ambē* of the HIPPOKRATIC CORPUS, JOINTS as like a door's bolt pin (Zn.7 von Staden) and explained the *kammaron* of the HIPPOKRATIC CORPUS (*Places in Man* 27) as what the Dorians in Italy called hemlock (*kōneion*) (Zn.8 von Staden). CAELIUS AURELIANUS refered to the "cassidony" drink for colon complaints of some Zēnōn, perhaps ours (*Chron.* 4.99 [*CML* 6.1.2, p. 830]).

von Staden (1989) 501–505; *OCD3* 1635 (#7), *Idem; NP* 12/2.752 (#9), V. Nutton.

GLIM

Zēnōn (of Athens?) (*ca* **80 – 120 CE**)

Among the guests discussing if food from the sea is better than food from the land in PLUTARCH'S *Table Talk* 4.4 (667C–669E). Together with KRATŌN (OF ATHENS?), he deemed fish "lighter," i.e., easier to digest, for sick people (4.4.3, 669C). Uncited elsewhere, he has been identified with the pharmacologist Zēnōn of Athens contemporary with some teacher of GALĒN, according to pseudo-Galēn, *De medicinis expertis* 10 (Chartier 1639: v. 10, p. 568; *cf.* Wickersheimer 1922).

RE 10A (1972) 146 (#14) F. Kudlien; *KP* 5 (1975) 1506–1507 (#13), J. Kollesch.

Alain Touwaide

Zēnōn of Kition (*ca* **305 –** *ca* **263 BCE**)

Founder of **Stoicism**. Zēnōn was born *ca* 335 BCE to a Phoenician family in Cyprus and came to Athens in 313 BCE. He studied philosophy under Kratēs the Cynic, POLEMŌN and XENOKRATĒS the **Academics**, and Stilpōn the Megarian before he began to give discourses at the so-called *Stoa poikilē*, the painted colonnade which gave **Stoicism** its name.

DIOGENĒS LAËRTIOS (7.4) attributes 20 titles to him, none of which survives. Zēnōn's system was to be considerably elaborated by his successors in the school, but the basic and most fundamental doctrines of **Stoicism** can often be traced back to Zēnōn. He emphasized the interdependence of physics, ethics, and logic, although as with later **Stoics**, ethics seems to have been his ultimate priority. Zēnōn's physics emphasized a unified ***kosmos*** whose substance is itself divine. Strict laws of causation meant that the ***kosmos*** was predestined but this did not absolve humans of moral responsibility, as highlighted by a story told in Diogenēs Laërtios (7.23): Zēnōn was beating a slave for stealing a loaf of bread. When the slave protested (sarcastically) that he was not guilty, for he had been fated to steal the bread, Zēnōn countered that this was true, as was the fact that the slave was fated also to be beaten.

The ***kosmos*** is unified and finite, surrounded by void. From the fundamental interconnectedness of the ***kosmos***, Zēnōn argued for the efficacy of divination. He is responsible for the original **Stoic** division into active and passive principles as the basis for physical explanation, as well as the doctrines of the interrelation of the four Aristotelian elements with ***pneuma***, and of the periodic conflagration of the ***kosmos***.

Ed.: *SVF* 1.1–72.
A. Graeser, *Zenon von Kition* (1975).

Daryn Lehoux

Zēnōn of Laodikeia (250 BCE? – 80 CE)

Greek physician, considered **Hērophilean** (Kudlien), but without evidence (Kollesch; von Staden 1989: 504–505, n.19). ANDROMAKHOS, in GALĒN *Antid.* 2.9 (14.163 K.), quotes his multi-ingredient theriac compounded from cardamom, ***herpullos***, parsley, bryony root, clover seed, anise, fennel root and seeds, birthwort, bitter vetch, ***opopanax***, in equal measures, beaten individually, mixed and administered with sour wine, then made into 3–obol pills, dried in the shade, one pill given every night in conjunction with a regimen of induced vomiting. ASKLĒPIADĒS PHARM., in Galēn *Antid.* 2.11 (14.171 K.), preserves his treatment for ***hudrophobia***, useful against any kind of venomous bite and compounded from much the same ingredients. PHILOUMENOS 10 (*CMG* 10.1.1, pp.14.25–15.6]) eschews Zēnōn's lengthy antidote, but records his plaster for healing venomous bites. PLINY (22.90) reports that a Zēnōn, possibly identifiable with our pharmacologist, recommended *soncus* root against strangury. The Zēnōn who prepared liquid colon remedies (CAELIUS AURELIANUS *Chron.* 4.99 = *CML* 6.1.2, p. 830.13–14) may, likewise, be our man.

RE 10A (1972) 146 (#13), Fr. Kudlien; *KP* 5.1506 (#2), J. Kollesch; *NP* 12/2.754 (#13), V. Nutton.

Alain Touwaide

Zēnōn of Sidōn (130 – 70 BCE)

Epicurean philosopher and **scholarch** at Athens. He lectured, and authored works on many different topics. He criticized the foundations of Euclidean geometry, and argued with the **Stoics** about whether inferences from individual cases can lead to knowledge. None of his writings has survived, but some of the works of his student, PHILODĒMOS OF GADARA, show his influence, including *On Signs* and *On Plain Speaking*.

DSB 14.612–613, K. von Fritz; A. Angeli and M. Colaizzo, "I frammenti di Zenone Sidonio," *CrErc* 9 (1979) 47–133; *OCD3* 1635, D. Obbink; *ECP* 584, D. Clay; *NP* 12/2.752–753 (#10), M. Erler.

Walter G. Englert

Zēnophilos (100 BCE – 360 CE)

Physician whose antidote for inflamed bladders and kidney stones comprised cassia, *sarxiphagos, betonikē, kuperos*, parsley, ***kostos***, chaste-tree, roasted linseed, ***malabathron***, spikenard, European wild ginger, dittany, laurel-berry, basil- and celery-seed, pine-nut, ginger, and honey, administered with honeyed or golden (*khrusattikos*) wine: OREIBASIOS, *Syn.* 3.197 (*CMG* 6.3, p. 116) = AËTIOS OF AMIDA 11.13 (p. 610 Cornarius, reading *XENOPHILVS*). The name, attested from *ca* 110 BCE (*LGPN* 1.199), and cited more frequently in the 2nd/3rd cc. CE (*LGPN* 2.193, 3A.187), is probably not Christian.

RE 10A (1972) 220, Fr. Kudlien.

GLIM

Zēnothemis (340 – 260 BCE)

Wrote a ***periplous*** of the known world in elegiacs, of which one distich is quoted by Tzetzēs, *Chil.* 7.675–677. He is also cited by PLINY for mineral products (37.34 amber, 86–88 Indian sardonyx, 90 Indian onyx, 134 Carmanian ceraunia), and by AELIANUS, *NA* 17.30 for cattle fed on live fish. For the rare name, *cf. LGPN* 1.194 (Dēlos, 297 BCE; Samos 240–220 BCE).

NP 12/2.756–757, E. Bowie.

PTK

Zeuxippos (225 – 175 BCE?)

Addressee of ARCHIMĒDĒS' lost treatises *On Numbers* and *Of Balances* (2.216 H.); mentioned in *Sand Reckoner* (2.236 H.).

Netz (1997) #118; *Idem, Works of Archimedes* (2004) 12.

GLIM

Zeuxis (Empir.) (200 – 100 BCE)

Empiricist physician, certainly prior to HĒRAKLEIDĒS OF TARAS (GALĒN, *Hipp. Epid.: CMG* 5.10.2.2, p.3). His identification with the skeptical philosopher Zeuxis mentioned by DIOGENĒS LAËRTIOS 9.106 (implying a date in the 1st c. CE) is groundless. He wrote commentaries on all treatises of the HIPPOKRATIC CORPUS that were regarded as authentic (certain: *Epid.* 2–3, 6, *De locis, De off.med., Prorrh.*; doubtful: *De hum.*), suggesting glosses, variants, and interpretations, and polemizing against previous commentators (**Hērophileans**, GLAUKIAS). He also resumed the controversy over the attribution to HIPPOKRATĒS of the marks contained in an Alexandrine copy of *Epidem.* 3, that had been advanced by the **Hērophilean** ZĒNŌN and contested by the **Empiricists** beginning with APOLLŌNIOS OF ANTIOCH. Galēn knew his commentaries on *Epid.* 3 and 6: he complained that they were difficult to find (*CMG* 5.10.2, p. 1 = 17A.605 K.).

Ed.: Deichgräber (1930) 209 (fragments), 263.
RE 10A (1972) 386–387 (#7), F. Kudlien; *KP* 5.1527 (#2), J. Kollesch; Smith (1979) 219–222; Manetti and Roselli (1994) 1594–1597; *OCD3* 1639, H. von Staden; *NP* 12/2.794 (#3), A. Touwaide; Ihm (2002) #264–271.

Fabio Stok

Zeuxis "the Hērophilean" (*ca* 80 – 10 BCE)

STRABŌN 12.8.20 notes that ". . .in my own time. . .," a Zeuxis had established a "large **Hērophilean** teaching center of medicine" at the Temple of Mēn Karou, located between Laodikeia and Karura in Phrugia, western Asia Minor. Strabōn continues to say that ALEXANDER PHILALĒTHĒS succeeded Zeuxis. Kudlien posits a "Zeuxis the Elder from Taras," but the evidence is shaky, and a recipe for the treatment of ***leikhēn*** ascribed to Zeuxis (GALĒN, *CMLoc* 5.3 [12.834 K.]) probably belongs to the homonymous **Empiricist** physician, not the **Hērophilean**. Nonetheless, a Zeuxis was named on two bronze coins issued in Laodikeia, after 27 BCE (the obverse carries *Sebastos*, Greek for *Augustus*), one of which displays a caduceus on the reverse. Nothing is known regarding Zeuxis' medical contributions, but one can suppose that his instruction at Mēn Karou included such typical **Hērophilean** topics as pulse lore, obstetrics and gynecology, and the physiology of reproduction.

RE 10A (1972) 387, Fr. Kudlien; J. Benedum, "Zeuxis Philalethes und die Schule der Herophileer in Menos Kome," *Gesnerus* 31 (1974) 221–236; *RE* S.15 (1978) 306–308, *idem* ["Philalethes"]; von Staden (1989) 529–531; Dueck (2000) 142.

John Scarborough

Zīg (Royal Tables) (450, 555/556, and *ca* 635 – 650 CE)

The Pahlavi *Zīg ī Šahryārān* (Arabic: *Zīǰ al-šāh*), three different versions of the Sasanian royal astronomical tables. The first was calculated in 450, according to Ibn Yūnis, under Yezdegird II (438–457 CE). Its mathematical parameters were probably derived from the Sanskrit PAITĀMAHASIDDHĀNTA.

Al-Hāšimī (*Kitāb al-zīǰāt, ca* 875) preserves a statement of Māšā'allāh that King Xusraw I ordered an astronomical meeting, where the Pahlavi version of PTOLEMY's *Syntaxis* was compared with the Pahlavi translation of the *Old Sūryasiddhānta*; this second set of tables was named *Zīg ī Arkand* (probably the Pahlavi translation of Sanskrit *ahargaṇa*: "series of days, calculated term"). Indian parameters were curiously preferred to the Ptolemaic ones. Al-Bīrūnī in his *al-Qānūn al-Mas'ūdī* (3.1473–1474) confirms that the meeting occurred in the 25th year of Xusraw (555–556).

Māšā'allāh wrote that, under Yezdegird III (632–652), a third set of tables was compiled. This *Zīg* was named "Trifold," because it utilized only three *kardaja* (Sanskrit *kramajyā*: "outstretched cord, sinus, etc."). Manuščihr's second *Epistle* (2.2.9–11, a 9th c. Pahlavi text), confirms the promiscuous use of different astronomical tables, such as the *Zīg of the Persian King* (or *Zīg ī Šahryārān*), the *Zīg of the Indians* (or *Zīg ī Hindūg*), and the *Zīg of Ptolemy* (or *Zīg ī Ptalamaius*). In particular the last redaction of the Sasanian *Zīg* had an enormous impact on subsequent sets of astronomical tables calculated by Arabo-Islamic and mediaeval astronomers.

Bailey (1943; 1971) 80; al-Bīrūnī, *Kitāb al-Qānūn al-Mas'ūdī* 3 vv. (1954–1956); E.S. Kennedy, *A Survey of Islamic Astronomical Tables* (1956); *GAS* 6 (1978) 106–111, 115 and 7 (1979) 102–108; 'Alī b. S. al-Hāshimī, *The Book of the Reasons behind Astronomical Tables (Kitāb fī 'ilal al-zīǰāt)*: translation by F.I. Haddad and E.S. Kennedy and a commentary by D.E. Pingree and E.S. Kennedy (1981), 95–95[R], 212–213; Pingree (1989) 238–239; Antonio Panaino, *Tessere il Cielo* (1998).

Antonio Panaino

Zōilos (of Cyprus?) (*ca* 305 BCE)

According to PLUTARCH, *Dēmētrios* 21.4–5, an engineer who made in Cyprus two *thōrakes* of extraordinarily hard steel (*sidēroi*).

(*)

PTK

Zōilos of Macedon (15 – 75 CE)

Although PLINY 1.*ind.*12–13 cites him as an authority on trees, ANDROMAKHOS, in GALĒN *CMLoc* 3.1 (12.632–633 K.), cites him as an oculist, and records an earache remedy, whereas ASKLĒPIADĒS PHARM., in Galēn *CMLoc* 4.7, cites several collyria, including one "from PACCIUS" (12.752), a "green" (12.763–764), and the *Nardinon* (12.771–772), which remained in use throughout antiquity: CASSIUS FELIX 29.13 (*CUF*, p. 63), AËTIOS OF AMIDA 7.117 (*CMG* 8.2, pp. 392–393), and ALEXANDER OF TRALLEIS (2.39–41 Puschm.). Besides nard, it contained acacia, aloes, antimony, **calamine**, saffron, ginger, ***malabathron***, myrrh, opium, ***psimuthion***, etc., in rainwater. Asklēpiadēs, in Galēn *Antid.* 2.12 (14.178–179), furthermore records his remedy for scorpion stings, also used by EPAPHRODITOS OF CARTHAGE.

NP 12/2.826 (#7), V. Nutton.

PTK

Zōpuros (Geog.) (250 – 120 BCE)

Wrote a geographical work, *On Rivers*, cited by the grammarian Harpokratiōn and by ALEXANDER OF MILĒTOS in STEPHANOS OF BUZANTION. Historical fragments of various kinds are attributed to the common name Zōpuros: IŌANNĒS "LYDUS," *Mens.* 4.150 (p. 168 Wu.) on early Rome, and Marcellinus, *Vit. Thuc.* on THUCYDIDES' death.

NP 12/2.836 (#7), H.A. Gärtner.

PTK

Zōpuros (Physiogn.) (440 – 400 BCE?)

Physiognomist known only from an anecdote about assessing Sōcratēs by physiognomic inference: "Stupid is Sōcratēs and dull, because he has no hollows at the joint of the collarbones, but these parts are blocked and stopped up; besides, he is a womanizer." At the audience's laughter, Sōcratēs defended Zōpuros' analysis, saying that this was indeed his natural inclination, but that by his intellect he had rid himself of it (CICERO *de Fato* 10; *cf.* *Tusc.* 4.80). This anecdote presumably stems from the dialogue *Zōpuros* by Phaidōn of Ēlis (DIOGENĒS LAËRTIOS 2.105) and probably served to illustrate the current theme in Socratic writings about the discrepancy between the appearance of the body and the nature of the soul (*cf.* ANTISTHENĒS). Another version of the anecdote calls Zōpuros a "wise man from Syria" and has him prophesy an unnatural death to Sōcratēs (D.L. 2.45).

RE 10A (1972) 768–769 (#3), K. Ziegler; *SSR* 1.491–492 (*Phaedon fr.*8–11) and 4.115–127; G. Boys-Stone, "Physiognomy and Ancient Psychological Theory. I. The Circle of Socrates (1): Phaedo of Elis," in Swain (2007) 22–33.

Sabine Vogt

Zōpuros of Alexandria (130 – 70 BCE)

Surgeon and pharmacologist, active in Alexandria, teacher of the **Empiricist** Apollōnios of Kition (whence we infer that he belonged to the **Empiricist** "school"). He kept up a correspondence with Mithradatēs VI: he sent the king an antidote suggesting that he try it on some condemned men (Galēn *Antid.* 2.8 [14.150 K.]; Scribonius Largus 169). He prepared an antidote, called *Ambrosia*, for one Ptolemy, perhaps XII Auletēs (Galēn *Antid.* 2.17 [14.205 K.]; Celsus 5.23.2). Further pharmaceutical prescriptions, probably from a work entitled *On simple remedies*, are attested by Pliny (24.87), Oreibasios (several remedies), Aëtios of Amida, and Dioskouridēs. The remedy called *zopyrium* from the name of its inventor, mentioned by Caelius Aurelianus *Acute* 3.47 (*CML* 6.1, p. 320), *Chron.* 2.210, 3.58, 5.118 (pp. 672, 712–714, 924), is perhaps named after him.

Ed.: Deichgräber (1930) 21, 205–206 (fragments), 261–262.
RE 10A (1972) 771–772 (#15), J. Kollesch; *NP* 12/2.836 (#8), V. Nutton; *AML* 938–939, G. Marasco.

Fabio Stok

Zōpuros of Gortuna (*ca* 20 – 55 CE)

Physician, Scribonius Largus' guest-friend during Zōpuros' ambassadorship to Rome (*Comp.* 172). Scribonius included one of Zōpuros' antidotes, now lost (*Comp.* 169, and index).

RE 10A (1972) 772 (#17), J. Kollesch.

GLIM

Zōpuros of Taras (220 – 210 BCE?)

Bitōn records two tension-catapults constructed by Zōpuros, whom Diels (1930: 22–23) identified with an otherwise unknown **Pythagorean** mentioned by Iamblikhos (*VP* 267), and dated to *ca* 360 BCE. But the name is very common (*LGPN*), and there is no need to make the identification. Sometime before 170 BCE, at Milētos, Zōpuros designed a mid-sized arrow-shooter (*gastraphetēs*: Bitōn pp. 61–64), perhaps in preparation for an attack expected from Attalos I in 218 BCE (Polubios 5.77.2–9). Zōpuros also designed at Italian Cumae a smaller device, the "mountain *gastraphetēs*" (Bitōn, pp. 65–67), probably for its emergency defense against Hannibal in 215 BCE (Livy 23.35–37). *Cf.* Hērakleidēs of Taras (Mech.), and other contemporary mechanics such as Numphodōros, Pausistratos of Rhodes, Philōn of Buzantion, and Truphōn of Alexandria, who innovatively redeployed old technology.

RE S.15 (1978) 1556 (#19a), E. Fischer.

PTK

Zoroaster, pseudo (500 – 300 BCE)

Legendary character credited spuriously with numerous treatises, some of which treated astrology, because of the renown of the Iranian *Magi*, frequently associated or confused with the Chaldaeans. According to Pliny (7.72) "magic" originated from Zoroaster's medicine. The legend of Zoroaster was developed in the Iranian (and later Islamic) framework, as well as among Classical and Western authors: for example the late-Byzantine author Plēthōn attributed the *Oracula Chaldaica* to Zoroaster.

Various astrological works in Arabic were attributed to Zoroaster. E.g., an astrological compilation of HERMETIC origin was translated into Pahlavi at the time of Xusraw I; the Greeks, and later the Arabs (and Persians), ascribed it to Zoroaster. This text was corrected and revised by Māhānkard *ca* 637, then translated into Arabic by Sa'īd ibn Xurāsānxurrah (*ca* 747/754) as *Kitāb al-mawālīd.* This Arabic text, incorporating many Pahlavi terms and loanwords, reveals a Greek foundation, intermingled with Indian and Iranian traditions and doctrines, although some elements could show a Ḥarrānian influence. Consequently, Pingree suspects that a Ḥarrānian native composed the Pahlavi version from a Greek original of the 3rd c. CE, which Māhānkard deeply revised.

Theophilos of Edessa (8th c.) read and translated Sasanian astrological material (whether in Pahlavi, Arabic, Syrian, or Greek), e.g., letters attributed to Zoroaster.

C. Clemen, *Fontes Historiae Religionis Persicae* (1920); V. Stegemann, "Astrologische Zarathustra-Fragmente bei den arabischen Astrologen Abū'l Ḥasan Alī b. abī'r-Riǧāl (11. Jh.)," *Orientalia* ns 6 (1937) 317–336; D.E. Pingree, *The Thousands of Abū Ma'shar* (1968) 7, 10, 22, 130; *Idem* (1978) 445; *GAS* 7 (1979) 81–86; C.M. Woodhouse, *George Gemistos Plethon* (1986); Pingree (1989) 227–239; P. Kunitzsch, "The Chapter on the Fixed Stars in Zarādusht's *Kitāb al-mawālīd*," *Zeitschrift für Geschichte der arabisch-islamischen Wissenschaften* 8 (1993) 241–249; Antonio Panaino, "Sopravvivenze del culto iranico della stella Sirio nel *Kitāb al-mawālīd* di Zarādušt ed altre questioni di uranografia sasanide," in E. Acquaro, ed., *Alle soglie della Classicità. Il Mediterraneo tra tradizione e innovazione. Studi in onore di S. Moscati* (1996) 1.343–354.

Antonio Panaino

Zōsimos (Med.) (10 BCE – 95 CE)

ASKLĒPIADĒS PHARM., in GALĒN *CMLoc* 4.8 (12.753 K.), records his collyrium, compounded from **calamine** roasted and quenched in Italian wine, plus acacia, aloes, antimony, copper oxide, saffron, myrrh, Indian nard, and opium, in gum and rainwater. OREIBASIOS, *Ecl. Med.* 69.6 (*CMG* 6.2.2, p. 232), records his ointment for tremors, a carefully prepared mixture of ***euphorbia*** (*cf.* IOUBA), marsh-salt (DIOSKOURIDĒS 5.119), pine-resin, natron, and ***opopanax***, in olive oil and beeswax; the recipe is praised and repeated by PAULOS OF AIGINA, 3.21 (*CMG* 9.1, p. 170), 7.19.16 (*CMG* 9.2, p. 378).

RE 10A (1972) 790 (#5), Fr. Kudlien.

PTK

Zōsimos of Panōpolis (*ca* 250 – 300 CE)

Earliest Hellenistic alchemist who wrote without a pseudonym and whose writings survive in any appreciable number. He was a prolific commentator on the works of previous alchemists and in his writings are found a blend of practical laboratory instructions, lore concerning the (pseudo-)history and mythology of alchemy and mystical and religious speculations of a Gnostic and Hermetic character. He was born at Panōpolis in the Thēbaïd region of Upper Egypt and was, perhaps, resident at Alexandria. He composed a large number of alchemical treatises, many of which addressed to THEOSEBEIA, and it is from these treatises that we have much of our knowledge of the other early alchemists and their works. Despite his fame and prominent position in alchemical history, almost nothing is known about his life and because of the allegorical and secretive style of the alchemists, little can be said with certainty about his alchemical doctrines.

Phōtios calls him a Thēban of Panōpolis and says that his writings were discussed in a work, of unknown title and author, which tried to prove that pagan intellectuals of all lands proclaimed Christian dogma in advance (*Bibl.* 170). The *Souda*'s short entry on Zōsimos (Z-168) calls him a philosopher of Alexandria, perhaps indicating, against the general opinion that he was a Panōpolite, that he was a resident of the mētropolis. The *Souda* says that Zōsimos wrote chemical works in 28 (*sic*) books, arranged alphabetically, addressed to his "sister" Theosebeia, and called by some the *Kheirokmēta* (*Things Wrought By Hand*; elsewhere *Kheirotmēta*). Although Zōsimos himself mentions a work of his thus titled and addressed to Theosebeia (Mertens 1995, §4.2), and treatises entitled with letters of the alphabet survive and are referred to in his extant works, it is difficult to judge which, if any, of these texts were part of the *Kheirokmēta*. Finally, the *Souda* says that he wrote a biography of PLATO, entirely lost.

The date of Zōsimos's lifetime is broadly bounded by two points: a) his citation of IULIUS AFRICANUS (*CAAG* 2.169) provides a *terminus post quem* in the first half of the 3rd c. CE, while b) his reference to the Serapeion at Alexandria as still existent (Mertens 1995, §1.8) allows a *terminus ante quem* at the destruction of that temple in 391 CE. Thus, most scholars date him to the late 3rd or early 4th c., although Hammer Jensen (1921: 99) dated him to *ca* 500 CE. Letrouit (1995: 46) would tighten these boundaries, noting that Zōsimos' veiled anti-Manichaean polemic (Mertens 1995, §1.14) would only have been appropriate between the introduction of Manichaeism into Egypt (*ca* 268 CE) and the death of Mani (278 CE).

Zōsimos was a fervent follower of MARIA whose descriptions of furnaces and other chemical apparatus he elaborates and whose alchemical maxims he preserves (Mertens 1995: CXIII-CLXIX). Likewise he followed HERMĒS TRISMEGISTOS, AGATHODAIMŌN, PSEUDO-DĒMOKRITOS and other early alchemical authorities. Aside from discussions of earlier alchemical works and their instruments, ingredients and procedures, Zōsimos' writings also occasionally took the form of allegorical dream-visions (Mertens 1995, §10–12), discussions of material similar to that found in the *Corpus Hermeticum* (e.g. in Mertens 1995, §1) and spiritual advice to Theosebeia (Festugière 1950: 366–368).

M. Berthelot and C. E. Ruelle (*CAAG* 2.107–252) edited his extant Greek works, but the inclusion in this edition of texts full of later interpolations, texts that are in fact by later authors merely citing Zōsimos, as well as numerous errors due to the obscurity of the subject matter and problems surrounding the manuscript tradition, have rendered the edition untrustworthy. Mertens is re-editing Zōsimos' Greek works; see also Letrouit (1995: 22–37) for a list of his known Greek works and discussion of those falsely attributed to him in the *CAAG*.

Texts attributed to Zōsimos are extant in Syriac (partial paraphrased translation in Berthelot and Duval 1893: 203–266) and Arabic, much of which appears to be authentic and some of which, in the case of the Arabic, are translations of extant Greek texts (Hallum 2008: 114–192). However, the gnomic sayings preserved in early Arabic literature and mentioned by Ullmann (1972: 160–161) do not derive from Zōsimos (Hallum 2008: 34–87).

Ed.: Mertens (1995) 1–49.

M. Berthelot and R. Duval, *La Chimie au moyen âge* 2: *L'achimie Syriaque* (1893); I. Hammer Jensen, *Die älteste Alchymie* (1921); *DSB* 14.631–632, M. Plessner; M. Mertens, "Project for a New Edition of Zosimus of Panopolis," in Z.R.W.M. von Martels, ed., *Alchemy Revisited* (1990) 121–126; *Dictionary of Gnosis and Western Esotericism* 2 (2005) 1183–1186, A. de Jong; *NDSB* 7.405–408, M. Mertens; Bink Hallum, *Zosimus Arabus: the Arabic/Islamic Reception of Zosimos of Panopolis* (Diss. London, 2008).

Bink Hallum

GAZETTEER

We list here all 290 or more sites and all 35 or more regions from which ca *1000 ancient scientists are attested or considered to have originated (when two or three homonyms came from the same place, that is marked with a parenthetical Arabic numeral "2" or "3" – see Alexandria, e.g.). A few other sites mentioned in the entries are also included. Identifications of ancient with modern sites are often controversial, and we have simply listed what we believe to be the scholarly consensus (indicating disagreements where we found them); a few sites have not yet been located. Because modern names are subject to change, we have given the latitude and longitude of each site. Moreover, about one tenth of the sites are ambiguous (marked by %), since two or more by the same name existed: often such sites were distinguished in antiquity by some epithet or localization (which may nevertheless have been lost in transmission), the oldest or largest site being sometimes unmarked: see esp. Antioch, Apameia, Hērakleia, Laodikeia, and Neapolis. In some cases, disambiguation is not certain, and we index such entries first, indicating the possibilities. Finally, where the attribution of a place of origin to one of the scientists is itself uncertain, the scientist's name is marked with "(?)".*

Notes

1) *Regions are underlined, and include a list of their cities in this gazetteer, plus a list of scientists from that region for whom no city is known. (Sufficiently small islands are considered sites.)*
2) *The historical sketches cover the period represented by the relevant entries; thus the later history of, e.g., Arados, Gadēs, Knōssos, and Nola is omitted.*
3) *Certain turning points are emphasized: foundation, autonomy or its loss, conquests by non-Greeks (esp. by* ***Rome****), and raids in the 3rd c. by the Goths (256–277: see* ***Argos****,* ***Boiōtia****,* ***Ephesos****,* ***Kilikia****,* ***Nikaia (Bithunia)****,* ***Nikomēdeia****,* ***Pergamon****,* ***Sidē****, and* ***Trapezous****), or Heruli (267: see* ***Athens****,* ***Buzantion****,* ***Corinth****, and* ***Sparta****), and invasions by the Vandals in the early 5th c. (****Burdigala****,* ***Ēpeiros****,* ***Hispalis****, and esp.* ***Africa*** *and* ***Mauretania****, within which see* ***Caesarea****,* ***Carthage****,* ***Cirta****,* ***Hippo Regius****, and* ***Sicca Veneria****).*

Abdēra (mod. Avdira; 40°57' N, 24°59' E): coastal city of **Thrakē** opposite **Thasos**, founded mid-7th c. BCE from **Klazomenai**, augmented with settlers from **Teōs** in 544 BCE (Hērodotos 1.168), allied with **Athens** in the 5th c. BCE, sacked in 350 BCE (by Philip II of **Macedon**) and 170 BCE (by Eumenēs II of **Pergamon**); small thereafter. *PECS* 3–4, D. Lazarides; *OCD3* 1, J.M.R. Cormack and N.G.L. Hammond; *BAGRW* 51-D3; *BNP* 1 (2002) 16 (#1), I. von Bredow.

Biōn, Dēmokritos, Diokleidēs, Hekataios, Leukippos, Prōtagoras.

Abudos (mod. Mal Tepe near Naara Point; 40°12' N, 26°23' E): Musian city on the Hellespont, between **Sigeion** and **Lampsakos**, sacked by Philip V of **Macedon**

(Polubios, Book 16, *frr.*29–34), and made a free city by **Rome** (Livy 33.30). *PECS* 5, G.E. Bean; *OCD3* 1–2, St. Mitchell; *BAGRW* 51-G4; *BNP* 1 (2002) 38 (#1), E. Schwertheim. (Contrast the Egyptian homonym, *BAGRW* 77-F4.)

Alkamenēs, Isidōros.

Africa: the northern portion of the modern African continent excepting **Egypt**, i.e., the coastal region from the western edge of the Nile Delta to Cape Delgado; later the Roman province *Africa* (initially approximately modern Tunisia and eastern coastal Algeria, later expanded eastwards and westwards), whose capital was **Utica**. Sometimes distinguished from, or at other times including, **Libya** and **Mauretania**. Taken by the Vandals 430 CE; retaken by Belisarius for **Buzantion** 534 CE. *OCD3* 33, J.M. Reynolds; *BNP* 1 (2002) 291–300, W. Huß *et al.*

Sites: **Auzia**, **Caesarea (Mauretania)**, **Carthage**, **Cirta**, **Hippo Regius**, **Kurēnē**, **Madaurus**, **Sicca Veneria**, **Utica**.

People: Marius Victorinus, Marcianus (?), Melitianus, Muscio/Mustio (?), Probinus, Macrobius Ambrosius Theodosius (?).

Agurion (mod. Agira; 37°39' N, 14°31' E): inland city of **Sicily** west of **Centuripae**, much built up by Timoleōn *ca* 335 BCE. *PECS* 18–19, M. Bell; *BAGRW* 47-F3; *BNP* 1 (2002) 398, G. Makris.

Diodōros.

% Aigai (mod. Yuntdağıköseler; 38°51' N, 27°12' E): east of **Kumē** and south of **Pergamon**, which controlled Aigai from 218 BCE; ravaged by Prousias II (*ca* 155 BCE). *PECS* 19, G.E. Bean; *OCD3* 16, N.G.L. Hammond; *BAGRW* 56-E4. Several homonymous sites exist, *cf.* Stephanos of Buzantion, s.v., from one of which Dionusios may have come: esp. **(A)** in **Macedon** (mod. Vergina, *BAGRW* 50-B4, small after 168 BCE; gone after 1st c. CE); **(B)** on Euboia (mod. Politika Kafkala, *BAGRW* 55-F3); **(C)** the Peloponnesian Aigai (mod. Akrata, *BAGRW* 58-C1); and **(D)** Aigai of **Kilikia**, mod. Yumurtalık, *BAGRW* 67-B3).

Dionusios (?), Pollēs.

Aigina (mod. Aigina; 37°45' N, 23°20' E): island, maritime trade center (whose monetary and metrological system were standards in commerce) in the 7th–6th cc. BCE; opposed and finally overwhelmed by neighboring **Athens** 488–431 BCE. Prosperous from Hellenistic to early Byzantine times. *PECS* 19–21, B. Conticello; *OCD3* 17, S. Hornblower; *BAGRW* 58-E2; *BNP* 1 (2002) 192–194, H. Kalcyk.

Paulos, Petrōn.

Aizanoi (mod. Çavdarhisar; 39°12' N, 29°37' E): well east of **Pergamon**, and well south of **Prousias**, in Phrugia; 184 BCE taken by Eumenēs II of **Pergamon** (from Prousias I of **Bithunia**); then under **Rome** from 133 BCE. Prosperous esp. in the 2nd c. CE. Strabōn 12.8.11; *RE* 1.1 (1893) 1131–1132, G. Hirschfeld; *PECS* 16, R. Naumann; *BAGRW* 62-C3.

Apollinarios.

Akhmim ⇒ Panōpolis

Akragas (mod. Agrigento; 37°19' N, 13°35' E): founded on south-west coast of **Sicily** *ca* 582 BCE from **Gela**, prosperous and democratic in the 5th c. BCE; sacked by **Carthage** in 406 BCE. Restored by Timoleōn *ca* 335 BCE; taken by **Rome** in 210 BCE, who enslaved the population, and resettled the city *ca* 195 BCE with Sicani. Plundered by Verres, 73–71 BCE. *PECS* 23–26, P. Orlandini; *OCD3* 9, A.G. Woodhead and R.J.A. Wilson; *BAGRW* 47-D4; *BNP* 1 (2002) 110–111, G. Manganaro.

AKRŌN, EMPEDOKLĒS, NIKŌN, PHAIAX (?).

Alabanda (mod. Araphisar; 37°38' N, 27°57' E): old Karian city upstream from **Tralleis**; sacked by Philip V of **Macedon** *ca* 200 BCE (POLUBIOS, Book 16, *fr.*24), and allied with **Rome** by *ca* 170 BCE. *PECS* 28–29, G.E. Bean; *OCD3* 49, W.M. Calder and S. Sherwin-White; *BAGRW* 61-F2; *BNP* 1 (2002) 418, H. Kaletsch.

HERMOGENĒS.

Alexandria "near Egypt" (31°12' N, 29°55' E): founded by Alexander of **Macedon**, 332/331 BCE; under Ptolemy I became the capital of the Ptolemaic Empire, and a center of commerce. Ptolemy placed his library in a shrine to the Muses, the "Mouseion" (with its scholars nearby), and his successors vigorously augmented the collection, by confiscating, copying, and translating. Ptolemy VIII expelled the librarian and scholars in 145/144 BCE (Ath., *Deipn.* 4 [83]), after which the librarianship was a sinecure for courtiers; the scholars turned to the systematic study and criticism of existing literature. The acquisitions policy had encouraged forgeries, and debates about authenticity occupied them. Much damaged by CAESAR's conquest, 49–48 BCE; and in the Jewish revolt ("Kitos" War) of 115–117 CE (*cf.* **Cyprus**, **Edessa**, **Kurēnē**, and **Nisibis**). Again damaged in wars and riots of the 3rd c. CE; later Christian riots destroyed more. Taken by the Sasanians 618–628 CE. *PECS* 36–38, S. Shenouda; *ODB* 60–61, P. Grossman and L.S.B. MacCoull; *OCD3* 61–62, D.W. Rathbone; *BAGRW* 74-B2; *BNP* 1 (2002) 496–498, K. Jansen-Winkeln; *EJ2* 1.632, A. Tcherikover. Some of these about 80 people may have been born or educated outside of Alexandria; *cf.* **Athens**.

ADAMANTIOS IOUDAIOS, AELIUS PROMOTUS (?), AGAPIOS, AGATHODAIMŌN, ALUPIOS, AMARANTOS, AMMŌNIOS (2), ANTIGONOS, APOLLŌNIOS, ASKLĒPIODOTOS, CLAUDIUS CLAUDIANUS, DĒMĒTRIOS, DIDUMOS (3), DIODŌROS (ASTRON.)(?), DIONUSIOS, DIONUSIOS PERIĒGĒTĒS, DIOPHANTOS, DIOSKOURIDĒS, EUCLID, EUDĒMOS, EUDŌROS, GORGIAS, HARPOKRĀS, HARPOKRATIŌN, HĒLIODŌROS (2), HĒRAKLEIANOS, HERMODŌROS, HĒRŌN, HIERIOS (?), HUPATIA, HUPSIKLĒS, IAKŌBOS (?), IŌANNĒS, IŌANNĒS PHILOPONOS, IULIANUS (?), KHRUSERMOS, KHRUSĒS, KLEOPATRA, KTĒSIBIOS, KURILLOS (?), LEŌNIDAS (2), MARINOS, MEGETHIŌN (?), MEGETHIOS, MENELAOS, MĒTRODŌROS, OLUMNIOS, OLUMPIODŌROS (2?), OLUMPOS, PALLADIOS, PAMPHILOS, PANKRATĒS, PAPPOS, PARMENISKOS, PAULOS, PHILŌN, PHILOUMENOS, PHILOXENOS, POTAMŌN, PTOLEMY, PUTHAGORAS (?), SELEUKOS, SERAPIŌN (2), SŌSTRATOS, SŌTIŌN, SPEUSIPPOS, STEPHANOS (2?), SYRIANUS, THEODOS, THEŌN (3), THEŌN *ARKHIATROS*, THEŌN (STOIC), TIMAGENĒS, TIMOKHARIS (?), TRUPHŌN, ULPIANUS (?), ZŌPUROS.

Alexandria Troas (mod. Eskistanbul/Eski Stambul ***or*** Dalyanköy; 39°46' N, 26°09' E): coastal city opposite **Tenedos**, built by Antigonos *ca* 310–305 BCE as Antigoneia and soon renamed by LUSIMAKHOS to Alexandria. Built up by AUGUSTUS, and in the 2nd c. CE. *PECS* 39, C. Bayburtluoğlu; *OCD3* 62, St. Mitchell; *BAGRW* 56-C2; *BNP* 1 (2002) 498 (#2), E. Schwertheim.

HĒGĒSIANAX, QUINTILII.

Amaseia (mod. Amasya; 40°39' N, 35°50' E): inland city of Pontos, founded *ca* 300 BCE, and the kingdom's capital until 183 BCE, when the capital became **Sinōpē**. An important city of the succeeding client kingdoms; the nearest port was **Amisos**. *PECS* 47, D.R. Wilson; *OCD3* 69, St. Mitchell; *BAGRW* 87-A4; *BNP* 1 (2002) 560, E. Olshausen.

STRABŌN.

Amida (mod. Diyarbakır; 37°59' N, 40°13' E): city at the upper navigable limit of the

Tigris, north of **Rešʿaina** and **Constantia**, west-south-west of **Artemita**. Under **Rome** from 66 BCE (*cf.* **Syria**); fortified against the bordering Sasanians by Constantius II *ca* 345 CE; taken by the Sasanians 359–363 CE. Monastic center in the 5th c.; taken by Sasanians and then successfully and devastatingly besieged by **Rome** 503–506 CE; under frequent attack by Persians and Huns until 562 CE; suffered from the plague of 542 CE. *PECS* 49, R.P. Harper; S.A. Harvey, *Asceticism and Society in Crisis* (1990) 57–65; *ODB* 77, M.M. Mango; *BAGRW* 89-C3; *BNP* 1 (2002) 580–581, J. Pahlitzsch.

AËTIOS.

Amisos (mod. Samsun; 41°17' N, 36°20' E): Greek colony on south coast of Black Sea, east of **Sinōpē** and west of **Trapezous**, founded in the mid-8th c. BCE (at the terminus of a trade route), restored to democracy by Alexander of **Macedon**, and then serving as port to **Amaseia**; expanded by MITHRADATĒS VI, destroyed and restored by Lucullus 71 BCE; survived a siege 48–47 BCE and a tyrant 36–31 BCE. *PECS* 49, D.R. Wilson; *OCD3* 72, St. Mitchell; *BAGRW* 87-B3; *BNP* 1 (2002) 581–582, E. Olshausen.

DĒMĒTRIOS, HUPSIKRATĒS.

Amiternum (mod. S. Vittorino; 42°24' N, 13°19' E): south-east of **Spoletium**, north-west of **Sulmo**, traditional capital of the Sabines, fully Roman by the 2nd c. BCE. *PECS* 49–50, E.T. Salmon; *BAGRW* 42-E4; *BNP* 1 (2002) 582, G. Uggeri.

SALLUSTIUS CRISPUS.

Amphipolis (mod. Amphipolis; 40°49' N, 23°51' E): coastal city of **Thrakē**, 4 km north of the estuary of the Strumon river (north of **Stageira**). Colonized from **Athens** by Hagnōn, son of Nikias, in 437–436 BCE, in turn capital of the Edones and **Macedon**, then serving as Alexander's chief mint. It remained an important way-station under the Romans on the *Via Egnatia*. *PECS* 51–52, D. Lazarides; *OCD3* 76, J.M.R. Cormack and N.G.L. Hammond; *BAGRW* 51-B3; *BNP* 1 (2002) 605, R.M. Errington. (Contrast the Syrian homonym, mod. Jebel Khaled; *BAGRW* 67-G4.)

EUPHRŌNIOS.

Amphissa (mod. Salona/Amfissa; 38°32' N, 22°22' E): largest city of west **Lokris**, west of **Khairōneia**; from 196 BCE allied with Aitolia, often in conflict with Delphi; many Aitolians settled here after Actium 31 BCE, and the city thereafter claimed to be Aitolian. *PECS* 993, L. Lerat; *OCD3* 76, W.M. Murray; *BAGRW* 55-C3; *BNP* 1 (2002) 606, G. Daverio Rocchi.

PHILŌTAS.

Anazarbos (mod. Anavarza Kalesi; 37°15' N, 35°52' E): city of **Kilikia** in the valley of the Puramos river, near **Mallos** and **Tarsos**; refounded 19 BCE by AUGUSTUS. *PECS* 53–54, M. Gough; *BAGRW* 67-B2.

DIOSKOURIDĒS, OPPIANUS.

Antinoeia/Antinoopolis (mod. Sheik ʿIbada; 27°49' N, 30°53' E): city of Upper **Egypt**, on the east bank of the Nile, across from Hermopolis Magna, downstream of **Panōpolis** and upstream of **Oxyrhynchos**; founded by Hadrian 130 CE, where his beloved Antinoos had drowned; capital of the Thebaid nome under Diocletian; a Christian bishopric from the 3rd c. STEPHANOS OF BUZANTION, s.v.; *RE* 1.2 (1894) 2442 (#2), R. Pietschmann; *PECS* 60, S. Shenouda; *OCD3* 106, W.E.H. Cockle; *BAGRW* 77-D1; *BNP* 1 (2002) 756, R. Grieshammer.

PEITHŌN (?), SERENUS.

% Antioch/Antiokheia: many cities were founded under this name in regions controlled by the Seleukids, from one of which these men may have come, if they are not from

Antioch on the Orontes: for example, (**A**) Antioch Epiphanēs on the Maiandros (mod. Aliagaçiftligi near Azizabat, *BAGRW* 61-H2/65-A2, Stephanos of Buzantion, s.v., #2); (**B**) Antioch of **Pisidia** (near mod. Yalvaç, *OCD3* 107, St. Mitchell; *BAGRW* 62-F5, Stephanos of Buzantion, s.v., #4); (**C**) Antioch of **Kilikia**, on Saros, formerly Adana (*BAGRW* 66-G3); and (**D**) Antioch of **Kilikia**, on Kragos (mod. Endiegney/Güney Köy; *BAGRW* 66-A4).

Apollōnios Biblas, Isidōros, Kárpos, Serapiōn.

Antioch ⇒ Arados

Antioch ⇒ Constantia

Antioch ⇒ Edessa

Antioch ⇒ Kharax

Antioch ⇒ Nisibis

Antioch ⇒ Tarsos

Antioch on the Orontes (mod. Antakya; 36°12' N, 36°09' E): founded by Antigonos 307 BCE as Antigoneia, displaced 300 BCE and renamed by Seleukos I (in honor of his father Antiokhos), and made the capital of the empire; expanded *ca* 235 and 170 BCE. After the conquest by **Rome**, 64 BCE, became the capital of the province *Syria* (*cf.* **Syria**). Frequently damaged by earthquakes (140 BCE, 37 CE, 115 CE); taken by the Sasanians 256 and 260 CE. Christian center from the 1st c., and school of literal-historical interpretation of the Christian scriptures. Capital of *Coele Syria* from 350 CE; of all *Syria* from 415 CE. After a fire (525 CE), earthquake (526), and Sasanian conquest (540), thoroughly rebuilt by Justinian. *PECS* 61–63, J. Lassus; *ODB* 113–116, M.M. Mango; *OCD3* 107, A.H.M. Jones *et al.*; *BAGRW* 67-C4; *BNP* 1 (2002) 757–758, A.-M. Wittke, 758–759, T. Leisten; *EJ2* 2.201–202, A. Haim and D. Kushner.

Alupios, Antiokhos VIII, Iōannēs "Khrusostomos," Krateros, Libanios, Rufinus, Vettius Valens.

% Apameia: many homonymous cities were founded in regions controlled by the Seleukids, from one of which these men may have come, if not from **Apameia on the Orontes**: for example, Apameia Kibotos, also known as Kelainai (mod. Dinar; *BAGRW* 65-D1); and Apameia on the Euphratēs (mod. Keskince, *BAGRW* 67-F2); *cf.* Stephanos of Buzantion, s.v.

Damostratos, Dēmētrios.

Apameia ⇒ Murleia

Apameia on the Orontes (mod. Qalaat el-Moudiq/Qalʿat al-Mudiq; 35°25' N, 36°24' E): founded 300 BCE by Seleukos I as Pella (in honor of Macedonian **Pella**), on the site of Pharnaka (south and upstream of **Antioch on the Orontes**, north of **Emesa**). Soon renamed, it served as the Seleukid treasury and horse-breeding center (Strabōn 16.2.10). The fortress was destroyed by Pompey 64 BCE; built up under **Rome** in the 1st c. CE;

became a Christian center (*cf.* **Syria**). *PECS* 66–67, J.-P. Rey-Coquais; *OCD3* 118, A.H.M. Jones and S. Sherwin-White; *BAGRW* 68-B3; *BNP* 1 (2002) 817 (#3), J. Oelsner.

Arkhigenēs, Noumēnios, Oppianus, Poseidōnios.

Aphrodisias of Kilikia (east of mod. Yeşilovacık; 36°12' N, 33°42' E): coastal city, south-west of **Seleukeia (Kilikia)**, attested by pseudo-Skulax, 40; little or no history is known. Stephanos of Buzantion, s.v. (listing ten homonyms, e.g. in **Thrakē**, *BAGRW* 51-H3); *RE* 1.2 (1894) 2726 (#1), A. Wilhelm; *BAGRW* 66-D4.

Menemakhos, Xenokratēs.

Aphrodisias of Karia (mod. Geyre; 37°42' N, 28°43' E): on a tributary of the Maiandros (not far from **Hērakleia Salbakē**, and upstream from **Tralleis**). The pre-Greek name Ninoē probably refers to a fertility-goddess; the name Aphrodisias is first attested in the 3rd c. BCE. Close relations with **Rome** began with Sulla (devoted to Venus), and continue under rulers from Caesar to the end of the 3rd c. CE. The temple of Aphrodite was made into a church in the 5th c. CE. *PECS* 68–70, K. Erim; *OCD3* 119–120, J.M. Reynolds; *BAGRW* 65-A2; *BNP* 1 (2002) 828–829, H. Kaletsch.

Adrastos, Alexander, Apollōnios.

Apollōnia Pontikē (mod. Sozopol; 42°25' N, 27°42' E): colony on west coast of Black Sea, founded by **Milētos** *ca* 610 BCE; south of **Kallatis** and about a day's sail north of the Bosporos. Traded and allied with **Athens** in the 5th c. BCE, but declined thereafter; allied with Mithradatēs VI, and plundered by Lucullus 71 BCE. *PECS* 72–73, A. Frova; *OCD3* 124, Max Cary and N.G.L. Hammond; *BAGRW* 22-E6; *BNP* 1 (2002) 865–866 (#2), I. von Bredow.

Diogenēs.

Aquitania: south-west France, approximately the Pays Basque (*cf.* Caesar, *BG* 1.1). *OCD3* 134, J.F. Drinkwater.

Sites: **Burdigala, Vasates, Vesunna**.

People: Victorius.

Arados (mod. er-Rouad/Arwad; 34°51' N, 35°52' E): millennium-old trading center on a small island near the Syrian coast (north of **Bublos**, and west of **Emesa**), head of a Phoenician trading alliance, conquered by Alexander of **Macedon** 332 BCE; autonomous under the Seleukids, and known as Antioch Pieria (Stephanos of Buzantion, s.v., #7). *PECS* 82, J.-P. Rey-Coquais; *OCD3* 135, J.-F. Salles; *BAGRW* 68-A4; *BNP* 1 (2002) 948, M. Köckert.

Kallias.

Arelate (mod. Arles; 43°41' N, 04°38' E): colony of **Phōkaia** founded as Thelinē in the 6th c., destroyed by Ligurians 535 BCE, and revived in the 4th c. BCE; under **Rome** became one of the chief cities of the province *Gallia Narbonensis* (see **Narbo**). Canals dug by Marius (104 BCE) repaired its access to the sea; sided with Caesar in 49 BCE against **Massalia**. Built up by Augustus, and again by Constantine, when it became a Christian center; occupied by Visigoths *ca* 480 CE. *PECS* 87–88, R. Amy; *OCD3* 151, J.F. Drinkwater; *BAGRW* 15-D2; *BNP* 1 (2002) 1044–1045, Y. Lafond.

Fauorinus, Hilarius.

Argos in Akarnania/Amphilokhia (mod. Loutron; 38°57' N, 21°12' E): archaic city on the Ambrakian Gulf, across from **Nikopolis**; allied with **Athens** in 430 BCE. *PECS* 91, M.H. McAllister; *BAGRW* 54-D4.

Euēnōr.

Argos (mod. Argos; 37°37' N, 22°44' E): pre-Hellenic city in the valley of the Inakhos not far inland, and south of **Corinth** and **Sikuōn**. Prosperous in Mycenaean times, rival of **Sparta** from early times, responsible for the destruction of **Asinē**, neutral in the Persian Wars. Democratic from *ca* 460 BCE and allied with **Athens**. Variously ruled in the Hellenistic period; devastated by the Goths 267 and by Alaric 395 CE. (Homonymous sites existed, all distinguished by some epithet: e.g., **Akarnanian Argos**, and Argos Hippion, *BAGRW* 45-C1.) *PECS* 90–91, J.F. Bommelaer; *OCD3* 155, R.A. Tomlinson and A.J.S. Spawforth; *BAGRW* 58-D2; *BNP* 1 (2002) 1070–1073, Y. Lafond.

HARPOKRATIŌN, KAIKALOS, LOBŌN, PANKRATĒS, POLUKLEITOS, SŌKRATĒS.

Arpinum (mod. Arpino; 41°30' N, 13°37' E): town of the Volsci, a few kilometers off the *Via Latina* (south-east of **Praeneste** and north-west of **Beneventum**), and under **Rome** since 305 BCE; granted full Roman citizenship in 188 BCE, and made a *municipium* in 90 BCE. *PECS* 95, D.C. Scavone; *OCD3* 175, E.T. Salmon and D.W.R. Ridgeway; *BAGRW* 44-E2; *BNP* 2 (2003) 20, G. Uggeri.

M. TULLIUS CICERO, Q. TULLIUS CICERO.

Artemita (mod. Edremit; 38°25' N, 43°15' E): "noteworthy" city on the east shore of Lake Thospitis (mod. Lake Van), across from **Xoren**, east-north-east on the road from **Amida**, and north-east of **Nisibis**: STRABŌN 16.1.17, PLINY 6.117. *RE* 2.2 (1896) 1443 (#1), S. Fraenkel; *BAGRW* 89-F2.

APOLLODŌROS.

% Asinē (Argive: mod. Tolon/Asine; 37°33' N, 22°52' E; Messēnian: mod. Koroni; 36°48' N, 21° 57' E): the ancient city, south-east of **Argos**, and directly south of **Corinth**, was subjugated by **Argos** *ca* 8th c. BCE, the refugees being settled by **Sparta** at Messēnian Asinē. It does not seem possible to determine from which Asinē Theodōros came. STEPHANOS OF BUZANTION, s.v.; *RE* 2.2 (1896) 1582 (#2), E. Oberhummer; *PECS* 100–101, P. Aström; *OCD3* 191, R.A. Tomlinson; *BAGRW* 58-D2 (Argive), B4 (Messēnian). (Contrast Asinē of **Lakōnika**, mod. Skoutari, *BAGRW* 58-D2.)

THEODŌROS.

Askalon (mod. Ashkelon; 31°04' N, 34°34' E): ancient coastal city, north-north-east of **Gaza** and west of **Eleutheropolis**; alternately Egyptian and Assyrian, and never Jewish; under the Persians, a dependency of **Turos**; after the conquest of Alexander of **Macedon**, alternately Ptolemaic and Seleukid. Independent from 104 BCE; a banking and commercial center, devoted to the worship of the fish-goddess Atargatis and resistant to Christianity. *PECS* 98–99, A. Negev; *BAGRW* 70-F2; *BNP* 2 (2003) 92, M. Köckert; *EJ2* 2.567–568, M. Avi-Yonah and Sh. Gibson.

EUTOKIOS, IULIANUS, ULPIANUS.

Askra (mod. Episkopi?; 38°19' N, 23°05' E): small town of **Boiōtia**, between **Khairōneia** and **Plataia**, and west of **Thēbai**. *PECS* 101, P. Roesch; *BAGRW* 55-E4; *BNP* 2 (2003) 107, K. Freitag.

HĒSIOD.

Assos (mod. Beyramkale; 39°29' N, 26°20' E): coastal colony of **Mēthumna** (7th c. BCE?) in the **Troas**, allied with **Athens** in the 5th c. BCE, and stronghold of the rebellious satrap Ariobazarnēs 366 BCE.; in the later 4th c., ruled by tyrants and visited by ARISTOTLE. Prosperous from shipping; came under the rule of **Pergamon** 227/6 BCE, and thence to **Rome** in 133 BCE. *PECS* 104–105, H.S. Robinson; *OCD3* 194–195, St. Mitchell; *BAGRW* 56-C3; *BNP* 2 (2003) 184–185, E. Schwertheim.

KLEANTHĒS.

Astupalaia (mod. Astupalaia; 36°33' N, 26°21' E): Aegean island, colonized by **Megara**, displacing native Karians; allied with **Athens** in the 5th c. BCE, and with the Ptolemies in the Hellenistic era. Prosperous from fishing and trade; under **Rome** (by 105 BCE), variously a *ciuitas libera* or *foederata*. *PECS* 105–106, G.S. Korrès; *BAGRW* 61-C4; *BNP* 2 (2003) 214, H. Kalcyk.

ONĒSIKRITOS.

Athens/Athēnai (mod. Athens; 37°58' N, 23°43' E): ancient site in a mountain-fringed coastal plain, prosperous in the 14th c. BCE, aristocratic in the 6th c. BCE, adopted democracy in 508 BCE, a target of the Persian invasions of 490 and 480 BCE. Founded (478 BCE), and soon dominated, the League of **Dēlos**, leading to conflict with **Sparta**, and the Peloponnesian War (431–404 BCE). A revived Athenian League (from 378 BCE) resulted in the revolt of allies 357–355 BCE. Opposed Philip II of **Macedon**; defeated at **Khairōneia** 338 BCE; mostly dominated by **Macedon** thereafter until *ca* 230 BCE. Site of PLATO's **Academy** (from *ca* 360 BCE), ARISTOTLE's Lyceum (**Peripatos**; from *ca* 335 BCE); ZĒNŌN's **Stoa** (from *ca* 310 BCE), and EPICURUS' **Garden** (from *ca* 305 BCE). Friendly or even allied with **Rome** from 229 BCE, rewarded with control of **Dēlos** 166 BCE. Supported MITHRADATĒS VI in 88 BCE; sacked by Sulla 86 BCE. Built up by Hadrian; sacked by the Heruli (267 CE); resisted Alaric 396 CE; built up in the 5th c. CE; sacked by Slavs 582 CE. *PECS* 106–110, J. Travlos; *ODB* 221–223, T.E. Gregory and N.P. Ševčenko; *OCD3* 203–205, A.J.S. Spawforth; *BAGRW* 58-F2; *BNP* 2 (2003) 253–280, H.R. Goette and K.-W. Welwei. Some of these *ca* 50 scientists may have originated elsewhere, but settled in Athens; *cf.* **Alexandria**.

AISKHINĒS, AKESIAS, AMMŌNIOS ANNIUS (?), AMPHILOKHOS, ANDREAS (?), ANDROTIŌN, ANTIOKHOS, ANTIPHŌN, ANTISTHENĒS, APOLLODŌROS, PSEUDO-APOLLODŌROS, APOLLŌNIOS, ARISTANDROS, ARKHELAOS, BROMIOS, DAMŌN, DŌROTHEOS, EPIMAKHOS, EUDĒMOS, EUKTEMŌN, EUPHRŌNIOS, EUTHUDĒMOS, HAGNODIKĒ, HĒLIODŌROS, HIPPONIKOS (?), IKTINOS, KALLIKRATĒS (ARCH.) (?), KHAIREAS, KHAIRESTEOS, KLEIDĒMOS, KRATISTOS (?), KRATŌN (?), KRATULOS, KRITIAS, LEŌN, METŌN, MNĒSARKHOS, MNĒSITHEOS, NEOKLEIDĒS (?), NIKĒRATOS (?), NIKĒTĒS (?), NIKOMAKHOS, PHAIDROS (?), PHILEAS, PLATO, PLOUTARKHOS, POLEMŌN, SILANIŌN, SIMŌN, SPEUSIPPOS, STEPHANOS (2?), THEAITĒTOS, THUCYDIDĒS, XENOPHŌN, ZĒNODŌROS (?).

Atrax (near mod. Aliphaka; 39°34' N, 22°13' E): city of **Thessalia** on the Peneios upstream (west-south-west) from **Larissa**; resisted the army of **Rome** in 198 BCE, and then (under **Rome**) the army of Antiokhos III in 191 BCE. *PECS* 110–111, T.S. MacKay; *BAGRW* 55-C1; *BNP* 2 (2003) 298, H. Kramolisch.

AGATHOKLĒS.

Attaleia (mod. Antalya; 36°54' N, 30°40' E): coastal city, east of **Phasēlis**, and west of **Pergē**, founded before 150 BCE, by Attalos II, and left free in 133 BCE; under **Rome** from 77 BCE; greatly restored by Hadrian 130 CE. *PECS* 111, G.E. Bean; *OCD3* 211, St. Mitchell; *BAGRW* 65-E4; *BNP* 2 (2003) 302, W. Martini. (Contrast the small Musian town, near mod. Selçikli, *BAGRW* 56-F3.)

ATHĒNAIOS (?).

Auzia (mod. Sour el-Ghozlane; 36°09' N, 03°41' E): inland city of **Mauretania**, some 150 km east-south-east from **Caesarea**; a *municipium* after *ca* 200 CE. *BAGRW* 30-G4; *BNP* 2 (2003) 421, W. Huß.

GARGILIUS MARTIALIS (?).

Babylōn (near mod. Hilleh; 32°28' N, 44°25' E): ancient cultural and religious center, on east bank of the Euphratēs, claimed to be the center of the ***kosmos***; made magnificent by Nebuchadnezzar II (604–562 BCE); selected as capital by Alexander of **Macedon**, and built up by Antiokhos I; declined from 275 BCE; taken by Parthians 129 BCE. *OCD3* 228, A.T.L. Kuhrt; *BAGRW* 91-F5; *BNP* 2 (2003) 441–442, S. Maul; *EJ2* 3.23–24, D.C. Snell.

BĒROSSOS, DIOGENĒS, KIDĒNAS, NABURIANOS, SERGIUS, ZAKHALIAS.

Babylōn of Egypt (mod. Fosta/Fustat/Misr al Qadimah, south edge of Cairo; 30°01' N, 31°14' E): south of **Hēliopolis**, where Necho's canal leaves the Nile to the Red Sea; fortified legionary camp under AUGUSTUS. *RE* 2.2 (1896) 2699–2700 (#2), H. Sethe; *BAGRW* 74-E4.

TEUKROS.

Bargulia (on mod. Varvil Bay, S of mod Güllük; 37°14' N, 27°36' E): Karian coastal town between **Iasos** and **Karuanda**, and dedicated to Artemis; allied with **Athens** in the 5th c. BCE; allied with **Iasos** in the 3rd c. BCE (POLUBIOS Book 16, *fr.*12); winter headquarters of Philip V of **Macedon** in 201/200 BCE; declared free by **Rome** in 196 BCE, and an ally in 133 BCE. *PECS* 143, G.E. Bean; *BAGRW* 61-F3; *BNP* 2 (2003) 509–510, H. Kaletsch.

PRŌTARKHOS.

Ben(e)ventum (mod. Benevento; 41°08' N, 14°47' E): city of **Campania**, near the confluence of the Cabre and Sabato, on the *Via Appia*, north-east of **Nola**, and south-east of **Arpinum**; renamed from Maleventum after the nearby victory over PURRHOS, 275 BCE. Under Latin law from 268 BCE, then a *municipium* from 89 BCE; *colonia* from 42 BCE. The northern terminus of the *Via Traiana*. *PECS* 149, E.T. Salmon; *OCD3* 239, *Idem* and T.W. Potter; *BAGRW* 44-G3; *BNP* 2 (2003) 598, M. Buonocore.

LITORIUS.

Bērutos (mod. Beirut; 33°53' N, 35°31' E): ancient port city, north of and allied with **Sidōn**; annexed by Antiokhos III after he defeated Philip V of **Macedon** (*ca* 200 BCE); sacked *ca* 140 BCE. Under **Rome** a *colonia* from 14 BCE, and became an administrative center (residence of King Herod and successors). From the 3rd c. CE site of an academy and a school of law; by *ca* 400 CE the most important Phoenician city. Earthquakes of 347/8 and 501/2 CE damaged the city; destroyed by the earthquake and tsunami of 551 CE (*cf.* **Boiōtia**); rebuilt by Justinian. *PECS* 152, J.-P. Rey-Coquais; *ODB* 284–285, M.M. Mango; *OCD3* 240, A.H.M. Jones *et al.; BAGRW* 69-C2; *BNP* 2 (2003) 610–611, U. Finkbeiner and T. Leisten.

ANATOLIOS, "BĒRUTIOS" (?), PAMPHILOS, STRATŌN, TAUROS.

Bērutos (*unlocated*): inland site distinct from the coastal city: *Souda* E-3045; STEPHANOS OF BUZANTION, s.v., describes it as in "Arabia" and called Diospolis (contrast the Diospolis of Judaea, *BAGRW* 70-F2).

HERMIPPOS.

Bithunia: a mountainous yet fertile and economically rich region of north-west Anatolia, from the peninsula of **Khalkēdōn** eastward to **Paphlagonia**, and southward along the Propontis to Mt. Olumpos in Musia/Mysia. *OCD3* 244–245, T.R.S. Broughton and St. Mitchell.

Sites: **Murleia, Nikaia, Nikomēdeia, Prousias**.

People: AGRIPPA, IOLLAS, ISKHOMAKHOS, KALLIMAKHOS (?), ŌRIŌN, THEODOSIOS (?), NIKOMĒDĒS IV.

Boiōtia: fertile and prosperous region of central Greece, north of Attica, and between west and east **Lokris**; a populous *amphiktionic* (temple-centered) league *ca* 525–480 BCE. Under **Athens** 457–447 BCE; federal state led by **Thēbai** 447 BCE and allied with **Sparta**; briefly allied with **Athens** (395–386) and then independent from 379 BCE, and defeated **Sparta** 371 BCE. Under **Macedon** from 362 BCE; after Alexander, the reconstituted Boiōtian League was independent and powerful until under **Rome**, 146 BCE. Ravaged by Goths in the late 3rd c. CE, and devastated by the earthquake of 551 CE (Prokopios, *Goth.* 4.25). *OCD3* 246–247, J. Buckler and A.J.S. Spawforth; *BNP* 2 (2003) 695–699, P. Funke and K. Savvidis.

Sites: **Askra, Eruthrai (?), Khairōneia, Plataia, Thēbai**.

People: HEIRODOTOS (?).

Bordeaux ⇒ Burdigala

Borissos (mod. Sofular?; 38°20' N, 34°27' E): near **Nazianzos**, north of **Tuana** and west-south-west of **Caesarea (Kappadokia)**. *RE* S.1 (1903) 256, W. Ruge; *BAGRW* 63-E4.

PHILOSTORGIOS.

Borusthenēs (mod. Berezan Is.; 46°36' N, 31°25' E): on an island in the Borusthenēs (Dnieper) estuary, first Greek colony on the north shore of the Black Sea, founded *ca* 650 BCE, and prosperous in the 6th–5th centuries BCE; after which small or abandoned. *PECS* 150, M.L. Bernhard and K. Sztetyłło; *BAGRW* 23-E2.

SPHAIROS.

Britain/Britannia: The island west of the European mainland whose inhabitants shared a long history and culture with northern Gaul. Raided by CAESAR (55–54 BCE: *BG* 5.11–23), then annexed by **Rome** in 43 CE under Claudius. Invaded by Scots and others beginning 360 CE; **Rome** officially withdrew 410 CE. *OCD3* 261–263, M.J. Millett; *BNP* 2 (2003) 774–783, M. Todd.

Sites: (none).

People: GILDAS.

Bublos (mod. Ǧubail/Jebeil/Jbeil; 34°07' N, 35°39' E): ancient port, north of **Bērutos**, conquered by Alexander of **Macedon**, after which declined. Rebuilt by King Herod; later famous for its cult of Adonis. *PECS* 176, J.-P. Rey-Coquais; *OCD3* 266, J. Boardman and J.-F. Salles; *BAGRW* 68-A5; *BNP* 2 (2003) 842, U. Finkbeiner.

PHILŌN.

Burdigala (mod. Bordeaux; 44°50' N, 00°35' W): estuarine port in Atlantic trade, just above confluence of Garonne and Dordogne, 60 km north-west of **Vasates** and west of **Vesunna**. Founded in the 3rd c. BCE; became a *municipium* under the Flavians, and the capital of **Aquitania** in the 2nd c. CE. Christianized in the 3rd c. CE; a university town *ca* 400 CE. Taken by the Vandals 409 CE; then by the Visigoths 419 CE. *PECS* 172, R. Étienne; *OCD3* 265, C.E. Stevens and J.F. Drinkwater; *BAGRW* 14-E4; *BNP* 2 (2003) 824–825, E. Frezouls.

EUTROPIUS, MARCELLUS, SIBURIUS.

Buzantion (mod. Istambul; 41°01' N, 28°59' E): at the southern mouth of the Bosporos, on the western shore, between the Propontis and the Golden Horn, a natural harbor. Founded by **Megara** in the 7th c. BCE, under Persian control by the late 6th, then allied with **Athens** against Philip II of **Macedon**. Sided with **Rome** by 200 BCE; razed by Septimus Seuerus 196 CE; sacked by the Heruli 267 CE. Refounded in 330 CE by Constantine as the "New Rome," and subsequently greatly expanded, including the building of a university; attracted many immigrants. *PECS* 177–179, W.L. MacDonald; *ODB* 344–345, C. Mango and A. Kazhdan; *OCD3* 266, A.J. Graham and St. Mitchell; *BAGRW* 52-D2; *BNP* 2 (2003) 846–858, A. Effenberger.

Aglaias, Aristophanēs, Dionusios, Epigenēs, Hēgētōr, Leōnidas, Maximus (?), Mētrodōros, Philōn, Stephanos, Themistios.

C- ⇒ K-

Caesarea (Kappadokia) (mod. Kayseri; 38°43' N, 35°28' E): on the north foot of Mt. Argaios (Erciyes Dağ), originally "Mazaka" and the capital of the Kappadokian kingdom; renamed Caesarea 10 BCE and the provincial capital from 17 BCE. Taken by the Sasanians 260 CE; Christian riots *ca* 360 CE destroyed temples; deprived of municipal status by Julian; later a military center with weapon and textile factories. *PECS* 182, R.P. Harper; *ODB* 363–364, C.F.W. Foss; *OCD3* 272, A.H.M. Jones *et al.; BAGRW* 63-G3/64-A3; *BNP* 2 (2003) 916–917, K. Strobel and T. Leisten.
Basil.

Caesarea (Mauretania) (mod. Cherchel; 36°36' N, 02°11'E): Phoenician or Punic emporium named "Iol"; renamed by Iouba II in honor of Augustus (Strabōn 17.3.12), and much built up; capital of the province from *ca* 45 CE. Sacked *ca* 370 CE; again in 429 CE by Vandals. *PECS* 413–414, J. Lassus; *OCD3* 272, T.W. Potter; *BAGRW* 30-D3; *BNP* 2 (2003) 917–918, W. Huß.
Priscianus.

Caesarodunum (mod. Tours; 47°24' N, 00°41' E): on the south bank of the Loire and the road from **Bordeaux**, founded in the 1st c. CE; destroyed by Germans 275 CE, soon moved and rebuilt; from 374 CE the provincial capital and a Christian bishopric. *PECS* 182–183, C. Lelong; *BAGRW* 14-F1; *BNP* 2 (2003) 920, E. Olshausen; *EJ2* 20.70, B. Blumenkranz and D. Weinberg.
Gregory.

Campania: mountain-fringed fertile coastal plain of west-central **Italy**, north of **Lucania**, south of Latium (the region of **Rome**); the site of many harbors, and of Mt. Vesuvius. Etruscan and Greek colonies 6th–5th centuries BCE; under **Rome** from the late 4th c. BCE, along the *Via Appia* (built 312 BCE). *OCD3* 283, N. Purcell; *BNP* 2 (2003) 1024–1026, U. Pappalardo.
Sites: **Beneventum, Neapolis, Nola, Puteoli**.
People: Sunerōs (?), Vitruuius (?).

Campi Macri (mod. Val di Montirone near Magreta; 44°36' N, 10°48' E): south across the Po river from **Verona**, and west-north-west of **Ravenna**; market-town (Varro, *RR* 2.*pr*.6). *RE* 14.1 (1930) 162, H. Philipp; *BAGRW* 39-H4.
Turranius (?).

Carthage/Karkhēdōn (across Lake Tunis from mod. Tunis; 36°51' N, 10°19' E): emporium founded *ca* 813 BCE by Phoenicians from **Turos**, as "New Town," east of **Utica**. Prospered and, with the decline of **Turos**, became capital of its own maritime trading empire (with many daughter colonies around the western Mediterranean, esp. in **Spain** and **Sicily**), until attacked (3rd c. BCE) and destroyed (146 BCE) by **Rome**. Refounded 44 BCE on Caesar's plans as a Roman colony, and *ca* 40 BCE capital of the province of *Africa*; became the second city of the western Roman Empire, and an intellectual center; a Christian bishopric from the 3rd c. CE. Captured by Vandals 430 CE, recaptured by Belisarius for **Buzantion** 534 CE. *PECS* 201–202, A. Ennabli; *OCD3* 295–296, W.N. Weech *et al.; BAGRW* 32-F3; *BNP* 2 (2003) 1130–1136, T. Leisten and H.G. Niemeyer; *EJ2* 4.499, U. Rappoport.

CAPELLA, EPAPHRODITOS, HANNO, KERAS, KHARŌN, KLEITOMAKHOS, NEMESIANUS, TEUKROS.

Catina ⇒ Katanē

Centuripae (mod. Centuripe; 37°37' N, 14°44' E): Siculan town east of **Agurion** and west of **Katanē**; ruled by Greeks, repeatedly rebelled against **Surakousai**, and affiliated with **Athens** 414–413 BCE (THUCYDIDĒS 6.94.3). Allied with **Rome** 263 BCE against **Carthage**, and a *ciuitas libera atque immunis* from 241 BCE. In CICERO's time, the richest city in **Sicily**, but reduced by civil war; built up by AUGUSTUS. *PECS* 213, P. Deussen; *BAGRW* 47-F3; *BNP* 3 (2003) 128, R. Patané.

APULEIUS CELSUS.

Cirta (mod. Constantine; 36°22' N, 06°37' E): inland western Numidian city on the Ampsaga river, influenced by **Carthage** from *ca* 250 BCE; part of **Mauretania** 106–46 BCE; rewarded by CAESAR and AUGUSTUS; made provincial capital *ca* 120 CE. Destroyed *ca* 310 CE; rebuilt and renamed by Constantine; under the Vandals 430–534 CE. *PECS* 224–225, P.-A. Février; *OCD3* 333, W.N. Weech *et al.*; *BAGRW* 31-F4; *BNP* 3 (2003) 364–365, W. Huß and H.G. Niemeyer.

CASSIUS FELIX.

Commum ⇒ Nouum Comum

Constantia (Osrhoēnē) (mod. Viranşehir; 37°14' N, 39°45' E): founded as **Antioch** by Nikanōr (PLINY 6.117); later variously renamed; located south of **Amida** and north of **Reš'aina**. *RE* 1.2 (1894) 2445 (#9), S. Fraenkel; *BAGRW* 89-B3. (Many other places were also called "Constantia"; *cf.*, e.g., **Salamis**.)

PETROS (?).

Constantinople ⇒ Buzantion

Corcyra ⇒ Kerkura

Corduba (mod. Córdoba; 37°53'N, 04°46' W): native (Tartessian?) city on the west bank of the Gualdalquivir river, navigable down to the sea (where **Hispalis** is found); refounded under **Rome** 152 BCE; made a *colonia ca* 46 BCE; sacked by CAESAR 45 BCE, and resettled with veterans by AUGUSTUS. Capital of the province *Hispania Baetica*, and prosperous from agriculture and mining; an intellectual center. *PECS* 239–240, J.M. Roldán; *OCD3* 389, S.J. Keay; *BAGRW* 26-F4; *BNP* 3 (2003) 786–788, P. Barceló.

ANNAEUS LUCANUS; ANNAEUS SENECA.

Corinth/Korinthos (mod. Corinth; 37°56' N, 22°56' E): city (with pre-Greek name, spelled Qorinthos in archaic times) just south of the isthmus between the Corinthian and Saronic gulfs, at the intersection of major roads, became prosperous *ca* 725 BCE, and eventually the Greek city with the largest area. Allied with **Sparta** in the 5th c. BCE, and afterward with **Athens** or **Macedon**, becoming the center of the final resistance to **Rome**, and thus destroyed in 146 BCE. Refounded by CAESAR 44 BCE, rapidly regained its position as a trade center. Devastated 267 CE by the Heruli, became the ecclesiastical and administrative center of Greece in the 4th c. CE. *PECS* 240–243, H.S. Robinson;

OCD3 390–391, J.B. Salmon; *BAGRW* 58-D2; *BNP* 3 (2003) 797–804, Y. Lafond and E. Wirbelauer.

DIONUSIOS, EUPHRANŌR, NUMISIANUS, POSEIDŌNIOS.

Crete/Krētē: largest Greek island (fifth-largest Mediterranean island), on important sea-routes linking Greece to **Cyprus** and **Egypt**; the Greek inhabitants in the historic period were predominantly Dorian, living in many small towns with constitutions resembling **Sparta**'s. Internal conflict was led by **Gortuna** and **Knōssos**, and influenced by **Pergamon**, the Ptolemies, and the Seleukids; inhabitants often served as mercenaries abroad. Philip V of **Macedon** encouraged piratical activity to subvert **Rhodes**. Subdued by Caecilius and annexed by **Rome** in 69–67 BCE. *OCD3* 408–409, W.A. Laidlaw *et al.; BAGWR* 60; *BNP* 3 (2003) 934–939, J. Niehoff.

Sites: **Gortuna, Khersonēsos, Knōssos**.

People: ANDROMAKHOS (FATHER AND SON), EPIKLĒS, EPIMENIDĒS, FLAUIANUS, HĒRAKLEIDĒS "KRITIKOS" (?), NEARKHOS, PARISINUS MEDICUS.

Croton ⇒ Krotōn

Cyprus/Kupros: third largest Mediterranean island, dry but fertile, forested in antiquity, rich in copper and salt; under Greek influence from *ca* 1400 BCE, and colonized from *ca* 1200 BCE by Peloponnesians; under **Egypt** from *ca* 600 BCE. Composed of many small kingdoms, which were subjected to the Ptolemies from 294 BCE until annexed by **Rome** 58 BCE. The Jewish revolt ("Kitos" War) in 115–117 CE destroyed **Salamis** (*cf.* **Alexandria, Edessa, Kurēnē**, and **Nisibis**). *OCD3* 419–420, H.W. Catling; *BAGRW* 72; *BNP* 3 (2003) 1075–1081, A. Berger *et al.; EJ2* 5.347–348, B. Oded and L. Roth.

Sites: **Kition, Salamis, Soloi**.

People: ABDARAXOS (?), APOLLŌNIDĒS, DIAGORAS, GEŌRGIOS, NIKAGORAS, STUPPAX, SUENNESIS, ZŌILOS (?).

Cyzicus ⇒ Kuzikos

Daldis (near mod. Yunuslar; 38°43' N, 28°06' E): hill-town of **Ludia** north across the Hermos river from **Sardēs**, north-east of the lake; founded in the Roman period. *RE* 4.2 (1901) 2021, L. Bürchner; *TAM* 5.1, pp. 200–202; *BAGRW* 56-G4.

ARTEMIDŌROS.

Damaskos (mod. Dimashq; 33°30' N, 36°17' E): ancient city on the Syrian plateau, between the mountains and the desert, *ca* 90 km east of **Sidōn**; a great city of the Persian Empire, conquered for Alexander of **Macedon** by Parmeniōn in 332 BCE (*cf.* **Syria**). At first ruled by the Ptolemies, then by the Seleukids (made a capital in 111 BCE), and then by **Petra** from 86 BCE. Taken by Pompey for **Rome** in 66 BCE, reverted to **Petra** 37–54 CE. Fortified by Diocletian against the Sasanians; provided with a church dedicated to John the Baptist by Theodosius I. *PECS* 256–257, J.-P. Rey-Coquais; *ODB* 580, M.M. Mango; *OCD3* 427, A.H.M. Jones *et al.; BAGRW* 69-D2; *BNP* 4 (2004) 54–57, H. Klengel and T. Leisten.

APOLLODŌROS, DAMASKIOS, HĒSUKHIOS, IAKŌBOS (?), NIKOLAOS, PAUSANIAS.

Daorsi (mod. Stolac; 43°05' N, 17°58' E): Dalmatian region and town, south-east of **Salona**, and north-west of **Durrakhion**; *municipium* by the early 2nd c. CE. *BAGRW* 20-E7; *BNP* 4 (2004) 78–79, M. Šašel Kos.

DOARIUS (?).

Dēlos (mod. Dēlos; 37°24' N, 25°16' E): Aegean island, north of **Paros** and east of **Suros**, early religious center (dedicated to Apollo, Leto, and Artemis), from *ca* 550 BCE controlled by **Athens**. From *ca* 310 BCE the prosperous center of an island league (including adjacent Rheneia, plus **Naxos**, **Lesbos**, **Rhodes**, and **Suros**), which had existed in some prior form, and then came under the Ptolemies 286 BCE; apparently dissolved *ca* 250 BCE and re-formed 188 BCE under **Rhodes**. Declared a "free port" by **Rome** in 166 BCE and given to **Athens**, made a center of the slave trade. Sacked by MITHRADATĒS VI in 88 BCE, and by pirates in 69 BCE, and thereafter small. *PECS* 261–264, P. Bruneau; *OCD3* 443–444, R.W.V. Catling; *BAGRW* 61-A3; *BNP* 4 (2004) 210–215, H. Kaletsch.

ANTIPHANĒS, SĒMOS, TLĒPOLEMOS (?).

% **Dianium**: two sites: **(A)** Spanish coastal town (mod. Denia, Spain; 38°50' N, 00°06' W), founded before *ca* 400 BCE by traders from **Massalia** as Hēmeroskopeion, and renamed when under **Rome** to its pre-Greek title (for which POSEIDŌNIOS gives "Artemisia"); **(B)** Italian isle (Giannutri, Italy; 42°20' N, 11°08' E), also named Artemisia by Greeks. *RE* 5.1 (1903) 340–341, E. Hübner (Spain); *PECS* 272, G. Monaco (Italy); *BAGRW* 27-F3 (Spain), 41-E5/42-A4 (Italy); *BNP* 4 (2004) 360, G. Uggeri (Italy).

ARTEMISIUS.

Dikaiarkhia ⇒ Puteoli

% **Diospolis** (mod. Karnak/Luxor; 25°41' N, 32°39' E): Diospolis "the great," also known to Greeks as "Egyptian **Thēbai**" (Thebes), the Egyptian center of the worship of Amun (identified with Zeus, hence "Diospolis"), well upstream from **Panōpolis**. The ancient Waset, occasional capital of **Egypt**, visited by HĒRODOTOS (2.143), which retained its importance under the Ptolemies. Center of revolts in 206, *ca* 130, and 88–85 BCE. Declined from mid-1st c. BCE (DIODŌROS OF SICILY 1.46–49; STRABŌN 17.1.46), and sacked by AELIUS GALLUS in 30/29 BCE. *PECS* 904, S. Shenouda; *OCD3* 1496, J.G. Milne and A.J.S. Spawforth; *BAGRW* 80-B2; *BNP* 4 (2004) 522, R. Grieshammer; *NP* 12/1.277–282, J. Quack. Several other smaller places were also called Diospolis (*BAGRW* 74-F2 ["Kato"], 77-G4 ["Mikra"]), from one of which Anoubiōn might have come (Hephaistiōn and Ōdapsos are from this one).

ANOUBIŌN, HEPHAISTIŌN, ŌDAPSOS, OLUMPIODŌROS.

Durrakhion/Dyrr(h)achium (mod. Durrës [formerly Durazzo]; 41°19' N, 19°27' E): coastal city of southern Illyria, founded as Epidamnos *ca* 626 BCE from **Corinth** and **Kerkura**, from the 5th c. also called Durrakhion, a name which became usual under **Rome**, from 229 BCE. In the 2nd c. BCE, the *Via Egnatia* was built from Durrakhion to Thessalonikē (east of **Pella** on the coast). A *ciuitas libera* in the late republic, Durrakhion had a library, and was strategically important from the Roman Civil Wars to the Byzantine period. *PECS* 311, P.C. Sestieri; *OCD3* 499–500, M. Cary and N.G.L. Hammond; *BAGRW* 49-B2; *BNP* 4 (2004) 760–762, D. Strauch and E. Wirbelauer.

AUFIDIUS (?), BITHUS, PHILŌNIDĒS.

Edessa (mod. Şanlıurfa or Urfa; 37°09' N, 38°48' E): ancient Syrian city Urhai (whence the region Osrhoēnē), renamed by Alexander of **Macedon** or Seleukos I in honor of the Macedonian city Edessa (west of **Pella**); also known as **Antioch** on the Kallirhoē (STEPHANOS OF BUZANTION, s.v., #8). Capital of an independent Aramaic/Syriac-speaking kingdom *ca* 130 BCE to *ca* 242 CE, which was under the control of **Rome** from *ca* 63 BCE (*cf.* **Syria**). The Jewish revolt ("Kitos" War) in 115–117 CE led to its sack (*cf.* **Alexandria**, **Cyprus**, **Kurēnē**, and **Nisibis**). Officially Christian by *ca* 200 CE, and a center of Syriac culture and Christian study; later a stronghold against the Sasanians; rebuilt by Justin I (*ca* 525 CE). *PECS* 61, J.B. Segal; *ODB* 676, M.M. Mango; *OCD3* 505, E.W. Gray and A.T.L. Kuhrt; *BAGRW* 67-H2; *BNP* 4 (2004) 802–803, T. Leisten; *EJ2* 6.146, I. Gafni and Sh. Gibson. (BARDAISAN is known to have been from the Syrian Edessa, and the Macedonian city is unlikely to be KUROS' place of origin: *PECS* 292–293, Ph.M. Petsas; *BAGRW* 50-B3; *BNP* 4 (2004) 802, R.M. Errington and E. Wirbelauer.)

BARDAISAN, KUROS.

Egypt/Aiguptos: the Nile valley and its Delta, explicitly excluding **Alexandria "near Egypt"**; unified under the Pharaohs from *ca* 3200 BCE, capital at **Memphis**; prosperous in the 2nd millennium BCE; exerted influence or rule over **Cyprus**, Palestine, and Nubia (maximum extent perhaps under Rameses II, 1279–1213 BCE). Mostly under Libyan, Nubian, or Assyrian control in the 8th–7th centuries BCE; ruled by Persia 525–403 BCE and again 343–332 BCE. Ruled by the Ptolemies (a Greek dynasty founded by Ptolemy), from 305 to 30 BCE (capital at **Alexandria**; the last ruler being KLEOPATRA VII). Controlled **Kurēnē** from *ca* 300 BCE to 96 BCE, **Cyprus** from 294 BCE to 58 BCE, much of Palestine to 198 BCE (see **Askalon**, **Damaskos**, **Gaza**, **Seleukeia Pieria**, **Sidōn**, and **Turos**), as well as other sites and regions (e.g., **Dēlos**, **Ephesos**, **Knidos**, **Kōs**, **Lukia**, **Naxos**, **Mēthumna**, **Samos**, **Sidē**, and **Suros**). In 168 BCE, intervention by **Rome** arrested Antiokhos IV's invasion near **Alexandria**. CAESAR conquered **Alexandria** 49 BCE; the whole of Egypt was under **Rome** from 30 BCE. *OCD3* 510–511, A.B. Lloyd, 511–512, D.J. Thompson, 512, D.W. Rathbone; *BNP* 4 (2004) 844–853, S.J. Seidlmayer and K. Jansen-Winkeln.

Sites: **Antinoopolis**, **Diospolis**, **Hēliopolis**, **Hērakleopolis**, **Kurtos**, **Memphis**, **Mendēs**, **Naukratis**, **Oasis**, **Oxyrhynchos**, **Panōpolis**, **Pēlousion**, **Philadelpheia**, **Saïs**, **Sebennutos**, **Thēbai**.

People: ARKHELAOS, ASAMŌN (?), BLATAUSIS, CYNCHRIS, HĒRAISKOS, HERMŌN, KHAIRĒMŌN, KŌMARIOS, NAUKRATITĒS (?), NEILAMMŌN (?), NEILOS, NINUAS, PHILIPPOS, PLŌTINOS, SEUERUS IATROS. (?), THEOPHULAKTOS.

Elaious (east of mod. Tekke Burnu; 40°03' N, 26°13' E): most probably, the city at the southern tip of the Thrakian **Khersonēsos**, opposite **Sigeion**, thus explaining Menekratēs' *Guide to the Hellespont*. Elaious was allied with **Athens** in the 5th–4th centuries BCE, and used by Alexander of **Macedon** as the base for his Asian campaign in 334 BCE. Less likely is coastal Aiolian Elaia (mod. Kazıkbağları), *BAGRW* 56-E4, south of **Pergamon**. (It is unlikely that Menekratēs is from the Attic deme Elaious, *BAGRW* 59-B2; or the Peloponnesian village Elaious, south of **Argos**, *BAGRW* 58-D2.) *BAGRW* 51-G4; *BNP* 4 (2004) 883, H. Lohmann.

MENEKRATĒS.

Elea (mod. Castellamare di Velia; 40°10' N, 15°09' E): coastal colony of **Phōkaia** in **Lucania** founded *ca* 540 BCE, from the 3rd c. BCE allied with **Rome**, and a *municipium* from 88 BCE. Remained Greek-speaking under the Roman Empire, but declined in prosperity, because the river silted up. *PECS* 295–296, L. Richardson, Jr.; *OCD3* 516, H.K. Lomas; *BAGRW* 46-B1; *NP* 12/1.1164–1165, A. Muggia.

PARMENIDĒS, ZĒNŌ.

Eleusis (mod. Eleusis; 38°02' N, 23°32' E): coastal town west of **Athens**, and center of the cult of Demeter; refortified *ca* 375 BCE; sacked 170 BCE and by Alaric 395 CE, from the latter of which, plus Christian depredations, it never recovered. *PECS* 296–298, G.E. Mylonas; *OCD3* 520, K. Clinton; *BAGRW* 59-B2; *BNP* 4 (2004) 913–917, J. Niehoff.

PHILŌN.

Eleutheropolis (mod. Bet Guvrin; 31°36' N, 34°53' E): previously named Baitogabra (Iosephus, *Bell. Iud.* 4.447; PTOLEMY, *Geog.* 5.16.6), on the road between **Askalon** and **Jerusalem**, conquered 64 BCE by Pompey for **Rome**; renamed "Eleutheropolis" by the emperor Septimius Seuerus *ca* 200 CE, and its inhabitants made Roman citizens; a Christian bishopric from before 325 CE. *RE* 5.2 (1905) 2353–2354, I. Benzinger; *PECS* 298, A. Negev; *BAGRW* 70-F2.

EPIPHANIOS.

Ēlis (mod. Ēlis or Palaiopolis; 37°53' N, 21°23' E): city of western Peloponnesos founded 472 BCE on the Peneios river, prosperous until the 3rd c. CE, when it declined. *PECS* 299–300, N. Yalouris; *OCD3* 521, T.J. Dunbabin *et al.; BAGRW* 58-A2; *BNP* 4 (2004) 921–924 (#1–2), Y. Lafond. (Contrast the small and unlocated Arkadian town, cited by STEPHANOS OF BUZANTION, s.v.)

AIGIMIOS, HIPPASIOS, HIPPIAS.

Emesa (mod. Homs; 34°44' N, 36°43' E): ancient inland city on the Orontes river, south of **Apameia** and north of **Damaskos**, a client kingdom of **Rome** from *ca* 50 BCE to Domitian who suppressed the kingship, and the center of worship of the Syrian sun-god, El-Gabal (*cf.* **Syria**). Prospered under Emperor Elagabalus (born here and a priest of El-Gabal), from 218 CE, and afterward. *PECS* 302, J.-P. Rey-Coquais; *ODB* 690, M.M. Mango; *OCD3* 523, J.F. Healey; *BAGRW* 68-C4; *BNP* 4 (2004) 940–941, C. Colpe; *EJ2* 6.394, L. Roth and A. Shmuelevitz.

MAGNUS, NEMESIOS, VLPIANUS.

Ēpeiros: north-western Greece from Gulf of Ambrakia to **Nikopolis**, south of **Illyria**, west of **Macedon**; some of the native peoples were under Greek influence, and a few allied with **Athens** or **Macedon** from the 4th c. BCE. The first king of all Ēpeiros was PURRHOS (297 BCE); in the late 3rd c. BCE, allied with **Macedon** against **Rome**; under **Rome** from 167 BCE. Divided into two provinces by Diocletian 297 CE; pillaged by the Vandals under Alaric 395–397 CE. *OCD3* 546–547, N.G.L. Hammond; *BNP* 4 (2004) 1123–1127, D. Strauch and J. Niehoff.

Sites: **Nikopolis (?)**.

People: MAXIMUS (?), PHILAGRIOS, PURRHOS.

Ephesos (mod. Selçuk; 37°56' N, 27°20' E): Greek colony (of **Athens**?) founded in the 10th c. BCE on a Karian site, in the delta of the Kaüstros/Cayster river, and devoted to the worship of Artemis. In the mid-6th c. BCE, Kroisos/Croesus moved the town inland, as did again LUSIMAKHOS *ca* 290 BCE. Allied with **Athens** in the 5th c. BCE; under Seleukid control 281–246 and 196–188 BCE; Ptolemaic 246–196 BCE; from 188 BCE under **Pergamon**. Under **Rome** from 133 BCE, as a *ciuitas libera ac foederata*; allied with MITHRADATĒS VI 89–84 BCE. Became a large and prosperous city, including e.g. the library of Celsus (135 CE); withstood siege by the Goths 268 CE. *PECS* 306–310, V. Mistopoulou-Leon; *OCD3* 528, W.M. Calder *et al.; BAGRW* 61-E2; *BNP* 4 (2004) 1024–1030, P. Scherrer and E. Wirbelauer.

ALEXANDER, ARTEMIDŌROS, BALBILLOS (?), DIODŌROS, DIONUSIOS, HĒRAKLEIDĒS, HĒRAKLEITOS, IMBRASIOS, MAGNUS, MENEKRATĒS, OFELLIUS (?), PAIŌNIOS, RUFUS, SŌRANOS, XENOKRATĒS.

Eresos (mod. Skala; 37°10' N, 25°56' E): on south-west part of **Lesbos** island, occupied from Archaic times, perhaps Sappho's birthplace. The harbor was small, strengthened by **Mutilēnē** 428 BCE. (See **Lesbos**.) *PECS* 502–503, M. Paraskevaïdis; *OCD3* 555, D.G.J. Shipley; *BAGRW* 56-B3; *BNP* 5 (2004) 22, H. Sonnabend.

Phainias, Theophrastos.

Eretria (mod. Eretria; 38°24' N, 23°48' E): on south coast of Euboia, near **Khalkis**, with whom long at odds; destroyed by Darius 490 BCE, and allied with **Athens** in the 5th c. BCE. Here Menedēmos had his school of philosophy, *ca* 300 BCE. Sacked by **Rome** 198 BCE, and in the war with Mithradatēs VI, after which abandoned. *PECS* 315–317, T.W. Jacobsen; *OCD3* 555, W.A. Laidlaw and S. Hornblower; *BAGRW* 59-C1; *BNP* 5 (2004) 23–24 (#1), P. Ducrey. (Contrast the small town of Phthiotis, *BAGRW* 55-D2.)

Diodōros.

% **Eruthrai** (mod. Ildır; 38°23' N, 26°29' E): coastal city opposite **Khios**; allied with **Athens** in the 5th c. BCE; then variously allied (including with **Rome** and **Pergamon**). *PECS* 317, E. Akurgal; *OCD3* 557, G.E. Bean *et al.*; *BAGRW* 56-C5; *BNP* 5 (2004) 54–55 (#2), H. Engelmann and E. Olshausen. (Compare many other sites with the same or similar names, Stephanos of Buzantion, s.v., such as in **Boiōtia**, mod. Darimari, *BAGRW* 55-E4/59-A2; in west **Lokris**, mod. Monastiraki, *BAGRW* 55-B4; or in Malis, mod. Phrantzi, *BAGRW* 55-C3.)

Diognētos (?), Hērakleidēs.

Etruria: homeland of the Etruscans, modern Tuscany; conquered by **Rome** in the 4th–2nd centuries BCE, the native culture had essentially disappeared by the mid-1st c. CE. *BNP* 5 (2004) 93–102, G. Camporeale.

Sites: **Volsinii**.

People: Amelios Gentilianus.

Firmum Picenum (mod. Fermo; 43°10' N, 13°43' E): north-east of **Spoletium**, 8 km inland from the Adriatic coast, occupied from the early Iron Age, under **Rome** from 264 BCE, as a *colonia*; sided with **Rome** against Hannibal (215 BCE) and in the Social Wars (90–89 BCE). Sided with Caesar in 48 BCE and with Octavian in 44 BCE. Declined in imperial times. *PECS* 329, L. Richardson, Jr.; *BAGRW* 42-F2; *BNP* 5 (2004) 435–436, G. Uggeri.

Tarutius.

Forum Iulii (mod. Fréjus; 43°26' N, 06°44' E): harbor town in *Gallia Narbonensis*, west of **Nikaia**, east of **Massalia**, near the mouth of the Argens river. Founded probably by Caesar 49 BCE; made an important naval base by Augustus after 31 BCE. *PECS* 335–336, C. Goudineau; *OCD3* 607, C.E. Stevens and J.F. Drinkwater; *BAGRW* 16-C3; *BNP* 5 (2004) 531 (#IV.4), E. Olshausen. (Contrast the Forum Iulii north of Aquileia, *BAGRW* 19-F3.)

Iulius Graecinus.

Gabala (mod. Jableh; 35° 22' N, 35° 55' E): old coastal town, with good harbor, a little south of **Laodikeia on the Sea** and north of **Arados**; Phoenician settlement under **Arados**. The 5th c. bishop Theodōrētos of Kurrhos declared it a "charming little town." *RE* 7.1 (1910) 415 (#5), I. Benzinger; *PECS* 340, J.-P. Rey-Coquais; *BAGRW* 68-A3.

Seuerianus.

Gadara (mod. Umm Qais; 32°39' N, 35°41' E): south-east of the Sea of Galilee; under the Seleukids from 198 BCE, as **Antioch** or **Seleukeia**, then renamed for the Macedonian city (Polubios 5.71.3); freed by Pompey in 64 BCE, the city was rebuilt. Presented by Augustus to Herod (*cf.* **Gaza**), upon whose death the city was annexed to the Roman province of *Syria* (*cf.* **Syria**). *PECS* 341, A. Negev; *BAGRW* 69-C4; *BNP* 5 (2004) 635, T. Leisten; *EJ2* 7.331, M. Avi-Yonah and Sh. Gibson.

Meleagros (?), Philodēmos, Philōn, Theodōros.

Gadēs (mod. Cádiz; 36°32' N, 06°18' W): originally a small island enlarged by silting and joined to the mainland by a bridge, on the coast south of **Hispalis**. Founded by Phoenicians, *ca* 800 BCE; allied with **Rome** in 206 BCE, as a *ciuitas libera ac foederata*. CAESAR granted inhabitants citizenship and placed the city in the province of *Baetica*. In the early empire, second in size only to **Rome** (STRABŌN 3.5.3). *PECS* 341–342, J.M. Blázquez; *OCD3* 618, S.J. Keay; *BAGRW* 26-D5; *BNP* 5 (2004) 635–637, P. Barceló and H.G. Niemeyer.

IUNIUS COLUMELA, MODERATUS.

Gaza (mod. Gaza; 31°30' N, 34°27' E): regional base for Egyptian operations (*cf.* HĒRODOTOS 2.159); resisted Alexander of **Macedon** 332 BCE, who enslaved the population; Ptolemaic outpost to 198 BCE; devastated by Alexander Yannai of Israel *ca* 97 BCE. Rebuilt and prosperous under **Rome** from 58 BCE, as the Mediterranean entrepôt of Arabian trade; granted to Herod in 30 BCE (*cf.* **Gadara**), upon whose death annexed to the Roman province of *Syria* (*cf.* **Syria**). Long remained a flourishing center of Greek culture and paganism, famous for its school for rhetors. *PECS* 345–346, A. Negev; *ODB* 825, G. Vikan *et al.; OCD3* 627, E.W. Gray and J.-F. Salles; *BAGRW* 70-E2; *BNP* 5 (2004) 715–716, E.A. Knauf and T. Leisten; *EJ2* 7.398–400, M. Avi-Yonah and Sh. Gibson.

AINEIAS, TIMOTHEOS.

Gela (mod. Gela, formerly Terranova; 37°04' N, 14°15' E): on south coast of **Sicily**, founded 688 BCE as colony (displacing native Sikani) from **Crete** and **Rhodes**; in turn founded **Akragas** *ca* 582 BCE. Gela's tyrant Hippokratēs conquered much of **Sicily**, including **Leontinoi** (495 BCE), and **Messēnē** (493 BCE); his successor Gelōn took **Surakousai** 484 BCE, and made that his capital, depopulating Gela. Repopulated from 466, the city of Aeschylus' death 456 BCE; allied with **Surakousai** against **Athens** 427–424 and 415–413 BCE. *PECS* 346–347, P. Orlandini; *OCD3* 627, A.G. Woodhead and R.J.A. Wilson; *BAGRW* 47-E4; *BNP* 5 (2004) 721–723, D. Palermo and E. Olshausen.

PAUSANIAS.

Gerasa (mod. Jerash, about 50 km north of Amman; 32°16' N, 35°53' E): ancient city, east of **Neapolis (Samaria)**, south-east of **Gadara**; probably refounded as a Greek city, **Antioch** on the Khrusorrhoas river, by Antiokhos IV (175–164 BCE). Captured by Alexander Yannai of Israel *ca* 100 BCE, annexed by **Rome** to the province of *Syria* in 63 BCE (*cf.* **Syria**). On the caravan route, prospered under Trajan who annexed **Petra**, but declined in the 3rd c. CE; revived under Justinian. *PECS* 348–349, W.L. MacDonald; *OCD3* 633, J.F. Healey; *BAGRW* 69-C5; *BNP* 5 (2004) 791–792, T. Leisten.

NIKOMAKHOS.

Gortuna (mod. Ag. Deka/Gortuna/Kainourgiou; 35°04' N, 24°56' E): in central **Crete**, occupied from prehistoric times; Greek from the 7th c BCE. By the 3rd c. BCE, grew in power and territory, acquiring harbors at Matala and Lebena; the foremost city of **Crete** in the 2nd c. BCE. Siding with the Romans after Q. Caecilius Metellus captured **Knōssos**, it was made the capital of the Roman province *Creta*. *PECS* 362–363, K. Branigan; *OCD3* 643, V. Ehrenberg *et al.; BAGRW* 60-C2; *BNP* 5 (2004) 942–944, H. Sonnabend. (*Cf.* also the Arkadian Gortuna, *BAGRW* 58-C2, with a sanctuary of Asklēpios: Paus. 8.28.1–3.)

TRUPHŌN, ZŌPUROS.

Hadrumetum (mod. Sousse; 35°50' N, 10°38' E): Phoenician colony founded in the 9th c. BCE, *ca* 100 km south of **Carthage**. Surrendered to Agathoklēs of **Surakousai** 310 BCE, Hannibal's base 203–202 BCE. Allied with **Rome** 146 BCE; opposed CAESAR 46 BCE; became prosperous under the empire, and made a *colonia* by Trajan. *PECS* 372, A. Ennabli; *OCD3* 663–664, W.N. Weech *et al.; BAGRW* 33-G1; *BNP* 5 (2004) 1088–1089, W. Huß.

CLODIUS ALBINUS.

Halikarnassos (mod. Bodrum; 37°01' N, 27°25' E): Karian city controlling the sea route between **Kōs** and western Anatolia. Colonized by Troizen *ca* 900 BCE, culturally Ionic during the Classical era. Allied with **Athens** in the 5th c. BCE, and served as an Athenian naval station after 412 BCE. Maussōllos made the site his capital, *ca* 370 BCE, and engaged in extensive and spectacular rebuilding. Besieged and captured by Alexander of **Macedon** 334 BCE; under **Rome** from 129 BCE. *PECS* 375–376, G.E. Bean; *OCD3* 664, J.M. Cook and S. Hornblower; *BAGRW* 61-E3; *BNP* 5 (2004) 1110–1113, H. Kaletsch and C. Höcker.

DIONUSIOS (?), HĒRODOTOS, SKULAX.

Hēliopolis (Egypt) (mod. Matariya at north edge of Cairo; 30°08' N, 31°18' E): ancient city east of the Nile and at the Delta's apex, north of **Diospolis**; center of the worship of the Egyptian sun-god Raʿ, and center of wisdom (HĒRODOTOS 2.3, 2.59); although mostly deserted by the 1st c. BCE (STRABŌN 17.1.29). *BAGRW* 74-E4; *BNP* 6 (2005) 76, S.J. Seidlmayer. (Contrast the homonym north of **Damaskos**, mod. Ba'albek, *BAGRW* 69-D1; *BNP* 2 [2003] 439–440, T. Leisten.)

ANTIMAKHOS, DŌROTHEOS, PETEËSIS.

% Hērakleia: many homonymous sites existed (*cf.* STEPHANOS OF BUZANTION, s.v., listing 23), from any of which these men may have come, if not from **Hērakleia Pontikē** or **Hērakleia Salbakē**. Compare, e.g.: **(A)** in **Lucania** (near mod. Policoro; *PECS* 384, R.R. Holloway; *OCD3* 684, T.R.S. Broughton and St. Mitchell; *BAGRW* 46-E1; *BNP* 6 [2005] 154–155 [#10], G. Camassa); **(B)** Hērakleia Minoa in **Sicily** (mod. Eraclea Minoa; *PECS* 385–386, P. Orlandini; *BAGRW* 47-C4; *BNP* 6 [2005] 153–154 [#9], G. Falco); **(C)** Hērakleia Lunkestis in **Illyria** (near mod. Bitola; *PECS* 385, J. Wiseman; *BAGRW* 49-D2; *BNP* 6 [2005] 150–151 [#2], R.M. Errington); **(D)** Hērakleia Trakhis (*PECS* 386, Y. Béquignon; *OCD3* 684, B. Helly; *BAGRW* 55-C3; *BNP* 6 [2005] 150 [#1], H. Kramolisch); **(E)** Hērakleia Idalē (near mod. Ayvalık; *BAGRW* 56-D3); **(F)** Peloponnesian Hērakleia (mod. Brouma; *BAGRW* 58-B2); **(G)** Hērakleia on Latmos (mod. Kapıkırı; *PECS* 384–385, W.L. MacDonald; *OCD3* 684, A.J.S. Spawforth; *BAGRW* 61-F2; *BNP* 6 [2005] 151 [#5], A. Peschlow-Bindokat); and **(H)** Syrian Hērakleia (mod. Ras Ihn Hani; *BAGRW* 68-A2).

EPIKRATĒS, MENANDROS, NOUMĒNIOS, PHILŌN.

Hērakleia Pontikē/Hērakleia on the Pontos (mod. Ereğli/Karadeniz Ereğli; 41°17' N, 31°25' E): on south coast of Black Sea, west of **Sinōpē** and east of **Kallatis**, colony of **Megara** (the natives subjugated as serfs) that founded several colonies of its own; originally primarily agricultural, after *ca* 450 BCE trade grew. Prosperous in the 4th c. BCE; a kingdom from 305 BCE, after which the export economy declined; LUSIMAKHOS suppressed the kingship in 284 BCE. Allied with **Rome** from 188 BCE; supported MITHRADATĒS VI in 73–70 BCE, then sacked by **Rome**, which made it a *colonia* from 45 BCE. *PECS* 383, D.R. Wilson; *OCD3* 684, T.R.S. Broughton and St. Mitchell; *BAGRW* 86-B2; *BNP* 6 (2005) 152–153 (#7), K. Strobel.

AMUNTAS, BRUSŌN, HĒRAKLEIDĒS, HĒRAKLEIDĒS (JUNIOR), MARCIANUS, NUMPHIS, XENAGORAS.

Hērakleia Salbakē (mod. Vakif; 37°37' N, 28°59' E): south-east of **Aphrodisias in Karia**; founded before 30 BCE. *BAGRW* 65-A2; *BNP* 6 (2005) 151–152 (#6), H. Kaletsch.

ARKHELAOS, KRITŌN.

Hērakleopolis in Egypt (the "Greater"; mod. Ihnasiyah al-Madinah, 29°05' N, 30°56' E): on the west bank of the Nile near the entrance to the Fayum, between **Oxyrhynchos** and **Memphis**, home to the Pharaohs of the 9th–10th dynasties. Two other cities by this name existed in **Egypt**: **(A)** the "Lesser" (Tell Ayid, 31°02' N, 32°12' E), in the Delta west and upstream of **Pēlousion** (*BAGRW* 74-G2); and **(B)** at the Kanopic mouth of the Nile, i.e., near **Alexandria**. STEPHANOS OF BUZANTION, s.v. (listing the "Greater" as Theophanēs' home); *RE* 8.1 (1912) 515, H. Grapow; *EAAE* 368–370, F. Gomaà; *BAGRW* 75-D2; *BNP* 6 (2005) 156, S.J. Seidlmayer.

THEOPHANĒS.

% Hierapolis: any of several homonymous sites could be Aigeias' home, *cf.* STEPHANOS OF BUZANTION, s.v.: **(A)** Phrugian Hierapolis, destroyed mid-5th c. and not rebuilt (mod. Koçhisar; *ODB* 928, C.F.W. Foss; *BAGRW* 62-D5); **(B)** Hierapolis Comana (mod. Şar; *BAGRW* 64-C4); **(C)** Lukian Hierapolis in the Lukos valley, near **Laodikeia** ad Lycum, with an oracle of Apollo (mod. Pamukkale; *BAGRW* 65-B2; *BNP* 6 [2005] 302, Th. Drew-Bear); **(D)** the Hierapolis later known as **Kastabala**; or **(E)** Syrian Hierapolis, "Bambukē," located on the Euphratēs and dependent upon **Antioch on the Orontes**; a military center in Byzantine times (mod. Membidj; *ODB* 928–929, M.M. Mango; *BAGRW* 67-F3; *BNP* 6 [2005] 482–483, T. Leisten).

AIGEIAS.

Himera (mod. Himera; 37°58' N, 13°49' E): founded by Zanklē (see **Messēnē**) 649 BCE on the north coast of **Sicily**, prosperous; allied with **Surakousai** against **Athens**; attacked by **Carthage** 480 BCE, then destroyed by **Carthage** 409 BCE, and never rebuilt. *PECS* 393, N. Bonacasa; *OCD3* 707, A.G. Woodman and R.J.A. Wilson; *BAGRW* 47-D3; *BNP* 6 (2005) 327–328, G. Falco.

PETRŌN, PHAIAX (?).

Hippo Regius (south of mod. Annaba; 36°54' N, 07°46' E): north African seaport on a deep bay, between **Caesarea** and **Utica**, used by Carthaginians from 5th c. BCE, where Scipio Africanus' deputy C. Laelius landed in 205 BCE, and the Pompeian fleet was captured in 46 BCE. By the 2nd c. CE, a *colonia* serving as base for one of the three *legati* of the proconsul of **Africa**. AUGUSTINE died during the Vandal siege. *PECS* 394–396, J. Lassus; *OCD3* 709–710, B.H. Warmington and R.J.A. Wilson; *BAGRW* 31-H3; *BNP* 6 (2005) 347, W. Huß. (Contrast the Hippo north-west of **Utica**, at mod. Bizerte.)

AURELIUS AUGUSTINUS.

Hispalis (mod. Sevilla; 37°23' N, 05°59' W): founded in the 8th c. BCE, at the head of the estuary of the Guadalquivir river (downstream from **Corduba**, and north of **Gadēs**), an ancient shipyard and commercial center (oil and metals) mentioned by CAESAR, who established there a modest veteran colony, reinforced under Otho. A Christian bishopric from 4th c. CE, later metropolitan; taken by the Vandals 428 CE; changed hands repeatedly; taken by the Visigoths 567 CE. *OCD3* 712, S.J. Keay; *BAGRW* 26-E4; *BNP* 6 (2005) 384, P. Barceló; *EJ2* 18.325–326, H. Beinart and Y.T. Assis.

ISIDORUS.

Hispania ⇒ Spain

Huampolis (mod. Exarkhos; 38°35' N, 22°57' E): city of eastern Phokis, north of Lake Kopais, south-west of **Opous**, and north-east of **Khairōneia**. *BAGRW* 55-D3; *BNP* 6 (2005) 592–593, G. Daverio Rocchi.

PHILŌN.

Iasos (mod. Asınkalesi; 37°17' N, 27°36' E): on a small peninsula of Karia, south of **Ephesos**, traditionally thought a colony of **Argos** and allied with Bargulia in the 3rd c. BCE (POLUBIOS

Book 16, *fr.*12). The population was mixed, but the culture was fully Greek. Probably an Ionian ally against Darius, and then allied with **Athens** in the 5th c. BCE, sacked by **Sparta** 412 BCE, destroyed by Lysander 405 BCE, rebuilt with help from **Knidos**. By 125 BCE under **Rome** (in the province of *Asia*), prosperous into the 2nd c. CE. *PECS* 401–402, C. Laviosa; *OCD3* 744, A.J.S. Spawforth and Ch. Roueché; *BAGRW* 61-F3; *BNP* 6 (2005) 687–688, H. Kaletsch.

DIODŌROS, LUKŌN.

Ilion (mod. Hisarlık; 39°58' N, 26°14' E): the traditional site of HOMER's Troy, refounded by Aiolians *ca* 700 BCE. Xerxēs, Alexander of **Macedon**, and Antiokhos III (192 BCE) prayed for victory at the temple of Athena. Under **Pergamon** 227–133 BCE. Destroyed by Flauius Fimbria of **Rome** 85 BCE; later rebuilt and patronized by Roman emperors. *BAGRW* 56-C2; *NP* 12/1.852–857, D. Mannsperger.

POLEMŌN.

Illyria: loosely-defined region, the eastern Adriatic coast north of **Ēpeiros**; subjected to or influenced by **Macedon** in the 4th–3rd centuries BCE; under **Rome** from 167 BCE. *OCD3* 747, J.J. Wilkes.

Sites: **Durrakhion, Salona**.

People: GENTHIOS, SARKEUTHITĒS (?).

Indos (Lukia) (mod. Çukurhisar?; approx. 36°40' N, 28°20' E): a river, on which a fort Thabusion (Livy 38.14.2); perhaps Isinda, **Lindos**, or some other city is meant. *BAGRW* 65-A4; *BNP* 6 (2005) 794, E. Olshausen.

ARBINAS.

Ioulis on Keōs (mod. Khora; 37°40' N, 26°19' E): nearest of the Kuklades/Cyclades islands to Attika, Ioulis the chief town. Prosperous in the late 6th to early 5th c. BCE; allied with **Athens** in the 5th c. BCE. *PECS* 446–447, J.L. Caskey; *OCD3* 311, R.W.V. Catling; *BAGRW* 58-G2; *BNP* 3 (2003) 129–130, H. Kaletsch and Ernst Meyer. (Since Aiskhulidēs is only attested to be from the island, he may well have been from one of the smaller cities, e.g., Karthaia.)

AISKHULIDĒS, ARISTŌN, ERASISTRATOS, KLEOPHANTOS.

Italy/Italia: approximately the mainland of the modern country; included many smaller regions, esp. **Etruria**, **Lucania**, and **Transpadana**; most of the cities of the south were Hellenized or even Greek foundations; north of **Rome** and **Etruria** dominated by the Etruscans; the Po Valley by Celts. **Rome** methodically subjugated the Italic peoples of the central peninsula from the 6th c. BCE; **Etruria** from the early 4th c. BCE; and most of the Greek cities of the south Italy during 270–212 BCE. Revolts of the early 1st c. BCE resulted in extensions of Roman citizenship throughout Italy. Declined in importance from *ca* 100 CE, and esp. after *ca* 200 CE; Diocletian/Maximian relocated the capital to **Mediolanum**, 293 CE. In the aftermath of the Gothic invasion under Alaric 401 CE, Honorius moved the capital to **Ravenna**, *ca* 403 CE; Alaric sacked **Rome**, 410 CE. Under Gothic rule from 488 CE; partly reconquered by **Buzantion** 536 CE; invaded by Lombards 568 CE. *OCD3* 773–774, E.T. Salmon and T.W. Potter; *BNP* 6 (2005) 994–1101, G. Uggeri and J. Niehoff.

Sites: **Amiternum**, **Arpinum**, **Beneventum**, **Campi Macri**, **Elea**, **Firmum Picenum**, **Krotōn**, **Lokroi Epizephurioi**, **Mantua**, **Mediolanum**, **Medma**, **Metapontion**, **Neapolis**, **Nola**, **Nouum Comum**, **Ostia**, **Poseidōnia**, **Praeneste**, **Puteoli**, **Ravenna**, **Reate**, **Rhēgion**, **Rome**, **Rudiae**, **Scylletium**, **Spoletium**, **Subaris**, **Sulmo**, **Taras**, **Tusculum**, **Verona**, **Volsinii**.

People: MAMERKOS (?), PAULOS (?).

Itukē ⇒ Utica

Jerusalem (mod. Jerusalem; 31°47' N, 35°13' E): ancient Jewish city; Antiokhos IV's efforts to impose the worship of Zeus in 167 BCE incited revolt, resulting in an independent Jewish kingdom. Besieged by Pompey in 63 BCE, and Hellenized under Herod. After revolts in the 40s and 60s, FLAUIUS TITUS sacked the city and destroyed the temple. Refounded as a military colony by Hadrian after the revolt of 135 CE, the wall rebuilt by Diocletian; heavily Christianized by Constantine (325 CE) and later. *PECS* 12–13 (s.v. Aelia Capitolina), K.M. Kenyon; *ODB* 1033–1036, K.G. Holum and G. Vikan; *OCD3* 794–795, T. Rajak; *BAGRW* 70-G2; *BNP* 6 (2005) 1169–1177, K. Bieberstein; *EJ2* 11.143–153, Sh. Gibson *et al.*

KURILLOS (?).

Kalkhēdōn ⇒ Khalkēdōn

Kallatis (mod. Mangalia; 43°49' N, 28°35' E): fertile grain-land, colonized from and west of **Hērakleia Pontikē** *ca* 6th c. BCE. Supplied grain to **Athens** in the 4th c. BCE. Heavily taxed by LUSIMAKHOS whom the citizens resisted in 313 and 310 BCE. Then under the Skuthians sided with MITHRADATĒS VI against **Rome**, but became a *ciuitas*, suffered from numerous invasions, but enjoyed revival under Diocletian and reconstruction in the Byzantine era. *PECS* 431–432, D. Adamesteanu; *BAGRW* 22-F5; *BNP* 2 (2003) 959, J. Burian.

DĒMĒTRIOS, HĒRAKLEIDĒS, SATUROS.

Kapitthaka (mod. Kaïtha 20 km east of Ujjain; 23°14' N, 76°01' E): small town near the prime meridian for Hindu geography; under the Śakas from the 2nd–4th centuries CE; center of learning in the 6th–7th centuries CE (Brahmagupta also worked here). J. Schwartzberg *et al.*, *A Historical Atlas of South Asia*, 2nd ed. (1992) grid D5 on maps 18–21, 25–26, and pp. 171–172.

VARĀHAMIHIRA.

Kappadokia: north of Tauros mountains, east of Halus river, west of Armenia; did not resist Alexander of **Macedon**; then influenced or dominated by **Macedon** and the Seleukids; declared free by **Rome** 96 BCE. Made a province of **Rome** from 18 CE, capital **Caesarea**; in 371 CE, split into two provinces, capitals at **Caesarea** and **Tuana**. *OCD3* 288–289, T.R.S. Broughton and A.J.S. Spawforth.

Sites: **Caesarea, Borissos, Kastabala, Laranda, Nazianzos, Nussa, Tuana**.

People: ACILIUS HYGINUS, ARETAIOS, ARKHELAOS, CAESARIUS, HĒRĀS.

Kardia (mod. Baklaburnu or Bakla Liman; 40°32' N, 26°45' E): on the north face of the Thrakian **Khersonēsos**, north-east of **Lampsakos**, founded in 7th c. BCE by **Milētos** and **Klazomenai**, colonized by **Athens** (under the elder Miltiadēs) in the 6th c. BCE, abandoned to Persia in 493 BCE but restored to Athenian hegemony by mid-5th c. Allied with Philip II of **Macedon** in 352 BCE, destroyed by LUSIMAKHOS *ca* 309 BCE, but rebuilt (as Lusimakheia) and became the largest city of the region by the 1st c. BCE. *OCD3* 290, E.N. Borza; *BAGRW* 51-H3; *BNP* 2 (2003) 1093–1094, I. von Bredow.

HIERŌNUMOS.

Karkhēdōn ⇒ Carthage

Karuanda (mod. Salih Adası; 37°09' N, 27°31' E): an island with sheltered anchorage, between **Mundos** and **Bargulia**, and north of **Halikarnassos**. Early in the 3rd c. BCE, the population migrated to a new mainland site (south of mod. Güllük Körfezi; 37°07' N, 27°20' E; Strabōn 14.2.20), becoming citizens of **Mundos** and effectively replacing that town. *PECS* 798, G. Webster; *BAGRW* 61-F3; *BNP* 2 (2003) 1147–1148, H. Kaletsch.
Skulax, Skulax (pseudo) (?).

Karustos (mod. Karustos; 38°01' N, 24°25' E): coastal city at the south end of Euboia, opposed the Persian invasion of 490 BCE, but aided Xerxēs in 480 BCE; then allied with **Athens**. In the 3rd c. BCE, variously a member of the Euboian League, or allied with **Macedon**; taken by **Rome** 196 BCE. *PECS* 438, M.B. Wallace; *OCD3* 563, W.A. Laidlaw *et al.; BAGRW* 58-G1; *BNP* 2 (2003) 1149–1150, H. Kaletsch.
Andreas, Antigonos, Dioklēs.

Kassandreia (mod. Nea Potidaia; 40°12' N, 23°20' E): originally Poteidaia, a colony of **Corinth** founded *ca* 600 BCE on the western peninsula of the **Khalkidikē**, near **Olunthos**. After the revolt in 432 BCE, the inhabitants exiled and the site resettled by **Athens**; restored to its former inhabitants 404 BCE. Taken by Philip II of **Macedon** 356 BCE, and the inhabitants enslaved; replaced with refugees from **Olunthos**. Kassandros/Cassander refounded the city in his own honor 316 BCE, and it remained part of the kingdom of **Macedon** until its end, 168 BCE. *PECS* 733–734, J.A. Alexander; *OCD3* 1235, N.G.L. Hammond; *BAGRW* 50-D4; *NP* 10.230–232, M. Zahrnt.
Aristoboulos, Artemōn.

Kastabala (near mod. Kırmıtlı north-west of Osmaniye; 37°10' N, 36°08' E): city in **Kappadokia** or **Kilikia**, east of **Anazarbos** and north of **Antioch on the Orontes**; known as **Hierapolis** from *ca* 170 BCE to Roman times; under the kingdom of **Kappadokia** 63 BCE to 17 CE; a Christian bishopric in late antiquity. Strabōn 12.1.4, 12.2.7; *BAGRW* 67-C2; *BNP* 2 (2003) 1175, H. Täuber.
Celsinus.

Katanē/Catina (mod. Catania; 37°31' N, 15°04' E): coastal city south-east of Mt. Aetna on **Sicily**, north of **Surakousai**; colonized in 729 BCE by citizens of **Khalkis** previously settled on **Naxos**. Controlled by **Surakuosai** in the early 5th c. to 461 BCE (the population being temporarily removed to **Leontinoi**); served as a base for maneuvers by **Athens** in 415–413 BCE. Conquered by **Rome** in 263 BCE, it became a *ciuitas*. Under Augustus, it enjoyed extensive rebuilding and became a *colonia*, gradually acquiring importance retained into the Byzantine era. *PECS* 442–443, G. Rizza; *OCD3* 302–303, A.G. Woodhead and R.J.A. Wilson; *BAGRW* 47-G3; *BNP* 3 (2003) 8–9, G. Falco and K. Ziegler.
Philōnidēs.

Kaunos (mod. Dalyan; 36°50' N, 28°37' E): prosperous old Karian coastal town, south-east of **Halikarnassos**, and north-east of **Knidos**; under Persian hegemony in the 6th c. BCE, under **Athens** in the 5th, then under Maussōllos' influence in the 4th. Under various Hellenistic kings until **Rhodes** purchased the city for 200 talents in 191 BCE (Polubios Book 30, *fr.*31.6). Freed by **Rome** in 167 BCE, but returned to **Rhodes** briefly. *PECS* 443–444, G.E. Bean; *OCD3* 305, S. Hornblower; *BAGRW* 65-A4; *BNP* 3 (2003) 39–40 (#2), H. Kaletsch.
Dionusodōros (?).

Keōs ⇒ Ioulis

Kerkura (mod. Kerkura, i.e., "Corfu"; 39°40' N, 19°45' E): island in Ionian sea off western coast of **Ēpeiros**, occupied from Neolithic era. In 734 BCE, **Corinth** established a colony, which then opposed the metropolis, fighting a naval battle in *ca* 660 BCE, and allying with **Athens**. Was (and remains) an important port for sailing between Greece and **Italy**, disputed by various Hellenistic potentates, coming under **Rome** in 189 BCE. Backed the losing sides in the Roman Civil Wars. *PECS* 449–450, L.V. Borrelli; *OCD3* 389, W.M. Murray; *BAGRW* 54-A2; *BNP* 3 (2003) 783–786, D. Strauch.

Apollodōros, Drakōn, Eumakhos.

Khairōneia/Chairōneia (mod. Kapraina/Khaironia; 38°30' N, 22°51' E): westernmost city of **Boiōtia** (east of Delphi, and across Lake Kopais from **Thēbai**), member of the Boiōtian League 424–146 BCE; site of two defeats: of the Greeks by Philip II of **Macedon** 338 BCE, and of Mithradatēs VI by Sulla of **Rome** 86 BCE. Destroyed by the earthquake of 551 CE (*cf.* **Bērutos**). *PECS* 215–216, P. Roesch; *OCD3* 315, J. Buckler; *BAGRW* 55-D4; *BNP* 3 (2003) 176–177, P. Funke.

Plutarch.

Khaldea/Chaldea: south and coastal end of Mesopotamia, and the region around the head of the Persian Gulf; next to or possibly including the region around **Babylōn**. *Cf.* Bērossos. *BNP* 3 (2003) 182–183, J. Oelsner.

Sites: **Kharax "Spasinou," Nehardea**, **Seleukeia**.

People: Dōrotheos.

Khalkēdōn (mod. Kadıköy; 40°59' N, 29°02' E): on the eastern bank of the Bosporos, occupied by Phoenicians and Thrakians, colony founded by **Megara** in 685 BCE. Allied with **Athens** in the 5th c. BCE, fell to Persia 387 BCE, freed by Alexander of **Macedon**. Sought alliance with **Rome** in 220s BCE; resisted siege of Mithradatēs VI in 73 or 74 BCE. *PECS* 216, G.E. Bean; *OCD3* 315, A.J. Graham and St. Mitchell; *BAGRW* 52-E3.

Dioklēs, Hērophilos, Xenokratēs.

Khalkidikē: the three mountainous peninsular "fingers" extending from the coast of **Macedon/Thrakē**; Greek settlers from **Khalkis** founded some 30 colonies here (displacing or subjecting the natives); more were founded from **Eretria**. Allied with **Athens** in the mid-5th c. BCE, revolted 432 BCE, and founded an autonomous confederacy with its capital at **Olunthos** (Thucydidēs 1.58); by 380 BCE the confederacy controlled **Pella**. Instigated by **Sparta**, Philip II of **Macedon** destroyed **Olunthos** and the confederacy in 348 BCE; thence until 168 BCE a semi-autonomous region within the kingdom of **Macedon**. *OCD3* 315–316, C.F. Edson and N.G.L. Hammond; *BNP* 3 (2003) 179–180, M. Zahrnt.

Sites: **Kassandreia**, **Mendē**, **Olunthos**, **Stageira**.

People: Xenophilos.

Khalkis (mod. Khalkis, formerly Negroponte; 38°28' N, 23°36' E): ancient city in Euboia, metalworking center controlling the Euripos channel at its narrowest. Colonized **Italy** and **Sicily** in the 8th c. BCE and the north Aegean in the 7th. Opposed Xerxēs; a tributary ally of **Athens** from 446–411 BCE; in the 4th c. BCE garrisoned by Philip II of **Macedon**. Under **Rome** in 197 BCE, partly destroyed in 146 BCE for supporting the Akhaean confederacy against **Rome**. Prosperous under **Rome**; refortified in the 6th c. CE (Prokopios *Aed.* 4.3.19). *PECS* 216–217, M.H. McAllister; *OCD3* 316, W.A. Laidlaw *et al.*; *BAGRW* 55-F4; *BNP* 3 (2003) 181–182, H. Kaletsch. (Although Ath., *Deipn.* 11 [502b], and Stephanos

OF BUZANTION, s.v., refer to a Khalkis of **Thrakē**, no such site has been found, and M. Zahrnt, *Olynth und die Chalkidier* [1971] 253 argues that it never existed.)

EUPHORIŌN, KRATĒS.

Khalkis ad Belum (Syria) (mod. Qinnesrin; 36°00' N, 37°00' E): founded in Seleukid times on a site formerly (and now) known as "eagle's nest," east of the Orontes valley, and east on the road from **Antioch**, north-east on the road from **Apameia** (*cf.* **Syria**). Base for the revolt of Diodotos Truphōn 145 BCE (DIODŌROS OF SICILY, Book 33, *fr.*4a); ravaged by the Sasanian king Shapur I, 252 CE; well-known in Byzantine times. *RE* 3.2 (1899) 2090–2091 (#14), I. Benzinger; *ODB* 406, M.M. Mango; *BAGRW* 68-C2.

IAMBLIKHOS.

Kharax "Spasinou" (mod. Abadan; 30°21' N, 48°17' E): at the head of the Persian gulf, founded by Alexander of **Macedon** as Alexandria, destroyed by floods, and rebuilt by Antiokhos IV (175–164 BCE) as **Antioch**. Under Seleukid rule, then semi-autonomous but under Nabataean influence, after 106 CE under Parthian hegemony, significant center of land and maritime Arab trade. *PECS* 60, D.N. Wilber; *BAGRW* 93-D3; *BNP* 3 (2003) 191, J. Oelsner. (Contrast the Kharax near **Apameia** in Syria, and Kharax in the Crimea.)

ISIDŌROS.

% Khersonēsos: besides the city (next), several homonymous sites and regions could also be Arkhelaos' home: **(A)** city and region on the north shore of the Black Sea (3 km west of mod. Sebastopol; 44°36' N, 33°29' E), (re)founded 421 BCE by **Hērakleia Pontikē**; allied with MITHRADATĒS VI, *ca* 115 BCE; independent until the 3rd c. CE (*PECS* 221–222, M.L. Bernhard and Z. Sztetyłło; *OCD3* 320–321, D.C. Braund; *BAGRW* 23-G4; *BNP* 3 [2003] 214–215 [#3], I. von Bredow *et al.*): for the peninsula on which this city was sited, see 213–214 (#2), I. von Bredow; **(B)** the peninsula adjacent to the Hellespont (*OCD3* 320, E.N. Borza; *BAGRW* 51-G/H4; *BNP* 3 [2003] 213 [#1], I. von Bredow): including the sites **Elaious** and **Kardia; (C)** peninsular region of south-west Karia, *BAGRW* 61-G4, probably under Ptolemaic control in the era of Arkhelaos.

ARKHELAOS.

Khersonēsos (mod. Khersonisos 26 km east of Iraklion; 35°18' N, 25°22' E): city of **Crete**, autonomous in the 4th–3rd centuries BCE; allied with **Knōssos** in the 3rd c.; allied with **Pergamon** in 183 BCE. *RE* 3.2 (1899) 2251–2252 (#4), L. Bürchner; *PECS* 221, D.J. Blackman; *BAGRW* 60-D2.

PHILŌNIDĒS.

Khios (mod. Khios; 38°24' N, 26°01' E): piney eastern Aegean island, wealthy commercial and industrial center. Colonized from Euboia in the 9th c. BCE, supported eastern powers in the 6th c. BCE, allied with **Athens** from 477 BCE. From 412 BCE allied with **Sparta**; in the 4th c. BCE allied with **Athens**, then under Maussōllos. Largely independent after Alexander of **Macedon**, pro-Roman from 190 BCE, captured by Sulla for **Rome** in 86 BCE, a *ciuitas libera* until Vespasian. *PECS* 715–716, G.B. Montanari; *OCD3* 323, D.G.J. Shipley; *BAGRW* 56-C5; *BNP* 3 (2003) 232, H. Kaletsch.

AGATHOKLĒS, ARISTŌN, EUNOMOS, GLAUKOS, HEKATŌNUMOS, HIPPOKRATĒS, HOMER (?), IŌN, MĒTRODŌROS, OINOPIDĒS, SKUMNOS.

Khoren ⇒ Xoren

Kilikia: possibly Semitic native population, ruled by the Seleukids from **Antioch** in the 3rd c. BCE; accused of piracy in the 2nd c. BCE. Taken by **Rome** 102 BCE and made a province; autonomous 44 BCE; again a province of **Rome** from 72 CE. Prosperous and densely populated; ravaged by Goths 276–277 CE; Diocletian 297 CE divided the province into western ("Rough") and eastern ("Plain"), the latter further divided *ca* 400 CE, with capitals at **Tarsos** and **Anazarbos**; declined after 6th c. *ODB* 462–463, C.F.W. Foss; *OCD3* 330–331, G.E. Bean and St. Mitchell; *BNP* 3 (2003) 329–331, H. Täuber and A. Berger.

Sites: **Anazarbos, Kastabala, Mallos, Seleukeia, Soloi, Tarsos, Tuana**.

People: ARIOBAZARNĒS, SIMPLICIUS, ZĒMARKHOS.

Kios ⇒ Prousias

Kition (mod. Larnaka; 34°55' N, 33°38' E): port town on south-west coast of **Cyprus** with a natural harbor, founded *ca* 1500 BCE by Mycenaeans, a center of Phoenician trade from the 9th c., and under Phoenician hegemony from 479 to 312 BCE, when ceded to Ptolemy I. *PECS* 456–458, K. Nicolaou; *BAGRW* 72-D3; *BNP* 3 (2003) 368–369, R. Senff.

APOLLODŌROS, APOLLŌNIOS, ZĒNŌN.

Klazomenai (mod. Kilizman; 38°22' N, 26°53' E): island joined to mainland on the south shore of the Gulf of **Smurna**. Ionian settlers colonized the mainland but moved to the island *ca* 500 BCE. Under Persian control until allied with Athens in the 5th c. BCE, surrendered to Persia 386 BCE, freed by Alexander of **Macedon**, granted immunity by **Rome** 188 BCE. *PECS* 458, G.E. Bean; *OCD3* 343, *Idem* and S. Sherwin-White; *BAGRW* 56-D5; *BNP* 3 (2003) 411–412, K. Ziegler and H. Engelmann.

ANAXAGORAS, APSURTOS (?), ARTEMŌN.

Knidos (new) (mod. Tekir; 36°41' N, 27°22' E): the inhabitants of **Knidos (old)** moved in 360 BCE to a better site at the tip of the peninsula. Under Ptolemaic control in 3rd c. BCE; under **Rhodes** in 2nd c. BCE. *PECS* 459, I. Love; *OCD3* 354, J.M. Cook and S. Sherwin-White. *BAGRW* 61-E4; *BNP* 3 (2003) 489–490, H. Kaletsch.

AGATHARKHIDĒS, ARISTEIDĒS (?), ARISTOGENĒS, DIDUMOS, KHRUSIPPOS (II), SŌSTRATOS.

Knidos (old) (mod. Datça; 36°45' N, 27°40' E): Dorian foundation in Karia, on long peninsula in Gulf of **Kōs**, south-east of **Halikarnassos**, perhaps a colony of **Sparta**; yielded to the Persians after 546. Allied with **Athens** in the 5th c. BCE, but supported **Sparta** after 413. Famous for a medical school, fine wine, and Praxitelēs' statue of Aphrodite. See **Knidos (new)**; *BAGRW* 61-F4.

EUDOXOS, EURUPHŌN, HĒRODIKOS, KHRUSIPPOS (I), KTĒSIAS.

Knōsos/Knōssos (mod. Makruteikhos; 35°18' N, 25°10' E): flourishing Greek town in the 9th–6th centuries BCE, then again in the 4th c. Resisted **Rome**, but after 27 BCE, a *colonia*, receiving settlers probably from Capua. *PECS* 459–460, K. Branigan; *OCD3* 354, L.F. Nixon and S.R.F. Price; *BAGRW* 60-D2; *BNP* 7 (2005) 73–74, H. Sonnabend.

AINESIDĒMOS, KHERSIPHRŌN, METAGENĒS.

Kolophōn (mod. Degirmendere; 38°07' N, 27°08' E): fertile coastal site, good port, famous for horses. Under Lydians, then Persians, until Alexander of **Macedon**. Resisted LUSIMAKHOS who deported inhabitants to **Ephesos**; they returned after his death (281 BCE), but the town did not fully recover; resisted Antiokhos III's circumvallation (Livy 37.26.5–8, 28.4, 31.3). *PECS* 233, W.L. MacDonald; *BAGRW* 61-E1; *BNP* 3 (2003) 578–579 (#1), K. Ziegler.

DAMIS, DEINŌN, HERMOTIMOS, NIKANDROS, XENOPHANĒS.

Kōs (mod. Kōs; 36°51' N, 27°14' E): south-east Aegean island, on shipping routes from the

Aegean to the east, colonized by Dorians (*ca* 900 BCE). Under Persians from *ca* 700 BCE, allied with **Athens** in the 5th–4th centuries BCE, under Alexander of **Macedon** from 336 BCE, then Ptolemaic from 309 to 260 BCE. Occupied by MITHRADATĒS VI 88 BCE, given immunity by Claudius 53 CE, destroyed by an earthquake 554 CE. *PECS* 465–466, V. Tusa; *OCD3* 403–404, W.A. Laidlaw and S. Sherwin-White; *BAGRW* 61-E4; *BNP* 3 (2003) 856–859, H. Sonnabend.

ANTONINUS, DEXIPPOS, DRAKŌN, HIPPOKRATĒS, LUSIMAKHOS, PHILETAS/PHILINOS, PHILIPPOS, PHULOTIMOS, PRAXAGORAS, SIMOS, SŌRANOS, THESSALOS, XENOKRITOS, XENOPHŌN (?).

Krētē ⇒ **Crete**

Krotōn (mod. Crotone; 39°05' N, 17°07' E): coastal south Italian city founded 710 BCE (STRABŌN 6.1.12), flourishing into the 6th c., destroyed by **Subaris** 510 BCE, but revived, allying with **Rome** against Hannibal, made a *colonia* 194 BCE, after which declined. *PECS* 470–471, W.D.E. Coulson; *OCD3* 411, H.K. Lomas; *BAGRW* 46-F3; *BNP* 3 (2003) 959–960, A. Muggia.

ALKMAIŌN, DĒMOKEDĒS, EURUTOS (?), HIPPŌN, IORDANES, NEOKLĒS, PHILOLAOS, XOUTHOS (?).

Kumē (Aiolian) (mod. Çakmaklı; 38°45' N, 26°56' E): coastal Aiolian city between two river-mouths, north of **Phōkaia** and south of **Pitanē**. Subjected to rule by Persia, **Athens**, Seleukids, **Pergamon**, and **Rome**. Devastated by the earthquake of 17 CE (compare **Magnesia on Sipulos, Philadelpheia [Ludia], Sardēs**), rebuilt with Tiberius' aid. *PECS* 472–473, G.E. Bean; *OCD3* 418, D.E.W. Wormell and St. Mitchell; *BAGRW* 56-D4; *BNP* 3 (2003) 1050 (#3), H. Kaletsch. (Contrast the alleged Euboian site, and the city of **Italy**, *BAGRW* 44-F4.)

EPHOROS.

Kurēnē (mod. Ain Shahat, Grennah; 32°49' N, 21°51' E): Theran colony in **Libya**, founded 630 BCE, annexed by Ptolemy I *ca* 300 BCE, not without subsequent struggles. Influenced by **Rome** from 163 BCE, semi-autonomous from 96 BCE, became a province under **Rome** in 74 BCE, devastated during Jewish revolt ("Kitos" War) in 115–117 CE (*cf.* **Alexandria**, **Cyprus**, **Edessa**, and **Nisibis**), but restored and repopulated by Hadrian. Suffered in 3rd c. CE from attacks from indigenous peoples, earthquake of 262, and encroachments of the desert. Divided into two provinces by Diocletian; refortified by Anastasios and Justinian in response to raids by Asturiani. *PECS* 253–255, D. White; *ODB* 570–571, R.B. Hitchner; *OCD3* 421–423, J.M. Reynolds; *BAGRW* 38-C1; *BNP* 4 (2004) 6–10, W. Huß; *EJ2* 5.349–350, I. Gafni.

APELLĀS, ARISTIPPOS, DAMŌN, DIONUSIOS, ERATOSTHENĒS, KALLIMAKHOS, KALLIMAKHOS JR., KARNEADĒS, LAKUDĒS, NIKOTELĒS, OPHELLĀS, PHILOSTEPHANOS, PRŌROS, PTOLEMAIOS, PTOLEMAÏS, SUNESIOS, THEODŌROS.

Kurrhos (mod. Aravissos; 40°50' N, 22°18' E): small city in **Macedon**, east of Edessa and west of **Pella**. *PECS* 473, J.-P. Rey-Coquais; *BAGRW* 50-B3; *BNP* 4 (2004) 15 (#1), R.M. Errington. (Contrast the Syrian Kurrhos, also known as Hagioupolis, mod. Nebi Ouri; *BAGRW* 67-D3.)

ANDRONIKOS.

Kurtos (*unlocated*): explained by STEPHANOS OF BUZANTION, s.v., as an Egyptian city "in the interior" (*sc.*, upper **Egypt**?).

Dionusios.

Kusumapura or Pāṭaliputra (mod. Patna; 25°36' N, 85°08' E): on the south bank of the Ganges, founded 490 BCE; center of Buddhist study and practice. Named "Palibothra" and described as of great size by Megasthenēs (although Ranajit Pal, *Non-Jonesian Indology and Alexander* [New Delhi 2002] has argued that Palibothra is mod. Kohnouj, south-west of mod. Jiroft/Sabzvaran, 28°40' N, 57°44' E). Kusumapura was built up by Ashoka as his imperial capital 273 BCE; eastern limit of conquests by Menandros king of Taxila (reigned *ca* 170–130 BCE); visited by Chinese scholars, esp. Fa Hsien, in the 4th–5th centuries CE. J. Schwartzberg *et al.*, *A Historical Atlas of South Asia*, 2nd ed. (1992) grid G4 on maps 18–20, and pp. 170–174, 179, 183.

Āryabhaṭa.

Kuthēra (mod. Kithira; 36°10' N, 23°00' E): island off Cape Malea, south of Peloponnesos, famous for murex (for purple dye), under **Argos**, and then seized by **Sparta** 550 BCE; changing hands several times in the late 5th and early 4th centuries BCE. Prosperous enough to mint its own coins in the 3rd c. BCE. Part of the non-Spartan Lakōnian League in the 2nd c. BCE; granted by Augustus to Iulius Eurycles of **Sparta** 21 BCE; held by the family until returned to **Sparta** by Hadrian. *PECS* 473, M.G. Picozzi; *OCD3* 423, W.G. Forrest and A.J.S. Spawforth; *BAGRW* 58–inset; *BNP* 4 (2004) 23–25, H.-D. Blume.

Ptolemaios.

Kuthnos (mod. Kuthnos; 37°23' N, 24°25' E): Aegean island south of **Keōs**, west of **Suros**, inhabited from early times, its constitution praised by Aristotle. *BAGRW* 58-G3; *BNP* 4 (2004) 25–26, H. Kaletsch.

Kudias (?).

Kuzikos/Cyzicus (mod. Belkis; 40°24' N, 27°53' E): city on the isthmus where Arktonessos joins the mainland, east of **Parion** and west of **Murleia**; provided with good harbors, and center of trade (its coins being widely standard); earliest Greek colony in the Propontis (Strabōn 14.1.6), founded 756 BCE possibly by **Corinth**, refounded 675 by **Milētos**. Alternately allied with **Athens** and **Sparta** in the 5th c. BCE; under the Persians 387–334 BCE; under **Pergamon** in the 2nd c. BCE until 133, when under **Rome**. Withstood a siege by Mithradatēs VI 74 BCE. *PECS* 473–474, E. Akurgal; *OCD3* 423–425, T.R.S. Broughton and St. Mitchell; *BAGRW* 52-B4; *BNP* 4 (2004) 26–28, Thos. Drew-Bear.

Adrastos, Apollodōros, Athēnaios, Eudoxos, Helikōn, Kallippos, Kleōn (?), Mnēsitheos, Neanthēs, Polemarkhos, Teukros.

Lakōnika: the south-eastern fifth of the Peloponessos, ruled by **Sparta**. *OCD3* 810–811, R.W.V. Catling; *BNP* 7 (2005) 148–150, Y. Lafond.

Sites: **Megalopolis**, **Stumphalos**, **Sparta**.

People: Arkhagathos, Dēmētrios.

Lampsakos (mod. Lapseki; 40°21' N, 26°41' E): colony of **Phōkaia** in northern **Troas**, at the entrance to the Hellespont from the Propontis, and generally prosperous. Allied with **Athens** in the 5th c. BCE, under **Sparta** 405–386 BCE; then again Athenian or Persian by turns, until conquered by Alexander of **Macedon** 334 BCE. Changed alliance several times, then from 281 BCE to the Seleukids, and from 227/226 BCE to **Pergamon**. Largely autonomous from 190–129 BCE, when came under **Rome**. *PECS* 480, T.S. MacKay; *OCD3* 813, D.E.W. Wormell and St. Mitchell; *BAGRW* 51-H4; *BNP* 7 (2005) 190–191, E. Schwertheim.

Idomeneus, Kōlōtēs, Mētrodōros, Poluainos, Stratōn, Xenophōn.

% Laodikeia: many cities were founded in regions controlled by the Seleukids under this name, including those listed separately below and: **(A)** Katakekaumēnē, between Phrugia and Lukaonia (mod. Ladik, *BAGRW* 63-A4); **(B)** on the Orontes or "ad Libanum" (mod. Tell Nebi Mend, *BAGRW* 68-C4); **(C)** Pontikē (mod. Gökçeyazı; *BAGRW* 87-A4); and **(D)** Median (mod. Nihavand; *BAGRW* 92-D2).

Apellās, Apollōnios, Diphilos, Iulianus, Papias, Zēnōn.

Laodikeia on the Sea (Syria) (mod. Ladhiqiyah; 35°31' N, 35°47' E): one of the four cities founded in Syria by Seleukos I *ca* 300 BCE, on an old Phoenician site, north of **Arados**, south of **Seleukeia "Pieria"**; prosperous and prominent by *ca* 150 BCE, conquered by Pompey (64 BCE), declared free by **Rome** *ca* 45 BCE. Sacked by Pescennius Niger (193/4 CE), and restored by Septimius Seuerus who made it a *colonia*. Prosperous center of linen and book production; damaged by earthquakes in 494 and 555 CE. *PECS* 482, J.-P. Rey-Coquais; *ODB* 1178, M.M. Mango; *BAGRW* 68-A2; *BNP* 7 (2005) 232–233 (#1), J. Gerber.

Philōnidēs, Proklos, Themisōn.

Laodikeia (Phrugia) on the Lukos (mod. Eski Hisar, 6 km north of Denizli; 3749' N, 2907' E): above the confluence of the Lukos and Maiandros rivers, north of **Hērakleia Salbakē**, east of Antioch on the Maiandros, and across the valley from one **Hierapolis**. Prosperous trade center, refounded *ca* 260 BCE by Antiokhos II in honor of his wife Laodikē. Antiokhos III deported Jews here from **Babylōn**, *ca* 200 BCE: Iosephus, *Ant. Iud.* 12.3.4. Under **Pergamon** by 188 BCE; then under **Rome** from 133 BCE. The medical school at **Mēn Karou** was nearby. Seriously damaged in the earthquake of 60 CE, restored without outside support. An early center of Christianity, and of textile production. *PECS* 481–482, G.E. Bean; *ODB* 1177, C.F.W. Foss; *OCD3* 815, W.M. Calder and S. Sherwin-White; *BAGRW* 65-B2; *BNP* 7 (2005) 234 (#4), K. Belke; *EJ2* 12.487–488, U. Rappaport.

Alexander, Anatolios, Polemōn, Theodās.

Laranda (mod. Karaman; 37°10' N, 33°13' E): chief town of Lukaonia, south-west along the road from **Tuana**, and north-north-west of **Seleukeia on the Kalukadnos**; the region was under the Seleukids 280–189 BCE, then under **Pergamon** to 133 BCE, then under **Rome**; at some point part of the province of *Cilicia*, and in the province of *Galatia* from 25 BCE; much of Lukaonia was again in *Cilicia* from *ca* 150 CE, but Laranda is cited by Ptolemy, *Geog.* 5.6.16, as in **Kappadokia**. *BAGRW* 66-C2; *BNP* 7 (2005) 244, K. Belke.

Nestōr.

% Laris(s)a: besides the two listed below, there were numerous sites named Larissa: e.g., the Larisa of Phthiotis, *BAGRW* 55-C1, or the Larissa of **Ludia**, *BAGRW* 56-F5, and see Strabōn 9.5.19, who lists a dozen or more sites. The most likely in Byzantine times may well be that in **Thessalia**.

Damianos, Hēliodōros.

Larissa of Syria (mod. Shaizar/Sizara; 35°32' N, 36°35' E): ancient city on the Orontes, north of **Emesa**, upstream from, and dependent upon, **Apameia**; originally named Sezar/Sidzara, and again usually called Sezar from the 1st c. CE. Strabōn 16.2.10; Stephanos of Buzantion, s.v. Larisa; *RE* 12.1 (1924) 873 (#12), L.A. Moritz; *BAGRW* 68-C3.

Domninos.

Larissa of Thessalia (mod. Larisa; 39°38' N, 22°25' E): most important town of **Thessalia**, on right bank of the Peneios and roughly in the center of the eastern Thessalian plain, occupied from prehistoric times. Flourishing artistic center in the 5th c. BCE weakened by internal discord, under **Macedon** from 344–196 BCE, when it became the capital of the Thessalian League under **Rome**, flourishing well into the Roman Empire, sacked by Ostrogoths in the late 5th c. and rebuilt by Justinian. *PECS* 485, T.S. MacKay; *ODB* 1180,

A. Kazhdan; *OCD3* 816, B. Helly; *BAGRW* 55-C1; *BNP* 7 (2005) 251–253 (#3), H. Kramolisch and E. Wirbelauer; *EJ2* 12.494, S. Marcus and Y. Kerem.

Anaxilaos, Polukleitos.

Lēmnos (mod. Lēmnos; 39°55' N, 25°15' E): northern Aegean volcanic island, whose native language resembled Etruscan; conquered by **Athens** *ca* 505 BCE (Hērodotos 6.137–140). The main cities were Murina on the west coast and Hephaistia on the north. **Rome** confirmed Athenian hegemony 167 BCE, which endured until *ca* 200 CE. *PECS* 496–497, J. Boardman; *OCD3* 842–843, E.N. Borza; *BAGRW* 56-A2; *BNP* 7 (2005) 382–384, H. Kaletsch and E. Meyer.

Apollodōros, Philostratos.

Leontinoi (mod. Lentini; 37°17' N, 15°00' E): colony of **Naxos** 729 BCE, in south-east **Sicily**, which forcibly expelled the native Sikels. Autonomous until conquered by Hippokratēs of **Gela** 495 BCE; prosperous in the 5th c.; allied with **Athens** 433–424 BCE (and 415 BCE), occupied by **Surakousai** 422 BCE, which then mostly dominated Leontinoi until conquest by **Rome** 215 BCE. *PECS* 497–498, G. Rizza; *OCD3* 844, A.G. Woodhead and R.J.A. Wilson; *BAGRW* 47-G4; *BNP* 7 (2005) 405–406, S.D. Spina.

Gorgias.

Lesbos: large north-eastern Aegean island, colonized nearby mainland **Sigeion**; dominated by Persia from 545 BCE. Allied with **Athens** in the 5th–4th centuries BCE; unsuccessfully revolted from **Athens** 428 and 412 BCE, allied with **Sparta** 405 BCE, but restored to Athenian rule by 390 BCE. Friendly with Persia until 334 BCE, when taken by Alexander of **Macedon**. Member of the island-league centered at **Dēlos** (q.v.); friendly with **Rome** beginning 200 BCE (which destroyed Antissa 167 BCE), allied after 129 BCE. *PECS* 502–503, M. Paraskevaïdis; *OCD3* 845, D.G.J. Shipley; *BAGRW ca* 56-C3; *BNP* 7 (2005) 429–431, H. Sonnabend.

Sites: **Eresos**, **Mēthumna**, **Mutilēnē**.

People: Salpē (?).

Libya/Libua: name applied ambiguously and variously to northern Africa west of **Egypt** (i.e., approximately the modern nation of Libya plus western Egypt), or the entire north African coastal zone, or the entire African continent. By Roman convention, it was the administrative district west of **Alexandria** to a point west of **Kurēnē**. Ancient sources emphasize the nomadic lifestyle of the indigenous peoples, but coastal areas supported Phoenician, Punic, Greek, and Roman settlements with mixed populations. *OCD3* 855–856, J.M. Reynolds; *BNP* 7 (2005) 515–516, K. Zimmermann.

Sites: **Kurēnē**.

People: Fauilla, Kleomenēs, Theokhrēstos (?).

Lindos (mod. Lindos; 36°05' N, 28°05' E): on **Rhodes** island, allied with **Athens** in the 5th c. BCE until 411; autonomous until creation of federal Rhodian state 408/7 BCE. *PECS* 756–757, R.E. Wycherley; *OCD3* 862–863, E.E. Rice; *BAGRW* 60-G3; *BNP* 7 (2005) 609–612, H. Sonnabend.

Kleomenēs (?), Panaitios.

Lokroi Epizephurioi (mod. Locri; 38°14' N, 16°16' E): coastal city in south **Italy** founded in the early 7th c. BCE by colonists from **Lokris**. Notable for an early written legal code and its restrictive hereditary oligarchy. Allied with **Surakousai** against **Rhēgion** and **Athens** in the 5th c. BCE; captured by Scipio Africanus for **Rome** 205 BCE. *PECS* 523–524, F.P. Badoni; *OCD3* 879, H.K. Lomas; *BAGRW* 46-D5; *BNP* 7 (2005) 774–778, D. Musti and L.D. Morenz.

Philistiōn.

% Lokris: two regions of Greece ("West," on north coast of Gulf of Corinth, and "East," across the mountains and facing Euboia). East Lokris was federal in the 5th c. BCE, capital at **Opous** and allied with **Sparta**; later often dominated by **Macedon**, autonomous again after 167 BCE. West Locris was also federal, by the 4th c. BCE, more loosely organized. *OCD3* 879–880, J. Buckler; *BNP* 7 (2005) 769–774, G. Daverio Rocchi.

Sites: **Amphissa** (West Lokris), **Opous** (East Lokris).

People: TIMAIOS (?).

Lucania: in south **Italy**, north of Bruttium, west of Apulia, and south of **Campania**; allied with **Rome** 326–317 BCE; subjugated by **Rome** 298 BCE; in revolt against **Rome** and allied with PURRHOS 281–272 BCE; from 206 BCE thoroughly under **Rome**. *OCD3* 886, H.K. Lomas; *BNP* 7 (2005) 826–828, M. Lombardo.

Sites: **Elea**, **Metapontion**, **Poseidōnia**, **Subaris**.

People: AISARA, OCELLUS.

Ludia/Lydia: independent kingdom on the west coast of Anatolia, subjected to Persia 546 BCE, then to Alexander of Macedon 334 BCE (some coastal cities excepted); under the Seleukids until 189 BCE; under **Pergamon** until 133 BCE; under **Rome** as part of the province of *Asia*. Reconstituted as a province in its own right by Diocletian 297 CE, capital **Sardēs**; devastated by Goths 399 CE. *OCD3* 898, W.M. Calder *et al.*; *BNP* 8 (2006) 2–11, H. Kaletsch.

Sites: **Daldis**, **Ephesos**, **Magnesia (2)**, **Milētos**, **Philadelpheia**, **Sardēs**, **Smurna**, **Tralleis**.

People: PERIKLĒS, PRISCIANUS.

Lugdunum (mod. Lyon; 45°46' N, 04°50' E): at the confluence of the Saône and Rhône, on the common border of the three parts of Gaul (*cf.* CAESAR, *BG* 1.1.1), founded as a *colonia* of **Rome** on the site of a Gallic hill-fort, in 43 BCE.; its site made it the prosperous hub of AGRIPPA's road system. In 15 BCE it became an imperial mint; in 12 BCE the seat of the provincial *concilium*; damaged by a fire in 66 CE; early Christian site. *PECS* 528–531, M. Leglay; *OCD3* 891–892, A.L.F. Rivet and J.F. Drinkwater; *BAGRW* 17-D2; *BNP* 7 (2005) 876–878, Y. Lafond and M. Leglay.

ABASKANTOS.

Lukaia (*unlocated*): town in north-west Arkadia (i.e., not far from **Megalopolis** and west of **Stumphalos**); resisted its incorporation into **Megalopolis**, and continued to exist in some form. Paus. 8.27.4–5; *RE* 13.2 (1927) 2229–2231 (#1), Ernst Meyer.

ALEXANDER.

Lukia/Lycia: mountainous region of south-west Anatolia, around the Xanthos river and the homonymous city (mod. Kınık); taken by Persia 546 BCE, then under Maussōllos and Alexander of **Macedon**; Ptolemaic *ca* 300–197 BCE, when taken by Antiokhos III; under **Rhodes** 189–169 BCE. Autonomous or dominated by **Rome** until made a province in 43 CE (joined with *Pamphylia* from 74 CE); prosperous region under **Rome**. Separated from *Pamphylia* by Constantine; prosperous in the late 6th c. *ODB* 1257–1258, C.F.W. Foss; *OCD3* 894–895, St. Mitchell; *BNP* 7 (2005) 916–920, M. Zimmermann.

Sites: **Indos**, **Oinoanda**, **Patara**, **Phaselis**, **Tlōs**.

People: CAPITO, DIOPHANTOS, PROKLOS.

Lydia ⇒ **Ludia**

Macedon: mountainous region between the Balkans and Greece, sloping down to the sea near **Khalkidikē**, between **Thessalia** and **Thrakē**, a nexus of important land-routes.

Occupied by Persia 512–479 BCE; united by Philip II from *ca* 360 BCE, who dominated Greece after the battle of **Khairōneia** 338 BCE. His son and successor Alexander conquered Persia including **Egypt**, 336–323 BCE. After Alexander's death, the kingdom continued to rule northern Greece until **Rome** opposed Philip V (211 BCE), and then confined **Macedon** to its old borders from 197 BCE; **Rome** abolished the kingdom and made the territory into four provinces 167 BCE. *OCD3* 904–905, N.G.L. Hammond; *BNP* 8 (2006) 57–72, R.M. Errington and L. Duridanov.

Sites: **Kurrhos**, **Pella**, **Stoboi**.

People: ARRABAIOS (?), LUKOS, LUSIMAKHOS, MAËS TITIANUS, PATROKLĒS, PHILIPPOS, POSEIDŌNIOS, THEODŌROS (?), ZŌILOS.

Madaurus (mod. M'Daourouch; 36°05' N, 07°49' E): inland Numidian city founded 3rd c. BCE, east of **Cirta** and west of **Sicca Veneria**; part of **Mauretania** from 106 BCE, then under **Rome** from 46 BCE, and in the province of *Africa*; made a *colonia* of veterans *ca* 70–95 CE. Although small, an intellectual hub with several schools; AUGUSTINE was partly educated here. *PECS* 541–542, G. Souville; *OCD3* 907, W.N. Weech *et al.; BAGRW* 32-A4; *BNP* 8 (2006) 106–107, W. Huß.

APULEIUS.

% Magnesia on Maiandros (mod. Tekin 3 km south of Ortaklar; 37°51' N, 27°32' E): colony founded from the region Magnesia, in **Ludia** near **Ephesos, Priēnē**, and **Tralleis**; refounded *ca* 400 BCE; sided with MITHRADATĒS VI against **Rome**, but made a free city 84 BCE. STEPHANOS OF BUZANTION, s.v. (knows only this one); *PECS* 544, G.E. Bean; *OCD3* 912, W.M. Calder *et al.; BAGRW* 61-F2; *BNP* 8 (2006) 173 (#2), W. Blümel. It does not seem possible to determine from which Magnesia these men came:

DIOKLĒS, KHARŌN, LASOS, MAKARIOS, PURRHOS, SIMŌN, THEUDIOS.

Magnesia on Sipulos (mod. Manisa; 38°37' N, 27°26' E): colony founded from the region Magnesia, in **Ludia** at the nexus of roads from the interior and the Propontis to **Smurna**. Frequently changed hands in the 3rd c. BCE. The site of the decisive Roman victory over Antiokhos III in 189 BCE; although under **Rome** only from 133 BCE. Reconstructed after 17 CE earthquake (compare **Kumē, Philadelpheia [Ludia], Sardēs**), an important late Byzantine military and political center. *PECS* 544–545, W.L. MacDonald; *OCD3* 912, J.F. Lazenby; *BAGRW* 56-E4; *BNP* 8 (2006) 173–174 (#3), H. Kaletsch.

HĒGĒSIAS.

Mallos (mod. Kızıltahta, north of Karataş; 36°45' N, 35°29' E): old city of **Kilikia** at the mouth of the Puramos river. After conquest by Alexander of **Macedon**, under the Seleukids in the 3rd–2nd centuries BCE; briefly independent in the 2nd c. Settled with ex-pirates by Pompey 67 BCE, accumulated honorific titles in rivalry with neighbors, especially **Tarsos**. *PECS* 547, M. Gough; *BAGRW* 66-G3; *BNP* 8 (2006) 204, M.H. Sayar.

ARISTOPHANĒS, KRATĒS, NIKIAS, PHILISTEIDĒS, ZĒNODOTOS (?).

Mantua (mod. Mantova/Mantua; 45°10' N, 10°48' E): Etruscan settlement on Mincius river in Cisalpine Gaul, south-west on road from **Verona**. Occupied by Gauls, then from 3rd c. BCE by **Rome**, which made the city a *colonia* and then a *municipium* with Latin rights *ca* 80 BCE; made a *ciuitas* 49 BCE. *PECS* 550, D.C. Scavone; *OCD3* 919, E.T. Salmon and D.W.R. Ridgeway; *BAGRW* 39-H3; *BNP* 8 (2006) 261–262, A. Sartori. (Contrast the Spanish Mantua, *BAGRW* 24-G4.)

VERGIL.

Massalia (mod. Marseilles; 43°17' N, 05°24' E): coastal city, east of the Rhône delta, colonized *ca* 600 BCE from **Phōkaia**, an excellent harbor and commercial hub, founded many of its own colonies; had a stable aristocratic constitution. Allied with **Rome** against Hannibal, 215 BCE; aggression by local Gauls resulted in submission to **Rome**, 125–121 BCE, becoming the leading free city of the new province. Taken by Caesar 49 BCE, who promoted **Narbo** in its place. *PECS* 557–558, F. Salviat; *OCD3* 935, A.L.F. Rivet, J.F. Drinkwater; *BAGRW* 15-E3; *BNP* 8 (2006) 441–445, Y. Lafond.

EUTHUMENĒS, KHARMĒS, KRINAS, MASSILIOT PERIPLOUS, PUTHEAS.

Mauretania: coastal north **Africa**, west of **Cirta**, dry and rocky (approximately the coasts of modern Morocco and Algeria). In the 8th–7th centuries BCE Phoenicians established coastal emporia; native kingdoms are recorded by the late 3rd c. BCE. King Bocchus I assisted **Rome** 106 BCE, and thus acquired western Numidia (including **Cirta** and **Madaurus**). His sons Bocchus (II) and Bogud ruled jointly, and supported CAESAR, 49–44 BCE; Bogud died fighting for Antony, whereas Bocchus II bequeathed the kingdom to Octavian (33 BCE). AUGUSTUS in turn appointed IOUBA II client king (25 BCE). After Caligula's murder (in 40 CE) of the son of Iouba II, Claudius (*ca* 42 CE) constituted two provinces (*Mauretania Tingitana* and *Mauretania Caesariensis*), overseen by procurators from Tingis and **Caesarea.** **Rome** founded several *coloniae*, and native cavalry served as *auxilia*. Much land remained under native rulers, and the late 3rd to 4th c. saw serious rebellions. Under the Vandals 430–534 CE. *OCD3* 939, W.N. Weech and R.J.A. Wilson; *BNP* 8 (2006) 493–496, W. Huß.

Sites: **Auzia**, **Caesarea**, **Cirta**, **Madaurus**.

People: IOUBA II.

Mediolanum (mod. Milano; 45°28' N, 09°10' E): at the junction of prehistoric roads from the plains and Alps, founded *ca* 396 BCE by the Insubres, under **Rome** by 194 BCE as a *municipium* and eventually a *colonia*. A center of the applied arts, and the site of a Roman mint, the city became the western Roman capital under Diocletian/Maximian (293 CE), until it was moved to **Ravenna**, *ca* 403 CE. *PECS* 561, M. Mirabella Roberti; *OCD3* 949–950, E.T. Salmon and T.W. Potter; *BAGRW* 39-E3; *BNP* 8 (2006) 583–584, C. Heucke; *EJ2* 14.231, A. Milano and S. Rocca.

AMBROSIUS, FAUONIUS, MALLIUS THEODORUS.

Medma (mod. Rosarno; 38°30' N, 15°59' E): also known as Mesma and Medmē, colony established from **Lokroi Epizephurioi** in 7th c. BCE, allied with **Lokroi** against **Krotōn** *ca* 500 BCE; opposed **Lokroi** 422 BCE; benefited by **Surakousai** 396 BCE. *PECS* 563–564, R. Holloway; *OCD3* 950, H.K. Lomas; *BAGRW* 46-C5; *BNP* 8 (2006) 589, M. Lombardo.

PHILIPPOS.

Megalopolis (mod. Megalopolis/Sinanou; 37°24' N, 22°08' E): established *ca* 370 BCE, by five Arkadian ***poleis***, north-west of **Sparta**, on the Alpheios river (a travel-route between **Lakōnika** and Arkadia). Opposed and conquered by **Sparta**; rebuilt 223 BCE. Influential from the 4th–2nd cc. BCE, supporting **Macedon**, declining under **Rome** but existing into late antiquity. *OCD3* 950–951, J. Roy; *BAGRW* 58-C3; *BNP* 8 (2006) 596–598, Y. Lafond and E. Meyer.

POLUBIOS.

Megara (mod. Megara; 38°00' N, 23°20' E): coastal city between **Athens** and **Corinth**, a vigorous center of colonization (**Buzantion, Khalkēdōn, Hērakleia Pontikē**), alternately allied or in contention with **Athens** in the 7th–6th centuries BCE; allied with **Sparta** in the 5th c. BCE. *PECS* 565, W.R. Biers; *OCD3* 951, J.B. Salmon; *BAGRW* 58-E2; *BNP* 8 (2006) 599–603, K. Freitag. (Contrast their colony Megara Hublaia on the east coast of **Sicily**, *BAGRW* 47-G4, as well as the unlocated site in Syria mentioned by STRABŌN 16.2.10.)

AGĒSIAS, EUPALINOS, HĒRODIKOS.

Memphis (mod. Mit Rahina; 29°51' N, 31°15' E): on west bank of the Nile, about 30 km upstream of the apex of the Delta; first capital of united Upper and Lower **Egypt** (Hērodotos 2.99). Here Alexander of **Macedon** celebrated a Greek-style victory in 332 BCE, and Ptolemy V was crowned in the temple of Ptah according to Egyptian rites (196 BCE). *PECS* 571, S. Shenouda; *OCD3* 955, D.J. Thompson; *EAAE* 488–490, D. Jeffreys; *BAGRW* 75-E1; *BNP* 8 (2006) 654–656, K. Jansen-Winklen.

Apollōnios, Isidōros.

Mēn Karou (near mod. Gereli; 37°55' N, 28°55' E): on the Lukos river, upstream from **Antioch** on the Maiandros, and downstream of **Laodikeia on the Lukos**; site of a temple to Mēn, an Anatolian horned god of healing and fertility; here was a **Hērophilean** medical school: Strabōn 12.8.20. *OCD3* 955–956, R.L. Gordon; *BAGRW* 65-A2; *BNP* 8 (2006) 656–658, G. Petzl.

See: Alexander of Laodikeia, Aristoxenos (Hēroph.), Dēmosthenēs (Hēroph.), Hikesios of Smurna, Zeuxis (Hēroph.).

Mendē (mod. Kalandra; 39°58' N, 23°24' E): coastal colony of Euboian **Eretria** on the western peninsula of the **Khalkidikē**, south of **Poteidaia**; declined in importance after founding of **Kassandreia**. *PECS* 572, S.G. Miller; *OCD3* 957–958, S. Hornblower; *BAGRW* 51-A5; *BNP* 8 (2006) 670–671, M. Zahrnt.

Polukritos.

Mendēs (mod. Tell el-Ruba; 30°57' N, 31°31' E): ancient city in the north-east of the Delta of the Nile, east of **Sebennutos**, and west of **Pēlousion**; flourished under the Ptolemies, declined under **Rome**. *OCD3* 958, D.J. Thompson; *EAAE* 497–498, D. Hansen; *BAGRW* 74-F3; *BNP* 8 (2006) 671, S.J. Seidlmayer.

Bōlos, Ōros, Thrasullos (?).

Messēnē (Sicily) (mod. Messina; 38°11' N, 15°33' E): at the north-east corner of **Sicily**, on the straits of Messina, colonized in the 8th c. BCE (as "Zanklē") by Cumae and Euboia; founded **Himera**. Occupied by **Rhēgion** and renamed after the Peloponnesian region *ca* 480 BCE (Thucydidēs 6.4.5–6). Independent from **Rhēgion** and allied with **Surakousai** in the late 5th c. Taken and sacked by **Carthage** 396 BCE, liberated by **Surakousai** 393 BCE; under whom until 354 BCE, then again from 337 BCE (with brief interruptions). Unemployed mercenaries from **Campania** seized the town 288 BCE and in 264 BCE precipitated war between **Carthage** and **Rome**; after 241 BCE under **Rome** as Messana, a prosperous *ciuitas foederata*. Base of operations for Pompey's war against Caesar. *PECS* 998–999, G. Scibona; *OCD3* 963, H.K. Lomas; *BAGRW* 47-H2; *BNP* 8 (2006) 752–753 (#1), H. Sonnabend. (To be distinguished from the Peloponnesian Messēnē, *BAGRW* 58-B3; *BNP* 8 [2006] 762–765 [#2], Y. Lafond.)

Aristoklēs, Dikaiarkhos.

Metapontion (mod. Metaponto; 40°22' N, 16°48' E): coastal city of **Lucania** at two river mouths, colonized from Akhaia; a prosperous site in rivalry with **Taras**; burial site of Pythagoras. Supported **Athens** against **Sicily** in 413 BCE; dominated by **Taras** from 370 BCE until captured by Kleonumos of **Sparta**. *PECS* 574–575, R. Holloway; *OCD3* 968, H.K. Lomas; *BAGRW* 45-E4; *BNP* 8 (2006) 792–793, A. Muggia.

Eurutus (?), Hippasos, Timotheos.

Mēthumna (mod. Mithimna/Molyvos; 39°22' N, 26°10' E): on north coast of **Lesbos** island, antagonist of **Mutilēnē**; involved in founding **Assos**. (See **Lesbos**.) *PECS* 502–503 (s.v. Lesbos), M. Paraskevaïdis; *OCD3* 969, D.G.J. Shipley; *BAGRW* 56-C3; *BNP* 8 (2006) 806–807, H. Sonnabend.

MATRIKETAS, MURSILOS.

Milan ⇒ Mediolanum

Milētos (mod. Balat; 37°31' N, 27°17' E): ancient Hittite city, then large coastal Ionian city, prosperous commercial center and hub of colonization. Instigated revolt against Persia in 499 BCE, sacked in 494 BCE. Allied with **Athens** in the 5th c. BCE, but revolted in 412 and became a naval base for **Sparta**. Then under Persia until liberated by Alexander of **Macedon**. Diplomatically active and of varying alliance in the 3rd c., until subjected to **Pergamon** in the 2nd c.; under **Rome** from 133 BCE. *PECS* 578–582, G. Kleiner; *OCD3* 980, P.N. Ure *et al.*; *BAGRW* 61-E2; *BNP* 8 (2006) 885–895, J. Cobet and F. Starke; *EJ2* 14.232, I. Gafni. (Contrast the Cretan Milatos, *BAGRW* 60-E2.)

AGATHOKLĒS, ALEXANDER, ANAXIMANDROS, ANAXIMENĒS, ARISTAGORAS (?), ARISTOPHANĒS, BAKKHEIOS, DAPHNIS, DĒMODAMAS, DIONUSIOS, DIONUSIOS (?), HEKATAIOS, ISIDŌROS (2), LEUKIPPOS, MNASEAS, NIKIAS, OLUMPIAKOS, THALĒS.

Mulasa (mod. Milas; 37°19' N, 27°47' E): Karian city east of **Iasos** and west of **Stratonikeia**; allied with **Athens** 450–440 BCE, then under Persia until 360 BCE when under Maussōllos. Taken by Alexander of **Macedon** 334 BCE, Ptolemaic *ca* 300–250 BCE, then under Seleukids by whom declared free; autonomous and variously allied until favored **Rome** 190 BCE, by whom again declared free. *PECS* 601–602, G.E. Bean; *BAGRW* 61-F3; *BNP* 9 (2006) 407–409, H. Kaletsch.

KUDIAS.

Mundos (mod. Gümüslük; 37°02' N, 27°14' E): coastal Karian city, west of **Halikarnassos**, resisted Alexander of **Macedon** for a year (334–333 BCE); naval harbor for Ptolemy in 308 BCE; under **Rome** from 133 BCE; naval harbor again in 43 BCE. *BAGRW* 61-E3; *BNP* 9 (2006) 410–411, H. Kaletsch.

ALEXANDER, APOLLŌNIOS.

Murleia (mod. Mudanya; 40°23' N, 28°53' E): colony founded on the Propontis by **Kolophōn**, as "Brulleion"; allied with **Athens** in the late 5th c. BCE; known as Murleia by 330 BCE. Taken by Philip V of **Macedon** in 202 BCE; absorbed by the kingdom of **Bithunia** and renamed "**Apameia**" either *ca* 170 BCE for the wife of Prousias II, or else *ca* 140 BCE by Nikomēdēs II. Refounded by AUGUSTUS *ca* 25 CE. STEPHANOS OF BUZANTION, s.v.; *BAGRW* 52-D4; *BNP* 1 (2002) 817 (#1), K. Strobel.

ASKLĒPIADĒS.

Mutilēnē (mod. Mutilēnē/Mitilini; 39°06' N, 26°33' E): on the east coast of **Lesbos**, the most important ***polis*** of the island, wealthy trade center. Supported MITHRADATĒS VI 88 BCE, and sacked by **Rome** 79 BCE; freed by Pompey; prospered into the late Roman era. *PECS* 502–503 (s.v. Lesbos), M. Paraskevaïdis; *OCD3* 1020, D.G.J. Shipley and Ch. Roueché; *BAGRW* 56-D3; *BNP* 9 (2006) 471–474, H. Sonnabend.

ARISTOTELĒS, HERMARKHOS, THEOPHANĒS.

Narbo (mod. Narbonne; 43°11' N, 03°00' E): city on right bank of the Atrax river, not far from **Arelate** and **Massalia**; capital of a Celtic kingdom; refounded 118 BCE as first *colonia* of **Rome** in the new eponymous province, *Gallia Narbonensis*; on the *Via Domitia*. Promoted by CAESAR during the Roman civil war in preference to **Massalia**. *PECS* 607–608, M. Gayraud and Y. Solier; *OCD3* 1026, J.F. Drinkwater; *BNP* 9 (2006) 504–505, Y. Lafond and E. Olshausen.

TERENTIUS VARRO.

Naukratis (mod. Kom Gaief/Geʿif; 30°54' N, 30°35' E): in the western Nile Delta, west of **Saïs** and upstream from **Alexandria**; founded by Greeks as a trading center under the authority of the Pharaohs (Hērodotos 2.178). Declined after foundation of **Alexandria**. *PECS* 609–610, S. Shenouda; *EAAE* 561–564, A. Leonard, Jr.; *BAGRW* 74-D3; *BNP* 9 (2006) 538–539, A. Möller.

Naukratitēs (?), Staphulos.

Naxos (mod. Naxos; 37°05' N, 25°28' E): Aegean island, just east of and opposed to **Paros**; said to have been colonized by **Athens**. Prosperous from the 10th c. BCE, and chief island of the Cyclades in the 6th c. BCE. Resisted Persia *ca* 500 BCE; taken by Persia 490 BCE; allied with or subjected to **Athens** in the 5th c. BCE. In a revived Athenian League from 376 BCE; then member of the league of islands centered on **Dēlos** (q.v.). *PECS* 611–612, N.M. Kontoleon; *OCD3* 1031 (#1), R.W.V. Catling; *BAGRW* 61-A3; *BNP* 9 (2006) 571–574, H. Sonnabend. (Contrast the Sicilian city destroyed in 403 BCE, *BAGRW* 47-G3.)

Kritōn, Leōnidas, Xanitēs (?).

Nazianzos (mod. Bekarlar; 38°23' N, 34°26' E): near **Borissos**, north of **Tuana** and west-south-west of **Caesarea (Kappadokia)**. Minor station on the highway from Anatolia to Palestine; a Christian bishopric by 325 CE. *RE* 16.2 (1935) 2099–2101, W. Ruge; *ODB* 1445–1446, C.F.W. Foss; *BAGRW* 63-E4; *BNP* 9 (2006) 576, K. Strobel.

Caesarius, Gregory.

% Neapolis: many homonymous sites in all areas, including especially: **(A)** mod. Nabeul, Tunisia (*BAGRW* 32-G4); **(B)** mod. Lebda, Libya (also known as Leptis Magna, *BAGRW* 35-G2); **(C)** Sardinian (mod. S. Maria di Nabui, *BAGRW* 48-A3); **(D)** mod. Izvor-Gromadje, Macedonia (*BAGRW* 49-D2); **(E)** mod. Kavalla, Greece (*BAGRW* 51-C3); **(F)** mod. Inebolu, Turkey (*BAGRW* 61-G2); **(G)** mod. Kiyakdede, Turkey (*BAGRW* 65-F2); **(H)** mod. Lemesos, Cyprus (*BAGRW* 72-C3); and **(I)** mod. Vezirköprü, Turkey (*BAGRW* 87-A3).

Diōn.

Neapolis (Italy) (mod. Napoli; 40°50' N, 14°15' E): port city near Mt. Vesuvius, founded from Italian Cumae *ca* 650 BCE, the principal Greek city in **Campania**, and center of Hellenic culture. Allied with **Rome** against Purrhos and Hannibal, made a *municipium* 89 BCE; sacked by Sulla 82 BCE; but recovered to become a fashionable resort town, with Greek culture persisting until the 3rd c. CE. *PECS* 614–615, W.D.E. Coulton; *OCD3* 1031–1032, H.K. Lomas; *BAGRW* 44-F4; *BNP* 9 (2006) 580–581 (#2), A. Muggia.

Clodius, Gaius, Lukos.

Neapolis (Samaria) (mod. Nablus/Shechem; 32°13' N, 35°16' E): ancient city Shechem, north of **Jerusalem** and west of **Gerasa**, refounded as Neapolis by Vespasian, 73 CE. Made a *colonia* by Philip the Arab (*ca* 250 CE); a Christian bishopric in Byzantine times; center of Samaritan revolt 484 CE against the building of a Christian church on their holy mountain; another revolt in 529 CE caused Justinian to build a city-wall. *ODB* 1447–1448, M.M. Mango; *BAGRW* 69-B5; *BNP* 9 (2006) 583 (#11), J. Pahlitzsch.

Marinos.

Nehardea (upstream from mod. Sippar; 33°10' N, 44°08' E): at the confluence of the Euphratēs and Malka rivers, north of **Babylōn** and west of **Seleukeia on Tigris**; settled by Jews exiled in the early 6th c. BCE (Iosephus *Ant. Iud.* 18.311); in the 2nd c. CE site of a Jewish academy, destroyed in 282 CE. *BAGRW* 91-F4; *BNP* 9 (2006) 475–476, K. Kessler; *EJ2* 15.59–60, Y.D. Gilat.

Samuel.

Nemrut Mt. (*ancient name unknown*; 37°59' N, 38°45' E): mountain in eastern Anatolia (north-east of mod. Adıyaman and north of Kahta), near whose peak Antiokhos I of Kommagēnē set up a *Hierothesion*. *PECS* 618–619, J.H. Young; *OCD3* 1034, A.J.S. Spawforth; *BAGRW* 67-H1; *BNP* 9 (2006) 633, J. Wagner.

"Lion Horoscope."

Nikaia (Bithunia) (mod. İznik; 40°26' N, 29°43' E): founded by Antigonos as Antigoneia; refounded by Lusimakhos as Nikaia 301 BCE; absorbed by the kingdom of **Bithunia** 282 BCE; bequeathed to **Rome** by Nikomēdēs IV in 74 BCE. Raided by Goths 258 CE. *PECS* 622–623, N. Bonacasa; *ODB* 1463–1464, C.F.W. Foss; *OCD3* 1040 (#1), O.A.W. Dilke and St. Mitchell; *BAGRW* 52-F4; *BNP* 9 (2006) 701–702 (#5), K. Strobel.

Hipparkhos.

% Nikaia (Gaul) (mod. Nice; 43°42' N, 07°16' E): founded by **Massalia** *ca* 350 BCE; assisted by **Rome** 154 BCE; faded in contrast to the nearby Roman city of Cemenelum, founded by Augustus, which prospered in the 2nd–3rd centuries CE. *PECS* 211 (s.v. Cemenelum), C. Goudineau; *OCD3* 1040 (#2), O.A.W. Dilke and J.F. Drinkwater; *BAGRW* 16-D2; *BNP* 3 (2003) 99, G. Mennella. All of these men may have come from this Nikaia, but are perhaps more likely to have come from **Nikaia of Bithunia**.

Antigonos, Asklēpiodotos (?), Diophanēs, Isigonos, Nikias, Prōtagoras, Sporos.

Nikomēdeia (mod. İzmit; 40°46' N, 29°55' E): founded 262 BCE by Nikomēdēs I, as the capital of his kingdom, **Bithunia**, taken by Mithradatēs VI 89 BCE; plundered 85 BCE, bequeathed to **Rome** 74 BCE, but again taken by Mithradatēs 73 BCE. Damaged by an earthquake 120 CE; raided by the Goths 258 CE; adorned by Diocletian; here Galerius issued the edict of toleration 311 CE. Libanios taught here 344–348 CE. *OCD3* 1043, T.R.S. Broughton and St. Mitchell; *PECS* 623–624, W.L. MacDonald; *BAGRW* 52-F3; *BNP* 9 (2006) 737–739, E. Wirbelauer and K. Strobel.

Apsurtos (?), Flauius Arrianus, Mēnodotos.

% Nikopolis: many homonymous sites existed, founded at various times: **(A)** mod. İslahiye, Turkey (*BAGRW* 67-D2, founded by Alexander of **Macedon**, according to Stephanos of Buzantion, s.v. Issos); **(B)** mod. Yeşilyayla, Turkey (Stephanos of Buzantion, s.v. #3; *BAGRW* 87-D4: south-west of **Trapezous**, founded *ca* 300 BCE by Seleukos I); **(C)** "Emmaus" (mod. Imwas, Israel, *BAGRW* 70-F2, from *ca* 220 BCE); **(D)** Issos (mod. Yeşil Hüyük, Turkey, *BAGRW* 67-C3); **(E)** of **Ēpeiros**, the best-known and most likely unmarked referent in Byzantine times (mod. Palaio-Preveza, Greece; Stephanos of Buzantion, s.v. #1; *BAGRW* 54-C3: founded by Augustus 30 BCE); **(F)** ad Istrum (mod. Nikiup, Bulgaria; *BAGRW* 22-C5, founded 102 CE); and **(G)** ad Nestum (mod. Gârmen, Bulgaria, *BAGRW* 51-B2, founded 106 CE). *OCD3* 1043–1044, St. Mitchell *et al.; BNP* 9 (2006) 741–744, (various authors).

Theomnēstos.

Nisibis (mod. Nusaybin; 37°04' N, 41°13' E): old Assyrian town, known in Seleukid times as **Antioch in Mugdonia** (Polubios 5.51; Stephanos of Buzantion, s.v., #3); east of **Constantia**, and east-north-east of **Reš'aina**. From 80 BCE under the Parthians; Jewish presence from the 1st c. CE; taken by **Rome** 115 CE, and then a focus of the Jewish revolt ("Kitos" War) in 115–117 CE (*cf.* **Alexandria**, **Cyprus**, **Edessa**, and **Kurēnē**). Taken by the Parthians and retaken by **Rome** in 194 CE; Sasanian 260–298 CE; a Christian bishopric and school of Christian theology from *ca* 300; returned to Sasanian control 363 CE by treaty, whereupon Jovian ordered the Christian population relocated to **Amida**. The east-Syrian Christian school returned here upon being exiled from **Edessa** 489 CE; border city engaging in trade from 540 CE; declined in the 7th c. *ODB* 1488, M.M. Mango; *OCD3* 1046, J. Whatmough; *BAGRW* 87-D3; *BNP* 9 (2006) 777–779, K. Kessler; A.H. Becker, *Fear of God and the Beginning of Wisdom* (2006) 41–97, 197–203; *EJ2* 15.276, E. Ashtor and M. Beer.

APOLLOPHANĒS, MAGNUS, SEVERUS SEBOKHT.

Nola (mod. Nola; 40°56' N, 14°32' E): old city in **Campania**, perhaps founded by **Khalkis** or by Etruscans, on the other side of Mt. Vesuuius from **Neapolis** (with whom early relations were good), and south-west of **Beneventum**; under **Rome** from 312 BCE; resisted Hannibal in 215 and 212 BCE. *PECS* 627–628, L. Richardson, Jr.; *OCD3* 1047, H.K. Lomas; *BAGRW* 44-G4; *BNP* 9 (2006) 790–791, E. Olshausen and V. Sauer.

MINIUS PERCENNIUS.

Nouum Comum (mod. Como; 45°49' N, 09°05' E): at south end of Como lake, *ca* 45 km north of **Mediolanum**, under **Rome** from the early 1st c. BCE; moved to the new site by CAESAR. *PECS* 234–235, M. Mirabella Roberti; *OCD3* 375, J.B. Ward-Perkins and T.W. Potter; *BAGRW* 39-E2; *BNP* 3 (2003) 678–679, A. Sartori.

PLINIUS SECUNDUS.

Nusa (mod. Sultanhisar; 37°53' N, 28°09' E): Karian city at the north edge of the plain of the Maiandros river, east of **Tralleis**, and west of **Antioch** on the Maiandros; Seleukid in the 3rd c. BCE; prosperous under **Rome** from 88 BCE. STEPHANOS OF BUZANTION, s.v., lists ten homonyms; *PECS* 636–637, G.E. Bean; *BAGRW* 61-G2; *BNP* 9 (2006) 930–931, H. Kaletsch.

IASŌN, SŌSTRATOS.

Nussa (mod. Nevşehir?; 38°37' N, 34°43' E): old city of **Kappadokia**, near **Caesarea**; no history before the middle-Byzantine period seems to be known. *RE* 17.2 (1937) 1662, W. Ruge; *ODB* 1506–1507, C.F.W. Foss; *BAGRW* 63-D3; *BNP* 9 (2006) 931, K. Strobel.

GREGORY.

Oasis (mod. al-Khargah; *ca* 25°15'N, 30°35' N): region of **Egypt** 200 km west of the far upper Nile, an "island" (its meaning in Egyptian) in the desert. *BAGRW* 79-B2. (Contrast the older Oasis Parva, mod. Bahariya Oasis, *BAGRW* 73-F5.)

APIŌN.

Oinoanda (mod. İncealiler; 36°48' N, 29°34' E): city in **Lukia**, a colony of Termessos, founded *ca* 200 BCE. *PECS* 640–641, G.E. Bean; *OCD3* 1062, St. Mitchell; *BAGRW* 65-C4; *BNP* 10 (2007) 51, H. Elton.

DIOGENĒS.

Olunthos (mod. Olinthos/Nea Olunthos; 40°17' N, 23°21' E): coastal city of **Khalkidikē** near **Poteidaia**, occupied by Thrakians until expelled by the Persians in favor of Greeks (HĒRODOTOS 8.127). Fearing **Sparta**, allied with Philip II of **Macedon** in 357, but fearing his power intrigued with **Athens**, and was destroyed in 348 BCE. *PECS* 651–652, J.W. Graham; *OCD3* 1067, N.G.L. Hammond; *BAGRW* 50-D4; *BNP* 10 (2007) 119–120, M. Zahrnt.

KALLISTHENĒS.

Opous (mod. Atalandi; 38°49' N, 23°00' E): coastal city and capital of East **Lokris**, north-east of **Huampolis**, opposite Euboia; allied with **Sparta** in the 5th c. BCE. *BAGRW* 55-E3; *BNP* 10 (2007) 178–179, G. Daverio Rocchi.

PHILIPPOS.

Ostia (mod. Ostia; 41°45' N, 12°18' E): port city of **Rome**, at the mouths (*ostia*) of the Tiber river; razed by Marius 87 BCE, then by pirates 68 BCE; rebuilt by CICERO, Tiberius,

Claudius, Trajan, and most extensively by Hadrian. *PECS* 658–661, R. Meiggs; *OCD3* 1081–1082, N. Purcell; *BAGRW* 43-B2; *BNP* 10 (2007) 280–281, 283–286, G. Uggeri and H.-M. von Kaenel.
MELIOR (?).

Oxyrhynchos (mod. Sandafa el-Far; 28°32' N, 30°40' E): west of the Nile, upstream from **Philadelpheia** and downstream from **Antinoopolis**, ancient capital of the Scepter Nome; site of rich deposit of Greek papyri. *PECS* 663, S. Shenouda; *OCD3* 1088, W.E.H. Cockle; *EAAE* 594–595, F. Gomaà; *BAGRW* 75-D3; *BNP* 10 (2007) 312–313, Jo. Quack.
HUBRISTĒS.

Panion (near mod. Şarköy; 40°37' N, 27°07' E): on the north shore of the Propontis, north across the sea from **Prokonessos**; scarcely attested before the middle-Byzantine period. *Souda* Pi-202; *RE* 18.3 (1949) 601, Johanna Schmidt; *BAGRW* 52-A3.
PRISKOS.

Panōpolis (mod. Akhmim; 26°34' N, 31°45' E): city in **Egypt** on the east bank of the Nile, well upstream of **Antinoopolis**; weaving and stone-cutting center, and early Christian center. HĒRODOTOS 2.156; STRABŌN 17.1.41; *BAGRW* 77-F3; *BNP* 10 (2007) 455–456, K. Jansen-Winkeln.
P. AKHMIM, ZŌSIMOS.

Parion (mod. Kemer; 40°25' N, 27°04' E): port in **Troas** at the north end of the Hellespont near **Lampsakos**, founded *ca* 700 BCE; allied with **Athens** in the 5th c. BCE; taken by LUSIMAKHOS 302 BCE, under **Pergamon** from 227/226 BCE. Under **Rome** from 133 BCE; made a *colonia* by AUGUSTUS. *PECS* 676, N. Bonacasa; *OCD3* 1113, E.N. Borza; *BAGRW* 52-A4; *BNP* 10 (2007) 536, P. Frisch.
ARTEMIDŌROS.

Paros (mod. Paros; 37°05' N, 25°09' E): Aegean island just west of, and opposed to, **Naxos**, colonized **Thasos**; under Persians 489 BCE, then allied with **Athens**, varying between a democratic and oligarchic constitution. Variously allied in the 3rd c. BCE, although never in the island league (see **Dēlos**). *PECS* 677–679, N.M. Kontoleon; *OCD3* 1116, R.W.V. Catling; *BAGRW* 61-A3; *BNP* 10 (2007) 553–555, H. Sonnabend.
KHARETIDĒS, SATUROS, THUMARIDAS (?).

Pāṭaliputra ⇒ Kusumapura

Patara (mod. Gelemiş; 36°16' N, 29°19' E): chief port city in **Lukia**, near mouth of the Xanthos river, a chief temple and oracle of Apollo. Taken by Alexander of **Macedon** 334 BCE; occupied by Antigonos (315 BCE) and Dēmētrios Poliorkētēs (304 BCE); then Ptolemaic until 197 BCE when under the Seleukids. Under **Rhodes** from 188 BCE; under **Rome** from 168 BCE. *PECS* 679–680, G.E. Bean; *OCD3* 1121, St. Mitchell; *BAGRW* 65-B5; *BNP* 10 (2007) 594–595, C. Marek.
MNASEAS.

Pella (mod. Arkhaia Pella; 40°46' N, 22°31' E): large city of **Macedon**, on the Ludias river (near the Thermaic Gulf), made the capital by Arkhelaos (413–399 BCE), birthplace of Alexander. Declined as Thermaic Gulf silted up; insignificant after 168 BCE. *RE* 19.1 (1937) 341–348 (#3), E. Oberhummer (with map by A. Struck); *PECS* 685–686, Ph.M. Petsas; *OCD3* 1132, R.A. Tomlinson; *BAGRW* 50-C3; *BNP* 10 (2007) 698–699, R.M. Errington. (Contrast the Pella in **Samaria**, mod. Tabaqat Fahl, *BAGRW* 69-C5, and the first name of **Apameia on the Orontes**.)

POSEIDIPPOS.

Pēlousion (mod. Tell el-Farama; 31°03' N, 32°36' E): ancient Egyptian city and border fort, at the north-east corner of the Nile Delta; point of attack upon **Egypt** for the Persians (342 BCE) and Alexander of **Macedon** (332 BCE). Ptolemaic until taken by Antiokhos IV (170 BCE); thrice taken by **Rome** in the 1st c. BCE. *OCD3* 1134–1135, D.J. Thompson; *BAGRW* 74-H2; *BNP* 10 (2007) 716–717, K. Jansen-Winkeln.

DŌSITHEOS, LAMPŌN, PRŌTĀS.

Pergamon (mod. Bergama; 39°07' N, 27°11' E): old city, whose regent Philētairos defected from LUSIMAKHOS in 282 BCE (in favor of Seleukos I) and defended the city from Galatian invaders, 278–276 BCE; his successor Eumenēs I revolted from the Seleukids, 263 BCE, and Eumenēs' son Attalos I made the city the capital of his newly-declared kingdom, which patronized arts and sciences. Attalos I also opposed **Macedon** from 220 BCE, and allied with **Rhodes** from 201 BCE, and then also **Rome** from 197 BCE. Eumenēs II around 190 BCE founded here a library in competition with **Alexandria**. ATTALOS III bequethed the kingdom to **Rome**, 133 BCE. Punished by **Rome** for supporting MITHRADATĒS VI (88–85 BCE). Prosperous and prestigious center of Roman imperial ruler-cult. Attacked by Goths, *ca* 265 CE, but remained an important intellectual center, where the emperor Julian studied philosophy. *PECS* 688–692, J. Schäfer; *ODB* 1628, C.F.W. Foss; *OCD3* 1138–1139, A.J.S. Spawforth and Ch. Roueché; *BAGRW* 56-E3; *BNP* 10 (2007) 754–772, W. Radt and W. Eder.

AISKHRIŌN, APOLLŌNIOS (2), ATTALOS, BITŌN (?), DĒMĒTRIOS KHLŌROS (?), EPIKOUROS, GALĒN, MENANDROS (?), MENIPPOS, NIKŌN, OREIBASIOS, PHILIPPOS (?), PHILISTIŌN, STRATONIKOS.

Pergē (mod. Murtana; 36°58' N, 30°51' E): ancient coastal site east of **Attaleia**, Hellenized by **Rhodes** from the 7th c. BCE; welcomed Alexander of **Macedon**, then under the Seleukids; under **Pergamon** from 188 BCE; plundered by Verres in the 1st c. BCE. *PECS* 692–693, G.E. Bean; *OCD3* 1139, *Idem* and St. Mitchell; *BAGRW* 65-E4; *BNP* 10 (2007) 773–775, W. Martini.

APOLLŌNIOS, ARTEMIDŌROS.

Perinthos (mod. Ereğli/Marmaraereğli; 40°58' N, 27°57' E): founded by **Samos** 602 BCE on the coast of the Propontis in **Thrakē**, west of **Buzantion**; taken by the Persians, allied with **Athens** in the 5th–4th centuries BCE; allied with Philip II of **Macedon** 355 BCE; after the conquests of Alexander of **Macedon**, variously allied or free. *OCD3* 1140, E.N. Borza; *BAGRW* 52-B3; *BNP* 10 (2007) 785–786, I. von Bredow.

HESTIAIOS.

Petra (mod. Wadi Musa; 30°20' N, 35°27' E): south-east of **Gaza**, south of the Dead Sea, capital of the Nabataean kingdom from at least 312 BCE until 106 CE when **Rome** transferred the provincial capital to Bostra, but continued as a trade center. Devoted to its native god, Dusares, and to Isis; early Christian center. Declined *ca* 150–250 CE, as Arabian trade shifted north. Survived an earthquake 363 CE, which damaged the aqueducts; another *ca* 415 CE. In Byzantine times, *metropolis* of the province *Palaestina Tertia*. *PECS* 694–695, J.-P. Rey-Coquais; *ODB* 1642–1643, W.E. Kaegi and A. Kazhdan; *OCD3* 1149, J.F. Healey; *BAGRW* 70-G5/71-A5; R.G. Hoyland, *Arabia and the Arabs* (2001) 70–74; *BNP* 10 (2007) 869–871, T. Leisten; *EJ2* 16.17–18, M. Avi-Yonah and Sh. Gibson.

GESSIOS.

Phasēlis (mod. Tekirova; 36°31' N, 30°33' E): coastal city at the east edge of **Lukia**, founded by **Lindos** *ca* 690 BCE, and then assisted in founding **Naukratis**. Under the

Persians to 469 BCE, then allied with **Athens**; autonomous in the 4th c. BCE; welcomed Alexander of **Macedon**, then Ptolemaic in the 3rd c.; under **Rhodes** from 188 BCE. Occupied by pirates *ca* 100 BCE, and destroyed by Seruilius 77 BCE; rebuilt and prosperous in the 1st–2nd cc. CE. *PECS* 700–701, G.E. Bean; *OCD3* 1156, *Idem* and St. Mitchell; *BAGRW* 65-E4; *BNP* 10 (2007) 939–940, A. Thomsen.

KRITOLAOS, MNĒSIMAKHOS.

% Philadelpheia: several homonymous sites could have been Dionusios' home, if not one of the two listed separately below: Isaurian/Kilikian (mod. İmşi Ören; *BAGRW* 66-C3, Roman and later); and mod. 'Amman, Jordan (*PECS* 703–704, J.-P. Rey-Coquais; *BAGRW* 71-B2, Hellenistic and later). *Cf.* STEPHANOS OF BUZANTION, s.v.

DIONUSIOS.

Philadelpheia (Egypt) (mod. Gerza; 29°25' N, 31°12' E): in the north-east of the Fayum, downstream from **Oxyrhynchos**, upstream from **Memphis**; founded as military colony by Ptolemy II Philadelphos *ca* 270 BCE; existed until the 4th or 5th c. CE. *OCD3* 1158 (#1), W.E.H. Cockle; *BAGRW* 75-E2; *BNP* 11 (2007) 7–8 (#4), K. Jansen-Winkeln.

MAGNUS.

Philadelpheia (Ludia) (mod. Alaşehir; 38°21' N, 28°31' E): east of **Sardēs**, founded by Attalos II Philadelphos *ca* 150 BCE; destroyed by the earthquake of 17 CE (*cf.* **Kumē, Magnesia on Sipulos, Sardēs**), and rebuilt. Called "little **Athens**" in the 6th c. *PECS* 703, T.S. MacKay; *ODB* 1648–1649, C.F.W. Foss; *OCD3* 1158 (#2), G.E. Bean and A.J.S. Spawforth; *BAGRW* 56-H5; *BNP* 11 (2007) 7 (#1), E. Olshausen.

IŌANNĒS.

Phleious (mod. Phleious, *ca* 2 km north of Nemea; 37°50' N, 22°42' E): in the Asopos valley (north-eastern Peloponnesos) and south-west of **Corinth**; ally of **Sparta**, against the Persians at **Plataia** 480 BCE (HĒRODOTOS 9.28.4); site of a **Pythagorean** school (DIOGENĒS LAËRTIOS 8.46) and the setting of PLATO's dialogue *Phaidōn*. Under Macedon in the early 3rd c. BCE; allied with the Akhaian League from 228 BCE. *PECS* 707–708, W.R. Biers; *BAGRW* 58-D2; *BNP* 11 (2007) 134–135, Y. Lafond.

EKHEKRATĒS, TIMŌN.

Phōkaia (mod. Foça; 38°40' N, 26°45' E): most northerly Ionian city in Anatolia, coastal site with poor land, early inhabitants were maritime traders closely connected with Tartessos (HĒRODOTOS 1.163). Their colonizing efforts in **Spain** and France were especially vigorous. Besieged by Persia in 546 BCE, many inhabitants emigrated, settling at **Elea**. Allied with **Athens** in the 5th c. BCE. *PECS* 708–709, E. Akurgal; *OCD3* 1172–1173, D.E.W. Wormell and A.J.S. Spawforth; *BAGRW* 56-D4; *BNP* 11 (2007) 137–138, Ö. Özyiğit.

THEODŌROS.

Pisidia: fertile mountainous land north of **Lukia** and south-east of **Ludia**, independent of Persia and of Greeks until loosely under **Pergamon** after 188 BCE; likewise loosely under **Rome** from 101 BCE, and part of the province of *Galatia* from 25 BCE. Christianized early; made into its own province in the early 4th c. CE, with Pisidian **Antioch** as its capital; remote and chaotic, often in revolt; suffered from an earthquake in 518 CE (*cf.* **Stoboi**), and the plague of the 540s CE. *ODB* 1680, C.F.W. Foss; *OCD3* 1186, St. Mitchell; *BNP* 11 (2007) 294–295, H. Brandt.

Sites: (none).

People: GEŌRGIOS.

Pitanē (mod. Çandarlı; 38°56' N, 26°56' E): coastal Aiolian city north of **Kumē**, allied with **Athens** in the 5th c. BCE; besieged in vain by Parmeniōn 336 BCE; free city within kingdom of **Pergamon** until 133 BCE, when it came under **Rome**. MITHRADATĒS VI was besieged here by Flauius Fimbria 85 BCE. *PECS* 715, E. Akurgal; *BAGRW* 56-D4; *BNP* 11 (2007) 303, E. Olshausen.

APOLLŌNIOS, AUTOLUKOS.

Plataia (mod. Kokkla; 38°13' N, 23°16' E): in southern **Boiōtia** between Mt. Kithairon and the Asopos river, not far from **Athens**, often under control of **Thēbai**. Best known as the site of the final battle against Persia, 479 BCE. Destroyed by **Sparta** and **Thēbai** 427 BCE, after a two-year siege. Rebuilt *ca* 380 BCE with help from **Sparta**, again destroyed by **Thēbai** 374/373 BCE; once more rebuilt, by Philip II of **Macedon** (*ca* 350 BCE). *PECS* 717, N. Bonacasa; *OCD3* 1189, J. Buckler and A.J.S. Spawforth; *BAGRW* 55-E4; *BNP* 11 (2007) 336–337, K. Freitag.

ARISTOPHILOS, DAIMAKHOS.

Pleuron (mod. Kato Retsina; 38°26' N, 21°25' E): ancient Aitolian city between the Akhelos and Euonos rivers, west of **Amphissa**. After destruction by Dēmētrios II in 230 BCE, the inhabitants moved 1.5 km northward to a fortified site on Mt. Arakunthos. *PECS* 717–718, N. Bonacasa; *OCD3* 1197, W.M. Murray; *BAGRW* 55-A4; *BNP* 11 (2007) 382–383, K. Freitag.

ALEXANDER.

Poseidōnia (mod. Paestum; 40°25' N, 15°00' E): coastal city of **Lucania** founded from **Subaris**, *ca* 600 BCE, south-east of **Neapolis**, it prospered in the 6th c. Taken over by natives *ca* 410 BCE, sided with PURRHOS against **Rome**, then under **Rome** as a Latin *colonia* from 273 BCE, and a *municipium* from 88 BCE. *PECS* 663–665, W.D.E. Coulson; *OCD3* 1091, H.K. Lomas; *BAGRW* 45-B4; *BNP* 11 (2007) 678–682, M. Lesky.

SIMOS.

Poteidaia ⇒ Kassandreia

Praeneste (mod. Palestrina; 41°50' N, 12°53' E): Etruscan-influenced site in the Apennines 40 km east-south-east of **Rome**, east of **Tusculum**, *colonia* from *ca* 500 BCE, but opposed **Rome** in the 4th c. BCE. After 90 BCE, a Roman *municipium*, supported Marius, sacked by Sulla, rebuilt and colonized with veterans 82 BCE. In imperial times, much visited by those consulting the oracle. *PECS* 735–736, L. Richardson, Jr.; *OCD3* 1239, E.T. Salmon and T.W. Potter; *BAGRW* 43-D2; *BNP* 11 (2007) 764–765, M.M. Morciano.

AELIANUS.

Priēnē (mod. Güllübahçe; 37°40' N, 27°18' E): Ionian city at the mouth of the Maiandros river, under Persians 546–499 BCE; revolted and suppressed; sometimes allied with **Athens** in the 5th c. BCE. Refounded at the foot of Mt. Mukalē in the 4th c. BCE, due to silting in the gulf. Territory ravaged by the Celts 277 BCE. Under **Rome** from 129 BCE, after which further silting of the Maiandros river depressed prosperity. *PECS* 737–739, G.E. Bean; *OCD3* 1245, J.M. Cook and A.J.S. Spawforth; *BAGRW* 61-E2; *BNP* 11 (2007) 832–837, Fr. Rumscheid.

DIODŌROS, MANDROLUTOS, MENANDROS, PUTHIOS.

Prokonessos (mod. Marmara; 40°37' N, 27°37' E): largest island of the Propontis, colony founded by **Milētos** *ca* 675 BCE; revolted from Persians 499 BCE, and sacked; allied with **Athens** in the 5th c. BCE; taken by **Kuzikos** *ca* 360 BCE, and the population deported. *BAGRW* 52-B3; *BNP* 11 (2007) 918, E. Olshausen and V. Sauer.

DEINOSTRATOS, MENAIKHMOS.

Prousias "ad Mare" (mod. Gemlik; 40°26' N, 29°09' E): coastal city on the eastern Propontis, the ancient Kios, colonized by **Milētos**, destroyed by Philip V of **Macedon**; rebuilt and renamed by Prousias I of **Bithunia**, *ca* 190 BCE; became prosperous through trade. *BAGRW* 52-E4; *BNP* 3 (2003) 370–371 (#1), H. Kaletsch and F.K. Dörner. (Contrast Prousias ad Hypium, *BAGRW* 86-B3, mod. Konuralp.)
APOLLŌNIOS, APSURTOS (?), ASKLĒPIADĒS.

Puteoli (mod. Pozzuoli; 40°49' N, 14°07' E): coastal city of **Campania**, north of **Neapolis**, settled by refugees from **Samos** *ca* 520 BCE, prosperous through trade but politically dependant upon nearby Cumae until conquered by the Samnites, 421 BCE. Successfully resisted Hannibal in 215; received a Roman customs station in 199 and maritime colony in 194. Its proximity to the *Via Appia* rendered its port preferable to **Neapolis**', and it became a favored resort town in the early Roman Empire. *PECS* 743–744, H. Comfort; *OCD3* 1280–1281, H.K. Lomas; *BAGRW* 44-F4; *NP* 10.606–608, M.I. Gulletta and D. Steuernagel.
AMBROSIOS RUSTICUS, PUBLIUS.

Ravenna (mod. Ravenna; 44°25' N, 12°12' E): city in the southern marshes of the Po delta, allied with **Rome** by 1st c. BCE, with full Roman status in 49 BCE, the base of AUGUSTUS' Adriatic fleet. Honorius transferred the imperial court here *ca* 403 CE, seeking the security of the marshes, whereupon the city's fortunes blossomed, continuing as the splendid capital of successive barbarian kingdoms until 455 CE; revived under Odoacer from 476 CE; in Byzantine hands after 568 CE. *PECS* 751, D.C. Scavone; *ODB* 1773–1775, T.S. Brown and D. Kinney; *OCD3* 1294, B.R. Ward-Perkins; *BAGRW* 40-C4; *NP* 10.796–800, C. Heucke and Al. Berger; *EJ2* 17.120, A. Toaff and S. Rocca.
AGNELLUS, RAVENNA COSMOGRAPHY.

Reate (mod. Rieti; 42°24' N, 12°52' E): small Sabine town on Velino river, west of **Amiternum**, north-east of **Rome**, south of **Spoletium**; under **Rome** from 3rd c. BCE; *municipium* from 27 BCE. *PECS* 751–752, L. Richardson, Jr.; *OCD3* 1294, E.T. Salmon and T.W. Potter; *BAGRW* 42-D4; *NP* 10.802, M.M. Morciano.
TERENTIUS VARRO.

Reš'aina (mod. Tell Fakhariya, near Ras al-'Ain; 36°49' N, 40°02' E): east-south-east of **Edessa**, south of **Amida** and of **Constantia**, west-south-west of **Nisibis** (*cf.* **Syria**); old Assyrian city (whose name means "head of the stream"), made a *colonia* under Septimius Seuerus (*ca* 200 CE); Theodosius I built it up and renamed it Theodosiopolis (381–383 CE). *RE* 1A.1 (1914) 618–619, F.H. Weissbach; *BAGRW* 89-C4.
SERGIUS.

Rhēgion (mod. Reggio di Calabria; 38°07' N, 15°40' E): southern Italian town colonized from **Khalkis** in 720 BCE, dominated the Straits of Messina (see **Messēnē**), allied with **Lokroi** against **Krotōn** in the mid-6th c. BCE; destroyed by Dionusios I of **Surakousai** 387 BCE; rebuilt 359/358 BCE; allied with **Rome** against PURRHOS in 280 BCE, and garrisoned by **Rome**'s Campanian allies, remaining loyal in the 2nd–1st centuries BCE. *PECS* 753–754, A. de Franciscis; *OCD3* 1312, H.K. Lomas; *BAGRW* 46-C5; *NP* 10.951–952, A. Muggia.
LUKOS, PHILIPPOS XĒROS (?), THEAGENĒS.

Rhodes/Rhodos (mod. Rhodos; 36°10' N, 28°00' E): large island in the south-east Aegean, center of trade from early times; submitted to Persia in 490 BCE, allied with **Athens** in the 5th c. BCE. The three Rhodian ***poleis*** (**Lindos**, Ialusos, Kamiros) revolted from **Athens** 412/411 in favor of **Sparta**. They founded a federal state 408/7 to facilitate commercial interests. Revolted from **Sparta** 395, allied with **Athens** 378–357 BCE, after which under Maussōllos. Allied with Alexander of **Macedon**; then resisted Dēmētrios Poliorkētēs' siege 305–304 BCE (the colossus of their patron god Hēlios was built in thanks, early 3rd c. BCE). Major sea-power in the 3rd c. BCE (*cf.* **Dēlos**); survived the earthquake of 226 BCE, which destroyed the colossus. Cooperated with **Rome** against Philip V and Antiokhos III, for which service acquired territory in Karia and **Lukia** (188 BCE). But **Rome** declared **Dēlos** a "free port" 166 BCE, thus ending Rhodes' leading role in trade; and alliance with **Rome** 164 ended Rhodes' political independence. Rhodes, nonetheless, remained a prosperous cultural center, where CICERO studied, and Tiberius spent his self-imposed exile. *PECS* 755–758, R.E. Wycherley; *OCD3* 1315–1316, C.B. Mee and E.E. Rice; *BAGRW* 60-G3; *NP* 10.996–998, H. Sonnabend. *Cf.* **Lindos**.

AGĒSISTRATOS (?), ANDRONIKOS, ANTISTHENĒS (?), ATHĒNODŌROS (?), ATTALOS, BAKŌRIS, DIOGNĒTOS, DIONUSIOS, EPIGENĒS, EUDĒMOS, EUDOXOS, GEMINUS (?), HIERŌNUMOS, KALLIXEINOS, KESKINTOS, PAUSISTRATOS, PUTHIŌN, SŌSIKRATĒS, TIMOSTHENĒS.

Rhodiapolis (10 km north of mod. Kumluca; 36°22' N, 30°18' E): city in south-east **Lukia** founded from **Rhodes** in the 4th c. BCE; in Lukian federation from 167 BCE. *PECS* 758, G.E. Bean; *OCD3* 894–895, St. Mitchell, and 1069, A.H.M. Jones and St. Mitchell; *BAGRW* 65-D5; *NP* 10.994, A. Thomsen.

HĒRAKLEITOS.

Rome (mod. Roma; 41°54' N, 12°30' E): city-state originally built on seven hills in the valley of the Tiber, the largest river in central **Italy**, with an easy crossing at Tiber island and navigable down to the sea. Settled from *ca* 1400 BCE, developed in the 6th c. BCE under influence from **Etruria**, adapting Etruscan building and hydraulic engineering. Methodically took control of **Italy**; central **Italy** and **Etruria** from the 4th c. BCE, and most of the Greek cities of south **Italy** being subjected 270–212 BCE. Revolts of the early 1st c. BCE in **Italy** resulted in extensions of Roman citizenship throughout **Italy**. Imperial capital from 30 BCE, and thus a center of intellectual practice; Christian center from *ca* 100 CE. Roman citizenship granted to all free citizens of the empire in 212 CE; Diocletian/Maximian moved the imperial capital to **Mediolanum** 293 CE, after which Rome declined. Sacked in the second Gothic invasion under Alaric 410 CE. *PECS* 763–771, E. Nash; *OCD3* 1322–1327, T.J. Cornell, 1327–1331, G.P. Burton, 1331–1334, F.F. Matthews, 1334–1335, I.A. Richmond *et al.; BAGRW* 44-B2; 10.1050–1077, W. Eder.

AMMŌN, ANDRŌN, APHTHŌNIOS, EUKLEIDĒS, HIPPOLUTOS, CAESAR (?), AUGUSTUS, MACHARIUS (?), PHILIPPOS (?), VITRUUIUS (?).

Rudiae (south-west of mod. Lecce; 40°21' N, 18°10' E): Messapian city prosperous in the 5th–3rd cc. BCE; a *municipium* under **Rome**. *PECS* 774, F.G. LoPorto; *OCD3* 1337, H.K. Lomas; *BAGRW* 45-H4; *NP* 10.1149, M. Lombardo.

ENNIUS.

Saïs (mod. Sa el-Hağar; 30°58' N, 30°46' E): city of **Egypt** in the western Delta of the Nile, west of **Sebennutos** and east of **Naukratis**, capital of the 24th dynasty (*ca* 730 BCE) and 26th dynasty (672–525 BCE). In STRABŌN's time (17.1.18) again an important city; a Christian bishopric from *ca* 325 CE. *BAGRW* 74-D3; *NP* 10.1234, K. Jansen-Winkeln.

P. HIBEH 27 (?), PHIMENAS.

Salamis (north-west of mod. Ammokhostos; 35°11' N, 33°54' E): old city on east coast of **Cyprus**, seat of the governor under the Ptolemies, and the second city of the island under **Rome**. The Jewish revolt ("Kitos" War) in 115–117 CE destroyed Salamis (*cf.* **Alexandria**,

Edessa, **Kurēnē**, and **Nisibis**). After several earthquakes (332, 342, 352 CE), rebuilt as Constantia by Constantius II (337–361 CE) and made capital of the island. *PECS* 794–796, K. Nicolaou; *OCD3* 1347 (#2), H.W. Catling; *BAGRW* 72-D2; *NP* 10.1243–1244, R. Senff.

EPIPHANIOS.

Salona (mod. Solin; 43°32' N, 16°29' E): coastal Dalmatian city near Split, taken by **Rome** 78/77 BCE, and established as a *colonia ca* 45 BCE; after 9 CE became provincial capital of Dalmatia and the center of the Romano-Dalmatian road network. Walls repaired 170 CE; a Christian center from the 2nd c. CE. In 305 CE Diocletian retired to a villa on the coast just a few miles away, and Salona became prosperous; a Christian bishopric from *ca* 350 CE. Under Ostrogoths in the 5th c.; reconquered by Justinian 537 CE. *PECS* 799, M. Zaninović; *ODB* 1832, A. Kazhdan; *OCD3* 1350, J.J. Wilkes; *BAGRW* 20-D6; *NP* 10.1264–1265, U. Fellmeth.

PELAGONIUS.

Samaria (mod. Sebastiya; 32°17' N, 35°12' E): ancient capital of the region Samaria, north of **Jerusalem** and south-west of **Gerasa**; taken by Alexander of **Macedon** 332 BCE; destroyed by the Jewish king Hyrcanus in 108 BCE; under **Rome** from 63 BCE, and rebuilt in 57 BCE. AUGUSTUS gave the city to King Herod, who renamed it Sebastē; destroyed in the revolt of 66–70 CE. *PECS* 800, A. Negev; *OCD3* 1350–1351, T. Rajak; *BAGRW* 69-B5; *NP* 11.1–2, R. Liwak.

RUFUS.

Samos (mod. Samos; 37°44' N, 26°50' E): large Aegean island, whose ancient capital Samos was at mod. Pythagorion/Tigani; Ionic Greek from *ca* 1000 BCE; center of trade from *ca* 700 BCE; involved in colonizing **Perinthos** and **Naukratis**. Under the Persians after Polukratēs' execution 525 BCE; allied with or subjected to **Athens** in the 5th–4th centuries BCE (although besieged by **Athens** 441–439 BCE), until taken by Alexander of **Macedon**, after which many changes of alliance. Ptolemaic 246–197 BCE, then after an attack by Philip V of **Macedon** under **Rhodes**; made autonomous by **Rome** 188 BCE, but still influenced by **Rhodes**. Under **Rome** from 129 BCE as part of the province of *Asia*; plundered by Verres *ca* 80 BCE; made free by AUGUSTUS; prosperous in the early empire. *PECS* 802–803, L. Vlad Borelli; *OCD3* 1351, D.G.J. Shipley; *BAGRW* 61-D2; *NP* 11.17–23, H. Sonnabend.

AËTHLIOS, AGATHARKHIDĒS, AGATHARKHOS, AGATHŌN, ARISTAIOS, ARISTARKHOS, ARISTEIDĒS, DIODŌROS, DIONUSIOS, DOURIS, EPICURUS, KONŌN, MANDROKLĒS, MELISSOS, NIKANŌR, PAUSIMAKHOS, PHOKOS, PROMATHOS, PUTHOKLĒS, PYTHAGORAS, RHOIKOS, THEODŌROS.

Sardēs/Sardeis (mod. Sartmahmut west of Salihli; 38°30' N, 28°03' E): east inland from **Smurna**, ancient city under Greek influence from the 12th c. BCE; capital of **Ludia** *ca* 650–547 BCE; then Persian until taken by Alexander of **Macedon** 334 BCE. Taken by Seleukos I 282 BCE; destroyed and rebuilt by Antiokhos II 213 BCE; then under **Pergamon** from *ca* 180 BCE; under **Rome** from 133 BCE. Damaged in the earthquake of 17 CE (*cf.* **Kumē**, **Magnesia on Sipulos**, **Philadelpheia [Ludia]**), and rebuilt. *PECS* 808–810, J.A. Scott and G.M.A. Hanfmann; *OCD3* 1356–1357, W.M. Calder *et al.*; *BAGRW* 56-G5; *NP* 11.54–65, H. Kaletsch; *EJ2* 18.53–54, anon.

EUNAPIOS, THRASUMAKHOS, XANTHOS (?).

Sarnaka (*unlocated*): perhaps in the **Troas**, since PLINY 5.126 refers to it as an inland Musian city; thus probably under **Pergamon** 227–133 BCE. *RE* 2A.1 (1921) 29, L. Bürchner.

MELAMPOUS.

Scylletium/Skullētion (mod. Roccalletta; 38°49' N, 16°36' E): on the Golfo di Squillace, on the border between **Krotōn** and **Lokroi**, later called Skulakion/Scylacium (STRABŌN 6.1.0), a colony of **Athens**, and perhaps allied with the metropolis in the 5th c. BCE (*cf.* DIODŌROS OF SICILY 13.3); under **Rome** and a *colonia* 199 BCE (Livy 32.7); augmented in 124 BCE and again under Nerua. *RE* 2A.1 (1921) 920–923, H. Philipp; *BAGRW* 46-E4.

CASSIODORUS SENATOR.

Sebennutos (mod. Sammanud; 30°58' N, 31°15' E): city of **Egypt**, in the central Nile Delta, west of **Mendēs** and east of **Saïs**, a local capital from the 9th c. BCE, home city of the Pharaohs of the 30th dynasty (380–342 BCE). *BAGRW* 74-E3; *NP* 11.313, K. Jansen-Winkeln.

MANETHŌN.

Seleukeia ⇒ Tralleis

Seleukeia "Pieria" (mod. Kapısuyu; 36°08' N, 35°56' E): many cities were founded in regions controlled by the Seleukids under the name Seleukeia (mostly in modern Turkey), but POLUBIOS (5.56, 58–59) makes it clear that Apollophanēs came from this one (*cf.* **Syria**). Founded *ca* 300 BCE by Seleukos I as capital and harbor of **Antioch on the Orontes** (*ca* 5 km north of the mouth). Under the Ptolemies 246–219 BCE. *PECS* 822, J.-P. Rey-Coquais; *OCD3* 1380, A.H.M. Jones *et al.; BAGRW* 67-B4; *NP* 11.356–357 (#2), J. Wagner.

APOLLOPHANĒS.

Seleukeia on the Kalukadnos (Kilikia) (mod. Silifke; 36°22' N, 33°55' E): coastal city founded by a Seleukid king; north-east of **Aphrodisias (Kilikia)** and south-west of **Soloi (Kilikia)**; no history appears to be known. *RE* 2A.1 (1921) 1203–1204 (#5), W. Ruge; *PECS* 821–822, T.S. MacKay; *BAGRW* 66-D4; *NP* 11.357 (#5), F. Hild.

XENARKHOS.

Seleukeia on the Tigris (mod. Tell ʿUmar, *ca* 35 km south of Baghdad; 33°10' N, 44°35' E): on west bank of the Tigris, at confluence of Malka river, *ca* 60 km north of **Babylōn**. Founded *ca* 305 BCE, large trade center; opposite to the Parthian capital Ktēsiphōn, and conquered by them in 141 BCE; grew to become one of the largest cities of the period. *RE* 2A.1 (1921) 1149–1184 (#1), M. Streck; *PECS* 822, D.N. Wilber; *BAGRW* 91-F4; *NP* 11.355–356 (#1), H.J. Nissen.

APOLLODŌROS, SELEUKOS.

Selumbria (mod. Silivri; 41°05' N, 28°15' E): city on the coast of Propontis in **Thrakē**, 60 km west of **Buzantion**; colonized by **Megara** *ca* 680 BCE; sacked by the Persians 493 BCE; then allied with **Athens**. *BAGRW* 52-C2; *NP* 11.373–374, Al. Berger.

HĒRODIKOS.

Sevilla ⇒ Hispalis

Shirak: region of Armenia, known as Sirakēnē to Greeks, in which lay Ani (1 km south of mod. Ocaklı; 40°30'N, 43°34' E), on the west bank of the Akhuryan river; Ani became the capital of Greater Armenia (962–1055 CE). R.H. Hewsen and C.C. Salvatico, *Armenia: A Historical Atlas* (2001) 55-C4/D5, 91-C4, 262-B1/2; *BAGRW* 88-B4.

ANANIA.

Sicca Veneria (mod. Le Kef; 36°11' N, 08°43' E): on the inland road between **Carthage** and **Cirta**; under **Carthage** from 241 BCE (POLUBIOS 1.66–67) to 201 BCE. Early *colonia* of **Rome** in **Africa**, temporarily part of **Mauretania**, and possibly briefly the capital of the province *Africa Noua*. Episcopal see from the mid-3rd c. CE; substantially fortified in the 4th–5th centuries CE; under the Vandals 430–534 CE. *OCD3* 1401, B.H. Warmington and R.J.A. Wilson; *PECS* 834, A. Ennabli; *BAGRW* 32-C4; *NP* 11.502–503, W. Huß.

CAELIUS AURELIANUS, LACTANTIUS.

Sicily/Sikelia/Sicilia: largest Mediterranean island; inhabited by three indigenous peoples: Sikeli (east), Sikani (central), Elumi (west), with influence from **Italy** in the north-east. Colonized by Greeks and Phoenicians in the 8th c. BCE, its cities prospered through trade, and from its fertility. **Carthage** attacked **Himera**, 480 BCE; **Athens** twice attempted to dominate the island, 427–424 and 415–413 BCE; **Carthage** again attempted to establish dominance from 409 BCE (now destroying **Himera**) to 405 BCE. Agathoklēs of **Surakousai** ruled much of the island 317–289 BCE; and **Carthage** and **Rome** warred for 20 years (264–241 BCE), after which all but **Surakousai** was a Roman province. After **Surakousai** sided with Hannibal in 215 BCE, and was taken by **Rome** 212 BCE, the whole island was under **Rome**. Although **Rome** generally respected the autonomy of local ***poleis***, Verres (73–71 BCE) notoriously exploited Sicily's wealth. AUGUSTUS founded *coloniae* at **Katanē**, **Surakousai**, **Tauromenion**, and elsewhere. Sicily continued to prosper well into the Roman imperial period. *OCD3* 1401–1403, A. Momigliano *et al.; NP* 11.505–512, E. Olshausen and G. Falco, 512–513, I. Toral-Niehoff.

Sites: **Agurion**, **Akragas**, **Centuripae**, **Gela**, **Himera**, **Katanē**, **Leontinoi**, **Surakousai**, **Tauromenion**.

People: AUFIDIUS, CAYSTRIUS, EUPHĒMIOS, EURUŌDĒS, ORTHŌN, SCRIBONIUS.

Sidē (mod. Selimiye; 36°46' N, 31°23' E): city on Pamphulian coast, east of **Pergē** and west of **Aphrodisias** (**Kilikia**), long occupied, colonized by **Kumē** in the 7th or 6th c. BCE. The city's good harbor long rendered it a commercial hub of the eastern Mediterranean, a center of the Anatolian slave trade, and a naval base. Under Persians until Alexander of **Macedon**; Ptolemaic 301–218 BCE, then Seleukid until 188 BCE; then autonomous until 102 BCE. Provided quarter and markets to pirates from **Kilikia**. The city was under **Rome** from 78 BCE, but semi-autonomous until Hadrian, when it became the capital of the province of *Pamphylia*. Attacked by Goths 268 CE, after which defenses were augmented; poor in the 4th c., and again prosperous in the 5th–6th centuries. *PECS* 835–836, G.E. Bean; *ODB* 1892, C.F.W. Foss; *OCD3* 1404, G.E. Bean and S. Hornblower; *BAGRW* 65-F4; *NP* 11.517–519, W. Martini. (Contrast Malian Sidē, *BAGRW* 55-C3, Sidē of **Lakōnika**, *BAGRW* 58-E5, and Pontic Sidē, *BAGRW* 87-C3.)

ARTEMIDŌROS, MARCELLUS, MNĒMŌN, TRIBONIANUS.

Sidōn (mod. Saida; 33°34' N, 35°24' E): ancient fishing port of Phoenicia, south of **Bērutos** and north of **Turos**; founder of the colony **Kition**; prosperous until 677 BCE, when destroyed by Esarhaddon of Assyria. Rebuilt and autonomous under the Persians; destroyed 351 BCE and its people deported, after a failed revolt. Submitted to Alexander of **Macedon** 322 BCE; republic from *ca* 250 BCE; regional center of Ptolemaic operations from 218 BCE; surrendered to Antiokhos III 198 BCE; again autonomous from 111 BCE. Under **Rome** from 20 BCE. *PECS* 837, J.P. Rey-Coquais; *OCD3* 1404, A.H.M. Jones and J.-F. Salles; *BAGRW* 69-B2; *NP* 11.520–521, R. Liwak and J. Wagner; *EJ2* 18.549–551, E. Ashtor and Sh. Gibson.

BOËTHOS (2), DŌROTHEOS, MEGĒS, PASIKRATĒS (?), ZĒNŌN.

Sigeion (mod. Yenişehir near Kumkale; 40°00' N, 26°12' E): coastal city in the **Troas** near **Ilion**, founded by **Mutilēnē**, allied with **Athens** in the 5th c. BCE. *OCD3* 1406, S. Hornblower; *BAGRW* 56-C2; *NP* 11.535–536, E. Schwertheim.

DAMASTĒS.

Sikuōn (mod. Vasilikon; 37°59' N, 22°44' E): Dorian coastal city west of **Corinth** and straight north of **Argos**, which dominated it through *ca* 650 BCE, after which linked with **Athens** until *ca* 550 BCE; then allied with **Sparta**. Under **Thēbai of Boiōtia** 371–251 BCE (moved 303 BCE by Dēmētrios Poliorkētēs to higher ground), after which in the Akhaian League; enriched after the destruction of **Corinth** 146 BCE, then declined after **Corinth** re-founded 46 BCE. *PECS* 839–840, R. Stroud; *OCD3* 1403–1404, J.B. Salmon; *BAGRW* 58-D2; *NP* 11.543–544, Y. Lafond and E. Olshausen.

ARISTARKHOS, ERASISTRATOS, HĒRAKLEITOS, POLUKLEITOS.

Sinōpē (mod. Sinop; 42°02' N, 35°10' E): on the south shore of the Black Sea, west of **Amisos** and east of **Hērakleia Pontikē**, with two good harbors. The earliest colony of **Milētos** founded on the Black Sea (late 7th c. BCE), commanded regional maritime trade and established many of its own colonies. Became the capital of the kingdom of Pontos 183 BCE (moved from **Amaseia**), then made a *colonia* under **Rome** by CAESAR, remaining prosperous into the 3rd c. CE. Source of tuna, timber, and ***Sinōpian earth***. *PECS* 842, E. Akurgal; *OCD3* 1412, T.R.S. Broughton and St. Mitchell; *BAGRW* 87-A2; *NP* 11.585–586, C. Marek.

MITHRADATĒS VI.

Siphnos (mod. Siphnos; 36°59' N, 24°40' E): Aegean island west of **Paros** and south-south-east of **Kuthnos**, with silver mines active from the 3rd millennium BCE; prosperous in the 6th c. BCE, until plundered by exiles from **Samos** 525 BCE. Allied with **Athens** against Persia; in decline from the 4th c. BCE, as mines were exhausted. *PECS* 842, M.B. Wallace; *OCD3* 1412–1413, R.W.V. Catling; *BAGRW* 60-C4; *NP* 11.589–590, A. Külzer.

DIPHILOS.

Skēpsis (mod. Kurşunlu Tepe; 40°08' N, 26°29' E): Greek city of **Troas**, augmented with settlers from **Milētos**, on the shore of the Hellespont, and north of **Assos**. STRABŌN 13.1.54 records the tale that ARISTOTLE's library was hidden here for a century (*ca* 190–90 BCE). *OCD3* 1362, J.M. Cook and S. Hornblower; *BAGRW* 56-D2; *NP* 11.611, E. Schwertheim.

MĒTRODŌROS.

Smurna/Zmurna, Smurnē/Zmurnē (mod. Izmir; 38°25' N, 27°09' E): coastal colony of **Kolophōn** at the head of the Hermaic Gulf, near the Maiandros valley; most prosperous from 650 to 600 BCE, when sacked by Alyattēs of **Ludia**. Refounded on new site *ca* 300 BCE; praised by STRABŌN 14.1.37 as a center of art and science, with a library (but no sewers); restored by M. Aurelius after the earthquake of 178 CE. *PECS* 847–848, E. Akurgal; *OCD3* 1417, W.M. Calder *et al.; BAGRW* 56-E5; *NP* 11.661–663, G. Petzl and Al. Berger.

ALBINUS, HERMIPPOS, HERMOGENĒS, HIKESIOS, HOMER (?), MARKIŌN, MĒNODŌROS, MIKIŌN, PELOPS, SATUROS, SOLŌN, THEŌN.

% Soloi: seemingly impossible to determine from which of the two homonymous sites these men came:

ARISTOMAKHOS, THEODŌROS.

Soloi (Kilikia) (mod. Viranşehir west of Mersin; 36°48' N, 34°37' E): near **Tarsos**, founded

7th c. BCE as a colony of **Lindos**; sided with the Persians against Alexander of **Macedon**. Ravaged and depopulated by Tigranēs of Armenia during the war of MITHRADATĒS VI; repopulated (with ex-pirates) and rebuilt by Pompey, after whom it was then renamed. STEPHANOS OF BUZANTION, s.v., lists only this one; *PECS* 851, M. Gough; *BAGRW* 66-F3; *NP* 11.704 (#2), F. Hild.

ARATOS, CHRYSIPPUS, KRANTŌR.

Soloi (Cyprus) (mod. Karavostasi/Potamos tou Kambou; 35°09' N, 32°50' E): near the southernmost point of the bay of Morphou, one of the chief kingdoms of **Cyprus**, and taken by the Persians 499 BCE; allied with Alexander of **Macedon**, and then controlled by the Ptolemies, who abolished the kingship. *PECS* 850–851, K. Nicolaou; *BAGRW* 72-B2; *NP* 11.703–704 (#1), R. Senff.

BIŌN, HIERŌN, KLEARKHOS.

Spain/Hispania: comprising much of the Iberian peninsula, with abundant natural resources; colonized by Phoenicians (who founded **Gadēs**) and Greeks (from **Phōkaia**), and culturally influenced by Celtic migrations (from *ca* 500 BCE). Territory in south-east Spain was taken by **Carthage** 237–218 BCE, leading to war with **Rome** 218–206 BCE, after which Mediterranean coastal strips were constituted as provinces (*Hispania Citerior* and *Hispania Ulterior*); **Rome** extended holdings westwards in wars of 155–133 BCE. AUGUSTUS' conquest of the remaining north-west corner (26–19 BCE) resulted in the creation of the new province *Lusitania*, the expansion of *Hispania Citerior* as *Hispania Tarraconensis*, and the renaming of *Hispania Ulterior* as *Hispania Baetica*. The provinces remained prosperous through the end of the 2nd c. CE, the origin of many notable Romans: ANNAEUS SENECA, ANNAEUS LUCANUS, COLUMELLA, Quintilian, Martial, Trajan, Hadrian, and Marcus Aurelius. *OCD3* 1429–1430, S.J. Keay; *BNP* 6 (2005) 384–395, P. Barceló *et al.*

Sites: **Corduba, Gadēs, Dianium, Hispalis, Tingentera**.

People: EGNATIUS (?), TURRANIUS.

Sparta (mod. Sparti; 37°04' N, 22°26' E): early Dorian city in fertile valley (Eurotas river), ruled by a pair of kings; often opposed to **Athens**; free city under **Rome** from 196 BCE; plundered by the Heruli 267 CE, and destroyed by Alaric 395 CE. *PECS* 855–856, P. Cartledge; *OCD3* 1431–1433, P. Cartledge *et al.*; *BAGRW* 58-C3; *NP* 11.784–795, K.-H. Welwei.

AGATHINOS.

Spoletium (mod. Spoleto; 42°44' N, 12°44' E): Umbrian town, north-west of **Amiternum**, north of **Reate**, east of **Volsinii**, and south-west of **Firmum Picenum**, a *colonia* under **Rome** from 241 BCE, *municipium* by 90 BCE. *PECS* 858, L. Richardson, Jr.; *OCD3* 1436, E.T. Salmon and T.W. Potter; *BAGRW* 42-D3; *NP* 11.834, G. Uggeri.

MAECENAS MELISSUS.

Stageira (near mod. Nizvoro; 40°32' N, 23°45' E): small coast city in northern **Khalkidike**; a colony of Andros, destroyed by Philip II 349, but restored by Alexander of **Macedon** at ARISTOTLE's petition (PLUTARCH, *Alex.* 7.3). *OCD3* 1437, S. Hornblower; *BAGRW* 51-B3; *NP* 11.914, M. Zahrnt.

ARISTOTLE, NIKOMAKHOS.

Stoboi (mod. Gradska; 41°05' N, 21°57' E): Paionian city at juncture of Črna and Vardar rivers; under **Rome** from 168 BCE, and a *municipium* from 69 CE. Prosperous thence through 479 CE, when raided by Goths under Theodoric; devastated by an earthquake 518 CE (*cf.* **Pisidia**), after which declined. *PECS* 859–860, J. Wiseman; *ODB* 1958, A. Kazhdan; *OCD3* 1445–1446, J.J. Wilkes; *BAGRW* 50-A1; *NP* 11.1010, R.M. Errington.

Iōannēs.

Stratonikeia (mod. Eskihisar; 37°19' N, 28°04' E): east of **Mulasa** and south of **Tralleis**, founded *ca* 270 BCE on the site of the Karian city Idrias/Edrias; variously controlled by the Seleukids or **Rhodes** until 167 BCE, when declared free by **Rome**. Participated in the revolt against **Rome** 133–130 BCE; allied with Mithradatēs VI 88 BCE; again declared free by **Rome** 81 BCE, and resisted the Parthian siege 40 BCE. Prosperous until at least the mid-3rd c. CE. *PECS* 861, G.E. Bean; *OCD3* 1449, *Idem* and S. Sherwin-White; *BAGRW* 61-G3; *NP* 11.1046–1047 (#1, 2), H. Kaletsch. (Contrast Stratonikeia of **Ludia**, at mod. Siledik, east on the Kaikos from **Pergamon**, under whom from 188 BCE, and renamed Hadrianopolis 123 CE: *BAGRW* 56-F3.)

Menestheus.

Stumphalos (mod. Stumphalia; 37°50' N, 22°27' E): north-east Arkadia, west of **Phleious**, south-west of **Sikuōn**, strategically important to **Sparta**, member of the Arkadian League in the 4th c. BCE. *PECS* 861, W.F. Wyatt, Jr.; *OCD3* 1449–1450, J. Roy; *BAGRW* 58-C2; *NP* 11.1062, Kl. Tausend.

Aineias.

Subaris/Sybaris (mod. Sibari; 39°47' N, 16°19' E): city of **Lucania**, founded by Akhaeans and Troizenians *ca* 720 BCE, expanded rapidly through exploitation of agricultural resources and Etruscan commercial connections; its wealth and luxury a literary topos; defeated and razed by **Krotōn**, 510 BCE; its exiled inhabitants founded a new Subaris on the Traente river. *PECS* 869–870, F. Rainey; *OCD3* 1458–1459, H.K. Lomas; *BAGRW* 46-D2; *NP* 11.1124–1125 (#4), A. Muggia.

Menestōr.

Sulmo (mod. Sulmona; 42°02' N, 13°56' E): principal town (with Corfinium and Superaequum) of the Paeligni, near the confluence of two tributaries to the Aternum (which debouches into the Adriatic), south-east of **Amiternum**, strategically on the road leading to Samnium. Allied with **Rome** since 305 BCE (Livy 9.45); resisted Hannibal 211; supported Caesar in the Roman Civil Wars. Modest *municipium* from 89 BCE, retaining its importance into the medieval period. *PECS* 867–868, A. La Regina; *OCD3* 1454, E.T. Salmon and T.W. Potter; *BAGRW* 44-E1; *NP* 11.1095, G. Uggeri.

Ovid.

Surakousai/Syracuse (mod. Siracusa; 37°05' N, 15°17' E): colony of **Corinth**, *ca* 734 BCE; prosperous city with a growing empire; Hierōn I (478–466 BCE) patronized Aeschylus, Simōnidēs, and Pindar. Resisted **Athens**' invasions 427–424 and 415–413 BCE. Agathoklēs of Surakousai ruled much of **Sicily** 317–289 BCE; under Hierōn II (270–216 BCE), the city prospered on trade throughout the Mediterranean, supporting an active building program, and became again a significant artistic and intellectual center. In 215, Hierōn's successor Hierōnumos favored **Carthage**, prompting a siege by **Rome**, in which Archimēdēs acted prominently. The city fell to **Rome**, but became a center of provincial government; suffered under Verres (73 BCE); received a colony under Augustus. *PECS* 871–874, G. Voza; *OCD3* 1463–1464, A.G. Woodhead and R.J.A. Wilson; *BAGRW* 47-G4; *NP* 12/2.1159–1172, I. Toral-Niehoff.

Antiokhos, Archimēdēs, Ekphantos, Epikharmos, Hermodōros, Hierōn II, Hiketas, Kleandros, Kleōn, Menekratēs, Numphodōros, Pheidias, Skopinas.

Suros (mod. Suros; 37°27' N, 24°54' E): Aegean island, east of **Kuthnos** and west of **Dēlos**, inhabited from the 3rd millennium BCE, and Ionic from *ca* 1000 BCE; colonized by, and allied with, **Athens** in the 5th c. BCE. Member of the island-league centered at **Dēlos** (q.v.). *PECS* 874, M.G. Picozzi; *BAGRW* 60-A5; *NP* 11.1187, A. Külzer.

PHEREKUDĒS.

Syracuse ⇒ Surakousai

Syria: approximately corresponding to modern Lebanon plus coastal Syria (the ancient borders were unstable), and sometimes including areas of Phoenicia; taken by Alexander of **Macedon** 333/332 BCE, and a core area of the Seleukid Empire, the coastal cities being disputed with the Ptolemaic Empire, until taken by Pompey for **Rome** 64 BCE. The cities were Hellenized, but the rural areas remained Aramaic-speaking. Enlarged by the absorption of client kingdoms, then divided into two provinces by Septimius Seuerus (194 CE), and into four by Diocletian. Afflicted by the Palmyrene revolt (260–273 CE), attacks by Sasanians, the plague, earthquakes, and Christian conflicts. *OCD3* 1464–1465, A.H.M. Jones *et al.; BAGRW* 68 and 69; *NP* 11.1170–1181, H. Klengel and E.M. Ruprechtsberger.

Sites: **Amida**, **Antioch**, **Apameia**, **Arados**, **Bērutos**, **Bublos**, **Constantia**, **Damaskos**, **Edessa**, **Emesa**, **Gabala**, **Gadara**, **Gaza**, **Gerasa**, **Khalkis**, **Larissa**, **Nisibis**, **Rešʿaina**, **Samaria**, **Seleukeia Pieria**, **Sidōn**, **Tyre**.

People: IAMBLIKHOS (GEOG.).

Tanagra (mod. Tanagra; 38°19' N, 22°32' E): town in **Boiōtia** on the Asopos river, west of **Eleusis** and east of Boiōtian **Thēbai**; occupied by Sparta 386 to *ca* 373 BCE; under **Rome** 145 BCE, which made Tanagra a *ciuitas libera et immunis*, after which prosperous. *PECS* 876–877, P. Roesch; *OCD3* 1472–1473, J. Buckler; *BAGRW* 59-B1; *NP* 12/1.6–7, M. Fell.

BAKKHEIOS.

Taras/Tarentum (mod. Taranto; 40°28' N, 17°14' E): Apulian town, colonized by **Sparta** in the 8th c. BCE; became prominent in the 5th c. as **Krotōn** declined; fell to **Rome** after PURRHOS' defeat, becoming an ally 270 BCE; revolted 213–209 BCE under Hannibal. *PECS* 878–880, W.D.E. Coulson; *OCD3* 1473–1474, H.K. Lomas; *BAGRW* 45-F4; *NP* 12/1.20–22, A. Muggia.

APOLLODŌROS, ARISTOXENOS, ARKHUTAS, EURUTOS (?), GLAUKIAS, HĒRAKLEIDĒS (2), IKKOS, KLEINIAS, ZEUXIS, ZŌPUROS.

Tarsos (mod. Tarsos; 36°55' N, 34°54' E): near **Soloi** (**Kilikia**), capital of the Kilikian kings, the Persian satraps, and the Roman province of *Cilicia* (from 72 CE); under Antiokhos IV (175–163 BCE) renamed **Antioch** on the Kudnos; a commercial hub known for its linen, a philosophical center, the birthplace of the Christian epistolographer Paul; and the base for Caracalla's Parthian war (216 CE). *PECS* 883–884, M. Gough; *OCD3* 1476, A.H.M. Jones and St. Mitchell; *BAGRW* 66-F3; *NP* 12/1.37–38, Fr. Hild. (Contrast the Roman-period town Tarsos of **Bithunia**, *BAGRW* 52-G3.)

ANTIPATROS, APOLLŌNIOS, AREIOS, ARISTARKHOS, ARKHEDĒMOS, ATHĒNAIOS (?), ATHĒNODŌROS, DIODŌROS, DIOGENĒS, LEUKIOS (?), MAGNUS, PHILŌN, SELEUKOS.

Tauromenion (mod. Taormina; 37°51' N, 15°17' E): coastal city in eastern **Sicily**, north of **Katanē**; established by Himilkōn of **Carthage** in 396/5 BCE on a site already Hellenized; refounded in 392 as a Greek city; supported PURRHOS; dominated by **Surakousai**; submitted to **Rome** when HIERŌN II died. *PECS* 886–887, M. Bell; *OCD3* 1477, A.G. Woodhead and R.J.A. Wilson; *BAGRW* 47-G3; *NP* 12/1.57–58, M.C. Lentini and K. Meister.

TIMAIOS.

Tenedos (mod. Bozcaada; 39°50' N, 26°04' E): small island close to the shore, opposite **Alexandria Troas**; settled from **Lesbos**; allied with **Athens** in the 5th c. BCE. *OCD3* 1483, E. Kearns; *BAGRW* 56-C2; *NP* 12/1.134–135, A. Külzer. (Contrast the Hellenistic Pamphulian town Tenedos, *RE* 5A.1 [1934] 498–499 [#2], W. Ruge; *BAGRW* 65-E4.)

ANDROITAS, KLEOSTRATOS, PHASITAS.

Teōs (mod. Sığacık; 38°12' N, 26°47' E): south of **Smurna** and north of **Ephesos**, on an isthmus, with a north and south harbor; foundation-legends connect it with **Boiōtia** and with **Athens**; in 544 BCE, the inhabitants migrated to **Abdēra**, though many soon returned; allied with **Athens** in the 5th c. BCE. *PECS* 893–894, G.E. Bean; *OCD3* 1483, *Idem* and S. Hornblower; *BAGRW* 56-D5; *NP* 12/1.137–138, W. Blümel and E. Olshausen.

ANDRŌN, NAUSIPHANĒS, SKUTHINOS.

Thasos (mod. Thasos; 40°43' N, 24°46' E): island, opposite **Abdēra**, colonized from **Paros** *ca* 680 BCE (displacing the native Thrakians), and prosperous from its gold-mines. Surrendered to the Persians and then allied with or subject to **Athens** from 477 BCE; remained prosperous then and under **Macedon**; declared free by **Rome** in 197 BCE. *PECS* 903, E. Vanderpool; *OCD3* 1492, E.N. Borza; *BAGRW* 51-D3; *NP* 12/1.244–246, A. Külzer.

ANAXIPOLIS, ANDROSTHENĒS, APELLĒS (?), ARISTOGENĒS, EUAGŌN, LEŌDAMAS, PHANOKRITOS (?), PHILISKOS, PUTHIŌN, STĒSIMBROTOS, THRASUALKĒS.

% Thēbai: seemingly impossible to determine whether these scientists came from one of the two sites listed separately below, or else from the Karian Thēbai (*BAGRW* 61-E2) or the Thēbai of Phthiotis (*BAGRW* 55-D2, mod. Akitsi):

EUMĒLOS, HIERAX, LUPUS (?), OLUMPIAS.

Thēbai of Boiōtia (mod. Thivai; 38°19' N, 23°19' E): large and storied settlement since *ca* 1400 BCE; north of **Plataia** and south-west of **Khalkis**; from the late 6th c. BCE sought to dominate all of **Boiōtia**; allied with **Sparta** in 457 BCE; formed Boiōtian confederacy in 447, and attacked **Plataia**; after 404 BCE, opposed **Sparta** and defeated her at Leuktra in 371 BCE; opposed Philip II of **Macedon** at **Khaironeia**, 338 BCE. Destroyed after failed revolt against Alexander of **Macedon**. *PECS* 904–906, P. Roesch; *OCD3* 1495–1496, J. Buckler; *BAGRW* 55-E4; *NP* 12/1.283–288 (#2), M. Fell.

ARISTOGEITŌN (?), LINOS (?).

Thēbai of Egypt ⇒ Diospolis Magna

Thessalia/Thessaly: mostly land-locked region of northern Greece, consisting of two large plains, enclosed by mountains, including Olumpos/Olympus, Ossa, and Pelion. Fertile land, famous for horses, notorious for witches, and a center of the Orphic cult, with distinctive but Greek-influenced culture. Militarily dominated central Greece until *ca* 600 BCE. Internal discord facilitated the intervention of Philip II of **Macedon**, 353–352 BCE; allied with Alexander of **Macedon**, then opposed **Macedon** 323–322 BCE. Partly controlled by **Macedon** in the 3rd c. BCE; declared free, but controlled, by **Rome** in 196 BCE; enlarged by **Rome** in 146 BCE. *OCD3* 1511–1512, B. Helly; *NP* 12/1.446–451, H. Beck.

Sites: **Atrax**, **Larissa**.

People: BILLAROS (?), KINEAS, NIKŌNIDĒS, POLUĪDOS.

Thrakē/Thrace: on the northern edge of **Macedon**, with varying and unclear borders: e.g., the 5th c. BCE kingdom of the Odrusai extended from the Danube to the Hellespont;

generally included the Thrakian **Khersonēsos**. Greeks considered Thrakians primitive and colonized coastal areas; but natives resisted Hellenization and apparently lived in villages until the Roman period. Subdued by Persia *ca* 516 BCE; allied with **Athens** against **Macedon** 429 BCE. Invaded by Philip II of **Macedon** 346 and 342 BCE; ruled by Alexander of **Macedon**, after whose death acquired by LUSIMAKHOS who founded Lusimakheia 308 BCE, after which under **Macedon** until 168 BCE. Under ever-increasing influence by **Rome** 149 BCE to 46 CE, when it became a province. Hardly urbanized; frequently raided from the north, esp. from the 4th c. CE. *ODB* 2079–2080, T.E. Gregory; *OCD3* 1514–1515, J.M.R. Cormack and J.J. Wilkes; *NP* 12/1.478–485, I. von Bredow and J. Niehoff.

Sites: **Abdēra**, **Abudos**, **Mendē**, **Elaious**, **Kardia**, **Stageira**.

People: MARSINUS.

Tingentera (mod. Algeciras; 36°08'N, 05°27'W): small coastal city, opposite the Pillars of Hēraklēs (Gibraltar), east of **Gadēs**; founded by colonists from **Mauretania**, perhaps Tingis (mod. Tangiers, 35°46'N, 05°48'W), south-west across the strait. *RE* 6A.2 (1937) 1383–1384, A. Schulten; *BAGRW* 26-E5.

POMPONIUS MELA.

Tlōs (mod. Yaka east of Düver; 36°33' N, 29°26' E): ancient and major city of **Lukia** on the east bank of the Xanthos river, north of **Patara** and south of **Oinoanda**. In the Lukian League from 168 BCE, with a democratic constitution; by *ca* 100 BCE one of the six leading cities of the federation (ARTEMIDŌROS OF EPHESOS in STRABŌN 14.3.3); the best-known city of **Lukia**, according to PLINY 5.101. *PECS* 927, G.E. Bean; *BAGRW* 65-B4; *NP* 12/1.637, W. Hailer.

ANTIOKHIS.

Tours ⇒ Caesarodunum

Tralleis (mod. Aydın; 37°51' N, 27°51' E): prosperous city in **Ludia** on north bank of Maiandros river valley; under Maussōllos in the mid-4th c.; under the Seleukids called **Seleukeia**; renamed Caesarea by AUGUSTUS who restored the city after an earthquake; flourished into late antiquity, a hub of monophysite activity; monumental aqueduct built in the 4th c. *PECS* 931, G.E. Bean; *ODB* 2103–2104, C.F.W. Foss; *OCD3* 1544–1545, W.M. Calder *et al.*; *BAGRW* 61-F2; *NP* 12/1.750–751, H. Kaletsch.

ALEXANDER, ANTHĒMIOS, ASINIUS POLLIO, ASKLĒPIOS, IULIANUS, MĒTRODŌROS, PHLEGŌN, PRŌTARKHOS, STEPHANOS, THESSALOS.

Transpadana: the region north of the Padus (Po) river, inhabited by Celts; the whole valley was conquered by **Rome** 222–218 and colonized 191–187 BCE; Transpadana itself remained inhabited mainly by Celts. Granted citizenship by CAESAR in 49 BCE, by which time it had become a prosperous region (STRABŌN 5.1.12). *OCD3* 1546, anon.; *NP* 12/1.756, A. Sartori.

Sites: **Mantua**, **Mediolanum**, **Nouum Comum**, **Verona**.

People: CORNELIUS NEPOS, PLINIUS SECUNDUS.

Trapezous (mod. Trabzon; 41°00' N, 39°44' E): on south-east coast of Black Sea, founded as a trading-post in 756 BCE by (and east of) **Sinōpē**. Annexed by **Rome** in 64 CE as a free city. As the nearest Black Sea port to the upper Euphrates river, its importance grew under Hadrian who improved the harbor. Sacked by Goths in 256 CE, but persisted as a garrison and then grew in early Byzantine times. *PECS* 932, D.R. Wilson; *OCD3* 1547, T.R.S. Broughton and St. Mitchell; *BAGRW* 87-E4; *NP* 12/1.763–764, E. Olshausen.

TUKHIKOS.

Trebizond ⇒ Trapezous

Troas: mountainous peninsula in north-west Anatolia, on the Hellespont; under Persia from the end of the 5th c. BCE; taken by Alexander of **Macedon**; variously ruled until under **Pergamon** 227 BCE; under **Rome** from 133 BCE. *OCD3* 1555, D.E.W. Wormell and St. Mitchell; *NP* 12/1.848–850, E. Schwertheim.
Sites: **Alexandria, Assos, Ilion, Lampsakos, Parion, Sarnaka (?), Sigeion**.
People: LUKŌN.
Tuana (mod. Kemer Hisar, five km south of Niğde; 37°55' N, 34°40' E): ancient Hittite city of **Kappadokia**, north of **Tarsos** and **Soloi (Kilikia)**. Large and prosperous *ca* 400 BCE (XENOPHŌN, *Anab.* 1.2.20); allied with **Rome** from 168 BCE; made a province 17 CE; made a *colonia* by Caracalla (*ca* 215 CE). STRABŌN 12.2.7; *PECS* 942, R.P. Harper; *BAGRW* 66-F1; *NP* 12/1.936, K. Strobel.
APOLLŌNIOS, PHILŌN.
Turos/Tyre (mod. es-Sur; 33°16' N, 35°13' E): prosperous port in Phoenicia, south of **Sidōn**; founded the colonies **Utica** and **Carthage**. Besieged for seven months by Alexander of **Macedon**, and destroyed; under the Ptolemies until 274 BCE, when it became a republic; taken by the Seleukids 200 BCE; again autonomous from 126 BCE. Taken by Pompey for **Rome** 67 BCE, but freed; then under **Rome** from 20 BCE. Prosperous throughout the Roman period (esp. through purple dye and glass; a terminus of the silk trade); capital of *Syria Phoenice* from 198 CE (*cf.* **Syria**); episcopal see by *ca* 300 CE; monophysite in the 6th c. (synod of 514). *PECS* 944, W.L. MacDonald; *ODB* 2134, M.M. Mango; *OCD3* 1568, A.H.M. Jones *et al.; BAGRW* 69-B3; *NP* 12/1.951–955, H. Sader; *EJ2* 20.218–219, E. Ashtor. (Contrast the Peloponnesian town, *BAGRW* 8-D3, and the inland town, *BAGRW* 71-B2.)
ANTIPATROS, MARINOS, PEPHRASMENOS, PORPHURIOS.
Tusculum (mod. Tuscolo; 41°48' N, 12°42' E): ancient and possibly Etruscan town on the slopes of Mt. Albanus, *ca* 25 km south-east of **Rome**, west of **Praeneste**, and first *municipium* created by **Rome** 381 BCE; by 80 BCE a *colonia*. *PECS* 941–942, B. Goss; *OCD3* 1565, E.T. Salmon and T.W. Potter; *BAGRW* 43-C2; *NP* 12/1.931–932, M.M. Morciano and J.W. Mayer.
PORCIUS CATO, VIBIUS RUFINUS.

Tyre ⇒ Turos

Utica/Itukē (mod. Utique; 37°03' N, 10°02' E): at mouth of the Bagradas river, north-west of **Carthage**, occupied from 8th c., and traditionally the earliest Phoenician colony in **Africa**. Second only to **Carthage**, which she opposed in 149, rewarded by **Rome** with land and the status of a free city; and in 146, became the capital of Roman province *Africa Noua*. *PECS* 949–950, A. Ennabli; *OCD3* 1575, W.N. Weech *et al.; BAGRW* 32-F2; *NP* 12/1.1067–1068, W. Huß.
DIONUSIOS.
Vasates (mod. Bazas; 44°26' N, 00°13' W): 60 km south-east of **Burdigala**, and south-west of **Vesunna**; from the 3rd c. BCE the capital of the homonymous tribe of **Aquitania**; flourished after the provincial reorganization by Diocletian 297 CE. *BAGRW* 14-E4; *NP* 12/1.1146–1147, M. Polfer.

Iulius Ausonius.

Verona (mod. Verona; 45°26' N, 10°59' E): town of north **Italy**, on the Adige river; described as large and important by Strabōn (5.1.6) and Martial (14.195.1); a *colonia* of **Rome** by 69 CE. *PECS* 968–969, B. Forlati Tamaro; *OCD3* 1588, E.T. Salmon and T.W. Potter; *BAGRW* 39-H3; *NP* 12/2.77–78, E. Buchi.

Aemilius Macer.

Vesunna (mod. Périgueux; 45°11' N, 00°43' E): hill-fort of the Petrucorii (a federation of Gallic tribes that supported Vercingetorix), north-east of **Vasates** and east of **Burdigala**; the city was moved down into the valley of the Isle river after conquest by **Rome** in 51 BCE; incorporated into **Aquitania** by Augustus in 27 BCE; flourished until the Germanic invasions in 276 CE. *RE* 19.2 (1938) 1306–1318, P. Goessler; *RE* 8A.2 (1958) 1800–1801, F.M. Heichelheim; *PECS* 972–973, A. Blondy; *BAGRW* 14-F3.

Anthedius.

Vocontii: (people from) the mountainous region *ca* 60 km north-east of **Arelate**, who occupied the western foothills of the Alps from 3rd c. BCE (corresponding to south-eastern France north of the coast). *OCD3* 1610, A.L.F. Rivet and J.F. Drinkwater; *BAGRW* 15-E1; *NP* 12/2.275–276, Ch. Winkle.

Sites: (none).

People: Pompeius Trogus.

Volsinii/Velzna (mod. Bolsena; 42°39' N, 11°59' E): the inhabitants of the Etruscan city Volsinii Veteres (mod. Orvieto on the upper Tiber, *BAGRW* 42-C3) were resettled by **Rome** in this city of **Etruria** (on Lake Bolsena), west of **Spoletium**, after their rebellion in 294 BCE. *PECS* 657, L. Richardson, Jr.; *OCD3* 1612, D.W.R. Ridgeway; *BAGRW* 42-B3; *NP* 12/2.314–315, G. Camporeale.

Auienus.

Xoren (near mod. Muş: either near Kanlar, west-north-west of Muş, 38°45' N, 41°22' E ***or*** near Laçkam, west-south-west of Muş, 39°00' N, 41°83' E): in the Armenian district of Taraun/Taron, west of Lake Thospitis (mod. Lake Van), west across the lake from **Artemita**, north-east of **Amida**, north of **Nisibis**; later called Khoronk. *RE* S.10 (1965) 789.36–38, E. Polaschek; *BAGRW* 89-D2; R.H. Hewsen and C.C. Salvatico, *Armenia: A Historical Atlas* (2001) 27-C3/D4, 48-A5, 194-A3 ***or*** 194A/B2; A.J. Hacikyan, *Heritage of Armenian Literature* 1 (2001) 306.

Moses.

Zakunthos (mod. Zakunthos; 37°48' N, 20°45' E): southernmost Ionian island, off the coast of **Ēlis**; mentioned by Homer (*Iliad* 2.634), settled by Akhaians or Arkadians; a fleet station for **Athens** 430–405 BCE; allied with **Sparta** after the war; supported Philip II and Alexander of **Macedon**; conquered by **Rome** 211 BCE, under whom it prospered. *OCD3* 1633, W.M. Murray; *BAGRW* 54–inset; *NP* 12/2.686–689, D. Strauch.

Puthagoras.

Zmurna/Zmurnē ⇒ Smurna

GLOSSARY

These Greek terms were in some cases used over many centuries, and their meanings shifted; we give here a basic definition, but in each case consultation of relevant dictionaries and other reference works is recommended. Terms are entered here if used thrice or more within the encyclopedia, and if their translation is ambiguous or complex. Within a glossary-entry, cross-references to main entries indicate those which provide further documentation; following each glossary-entry is a list of main-entries in which the relevant term is found. We provide a representative not complete bibliography.

We owe a debt of gratitude to 14 contributors who assisted in composing this Glossary: Bernard, Fischer, Hallum, Hellmann, Jones, Karamanolis, McCabe, Opsomer, Rochberg, Scarborough, Touwaide, Tybjerg, Wilson, and Zhmud. See also the MANUSCRIPTS, immediately following the Glossary.

Academy PLATO's school in Athens, closed 529 CE. Scholars generally divide its history in three phases, with "Middle" commencing with EUDŌROS OF ALEXANDRIA, and "Late" or "Neo" commencing with PLŌTINOS. An emphasis on, or at least admiration for, mathematics and astronomy persisted. No actual institution existed from *ca* 88 BCE to *ca* 410 CE; we nevertheless list here both all those attested to have been members or adherents of the school. Alan Cameron, "The last days of the Academy at Athens," *PCPS* 195 (1969) 7–29; J. Glucker, *Antiochus and the late academy* (1978); *BNP* 1 (2002) 41–46, Th.A. Szlezák.

Entries on Academics: AELIANUS (2), AGAPIOS (?), ALBINUS OF SMURNA, ALKINOOS, AMELIUS, AMMŌNIOS ANNIUS, AMMŌNIOS OF ALEXANDRIA (SON OF HERMEIAS), APULEIUS OF MADAURUS, ARISTEIDĒS QUINTILIANUS, ARISTIPPOS, ASKLĒPIOS OF TRALLEIS, ATHĒNAIOS OF KUZIKOS, ATTICUS, BASIL OF CAESAREA, BOËTHIUS, CALCIDIUS, CASSIUS LONGINUS, MARTIANUS CAPELLA, DAMASKIOS, DAMIANOS, DĒMĒTRIOS (MATH. II), DERKULLIDĒS, DIONUSODŌROS, EUDŌROS OF ALEXANDRIA, EUDOXOS OF KNIDOS, GAIUS (PLATON.), HARPOKRATIŌN OF ARGOS, HĒRAISKOS, HĒRAKLEIDĒS PONTIKOS, IAMBLIKHOS OF KHALKIS, IŌANNĒS PHILOPONOS, KARNEADĒS, KRANTŌR, KRONIOS, MARINOS OF NEAPOLIS, MARIUS VICTORINUS, NOUMĒNIOS OF APAMEIA, OFELLIUS, OLUMPIODŌROS, P. BEROL. 9782, PERIKLĒS OF LUDIA, PHILIPPOS OF OPOUS, PHILŌN OF ALEXANDRIA, PLATO, PLOUTARKHOS SON OF NESTORIOS, PLUTARCH, POLEMŌN OF ATHENS, PORPHURIOS OF TYRE, PRISCIANUS OF LUDIA, PROKLOS OF LUKIA, PTOLEMAIOS (PLAT.), CASSIODORUS SENATOR, SIMPLICIUS, SŌKRATĒS JUNIOR, SPEUSIPPOS OF ATHENS, STEPHANOS OF ALEXANDRIA, SYRIANUS, TAUROS, THEMISTIOS, THEODŌROS OF ASINĒ, THEODŌROS OF SOLOI, MACROBIUS THEODOSIUS, THEOL. ARITHM., THEUDIOS, THRASULLOS, TIBERIANUS, XENOKRATĒS OF KHALKĒDŌN.

See also: ADRASTOS OF APHRODISIAS, AISARA, ALEXANDER OF APHRODISIAS, AMBROSE, ANTIPATROS OF TARSOS, PSEUDO-APULEIUS, ARCHIMĒDĒS, ARISTARKHOS OF SAMOS, ARISTOTLE, CHRYSIPPUS, DIKAIARKHOS, DIOGENĒS LAËRTIOS, ERATOSTHENĒS, GALĒN, IULIUS FIRMICUS, IAKŌBOS PSUKHRESTOS, IAMBLIKHOS (ALCH.), IŌANNĒS "LYDUS," IŌANNĒS OF STOBOI, LONDINIENSIS MEDICUS, ON MELISSOS, XENOPHANĒS, AND GORGIAS, MNĒSITHEOS OF ATHENS, OLUMPIODŌROS OF THĒBAI, OREIBASIOS, PHILŌNIDĒS OF LAODIKEIA, TERENTIUS VARRO OF REATE, THEOPHRASTOS, THEOPHULAKTOS, TIMAIOS OF LOKRIS, M. TULLIUS CICERO, ZĒNŌN OF KITION.

acopum (pl. ***acopa***) Latin for ***akopon***.

aigilōps a kind of eye-ulcer (CELSUS 7.7; GALĒN *UP* 10.10 [3.808 K.]); to be distinguished from the homonymous plant (THEOPHRASTOS *CP* 5.15.5; DIOSKOURIDĒS 4.137).

See also: LAMPŌN, MENELAOS (PHARM.).

aithēr originally the clear air above the clouds and mist: HĒSIOD *Theog.* 125; HOMER, *Iliad* 2.455–458; PLATO, *Kratulos* 410b. Perhaps already with the **Pythagoreans** is considered as a distinct cosmic element (PHILOLAOS DK 12, whose authenticity Huffmann 1993: 392–395 doubts). Plato and the early **Academics** appear to consider it as the second most important element after fire (*Phaedo* 109b9, 111b2, *Timaeus* 58d1, EPINOMIS 984b–e), while ARISTOTLE takes it to be an indestructible element responsible for the eternal existence and circular movement of the planets, *De Caelo* 1.3 (268b10–270b31), 2.6 (288a13–288b35); ON THE KOSMOS 2 (392a5–b4); a point disputed within the **Peripatos**, esp. by XENARKHOS. *BNP* 1 (2002) 269–270, F. Graf.

See also: ANTIPATROS OF TYRE, ATTICUS, BOËTHOS OF SIDŌN (STOIC), EMPEDOKLĒS, HĒRAKLEIDĒS OF ATHENS, KLEARKHOS, ON THE KOSMOS, PHILISTIŌN, PTOLEMY, STEPHANOS OF ALEXANDRIA, TAUROS, THEOPHRASTOS.

***akopon* (Lat.: *acopum*)** an anodyne poultice, plaster, or potion, frequently used as a dermal anesthetic in surgery or in the treatment of wounds, e.g. those sustained by gladiators and soldiers. CELSUS 4.31.8, 5.24.1; SCRIBONIUS LARGUS 156, 161, 206, 254, 268 *et al.*; PLINY 23.89; GALĒN, *CMGen* 7.11–13 (13.1005–1039 K.).

See also: ANTIMAKHOS (PHARM.), APHRODĀS, APHRODISIS, AQUILA SECUNDILLA, CASTUS, DOMITIUS NIGRINUS, HALIEUS, IULIUS AGRIPPA, IULIUS SECUNDUS, KLUTOS, MARCIANUS OF AFRICA, ORFITUS, ŌRIŌN, PHILOKLĒS, QUADRATUS, SERTORIUS CLEMENS.

alkuoneion one of several types of "bastard corals" (zoophytes) generally native to the Black, Mediterranean, and Red Seas, employed as external medicines (usually reduced to ash or powdered) for skin-cleaning (including depilation) and tooth-whiteners, and internally as a diuretic: HIPPOKRATĒS, *Mul.* 1.106 (8.230 Littré); DIOSKOURIDĒS, *MM* 5.118; GALĒN, *Simples* 11.2.3 (12.370–371 K.). J. Berendes (*Pedanios Dioskurides aus Anazarbos Arzneimittellehre* [1902] p. 541) identifies the five sorts listed by Dioskouridēs: *Alcyoneum cortoneum* Pall., *A. papillosum* Pall., *A. palmatum* Pall., *Spongia panicea* Pall., and *A. ficus* Pall. To be distinguished from *kouralion*, "coral" (Dioskouridēs, *MM* 5.121).

See also: EIRĒNAIOS, PHAIDROS, SŌKRATIŌN.

ami or ***ammi*** Indian species transplanted to Egypt, variously known as ajowan (thus "Ajowan oil"), ajwain, Bishop's weed, sometimes Ammi Copticum; identified as *Carum Copticum* Benth. and Hook: DIOSKOURIDĒS, *MM* 3.62; PLINY 20.163; GALĒN, *Simples*

Ammi, Mount Athos Ω 75, f.19^{V}

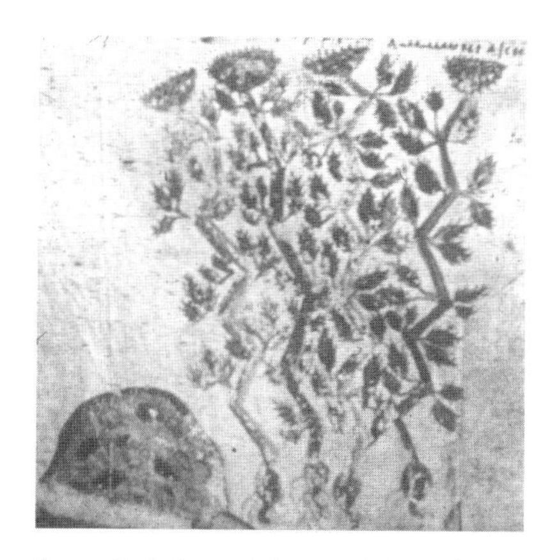

Ammōniakon, Mount Athos Ω 75, f.21^{R}

Amōmon, Mount Athos Ω 75, f.21^{V}

6.1.28 (11.824 K.); Usher 1974: 126; Durling 1993: 36. (Closely related to *Carum carvi* L., caraway: Usher 1974: 126; Stuart 1979: 167; Evans 1996: 263–264; Wichtl 2004: 116–118.)

See also: Akhillās, Marcellus (Pharm.), Nikēratos, Parisinus medicus, Pankharios, Theodos.

ammōniakon gum-exudate from the giant fennel, *Ferula marmarica* Asch. and Taub, native to north Africa, and named from the oracle of Ammōn in Egypt: Dioskouridēs, *MM* 3.84; Pliny 12.107; Galēn, *Simples* 6.1.37 (11.828 K.); Usher 1974: 253; André 1985: 116; *BNP* 1 (2002) 587, C. Hünemörder (*Ferula tingitana* L.).

See also: Alkimiōn, Amuthaōn, Andronikos (Med.), Aphros, Apollophanēs of Seleukeia, Aristoklēs, Attalos III, Diophantos of Lukia, Euboulos (Pharm.), Euelpidēs, Euthukleos, Harpalos (Pharm.), Harpokrās, Hierax, Ioudaios, Isidōros of Memphis, Khalkideus, Kleophantos (Pharm.), Lampōn, Leukios, Lusias, Marcianus of Africa, Megēs, Menelaos (Pharm.), Moskhos, Nikomēdēs IV, Onētidēs/Onētōr, Ōriōn, Parisinus medicus, Phaidros, Proëkhios, Puramos, Sarkeuthitēs, Sōkratiōn.

amōmon tropical Asian species, similar to ginger, often called the Nepal cardamom, *Amomum subulatum* Roxb. Generally imported into the Mediterranean world from India: Theophrastos *HP* 9.7.2; Dioskouridēs, *MM* 1.15; Galēn, *Simples* 6.1.38 (11.828 K.); Miller 1969: 67–69; Durling 1993: 40; *BNP* 1 (2002) 593, C. Hünemörder.

See also: Aphrodās, Celer, Kleopatra, Magnus Arkhiatros, Marcellinus (Pharm.), Pamphilos of Bērutos.

***anthrax* (Lat.: *carbunculus*; "glowing coal")** infected wound characterized by inflammation and swelling, leading to fever: *cf.* Hippokratic Corpus, Epidemics 2.1 (5.72 Littré); Pliny 26.5–6; Galēn, *CMGen* 5.15 (13.854–855 K.).

See also: Aelius Promotus, Apiōn, Apollōnios Ophis-Thēr, Kharmēs, Pasiōn, Polueidēs, Saturos of Smurna, Telamōn.

aphronitron (also ***nitrou aphros***) froth or efflorescence of natron: Dioskouridēs, *MM* 5.113; Pliny 31.112–113 (Lydian is best, then Egyptian); Galēn, *Simples* 9.3.5 (12.212–213 K.).

See also: Castus, Halieus, Kleophantos (Pharm.), Marcianus of Africa, Menemakhos, Polustomos, Sōkratiōn.

artēriakē ("<medicine for> the windpipe") drugs often for treatment of asthma, but also cough lozenges enabling a patient to breathe more freely: Durling 1993: 79.

See also: Aphrodās, Kharixenēs, Lusias, Marinos (Med.), Moskhiōn (Pharm.).

asēmos in alchemy, a silver or silver-like alloy, often an alloy of copper and silver.

Pseudo-Dēmokritos (Alch.), P. Holmiensis, P. Oxy. 3.467, Phimenas.

Asklēpiadeans medical school founded by Asklēpiadēs of Bithunia, and lasting until *ca* 100 CE, adhering to their founder's theory of disease.

Entries on Asklēpiadeans (most by Scarborough): Antiokhos Paccius, Areios of Tarsos, Asklēpiadēs of Bithunia, Aufidius, Damonikos, Diodotos (Pharm.), Erōtianos, Khrusippos, Mētrodōros (Pharm.), Moskhiōn (Pharm.), Nikēratos, Philōnidēs of Catina, Philōnidēs of Durrakhion, Samithra.

atomism the theory promulgated first by Leukippos and Dēmokritos, then revised by Epicurus (see **Epicurean**), that the ***kosmos*** is composed ultimately of atoms ("indivisibles") and void; *HWPhil* 1 (1971) 606–611, A.G.M. van Melsen. For entries on atomists, see **Epicureans**.

See also: Aristotelian Corpus On Indivisible Lines, Dēmokritos, Diogenēs Laërtios, Ekphantos, Empedoklēs, Epicurus, Galēn, Hērakleidēs of Hērakleia Pontikē Junior, Hierōnumos of Rhodes, Karneadēs, Leukippos, Mētrodōros of Khios, Nausiphanēs, Themisōn.

***bdellion* (Lat.: *bdellium*)** aromatic gum of the mukul myrrh tree (*Commiphora mukul* Engl.): Dioskouridēs, *MM* 1.67; Galēn, *Simples* 6.2.6 (11.849–850 K.); Durling 1993: 90; Langenheim 2003: 371–372.

See also: Amuthaōn, Emboularkhos, Iulius Agrippa, Iustus (Pharm.), Kallinikos (Pharm.), Klutos, Kosmos, Leukios, Lusias, Mēnās, Nikolaos (Pharm.), Parisinus medicus, Pamphilos of Bērutos, Poluarkhos, Sarkeuthitēs.

calamine (Grk. *kadmeia*) a zinc carbonate, oxide, or silicate; *cf.* Dioskouridēs, *MM* 5.74; Galēn, *Simples* 9.3.11 (12.219–221 K.).

See also: Ammōnios of Alexandria, Anthaios, Apollinarios (Pharm.), Areios of Tarsos, Asterios, Attalos III, Bithus of Durrakhion, Candidus, Diagoras, Euelpidēs, Euphranōr (Pharm.), Harpokratiōn, Hermeias (Ophthalm.), Hermophilos, Iōannēs Archpriest, Kuros, Leukios, Lunkeus, Manethōn (Pharm.), Naukratitēs med., Neilammōn, Nikētēs, Nonnos, Philōtas, Proëkhios, Prōtās, Ptolemaios (Pharm.), Puramos, Sergius of Babylōn, Sōsandros (Pharm.), Thamuros, Theophilos (Pharm.), Zōilos of Macedon, Zōsimos (Med.).

collyrium stamps steatite stamps (letters are in mirror-image) used to "sign" blocks of dried paste, which in use would be grated into water to make an eye-salve: Grotefend (1867); Espérandieu (1906); Voinot (1981–1982); Marganne (1997); Voinot (1999).

See also: Atimētos, Dionusodōros, Seuerus Ophthalm.

daimōn, -ones initially God or the divine, but later being(s) intermediate between the gods and humans (Hēsiod *Erga* 122–126, 314; Plato *Symp.* 202e–203a, *Polit.* 271d–e, *Tim.* 41a–d), usually imagined as non-corporeal, but sometimes said to be made of ***aithēr*** (Epinomis 984d–985b); Diogenēs Laërtios 7.151 (Chrysippus and Zēnōn); *HWPhil* 2 (1972) 1–4, F.P. Hager; *OCD3* 426, H.S. Versnel; *BNP* 4 (2004) 275–279, P. Habermehl.

See also: Apuleius of Madaurus, Neilos, Saloustios, Xenokratēs of Khalkēdōn.

decan in Pharaonic Egyptian astronomy, one of a series of 36 constellations used for time-reckoning (first known from Egyptian coffin lids of *ca* 2100–1800 BCE), parallel to and south of the ecliptic, rising once every ten days of the civil calendar, providing a system of marking consecutive hours of night by their successive risings. After transmission of the zodiac to Egypt in the Ptolemaic period, the decans were adjusted to the new system, whereby they came to represent one-third of a zodiacal sign.

See also: Hermēs Trismegistos, Pahlavi (Translations), Salmeskhoiniaka, Sōkratēs (Lith.), Teukros of Babylōn.

dēmiourgos "Demiurge," originally "craftsman in public service" (*OCD3* 451, F.W. Walbank and P.J. Rhodes), then the divine organizer of the ***kosmos***, the chief god of Plato's pantheon, who models the world to eternal patterns, the Forms: *Tim* 28–30; *HWPhil* 2 (1972) 49–50, W. Ullmann; *BNP* 4 (2004) 261–263, M. Baltes.

See also (most Neo-**Platonists** employ the term; here we list only entries that mention it explicitly): Aristotle, Cassius Longinus, Dēmokritos (Neo-Plat.), Harpokratiōn of Argos, Iamblikhos of Khalkis, Iōannēs Archpriest, Leontios, Plōtinos, Theol. Arithm., Tiberianus, Timaios of Lokris.

diaphorētikē perspirant compound, intended to extract the cold wet **humor** (phlegm): Durling 1993: 128–130.

See also: Candidus, Ktēsiphōn, Mnaseas, Pollēs (Med.), Sertorius Clemens, Silo.

dioptra optical instrument for sighting, surveying, and star-gazing, consisting in essence of a tube or other framework, used as a sighting guide, and possibly devices to indicate azimuth and level: Campbell (2000); *BNP* 4 (2004) 513–514, E. Olshausen and V. Sauer.

See also: Bitōn, Dēmokleitos, Eratosthenēs, Hērōn of Alexandria, Kleoxenēs, P. Oslo.73.

diphruges metallic-yellowish iron ore, either chalcopyrites or pyrites: Dioskouridēs, *MM* 5.103, 125; Durling 1993: 132–133.

See also: Apollōnios of Tarsos, Khalkideus.

Dogmatists see **Rationalists**.

dropsy (Grk.: *hudrōps*) a disease characterized by edema, of varying etiology (i.e., in modern medicine would be a symptom, not a diagnosis); *cf.* Hippokratic Corpus *Affections* 22 (6.232–234 Littré), *Internal Affections* 22–26 (7.220–236 Littré). Varieties included *anasarka*, Galēn *Loc. Aff.* 5.7 (8.353 K.) and *askites*, Galēn, *Caus. Sympt.* 3.3 (7.224 K.).

See also: Akhillās, Antiokhis, Apollōnios of Memphis, Brenitus, Cornelius (Pharm.), Celsus, Dēmētrios of Apameia, Erasistratos of Ioulis, Eugamios, pseudo-Galēn Definitiones Medicinales, Iamblikhos of Constantinople, Khrusermos, Khrusippos of Knidos (I, II), Kleophantos (Pharm.), Marcellus of Bordeaux, Mētrodōra, Ōriōn, Philagrios, Proklos (Method.), Ptolemaios (Erasi.), Sergius of Resʿaina, Sōkratēs (Med.), Tharseas, Thessalos of Tralleis.

duspnoia "difficulty in breathing" or "shortness of breath" – more a symptom than a diagnosis: Hippokratic Corpus Aphorisms 3.31; Pliny 23.48, 92; Galēn *Sympt. Caus.* 1.7 (7.137 K.); pseudo-Galēn, Def. Med. 262 (19.420 K.); but associated in Stephanos of Athens, *In Hipp. Aphor.* (*CMG* 11.1.3.2, pp. 196–197), especially with the elderly.

See also: Aelius Promotus, Alkimiōn, Aristarkhos of Tarsos, Kharixenēs, Podanitēs.

ecliptic circle the great circle inclined relative to the axis of the daily rotation and along which the Sun moves; the Moon and planets move near it. See Hipparkhos, Oinopidēs, and Ptolemy; *cf. BNP* 4 (2004) 792–794, W. Hübner.

See also: Autolukos, Babylonian Astronomy, Eratosthenēs, Eudoxos of Knidos, Geminos, Hupsiklēs, Naburianos, Terentius Varro, Teukros of Babylōn, Theodosios of Bithunia.

elephantiasis* or *elephas usually translated "leprosy," but the terms encompass at least two afflictions with often-loathsome symptoms. Aretaios 4.13 (*CMG* 2, pp. 85–90), describes what is most likely leprosy (also, he says, called the "lion-disease," and other names): the patient presents gradually increasing signs including: foul breath, muddy urine, dryish and cracked tumorous swellings, rough and thick like elephant-hide, enlarged veins, increasing hair-loss over the whole body, tongue covered with pellets resembling hailstones, fingers and toes encrusted with ***leikhenes***, pruritic ulcers under the ears, nose with rough black pustules, finally body-wide pruritic ulcers and nodules that forecast the decay of fingers, toes, then penis and testicles (if male), nose, feet and hands, which detach, leaving open, malodorous and large ulcers, especially on the legs. Still the patient does not die: the disease "has a very long life, like the animal, the elephant" (4.13.17 [p. 89]); *cf.* Aëtios of Amida 13.120 (pp. 730–731 Cornarius), and Celsus 3.25. Currently named *Hansen's Disease*, leprosy is a chronic infection by *Mycobacterium leprae*, with symptoms fairly similar to Aretaios' description. The ancients (incl. Aretaios) sometimes included *satyriasis* and *boubones* among the signs, suggesting *Bancroft's filariasis* (often termed "elephantiasis"; a mosquito-borne lymphatic infestation of *Wucheria bancrofti* (Cobbald) Seurat), dramatically displayed in the male by hugely swollen testicles, as well as "elephant-like" and overly swollen legs; Aëtios' cure for this *elephas* was castration (*ibid.*, pp. 732–733). The Hippokratic Corpus, *Regimen in Acute Diseases* (*App.*) 2 (2.398 Littré), on Galēn's reading (4.15 [15.758–761 K.]), describes a similar affection, without naming it *elephas*. See: J. Scarborough, *Medical and Biological Terminologies* 2nd ed. (1998) 50–51. Grmek (1989) 165–171 argues that *leprosy* did not afflict the Greco-Roman world before *ca* 250 BCE; *cf.* also *BNP* 7 (2005) 417–418, V. Nutton.

Athēnodōros (Med.), Celsus, pseudo-Dēmokritos (Med.), Hierōn (Veterin.), P. Mil. Vogl. 1.15, Philōn of Huampolis, Stratōn (Erasi.).

Empiricists sect of medical practice founded *ca* 250 BCE by Philinos of Kōs; they rejected all theorizing about the "hidden conditions" of the body, and asserted doctors could obtain all needed knowledge through careful observation, recording similar sets of symptoms and antecedent circumstances, and subsequent empirical determination of effective and ineffective treatments for each syndrome. See esp. Celsus, Epicurus,

Galēn, Hēgētōr, Hērakleidēs of Eruthrai, Hērakleidēs of Taras, Pelops, and Quintus. The school ended *ca* 200 CE (*cf.* Marcellus of Bordeaux); *BNP* 4 (2004) 953–954, V. Nutton.

Entries on Empiricists (most by Stok): Aiskhriōn of Pergamon, Akrōn, Apollōnios of Antioch and Son, Apollōnios of Kition, Arkhibios, Cassius (Med.), Celsus, Dionusios of Aigai, Epikouros of Pergamon, Glaukias, Hērakleidēs of Taras, Kalliklēs, Lukos of Macedon, Lukos of Neapolis, Mēnodotos of Nikomēdeia, Philinos of Kōs, Philippos (of Pergamon?), Ptolemaios of Kurēnē, Serapiōn of Alexandria, Sextus of Apollōnia, Theodās, Theodosios, Zeuxis, Zōpuros of Alexandria.

See also: Agathinos, Artemidōros of Parion, pseudo-Galēn Introductio, Hērodotos (of Tarsos?), Hikesios of Smurna, Leōnidas of Alexandria, Themisōn, Zeuxis (Hēroph.).

ephemeris (pl. ephemerides) astronomical table(s) listing calculated positions of the Moon and planets in a calendrical framework.

Babylonian Astronomy, Clodius Tuscus, Iōannēs "Lydus," Kidēnas, Krinas, Naburianos, Terentius Varro of Narbo.

Epicurean follower of the teachings of Epicurus and member of his school, the **Garden**; members of the school retained his theory of **atomism**, as well as his insistence on alternate explanations, all equally tenable so long as none divine. The school is hardly attested after *ca* 200 CE. By that era they had long since acquired an undeserved reputation as hedonists; therefore, and because the standard model of late antiquity was compounded from **Peripatetic**, **Platonist**, and **Stoic** elements, Epicurean works were lost in greater proportion.

Entries on Epicureans (most by Englert): Apollodōros of Athens, Artemōn, Basileidēs, Dēmētrios of Lakōnika, Diogenēs of Oinoanda, Diogenēs of Tarsos, Epicurus, Hermarkhos, Idomeneus, Kōlōtēs, Lucretius, Mētrodōros of Lampsakos, Philodēmos, Philōnidēs of Laodikeia, Poluainos, Polustratos, Prōtarkhos of Bargulia, Timagoras, Zēnōn of Sidōn.

See also: Apollōnios of Antioch and Son, Aristoklēs of Messēnē, Galēn, Hērakleitos of Rhodiapolis, Hērōn of Alexandria, Hieroklēs of Alexandria, Karneadēs, Lusimakhos of Kōs, Manilius, P. Berol. 9782, Plutarch, Rabirius, Thessalos of Tralleis, Vergilius, Zēnodōros.

epicycle a uniform circular motion, the center of which is carried uniformly in a circular path around the observer: Apollōnios of Pergē; Hipparkhos; and Ptolemy.

See also: Adrastos of Aphrodisias, Āryabhaṭa, Derkullidēs, On Kosmos, P. Mich. 3.149, Paitāmahasiddhānta.

Erasistrateans medical school founded by Erasistratos, still in existence at the time of Galēn, but apparently moribund thereafter. The members of the school continued to deny the worth of phlebotomy, and to promote Erasistratos' theories of anatomy and physiology.

Entries on Erasistrateans: Antigenēs, Apollōnios of Memphis, Apollophanēs of Seleukeia, Artemidōros of Sidē, Athēniōn, Erasistratos, Hērakleidēs of Ephesos, Hermogenēs of Smurna, Hikesios of Smurna, Kharidēmos (?), Khrusippos of Knidos (II), Kleophantos of Keōs (?), Martialius/Martianus, Mēdeios, Mēnodōros (?), Miltiadēs (?), Ptolemaios (Erasi.), Stratōn (Erasi.).

erusipelas "red-skin", a condition involving rough red skin, often described or prescribed for, and perhaps sometimes the same as the modern erysipelas, an acute streptococcal skin infection, with inflammation, and usually involving subcutaneous fat. Galēn *Diff. Dis.* 5.2 (6.849 K.); Galēn *MM* 14.1 (10.946–949 K.); pseudo-Galēn, Def. Med. 383 (19.441 K.); Johnston 2006: 54–55.
See also: Aelius Promotus, Agathoklēs (Med.), pseudo-Dēmokritos (Agric.), Diophantos of Lukia.

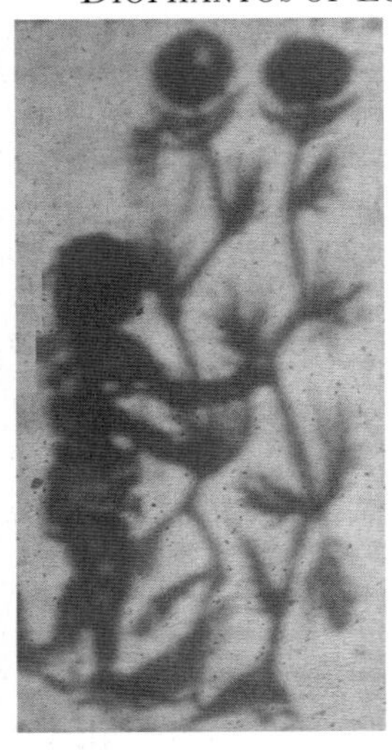

Euphorbia, Mount Athos Ω 75, f.48R

euphorbia plant discovered by Iouba and named for Euphorbos, the Mediterranean bush, *Euphorbia resinifera* Berg., whose resin was commonly used in medicine: Dioskouridēs, *MM* 3.82.1; Pliny 5.16, 25.77–79; Galēn, *Simples* 6.5.24 (11.879 K.); *BNP* 5 (2004) 181, C. Hünemörder.
See also: Abaskantos, Aineios, Antonius Musa, Aquila Secundilla, Diogenēs (Pharm.), Domitius Nigrinus, Eutonios, Hikesios of Smurna, Iulius Agrippa, Iulius Secundus, Lingōn, Logadios, Marcianus (of Africa?), Pamphilos of Alexandria, Philōn of Tarsos, Parisinus medicus, Sōkratēs (Med.), Timokleanos, Zōsimos (Med.).

galbanum (Grk. *khalbanē*) Indian Kasnib-resin, probably *Ferula galbaniflua* Boiss. and Buhse: Theophrastos, *HP* 9.7.2, 9.9.2; Dioskouridēs, *MM* 3.83; Pliny 12.107; Galēn, *Simples* 8.22.1 (12.153 K.); Usher 1974: 253; André 1985: 116; Durling 1993: 335; Evans 1996: 503; Langenheim 2003: 415.
See also: Alkimiōn, Amuthaōn, Andreas of Karustos, Antonius "root-cutter," Aphthonios, Apuleius Celsus, Aristoklēs, Aristokratēs, Arrabaios, Damonikos, Epigonos, Euangeus, Eugeneia, Euthukleos, Halieus, Hekataios (Pharm.), Hermōn, Hierax, Iulianus (of Alexandria?), Khalkideus, Kleophantos (Pharm.), Klutos, Leukios, Lunkeus, Lusias, Makhairiōn, Marcianus of Africa, Menoitas, Minucianus, Moskhos, Nikētēs, Nikostratos (Pharm.), Philoklēs, Philokratēs, Proëkhios, Sertorius Clemens.

Garden Epicurus' school in Athens: see **Epicurean**.

genethlialogy the composition and interpretation of birth horoscopes; contrast ***katarkhic* astrology**.
Antiokhos of Athens, Babylonian Astronomy, Sphujidhvaja, Varāhamihira, Wuzurgmihr.

glanders a highly contagious and often fatal bacterial infection of the lungs and upper respiratory tract of horses and other equids (capable of infecting humans and other animals as well).
Aemilius Hispanus, Agathotukhos, Caystrius, Litorius, Nephōn, Pelagonius.

glaukion latex or "juice" of a species of horned poppy (*Glaucium flavum* Crantz), often used as a purgative, light sedative, or adulterant to the latex of the opium poppy: Dioskouridēs, *MM* 3.86, 4.65.5; Galēn, *Simples* 6.3.5 (11.857 K.); Usher 1974: 275; André 1985: 111; Durling 1993: 103; Scarborough (1995).

See also: Hērōn (Med.), Poludeukēs, Sergius of Babylōn.

glaukōma eye-disease characterized by a bluish-grey opacity in the crystalline lens, distantly similar to the modern "glaucoma" (which designates an eye-disease characterized by intraocular pressure, with hollowing out and atrophy of the optic nerve, producing defects in the field of vision): Rufus of Ephesos, in Paulos of Aigina 3.22.30 (*CMG* 9.1, pp. 184–185); J. Hirschberg, *Wörterbuch der Augenheilkunde* (1887) 34–37.

See also: P. Ashmolean Library, P. Ross.Georg. 1.20.

grammatikos lit. "lettered," but as used first by **Stoics** from the 3rd c. BCE, teacher of language (including grammar) and literature, in general the rudiments of literate studies; R.A. Kaster, *Guardians of Language: The Grammarian and Society in Late Antiquity* (1988).

Akhilleus, Alexander of Pleuron, Anthēmios, Iōannēs Philoponos, Mētrodōros of Tralleis, Proklos of Lukia, Sēmos, Syrianus.

hedrikē soothing ointment for the "seat" (buttocks, anus, and perineum): Galēn, *CMLoc* 9.6 (13.306–312 K.); Durling 1993: 140–141.

See also: Ambrosios Rusticus, Kleophantos (Pharm.), Nikomēdēs IV, Nikostratos, Ōros, Samithra, Turannos, Xanitēs.

hēgemōn or ***hēgemonikon*** commanding faculty of a conscious being, especially a human – its location was much debated, the brain and heart being the two leading candidates; *cf.* Hippokratic Corpus Sacred Disease; Plato, *Tim.* 41c; Aristotle, *EN* 3.3.17–18 (1113a2–9); Hērophilos; Chrysippus; Ptolemy *Krit. Hegem.*; Alexander of Aphrodisias *De Anima*; Porphurios *ad Gaurum*, etc.; *HWPhil* 3 (1974) 1030–1031, Th. Kobusch.

See also: Antipatros of Tyre, Arkhedēmos, Athēnaios of Attaleia, Diogenēs of Babylōn, Dioklēs of Karustos, Kleanthēs, Marinos (Med.), Praxagoras, Ptolemy.

helepolis giant moveable siege-tower, on whose top level, and sometimes on whose intermediate levels, catapults and/or ballistae were mounted; when the tower reached the walls of the besieged city, ladders or boarding bridges were extended for assault. *BNP* 6 (2005) 67–68, L. Burckhardt.

Athēnaios Mech., Bitōn, Diognētos of Rhodes, Diokleidēs, Epimakhos, Kallias, Poseidōnios of Macedon.

Hērophileans medical school founded by Hērophilos, which persisted through the mid-1st c. CE. The members of the school continued to be interested in anatomy, pulse-lore, and medical doxography. See esp. von Staden (1989) and (1999).

Entries on Hērophileans: Alexander of Laodikeia, Andreas of Karustos, Apollōnios of Alexandria, Aristoxenos (Hēroph.), Bakkheios of Tanagra, Dēmētrios of Apameia, Dēmosthenēs Philalēthēs, Dioskouridēs Phakas, Gaius (Hēroph.), Hēgētōr (Hēroph.), Hērakleidēs of Eruthrai, Hērophilos, Kallianax, Kallimakhos of Bithunia, Khrusermos, Kudias (Hēroph.), Mantias (Hēroph.), Speusippos of Alexandria, Zēnōn (Hēroph.), Zēnōn of Laodikeia (?), Zeuxis (Hēroph.).

See also: Apollōnios of Antioch and Son, Apollōnios of Kition, Cassius Iatros., Hērakleidēs of Taras, Hermogenēs of Smurna, Kleophantos of Keōs, Londiniensis medicus, Ptolemaios of Kurēnē, Serapiōn of Alexandria.

herpullos the creeping thyme (*Thymus serpyllum* L.), a decoction of whose flowers and leaves was used as a heating, bitter tonic, and in the relief of severe coughs:

Dioskouridēs, *MM* 3.38, 41; Galēn, *Simpl.* 6.5.20 (11.877 K.); Usher 1974: 576; André 1985: 122.

See also: Epaphroditos of Carthage, Zēnōn of Laodikeia.

hiera a drastic cathartic, emetic, or purgative, combining harsh simples with ingredients characterized by mild, sweetish, or aromatic properties: Iustus in Oreibasios, *Coll.* 8.47.8 (*CMG* 6.1.1, p. 306) and Aëtios of Amida 3.114–118 (*CMG* 8.1, pp. 303–306).

See also: Antiokhos Paccius, Iustus (Pharm.), Logadios.

homoiomerous (parts) those parts of a body composed of a macroscopically homogenous substance, thus blood, bone, tendon, etc. See Aristotle, *De Caelo* 3.4 (302b10–303a3), *Meteor.* 4.10 (388a10–389a24), 4.12 (389b24–390b21), *GC* 1.1 (314a16–314b1), *PA* 2.1–2 (646a8–648a36), 2.4–9 (650b13–655b27); Galēn, *Elementis* 1.2 (1.425–426 K.); *HWPhil* 3 (1974) 1179–1180, A. Lumpe.

See also: Anaxagoras, Athēnaios of Attaleia, Marinos (Med.).

***hudrophobia* (Lat.: *rabies*)** although prehistoric, and despite references in Homer, *Iliad* 8.299, Xēnophōn, *Anabasis* 5.7.26, and Aristotle, *HA* 7(8).22 (604a5–13), the last denying human transmission, rabies does not enter Greek medical literature until Andreas, who first names it (Caelius Aurelianus, *Acute* 3.98 [*CML* 6.1.1, p. 350]), and Dēmētrios of Apameia (*ibid.* 3.106 [p. 354]), who describes it as chronic. Most securely-datable works on rabies are 1st c. BCE or later: Artemidōros of Sidē (*ibid.* 3.118 [p. 362]), Artorius (*ibid.* 3.104, 113 [pp. 354, 358]), Gaius the Hērophilean (*ibid.* 3.113–114 [p. 360]), Kharidēmos (*ibid.* 3.118–119 [p. 362]), etc.; *cf.* Athēnodōros (Med.). Rabies usually gives symptoms within a few days (fever, headache, malaise), and the "furious" form (encephalitis) occurs in about 80% of cases, the "dumb" form (paralysis) in roughly 20%. True rabies manifests an increasing restlessness, confusion, agitation, weird behaviors, reported hallucinations, inability to sleep, excessive salivation, and – the classic symptom – painful spasms of the laryngeal and pharyngeal muscles, thus any attempt to drink a liquid becomes *hudrophobia.* Death occurs usually in three to ten days after symptoms begin. But most individuals bitten by dogs or other animals would not contract rabies.

See also: Anthaios Sextilius, Aphrodās, Apollōnios Claudius, Apuleius Celsus, Bathullos, pseudo-Dēmokritos (Med.), Eudēmos (Method.), Hērās, Kratēs (Med.), Kratippos, Laïs, Lukos of Macedon, Makhairiōn, Menelaos (Pharm.), Menippos, Nemesianus, Nikēratos, Nikostratos (Pharm.), Pelops, Philagrios, Poseidōnios (Med. II), Rufus of Ephesos, Salpē, Theodorus Priscianus, Zēnōn of Laodikeia.

humor (Grk. *khumos*; Lat. *umor*) one of a small number of "elemental" fluids thought to constitute living beings; there were canonically four of these (blood, bile, phlegm, and black bile), but some systems propounded fewer or more. See Hippokratic Corpus (Anatomy and Physiology, On the Sacred Disease, and Nosology); Polubos; Dioklēs of Karustos; Praxagoras; Galēn; *HWPhil* 3 (1974) 1232–1234, W. Preisendanz; *OCD3* 733, J.T. Vallance; *BNP* 6 (2005) 571–572, V. Nutton.

See also: Alexander of Tralleis, Antiphōn, Aristoxenos (Hēroph.), Arkhigenēs, Celsus, Dexippos, Dioskouridēs Phakas, Erasistratos of Ioulis, Erasistratos of Sikuōn, pseudo-Galēn Definitiones Medicinales, pseudo-Galēn Introductio, Harpokratiōn, Hērophilos, Hippokratēs of Kōs, Hippokratic Corpus (Airs Waters Places, Ancient Medicine, Aphoristic

WORKS, EPIDEMICS, GYNECOLOGY, REGIMEN), KTĒSIPHŌN, ISIDORE OF HISPALIS (SEVILLE), MARCELLUS OF BORDEAUX, MNĒSITHEOS OF ALEXANDRIA, PARISINUS MEDICUS, PHASITAS, PLEISTONIKOS, POSEIDŌNIOS (MED. II), DE QUATERNIONIBUS, SABINUS (MED.), STRATŌN (ERASI.), STRATONIKOS OF PERGAMON, THRASUMAKHOS, VINDICIANUS.

***hupostasis* (Lat.: *substantia*)** the substantial nature, underlying reality, or actual existence of a thing in the world. The earlier medical meaning was approximately "precipitate" (that which settles to the bottom, as in urine): HIPPOKRATIC CORPUS APHORISMS 4.79 (4.530 Littré); ARISTOTLE, *PA* 2 (647b28), *Meteor.* 2 (358a8); THEOPHRASTOS *CP* 6.7.1–4; and POLUBIOS Book 34, *fr.*9.10–11; but later came to mean "what persists," "what is real." Neo-**Platonists** used the term to refer to substantial cause and to levels of reality: F. Romano and D.P. Taormina, edd., *Hyparxis e hypostasis nel neoplatonismo* (1994). In Christian usage, the term came to mean one "person" of the Trinity: each of the three divine persons is entirely God, not part of God. See H. Dörrie, *Hupostasis: Wort- und Bedeutungsgeschichte = Nachrichten der Akademie der Wissenschaften in Göttingen* (1955) #3, pp. 35–92, and *HWPhil* 3 (1974) 1255–1259, B. Studer; *BNP* 6 (2005) 644–645, S. Meyer-Schwelling.

See also (most Neo-**Platonists** employ the term; here we list only entries that mention it explicitly): AMELIUS GENTILIANUS, DĒMOKRITOS (NEO-PLAT.), MARIUS VICTORINUS, NOUMĒNIOS OF APAMEIA, PLŌTINOS, PORPHURIOS OF TYRE.

iatromathematics attempts to predict the outcomes of diseases based on numerology or astrology, usually based on astrological numbers derived from facts about the patient and the disease. Although the HIPPOKRATIC CORPUS, *Diseases* 4, discusses the diagnostic meaning of dreams about planets, iatromathematics *per se* begins *ca* 200 BCE. *BNP* 6 (2005) 691–692, A. Touwaide.

AMMŌN (ASTROL.), HERMĒS TRISMEGISTOS, IMBRASIOS, KRINAS, PANKHARIOS, PRŌTAGORAS OF NIKAIA.

***ikhthuokolla* ("fish-glue"; isinglass)** gelatin usually derived from the swim-bladders of freshwater fish, esp. sturgeons; it dried clear, hard, and airtight. Hellenistic and Roman surgeons valued high-quality fish-glue for sealing hard-to-close wounds, and joining fractures (*cf.* DIOSKOURIDĒS, *MM* 3.88; PLINY 32.73); a use that continued well into the 20th c.: Wood and LaWall (1926) 1338–1339. It is still used as an adhesive, in the clarification of some beers and wines, and in repairing books: T. Petukhova, "Potential application of Isinglass Adhesive for Paper Conservation," *Book and Paper Group Annual* 8 (1989) 58–61. Pliny 32.84–85 describes its use as a wrinkle-remover.

See also: DIONUSIOS OF SAMOS, EUANGEUS, PHILŌTAS.

Indian buckthorn (*lukion Indikon*) astringent liquid, used to treat diarrhea and sore throats; produced by digesting the wood of *Acacia catechu* Willd. in hot water: SCRIBONIUS LARGUS 19 and 142; DIOSKOURIDĒS, *MM* 1.100.4 ("It is reported that Indian *lukion* is made from a bush called *lonkhitis*."); PLINY 12.31, 24.125–126; GALĒN, *Simpl.* 7.11.20 (12.63–64 K.); Usher 1974: 12; Stuart 1979: 142; André 1985: 149 (#2); it was first imported from India after the time of Hippalos: Casson 1989: 192–193; Evans 1996: 230. Known as "Cutch," "Black Catechu," or "Pegu Catechu," it was used for tanning (thus the original "khaki"), and has high concentrations of polyphenols and tannins, explaining its astringency. *Cf.* below ***lukion***.

See also: AMMŌNIOS OF ALEXANDRIA, CORNELIUS (PHARM.), GLAUKŌN (MED.), HARPOKRATIŌN, HERMOLAOS (PHARM.), NIKOSTRATOS (PHARM.), POSEIDŌNIOS (MED. II).

katarkhic* astrology** offered judgment about auspicious and inauspicious times for various activities, and about the outcomes of actions taken at given times; contrast ***genethlialogy.

Ammōn (Astrol.), Dōrotheos of Sidōn, Iulianus of Laodikeia, Maximus, Serapiōn of Alexandria, Sphujidhvaja.

***khalkanthes*/*-on* (Lat. *flos aeris*)** "flower of copper," probably usually the bluish copper-sulfate weathering product of many copper-ores, or a solution thereof, often used in pharmacy and alchemy; *cf.* Dioskouridēs, *MM* 5.98; Galēn, *Simples* 9.3.34 (12.238–241 K.). The Latin term varied, Pliny 34.107 using *aeris flos*, and others *atramentum sutorium*, *cf.* Celsus 5.1.

See also: Akhillās, Aphrodās, Apiōn, Apollōnios of Tarsos, Artemidōros of Pergē, Atimētos, Axios, Bithus, Diomēdēs (Pharm.), Hierax, Kleoboulos, Pasiōn, Petasios, Polueidēs, Proëkhios, Terentius Valens, Threptos, Timaios (Pharm.).

khalkitis copper ore from Cyprus, likely a copper sulfide: Dioskouridēs, *MM* 5.99; Galēn, *Simples* 9.3.21 (12.226–229 K.) found above ***sōru*** and below ***misu***, and 35 (12.241–242 K.); Antullos in Paulos of Aigina, 7.24.11 (*CMG* 9.2, p. 398); Durling 1993: 336.

See also: Amphiōn, Aphrodās, Apollōnios Tarsos, Bithus, Euphranōr (Pharm.), Hermās thēriakos, Hermogenēs of Smurna, Iulianus (of Alexandria?), Leukios, Prōtās, Ptolemaios (Pharm.), Samithra, Sōkratiōn, Sōsandros (Pharm.), Turpillianus.

kosmos what is ordered (from Grk. *kosmein* to set in order), probably at first the ***polis*** or state, *cf.* Hērodotos 1.65.4 (plus 2.52), Thucydidēs 8.48.4; then the whole universe, considered as a system organized or ordered, by God or Nature. Compare Anaximander; Hērakleitos; Empedoklēs; Dēmokritos; Diogenēs of Apollōnia; Plato; Aristotle; Zēnōn of Kition; Epicurus; *HWPhil* 4 (1976) 1167–1176, R. Ebert.

See also: Anania, Anthēmios, Antipatros of Tyre, Apollodōros of Seleukeia, Apollophanēs of Nisibis, Apuleius of Madaurus, Arkhedēmos, Artemidōros of Parion, Asklēpiadēs of Murleia, Athēnaios of Attaleia, Atticus, Bardaisan, Basil of Caesarea, Boëthos of Sidōn, Chrysippus, Diodōros (Astron.), Diodōros of Tarsos, Derveni papyrus, Epidikos, Geōrgios of Pisidia, Gregory of Nussa, Harpokratiōn of Argos, Hellenizing School, Hēraiskos, Hippasos, Hippokratic Corpus (Anatomy, Airs Waters Places, Regimen), Kleanthēs, On Kosmos, Krantōr, Leukippos, Mētrodōros of Khios, Oinopidēs, Panaitios of Rhodes, Petrōn of Himera, Pherekudēs, Philolaos, Poseidōnios of Apameia, Ptolemy, Pythagoras, De Quaternionibus, Seleukos of Seleukeia, Stephanos of Alexandria, Severus Sebokht, Stratōn of Lampsakos, M. Tullius Cicero, Vergil, Xenokratēs of Khalkēdōn.

kostos a prominent Indian aromatic plant, commonly imported to the Mediterranean; probably *Saussurea lappa* Clarke: Theophrastos *HP* 9.7.3; Dioskouridēs, *MM* 1.16; Pliny 12.41; Galēn, *Simples* 7.10.45 (12.40–41 K.); Miller 1969: 84–87; André 1985: 76; Durling 1993: 210; *OCD3* 405, D.T. Potts.

See also: Anthimus, Aphrodās, Apuleius Celsus, Aristoklēs, Blastos, Bouphantos, Dioskoros (Pharm.), Dōsitheos (Pharm.), Eugeneia, Euskhēmos,

HERMŌN, IUSTINUS (PHARM.), KLEOPHANTOS (PHARM.), KOSMOS, KURILLOS, MAGNUS *ARKHIATROS*, MARCIANUS OF AFRICA, MELĒTOS, MITHRADATĒS VI, NEARKHOS, NIKOSTRATOS (PHARM.), PAMPHILOS OF ALEXANDRIA, PAMPHILOS OF BĒRUTOS, PHILIPPOS OF MACEDON, ROMULA, THEOPOMPOS, ZENOPHILOS.

krisis the decision-point in the course of a disease, usually accompanied by diagnostic signs on the basis of which doctors attempted to predict the final resolution (health or death); see HIPPOKRATIC CORPUS APHORISMS; DIOKLĒS OF KARUSTOS.

leikhēn skin-disease, named after lichen, and characterized by peeling and scaliness (perhaps similar to modern psoriasis): Durling 1993: 219.

IUNIUS CRISPUS, PAMPHILOS OF ALEXANDRIA, SŌKRATIŌN, ZEUXIS (HĒROPH.).

leukōma eye-disease, characterized by dense opacity of the cornea, or, more generally "a white spot on the eye," DIOSKOURIDĒS, *MM* 3.84.3 (***ammōniakon*** clears away *leukōmata*); PSEUDO-GALĒN, *Introd.* 16 (14.775 K.); AËTIOS OF AMIDA 7.38 (*CMG* 8.2, p. 290): "All scars on the iris of the eye appear white because the cornea is thicker [on account of the scar] and the blacker color from inside is unable to shine through it; those [scars] that are [like] papules are almost all white, those that are smooth [or level] are less white. . ."; J. Hirschberg, *Wörterbuch der Augenheilkunde* (1887) 51, 62–65. (Modern ophthalmology uses "leukoma" only as a general term to describe a white density of the cornea, and occasionally one sees in the professional literature an "adherent leukoma," i.e. a healed scarring of the cornea to which a portion of the iris is attached.)

See also: HĒLIADĒS, PETROS.

litharge (Grk.: *litharguros* "stone of silver") the residue from the cupellation of silver out of galena (lead sulfide); it is primarily lead monoxide, which also occurs as a mineral: DIOSKOURIDĒS, *MM* 5.87; GALĒN, *Simples* 9.3.17 (12.224–225 K.).

See also: AELIUS PROMOTUS, AMBROSIOS RUSTICUS, AMPHIŌN, ANDREAS OF KARUSTOS, ANTONIUS "ROOT-CUTTER," APHRODĀS, APOLLODŌROS *THĒRIAKOS*, APOLLŌNIOS CLAUDIUS, APOLLŌNIOS OF TARSOS, ATTALOS III, BATHULLOS, CASTUS, DAMONIKOS, DIOGENĒS (PHARM.), DIOPHANTOS OF LUKIA, EUBOULOS (PHARM.), HALIEUS, HIERAX, HIKESIOS OF SMURNA, IULIANUS (OF ALEXANDRIA?), IUNIA/IOUNIAS, KHALKIDEUS, KLEŌN OF KUZIKOS, KLEOPHANTOS (PHARM.), KUROS, LEUKIOS, MARCELLUS (PHARM.), MEGĒS, MENOITAS, MNASEAS, MOSKHIŌN (PHARM.), PASIŌN, PHILOKRATĒS, SAMITHRA, TELEPHANĒS, TELAMŌN, THEOTROPOS, TURPILLIANUS.

lukion "common" or "dyer's buckthorn," one of a number of species in the shrub genus *Rhamnus* (of over 100 species world-wide): e.g. *R. catharticus* L., *R. frangula* L., *R. infectorius* L., *R. lycoïdes* Boiss., *R. petiolaris* Boiss., *R. punctata* Boiss. SCRIBONIUS LARGUS 142; DIOSKOURIDĒS, *MM* 1.100.1–4 (the most complete description of medical uses, especially as an astringent for discharges, but also as a yellow dye: 1.100.3); PLINY 12.30–32 and 24.125–127; GALĒN *Simples* 7.11.20 (12.63–64 K.); Usher 1974: 501–502; Stuart 1979: 251; L. Boulos, *Medicinal Plants of North Africa* (1983) 151; André 1985: 149 (#1); Casson 1989: 192–193; Durling 1993: 226. (*R. catharticus* yields a purgative syrup from the berries, a yellow dye from the bark, and a hard yellow wood; the bark of *R. frangula* produces a laxative, primarily the glycoside *frangulin*, and a hard wood; and *R. infectoris* has berries that were an important source of a yellow dye.) Compare above **Indian buckthorn**.

See also: DŌSITHEOS (PHARM.), EUELPIDĒS, LAODIKOS.

Lyceum (Grk. Lukeion) alternate name for the **Peripatos**.

Malabathron, Mount Athos Ω 75, f.6R

malabathron or ***malobat(h)rum*** identified as either *Cinnamomum tamala* Nees., or *C. iners* Blume, or *C. zeylanicum* Blume, or other related species, *malabathron* is either the leaf of the "true" cinnamon (generally labeled as simply *C. verum*), from which a distilled oil is extracted, or else the Chinese patchouli-plant, *Pogostemon patchouli* Lab. [or Pellet.], which also yields a steam-distilled oil. STRABŌN 15.1.57; POMPONIUS MELA 1.2.11; DIOSKOURIDĒS, *MM* 1.12.1–2 (*malabathron* is said to be an "Indian spikenard," which he says is wrong, since one is easily misled by the "similarity of its scent to that of Cretan spikenard"); PLINY 12.129; PTOLEMY *Geography* 7.2.15–16; Miller 1969: 74–77 (inclining against a Chinese or Philippine source); Usher 1974: 473 (patchouli, ". . .often associated with fabrics from the East, where the plant is used as an insect repellant in cloth"); M.G. Raschke, "New Studies in Roman Commerce with the East," *ANRW* 2.9.2 (1978) 604–1361 (see "The Spice Trade," pp. 650–655 with nn. 1002–1127: extreme skepticism about all "Far Eastern spices," except for pepper, in the classical world); J.W. Purseglove *et al., Spices* 2 vv. (1981) 1.100–173 ("Cinnamon and Cassia", esp. 126–129: "Cinnamon leaf oil (from *C. verum*)"); André 1985: 151–152 (favoring patchouli); Casson 1989: 241 (reasonably suggesting that it was imported into the Mediterranean world by the Augustan era, probably from Sri Lanka and southern India, perhaps from points further north and east); Durling 1993: 283.

See also: AGAPĒTÓS, ANASTASIOS, AQUILA SECUNDILLA, AURELIUS (PHARM.), BLASTOS, DŌSITHEOS (PHARM.), EMBOULARKHOS, FLAUIUS "THE BOXER," IULIUS AGRIPPA, KOSMOS, KURILLOS, MAGNUS *ARKHIATROS*, MARCELLUS (PHARM.), NIKOSTRATOS (PHARM.), PAMPHILOS OF BĒRUTOS, POLUARKHOS, PTOLEMAIOS (PHARM.), ZĒNOPHILOS, ZŌILOS OF MACEDON.

mastic(h) (Grk.: *mastikhē*) brittle, pale-yellow, clear gum of some *Pistacia* sp., often chewed, *cf.* ***terebinth***. DIOSKOURIDĒS, *MM* 1.70; GALĒN, *Simples* 7.12.6 (12.68–69 K.); Langenheim 2003: 385–390; *BNP* 8 (2006) 451, R. Hurschmann (who identifies as *Pistacia lentiscus* L.). (Mastic is still produced on Khios.)

See also: APOLLOPHANĒS OF SELEUKEIA, APULEIUS CELSUS, AURELIUS (PHARM.), BLASTOS, DŌSITHEOS (PHARM.), KLEOPHANTOS (PHARM.), LAMPŌN, PTOLEMAIOS (MED.), THAÏS, THEOPOMPOS, TIMOKLEANOS.

melothesia system of correspondences between various parts of the human body and the 12 zodiacal constellations in connection with the planets, which were held to govern the specific body parts.

DĒMĒTRIOS (ASTROL.), HERMĒS TRISMEGISTOS, HIPPOKRATIC CORPUS (PAHLAVI), MELAMPOUS, P. MICH. 3.149, PAHLAVI (TRANSLATIONS), DE PLANETIS.

Methodists sect of medical practice (so labeled from their "method" of grouping diseases into either "chronic" or "acute" classes) founded either by THEMISŌN or by THESSALOS OF TRALLEIS; they sometimes claimed an intellectual heritage stemming from Asklēpiadēs of Bithunia (some practitioners do show a tendency to **atomism**). They taught that a diseased body exhibited one or the other of two "states contrary to nature," morbidities determined from direct observation, and termed "communities": one was "tightness" or "constriction," and the other was "looseness" or "flowing" (older translation, "flux"), with a third intermediate state occasionally allowed, the "mixed state" (sc. of the "loose" and the "tight"). These states were said to directly indicate the appropriate therapies. Many Methodists indulged in anatomy and dissection of animals, surgery, and precision in pharmacology (e.g. Sōranos of Ephesos, and the lengthy catalogue of physicians retailed by Caelius Aurelianus), so that any practical limitations by Methodist doctrines were, in practice, quite flaccid. Edelstein (1935/1967); Frede (1982); Pigeaud (1991); Tecusan (2004) 7–21; *BNP* 8 (2006) 801–802, A. Touwaide.

Entries on Methodists: Antipatros (Method.), Apollōnidēs of Cyprus, Auidianus, Caelius Aurelianus, Dionusios (Method.), Eudēmos, Iulianus (of Alexandria?), Megēs, Menemakhos, Mnaseas, Modius Asiaticus, Olumpiakos, Philōn (Method.), Proklos (Method.), Rhēginos, Sōranos of Ephesos, Themisōn, Thessalos of Tralleis.

See also: Agathinos, Antiokhos VIII, Antipatros (Pharm.), Aristoklēs, Cassius iatros., Celsus, Galēn, pseudo-Galēn Introductio, Hērodotos (of Tarsos?), Iulius Secundus, Leōnidas of Alexandria, Lusias, Magnus of Ephesos, Mulomedicina Chironis, Parisinus medicus, P. Turner. 14, Philōn of Huampolis, Philoumenos.

misu copper ore from Cyprus, probably the copper-sulfide ore chalcopyrite, found at the highest levels, above ***khalkitis***: Dioskouridēs, *MM* 5.100; Galēn, *Simples* 9.3.21 (12.226–229 K.).

See also: Akhillās, Andrōn (Pharm.), Aphrodās, Arkhagathos, Bithus, Eugeneia, Euphranōr (Pharm.), Hērakleidēs of Ephesos, Hierax, Kleōn of Kuzikos (Med.), Leukios, Proëkhios, Ptolemaios (Pharm.), Puramos, Samithra, Sōkratiōn, Sōsandros (Pharm.), Terentius Valens, Theophilos (Pharm.).

neuron (pl. neura) means both nerve and tendon, and from the latter also bow-string: Aristotle, *HA* 3.5 (515a27–515b25); Galēn, *PHP* 1.9.1–2 (*CMG* 5.4.1.2, v.1, pp. 94–95): "There are three structures similar to each other in bodily form but quite different in action and use. One is called 'nerve' (*neuron*), another 'ligament' (*sundesmos*), and the third 'tendon' (*tenon*). A nerve in every case grows from the brain or spinal cord and conveys sensation or motion or both to the parts to which it is attached. A ligament is without sensation; its use is expressed by its name. Finally a tendon is the nerve-like termination of a muscle, the product of ligament and nerve . . . (trans. De Lacy), and "All three classes are white, bloodless and solid, and all are separable into straight fibres except the very hard ligaments . . ." (1.9.8, *ibid.* pp. 96–97); Solmsen 1961; von Staden 1989: 155–161, 247–259.

See also: pseudo-Dēmokritos (Med.), Hērophilos, Philōn of Tarsos, Suennesis.

neusis the construction of a line segment such that its endpoints lie on given straight or curved lines, and such that the line segment (produced if necessary) passes through a given point; *cf.* Apollōnios of Pergē. *BNP* 8 (2006) 690, M. Folkerts.

See also: Hērakleitos (Math.), Hippokratēs of Khios, Nikomēdēs.

oikoumenē the whole known inhabited portion of the Earth's sphere (or sometimes just the known portion of the inhabited world), as distinguished from Ocean and from hypothetical trans-Oceanic landmasses: *HWPhil* 6 (1984) 1174–1177, S. Seigfried (Greco-Roman and Byzantine); *BNP* 10 (2007) 73–75, T. Schmitt.

See also: Aethicus Ister, Agathēmeros, Agathodaimōn of Alexandria, Artemidōros of Ephesos, Dionusios of Alexandria Periēgētēs, Eratosthenēs, Hērodotos, Isidōros of Kharax, Kratēs of Mallos, Lukos of Rhēgion, Marinos of Tyre, Pappos of Alexandria, Pausanias of Damaskos, Polubios, Prōtagoras (Geog.), Strabōn, Summaria Rationis Geographiae, Timaios of Tauromenion, Vergilius.

oktaetēris an eight-year calendar cycle, attempting to coordinate lunar months with a solar year by intercalating a month at intervals (the sequence of intercalations varied depending on the lengths assigned to the months and the year); see Eudoxos of Knidos, Harpalos, Kleostratos, Metōn. *BNP* 2 (2003) 451, J. Rüpke.

See also: Dōsitheos of Pēlousion, Eratosthenēs, Kritōn of Naxos, Menestratos (I, II), Nautelēs, P. Parisinus Graecus 1.

omphakion juice or paste made from unripe grapes and/or olives: Dioskouridēs, *MM* 3.7, 5.5; Durling 1993: 251.

See also: Ptolemaios (Med.), Ptolemaios (Pharm.), Theophilos (Pharm.).

opopanax the bitter flammable gum of ***panax***: Dioskouridēs, *MM* 3.48; Durling 1993: 254.

See also: Alkimiōn, Amuthaōn, Antiokhos VIII, Aristophanēs (Pharm.), Epaphroditos of Carthage, Euangeus, Euboulos (Pharm.), Iakōbos Psukhrestos, Isidōros of Memphis, Iustus (Pharm.), Kleophantos (Pharm.), Laodikos, Leukios, Lusias, Makhairiōn, Marcellus (Pharm.), Menippos, Serenus (Pharm.), Sōkratēs (Med.), Sōsikratēs (Pharm.), Zēnōn of Laodikeia, Zōsimos (Med.).

orthopnoia disease characterized by being able to breathe only in an upright position: Hippokratic Corpus, Prognosis 23 (2.176 Littré); Hippokratic Corpus, Sacred Disease 9 (6.370 Littré); Galēn *Loc. Aff.* 2.5 (8.120–121 K.), *CMLoc* 7.6 (13.105–106 K.).

See also: Andronikos (Med.), Kleomenēs, Onētidēs/Onētōr, Secundus, Sōsikratēs (Pharm.).

***oxumel* (Grk.: *oxumeli*, "sharp-honey")** honey boiled down in vinegar, used as a remedy, or a base for compound medicines. Hippokratic Corpus *Regimen in Acute Diseases* 58–60 (2.348–358 Littré); Dioskouridēs, *MM* 5.14; Durling 1993: 253.

See also: Diophantos of Lukia, Hērās.

panax numerous plants were given this name, esp. *Opopanax hispidus* Friv. or Grisb. ("Hercules' woundwort"), *Opopanax chironium* (L.) Koch. ("sweet myrrh"), and *Inula helenium* L.: Theophrastos, *HP* 9.11.1–4; Dioskouridēs, *MM* 3.48 ("Hercules"), 3.50 ("Chiron's"), etc.; Galēn, *Simples* 9.16.3–5 (12.94–95 K.); Usher 1974: 318, 424; André 1985: 186–187 (esp. #1, 5); Langenheim 2003: 97, 416.

See also: Eugērasia, Harpokrās, Khrusermos, Mēdeios, Olumpos.

paradoxon an event or observation contrary to reason; collected by **Stoics** and others to show the incomprehensibility of nature. *Cf. HWPhil* 7 (1989) 81–84, P. Probst.

Agathoklēs of Milētos, Amōmētos, Ampelius, Vindonius Anatolios, pseudo-Aristotle De Mirabilibus Auscultationibus, Cornelius Tacitus, Dionusios

OF BUZANTION, MURSILOS, NIKOLAOS OF DAMASKOS, PHLEGŌN, POMPONIUS MELA, SŌTIŌN.

parapēgma astronomical calendar of fixed-star phases and solar positions, indicating expected weather, and noting fixed-star or constellation phases (such as first morning rising or last evening setting), and solstices and equinoxes. Often constructed as a slab of stone with peg-holes for each day of the year, with a peg being moved from hole to hole for each day of the year; examples are also recorded on papyri and in MSS. *Cf.* DŌSITHEOS OF PĒLOUSION; EUKTĒMŌN; GEMINUS; PTOLEMY; *BNP* 10 (2007) 519–520, J. Rüpke; D. Lehoux, *Astronomy, weather, and calendars in the ancient world: parapegmata and related texts in classical and Near Eastern societies* (2007).

See also: CLODIUS TUSCUS, DĒMOKRITOS, EUDOXOS OF KNIDOS, HIPPARKHOS OF NIKAIA, IŌANNĒS "LYDUS," KALLIPPOS, KALYAṆA, KONŌN, METŌN, MĒTRODŌROS (ASTRON. I), PHILIPPOS OF OPOUS, SŌSIGENĒS (I).

pastille see ***trokhiskos***.

periēgēsis description of, or guide-book to, a region of the ***oikoumenē***. *RE* 19.1 (1937) 725–742, H. Bischoff; *BNP* 10 (2007) 783, E. Olshausen.

ASKLĒPIADĒS OF MURLEIA, CORNELIUS TACITUS, DIONUSIOS OF ALEXANDRIA (PERIĒGĒTĒS), DIONUSIOS OF CORINTH, DIONUSIOS OF RHODES, GLAUKOS (GEOG. I), HEKATAIOS OF MILĒTOS, HERMEIAS (GEOG.), KALLIXEINOS, MĒTRODŌROS OF SKĒPSIS, MNASEAS OF PATARA, POLEMŌN OF ILION, SKUMNOS OF KHIOS, SŌKRATĒS OF ARGOS, THEOPHILOS (GEOG.).

periodos description of a (possibly notional) trip around a region, or the whole, of the ***oikoumenē*** or even of the whole world.

PSEUDO-APOLLODŌROS OF ATHENS, HEKATAIOS OF MILĒTOS, MENEKRATĒS OF ELAIOUS, PHILEAS, SĒMOS, THUCYDIDĒS.

Peripatos ARISTOTLE's school in Athens, very active scientifically through the 3rd c. BCE, and surviving until *ca* 200 CE. See esp. THEOPHRASTOS, STRATŌN, DIKAIARKHOS, EUDĒMOS OF RHODES, and ALEXANDER OF APHRODISIAS.

Entries on Peripatetics: ADRASTOS OF APHRODISIAS, AGATHARKHIDĒS OF KNIDOS, ANDRONIKOS OF RHODES, ARISTOKLĒS OF MESSĒNĒ, ARISTŌN OF IOULIS, ARISTŌN OF KHIOS, ARISTOTELĒS OF MUTILĒNĒ, ARISTOTELIAN CORPUS (12 entries), ARISTOXENOS OF TARAS, ASPASIOS, BOËTHOS OF SIDŌN, HĒRAKLEIDĒS OF KALLATIS, HERMINOS, HERMIPPOS (OF SMURNA?), HIERŌNUMOS OF RHODES, KRITOLAOS, LUKŌN OF TROAS, MENŌN, ON MELISSOS, XENOPHANĒS, AND GORGIAS, NIKOLAOS OF DAMASKOS, PHYSIOGNOMISTA LATINUS, SŌSIGENĒS (II), SOTIŌN OF ALEXANDRIA, XENARKHOS.

See also: AGATHARKHIDĒS OF SAMOS, ALKINOOS, ANANIA, ANATOLIOS OF LAODIKEIA, APOLLODŌROS *THĒRIAKOS*, APULEIUS OF MADAURUS, AREIOS DIDUMOS, ARISTOPHANĒS OF BUZANTION, ARKHUTAS OF TARAS, CALCIDIUS, CASSIUS IATROS., DIODOTOS (ASTR. II), DIOGENĒS LAËRTIOS, DIOKLĒS OF KARUSTOS, ERASISTRATOS OF IOULIS, ERATOSTHENĒS, GALĒN, HARPOKRATIŌN OF ARGOS, HĒRAKLEIDĒS PONTIKOS, HIPPOKRATIC CORPUS PROTREPTIC WORKS, LONDINIENSIS MEDICUS, MĒNODOTOS OF NIKOMĒDEIA, OCELLUS, PLŌTINOS, PLUTARCH, SIMPLICIUS, THEMISTIOS, XENOKRATĒS OF KHALKĒDŌN.

periplous* (Lat.: *periplus*)** a voyage around a shore (of an island or continent), or the description thereof; contrast ***periēgēsis and ***periodos***. *RE* 19.1 (1937) 841–850 (#2), F. Gisinger; *OCD3* 1141–1142, N. Purcell; *BNP* 10 (2007) 799–801, J. Burian.

Agathōn of Samos, Alexander (Geog.), Alexander of Mundos, Anaxikratēs, Androitas, Androsthenēs, Apollōnidēs, Artemidōros of Ephesos, Auienus, Bakōris, Bōtthaios, Cornelius Tacitus, Damastēs of Sigeion, Dionusios of Buzantion, Dionusios son of Kalliphōn, Eudoxos of Knidos, Eudoxos of Rhodes, Euthumenēs, Hekataios of Milētos, Hērakleidēs "Kritikos," Himilkōn, Itineraries, Kallisthenēs, Kharōn of Carthage, Kleoboulos, Ktēsias, Marcianus of Hērakleia, Massilot Periplus, Menippos of Pergamon, Mnaseas of Patara, Numphis, Numphodōros of Surakousai, Ophellās, Pausimakhos, Periplus Maris Erythraei, Periplus Ponti Euxini, Sallustius Crispus, Sebosus Statius, pseudo-Skulax, Skulax of Karuanda, Sōsandros (Geog.), Stadiasmus Maris Magni, Terentius Varro, Thucydidēs, Timagenēs, Timagētos, Timosthenēs, Xenophōn of Lampsakos, Zenothemis.

pessary medicated vaginal suppository; the mode of abortion prohibited in the Hippokratic Corpus Oath.

See also: Arsenios, Boëthos (Med.), Celsus, Hippokratic Corpus Gynecology, Olumpias, Ōros, Pleistonikos.

phthisis "wasting," i.e., usually pneumonial tuberculosis (of which a common symptom is bloody sputum), very common in antiquity (Hippokratic Corpus, *Diseases* 2.48–50 [7.72–78 Littré]), and into the modern period, *cf.* Grmek 1989: 177–197.

Abaskantos, Amuthaōn, Dioklēs of Khalkēdōn, Euruphōn, Hērodikos of Knidos, Hippokratēs of Kōs, Krateros, Mēnodōros, Nikēratos, Ōrigeneia, Philagrios, Philippos of Rome, Prutanis, Ptolemaios (Pharm.).

phusis "nature" or everything that comes to be; frequent as the subject of writing on science: Aristotle, *Physics* 2.1 (192b8–193b21); *HWPhil* 7 (1989) 967–971, L. Deitz (on *phusis* and *nomos*).

See also: Aisara, Antiphōn, Carmen Astrologicum, Epidikos, Hippokratic Corpus Airs Waters Places, Kleidēmos, Lucretius, Mētrodōros of Khios, Nemesios, Polemōn of Athens, Physiologus.

Platonism see **Academy**.

pneuma originally "breath" or the "innate spirit" (Aristotle, *Motu Anim.* 10 [703a4–703b2]), later the active principle of **Stoic** cosmology. *OCD3* 1202, J.T. Vallance; *NP* 9 (2000) 1181–1182, T. Tieleman.

See also: Agathinos, Antipatros of Tarsos, Apollōnios of Memphis, Aretaios, Aristotelian HA 10, Aristotelian On Breath, Athēnaios of Attaleia, Chrysippus, Dioklēs of Karustos, Erasistratos of Ioulis, Gorgias of Alexandria, Hērophilos, Kleanthēs, Ktēsibios, Magnus of Ephesos, Mnēsarkhos, Mnēsitheos of Alexandria, P. Hibeh (Ophthalm.), P. Mil. Vogl. I.14, Pelops, Philistiōn of Lokri, Poseidōnios of Apameia, Praxagoras, Stratōn of Lampsakos, Theophrastos, Zēnōn of Kition.

Pneumaticists sect of medical practice founded by Athēnaios of Attaleia, whose primary explanatory principle was the ***pneuma*** of the **Stoa**. The human physiological system was made of ***pneuma*** and four elements in an equilibrium (*eukrasia*). ***Pneuma*** circulated through the cardio-vascular system, hence their strong interest in sphygmology, with elaborate classifying of pulses. A disturbance of the equilibrium (*duskrasia*) caused disease(s), a sign of which was fever. Therapy consisted partly in evacuating an excess of ***pneuma*** or of one of the four physiological elements. See esp. Agathinos,

Arkhigenēs, Aretaios. *OCD3* 1202–1203, J.T. Vallance; *NP* 9 (2000) 1183–1184, V. Nutton.

Entries on Pneumaticists (most by Touwaide): Agathinos, Antullos, Apollōnios of Pergamon, Aretaios, Arkhigenēs of Apameia, Athēnaios of Attaleia, pseudo-Galēn Definitiones Medicinales, pseudo-Galēn Introductio, pseudo-Galēn De Pulsibus, Hēliodōros of Alexandria, Hēraklās, Hērodotos (of Tarsos?), Leōnidas of Alexandria, Magnus of Ephesos, Philippos of Rome, Theodōros (of Macedon?).

See also: Hermogenēs of Smurna, Hippokratic Corpus Heart, Philoumenos, Philagrios.

polis (pl. ***poleis***) city-state, generally self-governing in principle; see esp. Aristotle, *Politics* 3.9 (1280 b40–1281 a4). *Cf. HWPhil* 7 (1989) 1031–1034, W. Nippel; *OCD3* 1205–1206, O. Murray; *NP* 10 (2001) 22–26, K.-W. Welwei and P.J. Rhodes.

See also: Aineias Tacticus, Hērakleidēs "Kritikos."

pompholux "bubbles," i.e., zinc oxide deposited on the interior chimney walls of a refinery furnace: Dioskouridēs, *MM* 5.75; Galēn, *Simples* 9.3.25 (12.234–235 K.).

See also: Athēnippos, Atimētos, Diomēdēs (Pharm.), Hermolaos (Pharm.), Kleōn (Med.), Manethōn, Neilammōn, Sergius of Babylōn, Stratōn of Bērutos, Suneròs.

psimuthion (Lat. ***cerussa***) "white lead," i.e., lead acetate or carbonate: Theophrastos, *Lapid.* 55–56; Dioskouridēs, *MM* 5.88; Galēn, *Simples* 9.3.39 (12.243–244 K.); Antullos in Paulos of Aigina, 7.24.11 (*CMG* 9.2, p. 398). Known to be poisonous, see Asklēpiadēs in Galēn *Antid.* 2.7 (14.144–146 K.).

See also: Aelius Promotus, Ambrosios Rusticus, Ammōnios of Alexandria, Andreas of Karustos, Apollinarios (Pharm.), Apollōnios Claudius, Arkhagathos, Artemōn (Med.), Asterios, Atimētos, Attalos III, Bathullos, Bolās, Candidus, Castus, Diomēdēs (Pharm.), Diophantos of Lukia, Euelpidēs, Gennadios, Hermōn, Hierax, Iunia/Iounias, Kleophantos (Pharm.), Kuros, Marcellus (Pharm.), Megēs, Moskhiōn (Pharm.), Neilammōn, Olumpos, Philōn of Tarsos, Polustomos, Prōtās, Samithra, Telamōn, Telephanēs, Timaios (Pharm.), Zōilos of Macedon.

pterugeion eye-disease characterized by triangular discoloration of the eye reaching from the inner angle to the pupil, i.e., the modern pterygium; *cf.* P. Aberdeen 11 and P. Ross.Georg.1.20.

purethron the plant pellitory, *Anacyclus pyrethrum* DC., typically north African, but also from Spain and Syria; a carminative and toothache-reliever. See Nikandros, *Thēr.* 683; Celsus 5.4, 5.8; Dioskouridēs, *MM* 3.73, 5.42; Scribonius Largus 9; Galēn *Simples* 8.16.41 (12.110 K.); Usher 1974: 43; André 1985: 212; Durling 1993: 279–280.

See also: Aristokratēs, Eutonios, Hikesios of Smurna, Ianuarinus, Iollas, Lingōn, Philōn of tarsos, Theodorus Priscianus, Timokratēs.

Pythagoreans followers of Pythagoras, politically organized *ca* 510–450 BCE; afterwards organized mostly intellectually through teacher-student connections (no **scholarchs** are attested). Around 350 BCE this succession ceased, but around 200 BCE pseudo-Pythagorean writings, signed with the name of Pythagoras and historical or invented Pythagoreans, began to appear and were fabricated until *ca* 100 CE. The authors of these works usually relied on **Academic** and **Peripatetic** interpretations. A revival of the Pythagorean movement, designated by modern scholars as

Neo-Pythagoreanism, occurred by *ca* 50 BCE (*cf.* e.g. NIGIDIUS). The pseudonymous tracts were gradually replaced by the writings of those who saw themselves as followers of a Platonized Pythagoras, but wrote in their own names. With few exceptions, all known Neo-Pythagoreans were **Platonists** (e.g., EUDŌROS OF ALEXANDRIA, MODERATUS OF GADĒS, NIKOMAKHOS OF GERASA, NOUMĒNIOS OF APAMEIA). In the late 3rd c. CE, PORPHURIOS OF TYRE and especially IAMBLIKHOS OF KHALKIS present a fusion or synthesis of Neo-**Platonism** and Neo-Pythagoreanism. Some of the original Pythagoreans (ARKHUTAS, HIPPASOS, PHILOLAOS, PYTHAGORAS, THEODŌROS OF KURĒNĒ), as well as some of the Neo-Pythagoreans (EUPHRANŌR, HIPPOKRATIC CORPUS SEVENS, TIMAIOS OF LOKRIS), focused on the role of number in the ***kosmos***.

Entries on Pythagoreans (most by Zhmud [early] or Centrone ["neo-"]): AISARA, ALKMAIŌN, ANAXILAOS, ANDROKUDĒS, ARISTOMBROTOS, ARKHUTAS OF TARAS, BŌLOS, DĒMĒTRIOS (PYTHAG.), DEMOKEDĒS, DIDUMOS (MUSIC) (?), EKPHANTOS, ENNIUS, EPIKHARMOS (?), EUBOULIDĒS (?), EUPHRANŌR (MUSIC), EUPHRANŌR (PYTHAG.), EURUTOS, HĒSUKHIOS (?), HIKETAS, HIPPASOS, HIPPOKRATIC CORPUS OATH, HIPPŌN, IKKOS, LEONTIOS, LUKŌN OF IASOS, MENESTŌR OF SUBARIS, MODERATUS OF GADĒS, MUIA, MUŌNIDĒS, NIGIDIUS FIGULUS, NIKOMAKHOS OF GERASA, OCELLUS, PSEUDO-ORPHEUS (ASTROL.), PAETUS, PHILOLAOS, PYTHAGORAS, DE QUATERNIONIBUS, SEXTIUS, TERENTIUS VARRO, THEODŌROS OF KURĒNĒ, THUMARIDAS, XENOPHILOS, XOUTHOS.

See also: AELIANUS (MUSIC), AKHILLEUS, ALEXANDER OF EPHESOS, AMMŌNIOS ANNIUS, APOLLODŌROS OF KUZIKOS, ARISTOXENOS OF TARAS, ATTIUS, CALCIDIUS, DAMASTĒS (MED.), DĒMĒTRIOS (MUSIC), EMPEDOKLĒS, EPAPHRODITOS AND VITRUUIUS RUFUS, ERATOKLĒS, ERATOSTHENĒS, EUCLIDEAN SECTIO CANONIS, EUDŌROS OF ALEXANDRIA, FAUONIUS EULOGIUS, GAUDENTIUS, HĒRAKLEIDĒS OF HĒRAKLEIA PONTIKĒ JUNIOR, IAMBLIKHOS OF KHALKIS, IŌANNĒS "LYDUS," IŌANNĒS OF STOBOI, KURANIDES, NOUMĒNIOS OF APAMEIA, OINOPIDĒS, PANAITIOS THE YOUNGER, PAPPOS OF ALEXANDRIA, P. GEN. INV. 259, PHEREKUDĒS, PHILIPPOS OF OPOUS, PLUTARCH (MUSIC), PORPHURIOS OF TYRE, PROKLOS OF LUKIA, PTOLEMAÏS, PTOLEMY, SALIMACHUS, SPEUSIPPOS OF ATHENS, SYRIANUS, MACROBIUS THEODOSIUS, THEAITĒTOS, THEOL. ARITH., THRASULLOS, VERGILIUS, ZŌPUROS OF TARAS.

rabies see ***hudrophobia***.

Rationalists somewhat contrived sect of medical theory, also known as "Dogmatists" (*cf.* CELSUS; GALĒN); a doctor was labeled "Rationalist" if s/he believed in a theoretical basis for medicine, and insisted etiology was a basis for treatment; contrast the **Empiricists** and the **Methodists**; the **Pneumaticists** could be considered a kind of Rationalist.

See also: ARKHIBIOS, CASSIUS, CELSUS, DIOKLĒS OF KARUSTOS, GALĒN, PSEUDO-GALĒN INTRODUCTIO, KALLIMAKHOS OF BITHUNIA, MĒNODOTOS OF NIKOMĒDEIA, MNĒSITHEOS OF ATHENS, PELOPS, PHILIPPOS (OF PERGAMON?), P. BEROL. 9782, THESSALOS OF KŌS.

sagapēnon probably *Ferula persica* Willd.: DIOSKOURIDĒS, *MM* 3.81, 5.42; GALĒN, *Simples* 8.18.1 (12.117 K.); Miller 1969: 100; Usher 1974: 253; André 1985: 223; Durling 1993: 286; Langenheim 2003: 416.

See also: ARISTOKRATĒS, HARPALOS (PHARM.), TELEPHANĒS.

Sagapēnon, Mount Athos Ω 75, F. 142[R]

sambukē scaling ladder, with mechanical elevator (winch or screw), often used on ships; *cf.* BITŌN, DAMIS OF KOLOPHŌN, HĒRAKLEIDĒS OF TARAS, and MOSKHIŌN (MECH.).

sarkokolla an *Astragalus* species, used in styptic or clotting ointments: DIOSKOURIDĒS, *MM* 3.85; GALĒN, *Simples* 8.18.4 (12.118 K.); Wood and LaWall 1926: 1463; Miller 1969: 101; Usher 1974: 68; André 1985: 227 (identifies as *A. fasciculifolius* Boiss.); Durling 1993: 287 (follows André).

See also: HĒRŌN (MED.), POLUDEUKĒS, SERGIUS OF BABYLŌN.

scholarch: head of a school, such as the **Academy**, **Peripatos**, **Stoa**, or **Garden**; elected usually for a life-term.

Entries on scholarchs: ALEXANDER OF APHRODISIAS, ANDRONIKOS OF RHODES, ANTIPATROS OF TARSOS, APOLLODŌROS OF ATHENS, ARISTŌN OF IOULIS, ARISTOTLE, ARKHEDĒMOS OF TARSOS, BASILEIDĒS, BOËTHOS OF SIDŌN (PERIP.), CHRYSIPPUS, DIOGENĒS OF BABYLŌN, EPICURUS, HĒRAKLEIDĒS PONTIKOS, HERMARKHOS, HIERŌNUMOS OF RHODES, IASŌN, IDOMENEUS, KARNEADĒS, KLEANTHĒS, KLEITOMAKHOS, KRITOLAOS, LAKUDĒS, LUKŌN OF TROAS, MARINOS OF NEAPOLIS, MĒTRODŌROS OF LAMPSAKOS (?), MNĒSARKHOS, PANAITIOS OF RHODES, PLATO, PLŌTINOS, POLEMŌN OF ATHENS, POLUAINOS (?), POLUSTRATOS, POSEIDŌNIOS OF APAMEIA, PROKLOS OF LUKIA, SPEUSIPPOS OF ATHENS, STRATŌN OF LAMPSAKOS, SYRIANUS, THEOPHRASTOS, XENOKRATĒS OF KHALKĒDŌN, ZĒNŌN OF KITION, ZĒNŌN OF SIDŌN.

See also: AISARA, ALEXANDER OF APHRODISIAS, EUDĒMOS OF RHODES.

shelf-fungus (Grk. *agarikon*) any of several tree-fungi, probably *Fomes officinales* Bresadola: DIOSKOURIDĒS, *MM* 3.1; GALĒN, *Simples* 6.1.5 (11.813–814 K.); Durling 1993: 1; G. Maggiuli, *Nomenclatura Micologica Latina* (1977).

See also: AMARANTOS, ANASTASIOS, DOARIUS, DŌSITHEOS (PHARM.), IUSTUS (PHARM.), PROKLOS THE METHODIST.

***silphion* (Lat.: *silphium* and *laserpicium*)** a *Ferula* species, now apparently extinct due to a supposed over-harvesting, native to **Kurēnē**, whose juice was widely-used in pharmacy: DIOSKOURIDĒS, *MM* 3.80; GALĒN, *Simples* 8.18.16 (12.123 K.); *NP* 11 (2001) 561, C. Hünemörder; Touwaide (2006).

See also: ARISTOKRATĒS, BOUPHANTOS, CLODIUS (ASKLĒP.), ISIDŌROS OF MEMPHIS, MARCELLUS (PHARM.), MARCIANUS OF AFRICA, POLLĒS (MED.), POSEIDŌNIOS (MED. II), TRUPHŌN OF GORTUN.

Sinōpian earth/ocher ruddle (red iron oxide) found near Sinōpē (see ***Gazetteer***), and used medicinally: THEOPHRASTOS, *Lapid.* 52; DIOSKOURIDĒS, *MM* 5.96.

See also: AKHAIOS, APHRODĀS, HALIEUS, HIERAX, LEUKIOS, MAGNUS OF PHILADELPHEIA.

skordion a *Teucrium* species, "water germander", astringent and bitter: DIOSKOURIDĒS, *MM* 3.111; GALĒN, *Simples* 8.18.25 (12.125–126 K.); Usher 1974: 572–573; André 1985: 231 (#1); Durling 1993: 293 (follows André).

See also: APELLĒS (OF THASOS?), ARISTOKLĒS, EPAPHRODITOS OF CARTHAGE, KŌDIOS TOUKOS, MITHRADATĒS VI.

skotodiniē/skotodinos/skotōma dizziness with darkening of vision, a commonly-cited symptom: HIPPOKRATIC CORPUS, *Affections* 2 (6.210 Littré), *Diseases* 2.4, 15, 18 (7.12, 28, 32 Littré = *CUF* v. 10.2, ed. Jouanna, pp. 136, 149, 152), *Epidemics* 7.84.4 (5.442 Littré = *CUF* v. 4.3, ed. Jouanna, p. 99); THEOPHRASTOS, *Dizziness*; GALĒN *Loc. Aff.* 3 (8.201 K.), *Sanit.* 5.10.14, 6.9.22, 6.12.1 (*CMG* 5.4.2, pp. 136, 149, 152); PSEUDO-GALĒN, DEFINITIONS 251 (19.417 K.). Contrast the modern scotoma, an area of defective visual acuity and *cf.* modern vertigo.

See also: PTOLEMAIOS (PHARM.), POSEIDŌNIOS (MED. II).

sōri or ***soru*** a copper ore from Cyprus, found in the lowest levels, below ***khalkitis***: DIOSKOURIDĒS, *MM* 5.74, 102; GALĒN, *Simples* 9.3.21 (12.226–229 K.); ANTULLOS in PAULOS OF AIGINA, 7.24.11 (*CMG* 9.2, p. 398); Durling 1993: 309.

spodion or ***spodos*** "ash," meaning copper oxide: HIPPOKRATĒS, *Diseases of Women*, 1.103–104; DIOSKOURIDĒS, *MM* 5.75; GALĒN, *Simples* 9.3.25 (12.234–235 K.).

See also: KLEŌN OF KUZIKOS (MED.), NIKĒTĒS, PHILŌN OF TARSOS, STRATŌN (ERASI.).

staphis/staphis agria a Delphinium (Larkspur) species, probably *Delphinium staphisagria* L., whose seeds are the source of an insecticide: DIOSKOURIDĒS, *MM* 2.159, 4.152 (non-"wild": 5.3); PLINY 23.17–18; Wood and LaWall 1926: 1027–1028; Usher 1974: 202; André 1985: 248; Durling 1993: 298 (follows André).

See also: ABASKANTOS, EUTONIOS, PRUTANIS.

Stoa/Stoic ZĒNŌN's school in Athens, originally a colonnaded porch (*stoa*) where he taught; the doctrines of Stoicism were largely influenced by CHRYSIPPUS OF SOLOI and POSEIDŌNIOS OF APAMEIA. The school seems not to have persisted after *ca* 200 CE.

Entries on Stoics (many by Lehoux): AEFICIANUS, AETNA, ANNAEUS LUCANUS, ANNAEUS SENECA, ANTIPATROS OF TARSOS, ANTIPATROS OF TYRE, APOLLODŌROS OF SELEUKEIA, APOLLOPHANĒS OF NISIBIS, ARATOS, ARISTOKREŌN (?), ARKHEDĒMOS OF TARSOS, BOËTHOS OF SIDŌN (STOIC), CHRYSIPPUS, DIODOTOS (ASTR. I) (?), DIOGENĒS OF BABYLŌN, DIONUSIOS OF KURĒNĒ, GEMINUS, HĒLIODŌROS (STOIC), HĒRAKLEIDĒS OF ATHENS, HIEROKLĒS OF ALEXANDRIA, IASŌN, KHAIRĒMŌN, KLEANTHĒS, KLEOMĒDĒS, MANILIUS, MĒDIOS, MNĒSARKHOS, OPPIANUS OF KILIKIA, PANAITIOS OF RHODES, POSEIDŌNIOS OF APAMEIA, SIMMIAS (STOIC), SPHAIROS OF BORUSTHENĒS, STRABŌN OF AMASEIA, THEŌN OF ALEXANDRIA (STOIC), ZĒNODOTOS (OF MALLOS?) (?), ZĒNŌN OF KITION.

See also: AELIANUS OF PRAENESTE, AGATHINOS, AGENNIUS URBICUS, AKHILLEUS, ANDRONIKOS OF RHODES, ARCHIMĒDĒS, AREIOS DIDUMOS, ARISTŌN OF IOULIS, ARISTOTLE OF MUTILĒNĒ, ARTEMIDŌROS OF DALDIS, ATHĒNAIOS OF ATTALEIA, BOËTHIOS OF SIDŌN (PERIP.), CASSIUS LONGINUS, DERVENI PAPYRUS, DIODŌROS (ASTRON.), DIOGENĒS LAËRTIOS, DIOKLĒS OF MAGNESIA, DIONUSIOS SON OF KALLIPHŌN, GALĒN, GEŌRGIOS OF PISIDIA, HĒROPHILOS, IULIUS FIRMICUS, HOMER, IMBRASIOS, KARNEADĒS, LONDINIENSIS MEDICUS, ON MELISSOS, XENOPHANĒS, AND GORGIAS, P. BEROL. 9782, PANAITIOS THE YOUNGER, PERIKLĒS OF LUDIA, PHILŌN OF ALEXANDRIA, PHILŌNIDĒS OF LAODIKEIA, PLŌTINOS, PLUTARCH, POLEMŌN OF ATHENS, POSEIDIPPOS, POSEIDŌNIOS (MED. I), SEXTIUS, SKULAX OF HALIKARNASSOS, SUDINĒS, TAUROS, TIBERIANUS, M. TULLIUS CICERO, VERGILIUS, VICELLIUS, ZĒNŌN OF SIDŌN.

sturax **(Lat.: *storax*)** the resin of *Styrax officinalis* L., an effective expectorant, also known as benzoin resin. Nowadays derived from the Sumatran species, the Greco-Roman *storax* was and is native to the eastern Mediterranean. Dioskouridēs, *MM* 1.66; André 1985: 252; Durling 1993: 302; *NP* 11 (2001) 1063, C. Hünemörder.

Emboularkhos, Eruthrios, Kleophantos (Pharm.), Klutos, Leukios, Lukomēdēs, Lusias, Nikēratos, Pamphilos of Bērutos, Platōn (Pharm.), Prutanis, Romula, Sōsagoras, Theopompos, Theosebios, Timokleanos.

sumpatheia the (**Stoic**) idea that beings in the world have connections with one another through which they can influence one another, since all things are linked together in a single system maintained by reason: *HWPhil* 10 (1998) 751–756, M. Kranz and P. Probst.

Akhaios, Androtiōn, Vindonius Anatolios, Aristotelian Corpus Problems, Arkhelaos of Khersonesos, Bōlos, Bothros, pseudo-Dēmokritos (Agric.), Geōrgios of Pisidia, Harpokratiōn of Alexandria, Hermēs Trismegistos, Kallikratēs (Astrol.), Kuranides, Iulius Firmicus, Nepualios, Orestinos, Petosiris, Pollēs of Aigai, Salpē, Stephanos of Alexandria (Alch.), Timotheos of Gaza.

sunankhē ("choking") acute inflammation of the throat or tonsils, as in whooping cough, the flu, or other illnesses; in some Latin texts *angina*; sometimes rendered into English as "quinsy" (from the name for the "more serious" kind, *kunankhē*). Aretaios 1.8 (*CMG* 2 [1958] 7–9) defines it as "inflammation (*phlegmonē*) of the respiratory parts, or . . . a pathology of the ***pneuma*** alone . . . The parts affected are the tonsils, epiglottis, pharynx, uvula, and uppermost part of the trachea." The "more serious" kind occurs when the inflammation spreads to the tongue, causing it to protrude (hence the name), and resulting in suffocation.

See also: Caelius Aurelianus, Marcianus of Africa, Orpheus (Med.), Scribonius Largus, Terentius Valens.

sympathy ⇒ *sumpatheia*

terebinth **(Grk. *terminthos*)** small Mediterranean tree or bush, a *Pistacia* species, whose sap was distilled (*cf.* **mastic**): Dioskouridēs, *MM* 1.71; Galēn, *Simples* 8.19.1 (12.137–138 K.); André 1985: 256; *NP* 12/1 (2002) 140–141, C. Hünemörder (who identifies as *Pistacia terebinthus* L.).

See also: Akholios, Amuntas (Geog.), Amuthaōn, Antiokhis, Aphrodisis, Aphthonios, Aquila Secundilla, Arkhagathos, Aristogenēs of Knidos, Aristoklēs, Arrabaios, Attalos III, Boëthos (Med.), Candidus, Castus, Damonikos, Diogenēs, Diophantos of Lukia, Dioskouridēs Phakas, Eruthrios, Eugeneia, Euangeus, Euboulos (Pharm.), Eutonios, Fauilla, Halieus, Harpalos (Pharm.), Harpokrās, Heirodotos, Hekataios (Pharm.), Hērakleidēs of Ephesos, Hermōn, Iakōbos Psukhrestos, Iulius Agrippa, Kallinikos, Kleophantos (Pharm.), Klutos, Ktēsiphōn, Leukios, Makhairiōn, Megēs, Menoitas, Minucianus, Murōn, Olumpos, Ōriōn, Ōros, Pollēs (Med.), Polustomos, Proëkhios, Samithra, Sarkeuthitēs, Truphōn of Gortun.

tetanos **(Lat.: *tetanus*)** the arched, stiffened back, locked jaws, inability to swallow, and frothy saliva of the modern "tetanus" or "lockjaw" is caused by an acute poisoning from a neurotoxin produced by the widely-distributed *Clostridium tetani*; what is generally regarded as the same disease is found in the Hippokratic Corpus, Epidemics 5.47 (5.234 Littré), 5.95 (5.254–256 Littré). The infection is invariably fatal in 5–10 days if untreated (prompt modern treatment reduces the death-rate to *ca* 30%). The

Hippokratic Corpus, Heart 12 (9.92 Littré), and Celsus 4.6.1 indicate that patients who managed to endure to the fifth day would live. Celsus records the Greek distinction between reverse (*opisthotonos*) and forward (*emprosthotonos*) spinal flexion (*cf.* Hippokratic Corpus, Diseases 3.12 [7.132 Littré], *Internal Affections* 52–54 [7.298–302 Littré]; Caelius Aurelianus, *Acute* 3.61 [p. 339 Drabkin; *CML* 6.1.1, p. 328]).

See also: Clodius (Asklēp.), Hippokratic Corpus Nosology, Pelops.

***trokhiskos* (Lat. *pastilla*)** "pill," often made up as a way to store a remedy, which in use would be dissolved (in wine or water) and applied, to eye or skin, etc.

Andrōn (Pharm.), Aristarkhos of Tarsos, Apollōnios of Memphis, pseudo-Apuleius, Athēniōn, Bithus, Cornelius (Pharm.), Celsus, Faustinus, Grēgorios (Veterin.), Harpokratiōn, Hērakleidēs of Ephesos, Isidōros of Antioch, Khrusermos, Kosmos, Magnus of Philadelpheia, Menestheus, Nikēratos, Pasiōn, Philoumenos, Proklos (Method.).

verdigris (Grk. *ios*) copper acetate prepared by steeping copper in vinegar: Theophrastos, *Lapid.* 57; Dioskouridēs, *MM* 5.79; Pliny 34.110–111; Galēn, *Simples* 9.3.10 (12.218–219 K.); Antullos in Paulos of Aigina, 7.24.11 (*CMG* 9.2, p. 398).

See also: Aphrodās, Aphros, Apollophanēs of Seluekeia, Areios of Tarsos, Castus, Diomēdēs (Pharm.), Dionusios of Samos, Epigonos, Epikouros (Pharm.), Euangeus, Euboulos (Pharm.), Halieus, Hermophilos, Hierax, Hikesios of Smurna, Isidōros of Memphis, Leukios, Lunkeus, Menoitas, Nikolaos (Pharm.), Olumpias, Olumpos, Pasiōn, Phaidros, Philōtas, Pollēs (Med.), Proëkhios, Sertorius Clemens, Theophilos (Pharm.), Truphōn of Gortun.

Manuscripts

Plant representations in classical botanical treatises are known only from MSS of Dioskouridēs' *De materia medica* (Greek, Latin and Arabic), and from fragmentary Egyptian pharmacological papyri (*cf.* Papyri). Numerous copies of illustrated late-antique pharmacological treatises in Latin (mainly on therapeutic plants) are also preserved. If their plant representations (still little studied despite Collins, probably because of their sheer quantity) descend from earlier prototypes as hypothesized, they might complement those from Greek MSS, as also do those in Arabic translations of Dioskouridēs.

While only one Latin illustrated copy of Dioskouridēs' text is known (Munich, Bayerische Staatsbibliothek, CLM 337), extant are a handful of Arabic copies (Grube) and almost 30 Greek codices ranging from early 6th to late 16th centuries, first studied in the early 19th c. (Millin; *cf.* Choulant). The most ancient of them, the "Vienna Dioscorides" (Vienna, Österreichische Nationalbibliothek, *medicus graecus* 1), was fully reproduced in a monumental work that remains a milestone in the historiography of ancient botanical illustration (Premerstein *et al.*). Despite several subsequent studies, including new and high-quality facsimiles (Gerstinger 1970), there is so far no comprehensive analysis, thus leaving open the origin of these illustrations (whether from Dioskouridēs or added later), their tradition (linkages between MSS or groups of MSS, and the correspondence between textual and iconic traditions), and their function and relation with the text (including the fundamental question of schematism vs. realism).

Greek codices of Dioskouridēs have been traditionally evaluated on the basis of their antiquity and artistic quality (hence the focus on the Vienna codex), but can be approached more appropriately from scientific and iconic viewpoints, with due consideration of Arabic copies. Possibly the most ancient set of pictures is that of MS Paris, BNF, *graecus* 2179 (9th c., southern Italy or Syria-Palestine), to which the early 13th c. codices of Istanbul, Suleymaniye Kütüphanesi, Ayasofia 3702 and 3703 are very close (Touwaide). Though relatively recent, the latter two probably reproduced a 9th c. model that, in turn, copied carefully the 9th c. or even earlier Greek codex used to translate Dioskouridēs' treatise into Arabic. None of these three MSS is complete.

The Vienna codex is usually paired with the parchment MS of Naples, Biblioteca Nazionale, *ex Vindobonensis graecus* 1 (7th c.). This codex presents the text in the layout of papyrus rolls, i.e., in two columns on the pages, and illustrations atop the columns. It is therefore deemed a close copy of the most ancient form of *De materia medica* illustrations. The Vienna and Naples MSS are traditionally considered copies of the same ancestor, which, given the realistic aspect of the pictures in the Vienna and Naples books, is believed closest to the most ancient form of *De materia medica* illustrations. Nevertheless, a systematic comparison of these two with all other illustrated MSS suggests their common model might have reinterpreted in a realistic way such pictures as those of Paris *graecus* 2179, as did also the Vienna codex thus adding a further layer of realism. (Their naturalism probably explains why the Vienna representations are often considered close to the original form.) This set of pictures, too, is incomplete, as the text of the two MSS is a *selection* from Dioskouridēs.

Later MSS generally reproduced the illustrations of the two groups above more or less deftly and can be divided into two major categories, for each of which the major items are given. The New York and Athos Dioskouridēs (respectively New York, Pierpont Library, M 652, 10th c., and Athos, Megisti Lavra, Ω 75, 11th c.) reproduce the text and the pictures of the Vienna-Naples group. However, they also add the text missing in these two codices, taking it from the full recension (represented by Paris *graecus* 2179). The pictures accompanying these parts of the text in the New York and Athos volumes do not correspond to those in the Paris codex, and seem to have been created by the artists of the two MSS, who probably lacked models. While the newly created tables in the New York codex are highly schematic, those of the Athos are much more realistic, and also include eastern drugs missing not only in the Vienna and Naples volumes, but also in the New York codex. Significantly, one of these illustrations corresponds closely to its equivalent in an Arabic copy of Dioskouridēs.

The second category is formed of late illustrated Dioskouridēs MSS, all of which descend from the *Parisinus graecus* 2183 (mid-14th c.), wherein the illustrations of several models have been meshed, just as the text itself, which results from the collation of all previous versions of Dioskouridēs' *Materia medica*. This was the set of illustrations that was first known in the Renaissance.

A.-L. Millin, "Observations Sur les Manuscrits de Dioscorides qui sont conservés à la Bibliothèque nationale," *Magasin Encyclopédique* 2 (1802) 152–162; L. Choulant, "Ueber die Handschriften des Dioskorides," *Archiv für die zeichnenden Künste* 1 (1855) 56–62; A. de Premerstein, C. Wessely, and I. Mantuani, *De codicis Dioscuridei Aniciae Iulianae, nunc Vindobonensis Med. Gr. 1 historia, forma, scriptura, picturis* (1906); E. Grube, "Materialen zum Dioscurides Arabicus," in *Aus der Welt der Islamischen Kunst: Festschrift für Ernst Kühnel* (1959)

163–194; H. Grabe-Alpers, *Spätantike Bilder aus der Welt des Arztes: Medizinische Bilderhandschriften der Spätantike und ihre mittelalterliche Überlieferung* (1977); Alain Touwaide, *Farmacopea araba medievale: Codice Ayasofia 3703*, 4 vv. (1992–1993); M. Collins, *Medieval Herbals: The Illustrative Traditions* (2000).

Alain Touwaide

TIME-LINE

1 Most (97%) of the entries possess a date-range, either the *termini post* and *ante* of actual working lifespan (e.g., Aristotle), or else the *termini post* and *ante* within which the author was active or the work created (e.g., the "Anonymous Londiniensis" here filed as "Londiniensis medicus"). There are 30 entries for which only one terminus is provided; and 12 entries for which no date at all is given (these are listed after the "Time-Line"). Moreover, 18 entries whose date-range (or *akmē*) places them after our *terminus* of 650 CE, but included in *EANS* either because previously assigned to our period or else helpful to clarify other entries, are also placed after the "Time-Line."

2 We eschew as far as possible the ill-defined concept of the "*akmē*" introduced by Apollodōros of Athens (and Latinized by Jerome as "*floruit*"), although in 70 cases where we have but one date that is all we can give (e.g., Aristōn of Keōs, Arkhagathos, Astrologos of 379, etc.). Some of those demonstrate the defect of using a *floruit*, since the sole date known is the death-date (e.g., Aemilius Macer, Ploutarkhos of Athens, and Theaitētos).

3 Note that the dates given for papyri are usually the papyrological date, not the date of the work; but see P. Berol. 9782, P. Hibeh 1.27, P. Hibeh 2.187, P. Oxy. 3.470, P. Oxy. 13.1609, and P. Parisinus graecus 1.

4 About half the entries have date-ranges much wider than the "maximum likely" working lifetime (which we take to be 50 years); these are placed in the right-hand column of the "Time-Line." However, 33 of the 1,014 entries with a "wide" date-range represent a known actual working lifetime, and those are treated as (i.e., categorized with) the other 962 that have a narrow range or an *akmē*: Alexander of Tralleis, Anaxagoras, Anthēmios, Antisthenēs of Athens, Antonius Castor, Arkhelaos of Kappadokia, Ausonius, Cornelius Nepos, Damaskios, Dēmokritos of Abdēra, Diodōros of Sicily, Galēn, Gorgias of Leontinoi, Hierōn II, Hierōnumos of Kardia, Hippokratēs of Kōs, Iōannēs of Alexandria (Philoponos), Kleanthēs, Lukōn of Troas, Lusimakhos of Macedon, Mithradatēs VI, Polemōn of Athens, Polubios, Poseidōnios of Apameia, Proklos of Lukia, Cassiodorus Senator, Strabōn, Terentius Varro of Reate, Theophrastos, Timaios of Tauromenion, Xenarkhos, Xenokratēs of Khalkēdōn, Xenophanēs.

5 In addition to the exclusions noted in # 1 above (entries with unknown, late, or partial date-ranges), 61 entries of the remaining 981 "wide" entries are so uncertainly dated, i.e., have *termini* so wide (525 years or more), that they cannot be meaningfully included in the "Time-Line" and are also listed afterwards.

6 In grouping the 995 net total "narrow" entries into 37 clusters of 35 years each (a notional "generation"), as well as in grouping the 920 net total "wide" entries into

14 clusters of 105 years, we are well-aware of the "sometimes-deceptive effects of aggregation" (E.R. Tufte, *Visual explanations: images and quantities, evidence and narrative* [1997] 35) – increments much smaller could cause possibly-misleading multiplication of the names, whereas increments much larger might falsely suggest synchronizations.

7 Four encyclopedia lemmata generate multiple entries in this index, either because they represent multiple authors or works: Apollōnios Biblas and Son (two narrow); Aristiōn (Mech.) (one narrow, one wide); and Zīg (two *akmē*, one narrow), or else because two disjoint date-ranges are suggested: Komerios (two wide); see also Hermolaos (Geog.) and Zēnariōn, for which two disjoint date-ranges are suggested, with the later one being after our terminus.

8 Finally, two kinds of entries are not indexed here at all:

  a eight on schools or collections, which extend over many centuries, and represent no single work: Arabic Translations; Aristotelian Corpus in Pahlavi; Babylonian Astronomy; Demotic Texts; Hellenizing School (Armenian); Hippokratic Corpus in Pahlavi; Pahlavi translations; and "Papyri" (the entry introducing all the individual papyrological entries).

  b four non-existent people: Sextus of Apollōnia, Salimachus, Silimachus, and Sōsandros (Vet.); moreover, Asklēpiadēs Titiensis and Auidianus should perhaps be likewise omitted.

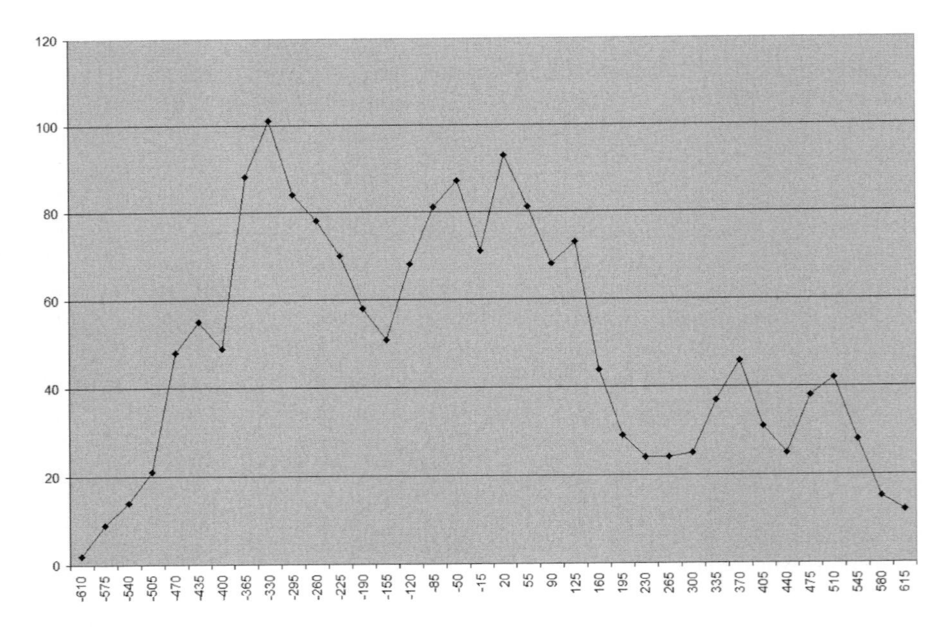

Number of scientists (with "narrow" date-ranges) per generation

All entries with "narrow" date-ranges are plotted above; the smaller fluctuations may not be significant, but the large rise (500 BCE to 330 BCE) and the high level (through *ca* 100 CE) surely are, as is also the precipitous fall after Hadrian (*ca* 140 CE). (The "late-Hellenistic" dip, of the 2nd c. BCE, may be significant.) Below we also plot all the entries with "wide" ranges, where the counts per "long" century (105 years) are weighted (lower for wider ranges):

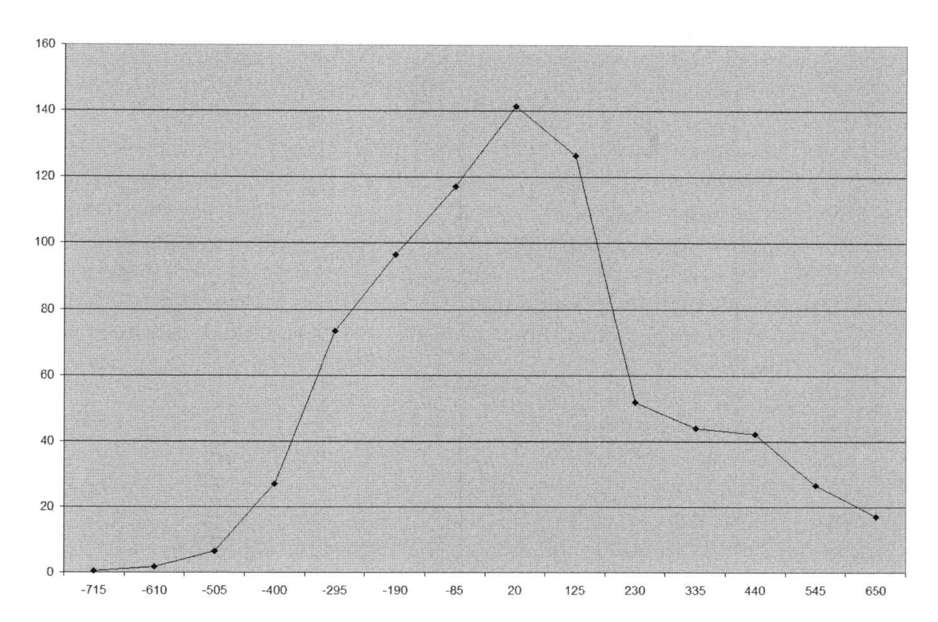

Number of scientists (with "wide" date-ranges) per century

The same general trend can be observed, a strong rise from *ca* 500 BCE, a high level from *ca* 300 BCE to *ca* 100 CE, and a precipitous drop *ca* 150 CE. (In principle one could combine these two graphs, but sufficient complexities would arise regarding the widths of the intervals, and no significant new results would be expected.) As argued in the *Introduction*, pp. 7–8, the decline in science around 150 CE was due to a shift of the political paradigm, the centralization of power and the loss of autonomy (*cf.* also P.T. Keyser, "Roman Science," in A. Barchiesi and W. Scheidel, *Oxford Handbook of Roman Studies* [2008: forthcoming]). This shift is from what Jane Jacobs, *Systems of Survival: A Dialogue on the Moral Foundations of Commerce and Politics* (1992), has called the "commercial" syndrome to the "extractive" syndrome. The latter is characterized by adherence to tradition, respect for hierarchy, honor, loyalty, obedience, and ostentatious acts of patronage, a predilection for military prowess and solutions, and rejection of investment and trade. In contrast the "commercial" syndrome encourages trade and investment, but eschews force, values thrift and industry, and respects invention, dissent, honesty, and cooperation. Thus, one expects to find evidence of a significant decline in trade around 150 CE, and indeed shipwreck evidence strongly confirms that: A.J. Parker, *Ancient Shipwrecks of the Mediterranean and the Roman Provinces* (1992) fig. 3 and 5, pp. 549 and 551; data pp. 10–14.

The table on the following pages (940–986) shows the scientists whose date-ranges are plotted in the two graphs above: narrow date-ranges on the left and wide date-ranges on the right.

Dates:	***Names:***	***Dates (Wide):***	***Names:***
750–700 BCE	Homer	820–715 BCE	Hēsiod
		715–610 BCE	Epimenidēs, Glaukos of Khios, Hēsiod
610–575 BCE	Anaximandros, Mamerkos	610–505 BCE	Dēmokedēs of Krotōn, Epimenidēs, Glaukos of Khios, Kleostratos, Mandrolutos, Massiliot Periplous, Pythagoras, Thalēs, Theodōros of Samos
575–540 BCE	(9) Anaximandros, Anaximenēs, Eupalinos, Euthumenēs, Khersiphrōn, Mamerkos, Metagenēs, Pherekudēs, Rhoikos		
540–505 BCE	(14) Anaximenēs, Eupalinos, Euthumenēs, Hekataios of Milētos, Hērakleitos of Ephesos, Himilkōn, Hippasos, Khersiphrōn, Mandroklēs, Metagenēs, Rhoikos, Skulax of Karuanda, Theagenēs, Xenophanēs		
505–470 BCE	(21) Alkmaiōn of Krotōn, Anaxagoras, Epikharmos, Eupalinos, Hanno of Carthage, Hekataios of Milētos, Hērakleitos of Ephesos, Himilkōn, Hippasos, Kleoitas, Mandroklēs, Melissos, Metagenēs, Parmenidēs, Phaiax, Puthagoras of Zakunthos, Rhoikos, Skulax of Karuanda, Theagenēs, Xanthos, Xenophanēs	505–400 BCE	Abas, Alkamenēs, Andrōn (Math), Apollodōros of Lēmnos, Bakōris of Rhodes, Dēmokedēs of Krotōn, Dēmophilos, Eratoklēs, Euruphōn, Harpalos (Astron.), Hērakleitos (Math.), Hērodikos of Selumbria, Hippokratic Corpus (Anatomy and Physiology, Aphoristic Works, Epidēmiai, Gynecological Works, Nosological Works, Prognostic Works, Protreptic Works, Regimen, Sevens, Surgery), Iktinos, Keras, Kharetidēs, Kleophanēs, Kleostratos, Massiliot Periplous, Melampous of Sarnaka, Milōn,
470–435 BCE	(48) Agatharkhos of Samos, Akrōn of Akragas, Anaxagoras, Antiphōn of Athens, Aristeidēs (Mech.), Aristōn (I), Arkhelaos of Athens, Artemōn of Klazomenai, Damastēs of Sigeion, Damōn of Athens,		

Dates:	*Names:*	*Dates (Wide):*	*Names:*
	Dēmokritos of Abdēra, Diogenēs of Apollōnia, Dionusios of Milētos, Empedoklēs, Euktēmōn, Gorgias of Leontinoi, Hērodikos of Knidos, Hērodotos of Halikarnassos, Hippias, Hippokratēs of Khios, Hippokratēs of Kōs, Hippōn of Krotōn, Ikkos, Iōn, Kallikratēs (Arch.), Karpiōn, Kleoitas, Leophanēs, Leukippos, Melissos, Menestōr, Metōn, Oinopidēs, Parmenidēs, Pausanias of Gela, Pausimakhos, Petrōn of Himera, Phaeinos, Phileas, Polukleitos of Argos, Promathos, Prōtagoras of Abdēra, Puthagoras of Zakunthos, Stēsimbrotos, Stuppax, Xanthos, Zēnō of Elea, Zōpuros (Physiog.)	505–400 BCE	Pephrasmenos, Petrōn of Aigina, Phaōn, Plentiphanēs, Polubos, Puthagoras (Med.), Puthoklēs, Pythagoras, Simōn of Athens, Skuthinos, Sminthēs, Suennesis, Theodōros of Kurēnē, Theokudēs, Thessalos of Kōs, Thrasumakhos, Timotheos of Metapontion, Xen(okh)arēs, Xouthos, Zoroaster (pseudo)
435–400 BCE	(55) Agatharkhos of Samos, Aiskhulos, Akrōn of Akragas, Anaxagoras, Antiokhos of Surakousai, Antiphōn of Athens, Antisthenēs of Athens, Aristeidēs (Mech.), Aristōn (I), Artemōn of Klazomenai, Damastēs of Sigeion, Damōn of Athens, Dēmokritos of Abdēra, Diogenēs of Apollōnia, Dionusios of Milētos, Dissoi Logoi, Empedoklēs, Euktēmōn, Gorgias of Leontinoi, Hērodikos of Knidos,		

Dates:	***Names:***	***Dates (Wide):***	***Names:***
	Hērodotos of Halikarnassos, Hippias, Hippokratēs of Khios, Hippokratēs of Kōs, Hippokratic Corpus (AWP, Ancient Medicine, Head Wounds, Sacred Disease), Hippōn of Krotōn, Iōn, Kallikratēs (Arch.), Karpiōn, Kratulos, Kritias, Ktēsias, Leophanēs, Leukippos, Melissos, Metōn, Nikomakhos of Stageira, Oinopidēs, Pausanias of Gela, Pausimakhos, Petrōn of Himera, Phaeinos, Phileas, Philolaos, Polukleitos of Argos, Promathos, Prōtagoras of Abdēra, Stēsimbrotos, Stuppax, Thucydidēs, Zēnō of Elea, Zōpuros (Physiog.)		
400–365 BCE	(50) Aineias Tacticus, Androtiōn of Athens, Antisthenēs of Athens, Aristagoras (of Milētos?), Arkhutas of Taras, Athēnaios of Kuzikos, Brusōn, Dēmokritos of Abdēra, Dexippos, Diodōros of Eretria, Dissoi Logoi, Drakōn of Kōs, Ekhekratēs, Ekphantos, Euphranōr (Music), Eurutos, Gorgias of Leontinoi, Helikōn, Hērakleodōros, Hiketas, Hippokratēs of Kōs, Hippokratic Corpus (Ancient Medicine, Head Wounds),	400–295 BCE	Abas, Abdaraxos, Adeimantos, Aëthlios of Samos, Agathoklēs of Atrax, Agathoklēs of Khios, Agathōn of Samos, Aiskhinēs, Aiskhriōn, Aiskhulidēs, Akesias, Alexander (Geog.), Alexias, Alkamenēs, Amphilokhos, Amuntas (Geog.), Anakreōn (Astron.), Anaxipolis, Androitas, Andrōn (Math.), Antigonos of Kumē, Antiphanēs of Dēlos, Apollodōros (Med.), Apollodōros of Kition, Apollodōros of Lēmnos,

Dates:	***Names:***	***Dates (Wide):***	***Names:***
	Khrusippos of Knidos (Med. I), Kleidēmos, Kleinias, Ktēsias, Kudias (of Kuthnos?), Leōdamas, Leōn, Loxos, Mētrodōros of Khios, Mnēsitheos of Athens, Neokleidēs, Nikagoras, Nikomakhos of Stageira, Philistiōn of Lokroi, Plato, Polukritos, Prōros, Putheos of Priēnē, Sōkratēs (junior), Speusippos of Athens, Theaitētos, Theodōros of Phokaia, Timagētos, Xenokratēs of Khalkēdōn, Xenophilos, Xenophōn of Athens, Zēnodotos (Math.)		Apollodōros of Taras, Apollōnios of Pergamon (Agric.), Apollōnios of Pitanē, Aristaios, Aristanax, Aristeidēs of Samos, Aristomakhos of Soloi, Aristombrotos, Aristomenēs, Aristophanēs of Mallos, Aristophilos of Plataia, Aristotelian Corpus (Colors, HA 10, Mēkhanika, Sounds, Winds), Arkhelaos (Geog.), Arkhidēmos, Arkhutas, Astunomos, Athēnagoras (Agric.), Bakkheios of Milētos, Bakōris of Rhodes, Basilis, Biōn of Abdēra, Biōn of Soloi, Botrus, Bōtthaios, Dadis, Daliōn (Med.), Damigerōn, Dēmarkhos, Derveni papyrus, Dieukhēs, Diodōros of Priēnē, Dioklēs of Karustos, Dionusios of Alexandria (Mech.), Dionusodōros (Pharm.), Diophantos (Geog.), Dōrotheos of Athens, Douris of Samos, Epigenēs (Med.), Epikratēs of Hērakleia, Eratoklēs, Euagōn of Thasos, Euboulos (Agric.), Eudēmos of Athens, Euēnōr of Argos, Eunomos of Khios, Euphrōnios of Amphipolis, Euphrōnios of Athens, Euphutōn, Euruōdēs, Euthudēmos
365–330 BCE	(88) Aineias Tacticus, Amphinomos, Amuntas of Hērakleia Pontikē, Androkudēs (Med.), Androōn of Teōs, Androtiōn of Athens, Aristagoras (of Milētos?), Aristeidēs (of Knidos?), Aristoboulos of Kassandreia, Aristotelian Corpus Flood of the Nile, Aristotle, Aristoxenos of Taras, Arkhutas of Taras, Athēnagoras son of Arimnēstos, Athēnaios of Kuzikos, Baitōn, Brusōn, Deinōn of Kolophōn, Deinostratos of Prokonessos, Dexippos, Dikaiarkhos, Diodōros of Eretria, Diognētos (of Eruthrai?), Dionusios (Med.), Drakōn of Kōs, Ekhekratēs, Ekphantos, Ephoros, Eudoxos of Knidos, Euphranōr		

Dates:	***Names:***	***Dates (Wide):***	***Names:***
	(Music), Helikōn, Hērakleidēs of Hērakleia Pontikē, Hērakleodōros, Hermodōros of Surakousai, Hermotimos, Hestiaios, Hiketas, Kallippos, Kallisthenēs, Khrusippos of Knidos (Med. I), Kleidēmos, Kleinias, Kratēs of Khalkis, Kudias (of Kuthnos?), Leōdamas, Leōn, Leōnidas of Naxos, Loxos, Lusimakhos of Macedon, Menaikhmos, Menekratēs of Surakousai, Menōn, Mētrodōros of Khios, Mnēsitheos of Athens, Nausiphanēs, Neokleidēs, Nikagoras, Paiōnios, Periandros, Phainias of Eresos, Philippos of Opous, Philistiōn of Lokroi, Philōn of Eleusis, Philōnidēs of Crete, Plato, Polemarkhos, Polemōn of Athens, Poluidos of Thessalia, Polukritos, Poseidōnios of Macedon, Prōros, Putheos of Priēnē, Saturos of Paros, Silaniōn of Athens, Simos of Poseidōnia, Skulax of Karuanda (pseudo), Sōkratēs (junior), Speusippos of Athens, Stratōn (Med.), Theophrastos, Theudios, Timagētos, Timaios of Tauromenion, Xenokratēs of Khalkēdōn, Xenophilos, Xenophōn of Athens, Zēnodotos (Math.)		of Athens, Glaukidēs, Glaukos (Geog. II), Hēgēsidēmos, Hēgētōr of Buzantion, Hērakleitos (Math), Hermarkhos of Mutilēnē, Hermās *thēriakos*, Hikesios (Agric.), Hippokratic Corpus (Anatomy and Physiology, Aphoristic Works, Epidēmiai, Gynecological Works, Heart, Nosological Works, Oath, Prognostic Works, Protreptic Works, Regimen, Sevens, Surgery), Huriadas, Kaikalos of Argos, Kallimakhos of Kurēnē, Kallistratos, Kēphisophōn, Keras, Khaireas, Khairesteos, Kharetidēs, Kharitōn, Kharmandros, Kharōn of Carthage, Kleōn of Surakousai, Kleophanēs, Kleëmporos, Kommiadēs, Krantōr of Soloi, Kratēs (Agric.), Kritōn of Naxos, Ktēsiphōn, Leōnidas (Geog.), Lukōn of Iasos, Lukos of Rhēgion, Lusimakhos, Massiliot Periplous, Melampous, Melampous of Sarnaka, On Melissos Xenophanēs and Gorgias, Menandros of Hērakleia, Menandros of Priēnē, Menekratēs of Ephesos, Menekritos, Menestratos I, Menestratos II, Milōn, Mnēsidēs, Mnēsimakhos, Monās, Mousaios, Nautelēs, Neanthēs of

Dates:	***Names:***	***Dates (Wide):***	***Names:***
330–295 BCE	(101) Aigimios, Amphinomos, Amuntas of Hērakleia Pontikē, Anaxikratēs (of Rhodes?), Androkudēs (Med.), Andrōn of Teōs, Androsthenēs of Thasos, Aristeidēs (of Knidos?), Aristoboulos of Kassandreia, Aristokreōn, Aristotelian Corpus (Flood of the Nile, Indivisible Lines, Physiognomy), Aristotle, Aristoxenos of Taras, Aristullos, Autolukos of Pitanē, Baitōn, Bērossos, Bromios, Daliōn (Geog.), Daphnis of Milētos, Deinostratos of Prokonessos, Dēmētrios (of Athens?), Diadēs, Dikaiarkhos, Diodōros of Iasos, Diognētos (of Eruthrai?), Diognētos of Rhodes, Dionusios (Med.), Dionusios son of Oxumakhos, Diphilos of Siphnos, Epicurus, Epimakhos of Athens, Euclid, Euclidean Sectio Canonis, Eudēmos of Rhodes, Euphranōr (Arch.), Hēgēsias of Magnesia, Hekataios of Abdēra, Hērakleidēs of Hērakleia Pontikē, Hērakleodōros, Hermodōros of Surakousai, Hermotimos, Hestiaios, Hierōn of Soloi, Hierōnumos of Kardia, Idomeneus, Kallias, Kallippos, Kallisthenēs, Kharias, Kineas,		Kuzikos, Neoklēs, Neoptolemos, Ninuas, Olumpias, Olumpikos (Lith.), Onētidēs/Onētōr, Ōros, Pankratēs of Argos, Papias, P. Hibeh 2.187, P. Louvre inv. 7733, P. Oxy. 13.1609, P. Ryl. III.531, Pephrasmenos, Persis, Phanokritos, Phaōn, Phasitas, Philippos of Medma, Philiskos of Thasos, Philomēlos, Phulotimos, Pleistonikos, Plentiphanēs, Pollis (Arch.), Polubos, Polukleitos of Larissa, Poseidōnios of Corinth, Potamōn (Pharm.), Puthagoras (Med.), Puthiōn of Rhodes, Puthios, Puthoklēs, Puthoklēs of Samos, Saturos (Lithika), Sēmos of Dēlos, Silēnos, Simōn of Magnesia, Simos of Kōs, Skuthinos, Sminthēs, Sōkratēs of Argos, Sōranos of Kōs, Sōsikratēs of Rhodes, Sōsimenēs, Sōteira, Suennesis, Teukros of Carthage, Theodōros of Soloi (Kilikia), Theokudēs, Theomenēs, Theophilos (Agric.), Theoxenos, Thessalos of Kōs, Thrasuandros, Thrasuas, Thrasumakhos, Timaris, Timaristos, Timokharis, Tlēpolemos, Xen(okh)arēs, Xenagoras (of Hērakleia Pontikē?), Xenophōn (of Kōs?),

Dates:	***Names:***	***Dates (Wide):***	***Names:***
	Klearkhos, Kratēs of Khalkis, Lasos, Leōnidas of Naxos, Lusimakhos of Macedon, Mēdeios, Megasthenēs, Menaikhmos, Menekratēs of Elaious, Menōn, Mētrodōros of Lampsakos, Mursilos, Nausiphanēs, Nearkhos, Nikias of Milētos, Nikomakhos of Athens, Onēsikritos, Ophellās, Orthagoras, Paiōnios, P. Hibeh, P. Hibeh 1.27, Parmeniōn, Patroklēs, Phainias of Eresos, Pheidias, Philōn (Geog.), Philōn of Eleusis, Philōnidēs of Crete, Polemōn of Athens, Poluainos, Poluidos of Thessalia, Poseidōnios of Macedon, Praxagoras, Putheas of Massalia, Silaniōn of Athens, Simōnidēs (Geog.), Sōstratos of Knidos, Sōtakos, Stratōn (Med.), Theophrastos, Theophrastos (pseudo), Theudios, Timaios of Tauromenion, Xenokratēs of Khalkēdōn, Xenophilos, Zēnōn of Kition, Zōilos (of Cyprus?)		Xouthos, Zēnothemis, Zoroaster (pseudo)
295–260 BCE	(84) Alexander of Pleuron, Amōmētos, Androsthenēs of Thasos, Antigonos of Karustos, Apeimantos, Apollodōros the *thēriakos*, Aratos, Aristarkhos of Samos, Aristokreōn, Aristotelian	295–190 BCE	Abdaraxos, Adeimantos, Aëthlios of Samos, Agatharkhidēs of Knidos, Agatharkhidēs of Samos, Agathoklēs of Atrax, Agathoklēs of Khios, Agathoklēs of Milētos, Agathōn of

Dates:	***Names:***	***Dates (Wide):***	***Names:***
	Corpus (Breath, Physiognomy, Problems), Aristullos, Bērossos, Daimakhos of Plataia, Daliōn (Geog.), Damas, Dēmētrios (of Athens?), Dēmodamas, Dikaiarkhos, Diodōros of Iasos, Dionusios (Astron.), Dionusios (Geog.), Dionusios of Ephesos, Dionusios son of Oxumakhos, Diphilos of Siphnos, Epicurus, Euclid, Euclidean Sectio Canonis, Eudēmos of Alexandria, Eudēmos of Rhodes, Hagnodikē, Hēgēsias of Magnesia, Hekataios of Abdēra, Hērakleidēs "Kritikos," Hērophilos, Hierōn II, Hierōnumos of Kardia, Hipponikos, Idomeneus, Kallianax, Khrusippos of Knidos (Med. II), Kineas, Kleanthēs, Klearkhos, Kleophantos, Kōlōtēs, Ktēsibios, Lasos, Lukōn of Troas, Lusimakhos of Macedon, Manethōn of Sebennutos, Mēdeios, Megasthenēs, Mētrodōros of Lampsakos, Mursilos, Nikias of Milētos, Noumēnios of Hērakleia, Numphis, Ophiōn, P. Hibeh, P. Hibeh 1.27, Parmeniōn, Patroklēs, Pheidias, Philōn (Geog.), Polemōn of Athens, Poluainos, Polustratos, Poseidippos, Praxagoras, Purgotelēs, Purrhos of		Samos, Agēsias, Aiskhinēs, Aiskhriōn, Aiskhulidēs, Akesias, Akhaios, Alexander (Geog.), Alexander of Lukaia, Alexias, Alkimakhos, Amphilokhos, Amphiōn, Amuntas (Geog.), Anakreōn (Astron.), Anaxipolis, Androitas, Andrōn (Pharm.), Andronikos (Med.), Antigonos (Med.), Antigonos of Kumē, Antipatros (of Tarsos?), Apellēs (of Thasos?), Apellis, Apollodōros (Med.), Apollodōros of Kition, Apollodōros of Taras, Apollōnios Glaukos, Apollōnios of Aphrodisias, Apollōnios of Pergamon (Agric.), Apollōnios of Pitanē, Apollōnios of Tarsos, Apollōnios "Ophis," Apollophanēs of Nisibis, Aristaios, Aristanax, Aristandros of Athens, Aristeidēs (Paradox.), Aristeidēs of Samos, Aristoboulos, Aristodēmos, Aristolaos, Aristomakhos of Soloi, Aristombrotos, Aristomenēs, Aristophanēs, Aristophanēs of Mallos, Aristophilos of Plataia, Aristotelian Corpus (Colors, HA 10, Mēkhanika, Sounds, Winds), Aristotle (pseudo: Mirab.), Arkhelaos

Dates:	***Names:***	***Dates (Wide):***	***Names:***
	Epeiros, Puthagoras (of Alexandria?), Simōnidēs (Geog.), Sōstratos of Knidos, Sōtakos, Stratōn (Erasistratean), Stratōn of Lampsakos, Theophrastos, Timaios of Tauromenion, Timōn of Phleious, Timosthenēs, Zēnōn of Kition		(Geog.), Arkhelaos of Khersonēsos, Arkhestratos, Arkhidēmos, Arkhutas, Arrabaios, Artemōn (Epicurean), Artemōn of Kassandreia, Asklatiōn (Med.), Asklēpios (Med.), Aspasios (Pharm.), Astunomos, Athēnagoras (Agric.), Athēnodōros (of Rhodes?), Bakkheios of Milētos, Basilis, Biōn of Abdēra, Biōn of Soloi, Boëthos (Med.), Bōlos, Botrus, Dadis, Daliōn (Med.), Damigerōn, Damōn (Geog.), Dēïleōn, Dēmarkhos, Dēmētrios (Pythag.), Dēmētrios of Alexandria, Dēmētrios of Apameia, Dēmētrios of Kallatis, Dēmoklēs, Dēmokritos (pseudo: Agric.), Dēmokritos (pseudo: Alch.), Dēmokritos (pseudo: Lith.), Derkullos, Didumos of Knidos, Dieukhēs, Diodōros of Priēnē, Diokleidēs of Abdēra, Dioklēs of Khalkēdōn, Diomēdēs, Dionusios (of Halikarnassos?), Dionusios of Alexandria (Mech.), Dionusios of Corinth, Dionusios of Rhodes, Dionusios of Samos, Dionusios: son of Diogenēs, Dionusodōros (Pharm.), Diophantos (Geog.), Diphilos, Dōriōn (Mech.), Dōrotheos of Athens, Dōrotheos of
260–225 BCE	(78) Alexander of Pleuron, Amōmētos, Andreas of Karustos, Antigenēs, Antigonos of Karustos, Apeimantos, Apollodōros the *thēriakos*, Apollōnios of Memphis, Aratos, Archimēdēs, Aristogenēs of Knidos, Aristokreōn, Aristōn of Ioulis, Aristophanēs of Buzantion, Aristotelian Corpus (Breath, Problems), Aristotheros, Bakkheios of Tanagra, Chrysippus of Soloi, Damas, Dēmētrios of Amisos, Dionusios (Geog.), Dionusios of Ephesos, Dionusios son of Oxumakhos, Diphilos of Siphnos, Dōsitheos of Pēlousion, Erasistratos of Keōs, Eratosthenēs, Eudēmos of Alexandria, Hēgēsias of Magnesia, Hērakleidēs "Kritikos," Hermippos (of Smurna?), Hērodotos (Mech.), Hierōn II, Hierōnumos of Kardia, Hierōnumos of Rhodes, Hipponikos, Kallianax, Kallimakhos Jr. of Kurēnē, Khrusippos of Knidos (Med. II),		

Dates:	Names:	Dates (Wide):	Names:
	Kleanthēs, Kleophantos, Kōlōtēs, Konōn, Ktēsibios, Lakudēs, Lasos, Leptinēs I, Lukōn of Troas, Mnēmōn of Sidē, Mursilos, Neileus, Nikias of Milētos, Nikotelēs, Noumēnios of Hērakleia, Numphis, Numphodōros, Numphodōros of Surakousai, Ophiōn, P. Hibeh (Ophth.), Pheidias, Philōn (Geog.), Philōn of Buzantion, Philostephanos, Polustratos, Poseidippos, Puthiōn of Thasos, Serapiōn of Alexandria (Empir.), Simmias (of Macedon?), Simmias son of Mēdios, Simōnidēs (Geog.), Sōstratos of Knidos, Sphairos, Stratōn (Erasistratean), Sudinēs, Thrasudaios, Timōn of Phleious, Timosthenēs		Hēliopolis, Dōrotheos of Khaldaea, Douris of Samos, "Dtrums," Eirēnaios, Epigenēs of Rhodes, Epigonos, Epikouros (Pharm.), Epikratēs of Hērakleia, Erasistratos of Sikuōn, Euagōn of Thasos, Euainetos, Euangeus, Euboulidēs, Euboulos (Agric.), Euboulos (Pharm.), Eudēmos the elder, Eudikos, Eudoxos of Rhodes, Euēnōr of Argos, Euēnos, Euhēmeros (Pharm.), Euphoriōn of Khalkis, Euphranōr (Pharm.), Euphrōnios of Amphipolis, Euphrōnios of Athens, Euphutōn, Euruōdēs, Euthudēmos of Athens, Euthukleos, Gennadios, Glaukidēs, Glaukos (Geog. I), Glaukos (Geog. II), Glukōn, Halieus, Harpokrās, Hēgēsidēmos, Hēgētōr (Med.), Hēgētōr of Buzantion, Hekataios (Pharm.), Hēliodōros of Athens, Hērakleitos (Math.), Hērakleitos of Sikuōn, Hermarkhos of Mutilēnē, Hermās *thēriakos*, Hermeias (Ophthalm.), Hierax of Thēbai, Hikatidas, Hikesios (Agric.), Hippokratic Corpus (Heart, Oath, Protreptic Works, Sevens), Idios, Ioudaios, Iriōn, Isis:
225–190 BCE	(70) Andreas of Karustos, Antigenēs, Apollōnios (of Alexandria), Apollōnios of Antioch, Apollōnios of Memphis, Apollōnios of Pergē, Apollophanēs of Seleukeia, Archimēdēs, Aristiōn (father: Mech.), Aristippos of Kurēnē, Aristōn of Ioulis, Aristophanēs of Buzantion, Arkhagathos of Lakōnika, Bakkheios of Tanagra, Basileidēs, Chrysippus of Soloi, Damastēs, Damis of Kolophōn, Damōn of		

Dates:	***Names:***	***Dates (Wide):***	***Names:***
	Kurēnē, Dēmokleitos, Diagoras, Diogenēs of Babylōn, Dioklēs, Dionusodōros (of Kaunos?), Dōsitheos of Pēlousion, Ennius, Eratosthenēs, Glaukias of Taras, Hēgēsianax, Heirodotos, Hērakleidēs of Taras (Mech.), Hermippos (of Smurna?), Hermogenēs of Alabanda, Hērodotos (Mech.), Hierōn II, Hippobotos, Kharōn of Magnesia, Kleoxenos, Konōn, Lakudēs, Minius Percennius, Mnaseas of Patara, Mnēmōn of Sidē, Mnēsitheos of Kuzikos, Moskhiōn (Mech.), Naukratēs, Neileus, Nikomēdēs, Nikotelēs, Numphodōros, Numphodōros of Surakousai, P. Parisinus graecus 1, Pausistratos, Philōn of Buzantion, Philostephanos, Prōtarkhos (Mech. and Pharm.), Saturos of Kallatis, Serapiōn of Alexandria (Empir.), Simmias (of Macedon?), Simmias son of Mēdios, Skopinas, Sōtiōn of Alexandria, Sphairos, Thrasudaios, Truphōn of Alexandria, Xenagoras son of Eumēlos, Zēnodōros, Zēnōn (Med.), Zeuxippos, Zōpuros of Taras		pseudo (Pharm.), Kalliklēs, Kallimakhos (of Bithunia), Kallimakhos of Kurēnē, Kalliphanēs, Kallistratos, Kallixeinos, Kēphisophōn, Khaireas, Khairesteos, Khalkideus, Kharitōn, Kharmandros, Kharōn of Carthage, Khios, Khrusippos (Agric.), Kimōn, Kleoboulos (Pharm.), Kleōn of Surakousai, Kleëmporos, Kloniakos, Klutos, Kōdios Toukos, Kommiadēs, Krantōr of Soloi, Kratēs (Agric.), Kritōn of Naxos, Ktēsiphōn, Kudias, Laodikos, Leōnidas (Geog.), Leptinēs II, Linos (pseudo), Lobōn, Lukōn of Iasos, Lukos of Rhēgion, Lunkeus, Lusimakhos, Lusimakhos of Kōs, Melampous, Melampous of Sarnaka, On Melissos Xenophanēs and Gorgias, Melitōn, Menandros of Hērakleia, Menandros of Priēnē, Menekratēs of Ephesos, Menekritos, Menestratos I, Menestratos II, Mēnodotos (Astr.), Menoitas, Mētrodōros son of Epikharmos (pseudo), Milēsios, Miltiadēs, Mnēsidēmos, Mnēsidēs, Mnēsimakhos, Molpis, Monās, Mousaios, Muia: pseudo, Murōn, Naukratitēs medicus, Nautelēs, Neanthēs of

Dates:	***Names:***	***Dates (Wide):***	***Names:***
			Kuzikos, Neoklēs, Neoptolemos, Nikētēs (of Athens?), Nikomakhos (Pharm.), Nikomēdēs (Hērakleitean), Ocellus, Olumpias, Olumpikos (Lith.), Olumpionikos, Onētidēs/Onētōr, Ōriōn of Bithunia, Ōros, Orpheus (pseudo: Astrol.), Orpheus (pseudo: Med.), Pankratēs of Argos, Pantainos, Papias, P. Ashmolean Library, P. Fayumensis, P. Hibeh 2.187, P. Louvre inv. 7733, P. Oxy. 13.1609, P. Ryl. III.531, Pasiōn, Pausanias "Hērakleiteios," Perigenēs, Perseus, Persis, Petrikhos, Phaidros, Phanias, Phanokritos, Philinos of Kōs, Philippos of Medma, Philiskos of Thasos, Philoklēs, Philokratēs, Philomēlos, Philōn (Meteor.), Philōn of Hērakleia, Philōn of Tuana, Philōnidēs of Laodikeia, Phulotimos, Pleistonikos, Plentiphanēs, Podanitēs, Politēs, Pollis (Arch.), Polueidēs, Polustomos, Poseidōnios of Corinth, Potamōn (Pharm.), Prutanis, Ptolemaios (Erasi.), Ptolemaios (Med.), Puramos, Purrhos of Magnesia, Puthagoras (Med.), Puthiōn of Rhodes, Puthios, Puthoklēs of Samos, Salmeskhoiniaka,

Dates:	*Names:*	*Dates (Wide):*	*Names:*
			Sarkeuthitēs, Saturos (Lithika), Seleukos of Tarsos, Sēmos of Dēlos, Sergius of Babylōn, Silēnos, Simōn of Magnesia, Simos of Kōs, Sminthēs, Sōkratēs of Argos, Sōkratiōn, Solōn, Sōranos of Kōs, Sōsagoras, Sōsandros (Pharm.), Sōsikratēs, Sōsikratēs of Rhodes, Sōsimenēs, Sōteira, Sōtiōn, Speusippos of Alexandria, Staphulos, Sunerōs, Telamōn, Telephanēs, Teukros of Carthage, Thamuros, Theanō (pseudo), Theodōros of Soloi (Kilikia), Theodosios (of Bithunia), Theokhrēstos, Theokritos, Theokudēs, Theomenēs, Theophilos (Agric.), Theophilos (Lithika), Theoxenos, Thrasuandros, Thrasuas, Timagoras, Timaios (Pharm.), Timaris, Timaristos, Timokharis, Timōn, Timotheos, Tlēpolemos, Xanitēs, Xen(okh)arēs, Xenagoras (of Hērakleia Pontikē?), Xenophōn (of Kōs?), Zēnōn of Laodikeia, Zēnothemis, Zeuxis (Empir.), Zōpuros (Geog.)
190–155 BCE	(58) Antipatros of Tarsos, Apollodōros of Kerkura, Apollodōros of Seleukeia, Apollōnios (of Alexandria), Apollōnios of Antioch, Apollōnios of	190–85 BCE	Abdaraxos, Abram, Adeimantos, Aemilius Hispanus, Agatharkhidēs of Knidos, Agatharkhidēs of Samos, Agathoklēs of Atrax,

Dates:	***Names:***	***Dates (Wide):***	***Names:***
	Pergē, Apollōnios "Biblas," Apollophanēs of Seleukeia, Aristiōn (father: Mech.), Aristippos of Kurēnē, Aristophanēs of Buzantion, Arkhedēmos of Tarsos, Basileidēs, Bitōn, Boëthos of Sidōn (Stoic), Damastēs, Damis of Kolophōn, Damōn of Kurēnē, Dēmokleitos, Diagoras, Diogenēs of Babylōn, Dioklēs, Dionusios of Kurēnē, Ennius, Fuluius Nobilior, Genthios, Glaukias of Taras, Hēgēsianax, Heirodotos, Hermogenēs of Alabanda, Hērodotos (Mech.), Hippobotos, Isidōros of Abudos, Kleoxenos, Kratēs of Mallos, Kritolaos, Menandros (of Pergamon?), Minius Percennius, Mnaseas of Patara, Mnēsitheos of Kuzikos, Moskhiōn (Mech.), Naukratēs, P. Parisinus graecus 1, Pasikratēs, Polemōn of Ilion, Polubios, Porcius Cato, Prōtarkhos (Mech. and Pharm.), Saturos of Kallatis, Seleukos of Seleukeia, Skopinas, Skumnos, Sōtiōn of Alexandria, Sulpicius Gallus, Xenagoras son of Eumēlos, Zēnodotos (of Mallos?), Zēnōn (Med.), Zeuxippos		Agathoklēs of Khios, Agathoklēs of Milētos, Agathōn of Samos, Agēsias, Aisara, Aiskhinēs, Aiskhriōn, Aiskhulidēs, Akhaios, Akhillās, Akhinapolos, Alexander (Geog.), Alexander of Lukaia, Alkimakhos, Alkimiōn, Ammōn (Astrol.), Amphilokhos, Amphiōn, Amuthaōn, Anakreōn (Astron.), Anaxipolis, Androkudēs (Pythag.), Andrōn (Pharm.), Andronikos (Med.), Andronikos of Rhodes, Antigonos (Med.), Antigonos of Kumē, Antipatros (of Tarsos?), Antipatros of Tyre, Antisthenēs (of Rhodes), Antonius "root-cutter," Apellēs (of Thasos?), Apellis, Aphrodās, Apios Phaskos, Apollodōros (Med.), Apollodōros Dēmokritean, Apollodōros of Artemita, Apollodōros of Kitiōn, Apollodōros of Taras, Apollōnios Glaukos, Apollōnios of Athens, Apollōnios of Pergamon (Agric.), Apollōnios of Pitanē, Apollōnios of Tarsos, Apollōnios "Ophis," Arbinas of Indos, Areios Didumos, Ariobarzanēs, Aristanax, Aristandros of Athens, Aristeidēs (Paradox.), Aristeidēs of Samos,
155–120 BCE	(51) Acilius, Andronikos of Kurrhos,		

Dates:	***Names:***	***Dates (Wide):***	***Names:***
	Ankhialos, Antipatros of Tarsos, Apollodōos of Athens Apollodōos of Kerkura, Apollodōros of Seleukeia, Apollōnios (Paradoxographer), Apollōnios (of Alexandria), Apollōnios of Antioch, Apollōnios "Biblas," Arkhedēmos of Tarsos, Attalos III, Attalos of Rhodes, Boëthos of Sidōn (Stoic), Damastēs, Diogenēs of Babylōn, Diogenēs of Tarsos, Dionusios of Kurēnē, Hērakleidēs of Kallatis, Hermogenēs of Alabanda, Hipparkhos of Nikaia, Hupsiklēs, Iollas, Iunius Silanus, Kassandros, Kleitomakhos, Kratēs (Geom.), Kratēs of Mallos, Kritolaos, Mantias (Hēroph.), Minius Percennius, Nikandros of Kolophōn, Panaitios of Rhodes, Pasikratēs, Pausanias of Damaskos, Petosiris, Polubios, Porcius Cato, Prōtarkhos of Bargulia, Ptolemaios of Kurēnē, Saturos of Kallatis, Seleukos of Seleukeia, Skopinas, Skulax of Halikarnassos, Skumnos, Sōtiōn of Alexandria, Sulpicius Gallus, Trebius Niger, Zēnodotos (of Mallos?), Zēnōn (Med.)		Aristiōn: grandson (Mech.), Aristoboulos, Aristodēmos, Aristoklēs, Aristolaos, Aristomakhos of Soloi, Aristombrotos, Aristomenēs, Aristophanēs, Aristophanēs of Mallos, Aristotle (pseudo: Mirab.), Arkhebios/Arkesios, Arkhelaos (Geog.), Arkhelaos of Khersonēsos, Arkhestratos, Arkhutas, Arrabaios, Artemidōros of Sidē, Artemōn (Epicurean), Artemōn of Kassandreia, Asklatiōn (Med.), Asklēpios (Med.), Aspasios (Pharm.), Astunomos, Athēnagoras (Agric.), Athēnippos, Athēnodōros (of Rhodes?), Azanitēs, Bakkheios of Milētos, Basilis, Bathullos, Biōn Caecilius, Boëthos (Med.), Boëthos of Sidōn (Perip.), Bōlos of Mendes, Boutoridas, Brenitus, Caecilius "Medicus," Campestris, Cornelius, Cornelius Bocchus, Dadis, Daliōn (Med.), Damigerōn, Damōn (Geog.), Dasius, Dēïleōn, Dēmarkhos, Dēmētrios (Geog.), Dēmētrios (Pythag.), Dēmētrios of Alexandria, Dēmētrios of Apameia, Dēmētrios of Kallatis, Dēmētrios of Lakōnika, Dēmētrios "physicus," Dēmoklēs, Dēmokritos (pseudo:
120–85 BCE	(68) Adrastos of Kuzikos, Agesistratos, Ainesidēmos, Andrias, Ankhialos, "Antikythera		

Dates:	*Names:*	*Dates (Wide):*	*Names:*
	Device," Antiokhis of Tlōs, Antiokhos VIII, Apollodōros of Athens, Apollōnidēs, Apollōnios (Paradoxographer), Apollōnios of Kition, Apollōnios of Mundos, Aristōn of Khios, Artemidōros of Ephesos, Asklēpiadēs Titiensis, Asklēpiadēs of Bithunia, Asklēpiadēs of Murleia, Aufidius of Sicily, Billaros, Dēmētrios Khlōros, Diodotos (Astr. I), Diogenēs of Tarsos, Diōn of Neapolis, Dionusios of Kurēnē, Dionusios of Utica, Dionusios son of Kalliphōn, Egnatius, Epainetos, Eudoxos of Kuzikos, Fufi(ci)us, Gorgias of Alexandria, Hērakleidēs of Kallatis, Hērakleidēs of Taras (Med.), Hērōn (Med.), Hikesios of Smurna, Hubristēs, Hupsiklēs, Iollas, Kassandros, Kleitomakhos, Krateuas, "Lion Horoscope," Mantias (Hēroph.), Mētrodōros of Skēpsis, Mithradatēs VI, Mnaseas of Milētos, Mnēsarkhos, Nikandros of Kolophōn, Nikomēdēs IV, Orestinos, Panaitios of Rhodes, P. bibl. univ. Giss. IV.44, Parmeniskos, Pausanias of Damaskos, Petosiris, Philoxenos, Polubios, Poseidōnios of Apameia, Ptolemaios of Kurēnē,		Agric.), Dēmokritos (pseudo: Alch.), Dēmokritos (pseudo: Lith.), Dēmokritos (pseudo: Med.), Dēmokritos (pseudo: Pharm.), Dēmotelēs, Derkullos, Dexios, Didumos of Knidos, Diodōros (Astron.), Diodōros (Empir.), Diodōros of Priēnē, Diodōros of Samos, Dioklēs of Khalkēdōn, Dioklēs of Magnesia, Diomēdēs, Diōn (Med.), Dionusios (of Halikarnassos?), Dionusios of Corinth, Dionusios of Kurtos, Dionusios of Philadelpheia, Dionusios of Rhodes, Dionusios of Samos, Dionusios: Sallustius, Dionusios son of Diogenēs, Dionusodōros (Pharm.), Diophantos (Geog.), Diophil–, Dioskoros (Pharm.), Diphilos, Dōriōn (Biol.), Dōriōn (Mech.), Dōrotheos of Athens, Dōrotheos of Hēliopolis, Dōrotheos of Khaldaea, "Dtrums," Eirēnaios, Elephantinē/ Elephantis, Emeritus (Hemeritos), Epagathos, Epainetēs, Epidauros, Epigenēs of Buzantion, Epigenēs of Rhodes, Epigonos, Epiklēs of Crete, Epikouros (Pharm.), Epikratēs of

Dates:	*Names:*	*Dates (Wide):*	*Names:*
	Septimius, Serapiōn of Antioch, Skulax of Halikarnassos, Sōstratos of Nusa, Themisōn, Theodotos, Theophanēs of Mutilēnē, Xenophōn of Lampsakos		Hērakleia, Erasistratos of Sikuōn, Euagōn of Thasos, Euainetos, Euangeus, Euboulidēs, Euboulos (Agric.), Euboulos (Pharm.), Eudēmos the elder, Eudikos, Euēnos, Eugeneia, Eugērasia, Euhēmeros (Pharm.), Eukleidēs "Palatianus," Euphranōr (Pharm.), Euphranōr (Pythag.), Euphrōnios of Amphipolis, Euphrōnios of Athens, Euphutōn, Euruōdēs, Euskhēmos, Euthudēmos of Athens, Euthukleos, Faustinus, Fronto (Astrol.), Geminos, Gennadios, Glaukidēs, Glaukōn/Glaukos, Glaukos (Geog. I), Glaukos (Geog. II), Glukōn, Granius, Halieus, Harpalos (Pharm.), Harpokrās, Hēgēsidēmos, Hēgētōr (Med.), Hēgētōr of Buzantion, Hekataios (Pharm.), Hēliodōros of Athens, Hērakleitos (Math.), Hērakleitos of Sikuōn, Hermās *thēriakos*, Hermeias (Astrol.), Hermeias (Ophthalm.), Hermēs Trismegistos (pseudo), Hermōn of Egypt, Hermophilos, Hierax of Thēbai, Hikatidas, Hikesios (Agric.), Hippokratic Corpus (Oath, Protreptic Works, Sevens), Hostilius Saserna and son,

Dates:	Names:	Dates (Wide):	Names:
			Hugiēnos, Idios, Ioudaios, Iriōn, Isis: pseudo (Pharm.), Kalliklēs, Kalliphanēs, Kallistratos, Kallixeinos, Karneadēs, Kēphisophōn, Keskintos: Inscription of, Khaireas, Khairesteos, Khalkideus, Khariklēs, Kharitōn, Kharōn of Carthage, Khios, Khrusanthos, Khrusippos (Agric.), Khrusippos (Med.), Kidēnas, Kimōn, Kleoboulos (Pharm.), Kleōn (of Kuzikos?), Kleëmporos, Kloniakos, Klutos, Kōdios Toukos, Kommiadēs, Korē Kosmou, Kratēs (Agric.), Kratippos, Kratōn, Ktēsiphōn, Kudias, Laïs, Lampōn, Laodikos, Leōnidas (Geog.), Leōnidas of Buzantion, Leptinēs II, Licinius Atticus, Linos (pseudo), Litorius, Lobōn, Lukomēdēs, Lukos of Neapolis, Lunkeus, Lupus, Lusias, Lusimakhos, Lusimakhos of Kōs, Magnus of Philadelpheia, Makhairiōn, Maria, Melampous of Sarnaka, Melissos Xenophanēs and Gorgias, Melitōn, Menandros of Hērakleia, Menandros of Priēnē, Menekritos, Menelaos (Pharm.), Menestheus, Menestratos II, Menippos, Mēnodotos (Astr.), Menoitas,

Dates:	***Names:***	***Dates (Wide):***	***Names:***
			Mēnophilos, Mētrodōros (Astr. I), Mētrodōros (Pharm.), Mētrodōros of Buzantion, Mētrodōros son of Epikharmos (pseudo), Mikiōn, Milēsios, Miltiadēs, Mnēsidēmos, Mnēsidēs, Molpis, Moskhiōn (Pharm.), Mousaios, Muia: pseudo, Muōnidēs, Murōn, Naukratitēs medicus, Neanthēs of Kuzikos, Neoklēs, Neoptolemos, Nikētēs (of Athens?), Nikias of Mallos, Nikias of Nikaia, Nikolaos (Pharm.), Nikomakhos (Pharm.), Nikomēdēs (Hērakleitean), Ocellus, Olumpias, Olumpikos (Lith.), Olumpionikos, Onētidēs/Onētōr, Oppius, Ōrigeneia, Ōriōn of Bithunia, Ōros, Orpheus (pseudo: Astrol.), Orpheus (pseudo: Med.), Orthōn, Panaitios Jr., Pankratēs of Argos, Pantainos, Papias, P. Ashmolean Library, P. Fayumensis, P. Lit. Lond. 167, P. Osloensis 73, P. Oxy. 13.1609, P. Oxy. 15.1796, Pasiōn, Pausanias "Hērakleiteios," Paxamos, Perigenēs, Periklēs, Perseus, Persis, Petrikhos, Phaidros, Phanias, Philippos (Astron.), Philippos of Macedon, Philiskos of Thasos,

Dates:	***Names:***	***Dates (Wide):***	***Names:***
			Philistidēs of Mallos, Philokalos, Philoklēs, Philokratēs, Philomēlos, Philōn (Meteor.), Philōn of Tuana, Philōnidēs of Laodikeia, Phoibos Ulpius, Platusēmos, Plentiphanēs, Podanitēs, Politēs, Pollis (Arch.), Polueidēs, Polustomos, Poseidōnios of Corinth, Potamōn (Pharm.), Primiōn, Proëkhios, Prōtagoras of Nikaia, Prōtarkhos of Tralleis, Prōtās, Proxenos, Prutanis, Ptolemaios (Erasi.), Ptolemaios (Med.), Puramos, Purrhos of Magnesia, Puthagoras (Med.), Puthiōn of Rhodes, Puthios, Puthoklēs of Samos, Quadratus, Rabirius, Salmeskhoiniaka, Salpē, Samithra, Sarkeuthitēs, Saturos (Lithika), Seleukos of Tarsos, Sēmos of Dēlos, Serapiōn of Alexandria (Astrol.), Sergius of Babylōn, Silēnos, Silo, Simos of Kōs, Sōkratēs of Argos, Sōkratiōn, Solōn, Sōranos of Kōs, Sōsagoras, Sōsandros (Geog.), Sōsandros (Pharm.), Sōsikratēs, Sōsikratēs of Rhodes, Sōsimenēs, Sōteira, Sōtiōn, Spendousa, Speusippos of Alexandria, Staphulos, Stratōn of Bērutos, Sunerōs, Tektōn,

Dates:	***Names:***	***Dates (Wide):***	***Names:***
			Telamōn, Telephanēs, Teukros of Carthage, Thamuros, Tharseas, Theanō (pseudo), Theodōros of Soloi (Kilikia), Theodosios (of Bithunia), Theokhrēstos, Theokritos, Theokudēs, Theomenēs, Theophilos (Agric.), Theophilos (Geog.), Theophilos (Lithika), Theopompos, Theoxenos, Thrasuandros, Threptos, Timagoras, Timaios (Pharm.), Timaios of Lokris (pseudo), Timaris, Timaristos, Timōn, Timotheos, Tlēpolemos, Turannos, Vicellius, Xanitēs, Xen(okh)arēs, Zakhalias, Zēnōn of Laodikeia, Zēnōn of Sidōn, Zēnophilos, Zeuxis (Empir.), Zōpuros (Geog.), Zōpuros of Alexandria
85–50 BCE	(80) Adrastos of Kuzikos, Agesistratos, Ainesidēmos, Alexander of Ephesos, Alexander of Milētos, Andrias, Antigonos of Alexandria, Antiokhis of Tlōs, Apollōnidēs, Apollōnios of Kition, Apollōnios of Mundos, Aristōn of Khios, Arkhelaos (Lithika), Artemidōros of Parion, Artemidōros of Pergē, Artorius, Asklēpiadēs of Murleia, Athēnodōros of Tarsos, Aufidius of Sicily, Billaros, Cornelius Nepos,	85 BCE–20 CE	Abaskantos, Abdaraxos, Abram, Adeimantos, Aemilius Hispanus, Aëtios, Aetna, Agatharkhidēs of Samos, Agathoklēs, Agathoklēs of Atrax, Agathoklēs of Milētos, Agathōn of Samos, Agēsias, Aineios, Aisara, Aiskhinēs, Aiskhulidēs, Akhaios, Akhillās, Akhinapolos, Alexander (Geog.), Alexander of Lukaia, Alkimakhos, Alkimiōn, Amarantos, Ammōn (Astrol.), Amphiōn,

Dates:	*Names:*	*Dates (Wide):*	*Names:*
	Dēmētrios Khlōros, Diodōros of Sicily, Diodotos (Astr. I), Diōn of Neapolis, Dionusios (of Milētos?), Diophanēs of Nikaia, Dioskouridēs Phakas, Egnatius, Eudōros of Alexandria, Fufi(ci)us, Gorgias of Alexandria, Hērakleidēs of Ephesos, Hērakleidēs of Taras (Med.), Hērōn (Med.), Hikesios of Smurna, Iulius Caesar, Khrusermos, Kleopatra, Krateuas, Licinius Caluus, "Lion Horoscope," Lucretius Carus, Mamilius Sura, Mēnodōros, Mētrodōros of Skēpsis, Mithradatēs VI, Mnaseas of Milētos, Naburianos, Nigidius Figulus, Nikomēdēs IV, Nikōn of Akragas, Nikōnidēs, Orestinos, P. bibl. univ. Giss. IV.44, Parmeniskos, Philodēmos, Philōnidēs of Durrakhion, Pompeius Lenaeus, Poseidōnios (Med. I), Poseidōnios of Apameia, Ptolemaios of Kurēnē, Sallustius (Cn.), Septimius, Serapiōn of Antioch, Sōsigenēs (I), Sōstratos of Nusa, Sueius, Tarutius, Terentius Varro of Narbo, Terentius Varro of Reate, Themisōn, Theodotos, Theophanēs of Mutilēnē, Timagenēs, Tremellius Scrofa, Tullius Cicero (M.), Tullius Cicero (Q.), Xenophōn of Lampsakos		Amuthaōn, Androkudēs (Pythag.), Andrōn (Pharm.), Andronikos (Med.), Andronikos of Rhodes, Anoubiōn of Diospolis, Antigonos (Med.), Antimakhos (Pharm.), Antiokhos of Athens, Antipatros (Pharm.), Antipatros of Tyre, Antisthenēs (of Rhodes), Antoninus of Kōs, Antonius "root-cutter," Apellēs (of Thasos?), Aphrodās, Aphrodisis, Apios Phaskos, Apollodōros Dēmokritean, Apollodōros of Artemita, Apollodōros of Athens (pseudo), Apollodōros of Kition, Apollodōros of Taras, Apollōnios Glaukos, Apollōnios of Alexandria ("Mus"), Apollōnios of Athens, Apollōnios of Pergamon (Med.), Apollōnios of Pitanē, Apollōnios of Prousias, Apollōnios of Tarsos, Apollōnios "Ophis," Aquila Secundilla, Arbinas of Indos, Areios Didumos, Ariobarzanēs, Aristanax, Aristarkhos of Sikuōn, Aristeidēs (Paradox.), Aristeidēs of Samos, Aristiōn: grandson (Mech.), Aristoboulos, Aristodēmos, Aristoklēs, Aristoklēs of Messēnē, Aristolaos, Aristomakhos of Soloi, Aristombrotos,

Dates:	*Names:*	*Dates (Wide):*	*Names:*
50–15 BCE	(87) Aelius Gallus, Aemilius Macer, Africanus (Pharm.), Alexander of Ephesos, Alexander of Laodikeia, Alexander of Milētos, Ambiuius, Ammōnios of Alexandria, Anaxilaos of Larissa, Antigonos of Alexandria, Antiokhos (Paccius), Antonius Musa, Aristōn (II), Arkhelaos (Lithika), Arkhelaos of Kappadokia, Artemōn (Med.), Artorius, Asinius Pollio, Athēnaios Mechanicus, Athēniōn (of Athens?), Athēnodōros of Tarsos, Castricius, Clodius Tuscus, Cornelius Nepos, Diodōros of Sicily, Dionusios (of Milētos?), Diophantos of Lukia, Dioskouridēs Phakas, Eudōros of Alexandria, Euelpistos (Terentius), Euphorbos, Florus, Fonteius Capito, Grattius Faliscus, Hērās, Hupsikratēs, Iouba II, Isidōros of Kharax, Iulius Caesar, Iulius Caesar (Augustus), Iulius Hyginus, Khrusermos, Kleopatra, Krateros, Licinius Caluus, Maecenas Licinius, Maecenas Melissus, Matius Caluena, Melētos, Mēnodōros, Mnaseas of Milētos, Naburianos, Nigidius Figulus, Nikolaos of Damaskos, Nikōn of Akragas, Olumpos of Alexandria, Ouidius Naso, Philodēmos, Pompeius		Aristophanēs, Aristotle (pseudo: Mirab.), Arkhebios/Arkesios, Arkhelaos (Geog.), Arrabaios, Artemidōros of Sidē, Asklatiōn (Med.), Asklēpiodotos (of Nikaia?), Asklēpios (Med.), Aspasios (Pharm.), Athēnippos, Athēnodōros (of Rhodes?), Attius, Axios, Azanitēs, Bathullos, Biōn Caecilius, Bithus of Durrakhion, Blastos, Boëthos (Med.), Boëthos of Sidōn (Perip.), Boutoridas, Brenitus, Caecilius "Medicus," Caesennius, Campestris, Candidus, Celer, Clodius (Asklēpiadean), Clodius of Naples, Cornelius, Cornelius Bocchus, Daliōn (Med.), Damigerōn, Damōn (Geog.), Damostratos/Dēmostratos, Dasius, Dēïleōn, Dēmarkhos, Dēmētrios (Geog.), Dēmētrios of Alexandria, Dēmētrios of Lakōnika, Dēmētrios "physicus," Dēmoklēs, Dēmokritos (pseudo: Agric.), Dēmokritos (pseudo: Alch.), Dēmokritos (pseudo: Lith.), Dēmokritos (pseudo: Med.), Dēmokritos (pseudo: Pharm.), Dēmosthenēs Philalēthēs, Dēmotelēs, Derkullidēs, Derkullos, Dexios, Didumos of Alexandria

Dates:	***Names:***	***Dates (Wide):***	***Names:***
	Lenaeus, Pompeius Trogus, Poseidōnios (Med. I), Potamōn of Alexandria, Puthiōn (Pharm.), Sabinius Tiro, Sallustius (Cn.), Sallustius Crispus, Sōsigenēs (I), Strabōn, Sueius, Tarutius, Terentius Varro of Narbo, Terentius Varro of Reate, Themisōn, Theodōros of Gadara, Theophanēs of Mutilēnē, Timagenēs, Tullius Cicero (M.), Tullius Cicero (Q.), Turranius, Valerius Messalla Potitus, Valgius Rufus, Velchionius, Vergilius, Vipsanius Agrippa, Vitruvius Pollio, Xenarkhos		(Metrol.), Didumos of Knidos, Diodōros (Astron.), Diodōros (Empir.), Diodōros of Samos, Diogas, Diogenēs (Geog.), Dioklēs of Khalkēdōn, Dioklēs of Magnesia, Diomēdēs, Diōn (Med.), Dionusios (of Halikarnassos?), Dionusios of Corinth, Dionusios of Kurtos, Dionusios of Philadelpheia, Dionusios of Rhodes, Dionusios of Samos, Dionusios: Sallustius, Dionusodōros (Pharm.), Diophil–, Dioskoros (Geog.), Dioskoros (Pharm.), Diphilos, Diphilos of Laodikeia, Domitius Nigrinus, Dōriōn (Biol.), Dōriōn (Mech.), Dōrotheos of Athens, Dōrotheos of Hēliopolis, Dōrotheos of Khaldaea, "Dtrums," Eirēnaios, Elephantinē/Elephantis, Emeritus (Hemeritos), Epagathos, Epainetēs, Epidauros, Epigenēs of Buzantion, Epigonos, Epiklēs of Crete, Epikouros (Pharm.), Epikratēs of Hērakleia, Erasistratos of Sikuōn, Euainetos, Euangeus, Euboulidēs, Euboulos (Pharm.), Eudēmos the elder, Eudikos, Euēnos, Eugeneia, Eugērasia, Euhēmeros (Pharm.), Eukleidēs "Palatianus,"
15 BCE – 20 CE	(72) Aelius Gallus, Alexander of Laodikeia, Alexander of Mundos, Ambiuius, Ammōnios of Alexandria, Antiokhos (Paccius), Antonius Castor, Aristōn (II), Arkhelaos (Lithika), Arkhelaos of Kappadokia, Artemōn (Med.), Asinius Pollio, Athēniōn (of Athens?), Atimētos, Caepio, Cassius, Cloatius Verus, Clodius Tuscus, Cornelius Celsus, Diodotos (Pharm.), Diogenēs (Pharm.), Diophantos of Lukia, Euelpidēs, Euelpistos (Terentius), Eunomos Asklēpiadean, Florus, Grattius Faliscus, Hēliodōros (Stoic), Hērās, Hupsikratēs, Iouba II,		

Dates:	***Names:***	***Dates (Wide):***	***Names:***
	Isidōros of Kharax, Iulius Atticus, Iulius Bassus, Iulius Caesar (Augustus), Iulius Caesar (Germanicus), Iulius Hyginus, Kárpos, Maecenas Licinius, Maecenas Melissus, Manilius, Mantias (Alch.), Marcianus (of Africa?), Matius Caluena, Megēs, Melētos, Menekratēs (Claudius), Nikēratos, Nikolaos of Damaskos, Ouidius Naso, Petronius Musa, Philemōn, Philōn of Alexandria, Philōn of Tarsos, Philōnidēs of Catina, Pompeius Trogus, Potamōn of Alexandria, Sabinius Tiro, Sabinus (Agric.), Sallustius Mopseates, Seleukos of Alexandria, Sextius, Strabōn, Theodōros of Gadara, Theōn of Alexandria (Stoic), Thrasullos, Truphōn of Gortun, Turranius, Turranius Gracilis, Valgius Rufus, Vipsanius Agrippa, Xenarkhos		Eumakhos, Euphranōr (Pharm.), Euphranōr (Pythag.), Euruōdēs, Euskhēmos, Euthudēmos of Athens, Euthukleos, Fauilla, Faustinus, Firmius, Flauianus of Crete, Flauius Clemens, Flauius "the boxer," Fronto (Astrol.), Gaius of Neapolis, Gemellus, Geminos, Gennadios, Glaukidēs, Glaukōn/ Glaukos, Glaukos (Geog. II), Glukōn, Granius, Halieus, Harpalos (Pharm.), Harpokrās, Harpokratiōn (Pharm.), Hēgēsidēmos, Hēgētōr (Med.), Hēgētōr of Buzantion, Hekataios (Pharm.), Hēliodōros of Athens, Hērakleidēs Pontikos of Hērakleia Pontikē (Junior), Hērakleidēs of Eruthrai, Hērakleitos of Sikuōn, Hermās *thēriakos*, Hermeias (Astrol.), Hermeias (Ophthalm.), Hermēs Trismegistos (pseudo), Hermōn of Egypt, Hermophilos, Hierax of Thēbai, Hikatidas, Hikesios (Agric.), Hippokratic Corpus: Sevens, Hostilius Saserna and sōn, Hugiēnos, Hulas, Iamblikhos (Geog.), Iasōn of Nusa, Idios, Ioudaios, Iriōn, Isigonos of Nikaia, Isis: pseudo (Pharm.), Iskhomakhos, Iulius Agrippa, Iulius Secundus,

Dates:	***Names:***	***Dates (Wide):***	***Names:***
			Iunia/Iounias, Iustinus (Pharm.), Iustus the Pharmacologist, On the Kosmos, Kalliklēs, Kalliphanēs, Kallistratos, Kēphisophōn, Keskintos: Inscription of, Khalkideus, Kharidēmos, Khariklēs, Kharitōn, Kharixenēs, Khios, Khrusanthos, Khrusippos (Agric.), Khrusippos (Med.), Kidēnas, Kimōn, Kleoboulos (Geog.), Kleoboulos (Pharm.), Kleomēdēs, Kleōn (of Kuzikos?), Kleophantos, Kleëmporos, Kloniakos, Klutos, Kōdios Toukos, Komerios, Kommiadēs, Korē Kosmou, Kratēs (Med.), Kratippos, Kratōn, Kritodēmos, Ktēsiphōn, Laïs, Lampōn, Laodikos, Leōnidas of Buzantion, Lepidianus, Licinius Atticus, Lingōn, Litorius, Lobōn, Logadios, Lukomēdēs, Lukos of Neapolis, Lunkeus, Lupus, Lusias, Lusimakhos of Kōs, Magnus of Philadelpheia, Magnus of Tarsos, Makhairiōn, Marcellinus (Pharm.), Maria, Markiōn, Melampous of Sarnaka, Melissos Xenophanēs and Gorgias, Melitōn, Menekritos, Menelaos (Pharm.), Menestheus, Menippos, Menippos of Pergamon, Menenius Rufus, Mēnodotos (Astr.),

Dates:	*Names:*	*Dates (Wide):*	*Names:*
			Menoitas, Mēnophilos, Mētrodōros (Arch.), Mētrodōros (Astr. I), Mētrodōros (Astr. II), Mētrodōros (Pharm.), Mētrodōros of Buzantion, Mētrodōros son of Epikharmos (pseudo), Mikiōn, Milēsios, Miltiadēs, Minucianus, Mnēsidēs, Molpis, Moskhiōn (Pharm.), Mousaios, Muia: pseudo, Muōnidēs, Murōn, Naukratitēs medicus, Neanthēs of Kuzikos, Nearkhos, Neoklēs, Neoptolemos, Nikētēs (of Athens?), Nikias of Mallos, Nikias of Nikaia, Nikolaos (Pharm.), Nikomakhos (Pharm.), Nikomēdēs (Hērakleitean), Ocellus, Olumpias, Olumpikos (Lith.), Olumpionikos, Onēsidēmos, Onētidēs/ Onētōr, Oppius, Orfitus, Ōrigeneia, Ōriōn of Bithunia, Ōros, Orpheus (pseudo: Astrol.), Orpheus (pseudo: Med.), Orthōn, Ostanēs (pseudo), Paconius, Panaitios Jr., Pantainos, Papias, Papirius Fabianus, P. Berol. 9782, P. Lit. Lond. 167, P. Mich. 3.148, P. Osloensis 73, P. Oxy. 13.1609, P. Oxy. 15.1796, Paradox. Vaticanus, Pasiōn, Patroklos, Pausanias "Hērakleiteios," Paxamos, Pelops (Med.), Perigenēs,

Dates:	*Names:*	*Dates (Wide):*	*Names:*
			Periklēs, Perseus, Phaidros, Phanias, Philippos of Macedon, Philiskos of Thasos, Philistidēs of Mallos, Philokalos, Philoklēs, Philokratēs, Philomēlos, Philōn (Meteor.), Philōn of Tuana, Philōtas, Phoibos Ulpius, Phulakos, Platōn (Pharm.), Platusēmos, Podanitēs, Politēs, Pollis (Arch.), Poluarkhos, Polueidēs, Polustomos, Poseidōnios of Corinth, Potamōn (Pharm.), Primiōn, Proëkhios, Proklos (Methodist), Prōtagoras of Nikaia, Prōtarkhos of Tralleis, Prōtās, Proxenos, Prutanis, Ptolemaios (Erasi.), Ptolemaios (Med.), Ptolemaïs of Kurēnē, Puramos, Purrhos of Magnesia, Puthagoras (Med.), Puthagoras (pseudo: Astrol.), Puthios, Puthoklēs of Samos, Quadratus, Rabirius, Ripalus, Salpē, Samithra, Sardonius, Sarkeuthitēs, Saturos (Lithika), Sebosus Statius, Serapiōn of Alexandria (Astrol.), Sergius of Babylōn, Sertorius Clemens, Silēnos, Silo, Simos of Kōs, Sōkratēs (Med.), Sōkratēs of Argos, Sōkratiōn, Solōn, Sōranos of Kōs, Sōsagoras, Sōsandros (Geog.), Sōsandros (Pharm.),

Dates:	Names:	Dates (Wide):	Names:
			Sōsikratēs, Sōsimenēs, Sōstratos of Alexandria, Sōteira, Sōtiōn, Spendousa, Speusippos of Alexandria, Staphulos, Stratōn of Bērutos, Sunerōs, Tektōn, Telamōn, Telephanēs, Teukros of Carthage, Teukros of Kuzikos, Thamuros, Theanō (pseudo), Theodōrētos, Theodōros of Soloi (Kilikia), Theodosios (of Bithunia), Theokhrēstos, Theokritos, Theokudēs, Theomenēs, Theophilos (Geog.), Theophilos (Lithika), Theopompos, Theoxenos, Theudās, Thrasuandros, Threptos, Timaios (Astrol.), Timaios (Pharm.), Timaios of Lokris (pseudo), Timaristos, Timokleanos, Timokratēs, Timōn, Timotheos, Tlēpolemos, Turannos, Turpillianus, Valerius Paulinus, Vicellius, Xanitēs, Xen(okh)arēs, Zakhalias, Zēnōn of Laodikeia, Zēnōn of Sidōn, Zēnophilos, Zeuxis (Hēroph.), Zōilos of Macedon, Zōpuros of Alexandria, Zōsimos (Med.)
20–55 CE	(94) Acilius Hyginus, Agathinos, Aglaias, Alexander of Laodikeia, Alexander of Mundos, Alfius Flauus, Alkōn, Ambrosios	20–125 CE	Abaskantos, Abram, Adeimantos, Adrastos of Aphrodisias, Aemilius Hispanus, Aëtios, Aetna, Agatharkhidēs of Samos,

Dates:	***Names:***	***Dates (Wide):***	***Names:***
	Rusticus of Puteoli, Ammōnios (Annius), Andromakhos of Crete (Elder), Annaeus Seneca, Anthaios (Sextilius), Antonius Castor, Apiōn of Oasis, Apollōnios (Claudius), Apuleius Celsus, Areios of Tarsos, Aristarkhos of Tarsos, Aristokratēs, Aristoxenos, Arkhibios, Artemōn (Med.), Athēnaios of Attaleia, Athēnodōros (Med.), Atimētos, Balbillos, Caepio, Cassius, Castus, Cornelius Celsus, Cornelius Valerianus, Damonikos (Claudius), Diodotos (Pharm.), Diogenēs (Pharm.), Dionusios (Meth.), Dioskouridēs of Anazarbos, Dōrotheos of Sidōn, Eudēmos (Method.), Euelpidēs, Eunomos Asklēpiadean, Hēliodōros (Stoic), Hermogenēs of Smurna, Iouba II, Isidōros of Antioch, Iulius Atticus, Iulius Bassus, Iulius Graecinus, Iunius Moderatus Columella, Khairēmōn, Kharmēs, Koiranos, Krinas, Kárpos, Leōnidas of Alexandria (Astron.), Leukios, Magnus of Ephesos, Manilius, Marcellus (Pharm.), Megēs, Melētos, Menekratēs (Claudius), Mnaseas (Method.), Moderatus, Nikēratos,		Agathoklēs, Agathoklēs of Atrax, Agathoklēs of Milētos, Agēsias, Aineios, Aisara, Aiskhinēs, Aiskhulidēs, Akhaios, Akhillās, Alexander (Geog.), Alkimiōn, Alkinoos, Amarantos, Ammōn (Astrol.), Amphiōn, Amuthaōn, Andronikos (Med.), Anoubiōn of Diospolis, Antimakhos (Pharm.), Antiokhos of Athens, Antipatros (Methodist), Antipatros (Pharm.), Antoninus of Kōs, Antonius "root-cutter," Antullos, Aphrodās, Aphrodisis, Apios Phaskos, Apollinarios of Aizanoi, Apollodōros of Kition, Apollodōros of Taras, Apollōnios Glaukos, Apollōnios of Alexandria ("Mus"), Apollōnios of Pergamon (Med.), Apollōnios of Pitanē, Apollōnios of Prousias, Apollōnios of Tarsos, Aquila Secundilla, Arbinas, Areios Didumos, Aristanax, Aristarkhos of Sikuōn, Aristoboulos, Aristodēmos, Aristoklēs, Aristoklēs of Messēnē, Aristolaos, Aristombrotos, Aristophanēs, Aristotle (pseudo: Mirab.), Arkhelaos (Veterin.), Arkhelaos of Hērakleia Salbakē, Arrabaios, Asklatiōn (Astrol.), Asklatiōn (Med.),

Dates:	***Names:***	***Dates (Wide):***	***Names:***
	Nikostratos (Pharm.), Ofellius Laetus, Paetus, P. Vindob. 19996, Periplus Maris Erythraei, Petronius Musa, Philemōn, Philippos of Rome, Philōn of Alexandria, Philōn of Huampolis, Philōn of Tarsos, Philōnidēs of Catina, Pliny, Pomponius Mela, Sallustius Mopseates, Scribonius Largus, Seleukos of Alexandria, Seuerus the Ophthalmologist, Sextilius Paconianus, Sextius, Sextius Niger, Strabōn, Terentius Valens, Thessalos of Tralleis, Thrasullos, Vettius Valens (Med.), Vibius Rufinus, Xenokratēs of Aphrodisias, Xenokratēs of Ephesos, Zōpuros of Gortuna		Asklēpiodotos (of Nikaia?), Aspasios (Pharm.), Athēnippos, Athēnodōros (of Rhodes?), Attius, Axios, Biōn Caecilius, Bithus of Durrakhion, Blastos, Boëthos (Med.), Brenitus, Caecilius "Medicus," Caesennius, Campestris, Candidus, Carmen Astrologicum, Celer, Clodius (Asklēpiadean), Clodius of Naples, Cornelius, Cornelius Bocchus, Daliōn (Med.), Damigerōn, Damōn (Geog.), Dasius, Dēïleōn, Dēmarkhos, Dēmētrios (Geog.), Dēmētrios of Alexandria, Dēmokritos (pseudo: Alch.), Dēmokritos (pseudo: Lith.), Dēmosthenēs Philalēthēs, Derkullidēs, Derkullos, Dexios, Didumos of Alexandria (Metrol.), Didumos of Knidos, Diodōros (Astron.), Diodōros of Samos, Diogas, Diogenēs (Geog.), Diogenēs of Oinoanda, Dioklēs of Khalkēdōn, Diomēdēs, Diōn (Med.), Dionusios (Lithika), Dionusios (of Halikarnassos?), Dionusios of Buzantion, Dionusios of Corinth, Dionusios of Kurtos, Dionusios of Rhodes, Dionusios of Samos, Dionusios: Sallustius,
55–90 CE	(81) Agathinos, Aglaias, Alfius Flauus, Ambrosios Rusticus of Puteoli, Ammōnios (Annius), Andromakhos of Crete (Elder), Andromakhos of Crete (Younger), Annaeus Lucanus, Annaeus Seneca, Anthaios (Sextilius), Antonius Castor, Apollōnios (Claudius), Areios of Tarsos, Aristarkhos of Tarsos, Aristogeitōn, Aristokratēs, Arkhibios, Asarubas, Athēnaios of Attaleia, Athēnodōros (Med.), Balbillos, Castus,		

Dates:	***Names:***	***Dates (Wide):***	***Names:***
	Damokratēs (Seruilius), Damonikos (Claudius), Didumos (Music), Dionusios (Meth.), Dioskouridēs of Anazarbos, Dōrotheos of Sidōn, Drakōn of Kerkura, Erōtianos, Flauius Vespasianus, Gaius (Hēroph.), Hēliodōros of Alexandria (Pneum.), Hermogenēs of Smurna, Hērodotos (Pneum.), Hērōn of Alexandria, Isidōros of Antioch, Iunius Crispus, Iunius Moderatus Columella, Khairēmōn, Kharmēs, Koiranos, Kosmos, Kratōn of Athens, Kritōn, Leōnidas of Alexandria (Astron.), Leōnidas of Alexandria (Pneum.), Leukios, Licinius Mucianus, Londiniensis medicus, Magistrianus, Magnus of Ephesos, Marcellus (Pharm.), Marinos (Med.), Mnaseas (Methodist), Moderatus, Nikostratos (Pharm.), Ofellius Laetus, Paetus, Pamphilos of Alexandria, P. Iandanae 85, P. Oxy. 3.467, P. Vindob. 19996, Periplus Maris Erythraei, Pharnax, Philippos of Rome, Philōn of Huampolis, Pliny, Plutarch, Pollēs of Aigai (pseudo), Pomponius Bassus, Pomponius Mela, Ptolemaios (Pharm.), Publius of Puteoli, Rufus		Dionusodōros (Maecius Seuerus), Dionusodōros (Pharm.), Diophil–, Dioskoros (Geog.), Dioskoros (Pharm.), Dioskouridēs (Metrol.), Diphilos of Laodikeia, Domitius Nigrinus, Dōrotheos of Athens, Dōrotheos of Hēliopolis, Dōrotheos of Khaldaea, Eirēnaios, Elephantinē/Elephantis, Emeritus (Hemeritos), Epagathos, Epainetēs, Epaphroditos of Carthage, Epidauros, Epikouros (Pharm.), Epikouros of Pergamon, Erasistratos of Sikuōn, Esdras, Euainetos, Euangeus, Euboulidēs, Euboulos (Pharm.), Eudikos, Euēnos, Eugeneia, Eugērasia, Euhēmeros (Pharm.), Eukleidēs "Palatianus," Eumakhos, Euphranōr (Pharm.), Euruōdēs, Euskhēmos, Euthukleos, Fauilla, Faustinus, Firmius, Flauianus of Crete, Flauius Clemens, Flauius "the boxer," Fronto (Agric.), Fronto (Astrol.), Gaius of Neapolis, Galēn (pseudo: Hist. Phil.), Gemellus, Gennadios, Glaukidēs, Glaukōn/Glaukos, Glaukos (Geog. II), Granius, Habrōn, Harpalos (Pharm.), Harpokrās, Harpokratiōn

Dates:	**Names:**	**Dates (Wide):**	**Names:**
	of Ephesos, Theotropos, Thessalos of Tralleis, Vibius Rufinus, Xenokratēs of Aphrodisias, Xenokratēs of Ephesos, Zēnōn (of Athens?)		(Pharm.), Harpokratiōn of Alexandria, Hēgēsidēmos, Hekataios (Pharm.), Hekatōnumos of Khios, Hēliodōros of Athens, Hērakleidēs Pontikos of Hērakleia Pontikē (Junior), Hērakleidēs of Eruthrai, Hērakleitos of Rhodiapolis, Hērakleitos of Sikuōn, Hermās *thēriakos*, Hermeias (Astrol.), Hermeias (Math.), Hermeias (Ophthalm.), Hermēs Trismegistos (pseudo), Hermippos of Bērutos, Hermophilos, Hērōnas, Hierax of Thēbai, Hikatidas, Hikesios (Agric.), Hulas, Hyginus Gromaticus, Iamblikhos (Geog.), Idios, Imbrasios (Paradox.), Ioudaios, Iriōn, Isigonos of Nikaia, Iskhomakhos, Iulianus of Tralleis, Iulius Agrippa, Iulius Secundus, Iunia/Iounias, Iustinus (Pharm.), Iustus the Pharmacologist, Kalliklēs, Kallikratēs (Astrol.), Kallinikos (Pharm.), Kalliphanēs, Kēphisophōn, Khalkideus, Kharidēmos, Khariklēs, Kharitōn, Kharixenēs, Khios, Khrusanthos, Khrusippos (Med.), Kimōn, Kleoboulos (Geog.), Kleoboulos (Pharm.), Kleomēdēs, Kleoneidēs,
90–125 CE	(68) Agrippa of Bithunia, Aiskhriōn of Pergamon, Andrōn of Rome, Apollodōros of Damaskos, Arkhigenēs, Artemidōros Capito, Asklēpiadēs Pharmakiōn, Aspasios (Perip.), Athēnodōros (Med.), Balbus, Calpurnius Piso (I), Cornelius Tacitus, Dioskouridēs of Alexandria, Dōrotheos of Sidōn, Drakōn of Kerkura, Fauorinus, Flauius Arrian, Gaius (Platonist), Galēn (pseudo: Def. Med.), Hēliodōros of Alexandria (Pneum.), Hēraklās, Hērodotos (Pneum.), Hieroklēs of Alexandria, Hyginus (Agrimensor), Iulius Frontinus, Kosmos, Kratōn of Athens, Kritōn, Leōnidas of Alexandria (Pneum.), Londiniensis medicus, Magistrianus, Magnus *arkhiatros*, Magnus of Ephesos, Manethōn (Astrol.), Marinos (Med.), Marinos of Tyre, Menelaos of Alexandria, Mēnodotos of Nikomēdeia, Nikanōr, Nikomakhos of Gerasa, Nikōn of Pergamon,		

Dates:	*Names:*	*Dates (Wide):*	*Names:*
	Ofellius Laetus, Orpheus (pseudo: Lithika), Paetus, P. Iandanae 85, P. Oslo. 72, P. Oxy. 3.467, P. Vindob. 19996, Pharnax, Philippos of Rome, Philōn of Bublos, Phlegōn, Pitenius, Plutarch, Polemōn of Laodikeia, Pollēs of Aigai (pseudo), Pompeius Sabinus, Pomponius Bassus, Ptolemaios of Kuthēra, Quintus, Rufus of Ephesos, Rufus of Samaria, Sabinus (Med.), Sōranos of Ephesos, Stratonikos of Pergamon, Theodās of Laodikeia, Theōn of Smurna, Zēnōn (of Athens?)		Kleophantos, Kleëmporos, Kloniakos, Kōdios Toukos, Komerios, Kommiadēs, Korē Kosmou, Kratēs (Med.), Kratippos, Kratōn, Kritodēmos, Ktēsiphōn, Kuranides, Kuros, Laïs, Lampōn, Laodikos, Lepidianus, Licinius Atticus, Lingōn, Litorius, Lobōn, Logadios, Lukomēdēs, Lunkeus, Lupus, Lusias, Maecius Aelianus, Maēs Titianus, Magnus of Philadelpheia, Magnus of Tarsos, Makhairiōn, Marcellinus (Pharm.), Maria, Markiōn, Melitōn, Menekritos, Menelaos (Pharm.), Menemakhos, Menestheus, Menippos, Menenius Rufus, Mēnodotos (Astr.), Mēnophilos, Mētrodōra, Mētrodōros (Arch.), Mētrodōros (Astr. II), Mētrodōros (Pharm.), Mētrodōros son of Epikharmos (pseudo), Milēsios, Miltiadēs, Minucianus, Mnēsidēs, Modius Asiaticus, Moskhiōn (Pharm.), Mousaios, Muia: pseudo, Murōn, Naukratitēs medicus, Nearkhos, Nepualios, Nikētēs (of Athens?), Nikias of Mallos, Nikolaos (Pharm.), Nikomakhos (Pharm.), Olumpiakos, Olumpias, Olumpikos

Dates:	Names:	Dates (Wide):	Names:
			(Lith.), Olumpionikos, Onēsidēmos, Onētidēs/Onētōr, Onētōr, Orfitus, Ōrigeneia, Ōriōn of Bithunia, Ōros, Orpheus (pseudo: Astrol.), Orpheus (pseudo: Med.), Orthōn, Ostanēs (pseudo), Pammenēs (Alch.), Panaitios Jr., Pantainos, Papias, Papirius Fabianus, P. Aberdeen 11, P. Ayer, P. Berol. 9782, P. Geneva inv. 259, P. Lit. Lond. 167, P. London 98, P. Mich. 3.148, P. Mich. 3.149, P. Mil. Vogl. I.14, P. Mil. Vogl. I.15, P. Osloensis 73, P. Oxy. 13.1609, P. Oxy. 15.1796, P. Tebtunis 679, Paradoxographus Florentinus, Paradoxographus Vaticanus, Parisinus medicus, Paulos (of Italy), Pausēris, Pēbikhios, Pelops (Med.), Perigenēs, Periklēs, Phaidros, Phanias, Philaretos (Alch.), Philippos of Egypt, Philistidēs of Mallos, Philokalos, Philoklēs, Philokratēs, Philomēlos, Philōn (Meteor.), Philōn (Meth.), Philōn of Tuana, Philōtas, Phoibos Ulpius, Phulakos, Physiologos, Platōn (Pharm.), Platusēmos, Podanitēs, Politēs, Pollēs (Med.), Poluarkhos, Polueidēs, Polustomos, Poseidōnios of Corinth, Potamōn (Pharm.),

Dates:	***Names:***	***Dates (Wide):***	***Names:***
			Praecepta Salubria, Primiōn, Proëkhios, Proklos (Methodist), Prōtagoras of Nikaia, Prōtās, Prutanis, Ptolemaios (Erasi.), Ptolemaios (Med.), Ptolemaios Platonikos, Ptolemaïs of Kurēnē, Puramos, Purrhos of Magnesia, Puthagoras (pseudo: Astrol.), Puthios, Puthoklēs of Samos, Quadratus, Rabirius, Rhēginos, Ripalus, Salpē, Samithra, Sardonius, Sarkeuthitēs, Saturos (Lithika), Sebosus Statius, Serapiōn of Alexandria (Astrol.), Serenus (Pharm.), Sergius of Babylōn, Sertorius Clemens, Sextus Empiricus, Siculus Flaccus, Silo, Sōkratēs (Lithika), Sōkratēs (Med.), Sōkratiōn, Solōn, Sōranos of Kōs, Sōsagoras, Sōsandros (Pharm.), Sōsikratēs, Sōsimenēs, Sōteira, Sōtiōn, Spendousa, Stratōn of Bērutos, Sunerōs, Telamōn, Telephanēs, Teukros of Egyptian Babylōn, Thamuros, Theanō (pseudo), Theodōrētos, Theodōros (of Macedon?), Theodōros of Soloi (Kilikia), Theokhrēstos, Theokritos, Theomenēs, Theophilos (Geog.), Theophilos (Lithika),

Dates:	***Names:***	***Dates (Wide):***	***Names:***
			Theopompos, Theosebios, Theoxenos, Theudās, Threptos, Timaios (Astrol.), Timaios (Pharm.), Timaios of Lokris (pseudo), Timaristos, Timokleanos, Timokratēs, Timōn, Timotheos, Turannos, Turpillianus, Valerius Paulinus, Vicellius, Xanitēs, Zēnariōn, Zēnōn of Laodikeia, Zēnophilos, Zōilos of Macedon, Zōsimos (Med.)
125–160 CE	(73) Aeficianus, Aelius Promotus, Aiskhriōn of Pergamon, Albinus of Smurna, Alexander of Aphrodisias (pseudo), Andrōn of Rome, Antigonos of Nikaia, Apollōnidēs of Cyprus, Apuleius of Madaurus, Aretaios, Artemidōros Capito, Artemidōros of Daldis, Aspasios (Perip.), Attalos (Med.), Atticus, Aurelius (Pharm.), Calpurnius Piso (I), Dionusios of Alexandria (Geog.), Fauorinus, Flauius Arrian, Gaius (Platonist), Galēn, Galēn (pseudo: Def. Med.), Galēn (pseudo: Introductio), Hēraklās, Hērakleianos of Alexandria, Hērakleidēs of Athens, Iulianus (of Alexandria?), Iulius Titianus, Kallimorphos,	125–230 CE	Abram, Adrastos of Aphrodisias, Aelianus "the Platonist," Aemilius Hispanus, Africanus (Metrol.), Agapētós, Agathoklēs of Atrax, Agēsias, Aiskhulidēs, Akhilleus, Alkinoos, Ammōn (Astrol.), Anastasios, Antiokhos of Athens, Antipatros (Methodist), Antullos, Apellās of Laodikeia, Apollinarios of Aizanoi, Apollinarios (Pharm.), Apollōnios of Laodikeia, Apollōnios of Pergamon (Med.), Apsurtos, Areios Didumos, Aristodēmos, Aristotle (pseudo: Mirab.), Arkadios, Arkhelaos (Med.), Arkhelaos (Veterin.), Arruntius Celsus, Artemisius Dianio, Asklatiōn (Astrol.), Auidianus,

Dates:	***Names:***	***Dates (Wide):***	***Names:***
	Kronios, Lukos of Macedon, Magnus *arkhiatros*, Manethōn (Astrol.), Marcellinus (Med.), Marcellus of Sidē, Martialius, Melior, Mēnodotos of Nikomēdeia, Mētrodōros of Alexandria, Nikanōr, Nikomakhos of Gerasa, Nikōn of Pergamon, Noumēnios of Apameia, Numisianus, Orpheus (pseudo: Lithika), Pankratēs of Alexandria, P. Oslo. 72, P. Ross. Georg. 1.20, P. Strassbourg Inv. Gr. 90, P. Turner. 14, Pelops of Smurna, Philōn of Bublos, Philoumenos, Phlegōn, Plutarch (Music), Polemōn of Laodikeia, Ptolemy, Quintilii, Quintus, Saturos of Smurna, Simmias the Stoic, Sōranos of Ephesos, Stratonikos of Pergamon, Suros, Tauros of Bērutos, Theodās of Laodikeia, Theōn (Astr.), Theōn of Alexandria (Med. I), Theōn of Smurna, Vettius Valens of Antioch, Volusius Maecianus, Yavaneśvara		Book of Assumptions, Campestris, Carmen Astrologicum, Clodius of Naples, Damigerōn, Dēmokritos (Neo-Plat.), Dēmokritos (pseudo: Alch.), Diodōros (Astron.), Diogenēs Laërtios, Diogenēs of Oinoanda, Dionusios (Lithika), Dionusios (of Halikarnassos?), Dionusios of Aigai, Dionusios of Buzantion, Dionusios of Rhodes, Dionusodōros (Maecius Seuerus), Dioskouridēs (Metrol.), Diphilos of Laodikeia, Dulcitius, Emeritus (Hemeritos), Epaphroditos and Vitruuius Rufus, Epikouros of Pergamon, Erasistratos (Astrol.), Esdras, Euboulidēs, Euhēmeros/Himerios, Eutychianus, Fronto (Agric.), Fronto (Astrol.), Galēn (pseudo: Hist. Phil.), Galēn (pseudo: Pulsibus), Gaudentius, Glaukos (Geog. II), Grēgorios (Pharm.), Habrōn, Harpokratiōn of Alexandria, Hekatōnumos (?) of Khios, Hērakleitos of Rhodiapolis, De Herbis, Hermās *thēriakos*, Hermeias (Astrol.), Hermeias (Doxogr.), Hermēs Trismegistos (pseudo), Hermippos of Bērutos, Hērōnas, Hulas,
160–195 CE	(44) Aelius Promotus, Albinus of Smurna, Alexander of Aphrodisias (pseudo), Ampelius, Amuntianos, Andro… of Rome, Antigonos of Nikaia, Antonius, Apuleius of Madaurus, Aretaios,		

Dates:	Names:	Dates (Wide):	Names:
	Aristotle of Mutilēnē, Attalos (Med.), Atticus, Aurelius (Pharm.), Bardaisan, Calpurnius Piso (II), Censorinus (I), Diodotos (Astr. II), Euphrates, Flauius Arrian, Galēn, Galēn (pseudo: Introductio), Harpokratiōn of Argos, Hērakleidēs of Athens, Herminos, Isis (pseudo: Alch.), Iulius Africanus, Iulius Titianus, Iustus the Ophthalmologist, Kallimorphos, Kronios, Martialius, Mētrodōros of Alexandria, Noumēnios of Apameia, Oppianus of Kilikia, P. Strassbourg Inv. Gr. 90, P. Turner. 14, Philistiōn of Pergamon, Philostratos, Philoumenos, Quintilii, Serenus Sammonicus, Vettius Valens of Antioch, Volusius Maecianus		Hyginus Gromaticus, Iamblikhos (Geog.), Imbrasios (Paradox.), Iulianus of Tralleis, Iunius Nipsus, Kalliklēs, Kallikratēs (Astrol.), Kleomēdēs, Kleoneidēs, Korē Kosmou, Kuranides, Kuros, Largius, Leontinos (Agric.), Lepidianus, Litorius, Lobōn, Logadios, Lupus, Maecius Aelianus, Maria, Mētrodōra, Mētrodōros (Astr. II), Muia: pseudo, Nepualios, Nonnos, Ōdapsos, Olumpiakos, Onētōr, Orpheus (pseudo: Astrol.), Pammenēs (Alch.), Panaitios Jr., Pankharios, P. Aberdeen 11, P. Ayer, P. Berol. 9782, P. Cairo Crawford 1, P. London 98, P. Lund I.7, P. Michiganensis 3.149, P. Mil. Vogl. I.14, P. Mil. Vogl. I.15, P. Ryl. III.529, P. Tebtunis 679, Paradoxographus Florentinus, Paradoxographus Palatinus, Paradoxographus Vaticanus, Paraphrasis eis ta Oppianou Halieutika, Parisinus medicus, Paulos (of Italy), Pausēris, Pēbikhios, Periklēs, Philaretos (Alch.), Philippos (of Pergamon?), Philippos of Egypt, Philōn (Meteor.), Philōn of Gadara,
195–230 CE	(29) Aelianus of Praeneste, Alexander of Aphrodisias, Alexander of Aphrodisias (pseudo), Aristotle of Mutilēnē, Arrianos, Artemidōros (Astron.), Atticus, Aurelius (Pharm.), Bardaisan, Calpurnius Piso (II), Cassius Iatrosophist, Censorinus (I), Clodius Albinus, Diodotos (Astr. II), Florentinus, Galēn, Gargilius Martialis, Harpokratiōn of Argos, Hippolutos, Hyginus (pseudo), Isis (pseudo:		

Dates:	***Names:***	***Dates (Wide):***	***Names:***
	Alch.), Iulius Africanus, Medicina Plinii, Nestōr, Oppianus of Apameia, P. Turner. 14, Philostratos, Serenus Sammonicus, Serenus of Antinoeia		Phoibos Ulpius, Physiologos, Planetis, Platusēmos, Politēs, Pollēs (Med.), Poseidōnios of Corinth, Praecepta Salubria, Proclianus, Proëkhios, Proklos of Laodikeia, Prōtagoras, Prōtagoras of Nikaia, Prothlius, Ptolemaios Platonikos, Puthagoras (pseudo: Astrol.), Puthoklēs of Samos, Quaternionibus, Rhēginos, Romula, Sardonius, Secundus, Serapiōn (Astron.), Serenus (Pharm.), Sextus Empiricus, Siculus Flaccus, Sōkratēs (Lithika), Sōsigenēs (II), Sporos, Stadiasmus Maris Magni, Theodōrētos, Theodōros (of Macedon?), Theodosios (Empir.), Theopompos, Theosebios, Thrasubulus, Tiberius, Timokleanos, Vicellius, Zēnophilos
230–265 CE	(24) Aelianus of Praeneste, Amelius, Anatolios of Laodikeia, Cassius Iatrosophist, Cassius Longinus, Censorinus (II), Dēmētrios (Math.), Dionusios (of Alexandria?), Diophantos of Alexandria, Florentinus, Gargilius Martialis, Hippolutos, Iulius Africanus, Iulius Solinus, Medicina Plinii, Mēdios (Stoic), Neilos, P. Rylandensis 27, PSI inv. 3011, Plōtinos,	230–335 CE	Aelianus "the Platonist," Aemilius Hispanus, Africanus (Metrol.), Agapētós, Akhilleus, Alupios, Ammōn (Astrol.), Anastasios, Antiokhos of Athens, Antullos, Apellās of Laodikeia, Aphthonios, Apollinarios (Pharm.), Apollōnios of Laodikeia, Apsurtos, Aristeidēs Quintilianus, Arkadios, Arkhelaos (Med.), Arkhelaos (Veterin.), Arruntius Celsus,

Dates:	***Names:***	***Dates (Wide):***	***Names:***
	Porphuriosof Tyre, Samuel, Theosebeia, Zōsimos of Panopolis		Arsenios, Artemisius Dianio, Asklatiōn (Astrol.), Auidianus, Bakkheios Gerōn, Book of Assumptions, Campestris, Carmen Astrologicum, Carmen de ponderibus et mensuris, Carminius, Celsinus of Kastabala, Constantinus, Dēmokritos (Neo-Plat.), Dēmokritos (pseudo: Alch.), Diodōros (Astron.), Diogenēs Laërtios, Dionusios (of Halikarnassos?), Dionusios of Aigai, Dioskoros (Alch.), Doarios, Dulcitius, Emeritus (Hemeritos), Epaphroditos and Vitruuius Rufus, Erasistratos (Astrol.), Esdras, Euboulidēs, Eugenios (Alch.), Euhēmeros/Himerios, Euteknios, Eutychianus, Fronto (Agric.), Fronto (Astrol.), Fullonius Saturninus, Galēn (pseudo: An Animal), Galēn (pseudo: Hist. Phil.), Galēn (pseudo: Pulsibus), Gaudentius, Grēgorios (Pharm.), Hekatōnumos (?) of Khios, De Herbis, Hermeias (Doxogr.), Hermēs Trismegistos (pseudo), Hērōnas, Hierios, Hieroklēs (Veterin.), Hulas, Hyginus Gromaticus, Iamblikhos (Geog.), Iamblikhos of Constantinople,
265–300 CE	(24) Amelius, Anatolios of Laodikeia, Cassius Longinus, Dēmētrios (Math.), Fauentinus (Cetius), Flauius (Med.-poet), Gargilius Martialis, Lactantius, Mēdios (Stoic), Megethiōn, Mulomedicina Chironis, Neilos, Nemesianus, Pamphilos of Bērutos, Pandrosion, Pappos of Alexandria, P. Rylandensis 27, PSI inv. 3011, Plōtinos, Plutarch (pseudo: Rivers), Porphurios of Tyre, Sphujidhvaja, Theosebeia, Zōsimos of Panopolis		
300–335 CE	(25) Adamantios, Albinus (Encyclo.), Anatolios of Bērutos, Ausonius (Iulius), Fauentinus (Cetius), Firmicus Maternus (Iulius), Flauius (Med.-poet), Hermodōros of Alexandria, Iamblikhos of Khalkis, Iulianus Imp., Lactantius, Megethiōn, Mīnarāja, Mulomedicina Chironis, Pamphilos of Bērutos, Pandrosion, Pappos of Alexandria, P. Leidensis V, Peutinger Map, Philagrios, Plutarch (pseudo: Rivers), Porphurios of Tyre, Theol. Arith., Tiberianus, Ulpianus of Emesa		

Dates:	***Names:***	***Dates (Wide):***	***Names:***
			Ianuarinus, Imbrasios (Paradox.), Isidōros (Alch.), Iulianus Vertacus, Iulius Honorius, Iunius Nipsus, Korē Kosmou, Kuros, Largius, Leontinos (Agric.), Lepidianus, Libanios of Antioch, Litorius, Logadios, Loukās (pseudo: Alch.), Lupus, Magnus of Emesa, Makarios of Magnesia, Marcellus (Geog.), Marcianus of Hērakleia, Maria, Maximianus, Maximus, Mētrodōra, Mētrodōros (Astr. II), Nonnos, Ōdapsos, Pammenēs (Alch.), Panaitios Jr., Pankharios, P. Cairo Crawford 1, P. Holmiensis, P. Leidensis X, P. Lund I.7, P. Ryl. III.529, Paradoxographus Palatinus, Paraphrasis eis ta Oppianou Halieutika, Paterios, Paulos (of Italy), Pausēris, Pēbikhios, Peithōn, Pelagios, Periklēs, Petasios (pseudo), Philaretos (Alch.), Philōn of Gadara, Phimenas, Phoibos Ulpius, Physiologos, Planetis, Platusēmos, Pollēs (Med.), Pontica, Praecepta Salubria, Priscianus, Proclianus, Proëkhios, Proklos of Laodikeia, Prōtagoras, Prōtagoras of Nikaia, Prothlius, Ptolemaios, Platonikos, Quaternionibus, Remmius Fauinus,

Dates:	***Names:***	***Dates (Wide):***	***Names:***
			Romula, Sardonius, Secundus, Serapiōn (Astron.), Serenus (Pharm.), Sporos, Stadiasmus Maris Magni, Sunesios, Theodōrētos, Theodōros of Asinē, Theomnēstos of Nikopolis, Theōn of Alexandria (Med. II), Theopompos, Theosebios, Thrasubulus, Tiberius, Timokleanos, Vibius Sequester, Zēnophilos
335–370 CE	(37) Adamantios, Albinus (Encyclo.), Alupios of Antioch, Anatolios of Bērutos, Andreas (of Athens?), Arbitio, Auienus, Ausonius (Iulius), Basil of Caesarea, De Rebus Bellicis, Caesarius of Nazianzos, Dardanos, Diodōros of Tarsos, Epiphanios of Salamis, Eutropius of Bordeaux, Expositio totius mundi, Firmicus Maternus (Iulius), Hēliodōros (Astrol.), Hermodōros of Alexandria, Innocentius, Iulianus Imp., Magnus of Nisibis, Marius Victorinus, Oreibasios, P. Johnson, P. Leidensis V, P. Mich. 17.758, Paulos of Alexandria, Pelagonius, Philagrios, Physiognomista Latinus, Saloustios, Siburius, Themistios, Theodorus Priscianus, Theol. Arith., Theōn of Alexandria (Astr.)	335–440 CE	Aemilianus (Palladius), Aemilius Hispanus, Agapētós, Agathēmeros son of Orthōn, Akholios, Alexander (Med.), Alexander Sophistēs, Alupios, Ammōn (Astrol.), Anastasios, Apellās of Laodikeia, Aphthonios, Apollōnios of Laodikeia, Apsurtos, Arkadios, Arkhelaos (Med.), Arruntius Celsus, Arsenios, Artemisius Dianio, Asklatiōn (Astrol.), Athēnagoras (Med.), Auidianus, Bakkheios Gerōn, "Bērutios," Book of Assumptions, Campestris, Carmen Astrologicum, Carmen de ponderibus et mensuris, Carminius, Celsinus of Kastabala, Constantinus, Damianos of Larissa, Didumos of Alexandria (Agric.), Dimensuratio and

Dates:	***Names:***	***Dates (Wide):***	***Names:***
370–405 CE	(46) Agennius Urbicus, Alupios of Antioch, Ambrose (Ambrosius), Ammōn (Metrol.), Arbitio, Astrologos of 379, Auienus, Aurelius Augustinus, Ausonius (Iulius), Basil of Caesarea, De Rebus Bellicis, Calcidius, Cassius Felix, Claudian, Dardanos, Diodōros of Tarsos, Epiphanios of Salamis, Eunapios, Eusebius son of Theodorus, Eutropius of Bordeaux, Fauonius Eulogius, Gregory of Nazianzos, Gregory of Nussa, Hupatia, Iōannēs of Antioch ("Chrysostom"), Iōannēs of Stoboi, Iōnikos, Macharius, Magnus of Nisibis, Mallius Theodorus, Marcellus of Bordeaux, Oreibasios, Orosius, P. Johnson, Paulos of Alexandria, Pelagonius, Philostorgios, Physiognomista Latinus, Placitus Papyriensis, Poseidōnios (Med. II), Rufinos of Antioch, Siburius, Sunesios of Kurēnē, Themistios, Theodorus Priscianus, Theōn of Alexandria (Astr.)		Diuisio, Diodōros (Metrol.), Dioskoros (Alch.), Doarios, Dulcitius, Emeritus (Hemeritos), Eruthrios, Esdras, Euax, Eugenios (Alch.), Euhēmeros/Himerios, Euteknios, Eutychianus, Fronto (Agric.), Fronto (Astrol.), Fullonius Saturninus, Galēn (pseudo: Hist. Phil.), Galēn (pseudo: Pulsibus), Gaudentius, Grēgorios (Pharm.), Hēliodōros of Larissa, Hermēs Trismegistos (pseudo), Hērōnas, Hēsukhios, Hierios, Hieroklēs (Veterin.), Hulas, Iamblikhos (Geog.), Iamblikhos of Constantinople, Ianuarinus, Imbrasios (Paradox.), Iōannēs Iatrosophist, Isidōros (Alch.), Iulianus Vertacus, Iulius Honorius, Iunius Nipsus, Korē Kosmou, Kratistos, Kurillos, Kuros, Largius, Lepidianus, Libanios (Geog.), Libanios of Antioch, Litorius, Logadios, Loukās (pseudo: Alch.), Loukās (pseudo: Med.), Lupus, Magnus of Emesa, Maiorianus, Makarios of Magnesia, Marcellus (Geog.), Marcellus (Mech.), Marcianus of Hērakleia, Maximianus, Maximinus, Maximus, Melitianus, Mēnās,
405–440 CE	(32) Adamantios of Alexandria, Agennius Urbicus, Aurelius Augustinus, Caelius Aurelianus, Capella		

Dates:	***Names:***	***Dates (Wide):***	***Names:***
	(Martianus), Cassius Felix, Dardanos, Domninos, Eunapios, Fauonius Eulogius, Gamaliel VI, Hephaistiōn, Hērōn (Math.), Hilarius, Hupatia, Iōannēs of Antioch ("Chrysostom"), Iōannēs of Stoboi, Marcellus of Bordeaux, Olumpiodōros of Thēbai, Orosius, Paitāmahasiddhānta, Periklēs, Philostorgios, Placitus Papyriensis, Ploutarkhos of Athens, Poseidōnios (Med. II), Proklos of Lukia, Seuerianus, Sunesios of Kurēnē, Syrianus, Theodōros (Mech.), Theodosius (Macrobius)		Mētrodōra, Moses of Xoren, Nemesios, Nonnos, Ōdapsos, Olumnios, Pankharios, P. Holmiensis, P. Laur. Inv. 68, P. Leidensis X, P. Lund I.7, Paraphrasis eis ta Oppianou Halieutika, Paterios, Paulos (of Italy), Peithōn, Pelagios, Petasios (pseudo), Phimenas, Phoibos Ulpius, Physiologos, Platusēmos, Pollēs (Med.), Pontica, Porphurios (Geog.), Praecepta Salubria, Priscianus, Probinus, Proclianus, Proëkhios, Proklos of Laodikeia, Prōtagoras of Nikaia, Prothlius, Quirinus, Remmius Fauinus, Romula, Sardonius, Secundus, Serapiōn (Astron.), Serenus (Pharm.), Sunesios, Theodōros of Asinē, Theomnēstos of Nikopolis, Theōn of Alexandria (Med. II), Thrasubulus, Tiberius, Timokleanos, Vibius Sequester, Vindicianus, Zēnophilos
440–475 CE	(25) Agapios of Alexandria, Ammōnios of Alexandria (Neo-Plat.), Anthedius, Asklēpiodotos of Alexandria, Caelius Aurelianus, Cassius Felix, Domninos, Domnus, Hephaistiōn, Hērōn (Math.), Hilarius, Iakōbos Psukhrestos, Marinos of	440–545 CE	Aemilianus (Palladius), Aethicus (pseudo), Agapētós, Agathēmeros son of Orthōn, Aineias of Gaza, Akholios, Alexander (Med.), Alexander Sophistēs, Anastasios, Anonymous Alchemist "Christianus," Arkadios,

Dates:	**Names:**	**Dates (Wide):**	**Names:**
	Neapolis, Mustio, Periklēs, Petros, Physica Plinii, Placitus Papyriensis, Priskos, Proklos of Lukia, Prolegomena to Ptolemy's Suntaxis, Theodōros (Mech.), Vegetius, Victorius, Zīg		Arkhelaos (Med.), Asklatiōn (Astrol.), Athēnagoras (Med.), Auidianus, "Bērutios," Book of Assumptions, Carmen Astrologicum, Carmen de ponderibus et mensuris, Cassianus Bassus, Damianos of Larissa, Didumos of Alexandria (Agric.), Dimensuratio and Diuisio, Doarios, Eruthrios, Esdras, Euax, Eugenios (Alch.), Eusebius (pseudo), Euteknios, Fronto (Agric.), Fullonius Saturninus, Galēn (pseudo: Pulsibus), Grēgorios (Pharm.), Heldebald, Hēliodōros of Larissa, Hērōnas, Hēsukhios, Hieroklēs (Geog.), Iamblikhos (Geog.), Iamblikhos of Constantinople, Imbrasios (Paradox.), Iōannēs Iatrosophist, Iōannēs of Alexandria, Isidōros (Alch.), Isidōros of Milētos' student, Isidōros the Younger, Iulianus Vertacus, Iulius Honorius, Kratistos, Kurillos, Kuros, Libanios (Geog.), Libanios of Antioch, Logadios, Loukās (pseudo: Alch.), Loukās (pseudo: Med.), Maiorianus, Marcellus (Mech.), Marcomir, Marsinus, Maximianus, Maximinus, Melitianus, Mēnās, Moses of Xoren, Nonnos, Olumnios,
475–510 CE	(39) Aëtios of Amida, Agapios of Alexandria, Ammōnios of Alexandria (Neo-Plat.), Anthēmios, Anthimus, Apuleius (pseudo: Herbarius), Āryabhaṭa, Asklēpiodotos of Alexandria, Athanarid, Boëthius, Capito, Castorius, Damaskios, Domnus, Gessios, Hēliodōros of Alexandria (Astron.), Hēraiskos, Isidōros of Milētos, Iuliana, Iulianus of Laodikeia, Khrusēs of Alexandria, Lollianus, Marianus, Marinos of Neapolis, Ouranios, Periklēs, Physica Plinii, Priscianus of Caesarea, Priskos, Proklos of Lukia, Prolegomena to Ptolemy's Suntaxis, Cassiodorus Senator, Sergius of Rešʿaina, Seuerus Iatrosophista, Stephanos of Tralleis, Theodōros (Mech.), Timotheos of Gaza, Ulpianus, Urbicius		
510–545 CE	(43) Aëtios of Amida, Aganis, Ammōnios of Alexandria (Neo-Plat.), Anthēmios, Anthimus, Apuleius (pseudo:		

Dates:	***Names:***	***Dates (Wide):***	***Names:***
	Herbarius), Asklēpios (Pharm.), Asklēpios of Tralleis (Math.), Boethius, Capito, Castorius, Damaskios, Eutokios, Gessios, Gildas, Hermolaos (Geog.), Iōannēs of Alexandria (Philoponos), Iōannēs of Philadelpheia ("Lydus"), Isidōros of Milētos, Iuliana, Iulianos of Askalōn, Iulianus (Pharm.), Khrusēs of Alexandria, Kosmās, Lollianus, Marianus, Megethios, Nonnosos, Olumpiodōros of Alexandria, Olumpiodōros of Alexandria (Alch.), Ouranios, Priscian of Ludia, Priscianus of Caesarea, Cassiodorus Senator, Sergius of Rešʿaina, Seuerus Iatrosophista, Simplicius, Stephanos of Buzantion, Stephanos of Tralleis, Theodore pupil of Sergius, Tribonianus, Urbicius, Wuzurgmihr		Palladios, P. Akhmim, P. Laur. Inv. 68, Paraphrasis eis ta Oppianou Halieutika, Pelagios, Porphurios (Geog.), Probinus, Proklos of Laodikeia, Quirinus, Serenus (Pharm.), Stephanos of Athens, Theomnēstos of Nikopolis, Theōn of Alexandria (Med. II), Thrasubulus, Tiberius, Tukhikos, Vibius Sequester
545–580 CE	(28) Aganis, Alexander of Tralleis, Anthēmios, Asklēpios (Pharm.), Asklēpios of Tralleis (Math), Aëtios of Amida, Burzoy, Capito, Gildas, Gregory of Tours, Hermolaos (Geog.), Iōannēs of Alexandria (Philoponos), Iōannēs of Philadelpheia ("Lydus"), Iordanes, Isidōros of Milētos, Kosmās,	545–650 CE	Aethicus Ister, Agapētós, Agathēmeros son of Orthōn, Alexander (Med.), Alexander Sophistēs, Anania of Shirak, Anonymous Alchemist Philosopher, Anonymous Alchemist "Christianus," Arkhelaos (Med.), Athēnagoras (Med.), Auidianus, Book of Assumptions, Cassianus Bassus,

Dates:	***Names:***	***Dates (Wide):***	***Names:***
	Mētrodōros of Tralleis, Mucianus, Olumpiodōros of Alexandria, Olumpiodōros of Alexandria (Alch.), Periplus Ponti Euxini, Cassiodorus Senator, Stephanos of Buzantion, Tribonianus of Sidē, Varāhamihira, Wuzurgmihr, Zēmarkhos, Zīg		Damianos of Larissa, Elias (pseudo), Eruthrios, Eugenios (Alch.), Eusebius (pseudo), Galēn (pseudo: Pulsibus), Ḥarith ibn-Kalada, Heldebald, Hēliodōros of Larissa, Hierophilos Sophistēs, Imbrasios (Paradox.), Iōannēs Iatrosophist, Iōannēs of Alexandria, Isidōros of Milētos' student, Isidōros the Younger, Iustinianus Imp., Komerios, Loukās (pseudo: Alch.), Loukās (pseudo: Med.), Marcomir, Marsinus, Maximianus, Maximinus, Moses, Olumnios, Palladios, Pappos (II), P. Akhmim, Ravenna Cosmography, Rhetorios, Stephanos of Alexandria, Stephanos of Alexandria (Alch.), Stephanos of Athens, Theomnēstos of Nikopolis, Tukhikos
580–615 CE	(16) Agnellus, Ahrun, Alexander of Tralleis, Burzoy, Geōrgios of Cyprus, Geōrgios of Pisidia, Gregory of Tours, Hērakleios Imp., Isidorus of Hispalis (Seville), Leontios (Astron.), Mētrodōros of Tralleis, Mucianus, Paulos (Music), Periplus Ponti Euxini, Cassiodorus Senator, Theophulaktos		
615–650 CE	(13) Abiyūn, Ahrun, Anqīlāwas, Geōrgios of Cyprus, Geōrgios of Pisidia, Hērakleios Imp., Isidorus of Hispalis (Seville), Leontios (Astron.), Paulos (Music), Paulos of Aigina, Severus Sebokht, Theophulaktos, Zīg		

These 18 entries have date-ranges that place them after our terminus; the three marked * are included solely because of their relation to other entries, whereas those not so marked have been dated to within our range by some scholars; *cf.* also Hermolaos (Geog.) and Zēnariōn, who may belong here:

Aethicus, pseudo *
Damaskēnos
Eleutheros
Euphēmios of Sicily
Expositio geographiae
Geōponika *

Geōponika in Pahlavi *
Hēliodōros (pseudo?)
Hupatos
Iōannēs Archpriest
Iōannēs Esdras
Iōannēs Iakōbos
Iōannēs Matthaios
Iōannēs of Antioch (*arkhiatros*)
Nikomēdēs Iatrosophist
Okianos
Philaretos (Med.)
Philippos Xēros

These 12 entries are not assigned any date, because the evidence for them is based on middle- or late-Byzantine sources, and internal evidence is not decisive:

Agathosthenēs
Ambrosios Sophistēs
Anthemustiōn
Antimakhos of Hēliopolis
Apollōnios of Tuana, pseudo
Aristogenēs of Thasos
Asamōn
Dēmētrios (Astrol.)
Epaphroditos (Meteor.)
Epiphanios (Meteor.)
Nephōn
Philogenēs

These 30 entries have only a single terminus (five *post* and 25 *ante*); those marked with * have only a *terminus post*, or else only a *terminus ante* late enough that their actual date may be outside our date-range:

Agathodaimōn of Alexandria *
Anakreōn (Pharm.) *
Arkhedēmos (Veterin.)
Auxanōn
Blatausis *
Book of the Signs of the Zodiac *
Caystrius
Dēmētrios (Music)
Erukinos
Eumēlos of Thēbai
Grēgorios (Veterin.)
Helenos *
Hipparkhos (Veterin.)
Hippasios of Ēlis
Hippokratēs (Veterin.)
Itineraries
Kalyāṇa
Kleomenēs the Libyan
Markianos *
Matriketas
P. Florentinus
Phokos of Samos
Sophar/Sōphar *
Sornatius
Stratonikos (Veterin.)
Summaria rationis geographiae *
Theodos of Alexandria
Theophanēs of Hērakleopolis
Thrasualkēs
Zarathuštra

These 61 entries have date-ranges too wide (525 years or more) to warrant entering them into the "Time-Line" above:

Agathodaimōn (pseudo)
Agathotukhos
Aigeias of Hierapolis
Amuntas (Med.)
Anaxilaïdēs
Andronikos (Paradox.)
Apellās of Kurēnē
Aphros
Apollodōros of Kuzikos
Asaf ha-Rofe
Aspasia
Asterios
Astrampsukhos
Bakkhulidios

Bolās
Bothros
Bouphantos
Claudianus (Alch.)
Diodōros of Ephesos
Dionidēs
Dōsitheos (Pharm.)
Emboularkhos
Epidikos
Epiphanēs
Eugamios
Eutonios
Hēliadēs
Hermeias (Geog.)
Hermolaos (Pharm.)
Hierōn (Veterin.)
Hipposiadēs
Iamblikhos (Alch.)
Imbrasios of Ephesos
Isidōros of Memphis
Khēmēs or Khumēs
Kleandros
Magnēs or Magnus
Manethōn (Pharm.)
Marpēssos
Meleagros
Menandros Iatrosophist
Minuēs
Neilammōn
Nikolaos (Math)
Numius
P. Oxy. 3.470
Pammenēs (Biol.)
Penthesileus
Philippos of Kōs
Poludeukēs
Porphurios (Med.)
Prosdokhos
Sandarius
Simōnidēs (Biol.)
Thaïs
Theoklēs
Theomnēstos
Theophilos (Pharm.)
Theophilos son of Theogenēs
Thumaridas
Trophilos

TOPICS

We offer here an index of the entries by "topic" using modern categories, which do not always map neatly onto ancient categories, but which are good (and even necessary) for us to think with. The categories are similar to those of the chapters of our earlier book, Irby-Massie and Keyser (2002). Of course, many authors, especially those known as philosophers, will appear in multiple categories.

Agriculture/Agronomy (102) (authors and writings on the methods and practice of farming; contrast next, **Agrimensores**); most entries by Marsilio, Rodgers, or Thibodeau:

Aemilianus, Palladius
Agathoklēs of Khios
Aiskhriōn
Aiskhulidēs
Ambiuius
Amphilokhos
Anatolios (Vindonios)
Anaxipolis
Androtiōn
Antigonos of Kumē
Apollodōros of Lēmnos
Apollōnios of Pergamon
Aristandros
Aristomakhos
Aristomenēs
Aristophanēs of Mallos
Arkhelaos of Kappadokia
Arkhutas
Arrianus
Athēnagoras
Attalos of Pergamon
Attius
Bakkheios of Milētos
Biōn of Soloi
Caepio
Caesennius
Cassianus Bassus
Castricius
Cloatius Verus
Clodius Albinus
A. Cornelius Celsus
Dadis
Deinōn
Dēmokritos, pseudo
Didumos of Alexandria
Diodōros of Priēnē
Dionusios of Utica
Diophanēs
Epigenēs of Rhodes
Euagōn
Euboulos
Euphrōnios of Amphipolis
Euphrōnios of Athens
Euphutōn
Firmius
Florentinus
Fronto
Gargilius
Geōponika
Hēgēsias
Hēsiod
Hierōn II of Syracuse

Hikesios
Hostilius Saserna & son
Iulius Atticus
L. Iulius Graecinus
C. Iulius Hyginus
L. Iunius Columella
Iunius Silanus
Khaireas
Khairisteos
Kharetidēs
Khrusippos
Kleidēmos
Kommiadēs/Kosmiadēs
Kratēs
Leontinos
Leophanēs
Lusimakhos
Maecenas Licinius
Maecenas Melissus
Mamilius Sura
C. Matius Caluenus
Menandros of Hērakleia
Menandros of Priēnē
Menekratēs of Ephesos
Menestratos (II)
Minius Percennius
Mnaseas of Milētos
Neoptolemos
Nestōr
Oppius
P. Hibeh 2.187
Paxamos
Persis
Philiskos
Plentiphanēs
Pompeius Lenaeus
M. Porcius Cato
Puthiōn of Rhodes
Puthoklēs of Samos
Sex. Quinctilii
Sabinius Tiro
M. Sueius
M. Terentius Varro
Theophilos
Trebius Niger
Cn. Tremelius Scrofa
Turranius
Turranius Gracilis
M. Valerius Messalla Potitus
P. Vergilius Maro

Agrimensores (8) (authors and writings on the measuring of land and surveying, primarily Latins); most entries by Campbell, Guillaumin, and Roth Congès:

Balbus
Epaphroditos and Vitruuius Rufus
Hērōn of Alexandria
Hyginus (Agrimensor)
Hyginus Gromaticus
pseudo-Hyginus
Innocentius
Iunius Nipsus
Siculus Flaccus

Alchemy (56) (authors and writings on the theory and method of material transformation; *cf.* **Lithika**); most entries by Hallum or Viano:

Agathodaimōn, pseudo
Anaxilaos of Larissa
Anonymous Alchemist "Christianus"
Anonymous Alchemist Philosopher
Attalos III of Pergamon
Bōlos of Mendēs
Claudianus
Dēmokritos, pseudo
Dionusios of Corinth
Dioskoros
Egnatius
Eugenios
Hēliodōros, pseudo
Hērakleios Imp., pseudo
"Hermēs Trismegistos"
Iamblikhos
Iōannēs Archpriest
Isidōros

Isis, pseudo
Iuliana Anicia
Iulianus Imp., pseudo
Sex. Iulius Africanus
Iustinianus Imp., pseudo
Khēmēs
Kleopatra VII
Kōmarios
Kudias of Kuthnos
Loukās
Maria
Mo(u)sēs
Neilos
Olumpiodōros
Ostanēs, pseudo
Pammenēs
Pappos II
P. Florentinus
P. Holmiensis
P. Iandanae 85
P. Leidensis V
P. Leidensis X
P. Oxy. 3.467
Pausēris
Paxamos
Pēbikhios
Pelagios
Peteēsis/Petasios
Philaretos
Phimenas
Sophar
Stephanos
Sunesios
Teukros of Kuzikos
Theophilos son of Theogenēs
Theosebeia
Zōilos of Cyprus
Zōsimos of Panopolis

Architecture (44) (authors and writings on the theory and method of construction, often including mathematical, mechanical, or other analyses); many entries by Howe, Kourelis, Miles, or Pfaff:

Andronikos of Kurrhos
Anthēmios
Arkesios
Daphnis
Dēmophilos
Epaphroditos
Eupalinos
Euphranōr of Corinth
M. Cetius Fauentinus
Fuficius
Hermogenēs of Alabanda
pseudo-Hyginus
Iktinos
Isidōros of Milētos (three men)
Iulianus of Askalon
Sex. Iulius Frontinus
Kallikratēs
Karpiōn
Khersiphrōn
Khrusēs
Leōnidēs of Naxos
Mandroklēs
Melampous of Sarnaka
Metagenēs
Mētrodōros
Nikōn of Pergamon
Paiōnios
Parmeniōn
Philōn of Eleusis
Pollis
Puthios of Priēnē
Rhoikos
Rufinus
Saturos of Paros
P. Septimius
Silēnos
Sōstratos of Knidos
Theodōros of Phokaia
Theodōros of Samos
Theokudēs
Vitruuius
Xēn(okh)arēs

Astrology (96) (authors and writings on the positions and effects of the "stars," based on a theory of affinity between them and the central Earth, supported by observable effects in the case of the Sun and Moon); *cf.* **Astronomy**. Many entries by Jones, Lehoux and Rochberg:

Abram
Akhinapolos
Ammōn
Andreas of Athens
Ankhialos
Anoubiōn
Antigonos of Nikaia
Antiokhos of Athens
Antipatros
Apollōnios of Mundos
Asklatiōn
Astrologos of 379
Attius
Balbillos
Bardaisan
Bērossos
Bōlos of Mendēs
Book of the Zodiac
Bothros
Calpurnius Piso (I)
Campestris
Carmen astrologicum
Censorinus (II)
Dēmētrios
Diodōros of Tarsos
Dōrotheos of Sidōn
Epigenēs of Buzantion
Erasistratos
Euax
Iulius Firmicus Maternus
Fonteius Capito
Fronto
Fullonius
Hēliodōros
Hephaistiōn
Hermeias
"Hermēs Trismegistos"
Iamblikhos
Imbrasios of Ephesos
Iōannēs of Philadelpheia
Iulianus of Laodikeia
Kallikratēs
Kárpos
Kassandros
Khairēmōn
Krinas
Kritodēmos
Leptinēs (I)
"Lion Horoscope"
Macharius
Manethōn
M. Manilius
Maximus
Melampous
Mīnarāja
Neilos
P. Nigidius Figulus
Ōdapsos of Thebes
Orpheus, pseudo
Pankharios
P. Londinensis 98
P. Michiganensis 3.148
P. Michiganensis 3.149
Paulos of Alexandria
Petosiris
Petros
T. Pitenius
De Planetis
Pollēs of Aigai
Prōtagoras of Nikaia
Ptolemy
Puthagoras, pseudo
De Quaternionibus
Rhētorios
Salmeskhoinaka
Serapiōn of Alexandria
Sextus Empiricus
Skulax of Halikarnassos
Sphujidhvaja
Stephanos
L. Tarutius
Teukros of Seleukeia
Thessalos of Tralleis
Thrasubulus
Thrasullos
Timaios

Tribonianus
Varāhamihira
Vettius Valens of Antioch
Vicellius
Wuzurgmihr
Yavaneśvara
Zarathuštra
Zēnarion
Zīg
Zoroaster

Astronomy (161) (authors and writings on the motion and nature of the "stars," often very mathematical, and also often descriptive), *cf.* also **Cosmology** (it is not always possible unambiguously to distinguish cosmology from astronomy); *cf.* also **Astrology**; *cf.* also **Meteōrologika**. Many entries by Bowen, Cusset, Jones, Lehoux, and Mendell:

Abiyūn al-Biṭriq
Adrastos of Kuzikos
Agrippa
Aiskhulos
Akhilleus
Alexander of Ephesos
Alexander of Lukaia
Alexander of Pleuron
Alkinoos
Ammōnios, son of Hermeias
Ammōnios of Alexandria, M. Annius
Anakreōn
Anania of Shirak
Anaxagoras
Andreas of Athens
Andrias
Anthedius
Apollinarios
Apollōnios of Laodikeia
Apuleius of Madaurus
Aratos of Soloi
Aristarkhos of Samos
Aristotheros
Aristullos
Artemidōros
Artemidōros of Parion
Āryabhaṭa
Attalos of Rhodes
Atticus
Auienus
Autolukos
Billaros
Biōn of Abdēra
Boethius
Calcidius
Censorinus (II)
Derkullidēs
Didumos of Knidos
Diodōros of Alexandria
Diodōros of Tarsos
Diodotos (I)
Diodotos (II)
Diogenēs of Oinoanda
Diogenēs of Tarsos
Diōn of Neapolis
Dionusios
Diophil-
Dōsitheos of Pēlousion
Egnatius
Ekphantos
Eratosthenēs
Euainetos
Euclid
Eudēmos of Rhodes
Eudoxos of Knidos
Euktēmōn
Eutokios
Fauonius
T. Flauius Vespasianus
Fuluius
Geminus
Gregory of Tours
Harpalos
Hēgēsianax
Helikōn
Hēliodōros
Hēliodōros of Alexandria
Hērakleidēs of Hērakleia Pontikē
Herminos
Hiketas
Hipparkhos of Nikaia
Hippokratēs of Khios

Hupatia
Hupsiklēs
Iōn
Iulianus Vertacus
Germ. Iulius Caesar
C. Iulius Hyginus
L. Iunius Columella
Kallippos of Kuzikos
Kalyāṇa
Keskintos Inscription
Kharmandros
Kidēnas
Kleomēdēs
Kleostratos of Tenedos
Konōn
Kritōn of Naxos
Lasos
Leōnidas of Alexandria
Leontios
Leptinēs (II)
Linos
Mallius
Mandrolutos
Matriketas
Menelaos of Alexandria
Menestratos (I)
Mēnodotos
Metōn
Mētrodōros (I)
Mētrodōros (II)
Naburianos
Nautelēs
Nearkhos of Crete
P. Nigidius Figulus
Oinopidēs
P. Ouidius Naso
Paitāmahasīddhānta
Pappos of Alexandria
P. Hibeh 1.27
P. Osloensis 73
P. Oxy. 3.470
P. Paris. Gr. 1
P. Rylandensis 27
Parmeniōn
Parmeniskos
Patroklēs
Phaeinos
Pheidias
Philippos
Philippos of Opous
Philolaos
Phokos
Plato
Plutarch
Polemarkhos
Priscianus of Caesarea
Priscianus of Ludia
Ptolemy
Purrhos of Magnesia
Sabinus
Samuel
Seleukos of Seleukeia
Serapiōn
Seuerus Sebokht
L. Sextilius Paconianus
Skopinas
Sminthēs
Sōsigenēs (I)
Sōsigenēs (II)
Stephanos
Sudinēs
C. Sulpicius Gallus
M. Terentius Varro
Theodōros of Kurēnē
Theodosios (of Bithunia)
Theōn
Theōn of Alexandria
Theōn of Smurna
Thrasullos
Timokharis
Timotheos
Tukhikos
M. Tullius Cicero
Q. Tullius Cicero
Varāhamihira
Victorius
Xenarkhos
Zēnodōros
Zēnodotos of Mallos

Biology (101) (writings on botany and zoology, treating animals *in se*, not as objects of farming, for which see **Agriculture**, nor as sources of *materia medica*, for which see **Pharmacy**); many entries by de Stefani, Meliadò, and Zucker:

Adeimantos
Aelianus of Praeneste
Aemilius Macer
Agathoklēs of Atrax
Agathōn of Samos
Alexander of Mundos
Alfius Flauus
Alkimakhos
Ambrosius of Milan
Amuntianos
Annaeus Lucanus
Antonius Castor
Apellēs of Thasos
Apollodōros Dēmokritean
Apollodōros of Alexandria
Aristodēmos
Aristophanēs of Buzantion
Aristotelian Corpus, Breath
Aristotelian Corpus, Problems
Aristotle
Arkhelaos of Khersonēsos
Bōlos of Mendēs
Calpurnius Piso (II)
Claudius Claudianus
Cornelius Valerianus
Damostratos
Dēmētrios "physicus"
Dionusios of Philadelphia
Dōriōn
Empedoklēs
Epainetēs
Epainetos
Eudēmos of Rhodes
Euteknios
Geōrgios of Pisidia
Grattius
Hēgēsidēmos
Hieroklēs of Alexandria
Iouba
Iulius Africanus
Kaikalos
Kallisthenēs of Olunthos
Khrusippos of Knidos
Klearkhos
Krateuas
Ktēsias
Lactantius
Leōnidas of Buzantion
Lukōn of Iasos
Lukōn of Troas (?)
Maecenas Melissus
Marcellus of Sidē
Marianus
Menekratēs of Ephesos
Menestōr
Mētrodōros of Buzantion
Nemesianus
Nemesios
Neoptolemos
Nepualios
Nestōr
Nikandros
Nikolaos of Damaskos
Nikomakhos of Stageira
Noumēnios of Hērakleia
Oppianus of Anazarbos
Oppianus of Apameia
P. Nigidius Figulus
P. Ouidius Naso
P. Oxy. 15.1796
Pammenēs
Pankratēs of Alexandria
Pankratēs of Argos
Papirius Fabianus
Petrikhos
Phainias
Philolaos
Philōn of Alexandria
Philoumenos
Physiologus
Plato
Plutarch
Polukleitos of Larissa
Pompeius Trogus
Poseidōnios of Corinth
Ptolemaios of Kuthēra
Seleukos of Tarsos
Simōn of Athens

Simōnidēs
Sornatius
Sōstratos of Alexandria
T. Lucretius Carus
Theoklēs
Theomenēs
Theophanēs of Hērakleiopolis
Theophrastos
Timotheos of Gaza
Trebius Niger
C. Vibius Rufinus
Xenokratēs of Aphrodisias
Xenophōn of Athens

Cosmology (65) (authors and writings on the nature and structure of the ***kosmos*** (*cf.* the Glossary) as a whole, including the "world-soul" and often touching on **Astronomy**, **Biology**, or other topics); most philosophers of the **Academy**, **Peripatos**, and **Stoa**, plus **Epicureans** and **Pythagoreans**, addressed cosmology, and those lists (in the Glossary) should also be consulted:

Alkinoos
Ampelius
Anaxagoras
Anaximandros
Anaximenēs
Apollodōros of Seleukeia
Arkhedēmos of Tarsos
Arkhelaos of Athens
Artemidōros of Parion
Atticus
Bardaisan
Boethius
Calcidius
Censorinus (II)
Chrysippus
Dēmokritos of Abdēra
Derveni Papyrus
Diodōros of Tarsos
Diogenēs of Apollonia
Egnatius
Empedoklēs
Epidikos
Geōrgios of Pisidia
Hērakleitos of Ephesos
Hēraiskos
Hippokratic Corpus, Regimen
Hippokratic Corpus, Sevens
Iōannēs Philoponos
Iōn
Iulianus of Tralleis
Kleanthēs
Koiranos
Korē Kosmou
On the Kosmos
Kratulos
Kritias
Leukippos
Linos
Mallius
Mandrolutos
Melissos of Samos
Mētrodōros of Khios
Nausiphanēs
Nikomakhos of Athens
Nikomēdēs (Hērakleitean)
Papirius Fabianus
Parmenidēs
Pausanias (Hērakleitean)
Petrōn of Himera
Pherekudēs
Philōn of Alexandria
Potamōn of Alexandria
Prōtagoras of Abdēra
Cn. Sallustius
Saloustios
Simplicius
Skuthinos
Stēsimbrotos
Thalēs
Theagenēs
Timaios of Lokris
M. Tullius Cicero
Xenophanēs of Kolophon
Xouthos
Zēnō of Elea

Doxography (38) (authors and writings surveying the teachings of earlier thinkers), although they make no original contribution, serve (like the *EANS* itself) to gather and organize contributions, and to show the range of answers and questions; typically written by Greeks in contrast to the **Encyclopedia**, typically written by Latins (most entries by Mejer):

Aëtios
Alexander of Milētos
Anaxilaïdēs
Antisthenēs of Rhodes
Areios Didumos
Aristippos
Aristoklēs of Messēnē
Athēnodōros (of Rhodes?)
Celsinus
Damas
Damōn of Kurēnē
Diodōros of Ephesos
Diodōros of Eretria
Diogenēs, Laertios
Dioklēs of Magnesia
Eunapios
Fauorinus
Galēn (pseudo), Hist. Philos.
Hērakleidēs of Kallatis
Hermeias
Hermippos (of Smurna?)
Hippobotos
Hippolutos
Iasōn
Idomeneus
Iōannēs of Stoboi
Lobōn
Manethōn of Sebennutos
Meleagros
Minuēs
Neanthēs
Nikias of Nikaia
Ofellius Laetus
Phanokritos
Saturos of Kallatis
Seleukos of Alexandria
Sōsikratēs of Rhodes
Sōtiōn of Alexandria

Encyclopedia (13) (work of systematic knowledge collection, distinguished from **Doxography** by being focused not on the thinkers or the questions, but on the answers, as perceived at the time of composition), typically written by Latins, instead of **Doxography**, typically written by Greeks (the two Greek encyclopedias here are marked with *):

Albinus
Ampelius
Anicius Manlius Boethius
Martianus Capella
Censorinus (II)
Cornelius Celsus
* Geōponika
Isidorus of Hispalis (Seville)
* Sex. Iulius Africanus
C. Plinius Secundus
Cassiodorus Senator
M. Terentius Varro of Reate
Macrobius Ambrosius Theodosius

Geography (246) (works and authors on the description of the Earth, or large portions thereof, as well as mathematical geography, latitudes and longitudes); some works of geography skirt close to **Paradoxography**, and writers of **Astronomy** and **Astrology** often treat geography. Many entries by Dognini, Dueck, Kaplan, Kuelzer, and Lozovsky:

Acilius
Aethicus Ister
Aethicus, pseudo
Aëthlios
Agatharkhidēs of Knidos
Agathēmeros

Agathodaimōn of Alexandria
Agathoklēs of Milētos
Agathōn of Samos
Alexander
Alexander of Ephesos
Alexander of Milētos
Alexander of Mundos
Alupios of Antioch
Amōmētos
Ampelius
Amuntas
Anania of Shirak
Anaxikratēs
Androitas
Andrōn of Teōs
Androsthenēs
Antiokhos of Surakousai
Apellās of Kurēnē
Apollodōros of Artamita
Apollodōros of Athens, pseudo
Apollodōros of Kerkura
Apollōnidēs
Apollōnios of Aphrodisias
Arbitio
Aristagoras
Aristarkhos of Sikuōn
Aristoboulos of Kassandreia
Aristokreōn
Aristōn of Khios
Aristotelian Corpus, *Flood of the Nile*
Arkhelaos
Artemidōros of Ephesos
Asamōn
Asklēpiadēs of Murleia
Astunomos
Athanarid
Athēnagoras
Athēnodōros of Tarsos
Auienus
Baitōn
Bakōris
Basilis
"Bērutios"
Biōn of Abdēra
Biōn of Soloi
Blatausis
Bōtthaios
Boutoridas
Capito of Lukia
Carminius
Castorius
Cornelius Nepos
Cornelius Tacitus
Daimakhos
Daliōn
Damastēs of Sigeion
Damōn
Dēmētrios
Dēmētrios of Kallatis
Dēmodamas
Dēmotelēs
Dikaiarkhos
Dimensuratio/*Diuisio*
Diodōros of Samos
Diodōros of Sicily
Diogenēs
Diognētos of Eruthrai
Dionusios
Dionusios of Alexandria, Periēgētēs
Dionusios of Buzantion
Dionusios of Corinth
Dionusios of Milētos
Dionusios of Rhodes
Dionusios, son of Diogenēs
Dionusios, son of Kalliphōn
Diophantos
Dioskoros
Douris
Ephoros
Eratosthenēs
Eudōros of Alexandria
Eudoxos of Knidos
Eudoxos of Kuzikos
Eudoxos of Rhodes
Euktēmōn
Euthumenēs
Expositio geographiae
Expositio totius mundi
Flauius Arrianus
Geōrgios of Cyprus
Gildas
Glaukos (I)
Glaukos (II)
Hanno

Priscianus of Caesarea
Priskos
Probinus
Promathos
Prōtagoras
Prōtarkhos of Tralleis
Ptolemy of Alexandria
Puthagoras of Alexandria
Putheas
Ravenna Cosmography
C. Sallustius Crispus
Sardonius
Sebosus Statius
Seleukos of Seleukeia
Sēmos
Serapiōn of Antioch
Simmias
Simōnidēs
Skulax of Karuanda
Skulax of Karuanda, pseudo
Skumnos
Sōkratēs of Argos
Sōsandros
Sōsikratēs of Rhodes
Stadiasmus Maris Magni
Staphulos
Stephanos of Buzantion
Strabōn
Summaria rationis geographiae
M. Terentius Varro
Theodōros of Gadara
Theophanēs of Mutilēnē
Theophilos
Theophulaktos
Thrasualkēs
Thucydidēs
Timagenēs
Timagētos
Timaios of Tauromenion
Timosthenēs
Turranius Gracilis
Vibius Sequester
M. Vipsanius Agrippa
Ulpianus of Emesa
Xanthos
Xenagoras of Hērakleia
Xenagoras, son of Eumēlos
Xenophōn of Athens
Xenophōn of Lampsakos
Zēmarkhos
Zēnothemis
Zōpuros

Harmonics (54) (authors and works on the **Mathematics** of music, i.e., such topics as concords and octaves, often now called "music theory"); most entries by Creese, Mathiesen, or Rocconi:

Adrastos of Aphrodisias
Aelianus
Alupios of Alexandria
Aristidēs, Quintilianus
Aristotelian Corpus, Problems
Aristotelian Corpus, Sounds
Aristoxenos of Taras
Arkhestratos
Arkhutas of Taras
Artemōn of Kassandreia
Athēnodōros (of Rhodes?)
Aurelius Augustinus
Bakkheios Gerōn
Calcidius
Censorinus (II)
Damōn of Athens
Dēmētrios
Didumos
Diogenēs of Babylōn
Dionusios of Halikarnassos
Eratoklēs
Eratosthenēs
Euboulidēs
Euclidean Sectio Canonis
Euphranōr (Music)
Euphranōr (Pythag.)
Fauonius
Gaudentius
Hērakleidēs of Hērakleia Pontikē
Hērakleidēs of Hērakleia Pontikē, junior
Hippasos
Hupatia

Kleoneidēs
Mucianus
Muōnidēs
Nausiphanēs
Nikomakhos of Gerasa
Panaitios
Paulos
Philodēmos
Philolaos
Plutarch, Music
Porphurios of Tyre
Ptolemaïs of Kurēnē
Ptolemy
Puthagoras of Zakunthos
Pythagoras
Sextus Empiricus
Simos of Poseidōnia
M. Terentius Varro
Theodōros of Kurēnē
Theōn of Smurna
Thrasullos
Xenophilos

Lithika (36) (works and authors on the nature and properties of stones or other substance dug from the earth; **Alchemy** often makes use of such substances, or tries to imitate them, **Paradoxography** often writes about them, and **Pharmacy** regularly employs them, so no rigid distinction can be made); most entries by Amato:

Agatharkhides of Samos
Aristoboulos
Arkhelaos
Asarubas/Hasdrubas
Claudius Claudianus
Cornelius Bocchus
Damigerōn
Dēmokritos, pseudo
Derkullos
Dionusios
Dionusios of Alexandria, Periēgētēs
Dōrotheos of Khaldea
Drakōn of Kerkura
Epiphanios
Euax
Hērakleitos of Sikuōn
Iouba
P. Nigidius Figulus
Nikanōr
Nikias of Mallos
Olumpikos
Orpheus, pseudo
Ostanēs, pseudo
Petosiris
Philostratos
Poseidippos
Saturos
Sōkratēs
Sotakos
Sudinēs
Theophilos
Theophrastos of Eresos
Timaris
Xenokratēs of Ephesos
Zakhalias
Zēnothemis

Mathematics (164) (authors and works on arithmetic and geometry, as well as on "logistics," i.e., calculation; distinguished from **Metrology** on the one hand, and from mathematical **Astronomy** and **Geography** on the other); **Academics** *per se* are not listed here (many if not all were concerned with some aspect of mathematics), but agrimensores and even numerology are included; many entries by Bernard, Campbell, Jones, Lehoux, and Mueller:

Aganis
Aigeias
Ammōnios of Alexandria, M. Annius
Ammōnios, son of Hermeias
Amphinomos
Amuntas of Hērakleia Pontikē
Anania of Shirak
Anatolios of Laodikeia

Androkudēs (Pythag.)
Andrōn of Rome
Anthēmios
Antiphōn of Athens
Apollodōros of Kuzikos
Apollōnios of Pergē
Archimēdēs
Aristaios
Aristotelian Corpus, Indivisible Lines
Aristotelian Corpus, Problems
Arkadios
Arkhutas of Taras
Asklēpios of Tralleis
Athēnaios of Kuzikos
Autolukos
Balbus
Boëthius
Book of Assumptions of Aqaṭun
Brusōn
Calcidius
Censorinus (II)
Deinostratos
Dēmētrios
Dēmētrios of Alexandria
Dēmētrios of Amisos
Dēmētrios (of Athens?)
Dēmētrios of Lakōnika
Dēmokritos of Abdēra
Derkullidēs
Dioklēs
Dionusios of Alexandria
Dionusios of Kurēnē
Dionusodōros (of Kaunos)
Diophantos of Alexandria
Domninos
Dōsitheos of Pēlousion
Eratosthenēs
Erukinos
Euboulidēs
Euclid
Euclidean Sectio Canonis
Eudēmos of Rhodes
Eudoxos of Knidos
Euphoriōn
Eurutos
Eutokios
Fauonius
Helikōn
Hērakleitos
Hermeias
Hermippos of Bērutos
Hermodōros of Alexandria
Hermodōros of Surakousai
Hermotimos
Hērōn
Hērōn of Alexandria
Hērōnas
Hierios
Hipparkhos of Nikaia
Hippasos
Hippias
Hippokratēs of Khios
Hipponikos
Hupatia
Hupsiklēs
Hyginus
Hyginus Gromaticus
Hyginus (pseudo)
Isidōros of Milētos
Isidōros of Milētos' student
Iulianus Vertacus
M. Iunius Nipsus
Kárpos
Kharmandros
Kleinias
Konōn
Kratēs
Kratistos
Leōn
Magnēs
Mamerkos
Marinos of Neapolis
Megethiōn
Melior
Menaikhmos
Menelaos of Alexandria
Metōn
Mētrodōros of Tralleis
Moderatus
Naukratēs
Nausiphanēs
Neokleidēs
Nikolaos
Nikomakhos of Gerasa

Nikomēdēs
Nikōn of Pergamon
Nikotelēs
Oinopidēs
Onētōr
P. Akhmīm
P. Ayer
P. Gen. inv. 259
P. Osloensis 73
P. Vindob. 19996
Panaitios Jr.
Pandrosion
Pappos of Alexandria
Paterios
Peithōn
Periklēs
Perseus
Philippos of Opous
Philolaos
Philōn of Gadara
Philōn of Tuana
Philōnidēs of Laodikeia
Poluainos
Proklos of Laodikeia
Proklos of Lukia
Prolegomena to Ptolemy's Suntaxis
Prōros
Prōtarkhos of Bargulia
Ptolemy of Alexandria
Puthiōn of Thasos
Pythagoras of Samos
Serapiōn of Antioch
Serenus of Antinoeia
Seuerus Sebokht
Sextus Empiricus
Siculus Flaccus
Simos of Poseidōnia
Sōkratēs (junior)
Speusippos of Athens
Sporos
Syrianus
M. Terentius Varro of Reate
Thalēs
Theaitētos
Theodōros of Kurēnē
Theodōros of Soloi
Theodosios (of Bithunia)
Theologumena arithmeticae
Theōn of Alexandria
Theōn of Smurna
Theudios
Thrasudaios
Thumaridas
Tukhikos
Ulpianus of Alexandria
Victorius
Xenagoras, son of Eumēlos
Xenokratēs of Khalkēdōn
Zēnodōros
Zēnodotos
Zēnōn of Sidōn
Zeuxippos

Mechanics (72) (authors and writings on the construction and operation of mechanical devices, often war-machines, in works entitled *Poliorkētika* or *Belopoiika*, or else automata, but also including pneumatic devices, as well as the theory of motions), a topic not always clearly distinct from **Architecture**; many entries by Tybjerg:

Abdaraxos
Agēsistratos
Aineias of Stumphalos
Andronikos of Kurrhos
Apellis
Apollodōros of Damaskos
Apollōnios of Athens
Archimēdēs
Aristiōn, grandfather & grandson
Aristotelian Corpus, Mēkhanika
Artemōn of Klazomenai
Athēnaios
Bitōn
Bromios
Daimakhos
Damis
De Rebus Bellicis
Dēmoklēs
Diadēs
Diognētos of Rhodes

Diokleidēs
Dionusios of Alexandria
Diphilos
Dōriōn
Epikratēs
Epimakhos
Euclid
Harpalos
Hēgētōr of Buzantion
Hērakleidēs of Taras
Hērodotos
Hērōn of Alexandria
Isidōros of Abudos
Kallias
Kallistratos
Kallixeinos of Rhodes
Kárpos
Keras
Kharias
Kharōn of Magnesia
Kineas
Kratēs of Khalkis
Ktēsibios
Leontios
Marcellus
Moskhiōn
Neileus
Nikōnidēs
Numphodōros
Paconius
Pappos of Alexandria
Pasikratēs
Pausistratos
Pephrasmenos
Perigenēs
Phaiax
Philōn of Buzantion
Poluīdos
Poseidōnios of Macedon
Prōtarkhos
Purgotelēs
Purrhos of Ēpeiros
Quirinus
Skopinas
Stuppax
Tektōn
Teukros of Carthage
Theodōros
Truphōn of Alexandria
Urbicius
Vitruuius
Zōpuros of Taras

Medicine (420) (writers on medicine, including commentators on Hippokratēs, and all members of medical schools, **Empiricist**, **Hērophilean**, **Erasistratean**, **Asklēpiadean**, **Methodist**, and **Pneumaticist**, as noted, but excluding people known only for **Pharmacy**; some tracts of **Astrology** touch on medicine, as do some works of **Cosmology**); many authors are ambiguous, classifiable as both medical or pharmaceutical:

Abas (Aias)
Adamantios of Alexandria
Aeficianus
Aelius Promotus
Aëtios of Amida
Africanus
Agapios
Agathinos (Pneum.)
Agathoklēs
Aglaias
Agnellus
Ahrun
Aigimios
Aiskhinēs of Athens
Aiskhriōn of Pergamon (Emp.)
Akesias
Akrōn
Alexander
Alexander, Sophistēs
Alexander of Laodikeia (Hēr.)
Alexander of Tralleis
Alkamenēs
Alkinoos
Alkmaiōn
Alkōn
Ammōnios of Alexandria

Amuntās
Andreas of Karustos (Hēr.)
Androkudēs
Anqīlāwas
Anthimius
Antigenēs (Era.)
Antigonos of Alexandria
Antiokhos Paccius (Askl.)
Antipatros (Meth.)
Antiphanēs of Dēlos
Antonius
Antonius Musa
Antullos (Pneum.)
Apeimantos
Apollodōros
Apollōnidēs of Cyprus (Meth.)
Apollōnios Glaukos
Apollōnios Ophis
Apollōnios of Alexandria (Hēr.)
Apollōnios of Antioch & Son
Apollōnios of Kition (Emp.)
Apollōnios of Memphis (Era.)
Apollōnios of Pergamon (Pneum.)
Apollōnios of Pitanē
Apollōnios of Prousias
Apollōnios of Tuana, pseudo
Apollophanēs of Seleukeia (Era.)
Areios of Tarsos (Askl.)
Aretaios (Pneum.)
Aristanax
Aristogenēs of Knidos
Aristogenēs of Thasos
Aristōn (I)
Aristotelian Corpus, Hist. Animals 10
Aristotelian Corpus, Problems
Aristotle of Stageira
Aristoxenos (Hēr.)
Arkhagathos
Arkhelaos
Arkhibios (Emp.)
Arkhidēmos
Arkhigenēs of Apameia (Pneum.)
Arsenios
Artemidōros, Capito
Artemidōros of Pergē (?)
Artemidōros of Sidē (Era.)
Artorius
Asaph
Asklatiōn
Asklēpiadēs Titiensis
Asklēpiadēs of Bithunia
Asklēpiadēs of Prousias
Asklēpios
Aspasia
Athēnaios of Attaleia (Pneum.)
Athēniōn (Era.)
Athēnodōros
Attalos
Aufidius (Askl.)
Auidianus (Meth.)
Ausonius
Bakkheios of Tanagra (Hēr.)
Basileios
Biōn, Caecilius
Bothros
Burzoy
Caecilius
Caelius Aurelianus (Meth.)
Caesarius
Cassius (Emp.)
Cassius Felix
Cassius Iatrosophist
Clodius of Neapolis
Constantinus
A. Cornelius Celsus (Emp.)
Daliōn
Damastēs
Damonikos (Askl.)
Dēmarkhos
Dēmētrios
Dēmētrios "Khlōros"
Dēmētrios of Apameia (Hēr.)
Dēmokedēs
Dēmokritos, pseudo
Dēmosthenēs Philalēthēs (Hēr.)
Dexippos
Diagoras
Didumos of Alexandria (II)
Dieukhēs
Diodōros (Emp.)
Diodotos (Askl.)
Dioklēs of Karustos
Diōn
Dionidēs

Theōn of Alexandria (I)
Theōn of Alexandria (II)
Thessalos of Kōs
Thessalos of Tralleis (Meth.)
Thrasuas
Thrasumakhos
Timotheos of Metapontion
Truphōn of Gortun
Vettius Valens
Vindicianus
Xenokratēs of Aphrodisias
Xenokritos of Kōs
Xenophōn of Kōs
Zēnōn (Hēr.)
Zēnōn (of Athens?)
Zeuxis (Emp.)
Zeuxis (Hēr.)
Zōpuros of Alexandria (Emp.)

Meteōrologika (61) (authors and writings on "things on high," the literal meaning of the term: at an early date included all that was later filed under **Astronomy**, and perhaps only separated by Aristotle and later writers); here, includes works only on matters not included in **Astronomy**, thus, what we would call optical effects in the atmosphere (meteors, rainbows, and the like, thus being hard to distinguish strictly from early **Astrology**), plus, consonant with the ancient sense, discussions about earthquakes, tides, volcanoes, and the like (thus being hard to distinguish strictly from **Geography**):

Aetna
Alexander of Aphrodisias
Anaxagoras
Anaximandros
Anaximenēs
Annaeus Seneca
Apollōnios of Mundos
Apuleius of Madaurus
Aratos
Aristombrotos
Aristotelian Corpus, On the Nile
Aristotelian Corpus, Problems
Aristotelian Corpus, Winds
Aristotle
Arkhedēmos of Tarsos
Artemidōros of Parion
Asklēpiodotos of Nikaia
Athēnodōros of Tarsos
Boëthos of Sidōn (Stoic)
Bōlos of Mendēs
Campestris
Claudius Claudianus
Clodius Tuscus
Damaskios
Dēmoklēs
Dikaiarkhos
Diodotos (Astr. II)
Diogenēs of Oinoanda
Dionusios of Corinth
Dōsitheos of Pēlousion
Epaphroditos
L. Flauius Arrianus
Geminus
Hippokratēs of Khios
Iōannēs "Lydus"
Iōannēs Philoponos
Kallisthenēs
T. Lucretius Carus
Mallius
Mētrodōros (I)
Milōn
Nearkhos of Crete
Nikanōr
Nikolaos of Damaskos
Ofellius Laetus
Olumpiodōros
Papirius Fabianus
Philippos of Medma
Philippos of Opous
Philōn
Plutarch
Poseidōnios of Apameia
Promathos
C. Sallustius Crispus
Seleukos of Seleukeia
Suros
P. Terentius Varro
Thalēs

Theophrastos
Theophrastos, pseudo
Vicellius

Metrology (29) (authors and works on systems of units and their relations, whether of length, area, volume, or weight); most entries by de Nardis:

Adamantios Ioudaios
Africanus
Agennius Urbicus
Ammōn of Rome
Arruntius Celsus
Balbus
Carmen de Ponderibus
Dardanios
Didumos of Alexandria
Diodōros
Dioskouridēs
Epiphanios
Eusebios (pseudo)
Hyginus
Hyginus Gromaticus
Hyginus, pseudo
Innocentius
Sex. Iulius Africanus
Sex. Iulius Frontinus
M. Iunius Nipsus
Kleopatra
Kurillos
Polukleitos of Argos (?)
Priscianus of Caesarea
Remmius Fauinus
Siculus Flaccus
Silaniōn (?)
Victorius
L. Volusius Maecianus

Optics (32) (authors and writings on geometrical optics, i.e., of perspective, surveying, and mirrors; refraction in a very few cases; we include here writings on color); many writers on optics are also writers on **Mathematics**, and discussions of optical phenomena often became discussions of **Meteōrologika** (writers on proportion and symmetry are filed with **Architecture**):

Agatharkhos of Samos
Anthēmios
Apollodōros of Seleukeia
Aristombrotos
Aristotelian Corpus, Colors
Aristotle
Censorinus (I)
Damianos
Dēmokleitos
Dēmokritos of Abdēra
Dioklēs
"Dtrums"
Empedoklēs
Epicurus
Hēliodōros of Larissa
Hērōn of Alexandria
Hestiaios
Hierōnumos of Rhodes
Hippokratēs of Khios
Kleandros
Kleoxenos
P. Hibeh
P. Louvrensis 7723
P. Osloensis 73
P. Oxy. 13.1609
Philippos of Opous
Plato
Plutarch
Ptolemy of Alexandria
Puthiōn of Thasos
Theophrastos
Zēnodōros (?)

Paradoxography (61) (authors and writings on ***paradoxa*** of the natural world, embodying collections of alleged observations serving to call into question or limit the comprehensibility of the world, or else to manifest its astonishing variety and power); many works of Hellenistic **Biology** or **Geography** and some of **Alchemy**, **Lithika**, **Pharmacy**, or **Medicine** incorporate or resemble paradoxography. Most entries by Guido Schepens, Jan Bollansée, and Karen Haegemans:

Agathosthenēs
Agēsias of Megara
Alexander of Mundos
Amōmētos
Anaxilaos of Larissa
Andronikos
Antigonos of Karustos
Apollōnios
Aristeidēs
Aristotle, pseudo, Mirab. Ausc.
Arkhelaos of Khersonēsos
Bōlos of Mendēs
Boutoridas
Damōn (Geog.)
Dēmētrios (Geog.)
Dēmētrios "physicus"
Demotelēs
Diophanēs
Dōrotheos of Hēliopolis
Eudikos
Eudoxos of Rhodes
Geōrgios of Pisidia
Granius
Habrōn
Hēgēsias of Magnesia
Hēgēsidēmos
Hēliodōros of Athens
Imbrasios
Isigonos
Kallimakhos of Kurēnē
Kallimakhos of Kurēnē, Jr.
Kalliphanēs
Kallixeinos
C. Licinius Mucianus
Mursilos
Nepualios
Nestōr
Nikagoras
Nikolaos of Damaskos
Numphodōros of Surakousai
Ofellius
Paradoxographus Florentinus
Paradoxographus Palatinus
Paradoxographus Vaticanus
Philōn of Hērakleia
Philostephanos
Philostratos
Phlegōn
Plutarch, pseudo, de Fluuiis
Polemōn of Ilion
Politēs
Polukritos
Prōtagoras (Geog.)
Ptolemaios of Kuthēra
Sornatius
Sōstratos of Nusa
Sōtiōn
Teukros of Kuzikos
Theokhrēstos
Theophulaktos
Trophilos

Pharmacy (500) (writers and works on the compounding of remedies; some writers of **Medicine** also wrote pharmacy, and writings in this category often touch upon **Biology** or **Lithika**); we list writers of **Veterinary** recipes separately; many authors are ambiguous, classifiable as both medical or pharmaceutical:

Abaskantos
Acilius Hyginus
Adamantios of Alexandria
Aelius Gallus
Aelius Promotus
Aemilius Macer
Africanus
Agapētós
Agathinos
Agathoklēs

Castus
Celer
Cornelius
Daliōn
Damaskēnos
Damokratēs
Damonikos
Dasius
Dēïleōn
Dēmokritos, pseudo
Dēmosthenēs Philalēthēs
Dexios
Diodotos
Diogas
Diogenēs
Dioklēs of Khalkēdōn
Diomēdēs
Dionusios of Kurtos
Dionusios (of Milētos?)
Dionusios of Samos
Dionusodōros
Diophantos of Lukia
Dioskoros
Dioskouridēs of Anazarbos
Doarios
Domitius Nigrinus
Dōrotheos of Athens
Dōrotheos of Hēliopolis
Dōsitheos
Dulcitius
Eirēnaios
Elephantis
Eleutheros
Emboularkhos
Epagathos
Epainetēs
Epaphroditos of Carthage
Epidauros
Epigonos
Epikouros
Epimenidēs
Erasistratos of Sikuōn
Eruthrios
Euangeus
Euboulos
Eudēmos (Meth.)
Eudēmos of Athens
Euelpidēs
Euēnos
Eugamios
Eugeneia
Eugērasia
Euhēmeros
Eukleidēs
Eumakhos
Eunomos
Eunomos of Khios
Euphēmios
Euphorbos
Euphranōr
Euphratēs
Eusebius son of Theodorus
Euskhēmos
Euthukleos
Eutonios
Eutychianus
Fauilla
Faustinus
Flauianus
Flauius (Poet)
Flauius “the boxer”
Flauius Clemens
T. Flauius Vespasianus
Florus
Gaius of Neapolis
Galēn
Gemellus
Gennadios
Genthios
Geōrgios of Pisidia
Glaukias
Glaukōn
Glukōn
Grēgorios
Halieus
Harpalos
Harpokrās
Harpokratiōn
Harpokratiōn of Alexandria
Heirodotos
Hekataios
Hekatōnumos
Hēliadēs
Hērakleidēs of Ephesos

Hērakleidēs of Taras
Hērās
<u>De Herbis</u>
Hermās
Hermeias
Hermogenēs of Smurna
Hermolaos
Hermōn
Hermophilos
Hērodotos (of Tarsos?)
Hērōn
Hērophilos
Hierax
Hikesios of Smurna
Hubristēs
Hugiēnos
Iamblikhos of Constantinople
Ianuarinus
Idios
Iollas
Ioudaios
Iriōn
Isidōros of Antioch
Isidōros of Memphis
Isis, pseudo
Iulianus the deacon
Iulianus (of Alexandria?)
Iulius Agrippa
Iulius Bassus
Iulius Secundus
Iunia/Iounias
Iunius Crispus
Iustinus
Iustus
Kallimakhos of Bithunia
Kallinikos
Kēphisophōn
Khalkideus
Khariklēs
Kharitōn
Kharixenēs
Kharmēs
Khrusanthos
Khrusermos
Khrusippos
Khrusippos of Knidos (I)
Kimōn
Kleëmporos
Kleoboulos
Kleōn of Kuzikos?
Kleophantos
Kloniakos
Klutos
Kōdios Toukos
Kosmos
Krateros
Krateuas
Kratippos
Kratōn
Kritōn of Hērakleia Salbakē
Ktēsiphōn
<u>Kuranides</u>
Kurillos
Kuros
Laïs
Lampōn
Laodikos
Lepidianus
Leukios
Licinius Atticus
Lingōn
Logadios
Loukās
Lukomēdēs
Lukos of Macedon
Lukos of Neapolis
Lunkeus
Lusias
Lusimakhos of Macedon
Maecius Aelianus
Magistrianus
Magnus *arkhiatros*
Magnus of Ephesos
Magnus of Philadelpheia
Magnus of Tarsos
Maiorianus
Makhairiōn
Manethōn
Mantias
Marcellinus
Marcellus
Marcellus of Bordeaux
Marcianus of Africa
Marinos of Alexandria

Markiōn
Marsinus
Maximianus
Mēdeios
Medicina Plinii
Megēs
Melētos
Melitōn
Menandros (of Pergamon?)
Mēnās
Menekratēs Claudius
Menelaos
Menemakhos
Menenius Rufus
Menestheus
Menippos
Mēnodōros of Smurna
Menoitas
Mēnophilos
Mētrodōros (Pharm.)
Mikiōn of Smurna
Minucianus
Mithradatēs VI
Mnaseas
Mnēsidēmos
Mnēsidēs
Moskhiōn
Mousaios
Murōn
Naukratitēs medicus
Nearkhos
Neilammōn
Nikēratos
Nikētēs
Nikolaos
Nikomakhos
Nikomēdēs of Bithunia
Nikostratos
Nonnos
Noumēnios of Hērakleia
Numius
Numphodōros
Okianos
Olumpiakos
Olumpias
Olumpionikos
Onēsidēmos
Onētidēs
Ophiōn
Orestinos
Orfitus
Ōrigeneia
Ōriōn
Ōros
Orpheus, pseudo
Orthōn
Ostanēs, pseudo
Pamphilos of Alexandria
Pantainos
P. Johnson
P. Laur. Inv. 68
P. Mich. 17.758
P. Ryl. III.531
PSI inv.3011
P. Strassbourg Inv. Gr. 90
P. Tebtun. 679
Pasiōn
Patroklos
Paulos (of Italy?)
Pelops
Pelops of Smurna
Perigenēs
Petrikhos
Petrōnios Musa
Petros
Phaidros
Phanias
Pharnax
Philippos of Macedon
Philippos of Rome
Philokalos
Philoklēs
Philokratēs
Philōn
Philōn of Huampolis
Philōn of Tarsos
Philōnidēs of Catina
Philōnidēs of Durrakhion
Philōtas
Philoumenos
Philoxenos
Phulakos
Physica Plinii
Sex. Placitus Papyriensis

Timōn
Tlēpolemos
Truphōn of Gortun
Turannos
Turpillianus
Valerius Paulinus
C. Valgius Rufus
Velchionius
Xanitēs
Xenokratēs of Aphrodisias
Zēnōn (Hēr.)
Zēnōn (of Athens?)
Zēnōn of Laodikeia
Zēnophilos
Zeuxis (Hēr.)
Zōilos of Macedon
Zōpuros of Alexandria
Zōpuros of Gortuna
Zōsimos

Physiognomy (10) (authors and works on the correlation between the body and the mind, traits of the latter thought to be revealed by observations on the former); most entries by Sabine Vogt:

Adamantios
Antisthenēs of Athens
Aristotelian Corpus, Physiognomy
Loxos
Melampous
Physiognomista Latinus
Polemōn of Laodikeia
Polukleitos of Argos (?)
Silaniōn (?)
Zōpuros

Psychology (48) (authors and works on the physical nature of the soul, or its structure, including on the location of the ***hēgemōn*** – not on ethics or on character; including works on interpretation of dreams); an admittedly fuzzy category, but a well-established part of the discourse especially of the **Academy**, but also of the **Peripatos, Stoics**, and **Epicureans**; medical writers on mental illness are also included here, although the distinction between "mental" and "physical" illness was itself debatable in ancient **Medicine** (note that items on the "world-soul" are filed under **Cosmology**):

Aisara
Alexander of Aphrodisias
Alkinoos
Antipatros of Tyre
Antonius
Apollophanēs of Nisibis
Aretaios
Aristotle
Artemidōros of Daldis
Astrampsukhos
Athēnaios of Attaleia
Boëthos of Sidōn (Peripatetic or Stoic?)
Cassius Longinus
Chrysippus
Damōn of Athens
Dēmokritos of Abdēra
Dikaiarkhos
Diogenēs of Babylōn
Diogenēs of Oinoanda
Dioklēs of Karustos
Dissoi Logoi
Ekhekratēs
Epicurus
Galēn
Hērophilos
Hieroklēs of Alexandria
Iamblikhos of Khalkis
Iōannēs of Stoboi
Karneadēs
Kleanthēs
T. Lucretius Carus
Marinos (Med.)
Marinos of Neapolis
Melampous
Mnēsarkhos of Athens
Noumēnios of Apameia

Plato
Plōtinos
Porphurios of Tyre
Poseidōnios of Apameia
Poseidōnios (Med. II)
Praxagoras
Priscianus of Ludia
Ptolemy
Stratōn
Sunesios of Kurēnē
Syrianus
Theophrastos of Eresos

Veterinary Medicine (41) (authors and writings on **Medicine** or **Pharmacy** as applied to animals; distinguished from **Biology** in being practical application, just as is **Pharmacy**); most entries by Anne McCabe:

Aemilius Hispanus
Agathotukhos
Ambrosios Sophistēs
Anthemustiōn
Apellās of Laodikeia
Apsurtos
Arkhedēmos
Arkhelaos
Astrampsukhos
Auxanōn
Caystrius
Emeritus
Epikharmos
Euboulos
Euhēmeros
Eumēlos
Flauius Arrianus
Geōponika
Grēgorios
Helenos
Hērakleidēs of Taras (?)
Hieroklēs
Hierōn
Hipparkhos
Hippasios
Hippokratēs, junior
Iulius Africanus
Kleomenēs
Litorius
Mulomedicina Chironis
Nephōn
Pelagonius
Phoibos
Secundus
Simōn of Athens
Sōsandros
Stratonikos
Theomnēstos of Nikopolis
Tiberius
Vegetius Renatus
Xenophōn of Athens

INDICES

Names by Ethnicity

Armenian: some works of Greek science, and many theological works, were translated into Armenian; see esp. Hellenizing School (Armenian) (most entries by Edward G. Mathews, Jr.):

Anania of Shirak
Andreas (of Athens?)
Moses of Xoren
Philōn of Alexandria
Tukhikos of Trebizond

Celtic names appear occasionally, though there is no evidence of scientific writing in ancient Gaulish or other Celtic language:

Brenitus
Gildas
Lingōn
(Pompeius) Trogus

Egyptians (59): by name, ancestry, or origin (see esp. Demotic Scientific Texts) form the second-largest ethnic group (after Latins) herein – Greeks from Alexandria excluded and only those papyri of clearly Egyptian orientation included:

Aganis
Ammōn(ios) (five men)
Anoubiōn
Apiōn
Arkhelaos
Asamōn
Baitōn
Blatausis
Bōlos
Boutoridas
Cynchris
Dōrotheos of Hēliopolis
Hēphaistiōn of Thēbai
Hēraiskos
Hermēs, pseudo
Hermōn
Isis, pseudo (two women)
Khairēmōn
Kōmarios
Kuranides
Magnus of Philadelpheia
Manethōn (two men)
Naukratitēs medicus
Neilammōn
Neilos
Nephōn

Ninuas
Olumpiodōros of Thēbai
Ōros
Pammenēs
P. Hibeh. 1.27
P. Holm.
P. Leid.
P. London 98
P. Oxy. 3.470
P. Oxy. 15.1796
Pausēris
Paxamos
Pēbikhios
Petosiris
Philippos of Egypt
Phimenas of Saïs
Plōtinos
Praecepta Salubria
Prōtās
Salmeskhoiniaka
Seuerus Iatrosophist
Teukros of Babylōn (Egypt)
Theophanēs of Hērakleopolis (Egypt)
Theophilos son of Theogenēs
Theophulaktos
Thrasullos of Mendē
Zōsimos of Panopolis

Gothic: east Germanic language, with a tradition of Scandinavian origin, attested mostly in the 4th-6th centuries CE:

Athanarid
Heldebald
Iordanes
Marcomir

Latin:

Many ancient cultures adopted Greek science to some degree, none more so than the Latins. More works of science survive in Latin than in any other non-Greek language, only a small number of which are mere translations: **Index A** (188 entries, over 9%).

In addition, people whose ethnicity appears to be Latin (based on bearing a wholly Latin name) wrote works of science in Greek, in numbers larger than for any other non-Greek group: **Index B** (154 names). In the latter list, up to 71 of the people named *may* have written in Latin, but we lack evidence that would allow certainty; moreover, the later the date the more likely that a "Latin" name was no longer exclusively Latin.

A) Writers and works in Latin (188)

Known, by testimony or because extant, to have been written in Latin.

Aemilianus (Palladius)
Aemilius Macer of Verona
Aemilius Spanus
Aethicus (pseudo)
Aethicus Ister
Aetna
Agennius Urbicus
Agnellus
Albinus
Alfius Flauus
Ambiuius
Ambrosius of Milan
Ampelius
Annaeus Lucanus
Annaeus Seneca
Anthedius of Vesunnici
Anthimius
Antonius Castor
Apuleius of Madaurus
Apuleius (pseudo)
Arbitio
Arruntius Celsus
Artemisius of Dianium
Attius
Auienus (Rufius Festus)
Aurelius Augustinus

Maecenas Melissus
Mallius Theodorus
Mamilius Sura
Manilius
Marcellus of Bordeaux
Marius Victorinus
Matius Caluenus
Maximinus
Medicina Plinii
Melior
Melitianus
Mucianus
Mulomedicina Chironis
Mustio

Nemesianus of Carthage
Nigidius Figulus

Oppius
Orosius of Bracara
Ouidius Naso

Paconius
Papirius Fabianus
Pelagonius of Salona
Peutinger Map
Physica Plinii
Physiognomista Latinus
Plinius Secundus
Pompeius Lenaeus
Pomponius Mela
Pontica
Porcius Cato
Priscianus of Caesarea
Probinus
Proclianus

Rabirius
Ravenna Cosmography
Remmius Fauinus

Sabinius Tiro
Sabinus (Poet)
Sallustius (Cn.)
Sallustius Crispus
Sardonius
Scribonius Largus
Sebosus Statius
Secundus
Cassiodorus Senator
Septimius
Serenus Sammonicus
Sextilius Paconianus
Siburius
Siculus Flaccus
Sornatius
Sueius
Sulpicius Gallus

Terentius Varro of Reate
Terentius Varro of Narbo
Theodorus Priscianus
Macrobius Ambrosius Theodosius
Thrasubulus
Tiberianus
Trebius Niger
Tremelius Scrofa
Tribonianus of Side
M. Tullius Cicero
Q. Tullius Cicero
Turranius
Turranius Gracilis

Valerius Messalla Potitus
Valgius Rufus
Vegetius Renatus
Vergilius Maro
Vibius Rufinus
Vibius Sequester
Vicellius
Victorius of Aquitania
Vindicianus
Vipsanius Agrippa
Vitruuius
Volusius Maecianus

B) Onomastically-Latin Writers (154)

Includes all writers whose names are wholly Latinate, and are not in list "A"; many of these may have thought of themselves not as "Latin" but as Greeks with a name of Latin origin, and only four are *known* to have been Latins (marked *). The 83 authors who certainly wrote in Greek are marked **(G)**; some of the other 71 *may* have written in Latin, though most, 55, are pharmacists quoted in Greek.

* Acilius **(G)**
Aeficianus **(G)**
Aelianus (Platonist) **(G)**
Aelianus of Praeneste **(G)**
* Aelius Gallus **(G)**
Aelius Promotus of Alexandria **(G)**
Africanus (Metrol.) **(G)**
Africanus (Pharm.)
Agrippa of Bithunia **(G)**
Albinus of Smurna **(G)**
Antoninus of Kōs (Pharm.)
Antonius (Atomist) **(G)**
Antonius "root-cutter" (Pharm.)
Antonius Musa **(G)**
Apuleius Celsus of Centuripae
Aquila Secundilla (Pharm.)
Arrianus **(G)**
Artorius
Asinius Pollio **(G)**
Atticus **(G)**
Aufidius of Sicily (Pharm.)
Auidianus
Aurelius (Pharm.)
Axios (if "Axius") (Pharm.)

Blastus (Pharm.)

Caesarius of Kappadokia **(G)**
Candidus (Pharm.)
Capito of Lukia **(G)**
Cassianus Bassus **(G)**
Cassius Iatrosophist **(G)**
Cassius Longinus **(G)**
Castus (Pharm.)
Celer (Pharm.)
Celsinus **(G)**
Censorinus (I) **(G)**
Claudianus **(G)**
Clodius (Med.)
Clodius of Naples **(G)**
Constantinus **(G)**
Cornelius (Pharm.)

Domitius Nigrinus (Pharm.)

Emeritus **(G)**

Fauilla (Pharm.)
Fauorinus of Arelatum **(G)**
Faustinus (Pharm.)
Flauianus (Pharm.)
Flauius "the boxer" (Pharm.)
Flauius Arrianus **(G)**
Flauius Clemens (Pharm.)
Florentinus **(G)**
Florus (Pharm.)
Fronto (Agric.)

Gaius (Platonist) **(G)**
Gaius (Hērophilean) **(G)**
Gaius of Naples **(G)**
Gaudentius **(G)**
Gemellus (Pharm.)
Geminus of Rhodes **(G)**

Ianuarinus (Pharm.)
Iuliana Anicia **(G)**
Iulianus (Pharm.)
Iulianus, Emperor (pseudo) **(G)**
Iulianus (of Alexandria?) **(G)**
Iulianus of Askalon **(G)**
Iulianus of Laodikeia **(G)**
Iulianus of Tralleis **(G)**
Iulius Africanus **(G)**
Iulius Agrippa (Pharm.)
Iulius Bassus (Pharm.)
Iulius Secundus (Pharm.)
Iunia (if not "Iounias") (Pharm.)
Iunius Crispus (Pharm.)
Iustinianus Imp., ps. **(G)**
Iustinus (Pharm.)
Iustus (Oculist)
Iustus (Pharm.)**(G)**

Lepidianus (Pharm.)
Licinius Atticus (Pharm.)

Ariobarzanēs
Astrampsukhos
Azanitēs
Burzoy
Geōponika, in Pahlavi
Hippokratic Corpus, in Pahlavi
Ostanēs, pseudo
Pharnax
Sophar
Wuzurgmihr
Zarathuštra
Zīg (Royal Tables)
Zōroaster, pseudo

Sanskrit: Indo-European language (and script) of north India, ancestor of most modern Indian languages (entries by Plofker & Knudsen):

Āryabhaṭa
Kalyāṇa
Mīnarāja
Paitāmahasīddhānta
Sphujidhvaja
Varāhamihira
Yavaneśvara

Semitic (34)

Arabic (11): many works of Greek science were translated into Arabic, and some survive only or primarily thus, see esp. ARABIC, TRANSLATIONS (most entries by Kevin van Bladel):

Abiyūn
Aganis
Anqīlāwas
Apollōnios of Tuana, pseudo
Ahrun ibn-Aʿyan al-Qass
Book of Assumptions by Aqāṭun
"Dtrums"
Epaphroditos (meteor.)
Euax
al-Ḥarith ibn-Kalada
Paulos (music)

Babylonian: language and (cuneiform) writing of a people dwelling in Mesopotamia, whose work in astronomy greatly influenced Greeks (most entries by Francesca Rochberg):

Babylonian Astronomy
Bērossos
Kidēnas
Naburianos
Sudinēs

Hebrew: language of the Jews, closely related to Punic, also related to Aramaic (all entries by A.Y. Reed):

Asaf ben Berekhiah
Gamaliel VI
Samuel of Nehardea

Mandaic: an eastern Aramaic language, closely related to Syriac, but whose script includes vowels (entry by Siam Bhayro):

Book of the Signs of the Zodiac

Punic: language of the Phoenicians and Carthaginians, closely related to Hebrew:

Asarubas
Hanno
Hasdrubal
Himilkōn

Syriac: an eastern Middle Aramaic dialect strongly associated with Christianity in the Orient (entries by Siam Bhayro):

Bardaisan of Edessa
Sergius of Reš'aina
Severus Sebokht of Nisibis
Theodore, pupil of Sergius

Other Semitic names:

Abdaraxos
Iamblikhos
Libanios
Malkh-
Ninuas
Nonnosos

Thrakian names appear in Greek culture from an early period, until late antiquity.

Bithus
Seuthēs
Siburius (?)
Thamuros

Other: 11 names of miscellaneous or uncertain ethnicity:

Arbinas (Lukian)
Balbillos
Bocchus (Numidian)
Bolās
Dulcitius
Genthios (Illyrian)
Gessios (Petraean?)
Iouba (Numidian)
Logadios (Celtic?)
Suennesis (Kilikian)
Xanthos (Ludian)

Unattested names (53): i.e., attested solely for the person in *EANS*; some are probably corrupt. Compare on the one hand these eight attested but unusual names: Epidauros, Hikatidas, Idios, Maēs, Monās, Phulakos, Proëkhios, and Sardonius; and on the other these 13 names which are almost certainly to be emended and removed: Apios Phaskos, Aphros, Asklēpiadēs Titiensis, Auidianus, Kleophanēs, Kōdios Toukos, Numius, Plentiphanēs, Salimachus, Sarkeuthitēs, Sextus of Apollōnia, Silimachus, and Trophilos. Of the 52 names listed below, ten (marked *) are reliably transmitted, and some are likely to become attested with further epigraphic or papyrological discoveries:

Abdaraxos
Akhinapolos
Baitōn (Egyptian?)
Bakkhulidios
Bakōris (Egyptian?)
* Balbillos
Bothros
Bōtthaios
Boutoridas (Egyptian?)
* Damigerōn
Dēmokleitos
Doarios (Illyrian?)
Emboularkhos
Euangeus (not a name?)
* Euax
Eugērasia
Euruōdēs
Fauilla

Hagnodikē
Hipposiadēs
Huriadas
Iounias (⇒ Iunia)
Iriōn
Kaikalos
Kloniakos (not a name?)
Kommiadēs (⇒ Kosmiadēs)
Lingōn
* Logadios
Macharius
Magistrianus
Melitianus
Minuēs
Nepualios
* Nonnosos
* Ōdapsos
Okianos
Olumnios
Olumpionikos
Pephrasmenos
* Phokos
Platusēmos
Podanitēs
Polustomos
Prothlius
* Salpē
Samithra
Sandarius
Skopinas
* Stuppax
Theotropos
Timokleanos
Velchionius (Etruscan)
Xanitēs (not a name?)

Names by Category

Female writers (30 or under 2%): rare, but less so than in any other field of ancient literature; note that several of these names are restored as feminine from transmitted masculine names. Many other obscure names may well conceal female writers, since scribes notoriously masculinize names (*cf.* perhaps Aigeias, Arsenios, Diophil-, Eugamios, Faustinus, Laodikos, Marpēssos, Marsinus, Penthesileus, Prosdokhos, and Zēnariōn); and several obscure names may be female (e.g., Samithra).

Aisara of Lucania
Antiokhis of Tlōs
Aquila Secundilla
Aspasia
Elephantinē
Eugeneia
Eugērasia
Hagnodikē of Athens
Hupatia of Alexandria
Isis, pseudo (alch.)
Isis, pseudo (pharm.)
Iuliana Anicia
Iunia/Iounias
Kleopatra VII of Alexandria
Laïs
Maria
Mētrodōra
Muia
Olumpias of Thebes
Ōrigeneia
Pandrosion
Persis
Ptolemaïs of Kurēnē
Romula
Salpē
Samithra ?
Sōteira
Thaïs
Theanō
Timaris

Monotheists (82 or 4%): although most writers of ancient science were polytheists, some were monotheists; these **20** figures are identified more or less certainly as Jewish:

Abram
Adamantios of Alexandria
Ahrun
Asaf ben Berekhiah
Domninos of Larissa
Domnus
Dōsitheos of Pelousion
Gamaliel VI
Ioudaios
Iulius Agrippa
Maēs Titianus
Maria
Mousēs
Nonnosos
Philōn of Alexandria
Rufus of Samaria
Samuel of Nehardea
Theodos of Alexandria
Theophilos son of Theogenēs
Zakhalias of Babylōn

These **62** figures are identified more or less certainly as Christian:

Agapētós
Aineias of Gaza
Anonymous Alchemist "Christianus"
Ambrose
Anania of Shirak
Anastasios
Anatolios
Arsenios
Aurelius Augustinus
Bardaisan of Edessa
Basileios of Caesarea
Caesarius of Nazianzos
Castorius
Elias (pseudo)
Epiphanios of Salamis
Eusebius son of Theodorus
Fauonius Eulogius
Geōrgios of Cyprus
Geōrgios of Pisidia
Gregory of Nazianzos
Gregory of Nussa
Gregory of Tours
Hermeias (doxogr.)
Hermolaos (geog.)
Hilarius of Arles
Hippolutos of Rome
Innocentius
Iōannēs of Alexandria, *iatrosophistēs*
Iōannēs of Alexandria, Philoponos
Iōannēs of Antioch, Chrusostomos
Isidorus of Hispalis (Seville)
Iulianus (Pharm.)
Iulius Africanus
Iulius Firmicus Maternus
Iustinianus, Imp. (pseudo)
Kosmās Indikopleustēs
Kurillos
Lactantius
Macharius
Mallius Theodorus
Marcellus (Empiricus)
Marius Victorinus
Moses of Khoren
Mucianus
Nemesios of Emesa
Pamphilos of Bērutos
Paulus Orosius
Petros of Constantia
Philostorgios
Physiologus
Praecepta Salubria
Priscianus of Caesarea
Ravenna Cosmography
Cassiodorus Senator
Sergius of Rešʿaina
Seuerus Sebokht
Stephanos of Alexandria
Stephanos of Buzantion
Sunesios of Kurēnē
Theodore (student of Sergius)
Theophulaktos
Timotheus of Gaza

Poets (119 or almost 6%): from the beginnings, Greek science (and its Latin offspring) included works in verse, meant to more memorable or pleasant, while remaining instructive;

rarely or never was original research presented in this way, and notably no works in verse on architecture, mechanics, music, or veterinary medicine are attested. (In addition, these 13 scientific writers composed non-scientific verse: Bardaisan, Diogenēs of Tarsos, Gregory of Nazianzos, Iōn, Iōnikos, Kallimakhos of Kurēnē, Kritias, Licinius Caluus, Philodēmos, Philostephanos of Kurēnē, Sunesios of Kurēnē, Valgius, and perhaps Meleagros.)

Aemilianus
Aemilius Macer
Aetna
Aglaias
Alexander of Ephesos
Alexander of Pleuron
Ammōn (Astrol.)
Anakreōn (Astron.)
Andromakhos of Crete (Elder)
Annaeus Lucanus
Annaeus Seneca
Anoubiōn
Antimakhos of Hēliopolis
Apollodōros of Athens (pseudo)
Apollodōros of Kuzikos
Aratos
Arkhelaos of Khersonēsos
Arrianus
Artemidōros (Astron. I) (?)
Auienus
Calpurnius Piso (I)
Carmen Astrologicum
Carmen de Ponderibus
Claudius Claudianus
Damokratēs
Diodotos (Pharm.)
Dionusios son of Kalliphōn
Dionusios of Alexandria, Periēgētēs
Dionusios of Corinth
Dionusios of Philadelpheia
Diophil-
Dōrotheos of Athens
Dōrotheos of Sidōn
Egnatius
Empedoklēs
Ennius
Epikharmos
Epimenidēs
Eratosthenēs
Eudēmos (?)
Euphoriōn
Flauius (?)
Geōrgios of Pisidia
Grattius
Hēgēsianax
Hēliodōros (Alch.), pseudo
Hēliodōros of Athens
Hērakleitos of Rhodiapolis
De Herbis
Hēsiod
Hilarius
Homer
Germ. Iulius Caesar
C. Iulius Caesar Octauianus (?)
Iunius Columella
Kaikalos
Kleostratos
Lactantius
Leōnidas of Alexandria (I)
Linos
Lucretius
Magnus of Nisibis (?)
Mallius Theodorus
Manethōn (Astrol.)
Manilius
Marcellus of Bordeaux
Marcellus of Sidē
Marianus
Maximus
Menekratēs of Ephesos
Nemesianus
Nestōr
Nikandros
Nikias of Milētos
Noumēnios of Hērakleia
Oppianus of Anazarbos
Oppianus of Apameia
Ōros
Orpheus, pseudo (Astrol.)
Orpheus, pseudo (Lith.)
Ouidius ("Ovid")
Pankratēs of Alexandria

Pankratēs of Argos
P. Oxy. 15.1796
P. Parisinus Graecus 1
Parmenidēs
Pausanias of Damaskos
Periandros (?)
Petrikhos
Philōn of Tarsos
Phokos
Polukritos
Pontica
Poseidippos
Poseidōnios of Corinth
Praecepta Salubria
Priscianus of Caesarea
Ptolemaios of Kuthēra
Remmius Fauinus
Rufus of Ephesos
Sabinus
Cn. Sallustius
Saturos (Lith.)
Q. Serenus
Sextilius Paconianus
Skuthinos
Sminthēs
Sphujidhvaja
Sueius
Terentius Varro Atax
Tiberianus
Timaris
Timaristos
Tribonianus
M. Tullius Cicero
Q. Tullius Cicero
Vergilius
Xenophanēs
Zēnothemis

Rulers (24 kings, queens, and tyrants, plus republican consuls and Roman emperors): rulers sometimes turned from politics to science, or presented themselves as wise through production of scholarship:

Antiokhos VIII
Arkhelaos of Kappadokia
Attalos III
T. Flauius Vespasianus ("Titus")
Fuluius Nobilior
Genthios
Hērakleios, pseudo
Hierōn II
Iouba II
Iulianus, pseudo
C. Iulius Caesar
C. Iulius Octauianus
Iustinianus, pseudo
Kleopatra
Laodikos
Lusimakhos of Macedon
Mithradatēs
Nikomēdēs
Ophellās
Porcius Cato
Purrhos of Ēpeiros
Sphujidhvaja
Sulpicius Gallus
M. Tullius Cicero

Scholarship

Textual criticism: **105** emendations are discussed or proposed within:

Abaskantos
Agēsias
Akhaios
Akhinapolos
Albinus of Smurna
Alkinoos
Alkōn
Anthaios
Apeimantos
Aphrodisis

Aphros
Apios Phaskos
Apollōnios, Claudius
Arbinas
Aristiōn
Arkhebios
Arrabaios
Asarubas
Aspasios (Pharm.)
Atimētos
Attalos (Med.)
Auidianus
Axios
Azanitēs
Bathullos
Bōtthaios
Brenitus
Castus
Doarios
Epagathos
Emboularkhos
Euangeus
Euelpidēs
Euphrōnios of Athens
Euruōdēs
Fauilla
Flauius
Florus
Fonteius Capito/Fronto (Astrol.)
Galēn, pseudo, An Animal
Habrōn
Harpokratiōn
Hēgēsias
Heirodotos
Hekatōnumos
Hērakleianos of Alexandria
Hērakleidēs "Kritikos"
Hērās of Kappadokia
Hipposiadēs
Huriadas
Idios
Iriōn
Iunia/Iounias
Kaikalos
Khairesteos of Athens
Kh(o)ios
Kleophanēs
Kloniakos
Klutos
Kōdios Toukos
Kommiadēs
Ktēsiphōn
Laodikos
Leontinos
Leukios
Lupus
Maecius Aelianus
Mamerkos
Marsinus
Mēnās
Menekratēs Claudius
Menekritos
Menenius Rufus
Menestheus
Mēnodotos of Nikomedia
Mnēsidēs
Moskhiōn (Pharm.)
Nikētēs
Numius
Olumpionikos
Onētidēs/Onētōr
Orthōn
Pasiōn
Paulos (of Italy)
Paxamos
Plentiphanēs
Prōtās of Pelousion
Prothlius
Purgotelēs
Salimachus
Samithra
Sarkeuthitēs
Serenus (Pharm.)
Sextus of Apollōnia
Silimachus
Silo
Simmias the Stoic
Sōkratēs (Lith.)
Trophilos
Valerius Messalla Potitus
Velchionius
Xanitēs
Xen(okh)arēs
Xouthos
Zēnophilos

New to scholarship (276, over one-eighth): many of the entries are new to scholarship, or effectively new (e.g., discussed in one article and then forgotten); the proportion rises from about 1/12th early in the English alphabet to over 1/7th at the end. These **155** names (excluding anonymi and entries representing portions of the Aristotelian or Hippokratic corpus) are not cited in any modern reference work at all, so far as we have been able to determine (see also Nutton 1985; Netz 1997; Parker 1997):

Adeimantos
Aemilius Hispanus
Andrias
Andronikos (Paradox.)
Apellis
Aphthonios of Rome
Aquila Secundilla
Arbitio
Aristarkhos of Sikuōn
Aristogenēs of Thasos
Aristophanēs
Attius
Auxanōn
Bakōris of Rhodes
Bithus of Durrakhion
Blastos
Book of Assumptions by Aqāṭun
Bouphantos
Bromios
Caystrius of Sicily
Celer
Claudianus (Alch.)
Clodius of Naples
Damis of Kolophōn
Damōn (Geog.)
Damōn of Kurēnē
Didumos of Knidos
Diogas
Dionidēs
Dionusios of Rhodes
Diophil-
Dioskoros (Alch.)
Dioskoros (Pharm.)
Dulcitius
Emeritus (Vet.)
Epikratēs of Hērakleia
Epiphanēs
Erasistratos (Astrol.)
Erasistratos of Sikuōn
Eruthrios
Euainetos
Eudikos
Euhēmeros (Vet.)
Grēgorios (Pharm.)
Grēgorios (Vet.)
Hekatōnumos of Khios
Helenos (Vet.)
Hēliodōros (Alch.)
Hērakleodōros
Hierōn (Vet.)
Hipparkhos (Vet.)
Hipposiadēs
Hulas
Huriadas
Iamblikhos (Geog.)
Iamblikhos (Alch.)
Imbrasios of Ephesos
Iounias
Iriōn
Kallias of Arados
Kallikratēs (Astrol.)
Kallistratos
Kēphisophōn
Keras of Carthage
Kharōn of Magnesia
Khēmēs
Khios
Kleandros of Surakousai
Kleoxenos
Klutos
Kōmerios
Kratēs (Med.)
Leontinos
Lepidianus
Libanios (Geog.)
Litorius of Beneventum
Lollianus
Loxos
Lunkeus
Lupus

Mantias
Maria
Marpēssos
Matriketas of Methumna
Maximinus
Meleagros
Melior
Melitianus
Menestratos (I)
Mēnodotos (Astr.)
Monās
Mousaios "the boxer"
Mo(u)ses
Muōnidēs
Murōn
Nautelēs
Neilammōn
Neilos (Alch.)
Nephōn
Nikolaos (Math.)
Nonnos
Ofellius
Oppius
Pausēris
Pelagios
Pelops (Med.)
Penthesileus
Phanias
Phasitas
Philaretos
Philōn (Meteor.)
Phimenas of Sais
Phoibos Ulpius
Pitenius
Platusēmos
Politēs
Pollēs (Med.)
Poludeukēs
Porphurios (Geog.)
Poseidōnios (Med. I)
Poseidōnios of Macedon
Priscianus
Probinus
Prothlius
Purrhos of Magnesia
Puthagoras of Zakunthos
Puthiōn of Rhodes
Romula
Sandarius
Sardonius
Seuthēs
Simmias the Stoic
Simōnidēs (Biol.)
Sophar
Stratonikos (Vet.)
Tektōn
Teukros of Carthage
Thais
Theoklēs
Theomenēs
Theomnēstos (Pharm.)
Theophanēs of Herakleiopolis
Theophilos son of Theogenēs
Theophrastos (Alch.)
Theosebeia
Theudās
Thrasumakhos of Sardis
Tiberius (Vet.)
Timaristos
Timokleanos
Timōn
Timotheos
Tukhikos of Trebizond
Velchionius
Zōilos (of Cyprus)

A further **121** names (all medical) are not cited in any reference work since Fabricius (1726), as far as we have been able to determine:

Akholios
Amphiōn
Apeimantos
Aphrodisis
Apollodōros Demokritean
Arbinas of Indos
Ariobarzanēs
Aristogeitōn
Aristolaos
Arrabaios (of Macedon)
Aspasia
Aspasios

Telephanēs
Terentius Valens
Thamuros
Theopompos
Theotropos
Theoxenos
Thrasuandros
Thrasuas
Threptos
Timaios (Pharm.)
Tlēpolemos
Turannos
Turpillianus
Valerius Paulinus
Xanitēs

Often-Cited Non-Scientists (Greeks and Byzantines)

Aeschylus: Athenian tragedian, active from *ca* 500 BCE, d. *ca* 456 BCE (for the name *cf.* the entry on Aiskhulos).

Agathoklēs of Surakousai: tyrant of the city from 317 BCE, and king of Sicily from 304 BCE; d. 289 BCE.

Alexander III "the Great": king of Macedon (336–323 BCE) and conqueror of the Persian Empire (333–323 BCE); founder of Alexandria "near Egypt" (332 BCE).

Antigonos: name of several rulers, esp. (1) Monophthalmos ("one-eyed"), father of Dēmētrios Poliorkētēs ("besieger"), governor and general in Asia Minor and Mesopotamia from 333 BCE, declared himself king in succession to Alexander 306 BCE, but was defeated and killed 301 BCE; (2) Gonatas, son of Dēmētrios Poliorkētēs, defeated the Gauls and claimed the throne of Macedon 277 BCE, soon ruling over much of Greece, d. 239 BCE.

Antiokhos: name of many rulers of the Seleukid Empire (and of other kingdoms), esp. the first, "Sōtēr" (reigned 281–261 BCE), the son of Seleukos I; and the third, "Great" (reigned 222–187 BCE).

Apollōnios of Rhodes: poet, student of Kallimakhos (q.v.), active *ca* 270–240 BCE, head of the library in Alexandria around 245 BCE, wrote the epic *Argonautica* and other works.

Aristophanēs: Athenian comedic playwright, active *ca* 425–385 BCE (for the name, *cf.* the three entries on men named Aristophanēs).

Athēnaios of Naukratis: composed the *Deipnosophists, ca* 200–230 CE, in which various characters (including Galēn) converse over dinner, mainly about food: the work is a mine of fragments of lost authors.

Clement of Alexandria: Christian theologian and **Platonist**, d. *ca* 215 CE; teacher of Origen.

Constantine: name of 11 emperors of the Byzantine Empire, the first of the name (reigned 306–337 CE) being known for legalizing Christianity (313 CE) and for re-founding Buzantion (Byzantium) as Constantinople (330 CE); the seventh of the name (d. 959 CE), Porphurogennētos ("purple-born"), being known for scholarship. (For the name, *cf.* the entry on the pharmacist Constantinus.)

Dēmētrios Poliorkētēs: son of Antigonos Monophthalmos, known as the "besieger" (not conqueror), ruled Macedon 294 – 288 BCE; father of Antigonos Gonatas.

Dēmētrios of Magnesia: wrote a book on homonymous individuals (*ca* 50 BCE), much used by Diogenēs Laërtios; he was a friend of Cicero's friend Atticus.

Dio Cassius: historian from Nikaia of Bithunia who wrote a history of the Roman Empire to 229 CE, and held various Roman offices *ca* 185–205 CE.

Euripidēs: Athenian tragedian, active from *ca* 455 BCE, d. 406 BCE.

Iosephus (T. Flauius Iosephus; Greek: Iōsēpos): participant in the Jewish rebellion (66–67 CE), surrendered to the Romans, became a Roman citizen and assisted them against the Jews; composed an account of the revolt (published *ca* 75 CE) and a history of the Jews (*ca* 95 CE).

Isokratēs ("Isocrates"): Athenian orator and teacher, active 392 BCE until his death 338 BCE.

Julian: emperor of the Byzantine Empire (ruled in the Western Empire from 355 CE, reigned as emperor 361–363 CE); attempted to restore paganism as the dominant religion; killed in battle against the Persians. (For the name, *cf.* the seven entries on men named Iulianus.)

Justinian: emperor of the Byzantine Empire (reigned 527–565 CE); contrast the later emperor Justinian II (reigned 685–695 and from 705–711 CE).

Kassandros ("Cassander"): son of Antipatros (Alexander's regent in Macedon), ruler of Greece and Macedon (from *ca* 316 BCE), then king of Macedon (reigned 305–297 BCE).

Lucian of Samosata: writer of essays and stories, often satirical, from the Roman province of *Syria*, d. *ca* 180 CE or later.

Lykophrōn of Khalkis: adoptive son of Lukos of Rhēgion (q.v.) and writer of tragedies and other poetry, active in Alexandria *ca* 285–245 BCE.

Maussōllos ("Mausolus"): Karian dynast who ruled, jointly with his sister-wife Artemisia, from Mulasa and then Halikarnassos (377–353 BCE); his tomb was regarded as a wonder.

Origen: Christian theologian and **Platonist**, d. *ca* 254 CE, student of Clement of Alexandria, and author of many works (for the feminine form of the name, *cf.* the entry on Ōrigeneia).

Periklēs ("Pericles"): Athenian politician and general, in power 461–429 BCE, sponsor of building and military programs. (For the name, *cf.* the two entries on men named Periklēs.)

Phōtios: Christian patriarch of Constantinople (858–867 and 877–886 CE), and productive scholar, whose *Library* (*Bibliotheca*) preserves summaries of many lost works; d. 893 CE.

Prokopios of Caesarea: historian of the reign of Justinian, and legal advisor to Justinian's general Belisarius; wrote on the wars of Justinian, the buildings of Justinian, and a "Secret History" (reporting scandals concerning Justinian).

Ptolemy: name of most of the rulers of the Ptolemaic Empire (Egypt and other possessions), esp. the first of the name, Sōtēr (reigned 323–283 BCE), who had been a general under Alexander; his son the second, Philadelphos (reigned 283–246 BCE); the third, son of the second, Euergetēs (reigned 246–221 BCE); and the fourth, son of the third, Philopatōr (reigned 221–205 BCE). For the name, *cf.* the six entries on men named Ptolemaios.

Pyrrho of Ēlis: philosopher active *ca* 330 – 270 BCE, teacher of Timōn of Phleious (q.v.), and "founder" of ancient skepticism.

Seleukos: name of many rulers of the Seleukid Empire, esp. the first, Nikatōr (reigned 312–281 BCE), who had been a governor under Alexander, and who was the father of Antiokhos I; and the second (reigned 246–225 BCE), father of Antiokhos III. (*Cf.* also three entries on men named Seleukos.)

Sōcratēs: character in Plato's dialogues, based upon the historical figure also known from Xenophōn of Athens (q.v.) and Aristophanēs; for the name, *cf.* the four entries on men named Sōkratēs.

Sophoklēs: Athenian tragedian, active from *ca* 470 BCE, d. 406 BCE.

Theokritos: poet active *ca* 275–260 BCE, some of whose works were addressed to Hierōn of Surakousai (q.v.). (For the name, *cf.* the entry on the pharmacist Theokritos).

Tzetzēs: from Georgia, worked in Constantinople, d. 1180 CE, writer of letters, poetry, and commentaries.

INDEX OF PLANTS

All plant species cited within the encyclopedia (and not already explained in the ***Glossary***) are indexed here, with their Greek (or Latin) name, and modern binomial name when known; we also include plant products (e.g., beer, wine). Some plants listed here (e.g., aloe, cassia/cinnamon, nard, or wormwood) could perhaps have been explained in the ***Glossary***, and the following 26 plants and plant-products are found there: *alkuoneion, ammi, ammōniakon, amōmon*, bdellium, *euphorbia*, galbanum, *glaukion, herpullos*, Indian Buckthorn, *kostos, lukion, malabathron*, mastic, *omphakion, opopanax, panax, purethron, sagapēnon, sarkokolla*, shelf-fungus, *silphion, skordion, staphis, sturax*, and *terebinth*. All binomial names of plants are listed at the end of this index with a cross-reference to the indexed name (whether in this index or in the glossary).

The raw materials for this index were created by Ian Lockey, and John Scarborough identified many plants. Paul T. Keyser is responsible for its final form.

Acacia (Greek: *akakia; Acacia nilotica* (L.) Willd. *ex* Delile or *Acacia arabica* Lam.): Amarantos, Anthaios Sextilius, Apollinarios (Pharm.), Apollophanēs of Seleukeia "Pieria," Areios of Tarsos, Bolās, Dasius, Dēmētrios (Geog.), Dioklēs of Khalkēdōn, Diōn (Med.), Euboulos (Pharm.), Gennadios, Harpokratiōn, Hermolaos (Pharm.), Kurillos, Kuros, Leukios, Mantias (Heroph.), Naukratitēs medicus, Neilammōn, Nikolaos (Pharm.), Olumpionikos, Olumpos, Parisinus medicus, Petrōnios Musa, Philippos of Macedon, Philōtas, Sandarius/Sardacius, Sergius of Babylōn, Stratōn of Bērutos, Terentius Valens, Themisōn of Laodikeia, Theophilos (Pharm.), Zōilos of Macedon, Zōsimos (Med.).

Acanthus, Egyptian (Greek: *akantha*; an *Acacia* species, probably the same as **Acacia**): Stephanos of Tralleis.

Aconite (Greek: *akoniton; Aconitum Anthora* L.): Aelius Promotus, Nikandros of Kolophōn, Praecepta Salubria.

Aethiopis (Latin; unidentified): pseudo-Dēmokritos (Pharmacy).

Agrimony (Greek: *eupatorion; Agrimonia eupatoria* L.): Mithradatēs VI, Nearkhos.

Akakallis: (see **Narcissus**)

Akanthis: (see ***Ērigerōn***)

Akinos (Greek; probably *Acinos rotundifolius* Pers. syn. *Calamintha graveolens* (Bieb.) Benth.): Andrōn.

Alfalfa (Greek: *poa Mēdikē; Medicago sativa* L.): Amphilokhos of Athens.

Alkibiadion (Greek; perhaps *Anchusa tinctoria* L.): Polueidēs.

Almond (Greek: *amugdalē; Prunus dulcis* (Mill.) D.A.Webb): Gargilius Martialis, Menemakhos, Mēnophilos, Ōriōn, Papyrus Hibeh 2.187, Papyrus Ryl. III.531, Praecepta Salubria.

Aloe (Greek: *aloē; Aloe perryi* Baker ["best" or "Indian": *] and *Aloe vera* L.): Agapētós, Akhillās, *Alkimiōn, Ammōnios of Alexandria, Anastasios, Aphrodās, Apollinarios (Pharm.), Attalos III of Pergamon, Blastos, Cornelius, Damonikos, Dōsitheos (Pharm.), Epigonos, Epikouros, Euangeus, Glaukōn/Glaukos (Med.), Harpokratiōn, Hermeias (Ophthalm.), Hermolaos (Pharm.), Hermōn, *Iulianus (of Alexandria?), Iustus the Pharmacologist, Kleōn (of Kuzikos?), Kratōn (Pharm.), Kurillos, Kuros, *Lampōn, Leukios, Logadios, Lunkeus, Mantias (Heroph.), Menestheus, Nikolaos (Pharm.), *Olumpionikos, Parisinus medicus, Pasiōn, Polueidēs, Poseidōnios (Med. II), Sextius Niger, Terentius Valens, Theodōrētos, Threptos, Zōilos of Macedon, Zōsimos (Med.).

Alussos (Greek; perhaps *Farsetia clypeata* R.Br. or *Sideritis romana* L.?): Antoninus of Kōs.

Amellus (Latin; *Aster amellus* L.): Vergilius.

Amurca (Latin; *cf.* Olive): Porcius Cato.

Anise (Greek: *anison; Pimpinella anisum* L.): Aelius Gallus, Ambrosios of Puteoli, Andronikos (Pharm.), Apuleius Celsus, Bouphantos, Daliōn (Med.), Flauianus of Crete, Iamblikhos of Constantinople, Khrusermos, Marcellinus (Pharm.), Marcellus (Pharm.), Olumpos, Pankharios, Pasikratēs, Sōsimenēs, Tlēpolemos, Zēnōn of Laodikeia.

Anonymus (Latin; unidentified): Aristogeitōn.

Apple (Greek: *mēlea, mēlon; Malus domestica* Borkh.): Androtiōn, Cloatius Verus, Gargilius Martialis, Matius Caluena.

Apple, Persian: (see **Citron**)

Aristolokhia: (see **Birthwort**)

Arugula (Greek: *euzōmon; Eruca sativa* Lam. or Mill.): Fronto (Agric.), Iamblikhos of Constantinople, Nearkhos.

Aspalathos (Greek; perhaps *Alhagi maurorum* (L.) Medik. or *Calicotome villosa* (Poir.) Link.): Antimakhos.

Asparagus (Greek *asparagos* and *aspharagos; Asparagus officinalis* L.): Khrusippos of Knidos (I), Nikēratos.

Asphodel (Greek: *asphodelos; Asphodelus ramosus* Willd. or *Asphodelus albus* Willd.): Dionusios (Methodist), Dionusios of Utica, Khrusermos, Simos of Kōs, Sōkratiōn.

Autumn Crocus (Greek: *hermodaktulos; Colchicum autumnale* L.; *cf.* perhaps ***Ephēmēron***): Alexander of Tralleis, Iakōbos Psukhrestos.

Balsam (Greek: *balsamon; Commiphora opobalsamum* L.): Aphrodisis, Epidauros, Iulius Secundus, Kleophantos, Lampōn, Proëkhios, Terentius Valens, Theodōrētos.

Banana (description in Strabōn 15.1.21; cultivar derived from *Musa acuminata* Colla 1820): Aristoboulos of Kassandreia.

Banyan tree (description in Strabōn 15.1.21; *Ficus benghalensis* L.): Aristoboulos of Kassandreia.

Barley (Greek: *krithē*; Latin: *hordeum; Hordeum vulgare* L.; *cf.* **Beer**): Anthimus, Eutonios, Hippokratic Corpus Regimen, Khrusermos, Lusias, Magistrianus, Pollēs, Proëkhios, Turranius Gracilis.

Basil (Greek: *ōkimon; Ocimum basilicum* L.): Andrōn, Diodōros (Empir.), Khrusippos of Knidos (I), Sabinius Tiro, Zēnophilos.

Bean, Broad: (see **Bean, Fava**)

Bean, Fava (Greek: *kuamos; Vicia faba* L.): Dulcitius, Mamilius Sura, Phainias, Vergilius.

Bear-Berry (Greek: *arkou staphulos; Arctostaphylos uva-ursi* (L.) Spreng.): Arrabaios.

Beech (Greek: *oxuē; Fagus sylvatica* L. and *Fagus orientalis* Lipsky): Menekratēs of Elaious.

Beer (Greek: *zumē; cf.* **Barley**): Anthimus, Glaukos (Geog. II), Praecepta Salubria.

Beet (Greek: *teutlon; Beta vulgaris* L.): Menandros (of Pergamon?), Seuerus Iatrosophista.

Ben-Nut (Greek: *murobalanos; Moringa arabica* Pers. or perhaps *Moringa oleifera* Lam.): Ianuarinus.

Betonikē (Greek; perhaps *Stachys officinalis* L. ["betony"] or *Cochlearia anglica* L. or *Rumex aquaticus* L.?): Zēnophilos.

Birthwort (Greek: *aristolokhia; Aristolochia clematitis* L.): Abaskantos, Andrōn, Antiokhos Paccius, Arbinas, Bithus, Epaphroditos of Carthage, Epigonos, Euangeus, Harpokratiōn, Hierax, Iriōn, Iulianus (of Alexandria?), Iustus the Pharmacologist, Khalkideus, Lepidianus, Makhairiōn, Philokratēs, Philōtas, Philoxenos, Prothlius/Protlius, Terentius Valens, Threptos, Zēnōn of Laodikeia.

Birthwort, Long (*Aristolochia longa* L.): Proklos the Methodist.

Blackberry (Greek: *batos; Rubus fruticosus* L.): Botrus.

Boukeras: (see **Ginger**)

Boupleuron (Greek; unidentified, perhaps *Bupleurum fruticosum* L.): Glaukōn/Glaukos (Med.).

Bran: (see **Wheat**)

Broom (Greek: *hupokistis* or *hupokustis*; a *Cytinus* species, probably *Cytinus hypocistis* L.): Dioklēs of Khalkēdōn, Hikesios of Smurna, Nikēratos (of Athens?), Theosebios.

Bryony (Greek: *bruōnia*; probably *Bryonia dioica* Jacq.): Alexander Sophistēs, Eugērasia, Zēnōn of Laodikeia.

Bulapathum (Latin; unidentified, perhaps *Rumex scutatus* L.): Solōn.

Bur-parsley, Small: (see ***Kaukalis***)

Butcher's-broom (Greek: *kentromurrinē, oxumurrinē,* and *muakanthos; Ruscus aculeatus* L.): Amarantos, Antonius Castor.

Cabbage (Greek: *krambē* and *rhaphanos; Brassica oleracea* L.): Androkudēs (Med.), Iulius Bassus, Khrusippos of Knidos (I), Mnēsitheos of Kuzikos, Nestōr, Porcius Cato.

Calamint (Latin: *calamintha*; one of several *Calamintha* Mill. species; *cf.* **Mint**): Cornelius, Kharitōn, pseudo-Orpheus (Med.), Pollēs.

Camel-Grass (Latin: *schoenus; Cymbopogon schoenanthus* L.): Apuleius Celsus.

Caper (Greek: *kapparis; Capparis spinosa* L.): Erasistratos of Sikuōn.

Cardamom (Greek: *kardamōmon; Elettaria cardamomum* (L.) Maton): Amarantos, Celer the Centurion, Harpalos (Pharm.), Harpokrās, Ianuarinus, Iulius Agrippa, Kosmos, Marcellus (Pharm.), Marcianus (of Africa?), Olumpos, Pamphilos of Alexandria, Parisinus medicus, Poluarkhos, Proxenos, Sandarius/Sardacius, Zēnōn of Laodikeia.

Carrot (Greek: *daukos; Daucus carota* L.): Ambrosios of Puteoli, Aristoklēs, Khrusermos, Kleophantos of Keōs, pseudo-Orpheus (Med.), Pasikratēs, Petrōnios Musa.

Carrot, Cretan (*Athamanta cretensis* L.): Eugērasia, Khariklēs, Lingōn.

Carrot, "Deadly": (see ***Thapsia***)

Cassia (Greek: *kasia; Cinnamomum aromaticum* Nees; *cf.* **Cinnamon**): Akhillās, Amarantos, Aphrodās, Apuleius Celsus, Aristoklēs, Asterios, Blastos, Celer the Centurion, Dioskoros (Pharm.), Emboularkhos, Eudēmos "the Elder," Hermōn, Iulius Agrippa, Iustinus (Pharm.), Iustus the Pharmacologist, Kleopatra, Kleophantos, Kosmos, Kratōn (Pharm.), Lampōn, Lusias, Melētos, Mēnās, Mithradatēs VI, Pamphilos of Bērutos,

Pasikratēs, Philippos of Macedon, Poluarkhos, Ptolemaios (Pharm.), Rufus of Ephesos, Stratōn (Erasistratean), Theodōrētos, Zēnophilos.

Cassidony (Greek: *stoikhas; Lavandula stoechas* L.): Euskhēmos, Zēnōn (Med.).

Castor (Greek: *kiki; Ricinus communis* L.): Ōros, Serenus (Pharm.).

Cedar (Greek: *kedrelatē* [tree] and *kedris* [its berry]; *Cedrus libani* L. or *Juniperus oxycedrus* L.; *cf.* **Juniper**): Apios Phaskos, Bothros, Eugērasia, Khrusermos, Nikias of Milētos, Orestinos.

Celery (Greek: *selinon; Apium graveolens* L.; *cf.* **Horse-celery** and **Parsley**): Aelius Gallus, Ambrosios of Puteoli, Anthimus, Apuleius Celsus, Diōn (Med.), Khariklēs, Kurillos, Marcellinus (Pharm.), Marcellus (Pharm.), Philōnidēs of Catina, Timokratēs, Zēnophilos.

Centaury (Greek: *kentaureion*; identifications vary: *Centaurea centaurium* L. ["greater"] or *Centaurium umbellatum* Gilib. syn. *Erythraea centaurium* L. ["lesser" or "white"]): Diophantos of Lukia, Seuerus Iatrosophista.

Centaury, White (*Centaurium umbellatum* Gilib.): Proklos the Methodist.

Chameleon, Black (Greek: *khamaileōn melas; Cardopatium corymbosum* L.): Castus, Epainetēs.

Chameleon, White (Greek: *khamaileōn leukos; Atractylis gummifera* L.): Hierax, Hikesios of Smurna.

Chamomile (Greek: *khamaimēlon* ["ground apple"]; *Matricaria chamomilla* L. syn. *M. recutita* L.): Iakōbos Psukhrestos, Pollēs.

Chaste-Tree (Greek: *agnos* or *lugos; Vitex agnus-castus* L.): Dioskouridēs of Anazarbos, Zēnophilos.

Cherry, Ground (Greek: *khamaikerasos*; perhaps *Prunus prostrata* Labill.): Asklēpiadēs of Murleia.

Chestnut (Greek: *kastana* and *kastanea; Castanea sativa* Mill.): Palladius Rutilius Taurus Aemilianus, Gargilius Martialis, Oppius.

Chickpea (Greek: *erebinthos; Cicer arietinum* L.): Andreas of Karustos, Nikēratos, Phainias.

Chicory (Greek: *intubion* and *intubos*; Latin: *seris; Cichorium endivia* L. and *Cichorium intybus* L. ["wild"]): Petrōnios Musa, Praecepta Salubria.

Cinnamon (Greek: *kinnamōmon; Cinnamomum verum* J. Presl syn. *C. zeylanicum* Nees; *cf.* **Cassia**): Adeimantos, Aelius Gallus, Anastasios, Antiokhos Paccius, Aphrodās, Apuleius Celsus, Aristoklēs, Emboularkhos, Hermōn, Iulius Africanus, Iulius Agrippa, Iustinus (Pharm.), Iustus the Pharmacologist, Kleophantos, Lampōn, Logadios, Magnus *arkhiatros*, Marcianus (of Africa?), Mithradatēs VI, Nikostratos (Pharm.), Pasikratēs, Philippos of Macedon, Pomponius Bassus, Praecepta Salubria, Quintus, Ripalus, Romula, Sandarius/Sardacius.

Cinnamon-wood (some *Cinnamomum* species): Poluarkhos.

Citron (Greek: *mēlea Persikē* ["Persian apple"]; *Citrus medica* L.): Africanus (pharm.), Oppius.

Clove (Greek: *karuophullon; Syzygium aromaticum* (L.) Merrill & Perry): Alexander of Tralleis, Anthimus, Eruthrios, Theodōrētos.

Clover (Greek: *triphullos; Trifolium pratense* L., *Tr. fragiferum* L., and other species): Antiokhos VIII Philomētōr, Apollōnios Claudius, Dionusios (Methodist), Epaphroditos of Carthage, Hippokratic Corpus Surgery, Simos of Kōs, Terentius Valens, Zēnōn of Laodikeia.

Colocynth (Greek: *kolokunthē; Lagenaria vulgaris* L.; *cf.* **Gourd, Squash**): Seuerus Iatrosophista.

Colts-foot (Latin: *tussicularis* [and other names]; *Tussilago farfara* L.): Lusias.

Condrion (Latin; unidentified): Dōrotheos of Athens.
Coriander (Greek: *koriandron* and *koriannon; Coriandrum sativum* L.): Kurillos, Marcianus (of Africa?).
Crocus: (see **Autumn Crocus** or else **Saffron**)
Cucumber (Greek: *sikuos; Cucumis sativus* L.): Ambrosios of Puteoli.
Cucumber, Wild: (see **Squirting Cucumber**)
Cumin (Greek: *kuminon; Cuminum cyminum* L.): Eirēnaios, Khrusippos of Knidos (I), Serenus (Pharm.).
Cumin, "Ethiopian" (probably the same as ***ammi***, for which see the Glossary): Andronikos (Pharm.), Iulianus (Pharm.), Marcianus (of Africa?), Philippos of Macedon.
Cyclamen (Greek: *kuklaminos; Cyclamen graecum* Link, and other species): Diogenēs (Pharm.), Eruthrios, Papyrus Laur. Inv. 68, Papyrus Oxyrhynchos 15.1796 (*De Plantiis Aegyptiis*).
Cypress (Greek: *kuparissos; Cupressus sempervirens* L.): Minius Percennius, Philōn of Huampolis.

Damson-plum (Greek: *damaskēnē; Prunus domestica* L.): Pamphilos of Alexandria.
Date (Greek: *phoinikobalanos* [fruit] and *phoenix* [tree]; *Phoenix dactylifera* L.): Euēnos, Kurillos, Lactantius, Olumpos, Prutanis, Stephanos of Tralleis.
Dill (Greek: *anēthon; Anethum graveolens* L.): Daliōn.
Dittany (Greek: *diktamnon; Origanum dictamnus* L.): Apellēs (of Thasos?), Diophantos of Lukia, Eutonios, Phulakos, Stratōn (Erasistratean), Zēnophilos.
Dodder (Greek: *epithumon; Cuscuta epithymum* L.): Theodōrētos.
Dropwort (Greek: *oinanthē*; formerly *Spiraea filipendula* L., now *Filipendula vulgaris* Moench): Apollōnios of Alexandria "Mus," Euēnos.

Elecampane (Greek: *helenion; Inula helenium* L.): Hikesios of Smurna, Iamblikhos of Constantinople, Marcellus (Pharm.), Nearkhos, Papyrus Johnson (Antinoensis).
Elelisphakos (Greek; *Salvia libanotica* Boiss. & Gaill. syn. *Salvia triloba* L.f.): Diōn (Med.), pseudo-Orpheus (Med.).
Elm (Greek: *ptelea*; Latin: *ulmus; Ulmus glabra* Huds., *Ulmus minor* Mill., and *Ulmus procera* Salisb.): Iulius Atticus.
Endive: (see **Chicory**)
Ephēmeron (Greek; unidentified, perhaps the **Autumn Crocus**): Stratōn (Erasistratean).
Ērigerōn (Greek; *Senecio vulgaris* L., "groundsel"): Kallimakhos (of Bithunia).
Eryngo (Greek: *ērungion; Eryngium campestre* L.): Amarantos, Nearkhos, Pollēs.
Eupatoria: (see **Agrimony**)

Fennel (Greek: *marathon; Foeniculum vulgare* Mill.; *cf.* ***Hippomarathron***): Anthimus, Isidōros of Memphis, Kurillos, Marcianus (of Africa?), Nikēratos, Petrikhos, Pomponius Bassus, Tlēpolemos, Zēnōn of Laodikeia.
Fennel, Libyan (*Ferula marmarica* L.): Nikēratos, Truphōn of Gortun.
Fenugreek (Greek: *tēlis; Trigonella foenum-graecum* L.): Andreas of Karustos, Harpokrās, Marcianus (of Africa?), Pantainos, Seuerus the Ophthalmologist, Timōn.
Fern (Greek: *pteris* and *thēlupterion; Pteris aquilina* L. or *Dryopteris filix-mas* (L.) Schott.; *cf.* **Maidenhair**): Alexander of Tralleis, Phainias.
Ferula (Greek: *narthēx; Ferula communis* L.): Antonius Castor.
Fig (Greek: *erineos* and *sukē; Ficus carica* L.): Androtiōn, Chrysippus of Soloi, Cloatius Verus,

Eirēnaios, Eleutheros, Lusias, Mithradatēs VI, Moskhiōn (Pharm.), Orestinos, Praecepta Salubria, Truphōn of Gortun.

Fir (Greek: *elatē; Abies cephalonica* L.): Lampōn.

Flax (Greek: *linon* and *linospermon; Linum usitatissimum* L.): Hēliadēs, Iunia/Iounias, Ōrigeneia, Zēnophilos.

Fleabane: (see ***Konuza***)

Flour: (see **Wheat**)

Frankincense (Greek: *libanos* and *libanōtos; Boswellia sacra* Flueck.): Aelius Gallus, Akhillās, Alkimiōn, Amarantos, Amuthaōn, Aphrodās, Apollophanēs of Seleukeia, Attalos III of Pergamon, Aurelius, Bithus, Blastos, Damonikos, Dēïleōn, Diomēdēs, Diōn (Med.), Dionusios of Samos, Diophantos of Lukia, Dōsitheos (Pharm.), Epidauros, Euangeus, Halieus, Harpalos (Pharm.), Harpokratiōn, Hekataios (Pharm.), Hērakleidēs of Ephesos, Hermeias (Ophthalm.), Hermogenēs of Smurna, Hierax, Hikesios of Smurna, Ioudaios, Iulianus (of Alexandria?), Iulius Bassus, Iunia/Iounias, Kimōn, Kleoboulos (Pharm.), Kleōn (of Kuzikos?), Kleophantos, Klutos, Krateros, Kurillos, Kuros, Leukios, Lusias, Manethōn (Pharm.), Menemakhos, Menoitas/Menoitios, Mithradatēs VI, Mnaseas (Method.), Moskhiōn (Pharm.), Murōn, Nikēratos, Olumpiakos, Olumpionikos, Parisinus medicus, Pasiōn, Patroklos, Petrōnios Musa, Philoklēs, Philōtas, Philoxenos, Primiōn, Puthiōn (Pharm.), Rufus of Ephesos, Sōkratiōn, Solōn, Telephanēs, Thaïs, Theudās "Sarkophagos", Threptos, Timaios (Pharm.), Truphōn of Gortun.

Fungus: (see **Mushroom**)

Garlic (Greek: *skorodon; Allium sativum* L.): Menandros (of Pergamon?), Praecepta Salubria.

Gentian (Greek: *gentianē; Gentiana lutea* L. and *G. purpurea* L.): Abaskantos, Amarantos, Anastasios, Aquila Secundilla, Arbinas, Aspasia, Brenitus, Doarios, Genthios, Iustus the Pharmacologist, Kallinikos, Kratippos, Lepidianus, Logadios, Nearkhos, Nikostratos (Pharm.), Ōrigeneia.

Gentian, Great/Yellow (*Gentiana lutea* L.): Proklos the Methodist.

Germander (Greek: *khamaidrus* ["ground oak"]; *Teucrium chaemaedrys* L.): Abaskantos, Epaphroditos of Carthage, Iustus the Pharmacologist, Lepidianus, Proklos the Methodist.

Ginger (Greek: *zingiberi; Zingiber officinale* Roscoe): Amarantos, Arbinas, Bouphantos, Cassius, Cornelius, Diophantos of Lukia, pseudo-Elias (pseudo-David), Kleophantos, Marcellus (Pharm.), Menestheus, Nikostratos (Pharm.), pseudo-Orpheus (Med.), Pankharios, Poseidōnios (Med. II), Sandarius/Sardacius, Sōkratēs (Med.), Zēnophilos, Zōilos of Macedon.

Ginger, European Wild (Greek: *asaron; Asarum europaeum* L.): Euskhēmos, Nikēratos, Pasikratēs, Zēnophilos.

Gladiolus (Greek: *xiphion; Gladiolus segetum* Gawler): Minucianus.

Gourd (Greek: *kolokunthis; Citrullus colocynthis* (L.) Schrad.; *cf.* **Colocynth, Squash**): Iustus the Pharmacologist, Khrusippos of Knidos (I).

Grape: (see **Vine (Grape)**)

Halikababon (Greek; probably *Physalis alkekengi* L., "winter cherry"): Hierax, Timaristos.

Hart's Tongue (Greek: *skolopendrion; Asplenium scolopendrium* L.): Nearkhos.

Hartwort (Greek: *seseli; Tordylium officinale* L.): Cornelius, Khrusermos, Pankharios.

Hartwort, Massilian (*Seseli tortuosum* L.): Hubristēs.

Hazelwort: (see **Ginger, European Wild**)

Heath (Greek: *ereikē; Erica arborea* L.): Aspasios (Pharm.), Hermās *theriakos*, Petrōnios Musa.

Helenion: (see **Elecampane**)

Hellebore (Greek: *elleboros*; see below for the two kinds): Agathinos of Sparta, Arkhigenēs of Apameia, pseudo-Dēmokritos (Medicine), Dieukhēs, Eudēmos (Methodist), Eudēmos of Athens, Eunomos of Khios, Hippokratic Corpus Surgery, Mnēsitheos of Athens, Mnēsitheos of Kuzikos, Phulotimos, Pleistonikos, Poseidōnios (Med. II).

Hellebore, black (*Helleborus niger* L.): Apios Phaskos, Diophantos of Lukia, Iustus the Pharmacologist, Logadios, Serenus (Pharm.).

Hellebore, white (Greek: *elleboros leukos* and *karpason; Veratrum album* L.): Asklēpiodotos of Alexandria, Axios, Moskhiōn (Pharm.), Orthōn, Philōnidēs of Catina, Ptolemaios (Pharm.), Sōranos of Ephesos.

Hemlock (Greek: *kōneion; Conium maculatum* L.): Aelius Promotus, Epainetēs, Erasistratos of Sikuōn, Euskhēmos, Iulius Bassus, Kallinikos, Nikandros of Kolophōn, Zēnōn (Med.).

Hemp (Greek: *kannabis; Cannabis sativa* L.): Glaukos (Geog. II).

Henbane (Greek: *huoskuamos* ["pig-bean"]; *Hyoscyamus niger* L.): Abaskantos, Aelius Gallus, Aelius Promotus, Akhaios, Antiokhos Paccius, Antonius "Root-Cutter," Aphrodās, Apollodōros the *thēriakos*, Apuleius Celsus, Dasius, Epainetēs, Erasistratos of Sikuōn, Flauianus of Crete, Florus, Hermās *theriakos*, Iulius Bassus, Kallinikos, Khariklēs, Kharixenēs, Krateros, Leukios, Lingōn, Lukomēdēs, Mantias (Heroph.), Marcellinus (Pharm.), Mnēsidēs, Nikēratos, Ōrigeneia, Pamphilos of Alexandria, Paulos (of Italy), Philōn of Tarsos, Philoxenos, Proxenos, Prutanis, Ptolemaios (Pharm.), Quintus, Ripalus, Rufus of Ephesos, Silo, Sōsagoras, Terentius Valens.

Henna (Greek: *kupros* or *kuprinon; Lawsonia inermis* L.): Amuthaōn, Aphrodisis, Khrusippos (Med.), Kleophantos, Petrōnios Musa.

Hippomarathron (*Prangos ferulacea* (L.) Lindl. or *Cachrys ferulacea* (L.) Calest.; *cf.* **Fennel**): Mikiōn.

Hog-fennel: (see **Sulfurwort**)

Holarrhena (Greek: *xulomaker; Holarrhena antidysenterica* Wall.): Castus.

Horehound (Greek: *prasion; Marrubium vulgare* L.): Antonius Castor, Kōdios Toukos, Nikēratos, Stratōn (Erasistratean).

Horse-celery (Greek: *hipposelinon; Smyrnium olusatrum* L.; *cf.* ***Smurneion***): Anakreōn (Pharm.).

Horseheal: (see **Elecampane**)

Hupokistis (see: **Broom**)

Hyacinth (Greek: *bolbos; Muscari comosum* Mill.): Daliōn.

Hyssop (Greek: *hussōpos; Hyssopus officinalis* L.): Akholios, Diophantos of Lukia.

Iberis (Latin; *Lepidium graminifolium* L.): Damokratēs.

Iris (Greek: *iris*; some *Iris* species, often *Iris pseudacorus* L. or *Iris pallida* Lam.): Aphrodisis, Apollophanēs of Seleukeia, Eugērasia, Harpalos (Pharm.), Harpokrās, Hekataios (Pharm.), Khrusermos, Lusias, Onētidēs/Onētōr, Philokratēs, Stephanos of Tralleis.

Iris, Illyrian (Greek: *iris Illurikē; Iris pallida* Lam.): Abaskantos, Bithus, Castus, Diophantos of Lukia, Euskhēmos, Leukios, Melētos, Pamphilos of Alexandria, Timokratēs, Truphōn of Gortun.

Iris, Yellow (Greek: *akoron; Iris pseudacorus* L.): Amarantos, Aristoklēs, Diophantos of Lukia.
Iskhas (Greek; *Euphorbia apios* L.): Apollōnios "Mus."
Ivy (Greek: *kissos; Hedera helix* L.): Harpokratiōn of Alexandria.

Juniper (Greek: *arkeuthos* [tree] and *arkeuthis* [its berry]; Latin *iuniper; Juniperus communis* L.; *cf.* **Cedar**): Cornelius, Flauius "the boxer," Nearkhos, Pankharios.
Juniper, Savin (Greek: *brathu*; Latin: *sabina; Juniperus sabina* L.): Antimakhos, Eugamios.

Kammaron: (see **Hemlock**)
Kangkhru (Greek; probably *Cachrys libanotis* (L.) Koch or *Lecokia cretica* Lam.): Hubristēs.
Kaukalis (Greek; perhaps *Tordylium apulum* L., *Caucalis playcarpos* L., or *Caucalis grandiflora* L. syn. *Orlaya grandiflora* (L.) Hoffm.): Khrusippos of Knidos (I), Petrikhos.
Khamaidrus: (see **Germander**)
Khloeron sisumbron (Greek; unidentified): Nikias of Milētos.
Konuza (Greek; either *Inula viscosa* Aiton ["fleabane"] or *Inula graveolens* Desf.): Pollēs.
Krotalon: (see **Narcissus**)
Kuperos (Greek; *Cyperus rotundus* L.): Aurelius, Poluarkhos, Zēnophilos.

Labdanum: (see ***Ladanon***)
Ladanon (Greek; *Cistus creticus* L. or *Cistus cyprius* L. or another): Hērās, Rufus of Ephesos.
Larch (Latin: *larix; Larix decidua* Mill.): Vitruuius.
Lathuris (Greek; perhaps *Euphorbia lathyris* L.): Aphrodās.
Laurel (Greek: *daphnē; Laurus nobilis* L.; *cf.* **Mustax**): Eugeneia, Khariklēs, Kurillos, Ōriōn, Sertorius Clemens, Zēnophilos.
Lavender Cotton (Greek: *abrotonon; Santolina chamaecyparissus* L.): Nikēratos.
Leek (Greek: *prason; Allium ampeloprasum* var. *porrum* (L.) J.Gay): Dionusios of Utica, Seuerus Iatrosophista.
Lees (Greek: *trux; cf.* **Wine**): Erasistratos of Sikuōn, Euelpidēs, Manethōn (Pharm.), Menelaos (Pharm.), Minucianus.
Lentil (Greek: *phakos; Lens culinaris* Medik.): Hippokratic Corpus Surgery, Praecepta Salubria.
Leopard's Bane (Greek: *skorpios; Doronicum pardalianches* Jacq. or similar species): Dioskouridēs of Anazarbos, Epainetēs.
Lettuce (Greek: *thridax; Lactuca sativa* L.): Iakōbos Psukhrestos, Praecepta Salubria.
Libanōtis (Greek; either ***Kangkhru*** or **Rosemary**): Antimakhos, Sertorius Clemens.
Licorice (Greek: *glukurrhiza* and *glukeia rhiza; Glycyrhiza glabra* L.): Amarantos, Apollōnios Claudius, Dioskoros (Pharm.), Eugeneia, Iakōbos Psukhrestos, Lusias, Nikostratos (Pharm.), Olumpos, Ōrigeneia, Stephanos of Tralleis.
Ligusticum: (see **Lovage**)
Lily (Greek: *sousinon; Lilium candidum* L.): Stratōn (Erasistratean).
Linseed: (see **Flax**)
***Lōtos* Tree, Libyan** (Greek; *Ziziphus lotus* Willd.): Marcianus (of Africa?).
Lotus Flower (Greek: *lōtos; Nymphaea caerulea* Sav. [blue] and *Nymphaea zenkeri* L. [red]): Pankratēs of Alexandria.
Lovage (Latin: *ligusticum; Levisticum officinale* Koch): Cornelius, Pankharios.
Lusimakhia (Greek; *Lythrum salicaria* L.): Lusimakhos of Macedon.

Madder (Greek: *eruthrodanon; Rubia tinctorum* L.): Nearkhos, Thaïs.
Maidenhair (Greek: *adianton* and *polutrikhon; Adiantum capillus-veneris* L. or *Asplenium adiantum-nigrum* L.; *cf.* **Fern**): Hērās, Orestinos.
Mallow (Greek: *malakhē; Malva silvestris* L.): Olumpias, Papyrus Laur. Inv. 68.
Mandrake (Greek: *mandragoras; Mandragora officinarum* L.): Abaskantos, Aelius Promotus, Alexander Sophistēs, Antiokhos Paccius, pseudo-Apuleius, Arbinas, Aristoklēs, Dioskouridēs of Anazarbos, Domitius Nigrinus, Epainetēs, Erasistratos of Sikuōn, Euskhēmos, Flauianus of Crete, Florus, Hērōn (Med.), Iulius Bassus, Kharixenēs, Krateros, Moskhiōn (Pharm.), Nikolaos (Pharm.), Nikostratos (Pharm.), Philoxenos, Ptolemaios (Pharm.), Rufus of Ephesos, Silo, Terentius Valens.
Marjoram (Greek: *sampsukhon; Origanum majorana* L.): Antimakhos, pseudo-Dēmokritos (Medicine), Marcellus (Pharm.), Numius, Philoklēs, Pollēs.
Meadow Saffron: (see **Autumn Crocus**)
Melilot (Greek: *melilōton; Melilotus officinalis* (L.) Pall. or perhaps *Trigonella corniculata* L.): Euēnos.
Mēon (Greek; *Meum athamanticum* Jacq.): Aristoklēs (Pharm.).
Millet (Greek: *kenkhros; Panicum miliaceum* L.): Glaukos (Geog. II).
Mint (Greek: *hēduosmon* and *minthē*; various *Mentha* L. species; *cf.* **Calamint**): Nikomēdēs IV of Bithunia, Sabinius Tiro, Seuerus Iatrosophista, Timokratēs.
Mistletoe (Greek: *ixos; Loranthus europaeus* Jacq. and *Viscum album* L. [rarely on oak]): Apollodōros of Kition, Apollodōros of Taras, Apollophanēs of Seleukeia.
Mistletoe, Oak: Idios, Minucianus, Nikomēdēs IV of Bithunia, Proëkhios.
Mithridatia (Latin; perhaps *Erythronium dens-canis* L.): Krateuas, Mithradatēs VI.
Moon-trefoil (Greek: *kutisos; Medicago arborea* L.): Amphilokhos of Athens, Aristomakhos of Soloi.
Mullein (Latin: *bugillo; Verbascum thapsus* L.): Iustus the Pharmacologist.
Mushroom (Greek: *mukēs*): Epainetēs, Euteknios, Phainias.
"Musian": (see **Beech**)
Must (Greek: *gleukinon* and *gluku*; Latin: *sapa*; pressed grapes: *cf.* **Vine**): Antimakhos, Aphrodisis, Harpalos (Pharm.), Harpokratiōn, Kimōn, Laodikos, Lusias, Papyrus Laur. Inv. 68, Prutanis, Timōn.
Mustard (Greek: *napu* and *sinapi*; Latin: *sinapi; Brassica nigra* Koch and *Sinapis alba* L.): Ianuarinus, Pantainos, Podanitēs.
Mustax (Latin; type of **Laurel**): Pompeius Lenaeus.
Myrrh (Greek: *smurnē; Commiphora myrrha* Arn.): Abaskantos, Aelius Gallus, Agapētós, Akhillās, Alkimiōn, Ambrosios of Puteoli, Amuthaōn, Andrōn, Andronikos (Pharm.), Anthaios Sextilius, Antiokhos Paccius, Aphrodās, Aphrodisis, Aphros, Apollinarios (Pharm.), Apollōnios Claudius, Apollōnios of Alexandria "Mus," Apollophanēs of Seleukeia, Apuleius Celsus, pseudo-Apuleius, Aquila Secundilla, Aristoklēs, Aristokratēs, Arkhelaos (of Hērakleia Salbakē?), Arrabaios, Artemōn (Med.), Asterios, Athēniōn, Attalos III of Pergamon, Bathullos, Blastos, Candidus, Castus, Celer the Centurion, Cornelius, Damonikos, Dasius, Diagoras of Cyprus, Diomēdēs, Diōn (Med.), Dionusios of Samos, Diophantos of Lukia, Emboularkhos, Epaphroditos of Carthage, Epigonos, Epikouros, Euangeus, Eudēmos "the Elder," Euelpidēs, Euēnos, Eugamios, Eugērasia, Euskhēmos, Florus, Gennadios, Glaukōn/Glaukos (Med.), Hagnodikē, Harpalos (Pharm.), Harpokrās, Harpokratiōn, Hērās, Hermeias (Ophthalm.), Hermolaos (Pharm.), Hermōn, Hierax, Hippokratic Corpus Surgery,

Iulianus (Pharm.), Iulianus (of Alexandria?), Iulius Agrippa, Iulius Bassus, Iulius Secundus, Iustinus (Pharm.), Iustus the Pharmacologist, Khrusermos, Kimōn, Kleoboulos (Pharm.), Kleōn (of Kuzikos?), Kleophantos, Klutos, Kosmos, Krateros, Kratippos, Kratōn (Pharm.), Kurillos, Lampōn, Logadios, Lunkeus, Lusias, Makhairiōn, Marcellinus (Pharm.), Marcellus (Pharm.), Melētos, Mēnās, Menemakhos, Mēnophilos, Mithradatēs VI, Naukratitēs medicus, Nikēratos, Nikētēs, Nikolaos (Pharm.), Nikostratos (Pharm.), Olumpiakos, Olumpionikos, Olumpos, Onētidēs/Onētōr, Ōrigeneia, Pamphilos of Alexandria, Pasiōn, Patroklos, Petrōnios Musa, Philōnidēs of Catina, Philōtas, Platōn, Polueidēs, Proxenos, Prutanis, Ptolemaios (Pharm.), Puramos, Ripalus, Rufus of Ephesos, Sōkratiōn, Solōn, Sōsikratēs, Stratōn of Bērutos, Sunerōs, Telephanēs, Terentius Valens, Thaïs, Theophilos (Pharm.), Theosebios, Threptos, Timaios (Pharm.), Timokratēs, Zōilos of Macedon, Zōsimos (Med.).

Myrtle (Greek: *murrinē* and *murton; Myrtus communis* L.): Androtiōn, Aristolaos, Aspasios (Pharm.), Deinōn of Kolophōn, Dōsitheos (Pharm.), Euboulos (Pharm.), Eunēnos, Euphranōr, Flauius "the boxer," Harpokrās, Hērās, Idios, Kimōn, Kleophantos, Kurillos, Moskhiōn (Pharm.), Murōn, Nikomēdēs IV of Bithunia, Romula, Sertorius Clemens.

Narcissus (Greek: *narkissos*; various *Narcissus* L. species; *cf.* ***Akakallis*** and ***Krotalon***): Eumakhos of Kerkura.

Nard (Greek: *nardos*; see below for kinds): Aphrodās, Aristoklēs, Celer the Centurion, Dioskoros (Pharm.), Eruthrios, Eudēmos "the Elder," Harpalos (Pharm.), Hermōn, Kleophantos, Kratōn (Pharm.), Naukratitēs medicus, Prutanis, Ripalus, Zōilos of Macedon.

Nard, Celtic (Greek: *nardos Keltikē; Valeriana celtica* L.): Marcianus (of Africa?), Olumpos.

Nard, Indian (also known as **Spikenard**): Abaskantos, Akhillās, Ammōnios of Alexandria, Anthaios Sextilius, Aphrodās, Aristoklēs, Atimētos, Blastos, Brenitus, Diagoras of Cyprus, Diōn (Med.), Euskhēmos, Hubristēs, Iulius Agrippa, Lukomēdēs, Magnus *arkhiatros*, Marcianus (of Africa?), Melētos, Pasikratēs, Poluarkhos, Theophilos (Pharm.), Zōsimos (Med.).

Nard, Pontic (Greek: *karpēsion; Valeriana dioscoridis* Sibth.): Quintus.

Nard, Pontic (Greek: *phou Pontikon; Valeriana phu* L.): Amarantos, Diophantos of Lukia.

Nard, Syrian: Apuleius Celsus.

Nasturcium (Latin; *Lepidium sativum* L.): Iamblikhos of Constantinople, Marcellus (Pharm.).

Nettle (Greek: *akalēphē* and *knidē; Urtica dioica* L.): Andreas of Karustos, Ianuarinus, Khrusermos, Ōrigeneia, Phainias.

Nightshade: (see ***Strukhnos***)

Oak (Greek: *drus*; any of *Quercus cerris* L., *Quercus faginea* Lam., *Quercus pontica* K.Koch, *Quercus pubescens* Willd., or esp. *Q. robur* L.): Amuntas (Geog.), Fronto (Agric.), Pherekudēs, Poseidōnios of Macedon.

Oak, Cork (Greek: *phellos; Quercus suber* L.): Thamuros.

Oak-gall (Greek: *kēkis*; grows on the leaves): Andrōn, Antimakhos, Aristoklēs, Bithus, Botrus, Hagnodikē, Harpokratiōn, Hippokratic Corpus Surgery, Iulianus (of Alexandria?), Kleoboulos (Pharm.), Kurillos, Leukios, Onētidēs/Onētōr, Papyrus Ryl.

III.531, Primiōn, Ptolemaios (Med.), Scribonius Largus, Terentius Valens, Theodorus Priscianus, Threptos, Timaios (Pharm.).

Oak, Winter (Latin: *aesculus*; probably *Quercus petraea* (Mattuschka) Liebl.): Palladius Rutilius Taurus Aemilianus.

Oinanthe: (see **Dropwort**)

Ōkhros (Greek; *Lathyrus ochrus* (L.) DC.): Phainias.

Oleander (Latin: *ther(i)onarca; Nerium oleander* L.): pseudo-Dēmokritos (Pharm.).

Olive (Greek: *elaa; Olea europaea* L.): Alkimiōn, Amphiōn, Vindonius Anatolios, Androtiōn, Attalos (Med.), Attalos III of Pergamon, Bathullos, Castus, Damigerōn, Damonikos, Diogenēs (Pharm.), Dioklēs of Karustos, Diophantos of Lukia, Dioskouridēs (Metrology), Eleutheros, Epigonos, Euangeus, Euelpistos, Eutonios, Florentinus, *Geoponika*, Halieus, Harpalos (Pharm.), Hermogenēs of Smurna, Hierax, Hikesios of Smurna, Iakōbos Psukhrestos, Ioudaios, Iulianus (of Alexandria?), Iunia/Iounias, Khalkideus, Kloniakos, Kōdios Toukos, Ktēsiphōn, Leukios, Megēs, Menelaos (Pharm.), Menoitas/Menoitios, Minucianus, Mnaseas (Method.), Olumpos, Ōriōn, Papyrus Turner. 14, Pasiōn, Philagrios, Philoklēs, Philōtas, Plutarch, Polustomos, Porcius Cato, Primiōn, Prothlius/Protlius, Puthiōn (Pharm.), Samithra/Tanitros, Sertorius Clemens, Sōkratiōn, Sōranos of Ephesos, Sunerōs, Telamōn, Telephanēs, Theophulaktos, Timokratēs, Truphōn of Gortun, Zōsimos (Med.).

Onothuris (Latin; *Epilobium angustifolium* L.): pseudo-Dēmokritos (Pharm.).

Opium (Greek: *opion; cf.* **Poppy**): Abaskantos, Akhaios, Akhillās, Anthaios Sextilius, Aphrodās, Aphros, Apollinarios (Pharm.), Aspasios (Pharm.), Asterios, Athēnippos, Atimētos, Bolās, Brenitus, Candidus, Cornelius, Diagoras of Cyprus, Dioklēs of Khalkēdōn, Diomēdēs, Diōn (Med.), Dōsitheos (Pharm.), Eugeneia, Euhēmeros, Euskhēmos, Florus, Gennadios, Glaukōn/Glaukos (Med.), Harpalos (Pharm.), Harpokrās, Harpokratiōn, Hermeias (Ophthalm.), Hermolaos (Pharm.), Hērōn (Med.), Hierax, Khariklēs, Kharixenēs, Kleophantos, Krateros, Leukios, Lingōn, Lukomēdēs, Lunkeus, Mnēsidēs, Moskhiōn (Pharm.), Naukratitēs medicus, Neilammōn, Nikētēs, Nikolaos (Pharm.), Philōn of Tarsos, Philōnidēs of Catina, Philoxenos, Poludeukēs, Pomponius Bassus, Proëkhios, Prōtās, Proxenos, Prutanis, Ptolemaios (Pharm.), Puramos, Quintus, Sergius of Babylōn, Silo, Sōkratēs (Med.), Solōn, Stratōn of Bērutos, Sunerōs, Terentius Valens, Theophilos (Pharm.), Theosebios, Zōilos of Macedon, Zōsimos (Med.).

Orache (Greek: *atraphaxus; Atriplex hortensis* L.): Dionusios (Methodist), Dionusios of Utica, Solōn.

Orchil (Greek: *phukos; Roc(c)ella tinctoria* (L.) de Cand.): Iouba, Thaïs.

Oregano (Greek: *origanon; Origanum vulgare* L.): Petrikhos, Praecepta Salubria.

Papyrus (Greek: *papuros* and *khartos; Cyperus papyrus* L.): Apellēs (of Thasos?), Aristolaos, Idios, Kleoboulos (Pharm.), Priscianus, Ptolemaios (Pharm.), Puthios, Thamuros.

Parsley (Greek: *petroselinon* ["rock celery"]; *Petroselinum crispum* (Mill.) Nyman *ex* A.W. Hill; *cf.* **Celery**): Amarantos, Ambrosios of Puteoli, Anastasios, Aristoklēs, Aspasios (Pharm.), Cornelius, Daliōn, Dionusios (Med.), Dionusios (Methodist), Dionusios of Utica, Doarios, Eugērasia, Harpokrās, Iulianus (Pharm.), Khrusippos of Knidos (I), Marcellus (Pharm.), Mithradatēs VI, Proklos the Methodist, Zēnōn of Laodikeia, Zēnophilos.

Parsley, Sardinian (described in fr. 2.10 M.; unidentified, perhaps *Ranunculus sardous* Crantz): Sallustius Crispus.

Parsley, Stone (Greek: *sinōn; Sison amomum* L.): Terentius Valens.

Parsnip (Latin: *siser; Pastinacea sativa* L.): Ophiōn.

Peach (Latin: *persicus; Prunus persica* (L.) Batsch): Gargilius Martialis.

Pear (Greek: *apios* and *apion; Pyrus communis* L.): Aëthlios of Samos, Aiskhulidēs of Keōs, Androtiōn, Apios Phaskos, Cloatius Verus, Turranius, Vergilius.

Pellitory: (see ***Purethron***)

Pennyroyal (Greek: *blēkhōn* and *glēkhōn; Mentha pulegium* L.): Akholios, Anthimus, Cornelius, Iustus the Pharmacologist, Kurillos, Papyrus Laur. Inv. 68, Serenus (Pharm.), Timokratēs.

Peony (Greek: *glukusidē* and *paiōnia; Paeonia officinalis* L.): Agapētós, Kuranides.

Peplis (Latin; *Euphorbia peplis* L.): Mētrodōros (Pharm.).

Pepper (Greek: *peperi; Piper nigrum* L.): Abaskantos, Aineios (of Kōs?), Akhillās, Akholios, Amuthaōn, Anthimus, Arbinas, Aristoklēs, Athēniōn, Bouphantos, Cornelius, Dasius, Dexios, Diogenēs (Pharm.), pseudo-Elias (pseudo-David), Epidauros, Euelpidēs, Eugērasia, Euskhēmos, Harpalos (Pharm.), Hubristēs, Ianuarinus, Iulianus (Pharm.), Iulius Agrippa, Iustus the Pharmacologist, Kallinikos, Khrusippos (Med.), Khrusippos of Knidos (I), Kosmos, Leukios, Logadios, Lusias, Magnus of Tarsos, Marcellus (Pharm.), Marcianus (of Africa?), Mēnophilos, Mithradatēs VI, Nearkhos, Nikostratos (Pharm.), Pamphilos of Alexandria, Philippos of Macedon, Pollēs, Ptolemaios (Pharm.), Sōsikratēs, Sunerōs, Timokleanos.

Pepper, Black (*Piper nigrum* L.): Acilius Hyginus, Aristokratēs, Atimētos, Flauianus of Crete, Iulius Bassus, Iustus the Pharmacologist, Ripalus, Timokratēs.

Pepper, Long (*Piper longum* L.): Apellēs (of Thasos?), Apuleius Celsus, Diophantos of Lukia, Eugeneia, Harpalos (Pharm.), Nikostratos (Pharm.), pseudo-Orpheus (Med.), Pomponius Bassus.

Pepper, White (*Piper nigrum* L.): Abaskantos, Acilius Hyginus, Amarantos, Antigonos (Med.), Antiokhos Paccius, Apellēs (of Thasos?), Apollōnios Claudius, Apuleius Celsus, Aquila Secundilla, Artemōn (Med.), Athēnippos, Attalos III of Pergamon, Brenitus, Diophantos of Lukia, Euelpidēs, Eugeneia, Fauilla of Libya, Flauianus of Crete, Harpalos (Pharm.), Hermophilos, Iulius Bassus, Iustus the Pharmacologist, Khariklēs, Kratippos, Mantias (Heroph.), Nikēratos, Nikostratos (Pharm.), Orthōn, Pamphilos of Alexandria, Patroklos, Philōn of Tarsos, Pomponius Bassus, Proëkhios, Prōtās, Proxenos, Ripalus, Rufus of Ephesos, Silo, Terentius Valens, Timokratēs.

Perdikias (Greek; probably *Convolvulus arvenis* L.): Timokratēs.

Peridexion (Greek; unidentified): Physiologos.

Persea (Greek; *Mimusops Schimperi* L.) Papyrus Oxyrhynchos 15.1796 (*De Plantiis Aegyptiis*).

Phaulia (Greek): Glaukidēs.

Philadelphum (Latin; unidentified): Apollodōros of Artemita.

Phlommos: (see **Elecampane**)

Pimpernel (Greek: *anagallis; Anagallis arvensis* L.): Agapētós, Epaphroditos of Carthage, Leukios.

Pine (Greek: *peukē* and *pitus*; usually *Pinus brutia* Tenore, *Pinus halepensis* Mill., or *Pinus pinea* L.; *cf.* **Pitch**): Aphthonios, Castus, Euelpistos, Hikesios of Smurna, Khalkideus, Leōnidas (Geog.), Moskhiōn (Pharm.), Murōn, Pasiōn, Philōn of Huampolis, Ptolemaios (Pharm.), Telamōn, Zōsimos (Med.).

Pine, Bruttian (*Pinus laricio* Poir.): Iulianus (of Alexandria?), Menippos, Puthiōn (Pharm.), Truphōn of Gortun.

Pine-nut (Greek: *purēn* and *strobilon; Pinus pinea* L.): Kratippos, Zēnophilos.

Piperitis (Latin; probably *Polygonum hydropiper* L. syn. *Persicaria hydropiper* (L.) Spach): Antonius Castor.

Pistachio (Greek: *pistakion; Pistacia vera* L.): Paxamos.

Pitch (Greek: *pissa* and *pitta*; Latin: *pix*; the **Resin** from **Pine**): Aineios (of Kōs), Aristoklēs, Aristophanēs, Castus, Damonikos, Diogenēs (Pharm.), Hekataios (Pharm.), Iriōn, Iulianus (of Alexandria?), Khrusippos of Knidos (I), Kurillos, Menippos, Murōn, Puthiōn (Pharm.), Truphōn of Alexandria.

Plantago (Latin): (see **Plantain**)

Plantain (Greek: *arnoglōssos; Plantago major* L.): pseudo-Apuleius, Nikēratos, Philippos of Macedon.

Polion (Greek; *Teucrium polium* L.): Apellēs of Thasos, Nearkhos.

Pomegranate (Greek: *rhoa; Punica granatum* L.): Androtiōn, Aspasios (Pharm.), Mantias (Heroph.), Papyrus Ryl. III.531, Ptolemaios (Med.), Sandarius/Sardacius, Sphairos.

Pomegranate-flower (Greek: *balaustion* and *kutinos*): Akhaios, Amarantos, Andrōn, Dioklēs of Khalkēdōn, Epagathos, Harpokratiōn, Lukomēdēs, Mantias (Heroph.), Nikēratos, Polueidēs, Terentius Valens.

Pomegranate-peel (Greek: *sidion*): Bithus, Harpokratiōn, Mēnophilos, Primiōn, Terentius Valens, Threptos.

Poplar (Latin: *populus; Populus alba* L. or *Populus nigra* L.): Theomenēs.

Poppy (Greek: *mēkōn*, and *mēkōnion* or *opos mēkōnos; Papaver somniferum* L.; *cf.* **Opium**): Alexander of Tralleis, Ambrosios of Puteoli, Andreas of Karustos, Antiokhos Paccius, Apollophanēs of Seleukeia, Apuleius Celsus, Areios of Tarsos, Aristokratēs, Dēmosthenēs, Dioskouridēs of Anazarbos, Epainetēs, Eudēmos "the Elder," Euelpidēs, Flauianus of Crete, Iulius Bassus, Laodikos, Mnēsidēmos, Nikēratos, Olumpionikos, Petrōnios Musa, Philagrios, Philōn of Tarsos, Philōtas, Ripalus, Rufus of Ephesos, Seuerus Iatrosophista, Sōranos of Ephesos, Sōsagoras, Terentius Valens, Themisōn of Laodikeia.

Potamogiton (Latin; perhaps *Hippuris vulgaris* L. or *Potamogeton natans* L.): Antonius Castor.

Psalakanthē (Greek; unidentified): Ptolemaios of Kuthera.

Pseudo-Mastic: (see **Chameleon, White**)

Puritis: (See ***Purethron*** in the Glossary)

Quince (Greek: *kudōnion*; Latin: *cotonea* and *cydonea; Cydonia vulgaris* L. syn. *Cydonia oblonga* Mill.): Gargilius Martialis, Glaukidēs, Philagrios, Sandarius/Sardacius.

Radish (Greek: *rhaphanis; Raphanus sativus* L.): Apollodōros of Kition, Apollodōros of Taras, Aristomakhos of Soloi, pseudo-Dēmokritos (Pharm.), Eleutheros, Mēdeios, Moskhiōn (Pharm.), Pleistonikos, Sōranos of Ephesos.

Ragged Robin (Greek: *lukhnis; Lychnis flos-cuculi* L.): Derkullos.

Rape: (see **Turnip, wild**)

Reed (Greek: *kalamos; Arundo* L. species): Orestinos, Theophrastos of Eresos.

Reed, aromatic (Greek: *kalamos; Acorus calamus* L.): Lampōn.

Resin (Greek: *rhētinē*; Latin: *resina; cf.* **Cedar, Chameleon (White)**, ***Euphorbia***,

Frankincense, Galbanum, *Ladanon*, Myrrh, Pine, Scammony, *Sturax*, Terebinth): Alkimiōn, Aristoklēs, Dēïleōn, Dexios, Epidauros, Epigonos, Euelpistos, Euphranōr (Pharm.), Hērakleidēs of Ephesos, Ioudaios, Iunia/Iounias, Kleophantos, Kloniakos, Nikolaos (Pharm.), Nikomēdēs IV of Bithunia, Poluarkhos, Prothlius/Protlius, Sertorius Clemens, Sōsagoras, Truphōn of Gortun.

Rhamnos (Greek; a *Rhamnus* species, probably *Rhamnus cathartica* L.): Kōdios Toukos.

Rhubarb (Greek: *rhā; Rheum officinale* L.): Amarantos, Aristoklēs, Arkhelaos (of Hērakleia Salbakē?).

Rhubarb, Pontic: Mantias (Heroph.), Pasikratēs, Theodōrētos.

Rice (Greek: *oruza; Oryza sativa* L.): Aristoboulos of Kassandreia.

Rose (Greek: *rhodon*; usually *Rosa canina* L., *Rosa gallica* L., or *Rosa sempervirens* L.): Aelius Gallus, Amarantos, Andreas of Karustos, Aphrodās, Arkhelaos (of Hērakleia Salbakē?), Caepio, Diagoras of Cyprus, Dōsitheos (Pharm.), Emboularkhos, Euēnos, Flauius "the boxer," Florus, Glaukōn/Glaukos (Med.), Khariklēs, Kleophantos, Kratippos, Leukios, Lukomēdēs, Marcellinus (Pharm.), Melētos, Naukratitēs medicus, Olumpos, Pamphilos of Bērutos, Philōtas, Poluarkhos, Poludeukēs, Sandarius/Sardacius, Stephanos of Tralleis, Terentius Valens, Themisōn of Laodikeia, Theopompos.

Rose-oil: Apellēs (of Thasos), Aphrodisis, Dēmosthenēs, Diogenēs (Pharm.), Eruthrios, Harpalos (Pharm.), Heirodotos (of Boiotia?), Khariklēs, Kleōn (of Kuzikos?), Mantias (Heroph.), Marcellus (Pharm.), Nikomēdēs IV of Bithunia, Ōros, pseudo-Orpheus (Med.), Philagrios, Prutanis, Stratōn (Erasistratean), Themisōn of Laodikeia.

Rose-water: Aphrodās, Idios, Seuerus Iatrosophista, Xanitēs.

Rosemary (Greek: *libanōtis; Rosmarinus officinalis* L.; *cf.* ***Libanōtis***): Nikēratos, Sandarius/Sardacius.

Rue (Greek: *pēganon; Ruta graveolens* L.): Apellēs (of Thasos?), Apollōnios of Alexandria "Mus," Arbinas, Diophantos of Lukia, Faustinus, Hubristēs, Iollas, Kōdios Toukos, Lepidianus, Marcianus (of Africa?), Mithradatēs VI, Nikomēdēs IV of Bithunia, Pantainos, Papyrus Michiganensis 17.758, Sabinius Tiro, Serenus (Pharm.), Sertorius Clemens, Sōsikratēs, Terentius Valens.

Rush (Greek: *skhoinos*; species of *Juncus* L. or of *Scirpus* L.): Lampōn, Melētos, Theodōrētos.

Saffron (Greek: *krokos; Crocus sativus* L.): Abaskantos, Aelius Gallus, Agapētós, Ammōnios of Alexandria, Anthaios Sextilius, Antiokhos Paccius, Aphrodās, Apollinarios (Pharm.), Apollōnios Claudius, Apollōnios of Memphis, Apuleius Celsus, Aristoklēs, Aristokratēs, Artemōn (Med.), Atimētos, Attalos III of Pergamon, Blastos, Bolās, Candidus, Celer the Centurion, Damonikos, Dasius, Dēmosthenēs, Diagoras of Cyprus, Diomēdēs, Diophantos of Lukia, Dioskoros (Pharm.), Dōsitheos (Pharm.), Emboularkhos, Eruthrios, Eudēmos "the Elder," Euelpidēs, Euēnos, Eugeneia, Euhēmeros, Florus, Glaukōn/Glaukos (Med.), Harpalos (Pharm.), Harpokrās, Harpokratiōn, Hērās, Hermeias (Ophthalm.), Hermolaos (Pharm.), Hermōn, Hērōn (Med.), Hierax, Isidōros of Antioch, Iulianus (Pharm.), Iustinus (Pharm.), Iustus the Pharmacologist, Kallinikos, Kimōn, Kleōn (of Kuzikos?), Kleophantos, Kosmos, Krateros, Kratippos, Lampōn, Leukios, Lingōn, Lukomēdēs, Lusias, Mantias (Heroph.), Marcellinus (Pharm.), Marinos (Med.), Melētos, Mēnās, Menemakhos, Menestheus, Mēnophilos, Mithradatēs VI, Naukratitēs medicus, Nikēratos, Nikostratos (Pharm.), Olumpiakos, Olumpionikos, Olumpos, Ōrigeneia, Pamphilos of Alexandria, Pasikratēs, Philippos of Macedon, Philōn of Tarsos, Philōnidēs of Catina, Philōtas, Poluarkhos, Poludeukēs,

Proëkhios, Prōtās, Proxenos, Prutanis, Ptolemaios (Pharm.), Puramos, Romula, Rufus of Ephesos, Sergius of Babylōn, Solōn, Stratōn of Bērutos, Suncrōs, Terentius Valens, Thaïs, Theodōrētos, Theophilos (Pharm.), Zōilos of Macedon, Zōsimos (Med.).

Sage (Greek: *sphakos; Salvia officinalis* L. syn. *Salvia cretica* L.): pseudo-Dēmokritos (Pharm.), Sertorius Clemens.

Sampsukhon: (see **Marjoram**)

Sarxiphagos (Greek; probably *Pimpinella saxifraga* L.): Zēnophilos.

Savory (Latin: *cunila; Satureja hortensis* L.): Sabinius Tiro.

Scammony (Greek: *skammōnia; Convolvulus scammonia* L.): Mnaseas (Method.), Serenus (Pharm.).

Scordion: (see ***Scordotis***)

Scordotis (Latin; *Teucrium scordium* L. and *Teucrium scorodonia* L.): Pompeius Lenaeus.

Shepherd's-purse (Greek: *thlaspis; Capsella bursa-pastoris* (L.) Medik.): Amarantos.

Smurneion (Greek; *Smyrnium perfoliatum* L.; *cf.* **Horse-celery**): Anakreōn (Pharm.).

Sonkhos (Greek; *Sonchus arvensis* L., *Sonchus asper* (L.) Hill, and *Sonchus oleraceus* L.): Kleëmporos.

Sorrel (Latin: *rumex; Rumex acetosa* L.): Iustus the Pharmacologist.

Sphondulion (Greek; *Heracleum sphondylium* L.): Aristokratēs, Kharitōn.

Spikenard (Greek: *nardostakhus; Nardostachys grandiflora* DC. syn. *Nardostachys jatamansi* DC.; *cf.* **Nard**): Agapētós, Anastasios, Anthimus, Antiokhos Paccius, Asterios, Damonikos, Emboularkhos, Eruthrios, Harpokrās, Hermolaos (Pharm.), Iulianus (Pharm.), Iulius Bassus, Iustinus (Pharm.), Kuros, Mantias (Heroph.), Mēnās, Olumpiakos, Olumpos, pseudo-Orpheus (Med.), Pamphilos of Bērutos, Philōn of Tarsos, Philōnidēs of Catina, Romula, Rufus of Ephesos, Stratōn (Erasistratean), Terentius Valens, Theodōrētos, Timokleanos, Zēnophilos.

Spurge: (see ***Euphorbia***)

Spurge-olive (Greek: *khamelaia; Daphne oleoides* Schreb.): Sextius Niger.

Spurge, Petty (Greek: *tithumallos; Euphorbia peplus* L.): Mikiōn.

Spurge, Sea (Greek: *tithumallos paralias; Euphorbia paralias* L.): Andreas of Karustos.

Squash (Greek: *kolokunthos; cf.* **Colocynth, Gourd**): Mēnodōros of Smurna.

Squill (Greek: *skilla/skillēs* or *skhinon; Urginea maritima* (L.) Baker): Amuntas (Geog.): Aristogenēs of Knidos, Cornelius, Diophantos of Lukia, Epimenidēs, Eugērasia, Iustus the Pharmacologist, Khrusermos, Logadios, Marcellus (Pharm.), Puthagoras (Med.), Truphōn of Gortun.

Squirting Cucumber (Greek: *sikuos agrios* ["wild cucumber"] and *elatērion; Ecballium elaterium* (L.) A.Rich.): Apios Phaskos, Damaskēnos, Eutonios, Minucianus, Philoklēs.

St. John's Wort (Greek: *huperikon; Hypericum perforatum* L. and *Hypericum crispum* L.): Amarantos, Hubristēs, Kurillos, Proklos the Methodist.

Strouthia (Greek): Glaukidēs.

Strukhnos (Greek; probably a plant of the nightshade family, one of *Solanum nigrum* L., *Atropa belladonna* L., or *Withania somnifera* Dun.): pseudo-Orpheus (Med.).

Sulfurwort (Greek: *peukedanon; Peucedanum officinale* L.): Khariklēs, Nikomēdēs IV of Bithunia, Sōsagoras.

Sumac (Greek: *rhous eruthros; Rhus coriaria* L.): Melētos.

Sumphuton: (see **Elecampane**)

Sycamore (Greek: *sukaminon; Ficus sycomorus* L.): Papyrus Oxyrhynchos 15.1796 (*De Plantiis Aegyptiis*), Physiologos.
Symphytum (Latin; *Symphytum bulbosum* Schimp.): Sandarius/Sardacius.

Thapsia (Greek; *Thapsia garganica* L.): Dionusodōros (Pharm.), Epidauros, Eutonios, Sōkratēs (Med.).
Ther(i)onarca (Latin; unidentified; contrast **Oleander**): pseudo-Dēmokritos (Pharmacy).
Thistle (Greek: *krission*; perhaps *Carduus pycnocephalus* L. or *Carduus tenuiflorus* Curtis): Andreas of Karustos, Iamblikhos of Constantinople, Khaireas.
Thyme (Greek: *thumon; Thymus vulgaris* L.): Cornelius, Iustus the Pharmacologist, Marcellus (Pharm.).
Tithumallos: (see **Spurge, Petty** and **Spurge, Sea**)
Tragacanth (Greek: *tragacantha; Astragalus gummifer* Labill.): Apollinarios (Pharm.), Apollōnios Claudius, Apuleius Celsus, Hērōn (Med.), Iakōbos Psukhrestos, Khrusermos, Mantias (Heroph.), Marinos (Med.), Neilammōn.
Tree Cotton (Greek: *dendron eriophoron; Gossypium arboreum* L.): Androsthenēs of Thasos.
Tree-Heath: (see **Heath**)
Turnip (Greek: *gongulis; Brassica rapa* L.): pseudo-Dēmokritos (Pharm.), Dionusios (Methodist).
Turnip, wild (Greek: *bounias; Brassica napus* L.): Amarantos, Dionusios of Utica.

Umbellifer (Greek: *petasōdē*): Phainias.
Valerian (Greek: *phou; Valeriana phu* L.; *cf.* **Nard, Pontic**): Proklos the Methodist.
Vervain (Latin: *uerbena; Verbena officinalis* L.): Iustus the Pharmacologist.
Vetch (Greek: *bikion*; Latin: *uicia; Vicia sativa* L.): Mamilius Sura.
Vetch, Bitter (Greek: *orobos*; Latin: *eruus; Vicia ervilia* (L.) Willd.): Antiokhos VIII Philomētōr, Eugērasia, Khrusermos, Terentius Valens, Zēnōn of Laodikeia.
Vine (Grape) (Greek: *ampelos; Vitis vinifera* L.; *cf.* **Lees, Must, Vinegar, Wine**): Palladius Rutilius Taurus Aemilianus, Vindonius Anatolios, Aristolaos, Cornelius Valerianus, Dēmosthenēs, Eusebius son of Theodorus, Florentinus, Fronto (Agric.), Hēsiod of Askra, Hostilius Saserna and son, Iulius Atticus, Iulius Graecinus, Iulius Hyginus, Kleidēmos, Kuranides, Nestōr, Olumpos, Pamphilos of Alexandria, Papyrus Hibeh 2.187, Petrōnios Musa, Porcius Cato, Tremellius Scrofa, Vergilius.
Vinegar (*cf.* **Vine**): Alkimiōn, Amphiōn, Andreas of Karustos, Apiōn of Oasis, Aristoklēs, Aristophanēs, Artemidōros of Pergē, Attalos III of Pergamon, Botrus, Castus, Damonikos, Dēïleōn, Dionusios Sallustius, Dionusios of Samos, Dionusodōros (Pharm.), Dioskouridēs of Anazarbos, Eirēnaios, Epigonos, Erasistratos of Sikuōn, Euangeus, Eugērasia, Halieus, Harpalos (Pharm.), Harpokrās, Hērakleidēs of Ephesos, Hierax, Hikesios of Smurna, Hippokratic Corpus Regimen, Ianuarinus, Ioudaios, Khariklēs, Kimōn, Kleōn (of Kuzikos?), Kleophantos, Kloniakos, Kurillos, Kuros, Lampōn, Leukios, Marcellus (Pharm.), Megēs, Menemakhos, Menippos, Mēnophilos, Moskhiōn (Pharm.), Nikēratos, Pasiōn, Philagrios, Philōtas, Puramos, Puthiōn (Pharm.), Serenus (Pharm.), Sōkratēs (Med.), Sōkratiōn, Sōsimenēs, Sunerōs, Theoxenos, Tlēpolemos, Truphōn of Gortun.

Walnut (Greek: *karua Persikē; Juglans regia* L.): Amuntas (Geog.), Apollōnios of Alexandria "Mus," Damonikos, Mithradatēs VI, Sueius.

Wheat (Greek: *puros*, or simply *amulos* and *sitos* ["flour"]; *Triticum vulgare* L.): Apollinarios (Pharm.), Iakōbos Psukhrestos, Iunia/Iounias, Nikēratos, Papyrus Michiganensis 17.758, Sōkratiōn, Stephanos of Tralleis.
Willow (Greek: *itea; Salix alba* L.): Dioskouridēs of Anazarbos, Lusimakhos of Macedon.
Winter Cherry: (see ***Halikababon***)
Wine (*cf.* **Vine**): Abaskantos, Aelius Gallus, Amarantos, Ambiuius, Androkudēs, Andrōn (Pharm.), Antigonos (Med.), Antimakhos, Antonius Musa, Apollodōros (Med.), Apollodōros the *thēriakos*, Apollōnios Claudius, Apollophanēs of Seleukeia, Apuleius Celsus, Aristolaos, Aristomakhos of Soloi, Aristotelian Corpus Problems, Artemōn, Asklēpiadēs of Bithunia, Aurelius, Biōn Caecilius, Bithus, Blastos, Chrysippus of Soloi, Cornelius, Daliōn (Med.), Damigerōn, Demotic Scientific Texts, Dexios, Dioklēs of Karustos, Dioskouridēs (Metrology), Dioskouridēs of Anazarbos, Epagathos, Epaphroditos of Carthage, Eruthrios, Euboulos (Pharm.), Euelpidēs, Euēnos, Eugamios, Eugērasia, Euhēmēros, Euphranōr, Euphrōnios of Athens, Eusebius son of Theodorus, Faustinus, Flauius "the boxer," Florus, Fronto (Agric.), Gemellus, Geōponika, Glaukōn/Glaukos (Med.), Glaukos (Geog. II), Harpalos (Pharm.), Hērās, Hermeias (Ophthalm.), Hermogenēs of Smurna, Hikesios, Hippokratic Corpus Regimen, Iakōbos Psukhrestos, Idios, Isidōros of Antioch, Khaireas, Kharitōn, Khrusermos, Khrusippos of Knidos (I), Khrusippos of Knidos (II), Kimōn, Kleophantos, Kleophantos of Keōs, Kōdios Toukos, Kommiadēs, Krateros, Kratippos, Kratōn (Pharm.), Licinius Mucianus, Lusias, Maecenas Licinius, Maiorianus, Mantias (Heroph.), Marsinus of Thrake, Matius Caluena, Menestheus, Mnaseas (Method.), Mnēsitheos of Athens, Moskhiōn (Pharm.), Nikēratos, Nikolaos (Pharm.), Nikomēdēs IV of Bithunia, Olumpos, Oppianus of Apameia, Pamphilos of Alexandria, Pantainos, Papyrus Laur. Inv. 68, Papyrus Ryl. III.531, Pasiōn, Petron(as) of Aigina, Petrōnios Musa, Phainias, Philagrios, Philippos of Macedon, Philoklēs, Philōnidēs of Durrakhion, Philōtas, Pleistonikos, Polueidēs, Porcius Cato, Praxagoras, Ptolemaios (Pharm.), Ripalus, Simos of Kōs, Solōn, Sōsandros (Pharm.), Sōsikratēs, Sunerōs, Thamuros, Themisōn of Laodikeia, Theophilos (Pharm.), Theosebios, Valerius Messalla Potitus, Vergilius, Zēnōn of Laodikeia, Zēnophilos, Zōsimos (Med.).
Wormwood (Greek: *abrotonon; Artemisia abrotanum* L.): Diophantos of Lukia, Kleophantos, Nikēratos, Praecepta Salubria, Serenus (Pharm.), Seuerus Iatrosophista.
Wormwood, absinthe (Greek: *apsinthinon; Artemisia absinthium* L.): Eruthrios, Praecepta Salubria.
Wormwood, tree (Greek: *artemisia; Artemisia arborescens* L.): Stratōn (Erasistratean).

Abies cephalonica L.	Fir
Acacia arabica Lam. and *A. nilotica* (L.) Willd. *ex* Delile	Acacia
Acacia catechu Willd.	**Indian buckthorn**
Acinos rotundifolius Pers.	*Akinos*
Aconitum Anthora L.	Aconite
Acorus calamus L.	Reed, aromatic
Adiantum capillus-veneris L.	Maidenhair
Agrimonia eupatoria L.	Agrimony
Alcyoneum cortoneum Pall., *A. papillosum* Pall., *A. palmatum* Pall., and *A. ficus* Pall.	***Alkuoneion***
Alhagi maurorum (L.) Medik.	Aspalathos

Allium ampeloprasum var. *porrum* (L.) J.Gay	Leek
Allium sativum L.	Garlic
Aloe perryi Baker and *Aloe vera* L.	Aloe
Amomum subulatum Roxb.	***Amōmon***
Anacyclus pyrethrum DC.	***Purethron***
Anagallis arvensis L.	Pimpernel
Anchusa tinctoria L.	*Alkibiadion*
Anethum graveolens L.	Dill
Apium graveolens L.	Celery
Arctostaphylos uva-ursi (L.) Spreng.	Bear-Berry
Aristolochia clematitis L.	Birthwort
Aristolochia longa L.	Birthwort, Long
Artemisia abrotanum L.	Wormwood
Artemisia absinthium L.	Wormwood, absinthe
Artemisia arborescens L.	Wormwood, tree
Arundo L. species	Reed
Asarum europaeum L.	Ginger, European Wild
Asparagus officinalis L.	Asparagus
Asphodelus ramosus Willd. or *A. albus* Willd.	Asphodel
Asplenium adiantum-nigrum L.	Maidenhair
Asplenium scolopendrium L.	Hart's Tongue
Aster amellus L.	*Amellus*
Astragalus fasciculifolius Boiss. (or other *Astragalus* L. species)	***Sarkokolla***
Astragalus gummifer Labill.	Tragacanth
Athamanta cretensis L.	Carrot, Cretan
Atractylis gummifera L.	Chameleon, White
Atriplex hortensis L.	Orache
Atropa belladonna L.	*Strukhnos*
Beta vulgaris L.	Beet
Boswellia sacra Flueck.	Frankincense
Brassica napus L.	Turnip, wild
Brassica nigra Koch	Mustard
Brassica oleracea L.	Cabbage
Brassica rapa L.	Turnip
Bryonia dioica Jacq.	Bryony
Bupleurum fruticosum L.	*Boupleuron*
Cachrys ferulacea (L.) Calest.	*Hippomarathron*
Cachrys libanotis (L.) Koch	*Kangkhru*
Calamintha graveolens (Bieb.) Benth.	*Akinos*
Calamintha Mill. species	Calamint
Calicotome villosa (Poir.) Link.	*Aspalathos*
Cannabis sativa L.	Hemp
Capparis spinosa L.	Caper
Capsella bursa-pastoris (L.) Medik.	Shepherd's-purse
Cardopatium corymbosum L.	Chameleon, Black
Carduus pycnocephalus L. or *C. tenuiflorus* Curtis	Thistle
Carum Copticum Benth. and Hook	***Ammi***

Castanea sativa Mill.	Chestnut
Caucalis playcarpos L. or *C. grandiflora* L.	*Kaukalis*
Cedrus libani L.	Cedar
Centaurea centaurium L.	Centaury
Centaurium umbellatum Gilib.	(White) Centaury
Cicer arietinum L.	Chickpea
Cichorium endivia L. and *C. intybus* L.	Chicory
Cinnamomum aromaticum Nees	Cassia
Cinnamomum tamala Nees, or *C. iners* Blume, or *C. zeylanicum* Nees	***Malabathron***
Cinnamomum verum J. Presl syn. *C. zeylanicum* Nees	Cinnamon
Cistus creticus L. or *C. cyprius* L.	*Ladanon*
Citrullus colocynthis (L.) Schrad.	Gourd
Citrus medica L.	Citron
Cochlearia anglica L.	*Betonikē*
Colchicum autumnale L.	Autumn Crocus
Commiphora mukul Engl.	***Bdellion***
Commiphora myrrha Arn.	Myrrh
Commiphora opobalsamum L.	Balsam
Conium maculatum L.	Hemlock
Convolvulus arvenis L.	*Perdikias*
Convolvulus scammonia L.	Scammony
Coriandrum sativum L.	Coriander
Crocus sativus L.	Saffron
Cucumis sativus L.	Cucumber
Cuminum cyminum L.	Cumin
Cupressus sempervirens L.	Cypress
Cuscuta epithymum L.	Dodder
Cyclamen graecum Link	Cyclamen
Cydonia vulgaris L. syn. *C. oblonga* Mill.	Quince
Cymbopogon schoenanthus L.	Camel-Grass
Cyperus papyrus L.	Papyrus
Cyperus rotundus L.	*Kuperos*
Cytinus hypocistis L.	Broom
Daphne oleoides Schreb.	Spurge-olive
Daucus carota L.	Carrot
Delphinium staphisagria L.	***Staphis***
Doronicum pardalianches Jacq.	Leopard's Bane
Dryopteris filix-mas (L.) Schott.	Fern
Ecballium elaterium (L.) A.Rich.	Squirting Cucumber
Elettaria cardamomum (L.) Maton	Cardamom
Epilobium angustifolium L.	*Onothuris*
Erica arborea L.	Heath
Eruca sativa Lam. or Mill.	Arugula
Eryngium campestre L.	Eryngo
Erythraea centaurium L.	Centaury
Erythronium dens-canis L.	*Mithridatia*
Euphorbia apios L.	*Iskhas*

Euphorbia lathyris L.	*Lathuris*
Euphorbia paralias L.	Spurge, Sea
Euphorbia peplis L.	*Peplis*
Euphorbia peplus L.	Spurge, Petty
Euphorbia resinifera Berg.	***Euphorbia***
Fagus sylvatica L. and *F. orientalis* Lipsky	Beech
Farsetia clypeata R.Br.	*Alussos*
Ferula communis L.	Ferula
Ferula galbaniflua Boiss. and Buhse	***Galbanum***
Ferula marmarica Asch. and Taub	***Ammōniakon***
Ferula marmarica L.	Fennel, Libyan
Ferula persica Willd.	***Sagapēnon***
Ficus benghalensis L.	Banyan tree
Ficus carica L.	Fig
Ficus sycomorus L.	Sycamore
Filipendula vulgaris Moench	Dropwort
Foeniculum vulgare Mill.	Fennel
Fomes officinales Bresadola	**shelf-fungus**
Gentiana lutea L.	Gentian, Great/Yellow
Gentiana lutea L. and *G. purpurea* L.	Gentian
Gladiolus segetum Gawler	Gladiolus
Glaucium flavum Crantz	***Glaukion***
Glycyrhiza glabra L.	Licorice
Gossypium arboreum L.	Tree Cotton
Hedera helix L.	Ivy
Helleborus niger L.	Hellebore, black
Heracleum sphondylium L.	*Sphondulion*
Hippuris vulgaris L.	*Potamogiton*
Holarrhena antidysenterica Wall.	*Holarrhena*
Hordeum vulgare L.	Barley
Hyoscyamus niger L.	Henbane
Hypericum perforatum L. and *H. crispum* L.	St. John's Wort
Hyssopus officinalis L.	Hyssop
Inula helenium L.	***Panax***
Inula helenium L.	Elecampane
Inula viscosa Aiton or *I. graveolens* Desf.	*Konuza*
Iris pallida Lam.	Iris, Illyrian
Iris pseudacorus L.	Iris, Yellow
Juglans regia L.	Walnut
Juncus L. species	Rush
Juniperus communis L.	Juniper
Juniperus oxycedrus L.	Cedar
Juniperus sabina L.	Juniper, Savin
Lactuca sativa L.	Lettuce
Lagenaria vulgaris L.	Colocynth
Larix decidua Mill.	Larch
Lathyrus ochrus (L.) DC.	*Ōkhros*

Laurus nobilis L.	Laurel
Lavandula stoechas L.	Cassidony
Lawsonia inermis L.	Henna
Lecokia cretica Lam.	*Kangkhru*
Lens culinaris Medik.	Lentil
Lepidium graminifolium L.	*Iberis*
Lepidium sativum L.	*Nasturcium*
Levisticum officinale Koch	Lovage
Lilium candidum L.	Lily
Linum usitatissimum L.	Flax
Loranthus europaeus Jacq.	Mistletoe
Lychnis flos-cuculi L.	Ragged Robin
Lythrum salicaria L.	*Lusimakhia*
Malus domestica Borkh.	Apple
Malva silvestris L.	Mallow
Mandragora officinarum L.	Mandrake
Marrubium vulgare L.	Horehound
Matricaria chamomilla L. syn. *M. recutita* L.	Chamomile
Medicago arborea L.	Moon-trefoil
Medicago sativa L.	Alfalfa
Melilotus officinalis (L.) Pall.	Melilot
Mentha L. species	Mint
Mentha pulegium L.	Pennyroyal
Meum athamanticum Jacq.	*Mēon*
Mimusops Schimperi L.	*Persea*
Moringa arabica Pers. or *M. oleifera* Lam.	Ben-Nut
Musa acuminata Colla 1820	Banana
Muscari comosum Mill.	Hyacinth
Myrtus communis L.	Myrtle
Narcissus L. species	Narcissus
Nardostachys grandiflora DC. syn. *N. jatamansi* DC.	Spikenard
Nerium oleander L.	Oleander
Nymphaea caerulea Sav. and *N. zenkeri* L.	Lotus Flower
Ocimum basilicum L.	Basil
Olea europaea L.	Olive
Opopanax hispidus Friv. or Grisb. ("Hercules' woundwort"), *O. chironium* (L.) Koch. ("sweet myrrh")	***Panax***
Origanum dictamnus L.	Dittany
Origanum majorana L.	Marjoram
Origanum vulgare L.	Oregano
Orlaya grandiflora (L.) Hoffm.	*Kaukalis*
Oryza sativa L.	Rice
Paeonia officinalis L.	Peony
Panicum miliaceum L.	Millet
Papaver somniferum L.	Poppy
Pastinacea sativa L.	Parsnip
Persicaria hydropiper (L.) Spach	*Piperitis*

Petroselinum crispum (Mill.) Nyman *ex* A.W. Hill	Parsley
Peucedanum officinale L.	Sulfurwort
Phoenix dactylifera L.	Date
Physalis alkekengi L.	*Halikababon*
Pimpinella anisum L.	Anise
Pimpinella saxifraga L.	*Sarxiphagos*
Pinus brutia Tenore, *P. halepensis* Mill., or *P. pinea* L.	Pine
Pinus laricio Poir.	Pine, Bruttian
Pinus pinea L.	Pine-nut
Piper longum L.	Pepper, Long
Piper nigrum L.	Pepper (Black & White)
Pistacia lentiscus L. (or other *Pistacia* L. species)	**Mastic(h)**
Pistacia terebinthus L. (or other *Pistacia* L. species)	***Terebinth***
Pistacia vera L.	Pistachio
Plantago major L.	Plantain
Polygonum hydropiper L.	*Piperitis*
Populus alba L. or *P. nigra* L.	Poplar
Potamogeton natans L.	*Potamogiton*
Prangos ferulacea (L.) Lindl.	*Hippomarathron*
Prunus domestica L.	Damson-plum
Prunus dulcis (Mill.) D.A.Webb	Almond
Prunus persica (L.) Batsch	Peach
Prunus prostrata Labill.	Cherry, Ground
Pteris aquilina L.	Fern
Punica granatum L.	Pomegranate
Pyrus communis L.	Pear
Quercus L. species, esp. *Q. robur* L.	Oak
Quercus petraea (Mattuschka) Liebl.	Oak, Winter
Quercus suber L.	Oak, Cork
Ranunculus sardous Crantz	Parsley, Sardinian
Raphanus sativus L.	Radish
Rhamnus cathartica L.	*Rhamnos*
Rhamnus L. species (e.g. *R. catharticus* L., *R. frangula* L., *R. infectorius* L., *R. lycoïdes* Boiss., *R. petiolaris* Boiss., *R. punctata* Boiss.)	***Lukion***
Rheum officinale L.	Rhubarb
Rhus coriaria L.	Sumac
Ricinus communis L.	Castor
Roc(c)ella tinctoria (L.) de Cand.	Orchil
Rosa canina L., *R. gallica* L., or *R. sempervirens* L.	Rose
Rosmarinus officinalis L.	Rosemary
Rubia tinctorum L.	Madder
Rubus fruticosus L.	Blackberry
Rumex acetosa L.	Sorrel
Rumex aquaticus L.	*Betonikē*
Rumex scutatus L.	*Bulapathum*
Ruscus aculeatus L.	Butcher's-broom
Ruta graveolens L.	Rue

Salix alba L.	Willow
Salvia libanotica Boiss. & Gaill.	*Elelisphakos*
Salvia officinalis L. syn. *S. cretica* L.	Sage
Salvia triloba L.f.	*Elelisphakos*
Santolina chamaecyparissus L.	Lavender Cotton
Satureja hortensis L.	Savory
Saussurea lappa Clarke	***Kostos***
Scirpus L. species	Rush
Senecio vulgaris L.	*Ērigerōn*
Seseli tortuosum L.	Hartwort, Massilian
Sideritis romana L.	*Alussos*
Sinapis alba L.	Mustard
Sison amomum L.	Parsley, Stone
Smyrnium olusatrum L.	Horse-celery
Smyrnium perfoliatum L.	*Smurneion*
Solanum nigrum L.	*Strukhnos*
Sonchus arvensis L., *S. asper* (L.) Hill, and *S. oleraceus* L.	*Sonkhos*
Spiraea filipendula L.	Dropwort
Spongia panicea Pall.	***Alkuoneion***
Stachys officinalis L.	*Betonikē*
Styrax officinalis L.	***Sturax***
Symphytum bulbosum Schimp.	*Symphytum*
Syzygium aromaticum (L.) Merrill & Perry	Clove
Teucrium chaemaedrys L.	Germander
Teucrium L. species	***Skordion***
Teucrium polium L.	*Polion*
Teucrium scordium L. and *T. scorodonia* L.	*Scordotis*
Thapsia garganica L.	*Thapsia*
Thymus serpyllum L.	***Herpullos***
Thymus vulgaris L.	Thyme
Tordylium apulum L.	*Kaukalis*
Tordylium officinale L.	Hartwort
Trifolium pratense L. and *Tr. fragiferum* L.	Clover
Trigonella corniculata L.	Melilot
Trigonella foenum-graecum L.	Fenugreek
Triticum vulgare L.	Wheat
Tussilago farfara L.	Colts-foot
Ulmus glabra Huds., *U. minor* Mill., & *U. procera* Salisb.	Elm
Urginea maritima (L.) Baker	Squill
Urtica dioica L.	Nettle
Valeriana celtica L.	Nard, Celtic
Valeriana dioscoridis Sibth.	Nard, Pontic
Valeriana phu L.	Nard (Pontic) & Valerian
Veratrum album L.	Hellebore, white
Verbascum thapsus L.	Mullein
Verbena officinalis L.	Vervain

Vicia ervilia (L.) Willd.	Vetch, bitter
Vicia faba L.	Bean, Fava
Vicia sativa L.	Vetch
Viscum album L.	Mistletoe
Vitex agnus-castus L.	Chaste-Tree
Vitis vinifera L.	Vine (Grape)
Withania somnifera Dun.	*Strukhnos*
Zingiber officinale Roscoe	Ginger
Ziziphus lotus Willd.	*Lōtos* Tree, Libyan